Textbook of Forest Science

Asim Kumar Mandal • Abel Nicodemus
Editors

Textbook of Forest Science

Springer

Editors
Asim Kumar Mandal
Ramakrishna Mission Vivekananda
Educational and Research Institute
Ranchi, Jharkhand, India

Abel Nicodemus
ICFRE-Institute of Forest Genetics and Tree Breeding
Coimbatore, Tamil Nadu, India

ISBN 978-981-97-8288-8 ISBN 978-981-97-8289-5 (eBook)
https://doi.org/10.1007/978-981-97-8289-5

This Springer imprint is published by the registered company Springer Nature Singapore Pte Ltd.
The registered company address is: 152 Beach Road, #21-01/04 Gateway East, Singapore 189721, Singapore

Preface

Forests have nurtured humans right from their origin and have shaped their course of evolution in many ways. Even after moving to settled agriculture, people continued to depend on forests for their livelihood and possessed a deep understanding of the delicate relationship between farming and forests. With the population increase, the demand for more products from the forests resulted in the emergence of forestry as a scientific discipline. Till recently deriving maximum output in terms of timber and other products has been the main focus of forestry. However, the science of forestry is receiving more attention from researchers, politicians, and the public than ever before due to growing awareness on the vital importance of forests in the amelioration of the world's environment. Forest science has also gained global importance in the context of climate change since forests are regarded as an important sink for sequestering carbon. They offer several tangible and intangible benefits, which are essential for human survival. Currently, the main focus of attention of national governments and international organizations is to evolve efficient methods of managing forests for human well-being and to conserve them for posterity.

Forestry is an interdisciplinary subject drawing inputs from almost all branches of science and engineering. Often students and others interested in forestry are daunted with searching the vast literature scattered in many sources to get the required information. This book is an attempt to present the latest developments in all forestry disciplines in a single volume. It is essentially designed as a text book for undergraduate and postgraduate students providing an in-depth reading of the traditional and emerging areas of forest science. It will also help them to readily identify knowledge gaps to which the future research and development efforts have to be channelized. The increased consciousness on environmental protection and ecological conservation in the context of changing climatic conditions have widened the stakeholder base of forestry worldwide. This volume will also serve as a reference book for professional researchers, teachers, practicing foresters, policy planners and anyone interested in forestry science.

There are 39 chapters grouped into seven sections that include Forests, Policy and Research; Forest Management and Conservation; Silviculture, Resource Assessment and Sustainability; Production Forestry; Tree Improvement and Biotechnology; Forest Protection; and Forest Products and Utilization. Each of these sections has three to nine chapters covering classical and contemporary areas of forest science written by eminent experts. The authors of the chapters have taken every effort to present a reappraisal of theory and practice, applications and future scope of the selected areas of forest science. While the chapters dealing with current status and traditional aspects of different areas of forestry provide the necessary foundation of the subject, the recent advances like biotechnology and remote sensing provide an outlook for future development.

We thank all contributors of chapters for their efforts in bringing out the most comprehensive knowledge and information on their area of expertise. We appreciate their cooperation during editing and taking up of revisions to make each chapter concise and to ensure that they are in harmony with other chapters of the book. We wish to place on record our special thanks to Mohua Mandal, Animesh Sinha, C. Narayanan and Lokesh Kumar Singh for their sincere technical help as and when required. We thank the publisher, Springer Nature, for opportunity given to us to compile this volume and for all the support received during the preparation and publishing of the book.

We hope that the vast range of topics address the needs of anyone seeking the latest knowledge on different areas of forestry. We would consider that our efforts have borne fruit if the readers find the book useful and easy to read. Any feedback and suggestions from the readers are welcome to improve the future editions.

Ranchi, Jharkhand, India Asim Kumar Mandal
Coimbatore, Tamil Nadu, India Abel Nicodemus

Contents

About the Editors

Asim Kumar Mandal an Agriculture graduate with Honours from Visva-Bharati, Santiniketan. He obtained his M.Sc. (Genetics) followed by Ph.D. (Genetics) from Indian Agricultural Research Institute, New Delhi. He did his Post-Doctoral research in Forest Genetics from the University of Edinburgh, UK. He received advanced training in the management of R & D organization from Administrative Staff College of India, Hyderabad.

Dr. Mandal began his professional career as Assistant Professor at the Institute of Agriculture, Visva-Bharati, Sriniketan. He moved to Indian Council of Forestry Research and Education (ICFRE), Dehradun as Senior Scientist (Forest Genetics) and worked at the Institute of Forest Genetics and Tree Breeding , Coimbatore; Tropical Forest Research Institute, Jabalpur; and Forest Research Institute, Dehradun, in different capacities from 1990 to 2004. Apart from research and teaching, Dr. Mandal provided academic and administrative leadership in all the above institutes as Head of Division. He researched in the areas of genetic improvement of timber yielding species, mainly teak and medicinal plants. He joined as Director of Tropical Forest Research Institute, ICFRE, Jabalpur, in January 2005 and led the institute as its Chief Executive for five and half years.

He has travelled abroad extensively to attend various professional programs. He is the first National Coordinator of Asia Pacific Forest Genetic Resources Program and represented ICFRE as Member in the Board of International Union of Forest Research Organization (IUFRO). He has served in different national organizations, including ICAR, DBT, MoEF, ICFRE, FDCM, and Central Silk Board as Member/Chairperson of different committees. He is Member and Fellow of a number of academic bodies.

Dr. Mandal has teaching experience at UG, PG and post PG levels. He is a recipient of national and international fellowships for study and research and also recipient of two national awards. He has successfully guided a number of students for their doctoral program and has executed a number of research projects funded by national and international funding agencies. He has released new improved varieties of medicinal plants and is credited with one Indian patent. He has 143 publications including books, conference proceedings, technical bulletins, book chapters and research papers in national and international peer-reviewed journals. He has chaired/co-chaired a number of technical sessions in national and international conferences. On retirement in 2012, he served R. K. M. Vivekananda Educational and Research Institute, Ranchi, as Honorary Professor for five years.

Abel Nicodemus works at the Division of Genetics and Tree Improvement, ICFRE-Institute of Forest Genetics and Tree Breeding, Coimbatore as Scientist G (Principal Scientist). He has been working in the field of forest genetics and tree breeding for over three decades. His broad areas of research are breeding of short rotation trees for fast growth and higher wood production, varietal development and commercialization and domestication and conservation of timber yielding trees. He has released new clones of Casuarina species and their hybrids possessing high adaptability and fast growth. These new varieties are widely planted throughout the country and substantially increased the wood production. He works in close partnership with Australian Tree Seed Centre, CSIRO and paper industries in India to increase the productivity of industrial farm forestry plantations and to secure pulpwood raw material requirements.

Dr. Nicodemus functions as the Deputy Coordinator for the IUFRO Working Party 20.08.02 on Improvement and Culture of Nitrogen fixing trees. He has published two books and edited three volumes in association with international experts. He has guided doctoral students and published around 50 research papers in peer-reviewed journals.

Part I

Forests, Policy and Research

Forest, Forestry, and State of Forests

1

Animesh Sinha

Abstract

Forest is the sole warehouse of natural resources to the mankind providing food, fuel, and shelter since the dawn of human civilization. Forests play the most significant role in sustaining the balance of ecosystem on this planet. However, industrialization and urbanization have always threatened the forestlands resulting in the rapid depletion of its cover round the globe. India's overall forest cover accounts for 23.60% of its entire geographic area and is divided into 16 major forest types. India is one of the few countries in the world where forest cover has increased in the last 10 years. However, the country still faces several challenges in the sustainable management of its forests, leading to poor production and productivity from this sector. Several legislative frameworks and different policies to generate mass awareness are being implemented to conserve the Indian forests. This chapter will provide a detailed insight to the forests of India with respect to the global scenario.

Keywords

Forestry · Tropical forest · Reserved forest · Forest cover · National parks

1.1 Forest

A "forest" is defined as any area of land larger than 0.5 hectare (ha) that has trees higher than 5 m with a canopy cover >10%, or trees that can naturally attain these thresholds (FAO 2006). It excludes land used primarily for urban or agricultural purposes. Nevertheless, the Forest Survey of India (FSI) defines "Forest Cover" as "an area larger than 1 ha in size with a tree canopy density of 10% or above, regardless of legal status or species composition." On the other hand, the term "Recorded Forest Area" refers to any "areas that have been notified as forest under any Government Act or Rule, or that have been documented as forest in the Government records." There may or may not be forest cover in the recorded orest area (ISFR 2021). There is overlap between recorded forest area and forest cover, but not coterminous relationship. The recorded forest area can be classified as:

- Reserved Forest (RF): A fully protected area (PA) declared under the terms of Indian Forest Acts or State Forest Acts. Unless otherwise approved, all activities like hunting, grazing, and so forth are forbidden in RFs.
- Protected Forest (PF): An area with limited protection that has been recognized under the Indian Forest Acts or State Forest Acts. In PFs, permit to activities as hunting and grazing are occasionally granted to populations living on the forest's outskirts that rely largely or entirely on forest resources or their derivatives for livelihood.
- Unclassified Forest (UF): A forest area that is not classified as reserved or protected forest. State-by-state ownership statuses of these forests differ.

Currently, India's total recorded forest area is around 23.6% of the country's entire geographical area, with the areas under RF, PF, and UF varying in proportion by 4:2:1 (Table 1.1).

1.1.1 Role of Forests

Forests have been an integral part of the human ecosystem since ancient times. Forests, nature's greatest gift to mankind, play a vital role in human life. Apart from providing shelter and protection to a large number of living beings, they have been a major source of food, fodder, fiber, wood, and many other products. They have enriched human life in diverse ways, both materially and psychologically.

A. Sinha (✉)
ICFRE - Institute of Forest Productivity, Ranchi, India

A. K. Mandal, A. Nicodemus (eds.), *Textbook of Forest Science*, https://doi.org/10.1007/978-981-97-8289-5_1

Table 1.1 Recorded forest areas in India

Categories of forest area	Area in '000 sq km
Reserved forest	442.3
Protected forest	212.3
Unclassified forest	120.7
Total recorded forest area	775.3
Geographical area	3287.5
% of Geographical area	23.6

Source: ISFR (2021)

Table 1.2 Carbon content (Mt) by component in India's forests between 1995 and 2021

Carbon	1995	2021	Incremental change
In biomass	2692.47	3038.80	346.33
In soil (Including dead wood and litter)	3552.30	4165.20	612.9
Total	6244.78	7204.00	959.23

1.1.1.1 Ethical

Forests are deeply intertwined with religion and cultural manifests of India. Most types of education were taught in ancient India in the vicinity of forests. Rabindranath Tagore, sometimes referred to as "Gurudev," is an acclaimed person in the field of modern educational theory and is well known for his contributions made both in India and beyond. One of the most remarkable aspects of his educational philosophy was his belief that education was the most effective instrument for modernizing India. At Santiniketan, Tagore founded his school, Brahmachary-Ashram, in 1901. Later, this institution evolved into Visva-Bharati University. The Vedas and Upanishadas, which comprise India's national inheritance, left their mark on Tagore's experimentation. Tagore intended to introduce a semblance of ancient ashramic teachings into contemporary India, where students and teachers would cohabit as members of a family. Learning would occur in a cooperative atmosphere involving both the environment and other people. The child's individuality would flourish as a result of exposure to music, literature, and the arts. The child would physically assist in daily tasks, much as how the disciple and other family members of the Guru used to feed the animals, milch the cow, and gather seasonal produce from the forest in the past. Cleaning the jungle, planting trees, and maintaining the environment all fell within the student's purview. According to Bhattacharya (2013), all of these activities would afford the child both mental and physical independence, enabling them to learn impulsively, passionately, and effortlessly. With so many outdoor recreational opportunities, forests promote healthy, active lifestyles. By connecting with others and their surroundings, they also improve their mental and spiritual health.

1.1.1.2 Ecological

Numerous plant and animal species are protected and maintained by the habitat that forests offer. Forests modulate hydrologic flow. Soil erosion is decreased by plants because they create a protective cover that decreases the force of rains on the soil. Soil anchoring is aided by root systems. They offer shade, keeping the soil from drying up too much enhancing the soil's ability to retain moisture. Forests are essential for conserving soil and water, and reducing the siltation of reservoirs. A healthy watershed depends heavily on the forests. Rivers that start in forested areas transport organic materials from the forest to downstream, sustaining a diversity of fish and aquatic life.

The multitude of different kinds and variations of plants, animals, and microbes may be found in these Indian habitats. A variety of ecosystem services, such as pollination, water supply, and genetic resources, that are necessary for agricultural systems' stability and nutritional security are also provided by biodiversity, forests, and trees outside of forests. Additionally, they have a significant impact on reducing pollutants in the environment, droughts, floods, and desertification.

By capturing atmospheric carbon dioxide and reducing the greenhouse effect, trees replenish the oxygen availability in the atmosphere through photosynthesis. Results show that between 1995 and 2021, the country's forest carbon stocks increased by 36.9 metric ton (Mt) of carbon per year, from 6244.8 to 7204.0 Mt (Table 1.2). This rise is attributed to conservation policies. The country's annual carbon dioxide removal from the forests was sufficient to offset 9.3% of its emissions in 2000. In COP 27 (Conference of the Parties), India committed to developing additional carbon sinks of 2.5–3.0 billion tons of CO_2 equivalent through increased forest cover by 2030. Forests play an important part in the country's carbon sequestration, which can fetch a handsome amount of foreign currency in exchange of trading carbon credits in the international markets (Gorain et al. 2021). Countries such as India, which have large forests and a large rural population reliant on forest products and services for subsistence, face a dilemma. With approximately 170,000 settlements categorized as forest fringe villages, around

300 million residents rely heavily on the surrounding forests (PIB 2023). As a result, people's reliance on the industry must be lessened by making alternate sources of forest and tree-based goods available to them. The potential for India's forests and tree cover to store carbon may be increased by the National Mission for a Green India, which will also provide LPG (liquified petroleum gas), biogas plants, and energy-efficient stoves to mountain dwellers living in the Western Ghats and the Himalayan area, enhancing forest protection legislation.

In the world's subtropics and tropics, mangroves are the only tropical coastal forests found where land meets the ocean (Alongi 2002). Mangroves are extremely prolific ecosystems with enormous ecological significance. They safeguard and maintain coastal zones while also nourishing and nurturing coastal water. The Sundarbans, located in the delta of the rivers Ganga, Brahmaputra, and Meghna, are the biggest coastal wetland in the world. They are made up of mangrove forests that span an area of around one million hectares. About 60% of this mangrove tract is shared by Bangladesh, and the remaining 40% by India. The climate is subtropical monsoon, with annual rainfall ranging from 1600 to 1800 mm and numerous cyclonic storms. Following the 2004 Indian Ocean tsunami, one of the most devastating natural catastrophes in recent memory, the role of coastal forests in reducing the effects of tsunamis suddenly became an important topic of discussion. There is substantial evidence that coastal vegetation may diminish the power, depth, and velocity of a tsunami, decreasing property damage and loss of life. Several scientific studies, surveys, and anecdotes carried out in different countries of South East Asia proved that the degree of damage by tsunamis in coastal region is significantly correlated with the absence of coastal forests there (Forbes and Broadhead 2008).

1.1.1.3 Economical

A component of sustainable forest management is the forestry sector's economic contribution to the country; data on this aspect are required to track advancements of the country. A consistent and internationally comparable dataset on the contribution of the forestry industry to national economies is made available by the Food and Agriculture Organization of the United Nations (FAO) for each and every nation and territory in the world. Three measures are often used to quantify the forestry industry's contribution to national economics: employment in the sector, gross domestic product (GDP) contribution, and trade contribution. Furthermore, although the economic appraisal of the forest sector is being conducted, the ecological services that forests provide are sometimes overlooked and underappreciated.

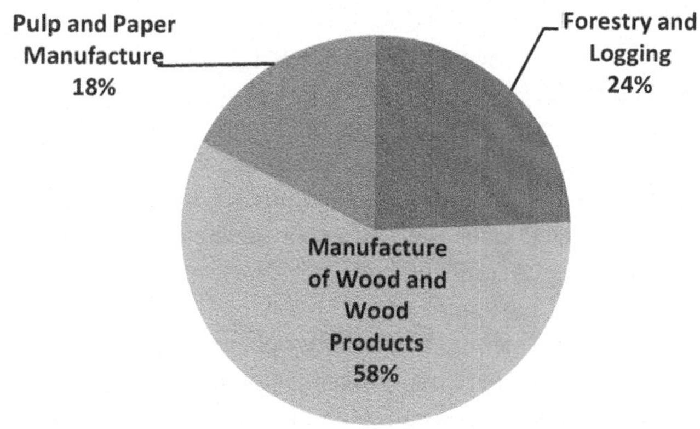

Fig. 1.1 Distribution of employment in forest sector across the world

Employment

According to a recent FAO and ILO (International Labour Organization) study by including 185 nations and 99% of global forest lands, 33.3 million people were engaged in the forest sector globally from 2017 to 2019, accounting for barely 1% of total global employment (Lippe et al. 2022). Between 2011–2013 and 2017–2019, the overall number with respect to (w.r.t) employment associated with forests faced a decrease of 15% from 39.5 million. Increased productivity, such as automation, and improved forest management methods may explain the declining level of forest-related jobs. According to Lippe et al. (2022), the production of wood and wood products engaged around 19.4 million people, while other subsectors such as logging and pulp-paper manufacturing generated 8.0 and 5.9 million employment in this sector, respectively (Fig. 1.1). As per Indian scenario, around 1.6 million people were engaged in forest sector, which is only 0.1% of the total population of the country (Li et al. 2019).

Contribution to GDP (Value-Added)

The overall contribution of forestry sector to world GDP has been decreasing continuously over time due to the rapid expansion of global economy. During the period of 2000–2011, the global GDP has increased around 40%, whereas the total value-added in forestry sector has seen an uplift of only 3%. The forestry industry generated a total value-added of US$ 606 billion in 2011 (Lebedys and Li 2014). Li et al. (2019) assessed that forest sector in India has contributed a total of US$ 44.4 billion total value-added to the country's GDP, which includes US$ 23.4 billion, US$ 8.1 billion, and US$ 12.9 billion as direct, indirect, and induced value-added, respectively.

Trade

Global forest product exports totaled $421 billion in 2011 (Lebedys and Li 2014). Despite fast increase in international trading, growth in trading in forest products is relatively slower than growth in other commercial items. Approximately 90% of the worldwide commerce in forest products was conducted in the top four exporting areas, which include the growing Asia-Pacific region, Northern America, and Western and Eastern Europe.

Northern America, Western Europe, and the industrialized Asia-Pacific area are major manufacturers and consumers of forest products, making up 60% of exports, 23% of employment, and 50% of value-added. If the forestry industry is only considered in terms of its contribution to GDP and employment opportunities, it is not a significant economic sector globally. However, contribution of the sector is undervalued both in economic and social terms as majority of times indirect use values are either not captured or underestimated (Gorain and Malakar 2020). In many countries, individuals gather fuel wood or non-timber forest products (NTFPs) for sustenance, which is usually not included as a contribution to the nation's GDP. It is challenging to monitor and quantify these activities on a national or regional level, even when these goods are gathered for commercial purposes. A research that evaluated the prospective financial rewards from collecting wood in Peru's Amazonian rain forest to those from using non-timber forest products (wild fruits and latex) amply demonstrated the superiority of NTFPs over timber (Peters et al. 1989). Compared to the merchantable timber production of US$ 11.49/ha/year, the sustainable fruit and latex output was predicted to be US$ 422/ha/year.

In India, the gross value contributed from forestry and logging was 1.96 trillion Indian rupees for the 2020 fiscal year (https://www.statista.com/statistics/911982/india-economic-contribution-of-forestry-and-logging-indistry/). Since the fiscal year 2013, there has been an upward tendency in this sector's contribution (Fig. 1.2). India produces a lot of food, medicine, fuel, lumber, and fodder from its forests. The ever-growing need for timber is met by forests. In the contemporary economy, NTFPs are enormously significant. Wild food goods, certain essential oils, gums, resins, turpentines, tans, colors, bamboo, fibers, tendu leaves, and so on are examples of such products. These items serve as the foundational raw materials for the phytochemical and phytopharmaceutical industries. Additionally, forests supply cattle and other livestock with feed. Certain plants have considerable feed value in their leaves and twigs. Even at crucial times, multipurpose trees such as *Grewia optiva*, *Leucaena leucocephala*, *Bauhinia variegata*, and *Morus serrata* have the capacity to provide animals with high-quality fodder when harvested at the ideal stage of maturity (Navale et al. 2022). The NTFP sector is one of India's major unorganized sectors, with a dependent population of over 275 million and an annual revenue of more than Rs. 6000 crores (Kanwal 2014). According to Shivaprasad and Chandrashekar (2014), this sector accounts for almost 70% of the exports of forest-based products and 50% of the forest revenue received by the Indian government.

1.1.1.4 Social

Forests and trees contribute to food security and nutrition in a variety of ways. Diets of individuals living in heterogeneous environments are more diverse and forests are capable of providing food of higher nutritional quality abundantly. Forests can act as a major source in times of food scarcity. Forests and tree-based agriculture systems support the lives

Fig. 1.2 Gross value-added from the forestry and logging industry in India

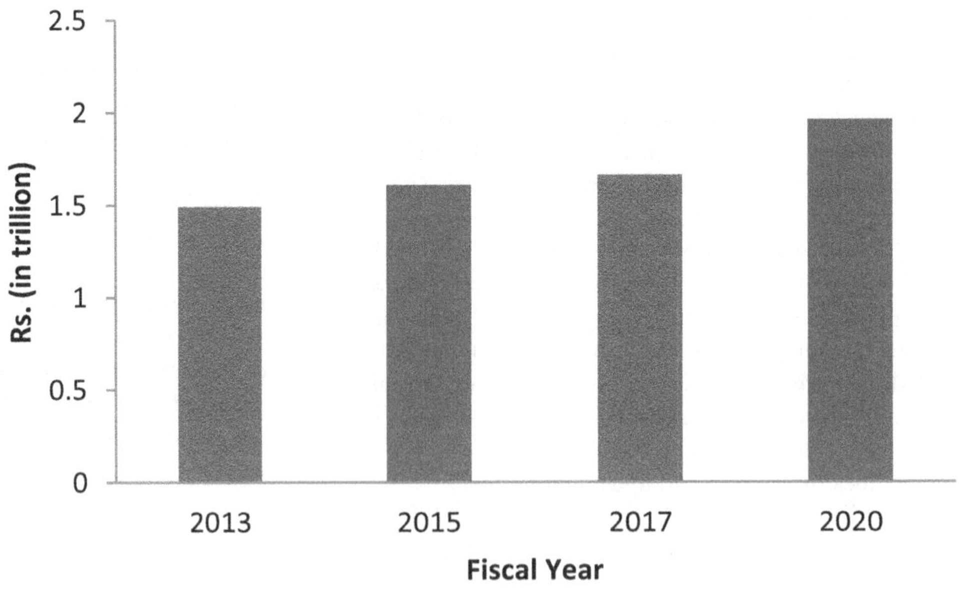

of an estimated one billion people worldwide (Sunderland et al. 2013).

Forest dwellers have several alternatives for self-employment. According to the International Labour Organization, 1 ha of forest planting requires about 630 man-days from nursery to harvesting. Approximately 70% of the funding allotted to plantings or afforestation is spent on direct pay for workers, with the other 30% going toward the procurement of seeds, planting supplies, equipment, and so on. It is worth noting that in India, women and tribals make up half of the workforce on forest plantations. Rural women spend 70–80% of their working days collecting NTFPs, fuel, and feed. NTFP-related activities offer employment during lulls in the economy as well as a buffer against risk and family crises (Singhal et al. 2003). Furthermore, forests serve as popular spots for leisure and recreation and also enhance the area's scenic appeal.

1.1.2 Forest Types

Champion initiated the first endeavor to categorize India's forests in 1936, and Seth amended it in 1968 (Champion and Seth 1968). Employing temperature and precipitation data, they categorized the Indian forest into 16 type groups (climatic kinds), 6 primary forest groups, and more than 200 subgroup types (Table 1.3). The lack of geographic information system (GIS) and remote sensing data at the time, however, likely made it impossible to create maps depicting the distribution of different forest types.

The first-ever "*Forest Type Atlas of India*" was published in 2011 by the Forest Survey of India (FSI), which carried out a nationwide mapping project of forest types from 2005 to 2010. The Champion and Seth categorization was used to create the 1:50,000 scale forest type maps. Table 1.4 shows the area statistics for the type groupings according to ISFR (2021). Group 5 consists of tropical dry deciduous forest (280,547 sq km), followed by Group 3, which consists of tropical wet deciduous forest (131,805 sq km). These two groups together encompass around 58% of India's forests.

Despite being largely accepted at the time, Champion and Seth's (1968) categorization was later revealed to have certain shortcomings. Champion's categorization was largely for forest usage, especially lumber extraction. Over time, the focus of Indian forestry shifted to forest conservation, with a renewed emphasis on the significance of forests in environment amelioration and climate change mitigation/adaptation. Because of the classification into over 200 subgroups, forest managers encountered extra intricacy. The Indian Council of Forestry Research and Education (ICFRE) sought to simplify this categorization even further by ground-proofing, more surveying, and reclassification measures. ICFRE (2013) established 10 primary groupings and 44 sub-groups (Table 1.5) to decrease redundancy, overlap, and to make it easier for practical foresters to work in the field.

Table 1.3 Forest types in India—Champion and Seth classification

Major group	Group	Major group	Group
A. Moist tropical forest	1. Tropical wet evergreen forest	C. Montane subtropical forest	8. Subtropical broad-leaved hill forest
	2. Tropical semi-evergreen forest		9. Subtropical pine forest
	3. Tropical moist deciduous forest		10. Subtropical dry evergreen forest
	4. Littoral and swamp forest	D. Montane temperate forest	11. Montane wet temperate forest
B. Dry tropical forest.	5. Tropical dry deciduous forest		12. Himalayan moist temperate forest
	6. Tropical thorn forest		13. Himalayan dry temperate forest
	7. Tropical dry evergreen forest.	E. Subalpine forest	14. Subalpine forest
		F. Alpine scrub	15. Moist alpine scrub
			16. Dry alpine scrub

Table 1.4 Area under different forest type groups in India as per ISFR (2021)

Type group	Area in sq km	% of Grand total	Type group	Area in sq km	% of Grand total
Group 1	19,572	2.74	Group 10	173	0.02
Group 2	69,195	9.69	Group 11	20,185	2.83
Group 3	131,805	18.47	Group 12	28,727	4.02
Group 4	5478	0.77	Group 13	4255	0.60
Group 5	280,547	39.30	Group 14	12,672	1.78
Group 6	13,259	1.86	Group 15	652	0.09
Group 7	835	0.12	Group 16	2396	0.34
Group 8	31,015	4.35	Plantation	75,221	10.54
Group 9	17,801	2.49	Total	713,788	100.00

Table 1.5 Forest type groups in India as per ICFRE (2013)

Major group	Subgroup	Major group	Subgroup
I. Tropical wet evergreen forests	IA. Andaman tropical evergreen forest	VI. Dry tropical thorn and scrub type	VIA. Dry thorn forest
	IB. West coast tropical evergreen forest		VIB. Desert thorn and ravine forest
	IC. Assam Valley tropical evergreen forest		VIC. Dry scrubs
	ID. Cachar tropical evergreen forest		VID. Dry evergreen forest
II. Tropical wet semi-evergreen forests	IIA. Andaman semi-evergreen forest	VII. Southern montane broadleaved forests	VIIA. Southern montane broadleaved forest
	IIB. West Coast semi-evergreen forest		VIIB. Southern montane broadleaved wet forest
	IIC. Assam valley semi-evergreen forest	VIII. Northern subtropical broadleaved and coniferous forests	VIIIA. East Himalayan subtropical broadleaved hill forest
	IID. Cachar semi-evergreen forest		VIIIB. Khasi wet hill forest
	IIE. Moist bamboo forest		VIIIC. Assam subtropical pine forest
III. Tropical moist deciduous forests	IIIA. Moist teak forest		VIIID. Himalayan subtropical pine forest
	IIIB. Mixed deciduous forest	IX. Himalayan temperate forests	IXA. East Himalayan broadleaved temperate forest
	IIIC. Moist sal forest		IXB. East Himalayan mixed temperate coniferous forest
	IIID. Moist peninsular sal		IXC. West Himalayan broadleaved temperate forest
	IIIE. Odisha moist deciduous forest		IXD. West Himalayan deodar forest
IV. Littoral and swamp forests	IVA. Littoral forest		IXE. West Himalayan mixed coniferous forest
	IVB. Tidal mangroves		IXF. West Himalayan dry temperate coniferous forest
	IVC. Tidal *Heritiera* forest		IXG. East Himalayan dry temperate coniferous forest
	IVD. Tropical fresh water *Myristica* swamps	X. Alpine forests	X A. Trans-Himalayan alpine forest
	IVE. Tropical fresh water seasonal swamp		XB. West Himalayan alpine forest
V. Dry deciduous type	VA. Dry teak bearing forest		XC. Himalayan *Rhododendron*
	VB. Red sander forest		
	VC. Dry mix deciduous forest		
	VD. Dry sal bearing forest		
	VE. Dry grassland		

1.2 Forestry

Forestry is defined as the science or practice of planting, maintaining, utilizing, preserving, and caring for forests and their related resources for human and environmental purposes. Forestry has various branches, and just a few of the significant ones are mentioned here.

1.2.1 Silviculture and Silvicultural Systems

The branch of forestry known as silviculture is concerned with the establishment, maintenance, and propagation of timber stands. It is the process of growing forest trees. The principal objectives of silviculture are (1) production of higher timber volume per unit area, (2) production of superior quality timber, (3) reduction of rotation age of species, (4) large-scale production of economically valuable species,

(5) afforestation, (6) reforestation, and (7) introduction of exotics.

The collection of silvicultural practices used to cultivate, harvest, and replace the crops that make up a forest with new, distinctively structured crops is known as a silvicultural system. The two main categories of silvicultural systems are coppice forest systems and high forest systems. In high forest systems, where rotation is often lengthy, regeneration originates from seedlings, whether naturally occurring or artificially created. In coppice forest systems, coppice is the source of regeneration and has a brief rotation.

1.2.2 Forest Mensuration

It deals with determining the shape, age, height, girth, weight, volume, and increment of individual trees, stands, or entire forests, either standing or after they have been felled. Forest mensuration aids in estimating the worth of any forest output for sale. It is necessary to quantify the number of standing trees in a forest and their yearly growth in order to use forest products sustainably. Forest mensuration provides crucial basic statistical data for future forest management planning.

1.2.3 Forest Resource Utilization

This field of forestry is concerned with the collection, transformation, disposal, and utilization of forest products. Timber, poles, and firewood are the most important forest products. Fruits, leaves, bark, fiber, colors, oils, essential oils, gum, resins, animal products (lac, honey, wax, silk, ivory, etc.), mineral goods, medicinal plant items, and so on are examples of minor forest production. These are also known as non-timber forest products (NTFPs) or non-wood forest products (NWFPs).

1.2.4 Wood Technology

The hard, fibrous, lignin-rich material beneath a tree's bark is called wood. The main components of wood are the isomeric carbohydrates cellulose and lignin. According to Perlin (1991), wood is hailed as the "unsung hero" of the revolution in technology that has propelled humanity from a society of stone and bone to the modern day. Wood technology deals with mechanical performance of wood and wood products, durability of wood, drying of wood, high performance in wood coating, environmental optimization of wood protection, innovative timber and composite elements, analytical techniques for biorefineries, etc.

1.2.5 Forest Policy and Legislation

The National Forest Policy (NFP) of 1988 placed a strong emphasis on outside-the-forest tree planting as well as the conservation of forests via public participation. In 2018, a draft national forest policy was developed. The policy's overarching goal is to protect people's ecological security and means of subsistence for both the current and future generations by managing forests sustainably to maintain the flow of ecosystem services (http://www.indiaenvironmentportal.org.in/content/452833/draft-national-forest-policy-2018/). For improved management of the nation's forests, the National Community Forest Management Mission (CFM) and National Board of Forestry (NBF), two national-level organizations, are suggested to be established under the draft policy. In order to carry out afforestation and reforestation activities in degraded forest regions, forest lands available with forest development corporations, and outside forests, public–private involvement models may be established.

1.2.6 Forest Management

The nation's approach to managing its forests has evolved over time, moving from production to protection to conservation-oriented forestry. The first National Forest Policy was formulated in 1894, marking the beginning of scientific forestry planning in India. Because forests are renewable resources, they may be preserved for all time with careful planning and management. Demands generated by population growth change the requirement for forest management and planning throughout time (Kumar et al. 2020). India saw massive tree-felling between the sixteenth and eighteenth centuries in order to make way for army movements, hunt animals, and expand agricultural fields. Deforestation escalated in the mid-nineteenth century due to the East India Company's construction of transportation, railroads, and electrical lines, as well as for ship building.

For the Pegu Yomo forests in Myanmar, Sir Dietrich Brandis developed the first forest management plan in 1860 (Singh 2007). He used the Linear Valuation Survey to assess the rising stock of forests for the first time, and he extensively counted rings to determine the growth rates of teak. When the nation's first forest policy was established in 1864, it may be said that organized forestry began. The current Indian Forest

Service was established in 1966 to oversee forests following a century of its predecessors.

1.2.7 Working Plan

Working plans have been the primary tool used in Indian forest management for about 150 years in order to maintain a sustainable yield (Kumar et al. 2020). The State and Union Territory Governments have been given primary authority for the management and control of forests. Nonetheless, the Central Government sets general policies and keeps an eye on the forest sector. With the exception of wildlife divisions, the territorial organization consists of 197 circles, 788 divisions, 4706 ranges, 11,685 blocks, and 43,884 beats (FSRI 2010). Using GIS and data from remote sensing, state forest departments develop work plans for the purpose of managing forests scientifically.

1.2.8 Forestry in Basic Education

By promoting good and sustainable forest management skills among elementary school students and, through forestry extension, among youth and adults, there is a great chance to enhance the effects of the changes at higher levels of education in forestry. Primary school students in underdeveloped nations are becoming strong proponents of plant and animal conservation. In a relatively short amount of time, school children's influence on parents, neighborhoods, and communities has proven quite effective in changing cultural beliefs. Examples of environmental education programs that provide forestry education and training to elementary school students include those run by India's National Council of Educational Research and Training (NCERT) for primary and secondary schools (Gasperini 2001). First introduced in the National Policy on Education (1986), the program's core tenet is that environmental conservation is a value that should be ingrained in the curriculum at every level of the educational cycle. At the end of 2000, NCERT created a national curriculum framework that outlined the anticipated learning results for students across the board. According to the framework, environmental studies is one of the required subjects for elementary school students' official curricula and is given 20% of class time. As students go through the educational system, environmental education as a distinct topic is phased out in favor of integrated science, which requires students to reflect on observations made about the environment, such as the balance of nature and population impacts on natural resources.

The ICFRE signed Memorandums of Understanding (MoUs) with Navodaya Vidyalaya Samiti (NVS) and Kendriya Vidyalaya Sangathan (KVS) in 2018 to begin the "Prakriti" program. The program's objectives are to increase awareness on forests and environment, and spark interest in preserving a balanced ecosystem among NVS and KVS students, and raise knowledge of forests. Scientists from ICFRE are educating students on current forestry research through lectures and interactive workshops. For practical experiences, field trips and laboratory visits by educators and students from NVS and KVS schools are also planned to ICFRE institutes.

1.3 State of Forests

1.3.1 Forest Cover and Forest Plantation

The worldwide forest cover is 4059 m ha, i.e., 30.8% of land globally. China, Brazil, Canada, the United States, and the Russian Federation are the five nations that account for over 50% of the planet's forests (FRA 2020). The distributions of global forests in major countries and in different climatic regions are depicted in Figs. 1.3 and 1.4, respectively. In Fig. 1.3, top five countries on the basis of forest area and India are depicted. India occupies the tenth position with respect to forest area in the world. The world's biggest geographical zone with forests is found in tropical regions, or the land areas roughly defined by the tropics of Cancer and Capricorn. Tropical forests are commonly assumed to be evergreen rainforests and moist forests; nevertheless, rainforest accounts for only up to 60% of tropical forest. Seasonally dry tropical forests, mangroves, tropical freshwater swamp forests, tropical coniferous forests, and savanna forests are among the many diverse forest types that make up the remaining tropical forests. The boreal climate zone has the world's second largest forest acreage and is situated in the Northern Hemisphere, roughly between 50°N and 70°N latitude. A subarctic climate is found in the boreal region. In Europe and Asia, boreal forests are frequently referred to as

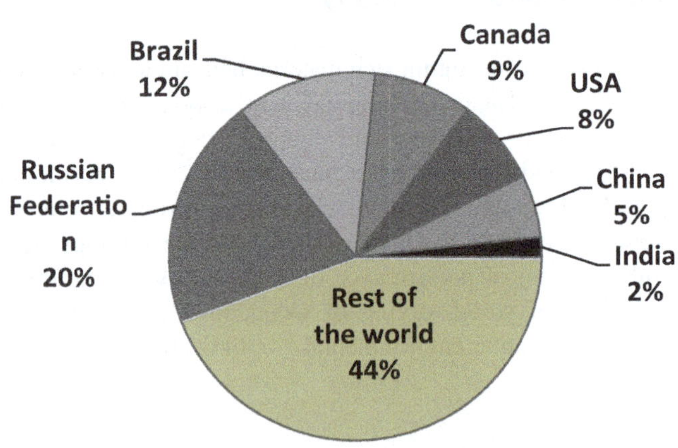

Fig. 1.3 Distribution of global forests in different countries

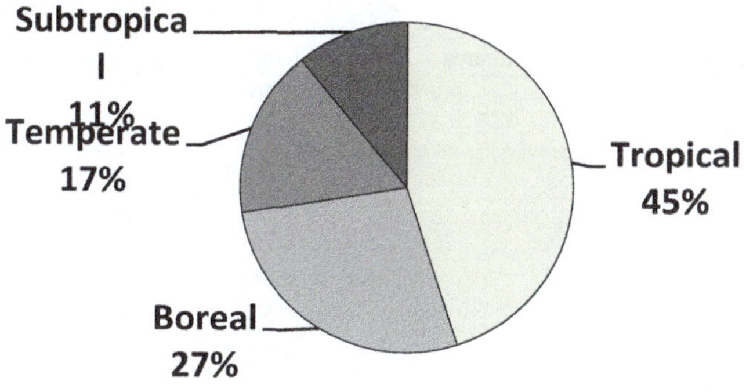

Fig. 1.4 Distribution of global forests in different climatic regions

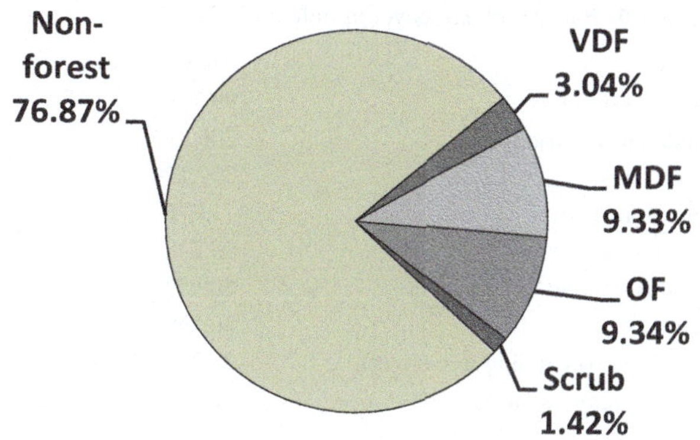

Fig. 1.5 Relative compositions of forest cover in India

taiga. Since these ecosystems are inherently susceptible to global warming, boreal forests are predicted to be particularly vulnerable and sensitive to climate change.

In 1987, the Forest Survey of India (FSI) in collaboration with the National Remote Sensing Agency conducted the first assessment of the nation's genuine forest cover. Since then, the forest cover has been monitored every 2 years using satellite imagery. However, the mapping exercise based on remote sensing has several drawbacks. Some of the constraints are as follows:

- Land cover characteristics on the ground with geometric dimensions <23.5 m are invisible due to the 23.5 m resolution of the Linear Imaging and Self Scanning Sensor (LISS) - III sensor data.
- The presence of agricultural crops like cotton, sugarcane, and others next to forests, as well as weeds like lantana, results in the mixing of spectral signatures.
- Insufficient reflection from newly planted areas and trees with reduced foliage or chlorophyll pigments.
- Atmospheric distortions, etc.

India has 7.14 lakh sq km, or 71.4 m ha, of total forest cover, making up 21.71% of the country's total land area (ISFR 2021). Three canopy density classifications are used to categorize forest cover in the country, i.e., very dense forest (VDF) with over 70% canopy density, moderately dense forest (MDF) with 40–70% canopy density, and open forest (OF) with 10–40% canopy density.

Figure 1.5 depicts the relative proportions of forest cover in different classes, scrub, and non-forest (including tree cover) regions. Any patches of trees that are <1 ha and located outside of a registered forest area (RFA) are considered to be part of the tree cover. Despite their apparent similarity, trees outside forest (TOF) and tree cover are two distinct concepts according to the FSI's evaluation. TOF refers to any trees that grow outside of RFA, regardless of patch size, which can be >1 ha. Consequently, tree cover falls under the TOF subset. A sampling-based approach is utilized to estimate tree cover, while high-resolution satellite data are employed for stratification. TOF is heavily reliant on tree cover. The country's tree cover and TOF are predicted to be 95,748 sq km and 292,907 sq km, respectively. The overall forest and tree cover of the nation is 809,537 sq km, accounting for 24.62% of the country's geographical area (Table 1.6). With 77,493 sq km, Madhya Pradesh has the nation's highest forest cover, which is followed by Arunachal Pradesh (66,431 sq km) and Chhattisgarh (55,717 sq km) (Table 1.7). On the other hand, of all the states and union territories, Mizoram (84.5%) and Lakshadweep (90.3%) have the greatest percentages of forest cover overall.

India is one of the few nations in the world with an observed growth in the amount of forest cover with a continuous upward trend (Fig. 1.6). Acacia, eucalyptus, teak, casuarina, poplar, pine, and bamboo plantations are the most commonly grown in the country. The large-scale plantations of exotics like acacias, eucalyptus, and casuarina and plantation of teak beyond its natural distribution have been undertaken. The plantations have reduced anthropogenic pressure to a considerable extent on the natural forests. However, the plantations in the natural forests have altered the floristic composition and ecological function of these forests.

1.3.2 Is the Target of Forest Cover Achievable?

India has continuously taken several actions to safeguard its natural resources, replenish its forest cover, and progress toward having one-third of its territory covered by forests. Progress has been slower than anticipated due to a number of factors, including its enormous and expanding population, the poverty of a sizable portion of the populace, the growing demand for forest products, a lack of funding, and pressure on land from its many developmental demands. The National Forest Policy (NFP) of 1952 advocated for a 33% forest cover; however, in hilly and mountainous areas, the percentage is 60%. The NFP 1988 placed emphasis on this goal, aiming for

Table 1.6 Forest and tree cover in India in 2021

Class	Area (sq km)	Percentage of geographical area	Change (%) w.r.t data in 2019
Very dense forest	99,779	3.04	0.50
Moderately dense forest	306,890	9.33	−0.51
Open forest	307,120	9.34	0.86
Total forest cover	713,789	21.71	0.22
Tree cover	95,748	2.91	0.76
Total forest and tree cover	809,537	24.62	0.28
Scrub	46,539	1.42	0.52
Non-forest (including tree cover)	2,527,141	76.87	−0.07
Total geographic area	3,287,469	100.00	

Table 1.7 State-wise forest cover in India in 2021 (as per ISFR 2021)

S. no.	State/UT	Total forest cover (sq km)	% of geographical area	Forest cover change (%) w.r.t 2019	Scrub (sq km)	Tree cover (sq km)	Trees outside forest (sq km)
1	Andhra Pradesh	29,784	18.28	2.22	8276	4679	10,224
2	Arunachal Pradesh	66,431	79.33	−0.39	797	1001	8752
3	Assam	28,312	36.09	−0.05	228	1630	9939
4	Bihar	7381	7.84	1.03	236	2341	4891
5	Chhattisgarh	55,717	41.21	0.19	615	5355	18,605
6	Delhi	195	13.15	−0.23	0.38	147	283
7	Goa	2244	60.62	0.31	0.00	244	1268
8	Gujarat	14,926	7.61	0.46	2828	5489	10,580
9	Haryana	1603	3.63	0.06	159	1425	2654
10	Himachal Pradesh	15,443	27.73	0.06	322	675	5474
11	Jharkhand	23,721	29.76	0.47	584	2867	14,306
12	Karnataka	38,730	20.19	0.40	4611	7494	23,676
13	Kerala	21,253	54.70	0.52	30	2820	14,394
14	Madhya Pradesh	77,493	25.14	0.01	5457	8054	20,775
15	Maharashtra	50,798	16.51	0.04	4247	12,108	26,866
16	Manipur	16,598	74.34	−1.18	1215	169	1934
17	Meghalaya	17,046	76.00	−0.43	663	698	2946
18	Mizoram	17,820	84.53	−1.03	1	444	708
19	Nagaland	12,251	73.90	−1.88	824	365	3988
20	Odisha	52,156	33.50	1.04	4924	5004	24,474
21	Punjab	1847	3.67	−0.11	34	1138	2197
22	Rajasthan	16,655	4.87	0.15	4809	8733	12,828
23	Sikkim	3341	47.08	−0.03	296	39	1326
24	Tamil Nadu	26,419	20.31	0.21	758	4424	13,312
25	Telangana	21,214	18.93	3.07	2911	2848	5366
26	Tripura	7722	73.64	−0.05	33	228	2517
27	Uttar Pradesh	14,818	6.15	0.08	563	7421	13,096
28	Uttarakhand	24,305	45.44	0.01	392	1001	8521
29	West Bengal	16,832	18.96	−0.41	156	2349	12,069
30	A & N Islands	6744	81.75	0.01	1	23	559
31	Chandigarh	23	20.07	3.86	0.38	15	29
32	Dadra & Nagar Haveli and Daman & Diu	228	37.83	0.04	4.85	32	100
33	Jammu & Kashmir	21,387	39.15	0.14	284	3511	11,722
34	Ladakh	2272	1.35	0.80	279	954	2426
35	Lakshadweep	27	90.33	0.00	0.00	0.05	27
36	Puducherry	53	10.88	1.70	0.00	23	75
	Total	**713,789**	**21.71**	**0.22**	**46,539**	**95,748**	**292,907**

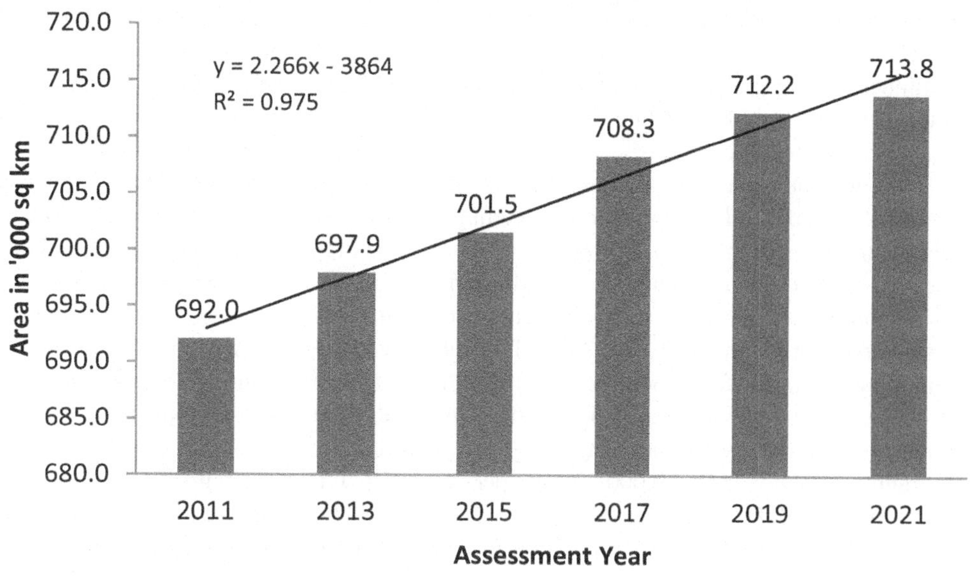

Fig. 1.6 Trend in forest cover in India during 2011–2021

Table 1.8 Number of Indian states/union territories meeting the criteria of forest cover

Sl. no.	Classification on the basis of target achievement	2011	2021
1	States/UTs meeting the criteria of 33% forest cover	15	17
2	States/UTs can meet the criteria of 33% forest cover provided entire wasteland is converted to green cover	3	2
3	States/UTs not meeting the criteria of 33% forest cover	17	17
	Total	35[a]	36[b]

[a]Number represents Ladakh as a part of Jammu and Kashmir State and Dadra & Nagar Haveli as a separate UT from Daman & Diu
[b]Number representing creation of Telangana and Ladakh

33% of the total area to be covered by forests or trees and 67% of the area to be covered in hills and mountains.

Joshi et al. (2010) evaluated the possibility of reaching the mandated policy objectives given the current state of land use practices by analyzing changes in the state's forest cover. Only a few Indian states adhered to the mandated policy, whereas the majority of states did not. Three states (Jharkhand, Odisha, and Rajasthan) have the potential to attain the 33% target as long as the transition to tree cover is not hampered and appropriate resources are accessible on the state level. The Planning Commission has placed an emphasis on adding trees outside of forests and other natural habitats in the XI Five-Year Plan (2007–2012) to increase forest cover. Of India's entire wasteland (17.45% of the country's geographical area), only 70% can be put to better use by boosting the amount of greenery (Mathur and Sachdeva 2003). India's green cover may further be increased by planting trees in farmlands (5 m ha) and forest areas (5 m ha) as part of the Green India Mission's 2010–2020 program. The situation regarding reaching the aim of 33% forest cover during the past 10 years has remained mostly unchanged (Table 1.8). During this time, only Odisha has achieved the target.

Degradation of forests in India is mostly caused by overgrazing and firewood collecting. It is necessary to establish a framework for collaboration at the local, regional, and national levels between the forestry industry and other sectors, particularly the animal husbandry sector for the management of grazing and fodder production and the energy sector for the management of rural energy requirements. The Participatory/Joint Forest Management (JFM) initiative is an excellent strategy for reducing forest degradation. JFM policies that are well designed can help to preserve and enhance community livelihoods, prevent deforestation, and restore forest cover and density.

Forestry research must shift its focus from conventional silviculture to present requirements in community-based forestry, environmental ethics, ecological economics, urban forestry, and other fields. Aside from afforestation and replanting in all denuded and degraded forest areas, promotion of trees outside forests and urban greens should be undertaken on a mission manner in order to achieve the aim of Indian forest and tree cover.

1.3.3 Production of Forest Resources

With an annual yield of around ten million m^3, India's natural forests continued to be the primary source of lumber until the 1970s. In 1998, the Indian Supreme Court issued an order prohibiting the cutting of trees from natural forests without the Central Government's prior permission. The yearly yield of timber coming from state forests decreased to around three million m^3. The majority of this output might be due to road expansion-related tree cutting.

India has the capacity to produce 6167.50 million m^3 of wood (ISFR 2021). The prohibition on cutting down forest trees contributes to the growth of trees outside of forests for the production of industrial timber. An estimated 58.5 million m^3 of commercial wood are produced nationwide from trees that are not part of forests (ICFRE 2016). India liberalized the importation of wood and wood products in 1996 in response to the growing focus on forest protection. Since then, the import of wood and wood products has grown significantly, averaging around INR 20,000 crores per year till FY 2020–2021 (IPIRTI 2021).

Wood is utilized for fuels in addition to being used for lumber. However, the majority of its manufacturing, transportation, and consumption have stayed inside the informal economy. Determining the amount of fuel wood produced in forests is an extremely challenging task. However, it has been projected to reach 52 million m^3 per year (FSI 2009). Forests in India fulfill about 40% of the country's energy demands and around 30% of the cattle population's feed needs. It is estimated that around 270 Mt of fuel wood, 280 Mt of fodder, and innumerable NTFPs are harvested from the forest each year (Singhal et al. 2003). In India, over 800 different types of non-timber forest products (NTFPs) are collected, but reliable information about their yearly output is lacking. Tendu leaf, bamboo, resins, and other NTFPs are among the most common. According to Bag et al. (2010), the states in Central India account for over 70% of all NTFP collection in India. According to Shivaprasad and Chandrashekar (2014), the gross value of non-timber products and services derived from India's tropical deciduous forests ranged from \$219 to \$357 per hectare annually.

1.3.4 Productivity of Forest

The highest timber yield that a forest can produce in a specific amount of time is a key indicator of site quality, which determines how productive a forest may be (Katwal et al. 2003). The quality of the site influences not only the yield of seeds but also the production of wood and other non-timber forest products. It is well understood that trees grown in fertile soil blossom more than trees grown in poor soil. Productivity for a particular species or forest type depends on habitat elements such as soil, nutrient condition, rainfall, and silvicultural management techniques. Forest productivity can be quantified in terms of mean annual increment (MAI). Specific elements that affect productivity include the amount of trash removed or retained, the condition of the ground vegetation, the density of the stock that will support future growth, and protection from substances that inhibit growth. The estimated productivity of these forests, using Paterson's index and taking into account rainfall, soil conditions, and length of growth period, would vary from 1.35 m^3/ha/year in the dry regions of India to 7.66 m^3/ha/year in the moist and wet regions, while rest forests of the country produce between 3 and 4 m^3/ha/year (MoEF 1999).

Aside from site quality, the genetic makeup of seed or planting material has a significant influence in enhancing forest production, particularly in planted forests. In India, tree improvement initiatives began in the 1960s. The Indian Council of Forestry Research and Education (ICFRE), Dehradun, was formed in 1987 as an apex organization under the aegis of the Ministry of Environment, Forest and Climate Change (MoEFCC), with one of its objectives being to boost forest production. To boost productivity, the following actions are typically taken:

- Selection of suitable species for the site
- Selection of superior trees, their evaluation, and release as varieties or clones
- Creating seed production sites, seedling seed orchards, and clonal seed orchards will enable the production of high-quality seeds
- Development of macro- and micro-propagation technique for mass multiplication

Aside from ICFRE and State Forest Department (SFD) research institutes, commercial sectors, particularly ITC Bhadrachalam, J.K. Paper Mills, Ballarpur Industries Ltd., Mysore Paper Mills, and WIMCO Rudrapur, have played an important role in tree development initiatives, particularly in promoting clonal forestry in India. Annual increment of growth of Indian forests is about 0.8 m^3/ha/year, which is less than the global average of 2.0 m^3/ha/year (ICFRE 2016). Despite commendable work carried out on tree improvement of large number of tree species, unimproved seed of various tree species is still being used for planting programs. This is mostly due to a scarcity of sufficient quantities of high-quality seeds and vegetative propagules. Refinement/upscaling of protocols/process developed is essential to reap the benefit of tree improvement works. Conventional tree improvement tactics must be combined with high-end molecular breeding technologies such as genomic selection and transgenics. A holistic tree improvement program on priority species should be implemented, without discontinuity, by involving multi-institute/organizations at national level as All India Coordinated Research Programmes (AICRPs).

1.4 Conservation of Forests, Wildlife, and Biodiversity

The nation boasts an immense range of ecological diversity, spanning from sea level to the world's highest mountain ranges; hot and dry conditions in the northwest to cold and dry conditions in the trans-Himalayan region; tropical wet evergreen forests in North East India and the Western Ghats; Sundarbans mangroves; and freshwater aquatic to marine ecosystems (Sharma and Singh 2000). Figure 1.7 illustrates the species diversity in India compared to the world. India is home to 47,513 plant species, including viruses, bacteria, algae, fungus, and lichens (Singh and Dash 2014), comprising 10.2% of the world's flora. Nonetheless, of the 15.66 lakh species worldwide, 101,167 species are known to exist in India, accounting for as much as 6.5% of all animal species (Venkataraman et al. 2020). Ten bio-geographical zones have been identified by the Wildlife Institute of India based on the abundance of plants and animals throughout the nation. Biogeography is the study of the distribution of species, organisms, and ecosystems across geographical territory and course of geological time. According to Kumar et al. (2012), India is ranked tenth globally and fourth in Asia among the top 17 mega-diverse nations. The following are the ten biogeographic zones: (1) Trans-Himalayan, (2) Himalayan, (3) Desert, (4) Semi-arid, (5) Western Ghats, (6) Deccan plateau, (7) Gangetic plain, (8) Northeast, (9) Coastal, and (10) Islands. There are 27 biogeographic provinces that comprise the additional divisions within each zone. Among these ten zones, the northeastern zone may be most abundant in terms of varied groups, species, and indigenous flora and fauna.

India has implemented the strict laws and regulations needed for the efficient protection of forests, animals, and biodiversity. The following section lists various legislations' main features along with how they affect Indian forestry.

1.4.1 Wildlife Protection Act 1972

In order to safeguard the nation's ecological and environmental security, the Indian Government passed this Act in 1972 to preserve wild animals, birds, and plants as well as items related, auxiliary, or incidental to them. The Act's primary goals are (1) to outlaw hunting; (2) to protect and maintain animal habitats; (3) creation of protected areas (PAs); (4) supervision and oversight of the trade in wildlife parts and commodities; and (5) zoo management.

The Act has been amended many times since then. Some of the important amendments are:

- Creation of a Central Zoo Authority 1991
- Constitution of National Board of Wildlife and Restructuring of State Wildlife Board 2002
- Creation of National Tiger Conservation Authority (NTCA) and Wildlife Crime Control Bureau 2006
- Creation of Rhino Conservation Authority 2017

1.4.2 Forest Conservation Act (FCA) 1980

The Government of India adopted this Act to regulate the conversion of forest land for non-forestry activities such as road and dam building, mining, transmission lines, defense objectives, and so on. After reviewing the feasibility, the Central Government authorizes the project if it is deemed acceptable and imposes specific requirements on the user agencies, such as compensatory afforestation, watershed treatment, safety zone development, and so on. An equivalent

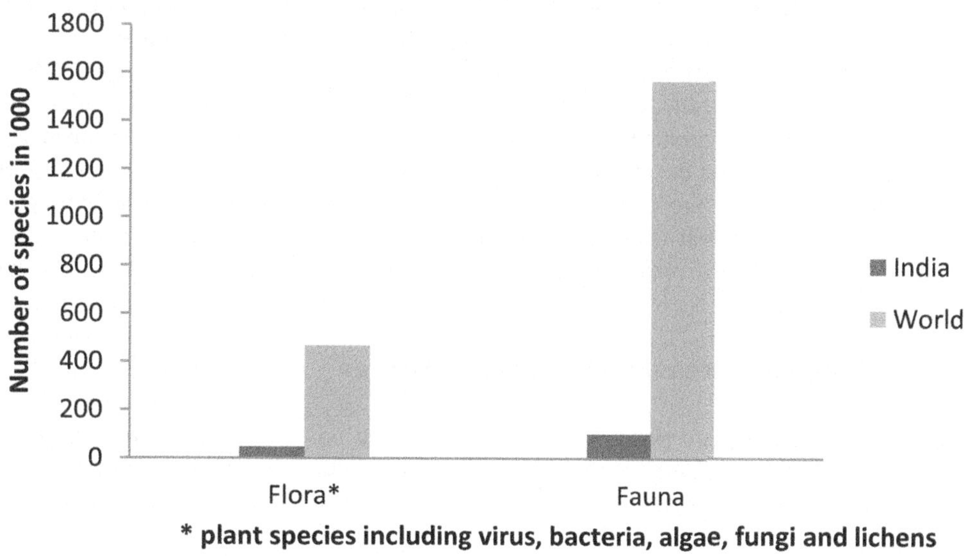

Fig. 1.7 Diversity of flora and fauna species in India and throughout the world

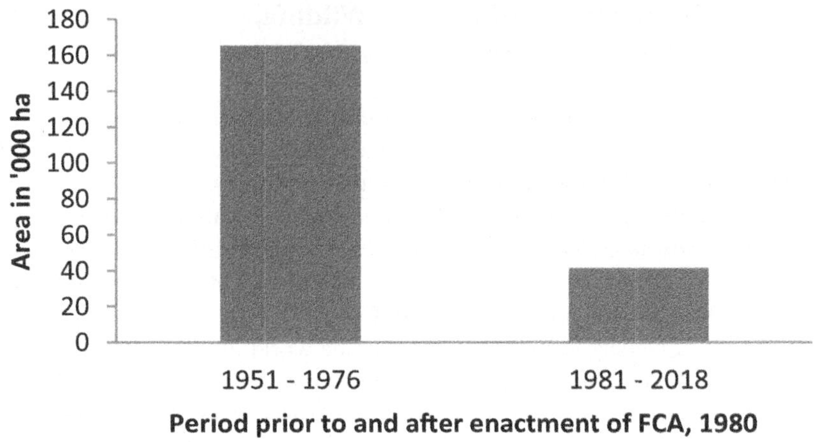

Fig. 1.8 Average annual diversion of forest land

amount of revenue land or non-forest land, or twice the area of degraded forest land, must be utilized for compensating afforestation when a portion of forest land is diverted for non-forestry uses. This property that has been forested is designated as Reserved or Protected Forests (RF/PF). Prior to the implementation of the FCA 1980, the conversion of forest land for non-forestry activities was 4.13 m ha during a 25-year period (1951–1976), and after implementation it amounted to 1.53 m ha over nearly four decades (1981–2018) (eGreenWatch 2020). Following enactment, Arunachal Pradesh (0.34 m ha), Madhya Pradesh (0.26 m ha), and Karnataka (0.11 m ha) are the top three states with the highest areas of forest land diversion. In contrast, Lakshadweep, Nagaland, and Pondicherry have no conversion of forest land. Prior to and during the Act, the average yearly diversion of forest area was 165,200 ha over 25 years and 41,351 ha over 37 years, respectively (Fig. 1.8), illustrating how the Act's enforcement lowered the pace of diversion by up to 75%.

The complete process of tracking plans for the diversion of forest land has been automated. This includes the ability to submit new proposals online, modify and update proposal information, and view the status of proposals at each step of the workflow. A web-based application known as PARIVESH (Pro Active and Responsive facilitation by Interactive and Virtuous Environmental Single-window Hub) has been created to facilitate the online submission and tracking of proposals made by proponents seeking clearance from district, state, and federal authorities for the environment, forest, wildlife, and conservation zones.

1.4.3 Protected Areas and Biosphere Reserves

Protected areas (PAs) are places where there is little or no human habitation or resource exploitation. Article 8 of the Convention on Biological Diversity (CBD) and the 2004 adoption of the Program of Work on Protected Areas (PoWPA) both serve as international declarations of the importance of PAs from national governments. During the CBD Conference of the Parties in 2010 in Japan, the Aichi Biodiversity goals were adopted. Aiming to preserve 10% of coastal and marine regions and at least 17% of terrestrial and inland water areas as PAs by 2020 is one of the 20 objectives. PAs currently encompass 16.64% of the world's terrestrial and inland waterways, as well as 7.74% of coastal and ocean regions (UNEP-WCMC and IUCN 2020). With 4.85 million km^2 (24%) of area protected, Latin America and the Caribbean have the most comprehensive coverage at the regional level. The Sustainable Development Goals (SDGs) and the 2030 Agenda for Sustainable Development were endorsed by the United Nations in 2015. Aichi Biodiversity Target 11's worldwide protected areas play a significant role in accomplishing SDGs 14 and 15.

Table 1.9 lays out India's network of 998 PAs, which as of January 2023 covered just 5.28% of the country's total land area. PAs are divided into four groups: National Park, Wildlife Sanctuary, Conservation Reserve, and Community Reserve. These groups are distinguished by the degree of protection and utilization of the resources found within each PA. Founded in 1936 to save the Bengal Tiger, which is currently endangered, Jim Corbett National Park is the first national park in India. Nonetheless, the Wildlife Protection Act of 1972 was amended in 2003, leading to the establishment of Community Reserves and Conservation Reserves. A community reserve may be established by the state government on any private or community land that is not included in a National Park, Sanctuary, or Conservation Reserve, provided that the community or an individual has committed to protect wildlife and its environment. The state government may designate any land it owns as a conservation reserve to preserve landscapes, seascapes, flora and fauna, and their habitat after consulting the local communities. This declaration applies especially to areas bordering national parks and sanctuaries and areas connecting protected areas. In addition to allowing local communities to be actively involved in

Table 1.9 Number and total area of protected areas in India

Sl. no.	Category	Number	Area (km^2)	Coverage (%) of country
1	National parks	106	44,403	1.35
2	Wildlife sanctuaries	567	122,565	3.73
3	Conservation reserves	105	5207	0.16
4	Community reserves	220	1455	0.04
	Total	998	173,630	5.25

Source: National Wildlife Database, Wildlife Institute of India (January 2023) (*wiienvis.nic.in*)

planning and conservation, conservation reserves offer a legally permissive environment for the creation of buffer areas and corridors for animal management. Tiruvidaimarudur Conservation Reserve became operational in 2005 as the country's first conservation reserve.

In addition to the Pas' network system that was already in place, biosphere reserves (BRs) were first announced in India in 1986 with the goal of providing a broader platform for the protection of the full spectrum of biodiversity and their ecological bases. It attempts to strike a balance between the preservation of the natural environment, concomitant cultural values, and economic and social growth. India has notified 18 BRs thus far.

1.4.4 Endangered Wildlife Species-Specific Conservation Efforts

India is making special efforts to save and preserve several species, including vultures, hangul, Ganges dolphin, tiger, elephant, snow leopard, and Indian (one horn) rhino. A few important projects are briefly detailed below.

1.4.4.1 Tiger Conservation

The tiger (*Panthera tigris*) plays a crucial role in maintaining and regulating ecological processes as one of the top predators. To sustain long-term healthy populations, tigers require huge undisturbed stretches of habitat with plenty of prey, thus serving as an umbrella species for the bulk of eco-regions on the Indian subcontinent. The main goal of "Project Tiger" is to conserve the environment as a whole; in addition to tigers, all other wild creatures in the project regions must also be multiplied. "Project Tiger" was begun in 1973, with nine Tiger Reserves (covering approximately 18,278 km^2). Among the 13 Tiger Range nations in the globe, India has the most tigers and their source areas. Since 2006, the National Tiger Conservation Authority (NTCA) has been in place to improve tiger conservation. It now encompasses 53 Tiger Reserves, or over 75,797 sq km, representing roughly 2.31% of India's total land area (NTCA 2022). The national tiger assessment conducted in India is the world's largest study of biodiversity. Since 2006, tiger numbers in India have been monitored every 4 years. The fourth cycle of assessment was conducted in 2018 and 2019, utilizing cutting-edge technologies such as automated segregation of camera trap photographs of species through the use of artificial intelligence and neural network models, as well as the electronic collection of primary field data through mobile phone applications. According to Jhala et al. (2019), the total estimated tiger population in India is 2967 (range 2603–3346). The yearly funding allotment for Project Tiger used to range from Rs. 300 to 350 crores.

1.4.4.2 Project Elephant (*Elephas maximus indicus*)

This was started in 1991–1992 to safeguard elephants, their habitat, and their corridors; it also looked after the well-being of elephants kept in captivity and addressed issues of conflict between humans and animals. Elephants are found in sixteen states. Its population was estimated to be 29,964 in the latest census performed in 2017 (PIB 2022). So far 33 Elephant Reserves have been notified (WIIENVIS 2022). Annual allocation of fund for Project Elephant varies from Rs. 30 to 35 crores.

1.4.4.3 Snow Leopard (*Panthera uncia*)

It is found across the majority of Central Asia's and the Himalaya's high altitudes (Jackson 1996). Anon. (2008) estimates that the approximate number of snow leopards worldwide is 7400, with an estimated 400–700 individuals living in India's 5 Himalayan states. Given the availability of robust density estimates, a revised estimate of the number of snow leopards in India utilizing better technique is probably 516 (range 238–1039) (Bhatnagar et al. 2016).

1.4.5 Are the Conservation Efforts by India Sufficient?

Researchers at Yale and Columbia universities in New York have found that India is rated the last out of 180 nations in the 2022 Environmental Performance Index (EPI) (https://epi.yale.edu/epi-results/2022/component/epi). The EPI report, which ranks 180 nations on 40 performance indicators across 11 issue areas spanning environmental health and ecosystem vitality, has now emerged as the leading metrics framework for international environmental policy analysis. Over the past 10 years, India has been struggling to perform effectively on a number of environmental challenges; as a result, it has only received the lowest EPI score of 18.90. India ranks 179th in the world for habitat and biodiversity, mostly due to its inability to fully use the conservation potential of its protected areas, particularly with regard to maritime habitats. The fact that India ranks 165th in the world for reducing climate change may be the most significant. Resilient forests are necessary to guarantee the sustainability of REDD-Plus (resolution to reduce emissions from deforestation and forest degradation) initiatives since climate change is a continuous phenomenon that directly affects the existence and health of species and ecosystems.

1.4.5.1 REDD-Plus

The resolution to reduce emissions from deforestation and forest degradation (REDD) in poor countries was approved by the Conference of the Parties (COP) to the United Nations Framework Convention on Climate Change (UNFCCC) at its 13th session in Bali in December 2007. REDD-Plus activities, which include lowering emissions from deforestation and forest degradation, conserving forest carbon stocks, managing forests sustainably, and increasing forest carbon stocks, are ways that developing country Parties can contribute to greenhouse gas (GHG) mitigation efforts in the forest sector. This decision was made by the UNFCCC COP during its 16th session. For REDD-Plus to be implemented, the following four essential components must be created:

- National Strategy or Action Plan
- Forest Reference Level/Forest Reference Emission Level
- Safeguards Information System
- National Forest Monitoring System

Meeting the objectives set forth in the 2015 Paris Climate Change Agreement requires continuous reductions in emissions of all greenhouse gases, and the 2022 EPI indicates that just two nations (Denmark and the United Kingdom) are decarbonizing swiftly enough to reach these climate change targets. Policies and programs may be designed to address poverty, biodiversity loss, and climate change all at the same time. India must demonstrate enduring dedication and well-designed initiatives to safeguard public health, preserve natural resources, and lower greenhouse gas (GHG) emissions if it hopes to rank among the top-scoring nations.

1.5 Disturbance

Numerous disruptions can negatively impact a forest's health and vitality as well as its capacity to offer a complete spectrum of commodities and ecosystem services. Disturbances alter the structure, function, and composition of forests. Abiotic phenomena like fires, floods, ice storms, and destructive wind events are examples of natural disturbances, as are biotic events like bug outbreaks and invading viruses. Here we address only two disruptions.

1.5.1 Invasive Species

There is rising concern across the world about the possible implications of invasive species on native species, animal habitats, and services provided by ecosystems. However, there has not been much focus on invasive plant species in Indian Protected Areas. Twenty-four plant species (Table 1.10) are reported to appear in the PAs of India (Hiremath and Sundaram 2013), based on the list of 100 invasive species worldwide (Lowe et al. 2000). These are a subset of the 225 foreign plant species that Khuroo et al. (2012) identified as invasive in India. An additional 134 species were identified by them as naturalized, although they still had the potential to spread quickly.

Like other parts of the world, invading species have found their way to India through a number of channels. The majority of foreign plant species that are known to be invasive in India's PAs were initially brought there as garden ornamentals. According to Hiremath and Sundaram (2013), deliberate introduction of foreign species was also done to fulfill fuel-wood requirements, stop the spread of the desert, and cultivate for commercial purposes.

India still lacks particular laws to control and screen the import of potentially invasive species, despite the increased global awareness of alien species invasions. A coordinated study initiative on "Invasive alien plants in Himalayan states:

Table 1.10 Invasive species found in protected areas of India

Sl. no.	Species	Sl. no.	Species	Sl. no.	Species
1	*Acacia mearnsii*	9	*Cytisus scoparius*	17	*Mikania micrantha*
2	*Ageratina* spp.	10	*Hyptis suaveolens*	18	*Mimosa invisa*
3	*Ageratum conyzoides*	11	*Imperata cylindrica*	19	*Opuntia stricta*
4	*Ageratum houstonianum*	12	*Ipomoea carnea*	20	*Parthenium hysterophorus*
5	*Arundo donax*	13	*Kappaphycus alvarezii*	21	*Polygonum polystachyum*
6	*Chromolaena odorata*	14	*Lantana camara*	22	*Prosopis juliflora*
7	*Clidemia hirta*	15	*Leucaena leucocephala*	23	*Sphagneticola trilobata*
8	*Coffea* sp.	16	*Merremia peltata*	24	*Ulex europaeus*

status, ecological impact and management," encompassing all 12 hill states of the Indian Himalayas, was launched by the MoEFCC in 2017 after realizing the gravity of the problem.

1.5.2 Forest Fire

Forest fires are not abnormal. Low-intensity fires prevent dead trees, senescent leaves and twigs, tree needles, dead grass, and dense undergrowth from naturally accumulating on the forest floor each year. Fire has therefore influenced the distribution of species and the patterns of flora in the woods. Not every forest experiences fire in the same way. However, depending on the plant type and climate, the same fire that is good for one ecosystem might be terrible for another. Tropical rain forests are least impacted by fires because they are engulfed in fog and constantly soaked by mists and rainfall. Because of the considerable precipitation, moisture, and very high humidity in temperate deciduous forests, fire occurs very seldom and causes minimal damage. Whereas, coniferous forests and evergreen forests with broad leaves in dry regions are often more prone to fire.

In India, forest fires are a major source of damage and forest loss (Satendra and Kaushik 2014). The frequency of forest fires and the resulting damage they cause are rising daily in tandem with the growth in population. The forest ecology is being affected by the fire in a number of ways. In addition to directly harming the forest trees, the fire has a negative impact on soil erosion, wildlife, microclimate, and forest regeneration. Approximately 2.56 lakh hectares of land in India have been affected by forest fires in 2019. Under the Forest Fires Prevention and Management Scheme, states earned around Rs. 45–50 billion per year from 2017 to 2019 (PTI 2020).

1.6 Conclusion

With 8.1 billion people on the earth and a total forest area of 4.06 billion hectares, there are 0.50 ha of forest for every person. Nonetheless, there is just 0.05 ha of usable forest land per person in India as of now. Deforestation has resulted in the loss of forest cover globally, but the pace of forest loss has significantly decreased. Due to a decline in deforestation in certain countries and increases in the amount of forest in others as a consequence of afforestation and natural forest expansion, the rate of net loss of forest declined significantly between 1990 and 2020 (FRA 2020). The main goal of forest management is the sustainable harvesting of timber and other non-timber forest products. Additionally, forests are also allocated for the preservation of biodiversity, the preservation of the soil and water, and social services including tourism, recreation, education, and the preservation of religious and cultural institutions.

India's forests have been progressively disappearing due to accelerated urbanization and the necessity to build homes and workplaces for the growing population. Among the issues include ongoing deforestation and degradation, rising demand for forest products, and legislative and policy barriers to accomplishing the objectives of forest policies. Therefore, it is necessary to harmonize the numerous contradictory policy announcements. Communities that depend on forests must also show a great deal of commitment and collaboration. Collaboration and resource sharing across academic and research institutions, communities, non-governmental organizations, and other government agencies are essential.

Lessons Learnt

1. Significance of forests from an ethical, ecological, economic, and social perspective
2. Classification of forests in India
3. Introduction to major branches of forestry science
4. State of forests in India with respect to global scenario
5. Production and productivity trend of forest sector in India
6. Legislative frameworks for conservation of flora and fauna in Indian forests
7. Critical impediments to Indian forest management

Key Questions

1. Describe the major roles that Indian forests serve to the country in achieving the SDGs.
2. What is the significance of forestry to be included as a subject in the curriculum of basic education?
3. Elaborate the trends of Indian forests in light of forest cover, plantation, production, and productivity. What measures should be taken to meet the target of 33% forest cover?
4. Enumerate the major conservation approaches and the challenges faced in Indian forestry sector.
5. Describe how forest fires and invasive species are posing great threats to the native biodiversity in Indian forests.

References

Alongi DM (2002) Present state and future of the world's mangrove forests. Environ Conserv 29:331–349

Anon. (2008) The project snow Leopard. Ministry of Environment and Forests, Government of India, New Delhi

Bag H, Ojha N, Rath B (2010) NTFP policy regime after FRA: a study in select states of India. Regional Centre for Development Corporation, Bhubaneswar

Bhatnagar YV, Mathur VB, Sathyakumar S, Ghoshal A, Sharma RK, Bijoor A, Raghunath R, Timbadia R, Lal P (2016) South Asia: India. In: Nyhus PJ, McCarthy T, Mallon D (eds) Snow Leopards.

Biodiversity of the world: conservation from genes to landscapes. Academic Press, Elsevier, Amsterdam
Bhattacharya A (2013) Rabindranath Tagore and Mahatma Gandhi: their thoughts on education from a postcolonial perspective. Postcolon Dir Educ 2(1):100–144
Champion HG, Seth SK (1968) A revised survey of the forest types of India. Manager of Publications, Government of India, New Delhi
eGreenWatch (2020) FCA projects. eGreenWatch. https://egreenwatch.nic.in/FCAProjects. Accessed 24 Apr 2021
FAO (2006) Global forest resources assessment 2005. FAO, Rome, Italy
Forbes K, Broadhead J (2008) The role of coastal forests in the mitigation of Tsunami impacts. FAO, Bangkok
FRA (2020) Forest resources assessment 2020. FAO, Rome
FSI (2009) Country report: India for FAO global forest resource assessment 2010. Forest Survey of India, Ministry of Environment and Forests, Government of India, Dehradun
FSRI (2010) Forest sector report India 2010. Indian Council of Forestry Research and Education, Dehradun
Gasperini L (2001) Forestry education for the future: expanding the scope. In: Moroccos R (ed) Expert consultation on forestry education. FAO, Rome. http://www.fao.org/3/y2993e/y2993e18.htm. Accessed 7 Jun 2020
Gorain S, Malakar A (2020) Status, trends and valuation methodologies of forestry sector in India. Curr J Appl Sci Technol 39(39):49–56. https://doi.org/10.9734/CJAST/2020/v39i3931103
Gorain S, Malakar A, Chanda S (2021) An analysis of carbon market and carbon credits in India. Asian J Agric Ext Econ Sociol 39(2): 40–49. https://doi.org/10.9734/ajaees/2021/v39i230528
Hiremath AJ, Sundaram B (2013) Invasive plant species in Indian protected areas: conserving biodiversity in cultural landscapes. In: Foxcroft LC, Pyšek P, Richardson DM, Genovesi P (eds) Plant invasions in protected areas: patterns, problems and challenges, Invading nature - Springer series in invasion ecology, vol 7. Springer
ICFRE (2013) Forest types of India: revised. Indian Council of Forestry Research and Education, Dehradun
ICFRE (2016) Advances in tree improvement and forest genetic resources conservation and management. In: Kumar A, Ashwani TS, Rathore K, Palanisamy K, Viswanath S (eds) ICFRE state of knowledge series III. Indian Council of Forestry Research and Education, Dehradun
IPIRTI (2021) IPIRTI bulletin volume III July. Indian Plywood Industries Research & Training Institute, Bengaluru
ISFR (2021) India state of forest report 2021. Forest Survey of India, Ministry of Environment, Forests and Climate Change, Dehradun
Jackson RM (1996) Home range, movements and habitat use of snow leopard (*Uncia uncia*) in Nepal. Ph.D. thesis, University of London, UK
Jhala YV, Qureshi Q, Nayak AK (2019) Status of tigers, co-predators and prey in India 2018. Summary report. National Tiger Conservation Authority, Government of India, New Delhi & Wildlife Institute of India, Dehradun
Joshi AK, Pant P, Kumar P, Giriraj A, Joshi PK (2010) National forest policy in India: critique of targets and implementation. Small-scale For 10:83–96
Kanwal KS (2014) Non timber forest products (NTFP) as a tool for sustainable socio-economic development of community. The Arunachal Times 26(58)
Katwal RPS, Srivastava PK, Kumar S, Jeeva V (2003) State of forest genetic resources conservation and management in India. In: Forest genetic resources working papers, working paper FGR/65E. Forest Resources Development Service, Forest Resources Division. FAO, Rome, 36p
Khuroo AA, Reshi ZA, Malik AH, Weber E, Rashid I, Dar GH (2012) Alien flora of India: taxonomic composition, invasion status and biogeographic affiliations. Biol Invasions 14:99–113
Kumar M, Singh MP, Singh H, Dhakate PM, Ravindranath NH (2020) Forest working plan for the sustainable management of forest and biodiversity in India. J Sustain For 39(1):1–22
Kumar NK, Raghunath TP, Jayaraj RSC, Anandalakshmi R, Warrier RR (2012) State of forest genetic resources in India: a country report. Institute of Forest Genetics and Tree Breeding, Indian Council of Forestry Research and Education, Coimbatore
Lebedys A, Li Y (2014) Contribution of the forestry sector to national economies, 1990–2011. In: Forest finance working paper FSFM/ACC/09. FAO, Rome
Li Y, Mei B, Linhares-Juvenal T (2019) The economic contribution of the world's forest sector. Forest Policy Econ 100:236–253. https://doi.org/10.1016/j.forpol.2019.01.004
Lippe RS, Schweinle J, Cui S, Gurbuzer Y, Katajamäki W, Villarreal-Fuentes M, Walter S (2022) Contribution of the forest sector to total employment in national economies - estimating the number of people employed in the forest sector. FAO and ILO, Rome and Geneva. https://doi.org/10.4060/cc2438en
Lowe S, Browne M, Boudielas S (2000) 100 of the world's worst invasive alien species. A selection from the Global Invasive Species Database. The Invasive Species Specialist Group (ISSG) of the World Conservation Union (IUCN), Gland
Mathur AS, Sachdeva AR (2003) Towards an economic approach to sustainable development. Planning Commission, GoI, New Delhi
MoEF (1999) State of environment -state of the environment report-India, vol I. Ministry of Environment and Forests, New Delhi
Navale MR, Bhardwaj DR, Bishist R, Thakur CL, Sharma S, Sharma P, Kumar D, Probo M (2022) Seasonal variations in the nutritive value of fifteen multipurpose fodder tree species: A case study of north-western Himalayan mid-hills. PLoS One 17(10):e0276689. https://doi.org/10.1371/journal.pone.0276689
NTCA (2022) Tiger reserve list. National Tiger Conservation Authority. https://ntca.gov.in/tiger-reserves/#tiger-reserves-2. Accessed 23 Jan 2022
Perlin J (1991) A forest journey. The role of wood in the development of civilization. Harvard University Press, Cambridge. Originally published by W.W. Norton, New York 1989
Peters CM, Gentry AH, Mendelsohn RO (1989) Valuation of an Amazonian rainforest. Nature 339:655–656
PIB (2022) World Elephant Day – 2022 celebrated at Periyar. Press Information Bureau, Government of India, on August 12, 2022. https://pib.gov.in/Pressreleaseshare.aspx?PRID=1851260. Accessed 24 Jan 2023
PIB (2023) Involvement of tribal communities in protection, conservation, and management of Forest Development. Press Information Bureau, Government of India, December 18, 2023. https://pib.gov.in/PressReleseDetailm.aspx?PRID=1987759#:~:text=As%20per%20the%20Census%202011,referred%20as%20forest%20fringe%20villages. Accessed 19 Dec 2023
PTI (2020) Over 2.56 lakh hectare land in India affected by forest fire in 2019. The Week, on September 24, 2020. https://www.theweek.in/news/sci-tech/2020/09/24/Over-2-56-lakh-hectare-land-in-India-affected-by-forest-fire-in-2019-Govt.html. Accessed 24 Apr 2021
Satendra, Kaushik AD (2014) Forest fire disaster management. National Institute of Disaster Management, Ministry of Home Affairs, New Delhi
Sharma JR, Singh DK (2000) Status of plant diversity in India: an overview. In: Roy PS, Singh S, Toxopeus AG (eds) Biodiversity and environment. Remote sensing and geographic information system perspectives. Indian Institute of Remote Sensing, National Remote Sensing Agency, Dehradun
Shivaprasad TM, Chandrashekar HM (2014) Impact of new forest policies on collection and marketing of minor forest produce in Karnataka, with special reference to LAMPS. IOSR J Humanit Soc Sci 19(11):9–14

Singh KD (2007) Updating forest planning system in India. Indian Forester 133(1a):10–20

Singh P, Dash SS (2014) Plant discoveries 2013 – new genera, species and new records. Botanical Survey of India, Kolkata

Singhal RM, Kumar S, Jeeva V (2003) Forests and forestry research in India. Trop Ecol 44(1):55–61

Sunderland T, Powell B, Ickowitz A, Foli S, Pinedo-Vasquez M, Nasi R, Padoch C (2013) Food security and nutrition: the role of forests. Discussion paper. CIFOR, Bogor

UNEP-WCMC, IUCN (2020) Protected Planet Report 2020. UNEP-WCMC and IUCN, Cambridge, UK and Gland, Switzerland

Venkataraman K, Sharma G, Banerjee D (2020) Faunal diversity of India. In: Dar G, Khuroo A (eds) Biodiversity of the Himalaya: Jammu and Kashmir State. Topics in biodiversity and conservation, vol 18. Springer, Singapore, pp 71–92

WIIENVIS (2022) Elephant reserves. ENVIS Centre on Wildlife & Protected Areas. http://www.wiienvis.nic.in/Database/ElephantReserves_8226.aspx. Accessed 24 Jan 2023

Forest Policy and Forest Acts

2

Asim Kumar Mandal and P. Bhattacharya

Abstract

Forests are most invaluable natural resources of any country. Indian forests are rich in floral and faunal diversity. Indian forests suffered most at the hands of foreign rulers who pursued an exploitative approach to the forests. Alarmed by the rapid deterioration of forests that threatened timber supply, the British formulated the first forest policy in 1894 to improve the condition of forests. Post-independence, India enacted two forest policies, one in 1952 and the other in 1988. Both the policies set target to have 33% forest cover. Many schemes including special initiatives were taken up to achieve the set target but there is still long way to go. This chapter discusses the above aspects and other important issues concerning Indian forests.

Keywords

Forest policy · Forest Act · Afforestation · Forest conservation · Ecological balance

2.1 Introduction

Human settlement preceded forest clearance, a tradition that is still continuing. Whenever there is requirement for additional land for expansion of agriculture, road–rail network development, mining, etc., it is forest that is cleared. Therefore, the forest remains vulnerable forever. Historically, the Indian forest suffered in the hands of native and foreign rulers except during the period of Maurya kings who took special interest in protecting forests and wildlife. The forest was classified and set aside for different purposes including grazing, supply of forest produce, and religious needs.

In the early days of British settlement, they considered forests as the main hindrance to the expansion of agriculture. The ecological and environmental impacts of forests were never given due to importance by the British. Most of the requirements of the government and the people were met from the forests without any restrictions. The two most important and versatile timber species of Indian forest are *Tectona grandis* (teak) and *Shorea robusta* (sal). The British exploited the best forest areas of both the species for making railway sleeper for expansion of railway network and ship building for the Royal Navy. At some point of time the British Empire transferred large parts of forest areas to landlords and large farmers of Bengal and Punjab. In the central provinces also most of the forest areas were sold out barring a few forest tracts, which were kept for meeting the requirements of public work. Shortage of forest products for railway and Royal Navy projects as well as meeting local requirements became acute.

Realizing the overexploitation of Indian forests, the British in 1864 took the first step toward reversing this trend. The British decided in favor of Sir Dietrich Brandis, a forester with prior experience of managing forests in Pegu, Myanmar, to manage forests in India as the IGF (Inspector General of Forests). The Pegu forests supplied teak for about 100 years but suffered setback due to overexploitation. As IGF, Brandis suggested the British government to manage and administer forests scientifically and technically. It was Brandis who laid the foundation of forestry administration in the country. Successful implementation of forestry organization and administration led to many changes, including initiation of social forestry, joint forest management (JFM), people's need for forest-based products, awareness about forests and wildlife conservation, etc. All these benefitted local people.

To reverse the trend of overexploitation, a number of initiatives were taken up including providing protection to a few selected species, introduction of improved methods of

A. K. Mandal (✉)
Ramakrishna Mission Vivekananda Educational and Research Institute, Ranchi, Jharkhand, India

P. Bhattacharya
Guru Govind Singh Indraprastha University, New Delhi, India

A. K. Mandal, A. Nicodemus (eds.), *Textbook of Forest Science*, https://doi.org/10.1007/978-981-97-8289-5_2

felling and extraction of trees, and establishment of timber supply syndicate in 1796 in Malabar. Unfortunately, the syndicate did not last long and none of the above methods could improve the condition of forests. The commission put restriction on the felling of teak trees having growth <21 inches. In the meantime, there was shortage of oak in England for the Royal Navy. So the Court of Directors in 1805 decided to seek clarification to what extent teak can be supplied to the Royal Navy. A forest committee was set up that submitted its report with the following suggestions:

1. The forests near the roads and towns had been severely exploited.
2. The capacity of forests to supply exploitable timber was overemphasized.
3. Providing protection to forests would result in rehabilitation of degraded forests.
4. Roads may be constructed to exploit forests at interior areas.

Following the above, a special officer having fairly good knowledge of the local people and their language was appointed. The main aim was to ensure protection and application of improved production techniques for teak and final product transferred to the East India Company. The Company appointed Captain Watson, a police officer, to look after the administration of Madras forest conservancy. He successfully did away with private rights in the forest and established timber monopoly in the Malabar region. This resulted in regular supply of timber for ship building. But due to growing discontent and anger, among the traders, the Governor of Madras abolished the post of Conservator in 1823. The landlords got back their forests but unrestricted harvesting again deteriorated the general condition of the forest in the absence of any effort toward their regeneration. In a short period of time when the supply of good-quality timbers declined, the Royal Navy suggested reopening of conservatorship. The Madras government reviewing the situation in 1838 gave the responsibility of teak to a revenue officer and not to an independent authority as done earlier. Following this, the Court of Directors in 1842 took a novel decision and suggested to take up teak plantation for regular supply of teak timbers.

The first successful teak plantation was raised by the then Collector of Malabar, Conoly, in 1842. He first set up a workable local forest department and framed simple rules for plantation and management. The teak plantation set up by him in Nilambur, Kerala, is still retained and is known as Conoly's plantation.

Following this development, the government of Bombay and the government of Madras Presidency in 1847 appointed Gibson and Cleghorn as the Conservator of Forest, respectively. Both Gibson and Cleghorn suggested that the government should exercise and claim proprietary rights over all disputed private forests, shifting cultivation should be banned in the hills, and strict conservation measures should be put in place. Unfortunately, the government did not pay much attention to the above suggestions and the forest conservancy did not flourish to discharge effective function.

Taking clue from the report submitted by McClelland in 1854 in connection with the teak forests at Pegu, Rangoon, the initial guidelines prepared took the form of Forest Policy 1855. Brandis started enumeration of forest trees and introduced a system of preparing a management plan, popularly known as forest working plan. In addition, Brandis introduced a system of forest improvement and protection. He involved people to plant teak under shifting cultivation and made a beginning of artificial regeneration.

Soon a beginning was made to establish formal forest department in India, and in various provinces, forest departments were established between 1847 and 1868. Established forest departments were placed under a Conservator. But with the increase in work load, departments were divided into "circles" and each circle was placed under a Conservator.

Shortage of trained foresters was felt in the department. So, people with interest to work in the forest department were invited. Many people joined and many proved to be fruitful to the department and did outstanding work.

In another development, Lord Dalhousie, Governor General of India, issued a memorandum in 1855. This memorandum was based on the report of McClelland who suggested imposition of restriction to check exploitation of forests by private parties. With this Charter of Indian Forestry, other policy developments followed at different periods of time. Each policy was framed to take care of certain aspects of forestry and environmental issues important at that time.

2.2 Forest Policy and Forest Acts

Forest and forest ecosystems are most vital aspects of any nation. Forest has very high commercial value as well as it improves quality of life. So even the British thought of having a forest policy for maintaining healthy forests. A forest policy consists of laws and regulations and basically includes general guidelines for the government and the people. It gives direction to efficiently protect and manage forests. Forest policies help in maintaining healthy forest ecosystem needed for improving quality of life. Acts means behaving in the manner specified or the process of doing right.

To get a brief but clear idea how the forest policy developed in India, an attempt has been made at the following paragraphs to discuss management intervention and policy development.

2.2.1 Forest Policy Prior to British Rule

Forests were considered as common man's property. There were no rules in writing. Interpersonal faith was the basis of management and sharing of forest resources. Kings, emperors, and even landlords followed these unwritten principles religiously. Sometimes they might have converted forest lands for other purposes including as game reserves, but they never put any restrictions to common people to use forests and forest resources. Pretzsch (2003) observed that symbolic and religious functions of trees complemented the institutional framework.

During the period of Mauryan Empire, forest was classified on the basis of need and use by the people: reserve forest for the state and general forests for public use (Dwivedi 1980). It was the responsibility of the Panchayat or the local community-level institution to maintain all natural resources at village level. A Panchayat is an administrative unit at village level. It is constituted of five senior persons who look after the general village affairs. Guha (1989) noted that the important duties of Panchayat are settlement of conflicts among villagers for water and land.

Borgoyary et al. (2005) gave a detailed account of forest land during the Mughal rule. They observed that the Mughal Empire did not impose any tax on forest land. State had direct claim on hunting reserves. For local people, hunting was banned but they were allowed to gather fallen tree branches, leaves, etc. for own use. Borgoyary et al. (2005) further observed that outside Mughal jurisdiction, local people used to manage local resources using local knowledge.

2.2.2 Pre-independence Forest Policy

Revenue generation was the main objective of the British rulers. They were interested in species like teak, sal, and deodar, which produced best-quality timber for the Royal Navy, carriages, and sleepers for railway track. The British claimed full control of forests producing the above species in abundance. The forests with mixed species and rich in biodiversity were treated as unproductive and cleared of vegetation for expansion of agriculture. So, the forest policies were aimed, as rightly pointed out by Gadgil and Guha (1992), to have total control on forest resources. The British started destroying the forests with the implementation of the Forest Act, 1865. Forest resource exploitation and mass destruction affected agriculture, hunter gatherer activities, and ecological state of forests. So, negative economic, social, and ecological effects were very prominent during the British rule (Sinha 2016).

During pre-independence period, three forest acts, namely Forest Acts 1865, 1878, and 1927 and a policy (Forest Policy 1894) were promulgated. The Forest Act 1927 is still in vogue.

2.2.2.1 Forest Act, 1865

This forest act passed and implemented in 1865 included two categories of forests, reserved and protective forests, which the forest department under British raj managed and protected for extraction of timbers. The extraction was mainly from reserved forests. Mass clearance of forests started with the implementation of this Act. The local people used to collect materials for their own use. There was existence of private forests also. These were under individual control and were not controlled by this Act. The British imposed higher tax on the landlords who cleared forests for expansion of agriculture. Gross violation of traditional rights to people living in forest fringe areas started around this time by depriving them of their livelihoods.

2.2.2.2 Forest Act, 1878

Unlike the 1865 Forest Act, the 1878 Forest Act classified three types of forest, namely reserved, protected, and village forests. The village forest was not under the control of British government. Biswas (2002) rightly pointed out that British understood the intimate relation of village people with the nearby forests. Entry to other two kinds of forests was banned including grazing. Provisions were kept for imposition of fine and imprisonment on the offenders. The British banned collection of forest produce by the local people. This led to conflicts between local people and the personnel managing forest affairs. The people resorted to both violent and non-violent means with limited success. Due to such movement the British administration decided to go for community-based management of forest, viz. Van Panchayats in Uttar Pradesh and Forest Cooperatives in Himachal Pradesh (Guha 1983).

2.2.2.3 Forest Policy, 1894

The Forest Policy 1894 was prepared after studying the report submitted by Dr. Voelcar. This policy classified Indian forests into four types: (1) forests that require protection, (2) timber forests for revenue generation, (3) minor forests of insignificant market value, and (4) forest lands with pasture value (Forest Policy 1894). The last two categories of forests sustained the livelihood of local people.

The British did not like the concept of shifting cultivation or jhum cultivation. Jhum cultivation was widely practiced in the northeastern part of India as well as by other tribal groups in other parts of India. In jhum cultivation an area inside the forest is burnt and cleared of debris for cultivation of crops till it produces good crops or remains productive. The area is abandoned after shifting to another plot of land and in the process the previous land regenerates. The main reason forest officials acted against jhum cultivation is not because of effect of burning on forest ecology and environment but because of loss of revenue from sale of timber trees. Though the British took severe steps against jhum cultivation, it could not be stopped totally.

Kant et al. (1991) in his study observed that local people used to meet their needs for fuelwood, small timber, and/or fodder from minor forests. Apart from the tribal groups, the common people involved in agriculture also suffered as there was restriction on collection of a number of resources including fuel, fodder, leaves, and small timbers. Common people continued to suffer at the hands of British as many timber and non-timber products were made under the full control of the government. Even for grazing, a fee was levied. Many areas were clear felled and replanted with single species. This led to loss in biodiversity and important forest products needed by the local people.

Apart from people of agricultural classes, the Indian artisan industry also suffered at the hands of colonial management. The forest managers imposed restrictions on the raw material from forest needed by the artisan industry. Bamboo was replaced by timber species. Tasar silk industry also suffered because of duties levied on them for collection of cocoons from forest. The other industries like charcoal-based iron manufacturing also suffered due to very high tax and irregular supply of charcoal. So, it was clear from the above that the 1894 forest policy benefitted the British by earning revenue through timber sale.

2.2.2.4 Forest Act, 1927

This Act went through several changes since its drafting in 1865 and was finally accepted in 1927. The Act provided for stringent measures to take control of forests by the government (Samra et al. 2005).

This forest Act was very stringent and imposed more restrictions on the rights of people. It provided for taking control of private forests. Forest department was given total rights over forest produce and no rights given to others. This Act entrusted enormous power to forest officers in decision-making and taking any action for the interest of Indian forests. There were prohibitions to the entry in the forest for the purpose of collecting minor products, fishing, etc. Forest officials had the power to arrest any offenders. The Act successfully suppressed the rights of people dependent on forests for their livelihood. The Forest Act, 1927 faithfully served the interest of the imperial government. This Act is still in vogue.

2.2.3 Post-independence Forest Policies and Acts

Post-independence government of India enacted two forest policies, in 1952 and 1988, and three Acts, in 1972, 1980, and 2006. Forest Rights Act 2006 will not be discussed here as a separate chapter in this book will exclusively discuss this Act.

2.2.3.1 National Forest Policy, 1952

Following independence, dire need was felt to raise revenue. Forest was considered an important source for revenue generation. It was stated in the policy that villagers will not have any right to collect resources from the forest (Mishra 2002). Singh and Khare (1993) and Kumar (2000) critically commented that the local people must be satisfied with minimum forest produce. So, it is very clear that though the right to minor forest products was granted in the colonial forest policy, the 1952 forest policy totally omitted this right.

As has been done in earlier forest Acts and policies, the 1952 policy classified forest into four categories, of which two are national forests and protection forests. The 1952 policy prescribed for the first time to have 33% forest cover of the geographical area in the plain and 60% in the hill areas. The policy enlisted a number of vital national needs. On the issue of relation between forestry and agriculture, the 1952 forest policy made it very clear that forestry has no inherent right to best lands, which should be diverted for crop cultivation (GOI 1952, para 8).

The policy did not suggest any change for the control of forests by the state. There was increase in forest area due to nationalization of forests under the control of landlords and in princely states. The imperial forest department became the state forest department managed by Indian Forest Service.

The National Commission on Agriculture prescribed a number of measures to increase productivity of forests by taking up large-scale plantation of commercially important species and use of improved planting materials. Production forestry gained momentum in the 1960s with the establishment of 16 Forest Development Corporations (FDC) in different states, which undertook reforestation work in normally harvested areas. Enough funds were provided to these Corporations.

The forest departments converted forests with mixed species into forest with species of high value. This situation led to breakdown of ecological balance. The 1952 forest policy did not produce expected result due to consideration of the fact that forest is a state subject (GOI 1952).

2.2.3.2 National Forest Policy, 1988

The National Forest Policy, 1988 is presently in force. This policy is a significant departure in respect of concept and content from all earlier policies.

The forest department was a part of Agriculture Ministry till 1985. Later it came under the jurisdiction of Ministry of Environment and Forests. This prompted the ministry to promulgate the National Forest Policy, 1988. A radicle change in the planning process over the earlier one was evident in this policy. This policy and a circular opened the scope for participation by local communities in forestry that led to JFM (joint forest management) in 1990. It enabled the establishment of village institutions to look after the forest resources from the designated forest areas. It was observed that by June 1991 as many as 25 states moved resolutions that enabled local people in managing forests jointly with the forest department.

Collection of minor forest produce, small timbers, fuel wood, etc. is an important source of livelihood for people living in the vicinity of forest areas. Their rights were recognized. Preserving biodiversity and up-keeping the natural resources were given due importance.

That the country should have 33% forest cover of the total land mass was first suggested in the 1952 forest policy. The same suggestion was made again in the 1988 forest policy. The policy assumed that forest to this extent will have positive effects on controlling soil erosion, degradation of land mass, stabilizing already fragile ecosystem, and bringing about environmental stability. The 1988 forest policy suggested to take up large-scale afforestation in forest areas and in degraded unproductive lands outside forest areas. Enhancing production and productivity both in forest areas and degraded areas was another aim of the policy to achieve. The policy laid emphasis on good forestry education, research and extension for development of suitable technology, and their transfer to the field.

As per the latest forest assessment report (ISFR 2021), the country has 23.6% forest cover. ISFR (2021) also reported 80,957 sq km of forest and tree cover in the country. Analysis done by Forest Survey of India (ISFR 2021) indicates that only six states have more than 33% forest area. Whereas another seven states will have 33% forest cover in near future. As many as a dozen states probably will not make the target individually.

To achieve the national target of forest cover, the Government of India launched a National Forestry Action Programme for 20 years with provisions for capacity building, technology transfer, and adequate financial outlay (MoEF 2007).

A number of initiatives to increase the green cover of the country were taken up in different times. A few of them are social forestry programs, agroforestry programs, planting trees outside forests, and setting up of a number of agencies including National Afforestation and Ecodevelopment Board (NAEB) and National Wasteland Development Board (NWDB) to oversee forestry plantation activities. These schemes have been effective in increasing forest and tree cover. Excessive firewood collection and grazing are major causes of forest degradation. Joint forest management seems to be an effective tool in checking forest degradation.

In the process of assessing the forest cover of the country, generally the tree component in the forest and outside forest areas is assessed. Apart from tree component, there are a number of other non-tree natural ecosystems including grasslands, deserts, and riparian. Forest Survey of India in their biannual survey now includes all the above components in assessing green cover of the country.

The national goal of achieving 33% forest cover is sometimes difficult to comprehend when it is treated as an absolute figure. A wiser approach would be to look at this from the point of view of density and quality of forests, including regenerating capacity.

2.2.3.3 Forest Policy, 2018 (Draft)

The principles and guidelines as laid down in the 1988 forest policy govern Indian forests. An update is offing for more than five years now but it is yet to be finalized. The draft policy proposed safeguarding livelihood security of people and ecological balance through balanced management of forests. The draft policy also emphasized mitigation of climate change effects.

The draft policy is withheld due to heavy criticism from all fronts on the following grounds: (1) Clear-cut policies on sustainable management of forests are missing; (2) No clear definition of forests accepted nationally has been provided; (3) No policy has been prescribed on inviolate forests (forests that are to be left untouched); and (4) Suggested to involve private players. On the issues of community participation, JFM, and Forest Rights Act 2006, the draft policy has moved backward. The draft policy is under active consultation process for suitable changes by the government and will be placed for consideration by the Parliament.

2.2.3.4 Wildlife (Protection) Act, 1972

Realization of the fact that game hunting reduced the wildlife population of India led to the promulgation of an Act known as the Wildlife (Protection) Act, 1972. Main objective of the Act is to ensure protection of wild animals. The major provisions of this Act are providing protection to wildlife and effective management of their habitats; protected area establishment; prohibition of hunting; control of illegal trade of animals, their parts, and products; establishment of zoos and their management; and restricted timber extraction.

Among others, Central Zoo Authority, Wildlife Crime Control Bureau, National Tiger Conservation Authority, etc. were established. The other important establishments are conservation reserves, tiger reserves, national parks, sanctuaries, etc. The Act annexed four schedules of animals depending on their importance. The Act is still in vogue with wide coverage for overseeing the welfare of wildlife.

2.2.3.5 Forest Conservation Act, 1980

Rampant use of forest land for purposes other than forestry like construction of road, rail networks, mining, transmission lines, and other infrastructure projects became a matter of concern. So, after careful consideration, the Government of India enacted and promulgated the Forest (Conservation) Act, 1980 for effecting control of diversion. Main Provisions of the Act are the following:

(a) Imposition of restrictions for non-forest purpose
(b) Restriction on leasing out of forest land to industries and individuals
(c) Restriction on the clear felling of trees
(d) Prior approval of the central government for diversion

On examination of a proposal for diversion of forest land if found suitable, the central government stipulates conditions like compensatory afforestation, safety zone creation, and catchment area treatment for the user agencies. Effect of this Act is spectacular in respect of diversion of forest land. About 4.13 m ha of forest areas were diverted during the period 1951–1976 compared to only about 1.53 m ha for the period 1981–2018 (e-Green Watch 2020). No diversion of forest land has been recorded in Nagaland, Puducherry, and Lakshadweep.

The Forest Conservation Act, 1980 played an important role in reducing diversion of forest areas significantly. But the Parliament passed the Forest (Conservation) Amendment Act, in 2023. One of the reasons given for the amended Act was to achieve net zero emission by 2070. However, other contents of the proposal were far from the original objective of reducing diversion of forest lands.

The amendment is widely feared to reverse the gains made in conservation over the past 40 years as it exempts large forest areas from the purview of the original 1980 Act. The amendment emphasized that forests falling within 100 km of international border will not require forest clearance for implementation of key projects like hydro-power projects, and highways. India has about 15,000 km of international border. Such a vast area houses a number of important ecosystems including rain forests, wetlands, deserts, grassland, and many wildlife, which remained virgin till date but will now be vulnerable.

The justification on strategic grounds is appreciable. But even today a large mass of resource-poor people depends on forest resources for their survival. Time will reveal the effect of this landmark amendment on the lives of forest-dependent people and preservation of forest, wildlife, and natural heritage for the posterity.

2.3 Conclusion

India is endowed with astonishing natural generosity. Indian forest management bears the testimony of colonial legacy. The colonialists did everything to extract forest resources for expansion of royal interest. This severely affected the Indian forests, wildlife, and the ecology. Forest laws were found to be more stringent in the Independent India (than British India) as the various rights and benefits enjoyed by the resource-poor people living in the forest fringe areas were withdrawn considerably. The 1988 forest policy brought some kind of relief and their role in joint management of forests was recognized. Though forest policies of Independent India had many progressive ideas, they were not able to achieve the magical target of one-third of forest cover of its geographical area. At present, this remains to be illusive. Forests are most valuable and indispensable resources not only to human beings but also for all other living organisms. Maintaining a right ecological balance is therefore important for the future generations and for the continuity of forests as well.

Lessons Learnt

- Negative social and ecological effects on Indian forests were marked during the colonial period.
- Implementation of the 1865 forest Act catalyzed mass forest clearance.
- Forest policies of Government of India came out in 1952 and 1988.
- The role of local people living in forest fringes in management of forest resources was recognized.
- Maintaining ecological balance is must for sustainability of healthy ecosystem for the future generation.

Key Questions

1. What are the forest policies and forest Acts promulgated during British India? List out five salient features of the Forest Act, 1865.
2. Critically compare different provisions of the forest policies of 1952 and 1988.
3. List out the objectives as outlined in the 1988 forest policy of India.
4. Forest Conservation Act, 1980 effectively reduced diversion of forest areas for other uses. Discuss.
5. Describe the conditions of forests during British India.

References

Biswas A (2002) Forests and people of India: special notes on West Bengal. In: Basu S (ed) Changing environmental scenario of Indian sub-continent. ACB Publications, Kolkata

Borgoyary M, Saigal S, Peters N (2005) Participatory forest management in India: a review of policies and implementation. Working paper no. 1, Overseas Development Group, University of East Anglia, Norwich, UK

Dwivedi AP (1980) Forestry in India. Jugal Kishore and Co., New Delhi

e Green Watch (2020) FCA projects. E Green Watch. https://egreenwatch.nic.in/FCAProjects

Gadgil M, Guha R (1992) This fissured land: an ecological history of India. Oxford University Press, New Delhi

Government of India (1952) National forest policy of (NFP) 1952, New Delhi
Guha R (1983) Forestry in British and post-British India: a historical analysis. Economic and Political Weekly, 29 October 1882–1896
Guha R (1989) The unquiet woods. Oxford University Press, New Delhi
ISFR (2021) India State of Forest Report 2021. Forest Survey of India, Ministry of Environment, Forests and Climate Change, Dehradun
Kant S, Singh NM, Singh KK (1991) Community based forest management system: case studies from Orissa. ISO/Swedforest Indian Institute of Forest Management and SIDA, Bhopal
Kumar N (June 2000) All is not green with JFM in India. Forest, Trees and People, No. 42
Mishra T (2002) Forests of Bengal. Paschimbanga Vigyan Mancha, Kolkata
MoEF (2007) Interim country report – India for United Nations Forum on Forest. Ministry of Environment and Forest, Government of India. http://envfor.nic.in/nfap/Unff2.pdf
Pretzsch J (2003) Forest related rural livelihood strategies in national and global development. Paper presented at the international conference on rural livelihoods, forests and biodiversity, 19–23 May, 2003, Bonn, Germany
Samra JS et al (2005) Water harvesting and recycling – Indian experiences. Central Soil and Water Conservation Research Centre, Dehradun
Singh S, Khare A (1993) People's participation in forest management. Commonw For Rev 72:279–283
Sinha S (2016) Forests in India: a review of colonial ordeal. Int J Res Humanit Soc Stud 3:40–43

3 Forests, Tribals and Forest Rights Act

Sanjay Upadhyay

Abstract

This chapter discusses one of the most cardinal legal instruments that has been enacted post-independence on the rights of the forest-dwelling communities, which had not been recognised and vested hitherto, namely the Scheduled Tribes and Other Traditional Forest Dwellers (Recognition of Forest Rights) Act, 2006.

The chapter introduces the basic framework and provides an eagle's eye view of the legislation. It then focusses on the often-misunderstood concept of who is the beneficiary under the Act and explains who is a 'Forest Dwelling Scheduled Tribe' (FDST) and who is an 'Other Traditional Forest Dweller' (OTFD). After establishing the beneficiaries under the Act, the chapter attempts to throw light on the most important aspect, that is, the rights that are being recognised and vested under the legislation, namely individual, community and community forest resource (CFR) rights.

Any legislation of such a nature that attempts to undo historical injustice will be hollow if it is not backed by a robust process that ensures the recognition and vesting of such rights, which is the next focus of the chapter. The section also dwells on novel concepts such as Critical Wildlife Habitats, as well as the empowered duties of the FDSTs and OTFDs, which highlight the fact that the legislation also has a conservation element embedded in it.

The next significant aspect of the chapter is the authorities that have been established under the legislation, and it is seen that it is for the first time the unit of governance, i.e. the Gram Sabha, has been given a statutory authority status. Further, a four-pronged approach is being implemented for ensuring the recognition and vesting of forest rights, wherein representatives from all important line departments namely the Revenue, Tribal, Forest and Panchayati Raj Departments are being included to ensure that the said process is seamless and complete.

The chapter also discusses the role of the nodal agency, namely the Ministry of Tribal Affairs, and its interaction with the Ministry of Environment, Forests and Climate Change.

The Act also has its set of challenges in its implementation and understanding, which is the next focus of the chapter. The chapter concludes with a brief summary of the lessons learnt and leaves some unanswered questions to ponder over.

Keywords

FDST · OTFD · FRA · PESA · Forest land

3.1 Introduction

The connect between Forest Governance and Scheduled Area Governance is one of the significant aspects of Natural Resource Governance Framework in India. The enactment of Provisions of Panchayats (Extension to the Scheduled Areas) Act, 1996 (PESA) and the Scheduled Tribes and Other Traditional Forest Dwellers (Recognition of Forest Rights) Act, 2006 (hereinafter Forest Rights Act, 2006 or FRA) were some of the landmark developments in strengthening this connect. Particularly, the Forest Rights Act of 2006 has heralded a new chapter in forest governance by recognising and securing traditional rights of the forest-dependent communities that were hitherto ignored, restricted or prohibited during the process of reservation.

Tribal history is often a history of their changing relationship with the nature and forest. The traditional practices for conservation and sustainable management of natural resources of indigenous tribes of Indian States highlight the predominant dependence of tribals on forests for sustenance. The Constitution of India made the first comprehensive attempt at securing the uniqueness of the tribal cultures in

S. Upadhyay (✉)
Supreme Court of India, New Delhi, India
e-mail: su@chambersofsanjayupadhyay.com

A. K. Mandal, A. Nicodemus (eds.), *Textbook of Forest Science*, https://doi.org/10.1007/978-981-97-8289-5_3

India through 'Scheduled Areas', thanks to the efforts of certain tribal advocates such as Shri Jaipal Singh, Yudhishthir Mishra, among others. Thus, Scheduled Areas were demarcated in the states of Andhra Pradesh, Telangana, Himachal Pradesh, Orissa, Gujarat, Rajasthan, Maharashtra, Chhattisgarh, Madhya Pradesh and Jharkhand, by the order of the President and were accorded a special status for administration.[1] For effective decentralised governance, Tribal Advisory Councils have been created in all the States where there is a sizeable population of tribals to advise on matters pertaining to the welfare and advancement of Scheduled Tribes (STs) and, more importantly, for the administration of these regions.[2] The Governor has been empowered to make regulations for the peace and good governance, especially on matters related to land transfer by or among members of Scheduled Tribe, regulating allotment of land to members of Scheduled Tribes and regulating money lending with the objective of safeguarding the marginalised from exploitation.[3] The Governor can also regulate the operation of any Act of the Parliament in the Scheduled Areas and allow/disallow it with modifications to suit the context of the Scheduled Areas.[4]

The special governance structure created under the Scheduled Areas for Tribal Communities is not only carried forward by the FRA, but also it goes a step further in recognising rights of other traditional forest dwellers. The FRA recognises the powers of vesting with the local communities, through Panchayat Bodies, to manage their natural resources in accordance with their traditional customary practices (Box 3.1).[5] The years preceding FRA witnessed a slew of measures from the Ministry of Environment and Forests (MoEF), which categorised forest dwellers as encroachers, and issued detailed Guidelines on restriction of regularisation of encroachment activities by State Governments post the enactment of the Forest (Conservation) Act, 1980. It was not until 2004, that Ministry of Environment and Forests explicitly recognised the exclusionary nature of prevailing forest legislations in incorporating and securing the traditional rights of tribals and stopped referring to them as encroachers. In this background, FRA is considered to be a milestone in facilitating the forest tribal interface as it aims to secure the traditional rights of the tribal communities as well as bona fide 'other traditional forest dwellers'. It recognises those rights that have not been recorded during the process of reservation of forests and by securing tenure, it aims at strengthening the conservation regime while ensuring a dignified livelihood source. The FRA provides in detail a framework of rights and responsibilities, authorities and duties to protect, regenerate, conserve and manage community forest resources (CFRs).[6]

Box 3.1 Scheduled Areas in FRA

The FRA recognises the uniqueness of scheduled areas in its frame. In fact, it starts with the basic unit of administration that is the village and recognises the unique village concept under the PESA that is a hamlet or a group of hamlets that is bound by customary practices, traditions and social norms rather than the conventional revenue village concept that is based on population.[7] In other words, it is the hamlet-based Gram Sabha and consequently the Forest Rights Committee that initiates the process of recognition and vesting of forest rights in scheduled areas.

The FRA also recognises traditional rights and gives precedence to customary law with regard to community forest resource management, which is in many ways unique to the scheduled areas where customs dictate the day-to-day living of scheduled tribes. The FRA also recognises the record of rights, which are traditionally recognised by any state and, in this context, it implicitly imbibes the record of rights system under the tenancy, especially the record of rights within such laws.

In January 2008, the Scheduled Tribes and Other Traditional Forest Dwellers (Recognition of Forest Rights) Act, 2006 (Rules) was notified. To facilitate the implementation of FRA, the Rules elaborate upon the roles, functions and duties of the Gram Sabha, State District Level Committees, District Level Committee (DLC) and State Level Monitoring Committee (SLMC) vis-a-vis the recognition and vesting of rights of the tribals. The Rules have also attempted to clarify the process of recognition by laying down detailed steps and the accompanying documents required to be submitted by claimants for verification of claims and completing the recognition of rights.

However, the operational arrangements for effective implementation of these two important Statutes (FRA and PESA) and the Rules are still not well understood. The Ministry of Tribal Affairs (MoTA) has issued more than 200 instruments for clarifying the implementation of FRA, the most recent being the September 2023 Guidelines on

[1]Article 244 of the Constitution of India, 1950.

[2]Part B of the Fifth Schedule of the Constitution of India, 1950.

[3]Part B of the Fifth Schedule of the Constitution of India, 1950.

[4]Ibid.

[5]ELDF, 'Towards Creating a Model Forest and Scheduled Area Governance in Chhattisgarh: A Manual on Forest Rights Act and PESA', 2012.

[6]Ibid.

[7]Section 4(b) of The Provisions of Panchayats (Extension to the Scheduled Areas) Act, 1996.

Conservation, Management and Sustainable Use of Community Forest Resources.

The Ministry of Panchayati Raj along with the States has also developed several operational arrangements for effective implementation of PESA. What is essential is to understand the context and the intent with which the said arrangements have been initiated. The language of the statute or the clarificatory instruments issued thereunder often become a barrier in effective implementation. The ground-level experience of the above two Statutes suggests that the knowledge of the law, the rights and duties it creates and the operational framework, is inadequate, especially in the tribal-dominated areas.

Therefore, it is essential to understand the Statute and particularly how it aims to facilitate the connect between forest governance and scheduled area governance. This becomes all the more crucial as many believe that the abovementioned Statutes could serve as the most important tools to overcome social unrest and the economic backwardness often associated with tribal-dominated areas in India.

3.2 The Discussion on the Act, Rules and Guidelines

The Forest Rights Act, 2006 is a landmark legislation based on the following cardinal principles:

- The aim of the Act is to address the historical injustices meted out to Forest-Dwelling Scheduled Tribes and Other Traditional Forest Dwellers by recognition and vesting of such forest rights, which could not be recorded.[8]
- The Act provides a framework for recording of rights including nature of evidence.[9]
- The Act also imposes responsibilities on the beneficiaries/ rights holders under the Act and authority for sustainable use and conservation of biodiversity.[10] It mandates that the conservation regime shall be strengthened while ensuring livelihood security.[11]
- The Act further aims to secure the tenure and traditional rights of the 'Forest-Dwelling Scheduled Tribes' and 'Other Traditional Forest Dwellers' on forest lands to address the long-standing insecurity of tenurial and access rights.[12]

Keeping in mind the above-mentioned cardinal principles, the Forest Rights Act, 2006 is divided into six chapters, covering the extent, applicability and definitions in the preliminary chapter, followed by a chapter enumerating the provisions on individual and community forest rights of Forest-Dwelling Scheduled Tribes and Other Traditional Forest Dwellers. The next chapter discusses the provisions pertaining to the recognition and vesting of forest rights in Forest-Dwelling Scheduled Tribes and Other Traditional Forest Dwellers, which is followed by a chapter on statutory authorities and the procedure for vesting of forest rights. The final two chapters deal with offences and penalties and miscellaneous provisions.

The Act has been named as the Scheduled Tribes and Other Traditional Forest Dwellers (Recognition of Forest Rights) Act, 2006 and extends to the whole of India.[13] It came into force on 2 January 2007 in pursuance of the Notification in the official gazette by the Central Government.

The Act defines some key terms that are integral to the basic structure of the Act. It is safe to presume that terms such as community forest resource, Critical Wildlife Habitats (CWHs), Forest-Dwelling Scheduled Tribes, forest land, forest rights,[14] forest villages, Gram Sabha, habitat,[15] other traditional forest dwellers, among others, that pave the way for the recognition and vesting of forest rights, have not been envisaged or defined in any other legislation, especially for the purposes of implementation in scheduled areas (Box 3.2).

Box 3.2 What Are Forest Lands?
Forest land is '*land of any description falling within any forest area and includes unclassified forests, undemarcated forests, existing or deemed forest, protected forest, reserved forest, sanctuary, national park. Different levels of individual and community forest rights are granted in all these lands to local communities*' (Section 2(d) of FRA, 2006).

The following sections will now discuss in detail the forest rights as laid down in the Act along with the process for their recognition, vesting and restoration, and also detail the authorities responsible for the same.

[8]Preamble of the FRA, 2006.

[9]See Section 6 of the FRA, 2006.

[10]See Section 5 of the FRA, 2006; read with the Preamble of the Act.

[11]Ibid.

[12]Preamble of the FRA, 2006.

[13]Section 1 of the FRA, 2006.

[14]Section 2(e) FRA, 2006.

[15]Section 2(h) FRA, 2006.

3.2.1 Beneficiaries Under the Forest Rights Act, 2006

3.2.1.1 Forest-Dwelling Scheduled Tribes[16] and Other Traditional Forest Dwellers[17]

The FRA, 2006 delineates two categories of persons/community as beneficiaries under the Act, namely the Forest-Dwelling Scheduled Tribes (FDST) and Other Traditional Forest Dwellers (OTFD). FDSTs are the members or community of the Scheduled Tribes who primarily reside in and who depend on the forests or forest lands for 'bona fide livelihood needs'[18] and include the Scheduled Tribe pastoralist communities. The definition of OTFD is more restrictive and it includes members or community who have been primarily residing and depending on forest land for their bona fide livelihood needs for at least three generations prior to the 13th day of December 2005 (Box 3.3).

> **Box 3.3 Who All Are Included in the Category of OTFDs?**
> *Any person or community who has been residing and depending on the forest or forest land for* bona fide *livelihood needs for a period of 75 years prior to 13.12.2005. It is not necessary for the person to belong to any tribe or caste to be eligible for the category of 'other traditional forest dwellers'* (S. 2(o) of the FRA, 2006).

3.2.2 Forest Rights Vested with FDSTs and OTFDs

The Forest Rights Act, 2006 has introduced two categories of rights, namely Individual Forest Rights[19] and Community Forest Rights,[20] for FDSTs and OTFDs.

3.2.2.1 Individual Rights

The Forest Rights Act, 2006 lays down four categories of individual rights of habitation and self-cultivation, rights over disputed land, right for conversion of pattas or leases on forest lands to titles[21] and right to in situ rehabilitation for FDSTs and OTFDs.

1. *What is meant by the right to dwell in a forest land and to cultivate forest land?*[22]
   This specific provision allows FDSTs and OTFDs to hold Forest land for self-cultivation or habitation or both either under individual or common occupation. The Amendment Rules of 2012 provide for another pragmatic aspect regarding the extent of land that shall be recognised for self-cultivation. It has explained that the '*land rights for self-cultivation recognised under clause (a) of sub section (1) of section 3 shall be, within the specified limit, including the forest lands used for allied activities ancillary to cultivation, such as, for keeping cattle, for winnowing and other post-harvest activities, rotational fallows, tree crops and storage of produce*'.[23] The 2012 Amendment makes it clear that the recognition of land for cultivation is not limited to the actual area under cultivation but also includes the lands that are used for ancillary purposes as those mentioned above.
   It is important to mention here that with regard to the forest rights as mentioned above, the land should be under the occupations of an individual or family or community on the date of commencement of FRA, shall be restricted to the area under 'actual occupation' and shall in no case exceed an area of 4 ha.[24]
2. *What are the individual rights over disputed land*
   In case of a dispute over custody of a particular land between forest department, revenue department and any other department or even between two States, there is no restriction on FDSTs and OTFDs from exercising their right or claim over such land.[25]
3. *Conversion of leases or pattas over forest land into permanent titles*
   By virtue of this right, the pattas or leases or grants on forest lands, which were issued prior to the enactment of the FRA, can be converted into permanent titles.[26]
4. *Right to rehabilitation when displaced or evicted from forest land*[27]
   Right holders under FRA, 2006 can also claim in situ rehabilitation, alternative land where they have been illegally evicted, displaced or have not received their legal entitlement prior to 13 December 2005. The monitoring of such claims is undertaken by the State Level Monitoring Committee (SLMC).[28]
   The right to land is one of the crucial forest rights recognised and vested under the Act to FDSTs and

---

[16] Section 2(c) FRA, 2006.

[17] Section 2(o) FRA, 2006.

[18] Rule 2(b) of the FRA Rules, 2008 clarifies that bona fide livelihood needs means '*fulfillment of livelihood needs of self and family through exercise of any of the rights specified in Sub-section (1) of Section 3 of the Act and includes sale of surplus produce arising out of exercise of such rights*'.

[19] As enlisted in Clauses (a), (f), (g) and (m) of Sub-section (1) of Section 3.

[20] The Amendment Rules of 2012 have categorically specified what would be community rights. They include rights listed in Clauses (b), (c), (d), (e), (h), (i), (j), (k) and (l) of Sub-section (1) of Section 3.

[21] Section 3(1)(g) of the FRA, 2006.

[22] Section 3(1)(a) of the FRA, 2006.

[23] Rule 12A (8) of the Amendment Rules, 2012.

[24] See Section 4(6) of the FRA, 2006 as well as Guidelines of 2012.

[25] Section 3(1)(f) FRA, 2006.

[26] Section 3(1)(g) FRA, 2006.

[27] Section 3(1)(m) FRA, 2006.

[28] See Rule 10 (f).

OTFDs provided they establish that they were displaced from their dwelling and cultivation without land compensation due to State development interventions, and where the land has not been used for the purpose for which it was acquired within 5 years of the said acquisition.[29]

3.2.2.2 Community Forest Resource[30]

The 'community forest resources' (CFRs) have been defined to mean '*customary common forest land within the traditional or customary boundaries of the village or seasonal use of landscape in the case of pastoral communities, including reserved forests, protected forests and protected areas such as Sanctuaries and National Parks to which the community had traditional access*'. It also includes the right to protect, regenerate or conserve or manage any community forest resource that they have been traditionally protecting and conserving for sustainable use.[31]

The concept of community forest resource is one of the most important rights recognised under FRA from both the rights and resources perspective. The law mandates that those areas that have been traditionally managed and conserved or protected, there should not be just a responsibility but a right to conserve the same. And in fact, a title deed needs to be conferred formalising such a right (as per the September 2012 Amendment in the Rules).

It is this community forest resource where the Gram Sabhas as well as similar village institutions have been empowered to protect and conserve such CFRs as a matter of duty.[32] What is crucial in this concept is the process of delineation of the community forest resource, with the real challenge being defining what is customary common forest land within traditional or customary boundary and the consequent conservation and management plan that is now a mandate under the September 2012 Amendment Rules. There should be complete clarity on these terms and their understanding in operational terms. The concept of CFR not only includes the seasonal use of landscape by pastoral communities but also extends to all types of forests including reserved forests, protected forests and protected areas such as National Parks and Sanctuaries to which the communities had traditional access.[33]

Community Rights

The Forest Rights Act, 2006 recognises and vests communities with nine different community rights (Box 3.4).

Box 3.4 What Are Community Rights?

The September 2012 Rules, for the first time, clarified what are community rights. It lists out the rights in Clauses (b), (c), (d), (e), (h), (i,) (j), (k), and (l) of Sub-section (1) of Section 3 of the FRA. In simple terms it means community rights such as Nistar, Minor Forest Produce (MFP), water bodies including fish, grazing, rights of pastoralist, community rights of Particularly Vulnerable Tribal Groups (PVTGs), forest villages to revenue villages, rights over community forest resource, and those recognised under other state laws or customary laws. Community intellectual property and right to biodiversity and other traditional rights are also now clarified as community rights under FRA. The FDST and OTFD can claim them as 'Community Rights'.

Public Rights over Public Forest Lands

The FRA recognises and vests community rights such as Nistar rights, which simply means public rights over public lands and it includes the community rights used in erstwhile Princely States, Zamindari or any intermediary regimes.[34] The community is entitled to rights such as fishing, grazing (both settled and transhumant), traditional seasonal use of resource by the nomadic or pastoralist communities and any other entitlements granted to the community.[35] The District Level Committee is the nodal authority for facilitating filing of claims by the pastoralists, transhumant and nomadic communities in the prescribed format before the concerned Gram Sabhas.[36]

Minor Forest Produce

Minor forest produce is the life blood of a tribal society and is used to meet every household need such as for food, fodder, fuel wood, medicines, festivals, building material, among others.[37] In this context, the FRA 2006 has entitled the FDSTs and OTFD with the right to own, access to collect, use and dispose the minor forest produce, which has been traditionally collected within or outside village boundaries.[38]

Minor forest produce includes '*all non-timber forest produce of plant origin including bamboo, brushwood, stumps, cane, tussar cocoon, honey, wax, lac, tendu or kendu leaves, medicinal plants and herbs, roots, tubers and the like*'.[39] The

[29] See Section 4(8) of the FRA, 2006.

[30] Section 2(a) FRA, 2006.

[31] Section 3 (1)(i) FRA, 2006.

[32] Section 5, FRA.

[33] ELDF, 'Towards Creating a Model Forest and Scheduled Area Governance in Chhattisgarh: A Manual on Forest Rights Act and PESA', 2012.

[34] Section 3(1)(b) of the FRA, 2006.

[35] Ibid.

[36] Section 3(1)(d) of the FRA, 2006.

[37] ELDF, 'Towards Creating a Model Forest and Scheduled Area Governance in Chhattisgarh, A Manual on Forest Rights Act and PESA', 2012.

[38] See Section 3(1)(c) of the FRA, 2006.

[39] Section 2(i) of the FRA, 2006.

phrase '*and the like*' in the definition of Minor Forest Produce clarifies that the said definition is inclusive and more varieties of plants can be added depending on their usage by the community. It is also important to highlight that the right to dispose the minor forest produce includes the right to sell as well as individual or collective processing, storage, value addition, and transportation within and outside forest area through appropriate means of transport for use of such produce or sale.[40] For finality in the number of resources that could constitute Minor Forest Produce, a list must be prepared in consultation with the Forest Department and a formal title to be obtained for the same.

To exercise the right to collect Minor Forest Produce, the FDSTs and OTFDs can enter the forest to collect the produce, use or sell it themselves or through cooperatives or associations or federations for livelihood.[41]

The Committee constituted under the Forest Rights Rules, 2007 and authorised by the Gram Sabha shall modify the transit permit for transportation of Minor Forest Produce. However, the procedural requirement will not impede or restrict the exercise of the right of collection or disposal of minor forest produce.[42] Further, the collection shall be free of all royalties, fees or any other charges.[43]

Water Bodies, Grazing and Nomadic Rights over Seasonal Resource

The Forest Rights Act, 2006 recognises and vests the right to entitlement or use resources such as fish and other products of water bodies, grazing (both settled and transhumant) and traditional seasonal resource access to nomadic or pastoralist communities.[44] For instance, the traditional systems of 'chuan' and other water bodies need to be identified and linked with such forest rights in the State of Jharkhand. Similarly, the identification of the nomadic tribes or pastoralists also has to be carried out and linked with the record of right system in Jharkhand.

Right of Habitat for Particularly Vulnerable Tribal Groups and Pre-Agricultural Communities

The Forest Rights Act, 2006 lays down that if Particularly Vulnerable Tribal Groups (PVTGs) such as Juang, Chenchu, or Baiga, and pre-agricultural communities such as Jhumias or other traditional forest dwellers have been cohabiting in a forest land, they can claim the right to their habitat, even if it falls inside a reserved forest or a protected forest.[45] However, there were not many claims being filed by the PVTGs, which led to 2012 Amendment to specifically mandate the District Level Committee to ensure that all PVTGs receive habitat rights, in consultation with the concerned traditional institutions of PVTGs, and their claims for habitat rights are filed before the concerned Gram Sabhas, wherever necessary by recognising floating nature of their Gram Sabhas.

In view of the differential vulnerability of PVTGs, the Amendment Rules of 2012 further mandated the District Level Committee to ensure that all PVTGs receive habitat rights, in consultation with the concerned traditional institutions of PVTGs, and their claims for habitat rights are filed before the concerned Gram Sabhas, wherever necessary by recognising floating nature of their Gram Sabhas.[46]

Right of Conversion of Forest Villages into Revenue Villages

The Forest Rights Act, 2006 ensures that old settlements, forest villages, old habitations, un-surveyed villages or any other village in forest if not recorded in revenue records of the government can now be converted into revenue villages under the Forest Rights Act (Box 3.5).[47] The conversion to revenue villages under the FRA, 2006 has certain advantages.

Box 3.5 Benefits of Conversion of Forest Villages into Revenue Villages

Although there is some form of recognition of forest villages by the Forest Department, certain revenue benefits including developmental schemes for education, irrigation and sanitation initiated by the government periodically does not accrue to such villages as they are not technically under the revenue department. This has caused numerous conflicts at the ground level. Thus, an urgent need was felt to bring such forest villages under a legal definition either under the Indian Forest Act or recognised as revenue villages depending on the circumstances. The challenge is to ensure that such conversion does not lead to honeycombing of forests.

Forest villages were created by the forest department to obtain a sufficient and continuous supply of labour for forestry operations such as forest protection and other works of improvement. An agreement was entered into between the residents of the village and the forest department, which would define the roles and responsibilities of the members of the village. The residents would receive several rights to fulfil their livelihood needs in lieu of their services.

[40] ELDF, 'Towards Creating a Model Forest and Scheduled Area Governance in Chhattisgarh: A Manual on Forest Rights Act and PESA', 2012; See Section 3(1)(d) of the FRA, 2006.

[41] See Amendment Rules of 2012.

[42] Rule 2(2) of the Forest Rights Rules, 2007.

[43] Rule 2(3) of the Forest Rights Rules, 2007.

[44] See Section 3(1)(d) of the FRA, 2006.

[45] See Section 3(1)(e) of the FRA, 2006.

[46] See Rule 12-B of the Forest Rights Rules 2007.

[47] See Section 3(1)(h) of the FRA, 2006.

These rights could be land for cultivation, which were to be held as service jagir and could not be transferred by sale, lease or mortgage. However, the residents did not have a title over the land and it could be taken back at the discretion of the forest department. No rent was charged on such lands. Besides, free grazing was allowed in the forest area for a reasonable number of cattle as determined by the forest department.[48] The residents were also allowed to collect forest produce such as timber, fruits, flowers, grass and bamboo for household and agricultural needs. A separate register was maintained by the forest department giving a description of the village, its boundaries, number of cultivators admitted and objective of the establishment of the village among others. A sketch map of the village was also prepared.[49]

Right to Protect, Regenerate or Conserve or Manage Any Community Forest Resource

The Forest Rights Act, 2006 also recognises and vests the right to protect, regenerate or conserve or manage any community forest resource that they have been traditionally protecting and conserving for sustainable use (Box 3.6).[50]

Box 3.6 What Are Community Forest Resource Rights?
Community forest resource[51] means customary common forest land within the traditional or customary boundaries of a village or seasonal use of landscape in the case of pastoral communities, including reserved forests, and protected forests such as sanctuaries and national parks to which the community had traditional access.

This would therefore mean that these are only those areas where communities can demonstrate their traditional access overlapped with legal categories like reserved forests, protected forests and protected areas such as Sanctuaries and National Parks under various legislations that would qualify to be community forest resource.[52]

It is a right to protect, regenerate or conserve or manage any community forest resource that they have been traditionally protecting and conserving for sustainable use. Thus far the Rules were silent on how to secure this right in the form of a title. The Amendment Rules of 2012 now adds another form under which such claim to CFR may not only be claimed but also a title is given to such community. A process has also been delineated in this regard. Further the Amendment Rules also now correct a long-standing typographical error where evidence of such community forest resource was wrongly written as evidence for community forest rights. This also means that to delineate community forest resource, a set of evidences can now be used for a legitimate claim on community forest resource that the community has been traditionally protecting.

It is this community forest resource where the Gram Sabhas as well as similar village institutions have been empowered to protect and conserve such CFRs as a matter of duty.[53] What is crucial in this concept is the process of delineation of the community forest resource. The real challenge being defining what is customary common forest land within traditional or customary boundary and the consequent conservation and management plan that is now a mandate under the September 2012 Amendment Rules. There should be complete clarity on these terms and their understanding in operational terms. The concept of CFR also includes seasonal use of landscape in the case of pastoral communities.

Rights That Are Recognised Under State Laws or Autonomous District Council or Autonomous Regional Council

This includes rights that are recognised under any State law or laws of any Autonomous District Council or Autonomous Regional Council or that are accepted as rights of tribals under any traditional or customary law of the concerned tribes of any State.[54] This primarily deals with those rights that have been recognised and also where the community has de facto control over such community resources particularly in the northeastern States of India and more specifically in the special states such as Nagaland, Arunachal Pradesh and Sixth Schedule Areas.

Rights of Access to Biodiversity and Community Right to Intellectual Property and Traditional Knowledge Related to Biodiversity and Cultural Diversity

Tribal communities possess rich culture, long-standing customs, practices and traditional wisdom referred to as their traditional knowledge. In many cases, this traditional knowledge has been orally passed for generations from person to person often through stories, legends, folklore, rituals, songs and even laws.[55] For instance, traditional knowledge may include knowledge about the medicinal properties of a particular species of plant or a unique farming practice that

---

[48] ELDF, 'Towards Creating a Model Forest and Scheduled Area Governance in Madhya Pradesh: A Manual on Forest Rights Act and PESA', 2012; See Section 3(1)(d) of the FRA, 2006.

[49] Ibid.

[50] See Section 3(1)(i) of the FRA, 2006.

[51] See Section 2(a) of the FRA, 2006.

[52] ELDF, 'Towards Creating a Model Forest and Scheduled Area Governance in Chhattisgarh: A Manual on Forest Rights Act and PESA', 2012; See Section 3(1)(d) of the FRA, 2006.

[53] See Section 5 of the FRA, 2006.

[54] See Section 3(1)(j) of the FRA, 2006.

[55] See Section 3(1)(k) of the FRA, 2006.

increases crop yield and that is exclusively possessed by the members of a particular tribal community as part of their culture. Often this time-honoured wisdom of the local communities is accessed without permission by private companies, individuals who unfairly exploit the knowledge to earn huge profits without sharing benefits with the local communities who have originally held that knowledge.

However, the Forest Rights Act provides a mechanism to secure control of the community over their traditional knowledge by claiming it as a Community Intellectual Property under the Act. Once this is done, no one can access and use the knowledge without the permission of the community.

Any Other Traditional Right Not Mentioned

It refers to any other traditional right customarily enjoyed by the Forest-Dwelling Scheduled Tribes or other traditional forest dwellers, as the case may be, which are not mentioned above. However, in the exercise of these traditional rights, Forest Rights Act does not allow hunting or trapping or extracting a part of the body of any species of wild animal. Such a clause is always added where the legislature ordinarily cannot envisage the entire gamut of rights and thus adds such a clause called as a 'residuary clause'.[56] This has been very creatively used, for example in some states to ensure the community forest resource title in the absence of such a provision prior to the Amendment Rules of 2012.

3.3 The Process of Recognition and Vesting of Forest Rights in Forest-Dwelling Scheduled Tribes and Other Traditional Forest Dwellers[57]

The Forest Rights Act, 2006 has laid down a detailed process of recognising and vesting of forest rights for its beneficiaries, which is discussed below.

Chapter III of the Forest Rights Act, 2006 provides explicit statutory expression for recognition, restoration and vesting of forest rights in FDSTs and OTFDs. By imputing the authorities with a legally binding mandate to recognise, restore and vest the rights of the Forest-Dwelling Scheduled Tribes and Other Traditional Forest Dwellers, it provides teeth to the objectives of the Act, ensuring they do not remain redundant on paper. The legal analysis of the process of recognition and vesting also draws references from the Forest Rights Rules of 2007, which were subsequently amended in 2012, to extensively and holistically record the substantive and procedural legal mandate for recognition and vesting of rights.[58]

[56] See Section 3(1)(l) of the FRA, 2006.

[57] Chapter III, Sections 4 and 5 of the FRA, 2006.

[58] As stated under Section 3 of the FRA.

3.3.1 Recognition and Vesting of Rights: The Mandate

The Forest Rights Act, 2006 provides[59] for mandatory recognition and vesting of rights laid down in Section 3 of the Act for Forest-Dwelling Scheduled Tribes and Other Traditional Forest Dwellers. It is preceded by a non-obstante clause, which implies that the forest rights in FDSTs and OTFDs were recognised and vested on the date of the enforcement of the legislation, i.e. 31 December 2007, and were not subject to any conditions. After such recognition and vesting, all that remained was the process for determining the extent of the right and the subsequent grant of title, for which a detailed process has been envisaged in the later sections read with the Rules, 2007 and the Amendment of 2012.

However, the Act itself provides for fulfilment of two pre-requisites, which are highlighted on reading Section 4 (3) of the Act together with Section 4(6). First, there should be occupation by the FDSTs and OTFDs before the cut-off date of 13 December 2005, and second, there should be continued occupation on the commencement of the Act as on 31 December 2007.[60] In addition to the two conditions, the OTFDs also have to prove that they have been resident on the said land/habitat over a period of three generations as on 13 December 2005.

The Act further subjects the recognition and vesting of right to hold and live in the forest land under the individual or common occupation for habitation or for self-cultivation for livelihood by member/members of FDST and OTFD, to land that has been under actual occupation and restricted to the ceiling limit of 4 ha.[61]

The forest rights recognised and vested in the forest dwellers would include those who were displaced from their dwelling and cultivation due to state developmental activities without land compensation, provided the land was not utilised by the State Government for the purpose for which it was acquired for 5 years of its acquisition.[62] State Level Monitoring Committee has been tasked to ensure right to land is recognised and vested for such forest dwellers.[63]

Further, the rights recognised under FRA are free of all encumbrances and procedural requirements as would otherwise be imposed for utilisation of forest land under the Forest Conservation Act of 1980 such as payment of 'Net Present

[59] Under Section 4(1) of the FRA, 2006.

[60] Letter dated 6 May 2010 from the Ministry of Tribal Affairs to the Government of Uttar Pradesh.

[61] Section 4(6) of the Forest Rights Act, 2006; Clause i. (j) of the Guidelines on the implementation of FRA, dated 12 July 2012.

[62] Section 4(8) of the Forest Rights Act, 2006.

[63] Clause v (e) of the Guidelines on the implementation of FRA, dated 12 July 2012.

Value' and 'compensatory afforestation' for diversion of forest area for any other activity, among others.[64]

In fact, until the process for recognition and vesting of rights, including individual[65] and community rights,[66] is completed, the beneficiary has complete protection from being evicted from the land under his occupation.[67] This would ensure that on recognition of right, the beneficiary can claim compensation for resettlement when an area is declared as a Critical Wildlife Habitat (CWH) or diverted for carrying out some other development activity.

Where a forest dweller has been evicted without completion of the recognition and vesting of rights, the District Level Committees may bring such cases to the notice of State Level Monitoring Committee for taking requisite actions against the violation.

3.3.2 Critical Wildlife Habitats[68]

After forest rights are recognised and vested as mandated in Section 4(1) of the Act and the process of recognition and vesting completed as per Section 6 of the Act, such forest rights may subsequently be modified or resettled for creating inviolate areas called Critical Wildlife Habitats for wildlife conservation (Box 3.7). Such areas are created as an exception to the implementation of the Forest Rights Act, 2006.

The Act envisages[69] six steps for modification/restriction of rights in areas that have been declared as Critical Wildlife Habitats under the Act (Fig. 3.1). However, the resettlement or modification of rights of STs and OTFDs will only be allowed if all the conditions specified in the Act are fulfilled.

Box 3.7 Identification and Declaration of a Critical Wildlife Habitat (CWH)

The key authority in identifying and notifying a CWH is the Ministry of Environment and Forests and Climate Change (MoEF&CC) through the following process:

- MoEF&CC shall identify areas within National Parks and Sanctuaries wherein it is specifically and clearly established, case by case, through scientific and objective criteria that such areas are required to be kept as inviolate for the purpose of wildlife conservation.

(continued)

Box 3.7 (continued)

- MoEF&CC is mandated to constitute an expert committee to initiate field survey through an open process of consultation with all stakeholders.
- After an area is made inviolate within a National Park or a Sanctuary and rehabilitation proceedings have been completed, the State Government cannot divert the said Critical Wildlife Habitat for any other purpose whatsoever [Proviso to Section 4(2)].

The pre-requisites for creation of Critical Wildlife Habitat requires that the process of recognition and vesting of rights has been completed[70] in all the areas under consideration. The State Government or its agencies in exercise of their powers under Wildlife Protection Act, 1972 has to establish that the activities or impact of the presence of holders of rights threaten the existence of wild animals and their habitat. At the same time, all other reasonable options of co-existence of holders of rights and wildlife, among other reasonable options, have been deliberated upon and ruled out by the State Government.

Once the area has been considered as inviolate for wildlife conservation, a resettlement or alternative package has to be prepared for providing secure livelihood and any other requirements of the affected individuals and communities in the proposed Critical Wildlife Habitat. The Gram Sabha will consider resettlement packages and has to give it free informed consent in writing regarding the same.[71] The implementation of the resettlement package will be monitored by the State Level Monitoring Committee.[72] Further, resettlement will take place only if all the facilities and allocation of land at the proposed site of relocation are complete as per the resettlement package that was promised.

If either one of the above-mentioned conditions (Fig. 3.1) remains absent, there cannot be any restriction or modification of the rights of STs and OTFDs even in Critical Wildlife Habitats. The Ministry had released Guidelines for Critical Wildlife Habitats in 2007 and subsequently in 2018. However, the same has not been enforced yet by the Centre or by the States. Thus, the provision remains redundant on paper. However, often a parallel has been drawn between Critical Wildlife Habitats and Tiger Reserves, although the latter was given statutory protection much later in 2006 under Wildlife Protection Act, 1972 and will not be governed by the regime

[64] Section 4(7) of the Forest Rights Act, 2006.

[65] Rule 12A of the Forest Rights Rules, 2007.

[66] Rules 12B of the Forest Rights Rules, 2007.

[67] Section 4(5) of the Forest Rights Act, 2006.

[68] Section 4(2) of the Forest Rights Act, 2006.

[69] Ibid.

[70] As specified under Section 6 of the Forest Rights Act, 2006.

[71] To be r/w Rule 4(1)(d) of the Forest Right Rules of 2007, which states the functions of Gram Sabha and includes consideration of resettlement packages under s. 4(2)(e) and passing resolutions regarding the same.

[72] Rule 10 (e) of the Forest Right Rules of 2007.

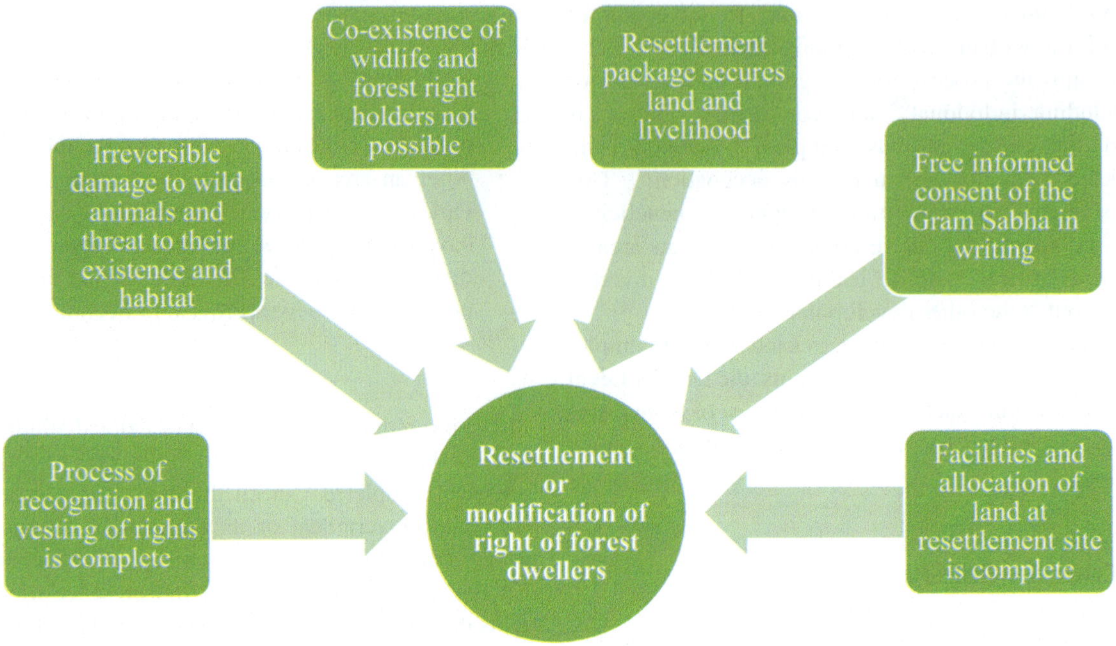

Fig. 3.1 Pre-requisites for creation of Critical Wildlife Habitats

prescribed for critical wildlife areas under the Forest Rights Act.

3.3.3 Transfer of Title of Forest Right Holder Under the Scheduled Tribes and Other Traditional Forest Dweller Act, 2006

The title granted under FRA is heritable and cannot be alienated or transferred to another party.[73] The objective of putting restriction of the transfer of titles is to ensure third parties do not undermine the process of recognition of rights through extraneous means of getting the title transferred in their favour. The title cannot be pledged with a bank for receiving loans.

3.3.4 Duties of Forest Dwellers

Any forest-related legislation would be incomplete without according the users of the forest resource or land the duty to continue conservation or ameliorative measures for diverted forest land and resource. The forest right holders also have not been provided a blanket exemption by virtue of their rights. All forest right holders along with the Gram Sabha and other village-level institutions have corollary duties towards continued conservation of forest, wildlife, biodiversity and water resources in the land over which they exercise their rights (Box 3.8).

Box 3.8 Duties Under Section 5 of the FRA

(a) Protect the Wildlife, Forest and Biodiversity
(b) Adequate protection of adjoining catchment area, water sources and other ecological sensitive areas
(c) Preservation of habitat of the forest-dwelling communities from destructive practices affecting their culture and heritage
(d) Ensure compliance with decisions of the Gram Sabha regulating access to community forest resources and restricting activities that adversely affect wildlife, forest and biodiversity

The Gram Sabha, to effectively exercise its duties,[74] has been empowered to create separate Committees with exclusive mandate towards protection of wildlife, forest and biodiversity.[75] The said constituted Committee would also prepare a conservation plan for management of community forest resources (CFRs) to facilitate sustainable and equitable management of such CFRs for the benefit of FDSTs and OTFDs. The conservation and management plan has to be integrated

[73] Section 4(4) of the Forest Rights Act, 2006.

[74] Under Rule 5 of the FRA Rules, 2008.

[75] Rule 4(1)(e) of the Scheduled Tribe and Other Traditional Forest Dwellers (Recognition of Forest Rights) Rules, 2007.

with micro plans or working plans or management plans of forest department, with modifications as considered necessary by the Committee.[76] In 2012, the Amendment to the Forest Rights Rules also made it mandatory for the Gram Sabha to approve all decisions of the Committee pertaining to issue of transit permits, use of income from sale of produce or modification of such management plans (Table 3.1).

3.4 Nodal Agency for the Act

Section 11 of the FRA states that the Ministry of Central Government dealing with Tribal Affairs or any officer or authority authorised by the Central Government on this behalf shall be the nodal agency for the implementation of the provisions of the Act. By virtue of such a provision, the Ministry of Tribal Affairs (MoTA) becomes the nodal agency for the implementation of the provisions of the Act and any notifications, clarifications etc. that are to be issued regarding the Act or its provisions shall be done by the MoTA.

Exception: The notification of a Critical Wildlife Habitat (CWH) is issued by the Ministry of Environment and Forests after a process of consultation by an expert committee, which includes experts from the locality including a representative of MoTA.

3.5 Procedure for Vesting of Forest Rights

Procedure for vesting of forest rights is presented in Flow Chart 3.1.

3.6 Some Myths and Perceptions

There are several myths and perceptions around FRA and its implementation. The sceptics declared that the Act hands over 4 ha of forest land to every tribal family! The answer is an emphatic No. The recognition is about existing occupation up to a maximum of 4 ha, to be validated in a four-level scrutiny process where every concerned department, including forest, tribal, revenue and panchayat, would be involved in verifying the veracity of the claims based on sound evidence.

The question that still remains is whether there is a rationale of 4 ha. Clearly, this magical figure of 4 ha was a bargain between 2.5 ha as envisaged in the first draft of the Act on the basis of existing Forest Village Rules in various states to 'as is where is basis' advocated by the Joint Parliamentary Committee. Clearly, there is no scientific or legal basis of 4 ha.

Another crucial debate is the cut-off date! This Act has no relevance to the cut-off date, be it 1980 when the Forest Conservation Act came into being or December 2005, which is mandated by the current Act. Historical injustice has to be proved by historical records. Clearly, 13 December 2005 is not a historical date and neither is 1980.

A popular myth is whether this Act is about regularisation of encroachment. This Act is not about regularisation of encroachment; in fact, it is about recording unrecorded rights and following a due process. Every Wildlife Protection and Forest Act in the country lays down a due process of settlement of bona fide claimants. The Forest Rights Act attempts to better the existing process. Another myth is that several hundred thousand hectares will be transferred to tribals and ownership will be given by the Forest Rights Act. The Act nowhere mentions the word ownership (except for minor forest produce) or forest land transfer. In fact, it only secures tenure on forest land to those who have been residing since generations.

There is also this mistrust by sceptics that the Gram Sabha, the lowest unit of governance, cannot be given this important task of initiating the recognition process for they lack capacity. The questions to such sceptics are 'What are the other institutional models available at the village level? Why are we so reluctant to rely on the wisdom of the village (the ordinary *Gram or Gaon)*?' It is time they have their chance and say!

3.7 Some Facts That Cannot Be Ignored

The FRA also is premised on some basic facts that cannot be ignored. The process of reservation under Indian Forest Act (IFA) or the state forest Acts has been faulty and not well documented. The settlement process under the IFA as well as Wildlife Protection Act has been far from satisfactory and the Supreme Court of India has been struggling to set this right since 1995 in the ongoing *Center for Environment Law—World Wide Fund for Nature* versus *Union of India* case. Therefore, the process of recognition and vesting of forest rights must be given its due, sans these myths and perceptions and in the backdrop of the facts on the ground.

3.8 Preliminary Lessons from the Field: More Questions than Answers

Several field visits across central and eastern and northeastern belts have revealed that when it comes to Forest Rights Act implementation, we have lost ourselves in only 1 out of the 13 set of rights that is envisaged under the law, i.e. land recognition of up to 4 ha. Clearly, the obsession with numbers has to go. Data from the MoTA show that individual

[76]Rule 4(1)(f) of the Scheduled Tribe and Other Traditional Forest Dwellers (Recognition of Forest Rights) Rules, 2007.

Table 3.1 Authorities under the FRA, their powers and functions

S. no.	Authority under the FRA	Covered u/s	Composition of the authority	Powers and functions under the Act, 2006	Powers and functions under the Rules, 2007 and Guidelines 2012	Remarks
1.	**Gram Sabha** (defined u/s 2(g) of the FRA to mean a village assembly convened by the Gram Panchayat that shall consist of all adult members of a village and in cases of states having no Panchayats, Padas, Tolas and other traditional village institutions and elected village communities, with full and unrestricted participation of women)	Section 6 (1)	All adult members of the village, with full and unrestricted participation of women	1. **Initiation of the process for vesting of forest rights** Gram Sabha initiates the process for determining the nature and extent of individual or community forest rights or both within the local limits of its jurisdiction under this Act. This is done by receiving, consolidating and verifying claims and preparing a map delineating the area of each recommended claim [Section 6(1), also Rule 4 and Para iv.c of Guidelines] 2. **Preparation of list of claimants** The Gram Sabha is also mandated under the Act to prepare a list of claimants of forest rights and maintain a register containing such details of claimants and their claims as the Central Government may by order determine [Section 6(1), also Rule 4] 3. **Passing a resolution on the claims presented** The Gram Sabha then passes a resolution on claims on forest rights after giving reasonable opportunity to interested persons and authorities concerned and forwards the same to the Sub-Divisional Level Committee (SDLC) [Section 6(1), also Rule 4] 4. **Regulating access to community forest resources** Gram Sabha is empowered to regulate access to community forest resources and stop any activity that adversely affects the	1. **Resolutions on resettlement packages** The Gram Sabha also considers resettlement packages under clause (e) of sub-section (2) of Section 4 of the Act and passes appropriate resolutions 2. **Demarcation of boundaries for community forest resource** Gram Sabha to demarcate the boundaries of the community forest resource for the purposes of filing claims for the recognition of forest rights [Para iv.c of the Guidelines of 2012] 3. **Constitution of committees from among its members** The Gram Sabha also constitutes Committees for the protection of wildlife, forest and biodiversity, from among its members, in order to carry out the provisions of Section 5 of the Act [Rule 4(1)(e) and r/w para iv.e of the Guidelines of 2012] 4. **Modification of Conservation and Management Plan** The Gram Sabha has to approve any modification to the Conservation and Management Plan prepared by the Committee for protection of wildlife, forest and biodiversity [Rule 4(1)(g)] 5. **Creation of the Forest Rights Committee (FRC)** The Rules mandate the creation of a specialised committee of not <10 but not exceeding 15 persons	The Forest Rights Committee (FRC) is a subset of the Gram Sabha, which has been constituted under the Rules of 2007 with the objective of assisting the Gram Sabha in its function to receive, acknowledge and retain the claims in the specified form and evidence in support of such claims [Rule 11(2) of the Rules of 2007] For diversion of forest land for non-forest purposes under the Forest (Conservation) Act of 1980, the MoEF letter dated 3 August 2009 encloses a list of evidences required to show compliance with the provisions of the Forest Rights Act, 2006, which includes letter from the concerned Gram Sabha stating their consent for the proposed diversion and compensatory, ameliorative measures required, or written consent or rejection of the Gram Sabha to the proposal for diversion of forest land

(continued)

Table 3.1 (continued)

S. no.	Authority under the FRA	Covered u/s	Composition of the authority	Powers and functions under the Act, 2006	Powers and functions under the Rules, 2007 and Guidelines 2012	Remarks
				wild animals, forest and bio-diversity [Section 5(d), also Rule 4(1)(e)]	as members that the Gram Sabha elects from among its members, called the Forest Rights Committee 6. **Institutional set up of the FRC** The Forest Rights Committee decides upon a Chairperson and a Secretary from among its members and intimates the Sub-Divisional Level Committee (SDLC) (Rule 3(2)) 7. **Composition of the FRC** The Rules mandate that two-third members of the Committee shall be from the Scheduled Tribes Not less than one-third of such members shall be women 8. **Raising Awareness regarding the provisions of the Act and Rules** The Guidelines require the Gram Sabha to be actively involved in the task of raising awareness regarding the provisions of the Act and Rules, especially the process of filing petitions [Para vi. c]	
2.	**The Sub-Divisional Level Committee (SDLC)**	Section 6 (1) of the Act	SDLC, which is constituted by the State Government, has the following members: (a) Sub-Divisional Officer or equivalent officer: Chairperson (b) Forest Officer in charge of a Sub-division or equivalent officer: Member (c) Three members of the Block- or Tehsil-level Panchayats to be nominated by the District Panchayat of	1. **Examination of the resolutions passed by the Gram Sabha** The SDLC has been mandated by the Act to examine the resolutions passed by the Gram Sabha [Section 6(4)] 2. **Preparation of the records of forest rights** [Section 6(4)] 3. The SDLC then forwards it to the District Level Committee, with their recommendations* on the veracity of the	Rule 6 grants the following powers and functions to the SDLC: 1. **Providing information to the Gram Sabhas about their duties** The Rules mandate the SDLC to provide information to each Gram Sabha about their duties and duties of holder of forest rights and others towards protection of wildlife, forest and biodiversity, with	*The SDLC does not have the power to take decisions on the claims. Also, the recommendations of the SDLC are not binding on the DLC, which means that the DLC can reject the recommendations of the SDLC

(continued)

Table 3.1 (continued)

S. no.	Authority under the FRA	Covered u/s	Composition of the authority	Powers and functions under the Act, 2006	Powers and functions under the Rules, 2007 and Guidelines 2012	Remarks
			whom at least two shall be from the Scheduled Tribes, preferably those who are forest dwellers or who belong to the primitive tribal groups, and where there are no Scheduled Tribes, two members who are preferably other traditional forest dwellers, and one shall be a woman member; or in areas covered under the Sixth Schedule to the Constitution, three members nominated by the Autonomous District Council or Regional Council or other appropriate zonal level, of whom at least one shall be a woman member (d) An officer of the Tribal Welfare Department in-charge of the Sub-division or where such officer is not available the officer in-charge of the tribal affairs (Rule 5) (Total members: 6)	claim for a final decision 4. **Appellate Authority** The SDLC hears petitions from persons, including State agencies, aggrieved by the resolutions of the Gram Sabhas (petition to be filed within 60 days and opportunity of being heard would be provided) (Section 6 (2) proviso)	reference to critical flora and fauna that need to be conserved and protected 2. **Providing forest and revenue maps to the Gram Sabha** The SDLC also provides forest and revenue maps and electoral rolls to the Gram Sabha or the Forest Rights Committee 3. Collation of all the resolutions of the concerned Gram Sabhas 4. Consolidation of maps and details provided by the Gram Sabhas 5. Examination of the resolutions and the maps of the Gram Sabhas to ascertain the veracity of the claims 6. **Adjudicatory powers** The SDLC also hears and adjudicates disputes between Gram Sabhas on the nature and extent of any forest rights 7. **Co-ordination with other SDLCs** The SDLCs shall also co-ordinate with other Sub- Divisional Level Committees for inter-sub-divisional claims 8. **Preparation of Block/Tehsil-wise record of rights** The SDLC prepares Block- or Tehsil-wise draft record of proposed forest rights after reconciliation of government records 9. **Raising awareness about the Act and the Rules** The SDLC shall also raise awareness among forest dwellers about the objectives and procedures laid down under the Act and in the rules	

(continued)

Table 3.1 (continued)

S. no.	Authority under the FRA	Covered u/s	Composition of the authority	Powers and functions under the Act, 2006	Powers and functions under the Rules, 2007 and Guidelines 2012	Remarks
					10. Ensure easy and free availability of proforma of claims to the claimants 11. Ensure that the Gram Sabha meetings are conducted in free, open and fair manner with requisite quorum	
3.	**The District Level Committee (DLC)**	Section 6 (5) of the Act	DLC is constituted by the State Government with the following members: (a) District Collector or Deputy Commissioner: Chairperson (b) Concerned Divisional Forest Officer or concerned Deputy Conservator of Forest: Member (c) Three members of the district panchayat to be nominated by the district panchayat, of whom at least two shall be from the Scheduled Tribes, preferably those who are forest dwellers or who belong to members of the primitive tribal groups, and where there are no Scheduled Tribes, two members who are preferably other traditional forest dwellers, and one shall be a woman member; or in areas covered under the Sixth Schedule to the Constitution, three members nominated by the Autonomous District Council or Regional Council of whom at least one shall be a woman member (d) An officer of the Tribal Welfare Department in-charge of the district or where such officer is not available, the officer in charge of the tribal	1. **Final authority on the status of the claims** Section 6(5) authorises the DLC to consider and finally approve the claims and record of forest rights prepared by the SDLC 2. **Final Appellate Authority regarding Status of Claims** The DLC is the final Appellate Authority to hear petitions from persons aggrieved by the orders of the SDLC	A. **Rule 8 enumerates the following functions of the DLC:** 1. Ensure that the requisite information under Clause (b) of Rule 6 has been provided to Gram Sabha or Forest Rights Committee 2. Examine whether all claims, especially those of primitive tribal groups, pastoralists and nomadic tribes, have been addressed keeping in mind the objectives of the Act 3. Co-ordinate with other districts regarding inter-district claims 4. Issue directions for incorporation of the forest rights in the relevant government records including record of rights 5. Ensure publication of the record of forest rights as may be finalised 6. Ensure that a certified copy of the record of forest rights and title under the Act as specified in Annexures II, III and IV of the FRA is provided to the concerned claimant and the Gram Sabha respectively B. **Paragraph iii of the Guidelines provide that DLCs should:** 1. Provide a record of prior recorded Nistar	–

(continued)

Table 3.1 (continued)

S. no.	Authority under the FRA	Covered u/s	Composition of the authority	Powers and functions under the Act, 2006	Powers and functions under the Rules, 2007 and Guidelines 2012	Remarks
			affairs (Total members: 6)		or other traditional community rights to the Gram Sabha. Further, it has to ensure that there are valid reasons for rejection of claims for usufructuary rights, and the same are recorded in writing 2. Facilitate filing of claims for pastoralists in Gram Sabha 3. Play a pro-active role for ensuring PTGs (now PVTGs) receive habitat rights and their claims are filed before the Gram Sabha	
4.	**The State Level Monitoring Committee (SLMC)**	Section 6 (8) of the Act	SLMC is constituted by the State Government with the following members, namely: (a) Chief Secretary: Chairperson (b) Secretary, Revenue Department: Member (c) Secretary, Tribal or Social Welfare Department: Member (d) Secretary, Forest Department: Member (e) Secretary, Panchayati Raj: Member (f) Principal Chief Conservator of Forests: Member (g) Three Scheduled Tribes member of the Tribes Advisory Council, to be nominated by the Chairperson of the Tribes Advisory Council, and where there is no Tribes Advisory Council, three Scheduled Tribes members to be nominated by the State Government (h) Commissioner, Tribal Welfare or equivalent who shall be the Member: Secretary (Total members: 10)	1. **Monitoring the process of right recognition** Section 6(8) mandates the SLMC to monitoring the process of recognition and vesting of forest rights in the State 2. **Co-ordination with the Nodal Agency** The SLMC shall be the contacting point for the nodal agency for matters related to the Act, which in this case is the Ministry of Tribal Affairs, and to submit to the nodal agency such returns and reports as may be called for by that agency	A. **Rule 10 envisages the functions of the SLMC as follows**: 1. Devise criteria and indicators for monitoring the process of recognition 2. Meet at least once in three months to monitor the process of recognition, verification and vesting of forest rights 3. Consider and address the field-level problems, and furnish a quarterly report in the format appended as Annexure V to these rules, to the Central Government on their assessment regarding the status of claims, the compliance with the steps required under the Act, details of claims approved, reasons for rejection, if any, and the status of pending claims 4. Take appropriate actions against the concerned authorities under the Act on receipt of a notice as mentioned in Section 8 of the Act, (cognisance of offences: Prior notice to be given to SLMC before initiating any action	–

(continued)

Table 3.1 (continued)

S. no.	Authority under the FRA	Covered u/s	Composition of the authority	Powers and functions under the Act, 2006	Powers and functions under the Rules, 2007 and Guidelines 2012	Remarks
					under the Act) [r/w para vi. d of the Guidelines] 5. Monitor resettlement under sub-section (2) of Section 4 of the Act (modification and resettlement of forest rights in Critical Wildlife Habitats) B. **Monitor compliance for rehabilitation under S. 3(1)(m) and 4(8) of the Act.** The Guidelines state that the SLMC should monitor compliance with in situ rehabilitation in cases where they have been illegally evicted or displaced from forest land without receiving legal entitlement to rehabilitation and when they are displaced without land compensations due to state development interventions [para v. e of the Guidelines of 2012] C. **Take action for evictions without settlement of rights** SLMC to take appropriate action against violation of the provisions contained in Section 4(5) of the Act, which restrict evictions or removal of FDSTs and OTFDs unless the process of recognition and verification is complete	

right, especially recognising lands for agriculture, is a preferred claim than CoFR or CFR. The shortcoming is that there are no separate State instructions on recognition or delineation process especially for community forest resource and/or Community Forest Rights. More recently after the passage of the September 2012 Rules there is a visible shift. Another key lesson is distinguishing community forest resource and Community Rights at the operational level, which is a must, which is still not visible in most sites.

Flow Chart 3.1 Vesting of forest rights

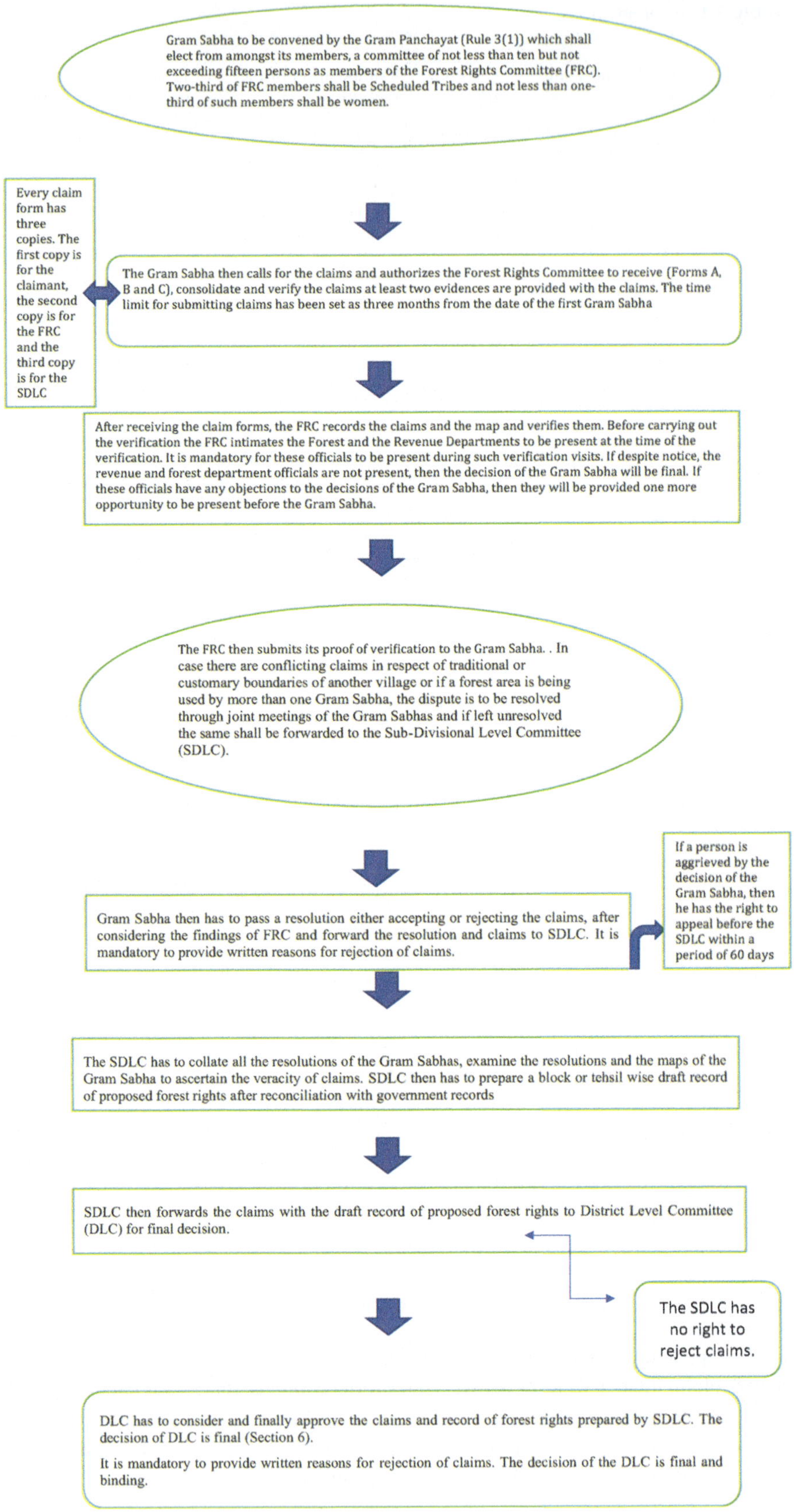

3.9 Points to Ponder and Way Forward

It is time to identify the real threats! Diversion of forest land and fragmentation of habitat are a reality, which is a consequence of the rush for the current growth rate. The unholy nexus of some executives–politicians–corporates has to be identified and exposed. New criteria for diversion of various categories of forest land and their livelihood impacts on dependent communities, especially in the light of new instruments such as Forest Rights Act, have to be developed. Can we go for Wasteland first? Further, strict monitoring of diverted land and whether they conform to the general and special conditions based on which they were granted diversion permissions are key. There is also a need for reversal of land transfer processes, especially when it has not been used for the purposes for which it was diverted or acquired. Last but not the least, it needs to be recognised that no right is absolute and it needs to be commensurate with duties and responsibilities.

3.10 Concluding Remarks: Key Issues and Challenges

The FRA, 2006, despite its objectives and scope, has always been in the eye of storm, with the Central and the State Governments either selectively implementing the Act, or implementing it in contravention of the mandate of the Act. Thus, there have always been two extreme positions on the Act. One holds that historical injustice has been done to forest dwellers while the other believes that the last forest tracts will be handed over to the people who cannot co-exist with wildlife, especially tigers.

As has been elucidated in the previous sections, the FRA, 2006 conceptualises a process of recognition and vesting of rights and acts as a one-law army to remedy historical injustices to the FDSTs and OTFDs. It is for this reason that one often, albeit mistakenly, adjudges the success of FRA, 2006 based on the number of rights that have been recognised and vested after a scrutiny. However, the success of the process of recognition and vesting of rights is determined by how effectively the regulatory, institutional authorities implement the mandate envisaged under the FRA, 2006 to ensure that it is not beyond the understanding of the lay person. Therefore, one needs to move beyond the number game to understand the impediments faced in implementation of FRA, 2006.

At this juncture, it would be prudent to talk about the latest developments on the implementation of the Act, which has far-reaching consequences, to say the least. The Honourable Supreme Court in the matter titled 'Wildlife First & Ors *vs* Union of India & Ors' (WP (C) No. 50/2008) first directed[77] the States to carry out evictions of those persons whose claims have been rejected. Before this, the States had been directed to file the status of the claims vis-à-vis those that had been granted and those that had been rejected. The Honourable Supreme Court interpreted the rejected claimants as 'encroachers', a term that has neither been mentioned nor defined in the FRA, 2006 or any forest law for that matter. The above-mentioned Order sent shock waves across the country and over one million 'tribals' and 'other traditional forest dwellers' were at the brink of being evicted from their habitats and livelihoods. At the time, the eviction Order was being seen as a response to the Central and the State Governments having failed to defend a law that protected the rights of these communities.

The Order dated 13 February 2019 was however stayed,[78] albeit temporarily, when the Honourable Court was appraised of the fact that even though the State Governments had filed their data on the status of claims including the rejected claims, the reasons and the procedure adopted for the rejections had not been stated adequately. It was stated that there was nothing on record to show as to who had rejected the claims and under which provision of the law were the evictions to be carried out, along with the competent authority for the same. There were several other lacunae in the rejection process that were pointed out, which have been summarised as following:

1. In most of the matters, tribals have not been served with the rejection orders of their claims
2. It is also not clear whether the three-tier Monitoring Committee constituted under the FRA, 2006 has supervised the rejections
3. Whether the tribals were given opportunity to adduce evidence
4. If yes, to what extent and whether reasoned orders have been passed regarding rejection of the claims

Even though the evictions have been temporarily skirted, there are huge gaps, apart from those mentioned above, that still plague the implementation of the FRA. They are, to name a few:

1. Discrepancy regarding the details of claims filed, rejected and granted between the Ministry of Tribal Affairs and the State Department
2. Rejections by SDLC, or other department(s), even though not mandated under the law to reject

[77] Vide Order dated 13 February 2019.

[78] By the Ld. Solicitor General on 28 February 2019. The Stay Order continues till date 19 January 2022, after having been extended vide Order dated 22 January 2020.

3. Mismatch between Gram Sabha Resolutions and DLC Certificates
4. Extent of cultivation reduced without giving any reasons
5. PVTG habitat rights not granted—despite the requirement of Suo motto enquiry under the Guidelines of 2012
6. The issue of procedural gap wherein Gram Sabha is not being empowered to fulfil the mandate of the FRA
7. Whether appeals made and pending, and yet rejected

What we also need to answer is whether mere granting of claims is enough? The FRA Rules were amended in 2012 wherein a new Rule was inserted (Rule 16), which places a duty on the State Governments to provide post-claim support to the forest right holders. For this, the State Governments have to ensure that all government schemes including those relating to land improvement, land productivity, basic amenities and other livelihood measures are provided to such right holders and communities whose rights have been recognised and vested under the Act. A robust post-claim strategy is essential to make the land productive so that communities get a more secured livelihood from their own land rather than being totally dependent on adjoining forests. It is clear from field experiences that mere title recognition without adequate handholding and support is insufficient. Support is required from government departments and agencies responsible for supporting agriculture, conservation and sustainable management of forest to help the communities in using, conserving and managing both the recognised land and the community forest resource, the title of which is now vested with them.

One thing is certain, that the FRA, 2006 is a unique law with a unique framework for implementation where, perhaps for the first time, four key line agencies have been legally mandated to scrutinise the dependence, the extent, and the post-claim support and, most importantly, secure the tenure on the land on which FDSTs and OTFDs have been historically dependent upon. The implementation of this law, therefore, is nothing short of a social engineering marvel. However, only time will tell whether this social engineering experiment created history or the experiment was confined to history!

Lessons Learnt

After having established the details of the FRA, let us summarise in a nutshell the lessons learnt vis-à-vis the implementation of the Act.

- There exists a major discrepancy regarding the data on the details of claims filed, rejected and granted between the Ministry of Tribal Affairs and the State Departments.
- More often than not, the reasons for rejection of claims are not mentioned. It has also been seen that the SDLC is rejecting claims, even though it is not mandated under the law to reject such claims.
- There is a huge mismatch between Gram Sabha Resolutions and DLC Certificates.
- There is an absolute lack of communication of rejected claims to the claimant.
- The extent of cultivation is reduced without giving any reasons.
- PVTG Habitat rights are not being granted, despite the requirement of Suo motto enquiry under the Guidelines of 2012.
- The issue of Critical Wildlife Habitat remains unresolved.
- The capacity of the Gram Sabha to prepare conservation and management plan for conservation and sustainable use is still a distant goal sans few exceptions.
- Little or no understanding of which of the Community Forest Rights are being exercised and how do they relate to the legal provisions!
- The confusion between community rights and development rights still exists.
- Delineation of CFR, the operational mechanism, is missing and communities are confused between traditional boundaries and legal boundaries.
- There is no clear linkage of existing Joint Forest Management/Participatory Forest Management/Community Forest Management/Eco Development Committee institutions or processes with CFR delineation process of Forest Protection Committees. It is hoped now that the new provision on aligning the above with the conservation and management plan may help.
- Inter-village and inter-panchayat claims that overlap and disputes resolution mechanisms still not well developed, including on community forest resource rights and community forest rights.
- There are no clear guidelines to First Appellate Bodies such as Sub-Division Level Committee and also to the District Level Committee on grounds of rejection or approval including reasoned orders—Now reasoned orders are a must.

Key Questions

1. Has the Ministry of Tribal Affairs been able to handhold, monitor and regulate the implementation of the FRA effectively?
2. Why is the Forest Department opposed to the FRA? Are the rights under FRA a threat to the forests and its wildlife?
3. There is overlapping of provisions of the Forest Rights Act (Critical Wildlife Habitat[79]) and Amendments in the Wild Life (Protection) Act of 1972 (Critical Tiger

[79] Section 2(b), Scheduled Tribes and Other Traditional Forest Dwellers (Recognition of Forest Rights) Act, 2006.

Habitat[80]). In case of a conflict between the two legislations, the question is as to which would prevail over which category and how each of these categories would be made inviolate? Also, how these new and seemingly conflicting categories need to be understood to iron the rough edges and ensure harmony between the two equally significant pieces of legislations?

4. Whether appeals made and pending, and yet rejected?

[80] Section 38V(5), The Wild Life (Protection) Act 1972.

Forestry Research, Education, and Extension in India

4

H. S. Ginwal and Kirti Chamling Rai

Abstract

Forest is the second largest land use in India next to agriculture. Protection and management of the forests in India have, at different points in the nation's history, been under a set of rules and regulations by different dynasties. Forestry research in India is one of the oldest in the tropical world. Indian forestry research was looked upon as a model for other countries in the region. When Forest Departments were established in the several British provinces during the early 1960s, scientific forestry got its start in India. Dietrich Brandis's appointment as the Inspector General of Forests in 1864 marked the start of the scientific era in forest management. With the advent of scientific management of forests, it became clear that organized research was necessary to facilitate proper conservation and utilization of forests in India. Dehradun has been the center of forestry education since 1884, when Imperial Forest College was established, to train rangers and foresters. It later came to be known as the Imperial Forest Research Institute, which was set up in 1906 by the Government of India for preparing a professional forestry course of high standard for the officers. Forestry research in Imperial Forest Research Institute during the pre-Independence period had two broad disciplines, viz. biological research and forest products research. The biological research discipline had silviculture, botany, and entomology as its various branches, which conducted experimental silviculture and statistical research, preparation of forest floras, and studies on forest insect pest. The Indian Council of Forestry Research and Education, or ICFRE, was established in 1986 as a national umbrella organization to handle the needs of the nation for forestry research, teaching, and extension. Several significant accomplishments have been made since organized forestry research in India first started. These include introducing novel silvicultural techniques, improving planting stock production methods, managing diseases and pests, introducing high-yielding clones, and introducing exotic species to increase forest productivity.

The National Commission on Agriculture (NCA 1976), a division of the Ministry of Agriculture, proposed the establishment of forestry education in India in 1976 with the goal of improving forest conservation, managing natural resources, and safeguarding the environment. This proposal was later included in the National Forest Policy of India (1988). India first launched university-level forestry education at Solan, Himachal Pradesh, in 1976. Following this, forestry programs were introduced by numerous agricultural universities in 1985 under the auspices of the Indian Council of Agricultural Research (ICAR) system (ICAR 2009). The nation's knowledge of biodiversity, ecology, and shared ground management can be strengthened, diversified, and integrated with the support of forestry education. Similarly, ICFRE has been periodically formulating and updating the forestry extension strategy for the country. Recently the "Extension Strategies in Forestry Research" was formulated for the country by ICFRE in 2011.

Adopting comprehensive, cross-sectoral methods to research, teaching, and extension is necessary in light of the rising environmental issues and rising global expectations from forests in the framework of the sustainable development goals. The nation has a largely good institutional framework in place to handle concerns related to forestry research and teaching under the specialized missions of already-existing institutes. To guarantee sufficient coverage of forestry concerns and to prevent duplication or overlap of efforts, these institutions' intersectoral coordination and networking must be improved. The historical development of scientific forestry is covered in this chapter, along with the main concerns of forestry research, teaching, and extension in India.

H. S. Ginwal (✉) · K. C. Rai
Forest Research Institute, Dehradun, Uttarakhand, India

A. K. Mandal, A. Nicodemus (eds.), *Textbook of Forest Science*, https://doi.org/10.1007/978-981-97-8289-5_4

Keywords

Inspector General of Forests · Imperial Forest College · National Commission on Agriculture · Van Vigyan Kendra

4.1 Historical Perspective

Historically, India has always been famous for its vast wealth of forests. The country had extensive tracts of wooded land till the eighteenth century. Brandis in his famous book on Indian Forestry noted that the total area of the British Indian Empire was 1,560,000 square miles with the number of indigenous trees exceeding 1200 and 120 species of bamboo. He also mentioned the important feature of traditional forestry in India of setting up sacred groves that were numerous in Garo and Khasi Hills, the Devara Kadus of Coorg, and the hill ranges of Salem district (Brandis 1897). During the early British Rule, the Indian forest went through a massive unprecedented destruction. According to Stebbing (1922), an imperial history of Indian woods, the British during this era were not familiar with the fundamentals of forestry science. Everyone agreed that, as far as the materials needed from the forest were concerned, the forest was boundless. As a result, the rate of forest degradation increased throughout the first few decades of British rule in India and was unchecked. Only when the government failed to provide local needs, such as lumber for shipbuilding, did it realize how bad things really were.

When Forest Research Institute (FRI), Dehradun, commemorated the centennial of the Indian Forest Department in 1961 with a paper providing a comprehensive history of forests prior to this time, scientific forestry reached its centennial (FRI 1961). The majority of the country was covered by forest, and most people lived there as forest dwellers, according to the journal. As agricultural production increased in search of food and pasture for cattle, the forests gradually began to disappear. Accessible woodlands suffered during the early years of British rule because valuable timber trees were cut down on a vast scale. The British realized sooner rather than later that they would need lumber for the Navy and the Empire, and that this would necessitate some sort of disruption. A commissioner was appointed in the Malabar region in 1800 to investigate the supply of teak. Watson was appointed Commissioner of Forests by the Madras government in 1806, with the responsibility of overseeing the production of teak and other timber fit for shipbuilding.

When Brandis was appointed as the Inspector General of Forests in India in 1864, the country entered a new era of scientific forest management. Nonetheless, the first known attempt to enhance and restore the nation's woods was made in 1842 when Conolly, the Collector of Malabar at the time, established the first teak plantations at Nilambur with the help of an Indian, Chathu Menon, a Sub-conservator of woods. The plantations had great success and gained international renown over time (Shanmuganathan 1997). The Directors of the East India Company realized that the south Indian forests needed to be improved as a result of Conolly's efforts. By inventing the Linear Valuation Survey, Brandis calculated the current increasing stock in the forest for the first time in relation to their size classes. Comprehensive ring counting was used to assess the rates of growth of teak. Even though Brandis started the process of creating a working plan for a few valuable woods, it moved very slowly, and, as a result, just 0.0282 million hectares of government-owned forests were under scientific management.

Similarly, the National Commission on Agriculture (NCA 1976) describes the beginning of scientific forestry in the early 60s of the last century with the creation of Forest Departments in the various British provinces. The newly enacted Indian Forest Act, 1865 required the forest officers to inspect tree-clad lands and all hilly regions to demarcate them followed by survey and mapping of suitable areas for "reserved" or "protected" forests. Following this the National Forest Policy of 1894, laid down some general principles on which the forests were to be managed. The state forests were to be administered with the sole objective of public benefit. Therefore, the rights and privileges of the inhabitants residing in the forest fringes were of the utmost concern for the conservation and preservation of the forest.

Since time immemorial, for the village dwellers all over the country, the forests had always provided for their daily needs, be it the fuel with which to cook their food or the small timbers for building shelter and other farm equipment to cultivate their land, etc. According to C.C. Wilson, a retired Chief Conservator of Forests and member of the Indian Forest Service (IFS), when the decrease in the forest cover was first noticed in 1861, scientific forest management became popular and forest conservation gained traction. The Governor General at the time, Lord Dalhousie, issued a memo on August 3, 1855, which became known as the "Charter of the Indian Forest" according to E.P. Stebbing. It laid out the principles for managing forests, which were to be treated as state property over which no community or individual had any rights. As a result, the Forest Act of 1865 was passed, establishing the Forest Department and governing the settlement, demarcation, management, and protection of forests (FRI 1961). But it did not take long to see that, in order to address the more pressing problems facing government and forest officials, a new and more extensive Forest Act was required. In 1869, Brandis drafted a measure that Baden Powell amended upon in 1874. A final version was then developed and passed as the Indian Forest Act of 1878. This is perhaps the most significant event in more than

100 years of scientific forestry in India, as this Act, though modified in 1927 by the Central Government and amended by several states, continues to be the central piece of legislation governing forest management.

There was lot of opposition to the newly formed policies of the Government by the forest dwellers, who had lived in these forests for centuries. Forests being an integral part of their lives were managed and owned by these communities, mainly tribal people. The colonial forest policies made them an outsider in their own land. Their customs were outlawed everywhere they could be found to be counterproductive to science. The tribal people rebelled against the colonial government's encroachment on their territory as a result. The movements for civil disobedience and non-cooperation are two major points of criticism to modern forestry. The Indian National Congress issued a number of resolutions that effectively expressed the public's disapproval of the unfair forest laws and the cruel forest management practices (Stebbing 1926). The National Forest Policy Resolution (1952) of the Government of India (GOI) essentially reaffirmed the methods and objectives of forest management that had been established by the British during the preceding 100 years or so, refusing to acknowledge the inefficiencies and injustice inherent in the forest policy. Much of the credit of framing the new policy is given to M.D. Chaturvedi, the first Indian Inspector General of India (1949–1954) (FRI 1961).

4.2 Inception of Scientific Forestry in India

Dietrich Brandis, the first Inspector General of Forests in India, established the Indian Forest Department in 1864, which gave rise to the idea of scientific forestry. In order for the new department to operate efficiently, legislation was required, and in 1865 the first Forest Act was approved. This Act allowed the government to regulate the use of forest resources through legislation, but it did not maximize resource optimization. Reserved woods, protected forests, and village forests were the three categories into which the forests were divided by the more detailed second Forest Act of 1878. This classification would not have been possible but for Dietrich Brandis, who drafted the first Forest Acts and initiated control over the forest resources as well as extended research activities in the highly heterogeneous forest areas in India. Further in 1894 when the forest map was clearer to the Government, the classification for protection became more effective. The forests that were to be preserved on climatic ground, forests to be protected to ensure supply of minor forest products, and forests to be used as pasture lands were divided.

Initially the interest in forestry research in India was more concerned with revenue extractions as large timber woods were required for the railway companies than developing forestry as a new area of scientific research. Till the 1850s, the colonial government through various sporadic proclamations and regulatory rules had asserted its sovereignty over the forests. For the sake of forest management, the majority of easily accessible woods with "valuable" trees like teak, sandal, etc. were taken over. But when the railroads arrived in 1853, everything suddenly changed. The foundation of colonial rule was the railway system, which required wood for sleepers to create the lines and as fuel for the locomotives. Right up to the end of the nineteenth century, firewood served as the primary fuel for railroads. Control over all of India's forests became essential to satisfy railroad needs. Due to the current policies and programs for forest conservation, the supply of timber was not guaranteed until 1857. The various railway firms' substantial sleeper requirements made the topic of forest conservation a crucial administrative issue. Around this crucial time, Forestry Research Institute and College in Dehradun was set up by the government. Initially, the institute was the place where forest rangers and other foresters were trained. Later, FRI and College became the ground where forest research became associated with forest education with the goals of research, maintenance, and improvement of forest estates; methods of improved utilization of the expensive resources of timber; education of young men in order to equip them to work and administer the work, etc. The institution did not concern itself for the commercial advancement of any particular material or product but being devoted to applied research, economic considerations naturally governed all its investigations. Further to ensure the sustainability of the work, the research work was divided into five heads: (1) Silviculture, (2) Forest Entomology, (3) Forest Utilization, (4) Forest Botany, and (5) Forest Chemistry. Therefore, in the first phase the idea of scientific forestry, which in theory aimed to protect Indian forests, was in reality implemented to extend the research activities to have a proper extraction of usable trees and other forest products, and here lies the true dichotomy of forest research in theory and forest preservation system in practice.

The inception of scientific forestry led to a phenomenal growth in forest revenue. The revenue for India during the year 1937–1938 was Rs. 30.19 million, which increased to Rs. 124.37 million at the end of WWII in 1944–1945. The revenue generated from the forest on the eve of Independence in 1946–1947, was Rs. 104.80 million. During the first 20 years after the enactment of the Forest Act, the total area of government forest was raised to over 100,000 square miles, from a total area of 14,000 sq. miles. This can be considered as the greatest achievements of the Forest Department (Stebbing 1923).

Forestry as a science was already prevalent in Europe with countries such as Germany and France having established private forestry schools in the early seventeenth century

(Chaney et al. 1998). After the colonization of India by the British, the changes in the general policy of resource use, most notably forest, led to the formation of the Forest Department in 1864 (Lawbuary 1999). With the advent of scientific management of forests, it became clear that organized research was necessary to facilitate proper conservation and utilization of forests in India. Management of forest in the colonial past was mainly aimed at consolidation of forests through various enactments and their conservation, besides utilization for various developments and needs of war. Dehradun has been the center of forestry education since 1878 when the Government of the North-West Province laid the cornerstone for a Forest School to train rangers and foresters. Only Europeans were initially eligible to apply for gazetted positions, and Indians were not allowed to join the provincial gazetted service until after 1912. When the school was given to the Government of India in 1884, it was called the Imperial Forest College. Later on, it gained recognition as the Forest Research Institute for providing officers with a top-notch professional forestry training. After the Indian Agricultural Research Institute, which was founded in 1901, it was historically the second central research institute to be formed by the Indian government (Ranganathan 1950). It was established in 1906. The establishment of the Forest Research Institute in Dehradun provided a significant boost to the forestry sector's efforts to maximize industrial growth. In November 1929, the Institute moved into a new £750,000 structure, which was dedicated by Viceroy Lord Irwin at the time.

4.3 Forestry Research

India is home to some of the oldest tropical forestry research projects worldwide. Other nations in the region looked to Indian forestry research as a model. Among the notable accomplishments were guidelines for creating working plans for a range of high and coppice forests, yield regulations, yield and volume tables, thinning techniques, and similar things (Singhal et al. 2003). Prominent publications like Troup's *Silviculture of Indian Trees*, Beeson's *Forest Insects*, and Kanji Lal's *Forest Floras*, as well as by Talbot Brandis, Mooney, and others attest to the high caliber and extreme dedication of Indian forestry study (Singhal et al. 2003).

Forestry research in Imperial Forest Research Institute during the pre-Independence period had two broad disciplines, viz. biological research and forest products research. The biological research discipline had silviculture, botany, and entomology as its various branches that conducted experimental silviculture and statistical research, preparation of forest floras, and studies on forest insect pest. Forest products research had timber mechanics, cellulose and paper, composite wood, wood seasoning, wood preservation, wood working, and chemistry of forest products as its branches. The forest products research was given a priority so as to meet the demands of supply during the First and Second World Wars, which worked with government departments (railways, defense) as well as forest-based industries to provide suggestions related to forest utilization. After Independence, the research facilities were broadened to encompass the increasing research requirements in the country for the proper implementation of forest development programs. Subsequently some research sections were made independent branches and a few branches were added, viz. Forest Ecology (1948), Publicity and Liaison (1950), Statistics (1950), Timber Engineering (1953), Forest Genetics (1957), and Plant Physiology (1958) (Singh 2017). For a long time, FRI was the only research center catering to all the State Forest Departments (SFDs) in the country. Later it was considered necessary to establish regional research centers (RRCs) of FRI to cope with the demands of the SFDs. The RRCs would supplement the research work of the FRI by focusing on the local problems. With the establishment of the Indian Council of Forestry Research and Education (ICFRE) and the transfer of responsibility for forest research, education, and training to this Council, a comprehensive reorganization of forestry research across the nation was initiated in 1986. Along with the Forest Research Institute, the RRCs were elevated to research institutions under the ICFRE, each with a distinct scientific and administrative mandate.

The forestry research suffered in the mid-war period when the demand for timber increased. This was the phase when the development of homogeneous forest came into prevalence and it affected the forests of India, which was heterogenous with a large number of species yet to be explored at the time. Single-variety plantation with a high return within a short period of time became the utmost aim of the research activities. The uncertain political future also affected the foresters in this time. Many of them, like Harry Champion, who was involved in the Indian forest service as silviculturist for long, left their job and returned to their home country. In addition to this, the traditional knowledge of the people for protecting the forest was wholly neglected and the concept of Joint Forest Management program, which was propagated by Sir Dietrich Brandis much earlier, was completely ignored.

4.3.1 Forestry Research After Independence

After the liberation of India, forestry research was still largely neglected in the initial years. In fact, the forest department was controlled by the Ministry of Agriculture just after the Independence. Some significant moves were made to reorganize the nation's forestry research and education in the mid-1980s, keeping this in mind as well as the new tendencies in forest management. To satisfy the expectations of the State Forest Departments, regional research centers of FRI were established in various regions. At a conference in 1956, the Central Board of Forestry (CBF) Expert Committee advocated the creation of Regional Research Centers (RRCs) to concentrate on local problems and support the FRI's research efforts (Expert Committee 1956:61–62; GoI 1957: 30). The purpose was to have one RRC in each region, i.e., the central, eastern, and southern regions, with FRI handling the research work of the northern region. Out of the four RRCs, two were established in southern region in Bangalore and Coimbatore. The RRC in Bangalore was established in 1958 and had its origins in the Forest Research Laboratory started by the erstwhile Mysore State in 1938. In Coimbatore, the RRC was established in 1959 by adding a biological research wing to the Southern Forest Rangers College (Singh 2017). Two RRCs, one each at Jabalpur and Burnihat, were established to meet the research requirements of the central and eastern Indian states, respectively.

In addition, a number of forestry and associated institutions were founded or restructured, and they are crucial to the effective management of the resources related to forests and animals. Below is a quick summary of significant national-level organizations.

4.3.2 Indian Council of Forestry Research and Education (ICFRE)

In order to handle the nation's demands for forestry research, education, and extension, the Indian Council of Forestry Research and Education (ICFRE) was established in 1986 (MoEF 1986). Under the former Ministry of Environment and Forests (MoEF), the ICFRE was established as an autonomous council in 1991 and registered as a society under the Societies Registration Act, 1860. ICFRE is currently the highest authority in the national forestry research system, promoting and carrying out need-based forestry research and extension. It has its headquarters in Dehradun. With its five centers spread across various bio-geographical regions of the nation and its nine regional institutes, the Council has a widespread presence throughout India (Table 4.1). The research institutes at the time of their creation framed mandates within the purview of which they could formulate their research programs.

The Council's goals are to conduct, support, encourage, and organize forestry research and its applications as well as education. It also hopes to create and maintain a national library and information center for allied sciences and forestry research, serve as a clearinghouse for general information and research, and offer consulting services in the areas of allied sciences, education, training, and forestry research. The ICFRE Society oversees the Council's operations. The Council's Board of Governors (BOG) is in charge of planning, managing, and funding it. The Council's chief executive is the Director General (DG), ICFRE. The IFS and SFS (State Forest Service) officers joined on deputation for specific tenures, while the permanent cadre of scientists and technical workers comprise the scientific, technical, and administrative personnel of the Council. Through the operationalization of the National Forestry Action Programme (NFAP) and National Forestry Research Plan (NFRP), the Council hopes to contribute to initiatives aimed at increasing the forest and tree cover as well as improving the productivity of forests (ICFRE 2000). In

Table 4.1 List of research institutes and advanced centers of ICFRE

S. no.	Name of institute/center	Location
Research institute		
1.	Arid Forest Research Institute (AFRI)	Jodhpur
2.	Forest Research Institute (FRI)	Dehradun
3.	Himalayan Forest Research Institute (HFRI)	Shimla
4.	Institute of Forest Biodiversity (IFB)	Hyderabad
5.	Institute of Forest Genetics and Tree Breeding (IFGTB)	Coimbatore
6.	Institute of Forest Productivity (IFP)	Ranchi
7.	Institute of Wood Science and Technology (IWST)	Bengaluru
8.	Rain Forest Research Institute (RFRI)	Jorhat
9.	Tropical Forest Research Institute (TFRI)	Jabalpur
Advanced center		
1.	Advanced Research Centre for Bamboo and Rattans (ARCBR)	Aizawl
2.	Centre for Forest Livelihood and Extension (CFLE)	Agartala
3.	Centre for Forestry Research and Human Resource Development (CFRHRD)	Chhindwara
4.	Centre for Social Forestry and Eco-Rehabilitation (CSFER)	Allahabad
5.	Forest Research Centre for Coastal Ecosystem	Visakhapatnam

addition, the Council earns its own funds through patent registration, externally funded programs, user agency consultations, and research projects.

Numerous significant accomplishments have been made since organized forestry research in India first got underway. These include the introduction of high-yielding clones and exotic species suitable for various physiographic zones of the nation to meet the growing needs of the wood-based industries for raw materials, as well as to increase the amount of forest cover and tree cover. They also include the scientific management of non-timber forest products (NTFPs) to address the livelihood concerns of the communities that live near the forest and in the forest, the value addition of NTFPs, and the endeavor to bring research inputs from the laboratory to the field. The major research outputs of ICFRE and state forest research organizations include silvicultural practices for a large number of species, agroforestry models, tree improvement work in general and of multipurpose tree species in particular, improved methods of resin tapping, wood seasoning and ammonia plasticization technique for bending of wood for commercial uses, development of cost-effective wood preservatives, extraction processes of chemicals from woods, development of bio-pesticides and natural dyes, technology for reclamation of degraded lands, development of machines and tools for implementation of various forestry operations including firefighting tools and implements, evaluation of physical and mechanical properties and identification of timber of various species, and preparation of volume, stand, and weight tables for different species.

4.3.2.1 Thrust Areas of ICFRE

Thrust areas and themes identified by ICFRE (ICFRE 2013) for categorization of various projects are presented in Table 4.2.

4.3.3 Indian Institute of Forest Management (IIFM)

Indian Institute of Forest Management (IIFM), Bhopal, was established in 1982 to undertake research on forest-management-related issues, which was not handled by Forest Research Institute, Dehradun. The objectives of the institute are to provide training in management and related subjects to persons and personnel of forest-related industries with a view to equipping them to practice the art and profession of management, in a most effective and efficient manner, prepare outstanding talented persons for careers leading to management responsibilities in forestry and forest-related systems, provide up-to-date information on forest management, and carry out research in matters concerning the use of management and allied techniques and methods conducive to the development of forestry.

4.3.4 Wildlife Institute of India (WII)

The Wildlife Institute of India (WII) had to be established in 1982 in order to properly handle research pertaining to wildlife. It was given autonomy in April 1986 and given the responsibility of training a group of scientists for nature conservation, qualified wildlife managers, and field biologists who could gather data on wildlife and protected areas for management purposes. It provides guidance to the government on issues pertaining to wildlife resource management and conservation. Primary sources of scientific data, field sites, are the focus of research programs carried out by WII around the nation. In addition to offering a 2-year postgraduate (PG) program on wildlife management, it also offers a nine-month postgraduate diploma program on wildlife management for forest officers, which is open to applicants from overseas. It also organizes three-month certificate course on wildlife management.

4.3.5 The Forest Survey of India (FSI)

The Forest Survey of India (FSI) was established in June 1981 with the intention of providing data for national planning, conservation, and management of environmental preservation as well as for the purpose of routinely monitoring the changing state of land and forest resources. In 1986, following an evaluation of its operations, the FSI underwent a reorganization. Every 2 years, FSI prepares a comprehensive State of Forest Report (SFR) that includes a national forest vegetation map. It also gathers, stores, and makes available the data related to forestry that are needed for planning and designing methods related to forest surveys. Finally, it provides training on contemporary forest survey inventory systems and supports and supervises the inventory work carried out by State Forest Departments (SFDs). The Director is in charge of the FSI, which has its main office in Dehradun. In addition to IFS officers and Indian Statistical Service officers appointed through deputation, the administrative and technical staff consists of a permanent cadre of technical and administrative workers. FSI has four regional offices: one each in Bangalore, Kolkata, Nagpur, and Shimla. Organizing fieldwork to gather inventory data, interpreting aerial photos, and carrying out studies on wood use are some of the main responsibilities of regional offices. The MoEF has direct administrative authority over the FSI.

In the formative years, research in the field of forestry was conducted at FRI and other institutes under the directive of ICFRE. Later on, the forestry research started to be undertaken in several state institutes and State Agricultural Universities (SAUs) of ICAR. The researches that were required to be conducted in laboratories were done by these institutes and SAUs, which are otherwise difficult to be

Table 4.2 Thrust areas and themes for categorization of research projects

Thrust areas	Themes
A. Research	
1. Managing forest and forest products for livelihood support and economic growth 2. Biodiversity conservation and ecological security 3. Forests and climate change 4. Forest genetic resources management and tree improvement	1. Silviculture 2. Social forestry, agroforestry, farm forestry 3. Sustainable forest management (SFM) 4. Forest economics 5. Forest biometrics and yield modeling 6. Participatory forest management 7. Wood science and technology 8. Chemistry of forest products 9. Wood based industries 10. NTFPs resource development 11. Bio-prospecting and bio-piracy 12. Seed science and technology 13. Forest certification 14. Forest hydrology 15. Food security 16. Bio-fuels and Bio-energy 17. Integrated pest and disease management 18. Application of microbes in forestry 19. Weeds and invasive species 20. Forest fire and grazing 21. Bio-informatics and geo-informatics 22. Policy and legal issues 23. Biodiversity conservation 24. Forest botany 25. Ethnic and traditional knowledge systems 26. Forest soils and land reclamation 27. Wetland and marine ecology 28. Watershed management 29. Climate change and forests 30. Forest ecology 31. Conservation of forest genetic resources 32. Tree improvement 33. Vegetative propagation 34. Biotechnology 35. Environment management
B. Forestry education	
Forestry education and policy research to meet emerging challenges	1. Improving formal forestry education 2. Accreditation of universities 3. Networking forestry education with research and extension 4. Capacity building of scientific and management cadre
C. Extension	
Forestry extension for taking research to people	1. Collection, compilation, and publication of forestry reports/journals 2. Dissemination of developed technologies 3. Evolving and coordinating comprehensive extension strategies in forestry research 4. Consultancy services

performed in by the State Forest Departments (SFD) (Singh 2017). Some of the major research institutes of ICAR involved in forestry research include: (1) Indian Grassland and Fodder Research Institute (IGFRI), Jhansi; (2) Central Agroforestry Research Institute (CAFRI), Jhansi; (3) Central Arid Zone Research Institute (CAZRI), Jodhpur; (4) Central Soil Salinity Research Institute (CSSRI), Karnal; (5) Central Soil and Water Conservation Research and Training Institute, Dehradun; (6) ICAR Research Complex for North Eastern Hills Region (ICAR-RC-NEH), Shillong; (7) Vivekananda Parvatiya Krishi Anusandhan Sansthan, Almora; (8) Central Research Institute for Dryland Agriculture (CRIDA), Hyderabad; and (9) Central Agricultural Research Institute (CARI), Port Blair. Apart from the institutions under the ICFRE and ICAR, a number of other government and semi-government organizations are directly or indirectly involved in forestry research. Kerala Forest Research Institute, Peechi, and State Forest Research Institute, Jabalpur, are the two State Government institutes that have contributed significantly to research in tropical forestry and biodiversity conservation. There are still some other institutes like Botanical Survey of India (BSI) or the Zoological Survey of India (ZSI) that conduct research on forestry-related issues to generate information to supplement the work done by ICFRE institutes.

Private sector forestry research includes some paper mills with R&D departments that conduct research to meet their specific needs, such as Andhra Paper Limited, Ballarpur Industries, ITC Bhadrachalam Paper Boards, JK Paper Mills, Tamil Nadu Newsprint and Papers Limited, and West Coast Paper Mills. Research on forestry and agroforestry is also conducted by several non-governmental organizations (NGOs).

4.3.6 Forestry Education in India

In 1878, a college of forest rangers was established in Dehradun to provide a two-year professional forestry training program. This marked the beginning of forestry education in India (Arunachalam and Pandey 2019). As part of the third National Forest Policy of India, 1988, the National Commission on Agriculture (NCA), housed within the Ministry of Agriculture, Government of India, recommended forestry education in order to improve forest conservation, manage natural resources, and safeguard the environment (Dhaka and Choudhari 2018; Arunachalam and Pandey 2019). The National Forest Policy (1952) should be revised, and forest science should be taught in agriculture and other universities, according to the recommendation. The NCA also suggested that an organization be founded to advance forest science and education in India. According to the National Forest Policy of 1988, forestry was acknowledged as a professional and scientific field. Prior to this, forestry education was governed by the Indian Council of Agricultural Research (ICAR), which began in 1985, and forestry education started in about 14 State Agricultural Universities (SAUs) throughout the nation (Arunachalam and Pandey 2019).

In terms of forestry education, 31 universities currently provide Undergraduate (UG), Postgraduate (PG), and Doctor of Philosophy (PhD) programs in forestry. An estimated 1500 graduates in forestry from various Indian states graduate on average each year (Arunachalam and Pandey 2019). The Forest Research Institute, out of all the institutes, has come to represent forestry research and education in India. On the advice of the University Grants Commission (UGC), Ministry of Human Resource Development, Government of India, FRI was designated as a deemed university in 1991. In order to enhance India's forestry workforce, FRI, Dehradun, launched PhD and postgraduate programs in forestry and related fields.

The lack of adequate funding and people with the necessary educational credentials has negatively impacted India's university forestry education system. It is necessary to design new courses in the framework of sustainable development in a way that is socially, economically, and environmentally sound. In order to improve the planning of the forestry education system in India, new views on forestry education should be pursued in order to eliminate the discrepancies between the current system and national requirements of forestry professionals/scientists and teachers (Quli et al. 2013).

4.3.7 Professional Forestry Training

Forestry training and education was kept directly under the MoEF and was split into two parts, i.e., training for IFS and SFS, and training of forest rangers (Singh 2017). National-level organizations took up the responsibility to train and educate personnel and people to efficiently manage and handle problems related to forest and wildlife resources. In 1938, Indian Forest College was founded in Dehradun with the purpose of training Forest Service officers. In May 1987, this college underwent an overhaul and was renamed the Indira Gandhi National Forest Academy (IGNFA), serving as a staff college for the IFS under the Ministry of Environment and Forests' administrative supervision. The Academy acts as the top institution for forestry personnel capacity-building and trains a cadre of capable forest officers to manage India's forest resources. For officers at different levels of seniority, the Academy arranges forestry professional introduction training courses, skills-upgradation courses for promoted IFS officials, and several regular in-service refresher courses, training workshops, and seminars. The Academy has hosted several international trainers and trainees who have completed professional courses. Additionally, the Academy is strengthening the forestry industry's capability for knowledge management.

The MoEF served as the cadre-governing authority when the Indian Forest Service (IFS) was established in July 1966 as an All-India Service pursuant to the All-India Services (AIS) Act, 1951. The Forest Department is organized into four hierarchical levels: field executive staff (which includes deputy range officers, foresters, and forest guards), SFS, FROs, and IFS. The Union Public Service Commission's (UPSC) annual all-India competitive examination serves as the foundation for the IFS recruiting process. After being chosen, the officers attend the IGNFA in Dehradun for professional training before being assigned to states or joint cadres in accordance with the shared policies for the three AIS. Every 5 years, the GOI reviews each cadre's strength and makeup after consulting with the relevant states.

Under the direct administrative jurisdiction of the MoEF, the Directorate of Forest Education, housed on the FRI campus, is in charge of providing induction training for direct recruit officers of the State Forest Service (SFS) and Forest Range Officers (FROs). For the SFS officers, Forest Range Officers (FROs), and frontline staff (Deputy FROs, foresters, and forest guards), the Directorate also arranges short-term in-service training courses, training workshops, and seminars at the colleges under its jurisdiction and in the State Forest Training Schools (SFTSs) across the nation. The Directorate supports State Government efforts to enhance the capabilities of frontline forest workers. Currently, the Directorate of Forest Education oversees one Central Academy of Forest Education (CAFÉ) at Kurseong, West Bengal, and three Central Academies for State Forest Service (CASFoS) at Dehradun, Coimbatore, and Burnihat. Furthermore, the Directorate possesses technical supervision over the schools of the Rangers.

In the beginning, from 1971–1973 to 1973–1975, training for State Forest Service officers was done in tandem with training for Indian Forest Service probationers. Subsequently, additional courses were arranged separately for Indian Forest Service probationers starting in 1976–1978. A distinct college for the training of State Forest Service officers was established in May 1976 at Burnihat, Assam, following the division of the State Forest Service and Indian Forest Service classes. In January 1980, a second State Forest Service college was established at Coimbatore to meet the growing need for State Forest Service officers who were trained. The third State Forest Service college was subsequently established in May 1981 in Dehradun. The Central Academy for State Forest Service (CASFoS) is the current name for these colleges.

The Directorate also has eight accredited institutions operating under it, viz. (1) Forest Training Institute and Rangers College (FTI&RC), Sundernagar, Himachal Pradesh; (2) Gujarat Forest Rangers College (GFRC), Rajpipla, Gujarat; (3) Karnataka State Forest Academy (KSFA), Gungaragatti, Dharwad; (4) Kundal Academy of Development, Administration and Management (KADAM), Kundal, Maharashtra; (5) Odisha Forest Rangers College (OFRC), Angul; (6) Tamil Nadu Forest Academy (TNFA), Coimbatore; (7) Telangana State Forest Academy (TSFA), Hyderabad; and (8) Uttarakhand Forestry Training Academy (UFTA), Haldwani, Uttarakhand. All these institutes conduct induction training of Forest Guards and Range Forest Officers and refresher courses to in-service forest personnel.

One of the top civil services in the states and union territories (UTs), the State Forest Service is second in command. Officers are recruited through their respective State Public Service Commissions (PSC) in accordance with state-specific recruitment regulations, and roughly half of them are promoted from the FRO grade. The SFS colleges are in charge of organizing the professional induction training. Being a feeder service for IFS, SFS fills one-third of the positions in the IFS in any state or UT cadre through promotion from SFS through seniority-cum-merit selection. The SFS officers oversee territorial or functional forest divisions as Divisional Forest Officers (DFOs), or they are posted as assistant conservator of forest (ACFs) on various functional postings in the SFDs.

4.3.8 Forestry Extension

Extension is regarded as a kind of instruction that is not formal. Techniques of forestry extension spread technology and essential knowledge to the intended target audiences. Extension, in its broadest definition, refers to a method of education that links, informs, and persuades individuals. Information sharing between farmers and other resource users, researchers, administrators, managers, and leaders is made easier by it. Extension efforts have a wide range of goals, which are determined by the goals of the project or program. Project-specific objectives include things like establishing tree nurseries, overseeing block plantations, and promoting agroforestry practices, among other things. On the other hand, the general goals of extension systems can be summed up as follows: sharing knowledge, offering technical forestry skills, promoting local involvement, and achieving a project's tangible goals. Forestry extension is an important vehicle for expansion of forest resources in the country, and a tool for forest resources conservation and development.

In order to provide technology packages, including suitable models, to the target groups—farmers, State Forest Departments, industries, craftsmen, craftspeople, etc.—the ICFRE maintains a fully functional Directorate of Extension. The Directorate develops comprehensive extension strategies and oversees the numerous extension initiatives of ICFRE institutions and centers. Additionally, it offers consulting services in areas such as environment management. The Directorate is tasked with transferring technology through a variety of publications, including yearly reports, the network of Demo Villages (DVs) and Van Vigyan Kendras (VVKs), the organization of Tree Growers' Melas (TGMs) and the networking of VVKs with Krishi Vigyan Kendras (KVKs) of ICAR, and the development of technology demonstration centers.

The nation's forestry extension strategy has been developed and updated on a regular basis by the Directorate of Extension at ICFRE. The Directorate previously produced papers for ICFRE such as the "Forestry Research Extension Programme" in 2000 and the "Extension Strategies in Forestry Research" in 2011. These initiatives sprang from discussions based on real-world forestry extension experiences. ICFRE has also periodically developed extension programs like Tree Grower Mela and Direct to Consumer to address the issues in the forestry extension sector.

ICFRE recently created an extension strategy and action plan for the nation based on technology, which will be implemented over the course of 5 years, from 2018 to 2023.

In addition, the extension divisions of the regional ICFRE institutes act as a liaison between the organization and stakeholders and end users, bolstering the findings of the research that ICFRE scientists conduct on many forestry-related topics. As part of their extension activities, the divisions have been organizing and carrying out short-term training programs as well as specially designed courses for various stakeholders. A number of exhibitions have also been planned to raise public awareness and highlight the division's accomplishments. ICFRE institutes have set up Van Vigyan Kendras (VVKs) and Demo Villages under the extension program to assist user groups in States and UTs in accessing the most recent technical know-how.

Since the Green Revolution, India's agricultural output has increased several times, resulting in agricultural self-sufficiency. It was a coordinated effort involving support for extension, research, and policy. In the forestry sector, comparable advancements are still pending. Despite the greatest efforts of the ICFRE institutions, the forest department, non-governmental organizations, and forestry universities, the pace of technology transfer and information penetration among users remains slow. The primary cause of this seems to be a shortage of skilled labor and ongoing financial support. When compared to the size of the clientele that needs their services, the number of forestry extension employees is woefully inadequate. In order for forest managers to tackle the current situation, better data, analysis, and research are needed. The spread of the services has long been limited by inadequate funding for extension programs. It is necessary to reinforce the extension and information-dissemination system, which includes resources and techniques for sustainable development and conservation, farming and management, harvesting, and processing forest products. The needs and interests of the populace must be the foundation of extension activities. Micro programs must be created in accordance with needs that vary among people, communities, states, and blocks. It is usually preferable to use a participatory method to extend forestry projects that guarantee people's active engagement. For the successful, timely, and economical transfer of technology to users, forestry extension is essential. Institutional capacity-building and continuous extension training programs for forest department personnel, industries, and other stakeholders are needed for better forestry extension service.

4.4 Way Forward

India is blessed with a wide variety of forest types, from tropical thorn forests in central and western India to tropical wet evergreen forests in the northeast. Despite being the second most extensive land use in India after agriculture, accounting for roughly 23.57% of the country's total land area (as of 2001), forestry's contribution to GDP is negligible (only 1.1%). An estimated 41% of the nation's forest cover has seen some level of degradation. According to MoEF&CC (2007), up to 78% of the forest area is prone to excessive grazing, and 50% of the forest area is vulnerable to forest fires. One of the main causes of forest degradation in the majority of the nation is a mismatch between the supply and demand for forest products. The demand for fuel wood and lumber in the country is far more than what can be sustained. India has established a voluntary goal to achieve 12 National Biodiversity Targets (NBTs), attain land degradation neutrality on 26 million hectares of land, and trap 2.5 billion tons of carbon dioxide through increased forest and tree cover by 2030. Forest resources are under tremendous strain for agriculture output, forest use, water utilization, and other environmental services due to the expanding population and rising societal expectations. As a result, the forests' degradation and loss of the products and services they once provided continue. Adopting comprehensive, cross-sectoral methods to research, teaching, and extension is necessary in light of the rising environmental issues and rising global expectations from forests in the framework of the sustainable development goals. To meet demand, stress areas must be cultivated and the amount of forest produce produced from currently accessible land must be increased. Under these circumstances the research should be linked with the ongoing programs on tree management.

A few of the top research areas that need immediate attention are: creating models for farm forestry and agroforestry; creating productive and adaptable varieties; managing forest fires; involving the community in Joint Forest Management; revitalizing and maintaining India's Preservation Plots; quantifying the ecosystem services provided by forests; studying regeneration in various forest types; studying hydrology in forests; studying the eradication of exotic invasive species; and forestry extension. It is necessary to conduct research and incorporate findings from forest management into understanding how climate change affects the structure and functions of forests. For India's forestry research and education needs, research and development funds must be established.

In order to create high-quality planting material and transmit technology, relationships between research institutes, state forest departments, and the business sector must be established. Services for forestry extension must be improved or enhanced. These services ought to be focused on assisting rural communities in expressing their needs, environmental issues, and opinions about potential fixes for problems that have been identified. The nation has a largely good institutional framework in place to handle concerns related to forestry research and teaching under the specialized missions of already-existing institutes. To guarantee proper coverage of forestry concerns and to prevent duplication or overlap of efforts, these institutions' inter-sectoral coordination and networking must be improved. Additionally, it is necessary to improve human resources and increase the capacity of forestry employees.

Lessons Learnt

- In 1864, the Imperial Forest Department was established by the British India Government. Dietrich Brandis, a German forest officer, was named Inspector General of Forests in 1866. The Imperial Forest Service was established in 1867 after it was realized that a first-rate forest service was required to oversee the nation's diverse natural resources and coordinate the operations of the Imperial Forest Department.
- In order for the new department to operate efficiently, legislation was required, and the first Forest Act was approved in 1865. Reserved woods, protected forests, and village forests were the three categories into which the forests were divided by the more detailed second Forest Act of 1878.
- Since 1878, when the North-West Province government set the groundwork for a Forest School to train rangers and foresters, Dehradun has served as the hub for forestry education. In 1884, the school changed its name to Imperial Forest College. It later came to be known as the Forest Research Institute for conducting professional forestry course for the officers. After the Indian Agricultural Research Institute, which was founded in 1901, it was historically the second central research institute to be formed by the Indian government in 1906.
- The Indian Council of Forestry Research and Education, or ICFRE, was established in 1986 as a national umbrella organization to handle the needs of the nation for forestry research, education, and extension. ICFRE, which has its headquarters in Dehradun, is currently the highest authority in the country's forestry research system, promoting and carrying out need-based forestry research extension. With its five centers spread throughout various bio-geographical regions of the nation and its nine regional institutes, the Council has a widespread presence throughout India.
- In order to improve forest conservation, manage natural resources, and safeguard the environment, the Ministry of Agriculture's National Commission on Agriculture (NCA 1976) recommended starting forestry education in India. This idea was duly included in the National Forest Policy of India, 1988.
- The Indian Forest Service was created in July 1966, as an All-India Service under the All-India Services (AIS) Act 1951, with the MoEF becoming the cadre-controlling authority.
- The State Forest Service is the second-in-command among the top civil services in the states and UT. May 1976 saw the opening of a college in Burnihat, Assam, to train State Forest Service staff. In January 1980, a second State Forest Service college was established at Coimbatore to meet the growing need for State Forest Service officers who were trained. In May 1981, the third State Forest Service College was subsequently established in Dehradun. The Central Academy for State Forest Service (CASFoS) is the current name for these colleges.
- ICFRE has been developing and revising the nation's forestry extension strategy on a regular basis. "Extension Strategies in Forestry Research" were developed by ICFRE for the nation in 2011.

Key Questions

1. What do you understand by Scientific Forestry? Who started Scientific Forestry in India? Describe its importance to Indian context.
2. When was the first Indian Forest Act enacted, and what were its main features?
3. When was the Forest Research Institute and colleges established in India? What were the areas initially it started working on?
4. What do you understand by ICFRE? What are its main objectives, role, and functions in the country?
5. Describe the structure and function of professional forestry training in India.

References

Arunachalam A, Pandey L (2019) Transforming forestry education for better job prospects. Curr Sci 117(5):748–752. https://doi.org/10.18520/cs/v117/i5/748-752

Brandis D (1897) Indian forestry. Oriental Institute, Woking

Chaney WR, Pope PE, Edlin HL (1998) Encyclopaedia Britannica, s.v. "Internet." Encyclopaedia Britannica, Chicago. https://www.britannica.com/science/forestry. Accessed 8 June 2020

Dhaka RK, Choudhari CK (2018) Forestry education in India: objectives, needs, current status and recommendations. Pharma Innov J 7(12):320–324

FRI (1961) One hundred years of Indian forestry, vol II. Forest Research Institute, Dehradun

ICAR (2009) New and restructured post-graduate curricula and syllabi of forestry. Indian Council of Agricultural Research, New Delhi

ICFRE (2000) ICFRE National Forestry Research Plan, vol 1. Indian Council of Forestry Research and Education, Dehradun
ICFRE (2013) ICFRE -Vision 2040. ICFRE Publication. Publisher - Director General, Indian Council of Forestry Research and Education, Dehradun, India.
Lawbuary J (1999) Reclaiming the forests. People's participation in forest management, East India. B.Sc. thesis, Human Environmental Science, King's College, London
MoEF (1986) Reorganization of forestry research, education and training. Resolution no. 15-48/86-RT dated 31 December 1986. Ministry of Environment and Forests, Government of India, New Delhi
MoEF&CC (2007) National Biodiversity Action Plan. Ministry of Environment and Forests and Climate Change, Government of India
NCA (1976) National Commission on Agriculture, vol IX. Government of India, New Delhi, 350 p
Quli SMS, Sivaji V, Srivastava KK (2013) Forestry education in India: retrospect, prospect and recommendations for improvement. Indian Forester 139(4):316–323
Ranganathan CR (1950) Forest research and education in India. Unasylva 4(3). https://www.fao.org/3/x5356e/x5356e04.htm#forest%20research%20and%20education%20in%20india. Accessed 27 Jan 2021
Shanmuganathan K (1997) Nilambur teak—the history and a resume of early planting activities. In: Basha SC, Mohanan C, Sankar S (eds) Proceedings of international teak symposium. Kerala Forest Department and Kerala Forest Research Institute, pp 226–31
Singh RV (2017) Forestry research in India: evolution, expansion and future challenges. D.K. Printworld (P) Ltd, New Delhi, isbn:9788124608777
Singhal RM, Kumar S, Jeeva V (2003) Forest and forestry research in India. Trop Ecol 44(1):55–61
Stebbing EP (1922) The forests of India, vol I. London, pp 62–63
Stebbing EP (1923) The forests of India, vol II. London, pp 8–10
Stebbing EP (1926) The forests of India, vol III. London, pp 257–260

Taxonomy: Importance, Relevance and Application

5

C. Kunhikannan, M. V. Anju, and R. Prasanna

Abstract

Taxonomy is the science of classifying and identifying an organism using scientific names. The Swedish Botanist Carolus von Linnaeus brought out a system for the development of plant taxonomy. The development of taxonomy was supported by new discoveries in a variety of scientific fields, and it is crucial to regularly update our understanding of taxonomy and field botany in order to effectively communicate research findings to other scientific fields, such as medicine, agriculture and forestry. India possesses an abundance of 55,387 plant species (22,108 species of Angiosperms, 83 Gymnosperms, 1319 Pteridophytes, 2819 Bryophytes, 3044 Lichens, 15,701 Fungi, 9035 Algae and 1278 microbes) accounting for 14.66% of the global flora. India is home to about 27% of endemic plant species. Documentation requirements such as the People's Biodiversity Register coming under Biological Diversity Act of 2002 and Report on Biological Environment required for Environmental Impact Assessments (EIAs) also have a role in taxonomy expertise.

Plant classification is the process of grouping plants together. It entails recognising and characterising a species. The phylogenetic classification system, which is according to evolutionary importance of plants, evolved from a basic artificial system (based on the morphological characteristics of the plant species) through a natural system. This chapter examines a variety of classification schemes, including natural, artificial and phylogenetic systems of classification. Bentham and Hooker's classification system and the most recent Angiosperm Phylogeny Group IV (APG IV) classification are used as examples.

Botanical nomenclature refers to the scientific method of naming plants. Carolus Linnaeus proposed a binomial nomenclature system. Every species is assigned a two-worded name: the first word is the genus name and the other is species epithet. The different principles and rules of plant nomenclature are explained, as well as the notable elements of both the International Code of Botanical Nomenclature (ICBN) and International Code of Nomenclature (ICN) for algae, fungi and plants. The accurate identification of a species is critical, because communication is only possible when the name is correct. The process of linking an unknown organism with a recognised one is known as identification. It is done by consulting a taxonomist or botanical literature, particularly flora and herbaria. This chapter also includes information on a dichotomous key for easy identification and a synopsis of Herbarium preparation processes.

Keywords

Taxonomy · Binomial nomenclature · Phylogenetic system · APG classification · Herbarium

C. Kunhikannan (✉) · M. V. Anju · R. Prasanna
Institute of Forest Genetics and Tree Breeding, R S Puram, Coimbatore, Tamil Nadu, India

5.1 Introduction

Taxonomy is a field of science focused with the systematic organisation of distinct species; in other words, it is the study of classification and identifying plants using scientific names, and it is concerned with the principles guiding an organism's categorisation. Augustin Pyramus de Condolle, a Swedish botanist, invented the term Taxonomy in 1813, which combines two Greek terms, *taxis* meaning arrangements and *nomos* meaning laws or knowledge. Plant taxonomy is also known as systematic botany. The three most significant components of taxonomy are classification, identification, and description and providing names to the plants. Knowledge of taxonomy depended on shape and structure of organisms. Different organism groupings are classed differently. Plant classification is distinct from microbiological and animal classification. Plant observation began when humans began using them for livelihood needs like food, housing, clothing, medicines and so on. Sufficient knowledge was

A. K. Mandal, A. Nicodemus (eds.), *Textbook of Forest Science*, https://doi.org/10.1007/978-981-97-8289-5_5

obtained about agricultural, medicinal and other plants that were essential for day-to-day lives of humans in ancient times, even before the beginning of the Christian period, making botany one of the earliest sciences. Egyptians (4000 BC) used images to record a lot of information about crop plants and fruit trees. The Chinese had mastered plant cultivation by 2500 BC. The Atharvaveda, which was composed in India around 2000 BC, included a plethora of knowledge on several medicinal plants and their purposes. Botany as a discipline was founded by ancient Greek philosophers and Indian herbalists.

In his work *The Historia Plantarum*, Theophrastus (370–287 BC) detailed the external and internal characteristics of 480 different species of plants. Botany experiments began in the sixteenth and seventeenth centuries. The *Hortus Malabaricus* by Hendrik Adriaan van Rheede, published between 1678 and 1693, is the first record of printed work on the flora of Asia and the tropics, which contains ethno-medicinal records of Malabar Coast of India, presently the states of Kerala and Karnataka. The botanical descriptions of plants in their natural environments had gained the significance of therapeutic herbs, and it is the age of herbalists. Taxonomy advanced during the eighteenth century. Carolus von Linnaeus (1753), a Swedish botanist, was instrumental in the establishment of plant classification.

5.2 Importance of Plant Taxonomy

Correct plant identification has an impact on all branches of botany, either directly or indirectly. It is critical to identify the existing organisms and their arrays. Only when a species' name is right can it be communicated with. As a result, categorisation is thought to be critical for segregation and sorting (Stace 1991). The primary goal of plant taxonomy is to characterise plants and organise them in a systematic order using a classification system based on their hierarchical order. The first stage in surveying vegetation and estimating phytodiversity is taxonomic identification of plants (Fralish and Franklin 2002). Plant taxonomy plays an essential role for biodiversity conservation through the selection of designated areas and management plans, and the European Commission has identified it as a crucial field in the development of mitigation methods to safeguard the environment (European Commission 1996). The essential condition for every location to be protected for biodiversity is recognition and well-defined species identification. Knowledge of a species' presence and composition is required for the species' preservation, protection and conservation (Williams and Humphries 1994; Hollingsworth 2003).

Taxonomy is simply the classification, naming and identification of living things. It will, however, aid in the classification of species based on morphological characteristics.

- Taxonomy will aid in determining the number of creatures on the planet. So currently, millions of plant and animal organisms were discovered, named and categorised.
- The primary goal of taxonomy is to scientifically categorise living entities for greater understanding.
- It will provide information on the area's flora and fauna, which will aid in distinguishing indigenous species.
- Taxonomy is composed of three fundamental components: identification, nomenclature and classification.

Because of the increasing importance of biodiversity, there is a rising demand for additional taxonomists. The Conference of Parties (COP) to the Convention of Biological Diversity (CBD) cited a shortage of taxonomists as an impediment to implementing convention decisions at the national and international levels. One out of a million species has been lost every year during the past half a billion years. Today's yearly extinction rate is 1000–10,000 times quicker. Our efforts to protect biodiversity will be considerably easier if we understand the fundamental units of species and their interactions. Taxonomy facilitates the finding and identification of these fundamental elements and their interactions.

India has a total of 54,733 plant species, with the addition of 253 species and intra-specific categories of flowering plants, pteridophytes, bryophytes, fungi, lichens, algae and microbes to the Indian flora during the year 2020 (BSI 2021), representing up to 14% of the world flora (BSI 2023). About 28% of the plants that exist in India are endemic to the country. These resources include wild relatives of many agricultural species, timber and pulp producing species, non-wood forest products, and medicinal plants. Correct species identification is critical for optimal use and conservation of these resources. The CBD recognised the problem of a dramatic reduction in taxonomic competence and the challenges experienced in correctly identifying plants and animals and offered a push to revitalise and develop taxonomy.

As a CBD signatory, India also launched measures to improve taxonomy knowledge by launching a plan, the 'All India Coordinated Project on Taxonomy (AICOPTAX)'. The initiative aims to improve the country's capacity for inventorying, monitoring, preserving and using biodiversity, as well as to develop leadership in the area of taxonomy at the regional and global levels. Recent documentation requirements, such as the People's Biodiversity Register mandated by the Biological Diversity Act of 2002, concerning biological environment as part of Environmental Impact Assessment (EIA), look forward to expertise in Taxonomy, which is also linked to various fields of science such as medicine, agriculture, forestry and so on (Fig. 5.1).

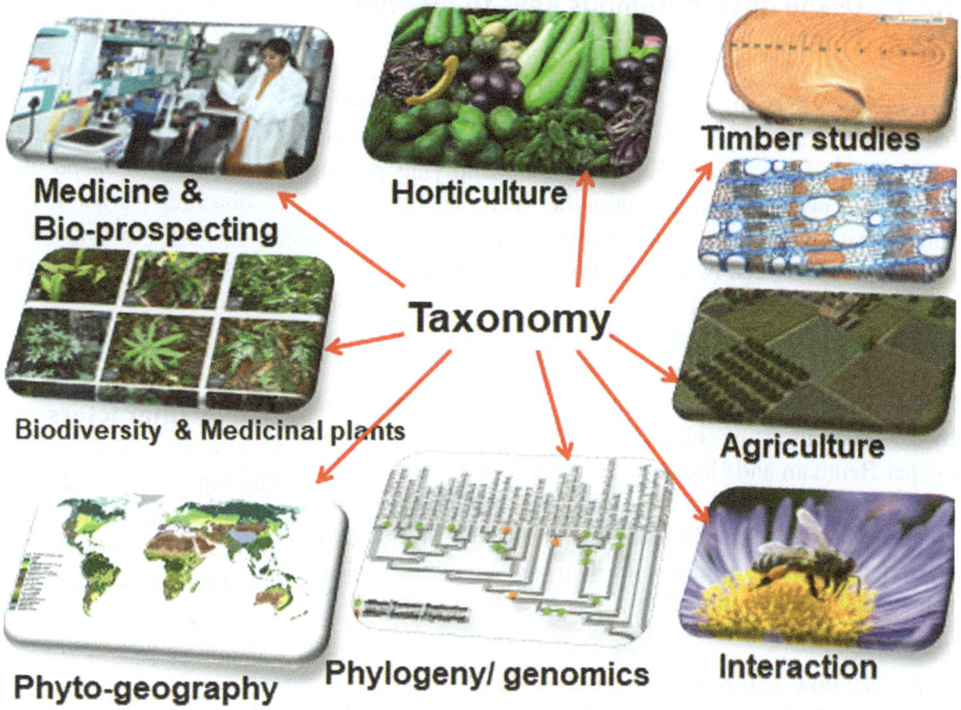

Fig. 5.1 Relation of taxonomy with other branches of science

The process of linking an unknown organism with a recognised one is known as identification. Identification is the evaluation of whether or not a taxon is comparable to one already reported (Box 5.1). For example, if a person accidentally comes in contact with a plant leaf in the forest and experiences itching and feverish symptoms within a few minutes, he can identify the plant by comparing it to recognised plants with similar features. It is possible to do so by examining the botanical literature, particularly flora, or by referring to herbaria, or it may be readily recognised based on the description of the taxa, botanical keys supplied in the flora. Otherwise, they do a thorough investigation to determine the relationship, categorise, and identify it as a new species. Naming has to conform to nomenclature regulations. Herbarium is another source of identification. To identify any plant specimen, the scholar or student should be well-versed in plant morphology.

Box 5.1 Tips for Identification in the Field

In field condition, one will always get clues for identification. For example, if someone discovers a few orange-red matured leaves on an evergreen forest floor, they may be sceptical of the presence of *Elaeocarpus* species nearby. Similarly, if one sees yellow dried leaves, a *Symplocos* species may be present. Certain fragrances also signal the presence of certain species in the forest. The presence of particular butterflies also aids in the identification of specific

(continued)

Box 5.1 (continued)

plant species; for example, the presence of the Southern birdwing butterfly indicates the presence of plants of the Aristolochiaceae family.

5.2.1 Questions Generally Asked to Collect Information About a Plant

1. What is the overall form of a plant?
   - Herbaceous/shrubs/trees, broadleaved, evergreen/deciduous?
2. Which characteristics do individual plant parts have?
   - Bark features: colour/texture, spines/thorns, presence of sap/latex, its colour, etc.
   - Leaf type: simple/compound/pinnate or palmate, etc.
   - Leaf arrangement: alternate, opposite, opposite-decussate, whorled, etc.
   - Types of venations: parallel/reticulate, etc.
   - Leaf margins: lobed, wavy, serrate, dentate, crenate, etc.
   - Flower: type/colour, number of floral parts, etc.
   - Fruit/seed: type/colour, number of seeds, etc.
3. What are the associated species?
4. What is the habitat of the plant?
   - Type of habitat: marshy land/dry rocks/grassland/forest/sea shore, etc.
5. What are the uses (if known)?
   - Edible/fodder, medicinal/timber/cultural uses, etc.

5.2.2 Usage of Dichotomous Keys to Identify Plants

Taxonomic keys: It is a written method of assisting individuals in identifying an unfamiliar plant. Dichotomous (forking) keys are widely consulted for identifying plants. These keys provide users an option between two characters, and perfect identification is attainable by making a sequence of selections between two contrasting characters.

The key is made up of couplets, which are options between two opposing characters. Using such keys, any flora of a certain place can aid in identification.

Following are certain keys to distinguish families/genera by using contrasting characters with certain exceptions (as per Bentham and Hooker's System) (Table 5.1).

We need to identify a plant, read the description in flora and confirm for further details. Herbarium specimens also can be referred to establish identity or the help of taxonomy specialists can be sought for confirmation of a species (Box 5.2).

Box 5.2 How a Student Can Identify a Plant?

If a student receives a plant, he should examine it and compare it to plants with comparable characteristics that he is familiar with. If he does not know the species' name, he must gather information from existing resources such as local available flora or herbaria, or consult with experts such as Vaidyas or medicinal plant collectors to learn its local name or consult with experts such as college/university professors/scientists with taxonomic knowledge to identify it. If he obtains some information about its group or family, then he must describe it in complete detail in botanical descriptions and confer with any plant group specialists accessible in the locality, region or nation, and then to the world. He will be happy if he gets a name. If specialists propose or opine that it may be a new species, he must describe it in detail, determine its relationship and location in the available taxonomic group or families, identify closest relatives, place it in a suitable family and genus, and publish it as a new species. While publishing, he must follow the Code of Nomenclature. The code requires that herbarium specimens be created, a type specimen be designated, deposited in major herbaria, and the name be published in a reputed journal. For plant identification, there are several online Apps, groups and websites available. Different images for the same species may appear on a website or blog. However, ultimate confirmation must be obtained by visiting a knowledgeable specialist or an actual herbarium. The herbarium of Botanical Survey of India including its regional herbaria located

(continued)

Box 5.2 (continued)

throughout the country, and various research institutes/universities/colleges also can be utilised.

A taxonomy student should be a keen observer. He or she must spend time in nature. Only an inquisitive observer learner may become a taxonomy specialist. To sustain competence, one must consistently study, preserve plant knowledge and share that information with others.

5.2.3 How to Construct Keys Using Contrasting Characters

The following are six simple examples of artificial keys for species identification. Following a thorough examination of the plants, the principal contrasting characteristics have to be noted on an index card.

Using these characters, keys can be made like the examples given below.

1a.	Plants erect	2	
1b.	Plants not erect	3	
2a.	Stem woody	4	
2b.	Stem not woody	5	
3a.	Climbing plants with latex on stem and leaves; flowers yellow	**Allamanda**	
3b.	Climbing plants without latex on the stem and leaves; flowers white	**Jasminum**	
4a.	Stem green with nodes and internodes, leaves grass like	**Bambusa**	
4b.	Stem grey brown, nodes and internodes not prominent, leaves not grass like	**Mangifera**	
5a.	Leaves with aromatic smell	**Ocimum**	
5b.	Leaves without aromatic smell	**Amaranthus**	

Table 5.1 Key characters for identification of family and genera

Key characters	Family or Genera
Dicots	
Petals free (Polypetalous)	
Thalamus long, swollen, conical; aggregate fruit; ruminate endosperm	Annonaceae
Coloured petals; sap yellow/milky; mostly herbs	Papaveraceae
Coloured petals, numerous in several whorls; mostly aquatic	Nymphaeaceae
Petals-4, free; stamens-6 (4 + 2), free; fruit with an internal partition; mostly herbs	Brassicaceae Capparidaceae
Leaves gland dotted, intramarginal vein absent; fruit hesperidium	Rutaceae
Leaves gland dotted, intramarginal vein present; fruit not hesperidium	Myrtaceae
Filaments of anthers united into a tube; epicalyx present	Malvaceae (except *Sida* and *Abutilon*)
Filaments of anthers united, less than 15; epicalyx absent	Sterculiaceae
Filaments of anthers free, numerous	Tiliaceae
Papilionaceous flower; fruit legume, dehiscent	Fabaceae
Caesalpinaceous flower; fruit indehiscent	Caesalpiniaceae
Flowers in head or condensed spike; fruit lomentum, indehiscent	Mimosaceae
Plants prickly; stipules adnate; petals coloured	Rosaceae
Plants mostly tendrillar; nodes swollen, jointed; fruit berry	Vitaceae
Stems angular; tendrillar climbers; unisexual; epigynous; stamens filaments and anthers fused (synandry); fruit pepo	Cucurbitaceae
Mostly trees, glands on the base of the leaf (except *Terminalia chebula)*	*Terminalia* sp.
Mostly scented herbs; flowers mostly in umbels	Apiaceae
Stems fleshy, jointed, bearing large thorns and small spines; xerophytic	Cactaceae
Petals united (Gamopetalous)	
Mostly herbaceous; flowers in heads; syngenesious (anthers united) stamens	Asteraceae
Leaves opposite, intra- or inter-petiolar stipules, epigynous	Rubiaceae
Herbs; calyx covered with stalked glands	Plumbaginaceae
Trees or shrubs; leaves often gland dotted; absence of stalked glands on the calyx	Myrsinaceae
Plants with milky latex; pollinia absent; carpels free	Apocynaceae
Plants with milky latex; presence of staminal corona; pollen grains in pollinia; comose seeds	Asclepiadaceae
Presence of persistent calyx, extra-axillary inflorescence, sagittate anther	Solanaceae
Leaves with cystoliths; stamens-4, didynamous; seeds on retinacula	Acanthaceae
Herbs; stems quadrangular, aromatic; inflorescence verticillaster; corolla bilabiate; gynobasic; fruit 4-lobed	Lamiaceae
Mostly trees or shrubs; stems quadrangular, non-aromatic, inflorescence not verticillaster; fruits not lobed	Verbenaceae
No distinction of sepals or petals (Monochlamydeous)	
Mostly herbs; flowers in spikes; basal placentation; fruit achene	Amaranthaceae
Plants with milky latex; inflorescence cyathia or catkin; ovary 3-celled; seeds carunculate	Euphorbiaceae
Plants with milky latex; inflorescence hypanthodium	Moraceae
Plants with watery sap, mostly covered with stinging hairs; inflorescence not hypanthodium	Urticaceae
Leafless herbs; root parasitic herbs; flowers in crowded heads	Balanophoraceae
Leafless trees, stems jointed; non-parasitic; flowers in spikes	Casuarinaceae
Monocots	
Mostly aquatic herbs; unisexual	Hydrocharitaceae
Mostly epiphytes; velamen roots; presence of labellum, rostellum	Orchidaceae
Climbing shrubs; leaves reticulate; capsule 3-winged	Dioscoreaceae
Root stock rhizome or bulb; flowers in umbels	Liliaceae
Trees; flabellate-pinnatisect leaves; flowers in compound spadix	Arecaceae
Rhizomatous or tuberous herbs; flowers in simple spadix	Araceae
Stems 3-angled; leaves 3-ranked; flowers in spikes; fruit nut like	Cyperaceae
Stems round; leaves 2-ranked; flowers in compound spike; fruit caryopsis	Poaceae

5.3 Plant Classification

Plant classification is the process of categorising plants into various groups. It entails characterising (assigning traits to a taxon) and identifying (figuring out that an unknown element corresponds to an established one) a species. The ultimate objective is to arrange plants in a logical order according to similarities between them. The plants that are possessing similar characters are treated as related and grouped together. Another purpose of classification is to discover the evolutionary relationships between diverse plant groups. The number of shared traits decreases as we progress from species to family to kingdom, with species sharing the most basic commonalities and creatures in the same kingdom sharing the fewest. Bessey (1897) has given a thorough link between phylogeny and taxonomy in *Botanical Gazette*.

The categorisation method utilised morphological characteristics up to the eighteenth century. This traditional technique is called artificial system (Alpha taxonomy). Evidence from other fields of study, notably phylogenetic categorisation, was also considered in subsequent years. In recent years, a multidisciplinary approach to taxonomic examination has gained prominence. DNA information from plant species is also regarded a significant criterion and has become widely acknowledged as an important element of the categorisation process at various stages (Stace 2005). Data from field investigations, experimental gardens, laboratories, herbariums, libraries and computer use have resulted in various alterations in the old approach to current taxonomic approaches.

5.3.1 Types of Classification

Plant taxonomy has a long history that is intertwined with human evolution. When humans lived a hunter-gatherer lifestyle, they attempted to categorise plants based on their usefulness, such as food plants, fuel plants, medicinal plants and so on. They gradually began categorising plants based on their requirements. It is vital to highlight that the evolution of current categorisation systems began with ancient Greek thinkers and naturalists. The early categorisation systems were quite simple and based on only a few traits, with emphasis placed on vegetative characteristics. Floral features being stable in nature were given more weight in later classification systems.

Following are the major types of systems for plant classification:

1. Artificial systems (based on the morphological characters of the plant species)
2. Natural systems (based on the reproductive parts of the plants)
3. Phylogenetic systems (based on the evolutionary significance of plants)

5.3.1.1 Artificial System

It was centred solely on one or two superficial characters. The Swedish Scientist Carolus Linnaeus published the famous book *Species Plantarum* in 1753 in which descriptions for 7300 species were provided. Based on staminal characters like number of stamens, length and union and other characters the plants were placed in 24 categories. As a result, this system is also referred to as the sexual classification system. Linnaeus' classification recognised the importance of floral features, which brought wide acceptability. The fundamental disadvantage of this approach was that completely different plants were kept in same group, while highly related plants were kept in different groups. For example, plants coming under the Monocotyledon, Zingiberaceae, and the Dicotyledon, Anacardiaceae, have been combined into a single group named Monandria since they have a single stamen. Another flaw in this method was that natural or phylogenetic references between plant groups were not prioritised.

5.3.1.2 Natural System

Plants are grouped in this system based on their natural affinities. Several characteristics were considered in the natural system. It is mostly based on information gathered during first-hand plant observation. George Bentham and Joseph Dalton Hooker, two British botanists, created the significant natural system of seed plant classification. This categorisation aids in determining the link between different plant groups. This method is still used in a number of herbaria in India. However, it makes no attempt to uncover evolutionary links between distinct plant families.

5.3.2 Benthem and Hooker System of Plant Classification

Benthem and Hooker devised the method of seed plant classification, which is the well-established natural categorisation system and was published in the *Genera Plantarum* (1862–1883). They described 97,205 seed plant species ranging from Ranunculaceae to Gramineae, consisting of 7569 genera and 202 families. This also includes three Gymnosperm families.

The genera were determined through natural affinities; the concept was pre-Darwinian. In this approach, all seed-bearing plants were placed in three groups: Dicotyledons

(165 families), Gymnosperms (3 families), and Monocotyledons (34 families).

Class 1: Dicotyledons: Plant seeds have two cotyledons, leaves with reticulate venation and flowers are pentamerous or tetramerous in nature. It is classified into three subclasses and consists of 14 series, 25 orders and 165 families, viz. Polypetalae, Gamopetalae and Monochlamydeae.

1. **Subclass—Polypetalae**: Plants in this subclass can be identified by the presence of separate sepals and free petals. It is divided into three series:
   (a) *Thalamiflorae*, with hypogynous flowers, many stamens and no disc, is composed of six orders: Ranales, Parietales, Polygalineae, Caryophyllineae, Guttiferales, and Malvales.
   (b) *Disciflorae* having hypogynous flowers with a disc under the ovaries (four orders under the group are Geraniales, Olacales, Celastrales, and Sapindales).
   (c) *Calyciflorae* has Perigynous or epigynous flowers with five orders such as Rosales, Myrtales, Passiflorales, Ficoidales, and Umbellales.
2. **Subclass—Gamopetalae**: Plants in this category have separate sepals and petals, and the petals are united. It is divided into three series:
   (a) *Inferae*: Plants with inferior ovary and having three orders, namely Rubiales, Asterales, and Campanales.
   (b) *Heteromerae*: Members have a superior ovary, one or two whorls of stamens and more than two carpels. This series has three orders, viz. Ebenales, Primulales, and Ericales.
   (c) *Bicarpellatae*: Plants in this family have a superior ovary, stamens in one whorl and two carpels. It is divided into four orders, namely Gentianales, Polymoniales, Lamiales, and Personales.
3. **Subclass—Monochlamydeae**: Plants in this subclass have apetalous blooms, meaning the perianth is either absent or sepals and petals are not differentiated. There are eight series in it, which are listed below.
   (a) Curvembryeae: Plants usually have coiled embryo and one ovule
   (b) Multiovulateaquaticae: Aquatic plants with many ovules
   (c) Multiovulateterrestress: Terrestrial plants with many ovules
   (d) Microembryeae: Plants with minute embryo
   (e) Daphnales: Plants having one carpel and one ovule
   (f) Achlamydosporae: The ovary is inferior, unilocular with one to three ovules
   (g) Unisexuales: Flowers are unisexual
   (h) Ordinesanomali: Plants of uncertain relationship

Class 2: Gymnospermae: The plants with naked ovules. The class consists of three families, viz. Cycadaceae, Gnetaceae, and Pinaceae.

Class 3: Monocotyledons: Trimerous flowers and leaves with parallel venation and seeds with one cotyledon. This class is divided into 7 series and 36 families.

1. Microspermae: Plants in this family have inferior ovary with very little seeds.
2. Epigynae: This plant family has a smaller ovary and bigger seeds.
3. Coronarieae: This plant family has a coloured perianth, a superior ovary and joined carpels.
4. Calycinae plants feature a green perianth, a superior ovary and joined carpels.
5. Nudiflorae: Plants lack perianth and have superior ovary.
6. Apocarpae: Plants with superior ovary and more than one free carpel.
7. Members of the Glumaceae have blooms wrapped in glumes and a shortened perianth with a superior ovary.

5.3.2.1 Advantages of the Bentham and Hooker System of Classification

- Every species' description is equitably precise and detailed, as it has been recorded by means of the original specimens or from the herbarium sheets.
- This technique is extremely practical and may help students of systematic botany identify species quickly.
- The system provides a geographical overview of species and genera, and generic definitions are extensive, precise and are founded on observations made first-hand.
- Large genera include subgenera having their own group of species.
- Ranales are widely regarded as the most basic angiosperms, however dicots begin with this order in this system.
- The dicots are placed before monocots following historical tendencies.
- It is also natural to place series Disciflorae between Thalamiflorae and Calyciflorae.
- Because petal union is considered an advanced trait above the free state, the positioning of gamopetalae after polypetalae is appropriate.

5.3.2.2 Demerits of Bentham and Hooker System of Classification

- It is unusual to place gymnosperms between dicots and monocots.
- It does not appear natural to place monochlamydeae after gamopetalae.

- The monochlamydeae subclass is extremely artificial.
- Some monochlamydeae families, such as Chenopodiaceae and Caryophyllaceae, are closely linked to polypetalae groups. Those species showing relatedness are placed far apart, while others are put near together.
- Treating an advanced family like Orchidaceae as primitive and placing it at the top of monocotyledons is odd.
- Compositae (Asteraceae) has also been listed at the top place under gamopetalae making it unnatural.
- The families Liliaceae and Amaryllidaceae that show relationship were separated based only on ovary characteristics.
- Phylogenetic relations are not considered.

Bentham and Hooker's system of classification is summarised below (Fig. 5.2):

5.3.3 Phylogenetic System of Classification

This classification method is based on the evolutionary sequence and genetic links between various taxa. Furthermore, it took into account as many taxonomic features as accessible. Theory of origin by Charles Darwin inspired the development of phylogenetic systems. In their work *Die Naturalichen Pflanzen Familien*, Adolf Engler (1844–1930) and Karl Prantl (1849–1893) created a phylogenetic system of classification. Floral characteristics like a solitary whorl with perianth or a lack of perianth, as well as anemophilous unisexual flowers were regarded as less evolved under this paradigm, as opposed to a pair of whorls of perianth and bisexual flowers pollinated by the insects. They have described 288 families in this classification. Plants from the dicotyledon Asteraceae and the monocotyledon Orchidaceae were highly valued, according to them.

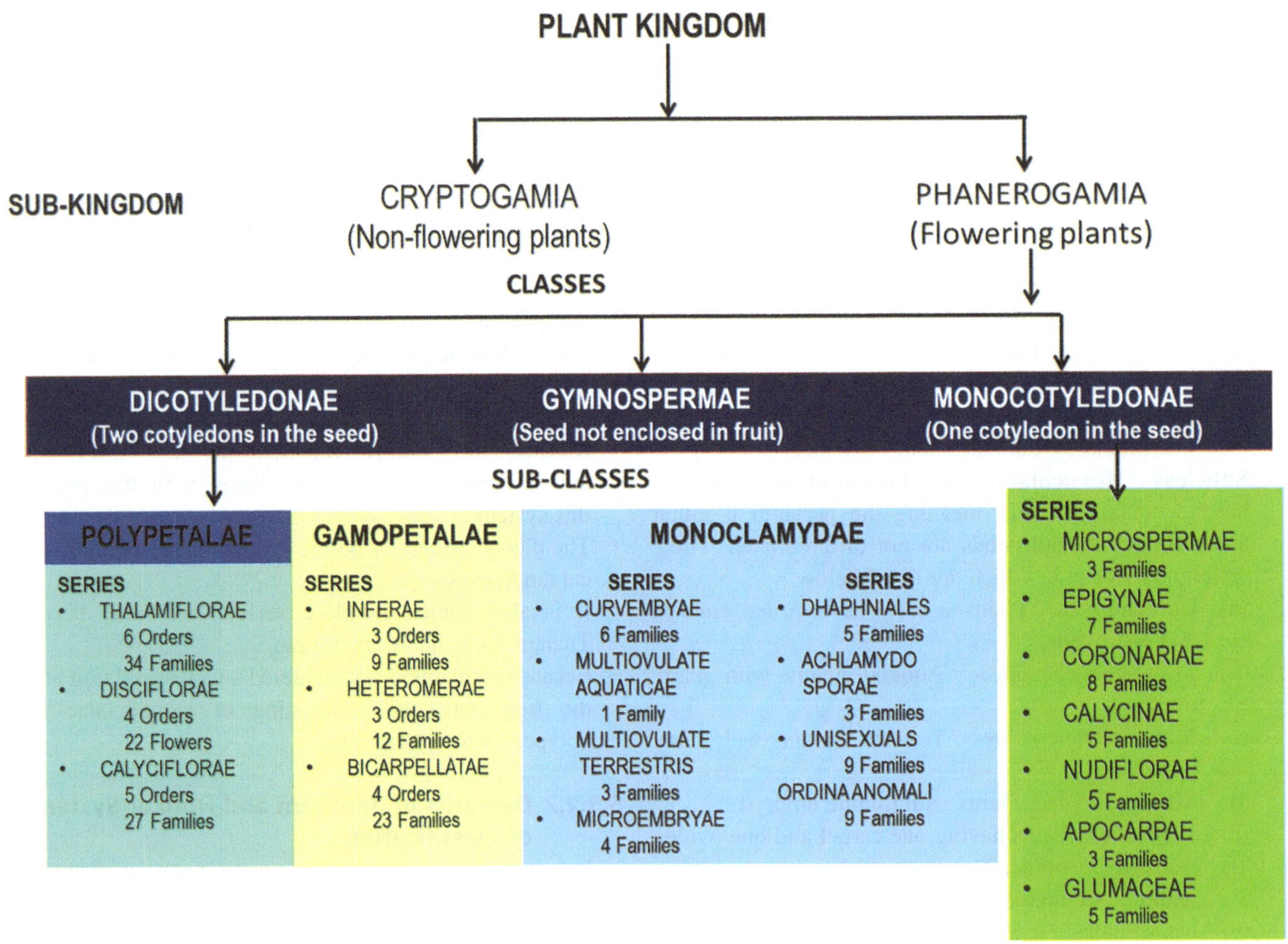

Fig. 5.2 Bentham and Hooker's system of classification

Charles Edwin Bessey (1845–1915), an American and disciple of Asa Grey, attempted a categorisation scheme comparable to those of Bentham and Hooker. During 1883, he presented a system. The ultimate and comprehensive system was published in 1915. He classified angiosperms (Anthophyta) as Oppositifoliae (Dicots) and Alternifoliae (Monocots). Bentham and Hooker's cohorts were referred to as orders in this system, while natural orders were referred to as families.

John Hutchinson (1874–1972) devised a phylogenetic system, which was published in *Families of Flowering Plants* in two volumes (1926, 1934) and amended in 1948. In this system, he has included 328 families and emphasised the woody and herbaceous natures of plants and so classified dicots as Lignosae (Woody) and Herbaceae (Herbaceous). Hutchinson considered the woody character was basic.

Armen Takhtajan (1910–2009), a Russian botanist of the Soviet Academy of Sciences, Leningrad, USSR, created a superior phylogenetic approach by considering every existing data from different fields of botany in 1954, which was later improved in 1969 and 1980. Magnoliophyta replaced angiosperms, Magnoliopsida replaced dicots and Liliopsida replaced monocots. He employed a Super Order extra rank between Subclass and Order with the ending 'anae'. In this categorisation, he has described 592 families.

Arthur Cronquist (1911), New York Botanical Garden, created a similar method and published it in the book *Evolution and Classification of Flowering Plants* during 1968. This was eventually renamed *An Integrated System of Classification of Flowering Plants* in 1981. Cronquist considered pteridosperms to be the most likely predecessors of angiosperms. Cronquist (1981), like Takhtajan, adopted the word Magnoliophyta for angiosperms, which have been separated into Liliopsida and Magnoliopsida, which are analogous to conventional monocotyledons and dicotyledons, respectively. These were then subdivided into 11 subclasses, 6 in the Magnoliopsida (Caryophyllidae, Magnoliidae, Hamamaelide, Dilleniidae, Rosidae, and Asteridae, with 64 orders, 318 families and with about 165,000 species) and 5 in Liliopsida (Arecidae, Alismatidae, Commelinidae, Zingiberidae, and Liliidae, with 19 orders, 65 families and about 50,000 species).

The Magnoliidae was regarded to be the base complex, from which the next five subclasses were formed individually. The Hamamelidae are primarily pollinated forms with reduced and apetalous blooms, similar to the Amentiferae. Caryophyllales and their near relatives were grouped together as Caryophyllidae. The Rosidae and Dilleniidae were thought to be parallel groupings because they both had multiple stamens that grow centripetally in the former and centrifugally in the latter. The Asteridae descended from the Rosidae and included the most advanced group, such as higher gamapetalous families with as many stamens as corolla lobes. The Liliopsida was thought to be evolved from hydrophytic predecessors and sharing affinities with present Nymphaeales. The five Liliopsida subclasses are said to have ecological affinity. The Alismatidae are mostly aquatic, whereas the rest are mostly land plants. The Arecidae are often arborescent plants with large leaves and many tiny flowers supported by a conspicuous spathe. The Commelinidae and Liliidae, which are likely descendants of a single ancestor, have different and parallel evolutionary histories.

5.4 APG Classification

During the 1990s, tremendous breakthroughs in molecular research with regard to DNA sequences, notably from the plastid gene rbcL, gave fresh and valuable data that influenced angiosperm classification schemes and led to the rearrangement of numerous families. It is an ordinal classification proposed by the Angiosperm Phylogeny Group (APG), an international group composed of plant taxonomists from the Royal Swedish Academy of Sciences, Upsala University, Sweden; Royal Botanical Garden, UK; University of Maryland, College Park, USA; and Florida, Gainesville and Missouri Botanical Garden, USA. The APG method combines substantial DNA sequence data with morphological data (for example, the use of pollen characteristics to delimit eudicots—the three-pored pollen carrying flowering plants). The APG method organised families and orders into informal groups that represented significant branches in the flowering plant phylogenetic tree. The Monocots, Eudicots, Rosids, and Asterids were four such significant groups. The APG System is now the most widely used for categorisation of flowering plants.

This system, established in 1998, was modified in 2003, 2009 and 2016. The APG method relies profoundly on morphology, anatomy, embryology, phytochemistry and comprehensive DNA sequence data, as well as molecular investigations with reference to DNA sequences. The method is according to the good phylogenetic idea of classifying taxa according to confirmed monophyly. The old angiosperm classification was left behind, and other monocot species were placed between the eudicots and primitive angiosperms. The monocots were divided into two categories: commelinids and the remainder of the monocots. These two families come after former angiosperms.

The APG 1988 approach recognised 462 families in 40 monophyletic orders that were further categorised into a few higher groups that are monophyletic like monocots, eudicots, rosids, asterids and so on. This system has 81 unplaced families (Fig. 5.3). Angiosperms are divided into three groups: monocots, eudicots, and four unclassified orders: Ceratophyllales, Laurales, Magnoniales, and Piperales. Acorales, Alismatales, Asparagales, Dioscoreals, Liliales, Pandanales, Arecales, Poales, Commelinales and

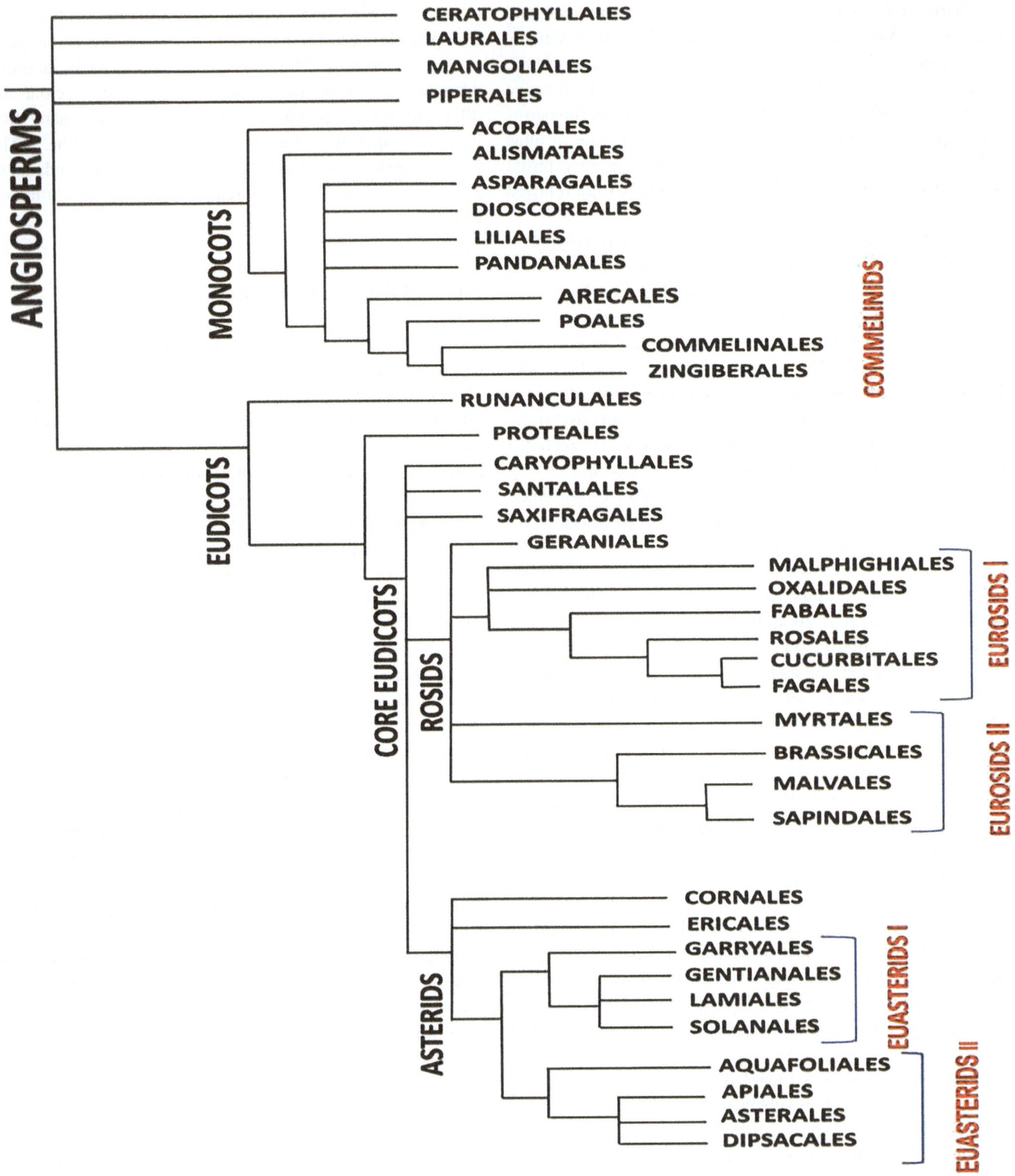

Fig. 5.3 Phylogenetic inter-relationships among different orders of APG classification. (Reproduced from APG 1998)

Zingiberales are examples of monocot orders. The final four orders are classified as Commelinoids. Eudicots are made up of Core-eudicots and two unclassified orders: Proteals and Rananculales. There are two groups of Core-eudicots, Rosids and Asterids, and three unclassified orders, Caryophyllales, Santalales, and Saxifragales. Eurosids I (Malpigiales, Fabales, Rosales, Cucurbitales and Fagales), Eurosids II (Mirtales, Brassicales, Malvales and Sapindales), and one unclassified order Geraniales are the three groupings of Rosids. The Asterids are divided into two groups: Euasterids I (Garryales, Gentianales, Lamiales and Solanales) and Euasterids II (Aquifoliales, Apiales, Asterales and Dispacales), plus two unclassified orders (Cornales and Ericales).

5.5 APG II

APG II recognised 457 families classified under 45 orders (Fig. 5.4); 44 of the 45 orders are divided into 11 informal groups. In contrast to APG 1998, which had 81 unplaced families, this number has been decreased to 40 in APG II. Angiosperms are categorised as Eudicots, Monocots and a few unplaced orders (Austrobaileyales) and families (Amborellaceae, Nymphyaceae, Chloranthaceae) under this system. Four orders, Canellales, Piperales, Magnoliales, and Laurales, which are classified as Magnolids, are also included in the unclassified ones.

5.6 APG III (2009)

APG III (Fig. 5.5) was published in the year 2009 by eight scientists. The updated edition was directed by Prof. P.F. Stevens. According to this approach, all plant species were classified into 413 families (with the exception of 10 families such as Ceratophyllaceae and Boraginaceae), which were assigned to 59 orders. Except for 5 orders, which are not classified, these orders were divided into 11 clades. Four of them are listed at the beginning, with one (Ceratophyllaceae) placed after monocots. This classification placed five taxa (two families, Apodanthaceae and Cynomoriaceae, and three genera, Gumillea, Petenaea, and Nicobariodendron) in an ambiguous position.

5.7 APG IV (2016)

This updated categorisation system was released in March of 2016. Angiosperm species were classified into 64 orders and 416 families. Unplaced groups were restricted to seven genera in this method. A poster with illustrations is available at http://www.plantgateway.com/download/2124/. The categorisation scheme's abstract is provided below:

5.7.1 Basal Angiosperms

Ana Grade: Amborellales/Nymphaeales/Austrobaileyales
Independent lineage: Chloranthales

5.7.2 Mesangiosperms

1. Magnolids: 4 orders
2. Monocot (11 orders): Commelinds (4 orders, now merged)
3. Probable sister of eudicots: Ceratophyllales
4. Eudicots (43 orders): 4 orders
   Core eudicots (39 orders)
      Order 1: Dilleniales
   Superrosids (18 orders)
      Order 1: Saxifragales
   Roside (17 orders)
      Order 1: Vitales
   Fabids: 8 orders
   Malvids: 8 orders

Superasterids (20 orders)
   Order 3: Berberidales/Santalales/Caryophyllales
Asterids (17 orders)
   Order 2: Ericales/Coronales
Lamids: 8 orders
Campanulids: 7 orders

5.7.3 Family Delimitations in APG with Respect to Earlier Classification Systems

- Earlier family Asclepiadaceae is merged with Apocynaceae
- Bombacaceae, Sterculiaceae and Tiliaceae are merged with Malvaceae
- Brassicaceae *s.l.* split into three families, viz. Brassicaceae *s.s.*, Capparaceae and Cleomaceae
- Chenopodiaceae is merged with Amaranthaceae
- Euphorbiaceae *s.l.* is split into Euphorbiaceae *s.s.*, Phyllanthaceae, Picrodendraceae and Putranjivaceae
- Fabaceae comprises three subfamilies: Fabaideae, Ceasalpinioideae and Mimosoideae
- Liliaceae *s.l.* is split into 14 families and many taxa are transferred to Asparagaceae
- Molluginaceae and Gisekiaceae are recognised separately from Aizoaceae

5.8 Botanical Nomenclature

Botanical nomenclature is the scientific method of giving names to the plants. Plant naming is important in determining their identification and connection. Around the mid-eighteenth century, plant names were mostly polynomials, or a succession of words that sounded like a description of the plant. Caryophyllum is a good illustration

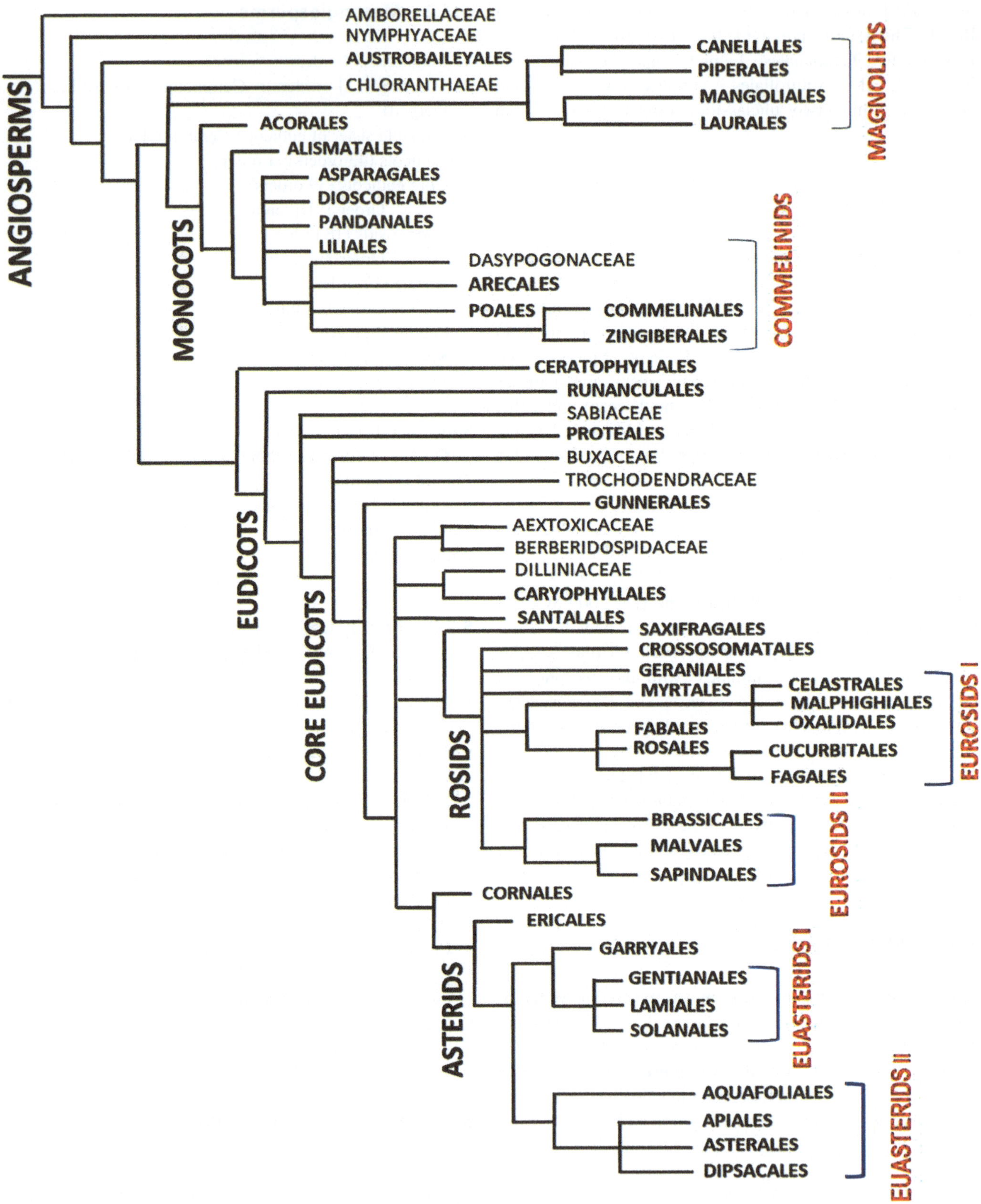

Fig. 5.4 Phylogenetic inter-relationships among different orders and families of APG-II classification. (Reproduced from APG 2003)

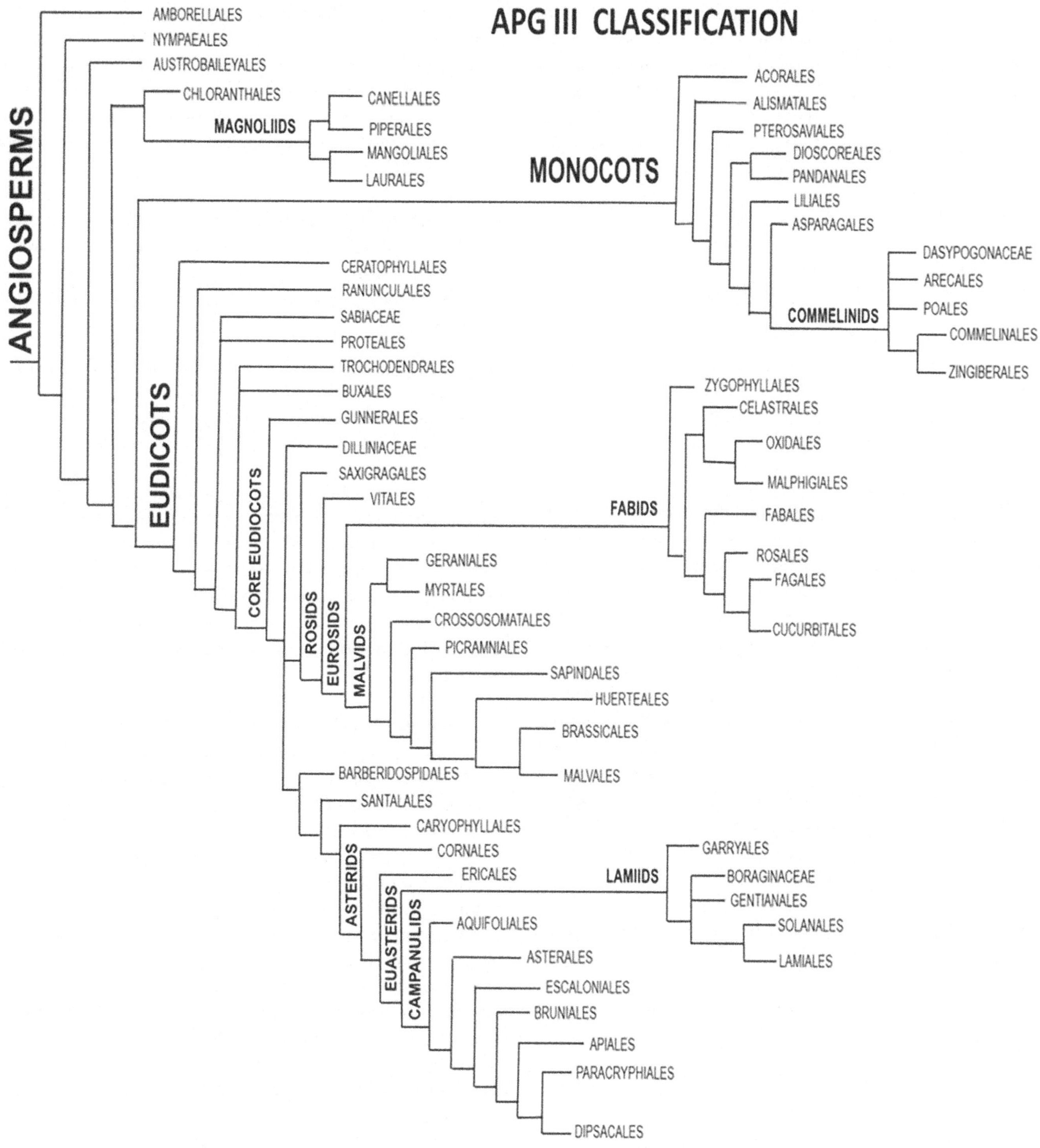

Fig. 5.5 Phylogenetic inter-relationships among different orders and families of APG III classification. (Reproduced from APG 2009)

of this. *Caryophyllum saxatilis folis gramineus umbellatis corymbis* means 'Caryophyllum growing on rocks with grass-like leaves and an umbellate corymbose inflorescence'. The binomial system was created by Gaspard Bauhinas in 1623, but it was Carolus Linnaeus who improved it by employing it in his book *Species Plantarum*. Linnaeus utilised this methodology to meticulously categorise plants in the 1700s. Linnaeus provided several aphorisms and announcements for naming plants in his *FundamentaBotanica* (1736) and *CiticaBotanica* (1737), and expanded on them later in *PhilosophiaBotanica* (1751), marking the true foundation of a good system of nomenclature. There were no acknowledged guidelines for naming plants until the eighteenth century, and botanists and herbalists did it their own way. Linnaeus's monumental book *Species Plantarum*, published in 1753, gathered plant names from all over the world under the binomial system of classification. It introduced the Binomial System of Nomenclature, which was an important turning point in the history of plant taxonomy.

Taxonomic group	Suffix
Kingdom	-a, ae
Division	-ta, ota
Class	-opsida
Order	-eles
Sub order	-ineae
Family	-aceae
Sub family	-oideae
Tribe	-ceae
Sub tribe	-ineae

This system is still in use around the world. The language employed in botanical naming is mostly Latin, with Greek and a few other 'dead languages'. These language's vocabulary and meanings are unlikely to change all over time.

A species' name has two words in binomial nomenclature. *Cocos nucifera*, for example, is the binomial nomenclature of the coconut tree. The term 'Cocos' is the genus, and 'nucifera' is specific epithet. The plant's name is made up of the two terms together. Because the binomial is a binary name, no two plants may have the same combination of generic and particular names since Linnaeus's time.

The International Code of Botanical Nomenclature (ICBN), which is a set of regulations and suggestions for naming plants, governs the official botanical names given to plants. The International Botanical Congress (IBC) is the only body with the authority to amend the ICBN (Swain 1963). The ICBN has established specific standards that must be rigorously obeyed, such as the suffixes of taxonomic groupings.

Old name	Alternatives
Palmae	Arecaceae
Cruciferae	Brassicaceae
Graminae	Poaceae
Leguminosae	Fabaceae
Guttiferae	Clusiaceae
Umbelliferae	Apiaceae
Compositae	Asteraceae
Labiatae	Lamiaceae

According to Article 18 of the ICBN, the family name must conclude with the suffix '-ceae'. However, due to long-term usage, eight families were permitted to keep their original names, and new replacements were also offered in accordance with the laws.

A family is a collection of plants that share similar traits, particularly reproductive organs such as flowers, fruits and seeds, which are used to distinguish them. A family might have as little as one species or as many as a 100. As an example, Ginkgoaceae consists of only one genus *Ginkgo biloba* whereas Rosaceae includes more than 100 genera (*Malus, Spiraea, Rosa*).

- A genus (plural genera) is a plant group that is closely related and identifiable by comparable traits (flowers, fruit, stem or roots) and genetic affinity. The genus name is generally a capitalised word that might denote one of the following:

A plant's appearance:	*Hemerocallius* (day and beauty)
Medicinal qualities:	*Pulmonaria* (lungwort)
Resemblance to body parts:	*Hepatic* (liver)
In honour of a person:	*Aenhenrya* (A N Henry)

The word that follows genus in a scientific name of the plant is the specific epithet that is not capitalised and usually adjective in nature like size (big size, *grandis*), colour (red colour, *rubra*), leaf form (heart shape, *cordifolia*), growth habit (*erecta*), in honour of a person (*linnae, roxburghiana, rheedei)* and use of the plant (*officinalis* is used in medicine). The specific epithet may contain information about the plant, for example:

- *Cotoneaster horizontalis* (horizontal branching)
- *Tectona grandis* (Large size of the tree)
- *Phyllanthus rheedei* (commemorating Van Rheede)
- *Madhuca longifolia* (long leaves)
- *Tephrosia purpurea* (having purple flowers)

The binomial system of naming is based on treating a population of individuals within a species capable of interbreeding. Scientific names of species should always be underlined or italicised. A scientific name is regarded as complete only when it ends with the person's name who described or named it first, a practice known as author citation. Paddy, for example, has the entire binomial *Oryza sativa* Linnaeus or *Oryza sativa* L. A species can be further subdivided into subspecies (subsp.), varieties (var.), cultivated varieties or horticultural varieties, and hybrids.

Plant naming is governed by a collection of rules and guidelines known as the International Code of Botanical Nomenclature (ICBN), which is now known as the International Code of Nomenclature (ICN). The International Botanical Congress (IBC) can only amend the ICBN. Following Linnaeus, the efforts of various taxonomists and the botanical congress resulted in a robust Code of Botanical Nomenclature. Their most significant contributions are listed below.

de Candolle Augustin (1813): Following Linnaeus, he wrote his most important book, *Theorieelementaire de la* *botanique*, in which he regarded plant anatomy and morphology to be key aspects in categorisation.

Steudel (1821): Plant names with synonyms are listed in *Nomenclatorbotanicus*. More than 3300 genera and almost 40,000 species were classified alphabetically.

Paris Code (1867): In Paris, the first International Botanical Congress was held to standardise and define Botanical Nomenclature. Alphonse de Candolle proposed Lois de la nomenclature botanique based on his father's *Theorieelementaire de la boatnique* (1813). '*Laws of Botanical Nomenclature*', published in 1867, was divided into three parts and had 68 articles. Priority rules, author citation, legitimate publishing and name approval and rejection were all established.

Kew Rule: When modifying a plant's generic name, the oldest specific epithet was not kept.

Rochester Code (1892): N.L. Britton and a group of botanists met in Rochester, New York, to discuss the American Association for the Advancement of Science. This meeting determined that each species must have a herbarium specimen in order to be designated as a Nomenclatural Type and recognised binomial nomenclature.

Vienna Code (1906) (The Third International Botanical Congress): Accepted Linnaeus' *Species Plantarum* (1753) as the beginning point for Nomenclature; generic description and generic names were preserved; Latin diagnosis, priority based on publication date.

American Code (1907): Type concept was accepted.

Brussels Code (1912) (Fourth International Botanical Congress): Priority of names and type concept was accepted.

Cambridge Code (1935) (Fifth International Botanical Congress 1930): T.A. Sprague and M.L. Green made every effort to establish norms and standards for plant naming and embraced the type idea as 'the application of names of taxonomic groups is determined by nomenclatural types'.

Amsterdam Code (1947) (Sixth International Botanical Congress): All plant diagnoses must be in Latin if published after 1 January 1935 rather than 1 January 1932. The Rule was broken down into Chapters and Sections, which were further separated into Articles and Recommendations. There is also a list of NominaGenericaCoservandawas.

Stockholm Code (1952) (Seventh International Botanical Congress): Minor adjustments and additions were enacted. The term Taxon (plural Taxa) was coined.

Paris Code (1956) (Eighth International Botanical Congress): The Code should be published in English, French and German, with the preamble and principle separated from the Rules and Recommendations. Appendix III (Nomina generic conservandaetrejicienda) has been revised and expanded. Appendix IV (Determination of Types) was reproduced in adherence with the Nomenclature Section's judgement in Paris.

Montreal Code (1961) (Ninth International Botanical Congress): Appendix II (Special provisions respecting fossil plants) was included with the introduction of Nominafamiliarumconservandafor Angiospermae.

Edinburgh Code (1966) (Tenth International Botanical Congress): The Congress established a committee to create a glossary of technical words used in the Code, which resulted in 'An annotated glossary of botanical nomenclature' (McVaugh et al. 1968).

Seattle Code (1972) (11th International Botanical Congress): Articles 22 and 26 have been improved, and the term autonym (automatically established names) has been added.

Leningrad Code (1978) (12th International Botanical Congress): The present ICBN system was adopted based on the resolution of this conference. Article 69 has been amended to read: 'A name must be rejected if it has been widely and persistently used for a taxon that does not belong to its type'. The rejected name will be added to a list of Nominarejicienda (Art. 69.1). The Code was outlined by the Congress in 1978 in Leningrad as follows:

This Code aspires for a consistent way of naming taxonomic categories, avoiding and rejecting names that may create mistake or ambiguity, and avoiding the development of names that are unnecessary.

1. Despite their unquestionable importance, the code viewed exact grammatical accuracy, regularity or euphony of names, more or less prevalent tradition, care for individuals and so on to be somewhat incidental.
2. The principles serve as the foundation for the botanical nomenclature system.
3. The detailed provisions are split into Rules and Recommendations, as outlined in the Articles. To explain the rules and guidelines, examples are included.
4. The Rules' purpose is to put the past nomenclature in order and to plan for the future; names that violate a rule cannot be kept.
5. The Recommendations deal with Secondary concerns, with the goal of achieving better uniformity and clarity, particularly in future nomenclature; names that contradict a suggestion cannot be used.
6. The provisions governing the amendment of this code include its final division.
7. With the exception of bacteria, the Rules and Recommendations apply to all species classified as plants (including fungus), whether recent* or fossil.
8. The only valid reason for altering a name is either a more comprehensive understanding of the facts as a consequence of adequate taxonomic study or the need of abandoning a rule-breaking nomenclature.
9. Established custom is observed in the absence of a relevant regulation or when the effects of rules are uncertain.

5.8.1 Some Salient Features of ICBN

1. A singular noun is the general name. Always write the initial letter of the generic name in capital letters. The particular epithet is always written in tiny characters and is an adjective. It derives from a variety of origins and may consist of one or two words, such as *Oryza sativa*.
2. The name should be brief, accurate and simple to say.
3. The binomial names are italicised or underlined. The generic name and specific epithet are both highlighted, for example, *Abutilon neilgherrense* or Abutilon neilgherrense.
4. When a new name is awarded to a plant, the herbarium specimen preparation of the same plant, along with its previous description, must be retained in any recognised herbaria. The specimens are referred to as type specimens.
5. The person who first publishes the description of a plant or gives it a new name is regarded the author. After the specific epithet, the plant name should include the author's name or an abbreviated name. This is known as author citation. For prominent taxonomists, abbreviations were created. Linnaeus' name was shortened to L. or Linn., Robert Brown to R.Br., and Sir Joseph Dalton Hooker to Hook, for example, Malva sylvestris Linn.

Sydney Code (1981), Australia (13th International Botanical Congress): Official versions of the code in English, French and German (with the English version taking precedence in the event of a dispute); conservation procedure (and rejection) extended to species names 'of major economic importance'; fungi starting date restored to 1753 with sanctioned name status established; types of genera and higher categories to become types of species (i.e. the taxa themselves are no longer types, only specimens or illustrations).

Berlin Code (1987), Germany (14th International Botanical Congress): The official version of the code is only available in (British) English, subsequent translations are available in French, German and Japanese.

Tokyo Code (1993), Japan (15th International Botanical Congress): Plant names are being registered; the nomenclature code is being reorganised extensively. The official version of the code is only available in (British) English, Chinese, French, German, Italian, Japanese and Russian; Slovak translations followed.

St. Louis Code (1999), USA (16th International Botanical Congress): Refinement of type standards; pictures as types were mainly prohibited beginning in 1958; morphotaxa for fossils. The Bio code and the registration of plant names were among the proposals that were defeated.

Vienna Code (2005), Austria (17th International Botanical Congress): Morphotaxa and regular taxa for fossils; pictures as types mainly prohibited since 2007; glossary added to nomenclature code.

Melbourne Code (2011), Australia: Electronic publication is authorised; registration of fungal names is permitted; English or Latin descriptions are permitted beginning in 2012; and the categories of amorph and teleomorphs (for fungi) and morphotaxa (for fossils) are removed.

The International Botanical Congress gathered in Paris during August 1867 and worked to standardise and legislate nomenclatural procedures. After receiving comments and recommendations (Amendments), it was agreed to meet the International Botanical Congress once every 5 years for improvements.

5.8.2 International Code of Nomenclature (ICN) for Algae, Fungi and Plants

It was determined during the 2011 International Botanical Congress in Melbourne, Australia, to include algae, fungi and plants within the standards of nomenclature. The International Code of Nomenclature for Algae, Fungi and Plants (ICN) superseded the International Code of Botanical Nomenclature (ICBN). It is unaffected by other codes such as:

- Animals: International Code of Zoological Nomenclature (ICZN)
- Bacteria: International Code of Nomenclature of Bacteria (ICNB)
- Cultivated plants: International Code of Nomenclature of Cultivated Plants (ICNCP)
- Plant associations: International Code of Phytosociological Nomenclature
- Viruses: International Committee on Taxonomy of Virus (ICTV)
- The term 'botanical' was deceptive, implying that the code only applied to green plants and excluded fungus and varied algae lineages. The term International Code for Botanical Nomenclature was changed to International Code of Nomenclature (ICN) for Algae, Fungi and Plants.

5.8.3 Electronic Publication of Nomenclatural Acts (Permitted from 1 January 2012)

The Nomenclature section unanimously endorsed a number of suggestions made by the Special Committee on Electronic Publication, which was established by the Vienna Congress in 2005 (Chapman et al. 2010). This implies that the new names of plant fungi and algae (as well as type designations) will no longer need to exist in the printed content in order to be effectively published—effective publication being key condition for the code of approval of any nomenclatural act. Publishing online in Portable Document Format (PDF) in a publication with an International Standard Serial Number (ISSN) or International Standard Book Number (ISBN) will be accepted as an alternative.

5.8.4 Latin and English

Except for names of fossils, it is presently essential to submit a description and/or diagnosis in Latin in order to legitimately publish the name of a new taxon, e.g. a species. The Nomenclature section changed this such that the description and/or diagnosis must be in either English or Latin for the names issued on or after 1 January 2012. This will apply to the names of new taxa in all the Code's groupings. It is already required for new fossil taxa names published on or after 1 January 1996.

The code is divided into three major divisions:

1. **Principles (1–6)**
   (a) Botanical Nomenclature is independent of ICZN, ICBN and ICTV.
   (b) The application of names of taxonomic groups is determined by means of nomenclatural types.
   (c) The nomenclature of a taxonomic group is based upon priority of publication.
   (d) Each taxonomic group with peculiar circumscription, position and rank can bear only one correct name, the earliest that is in accordance with rules, except in specified cases.
   (e) Scientific names of taxonomic groups are treated as Latin regardless of their derivation.
   (f) The rules of nomenclature are retroactive unless expressly limited.
2. **Rules/Articles and Recommendations (1–75)**
   The Rules or Articles contain precise prescriptions for restoring historical nomenclature and planning for the future; names that violate a guideline cannot be retained.
3. **Provisions for Modification of the Code**
   The Recommendations handle minor problems. Their goal is to achieve better uniformity and clarity, particularly in future nomenclature. Recommendations are frequently practical applications of Rules. They have no bearing on the authenticity of a name, yet it would be foolish to ignore them.

5.8.5 Nomenclature Criteria

Give the accurate name of the plants along with the author citation by referring to the most recent literature such as floras/monographs/revisions published and websites such as http://www.theplantlist.org, http://www.ipni.org/ or http://www.tropicos.org/.

To enable standard reference by scientists all around the world, the authors' names are also shortened or quoted as provided in 'Authors of Plant Names', Royal Botanical Garden, Kew London (1992).

5.8.6 Definitions of Terms

1. **Valid publication of a name**
   - A legitimately published name is one that has been published in conformity with Articles 32–45 of the ICBN. A taxon's name has no legal standing under the 'Code' until it is officially published. Only a lawfully published name can be considered legitimate or illegitimate.
   - The term 'invalid' does not occur in the Code; it typically refers to a name that has not been properly issued. However, this phrase is frequently incorrectly interpreted to mean illegitimate.
2. **Legitimate name**
   - A legally published name or epithet that follows all of the criteria.
3. **Illegitimate name**
   - A legally published name or epithet that violates one or more Code Articles. Superfluous names, later homonyms, and tautonyms are examples of popular illegitimate names.
4. **NominaConservanda**
   - The term NominaConservanda (nom. cons.) was first used at the International Botanical Congress in Vienna (1905). This was used to avoid unfavourable modifications in the names of families and genera caused by the rigid implementation of the priority principle. These names have been preserved because of their long use in botanical journals and for the sake of nomenclatural stability.
   - Appendix II: The Code consists of a list of conserved families (NominaFamiliarumConservanda).
   - Appendix III: The Code includes a list of preserved and rejected genus names (NominaGenericaConservandaetRejicienda). For example, Pleurolobus (1812), Desmodium (1813), Meibomia (1759).
5. **Rejected names (*Nominarejicienda*)**
   A name must be rejected if it has been extensively and repeatedly used for a taxon other than its type, i.e. a name that is used in several contexts and has long been a cause of mistake or confusion.
6. **Correct name**
   A legal name required by the regulations for a taxon having a certain circumscription, position and rank. According to different taxonomic definitions, a valid name may be right or wrong. For example, the generic

name *Hibiscus* L. is legal since it follows the standards. The generic name *Abelmoschus Medicus* is the same. When the genera are maintained distinct, both generic names are accurate. Augustin de Candolle combined *Abelmoschus Medicus* and *Hibiscus* L. into a single genus, and when this approach was adopted, as it was in the majority of previous Indian floras, *Hibiscus* L. was the sole accurate name with this circumscription. However, K. Schumann and numerous other monographers re-established the genus *Abelmoschus Medicus*. As a result, the real generic name *Abelmoschus Medicus* may be valid or inaccurate depending on taxonomic concepts.

7. **Synonym**

Synonyms are names that are different for the same taxonomic group or taxonomic taxon. In most cases, a synonym refers to a following synonym or synonymous phrase that will be rejected. There are two kinds of synonyms: Nomenclatural Synonyms and Taxonomic synonyms.

- Nomenclatural Synonyms or Homotypic Synonyms are the names based on the same type.

  For example:

  Miliusa tomentosa (Roxb.) Sinclair, *Saccopetalum tomentosum* (Roxb.) Hook.f. & Thoms., and *Uvaria tomentosa* Roxb. are all names derived from the same type of specimen and so are nomenclatural synonyms.

- *Basionym*

  For *Miliusa tomentosa* (Roxb.) Sinclair, the basionym is *Uvaria tomentosa* Roxb. Pl. Cor. T. 35, 1975, and not *Saccopetalum tomentosum* (Roxb.) Hook.f. & Thomson Fl. Ind. 152, 1855 & Hook.f. Fl. Brit. India 1: 88, 1872.

- Taxonomic Synonyms, also known as Heterotypic Synonyms, are names based on different types that are thought to belong to the same taxon.

  Example: *Hibiscus vitifolius* L. (1753) and *Hibiscus obtusifolius* Willd. (1801) are based on two separate type specimens (Linnaeus n. 265 and Klein s. n. [B-W n. 12899], respectively). On enquiry, both the types are found to be identical, thus completely dependent on taxonomic judgement.

  Hence *Hibiscus obtusifolius* Willd. is a taxonomic synonym of *Hibiscus vitifolius* L.

- *Homonym*

  The International Code of Botanical Nomenclature states that the earliest published homonym of two or more is to be utilised; a later homonym is 'illegitimate' and not to be used unless conserved. As an example:

  In the Leguminosae, the later homonym Myroxylon L.f. (1782) is conserved against the older homonym Myroxylon J.R. Forst. & G. Forst. (1775) in the Flacourtiaceae.

  Names that are similar enough to be mistaken are deemed homonymous under the botanical code (Article 53.3).

  For example: Astrostemma Benth. (1880) is an illegitimate homonym of *Asterostemma* Decne. (1838).

  A name with same spelling like a validly published name for a taxon of the same rank based on a different type.

  For example: *Jatropha heterophylla* Steudel, Nom. Bot. ed 2.1:799, 1840, and *Jatropha heterophylla* Heyne ex Hook.f. Fl. Brit. India 5:382, 1887.

  The *later homonym***,** i.e. *Jatropha heterophylla* Heyne ex Hook.f., is *illegitimate* and is replaced by the *new name—Jatropha heynei* Balakr. (Bull. Bot. Surv. India 3:41, 1961).

- *Tautonym*

  A binary name in which specific epithet identically replicates the generic name, either with or without the written sign, e.g. *Malus malus.*

  In botanical nomenclature, tautonyms are considered *illegitimate.* However, names such as *Cajanus cajan* and *Bambusa bambos* are not called as tautonyms as the specific epithets *do not repeat* exactly the generic names.

- *Autonym*

  A name that is automatically assigned to a taxon as a result of proposal of another taxon. For example, recognising a subspecies or variety that lacks the type species automatically establishes a subspecies or variety that does possess the type. As a result, at the infraspecific level, the last epithet of an autonym is the specific name unaltered but without mention of the author's name, e.g. *Abutilon indicum* (L.) Sweet subsp. *indicum.*

  The publication of the name *Crotalaria willdenowiana* subsp. *glabrifoliolata Ellis* in 1965 immediately created the name of another subspecies *Crotalaria willdenowiana* DC. subsp. *willdenowiana*, the type of which is that of the same subspecies *Crotalaria willdenowiana* DC. subsp. *willdenowiana*, the type of which is that of the same subspecies *Crotalaria willdenowiana* DC.

8. **Comb. nov. (Combinatio nova)**

Comb. nov. means *new combination*, a *new name framed from a previously existing epithet created by the type of the basionym.*

When a taxon of lower rank than genus is shifted to another genus or species, with or without change of rank, based on taxonomic decision, the original epithet if legitimate (basionym) is retained and the author of the basionym is retained in parenthesis followed by the name of the author who effected the change (the author of the new combination). For example, *Miliusa tomentosa* (Roxb.) Sinclair is a new combination based on *Uvaria tomentosa* Roxb.

When the rank of the taxon is modified during the transfer and its epithet from the name in the former rank has been retained, the phrase *stat. nov.* (*status novus*) is used to describe the new combination. For example: *Abutilon indicum* (L.). Miq. Sweet subsp. *albescens* Borgs. is a new subspecific status (stat. nov.) given for the species *Abutilon albescens* Miq.

9. **Nom. nov. (nomen novum)**

   Nom. nov. means a *new name*. A new name is usually an *alternate to replace an older name, designated by the type of the older name*. This is often shown as such only by the citation of the older name in the synonomy (replaced synonym). *A new name is given only when no legitimate name is available.*

   Tabernaemontana gamblei Subr. & Henry is a new name for *Ervatamia caudata* Gamble. *Ervatamia caudata* Gamble. when shifted to the genus *Tabernaemontana* must be called *Tabernaemontana gamblei* Subr. & Henry (1972) because of existing of the name *Tabernaemontana caudata* Merr. (1909), denoting to a different species. In other words, the epithet *caudata* is already available in the genus *Tabernaemontana*.

10. **Latin/English diagnosis**

    A diagnosis is a concise description of characteristics that distinguishes a taxon from its closest relatives. A species' or infraspecific taxon's diagnosis might be as simple as stating one or two traits that other species in the same genus lack. Earlier Latin diagnosis was compulsory, but after Melbourne Code (2011) this was relaxed.

5.9 Herbarium

A collection of preserved plants serves as voucher specimens for identifying an unknown plant. Herbarium is essentially a warehouse of dried and pressed plant specimens that are methodically categorised based on generally recognised classification systems and made available for future scientific research or evaluation of all material of a given taxon. Most of Indian herbaria, such as the Herbarium of the Botanical Survey of India, use the Bentham and Hooker categorisation system.

A plant specimen is deposited in a recognised herbarium whenever a new name for a plant is proposed. They serve as type specimens. The family name is always based on the type genus. The type specimens are treated as the most important accession in any herbarium and are accorded the maximum care and protection. Such specimens are usually lodged in fire-resistant cabinets.

The life of herbarium specimens can be prolonged if they are handled well by the users. To preserve specimens against fungus and insects, precautions should be taken. It is usually preferable to employ pesticides that repel insects from herbarium specimens. A label is always attached to the herbarium. It contains information such as the botanical name, family, habit, location and date of collection, as well as the details of specimen collector and a brief comment.

5.9.1 Importance of Herbarium

Herbarium records are vital and verifiable records of a species at a certain location and time. They are regarded as the primary instrument for plant categorisation and naming. Dried plant collections are an essential part of plant taxonomy since they aid in identification, morphological research, regional distribution, ecology and other aspects. The Italian Luca Ghini (1490–1556) was the first to use this method, stating that herbaria are the foundation of the scientific method and approach to plant taxonomy (Funk et al. 2005). The following are the primary purposes and applications of herbarium:

- A herbarium serves as a repository of knowledge on the flora of a region, location or country, as well as a data storage facility for plant information.
- The type specimen aids in precise plant identification.
- It supplies material for taxonomic, phytogeographical, palenological, phylogenetic, and anatomical research.
- Taxonomy has placed a strong emphasis on typical pollen characteristics. Even after over 200 years of storage, the pollen's morphological characteristics remain constant.
- It is extremely valuable in the study of cytology, DNA structure, numerical taxonomy and chemotaxonomy. It serves as a repository for gene pool research.

5.9.2 Herbarium Techniques

The following are the steps taken up while preparing herbarium specimens:

1. Collection
2. Poisoning and drying
3. Mounting and stitching
4. Labelling
5. Deposition

1. *Collection*
   Plant gathering is a serious business. In most cases, a twig with flowers must be obtained for the specimen. The specimen's part should have evident phyllotaxy and a branching system. Herbs and small shrubs up to 2–3 ft. tall should be picked, and lateral branches and some leaves may be cut to minimise bulk if necessary. Grasses, algae, bryophytes and ferns should be collected whole. To depict the layout of large shrubs and trees, a twig or tiny branch with three or more nearby leaves, or at least leaf bases, should be used. It must produce flowers and/or fruits. Plants having distinct male and female flowers, whether monoecious or dioecious, should be sought after for collection with both male and female flowers.
2. *Poisoning and Drying*
   To avoid pest and insect attacks, the specimen must be poisoned after collection from the field. There are various treatment options available to keep the specimen pest-free indefinitely. The specimens are poisoned by immersing them in an alcoholic solution of mercuric chloride and then dried between the folds of old newspapers or blotting sheets within a press. A plant press is made up of two boards tied by straps that assist tighten the newspapers or blotting sheets with specimens between the boards. When the weather is nice and dry, the press can be kept outside to dry. The sheets must be changed at regular intervals until the plants are thoroughly dry. The drying of plant specimens is a time-consuming operation. Folded leaves or flower components can be rearranged during pressing to properly position the specimen. At the end of the drying process, any remaining dirt is manually removed using a needle or scissors.
3. *Mounting and Stitching*
   Finally, the pressed and dried specimens are glued or mounted on a herbarium sheet. A typical herbarium sheet is 29 cm by 41.5 cm and is made of thick handmade paper or a card sheet. The paper used to mount specimens should be completely dry. The sheet should be sufficiently firm to prevent damage during specimen handling. The goal of 'mounting' plant specimens is to provide them with a sturdy physical support that allows for normal handling with little harm. A variety of methods are used to secure the specimens to the sheet. Some quick and easy ways to secure the specimens are to use sticky linen, paper or cellophane strips.
4. *Labelling*
   The label is positioned in the bottom right corner of the herbarium sheet, and the relevant information is recorded or printed directly on the sheet or on the paper slips that are glued on the sheets. It is best if the information is typed on the label. Handwriting should be done in permanent ink. Ball pens should never be used since the ink frequently spreads after few years. The size of the standard herbarium label is 4 × 6 inches or it may vary.
5. *Deposition*
   Deposition is the process by which mounted, tagged and treated specimens are eventually placed into a herbarium, where they are carefully preserved and maintained without harm. Specimens in each herbarium are organised alphabetically by family, genus and species, using a specific classification system.

5.10 Virtual Herbarium

A virtual herbarium is an Internet-accessible digital database that contains photographs of herbarium specimens as well as the descriptive label. It is an important improvement in the usage of herbarium, connecting biological specimens with digital herbarium over the Internet. It assists researchers in integrating comprehensive specimen data, as well as resource production and retrieval. Even though a virtual herbarium cannot exist without an actual herbarium, it serves a more beneficial function than a physical herbarium. Many large organisations, like the New York Botanical Garden (KE EMu), the Fairchild Tropical Garden (E-FTG), the Australian Virtual Herbarium (AVH) and the Royal Botanical Gardens Kew (ePIC) are creating searchable databases similar to Virtual herbaria.

5.10.1 Advantages of Virtual Herbarium

1. The user does not need to touch physical specimens because photos are provided electronically, preventing specimen damage to a considerable extent.
2. The time required in sorting out the specimen of a collector or a nation in a physical herbarium will be more than that required in a virtual herbarium. The virtual herbarium on the other hand can achieve this in seconds.
3. Traditional herbarium may only have a few visits in a particular period, while digital herbarium capabilities are referred to by a huge number of users from all over the world.
4. It takes a long time for a researcher to obtain essential information from preserved plant specimens in a physical herbarium. Virtual herbarium on the other hand provides descriptive features such as geographical distribution, pictures, drawings, manuscripts, published work, microscopic preparations, gene sequences and nomenclature via easily accessible hyperlinks.
5. A physical herbarium can only provide its own specimens for research, but portals of large virtual herbaria allow users to browse many virtual herbaria at the same time.

6. Thousands of specimens and precious holotypes kept in the Berlin Herbarium were lost during World War II. However, digitised pictures of herbarium can be preserved on several computers in separate places, allowing them to be retrieved from any one source even if others are destroyed.
7. Because the majority of research projects do not require physical specimens, computerised pictures may be used, saving time and money on transportation of genuine specimens.

5.11 Conclusion

Taxonomy is an ever-changing subject. Development in current fields of sciences gives evidence for the advancement of taxonomy, while frequent updating of information on taxonomy and field botany is required to help other branches of sciences in effectively and correctly communicating research results. The CBD recognised the sharp decline in taxonomic expertise and the difficulties confronting biological sciences, and India, as a signatory, launched the All India Coordinated Project on Taxonomy (AICOPTAX) with the goal of improving the country's capabilities for inventorying, monitoring, conserving and utilising biodiversity, as well as establishing leadership in the field of taxonomy at the regional, national and global levels. If we have enough time to spend with nature, taxonomy is a simple topic. An inquisitive observer can build competence, and current devices such as mobile phones, laptop computers and tablets can be efficiently utilised to record images of species and their life processes, and many web resources can also be used to develop taxonomic expertise.

Lessons Learnt

- Various classification systems with historical views
- Taxonomy applications and current significance
- Finer details of Bentham and Hooker's plant classification system, its virtues and demerits
- Details of the APG plant classification system (APG I, II, III and IV)
- Binomial nomenclature
- Key properties of various families and genera
- Creating a key for species identification
- Various concepts and laws of plant naming with examples
- Herbarium preparation and maintenance methods, as well as their significance

Key Questions

1. What is taxonomy? Explain the binomial nomenclature.
2. What is plant classification? Explain the Bentham and Hooker's system of plant classification.
3. Explain the APG system of plant classification.
4. What is binomial nomenclature? Explain with examples.
5. Describe salient features of International Code of Botanical Nomenclature (ICBN).

References

APG (Angiosperm Phylogeny Group) (1998) An ordinal classification for the families of flowering plants. Ann Mo Bot Gard 85:531–553

APG II (2003) An update of the angiosperm phylogeny group classification for the orders and families of flowering plants: APG II. Bot J Linn Soc 141:399–436

APG III (2009) An update of the angiosperm phylogeny group classification for the orders and families of flowering plants. APG III. Bot J Linn Soc 161:105–121

APG IV (2016) An update of the angiosperm phylogeny group classification for the orders and families of flowering plants: APG IV. Bot J Linn Soc 181:1–20

Bessey CE (1897) Phylogeny and taxonomy of the angiosperms. Bot Gaz 24(3):145–178

BSI (2023) Plant discoveries 2022. Botanical Survey of India, Kolkata

Chapman AD, Turland NJ, Watson MF (2010) Report of the special committee on electronic publication. Taxon 59(6):1853–1862

de Candolle AP (1813) Théorieélémentaire de la botanique (Elementary theory of botany). Déterville, Paris

European Commission (1996) Interpretation manual of European Union habitats. Version EUR 15. European Commission DG XI. European Commission, Brussels

Fralish JS, Franklin SB (2002) Taxonomy and ecology of woody plants in north American forests (excluding Mexico and subtropical Florida). Wiley, Hoboken

Funk VA, Hoch PC, Prather LA, Wagner WL (2005) The importance of vouchers. Taxon 54:127–129

Hollingsworth PM (2003) Taxonomic complexity, population genetics and plant conservation in Scotland. Bot J Scotl 55(1):55–63

McVaugh R, Ross R, Stafleu FA (1968) Annotated glossary of botanical nomenclature. Regnum Veg 56:1–31

Stace CA (1991) Plant taxonomy and biosystematics. Cambridge University, Cambridge

Stace CA (2005) Plant taxonomy and biosystematics—does DNA provide all the answers? Taxon 54(4):999–1007

Steudel (1821) Nomenclator botanicus—indexed plant names with synonym

Swain T (ed) (1963) Chemical plant taxonomy. Academic, London

Williams PH, Humphries CJ (1994) Biodiversity, taxonomic relatedness and endemism in conservation. In: Forey PL, Humphries CJ, Vane-Wright RI (eds) Systematic and conservation evaluation. Clarendon Press, Oxford, pp 269–287

Statistical Methods in Forestry Research

6

Rajiv Pandey and Abhishek K. Verma

Abstract

The result of a random experiment provides one outcome out of all possible outcomes. Statistical methods provide solution for such experiments for precise decision making under uncertainty. The decision making under uncertainty requires three principal stages where experimenter should take adequate precautions. First precaution requires appropriate application of statistical techniques for sample selection and subsequently data collection for the required variables from the specified selected units. Second precaution requires application of appropriate methods for data analysis and presentation of the analysis for achieving the objectives. Third one requires appropriate interpretation of the data analysis. This chapter discusses all the three precautions along with suitable examples of from forestry experiments using basic statistical methods. Introductory remark about the forestry experiments along with basics of statistics, that is, variable and measurement scale is elaborated with examples. Second section discusses the data collection protocol using suitable examples from forestry field for the basic experiments, that is, sampling methods and designing of the experiments. Third section deals with data analysis for forestry experiments along with suitable examples. Data analysis includes the representation of data, that is, qualitative and quantitative method of data representation. The inferential examples are included for testing of the population(s) besides modeling of the cause and effect relationship. Each example is elaborated and interpreted for precise information of the experiment as per the data analysis.

Keywords

Sampling variation · Sample survey · Experimental design · Analysis of variance · Regression

6.1 Introduction

Forestry is a conglomeration of art and science dealing the management of forests for ensuring sustained supply of ecosystem services (Kimmins 2002). With time, the focus of forestry research has been shifted from species-specific and production to forest ecosystem and the emergent properties for sustainable flow of ecosystem services. Precisely, the research in forestry has shifted the focus from yield-based forestry towards more interdisciplinary dimensions of forestry research. In this discourse, forests are ecologically important and dynamic system leading to a paradigm shift to ecosystem-based management integrating the fabrics of ecological, social and economic objectives. Currently, the ecosystem-based researches are pronounced by evaluating functions and processes of the forests using functional traits, and also ensuring the sustainable flow of ecosystem services.

The researches for achieving the objective about forest functioning and management require experimentation following scientific methods so that the required knowledge for the forest may be obtained on various spatial and temporal scales. Scientific methods start with a problem followed by a procedure for data collection and analysis leading to drawing of a valid conclusion, that is, generation of knowledge. Moreover, to be parsimony as per Occam's razor principle, it is essentially required that simpler explanation, that is, knowledge for any phenomenon, should be preferred over complex explanation, if all other things being equal. The generation of knowledge, say explanation of functioning or processes of the forest, is based on inductive and deductive approaches. In inductive reasoning, an individual instance supports the truth of a generalization, contrary to deductive

R. Pandey (✉)
Indian Council of Forestry Research and Education, Dehradun, India
e-mail: pandeyr@icfre.org

A. K. Verma
Indian Council of Forestry Research and Education, Dehradun, India

Forest Ecology and Climate Change Division, Forest Research Institute, Dehradun, India

A. K. Mandal, A. Nicodemus (eds.), *Textbook of Forest Science*, https://doi.org/10.1007/978-981-97-8289-5_6

reasoning, where a conclusion is logically drawn based on the premises.

The deduction approach revolves around hypothesis testing through collecting and analyzing the data for testing the hypothesis/theory. Deductive approaches begin with the knowledge based on findings on the basis of reasoning and logic to evolve theories and assumptions from general or universal whether conclusion is based on generally accepted statements or facts. Inductive approach begins with question, followed by data collection and essentially data analysis highlighting the truth. In inductive approach, research study explored data for regularities, patterns, and themes and formed generalizations and subsequently a theory. Inductive generalization requires a large number of observations from very different conditions with the condition that none of the observation statements contradict the general hypothesis.

Major purposes in planning for the collection of data through experimentations,

- **Validity** ensures the validity of the conclusions, which is logically attainable through investigation and free from personal bias
- **Precision** ensures that the variation due to uncontrolled factors (errors) in the experimentation should be contained at a low level
- **Coverage** ensures the wide applicability of obtained result from the experiment

Experimentation in forestry is complex due to the interactions of various uncontrolled factors leading to produce random variation. The random variation in the forestry experimentation requires the application of a scientific discipline, which is designed to evaluate the uncontrolled variation and is not deterministic rather probabilistic in nature, that is, occurrences cannot be predicted well in advance. In 1948, Anscombe, a statistician, discussed that in random experimentation repeated observations of same or similar events do not agree with another exactly, and the researcher cannot make the agreement perfect by taking any reasonable precautions to keep conditions constant; for example, the growth of neighboring trees receiving similar treatments is not equal. It may happen that the absolute performance of treatments varies erratically, though the relative performance of two treatments may be fairly consistent. In such cases, it may be possible to state that in similar circumstances one treatment gives substantially better response than another in relative terms, even when we cannot state at all exactly what results either will give. In such fields of experimentation, experiments are likely to be comparative, and only secondary interest is attached to absolute performance of treatments. In forestry, such comparative experiments are numerous, namely, comparison of carbon sequestration potential between natural forest with plantation, comparison of performance of different clones in a locality, comparison of performance of preservatives on wood pieces of different species, comparison of performance of insecticides on different insect, testing of chemical characteristics of plant parts of medicinal plants, selection of wood for box making (packaging purposes) in terms of strength and economics, etc. However, the value of performance may be estimated on average basis based on all the collected values/observation and thus the experiments are estimative. The estimation and comparison require random experimentations and are a prerequisite for the solution of any scientific problem under consideration and the two aspects, that is, estimation and comparison are dealt by statistics.

The homogeneity in experimental materials is a prerequisite of the comparative experiments. The homogeneity should be ensured by addressing the factors which can cause the variation. In forestry, following factors caused variation in the response or in experimental materials.

- **Developmental variation** accounts variation in the growth and development among units, for example, a mature tree is taller than a young seedling of the same species. This sort of variation can be easily eliminated by choosing similar units from the population; for example, trees of same age are required in the layout for the experiment.
- **Environmental variation** accounts variation in the environment among the units, for example, a tree on a good site is gregarious than that of a tree on poor site having same age and of same species. Various environmental conditions may vary between sites, such as differences in how clones behave at different latitudes and altitudes. There can also be macro-variations within a site, such as a fertility gradient from one corner of a forest to another, and micro-variations within a site, such as changes in nutrient availability near neighboring trees or locations. Environment variation cannot be eliminated completely but can be reduced by taking trees to be compared on the same sites and designing tactfully.
- **Genetical variation** accounts variation in the genetic structure and behavior; for example, one resin producing tree may produce more resin than another of the same species and age growing on the same site. Genetical variation may be reduced through the knowledge of the subject
- The above three types of variations in combination are known as phenotypic variation. The variations are due to the experimental materials being used in experimentation.

(continued)

- **Sampling variation** is due to the study of a representative subset of the units rather than evaluation of all units.
- **Experimental error** originates due to inaccuracies of assessment/ measurement/data recording.
- The sampling variation and the experimental error in combination are termed as the residual variation, that is, the variation, which is remaining, when genetic and the environmental variation have been identified and addressed already.

Statistical procedures are dealt with scientific approaches, and drawing and generalizing interpretation of the random experiments are art. Statistics supports to derive the inferences and conclusions from experiments having random variation by suitably designing as per the objectives, available conditions, and resources for collecting the information. The information collected from the experiment is subjected to statistical analysis as per the objectives of experiment and interpretation of results. For example, if a researcher considers to evaluate water stress impacts on poplar (*Populus* sp.) clones, he has to lay an experiment, which facilitates similar conditions to respond against the external treatment to all the poplar clones so that chances of being in advantage to any clones are refuted. The data collected from such an unbiased experiment, once analyzed suitably, may yield valid inferences about the poplar clones for the posed objective. Statistics provides a kit of tools to deal with random experiment only and the inferences drawn therefore are generalized for the target population for which the observations are obtained from the representative sample. The three steps required to obtain valid inferences from the random experiments are (1) data collection method, (2) data analysis methods, and (3) techniques for proper interpretation of results. The three steps are explained in details in the following section with some examples.

6.2 Methods of Data Collection

Data, that is, information, may be retrieved either from the existing information like productivity of different patches of a forests or can be constructed (manufactured) as growth of eucalyptus species under various growth regulators, that is, NAA, IBA. The two types of information, that is, the existing and manufactured information, require different approaches for the data collection. The existing information can be collected through sampling design and the manufactured information as per the researcher's requirement can be collected through experimental design (Chart 6.1). Precisely, sampling design aims to draw valid conclusions on the basis of representative samples from the naturally existing population through proper scientific technique, such as pH of soil under different tree species in a natural forest. On the other hand, in experimental design no such population exists in its own way, rather the data has to be manufactured as per the problem by proper experimentation to answer the posed problem. In an experimental design, the aim is to study the variation produced in response to purposively introduced variation through factors or applied treatments, such as to evaluate effect of different media for plant growth under natural condition. For example, the media, that is, factors or treatments, are purposely introduced to create the variation in response as per the objective. Creation of controlled conditions is the main characteristic feature of the design of experiment.

- An **observation** is implicitly coded information, for example—the height of deodar tree is more than oak tree.
- A **datum** (plural is data) is explicitly coded information, for example—the height of deodar tree is 27.9 m.
- A **variable** is a characteristic feature of the unit under investigation, which varies with respect to time or space or units or for any combination of the three.
- A **qualitative variable** represents the information of the variable in terms of quality (non-numeric), for example, the leaf type of various plant species.
- A **quantitative variable** represents the values of the variable in terms of quantity (numeric), for example, the pH of a forest patch is 6.7.
- A **univariate** considers only one variable at a time.
- A **bivariate** considers two correlated variables simultaneously.
- A **multivariate** considers more than two variables simultaneously.

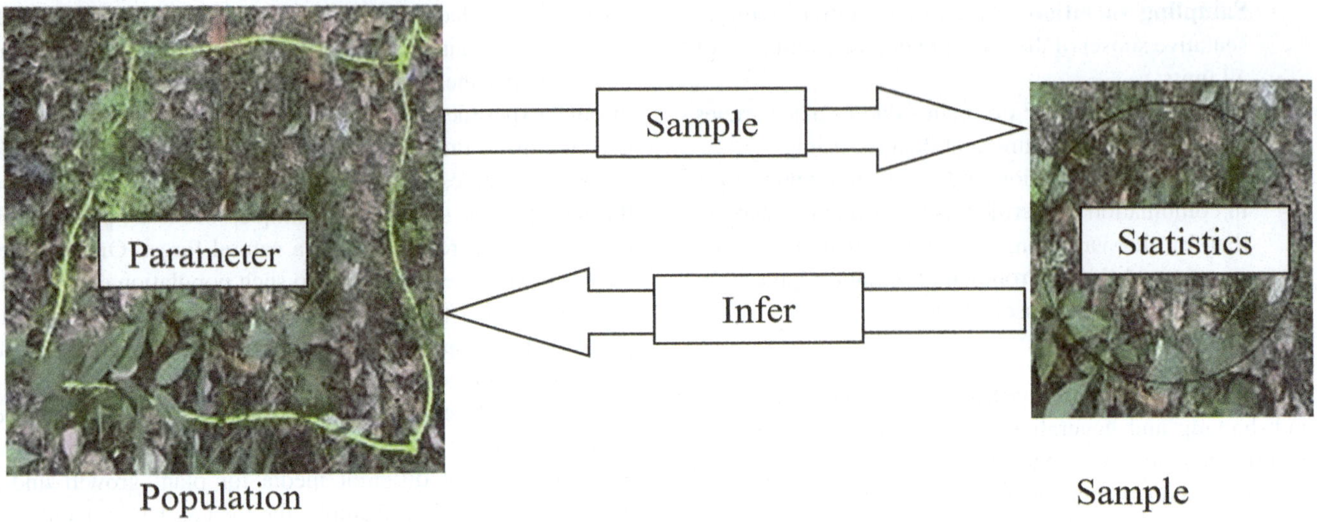

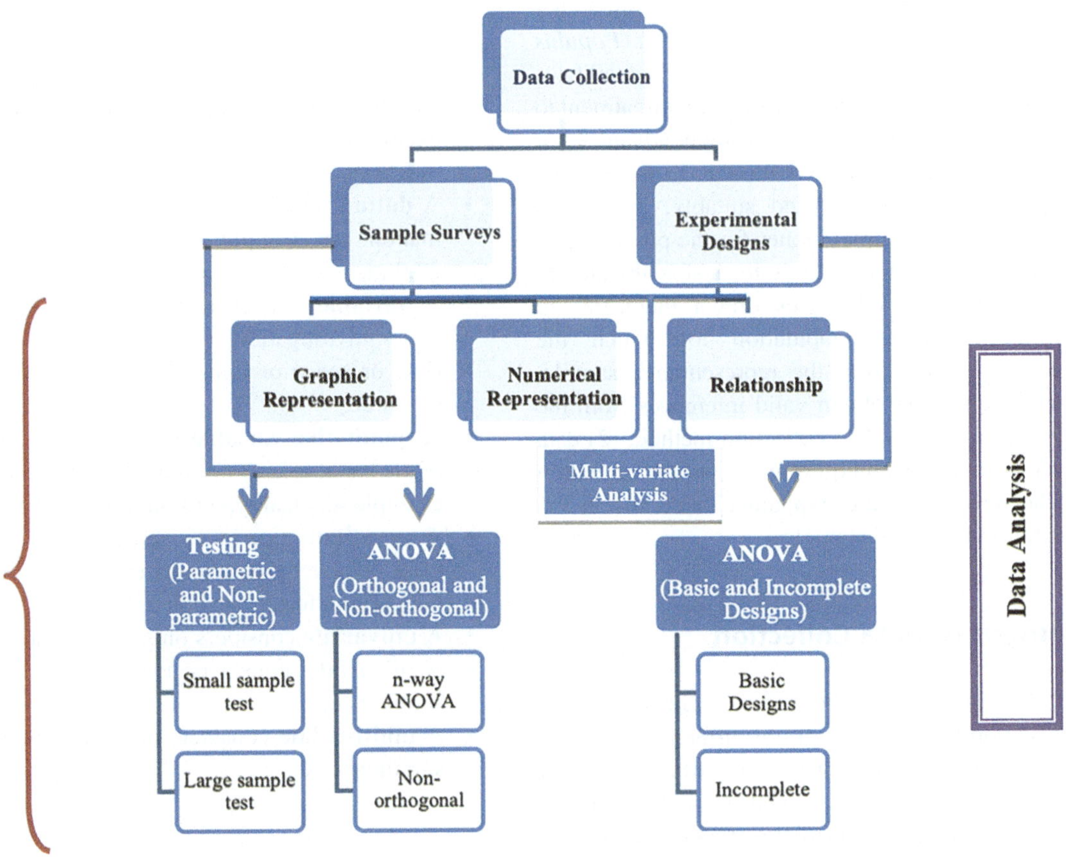

Chart 6.1 Flowchart of materials in the document

6.2.1 Sample Survey

Sample survey deals with the selection of representative samples through suitable sample drawing protocol for the units of the population under evaluation. The data for the selected units would be analyzed for drawing the inferences about the population. The representative sample means a subset (or part) of the units from the population (which is a collection of all units as per objective) containing all sorts of variation present in the population. Precisely, the sample must contain all sort of variability present in the population. The sampling is easy, resource saving, and more scientific. The sampling process includes two types of error. Sampling error exists due to the consideration of only a part of the population, as some of population units are not part of the sample. Non-sampling error exists due to use of faulty instruments, data recording, data digitization, and measurements.

Characteristic	Parameter	Statistics
Source (origin)	Population (all units)	Sample (representative units)
Notation (representation)	Greek (as μ for population mean)	Roman (as x bar for sample mean)
Fixed value	Yes, as it is based on all units	No (changes with sample)
Computation	No	Yes (for each sample, estimation is being made)

- A **unit** is the smallest element on which a researcher wishes to measure the single required information, for example, leaf area of one leaf for a single species.
- A **population or universe** is the collection of all units, which is under investigation by the researcher.
- A **census** is an evaluation of all the units in the population; for example, if researcher is estimating the biomass carbon of trees of a natural forest, for census, researcher would measure biomass carbon of each and every tree of the natural forests.
- A **sampling frame** is used for selection of the units for the sample surveys and contains the list of all the units of the population.
- A **sample** is a representative subset (part of the population) of the units in the population. The representative subset considers all sort of variation present in the units.
- A **population parameter** is a value, which is based on the measurements of all the units in the population for the variable under investigation.
- A **statistics** is a value, which is based on the measurements of subset (part of population) of the units in the population, that is,. the units of the sample for the variable under investigation. The statistics estimated from the sample is representing population parameters.
- A **sampling unit** is smallest units, from which a single observation is drawn.

(continued)

Sampling for any experimentation revolves around presampling and sampling choices. Presampling choices include the nature of the study, that is, exploratory, descriptive, or conformity and variables to be measured along with its measurement protocol for achieving the objectives from the defined target population. Sampling choices include the listing of the target population, that is, sampling frame, so that units can be drawn as per sampling design along with the acceptable (tolerable) error in the estimate which the researchers expect; and sampling techniques and number of units drawn in the sample, that is, sample size. The sample drawing is of two types: (1) drawing as per the choice of the researcher (non-probability sampling) and (2) drawing by any scientific procedure say based on probabilistic phenomenon (probability sampling).

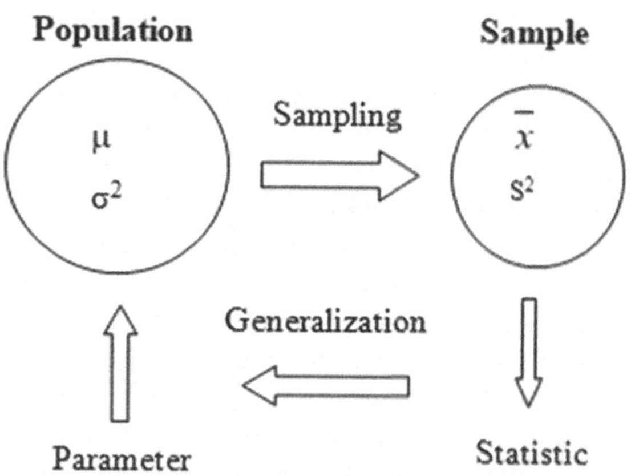

Non-probability sampling selects the units based on either convenience or based on choices of the researcher, for example, selection of healthy seedlings from nursery; selection of forest patches of natural forests near to road. The basic limitation of the sampling design is consideration of only part of units of the population rather than consideration of all units of the population for drawing the samples. Therefore, the generalization of the result based on such sample is not possible and rather limited for the intended use. Several types of non-probability sample designs exist as convenience sampling (samples drawn at the convenience of the researcher), quota sampling (samples drawn as per quota, i.e., specific number of certain types of units to be surveyed),

Table 6.1 Comparison between the lottery and random number table

Procedure	Lottery method	Random number table
Population size number of all units in the population)	*N*	*N*
Sample size (number of units in the sample)	*N*	*n*
Sampling frame (listing of all units in the population)	Required	Required
Assigning number to the units of the population	Number 1 to *N* will be allocated for the *N* units of the population	Number 1 to *N* will be allocated for the *N* units of the population
Process for selection of units	Allocated number of units is marked on *N* number of identical slips of the paper. All the *N* number of slips would be placed in a box	Not required
Selection procedure for *n* units	One by one drawing of slips after through reshuffling the box should be taken place for "*n*" times or more than "*n*" times to derive a sample of size of "*n*" units	A k-digit random number is selected from any point in the random number table ("*k*" is the number of digits for the number of the last population units). If the number selected from the table is less than or equal to "*N*," then the unit allocated the number would be selected. If number selected is higher than "*N*," the number will be discarded
Number of draw	"*n*" draw	May be more than "*n*" draw
With replacement (WR) and without replacement (WoR)	Both	Both

judgment sampling (samples drawn based on personal judgment), and snowball samples (selection of additional sampling unit is based on referrals from the initial units).

Probability sampling selects the units either based on equal or non-equal probability, for example, for estimating biomass carbon in a natural forest, compartment (smallest management units) number 8 and 13 may be drawn through a random process out of total 17 numbers of compartments in the natural forest. In this sampling, all units are considered while drawing the units from the population and therefore, the results derived from the sample can be generalized to whole population. A sample is random, provided it is drawn by such machinery, that in the long run any one individual of the population would get selected in sample as often as any other, according to Peirce. The probabilistic sampling design depends on the objective of the experiments and availability of the resources and several sampling designs are elaborated in the literature. Moreover, several methods for the drawing of samples through random process are available.

The simplest random process is age old lottery method (Table 6.1). In lottery method, "*N*" number of paper pieces is made for a population containing "*N*" units. Each population unit is allocated and assigned a number, and the number is marked on the slips of paper of the same shape, size, and color. The marked slips of paper are folded and kept in a box for drawing a sample. Now suppose the researcher attempts to draw a "*n*" number of units from the population; therefore, the researcher will have to make "*n*" draw one by one either by replacing or not replacing the drawn slips in the box in subsequent draw. The unit as per the marked number on the drawn slip would be selected in the sample. The sample selected by replacing the drawn slips in the box in subsequent draw is known as sample with replacement (SWR). In this, after the "*n*" drawing, the researcher gets "*n*" unit which forms the desired sample. The sample may and may not be having distinct units in the sample. However, if the drawn slips are not replaced in the box in subsequent draw, then all the units will be distinct and the sample is termed as sample without replacement (SWoR). Suppose, the researcher wishes to select ten Shisham (*Dalbergia sissoo*) trees through a random process from a plantation of 150 trees, he has to marked each tree and the mark would be noted down on 150 distinct slips having all similarity. The marked 150 folded slips would be placed in a box and the box will be thoroughly reshuffled. Thereafter, the researcher would pick ten slips one by one without replacement and the trees having the number as marked on the slips would form the random sample of ten trees. The lottery method is not applicable for the population which has infinite number of units due to practicality.

Another random process can be obtained through the use of random number table (Table 6.1). Random number table is arrangement of numbers in rows and columns in such a fashion that each of the digits 0, 1, 2, . . ., 9 appears independently and approximately with the same frequency across the table. Drawing samples through random number table process is most practical, easy, and inexpensive. Various "Random Number Table" has been made and named as Fisher and Yates random number table, Tippett random number table, Kendall and Smith random number table, and Rand

Corporation random number table. The researcher can use any one table as per his convenience. In random number table method, each "N" number unit of the population is allocated and assigned a number from 1 to "N" and a number from an identified random number table is selected from the k-digit random number from any point in the random number table. The "k" is the number of digits in the number of the last population units; for example, if in a population, there are 1,15,325 number of trees, then N is 1,15,325 and "k" is 6. Suppose, if a researcher wants to select a sample for a specified sample size from a population of size N (<99), then any two digit random number from anywhere in the random number table can be selected from (00–99) from the random number table. If the selected number from the table is less than or equal to N, the unit matches with the selected number of the table and is included in the desired sample; otherwise, the selected number of table is discarded, and a new number from table is selected. The next number not exceeding N may be selected systematically either moving horizontally, vertically or any other direction as per the choice of the researcher. The process is repeated till the researcher gets the desired number of units for the sample. In this process also, sample is drawn with or without replacement depending upon whether the selected number is being considered or not in next draws.

Random sampling attaches a selection probability to each and every unit in the population to be included in the sample and refers to random manner selection of samples from the population; for example, if researcher wants to evaluate the productivity of a natural forest containing 12 compartments through random sampling, he may select any three (sample size) compartments through lottery method or random number table approach. Following are random sampling approaches being extensively used in the forestry experimentation.

- Simple random sampling
- Stratified random sampling
- Systematic sampling
- Cluster sampling

Simple Random Sampling: In simple random sampling (SRS), each and every unit in the population has equal and independent chance of being selected in the sample. The information may be collected from the selected units of the sample and used for the analysis. SRS is applicable only if all the units of the population are more or less similar with respect to the objective of the investigation. For example, the researcher wishes to estimate litter fall in a natural forest. In this, the forest area will be divided into small patches containing some trees, say "N" units. The researcher inspected all units and found that more or less all "N" units (patch) of forest are having more or less same number of trees with similar growth and development. The listing of all "N" units, that is, patches is sampling frame. Selection of a sample of size of "n" units from a population containing "N" units through a random process requires allocation of a number 1 to N to the units of the population. The Tippett random number table is used for selection of "n" units from the population by selecting numbers from the table. The "n" selected units, that is, patches from the population will be explored for collection of litter from the patches. The collected data will be subjected for the desired analysis as per objective. The process is exemplified below for better understanding.

Patches of the forest—population ($N = 25$) (Sampling frame)					Selected numbers by random number table				Selected patches from the patches of the forests—Sample ($n = 5$)
1	**2**	**3**	**4**	**5**					**6** | **9** | **12** | **18** | **19**
6	**7**	**8**	**9**	**10**	6			9	The 5 at random selected patches (sample size) from 25 patches (population size) of the population constitute the sample. These five patches will be evaluated for data collection
11	**12**	**13**	**14**	**15**		12			
16	**17**	**18**	**19**	**20**			18	19	
21	**22**	**23**	**24**	**25**					

Note: It is assumed that patches of forests (represented by number 1–25) are more or less similar. This is pre-requisite for SRS

Stratified Random Sampling: Stratified sampling is a fit case for the units of the population where variation across the units of the population is apparent. If the units of the population are heterogeneous for the variable under investigation, stratified random sample may be applicable for collection of sample. In this sampling, first the heterogeneous population is classified into different strata (subpopulation) in such a manner, so that the units of the different strata are homogeneous within themselves. The strata are designed in such a manner that they do not overlap each other. Now, the sample size for each stratum is decided either through equal allocation of units from each stratum (termed as equal allocation procedure) or based on proportion of units of the strata with respect to the total units of the population (termed as proportional allocation). A sample as per the size for the stratum is selected through random process and the same process is followed for all the strata of the population. The collection of all units from each stratum constitutes the stratified random samples. The information may be collected from the selected units of the sample and used for the analysis.

For example, the researcher wishes to estimate soil organic carbon (SOC) in a natural forest containing dominated species from two leaf habits, that is, deciduous and conifer in different patches. The SOC is dependent on the tree species, and therefore the species under two leaf habits would differ for their SOC stock. Therefore, the forest has to be stratified in two strata one for each leaf habits. In this forest area will be divided into small patches containing different tree species of the two leaf habits, say "N" units, and out of N, conifer species dominates in N_1 units and deciduous species dominates in N_2, namely, $N = N_1 + N_2$. The listing of all "N" units, that is, patches constitutes sampling frame. Selection of a sample of size of "n" units from a population containing "N" units through random process in the sampling requires to select a sample of "n_1" units from conifer and "n_2" from deciduous species such that $n = n_1 + n_2$. The Tippett random number table is used for selection of "n_1" and "n_2" units from the two strata by selecting numbers from the table. The "$n = n_1 + n_2$" selected units, that is, patches from the population will be explored for collection of SOC from the patches under the two strata. The collected data will be subjected for the desired analysis as per objective. The process is exemplified below for better understanding.

Patches of the forest—population ($N = 15 + 10 = 25$) (Sampling frame)					Selected numbers by random number table					Selected patches from the patches of the forests—Sample ($n = 3 + 2 = 5$) proportional allocation
1	**2**	**3**	**4**	**5**				4		**4** **7** **16** **20** **23**
6	**7**	**8**	**9**	**10**		7				The 5 at random selected patches (sample size); 3 from deciduous and 2 from conifers from 25 patches (population size) of the population constitute the sample as per proportional allocation. These five patches will be evaluated for data collection
11	**12**	**13**	**14**	**15**						
16	**17**	**18**	**19**	**20**	16				20	
21	**22**	**23**	**24**	**25**			23			
Deciduous				Conifer	Deciduous				Conifer	

Note: Due to the heterogeneity in terms of the leaf habits (contribute differentially for SOC), the forest patches (represented by number 1–25) has been divided into two strata as per the dominance of the conifer or deciduous tree species. It is assumed that patches of forests are more or less similar within the strata

Systematic Random Sampling: Systematic sampling is a fit case for the population, where sampling frame cannot be created due to various reasons such as listing of all trees in a large natural forest. The sampling starts with selection of the first unit through a random process and the remaining units are selected as per sample size according to some pre-defined system. In this sampling interval (k) has been defined as the integer of the ratio of population size and sample size, that is, $k = N/n$. The first unit is selected by random process from the first k units and subsequent units are selected by adding the sampling interval with the number of the first selected units, that is, every kth units, starting from the first selected unit is included in the sample.

For example, the researcher wishes to estimate crown size of trees in a huge natural forest containing different species. In this forest, counting or listing of tree is not available rather not possible also; however, roughly, 1,50,000 trees are present in the forests. Therefore, the population size is 1,50,000 = N and the researcher wishes to evaluate 150 trees, that is, sample size is 150 = n. The sample size is estimated through scientific protocol. In systematic sampling first, sampling interval has to be estimated which will be k = 150,000/150 = 1000. Now, the researcher will select the first unit through random process by selecting a number from random number table from the first 1000 trees of the forest. Say, the first tree was selected as 577 and then the remaining trees may be selected as 577 + 1000 = 1577th tree; 1577 + 1000 = 2577th tree and the process will continue till all the 150 trees required for the sample is collected. The selected trees will be explored for the data and the collected data will be subjected for the desired analysis as per objective. The process is exemplified below for better understanding.

Patches of the forest—population ($N = 25$) (Sampling frame)					Selected first number by random number table ($k = 25/5 = 5$)				Selected patches from the patches of the forests—Sample ($n = 5$)
1	**2**	**3**	**4**	**5**				4	**4** **9** **14** **19** **24**
6	**7**	**8**	**9**	**10**				9	The 5 selected trees (sample size) with the first as fourth tree as per the selected random number; and remaining four as per sampling interval (5) constitute the sample. These five trees will be evaluated for data collection
11	**12**	**13**	**14**	**15**				14	
16	**17**	**18**	**19**	**20**				19	
21	**22**	**23**	**24**	**25**				24	

Note: Due to the non-availability of listing of the trees, systematic sampling is adopted by obtaining the first unit as fourth tree and subsequent trees through systematic approach

Cluster Random Sampling: Cluster random sampling is preferred for large-scale surveys where the sampling frame for the sample selection is not available. In this sampling, the population is divided into smaller groups based on adjacent units and termed as clusters. The cluster comprises of adjacent units without overlapping of units and generally decided based on pre-existing units. The units within the cluster should be as diverse as possible and all clusters together should cover the entire population. Ideally, the cluster sampling would be highly precise, if cluster truly represents the entire population. However, since clusters are in general naturally occurring groups, they are homogeneous than the population as a whole. The sampling is made by randomly selecting some clusters from all the clusters following random selection procedure. Cluster sampling is time and resource efficient, especially for units with widely geographically spread.

Patches of the forest—Population ($N = 30$) (Sampling frame) Cluster 5 (Each containing six units)					Selected two clusters by random number table (Cluster—2)					Selected trees for biomass estimation (2 clusters each with six trees)				
1	2	3	4	5	1	2				4	9	14	19	24
6	7	8	9	10	6	7				Two clusters have been selected i.e. all 12 units have been considered for data collection				
11	12	13	14	15	11	12								
16	17	18	19	20										
21	22	23	24	25			23	24	25					
26	27	28	29	30			28	29	30					

For example, the researcher wishes to estimate the biomass of trees in a large natural forest containing different species. In this forest, counting or listing of trees is not available easily. However, a group of trees may be constituted as clusters. Say for example, a forest contains 30 trees and researcher can constitute five clusters each containing six trees. Now out of five clusters, researcher can select two clusters at random and evaluate each and every units of the cluster following random sample selection procedure. The selected trees will be explored for the data and the collected data will be subjected for the desired analysis as per objective. The process is exemplified below for better understanding.

6.2.2 Design of Experiments

Finney explains that the design of an experiment includes the observations selected for comparison; the specification of the plots for the application of the treatments; the rules by which the treatments are to be allocated to plots; and the specification of the measurements or other records to be made on each plot. The effective and valid design requires three basic principles: randomization, replication, and local control.

Randomization is a "physical basis of the validity of the test or trial" achieved through allocating the treatments to the experimental units in such a fashion, that is, random process, which ensures independency of observation from each experimental units. Cochran and Cox (1957) define "randomisation is somewhat analogous to insurance, in that it is a precaution against disturbances that may or may not occur and that may or may not be serious, if they do occur." The random allocation safeguards the experiment from the continuous influence of extraneous factors.

- An **experiment** is a controlled application, introduction, or intervention of some treatments on a set of individual sampling unit.
- An **experimental unit** is the subject or material or field on which the external treatment or stimulus is applied. These experimental units are the subset of the population units, on which the conclusion is generalized. For example, in disease evaluation of a small nursery, all seedlings of the nursery may be considered as experimental material.

(continued)

- A **plot** is the smallest unit on which a treatment is applied and from which single observation is recorded. A plot may be an area of land on which a crop is grown, may be a piece of animal tissue, a forest stand, a tree, a group of trees, a branch of tree, etc.
- A **block** is a group of homogeneous plots used in the layout of the design and on which various treatments are applied.
- A **treatment** is in a generic sense indication of the stimulus, and is intentionally introduced variation causing elements to be tested in experimentation. For example, different insecticides are treatment as they are being applied to control the insect infestation.
- A **factor** is a well-defined and designed source of explainable variability introduced in the experimentation and has various doses, termed as levels of the factor. For example, to evaluate the effect of moisture content on seed germination, different levels of moisture content say 30, 40, 50% are three levels of the factor "Moisture."
- A **level** is distinguishable different qualitatively or quantitatively states of a factor. A factor whose levels differs only in a nominal manner is a qualitative factor, while a factor whose levels differ only in quantitative ways is a quantitative factor.
- Selection of levels or does is determined either theoretically or as per the experience of researcher.
- A **crossed classification** accounts multifactor experiments and ensures that each level of one factor is observed at all levels of the other factor(s).
- Ross advocates that when experiments involve many factors at many levels and where there is access to an almost unlimited choice of land, so that uniform sites may be chosen, then larger blocks may be supported.
- A **yield** is any quantitative measure obtained from a plot against the introduced treatments. For example, height of plants due to different treatments on yield for that experiment.
- A **control group for comparison** is determined as per the objective of the experiment. Control may be "no application of treatment" or "a well-established standard" or as per the choice of the researcher.
- A **layout** is the allocation of treatments on the experimental units or plots.

Replication is repetition of the treatment to facilitate increased accuracy of estimates of treatment effect. Replications are necessary to eliminate the factors of variability and to estimate the experimental error, in order to increase the precision of the experiment. The number of replications depends upon the permissible expenditure and the required degree of precision, that is, the size of true differences that the experiment is to show as significant. One of the conventional criteria for determining number of replications is to consider replications which ensure at least 12 degrees of freedom (df), or preferably more for the error source of variation in the analysis of variance (ANOVA) of the design applied as below 12 df, the value of "*F*"- statistic is not stable, and test becomes sensitive.

Local control is a tool to control error and ensure reduction in variance of the estimates of the treatment effects and can be achieved by making the experimental units homogeneous or formation of several homogeneous groups. The basic principle underlying local control is that the experimental units should be as similar as possible before the introduction of treatment. Another importance of local control is consistency in the data recording process through using the same measurement devices (instruments), background conditions (aspects, slopes), and unbiased measurements of the data.

Types of design: The design is classified broadly into two types, that is, orthogonal designs and non-orthogonal designs. Orthogonal design is a design in which all the levels of a factor occur with every level of other factor. Orthogonality ensures that the effect of a factor is not entangled with the effect of other factors. There are three basic designs, which are commonly applicable in forestry experiments depending upon the objectives of the experiment and the nature of the experimental materials.

- Completely randomized design (CRD)
- Randomized block design (RBD)
- Latin square designs (LSD)

Non-orthogonal designs are of several types. Here, we would only discuss about the incomplete block design.

CRD is suited for the comparing of the treatments to be applied on the homogeneous experimental material. Lab conditions and nursery conditions, where the conditions are similar for all the similar experimental material, are the fit case for CRD. For example, if the researcher wishes to compare the growth of seedlings of four species growing in a nursery, it can be considered as homogeneous condition and is suitable for CRD. In such random allocation, the whole experimental material is homogenous and the differences to be observed in the response are completely due to the actual differences of the effect of treatments that is the species in this experiment. The principle of local control is not taken into consideration in CRD, due to the homogeneous nature of experimental units.

Layout of the CRD for data collection: In this arrangement, the experimental material is divided into N number of plots or pots (in nursery trial) or group of pots, such that for the v treatments each replicated r times follows $v \times r = N$. For random allocation of the treatments with their replications, the plots are marked as 1, 2, …, N and the treatments are allocated in the plots using random process, that is, through random number table. Unequal replication is also possible in the design and is applicable if some treatments have more variance than others or has greater interest than other treatments.

T2R3	T6R4	T5R1	T3R3	T4R2	T6R3
T4R3	T3R1	T3R2	T2R4	T1R1	T4R4
T5R3	T1R2	T2R1	T1R3	T3R4	T5R4
T1R4	T4R1	T6R2	T5R2	T6R1	T2R2

Note: CRD layout for six treatments (T) replicated (R) four times, where TiRj means *i*th treatment at *j*th replication

Application of CRD: This design can be extensively used in forestry research for:

1. Comparative germination tests in nursery or lab
   (a) Method of storage of seeds for species
   (b) Pre-sowing treatments to stimulate germination for species
   (c) Suitable germination media for species
   (d) Germination of seeds of some species
2. Comparative nursery trials
   (a) Soil preparation experiments
   (b) Nursery manuring experiments
   (c) Shading and sheltering experiments
   (d) Nursery weeding experiments
   (e) Watering experiments
3. Planting experiments in nursery
   (a) Entire plants of different ages and sizes
   (b) Storage and transplant of plants
   (c) Drought and frost stress

Homogeneity cannot be ensured for large experimental materials and therefore blocking of experimental units, which provides homogeneous groups of units and treated as replication is an appropriate way for designing the experiments. In RBD, random allocation of each treatment is made in each block. It is assumed that though the results differ in different blocks, the relative effects of treatments are same in all the blocks apart from experimental error, that is, there is no interaction between treatments and blocks. In simple words, the differences among treatments are about the same in all the blocks. Due to this fact, it is not advisable to have blocks that differ greatly between each other; such condition may cause interaction with treatment. Interaction, if exists, must be eliminated from the experiment, as the large interaction will provide misleading result of the experiment.

Layout of the RBD for data collection: In RBD, all the "*k*"(say) treatments to be evaluated are applied randomly in all blocks (say *r*), in a total "*kr*" number of experimental units, where each of "*r*" blocks is subdivided into "*k*" plots. In simple RBD, number of replications for all the treatments is equal. The treatments are allocated to the plots of a block in a random manner and a separate randomization is considered for allocating treatments in each block.

Block 1	Block 2	Block 3	Block 4	Block 5	Block 6
T4	T2	T1	T3	T4	T2
T2	T1	T4	T2	T1	T3
T1	T3	T3	T4	T2	T4
T3	T4	T2	T1	T3	T1

Note: RBD layout for four treatments (T) replicated (R) (shown as block) six times, where T_i represents *i*th treatment

Application of RBD: RBD is extensively used in forestry research for:

1. Comparative germination tests
   (a) Pre-sowing treatments in field to stimulate germination
   (b) Tests of germination media for field experiment.
   (c) Depth of sowing of seeds in field
2. Comparative field trials
   (a) Nursery manuring experiments
   (b) Shading and sheltering experiments
   (c) Weeding experiments
   (d) Watering experiments
   (e) Spacing of transplants
3. Investigations on planting out experiments
   (a) Entire plants of different ages and sizes in natural forest
   (b) Partial pruning of plants in root or shading effect
   (c) Root and shoot pruned plants (stumps) of different dimensions
   (d) Storage and transplant of plants after removal from nursery
   (e) Frequency, date and type of weeding
   (f) Protection methods for drought and frost
   (g) Effect of training and pruning
4. Experiments on single trees on established crops
   (a) Coppicing trials
   (b) Resin and gum tapping experiments
   (c) Pruning trials
5. Experiments on plots of established crops
   (a) Weeding experiments
   (b) Cutting back and burning experiments
   (c) Coppice thinning experiments
   (d) Root competition experiments
   (e) Cover crop experiments
   (f) Crop thinning experiments

Note: Testing the block effect is not strictly essential, but the ratio of the block sum of squares to the error mean sum of squares provides an indication of the precision gained through blocking. A large block mean sum of squares indicates that the design is more sensitive than a Completely Randomized Design (CRD). However, equal block and error mean sums of squares indicate that blocking does not result in any improvement in precision.

Latin square design (LSD) accounts for the two directions variability in the experimental material. In LSD, the plots are arranged in the form of a square, so that the number of rows and columns are equal to number of treatments. Each treatment should be allocated in such a way that the treatment should occur only one time in each row and only one time in each column. For example, if a researcher wishes to compare the growth of seedling under four types of fertilizer for four provinces, each with four different ages, then LSD is the appropriate design to study the effect of different treatments. This arrangement eliminates variations due to different provenance and different ages, and provides an estimate of the true effect of fertilizers. Generally, a Latin square design is applicable for 4–12 treatments. The basic assumption one has to take care while using LSD is that the two sources of variability do not interact and they should act independently without affecting to each other.

Layout of the LSD for data collection: A Latin square of an arrangement of ν symbols, known as order of Latin square, arranged in ν^2 cells in ν rows and ν columns, such that every symbol occurs only once in each row and once in each column. A Latin square is in the standard form, if the symbols in the first row and first column are in the natural order.

Block 1	Block 2	Block 3	Block 4
T1	T2	T3	T4
T4	T1	T2	T3
T3	T4	T1	T2
T2	T3	T4	T1

Note: LSD (standard) layout for four treatments (T) and replicated ® four times

Applications of LSD: LSD is used in the following forestry experiments:

1. Comparative germination tests
   (a) Five pre-sowing treatments for five species in field for germination trial
   (b) Four germination media for four humidity in field for growth
   (c) Six depth of sowing with six planting media for survival of seedlings

2. Comparative trials
   (a) Coppicing trials for five height with five growth regulators
   (b) Pruning trials for four directions and four volume of pruning

In designing experiment, sometimes, the number of treatments is more than the number of plots in a block; hence, the all treatment cannot be accommodated in the block, that is, block is incomplete and such a design termed as incomplete blocks is called an incomplete block design. An incomplete block design is defined as balanced incomplete block design (BIBD) for v treatments and b blocks each of size k ($v > k$) such that each treatments is replicated r times and each pair of treatments should occur in same number of times. BIBD is applied, if all treatment comparisons have equal importance as BIBD ensures equal precisions of the estimates of all pairs of treatment effects. The situation happens when a large number of treatments are required to be evaluated; for example, researcher is interested to compare 40 number of provenances of Shisam (*Dalbergia sissoo*) at a location; for the experiment, 40 plots of same size is accommodated in each block, thereby the homogeneity among the plots of a block is difficult to maintain, rather leading to large variability within the blocks thereby making the design imprecise.

Precisely, a BIBD is an arrangement of v treatments in b blocks of k plots each ($k < v$) in following manner,

1. Each treatment occurs only once in r blocks and
2. Each pair of treatments occurs together in λ blocks.

The integers b (number of blocks), v (number of treatments), r (number of replications for each treatment), k (number of plots in each block), and λ (number of blocks in which each pair of treatments occurs or number of times any two treatments occur together in a block) are parameters of the BIBD. The following parametric relations are necessarily required for the existence of a BIBD.

1. $vr = bk$
2. $\lambda\,(v-1) = r\,(k-1)$
3. $b \geq v$ (Fisher's Inequality)

Incomplete block designs are categories in many types as balanced incomplete block design and partially balanced incomplete block designs, which contain a very broad class of designs.

Layout of the IBD for data collection: The randomization of treatments to experimental units due to incomplete block includes randomization of the treatment label, randomization of the blocks, randomization of the treatments within each block, and a separate randomization in each block.

Applications of IBD: IBD can be applied in various forestry experiments where performance of a large number of treatments is evaluated.

1. Provenance, progeny, and clonal trials
2. Carbon capturing potential of various evergreen trees

6.3 Methods of Data Analysis

Data collected from the field is in raw form and needs to be summarized, processed, and analyzed to derive the inherent information. In following section, data summarization and analysis are discussed.

6.3.1 Data Presentation

Data and information can be presented in textual, tabular, and graphical forms. Text, tables, and graphs are communication tools for data and information by making them easy to understand, attractive, and to efficiently comprehend large and complex information. Data presentation should focus on simplicity of presentation and choice of presentation should be based on data format, measurement scale, method of analysis, and objective of the experiment.

Meaningful information from raw data can be obtained by summarizing the information through various means depending upon the data, data format, and objective of investigation. Summarization of quantitative information starts with arranging the numeric data in ascending or descending order and known as array; however, qualitative information can be summarized by classifying the data into different categories; for example, leaf margin of the tree species can be classified into various categories as entire leaf, lobed leaf, toothed leaf, and parted leaf. Other methods of visual representation are discussed below.

- The **raw data** are collected directly from field and have not been processed and organized by any means.
- The **primary data** are collected from the units or individuals directly by the researcher for his research purposes. These data are used for the first time for his research.

(continued)

- The **secondary data** are collected by other persons or agency for their defined purposes; however, they are used for other than the primary purposes.
- A **frequency** is number of total individuals in a category.
- A **frequency distribution** is a tabular presentation of summary of frequency (or number) of individuals in different non-overlapping classes or categories as per the information or data.
- The **relative frequency** of a class is the ratio (fraction) or proportion of the total frequency of the units (items) belonging to the class.
- A **relative frequency distribution** is a tabular presentation of the summary of relative frequency for each class.
- The **percent frequency** is the product of relative frequency with 100.
- A **percent frequency distribution** is a tabular presentation of the summary of percent frequency for each class.
- A **comparison chart** is used to compare one or more data sets.
- A **relationship chart** is used to show a relation between two or more variables.
- A **composition chart** is used to display parts of a whole and change over time.
- A **distribution chart** is used to show how variables are distributed across the categories.

Data presentation can be of two types, that is, graphical and tabular presentation and numeric presentation.

Numerous approaches may be followed for representation of the data. The representation depends on the data types and objective of the research.

Summarization by text presentation: Text explains findings, outlines trends, and provides contextual information and also express the interpretation of results. Text can be used for comparing or discussing two or three scenarios at a certain time point; for example, the productivity of evergreen species is more than the productivity of conifer species in temperate forests. Moreover, data presentation of long list of information in text is difficult to grasp than tables and graphs.

Summarization by table presentation: A table is a representation of information in rows and columns and is used for representing both quantitative and qualitative information and most suited once researcher is interested to provide equal attention for all information (Table 6.2). Tables summarize accurate information and are useful for comparing quantitative information of variables. However, the interpretation of information in table requires more efforts than the graphs and is also difficult to evaluate the trends.

Heat map has better visualization of information than tables by using colors to the background of cells (Table 6.2). The colors or color saturation in the cell may be adjusted for precise communication of the information, and it facilitates the readers to quickly identify the interested information.

Summarization by graphic presentation: A graph is a way to represent the information in a visual form as line, bar, and picture which facilitate comparison, reveal trends over time and relationships. Graphs facilitate to simplify complex information using visual tools and presenting data patterns, and are also useful for summarizing, explaining, or exploring quantitative data.

Bar graph: A bar graph represents and compares values, frequencies, or other statistical parameters of a qualitative variable containing discrete categories or groups of a single information. The height or length of a bar represents the value (or frequency) of information in a category. A bar graph is also used to represent information in a grouped or subdivided bar format in cases of more than one information contained in each category; such graph is termed as multiple bar chart (Fig. 6.1a, b). A bar chart may also be used to represent the sum of each category or percent of each category through the stacks and termed as stacked bar chart or percent bar chart (Fig. 6.2a, b; Table 6.3).

Pie chart: A pie chart is two-dimensional circular representation of nominal data distributed in several categories, showing the proportion of category representing the segment (Fig. 6.3). The area of segment is proportional to the angles of corresponding category and is calculated by multiplying the relative frequency with 360.

Table 6.2 Example of a tabular and heat map presentation of diversity indices of some forest types

Table			Heat map		
Forest type	Shannon-Weaver index	Simpson dominance	Forest type	Shannon-Weaver index	Simpson dominance
Banj Oak Forest	3.32	0.95	Banj Oak Forest	3.32	0.95
Dry Deodar Forest	2.7	0.92	Dry Deodar Forest	2.7	0.92
Moist Temperate Deciduous Forest	2.92	0.94	Moist Temperate Deciduous Forest	2.92	0.94
Moru Oak Forest	3.73	0.97	Moru Oak Forest	3.73	0.97

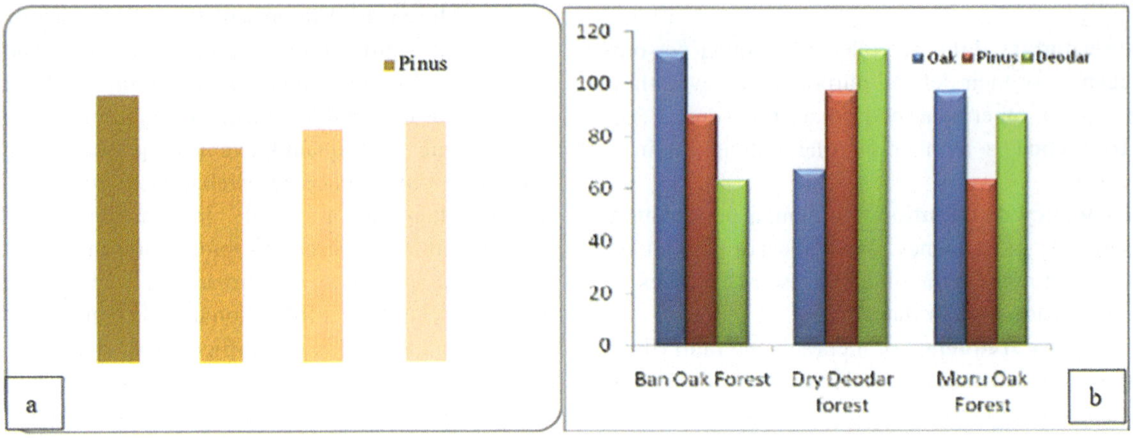

Fig. 6.1 (**a**) Simple bar chart and (**b**) multiple bar chart for number of individuals of species

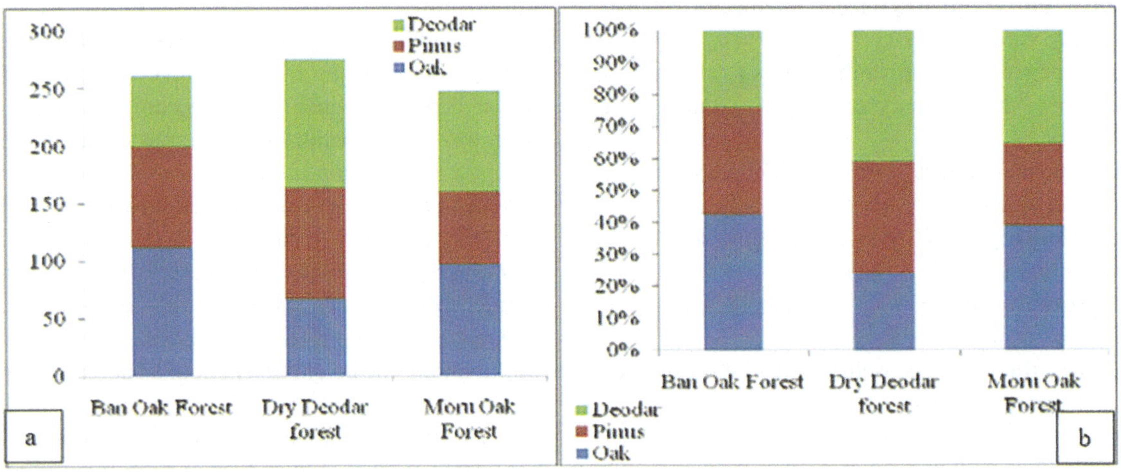

Fig. 6.2 (**a**) Stacked bar chart and (**b**) percent bar chart for number of individuals of species

Table 6.3 Suitability of types of charts along with purpose

Purpose	Nature	Category	Chart type
Composition	Item (Frequency)	One or two per items	Bar chart
	Over time (Time series)	Many periods	Circular area/line chart
		Few periods	Bar/line chart
Relationship	If exits	Two variables	Scatter plot
Distribution	Item	One	Bar/line histogram
		Two	Scatter chart
Comparison	Dynamic	Only relative differences	Percent stacked bar chart
		Relative and absolute differences both	Stacked bar chart
	Static	Simple share of total	Pie chart
		Accumulation	Waterfall chart
		Components	Percent stacked bar chart with subcomponents

Box and whisker chart: A box and whisker chart (Fig. 6.4) is a graphical display of single information and consists of boxes representing inter-quartile range (one to three), the mean and median of the data, and whiskers are presented as lines outside of the boxes and reflecting the highest and lowest values of the data. Besides, box plot also presents individual points outside box, which are outliers (Fig. 6.5). The relative location of the median within the box provides information about the skewness and the distance between the two ends of the box is inter-quartile range.

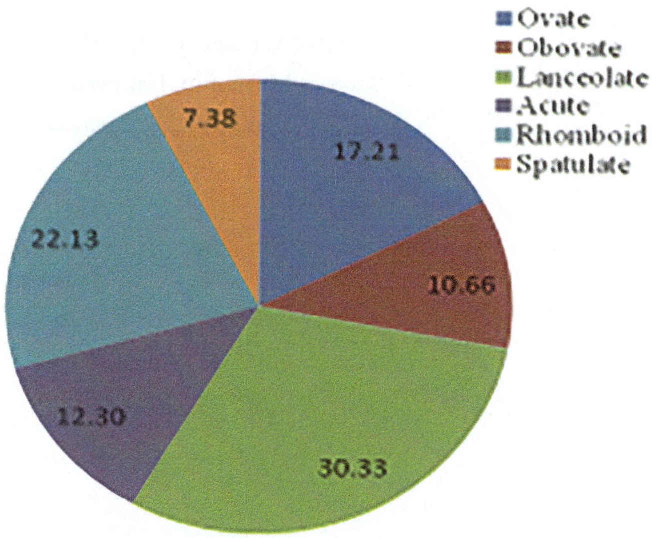

Fig. 6.3 Pie chart representing the proportion of leaf shapes of various shrubs in a natural forest

Line plot with whiskers: A line plot represents time-series information such as daily or monthly precipitation and annual productivity of the forests (Fig. 6.6). Line plot is useful for evaluating trends and patterns for time-series data. For example, line chart may be made for the data, measured over the progression of a continuous variable, such as distance. Line plot can contain a multiple data sets on a single line graph to compare and analyze patterns across different data sets.

Scatter plot: A scatter plot is the simplest way to evaluate the relationship (association) by plotting the pairwise data of two variables on the two-dimensional graph paper (Fig. 6.7). Scatter plot provides an indication about the nature of relationship between the two correlated variables; for example, the relationship between the infected area of leaves against weight loss with following data can be elucidated by scatter plot.

Weight loss (g)	12	13	16	17	18	10	11	14	9	9	15	16
Infected area (mm^2)	3.5	4.5	5.5	6.5	6.5	3.5	4.5	4.5	2.5	2.1	5.6	6.2

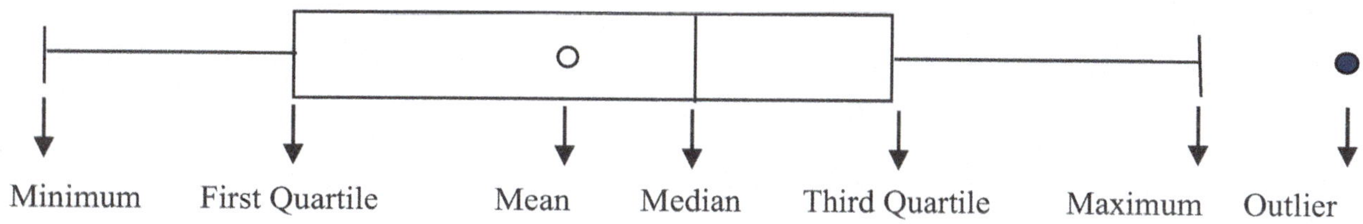

Fig. 6.4 A box plot with whiskers

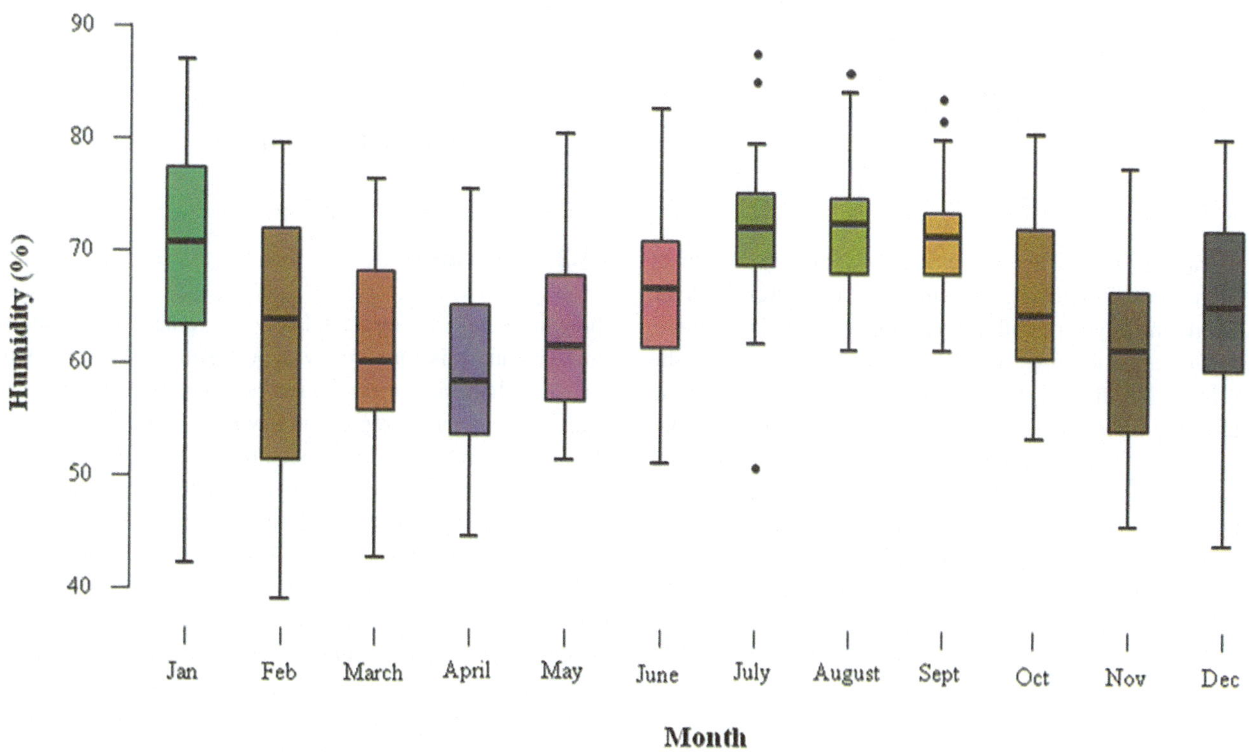

Fig. 6.5 Box plot of monthly humidity (%)

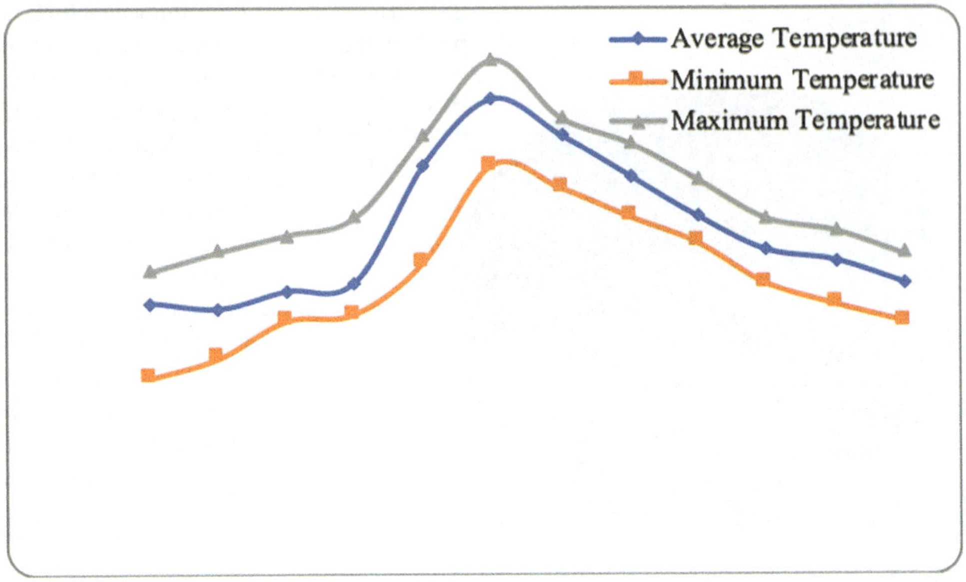

Fig. 6.6 Line plot for monthly temperature (°C)

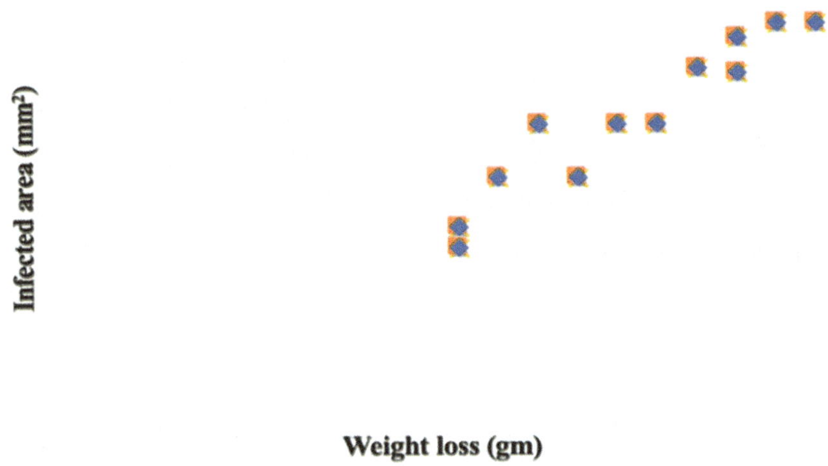

Fig. 6.7 Scatter plot between infected area with weight loss

The raw data can be represented by summary measures through numerical entities. The summary measures describe the data in precise form with the support of numerical values. Main summary measures include measures representing the center or middle most values or most common value and the spread of the data along the central value. The two summary measures are termed as measures of central tendency or measures of averages and measure of dispersion or measures of variance. These summary measures describe charecteristics of the data set in numerical terms and dealt under the theme of Descriptive Statistics in statistical literature. The following section deals with the summary measures.

6.3.2 Averages

Three measures of central tendency or averages are common in statistical analysis for forestry experiments and are known as mode, median, and mean. Each measure of the central tendency is designed to represent a typical score. The choice of the use of the measure depends on the shape of the distribution (normal vs. skewed) and the measurement scale of the variable (nominal, ordinal, interval and ratio) (Table 6.4).

Mode: Mode represents the most frequent value in a data set and is typically associated with nominal variables. For example, in conifer forests, the most common leaf shape is needle-like, indicating that the predominant leaf shape in such forests is needle-like. Similarly, in eucalyptus species, the bole of trees of the species are generally straight and clean, which suggests that the bole of trees of the eucalyptus species are straight and clean.

Table 6.4 Suitability of numerical measures with purpose

Measure	Purpose (Scale)	Description	Interpretation
Mean	Middle value (ratio scale)	Data value lies above and below to the middle value	Measure of central value
Mode	Middle most (nominal scale)	The value occurs most times	Measure of central item
Median	Middle point (ordinal scale)	The midpoint—half above and half below	Measure of central point
Standard deviation	Spread in a sample (ratio scale)	Variation in a sample along mean	Spreadness of the data with respect to mean of a sample
Standard error	Spread in population (ratio scale)	Variation in a population along mean	Spreadness of the data with respect to mean of a population

Median: Median is the middle point of the data arranged in an order. Any data set can be said to be ordered if the values of the data is arranged in either descending or ascending order. Median is generally dealt with the ordinal variable, for example, the midpoint of a forest.

Mean: Mean is the numerical measure based on all the observations of the variable and representing the central value of the data. Mean is estimated by the ratio of sum of all the observations and the number of observations. Mean is also used to predict the values for other items.

The precision of mean in general and for prediction needs to be evaluated. This can be achieved by evaluating the dispersion (spread out or scatteredness) of data on each side of the center. Three measures of dispersion or variation are commonly used in statistical analysis for forestry experiments as range, standard deviation (SD), and coefficient of variation. Each measure of dispersion is designed to represent a typical score and the choice of the use of the measure depends on the shape of the distribution (normal vs. skewed), and the measurement scale of the variable (nominal, ordinal, interval and ratio) (Table 6.4).

Range: Range is estimated by the difference between the largest and the smallest values of the variable in the distribution. If $x_1, x_2, \ldots, x_n$ are the values of observations of the variable, then range (R) of the variable is estimated as follows:

$$R(x_1, x_2, \ldots, x_n) = \max\{x_1, x_2, \ldots, x_n\} - \min(x_1, x_2, \ldots, x_n\}$$

Mean Deviation: The mean deviation is a measure of dispersion and is an average of absolute deviations of individual observations from the central value (mean, median, and mode) of a series. If $x_1, x_2, \ldots, x_n$ are the values of observations, then mean deviation of the variable is estimated as follows:

$$\text{Mean deviation} = \frac{\sum |x_i - \bar{x}|}{n}$$

Standard Deviation: Standard deviation (SD) is the positive square root of the mean-square deviations of the observations from their arithmetic mean. The square of SD is variance and estimated as below for the observation $x_1, x_2, \ldots, x_n$

$$\text{Standard deviation(SD)} = \sqrt{\frac{\sum_{i=1}^{n} (x_i - \bar{x})^2}{n}}$$

Standard Error: Standard error (SE) is the ratio of SD to the square root of the size of sample. SE measures the likely discrepancy between the estimated sample mean compared to the population mean.

$$\text{Standard error(SE)} = \frac{\text{SD}}{\sqrt{n}}$$

Standard error is a measure of spread and provides the precision of a sample mean by measuring the sample-to-sample variability of the sample means. SE decreases with increase in sample size. If analysis is based on sampling, which is generally applicable in research, it is always advisable to use standard error.

Coefficient of Variation: A coefficient of variation (CV) is a measure of dispersion and is a ratio of the standard deviation to the mean of the distribution. CV is spread as a proportion of its mean. CV is a dimensionless number that can be used to compare the amount of variance between populations with different means.

$$\text{CV} = \frac{s_x}{\bar{x}}$$

Example: Following are the leaf area (in cm^2) of 25 leaves of *Abies pindrow* in a temperate forest of Mussoorie Forest Division, Uttarakhand. Use this data to estimate all relevant numerical measures.

21.6	22.4	19.9	18.9	22.5	21.5	24.7	17.8	27.6	20.6	21.3	23.3	19.6
24.2	25.7	18.7	20.8	19.6	21.4	20.8	21.9	26.7	19.6	20.4	25.5	

Solution: $n = 25$; Maximum observation = 27.6; Minimum observation = 17.8

Data processing of leaf area for 25 leaves for estimation of averages and dispersion

Sl. no.	Leaf area (x)	$x_i - \bar{x}$	$\lvert x_i - \bar{x} \rvert$	$(x_i - \bar{x})^2$
1	17.8	−4.08	4.08	16.6464
2	18.7	−3.18	3.18	10.1124
3	18.9	−2.98	2.98	8.8804
4	19.6	−2.28	2.28	5.1984
5	19.6	−2.28	2.28	5.1984
6	19.6	−2.28	2.28	5.1984
7	19.9	−1.98	1.98	3.9204
8	20.4	−1.48	1.48	2.1904
9	20.6	−1.28	1.28	1.6384
10	20.8	−1.08	1.08	1.1664
11	20.8	−1.08	1.08	1.1664
12	21.3	−0.58	0.58	0.3364
13	21.4	−0.48	0.48	0.2304
14	21.5	−0.38	0.38	0.1444
15	21.6	−0.28	0.28	0.0784
16	21.9	0.02	0.02	0.0004
17	22.4	0.52	0.52	0.2704
18	22.5	0.62	0.62	0.3844
19	23.3	1.42	1.42	2.0164
20	24.2	2.32	2.32	5.3824
21	24.7	2.82	2.82	7.9524
22	25.5	3.62	3.62	13.1044
23	25.7	3.82	3.82	14.5924
24	26.7	4.82	4.82	23.2324
25	27.6	5.72	5.72	32.7184
Total	547	0	51.4	161.76
Mean	21.88	0	2.056	6.4704

Mean = $(\Sigma X_i)/N$ = 547/25 = 21.88

Median = (Total number of observation + 1)/2 = (25 + 1)/2 = 13th observation (ordered). Therefore, the 13th observation will be median, that is, 21.4.

Mode is most occurring value, which is 19.6 as it is occurring three times (the most).

Range = 27.6 − 17.8 = 9.8

Mean deviation $= \frac{\sum |x_i - \bar{x}|}{n} = 51.4/25 = 2.06$

Standard deviation $= \sqrt{\frac{\sum_{i=1}^{n} (x_i - \bar{x})^2}{n}} = \sqrt{\frac{161.76}{25}} = \sqrt{6.47} = 2.54$

Coefficient of variation $= \frac{s}{\bar{x}} = (2.54/21.88) * 100 = 11.61$

Result Presentation

Mean ± SD	(Min, Max)	CV
21.88 ± 2.54	(17.8 – 27.6)	11.61

SE should be preferred over SD; Min is minimum value and Max is maximum value in the data set

Interpretation: The leaf area of leaves of *Abies pindrow* in temperate forests of Mussoorie Forest Division, Uttarakhand was 21.88 cm^2 with 12% variability.

Note: It is advisable, in general, to estimate the measure of central tendency (especially mean) and standard deviation or standard error for the collected data in all cases until any other measures are prescribed. The estimate forms summary numerical measures for the data.

6.4 Relationship Estimation

Relationship can be evaluated between two variables (bi-variate). Many instances in forestry, the relationship evaluation is critical. For example, the relationship between following variables

- Water stress versus biochemical changes in plants
- Leaf area versus transpiration rate
- Biomass of seed versus oil yield
- Tensile strength versus fiber length
- Soil microbial biomass carbon versus specific leaf area in temperate forests
- Leaf biomass versus leaf C/N
- Leaf N/P versus stomatal conductance in the leaves
- Leaf carbon versus intercellular CO_2 concentration in the evergreen species
- Soil microbial biomass versus soil moisture in temperate forest
- Fragmentation versus species richness in disturbed forest

The relationship between two or more variables can be assessed by correlation (between two interdependent variables—linear relation only) and regression (dependent sort variables—two or more variables—linear and non-linear relation both can be measured).

Linear relationship accounts a relation in which the successive increment in one variable is more or less similar due to each unit change in the other variable. The differential increments in successive unit change demonstrate in non-linear relationship. In forestry, most of the relationships are non-linear in general.

6.4.1 Types of Assessment of Relationship

The relationship can be assessed by the two approaches depending upon the data and objective of the research.

- Linear relationship—correlation
- Linear and non-linear relationship—regression

Linear relationship can be measured or estimated based on following approaches depending upon the data and objective of the research.

Scatter Plot: The simplest way to evaluate the relationship between two variables can be adjudged by plotting the pairwise data of two variables on the two-dimensional graph paper, that is, scatter plot between two variables. The plot provides an information about the nature of relationship between the two variables; however, it does not provide any numerical measure. Scatter plot may provide a clear and precise measure of relationship for comparative purposes.

Example: Evaluate the relationship between temperature (°C) and infected area (cm^2) of leaves for 16 leaves in a laboratory experiment with the following data.

Temperature	12	13	16	17	18	10	11	14	9	11	12	13	17	18	13	16
Infected area	3.5	4.5	5.5	6.5	6.5	3.5	4.5	4.5	2.5	2.7	4.5	5.8	6.5	6.8	4.4	6.1

Solution: The relationship between two variables may be evaluated based on scatter plot.

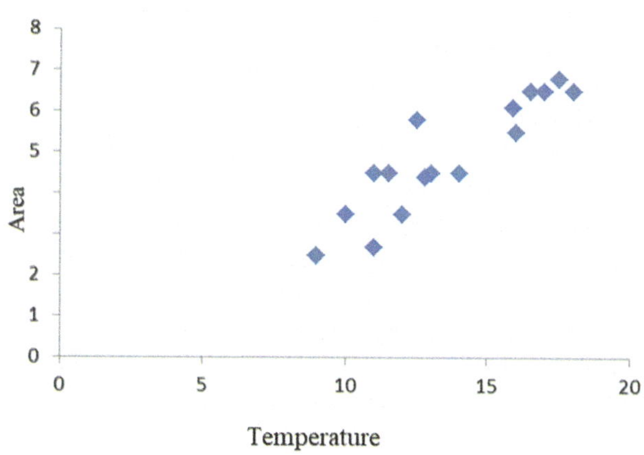

Interpretation: The scatter plot for the data sets shows that there is a positive relationship between temperature and infected area. Precisely, it shows that as temperature increases, the infection in the leaf area also increases in the data range.

Covariance—Numerical Measure of Relationship: Covariance quantifies the degree of linear association (relationship) between the two variables and is a measure of the corresponding change in one variable due to the change in the second variable.

Let the observations of the variable X is $x_1, x_2, x_3, \ldots, x_n$; and the variable Y is $y_1, y_2, y_3, \ldots, y_n$. Then the covariance is

$$\mathrm{cov}(x, y) = S_{xy} = \frac{1}{n}\sum_{i=1}^{n}(x_i - \bar{x})(y_i - \bar{y})$$

The value of cov (x, y) depends on the total number of observations due to the $(1/n)$ factor in the formula and determines the direction of relationship.

- cov(X,Y) < 0 X and Y are inversely correlated
- cov(X,Y) > 0 X and Y are positively correlated
- cov(X,Y) = 0 X and Y are independent

Covariance is an absolute measure and not a relative measure, which restricts its use. It has unit and measures for linear relationship only. Moreover, if two variables are independent, then the covariance between the two variables will be zero, but vice versa is not true.

Correlation for continuous data or Karl Person Correlation Co-efficient: Sir Francis Galton proposed the concept of correlation in 1894, and described mathematically by Karl Pearson in 1896. Correlation measures intensity or degree of linear relationship between two continuous variables by correlation coefficient, which is a unit less quantity having a range between −1 to 1.

Correlation is quantified by Karl Pearson correlation coefficient, also referred as product-moment correlation coefficient. Let the observations of the variable X is $x_1, x_2, x_3, \ldots, x_n$; and the variable Y is $y_1, y_2, y_3, \ldots, y_n$. Then the correlation coefficient formula for measuring a numerical linear relationship can be defined as

$$r = \frac{\mathrm{cov}(x, y)}{\sqrt{V(x)}\sqrt{V(y)}} = \frac{\sum_{i=1}^{n}(x_i - \bar{x})(y_i - \bar{y})}{\sqrt{\sum_{i=1}^{n}(x_i - \bar{x})^2}\sqrt{\sum_{i=1}^{n}(y_i - \bar{y})^2}}$$

Assumptions of correlation

- The two variables should be linearly related.
- Each of the variables is being affected by a many causes, so that the variables should follow normal distribution.
- The cause and effect relationships should be existed between different forces responsible for changes in the variables.

Example: Evaluate the relationship between temperature and infected area of leaves for 16 leaves in a laboratory with the data from the above example.

Solution: The various estimates are as follows:

Variable	Mean	Variance	Co-variance	Co-relation
Temperature	13.61	7.66	3.61	0.90
Infected area	4.89	1.83		

Interpretation: The covariance shows positive relationship and correlation showed a strong positive linear relationship, that is, increase in infected area is proportional to the increase in temperature.

Correlation for other than continuous data or Spearman's correlation coefficient: The correlation coefficient for qualitative data is measured by Spearman's rank correlation coefficient (proposed by Charles Spearman) and represented by Spearman's rho (ρ). Rho is a non-parametric measure of statistical relationship (linear) between two variables with qualitative data. Rho measures the relationship between two variables using a monotonic function. Monotonic function is a function between ordered sets that preserves or reverses the given order.

Let the n raw scores (characteristic) of the qualitative variable is assigned ranks as $x_1, x_2, x_3, \ldots, x_n$ for X and $y_1, y_2, y_3, \ldots, y_n$ for Y variable. Then the differences $d_i = x_i - y_i$ between the ranks of each observation on the two variables are calculated; if no tied rank is observed, then the correlation coefficient rho (ρ) for measuring a numerical linear relationship can be defined as

$$\rho = 1 - \frac{6\sum d^2}{n(n^2 - 1)}$$

The rho (ρ) is estimated once, one or both of the relevant variables are ordinal; thus rho (ρ) relies on ranks rather than actual observations. Rho (ρ) is less sensitive to outlying values than Pearson's correlation coefficient, and it measures the strength of linear relationship.

General Remarks About Correlation Coefficients: Following are the important points about the correlation coefficients and generally considered for both the approaches of estimation of correlation coefficients.

- Correlation coefficients measure and describe the direction and strength of an association between variables.
- Correlation does not imply causality and therefore inferring causation based on correlation coefficient must be avoided.
- Correlation is unsuited for analyses of agreement between two measurements by two different instruments.
- Scatter plots must be used to evaluate the relationship, as correlation does not adequately describe nonlinear or non-monotonic relationships.
- Low correlation values do not categorically refute existence of relationship rather if theory supports, chances might exit for non-linear relationship between the two variables.
- Correlation is not a robust measure. An unusual or freak observation of a data may have a strong impact on correlation coefficients.

Correlation cannot measure the cause and effect relationship, and therefore manipulation in the independent variable and observing the effect on dependent variable require a different approach to measure the relationship. The approach to define the relationship between dependent and independent variables or cause and effect relationship can be modeled by regression analysis. Regression analysis can measure both the dependent and independent relationship. For example, forest functioning can be modeled by diversity and traits of the species.

Forest functioning = f (diversity, soil traits, plant traits)

If regression is estimated linear relationship for a single independent variable, then the regression is known as simple regression analysis, for example, biomass = f (diameter). More than one independent variables accounting linear relationship is known as linear multiple regression analysis for example,

Wood strength = f (fiber length, wood density);

Habitat suitability = f (food availability, biodiversity, distance to water source);

Disease spread = f (temperature, humidity, defense mechanism of plant).

The variable which affects the response variable (dependent) is known as independent variable. Therefore in the above example, forest functioning is depending (dependent) on the diversity and traits; therefore, these (diversity and traits) are known as independent variables. The above example can be demonstrated as follows:

Carbon in forest (C) = f (diversity (I), height (H)

Mathematically, a model can be written as

$$Y = \beta_0 + \beta_1 X + \varepsilon$$

Specifically, say for a forest, Y (the dependent variable) may be carbon (C); X may be height (H); and β_0 and β_1 are model coefficient; and ε is error. Therefore, the model can be written as

$$C = \beta_0 + \beta_1 H + \varepsilon$$

Precisely, the statistical model has deterministic model and random (error term). This happens as the response of any phenomenon does not remain same in spite of similar conditions. This nature of experimentation is known as random behavior and occurred as all experimental conditions cannot be ensured exactly perfect due to intrinsic nature of living materials. These two interwoven components in biological systems describe the following:

- A deterministic model describes and measures the deterministic (measurable/defined) biological process and
- A stochastic (random) component describes and measures the random variation, which influences the biological process, which is being modelled.

The deterministic component typically, in general, forms the biological process of interest. The deterministic model is easy to understand and may be relatively straightforward. For example, an exponential population growth model may include a description of multiple populations, such as density independent model or coupled predator–prey models; or may explicitly describe within-population structure, such as density-dependent model or size-structured dynamics. The stochastic component(s) of a statistical model are described by probability distributions based on a wide range of functions. The stochastic component of a statistical model is often an "error" term, that is, variation around a deterministic model.

The functional relationship can be used for two broader purposes, that is, (1) Prediction for example, predict biomass for the given diameter of a tree, predict the growth of a tree in a stand. (2) Quantifying casualty, that is, determine factors that relate to the variable to be predicted, for example, species richness as a function of latitude, altitude, and rainfall. In prediction, the goal is to develop a functional form (equation), that is termed as model, that enables predictions about the dependent variable based on the observed values of the independent variables. In a causal analysis, the independent variables are regarded as causes of the dependent variable. The aim of the study is to determine whether a particular independent variable really has impact on the dependent variable, and thus facilitate to estimate the magnitude of that effect, if any.

Regression: Regression explains the variation in a dependent variable due to independent variables by using the variation in independent variables. Regression is, therefore, an explanation of causation by estimating an appropriate functional relationship between variables. The functional form is decided based on either understanding of their relationship (theoretical construct) or through analysis (data); sometime hit and trial approaches are also followed. If in the estimated model the independent variable(s) sufficiently explain the variation in the dependent variable, the model can be used for prediction. Regression is the measure of the average relationship between two or more variables in terms of the original units of the data.

- Biomass carbon = f (wood density)
- Habitat = f (food availability, environment)
- Oil yield = f (biomass, chemical composition)
- Disease infection = f (plant vigor, climatic parameters)
- Ecosystem functioning = f (leave trait, soil trait)

Regression provides (1) description of system (relationship definition), that is, the functional relationship between two sets of variable (dependent and independent) reflects the type of relationship between the two sets of variables; (2) inference about the system (response mechanism), that is, the functional relationship portray the behavior of the system by utilizing the direction and response due to changes in independent variables; and (3) prediction (forecasting) for future, that is, future response.

Fundamentally, regression approach requires three principal steps. (1) *Model selection (Construction and Development),* which identifies the most parsimonious set of predictors that explain the most variation in the independent (response) variable along with the model functional relation; (2) *Evaluation of Assumptions (Testing),* which ensures the fulfillment of the assumptions of the regression model; and (3) *Model validation,* which ascertains the validity of the model for precise prediction. Regression is based on least square method, which is based on minimization of error.

Various model form and equations for one independent variable (x)

Model form	Equation	Graph	Symbol
Linear	$y = a + bx$	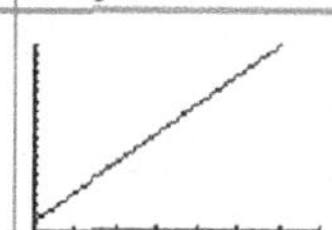	a and b are coefficients
Logarithmic	$y = a + b\ln x$	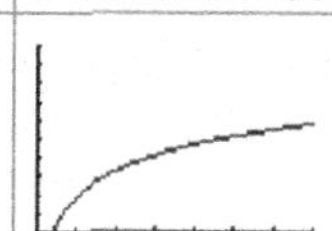	a and b are coefficients
Logarithmic	$y = a \ln x$	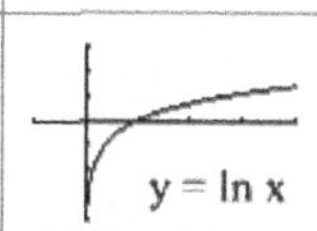	a is coefficient
Exponential	$y = ae^{x}$	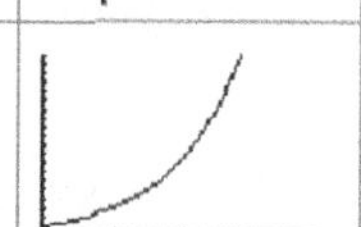	a is coefficient
Exponential	$y = ae^{-x}$	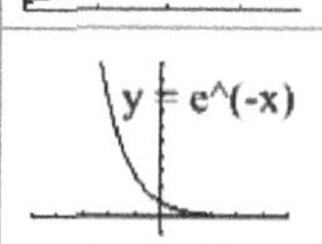	a is coefficient
Power	$y = ax^{b}$	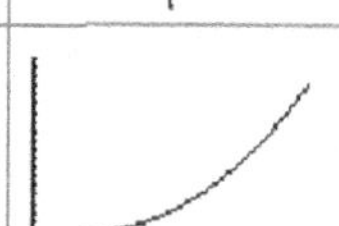	a is coefficient
Quadratic	$y = ax^2 + bx + c$	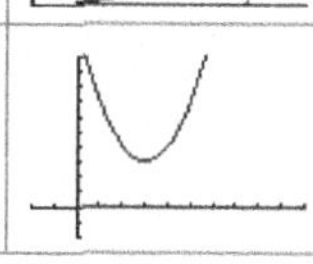	a, b and c are coefficients

(continued)

Model form	Equation	Graph	Symbol
Cubic	$y = ax^3 + bx^3 + cx + d$	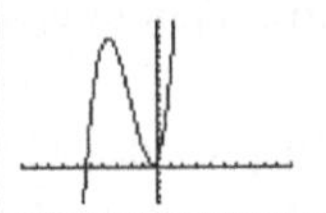	a, b, c and d are coefficients.
Logistic	$y = c/(1 + ae^{-bx})$	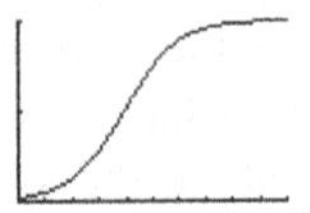	a and b are coefficients
Sinusoidal	$y = a \sin(bx + c) + d$		a, b and c are coefficients

General Remarks About Regression: Following are important points, which must be considered while using regression.

- Evaluate the choice of the model for its realistic nature.
- Fitting the data well is no guarantee for good predictive performance.
- Past experience with similar data may facilitate in model form selection.
- Theoretical understanding may also guide for selection of model form.
- Do not use a model which can predict unrealistic values.
- Avoid complex models for small data sets.
- Validation of model may be approached with new data.
- The SE and p-values do not account for model uncertainty rather assume that the model is fixed.
- "The best choice (of a model) depends on the set of data being analyzed and requires an exercise in judgment, not just computation" ("Modeling the US Population" by Shelly Gordon).

Various model selection criterions

Criterion	Formula	Symbol description
Residual sum of square (RSS)	$\text{RSS} = \sum_{i=1}^{n} (y_i - f(x_i))^2$	y_i is observed values of dependent variable $f(x_i)$ is predicted values of dependent variable
Residual mean sum of square (RMS)	$\text{RMS} = \text{RSS}/(n-k)$	n = the number of observations; k = the number of free parameters to be estimated
Coefficient of determination (R^2)	$R^2 = \frac{\text{SSR}}{\text{SST}}$	SSR—sum of square due to regression SST—total sum of square
Adjusted R^2 (Adj. R^2)	$\overline{R}^2 = 1 - \frac{n-1}{n-k}(1 - R^2)$	N = number of observations K = number of independent variables
Mallow's Cp	$Cp = \frac{\text{RSS}_p}{\text{RMS}_{\text{FULL}}} + 2p - n$	RSS is the residual sum of square of the model, p is the number of regression coefficients in the model, and RMS_{FULL} is the estimate for σ using the full model

(continued)

Criterion	Formula	Symbol description
Akaike's Information criterion (AIC)	$\text{AIC} = 2k + 2 \ln (L)$	k is the number of parameters in the model, and L is the likelihood function
Schwarz's BIC	$\text{BIC} = 2^* \ln (L) + k \ln (n)$	n = the number of observations k = the number of free parameters to be estimated L = the maximized value of the likelihood function for the estimated model

Example: Establish a linear relationship between leaf weight (Y) in gm with leaf area (X) in cm^2 for 16 leaves of *Poplus deltoides*.

Solution: Sample size $n = 16$

The linear relationship can be adjudged by $Y = a + bX$ (Equation of straight line); where a is intercept and b is coefficient and needs to be estimated. The formula for

$$b = \frac{\sum_{i=1}^{n} (y_i - \overline{y}) * (x_i - \overline{x})}{\sum_{i=1}^{n} (x_i - \overline{x})^2}$$

$$a = \overline{x} - b^* \overline{y}$$

$$b = \frac{\sum_{i=1}^{n} (y_i - \overline{y}) * (x_i - \overline{x})}{\sum_{i=1}^{n} (x_i - \overline{x})^2} = \frac{0.42}{18.43} = 0.02$$

$$a = \overline{x} - b^* \overline{y} = 4.48 - 0.02^* 8.48 = 4.31$$

Observation	Y	X	$(y_i - \overline{y})$	$(x_i - \overline{x})$	$(y_i - \overline{y})^2$	$(x_i - \overline{x})^2$	$(x_i - \overline{x}) * (y_i - \overline{y})$
1	2.50	8.60	−1.98	0.13	3.93	0.02	−0.25
2	4.60	9.20	0.12	0.73	0.01	0.53	0.09
3	3.50	8.50	−0.98	0.03	0.96	0.00	−0.02
4	3.20	7.50	−1.28	−0.98	1.64	0.95	1.25
5	4.50	9.20	0.02	0.73	0.00	0.53	0.01
6	4.50	7.90	0.02	−0.57	0.00	0.33	−0.01
7	3.50	8.10	−0.98	−0.38	0.96	0.14	0.37
8	4.50	8.60	0.02	0.13	0.00	0.02	0.00
9	2.90	7.40	−1.58	−1.08	2.50	1.16	1.70
10	5.10	9.20	0.62	0.73	0.38	0.53	0.45
11	4.50	8.80	0.02	0.33	0.00	0.11	0.01
12	2.50	7.90	−1.98	−0.57	3.93	0.33	1.14
13	4.60	9.10	0.12	0.63	0.01	0.39	0.07
14	8.50	10.50	4.02	2.03	16.15	4.10	8.14
15	4.10	9.50	−0.38	1.03	0.15	1.05	−0.39
16	8.70	5.60	4.22	−2.88	17.80	8.27	−12.13
Total	71.68	135.68	0.00	0.00	48.42	18.43	0.42
Mean	4.48	8.48	0.00	0.00	3.03	1.15	0.03

Thus, the equation is

$Y = 4.31 + 0.02X$

Therefore, the independent variable, that is, leaf area is positively related to the dependent variable, that is, weight.

Interpretation: The equation clearly demonstrates that the leaf weight is linearly related to leaf area and unit increase in leaf area will contribute 4.31 g leaf weight within the domain of the data.

6.5 Hypothesis Testing

Statistical methods are applied for an exploratory sense, during the start of an investigation, when questions are imprecise and needs more critical thinking. This exploratory stage is a creative part of scientific analysis, since it provides an understanding about the causes, which can be formulated as a research hypothesis and causal models. In comparative experiments, the interest lies with comparison of treatments rather than their absolute performances. In forestry, such comparative experiments are numerous, namely assessment of differences in twin's behaviors in different environment, assessment of genetic fraction of the total variation in a strain having high fraction than reported value, comparison between different progeny of a species in respect of variation between progenies. However, for the comparative purposes, the experimental materials should be homogeneous so that no other sort of variation may creep into the real performance of the treatments. In forestry experimentation, due to intrinsic and extrinsic variations, any comparison of samples, merely based on observing the numerical magnitudes of the characteristic under consideration, may be misleading. It is, therefore, quite vexing to decide whether the difference observed is real, that is, as consequence of treatments applied rather a result of the sampling fluctuations. Inferential statistics, a branch of statistics, deals with the testing of hypothesis. Testing of hypothesis deduces the significant difference between the two situations based on the standard test for the situation. These test statistics are based on statistical distribution and requirements of the problem under investigation.

- A **hypothesis** is statements about population parameters.
- A **null hypothesis** is no difference statement about population and contains an equality statement as $\geq$, $=$ or $\leq$.
- An **alternative hypothesis** contains a statement of inequality such as $<$, $\neq$ or $>$.

(continued)

- A **type I error** occurs once researcher fails to accept the null hypothesis when the null hypothesis is actually true.
- A **type II error** occurs once researcher does not reject the null hypothesis although the null hypothesis is false.
- A **level of significance** is the maximum probability of type I error.
- A **critical region** corresponds to the test statistics, which amounts to rejection of null hypothesis.

Hypothesis testing facilitates inference about population by setting up of hypotheses and thereof testing the claim. In each problem, the problem is simplified into two competing claims formulated as hypotheses, that is, null hypothesis (H_0) against the alternative hypothesis (H_1), and researcher will choose one based on evidence, that is, based on the data. These two competing claims are not treated on an equal basis; rather, null hypothesis is favored due to statistical requirement. We have two common situations:

1. The experiment has been carried out to disprove or reject the claim. Priority will be attached to "null hypothesis" so that rejection of null hypothesis is not possible unless sufficiently strong evidence exists against the hypothesis, for example, there is no difference in tissue width between poplar and teak wood against there is difference in tissue width between poplar and teak wood.
2. If one of the two hypotheses is "simpler," the priority will be attached to the simpler, to avoid a more "complicated" theory unless there is sufficient evidence against the simpler one, for example, a "simpler" claim may be as there is no difference in wood strength between poplar and melia wood than there is a difference.

Steps in a Hypothesis Testing

1. Formulation of hypothesis: Stating the null (H_0) and the alternate (H_1) hypothesis
2. [Facilitate for decision about the single versus two-tail test]
3. Set the level of significance (denoted by α).
4. Identify the sampling distribution
5. Calculation of the test statistic
6. Compare the test statistics with standard and make decision

Hypotheses framing: A statistical hypothesis is actually the presumption based on past experience, or any desired value which researcher desires or is likely to arrive at. Null hypothesis asserts no differences between the two treatments and the observed difference is purely due to chance only. For example, in a comparative experiment, two media for seedling growth in nursery may be tested with the following hypothesis:

H_0: The two growth media have similar effect, on average ($\mu = \mu_0$);

H_1: The two growth media differ in their effects, on average ($\mu \neq \mu_0$)

or

H_1: Media "A" is better than the media "B," on average ($\mu > \mu_0$).

The decision for single tailed and two tailed tests are evaluated based on the alternative hypothesis. If alternate hypothesis has one option, that is, unidirectional, then one-tailed test will be applied; otherwise, for more than one options, that is, bi-directional alternate hypothesis, two tailed test will be used. A one-tailed test is appropriate only when previous data, physical limitations, practical experience, or common sense are in favor that a difference is directional. For example, for the experiment of testing the survival of pine seeds under the temperate environment, it may be hypothesized that survival of pine seeds is less than 65%, then it may be a fit case of one-tail test with the hypotheses as H_0: $\mu = 65$, against H_1: $\mu < 65$. In another experiment, the researcher wishes to test the average primary branch angle of *Acacia nilotica* wood is 25^0 and the hypothesis may be as H_0: $\mu = 25^0$, against H_1: $\mu \neq 25^0$, which is a case of two-tailed test.

Significance level: For example, in a trial of testing growth of a new clone of poplar, the null hypothesis might be that the new clone is no better than the best poplar clone in terms of growth. A type I error may occur, if researcher concludes that the two clones are different, whereas the two clones are same. However, a type II error occurs if researcher concludes that the two clones are similar, whereas the new clone has better growth than the best clone.

A type I error is often more serious, therefore more important to avoid, than a type II error. The hypothesis test procedure is, therefore, made balanced so that there is a guaranteed "low" probability attached for rejecting the null hypothesis wrongly. The probability should be based on the nature and objective of investigation and the precision required for the result. The probability of a type II error is generally unknown, and denoted by β. Type II error may arise when the sample may small enough to unable to identify the falseness of the null hypothesis (especially if the truth is very close to hypothesis). For any given set of data to compare through statistical tests, type I and type II errors are always present and cannot be completely removed. The two errors (Type I and Type II) are related inversely, that is, the lower the risk of one, the higher the risk of the other.

Sampling distribution and calculation of test statistics: Sampling distribution determines the test statistic of posed problem assuming that the H_0 (equality condition) is true and the experiment is repeated for an infinite number of times. Test statistics is estimated based on the sample data through some defined function under certain condition and used to decide the support or rejection of null hypothesis.

Make decision: If the p-value is less than α (the level of significance), the experimenter would reject H_0. If the p-value is greater *than* α, the experimenter would fail to reject H_0. The final conclusion of the test is always reported in terms of the null hypothesis. The conclusion of the test should be either "Reject H_0 in *favour* of H_1" or "Do not reject H_0." The experimenter should never conclude "Reject H_1" or even "Accept H_1." "Do not reject H_0" necessarily mean that there is not sufficient evidence against H_0 in favor of H_1. Rejecting the null hypothesis then suggests that the alternative hypothesis may be true.

The sample space is partitioned into two regions for the test statistic. One region termed as critical region or rejection region accounts for rejection of the null hypothesis H_0; however, the other region accounts the non-rejection of H_0 (the acceptance region). If the observed value of the test statistic falls in the critical region, then the conclusion would be states as H_0 does not have ample evidence in favor, and if it does not fall in the critical region, then H_0 has ample (sufficient) support in favor. This threshold point is known as critical value for a hypothesis test. The level of significance for the critical value may be more than 0.05 depending upon the conservative and demand/expect for a high degree of confidence for 0.10.

If the claim is the null hypothesis, then the claim will be either rejected or determined that there is not enough evidence to reject the claim. If the claim is the alternative hypothesis, then the claim will be either supported or determined that there is not enough evidence to support the claim (Box 6.1).

Calculate the p-value for the test statistic: The probability value is termed as p-value of a statistical hypothesis test. P-value is the probability of getting a value of the test statistic as extreme or more extreme than that observed by chance alone, if the null hypothesis is true. The p-value is the probability of wrongly rejecting the null hypothesis, while the null hypothesis is in fact true. The p-value is compared with the actual significance level of test and; the result is significant if p-value is smaller than significance level. Small p-value shows that the difference is quite unlikely to be caused by random sampling rather, actually, the populations have different mean, that is, if the null hypothesis is to be rejected at the 5% significance level, that is, "$p < 0.05$." The p-value provides a measure of the strength of evidence for rejecting the null hypothesis, rather than simply concluding rejection or not rejection H_0.

The p-value of 0.04 means that there is a 4% chance of observing a difference between the estimated and test value as large as researcher observed even if the two population means are identical. Precisely, it can be said that random sampling from identical populations would lead to a difference smaller than observed in 96% of experiments and larger than observed in 4% of experiments.

Statistical significance: A result is statistically significant when the result would be surprising if the populations were really identical. Statistically significant results have two possible explanations. (1) The populations are identical, so there is actually no difference. By chance, researcher obtained larger values in one population than the other population. (2) The populations are actually different, so conclusion is correct.

Biological relevance in designing the experiment is critical for statistical analysis. Biological relevance and statistical significance are not necessarily linked. However, a broad consensus in modern applied statistics exists that the statistical significance on its own cannot completely summarize all the information from the data and therefore the biological relevance of any change observed should be of primary importance in the experimentation rather than the statistical significance (EFSA Scientific Committee 2011).

6.5.1 Testing of Hypothesis for Various Situations

Numerous tests are available depending upon the data types and objectives (Table 6.5).

t-test for testing the mean of one sample: One sample hypotheses is useful, when researcher is interested to know the effect of a treatment is different than the known yardstick or assumed value, for example, soil pH under forest patches in comparison to the pH in adjacent soil; or amount of photosynthesis under dark conditions in comparison with light condition.

Suppose a random sample $x_1, \ldots, x_n$ of size n ($n \geq 2$) is collected from a normal population whose variance σ^2 is unknown and researcher wishes to test whether the sample belongs to the population or not, that is,

$H_0{:}\mu = \mu_0$, that is, The estimated mean differs with population mean

$H_1{:}\ \mu \neq \mu_0$

Test statistic:

$$t = \frac{\overline{x} - \mu_0}{s/\sqrt{n}} \sim t_{n-1}$$

where $\overline{x} = \frac{1}{n}\sum_{i=1}^{n} x_i$ is sample mean, $s^2 = \frac{1}{n}\sum_{i=1}^{n}(x_i - \overline{x})^2$ is variance.

The computed value is compared with the tabulated value at desired levels of significance and $(n-1)$ df for inference.

Example: A random sample of 10 trees from a natural stand of chir pine had the following girths in cm.

7.5, 8.8, 11.0, 12.0, 9.8, 9.0, 10.0, 10.5, 8.5, and 9.5.

Do these trees belong to the stand, in which, the mean girth of tree is 10.0 cm?

Table 6.5 Situation, tests with assumptions for small samples ($n \leq 30$)

Situation/condition	Appropriate test and assumption
To test the sample mean ($\overline{x}$) differs from hypothetical mean μ of population or, whether the sample has been actually drawn from the same population with known mean	One sample t-test 1. Parent population should be normal 2. The sample is drawn randomly. 3. The variance of the population is unknown
To test whether observed mean of two samples differ, or whether these two samples belong to the same population	Two samples independent t-test 1. Parent population should be normal 2. The populations' variances are equal but unknown 3. Two samples are random and independent
To test whether observed mean of two paired samples differ	Paired t-test 1. Samples must have some relationships with each other 2. Samples should be drawn from normal populations 3. Variances of two populations are equal but unknown
To test the equality of several means	F-test as Analysis of variance (ANOVA) 1. The observations are independent to each other 2. The parent population from which the observations have been drawn is normal 3. The various treatments and environment effects should be additive 4. The variances are homoscedastic and mutually uncorrelated
To test whether two independent samples have been drawn from same population having variance (σ^2), or whether the two independent estimates of the population variance are homogenous	F-test for test of equality of two variances 1. Parent population should be normal 2. The sample is drawn randomly

Solution: The question provides following information

$\mu = 10.0$ cm, $n = 10$, $\overline{x}$, s^2 are to be estimated from the given data.

H_o: $\mu = 10.0$ cm, that is, samples are collected from stand with mean girth of 10.0 cm.

H_1: $\mu \neq 10.0$ cm, that is, the sample does not belong to the stand.

Level of significance is 5% (say).

The test statistics:

$$t = \frac{\overline{x} - \mu_0}{s/\sqrt{n}} \sim t_{n-1}$$

$$s^2 = \frac{1}{n}\sum_{i=1}^{n}(x_i - \overline{x})^2 = \frac{15.324}{10} = 1.5$$

Therefore, $t = \frac{9.66 - 10.00}{\sqrt{1.50/10}} = \frac{-0.34}{\sqrt{0.150}} = -0.84$

The tabulated value of "'t" at 5% level, and 9 degrees of freedom is 2.26, that is, $t_{5\%}$, 9 df = 2.26. Thus, the calculated "'t" is less than tabulated "'t"; hence, there is ample evidence in favor of H_0 at 5% level of significance, that is, sample is selected from the stand with mean girth of 10.0 cm.

***t*-test for difference of means of two independent populations**: Let $\overline{x}_1$ and $\overline{x}_2$ be the sample means of two samples of sizes n_1 and n_2, respectively, from two normal populations with mean μ_1 andμ_2 and uniform unknown variance of the two populations as σ^2, for example, amount of moisture in soil in dense and open forest patches; productivity of two tree species under plantation; or amount of photosynthesis in two species.

$H_0 : \mu_1 - \mu_2 = 0$, that is, means of two populations are same.

Test statistic: $t = \frac{\overline{x}_1 - \overline{x}_2}{\sqrt{s^2\left(\frac{1}{n_1}+\frac{1}{n_2}\right)}} \sim t_{n_1+n_2-2}$

Since σ^2 is unknown, therefore, it is estimated from the sample:

Pooled sample variance $s^2 = \frac{\sum(x_{1i} - \overline{x}_1)^2 + \sum(x_{2j} - \overline{x}_2)^2}{(n_1+n_2-2)}$

Example: In a range experiment a group of 5 plots of same area treated with nitrogen at 20 kg/ha yielded grass productivity of 4.2, 3.9, 4.8, 6.0, and 4.1 kg, whereas a second similar group of 7 plots with same area treated with nitrogen at 40 kg/ha yielded productivity of 3.8, 4.2, 5.6, 6.4, 6.8, 6.9, and 6.2 kg. Test whether, the higher dose of nitrogen yield more grass productivity?

Solution: We have

$n_1 = 5, n_2 = 7, \overline{x_1} = 4.6, \overline{x_2} = 5.7, s^2 = (91.22 + 225.86)/(5 + 7 - 2) = 31.708$

The problem under study will be tested by two samples independent *t*-test against the following hypothesis: $H_0 : \mu_1 = \mu_2$,

$H_1 : \mu_1 < \mu_2$

$$t = \frac{\overline{x_1} - \overline{x_2}}{\sqrt{s^2\left(\frac{1}{n_1} + \frac{1}{n_2}\right)}}$$

$$= \frac{4.6 - 5.7}{\sqrt{31.708\left(\frac{1}{5} + \frac{1}{7}\right)}} - 0.52$$

The calculated "*t*" is less than tabulated "*t*," as 1.81 at 5% and 10 df. The researcher can conclude that there is no evidence against the null-hypothesis H_o: $\mu_1 = \mu_2$, that is, the yield from two nitrogen doses does not differ significantly. The yield of two samples may be regarded as equal and the observed difference is due to chance.

***t*- test for difference of means of paired samples**: Paired means that there is a relationship between the observations in the first sample and the second sample. Two samples are related if each member of one sample corresponds to a member of the other sample. Related samples are paired samples or matched samples. Paired *t*-test is used to test the difference of two normal population means when a sample is randomly selected from a population such that each member of the first sample must be paired with a member of the second sample of another population, for example, amount of moisture in soil before and after monsoon; body weights of mice before and after the nutrient supplement; or amount of photosynthesis under light and dark conditions.

H_0: $\mu_1 = \mu_2$' that is' Means of two populations are same.

H_1: $\mu_1 < \mu_2$

$$t = \frac{\overline{d} - \mu_d}{s_d/\sqrt{n}}.$$

$$\overline{d} = \frac{\sum d}{n}$$

The difference between a data pair, $d = x_1 - x_2$ and

$$s_d = \sqrt{\frac{n(\sum d^2) - (\sum d)^2}{n(n-1)}}$$

Example: An industrial unit claims that a preservative developed by them would perform better on the graveyard test than standard. The table shows number of the month first termite attack was observed on two sets containing six number of homogeneous logs. At $\alpha = 0.05$, is there enough evidence to conclude that the developed preservative is better than the standard?

Log	1	2	3	4	5	6
Month with standard	85	96	70	76	81	78
Month with preservative	88	85	89	86	92	89

Solution: We have,
Hypothesis
H_0: $\mu_d \leq 0$
H_a: $\mu_d > 0$ (Claim)

$$\overline{d} = \frac{\sum d}{n} = \frac{-43}{6} \approx -7.167$$

$$s_d = \sqrt{\frac{n(\sum d^2) - (\sum d)^2}{n(n-1)}} = \sqrt{\frac{6(833) - 1849}{6(5)}} \approx 10.245$$

The *t*-test

$$t = \frac{\overline{d} - \mu_d}{s_d/\sqrt{n}} = \frac{-7.167 - 0}{10.245/\sqrt{6}} \approx -1.714.$$

The calculated "*t*" is less than tabulated "*t*," at 5% and 5 df. The researcher can conclude that there is no evidence against the null-hypothesis, that is, there is not enough evidence to support the claim at the 5% level that the new preservative is better than the standard preservative.

***F*-test for testing of equality of two variances:** "*F*" is the ratio of two unbiased estimates of population variance. Let x_i, $i = 1,\ldots,n_1$ and x_j, $j = 1,\ldots,n_2$ be the two random samples with sizes n_1 and n_2 collected from two independent normal populations, that is, $N(\mu_1, \sigma_1^2)$ and $N(\mu_2, \sigma_2^2)$, respectively.

H_0: ${\sigma_1}^2 = {\sigma_2}^2 = \sigma^2$, that is, the two population variances are equal or homogeneous (homogeneity of variances).

H_1: ${\sigma_1}^2 \neq {\sigma_2}^2$

Test statistic: Assuming $s_1^2 > s_2^2$ (Required as calculated "*F*" is greater than unity due to the nature of "*F*" table which calculates the right tail, and having all values greater than or equal to 1).

Under H_0

$$F = \frac{s_1^2}{s_2^2} \sim F_{n_1-1, n_2-1}$$

s_1^2 and s_2^2 are the sample variances of the two samples.

$$s_1^2 = \frac{1}{n_1 - 1}\sum_{i=1}^{n_1}(x_i - \overline{x_1})^2, \quad s_2^2 = \frac{1}{n_2 - 1}\sum_{i=1}^{n_2}(x_i - \overline{x_2})^2$$

$$\overline{x_1} = \frac{1}{n_1}\sum_{i=1}^{n_1} x_i, \quad \overline{x_2} = \frac{1}{n_2}\sum_{j=1}^{n_2} x_j$$

This value will be compared with the table *F* value for obtaining significance.

Example: In an experiment, compare the stability of organism of bird (measured by the difference between length of largest tarsus with length of smallest tarsus) in polluted (Population 1) and non-polluted (Population 2) environment with the following data. Compare the environmental impact of pollution.

Population 1	3	5	4	3	4	5
Population 2	4	10	4	7	7	10

Solution: *F*-test for testing of equality of variances with the following hypothesis may be applicable.

H_0: ${\sigma_1}^2 = {\sigma_2}^2$. There is no difference in stability of organism between the polluted and non-polluted environment.

H_1: ${\sigma_1}^2 \neq {\sigma_2}^2$

Level of significance: 5% (Say)

Estimated values of mean and variance as per sample data are as follows:

Population 1	Mean = 4;	Variance = 0.8
Population 2	Mean = 7;	Variance = 7.2

Calculate *F*-test: $F = \frac{s_1^2}{s_2^2} = 7.2/0.8 = 9.0$

Calculated "*F*" is greater than tabulated "*F*" ($F_{0.05\ (2),\ 5,\ 5}$ = 7.15); hence, variances are significantly different, that is, there are ample evidence that there is instability in organism with polluted than non-polluted environment.

Examples for *t*-test for testing mean of one sample

1. Does the new forest management strategy based on the returned migrants (due to coronavirus disease 2019 [COVID-19]) skill based forest guarding requires less resources?
2. Do the forest fertilizers from the floor of forests improve the farm productivity?
3. Do the forest resources yield better result for COVID-19 management than the traditional approach?
4. Comparison of new clone of poplar with the most prominent clones (variable of interest—plant growth, productivity, health, quality)
5. Comparison of chemical composition between the two wild varieties of a medicinal plant
6. Do genetically modified plants more productive than the existing control plant

Examples for *t*-test for testing means of two independent samples

1. Comparison of strength of composite boards produced from two different production units
2. The economic performances of the two forest management regimes
3. The response of a disease under two temperatures
4. Fungal growth in two different environment or media
5. Comparison of two pruning methods for productivity improvement

Examples for paired t-test for testing means of paired samples

1. The change in resin yield before and after the application of chemical stimulants was paired with the same tree resin yield.
2. The change in number of insects count before and after the spray of an insecticide was paired with the insects on the same tree.
3. Comparison between monozygotic twins or clones of plants can be paired in an experiment.
4. Comparison between twins having one of the twins with a particular diet and the other twin with a normal diet.

Examples for F-test

1. The researcher might be interested in the difference between the variability of response of inbred lines (little genetic variation but environmental variation within population) versus an outbred populations (lots of genetic and environmental variation due to outside population too).
2. Comparison of vigor of inbred and outbred of a tree species (the vigor may be measured by gregariousness of trees).

Box 6.1 Decision Making

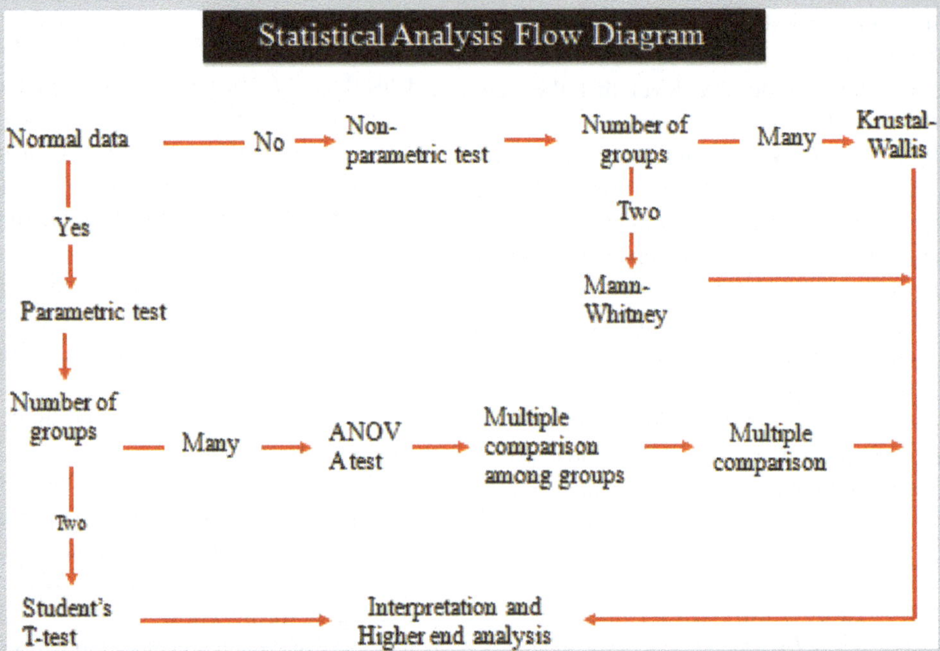

(continued)

Box 6.1 (continued)

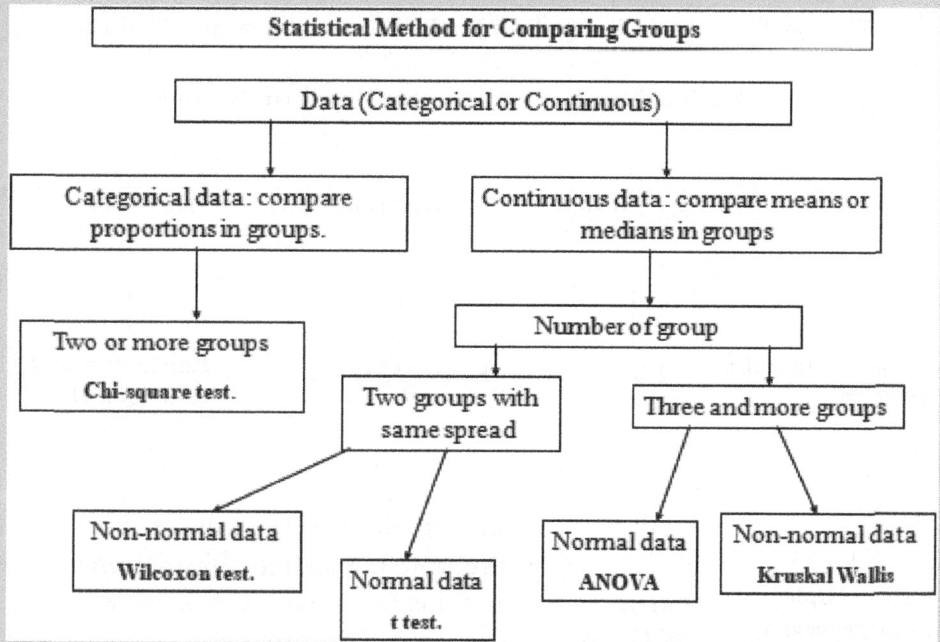

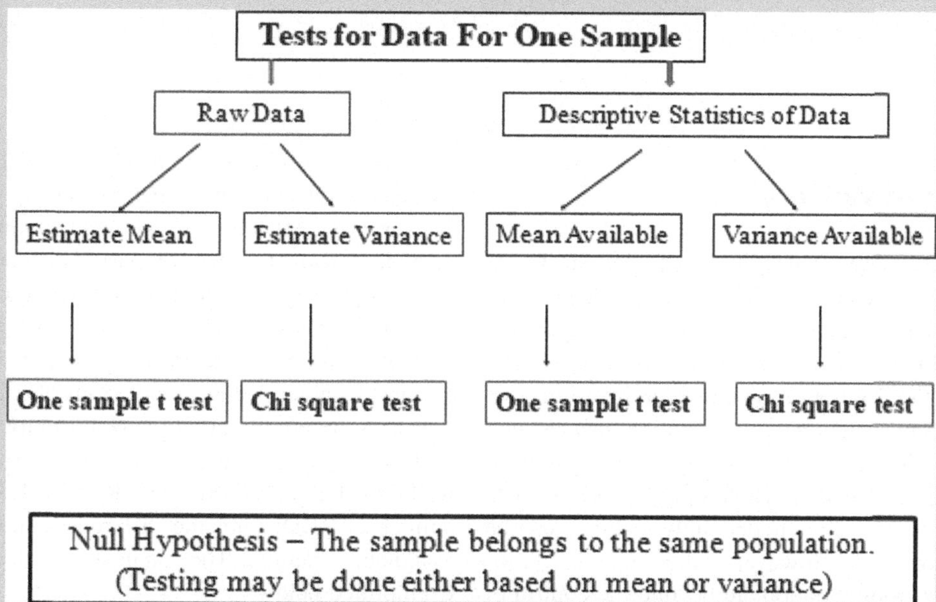

(continued)

Box 6.1 (continued)

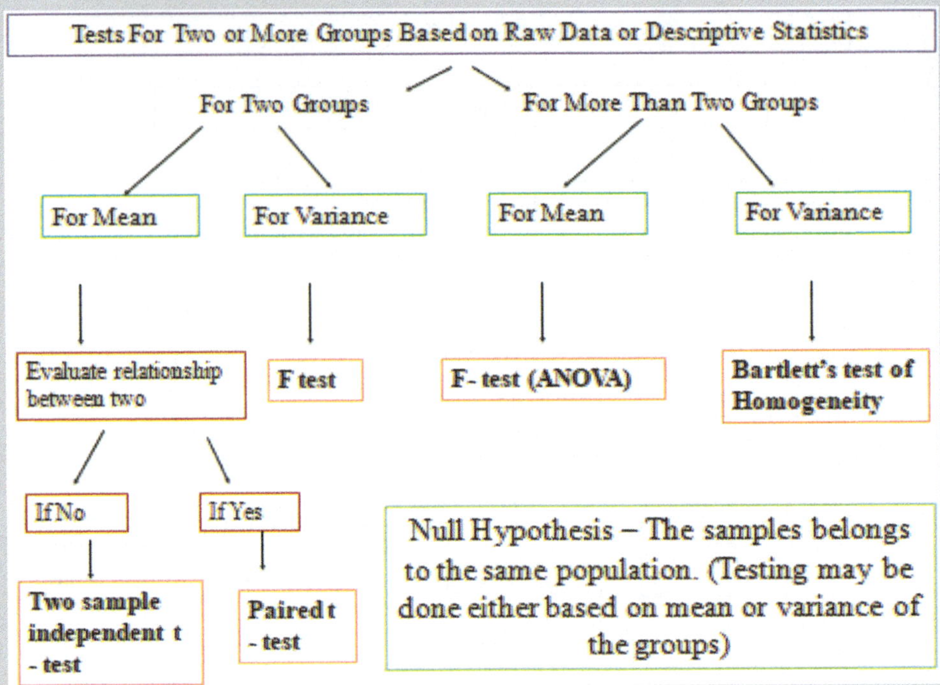

6.6 Analysis of Variance

Analysis of variance (ANOVA), introduced by Sir R. A. Fisher in 1925, has basic purpose to test the homogeneity of mean of treatments or treatment combinations by partitioning the total variation in the data into its various components. The total variability in the data set of random experiment may be due to the different components of known causes and unknown causes. The known causes, termed as assignable causes, are introduced in experimental units by researcher to see the responses of the factors under study and thus can be treated as controlled. However, the unknown or non-assignable causes of variation are beyond the control of researcher and present due to the nature of experimental units. For example, in forestry experiments, which have inherent variability, the non-assignable cause may be the fertility of two adjoining fields, which cannot be exactly same in spite of all similar conditions and assignable cause may be seed of various species, or, height of plants of various parents. ANOVA technique enables us to partition the total variation into its components, that is, known and unknown components. The variation due to non-assignable or unknown causes is known as the experimental error that is receiving of different yields (result) while applying same treatments on alike experimental units. ANOVA estimates the variance contributed by known and unknown factors and provides information for comparing the variance attributed to "known causes" with that of "unknown causes." The comparison is based on *F*-test, which enables us to test the homogeneity of several means, simultaneously. ANOVA is a robust technique, that is, violations of a few assumptions of ANOVA are not affecting the result in totality.

In the synthesis, we are only discussing the analysis of RBD, as the analysis of the CRD and LSD is more or less similar as RBD; only the differences lies in the factors of assignable causes. The analysis of the RBD is being exemplified below.

Example: The fresh weight of plants is noted for an experiment with six treatments in four randomized blocks. Based on the data reported below, compare the significance of treatments. Numbers within parenthesis indicate the treatments.

Replication	Fresh weight					
I	(1) 24.7	(3) 27.7	(2) 20.6	(4) 16.2	(5) 16.2	(6) 24.9
II	(3) 22.7	(2) 28.8	(1) 27.3	(4) 15.0	(6) 22.5	(5) 17.0
III	(6) 26.3	(4) 19.6	(1) 38.5	(3) 36.8	(2) 39.5	(5) 15.4
IV	(5) 17.7	(2) 31.0	(1) 28.5	(4) 14.1	(3) 34.9	(6) 22.6

Solution: The analysis of the above data is as follows:

Replication	Treatment 1	2	3	4	5	6	Total
1	24.7	20.6	27.7	16.2	16.2	24.9	130
2	27.3	28.8	22.7	15	17	22.5	113.3
3	38.5	39.5	36.8	19.6	15.4	26.3	176.1
4	28.5	31	34.9	14.1	17.7	22.6	148.8
T_1	119	119.9	122.1	64.9	66.3	96.3	388.5 (GT)

Step 1: Correction factor (CF) $= \frac{(GT)^2}{N} = \frac{34633.25}{24} =$ 14430.5

Step 2: Total SS $= S_T^2 = \Sigma\Sigma y_{ij}^2 - CF = 15780.76 - 14430.51 = 1350.25$

Step 3: SS due to testaments (SST)

$$S_T^2 = \frac{1}{4}\sum_{i=1}^{6} T_i^2 - C.F. = \frac{6132.81}{4} - 14430.51 = 901.19$$

Step 4: SS due to blocks (SSB) $= S_R^2 = \frac{1}{6}\sum_{j=1}^{4} B_j^2 - C.F. = \frac{87899.63}{6} - 14430.51 = 219.43$

Step 5: Error SS $= S_E^2 = \text{TSS} - \text{SST} - \text{SSB}$

$$= 1350.25 - 901.19 - 219.43 = 229.63$$

Step 6: Analysis of Variance table

Source of variance	df	SS	MSS	F (Cal.)
Treatment	5	901.19	180.24	$F_t = \frac{180.24}{15.31} = 11.8^*$
Block (replication)	3	219.43	73.14	$F_b = \frac{73.14}{15.31} = 4.7$
Error	15	229.63	15.31	
Total	23	1350.25		

Tabulated $F_{0.05\,(3,\,15)} = 5.42$ and $F_{0.05\,(5,\,15)} = 4.50$

ANOVA table reflects the insignificance of block variation and significance of treatment variation, that is, there is evidence against the null hypothesis H_0 for the treatment.

Step 7: For getting pair wise comparison, critical difference (CD) is estimated.

$$CD = \sqrt{\frac{2 \times EM.S.S}{r}} \times t_{0.05,15}$$

$$CD = \sqrt{\frac{2 \times 15.31}{4}} \times 2.131$$

$$= 2.7668 \times 2.31 = 5.90$$

Based on this analysis, it can be concluded that the treatment 3 is having maximum weight, which is statistically at par with treatment 1 and 2. However, the minimum weight is with treatment 4 which is statistically at par with treatment 5.

Interpretation: The ANOVA results indicate that the block effect (i.e., replication) is non-significant, while the treatments have significantly different effects on the fresh weight of the plants. The post-ANOVA technique, that is, LSD results that treatment 3 has similar effects on fresh weight of plants as treatment 1 and 2. Treatment 4 and 5 have least effect for weight gain and statistically similar, that is, the effect of treatment 4 and 5 is similar for fresh weight of plant.

6.7 Interpretation and Reporting of Data Analysis

This section deals with the interpretation of data analysis in precise manner for most of the methods. One hypothetical data has been considered for exemplification.

Example: Leaves of one evergreen (*Euonymus pendulous*) and one deciduous (*Pyrus pashia*) species were evaluated for chlorophyll content (mg g^{-1}) in the temperate forest in Mussoorie Forest Division. Following raw data were obtained. The various data analysis is being exemplified below for data representation, numerical measures, and *t*-test.

Data of chlorophyll content (CC) (mg g^{-1}) for *Euonymus pendulous* (EP) and *Pyrus pashia* (PP)

Sl. no.	EP	PP	Sl. no.	EP	PP	Sl. no.	EP	PP
1	12.8	12.3	10	11.9	11.1	19	12.5	12.2
2	14.7	12.3	11	12.2	12.8	20	13.6	14.5
3	13.9	11.5	12	13.2	11.9	21	12.9	13.7
4	12.9	13.7	13	12.9	12.7	22	14.9	12.3
5	13.6	13.8	14	14.4	13.2	23	14.7	11.9
6	12.6	11.5	15	15	12.4	24	15.8	11.7
7	11.9	10.9	16	14.9	13.7	25	12.4	12.6
8	14.2	14.1	17	13.3	12.3			
9	12.9	12.3	18	14.4	13.9			

Solution: The ordered data of both the series are reported below for data processing. Based on the arranged data, mean, median, and mode were estimated. The mean and mode can be estimated without ordering the data; however, for median the ordering is essentially required as per definition of the median.

Data processing of chlorophyll content (CC) for 25 leaves for estimation of averages and dispersion of *Euonymus pendulous* (EP) and *Pyrus pashia* (PP)

Si. no.	EP CC (x)	$x_i - -x$	$\lvert x_i - -x \rvert$	$(x_i - -x)^2$	PP CC (y)	$y_i - -y$	$\lvert y_i - -y \rvert$	$(y_i - -y)^2$
1	11.90	−1.64	1.64	2.69	10.90	−2.64	2.64	6.97
2	11.90	−1.64	1.64	2.69	11.10	−2.44	2.44	5.95
3	12.20	−1.34	1.34	1.80	11.50	−2.04	2.04	4.16
4	12.40	−1.14	1.14	1.30	11.50	−2.04	2.04	4.16
5	12.50	−1.04	1.04	1.08	11.70	−1.84	1.84	3.39
6	12.60	−0.94	0.94	0.88	11.90	−1.64	1.64	2.69
7	12.80	−0.74	0.74	0.55	11.90	−1.64	1.64	2.69
8	12.90	−0.64	0.64	0.41	12.20	−1.34	1.34	1.80
9	12.90	−0.64	0.64	0.41	12.30	−1.24	1.24	1.54

(continued)

Si. no.	EP CC (x)	$x_i - -x$	$\|x_i - -x\|$	$(x_i - -x)^2$	PP CC (y)	$y_i - -y$	$\|y_i - -y\|$	$(y_i - -y)^2$
10	12.90	−0.64	0.64	0.41	12.30	−1.24	1.24	1.54
11	12.90	−0.64	0.64	0.41	12.30	−1.24	1.24	1.54
12	13.20	−0.34	0.34	0.12	12.30	−1.24	1.24	1.54
13	13.30	−0.24	0.24	0.06	12.30	−1.24	1.24	1.54
14	13.60	0.06	0.06	0.00	12.40	−1.14	1.14	1.30
15	13.60	0.06	0.06	0.00	12.60	−0.94	0.94	0.88
16	13.90	0.36	0.36	0.13	12.70	−0.84	0.84	0.71
17	14.20	0.66	0.66	0.44	12.80	−0.74	0.74	0.55
18	14.40	0.86	0.86	0.74	13.20	−0.34	0.34	0.12
19	14.40	0.86	0.86	0.74	13.70	0.16	0.16	0.03
20	14.70	1.16	1.16	1.35	13.70	0.16	0.16	0.03
21	14.70	1.16	1.16	1.35	13.70	0.16	0.16	0.03
22	14.90	1.36	1.36	1.85	13.80	0.26	0.26	0.07
23	14.90	1.36	1.36	1.85	13.90	0.36	0.36	0.13
24	15.00	1.46	1.46	2.13	14.10	0.56	0.56	0.31
25	15.80	2.26	2.26	5.11	14.50	0.96	0.96	0.92

Averages and dispersion statistics for chlorophyll content (mg g^{-1}) for *Euonymus pendulous* (EP) and *Pyrus pashia* (PP).

Species	Size	Max	Min	ΣX_i	Mean $= (\Sigma X_i)/N$	Mode	Median	MD	Range	SD	SE	CV
EP (x)	25	15.80	11.90	**338.50**	**13.54**	12.90	13.30	0.93	3.90	1.07	0.21	7.88
PP (y)	25	14.50	10.90	**315.30**	**12.61**	12.30	12.30	1.14	3.60	1.34	0.27	10.59

Size—No of observation; Max—Maximum observation; Min—Minimum observation; MD—Mean deviation; SD—Standard deviation; SE—Standard error

Interpretation: Numerically, the data of chlorophyll ranges from below 11.90 to above 15.80 (mg g^{-1}) for EP and approximately 10.90 to above 14.50 (mg g^{-1}) in PP. Mean of chlorophyll is 13.54 and 12.61 (mg g^{-1}) for EP and PP, respectively, with low variability in CC for EP than PP as evident by SD and SE (Box plot). The low CV of EP demonstrates low variability rather high consistency in CC for EP than PP. The result supports that evergreen species (i.e., EP) containing leaves throughout the year have tendency to maintain chlorophyll level more consistent than deciduous species (i.e., PP) sheds leaf in fall season. The tendency to maintain the leaves throughout the year assists to maintain the chlorophyll level without much fluctuation in evergreen species than deciduous species.

Interpretation

The data of chlorophyll ranges from below 12 to above 15 (mg g^{-1}) for EP and approximately 11 to above 14 (mg g^{-1}) in PP. Box plot shows that upper range of chlorophyll is highly diverse than the lower range of CC in EP and PP.

Comparison between the CC of two species (Independent Samples *t*-test)

t-test result for comparison between *Euonymus pendulous* (EP) and *Pyrus pashia* (PP) for chlorophyll content (mg g^{-1}).

Species	N	Mean	SD	SE	*W* (*p*-value)	*F* (*p*-value)	Mean difference	SE of differences	df	*t*-value (*p*-value)
EP	25	13.54	1.07	0.21	0.95 (0.26)	0.74 (0.39)	0.93	0.29	48	3.17 (0.003)
PP	25	12.61	1.34	0.27	0.95 (0.31)					

W—Test of normality (Shapiro–Wilk test)

Interpretation

The *t*-test for testing the mean CC of EP and PP reflects the statistically significant difference between the two species with high CC in EP. Moreover, the data of CC for both species follows the assumptions of the t-test, that is, normally distributed as Shapiro–Wilk test is non-significant and the variance is also equal by *F*-test. The lower mean value of CC in deciduous species (PP) indicates that less chlorophyll is required to fulfill energy demand in comparison to evergreen species (EP). This experiment also put ups a question to study more about the chlorophyll use efficiency in different leaf types of species.

Presentation of data (Box plot)

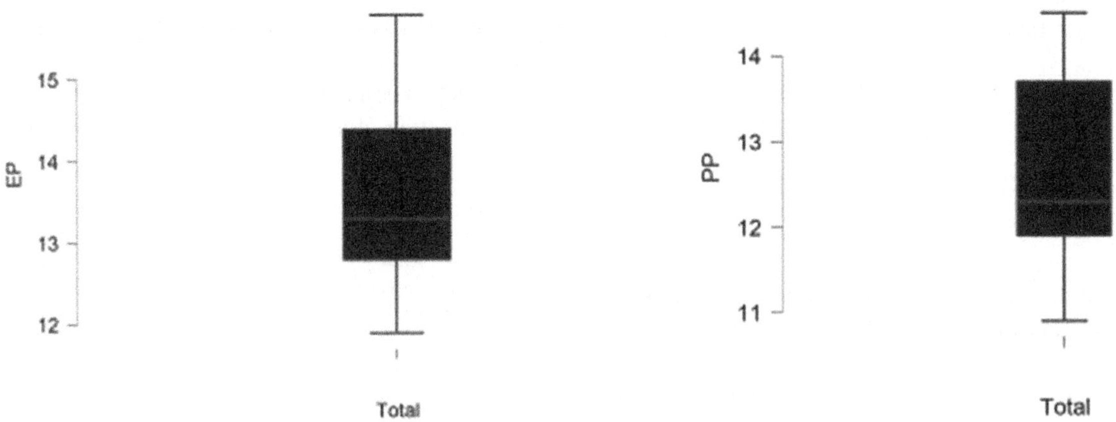

Box plot for chlorophyll content (mg g^{-1}) for *Euonymus pendulous* (EP) and *Pyrus pashia* (PP)

Lesson Learnt

- Statistical methods are prime for decision-making under uncertainty for any random experimentation. The application of statistical method requires basic understanding about the objectives of the experiment and how to apply the statistical method to achieve the objective. Wrong application of the statistical methods leads to poor information generation and wasteful experimentation.
- Details about the basic statistical methods along with their protocol, assumptions, and application with forestry experiments. Demonstrated the application of methods as per the requirement of the experiment leading to precise and unbiased results of the experiment. Attempted to fill the gap using examples from basic statistical methods. Demonstrated the representation of the result as per the data analysis protocol along with their interpretation. Suitable interpretation of the experiment is generally lacking and therefore, undermining the experimentation.

Key Questions

1. Elaborate the significance of the statistical methods in forestry experimentation.
2. Explain the examples of forestry experimentation where regression modeling is appropriate along with all the necessary protocol.
3. Elaborate the situations where two sample *t*-test can be applied for (1) independent samples and (2) paired samples.
4. Discuss the basic steps for conducting sampling in forestry experiments.
5. Define all possible variations for forestry experimentation.

References

Anscombe FJ (1948) The validity of comparative experiments. J R Stat Soc A 111(3):181–211

Cochran WG, Cox GM (1957) Experimental designs, 2nd edn. Wiley, New York

EFSA Scientific Committee (2011) Statistical significance and biological relevance. EFSA J 9(9):2372. https://doi.org/10.2903/j.efsa.2011.2372

Kimmins JH (2002) Future shock in forestry where have we come from; where are we going; is there a "right way" to manage forests? Lessons from Thoreau, Leopold, Toffler, Botkin and Nature. For Chron 78(2):263–271. https://doi.org/10.5558/tfc78263-2

Further Reading

Cochran WG (1977) Sampling techniques, 3rd edn. Wiley, New York

Das MN, Giri NC (1991) Design and analysis of experiments, 2nd edn. Wiley Eastern Ltd., New Delhi

Draper NR, Smith H (1998) Applied regression analysis, 3rd edn. Wiley, New Delhi

Fisher RA (1937) The design of experiments, 2nd edn. Oliver and Boyd, Edinburgh

Fisher RA (1938) Statistical methods for research workers, 7th edn. Oliver and Boyd, Edinburgh

Gupta SC, Kapoor VK (2014) Fundamentals of applied statistics. Sultan Chand & Sons, New Delhi

Gupta SC, Kapoor VK (2017) Fundamentals of mathematical statistics. Sultan Chand & Sons, New Delhi

Jayaraman K (1999) A statistical manual for forestry research. FAO, Bangkok

Polinko AD, Coupland K (2020) Paradigm shifts in forestry and forest research: a bibliometric analysis. Can J For Res 51:154. https://doi.org/10.1139/cjfr-2020-0311

Snedecor GW, Cochran WG (1989) Statistical methods, 8th edn. Wiley-Blackwell, New York

Part II

Forest Management and Conservation

Forest Biodiversity Conservation

7

Prodyut Bhattacharya

Abstract

Biodiversity is in a broad sense the total expression of life on the earth. India is one of the 12 mega-diversity country; it has two of the 18 biodiversity hotspots: Western Ghats and Eastern Himalayas. It covers about 7% of the global biodiversity. Biodiversity is essential component for the survival of all species, as each life form is linked to another in a fragile web of life. Habitat change has a high and continuing impact on the local biodiversity, whereas climate change impact is moderate, but it may have very rapid change on biodiversity as a whole, due to overall changes of the area. The systems that cause deforestation, fragmentation, and degradation are directly or indirectly cause the biodiversity loss. However, the most significant contributors to the decline of forest biodiversity are usually of human origin. The transformation of forests into agricultural areas, overgrazing, uncontrolled shifting cultivation, forest mismanagement, invasive alien flora and fauna, infrastructure expansion (such as road construction, hydro-electric projects, urban expansion), mining, oil exploitation, human-induced forest fires, pollution, and climate change collectively contribute to adverse effects and serve as causative factors for the loss of biodiversity in forest ecosystems. As the forests degrade, the status of the biological diversity also gets degraded. This degradation diminishes the resilience of forest ecosystems, rendering them less capable of adapting to evolving environmental conditions. This chapter analyzes the forest biodiversity condition, actions, reasons for the biodiversity loss, research methodology and biodiversity legislation, and conservation strategies associated with this global agenda.

Keywords

Biodiversity · CBD · Biocultural diversity · Biogeographic zone · Habitat fragmentation

7.1 Introduction

Forests are the preeminent habitats for biodiversity on land, as they harbor a vast majority of the earth's terrestrial species. Forests represent the apogee of biodiversity on land. Biodiversity is simply understood by the total expression of life on the earth. The rain forest ecosystem is among the oldest ecosystems on earth, having very rich biodiversity among other terrestrial ecosystems. Forests provide plenty of useful products for human beings like timber, fuelwood, pulpwood, fodder, meat, gum, essential oils, cash crops, and medicinal plants. They meet the livelihood needs for many millions of people globally. But only a fraction of species are known, documented, and examined so far for their value. Biodiversity, which constitutes the variety and variability of life forms on the planet, results from a cumulative evolutionary process spanning over 3.5 billion years. Its advantages to human societies are diverse, encompassing economic, ecological, cultural, social, spiritual, scientific, and aesthetic dimensions. The distribution of biodiversity across the globe is not uniform, as a select few countries, predominantly situated in tropical regions, contribute significantly to the global biodiversity. Specifically, between 60 and 70% of the global biodiversity is attributed to only a dozen nations. These nations are referred to as "Mega-diverse Nations."

The term "forest biological diversity" is comprehensive, encompassing all living organisms within forested areas and the ecological functions they fulfill. In this context, it extends beyond trees to include various lower plant forms such as moss, ferns, algae, fungi, and lichens, as well as animals and micro-organisms inhabiting forested regions, along with their

P. Bhattacharya (✉)
University School of Environment Management, GGS Indraprastha University, New Delhi, India

A. K. Mandal, A. Nicodemus (eds.), *Textbook of Forest Science*, https://doi.org/10.1007/978-981-97-8289-5_7

corresponding genetic diversity. Historically and presently, the predominant emphasis in conservation endeavors aimed at conserving biological diversity has been directed toward species, subspecies, and populations. Forests, along with their biodiversity, function as a crucial safety net for humanity, offering essential environmental services such as ensuring clean air, water cycle regulation, carbon sequestration, natural disaster mitigation, and livelihood support (FAO 2018).

The biological diversity of forests can be examined across various levels, encompassing the ecosystem, landscapes, species, populations, and genetic levels. The intricate interactions within and among these diverse levels constitute an integral aspect of biodiversity. Within biologically diverse forests, this complexity facilitates the gradual adaptation of organisms to diverse ecological conditions and the ongoing adjustments to micro-environmental changes, thereby sustaining essential ecosystem functions. At present, approximately 18% (726 M ha) of the worldwide forest area is actively managed for conservation, while around 30% (1.15 M ha) is dedicated for timber and non-wood forest products (FAO 2020).

The global forest area is diminishing by 12 million hectares annually, predominantly affecting tropical rainforest characterized by distinctive and diverse biodiversity. Encompassing around 30% of earth's land surface, forests play a pivotal role in supplying essential ecosystem goods and services. They serve as carbon reservoirs, offer habitat for a diverse array of species, and contribute to the preservation of land by mitigating degradation and preventing desertification. Worldwide, the biodiversity of forests faces imminent threats from rapid deforestation, fragmentation, degradation, forest fires, open grazing, hunting, and the introduction of invasive species from different habitats. Conserving the myriad species of existing biodiversity necessitates comprehensive, large-scale strategies implemented at the ecosystem and landscape levels (Franklin 1993).

Various endeavors have been undertaken to preserve global biodiversity through the safeguarding of species outside their native habitats, employing ex situ methods. Notably, economically significant trees are conserved in seed orchards and gene banks to safeguard their genetic diversity. But many forest species have short life seeds (recalcitrant seeds) that do not survive after storage as they have more oil content in their seeds, and some animal and plant species are hard to protect once isolated from their ecosystems. The soil biodiversity is significant for stabilizing and regulating the earth's climate. The invisible soil biota, integral to fostering healthy soil and, consequently, sustaining healthy flora and fauna, is experiencing a decline in genetic diversity attributed to the detrimental application of pesticides and other forms of soil pollution. Despite the magnitude of diversity in soil fauna surpassing that above ground, life within the soil is largely underestimated and overlooked in the realm of biodiversity conservation (Rawal et al. 2013).

Widespread human interventions in natural ecosystems in recent years have led to a decline in biodiversity. Acknowledging the inevitability of ongoing changes, it becomes imperative to make informed decisions about the species that are of critical importance and determine strategies for their preservation within their natural habitats. While the challenge lies in avoiding an exclusive focus on a select few commercially significant species, it is also important to recognize and analyze the interests of local populations who rely on forest resources, particularly indigenous groups, their 60–90% dependence revolves around local biodiversity. The current diversity observable in nature is the outcome of evolutionary processes spanning several billion years. This implies two significant aspects: first, the existing diversity encapsulates a reservoir of genetic information, and second, the evolutionary processes have transpired within diverse environmental conditions.

Tropical forests likely comprise more than 50% world's species, representing substantial natural capital. Variations in patterns arise depending on the specific indicator, whether related to mammals, insects, plants, and so forth. Brazil is considered the most bio-diverse country on the planet, where one-third of the total global species is present. Islands too have a significant role to play, often containing high endemic species evolved in a particular ecosystem and restricted to that area. For example, the recent assessment of IUCN (2022) reports Papua New Guinea, a most tree diverse island country having 10,973 described plant species including 2854 trees native to the country of which 1284 are endemic which show the unique biodiversity of the country. Trees and forests are vital to the livelihoods and well-being of Papua New Guinea's population, with approximately 80% residing in the country's rural areas (Barstow et al. 2022).

Biodiversity, as defined by Wilson in 1988, encompasses the variety and variability of life, along with the interrelated ecological processes and functions. The intricate ecological relationships, processes, and functions sustain the diversity of life, illustrating the interdependence among species, plants, animals, and their abiotic environment and highlighting the intricate nature of ecological complexities. The term "Biodiversity" was first coined by Lovejoy (1980). However, Rosen (1985) used the term "Biodiversity" in his publication and Edward Wilson (1986) popularized the concept in his book "The Diversity of Life" (1992).

Biological diversity encompasses the variety of earth's species, their genetic makeup, the ecosystems they inhabit, and the processes of energy flow and nutrient cycling that support life. Species diversity serves as a crucial indicator of overall biological diversity, with species richness showing an increase from polar regions to the equator. The well-documented high species density in tropical forested areas is highlighted in research by Reid and Miller (1989). Establishing protected forest areas is among the most effective strategies for conserving forest biodiversity. However, these areas must either meet a specific size criterion or form a well-designed network to ensure the continued effective operation of local forest ecosystems. The management of the forest surrounding the protected area (PA) is crucial to serve as a buffer zone. Moreover, the surrounding forests provide an opportunity for local communities to earn a livelihood without encroaching upon the protected forest.

7.2 History of Biodiversity Conservation

Over the previous 8000 years, over 45% of the earth's natural forest cover has been destroyed, with the majority of this loss occurring in the last century. According to the Food and Agriculture Organisation of the United Nations, deforestation claims an estimated 13 million hectares of the world's forests each year (FAO 2018).

Historical scriptures provide valuable insights that can be harnessed to cultivate beneficial relationships and services for humankind in conjunction with biodiversity. This knowledge is rooted in the co-evolutionary dynamics and interactions among all species. The biological diversity seen in forests is the result of thousands or even millions of years of evolutionary processes that have been shaped by ecological pressures including competition, disturbance, fire, and temperature. In addition, the richness of forest ecosystems encompassing both physical and biological components is a testament to nature's high levels of adaptation.

India, the seventh largest country globally, has an area of 329 M ha. Despite occupying only 2.4% of the global land area, the country is home to 7–8% of globally documented floral and faunal species. Recognized as a prominent center of crop diversity, India also serves as a habitat for crop wild relatives. With the terminology, "biodiversity conservation" research and study as such in true sense started in India three to four decades ago, but the floristic and faunal documentation and herbarium or museum specimen collection for record have started in the past century.

In India, a multitude of ecological and climatic conditions, combined with distinctive physical and cultural features, have fostered a remarkable array of habitats that support extensive biological diversity across all levels. While the precise count remains uncertain, it is evident that at least 10% of India's documented wild flora, and potentially a higher proportion of its wild fauna, are classified as threatened species. Many of these species are on the brink of extinction, primarily due to the loss of at least half of India's forests. There is a likelihood that numerous species, numbering in the hundreds or possibly thousands, have vanished without study or documentation due to our lack of awareness (Singh and Kushwaha 2008).

In the field of forest biological diversity, there is an increasing focus on the creation and improvement in forest indicators through "Criteria and Indicator" as an applied tool for developing criteria and indicators for sustainable forest management (SFM) evolved through the International Tropical Timber Organization (ITTO) (1991), Montreal Process (1995), Tarapoto Proposal (1995), Dry zone Africa Process (1995), Dry Forests in Asia Regional Initiative (1999), etc. These initiatives promoted regularly the development of internationally accepted tool for SFM through Criteria and Indicator, where biodiversity is considered as important criteria. In the same manner, Forest Stewardship Council (FSC) has given the significant importance of biodiversity in the international and national standards including HCV (high conservation value) forest for the certification of managed and unmanaged forest, natural and plantation forest. Indicators play a crucial role as informational tools, enabling the evaluation of the global status and trends of forest resources, including progress toward the Aichi Biodiversity Targets 2011–2020, for action by all countries and stakeholders to save biodiversity and increase its benefit for local communities.

7.3 Forest, Biodiversity, and Bio-cultural Diversity Linkages

About 60% of all angiospermic and gymnospermic species are found in rain forests, making forests unique among earth's biodiverse environments. Approximately 1300 kinds of forest plants are used for medical or cultural purposes in the Amazon basin alone, underscoring the significant value of forest biodiversity in providing food, medicine, fodder, and raw materials. Recently, meeting the sustainable development goals is closely tied to forest biodiversity. The preservation of ecological processes within distinct forest ecosystems hinges on maintaining their biological diversity.

Biodiversity encompasses both the quantity and diversity of organisms within a specific region (Wilson 1992). This definition extends to the variability found within and among species, as well as within and among ecosystems. Variety and variability of living organisms are assessed at the following three levels:

Species diversity: A number of different taxonomic units in a given region.

Genetic diversity: Species with inter population differences, genetically adapted to specific habitat conditions.

Ecological diversity: Refers to diverse habitats that support distinct flora and fauna.

Biodiversity serves as the foundation for ecosystem services, constituting the life support system for human existence. The Millennium Ecosystem Assessment (MEA) in 2005 categorizes ecosystem services into four groups:

1. *Provisioning services*, encompassing goods like timber, honey, water, medicines, fiber;
2. *Regulating services*, which include regulation of water, climate, flood and control of disease, soil erosion;
3. *Cultural services*, offering benefits in religious activity, knowledge, awareness; and
4. *Supporting services*, involving nutrient balance, N-fixation, soil formation, and plant growth. Biodiversity has two main types of values: Trade values and environmental or quality-of-life values (Mendelsohn 2001). Trade values represent the role of biodiversity as an input to economic sectors like farming, development, and health. Quality-of-life values underscore how biodiversity enhances our lives aesthetically and contributes to ecosystem services, even if not directly utilized in the economy. The extinction of certain species and populations, though not directly contributing to market values, could result in significant losses for nations. Native species are those that live the habitat or community from where they have originated, whereas exotic species are deliberately or accidentally introduced into other countries where they are considered as invasive, non-native, alien or introduced species.

The Convention on Biodiversity (CBD) aims to conserve and sustainably utilize forest biodiversity, emphasizing the just and equitable distribution of benefits derived from using forest genetic resources. As a crucial component of biodiversity targets, there is a focused effort to maintain and improve the resilience of biodiversity components to adapt to climate change. It is noteworthy that biodiversity in forests is not solely influenced by actions within forest management regimes, whether planned or unplanned. Policies and management measures in other sectors can often yield unintended, and at times intended, consequences for forests and their biodiversity (CBD 2018). The biocultural diversity is linked with the biodiversity and human culture, food, religious belief system, language, and other dependent and interrelated complex socioecological adaptive system of nature which people live in. This idea is a major paradigm shift in biodiversity conservation strategies. A few decades ago, the general consensus was that the most suitable way to protect biodiversity was to keep people out of natural reserves like national parks and sanctuaries. However, today, conservation strategies are changing to see indigenous people as potential allies in biodiversity protection. There are many examples across the globe of successful community conservation projects that empower local people to actively protect and enhance their local environment.

Each culture has developed ways of adapting to their way of living, drawing on nature for goods, services, inspiration, mythology, and much else besides the identity of flora and fauna. Biocultural conservation represents a forward-thinking approach that acknowledges and respects the inherent connections between nature and humanity. This approach constructs conservation programs rooted in cultural values and aligned with community priorities. Indigenous communities have preserved their cultural and spiritual heritage, embodying a harmonious coexistence with nature across centuries. The livelihoods of millions are intricately tied to the biodiversity of their surroundings, making biodiversity conservation a national priority evident in policies, acts, and other legislative measures.

7.4 Current Status of Biodiversity in India

India, one of the 12 mega-diverse nations, is home to two of the world's 18 biodiversity hotspots, the Eastern Himalayas and the Western Ghats. At about 7% of the world's biodiversity, India is home to 81,000 animal species, 45,000 plant species (6000 of which are endemic), 50% of the world's tiger population, and all the Asiatic lion population. The diverse edaphic, topographic, and climatic conditions have given rise to a broad spectrum of ecosystems, such as grasslands, forests, coastal area, wetlands, marine, and desert ecosystems. The mountainous region is spread across 100 M ha, while arid and semi-arid zones span more than 30 M ha, complemented by an approximately 8000 km-long coastline. Indian biodiversity is sustained by (1) the Palearctic Realm, which is located in the Himalayas, and the Malayan Realm, which is located in the remaining sub-continent; (2) five biomes, which include warm deserts and semi-deserts, alpine meadows, tropical humid forests, and tropical dry deciduous forests (including monsoon forests); (3) ten biogeographic zones and 27 biogeographic provinces.

With around 45,500 plant species, India ranks fourth in Asia and tenth internationally in terms of plant diversity, making up approximately 11% of the world's flora (Table 7.1). Similar to global patterns, numerous organisms, particularly in lower plant groups like bacteria, algae, fungi, lichens, and bryophytes, remain incompletely described. Comprehensive exploration of remote geographical areas is still needed to enhance our understanding of the full spectrum of plant diversity in the country.

Table 7.1 Status of species in major groups of plants and microorganisms

Plant groups	No. of species described in India	No. of species described in the world	% of India to the world
Angiosperms	17,527	2,50,000	7.0
Gymnosperms	67	650	10.3
Pteridophytes	1200	10,000	12.0
Bryophytes	2500	14,500	17.2
Algae	7175	40,000	17.9
Lichen	2223	13,500	16.4
Fungi	14,500	72,000	20.1
Virus and bacteria	850	8050	10.6

Source: BSI (2009)

Major Indian categories of plant species are as below:

Angiosperms: India has around 17,500 species (7% of the total globally known) angiosperms. This botanical richness is distributed across 247 families and 2984 genera. Several dominant plant families play a significant role in this diversity, with more than 500 species each: Poaceae (Grasses): 1291 species, Orchidaceae (Orchids): 1229 species, Leguminosae (Legumes): 1225 species, Rubiaceae: 616 species, Cyperaceae (Sedges): 545 species, Euphorbiaceae: 527 species, Acanthaceae: 510 species, Asteraceae-892.

Gymnosperms: Approximately 67 species are represented by Pinaceae, Cupressaceae, Ephedraceae, and Gnetaceae. The Gnetaceae family includes one genus Gnetum, Cycadaceae includes Cycas, Pinaceae includes Pine & Deodar which are available in Himalayan region, Western Ghats and Andaman & Nicobar Islands.

Pteridophytes: India hosts a diverse array of pteridophytes, with approximately 1200 species distributed across 204 genera, like Marsilea, Azolla, and Salvinia, Acrostichum. The distinctive flora of India is greatly enhanced by the approximately 17% of these species that are native to the country. Prominent families and genera in the pteridophytic flora of the Indian subcontinent include Polypodiaceae, Dryopteridaceae, Athyriaceae, Thelypteridaceae, and Selaginellaceae. These families and genera play a pivotal role in shaping India's diverse pteridophytic landscape.

Bryophytes: These are mosses found mainly in moist, clean, and pollution-free habitats of North-eastern and Himalayan region including Western Ghats, having 2500 species. Nineteen genera and 629 species are endemic to India. The Indian bryo-flora has 1576 species, followed by liverworts and hornworts (924 species).

Lichens: These represent fungal and algal symbiosis; they comprise 2200 species, considered to be environmental indicator species found in pollution free areas mostly in the undisturbed Himalayan and southern region (Western Ghats) of the country; endemic species make up 23% of the species.

Fungi: With a great diversity found in the Western Ghats and the Himalayan region, they are home to the majority of India's 14,500 species of fungi, which are divided into 2300 genera and 250 families including subclasses Deuteromycetes, Ascomycetes, and Basidiomycetes. About 3500 species of Indian fungi are endemic, with Cercospora being the biggest genus with 707 species, followed by Puccinia (328 species) and Phyllosticta (280 species).

Algae: This group is spread across a wide range of habitats, from freshwater and marine to terrestrial, and are represented by over 7100 species. The largest family is Chlorophyceae, with 4495 species.

The Indian region boasts an impressive 11,058 species of endemic flora, of which 6200 are flowering plants alone. High levels of endemism can be found in regions such as parts of Himalayan region, peninsular India, and the A&N Islands. The status of endemism in different plant groups of India is given in Table 7.2.

As per Nayar (1997), the Indian wet evergreen forests, covering an area of 51,249 sq. km (belonging to 1.5% India's land), are home to approximately 7000 species of flowering plants. This constitutes nearly 50% angiospermic flora of India. The remarkable diversity of tropical forests is exemplified by the Silent Valley in the Western Ghats, where a mere 90 sq. km area harbors 966 flowering plant species (Manilal 1988).

Although Himalaya being an active speciation zone, still new variety of plants develops there due to suitable habitat condition. Further Indian biodiversity provided impressive status of the country comprising 7 World Heritage Sites (natural, total 40, 32 cultural), 18 biosphere reserves (BRs; 2022), 75 Ramsar wetlands, 106 national parks (IUCN Category II), 565 animal sanctuaries (IUCN Category IV) covering 8.1 million ha accounting for 5–27% of the geographical area. India's rich biodiversity spans diverse habitats, encompassing dry deserts, cold deserts, alpine regions, temperate zones, tropical habitats, wetlands covering 4.1 million hectares, and mangroves. Varied ecological conditions (high rainfall area, coldest place on earth, etc.) also support such diversity. India's biodiversity thrives across varied altitudes, ranging from sea level to the world's highest mountain peaks. Geological evidence highlights the Gondwana connection as

Table 7.2 Endemism in different plant groups of India

Plant group	Total no. of species in India	Number of endemic species reported	Percentage
Algae	7175	1925	26.8
Lichens	2223	527	23.7
Fungi	14,500	3500	24.0
Bryophytes	2500	629	25.1
Pteridophytes	1200	193	16.0
Gymnosperms	67	07	14.9
Angiosperms	17,527	6200	35.3

Source: BSI (2009)

a pivotal factor in the evolutionary processes that shaped the origin and dispersion of flora and fauna region.

Biodiversity hotspots are areas of unique biodiversity with high rates of endemic or monotypic biota but are threatened and prone to anthropogenic disturbances (Myers et al. 1988). Thirty five world biodiversity hotspots host 44% of all plant species which include:

- Unique germplasm centers
- Exceptional scientific interest
- Unique/specialized forest communities (wet humid tropical forests, tropical swamps, sacred forests)

Four of the 35 global biodiversity hotspots are located in India: the Nicobar Islands, parts of the Sundaland hotspot; parts of the northeast India in the Indo-Burma hotspot; the Himalayan range includes the notheast India, Bhutan, and Nepal; and the Western Ghats along the India's western coast.

Biodiversity hotspots, characterized by high species richness and notable endemism, provide a unique environment showcasing the evolutionary process of speciation. However, these areas face severe threats (Myers et al. 2000). The Himalaya hotspot includes all mountains greater than 8000 m globally and features some of the deepest river gorges globally. The rich and diverse Indo-Burma hotspot covers most of North-eastern India. The Western Ghats–Sri Lanka hotspot, and further 400 km into Sri Lanka, preserves the remaining tropical moist evergreen forests (rainforest) and corresponding high endemic biodiversity even with high anthropogenic pressure.

7.5 Why Biodiversity Is Important for Sustainable Forest Management

Every species has its role to play in the ecosystem, so making species diversity is very important for future sustainability. Biodiversity is crucial for the survival of every species, as organisms are interconnected in a delicate web of life. In ecosystems, various species create a food chain linking producers to consumers, playing vital roles. The extensive diversity of wild species enhances the potential for generating new varieties and ensures sustainability in the future.

An environment with higher species abundance is often perceived as more stable than one with lower species abundance. The necessity of biodiversity becomes evident when considering our dependence on the environment. Human reliance on various plant species for diverse needs and dependence on different animal and microbial species underscore the critical role of biodiversity in sustaining our way of life.

In the past, forests were primarily seen as sources of timber production. However, in recent decades, there has been a shift in perception towards a more multi-functional approach, emphasizing ecosystem functions and biodiversity conservation. A wider understanding of the many roles that forests play is reflected in the recognition of forest functions and services, under sustainable forest management (SFM).

Biodiversity faces threats from habitat loss, resource overexploitation, climate changes, pollution, invasive species, diseases, hunting, among other factors. Given its economic, ethical, and aesthetic significance, conserving biodiversity is crucial. This involves biodiversity management and realizing the significance of safeguarding ecosystem across different spatial scales. Forest certification schemes like Forest Stewardship Council (FSC) and Programme for the Endorsement of Forest Certification (PEFC) keep high conservation value approach in their certification standard for biodiversity conservation in Forest Management Unit (FMU) (Harrison et al. 2022).

7.6 Biodiversity in Various Forest Ecosystems

In India, mountainous regions span across the two global hotspots, the Himalaya and the Western Ghats. In close observation, within forest ecosystem as such we can identify various micro-climatic zones, and each zone based on moisture, sunlight, temperature, slopes, and altitudes ecosystem will have very specific biodiversity components. Biological diversity in the context of forest management refers to the variety and variability in the different floral and faunal functional groups.

7.6.1 Himalayan Mountain Forest Biodiversity

The Indian Himalayan Mountain (IHM) range encompasses diverse ecological and cultural landscapes, contributing to a rich biodiversity. Majestic Himalayan Mountain system covers 5.3 lakh km^2 area and administratively IHM covers 10 states of India. Himalayan flora is exceptionally diverse, featuring numerous medicinal and aromatic species like Primula and Rhododendron, along with a significant presence of orchids. Notably, the region constitutes 71 endemic genera and 32% endemic species. More than 90% of the species in Berberidaceae and Saxifragaceae are endemic to the Himalayas (Rawal et al. 2013).

The Eastern Himalayas, Sikkim, and Arunachal Pradesh have reported a significant orchid varieties (MoEF 2009). Notably, Cordyceps sinensis (Kira jari), a rare and unique fungus, thrives in alpine meadows above 3500 m asl for 2 months. This fungus holds high medicinal value and is particularly expensive, with a price ranging from 5 to 8 lakh rupees per kilogram. It is extensively used in Chinese and Tibetan traditional medicines.

The Himalayan region is home to about 300 recorded mammal species, 12 of which are endemic to the region. Among the varied faunal elements, the region is home to wild chicken and large mammals like Zebu, Mithun, and Yak. Endemic species include the Golden Langur, the Himalayan Tahr, the Pygmy Hog, the Flying Squirrel. The region boasts around 1200 butterfly species and 979 bird species, including 15 endemics. Four endemic bird areas (EBAs) partially or entirely overlap with the Himalayan hotspot. Among reptiles (177 species), 49 species are endemic.

The Himalayan region is home to approximately 60 million people, comprising over 800 culturally distinct communities and more than 200 ethnic groups. The livelihoods of the inhabitants in this region are directly dependent on a diverse plant and animal resource base. The ecosystem services and bio-resources originating from this area not only support the local population but also extend their benefits to the plains. Additionally, these resources play a substantial role in ecological and food security at various regional scales.

The Himalaya has not been immune to the effects of human activity, especially development projects that change land use patterns and cause habitat loss and fragmentation, despite its remoteness and difficulty of access. Animals like tigers and rhinoceroses are seriously threatened by poaching, and the unsustainable extraction of medicinal herbs is a growing source of concern. Even with its abundance and diversity of natural resources, the area is still underdeveloped, and the pervasive poverty there may be a factor in the deterioration of the environment.

7.6.2 Western Ghats Forest Biodiversity

The mountain range that stretches from the Vindhya-Satpura hills to the southernmost point of India's west coast is known as the Western Ghats. Its varied habitats include forest types such as tropical wet evergreen, montane evergreen, moist deciduous, and more. The Shola forest habitat, which is home to numerous endemic species, is particularly distinctive to this area. It is found in the upper elevations of the Sahyadri mountain range.

The World Conservation Monitoring Centre (WCMC) has recognized this region as a crucial area for freshwater biodiversity. Its diverse topography, climate, and geology have immensely contributed to local habitat, hosting about 30% Indian flowering plants. The Nilgiri Biosphere Reserve (NBR), which spans three states, is the first biosphere reserve established in India. Interestingly, 1500 species covering 56 genera of flowering plants, with 63% of Indian evergreen forest tree species, are endemic to the Western Ghats.

In the Western Ghats, there are 14 known endemic mammal species. The mammalian fauna of this region is characterized by a dominance of insectivores, bats, and rodents. Among the 508 recorded bird species in the Western Ghats, 144 are aquatic, and 324 are terrestrial. Sixteen bird species are exclusive to the Western Ghats. Out of the total known fauna, 102 species are categorized as threatened, with mammals (30 species; 21.9%) and amphibians (52 species; 33.3%) being prominent among the threatened groups.

Until recently, the forests of the Western Ghats underwent selective logging, and extensive portions were converted into agricultural lands for monoculture plantations such as eucalyptus, tea, coffee, cocoa, cashew, oil palm, and teak. Additionally, the region has been impacted by road construction, tourism infrastructure (hotels, resorts), railways, and the development of reservoirs. The overlapping requirements of the locals residing around the forest area contribute to human–wildlife conflicts in the area.

Approximately 20% of the original forest cover in the Western Ghats remain relatively pristine. This pressure includes the collection of firewood and non-timber forest products (NTFPs) for subsistence. Common threats to the region involve mass tourism, grazing, mining, and forest fires. Widespread poverty persists, though recent economic development has improved conditions in the areas adjacent to forests, including protected areas (PAs).

7.6.3 Mangrove Biodiversity

The mangrove ecosystem is an intricate combination of plant, animal, and microbes, along with their abiotic environment, functioning as an integrated unit. This ecosystem serves as a connection between land and sea water ecosystems and is located in the inter-tidal zones of shores, creeks, marshes, mud-flats, estuaries, backwaters, including lagoons. Mangroves are recognized as highly productive and biodiverse ecosystems, playing crucial roles as habitats, spawning grounds, nurseries, and nutrient sources for numerous animals. These habitats support various endangered species, including mammals (tigers, deer, otters, and dolphins), reptiles (like crocodiles, iguanas, and snakes), amphibians, and birds (herons, egrets). Only a few plant families, such as Rhizophoraceae, Avicenniaceae, and Combretaceae, are morphologically adapted to sustain in salt water habitat and tidal effects.

In India, mangrove vegetation constitutes approximately 5% of the global mangrove area, extending to about 4900 sq. km along the coasts and union territories of the country. West Bengal boasts the highest mangrove cover (42.45%) in the nation, followed by Gujarat (23.66%), A&N Islands (12.39%), Odisha (Bhitarkanika), and Tamil Nadu (Pichavaram). Sundarbans of West Bengal is considered as marvels of biodiversity; they are considered as UNESCO world heritage site due to unique habitat of Royal Bengal Tiger. Now Sundarban is facing tremendous challenges due to repeated cyclonic event due to climate change impact. Vigorous endeavors are imperative for the restoration of mangroves to realize ecological and local livelihood benefits, with a specific emphasis on sustainable mangrove management and improving the ecology of the area for enhancing fish and crab production.

7.6.4 Desert, Arid and Semi-arid Lands Biodiversity

India's arid and semi-arid regions encompass 127.3 million hectares, constituting 38.8% of the total area, and are distributed across 10 states. The hot arid zone predominates in Rajasthan (60%), Gujarat (20%), Punjab and Haryana (9%), and Andhra Pradesh, Karnataka, and Maharashtra (10%). Meanwhile, the cold arid zones are situated in the Trans-Himalayan region of Jammu & Kashmir, Himachal Pradesh, Uttarakhand, and Sikkim, covering an area of 184,823 square kilometers or 5.62% of the total geographical area.

The Thar Desert, located in the hot and semi-arid region, is the seventh biggest desert globally and is recognized as the most challenging eco-region. This extensive eco-region is situated west of the Aravalli Range and is characterized by extreme climatic conditions. Rainfall is scarce, typically ranging between 100 and 150 mm. The flora of the Indian desert includes 682 species (87 families, 352 genera, 86 angiosperm families, a single gymnosperm family).

The Thar Desert exhibits a relatively higher degree of endemism in its plant species at 6.4%, surpassing the 3% endemism observed in the Sahara Desert. Notable endemic species in the Thar Desert include *Calligonum polygonoides* (Polygonaceae), *Prosopis cineraria* (Mimosaceae), *Tecomella undulata* (Bignoniaceae), *Cenchrus biflorus* (Poaceae), and *Suaeda fruticosa* (Chenopodiaceae). Many arid, semi-arid, and dry sub-humid regions in India are prone to desertification, drought, or considered wastelands. In arid Rajasthan, approximately 92% of the area is affected by desertification, with 76% experiencing wind erosion and 13% facing water erosion. In neighboring Gujarat, about 93% of the area is affected by desertification, and 39% is impacted by water erosion, affecting both agricultural and wild lands.

The landscape is home to a diverse range of species that have adapted to its challenging conditions. The mammals include 41 species, featuring the lion, leopard, and tiger. The region boasts magnificent grasslands and serves as a sanctuary for the Great Indian Bustard (GIB, *Choriotis nigriceps*), a charismatic bird facing global threats. Other mammals include blackbuck, desert cat, antelope, gazelle, striped hyena, grey mongoose, wild ass, chinkara, caracal, and desert fox. The blackbuck, a globally threatened species, finds refuge in this area. Among the 140 bird species recorded, the GIB stands out as a globally threatened species. Additionally, the region serves as a migration flyway for cranes and flamingos.

Degradation has taken its toll on a substantial portion of India's 329 million hectares of geographical area, impacting at least one-third of it. The arid regions, encompassing 49.5 million hectares, bear the brunt of degradation, with notable impact in the western part of Rajasthan, and in arid Gujarat (6.22 million hectares).

Desertification process is biggest threat for local biodiversity in the area. Cyclical droughts, intense winds, nutrient-poor sandy soils, and heightened human and livestock demands for food, fodder, and fuelwood contribute to the excessive exploitation of delicate resources. This leads to issues like erosion, waterlogging, salinity-alkalinity, and degradation of habitat of flora.

Recognizing the significance of local flora and fauna in dry area, various protected areas have been set up in the region. This encompasses national parks covering 3162 sq. km and wildlife sanctuaries spanning 12,914 sq. km. Furthermore, the Rann of Kutch, extending over 12,454 sq. km, has been designated as the largest biosphere reserve (BR) in India.

7.6.5 Urban Biodiversity

The rapid changes in various parameters of the urban environment pose both opportunities and challenges for the species inhabiting it. These factors influence population size, abundance, density, and the coexistence of diverse species, thereby shaping urban biodiversity. Urbanization is very rapid in some continents of the globe, the urban areas of the world is more than 50%, and such a high rate of urban expansion is one of the causes of biodiversity loss. Biodiversity is required by the city dwellers almost same manner as we see in rural areas. Urban green spaces (UGS) protect and maintain the diverse biodiversity of surroundings in their various habitats like green habitat, blue habitat, and brown habitat. UGSs serve as a biodiversity hotspot for significant bird and bee population (Aronson et al. 2017), promoting pollination and seed dissemination (Tian et al. 2020). Biodiversity in urban areas plays a crucial role in regulating atmospheric composition, temperature, hydrological cycles, soil formation, and nutrient cycling. Additionally, it contributes to essential services such as pollination, while urban green spaces provide recreational and cultural opportunities. These areas not only promote healthy living and community spirit but also act as green lungs, fostering biodiversity in the heart of the city. Urban areas require per capita 9 sq. m green space area. They are an indispensable component of cities which play an important role due to their site specific plants and animals in urban areas (Tian et al. 2020; Zhang and Ramírez 2019). People get benefited from UGSs in terms of physical and mental health because they are popular as venue for physical activities while also providing peacefulness for enhanced restfulness, mental health, and cognitive growth.

In New Delhi, the capital city of India, there are over 18,000 parks and gardens covering approximately 8000 hectares. These green spaces are significant sources of biodiversity, hosting a variety of plants and animals. Diversity is the key to safeguard any ecosystem. For example, in an urban park area 100 individual birds of three species are available, whereas, in another park only 50 birds of 12 different species available, the second area is having more biodiversity. The same is happening in urban area; only a few birds like pigeon, myna, and crow are mostly available in large numbers (Bhalla and Bhattacharya 2015). Habitat modification may restrict some species to small pockets of the urban environment. Gradual slow extinction of a few birds and other animals often go unnoticed by local people and their absence is felt only a long time after their disappearance. Various agencies, including the Municipal Corporation of Delhi (MCD), Forest Department, Delhi Development Authority (DDA), Public Works Department (PWD), Central Public Works Department (CPWD), and Delhi Cantonment Board (DCB), are involved in maintaining these areas. Residents Welfare Associations (RWAs) also actively participate in the upkeep of local neighborhood parks in collaboration with government agencies. To coordinate, manage, and monitor these urban green spaces, the government of the National Capital Territory (NCT) of Delhi established the "Delhi Parks and Gardens Society" in 2008 (Anand and Bhattacharya 2021).

The Indian State of Forest Report (2021) indicates that Delhi has approximately 195 square kilometers of forest cover inside or outside the recorded forest area. The city encompasses various types of urban green spaces, including reserved forests, institutional areas, roadside plantations, protected forests, society parks, neighborhood parks, tot lots, gardens, green belts, and avenue plantations. Common tree species found in these green spaces are Neem, Gulbakabali, Amaltas, Ashok, Shehtoot, Peepal, Australian acacia, Arjun, Jamun, Mango, Sissoo, Karanj, Gulmohar, Kali Siris, and Semul.

7.6.6 Urban Biodiversity Park

Establishment of city biodiversity parks as a new model of urban biodiversity conservation has been practiced in India for the past two decades. Implementing an in situ conservation strategy for urban green spaces and promoting the development of flora and fauna requires a scientific approach. In the NCT of Delhi, Aravalli Biodiversity Park and Yamuna Biodiversity Park were established in 2005 to change the areas occupied by *Lantana* and *Prosopis* (both are invasive species) and to conserve the ecosystem and bringing back the original native species of the area.

7.7 Biodiversity Measurement and Monitoring Study

Biodiversity indicators are essential for gauging the health and condition of biodiversity in a specific area. They encompass various aspects, including species diversity, rare and threatened species, habitat quality and extent, threat levels, compliance with regulations, ecosystem services, community engagement, and ecological resilience. These indicators help assess the effectiveness of conservation efforts and provide a comprehensive understanding of biodiversity status. In the planning process of forest management, for understanding the health of forest, biodiversity status could be a crucial parameter to study a forest area. For achieving desirable objectives for forest management, foresters should follow procedure based on current condition of the place, long-term interest of the government, and the local communities. Following aspects are relevant to take care of biodiversity planning:

- Make an inventory of mature trees, threatened areas, and natural forest areas and plantations.
- Survey of other natural areas (e.g., open grazing/community land and wetlands).
- Provision kept for protection of cultural values of local people.

- Understanding about connections within the forest management unit and across other landscape including grassland.

7.8 Measurements Through Permanent and Temporary Plots

Using the quadrat method facilitates the evaluation of biodiversity through parameters such as species richness, abundance, the Simpson index, and the Shannon-Wiener index. Species richness involves counting the number of species in a plot or quadrat. The Simpson and Shannon-Wiener indices consider both richness and the distribution of individuals among species. Calculating these indices requires information of species and their number of species and corresponding number of individuals across sites in various seasons.

7.8.1 Primary Analysis of Vegetation

The biodiversity is assessed through collection of the basic data with regard to vegetation in the sample plots laid under different management regimes. The data gathered is used to determining abundance, density, and frequency for each tree species using the below mentioned formulae (Curtis and McIntosh 1950).

7.8.1.1 Frequency

It is the spread of an individual species in an area and is expressed in terms of percentage occurrence. It is calculated by equation:

$$\text{Frequency} = \frac{\text{Number of sample plots in which species occurred}}{\text{Total number of sample plots studied}} \times 100$$

7.8.1.2 Density

Density represents the number of individuals of a species in a unit area. It represents the numerical strength of the species in the community.

$$\text{Density} = \frac{\text{Total number of individuals of a species in all plots}}{\text{Total number of sample plots studied}}$$

7.8.1.3 Abundance

Abundance is the measure of the number of individuals of a species per sampling unit of occurrence. It is calculated as below:

$$\text{Abundance} = \frac{\text{Total number of individuals of a species in all plots}}{\text{Total number of sample plots studied}}$$

7.8.1.4 Basal Area

Basal area pertains to the ground area covered with tree stems. It is a reliable indicator of the size, volume, or weight of a tree. Basal area offers insights into the relative dominance of larger and smaller trees within a community. This parameter holds significance in estimating the standing biomass in a given area and serves as a metric for productivity. The basal area is calculated by using formula:

$$\text{Basal area} = \frac{(\text{CBH})^2}{4\pi}$$

where CBH = circumference at breast height.

7.8.1.5 Importance Value Index

The Importance Value Index (IVI) of a species within a community provides insight into its relative importance in comparison to other species (Curtis 1959). This statistical measure offers a comprehensive assessment of a species' significance within the plant community. In calculation relative values are used instead of absolute values. The percentage values of relative density, relative frequency, and relative dominance for a species are summed, yielding the importance value index. IVI plays a pivotal role in determining the vegetation status and the importance of component species within a vegetation. It, thus, incorporates following three parameters:

$$\text{Relative density} = \frac{\text{Density of the species}}{\text{Total density of all species}} \times 100$$

$$\text{Relative frequency} = \frac{\text{Frequency of the species}}{\text{Total frequency of all species}} \times 100$$

$$\text{Relative dominance} = \frac{\text{Dominance of the species}}{\text{Total dominance of all species}} \times 100$$

$$\text{IVI} = \text{Relative density} + \text{Relative frequency} + \text{Relative dominance}$$

7.8.1.6 Species Diversity

Species diversity is a measure of the diversity within an ecological unit. It is calculated as:

1. **Shannon-Wiener index**: It is used to calculate the species diversity in the community and is represented by *H* (Shannon and Weaver 1949).

$$H = -\sum P_i \ln P_i$$

where, $P_i = n^I/N$. n^I is the number of individuals of the species and N is the total number of individuals of all species.

2. **Simpson's index (1949)**

$$D_s = \sum_{i=1}^{s} p_2$$

where, $p_i = n_i/N$ (proportional abundance of ith species), S = number of species, N_i = the total number of organisms of a particular species, N = the total number of organisms of all species.

7.8.1.7 Species Richness

Species richness is the number of species in a sample or habitat per unit area. Higher the value, greater the species richness. The indices used:

1. **Margalef's Index (1968)**

$$Da = \frac{s-1}{\ln(N)}$$

where S = total number of species, N = total number of individuals in the sample, ln = natural logarithm.

2. **Menhinick's Index (1964)**

$$Db = \frac{s}{\sqrt{N}}$$

where s = number of species, N = total number of individuals.

7.8.2 Geospatial Approach

Satellite imagery with extensive coverage introduces a new aspect to biodiversity assessment. Maps of vegetation types derived from satellite data can serve as crucial inputs for a biodiversity assessment at the landscape level. Various vegetation indices are used like NDVI (Normalized Difference Vegetation Index), to identify vegetation health status, along with other indices to analyze different land use and land cover changes and trends.

7.8.3 Temporal Assessment

Biodiversity monitoring, focusing on the temporal scale, can be approached in three ways: (1) through consistent surveys of permanent plots in various habitats and forest groups; (2) by seasonal monitoring of keystone and indicator species; and (3) by employing geo-spatial methods across expansive regions. The camera trap technique is utilized for repeated censuses of large mammals (e.g., bison, lion, leopard, rhino, tiger, antelope) despite limited accuracy, and further technological refinement is needed for improvement.

Monitoring and assessment of keystone species like Ficus sp. in the mixed miscellaneous forest of Madhya Pradesh, Chhattisgarh, Odisha or Oaks (Quercus sp.) in Himalaya area and Sal (Shorea robusta) in tropical moist deciduous forests. Monitoring of exotic weeds such as *Prosopis juliflora, Lantana camara, Euphatorium odoratum, Mimosa invasa*, and *Parthenium hysterophorus* are key to evaluating changes in local flora. To study biodiversity over time, it is essential to correlate the occurence and population size of the selected species with the biodiversity level.

7.8.4 Participatory Assessment

7.8.4.1 Participatory Assessment Monitoring and Learning (PALM)

It is one of the Participatory Rural Appraisal (PRA) methods used along with local inhabitants in natural forest areas, wherein villagers monitor the biodiversity of the Joint Forest Management (JFM) areas regarding plants and animals including bird species available in their forest areas. This provides first-hand information to local community about all landscape components and biodiversity highlighlighting how new regenration is restoring the native species in their area. Their indigenous knowledge always helps in conservation of local species in different season. They even do the yearly monitoring of biodiversity components, including water bodies which they utilize for their day-to-day needs. Villagers may also do the documentation of yearly monitoring data in local language place in sign board at the village meeting, so that all members of village understand about the biodiversity status and changes in their village. In such a monitoring, local community actively participates under the guidance of field experts, which is more effective than the monitoring by the outside consultants, with limited knowledge about local flora, fauna, and history of the forest area.

7.8.5 People's Biodiversity Registers

The Biodiversity Rules of 2004 have introduced an inventive method for cataloging biodiversity at the local level known as People's Biodiversity Register (PBR) or Panchayat Biodiversity Register. This initiative was initially launched by the Foundation for Revitalization of Local Health Traditions (FRLHT) in collaboration with Indian Institute of Science,

Bangalore (Utkarsh et al. 1999), this program has been implemented by various State Biodiversity Boards since 2010, resulting in nearly 10,520 PBRs by 2020. The knowledge of availability of species, seasonality, utilization, habitat of occurrence, practices, harvest methods and preservation, as well as local uses of bioresources is available in each panchayat are documented by involving local communities, school teachers, botanists, zoologists, and foresters (Saha and Bhattacharya 2011).

7.9 Approaches to Forest Biodiversity Conservation and Management

It is obvious that the natural forests have more biodiversity than the plantation forest. However, effectively managed forests have the potential to sustain biodiversity levels like those found in untouched forests, contributing to essential ecosystem services (Gibson et al. 2011). Unfortunately, the management and conservation of genetic resources and genetic diversity in productive forests are aspects frequently neglected in forest biodiversity conservation. The crucial initial step in establishing a biodiversity management plan is to identify and define biodiversity objectives, forming an integral component of both short-term and long-term forest planning processes.

Biodiversity conservation is a very complex process which demands scientific measures to improve the composition, diversity, and sustainability. Generally, biodiversity can be conserved in the following ways:

- In situ conservation
- Ex situ conservation

7.9.1 In Situ Conservation

In situ conservation of biodiversity involves the preservation of species within their natural habitats, maintaining and protecting the natural ecosystem. The in situ conservation has the following advantages.

In situ conservation proves to be a cost-effective and convenient method for preserving biodiversity. This approach allows the simultaneous conservation of a large number of living organisms. Being in their natural ecosystem, these organisms can evolve more effectively and readily adapt to diverse environmental conditions. Designated protected areas like national parks, wildlife sanctuaries, and biosphere reserves serve as important sites for in situ conservation efforts.

India has strategically employed its protected area network to manage natural resources for biodiversity conservation and the welfare of populations dependent on these resources. The country boasts a network of 770 protected areas (PAs), covering 1,62,365.49 km^2, equivalent to 5.2% of the total geographical area. This network includes 103 national parks, 544 wildlife sanctuaries, 46 community reserves, and 77 conservation reserves. Among these, there are 39 tiger reserves and 28 elephant reserves, and India hosts 6 World Heritage Sites under UNESCO's framework. Both scientific monitoring and traditional observations affirm the restoration of depleted natural resources and the sustained preservation of pristine ecological conditions. Notably, improved conservation effectiveness has yielded positive outcomes for various species. Tiger and elephant populations have shown an increase in recent years, and the conservation status of the Indian rhino has been upgraded from endangered to vulnerable.

India has six natural World Heritage Sites recognized for their "Outstanding Universal Values." These sites include Kaziranga National Park, Manas National Park, Keoladeo National Park, Nandadevi National Park, Sundarbans National Park, and the Western Ghats serial site. Additionally, several other natural sites in India are being tentatively assessed and evaluated for potential inclusion as World Heritage Sites. Furthermore, India has identified 12 Transboundary Protected Areas, and collaborative efforts through bilateral and/or multilateral cooperation have been initiated with neighboring nations in this regard.

7.9.1.1 National Parks

These are compact reserves managed by the government with clearly defined boundaries. Activities such as grazing, forestry, and habitat cultivation are strictly prohibited within these areas, for example, include Kanha National Park and Bandipur National Park.

7.9.1.2 Wildlife Sanctuary

In such areas, only certain activities like timber harvest, NTFP collection is allowed without disruption of the conservation efforts. These areas also attract tourists for recreational purposes. Table 7.3 shows the number and area covered by four categories of protected areas under Wildlife Protection Act (1972).

India has various areas, distinct from traditional protected areas, that fall under the definition of other effective area-based conservation methods (OECM) as per the CBD. These geographically defined regions, managed in diverse ways, aim to achieve positive and sustained long-term outcomes for in situ biodiversity conservation, including associated ecosystem functions, services, and relevant cultural, spiritual, socioeconomic values. These areas exist across different states and are conserved for diverse purposes, each governed by unique mechanisms.

Communities actively conserve community conserved areas (CCAs) for cultural, religious, and livelihood purposes,

Table 7.3 Number and area covered under different protected areas in India

Category	Number of reserves	Area (km^2)
National parks	106	44,402
Wildlife sanctuaries	573	123,762
Community reserves	220	1455
Conservation reserves	115	5548
Total protected areas	1014	1,75,167

employing customary practices. Specific legal/administrative arrangements in some states, such as the Village Council Act in Nagaland, the Panchayat Proclamation Act in Arunachal Pradesh, and the Van Panchayat Act in Uttarakhand, aid in protecting CCAs. Another significant category is "Sacred Groves," recognized by various names in different regions of India. With over 7000 sacred groves documented, many preserved in their pristine form through generations; experts estimate the actual number to be between 0.1 and 0.15 million.

7.9.1.3 Biosphere Reserves

Biosphere reserves, supported by UNESCO, are multipurpose areas that safeguard wildlife, the traditional knowledge of local inhabitants, and domesticated plants and animals. These reserves allow for both tourist and research activities. Currently, there are 18 designated biosphere reserves, covering an area of 89,531 km^2. Ten of these reserves are part of the World Network of Biosphere Reserves.

7.9.1.4 Sacred Groves and Community Reserves

Several Indian societies hold a deep reverence for nature and natural elements, considering them sacred and even worshiping them as deities. In alignment with these traditions, certain communities have designated forest patches as sacred groves, dedicated to ancestral spirits and deities. These groves are typically maintained in their natural state, representing a climax stage of vegetation succession. Estimates suggest that India may have thousands of sacred groves, covering a significant percentage of the country's geographical area. Various states, such as Meghalaya, Kerala, Maharashtra, Tamil Nadu, Uttarakhand, and Himachal Pradesh, host sacred groves with diverse floral and faunal biodiversity. The management and protection of these groves vary, with some being overseen by Gram Panchayats, communities, or clans, depending on the region (Malhotra 1998; Pushpangadan et al. 1998; Ramakrishnan 1998).

The preservation of sacred groves in India faces various challenges, including the erosion of cultural values, forest degradation from canopy opening, livestock grazing leading to invasive weed infestations, agricultural encroachment, and natural disasters. To address these issues and conserve sacred groves, potential strategies include conducting a comprehensive inventory of groves in all biogeographic zones; promoting mass education and revitalizing cultural values among local populations; developing site-specific conservation and restoration plans in collaboration with communities; and offering economic incentives to communities safeguarding and managing sacred groves, thereby ensuring the sustainability of these ecosystems for future generations.

The National AYUSH Mission and National Medicinal Plant Board (NMPB) have facilitated the establishment of 110 Medicinal Plants Conservation Areas (MPCAs), each averaging 200 hectares in size, covering 34 forest types. These areas serve as habitats for wild breeding populations including flagship threatened species. As part of the scheme's objectives to expand the cultivation, conservation, utilization, and knowledge advancement of medicinal plants, 57 MPCDAs have been established, covering an area of 10,702.15 hectares.

The wealth of biodiversity contributes to the utilization of nearly 8000 medicinal plants for health and various purposes in various AYUSH systems. The reliance on sourcing medicinal plants from the wild has decreased steadily, thanks to the implementation of various conservation and cultivation techniques since the 1990s.

7.9.2 Ex Situ Conservation

Ex situ conservation entails the breeding and upkeep of endangered species within artificial ecosystems, including zoos, nurseries, botanical gardens, gene banks, and more. This method reduces competition for resources like food, water, and space among organisms. In the realm of ex situ conservation, various national gene banks have been established for diverse forms of life, including plants, animals, insects, fish, and agriculturally-significant microorganisms. These gene banks collectively safeguard 366,933 unique accessions of plant genetic resources and 2517 microorganisms.

Ex situ conservation has the following advantages:

1. The animals are provided with a longer time and breeding activity.
2. The species bred in captivity can be reintroduced in the wild.
3. Genetic techniques can be used for the preservation of endangered species.

The National Bureau of Plant Genetic Resources (NBPGR), which was founded as the first national gene bank in India, is in charge of conserving both cultivated plants and their wild counterparts. The National Facility for Plant Tissue Culture Repository (NFPTCR) is engaged in cryo-preservation activities as well as in vitro culture of plants with ecological significance (Chandel 1996).

Numerous botanical gardens and arboreta play a vital role in the preservation of plant species across diverse categories for scientific documentation. The National Bio-resource Development Board (NBDB) established in 1999 is conferred the task of devising suitable strategies for conservation efforts for valuable biological resources.

7.10 Threats to Forest Biodiversity

The depletion of biodiversity is attributed to factors such as habitat loss, resource over-exploitation, climate change, pollution, invasive species, diseases, and hunting. Given the multitude of economic, ethical, and aesthetic benefits derived from biodiversity, its conservation becomes imperative. In addition to specific species, communities and systems as a whole are also facing degradation and decreased resilience. Every year, approximately ten million hectares of land are subjected to deforestation, leading to the substantial loss of forest biodiversity and the services provided by ecosystems (Gibson et al. 2011; FAO 2020).

7.10.1 Indian Biodiversity Loss

India's per capita forest area is approximately 0.064 hectares, significantly lower than the global average of 0.64 hectares. Around 78% of forest land in the country experiences intense grazing, and about ten million hectares are affected by shifting cultivation. Furthermore, extensive areas continue to be converted to developmental and infrastructure sites.

The unsustainable collection of fuelwood from forests has emerged as a significant issue, especially in the last decade, contributing to forest degradation and consequent biodiversity loss. The heightened biotic pressure, characterized by dense human and cattle populations, is a major factor contributing to the decline in forest cover. These factors not only lead to the loss of biodiversity but also contribute to forest fragmentation. This fragmentation causes disruption in the gene flow resulting in genetic impoverishment.

Forest timber plantations, featuring species like eucalyptus, teak, rosewood, conifers, and exotics such as rubber, acacia, alder, and silver oak pose significant threats to natural diversity. In addition, the biological invasive species like *Lantana, Mimosa,* and *Euphatorium* are notably endangering native biodiversity.

Approximately 77% of forests in the Western Ghats–Sri Lanka region and 75% in the Himalayas have been lost or degraded primarily due to human encroachment. Biodiversity depletion is also driven by trade in economically and medicinally valued species. The international trade encompasses a wide range, including live plants and animals, various food products, exotic leather, animal products, wooden musical instruments, timber, tourist curios, and medicines. The combination of high levels of exploitation, trade, and habitat loss can severely deplete populations of certain plant species and push them close to extinction. The following are some of the key causes of biodiversity loss.

7.10.1.1 Unsustainable Harvesting

A long list of biodiversity products available in forest areas such as NTFPs are often over harvested by the action of local people, traders and herbal industries. There are three ways of unsustainable harvesting: over harvesting (more amount of collection than it should be), pre-harvesting—in this case before the maturity of the product (fruit, roots, barks, etc.) they are harvested, faulty method of harvesting—mechanism of collection is not proper (e.g., cutting of whole branch for harvesting of Anola fruits from a tree) which result in death of the plant or limited produce in subsequent years. The direct harvesting and utilization of species, whether for timber, non-timber forest products (NTFPs), animal meat, or the wildlife trade, can result in significant negative impacts on their populations (Ticktin 2004; Shackleton et al. 2018). These impacts include the erosion of genetic diversity (Thomas et al. 2014; Chiriboga-Arroyo et al. 2021). Several highly sought-after timber species such as rosewood, red sanders, and agar trees are among the 76 species that are currently threatened. Overharvesting of animals for bushmeat and medicinal plants, as well as the pet trade in many countries, is a pervasive problem, particularly in tropical and subtropical forests.

7.10.1.2 Invasive and Alien Species

The term "invasive species" refers to non-native species that have the potential to cause sociocultural, economic, environmental, or health harm upon introduction. These species can have negative effects on new environments because they lack natural enemies, which normally maintain population balance, or because their new host trees are unable or unwilling to resist them. As a result, invasive species can have negative effects on forests and the related goods and services they support. Exotic species refer to those introduced from one geographic region to an area outside their natural range. Exotic species are deliberately or accidentally introduced into other countries where they may also be referred as invasive, non-native, alien or introduced species, such a group of species are very important to understand the biodiversity of the area. Worldwide, the economic cost of invasive species, encompassing losses in crop, pasture, and forest production, along with expenses for environmental damage and control efforts, is estimated to be USD 1.288 trillion over the last 50 years (Zenni et al. 2021). The characteristic which supports exotics for survival is high adaptability and high reproductive success to adopt a new environment. Exotics are usually fast growing and they compete for food and habitat with indigenous populations. They establish in disturbed ecosystem with niches or in early successional stage. They are usually not comfortable in the stable ecosystem with complex food webs. They have shorter life cycle, long survival capacity, high dispersal rate. Some of the common exotic species are Bilati Babool (*Prosopis juliflora*), water hyacinth (*Eichhornia crassipes*) native to Brazil, *Ipomoea fistulash* introduced from Mauritius, *Argemone mexicana* from Mexico, *Lantana camara* from North America, *Parthenium hysterophorous* from North America, *Eucalyptus* species from Australia.

Native species live in the habitat or community from where they have originated. In numerous instances, introduced species assimilate into the native flora or fauna without causing measurable harm. Over time, these new species may become naturalized during their journey. In the forest habitat, they compete for food and habitat, may predate on the indigenous populations, transmit parasites and diseases, and have greater viability and reproductive success. Invasive species introduced in an area can create genetic pollution as they often breed with native relatives to produce hybrids, altering the genetic pool irreversibly leading to loss of genetic diversity of the place. Therefore, naturally occurring species can be threatened by genetic pollution due to regeneration/breeding of wild species.

7.10.1.3 Forest Fire

Large uncontrolled forest fires have become a significant factor in the drastic degradation of forest bio-resources, leading to increased negative impacts globally in recent decades. Biodiversity, especially in dry deciduous forests, faces a major threat from the extensive destruction caused by these fires, resulting in significant losses of flora and fauna. Usually fire occurs in probability and tendency of spreading damage is quite high, from mid-March to mid-June. Most of the human-made fires are associated with acts like collection of mahua flowers from forest floor, sal seed collection, promoting better grass growth after the rains and increasing flush of tendu leaves. Therefore, it is essential to identify the setters of fire on dense forest land and these kinds of acts should be placed in the offenses under various laws. The results from the spatial statistical analysis in fire research show that fire affects first teak mixed dry deciduous forests and then comes bamboo, teak, sal mixed dry deciduous and finally mixed plantation vegetation types. The occurrence of fire due to the thick litter layer which serves as excellent fuel for fire to occur which is a typical characteristics in tropical dry deciduous forest.

In India, the role of forest fire management is not adequately acknowledged, and fires are still perceived as harmful to forest health. This perspective persists despite the globally accepted notion of prescribed fires and diverse fire regimes, recognizing their role in promoting biodiversity and fostering hardy timber species. While acknowledging potential adverse effects on timber trees and their regeneration in the short term, re-evaluation of the overall role of fires in the functioning of forest systems in India is required. Shifting or jhoom cultivation uses native seeds for cultivation, which is much prevalent in forest areas of north-east part of India, is also responsible for fire.

The forest fire is a natural part of the earth's ecosystem and has mostly negative impact on the local flora and fauna, though they also have a few positive impacts as well. The negative impacts include the habitat loss for animals; many endangered or RET species die just after fire happens in any area. It also damages soil profile, soil erosion take place and loss of plant biodiversity particularly lower plants like mosses, fern, fungi which perish soon after the fire incidence.

Uncontrolled burn can result in millions of dollars' worth damage to the habitation. Many species like pines, acacias as pioneer species come in the forests in large number. The ash left post fire puts nutrients in to the soil, may kill invasive species, allowing native biodiversity to regain territory. Fire indirectly promotes biological diversity and healthy ecosystem. It removes dry, dead wreaked trees and plants to create a more open space on the forest floor which allows sunlight for better growth.

7.10.1.4 Climate Change Impact on Biodiversity

The interconnection between climate change and biodiversity is evident, especially concerning forests. Forests serve as both carbon sinks and are susceptible to the impacts of climate change, making them a crucial consideration for various governments and organizations. Recent studies indicate a worldwide reduction in forest resilience due to climate change, particularly affecting tropical, arid, and temperate forests, leading to declines in primary productivity (Forzieri et al. 2022).

The Intergovernmental Panel on Climate Change (IPCC) says that the conservation and management of forests offer an effective means to significantly reduce emissions, with potential co-benefits for adaptation and sustainable development, namely, ways to reduce land degradation and forest conservation and restoration can enhance co-benefits. Furthermore, forests play a crucial role in climate change adaptation by retaining ecosystem services and providing livelihood options.

Climate change is expected to have an impact on forest biodiversity as well as the capacity of forests to provide habitat for species, protection of soil and water, and other ecosystem services. Certain forest ecosystems, including mangroves, boreal forests, tropical forests, cloud forests, and dry forests, are identified as particularly vulnerable.

Dry forests face potential negative impacts, given that their soils are susceptible to erosion. Drylands cover 41% of the earth's land area and support over two billion people. Human activities such as fire, grazing, agriculture, and firewood collection have significantly altered many dry forests, leading to degradation and desertification threats. Forests managed for diversity and multiple benefits, as well as primary forests with high biodiversity tend to be more resilient to disturbances. These ecosystems maintain healthy soils, provide natural habitats for forest biodiversity, and serve as a stable carbon store.

7.10.1.5 Habitat Fragmentation and Destruction of Biodiversity Corridor

In forests, large continuous habitats of species that get divided into smaller, scattered and isolated patches are called habitat fragmentation, which happens largely due to anthropogenic causes like construction of road, dam, long boundary walls, and other infrastructure. Fragmentation divides the animal and plant populations into smaller isolated groups with reduced size. The consequence is breaking up of free flow of biological association making the population vulnerable to extinction as it prevents gene flow among the fragmented patches.

Extreme land-use alteration like fragmentation has directly affected the persistency of biodiversity and ecological processes. Plantations could be the alternative to reestablish the corridor along with structural modification in the habitat which has been fragmented to develop biodiversity again in the same place. Plantation forests typically exhibit a uniform age structure and low species and genetic diversity. Over time, the biodiversity value of plantation stands tends to increase due to processes like under storey natural regeneration and colonization by epiphytes on trunks and branches. Large trees within plantations may also offer additional wildlife resources, such as nest sites. The potential of plantations to support biodiversity is limited by a number of factors, despite their regular management cycle. However, through suitable design and management practices, the biodiversity value of fragmented plantation sites can be significantly enhanced.

7.11 Biodiversity Conservation Policy, Legislation, and Acts

The formulation of policies and their proper implementation through different schemes, in the respective countries towards conservation of biodiversity has been in place. Globalization of the forest sector in many developed countries has primarily focused on the effects of using exotic plantation species and maintaining forest cover as a holistic consideration, rather than specifically improving biodiversity. In this context, careful policy analysis of multi-lateral trade agreements is necessary to ensure that the impacts on forests and biodiversity are minimized. Current forest certification standards keep the importance and safeguard the biodiversity, a high priority area in their mechanism of achieving sustainable forest management (SFM).

The Biological Diversity Act, 2002, was formulated by India in response to the CBD. The primary objectives of the act include the preservation of biological diversity, sustainable use of its components, and the fair and equitable sharing of benefits arising from the utilization of genetic resources. Sustainable use, as defined in section 2(o), refers to the utilization of components of biological diversity in a manner and at a rate that avoids the long-term decline of biodiversity, ensuring its potential to meet the needs and aspirations of present and future generations. The act envisages a three tier institutional structure for regulation of access to the biodiversity resources, comprising of:

1. National Biodiversity Authority (NBA)
2. State Biodiversity Boards (SBB)
3. Biodiversity Management Committees (BMC) at the local level

Many focal areas under biodiversity act which government follows through the above instruments particularly for the Intellectual Property Right (IPR), bio-utilization, research, transfer and exchange of bio-resources, and Access and Benefit Sharing (ABS).

Biodiversity Rules, 2004: Under the authority granted by Section 62 of the Biological Diversity Act, 2002, and superseding the National Biodiversity Authority Rules, 2003, the central government established new rules, which became effective on April 15, 2004. The Biodiversity Rules have following points developed for action:

- Procedure for seeking prior approval before applying for intellectual property protection
- Criteria for equitable benefit sharing
- Approval for settlement of dispute under section 50
- Safeguarding Traditional Knowledge (TK) linked to biological resources—requiring suitable legal and institutional mechanisms to acknowledge the rights of tribal communities over their traditional knowledge related to biological resources on the global stage.

Article 6 of the CBD necessitates the formulation of "National Biodiversity Strategy and Action Plans" (NBSAPs), a crucial document mandated for creation by individual countries to implement the CBD's goals through their national strategies and action plans. Over 190 nations have committed to enhancing efforts to incorporate biodiversity into the policies of various sectors such as forestry,

Table 7.4 India's legislations related to biodiversity conservation

Name of the legislation	Key features
Wildlife (Protection) Act, 1972	Addresses the safeguarding of wildlife and habitats and outlines provisions for the safeguarding of wild animals, birds, plants, and associated aspects, aiming to guarantee the ecological and environmental security of the nation
Forest (Conservation) Act, 1980	Designed for the conservation of forests and related matters
Biological Diversity Act, 2002	Enables the conservation of biological diversity, sustainable utilization of its components, and fair distribution of benefits arising from the utilization of biological resources, knowledge, and associated aspects
Biological Diversity Rules, 2004	Deals with internationalizing the Biological Diversity Act

fisheries, agriculture, tourism, and more. The Sharm El-Sheikh Declaration in 2018 expanded this commitment to additional sectors like energy, infrastructure, manufacturing, and processing. India's NBSAP aimed to integrate biodiversity issues under varied sectors and ensure democratic and equitable decision-making for access to biological resources and associated benefits. The NBSAP is a project of the Ministry of Environment and Forests (MoEF) of the Government of India, funded by the Global Environment Facility (GEF) through the United Nations Development Programme (UNDP). Table 7.4 presents a list of India's legislations on biodiversity conservation.

Following legislations of India further support biodiversity conservation:

- The National Working Plan Code, 2014 for Sustainable Management of Forest and Biodiversity in India incorporates strategies for managing invasive species to uphold and improve forest health and vitality.
- The National Assessment of Tigers, which is conducted every fourth year since 2006, is an integral part of evaluating co-predators, prey, and their habitats. Trained biologists and forest staff systematically sample all the forest patches within the tiger ranges across the country.
- The Plant Quarantine (Regulation of Import into India) Order, 2003, and the Sub-Mission on Plant Protection and Plant Quarantine play a crucial role in preventing the introduction of exotic pests, diseases, and weeds that may be introduced through the import of agricultural commodities or plant materials into India. These measures also fulfill obligations under the International Plant Protection Convention (IPCC).
- Plants, Fruits & Seeds (Regulation of Import in India) Order 1989 (PFS Order 1989): Regulates the import of plants, fruits, or seeds in India.
- The Destructive Insects and Pests Act, 1914, along with its amendments, serves the purpose of preventing the introduction into and transport from one state to another in India of any insects, fungus, or other pests that may pose a threat to forest plantations and agricultural crops.

7.12 Capacity Building in Biodiversity Conservation

The science of biodiversity is an emerging area for education in the field of environment. In the past three decades lots of skill and capacity building started in the subject of biodiversity. There are several universities started offering master's program in Biodiversity Conservation, Natural Resource Management, Wetland Management, Wildlife Management, Forestry, Sustainable Development, etc. and all these subjects have a great role to play. These subjects provide reasonable job opportunity for the future generation. Improving biodiversity outcomes can be achieved by supporting research and training in forest management and biodiversity conservation at higher learning institutions. Governments and educational institutions play a crucial role in mainstreaming biodiversity by developing national biodiversity databases and digital tools that incorporate local knowledge and citizen science. Leveraging innovative technologies can enhance forest law enforcement. Collaboration among government departments (BSI/ZSI), civil society organizations, private sector companies, and development partners is essential to build the capacity of forest managers and staff in integrating biodiversity conservation into mainstream forest management through best practice guidelines and training. At the international, national, and local levels, awareness development is a key strategy for biodiversity conservation. National governments make international commitments, translate them into National Biodiversity Strategy and Action Plans (NBSAPs), and integrate them into sectoral laws, regulations, and policies. The National ENVIS (Environment Information System) program of the Ministry of Environment, Forest and Climate Change (MoEF&CC) is used for management, research, training organizations, and awareness development for various stakeholders involved in biodiversity management.

Capacity building of biodiversity (plant, animal) identification, taxonomic study is very crucial for the correct nomenclature of plants and animals. We need to educate local community member, panchayat member, civil society organization about the significance of biodiversity for today and in the future for sustainable environment and livelihood development. For example, PBR preparation by local people. The People's Biodiversity Register (PBR) is anticipated to: (1) document, monitor, and offer information for the sustainable management of biodiversity resources; (2) encourage development that is friendly to biodiversity; (2) establish the claims of individuals and local communities regarding the knowledge of biodiversity resource uses; (4) preserve and promote the practical ecological knowledge of local communities for Ayurveda and Unani medicine, aligning with the goals of the CBD. This system is particularly crucial to safeguard the interests of India's indigenous people, who have been instrumental in conserving the nation's biodiversity.

7.13 Organization and Institution for Biodiversity Conservation

The National Biodiversity Authority (NBA) was formed in Chennai by the Government of India in 2003 to enforce the Biological Diversity Act, 2002. As a statutory body, NBA plays a facilitative, regulatory, and advisory role on behalf of the Indian Government, addressing issues related to the conservation and sustainable use of biological resources, along with the fair and equitable sharing of benefits from their utilization. State Biodiversity Boards (SBBs) advise State Governments on biodiversity conservation matters, while Biodiversity Management Committees (BMCs) at the local level promote conservation, sustainable use of flora and fauna, and document biological diversity. Since its inception, NBA has aided in the establishment of SBBs in 28 States and Union Territories, along with approximately 2,66,499 BMCs.

In many developing countries with lower incomes, protected areas often experience deforestation at similar rates to unprotected forests (Anderson and Mammides 2020; Gaveau et al. 2013; Geldmann et al. 2019). Unfortunately, protected areas have proven ineffective in mitigating the decline of wildlife driven by trade in tropical and subtropical forests (Cardoso et al. 2021; Morton et al. 2021). A review of 144 updated National Biodiversity Strategies and Action Plans (NBSAPs) by Whitehorn et al. (2019) revealed that 91% of NBSAPs acknowledge the contribution of biodiversity to the national economy, with 43% providing specific details.

Biodiversity offsets represent a novel approach wherein developers, either voluntarily or mandated, offset unavoidable biodiversity loss by preserving, enhancing, or creating equivalent habitats elsewhere. In essence, this strategy aims to foster a pro-development environment while concurrently safeguarding or enhancing biodiversity. Although biodiversity offsetting may not be directly applicable to developers in the forest sector, projects related to biodiversity offsetting or Bio-fin could potentially serve as a source of income for expanding and enhancing protected area management, or funding initiatives for forest restoration.

7.14 Conclusions

Biodiversity conservation emerged as a prominent global agenda with the ratification of the Convention on Biological Diversity in 1992 by numerous countries. The emphasis on preserving biological diversity necessitates a focus on the ecosystem level due to the vast number of species, a majority of which remain unidentified. Globally, about two billion people indirectly depend on biodiversity-based livelihood particularity in the developing countries. Biodiversity losses are irreversible, once a species of plant or animal goes extinct, it is lost forever and will never be seen again, which has to be understood by the society. Biodiversity conservation has been included in the education curriculum of primary to graduation level. Adopting an ecosystem approach is crucial for conserving processes and habitats, along with their constituent species. One widely employed landscape-level strategy is the establishment of a suitable system of habitat reserves, a practice observed globally.

The Paris Agreement on climate change (COP21) 2015 urges all parties to recognize "the importance of ensuring the integrity of all ecosystems, including oceans, and the protection of biodiversity, recognized by some cultures as Mother Earth." Before undertaking significant operations, forest managers should perform biodiversity risk assessments regularly. This process typically involves completing a straightforward checklist of considerations and identifying activities that pose elevated risks. Quantification of forest biodiversity should be an ongoing process in the management of forest system. The value and contribution of biodiversity to the productive sectors and ecosystem service sector have not been studied and realized properly. Participatory monitoring of biodiversity in managed forest may be introduced to understand after three decades of participatory forestry what kind of change has happened in local flora and fauna.

Our industries, development organizations, NGOs, and financial institutions must understand the importance of this topic for future sustainability, biodiversity research, and take care of local biodiversity where they operate, increasingly such action within the stakeholders coming under "Global Reporting Initiatives"(GRI) and Environment, Sustainability and Governance (ESG) which will help to offer private and public finance for biodiversity conservation.

Despite the implementation of favorable forest policies and robust regulatory frameworks, forest degradation and biodiversity loss persist globally due to escalating human population demands, land-use changes, and the spread of invasive alien species. Continuous monitoring of biodiversity loss and active engagement of people in large-scale conservation and rehabilitation efforts are essential. However, there is a concern that an exclusive focus on maximizing carbon sequestration may lead to prioritizing afforestation with a limited number of exotic species. The nexus between carbon storage and biodiversity is complex, and while there may be a correlation, it may not be guaranteed. Human-induced environmental changes may result in some native species becoming overly abundant, causing ecological imbalances. In regions where large predators are absent, prey populations can surge, leading to vegetation damage. Therefore, harvesting sustainable levels and employing scientifically sound methods should be determined through best available knowledge and adjusted through continuous monitoring.

It is needed to increase our engagement and cathartic action at scale and with urgency if it is required to reverse the biodiversity loss that has been witnessed around the globe over the past decades. There are only a few countries of the world who have the Biodiversity Act in place; India is most fortunate for that; we must cooperate with our government for biodiversity conservation and make it as a mass movement in our country. Developing biodiversity strategies through broad stakeholder consultation, including relevant line ministries, is crucial for promoting responsibility for biodiversity mainstreaming across government sectors. Allocating financial resources, supporting fund utilization, and capacity development are essential processes, with academic institutions playing a critical role in training and teaching. Community-based forest management has been shown to produce better biodiversity outcomes compared to top-down approaches by government agencies. Similarly, lands traditionally owned, managed, used, or occupied by indigenous peoples perform better in resisting deforestation than unprotected areas. Addressing biodiversity loss requires a multidimensional approach, including halting and reversing deforestation, combating illegal and unregulated forest activities. Recognizing the forest tenure of indigenous peoples and local communities, preventing the conversion of natural forests, reducing reliance on mono-specific forest plantations, ensuring sustainable management of economic species, managing and controlling invasive are the priority area to work. There is a need for investing on a large scale in research, knowledge dissemination, and capacity development activities in core forest biodiversity conservation areas.

Lessons Learnt

- Forests play a pivotal and complex role for biodiversity conservation by providing its habitat; condition of habitat dictates the quality and quantity of biodiversity status.
- Forest biodiversity extinction is a slow process which can be restored with appropriate scientific and legal measures.
- Biodiversity Act 2002 adopted by government of India has created significant scope for the conservation initiative and appropriate measures to manage bio-resource based businesses.
- India is a mega-diverse country, for managing its diversity appropriate monitoring and habitat conservation and restoration is need of the day.
- The ecosystem services of biodiversity are high, other than provisional services; cultural and religious services of biodiversity are also important.
- Maintaining ecological balance is essential for the sustainability of healthy ecosystem and is the need for future generation to sustain biodiversity.
- A large-scale environmental awareness, knowledge, and capacity building is required for the people to understand the significance of biodiversity.

Key Questions

1. What is the significance of forest biodiversity conservation for a country?
2. List out the strategies of in situ and ex situ methods in biodiversity conservation.
3. Analyze the causes of biodiversity threats and specify how to control those causes.
4. What are the international efforts for sustaining biodiversity hotspots of the world?
5. Describe the biodiversity related acts, polices, and legislation available in India to safe guard its forest biodiversity resources.

References

Anand, Bhattacharya P (2021) Urbanites' perceptions of green spaces and their roles in effective management: a survey based study from Delhi, India. J Environ Stud Sci 13(1):31–42. https://doi.org/10.1007/s13412-022-00799-1

Anderson E, Mammides C (2020) The role of protected areas in mitigating human impact in the world's last wilderness areas. Ambio 49(2):434–441. https://doi.org/10.1007/s13280-019-01213-x

Aronson MFJ, Lepczyk CA, Evans KL, Goddard MA, Marra PP, MacIvor JS, Nilon CH, Vargo TL (2017) Biodiversity in the city: key challenges for urban green space management. Front Ecol Environ 15(4):189–196. https://doi.org/10.1002/fee.1480

Barstow M, Jimbo T, Davies K (2022) Extinction risk to the endemic trees of Papua New Guinea. Plants People Planet 5:508. https://doi.org/10.1002/ppp3.10342

Bhalla P, Bhattacharya P (2015) Urban biodiversity and green spaces in Delhi: a case study of new settlement and Lutyens' Delhi. J Hum Ecol 52(1–2):83–96. https://doi.org/10.1080/09709274.2015.11906933

BSI (2009) Unpublished document of botanical survey of India. Kolkata, MoEF

Cardoso P, Amponsah-Mensah K, Barreiros JP, Bouhuys J, Cheung H, Davies A, Kumschick S et al (2021) Scientists' warning to humanity on illegal or unsustainable wildlife trade. Biol Conserv 263:109341. https://doi.org/10.1016/j.biocon.2021.109341

CBD (2018) Report of the conference of parties to the convention on biological diversity on its fourth meeting, Egypt. https://www.cbd.int/doc/c/1081/32db/e26e7d13794f5f011cc621ef/cop-14-14-en.pdf

Chandel KPS (1996) Biodiversity and strategies for plant germplasm conservation in India. Trop Ecol 37(1):21–29

Chiriboga-Arroyo F et al (2021) Genetic threats to the forest giants of the Amazon: habitat degradation effects on the socio economically important Brazil nut tree (*Bertholletia excelsa*). Plants People Planet 3(2):194–210. https://doi.org/10.1002/ppp3.10166

Curtis JT, McIntosh RP (1950) The interrelations of certain analytic and synthetic phytosociological characters. Ecology 31:434–455. https://doi.org/10.2307/1931497

Edward W (1986) Biophila—the human bond with other species. Harvard University Press, p 176. isbn:9780674074422

FAO (2018) The state of the world's forests 2018—Forest pathways to sustainable development. Rome, 118p

FAO (2020) Forest resources assessment 2020: main report. Rome

Forzieri G et al (2022) Emerging signals of declining forest resilience under climate change. Nature 608:534–539. https://doi.org/10.1038/s41586-022-04959-9

Franklin JF (1993) Preserving biodiversity: species, ecosystems, or landscapes? Ecol Appl 3(2):202–205

Gaveau DLA, Kshatriya M, Sheil D, Sloan S, Molidena E, Wijaya A, Wich S et al (2013) Reconciling forest conservation and logging in Indonesian Borneo. PLoS One 8(8):e69887. https://doi.org/10.1371/journal.pone.0069887

Geldmann J, Manica A, Burgess ND, Coad L, Balmford A (2019) A anthropogenic pressures. Proc Natl Acad Sci U S A 116(46): 23209–23215. https://doi.org/10.1073/pnas.1908221116

Gibson L, Lee TM, Koh LP, Brook BW, Gardner TA, Barlow J, Peres CA, Bradshaw CJA, Laurance WF, Lovejoy TE, Sodhi NS (2011) Primary forests are irreplaceable for sustaining tropical biodiversity. Nature 478(7369):378–381. https://doi.org/10.1038/nature10425

Harrison RD, Shono K, Gitz V, Meybeck A, Hofer T, Wertz-Kanounnikoff S (2022) Mainstreaming biodiversity in forestry. CIFOR. https://doi.org/10.4060/cc2229en

Malhotra KC (1998) Anthropological dimensions of sacred groves in India: an overview. In: Ramakrishnan PS, Saxena KG, Chandrasekhara UM (eds) Conserving the sacred for biodiversity management. UNESCO and Oxford-IBH Publishing, pp 423–438

Manilal KS (1988) Flora of silent valley tropical rain forests of India. The Mathrubhumi Press, Kerala

MEA (Millennium Ecosystem Assessment) (2005) Ecosystem and human well being: wetlands and water. World Resources Institute

Mendelsohn R (2001) Measurements of economic value of biodiversity. In: Levin SA (ed) Encyclopedia of biodiversity, vol 2. Academic, pp 285–290

Ministry of Environment and Forests, Government of India (2009) India's fourth national report to the convention on biological diversity. New Delhi. https://nbaindia.org/uploaded/Biodiversityindia/4th_report.pdf

Morton O, Scheffers BR, Haugaasen T, Edwards DP (2021) Impacts of wildlife trade on terrestrial biodiversity. Nat Ecol Evol 5(4): 540–548. https://doi.org/10.1038/s41559-021-01399-y

Myers N et al (1988) Threatened biotas: hot spots in tropical forests. Enviornmentalist 8(3):187–208

Myers N, Mittermeier R, Mittermeier C et al (2000) Biodiversity hotspots for conservation priorities. Nature 403:853–858. https://doi.org/10.1038/35002501

Nayar MP (1997) Biodiversity challenges in Kerala and science of conservation biology. In: Pushpangadan P, Nair KSS (eds) Biodiversity of tropical forests: the Kerala Scenario. STEC, Trivandrum, Kerala

Pushpangadan P et al (1998) Sacred groves of Kerala—a synthesis on the state of art of knowledge. In: Ramakrishnan PS, Saxena KG et al (eds) Conserving the sacred for biodiversity management. Oxford and IBH Publishing Co., New Delhi, pp 193–210

Ramakrishnan PS (1998) Conserving the sacred for biodiversity: the conceptual framework. In: Ramakrishnan PS, Saxena KG et al (eds) Conserving the sacred for biodiversity management. Oxford and IBH Publishing Co., New Delhi, pp 3–16

Rawal RS, Bhatt ID, Sekar KC, Nandi SK (eds) (2013) The Himalayan biodiversity: richness, representativeness, uniqueness and life-support values. GB Pant Institute of Himalayan Environment and Development (GBPIHED). https://gbpihed.gov.in/PDF/Publication/Himalayan_Biodiversity_2013_book.pdf

Reid WV, Miller KR (1989) Keeping options alive. World Resources Institute

Saha R, Bhattacharya P (2011) Biodiversity register and indigenous knowledge: a case study of Baigachak area in Dindori district of Madhya Pradesh. J Biodivers 2(2):127–140

Shackleton CM, Ticktin T, Cunningham AB (2018) Non-timber forest products as ecological and biocultural keystone species. Ecol Soc. 23(4). https://www.jstor.org/stable/26796885

Shannon CE, Weaver W (1949) The mathematical theory of communication. University of Illinois Press, Urbana

Singh JS, Kushwaha SPS (2008) Forest biodiversity and its conservation in India. Int For Rev 10(2):292–304. https://doi.org/10.1505/ifor.10.2.292

Thomas E, Jalonen R, Loo J, Boshier D, Gallo L, Cavers S, Bordács S, Smith P, Bozzano M (2014) Genetic considerations in ecosystem restoration using native tree species. For Ecol Manag 333:66–75. https://doi.org/10.1016/j.foreco.2014.07.015

Tian Y, Wu H, Zhang G, Wang L, Zheng D, Li S (2020) Perceptions of ecosystem services, disservices and willingness-to-pay for urban green space conservation. J Environ Manag 260:110140. https://doi.org/10.1016/j.jenvman.2020.110140

Ticktin T (2004) The ecological implications of harvesting non-timber forest products. J Appl Ecol 41(1):11–21. https://doi.org/10.1111/j.1365-2664.2004.00859.x

Utkarsh G, Madhav G, Sheshagiri Rao PR (1999) Intellectual property rights on biological resources: benefiting from biodiversity and people's knowledge. Curr Sci 77(11):1418–1434

Whitehorn PR, Navarro LM, Schröter M, Fernandez M, Rotllan-Puig X, Marques A (2019) Mainstreaming biodiversity: a review of national strategies. Biol Conserv 235:157–163. https://doi.org/10.1016/j.biocon.2019.04.016

Wilson EO (1988) The current state of biological diversity. In: Wilson EO, Peter FM (eds) Biodiversity. National Academy Press, Washington

Wilson EO (1992) The diversity of life. Cambridge

Zenni RD, Essl F, García-Berthou E, McDermott SM (2021) The economic costs of biological invasions around the world. NeoBiota 67:1. https://doi.org/10.3897/neobiota.67.69971

Zhang S, Ramírez FM (2019) Assessing and mapping ecosystem services to support urban green infrastructure: the case of Barcelona, Spain. Cities 92:59–70. https://doi.org/10.1016/j.cities.2019.03.016

Forest Management

8

A. Balasubramanian, C. N. Hari Prasath, S. Varadha Raj, and S. Suresh Ramanan

Abstract

Forest management is the operating tool for forest properties through scientific, technical, and economic principles. This chapter begins with the history of forest management in India from the British rule till the present draft national forest policy 2018. The 1988 forest policy has played a significant role in emphasizing the concept of ecological balance and environmental stability for future generations. Joint Forest Management (JFM) is an important criterion for determining the togetherness of people with the forest administration for the development and conservation of forests. In the JFM activity, the entire crux is taking care and control of the locals' livelihood improvement by providing employment opportunities through non timber forest products (NTFP) collection, etc. As per the directions of 1952 National Forest Policy, forest cover must be one-third of the country's total geographical area, which could be brought about by planting and afforestation schemes. The importance of forest and forest management is described in various five-year plans from 1951 to 2012 in eleven phases. The criticality and inevitability of forestry education for managing the forest in a scientific, technical, and economic perspective are also highlighted.

Keywords

Forest management · National Forest Policy · Tribal development · Five year plan · Centrally sponsored scheme

A. Balasubramanian (✉) · C. N. H. Prasath · S. V. Raj
Forest College and Research Institute, Mettupalayam, Coimbatore, Tamil Nadu, India

S. S. Ramanan
Forest College and Research Institute, Mettupalayam, Coimbatore, Tamil Nadu, India

Central Agroforestry Research Institute, Jhansi, India

8.1 Outlook of Forest Management

Management is an essential attribute for handling any type of resources, both material and workforce. Forest is a simple term representing the natural resource pool which includes trees, wild animals, grasslands, medicinal plants, and even some minerals. The increasing population has been a challenge in many sectors. It also heavily impacts forest management, where forest policy makers and managers are changing the management strategies to suit the requirement of efficiently utilizing the forest resources sustainably, besides ensuring that the benefits of forest reach all sects of people (Poffenberger 1990). This is a paradigm shift from conventional forest management. Overall there is no denying that there is a change in forest management. However, the traditional management still holds relevant even today.

Forest management is always pivotal to the scientific forestry approach. It combines the technical inputs from main branches of forestry like silviculture, economic botany, soil science, ecology, geology, pedology, botany, entomology, pathology, economics, and finance to achieve the target. The target may change depending on the managerial objective. Implicitly, forest management is not an academic discipline in and of itself; rather, it is the practical application of science, technology, and economics to a forest estate to accomplish specific goals, such as the production of wood, timber, and industrial raw materials as well as other forest products (Jodha 1986). One way that forests are managed varies based on whether they are used for leisure purposes or to produce minor forest products. Moreover, the management regimes and implementing the scientific principles for forest management vary depending on the forest types (AFC 1994). Thus, practical experience of forester or forest managers is invaluable but the fundamental of forest management has always been the same (Saxena 1996).

By literal definition, forest management is defined as "the scientific, technical and profitable rules of forestry." Another definition states "The management of a forest property

A. K. Mandal, A. Nicodemus (eds.), *Textbook of Forest Science*, https://doi.org/10.1007/978-981-97-8289-5_8

involves the utilization of both technical forestry concepts and business strategies". The use of technical forestry ideas and business practices in the management of a forest property is known as forest management. Irrespective of the definition, the economical part or utility purposes decides the managerial rules. Classical natural forests are traditionally managed for source of timber, but owing to increased concern on environmental benefits much of the natural forests across the globe are protected, that is, timber extraction is either banned or carried out sustainably (Gilmour and Fisher 1998). Alternate source of timber from man-made plantations is generated eventually and the area under plantation forestry/planted forest is showing increasing trend steadily, since 2000. Presently, Global Forest Resources Assessment 2020 by FAO (FRA 2020) explained that more than 130 million ha of land surface is under plantation forestry which accounts to be 3% of total global forest area.

8.1.1 Scope of Forest Management

Like managing any other enterprise, managing forests entails establishing and carrying out policy decisions to accomplish the owner's goals. Consequently, a plan of action is included in these judgments. Planning, which is the responsibility of the states and the center in the case of state-owned forests, is governed by the national and state forest policies as general principles (Poffenberger 1995). Professional forest managers create detailed management plans, which are carried out by staff members at the technical level (Murali et al. 2003). The forest manager must continually control the expanding stock to accomplish certain management goals; this involves determining "how much, when, where, and how to chop." The impact of forest management is enormous because it covers a wide range of topics. The most important ones include,

(a) *Control of growing stock, its structure, and composition*: Site adaptability, species selection/choice, produce harvesting, stand modification, regeneration, and protection (Hobley 1996).
(b) *Distribution and marketing of produce*: Communication and transportation, produce sales, marketing data, logging plans, and revenue.
(c) *Administration of forest property*: Forestry Commission, personnel management, keeping an eye on things and exercising control welfare and labor management, budgetary restraint and economic effectiveness, performing social duties, and keeping a record for future use.

8.2 Forest Management in India

India's forestry industry is one of the first in the world to be operated using scientific silvicultural principles codes. It just so happens that the industrial revolution in the West coincided with the establishment and application of forest management concepts in the middle of the eighteenth century (Saxena 1996). During the pre-independence era, the forests became valuable resources due to the rising need for raw materials and the necessity of expanding the railway network (Balooni 2002). Thus, at that time, forestry was production-oriented (Raju 1997; Gaikwad 1981). Nonetheless, the National Forest Policy of 1952 brought about the fundamental shift in thinking, moving away from production forestry and toward a focus on achieving the goals of both upholding ecological balance and optimally satisfying stakeholder requirements. (GoI 1952).

8.2.1 History of Forest Management in India

8.2.1.1 Forest Management Under British Rule

Natural resources are highly valued in India and have long been an essential component of the country's economy and culture (Stebbing 1926). Literature, politics, and religion from antiquity attest to the idea that humans have always been seen as a natural component of the environment rather than as something superior to it (Raju et al. 1993). The wealth of information regarding forest management regimes under British rule means that the only subjects covered in this part are the colonial approach to forest management and its policies affecting those who depend on forests (Ribbentrop 1989; Guha 1983; Balooni 2002).

It is commonly known that several Indian forests have been managed at various times throughout the country's history under a system of laws and guidelines created by various groups (Shepherd 1989; Saldanha 1996). Some of the self-initiated forest conservation organizations have persisted or been re-established in response to the urgent need to preserve community forests (Eggert 2013; Balooni 2002).

8.2.1.2 National Forest Policy, 1894

The expansion of agriculture, a significant source of income, and commercial interests were the focal points of the British government's forest strategy. The National Forest Policy of 1894, India's first official forest policy, contains specific documentation of these goals (Singh 1994; Desloges and Gauthier 1997). This policy stated that "wherever Forests which are the reservoirs of valuable timbers should be managed on commercial lines as a source of revenue to the States" and "there is an effective demand for cultivable land that can only be supplied by a forest area, the land should ordinarily be relinquished without hesitation" (Government of India 1894). This policy stated that the advancement of state interests was the only reason woods were managed under British administration (CIDA 1991). Few most important goals of the national Forest Policy, 1894 were,

- Managing forests to ensure the overall welfare of the nation.
- Ensuring enough forest cover to preserve the climate and physical environment while meeting people's needs.
- The forest strategy stipulated that permanent cultivation should take precedence over forestry, subject to these objectives.
- Local needs should be satisfied at non-competitive rates, taking into account revenue as a secondary priority.

For management purposes, the British administration divided the forests into four sections, as described in the National Forest Policy of 1894 (Kumar 2002). The first type of forests was thought to be necessary to shield farmed areas from harm brought on by landslides and hill torrents. These forests were typically found on hill slopes (Bhatt et al. 2014). They, therefore, played a conservation function that benefited plains agriculture (Straub et al. 2008; Bathla 1999; Haque 1998). The enormous reserves of priceless timber trees, such as *Tectona grandis, Shorea robusta*, and *Cedrus deodara*, comprised the second class of forests. To encourage the first two species' natural regeneration and the third species' artificial regeneration—which is motivated by commercial interests—forest management strategies were devised (Kolavalli 1994; Balooni 2002). However, to create techniques for *Shorea robusta* artificial regeneration, the taungya system was employed in several parts of northern and eastern India (Government of India 1976).

The British authority did not prioritize the use of forests to suit people's needs. People's requirements were met by the third class of forests, known as "minor forests," which only produced inferior lumber, fodder, or timber, and the fourth class, known as "pastures and grazing grounds," which were subject to restrictions (Kant et al. 1991). In addition, the policy recommended that "the constitution and preservation of forests and, to a greater or lesser extent, the regulation of rights and the restriction of privileges of users in those forest areas which may have previously been enjoyed by the inhabitants of the immediate neighborhood" be the fundamental principle to be observed (Government of India 1894). In summary, during colonial authority, the state's commercial interests overruled the interests of the people when it came to forests (Bardhan 1993; Karanth et al. 2008).

8.2.1.3 Indian Forest Act, 1927

Similarly, the people that relied on forests were also impacted by the British administration's enforcement of the Indian Forest Act, 1927. The majority of forests were put under public ownership in 1865 when the Indian Forest Act was originally enacted. It underwent more revisions in 1878 and was combined in 1927 (Government of India 1927). It is crucial to remember that the goals of forest management as stated in the (Draft) Forest Acts of 1865 and 1878 gave rise to the National Forest Policy of 1894 (ICFRE 2000). "An Act to combine the laws pertaining to forests, the transportation of forest products, and the duties imposed on timber and other forest products" was the description of the Indian Forest Act of 1927 (Government of India 1927). This act's content was organized into 13 chapters, each containing a profusion of guidelines, sanctions, and processes meant to strengthen government control over forests while undermining individuals' rights to use them (Lundmark 1995; Samra et al. 2002). Consequently, communities lost many of their customary rights over forests because of this act's facilitation of the state's control over forests. In other words, "privileges replaced people's rights to use forests, which were destroyed." This act strained village communities' long-standing symbiotic relationship with forests even more (Inform 2002; Balooni 2002).

Following their independence in 1947, the Indian states enacted the Forest Act of 1927. Later, the act underwent several revisions, the majority of which limited the local usage of woods (Guhathakurta and Bhatia 1995; Iyengar 1989). Moreover, the forest acts were enacted by the Indian states. For instance, the Orissa (State) Forest Act 1972 stated that areas designated for reservation shall not be subject to claims for shifting cultivation. "Forest offences as defined in the Indian Forest Act 1927 were re-categorized and tougher sanctions were provided" following independence (Balooni 2002).

8.2.1.4 Policy Setup on Forest Management After Independence

The necessity to reconsider the forest strategy arose considering the most recent developments, following India's 1947 independence from British rule. As a result after reviewing the forest policy, the Indian government unveiled the first National Forest Policy of Independent India in 1952. "India as a whole should aim at maintaining one-third of its entire land area under forest," according to the Policy. In essence, it stated that local communities should be permitted to meet their requirements, if they do so within justifiable bounds and bear in mind the national interest (Singh and Khare 1993; Kumar 2000). The policy also addressed issues including illegal felling and encroachment, restoring timber products, and combating fungal and other forest damage. It also emphasized how important it is to protect and maintain forests as a valuable national resource (Sundar 2000).

But none of the main goals of the National Forest Policy of 1952 were met. The policy's ecological component was mainly disregarded, which led to the widespread transfer of "low-value mixed forests" into "high value marketable plantations" (Joshi 1998). There were insufficient efforts made by the policy to tackle the problem of forest depletion, and no noteworthy initiatives were made to maintain forests. The 1952 forest policy's failure is also attributed in part to the fact that forests were a state topic (Government of India 1952).

The **National Commission on Agriculture (NCA), instituted in 1976**, was the next significant step toward forest protection. The commission suggested that a modification be made to the government's policy. It promoted switching from a "conservation-oriented" to a "productive forestry" approach, which is more dynamic (Baland and Platteau 1996; Kanbur 1992). In addition, it suggested social forestry programs for the states, emphasizing that the policy should be centered on maximizing forest productivity, avoiding erosion and denudation, maximizing the use of forest resources for goods and services, and increasing the possibility of creating jobs for the benefit of the country. In that same year, the government proposed a constitutional amendment putting woods on the concurrent list, enabling both the centralized and state governments to pursue laws relevant to them.

The **Forest (Conservation) Act, 1980** was passed by the central government in 1980. The act's main goal was to impede the development of forest land for uses other than forests. Rather than outright prohibiting such conversion, the act stipulated that it could only be carried out with the central government's assent. Additionally, it limited the removal of reserves from forests (Ballabh et al. 2002a, b). The purpose of the act was to elevate the concept of forest protection. The Forest Act lacked powers to prevent such property conversion (FCA, GoI 1980; Baland and Platteau 1999).

With the approval of the **National Forest Policy in 1988**, the government's perspective underwent a significant paradigm change. In contrast to the 1952 policy's use-oriented approach, this policy placed more emphasis on the environmental aspect of forests.

The aims of the policy were as follows:

- Preservation and restoration of ecological balance as needed to maintain environmental stability.
- Preserving the nation's genetic resources and natural heritage.
- Expanding the percentage of land covered by forests (33%), for example, through afforestation and social forestry initiatives.
- Enhancing the efficiency of forests to justify the needs of the region and the country.
- Organizing a large-scale public campaign to enhance and preserve the amount of forest cover.
- Economic benefit extraction will take a backseat to these main objectives.

In addition to the aforementioned goals, the policy made additional significant contributions. It was the first significant policy that addressed how the tribal people who reside in these forests were treated (GoI 1990). It was stated that the legitimate forest dwellers must receive all rights and concessions pertaining to the forest (i.e., tribals). Furthermore, since these populations depend on the forest for their livelihood, they must be intimately involved in its preservation, regrowth, and development. Their domestic needs should take precedence over the produce of the forests they live in, and the rights and concessions must likewise extend to the forest produce (Bahuguna 2001).

The policy stipulated that expert should thoroughly assess the costs and benefits of converting forest land for non-forest uses, taking into account ecological, social, and environmental factors. It should be mandatory for those converting such land to non-forest uses (such as mining and quarrying) to restore and reforest the area in accordance with local forestry practices. The policy also addressed the preservation of wildlife and the establishment of corridors between protected areas to preserve genetic continuity among artificially divided subpopulations of migratory wildlife (Saravanan 2008).

The **Biological Diversity Act of 2002** was another significant piece of legislation introduced at this time. It contained elements that, even if they were unrelated to forest protection, would eventually aid in forest conservation (It is noteworthy to note that these two laws also resulted from India's international obligations resulting from specific conventions that were in effect at the time). In addition, the 1988 policy's mention of wildlife conservation was given effect by amending the **Wildlife (Protection) Act, 1972**. Another significant legislation from this era that supported the preservation of forests was the **Environment Protection Act (1986)** (Ravindranath et al. 2000).

Many environmentalists highly valued the 1988 policy in light of all these significant systems and modifications that were implemented after it was first announced. (Government of India 2001a; Government of India 2001b). The National Forest Commission further supports this by stating in its 2006 report that,

- Scientific research should be prioritized in determining the ideal amount of forest cover.
- Amendment 1927 to the Forest Act.
- The Forest Department will execute the Environment Protection Act of 1986 and the Biological Diversity Act of 2002.

- Rearranging of species in accordance with the 1972 Wildlife Protection Act.
- The 1988 Policy remains unchanged.

The central government finally revealed the draft **National Forest Policy, 2018** in the year 2018. The recent issues that the nation are facing are taken care of by the draft policy. These issues include decreasing green cover, human-animal conflict, and climate change (Draft NFC 2018). The principal attributes of the policy are as follows,

- The objective is to employ scientific methods to ensure that forests occupy a minimum of one-third of India's total land area.
- Current policy is centered on climate change, as opposed to prior policies that addressed ecological balance and environmental stability.
- To tackle conflicts between humans and animals, it suggested forming quick response teams comprised of skilled and prepared individuals who can evaluate the harm and promptly provide aid to the affected parties.
- Wildlife observation and management.
- The involvement of both public and private sectors in afforestation and reforestation projects.
- Preserve forests by tightly controlling their diversion for non-forest uses.

Although the strategy aims to tackle fresh issues facing the nation, it has also drawn criticism from a significant portion of the populace (Arul 1998) for the reasons listed below,

- Incorporating private interests may result in the privatization of resources and the establishment of private forests.
- In view of the competitive demands for forest lands, there is no system in place to accomplish these goals.
- Regeneration is not as much of a focus as conservation and protection are.
- Does not address the controversial problem of using forest land for mining and other uses.

8.3 Joint Forest Management

8.3.1 Departmental Structure to the Evolution of JFM

According to Jodha (1997) and Kuperus (1987), the 1988 forest policy allows for participatory management of the forest or integrating people in management system. The Joint Forest Management (JFM) initiative got its name from a resolution that was approved by all state forest departments to allow participation (Arora 1994; Balooni 2002). There may be some similarities between the concepts of cooperative forest management, community forestry programs, and participatory forest management. (Rastogi 1995; Balooni 2002).

- The term "community forest management" (CFM) describes community-based initiatives aimed at ensuring the forest is used sustainably. The history of community-managed forests confirms that CFM is a reasonable guiding principle in forest management, despite its lack of a definition.

As previously said, the definition of CFM is challenging due to its diverse forms and circumstances; instead, it is easier to focus on how the concepts of community engagement, sustainability, and development interact. The method that administration was done was significantly altered by the policymakers. Nevertheless, everyone had to support the project that was created by someone (Arnold and Stewart 1991).

The Joint Forest Management is different in this situation. Communities and individuals are also involved in the project formulation process in JFM (MacKenzie 1991; Saberwal 2001; Gadgil and Guha 1992). In order to manage public forest areas sustainably and share benefits with others, the Joint Forest Management Program aims to create partnerships between state forest agencies and local community organizations (Balooni and Singh 2001).

According to Malhotra et al. (2006), JFM's primary objective is to ensure that forests are used sustainably so that local demands may be fairly met while environmental sustainability is preserved. The fundamental idea is that local forest-dependent women and men have the most stake in sustainable forest management (Balooni 2002). According to Arnold and Stewart (1991) in Community Forest Management, the government serves as passive supporter while the community takes the initiative and handles the resources. But in Participatory Forest Management, the government takes the lead and formulates projects and the community participates in it, sometimes as hired labor.

8.3.2 Joint Forest Management: An Inadequate Approach

Thus, a people's campaign to safeguard the nation's forest resources was suggested by the 1988 National Forest Policy (Balooni 2002). In June 1990, the central government of India followed suit and released a circular on Joint Forest Management, which addressed community forest conservation initiatives and operationalized participatory forest management (Gilmour et al. 2004). The explanation of the JFM circular states:

- The 1988 National Forest Policy envisions public involvement in the development of forests. Priority should be given to the needs of the tribal people and other villages who live in and around the forests for fuelwood, fodder, and small timber for building homes (Fernandes and Kulkarni 1983; Heinze et al. 2011). According to the policy statement, encouraging forest communities to identify with the growth and preservation of the forests that provide them with benefits is one of the fundamental components of effective forest management. The key components of the program recommended by the resolution of the Government of India (Appukuttannair and Stefanie 2002; Bhatt 1988) are as follows,
- The recipients will receive usufructs such as grasses, branch tips and lops, and non-timber forest products. A percentage of the money made from the sale of mature trees will be awarded to them upon the successful protection of forests (IIM 1994).
- Planting fruit trees, bushes, legumes, and grasses may be allowed in addition to trees that provide fuel, feed, and lumber for the village community.
- The forestlands that the village communities protect are not to be used for grazing. To encourage stall feeding, however, the people can trim and move grass for free.
- Usufructuary benefits may be lost if the territory is not protected against encroachment, grazing, and other threats.
- The advantages of people participating should only benefit the village communities, not businesses or other interests.

8.3.3 Policy Trends in Joint Forest Management

8.3.3.1 Policy Directives

The new National Forest Policy of 1988, which is the fundamental to focus the requirement of nation contribution in forest management and protection, is considered at the beginning of this section (Ballabh and Singh 1988; Bleeker et al. 2007). This strategy was developed in response to the nation's successful, if limited, implementation of a participatory forest management scheme (Kerr 1997; Saravanan 2008). The directives and guidelines on JFM issued by the Indian government are cited in this section.

Creating a People's Movement

The Indian Forestry Sector has undergone a decade-long transformation according to the National Forest Policy, 1988, the country's second forest policy following independence (Annexure 1). It is focused on both production and conservation. Maintaining environmental stability by protecting forests as a natural legacy is the major goal of this approach (Schabel and Palmer 1999). Additionally, it emphasizes the importance of significantly raising the nation's forest production and tree cover to meet needs on a national scale (Starfinger 1997; Commander 1986). The phrase "In order to accomplish the aforementioned goals and reduce the strain on current forests, a large-scale people's movement involving women should be established" was a unique aspect of this new program, nevertheless. It's an extreme change starting the former National Forest Policy, which was established in 1952 and called for public involvement in the preservation and growth of forests. A sign of "management change," or the transition from government-managed to community-managed forests, is the National Forest Policy (Ballabh et al. 2002a, b). In the last 10 years, the Indian government has periodically released directives and recommendations on JFM as a follow-up to the National Forest Policy of 1988 (Chatterjee 2003). This demonstrates the government's determination to mobilize a sizable public movement and promote involvement in forest management, as stated simply in the National Forest Policy (Desloges and Gauthier 1997).

Following this, the Forest (Conservation) Act was passed in 1980, requiring the approval of the central government before any forest land could be converted for uses other than reforestation (Angelsen and Wunder 2003). In some ways, the Forest (Conservation) Act of 1980 has assisted in preventing the conversion of forest land to purposes other than forests. The leap of transition of forest land for non-forest uses declined to roughly 22,625 ha yearly between 1981 and 1988 (ICFRE 2000), compared to 1,43,000 ha annually prior to 1980 (Balooni 1998; Mishra 1992). This act has made it easier to implement the JFM program on forest land in certain ways, as encroachment usually occurs on area that would otherwise be ideal for JFM management, usually on the edges of existing forests (Balooni and Ballabh 2000). Functional aspects of joint forest management in the country are depicted in Flowchart 8.1.

Expansion of JFM to Non-Forest Areas

In addition, the government established a "Standing Committee on JFM" in 1988 to examine how JFM programs were being implemented and what JFM arrangements were already in place in the nation (Ahuja 1998). According to Agarwal (2001a, b) and Balooni (1998), the principal objective of the committee was to provide guidance to the government regarding the operational aspects of JFM, encompassing institutional setups. It was also anticipated that the group will talk about ways to extend JFM outside of forests (Sarin 1998).

In addition to the forest land under the State Forest Departments' ownership and management, a sizable portion of India's land area (roughly ten million hectares) is dedicated to non-forest and non-agricultural uses, including

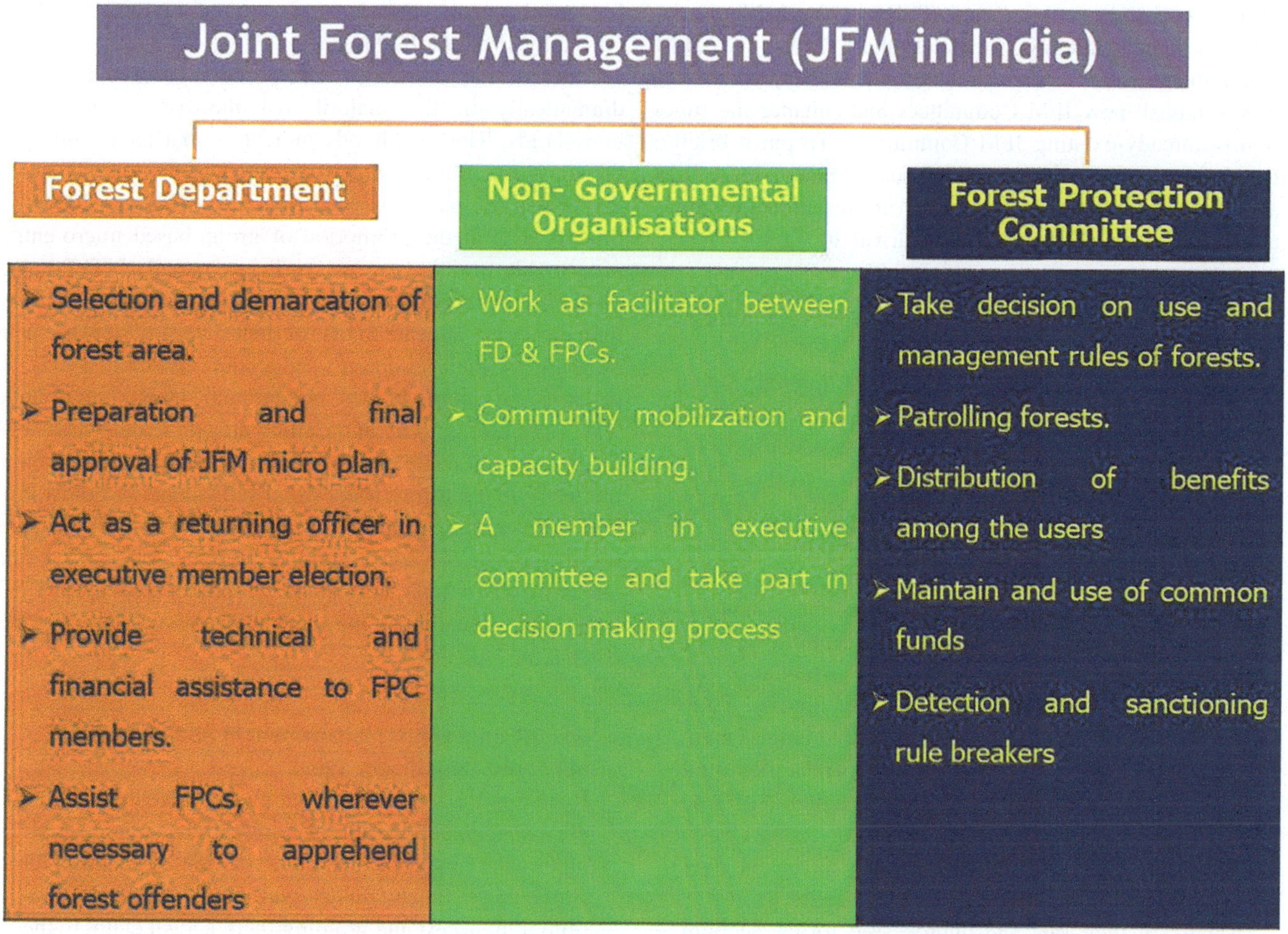

Flowchart 8.1 Functional aspects of joint forest management

permanent pastures, culturable wastelands, and other grazing areas (Mohanraj and Veenakumari 1996). These lands are sometimes referred to as "common property resources," yet they are legally owned by the revenue department and other government agencies. These areas are primarily considered "open access resources" (Balooni and Singh 2001). Despite being severely damaged due to "the tragedy of commons," these uncultivated lands nevertheless have the potential to help JFM spread throughout the nation. (Balooni 2002).

JFM in Afforestation Schemes

In keeping with the administration's focus on participatory forest management, the afforestation investments in the Five-Year Plans are being modified to take "people's involvement in project planning and execution" into account (Shrestha 1996; SPWD 1992). Following India's 1947 declaration of independence, the government introduced a number of Five Year Plans featuring specific budgetary allotments for the advancement of different industries. Between 1951 and 1956, the first Five-Year Plan was put into effect (Campbell 1992; CMA 1994).

In a recent move, the tenth Five-Year Plan's National Afforestation Program would be formulated using new operational instructions released by the Ministry of Environment and Forests (Agrawal 2001). To ensure "consistency in financial patterns and implementation mechanisms" and " prevent duplication of schemes with similar objectives," the forestry programs that were in place throughout the ninth plan have been integrated into the National Afforestation Program (Molder et al. 2014).

One of the key features of these recommendations is that all new centrally financed afforestation schemes would be implemented through a two-tier structure consisting of JFM Committees and Forest Development Agencies (FDAs) to allow for greater community involvement in planning and execution. There were 165 FDAs in operation in 21 Indian states as of July 2002. FDAs are recently established institutional entities that operate at the territorial or wildlife forest division level and are registered under the Societies Registration Act.

In addition to JFM Committees, already-existing village institutions will serve as the grassroots implementing agency to address village requirements (TERI 1992). A Memorandum of Understanding stipulates that FDAs and JFM Committees will collaborate (Agarwal 2001a, b). FDAs both establish new JFM Committees and enhance the function of already-existing JFM Committees. To put it briefly, the National Afforestation Program seeks to incorporate JFM as a fundamental component of all afforestation initiatives across the nation (Hobley 1996; Agarwal 2001a, b).

Box 8.1 Success of JFM and Non Timber Forest Products (NTFPs) (Bhatt et al. 2014)

The term "NTFPs" refers to all forest-derived goods other than lumber. Because of the forest's significance for the social, cultural, and economic sustenance of these communities, NTFPs are essential to their survival. Gathering non-timber forest products (NTFPs) such as anola, chironjee seeds, mahua flowers, tamarind, tendu leaves, sal leaves, and various seeds, among others, offers a supplementary food and financial source. NTFPs are important to JFM initiatives for several reasons. First of all, they are essential to the survival of populations that depend on forests. They meet fundamental needs, offer profitable employment during hard times, and augment wages and agricultural revenue. Second, when it comes to the amount of time required to produce sizable amounts of economically valuable output, NTFPs clearly have an advantage over timber. Third, non-timber forest products (NTFPs), usually in their raw and unprocessed forms, account for more over half of all forest revenue nationally.

JFM and Tribal Development

Most tribal people on the planet reside in remote areas or on the edges of forests. Every indigenous tribe has its own customs and way of life, and they rely heavily on the natural resources. Without causing any harm to the forest, they have been gathering resources from it (Locke 1999). They can secure their food and means of subsistence thanks to the forest. Approximately 100 million people live in forests, while an additional 275 million people live nearby and depend on forests for their livelihood. Approximately 370 million people rely on forest products for their direct or indirect livelihood, such as fuel wood, fodder, food, medicines, and other natural resource inputs (Box 8.1) that are then turned into financial outputs and sustenance the maintenance of forest dwellers (Agarwal 1997). According to the Indian State Forest Report 2019 (MoEF, GoI 2009), most India's forests are held by the state and span 78.29 million hectares, or 24.56% of the nation's total land area.

Important elements of the program's goal include enhancing the livelihoods of communities that live near forests, lowering rural poverty, and managing and conserving forests. Indian tribes make up a sizable portion of the forest residents (Westoby 1987; GoI 2000).

Due to JFM assistance, livelihood indicators improved dramatically in the majority of the tribal settlement's households. The livelihoods indicators that have improved the most are drinking water, livestock activities, education level, and employment and income (Roering 2004). It directly aids in the promotion of group-based micro-entrepreneurship initiatives as well as small NTFP-based businesses. Since JFM activities generate a steady income, they can freely allocate funds for their own skill development as well as their children's education (Saxena 1995). It supports the sample homes' educational status in an oblique way. Livestock activity in the study area increases when JFM guidelines are made freely available for gathering animal fodder from the forest. Most tribal settlements in Forest Protection Committee (FPC) construct submersible water tanks for irrigation and consumption. (Saxena 2000; FAO and Department for International Development 2001). JFM has been implemented in all the states with extraordinary success (Box 8.2).

Box 8.2 Joint Forest Management in Andhra Pradesh (AP Forest Department 1999)

Realizing the value of community involvement in forest management, the "Joint Forest Management" strategy was put into practice, involving the inclusion of all fringe forest inhabitants as partners in the administration of the forest and granting them a legal claim to the usufructs. The formation of Vana Samarakshana Samithis (VSS) by people residing on the edges of the forests was promoted to preserve and revitalize degraded forest regions.

A total of 6580 Vana Samarakshana Samithis (VSS) have been established. Of these, 16.58 lakh hectares of degraded forest area have been brought under JFM; of these, 5.38 lakh hectares have been treated thus far through regeneration work, which includes soil and moisture conservation and gap planting. Thirteen thousand VSS members, including six thousand women, are actively involved in JFM. Through VSS, 24,000 hectares of encroaching forestland have been rejuvenated.

The government's JFM ruling grants the VSS villagers 100% usufruct rights over forest products, subject to the requirement that the village community set aside funds needed for sustainable forest management and regeneration. Forest preservation and enhancement have been major initiatives of the Vana Samarakshana Samithis. The members of Samiti are

(continued)

Box 8.2 (continued)

working together to create forests by clearing weeds and shrubs to allow vital tree species to grow properly and meet the needs of the local population. They are utilizing microplans that they have designed for this purpose. Another VSS thrust area is village development, with an emphasis on bettering roads, drinking water quality, community hall construction, and the economic standing of women via supporting women's organizations, among other things. Additionally, the Samithis have been effective in uplifting and integrating tribal people into society.

Special Features of JFM in Andhra Pradesh

- Members of VSS have complete ownership rights over forest products.
- VSS receives half of the net revenue from beedi leaves that member collect.
- Members of the VSS receive 25% of the fines collected from forest offenses that they have arrested.
- By using the Joint Account System to transmit the monies, transparency in VSS activities is guaranteed.
- Investment decisions are based on a microplan that was created after consulting with VSS members.
- Women are required to join in the general body at a rate of 50%, and in the Managing Committee, the minimum is 30%. To conserve the money they receive from the Samithis, women have organized into thrift clubs.
- Every group of people, including SCs and STs, is included and encouraged to participate actively.
- NGOs are contributing to the enhancement of VSS by means of encouragement, expansion, instruction, and further assistance.
- It has been determined which medicinal plants grow in forest regions, and members of VSS are urged to cultivate and preserve these plants there.
- Obtaining resources from the World Bank, NABARD, Employment Assurance Scheme, and centrally supported programs to ensure the necessary monies are accessible.
- Effective collaboration across departments to ensure the growth of VSS villages into "Model villages".

Satellite images taken in 1996 and 1998 reveal that, under VSS management, the forest cover significantly improved in just two years.

(continued)

Box 8.2 (continued)

Andhra Pradesh's JFM program has been ranked as the finest in the nation by the World Bank.

Success stories from across the state show how community-driven JFM initiatives have raised productivity, raised the water table, created sufficient jobs, and improved socioeconomic conditions. In a paper titled "Jyoti se Jyoti jale Joint Forest Management: Forest conservation with people's participation: a saga of success," the A.P. Forest Department chronicled these. A few of these accomplishments are listed here. We appreciate permission to incorporate these research from A.P., Principal Chief Conservator of Forests.

From Scrub Jungle to Thick Forests

A 329-hectare degraded reserve forest is in **Kamdhan village, Mahbubnagar district**. The hamlet was chosen to implement JFM, and all houses joined the VSS after it was established. A PRA exercise was used to create a microplan for the growth of the forest. The degraded forest's vegetation and animals have improved because of increased protection and citizen participation. Five groups of women have formed and are saving money. The groups have benefited from a matching grant provided by the Forest Department. The organization raised 5000 nursery seedlings, which were utilized to fill up gaps in the forest and distributed to nearby villages for planting.

A regular nominal fee has been established for grazing, and the money collected is paid into the VSS corpus fund. The VSS has made NTFP marketing more regular. Members of the VSS can gather fruits such as custard apples for personal use, and the VSS auctions off any marketable excess. The corpus find account is where the sale proceeds are put. Clonal plantations of Seetaphal, Usiri, and Tamari have been cultivated. The VSS suggested grafting more clones and giving them to the locals. A percolation tank, a check dam that has stopped soil erosion, and a raised ground water table are examples of large-scale soil and moisture conservation projects undertaken.

Consistent gatherings, instruction, and exposure trips have raised awareness among VSS members. The VSS was determined to be the finest in the district in January 1998. In addition to preserving the forest, the VSS now actively participates in developmental initiatives, sends kids to school, and organizes into thrift groups. The VSS has developed into a strong institutional structure for the effective localized execution of government initiatives.

8.4 Forest Management for Tribal Welfare After 2000

8.4.1 Scheduled Tribes and Other Traditional Forest Dwellers (Recognition of Forest Rights) Act, 2006

The past 20 years have been eventful for conservation policy, with community conservation efforts and resource rights receiving more attention in global academic discourse, policy, and practice. In order to defend local communities' and Indigenous peoples' rights to forests and other natural resources, many nations have recently implemented modifications to forest tenure (Heinze et al. 2011). Reforms to forest tenure are taking place in the background of mounting evidence that truth--based tactics to preservation are important, especially when compared to the traditional conservation strategy that excludes rights and community involvement (Eggert 2013). These shifts are occurring in the midst of disagreements over how industrialization and the neo-liberal model of resource-based development have affected and continued to harm the lands, forests, and means of subsistence for local communities and Indigenous peoples around the globe (Jyotish 2008; Saravanan 2008).

The Forest Rights Act has sparked a heated debate about conservation since it went into effect in January 2008. One side represents the conservation orthodoxy, which maintains that rights and conservation cannot coexist (Prasad 2006).

8.4.1.1 Contribution of Forest Rights to Conservation

Distinct and regular contract over forest lands and forests is guaranteed by the Forest Rights Act. According to Malhotra and Bhattacharya (2011), the act acknowledges the geographical rights of Primitive Tribal Groups (PTGs), the rights of pre-agricultural communities, and the rights over communal management activities. Forest communities have utilized the act's provisions to assert a range of forest rights since it went into effect in 2008 (Singh and Quli 2011; Eggert 2013). In addition, the act gives privileges to populations ejected by state development actions, rights to revenue village conversion from forest villages, and rights to development amenities building like roads, schools, health centers, electricity, and irrigation in the forest villages (Chatterjee 2007; GOI 2009). The protection of such a wide range of rights related to conventional accessibility, use, and interactions reflects the strong preservation ethic of the many communities that the act wants to conserve (Balooni 2002).

Community conservation zones safeguard some of the most important habitats, such as wetlands, forests, coastal/estuarine ecosystems, and mangrove ecosystems. These ecosystems are vital to local communities' livelihoods and have high ecological value. These conservation efforts cover a number of rare, endangered, and vulnerable species of plants and animals, including migratory birds, black bucks, peacocks, tigers, elephants, and sea turtles. Despite government efforts to rid protected areas of human habitation, community conservation zones also encompass the majority of the state's tiger, elephant, and protected areas.

A foundation for these long-standing community conservation efforts on forest land including wetlands and mangroves to be legally recognized has been established by the Forest Rights Act's Sect. 3(1)(i) right to protect, conserve, regenerate, and manage community forest resources. Many local environmental groups have already claimed and obtained legal authority over their community forests since the Forest Rights Act's adoption (Sunderlin and Ba 2006; Sahu 2007).

8.4.1.2 Process to Support Social Institutions for Conservation

Gram Sabha, or village assembly, is empowered by the Forest Rights Act to start the process of determining rights, involves gathering, compiling, and examining the claims. Forest Rights Committees are the means by which the Gram Sabha carries out these functions. The Forest Rights Act stipulates that women, other populations that live in forests, and scheduled tribes must be adequately represented in the Gram Sabha and Forest Rights Committee, as well as have enough involvement. It takes a joint effort on the part of the affected villages and communities to determine the nature and scope of the rights to determine and claim them (Poffenberger and McGean 1996). The process of determining rights has contributed to the formation of the institutional framework for the management of community forest resources, the revitalization of community institutions, the creation of social capital, and the collective visualization of natural landscapes (Chatterjee 2007). It has also aided in resolving disputes and claim overlaps, addressing social justice and equity issues, and creating room for women to participate.

There are three steps in the recognition process. The process for determining rights and claims is first started by the Gram Sabha (Pathak 1994). Subsequently, a district-level team that reviews the claims and develops the top of forest rights is advised to review the draft record of forest rights prepared by the sub-district-level committee (Bleeker et al. 2007). Members of the Panchayati Raj institutions as well as government representatives from the departments of revenue, forest, and tribal welfare sit on the committees at the district and sub-district levels. The potential to add different authorities and features to the procedure of recognizing forest rights has been made possible by the involvement of individuals from several departments (Agarwal 1989).

Furthermore, Section 5 of the Act empowers Gram Sabha and village-level institutions, as well as holders of any forest right, to take action to protect collaborative forest resources and habitat areas from activities that endanger the forest, wildlife, or biodiversity. Communities employ the rights and empowered authority clauses to regulate external dangers that harm community woodlands and to bolster community conservation programs (Prasad 2006).

8.4.1.3 Issues of Operation and Suggestions for Ways Forward

The Forest Rights Act offers communities enormous opportunities to be recognized and respected for their customary rights and conservation actions. But the effectiveness of the above-discussed legal structure of rights and empowered local authorities in practice hinges mostly on how government agencies interpret and carry out the act, especially its sections regarding community rights (Agarwal and Narain 1991).

Additionally, unimpressive is the state of the nation. By May of 2015, out of almost 28,22,000 claims made, just 32% had been paid out. Less than 3% of the about 49,000 community claims that were filed were fulfilled by this. The claims and recognition status to date depict a depressing picture, given that there are approximately 1,70,400 forest edge communities and 275 million forest-dependent people in India (Pandey et al. 2011). It may be argued that the Gram Sabhas, who are empowered by the act to determine rights, lack the knowledge and resources required for the assessment and validation of claims, including documentation, maps, proof, and technical assistance (Molder et al. 2014).

8.5 Forest Management Under Five-Year Plans

8.5.1 The First Five-Year Plan (1951–56)

The First Five-Year Plan (1951–56) gave the growth of forests a lot of attention. The objectives were to assemble the developing request for timber and other forest products soon, to improve and expand the areas covered by forests, and to prepare for the long-term development of forest resources (Bhaskaran 1990). In addition to this, the income that forests produced was a significant factor in the management of forests. The New National Forest Policy of India, which updated the 1894 Old Forest Policy to meet modern economic needs, was a major step in this direction when it was announced in 1952 (Azra 2012). The strategy highlighted the protective and productive aspects of forests and recommended that one third of the land area (roughly 33%) be covered by forests as a desirable long-term goal. The First Five-Year Plan's Forest development initiatives in the states were primarily carryovers from post-war development plans that had been started between 1947 and 1951 (Review of the First Five Year Plan 1957).

Plans for plantations and afforestation, bettering forest communications, creating working plans, implementing modern intensive forest management systems, introducing modern logging techniques, defining newly acquired areas to restore them, enhancing growing stock, and fortifying forest administration were among them (Review of the First Five Year Plan 1957). Given the pressing need to preserve India's diverse animal history, wildlife conservation was viewed as an essential component of forest management (Emma 1998). Because of this, the Indian Board for Wildlife was established in 1952 and has since made significant contributions to the nation's efforts to preserve its wildlife (Anon 1961). Plans for forestry and wildlife conservation programs included a total of Rs, 7.64 crores in the budgets for the center and the states.

The First Five-Year Plan also highlighted how crucial woods are to the preservation of soil. As a result, the Ministry of Food and Agriculture established the Central Soil Conservation Board in 1953 with the goal of conducting research on soil and water conservation and supporting state-level soil conservation initiatives. In addition to funding several centrally managed projects, the board approved a total of Rs, 47 lakh for soil conservation initiatives (Review of the First Five Year Plan 1957).

8.5.2 Second Five-Year Plan (1956–1961)

Aiming primarily to expand upon the processes started in the First Plan, the Second Five-Year Plan (1956–1961) allowed for significant increases in the rate of forest development (Azra 2012). Apart from completing the tasks initiated by the First Five-Year Plan, the Second Plan's initiatives encompassed afforestation, enhancing the condition of the nation's less developed forest regions, and extension forestry; planting commercial and industrial species; advocating for approaches to increase the output and accessibility of timber and other forest products in the near future; preserving wildlife; enhancing the working conditions of laborers and employees in the forest sector; augmenting the quantity of forest research; boosting the availability of technical personnel; and offering central coordination and guidance in the execution of forest development schemes throughout the country (The Second Five Year Plan 1956–1961). The plan (in both the center and state budgets) allocated a total of Rs. 21.21 crores for the development of wildlife and forestry. A plan was established that included a budget of Rs. 20 crores for soil conservation work that was also to be done intensively (The Second Five Year Plan 1956–1961).

8.5.3 Third Five-Year Plan (1961–1966)

During the Third Plan (1961–1966), the nation's efforts to meet its long-term agricultural and industrial needs were stepped up, and measures to ensure a more cost-effective and effective use of the accessible forest products, including wood residues and inferior timber, were given special attention (Azra 2012). The immediate goals were to improve forest communications, enhance the use of preservation and seasoning procedures, and increase output through improved timber extraction techniques (Anon 2006). The plan states that "developing the country's forest wealth on a long-term basis is essential not only for meeting the needs of development and various industries, such as paper, rayon, plywood, matches, and tanning, but also to supply sufficient energy and timber for the rural areas" (The Third Five Year Plan 1961–66).

The plan comprised village and extension forestry, fast-growing species plantings, commercial and industrial plantations, and other significant forest development projects (Karan 1994). Although the plan recognized biodiversity conservation as a critical component of forest development, it also contained programs for the establishment of ten wildlife sanctuaries, five national parks, and five zoological parks (Azra 2012). The Third Five-Year Plan (1961–66) allocated a total of Rs. 6.7 crores for centrally funded schemes and an additional Rs. 45.85 crores for various programs of forest advancement in states and union territories. The plan also introduced several soil conservation initiatives, with funding of roughly Rs. 72 crores allocated for their implementation (The Third Five Year Plan 1961–66).

8.5.4 Fourth Five-Year Plan (1969–74)

The Forth Five-Year Plan (1969–74) outlined three primary goals for the field of forestry: raising forest productivity, connecting forest development to other forest-based businesses, and fostering the growth of forests as a means of bolstering the rural economy. Given the extensive need for wood and other forest products for industrial purposes, the plan called for measures to be made in order to construct large-scale plantations of valuable, swiftly developing species as well as species of commercial and economic value (Pandey et al. 2011). Achieving self-sufficiency in forest products was a key goal, particularly for large forest-based companies (Saigal 2005; Prasad 2006).

The strategy called for specialized training in ecology and conservation when it came to wildlife conservation. In response, the Forest Research Institute in Dehradun established specialized training programs in wildlife conservation, management, and research (The Fourth Five Year Plan 1969–74). In 1973, two additional centrally supported lists, Project Tiger and Development of National Parks and Sanctuaries, were also established, even though it was recognized that habitat protection and efficient management were essential for the preservation of the nation's abundant wildlife. The Fourth Plan's overall expenditure for forestry programs (including wildlife) in the federal government, the states, and the union territories was Rs. 89.42 crores. This amount included Rs. 1.39 crores for programs sponsored by the federal government (The Fourth Five Year Plan 1969–74).

8.5.5 Fifth Five-Year Plan (1974–1979)

The Fifth Five-Year Plan (1974–79) saw a change in emphasis from forestry focused on conservation to a vigorous producing forestry program. The main goal of the plan was to start clearing trees and planting large-scale artificial forests with institutional funding to increase the production of industrial timber and other forest products (Gupta 1998). Increasing the supply of fuel and lumber in rural regions by developing agricultural forestry and improving damaged forests was the second major goal (Giri 2010). It was also recommended to build a system of pertinent forest surveys in conjunction with an adequate information system in order to assess the expanding stocks, increments, and predicted enhancements by forest divisions, natural regions, and states (Stebbing 1982). But the plan also highlighted the nation's widespread deforestation, placed a strong focus on intense forest development, and made sure that forest areas would not be cleared for the expansion of agricultural lands (Kishwan et al. 2009). It was suggested that efforts be made to afforest equal areas out of wastelands to make up for the lost forest area. In terms of wildlife conservation, the strategy aimed to, on the one hand, bolster the administration of wildlife and, on the other, conduct frequent surveys and enumerations of different wildlife species. In nine designated areas, where commercial forestry operations would not be conducted, special measures were planned to conserve the tiger population (The Fifth Five Year Plan 1974–79). According to the plan, the center and state budgets would provisionally allot Rs. 208.84 crores for forestry and wildlife conservation programs (The Fifth Five Year Plan 1974–1979).

8.5.6 Sixth Five-Year Plan (1980–1985)

The Sixth Five-Year Plan (1980–1985) set forth the goal of the forestry sector programs, which included the preservation of the nation's current forest cover as well as the implementation of a national afforestation and social forestry initiative. A number of laws and organizations were established for the purpose of protecting the environment and forests throughout the plan period. For example, the primary goal of the 1980 Forest (Conservation) Act was to prevent the use of forest land for uses other than forestry (Rosencranz 1991).

Similarly, the Indian Institute of Forest Management was founded in Bhopal in March 1981 with the goal of efficiently implementing forestry sector activities via the use of contemporary management methods. Additionally, the National Commission on Agriculture's recommendations led to the creation of the Forest Survey of India (FSI) in June 1981, which was tasked with conducting yearly assessments of the nation's forest resources. Consequently, the Wildlife Institute of India was established in 1982–1983 as part of the plan with the goals of conducting research, disseminating information, and offering guidance on scientific wildlife management in addition to expanding scientific knowledge on wildlife reserves (The Approach Paper to the Seventh Five Year Plan 1985). The plan assigned Rs. 692.49 crores for forestry and wildlife management options in both the center and state budgets (The Sixth Five Year Plan 1980–1985).

8.5.7 Seventh Five-Year Plan (1985–1990)

The relevance and significance of the forestry industry for the nation's economic development in terms of ecosystem protection and the supply of diverse forest products were properly appreciated for the first time during the Seventh Five-Year Plan (1985–1990) (Prabhkar 1998). Reaching the goal of having one-third of the nation's land covered by trees became imperative, and every effort was made to meet the deadline (Joshi et al. 2011). In December 1988, a revised National Forest Policy was developed as part of the plan. It placed a strong emphasis on managing forests for conservation and addressing the needs of local communities, with commercial exploitation and revenue production coming in second (Saigal 2005). Additionally, the strategy promoted public involvement in forest management and protection. Thus, with the implementation of the JFM program in 1990, a devolved system of forest management was started in the Seventh Plan (Ballabh 1996; Reddy 1998). The goals and tactics outlined in the plan included protecting biological diversity in terms of flora and fauna, conserving ecologically fragile ecosystems, and meeting basic human needs for fuel wood, fodder, minor forest products, and small timber. Additionally, it made sure that forestry initiatives and the well-being of indigenous people and other historically forest-dependent populations were closely related. It did this by emphasizing ecological research and education in particular and by putting the Wildlife Action Plan into action to conserve wildlife (The Seventh Five Year Plan 1985). The Approach Paper to the Seventh Five-Year Plan (1985) states that the plan allotted Rs, 1859.1 crores for forestry and wildlife conservation projects in the states, the union territories, and the center.

Concerns about the state of forests persisted even after the Seventh Five-Year Plan's numerous initiatives for forest conservation were launched. The nation's soil and water resources have suffered greatly because of forest destruction and degradation (Kulkami 1989). An estimated 6000 million tons of high-quality soil containing vital nutrients were being washed into the ocean annually. The loss of topsoil, vegetation, uncontrolled surface runoff, and inadequate aquifer replenishment had a detrimental impact on society overall, and on tribal people (The Eighth Five Year Plan 1992–97).

8.5.8 Eighth Five-Year Plan (1992–1997)

The Eighth Plan also launched several programs and initiatives pertaining to the preservation and protection of forests in light of these issues. The National Afforestation and Eco-Development Board (NAEB) was founded in 1992 as part of the Eighth Plan's forestry conservation initiative. Under the NAEB, several afforestation projects were started to restore degraded forest regions that bordered forests (Bowonder 1986). In addition, the Ministry of Rural Development of the Government of India established a new Department of Wastelands Development in July 1992, tasked with developing non-forest wastelands. The program for developing national parks and sanctuaries for the conservation of biodiversity was also included in the plan (The Ninth Five Year Plan 1997–2002). A total of Rs. 4910 crores were proposed for forestry and wildlife conservation projects in the center and state budgets (The Ninth Five Year Plan 1997–2002).

8.5.9 Ninth Five-Year Plan (1997–2002)

The significance of forests in conserving the earth's life support system, upholding ecological balance, and attaining the objective of sustainable development was acknowledged by the Ninth Five-Year Plan (1997–2002). Nonetheless, the nation was also gravely concerned about the country's declining forested area as a result of population pressure and several development initiatives like industry, highways, irrigation, agriculture, and power projects (Leelakrishnan 1999). During the plan in 1999, a National Forestry Action Plan was also introduced to address significant concerns in the forestry sector (Balooni and Inoue 2009). Two viewpoints were used to analyze the significant concerns that had surfaced in the convention on biodiversity: national requirements and activities, as well as international negotiations' needs, commitments, and actions. It was widely acknowledged that comprehensive law was necessary, and works on creating the biodiversity conservation through people participations had begun (Chatterjee 2007; Nomani 2004). In the Ninth Plan, it was also suggested that the Ministry of Environment and Forests establish a distinct biodiversity cell (Panda and Das 2009). A number of noteworthy initiatives were included in the strategy for the protection of biodiversity, such as 52 new research projects related to biosphere reserves, the oversight and preservation of wetlands and mangroves, and eco-development projects in seven tiger reserves (The Ninth Five Year Plan 1997–2002). A total of Rs. 8189.09 crores were allocated in the plan's center and state budgets for the execution of various forestry and wildlife activities (The Ninth Five Year Plan 1997–2002).

8.5.10 Tenth Five-Year Plan (2002–2007)

The status of forests in maintaining the sphere's life support systems and attaining environmental and economic sustainability was underlined once again in the Tenth Five-Year Plan (2002–2007). However, the plan did voice concerns about the continued depreciation of forests from a social and economic perspective (Misra 2000). Despite massive resource flows and national concern, it was found that the potential of forests to reduce poverty, accomplish growth in the economy, and contribute towards both the global and local environment had not been fully realized. As a result, the plan called for promoting integrated forest protection, which incorporates all aspects of forest protection and sustainable forest management (Sahu 2007). Tenth Five-Year Plan's Approach Paper mandated that 25% of the land be covered by forests and trees by 2007 and 33% by 2012 (Pandey et al. 2011). The plan placed a strong emphasis on managing forests, wildlife, and rural development in a coordinated and balanced manner while incorporating the knowledge, abilities, and practices of the local population into conservation strategy, planning, and management (Mathur and Sachdeva 2003). According to Vedeld et al. (2004), more targeted actions were needed to preserve natural forests for the protection of ecosystems and to uphold the local and global forest values. In order to mitigate the negative effects of natural disasters, the plan included a number of initiatives related to forest protection, including the development of agroforestry, the strengthening and improvement of the protected area network, the creation of national parks and sanctuaries, the improvement of productivity on forest plantations, the expansion of technological advances and research to boost profitability and generate new products, the avoidance of poaching and the illegal trade in wild animals and plant species, and the development of shelter belt plantations. Protection of Wildlife outside Protected Areas (PWOPA) was a brand-new initiative proposed for the Tenth Plan (Thakur 1997; World Bank 2006). Furthermore, it was recommended that the National Afforestation Program (NAP) incorporate all of the afforestation activities of the National Afforestation and Eco-development Board (NAEB) (Sunderlin and Ba 2006). In order to protect watersheds, encourage the regeneration and eco-development of degraded forests and surrounding areas, and preserve natural resources by actively involving the public and halting land degradation, deforestation, and biodiversity loss, this 100% Centrally Sponsored Scheme (CSS) was launched in 2002–2003 (The Tenth Five Year Plan 2002). For the purpose of implementing various projects pertaining to the forestry and wildlife sectors in the union territories, states, and the center, a total of Rs. 14,344 crores were allotted by the Plan (The Ninth Five Year Plan 1997–07).

8.5.11 Eleventh Five-Year Plan (2007–2012)

The Eleventh Plan's approach to the development of the forestry sector was to foster an atmosphere that would enable sustainable wildlife and forestry management, with an emphasis on the socioeconomic goals (Malhotra and Bhattacharya 2011). As a result, the strategy started several initiatives to improve the state of green cover and grow forestry (MoEF 2006a, b; ICFRE 2010; GOI 2007).

In order to meet the goal of increasing the amount of forest cover by 5% of the total geographical area, the plan concentrated on restoring the condition of degraded forests through joint forest management committees, combining forests to livelihood and increasing the involvement of citizens through the National Afforestation Program (MoEF 2009; FAO 2006). The strategy called for the intense management of already-existing commercial plantations using inputs that were scientifically planned to achieve maximum output. The Plan (in the center and state budgets) included a provision for forestry and wildlife conservation programs totaling an estimated Rs. 15583.02 crores.

8.6 Forestry Education

Forestry education is the process of distilling teaching, learning, story, information, skill, training, values, habits, beliefs, and concentrated forestry-related research. For the benefit of people and the environment, forestry science is both an art and a science (traditional and applied), according to Dhaka and Choudhari (2018). Its main goals are the comprehension, production, management, preservation, and ethical exploitation of natural resources. "Students studying forest science are better equipped to manage natural resources in a way that is sustainable for the environment, society, and economy."

8.6.1 Recommendations for Forestry Education in India

There are fifteen parts and 69 chapters in the 1976 National Commission on Agricultural Report. Six chapters (Research and Education, Forest Policy, Production and Social Forestry, Minor Forest Produce, Forest Ecology and Wildlife Management, Forest Protection and Law, and Forest Planning) which make up Part IX, which studies forestry (NTGCF 1996). But in 1974, the National Commission on Agriculture received a comprehensive report from Central Board of Forestry, which had been founded in 1950. The National Commission on Agriculture (1976) made recommendations on the topic of forestry, emphasizing a few themes including planning, policy, education, and research (NCA 1976).

According to Paragraph 4.11 on Forestry Education of the National Forest Policy (1988), "Forestry should be recognized both as a scientific discipline and a profession" (NFP 1988). Universities that specialize in agriculture and other establishments dedicated to advancing forestry education should design curricula and courses that not only teach academic knowledge but also enable postgraduate research and professional competence, considering the labor demands of the country. Candidates for jobs in the State Forest Service and the Indian Forest Service should consider their professional and academic backgrounds in forestry. It is necessary to promote specialized and orientation courses for improving management abilities through in-service training, keeping in mind the most recent advancements in forestry and allied fields (Dhaka and Choudhari 2018).

Additionally, the National Forest Commission (2006) suggested using forestry graduates in Chap. 16 with the items listed as (NFC 2006): Article 301 provides that "Forest administration should take advantage of forestry education in the universities by at least giving preference in selection for the posts of forest officers." Article 302 provides that "B.-Sc. Forestry graduates produced by universities imparting forestry education should be recruited to forest rangers." Article 303 of the Constitution says that "Forestry should be recognized as a subject for competitive examinations in state and All India Administrative Services." Lastly, Article 307 states that "Specialization is a prerequisite in forestry to enable the service to fulfill its role in conserving the forest ecosystems and its biota, in extending forestry within and without existing forests, and in fulfilling the needs and aspirations of the people" and Article 309: The aforementioned study should also take into account any modifications that should be made to the recruitment guidelines for each of the specialized sub-cadres, as well as the length of training and education that must be provided to recruits who hold degrees in fields related to forestry and forestlands, such as botany, biology, zoology, ecology, forestry, ethology, and environmental sciences, as well as to recruits with other science degrees (Tewari 1993).

However, recruits with degrees in forestry-related disciplines must be given preference over those with science degrees unrelated to forestry and ecology (Bhat 2005).

8.6.2 Forestry Education and Forest Management in India

First offered at the university level in India, forestry education began with the M.Sc., in 1976 saw the forestry in Solan, Himachal Pradesh. In 1979, the B.Sc. Forestry program was established in Ranchi, Jharkhand. A lot of Agricultural Universities began offering B.Sc. The Indian Council of Agricultural Research (ICAR) system includes a forestry program (ICAR 2009). Today, professional degree programs in forestry are offered by 28 central and state agricultural universities.

Within agricultural universities, there are distinct forestry colleges; some combine forestry with horticulture colleges. However, following the establishment of the Indian Council of Forestry Research and Education (ICFRE) in 1988, the Forest Research Institute (FRI), Dehra Dun, was elevated to the status of Deemed University in December 1991. To increase the number of forestry professionals in India, PG and Ph.D. programs in forestry were offered at FRI, Dehra Dun. Despite this, there are a number of private colleges and universities as well as central and state universities that offer forestry instruction.

8.6.3 Current Status of Forestry Education in India

When compared to the ICFRE system, forestry education in India is currently operating under the ICAR system quite successfully. Out of the 51 central/state agricultural universities in the nation, 28 offer high-quality forestry education at the Undergraduate (UG), Postgraduate (PG), and Doctor of Philosophy (Ph.D.) levels, 40 years after the program's inception.

Under the ICAR system, the nation annually produces about 1500 forestry experts. Nonetheless, under the ICFRE system, FRI offers 20 Ph.D. and 80 PG degrees in forestry each year (Dhaka and Choudhari 2018).

8.7 Management of Forest for Timber and Biodiversity Conservation

8.7.1 Forest Management for Biodiversity and Wildlife Conservation

Biodiversity provides resources that support families, communities, countries, and future generations. It is an integral element of our everyday existence and way of life. Each species, including variations, is distinct and has a specific function to fulfill within an ecological niche. Because of human activity, the rate of extinction of species has grown rapidly during the previous few decades. Many species are declining, and ecosystems are being destroyed or fragmented. The biological variety of our planet is seriously threatened by habitat loss, degradation, and fragmentation. Given our reliance on food crops, medications, and other biological resources, these losses are irrevocable and a threat to our personal wellbeing (Prasad and Kant 2003).

Sustainable resource use is the goal of the Biodiversity Conservation and Management Plan, which entails scientific management of the natural wealth in relation to developmental activities that may have an impact on these resources. In order to preserve and manage the forest ecosystems in areas with diminishing natural resources, a plan known as Biodiversity Conservation and Management has been developed (Prasad 2006). The recommendations aim to support the recruitment of significant structural features and long-term stand level maintenance, including species diversity, wildlife, riparian areas and wetlands, special or unique habitats for floral and faunal wealth, coarse woody debris, and horizontal and vertical structural diversity.

Managing biodiversity is regarded as a challenging endeavor since it encompasses diversity on all scales, including genetic, species, and community. It is difficult to execute biodiversity conservation strategies in places like Arunachal Pradesh, northeastern India, where most of the population is tribal and views itself as an essential component of the forest ecosystem.

One step toward the conservation of the environment is the creation of a strategy for the management of biodiversity and the conservation of wildlife for a developmental project. Preventing further habitat degradation is one of the most effective and efficient ways to conserve biodiversity, as human activities such as road construction, urbanization, and agricultural expansion are thought to pose serious threats to wildlife and biodiversity. Reducing anthropogenic pressure, rehabilitating endangered species, in-situ strategy, and ex-situ strategy are the four strategies needed for biodiversity management.

8.7.1.1 Biodiversity Conservation and Wildlife Management Plan

The country requires economic expansion and vigor. But environmental and ecological preservation and economic growth are not mutually exclusive or alternatives. Both must go hand in hand to preserve or, better yet, restore natural ecosystems for the continuous flow of products and services that are essential to human subsistence. Each entrepreneur must establish a relationship with natural objects and dependent communities while being mindful of the ecological effects of their industry. This will ensure the development's sustainability and provide society with enduring joy.

The examination and assessment of habitat, regular monitoring of animal population status and vegetation cover, and identification of habitat elements that promote and hinder population expansion are all components of wildlife management. The promotion of welfare factors, the arrest of unfavorable factors, and the mitigation of limiting variables improve habitat carrying capacity and allow populations to reach the intrinsic equilibrium point of the species. The local public's involvement and support are needed for the conservation strategy to be successful and for the results to be sustained. This conservation plan is designed to lessen the impact on the flora and fauna in consideration of the aforementioned factors and in compliance with legal requirements (WMP 2017).

8.8 Overview

The vast industrialization and globalization have put forestry sector to many challenges particularly the role of natural forest and its management has made it to paradigm shift from production to conservation forestry. Accordingly, the management options and tools that are used for the forest regulations were also changed greatly. In India, the traditional forest management has made a great shift during eighteenth century by the introduction of scientific principles and silvicultural tools in managing and regulating natural forest. After realizing the greater exploitation of natural forest for harvesting tangible benefits, many of the forest management activities were geared up towards the promotion of planted forest and rejuvenation of natural forest. The era of post-independence also gave importance for developmental sectors at the cost of forest resources. During 1978, the forest was bought under the concurrent list which gave focus on conservation and the same was duly supported by Forest Conservation Act 1980 and National Forest Policy 1988. Both were insisting on prevention of converting natural forest area to non-forestry purposes. Such policy regulations made to think new forest management options which brought out new components in forest regulations like joint forest management, sustainable forest management, wildlife management, and biodiversity conservation. This new array of forest management has many challenges in terms of converting government into governance. Hitherto, more importance was not given in the government sector for promoting forest management activities compared to other sectors of the government. Due to policy level changes, many good developments have taken place in forestry sectors like tree outside forest, agroforestry, farm forestry, and industrial forestry. Again, new forest management policies need to be developed and implemented to suite the new requirements so that forest can be managed both sustainably and also useful for human well-being. Towards, this objective, forest management has many challenges to face like inter sectoral cooperation, people's participation, government policies and budget, forest education, and the role of forest in industrial utility. Hence, this is the time for making a new scenario of forest management both at nationally and global levels.

Lessons Learnt

- Like any other business, managing forests entails formulating and carrying out policy decisions to accomplish the owners' goals.
- In India, forest resources are highly valued and play a crucial role in both the economy and culture.
- Long-term development of the nation's forest resources is vital in order to meet industry needs and provide fuel and fodder for rural communities.
- Plantations of fast-growing species and large-scale commercial plantations are among the main initiatives for forest growth.
- A few new institutes have been developed to effectively implement activities in the forestry sector with current management technology.
- The industrialization and globalization have put forestry sector to many challenges particularly, the role of natural forest and its management have made a paradigm shift from production to conservation forestry.
- Due to policy level changes, many good developments have taken place in forestry sectors like tree outside forest, agroforestry, farm forestry, and industrial forestry.

Key Questions

1. Trace the history and the present scenario for enacting policies related to scientific management of forest.
2. Discuss the contribution of JFM in local/tribal development.
3. Enumerate and discuss the salient features of Five-Year Plan and its impact on forest management.
4. Describe why much emphasis is given to forestry education for the upliftment of the management of forests in India in a scientific way.
5. Elaborate and discuss the need for biodiversity and wildlife conservation and their scientific management.

References

A.P. Forest Department (1999) Jyoti se Jyoti Jale. Joint Forest Management: Forest Conservation with People's Participation; a saga of success

AFC (1994) Peoples Participation in forestry—joint forest management. Agriculture Finance Corporation Ltd., New Delhi

Agarwal A, Narain S (1991) Towards green villages. Centre for Science and Environment, New Delhi

Agarwal B (1989) Rural women, poverty and natural resources: sustenance, sustainability and struggle for change. Econ Polit Weekly XXIV 43:8–46

Agarwal B (1997) Environmental action, gender equity and women's participation. Dev Chang 28(1):1–43

Agarwal B (2001a) Participatory exclusions, community forestry, and gender: an analysis for South Asia and conceptual framework. World Dev 29(10):1623–1648

Agarwal B (2001b) Participatory exclusions, community forestry, and gender—an analysis for South Asia and a conceptual framework. World Dev 29:1623–1648

Agrawal A (2001) Common property institutions and sustainable governance of resources. World Dev 29(10):1649–1672

Ahuja V (1998) Land degradation, agricultural productivity and common property: evidence from Cote d'Ivoire. Environ Dev Econ 3(1): 7–34

Angelsen A, Wunder S (2003) Exploring the forest-poverty link: key concepts, issues and research implications. CIFOR Occas Pap No. 40

Anon. (1961) 100 Years of Indian Forestry (1861–1961). Forest Research Institute, Dehradun

Anon (2006) Handbook on international environment agreements: an Indian perspective. Ministry of Environment and Forests, Centre for Environmental Law, New Delhi

Appukuttannair D, Stefanie E (2002) JFM in India: assessment of performance and evaluation of impacts, vol 77. ZEF Discussion Papers on Development Policy

Arnold JEM, Stewart WC (1991) Common property resource management in India. Tropical Forestry Paper No. 24. Oxford Forestry Institute, University of Oxford

Arora D (1994) From state regulation to people's participation: case of forest management in India. Econ Polit Wkly 29(12):691–698

Arul MJ (1998) Participatory management of forests. IRMA Working Paper 119. Institute of Rural Management Anand, Anand

Azra B (2012) Environment and forest management in India. Department of Economics, Aligarh Muslim University, Aligarh, Uttar Pradesh, India

Bahuguna VK (2001) Production, protection and participation in forest management—an Indian perspective in achieving the balance. In: Proceedings of the 16th commonwealth forestry conference on forests in a changing landscape. Promaco Conventions Pty Ltd., Canning Bridge, Australia, pp 1–16

Baland J, Platteau J (1999) The ambiguous impact of inequality and on local resource management. World Dev 27(4)

Baland JM, Platteau JP (1996) Halting degradation of natural resources. Is there a role for rural communities? Clarendon Press, Oxford

Ballabh V, Balooni K, Dave S (2002a) Why local resource management institution decline: a comparative analysis of van (forest) panchayats and forest protection committees in India. World Dev 30(12): 2153–2167

Ballabh V (1996) Joint forest management in India-promise and prospects, Mimeo. Institute of Rural Management Anand, Anand

Ballabh, V. and Singh, K. 1988. Van (Forest) Panchayats in Uttar Pradesh hills—a critical analysis. IRMA Research Paper 2. Anand: Institute of Rural Management Anand

Ballabh V, Balooni K, Dave S (2002b) Why local resources management institutions decline—a comparative analysis of Van (Forest) Panchayats and forest protection committees in India. World Dev 30: 2153–2167

Balooni K (1998) Financing of afforestation of wastelands. Unpublished Ph.D. Dissertation. Department of Economics, Sardar Patel University, Vallabh Vidyanagar

Balooni K, Inoue M (2009) Joint forest management in India: the management change process. IIMB Manag Rev 21(1)

Balooni K, Singh K (2001) Tree plantations for restoration of degraded lands and greening of India—a case study of tree growers' cooperatives. Nat Res Forum 25(1):21–32

Balooni K, Ballabh V (2000) Managing village plantations through tree growers' co-operatives—emerging issues and policy implications. Agric Econ Res Rev 13(1):54–70

Balooni K (2002) Participatory forest management in India—an analysis of policy trends amid 'Management Change'. Policy Trend Rep 2002:88–113

Bardhan P (1993) Analytics of the institutions of informal cooperation in rural development. World Dev 21:633–639

Bathla S (1999) Externalities impinging on participatory forest management in India. Policy and joint forest management series-4. World Wild Fund for nature, India

Bhaskaran ST (1990) The rise of the environmental movement in India. Media J 37(2):13–16

Bhat KM (2005) Forest education and research in India: country report 1–2. pp. 185–192

Bhatt E (1988) Shramshakti, National Commission on self-employed women and women in the informal sector. Department of Women and Child Development, Government of India, New Delhi

Bhatt JD, Thaker NM, Sharma L (2014) Joint forest management in India: a case of Haryana. Summer Internship Soc 6(1):26–33

Bleeker W, Schmitz U, Ristow M (2007) Interspecific hybridisation between alien and native plant species in Germany and its consequences for native biodiversity. Biol Conserv 137(2):248–253

Bowonder B (1986) Environmental management problems in India. Environ Manag 10(5):599–609

Campbell JY (1992) Joint forest management in India. Soc Change 22(1):36–54

Chatterjee B (2007) Environmental laws—implementation problems and perspectives. Deep and Dedeep Publication, New Delhi

Chatterjee, M. 2003. Women in joint forest management: a case study from West Bengal. Technical Paper 4. IBRAD, Calcutta

CIDA (1991) Andhra Pradesh social forestry project. Canadian International Development Agency, New Delhi

CMA (1994) Joint forest management: selected case studies from Gujarat and West Bengal. Centre for Management in Agriculture, Indian Institute of Management, Ahmedabad (Draft for discussion)

Commander S (1986) Managing Indian forests: a case for the reform of property rights. Dev Policy Rev 4:325–344

Desloges C, Gauthier M (1997) Community forestry and forest resource conflicts—an overview. In: Proceedings of the XI World Forestry Congress on forestry for sustainable development: towards the 21st century, Ankara, Turkey, vol 5, pp 109–121

Dhaka RK, Choudhari C (2018) Forestry education in India: objectives, needs, current status and recommendations. Pharma Innov J 7(12): 320–324

Eggert K (2013) An unexpected pair: the Nazis and the environment. Senior Capstone Theses, Paper 9: pp. 1–31

Emma M (1998) After Chipko: from environment to region in Uttaranchal. J Peasant Stud 25(4):36–54

FAO and Department for International Development (2001) How forests can reduce poverty. FAO, Rome

FAO (2006) Global forest resources assessment 2005, progress towards sustainable forest management. FAO Forestry Paper No. 147, Rome

Fernandes W, Kulkarni S (1983) Towards a new forest policy: people's rights and environmental needs. Indian Social Institute, New Delhi

Gadgil M, Guha R (1992) This fissured land—an ecological history of India. Oxford University Press, New Delhi

Gaikwad V (1981) Community development in India. In: Community development: comparative case studies in India, the Republic of Korea, Mexico and Tanzania. Croom Helm, London

Gilmour D, Malla Y, Nurse M (2004) Linkages between community forestry and poverty. Regional Community Forestry Training Center for Asia and the Pacific, Bangkok

Gilmour DA, Fisher RJ (1998) Evolution in community forestry—contesting forest resources. In: Victor M, Land C, Bornemeier J (eds) Community forestry at a crossroads: reflections and future directions in the development of community forestry—Proceedings of an international seminar, held in Bangkok, 17–19 July 1997, RECOFTC Report 16. RECOFTC, Bangkok, pp 27–44

Giri DK (2010) Forest rights act: combining conservation with human development, 18th Commonwealth Forestry Conference. In: Edinburgh International Conference Centre, Scotland, 28th June–2nd July

GoI (1990) Involving village communities and voluntary agencies for regeneration of degraded forest lands (Letter no. 6-21/89-PP dated1 June 1990). Government of India, Ministry of Environment and Forests, New Delhi

GoI (2000) Guidelines for strengthening of joint forest management programme (Letter no. 22-8/2000-JFM (FPD) dated 21 February 2000). Government of India, Ministry of Environment and Forests, New Delhi

GOI (2007) Report of the working group on environment and environmental regulatory mechanisms in environment and forests for the eleventh five year plan (2007–2012). Planning Commission, Government of India, New Delhi

GOI (2009) A report of planning commission on the ninth five year plan (1997–2002). Government of India, New Delhi

Government of India (1894) Forest policy, 1894, circular no. 22-F, dated 19th October 1894. Available at http://www.rupfor.org/nat-scenario/FORESTPOLICY1894.rtf

Government of India (1927) The Indian Forest Act, 1927, 16 of 1927, dated 21st September 1927. Available at http://www.rupfor.org/nat-scenario/FORESTACT1927.rtf

Government of India (1952) Forest Policy, 1952, Resolution No. 13-1/52F, dated 12th May 1952. Board of Forestry, New Delhi

Government of India (2001a) Report of the task force on greening India for livelihood security and sustainable development. Planning Commission, New Delhi

Government of India (2001b) Approach paper to the tenth five year plan (2002–2007). Planning Commission, New Delhi

Guha R (1983) Forestry in British and post British India: a historical analysis. Econ Polit Wkly 18:1940–1947

Guhathakurta P, Bhatia A (1995) JFM in West Bengal. World Bank, New Delhi

Gupta S (1998) Natural resource management: need for 21st century. Sarup and Sons, New Delhi

Haque MS (1998) Financial viability and bankability of JFM projects in India. Indian Forester 124:487–494

Heinze E, Boch S, Fischer M, Hessenmöller D, Klenk B, Müller J, Prati D, Schulze ED, Seele C, Socher S, Halle S (2011) Habitat use of large ungulates in northeastern Germany in relation to forest management. For Ecol Manag 261(2):288–296

Hobley M (1996) Participatory forestry—the process of change in India and Nepal. Rural Development Forestry Study Guide 3. Overseas Development Institute, London

ICAR (2009) New and restructured post-graduate curricula and syllabi of Forestry. Indian Council of Agricultural Research, New Delhi

ICFRE (2000) Forestry Statistics India—2000. Indian Council of Forestry Research and Education, Dehradun

ICFRE (2010) Forest sector report India, Indian Council of Forestry Research and Education. Ministry of Environment and Forest, Government of India, New Delhi

IIM (1994) Study of joint forest management. Indian Institute of Management, Ahmedabad

Inform (2002) Headlines—current news on participatory forestry. Inform Bullet Participatory Forest Manage (Resource Unit for Participatory Forestry, Winrock International India) 2(3):2

Iyengar S (1989) Common property land resources in Gujarat—some findings about their size, status, and use. Econ Polit Wkly 24:A67–A77

Jodha NS (1986) Common property resources and the rural poor. Econ Polit Wkly 21(5):1169–1181

Jodha NS (1997) Management of common property resources in selected dry areas of India. In: Kerr JM, Marothia DK, Singh K, Ramasamy C, Bentley WR (eds) Natural resource economics—theory and application in India. Oxford & IBH Publishing Co., New Delhi and Calcutta, pp 339–361

Joshi A (1998) Progressive Bureaucracy: an Oxymoron? The Case of Joint Forest Management in India. Rural Development Forestry Network, Oversea Development Institute, London

Joshi AK, Pant P, Kumar A, Giriraj A, Joshi PK (2011) National Forest Policy of India: critique and Implementation. Small Scale Forest 10: 83–96

Jyotish P (2008) JFM, Deforestation and local people participation—a case study in West Bengal, India. Available at http://moef.nic.in accessed on 20th January 2012

Kanbur R (1992) Heterogeneity, distribution and cooperation in commons property management. Background paper for the 1992 world development report. World Bank

Kant S, Singh NM, Singh KK (1991) Community based forest management system: case studies from Orissa. ISO/Swedforest Indian Institute of Forest Management and SIDA, Bhopal

Karan PP (1994) Environmental movements in India. Geogr Rev 84(1): 32–41

Karanth KK, Kramer RA, Qian SS, Christensen NL Jr (2008) Examining conservation attitudes, perspectives, and challenges in India. Biol Conserv 141(9):2357–2367

Kerr JM (1997) Market failures in natural resource management. In: Kerr JM, Marothia DK, Singh K, Ramasamy C, Bentley WR (eds) Natural resource economics—theory and application in India. Oxford and IBH Publishing Co, New Delhi and Calcutta, pp 185–222

Kishwan J, Pandey R, Dadhwal VK (2009) India's forest and tree cover: contribution as a carbon sink. Technical Paper, Indian Council of Forestry Research and Education, India

Kolavalli S (1994) Collective action for Joint Forest Management. Paper presented at the National Workshop on Joint Forest Management at the Indian Institute of Management, Ahmedabad, August 25–26

Kulkami S (1989) Forests: law versus policy. Econ Polit Wkly 24(16): 859–862

Kumar N (2000) All is not green with JFM in India. Forest Tress and People, No 42, June

Kumar S (2002) Does participation in common pool resource management help poor? A social cost-benefit analysis of joint forest management in Jharkhand, India. World Dev 30(5):763–782

Kuperus S (1987) Participation in forestry. Paper at a workshop, Department of Forest Management, Wageningen Agricultural University

Leelakrishnan P (1999) Environmental law in India. Butterworths, Lexis Nexis

Locke C (1999) Constructing a gender policy for joint forest management in India. Dev Chang 30(2):265–285

Lundmark T (1995) The recreational entrance right in Germany. North European Environmental Law 1995

MacKenzie C (1991) The challenges of participatory forest management in Sri Lanka. In Community forestry at a crossroads: reflections and future directions in the development of community forestry. In: Victor M, Land C, Bornemeier J (eds) Proceedings of an International Seminar, held in Bangkok, 17–19 July 1997, RECOFTC Report 16. Bangkok, RECOFTC, pp 231–245

Malhotra KC et al (2006) Joint management of forest lands in West Bengal: a case study of Jamboni Range in West Midnapore District. Technical Paper No. 2, Indian Institute of Bio- Social Research and Development, Calcutta

Malhotra KC, Bhattacharya P (2011) Forest and livelihood-status paper (forest ecosystem). Research Unit for Livelihoods and Natural Resources, Centre for Economic and Social Studies, Hyderabad

Mathur AS, Sachdeva AS (2003) Towards an economic approach to sustainable forest development, working paper series, Paper No. 2/ 2003-PC. In: Perspective Planning Division, Planning Commission, Government of India

McGean B (ed) (1991) NGO Support Groups in Joint Forest Management: emerging lessons. Sustainable Forest Management Working Paper No. 13. Ford Foundation, New Delhi

Mishra VK (1992) The tree growers' co-operative societies and the village commons. Paper presented at The Workshop on Co-operatives in Natural Resource Management, held in Institute of Rural Management, Anand, 7th–11th December 1992

Misra B (2000) Sustainable development—problems and challenges of 21st century. IASSI Quart 19(2):15–31

MoEF (2006a) National Environment Policy 2006, Ministry of Environment and Forest, Government of India, New Delhi

MoEF (2006b) National Status Report on Forests and Forestry in India. Ministry of Environment and Forests, Government of India, New Delhi

MoEF (2009) India forestry outlook study. Ministry of Environment and Forest, Government of India, New Delhi

Mohanraj P, Veenakumari K (1996) Nomenclature, classification and the basis of the Schedules in the Indian Wildlife (Protection) Act, 1972. Curr Sci 70(6):428–432

Molder A, Streit M, Schmidt W (2014) When beech strikes back: how strict nature conservation reduces herb-layer diversity and productivity in Central European deciduous forests. For Ecol Manag 319: 51–61

Murali KS et al (2003) Institutional and policy issues of participatory forestry. Indian Exp Trop Ecol 44(1):73–84

National Commission on Agriculture (NCA) (1976) Report of the national commission on agriculture, Part IX, Forestry. Ministry of Agriculture and Cooperation, New Delhi

NFC (2006) National Forest Commission Report, India. Government of India, New Delhi

NFP (1988) National Forest Policy, India. Government of India, New Delhi

Nomani ZM (2004) Natural Resources Law and Policy. Uppal Publishing House, New Delhi

NTGCF (1996) Mid-term review, November 1996. National Tree Growers' Co-operative Federation Limited, Anand

Panda RK, Das CR (2009) Joint forest management towards sustainable development in Orrisa: a study. IUP J Governance Public Policy 4(1):22–34

Pandey A, Bhargava P, Negi M (2011) Introduction of JFM. Indian Forester 137(7)

Pathak A (1994) Contested domains—the state, peasants and forests in contemporary India. Sage Publications, New Delhi/Thousand Oaks/London

Poffenberger M, McGean B (1996) Village voices, forest choices: joint forest management in India. Oxford University Press, New Delhi

Poffenberger M (1990) Forest management partnerships: regenerating India's forests. Ford Foundation, New Delhi

Poffenberger M (1995) The resurgence of community forest management in the Jungle Mahals of West Bengal. In: Arnold AD, Guha R (eds) Nature, culture, imperialism—essays on the environmental history of South. Oxford University Press, New Delhi, pp 336–369

Prabhkar KV (1998) Social and community forestry. International Book Distributors, Dehradun

Prasad R (2006) Strategy for sustainable forest management in India, National Status. Report for ITTO Mission to India. Ministry of Environment and Forests, New Delhi

Prasad R, Kant S (2003) Institutions, forest management and sustainable human development—experiences from India. Environ Dev Sustain 5:353–367

Singh PK, Quli SM (2011) Economic valuation of nontimber forest products contribution in tribal livelihood in West Singhbhum District of Jharkhand. Indian Forester 137(11):18–28

Raju G (1997) Joint forest management—the dilemma of empowerment. IRMA Working Paper 109. Institute of Rural Management Anand, Anand

Raju G, Vaghela R, Raju MS (1993) Development of peoples institutions for management of forests. VIKSAT, Ahmedabad

Rastogi A (1995) Impact of culture on process of joint forest management in India. Ambio 24:253–255

Ravindranath NH, Murali KS, Malhotra KC (2000) Joint Forest Management and Community Forestry in India. Oxford and IBH Publishing, Calcutta

Reddy VR (1998) Environmental movements in India: some reflections. J Indian School Polit Econ 10(4):685–695

Review of the First Five Year Plan (1957) Planning Commission. Government of India, New Delhi

Ribbentrop B (1989) Forestry in British India. Indus Publishing Company, New Delhi

Roering HW (2004) Study on Forestry in Germany. Institute for Economics, Federal Research Centre for Forestry and Forest Products, Hamburg, Germany

Rosencranz A (1991) Environmental law and policy in India—cases, materials and statutes. Tripathi Private Limited, Bombay

Saberwal VK (2001) People, parks, and wildlife: towards coexistence, vol 14. Orient Longman Private Limited, New Delhi

Sahu G (2007) People's participation in environmental protection: a case study of Patancheru. Working Paper 180, Institute of Social and Economic Change

Saigal S (2005) Joint management of state forest lands: experiences from India. In: Garforth M, Mayers J (eds) Plantations, privatization, poverty and power: changing ownership and management of state forests. Earthscan, UK

Saldanha IM (1996) Colonialism and professionalism: a German forester in India. Environ History 2(2):195–219

Samra JS, Sharda VN, Sikka AK (2002) Water harvesting and recycling—Indian experiences. Central soil and Water Conservation Research Centre, Dehradun

Saravanan V (2008) Economic exploitation of forest resources in south India during the pre-Forest Act colonial era, 1793–1882. Int For Rev 10(1):65–73

Sarin M (1998) Grassroots priorities versus social responses—The dilemmas facing community forestry management in India. In: Community forestry at a crossroads: reflections and future directions in the development of community forestry. Proceedings of an International Seminar, Bangkok, 17–19 July 1997

Saxena NC (1995) Towards sustainable forestry in the U.P. Hills. ODA, New Delhi

Saxena NC (1996) I manage, you participate—the saga of participatory forest management in India (Draft report). Mussoorie, India: Centre for Sustainable Development, LBS National Academy of Administration and Bogor, Indonesia: Centre for International Forestry Research

Saxena R (2000) Joint forest management in Gujarat—Policy and managerial issues. IRMA Working Paper 149. Institute of Rural Management Anand, Anand

Schabel HG, Palmer SL (1999) The Dauerwald: its role in the restoration of natural forests. J For 97(11):20–25

Shepherd G (1989) Joint Forest Resources Management in India: legal constraints and opportunities relevant to the proposed Forestry and Environment Project in the Western Ghats. Report prepared for the ODA: Overseas Development Corporation, London

Shrestha ML (1996) Problems in developing private forestry in Nepal. Community Forestry Case Study Series CR002/96

Singh K (1994) Managing common pool resources—principles and case studies. Oxford University Press, New Delhi

Singh S, Khare A (1993) People's participation in forest management. Commonwealth Forest Rev 72:279–283

SPWD (1992) Joint forest management: constraints and opportunities. Society for Promotion of Wastelands Development, New Delhi

Starfinger U (1997) Introduction and naturalization of *Prunus serotina* in Central Europe. Plant invasions: studies from North America and Europe. Backhuys Publishers, Leiden, The Netherlands, pp 161–171

Stebbing ARO (1982) Hormesis—the stimulation of growth by low levels of inhibitors. Sci Total Environ 22:213–234

Stebbing EP (1926) The forests of India. John Lane, London

Straub E et al (2008) The German wildlife information system: population densities and development of European Hare (*Lepus europaeus* PALLAS) during 2002–2005 in Germany. Eur J Wildl Res 54(1): 142–147

Sundar N (2000) Unpacking the 'joint' in joint forest management. Dev Chang 31(1):255–279

Sunderlin WD, Ba HT (2006) Poverty alleviation and forests in Vietnam. Forest Policy Econ 8(4):386–396

TERI (1992) Impact of participatory forest management on the ecology of Shivalik Hills in Haryana State. JFM Series 2. Tata Energy Research Institute, New Delhi

Tewari DN (1993) Forestry research: India. FORSPA, Occasional Paper 13, FAO of UN, Bangkok

Thakur K (1997) Environmental protection law and policy in India. Deep and Deep Publications, New Delhi

The Approach Paper to the Seventh Five Year Plan (1985–90) Planning Commission. Government of India, New Delhi

The Eighth Five Year Plan (1992–97) Planning Commission. Government of India, New Delhi

The Eleventh Five Year Plan (2007–12) Planning Commission. Government of India, New Delhi

The Fifth Five Year Plan (1974–79) Planning Commission. Government of India, New Delhi

The Fourth Five Year Plan (1969–74) Planning Commission. Government of India, New Delhi
The Ninth Five Year Plan (1997–02) Planning Commission. Government of India, New Delhi
The Second Five Year Plan (1956–61) Planning Commission. Government of India, New Delhi
The Seventh Five Year Plan (1985–90) Planning Commission. Government of India, New Delhi
The Sixth Five Year Plan (1980–85) Planning Commission. Government of India, New Delhi
The Tenth Five Year Plan (2002–07) Planning Commission. Government of India, New Delhi
The Third Five Year Plan (1961–66) Planning Commission. Government of India, New Delhi
Vedeld P et al (2004) Counting on the environment: forest, incomes and the rural poor. Environment Economics Series No. 98. World Bank, Washington, DC, USA
Westoby J (1987) The purpose of forests: the follies of development. Basil Blackwell, Oxford
World Bank. (2006). India: Unlocking opportunities for forest dependent people in India. Report No. 34481, World Bank: South Asia Region, Washington DC

Forest Soils

9

S. K. Gupta, Pankaj Panwar, Rakesh Banyal, and Suresh Ramanan S

Abstract

The well-defined succession of natural horizons with a distinct O horizon or organic layer is the unique feature of the forest soils. Unlike agricultural soils, vegetation is present continuously with minimal disturbance in a forested landscape. The top layer of soil, called the O horizon, is split into three parts or strata: 1. Litter layer (L-layer), 2. Fragmented layer (F-layer), and 3. Humus layer (H-layer). The litter layer comprises the remnants of deceased flora and fauna. The fully decomposed organic substance is referred to as humus. The "H" layer is frequently overlooked due to its easily crumbled texture and significant mineral content. Humus is categorized into mor humus, mull humus, and duff mull humus based on its morphological attributes. The forest soils have better soil texture, soil structure, increased porosity, lower bulk density, good aeration, reduced soil temperature, and better water retention capacity. Coniferous forests exhibit lower soil pH levels compared to deciduous forests. A soil having a cation exchange capacity (CEC) value of more than 10 meq/100 g is regarded as high CEC, typically forest soils. The soil's biological processes include activities such as litter decomposition, fine root decomposition, mycorrhizal networks, nitrogen fixation, contributing to nutrient cycling.

In the field of forestry, the term used to indirectly refer to soil quality is site index (SI), which represents the average height of trees at a specific age. SI is widely utilized and valuable due to the direct correlation between height growth and volume growth, as well as the fact that height growth is not dependent on stocking. Fire negatively affects soil structure, reducing soil organic matter, porosity, and increasing pH levels. On the other hand, decomposition positively impacts soil organic matter, mineralization, and nutrient cycling. Climate change directly impacts soil quality and fertility, while also indirectly affecting forest soils through increased pest and disease incidence, as well as heightened fire occurrences that result in reduced carbon sequestration and forest degradation.

Forestry and agroforestry play a crucial role in utilizing trees for productive purposes and ecological rehabilitation due to their ability to thrive in challenging climates and soil conditions. For reclaiming saline and alkali soils, trees species like *Acacia auriculiformis, Azadirachta indica, Ailanthus excels, Casuarina equisetifolia, Prosopis cineraria, Acacia tortilis, Dalbergia sissoo,* and *Acacia nilotica* are the ideal ones.

Keywords

Soil forming process · Soil profile · Humus · Electrical conductivity · Nutrient cycling

9.1 Introduction

Soil is a unique and important natural resource and understanding forest soils is essential for a forester for effective and efficient forest management. Forest soils are characterized as soils formed under the impact of forest cover, and their characteristics are significantly shaped by the composition of the forest. A more appropriate definition of forest soil for foresters will be "Soil science focuses on the natural resource of soil on earth's surface, especially in areas covered by forests. It explores how the forest and soil interact, affecting

S. K. Gupta (✉)
Sher-e-Kashmir University of Agricultural Sciences and Technology of Jammu, Jammu, Jammu and Kashmir, India

P. Panwar
ICAR-Indian Institute of Soil and Water Conservation, Research Center, Chandigarh, Haryana, India

R. Banyal
ICAR-Central Soil Salinity Research Institute, Karnal, Haryana, India

S. Ramanan S
ICAR-Central Agroforestry Research Institute, Jhansi, Uttar Pradesh, India

A. K. Mandal, A. Nicodemus (eds.), *Textbook of Forest Science*, https://doi.org/10.1007/978-981-97-8289-5_9

Table 9.1 Features of forest soils as compared to cultivated soils

S. no	Forest soils	Cultivated soils
1	Forest soils have a well-defined succession of natural horizons	The soil horizons may get damaged due to human activity
2	The organic horizon is present	The organic horizon is absent
3	High and efficient nutrient cycling occurs naturally	Poor nutrient cycling, so the addition of nutrients in form of fertilizers is practiced
4	Rich and diverse microflora and fauna in the soil due to modification in the microclimate	Less diversity in the cultivated soil due to frequent disturbance
5	In a forested landscape, the soil temperature and moisture condition are more consistent.	The soil temperature and moisture levels vary depending on the specific crop being cultivated. Fallow and uncultivated lands will have high soil temperature and less soil moisture conditions
6	The concentration of organic matter is quite high	Limited concentration of organic material
7	Nutrient pumping from deep soil horizon occurs due to fine roots of trees	No nutrient pumping
8	Certain forest soils lack proper drainage, have steep slopes, and contain large boulders and stones	Mostly well-drained, flat, pulverized soils

Source: Binkley and Fisher (2013)

each other's formation and properties-covering physical, chemical, and biological aspects. The goal is to understand how these soil properties impact sustainable use for various forest outputs, such as wood, wildlife, water, and environmental well-being"(Osman 2013).

The distinctive quality of forest soils is influenced by the deep roots of trees, the addition of nutrients through litter and fine root decomposition, and the array of microorganisms found under the tree canopy. Yet, not every forest soil is fertile enough to sustain a diverse range of forest vegetation. In some places, forest soils are poor in soil fertility, steep-sloped, and stony, have low moisture, and even may comprise of saline or sodic soils (in mangrove forests). Despite high variability, still, these soils support forest vegetation. Table 9.1 outlines the distinctive characteristics of forest soil in comparison to cultivated soil.

9.2 Soil Forming Processes in Forests

The pedogenic processes are faster than geological processes in transforming lifeless parent material into fertile soil. These processes are intricate and ever-changing, encompassing biological and chemical reactions that occurs in a particular area. Occasionally, one process may counterbalance another, while in other instances, diverse processes collaborate to attain a common result. The functioning of these processes is influenced by various natural conditions, and the interplay of soil-forming factors in different environments dictates the specific occurrences of soil-forming processes (Binkley and Fisher 2013). The process involved in soil formation are as follows (Fig. 9.1):

- The soil gets water from rainfall and receives organic and mineral substances.
- The mentioned materials have been depleted.
- The soil undergoes alterations as minerals and organic substances undergo transformations within it.
- Translocation involves the movement of soil materials from one place to another within the soil. This process is typically categorized into different phases:
  - Solution movement (leaching), and.
  - Suspension movement of organic matter, hydrous oxides and clay.

9.2.1 Soil Profile

A soil profile refers to "A cut section of the soil that displays different layers, ranging from the top layer to the untouched parent material and these layers are called as horizons." In a hypothetical mineral soil profile, one can expect to find O, A, B, C, and R master horizons, along with all the potential sub-horizons (Fig. 9.2).

The organic horizon O is predominant in forest soils compared to agricultural soil. Unlike agricultural soils, vegetation is present continuously with minimal disturbance in a forested landscape. The time between forest disturbances lets organic soil layers (O horizon or O layer) form on top of soil. The thickness of the layer depends on the diversity of plants that grow in the forest. For instance, the moist deciduous forest has a thick O horizon compared to dry deciduous forest or thorn scrub. Also, animal waste, animal and plant debris, microorganisms all contribute to creation of the layer. The segregation of organic material within forest soils not only alters root distribution and vegetation growth but also significantly influences nutrient cycling. This impact extends to potassium immobilization while aiding in the gradual release

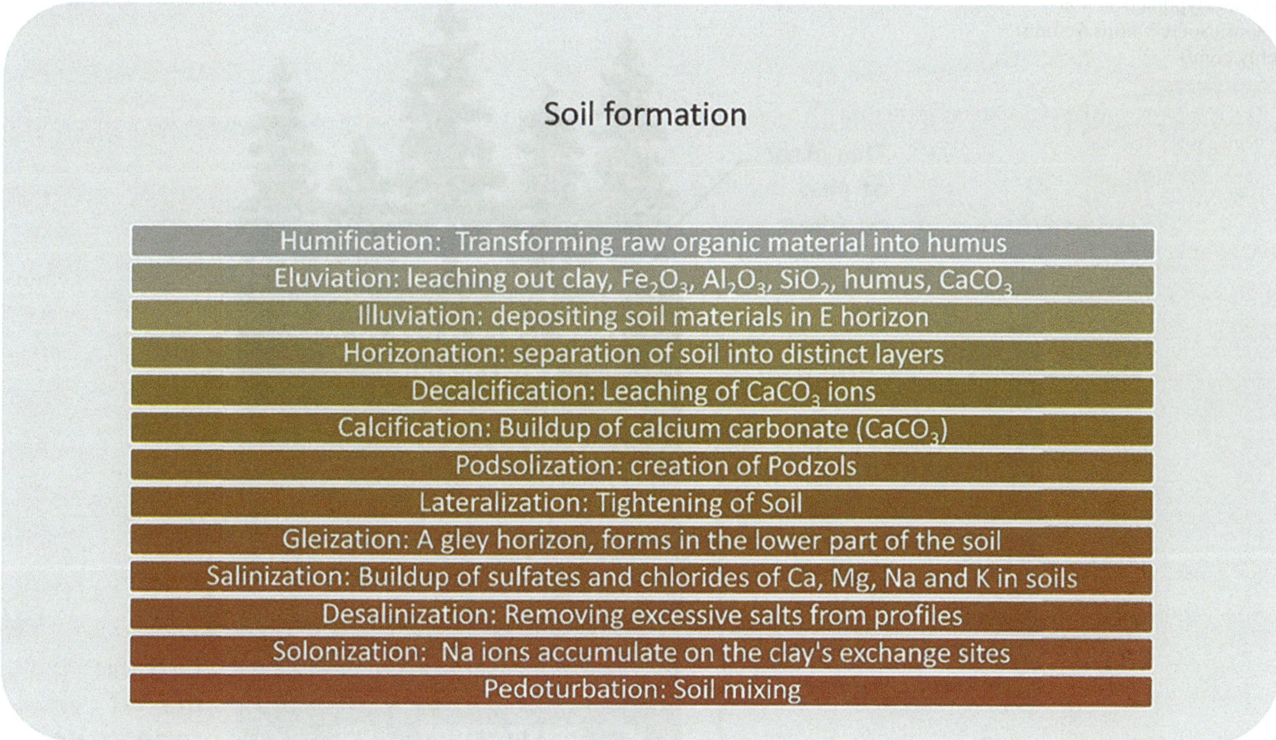

Fig. 9.1 Various soil forming process

of nitrogen and phosphorus. The organic substances combine with minerals and readily influence the soil aggregate formation, thus affecting the soil structure and soil productivity. The soil microorganisms thriving on this organic layer produce polysaccharide gums and the fungal hyphae network are the important components aiding soil aggregation. The organic matter increases the aeration, infiltration and permeability of soils. This layer is designated into three different layers or strata as under:

1. Litter layer (L-layer).
2. Fragmented layer (F-layer).
3. Humus layer (H-layer).

The litter layer is the remains of dead plants and animals. This layer still retains the structural integrity, i.e., decomposition has not yet started. In "F" layer, fragmentation and partial decomposition of the litter has started. The strength and stability of the litter's structure start to breakdown. Finally, in the "H" layer, the litter becomes decomposed amorphous organic matter. The completely decomposed organic matter is called humus. The "H" layer is commonly disregarded due to its friable, crumb-like structure, despite containing a notable proportion of mineral materials. It is the place where the carbon gets stored in the soil. The C: N ratio of humus is around 10–12:1, cation-exchange capacity (CEC) is 100–300 C mol(P^+)/kg soil, with a density of 1.3–1.5 Mg/m^3.

Humus is categorized by its morphological traits into three main types: mor humus, mull humus, and duff mull humus. It is the simplest classification possible, though more detailed information is available in Chertov and Nadporozhskaya (2018).

(a) Mor humus: Mor humus refers to the surface accumulation of organic residues, featuring a distinct organic layer separated from the underlying mineral horizon. Fungi play a significant role as the predominant microorganisms in this type of humus, contributing to its acidic nature.
(b) Mull humus: Mull humus refers to the complex blend of amorphous humus and mineral soil. It contains a higher proportion of organic substances combined with minerals like sodium, magnesium, calcium, and potassium. Bacteria and actinomycetes are the prevalent microorganisms in mull humus. Unlike mor humus, it doesn't sharply distinguish itself from mineral soils and typically showcases a granular or crumb-like structure. Mull humus often hosts large populations of earthworms.
(c) Duff mull humus: This humus type shares traits from both mull and mor humus. It forms due to ongoing

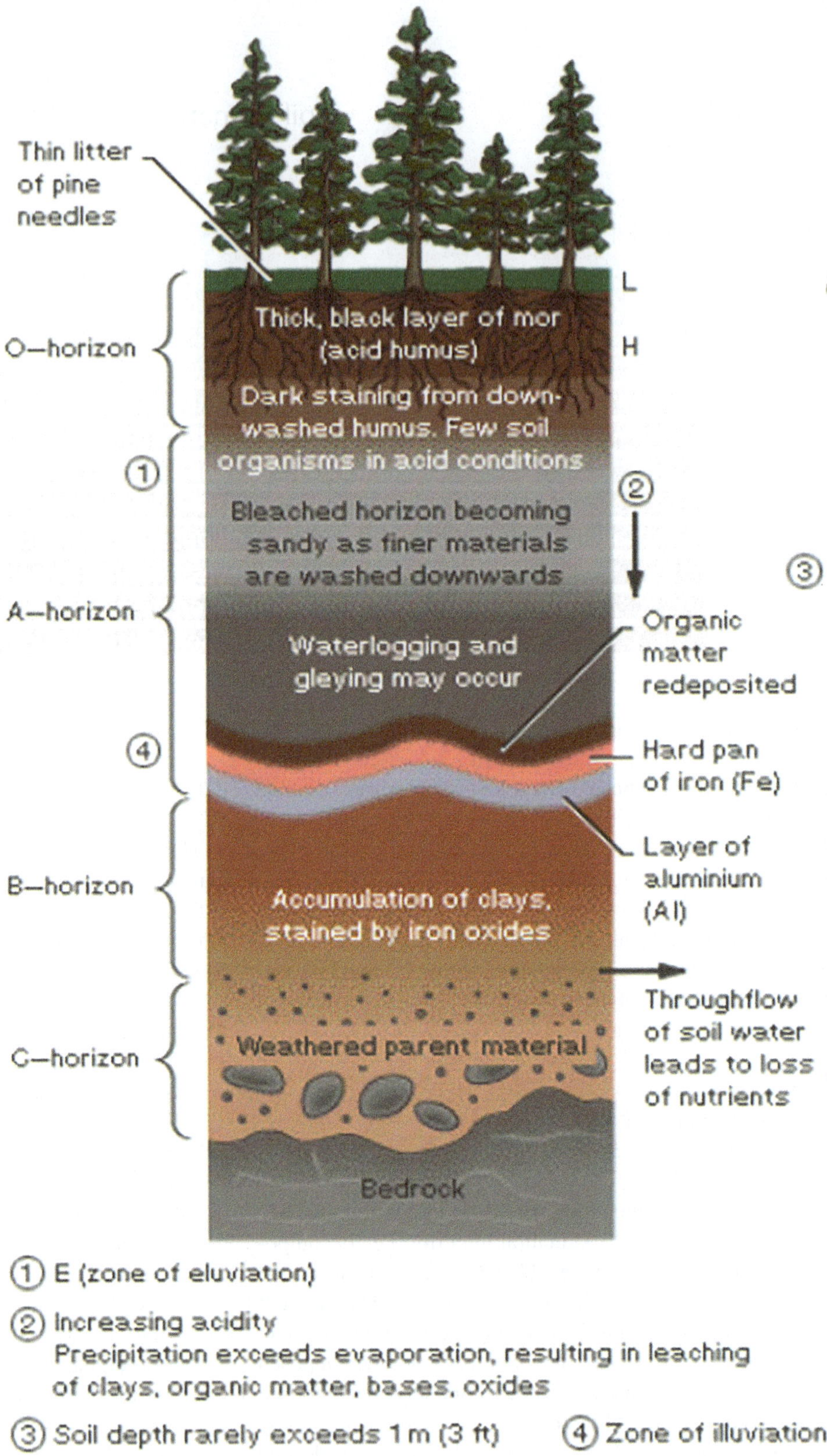

Fig. 9.2 Hypothetical soil horizons (Source: https://cforest.weebly.com/)

blending of mineral soil into the lower forest floor by ants or rodents, maintaining the organic horizons without destroying them.

The uppermost layer, eluviation horizon, is susceptible to leaching, leading to the leaching out of specific soil constituents. Horizon B, the illuviation horizon, is where these leached materials find redeposition. Horizon C signifies the parent material. An analysis of the soil vertical section is paramount as it encapsulates a historical narrative of soil-forming processes. This serves as the cornerstone for pedagogical inquiries and soil classification. Understanding the soil profile is instrumental in effectively harnessing soil resources.

In forest ecosystems, there's a consistent and substantial accumulation of organic material, fundamentally enriching the soil's physical, chemical, and biological characteristics.

9.3 Nature of Forest Soils

The properties inherent to soil play a pivotal role in determining the health and productivity of forests. The continued presence of vegetation has made forest soils unique compared to other land use. Disturbances (either natural or anthropogenic) inside the forests can alter the soil properties. Foresters generally prefer usage of the term "site quality" while discussing with forest soil productivity. Although there are fundamental distinctions in assessing the soil fertility and site quality, still site quality can be regarded as an indirect conation of soil properties as site quality is directly determined. The site index, which gauges the average height of dominant trees at a specific age within a stand, is considered a gauge of site quality. However, it may not accurately represent the genuine physical and chemical attributes of the forest soils.

The soil properties are categorized as physical, chemical, and biological which are influenced by several factors and understanding them becomes essential for the forest as well as plantation management.

9.3.1 Physical Properties

Plant growth purely depends upon the root architecture linked with water and nutrient uptake. Root spread and growth are often restricted by mechanical impedance, shallow bedrock, lack of aeration, low temperature, and sometimes even the presence of heavy metals or other toxic substances. Apart from these factors, root growth is hampered by soil compaction, hard pans. Holistically this can be comprehended as soil physical properties.

The key physical properties of soil consist of its texture, structure, porosity, density, aeration, temperature, and water retention capacity. These properties are determined by the size, distribution, and arrangement of particles within the soil matrix. Starting from the availability of water, water absorption, ease of root penetration, oxygen availability, and many other vital functions dependent on soil physical properties which act as indicators for soil fertility and productivity. Human activities such as logging, land operations, and shifting cultivation can notably modify the soil's physical properties. (Osman 2013).

9.3.1.1 Soil Texture

Soils are primarily composed of sand, silt, and clay, and their texture varies based on the proportions of these components. It is an innate factor of the soil—an outcome of weathering process and is difficult to alter. Different classifications for soil texture are provided by organizations such as the US Public Roads Administration, Massachusetts Institute of Technology, International System (Atterberg Classification), US Bureau of Soils, Indian Standards Institution, and British Standards Institute. Amongst these, the classification by the US Department of Agriculture presented in Table 9.2 is widely referred.

Table 9.2 Classification of soil texture

Texture of the soil	Sand	Silt	Clay	Textural class
Coarse (sandy soils)	86–100	0–14	0–10	Sand
	70–86	0–30	0–15	Loamy sand
Moderately coarse (loamy soils)	50–70	0–50	0–20	Sandy loam
Medium (loamy soils)	23–52	28–50	7–27	Loam
	20–50	74–88	0–27	Silty loam
	0–20	88–100	0–12	Silt
Moderately fine (loamy soils)	20–45	15–52	27–40	Clay loam
	45–80	0–28	20–35	Sandy clay loam
	0–20	40–73	27–40	Silty clay loam
Fine (clayey soils)	45–65	0–20	35–55	Sandy clay
	0–20	40–60	40–60	Silty clay
	0–45	0–40	40–100	Clay

(Source: http://www.fao.org/tempref/FI/CDrom/FAO_Training/FAO_Training/General/x6706e/x6706e06.htm)

9.3.1.2 Structure of Soil

It is the segregation of sand, silt, and clay particles, forming aggregates. Essentially, it's how primary and secondary particles are organized within the soil mass. This structure governs the occurrence of water and air in the soil. For seeds and roots to grow, they need ample air and oxygen. Bacterial activity also relies on the presence of air and water molecules in the soil. Soil particles can exist individually as grains or as aggregates—clusters of particles bound into granules or compound particles known as secondary particles. Sandy or silty soil mostly comprises single grains, whereas clayey soil tends to form aggregates. Individual particles are typically solid, while aggregates have a porous or spongy character. Soil lacking aggregates is considered structureless, whereas those with secondary particles are termed aggregate, granulated, or crumb structure.

9.3.1.3 Soil Density (Bulk and Particle Density)

Soil density is commonly described through two key concepts: particle density and bulk density.

Particle density: Particle density refers to the weight per unit volume of the solid part of soil. It's quantified in the metric system as megagrams per cubic meter (Mg m^{-3}). For instance, if 1 cubic meter of soil solids weighs 2.6 Mg, the particle density is 2.6 Mg m^{-3} (since 1 Mg = one million grams and 1 m^3 = one million cubic centimeters). Consequently, particle density can also be articulated as 2.6 g cm^{-3}. Typically, normal soils exhibit a particle density of 2.65 grams per cubic centimeter. When a considerable number of heavy minerals like magnetite, limonite, and hematite are present, the particle density tends to be higher. Conversely, as soil organic matter increases, particle density decreases. Particle density is also referred to as true density.

Bulk density: Bulk density signifies the weight of a specific soil volume, accounting for pore spaces, after drying in an oven. It consistently registers lower than particle density. Sandy soil typically displays a bulk density around 1.6 g cm^{-3}, while organic matter tends to be approximately 0.5. Finer-textured mineral soils generally exhibit reduced bulk density. This density inversely correlates with total pore space, offering a reliable gauge of soil porosity. Understanding bulk density is more pivotal than particle density in comprehending soil's physical characteristics.

Soils boasting low bulk density often provide favorable physical conditions. For instance, forest soils commonly showcase lower bulk density compared to deforested areas. Other physical attributes like soil temperature and water retention predominantly concern agricultural or cropped land utilization.

9.3.2 Chemical Characteristics of the Soil

Soil is a blend of organic, inorganic solids, solutes, liquids and gases. The inorganic solids present in soil comprise fine silicate clays, oxides, and hydroxides of elements like iron and aluminum. Additionally, substances such as lime, gypsum, phosphates, and various essential nutrient ions are part of this composition. All these characteristically influence soil chemical properties. Among the chemical properties of soil, the most frequently studied ones include pH, cation-exchange capacity (CEC), anion-exchange capacity (AEC), base saturation (BS) percentage, exchangeable sodium percentage (ESP), and electrical conductivity.

9.3.2.1 pH

Generally assessed chemical properties is pH, a concept introduced by Arrhenius. pH represents the negative logarithm of the reciprocal of H+ ion concentration in the soil solution, consisting of dissolved solutes like ions, ion pairs, ligands, and complexes. Soils are rated on a scale from 1 to 14, based on the amount of presence of hydrogen ions. A pH below 7 indicates acidity, while above 7 indicates alkalinity. For a heterogeneous medium like soil, pH values between 6.5 and 7.5 are considered fairly neutral. Lower pH values signify higher acidity.

Coarse-textured, highly weathered, and intensely leached soils tend to be acidic. Soil acidity can also increase due to the increase of organic substance and subsequent decomposition, leading to humus formation. Typically, soils in coniferous forests tend to have lower pH levels in contrast to those in deciduous forests. The pH level alone will not impact plant growth; however, pH value influences the nutrient availability which is vital for plant growth. In acidic soils, essential nutrients like phosphate, calcium, magnesium, potassium, and molybdenum become less available to plants, often leading to deficiency symptoms. Conversely, when soil pH exceeds 8, it becomes sodic or alkaline, characterized by a sodium adsorption ratio (SAR) of 15 or higher. Such soils face limitations in nutrient availability and are deficient in zinc, iron, phosphorus, and occasionally calcium, potassium, and magnesium. However, nitrogen scarcity becomes the primary concern in alkaline soils. Sodic soils might contain phytotoxic concentrations of boron. High bicarbonate levels primarily trigger iron deficiency, which can worsen in high soil moisture or anaerobic conditions.

9.3.2.2 Cation Exchange Capacity

Elements present within the soil solution exist in an ionic form that interacts with soil particles. This interaction is gauged as exchange capacity, which can be either cation

Table 9.3 Classification of soils based on ESP and EC level

Nature of soil	EC (dS m^{-1})	ESP	pH
Saline	>4.0	<15	<8.5
Sodic	<4.0	>15	>8.5
Saline-sodic	>4.0	>15	>8.5

exchange capacity or anion exchange capacity. Generally, most of the soils have negative charges and thus cation exchange capacity is commonly measured in soils. It is defined as the quantity of centimoles of charges per kilogram of soil. It primarily relies on the presence of clay minerals and humus content within the soil. CEC signifies the soil's ability to retain cations at particle surfaces. Similarly, for anions, it is expressed as anion exchange capacity. Predominately the soils have higher CEC due to their innate negative charges. Measuring CEC is important because vital plant nutrients like potassium, manganese, calcium, magnesium, and zinc are present as cations in the soil. CEC is also correlated to the pH and water holding capacity. A soil having a CEC value of more than 10 meq/100 g is regarded as high CEC, typically the forest soils. Indeed, activities like deforestation, forest fires, and other disturbances diminish the organic content of the soil, consequently reducing its CEC.

9.3.2.3 Exchangeable Sodium Percentage (ESP) and Electrical conductivity (EC)

Excess of sodium in the soils can hamper plant growth. Therefore, assessing the sodicity or alkalinity of the soils is vital. In agricultural land, measurement of EC and ESP is crucial as they decide on the nature of crops that can be cultivated. On contrary, measurement of EC or ESP in forest soils will not be so important. However, before taking up any afforestation and reforestation activity these are taken into account. It has a direct implication on microbial diversity thus affecting processes like decomposition, nitrification, denitrification, and many others. Depending on the ESP and EC levels, the soils are classified as given in Table 9.3.

9.3.3 Biological Properties

"Essentially, all life depends upon the soil ...There can be no life without soil and no soil without life; they have evolved together"- Charles E. Kellogg (van Elsas et al. 2019).

Apart from the mineral and organic component, the soil contains water and diverse organisms inside it. Soils function as living systems where plant roots intermingle with microbial and animal communities thriving within the soil.

The extensive variety of microbes and animals dwelling in soil forms the soil food web, crucial for recycling organic substance obtained from plants above ground. This network encompasses a multitude of members, predominantly microbes like bacteria and fungi, alongside diverse animal species of varying sizes within the soil. They are classified into size-based groups: Microfauna (1–1000 μm), including bacteria, fungi, protozoa, and nematodes; Mesofauna (100 μm to 2 mL), encompassing tardigrades, mites, microarthropods, and enchytraeids; and macrofauna (also called megafauna) (>2 mm), comprising earthworms, ants, termites, and millipedes.

These soil fauna occupy different trophic levels and are typically grouped based on their feeding habits within the soil food web. Some primarily consume microbes (microbial-feeders) or litter (detritivores), while others predominantly feed on plant roots (herbivores) or other animals (carnivores). Studies have highlighted the prevalence of omnivores in soil food webs, suggesting their tendency to feed across various trophic levels (Mikola and Setälä 1998). The simplest classification of soil animals based on their body size is tabulated in Table 9.4.

9.4 Nutrient Dynamics in Forest Soils

9.4.1 Nutrient Cycling

Understanding interactions between forest trees and soils is an essential pre-requisite for a higher and sustained or progressive yield of timber. In long-duration forest crops, assessing plant-available nutrients becomes complex due to challenges in correlating soil nutrient availability with their cycling processes. Various methods, including site classification, soil analysis, foliar analysis, have been employed to define nutrient availability and soil fertility, facilitating the assessment of nutrient cycling.

(i) *Site classification method*: The site classification method addresses soil fertility within forest lands. Originally developed for British Columbia, this system, adopted by the B.C. Forest Service utilizes an edatopic grid. It organizes sites based on soil fertility and moisture regimes for effective classification (Klinka et al. 1981; Lowe and Klinka 1981). One negative point of this system is the lack of an objective method to ascertain soil fertility at a specific site and time. It relies on classes determined by the mineralogical composition of parent material, forest floor, and ground cover vegetation. Although effective for classifying extensive land areas concerning wood production, its adaptability across regions or providing reliable

Table 9.4 Classification of soil animals

Group	Known species	Estimated species	% described
Microfauna			
Nematodes	20,000–25,000	1,000,000–10,000,000	0.2–2.5
Protists	21,000	7,000,000–70,000,000	0.03–0.3
Fungi	97,000	1,500,000–5,100,000	1.9–6.5
Bacteria	15,000	>1,000,000	<1.5
Mesofauna			
Mites	40,000	100,000	55
Collembolans	8500	50,000	17
Macrofauna			
Earthworms	7000	300,000	23
Ants	14,000	25,000–30,000	50–60
Termites	2700	31,000	87

Source: Orgiazzi et al. (2016)

information for various tree species on a particular site can be challenging.

(ii) *Soil analysis method*: The method of soil analysis refines the classification of forest land into distinct site classes. It involves assessing the soil characteristics at a specific location and establishing connections between anticipated tree growth and soil parameters. In addition to the site-specificity of this approach, it also includes soil physiochemical characters that are recognized as growth-limiting factors. Yet, the accuracy of this approach concerning individual nutrients is restricted due to the diversity found within soils.

(iii) *Foliar analysis*: Another method to assess nutrient availability involves analyzing the nutrient content present in foliage. Foliar analysis diagnoses the deficiency and also helps in the prediction of tree response with fertilizer applications. Due to numerous limitations in correlating foliar nutrient concentrations with tree growth or soil nutrients, it doesn't seem to serve as a reliable indicator of soil nutrient availability.

Thus, to have a fair idea on nutrient cycling a combination of all the methods is essential. Studies on nutrient cycling in forest ecosystem have various factors that are outlined below and need to be studied at both spatial and temporal scale.

a) **Nutrient uptake in the forest biomass**: The way nutrients are absorbed by forest biomass is comparable to other crops, yet the quantity accumulated in forest biomass typically ranges from one-quarter to one-third of the overall uptake (Miller 1981). Coniferous forests typically exhibit lower nutrient demands compared to hardwoods, although there's substantial variability within both categories. However, assessments of forest nutrient requirements are predominantly obtained from previous studies conducted on relatively homogeneous natural stands or single-species plantations.

b) **Nutrient input from atmospheric sources**: Several researchers have documented the atmospheric deposition of nitrogen in various forms, including nitrate, ammonium, and diverse organic nitrogen compounds. Researchers have estimated the overall (combining wet and dry forms) nitrogen quantities in bulk precipitation, ranging from approximately 3 kg N ha^{-1} $yr.^{-1}$ in unpolluted regions (McColl et al. 1982) to as much as 20 kg N ha^{-1} $yr.^{-1}$ in polluted areas (Likens et al. 1977). While these additions of nitrogen from the atmosphere might seem minor compared to the nutrient quantities within forest biomass (Cole and Rapp 1981) and forest soils (Mahendrappa 1980), over the average rotation age of forests (usually 60–80 years), even the lowest rates of annual nitrogen deposition can contribute significantly to the total nitrogen content in forest ecosystems. Lindberg et al. (1986) suggested that atmospheric deposition accounted for 40–100% of the nitrogen and sulfur needed for annual woody growth in a mixed hardwood forest.

c) **Nutrients cycled as through fall and stem flow**: Through fall and stem flow can contribute nutrient transfer to the soil, particularly the K and Mg (Parker 1983). Despite potassium's tendency to leach from foliage and its mobility within forest ecosystems, there is substantial evidence indicating that forest vegetation efficiently conserves potassium (Stone and Kszystniak 1977).

d) **Re-translocation within trees**: The movement of nutrients inside plants is crucial for their cycling and conservation. Before leaves fall, a significant amount of nutrients, especially nitrogen and phosphorus, is taken from the leaves and moved within the tree. Research by Miller (1984) suggests that 60–84% of the nutrients in foliage are re-distributed before the leaves drop. This re-distributed nutrient often meets 50–60% of the needs of trees like *Pinus nigra* in Scotland. Similar patterns are observed in *Pinus pinea* (Rapp et al. 1979). Hardwood trees tend to move more nitrogen before leaf fall compared to conifers, and this behavior seems to vary by tree species. The quantity of a nutrient moved within trees is influenced by how much of that nutrient is available in the soil. Trees in areas with limited nutrients tend to move more nutrients out of their leaves than species in nutrient-rich areas (Melillo and Gosz 1983).

e) **Litter fall**: The amount of litter produced in forests is closely linked to their basal area and age. Both deciduous and coniferous forests shed a substantial amount of leaf biomass in autumn and smaller amounts throughout the year. In conifers, about 40–60% of the total annual litter fall occurs in autumn (Perala and Alban 1982), while for hardwoods, over 80% happens in autumn (Morrison 1985). Nutrient cycling studies often overlook the yearly contribution of organic matter and nutrients from the decay of fine roots. Research by Vogt et al. (1982) estimated that in silver fir forests, the return of macronutrients to the soil from fine roots (including mycorrhizae) is 3–10 times greater annually compared to aboveground litter fall. Additionally, Kimmins and Hawkes (1978) found that mature white spruce-subalpine fir stands in British Columbia had significant fine root biomass, measuring 1870 kg ha^{-1} with 38 kg N ha^{-1} for overstory plants, and 7880 kg ha^{-1} with 139 kg N ha^{-1} for understory vegetation.

f) **Nutrient supply to the trees**: Nutrient uptake by trees comprises two stages: transporting nutrients to the root surface and then absorbing them into the tree. Certain elements, such as aluminum, chlorine, and sodium, are filtered out by plants, while they selectively absorb others like manganese, calcium, phosphorus, nitrogen, and potassium (Prenzel 1979). The primary constraint in nutrient uptake for plants lies in how quickly ions moves to the roots from the soil. Therefore, the abundance of roots plays a pivotal role in tree nourishment and nutrient circulation. Tree roots tend to explore deeper soil layers when the soil's physical and chemical properties are conducive to nutrient absorption (Sutton 1969).

g) **Leaching**: Forest nutrient uptake decreases when ions leach beyond the root-absorption zone. According to McColl and Cole (1968), cation leaching in forest soils is restricted by mobile soil anions' availability. Comparatively, leaching losses are insignificant in forest conditions when considering the overall nutrient cycling processes (Mahendrappa and Salonius 1982).

h) **Mycorrhiza**: The presence of mycorrhizal roots in low soil fertility conditions increase nutrient uptake by forest trees. In nutrient-poor and water-deprived soils, forest trees must allocate more photosynthate and energy to create and maintain an active root system (Keyes and Grier 1981).

i) **Weathering**: Weathering primarily produces base cations like calcium (Ca), magnesium (Mg), potassium (K), and sodium (Na), along with certain acid cations such as iron (Fe), aluminum (Al), and manganese (Mn). However, it generates very limited amounts of phosphorus (P). Depending on the parent material's characteristics, the estimated nutrient contributions (Ca, Mg, and K) from weathering range widely from under 10 to over 700 kg ha^{-1} $yr.^{-1}$ (Johnson et al. 1984). Even the smallest estimates of nutrient input from weathering over a rotation period of up to 100 years can be substantial in supplying nutrients.

j) **Nitrogen fixation**: Within the forest ecosystem, elemental nitrogen (N_2) is converted into plant-absorbable ammonium nitrogen through various processes. The estimated annual rates of nitrogen fixation by symbiotic or free-living organisms vary widely, ranging from 0.22 to over 50 kg N ha^{-1} $yr.^{-1}$, applicable to both hardwood and softwood stands. For some stands, the nitrogen needed can be fulfilled by the nitrogen fixed by microorganisms. However, soils experiencing high rates of nitrogen fixation often exhibit increased leaching of base cations due to elevated nitrification rates (Cole et al. 1967).

Other disturbances occurring in forest like forest fire and wild animals also affect nutrient cycling.

9.4.2 Soil and Its Relation to Stand Dynamics

Changes in stand structure and composition are shaped by disturbances, tolerance, and competition (Hart and Chen 2008; Rohne et al. 2012; St. Clair et al. 2013), as well as soil attributes. In forestry, soil quality is often indirectly measured through site index (SI), which represents the average tree height at a specific reference age. SI serves as an indicator of the long-term productivity of forest stands. It's commonly utilized in even-aged single-species stands (Skovsgard and Vanclay 2007) but also used in mixed-species scenarios (Anyomi et al. 2013). The direct correlation between stand height growth, volume growth, and the independence of height growth on stocking (Skovsgard and Vanclay 2007) makes SI a widely used and valuable factor to calculate the forest productivity. Additionally, Anyomi et al. (2015), in evaluating stand dynamics and soil relationships, highlighted soil texture as a major influencing factor ($R2 = 8\%$) among various soil parameters.

Broadben et al. (2014) conducted research in the Amazon forest and discovered notable associations between age of stand and various soil properties. They found a counter correlation value between age of stand and soil d13C ($R2 = 0.45$), while observing positive correlations with cation-exchange capacity (CEC) ($R2 = 0.31$) and soil nitrogen ($R2 = 0.34$). Interestingly, positive correlations of significance were discovered between stand age and the total soil mass at depths of 0–30 cm ($R2 = 0.55$), specifically within 10–20 cm ($R2 = 0.55$) and 20–30 cm ($R2 = 0.55$). However, there is no significant correlation in between 0 and 10 cm depth. Although soil % C did not relate to bulk density, a primitive negative impact was observed between soil organic substance (%) and bulk density ($R2 = 0.26$). In a study by Feldpausch et al. (2004) on young successional stands (14 years), they noticed an increase in soil nitrogen while phosphorus relocated from soil to biomass. This trend during succession indicates a decline in available soil phosphorus due to nutrient redistribution to vegetation, leading to soil phosphorus deficit (Markewitz et al. 2004). Soil nitrogen, on the other hand, showed potential increases due to symbiotic nitrogen fixation, particularly in early succession stages (Cleveland et al. 1999; Rastetter et al. 2001) and from atmospheric deposition (Holland et al. 1999). Interestingly, soil phosphorus did not exhibit variation concerning stand age, suggesting these alterations were unrelated to the succession process but exhibited a higher counter correlation with soil clay. Additionally, Silver et al. (2000) identified an increase in soil clay content led to decreased labile phosphorus, while total phosphorus increased, attributed to phosphorus easily forming complexes with exchangeable aluminum and iron.

9.4.3 Forest Disturbances

Among forest disturbances, forest fire is the main factor. Fires significantly deteriorate structure of the soil and diminish humus, leading to decreased porosity and elevated pH levels (Verma and Jayakumar 2012). They result in a decline in soil nutrients within forests due to processes like volatilization, ash transport, leaching, and erosion. These alterations amplify hydrophobicity, lower infiltration rates, and heighten runoff (Jiménez-Pinilla et al. 2016). Over the long term, fires might contribute to reduced nutrient availability by removing nutrients obtained from soil humus and microbial biomass via runoff and leaching from the ecosystem (Miesel et al. 2012).

9.4.4 Forest Litter and Soil Fertility

Different factors like stand age, density, and development stage significantly impact litter fall rates. In *Acacia auriculiformis* plantations, litter fall ranged from 12.2 to 14.4 Mg ha^{-1} $yr.^{-1}$ (Kumar and Deepu 1992), while Kunhamu et al. (1994) reported an annual litter fall of 12.9 Mg ha^{-1}. Comparatively, litter fall tends to be lower in agroforestry systems than in tropical forests. For instance, Sharma et al. (1997) mentioned values ranging from 1.5 to 2.9 Mg ha^{-1} $yr.^{-1}$ for leaf and twig components in agroforestry systems in Sikkim Himalayas, India. In humid central Kerala's silvopastoral systems, litter fall varies significantly among four multipurpose tree species, recording rates from 1.92 to 6.25 Mg ha^{-1} $yr.^{-1}$.

Nutrient recycling hinges on the decomposition of litter by soil organisms, a critical process defining soil quality and forest productivity (Wang et al. 2008). Rapid nutrient release might lead to losses through leaching or volatilization (Palma et al. 1998). Conversely, slow decomposition could inadequately supply nutrients to plant roots, thereby limiting plant growth and development (Montagnini and Jordan 2002). Hence, the rates of litter decomposition and nutrient release are vital factors for ecosystem functioning.

The decomposition of forest litter stands as a pivotal phase in the genesis of soil organic matter. This intricate process is subject to modulation by diverse factors encompassing soil composition, its physical and chemical attributes. Among these factors, texture emerges as a paramount influencer, exerting pronounced effects on nutrient and water dynamics, porosity, permeability, and surface area. The mineralization of organic nutrients and the equilibrium of carbon within forest ecosystems, as elucidated by Austin and Ballaré (2010), play a pivotal role in steering the nutrient cycle, aligning with the findings of Song et al. (2008), thereby enhancing their accessibility for optimal plant growth, in line with the insights provided by Pan et al. (2004). Moreover, the quality of litter substrate and its attendant rates of decomposition, as highlighted by Marínez-Yrzar et al. (2007), profoundly influence soil nutrient levels, consequently impacting forest productivity. In a reciprocal manner, the availability of soil nutrients assumes significance in dictating the pace of litter decomposition, as soils serve as the fount from which decomposers, such as soil-dwelling organisms, derive essential nutrients to sustain their life activities, an aspect underscored by the work of Swift et al. (1979). The rates of litter decomposition within mixed-species compositions often diverge significantly from the cumulative decomposition observed in single-species litter scenarios. Notably, mixed litter configurations exhibit accelerated decomposition rates, particularly evident when constituent species vary in their litter's nutrient concentrations, as observed in the study by Quested et al. (2002). Ball et al.'s investigation in 2009, focusing on a temperate forest, delved into a mixed-litter decomposition study involving four dominant tree species. Their findings underscored that species within a litter mixture possessing higher nutrient concentrations facilitated enhanced nutrient release, while those with lower nutrient concentrations tended to retain these vital elements. Similarly, the research by Yang et al. (2002) unveiled that the presence of mixed forests, specifically combining *Cunninghamia lanceolata* and *Tsoongiodendron odorum*, fostered a higher usable environment for recycling of nutrient. This was evidenced by a heightened annual litter fall nutrient return and subsequent nutrient release compared to pure forests. Furthermore, it's crucial to note that negative interactions might arise in litter types abundant in secondary compounds, such as tannins, directly impeding microbial growth and activity, thus affecting decomposition dynamics.

Adding fertilizer sometimes makes things break down faster (Hobbie and Vitousek 2000). When we add nitrogen, it can stop the breakdown of a tough material called lignin by stopping certain helpful enzymes. Nitrogen in the soil is the main thing that controls this, while phosphorus is usually in short supply in big forests. Inside a pile of fallen leaves and stuff (litter), calcium, nitrogen, and phosphorus break down pretty quickly, but it takes a few weeks or months (Devi and Yadava 2007).

The shade or exposure to sunlight also affects the decomposition of litter. The litter under the forest canopy is softer and disappears more quickly than leaves exposed to sunlight (Giller and Gadisch 1997).

The attribute of litter was also found to have influence on the decomposition rate. In general, the decomposition rate is high in species with extreme ash and nitrogen contents and minimum C/N ratios and lignin contents (Singh 1969). When the C:N ratio falls below a certain point (25:l), mineral N release to the soil increases (Berg and Staaf 1981). The initial lignin concentration of litter (Fogel and Cromack 1977) and litter N concentration (Berg and Staaf 1981) have been suggested as better predictors of decomposition rates than the C:N ratio. High lignin concentrations decrease the rate of decomposition of leaf and needle litter (Melillo et al. 1982).

9.5 Forest Soils and Climate Change

Climate change stands as a paramount concern in contemporary times, exerting a profound influence on forest ecosystems and the vitality of forest soils. Its dynamic shifts directly impact crucial facets like nutrient cycling, the respiration of ecosystems, and the storage of carbon within forests. The intricate relationship between soils and forests is multidimensional, with substantial repercussions on each other and the broader environment. Over millions of years, soils have served as the foundational support for trees and entire forest ecosystems. Their symbiotic interactions create and support environmental situations needed for agricultural production. These positive outcomes extend far-reaching benefits, ensuring the sustenance of a productive food system, the enhancement of rural livelihoods, and the preservation of a healthy environment amid the challenges posed by climate change. Forest ecosystems are subjected to change in climate which has already shown a myriad of consequences in forests and their soil. Over the ages, factors of climate change have more or less modified the bio-geochemical cycles of carbon, nitrogen, and phosphorous. This, in turn, has affected soil productivity having a direct effect on forests as a whole (Raison and Khanna 2011). Numerous scientific inquiries examining the nexus between climate change and vegetation dynamics have illuminated the distinctive correlations between specific climatic regimes and the constitution of plant communities or functional types (Walter 1985). Consequently, there's a logical inference that climatic variations will provoke substantial reconfigurations within forest ecosystems (IPCC 1996). Even conservative projections, envisioning a global temperature rise of merely 1–2 °C—substantially below recent forecasts—portend notable repercussions in terms of altered species composition, productivity, and biodiversity across diverse global ecosystems. These anticipated impacts bear direct relevance to communities reliant on forest resources for sustenance (Gitay et al. 2002). Insights gleaned from global models suggest that the amplified atmospheric CO_2 levels attributed to climate change are poised to augment the net primary productivity of forests. India, celebrated for its significant biodiversity and with forests spanning about 23% of its geographical expanse, is poised to experience discernible effects from climate change on its forested regions.

9.5.1 Climate Change: Role of Forest Soil

Carbon emissions play a pivotal role in driving climate change. Forests, in their multifaceted capacity, serve as substantial reservoirs for carbon. Of the global carbon stockpile totaling 650 billion tons, nearly one-third is sequestered within terrestrial ecosystems, predominantly in forests. Notably, forest soils account for approximately 45% of the carbon stored in the global forest biomass, with an additional 10% contained in deceased wood and leaf litter within these forested areas. Remarkably, the carbon storage capacity of forests parallels that of the earth's atmosphere. Methodically managed forests offer extensive ecosystem services, mitigating soil erosion, diminishing the risk of geological hazards like landslides and avalanches, and significantly contributing to the regulation of clean water sources and the maintenance of a balanced water cycle. Forest soils exhibit distinct characteristics due to their relatively low disturbance levels and high organic matter content, endowing them with substantial resilience against the climate change effect. Although the influence of change of climate on forest soils tend to manifest gradually, they accumulate progressively over time. On the other hand, industrial plantations are often at the behest of frequent disturbances from intensive management options (site preparation, fertilizer application, harvesting, slash burning, etc.). All these factors could contribute to the susceptibility or exposure of climate change but with an offer to manage emerging impacts through intensive plantation management practices. The average carbon sink capacity linked with agroforestry tree plantations in India is estimated to be 25 t C ha^{-1} from 96 million ha area despite substantial regional variation which found to regulate biomass production (Basu 2014).

Climate change directly affects soil moisture by altering temperature and rainfall patterns. Among the factors impacted by climate change in soils, the regulation of soil development processes significantly influences the breakdown of rocks and minerals. Extensive scientific evidence supports the correlation between specific temperature and rainfall conditions and various stages of soil development processes. The expected changes in climate variables will, directly and indirectly, affect the various developmental processes in the forest soils and all such perceived changes are summarized in Table 9.5 and Fig. 9.3 for understanding its basic processes with the interactive mechanism.

Climate change induces direct impacts like heightened temperatures, elevated atmospheric CO_2 concentrations, alterations in precipitation patterns, and increased occurrences of extreme climatic events such as heat waves, droughts, storms, and frosts. Indirectly, it leads to pest and pathogen outbreaks, frequent fire incidents, modifications in vegetation growth and species composition, and shifts in litter inputs. These multifaceted influences significantly affect various soil processes, directly linking to the health status of forest soil.

9.5.2 Direct Effect of Climate Change on Forest Soil

Temperature: The escalation of global temperature extremes, a consequence of advancing climate change, is well-documented (Hoegh-Guldberg et al. 2018). Elevated temperatures drive increased metabolic rates across various trophic levels within organisms (Walther et al. 2002). This heightened metabolic activity contributes to amplified forest productivity, a process intricately regulated by soil nutrient availability and precipitation patterns. (McMahon et al. 2010). At higher temperature P concentration in forest litter tends to decrease with the increase in N:P and C:P ratio, and leads to direct influence on organisms engaged in the decomposition process of litter (Yuan and Chen 2009). Rising temperature will result in increased soil respiration rate, loss of stored C, higher rate of N mineralization, and several subsurface mechanisms influence carbon dioxide (CO_2) fluxes. Soil respiration is intimately linked to the pace of chemical and biochemical reactions. Activities such as soil microbes and root respiration release more CO_2 and methane gases, contributing to greenhouse gas (GHG) emissions. Forests, however, can counterbalance some of these emissions through various biogeochemical processes, serving as a vital carbon sink to mitigate the impacts of climate change. Higher temperatures lead to increased carbon turnover in soils, limiting its availability for forest vegetation growth. While plants primarily acquire carbon through photosynthesis, soil carbon plays an equally

Table 9.5 The influence of climate change factors on soil processes

Climate variable(s)	Expected effects
Rising temperature	Declination of soil organic substances
	Decrease in a readily available organic particle pool
	Soil moisture decline
	Acceleration of mineralization rate
	Deterioration of soil structure
	Heightened soil respiration rate
Elevating CO_2 concentration	Rise in organic matter content in the soil
	Enhanced water use efficiency
	Greater carbon accessibility for soil microorganisms
	Heightened nutrient cycling speed
	Augmented soil moisture levels
	Increased surface runoff and erosion
Increasing rainfall	Nutrient losses through a higher rate of leaching
	Increased reduction of Fe and NO_3
	Higher volatilization losses of nitrogen
	Enhanced productivity in arid regions
Decrease in precipitation	Decrease in soil organic matter
	Increase in the rate of soil salinization
	Low nutrient availability for plants and crops

Source: Recreated and modified from Pareek (2017)

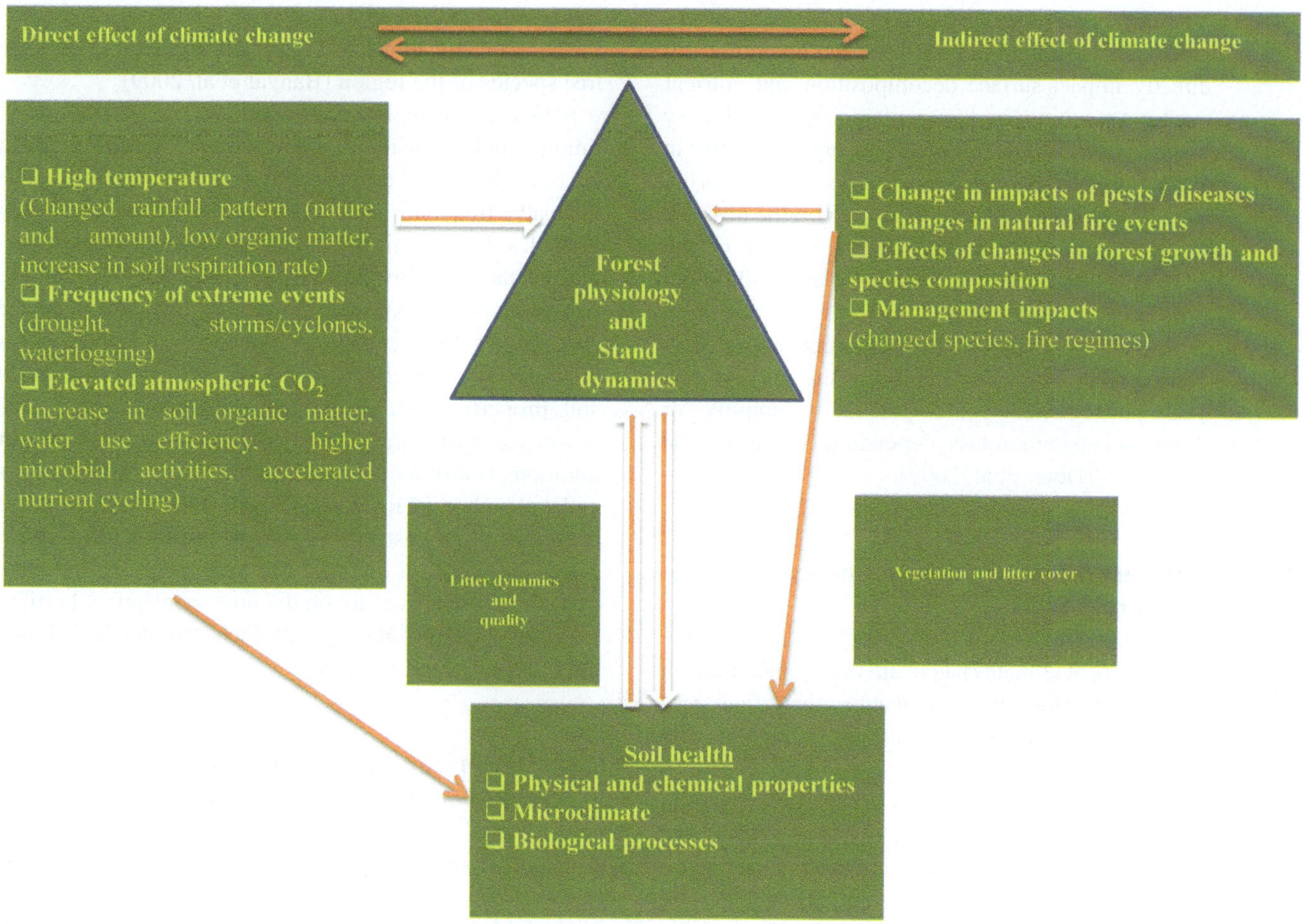

Fig. 9.3 Direct and indirect effects of climatic change on forest soil health
Source: Recreated and modified from Raison and Khanna (2011)

crucial role in sustaining plant growth and vitality. Soil carbon serves as a reservoir for water, nutrients, and an energy source for decomposing microbes. Elevated soil respiration rates result in the release of higher amounts of CO_2 into the atmosphere (Conant et al. 2008). A higher decomposition rate gives a higher amount of organic minerals with less sink of carbon, the source for higher biomass production in forests. Soil nitrogen availability will be higher in the prevalence of increased temperature which will give the higher carbon sequestration because of better and luxuriant growth of plant. On the other hand, N removed from the ecosystem will be there because of the priming effect of the excess nitrogen. Such losses from the ecosystem will give rise to other ecological problems of eutrophication in nearby water course and/or lowering of pH in soils (Galloway et al. 2003). Therefore, it is necessary to monitor the N losses from the forest ecosystems for assessing the influence of climate change concerning elevated soil respiration and temperature.

Precipitation pattern: Climate change is expected to directly affect the pattern of water availability, supported by various climate prediction models. Some models predict longer drought and wet periods, while others predict higher precipitation rate. However, it is still opaque and unclear to confirm that how such a scenario will be changed in the future. But, it is confirmed that climate change is expected to significantly alter the pattern of water availability which will result in the following soil concerns:

- High rainfall intensity and prolonged dry spells will result in a higher risk of soil erosion. Heavy precipitation can accelerate runoff and leaching because of the restricted buffering capacity of each type of soils to absorb and retain water. Contrary to this, the prolonged dry spell will decrease vegetation cover and put the soils at the risk of losing organic matter at a faster pace. Erosion at a large scale can also change the landscapes by losing stored carbon in forest areas and reducing plant productivity.

- Nutrient loss is the first victim of erosion due to change in precipitation pattern. Arid conditions will directly impact surface decomposition and nutrient cycling, resulting in lower plant biomass. Heavy precipitation can also deplete soil organic matter at a faster rate due to a higher pace of leaching (Nearing et al. 2004). But still the information on the possible changes in forest biomass pattern because of the increased rate of erosion are scanty and will be the subject of future research.
- Soil hydrophobicity, or water repellency, may become prevalent due to shifts in the microbiome. This phenomenon could potentially amplify or diminish soil carbon loss, depending on specific site conditions (Doerr et al. 2009).

9.5.3 Indirect Effect of Climate Change on Forest Soils

The indirect effects of climate change are tied to alterations in forest growing conditions and disturbances that affect ecosystem processes. These changes ultimately regulate the overall health of forest soils.

Change in forest growth, the input of organic matter and nutrients to soils: Forest growth is regulated by temperature variations which can increase or decrease the limiting factors existing in any site. In certain site-specific situations, increased atmospheric inputs of nutrients limit net primary productivity (NPP). It is already known that elevated CO_2 concentration is responsible for restricting nutrient availability to vegetation (Norby et al. 2010).

Dynamics of pests and diseases: The incidence of pests/diseases on forests is generally governed by (1) forest types, species composition; (2) climatic and edaphic factors; and (3) insects, pathogens, vertebrates. All these governing factors have strong interactions for deciding the forest stand health. Outbreaks can occur when all the stated factors positively promote pests and pathogens and negatively for host trees and/or forest stand and the reverse interactions for recovering the trees from such epidemics. By considering these actions is imperative for the management of pests and diseases, and it's essential for evaluating potential implications of climate change on forest health. Pathogens induce mortality across multiple native plant communities as large-scale mortality in young and mature trees of *Acacia nilotica* and *Dalbergia sissoo* was observed in south-western parts of Punjab, India. This was due to change in climatic factors (extremes in temperature and precipitation pattern), responsible for the weakening of plant system by causing physiological changes and resulted in the outbreak of pest and diseases and stress factors, causing threat to the existence of such valuable tree species of the region (Banyal et al. 2009).

Fire regimes: Fire regimes are regulated by the frequency, timing, and intensity of occurrences, impacting forest areas by inducing drought and reducing humidity. The intensity, frequency, and seasonal timing constitute key variables determining fire incidents. The combined interplay of these factors defines the specific fire outbreak to which forests are exposed. This fire regime, alongside site and species characteristics, influences greenhouse gas emissions and the effects of fire on nutrient cycles, forest soil properties, and ecosystem processes. Forest fires occurring more frequently and persisting for longer durations contribute to a reduction in forest carbon stocks and soil carbon (Raison and Kirschbaum 2008). Repeated occurrence of forest fires at shorter intervals has a direct effect on trees and underground vegetation on one hand and overall tree diversity on the other perceptive. Fertility may get regulated through all such changes and finally altering the forest soil health.

9.5.4 Understanding Soil Carbon and Sequestration in Forest Soils

Soil carbon storage is contingent upon inputs from vegetation, losses via decomposition, soil characteristics, and climatic variables such as temperature and precipitation. Forests contribute significantly to the soil organic substances sink via diverse sources like leaf litter, coarse woody material, roots, root exudates, and dissolved organics leached from plants and litter layers. These inputs undergo stabilization within the mineral soil matrix or biodegradation, releasing carbon dioxide (CO_2) into the atmosphere. The stabilization process is intricately tied to interactions among soil minerals, plants, soil organisms, and organic components, influenced by climate, stand management, and broader-scale factors. Interestingly, a large aboveground carbon stock doesn't necessarily correlate with substantial belowground carbon stocks, as heightened forest productivity can lead to increased nutrient pools, affecting organic matter decomposition and consequent carbon release into the atmosphere.

Contrarily, the decomposition process faces hindrance in conditions of asphyxiation (lack of oxygen) and low temperatures. The productivity both above and below ground is regulated by the mineralogical and chemical traits of the soil, coupled with drainage provisions. Clay soils typically harbor higher carbon stocks due to their increased surface area on clay particles. The adhesive nature of the mineral matrix can be linked to the origin of the clay's parent material. In the United States, carbon-rich forest soils (Andisols) in the Northwestern region stem from volcanic parent material, resulting in soils with elevated surface areas. These support high-quality mixed conifer hardwood forests, boasting the nation's highest quantities of carbon stored in the soil. Conversely, sandy soils exhibit lower carbon pool owing to their reduced adsorption area compared to clay soils (D'Amore and Kane 2020).

Soil carbon dynamics exhibit responsiveness to climate change, primarily mediated by the intricate relationships among biomass production, decomposition processes, and shifts in vegetation patterns influenced by climatic alterations. Temperature acts as a pivotal regulator, exerting both direct and indirect effects on the rates of vegetation growth and decomposition. Temperature fluctuations notably influence the release of mineral nutrients through weathering processes, contributing to carbon dioxide released during the breakdown of organic matter. This, however, presupposes that limiting factors such as oxygen, moisture, and vegetation inputs are not constraining. The rates of primary production and decomposition exhibit potential increases or decreases in response to temperature variations, contingent upon the nuanced growth and responsiveness of individual forest components, transcending the influence of temperature alone. Alterations in rainfall patterns engender two distinct extremes: drought and waterlogging. These extremes may arise from heightened erosion impacting stabilized mineral soil carbon or reduced decomposition attributed to low oxygen concentrations in waterlogged soils (Wuebbles et al. 2014). Scientific investigations suggest an anticipated transformation in forest vegetation patterns, with consequential impacts on growth and decomposition processes across global forest landscapes. Existing empirical studies, exemplified by Griffith et al. (2009) and Johnson et al. (2009), contribute to envisioning prospective scenarios the accumulation of carbon in the form of biomass and its storage in the soil across different forest types in diverse climatic conditions.

Under elevated CO_2 conditions, plants exhibit a shift in their metabolic allocation, decreasing the creation of nitrogen-rich metabolites while increasing the quantity of carbon-rich metabolites in root exudates (Tarnawski and Aragno 2006). This leads to heightened microbial activity, resulting in increased CO_2 production. The higher CO_2 levels impede the storing of carbon in soils, thereby impacting the ability to store carbon in the soil. Carbon isotope studies reveal that CO_2 production by roots and microorganisms in the rhizosphere is notably influenced by elevated CO_2 conditions during plant growth. Studies indicate that the increase in CO_2 respiration in the rhizosphere may surpass the growth enhancement in root biomass. For instance, Cheng and Johnson (1998) found that while plants exhibited a mere 15–26% increase in biomass under elevated CO_2 conditions, and carbon respiration surged by 56–74% compared to ambient CO_2 levels. Elevated atmospheric CO_2 concentrations often coincide with heightened mycorrhizal colonization due to increased plant demand for nutrients, leading to augmented carbon accumulation rates. Consequently, mycorrhizal biomass increases as plants allocate more photosynthates underground, driven by increased nutrient demands alongside carbon assimilation rates (Drigo et al. 2008). The drivers of global climate change, such as rising temperatures, alterations in rainfall patterns, elevated atmospheric CO_2, fire regimes, and nutrient enrichment, are intricately interconnected at various biological levels, affecting carbon, nitrogen, and phosphorus stoichiometry between producers and the soil biome (Fig. 9.4).

9.6 Forest Destruction, Fragmentation, REDD, and REDD$^+$ Approaches

Deforestation transpires when areas predominantly inhabited by naturally occurring tree species undergo conversion to fulfill specific human demands. It denotes the transition of forests into alternative land uses or a sustained reduction in tree canopy cover, often falling below the 10% threshold (FAO 2015). The cumulative effects of deforestation and forest degradation contribute to approximately 11% of carbon emissions, surpassing the global transportation sector and trailing only behind the energy sector. Primary drivers of deforestation encompass agriculture, unsustainable forest management, mining, infrastructure projects, and escalated incidences and intensity of fires. Infrastructure endeavors like road construction, while seemingly isolated, exert significant indirect impacts by opening up forests to settlers, initiating agricultural activities, and other operations. Degradation, a consequence of poor forest management and unsound extraction practices, often initiates a gradual and pervasive form of deforestation, metaphorically termed as "death by a thousand cuts." Forest fragmentation characterizes the transformation of continuous forest landscapes into fragmented patches interspersed with non-forested areas. It emerges from the intricate interplay between natural landscapes and escalating societal demands on land, leading to a patchwork of both natural and human-modified environments (FAO 2007).

It is now clear that the stabilization of global temperatures within 2 °C cannot be achieved without reducing emissions from the forest sector along with merging mitigation actions. According to the Global Forest Resources Assessment (FRA), global forest cover diminished by 3.16% between 1990 and 2015, currently standing at around 30.6% compared to 31.6% in 1990. The alarming rate of forest cover decline presents an imminent threat if left unchecked. Forecasts indicate that future demands on forest resources will intensify competition among nations, leading to an estimated annual loss of 18.7 million acres of forest cover

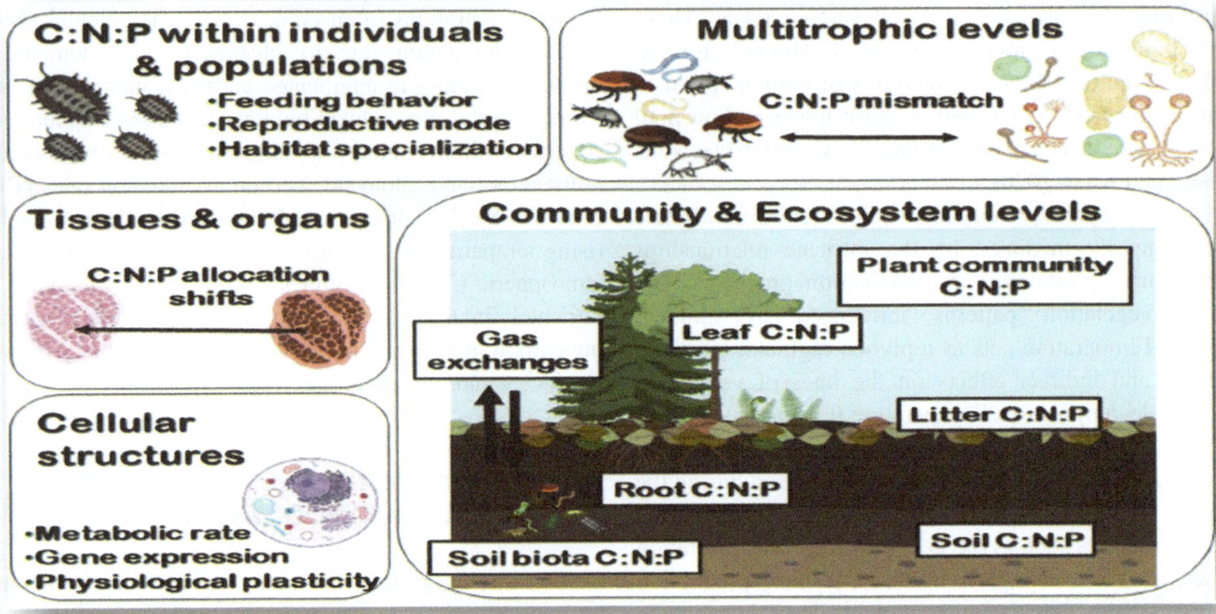

Fig. 9.4 Relationship of C, N, and P stoichiometry of primary producer and soil biota
Source: Maaroufi and Long (2020)

(Bradford 2018). Deforestation has multifaceted causes, with prominent contributors being forest conversion, wildfires, illegal and harvesting, unsustainable logging, fuelwood, mining, and the impact of climate change.

9.7 Problem Soils and Their Bio-Reclamation

Soils that limit crop production because of nutrient deficiency and abiotic stresses are classified as problem soils. These soils are always under the shadow of unfavorable soil properties and need special management for their proper utilization with productive functions Soils characterized by soluble salts, high groundwater tables, impermeable argillic horizons, erosion, low-fertility sandy composition, frequent coastal inundation, and laterites (Oxic soils) and acidic properties fall under the category of problematic soils. This classification encompasses vertisols, acid sulfate soils, saline-sodic soils, peat soils, and fine-textured alluvial soils as the commonly encountered problematic soil types. Physical and chemical characters are responsible for the formation of such soils. Physical factors are fluffy paddy soil, sandy soil, sub-soil hardpan, surface crusting, water logging, peat, and marshy soils, whereas the quantity and nature of salts govern the chemical factors. Fluffy paddy soils are developed due to continuous puddling operations for rice cultivation, resulting in structure less mass of soils due to breaking of soil aggregates. Sandy soils are the result of weathering process in rocks specifically granite, quartz and lime stone, poor in water holding capacity and low nutrients availability, making them unsuitable for agriculture and plantations activities. Hardpan soils are the feature of red soils due to illuviation of clay in association with oxides of Fe, Al, and $CaCO_3$. Roots could not penetrate the soils, limiting the nutrient uptake by the plants. Surface crusting in soils arises from the aggregation of colloidal oxides of iron and aluminum in alfisols, resulting in the binding of soil particles during moist conditions. In contrast, peat and marshy soils are products of organic matter accumulation, contributing to elevated levels of soil acidity. Such type of soils is commonly found in humid regions. Waterlogged soils occur due to excessive water content and inadequate aeration in the soil. Acid soils are created because of chemical problems in the soils. pH in such soils remains below 7.0 and even the range is much below up to 4.75 and 4.0 in extremely acidic conditions. The basic reasons for the formation of acidic soils are weathering of acidic parent materials (granite and sandstone), leaching of bases particularly in high rainfall zones. Temperate soils in pine forests are enriched with acid soils. Salt-affected soils, encompassing both saline and sodic/alkali types, result from a combination of natural and human-induced factors. Saline soils differentiated by the prevalence of neutral salts, including chlorides and sulfates of Na^{2+}, Ca^{2+}, and Mg^{2+}, exhibiting an electrical conductivity (EC) exceeding 4 dS/

m, a pH lower than 8.5 but not <7.0, and an exchangeable sodium percentage (ESP) below 15. These soils manifest an excess of soluble salts while maintaining small concentrations of exchangeable sodium, yielding favorable physical properties, flocculated soil structure, and high permeability akin to conventional soils. Saline soils can be categorized into three distinct groups: (a) inland saline soils, usually linked to a high water table and commonly occurring in irrigation command areas; (b) saline soils arising from the use of saline groundwater for irrigation, prevalent in dry and semi-dry areas; and (c) coastal saline soils. These are recognized by different regional names such as Shora, Khar, Thur, and Lavniya across various regions in the country.

The soils with a predominance of salts with alkaline in nature are termed sodic/alkali soils. These soils having electrical conductivity (EC) <4 dS/m, pH <8.2 and these soils are typically dominated by sodium carbonate (Na_2CO_3), sodium bicarbonate ($NaHCO_3$) with a kanker pan layer of $CaCO_3$. Physically, soil structure is highly dispersed with poor infiltration rate. Sodic soils exhibit degradation across several critical properties, including structural, chemical, nutritional, hydrological, and biological aspects and are represented by their compact, heavy nature, featuring a high bulk density and a silty clay loam texture categorized as Typic Natrustalf. They notably contain a higher proportion of sodium relative to other cations, both within the soil solution and on the exchange complex. In the Indo-Gangetic plain, sodic soils typically lack gypsum ($CaSO_4$, $2H_2O$) but contain an amorphous form of $CaCO_3$. This calcium carbonate variation increases with soil depth, often observed in concretions or forming an indurate bed, typically around 1 meter below the surface. The soils with pH > 8.2, ECe >4.0 dS m^{-1} and ESP >15 are designated as saline-alkali soils and are commonly known by different names across the country as *Usar, Reh, Kallar, Kshar*. These soils exhibit characteristics of both saline and alkaline soils, and appropriate reclamation and management strategies are implemented to optimize productivity.

Abnormal nutrient physiology in plants occurs because of ionic non-equilibrium due to high pH (>10) and ESP (>60). The elevated Na:Ca ratio and reduced C:N ratio in sodic soils create an inhospitable environment for sustaining vegetation. The growth of most crops is significantly hindered due to compromised physical conditions, disrupted nutrient availability, and suppressed biological activity. On these soils, deficiencies in certain micronutrients (Zn, Fe, Cu, Mn) alongside toxicity from other elements (Na, B, Mo) further impede the growth and development of crop plants. Sodic soils commonly exhibit poor water permeability, characterized by reduced hydraulic conductivity and infiltration rates. This limitation stems from pore space interlocking and soil compactness, which, in turn, hinder the root development of plants. The hostile conditions in sodic soils contribute to a restricted range of microbial populations and reduced diversity. Consequently, this environment leads to decreased litter decomposition and nutrient mineralization processes. Major problem soils are enlisted in Table 9.6 with key characteristics and major impediments.

Table 9.6 Major problem soils of India

S. No.	Problem soils	Key characteristics of diagnosis	Major impediments
1.	Clay soil	Higher clay particle	Soil saturated and compressed with little aeration
2.	Sandy	Greater coarse sand particles	It has a low fertility and less moisture retaining capacity. So vulnerable to erosion
3.	Acid Soil	Soil pH < 6.5	The presence of ferrous and aluminum content
4.	Salt affected		
	Saline soil (Solonchak)	ECe > 4.0 dS m^{-1} cations: Ca, Mg, Na and K anions: CO_3, HCO_3, Cl, SO_4	It has greater potential to hold osmotic pressure and it leads to the imbalance of the nutrient
	Sodic or alkali soil (Solonetz)	ESP > 15 Carbonates (CO_3^-) and bicarbonates (HCO_3^-) of Na salts. The amount of Cl^- and SO^-_4 is less	Degraded physical state or compromised physical condition. Toxicity of sodium, nutrient disproportion
	Saline-Sodic soil	ECe > 4.0 dS $m^{-1}$and ESP > 15	Elevated osmotic potential, degraded physical state, nutrient disproportion
5.	Calcareous soil	$CaCO_3$ > 5.0%	P and Fe deficiency
6.	Waterlogged soil	Water logging, poor permeability	Inadequate oxygenation with asphyxiation condition
7.	Compacted soil	Increased soil density	Inadequate oxygenation, restricted root penetration, water stagnation
8.	Impermeable soil	Low hydraulic conductivity and rate of infiltration	Inadequate oxygenation, water stagnation

9.7.1 Reclamation of Problematic Soils Through Forestry

Trees possess remarkable resilience to thrive in challenging climatic and soil conditions, making forestry and agroforestry pivotal in utilizing their capabilities for both productivity and ecological restoration. The deposition of tree litter significantly contributes to soil conservation by increasing soil organic substance, enhancing its physical properties, and simultaneously providing essential nutrients. In natural forest ecosystems, the carbon and nutrient cycles operate within a relatively closed system, characterized by extensive recycling and minimal external inputs and outputs. Reclamation forestry experiences have illustrated the efficacy of trees in enhancing fertility within problematic soils, showcasing their ability to rejuvenate and build up soil fertility. Trees have a dense network of fine and deep roots, high quality of litter, bind the soil to prevent soil erosion, ultimately keeping nutrients intact on soil surface thus, in all ways, trees promote more closed nutrient cycling. However, all trees are not having all the traits to be capable of soil reclamation. Certain tree species have been identified by Bhatt et al. (2017) as being perfect for recovering alkaline and saline soils. These kinds of trees are *A. auriculiformis, A. nilotica, A. tortilis, A. excelsa, P. cineraria C. equisetifolia, A. indica, and D. sissoo*.

Salt affected soils are a major challenge for rehabilitation. Several tree species got evaluation for their tolerance to salinity, particularly in dry and semi-dry areas. Through this assessment, specific species have emerged as particularly adept at thriving in saline conditions, surpassing others in their resilience. These standout species have been recommended as principal choices for cultivation in saline environments. The list of promising tree species suitable in saline conditions is illustrated in Table 9.7.

Alkali soils tend to support a limited natural vegetation spectrum, often comprised of only a select few species. Notably, *P. juliflora* has extensively established itself in these soils, forming dense clusters, notably on deserted lands alongside roadways and railway tracks. Other tree species, including *A. nilotica, S. oleoides, S. persica, C. decidua, C. septaria, and C. phlomidis*, rank among the important species that thrive in high-pH soils. Moreover, species like *A. leucophloea, A. eburnea, M. hamata, P. cineraria, B. monosperma, D. tomentosa, B. roxburghii, and M. emarginatus* are commonly found in slightly lower pH environments, extending up to 9.0 on the pH scale. Table 9.8 provides a list of recommended trees based on their tolerance to alkaline soils. The soils having higher pH sustain only for a few tree species with no fruit trees, especially in >10 pH soils. But, with the decrease in pH level more trees are accommodating with a wider horizon of usability in addition to reclamation benefits.

Salt affected water logging soil is another type of problem soils that need special management practices. Pandey et al. (2015) highlighted some of the prominent factors like heavy rainfall in some areas, poor drainage, excess and faulty irrigation methods, hardpan, shallow water table and seepage from the canal is a primary contributor to waterlogging issues. Arid and semiarid regions are the victim of the water logging problem due to canal irrigation without proper drainage options, which also results in secondary salinization. For managing such soils biodrainage is the only alternative approach which is effective, affordable, socially acceptable and environment friendly without causing any degradation to natural resources as experienced in some other available

Table 9.7 Relative salinity tolerance of trees

Tolerance range (ECe dS/m)	Agroforestry and forest/fruit trees/shrubs
Very high saline soils (20–30 dS/m)	Inland saline soils *P. juliflora, P. aculeata, S. oleoides, S. persica, A. farnesiana, S. fruiticosa, S. maritima, Haloxylon* spp., *Artiplex* spp., *T. articulata, T. troupi, T. ericoides* For coastal saline soils *A. marina, A. officinalis, R. apiculata, R. mucronata, B. gymnorrhiza, B. parviflora, C. tagala, A. ilicifolius, A. volubilis, A. corniculata, E. agallocha, H. fomes, H. littoralis, N. fruticans, S. alba, S. caseolaris, C. equisetifolia, B. asiatica, C. inerme, Pandanus* spp., *P. pinnata, C. ubcordata, T. catappa, C. inophyllum, F. retusa, M. littoralis, T. populnea*
Highly saline soils (14—20 dS/m)	*A. auriculiformis, A. nilotica, A. pennatula, A. tortilis, C. lanceolatus, C. equisetifolia, C. gluaca, C. obesa, E. camaldulensis, F. limonia, L. leucocephala, P. cineraria, Z. mauritiana*
Saline soils (10–14 dS/m)	*A. senegal, C. cunninghamiana, E. tereticornis, A. auriculiformis, A. cartibea, L. shannoni, P. pinnata, S. saman, T. arjuna*
Moderately saline soils (5.0–10 dS/m)	*A. procera, M. composita, M. dubia, D. sissoo, E. officinalis, G. ulmifolia, P. granatum, S. saman, A. deamii, A. guachapele, A. herbertsmithite, C. eriostachya, C. velutina, H. brasiletto, Salix* spp., *S. cumini, S. fruticosum, T. indica*
Less saline soils (<5.0 dS/m)	*A. auriculiformis, A. deamii, A. catechu, S. cumini, S. babylonica, M. azedarach, M. composita, M. Dubia, T. indica, A. lebbeck, L. leucocephala, Populus* spp., *A. indica, D. strictus, B. monosperma, T. arjuna, A. excelsa, D. sissoo, B. roxburghii*

Source: Banyal et al. (2017)

Table 9.8 Tree species recommended with relative tolerance for afforestation of alkali soils

Tolerance range (pH)	Agroforestry and forest/fruit trees/shrubs
Highly alkali soils >10.0	*P. juliflora, A. nilotica, T. articulata* (fruit trees are not recommended under this category)
Alkali soils 9.60–10.0	*E. tereticornis, C. decidua, P. dulce, P. alba, P. cineraria, C. equisetifolia, S. persica, S. oleoides, T. arjuna, C. carandus, P. guajava, Z. mauritiana, E. officinalis*
Moderately alkali soils 9.00–9.60	*C. rothii, A. lebbeck, C. siamea, P. pinnata, S. sesban, P. aculeata, D. sissoo, K. pinnata, B. monosperma, P. granatum, P. dactylifera, A. japota, T. indica, S. cuminii, F. limonia*
Less alkali soils 8.20–9.00	*G. robusta, A. indica, M. azedarach, L. leucocephala, H. binata, M. oleifera, P. deltoids, T. grandis, G. asiatica, A. marmelos, P. persica, P. communis, M. indica, M. alba, Ficus* spp., *S. laurifolium, V. vinifera*

Table 9.9 Tree species recommended with salinity and alkalinity tolerance for waterlogged areas

Tolerance range (pH)	Agroforestry and forest/fruit trees/shrubs
Very high saline soils (20–30 dS/m)	*A. farnesiana, P. juliflora, P. aculeata, T. aphylla*
Highly saline soils (14–20 dS/m)	*V. nilotica, A. pennatula, A. tortilis, C. lanceolatus, C. glauca, C. obesa, C. equisetifolia, E. camaldulensis, F. limonia, L. leucocephala, Z. jujube*
Saline soils (10–14 dS/m)	*Casuarina* spp., *E. tereticornis, T. arjuna*
Moderately saline soils (5.0–10 dS/m)	*A. lebbeck, D. sissoo, G. ulmifolia, P. pinnata, S. saman*
Less saline soils (<5.0 dS/m)	*A. auriculiformis, A. catechu, A. deamii, S. cumini, Salix* spp., *T. indica*
Highly alkali soils (>10.0 dS/m)	*Prosopis juliflora, Vachellia nilotica, Tamarix articulata*
Alkali soils (9.60–10.0 dS/m)	*E. tereticornis, P. alba, P. cineraria, C. equisetifolia, S. persica, S. oleoides, T. arjuna, Z. mauritiana*
Moderately alkali soils (9.00–9.60 dS/m)	*A. lebbeck, C. siamea, P. pinnata, D. sissoo, B. monosperma, S. cuminii, F. limonia*
Less alkali soils (8.20–9.00 dS/m)	*A. indica, M. azedarach, L. leucocephala, P. deltoides, Ficus* spp., *S. laurifolium*

Source: Dash et al. (2005), Banyal et al. (2017)

technologies (sub-surface drainage). The trees planted under bio-drainage plantation are listed in Table 9.9.

Banyal et al. (2019) showed that *Eucalyptus tereticornis* emerged as the prime choice for drawdown in the water besides soil reclamation in the saline places of trans Indo-Gangetic plains. Soil conditions improved in general and 0–30 cm profile (plough layer) in particular. There is a presence of consistent reduction in ECe (Soil salinity) underneath Eucalyptus trees with an increase in profile depth from 0 to 120 cm than the bare fallow area. The other essential nutrients like organic carbon, N and P were also higher underneath plantations which clearly showed the reclamation behavior of Eucalyptus in resource-poor waterlogged saline soils.

9.7.2 Problem Soils: Reclamation Mechanisms

Trees can play multifarious roles like the reclamation of problem soils, wood products and environmental impacts. Numerous research findings highlight the positive contributions of various tree species, particularly in problematic soils and those affected by salinity. In alkali soils, species such as A. *nilotica, T. articulata, and P. juliflora* have shown the power to improve salt-affected soil conditions. These species have demonstrated the deduction of exchangeable Na % and pH values while contributing to an increase in soil organic carbon content in the surface layer. The primary drivers of soil improvement include tree roots and leaf litter that settle on the soil surface. Over time, these factors have led to alterations in soil properties such as pH, exchangeable sodium, bulk density, porosity, and water holding capacity. Observations by Garg and Jain (1996) highlighted the alteration of physical and chemical properties in sodic wastelands, showcasing a decrease in pH and ESP while witnessing an increase in soil organic carbon, notably more pronounced under *D. sissoo* than *T. arjuna* plantations.

Leaf litter plays a crucial role in nutrient circulation, with the recycling order being N > Ca > Mg > K > P. Studies by Singh (1998) emphasized that *P. deltoides* clones contributed to soil amelioration through elevated levels of total nitrogen, exchangeable Ca and Mg contents. Further research by Singh

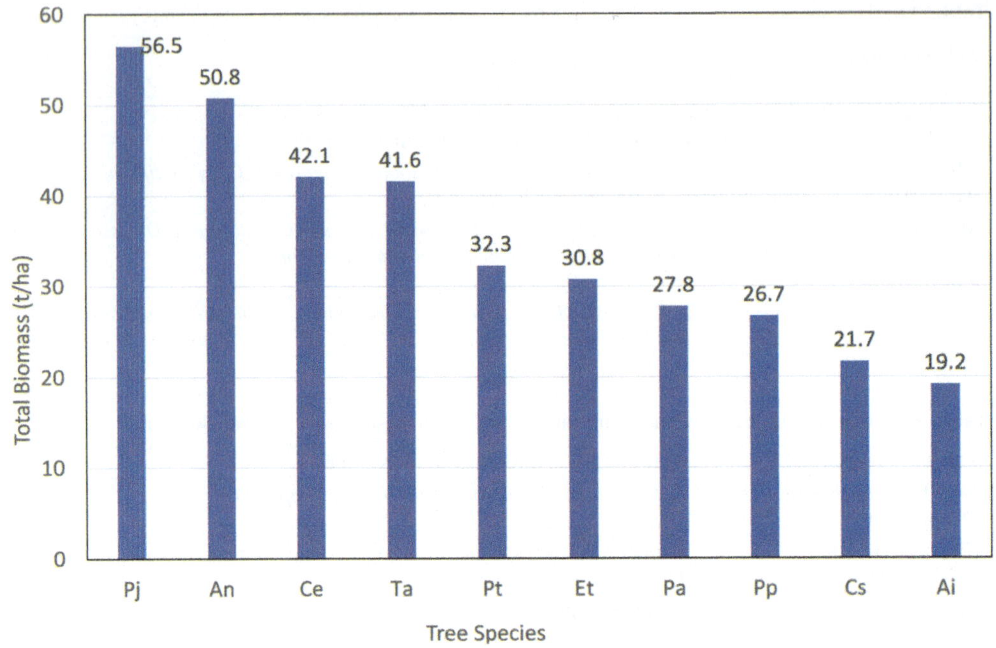

Fig. 9.5 Above ground biomass (t ha^{-1}) of *Pj P. juliflora*, *An A. nilotica*, *Ce Casuarina equisetifolia*, *Ta Terminalia arjuna*, *Pd Pithecellobium dulce*, *Et Eucalyptus tereticornis*, *Pa Prosopis alba*, *Pp Pongamia pinnata*, *Cs Cassia siamea*, *Ai Azadirachta indica*. Source: Singh et al. (2008)

et al. (2011) compared the impact of ten multipurpose tree species on growth, biomass yield, and physicochemical properties in alkali soils within the Indo-Gangetic alluvial plains. *C. equisetifolia* exhibited the highest survival rate (100%), followed by *T. arjuna, P. juliflora, P. pinnata*, and *P. dulce*. *C. equisetifolia* displayed the highest aboveground biomass production, followed by *A. nilotica* and *P. juliflora*. Overall, the specific tree species significantly improved alkali soils by reducing ESP and pH & elevating organic carbon content. Biomass production was noticed to follow the order: *E.tereticornis*>*A.nilotica*>*P.juliflora*>*C.equisetifolia*>*P.alba*>*P.dulce*>*T.arjuna*>*P.pinnata*>*A.indica*>*C.siamea*. (Fig. 9.5).

Substantial improvements were noticed in soil attributes under tree plantations, including a notable deduction in soil bulk density, declining from 1.57 to 1.21 mg m^{-3}. This reduction was accompanied by an increase in porosity, rising from 40.7 to 54.3%, and a significant enhancement in infiltration rate, escalating from 2.10 to 26.30 mm day^{-1}. Moreover, the water holding capacity experienced a remarkable increase, reaching up to 64.7% higher than in natural fallow conditions, particularly under *P. juliflora* tree plantations (Table 9.10).

Comparable enhancements were observed in soil parameters within the plantations compared to the natural bare area. These improvements included significant elevations in soil pH levels, reductions in electrical conductivity and substitutable sodium percentage, alongside notable increases in organic C content and the presence of nitrogen, phosphorus, and potassium (Table 9.11).

In saline soils, specific tree species including *E. tereticornis, M. composita, A. indica, D. sissoo,* and *T. arjuna* have demonstrated a capacity to reduce the electrical conductivity (EC) values compared to the initial soil EC levels (CSSRI 2018). In West Bengal, Banik et al. (2018) noted substantial improvements in soil properties, such as pH, EC, and cation exchange capacity (CEC), within plantation strips compared to non-vegetated fields. They observed variable enhancements in organic C, available NPK content in the topsoil under vegetation canopy compared to open sites. In Haryana, Bhojvaid and Timmer (1998) found that *P. juliflora* stands altered the microclimate and enhanced soil moisture status and exchangeable O, N, P, Ca, Mg and K. These alterations contributed significantly to the reclamation of sodic sites, resulting in decreased pH, electrical conductivity, and exchangeable sodium levels.

Similarly, Acacia nilotica contributed to organic matter and humus buildup after 10 years of tree growth, enhancing the soil's cation exchange capacity through the form of humus as an organic colloid. Under Dalbergia sissoo plantation, significant improvements were noted in soil properties, indicating a notable decrease in pH, electrical conductivity, and exchangeable sodium percentage, alongside increases in organic carbon, nitrogen, and available nutrients (Mishra 2002). Research by Bala et al. (2014) in the Indian desert's canal command waterlogged area observed higher soil organic carbon and electrical conductivity, along with increased nitrate levels in Eucalyptus rudis compared to Eucalyptus fastigata. Furthermore, in the natural succession of *P. juliflora, T. dioica*, and *S. munja* in response to

Table 9.10 Changes in hydraulic and physical properties of alkali soils under 10-year-old plantations

Tree species	Bulk density (g cm^{-3})	Soil porosity (%)	Water holding capacity (%)	Cumulative infiltration rate (mm day^{-1})
T. arjuna	1.49 ± 0.003	43.51 ± 1.32	41.10 ± 2.32	21.20 ± 3.25
A. indica	1.52 ± 0.002	42.60 ± 1.53	32.40 ± 2.12	21.70 ± 2.20
P. juliflora	1.39 ± 0.001	47.55 ± 1.52	43.20 ± 3.10	26.30 ± 3.12
P. pinnata	1.46 ± 0.003	44.65 ± 0.91	35.60 ± 1.53	24.30 ± 1.10
C. equisetifolia	1.31 ± 0.001	50.35 ± 1.15	38.20 ± 2.10	25.80 ± 2.10
P. alba	1.49 ± 0.010	43.75 ± 1.32	32.30 ± 2.32	20.00 ± 1.52
A. nilotica	1.43 ± 0.005	45.85 ± 1.45	42.20 ± 2.20	21.90 ± 1.40
E. tereticornis	1.44 ± 0.020	45.50 ± 1.50	41.30 ± 1.65	19.70 ± 0.92
P. dulce	1.41 ± 0.004	46.60 ± 1.62	36.20 ± 1.54	23.10 ± 2.12
C. siamea	1.47 ± 0.002	44.55 ± 1.34	36.20 ± 2.10	15.80 ± 3.10
Mean	1.43 ± 0.005	45.49 ± 1.36	37.87 ± 2.09	21.98 ± 2.08
Natural fallow	1.44 ± 0.003	42.05 ± 1.20	26.23 ± 1.63	11.80 ± 1.25
Initial	1.64 ± 0.020	35.50 ± 1.15	3.20 ± 2.450	4.0 ± 0.080
$LSD_{0.05}$	0.08	2.26	1.23	4.34

Source: Singh et al. (2011)

Table 9.11 Changes in soil chemical properties in alkali soils in 10-year-old plantations

Tree species	pH	EC_2 (dS m^{-1})	ESP	OC (%)	N (kg ha^{-1})	P (kg ha^{-1})	K (kg ha^{-1})	Na^+ (c mol kg^{-1})	Ca^{++} (c mol kg^{-1})
T. arjuna	9.80	0.39	55	0.35	122.6	62.5	509.2	1.55	11.3
A. indica	9.80	0.33	51	0.27	114.3	36.4	432.5	1.44	12.4
P. juliflora	9.50	0.30	46	0.43	138.5	76.3	610.4	1.10	14.2
P. pinnata	9.70	0.61	49	0.40	120.0	58.2	512.5	1.35	11.2
C. equisetifolia	10.0	0.68	66	0.36	131.4	64.3	540.3	1.21	10.3
P. alba	9.80	0.63	59	0.33	126.4	56.4	473.5	1.32	11.2
A. nilotica	9.70	0.77	51	0.35	126.3	62.6	522.4	1.42	10.2
E. tereticornis	9.80	0.86	57	0.24	106.5	36.4	410.4	1.50	8.65
P. dulce	9.90	0.70	60	0.27	121.5	56.2	482.5	1.48	9.52
C. siamea	10.0	0.69	66	0.26	116.1	41.5	416.5	1.46	8.63
Natural fallow	10.3	1.24	81	0.12	96.20	26.5	392.4	1.55	7.53
Initial	10.5	1.43	89	0.08	94.00	19.5	388.0	12.7	5.32

Source: Singh et al. (2011)

groundwater table recession, these species significantly contributed to further lowering the groundwater table and increasing productivity. Generally, forests have important turn in reclaiming problematic soils by enhancing their physical and chemical properties. While fruit species are typically sensitive to salt stress, certain species such as Z. mauritiana, Punica granatum, Syzygium cumini, Emblica officinalis, and Tamarindus indica demonstrated better performance in saline soils (Behera et al. 2015).

In saline soils, plants and crops employ three primary mechanisms for salt tolerance: salt exclusion, salt excretion, and salt accumulation (Fig. 9.6). These mechanisms allow plants to endure saline conditions and contribute to soil remediation. Transpiration have major play in limiting salt accumulation on the soil surface. By drawing soil moisture from lower layers, it maintains higher salt concentration beneath the surface and reduces it on the topsoil. Furthermore, tree canopies provide shade, slowing soil water evaporation and decreasing upward water flux containing capillary rise of saline water through capillary action to the soil surface. Deep-rooted trees aid in mitigating salinity. Their robust root systems penetrate the soil, enhancing water permeability and facilitating salt leaching. Fine roots has a significant turn in rehabilitating soil structure, pH balance, and water permeability. Roots of plant exude various compounds into the rhizosphere-amino acids, sugars, phenolics, and organic acids—essential for neutralizing salt in saline soils. Organic acids from decomposed leaf litter react with soil calcium carbonate, releasing calcium to replace sodium in the soil's exchange complex. As trees mature, the decomposition of abundant litter generates weak acids, like humic and fumic acids, which lower soil pH and electrical conductivity (EC).

Trees also act as bio-drains, extracting sodium through their root systems. The roots produce carbonic acid, solubilizing native $CaCO_3$ in these soils, leads to a decrease in substitutable Na content. Additionally, tree roots break through clay barriers, loosening subsoil, enhancing permeability, and aiding in downward sodium translocation. This

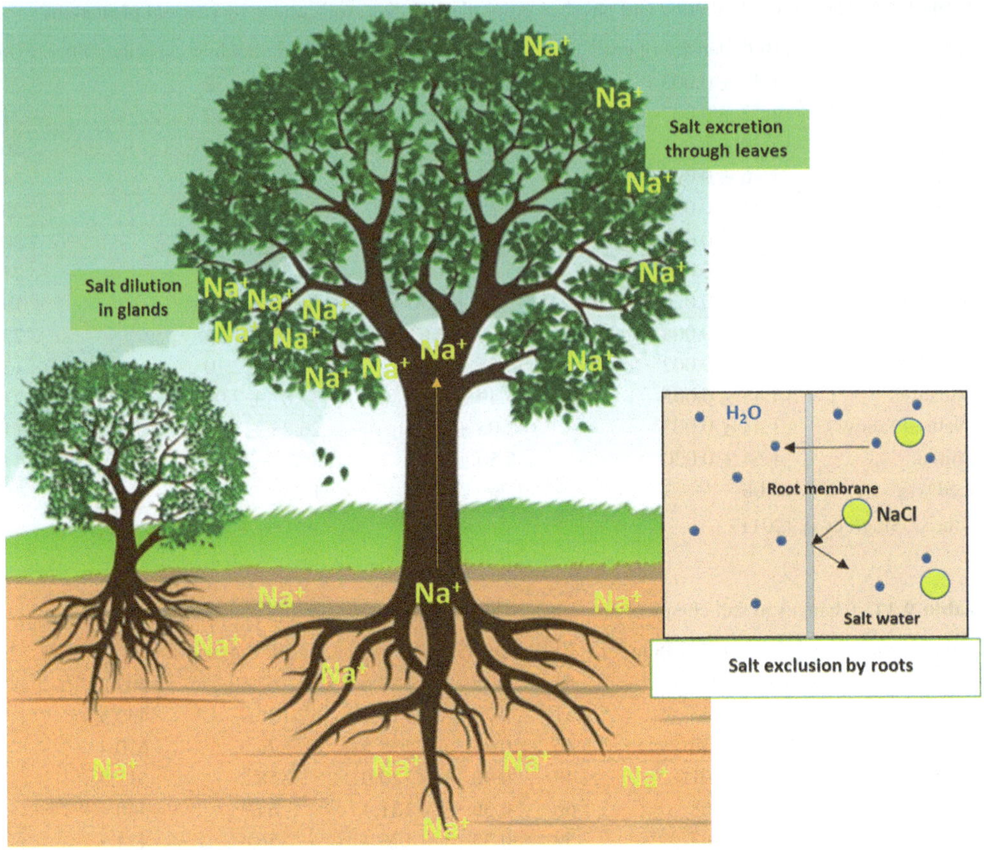

Fig. 9.6 Salt movement in soil and plant system in saline ecology. Source: Banyal et al. (2017)

process substantially enhances the physical attributes such as bulk density, porosity, soil moisture retention, and infiltration rates (Singh and Dagar 2005).

9.8 Conclusion

Understanding the soil is necessary for foresters to be successful in their forestry operations. More specifically, life depends upon the soil. However, anthropogenic factors including agricultural activities have changed the soil structure and its interactions. Therefore, there is demand for exclusive research on forest soils as it differs from agricultural soils *per se*. Forest soils are unique because the occurrence of tree species is regulated by the soil type and its properties and also, the tree species impact soil formation and its characteristics. Hence, understanding forest soils involves more than just considering productivity; it encompasses soil development. The forest floor, consisting of organic debris, forms a crucial part of a forest, often holding a substantial portion of its organic matter and nutrients both in quantity and composition. Fundamentally, the "O" horizon of forest soils makes the difference. Unlike agricultural soils, vegetation is present continuously with minimal disturbance in a forested landscape allowing the improvement of "O" horizon above the mineral soil surface. The extent of the "O" horizon is entirely contingent on the forest vegetation type, rendering the physical, chemical, and biological attributes of forest soils distinctive. Further, the nutrient cycling process is holistic in that only a meager quantity of nutrients gets leached from the system, that is, efficient nutrient cycling process occurs in the forest ecosystem. Soil quality acts as an indirect gauge of forest productivity, typically assessed through methods like site index. In the realm of climate change, direct soil impacts encompass heightened temperatures, increased atmospheric CO_2 levels, altered precipitation patterns, and more often climatic hazards like heat waves, droughts, storms, and frosts. Moreover, the indirect effects involve heightened cropping intensity due to amplified occurrences of pests and diseases. Carbon stored in forest soils is more stable compared to other land use owing to the slow decomposition and weathering process. Another important aspect of linkage between forest and soils is in the reclamation process of degraded and wastelands through afforestation or agroforestry activity. The exceptional adaptability of trees to thrive in adverse environmental and soil conditions underscores the pivotal role of forestry and agroforestry in productive applications that integrate ecological restoration and soil conservation.

Lessons Learnt

- Overview of soil and unique features of forest soils.
- Basics of forest soil-forming process and characteristics.
- Physical and chemical properties including the organic layer of forest soils.
- Soil biology and its biological properties.
- Stand management practices and their relation to soil properties.
- Implications of change of climate on forest soils.
- Problematic soils and their reclamation.

Key Questions

1. Process of eluviating iron and aluminum oxides, humus, and carbonates from the upper horizons and their accumulation in subsurface horizons is called as.
   (a) Lateralization; (b) Podsolization; (c) Decalcification; (d) Calcification
2. Why the C/N ratio is found to be more in 15–30 cm layer of forest soil? Explain.
3. The intricate mixture of amorphous humus and mineral soil is called ___________.
4. Tardigrades, mites, microarthropods, and enchytraeids are ________.
5. Soil nitrogen can be increased via N_2 fixation, commonly in the ______stages of succession.

References

Anyomi KA, Raulier F, Bergeron Y, Mailly D (2013) The predominance of stand composition and structure over direct climatic and site effects in explaining aspen (*Populus tremuloides* Michaux) site index within boreal and temperate forests of western Quebec, Canada. For Ecol Manag 302:390–403

Anyomi KA, Lorenzetti F, Bergeron Y, Leduc A (2015) Stand dynamics, humus type and water balance explain aspen long term productivity across Canada. Forests 6:416–432. https://doi.org/10.3390/f6020416

Austin AT, Ballaré CL (2010) Dual role of lignin in plant litter decomposition in terrestrial ecosystems. Proc Natl Acad Sci U S A 107(10): 4618–4622

Bala N, Singh G, Bohra NK, Limba NK, Baloch SR (2014) Biodrainage for restoration of canal command waterlogged area in Indian desert. Indian Forester 140(5):462–467

Ball BA, Bradford MA, Hunter MD (2009) Nitrogen and phosphorus release from mixed litter layers is lower than predicted from single species decay. Ecosystems 12(1):87–100

Banik M, Sarkar A, Ghatak P, Ray R, Patra SK (2018) Reclamation of waterlogged lowland in Indo-Gangetic alluvial plains using some biodrainage species. Int J Curr Microbiol Appl Sci 7(2):1028–1038

Banyal R, Kaushal P, Dogra AS (2009) Effect of climatic factors on survival of Kikar and Shisham in Punjab. Indian Forester 135(11): 1477–1486

Banyal R, Rajkumar K, M., Yadav, R.K. and Dagar, J.C. (2017) Agroforestry for rehabilitation and sustenance of saline ecologies. In: Dagar JC, Tewari VP (eds) Agroforestry: anecdotal to science. Springer Nature Singapore Pvt. Ltd., pp 413–454

Banyal, R., Bhardwaj, A.K., Singh, R.K., Gajender, KM, Aslam P, Jagdish C, Bhatia VK (2019) Impact of *Eucalyptus* plantations on waterlogged saline ecologies in Indo-Gangetic plains, Project Report, ICAR-CSSRI, Karnal, Haryana, p 84

Basu JP (2014) Agroforestry, climate change mitigation and livelihood security in India. New Zealand J For Sci 44(Suppl. 1):1–10

Behera L, Nayak MR, Dhiraji Patel D, Abhishek Mehta A, Sinha SK, Gunaga R (2015) Agroforestry practices for physiological amelioration of salt affected soils. J Plant Stress Physiol 1(1):13–18

Berg B, Staaf H (1981) Leaching accumulation and release of nitrogen in decomposing forest litter. In: Clark FE, Rosswall T (eds) Terrestrial nitrogen cycles: processes, ecosystem strategies, and management impacts. Ecological Bulletins, Stockholm, pp 163–178

Bhatt H, Husain M, Rathore JP, Sah VK (2017) Bioremediation of problematic soils through agroforestry practices. J Pharmacogn Phytochem 6(5):2044–2048

Bhojvaid PP, Timmer VR (1998) Soil dynamics in an age sequence of *Prosopis juliflora* planted for sodic soil restoration in India. For Ecol Manag 106(2–3):181–193

Binkley D, Fisher RF (2013) Ecology and management of forest soils, 5th edn. Wiley-Blackwell, Wiley Online Library

Bradford A (2018) Deforestation: facts, causes and effects. https://www.livescience.com/27692-deforestation.html

Broadbent EN, Almeyda Zambrano AM, Asner GP, Soriano M, Field CB et al (2014) Integrating stand and soil properties to understand foliar nutrient dynamics during forest succession following slash-and-burn agriculture in the Bolivian Amazon. PLoS One 9(2): e86042. https://doi.org/10.1371/journal.pone.0086042

Cheng WX, Johnson DW (1998) Elevated CO_2, rhizosphere processes and soil organic matter decomposition. Plant Soil 202(2):167–174

Chertov OG, Nadporozhskaya MA (2018) Humus forms in forest soils: concepts and classifications. Eurasian Soil Sci 51:1142–1153. https://doi.org/10.1134/S1064229318100022

Cleveland CC, Townsend AR, Schimel DS, Fisher H, Howarth RW et al (1999) Global patterns of terrestrial biological nitrogen fixation in natural ecosystems. Glob Biogeochem Cycles 13:623–645

Cole DW, Rapp M (1981) Elemental cycling in forest ecosystems. In: Reichle DE (ed) Dynamic properties of forest ecosystems, vol 23. Cambridge University Press, Cambridge, UK, pp 411–450

Cole DW, Gessel SP, Dice SF (1967) Distribution and cycling of nitrogen, phosphorus, potassium and calcium in a second growth Douglas-fir ecosystem. In: Young HE (ed) Primary productivity and mineral cycling in natural ecosystems. University of Maine Press, Orono, Maine, pp 197–213

Conant RT, Drijber RA, Haddix WL et al (2008) Sensitivity of organic matter decomposition to warming varies with its quality. Glob Chang Biol 14:868–877

CSSRI (2018) Annual report 2017–18. ICAR-Central Soil Salinity Research Institute, Karnal Haryana, India

D'Amore Davis, Kane Evan. (2020) Forest soil carbon and climate change. USDA Climate Change Resource Center, pp. 1–7

Dash CJ, Sarangi A, Singh AK, Dahiya S (2005) Bio-drainage: an alternate drainage technique to control waterlogging and salinity. J Soil Water Conserv 4(3&4):149–155

Devi AS, Yadava PS (2007) Wood and leaf litter decomposition of *Dipterocarpus tuberculatus* Roxb. in a tropical deciduous forest of Manipur, North East India. Curr Sci 93:243–246

Doerr SH, Shakesby RA, MacDonald LH (2009) Soil water repellancy: a key factor in post-fire erosion. In: Cerda A, Robichaud PR (eds) Fire effects on soils and restoration strategies. Science Publishers, Enfield

Drigo B, Kowalchuk GA, Johannes AV (2008) Climate change goes underground: Effects of elevated atmospheric CO_2 on microbial community structure and activities in the rhizosphere. Biol Fertil Soils 44(5):667–679

FAO (2007). http://www.fao.org/forestry/18222-045c26b711a976bb9d0d17386ee8f0e37.pdf
FAO (2015) Global forest resources assessment. FAO forestry paper No. 1. UN Food and Agriculture Organization, Rome
Feldpausch TR, Rondon MA, Fernandes EC, Riha SJ, Wandelli E (2004) Carbon and nutrient accumulation in secondary forests regenerating on pastures in central Amazonia. Ecol Appl 14:164–176
Fogel R, Cromack K (1977) Effect of habitat and substrate quality on Douglas fir litter decomposition in western Oregon. Can J Bot 55: 1632–1640
Galloway JN, Aber JD, Erisman JW et al (2003) The nitrogen Cascade. Bioscience 53:341–356
Garg VK, Jain RK (1996) Effect of fuelwood plantations on some properties of sodic wastelands. J Trop For Sci 9:194–205
Giller KE, Gadisch G (1997) Driven by nature: a sense of arrival or departure. CAB International, Wallingford, pp 393–399
Gitay H, Suarez A, Watson RT, Dokken DJ (2002) Climate change and biodiversity. Report of Intergovernmental Panel on Climate Change
Griffith RP, Madritch MD, Swanson AK (2009) The effects of topography on forest soil characteristics in the Oregon Cascade Mountains (USA): implications for the effects of climate change on soil properties. For Ecol Manag 257:1–7
Hart SA, Chen HYH (2008) Fire, logging and overstory affect understory abundance, diversity and composition in boreal forest. Ecol Monogr 78:123–140
Hobbie SE, Vitousek PM (2000) Nutrient limitation of decomposition in Hawaiian forests. Ecology 81(7):1867–1877
Hoegh-Guldberg O, Jacob D, Taylor M et al (2018) Impacts of 1.5 °C global warming on natural and human systems. In: Masson-Delmotte V, Zhai P, Pörtner HO (eds.) Global warming of 1.5 °C. An IPCC Special Report on the impacts of global warming of 1.5 °C above pre-industrial levels and related global greenhouse gas emission pathways, in the context of strengthening the global response to the threat of climate change, sustainable development, and efforts to eradicate. Intergovernmental Panel on Climate Change, Geneva, 211p
Holland EA, Dentener FJ, Braswell BH, Sulzman JM (1999) Contemporary and pre-industrial global reactive nitrogen budgets. Biogeochemistry 46:7–43
IPCC (1996) Climate change: impacts, adaptations and mitigation of climate change, scientific-technical analyses. Intergovernmental panel on climate change. Cambridge University Press, Cambridge, UK
Jiménez-Pinilla P, Lozano E, Mataix-Solera J, Arcenegui V, Jordán A, Zavala LM (2016) Temporal changes in soil water repellency after a forest fire in a Mediterranean calcareous soil: influence of ash and different vegetation type. Sci Total Environ 572:1252–1260
Johnson DW, Van Miegroet H, Cole DW, Richter DD (1984) Effects of acid precipitation and natural processes on cation leaching from four diverse forest ecosystems. In: Stone EL (ed) Sixth North American forest soils conference: forest soils and treatment impacts. Knoxville, Tenn, pp 247–263
Johnson K, Scatena FN, Johnson AH, Pan Y (2009) Controls on soil organic matter content within a northern hardwood forest. Geoderma 148:346–356
Keyes MR, Grier CC (1981) Above and below-ground net production in 40-year-old Douglas-fir stands on low and high productivity sites. Can J For Res 11:599–605
Kimmins JP, Hawkes BC (1978) Distribution and chemistry of fine roots in a white spruce-subalpine fir stand in British Columbia: implications for management. Can J For Res 8(3):265–279
Klinka K, Green RN, Trowbridge RL, Lowe LE (1981) Taxonomic classification of humus forms in ecosystems of British Columbia: first approximation. Land Management Report No. 8. British Columbia Ministry of Forests, Victoria, B.C. 54p
Kumar BM, Deepu JK (1992) Litter production and decomposition dynamics in moist deciduous forests of the Western Ghats in Peninsular India. For Ecol Manag 50:181–201
Kunhamu TK, Kumar BM, Assif PK, Jayadevan C (1994) Litter yield and decomposition under *Acacia auriculiformis*. Nitrogen Fixing Tree Res Rep 12:29–32
Likens GE, Bormann FH, Pierce RS, Eaton JS, Johnson NM (1977) Biogeochemistry of a forested ecosystem. Springer Verlag, New York, p 146
Lindberg SS, Lowett GM, Richter DD, Johnson DW (1986) Atmospheric deposition and canopy interactions of major ions in a forest. Science 231:141–145
Lowe LE, Klinka K (1981) Forest humus in the coastal western hemlock biogeoclimatic zone of British Columbia in relation to forest productivity and pedogenesis, vol No. 89. Research Note, B.C. Min. of Forests, Victoria, B.C. 83p
Maaroufi IN, De Jonathan R, Long. (2020) Global change impacts on forest soils: linkage between soil biota and carbon – nitrogen and phosphorus stoichiometry. Front Glob Change Clim 3:16
Mahendrappa MK (1980) Relationships between different estimates of mineralizable N in the organic materials under black spruce stands. Can J For Res l0(4):527–522
Mahendrappa MK, Salonius PO (1982) Nutrient dynamics and growth response in a fertilized black spruce stand. Soil Sci Soc Am J 46: 127–133
Marínez-Yrzar A, Núñez S, Búrquez A (2007) Leaf litter decomposition in a southern Sonoran desert ecosystem, northwestern Mexico: effects of habitat and litter quality. Acta Oecol 32(3):291–300
Markewitz D, Davidson E, Moutinho P, Nepstad D (2004) Nutrient loss and redistribution after forest clearing on a highly weather soil in Amazonia. Ecol Appl 14:S177–S199
McColl JG, Cole DW (1968) A mechanism of cation transport in forest soil. Northwest Sci 42:134–140
McColl JG, Monette LK, Bush DS (1982) Chemical characteristics of wet and dry atmospheric fallout in northern California. J Environ Qual 11:585–590
McMahon SM, Parker GG, Miller DR (2010) Evidence for a recent increase in forest growth. Proc Natl Acad Sci U S A 107:E86–E87. https://doi.org/10.1073/pnas.0912376107
Melillo JM, Gosz JR (1983) Interactions of Biogeochemical cycles in forest ecosystems. In: Bolin, Cook (eds) Volume on major biogeochemical cycles and their interactions. John Wiley and Sons, Toronto, Ontario. SCOPE 21, 177-222.Interactions of biogeochemical cycles in forest ecosystems
Melillo JM, Aber JD, Muratore JF (1982) Nitrogen and lignin control of hardwood leaf litter decomposition dynamics. Ecology 63:621–626
Miesel JR, Goebel PC, Corace RG, Hix DM, Kolka R, Palik B, Mladenoff D (2012) Fire effects on soils in Lake states forests: a compilation of published research to facilitate long-term investigations. Forests 3:1034–1070
Mikola J, Setälä H (1998) Interplay of omnivory, energy channels and C availability in a microbial-based soil food web. Biol Fertil Soils 28: 212–218. https://doi.org/10.1007/s003740050485
Miller HG (1981) Nutrient cycles in forest plantations, their change with age and the consequences for fertilizer practice. In: Proceedings of the Australian forest nutrition workshop: productivity in perpetuity, 10–14 August 1981
Miller HG (1984) Dynamics of nutrient cycling in plantation ecosystems. In: Bowen GD, Nambiar EKS (eds) Nutrition of plantation forestry. Academic Press, Toronto, Ontario
Mishra A (2002) Ameliorative effect of tree species on physical and chemical properties of sodic soils in Uttar Pradesh. Ph.D. Thesis. ICFRE, Dehradun
Montagnini F, Jordan CF (2002) Reciclaje de nutrientes. In: Guariguata MR, Kattan G (eds) Ecolog'ıa y Conservaci'on de Bosques Neotropicales. LUR, Cartago, Costa Rica, pp 167–191

Morrison IK (1985) Biomass and macroelements in a tolerant hardwood stand. Turkey Lakes Watershed, Ontario. In: Proc. Acid Rain For. Resour. Conf., 14–17 June 1983, Qubec City, Que

Nearing MA, Pruski FF, O'Neal MR (2004) Expected climate change impacts on soil erosion rates: a review. J Soil Water Conserv 59:43–50

Norby RJ, Warren JM, Iversen CM, Medlyn BE, McMurtrie RE (2010) CO_2 enhancement of forest productivity constrained by limited nitrogen availability. Proc Natl Acad Sci U S A 107(19):368–373

Orgiazzi A, Bardgett RD, Barrios E (2016) Global soil biodiversity atlas. European Commission

Osman KT (2013) Forest soils. Springer International Publishing, Cham. https://doi.org/10.1007/978-3-319-02541-4

Palma RM, Prause J, Fontanive AV, Jimenez MP (1998) Litter fall and litter decomposition in a forest of the Parque Chaqueno Argentino. For Ecol Manag 106(2–3):205–210

Pan KW, He J, Wu N (2004) Effect of forest litter on microenvironment conditions of forestland. Chin J Appl Ecol 15(1):153–158

Pandey DS, Singh SP, Singh G (2015) Underprivileged agriculture: retrospection and future prospects. In: Pandey GB (ed) Compendium of lectures on management of underprivileged agriculture. Pant University of Agriculture and Technology, Pantnagar, India. 311p

Pareek N (2017) Climate change impact on soils: adaptation and mitigation. MOJ Ecol Environ Sci 2(3):136–139

Parker GG (1983) Through fall and stemflow in the forest nutrient cycle. Adv Ecol Res 13:57–133

Perala DA, Alban DH (1982) Biomass nutrient distribution and litterfall in Populus, Pinus and Picea stands on two different soils in Minnesota. Plant Soil 64:111–192

Prenzel J (1979) Mass flow to the root system and mineral uptake of a beech stand calculated from 3-year field data. Plant Soil 51:3949

Quested HM, Press MC, Callaghan TV, Cornelissen JHC (2002) The hemiparasitic angiosperm *Bartsia alpina* has the potential to accelerate decomposition in sub-arctic communities. Oecologia 130(1): 88–95

Raison RJ, Khanna PK (2011) Possible impacts of climate change on forest soil health. In: Singh BP et al (eds) Soil health and climate change. Springer-Verlag, Berlin, Heidelberg

Raison RJ, Kirschbaum MUF (2008) Fire and the carbon balance of terrestrial ecosystems. Geophys Res Abstr 10:EGU2008-A-02946

Rapp M, Leclerc C, Lassaint P (1979) The nitrogen economy in a *Pinus pinea* L. stand. For Ecol Manag 2:221–231

Rastetter EB, Vitousek PM, Field C, Shaver GR, Herbert D et al (2001) Resource optimization and symbiotic nitrogen fixation. Ecosystems 4:369–388

Rohne B, Bigler C, Wunder J, Brang P, Bugmann H (2012) Fifty years of natural succession in Swiss forest reserves: changes in stand structure and mortality rates of oak and beech. J Veg Sci 23:892–905

Sharma R, Sharma E, Purohit AN (1997) Cardamom, mandarin and nitrogen fixing trees in agroforestry systems in India's Himalayan region. I Litterfall and decomposition. Agrofor Syst 35:239–253

Silver WL, Ostertag R, Lugo AE (2000) The potential for carbon sequestration through reforestation of abandoned tropical agricultural and pasture lands. Restor Ecol 8:394–407

Singh KP (1969) Studies in decomposition of leaf litter of important trees of tropical deciduous forests at Varanasi. Trop Ecol 10:292–311

Singh B (1998) Biomass production and nutrient dynamics in three clones of *Populus deltoides* planted on Indogangetic plains. Plant Soil 203(1):15–26

Singh G, Dagar (2005) Greening sodic soils: Bichhian model. In: Technical Bulletin 2/2005. Central Soil Salinity Research Institute, Karnal. 51p

Singh YP, Sharma DK, Singh G, Nayak AK, Mishra VK, Singh R (2008) Alternate land use management for sodic soils. Bull. No. 2. Central Soil Salinity Research Institute, Regional Research Station, Lucknow, Uttar Pradesh, 7p

Singh YP, Singh G, Sharma DK (2011) Ameliorative effect of multipurpose tree species grown on sodic soils of Indo-Gangetic alluvial plains of India. Arid Land Res Manag 25:55–74

Skovsgard JP, Vanclay JK (2007) Forest site productivity: a review of the evolution of dendrometric concepts for even-aged stands. Institute of chartered foresters. Forestry 81. https://doi.org/10.1093/forestry/cpm041

Song XZ, Jiang H, Zhang HL, Yu SQ, Zhou GM, Ma YD, Chang Scott X (2008) A review on the effects of global environment change on litter decomposition. Acta Ecol Sin 28(9):4414–4423

St-Clair SB, Cavard X, Bergeron Y (2013) The role of facilitation and competition in the development and resilience of aspen forests. For Ecol Manag 299:91–99

Stone EL, Kszystniak R (1977) Conservation of potassium in the *Pinus resinosa* ecosystem. Science 198:192–194

Sutton RF (1969) Form and development of earlier root systems. Commonwealth For. Bur Tech Commun No.7, 13lp

Swift MJ, Heal OW, Anderson JM (1979) Decomposition in terrestrial ecosystems. University of California Press, Berkeley

Tarnawski S, Aragno M (2006) The influence of elevated CO_2 on diversity, activity and bio-geochemical function of rhizosphere and soil bacterial communities. In: Long et al (eds) Managed ecosystems and CO_2 case studies, processes and perspectives, ecological studies series, vol 187. Springer, Berlin, pp 393–409

van Elsas JD, Trevors JT, Rosado AS, Nannipieri P (2019) Modern soil microbiology. CRC Press

Verma S, Jayakumar S (2012) Impact of forest fire on physical, chemical and biological properties of soil: a review. PIAEES 2(3):168–176

Vogt KA, Grier CC, Meier CE, Edmonds RL (1982) Mycorrhizal role in net primary production and nutrient cycling in *Abies amabilis* ecosystems in western Washington. Ecology 63:370–380

Walter H (1985) Vegetation systems of the earth and ecological systems of the geo-biosphere. Springer-Verlag, Berlin

Walther GR, Post E, Convey P, Menzel A, Parmesan C, Beebee TJC, Fromentin J-M, Hoegh-Guldberg O, Bairlein F (2002) Ecological responses to recent climate change. Nature 416:389–395

Wang Q, Wang S, Huang Y (2008) Comparisons of litterfall, litter decomposition and nutrient return in a monoculture *Cunninghamia lanceolata* and a mixed stand in southern China. For Ecol Manag 255(3–4):1210–1218

Wuebbles D, Meehl G, Hayhoe K et al (2014) CMIP5 Climate model analyses: climate extremes in the United States. Bullet Am Meteorol Soc 95(4):571–583

Yang YS, Chen GS, Guo JF, He ZM, Chen YX (2002) Litter decomposition and nutrient release in a mixed forest of *Cunninghamia lanceolata* and *Tsoongiodendron odorum*. Acta Phytoecol Sin 26(3):275–282

Yuan ZY, Chen HYH (2009) Global trends in senesced-leaf nitrogen and phosphorus. Glob Ecol Biogeogr 18:532–542

Forest Hydrology and Watershed Management: Current Perspectives

10

S. Shoba, D. Ravikumar, G. M. Devagiri, and Rajashekhar D. Barker

Abstract

The forests comprise 26% of the world's land surface area, encompassing distinct biotic communities which harbor rich biodiversity and provide diverse ecosystem services. The interaction between forest, water, and soil is multidimensional as forests provide canopy surfaces to intercept precipitation and allow evaporation back into the atmosphere. The mechanism of regulating the quantum of water reaching the forest floor as through fall and pull water from the soil for transpiration is the very basis of forest hydrology. The hydrological processes in forested watersheds are influenced by various environmental, and physiological factors such as precipitation, radiation, temperature, species type, leaf area, and extent and structure of forest ecosystems. Of late, the forest watersheds have been affected globally by anthropogenic activities such as forest harvesting, and conversion of forested landscapes for plantations and urbanization. As a result, hydrological processes in forested watersheds have been altered and these changes are expected to accelerate due to climate change phenomena. Hence, understanding the interaction of various environmental, physiological, and physical drivers is crucial for sustainable management of forested watersheds.

This chapter presents the basic concepts of hydrology and watershed management. The fundamental properties and principles of water energy relationships on earth are presented which will serve as a basis for understanding the hydrological processes such as circulation, water form transformations, and flow processes that occur in watersheds. The subsequent sections deal with hydrological components, hydrographs, soil erosion, soil loss measurement, watershed delineation using geospatial technology, and watershed management.

Keywords

Forest hydrology · Hydrological cycle · Hydrograph · Watershed · Watershed management

10.1 Introduction

Water is a valuable natural resource to sustain life on earth. Although water covers 70% of the surface of the earth, there is an uneven distribution of water which leads to conflict for its access. The global hydrological cycle amounts to 42,000 km^2/year of renewable water resources. Around 70.8% of earth's surface water is saline which amounts to 97.4% of the total water present on earth and the remaining 2.6% is freshwater. Out of the total freshwater available, about 77.23% is locked in the polar regions as icecaps and glaciers, 22.2% as underground aquifers, and 0.36% in rivers and lakes (Baumgartner and Reichel 1975)

10.1.1 Water Issues

The internal renewable water resources (IRWRs) from rivers and aquifers are currently depleting; the world water budget

S. Shoba (✉)
DSLD CHEFT, Devihosur, University of Horticultural Sciences, Bagalkot, Karnataka, India
e-mail: shobha.s@uhsbagalkot.edu.in

D. Ravikumar
DSLD CHEFT, Devihosur, University of Horticultural Sciences, Bagalkot, Karnataka, India

KVK, Hiriyur, KSNUAHS, Shivamogga, India

G. M. Devagiri
DSLD CHEFT, Devihosur, University of Horticultural Sciences, Bagalkot, Karnataka, India

College of Forestry, Ponnampet, KSNUAHS, Shivamogga, India

R. D. Barker
DSLD CHEFT, Devihosur, University of Horticultural Sciences, Bagalkot, Karnataka, India

Department of Agricultural Engineering, Hiriyur, KSNUAHS, Shivamogga, India

A. K. Mandal, A. Nicodemus (eds.), *Textbook of Forest Science*, https://doi.org/10.1007/978-981-97-8289-5_10

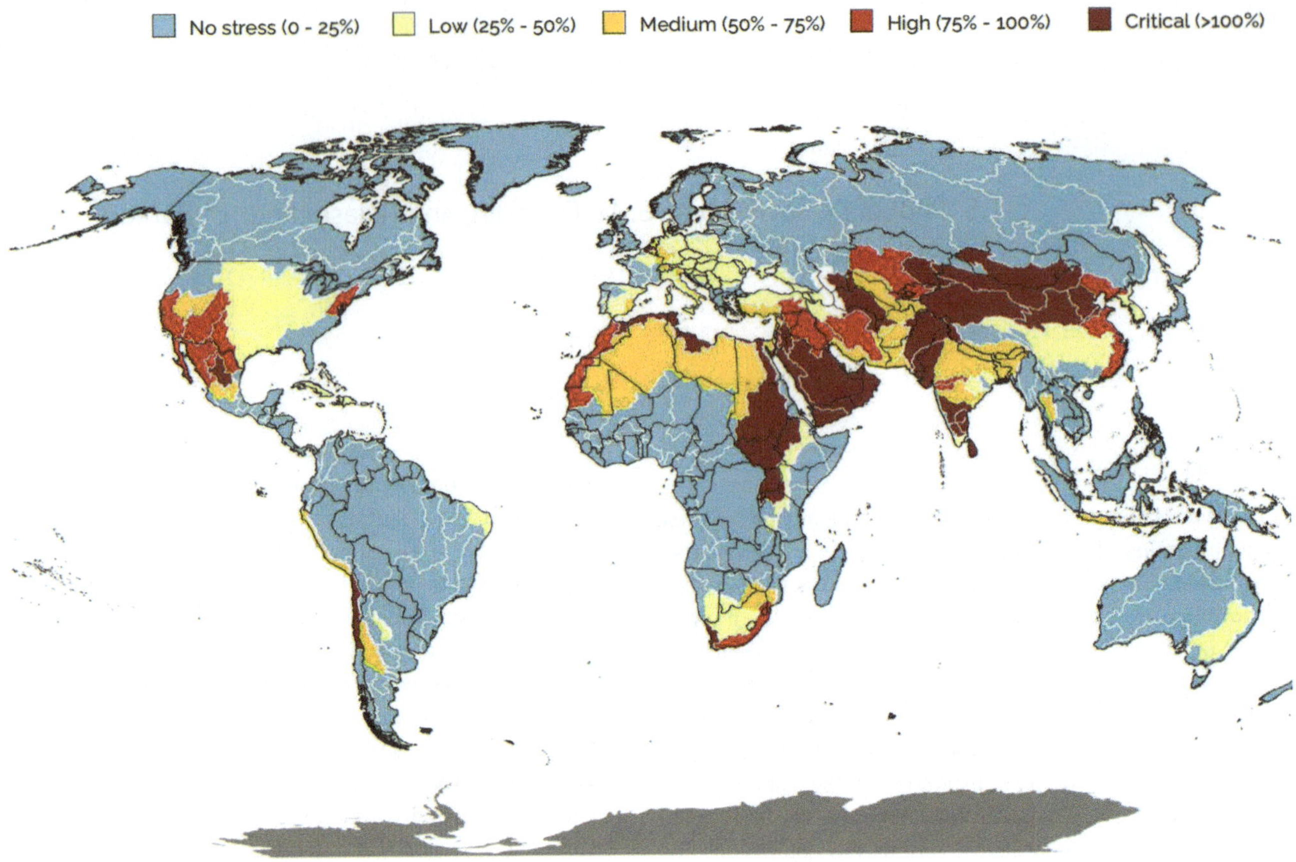

Fig. 10.1 Level of water stress of all sectors by major basin (Source: FAO & UN-Water 2021)

is currently under strain, amounting to 44,000 km^3/year, and withdrawals (all sectors) are 4000 km^3/year, which is 10% of the IRWRs (*www.fao.org*). The local impacts of physical water scarcity and freshwater pollution are spreading and accelerating. In many cases, the first sign of scarcity from increased withdrawals is falling groundwater levels as shown in Fig. 10.1.

The average withdrawal rate of more than 20% of renewable water resources is thought to stressed upon water resources in terms of physical water scarcity, with more than 40% considered "critical." Countries are already receding beyond crucial thresholds in some regions, most notably the Middle East, Northern Africa, and Central Asia. The way these pressures impact ecological systems is becoming more and more apparent. According to current estimates, almost 40% of rural populations worldwide reside in river basins with limited water supplies.

Water and forests both cover large portions of the earth and both are crucial to lives and the environment. The demands of a growing global population are mirrored in the degree of natural resource utilization. Therefore, the loss of forests and increasing human use of water are two important concerns in natural resources management (Cohen 2008). Certainly, the most crucial problems facing the twenty-first century are associated with water and forests.

We are concerned about water and forests not only as raw materials required by cultures and industries but also as key factors in the environment. Water and forests are not two independent natural resources; a close linkage exists between them. Consequently, the study of the interface between these two resources, called forest hydrology, has become an important field. It provides basic knowledge and foundations for watershed management—a discipline and skill for maintaining land productivity and protecting water resources.

Forest distributions enhance the forest–water relation significantly. Forests usually grow and develop in areas with annual precipitation of 500 mm or higher. These are also the areas suitable for certain agricultural activities. Approximately 30% of the area is covered by forests, while 60% of the runoff is produced by this 30% of forested land (Chang 2012). This means that much of the water that we use for drinking comes from forests. Any anthropogenic activity, development, and utilization of forested areas will inevitably destroy forest canopies and disturb forest floors to a certain degree. These actions may affect water quantity through their impacts on transpiration and canopy interception losses, infiltration rate, water-holding capacity, and overland flow velocity.

Water quality is impacted by several factors, including the loss of root system's capacity to bind the soil, increased overland flow, accelerated decomposition of organic materials, and direct raindrop impacts on mineral soils. All these factors may result in accelerating soil erosion and nutrient losses. All forests should be managed with their impacts on water resources in mind so that water resources can be properly protected and utilized. The proper scientific management of soil and water is of utmost importance to avoid the environmental and economic implication of soil degradation, afforestation, groundwater depletion, and climate change catastrophes. Thus, a basic understanding of hydrology is crucial for the effective management of forest watersheds.

10.1.2 Hydrology

The study of occurrence, circulation, and distribution of earth's water, its chemical and physical properties and interactions with the environment is known as hydrology. Hydrology is an interdisciplinary field since it includes aspects of meteorology, geology, statistics, chemistry, physics, and fluid mechanics. The following is the broad classification of hydrology.

(a) Scientific hydrology: It is concerned with the scholarly elements of hydrology.
(b) Applied/engineering hydrology: Applied/engineering hydrology deals with application aspects of hydrology such as water resources estimation, study of hydrological processes and their interactions and, water issues such as floods/drought and its combating strategies.

Based on the land use conditions, hydrology can be classified as rangeland hydrology, agricultural hydrology, forest hydrology, urban hydrology, wetland hydrology, and desert hydrology.

Forest hydrology is the study of hydrology in forest. It deals with forest in relation to all phases of water and the influence of forest management on the hydrological processes, soil erosion, sedimentation, nutrient losses, floods, and drought. Hydrological cycle has profound influence on geography, economics, sociology, agriculture, forestry, and politics. A scientific perspective and knowledge of forest hydrology are imperative for managing water resources in forest watersheds, designing, and running projects involving hydropower, irrigation, drainage, flood control, navigation, coastal works, salinity control, and recreational water use.

10.1.3 Hydrologic Cycle

Water is found in the atmosphere as clouds or vapor, and on earth's surface as ice, snow, or water in lakes, streams, and seas. It exists as groundwater and soil moisture below the surface. Except for deep groundwater, the whole earth's water supply is constantly moving due to various natural processes. The hydrological cycle is the movement of water across the surface or below the earth from the ocean and land surface to the air, and from the air to the land and back to the oceans as shown in Fig. 10.2.

The water present in the oceans evaporate due to the solar radiation. The clouds are formed due to the evaporation of

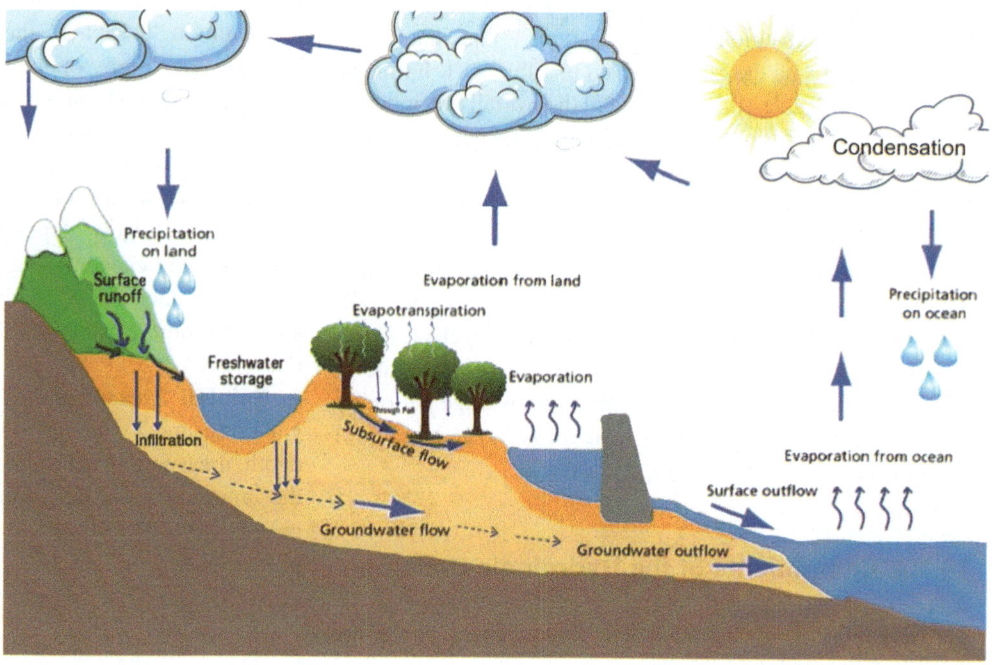

Fig. 10.2 Schematic diagram of the hydrologic cycle

Table 10.1 Annual water balance of earth calculated by different investigators

Land			Ocean			Globe	Reference
P_t	V	RO	P_t	V	RO	$P_t = V$	
800	485	315	1270	1400	130	1130	Korzoun (1978)

water from the oceans. A portion of the clouds is forced onto the land by winds, while the majority of the clouds condense and return to the oceans as rain. The clouds that are driven to the land surface, condense, and precipitate as rain, sleet, hail, and so on onto the land mass. Even while it falls, some of the precipitation may evaporate and return to the atmosphere. A certain portion of precipitation may descend to the ground surface after interception by plants, buildings, and other obstructions. A percentage of the water that falls on land infiltrates the earth's surface, increasing the soil's moisture content and reaching the groundwater body. Through transpiration, vegetation returns a fraction of the water from the subsurface to the atmosphere. After fulfilling the requirements for infiltration and evaporation, precipitation reaches the ground's surface, travels along the natural slope, passes through a system of gullies, streams, and rivers, and eventually reaches the ocean. After spending a significant amount of time below the surface, groundwater can emerge through springs and other outlets. Runoff is the part of precipitation that travels in different directions both above and below the earth's surface before it enters a stream channel. Runoff turns into stream flow after it gets into a stream channel.

The law of conservation is applicable to the hydrological system, as there exists a balance between the input and output components of a watershed system. The average annual balances of precipitation, P_t, vaporization, V and runoff, RO for land, ocean, and the globe is given in Table 10.1. It may be observed from Table 10.1 that for land the overall precipitation is greater than vaporization (V) and the difference appears as runoff to supplement the vaporization required in the ocean.

10.1.3.1 Energy Balance

The sun is the ultimate source of energy for the occurrence of the hydrological cycle, and the total amount of water available on earth is constant. The hydrological cycle requires energy in the form of solar radiation to change the state of water in the atmosphere and on earth. The average net radiation absorbed by the planet is 79 kcal/cm^2/year, and 84.5% of that energy is lost through evaporation.

10.1.4 Components of the Hydrologic Cycle

The essential components of the hydrological cycle are described hereunder:

10.1.4.1 Precipitation

The term precipitation refers to all forms of water derived from atmospheric vapor and deposited on the earth's surface. Water vapor in the atmosphere condenses to produce precipitation. Precipitation occurs in different forms, including rain, snow, hail, frost, and dew. Rain and snow are two of the main types of precipitation that contribute significantly to water content. However, the vegetation canopies prevent some portion of the rain and snow from reaching the earth's surface. The occurrence of rainfall necessitates a few fundamental requirements such as the presence of moisture in the atmosphere and an adequate number of nuclei to facilitate the condensation of water vapor into droplets smaller than 0.1 mm. Because the winds help the clouds move, they are crucial to the precipitation process. When the water droplets combine to produce larger drops, precipitation happens. Once the precipitation reaches the ground, it can move as surface runoff and a part may enter the soil as subsurface runoff or base flow. Some part of precipitation may again be lost to the atmosphere as evaporation, transpiration or evapotranspiration.

Rain: It is the most prevalent type of precipitation, with water droplets typically larger than 0.5 mm in diameter and its maximum size is about 6 mm. Any drop that is bigger than this usually fragments into tiny drops as it descends from the clouds. The various type of rainfall categorized based on intensity of rainfall are fog, mist, drizzle, light rain, moderate rain, heavy rain, excessive rain, and cloud burst (Humphreys 1940)

The canopies in the forest intercept the precipitation and act as a barrier for the precipitation to reaches the ground. Some portion of the intercepted precipitation reaches the ground through stems and some portion drips intercepted precipitation to reach the ground through the foliage, branches, or canopy openings. In forest watersheds, approximately 10–25% of annual precipitation is lost due to canopy interception depending upon evaporation potential, rainfall characteristics and vegetation. However, litter interception is much smaller than canopy interception. The amount of litter interception largely depends upon the thickness of litter, water-holding capacity, the frequency of wetting and evaporation rate. Generally, about 1–15% of annual precipitation and <50 mm/year are lost to litter interception (Helvey and Patric 1965).

Rainfall Measurement: Understanding the process for gathering and evaluating rainfall data is essential. Rainfall depth, measured in centimeters or millimeters, is a measure of the amount of rainfall falling on a level surface. The rain gauges are used to measure rainfall.

10.1.5 General Guidelines for the Location of Rain Gauges

Improper installation of the rain gauges may cause errors in the rainfall data. While locating the rain gauges, the fundamental guidelines need to be considered:

1. The opening of the rain gauge must be at least 75 cm above the ground surface.
2. The rain gauge should be placed in an open area free from obstructions.
3. The rain gauge should be installed away from any nearby trees or buildings and conveniently as twice as the height of obstruction.
4. The rain gauge should be placed on level ground and uneven terrain should be avoided.

The vertical depth of water that would accumulate in the rain gauge while on a level surface without any losses due to evaporation, infiltration, or seepage is considered as rainfall. Installing rain gauges in strategic places allows one to measure the amount of rainfall. Rain gauges can be categorized as non-recording and recording.

1. **Non-recording rain gauge**: The Symons' gauge as shown in Fig. 10.3 is the most prominent type of non-recording rain gauge that is widely used in India. It is a 12.7 cm diameter circular collecting area attached to a funnel. At a height of about 30.5 cm above the ground, the collector's rim is positioned in a horizontal plane. The rainfall is discharged from the funnel into a receiving vessel. The receiving vessel and funnel are stored in a metal container. Using a properly graduated measuring glass, the amount of water in the receiving vessel is determined with an accuracy of up to 0.1 mm.
2. **Recording rain gauge**
   When rain gauges are recorded, a continuous plot of precipitation versus time is created, which provides valuable data for hydrological storm analysis as it provides details on the amount and duration of rainfall. The following are some of the commonly used recording type rain gauges.

1. Tipping- bucket type
2. Weighing-bucket type
3. Natural siphon type

Fig. 10.3 Non-recording rain gauge (Symons' gauge)

10.1.6 Natural Siphon Type

The natural siphon type rain gauge also called as "float-type gauge" is used to measure the intensity, frequency, duration, and amount that affect the amount and rate of runoff. The natural siphon type rain gauge is shown in Fig. 10.4.

The Natural Siphon type rain gauge is made up of a galvanized iron cylinder measuring 22.5 cm in diameter and 60 cm height, funnel top. The funnel's opening connects into a circular brass inner tube. The cylinder collects the water that falls from the tube. A graph of rainfall against time is recorded on a record sheet using a pen and is carried by a clock-driven drum as part of the recording gear. The recording pen is secured to a rod that is attached to the inner cylinder's float. The float rises in tandem with the inking

Fig. 10.4 Natural siphon type rain gauge

pen, which captures the features of the storm, while water builds up in the cylinder. As soon as the cylinder is full, the inner cylinder and siphon tube are connected, the water in the cylinder is completely drained, and the float and inking pen return to their zero positions. Data from rain gauges is easier to analyze and comprehend when presented graphically.

Analysis of rainfall data: The rainfall data that a rain gauge station gathers is useful for various applications. The analysis of rainfall can be undertaken as follows.

1. **Mean annual rainfall**
   The average annual rainfall is recorded by using a rain gauge over a period of consecutive years. Mean monthly or mean weekly rainfalls can similarly be calculated.
2. **Rainfall intensity**
   The depth of rainfall received per unit time as expressed in inches per hour, millimeters per hour, or centimeters per hour is commonly referred as rainfall intensity. If rain gauges are not recording type, we can obtain only the daily average rainfall. The automatic rain gauges can calculate the rainfall intensity.
3. **Average depth of rainfall over an area.**
   The average rainfall depth of an area can be calculated using the following three methods.
   (a) Arithmetic mean
   (b) Thiessen method
   (c) Isohyetal method
   Arithmetic mean: It is obtained by dividing the total depth of rainfall recorded at all stations in the location by the total number of stations.
   Thiessen method: The location of the rain gauges is plotted on a map of the area and the stations are connected by means of straight lines as shown in Fig. 10.5. The perpendicular bisectors are constructed on each of the lines. Such that each of the rain gauge station is enclosed in a certain area. For example, station C is enclosed by the polygon $O_1.O_2$, O_7, O_8.

If P_1 P_2 P_3... are amount of rainfall recorded in each station; A_1, A_2, A_3... are the area of the polygons enclosing them. Then average rainfall P over the given area A is as given below

$$P = \frac{P_1A_1 + P_2A_2 + P_3\ A_3 + \ldots\ldots.}{A_1 + A_2 + A_3 + \ldots\ldots.}$$

Isohyetal method: A line connecting points of equal rainfall is called as isohyet. The location of the rain gauges stations is plotted and the depth of rainfall is recorded at each station. Isohyets are then drawn by interpolation as shown in Fig. 10.6.

$$P = \frac{a_{1\frac{(P_1+P_2)}{2}} + a_{2\frac{(P_2+P_3)}{2}} + \ldots.}{A}$$

where P_1, P_2.... are the value of isohyets; a_1, a_2 inter isohyetal areas; A is the total area

10.1.6.1 Evaporation

Evaporation is the process where water transforms from a liquid into a vapor state at the free water surface. Both the land surface and water bodies evaporate water during the

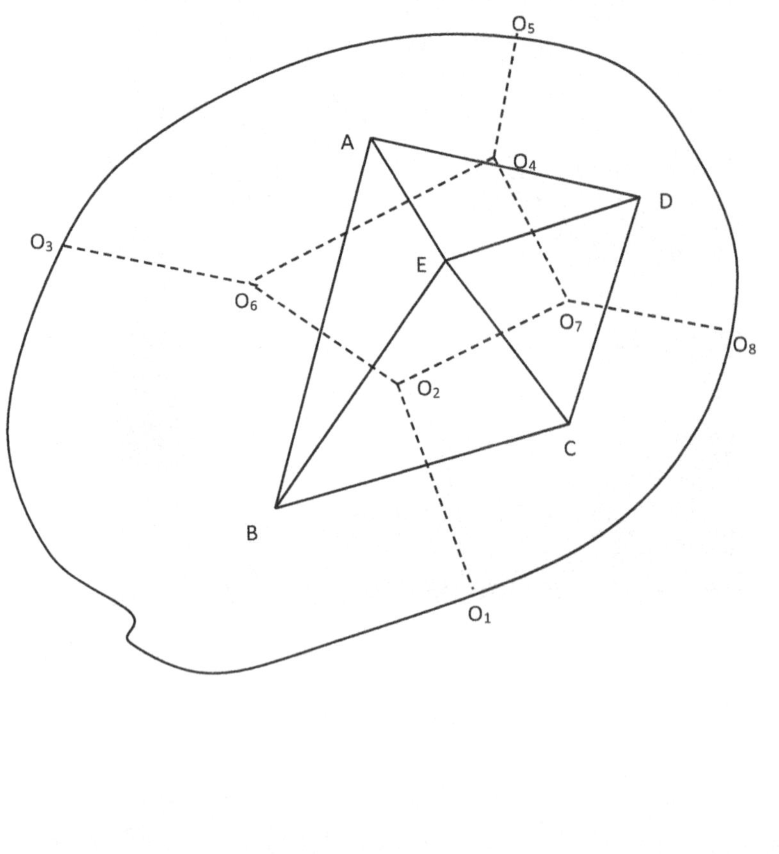

Fig. 10.5 Thiessen method

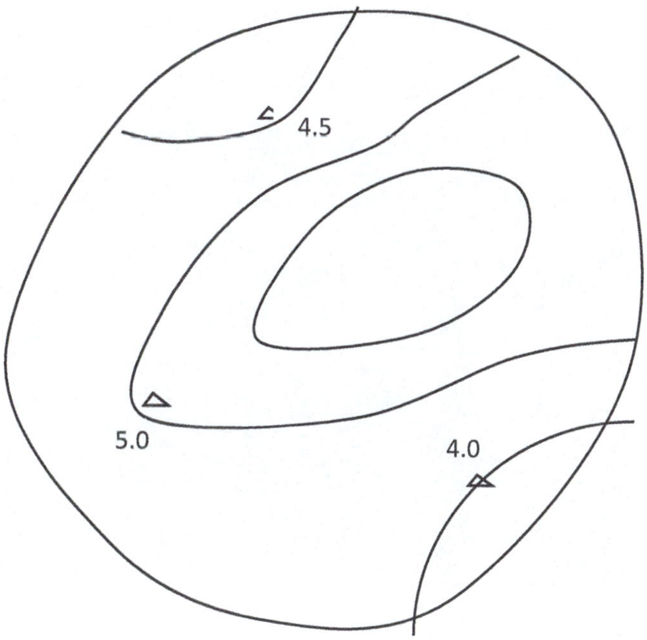

Fig. 10.6 Isohyetal method

hydrologic cycle. Evaporation is a heat and mass transfer process wherein about 585 cal of latent heat of vaporization is consumed for evaporating a gram of water. The vapor pressure at the water's surface and in the air above, air and water temperatures, wind patterns, atmospheric pressure, water quality, and the dimensions of the water bodies are some of the variables that influence the rate of evaporation.

In the case of forest watersheds, which are covered with litter floor evaporation occurs from 1% to 2% water surface of the watershed. A portion of the rainfall that is intercepted by obstructions evaporates as water vapor and reaches back into the atmosphere.

10.1.7 Estimation of Evaporation

The estimation of amount of water that has evaporated from the surface of water can be estimated using evaporimeters and analytical methods. The following formula can be used to determine the rate of evaporation.

$$E\ (\mathrm{g/sq.cm/s}) = \frac{-\rho_s - \rho_a}{r_a}$$

where,

$\rho_s - \rho_a$ is the vapor pressure gradient (g/cm^3)
r_a is aerial resistance to vapor diffusion (s/cm)

Fig. 10.7 Evaporation pan

10.1.8 Evaporimeter

Evaporimeters are open pans filled with water, allowing the water's evaporative loss to be recorded on a regular basis. It is made up of a shallow pan with water in it that is subject to the evaporative effects of the weather. The US Weather Bureau's Class A Standard Pan is shown in Fig. 10.7. The Class A Standard Pan has 1.21 m diameter, 25 cm depth, and is placed 15 cm above the ground.

Evaporimeters are simply open-air pans that hold water, allowing the water's evaporative loss to be regularly monitored. It consists of a shallow open pan filled with water and exposed to surface evaporation. The industry standard is the Class A Pan from the US Weather Bureau. Its dimensions are 1.21 m in diameter, 25 cm in depth, and 15 cm above the ground (Fig. 10.7). The readings taken from the pan (E_{pan}) need to be multiplied by a "Pan Coefficient" (K_{pan}) to obtain ETo.

$$\text{Thus}: \mathrm{ETo} = E_{\mathrm{pan}} \times K_{\mathrm{pan}}$$

K_{pan} ranges from 0.35 to 0.85 for the Class A evaporation pan, with an average of 0.70. An estimate can be made using the average value of 0.70 if the exact pan factor is unknown.

10.1.8.1 Infiltration

The movement of water from the soil's surface and penetrating downwards into the soil profile is known as infiltration. The term "infiltration rate," which is typically expressed as cm/h, refers to the speed at which water is permeating the soil's surface at any moment of time. Infiltration capacity is the highest rate of water absorption by soil at any instant of time. The rate of infiltration is high in the beginning, but it gradually decreases as water slowly seeps into the soil pores until it reaches a constant value at which the rate of infiltration equals the rate at which water is removed from the soil profile by gravity. The nearly constant rate attained after elapse of some time since the start of the precipitation is known as basic infiltration rate. The entire amount of water that percolates into the soil over a given amount of time is known as cumulative infiltration, also referred to as accumulated infiltration. The infiltration rate and the accumulated infiltration are the two aspects that are commonly used to evaluate the infiltration properties of soil. The cumulative infiltration and elapsed time are typically described by the following equation.

$$y = at^{\alpha} + b$$

where y = accumulated infiltration, cm; t = elapsed time in min; a, b and α = characteristic constants.

10.1.9 Factors Affecting Infiltration

The initial moisture content, soil surface condition, soil texture, structure, porosity, tillage, hydraulic conductivity of the soil profile, degree of soil swelling, organic matter, vegetative cover, length of irrigation or rainfall, etc. are the main factors that affect infiltration. Heavy-textured soils often have lower infiltration rates than light-textured soils. Grassland has higher infiltration rates than bare, uncultivated soil. The rate of penetration is significantly increased by the addition of organic waste.

10.1.10 Measurement of Infiltration

The following are the three commonly used methods for estimating infiltration in soil:

(a) Infiltrometer method
(b) Measurement of subsidence of free water in a large basin
(c) Estimation of accumulated infiltration from the water front advance data.

The measurement of infiltration using a double ring infiltrometer is shown in Fig. 10.8. The cylinders are made of 2 mm rolled steel and are typically about 25 cm deep. The inner cylinder, which is used to quantify infiltration, has a diameter of 30 cm and the outer cylinder is 60 cm in diameter which acts as a buffer pond. The cylinders are buried in the ground at a depth of roughly 10 cm. Every experiment is conducted with the cylinders installed at the same depth.

The infiltration process is greatly impacted by the type, texture, structure, permeability, and land use of the soil. Forest soils will have higher infiltration capacity as compared to similar soil types in urban watersheds.

Fig. 10.8 Double ring infiltrometer

10.1.10.1 Transpiration

Transpiration is a process of transferring water from the soil into the atmosphere in the form of water vapor by living plants. A substantial amount of the water that plants take up from the soil is lost through transpiration, and only a small percentage of 1% is retained by the plants for growth and development.

10.1.10.2 Runoff

Runoff is that portion of rainfall that flows as surface or subsurface runoff toward streams, lakes, or oceans. Only that water which enters the stream without seeping into the earth is considered as surface runoff. But when the word "runoff" is used in isolation, it usually refers to surface runoff. Surface and subsurface flows fall under the category of stream flow.

Rainfall is a major contributor for the generation of runoff. Interception is the portion of rainfall that is blocked from reaching the ground surface by obstructions such as trees, structures, and other objects. Depression storage (original detention) is that portion of rainfall that is held in surface depressions and eventually penetrates the soil or evaporates back to the atmosphere. The excess rainfall after satisfying these losses tends to travel across the land surface and drains into a channel or stream is referred to as "runoff." The various runoff pathways are depicted in Fig. 10.9. When rainfall intensity is less than soil infiltration capacity, rainwater is absorbed into the soil. When rain continues to saturate the soil, the infiltration capacity reduces thereby signaling the beginning of an overflow. The length and depth of overland flows are usually small. The overland flow continues to

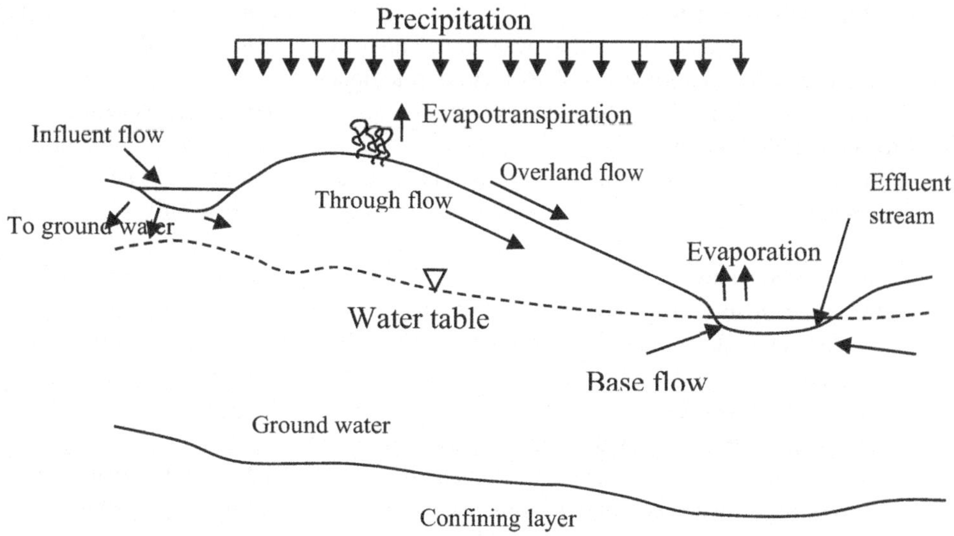

Fig. 10.9 Different routes of runoff

merge into bigger streams from smaller channel flows until it reaches the catchment outlet.

Interflow, through flow, storm seepage, subsurface flow, or quick return flow are terms used to describe a fraction of rainwater that returns to the surface after infiltrating and moving laterally through upper layers of the subsurface. The catchment's geological conditions determine how much interflow occurs. A considerable amount of the filtered water also passes through deep percolation to reach the groundwater table. The groundwater takes a complex and long course, eventually reaching the surface. The time lag between the entry of subsurface runoff into soil and the outflow from it is very large, may be months and years. This part of runoff is called groundwater runoff or groundwater flow.

The time delay between the onset of rainfall and runoff is used to categorize the runoff as follows.

1. Direct runoff: It is that part of the runoff which enters the stream immediately after the rainfall. Direct runoff includes surface runoff, prompt interflow, and rainfall on the surface of the stream itself.
2. Base flow: The delayed flow that reaches a stream essentially as groundwater flow is called base flow. Many times, delayed interflow is also included under this category.

10.1.11 Factors Affecting Runoff

The following two elements have the most impact on an area's runoff volume and rate:

A. Climatic factors.
B. Physiographical factors.

A. **Climate factors**

The climatic factors influencing runoff are discussed below.

1. **Types of precipitation**
   The type of precipitation has a profound influence on the runoff. The precipitation occurring in the form of rainfall will immediately tend to start as surface runoff based on the rainfall intensity while the precipitation received as snow will not result in surface runoff.
2. **Rainfall intensity**
   The infiltration rate of soil is dependent upon the rainfall intensity. If the rainfall intensity is more than the infiltration rate of soil, then runoff will be initiated with the onset of rainfall. In contrast, runoff begins later when rainfall is not as intense. High rainfall intensities hence result in increased runoff.
3. **Duration of rainfall**
   The rate of infiltration decreases with the length of rainfall. The duration of rainfall is directly correlated with the volume of runoff. Therefore, if the duration of the rainfall is prolonged, medium intensity rainfall even produces a significant amount of runoff.
4. **Rainfall distribution**
   Runoff from a watershed depends on rainfall distribution pattern. It is also expressed as "distribution coefficient" and defined as average ratio of maximum rainfall at a point to the average rainfall of watershed.
5. **The prevailing wind's direction**
   Peak low occurs when the drainage system and the predominant wind are pointing in the same direction. When a storm moves toward the stream's slope, it produces a higher peak faster than when it moves the other way.

6. **Other climate factors**
   The water losses from the watershed area are influenced by additional variables including temperature, wind speed, relative humidity, yearly rainfall, etc.

B. **Physiographic factors**
   It encompasses the following features of channels and watersheds:

1. **Size of watershed**
   In comparison to smaller watersheds, larger watersheds require more time to drain their runoff to the outlet.
2. **Shape of watershed**
   The shape of the watershed has a significant impact on runoff. Circular or fan-shaped watersheds have high rates of runoff when compared to other shapes because runoff from different points in the watershed is more likely to reach the outlet at similar times. High rates of runoff on small catchments of this shape are short-lived. For long, narrow (elongated) watersheds, tributaries join the main stream at intervals along its length. High flow rates from the downstream tributaries pass the gage before high flows from the upper tributaries arrive. Peak flow at the gage is less than for a fan-shaped catchment but persists for a longer time.
3. **Soil type**
   The infiltration rate varies with type of soil. Geology is significant for the formation of soil and physical characteristics of watersheds and the establishment of surface and subsurface flow systems of the hydrologic cycle. Geological features (type of rock formations, faults, etc.) and processes (such as the movement of glaciers more than 10,000 years ago) have helped to define the surface divide or ridge between watersheds, establish and control stream channel gradients, and the subsurface boundaries which control the movement of groundwater to surface streams. Groundwater flow boundaries do not always coincide with watershed divides and are more difficult to define.

10.1.12 Topographic characteristics

It includes those topographic features which affect the runoff such as slope, aspect, and land use.

10.1.12.1 Slope

Greater runoff occurs on undulating terrain than on level terrain because the slope gives runoff water more energy (velocity) and shorter infiltration time. Land slope affects the time of concentration and overland flow. More runoff from a sloping watershed is produced by increased runoff velocity, and vice versa.

10.1.12.2 Aspect

The compass direction that the land surface faces is termed aspect or orientation. The aspect can be important where there is only one major slope face. On large areas, there are a multitude of aspects and the average tends to balance out to no specific direction. The aspect affects soil water content, vegetation, and is also important for soil freezing. In the northern hemisphere, north-facing slopes receive less solar energy, soil freezes to greater depth, and the period of frost is longer.

10.1.12.3 Land Use

Compared to unprotected land surface, the vegetative cover of watersheds is likely to reduce storm runoff when soil water is significantly less than saturation. Vegetation improves the soil structure and depletes soil water, thus causing more infiltration and less surface runoff. Trees and deep-rooted crops usually consume more soil water by ET than shallow-rooted crops, especially in dry periods, and consequently their watersheds have less runoff. Bare soil surfaces where infiltration rates diminish rapidly have more surface runoff. This is because bare soils tend to have poor structure and are less permeable. Even in permeable soils such as sand, a common problem known as surface sealing often occurs shortly after rainfall begins. Surface sealing can be due to both physical and chemical processes. For example, in an experiment with a sandstone spoil material from a surface mine, chemical bonding of particles occurred shortly after rainfall. This results in the formation of a very thin, almost impervious layer at the surface. A seal due to physical processes occurs when fine soil particles are washed into larger surface cracks and pores. Soil and water conservation tillage practices that maintain high rates of infiltration could reduce surface runoff and those that reduce soil water evaporation could increase subsurface runoff. The effect of these practices on runoff depends somewhat on the soil water balance (the availability of excess soil water for surface or gravity flow).

10.1.12.4 Drainage Density

The ratio of the channel length to the entire watershed area is known as the watershed's drainage density. If the drainage density of a watershed is higher, then more runoff will be generated.

10.1.13 Methods for Estimation of Runoff

Peak runoff rate: It is the maximum rate of runoff per unit time. Peak runoff rate is essential to design soil and water conservation structures.

The common methods are:

1. Rational method
2. Curve number method (hydrologic soil cover complex number method)
3. Soil conservation service method
4. Empirical formula: (a) Dicken's formula, (b) Ryve's formula

1. Rational method: Rational method is used to predict the peak runoff rate (Q). The peak runoff may be calculated using the following equation.

$$Q = \frac{\mathrm{CIA}}{360}$$

where,

Q = Peak runoff rate, m^3/s; C = Runoff coefficient; I = Rainfall intensity (mm/h) for duration equal to time of concentration and for a given recurrence interval; A = Watershed area (ha)

10.1.14 Rainfall Intensity (I)

Rainfall intensity is defined as the rate of fall of precipitation, expressed in depth per time (mm/h)

$$I = P/T$$

where I = Rainfall intensity, mm/h; P = Amount of rainfall, mm; T = Duration of rainfall, h

The value of the intensity of rainfall (I) should be calculated for the period equal to the time of concentration of the catchment.

10.1.15 Time of Concentration

It is the time required for the runoff water to flow from the most remote point of the area to the outlet

$$t_c = 0.02L^{0.77}S^{-0.385}$$

t_c = Time of concentration, min; L = Length of channel reach, m; S = Average slope of channel reach, m/m

10.1.16 Runoff Coefficient

Runoff coefficient is the ratio between the peak runoff rate and intensity of rainfall. The values of runoff coefficient depend upon the land slopes and soil type. It ranges from 0.10 to 0.30 for sandy loam, 0.30–0.50 for clay and silt loam, 0.40–0.60 in silty clay in woodland (Subramanya 2008).

10.1.17 Empirical Formula

(a) **Dicken's formula**

$$Q = CA^{3/4}$$

where, Q = Peak runoff, m^3/s; A = Area of the watershed, km^2; C is a constant whose value varies widely between 2.8 and 5.6 for catchments in plains and between 14 and 28 for catchments in hills depending upon catchment characteristics. Dicken's formula is used for catchments in north and central India. For catchments in south India, Ryve's formula is preferred.

(b) **Ryve's formula**

$$Q = CA^{2/3}$$

where Q = Peak runoff, m^3/s; A = Area of the watershed, km^2; C is a constant whose value varies widely between 6.8 (for flat or plain catchments) and 42.4 for western coast region.

10.1.18 Runoff Measuring Devices

There are many devices for measuring water. Some of the commonly used water measuring devices are listed below.

1. Weirs
2. Parshall flumes
3. Orifice
4. Meter gates

The use of the above-mentioned device depends on the expected flow rates site conditions.

10.1.19 Hydrographs

A hydrograph is a graphical representation of the instantaneous runoff rate against time as shown in Fig. 10.10. The area under hydrograph gives the runoff volume. Runoff hydrographs are useful in designing structural measures and also to study the hydrologic effect of any catchment.

The response of a specific watershed to rainfall is called a hydrograph. It encompasses the surface runoff, interflow, and base flow. Identical storm in different catchments produces dissimilar hydrographs. The intricate relationships between different storms and catchments are remarkable. It is theoretically difficult to analyze hydrographs under such situations since there will be several peaked hydrographs. Therefore, hydrographs obtained from isolated storms are helpful in hydrograph studies.

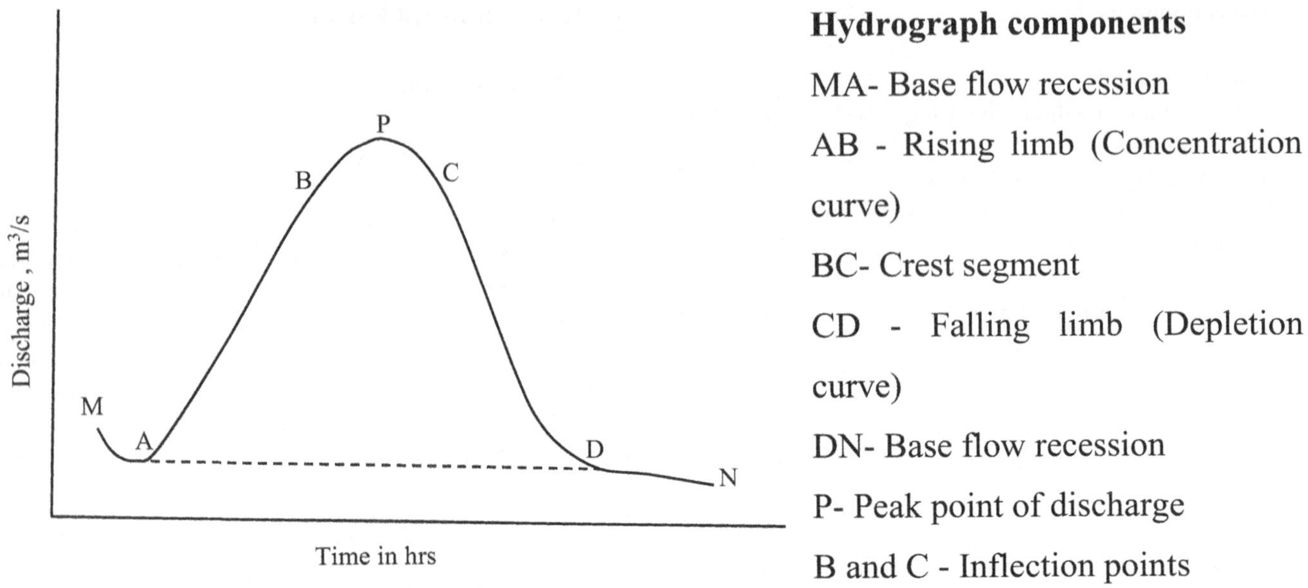

Fig. 10.10 Hydrograph. Hydrograph components: MA—Base flow recession; AB—Rising limb (Concentration curve); BC—Crest segment; CD—Falling limb (Depletion curve); DN—Base flow recession; P—Peak point of discharge; B and C—Inflection points

10.1.19.1 Factors Affecting Flood Hydrograph

The climatic and physiographic factors affect the shape of the hydrograph. The climatic factors influence the rising limb and the recession limb is influenced by catchment characteristics only.

Climatic Factors

1. Storm characteristics
2. Initial losses
3. Evapotranspiration

Physiographic Factors

1. Basin characteristics: Shape, size, slope, elevation, drainage density, etc.
2. Infiltration characteristics: Land use and cover, soil type and geological condition, lakes, swamps, etc.
3. Channel characteristics: Cross-section, roughness, and storage capacity.

Shape of the basin: The shape of the basin determines the time of concentration. Therefore, it affects the peak and the shape of the hydrograph. Fan-shaped catchment gives high peak and narrow hydrographs. Elongated catchment gives broad and low peaked hydrograph as shown in Fig. 10.11.

Size: The behavior of small basins is different from large basins. In small catchments, the overland flow is predominant over channel flow as rainfall intensity plays a key role in small catchments. In large channel flow, phase is predominant.

Slope: Large stream slope gives rise to quicker depletion of storage and hence steeper recession limbs and results in smaller time base. The basin slope is important in small catchment where the over land flow is relatively more important. In such cases, steeper slope of the catchment results in larger peak discharges.

Drainage density: Drainage density is defined as the ratio of the total channel length to the total drainage area. A large drainage density creates situation for quick disposal of runoff down the channel. This fast response will lead to pronounced peaked discharge. In basins of smaller drainage densities, the overland flow is predominant and the resulting hydrograph is squat with slowly rising limb as shown in Fig. 10.12.

10.1.20 Soil Erosion

The top soil is very important for sustaining the growth and development of vegetation which has profound influence on human beings and wildlife. When the top layer especially the top 30 cm is continuously exposed to the action of erosive agents such as water and wind, the top soil layer detaches and moves from one place to another.

Soil erosion: It is defined as detachment, transportation, and deposition of soil particles from one place to another place under influence of wind, water, or gravity forces. It is a three-phase phenomenon consisting of the detachment of individual soil particles from the soil mass and their transport by erosive agents such as running water and wind

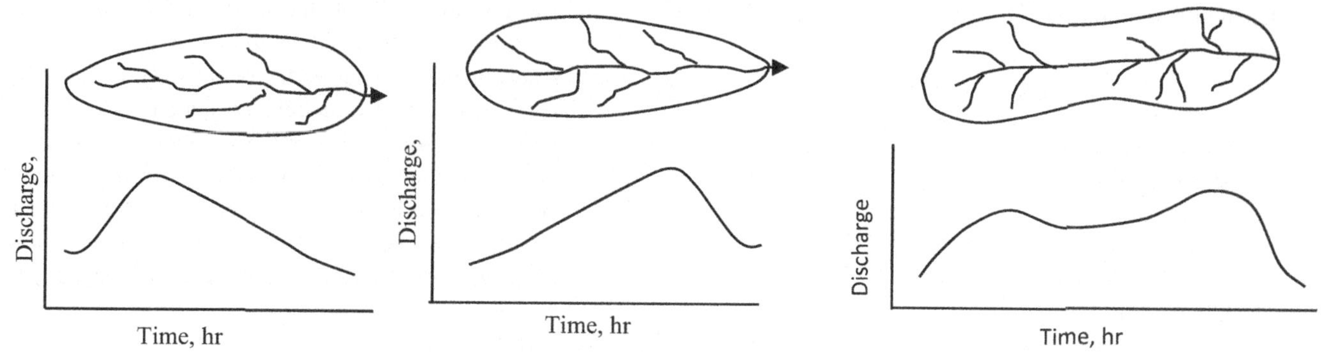

Fig. 10.11 Shape of the basin

Fig. 10.12 Shape of the basin

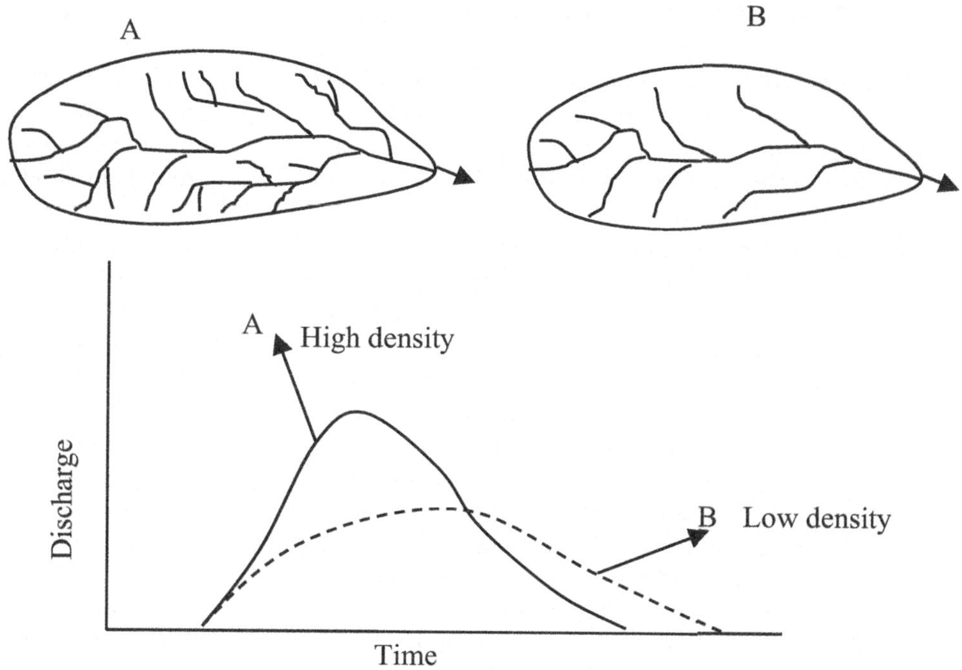

when energy to transport the particles is not available then third phase called deposition take place.

Erosivity: Erosivity can be defined as the potential ability of the rain to cause erosion. It is a function of the physical characteristics of the rainfall.

Erodibility: Erodibility is defined as the susceptibility (vulnerability) of the soil to erosion. It is a function of both the physical characteristics of the soil and the land management practices.

Soil erosion may be broadly classified into two groups:

1. Geological erosion
2. Accelerated erosion

1. **Geological erosion:** It is also called as natural or normal erosion. This process occurs due to the action of water, wind, gravity, and glaciers. The rate of geological erosion is slower as the soil loss is compensated by soil formation by natural weathering process. The vegetative cover like grasses, shrubs, tree canopy, and forest floor retards the transportation of soil material and control soil erosion. Even under the presence of vegetative cover, the occurrence of some certain amount of geological erosion is inevitable.
2. **Accelerated erosion:** When land is put under cultivation, natural equilibrium between soil formation and removal is disturbed. The removal of surface soil takes place at a much faster rate than it can be built up by the soil forming process. This is known as accelerated erosion or abnormal erosion and it is due to artificial causes. The rate of accelerated erosion is higher than the geological erosion. Accelerated erosion is one of the major reason for depletion of soil fertility in agricultural lands.

Depending upon the erosive agents, erosion is classified as follows.

1. Water erosion
   (a) Raindrop erosion
   (b) Sheet erosion
   (c) Rill erosion
   (d) Gully erosion
   (e) Stream bank erosion
   (f) Slip erosion
2. Wind erosion and
3. Sea side or coastal erosion

10.1.21 Types of Water Erosion and Their Effects

1. **Water erosion**: Water erosion is the process by which soil is removed from the exposed land surface by the force of rain drops and flowing water, including runoff from precipitation, melted ice, and snow. The different stages of water erosion are listed below.
   (a) **Raindrop erosion/Splash erosion**
   Raindrops falling on land surface cause detachment of soil particles and are subsequently transported with the flowing water. The soil splash is caused by the impact of falling raindrops. The rate of raindrops erosion is influenced by climate, soil properties, topography, and plant cover. The climatic factors like rainfall and temperature; soil properties like, erodibility and infiltration rate; topographic factors like length and degree of slope; plant cover factors such as barren, vegetative cover/residues contribute to initiation of raindrop/splash erosion.
   Raindrop splash is of major contributor to soil erosion. If raindrops strike on land covered with vegetation, the raindrop breakdowns into a spray of clear water which eventually moves into soil. On the contrary, when the raindrop strikes bare soil surface, considerable splashing occurs. The falling raindrops break down soil aggregates and detach soil particles from the soil mass; therefore, the fine soil particles get suspended to form a muddy splash. As the muddy water percolates into the soil, the finer particles form a thin film on the surface which leads to choking of the soil pores. These effects coupled with reduced infiltration cause higher surface runoff. The higher volumes of runoff carry more material and gradually remove the fine material from the soil rendering the soil less fertile sands and gravelly.
   The major effect of surface flow of water is to carry off the soil loosened by splash erosion. As the depth and concentration of the overland flow increase, it tends to form rills and then gullies. When this point is reached, the flowing water is not only transporting soil but is actually picking up added soil from the gullies. A good portion of the soil which is picked up in this manner is moved along the bottom of the stream as bed load. The droplet size and its velocity largely influence the raindrop/splash erosion as large drops may increase the sediment carrying capacity by 12-fold.
   (b) **Sheet erosion**
   A thin film of soil layer is detached and transported by the rainwater flowing on the land surface. This type of erosion is known as sheet erosion. Sheet erosion may not be evident significantly, but land subjected to sheet erosion loses a thin layer of top fertile soil every year.
   Sheet erosion is the removal of uniform layer of soil from the land surface by the action of rainfall and runoff. In sheet erosion, soil is removed from the surface in a thin layer or sheet of relatively uniform thickness. This type of erosion, though extremely harmful to the land, is usually so slow that the farmer/forester is not conscious of its existence. It is common on land having a gentle slope. It results in the uniform removal of the top soil with every hard rain. The land having loose, shallow top soil is most susceptible to sheet erosion. The occurrence of sheet erosion is not evident visually.
   Movement of soil by raindrop splash is the primary cause of sheet erosion. Raindrops cause the soil particles to be detached and the increased sediment reduces the infiltration rate by sealing the soil pores. As the water passes over a smooth slope, it flows along in a sheet of more or less uniform depth. In areas of unprotected sloping land, the sheet of water picks up soil particles and carries them down the hill. Sheet erosion has damaged millions of hectares of sloping land throughout India.
   (c) **Rill erosion**
   Rill erosion is the second stage of sheet erosion, during which the landscape is characterized by finger-like rills. Year after year, rills gradually expand in numbers, shape, and size, but they are normally smoothed out by regular farm operations. They get wider and deeper and affect crop production.
   Rill erosion is the formation of a series of small shallow branching channels. There is no clear demarcation between sheet erosion and rill erosion, but rill erosion is more distinct than sheet erosion.

The rills tend to expand due to concentrated runoff and if neglected further, rills grow into large gullies.

(d) **Gully erosion**

Gully erosion is seen as a more sophisticated form of rill erosion. When ignored, rills develop larger and eventually turn into gullies. Gullies can also be caused by runoff concentrating at a point on the agricultural lands. Gullies when not controlled expand year after year. Ravines are a form of extensive gullies.

Gully erosion is caused by higher concentration of runoff which results in the formation of channels of size 30 cm–10 m or more. Gully erosion is an advanced stage of rill erosion; any concentration of surface runoff is a potential source of gulling. Cattle paths, cart tracks, dead furrows, tillage furrows, or other small depressions down a slope favor concentration of water. Neglected rills get deepened and widened every year and begin to form gullies.

Within a few years an entire landscape may be filled with a network of gullies. Large areas of fertile lands thus get swallowed by the ever increasing gullies and are rendered uncultivable. Gully erosion is more spectacular than other types of erosion. The distinction between ravine, gully, and rills is that of size. A gully is too large to be filled by normal tillage practices.

A ravine is a deep narrow gorge. It is larger than a gully and is usually worn down by running water.

10.1.22 Principles of Gully Erosion

The rate of gully erosion depends mainly on the runoff producing characteristics of the watershed, the watershed area, soil characteristics, the alignment, size and shape of the gully, and the slope of the gully bed. The main processes in the development of a gully are waterfall erosion and channel erosion.

(a) **Waterfall erosion:** Water falling over the edge of a gully or bank of a ditch forms deep and rapidly expanding gullies. The falling water caves in and the soil so detached is transported from the site by the large volume of runoff and the waterfall moves upstream. As the over fall advances up the slope, its vertical height increases. In this manner, gullies often start in the bank of natural water courses that have been eroded to a great depth. They extend back into cultivated fields and grow deeper up the slope, often attaining depths of 15–18 m or more. As they extend backward and cross-lateral drainage ways, or natural depressions, waterfalls are sometimes formed on the sides of these depressions, and branch gullies develop. This branching may continue until a network of gullies cover the entire drainage area.

(b) **Channel erosion:** Channel erosion is essentially the scouring away of the soil by concentrated runoff as it flows over unprotected depressions. It may also be caused by the raindrop splash on unprotected soil in the gully. Gullies formed by channel erosion usually have sloping heads and sides and often called V-gullies. As the scouring continues, the gully becomes longer, deeper, and wider. The extension in length is much faster than the widening of the gully because a greater volume of runoff passes over the gully head than over the sides. The rate at which the gully deepens is very rapid on the upper part of the area, where the slopes are comparatively steep.

Channel erosion and waterfall erosion are commonly found in the same gully. The extension of the gully head is usually by waterfall erosion; while the scouring of bottom and sides (channel erosion) enlarges the depth and width. Gullies usually start with channel erosion. When an over fall develops at the head of the gully, the gully continues to develop by waterfall erosion.

The four stages of gully development are described hereunder:

Stage 1: Formation stage: In this stage, channel erosion is initiated due to downward scouring of the top soil.

Stage 2: Development stage: It consists of upstream movement of the gully head and gulley expansion.

Stage 3: Healing stage: Vegetation tends to grow in the gully.

Stage 4: Stabilization stage: In this stage, gully walls reaches a stable gradient and vegetative growth is observed on the gulley.

During the last two stages, the gully head progresses to the upper end of the watershed. The rate of runoff into the gully head decreases due to the reduction in the drainage area. The remaining portion of the runoff enters at different points along the length of the gully.

10.1.23 Classification of the Gullies

Gullies are classified as G1, G2, G3, and G4 depending upon the depth, breath, and side slopes of the gully. G1 gullies are very small up to 3 m deep, bed width <18 m with varying side slope; G2 gullies are small gullies up to 3 m depth, bed width >18 m, side slope between 8% and 15%; G3 are medium gullies of depth between 3 and 9 m, bed width not <18 m, sides uniformly sloping between 8% and 15% and G4 are deep and narrow gullies of size 3–9 m deep, bed width <18 m, varying side slope or depth >9 m, bed width and side slopes varying with mostly steep or even vertical having intricate and active branch gullies (Tejwani and Dhruva Narayana 1961). This classification was evolved after critically observing ravine areas along most of the river banks in Gujarat.

Gullies may also be classified as V or U shaped based on the shape of their cross-section. Both V- and U-shaped gullies may be found in the same channel.

As the gullies extend upstream, the most active portion is at the gully head or the upper end, whereas the most stable section is generally at the lower end. Active gullies are those gullies that continue to enlarge. They may be recognized by the presence of bare soil exposed on the side slopes.

Drainage channels or gullies convey runoff and sediment in a watershed from the upper reaches to the catchment outlet. Steep slope of a channel causes high runoff velocities with heavy sediment flow. Hence, reduced channel gradient can bring the runoff velocities within permissible limits.

(c) **Stream bank erosion:** The erosion of the stream banks by flowing water is known as stream bank erosion. In certain areas where the river or streams change course, stream banks are eroded in an accelerated rate. Stream bank erosion damages adjoining agricultural lands, highways, roads, and bridges.

(d) **Slip erosion:** Here the landslides and slips caused due to saturation of steep hills and slopes are referred to as slip erosion. Slip erosion disturbs the natural landscape and also may damage the highway and add additional sediment to stream and rivers.

10.1.24 Soil Loss Measurement

10.1.24.1 Revised Universal Soil Loss Equation (RUSLE)

The RUSLE, known as Revised Universal Soil Loss Equation is used for assessment and quantification of annual soil erosion rate and developing soil erosion risk map. Renard et al. (1996) enhanced and modified the Wischmeier and Smith (1965, 1978) universal soil loss equation (USLE) and developed RUSLE. RUSLE is the most widely used soil erosion models worldwide by researchers.

The RUSLE model is expressed as:

$$A = R^{*}\ K^{*}\ LS^{*}\ C^{*}\ P$$

where,

A = Computed average annual soil loss (tons/ha/y)
R = Rainfall-runoff erosivity factor
K = Soil erodibility factor
LS = Slope length-steepness factor
C = Cover management factor
P = Conservation practice factor

The L and S factors were calculated as two distinct parameters in the Universal Soil Loss Equation, but in the RUSLE, both components are consolidated into a single factor (the LS factor) and then computed.

Rainfall erosivity factor (*R*): The amount of rainfall erosion index units (EI30) at the study site over a specific time period is the factor R. A storm's rainfall erosion index unit (EI30) is described as

$$\mathrm{EI}_{30} = \frac{\mathrm{KE} \times I_{30}}{100}$$

where,

KE represents the storm's kinetic energy.
The formula for calculating the KE in metric tons/ha-cm is
KE = 210.3 + 89 log I, where I is the rainfall intensity in cm/h and
I_{30} is the storm's maximum intensity of rainfall for 30 min

The study period can be a week, month, season, or year. The storm EI_{30} values for that length of period are summed up. A summary of the storm EI30 values for that time frame is provided. Lines with equal EI30 values, sometimes called iso-erodent lines, are typically drawn for the region covered by the data stations for immediate use in USLE. Annual EI30 values are typically calculated from data available at various meteorological stations. Annual EI30 readings range from 250 to 500 all over the Karnataka state, except for the Western Ghats, which range from 500 to 1500. The state of Tamil Nadu has yearly EI30 readings between 300 and 700.

Soil erodibility factor (*K*): In the event of predicting and controlling soil erosion, the soil erodibility factor plays a crucial role. The soil erodibility factor (K) indicates the change in the soil per unit of applied external force of energy by reflecting how easily the soil is separated by splash during rainfall and by surface flow. This component explains how soil parameters affect soil loss during storm events and is linked to the combined effects of rainfall, runoff, and infiltration. The following methodology, as outlined by Wischmeier and Smith (1978), was used to compute the K factor utilizing data on inherent soil qualities (Parysow et al. 2003):

$$K{=}1.2917\{(2.1\times10^{-4}\mathrm{M}^{1.14}(12-a)+3.25(b-2)+2.5(c3/100$$

where M is equal to (percent silt + percent very fine sand) × (100 − percent clay)

"a" represents percent organic matter
"b" denotes the soil structure code (used in soil classification) and
"c" indicates the permeability class

The laboratory analysis provided data on the physical and chemical properties, which were used to estimate soil erodibility. The percentage of clay, very fine sand, silt, and weighted mean soil organic matter was computed and then averaged according to the area of each constituent soil series of a particular mapping unit.

Topographic factor (*LS*): The topographic factor (*LS*), which is generated by combining the two variables *L* and *S*, is typically

$$LS=\left(\frac{\lambda}{22.13}\right)^{m}\left(65.41\mathrm{Sin}^{2}\theta+4.56\ \mathrm{Sin}\theta+0.065\right)$$

where λ represents field slope length in meters; "m" indicates exponent factor (varying from 0.2 to 0.5); "θ" denotes angle of slope.

Crop management factor (C): The impact of multiple crop management techniques is combined to create this component. Table 10.2 provides factor C values for the areas around specific land use.

Supporting cultivation practice (P): This variable is the ratio of soil loss up and down the slope with a support practice to that with straight-row farming. Factor P is provided for a few support practices in Table 10.2.

Table 10.2 Crop cover management (*C*) and conservation practices (*P*) factor

S. no.	Land utilization type	*C* value	*P* value	CP factor
1	Agriculture	0.35	0.5	0.17
2	Plantation	0.38	0.3	0.11
3	Double crop	0.38	0.3	0.11
4	Scrub land	0.2	1	0.20
5	Waterbody, habitation	0	0	0

Source: Ravikumar (2020)

10.2 Watershed

The term watershed has been coined from water and shed originally to refer to the line that separate two river basins. This line usually occurs as a ridge line of high land where rainwater on one side flows into one stream and rainwater on the other side flows into a different stream. In the word watershed, the meaning of shed is "to part" divide or separate.

The word catchment is used for the area that catches the runoff and flows into a particular river or lake and the watershed is the term for the boundary of the catchment, that is, the ridge line or divide line that separates water on either side. In many countries including India, watershed is the popular term to denote the catchment area. The terms watershed, catchment, and drainage basin are synonymously used to denote the area that catches the runoff for a particular stream or river. Some people attribute certain differences between the terms watershed, catchment, and basin. Watershed is a common term to refer to any sort of area that catches the runoff. A basin is usually a big watershed, but the outlet must be sea or ocean, for example, the Indus basin and Kaveri basin.

"A watershed is any geographical area that is bounded by a ridge line and includes all of the land that contributes runoff to a common point or outlet." Another name for it is the ridge line. Alternatively, a "watershed" is a section of land that is enclosed by a natural ridge line and from which all surface runoff is directed toward a single drainage point. The size and shape of a watershed are determined by surface topography bound on all sides by a divide or ridgeline. The simplest method for defining surface watersheds is to locate the highest points on the land that drain into the body of surface water.

10.2.1 Concepts of Watershed Management

Vegetation, soil, and water are the three fundamental natural resources crucial for the existence of both humans and animals. On a watershed basis (a unit of management), these three interrelated resources can be managed collectively, conveniently, concurrently, and efficiently.

10.2.2 Watershed Delineation Using Geospatial Technologies

Delineating a watershed involves defining or illustrating the region enclosed by a ridge line, contributing runoff at a common point and dividing or separating it from the adjoining area.

Fig. 10.13 (**a**) Find the sink (the red pixel) that is made up of just one pixel. (**b**) The pixel in green should be assigned the smallest value among its eight neighboring pixels

(a)

58	52	55	53	56	58
55	40	42	45	51	55
48	33	35	33	48	52
33	23	15	27	25	38
17	17	17	22	17	12
12	10	15	18	16	14

(b)

58	52	55	53	56	58
55	40	42	45	51	55
48	33	35	33	48	52
33	23	17	27	25	38
17	17	17	22	17	12
12	10	15	18	16	14

Studying survey of India toposheets and conducting reconnaissance surveys can help define priority areas to some extent. On the other hand, this method is not precise and is time-consuming. Typically, 1:50,000 scale topographic sheets can be employed for this purpose.

Watershed-based delineation of priority regions is necessary because efficient soil conservation techniques require a comprehensive approach to watershed management. The size of the watershed that needs to be drawn must also fall between 10,000 and 20,000 hectares, as this is the range at which working plans for soil conservation can be easily and practically developed for small watersheds.

Watershed delineation using latest tools and different data is crucial for watershed management at micro-level. This section describes the watershed delineation using **digital elevation model** (DEM) and other ancillary data by geospatial technologies. The following are the conceptual overview of watershed delineation:

1. **DEM (Digital Elevation Model) data**
   It is a representation of the bare earth (bare ground) topographic surface of the earth exclusive of all surface objects, namely, trees and buildings. DEMs are digital representations of the earth's topographic surface.
2. **Sink**
   Local depressions in a DEM are eliminated by the Fill sinks operation, which creates flat areas in the output DEM in place of the local depressions. A cell or group of spatially related cells that cannot have its flow direction assigned to one of a flow direction raster's eight permissible values is referred to as a sink. This can happen when two cells flow into one another, forming a two-cell loop, or when every nearby cell is higher than the processing cell (Fig. 10.13a, b).
3. **Fill**
   It uses the analogues of several functions, such as focal flow, flow direction, sink, watershed, and zonal fill, to find and fill sinks. Until every sink inside the designated Z limit is filled, the function iterates. At the edges of the filled sections, sinks can form as they fill up; these are eliminated in the subsequent iteration. The sinks in the DEM data will be filled in this stage (Fig. 10.14). This stage will result in a depression-less DEM, which will serve as the basis for the next stages in the process (Planchon and Darboux 2002).
4. **Flow direction**
   Determining a surface's hydrologic properties requires knowing the flow direction from each cell in the raster data (DEM). The flow direction tool is used for this. This tool displays the direction of flow out of each cell as a raster after receiving a surface as input. Selecting the output drop raster option results in the creation of an output raster that displays a percentage-based ratio of the maximum elevation change from each cell along the flow direction to the path length between cell centers. All cells at the surface raster's edge will flow outward if the option to force all edge cells to flow outward is selected. The eight neighboring cells that flow could enter eight appropriate output directions associated with them. This method, which is sometimes called an eight-direction (D8) flow model, is based on one presented by Jenson and Domingue (1988) (Fig. 10.15).
5. **Flow accumulation**
   The calculation of accumulated flow involves summing the weights of all cells that drain into each downslope cell in the resulting raster. In cases where no weight raster is provided, a default weight of 1 is assigned to each cell. The number of cells flowing into each cell is represented in the output raster. The accompanying graphic's top-left image shows the flow direction from each cell, while the top-right image shows the number of cells that contribute to each cell (Fig. 10.16).
   Jenson and Domingue (1988) propose this technique for calculating accumulated flow from a DEM. In Tarboton

Fig. 10.14 An image of a sink profile taken both before and after fill

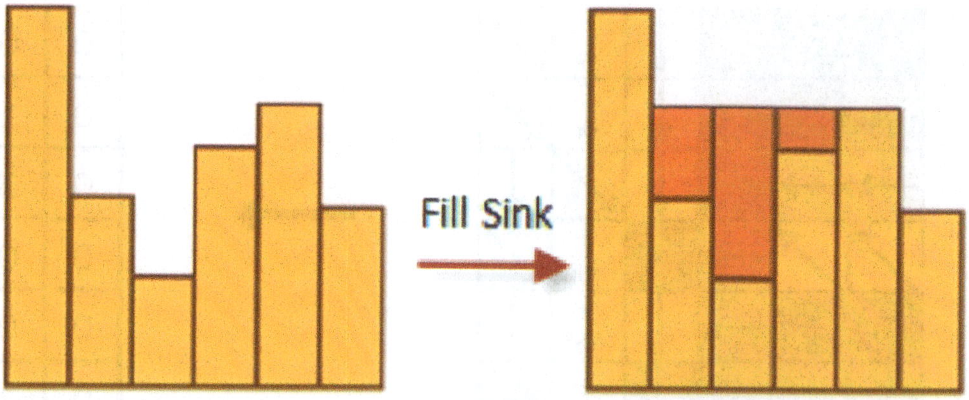

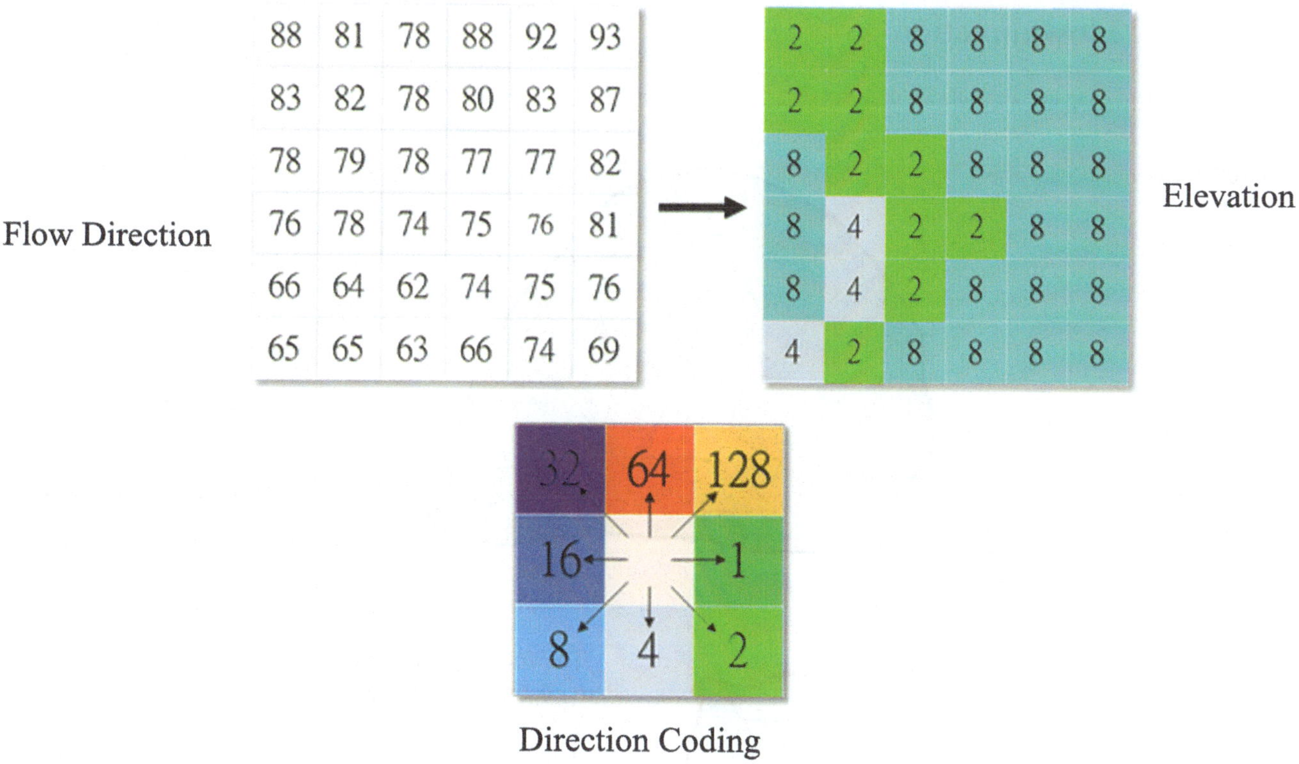

Fig. 10.15 The coding of the direction of flow

et al. (1991), an analytical approach to choosing a suitable threshold value for stream network delineation is described.

6. **Snap pour point**
The output raster of the basin analyst has great watershed extent, so, in hydrology analysis, sometimes we will need to split it to subsets. In this case, we need to learn the locations of the pour points of the subset before processing the analysis. In some study, users might hope to split to subsets by defining the pour points. However, the pour points defined by users might not fall on the reasonable positions (Fig. 10.17). With the snap pour point analyst, it helps to snap the pour points to locations with the greatest flow accumulation within the specified snap distance; so, the locations will have the greatest flow accumulation.
7. **Watershed**
Each watershed's value is derived from the source's value in the input raster or feature pour point data. The cell values will be utilized if the pour point is a raster dataset. The values will originate from the field specified if the pour point is a point feature dataset. The watershed delineation is based on snapped stream gage pour point (Fig. 10.18).

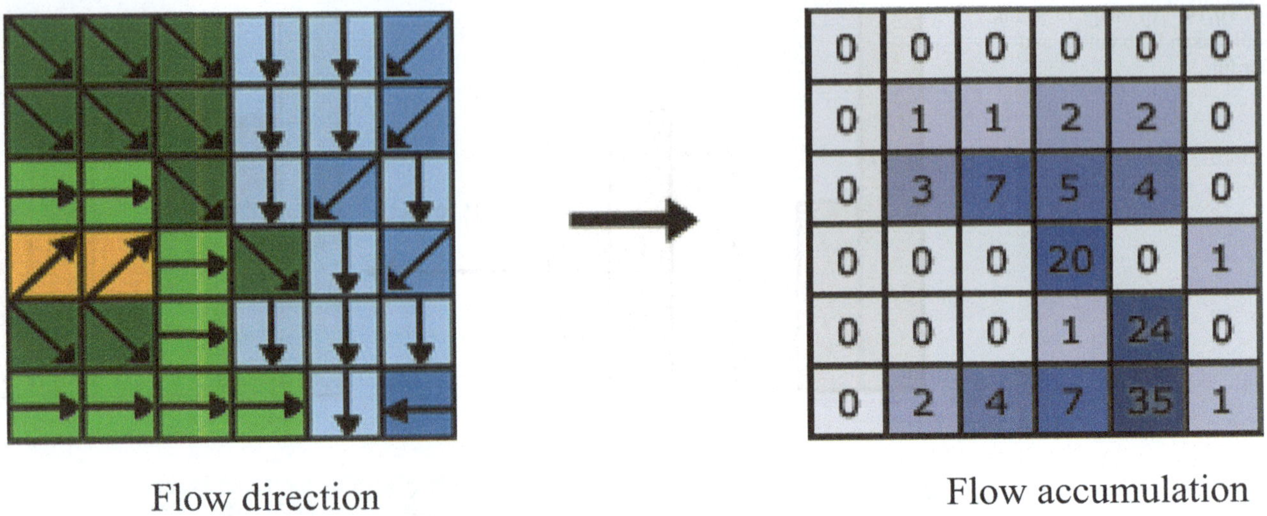

Fig. 10.16 Determining the accumulation of flow

Fig. 10.17 Determining the snap pour point

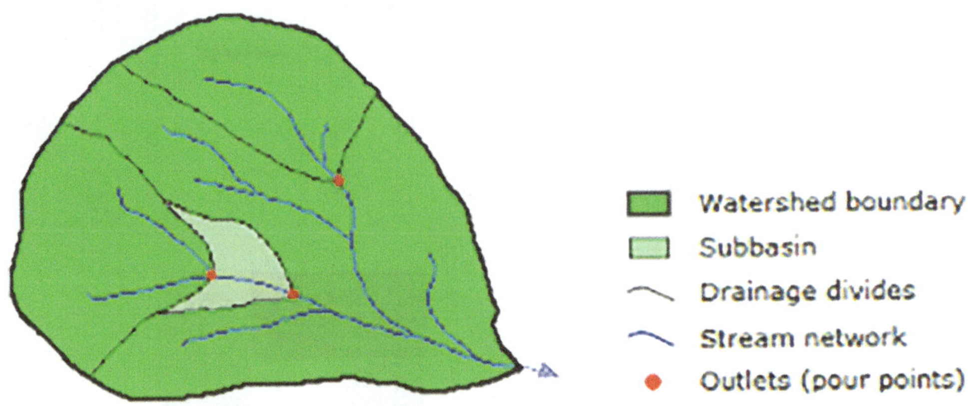

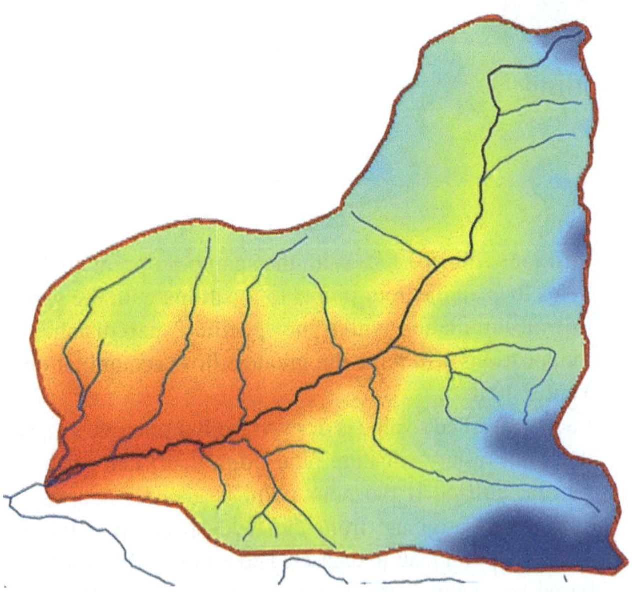

Fig. 10.18 Delineation of watershed boundary

The following is a description of the stages involved in demarcating a small watershed:

1. After identifying the major streams, divide the overall watershed into several sub-watersheds.
2. Further, split each sub-watershed into smaller areas, keeping care of the various tributaries and streams that flow through each sub-watershed.
3. Finally, further subdivide every small part of the watershed [size ranges of 10,000 to 20,000 ha, determined in step (2)].

10.2.3 Classification of Watershed

Based on its size and shape, watersheds are divided into different categories. Table 10.3 presents the watershed classification based on area.

Table 10.3 Classification of watershed based on area

Sl. no.	Watershed type	Area (ha)
1	Micro-watershed	0–10
2	Small-watershed	10–40
3	Mini-watershed	40–200
4	Sub-watershed	200–400
5	Macro-watershed	400–1000
6	River basin	Above 1000

10.2.4 Classification of Watershed Based on Shape

1. Square
2. Triangular
3. Rectangular
4. Oval
5. Fern leaf (elliptical)
6. Palm
7. Polygon
8. Circular (Fan shaped) and
9. Secateur shaped

10.2.4.1 Function of the Watershed

Receiving and disposing of incoming precipitation is the primary role of the watershed. This is the fundamental idea of conserving water and soil.

10.2.4.2 The morphological Features of the Watershed

Every watershed is unique and has many important characteristics that influence how it functions. There are seven such characteristics known.

1. Size (area)
2. Shape
3. Topography
4. Geology
5. Climate
6. Vegetation and
7. Land use.

1. **Size**
   The quantity of rainfall that is received, retained, and discharged (runoff) determines the size of the watershed. The larger the watershed, the higher would be the basin's channel and capacity for water storage. Geology, soil, climate, vegetation, and terrain are large watershed characteristics.
2. **Shape**
   A watershed might be square, triangle, rectangle, oval, palm, fern leaf, or any other shape. The length-to-width ratio, or shape index (form factor Ff = WB/Lb), is determined by the shape of the watershed and has a significant impact on runoff disposal. The larger the area of the watershed, the higher the time of concentration and more water will infiltrate, evaporate, or get utilized by the vegetation. When the length of the watershed is less than the width, the situation is reversed.
   Compactness coefficient (Cc)
   According to Gravelius (1914), the compactness coefficient (Cc) is the ratio of the basin's perimeter to its circular area's circumference, which equals the basin's area. The compactness coefficient depends purely on the slope and is not affected by the size of the watershed.
3. **Topography**
   (a) Soil loss and water disposal are influenced by the slope, degree, length, and uniformity of the slope. The slope's length and degree have an impact on water infiltration and time of concentration (T_c).
   (b) Drainage: Topography controls the flow of water. Topography determines the drainage density (length of all drainage channels divided by unit area), as well as the main outlet's size and dimensions. Concentration time is impacted by drainage pattern.
4. **Geology**
   Geological formations, rock types, and soil properties all have an impact on the amount of water erosion, the erodibility of channels and hill faces, and the creation of sediment. For instance, the dissolution rate of igneous rocks is slower compared to rocks such as phyllites and shales. The physical and chemical attributes of the soil, particularly soil texture, structure, and depth, play a crucial role in affecting water processes, including infiltration, storage, and runoff.
5. **Climate**
   The state of the climate affects watershed management and functioning in two ways.
   (a) Rainfall provides precipitation and all of its attributes, including strength, frequency, and volume.
   (b) Factors such as temperature, humidity, wind speed, and rainfall influence the characteristics of the soil and vegetation.
6. **Vegetation**
   Watershed functions such as infiltration, runoff production, water retention, sedimentation, and erosion are regulated by the type and the extent of vegetation present.

7. **Land use**
   Watershed behavior is primarily influenced by the kind, extent, and management of land use. The primary determinants of watershed behavior are the kind of land use, its extent, and its management.

10.2.5 Watershed Management

The watershed management emphasizes the judicious management of three fundamental resources, namely, soil, water, and vegetation, on a watershed basis, to accomplish specific objectives for the welfare of people. It involves both biological and engineering approaches for treating the land most properly.

10.2.5.1 Objective of Watershed Management

1. The process of producing food, fuel, and fodder
2. Pollution management
3. Reducing the exploitation of resources
4. Storage of water, prevention of flooding, and monitoring of sedimentation
5. Wild life protection
6. Erosion control and prevention of soil, degradation, and soil and water conservation
7. Employment generation through industrial development, dairy, fishery production
8. Recharging of groundwater
9. Recreational facility

10.2.5.2 Steps in Watershed Management

Watershed management encompasses preparation of alternative soil and water resource actions plans to address soil and water related issues. For this reason, information about problems of land, soil, water, and vegetation in the watershed is essential.

In order to have a practical solution to above problem, it is necessary to go through four phases for a full scale watershed management.

1. Recognition phase,
2. Restoration phase,
3. Protection phase and,
4. Improvement phase.

1. **Recognition phase**
   It involves following steps
   (a) Identification of the issue
   (b) An examination of the issue's root cause and impact
   (c) Development of alternative solutions of problem
   Many surveys, including those on soil, land capabilities, agronomy, forests, engineering, and socioeconomics, are used to gather the necessary data. This data is used to address and prioritize land treatment plans, identify issues with watersheds, and identify the causes and consequences of issues affecting both people and the land.
2. **Restoration phase**
   The restoration phase includes two main stages.
   (a) Finding the best treatment plans to problems identified
   (b) Implementation of the treatment plans to the problems of the land
   As per the treatment priorities, treatments are imposed to critical areas initially. After this, proper measures like biological and engineering measures are applied to all types of lands.
3. **Protection phase**
   The protection phase takes care of the overall health of the watershed and ensures normal functioning. The protection is against all factors that may cause deterioration of watershed condition.
4. **Improvement phase**
   The entire area is covered during this phase, which focuses on improving the watershed. To meet the goals of watershed management, emphasis is given to the agriculture production, forest management, forage production and management of pasture, as well as the socioeconomic conditions.

Causes of Watershed Deterioration

1. Degradation of land due to improper forestry, pasture management, and agriculture practices
2. Unscientific mining and quarrying
3. Faulty road alignment and construction
4. Industrialization
5. Forest fire
6. Apathy of the people

Effects of Watershed Deterioration

1. Less production from agriculture, forests, grass lands, etc.
2. Soil erosion
3. Rapid siltation of reservoirs, lakes and river beds
4. Less storage of water and lowering of water table
5. Poverty as a result of less food production

10.3 Sedimentation

Soil erosion, sediment transportation, and deposition within a watershed are examples of natural processes that are closely related to hydrologic processes. The success and economy of many water resources development activities in a watershed depend on the important parameters like soil and water conservation and reservoir sedimentation.

10.3.1 Erosion Process

The whole watershed, including the waterways, is eroded, when raindrops strike with a soil surface during a rainstorm event, their kinetic energy splits the soil aggregates, and separates the soil particles in the impact area. Surface runoff carries the separated particles. Deposition of the eroded material enrooted depending on the flow conditions, channel topography, geometry, etc. There are several ways that erosion occurs (which were covered in the section on soil erosion), including gully, sheet, channel, and rill erosions.

Integrated river basin development involves construction of storage reservoirs for the purpose of irrigation, power, and flood control. The loss of storage capacity caused by silting in these reservoirs is one of the difficulties with their design and maintenance. The deposition of sediments occurs when the transporting capacity of the stream reduces due to sudden entry into the still water of the reservoir. Silt entered into the reservoir settles there itself and after that it is economically not easy to remove it by artificial means.

While planning the reservoirs, it is very essential to make provisions for the storage of sediment. This given volume is referred to as dead storage. Live storage is the term for the storage capacity used for purposes like electricity and irrigation, etc. Live storage capacity decreases when the sediment depositing is increases and it occupies the live storage year after year and thereby live storage volume decrease (Fig. 10.19).

10.3.2 Sources of Sediments in Water Storage Structures

The sources of sediments in water storage structures vary from catchment to catchment; the various sources of sediments are given below:

- Sheet erosion from agriculture, forest, and waste lands
- Mass movements of soil from landslides, soil creep
- Gulling by concentrated runoff
- Stream channel erosion
- Floods
- Erosion caused by activities like construction of roads, rail, clearing of land for housing, and industries
- Mining and dumps left as waste

10.3.3 Factors Affecting Sedimentation

The quantity of sediment that is produced by erosion in a watershed depends upon the following factors:

- Climate—rainfall intensity, duration, temperature, etc.
- Soil characteristics—soil, type of soil, grain size, shape, porosity, etc.
- Topography—slope, orientation, degree, length, etc.
- Soil cover—vegetation /plant cover.
- Land use–agricultural practice, mining, roads, building, construction, etc.
(already discussed in the runoff and erosion section)

10.3.4 Definition

Gross erosion: Gross erosion is the sum of all erosions in a watershed.

Sediment yield: Total sediment out flow from a watershed at a reference section in a selected time interval is known as sediment yield. All sediment produced due to erosion in a watershed is not transported, out of it considerable temporary depositions takes place in various places and locations. The sediment yield is always less than the gross erosion.

Sediment delivery ratio: The ratio of sediment yield to gross erosion is known as sediment delivery ratio (SDR). The sediment yield of a watershed varies with the size of the contributing area.

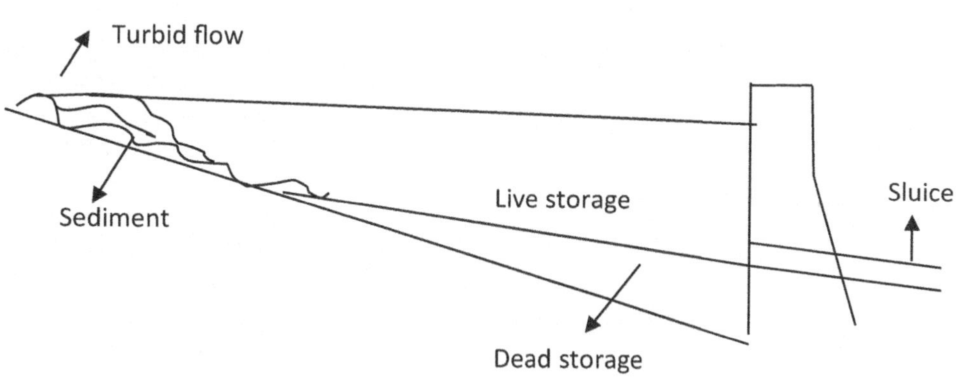

Fig. 10.19 Sediment deposition

10.3.5 Estimation of Sediment Flow Rate

The estimation of the sediment flow rates by a stream is essential to determine the extent of erosion in the catchment and for assessing the storage life of reservoir. Sediment observation stations are to be located along with discharge observations to estimate the sediment loads in stream or rivers. Sediment particles in stream are transported by three process, that is, suspension, saltation, and surface creep (similar to the transportation of soil particles in wind erosion). In case of sediment transport by streams, the saltation and surface creep are combined and are known as the bed load. The sediment transported in suspension is referred to as the suspended load as shown in Fig. 10.20. The total sediment load is determined by using both the bed load and suspended load. The bed load and suspended load are separately determined and added as the bed load moves at lower velocity than that of water.

10.3.6 Controlling Sedimentation of Reservoir

The life of a reservoir can be considerably prolonged if proper remedial measures against silting of reservoir are taken. The method proposed or tried for this purpose may be broadly divided into six groups.

- Selection of reservoir site
- Design of reservoir
- Control of sediment inflow
- Control of sediment deposition
- Removal of sediment deposits
- Watershed erosion control

Design aspects of the reservoir to be taken into consideration are provision of adequate capacity taking the expected sediment load and provision of sluices in the dam. Sluices eject only a part of the sediment settled very near to the dam.

Control of sediment inflow is achieved by constructing settling basins and propagating vegetative screens in the upstream part of the reservoir. Control of sediment deposition to some extent is achieved by allowing the passage of first inflow from the catchment out of the storage structure. This process ensures that the major part of the sediment does not get deposited in the storage structure. The sediment deposits that are settled down becomes economically impracticable to remove.

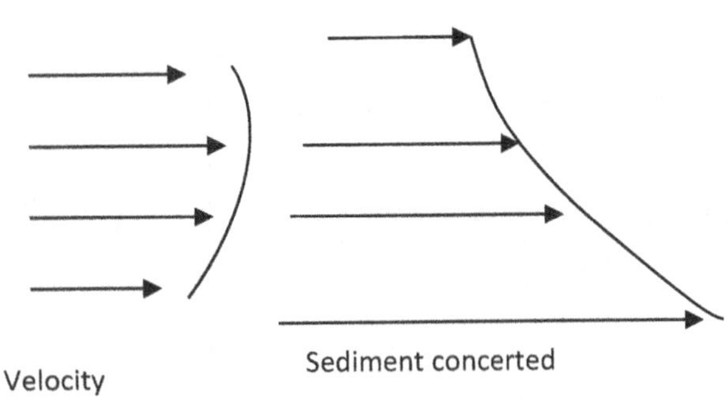

Fig. 10.20 Velocity distribution and sediment distribution in a straight natural stream

Watershed erosion control includes those measures which are effective in preventing the sedimentation into the reservoir. The various soil conservation measures for controlling sedimentation process are listed below.

- Contour trenching on barren hill slopes
- Afforestation
- Bench terracing of steep slopes
- Diversion channel
- Contour or graded bunding
- Gully protection
- Construction of small ponds
- Stream bank erosion control
- Road side stabilization
- Improved agricultural practices like contour cultivation, crop rotation, strip cropping
- Water spreading for ground recharge
- Control of land slide and land slips, etc.
- Pasture development
- Flood plain protection
- Reservoirs shore line protection, etc.

10.3.7 Recommended Soil and Water Conservation Measures for Watershed Management

The appropriate conservation structures recommended for any area is dependent on the slope percent, severity of erosion, amount of rainfall, land use, and soil type.

10.3.8 Greening of Watershed

As part of the greening program in the watersheds, it is envisaged to plant a variety of horticultural, forestry, and other tree species that are edible, economical and produce a lot of biomass which helps to restore the ecological balance in the watersheds. The lands that are suitable for greening program are non-arable lands (land capability classes V, VI, and VII) and also the lands that are not suitable or marginally suitable for growing annual and perennial crops. The method of planting these trees is given below.

It is recommended to open pits during the first week of March along the contour and heap the dugout soil on the lower side of the slope in order to harness the flowing water and facilitate weathering of soil in the pit. Exposure of soil in the pit also prevents spread of pests and diseases due to scorching sun rays. The pits should be filled with mixture of soil and organic manure during the second week of April and keep ready with sufficiently tall seedlings produced either in poly bags or in root trainer nurseries so that planting can be done during the second or third week of April depending on the rainfall.

The tree species suitable for the area considering rainfall, temperature, and adaptability is listed below (Table 10.4); waterlogged areas are recommended to be planted with species like Neral (*Syzygium cumini*) and Bamboo. Dry areas are to be planted with species like Honge, Bevu, and Seetaphal.

The different kinds of conservation structures are listed below:

1. Contour bunds
2. Graded bunds
3. Terraces
4. Trench cum bunds(TCB)/strengthening
5. Crescent bunds

The description of contour bunds, graded bunds and terraces are detailed below.

10.3.9 Contour Bund

In soil conservation practices, contouring, strip cropping, etc. are important mechanical practices employed to control soil erosion. Contour bunding works are carried out over wide areas in many parts of India (contour bunds are comparable to the narrow base level terraces).

Contour bund is a small embankment constructed on a contour. In soil conservation practice, contour bunding is the construction of small bunds across the slope of the land on a

Table 10.4 Tree species suitability in greening of watershed

Dry deciduous species			Temp (°C)	Rainfall (mm)
1.	Bevu	*Azadirachta indica*	21–32	400–1,200
2.	Tapasi	*Holoptelea integrifolia*	20–30	500–1000
3.	Seetaphal	*Anona squamosa*	20–40	400–1000
4.	Honge	*Pongamia pinnata*	20–50	500–2500
5.	Kamara	*Hardwikia binate*	25–35	400–1000
6.	Bage	*Albizzia lebbek*	20–45	500–1000
7.	Ficus	*Ficus bengalensis*	20–50	500–2500
8.	Sisso	*Dalbergia sissoo*	20–50	500–2000
9.	Ailanthus	*Ailanthus excelsa*	20–50	500–1000
10.	Hale	*Wrightia tinctoria*	25–45	500–1000
11.	Uded	*Stereospermum chelonoides*	25–45	500–2000
12.	Dhupa	*Boswellia serrata*	20–40	500–2000
13.	Nelli	*Emblica officinalis*	20–50	500–1500
14.	Honne	*Pterocarpus marsupium*	20–40	500–2000
Moist deciduous species				
15.	Teak	*Tectona grandis*	20–50	500–5000
16.	Nandi	*Lagerstroemia lanceolata*	20–40	500–4000
17.	Honne	*Pterocarpus marsupium*	20–40	500–3000
18.	Mathi	*Terminalia alata*	20–50	500–2000
19.	Shivane	*Gmelina arborea*	20–50	500–2000
20.	Kindal	*Terminalia paniculata*	20–40	500–1500
21.	Beete	*Dalbergia latifolia*	20–40	500–1500
22.	Tare	*Terminalia bellirica*	20–40	500–2000
23.	Bamboo	*Bambusa arundinacea*	20–40	500–2500
24.	Bamboo	*Dendrocalamus strictus*	20–40	500–2500
25.	Muthuga	*Butea monosperma*	20–40	400–1500
26.	Hippe	*Madhuca latifolia*	20–40	500–2000
27.	Sandal	*Santalum album*	20–50	400–1000
28.	Nelli	*Emblica officinalis*	20–40	500–2000
29.	Nerale	*Syzygium cumini*	20–40	500–2000
30.	Dhaman	*Grewia tiliifolia*	20–40	500–2000
31.	Kaval	*Careya arborea*	20–40	500–2000
32.	Harada	*Terminalia chebula*	20–40	500–2000

contour so that the long slope is cut into a series of small ones and a series of such bunds divide the area into strips and act as barriers to the flow of water, so they reduce the amount and velocity of the runoff. No cultivation is allowed on the bund like graded bunds. Under contour bunds an area of about 5% is lost for constructing bunds and is not available for cultivation/plantation. Bunds constructed along contour or with permissible deviations from contours are called contour bund.

Bunds serve two purposes. Bunds constructed along the periphery without reference to contour of a field are known as the peripheral bunds. Bunds constructed along the margins of roads, streams, or rivers are marginal bunds. Contour bunds serve as fences and in addition give some protection from water and wind erosion in areas of low rainfall. But in heavy rainfall areas having low infiltration rate they do more harm than good.

A contour bunding is recommended for semiarid areas where intensive farming is not feasible. Contour bunding is done in low rainfall areas and where soils are permeable. Contour bunds can be adopted on all types of relatively permeable soils. In rolling (slopes <6%) and flatter lands with scanty or erratic rainfall contour bunding is practiced. The main objective of contour bunds is to intercept the runoff, conserve moisture, as well as to reduce soil erosion. The cross-section of bunds differs according to slope, rainfall, and depth of soil.

10.3.10 Specifications for Contour Bunds

10.3.10.1 Spacing of Contour Bunds

Spacing is mainly dependent on the slope of the land. The distance between contour bunds may be measured either as the vertical interval (VI) or as the horizontal interval (HI) between them. Relationship between VI and HI is as follows

$$\mathrm{VI} = \frac{\mathrm{HI} \times S}{100}$$

where,

VI is vertical interval, HI is horizontal interval,
S is the percentage of slope

$$\mathrm{HI} = \frac{\mathrm{VI} \times 100}{S}$$

10.3.10.2 Ramser's Formula

$$\mathrm{VI} = 0.3\left[\frac{S}{3} + 2\right]$$

where

VI = Vertical interval, m,
S = Slope, %
(This formula can be applied for the soils with good infiltration rates and low rainfall)

10.3.10.3 USDA Formula

$$\mathrm{VI} = 0.3\left[\frac{S}{4} + 2\right]$$

(For soils with low infiltration rates and high rainfall)

These equations were developed for USA based on rainfall pattern, and also in the above equations the effects of climate and vegetative cover had not been taken into account. Slope was the only criterion considered in the above equation. M.P. Cox suggested the following equation considering climate and vegetative cover for Indian situation.

10.3.10.4 Cox Formula

$$\mathrm{VI} = (XS + Y)0.3$$

where, S = Slope %, X = Rainfall factor (i.e., 0.4–0.8), Y = Infiltration crop cover (1.0–2.0).

10.3.10.5 Cross-Section of Bund

The cross-section of bunds differs according to slope, rainfall, and depth of soil. The shape of the bund is usually trapezoidal. Cross-sectional area of the bund can be calculated as,

$$\text{Cross-section of bund} = \frac{(\text{Top width} + \text{Bottom width}) \times \text{Height}}{2}$$

The bund height is mainly dependent on the depth of impounding water. A freeboard of 20% of depth of water should be provided to avoid over topping

10.3.10.6 Length of Bund Per Hectare

Total length of bund required per hectare depends upon the vertical interval and slope.

Length of contour bund per ha = 10000/HI

$$\text{As} \quad \text{HI} = (\text{VI}/\text{S}) \times 100$$

$$L = \frac{10000 \times S}{100 \times \text{VI}}$$

$$L = \frac{100S}{\text{VI}}$$

10.3.10.7 Area Lost Due to Contour Bunding

When contour bunding is adopted, the structures occupy some area, which are not available for effective cultivation/plantation. The base width of bunds, slope, and the vertical interval affects the net area.

Area lost per ha due to bunding = $L \times b$ where, L = length of contour bund per ha. (m) and b = bottom width in m

$$\text{As } L = 100S/\text{VI}$$

$$\text{Area lost per ha due to bunding } (\text{m}^3) = \frac{100S \times b}{\text{VI}}$$

$$\text{So, percentage area lost per hectare due to bunding} = \boxed{\frac{S \times b}{\text{VI}}}$$

10.3.10.8 Benefits of Contour Bunding

1. Betters infiltration of water into the soil
2. Replenishes the underground storage
3. Controls the fertile soil erosion.
4. Reduces soil erosion by reducing velocity of flow of water.

10.3.10.9 Design Procedure of Contour Bund

The design of contour bund primarily consists of collecting basic information related to rainfall, runoff, infiltration characteristics, soil type and topography. The following steps are followed in the design of contour bunds.

1. Spacing of contour bunds (vertical interval)
2. Horizontal interval
3. Calculation of cross-sectional area of bund
   - Calculation of runoff (in cm)
   - Storage area required behind the bund
   - Calculation of storage capacity of the bund
   - Calculation of height of the bund (including 20% of free board), base width, top width, etc.

10.3.10.10 Calculations

1. Vertical interval, $VI = (XS + Y)\ 0.3$
   where, X = Rainfall factor; Y = infiltration and crop factor; S = slope factor
2. Horizontal interval, $\text{HI} = \frac{\text{VI} \times 100}{S}$
   where, VI = Vertical interval; S = land slope%
3. Calculation of runoff = RO (cm) = Rainfall (cm) − infiltration (cm)
4. Storage area $A = \frac{\text{RO}}{100} \times \text{HI}$
5. Calculation of storage capacity of bund by assuming d, slope of bund, slope of seepage line in terms of d and calculate depth of water to be stored
6. Calculate the height of bund $H = d + \frac{20\ d}{100}$
7. Base width $Bw = xd + yd$
   where, x = Slope of seepage; y = slope of bund and d = depth of water storage
8. Top width $(Tw) = Bw - yH + YH$
9. Cross-sectional area, $\text{CSA} = \frac{Tw + Bw}{2} \times H$

Example: Design a contour bund in medium soil having average land slope 2%. Rainfall expected is 650 mm. The infiltration capacity of the soil of the area is 40% of the rainfall absorbed in the field. Assume rainfall factor (X) – 0.6, infiltration and crop factor (Y) – 1.5, slope of seepage line to be 4:1 and bund slope to be 1.5:1.

Data given are as follows:

Rainfall 650 mm	Land slope 2%
$X = 0.6$	$Y = 1.5$

$$\begin{aligned}\text{Vertical Interval VI} &= (XS + Y)0.3\\ \text{VI} &= (0.6 \times 2 + 1.5)0.3\\ &= 0.81\,\text{m (say 0.8 m)}\end{aligned}$$

Horizontal interval (HI)

$$\text{HI} = \frac{\text{VI} \times 100}{S}$$

$$\text{HI} = \frac{0.8 \times 100}{2} = 40\ \text{m}$$

Cross-sectional area of bund

Calculation of runoff = RO (cm) = Rainfall (cm) − infiltration (cm)

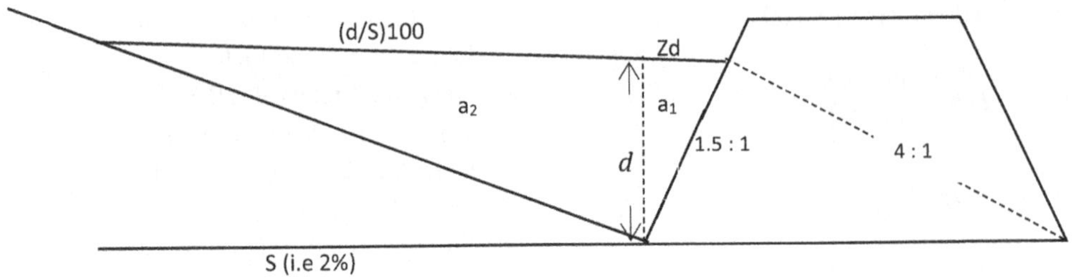

Rainfall 650 mm = 65 cm,
Consider infiltration = 40% of RF = 65 × 40/100 = 26 cm
Runoff = 65 − 26 = 39 cm
Storage area

$$A = \frac{\text{RO}}{100} \times \text{HI}$$

$$A = \frac{39}{100} \times 40 = 15.6 \text{ m}^2$$

Calculate the storage capacity of bund from figure
We know that $A = a_1 + a_2 = 15.6 \text{ m}^2$
Assuming the depth of water stored in the bund as d

$$a_1 = ½ \times 1.5d \times d, \text{ that is, } ½\, zd \times d = 0.75\, d^2$$

$$a_2 = \frac{1}{2}\left(\frac{d}{S} \times 100\right)d \quad a_2 = \frac{1}{2}\left(\frac{100}{2} \times d\right)d$$

$$a_2 = \frac{1}{2}(50d)\, d$$

$$a_2 = 25\, d^2$$

$$A = a_1 + a_2 = 0.75\, d^2 + 25\, d^2$$

$$15.6 = 25.75\, d^2$$

$$d^2 = \frac{15.6}{25.75} = 0.606$$

$$d = 0.246 = 0.25 \text{ m}$$

Consider a free board of 20% of d

$$H = 0.25 + \frac{20}{100} \times 0.25 = 0.3 \text{ m}$$

So,

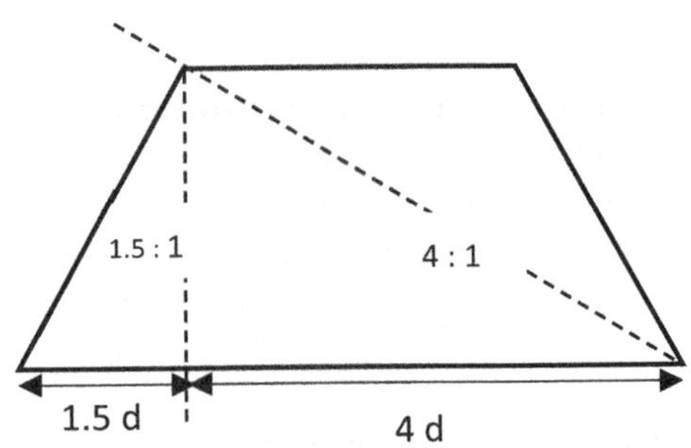

$$\begin{aligned} \text{Base width } &= zd + yd \\ &= 1.5d + 4d \\ &= 4(0.25) + 1.5(0.25) \\ &= 1 + 0.375 \\ &= 1.375 \text{ m} \quad \text{Say } 1.4 \text{ m} \end{aligned}$$

$$\begin{aligned} \text{Top width } &= Bw–(zH + zH) \quad \text{or} \quad Bw–2(zH) \\ &= 1.4–(1.5 \times 0.3 + 1.5 \times 0.3) \\ &= 1.4–0.9 \\ &= 0.5 \text{ m} \end{aligned}$$

$$\text{CSA} = \frac{Tw + Bw}{2} \times H = \frac{0.5 + 1.4}{2} \times 0.3 = 0.95 + 0.3$$
$$= 0.285 \text{ m}^2$$

Assuming a length of bund as 500 m

$$\begin{aligned} \text{Volumeofearthwork} &= \text{CSA} \times L \\ &= 0.285 \times 500 \\ &= 142.5 \text{ m}^3 \end{aligned}$$

If rate of earth work is Rs. 40/m^3
Then the cost will be = 142.5 × 40 = Rs. 5700

10.3.10.11 Graded Bunds

Graded bunds or graded terraces or channel terraces are laid along a predetermined longitudinal grade instead of along contour. They are used for the safe removal of excess runoff in high rainfall areas and regions where soil is relatively impervious.

Graded bunds consist of constructing wide and relatively shallow channels across the slope vary near the contour ridges and at critical intervals. These terraces act primarily as drainage channels for inducing and regulating the excess runoff water and drainage the same with mild and non-erosive velocity. The purpose of graded bunding is to make runoff water to trickle rather than to rush out.

It is adopted in areas where annual rainfall exceeds 80 cm and particularly in clayey soils even with lesser rainfall.

Specifications: VI and horizontal spacing is adopted as in case of contour bunds.

Grade: In general, a grade of 0.2–0.4% is provided depending on the soil type. In permeable soils, the grade may vary from 0% at upper end to 0.5% at the outlet ends. In case of impervious soils, it may started with 0.2% at upper end increased to 0.4% at the outlets.

Graded bunds may have uniform grade or variable grade.

Uniform graded bunds: Such bunds are suitable where the length of the bund is less and the discharge behind the bund or in the channel is not much and uniform grade is given throughout the length of the bund.

Variable graded bunds: These bunds are suitable where the length of bunds and discharge are more. Variable grade is providing in different sections of the bunds so that velocity of flow is within non-erosion limit.

The height of the bund should not be <45 cm. The grade of the bund should be sufficient to dispose off the water at non-erosion velocity. The top width varies from 30 to 90 cm depends on weighted bund. The base should be sufficiently wide so that the seepage line should not appear above the toe on the downstream side of bund. The graded bunds may be constructed with manual labor or with animal drawn buck scraper and with bulldozer.

10.3.10.12 Design of Graded Bund

1. Collect the basic information
2. Estimate the peak rate of runoff to be carried out by the bund. By rational formula

$$Q = 0.0276 \text{ CIA}$$

Where,
Q = Peak RO in m^3
C = Runoff coefficient (Ref. Table 10.5)
I = Maximum average RF (cm/h)
A = Watershed area (ha)

Time of concentration–
time required to water to flow from remote point to outlet

$$\text{Tc} = 0.0195 \text{L}^{0.77} S^{-0.385}$$

Where,
T_c = Time of concentration (min)
L= Maximum length of flow, S = average slope of area (m/m)
where L= HI + Bund length

3. Calculate cross-sectional area of bund

$$\textbf{Area} = \frac{\textbf{HI} \times \textbf{L}}{\textbf{10000}}$$

Table 10.5 Runoff coefficient values for different topographical conditions

	Runoff coefficient, C					
	Soil group A			Soil group B		
Slope	<2%	2–6%	>6%	<2%	2–6%	>6%
Forest	0.08	0.11	0.14	0.10	0.14	0.18
Meadow	0.14	0.22	0.30	0.20	0.28	0.37
Pasture	0.15	0.25	0.37	0.23	0.34	0.45
Farmland	0.14	0.18	0.22	0.16	0.21	0.28

4. Determine the top width, base and height of bund
   Assume depth of bund (say 50 cm)
5. Estimate the velocity of runoff behind the bund using Manning's formula

$$V = \frac{R^{\frac{2}{3}}\ S^{\frac{1}{2}}}{n}$$

where,
R is hydraulic radius (m),
n is constant (i.e., 0.018–0.04),
S is slope (%)

$$R = \frac{A}{P}$$

where,
$A = C/s$ area of bund (m^2),
P is wetted perimeter (m)

6. Observe the safe velocity of flow (calculated velocity should be < safe velocity as in tables)
7. Calculate the discharge capacity of the channel

$$Q = A\ V$$

where
Q—discharge in m^3/s,
A—c/s area (m^2),
V—Velocity (m/s)

8. Compare the discharge capacity of the channel and peak runoff rate capacity, that is, calculated discharge should be more than the peak runoff rate (step7 > step3)
9. Otherwise assuming *d* again repeat the above steps

10.3.10.13 Bench Terraces

Bench Terraces are the platform like constructions, which are constructed along the contour of the sloping land. This type of terrace is generally constructed on the land slope ranging from 6% to 33%. The hilly land is modified in the form of several steps, which intercept the flowing water on the soil surface. Bench terracing is one of the prominent mechanical soil conservation practices adopted by the farmers of India on sloping and undulating lands. Intensive farming coupled with bench terracing includes the construction of step like platforms along contours by half cutting and half filling. Original slope is converted into level fields and in case of sloping irrigated land, bench terracing aids in proper water management.

10.3.10.14 Types of Bench Terraces

Bench terraces are broadly categorized as given below:

1. Hill type bench terrace—Used for hilly areas with reverse grade towards the hill
2. Irrigated bench terraces—Level benches adopted under irrigated condition
3. Orchard bench terraces—Narrow width terraces for individual trees

Bench terraces are also classified depending upon the slope of benches as under (Ram Rao 1974). The different types of bench terraces are shown in Fig. 10.21.

1. Level bench terrace (for medium rainfall, deep, and highly permeable soil)
2. Inwardly sloping bench terraces (for heavy rainfall)
3. Outwardly sloping bench terraces (for low rainfall)
4. Puerto Rican or California type of terraces

1. **Level bench terrace:** Level bench terraces are adopted on slopes steeper than 6–7%. Level bench terraces are required in paddy growing areas on slopes as mild as 1%, to facilitate uniform impounding of water. These are also called as table top or paddy terraces.
2. **Inwardly sloping bench terraces:** Crops like potato are extremely susceptible to waterlogging. In this case, the benches are made with inward slope to quickly drain off the excess water. These are suited for steep slopes, to drain the excess runoff towards the inner side with a suitable grade along its length through a well stabilized vegetated waterway.
3. **Outwardly sloping bench terraces:** In places of low rainfall or shallow soils, these are used to reduce the existing steep slope to mild slope (say from 8% to 4%). These type of terraces are constructed on soils not having good permeability' provision of graded channels at lower end has to be kept to safely dispose of surplus water to some water way. In very permeable soils, a strong bund with spillway arrangement may take care of most of the rainfall events. While during heavy rainfall storms, the excess water may flow from one terrace to another. Usually dispose off this to some waterway at an earliest possible spot.

10.3.10.15 Design of Bench Terraces

The design of bench terrace consists of determining the following components.

(a) Type of bench terrace
(b) Terrace spacing or depth of cut
(c) Terrace width
(d) Terrace cross-section

Terrace spacing is generally expressed as the vertical interval between two terraces. The vertical interval is dependent upon the depth of cut

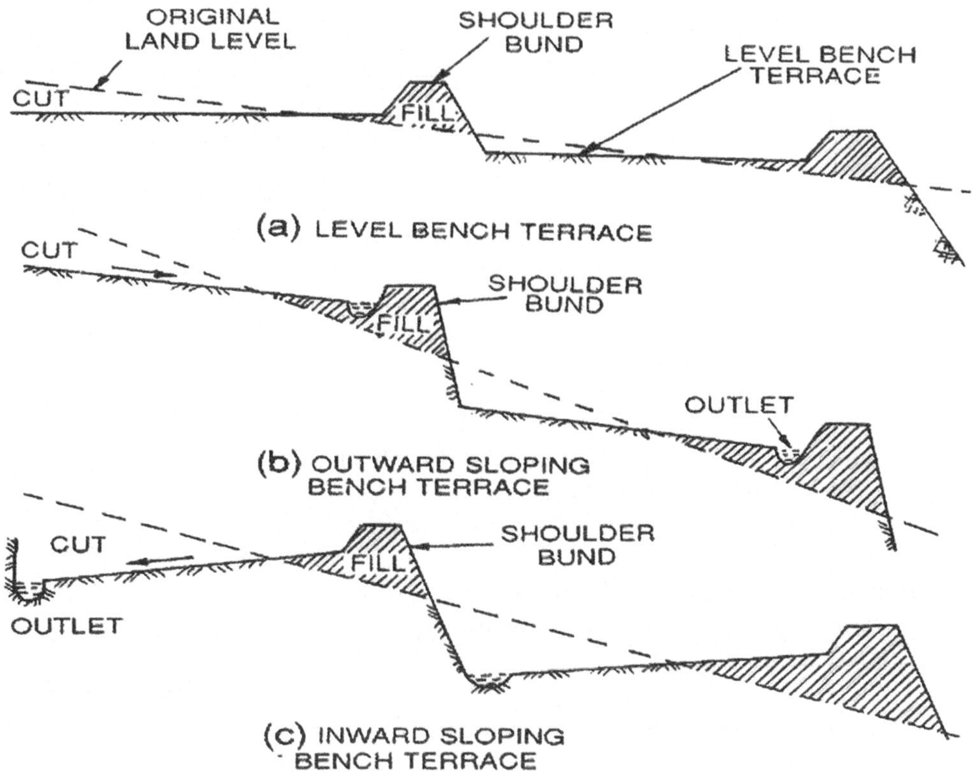

Fig. 10.21 Types of bench terraces

Depth of terrace (cut)

$$\mathrm{VI} = D = \frac{WS}{100}$$

where

W = width of terrace,
S = Slope
$D/2$ = Depth of cut

When batter slope is 1:1 $D = \frac{WS}{100 - S}$
When batter slope is ½:1 $D = \frac{2WS}{200 - S}$

10.3.10.16 Area Lost for Cultivation

When batter slope is 1:1 $a = \frac{S+200}{\frac{200}{S} + \frac{S}{100}}$
When batter slope is ½:1 $a = \frac{S+100}{\frac{200}{S} + \frac{S}{100}}$

$$\text{Length per hectare} = \frac{10000}{W}$$

Earth work excavation $\frac{1}{2} X \frac{W}{2} X \frac{D}{2}$

10.3.10.17 Design of Bench Terraces

Example Design a 150 m long bench terrace for a land having an average slope of 20%. The soil is clay loam. The terrace channel has a uniform grade of 0.5%. Maximum intensity of rainfall expected during the 10 year recurrence interval is 10 cm/h.

Solution
An inward slope type of terrace is selected (Fig. 10.22). The inner end of the terrace is 30 cm lower than the outer end.

Assume width of terrace as 4.5 m (as per established practice)

The slope of terrace face (riser) is 1:1
Shoulder bund side slope 1:1,
Bund height =30 cm, Bottom width 75 cm (as per practice)
Vertical interval between the bunds

$$D = \frac{WS}{100 - S} = \frac{4.5 \times 20}{100 - 20} = 1.125$$

We know that, $S = \frac{H}{L}$
Where, $L = 4.5 + 150 = 154.5$

$$H = 0.3 + (0.5/100)150 = 1.05 \text{ m}$$

$$S = \frac{1.05}{154.5} = 0.00679$$

So, Time of concentration

$$Tc = 0.0195\, L^{0.77} S^{-0.385}$$

$$Tc = 0.0195 \times 154.5^{0.77} 0.00679^{-0.385} = 6.4579 \text{ min}$$

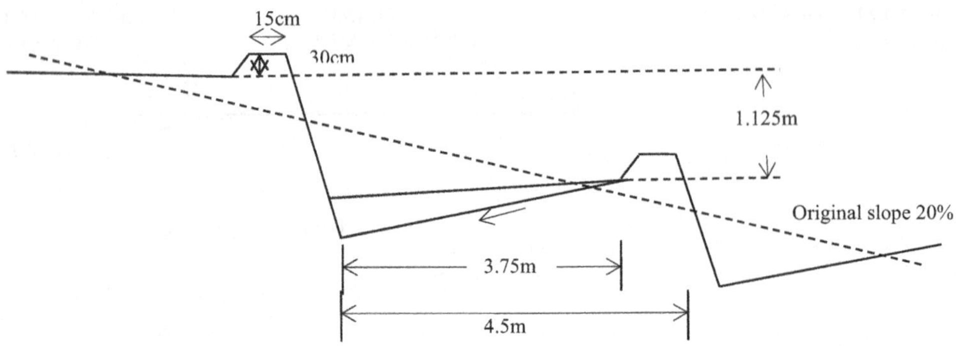

Fig. 10.22 Design of bench

Using Fig. 10.23 convert the 1 h intensity of 10 cm/h for a 6.5 min, intensity of rainfall is 24.75 cm

Peak runoff to be carried out by the channel is estimated by the rational formula

$Q = 0.0276$ CIA where, $C = 0.6$ (for clay loam)

$A = 150 \times 4.5 = 0.0675$ ha (watershed area)

$I = 24.75$ cm/h

$$\text{Designrunoffrate } Q = 0.0276 \times 0.6 \times 24.75 \times 0.0675 = 0.0276 \text{ cum/s}$$

$$a_1 = ½ \times 0.3 \times 0.3 = 0.045 \text{ m}^2$$

$$a_2 = ½ \times 3.75 \times 0.3 = 0.564 \text{ m}^2$$

$$a = a_1 + a_2 = 0.045 + 0.56 = 0.609 \text{ m}^2$$

Wetted perimeter (P) of water spread area is

$$S_1{}^2 = ac^2 = ab^2 + bc^2 \text{ (Using Pythagoras theorem)}$$

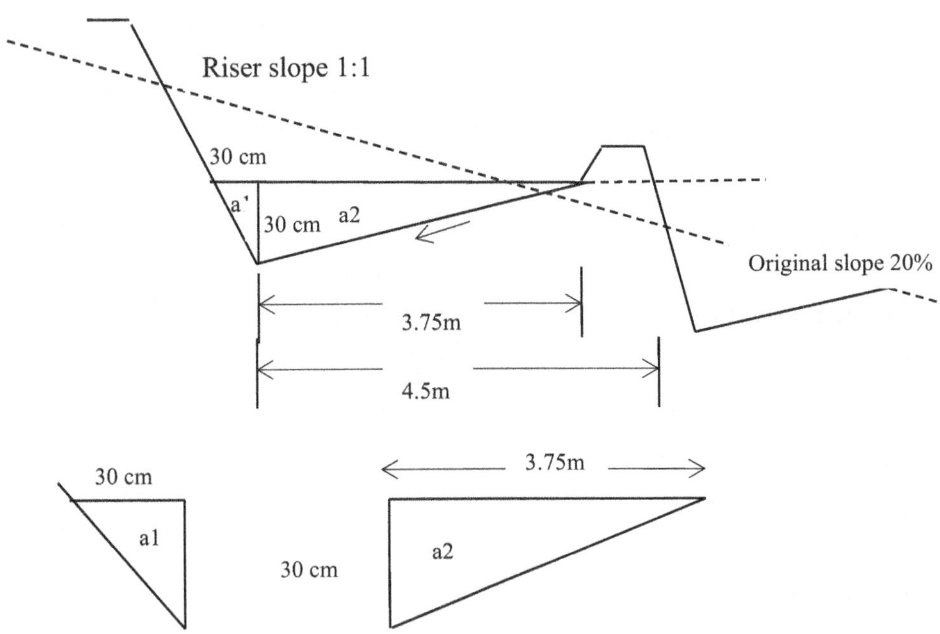

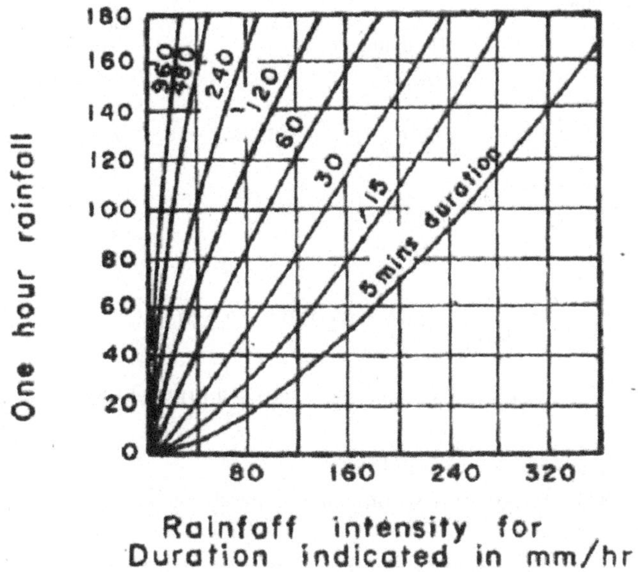

Fig. 10.23 Relation of one-hour rainfall intensity to intensity at other duration

$$S_1{}^2 = 0.3^2 + 0.3^2 = 0.09 + 0.09 = 0.18$$

$$S_1 = 0.42 \text{ m}$$

$$\text{Similarly}, S_2{}^2 = 0.3^2 + 3.75^2 = 14.153$$

$$S_2 = 3.761 \text{ m}$$

$$P = S_1 + S_2 = 0.42 + 3.761 = 4.18 \text{ m}$$

$$\text{Hydraulic radius } (R), R = \frac{A}{P} = \frac{0.609}{4.18} = 0.146 \text{ m}$$

Velocity of flow and carrying capacity of the channel is

Velocity using Manning's formula $V = \frac{R^{2/3} S^{1/2}}{n} = \frac{0.146^{2/3} 0.005^{1/2}}{0.04} = 0.497 \text{ m/s}$

The velocity is non-erosive

Therefore, carrying capacity

$$Q = \text{A V}$$

$$Q = 0.609 \times 0497 = 0.3028 \text{cum/s}$$

Hence the section has sufficient carrying capacity.

10.3.11 Grassed Waterways (Vegetated Outlets and Water Courses)

Grassed waterways are open channels protected with suitable grasses. These are constructed along the slope and act as outlet for the terraces or graded bunds. Waterways are vital water management channels for erosion control, excluding level terraces and contour bunds. These are used to carry natural runoff or to handle the discharge from mechanical structures such as terrace or bund, spillways of water storage structures, etc. Therefore, the vegetated outlets should be well established before handling the runoff from bunds, terraces, diversions channels, etc.

Design: The design of the grassed waterways is similar to the design of irrigation channels. The design of the grassed water consists in determining the size and shape of the waterway.

Size of waterways: The size of the waterway depends upon the expected runoff. A 10-year recurrence interval is used to calculate the maximum expected runoff to the waterway. As the catchment area of the waterway increases towards the outlet, the expected runoff is calculated for different reaches of the waterway and used for design purpose. The waterway is to be given greater cross-sectional area towards the outlet as the amount of water gradually increases towards the outlet. The cross-sectional area is calculated using the formula

$$a = Q/V$$

where a = area of cross-section, Q = expected maximum runoff, V = velocity of flow.

Shape of waterway: The shape of the waterway depends upon the field condition and the type of the construction equipment used. Commonly adopted shapes are trapezoidal, triangular, and parabolic. In course of time due to flow of water and sediment deposition, the waterway shape becomes nearer to parabolic shape. Usually parabolic and trapezoidal shapes are preferred considering the crossing of farm machinery equipment.

Velocity of flow: The permissible velocity in a grassed waterway depends on the type, condition, and density of vegetation and erosive characteristics of the soil. An average velocity of 1.5–2 m/s is used for design purpose.

Channel grade: Grassed waterways generally run down the slope and the channel grade is usually governed by the land slope. Channel slope should not exceed 10%, while it is normally desirable to keep the grade within 5% (A diversion ditch, however, constructed across the slope

Table 10.6 Calculation of area, wetted perimeter, hydraulic radius, and top width

Shape	Area	Wetted perimeter	Hydraulic radius	Top width
Trapezoidal	$bd + zd^2$	$b + 2d\sqrt{z^2+1}$	$\frac{bd+zd^2}{b+2d\sqrt{z^2+1}}$	$t = b + 2dz$ $T = b + 2Dz$
Parabolic	$\frac{2}{3}\ t\ d$	$t + \frac{8d^2}{3t}$	$\frac{t^2d}{1.5t^2+4d^2}$	$t = \frac{A}{0.67d}$ $T = t\left(\frac{D}{d}\right)2$
Triangular	zd^2	$2d\sqrt{z^2+1}$	$\frac{zd}{2d\sqrt{z^2+1}}$	$t = 2dz$ $T = \frac{D}{d}\ t$

for the purpose of intercepting surface runoff and dispose of it to safe outlet)

Channel dimensions: After the runoff, channel grade and design velocity have been determined; next step is to decide the channel dimensions. It is based on Manning's formula. The coefficient of roughness "*n*" assumed as 0.04. Side slope of channel should be 4:1 or flatter. About 15 cm freeboard should be provided to take care of the build up by sediment and variation in the value of the roughness coefficient.

Calculations of area (*A*), wetted perimeter (*P*), hydraulic radius (*R*), and top width (*t*, *T*) are given in Table 10.6.

Example Design a parabolic shaped grassed waterway to carry a flow of 3 m^3/s down a slope of 4%, an excellent stand of dub grass is to be maintained in the waterway.

Solution

Assume that the top width, $t = 4$ m and depth of flow, $d = 60$ cm

The cross-sectional area

$$a = \frac{2}{3}\ t\ d = 2/3 \times 4 \times 0.6 = 1.6\ \text{m}^2$$

Wetted perimeter,

$$P = t + \frac{8d^2}{3t} = 4 + \frac{8 \times 0.6^2}{3 \times 4} = 4.24\ \text{m}$$

$$\text{Hydraulic radius, } R = \frac{a}{p} = \frac{1.6}{4.24} = 0.377$$

Mean velocity, $v = \frac{1}{n} R^{2/3} S^{1/2} = \frac{1}{0.04} 0.377^{2/3} \times 0.04^{1/2} =$ 2.61 m/s

So, $Q = a\ v = 1.6 \times 2.61 = 4.17\ m^3/s$

The velocity is unsafe and the carrying capacity is excessive

Assume, $t = 4.15$ m, and $d = 47.5$ cm

Then, $a = 1.314\ m^2$ $p = 4.294$ m

And $R = 0.306$ m

$$\frac{1}{0.04} 0.306^{2/3} \times 0.04^{1/2} = 2.289\ \text{m/s}$$

$$Q = a \times v = 1.314 \times 2.289 = 3.007\ \text{m}^3/\text{s}$$

The section is suitable

The design dimensions of the parabolic channel are as follows,

Top width 4.15 m, depth of flow of water 47.5 cm

Free board 15 cm

10.4 Concluding Remarks

The natural resources like soil and water are imperative to life support system on earth. The increase in population, over exploitation of natural resources for human needs such as food, fodder, and fuel requirement coupled with unscientific management of vital natural resources has led to fragile and precariously balanced ecosystem. There exists a mutual relationship between water and forest as forests generate 60% of total runoff and a major share of our drinking water supply is from forest watersheds. The increasing degradation of these resources is affecting the stability and resilience of our ecosystems and the environment. The knowledge of hydrology is crucial to restore this ecological imbalance and manage degraded agricultural, forest and wastelands. The problem of degraded land and water management is multifaceted which necessitates multidimensional approach. This process aims to develop human resource in watershed development and management and generate awareness about the importance of sustainable utilization of natural resources.

The effective conservation of natural resources is aimed at obtaining optimum and sustained returns. Watershed is a hydrological unit which responds most effectively to various engineering, biological and cultural treatments. A right balance between socio-economic and environmental objectives is important in watershed management. The rising conflicts for water resources between different stakeholders mandate sustainable water management which is crucial for economic development and livelihood of the people. It would be very appropriate and timely to apply the Remote Sensing and Geographic Information System techniques for watershed management to mitigate natural disasters such as droughts, floods, and landslides.

Lessons Learnt

- Hydrology and hydrological cycle
- Essential components of hydrological cycle and their measurements
- Soil erosion, types, and their effect and soil loss measurement
- Watershed, its functions, and morphological characteristics
- Watershed delineation using geospatial technology
- Steps in watershed management
- Soil and water conservation measures for watershed management

Key Questions

1. What are the different components of hydrological cycle? Enlist the major threats.
2. What is the role of forest in watershed management?
3. What is the main problem with watershed management?
4. What is Integrated Watershed Management Program? How is it useful for sustainable management of natural resources?
5. What are the applications of geospatial technology for forest hydrology and watershed management?

References

Baumgartner FA, Reichel E, (1975) The world water balance: mean annual global, continental and maritime precipitation. Evaporation and Runoff, p 179

Chang M (2012) Forest hydrology: an introduction to water and forests. CRC Press

Cohen MJ (2008) Forest hydrology: an introduction to water and forests. Forest Sci 54(6):658

FAO (2022) The state of the world's land and water resources for food and agriculture – systems at breaking point. Main report. Rome

Gravelius H (1914) Flusskunde. Goschen Verlagshan dlug Berlin. En Zavoianu, I. 1985. Morphometry of drainage basins. Elsevier, Amsterdam

Helvey J, Patric JH (1965) Canopy and litter interception of rainfall by hardwoods of eastern United States. Water Resources Res 1(2): 193–206

Humphreys WJ (1940) Physics of the air, 3rd edn. McGraw-Hill, New York. (reprinted by Dover Publications, New York, 1964)

Jenson SK, Domingue JO (1988) Extracting topographic structure from digital elevation data for geographic information system analysis. Photogram Eng Remote Sensing 54(11):1593–1600

Parysow P, Wang G, Gertner G, Anderson AB (2003) Spatial uncertainty analysis for mapping soil erodibility based on joint sequential simulation. Catena 53(1):65–78

Planchon O, Darboux F (2002) A fast, simple and versatile algorithm to fill the depressions of digital elevation models. Catena 46(2–3): 159–176

Rao MR (1974) Soil Conservation in India. Indian Council of Agricultural Research

Ravikumar D (2020) Micro level land and water resource development plan for Koranahalli Subwatershed in Tarikere Taluk, Chikmagulur District Using Remote Sensing and GIS Techniques. Kuvempu University. Ph.D. Thesis, 34p

Renard KG, Foster GR, Weesies GA, McCool DK, Yoder DC (1996) Predicting soil erosion by water: a guide to conservation planning with the Revised Universal Soil Loss Equation (RUSLE). Agriculture Handbook, p. 400

Subramanya K (2008) Engineering hydrology. McGraw-Hill

Tarboton DG, Bras RL, Rodriguez-Iturbe I (1991) On the extraction of channel networks from digital elevation data. Hydrol Processes 5(1): 81–100

Tejwani KG, Dhruva Narayana VV (1961) Soil conservation survey and land use capability planning in the ravine lands of Gujarat. J Indian Soc of Soil Sci 9(4):233–244

Korzoun VI (1978) World water balance and water resources of the earth. UNESCO Press, Paris

Wischmeier WH, Smith DD (1965) Predicting rainfall-erosion losses from cropland east of the Rocky Mountains: Guide for selection of practices for soil and water conservation (No. 282). Agricultural Research Service, US Department of Agriculture

Wischmeier WH, Smith DD (1978) Predicting rainfall erosion losses: a guide to conservation planning (No. 537). Department of Agriculture, Science and Education Administration

11 Desert Ecology and Functional Aspects of Desert Ecosystems

G. Singh and Sachin Sharma

Abstract

Deserts are arid to extremely arid ecosystems on the earth and cover about one-fifth of the global land surface. The deserts cover large barren and abandoned lands dominated mostly by hills of sand, though they are also consisting of rocks and mountains. Such areas are characterised by <250 mm of rainfall annually with extremes of temperatures. Except Europe, almost every continent of the earth has deserts, which are situated near 23.4° N (the Tropic of Cancer) and 23.4° S (the Tropic of Capricorn) latitudes. Subtropical, cool coastal, rain shadow and cold winter are climatically delineated deserts, whereas trade wind, mid-latitude, rain shadow, coastal, monsoon, polar and montane are geographically demarcated deserts. Such wide differences in latitudinal, altitudinal, climatic, geomorphological and biological features of the deserts lead a diverse habitats and biological diversity and sustain peoples' livelihoods in the region. Circadian and seasonal extremities, intense solar radiation, low water availability, high salinity and wind and dust storms jointly affect the life in desert. While negative biotic interactions play a role in shaping the desert ecology, remarkable ways of adaptations allow life to sustain in these deserts. The deserts have a significant role in shaping animals' and humans' lives, and the surrounding environment and promote the formation and concentration of some of the essential minerals, and provide chemical-based adapting bioresources (animal and plant life) as a source of bioprospecting. These deserts are ideal places of living and for preserving human relics and remains as well. The desert sands also act as an important carbon sink on the earth. A range of diversity in habitats and isolated populations results in higher within-species diversity and is positively related to various functions of the region. However, the growing human population and corresponding anthropogenic activities coupled with climatic adversities are resulting in habitat degradation and productivity loss. Excessive removal of vegetation for fodder and fuelwood, altered land use, increased atmospheric carbon dioxide levels and global warming are leading factors to habitat loss and land degradation. Human-induced destruction of the environment results in biodiversity loss leading to a decline in ecosystem functions including water supply, soil functioning and climate change resilience. This chapter further describes the existing threats to the desert ecosystem, research and experimentations, developmental programmes and the way forward to restore and conserve the desert ecosystem and its biological treasures.

Keywords

Desert · Ecology · Ecosystem · Plants adaptation · Compensatory afforestation

11.1 Introduction

Desert ecology is described as interactions between biotic and abiotic components of the desert environment. Thus, the desert ecosystem is interaction between organisms, the surrounding atmosphere and other non-living factors that influence the living environment. The functioning of ecosystems on the earth depends primarily on the utilisation of energy and resources (water, minerals) and autotrophs as the main agents to utilise these resources (Hufnagel et al. 2018). Among various ecosystems based on the efficient utilisation of energy and resources, tropical rain forests are the most efficient, whereas deserts lie on the other end constituting a zone of struggle. Because of moisture constraints, these areas are referred to as ecosystems

G. Singh (✉)
Arid Forest Research Institute, Jodhpur, Rajasthan, India

Lucknow, India

S. Sharma
Arid Forest Research Institute, Jodhpur, Rajasthan, India

A. K. Mandal, A. Nicodemus (eds.), *Textbook of Forest Science*, https://doi.org/10.1007/978-981-97-8289-5_11

controlled by water availability—a prime resource required for the survival of the desert organisms, which depend primarily on the water balance, that is, water inflow and loss. The deserts are not in recent origin rather they have existed since glaciers covered large portions of the earth, that is, the great Ice Ages. The deserts are formed by global wind circulation patterns including trade winds, westerlies, the polar easterlies and the jet streams. Deserts occur from areas close to the poles and down to areas near the equator of the earth, hence they are not restricted by latitude, longitude, or elevation. However, most of the deserts are situated near 23.4° north and 23.4° south latitudes from the equator that is the Tropic of Cancer and the Tropic of Capricorn respectively.

All deserts have two common characteristics: (i) they are dry and (ii) they support a smaller number of plants and animals' lives (White et al. 2013). The deserts are not only hot but also cold in nature, where all moisture is frozen. The percentage of land classified as desert was only around 9% of the total earth's surface in 1882, but by 1950, arid and semi-arid areas together had occupied up to more than 23% global area (Grove 1974). Recent observations indicate that because of desertification—a process leading to desert-like conditions, the Sahara Desert in North Africa has advanced southward by another 1000 km (Zeng 2003; Kandji et al. 2006). The study by Thomas and Nigam (2018) indicates that the Sahara Desert has expanded by 11–18% depending on the season and by 10% when annual rainfall was considered for the period of 1902–2013.

11.1.1 Desert Formation

Despite the scarcity of geological evidence, researchers agree that some deserts have always been present. However, they were possibly smaller in size as compared to the deserts of the present time. The fossil record of a small species of hippopotamus found in the Arabian Desert, which covers the Arabian Peninsula and the east of North African regions, indicates that this region was once a wetland (Engel et al. 2017). Rock paintings from about 5000 years ago in the Sahara Desert of North Africa showing pictures of elephants, giraffes and herds of antelope also indicate their presence in the region (Jaroff 1997). The origins and dynamic properties of the desert conditions vary with regional conditions. For example, the processes of desertification in warm-temperate and sub-arid conditions are manifested with the leading role of anthropogenic factors, whereas the processes in tropical and subtropical areas are climate influenced (Zalibekov 2011; Weber et al. 2013). Geographer Peveril Meigs (in 1953) divided deserts of the world into three categories based on the amount of rainfall the desert receives. In this, extremely arid zone areas are without rainfall for at least 12 succeeding months, arid lands with a mean annual rainfall of <250 mm, and semi-arid lands with a mean annual rainfall of 250–500 mm.

Aridity index (AI) in one criterion for dividing the world into different climatic regions. The aridity index is the ratio of average annual precipitation and total annual potential evapotranspiration (PET). Accordingly, FAO (Food and Agriculture Organization of the United Nations) has classified the world's climate into hyper-arid, arid, semi-arid, dry sub-humid and humid) regions, which have AI values of <0.05, 0.05–0.2, 0.2–0.5, 0.5–0.65 and >0.65, respectively. Cold climate areas are the special case where the mean annual PET is <400 mm (FAO). Extremely arid and arid zones are considered as deserts, whereas semi-arid grasslands are largely referred to as grasslands or steppes. In the process of desertification, areas with aridity index of <0.5 have globally increased from 28.4% to 29.6% (Spinoni et al. 2013). At the same time, cold climate areas have decreased from 26.6% to 25.4% in the Northern Hemisphere. Comparing the aridity maps of the two periods (1951–1980 and 1981–2010), the areas which shifted most remarkably to lower aridity index values (towards more arid conditions) are Canada, Brazil, the Mediterranean region, Eastern Europe, most of Africa, the Middle East, Eastern parts of China, Borneo in Indonesia and Australia. At the regional/country level, a shift of one class towards a more arid region has been observed in the Alaska region of the USA, Alberta (Canada), Patagonia region of Argentina, Pernambuco of Brazil, western Peru, Spain, the Southern Sahara and North-Eastern Kalahari Desert regions, Rajasthan and Madhya Pradesh provinces in India, Mongolia, the Yang-Tze Basin of China and the North-Eastern and South-Western Australian coast regions (Spinoni et al. 2013). On the other hand, Central U.S., Paraguay and Northern Argentina, Scandinavia, Northern Australia and Western China have shifted to a wetter climate during 1981–2010 (Spinoni et al. 2013).

11.1.2 Kinds of Deserts

Global deserts can be divided into five types. These are subtropical, coastal, rain shadow, cold winter and polar deserts. The **subtropical deserts** are hot with 30 terrain and high rate of evaporation (Table 11.1). The intense solar radiation received in the tropical region lying between the line of Cancer and the line of Capricorn hits the land surface. This results in setting off the air currents bringing dry winds to the subtropical regions situated between 25 and 35° latitude in both the North and South of the equator. This affects precipitation negatively and makes the area deserts. The **cool coastal deserts** are situated in between the subtropical deserts, but the average temperature is much lower because of their position in the western part of the continents. Here the

Table 11.1 Various deserts, their area and distribution in the world

Desert name	Location	Area (million km^2)	Topographical features
Subtropical deserts			
Sahara	Covering the countries like Algeria, Chad, Egypt, Eritrea, Ethiopia, Libya, Mali, Mauritania, Morocco, Niger, Somalia, Tunisia and Western Sahara.	9.07	Topography is characterised by shallow basins, large depressions, gravel plains (70%), plateaus, mountains, sand sheets, dunes and sand seas (30%). It is the world's largest non-polar desert
Arabian Desert	It covers Saudi Arabia, Kuwait, Oman, Qatar, United Arab Emirates, Yemen and parts of Jordan and Iraq	2.59	Gravel plains and rocky highlands. Southern one -fourth area is the Empty Quarter or Rub' al-Khali desert, which is the largest unbroken sand hills in the world
Kalahari	Mostly the areas of Botswana, but the eastern third of Namibia and the northern Cape area of South Africa also represent it	0.57	Dominant landforms in this desert are sand sheets and longitudinal/linear dunes with occasional exposed bedrocks
Great Victoria	It occupies south- and west Australia	0.35	These deserts are dominated by gravel-covered terrain with a thin cover of grasses. However, extensive areas are undulating red sand plains and dune fields, some low rocky ridges with a high degree of laterite development. Simpson's straight is parallel sand dunes
Great Sandy	It occurs in the Northern Territory and Western Australia	0.27	
Gibson	Western Australia	0.18	
Simpson and Sturt Stony	It covers the eastern half of the Australian continent	0.18	
Thar	Parts of India and Pakistan	0.20	Rocky, sand and dune fields are the dominant landforms dotted by saline depressions
Coastal deserts			
Namib	It covers parts of Angola, Namibia and South Africa	0.034	Dominated by gravel plains
Atacama	Part of Chile in South America	0.14	It is dominated by salt depressions, sand and lava
Rain shadow deserts			
Great basin	Nevada, Oregon and Utah of USA (including Death valley area)	0.49	Dominated by basin and hill range topography. About 1% area is covered by sand dunes
Chihuahuan	Part of Mexico and the Southwestern part of USA	0.45	It is a shrub desert with large basins surrounded by terraces, mesas and hill ranges
Colorado Plateau	Arizona, Colorado, New Mexico, Utah and Wyoming parts of the USA	0.34	It is a desert region of the intermontane plateau dominated mostly by sedimentary rock and mesas including the Grand Canyon
Sonoran	Arizona and California of USA and the northern part of Mexico	0.31	It covers low elevated basins and plains bordered by parallel but widely scattered small hill ranges
Mojave	Arizona, Colorado, Nevada, Utah and California of USA	0.14	Represented by mountains and basin types of topography. Hill chains, dry lake beds and dunes rich in $CaCO_3$
Cold deserts			
Patagonian	Part of Argentina in South America	0.67	Dominated mostly by gravel plains, plateaus and basalt sheets
Kara-Kum	This desert is in Uzbekistan and Turkmenistan	0.35	About 90% area is grey-layered sand, which means 'black sand'.
Kyzyl-Kum	It covers parts of Uzbekistan, Turkmenistan and Kazakhstan	0.30	Sands and rock (red sand) dominate in this desert
Iranian	Mostly Iran	0.26	Dominated by salt, gravel and rock
Taklamakan	China and some parts of Kyrgyzstan	0.27	This desert is dominated by sand dunes and gravel
Gobi	It involves parts of China and Mongolia	1.30	Dominated by stony and sandy soil and steppes (the dry grasslands)
Polar deserts			
Arctic	It covers the Alaskan part of the USA, Canada, Greenland, Iceland, Norway, Sweden, Finland and Russia	13.99	It is dominated by snow and glaciers and exhibits a tundra climate
Antarctic	Antarctica	14.25	Dominated by ice, snow and bedrock

Source: Extracted from Alsharif et al. (2020) and Smith (2023)

dry inland conditions are induced by the upsurge of the deep cold ocean water and propelled by the ocean currents influencing air to rain on the ocean itself leaving dry air to move towards the coastal land areas. The Atacama and Namibian Deserts are examples of this kind. Some of the areas of the Atacama Desert are always under fog but without rain. Some deserts are also situated on the lee sides of the mountain ranges and are called **rain shadow deserts.** Because these deserts are deprived of ocean/sea-generated moisture. Death valley is the best example of this type of desert falling in the rain shadow of the Sierra Nevada Mountains of North America. **Cold winter deserts** are noticeable in high-temperature differences between seasons, where temperatures range from −12 °C in the winter season to 38 °C in the summer season. Because of the locking up of all moisture in the ice form, the polar regions are also considered as deserts, that is, **Polar deserts** covering parts of the Arctic and Antarctic regions (UCMP 2010).

The deserts are also classified based on geographical locations. Intense solar radiations in the trade winds belts of both northern and southern tropical regions cause heating as they move toward the equator. These dry winds disperse the cloud cover, resulting in more sunlight to reach and heat the terrain. As majority of deserts lie in the areas crossed by the trade winds and hence these deserts are called **trade wind deserts.** The Sahara desert in North Africa is the best example of trade wind desert. The second category is **mid-latitude deserts** that occur between 30 and 50° latitudes in subtropical high-pressure zones. These deserts are in general inland drainage basins distantly situated from the oceans and experience an extensive range of temperatures annually. Sonoran Desert is an example of this category. The **rain-shadow deserts** are developed as a result of prevention of moisture-rich clouds from reaching the areas on the lee side, or protected side of the tall mountain range. The Death valley in California and Nevada parts of the USA are examples of rain shadow because these are situated on the leeward side (rain shadow) of the Sierra Nevada Mountain range and are very hot and dry. The **coastal deserts** are situated on the western limits of the continents and near the Tropic of Cancer and the Tropic of Capricorn. In general, they are affected by cold ocean currents, which flow along the coast and are relatively complex because they are at the juncture of terrestrial, oceanic and atmospheric systems. The Atacama of South America and Namibian in Africa are examples of this type of desert. The other one is **monsoon deserts** in which monsoons develop in response to the temperature variations between landmass and the oceans. The Thar Desert in India and the Cholistan Desert in Pakistan are examples of monsoon deserts. **Polar deserts** are areas with annual precipitation of <250 mm and a mean temperature during the warmest month is <10 °C. In addition to these, **Montane deserts** are arid areas situated at a very high altitude. The most prominent examples of this kind of desert are Kunlun mountains in the Himalayas and the Tibetan Plateau.

11.1.3 Desert Characteristics

11.1.3.1 Cold Deserts

Cold deserts are arid regions with annual rainfall of <250 mm and are situated mainly in temperate and polar regions. Greenland, Antarctic and the Arctic zones of the world fall into this category of deserts. Cold winters and snowfall during the season and sometimes in summer are characteristics of these deserts. Average winter temperatures in these deserts range from −2 °C to 4 °C, whereas during summer the average temperature varies between 21 and 26 °C (Strahler and Strahler 1984). The summer season is short, moist and tolerably warm. Mean annual precipitation in these deserts ranges between 150 and 260 mm, and the rain occurs in April-May or autumn, depending on the regional variation. The soils of these deserts are heavy, moderately porous with varying extents of silt and salt, well-drained favouring leaching out of the salts. Plants are scattered, shrubby and deciduous in nature with spiny leaves. The height of this vegetation varies from 15 to 122 cm.

11.1.3.2 Hot Deserts

Such deserts exhibit extreme temperatures because atmospheric humidity in this region is extremely low and is unable to interfere with the incoming solar radiations. Average yearly temperature varies from 20 °C to 25 °C, whereas annual rainfall is very low and concentrated in a very short period. The rate of evaporation is greater than the precipitation. In some of the regions, rain falling evaporates before it reaches the ground surface. Deserts' soils are coarse in texture, shallow, rocky, gravely and well drained. It is without subsurface water. Vegetation growing in these deserts is mostly scrub and small trees with modified leaves for conserving available water. Some plants open their stomata only at night to minimise evaporation losses of water. Bushes like opuntia, false mesquite, agaves, some Euphorbias and brittlebush are some of the plants growing in such types of deserts.

11.1.3.3 Semi-arid Deserts

Semi-arid deserts are characterised by moderately long and dry summers. Rainfall is low and normally occurs in winter. The cool nights of these deserts reduce moisture loss by transpiration in plants and by sweating and breathing in animals, which is beneficial for these biotas inhabiting the region. Soils in these deserts are sandy in texture and include rock fragments and sometimes gravel. The soil of the mountainous slopes is shallow, rocky, gravely, and well drained. The soil of the bottomland is sandy and fine-textured. In these deserts, growing vegetations protect themselves with the help of modified spines that reduce transpiration loss of water.

Some plants have smooth silvery leaves that help the leaves to reflect the radiation received on them. *Larrea tridentata* (Creosote bush), *Franseria dumosa* and *Ficus deltoidea* (bursage), *Crataegus laevigata* (white thorn), *Dolichandra unguis-cati* (cat claw), *Prosopis juliflora* (mesquite), *Encelia farinosa* (brittlebush) and *Zizyphus jujuba* (ber) are some of the important vegetation of the region.

11.1.3.4 Coastal Deserts

Coastal deserts occur in moderately cool to warm areas like the afrotropical and neotropical realms. The average temperature in these deserts varies between 13 and 24 °C. The winter temperatures in these deserts decreased down to 5 °C or lower than this. Mean annual rainfall in these deserts varies from 80 to 130 mm. The Atacama Desert in Chili, South America falls in this category and is the driest desert on earth. It receives only <1 mm of annual rainfall and is even difficult to measure. The soil in the coastal desert areas is fine-textured with moderate soil content, porous and well drained. Thick and fleshy leaves or stems are characteristics of the vegetation of this region that can store water for future use after absorption during sufficient water availability. Some plants exhibit corrugated surfaces with longitudinal grooves and ridges. The *Atriplex* species (saltbush), *Fagopyrum esculentum* (buckwheat bush), *Coleogyne ramosissima* (black bush), *Leersia hexandra* (rice grass), *Tetradymia glabrata* (little leaf horse brush), *Salvia mellifera* (black sage), *Prosopis* species and cactus species are some of the plants found in the desert.

11.2 Landforms and Geomorphology

Various landforms of deserts interacting with climate and biotic factors determine the soil characteristics. Landforms directly affect the inflow and outflow of water and other materials in the form of run-off, infiltration, erosion, salt accumulation and water storage. Different landforms have different tendencies towards water and hence, they act as one of the agents for bringing heterogeneity in the habitat. The formation of these landforms is shaped by wind actions, which exposes the bed rocks resulting in the creation of different structures. They are important geomorphic structures and landforms of these fantastic biomes. Some of the wind-borne geomorphic features include dunes of different types and sizes, mesas and buttes. Other important geomorphic structures in the desert are playas, alluvial fans, bajada, pediments and inselbergs (Dorn 2018).

Vegetation cover, soil moisture and wind velocity have a significant bearing on wind erosion and sand drift. Sand dunes will not move or sand storms will not generate in the presence of ample vegetation and low wind velocity (Meng et al. 2018). Atacama, Libyan and Namib deserts are hyper-arid, whereas the Thar, Kalahari and Australia deserts are substantially moist in nature. Some of the deserts exhibit high-energy winds, whereas the others deserts have low velocity winds. This helps to explain variations in sand dune forms, and hence the presence or absence of wind erosion features results in dune formation and their types. Droughts and human activities make the Aeolian deposits more susceptible to wind erosion. The areas which are more sensitive to wind erosion are loess like those of China, desert margins where man interacts with the arable land, and the coastal dune fields. Five basic types of dunes observed across the globe are crescentic, linear, star, dome and parabolic. Their formations and movements depend upon the extent and direction of the wind energy. The **crescentic** dunes, which are known as barchans or transverse dunes, are usually wider than long. This type of dune moves much faster, that is, >100 m a year. The largest dune on the earth is in the Takla Makan desert in China having crest-to-crest width of >3 km. The **linear dunes** are long, straight, or slightly sinuous sand ridges. They are typically much longer than wide. The best example is >160 km long Simpson's straight, the linear and parallel sand dunes are the longest in the world and available in the Sturt Stony desert of Australia. The other type is **Star dunes,** which are pyramidal sand mounds having slip-face in many directions generally three or more arms that spread out from the highest centre of the sand mound (dune). About 500 m tall sand mound (star dune) has been observed in the Badain Jaran desert in China. The fourth type is **Dome dunes,** which are oval or circular sand mounds. Such dunes are generally without a slip face and generally increase in height. It is a rare dune type and occurs at the far upwind margins of the sand seas. The **parabolic dunes** are U-shaped (blowout or hairpins) sand mounds with convex noses in front (in wind direction) and trailed by elongated arms backside. The longest known parabolic sand dune was reported with a trailing arm of about 12 km long (Kollegger and Grunert 2017). Parabolic dunes also form from transverse or barchan dunes, when the dunes stay immobile for a period to support vegetation to grow on the dune. Some topographical features are presented in Fig. 11.1.

Among other wind-born features, mesas are wide and flat-topped erosional structures bearing steep slopes on all sides. Weak rock material is often withered by wind or water actions until a strong cap-rock material arrives to withstand the erosional process, thus constituting a 'Mesa'. Buttes are similar to the 'Mesas' but are larger and narrower thus forming a pillar-like structure standing alone or in groups. 'Playas' are dry lakes where the shallow water layer is lost because of high rates of evaporation in the region. These are often characterised by mud cracks and salt pans. Here problem of salinity remains a source of concern for ecologists. Alluvial fans are formed due to sediment deposition along the slope, typically at the mouth of shallow canyons when these

Fig. 11.1 Shifting sand dune near Bandra Village (left), the riverbed of Sukri River with isolated trees of *Prosopis juliflora* offering shade to the domestic animals (centre), and a bare dune near Miyanjhalar (right) in Barmer District of Rajasthan

alluvial fans merge to form a bajada, hence forming a wide alluvial rolling surface. The pediments are gently sloping low structures that often occur adjacent to the mesas, inselbergs, or mountains as a result of erosion of the mountains near their base. These are often covered by depositions leading to alluvial fans or bajadas. Inselbergs are residual uplands which survived the erosion process over the years and in other words these are isolated, rocky patches with steep slopes that rise above flat desert plains (Parsons and Abrahams 2009).

In addition, various other features like ephemeral watercourses are of immense importance as these streams provide conducive environments for riparian habitats to develop amid the desert. Shady slopes and rock crevices along with boulders provide hiding places or shelter for many species of invertebrates and vertebrates (Whitford 2002). Small sites among the rocky fields often support species which require relatively more moisture and low temperatures. Sedimentary rocks with terraced slopes support various vascular plants starting from some grasses to the occasional appearance of large trees. Variation in vegetation communities is a function of variable geomorphic habitats and their age, though variations in water availability also influence vegetation structure by regulating soil texture (Meng et al. 2018). The vegetation distribution in the bajadas of Sonoran Desert has also been affected by edaphic and topographic factors like slope, soil texture and soil salinity (Phillips and Mac Mahon 1978). The riparian vegetation along the ephemeral water courses provides habitat to many species which are absent in otherwise desert areas. For example, *Senegalia senegal* shows very good growth along the streams and water channels originating from isolated hills in the Thar Desert, India.

11.3 Soil and Water Conditions

Soils vary in their origin, structure and physical and chemical properties depending upon available rocks and the desert environments, which are frequently encountered by large diurnal changes in temperature (Rücknagel et al. 2013). The large variations in diurnal temperature affect the mechanical and physical weathering of rocks and wind-blown sands. The physical breakdown of the rocks results in somewhat large fragments, which are then gradually disintegrated into finer particles under chemical weathering. Existing flora in the region also plays a significant role in the process of soil formation by promoting disintegration and chemical weathering of rock-fragments and consequently soil organic matter enrichment by added litter and fine root turn over (Phillips et al. 2008). Soil types observed in hot desert areas are formed in an extreme environment leading to the development of non-flushing types of soils that accumulate soluble constituents like soluble salts, gypsum and carbonates (Khormali and Monger 2020). Thus, soils of this region are often limited by the availability of hardpan layer in subsurface layers affecting water-holding capacities of the soil and rooting depth. These hardpans consist of calcium carbonate aggregates or salicaceous materials, which occur continuously in 5–80 cm below the soil surface. However, metres of sand deposition over the ground—a general feature in sand drifting areas under Aeolian activity may increase the depth of the hard pan (Verheye 2006; Williams and Eldridge 2011). Processes of sand deposition in larger areas have a marked effect on the soil nutrient pools by way of increasing the accumulation of plant-available nitrogen on the organic crust. However, deposition, accumulation, or decomposition of plant material (litters) affects the availability of soil organic matter and thus soil fertility. However, the limited organic matter available in the soil is lost quickly under cultivation and cultural operations. Further, soils show rigorous weathering of minerals on formerly exposed surfaces and are well drained with a high rate of nutrient leaching (Chen et al. 2000; Sharma and Sharma 2011). Pavements, varnish and biocrusts are other important features in desert regions. In this, biocrusts help in fixing nitrogen, influence surface runoff, and protect soils from wind erosion with an additional role in the biomineralisation of carbonates (Khormali and Monger 2020).

Water availability has an important bearing on human settlements and well-being. It supports the two key terrestrial functions: production and recycling of nutrients. In desert regions, most of the water received through rainfall is lost by evapotranspiration, whereas groundwater recharge is negligible and confined only by seepage of water through the soil profile (Murray 2013). Surface runoff is negligible, and soil moisture storage and ground water recharge are highly variable, whereas withdrawal of groundwater far exceeds the rates of recharge in most of the desert areas (Jain et al. 2007; Mzezewa and van-Rensburg 2011). In north-west India, a simulation study indicates considerable increases in surface runoff by 2070–2099, with a mean change of 189 mm/year for 2 °C climate change (Murray 2013). However, rainfall is shown to have an important bearing on runoff generation, and the degree to which is also influenced by the quantity of surface runoff (Guhathakurta et al. 2011). In many areas, water availability for general use is also affected by salinity. The surface water available to the people living in the desert regions is confined to some large rivers in the region and originates in areas of higher elevation (Liniger et al. 1998). Some of these rivers are the Nile in the Sudan and Egypt in Africa, the Tigris in Iraq, the Indus in the Indo-Pakistan region, the Senegal and the Niger in African region and the Colorado in the western United States (IUCN 2008; FAO 2011). The scarcity of water in the region leads to the clustering of human settlements along these rivers' courses and/or around some water sources like springs, wells and oases. About 50% of the world's population is going to live in areas of high-water scarcity by 2030 under the existing climate change scenario. For this, adopting traditional methods of harvesting, conserving and transporting water, specialised land management techniques and devising structures to capture and retain precipitation, or to promote groundwater recharge may be important climate change adaptations in the desert regions.

11.4 The Indian Deserts

11.4.1 Cold Desert

Cold deserts in India are situated in the western Himalayas and range between Ladakh in north-western to the Kinnaur of Himachal Pradesh in the south-eastern. It comprises the Leh and Kargil Districts of the Ladakh region, Lahaul and Spiti Districts and a part of the Kinnaur District in Himachal Pradesh. This cold desert biome is characterised by harsh climatic conditions because of its location on the leeward side of the Himalayas. This rain-shadow area is challenging to the annual south eastern monsoon winds resulting in low rainfall and desert-like conditions. These deserts are situated at very high elevations ranging from 3000 to 6000 m above average sea level that adds to the coldness of the regional environment. Ladakh is located at an elevation of 2900 m to 5900 m above msl and temperature range of −30 °C to −70 °C resulting in the coldest region of the earth. This is almost barren region and the population is sparse that inhabits mostly along the river banks of Indus, Nubra, Changthang, Zanskar and Suru in different valleys (Butola et al. 2012; Raj and Sharma 2013). Annual precipitation is <50 mm most of which is received in the form of snowfall mainly in the winter season. In the region, afternoon hours are characterised by fast-blowing winds at 40–60 km h^{-1}. Thus, blizzards, snowstorms and avalanches are the common features in the region. Soils are gravely and sandy loam on the alluvial fans to silt clay loam in the Indus plain region. Loose sandy loam soil with a high percentage of gravel fraction, low water holding capacity, high bulk density and low fertility is because of the irregular distribution of the sparse vegetation and altitudinal gradient (Charan et al. 2013; Butola et al. 2012). Some soil patches exhibit alkaline conditions with organic matter concentrations of 0.08–2.98% (Singh and Gupta 1990). Soil water is in frozen condition during winters and very low during the summer months. This allows a very short growing period making the areas bare for most of the parts of the year. Water availability is low and comprises glacier-fed streams. Shifting of tree lines, migrating dunes and changes in soil moisture conditions are the indicators of a shift in precipitation pattern from winter to summer. These altogether have increased the process of desertification in this trans-Himalayan region (Sharma 2019). Sometimes, it leads to a very high and intense rainfall resulting in flash floods, cloud bursts, etc. Such events promote erosion of the fertile soil and demolition of agriculture lands, trees and forests in areas, which support sparse vegetation. This transforms fertile lands into barren and non-productive lands. A recent study indicates that about 80% of the glaciers in the region are smaller than one square kilometre in area, whereas these glaciers are critical to the people, whose drinking water supply is mainly met by glaciers and snow melt.

11.4.2 Thar Desert

The Indian desert—also known as 'Thar desert' encompasses 200,000 km^2 area. It occupies Rajasthan and extends partly up to Haryana and Punjab province in the north and northern Gujarat in south. Eastern Sindh province and the southeastern part of Punjab province in Pakistan are also part of this desert. The Cholistan desert near the Thar Desert spreads into the Punjab province of Pakistan. Three principal landforms in the Thar Desert are: (i) areas predominantly covered by sand, (ii) the plains area with isolated hills including the central dune and (iii) the semi-arid area in the west of Aravalli hill ranges. The slope of this desert is towards the west but unevenness in topography is mainly due to sand dunes of different heights. The sand dunes in the south region of the desert are up to 152 m in height. In the north region, the dunes are lower in height and rise to 16 m above the ground level. The climate of this desert is characterised by low and erratic rainfall, high diurnal and annual temperatures, low moisture and high wind speed (Pant and Hingane 1988). Summer months of March to June generally exhibit the maximum temperature, which varies from 40 to 45 °C, but reaches up to 51 °C occasionally. Night temperatures range between 20 and 29 °C. January is the coldest month when temperatures decrease to −2 °C at night. In western-, northern- and eastern Rajasthan, the occasional secondary western disturbances cause light rain, but increased wind speed causes a chilling effect during the winter months. The Thar Desert region receives <100,400 mm rainfall annually. Based on rainfall and edaphic factors, the Thar Desert is divided into four subdivisions (Sehgal et al. 1992; Sharma and Mehra 2009). These are (i) the transitional plain of Luni basin, comprising Pali, Jalore and parts of Barmer, Jodhpur, Ajmer, Sirohi and Nagaur Districts; (ii) transitional plain of inland drainage encompassing Sikar, Jhunjhunu and northern Nagaur District; (iii) the agriculturally rich district like SriGanganagar and Hanumangarh Districts; and (iv) arid western plain that includes Jaisalmer and parts of Barmer, Jodhpur, Bikaner and Churu Districts of Rajasthan.

11.5 Floristics of the Deserts

11.5.1 Cold Desert

Vegetation of the cold deserts is dominated by grasses, which tend to grow in clumps and are hence called bunchgrass. The terrain is also seen with shrubs and brushes among which the sagebrush is common in the United Kingdom. A unique two-leaved shrub *Welwitschia mirabilis* grows in the Namib desert in south-western Africa (Bombi et al. 2020). This species produces colourful cones and grows between 0.5 and 2 m in heights. The availability of trees is very low in these cold deserts. Some of them to mention are camel thorn acacia (*Acacia erioloba*) that grows in the Gobi Desert, a small and bushy sexual tree (*Haloxylon ammodendron*) grows in the Turkestan desert, Pistachio trees (*Pistacia vera*) are common in the Iranian desert, and tamarugo trees (*Prosopis tamarugo*) grows in Atacama Desert. *Prosopis tamarugo* produces an edible fruit used by livestock as well as human beings. Cactus species are not common in the cold deserts as they occur in the hot deserts. However, the giant cardon cactus (*Pachycereus pringlei)* is found growing in the Atacama Desert of South America. Several medicinal plant species of economic importance are also found in cold desert areas of Ladakh. Some of these species are sea buckthorn (*Hippophae rhamnoides* L.) and apricot (*Prunus armeniaca* L.). This region exhibits semi-desertic to desertic vegetation depending on the prevailing climatic condition. In some areas, the general desert conditions are changing because of the occasional escape of the monsoon winds that bring mild rainfall. Annual and perennial herbs and some stunted shrubs and bushes dominate in the alpine and high alpine zones of Ladakh (Kumar et al. 2016). The plants start growing at the beginning of summer when the melting of snow provides sufficient soil moisture. The vegetation is in full bloom in the month of August but starts disappearing by the end of September months. The mountain slopes, the meadows and the alpine grasslands provide a stunning view of flowers of barren mountains of cold deserts. *Aconitum violaceum*, *Aquilegia fragrans* and *Capparis spinosa* are some **alpine mesophyte**, the zone which exhibits high humidity and rainfall and is found in the Suru valley in the Ladakh region. The **oasitic vegetation** in the cold desert area of the western Himalayas represents a variety of exotic and indigenous species. Such vegetation grows near habitation, along water channels, streams, nallahs (drains), or in moist areas and Kargil and Leh regions are represented by such types of vegetation. Little or no rainfall, low humidity, extreme fluctuation of diurnal temperature and high-velocity winds are characteristics of the region involving **desert vegetation** zone, where maximum numbers of plant species show long roots and small leaves. *Salix elegans, S. alba, S. fragillis, S. sclerophylla, Populus alba, P. euphratica, P. ciliata, Juniperus polycarpos, J. communis, J. recurva, Astragalus* spp., *Artemesia* spp., *Myricaria bracteata, Hippophae rhamnoides, Tamarix gallica, Elaeagnus angustifolia, Caragana pygmaea, Atriplex crassifolia, Ephedra gerardiana, Rosa webbiana, Medicago sativa* and *Haloxylon thomsonii* are the woody perennials with direct relevance of the livelihoods of local population. Medicinal and aromatic plants of this region are *Aconitum rotundifolium, Arnebia euchroma, Ephedra gerardiana, Ferula jaeschkeana, Hyoscymus niger*, and have livelihood options for tribal communities in this desertic environment (Butola et al. 2012; Tewari and Kapoor 2013; Chauhan et al. 2020).

11.5.2 Hot Deserts

A diverse range of macrohabitats and corresponding vegetation in the desert are because of large topographical, geological, soil type, seasonal patterns of rainfall, fires, herbivore pressure and human management coupled with water scarcity in such regions (Goudie 2002; Iwara et al. 2011). Deserts are in general interrupted by varying sizes of streams, rivers and floodplains, which are considered to be more productive because of increased water availability as compared to the parts of the desert areas (Free et al. 2013). These dry riverbeds are beneficial in recharging ground waters help in the establishment of perennial vegetation and support many ecological functions (Arthington 2012; Steward et al. 2012). Based on such variations, one can notice four broad categories of adaptation in these drylands. These are:

1. **Drought escapers** are migrating animals that move in search of water and foods. Insects escape into the egg or pupal stage, whereas the plants escape some phenological stage (fruiting). All these biota waits until favourable weather condition returns.
2. **Drought evaders** are a group of plants like the salt bush, which develops deep and efficient root systems and reptiles that hibernate themselves belowground to avoid the heat or unfavourable climatic conditions.
3. **Drought resistors** are the group that includes cacti, which store water in plant system, and the animal-like camels that minimise water loss.
4. **Drought endurers** are groups including plants category that go for the dormant period and animals that go aestivate (e.g. frogs) occasionally.

Qureshi and Bhatti (2008) recognised five distinct habitats in Nara desert area in Pakistan based on topographical features and distinct vegetation composition. The habitats include: (i) top of the sand dune called crest; (ii) sloppy area/swale/flanks position; (iii) sandy plain or low-lying flat areas; (iv) water bodies or wetland habitats; and (v) saline-sodic areas. Likewise, Prakash (1994) has identified four major types of habitats in the Indian desert on the basis of flora and fauna. In this, the Thar Desert is represented by about 700 plant species. These species belong to 91 families and 385 genera (Shetty and Singh 1993; Sharma and Pandey 2010). In terms of number, the most dominant family is Poaceae followed by Cyperaceae with 125 and 35 species respectively. Among these species, 45 species are considered as rare and/or endangered (Fig. 11.2). In addition, gymnosperms (*Ephedra foliata*) and bryophytes (*Fumaria hygrometrica*, *Marcantia palmata*, and *Riccia robusta*) have also been reported in some of the areas of the Thar Desert (Choyal and Sharma 2011). Most of the grass species of this region are xerophilous. The grasslands comprise the species like *Eragrostis* species (*E. tremula* and *E. minor*), *Aristida adscensionis, Cenchrus biflorus, C. ciliaris, Cymbopogon* species, *Cyperus* species, *Eleusine compressa*, *Panicum* species, *Lasiurus sindicus, Aeluropus lagopoides* and *Sporobolus* species depending on the habitat types (Mares 1999; Singh 2005). Interestingly, some of the species act as indicator species for varying edaphic conditions in the region like dune, saline-alkaline, calcareous soil and rocky habitat. Electrical conductivity and soil phosphorus have been found to control the positive to negative associations between the plant species of the region, that is, positive interaction between *Blepharis sindica*—*Lasiurus sindicus* and *Lepidagathis cristata*—*Sonchus asper* for example (Mathur and Sundaramoorthy 2013). Plants of different habits include trees (*Prosopis cineraria*, *Tecomella undulata*, *Maytenus emarginata, Ziziphus mauritiana* etc.), shrubs species like *Acacia jacquemontii, Capparis decidua*, *Commiphora wightii*, *Euphorbia caducifolia*, *Grewia tenax*, *Ziziphus nummularia*, etc., the perennial herbs (*Tephrosia purpurea, Farsetia hamiltonii, Indigofera linnaei*, *Trianthema portulacastrum*, *Echinops echinatus*, *Convolvulus microphyllous, Solanum surattense* etc.), annual herbs (*Argemone mexicana, Alysicarpus monilifer, Fagonia cretica, Vernonia cinerea, Arnebia hispidissima*, *Heliotropium marifolium, Indigofera cordifolia*, *I. hochstetteri, Phyllanthus amaras, P. neruri*) and ephemerals (*Cleome viscosa*, *Polygala erioptera*, *Corchorus tridens, Tribulus terrestris,* etc.).

Fig. 11.2 *Ephedra foliata* as pendent on *Capparis decidua* (left), *Commiphora wightii* growing on a rocky site (centre) and **Chascanum** *marrubiifolium* growing on gypseous soil (right)

11.6 Adaptations and Biotic Interactions

11.6.1 Plants Adaptation and Survival

Plants of cold desert regions have many ways of adapting to dry, cold and salty climates. One way that these plants deal with the dryness is by losing some or all their leaves when there is dryness or very cold. Other plants are **halophytes** adapting to salty environments. Plants in cold desert areas, particularly high altitude exhibit several environmental, morphological and physiological adaptations. These adaptational changes help the plants to counteract the impact of the harsh climate prevailing in the Ladakh region. Some of them are as follows:

1. Extensive root system.
2. Changes in growth characteristics.
3. Changes in physiological characters.
4. Modified reproductive strategies.

Roots of most of the plant species are extensive and very deep so as to absorb ground water from the deeper soil layers. The extensive root systems are also useful in withstanding towards strong wings, and snow blizzards and in escaping the damage caused by subzero ambient winter temperatures as well. For example, the horizontally spreading roots of seabuckthorn (*Hippophae rhamnoides* L.) in Lahaul valley extend up to 2.5–6.8 m and generate 9–27 numbers root suckers per root (Singh and Dogra 1996). Herbs followed by some stunted and highly branched shrubs or bushes dominate the vegetation composition of the cold desert areas of the Ladakh region. The dwarf nature of the vegetation of this cold region is because of suppression of internodes due to slow growth and a very short growing period, that is, May to October. It is very much related to the extremely low temperature, high solar radiation and impedance in sap flow in the plants. In the case of physiological behaviour, the important characteristic of the plants of the high-altitude area is frost resistance. The freezing injury in plants is primarily caused by the formation of ice crystals within the plant cell. Adaptation to freezing temperature reduces ice crystal formation in these plants (Stegner et al. 2020). For successful survival of a species in such cold environmental conditions is an effective mode of propagation such as reproduction and perennation during adverse climatic conditions. Effective reproduction and dispersal of seeds enable the plants of cold desert to survive in the form of sparse vegetation. Reproduction in the plants of cold desert occurs both by sexual and vegetative methods.

In hot desert areas like that in the Thar Desert, most of the vegetation is confined to the area with water availability. Extreme cases of aridity that had led to the development of groves of various species including *P. cineraria*, *Capparis decidua*, *Z. rotundifolia*, Salvadora species were because of varying adaptations. Perennial plants of hot deserts remain inactive during the dry spells and become active after receiving rain during monsoon. Some annuals start germination by absorbing water immediately during the onset of rain and complete their life cycle in very short time. Some adaptational changes in plants of hot deserts are as follows:

1. Deep and extensive root systems.
2. Stem modification.
3. Leaf modification.
4. Physiological adaptations.

Desert plants grow in a water scares areas and hence exhibit a deep root system as a mechanism for survival. The roots grow deep in the soil profile and absorb soil water, which is then conveyed to the upper parts of the plants. The varieties of plants that adapt themselves by developing very long and deep root systems are known as phreatophytes. Mesquite (*Prosopis* species) tree or other species with roots more than 15 metre long is an example of phreatophyte (Singh et al. 2017). *Calligonum polygonoides*—a shrub of Indian desert spreads its root more than 6 times of its canopy as an adaptation and mines more than 700 tonnes of soils (Fig. 11.3). Some plants including cactus and Euphorbias have shallow rooting systems that extend radially to absorb as much soil moisture as possible, particularly at the time of seasonal rains. Some plants exhibit both radial and long taproot systems to maximise soil moisture absorption, that is, *Prosopis juliflora*. The horizontal roots of up to 20 m help absorb and store soil moisture after rainfall and perhaps even also dewfall for its best-fit survival. In a mechanism called 'hydraulic lift,' the deeply available soil water is lifted to top dry soil layers through the extensive deep roots, especially during night, and helps in improving water balance and nutrient cycling (Singh and Shukla 2011; Davies et al. 2012). This may also help to keep the shallow-growing lateral rootlets alive during dry spells and assist in quickly absorbing soil moisture after rainfall. The second most thoughtful characteristic of desert plants is their ability to store water in different organs like roots, stems and leaves.

Opuntia—a cactus species, accumulates water in the green, flattened, succulent stem known as phylloclade. These leaves like structures in these plants perform photosynthesis to make food and function in the same way as a typical leaf of regular plants do. *Welwitschitia mirabilis*—a brown colour plant of the Namibian desert, is adapted to strong solar radiation and temperature because of stem thickness (spreading on soil surface in about 20 feet) rather than its height and survives up to 2000 years. Sternberg and Shoshany (2001) observed a decrease in foliage to wood ratio from the periphery of the crown to the interior of the

Fig. 11.3 Extent of root distribution of *Calligonum polygonoides* (phog) shrub of Indian Thar Desert

trees and shrubs. In this, foliage biomass and water were found limited mainly to the top 30 cm of the crown in *Quercus calliprinos, Phillyre alatifolia, Pistacia lentiscus, Cistus creticus, Coridothymus capitatus* and *Sarcopoterium spinosum*. The plants of the desert region with small foliage and waxy surfaces were to reduce the rate of transpiration, which helps in conserving water in these species. In some xerophytic plants, the stomatal opening remained close during day time and open only at night. There are no true leaves in these plants rather the leaves are modified into prickly spines not only for reducing water loss but also for protecting themselves from herbivory. Many plant species in desert regions adapt themselves by developing thorns. To survive in the existing harsh environmental conditions, some plants develop specific adaptive traits. Plant adaptations for salinity tolerance are avoidance of saline habitats, modification in root morphology and retention of some toxic chemicals in vacuoles or granular compartments and osmotic corrections. Apart from the reverse flux from the leaves to the roots by phloem tissues may also help in diluting excessive salts, whereas the mechanisms related to seed preservation, protection and germination are adopted by the plants.

11.6.2 Biotic Interactions

Abiotic factors are more often held responsible as compared to biotic constituents for the determination of structure of a plant community. However, biotic interactions also play a major role in the determination of desert communities. These interactions are either positive or negative. Thus, presence or absence of a species can change the structure of existing communities (Allen et al. 2008). As compared to above-ground interactions, below-ground competition is more prevalent in deserts where the roots of some species are vast and robust and do not easily let other species establish and grow or are sometimes responsible for regular distribution in forested habitats. In contrast, some strong associations are also visible in desert areas where a group of 3–8 species are found associated and supporting their survival (Singh 2008a).

Because of the presence of a wide range of animals, animal husbandry as an important occupation in the desert regions, interactions between plants and animals are central as these interactions play significant roles in the evolutionary history of the region (Singh 2018). Some other biotic interactions observed to play important role in deserts are pollination and seed dispersal. However, studies on biotic interaction in deserts still have scope as the exact contribution of biotic and abiotic interactions is yet to be determined.

Net primary productivity depends mainly on water availability and autotrophic life in the area. However, studies in the Sonoran Desert reveal the productivity of different vegetation types that is affected by temperature or precipitation gradient, which in turn are influenced by topography of the region (Whittaker 1975). Likewise, salinity, edaphic conditions, nutrient deficiency and substrate instability are often held responsible for the low productivity of deserts. Productivity in seasonal plants varies temporally as well as spatially. Productivity in desert plants ranges from <100 to >2000 kg dry weight per hectare per year (Lathrop and Rowlands 1983). Underground production of desert plants is often more than the production above-ground and it may be six times or more in some of the plants (Holmgren and Brewster 1977). The nutrient most limiting in desert productivity is nitrogen as the primary production is reduced in the absence of nitrogen even during the wet months. Annuals also follow the same trend in case of nitrogen and phosphorus deficiencies. Conservation of nutrients within the plant organs and the ability of the plant to rapidly cycle the nutrients are the two adaptations which help desert plants to improve productivity (Freidel et al. 1980). Several other factors which may have an impact on nitrogen availability and productivity of desert ecosystems can be microbial biomass, decomposition rate, presence of nitrogen-fixing plants, soil fauna, excretion by animals, etc. (Singh et al. 2000; Tripathi et al. 2013).

11.7 Cultural and Scientific Wealth

11.7.1 Cultural and Economic Activities

Deserts of the world are inhabited by about 8% of the total population of the world. Humans inhabited deserts since ancient times and their livelihood largely depends on natural resources available there (Fig. 11.4). That in turn is governed by ecological aspects of the deserts. Despite harsh environmental conditions arid areas were known to support civilisation including Indus and Harappan that flourished during 3300–16,000 BCE (Dhir et al. 2017). The earliest inhabitants of the desert were hunters and gatherers and it was the culture and means of living for early humans. The lifestyle of early humans as hunters or gatherers was based on their habit of hunting animals and foraging for food. But during scarcity, these people used to migrate and start hunting at new sites hampering the ecological balance of the area, which further leads to an imbalance between biotic and abiotic resource utilisation. Knowledge of natural resources and climatic conditions proved helpful for survival in this kind of livelihood practice, but excessive hunting and gathering and harvesting from nature appeared injudicious. This type of livelihood practice has now been somewhat restricted because of the increased availability of alternative means of livelihood under technological advancement (Rowley-Conwy and Layton 2011).

Rainfall is a vital factor regulating agriculture, grazing, etc. but other parameters are high temperatures and strong winds. Although humans adapt to these conditions, they show behavioural, cultural and technological adaptations instead of morphological and physiological changes. Major livelihood practices in deserts were hunting and gathering in the past but with time domestic livestock herding and agriculture prevailed along with mining activities and tourism. Thus, the survival of humans in desert ecosystems primarily depends on an adequate supply of food, water and shelter.

Fig. 11.4 A village and a hamlet in Thar Desert region of India

Recent technological advancements have made the desert slightly comfortable. Industrialisation and an adequate supply of goods and services have improved the lifestyle of desert dwellers (Dhir et al. 2017). Besides, solar energy and wind energy utilisation in the region have brought some good changes and comfort in the lifestyle of the people. Along with the development and use of domestic crops and animals, early humans started herding and carrying out agricultural activities in varying proportions. In extreme conditions, the human population of deserts is regulated by changing climates, and technological advancements in the field of agriculture, irrigation and renewable resources. Many institutions including government and non-government are engaged in resource management and agricultural and horticultural activities, which are important for desert development (Singh 2017).

Pastoralism is a type of livelihood practice wherein animals are domesticated and utilised for various products such as milk, skins, wool and meat. Camel is the earliest domesticated animal in the hyper-arid deserts and is physiologically more advanced than other commonly domesticated animals in the region. In modern times, camels have been replaced by cattle (buffaloes also in the Thar Desert) because of their better market value. People have now shifted towards other livestock like cows, buffalos, sheep and goats as they provide better income. Animal husbandry is the second most important occupation in western Rajasthan. According to 2019 census, Rajasthan has livestock population of 56.80 million, which indicates a decrease in livestock population by 0.9 million as compared to the 2012 census. Census 2019 comprises 13.90 million cattle, 13.70 million buffaloes, 7.90 million sheep, 20.84 million goats and 0.29 million camels as compared to 13.33 million cattle, 13.0 million buffaloes, 9.08 million sheep, 21.66 million goats and 0.33 million camels in 2012 census. Goat and sheep combined contribute to about 60.8% of the total livestock population in Rajasthan. Mobility is important for successful pastoral activity in the deserts. Early pastoralists used to follow rainfall patterns and different grasslands that picked up growth in certain areas in some years depending on rainfall (Henschel et al. 2005). After a good rain, pastoralists and their herds used to gather in the area where sufficient rain offered good grazing ground (Kinahan 2016). Nowadays graziers rarely live in deserts rather use to move in other areas where the availability of fodder (grasses and herbs) is generally high and herd (gathering) their animals at the places to sell their animal products in urban markets (Fig. 11.5). Further, the population of buffaloes is increasing to meet the local as well as the urban milk requirements. However, the shrinking of rangelands and pastures and their grazing capacity on which domestic and wild animals depend are leading to overgrazing. Increased livestock numbers bring about overcrowding of grazing lands, and consequently trampling, reduced infiltration and accelerated surface runoff and soil erosion.

Agriculture has been central to the people of dry regions since this practice was started. A large chunk of the population depends on agriculture and animal husbandry as the two main occupations. Cropped area, cropping pattern and yield of agricultural crops are governed by soil texture and depth, terrain slopes, water availability, severity of natural hazards, rainfall pattern and socioeconomic conditions of the local population. The evolution of irrigated agriculture in different places depends on water availability and crop adaptability. Rainwater harvesting, drip and sprinkler methods of irrigation, mechanised farming, crop varieties and access to ground water are a few modifications that took place recently and these technologies provided better returns to desert dwellers (Singh et al. 2017). Paliwal families in Jaisalmer areas of western Rajasthan are accustomed to harvest rainfall on their

Fig. 11.5 Movement of herds of livestock in western Rajasthan. Traditional agriculture involving camel in ploughing

farmlands from the upland areas during the monsoon period and used to cultivate crops like wheat, gram and mustard during the Rabi season on the conserved moisture, what is called 'Khadin System' of cultivation (Fig. 11.6). Oases are also known for cultivating many types of crops such as dates, vegetables, cereals and cash crops. Other crops include salt-tolerant varieties of barley and fodder crops such as alfalfa. Many local cultivars of these crops have also been developed to suit the local conditions (Moore et al. 1994). The main agricultural crops of the Indian desert are cereals, oilseeds, pulses and others. Cereals include bajra (*Pennisetum typhoides*), jowar (*Sorghum bicolor*), maize (*Zea mays*), wheat (*Triticum aestivum*) and barley (*Hardenum vulgare*) in different locations and their yield vary considerably. Oilseed crops include sesame (*Sesamum indicum*), mustard (*Brassica nigra*), taramira (*Eruca sativa*) and castor (*Ricinus communis*). Pulses constitute about 10% of the total area under cultivation and include moong (*Vigna radiata*), moth (*V. aconitifolia*), cowpea (*V. unguiculata*) and gram (*Cicer arietinum*). Besides these, several other cash crops like cotton (*Gossypium* spp.), chilli (*Capsicum annum*), methi (*Trigonella foenum-graecum*), Isabgol (*Plantago ovata*), groundnut (*Arachis hypogea*) and cumin (*Cuminum cyminum*) are also cultivated in Thar Desert, which has become one of the highest groundnut-producing area.

Though deserts sustain relatively a small human population, but the environment conditions of the region make these people prosperous through increased availability of some non-renewable mineral resources. These mineral resources are exported to non-desert regions frequently and include gypsum, common salt, limestone and nitrate compounds of sodium and potassium. Numerous labours are also involved in excavation, loading, transportation and other related activities. However, the expansion of mining activities and disposal of large quantities of debris are posing a severe threat to the desert ecosystem. The impact of physical disturbances to the top soil during stripping, stock piling and reinstatements results in soil degradation by reduced soil structure, accelerated soil erosion and leaching, soil compaction, decrease in soil pH, heavy metals accumulation in the soils, depletion of soil organic matter, reduction in the soil available nutrients and cation exchange capacity, decreased microbial activities, soil fertility and productivity (Mensah 2015).

Fig. 11.6 Khadin system of cultivation: an ancient system of water conservation for crop cultivation with a dug well down side the slope

11.7.2 Centre of Recreation

Deserts also contribute extensively to culture and tradition that helps promote tourism in the region. Encampment, dune and hill-walking and sometimes fishing and hunting are very much popular in desert areas. In search of sunshine, warm weather, varying landscapes and remarkable flora and fauna are other attractions. Off-road vehicle for transport of tourists on the sand dunes like buggies and bikes or ordinary vehicles are becoming popular nowadays. Movies and books based on deserts have immense contributions to the rise of tourism in the region. The Thar Desert has become a recreational centre in view of desert festivals, which are celebrated with great zeal every year, particularly during the winter season. The people of the desert region use to dance and sing unforgettable songs of bravery (valour), romance and tragedy following dressed brilliantly in hued costumes. The fair of the desert includes puppeteers, acrobats and locally available folk performers. Camels are central to this festival, where the rich and colourful folk culture of Rajasthan is visible. The best example of such tourism activities is an influx of tourists during the winter season in the Sam sand dune area of Jaisalmer or another dune in the region making the desert a vibrant place of living (Tawade 2018; Fig. 11.7).

Nowadays, many desert areas like 'Thar Desert' have become common destinations for tourists. There may be chances of increased inflow of public because these deserts may work as rehabilitation centres for patients suffering from diseases like asthma or arthritis. Although tourism is going to be the main source of income generation for the desert dwellers in various deserts globally, but again there must be some regulations, which should be implemented stringently to protect these deserts from degradation. For example, the loosening of surface soils increases erosional processes by wind and water and leads to the degradation of land, while soil compaction reduces the rate of water infiltration. The negative effect of vehicle movement in desert areas has been observed in a study indicating that the tracks made by the single visit of the outdoor vehicles through lichen-rich areas remain visible for few decades in Namib desert (Seely 2004).

11.7.3 Act as Corridors

The trading convoys—called caravans, and migrant birds both use deserts as their corridors. These desert corridors connect one non-desert area as the starting point to another non-desertic region as the destination. While using the desert corridor, they use to link up with the biotic and a biotic environment of the desert. Earlier evidences indicate that traders of western countries like Arab and Afghanistan cross the Thar Desert to reach Delhi, Agra and other cities as the trade centre. Likewise, the Garamantes—an ancient people and descendent of Berber tribes or Saharan pastoralists that settled in the Fezan region of southwestern Libya about 3000 years ago, and used to control trans-Saharan trade (Nikita et al. 2012).

11.7.4 A Place of Research

Various experiments are being conducted in many desert regions of different countries to test science-driven rovers' operations to improve human–robot interactions in simulated conditions in deserts like those on the celestial bodies such as Moon and Mars (Fosum et al. 2018). India did nuclear explosion in the Thar Desert near Chandhan in the Jaisalmer District. Low levels of pollution, atmospheric humidity and

Fig. 11.7 Tourist influx to visit Sam sand dune and fossil reserved in Akal Fossil Park in Jaisalmer

air turbulence make the desert region suitable site for such experiments. Further, it is easy to find almost plain and unhindered (treeless) areas with reduced turbulence of air-flow and limited human interference in deserts. Indeed, some largest and most expensive astronomical instruments of the international astronomical community have also been placed on mountain tops of the deserts. This is because of reduced atmospheric layers between the telescope and the celestial object. For example, the very large telescope array (VLT) of the European Southern Observatory (ESO) is located on Cerro Paranal, the high area of the Atacama Desert in Chile. This consists of four individual telescopes that are equipped with a primary mirror measuring about 8.2 m diameters (Gregersen 2023). Likewise, the Southern African Large Telescope (SALT) is located near Sutherland—a town in the semi-desert region of the Karoo in South Africa and is a facility of the South African Astronomical Observatory, South Africa (Buckley et al. 2005). The SALT is operating since 2005, and other observatories are being scheduled for other desert sites.

11.7.5 A Place of Treasure of Paleontological Evidences

Deserts are characterised by low vegetative cover, shallow soils, high aridity and large areas under exposed rock, these all make the region better to work as a natural laboratory. Scanty rainfall affects chemicals leaching negatively and favours the preservation and discovery of fossils available in the region. These all together help in interpreting the evolutionary history of different biotic components. Akal Fossil Park near Jaisalmer in Rajasthan is indicative of the presence of gymnosperms in the Thar Desert. Paleontological records of pollen grains of the plants of the family Graminaceae, Chenopodiaceae, Amaranthaceae, Cyperaceae, etc., and evaporite minerals like gypsum and halite in the soil profiles indicate that this area remained brackish to saline even during the mid-Holocene, that is, Palaya region of Bap area of Jodhpur and the surrounding vegetation was dominated by grass (Deotare et al. 2004).

11.7.6 A Place of Physiological Adaptation

Societies in desert region have a long record of managing the impacts of climate-related events like droughts and floods of varying levels. By working on these records, additional adaptation measures may be adopted to reduce the adverse impacts of projected climate change and variability. Researches in dry regions reveal that physiological adjustment assists not only the plants but also the mammals and the birds to animate in an environment where water is scars and atmospheric temperature is high (Bhatnagar et al. 2011; Yan et al. 2013). But the adaptations in humans are solely culture and behaviour based. This is the main driving factor for the researchers to conduct research on the physiological responses of animals and plants to the surrounding environment. These studies may determine the potential of biota in familiarizing, rather than adapting to this stressful environment (Shkolnik et al. 1980). The types of adaptations acquired by humans and the livestock inhabiting the desert regions also compel the researchers to observe the damaging effects of extreme temperatures, blasting winds, chronic dehydration, and food scarcity on humans and the domesticated (like camels, sheep, goats and donkeys) as well as wild animals in the region (Singh 2018).

11.7.7 A Place of Ecological Research

Ecological processes in dry regions are governed by environmental conditions, particularly rainfall. Because of the relatively a smaller number of species and their population sizes, the desert ecosystem looks to be simpler and therefore easy to comprehend as compared to the other ecosystems, which are relatively complex in nature. These features inspire the researchers to use deserts as an outdoor laboratory. Likewise, the concept and theories devised in non-desert environmental conditions can also be tested in desert areas. Evolving through the specific mechanism of sectionalisation of limiting resources empowers many of the desert species to facilitate each other and coexist. This type of association leads to build up of diverse patches of biota in desert regions (Fig. 11.8). Since soil moisture is limited in top soil layer, the coexistence of many annual plant species is not due to drawing water from different soil layers and niche differentiation, rather facilitation by the associated species. Some species like *Peristrophe paniculata* grow well in association with *P. juliflora* and is because of shade and soil fertility improvement by the latter species (Singh and Shukla 2011). Despite low primary productivity in deserts, long food chains with several predators on top have also been observed in the region. Most of the primary produce in the desert are not consumed by the herbivores and converts in to plant litter, which does not decompose readily by the soil microorganisms because of low humidity in the atmosphere and soil as well. Therefore, most of the litter accumulates on the soil surface and is consumed by a large number of arthropods that leads to an increase in soil organic matter and soil fertility (Tripathi et al. 2013).

Fig. 11.8 Associations of *S. oleoides, Clerodendrum phlomidis* and *Cocculus pendulus* (left) and *Rivea hypocrateriformis, Capparis decidua* and *Lycium barbarum* (right) for best fit survival

11.8 Threats to Desert Ecosystem

Increased development activities are the biggest threats to deserts. However, changes in land use patterns in deserts reduce floral diversity in both hot (Singh et al. 2008) and cold desert regions (Oinam et al. 2005). Increasing the area under irrigation for the cultivation of *Arachis hypogaea* (ground nut) and *Ricinus communis* (castor) in the western region of Rajasthan enhances the economy of the villagers but simultaneous increase in the fuel wood consumption reduces the vegetation diversity and the natural habitats of some of the species like *C. polygonoides* and *L. sindicus* (Singh et al. 2008; Davies et al. 2012; Weber et al. 2013).

People are also heading towards the desert for recovery of their health particularly who suffer from various respiratory ailments leading to crowding in desert areas. Air conditioning has also made such an inroad that desert cities began to thrive. It is also fuelled by large quantities of water inputs from the dams on major rivers, that is, the Colorado in America or the Bhakhra dam on Sutlej in India. Water supply by these irrigation projects allows people to plant many species either in lawns or in their agriculture fields influencing microclimate. Many mysterious desert dwellers like *Coccidioides immitis* and *Coccidioides posadasii* have been found in arid and semi-arid alkaline soil and for them, harsh climatic conditions may be right for their growth. The spores of these fungi reach the lungs and cause infections of varying intensity like Valley Fever (Kollath et al. 2019). The increasing incidence of malaria in desert regions is a matter of great concern (Baeza et al. 2013)

An increase in irrigation in coastal agriculture makes the region to become much more prone to contamination by rising water tables due to the intrusion of salty seawater seeps into underground freshwater aquifers. Diversion and damming of large rivers passing through deserts to provide water to the nearby cities and agricultural farms cause downstream water supplies to dry up and will show a serious impact on flood plains (by increasing soil salinity) and river ecology. Salinisation in the Patan area of Gujarat due to the dam on the river Banas in Palanpur is the best example. Road construction, habitation expansion, other developmental works in the nearby areas, and mining are also common in the regions making the desert soil erosion prone. Many of the small hillocks are being removed in the process of making building materials. It may influence the hydrology of the area. Some areas have been put under irrigation using groundwater to cultivate cash crops, which are useful in the near term, but if water is not supplied in abundance to flow off the land and carry away the accumulated salts then those salts will build up, leading to salinisation of the agriculture land or region influencing vegetation composition and agricultural production (Singh et al. 2007; Ahmed et al. 2013).

Uncontrolled grazing over the carrying capacity of the areas can have severe impacts on floral abundance. The grazing animals are often placed in grazing lands in far greater numbers than the land normally supports. In addition to trampling and killing of desert plants, the high level of grazing (animals) alters the composition of indigenous plant communities and the biodiversity (Wuerthner 2023). The increasing population of grazing animals also poses forage competition with local wildlife and hence the greatest threat to the desert ecosystem (Wuerthner 2023).

Change in land uses also affects productivity as well as soil carbon stock. An estimate on woody perennials of traditional agroforestry in the Indian cold desert area indicates the addition of 400 tonnes of carbon through litter and is responsible for sequestering about 70,000 tonnes of carbon in the system (Kumar et al. 2009). A study conducted in western Rajasthan indicated the lowest carbon stock of 6.17 tonnes ha^{-1} in the soils of roadside, but it was almost close to the carbon stock in the soils of other land uses like sacred groves (Oran), grasslands (gauchar), agriculture and forest lands (Singh 2015). The highest level of carbon stock (10.08 tonnes ha^{-1}) was in the fallow land (Singh 2015). However, the effect of spatial variations on soil organic carbon stock is stronger than the effects of land use, though this variation is also related to rainfall and edaphic factors (Singh and Sharma 2017; Venkanna et al. 2014).

Changing landscape, soil and water conservation measures, water harvesting and storage, afforestation and grasslands development along with agriculture, horticulture and animal husbandry, all influence regional ecology and production system (Tewari and Kapoor 2013; Dhir et al. 2017). There may be chances of a change of species in each regional desert. A study conducted in the Chihuahua desert anticipates the replacement of about 50% of species of birds, mammals and butterflies by other species not later than the year 2055 (Peterson et al. 2002). Rising aridity will favour shrubs in grasslands resulting in a new stable ecosystem. Likely changes in plant community structure in many deserts are because of variations in the biochemical pathways of photosynthesis in some desert plant species and are responsive to elevated CO_2 as compared to the other species of the same community (Morgan et al. 2001). Some plant species introduced in restoration or other uses from other regions have also become invasive and threatening the native vegetation. Greening the Indian desert by the introduction of plants like *Prosopis juliflora* or ornamental plants like *Lantana camara* complex, both have become invasive in most of the land use systems. Likewise, the dry river beds in western Rajasthan are also occupied by *Tamarix* spp.

An increase in global temperature is another potential threat but hard to imagine in desert regions. The people of the Indian desert region in western Rajasthan have started realizing a rise in air temperature and heat waves (Singh 2018). As a function of climate change, a warming desert may be worse than some of the hot deserts. Even a small change in atmospheric temperature or rainfall can have significant impacts on agriculture and desert populations (Gobin et al. 2012; Hasan and Öztürk 2014). In some cases, global warming is predicted to increase the area of deserts under desertification, but it is not clear that the desert biota will be benefited from such changes. However, human activities like gathering of firewood or overgrazing of pastures and community lands are converting fragile grasslands into deserts, particularly in desert fringe area—a process called desertification.

The climate-induced increase in atmospheric temperature is increasing evapotranspiration (ET) in desert regions. High ET, low rainfall and severe droughts in combine reduce soil water—a limiting resource on which desert productivity depends. The expected increase in air temperature by 20% in the Thar Desert area will increase ET by 14.8% (Goyal 2004). Since all life forms depend on the plants either directly or indirectly, changes in the biological diversity of the region are inevitable. A study conducted in Rajasthan indicates that the standing biomass of the forests increases with an increase in rainfall, promoting soil carbon stock and litter production and showing positive correlation with species richness and diversity (Singh 2014). Less mobile animals and plant species of low dispersal ability may decline in numbers under the influence of global climate change (Croteau 2010). However, the species with dispersal rates faster than the expected rate of warming will extend their distribution ranges into the areas that had been relatively cooler earlier. Recent mortality in the state tree of Rajasthan, that is, Khejri (*Prosopis cineraria*) may also be linked with an increase in atmospheric humidity under irrigated agriculture and is a point of research.

Plants available in such desert areas are a potential source of new medicines. Global warming may affect some untapped plant species (bioresource) useful in extracting anti-cancer and anti-malarial drugs. Because of a decline in rainfall by 5–15%, the cold deserts like Australia's Great Victoria Desert, the Atacama and the Colorado Great Basin are becoming much more vulnerable. However, some deserts like Gobi and Thar are experiencing an increase in rainfall. There may be chances of an average increase in air temperature by 3 °C annually in some of the deserts. Furthermore, increase in winds that pick up sand grains off the dunes and deposit them in vegetated areas or agricultural lands, thus turning these arable lands into deserts.

11.9 Desert Development Programmes

Because of increasing threats of land degradation and living conditions in deserts and their margins, top priority programs for governments, international organisations and nongovernmental organisations include various activities under combating desertification. Agenda 21 of the United Nations specifically lists the steps to be taken to combat desertification (United Nations 1992). These are:

- Monitoring, modelling, and prediction of desertification for effectively combating desertification and sustaining economic development,
- Strengthening the knowledge base and developing information and monitoring systems for region prone to desertification and drought. This includes the economic and social aspects of desert ecosystems.

- Use of best fit and site-specific technology involving indigenous knowledge in natural resource management, control of land degradation such as sand dune stabilisation, intensification of agricultural production through tree integration focusing on indigenous species etc.
- Involvement of local community for effective long-term solution to desertification. Local people are required to be aware of the types of benefits and participate fully in the design and implementation of the programmes.
- Environmental education to reduce poverty and ensure food security among the poorest communities of these regions.

Major strategies in combating land degradation involve community-based approaches. This includes the promotion of activities that help in improving the economy and quality of living of the local people; generating awareness among the people about the environment and forests; drought management, preparedness and mitigation; research and development initiatives and interventions best suited and beneficial to the locals; and strengthening self-governance leading to empowerment of the local communities. Land treatment strategies cover: (i) prevention of land degradation, (ii) rehabilitation and restoration and (iii) reclamation. Prevention of land degradation includes policy interventions, protection, land and water management, protection of vegetation cover, encouraging mixed farming, blending indigenous knowledge with modern scientific techniques and promotion of alternative livelihood. Rehabilitation and restoration approaches involve various methods of afforestation and reforestation to restore these degraded lands, whereas reclamation includes reclamation of water-logged and saline-alkaline area, wastewater treatment and rehabilitation of mine spoils.

Various initiatives have been taken by the government of India and non-government organisations, research institutions, environmentalists, academia and the public in this direction. These include state policies and rules of different state governments including, Rajasthan, Punjab and Haryana by framing out biodiversity rules, state environment policies, state action plan on climate change, forest policies etc. Institutional works involve assessment of groundwater quality and status and management of biological diversity (Singh 2014, 2008b), vulnerability assessment of the region, strategies for wasteland development, a proposal for conservation of flood-prone catchment area, geohydrology and geomorphology, stratigraphic development of river basin etc. Some initiatives related to forestry and water resource conservation in Rajasthan regions include afforestation and reforestation through various national and international programmes. By now, about 0.4 million ha of lands have been treated under afforestation and reforestation using sand dune stabilisation (SDS), shelterbelt plantation, eco-restoration, silvipastoral, block plantation, canal side plantation (CSP), rehabilitation of degraded forest (RDF) under the programmes like Desert Development Programme (DDP), Combating Desertification Programme (CDP), Compensatory Afforestation and Management Planning Authority (CAMPA), Rajasthan Forestry and Biodiversity Project (RFBP), Forest Development Agency (FDA), Rajasthan Krishi Vikas Yojana (RKVY), National Bank for Agriculture and Rural Development (NABARD) and Border Area Development Project (BADP), Harit Rajasthan.

11.9.1 Compensatory Afforestation

In this program, about 57 plus species have been planted in Rajasthan. Frequently planted species are *Vachellia tortilis* (18.90%), *Senegalia catechu* (11.55%), *Z. mauritiana* (10.84%), *Vachellia leucophloea* (9.60%), *V. nilotica* (8.49%), *Dendrocalamus strictus* (8.24%), *Senegalia senegal* (6.02%), *Azadirachta indica* (2.94%), *Albizia lebbeck*, *Gmelina arborea*, *Pongamia pinnata*, *Prosopis cineraria*, *Commiphora wightii*, *Holoptelia integrifolia*, *Bombax ceiba*, *Terminalia indica*, *Salvadora oleoides*, *Emblica officinalis*, *Butea monosperma*, *Ficus religiosa*, *Dalbergia sissoo*, *Cassia fistula*, etc. Contributions of two indigenous species in desert region such as *P. cineraria* and *Tecomella undulata* plants in total plantations are only 0.33% and 0.46%, respectively (Table 11.2). Overall survival of plantations under CAMPA is 49.37%. Sixteen plant species showed >49.27% survival. *T. arjuna*, *S. cummini*, *Mitragyna parviflora*, *B. ceiba*, *Madhuca indica* etc. exhibited extremely low survival whereas species showing >49.4% survival are *S. catechu*, *S. senegal*, *V. tortilis*, *Ailanthus excelsa*, *A. indica*, *Anogeissus pendula*, *B. monosperma*, *C. wightii*, *Cordia gharaf*, *D. strictus*, *F. benghalensis*, *Manilkara hexandra*, *Moringa oleifera*, *W. tinctoria* and *Z. mauritiana*.

11.9.2 Research and Experimentation

Though there has been a large amount of research conducted on desertification, considering the climatic variations and its impacts in deserts and dry lands further research work may include: (1) dynamic monitoring of Aeolian desertification and environmental resources in the areas, and establishing a network for monitoring and surveillance all over the country covering hot and cold areas including inland and coastal dunes; (2) establishing a representative area in each zone for experimentation and data generation; (3) providing information and advisory services; and (4) conducting

Table 11.2 Different species and extent of their use in CAMPA plantation in Rajasthan

S. no.	Planted species	Total (%)	S. no.	Planted species	Total (%)
1	*Senegalia catechu*	13.7	22	*E. camaldulensis*	0.2
2	*Vachellia leucophloea*	10.1	23	*Ficus benghalensis*	0.0
3	*V. nilotica*	8.0	24	*F. religiosa*	0.0
4	*S. senegal*	6.2	25	*Gmelina arborea*	1.3
5	*V. tortilis*	15.3	26	*Holoptelea integrifolia*	5.3
6	*Ailanthus excelsa*	0.6	27	*Manilkara hexandra*	0.2
7	*Annona squamosa*	0.1	28	*Moringa oleifera*	0.0
8	*Albizia lebbeck*	1.0	29	*Madhuca indica*	0.3
9	*Azadirachta indica*	2.7	30	*Pithecelobium dulce*	0.6
10	*Balanites aegyptiaca*	0.2	31	*Mitragyna parviflora*	0.4
11	*Bambusa vulgaris*	0.0	32	*Pongamia pinnata*	1.8
12	*Aegle marmelos*	0.3	33	*Prosopis cineraria*	0.4
13	*Anogeissus pendula*	0.0	34	*Syzygium cumini*	0.6
14	*Bombax ceiba*	0.6	35	*Tamarindus indica*	0.1
15	*Butea monosperma*	0.9	36	*Tecomella undulata*	0.5
16	*Cassia fistula*	2.9	37	*Tectona grandis*	0.3
17	*Commiphora wightii*	0.0	38	*Terminalia arjuna*	0.6
18	*Cordia myxa*	0.1	39	*T. bellirica*	0.3
19	*Dalbergia sissoo*	1.4	40	*Wrightia tinctoria*	0.1
20	*D. strictus*	9.9	41	*Ziziphus mauritiana*	10.9
21	*Emblica officinalis*	1.7	42	Others	0.3

comprehensive research on the relationships of society, the economy and types of land degradation for an effective and integrated desertification control programme.

Use of vegetative barriers of tree sticks in wind erosion control favourably influences soil and water conservation and growth and production of pearlmillet. Strip cropping of legume crops with grass in 1:3 ratio reduces soil loss at 20.5 tonnes ha^{-1} than losses of 48 tonnes ha^{-1} in bare field. Estimated loss of soil organic carbon and total nitrogen are 44–48 kg ha^{-1} and 3.0–3.2 kg ha^{-1} respectively from the soils left in bare after sole cropping (Soni et al. 2007). In the contrary, promoting grasses and afforestation encourages sediment deposition, and provides a net gain in soil organic carbon and total nitrogen (Raji and Oguwole 2006). Planting 'Aloes' or thornless cactus (*Opuntia ficus-indica*) facilitates the effectiveness of grass reseeding for rangeland restoration, enhancing the benefits gained from the more common strategy of using the branch together. The major functions of this cactus are used as food, feed for livestock, income generation, environmental protection, fence, firewood, etc. (Gebretsadik et al. 2013). Small bushes and grasses are better sand binders as compared to the tree species. However, maintaining grasses has an associated problem of uncontrolled grazing on vulnerable sites covering dunes and sandy plains and most importantly is the drying of the grasses during the summer season, when maximum sand drift takes place. Bio-fencing of locally available non-palatable species is a cheaper option and more effective form of barrier. Most reactivated hairpin parabolic dunes in the zone of very high wind erosion are now being planted with trees and shrubs, especially *Vachellia tortilis* and *Calligonum polygonoides* (phog) and occasionally *Prosopis juliflora*. A combination of *C. polygonoides* and *Cassia angustifolia* (sonamukhi) has been found very effective in sand drift control and improving economic benefits (Singh and Rathod 2002). Erection of tree rows in the form of shelterbelts on boundaries of agricultural fields helps in reducing injuries to the newly planted seedlings caused by sandblasting, and hot and desiccating winds. Increasing tree cover on agricultural lands through agroforestry of varying kinds involving people also diversifies income generation opportunities and increases the yields of farm products. It enhances the socio-economic conditions of the farmers besides increasing tree cover in this hostile region. Overall environmental benefits of combating desertification are:

- Reduction in amount of atmospheric dust and sand that pollute the surrounding areas due to mild to severe sand storms.
- Provides benefits of carbon sequestration, biodiversity enhancement, improved soil and climate conditions and the economic value of associated forest products (Singh and Rathod 2002).
- Removal of CO_2 from the atmosphere and minimisation of GHG emissions from the soil by conserving and enhancing soil carbon stock.
- Positive effects of planted trees on micro-climate and decrease in soil erosion and sand drift.

11.10 Synthesis and Future Directions

A wide difference in latitudinal, altitudinal, climatic, geomorphological and biological features in deserts results in varying nature of habitats. Such variations are responsible for enjoyable biological diversity and survival of different species and sustain the livelihoods of the people living in these extreme but amazing areas. Changing life span, growth patterns, physiology and biochemical behaviour are the ways by which plants and animals adapt themselves towards the adversities of the environmental conditions of the world's deserts. Life forms like ephemerals, annuals and succulent and non-succulent perennials are important ways of adaptation among the plants in these deserts. Multiple sorts of stresses like diurnal and seasonal extremities, intense solar radiation and high temperatures, water scarcity, soil salinity and wind and dust storms all together influence the life in desert. Biotic stresses such as competitive interactions also play an important role in desert ecology. But remarkable ways of adaptations allow life to be sustained in the desert region. Deserts are usually low in species richness and diversity of the biota with respect to other favourable biomes of the world. However, diversity in habitats and isolated populations results in higher within-species diversity and is positively related to various functions of the desert ecosystem. However, recent trends indicate degradation in these habitats because of the increase in population of both livestock and human, and related anthropogenic activities and climatic adversities. Unsustainable livelihood practices like overutilisation of vegetation for fodder and fuelwood, along with altered land use, increased atmospheric CO_2 level and global warming are leading to habitat loss and land degradation. Their interactive outcome affects biodiversity, water supply, soil functioning and resilience and consequently brings a decline in ecosystem services and local economy.

A lot of researches have been conducted and programmes implemented to tackle the problems associated with desertification and improve the socio-economy of the desert people. For this, long-term planned restoration efforts are still required to check the process of desertification and secure the essential resources for the ecological and economic growth of the region. The process of conservation of deserts appeared inconsistent in the past and varied with countries and the availability of resources. In most of the cases, deserts were considered as wastelands and the only interest in deserts was mining or military testing.

The use of fast-growing exotic species *Vachellia tortilis* in Thar Desert regions appeared beneficial for the initial phases of sand drift control and stabilisation under effective protection and caring. However, there is a need to focus on indigenous species in replacing the exotics in a sequential manner. Because of the availability of pasturelands, bushlands, sacred groves etc., the deserts hold immense potential for its development into a rangeland. *Lasiurus sindicus*, *Cenchrus ciliaris*, *C. setigerus* and *Cymbopogon jwarancusa*, etc. are fodder grasses of high nutritive value and are well adapted to the Thar Desert region. These grasses can be utilised in restoring degraded community lands or pasturelands. Invasion of many alien species including *Prosopis juliflora* is also taking place in the deserts. These invasive species need to be eradicated and the land should be put into production and cattle rearing. However, large investment costs and technological interventions in the desert restoration process require documentation of the causal factors involving the local people especially the rural populations in the restoration process and reversing the process of land degradation/desertification also.

Lessons Learnt

- Origins and dynamic properties of desert-like conditions depend on region-wide conditions. While anthropogenic factors play a leading role in desertification in warm-temperate and sub-arid areas, the climatic factors influence desertification in tropical and subtropical zones. Wind action is important in shaping different landforms such as dunes of different shapes and sizes, mesas and buttes. Other important geomorphic structures are playas, alluvial fans, bajada, pediments and inselbergs. These landforms directly affect the inflow and outflow of water and other materials in the form of run-off, infiltration, erosion, salt accumulation, water storage etc. These altogether cause heterogeneity in the habitats supporting different flora and fauna adapted to the existing climatic conditions. Hence, deserts are not deserted despite many adversities in the region. This is the biome type, the researchers and conservationists do not want to expand, but they ensure that the deserts should not disappear because of coarse grain production, low biodiversity and extreme climatic conditions. The deserts are very sensitive, influenced easily by the climatic changes in the adjoining landscapes and present a risk of degradation as well to that.
- Deserts are home to 17% of the human population, livestock like camels, goats, antelope, kangaroo and plants like dates, figs, olives and other food crops. They are important places of mineral resources with 13 minerals concentrates, out of 15 minerals on the earth, well-preserved archaeological ruins (e.g. Village Kuldhara near Jaisalmer), generating renewable energy (both solar and wind energy), rich archaeological sources and tourist destination (attract millions of visitors every year). Deserts support unique biological profiles and species that exist in these environments, simply do not appear in other regions. Soil physicochemical properties like water content, organic carbon, salt contents and nutrients and more importantly human activities are the major forces

regulating plant communities' distribution and their diversity, structure and functions. This is observed through significant relationships between plant species diversity, organic carbon and nitrogen.

- Desert spaces are to be conserved and restored, and encroaching on them or unnecessarily altering them for other purposes should be avoided. Researchers and conservationists working in desert regions understand the importance of their conservation for cultural and ecological reasons. Restoration efforts made in the past have resulted in reestablishing land uses of desert ecosystems like tree integration in farmlands, shelterbelt plantation, stabilizing drifting sand dunes, development of community pastures, compensatory afforestation, etc. to make the respective land use more productive, resilient and their ability to supply higher quality ecosystem services. The desert ecosystems can be conserved and restored, if necessary, by promoting growth and or planting and seeding of native species, managing available water (both rain and groundwater), manipulating soil properties, controlling non-native species and effective management of already restored ecosystems following community-based approaches and activities to enhance livelihoods of the desert communities.

Key Questions

1. What are the factors responsible for desert formations and their geographical distribution?
2. How many types of deserts are there in the world and what are their characteristics?
3. What are the adaptational changes in plant communities and biotic interactions in desert environments?
4. In what way deserts are important and beneficial to human well beings?
5. What are the varying threats to the desert ecosystems?

References

Ahmad SS, Nawaz M, Khan SM (2013) Evaluating the effects of soil pH and moisture on the plant species of Changa Manga Forest using Van Dobben circles. Middle-East J Sci Res 17:1405–1411

Allen AR, Pockman WT, Restrepo C, Milne BT (2008) Allometry, growth and population regulation of the desert shrub *Larrea tridentata*. Funct Ecol 22:197–204

Alsharif W, Saad M, Hirt H (2020) Desert microbes for boosting sustainable agriculture in extreme environments. Front Microbiol 11. https://doi.org/10.3389/fmicb.2020.01666

Arthington AH (2012) Environmental flows: saving rivers in the third millennium. University of California Press, Berkeley and Los Angeles

Baeza A, Bouma MJ, Dhiman RC, Pascual M (2013) Long-lasting transition toward sustainable elimination of desert malaria under irrigation development. Proc Natl Acad Sci 110(37):15157–15162

Bhatnagar C, Meena SS, Pandey VN (2011) Habitat preference of *Trapelus agilis* Olivier in Thar Desert of Jaisalmer, India. Russian J Herpetol 19:207–211

Bombi P, Salvi D, Shuuya T, Vignoli L, Wassenaar T (2020) Very high extinction risk for *Welwitschia mirabilis* in the northern Namib Desert. J Arid Environ 190:104529. https://doi.org/10.1101/2020.05.05.078253v1.full.pdf

Buckley DAH, Charles PA, O'Donoghue D (2005) Science with the Southern African large telescope. South Afr J Sci 101:531–535

Butola JS, Malik AR, Baba JA (2012) Livelihood diversification in cold arid desert of Indian Himalaya: urgent need of transforming traditional agroforestry systems with emphasis on adoption of herbal farming. Int J Med Aromatic Plants 2(3):544–548

Charan G, Bharti VK, Jadhav SE, Kumar S, Acharya S, Kumar P, Gogoi D, Srivastava RB (2013) Altitudinal variations in soil physico-chemical properties at cold desert high altitude. J Soil Sci Plant Nutr 13(2):267–277

Chauhan A, Jishtu V, Thakur L, Dolma T (2020) Medicinal plants of the trans-Himalayan cold desert of Ladakh—a review. Int J Sci Environ Technol 9(2):239–253

Chen J, Blume HP, Beyer L (2000) Weathering of rocks induced by lichen colonization—a review. Catena 39:121–146

Choyal R, Sharma SK (2011) Studies on bryophytes of Thar Desert with particular reference to Ganganagar District, Rajasthan, India. Indian J Fundam Appl Life Sci 1:239–245

Croteau EK (2010) Causes and consequences of dispersal in plants and animals. Nat Educ Knowl 3(10):12

Davies J, Poulsen L, Schulte-Herbrüggen B, Mackinnon K, Crawhall N, Henwood WD, Dudley N, Smith J, Gudka M (2012) Conserving dryland biodiversity. xii +84p. IUCN (International Union for the Conservation of Nature), Nairobi, Kenya. http://www.unccd.int/Lists/SiteDocumentLibrary/Publications/drylands_bk_2.pdf

Deotare BC, Kajale MD, Rajaguru SN, Kusumgar S, Jull AJT, Donahue JD (2004) Palaeo-environmental history of Bap-Malar and Kanod playas of western Rajasthan, Thar desert. Abstracts, accessed on 20th December 2010. http://www.ias.ac.in/jess/abssep2004.html

Dhir RP, Joshi DC, Katju S (2017) Thar Desert in retrospect and prospect. Scientific Publisher, Jodhpur. 405p

Dorn RI (2018) Necrogeomorphology and the life expectancy of desert bedrock landforms. Prog Phys Geogr: Earth Environ 42(5):566–587

Engel M, Matter A, Parker AG, Parton A, Petraglia MD, Preston GW, Preusser F (2017) Lakes or wetlands? A comment on 'The middle Holocene climatic records from Arabia: reassessing lacustrine environments, shift of ITCZ in Arabian sea, and impacts of the southwest Indian and African monsoons. Glob Planet Change 148: 258–267

FAO (2011) Overcoming the crisis in the horn of Africa. Available at: http://www.fao.org/fileadmin/templates/horn_africa/docs/Overcoming_the_crisis_in_the_Horn_of_Africa.pdf. Accessed 5th of August 2015

Fossum KR, Danielsen BE, Aarseth W, Johnsen SO (2018) Technologies and human-robot collaboration for surface EVA exploration activities and training in European analogue environments. file:///C:/Users/HP1/Downloads/CORDIS_project_607346_fr.pdf. Accessed on 13th December 2023

Free CL, Baxter GS, Dickman CR, Leung LKP (2013) Resource pulses in desert river habitats: productivity-biodiversity hotspots, or mirages? PLoS One 8(10):e72690. https://doi.org/10.1371/journal.pone.0072690

Friedel MH, Cellier KM, Nicolson KP (1980) Nutrient deficiencies in central Australian semi-desert rangelands, with reference to decline in range condition. Aust Rangeland J 2:151–161

Gebretsadik G, Animut G, Tegegne F (2013) Assessment of the potential of cactus pear (*Opuntia ficusindica*) as livestock feed in Northern Ethiopia. Livestock Res Rural Dev 25(2) http://www.lrrd.org/lrrd25/2/moen25026.htm

Gobin A, Hai LT, Linh PH, Hens L, Ozer P, Hien LTT, Binh NT, Vinh PQ (2012) Impact of global climate change and desertification on the environment and society in Southern Centre of Vietnam (case study in Binh Thuan province). http://www.belspo.be/belspo/organisation/Publ/pub_ostc/BL/Rapp_BL03V28_en.pdf

Goudie AS (2002) Great warm deserts of the world: landscapes and evolution. Oxford University Press, Oxford

Goyal RK (2004) Sensitivity of evapotranspiration to global warming: a case study of arid zone of Rajasthan, India. Agric Water Manag 69: 1–11

Gregersen E (2023) Very large telescope. Encyclopedia Britannica, 2 Dec. 2023, https://www.britannica.com/topic/Very-Large-Telescope. Accessed 14 December 2023

Grove AT (1974) Desertification in the African environment. Afr Affairs 73:137–151

Guhathakurta P, Sreejith OP, Menon PA (2011) Impact of climate change on extreme rainfall events and flood risk in India. J Earth Syst Sci 120:359–373

Hasan BMD, Öztürk AB (2014) Desertification and its consequences in Turkey and the Middle East. Glob Climate Change Public Health 7: 293–305

Henschel JR, Burke A, Seely M (2005) Temporal and spatial variability of grass productivity in the central Namib Desert. African Study Monogra Suppl Issue 30:43–56

Holmgren RC, Brewster SF Jr (1977) Distribution of organic matter reserve in a desert shrub community. U.S. Dept. Agriculture Forest Service Research Paper INT-130: 15 p

Hufnagel L, Mics F, Pálinkás M, Homoródi R (2018) Introductory chapter: global aspects and scientific importance of desert ecological research. In: Hufnagel L (ed) Community and global ecology of deserts. Intech Open, pp 1–10

IUCN (2008) The nature of drylands: diverse ecosystem, diverse solutions. IUCN World Conservation Congress, Barcelona

Iwara AI, Gani BS, Njar GN, Deekor TN (2011) Influence of soil physico-chemical properties on the distribution of woody tree/shrub species in south-southern Nigeria. J Agric Sci 2:69–75

Jain SK, Agarwal PK, Singh VP (2007) Hydrology and water resources of India. Springer, Dordrecht, Netherlands. 1258 p

Jaroff L (1997) Etched in stone. Time. http://www.time.com/time/printout/0,8816,986446,00.html. Retrieved on 20th December 2010. Wildlife and humans tend to get equal billing in African rock art

Kandji ST, Verchot L, Mackensen J (2006) Climate change and variability in the Sahel region: impacts and adaptation strategies in the agricultural sector. UNEO and World Agroforestry Centre. https://www.worldagroforestry.org/publication/climate-change-and-variability-sahel-region-impacts-and-adaptation-strategies

Khormali F, Monger C (2020) Hot desert soils—global distribution and unique characteristics. Geoderma Reg 2020(23):e00330. https://doi.org/10.1016/j.geodrs.2020.e00330

Kinahan J (2016) Human responses to climatic variation in the Namib desert during the last 1000 years. Afr Archaeol Rev 33(2):183–203

Kollath DR, Miller KJ, Barker BM (2019) The mysterious desert dwellers: *Coccidioides immitis* and *Coccidioides posadasii*, causative fungal agents of coccidioidomycosis. Virulence 10(1):222–233

Kollegger M, Grunert T (2017) Major types of dunes: where and how they form. https://sedimentology.uconn.edu/2017/02/07/major-types-of-dunes-where-and-how-they-form/. Accessed on 11th December 2023

Kumar GP, Murkute A, Gupta S, Singh SB (2009) Carbon sequestration with special reference to agroforestry in cold deserts of Ladakh. Curr Sci 97(7):1063–1068

Kumar A, Mitra M, Adhikari BS, Rawat GS (2016) Flora of Niti Valley: a cold arid region of Nanda Devi Biosphere Reserve, Western Himalaya, India. Check List 12(1). https://doi.org/10.15560/12.1.1824

Lathrop EW, Rowlands PG (1983) Plant ecology in deserts: an overview. In: Webb RH, Wilshire HG (eds) Environmental effects of off-road vehicles. Springer-Verlag, New York

Liniger H, Weingarten R, Grosjean M (1998) Mountains of the world: water towers for the 21st century. Mountain Agenda, Centre for Development and Environment. Institute of Geography, University of Bern, Bern

Mares MA (1999) Encyclopedia of deserts. University of Oklahoma Press, USA

Mathur M, Sundaramoorthy S (2013) Interspecific associations of herbaceous vegetation in semi-arid Thar Desert, India. Range Manage Agrofor 34:26–32

Meng Z, Dang X, Gao Y et al (2018) Interactive effects of wind speed, vegetation coverage and soil moisture in controlling wind erosion in a temperate desert steppe, inner Mongolia of China. J Arid Land 10: 534–547. https://doi.org/10.1007/s40333-018-0059-1

Mensah AK (2015) Role of revegetation in restoring fertility of degraded mined soils in Ghana: a review. Int J Biodivers Conserv 7(2):57–80

Moore MR, Gollehon NR, Carey MB (1994) Multicrop production decisions in western irrigated agriculture: the role of water price. Am J Agric Econ 76(4):859–874

Morgan J, LeCain D, Mosier A, Milchunas D (2001) Elevated CO_2 enhances water relations and productivity and affects gas exchange in C_3 and C_4 grasses of the Colorado shortgrass steppe. Glob Change Biol 7:451–466

Murray SJ (2013) Present and future water resources in India: insights from satellite remote sensing and a dynamic global vegetation model. J Earth Syst Sci 122:1–13

Mzezewa J, van Rensburg LD (2011) Effects of tillage on runoff from a bare clayey soil on a semi-arid ecotype in the Limpopo province of South Africa. Water SA 37:165–172

Nikita E, Mattingly D, Lahr MM (2012) Sahara: Barrier or corridor? Nonmetric cranial traits and biological affinities of North African late Holocene populations. Am J Phys Anthropol 147(2):280–292

Oinam SS, Rawat YS, Khoiyangbam RS, Khwairakpam G, Kuniyal JC, Vishvakarma SCR (2005) Land use and land cover changes in Jahlma watershed of the Lahaul valley, India. J Mt Sci 2(2):129–136

Pant GB, Hingane LS (1988) Climatic changes in and around the Rajasthan desert during the 20th century. Int J Climatol 8:391–401

Parsons AJ, Abrahams AD (2009) Geomorphology of desert environments. In: Parsons AJ, Abrahams AD (eds) Geomorphology of desert environments. Springer, Dordrecht. https://doi.org/10.1007/978-1-4020-5719-9_1

Peterson AT, Ortega-Huerta MA, Bartley J, Sanchez-Cordero V, Soberon J, Buddemeier RH, Stockwell DRB (2002) Future projections for Mexican faunas under global climate change scenarios. Nature 416:626–629

Philips DL, Mac Mahon JA (1978) Gradient analysis of a Sonoran desert bajada. South Western Nat 23:669–680

Phillips JD, Turkington AV, Marion DA (2008) Weathering and vegetation effects in early stages of soil formation. Catena 72:21–28

Prakash I (1994) Ecology and conservation of the Indian Desert. In: Balakrishnan M, Borgstrom R, Bie SW (eds) Tropical ecosystems: a synthesis of tropical ecology and conservation. Oxford & IBH, New Delhi, pp 167–191

Qureshi R, Bhatti GR (2008) Diversity of micro-habitats and their plant resources I. Nara Desert, Pakistan. Pak J Bot 40(3):979–992

Raj A, Sharma P (2013) Is Ladakh a cold desert? Curr Sci 104(6): 687–688

Raji BA, Ogunwole JO (2006) Potential of carbon sequestration in the various land use in the sub-humid and semi-arid savannah of Nigeria: lessons from long term experiments. Int J Soil Sci 1:33–43

Rowley-Conwy P, Layton P (2011) Foraging and farming as niche construction: stable and unstable adaptations. Philos Trans R Soc Lond Biol Sci 366(1566):849–862

Rücknagel J, Götze P, Hofmann B, Christenm O, Marschall K (2013) The influence of soil gravel content on compaction behaviour and pre-compression stress. Geoderma 209–210:226–232

Seely M (2004) The Namib: natural history of an ancient desert. Desert Research Foundation of Namibia

Sehgal J, Mandal DK, Mandal C, Vadivelu S (1992) Agro-ecological regions of India. II Edition, Tech. Bull. No. 24, NBSS and LUP, 130p

Sharma DC (2019) Cold desert regions need special attention: Experts. https://www.downtoearth.org.in/news/climate-change/cold-desert-regions-need-special-attention-experts-66553

Sharma KK, Mehra SP (2009) The Thar of Rajasthan, India: ecology and conservation of a desert ecosystem. In: Shivaperuman C, Baqri QH, Ramaswamy G, Naseema M (eds) Faunal ecology and conservation of the Great Indian Desert. Springer, Berlin, Heidelberg, pp 1–6

Sharma KK, Pandey AK (2010) Phytosociological study of vegetation of some selected arid region of the Thar desert of Rajasthan, India. Curr World Environ 5:51–58

Sharma B, Sharma K (2011) Phytosociololgical studies on the vegetation of sand dunes and Sandy plains of Ajmer, Rajasthan. Curr Bot 2(10):1–6

Shetty BV, Singh V (1993) Flora of Rajasthan, vol III. Botanical Survey of India, Calcutta

Shkolnik A, Taylor CR, Finch V, Borut A (1980) Why do Bedouins wear black robes in hot deserts? Nature 283:373–375

Singh G (2005) Ecological and environmental assessment in the on-shore area of RJ-ON-90/2 block'. Technical Report AFRI, Jodhpur. Submitted to M/S Cairn Energy Pvt. India Ltd., Chennai

Singh G (2008a) Utilizing facilitative effects of *Euphorbia caducifolia* in rehabilitating degraded hills. Digital Library of the Commons, The Trustees of Indiana University. http://dlc.dlib.indiana.edu/archive/00003663/01/Degradedhills.pdf

Singh G (2008b) Baseline survey study on biological diversity in Mangala, Saraswati and Rageshwari areas of Rajasthan Hydro Carbon Project. Submitted to M/S Cairn Energy Pvt. India Ltd., New Delhi. 217p

Singh G (2014) Studies on carbon sequestration in different forest types of Rajasthan. Arid Forest Research Institute, Jodhpur. A project concluding report submitted to Indian Council of Forestry Research and Education, Dehradun

Singh G (2015) Studies on the effects of MPOWER programme on mitigation and adaptation towards climate change in western Rajasthan. AFRI Project concluding report submitted to ICFRE, Dehradun

Singh G (2017) Studies on the effects of MPOWER programme on mitigation and adaptation towards Climate Change in western Rajasthan-Part II. Project report submitted to MPOWER, Jodhpur, Rajasthan and ICFRE, Dehradun

Singh G (2018) Assessment of biological diversity and people perception for developmental plan and awareness generation in different community reserve areas in Jodhpur district. Project report submitted to State Forest Department, Jodhpur division, Jodhpur and ICFRE, Dehradun

Singh V, Dogra KK (1996) Characteristics, distribution, utilization, regeneration, biomass and nutritional values of Seabuckthorn (Hippophae). Indian Forester 122(6):486–491

Singh RP, Gupta MK (1990) Soil and vegetation study of Lahaul and Spiti Cold Desert of Western Himalayas. Indian Forester 116(10): 174–180

Singh G, Rathod TR (2002) Plant growth, biomass production and soil water dynamics in a shifting dune of Indian desert. For Ecol Manag 171:309–320

Singh G, Sharma R (2017) Effects of different land use changes and spatial variation in rainfall on soil properties and soil carbon storage in western Rajasthan, India. Ann Adv Agric Sci 1(2):43–53

Singh G, Shukla S (2011) Effects of *Prosopis juliflora* (DC.) tree on under canopy resources, diversity and productivity of herbaceous vegetation in Indian desert. Arid Land Res Land Manage 26:151–165

Singh G, Gupta GN, Kuppusamy V (2000) Seasonal variations in organic carbon and nutrient availability in arid zone agroforestry systems. Trop Ecol 41(1):17–23

Singh G, Bala N, Kumar P (2007) Challenges and strategies for sustainable development of Indira Gandhi Nahar area of Rajasthan. Revisiting Indira Gandhi Canal Command meeting held at CAZRI, Jodhpur on 15–16 December 2007

Singh G, Chouhan S, Rathod TR (2008) Vegetation diversity and socioeconomic profile relations in some selected villages of Indian desert. Indian Forester 134(6):744–756

Singh G, Singh B, Tomar UK, Sharma S (2017) A manual for dryland afforestation and management. Scientific Publisher (India), Jodhpur, p 512p

Smith JMB (2023) Desert. Encyclopedia Britannica, 13 Dec. 2023, https://www.britannica.com/science/desert. Accessed 21 December 2023

Soni ML, Yadava ND, Beniwal RK, Singh JP, Kumar S (2007) Production potential of arid legumes under grass based strip cropping system in arid rainfed condition of western Rajasthan. J Arid Legumes 4(1):9–11

Spinoni J, Micale F, Carrao H, Naumann G, Barbosa P, Vogt J (2013) Global and continental changes of arid areas using the FAO Aridity Index over the periods 1951–1980 and 1981–2010. Geophys Res Abstracts 2013:15. EGU2013-9262-1. http://adsabs.harvard.edu/abs/2013EGUGA.15.9262S (Accessed on 3rd December 2013)

Stegner M, Lackner B, Schafernolte T, Buchner O, Xiao N, Geirlinger N, Neuner G (2020) Winter nights during summer time: stress physiological response to ice and the facilitation of freezing cytorrhysis by elastic cell wall components in the leaves of a nival species. Int J Mol Sci 27(19):7042. https://doi.org/10.3390/ijms21197042

Sternberg M, Shoshany M (2001) Above ground biomass allocation and water content relationships in Mediterranean trees and shrubs in two climatological regions in Israel. Plant Ecol 157:171–179

Steward AL, von Schiller D, Tockner K, Marshall JC, Bunn SE (2012) When the river runs dry: human and ecological values of dry riverbeds. Front Ecol Environ 10:202–209

Strahler AN, Strahler AH (1984) Elements of physical geography. John Wiley & Sons, New York

Tawde P (2018) Cultural heritage of the thar desert and the tourist places in jaisalmer. https://www.veenaworld.com/blog/tourist-places-to-visit-in-jaisalmer (Retrieved on 20th April 2021)

Tewari VP, Kapoor KS (2013) Western Himalayan cold deserts: biodiversity, eco-restoration, ecological concerns and securities. Ann Arid Zone 52(3&4):223–230

Thomas N, Nigam S (2018) Twentieth-century climate change over Africa: seasonal hydroclimate trends and Sahara Desert expansion. J Clim 31(9):3349–3370

Tripathi G, Deora R, Singh G (2013) The influence of litter quality and micro-habitat on litter decomposition and soil properties in a silvopasture system. Acta Oecol 50:40–50

UCMP (2010) University of California museum of paleontology. Retrieved 2010-12-20. http://www.ucmp.berkeley.edu/about/index.php

Venkanna K, Mandal UK, Raju AJS, Sharma KL, Adake RV, Pushpanjali Reddy BS, Masane RN, Venkatravamma K, Babu BP (2014) Carbon stocks in major soil types and land-use systems in semiarid tropical region of southern India. Curr Sci 106:604–611

Verheye W (2006) Dry lands and desertification. In: Land use, land cover and soil sciences. http://www.eolss.net/Sample-Chapters/C19/E1-05-06.pdf (Accessed on 14th January 2015)

Weber K, Lange CA, Kellner K, Mishra D, Soni V (2013) Probing the process leading to desertification and its impact on plant biodiversity in the Thar Desert of Rajasthan, India. J Econ Nat Ecol World 4:8–20

White DH, Lubulwa G, Zuo H, Wint W, Slingenbergh J (2013) Agro-climatic classification systems for estimating the global distribution of livestock numbers and commodities. Environ Int. http://www.ergodd.zoo.ox.ac.uk/livatl2/reports/White_EI_final.doc

Whitford W (2002) Ecology of desert systems. Academic Press. 343 p

Whittaker RH (1975) Communities and ecosystems, 2nd edn. Macmillan Publishing Co., Inc., New York. 385p

Williams WJ, Eldridge DJ (2011) Deposition of sand over a cyanobacterial soil crust increases nitrogen bioavailability in a semi-arid woodland. Appl Soil Ecol 49:26–31

Wuerthne G (2023) Livestock impacts to desert regions. The Wildlife News. https://www.thewildlifenews.com/2023/01/19/livestock-impacts-to-desert-regions/#:~:text=Livestock%20production%20is%20one%20of,desertification%20and%20loss%20of%20biodiversity. Accessed on 15th December 2023

Yan E-R, Yang X-D, Chang SX, Wang X-H (2013) Plant trait-species abundance relationships vary with environmental properties in subtropical forests in eastern China. PLoS One 8(4):e61113. https://doi.org/10.1371/journal.pone.0061113

Zalibekov ZG (2011) The arid regions of the world and their dynamics in conditions of modern climatic warming. Arid Ecosyst 1:1–7

Zeng N (2003) Atmospheric science: drought in the Sahel. Science 302: 999–1000

Climate Change and Forest Sector in India

12

R. S. Rawat and V. R. S. Rawat

Abstract

The key issue behind the recent global climate change is a continuous increase in the concentration of greenhouse gases in the atmosphere. Emission of greenhouse gases during the time period from 2010 to 2019 was higher across the globe than the previous time periods in human history. The anthropogenic emission of greenhouse gases since 1850–1900 is responsible for warming of 1.1 °C. Taking note of the serious issues of global climate change, world governments in the year 1992 agreed on a global treaty called 'United Nations Framework Convention on Climate Change' for stabilising the concentration of greenhouse gases. Subsequently, Paris Agreement adopted in 2015 aimed '*to hold global average temperature increase to well below 2 °C and pursue efforts to limit temperature rise to 1.5 °C above pre-industrial levels*'. Countries made commitments under the Paris Agreement through their nationally determined contribution.

Protection of forested lands, improved forest management, restoration of degraded forests and other ecosystems have major potential to offset emission of greenhouse gases. As per the India State of Forest Reports, carbon stocks in India's forest increased from 6071 million tonnes in 1996 to 7204 million tonnes in 2021. India's third National Communication to the UNFCCC reveals that forests were removing 20% of total greenhouse gas emissions in 2019. Shifting of distribution ranges of many species would occur in future owing to the changing climate combined with land-use change. Significant changes in Himalayan forests with respect to their vegetation composition and distribution pattern at different altitudes were observed mainly due to changing climate. Ecosystem-based adaptation and nature-based solutions can enhance mitigation and adaptation benefits of forests towards climate change. Studies on climate change vulnerability, impacts, loss and damage, adaptation and long-term sustainability are priority areas for future research in forest sector. Inadequate financial and technical resources in the country are key barriers for conducting these studies and implementing suitable adaptation measures. A programme to reduce emission from deforestation and forest degradation in developing countries (REDD) was initiated in 2005 by UNFCCC primarily to address the causes of deforestation. However, a comprehensive policy approach put forth by India led to the inclusion of conservation of forest carbon stocks and sustainable management of forests in the agenda of REDD and since then it was upgraded to REDD+. In India, National REDD+ Strategy, Forest Reference Level and Safeguards Information System are already in place to facilitate implementation of REDD+ activities. Forest sector in India is witnessing a number of domestic initiatives to address the climate change issues. Eight national missions under the National Action Plan on Climate Change reflect India's vision of sustainable development along with meeting its climate change mitigation and adaptation objectives. The NDC target for forest sector of India is '*to create an additional carbon sink of 2.5–3 billion tonnes of* CO_2 *equivalent through additional forest and tree cover by 2030*'. In India, a strong policy framework is already in place for the protection and conservation of forests. The objective of the National Forest Policy of 1988 is 'to ensure ecological constancy and maintenance of ecological balance with an aim to bring one-third of geographical area of the country under forest and tree cover'. The Forest (Conservation) Act, 1980 was brought to check rampant diversion of forest lands for non-forest purposes and to regulate and control the land use changes in forests. With the implementation of this act, the rate of diversion of forest land for non-forest purposes has been radically reduced. The National Working Plan Code also attempts to include climate change and biodiversity

R. S. Rawat · V. R. S. Rawat (✉)
Biodiversity and Climate Change Division, Indian Council of Forestry Research and Education, Dehradun, Uttarakhand, India

A. K. Mandal, A. Nicodemus (eds.), *Textbook of Forest Science*, https://doi.org/10.1007/978-981-97-8289-5_12

conservation aspects through forest working plans in the country.

Keywords

Forest · Climate change · Adaptation · Mitigation · REDD+

12.1 Introduction

According to the Intergovernmental Panel on Climate Change, a change in the climate is one that is characterised by variations in the mean and/or variability of its features and lasts for several decades or more. Climate change refers to any long-term shift in climate due to human actions or natural variability (IPCC 2007). The United Nations (1992) described climate change as '*a change in climate which is attributed directly or indirectly to human activity and is additional to the natural climatic variability observed so far*'. The highest emission of greenhouse gas from human activity was reported globally in the past 10 years from 2010 to 2019. Since 2010, emissions of greenhouse gases have continued to grow reaching 59 ± 6.6 $GtCO_2$ eq. in 2019, but the average annual increase in the past 10 years (2010–19) was 1.3%, which was lower (2.1%) than in the time period of 2000–2009 (IPCC 2022a, b).

Rigorous analysis of all data and lines of evidence shows that most of the observed climate changes of the past several decades cannot be only explained just by natural causes but by considering anthropogenic ones. Anthropogenic activities have significantly disturbed the natural carbon cycle by extracting long-buried fossil fuels and burning them for energy needs, thus releasing CO_2 into the atmosphere.

The concentration of greenhouse gases increased since preindustrial time in the atmosphere such as CO_2 increased by about 40%, CH_4 increased by about 150% and N_2O increased by about 20%. The average concentration of CO_2 was measured from 316 ppm in 1959–400 ppm in 2013 at the Mauna Loa Observatory at Hawaii (Dlugokencky et al. 2014) and now its concentration reached 420 ppm. The same rates of increase have since been recorded at numerous other stations worldwide. Globally, the emission of greenhouse gases has been continuously increasing since the release of the fifth assessment report of IPCC. However, the average rate of emissions of greenhouse gases decreased from 2.4% (from 2000 to 2010) to 1.3% (from 2010 to 2019). Growth in population and gross domestic product are important key factors for increase in emission of greenhouse gases. Interaction of robust improvement in energy efficiency and placement of low-carbon technologies contributed to reduction in emission of greenhouse gases. However, emission reductions were not enough to offset the continued increase of emission at global level (UNEP 2018; IEA 2019; IPCC 2022a, b).

According to the sixth assessment report of IPCC, about 34% (20 $GtCO_2$ eq.) of total net anthropogenic greenhouse gas emissions emitted from energy supply sector, 24% (14 $GtCO_2$ eq.) from industrial sector, 22% (13 $GtCO_2$ eq.) from agriculture, forestry and other land use sector, 15% (8.7 $GtCO_2$ eq.) from transport sector and 6% (3.3 $GtCO_2$ eq.) from buildings sector. Land is a net sink of (−) 6.6 Gt CO_2 per year for the period from 2010 to 2019. The gross sink during this period was reported to be (−)12.5 $GtCO_2$ per year resulting from both anthropogenic changes and natural climate variability. Based on book-keeping models, net anthropogenic CO_2 emission from land use, land use change and forest, was estimated to +5.9 Gt CO_2 per year (IPCC 2022a, b).

The findings of the sixth assessment report of the IPCC reflect that emissions of greenhouse gases from human activities were responsible for approximately 1.1 °C of warming since 1850–1900 (IPCC 2021). It is expected that the average global temperature can reach to 1.5 °C of warming over the next 20 years. This warming is associated with rising atmospheric concentration of carbon dioxide. Increasing heat waves, longer warm seasons and shorter cold seasons are projected for 1.5 °C of warming while a warming of 2 °C will likely to result in heat extremes that would more often reach critical tolerance thresholds for agriculture and health. Apparently in some regions, every additional 0.5 °C of global warming has resulted in changes in intensity and frequency of droughts with more regions showing increases than decreases. In some regions with increasing global warming, increases in frequency and intensity of hydrological droughts become larger. Additional global warming even at 1.5 °C, will witness an increasing occurrence of some extreme climatic events (IPCC 2021).

Consequent upon the adoption of the Paris Agreement in 2015, the IPCC special report on Global Warming of 1.5 °C also concluded that due to anthropogenic activities, average global temperature has increased by about 1.0 °C when compared with pre-industrial levels (IPCC 2018). Extreme weather events have shown an increasing trend both in magnitude and in occurrence. At a warming of 1.5 °C, the effects of rise in temperature on society and ecosystems are much greater than expected in previous scientific reports. The overall impact at 2 °C will be disastrous for the developing countries. IPCC (2018) further emphasised to reduce CO_2 emissions by 45% from 2010 levels by 2030 and attaining net zero emissions by 2050 to keep temperature rise under 1.5 °C. A number of climate change impacts that could be avoided by limiting global warming to 1.5 °C compared to 2 °C have been highlighted in the report. Few instances include global sea level rise of 10 cm lower with global warming of 1.5 °C compared with 2 °C by 2100 (IPCC 2021).

12.2 Global Efforts Dealing with Climate Change

The Intergovernmental Panel on Climate Change (IPCC) was established by World Meteorological Organization and United Nations Environment Programme. It provides assessments reports of the scientific basis of climate change, its impacts and future risks and recommends options especially to the policy makers for adaptation and mitigation aspect of climate change at regular intervals. Assessment reports of IPCC provide a scientific basis for world governments to develop climate-related policies, and the findings of these assessment reports are also the basis for negotiations at the UNFCCC. The first assessment report was released by IPCC in 1990 noting that 'human activities were considerably increasing the concentration of greenhouse gases in the atmosphere'.

The Earth Summit in 1992 generated the overwhelming concern with the problem of climate change which led to the signing of the international treaty 'UNFCCC'. The UNFCCC was established with the objectives to stabilise greenhouse gas concentrations at levels that prevent dangerous anthropogenic interference with the climate system, to ensure that economic development proceeds in a sustainable manner and to prevent any threat to future food production. The convention enjoys near universal acceptance with 196 countries joining the treaty (UN 1992).

The Kyoto Protocol: In 1997, country Parties of UNFCCC approved an addition to the climate change treaty, called the Kyoto Protocol—legally binding measure which came into force in 2005. The first commitment period of the Protocol begun in 2008 and was to end in 2012. Parties to the UNFCCC attempted a strong multilateral framework agreement to ensure that there is no gap between the end of the Kyoto Protocol's first commitment period in 2012 and the entry into force of a future climate regime. With the enactment of the Kyoto Protocol, industrialised countries (Annex I countries of UNFCCC) were legally bound to reduce their greenhouse gas emission by 5.2% before 2012 with targets set for each nation based on their 1990 level. The Kyoto targets covered emissions of the six major greenhouse gases, namely carbon dioxide, methane, nitrous oxide, hydrofluorocarbons, perfluorocarbons and sulphur hexafluoride. With UNFCCC and Kyoto Protocol, carbon became a tradable commodity and carbon trading started happening under the compliance market supervised by UNFCCC. The Conference of Parties to the UNFCCC at its seventeenth session (COP 17) held in Durban in 2011 agreed to launch a new climate agreement for post-2020 and a second commitment period of Kyoto Protocol till 2020 with more ambitious emission reduction targets from Annex I countries.

Paris Agreement: The Paris Agreement was adopted in 2015 and aimed to limit the temperature increase to 1.5 °C above pre-industrial levels and keep the global average temperature rise well below 2 °C above pre-industrial levels. The agreement also intends to make finance flows consistent with a low greenhouse gas emissions and climate resilient path, and to strengthen nations' capacity to address the effects of climate change.

Encouraging progressively ambitious climate action from all country Parties on the basis of their voluntarily submitted Nationally Determined Contributions (NDCs) is the cornerstone of the Paris Agreement (Rajamani 2016). This requires that all Parties report regularly on their GHG emissions from all sources and their efforts on implementation of actions for NDCs. A global stocktake will take place every 5 years and the collective progress towards achieving the purpose of the agreement will be assessed (UNFCCC 2016). Mitigation goal under the Paris Agreement is to achieve a balance between anthropogenic emissions by sources and removals by sinks of greenhouse gases in the second half of this century. The Green Climate Fund (GCF) has been given an important role in supporting Paris Agreement goals.

The Paris Agreement *inter alia* has also shown strong commitment to reducing emissions from deforestation and forest degradation and conservation efforts leading to the enhancement of forest carbon stocks. The Paris Agreement recognises the role of forests as carbon sink for mitigation of climate change. Article 5 of the Paris Agreement encourages all Parties to act for the conservation and enhancement of sinks and reservoirs of greenhouse gases, including forests. The Agreement also encourages countries to implement and support activities related to reducing emissions from deforestation and forest degradation including conservation (REDD+) through results-based payments. Legitimising REDD+ activities under the Paris Agreement is a strong political signal for forest conservation-based mitigation actions. This will boost the confidence of developing forested countries to start working with REDD-plus strategy and readiness activities (Rawat and Singh 2016a, b).

Glasgow Climate Pact: Glasgow Climate Pact was adopted with an aim to make the 2020s as a decade of climate action and support. Country Parties agreed on, *inter alia* supporting efforts to build resilience to climate change, to curb greenhouse gas emissions and to provide the necessary finance for both. Parties reaffirmed their duty to fulfil the pledge of providing 100 billion dollars annually from developed to developing countries. Parties also agreed to work to close the gap between current emission reduction plans and what is needed to reduce emissions, so that the rise in the global average is limited to 1.5 °C. Nations also agreed on the transparent reporting of climate actions and support, rules related to market mechanisms and non-market approaches under the Agreement (UNFCCC 2021a).

Estimates of the global emissions outcome of current nationally submitted mitigation actions in the form of nationally determined contributions (NDCs) under the Paris Agreement by 2030, would lead to global greenhouse gas emissions of 52–58 billion tonnes of CO_2 eq. per year. Various pathways reflecting these NDC actions would not limit global warming to 1.5 °C, even if further supplemented by very ambitious emissions reductions after 2030. Avoiding the overshoot of GHGs and reliance on large-scale implementation of carbon dioxide removal technologies in the future can only be achieved if global CO_2 emissions start to decline well before 2030 (IPCC 2018).

12.3 Climate Change Projections for India

The Ministry of Environment and Forests, Government of India in its Second National Communication submitted to UNFCCC analysed climate change scenarios in India using the high-resolution regional climate model. Basic parameters like air temperature, rainfall and mean sea level pressure were analysed in order to obtain climatic projections for the end of this century. In order to generate an ensemble of future climate change scenarios for the Indian region, modelling simulations were carried out for the period from 1961 to 2098. The findings include no significant decrease in the monsoon rainfall in the future except in some parts of the southern peninsula. Its simulations for the year 2020, 2050 and 2080 indicated an all-round warming. According to model projection, the annual mean temperature increase will range from 3.5 °C to 4.3 °C by the end of this century (MoEF 2012).

Ministry of Earth Sciences, Government of India conducted climate change assessment in different parts of the country and brought a report titled Assessment of Climate Change over the Indian Region. The report covered various aspects like local climate change, variability of monsoon, variations in mean temperature, variability in precipitation, variation in greenhouse gas emissions, frequency of droughts and floods, impact of ocean warming, rise in sea level and melting of glacier (Krishnan et al. 2020). The salient features of this report are highlighted as under:

Changes in Temperature: Increase in the average temperature in India was reported to be around 0.7 °C during the time period of 1901–2018. This rise in temperature was largely due to emission of greenhouse gases. However, average temperature over India was projected to be risen by 4.4 °C comparative to the time period of 1976–2005 under the Representative Concentration Pathway 8.5 (RCP 8.5) scenario by the end of this century.

In the recent time period of 1986–2015, temperature of the warmest day and coldest night was risen by about 0.63 and 0.4 °C respectively. The frequencies of occurrence of warm days and warm nights are projected to be increased by 55% and 70%, respectively relative to the period of 1976–2005 by the end of this century. Magnification of heat stress is expected across India because of rise in surface temperature mainly over the Indo-Gangetic and Indus River basins.

Warming of Indian Ocean: Average surface temperature of the tropical Indian Ocean was risen by 1 °C during the time period of 1951–2015, which is distinctly higher than the global average of 0.7 °C for the same time period. The Indian Ocean had also exhibited an increasing trend in ocean heat content in the upper 700 m during the past six decades (1955–2015), while abrupt rise was noticed during the past two decades of 1998–2015. Further, sea surface temperature and ocean heat content are projected to be risen continuously in the Indian Ocean during this century.

Changes in Rainfall: A decline of 6% in the summer monsoon precipitation (June to September) was recorded from 1951 to 2015 with marked decreases over the Indo-Gangetic Plains and the Western Ghats. During the summer monsoon season, a shift towards more frequent dry spells (27% higher for the period 1981–2011 as compared to 1951–1980) and more intense wet spells were observed. Worldwide increased occurrence in the frequency of localised heavy rains has been observed in response to increased atmospheric moisture content. The frequency of extreme events of daily precipitation resulting in more than 150 mm of rainfall intensities per day increased by about 75% during the time period of 1950–2015 over central India. The anticipated reductions in future anthropogenic aerosol emissions coupled with persistent global warming, models (Coupled Model Inter-comparison Project Phase 5 CMIP5) projected an increase in the mean and variability of monsoon precipitation by the end of this century, together with considerable increases in daily precipitation extremes.

Droughts: During the past 6–7 decades, the overall decrease in seasonal summer monsoon rainfall has resulted in an increased tendency of droughts in the country. During the period of 1951–2016, there was a significant increase in the frequency and spatial extent of droughts mainly over central India, southwest coast, southern peninsula and north-eastern India where on an average more than 2 droughts per decade were observed during this period. Over the same period, the area affected by drought resulted in 1.3% increase per decade. Climate change modelling under the RCP8.5 scenario projected a high probability of increase in the frequency (>2 events/decade), intensity and area under drought conditions because of the increased variability of monsoon rains and increase in water vaporisation in a warmer atmosphere by the end of this century in the country.

Sea Level Rise: Globally, melting of ice and thermal expansion of ocean in response to atmospheric warming resulted in rise in sea levels. During the period of 1874–2004, sea level rise in the North Indian Ocean occurred at the rate of 1.06–1.75 mm/year. However, during the time period of 1993–2017, it was enhanced to 3.3 mm/year, which was quite comparable to the current rate of global mean sea-level rise. At the end of this century, sea level in the Northern Indian Ocean was projected to be increased by about 300 mm under the RCP4.5 scenario.

Tropical Cyclones: Over the North Indian Ocean basin, annual frequency of tropical cyclones had shown a significant reduction since the time period of 1951–2018. During the post-monsoon season, frequency of very severe cyclonic storms had increased significantly in the time period of 2000–2018. Rise in the intensity of tropical cyclones in the North Indian Ocean basin during the twenty-first century has also been projected by the climate projection models.

Changes in the Himalayas: During the period of 1951–2014, rise in temperature of about 1.3 °C was observed in the Hindu Kush Himalaya. In recent decades, the declining trend in snowfall and melting of glaciers were also observed in several areas of Hindu Kush Himalaya, while the high-elevation of the Karakoram ranges of Himalaya has experienced an increase in snowfall that contributed to glacier recharge in the region. The annual mean surface temperature was projected to be increased by about 5.2 °C under the RCP8.5 scenario over the Hindu Kush Himalaya by the end of this century. It was also projected (by the CMIP5 projections under the RCP8.5 scenario) increase in annual precipitation and decrease in snowfall over the Hindu Kush Himalayan region by the end of this century.

12.4 Impacts of Climate Change on Forests in India

Climate in India is strongly determined by its topographical and geographical features, particularly the Himalayas in the north, the seas in the east, west and south and desert in the west. During the winter, the Indian subcontinent is shielded from widespread incursions of cold, extratropical winds by the Himalayas and the Hindu Kush mountain system. Furthermore, the Himalayas and the Tibetan Plateau's seasonal warming creates a north-south temperature differential with respect to the tropical Indian Ocean, which is crucial for starting the summer monsoon circulation on a big scale. A peculiar feature of India's climate is the remarkably robust seasonal wind and precipitation cycle (Turner and Annamalai 2012).

The Indian summer monsoon has a significant role in the global climate. The Western Himalaya also receives significant amounts of precipitation from the western disturbances, which originate in the Mediterranean region, during the winter and early spring in addition to monsoonal rains (Dimri et al. 2015; Hunt et al. 2018). In addition, the Indian subcontinent is vulnerable to a variety of extreme weather events such as heat waves, tropical cyclones, thunderstorms, floods and droughts.

12.4.1 Climate Change Factor in Driving Forest Dynamics

Average temperature in India during the time period of 1901–2018 increased by about 0.7 °C (Krishnan et al. 2020). Furthermore, there was a general downward trend in rainfall, a decrease in seasonal mean rainfall and a rise in monsoon break days. Additionally, there was a rise in the frequency of intense precipitation events and a decrease in light and moderate rainfall occurrences (Christensen et al. 2013). The climate change trends in India and their impact on forests and vegetation formations are one of the most widely studied and discussed in recent years. There are many models and projections being developed and tested by many workers all over the country on the trends, and their impact on ecosystem processes, vegetation changes and other biophysical environments (MoEF 2010, 2012; Chaturvedi et al. 2008, 2011, 2012). MoEF (2010) while assessing the impact of climate change on different sectors concluded that by 2100, more than one-third of the forest cover will mostly convert from one type to another type of forest. Increase in net primary productivity in the Western Ghats by 23%, in the north-eastern region by 20%, in the Himalayas by 57% and in the coastal region by 31% is expected in the 2030s. The distribution of species tends to shift towards higher latitudes and altitudes as the earth warms. For each 1 °C of warming, tree ranges in the northern hemisphere have the potential to expand 100 km northward while southern boundaries retreat. This is a process which has been tracked since the last ice age (Davis 1989).

12.4.2 Patterns of Climate Change in India

Several assessments were made on different climate change scenarios in India and some are highlighted as under:

Patterns of Indian Summer Monsoon Rainfall: There are many studies and reports on the rainfall pattern at the all-India level. Based on the rainfall series constructed by Parthasarathy et al. (1994), it is found that, over a long period of time, especially when considering the all-India time scale, the monsoon rainfall is primarily random and devoid of any trend (Mooley and Parthasarathy 1984). However, on a spatial scale, some trend was noticed by Parthasarathy et al. (1994). Indian summer monsoon rainfall displays multi-decadal variations, in which there was clustering of wet or dry anomalies (Pant and Kumar 1997).

Projections of Climate Change in India: Earlier BIOME model (versions 3 and 4) was applied for making an assessment of possible climate change effects on Indian forests (Ravindranath et al. 2006). It was projected that under A2 and B2 scenarios, there would be shift in forest types in 77% and 68% of India's forested grids, respectively due to climate change. Using the current set of probable radiative forcing-based greenhouse gas emissions scenarios, climate models under the Coupled Model Inter-comparison Project showed that the Indian subcontinent will warm by 1–2 °C (compared to the 1961–1990 baseline) by 2030 under the more moderate Representative Concentration Pathway 4.5. Greater warming (2–3 °C) was expected in the northwest and Himalayan regions (Chaturvedi et al. 2012). Under RCP 4.5, projections for the end of the century range from 2 to 3 °C for the most parts of India, with the Himalayas experiencing a 5 °C increase, while under the extreme Representative Concentration Pathway 8.5 scenario country's temperatures might rise by 5 °C by the end of the century with Himalayan warming surpassing 7 °C (Chaturvedi et al. 2012). Models indicate that during the Indian summer monsoon, there will be an increase in inter-annual variability along with a rise in the mean and extremes of rainfall, despite recent downward trends in rainfall (Christensen et al. 2013). Precipitation was expected to rise by 6–14% overall later this century (Chaturvedi et al. 2012).

Analysis of Climate Data: ICFRE (2013) analysed the climatic data of rainfall and temperature for the period of 80 years (1930–2010) from 83 stations representing different forest types of the country. At the national scale, the analysis of climate data revealed an increase of 0.7 °C, 0.5 °C and 0.5 °C in mean maximum, mean minimum and average annual temperature from 1931 to 2010 respectively. Highest increase of 0.9 °C was observed in mean maximum and mean minimum for central India. In the southern region, mean maximum and mean minimum and annual mean temperatures increased by 0.9 °C, 0.4 °C and 0.6 °C respectively. In western India, mean maximum and mean minimum increased by 0.7 °C. In east and north east India, mean maximum temperature increased by 0.6 °C. In north India mean maximum, mean minimum and annual mean temperatures increased by 0.5 °C, 0.3 °C and 0.2 °C respectively.

At national level, total annual rainfall and number of annual rainy days decreased by 241 mm and 5 days, respectively. During this period rainfall in different regions reduced by 55 mm in western India to 1033 mm in north east India. In Northeast, this remarkable decrease is attributed mainly to a sharp decline in rainfall in Cherrapunjee once rated as the highest rainfall place in India. In Cherrapunjee rainfall decreased from 11,418 mm in 1931–60 to 3675 mm in 1990–2010 period, a net decrease of 7743 mm over the period. Rainy days also decreased from 3 in west to 10 in central India.

12.4.3 Climate Change Impacts on the Vegetation Composition

Migration of Species Towards Higher Elevations: Climate combined with patterns of land-use change can result in the contraction or shift in species ranges. Various studies have depicted that, under future climate scenarios, the suitable areas for many species would change, reducing or increasing in size or shifting in latitude and elevation (Dullinger et al. 2012; Niskanen et al. 2019). ICFRE (2013) study revealed that forest types described by Champion and Seth (1968) at different elevations have shown few significant changes with regard to their distributions. The blue pine found in the higher elevations up to 1700 m is now reported from in higher elevations up to 2000 m indicating the shift in the tree lines towards higher elevations. The change may be climate-induced due to the sites becoming warmer (ICFRE 2013).

Impact of Warming on Alpine Flora: Alpine zone is a unique vegetation type that occurs in the higher elevations of the Himalayas, which are regarded as biodiversity-rich hot spots. The alpine zone is the critical repository of biodiversity harbouring a high number of endemic and endangered species (Korner 2003; Viviroli and Weingartner 2004; Woodwell 2004). In Europe, studies of plant communities have indicated lower elevation dwarf-shrubs, sedges and meadows to have increased their cover at high altitudes at the expense of pioneer snow-bed communities. There are also studies reporting the upward migration of species and increase in species richness at summits during the last century (Walther et al. 2005). Singh et al. (2021) found that though evidences for the upslope movement of tree line at the study sites (Kashmir, Uttarakhand and Sikkim) are lacking, densification of tree line ecotone and advancement of krummholz species (*Rhododendron campanulatum*) into alpine meadows are noticeable. Both can be a serious threat to the diversity and integrity of alpine meadows. In Sikkim Himalaya, Telwala (2012) observed that Alpine plant life is proliferating and the warming of Sikkim Himalaya during the last century has resulted in upward migration of the range margin of the species.

Changing Himalayan Flora: Studies conducted by the Indian Council of Forestry Research and Education on the Himalayan flora indicated very significant changes in the vegetation composition and their distribution pattern (ICFRE 2013). The lower Himalayan forests which were occupied by the *Pinus roxburghii*, due to disturbances are now found either in low proportion or completely absent from many sites. The causes could be due to both anthropogenic and climate change-induced episodes. In the last 50 years, the disturbances induced by fire, grazing and lopping have affected the regeneration of broad-leaved species like oaks and at lower elevations broad-leaved species are missing. It is gratifying to note that some of the sites at higher elevations which were earlier dominated by the conifers are now being colonised by the oaks. The study concludes that this may be due to the sites becoming warmer during summer, which is favourable for the oaks. Moisture regimes in the moist temperate climate of the Western and Central Himalayas are showing changes in the species composition due to the availability of more moisture in some sites as a result of the melting of snow. Thus, a twin benefit of sites becoming warmer and moist is favourable for the broad-leaved species (ICFRE 2013).

Chitale and Bahera (2012) based on species distribution models predicted the geographic ranges of *Shorea robusta* (*sal*) from occurrence records and site-specific environmental data. The authors have (i) generated the 1960s scenario for *sal* on the basis of the existing published literature, (ii) confirmed the species occurrence data using satellite imagery for the period of 1972–75, (iii) run the Maxent species distribution model to predict the distribution for the year 2020 under climate change scenario SRES A1-B and (iv) validated the prediction using more than double the amount of species occurrence data gathered during the last decade (1998–2008). The model identified moisture as the prominent factor that would influence with >90% certainty the shift towards northern and eastern India.

North-East India: A few stations of north-Indian states have shown decreasing trends in rainfall. The change in the rainfall pattern in the north-eastern India has shown some significant changes in the species composition as well wherein more evergreen species are becoming dominant in the semi-evergreen and deciduous forest types. The increased wetness and reduction in the dry period also contributed for such a change. Absence of *sal* (*Shorea robusta*) in the Kalimpong from the East Himalayan forest type is one such occurrence, which is attributed to the canopy closure and appearance of semi-evergreen and evergreen species (ICFRE 2013).

Vegetation Changes in the Western Ghats: Similar to the Himalayas, the Western Ghats also have shown some evidences of vegetation changes at the species level. Though it is difficult to establish direct correlations with the climate change and the vegetation, the broader indications are available to relate the species composition in the most fragile ecosystems, which have a very narrow range of ecological amplitude. The *sholas* and the grasslands have shown the evidence of changing species composition (ICFRE 2013).

Changes Observed in Tropical West-Coast Evergreen Forest Type: The tropical west coast evergreen forest types according to the Champion and Seth's (1968) classification, predominantly have *Dipterocarpus indicus, Vateria indica, Cullinea* spp., *Poeciloneuron indicum, Lophopetalum wightianum*, etc. However, ICFRE (2013) study showed that predominant species of west-coast evergreen forests such as *Dipterocarpus indicus and Vateria indica* were not found in the study plots. The change in the species composition particularly the absence of *Vateria* and *Dipterocarpus* spp. can be related to reduction in the rainfall by nearly 300 mm and a temperature rise by 1 °C in Kozhikode, a representative station of southern Western Ghats where the low elevation west coast evergreen forests were found (ICFRE 2013).

Enlarging Sholas: *Sholas* are the hill forests found distributed on the slopes of Western Ghats mountains sometimes in the depressions and valleys surrounded by the tropical evergreen forests. In the Western Ghats, the top of the hills adjacent to *sholas* is completely devoid of vegetation and this formation was described by Ranganathan (1938) as the climatic climax formation due to repeated fires forming the grasslands. However, ICFRE (2013) study of the site showed the appearance of a few tree species indicating the change in the site conditions favouring the tree species particularly primary species to colonise. Thiruvananthapuram, Coonoor and other places in Kerala and Karnataka have recorded increased rainfall by 150 mm over the period of the last 80 years (1930–2010) which has favoured the species changes at higher elevations.

Semi-Evergreen Forest Type: These forests are generally found in the areas where the annual rainfall is limited to 1500 mm. Floristically the predominant species found are *Syzygium cumini, Dysoxylum malabaricum, Mangifera indica, Xylia xylocarpa, Terminalia tomentosa, Bambusa bamboo, Pterocarpus marsupium, Dalbergia latifolia,* etc. The lateritic secondary semi-evergreen forest type in Karnataka exhibited a slight change in the species composition wherein there are more evergreen species recorded than the moist deciduous type species indicating the change in the species composition due to progressive succession of semi-evergreen forests towards evergreen forest type (ICFRE 2013).

Tirunelveli Semi-Evergreen Forest Type: Many stations in Western Ghats have shown an increased trend of rainfall which is also in conformity with the rainfall analysis by Parthasarathy et al. (1994). Due to the increased wet period, semi-evergreen forest type is showing more evergreen species. The presence of *Hopea parviflora, Vateria indica* is an indication of the type turning into evergreen type. Representative stations, that is, Tuticorin, Thiruvananthapuram and Ooty have recorded increased rainfall. The change in the climate seems to have an impact on the transformation of the forest type (ICFRE 2013).

Das and Sukumar (2018) while studying long-term forest dynamics in tropical deciduous teak forests in Mudumalai Forest Dynamics Plot indicate that through its impact on species diversity, tree growth and mortality, fire frequency and its intensity, the amount and seasonality of precipitation play a vital role in determining the dynamics of forests. Based on the characteristics of the dominant species in these communities, the impact of drought and fire seems to vary.

The ICFRE (2013) study concluded that there appears to be a differential response of the vegetation at different locations mainly due to complex change factors, that is, climate, edaphic and anthropogenic. The present forest vegetation in different locations is differently getting affected by the temperature rise and differential intensity of rainfall as well as by the decreasing number of rainy days. Together with management interventions and biotic factors, the forests are responding and vegetation changes becoming evident. It is, therefore, of utmost importance that the forest managers take adaptation measures particularly dealing with forest composition, hydrology and rain conditions and also choose species carefully for the purpose of mitigation. Above all local community need to be made partners in these actions. This will ensure scientific as well as people-centric response to climate change.

Due to the high temporal variability in the dynamics of tropical forests and the fact that simultaneously several interacting global change factors like climate change, invasive species, habitat degradation and loss and nitrogen deposition etc. are affecting them, it is difficult to identify, infer and forecast the effects of changes driven by climate. Das and Sukumar (2018) suggested that persistent long-term monitoring studies are required to comprehend how forest ecosystems and the services they provide are changing as a result of climate change. The long-term ecological observatories programme of the Ministry of Environment, Forest and Climate Change, Government of India (MoEFCC 2015a, b) as a constituent activity of the Climate Change Action Programme of the country is an attempt to address these gaps. The fundamental objective of this network is to understand the biophysical and anthropogenic drivers of ecosystem change and their effects on social-ecological responses. The programme aims to understand the link between climate change and ecological processes by covering the wide range of diversity and complexity of representative landscapes of the country.

India being a tropical country with diverse climatic conditions, ecosystem and vegetation types are highly sensitive and vulnerable even to a smaller variation in the temperature and rainfall regimes. In this context, trends analysis of climate change in different climatic zones and their corresponding impacts on the forest types, sub-types and species composition is highly significant and valuable, to develop and to test the climate projections models. Further, the climate change impacts can also be regularly monitored at landscape, vegetation types and species level to mitigate the climate change impacts.

12.5 Vulnerability of Forest to Climate Change

Various studies have revealed consistent changes in the species with changing climate such as shifting to higher altitudes/latitudes, occurrence of early phenological events and occurrence of new invasive species. Modelling-based studies projected that the structure of food web and its stability favour the invading of species as a result of warming (Sentis et al. 2021). Phenology of exotic species responded more adaptively to warming than those of native species in simulated experiments (Zettlemoyer et al. 2019). Under long-term observations, the greater phenological responsiveness to warming was the reason behind the success of exotic species (IPCC 2022a).

Population extinctions driven by climate change are higher in the tropics (55%) than in temperate regions (IPCC 2022a). Such events are higher in freshwater (74%) than in marine (51%) or terrestrial (46%) habitats and higher in animals (50%) than in plants (39%) (IPCC 2022a). Since 1979, fire seasons have now prolonged in one-quarter of vegetated areas due to increasing temperature, aridity and drought. Consistent with climate change burned areas have been increased in the Amazon, the Arctic, Australia and parts of Africa and Asia (IPCC 2022a). Management practices such as prescribed burning, thinning and allowing few controlled forest fires where it is beneficial to the ecosystem can be introduced (Meigs et al. 2020).

Forests are under stress due to climatic and non-climatic factors, and studies were conducted by various researchers on vulnerability assessment of forests to climate change in India (Aaheim et al. 2011; Gopalakrishnan et al. 2011; Murthy et al. 2011; MoEF 2012; Sharma et al. 2013; Upgupta et al. 2015; Sharma et al. 2017; Kumar et al. 2019; Alam et al. 2022; Kanwar and Kuniyal 2022; Roshani et al. 2022; Thakur et al. 2022).

Based on climate projections from the Regional Climate Model of the Hadley Centre and the global dynamic vegetation model IBIS for A1B scenario, Gopalakrishnan et al. (2011) assessed the effects of projected climate change on forest ecosystems in India for both the short-term (2021–2050) and long-term (2071–2100) periods. Based on the dynamic global vegetation modelling, vulnerable forested regions of India have been identified to assist in planning adaptation interventions. Upgupta et al. (2015) assessed the vulnerability of forests to climate change in the state of Himachal Pradesh and prioritised the districts for vulnerability reduction under 'current climate' and 'future climate' scenarios. Identifying forests that are vulnerable to climate change will guide forest managers to prioritise the allocation of available resources and appropriate interventions in forest management practices restoring health and productivity of forests and also to plan climate change resilience in the long term.

Sharma et al. (2017) conducted a study on climate change driven vulnerability assessments for Indian forests and concluded that percentage of grid points under high and very high inherent vulnerability classes for Himalayan moist temperate forests, subalpine and alpine forests. Low inherent vulnerability of these forests has great significance because most of the forest biodiversity and endemicity are hosted in India by these forest types in the biodiversity hotspots in Western Ghats in peninsular India, the north Himalayan region and the northeast. Plantation forests show higher inherent vulnerability than natural forests. Roshani et al. (2022) conducted a systematic review from 1990 to 2019 and examined vulnerability of forests to climate change and its management practices. The issues that are currently being faced include the conversion of forest land into other land uses, fragmentation and disturbance of forests, and the impact of climate change on the forest ecosystem.

12.6 Forests for Climate Change Adaptation

Greenhouse gas emissions especially after the industrial revolution in the west and industrial development in developing economies in the recent past are the reason for current climate change. The scale of future change will largely depend on how and to what extent these man-made emissions of greenhouse gases are reduced. However, owing to historical emissions and inertia in the climate system, climate changes are unavoidable irrespective of climate change mitigation actions by the global community. Evidence of changes in extreme climatic events such as heatwaves, heavy precipitation, droughts, frequency and magnitude of tropical cyclones are increasing. The successive Assessment Reports of the IPCC presented clear evidence that the climate is changing which is mainly driven by emission of greenhouse gases (IPCC 2022b). Mitigation and adaptation to climate change are now being integrated into resource management strategies as well.

Adaptation to Climate Change: Adaptation is required to maintain the functioning of the natural and social systems which are vulnerable to changing climate. Adaptation can be distinguished as anticipatory or reactive, autonomous or planned. IPCC (2007) defined adaptation as '*adjustment in natural or human systems in response to actual or expected climatic stimuli or their effects, which moderates harm or exploits beneficial opportunities*'. In nature, biological adaptation is autonomous and reactive as organisms respond over time to changing conditions. In response to changes, people, on the other hand, may show autonomous adaptation, or plan for adaptation either in response to changes (Adger et al. 2007).

Structure and functioning of forest ecosystems are likely to be altered by the changes in climate as climate exerts a significant influence on the distribution, structure and ecology of forests. It was projected that climate change has adverse impacts on forest ecosystems and this in turn will affect the forest biodiversity, biomass and regeneration (Ravindranath et al. 2012; ICFRE 2013; Bahuguna et al. 2016). Forests support the livelihoods of about a billion people living in extreme poverty worldwide and provide employment for over 100 million people. More than 80% of the world's terrestrial biodiversity is housed in forests. Forests help protect watersheds that are critical for the supply of clean water to most of humanity. Climate change, however, poses enormous challenges for forests and people (Singh et al. 2015).

There is a close association between forests and community adaptation. Forests are known to play a key role in the adaptive capacity of local communities by providing various ecosystem goods and services, while actions of the community can enhance or reduce the adaptive capacity of forests to changing climate (Locatelli et al. 2011). There are inherent social limits to adaptation that are deeply rooted in the cultural values of the societies. Many societies have value systems rooted in their sense of belongingness to land and forests (Kant and Wu 2012).

Adaptive Strategies: Three types of adaptive strategies have been identified to deal with the climate change *viz*., incremental, anticipatory and transformational (Kates et al. 2012; Joyce et al. 2013; Stanturf 2015). Incremental adaptations are also known as no-regrets approaches. This kind of strategy provides adaptation to future conditions while at the same time, benefits are also realised under current climatic conditions. It composed of extensions of existing practices instituted in response to variations in climate and extreme events (Hobbs et al. 2011; Kates et al. 2012; Joyce et al. 2013). Many of the same techniques of incremental approaches may also be applied in anticipatory approaches, however, focussing on futuristic climate change adaptation (Alfaro et al. 2014).

Transformational adaptations are larger in scale or more intense than incremental adaptations. These approaches represent an effort to foresee or react to climate change proactively (Kates et al. 2012; Joyce et al. 2013). Such approaches anticipate larger shifts in climate, which may call for long-term major changes in management objectives or production systems. Aided migration of species well beyond their natural range of distribution, introducing species which are not native and restoring keystone species through genetic modification are also part of transformational adaptations (Davis et al. 2011; Pedlar et al. 2012; Williams and Dumroese 2013; Jacobs et al. 2013). The frequency and intensity of extreme climate events are expected to increase in future and transformational adaptation will likely to be associated with such events (Allen et al. 2010; Reichstein et al. 2013; Cai et al. 2014). The natural disasters are caused by long-term droughts, insect pest epidemics, forest fires and wind disturbances whether linked to climate change or not, all these calamities offer stimulus for restoration efforts (Stanturf et al. 2014; Van Aalst 2006). For an effective adaptation strategy, considerable planning and adaptive management need to be required (Joyce et al. 2009) especially in the advancement of extreme climatic events (Stanturf et al. 2007). The use of genetically diverse seed sources, planting material, placement of mixed systems in a systematic manner and promoting combination of tree species that are adapted to climate change could be a strategy for a less dramatic transformational approach (Prabhu et al. 2015).

Adaptation to Climate Change in Terrestrial and Freshwater Ecosystems: IPCC reports have highlighted adaptation issues for reducing climate change vulnerability of ecosystems and their services, and also acknowledging the need for both human-assisted and autonomous adaptation to safeguard species and ecosystems. The Fifth Assessment Report of IPCC recognised ecosystem-based adaptation, that is, adaptation for people which is based on improved environmental management, restoration and preservation as an emerging possibility (IPCC 2014). Ecosystem-based adaptations, conservation and similar other ideas were also integrated throughout the Special Report on Climate Change and Land (IPCC 2019). Natural systems react in different ways to changes in the climate and other environmental factors. Individual organisms can react to the changing environment through processes of growth, mobility and development. During the course of succeeding generations, genetic adaptation to shifting environmental conditions drives the evolution of species and populations. There is a limit on these natural processes which can preserve biodiversity and the advantages that humans get from nature (IPCC 2022a). Maintaining the sustainable flow of vital ecosystem services such as carbon sequestration and water supply also requires protection from exploitation, deforestation and forest degradation (Dinerstein et al. 2020; Portner et al. 2021).

Ecosystem-Based Adaptation: Ecosystem-based adaptation encompasses a variety of approaches such as the restoration of river and coastal systems to lower flood risks and enhance water quality, and the creation of natural green space inside cities to lower temperatures through the process of evaporative cooling and shade effect. Many other ideas such as ecosystem services, natural capital and disaster risk reduction are intimately related to ecosystem-based adaptation (IPCC 2014). Ecosystem-based adaptation can function at different levels, including local, catchment and regional. At the local level, lowering down of temperature by vegetation and managing microclimate can be achieved in number of ways. Such ecosystem-based adaptation techniques are now more widely gaining importance. Shade trees are a traditional method that can be currently used for climate change adaptation in both urban and rural areas. Likewise, in many places of the world, there is an increased risk of forest fires to both people and ecosystems. Apart from climate change, forest fire is also caused by past management practices. Restoring more natural fire regimes and scaling up better fire management practices can lower the fire-related hazards (IPCC 2022a).

Improving connectivity to promote the colonisation of new places is the most frequently mentioned strategy for species to adapt to changing climate. There are numerous ways to improve connectivity, varying from creating ‘stepping stones’ and ‘corridors’ to expand the habitat suitable for the species. Depending on the species and situation, different plans may work better in different situations (Keeley et al. 2018). Furthermore, habitat connectivity can help to make populations more resilient to extreme climatic conditions (Oliver et al. 2015). The benefits of climate change adaptation strategies and their effectiveness will become apparent in the coming years and decades which are inherently difficult to evaluate (Morecroft et al. 2019). However, the application of suitable measures for adaptation can lessen the vulnerability of many aspects of biodiversity to climate change.

Ecosystem-based adaptation is also known as a nature-based solution besides supporting sustainable development goals also contributing for the adaptation and mitigation of climate change. Inclusive decision-making and adaptive management pathways that produce climate-resilient systems supporting numerous sustainable development goals are essential for an effective nature-based climate change mitigation programme. Inadequately planned and poorly designed nature-based mitigation initiatives can have a number of detrimental effects, such as making other sectors compete for land and water resources, negatively affecting human well-being, and unable to provide long-term, sustainable mitigation goals (IUCN 2020). Nature-based solutions offer benefits of both adaptation and mitigation to climate change that help protect habitats and wild species. Nature-based solutions often contribute to other sustainable goals as well (Griscom et al. 2017; Keesstra et al. 2018; IPCC 2018, 2019; Lewis et al. 2019; Malhi et al. 2020; Seddon et al. 2020a, b). Nature-based solutions can vary from the protection of terrestrial, freshwater and marine ecosystems to the restoration of degraded ecosystems. These also include naturally regenerating ecosystems that are sustainably managed and provide food, fibre and energy (Box 12.1).

Conservation of indigenous biodiversity and, mitigation and adaptation to climate change can be achieved through agroecological techniques (Sinclair et al. 2019; Snapp et al. 2021). Climate change mitigation schemes based on nature-based solutions that are well conceived and applied have the potential to reduce greenhouse gases emissions by boosting carbon uptake in the natural ecosystems along with protecting and restoring biodiversity. Many species that are unable to thrive in habitats that have been degraded find refuge in intact natural forest ecosystems, which also serve as important storehouses of carbon. Such ecosystems have been cleared or deteriorated, or they are managed both sustainably and unsustainably in various places. Land degradation and deforestation still remain major contributors to global greenhouse gas emissions (IPCC 2022b). Highly effective nature-based solutions include protection of natural forests and sustainable management of forests that perpetually provide goods and services (Bauhus et al. 2009). Apart from offering a wide range of ecosystem goods and services, natural forests and sustainably managed forest that are rich in biodiversity play a crucial role in the adaptation and mitigation of climate change (Favero et al. 2020; Mackey et al. 2020). Forests play an important role in regulating the water cycle and improving water quality, increasing infiltration to ground water and reduction in surface runoff at local, regional and global scale (Bruijnzeel 2004; Zhou et al. 2015; Creed and van Noordwijk 2018; Alvarez-Garreton et al. 2019). Conservation of biodiversity, mitigation of climate change and provision for goods and services from forests can be improved through sustainable management of forests and protection of natural forests in an inclusive way (Bauhus et al. 2009; Nabuurs et al. 2013). Reforestation is one of the most useful and affordable methods of sequestering and storing carbon, which can also contribute to the preservation and recovery of biodiversity (Nabuurs et al. 2017; IPCC 2018; Smith et al. 2018; Cook-Patton et al. 2020; Cowie et al. 2021; Drever et al. 2021). This task can be accomplished by planting trees and shrubs or by letting them naturally colonise the area.

Box 12.1 Nature Based Adaptation: Mix Farming to Counter Rain Dependency

The indigenous farmers in Garhwal region, Uttarakhand State in India have, over the centuries built a rich base of agricultural biodiversity adapted to local agroclimatic conditions. The mixed cropping system developed is especially suitable for rainfed areas in high altitudes. The farming communities cultivate many species and varieties of legumes. A similar practice of diverse farming is also seen in lower altitudes of Garhwal. This system helps the farmers to withstand the uncertainties of weather, so that overall production is never zero at any point of time. The synergetic combination of many different species preserves soil fertility and land does not have to be kept fallow. The farmer is assured of at least one crop, even if major crop is failed because of pests, diseases or drought. The farmers of Garhwal region grow 126 varieties of rice, 8 varieties of wheat, 40 varieties of finger millet, 6 varieties of barnyard millet, 110 varieties of kidney beans, 7 varieties of horse gram, 8 varieties of traditional soybean and 10 varieties of French beans. Since all crops are not harvested at the same time, the diverse agriculture output helps to meet various household requirements over a long period. This agrobiodiversity is another crucial component of the region's sustainable food system. It enables the farmers to benefit from certain varieties even when there is damage to other crops.

Source: Centre for Science and Environment 2014

12.6.1 Forests and Climate Change Adaptation in India

Adaptation to climate change in forests is also characterised by long gestation periods that necessitate preparation, execution and oversight in advance. Adaptation to climate change in the forest sector is required to address current stresses such as causes of deforestation and forest degradation, forest fragmentation, cattle grazing and non-sustainable extraction etc. Ravindranath (2007) highlighted some strategies and practices to reduce the vulnerability of forests and plantations that include: (i) forest and biodiversity conservation, (ii) expansion of protected areas, (iii) sustainable logging and management of forests, (iii) mixed species forestry, (iv) planting in advance and facilitating natural migration of some species by transplanting and (v) adoption of sound silvicultural practices to lessen the impact of climate change.

Afforestation and reforestation are the main mitigation options. Agroforestry is one of the good choices for enhancing overall productivity and carbon sequestration of agriculture lands, and also improving resilience to the negative effects of climate change. It is anticipated that plantations for bioenergy purposes will be managed rigorously in order to maximise biomass productivity. Considering adaptation to climate change at the time of conceiving and operationalising of carbon forestry projects or projects under a new market mechanism of the Paris Agreement could be an additional co-benefit of carbon mitigation projects.

It is very important to evaluate the expected impacts of future climate change on forests and develop suitable adaptation strategies and actions which ensure conservation of biodiversity, safeguarding of the livelihoods of the communities depending on forests and assurance of roundwood production for industrial and commercial purposes. However, in some instances, it has been noted that some communities have taken initiatives in respect of adaptation measures. In the Himalayan region, farmers have started growing crops like cardamom, bamboo and fruit trees under agroforestry practices to ensure food security and additional income due to erratic rainfall during spring and summer seasons (Verma 1998). Similarly, in the north-eastern part of India, the age-old practice of soil and vegetation management has resulted in reduced erosion losses in the areas practicing shifting cultivation (Ramakrishnan 2007).

Forests in India are managed through a system of forest working plans which is the main guiding document for forest management in India. Forest working plans are prepared to promote sustainable management of forests on the principle of sound silviculture. The forest working plan code was revised in 2014 and further updated in 2023. The new revised working plan code aims at making forest management more people-centric by making provisions for goods and services to the local communities from the forests on a sustainable basis. The new working plan code also ensures biodiversity conservation, along with safeguarding the rights of forest-dwelling communities for non-timber forest products. The inclusion of the climate change mitigation and adaptation aspects are also important additions to the working plan code 2014 and 2023 (MoEFCC 2014a, 2023a).

12.6.2 Managing Forests for Climate Change Adaptation

Good silvicultural practices, including maintenance of optimum stocking levels and selection of trees which are best adapted to existing sites, should ensure that forests remain vigorous and relatively free of site and stand-related stress. These practices should help forests adapt to climate change. Practices which could help forests adapt to climate change include (Ciesla 1995):

- A shorter rotation length would reduce possibility of senescence related stresses and related hazard of damage by pest and diseases.
- Control of competition for available moisture, light and soil nutrients.
- Selection of species and provenances best adapted to site conditions.
- Properly scheduled thinning to maximise growth and increase resistance to damage from high winds, insect and diseases.
- Tree improvement programme to create planting stock from broad genetic base with high growth rates, better form and adapted to a varied range of site conditions.
- Create both in situ and ex situ reserves of important forest tree species with gene pools of wide variability that are capable of adapting to climate change.
- Reduce reliance on one or two tree species in afforestation and reforestation programmes. Instead include a number of species, where feasible in mixed species plantings, which are well adapted to local sites and climatic conditions and meet national needs for forest products and services.
- Protection from destructive effects of fire, pest and diseases.
- Periodic inventories and stand examination to provide the basis for silvicultural prescriptions and harvest scheduling.
- Accelerate timber salvage and fuel management programmes to reduce the hazard of wildfire in forests, especially those which have suffered from high levels of pest and disease damage or forest decline events.

- Design insect and disease monitoring programmes which are capable of capturing forest decline and in the active presence of new pests and diseases in addition to those which have historically caused losses.

12.7 Forest for Climate Change Mitigation

Various ecosystem services *viz.*, protecting water quality and quantity by retarding runoff, protecting biodiversity, sequestering carbon, wood and other non-timber forest products and aesthetic element in the landscape are being provided by forests (Heal 2000). Forests are important in the context of climate change mainly because of three reasons *viz.*, globally, deforestation and forest degradation contribute to about 9–11% of CO_2 emissions, provide a large climate change mitigation opportunity and forest ecosystems are expected to be adversely affected by the projected climate change in turn affecting biodiversity, biomass growth and forest regeneration (Ravindranath et al. 2012).

Mitigation and adaptation are two options for addressing the issue of climate change. Mitigation actions seek to address the causes of climate change while adaptation actions targeted to reduce the impacts of climate change. Mitigation strategies in the forests consist of reducing emissions from deforestation and forest degradation; enhancing carbon stocks in the forests and product replacement such as replacing fossil fuels with woods for energy and forest products in place of materials whose manufacturing process involves high emission of greenhouse gases. Mitigation actions are human efforts to reduce the emission or enhance the carbon sinks. Adaptation actions reduce the vulnerability of forests and forest-dependent communities to climate change (FAO 2012). Climate change mitigation and adaptation actions collectively contribute to stabilise the concentration of greenhouse gases in the atmosphere that would avoid dangerous anthropogenic intrusion into the climate system (IPCC 2014).

Forests play a key role in regulating the global carbon cycle as they account for a greater part of carbon exchange between the atmosphere and terrestrial biosphere thereby contributing to climate change mitigation (Pan et al. 2011). In recent decades, terrestrial carbon sink has enlarged; however, its size and location remain uncertain. Forest inventory data and long-term ecosystem carbon studies estimated a total forest sink of 2.4 ± 0.4 petagrams of carbon per year worldwide from 1990 to 2007 (Pan et al. 2011). Forest ecosystems store more than 80% of all terrestrial aboveground carbon and more than 70% of all soil organic carbon (Batjes 1996). Kohl et al. (2015) have indicated that world forest ecosystems remove nearly 2 petagrams per year of carbon from the atmosphere through net growth thus absorbing approximately 30% of anthropogenic CO_2 emissions.

Agriculture, forestry and other land use are a substantial net source of emissions which contribute about 23% of anthropogenic emissions of carbon dioxide, methane and nitrous oxide for the period of 2007–2016 (IPCC 2019). The IPCC (2022b) estimated that for the period of 2010–2019 the total net land-atmosphere flux of carbon dioxide on both managed and unmanaged lands provided a global net removal of CO_2 in the range of 6.6 ± 4.6 $GtCO_2$/year. This net removal of CO_2 consisted of two major components *viz.*, (i) net anthropogenic emissions from agriculture, forestry and other land use (5.9 ± 4.1 $GtCO_2$/year) which was mainly attributed to land cover change, including deforestation and plantations activities, and wood harvesting which accounted for about 13% of total net anthropogenic CO_2 emissions), and (ii) net removals because of non-anthropogenic processes (12.5 ± 3.2 $GtCO_2$/year) on managed and unmanaged lands (Fig. 12.1). The net effects of all anthropogenic and non-anthropogenic processes on land sector result in a net removal of CO_2 from the atmosphere. In proportion to total atmospheric emissions of CO_2, the net removal of CO_2 by managed and unmanaged land has generally increased over the last 60 years. Since around the middle of last century, the land sector has been a net sink for CO_2. However, earlier it was a net source of CO_2 mainly due to emissions from agriculture, forestry and other land use. All pathways considered by IPCC that overshoot the 1.5 °C target and require the use of technologies/techniques related to carbon dioxide removal such as plantation activities on degraded lands, carbon sequestration by soil, bioenergy with carbon capture and storage and technologies leading to direct carbon capture and storage from air (IPCC 2018).

As described earlier, mitigation-related programmes are the actions to control the anthropogenic factors responsible for climate change and their cumulative impacts, especially the emission of greenhouse gases and other pollutants like black carbon particles affecting the energy balance of the earth. Efforts towards enhancing carbon sinks such as greenhouse gas removal processes from the atmosphere also considered mitigation activities. Mitigation is part of a broader policy strategy that includes adaptation to climate change impacts as it lowers the anticipated effects of climate change as well as the risks of extreme impacts (IPCC 2014). Generally, climate change mitigation actions are aimed at limiting the extent and rate of long-term climate change. Mitigation actions are generally targeted towards reductions in anthropogenic emissions of greenhouse gases. Global as well as national policies on mitigation can further considerably reduce the projected risks of climate change (IPCC 2014).

Afforestation, reforestation and improved forest management are the most cost-effective options to reduce greenhouse

Fig. 12.1 Annual CO_2 fluxes from land sector (values are average of 2010–2019). Source: IPCC (2022b)

gas emissions from the forest sector. The value of key forest ecosystem services such as air quality improvement, flood mitigation, erosion control and landslide prevention, water filtration and local/regional precipitation cycles is now being recognised by many communities. Agroforestry is an example of a set of innovative trees growing practices along with agricultural crops designed to enhance overall productivity, along with the benefits of increased carbon sequestration. Such practices can also strengthen the capability of the system to cope with the adverse impacts of climate change. Possible benefits of these activities include mitigation and adaptation, ecosystem services, restoration of degraded land, biodiversity conservation and financial benefits from the sale of timber and non-timber forest products. Bioenergy can also play an important role in mitigation of climate change; however, there are issues of sustainability and the efficiency of bioenergy systems. There are diverse bioenergy technologies and a wide range of options available. Bioenergy-related options like improved cookstoves, small-scale biogas and biopower production technologies could reduce the emission of greenhouse gases and also improve livelihoods and health benefits (IPCC 2014).

Global Land Sector Mitigation Potential: IPCC reports indicated that agriculture, forests and other land use sector were responsible for total global net anthropogenic greenhouse gas emissions of 11.9 (±4.4) $GtCO_2$ eq./year on average for the period of 2010–2019 which is close to 21% of total global net anthropogenic greenhouse gas emissions from all sources (IPCC 2022a, b). Options for land-based climate change mitigation consist of various sustainable land management practices that reduce emissions of greenhouse gas and enhance carbon sequestration in forests, wetlands, grasslands, croplands and pasturelands. The land-based climate change mitigation measures are also known as nature-based solutions or natural climate solutions for well-being of human and biodiversity (IPCC 2022a, b). The technologies pertaining to net removal of greenhouse gases from the atmosphere and storing them in living or dead organic material or in geological formations are known as carbon dioxide removal/greenhouse gas removal/negative emissions technologies (Rogelj et al. 2018; Jia et al. 2019). Activities aiming at the reduction of deforestation and forest degradation were identified to have major potential for reducing supply-side emissions (0.4–5.8 $GtCO_2$ eq./year) followed by combined agriculture measures, 0.3–3.4 $GtCO_2$ eq./year (Jia et al. 2019). Afforestation/reforestation (0.5–10.1 $GtCO_2$ eq./year), soil carbon sequestration in croplands and grasslands (0.4–8.6 $GtCO_2$ eq./year) and bioenergy with carbon capture and storage (0.4–11.3 $GtCO_2$ eq./year) were among the activities identified to have the highest potential for carbon dioxide removal.

Globally land-based mitigation potential was estimated to be in the range of about 8–14 Gt CO_2eq. per year during the year 2020. Actions like forest protection, improved forest management and restoration of degraded forests and other ecosystems such as wetlands, savannas and grasslands were listed among various mitigation choices. These choices have the highest potential to reduce emissions and increase carbon sequestration to the tune of 7.3 Gt of CO_2 eq. per year. The second major share of mitigation potential of 4.1 Gt of CO_2 eq. per year was estimated for agriculture sector and this potential can be realised from soil carbon management in croplands and grasslands, agroforestry, biochar, rice cultivation, livestock and nutrient management (IPCC 2022a, b).

IPCC (2019) estimated a wide range of climate change mitigation potential of 0.5–10.1 Gt of CO_2 per year by 2050 from plantation activities (afforestation and reforestation). The global mitigation potential of 3.9 $GtCO_2$ per year for afforestation and reforestation activities and 1.7 $GtCO_2$ per year for improved forest management were estimated by the year 2050 with medium confidence (IPCC 2022a, b). Climate change impacts forest growth and composition and changes in disturbances regimes including forest fire will affect the mitigation potential of plantations (afforestation/reforestation) activities. Global initiatives like Bonn Challenge launched by the Government of Germany and IUCN in 2011, with an objective to bring 150 million hectares of degraded and deforested landscapes into restoration by 2020 and 350 million hectares by 2030, the UN Decade on Ecosystem Restoration (2021–2030) to prevent, halt and reverse the degradation of ecosystems on every continent and in every ocean, and 'One Trillion Trees Initiative' launched by the World Economic Forum in 2020 to grow, restore and conserve 1 trillion trees across the world by 2030 are responding to urgent issue of land degradation and climate change mitigation by raising new plantations. However, these initiatives are under intense discussions on the level of scale, their effectiveness and drawbacks of plantations for climate mitigation (Luyssaert et al. 2018; Bond et al. 2019; Anderegg et al. 2020; Heilmayr et al. 2020; Holl and Brancalion 2020).

Carbon Sequestration by Forests in India: Carbon sequestration by forests has attracted much interest as a mitigation approach, as it has been considered a relatively inexpensive means of addressing climate change. Forest management practices that increase carbon sequestration include afforestation/restoration of degraded forest lands, agroforestry, urban forestry, tree planting in rural landscapes, enhancement of forest carbon stocks and sequestration capacity through modification of forestry management practices.

National Forest policy, 1988 mandated to bring a minimum of one-third of the total land area of the country under forest and tree cover. Chaturvedi et al. (2010) analyse the mitigation potential of Forest Policy over the period of 2010–2030 under two scenarios *viz.*, (i) a rapid afforestation rate (achieving one-third goal of forest and tree cover by 2020) and (ii) a moderate afforestation rate (achieving one-third goal of forest and tree cover by 2030). Assessment using the generalised comprehensive mitigation-assessment process model shows that under the scenario of 2020, afforestation could mitigate 5.2 $GtCO_2$ in the time period of 2010–2030 as compared to 3.96 $GtCO_2$ under the scenario of 2030 in the same time period.

The total growing stock of India's forests and trees outside forests was estimated to be 6167.50 million m^3 which comprises 4388.15 million m^3 inside the forests and 1779.35 million m^3 outside the forests. There was an increase of 251.74 million m^3 in total growing stock of the country between two successive assessments (FSI 2019, 2021). The biomass carbon stock in India's forests was 7940 million tonnes during 1880, which was declined by nearly half over a period of hundred years in 1980 (Richards and Flint 1994). The carbon estimates also vary depending upon the methods of estimation used and the availability of the precise data for various components. Biomass and soil carbon stocks were estimated in the range of 8.58–9.57 $GtCO_2$ in 1986 (Ravindranath et al. 1997; Chhabra and Dadhwal 2004). Forest Survey of India made a detailed and comprehensive assessment of forest carbon stocks and changes in India's forests between two time periods, viz. 1994 and 2004 (FSI n.d.). According to this estimate, forest carbon stocks for the year 1994 and 2004 were 6071 and 6663 million tonnes respectively. During this period, India had successfully enhanced its forest carbon stocks by 10% and added 592 million tonnes of carbon in its forests. Since 2013, the Forest Survey of India added a new feature for assessment of forest carbon stocks for the states and union territories in its regular biennial India State of Forest Report. Total carbon stock in India's forests was estimated 6941 million tonnes in 2013 (FSI 2013), Fig. 12.2.

Plantations/Afforestation/Reforestation Activities in India: The first National Forest Policy of 1952 of India aimed at 33% of the geographical area of the country to be brought under forest and tree cover. Under this policy and to meet the demands of goods and services, plantations of various species were raised. Plantations undertaken until 1974–79 were about 3.32 million ha in the country. Afforestation and reforestation are being carried out under various government-sponsored programmes such as social forestry initiated in the early 1980s, programme on Joint Forest Management initiated in 1990, afforestation activities under National Afforestation and Eco-development Board programmes since 1992. National Afforestation Programme (NAP) of National Afforestation and Eco-development Board, one of the flagship schemes of the Government of India assists in the rehabilitation of degraded forests through the programmes of Joint Forest Management Committees. Plantation forestry activities have also been initiated by farmers, tree growers and industries. The flagship

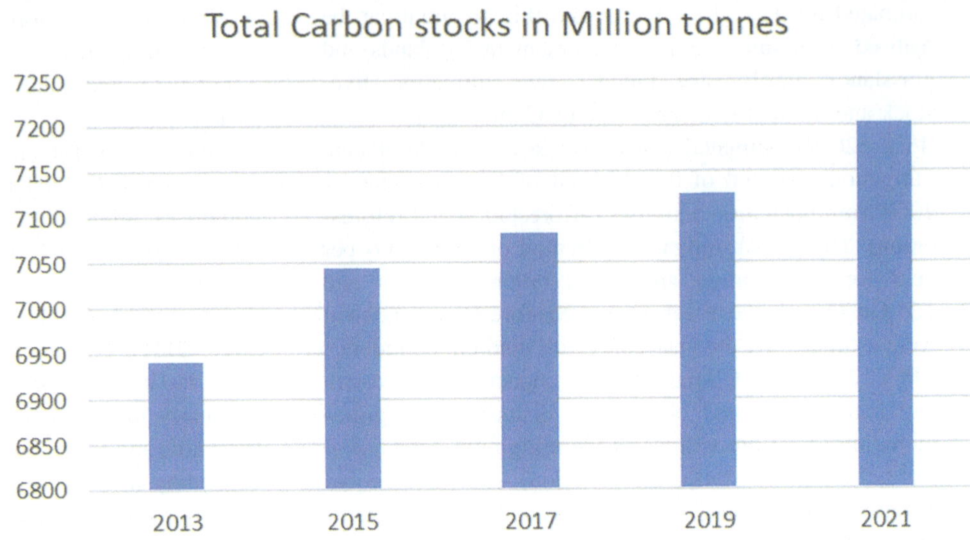

Fig. 12.2 Assessment of carbon stocks in India's forests

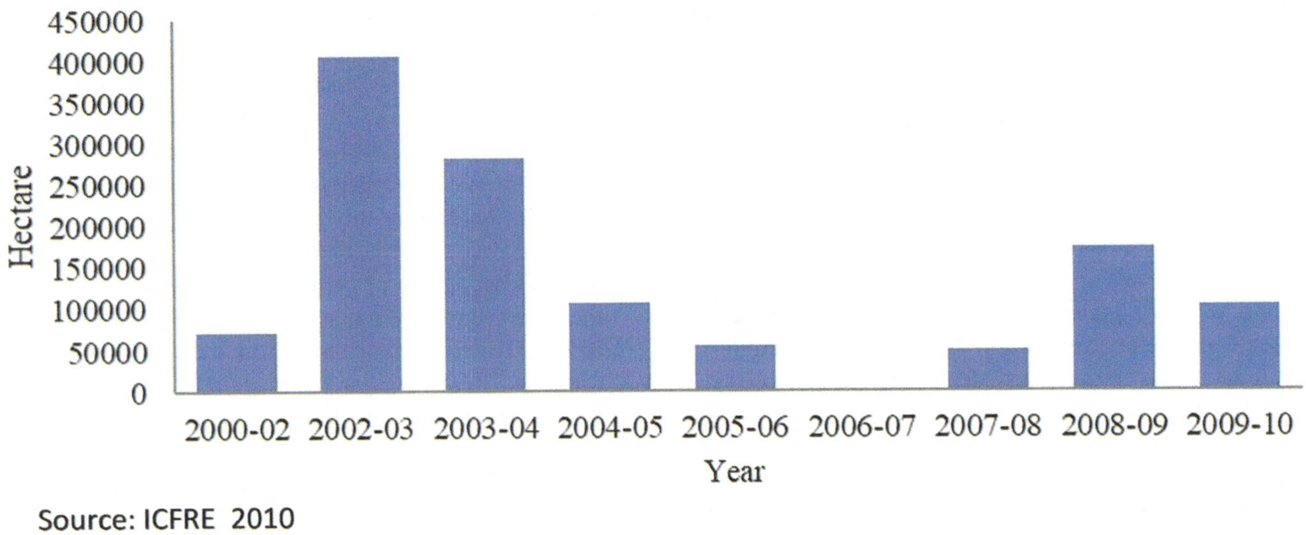

Fig. 12.3 Plantation carried out under NAP in India during successive years. Source: ICFRE (2010)

government initiative of creating the Compensatory Afforestation Fund Management and Planning Authority (CAMPA) is to accelerate activities for the preservation of natural forests, management of wildlife, infrastructure development and other associated works in the forest sector. These initiatives have contributed in increasing the annual rate of plantation to 1.50–1.80 million ha. The cumulative area afforested during the period 1980–2005, including community woodlots, farm forestry, avenue plantation and agroforestry, was about 34 million hectares at an average rate of 1.32 million hectares/year (Chaturvedi et al. 2008).

Forest Survey of India estimated that around 3.63 million ha of plantation and trees outside forests presently exist in the country which constitutes about 5.07% of country's total forest and tree cover (FSI 2011). Contiguous plantations in the forests by the state forest departments have enriched the forests of open (10–40% crown cover) and moderately dense categories (40–70% crown cover). The plantations have reduced anthropogenic pressure to a considerable extent on the natural forests and have also met both domestic and industrial demands besides meeting the rural energy demand. Details of plantation activities carried out under NAP in India during successive years are given in Fig. 12.3. Tree plantations outside forest land are also done by other departments of the State Government under Twenty Point Programme.

An assessment of the rate of carbon flux in plantation forests in India revealed that planted forests comprising of short-rotation tree species with regular leaf shedding patterns have more capacity for carbon sequestration in the litter as compared to species with annual or bimodal leaf shedding patterns (Raizada et al. 2003). Litters of fast-growing conifers may be slow in decomposition leading to accumulation on the forest floor, thus conifer forests are more prone to the risk of forest fire and eventually decline in ground floral diversity and productivity. Plantation forests comprising mixed tree species of exotic and native origin could be more efficient in carbon sequestration than monocultures.

Low Carbon Strategy for Forestry Sector in India: The Planning Commission of India (2014) highlighted various forestry actions for a low carbon strategy for the forestry sector in India. Forests within the protected areas are always preferred areas for saving and maintenance of forest carbon stocks because of the rigorous management of protected areas for wildlife and other specifically conserved forest areas and can be grouped under conservation forests. Forest areas which are subject to harvests in accordance with a prescription of forest working or management plans based on the sound principle of silviculture can be put under sustainably managed forests. Under the Bali Action Plan of UNFCCC, climate change mitigation actions in the context of REDD+ used the terms conservation and sustainable management of forests. Forest management practices of conservation forestry and sustainable management of forests would result in maintaining existing forest carbon stocks over a period of time. Such practices would also result in quantitative increment due to the natural growth process of conserved vegetation. Other options suggested for saving wood biomass are the use of improved cook stoves, which are more energy efficient and thus contribute towards forest conservation. Actions leading to an increase in the forest and tree cover improvement in carbon density and/or increment in various carbon pools in a given forest or woodlands can lead to the enhancement of forest carbon stocks. Tree planting activities (afforestation, reforestation, agroforestry and energy plantations for fuel wood and biodiesel) could contribute to the enhancement of forest carbon stocks. Estimates of carbon sequestration through different forestry mitigation options are given in Table 12.1.

Land Sector Contribution to National Greenhouse Gas Emission: Various government policies and programmes in India have transformed forests from net source to net sink of carbon dioxide. India's initial National Communication to UNFCCC reported that India's forests were net source of CO_2 for the reporting year 1994 which was 1.16% of total national emissions (MoEF 2004). India's forests were net sink of 17 % of total national emissions when India submitted its Second National Communication to the UNFCCC (MoEF 2012). In its first, second, third biennial update reports and third national communication to UNFCCC, Indian forest sector contributed in

Table 12.1 Carbon sequestration estimates of different forestry mitigation options

Option	Net carbon sequestration
Conservation and improvement of existing forests	
1. Protected areas	Avoiding GHG emissions from deforestation and forest degradation by conservation actions in protected areas. This action will annually add 47 $MtCO_2eq$ to forest carbon sink
2. Sustainable management of forests other than protected areas	The forest areas (53 Mha) under sustainable harvests are annually adding 62.0 Mt CO_2eq to the forest carbon sink
3. Improvement in forest and tree cover	This option aims to improve 1 million ha area of open forests and medium dense forests each, so that these forests transition to the next higher class. This transition can enhance the forest carbon sink to the tune of 7.3 $MtCO_2eq$ per year
4. Improved wood-burning cookstoves	By replacing traditional cook stoves with fuel-efficient models, the avoided emissions from the overuse of fuel wood in cooking stoves in rural regions (800 million people/160 million families) will contribute in increasing the carbon stocks in forests. This action would offset emissions equal to 58.2 $MtCO_2eq$
Afforestation	
5. Increase in forest and tree cover in forest fringe villages	This option can help in adding additional 17 million ha by creating additional forest in and around 170,000 forest fringe villages. Annually 1.7 million ha can be afforested/reforested on an average. This option can add 12.5 million tonnes of CO_2eq annually to the forest carbon sink
Wood products use management	
6. Harvested wood products management (as substitutes)	Carbon can be stored in wood products for a sufficiently long time. By encouraging wood products in building construction by substituting high energy-intensive products like cement and metallic door and window frames etc., 2 tonnes of CO_2eq can be saved for each cubic m of metal/cement concrete replaced
7. Replacement of office and domestic furniture using metals with wooden furniture	Currently about 35% of furniture used in homes and offices is made of metals and plastics. Substituting about 50% of the metals and plastics furniture every year by wood could replace about 1.5 million m^3 of metal and plastic furniture, resulting in additional sequestration of 3 million tonnes of CO_2eq annually

Source: Planning Commission (2014)

Table 12.2 Forestry sector (LULUCF) contribution to national greenhouse gas emission

S. No.	Study	Reference year	National emissions (million tonnes of CO_2 eq)	Forestry contribution to (as %) total national emission	Methodology followed for estimation of LULUCF emissions
1.	Initial National Communication	1994	1228.54	1.16 (net source)	Default values used
2.	Second National Communication	2004	1301.21	17.1 (net sink)	Tier 2 and 3 methodologies of IPCC used in combination
3.	1st Biennial Update Report	2010	1884.31	11.81 (net sink)	Combination of IPCC Tier 2 and 3 methodologies of IPCC used in combination
4.	2nd Biennial Update Report	2012	2,306.29	11.55 (net sink)	Tier 2 and 3 methodologies of IPCC used in combination
5.	3rd Biennial Update Report	2016	2,531,069	10.84 (net sink)	Tier 2 and 3 methodologies of IPCC used in combination
6.	Third National Communication	2019	26,46,556	20 (net sink)	Tier 2 and 3 methodologies of IPCC used in combination

Source: MoEFCC (2004, 2012, 2015a, 2018a, 2021, 2023b)

offsetting 11.81%, 11.55%, 10.81% and 20% of total national greenhouse gas emissions respectively (MoEFCC 2015a, 2018a, 2021, 2023b). Based on various national reports submitted to UNFCCC, it can be very well inferred that the forestry sector in India is now a net sink of CO_2 and contributing towards climate change mitigation in India. Forestry sector contribution to India's total greenhouse gas emissions is given in Table 12.2.

Forests are the largest carbon pool on earth and are a major source as well as sink of carbon. Thus, it has the potential to form a main component in mitigation and adaptation to climate change. Several land-based climate solutions are now available that can reduce and/or remove carbon emissions from the atmosphere. Afforestation and reforestation have the highest carbon removal potential followed by enhancing soil carbon and using bioenergy in combination with carbon capture and storage. The conservation focussed forest policies and afforestation/reforestation programmes in India have transformed India's forest from source to net sink of CO_2. Those progressive policies and programmes at national level have contributed to reducing CO_2 emissions, stabilising and improving forest carbon stocks and conserving the biodiversity.

12.8 REDD+ for Climate Change Mitigation and India's Readiness

Deforestation and forest degradation worldwide are occurring at alarming rates, which is a major contributing factor to greenhouse gas emissions and biodiversity loss. An estimated 420 million hectares of forests have been destroyed since 1990 as a result of land use conversion for different purposes. In the last 5-year time period of 2015–2020, an estimated annual deforestation rate was 10 million hactares, which was 12 million ha during the period of 2010–2015. Nonetheless, between 1990 and 2020, the rate of net forest loss significantly declined as a result of decreased deforestation in some countries and increases in forest cover in others as a result of plantations and expansion in area of forests. Globally annual rate of net forest loss came down from 7.8 million ha in the decade 1990–2000 to 5.2 million ha in 2000–2010 and 4.7 million ha in 2010–2020 (FAO 2020; FAO and UNEP 2020). Land use sector collectively known as agriculture, forestry and other land use sectors contributes about one-quarter of global greenhouse gas emissions. About 13% of total anthropogenic CO_2 was emitted from deforestation and peatland degradation alone. Reducing the rates of deforestation and forest degradation is one of the most reliable and successful ways to mitigate climate change, with significant global benefits (IPCC 2019). Duan and Tan (2019) have analysed that the forests in developing countries decreased from 1992 to 2015; however, this decline slowed down after 2004. In developing countries, the total forest cover decreased from 21.8 to 21.3 million sq km from the year 1992–2015 and 41.2–40.6% decrease in total forest coverage was recorded. South America recorded the largest reduction in forest area (505,100 sq km) accounting for approximately 85% of total forest loss.

12.8.1 Evolution of REDD+ at UNFCCC

The unique capability of forests to address climate change mitigation as well as adaptation by way of capturing and storing carbon has attracted the attention of the scientific community and policymakers since the beginning of climate change discussions. However, the adaptive capacity of forests is adversely impacted by changing climate regime by making them more vulnerable to climate change (IPCC 2022a, b). The capability of forests for large-scale reduction in greenhouse gases, as well as reducing emissions from deforestation and forest degradation has been projected as a cost-effective approach for mitigating climate change (Sathaye et al. 2007; Parker and Mitchell 2009). The Stern Review (2006) categorically mentioned that reduction in deforestation is the '*single largest opportunity for cost-effective and immediate reductions of carbon emissions*'.

Reducing emissions from deforestation and degradation (REDD) in developing countries was introduced for negotiations in 2005 for the first time at COP 11 of UNFCCC. The agenda of REDD came under argument and disparagement for its narrow focus covering only deforestation and forest degradation. India took the lead to correct the skewed focus of REDD with strong support of the like-minded countries where there was no large-scale deforestation and forest cover was increasing or stabilised as a result of their policies on forest conservation and sustainable management of forests (Rawat and Kishwan 2008; Rawat 2010).

Transformation of REDD to REDD+: India in COP 12 of UNFCCC in 2006 presented a new policy approach for the countries those have implemented strong conservation measures and regulations to be suitably compensated under the instrument of REDD for rewarding forest conservation and enhancement of forest carbon stocks (Rawat and Kishwan 2008). The Indian proposal aimed to compensate the countries for maintaining and increasing their forests as carbon pools as a result of conservation and improvement in their forest cover. India's proposal of inclusion of conservation and sustainable management of forest cover as a policy approach to REDD was finally recognised and given effect in the Bali Action Plan of COP 13 in 2007 which read as '*Policy approaches and positive incentives on issues relating to reducing emissions from deforestation and forest degradation in developing countries; and the role of conservation, sustainable management of forests and enhancement of forest carbon stocks in developing countries'*. The above paragraph of Bali Action Plan is collectively referred to as REDD+ (UNFCCC 2008).

UN negotiations leading to further implementation of REDD+ are briefly described below:

Cancun Agreements: In Cancun (Mexico), in the year 2010, Governments agreed to accelerate action to check emissions from deforestation and forest degradation in developing countries (MoEFCC 2014b). Cancun Agreements on REDD+ encourages developing countries to contribute to mitigation actions in the forest sector by undertaking the activities such as reducing emissions from deforestation, reducing emissions from forest degradation, conservation of forest carbon stocks, sustainable management of forest and enhancement of forest carbon stocks; in accordance with their respective capabilities and national circumstances. The Cancun Agreements include a number of guiding principles concerning safeguards for forest governance, social safeguards respecting the rights of indigenous peoples and local communities, and environmental safeguards for protection and conservation of biological diversity and ecosystem services.

Warsaw Framework for REDD+: The Warsaw Climate Change Conference (COP19 in 2013), countries agreed on a decision package, known as Warsaw Framework for REDD+. A series of agreements on methodological guidance, institutional arrangements and results-based finance were made.

REDD+ and Paris Agreement: Paris Agreement (UN 2015) also reaffirmed the role of REDD+ and lays out a framework for the conservation of carbon sinks, including forests, through schemes such as results-based payments for REDD+ actions. Although the broad rules and methodological guidance for REDD+ were already agreed upon under 'Warsaw Framework for REDD+' and other relevant COP decisions, legitimising and 'regulating' REDD+ activities under the Article 5 of the Paris Agreement was a strong political signal. Inclusion of REDD+ in Paris Agreement will further encourage the developing countries to continue with REDD+ strategy and readiness activities (Rawat and Singh 2016a, b).

12.8.2 India's Readiness for Implementation of REDD+ Activities

Developing countries willing to participate in REDD+ programmes are required to develop their (i) National REDD+ Strategy or Action plan, (ii) National REDD+ Reference level/Reference Emission Level, (iii) National Forest Monitoring System and, (iv) a Safeguard Information System. India already developed its Forest Reference Level for REDD+. Forest Reference Level submitted by India to UNFCCC is (−) 49.70 million tonnes of CO_2eq per year (UNFCCC 2018). India submitted its Safeguards Information System for implementation of REDD+ activities to UNFCCC in 2023.

Scope of REDD+ Implementation in India: About 1,73,000 villages are categorised as forest fringe villages in India with an estimated population of about 350 million with high dependence on forest resources for their livelihoods. In India, good chunk of forests is being managed under Joint Forest Management through more than 1,18,000 Joint Forest Management Committees. There is a lot of scope and opportunities for integrating REDD+ activities with the community-managed forests. Promoting and integrating REDD+ activities in the joint forest management activities to enhance carbon sequestration for meeting national targets of climate change mitigation and international commitments as a responsive member of the international community (Rawat and Kishwan 2008; Rawat et al. 2020). India's forest and tree cover besides meeting the goal of the National Forest Policy and livelihood needs of local communities are also crucial for meeting the country's Nationally Determined Contribution target under the Paris Agreement. Improvement in the health of natural forests of the country will need to be supplemented by focusing on trees outside forests to meet the India's Nationally Determined Contribution (NDC) target.

National REDD+ Strategy: National REDD+ Strategy 2018 of India builds upon existing national circumstances and aligned with the precepts of the National Forest Policy, 1988. The overarching objective of National REDD+ Strategy is to facilitate the implementation of REDD+ programme in the country in conformity with relevant decisions of UNFCCC, in particular the Cancun Agreements, Warsaw Framework for REDD+, Paris Agreement and the national legislative and policy framework for conservation and improvement of forests and the environment (MoEFCC 2018a, b).

As per the National REDD+ Strategy, National Governing Council for REDD+ (NGC-REDD+) has been established at national level to coordinate and guide REDD+ related actions. NGC-REDD+ shall be chaired by the Union Minister for Environment, Forest and Climate Change, Government of India. A National Designated Entity for REDD+ has also been established at the Ministry of Environment, Forest and Climate Change, Government of India to liaise with UNFCCC and state governments. The strategy delegates key responsibility for the implementation of REDD+ activities to the Forest Departments at the state level.

The National REDD+ Strategy focuses on creation of trained human resource to carry out forest-related measurements at all levels of execution of REDD+ activities. The National REDD+ Strategy also empowers youths as community foresters to lead the charge at the local level through building their capacity under the green skill development programme. Implementation of REDD+ activities will provide opportunities for additional jobs in the forest sector. Services of the community foresters can be utilised effectively in the implementation of the activities for climate change adaptation in forests such as assisted natural regeneration, soil and moisture conservation, silvicultural operations, maintaining forest nurseries and raising of quality planting stocks, and control of forest fires, pest and disease and invasive species.

Forest management in India is now transitioning to people centric. National REDD+ strategy proposes that civil society collaborates with National Institutions and State Forest Departments in organising capacity building programmes for the local communities including *Gram Sabha* and Joint Forest Management Committees. The strategy also provided a roadmap for addressing the drivers of deforestation and forest degradation and importance of safeguards for implementation of REDD+ activities.

REDD+ mechanism is primarily designed as climate change mitigation programmer under UNFCCC. However, an effective implementation of REDD+ programme will also provide opportunity for alternate income generation, livelihood security, climate resilience and social wellbeing. Forest ecosystem services and their continuous supply are now becoming increasingly important in the context of adaptation to climate change. India was active in shaping the evolution of the wide ranging and inclusive concept of REDD+ at the international level; however, more efforts need to be done domestically for the purposeful implementation of REDD+ activities.

12.9 Initiatives for Climate Adaptation and Mitigation in Forest Sector

Climate change is a global process but with local repercussions. India faces a bigger challenge in handling with the consequences of climate change than most other countries as local communities have large dependence on forest resources. India is a mega-biodiversity country where forests and tree cover account for 24.62% (80.95 million ha) of the geographical area (FSI 2021). Thus, it is very crucial to assess the likely impacts of projected climate change on forests and to develop adaptation strategies and action plan for conservation of biodiversity, safeguarding the livelihoods of the local communities, besides ensuring production of timber for commercial requirements. In India, forest policies and programmes are forest conservation centric and aim to enhance forest and tree covert through afforestation and reforestation thereby contributing in reduction of carbon emissions. Thus, forests in India are contributing positively in mitigating global climate change and promoting sustainable development (Rawat and Kishwan 2008). India is among the few developing countries whose forests are net sink of CO_2 and it is reported that forests capture almost 20% of the total greenhouse gas emissions (MoEFCC 2023b).

There are both external and domestic dimensions to India's policy on climate change which has been articulated through two key documents such as National Action Plan on Climate Change (NAPCC) and Nationally Determined Commitments submitted to the UNFCCC. The NAPCC incorporates India's vision of ecologically sustainable development and steps to be taken to implement it. It is based on the awareness that climate change action must proceed simultaneously on several intimately inter-related domains, such as energy, industry, agriculture, water, forests, urban spaces and the fragile Himalayan environment. The NAPCC acknowledged that climate change and energy security were two sides of the same coin; that India had to make a strategic shift from its current reliance on fossil fuels to a pattern of economic activity based progressively on renewable sources of energy such as solar energy and cleaner sources such as nuclear energy.

National Action Plan on Climate Change: India launched its National Action Plan on Climate Change way back in the year 2008. It identifies a number of measures in the form of eight national missions (National Solar Mission, National Mission for Enhanced Energy Efficiency, National Mission for Sustainable Habitat, National Water Mission, National Mission for Sustaining the Himalayan Ecosystem, National Mission for a Green India, National Missions for Sustainable Agriculture and National Mission on Strategic Knowledge for Climate Change) that simultaneously advance the country's development and climate change related objectives of adaptation and mitigation. The eight National Missions comprises multifaceted, long-term and integrated strategies for accomplishing India's primary goals in the context of climate change are the central component of the National Action Plan (MoEFCC 2014c).

The National Mission for a Green India is one of the key missions of NAPCC. The mission recognises that climate change will extremely affect and alter the distribution, type and quality of natural resources and associated livelihoods of the local communities in the country. The Mission recognises the impacts that the forests have on environmental well-being through mitigating climate change, food and water security, conservation of biodiversity and livelihood security of forest dependent communities.

The actions proposed under Green India Mission respond to climate change through combination of adaptation and mitigation strategies, which help in enhancing carbon sinks in sustainably managed forests and other ecosystems, adaptation of vulnerable species/ ecosystems to the changing climate and adaptation of forest dependent local communities in the face of climatic variability. The Mission would strengthen the management and restoration of forest ecosystems in reducing the vulnerability of poor communities to climate change related disasters. Protecting and enhancing ecosystem services through afforestation, reforestation, biodiversity conservation and restoration of pasturelands can help in improving the livelihoods of forest dependent communities and their adaptive capacity in view of the adverse impacts of climate change.

State Action Plans for Climate Change: Simultaneously after launch of National Action Plan on Climate Change, the Prime Minister's Council on Climate Change called upon state governments to prepare State Action Plans on Climate Change in line with NAPCC. Government requested states to priorities adaptation actions *vis-a-vis* state priorities and local level impacts. Forest sector is one of the major sectors in various State Action Plans of Climate Change (SAPCC). The objectives of SAPCCs are to support the NAPCC's vision while also being in line with the eight national missions. The impact of climate change, vulnerability assessment, adaptation, mitigation strategies, financial requirements and capacity building requirements to carry out the selected actions are all covered in detail by SAPCCs.

12.9.1 Nationally Determined Contribution (NDC)

First NDCs of India for achieving the objectives of the Paris Agreement were submitted to UNFCCC in October 2015 for achieving its targets by 2030. Later, India has submitted its updated NDCs in August 2022. In the updated NDC, India's NDC target for forest sector remain unchanged ie '*to create an additional carbon sink of 2.5–3 billion tonnes of* CO_2 *equivalent through additional forest and tree cover by 2030*' (MoEFCC 2022).

12.10 Mainstreaming Climate Change in Forest Management: Legal and Policy Framework

The country's natural resources including forests, wildlife and biodiversity are effectively protected by the policy and legislative framework that have developed in accordance with the provisions of the Constitution of India. This framework protects and upholds the rights of the local communities over lands and forest products while guiding and directing the sustainable management of forests to ensure biodiversity conservation (MoEFCC 2018b). The policies, legislations and regulations listed below provide strong commitments of the Government of India for conservation and protection of forests, wildlife and biodiversity.

Indian Forest Act, 1927: It consolidates the laws pertaining to forests, transportation of forest products, and levy imposed on timber and other forest products. Various provisions of this Act capable for enhancing quality and extent of forest and tree cover along with biodiversity conservation in the natural forests.

Wild Life (Protection) Act, 1972: The purpose of this Act is to protect wild animals, birds and plants. It works in perfect harmony with preserving biodiversity and keeping natural forests from being converted into plantations.

Forest (Conservation) Act, 1980: This is one of the most effective legislations controlling the conversion of forest lands to non-forestry purpose. This act has drastically reduced indiscriminate diversion of forest lands for non-forestry purposes, and regulated the land use changes in forests. Under this Act, Government of India is empowered to approve the diversion of forest land for non-forestry purpose. Enactment of this act contributed to strike a balance between conservation and developmental activities, thus it helps to enhance implementation of mitigation and adaptation actions in the country through conservation of natural forests.

Environment (Protection) Act, 1986: This act provides protection and improvement of **environment** and matters connected with it. It empowers the Union Government to establish authorities to control all kind of environmental pollution and to address specific environmental issues that are unique to various regions of the country.

National Forest Policy, 1988: The main objective of National Forest Policy, 1988 is to ensure environmental stability and maintenance of ecological balance. Another striking feature of the policy is to bring one-third of the country's geographical area under forest and tree cover.

Panchayat (Extension to Scheduled Areas) Act, 1996: This act was enacted to enable Tribal Self Rule in their areas. Through this Act the provisions of *Panchayats* were extended to the tribal areas of nine states that have Fifth Schedule Areas. Through PESA special powers especially for the management of natural resources were given to the *Gram Sabha*.

Biological Diversity Act, 2002: The Act provides for conservation of biological diversity, sustainable use of its components and fair and equitable benefit sharing from biological resources, knowledge and other related matters.

National Environment Policy, 2006: The objectives of the National Environment Policy, 2006 are to include conservation of critical environmental resources, intra-generational equity and livelihood security for the poor, inter-generational equity, integration of environmental concerns in economic and social development, efficiency in environmental resource use and enhancement of resources for environmental conservation.

Scheduled Tribes and Other Traditional Forest Dwellers (Recognition of Forest Rights) Act, 2006: The Act enjoins upon the local communities, to recognise their role in forest and **biodiversity** protection through sustainable management practices, which will yield long term benefits to them.

National Agroforestry Policy, 2014: The policy lays emphasis on the environmental contribution of agroforestry by halting deforestation, and enhancing carbon storage in trees, biodiversity, soil and water conservation. This policy contributes to the objectives of climate change mitigation by enhancing the area of trees outside forest and thus preventing pressure on natural forests.

12.10.1 Paradigm Shift in Forest Governance in India

In the year 1952, the first Forest Policy of independent India was formulated with the statement that village communities should not be allowed to use forests at the cost of national interest. The Forest Conservation Act of 1980 was the first shift in forest governance from commercialised use of forest to conservation centric objectives. With the promulgation of this Act, conversion of forest lands to non-forest use has been drastically reduced. The speed of diversion of forest land for non-forest purposes was close to 1,40,000 hectares per year from 1951 to 1980 which was drastically reduced to 24,000 ha per year (MoEFCC 2023b) during the period of 1980–2020 after the implementation of this Act (Fig. 12.4).

The National Forest Policy 1988 was a paradigm shift in the philosophy of forest governance in India. This policy differed from the previous policies of independent India. Earlier forests were considered merely as a source of revenue. The objective of the 1988 forest policy was to ensure that the rights of the forest dependent people are safeguarded. The ecological security was given priority over revenue earning from forest resources. The policy also stressed on the close relationship between the forests and tribal population. Current initiatives like National Afforestation Program, along with programmes in other sectors like agriculture and rural development are adding about 1 mha of forest and tree cover annually. This contributes to sequestering about 1 million tonne of carbon incrementally. Combined with the accumulation of biomass in the managed forests, protected areas and in tree cover outside forests, the total carbon service from India's forest is estimated at 138 mt CO_2 eq. every year (Kishwan et al. 2009).

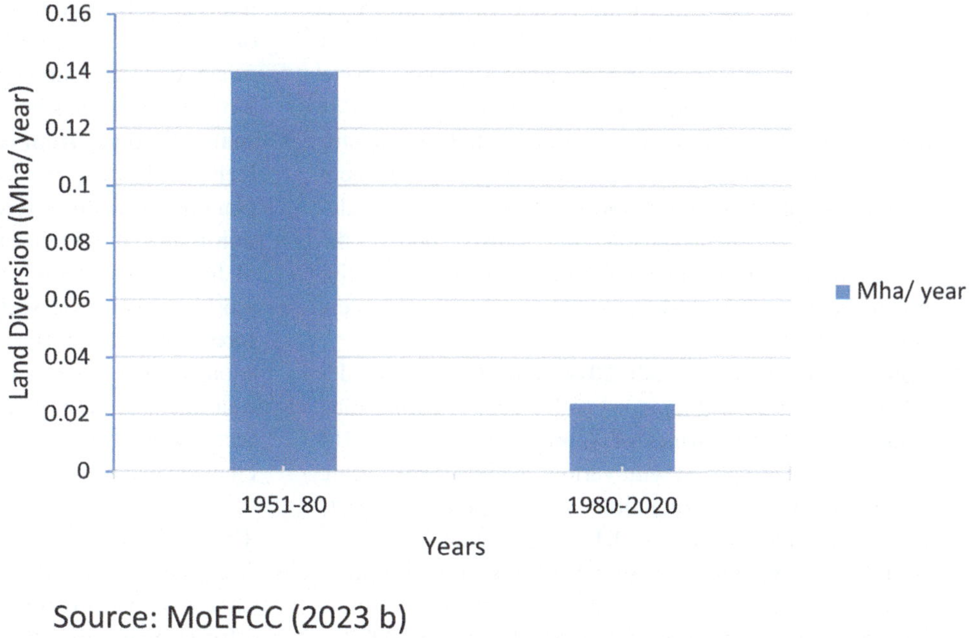

Fig. 12.4 Impact of Forest (Conservation) Act, 1980 on pace of diversion of forest land. Source: MoEFCC (2023b)

12.10.2 Mainstreaming Climate Change in Forest Management Plans in India

In India, scientific management of forests started in 1864 when Sir Dietrich Brandis, a German forester was appointed as first Inspector General of Forests. Following this, laws were passed, most notably the 1865 Forest Act, which was amended in 1878 to give more powers to the forest departments (MoEF 1999). Earlier objectives of forest management were primarily focussed towards revenue earning from forest resources; however, National Forest Policy, 1988 shifted attention of forests from merely a source of revenue to more people oriented by protecting the rights of the forest dependent people and ecological security of the nation as primary goal (MoEF 1988).

During the pre-independence period, creation of reserved and protected forests and the settlement of rights therein were the major task undertaken by the forest departments. Working plans were implemented for the majority of the exploitable areas and important tree species. Various silvicultural systems such as selection and selection-cum-improvement felling, conversion to uniform system, coppice with standard/reserves as well as the *taungya* system (agri-silviculture) involving clear felling and strip planting with short duration (3–5 years) of agriculture crops were also developed. These systems were all created to harvest and regenerate significant timber species in order to generate resources. All these systems were all designed to harvest and regenerate important timber species with prime objective of resource generation. In 1927, the previous Indian Forest Act was revised; however, many provisions of previous Act are still in force (Rawat and Kishwan 2008).

12.10.3 Forest Management Regime Supporting Climate Change Adaptation and Mitigation

Forest management in India is already aligned to implement various climate change related mitigation and adaptation aspects and as such it does not need change significant changes, in order to implement the programmes related to climate change. However, certain aspects of forest management will need to be strengthened by appropriate infrastructure inputs and capacity building to maximise the climate change related benefits out of these policies and programmes. These recommendations have been adequately described in the Green India Mission, India's NDC to UNFCCC (MoEFCC 2022) and National REDD+ Strategy (MoEFCC 2018b).

Integrating Climate Change Aspects in Forest Management Plan in India: The National Forest Policy, 1988 categorically emphasises that no forest should be allowed to be worked without a working plan that has been approved by the competent authority. All forests are to be managed sustainably under approved working plan/ scheme's prescriptions. The working plan shall be approved by the designated authority at the Ministry of Environment, Forest and Climate Change, which will also ensure its implementation. The working plans prepared as per the National Working Plan Code are generally valid for a period of 10 year.

At the national level, Forest Working Plan has been the main guiding document for forest planning ensuring scientific management of forests. Earlier different states of the country had been following their own working plan codes. However, in 1996 after the Supreme Court of India's intervention, all forest working plans were required to be approved by the Government of India as forests being brought in the concurrent list. The Ministry of Environment, Forest and Climate Change, Government of India adopted a uniform National Working Plan Code-2004 for the management of forests under the prescriptions of a working plan/scheme and to standardise the procedure. This code was later replaced with new codes in 2014 and 2023.

National Working Plan Code-2014: Working Plan Code-2004 was revised and replaced with a new Working Plan Code-2014. The doctrines of Working Plan Code-2014 were more or less tuned with various actions contributing to climate change mitigation and adaptation aspects including implementation of REDD+ activities.

National Working Plan Code-2023: Ministry of Environment, Forest and Climate Change, Government of India has released the 'National Working Plan Code-2023' in 2023 for scientific management of forests and evolving new approaches. While keeping the essential tenets of its predecessor, the National Working Plan Code-2023 will act as a guiding principle for State Forest Departments in preparation for the working plan for different forest divisions in the country. National Working Plan Code-2023 deals in detail with the essentials of forest management planning, incorporating the principles of sustainable management of forests. This includes extent and condition of forest and tree cover; maintenance, conservation and enhancement of biodiversity including wildlife, forest health and vitality, conservation and management of soil and water resources, enhancement of forest resource productivity, maintenance and enhancement of social, economic, cultural and spiritual benefits and providing the appropriate policy, legal and institutional framework. For the first time, the National Working Plan Code-2023 has prescribed state forest departments to engage in continuous data collection and its updating in a centralised database. The new working plan code of 2023 will help in generation of robust data base which will facilitate in developing various forest and climate change related programmes. The National Working Plan Code provides enough provisions to integrate issues related to climate change and conservation of biodiversity through the system of working plans.

Carbon Inventory: At national level, Forest Survey of India is primarily responsible for carrying out biomass and soil carbon inventory through its regular activity of national Forest Inventory assessment. In order to establish linkages between National Forest Inventory and working plans, the working plan code has adopted grid-based systematic sampling. Changes in carbon stocks in different carbon pools are estimated following the Good Practice Guidance of IPCC (2003). In working plans, for estimation of appropriate certainty, application of IPCC Tier 2 and Tier 3 methodology has been recommended.

Climate Change Impact, Adaptation and Mitigation-Related Issues: For effective forest management and planning to address forest related adaptation and mitigation issues, it is important to understand the vulnerability of forest ecosystems and the likely impacts. The revised working plan code seeks to increase the productivity of forests as well as improvement in forest quality and their capacity to sequester carbon through effective intervention and planning. The code, therefore, directly or indirectly deals with the various aspects of forests and climate change such as impacts, adaptation and mitigation.

The current system of forest management planning, policies and regulations and forestry practices in India are much apt for mainstreaming of climate change related aspects into forest management plans. Various polices, legislations, acts and regulations related to forests, biodiversity, wildlife and environment provide a robust statement on India's obligation towards protection of environment, biodiversity and at the same time addressing climate concerns. The National Forest Working Plan Code, objectives of Green India Mission, National Afforestation Programme, National REDD+ Strategy (MoEFCC 2018b) and National Forest Policy (1988) provide enough space for incorporating climate change mitigation aspects of forests while developing forest management plan. At the time of preparation of forest working plans, the linkage of climate change mitigation efforts through afforestation, reforestation, forest restoration and REDD+ with national forest inventory required to be strengthened so that various national programmes addressing climate change mitigation through forest-based actions can be achieved at the forest division level. The code also provides opportunity to create effective forest conservation and adaptation plans for climate change mitigation by the systematic and continuous observations of various parameters of forest management.

12.11 Way Forward

Despite all global efforts, increase in emission of greenhouse gas continues unabated. Concentrations of greenhouse gases are continuously increasing, reaching annual averages of 410 ppm for carbon dioxide, 1866 ppb for methane, and 332 ppb for nitrous oxide in 2019. The atmosphere, ocean, cryosphere and biosphere have all undergone extensive and swift changes. Anthropogenic activities are estimated to have caused 1.1 °C of global warming above pre-industrial levels (IPCC 2021) and likely to reach 1.5 °C between 2030 and

2052, if it continues at the current rate (IPCC 2018). Among the major global efforts to limit global warming, the Paris Agreement seeks to fortify the international response to the threat posed by climate change, by keeping the rise in the average global temperature to well below 2 °C with additional efforts for limiting it to 1.5 °C above pre-industrial levels.

12.11.1 Analysis of Current National Pledges Under NDC

The latest NDC Synthesis Report published by UNFCCC (UNFCCC 2021b) contains some worrying findings. When accounting the implementation NDCs submitted by 191 country Parties, the overall global greenhouse gas emission level in 2030 is predicted to be 16.3% higher as compared to 2010. According to IPCC special report on 1.5 °C (IPCC 2018), in order to limit temperature, rise to 1.5 °C goal, global net anthropogenic CO_2 emissions need to decrease by about 45% from the 2010 level by 2030, reaching net zero around 2050. To limit temperature rise to below 2 °C, CO_2 emissions need to decline by about 25% by the year 2030 from the 2010 level and reach net zero around 2070.

The NDC Synthesis Report (UNFCCC 2021b) illustrates that the new and updated NDCs are not sufficient enough to achieve the Paris Agreement goal on limiting global temperature. The Report suggest an urgent need either for a significant increase in the level of drive of countries NDCs between 2021 and 2030 or a significant over achievement of the latest NDCs, or a combination of both options, in order to keep warming well below 2 °C or limiting it to 1.5 °C. If emissions are not reduced by 2030, they will need to be substantially reduced thereafter to compensate for the slow start on the path to net zero emissions. IPCC (2018) in its Special Report on 1.5 °C identifies net zero CO_2 emissions as a prerequisite for halting warming at any level.

12.11.2 Achieving Net Zero Emission by Countries

Net-zero emissions is a state where the sum total of all anthropogenic emissions by sources and removals by sinks are zero. By now around 137 country Parties pledged to achieve net-zero emissions by the year 2050. Notably among them are the United States and the European Union set out its bloc-wide net-zero target for 2050, The United Kingdom has a legally binding net zero target by 2050.[1] China and Russia announced to be carbon-neutral by 2060. Other major European economies include France, Denmark, New Zealand and Hungary (all with 2050) and Sweden (by 2045). India also announced net-zero emissions by 2070. Other countries have amended their NDCs, proposed legislation, or stated their desire to achieve net-zero emissions

12.11.3 Carbon Dioxide Removal and Bioenergy With Carbon Capture and Storage

The recent series of IPCC reports (IPCC 2018, 2019, 2022a, b) reaffirms the importance of land restoration, afforestation and reforestation for mitigation of climate change. To limit global temperature, rise to 1.5 °C above pre-industrial levels extraordinary efforts in many sectors of economy particularly in energy and industrial would be required. Through various global efforts and with international cooperation it is possible to limit global warming by 1.5 °C as agreed by country Parties to UNFCCC under the Paris Agreement. Theoretically, it is feasible that carbon dioxide removal techniques to extract CO_2 from the atmosphere might help to limit temperature rise to 1.5 °C.

All pathways assessed by IPCC reach net CO_2 emission by around 2050, all of them also require some level of CO_2 removal measures or technologies and particularly in case of overshoot pathways where temperature exceeds 1.5 °C before 2100 then to bring it down to 1.5 °C. There are several possible options for carbon dioxide removal. The pathways assessed by the IPCC 1.5 °C report (IPCC 2018) mostly depended on bioenergy with carbon capture and storage, afforestation and reforestation. To achieve Paris temperature goals, 'net negative emissions' can only be realised if the removal of CO_2 from the atmosphere is greater than the amount entering the atmosphere. The quantity of carbon dioxide removal (CDR) required for this would be more than for stabilising the atmospheric concentration of CO_2 and thus global temperature at a certain level.

A variety of CDR techniques exist, each with varying potentials for producing negative emissions and also with diverse costs and adverse impacts. These techniques are at various stages of development, some of them are more conceptual than others. Bioenergy with carbon capture and storage is one example of CDR which is still in the demonstration phase. In this technique, a growing tree captures carbon dioxide from the atmosphere and then energy is produced by burning tree biomass. The carbon dioxide released during the process of burning is captured before it reaches the atmosphere and then stored in deep geological formations.

[1] https://commonslibrary.parliament.uk/global-net-zero-commitments/#:~:text=The%20UK%20has%20a%20legally,emissions%20by%2078%25%20by%202035.

12.11.3.1 Afforestation and Reforestation Activities for CDR

Tree plantation activities such as afforestation and reforestation also qualify for CDR because these plantations enhance natural sinks of carbon dioxide. Chemical processes are another category of CDR technique capturing CO_2 from the atmosphere and storing it for a long time. Another technique to store carbon out of the atmosphere is to turn waste plant residue into biochar and bury it in the earth for decades or perhaps centuries.

These aforesaid reports really bring important contributions from land-based activities including those that fit into CDR measures. Globally, ecosystems and people are affected by land degradation, which both causes and is influenced by climate change. Forest restoration as one of the key activities that can be taken within CDR measures and also bringing not only restoration of land but co-benefits of adaptation and mitigation are focus of the IPCC special report on climate change and land (IPCC 2019).

The large-scale implementation of CDR measures can have impacts that are yet not known specially if they are implemented on a large scale. These can have an impact on land, food security and local community at large and also biodiversity and water security. Level of scientific knowledge is yet not very well established but some of the CDR measures related to forest and other land use can have an important positive effect on land, biodiversity restoration of natural ecosystems and thus contributing to mitigation of climate change.

12.11.4 Forest-Based Mitigation Opportunities and Co-benefits

Forest-based mitigation options in national climate change mitigation programmes also contribute in achieving sustainable development goals as a co-benefit of these programmes. Opportunities for the restoration and rehabilitation of lands with potentially large co-benefits exist in areas where afforestation and reforestation activities take place on previously degraded lands. It further depends on whether restoration actions are carried out in plantations or natural forests. Thus, there are ample opportunities for adopting sustainable management of forests that can contribute to a reduction in atmospheric CO_2 through reducing deforestation and forest degradation, encouraging actions like forest conservation, forest restoration and enhancements of forest carbon stocks. Preserving carbon in harvested wood products for a longer period is also a potentially new opportunity for climate change mitigation with host of other co-benefits.

Most mitigation and adaptation options including REDD+ described earlier could contribute to income generation and livelihoods of the forest fringe communities as well. In bioenergy options, higher demand for woody biomass will tend to increase forest landowner revenue, while greater reliance on small-diameter biomass from faster-growing species may help provide a more continuous source of revenue than traditional timber production. Employment generation activity and income in rural and urban areas of developing countries through wood energy production and use also contribute to socio-economic development of the communities. Initiatives like Bonn Challenge and 'One Trillion Trees Initiative' have emerged to take advantage of opportunities for using the available land. Emerging technologies like bioenergy with carbon capture and storage can also be attempted through such initiatives.

12.11.5 Finance for Climate Change Actions

The Paris Agreement *inter alia* highlights the significance of equity and non-carbon benefits for sustainable development while supporting a broad range of policy measures such as REDD+, sustainable management of forests, joint mitigation and adaptation (UNFCCC 2016). In order to fully realise the potential of land-based mitigation efforts, a significant amount of financial input is also necessary. An investment of USD 0.7 billion/year has already been made in land-based mitigation programmes, however, as observed in IPCC Special Report 1.5 °C (IPCC 2018), there still remains a huge funding gap between these efforts and the scale of efforts required for 1.5 or 2.0 °C targets. Through forest-based activities CO_2 benefits up to 5.8 Gt per year can be achieved. However, the cost estimates of these actions could rise from USD 178 to USD 400 billion per year by 2050. The existing quantified attempts to reduce net emissions with trees and agricultural initiatives are important; however, in order to attain carbon sequestration levels aligned with high levels of mitigation, public, private sector and society will need to fast ramp up investments.

Natural climate solutions offer up to one-third of the solutions required to meet the climate change goals by 2030. To achieve the climate change, biodiversity and land degradation targets at global level, a financing gap of USD 4.1 trillion needs to be overcome by 2050. The current investments in nature-based solutions amount to USD 133 billion, most of which comes from public sources (UNEP 2021). UNEP (2021) calls for tripling the investments in nature-based solutions by 2030 and four times increase by 2050 from the current level. While more public money would help close some of the gap, the private sector's investment in nature-based solutions has to rise significantly.

12.11.6 Climate Change Priorities for Forest Sector

India has reason to be concerned about climate change as majority of population in India depends on livelihood in climate-sensitive sectors like agriculture and forestry. Forests in India have been historically used by its rural population and continue to be crucial for the livelihoods of at least 275 million rural people for collection of firewood, grazing of livestock, use of timber and bamboo for construction and non-timber forest products and a large variety of food and medicinal plants. The average mean annual increment of India's forests is merely 0.7 cubic meters per hectare per year which is significantly below the global average of 2.1 cubic meters (World Bank 2006). Although productivity from plantation activities is significantly high using modern inputs and technology, increased investments to improve forest productivity of natural forest are also necessary. A number of progressive national and state policies have been developed in recent past. Effective implementation of National Forest Working Plan Code-2023 (MOEFCC 2023a) and National REDD+ strategy are new initiatives to enhance forest productivity through effective management of natural forests.

India's development plans are designed to strike a balance on economic development and environment. The planning process in India is directed by the principles of sustainable development with emphasis on clean and green environment. India's National Action Plan on Climate Change incorporates India's vision of ecologically sustainable development and aligns the environmental and economic objectives. Land restoration opportunities and their climate change mitigation adaptation benefits have been identified through various policy, management and institutional instruments. The salient features of National Forest Policy (1988), National Agroforestry Policy (2014), National Working Plan Code 2023, Joint Forest Management; Green India Mission; Green Highways Policy (2015), National Bamboo Mission; Mahatma Gandhi National Rural Employment Act (2005); National Afforestation and Ecodevelopment Board and Compensatory Afforestation Fund Management and Planning Authority Act are a few examples. Concepts like carbon dioxide removal and bioenergy with carbon capture and storage are very naive in the Indian context. Additionally, there is a pressing need for investment planning and action for research in the new and developing fields of climate change in India's forestry sector.

To understand the state of forests, and their response to climate change, and to draw adaptation plans, a systematic long-term research, monitoring and modelling programme is required. The Ministry of Environment, Forest and Climate Change has initiated the Indian Long-Term Ecological Observatories (ILTEO) network with the primary goal of understanding the ecology of major biomes/social ecosystems at multiple spatial and temporal scales, through long-term and interdisciplinary research. Understanding the biophysical and anthropogenic drivers of ecosystem change and their effects on social-ecological responses is the core research premise of the Indian Long-Term Ecological Observatories network (MoEFCC 2015b).

Lessons Learnt

- Despite various global efforts, emissions of GHGs remain unabated resulting in warming of 1.1 °C since 1850–1900. The global temperature is expected to reach or exceed 1.5 °C by the year 2040 making it more challenging to fulfil the Paris Agreement's goal
- Forests play an important role in the context of climate change mainly because (i) deforestation and land degradation contribute to global CO_2 emissions, (ii) forests provide a large opportunity to mitigate climate change at a relatively lower cost and (iii) forest ecosystems are projected to be adversely affected by climate change, thereby affecting its biodiversity, biomass growth and regeneration of forests.
- Mitigation and adaptation are two crucial options to address climate change. Globally mitigation actions such as protection, improved forest management and restoration of forests and other degraded ecosystems offer a promising potential of reducing GHG emissions and/or sequestering carbon at 7.3 (3.9–13.1) $GtCO_2$ eq. per year.
- Forests are vulnerable to climate change and need adaptation efforts. Among the areas identified as emerging opportunities for these actions include ecosystem-based adaptation (EbA), adaptation for people, management and restoration of the natural ecosystems etc. EbA can function at several different levels, including local, catchment and regional. There are many situations where vegetation can control microclimates and reduce temperatures at the local level and some of these EbA techniques are more widely in use.
- Nature-based solutions besides supporting many sustainable development goals offer climate change adaptation and mitigation benefits. The use of shade trees is a traditional technique for climate change adaptation in both urban and rural scenarios. An all-inclusive decision-making process and adaptive management that produce climate-resilient systems are essential for effective nature-based climate change mitigation. Such an approach will also support many of the sustainable development goals.
- Along with offering a wide range of other forest ecosystem services, natural forests and other sustainably managed forest landscapes for biodiversity are vital for mitigating the effects of climate change and supporting adaptation as well. Natural forests in protected areas and sustainable management of other production forests can optimise the provision of the many forest ecosystem

services, along with climate mitigation and thus benefiting the society as a whole.

- Climate change is affecting the structure and functioning of forest ecosystems as well. Due to the high temporal variability in forest dynamics and the fact that several interacting global change factors such as changes in climate, degradation of habitats and invasive species, etc. are affecting them at the same time, as such it is difficult to identify, assign and forecast the consequences of climate-driven changes. Thus, in order to comprehend how climate change is affecting the ecosystems and the services that flow from them, Indian programme on long-term ecological observatories in the form of persistent long-term monitoring studies in various ecosystems is an attempt to develop a deep understanding of aforesaid issues.
- Government of India has initiated various programmes to address the issues related to climate change mitigation and adaptation at national level under the flagship programme of National Action Plan on Climate Change (NAPCC). Green India Mission is one such Mission under the NAPCC. Green India Mission has a target to increase the forest and tree cover by 5 million hectares and improve the quality of forest on another 5 million hectares along with providing livelihood support to the forest fringe village communities.
- These efforts have been further augmented by policies like National Agroforestry Policy, National REDD+ Strategy, Joint Forest Management, National Afforestation Programme, Namami Gange Programme and Green Highway Mission. These activities besides other environmental benefits will contribute to meeting India's NDC commitment of capturing 2.5–3 billion tonnes of additional CO_2 eq. through additional afforestation and reforestation activities.
- Forest policies targeted towards forest conservation and massive afforestation/ reforestation drives under various schemes/programmes have transformed India's forest from a net source to net sink of CO_2. Thus, forests in India are making a positive contribution to global climate change mitigation and in achieving various sustainable development goals.

Key Questions

1. Describe in brief global efforts to combat problem of climate change.
2. Forests are both source and sink of GHGs, explain. How much India's forests contribute to National GHG emissions?
3. What is ecosystem-based adaptation? How nature-based solutions could contribute to climate change adaptation?
4. What could be various forest management practices that can augment adaptation to climate change?
5. What is REDD+ programme under UNFCCC? What are the salient features of India's National REDD+ Strategy?
6. How current forest management in India can be attuned with climate change concerns?

References

Aaheim A, Gopalakrishnan R, Chaturvedi RK, Ravindranath NH, Sagadevan AD, Sharma N, Wei T (2011) A macroeconomic analysis of adaptation to climate change impacts on forests in India. Mitig Adapt Strat Glob Chang 16:229–245. https://doi.org/10.1007/s11027-010-9266-6

Adger W, Agrawala S, Mirza MMQ, Conde C, O'Brien K, Pulhin J, Pulwarty R, Smit B, Takahashi K (2007) Assessment of adaptation practices, options, constraints and capacity. In: Climate change 2007: impacts, adaptation and vulnerability. Contribution of working group II to the fourth assessment report of the intergovernmental panel on climate change, pp 717–743

Alam MK, Dasgupta S, Barua A, Ravindranath NH (2022) Assessing climate-relevant vulnerability of the Indian Himalayan Region (IHR): a district-level analysis. Nat Hazards. https://doi.org/10.1007/s11069-022-05233-x

Alfaro RI, Fady B, Vendramin GG, Dawson IK, Fleming RA, Saenz-Romero C, Lindig-Cisneros RA, Murdock T, Vinceti B, Navarro CM, Skroppa T, Baldinelli G, El-Kassaby YA, Loo J (2014) The role of forest genetic resources in responding to biotic and abiotic factors in the context of anthropogenic climate change. For Ecol Manage 333:76–87

Allen CD, Macalady AK, Chenchouni H, Bachelet D, McDowell N, Vennetier M, Kitzberger T, Rigling A, Breshears DD, Hogg EH (2010) A global overview of drought and heat–induced tree mortality reveals emerging climate change risks for forests. For Ecol Manage 259:660–684

Alvarez-Garreton C, Lara A, Boisier JP, Galleguillos M (2019) The impacts of native forests and forest plantation on water supply in Chile. Forests 10(6):473

Anderegg WRL, Trugman AT, Badgley G, Anderson CM, Bartuska A, Ciais P, Cullenward D, Field CB, Freeman J, Goetz SJ, Hicke JA, Huntzinger D, Jackson RB, Nickerson J, Pacala S, Randerson JT (2020) Climate-driven risks to the climate mitigation potential of forests. Science 368(6497):eaaz7005. https://doi.org/10.1126/science.aaz7005

Bahuguna VK, Swaminath MH, Tripathi S, Singh TP, Rawat VRS, Rawat RS (2016) Revisiting forest types of India. Int For Rev 18(2):135–145. https://doi.org/10.1505/146554816818966345

Batjes NH (1996) Total carbon and nitrogen in the soils of the world. Eur J Soil Sci 47(2):151–163. https://doi.org/10.1111/j.1365-2389.1996.tb01386.x

Bauhus J, Puettmann K, Messier C (2009) Silviculture for old growth attributes. For Ecol Manage 258(4):525–537. https://doi.org/10.1016/j.foreco.2009.01.053

Bond WJ, Stevens N, Midgley GF, Lehmann CER (2019) The trouble with trees: afforestation plans for Africa. Trends Ecol Evol 34(11):963–965. https://doi.org/10.1016/j.tree.2019.08.003

Bruijnzeel LA (2004) Hydrological functions of tropical forests: not seeing the soil for the trees? Agr Ecosyst Environ 104(1):185–228

Cai W, Borlace S, Lengaigne M, Van Rensch P, Collins M, Vecchi G, Timmermann A, Santoso A, McPhaden MJ, Wu L (2014) Increasing frequency of extreme El Nino events due to greenhouse warming. Nat Clim Change 4:111–116

Champion HG, Seth SK (1968) A revised survey of the forest types of India. The Manager of Publications, Delhi-6

Chaturvedi RK, Tiwari R, Ravindranath NH (2008) Climate change and forests in India. Int For Rev 10(2):256–268

Chaturvedi R, Gopalakrishnan R, Sukumar R, Ravindranath NH (2010) Carbon management in Indian forests: a policy analysis to assess mitigation potential. Carbon Manage 1(1):109–117

Chaturvedi RK, Gopalakrishnan R, Jayaraman M, Govindasamy B, Joshi NV, Sukumar R, Ravindranath NH (2011) Impact of climate change on Indian forests: a dynamic vegetation modeling approach. Mitig Adapt Strat Glob Chang 16:119–142. https://doi.org/10.1007/s11027-010-9257-7

Chaturvedi RK, Joshi J, Jayaraman M, Bala G, Ravindranath NH (2012) Multi-model climate change projections for India under representative concentration pathways. Curr Sci 103(7):791–802

Chhabra A, Dadhwal VK (2004) Assessment of major pools and fluxes of carbon in Indian forests. Clim Change 64(3):341–360

Chitale VS, Behera MD (2012) Can the distribution of sal (*Shorea robusta* Gaertn. f.) shift in the north eastern direction in India due to changing climate? Curr Sci 102(08):1126–1135

Christensen JH, Krishna Kumar K, Aldrian E, An S-I, Cavalcanti IFA et al (2013) Climate phenomena and their relevance for future regional climate change. In: Stocker TF, Qin D et al (eds) Climate change 2013: the physical science basis. Contribution of Working Group I to the Fifth Assessment Report of the Intergovernmental Panel on Climate Change. Cambridge University Press, Cambridge, UK, New York, NY

Ciesla WM (1995) Climate change and forest management: an overview. FAO Forestry Paper 126. FAO of United Nations, Rome

Cook-Patton SC et al (2020) Mapping carbon accumulation potential from global natural forest regrowth. Nature 585(7826):545–550. https://doi.org/10.1038/s41586-020-2686-x

Cowie et al (2021) Applying a science-based systems perspective to dispel misconceptions about climate effects of forest bioenergy. GCB Bioenergy 13(8):1210–1231

Creed IF, van Noordwijk M (2018) Forest and water on a changing planet: vulnerability, adaptation and governance opportunities: a global assessment report. IUFRO World Series 38. International Union of Forest Research Organizations. Available at: https://www.iufro.org/science/gfep/forests-and-water-panel/report/ (accessed 05/09/2022)

Das A, Sukumar R (2018) Climate change and forest dynamics: the importance of long-term studies. In: Bhatt JR, Das A, Shankar K (eds) Biodiversity and climate change: an Indian perspective, New Delhi, India. Ministry of Environment, Forest and Climate Change, Government of India

Davis MB (1989) Lags in vegetation response to global warming. Clim Change 15:75–82

Davis MA, Chew MK, Hobbs RJ, Lugo AE, Ewel JJ, Vermeij GJ, Brown JH, Rosenzweig ML, Gardener MR, Carroll SP (2011) Don't judge species on their origins. Nature 474:153–154

Dimri AP, Niyogi D, Barros AP, Ridley J, Mohanty UC, Yasunari T, Sikka DR (2015) Western disturbances: a review. Rev Geophys 15(2):225–246

Dinerstein E et al (2020) A "Global Safety Net" to reverse biodiversity loss and stabilize Earth's climate. Sci Adv 6. https://doi.org/10.1126/sciadv.abb2824

Dlugokencky EJ, Hall BD, Montzka SA, Dutton G, Muhle J, Elkins JW (2014) Global climate, atmospheric composition, long-lived greenhouse gases: "State of the Climate in 2013". Bull Am Meteorol Soc 95(7)

Drever CR et al (2021) Natural climate solutions for Canada. Sci Adv 7(23):eabd6034. https://doi.org/10.1126/sciadv.abd6034

Duan Q, Tan M (2019) Spatial and temporal variations of forest cover in developing countries. Sustainability 11:1517. https://doi.org/10.3390/su11061517

Dullinger S, Gattringer A, Thuiller W, Moser D, Zimmermann NE, Guisan A, Willner W, Plutzar C, Leitner M, Mang T, Caccianiga M, Dirnböck T, Ertl S, Fischer A, Lenoir J, Svenning JC, Psomas A, Schmatz DR, Silc U, Vittoz P, Hülber K (2012) Extinction debt of high-mountain plants under twenty-first-century climate change. Nat Clim Change 2:619–622

FAO (2012) FAO, forests and climate change. Working with countries to mitigate and adapt to climate change through sustainable forest management. FAO, Rome

FAO (2020) Global forest resources assessment 2020: main report. FAO, Rome. https://doi.org/10.4060/ca9825en

FAO and UNEP (2020) The state of the world's forests 2020. Forests, Biodiversity and People, Rome. https://doi.org/10.4060/ca8642en

Favero A, Daigneault A, Sohngen B (2020) Forests: carbon sequestration, biomass energy, or both? Sci Adv 6(13):eaay6792. https://doi.org/10.1126/sciadv.aay6792

FSI (2011) India state of forest report 2011. Forest Survey of India, Ministry of Environment and Forest, Government of India, Dehradun

FSI (2013) India state of forest report 2013. Forest Survey of India, Ministry of Environment and Forest, Government of India, Dehradun

FSI (2019) India state of forest report 2019. Forest Survey of India, Ministry of Environment, Forest and Climate Change, Government of India, Dehradun

FSI (2021) India state of forest report 2021. Forest Survey of India, Ministry of Environment Forest and Climate Change, Govt of India

FSI (n.d.) Carbon stock in India's forests. Forest Survey of India, Dehradun. 145 p

Gopalakrishnan R, Jayaraman M, Bala G, Ravindranath NH (2011) Climate change and Indian forests. Curr Sci 101:348–355

Griscom BW et al (2017) Natural climate solutions. Proc Natl Acad Sci 114(44):11645–11650. https://doi.org/10.1073/pnas.1710465114

Heal G (2000) Nature and the market place: capturing the value of ecosystem services. Island Press, Covelo, CA

Heilmayr R, Echeverría C, Lambin EF (2020) Impacts of Chilean forest subsidies on forest cover, carbon and biodiversity. Nat Sustain 3. https://doi.org/10.1038/s41893-020-0547-0

Hobbs RJ, Hallett LM, Ehrlich PR, Mooney HA (2011) Intervention ecology: Applying ecological science in the twenty–first century. BioScience 61:442–450

Holl KD, Brancalion PHS (2020) Tree planting is not a simple solution. Science 368. https://doi.org/10.1126/science.aba8232. https://unfccc.int/files/essential_background/convention/application/pdf/english_paris_agreement.pdf

Hunt KMR, Turner AG, Shaffrey LC (2018) The evolution, seasonality and impacts of western disturbances. Q J Roy Meteorol Soc 144: 278–290. https://doi.org/10.1002/qj.3200

ICFRE (2010) Forest sector report India. Indian Council of Forestry Research and Education, MoEF, GoI, Dehradun

ICFRE (2013) Forest types of India revisited. Indian Council of Forestry Research and Education, Dehradun

IEA (2019) Global energy and CO_2 status report. In: The latest trends in energy and emissions in 2018. Paris, France. https://iea.blob.core.windows.net/assets/23f9eb39-7493-4722-aced-61433cbffe10/Global_Energy_and_CO2_Status_Report_2018.pdf

IPCC (2003) Good practice guidance for land use, land-use change and forestry. IGES, IPCC National Greenhouse Gas Inventories Programme

IPCC (2007) Climate Change 2007: Mitigation of Climate Change: Contribution of Working Group III to the Fourth Assessment Report of the Intergovernmental Panel on Climate Change. (Eds. B. Metz, O.R. Davidson, P.R. Bosch, R. Dave, and L.A. Meyer). Cambridge University Press, Cambridge, UK

IPCC (2014) Transport. In: Edenhofer O, Pichs-Madruga R, Sokona Y, Farahani E, Kadner S, Seyboth K, Adler A, Baum I, Brunner S,

Eickemeier P, Kriemann B, Savolainen J, Schlömer S, von Stechow C, Zwickel T, Minx JC (eds) Climate change 2014: mitigation of climate change. Contribution of Working Group III to the Fifth Assessment Report of the Intergovernmental Panel on Climate Change. Cambridge University Press, Cambridge

IPCC 2018. Global warming of 1.5 °C. In: V. Masson-Delmotte, P. Zhai, H.O. Pörtner, D. Roberts, J. Skea, P.R. Shukla, A. Pirani, W. Moufouma-Okia, C. Péan, R. Pidcock, S. Connors, J.B.R. Matthews, Y. Chen, X. Zhou, M.I. Gomis, E. Lonnoy, T. Maycock, M. Tignor, T. Waterfield (eds) An IPCC special report on the impacts of global warming of 1.5 °C above pre-industrial levels and related global greenhouse gas emission pathway in the context of strengthening the global response to the threat of climate change, sustainable development, and efforts to eradicate poverty

IPCC (2019) Summary for policymakers. In: Shukla PR, Skea J, Calvo Buendia E, Masson-Delmotte V, Pörtner HO, Roberts DC, Zhai P, Slade R, Connors S, van Diemen R, Ferrat M, Haughey E, Luz S, Neogi S, Pathak M, Petzold J, Portugal Pereira J, Vyas P, Huntley E, Kissick K, Belkacemi M, Malley J (eds) Climate change and land: an IPCC special report on climate change, desertification, land degradation, sustainable land management, food security, and greenhouse gas fluxes in terrestrial ecosystems. Cambridge University Press, Cambridge

IPCC (2021) Summary for policymakers. In: Masson Delmotte V, Zhai P, Pirani A et al (eds) Climate change 2021: The physical science basis. Contribution of Working Group I to the Sixth Assessment Report of the Intergovernmental Panel on Climate Change. Cambridge University Press, Cambridge, United Kingdom and New York, NY, USA, pp 3–32. https://doi.org/10.1017/9781009157896.001

IPCC (2022a) Summary for policymakers. In: Pörtner HO, Roberts DC, Tignor M, Poloczanska ES, Mintenbeck K, Alegría A, Craig M, Langsdorf S, Löschke S, Möller V, Okem A, Rama B (eds) Climate change 2022: impacts, adaptation, and vulnerability. Contribution of Working Group II to the Sixth Assessment Report of the Intergovernmental Panel on Climate Change. Cambridge University Press

IPCC (2022b) Summary for policymakers. In: Shukla PR, Skea J, Slade R, Al Khourdajie A, van Diemen R, McCollum D, Pathak M, Some S, Vyas P, Fradera R, Belkacemi M, Hasija A, Lisboa G, Luz S, Malley J (eds) Climate change 2022: mitigation of climate change. Contribution of Working Group III to the Sixth Assessment Report of the Intergovernmental Panel on Climate Change. Cambridge University Press, Cambridge, UK and New York, NY, USA. https://doi.org/10.1017/9781009157926

IUCN (2020) IUCN Global standard for nature based solutions, 1st edn. Gland, Switzerland

Jacobs DF, Dalgleish HJ, Nelson CD (2013) A conceptual framework for restoration of threatened plants: the effective model of American chestnut (*Castanea dentata*) reintroduction. New Phytol 197:378–393

Jia G et al (2019) Land–climate interactions. In: Shukla PR et al (eds) Climate Change and Land: an IPCC special report on climate change, desertification, land degradation, sustainable land management, food security, and greenhouse gas fluxes in terrestrial ecosystems. IPCC

Joyce LA, Blate GM, McNulty SG, Millar CI, Moser S, Neilson RP, Peterson DL (2009) Managing for multiple resources under climate change: national forests. Environ Manag 44:1022–1032

Joyce LA, Briske DD, Brown JR, Polley HW, McCarl BA, Bailey DW (2013) Climate change and North American rangelands: assessment of mitigation and adaptation strategies. Rangel Ecol Manage 66: 512–528

Kant P, Wu S (2012) Should adaptation to climate change be given priority over mitigation in tropical forests? Carbon Manage 3:303–311

Kanwar N, Kuniyal JC (2022) Vulnerability assessment of forest ecosystems focusing on climate change, hazards and anthropogenic pressures in the cold desert of Kinnaur district, north western Indian Himalaya. J Earth Syst Sci 131:51. https://doi.org/10.1007/s12040-021-01775-z

Kates RW, Travis WR, Wilbanks TJ (2012) Transformational adaptation when incremental adaptations to climate change are insufficient. Proc Natl Acad Sci 109:7156–7161

Keeley AT, David D, Ackerly DD, Cameron DD, Heller NE, Huber PR, Schloss CA, James H, Thorne JH, Merenlender AM (2018) New concepts, models, and assessments of climate wise connectivity. Environ Res Lett 13(7):073002

Keesstra S, Nunes J, Agata Novara A, Finger D, Avelar D, Zahra Kalantari Z, ArtemiCerdà A (2018) The superior effect of nature based solutions in land management for enhancing ecosystem services. Sci Total Environ 610–611:997–1009

Kishwan J, Pandey R, Dadhwal VK (2009) India's forest and tree cover: contribution as a carbon sink. Technical Paper No. 130 ICFRE BL—23. Indian Council of Forestry Research and Education, MoEF, GoI, Dehradun

Kohl M, Lasco R, Cifuentes M, Jonsson O, Korhonen KT, Mundhenk P et al (2015) Changes in forest production, biomass and carbon: results from the 2015 UN FAO Global Forest Resource Assessment. For Ecol Manage 352:21–34. https://doi.org/10.1016/j.foreco.2015.05.036

Korner C (2003) Alpine plant life: functional plant ecology of high mountain ecosystems. Springer, Berlin

Krishnan, R., Sanjay, J., Gnanaseelan, C., Mujumdar, M.M., Kulkarni, A., Chakraborty, S. (eds.) (2020) Assessment of climate change over the Indian region, a report of the Ministry of Earth Sciences (MoES), Government of India, Springer Open.

Kumar M, Savita, Singh H, Pandey R, Singh MP, Ravindranath NH, Kalra N (2019) Assessing vulnerability of forest ecosystem in the Indian Western Himalayan region using trends of net primary productivity. Biodivers Conserv 28:2163–2182. https://doi.org/10.1007/s10531-018-1663-2

Lewis SL, Wheeler CE, Mitchard ETA, Koch A (2019) Restoring natural forests is the best way to remove atmospheric carbon. Nature 568(25):25–28

Locatelli B, Evans V, Wardell A, Andrade A, Vignola R (2011) Forests and climate change in Latin America: linking adaptation and mitigation. Forests 2:431–450

Luyssaert S, Marie G, Valade A, Chen YY, Djomo SN, Ryder J, Otto J, Naudts K, Lansø AS, Josefine Ghattas J, Matthew J, McGrath MJ (2018) Trade-offs in using European forests to meet climate objectives. Nature 562:259–262. https://doi.org/10.1038/s41586-018-0577-1

Mackey B et al (2020) Understanding the importance of primary tropical forest protection as a mitigation strategy. Mitig Adapt Strat Glob Chang 25(5):763–787. https://doi.org/10.1007/s11027-019-09891-4

Malhi Y, Franklin J, Seddon N, Solan M, Turner MG, Field CB, Knowlton N (2020) Climate change and ecosystems: threats, opportunities and solutions. Philos Trans Royal Soc B: Biol Sci 375(1794):20190104. https://doi.org/10.1098/rstb.2019.0104

Meigs GW, Dunn CJ, Parks SA, Krawchuk MA (2020) Influence of topography and fuels on fire refugia probability under varying fire weather. Can J For Res 50(7). https://doi.org/10.1139/cjfr-2019-0406

MoEF (1988) National forest policy. Ministry of Environment and Forests, Government of India, New Delhi

MoEF (1999) National forestry action programme—India, volume 1: Status of forestry in India. Ministry of Environment and Forests, Government of India, New Delhi, India

MoEF (2004) India initial national communication to UNFCCC. Ministry of Environment and Forests, Government of India

MoEF (2010) Climate change and India: a 4 × 4 Assessment—a sectoral and regional analysis for 2030s. INCCA Report # 2. INCCA: Indian Network for Climate Change Assessment. 160 p
MoEF (2012) India second national communication to the United Nations framework convention on climate change. Ministry of Environment and Forests, Government of India
MoEFCC (2014a) National working plan code—2014. Ministry of Environment, Forest and Climate Change, Government of India, New Delhi
MoEFCC (2014b) Reference document for REDD plus in India. Ministry of Environment and Forests, New Delhi. (Contributors: Jagdish Kishwan, N H Ravindranath, Rekha Pai, T P Singh, V.R.S. Rawat, Rajesh Kumar, Ruchi Badola, Madhu Verma, J V Sharma, Rajiv Pandey, A. Duraisamy, and Subhash Chandra) Available at: http://envfor.nic.in/content/reference-document-redd-india
MoEFCC (2014c) India's progress in combating climate change. Briefing Paper for UNFCCC COP 20 Lima, PERU, December, 2014, Ministry of Environment and Forest and Climate Change, Government of India, New Delhi
MoEFCC (2015a) India's first biennial update report to the United Nations Framework Convention on Climate Change. Ministry of Environment, Forest and Climate Change, Government of India
MoEFCC (2015b) Indian Long-Term Ecological Observatories (ILETO). Ministry of Environment, Forest and Climate Change, Government of India
MoEFCC (2018a) India: Second Biennial Update Report to the United Nations Framework Convention on Climate Change. Ministry of Environment, Forest and Climate Change, Government of India
MoEFCC (2018b) National REDD+ Strategy, India. Ministry of Environment, Forest and Climate Change, Government of India, https://redd.unfccc.int/files/india_national_redd__strategy.pdf
MoEFCC (2021) India: Third biennial updated report to UNFCCC. Ministry of Environment Forest and Climate Change, Government of India
MoEFCC (2022) India's updated first nationally determined contribution under Paris Agreement (2021–30). Ministry of Environment Forest and Climate Change, Government of India. Available at https://unfccc.int/documents/611411
MoEFCC (2023a) National working plan code—2023. Ministry of Environment, Forest and Climate Change, Government of India, New Delhi
MoEFCC (2023b) India: Third national communication and initial adaptation communication.
Mooley DA, Parthasarathy B (1984) Fluctuations in All-India summer monsoon rainfall during 1871–1978. Clim Change 6:287–301
Morecroft MD et al (2019) Measuring the success of climate change adaptation and mitigation in terrestrial ecosystems. Science 366(6471):aaw9256. https://doi.org/10.1126/science
Murthy IK, Tiwari R, Ravindranath NH (2011) Climate change and forests in India: adaptation opportunities and challenges. Mitig Adapt Strat Glob Chang 16:161–175. https://doi.org/10.1007/s11027-010-9261-yMoEF
Nabuurs GJ et al (2013) First signs of carbon sink saturation in European forest biomass. Nat Clim Change 3(9):792–796. https://doi.org/10.1038/nclimate1853
Nabuurs GJ et al (2017) By 2050 the mitigation effects of EU forests could nearly double through climate smart forestry. Forests 8(12): 484
Niskanen AKJ, Niittynen P, Aalto J, Väre H, Luoto M (2019) Lost at high latitudes: Arctic and endemic plants under threat as climate warms. Divers Distrib 25:809–821
Oliver TH et al (2015) Interacting effects of climate change and habitat fragmentation on drought-sensitive butterflies. Nat Clim Change 5(10):941
Pan Y, Birdsey RA, Fang J, Houghton R, Kauppi PE, Kurz WA, Phillips OL, Shvidenko A, Lewis SL, Canadell JG, Ciais P, Jackson RB, Pacala SW, McGuire AD, Piao S, Rautiainen A, Sitch S, Haye D (2011) A large and persistent carbon sink in the world's forests. Curr Forestry Rep 7:245–266. 26113 2011;333(6045):988LP-993LP
Pant GB, Rupa Kumar K (1997) Climates of South Asia. John Wiley & Sons, Chichester, p 320
Parker C, Mitchell A (2009) The little REDD+ Book: a guide to governmental and non-governmental proposals for reducing emissions from deforestation and forest degradation. Global Canopy Programme, Oxford
Parthasarathy B, Munot A, Kothawale DR (1994) All-India monthly and seasonal rainfall series 1887–1993. Theor Appl Climatol 49:217–224
Pedlar JH et al (2012) Placing forestry in the assisted migration debate. BioScience 62:835–842
Planning Commission (2014) The final report of the expert group on low carbon strategies for inclusive growth. Planning Commission, Government of India
Portner HO et al (2021) Scientific outcome of the IPBES-IPCC co-sponsored workshop on biodiversity and climate change. IPBES and IPCC., 256 p, Bonn, Germany. https://doi.org/10.5281/zenodo.4659158
Prabhu R, Barrios E, Bayala J, Diby L, Donovan J, Gyau A, Graudal L, Jamnadass R, Kahia J, Kehlenbeck K, Kindt R, Kouame C, McMullin S, van Noordwijk M, Shepherd K, Sinclair F, Vaast P, Vågen TG, Xu J (2015) Agroforestry: realizing the promise of an agroecological approach. In: FAO agroecology for food security and nutrition. Proceedings of the FAO International Symposium, Rome, pp 201–224
Raizada A, Parandiyal AK, Ghosh BN (2003) Estimation of carbon flux through litter fall in forest plantations of India. Indian For 129(7): 881–894
Rajamani L (2016) The 2015 Paris agreement: interplay between hard soft and non-obligations. J Environ Law 28(2):337–358. https://doi.org/10.1093/jel/eqw015
Ramakrishnan PS (2007) Traditional forest knowledge and sustainable forestry: a north-east India perspective. For Ecol Manage 249(1-2): 91–99
Ranganathan CR (1938) Studies in the ecology of the shola grassland vegetation of the Nilgiri Plateau. Indian For 64:523–541
Ravindranath NH (2007) Mitigation adaptation synergy in forest sector. Mitig Adapt Strat Glob Chang 12(5):843–853
Ravindranath NH, Somashekhar BS, Gadgil M (1997) Carbon flow in Indian Forests. Clim Change 35(3):297–320. ISSN 0165-0009
Ravindranath NH, Joshi NV, Sukumar R, Saxena A (2006) Impact of climate change on forests in India. Curr Sci 90:354–361
Ravindranath NH, Sharma N, Sagadevan A, Jayaraman M, Munsi M (2012) Climate change and its impacts on forests of Sikkim. In: Arrawatia ML, Tambe S (eds) Climate Change in Sikkim—Patterns, Impacts and Initiatives. Information and Public Relations Department, Government of Sikkim
Rawat VRS (2010) Reducing emissions from deforestation in developing countries (REDD) and REDD plus under the UNFCCC negotiations. Indian For 136(1):129–133
Rawat VRS, Kishwan J (2008) Forest conservation-based climate change mitigation approach for India. Int For Rev 10(2):269–280
Rawat VRS, Singh TP (2016a) Forestry sector contribution to India's INDC to UNFCCC. Indian For 142(7):711–712
Rawat VRS, Singh TP (2016b) Forests under Paris climate agreement. Indian For 142(5):513–514
Rawat VRS, Rawat RS, Karky BS, Bhattarai N (2020) REDD+ and Hindu Kush Himalayan countries. In: Rawat RS, Karky BS, Rawat VRS, Bhattarai N (eds) REDD+ Readiness in Hindu Kush Himalaya. Indian Council of Forestry Research and Education, Dehradun and International Centre for Integrated Mountain Development, Kathmandu, Nepal

Reichstein M, Bahn M, Ciais P, Frank D, Mahecha MD, Seneviratne SI, Zscheischler J, Beer C, Buchmann N, Frank DC, Papale D, Rammig A, Smith P, Thonicke K, van der Velde M, Vicca S, Walz A, Wattenbach M (2013) Climate extremes and the carbon cycle. Nature 500:287–295

Richards JF, Flint EP (1994) Historic land use and carbon estimates for South and Southeast Asia 1880–1980. ORNL/CDIAC-61 NDP-046, Oak Ridge National Laboratory, Tennessee, USA

Rogelj J, Popp A, Calvin KV, Luderer G, Emmerling J, Gernaat D, Fujimori S, Strefler J, Hasegawa T, Marangoni G, Krey V, Kriegler E, Riahi K, van Vuuren DP, Doelman J, Drouet L, Edmonds J, Fricko O, Harmsen M, Havlík P, Humpenöder F, Stehfest E, Tavoni M (2018) Scenarios towards limiting global mean temperature increase below 1.5 °C. Nat Clim Change 8:325–332

Roshani SH, Kumar P, Masroor M, Rahaman MH, Rehman S, Ahmed R, Sahana M (2022) Forest vulnerability to climate change: a review for future research framework. Forests 13:917. https://doi.org/10.3390/f13060917

Sathaye JA, Najam C, Cocklin T, Heller F, Lecocq J, Llanes-Regueiro J, Pan G, Petschel-Held S, Rayner J, Robinson R, Schaeffer Y, Sokona R, Swart H, Winkler (2007) Sustainable development and mitigation. In: Metz B, Davidson OR, Bosch PR, Dave R, Meyer LA (eds) Climate change 2007: mitigation. Contribution of Working Group III to the Fourth Assessment Report of the Intergovernmental Panel on Climate Change. Cambridge University Press, Cambridge, United Kingdom and New York, NY, USA

Seddon N et al (2020a) Global recognition of the importance of nature-based solutions to the impacts of climate change. Glob Sustain 3 (e15):1–12

Seddon N, Sengupta S, Garcia Espinosa M, Hauler I, Herr D, Rizvi AR (2020b) Nature-based solutions in nationally determined contributions. IUCN Library System. IUCN and University of Oxford

Sentis A, Montoya JM, Lurgi M (2021) Warming indirectly increases invasion success in food webs. Proc Royal Soc B: Biol Sci 288(1947):2020–2622. https://doi.org/10.1098/rspb.2020.2622

Sharma J, Chaturvedi R, Bala G, Ravindranath NH (2013) Challenges in vulnerability assessment of forests under climate change. Carbon Manage 4:403–411. https://doi.org/10.4155/cmt.13.35

Sharma J, Upgupta S, Jayaraman M, Chaturvedi RK, Bala G, Ravindranath NH (2017) Vulnerability of forests in India: a national scale assessment. Environ Manag 60:544–553. https://doi.org/10.1007/s00267-017-0894-4

Sinclair F et al (2019) The contribution of agroecological approaches to realizing climate-resilient agriculture. GCA, Rotterdam and Washington, DC. 46 p

Singh TP, Rawat VRS, Rawat RS (2015) Climate change mitigation and adaptation in forestry sector in India. Indian Council of Forestry Research and Education, Dehradun

Singh SP et al (2021) Indian Himalayan timberline ecotone in response to climate change—initial findings. Curr Sci 120(5):859–871

Smith et al (2018) Impacts on terrestrial biodiversity of moving from a 2 °C to a 1.5 °C target. Philos Trans Royal Soc A 376. https://doi.org/10.1098/rsta.2016.0456

Snapp S et al (2021) Agroecology and climate change—rapid evidence review: performance of agroecological approaches in low- and middle- income countries. CGIAR Research Program on Climate Change, Agriculture and Food Security. Available at: https://cgspace.cgiar.org/handle/10568/113487 (accessed 2021/09/08/03:49:15)

Stanturf JA (2015) Future landscapes: opportunities and challenges. New For 46:615–644

Stanturf JA, Goodrick SL, Outcalt KW (2007) Disturbance and coastal forests: a strategic approach to forest management in hurricane impact zones. For Ecol Manage 250:119–135

Stanturf JA, Palik BJ, Williams MI, Dumroese RK, Madsen P (2014) Forest restoration paradigms. J Sustain For 33:S161–S194

Stern N (2006) Stern review: the economics of climate change. Cambridge University Press, Cambridge, UK. https://doi.org/10.1017/CBO9780511817434

Telwala Y (2012) Climate change and alpine flora in Sikkim Himalaya. In: Climate change in Sikkim-patterns, impacts and initiatives. Information and Public Relations Department, Government of Sikkim, Sikkim, India. http://www.sikenvis.nic.in/writereaddata/7Chapter_Climate_Change_and_alpine_Flora_in_Sikkim_Himalaya.pdf

Thakur AK, Kumar R, Verma RK, Kumar P (2022) Analyzing environmental determinants of tree species distributions and regeneration pattern in Western Himalaya via National Forest Inventory of India. J Sustain For. https://doi.org/10.1080/10549811.2022.2123354

Turner AG, Annamalai H (2012) Climate change and the South Asian summer monsoon. Nat Clim Change 2:587–595. https://doi.org/10.1038/nclimate1495

UN (1992) United Nations framework convention on climate change. New York, USA (also available at https://unfccc.int/resource/docs/convkp/conveng.pdf)

UN (2015) Paris agreement. United Nations, New York

UNEP (2018) Emissions gap report 2018. United Nations Environment Programme, Nairobi, Kenya

UNEP (2021) Emissions gap report 2021: the heat is on—a world of climate promises not yet delivered—executive summary, Nairobi

UNFCCC (2008) Report of the conference of the parties on its 13th session, Bali, 3–15 Dec, 2007. unfccc.int/files/meetings/cop_13/application/pdf/cp_bali_action.pdf

UNFCCC (2011) Report of the conference of the parties on its 16th session, Cancun, Mexico, 29 Nov–10 Dec 2010. Addendum: Part Two: Action taken by the Conference of the Parties at its sixteenth session. Document FCCC/CP/2010/7/Add.1. Bonn, Germany (also available at https://unfccc.int/resource/docs/2010/cop16/eng/07a01.pdf)

UNFCCC (2016) Report of the conference of the parties on its twenty-first session, held in Paris from 30 Nov to 13 Dec 2015, United Nations Framework Convention on Climate Change, Document No. FCCC/CP/2015/10, United Nations

UNFCCC (2018) India's proposed submission on forest reference levels for REDD+, result based payments under UNFCCC, https://redd.unfccc.int/files/2018_frel_submission_india.pdf

UNFCCC (2021a) The Glasgow climate pact – key outcomes from COP26 | UNFCCC accessed on 3rd Sept 2022

UNFCCC (2021b) Nationally determined contributions under the Paris agreement. Revised synthesis report by the secretariat, UNFCCC Document No. FCCC/PA/CMA/2021/8/Rev.1 dated 25 Dec 2021

Upgupta S, Sharma J, Jayaraman M, Kumar V, Rabindranath NH (2015) Climate change impact and vulnerability assessment of forests in the Indian Western Himalayan region: a case study of Himachal Pradesh, India. Clim Risk Manag 10:63–76

Van Aalst MK (2006) The impacts of climate change on the risk of natural disasters. Disasters 30:5–18

Verma LR (1998) Forestry and agroforestry management practices. In: Indigenous knowledge for watershed management in upper north-west Himalayas of India. PWMTA Field Document No. 15, Participatory Watershed Management Training in Asia (PWMTA) Program GCP/RAS/161/NET, FAO, Kathmandu, Nepal. Available at: http://www.fao.org/docrep/x5672e/x5672e00.htm

Viviroli D, Weingartner R (2004) The hydrological significance of mountains: from regional to global scale. Hydrol Earth Syst Sci 8(6):1017–1030

Walther GR, Beibnera S, Burga CA (2005) Trends in the upward shift of alpine plants. J Veg Sci 16(5):541–548

Williams MI, Dumroese RK (2013) Preparing for climate change: forestry and assisted migration. J For 114:287–297
Woodwell GM (2004) Mountains: Top down. Ambio Suppl 13:35–38
World Bank (2006) India: Unlocking opportunities for forest-dependent people in India: main report: Volume I. No. 34481-IN. Agriculture and Rural Development Sector Unit, South Asia Region, The World Bank, New Delhi
Zettlemoyer MA, Schultheis EH, Lau JA (2019) Phenology in a warming world: differences between native and non-native plant species. Ecol Lett 22(8):1253–1263
Zhou G et al (2015) Global pattern for the effect of climate and land cover on water yield. Nat Commun 6:5918. https://doi.org/10.1038/ncomms6918

Community Participation with Special Reference to Joint Forest Management 13

M. R. Baloch

Abstract

The natural plant resources are the bases of life support system of human and other animals as they are primary producers in the food chain. Due to increased human and cattle populations and infrastructure development, the natural resources have greatly been depleted and now causing threats to the existence of human life. During the ancient eras (Indus Valley Civilization and Vedic period) when humans started settling near rivers and cleared forests for agriculture and habitations, the fauna and flora were plundered by them. Now the importance of their protection is realized to combat the threat of climate change which is adversely affecting all of us.

In 1970, the dense Sal forests of south Bengal were repeatedly damaged. The then Divisional Forest Officer (DFO) realized that forest cannot be protected only through forest field staff and by uses of forces. Forests can be protected with the help of tribal people and other local forest dwellers who are residing in and around forests since ages. The DFO held a series of meetings with villagers and tribals of Arabari forest area (Medinipore, West Bengal) and convinced them to perform duties with forest staff and also not to cut, damage and burn forests and kill animals. In turn, the DFO offered free collections of grasses, fallen leaves and twigs, other non-timber forest produced and allowed them to work in nursery, plantations and various construction works by replacing the contractors. Villagers gradually developed faith in forest administration and stopped illicit felling, encroachments, staff assaults and poaching etc. This approach has developed a sort of symbiotic relationships with forest field staff and villagers. Later, Government of West Bengal recognized this successful approach and termed it as Joint Forest Management (JFM). Government gradually announced various benefits for the tribals, namely usufructs rights and sharing of benefits accrued from Non-timber forest products (NTFP) and timber sales. JFM committees were formed involving men and woman. The concept of JFM was accepted in all States and Union Territories of India and also in abroad. The Government of West Bengal and other States and Government of India have issued many notifications and guidelines to make Joint Forest Management legally stronger. Due to JFM, community has realized the importance of forests and sustainable uses of forests which resulted in reduction in forest and wildlife crimes. It is a successful approach to protect and conserve the natural resources.

Keywords

Sacred grove · Khejri tree · Chipko Movement · JFM · Ecotourism

13.1 Management of Forest Resources: A Historical Background

The vegetation is the primary producer (able to synthesize food through water, minerals, and sun energy) in a food chain as it is the prime source of food energy, and this energy flows from primary consumers to secondary and then tertiary consumers in an ecosystem. Humans and all other living organisms survive only on the food energy produced by green vegetation. Thus, the importance of forest, wildlife and their ecosystem is very important to human beings since its existence. Humankind has consumed, conserved, worshipped, enjoyed, loved and protected forest and its associated wealth in a symbiotic relationship mode in all ages of its evolution. During the primitive jungle life of humans, the nature itself has full control over natural and biological resources and after settlements during civilization human gradually started the control over these resources for their shelter, food and other needs. With the development of civilization powerful tribes and *rajanyas*, kings and lords had controlled these resources. After industrialization due to

M. R. Baloch (✉)
Arid Forest Research Institute, Jodhpur, India

A. K. Mandal, A. Nicodemus (eds.), *Textbook of Forest Science*, https://doi.org/10.1007/978-981-97-8289-5_13

economic needs industries and the Governments took over full charge of these resources. These takeovers were initially as per minimum local needs and then kept on enhancing over a part or entire of an ecosystem. The tribals who were real owners of these resources were gradually stripped of their rights. When local communities (tribal and other forest dwellers) ultimately got alienated from the forests they felt that forests are not their properties and hence started destroying them for their daily uses. The age-old symbiotic relations got disconnected. Forest department or Governments tried best to protect and conserve these resources with sticks and guns but miserably failed. Ultimately, the concept of public or community participation emerged in the form of joint forest management in which usufructs rights of various extents were shared to regain the age-old symbiotic bonds of local communities with forest and wildlife. A brief historical account of forest, biodiversity and environment conservation through varied extents of community participations from ancient India to till now is discussed in this chapter.

13.1.1 Indus Valley Civilization (3300–1700 BC)

Aryans migrated to the subcontinent and settled in north and northwest and after clearing forests started agriculture and animal husbandry. Human culture and nature had symbiotic relationships. Religion the cult of fertility emerges, writing begins and animal husbandry, pastoral and agriculture economy and wide trading network develop. Aryans of Indus Valley Civilization (Harappa, Lothal, Kalibanga and Mohenjo-Daro) showed biodiversity awareness with the fact that finding of varied articles of flora (leaves, woods to burn bricks and mud pots) and fauna (peacocks, unicorn deer, elephants, bulls, tigers, etc.) during excavation. Deforestation, changing climate, migration and overpowering indigenous tribes, etc. started increasing.

13.1.2 Vedic Period (1500–600 BC)

Rig Veda and later other Vedas were composed. *Ayurveda* (*Atharva Veda*) the vegetation-based medicinal system (Ayurveda's God Dhanvantri, Sushruta Samhita by Sushruta) emerged. Aryan mixed with local tribes settled down in various parts of India, agriculture improved and epics composed. Nature worship and conservation was part of religion and culture in the society. Vedas described four types of living beings, namely *Andaj* (oviparity), *Jeevaj* (viviparity), *Swedj* (aquatic) and *Udbhij* (earth-borne vegetation). Vedas have the description of protection and uses of forest wealth. In *Ishopanishad* the stress was given on sustainable development and conservation of nature. In Ayurveda, the forests have been classified as *Atavi*, *Bipina*, *Gahana*, *Van* and *Aryani*. Several plants were considered sacred due to their food, aesthetic and medicinal values. Some of the plants were considered very significant due to their proximity to particular deities. In the Vedic seal, some animal and tree (*Pipal*) have been found depicted which suggest that these were being worshipped. Barks of some of the trees were used for tanning leather and wood to make fire by frictions. In the epic *Ramayana*, a beautiful portrait of wildlife, trees, grasslands, *Dandakarnya* forests have been narrated. Hanuman fetched the *Sanjivanee Buti* (herb) from the Himalayas to revive injured Laxmana shows the treasure of herbs in the Himalayas. In Vedic traditions, every habitat had the category of forest called *Mahavan*, *Tapovan* and *Shrivan*. *Mahavan* or great natural forest near village could be like protected forests of today (Kumar 2014). There was a system of reforestation of lands which had natural forest earlier but cleared for some reason. *Shrivan* or the prosperity or wealth forests indicates that the forest productivity was maintained inter alia, for timber, NTFPs, fodder, water, soil enrichment, agroforestry etc. The *Tapovan* or forest of religion was for shelter to sages in which no felling of trees or poaching of animal was allowed. In the ancient society, the life of a person was divided into fourfold *ashramas* in which the last one was *Vanprasthashram* in which an old person used to pass his remaining life in the forest. In the Vedic period the forests were looked by Panchayat, a group of five elders. Every household should have a cluster of five trees called *Panchavati* symbolizing the five basic *tatva* or elements that are earth, water, fire, air and sky or totality of every living being. In ancient Tamil *Sangam* period literature, the land was categorized into four ecotypes: the littoral, wet, grassland and rice fields. In the various sites of the Harappan civilization, wild animals were the main sources of food and other uses. In the Indus Valley sites the bones of hare, cheetal, jackal, one-horned rhinoceros, wild ass and elephants have been found. The seeds of some of the plants found have now been extinct from the area. The swamp deer and *barasingha* were found in the Baluchistan area till 300 BC. Taming of elephants, roosters, zebu cow and buffalo and herding of some of the animals along with farming were the landmark of evidence of that era. Himalaya is the great repository of herbs and medicinal plants so Vedic literature of *Dhanvantri*, *Sushruta*, *Charaka*, *Nagarjuna*, *Parasara*, *Valmiki* and numerous *rishis* and *munis* exemplify it. *Aranyakas* (forest works), *Upanishad/Brahdaranyaka* (great forest text) are very important from the forestry point of view. Smritis prescribe the importance, management and uses of forests. The faith and religion were used as a tool to protect nature and natural resources, besides a wide range of ethnobotany traditions. Tribals were nature worshipers and fully depend on forest and wildlife and nature.

13.2 Descriptions of Forest in Ancient Indian Texts

Jainism and Buddhism prescribed the path of ahimsa for all sorts of organism and creatures. Each Hindu deities was assigned various animals and birds as their *Vahanas*, the carriers which make the animals and birds sacred. The poet Kalidas in *Abhigyan Shakuntalam* mentioned the biodiversity and ecology of *Tapovan* the sacred groves dedicated to local deities. In Bhagavad Gita, Lord Krishna considered a *Banyan* tree and its branches with the world and all fauna and flora denotes the concept of community ecology. *Kautilya* or *Chanakya* (Chandragupta Mourya Dynasty, 321–297 BC) in *Arthashastra* showed the environmental awareness and the management of forests, gardens and orchards for recreational and revenue purpose and did their classifications (www.indianetzone.com/50/forests_ancient_india_forests.htm). He prescribed the penalties and punishment for injuries to living creatures. The current forestry and wildlife acts and rules have great similarities with the forms and contents of the *Arthshastra*. Forest directors, animal supervisors were appointed for the protection of wild animals, plants and domestic cattle. In non-agricultural lands, the killing of animals was prohibited. The village head was bound to prevent cruelty to animals. Fines were imposed for cutting of any part of a tree. Animal treatments were also prescribed. The King was supposed to protect forests, elephants, mines and water bodies. Forests were considered as important resources and should be used in a sustainable manner. As per *Arthshastra* the country was divided into forest (*aranya*), village area (*gramya*), hills (*parvata*), wet areas (*audak*), arid lands (*bhauma*), maidan (*sama*) and undulating land as (*viswa*). Forest-based factories were recommended for various forest types; *jangala* (open forest), *anupa* (marshy lands) and *sadharana* (common type lands) where different types of trees exist for timber and other uses.

Modern day's environmental laws show similarity with environmental issues mentioned in the *Arthshastra*. The traces of forestry management found in India in 300 BC during the reign of *Chandragupta Maurya* and later till Samrat Ashoka, a forest superintendent looked after the forest and wildlife in the State. Tree plantings were done along roads, community and camping sites. In Varahmihira's *Vrihat Samhita* (700 AD) the description of the technical relationship between irrigation tank and vegetation (tree) is the sort of a forestry research. The many trees were named which were to be planted over-irrigation embankment. The *Sanatan* culture believes that the whole world is a forest, so we must keep this intact. It gives the concept of sacred groves. *Neem*, *Shami* (*khejri*), *Karam*, *Mahua*, *Harsingar* or *Parijat* and *Tulsi* are worshipped all over India. As per belief, deities reside on plants during *Chaturmasa* (rainy season) so it forbids to fell tree during *Chaturmasa*. If very necessary to cut, then mantra is chanted and the tree spirit is urged for forgiveness. Cutting a tree after sunset is considered a big sin among some communities. In Kerala after childbirth a tree is allocated as *Nakshatra Vanam* or birth star. *Manusmiriti* (200 BC) described the whole organism kingdom into *Chara* (animals) and *achara* (plants). For saving of plants *Manusmriti* also describes various punishment for the offenders. *Manu* prohibited killing all type of fishes, carnivorous birds, village birds, web-footed birds, beak striking birds and one-hoofed animals, village pigs, solitary moving animals and unknown animals for food. He stated the killing of ass, horse, camel, deer, elephant, goat, snake and buffalo is a sin. *Charaka* and *Sushruta Samhitas* are ancient medical science books. Land was classified as per soil, climate and vegetation. In another post-Vedic text, *Surapala's Vrikhsayurveda* (the science of plant life) which describes soil, manure and watering, growth and health of vegetation. Its *bijottpati* chapter describes the forest ecosystem. Forests were named like *atavi*, *bipina*, *gahana*, *kanana*, *bana*, and *aranyani*. *Mahabharta* and *Ramayana* periods have different type of description of forest and animals. Lord Krishna was the biggest nature lover. He praised the *Vrindavan* which has vegetation like *kadamba*, *pipal*, *tamal* and *vata*. Lord Krishna compared the whole world with a banyan tree with numerous branches in which all creatures wander which reflect the concept of community ecology. Ramayana also describes *Dandakarnya* forest with the trees like *arjuna*, *jupapushpa*, *sala*, *ashoka*, *champa*, *sitaphal* and *chandan* (Bhattacharya 2014).

13.2.1 The Sacred Places, Trees and Groves

The sacred (Latin: sacer, means restricted or set off) is the object of veneration and admiration. The communities for common use or worship or protect or admire the places, plants, animals, natural objects etc. declared them traditionally as sacred or *pavitra* and protected them at any cost. *Oran* (*aranya*), *Gouchar*, *Devvan* etc. are the sacred places or groves which are example of beautiful community participation for conservation of natural resources during ancient eras. Mostly these are dedicated to some deities so are the most pious entities for all communities. Not only local but other communities are also bound to pay respect to such entities and protect them (Singh 2016).

13.2.1.1 Sacred Places

These can be springs, caves, meadows, forests, mountains, rivers and other water bodies. These sites are related to deities, legendry heroes of communities, religious, economic and environmental importance. Sacred places are viewed usefully from the natural, social, cultural, historical, intrinsic to extrinsic in values, uncontested and contested and protected and endangered. Yamunotri, Gangotri, Rishikesh and Haridwar (Uttarakhand), Kashi and Mathura (UP), Amarkantak and Chitrakoot (MP), Jaliyanvala and Amritsar (Punjab), Ajanta and Ellora caves (Maharashtra), Bodhgaya and Sarnath (Bihar), Pushkar, Khwaja Ajmer and Runecha or Ramdevra (Rajasthan), Dwarika and Somnath (Gujrat) and enumerable such places are examples from all over India.

13.2.1.2 Sacred Plants

Vegetation is a strong medium that connects living and non-living things in nature. Trees have played important roles in many of world's mythologies and religions and were given deep and sacred thoughts throughout the ages. Trees are symbols of physical and spiritual nourishment, transformation and liberation, sustenance and spiritual growth, union and fertility. They are with us right from birth to death. For example, a place to place and community to community, some trees are considered as the most sacred, like Deodar, Sal, Coconut, Rudraksha, Beil, Ashok, Kadam, Arjun, Pipal (*ashvtha*), Bargad, Khejri (*Prosopis cineraria*) Kalpavriksha (*Adansonia digitata*), Jal or Pilu (*Salvedora persica*), Kair (*Caprice decidua*), Neem, Karam (*Nauclea parviflora*), Mahua, Akand (*Calotropis* sp.), Aam, Kalam (*Mitragyna parviflora*), Christmas tree (*Araucaria* sp.) and Tulsi and many more species. Even Hindu religious festivals and pious days (*divas*) are named after some sacred trees like *Neem Saptami*, *Vat Savitri*, *Bilva Mangal*, *Kadli Vrita* and *Aonla Ekadashi*. Plant species are used for worshipping, naming of area or villages and clans etc.

13.2.1.3 Sacred Groves

It is a patch of forest or a clump of trees associated with another form of life and affords protection based on religious practice or faith. It is called by various names at different places like Oran (Aranya), Kenkari, Malvan and Yogmaya (Rajasthan), Dev van (HP), Devray (Maharashtra), Devkhand and Siddaryanam (Karnataka), Sarana (Bihar), Jaher (Jharkhand and Odisha), Ngwpui (Nagaland), Ki Law (Mizoram), Kavu (Kerala) and Pavithravana (Andhra Pradesh), Than, Madaico (Assam), Gumpha Jungle (Sikkim, Arunachal Pradesh), Swami shola (TN) and Harithan, Sabitrithan (WB). These can be on forest, revenue or private lands with a small patch to thousands of square km in size. These sacred groves provide services like soil moisture conservation, water, fodder, grasses, NTFP and timber supply apart from religious, recreational, cultural services. They sink carbon and are places of education, research, wildlife shelter, and conservation of biodiversity and ethnomedicinal importance. Nowadays many of these groves are under severe degradation due to overgrazing, mining, encroachment and other human greed and activities. These need to be conserved and rehabilitated with priority.

Around 365 Devvan have been reported from Kullu District (HP) alone (HFRI 2020). In 27 states of the country, a total of 18,669 sacred groves have been recorded (Singh 2016). In Rajasthan, 2804 numbers of orans distributed in 25 districts are documented. *Gouchar bhumi* is also equally sacred like Oran in Rajasthan. Actual numbers of sacred groves may be a few lakhs in the country, which are yet to be surveyed and recorded.

Ancient communities, along with plants and animals had given equal importance to wild animals, birds and even non-living objects or items in particular areas which are considered as sacred. Various birds and animals are considered as *vahanas* (carriers) of various Gods, Goddesses and deities in *Sanatan* philosophy, so they are worshipable. In Nepal during Diwali, Hindus worship crow and cows. Bishnoi of North India worship and protect blackbucks and Chinkaras. *Charan* (Bards) community at Karnimata Temple (Deshnokh, Bikaner) protects and worships rats. Nilgai (Bluebell, an antelope) is not killed as its name is matching with the holy cow the *Gaumata*. Naga (Cobra), Elephant, Tiger, many birds, reptiles, fish and many other living beings are also worshipped. All *Kshatriya* clans of North India (Rathore, Solanki, Chouhan, Bhati, Parmar or Panwar, Pratihar and Kachhawa) have one or more animals and birds (Lion, Tiger, Fish, Chinkara, Kite and Eagle, Dove etc.) in logos and symbols of their kingdom. They consider these animals as pious and protect them. Even at some places soil, stone and other non-living objects are prohibited by the local community to be damaged or collected. It means in our culture living and non-living components of an ecosystem are protected by a community which tradition is akin to the modern concept of biosphere reserve wherein everything is considered as heritage and protected. Thus, the old concept of sacred groves is the traditional way of protecting biological and other natural resources by the whole communities.

13.3 Forest Management: Post Ancient to Modern Era

13.3.1 Bishnois: The Great Green Warriors for Nature Conservation

The Bishnois, often referred to as Vishnois, are a vegetarian religious sect within Hinduism. They were established in Rajasthan's Thal (Indian Thar) desert and then extended to the northern Indian states of Haryana, Punjab, MP, and UP. The word 'Thar' is a mispronounced of the Marwari word 'Thal', which is the abbreviated version of the Sanskrit word 'Sthala', which denotes a location, an elevated point, or a specific area. The name Thal was formally changed to Thar because Sindhis and British people occasionally pronounce L as R. The 29 guiding principles of Vishnois are attributed to Guru Jambheshwar, who established Vishnoi or Bishnoi Panth (Sect) on Kartik Vadi 8, 1542 VS (1485 AD), while delivering a sermon to his pupils at Samarathal Dhora (sand dunes). Vishnois are the disciples of Guru Jambhoji, who was an ardent supporter of Vaishnvaite or Vaishnavi dharma. The term 'Bishnois' refers to the group of people who are believed to have been inspired by Guru Jambhoji to spread the concepts of Bees (20) plus Noi (9) or 29 principles. This vast community of environment lovers known as Bishnois or Vishnois have emerged in North India as a result of enthusiastic support of Guru Jambhoji, spanning all castes and creeds, who are willing to die for wild animals and verdant forests. His beliefs were recorded in a text known as Shabdwani, which is composed of 120 shabdas and is written in the Nagri alphabetical structure. Of his 29 ideals, nine deal with personal cleanliness and sanitation as well as preserving general health, seven deal with appropriate social behaviour and five deal with worshiping God. There are eight guidelines that have been established to protect biodiversity and promote humane treatment of animals. Of them, two more deep precepts pertain specifically to the preservation and safeguarding of forests and wildlife: first, love and compassion for all living things (Jeev Daya Palani); and second, refrain from felling the healthy trees (Roonkh Leela Ni Ghave) in order to preserve the ecosystem. These include protecting all living forms and outlawing the killing of animals and green trees.

13.3.2 Khejarli Massacre: The Supreme Sacrifice of *Amrita Devi* with 363 *Bishnois*

Amrita Devi Beniwal was a Bishnoi woman who, along with over 363 other Bishnois, perished in the Thal Desert while defending the most significant, most valuable and holy Khejri or Shami (*Prosopis cineraria*) trees. Early in the eighteenth century, the Marwari ruler of Jodhpur needed wood to heat lime for building the fort. The king, then, dispatched his warriors under the command of Hakim Girdhar Das Bhandari to cut down trees in the neighbouring village of Jalnadi (26 km south of Jodhpur city), which became known as Khejarli later on because of the town's abundance of dense forests of Khejri trees, that had been safeguarded for millennia by the Bishnois. However, the villagers and Amrita Devi Beniwal objected to the king's men after learning about it. She was threatened with death by the feudal army if she refused to support the felling of trees. She said she would consider it an insult to her religious beliefs and would sooner give her life to save the green trees, especially the Khejaris, without caring about the threat. According to the Indian lunar calendar, a 'black day' occurred in Indian history 291 years ago on September 11, 1730 AD, or 10 Bhadrapad Sudi, 1787 VS. The first person to be killed by the brutal royal army of wood cutters was Amrita Devi, along with her three tiny daughters, Asu, Ratni and Bhagu, as they hugged khejri trees in front of their house. Along with their mother, three daughters were quickly sacrificed. Prior to scarfication Amrita Devi said, 'Dam diya dag lage, tukro na deva dan, Shir saate roonkh rahe to bhi sasto jaan', which translates to, 'neither money will be paid nor a small donation will be given; if a tree is saved at the cost of one's head, still it must be considered as the worthiest'.

The word about the incident went viral. The Bishnois got together and summoned the 84 villages to come to discuss what they should do next. Since the ruling court was not satisfied with the ultimate sacrifice made by those four heads and the green tree felling kept going, it was decided that one Bishnoi volunteer would sacrifice their life in order to preserve the green trees. King extended his personal apologies to the people. That was the time he made the promises that hunting would never be allowed close to the Bishnois villages, that no khejri tree would ever be cut, and that they would never be asked to provide timber to the ruler again. The Jodhpur ruler, in recognition of the bravery of the Bishnoi people, issued a royal proclamation that stated: 'All hunts and cutting of green trees throughout the revenue borders of Bishnoi villages had been strictly prohibited'. This was the first widespread Chipko-kind Movement.

13.3.3 The Continuing of Tradition of Sacrifice for Trees and Wildlife

Before sacrifice of Khejarli village many smaller and random incidences of sacrifices for tree and wildlife had happened at various places including Ramsani, Jodhpur, 1604 AD and Tilvasani village, Jodhpur (Jetaram 2016). The contribution of various Vishnoi saints, for example, Sant Vilhoji (1532–1616 AD) in *Battis Akhari* has stated that '*jadi hiran sanhar, dekh sir dijiye*' means give your head if there is destruction of khejri tree and deer (Jetaram 2019). Guru Jambheshwar the reincarnation of Lord Vishnu told that *Vanaspati Vaso* (vegetation is the home) and '*hari kankari manap medi, tahan hamara vasa*', means the green kankeri tree is like a palace where is my residence (Jetaram 2019).

Bishnois have a deep affection for wild animals. Because of their protection, antelope (like chinkaras, blackbucks and blue bulls) and deer can be seen grazing in fields in Bishnoi-dominated areas, despite the fact that the state of Rajasthan, where the Bishnois are primarily found, has severe water shortages. The Bishnoi community began fierce protests in 2008 in response to the murder of black buck by a Bollywood celebrity and numerous other poachers.

Each of these 29 tenets prioritized protecting the environment. The sect has adhered to these principles religiously ever since. Numerous accounts exist of the Bishnoi people abusing hunters and poachers for trespassing into their territory. These people revere nature, and part of their sacred traditions includes protecting the environment, wildlife and plants. There are lots of trees and animals in the Bishnoi homeland, which is in the forests close to Jodhpur. This area has a greener landscape than other places, and the antelopes that inhabit it, especially the blackbuck and chinkara, don't fear people. In fact, they usually feed off the hands of the villagers who live close to these forests. When chinkaras and other wild animals graze their crops in the fields owned by the Bishnois, they are never chased away. Over local forest and police administrations in Rajasthan, the Bishnois provide greater protection to wildlife and trees. Akhil Bhartiya Jeev Raksha Bishnoi Sabha, the supreme body of Bishnois, is closely affiliated with organizations that prioritize the preservation of the environment, wildlife, and forests.

13.3.4 Chipko Andolan: Government's Initiatives for National Awards for Environment Protectors

The Ministry of Environment and Forests (MoEF) of the Government of India established national awards in honour of Amrita Devi and Gaura Devi, both revered figures in India's environmental conservation movement. These awards were instituted during the UN-declared International Year of Forests in 2011, aimed at promoting awareness of sustainable forest management, conservation and development. Amrita Devi, who sacrificed her life nearly 287 years ago in Khejarli village near Jodhpur while protecting the khejri tree, inspired Gaura Devi. Gaura Devi, in turn, played a pivotal role in the renowned Chipko Movement. This movement originated in 1973 when Gaura Devi led a group of courageous women in Reni village, Chamoli District of Uttarakhand (formerly Uttar Pradesh) to safeguard their trees. The Chipko Movement spearheaded by Gaura Devi became a benchmark for socio-ecological movements, influencing other forest regions in Himachal Pradesh, Rajasthan, Bihar and beyond. Gaura Devi's leadership in Uttarakhand inspired the Chipko Movement, which had a profound impact on environmental activism. This movement, with its emphasis on protecting trees and ecological balance, later inspired the Appiko Movement in Karnataka. The Appiko Movement successfully halted tree felling in the Western Ghats and Vindhyas, showcasing the far-reaching influence of grassroots environmental movements in India.

The Chipko Movement, initiated in 1973, emerged as a pivotal forest conservation movement in India, setting a precedent for non-violent protests that resonated globally. In a time when environmental movements were scarce in the developing world, Chipko's success garnered international attention, becoming a source of inspiration for numerous environmental initiatives. This non-violent movement contributed significantly to slowing deforestation, exposing vested interests, fostering ecological awareness and showcasing the collective power of people. While men held leadership roles, women were the backbone and mainstay of Chipko, bearing the brunt of rampant deforestation's impact. Women, profoundly affected by the diminishing resources like firewood, fodder, and water, played a crucial role in the movement. They not only highlighted the environmental consequences but also became primary stakeholders in afforestation efforts initiated by Chipko. In recognition of its achievements, the Chipko Movement was honoured with the Right to Livelihood Award in 1987. Employing Gandhian principles of Satyagraha, both male and female activists played vital roles in Chipko, with notable figures such as Gaura Devi, Sudeesha Devi, Bachni Devi and Chandi Prasad Bhatt contributing significantly to its success.

On March of 1973, loggers arrived in Gopeshwar, and after a few weeks, a significant encounter occurred in the village of Mandal on April 24, 1973. Approximately a hundred villagers and workers gathered, beating drums and shouting slogans, compelling the contractors and loggers to withdraw. This marked the initial confrontation of the movement. Eventually, the contract was revoked and granted to the Sangh instead. As part of their non-violent protest strategy, the Sangh adopted tree-hugging, known as Chipko. For their contributions to the modern Chipko Andolan, Chandi Prasad Bhatt received the Ramon Magsaysay Award in 1982, and Late Sri Sundarlal Bahuguna was honoured with the Padma Vibhushan in 2009, along with other accolades. However, the real Chipko Movement was started by Amrita Devi and her hundreds of Bishnoi followers who sacrificed their lives to save the khejri trees at Khejarli village near Jodhpur in 1730 AD. Therefore, the Bishnois are rightfully called the first environmentalist of India.

The Bishnoi manner of life includes some very humble facts and customs, which should cause us to reflect carefully. Some of them are stated below.

- Bishnois are kind to all living things and do not cut down green trees. The materials used to construct the Bishnoi neighbourhood are sourced locally, and great care is taken to ensure environmental sustainability.
- They don't chop down trees. All they gather is deadwood. Even a carpenter bides his time until the tree falls.
- They have a variety of animals that roam around their neighbourhoods, including deer, chinkaras, peacocks, blue bulls and black bucks. Many animals have survived this long because of the Bishnois and their dedication to protecting the natural world. Nowhere else would one find all types of fauna and flora in perfect harmony?

- Long before the idea of rain harvesting gained popularity, the Bishnois constructed water reservoirs that could catch and hold rainwater for use by humans and animals in order to combat severe drought and water scarcity.
- The cremation of deceased people is a religious custom that the Belarusians reject. They bury the dead to save firewood from being wasted.
- Despite being a sect of religion, the Bishnois uphold a strong foundation of respect for Mother Nature by consistently emphasizing kindness, affection and peace rather than needless rituals, idol worship, the caste system etc.
- They cook using cow dung cakes to reduce their reliance on green trees.
- One of the main principles of 'Amar Rakhave That' is to always give abandoned animals a place to live so they can do so with dignity for the remainder of their lives.

13.3.5 Community and Individual Sacrifice for Nature: Forest Martyrs Day

Around 300 years ago, courageous members of the 'Bishnoi' tribe in Rajasthan made a significant sacrifice to halt the widespread cutting of trees. On September 11, 1730 AD, 363 individuals from this tribe lost their lives. In acknowledgment of their selfless act, the Indian Government declared that starting from September 11, 2013, this date would be observed as 'Van Sahid Divas'. This day commemorates the Khejarli incident and honours all forest personnel who have given their lives in safeguarding forests and wildlife. To honour this historic event, the Indian Council of Forestry Research and Education (ICFRE), under the Government of India, erected a Van Shahid Smarak at the Forest Research Institute in Dehradun, Uttarakhand. Additionally, several states and union territories across the nation have also established Van Shahid Smarak, signifying a united effort to annually pay homage to these heroes of nature.

13.4 Nature Conservation During Pre- and Post-independent India

During the Mughal period, forests were designated as hunting reserves for the royals to pursue large game. They promoted the planting of trees along avenues and for practical gardening purposes. Moreover, forests were increasingly cleared for agricultural needs. There existed royal hunting grounds and grasslands where Falcons and Cheetahs were bred for entertainment. Elephants and cattle were received as gifts instead of monetary compensation. Special areas were preserved for the exclusive use of royalty. The British Indian Government introduced scientific forest management through the Imperial Forest Department and appointed Dr. Dietrich Brandis, a German Forest Officer, as the first Inspector General of Forests in 1866, marking the inception of organized forest management in India. The Imperial Forest Service, established in 1867, oversaw Provincial Forest, Executive, and Subordinate Services. Initially, officials of the Imperial Forest Service received training overseas. Later, from 1927, they were trained at the Imperial Forest Research Institute (IFRI) in Dehradun, subsequently renamed the Indian Forest College (IFC), and now known as the Indira Gandhi National Forest Academy (IGNFA).

The administration of forestry was transferred to the provincial list by the Government of India Act in 1935. The initial Indian Forest Policy in 1894 aimed for the scientific management of forests and timber production. The British focus was on maximizing economic gains for the Empire through the commercialization of timber, which restricted the rights of forest dwellers and rulers, ultimately resulting in extensive deforestation due to excessive logging. However, the government permitted community-based forest management, particularly in the Himalayas, through initiatives like Van Panchayat (forest villages) in some northern hill states. The Indian Forest Service was established in 1966 for improved forestry administration. In 1976, forest and wildlife matters were included in the Concurrent list of the Indian Constitution. Amendments were made to the Indian Forest Act of 1927 by both the Centre and States. Chapter III of this act provides the state the authority to grant governance rights to any village community over lands designated as village forests. The forest village community, when assigned a specific area, may access timber, forest products or grazing resources while being responsible for protecting and enhancing the designated forest. Several legislative acts in India, including the Scheduled Tribes and Other Traditional Forest Dwellers (Recognition of Forest Rights) Act of 2006, the Biodiversity Act of 2002, the Environment (Protection) Act of 1986, the Forest (Conservation) Act of 1980, and the Wildlife (Protection) Act of 1972, among others, were enacted to promote nature conservation and endorse the concept of Joint Forest Management (JFM). Subsequent Indian Forest Policies (IFPs) in 1952 and 1988 were introduced, and more recently, IFP 2018 was drafted, shifting objectives from production and revenue generation towards sustainable conservation.

13.5 Joint Forest Management (JFM): The Concept

13.5.1 The Arabari Experiment

In 1971–1972, Dr. Ajit Kumar Banerjee, then Divisional Forest Officer (Silvia-South Division), Medinipore, West Bengal initiated a pilot project (Socio-Economic Forest

Complex Plantation) at Arabari Research Range involving local people (11 forest fringe villages, 618 families) over 1270 ha of degraded forest of sal (*Shorea robusta*). Previously, despite strict enforcement of rules and duties, the poor tribal and other forest dwellers were illicitly cutting back the sal coppices, illegally collecting forest produce regularly for various purposes. Local people were earning from forest theft about Rs. 100,000 per year equivalent to 50,000 man days of employment in 1970. There were regular clashes between forest fringe people and forest front line staff. The DFO came to know the importance of the concept of the give-and-take policy and thought to unofficially involve these people in some protection activities and in turn, allow them to collect some minor forest produces (MFPs). Therefore, they were motivated through constant persuasion and in turn they were allowed to collect (free) dry leaves, fallen twigs, grass, fodder and small poles for household use on nominal prices and given all forestry works (nursey, afforestation, construction, watch and ward etc.) which earlier were generally given to outside contractor. Later, this pilot project got highlighted among senior forest officials, forest communities and politicians and was highly acclaimed.

A tall and dense sal coppices forest stood over 700 ha was found to generate around 2.3 lakh human days for the local people. The growing stock of 700 ha sal coppices were valued at Rs. 3 crores (against the expense of Rs. 12.10 lakhs) in 1986. This patch of sal forest is capable of generating an annual income of Rs. 23 lakhs in perpetuity. Forest Department of West Bengal, seeing the great success of Arabari Experiment, thought to regularize the above-mentioned activities through legal support so it can be replicated elsewhere in the State. Government of West Bengal approved the project and issued memo No 1118-FOR Dated 7.3.1987 granting usufructuary net benefits 25% of harvested yield to the involved beneficiaries and this led to the official recognition of JFM in India and the concept abroad.

The two special features in the participation were, the taking of all important management decisions by all participants and not by a committee or specific leadership alone; and the participants nominated representative handled the cash to pay wages. The post of nominated representative was rotational. This feature has brought the transparency in the working of the local forest department and villagers started listening to field staff and performing duties of protection with staff. Actually, previously there was contractor system in forestry works through which contractors were exploiting the tribals and they were not paid properly. Now, fringe people were interacting directly with staff as middlemen were removed (Banerjee 2014). Subsequently, many office orders have been issued for usufructuary sharing with JFMCs in respect to other forest produces. After success of JFM in West Bengal in 1972 (Fig. 13.1), Haryana (Sukhomajri village, 1976), and Rajasthan (Gopalpura village, 1986) and later in various states of the country participatory units came into existence to develop tribal area rural livelihoods and regenerate forests through joint management.

It is to state that well before the concept of jointly managing forests of Arabari in 1970, the novel idea of sharing of revenue with fringe people was tried out by the forest department. The sharing of revenue in the Tirap District forest of present Arunachal Pradesh with tribal chiefs and villagers experimented during as early as 1896. From their sharing the Nocte tribal people bought 40% shares of a local veneer mill. Similar product sharing was made with villagers in Zeiling province of Nagaland in 1960 (Banerjee 2014).

13.5.2 JFM: Background and Introduction

The old and new approach in forest wealth management can be understood clearly through comparison as given in Table 13.1.

Therefore, sustainable management should be equitable, interdisciplinary, alleviating poverty and building capacity, involving tribal, women, youth, elder, business, research and NGO and be on the decentralized decision making at various levels of participation.

13.6 JFM: Constitutional Provisions/Acts/Rules/Guidelines

The forestry sector falls within the Concurrent List of the Constitution, allowing both the Central and State Governments to create laws and regulations. Numerous articles within the constitution directly or indirectly focus on safeguarding forests, wildlife, the environment and natural resources. Article 14 ensures equal protection under the law, prohibiting discriminatory laws or rules. Article 21, often referred to as the Right to Life and Personal Liberty, protects individuals from being deprived of life or liberty except through established legal procedures. Article 19(1) grants equality to tribal communities for utilizing forest resources, ensuring the right to livelihood and residency in their domicile areas. Article 31A deals with laws pertaining to government acquisition of property rights, while Articles 39(B) and (C) emphasize fair distribution of community resources and preventing wealth concentration to the detriment of the common good. The Seventh Schedule includes Directive Principles that safeguard tribal interests. Article 48A obliges the State to strive for environmental protection, safeguarding forests, and wildlife. Article 51A (g) outlines fundamental duties, calling upon Indian citizens to protect and enhance the natural environment, including forests, lakes, rivers, and wildlife, and to show compassion for all living creatures.

Fig. 13.1 A degraded sal forest has been rejuvenated into a dense coppice forest through successful implementation of JFM in West Bengal

Table 13.1 Comparison between traditional and joint forest management

S. No.	Traditional FM (TFM)	Joint FM (JFM)
1	*Objectives*: Set by Forest Department	Participatory with public
2	*Strategy*: Controlling people	Facilitating, social fencing
3	*Approach*: Top down, macro, single technical package, hierarchical, sectorial and fixed	Bottom-up, micro-planning, multi-options, networking, holistic and flexible
4	*Fringe People*: Outsiders	Partners/insiders
5	*Management*: Centralized through large working and management plans	Participatory/Joint through micro-plans (participatory and rapid rural appraisal)
6	*Transparency*: Almost nil	Fully transparent
7	*Bias/Mandate*: Production forestry, Government's revenue, industry needs	Protection, usufructs sharing with people, sustainable conservation
8	*Species*: Single/monoculture. Loss to biodiversity and animal foods	Multiple species and NTFP. Biodiversity increases and the variety of foods
9	*Knowledge*: Foresters' wisdom	Emphasis on Indigenous system
10	*Results*: Alienation, confrontation, conflicts, with fringe people, degradation and deterioration in forest quality, extent and management	Co-operation, collaboration, conflicts resolution, forest and biodiversity conservation, gradation in sustainable management

Article 298 addresses the government's contractual liability. Leveraging the constitutional provisions mentioned above, a significant legislation, commonly known as the Forest Rights Act (FRA), was enacted in 2006. This act aims to empower not only tribal communities but also other forest dwellers, securing their rights within the forested areas.

13.6.1 Forest Rights Act (FRA) 2006

The enactment of the 'Scheduled Tribes and Other Traditional Forest Dwellers (Recognition of Forests Rights) Act, 2006' stands as a pivotal moment in forestry history. This legislation grants conditional ownership of forest lands to the indigenous inhabitants who have resided in and cultivated these lands for generations. They are no longer treated as outsiders but recognized as co-owners of the forests, with a primary responsibility for forest protection and conservation. The FRA 2006 solidifies the principles and objectives of Joint Forest Management (JFM) across the nation.

The aims of FRA 2006 are multifaceted:

- To rectify historical injustices faced by tribal and other forest-dwelling communities.
- To secure land tenure, livelihoods, and food security for these communities.
- To reinforce conservation efforts by delegating responsibilities and authorities to Forest Rights holders.

This Act empowers forest inhabitants to utilize forest resources for various daily needs, including sustenance, shelter, and socio-cultural requirements. Prior to this legislation, India's forest management policies did not acknowledge the symbiotic relationship between tribal communities and forests, evident in their dependence on and traditional wisdom regarding forest conservation.

The FRA 2006 encompasses individual rights like self-cultivation and habitation, as well as community rights such as grazing, fishing and access to water bodies within forests, including traditional seasonal resource access. Aligned with the National Forest Policy of 1988 and the Government of India's guidelines for managing degraded forest land under State Forest Departments (SFDs) with local community involvement, these departments have devised their resolutions on Joint Forest Management (JFM). By the early 1990s, almost all State Forest Departments had notified these resolutions, leading to the establishment of 118,000 Joint Forest Management Committees (JFMC) nationwide. These committees aim to rehabilitate and manage 22 million hectares of degraded forest lands.

Moreover, Eco Development Committees (EDC) have been formed by SFDs to engage fringe communities in Protected Areas (PA) management through community participation. Notably, this involves JFMCs in forest protection, scheme implementation, afforestation and reforestation efforts. The increase in forest cover, excluding Tree Outside Forest (TOF), from 638,804 km^2 (19.43%) in the India State of Forest Report (ISFR) 1989 to 708,273 km^2 (21.54%) in ISFR 2017, and the overall forest and tree cover, including Tree Outside Forest (TOF), at 24.59% according to ISFR (2019), highlights the significant role of Joint Forest Management (JFM) initiatives and intensive afforestation in expanding forest cover.

13.6.2 National Forest Policies (NFP)

Till now three NFPs have been formulated and the fourth and recent one (draft) is in the final stage of its approval. As per circumstances and contemporary needs the objectives have been kept on changing from revenue generation to sustainable conservation (Table 13.2).

Table 13.2 Comparison among different national forest policies

S. No.	National Forest Policy, 1894	National Forest Policy, 1952	National Forest Policy, 1988	National Forest Policy, 2018 (Draft)
1	Management of forests for the wellbeing of India (National interest)	A top-down approach to forest management	Meeting the local people's needs (NTFP) and encouraging their participation in the protection and management of forests (JFM)	Control over the community forest resources and JFM by launching a National Community Forest Management (CFM) Mission
2	The cultivation (agriculture) is before the forest (Food security)	Proposed to increase forest cover to 1/3 (33%) of the total Indian landmass	Reiterated 33% forests cover, outside forestry (social, agroforestry) and intensive afforestation. Protection forestry	Safeguarding ecological and livelihood security of people, sustainable management for the flow of ecosystem services. Hills and plains: 2/3rd forest and tree cover
3	Maximum revenue generation from forests (State's income)	Stress: Training, research, economic benefits, production forestry	Ecological balance environmental stability. Preserving the natural heritage, customary rights of tribal	Erosion check, water supply, biodiversity, NTFP, urban, agroforestry, wildlife, ecosystem services, research and education, management, climate change etc.

13.6.3 JFM: Government of India's Guidelines (1990)

The residents in and around forest areas are to be given primary rights to minor forest produces (MFPs). It is essential to encourage forest communities to actively participate in the development, management and protection of the forests from which they derive benefits. The Government of India, through the Ministry of Rural Development, endorsed the integration of watershed projects with Joint Forest Management (JFM) in 1998, encompassing initiatives like Drought Prone Areas Programme (DPAP), Desert Development Programme (DDP), Integrated Wastelands Development Programme (IWDP), and Employment Assurance Scheme (EAS).

In February 2000, the former Ministry of Environment and Forests (MoEF) issued a new set of guidelines. These guidelines included standardized nomenclature and legal support for Joint Forest Management Committees (JFMCs) nationwide. Moreover, the extension of JFM to ecologically valuable forest areas became a focal point, emphasizing the sustainable management of Non-Timber Forest Produce (NTFPs). These include

1. Putting them on the Society Registration Act registration list
   (a) The JFM Committee is a common name.
   (b) The state government and JFMCs will sign an MOU.
   (c) Every adult will be able to join
2. Specific threshold criteria were suggested for JFMCs in order to encourage the involvement of women
   (a) At least 50% of members overall are women.
   (b) At least 33% of the Executive Committee (EC) are women.
   (c) One member or at least one-third of the EC's members must be women.
   (d) President, Vice-President, and Secretary: Women should occupy at least one of these positions.

13.6.4 Extension of JFM Outside of Degraded Forests: JFM in Good Forest Areas

- Pilot programme proposed; to be reviewed and expanded later
- 20% revenue sharing maximum
- Minimum of 10 years of protection to be eligible for revenue sharing
- Guidance for microplan preparation provided for both new Work Plan and existing Work Plan areas
- Management generally in accordance with working plan, with JFM primarily for NTFP management

Participatory Approach: (Central Sponsored Scheme, CSS), 10th 5 Year Plan: National Afforestation Program (Forest Development Agencies (FDA), Self Help Group (SHG), MFP) and EDCs (Eco Dev Committees) in Protected Areas (PAs) (2005) for Forest and Wildlife Conservation and Eco-Development of fringe people. Previously, for PAs (National Parks, Sanctuaries, Biosphere Reserve, etc.) FPC was termed as EDC. Both have different usufruct rights and functions due to different levels of protection were required. In December 2002, additional guidelines were released regarding the establishment of a mechanism for dispute resolution with the Panchayat Raj organizations in order to secure their cooperation with forest management.

13.6.5 Resolution on Eco-Development Committee (EDC) by SFD, WB

The West Bengal government published guidelines for the creation of EDCs in PAs and (Sanctuaries and National Parks, etc.) on June 26, 1996, through resolution no. 3841-for/d/11m-795. DFO in consultation with local Panchayat Samiti will propose and approve the Constitution of EDC. EDC will have one executive committee for the overall management of EDC. EDCs will perform the duties and will assist forest field staff in the conservation and protection of forest and wildlife etc.

13.6.6 Usufructuary Benefits from PAs

1. Enhancing and managing the wildlife within a PA requires collecting and elimination of specific forest goods from the designated zone or zones. To such collection and removal, permission may be granted by the Chief Wildlife Warden. When collected by a government agency, members of the EDC will be entitled to an equal share of these forest products, free of royalty but subject to payment of collection costs:
   (a) A quarter of the poles derived from drift and overwood removal.
   (b) The entirety of the firewood, grasses and certain inedible fruits, pods, flowers, seeds, lovely fungi and leaves.
2. Upon demonstrating adequate forest protection for a minimum of a year, the recipient will be eligible to receive a 25% portion of government receipts related to transportation, tourism and related activities like photography within the protected area.

13.6.7 JFM Support Activities

Following entry point activities are carried out under JFM:

1. Agriculture development, irrigation, animal husbandry and improved breeds, school buildings.
2. Fencing, water, electricity, houses, watch towers, roads, bunds, culverts, etc. works.
3. Organization of medical camps, eye camps, etc.
4. Agriculture and forestry implement, and gas and chulha, etc. distribution.
5. Vocational trainings: mushrooms cultivation, making sal leaf plates, etc.
6. Pisciculture, weaving, sericulture, bee keeping, tailoring.
7. Abeer (brightly coloured powder), agarbatti (insence) sticks, pickle and murabba making, etc.
8. Agro and farm forestry and intercropping, dairy, piggery, pisciculture, poultry, duckery, goatery.
9. Development of marketing facilities, value addition of NTFP resources.
10. Formation of Self Help Group (SHG) and SHG of women exclusively.
11. Generating the corpus funds (in joint bank accounts) for common use by SHGs under JFMCs.
12. Organization of JFM related awareness camps and melas (fair).
13. Conducting of tours of one JFMC to other JFMC areas for showing the activities.

In WB to rejuvenate the JFMCs an annual mela called Bon Bandhab Utsab (van bandhu utsav) is organized for all JFMCs of one district in which senior forest officials, district administration, panchayat members, politicians, ministers, JFMCs members and other line departments participate for 2 days and interact with forest communities. NTFP and herbal items, decorative and flowering plants, cactus and orchids, sale of various forest-based handicraft articles prepared by JFMCs, forest, wildlife awareness and extension materials and soil and moisture conservation works, etc. displayed in the stalls. Cultural, musical and sports activities are also organized. Revenues accrued due to usufructuary rights are also distributed to JFMCs. Forestry and agriculture implements are also distributed to JFMCs. This mela (fair) has become an effective rejuvenating platform for JFMCs.

Some of the entry point activities done for JFMCs at Baikunthapur Division, Jalpaiguri, WB, are depicted below (Figs. 13.2, 13.3, 13.4, 13.5, and 13.6):

13.6.8 Intercropping on Forest Lands for Landless JFMC Members

As per land ownership surveys conducted in JFMCs (forest villages and fringe area villages) under Jalpaiguri Division, WB 20% of households were found landless, 35% had less than 0.5 acres, and 30% between 0.5 and 1.0 acre of land. Only 15% had land holdings above 1 acre (Fig. 13.7). This means, maximum households are either marginal or landless farmers. The forest department had initiated an intercropping scheme for these landless poor JFMC farmers to grow seasonal crops (except paddy, wheat, jute etc.) and vegetables over forest afforestation or clear-felled coupe sites as intercrops. Incentives in the form of lump-sum amount per ha were given to farmers to procure seeds, slips, manure etc. for intercrops. In sal Working Circles in some of the plantation models 4 m spacings (between lines of sal and miscellaneous species) were kept to enable farmers to grow 1 or 2 rows of crops for 3–5 years till sapling canopies are closed over inter line spacing (Fig. 13.8). Much needy tribal used to grow intercrops and used to sell in the nearby weekly *haat* (weekly market). In elephant and other wildlife depredation areas, they used to grow *Citronella slips* (lemon grass) as these are not grazed by any animals. Citronella leaves are harvested two to three times a year. Harvested, leaves are purchased by Silviculture Division for their citronella extraction units. Lemon grass oil is highly precious and there is huge demand in the market. All the benefits directly go to the JFMCs and Forest Department never used to share or recover anything from the harvests of these intercrops. The JFMCs were doubly benefitted out of this scheme as they also used to get full man-days against cleanings, maintaining and watching of plantations. Forest department used to get well protected and clean plantation sites due to enhanced interest of JFMCs. This scheme is still very popular in JFMC areas of North Bengal.

13.6.9 The Support and Sources of Funding for JFMC: Projects and Schemes

1. State Plan/Non-Plan budget and funds from line departments
2. Rashtriya Krishi Vikas Yojana (RKVY) and Forest Development Agencies (FDA)
3. Rural Infrastructure Development Fund (RIDF)
4. CSS: Integrated Forest Management (IFM), Grants-in-Aid (GIA)
5. Medicinal Plants Conservation Authority (MPCA) and NAP

Fig. 13.2 Digging of fish ponds for JFMCs by SFD

Fig. 13.3 Providing the poultry to JFMCs by SFD

Fig. 13.4 Distribution of various machineries

Fig. 13.5 Sal patta plate making

Fig. 13.6 Free medical camps for JFMCs

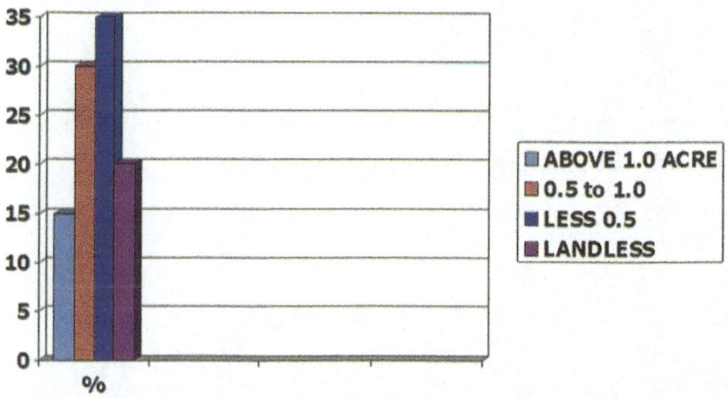

Fig. 13.7 Average agri-land holdings by JFMCs in Jalpaiguri Division, WB

6. Economically Weaker Section (EWS), a GoI Scheme
7. Mahatma Gandhi National Rural Employment Guarantee Act (MNREGA)
8. Japan International Cooperation Agency (JICA) and other foreign assistance
9. Compensatory Afforestation Management and Planning (CAMPA) funds
10. State Forest Development Corporation (SFDCs) and Tourism Department
11. NGO, local bodies and miscellaneous funding sources

13.6.10 National Level Monitoring of the JFM

A JFM cell was created in the Ministry of Environment and Forests (MoEF) in August 1998, to monitor the progress of JFM and to collect and compile the information on JFM. National Workshop on JFM was organized by Forest Research Institute, Dehradun during 27–28 June 2011. At the country level some important data have been compiled and published through proceeding which gives the latest status of JFM in India (Tables 13.3, 13.4, 13.5, and 13.6).

13.6.11 JFM and Wildlife Depredation

Forest villagers, tribal and forest fringe people are highly prone to depredation caused by wildlife. The habitations falling on elephants and other wildlife corridors and tea gardens are also victims of wildlife depredation. Mostly, they are the members of JFMCs and due to continuous threats and loss of life and properties, they are badly affected. They also develop some adverse sentiments for the wildlife as well as forest employees and workers and gradually get alienated from the JFM activities. In various parts of India, animal depredation is caused mainly by Elephants, Leopards, Tigers, Bisons, Rhinoceros, Wild boars, Bears, Blue bulls, monkeys, etc. In India around 80,000 incidences (18 States) of human-wildlife conflicts were reported during 2012–2013. Out of which crop and property damages were 73.4% followed by livestock injury or losses 20% and human injury and death cases 6.6% (Figs. 13.9 and 13.10).

To compensate the wildlife caused losses and to reduce the threats to some extent, MOEF&CC, Government of India had issued some guidelines in 2018 under 'Development of Wildlife habitats' for Project Tiger and Project Elephant which are funded through a Centrally Sponsored Schemes (CSS), and under which provision of ex-gratia for wildlife depredation has been made. MOEF&CC ordered ex-gratia Rs. 5 lakhs for human death or incapacitation, Rs. 2 lakhs for grievous injuries and Rs. 25,000 for minor injuries. For property and other losses, States/UTs can prescribe the compensations. Through order no. 1/123199/2021 dated 26/2/2021 Forest Department of West Bengal fixed

Fig. 13.8 Intercrops: potato and mustard in North Bengal

ex-gratia and compensation rates as follow: Human death case: Rs. 5 lakhs, permanent disability: Rs. 2 lakhs, permanent sight/hearing loss: Rs. 1 lakh, partial sight and hearing loss: Rs. 0.50 lakh, grievous injury needed hospitalization more than a week: Rs. 0.5 lakh plus free treatment in Government hospitals, and for less than week period Rs. 25,000 plus free treatment in Government hospital. For property (huts, livestock, crops etc.) separate categories and rates have been prescribed.

Different SFDs decide different categories and rates of compensation and these are revised from time to time. During depredations wildlife also gets injured or killed so losses are on both sides. During peak depredation periods (crop season, lean period-food and water shortage, breeding, migration, change in feeding taste and priority, winter and rainy periods; obstacle or disturbances in habitats or corridors etc.) wildlife depredations increase over the areas. To control or mitigate the problem forest department erect fencings, distribute search lights and crackers, construct the watch towers and deploy staff and JFMCs joint teams as wildlife squads for helping fringe people to ward off the wild animals. Through social media, radio, TV, newsletters, pamphlets, phones,

Table 13.3 Status of JFM in India—Number of JFM committees and forest area under JFM

States	JFM committees	Area under JFM (ha)	Number of families involved			
			GN	SC	ST	Total
A & N Islands	4	262	360	–	–	360
Andhra Pradesh	7718	1,519,000	492,000	395,000	551,000	1,438,000
Arunachal Pradesh	1013	100,377	–	–	33,048	33,048
Assam	1184	52,499	148,074	27,392	119,368	294,834
Bihar	682	462,333	91,606	80,586	39,482	211,674
Chhattisgarh	7887	3,319,000	313,000	190,000	614,000	1,117,000
Goa	26	10,000	–	–	336	336
Gujarat	3067	414,151	147,302	29,068	240,662	417,032
Haryana	2487	41,188	53,026	13,010	–	66,036
Himachal Pradesh	1023	205,056	190,000	65,000	8024	263,024
Jammu & Kashmir	4173	38,736	236,388	64,469	128,939	429,796
Jharkhand	9926	1,721,700	236,388	64,469	128,939	429,796
Karnataka	3848	808,020	185,290	55,480	32,035	272,805
Kerala	576	207,404	47,407	9708	21,386	78,501
Madhya Pradesh	15,228	6,687,390	645,000	255,000	800,000	1,700,000
Maharashtra	12,665	2,403,344	1,820,640	358,097	529,860	2,708,597
Manipur	665	166,767	7884	132	16,086	24,102
Meghalaya	285	17,245	–	–	39,210	39,210
Mizoram	613	55,990	–	57	80,628	80,685
Nagaland	951	42,929	–	–	159,587	159,587
Odisha	12,494	1,148,676	726,102	271,139	645,741	1,642,982
Punjab	1224	178,333	70,696	21,140	14	91,850
Rajasthan	5316	858,614	238,015	73,802	259,234	571,051
Sikkim	219	88,518	29,000	–	17,000	46,000
Tamil Nadu	3487	756,446	354,002	98,298	29,969	482,269
Tripura	920	241,138	19,561	14,643	45,241	79,445
Uttar Pradesh	3426	183,393	450,251	241,689	14,110	706,050
Uttarakhand	12,738	564,221	501,000	110,000	18,000	629,000
West Bengal	4368	646,084	225,309	159,425	120,415	505,149
Total	118,213	22,938,814	7,228,301	2,597,604	4,692,314	14,518,219

Source: Proceedings of National Workshop on JFM, Dehradun, 27–28 June 2011

meetings, and notices, fringe people, panchayat members and administration are informed about the depredation timings and areas. Fringe peoples are also trained in how to tackle the problem. Volunteers of JFMCs and paid members are involved to mitigate the depredation. By adoption of these measures the lives and properties of fringe people are saved and also losses were decreased. In the absence of JFMCs, there were severe and very tense conflicts between forest staff and villagers and in many cases forest field staff were *gheraod* (protest) and assaulted or even beaten to death and offices were burnt. Wild lives also were being injured, killed and poisoned.

13.6.12 JFM and Ecotourism

Ecotourism is a nature-based form of special travel defined by The International Ecotourism Society (TIES) in 1991 as '*responsible travel to natural areas which conserved the environment and sustain the wellbeing of local people*'. Sustainable tourism is defined as 'tourism that meets the needs of the present tourists and host regions while protecting and enhancing opportunity for the future'. Thus, it is an umbrella concept that would include ecotourism. The eco-tourism term was initially coined by Hector Ceballos Lacurian in 1983 to describe Nature-Based Travel. Ecotourism is based on three important principles: conservation (economic benefit will act as incentives to conserve and protect the natural resources), communities (community participation will empower them and help in achieving sustainable development) and interpretation (generating individual and mass environmental awareness). The Tourism Policy (2002) in India gives special thrust to rural tourism in small settlements, to advise the MOEF&CC and SFDs on initiating compulsory and voluntary regulations regarding ecotourism activities and to undertake specific programmes for national parks and sanctuaries. In June 2011 MoEF, GoI asked State/UT governments to frame ecotourism policies and accordingly almost all States have drafted their ecotourism policies for PAs and other forest sites. Many SFDs have created

Table 13.4 Percentage of forest area covered by JFM

States	Recorded forest area (ha)	Area under JFM (ha)	Forests covered by JFM (%)
A & N Islands	717,100	262	0.04
Andhra Pradesh	6,381,400	1,519,000	23.8
Arunachal Pradesh	5,154,000	100,377	1.95
Assam	2,683,200	52,499	1.96
Bihar	647,300	462,333	71.42
Chhattisgarh	5,977,200	3,319,000	55.53
Goa	122,400	10,000	8.17
Gujarat	1,892,700	414,151	21.88
Haryana	155,900	41,188	26.42
Himachal Pradesh	3,703,300	205,056	5.54
Jammu & Kashmir	2,023,000	38,736	1.91
Jharkhand	2,360,500	1,721,700	72.94
Karnataka	3,828,400	808,020	21.11
Kerala	1,126,500	207,404	18.41
Madhya Pradesh	9,468,900	6,687,390	70.62
Maharashtra	6,193,900	2,403,344	38.8
Manipur	1,741,800	166,767	9.57
Meghalaya	949,600	17,245	1.82
Mizoram	1,671,700	55,990	3.35
Nagaland	922,200	42,929	4.66
Odisha	5,813,600	1,148,676	19.76
Punjab	305,800	178,333	58.32
Rajasthan	3,263,900	858,614	26.31
Sikkim	584,100	88,518	15.15
Tamil Nadu	2,287,700	756,446	33.07
Tripura	629,400	241,138	38.31
Uttar Pradesh	1,658,300	183,393	11.06
Uttarakhand	3,465,100	564,221	16.28
West Bengal	1,187,900	646,084	54.39

Source: Proceedings of National Workshop on JFM, Dehradun, 27–28 June 2011

senior level posts for looking after exclusively the ecotourism activities, policy implementation, monitoring, coordination, management and extension.

13.6.13 Rajasthan Ecotourism Policy, 2020

Policy period: It is for 10 years or as may be decided by State Government.

Vision: To promote and expand access to exceptional, high-quality, low-impact ecotourism locations in Rajasthan that are focused on protecting the state's natural and cultural heritage, including its flora and fauna, creating high-quality educational opportunities and helping to create jobs locally, reduce poverty and advance conservation efforts.

The following are the Rajasthan Ecotourism Policy's primary goals:

1. Uniting all interested parties on a single platform to promote ecotourism
2. Ensure that all parties involved participate equally, effectively and actively
3. Sustainable ecotourism utilizing the Global Sustainable Tourism Criteria (GSTC)
4. Promoting ethical conduct both during and after their cooperation in efforts to conserve nature
5. Encourage sustainable utilization of biodiversity by creating revenue, employment and commercial opportunities
6. Equitably distribute the advantages of growing ecotourism to the nearby communities
7. Giving communities the chance to decide how best to present their cultural values
8. Preserve the state of Rajasthan's current biodiversity, ecosystems, culture and traditions

13.6.14 Strategy and Arrangements

The State Forest Department (SFD) will act as the central agency responsible for implementing the policy, utilizing the services of the Rajasthan Ecotourism Development Society (REDS), duly registered under the Rajasthan Society Registration Act, 1963. REDS is designated to support the

Table 13.5 Annual employment generation through JFM

States	No. of JFMC	Total no. of families	Annual employment generated in lakh man-days
Andhra Pradesh	7718	1,438,000	100.00
Arunachal Pradesh	1013	33,048	2.64
Assam	1184	52,499	4.88
Bihar	682	211,674	4.72
Chhattisgarh	7887	1,117,000	70.00
Gujarat	2195	417,032	182.76
Haryana	2487	66,036	7.05
Himachal Pradesh	1023	263,024	2.70
Jharkhand	9926	429,796	8.60
Karnataka	3848	272,805	74.90
Kerala	576	78,501	4.00
Maharashtra	12,665	2,709,000	91.37
Manipur	665	24,102	6.43
Meghalaya	285	39,210	16.04
Nagaland	951	159,587	2.60
Punjab	1224	91,850	8.00
Rajasthan	5316	571,051	51.35
Tamil Nadu	3487	482,269	16.99
Tripura	920	79,445	39.00
Uttar Pradesh	3426	706,050	17.23
West Bengal	4386	505,149	38.46
Total	71,864	9,747,128	749.72

Source: Proceedings of National Workshop on JFM, Dehradun, 27–28 June 2011

Table 13.6 Benefits from JFM

			Benefit in Rs. Lakhs				
States	No. of JFM committees	Total no. of families	Fuel wood	Fodder	NTFP	Others	Total
Andhra Pradesh	7718	1,438,000	1022.61	217.90	700.53	897.37	2838.41
Arunachal Pradesh	1013	33,048	1883.56	9198.00	117.90	44.25	11,243.71
Chhattisgarh	7887	1,117,000	14,400.00	24,000.00	93,050.00	800.00	132,250.00
Haryana	2487	66,036	94.60	78.48	20.98	0.27	194.33
Jharkhand	9926	429,796	8580.00	–	3124.00	–	11,704.00
Karnataka	3848	272,805	1628.58	–	–	1515.42	3144.00
Madhya Pradesh	15,228	1,700,000	–	–	–	2841.91	2841.91
Maharashtra	12,665	2,709,000	919.63	117.63	25.48	2.92	1065.66
Meghalaya	285	39,210	68.75	24.62	7.82	6.00	107.19
Nagaland	951	159,587	675.00	0.00	20.00	0.00	695.00
Punjab	1224	91,850	25.00	20.00	40.00	5.00	90.00
Rajasthan	5316	571,051	–	488.46	75.26	129.75	693.47
Tripura	920	79,445	4472.00	762.00	483.00	170.00	5887.00
Uttar Pradesh	3426	706,050	1152.02	791.18	50.33	53.22	2046.75
West Bengal	4368	505,149	10,000.00	3000.00	5000.00	0.00	18,000.00
Total	77,262	9,918,027	44,921.75	38,698.27	102,715.30	6466.11	192,801.43

Source: Proceedings of National Workshop on JFM, Dehradun, 27–28 June 2011

realization of the vision and objectives outlined in the Eco-tourism Policy, specifically focusing on forests, protected areas and natural areas beyond the jurisdiction of the Rajasthan Forest Department. REDS is tasked with promoting the development of ecotourism that harmoniously integrates nature-based and cultural attractions, ensuring environmental and socio-cultural sustainability. Functioning with a significant degree of autonomy, the society's ecotourism activities will be executed through the field units of the Forest Department. At the district level, the establishment of a District Ecotourism Committee (DLEC) is mandated, with the Collector serving as the Chairman and the District Forest

Fig. 13.9 Paddy field raiding by wild elephant herd

Officer (DCF) as the Member Secretary. The committee will include members from various relevant departments such as revenue, tourism, tribal development and rural development. Their collective role is to foster, administer and regulate ecotourism sites.

13.6.15 Strategy for Managing Ecotourism Sites

The Chief Wildlife Warden will create regulations governing the wildlife and recreational experiences at locations that fall under Protected Areas.

1. Locations under Project Tiger's (PT) authority and additional Protected Areas (PAs)
2. The Protected Areas' eco-sensitive zones (PAs)
3. Sites with distinctive natural features that are found in unclassified forest areas and reserve forests
4. Places where the forest department has administrative authority but are not situated on forest land
5. Locations under the administrative supervision of other government agencies

The initiative's primary beneficiary is the local community, and the government's Forest and Tourism departments are utilizing the partners' interdependencies to ensure the success of the conservation efforts.

13.6.16 Ecotourism Product Development, Promotion, Publicity and Marketing

The sites chosen for ecotourism will be distinct from others in terms of their aesthetic, natural, cultural and landscape qualities. These qualities will be acknowledged and transformed into an ecotourism offering that will establish the site's identity.

13.6.17 Sharing of Revenue Benefits

The local community that directly manages and maintains eco-tourism facilities receives 80% of earnings as compensation, with the remaining 20% going to the government as fees for facility entry and use. In addition, the Rajasthan Ecotourism Development Society (REDS) will have access to an

Fig. 13.10 Driving away to a wild loner tusker by farmers

extra 25% eco-development surcharge, which will be used to address issues related to local livelihood development, managing conflicts between humans and wildlife and conservation through eco-development.

13.6.18 Successful Ecotourism: A Case Study from Darjeeling, West Bengal

West Bengal also has got a draft ecotourism policy. Since the State is full of varied natural beauties (Himalayas to the ocean) ecotourism is highly developed all over state. Traditionally, Bengalese have the highest passion for traveling thus, all type of in-house tourists are abundantly available to utilize these eco-tourism opportunities.

As per the Annual Administrative Report for 2017–2018 WB State Forest Development Agency (WBSFDA) has developed a website for booking online ecotourism facilities (camps, huts, tents, park, tree houses, cottages, rooms, elephant ridings, trekking, FRH and other ecocentres under various forest divisions) (Fig. 13.11) and could get revenue of about Rs. 44 lakhs only from online bookings during the report period. WBFDC, Territorial, Wildlife and Functional Divisions also are engaged in ecotourism activities.

13.6.19 Ecotourism at Lamahatta and Around in Darjeeling: A Case Study

In March 2012 while passing through Lamahatta forest village the Chief Minister (CM) on seeing the eye-catching scenic natural beauty, asked accompanying forest officials to start eco-tourism activities at Lamahatta. DFO, Darjeeling Division took the eco-tourism works at war footing level and completed Lamahatta eco-park works which the CM inaugurated on 29th March 2013. The funding for this project was jointly arranged by SFD, Tourism and North Bengal Development Departments. Lamahatta is a small forest

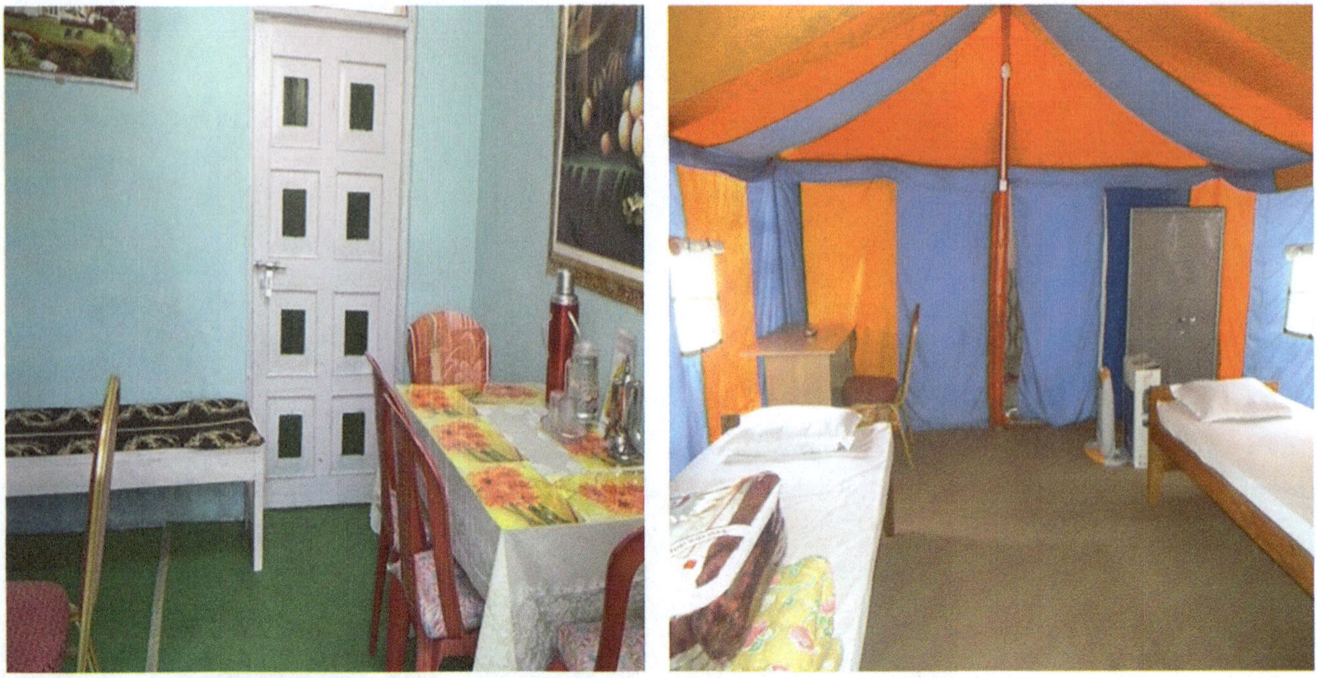

Fig. 13.11 Fully furnished room and tent houses at Lamahatta Home stays, Darjeeling Division, WB

busty on Darjeeling-Kalimpong and Sikkim road and has around 140 households inhabited mainly by multi-ethnic Buddhist communities and some Hindus. The Lamahatta forest village site selection for developing into a sustained ecotourism village had following important points which are essential to attract more and more tourists.

1. Lamahatta is only 23 km from Darjeeling town which is the world-famous hill station having beautiful rhododendron, dense forest valleys, world heritage Himalayan toy train, tea gardens of costliest Darjeeling tea brands, Buddhist monasteries, Himalayan zoological park and mighty snow-laden Kanchanzongha mountain chains. Thus, Lamahatta is very close to Queen of Hills which attracts lakhs of tourists every year.
2. Lamahatta (altitude 2077 m) is situated on the main road through which all tourists pass for visiting Kalimpong, Siliguri and Sikkim. On this route Tista Rangit River Sangam (lovers point), Tista River with bridges, Teesta dams, Latpanchar bird watchers, paradise point, FRHs, Cinchona plantations, many tea and orange gardens are spots of tourist attraction. Namchi hills of Sikkim, Tiger hills of Darjeeling can be clearly seen from Lamahatta.
3. Lamahatta has got picturesque landscape of Himalayan hillocks and cloud-filled valleys and snow clad Kunchenjunga mountain chains on the north and a series of very old tall pencil stemmed *Dhupi* (Japanese cedar, *Cryptomeria japonica*) trees along with beautifully developed eco-garden which has got many local and orchid flowers and is decorated with rows of praying flags.
4. A medicinal plants garden, watchtower, meditation centre, roof-top cafeteria, nature-trekking trail, a small natural pond at hillock top etc. are attractions which are available at the same location.
5. Nearby tourist sites are *Takdah*'s rare orchid species nursery and orange garden, rock climbing points, *Trinchule* organic village and *Peshok* Tea garden, *Senchal* Wild Life Sanctuary which can be visited by tourists during single trip.
6. Tourism Department funded 20 home stays at Chimney and Pokhariatar Forest Villages (JFMCs), community kitchen cum centre, watch tower, improvement of salamander ponds under Kurseong division also have been established during 2016–2018.

13.6.19.1Ecovillage: Implementation of Ideas

Many male members of Lamahatta were in the army and also working in nearby tea gardens as daily labourers and at the Takdah Cantonment area. Some were engaged in agricultural activities (mainly vegetables). Many youths either were working in other states or even in Arabian countries. Only women folk with aged persons and children are left in the village. Employment opportunities are very meagre.

DFO, Darjeeling and local RO with field staff had a series of meeting with forest villagers to peruse them for converting

Lamahatta as an eco-tourism centre. In 2012, a JFMC with 158 numbers was formed. To start with, jobless men and women were engaged in developing the eco-park and associated activities (flower garden, trekking trail works, construction of walls, towers and huts, cleaning of weeds, soil moisture works, nursery and watch ward works etc.). Later villagers also had given volunteer services in the development of eco-park and facilities. An entry fee of Rs. 10 per visitor was realized through a ticket counter by JFMC for the eco-park visit. At the village entry gate main information about eco-centre was displayed. JFMC members were trained by Darjeeling Division in hospitality and cooking services. Initially, 7 numbers of existing houses were converted into home stay facilities. Later 8 numbers of more home stays and 5 numbers of tent houses were added. Booking rates were Rs. 1000/- and 2000/- per day respectively for rooms and tents. For foods and other services, charges were additional to rents. As of today total 56 numbers of home stays have been developed under Takdah range. Out of which most of them have been funded by SFD (NBDB) with the initial financial help of Rs. 2.52 lakhs per household.

As there were no prescribed mechanism or direction from the department for sharing the revenues Division Office, Darjeeling had instructed to collect 25% of total revenue from the home stays and later use the same, 25% for the maintenance of the home stays and 50% is to be retained by the owner of the home stays.

For Lamahatta Ecotourism Centre, all the revenues collected are deposited in JFMC's joint account but for withdrawal, Member Secretary (Beat Officer) and JFMC President had to endorse the cheques. Every proposed work had to be passed by at least 60% of the concerned JFMC members by resolution.

13.6.20 Objective of Management

1. To provide livelihood opportunities to the local JFMC committee in a sustainable manner and involve the local community in participatory management and benefit-sharing.
2. Capacity building of local communities in planning, operating and managing ecotourism facilities.
3. To motivate and involve local JFMC committee in forest protection and conservation of nature and natural resources. Spreading nature awareness among the masses.
4. To generate Government revenue by optimum utilization of assets created.

13.6.21 Proposed Sharing of Revenue

- Sixty percent of net revenue generated shall be given to the concerned JFMC members as 'FPC share'.
- Twenty percent of the net income earned shall be deposited as a 'Government Revenue'.
- Twenty percent of the net income generated shall be kept as 'Maintenance Fund'.

13.6.22 Constitution of Management Committee

1. To manage Takdah Forest House (Old Club House) some management works that is catering, cleaning, watch and ward etc. have been assigned with local JFMC under Takdah Range. For a more systematic and transparent way and to impart a sense of accountability amongst all its stakeholders a 'Management Committee' shall be constituted with the following as its members:
   Range Officer, Takdah Range—Ex-officio Member Secretary
   Beat Officer, Takdah Beat-Member
   Two (2) nos. of JFMC members—Members
2. JFMC members shall be nominated by the concerned JFMC in its annual general meeting and shall act as members of the committee for one year only. The same member(s) cannot be a member of the committee for two consecutive years.

13.6.23 Duties of Management Committee

1. To regularly visit and monitor all the activities of Takdah Forest House and report to the Divisional Forest Officer, in writing, about any issue pertaining to the management of Takdah Forest House.
2. To call a meeting, ideally during the first week of each month, to decide on matters pertaining to the expenses that will be paid for out of the 'Maintenance Fund' and to present estimates to the DFO for final consent or confirmation.
3. Any other duty as may be decided by the Divisional Forest Officer.

13.6.24 Financial Management

1. The 'FPC Share' and 'Maintenance Fund' shall be deposited in two separate joint accounts operated jointly by the Range Officer, on behalf of the Forest Department and one of the member nominated by the concerned JFMC by a resolution signed by a majority of the members. The above two accounts shall be opened in any nationalized bank in consultation with the Divisional Forest Officer.
2. The 60% FPC share allotted to the JFMC members is a community fund which shall be utilized in a manner as decided by the concerned JFMC by taking resolution in Annual General Meeting signed by all the individual

members of the committee and subject to final approval of the Divisional Forest Officer.

3. The 20% amount kept as 'Maintenance Fund' shall be utilized for day-to-day maintenance of Takdah Forest House including the purchase of consumable articles etc. No fund shall be utilized from this account until and unless it is passed by taking resolution in a monthly meeting of 'Management Committee' and approved by the Divisional Forest Officer.
4. A separate cash book shall be maintained by the Range Officer, Takdah Range which shall reflect all the money receipts and expenditure done including details about FPC Share, Government Revenue, and Maintenance Fund.
5. That while maintaining accounts all the existing financial rules shall be invariably followed by the Range Officer.
6. All the vouchers should be countersigned by all the members of the Managing Committee.
7. That a monthly account shall be submitted to the DFO for final approval and record.

13.6.25 Engagement of Workers

1. Local JFMC members shall be engaged as a worker in Takdah Forest House wherever possible.
2. Selection of JFMC members shall be done strictly based on their educational qualification, experience, records and other important criteria as deem fit by the Divisional Forest Officer.
3. All such engagement of JFMC members shall be purely temporary and may be removed by the Divisional Forest Officer (DFO) on the ground of misbehaviour, incapacity, dishonesty, negligence in duty or any other reason detrimental to the safety and security of visitors and Government property. The decision taken by the DFO in this regard shall be final.
4. An undertaking shall be submitted by the concerned JFMC members, before his/her engagement as a worker, to the effect that he/she, in future, shall not claim to the government for any kind of employment, temporary or permanent, in the event of his/her removal as a worker of Takdah Forest House on any ground whatsoever.
5. To provide satisfactory services to the visitors a training in culinary and hospitality is provided to the local JFMC members who have been engaged as workers in Takdah Forest House.
6. Provision for uniform shall be made for cooks, house cleaners, security staff etc.

13.6.26 Methods of FRH Booking

- Booking shall be done through by the office of DFO and by Range Officer, Takdah Range as spot booking (information to DFO) in case of availability of room on the particular day.
- However, recently *Takdah* FRH is placed under WBFDC for all ecotourism purposes.

13.7 Economic Growth: Empowerment of Women

1. Maximum numbers of womenfolk took initiatives and put hard work to establish and to successfully run the ecotourism business. They got wages and profit money directly in their hands.
2. Some of homes stays have been created by enthusiastic villagers with their own money and money earned by them through ecotourism. Home stays are being run by women mainly.
3. They could easily sell their organic agri-products, handicrafts, tea and snacks.
4. Womenfolk accepted ecotourism as a miracle assignment for them. They got engaged in catering, mushroom cultivation, sale of pickles and groceries etc.
5. Some of the low paid persons left the works at Tea Gardens, other places and joined the women to assist them in home stay and other household works as they found an alternate source of income at their native place.
6. Some men returned from outside to work and invest in home stay occupation at home. Some have purchased vehicles for engaging in tourist travelling and guiding activities.
7. The village turned as global in ecotourism. They have their own website for booking as well as extension purposes. Sometimes the booking status becomes houseful as all the home stays are booked 5–6 months advance during peak tourism period.
8. Lamahatta ecotourism spot has found as the main tourist attraction in the list of important worth visit tourist sites under Darjeeling District and this place has been awarded the higher rankings by Trip advisor agency.
9. Lamahatta home stays have become almost self-sustained. For long-term success, the JFMC and SFD both will have to be more vigil, active, cooperative and will have to maintain a high standard of services and securities and satisfaction to the visitors.

13.7.1 Study on JFMC: National Scenario

The Energy Research Institute (TERI) had conducted a study on the effectiveness of JFMCs in the country and major findings are:

1. JFM scheme has been adopted by 22 States as on 2007.
2. Forest regeneration increased by 40%.
3. Pressure on forests, fires, illicit felling reduced by 20%.
4. Seventeen states had constituted 41,249 JFMCs and 11.24 m ha forest land covered in 14 States.
5. State wise usufructs shares to FPCs/VFCs are: Fuelwood 100% (AP, Gujarat, Odisha, MP), 50% (WB, AP), Kendu 50% (AP), Cashew 25% (WB), Timber 100% (AP, TN, Bihar), 50% (Odisha, Gujarat, Rajasthan), and 15% in WB. Entire sale proceeds given to FPC as revolving funds in HP, Haryana. The extent of items and usufruct sharing is modified time to time by state forest departments.

13.7.2 Strategies to Curb Illicit Felling: Reference to North Bengal Forest Division, WB

1. Regular FPC (JFMC) meeting at Beat, Range, Block and Divisional/District level.
2. Awareness through inter FPC trips, success stories, the importance of natural resources etc.
3. Distribution of leaflets and miking in areas and advertise through TV, newspapers etc.
4. Basic facility (arms, uniform, transport) and welfare measure (quarter, security, incentives) for staff.
5. Night camp in main outlet area/riverside where generally vehicles cannot ply.
6. Randomly patrolling on foot to avoid the indication that the patrol party is striking.
7. Strict, regular and effective supervision from Forester to DFO and even Circle Incharge.
8. Keeping personal informer (Intelligence) and maintaining high secrecy of informer and place.
9. Special patrolling (senior forest officials with FPC and Forest Protection Force) over prone forest areas.
10. Instant and immediate information collection by staff and communication to higher-ups.
11. Joint raid in difficult areas (populated areas, tea gardens) with local people/police etc.
12. Monsoon mobile camps (with boats, arms, life jacket etc.) in the rivers to check logs drifting.
13. Remuneration and appreciation to the staff and financial support to informers.
14. Keeping vigil on secret links between field staff and smugglers and poachers.
15. Taking local Panchayat leaders and politicians in confidence before any hard action taken.
16. Strong and legal backing to each case right from raids, seizures, arrests to courts.
17. Better liaison at all level with police, judiciary, politicians, local body members and JFMCs.
18. Regular workshops and training for frontline staff for support, updates on legal and other matters.

13.7.3 Conservation Efforts and Strategies

1. Minimize illicit felling/grazing and illegal entry in forests through fringe area.
2. Forest theft information is mostly from many JFMC members.
3. Cooperation and mutual confidence built up between the forest department and fringe public.
4. Gherao, assault on staff and firing incidences and other conflicts considerably reduced.
5. Small staff team could search, seize and arrest in JFMC areas without much oppose.
6. Regeneration of forest and coppice areas increased and fires incidences decreased.
7. Tendency for purchase of timbers from authorized depots that provide TP reference on bills.
8. Due to constant pressure from Forest Department, Panchayat and FPC members, the forest offenders started surrendering.
9. Illegal chain between smugglers, suppliers, purchasers, sawmill and furniture shop owners broken.
10. More and more JFMC members started participating in the joint patrolling with staff.

13.8 Threats and Challenges to JFM

1. Due to lack of full awareness, suitable alternate livelihood opportunities and due to acute poverty, some fringe forest dwellers are still engaged in anti-forest and wildlife conservation activities and remain out of mainstream of the JFMC concept.
2. Many JFMC members do not attend meetings, do not perform duties and play political biases and have the tendency of opposing the work and activities of JFMC.
3. Many JFMC members found cutting trees, engaged in poaching and illegal grazing, putting fires to jungles and stealing forest produces and having connections with outer forest offenders.
4. Some JFMC members are engaged in blackmailing, gheraoing, lodging minor or false complaints and apply pressuring tactics to field forest staff and demoralize them.

5. Some of JFMC members who found to be informers for timber mafias, also put field staff in dangerous situation during forest duties and raiding/seizing and searching/arresting operations.
6. Even in some cases they attack forest staff and injure or even kill them.
7. Womenfolk are found more and regular trespassers than men; hence they cause more damages to the forest. They cut back green coppices, newly raised saplings for firewood collections and sales.
8. Noticed that one of the JFMC members protect their own allotted forest well but clandestinely damage to the forest of nearby or other areas which have been allotted to other JFMCs.
9. Out of forest offenders arrested many were found to be JFMC members. In many raids and searches illegally collected timbers, split fire woods, other forest produces and poached wildlife are seized from the houses of JFMC members. Even at some places killed (by field staff or forest protection force in self-defence) forest offenders were found to be JFMC members.
10. It is seen after completion of seasonal forestry works, JFMC members are jobless and none of other Government agency provides jobs (inside the recorded forest areas wherein non-forestry activities are not allowed due to FCA, 1980) so for livelihood or for extra income they are engaged in forest-based illegal activities and also they migrate to other towns for earnings.
11. Some of JFMC members are so poverty-stricken that they cannot wait for revenue share or seasonal plantation works or free collections of some NTFPs, etc.
12. Some forest workers do not maintain regular intimacy with JMFC members, do not maintain transparency in works, do not pay needed attention towards the genuine grievances of the JFMC members. Such attitude and performance even by one or two staff may cause distrust and weakness in the JFM system.
13. Insufficient and delays in allotment and release of funds for JFMC works also affect the efforts.
14. Change in the policy and the priority of the existing or new Government towards JFM concept.
15. Shortage and transfer of staff, excessive load of field duties and lack of incentives to best performing staff and JFMC members also cause a gradual downgrade in the quality of works.

13.9 The Future of JFM

At present, in the absence of other more suitable alternates the JFM is the best concept to manage the natural resources thus this has also been adopted by some other countries. The sincerity, transparency, sharing of usufruct rights, responsibility of community towards their duties and mutual trust are the main keys to maintain the symbiotic relations between Forest Departments and JFMCs. Greed, irresponsibility and lack of the above-mentioned attributes shall cause deterioration in the sustainability of this concept.

The DFOs, ROs and JFMCs must continuously maintain the true spirit of the JFM concept. JFM represents a paradigm shift in the forest protection and regeneration method thus is a movement. Shabaz Ahmed in his book titled 'The Dilemma of People's Participation' has beautifully stated about the dilemma with the current participatory approaches as below:

1. The social welfare role of government is denied or disallowed.
2. Formalities in the activities at a village level cannot be conducive for social bonding.
3. Less importance on the long-term management of the large forest resources.
4. Requires complete overhauling of the existing administrative, managerial and sociocultural system without assuring benefits to either sides.
5. Put the JFMC members/villagers and resources at the mercy of market forces.

However, despite above-mentioned drawbacks, the present concept of participation has been strongly supported by the World Bank and emphasis was given to its all aided programmes.

Ahmed, while discussing the grey areas of participation stated ambiguity in the interpretation as there are high and low, direct and indirect, active and passive participation. An international association of public participation (IAPP) defines public participation, as any process that involves the public in problem-solving and decision making and that uses public input to make a better decision. The business concept of stakeholders (SFD is a stakeholder in the participation) is not suitable for huge natural resources where there is strong vested interest, and where strong decision taken required to ensure the fulfilment of the confide needs of the weaker sections of the society. The business enterprise is profit-making and social welfare or conservation of resources, not a primary concerns.

However, the JFM concept yet cannot equate with the said definition as the idea and goal is somewhat different in this type of participation. Full control of the public over resources may result in the violation of freedom and rights of individuals and the mandate of Government. In case of JFM, participation has been regularized by the Government thus final decision of Government will prevail in case of any dichotomy is created between participants and Government. If the ecotourism communities will have only business targets and do not have concern with forest and wildlife protection then they cannot be called as true joint participants. This is practically very difficult to maintain the true spirit of JFM concept for a longer period, thus it is a big challenge.

The word participation has been used with many terminologies with slightly different connotations, namely community participation, public or peoples' participation, participatory management, public–private participation (PPP) and joint management participation, etc. The common goal is to involve other stakeholders, institute agency, organization or individual body in an activity or action for a well-desired purpose; however, the extent and intensity of participation may vary in each case. Thus, the word JFM may not qualify the perspectives of different connotations of the concept of participation.

In the MP Forestry Project (second phase), the World Bank was not satisfied with the term JFM and gave an option of Community Forest Management (CFM). The bank argued that JFM was the initial stage of CFM and now transforms the relation between forest-dependent communities and SFD to a partnership that progressively empowers these communities to take lead roles of forest management and with SFD merely playing a need-based technical and facilitating roles. Government of MP did not agree with CFM concept of the World Bank so the project did not come up (Ahmed 2013).

Shabaz Ahmed also stated that another concept of PPP has cropped up in the 1990s wherein a public or private body purchase a service. The possible outcome of applying of PPP model in natural resource management (implement this approach in the regeneration and rehabilitation of degraded forests) may look appealing. Once a community is entrusted with ownership of a particular chunk of forestland it will be revenue earning entity and the matter of sustainability and ecosystem service to be of no immediate relevance. Legally and practically this PPP type participation in the management of fragile eco-resources will not be feasible and possible in country like India.

The JFM is theoretically run as envisaged in National Forest Policy 1988 and SFDs resolutions and there is no clear-cut role and responsibility at execution level. The World Bank's concept of CFM cannot be implemented in India whereas the same is being practised in France and Germany (Ahmed 2013). Thus, following of all concepts of participation at a place are not practical and feasible.

In nutshell, it is concluded that as of now, existing public participation in the form of JFM is the most suitable in developing country like India.

Lessons Learnt

- During early civilizations human had co-existence with forests, animals and wildlife and was completely dependent on these resources. Later, they started protecting and conserving the natural resources for shelter and sustenance. During Vedic period people started to worship some fauna and flora and classified the forests and animals, thus these became the part of their culture and religion. Many scriptures were written on the uses of medicinal plants for curing of various ailments and named this knowledge as the Ayurveda.
- Some of forest areas were set aside as Sacred Groves as Orans (Aranya), Gouchars and Dev vans. Illicit felling, grazing and other damages were finally prohibited. These were common resources of all nearby villages and trespassers were punished. Thus, the concept of community management of natural resources was existing since earlier times.
- In medieval and modern eras people raised voices against plundering of forest and poaching of animals. As many as 363 numbers of Bishnois of Rajasthan in eighteenth century AD sacrificed their lives to save Khejri (*Prosopis cineraria*) trees in Thar Desert of Jodhpur, and also they laid many sacrifices to protect Chinkaras and Blackbucks in the Rajasthan and adjoining states. Learning from khejarli massacre of Bishnois, Gaura Devi and other women started agitation against forest trees felling by contractors in Garhwal Himalayas, which was called the 'Chipko Movement'. This movement later spread pan India with different names.
- The programme initiated by the DFO (Medinipore, West Bengal) became the basis for Joint Forest Management involving local tribals and villagers by sharing of some benefits and rights. Seeing the success of JFM, it was started in entire West Bengal and then in India.
- Here are various constitutional provisions for the conservation and protection of biological natural resources. Thus, JFM concept fulfilled the aspirations of these provisions. FRA Act 2006 also has given some rights to tribals and forest dwelling communities due to which these people have started protecting and conserving the forest and wildlife. Thus, constitutional safeguards for protection and conservation of natural resources have been implemented through Joint Forest Management.
- Many entry point activities and JFM support activities have been funded by State Forest Department and Government of India. People are getting employments through forest department, so their direct dependencies on forest resources have been reduced. Due to ecotourism people are being made aware about the importance of forests, wildlife and environment. Now, forest resources are being used on sustainable basis in which poor tribals are being regularly benefitted and forest and wildlife related crimes have reduced drastically.

Key Questions

1. Define the Joint Forest Management (JFM) and briefly give an account of genesis of JFM.
2. Compare the concepts of Traditional Forest Management and Joint Forest Management.
3. Which major clauses in the Indian Constitution are devoted to the preservation and protection of the country's natural biodiversity?
4. Give a brief account of strengths and weakness of JFM.
5. How JFM helps in the empowerment of women in India?

References

Ahmed S (2013) The dilemma of peoples participation. LAP Lambert Academic Publishing

Banerjee AK (2014) 150 years of forestry (1864–2014). Directorate of Forest, Government of West Bengal, Kolkata

Bhattacharya S (2014) Forest and biodiversity conservation in ancient Indian culture: a review based old texts and archaeological evidences. Department of Environmental Studies, Rabindra Bharati University, Kolkata, India

HFRI (2020) Dev van ek prachin dharohar. HFRI, Shimla, Himachal Pradesh

ISFR (2019) India state of forest report 2019. Forest Survey of India, Ministry of Environment, Forests and Climate Change, Dehradun, India

Jetaram V (2016) Jambhabani Smarika, vol 3. JNV University, Jodhpur, Rajasthan

Jetaram V (2019) Jambhdhara Shodhpatrika. JNV University, Jodhpur, Rajasthan

Kumar BM (2014) Forestry in ancient India: some literary evidences on productive and protective aspects. College of Forestry, Kerala Agricultural University, Thrissur, Kerala

Singh G (2016) Sacred groves of Rajasthan, threats and management strategies. AFRI Publication, Jodhpur

Web References (URL)

https://www.indianetzone.com/50/forests_ancient_india_forests.htm

https://www.downtoearth.org.in/news/forests/private-funds-for-madhya-pradesh-forests-beware-of-legal-concerns-warn-experts-73906

https://www.ifs.nic.in Indian Forest Service, MOEF & CC, Govt of India

https://www.tribal.nic.in/FRA.aspx

https://www.moef.gov.in

https://www.0rissaaea.in/Journal/Jouranal_2018.pdf

https://docplayer.net/55574464-Conservation-across-landscapes-india-s-approaches-to-biodiversity-governance.html

https://www.dfe.gov.in

https://www.frienvis.nic.in

https://www.westbengalforest.gov.in/upload/actrules/jfm8.pdf

https://forest.rajasthan.gov.in/content/dam/raj/forest/ForestDepartment/PDFs/

Forest Invasive Species and Their Management 14

N. Roychoudhury

Abstract

Environmental alteration, increase of earth's temperature, different calamities like cyclonic disaster, biotic pressures, Invasive Alien Species (IAS), deforestation, pollution, rainfall pattern, landslides, cloudburst, thunderstorm etc. affecting the ecosystems and native species across the world. IAS is one of the serious warning to biodiversity and considered the main cause of habitat destruction.

An invasive species particularly of forest ecosystem is referred as Forest Invasive Species (FIS). FIS is directly related to the forestry and serious threat to forest cover. FIS can remove native species, food and forest cover. Human trade, tourism, transport and travel have enhanced the spread of FIS. For example, *Lantana* has invaded forest lands, *Parthenium* in agricultural and forestry areas and *Eupatorium* in mountain ecosystems, etc. FIS is highly significant in the national context. The bioclimatic conditions of India have been congenial for the alien species to invade in to the various landscapes and ecosystems. There is a lacuna in the knowledge on the biology, geographical distribution, genetic diversity and inter-crossing (hybridizing) potentials of the forest invasive alien species. The entry of eucalyptus gall insect, *Leptocybe invasa*, in different countries, including India has threatened *Eucalyptus* species throughout the world. The present chapter has contemplated and described succinctly the IAS with special emphasis on FIS in India. Further, it also highlights the impacts of FIS in forestry and forest cover, their prevention, current position, measures to control FIS, categorization of invasiveness, environmental impact assessment, possible management of FIS to save native forests as well as environment and finally the future scope of research.

Keywords

Invasive species · Alien plants · Prevention · Quarantine · Management

14.1 Introduction

After *Homo sapiens* appeared on earth, they soon began interaction with ecosystems. Their quest to colonize the planet, humans deliberately or inadvertently transported certain plants and animals with them. This human assisted relocation of species has continued throughout the history of mankind with varying consequences. However, the massive movement of species by rapidly accelerating human trade has breached the development of natural ecosystems that were unique and distinct hitherto, and resulted in an ever-increasing occurrence of alien/exotic species in agriculture, horticulture and forestry.

All non-native species are not harmful, since in addition to accidental dispersal of different harmful species, humans have also actively introduced a number of useful and beneficial species into new countries including India. A majority of alien species are benign and beneficial to human civilization like wheat and soybean and animals like fishes were introduced for human food (Gogoi and Babu 2006). The non-native (alien/exotic) species in the absence of their coevolved predators and parasites, become established, grow, adapt, multiply and spread, causing tremendous harm, often irreversible to the economy and in some cases to human health and are collectively known as 'invasive species' (Knowler and Barbier 2000). Many harmful exotic species were deliberately introduced. Horticultural varieties, namely Lantana and Water hyacinth and fishes introduced have exterminated native species.

N. Roychoudhury (✉)
ICFRE-Tropical Forest Research Institute, Jabalpur, India

A. K. Mandal, A. Nicodemus (eds.), *Textbook of Forest Science*, https://doi.org/10.1007/978-981-97-8289-5_14

14.2 Invasive Species

According to Oxford dictionary, the word *invasive* means tending to spread in a harmful or unpleasant way. An 'invasive species' is defined as a species that is not native or alien to the ecosystem and whose introduction causes or is likely to cause economic or environmental harm or harm to human health. Invasive species can be plants, animals and other organisms (e.g. microbes). Human actions are the primary cause of invasive species introductions. These harmful exotic species have grown successfully in a new area, increase quickly in number and disperse in ways that are harmful to human interests and natural systems are considered as Invasive Alien Species (IAS) (Goyal 2005). IAS is distinguished by following traits: rapid development, prolific reproduction, efficient dispersal capabilities and tolerance of a broad range of environmental conditions. Controlling invasive species requires complete information regarding its taxonomy, structural features, developmental biology, reproductive potentiality and seasonal history, including phytochemistry.

14.2.1 Special Features of Invasive Flora

The invasive flora has some distinguishing attributes, which are as follows:

1. Quick growth and early maturation.
2. Prolific seed production.
3. Reproduction by easily dispersed seeds and/or vegetative structures.
4. Long lived seeds or reproductive structures.
5. Dormant seeds ensure periodic germination.
6. Spread with crop seeds by natural agents and humans.
7. Produce biological toxins that suppress the growth of other plants.
8. Presence of spines or thorns that repel animals.
9. Parasitize other plants.
10. Seeds are same size and shape as crop seeds, makes cleaning difficult.
11. Roots or rhizomes with large food reserves.
12. No adverse environmental effects on survival and seed production.
13. High photosynthetic rate.
14. Drought tolerant.
15. Free of natural enemies.

14.2.2 Impact of Alien Invasive Flora

Food and fibre production has been threatened by invasive plants throughout the world. The effects of alien invasive are immense, insidious and usually irreversible, resulted in loss and degradation of habitats. Thus, the invasion by the alien plants is resulting in a shifting of species composition of diverse plant communities towards a preponderance of weedy, undesirable species.

14.2.3 Control of Invasive Species

The techniques for control of harmful exotic species are expensive and lengthy. But, complete eradication is must to stop regeneration. According to Reddy et al. (2008), three main methods for management of exotic weed species are as follows:

Mechanical method of control: This method is highly effective for management of harmful exotic species. This method includes removal of harmful exotic species through ploughing, scything, mowering, hoeing, cultivating, etc. By this technique the unwanted species are removed from the area.

Chemical method of control: Management by use of chemical is costly and successful in some cases. But, this technique is not desirable because chemicals are responsible for pollution.

Biological method of control: In most cases of alien pest invasions, biological control has proven the most effective promising and environmentally safe technique for the management of invasive species. Some examples for control of invasive species are release of natural enemy, water hyacinth predator weevil, *Neochetina* sp. for *Eichhornia*, *Lantanophaga pusillidactyla* predator insect for *Lantana camara*, *Zygogramma bicolorata* predator insect and *Cassia uniflora* plant species for *Parthenium hysterophorus*.

Critical alien plant ecological aspects that affect control methods are:

1. Sometimes native predators are not effective or very weak, predators need to be imported from the plant's homeland for control of invasive aliens.
2. There is a 'lag phase' existing for a long time after introduced ion of an alien plant species before an exponential phase (Baskin 2002). Some species that presently exhibit non-invasive but start to spread rapidly after some period.
3. Invasive alien plants cause retardation of natural succession and reforestation due to quick dense infestations.
4. Biological diversity reduces due to invasion of aliens and detracts from their primary mission.
5. Some shade tolerant alien species (*Cassia pumila*) can establish under forest canopies.

6. Some alien species, such as *Alternanthera philoxeroides*, *Eichhornia crassipes* and *Pistia stratiotes* are spreading riparian/aquatic ecosystem.
7. Alien species initially grow along highway then to disturbed habitats and riparian systems and finally to forest areas.
8. Alien plant species generally grow in disturbed habitats in cities and cause serious problems for urban forestry.
9. Some species, such as *Prosopis juliflora*, *Hyptis suaveolens* and *Lantana camara* have allelopathic effects that reduce growth and development of native species.
10. Successful growth and invasions of alien species lead to extinction of native species.

14.2.4 Flora of India

The quick increase in population of invasive species is homogenizing flora and fauna of the world (Mooney and Hobbs 2000). This is the major reason of world biodiversity loss. The infestation harmful exotic species is a biological pollution and one of the important components and major cause of floral change and species disappearance (Mooney and Drake 1987; Drake et al. 1989).

The serious problems caused by the infestation of harmful exotic species into ecosystems and the related harmful effects on native biodiversity are well known to taxonomists, ecologists and foresters (Reddy et al. 2008). Many exotic species that can able to remove, indigenous flora and fauna, affect nutrient and fire cycles and cause changes in the pattern of plant succession.

About 1,604,000 species have been described at the global level, reported by World Conservation Monitoring Centre (WCMC). India accounts for around 8% of the global biodiversity are found in India (http://www.bsienvis.nic.in/Database/Invasive_Alien_species_15896.aspx#). According to Nayar (1989), the number of flowering plant species endemic to India is 4900 out of a total of 15,000, that is, 33%. Hajra and Mudgal (1997) have reported 17,000 angiospermous species of India, out of which 5400 are endemics.

The flora found in India shows similarity with the flora of Indo-Malayan and Indo-Chinese region. According to Nayar (1977), 35% of Indian flora has south-east Asian and Malayan, 8% temperate, 1% steppe, 2% African and 5% Mediterranean-Iranian elements. Chatterjee (1940) has reported 6850 plant species as endemic to India. According to Reddy et al. (2008), 282 alien species have been found to be utilized for cultivation of food/forage/timber/ornamental/other economic purposes in our country introduced from Australia (33 species), Western Asia and Europe (17 species), Mediterranean Area and Africa (44 species) and Europe, America, Mexico, West Indies and Brazil (188 species).

14.2.5 Occurrence of Invasive Species

There are 45 scientific names belonging to the Kingdom Plantae (Table 14.1) and 51 scientific names belonging to the Kingdom Animalia (Table 14.2) found in the database of alien invasive species of India (www.ncbi.org.in). Reddy (2008) and Reddy et al. (2008) have documented a total of 173 invasive alien species of India, which includes 117 genera belonging to 44 families. The most serious invasive species of India are *Alternanthera philoxeroides*, *Cassia uniflora*, *Chromolaena odorata*, *Eichhornia crassipes*, *Lantana camara*, *Parthenium hysterophorus*, *Prosopis juliflora* and others.

Out of nearly 89,317 species of animals reported from India (Alfred 1998), during the year 2019, the Zoological Survey of India (ZSI) has prepared an inventory of 157 invasive species of animal origin (https://www.indiatoday.in/education-today/gk-current-affairs/story/zsi-alien-invasive-species-list-india-1108621-2017-12-19).

IAS belongs to plant or animal kingdom, but the scope of this chapter is restricted only to Indian forest invasive species (FIS) of plant and animal origin.

14.3 Forest Invasive Species

Forest invasive species (FIS) can be defined as those floras, fauna and microbes which are alien and not native and have been introduced deliberately to the forest and becomes an important pest in the area that warns forest ecosystems, habitats and native biodiversity. An invasive species particularly of forest ecosystem is referred to as FIS.

FIS are found in both plant and animal kingdom and practically in every ecosystem and almost all countries of the world. One of the key reasons for loss of forest biodiversity is infestation by FIS. Effects from FIS on biodiversity can be direct, indirect and cumulative.

FIS affect the ecosystem function and structure and effect at the level of communities or habitats as well as at the level of species in protected areas, as elsewhere. The influence of FIS is directly or indirectly related to livelihoods and poverty alleviation, through affecting ecosystem services or sustainable use of biodiversity or through impinging on cultural and heritage values. The information pertaining to this aspect is available at global level (Anon 2007a). A total of 326 IAS have recorded in 487 protected sites, located in 106 countries belonging to Asia, Africa, South and Central America,

Table 14.1 Alien invasive flora

Scientific name	Common name	Origin
Acacia farnesiana[IAS-6]	Babool	Australia
Acacia melanoxylon[IAS-6]	Australian blackwood	Australia
Alliaria petiolata[IAS-6]	Garlic mustard	Europe
Araucaria cunninghamii[IAS-8]	Australian hoop	Australia
Araucaria excelsa[IAS-8]	Norfolk island pine	Norfolk island (Pacific Ocean)
Carica papaya[IAS-9]	Papaya	Central America
Catesbaea spinosa[IAS-9]	Lily thorn	Cuba
Cereus peruvianus[IAS-9]	Peruvian apple	Argentina, Brazil and Peru
Chromolaena odorata[IAS-5]	Communist pacha	South and Central America
Chromolaena odorata[IAS-6]	Communist pacha	South and Central America
Cirsium arvense[IAS-6]	Canada thistle	Europe
Citrullus lanatus var. *lanatus*[IAS-9]	Watermelon	South Africa
Coffea arabica[IAS-4]	Arabian coffee	Ethiopia
Coffea robusta[IAS-4]	Robusta coffee	West Africa
Coffea syenophyla[IAS-4]	Bush coffee	Guinea and Sierra Leone
Combretum coccineum[IAS-9]	Combretum	Madagascar and Mauritius
Cryptomeria japonica[IAS-8]	Japanese cedar	Japan
Cryptostegia grandiflora[IAS-6]	Rubber vine	Madagascar
Cunninghamia sinensis[IAS-8]	China fir	China
Eichhornia crassipess[IAS-5]	Water hyacinth	South America
Elaeagnus umbellata[IAS-6]	Autumn olive	China, Korea and Japan
Epiphyllum macropterum[IAS-9]	Climbing cactus	Costa Rica and Panama
Eriocereus bonplandii[IAS-9]	The midnight lady	The Netherland
Eugenia uniflora[IAS-9]	Surinam cherry	Brazil
Hedychium flavescens[IAS-6]	Yellow ginger-lily	Himalayas
Lantana camara[IAS-5]	Lantana	Tropical America
Leucaena leucocephala [IAS-6]	Subabul	Mexico
Mesembryanthemum crystallinum[IAS-5]	Common iceplant	Africa
Mikania micrantha[IAS-5]	Mile-a-minute	Central and South America
Nopalea cochenillifera[IAS-9]	Cochineal cactus	Mexico
Opuntia stricta var. *dillenii*[IAS-9]	Erect prickly pear	Africa
Parthenium hysterophorus[IAS-5]	Congress grass	Tropical America
Passiflora edulis[IAS-9]	Passion fruit	South America
Passiflora foetida[IAS-9]	Wild water lemon	Tropical America
Passiflora laurifolia[IAS-9]	Water lemon	Tropical America
Passiflora raddiana[IAS-9]	Passionflower	Brazil
Passiflora stipulate[IAS-9]	Stinking passionflower	China
Pennisetum clandestinum[IAS-9]	Kikuyu grass	Africa
Pereskia grandiflora[IAS-9]	Rose cactus	Mexico and Brazil
Phalaris arundinacea[IAS-6]	Reed canary grass	North America
Pittosporum undulatum[IAS-6]	Sweet pittosporum	Australia
Polyscias guilfoylei[IAS-9]	Geranium aralia	Malaysia
Psidium guajava[IAS-6]	Common guava	Tropical America
Solanum viarum[IAS-6]	Tropical soda apple	Brazil and Argentina
Thauja orientalis[IAS-8]	White-Cedar	China and Korea

Source: www.ncbi.org.in

including Mexico and the Caribbean and Europe. Around 43 species of IAS are recorded from protected areas of Asia (https://www.iucngisd.org/gisd/). These species are *Acacia mearnsii*, *Acacia nilotica*, *Adhatoda vasica*, *Ageratum conizoides*, *Annona glabra*, *Aulacaspis yasumatsui*, *Axis axis*, *Bubalus bubalis*, *Cervus timorensis*, *Cervus unicolor*, *Chitala ornate*, *Chromolaena odorata*, *Cirsium vulgare*, *Clerodendron viscosum*, *Clidemia hirta*, *Dioscorea sanibariensis*, *Egeria densa*, *Eichhornia crassipes*, *Elephas maximus*, *Eupatorium* sp., *Imperata cylindrica*, *Lantana camara*, *Micania micrantha*, *Mikania* sp., *Mimosa diplotricha*, *Mimosa invisa*, *Mimosa pigra*, *Mimosa pudica*,

Table 14.2 Alien invasive fauna

Scientific name	Common name	Origin
Achatina fulica[IAS-3]	Giant African land snail	Africa
Amblygaster leiogaster[IAS-3]	Smooth-belly Sardinella	Central Pacific
Anoplolepsis gracilipes[IAS-6]	Yellow crazy ant	West Africa
Argulus foliacecus[IAS-3]	Common fish louse	Bangladesh
Aristichthys nobolis[IAS-2]	Big head carp	Siberia
Balanus amphitrite hawaiiensis[IAS-7]	Barnacles	Malay Archipelago and Persian Gulf
Balanus amphitrite var. *stutsburi*[IAS-7]	Striped barnacle	Africa
Barbonymus gonionotus[IAS-2]	Silver barb	Thailand, Myanmar
Barentsia ramose[IAS-7]	Ramos	Pacific, California, Belgium
Carassius auratus[IAS-2]	Goldfish	North America
Carassius auratus auratus[IAS-2]	Edible goldfish	China
Chimaera monostrosa[IAS-3]	Chimera	Japan
Cilicaea lateraillie[IAS-7]	Isopod	Indonesia, Philippines
Clarias gariepinus[IAS-2]	Catfish	Africa
Cookeolus boops[IAS-3]	Glasseye snapper	Atlantic
Ctenopharyngodon idellus[IAS-2]	Grass carp	China
Cyprinus carpio[IAS-6]	European carp	Western Asia
Cyprinus carpio carpio[IAS-2]	Common carp	Europe
Dinematichthys iluocoeteoides[IAS-3]	Yellow cuskeel	South America
Euglandina rosea[IAS-6]	Rosy wolfsnail	Latin America
Eusynstyela tincta[IAS-7]	Ascidian	Atlantic, Africa, Sri Lanka
Gambusia affinis[IAS-2]	Mosquitofish	Gulf of Mexico
Gambusia holobrooki[IAS-2]	Mosquitofish	South-eastern America
Heptranchias perlo[IAS-3]	Sharpnose sevengill shark	Atlantic ocean
Hypophthalmichtys molitrix[IAS-2]	Silver carp	America
Icerya purchasi[IAS-10]	Cottony cushion scale	Australia
Lyrodus medilobata[IAS-7]	Wood borer	New Zealand, Hawaiian islands
Macrorhamphosodes platycheilus[IAS-3]	Trumpetsnout spikefish	Africa
Mercierella enigmata[IAS-7]	Hydroid	Australia
Mytilopsissallei[IAS-7]	Black striped mussel	Atlantic Ocean
Notorynchus cepedianus[IAS-3]	Broadnose sevengill shark	Atlantic Ocean
Oncorhynchus mykiss[IAS-2]	Rainbow trout	Alaska
Oreochromis mossambicus[IAS-2]	Mozambique tilapia	Africa
Oreochromis mossambicus[IAS-3]	Mozambique tilapia	Africa
Oreochromis niloticus niloticus[IAS-2]	Nile tilapia	Africa
Osphronemus goramy[IAS-2]	Giant gourami	Java, Sumatra
Phallusia nigra[IAS-7]	Black sea squirt	Bermuda, Brazil
Poecilia reticulate[IAS-2]	Guppy	West Indies and South America
Rattus exulans[IAS-6]	Polynesian rat	Indonesia
Rhiniodon typus[IAS-3]	Giant whale shark	Atlantic ocean
Salmo trutta fario[IAS-2]	Brown trout	Europe and Asia
Salmo trutta trutta[IAS-2]	Brown trout	North America
Salvelinus fontinalis[IAS-2]	Brook trout	North America
Solenopsis geminata[IAS-6]	Tropical fire ant	Central America
Styela bicolor[IAS-7]	Ascidian	Australia
Tinca tinca[IAS-2]	Tench	Canada
Torpedo fairchildi[IAS-3]	Electricray	New Zealand
Torpedo fuscomaculata[IAS-3]	Black-spotted torpedo	Madagascar
Torpedo macneilli[IAS-3]	Shorttail torpedo	Australia
Urotrygon daviesi[IAS-3]	Giant stingarees	South America

Source: www.ncbi.org.in

Muntiacus muntjak, *Najas marinus*, *Nelumbo nucifera*, *Oncorhynchos mykiss*, *Pomacea* sp., *Opuntia dillennii*, *Parthenium hysterophorus*, *Procyon lotor*, *Prosopis juliflora*, *Salvinia molesta*, *Salvinia* sp., *Spartina* sp., *Sus scrofa*, *Tiliacora acuminata* and *Ulex europaeus*.

In India, invasive plant species in protected areas have given very little attention until recently. This is possibly due to main emphasis on wildlife protection, rather than on conservation of biodiversity and ecosystem functioning. Hiremath and Sundaram (2013) have reported 20 invasive alien forest plant species found in Indian protected areas, such as *Acacia mearnsii*, *Ageratina* spp., *Ageratum conyzoides*, *Ageratum houstonianum*, *Chromolaena odorata*, *Coffea Arabica*, *Coffea canephora*, *Cytisus scoparius*, *Hyptis suaveolens*, *Ipomoea carnea*, *Kappaphycus alvarezii*, *Lantana camara*, *Merremia peltata*, *Mikania micrantha*, *Mimosa invisa*, *Opuntia stricta*, *Parthenium hysterophorus*, *Polygonum polystachyum*, *Prosopis juliflora* and *Ulex europaeus*.

FIS are directly related to forestry. Many of the plant and insect species deliberately introduced in different countries and have become main threats to the forest biodiversity. Goyal (2005) has suggested that due to the importance of FIS, re-orienting the research priorities and strengthening of forestry research organizations is required.

Due to serious threat of FIS to forestry, the Asia Pacific Forestry Commission (APFC) has formally launched 'Asia Pacific Forest Invasive Species Network' (APFISN) at the 20th session of the APFC held at Nadi, Fiji, from April 17–21, 2004 with the following objectives:

- Among the member countries, mutual sharing of information on forest invasive species is necessary.
- Research, education and training opportunities need to be created to facilitate access to expertise.
- Research and management of FIS is necessary to strengthen capacities of the member countries.
- Coordination and cooperation among the member countries needs to be inceased and development of regional strategies for FIS is necessary.
- Awareness programme is necessary on FIS as an important issue throughout the Asia and Pacific region.

14.3.1 Occurrence of Forest Invasive Species in India

About 48 alien FIS have been identified (Figs. 14.1, 14.2 and 14.3) by the scientists of Indian Council of Forestry Research and Education (ICFRE), Dehradun, in different areas of the country, which are serious threat to the natural forest cover (Anon 2005a) (https://apfisn.net/wp-content/uploads/2018/07/India.pdf). Out of the identified 48 FIS, 29 species belong to plants (Table 14.3), 10 species belong to fungi (Table 14.4) and 9 species belong to insects (Table 14.5). Many FIS like *Lantana* and *Parthenium* are present and invading the forest areas (Roychoudhury and Sharma 2013; Roychoudhury et al. 2019). Shukla et al. (2009) have published a documentary list of 106 invasive alien plant species belonging to 77 genera, spread over 36 families, representing 10.48% of the flora in Achanakmar-Amarkantak biosphere reserve, located in Madhya Pradesh and Chhattisgarh.

Fig. 14.1 Podina ghas, *Lantana camara*

14.3.2 Effects of Forest Invasive Species in Forestry

Due to invasions of invasive species, forest can be seriously affected. Some exotic forestry flora that were planted for economic reasons have become invasive. Alien tree species have long been introduced intentionally in several countries for commercial forestry, erosion control or landscaping. Salt cedar, *Tamarix pentandra* was collected from central Asia to the south-west United States to control erosion along river banks. Similarly, *Lantana camara* was introduced long ago in India for ornamental purposes which have at present reached unmanageable proportions. The entry of alien invasive insect species of eucalyptus, *Eucalyptus* spp. has created significant trouble in nursery stage and Australian melaleuca, *Melaleuca quinquenervia*, which can be particularly harmful in context to ecology because its leaf litter secretes chemicals that do not allow other species to grow.

During the year 2002, an alien forest invasive insect species called blue gum chalcid wasp, *Leptocybe invasa*

Fig. 14.2 Vilayati babool, *Prosopis juliflora*

Fig. 14.3 Invasion of Amarbel, *Cuscuta reflexa* on *Ziziphus mauritiana* at Maharajpur Forest Range, Mandla Forest Division, Madhya Pradesh

Table 14.3 Forest invasive plants and weeds species

Scientific name	Common name	Origin
Acacia farnesiana	Babool	Australia
Acacia mearnsii	Black wattle	Mexico
Ageratina adenophora	Crofton weed	Mexico
Ageratum conyzoides	Podina ghas	Tropical America
Ageratum houstonianum	Podina ghas	Tropical America
Alternanthera philoxeroides	Alligator weed	Indonesia and Burma
Argemone mexicana	Mexican poppy	Central and tropical America
Chromolaena odorata	Communist pacha	South and Central America
Cytisus scoparius	Yellow broom	Europe
Eichornia crassipes	Water hyacinth	South America
Eupatorium odoratum	Siam weed	Central and South America
Galinsoga parviflora	Yellow weed	Neotropical
Imperata cylindrical	Thatch grass	Philippines
Ipomea carnea	Behaya	South America
Ipomea fistulosa	Besharam	Tropical America
Lantana camera	Lantana	Tropical America
Leucaena leucocephala	Subabul	Maxico
Mikania micrantha	Silver bullet	Neotropical
Mimosa pigra	–	Mexico
Parthenium hysterophorus	Congress grass	Tropical America
Prosopis chilensis	Vilayati kikkar	South and Central America
Prosopis juliflora	Vilayati babool	South America
Salvinia molesta	–	South America
Sida acuta	Snake's tongue	Pantropical
Sida orientalis	Paroquet bur	Pantropical
Solanum eleagnifolium	Silver leaf	South America
Solanum viarum	–	South America
Tridax procumbens	–	Tropical America
Ulex europaeus	Gorse	Europe

Source: Anon (2005a)

Table 14.4 Forest invasive fungi species

Scientific name	Common name	Origin
Cercospora pinidensiflorae	Needle cast of chir pines	Japan
Cryphonectria cubensis	Stem canker	Cuba
Diplodia pinea	Whorl canker	South Africa
Dothistroma pini	Red band	USSR
Fusarium moniliforme	–	Japan
Lophodermium pinastri	Needle cast	Mexico
Monochaetia unicornis	Canker of cupresses	USA
Oidium heveae	–	Brazil
Peniophora gigantean	–	New Zealand
Pseudoperonospora cubensis	–	Cuba

Source: Anon (2005a)

Fisher & LaSalle (Hymenoptera: Chalcidoidea: Eulophidae), has been reported from coastal Tamil Nadu, which is a new insect genus of minute wasp, solely responsible for gall formation in eucalyptus (Anon 2007b; Jacob et al. 2007). The epidemic infestation of this exotic pest in peninsular India has completely damaged the vast areas of this potential crop (Jacob 2009). Roychoudhury et al. (2007) have noticed this alien forest invasive species in forest nurseries of central India (Madhya Pradesh, Chhattisgarh and Maharashtra) during the month of April–May, 2007 and subsequently to young plantations of eucalyptus that caused serious damage like epidemic in large scale (Roychoudhury 2013, 2016a; Roychoudhury et al. 2016). Presently, this alien forest invasive insect species is inflicting eucalyptus in large scale all

Table 14.5 Forest invasive insect species

Scientific name	Common name	Origin
Aleurodicus dispersus	Spiraling white fly	Central America
Anoplolepis gracilipes	Yellow crazy ant	West Africa
Heteropsylla cubana	Jumping louse	Central and South America
Icerya purchasi	Cottny cushion scale	Australia
Pineus laevis	Adalgid	Europe
Quadraspidiotus perniciosus	San Jose scale	China
Solenopsis geminata	Fire ant	Central America
Teleonemia scrupulosa	Lantana lace bug	Mexico
Leptocybe invasa	Blue gum wasp	Australia

Source: Anon (2005a)

over the country (https://www.cabi.org/isc/datasheet/108923#tosummaryOfInvasiveness). Inventory of this insect has been prepared (Table 14.6).

FIS can replace native plants, eliminate food and forest area for wildlife and warn rare plant and animal species. These species can alter the functions of ecosystem and increase loss in forestry and increase costs of natural resource management.

FIS create a danger to the biological diversity and other human enterprise and human health. Some FIS are growing successfully in India to utilize for various purposes, such as medicine, furniture and composting.

14.3.3 Forest Invasive Species and Forest Cover

The total forest cover and total forest and tree area of the country as per assessment of the year 2019 are 712,249 and 807,276 km^2 constituting 21.67% and 24.56% respectively of the area of the country (Table 14.7) (Anon 2019). Of this, 99,278 km^2 (3.02%) is very dense forest, 308,472 km^2 (9.38%) is moderately dense forest, while 304,499 km^2 (9.26%) is open forest cover. The scrub forest accounts for 46,297 km^2 (1.41%).

Many of the alien FIS especially weeds require light for vigorous growth and cannot penetrate undisturbed natural forest areas. Many forests in India remain intact, much of this is now exploited to some extent and could be invaded by alien invasive weeds. Over 42.75% of the forest cover of India now only has a crown cover between 10% and 40%.

14.3.4 Prevention of Forest Invasive Species Introduction

The important steps towards prevention of FIS introduction are minimum soil disturbance, immediately re-vegetation of disturbed areas, use of certified 'weed free' seed for re-vegetation, early observation and removal, survey of invasive plant, include invasive plant management strategy in planning phase, education and awareness.

14.3.5 Current Status of Forest Invasive Species

At present there is no systematic national level initiative on techniques to be employed to monitor or control FIS. Most of the threats are dealt at the local/district level by the local governments. Likewise, the cost of preventing the invasion of FIS and the damages in economic terms needs to be evaluated.

14.3.6 Area Infested by Invasive Weeds

Invasions of invasive species are represented by the alien weeds entered accidentally or deliberately outside their native ranges. For example, introduction of *Parthenium hysterophorus*, an invasive weed of North American origin, invaded 14.25 million hectares of farm land alone in India. Similarly, *Prosopis juliflora* infested 1.8% in India. The world's potential 100 invasive species include *Lantana camara*, *Parthenium hysterophorus*, *Mikania micrantha* and *Chromolaena odorata*. All these species exist over a large area across continents. *L. camara* is an alarming invasive species in forest ecosystems of India. Out of 712,249 km^2 forest area in India, 20.04% come under the protected areas network of which 24.37% are national parks and 75.59% are wildlife sanctuaries. These protected areas are mainly wildlife habitats and include many diverse forest communities. Invasive alien species, such as *Lantana* and *Parthenium*, have infested many forest areas due to anthropogenic disturbances such as grazing and deforestation. Many of these alien weeds require light for vigorous growth and cannot penetrate undisturbed natural forest areas.

Table 14.6 Inventory of forest invasive species

Inventory	*Leptocybe invasa* Fisher & La Salle (Hymenoptera: Eulophidae) (Mendel et al. 2004)
Common name	Blue gum chalcid wasp
Origin and distribution	Australia Distributed to most Mediterranean countries, Africa and Asia, namely Algeria, France, Israel, Italy, Jordan, Morocco, Portugal, Spain, Turkey, Kenya, Syria, Tanzania, Uganda, Iran, Thailand, Vietnam and India (https://www.cabi.org/isc/datasheet/108923#tosummaryOfInvasiveness)
Associates	*Eucalyptus* spp.
Pathways of spread	Movement of host plants for planting can move the pest over long distances (Nyeko et al. 2007)
Threat levels	Severe in nurseries and young plantations (Hesami et al. 2005; Roychoudhury et al. 2016)
Major hosts	*Eucalyptus* species, such as *E. botryoides*, *E. bridgesiana*, *E. camaldulensis*, *E. globulus*, *E. gunnii*, *E. grandis*, *E. saligna*, *E. maidenii*, *E. robusta*, *E. tereticornis* and *E. viminalis* (Anon 2005b)
Management/eradication methods	Selection of resistant species/clones/provenances (Thu et al. 2009; Kulkarni 2010a; Nyeko et al. 2010) and use of potential biopesticides and minimum dose of less hazardous chemical pesticides (Roychoudhury 2013)
Rate of change in the habit under FIS	Spreading rapidly in the country (Jacob 2009; Roychoudhury 2016a)
Additional information	This species is commonly known as blue gum chalcid, a new genus of minute wasp, exclusively responsible for gall formation in different species of *Eucalyptus* (Mendel et al. 2004) In India, it was first noticed during the year 2002 from coastal Tamil Nadu and now it has spread over to peninsular India (Anon 2007b; Jacob et al. 2007) This alien invasive species has entered Madhya Pradesh of central India and massive infestations of gall maker observed in nurseries and plantations of *Eucalyptus*, during the month of May, 2007 (Roychoudhury et al. 2007) *L. invasa* causes galls on the mid-ribs, petioles and growing stems of new shoots both in nurseries and young plantations, including coppice shoots Studies have been conducted on morphology of gall insect and galls (Roychoudhury and Vaishy 2015), assessment of damage caused by gall insect in nurseries and plantations (Roychoudhury et al. 2016), development of galls and its effects on growth of seedlings (Roychoudhury 2018) Regarding the management aspects, Nyeko et al. (2007) have found that spraying of Dimethoate and Fenkil in nursery stage is highly effective against *L. invasa*. Kumari (2009) has reported that among the different insecticides tested both under nursery and field conditions, Imidacloprid recorded lowest adult emergence followed by Azadirachtin, Thiamethoxam and Acetamiprid, whereas Dimethoate and Profenophos have no significant effect in reduction of adult emergence. Spot application of Carbofuran 10 G at 0.5 and 1.0 g per plant at root zone, foliar spray of Methyl parathion 50 EC (1 mL/L) and Imidacloprid 17.8 SL (0.25 mL/L) have been found effective in reducing the number of fresh galls per 10 cm twig length and superior to botanicals (Javaregowda et al. 2010). Treatment of Carbofuran 3 G or Phorate 10 G at 1 g/plant in the soil, 45 days after transferring the seedlings to polyethylene bags followed by spray application of Dimethoate at 0.03% or Phosphamidon at 0.04% or Methyl-*o*-demeton at 0.025% or Acephate at 0.075% at 15 days interval starting after 1 month of granular application could be the effective strategy to check the infestation by the gall insect in eucalyptus nursery (Jhala et al. 2010). Kulkarni (2010b) has reported that the neonicotinoids, Thiacloprid 21.7% SL and Imidacloprid 17.8% SL are effective in reducing the gall development and enhance plant growth by reducing the adult emergence and oviposition Recently, an attempt has been made to examine the response of biopesticide, Spinosad 45% SC and chemical pesticides, Acetamiprid 20% SP, Dichlorvos 76% EC, Dimethoate 30% EC, Imidacloprid 17.8% SL and Monocrotophos 36% SL, against gall insect in nursery stage (Roychoudhury 2013). Results revealed that to combat gall wasp, spraying of biopesticide and chemical pesticides on seedlings of eucalyptus at the rate 0.05% for 6 months at 15 days interval, significantly ($p > 0.05$ to $p < 0.01$) increased the growth of seedlings and reduced the gall formation (Roychoudhury 2016b, 2017; Roychoudhury and Mishra 2018). The results suggest that spraying on seedlings needs to be started before reaching 1 month of age for management of gall wasp to save eucalyptus in nursery stage. On the basis of seedling growth and reduction in gall production, pesticides used in the present work could be arranged in descending order: Monocrotophos = Imidacloprid > Acetamiprid > Dimethoate = Dichlorvos = Spinosad

14.3.7 Forest Invasive Weeds in India

A vast majority of the weeds found in India too are exotic in origin. Some weeds such as *Lantana camara* and *Eichhormia crassipes* were deliberately and more and more have got entry into the country accidentally. The most important ones are being *Argemone mexicana*, *Salvinia molesta*, *Solanum elaeagnifolium*, *Tridax procumbens*, etc. The most conspicuous one, however, is *Parthenium hysterophorus* which is believed to have been introduced into the country through large scale wheat imports from the United States and Mexico in the 1950s. In the recent past, *Chromolaena odorata* (syn. *Eupatorium odoratum*) has become a serious menace in forest and plantation crops in Goa, Karnataka, Tamil Nadu and north-eastern states. It has also become a problem in Himachal Pradesh where this weed along with *Ageratum houstonianum* is squeezing the already limited grasslands. Similarly, another invasive weed, *Mikania micrantha*,

Table 14.7 Status of forest and tree cover of India

Class	Area (km^2)	Geographical area (%)
Forest cover		
Very dense forest (>70%)	99,278	3.02
Moderately dense forest (40–70%)	308,472	9.38
Open forest (10–40%)	304,499	9.26
Total forest cover[a]	**712,249**	**21.67**
Tree cover	95,027	2.89
Total forest and tree cover	**807,276**	**24.56**
Scrub (<10%)	46,297	1.41
Non-forest[b]	2,528,923	76.92
Total geographic area	**3,287,469**	**100.00**

Source: Anon (2019)
[a] Includes 4975 km^2 under mangroves cover
[b] Non-forest includes tree cover (percentage rounded off)

Fig. 14.4 Invasion of *Mimosa rubicaulis* in Kaziranga National Park, Assam

popularly known as 'Mile-a-minute', has generated serious problem in most of the plantation crops in Kerala. The bamboo plantations are fast disappearing being unable to withstand the competition from *Mikania*. All the vegetation, including tall trees of north-eastern states of India is covered with a thick canopy of *Mikania*. Most recently, *Mimosa rubicaulis*, a spiny weed has badly affected the world famous Kaziranga National Park in Assam state. It is seriously damaging the local vegetation of several herbivorous animals inhabiting the park, including the endangered single-horned rhinoceros (Fig. 14.4). Invasion of *Lantana* in teak forests is very common (Fig. 14.5), whereas invasion of *Parthenium* and *Chromolaena* is frequently noticed in sal forests of central India (Figs. 14.6 and 14.7).

14.4 Prevention and Management of Forest Invasive Weeds

14.4.1 Prevention

The most cost effective line of defence against invasive alien weeds is the prevention. To prevent invasive weeds, some methods are found effective, which do not allow to establish in an area where they are non-native. The first step in prevention is identification of the alien species and listing them under different categories or groups based on their invasive potential (Wittenberg 2000). Such lists can include:

Fig. 14.5 Invasion of *Lantana camera* in teak forests of Madhya Pradesh

Black lists: This includes species with highly destructive and their introduction should be strictly prohibited.

White lists: This list includes species with such a low probability of invasion and be allowed to introduce.

Grey lists: This includes the great majority of species whose invasiveness is unknown.

Major tools to prevent invasions are:

14.4.1.1 Interception

Implementation of rules and regulations at the border is necessary for risk analysis of species to be introduced. The tools useful in this process are:

Quarantine laws and regulations and prevention measures and their enforcement: International trade rules need to be establish for regulatory quarantine (sanitary and phytosanitary) measures to protect plants, animals and humans from damage due to pests and diseases. Laws need to be strictly enforced. Many countries have established quarantine laws for dealing with import of pest species of agriculture and forestry.

Risk assessment/risk analysis and environmental impact assessment: International procedures and instruments are available to measure pest and pathogen risk analysis. The risk assessment method is used for rating and ranking of invasive species.

Risk analysis method comprises of risk assessment, risk management and risk communication. Risk analysis is necessary not only for the economic implications but also environmental issues concerning the introductions. The risk assessment method begins with the proper identification of candidate invasive species and its pathways. The successful entry can be assessed through review of scientific literature, opinion of experts and qualitative and quantitative analysis. Some other factors can also be considered such as invasiveness, way of entry, establishment, rate of spread and economic and environmental impact.

Public education: Public education is extremely important and absolutely necessary for prohibition and control of harmful exotic flora and fauna. Public consciousness and support can greatly increase the success to protect and save biodiversity. Travelers do not have any knowledge of laws and regulations about invasive flora and fauna to prevent entry. Printed material like bulletin, brochure, poster and leaflet needs to be used for public awareness programme.

Treatment: Any suspected goods and packaging material infested with harmful non-indigenous organisms must be treated with ecofriendly way. Treatment should be done with environmentally safe pesticides applications, such as fumigation, biopesticde application, water immersion, heat and cold treatment, pressure or irradiation.

Early detection and rapid response: 'Weeds won't wait' is a truism. It is very difficult to stop the entry of harmful

Fig. 14.6 Invasion of *Parthenium hysterophorus* in core zone of Achanakmar-Amarkantak biosphere reserve, Chhattisgarh

exotic species. Only quick identification of invasive species is must to eradicate before they become spread widely and control becomes difficult technically. This early detection of exotic species is possible by regular periodical surveys and carry out this surveys by well trained personnel are necessary who are able to identify harmful exotic species of different taxonomic groups. Trainings can be imparted for farmers, forestry field staff, land surveyors, tour operators and the concerned public to learn about climate and well acquainted about the effects of harmful exotic. This education programme can include media promotion, displays and personal interactions.

Fig. 14.7 Invasion of *Chromolaena odorata* in buffer zone of Achanakmar-Amarkantak biosphere reserve, Madhya Pradesh

14.4.2 Management

The control of weeds involves four main strategies such as eradication, containment, control and mitigation.

Eradication: Eradication is the complete removal of weeds in an area. Sometimes prevention technique is not effective then eradication strategy is selected as feasible technique.

Containment: Containment is a specific form of management. The main aim is to control the spread of an alien species in a population of geographical range. The methods used for containment are similar to those described for prevention, eradication and control.

Control: Control aims for the long-term reduction in population density and abundance of exotic species.

Mitigation: If eradication, containment and control are not successful in reducing an invasive alien weed, the last resort is to 'live with' it in the best way and mitigate impacts on biodiversity. Mitigation is most commonly used in the preservation and protection of species becoming extinct.

Regarding the effective control and management of alien invasive weeds, cultural methods (e.g. crop rotation, revegetation, grazing and water level manipulation); physical methods (e.g. fences, equipment sanitation and electric dispersal barriers); removal (e.g. hand-removal and mechanical harvesting), the judicious use of environmentally safe chemical and biopesticides; release of selective biological control agents (such as host-specific predator/herbivore organisms); interference with reproduction (e.g. pheromone-baited traps and release of sterile males) and habitat management (e.g. prescribed burning and grazing) can be used. Sometimes integrated methods are used within an overall strategy. However, success mainly depends on the awareness, participation of public and all stakeholders to a large extent.

14.4.3 Consequences of Factors on Dynamics and Long-Term Ecological Impact on Forest Environmental Habitats

Undisturbed natural forest with closed canopy: Alien weeds are likely to be unimportant regionally as they cannot compete with native vegetation.

Natural forests, open canopy and forests with short fallow cycles: Alien weeds are likely to increase their ranges, if fallows are left for a longer period.

Intensively exploited forest habitats: In these circumstances, there is likely to be extensive range expansion of the weeds. This may either be through human aided dispersal and superior competitive ability and/or lack of co-evolved natural enemies. The alien invasive weeds provide soil stabilization and cover for some wildlife like tigers.

Alien invasive species also contribute a large fraction of biomass of early succession communities after shifting agriculture in northeastern India and, hence, reducing losses of nutrients in-run-off. However, almost complete monopolization of the habitat and a severe reduction in biodiversity may not be able to recover.

Plantations and agroforestry systems: As mentioned before, these are similar to the category of intensively exploited forest habitats. The evidence suggests that the alien weeds are expanding their range within these systems. There is a particular problem of the weed being spread by local communities. The cost there is on keeping the weeds below their economic thresholds (by weeding or other means).

14.4.4 Implications for Management

As the weed problem is complex and in view of the evidence for the likely range expansions and long-term negative impacts, management strategies need to be urgently developed.

The unwanted plants or weeds may consist of several subspecies/biotypes and these are invasive across a range of forest types, plantations and agricultural land throughout the region. This factor, together with a general lack of involvement of communities in tackling the problem, is mostly the reason why previous efforts focused on a single-component solution for control have either achieved only partial success at a local scale or failed altogether. A management strategy for these weeds must be based on a thorough understanding of these problems. Thus, some recent calls for action have emphasized for the development of an integrated weed management (IWM) project. An IWM project also needs to be community based and the solutions integrated into the overall joint forest management process, including forest rejuvenation and prevent soil erosion.

Weed impact studies need to focus on effects of the weeds on tree and flora regeneration as well as social problems. Ecological studies also need to identify major factors (natural and human) contributing to dispersal, colonization and competitive ability. This will help in predicting further spread of weeds and for the development of new management tactics (e.g. tactics to minimize spread by humans, encouragement of superior native plant competitors, etc.).

Various pilot schemes for the cultural control of weeds are implemented. Schemes have included green manure and composts, bio-gas production and tree planting. These schemes need to be objectively assessed to ascertain developmental potential and particularly adoption. A similar assessment also made of any control measure being used by communities on a local basis. Any current herbicide use also needs to be evaluated and needs for its future use carefully assessed within the framework of IWM.

Techniques of cultural control will be most appropriate for agricultural areas and local communities. However, given that the weeds mainly affect natural habitats and plantations, the basis of the IWM project needs to focus on low inputs and sustainable solutions. Thus, biocontrol techniques by using biopesticides and natural enemies (parasitoids and predators) can provide the best core component of an IWM project. Studies on the area of origin of the invasive weed populations may help facilitate the location of host-specific co-evolved natural enemies.

14.5 Future Scope

About 190 IAS including FIS are found in India. Reddy et al. (2008) have suggested that observation and surveillance of invasion can be done through seasonal species inventory (qualitative research) and phytosociological studies through mapping, GPS, remote sensing and digital data (quantitative research). All these studies involve active links between taxonomists, ecologists and forest managers to monitor and control.

Management of IAS needs good planning, understanding, correct methods, institutional and stakeholder supports, monitoring, evaluation, feedback and review as necessary. Adaptive management needs to be applied, which is very important to control IAS (Wittenberg and Cock 2001; Tu et al. 2001). The adaptive management for control of IAS as suggested by Tu et al. (2001) is mentioned hereunder. The same approach can be applied to all other taxa of IAS (Anon 2007a).

1. Establishment of conservation targets and goals.
2. Identification and prioritization of species/infestations that threaten targets and goals.
3. Assessment of control techniques.
4. Development and implementation of weed management plan.
5. Monitoring and assessment of impact management actions.
6. Review and modify, if necessary.

Thus, it is clear that there is an immediate need for documentation, identification through taxonomical studies, preparation of inventory of IAS including FIS and their digitalization; generate area/locality wise status information, extent of occurrence of FIS and assessment of their impact on population diversity in forest eco-system. National strategy, planning, action plan on FIS and compendium of FIS in India are necessary to identify all invasive species in the country and their concomitant feasible eco-friendly management and future control can ensure to save our native forest as well as environment.

Lessons Learnt

- Invasive Alien Species (IAS), particularly Forest Invasive Species (FIS), is one of the serious threats to forest biodiversity.
- FIS is directly involved to the forestry, a serious problem to forest cover that can eliminate plant and animal species of native origin.
- Many invasive species of flora and fauna have documented in India, belonging to different taxonomic groups.
- In India, 48 FIS have been found so far that include 29 plant species, 10 species of fungi and 9 insect species.
- Important FIS flora species in India include *Lantana camara*, *Parthenium hysterophorus*, *Eupatorium odoratum*, *Prosopis juliflora*, *Mikania micrantha* and *Chromolaena odorata*.

Key Questions

1. How to determine invasive potential of FIS for listing them under different categories?
2. Is there a complete inventory of FIS in India that includes microbes, plants and animals?
3. Describe an impact assessment study on FIS.
4. What are the important measures that can be considered to prevent invasions of FIS?
5. What are the feasible steps towards management of FIS?

Acknowledgements The author is thankful to Dr. Rajesh Kumar Mishra, Chief Technical Officer, Computer and Information Technology Section, ICFRE-Tropical Forest Research Institute, Jabalpur-482021, Madhya Pradesh, for necessary help in preparation of the manuscript.

References

Alfred JRB (1998) Faunal diversity in India: an overview. In: Alfred JRB, Das AK, Sanyal AK (eds) Faunal diversity in India. Zoological Survey of India, Kolkata, pp 1–9

Anonymous (2005a) Forest invasive species. Indian Council of Forestry Research and Education, Dehradun. 49 p. https://apfisn.net/wp-content/uploads/2018/07/India.pdf

Anonymous (2005b) Pest alert—blue gum chalcid. Tree Protect News 10:13

Anonymous (2007a) Invasive alien species and protected areas: a scoping report. http://www.issg.org/pdf/publications/gisp/resources/ias_protectedareas_scoping_i.pdf

Anonymous (2007b) Report on field visit-cun-meeting to discuss the strategy to be adopted and contain the insect gall problem in *Eucalyptus*, caused by *Leptocybe invasa* Fisher & La Salle, held on 17th April, 2007 at Chennai, organized by IFGTB, ICFRE, Coimbatore, Tamil Nadu

Anonymous (2019) India state of forest report 2019, vol I. Forest Survey of India, Ministry of Environment, Forest & Climate Change, Government of India, Dehradun. 183 p. http://164.100.117.97/WriteReadData/userfiles/ISFR2019%20Vol-I.pdf

Baskin Y (2002) A plague of rats and rubbervines: the growing threat of species invasions. Island Press, Washington, DC

Chatterjee D (1940) Studies on the endemic flora of India and Burma. J Asiat Soc Bengal 5:19–67

Drake JA, Mooney HA, Castri FD, Groves R, Kruger F, Rejmanek M, Williamson M (1989) Biological invasions: a global perspective. Wiley, New York

Gogoi AK, Babu MBBP (2006) Alien invasive weeds and their management. Vaniki Sandesh 30:22–32

Goyal AK (2005) Asia-Pacific Forest Invasive Species Network (APFISN). http://ifs.nic.in/rt/misc/apfisn/apfisn.htm

Hajra PK, Mudgal V (1997) Plant diversity hotspots in India—an overview. Botanical Survey of India, Kolkata

Hesami S, Alemansoor H, Seyedebrahimi S (2005) Report of *Leptocybe invasa* (Hym.: Eulophidae), gall wasp of *Eucalyptus camaldulensis* with notes on biology in Shiraz vicinity. J Ent Soc Iran 24:99–108

Hiremath AJ, Sundaram B (2013) Invasive plant species in Indian protected areas: conserving biodiversity in cultural landscapes. In: Foxcroft LC (ed) Plant invasions in protected areas: patterns, problems and challenges, Springer series in invasion ecology, vol 7. Springer, pp 241–266. https://doi.org/10.1007/978-94-007-7750-7_12

Jacob JP (2009) Invasive insect pest (*Leptocybe invasa*) in *Eucalyptus* plantations in India. ENVIS For Bull 9:65–70

Jacob JP, Devaraj R, Natarajan R (2007) Outbreak of the invasive gall-inducing wasp, *Leptocybe invasa*, on eucalypts in India. Invasives 8: 4–5

Javaregowda PST, Prabhu ST, Patil RS (2010) Evaluation of botanicals and synthetic insecticides against eucalyptus gall wasp, *Leptocybe invasa* (Hymenoptera: Eulophidae). Karnataka J Agric Sci 23:200–202

Jhala RC, Patel MG, Vaghela MN (2010) Effectiveness of insecticides against blue gum chalcid, *Leptocybe invasa* Fisher & La Salle (Hymenoptera: Eulophidae), infesting eucalyptus seedling in middle Gujarat, India. Karnataka J Agric Sci 23:84–86

Knowler D, Barbier E (2000) The economics of an invading species: a theoretical model and case study application. In: Perrings C, Williamson M, Dalmazzone S (eds) The economics of biological invasions. Edward Elgar Publishing, Cheltenham, UK, pp 70–93

Kulkarni HD (2010a) Screening eucalyptus clones against *Leptocube invasa* Fisher and Lasalle (Hymenoptera: Eulophodae). Karnataka J Agric Sci 23:87–90

Kulkarni H (2010b) Distribution, seasonal incidence and management of eucalyptus gall wasp, *Leptocybe invasa* Fisher and La Salle (Hymenoptera: Eulophidae) with special reference to native parasitoids. M.Sc. Thesis, University of Agricultural Sciences, Dharwad, Karnataka

Kumari NK (2009) Bioecology and management of eucalyptus gall wasp, *Leptocybe invasa* Fisher & La Salle (Hymenoptera: Eulophidae). M.Sc. Thesis, University of Agricultural Sciences, Dharwad, Karnataka

Mendel Z, Protasov A, Fisher N, La Salle J (2004) Taxonomy and biology of *Leptocybeinvasa* gen. & sp. n. (Hymenoptera: Eulophidae), and invasive gall inducer on *Eucalyptus*. Aust J Ent 43:101–113

Mooney HA, Drake JA (1987) The ecology of biological invasions. Environment 29:12

Mooney HA, Hobbs RJ (2000) Invasive species in a changing world. Island Press, Washington, DC

Nayar MP (1977) Changing patterns of the Indian Flora. Bull Bot Surv India 19:145–155

Nayar MP (1989) *In situ* conservation of wild flora resources. Bull Bot Surv India 29:319–333

Nyeko P, Mutitu EK, Day RK (2007) Farmer's knowledge, perceptions and management of the gall-forming wasp, *Leptocybe invasa* (Hymenoptera: Eulophidae) on *Eucalyptus* species in Uganda. Int J Pest Manage 53:111–119

Nyeko P, Mutitu KE, Otieno BO, Ngae GN, Day RK (2010) Variations in *Leptocybe invasa* (Hymenoptera: Eulophidae) population intensity and infestation on eucalyptus germplasms in Uganda and Kenya. Int J Pest Manage 56:137–144

Reddy CS (2008) Catalogue of invasive alien flora of India. Life Sci J 5: 84–89. http://www.lifesciencesite.com/lsj/life0502/16_life0502_84_89_Catalogue.pdf

Reddy CS, Bagyanarayana G, Reddy KN, Raju VS (2008) Invasive alien Flora of India. National Biological Information Infrastructure, USGS, USA

Roychoudhury N (2013) Damage assessment of gall making insect species of eucalyptus and its management by pesticides. Project Completion Report ID No.: 153/TFRI/2010/Ento-3(24), submitted to Indian Council of Forestry Research and Education (ICFRE), Dehradun, India

Roychoudhury N (2016a) Gall insect menace in *Eucalyptus*. Indian J Forestry 39:349–357

Roychoudhury N (2016b) Efficacy of chemical pesticides against *Eucalyptus* gall insect, *Leptocybe invasa* Fisher & La Salle (Hymenoptera: Eulophidae). Pestology 40:40–47

Roychoudhury N (2017) Evaluation of biopesticide, spinosad and neonicotinoids, acetamiprid and imidacloprid against *Eucalyptus* gall insect, *Leptocybe invasa* Fisher & La Salle. Pestology 41:23–32

Roychoudhury N (2018) Development of galls in *Eucalyptus* due to infestation of *Leptocybe invasa* and its effects on growth of seedlings. Indian J Forestry 41:79–83

Roychoudhury N, Mishra RK (2018) *Eucalyptus* gall insect, *Leptocybe invasa* Fisher & La Salle and its management by bio- and chemical pesticides—an update. Pestology 42:11–22

Roychoudhury N, Sharma R (2013) Invasive alien species with special emphasis on forest invasive species in India. In: Balu A, Jayaraj RSC, Regupathy A, Mohan V, Warrier RR, Raghunath TP, Krishnakumar N (eds) Forest health management. Institute of Forest Genetics and Tree Breeding, Coimbatore, Tamil Nadu, pp 343–353

Roychoudhury N, Vaishy N (2015) Morphology of *Leptocybe invasa* Fisher & La Salle and size of galls in Eucalyptus. J Trop Forestry 31: 78–84

Roychoudhury N, Chandra S, Joshi KC (2007) Infestation of Australian insect, *Leptocybe invasa*, on *Eucalyptus* in Madhya Pradesh. Vaniki Sandesh 31:13–15

Roychoudhury N, Chandra S, Singh RB (2016) Pest status of gall making insect, *Leptocybe invasa* Fisher & La Salle, in nurseries and plantations of Eucalyptus in Madhya Pradesh. Indian Forester 142:601–606

Roychoudhury N, Meshram PB, Mishra RK (2019) Invasive alien species. In: Cerificate course on Forest Entomology & Pest Control under Green Skill Development Programme (GSDP), Government of India, Ministry of Environment, Forest and Climate Change (MoEF&CC), New Delhi. Tropical Forest Research Institute, Jabalpur, Madhya Pradesh, pp 72–77

Shukla AN, Singh KP, Singh JS (2009) Invasive alien species of Achanakmar-Amarkantak biosphere reserve, central India. Proc Nat Acad Sci Sect B Biol Sci 79:384–392

Thu PQ, Dell B, Burges TI (2009) Susceptibility of 18 eucalypt species to the gall wasp *Leptocybe invasa* in the nursery and young plantations in Vietnam. Sci Asia 35:113–117. https://doi.org/10.2306/scienceasia1513-1874.2009.35.113

Tu M, Hurd C, Randall JM (2001) Weed control methods handbook. The Nature Conservancy. http://tncweeds.ucdavis.edu

Wittenberg R (2000) Best practices for the prevention and management of alien invasive species. Final report of GISP workshop on management and early warning systems, Kuala Lumpur, Malayasia, 22–27 Mar 1999

Wittenberg R, Cock MJW (2001) Invasive alien species: a toolkit of best prevention and management practices. CAB International, Wallingford, Oxon

Wildlife Management

15

Khursheed Ahmad and Amir Bhat

Abstract

Fundamental to a modern society, the essence of wildlife conservation relies in understanding and applying the principles of scientific wildlife management. A multidisciplinary subject, wildlife management encompasses scientific interpretation and logical reasoning. It involves various activities such as monitoring, protection, habitat restoration, and population control to maintain healthy ecosystems with blending of balancing basic human needs and requirements of wildlife. Wildlife conservation and management have recently gained prominence alongside persistent challenges like climate change, global warming, and health education. This has resulted in not only elevating wildlife as a scientific discipline but has also attracted students from all backgrounds who are eager to explore and understand its core principles, practices and benefits. Although wildlife management and conservation have roots in ancient and medieval times in India, it gained momentum during the British colonial era, which laid the foundation for scientific management practices for the conservation. Encouraged by these initial efforts, several national and international organizations emerged to conserve this invaluable natural resource.

In the past, studies on wildlife management and conservation were limited, focusing primarily on population estimation, habitat management and natural history observations. However, in the mid-twentieth century, these studies evolved to include scientific appraisal and wildlife legislation, supporting the development and expansion of protected area networks. The vast areas of community reserves, national parks, wildlife sanctuaries and conservation reserves collectively known as protected areas have experienced considerable progress since 1972. As of January 2023, the protected area network (PAN) spans an impressive area of the country's 173,629.52 km^2, representing approximately 5.28% total territorial area. This expansion demonstrates the continued commitment of legislative governments to conserving its diverse ecosystems and safeguarding the rich biodiversity found within them.

Advanced technologies including radio telemetry, satellite technology, camera/CCTV trapping, and drones have emerged as effective alternatives for collecting data on wildlife populations and monitoring. Such Internet-based technologies present numerous benefits, including higher accuracy, reduced costs, and reduced disturbance to animals. As technology continues to evolve, it is expected more advanced and sophisticated techniques will be developed to improve our understanding of wildlife, its populations and their respective habitats. Furthermore, India's liberalization policies and the advent of the new millennium opened doors to new avenues, especially in the IT sector, facilitating the access and exploration of the digital era in wildlife photography and filming.

Today, climate change and the growing human demand on freshwater, marine, and terrestrial resources are intricately linked to the preservation and its management of important wildlife habitats and their biodiversity. It is projected that one-third of the expected baseline biodiversity will be approximately declined by 2050 through climate change. Deforestation and urbanization are two significant contributors to species extinction, with an alarming rate of 36 million acres of natural forest being destroyed globally each year, as per the latest data from organizations like the World Wide Fund for Nature also known as WWF. These human activities have pushed over 16,000 species to the edge of extinction. The interaction of humans with wildlife has become increasingly complex primarily due to human-assisted change in the environment.

Wildlife management, as a permanent conservation policy, is essential to ensuring the long-term viability of

K. Ahmad (✉) · A. Bhat
Division of Wildlife Sciences, Sher-e-Kashmir University of Agricultural Sciences & Technology of Kashmir, Srinagar, Jammu and Kashmir, India

A. K. Mandal, A. Nicodemus (eds.), *Textbook of Forest Science*, https://doi.org/10.1007/978-981-97-8289-5_15

wildlife resources. Different underlying values shape conservation policies, with ecological, economic, and cultural considerations playing key roles. Such ideals can sometimes lead to ethical dilemmas and disagreements, highlighting the complexity of wildlife management and the need for careful decision-making. Furthermore, it is critical to acknowledge the historical evidence of human reliance on animal resources for survival, emphasizing the interdependence of human and wildlife existence.

Keywords

Wildlife · Territoriality · Home range · Corridor · Wetland

15.1 Introduction

The term "wildlife" lacks a universally accepted precise definition. It generally encompasses all living things existing outside direct human control. Free ranging animals, particularly non-domesticated ones, constituted wildlife in earlier definitions. In addition to this, wildlife was also referred to hunting of animals for sport. In recent times, wildlife can be succinctly defined as "animals, plants, fungi, and living organisms that dwell an area without the direct intervention of human beings." Multiple terminologies are used to address wildlife such as "wild plants and animals" and "wild living resources." The essence of the term "wildlife" is captured in a definition from the inaugural issue of *The Journal of Wildlife Management* (1937): "The practical ecology of all vertebrates and their plant and animal associates."

The term "management" distinctly suggests the exertion and application of human influence and manipulation. In the context of wildlife, management might also be termed as wildlife conservation because in some circumstances, the best management practices may be to leave wildlife alone. Nevertheless, wildlife management is closely connected with various other disciplines. Among them, forestry, range management and park administration are prominent. Therefore wildlife management can best be defined through Bolen and Robinson (2003) definition which states "ecological knowledge on populations of vertebrate animals and their plant and animal associates in a manner that strikes a balance between the needs of those populations and the needs of people." Ecological knowledge is applied in three basic management approaches, which serve as types of wildlife management:

1. **Preservation**: Permitting nature to unfold naturally without human interference.
2. **Direct management**: Involves actions such as trapping, shooting, poisoning, and stocking to control animal populations.
3. **Indirect management**: Involves altering key components of wildlife habitat, such as vegetation or water sources.

Wildlife management does not fall under the category of a fundamental science like physics, chemistry, zoology or botany, nor does it align with pure technology as seen in engineering. Wildlife management draws from these fields and entails an art that is integrated and used skillfully with logic and scientific interpretation.

Conservation, when applied to natural resources, is attributed to Gifford Pinchot, a forester in President Theodore Roosevelt's administration. Pinchot rejected terms like "*protection*" and "*preservation*," which, in his view, suggested keeping natural resources locked and unused. He believed these terms lacked a description for achieving sustained yield from forests, wildlife, and other resources. Consequently, he coined the term "conservation" in 1907, drawing from the Latin roots "*servare*" (to guard) and "*con*" (together). President Roosevelt promptly embraced Pinchot's term, making conservation a priority in the subsequent years of forest services.

15.1.1 Importance of Wildlife Management

The current vistas of education and teaching have changed throughout the globe. This means new dimensions of academic exercise has been incorporated to train the next generation. Dimensions like change in climate and warming of globe, health education, wildlife conservation and its management, policy and governance, etc. are some of the examples that have changed the academic scenario of the world. This transformation that has placed wildlife conservation and its management as one of the core focus has indirectly raised the scope and potential of wildlife as a science.

The attraction toward wildlife, its conservation and management have drawn many students of nature as well as general public to learn, understand and explore the habits, habitats, behavior and other attributes of wildlife. Besides this, wildlife appeals to the people as most popular inquisitiveness science where its exploration is dealt in-depth.

Planet Earth contains a vast number of animals, plants and other lower strata species. However, majority of them are either yet to be identified or those already identified are yet to be subjected to advanced studies which emphasize the importance of wildlife sciences. This approach has largely gained more importance in a country like India where such studies are scarcely documented. Thus data generation on wildlife research and focus on its management would add enormously to sciences of conservation of wildlife and its management.

15.1.2 Scope, Potential of Wildlife Management

Among other provisions of 1972 Act of Wildlife Protection (WPA), application of wildlife management and biological principles to maintain sustainable populations of wild fauna is envisaged. The application of management principles envisions the human intervention for resource protection as well as livelihood security of the indigenous people. This interventional management of wildlife resources vis-à-vis livelihood security assumes more importance in Indian context. First, protected area network (PAN) of India is vast and diverse, spread all over its states and union territories. Second, availability of limited expertise and human resource required for maintaining the sustainable population of wild species and third, enormous requirement of forest resources for populous and diverse population that directly or indirectly rely on forests.

In addition to this, recreational values of wildlife that drives people in millions to the National Parks and Wildlife Sanctuaries are contributing to the country's revenues and boosting the concepts like ecotourism. Several countries, besides India, have interesting wildlife tourist destinations. Chardonet et al. (2002) and Chundadwat et al. (2017) have compiled revenue generation contributions of several countries and found that ecotourism activities around the wildlife areas have a considerable contribution toward the overall GDP of countries including Kenya, Tanzania, South Africa, West and Central African countries, North and South American countries, Sri Lanka, Nepal and Italy.

15.2 Brief History of Wildlife Management

Up until the 1960s, wildlife management predominantly focused on game management, animal husbandry and population regulation of birds and mammals hunted for sport. While game management still remains a significant aspect of wildlife management, professionals now address a broader range of wild species and their issues.

Although the dynamics of wildlife management has evolved, yet its historical context remains pertinent to both its present and future. Wildlife management practices are intertwined within the domains of human ethics, culture, diverse perceptions, and relevant legislations. Narration of wildlife conservation started with understanding and recognizing the role of wild animals in diverse landscapes and thus the imperative need to safeguard these endangered species. These conservation efforts inspired many national and international organizations primarily established for the purpose of preserving valuable natural resources. International Council for Bird Preservation (ICBP) was the first such organization which was established in 1922. Establishment of International Union for Conservation of Nature and Natural Resources (IUCN) in 1948 became the first post-war organization funded by United Nations Educational, Scientific and Cultural Organization. After its successful establishment, IUCN formed the first fundraising organization, the International Wildfowl Research Bureau (IWRB) in 1961. In addition to these, Food and Agricultural Organization (FAO) of the United Nations, Smithsonian Institute and the Conservation Foundation both in Washington DC and the Fauna Preservation Society in London were some of the other notable conservation organizations formed in late sixties and early seventies. Zoologists Gessellschaft in Frankfurt, West Germany, has raised important funds for conservation, particularly for Africa.

The IUCN is dealing specifically with animals threatened with extinction, through collection of data on endangered species and to initiate action to prevent the extinction of such species. Maintaining a list of threatened mammals, birds, reptiles, amphibians, fishes, etc., called "The Red Data Book" and conservation expansion has produced positive results throughout the world.

Wildlife conservation in India has rich history and has been treated with cultural reverence. Hindu mythology, religious scriptures of Islam and Buddhism are some of the references which are full of love for wildlife and its conservation. The study of wildlife conservation in India may be divided into the following two groups:

1. **Pre-Sovereign India**
2. **Post-Sovereign India**

15.2.1 Pre-Sovereign India

People in ancient India recognized the animal rights to co-exist with humans. Based on Islamic literature, the reverence of certain animals and plants is highly acknowledged. Eighth century Sufi saint Aziz-ul Nasafi compares each genus of plants and animals to verses of Quran. Based on literature of Hinduism, the Vedas, Upanishads, and Puranas meticulously describe the wildlife. Buddhist literature has numerous tales and fables such as Buddhist Jataka tales. Wildlife in Ramayana mentions special relation of Ram and Laxman with wild animals, whereas Mahabharata mentions about Rishis and Munis conserving wild fauna such as deer and birds around their ashram.

Wildlife conservation in India dates to the period of Chandragupta Maurya in third century BC with the appointment of head of forest department that had to supervise the game laws laid down by the Arhtashastra of Kautilya. This is regarded as the first formal administrative document outlining the conservation strategies for wildlife.

Ashoka, the famous descendent of Chandragupta Maurya further championed the cause of wildlife conservation. His fondness for wildlife, which manifested in art and aesthetics, as exemplified by the Lion Capital of the Ashoka pillar, in which four lions were crafted with their back to each other atop the pillar. Ashoka sat up the first concept of wildlife sanctuaries.

Wildlife resources in India during medieval times especially around the Mughal period were vast. Historical data and records reveal that *rhinoceroses* were once distributed across the Gogra River and throughout Peshawar, whereas, wild elephants were abundant in the Karrah region (Allahabad). A distinctive form of hunting known as "*Gamargha Hunt*" was introduced during the time of Emperor Akbar, the grandson of Mughal dynasty's founder Babur. The hunt would entail dispatching hundreds of beaters with drums over a considerable distance to systematically drive all the wild animals in the region into a circumscribed area. After gathering the animals in this manner, the king and noblemen would then traverse the perimeter of the circle, engaging in the hunt with specialized weapons. During his period, wild animals especially tigers and leopards were bred in captivity specifically for hunting purposes in specific menageries in *Fatehpur* and *Agra*. Stringent laws forbidding the hunting of wild animals for sports were implemented during the reign of Jahangir, the son and successor of Akbar. Emperor's professionals, however, were granted special permission to hunt the wild animals involved in man–animal conflicts. Jahangir successfully bred hunting leopards and cheetahs in captivity, a significant achievement considering these species had nearly faced extinction during the reign of Akbar. This, however, didn't deter the royal hunting by the ruling family which continued without any interruption. This was evident in Jahangir's own memoir where the emperor himself documented hunting of 17,167 big and small game wild animals during a span of 36 years.

Wildlife began to decline around the period of Jahangir's death in 1627, same period when the British Empire was laying the foundation to establish itself in India. Almost all the laws framed to safeguard wildlife were overlooked. *Shikars* (hunting) were arranged to symbolize chivalry, bravery and a particular status symbol among the British. Even a regiment of *Shikaris* (hunters) comprising mainly of local tribes knowledgeable about the forests were maintained. This resulted in drastic decline of numerous wild animals who were driven towards their final existing refuge, exemplified by the sole surviving population of Asiatic lions in Gujarat and the one-horned rhinoceros in North-East India. Under these circumstances, India's first National Park, the Hailey National Park (now Jim Corbett National Park), was established in 1936, largely owing to the endeavors of Jim Corbett, a former hunter turned conservationist and naturalist. Historian Michael Lewis, in his work "*Inventing Global Ecology*," points out that by the time of initial decades of twentieth century, under British rule, Indian forests had also become the sites of some academic studies, primarily taxonomical in nature, concentrating on the collection and identification of species.

15.2.2 Post-Sovereign India

The unpretentious necessity to protect wildlife across the national level was realized post the independence of the country. Laws pertaining to wildlife conservation in India varied from state to state and in 1952, the Indian Wildlife Board was established to centralize all the rules and regulations related to wildlife conservation in India. Later in the year 1956, the Board passed a historical declaration that accorded all the game parks the status of a wildlife Sanctuary or a National Park. Wildlife conservation underwent an imperative shift from being solely the obligation of the respective governments and policymakers to becoming a concern of the general public at large. During the 1970s, two landmark actions were made that influenced the wildlife conservation for its sustainable future. The first being the implementation of the stringent Wildlife Protection Act in 1972, and the other was the initiation of Project Tiger in 1973, which became the largest wildlife conservation project of its time.

Besides the mainstream legislative measures, wildlife conservation also gained pace through large public movements. For their simplicity, non-violent approach and positive results, Chipko Andolan movement led by Sundar Lal Bahuguna, along with the wildlife conservation movement of the *Bishnois* in Rajasthan, gained worldwide recognition. To this day, they actively participate in the ongoing protection particularly of blackbucks and other wildlife present in their region. Widespread application of technology and science prevalent since 1990s has further supplemented the conservation efforts. Combined technological approaches including wildlife forensics, telemetry, remote sensing techniques, and geographic information systems were all integrated into the practice of wildlife conservation in India. Conservation has transformed from an informal practice offering arbitrary protection to wildlife to its current form as a disciplined art and science. This is prime time for evolving and expanding conservation beyond its current boundaries.

15.2.3 Evolution of Wildlife Management Studies

Early records containing diversity information about wildlife worldwide were first reported by Captain John Smith between 1580 and 1631 AD. However, such reports along with other during the first three centuries of wildlife studies failed the enumeration of wildlife data. Although subsistence was blamed to be sole cause of wildlife killing, sport killing by colonists especially the wealthy among them also led to the cause of wildlife studies during that era.

In 1600s William Bradford, Governor of Plymouth recognized the need of productive habitat to stop the decline of waterfowl population. In 1710, Massachusetts banned the activities that led to disturbance of the habitat conservation and prohibited the practice of fire lighting after studying its impact on the bird conservation.

In 1800s John James Audubon studied the feeding habits of ducks and swans along the shores of river banks in the USA. Increasing human population during the nineteenth century led to the expansion of agriculture which in turn impacted the breeding grounds of important bird species. For example, breeding ground spreading over 70 acres was drained due to Swamp Act of 1849. Later, timber cutting and illegal wildlife trade activities were studied to generate data on wildlife losses. Aldo Leopold laid the foundation for wildlife management in the Southwest Region, from 1909 to 1924.

In the mid-1930s, new regulations, organizations, and surveys were instituted to safeguard, assist, and assess wildlife populations. Airborne assessments and documentation of wildlife population also began around the same period. Notwithstanding the conservation efforts of wildlife in early twentieth century, it was after 1940s that federal and state agencies initiated scientific wildlife management and conservation in a bid to support the wildlife. Past emphasis of wildlife investigations by The United States Department of Agriculture (USDA) focused on the adverse impact of wildlife on human population. This led to the concept of human–animal conflict.

Post-1950s, legislation on wildlife management and its conservation focused on the sustained use of wildlife with strong emphasis on protected area network. By 1990s, concept on generating revenue through wildlife tourism popularly called as "Ecotourism" was incorporated by many developed nations in a bid to boost the economy and create awareness among the local populations.

The first scientific book on Indian birds was written by T.C. Jerdon, a British botanist, zoologist and physician, in 1862–1864 in two volumes, covering over 1000 species. In addition, T.C. Jerdon was also involved in documenting of *The Fauna of British India*, including Ceylon and Burma series. For the next 100 years, many books and research papers were written, mainly describing new species and their distribution, culminating in author Ali and Ripley's voluminous *Handbook of the Birds of India and Pakistan*, the first volume appearing in 1968 and the last in 1974. The year 1983 marked a crucial time for the Indian conservation movement. It was the year the Bombay Natural History Society (BNHS) celebrated its 100th birthday. Along with other centenary-celebration events, Indira Gandhi, Late Prime Minister of India, released "*A Pictorial Guide to the Birds of the Indian Subcontinent*" written by two doyens, Dr. Sálim Ali and Dr. S. Dillon Ripley. Birdwatchers and ornithologists long used to James Lee Peter's classification (used with some modification by Ali and Ripley in the Handbook) suddenly found an entirely new bird species sequence largely based on C.G. Sibleys's and B.L. Monroe Jr.'s work on *Distribution and Taxonomy of Birds of the World*, first published in 1990. The taxonomic sequence was based on genetics and other objective methods of assessment. Several English bird names were also changed. Post-1990s, global travel became more frequent, and it was felt that birders needed greater standardization of common names. Many birds were, after all, found in multiple countries, but were known by different English names. Understandably, this created confusion with itinerant, amateur birdwatchers. It eventually became a necessity to refer to them based on their distribution. India's liberalization policy of the 1990s and the beginning of a new millennium saw a boom in wealth, particularly within the IT sector, and access to photography equipment grew rapidly.

15.2.4 Values of Wildlife

The values of wildlife are diverse and sufficient enough to justify wildlife management. Although several classifications exist on the values of wildlife, still many authors point out discrepancies or ignorance by international community in classifying such values. Since wildlife possesses both direct (consumptive or productive) or indirect (nonconsumptive or existence) values, here they are classified in positive and negative values.

15.2.5 Positive Values

15.2.5.1 Recreational

Recreational values including activities in national parks and sanctuaries attract large number of people. Although it benefits the society, such values are hard to quantify. This form of wildlife value generates the tourism flow commonly termed as ecotourism. Countries like the USA, Canada, Kenya, Tanzania, South Africa, Latin countries, India, Sri-Lanka, and Italy are some of the notable ecotourism destinations.

In India, wildlife tourism is increasing and has generally spread across various national parks, sanctuaries, and tiger reserves. The common tourist flux destinations include Ranthambore, Sariska, Periyar, and Corbett Tiger Reserves.

15.2.5.2 Aesthetic

Despite the complexity in calculating or comparing their values, important intangible services have a significant role in wildlife conservation. The pure pleasure in watching the animals in wild, natural beauties, and artistic appreciation are popular among general masses and also serve as an important component with regard to wildlife conservation.

15.2.5.3 Scientific and Educational Values

The importance of wildlife in teaching and learning lies in the fact that among the young generations wildlife holds a unique attraction. Thus, when this attraction is blended with scientific knowledge, attributes, habits, behavior, and habitats of wildlife, it yields more pronounced impact on the scope and potential of wildlife sciences.

15.2.5.4 Ecological Values

Any extermination of wild animals from the ecosystem can result in irreversible ecological change which may spoil or disturb the natural functioning. The values like "keystone species" which indicate the profound influence of any living species on the ecosystem, underline the importance of wildlife to overall biological functioning of planet Earth. For example, in Sariska Tiger Reserve, leopards occupied the top predatory niche modifying the anti-predator strategies among wild ungulates and resulting in extinction of tigers.

15.2.5.5 Utilitarian Values

This value is also considered as a practical value of wildlife. The importance of wild animals in bio medical research is common. Animals like Rhesus monkey (*Macaca mulatta*) and Chimpanzee (*Pan troglodytes*) are usually used in such research. Susceptibility of any wild animal is also considered important like long-nosed Armadillo (*Dasypus novemcinctus*) which is highly and only susceptible animal to leprosy disease. Venomous snakes also help in making of antivenins that are important for curing snake bites.

15.2.5.6 Commercial Values

Wildlife largely contributes to commercial viability of several nations. In many countries, ecologically justified use of wildlife is a legitimate activity. Convention on International Trade in Endangered Species of Flora and Fauna generally known as CITES usually regulates such activities. As many as 175 countries including India are signatories to this convention. In North America, fur bearing species are commercially important. Game Ranching and Reindeer herding are important in Russia, the USA, and other Scandinavian countries. Deer farming is also reported from several countries as a viable commercial activity. Unless regulated through a sustainable system, commercial exploitation of wildlife poses a grave threat to the survival of several species including rhino, elephant, and musk deer.

15.2.6 Negative Values

15.2.6.1 Accidents

Encounters between large mammals and human beings lead to the damage to both humans and wild animals involved. Top carnivores, elephants, guars, bisons, bears threaten human lives and damage to the property. However, such conflicts are consequence of humans penetrating into wildlife habitats leading to the intrusion among them. Another accidental damage is the flock of birds that cause bird hit accidents as they get sucked into the jet engines causing fatality to both humans as well as birds.

15.2.6.2 Livestock and Crop Damage

Damage to agricultural and horticultural fields in India is not uncommon. Several wild animals including nilgai, blackbuck, wild boar, birds, rodents, and porcupine have caused immense considerable damage. These damages have led some states to categorize such animals as vermin's whose eradication is permitted in the larger interest of the society.

15.2.6.3 Disease Reservoirs

Wild animals cause a serious threat to human survival and general health because several animals act as reservoirs of deadly diseases and transmit them to human population. Diseases like rabies, anthrax, foot and mouth, Covid-19 and Ebola are transmitted to humans by animals. Periodic surveillance, disease surveys, and timely precautions can, however, prevent damage by such diseases. Number of incidences are available where a livestock disease has been transmitted to wild animals and threaten their population. For example in Gir National Park, Asiatic lions were badly affected by *canine distemper virus disease* and reduced their population drastically.

15.3 Management of Basic Requirements of Wildlife

Something that is required or necessary for general wellbeing of the wildlife is known as basic requirement. Basic wildlife needs are those that are required for complete physical, physiological, and mental wellbeing of wild flora and fauna. Basic requirements in different animals and plants differ greatly depending upon the species and its environment necessary for their survival. It includes food, water, cover, and space.

15.3.1 Food

Food is essential for wildlife survival. Animals that receive sufficient food and proper nutrition throughout their lives tend to grow larger and maintain better health compared to animals that undergo periods of poor nutrition or face nutritional deficiencies at any point in their lives. In general, wildlife in good condition exhibits higher reproduction rates, increased resistance to diseases, and better abilities to evade predators compared to animals in poor condition. However, nutritive deficiencies among the wild animals affect their birth and death rates, key for their overall survival and population regulation.

15.3.1.1 Food Availability

Seasonality and geographical location determine the availability of food with one area in a given season having abundant food while the other having critically short food supply. During the period of food scarcity, depletion of food resources and deterioration of food quality due to lignifications (hardening of plant cells) in food plants make them less digestible and palatable. Such nutritional stresses are common across different climatic regions (Yarrow 2009) and forest composition and variability (DeWalt et al. 2003). Food availability across the space and time also determines the habitat quality essential for survival of wildlife populations, for example, grizzly bears of West Central Alberta (Nielsen et al. 2010).

Nutritional needs of wild animals in their natural habitats are also driven by the availability of food both qualitatively and quantitatively. For instance, mice, voles etc. forms the diet of the coyotes for a large period of the year. However, due to the abundant availibility during summers, different insects, fruits and berries forms approximately 80% of coyotes diet. Similarly naturally granivore birds (seed eaters), for example, ruffed grouse, bobwhite quail, and turkey become insectivorous (insect eating) for optimal protein requirements during their breeding season. Because most animal matter is nutritionally complete and easy to digest, predators typically do not encounter issues with the quality of their diet. Unlike predators, herbivores exhibit distinct feeding patterns. These patterns are termed as food preferences, based on their prevalence in the diet relative to their availability in the environment. The ratio of utilization over availability serves as a valuable indicator of wildlife food preferences.

Another aspect to food dynamics in the wildlife is in understanding the difference between starvation (specifically lack of adequate caloric intake leading to serious health issues) and malnutrition (a broader concept encompassing various misbalancing of nutrition's that can arise from different causes). Wild animals perish due to starvation when they do not receive an adequate amount of food necessary for their survival. Carnivores usually succumb to death because they are unable to capture a sufficient amount of prey for sustenance. If an event has occurred that diminishes or eradicates the populations of small mammals (such as rabbits) in a particular area, a red fox residing in that region might face starvation. On the other hand, animals perish due to malnutrition due to inability of food location that otherwise satisfies their nutritional requirements, indicating a deficiency in food quality. Sometimes herbivores also suffer from malnutrition. Effectively managing food for herbivores involves aligning the animals' dietary habits and nutritional requirements with the available resources of the land. Next-generation young animals also are adversely affected especially with winter malnutrition, because newly born growing wild animals need more protein than the adults. Although milk serves as a primary source of protein for newly-born mammals, young carnivores frequently augment their protein requirements by consuming meat. Adequate milk supply is essential for young herbivores (Yarrow 2009).

Foods are categorized as **preferred** when they are more plentiful in an animal's diet than in the surrounding environment, **staple** if regularly consumed and meeting the nutritional requirements (considered the animal's second choice), **emergency** when consumed to address short-term nutritional needs, and **stuffers** when eaten out of necessity due to the absence of alternative options.

15.3.1.2 Food Management

Management of food is typically perceived as a two-step process, that is, determining food habits and supplying those foods. However, food management is not so simple activity. Food habits are determined by the digestive tract or animal dropping analysis. The amount and detail of food habits information should be tied to the objects of management; for example, local food analysis with the dietary needs of the animal determines the success of a reintroduction program. In some circumstances, food plants may be propagated for wildlife, for example, *Quercus* spp., for long-term acorn production and *Secale cereal* for short-term fodder production. Seasonality of food availability also determines the planting process. Thus with the combined information from habitat and wild animals, management can reasonably determine if the population is too large for the available food supply. The management interventions include reducing the animal population by (a) manipulating the population, (b) introduction of predators, (c) liberalization of hunting, live trapping, removal and relocation of animal populations, artificial feeding and habitat modification.

Eliminating plants which have high concentrations of nutritional value can be equally detrimental to a wild animal population as hunting them with the weapons. Conversely, a management plan should include managing the land in a way that promotes the flourishing of specific types of plants and plant communities, including those forming high-quality vegetation for animals (such as food plots), can enhance the health, quality, and abundance of local wildlife populations. Given that supplemental feeding of wildlife is frequently not economically viable, management endeavors should focus on proactively preventing nutritional issues before they arise. The most effective approach to warding off nutritional issues is to offer high-quality natural foods, which are achieved by managing the habitat (Yarrow 2009).

15.3.2 Cover

Cover refers to any component within an animal's environment that offers protection and contributes to the animal's survival or reproduction. It serves as a refuge against unfavorable weather conditions, such as during winter (thermal or winter cover), and offers protection from predators (screening or escape cover). In addition, wildlife also requires cover for activities such as nesting, breeding, rearing young, and loafing.

15.3.2.1 Types of Cover

Brood rearing cover: It is a cover of undergrowth vegetation including the shrubs. This cover is mostly used by the pheasants, quail, etc. This cover is found around the agricultural and other cultivated lands. **Nesting cover**: It is a cover of vegetation including grasses, low shrubs and thickets that is used for nesting sites. Such cover is mostly used by breeding bird population, for example, quail, grouse, songbirds. **Roosting cover**: It is a cover of open areas (cultivable as well as non-cultivable) with grassy vegetation characterized by strong winds, for example, quails. **Escape cover**: It is a cover of rock crevices, vine mats, grassy or branchy thickets, and hollow trees. Appropriate resting and escape cover is crucial for ground-nesting birds, rabbits, and other small game. However, most birds prefer living brush for cover. **Dusting cover**: It is a cover of open sandy areas mostly found in arid zones. Large mammal species use this cover for their daily activities. **Loafing cover**: Certain species of wild animals often engage in aimless activities, choosing secluded spots within their habitat to pass the time. A location that offers shade during the hot summers and sufficient protection from harsh winter winds can function as a suitable loafing cover. On the basis of cover, wildlife is categorized into two types:

1. **Selective wildlife animals**: Animals those are selective and particular for cover, for example, large carnivores and felines.
2. **Non-selective wildlife animals**: Animals that may adapt to two or more cover depending upon the availability.

Cover Management

Many animals including birds frequently visit the places where two habitats intersect each other. The apparent increase in wildlife at such places is known as edge effect. Habitat management of these edge effects on wildlife population usually has reported better results. Creation of edge appears beneficial for various wildlife species including waterfowl, woodcock, deer, and some songbirds. Species such as ovenbirds and spotted owls require large tracts of unbroken woodland habitat.

15.3.3 Water

While many wild animals can endure weeks without food, their ability to survive is typically limited to just a few days without access to water. Wild animals tend to inhabit areas relatively close to water sources, even in situations where food and cover might not be abundant. Animals depend on water for natural processes, for example, digestion and metabolism, as well as for regulating body temperature and eliminating metabolic wastes. Sources of water for wildlife are springs, creeks, farm ponds, food primarily of green plants, leaf dew, and a by-product of the body breaking down fats and starches.

The water requirements depend upon:

1. **Growth**: More growth means more requirement of water.
2. **Size**: Animals of larger size require more water for use than smaller animals, for example, elephants.
3. **Reproduction**: Water requirement during breeding is comparatively more.
4. **General body condition**: Overall body condition of the wild animals also determines its water use capacity.

15.3.3.1 Water Management

Ecological relationships concerning water are direct and indirect, including its physical force on the landscape and hence on wildlife habitat. Animals with specialized tolerances to water regimes may not thrive when they are introduced into areas with wetter or drier climates. A phenomenon known as "reservoir effect" greatly increases biological productivity after land is flooded, but its influence is determined by the continuous presence of water on land surface. In parts where water levels can be manipulated, wet-dry-wet regimes create several water depths, thereby exposing soil which establishes important food and cover for wildlife during different seasons. Vegetation and mud piled on edges of drainage holes create diversity and produce new ecological habitat conditions for various kinds of wildlife.

Water developments involving drainage, channelization, and large-scale impoundments usually produce undesirable effects on wildlife populations and their habitat. With planning, many kinds of water developments can benefit wildlife without interfering with their primary purpose.

15.3.4 Space

It is a behavioral requirement of wild animals and considered as a social response developed with the passage of time ensuring animals' welfare. Space requirement for a wild animal is specifically for **movement**, **lifesaving**, **mating** and for overall **survival** and **well-being**. Space requirements may vary depending on:

1. **Species**: Large animals especially carnivores, felines, hyenas require abundant space for the overall territorial control.
2. **Food quantity and quality**: Abundant and quality requirement of food will mean larger space exploration.
3. **Cover**: More cover area, more availability of food.
4. **Water**: Areas with abundant water supply will have more food availability.

In addition, some other factors influencing space requirements of wildlife include:

1. Animal size: Requirement of specie is more in large animals.
2. Animal's food preferences: Space requirement for food availability is more in carnivores than herbivores.
3. Ability to withstand and adapt crowded conditions.

15.4 Territoriality

Territoriality is the defense of an area from intrusion by other individuals. Territories are maintained when the resources gained are rewarding and defensible. Territoriality also arises in a group living animals like in red deer populations, when resources are scarce and unevenly distributed during the rut, the spatial utilization by females creates locations that are deemed valuable and worthy of defense. For ungulates, defense of an area is a low benefit–low cost mating strategy, favored under the conditions. In this context, the potential survival costs linked to alternative mating strategies are expected to be notably increased. Territoriality is an important and dynamic animal behavior which is governed together by the different ecological and animal behaviors; for example, Sacks et al. (1999) found that breeding populations of *Canis latrans*, a coyote's species found in the western part of the USA ecosystems, generally use mutually exclusive territories. Animal density, generally the group size in a habitat, also affects the territorial behavior Sparklin et al. (2010). It was noted that sounders exhibited nearly exclusive home ranges with entirely exclusive core areas, indicating that female feral pigs displayed territorial behavior at the sounder level rather than at the individual level. In parturition stages of white tailed deer species, it was observed that the mother protects the territory rigorously up-to a certain time of the fawn (Ozaga et al. 1982), whereas, social structure among various mammal classes also reflect in their behavior of territoriality and its defense (Pitt et al. 2003).

15.4.1 Types of Territory

1. Multipurpose territories: for example, breeding season, mating, nesting, feeding (songbirds).
2. Restricted to mating and nesting only: for example, red-winged blackbirds, tree swallows.
3. Restricted to mating only: for example, sage grouse.
4. Restricted to nesting only: for example, colonial water birds.
5. Restricted to non-breeding: for example, feeding in robins during winter.
6. Group activities: varied reasons: for example, in acorn woodpecker.
7. Restricted to roosting only: for example, mixed flocks such as starlings, blackbirds.

15.4.2 Model of Territoriality

A territoriality model was developed by Brown in 1964, which stated "in situations where resources are limited and the population is high, intense competition arises for food. The greater the demand for resources, the more pronounced the impact on population size." Intense competition necessitates heightened aggressiveness in securing essential resources. However, if maintaining territories comes with a prohibitively high economic cost, individuals may refrain from holding them.

Behavior that is generally utilized by animals to optimize their chances of survival and reproduction is called as aggressive behavior. Individuals within a population are often favored by such behaviors especially when their goals are consistent and easily accessible. Such behavior if consistent and easily accessible favors individuals within a population. For example, food supply that is generally mobile and transient cannot be defended; in such situations, no territorial system is evolved to defend it; and the territory, if present, may be restricted only to the nest and the area reachable by the parents on the nest. Instances of such behavior are observed in colonial seabirds, nomadic and socially feeding passerine species, as well as aerial feeders. In these species, achieving the objective of an increased or assured food supply is improbable through aggressive behavior. Conversely, when an individual relies on a relatively stable and clearly defined area for nesting, securing a food supply, and attracting a mate, this crucial space is usually defensible and transforms into what is commonly known as a classical territory.

The dependability of essential resources such as food, mates, mating locations, nests, or other necessities for reproduction or survival stands out as a crucial factor influencing the development of territorial behavior through natural selection. The concept of "dependability" encompasses considerations related to an individual's time and energy budgets, in addition to purely physical factors. Since intraspecific aggressiveness primarily stems from behavioral responses to competition for limited ecological resources, the primary factor driving increased aggressiveness through natural selection is competition. The intensity of competition is directly tied to population density and inversely related to the availability of the contested resources.

This relationship is complexly entwined with productivity, birth rates, mortality, and all ecological and species-specific characteristics influencing them. Excessive aggression in the absence of a scarcity of the contested resource would ultimately be detrimental. Thus, a delicate balance must be struck between the positive benefits of acquiring food, mates, nesting areas, family protection, etc., and the negative consequences of time intervals and energy expenditure, missed opportunities, and the threats of injury.

Determining where this balance lies in a particular species is influenced by a myriad of factors, including population density, physiological limitations, nest construction and site requirements, distance to food from the nest, developmental stage of young at birth, foraging time needed to raise young, clutch size, time required to protect young, potential mate's response to varying levels of aggressiveness, susceptibility to predators, migration, climate, weather, bird size, and richness of the food supply. In summary, it is argued that the type of territoriality that evolves in a species hinges on the specific resources with existence of competition and the magnitude of economically defensible ability, considering the balance between the advantages and disadvantages of such defense for individual members, not just the population as a whole.

15.4.3 Purposes of Territories

15.4.3.1 Territories in Relation to Reproduction

(a) **Preparation of nest sites**: For example, Nest holes (nuthatches, bluebirds, wrens); nest posts or nest tree snags (osprey); nest cliff ledges (sea birds, peregrine falcon). In this scenario, nest sites are carefully selected, and territories are established around them. In an experiment conducted by Sargent (1982), the primary factor influencing female nest-choice and reproductive success was the quality of the nest territory, specifically in terms of nest concealment by male sticklebacks. Conversely, the house sparrow exhibits colonial nesting behavior, and its territorial behavior is constrained to the defense of the nest site (Smith 1958).

(b) **Formation of pair**: Male birds, such as yellow-headed blackbirds and red-winged blackbirds, can attract females by advertising and showcasing a quality territory, possibly including a suitable nesting site. Migratory stonechats (*Saxicola torquata*) spend the winter in the northern Negev desert in Israel. Upon their arrival in October, most birds establish and defend territories throughout the winter, primarily in the form of heterosexual pairs (Gwinner et al. 1994). Solitarily breeding pairs *Coccothraustes coccothraustes* defend a small territory surrounding the nest for the whole of the nesting period (Mountfort 1956).

(c) **Opportunities for females to select best male**: In polygynous species, the quality of a territory and the quantity of females already settled can have a crucial role in influencing female mating decisions (Orians 1969).

15.4.3.2 Territories in Relation to Habitat

This set of behavioral conditions in wildlife allows the increased inbreeding within a population which serves as a means of readily and intricately adapting to a given local set of conditions (Gopal 2018).

15.4.3.3 Territory in Relation to Food

It facilitates animals to become more familiar in a habitat. It provides time and energy for mating and predation, food availability to younger ones, and resource partitioning between dominant and subordinate individuals.

15.4.3.4 Territories in Relation to Predation, Disease and Parasitism

(a) **Disease reduction and less predation**: This is likely an outcome of territorial behavior rather than a cause for it. It reduces the chances of disease transmission.

15.4.3.5 Territories in Relation to Numbers

(a) **Less aggression**: Once territories are established, birds establish a clear understanding of their relationships with one another, and territory maintenance is implemented, thereby aiding in the prevention of competition for resources.

(b) **Resource stability**: It helps in dispersal of animals, which result in the proper utilization of the habitat.

(c) **Population control**: Experiments conducted over 6 years with red grouse revealed that breeding stock size was influenced by territorial behavior (Watson and Jenkins 1968).

15.4.4 Size and Shape of Territory

The dimensions of territories exhibit variations influenced by individual characteristics, species attributes, and environmental factors. While most territories tend to be roughly circular, their shapes can vary. Linear territories are observed in stream bank feeders like kingfishers. Birds that primarily consume animal-based food maintain larger territories compared to those primarily feeding on plant-based resources. As such, there are both maximum and minimum sizes for territories, with the maximum size dictated by defensibility considerations and the minimum size determined by resource density.

15.4.5 Aggression and Territory

The level of aggressiveness in territorial defense varies, with a tendency to strongly defend a high-quality territory over weakly defending a poor one, as it may offer more advantages. Interspecific competition between different species is also observed. Typically, birds of one species tolerate individuals from other species in their territories and only expel those of the same species. For instance, a bald eagle and a great horned owl nested in the same tree in Florida with just a 1-m separation. Swallows often nest in the nests of eagles and hawks. Birds commonly drive away other birds from their territories if they are perceived as potential competitors or predators. For example, oystercatchers will chase gulls, ravens, and crows from their territories.

15.4.6 Territoriality Influence

Territorial behavior can exert a significant impact on populations, potentially compelling certain individuals into suboptimal habitats where breeding becomes challenging. This may result in increased mortality among adults and/or young individuals, or a decreased success rate of hatching/fledging. A study in Finland focusing on the willow warbler, which exhibits a preference for spruce-birch forests, revealed that during years of low species abundance, almost all breeding warbler pairs were located in the spruce-birch forest. In times of high abundance, the warblers were also found in other habitats, but their highest density persisted in the spruce-birch environment. Consequently, population densities tend to be much higher in suboptimal habitats during years of abundance. In such years, territorial numbers increased only marginally, and territory sizes slightly decreased, allowing more birds to fit into optimal habitats. Thus, territorial behavior can lead birds to occupy suboptimal yet adequate habitats during periods of abundance.

15.4.6.1 Home Range

Majority of animals are not nomadic and instead inhabit relatively restricted areas where they carry out their daily activities. These areas are referred to as home ranges. Burt (1943) defined a home range as "That area traversed by the individual in its normal activities of food gathering, mating, and caring for young."

15.4.6.2 Need of the Home Range for Animals

1. Avoiding or escaping predators
2. Maneuvering through its own environment
3. Providing more benefits than the incurring cost

15.4.6.3 Home Range Sensitivity

1. Year
2. Season
3. Lifetime

Animals can be acquainted with areas that they do not actively utilize. For instance, an Arctic fox (*Alopex lagopus*) may be familiar with expanses exceeding 100 km^2, while confining its actual usage to a smaller portion, approximately 25 km^2, within those areas.

15.4.6.4 Home Range Edges

Many animals have the peripheries of their home ranges as regions that they are familiar with but use sparingly. The precise boundaries of the home range may not be of significant concern to the animal, particularly if it predominantly spends the majority of its time in other areas. Except for certain territorial animals, the interior of an animal's home range is frequently more crucial both to the animal itself and to gaining insights into how the animal lives and the reasons behind its choice of habitat.

15.4.6.5 Estimators of Home Range

1. Utility distribution
2. Grid location
3. Minimum convex polygon
4. Fourier series
5. Circle and ellipse approach
6. Harmonic mean distribution
7. Fractal estimators
8. Kernel density estimators (Corsi and De Leeuw 2000)

15.4.6.6 Home Range Core

The core of a home range comprises two essential components. First, it must be utilized more intensively than the apparent clusters of high activity that result from a uniform and random use of space within the overall home range. Second, the core must not be exclusively determined by the overall size of the home range (Table 15.1).

Table 15.1 A comparison between home range and territory

Home range	Territory
Area where animal spends his/her time	An area which is defended by the animal which controls it
Different territories form a single home range	Complete control of an individual or its family
No exclusive right over the area, that is, home range	Exclusive right over the area, that is, territory
Home range can be shared with other species also	Exclusive and preferential use of the space by animal
Delineation of boundaries are easily recognized within an animal's home range	Identification of boundaries is difficult
Home range varies with food and other biological factors, for example, polar bear	Territory remains same of an animal through majority of animal life

15.4.7 Ecological Niche

The fundamental role of a living organism in an ecosystem, especially within its own habitat, is known as the ecological niche. This concept outlines how a species interacts with its environment. The ecological niche of a species is influenced by both biotic and abiotic factors, shaping the species' capacity to survive and thrive. Joseph Grinnell, a field biologist, expanded upon the basic notion of the niche. Biotic factors encompass elements such as food availability and the presence of predators, while abiotic factors involve temperature, landscape features, soil nutrients, light, and other non-living components.

15.4.7.1 Types of Ecological Niche

According to Ecologist Charles Elton's niche emphasized

1. Community similarity
2. Competition

Zoologist G. Evelyn Hutchinson introduced two forms of niche. First, the fundamental niche emphasizes the circumstances under which a species could exist without any ecological interactions. Second, the realized niche considers the population's existence in the presence of interactions, such as competition.

15.4.7.2 Importance of Ecological Niche

1. Ecological niches provide insights into the relationships between communities and environmental conditions, fitness, trait evolution, and predator–prey interactions within ecosystems.
2. The concept of ecological niches enables species to coexist in their environment by thriving under suitable conditions and playing distinctive roles.
3. Presence of ecological niche is crucial for biodiversity, ensuring a balanced ecosystem. Without them, biodiversity would diminish, leading to an imbalance in the ecosystem.

15.4.7.3 Competitive Exclusion Principle

To comprehend the coexistence of species, the Competitive Exclusion Principle is established, which declares that two species cannot coexist in the similar ecological niche due to competition for resources within a habitat. The competition within a niche prompts each species to specialize in different ways, avoiding the use of identical resources, or otherwise this might lead to the extinction of any of the competing species. This perspective aligns the natural selection concept. Two theories are commonly employed to address competitive exclusion.

R Theory: Coexistence of multiple species sharing the same resources is contingent upon niche differentiation. When resource density reaches its nadir, species populations most constrained by the resource are prone to competitive exclusion.

P Theory: Consumers can thrive at high densities by sharing common targets. Resource partitioning involves the separation of resource utilization. In simple terms, species can adapt their resource usage patterns to minimize depletion. This adaptive behavior facilitates the coexistence and evolutionary development of species.

1. An example of resource partitioning is observed in lizards like anoles. They utilize different parts of their overlapping habitats in distinct ways. Some anoles inhabit the forest floor, while others reside high in the canopy or along trunks and branches. Some may even venture into non-plant environments, such as deserts or coastal areas.
2. Dolphins and seals, though consuming similar species of fish, exhibit resource partitioning through differences in their home ranges. This spatial separation allows them to effectively utilize shared resources.
3. Darwin's finches exemplify resource partitioning through the specialization of their beak shapes over evolutionary time. This adaptation enables them to exploit resources in varied ways.

15.4.7.4 Niche Construction Theory

The niche construction theory elucidates how organisms modify their environments to better suit specific niches. Examples include constructing burrows, building nests, creating shade, and building beaver dams. Biologist John Odling-Smee first introduced the concept, asserting that niche construction should be viewed as an evolutionary process—a form of "ecological inheritance" passed on to descendants, rather than a genetic inheritance. Four core principles that underlie niche construction theory are as follows:

1. **Non-random modification**: Involves the non-random modification of the environment by a species, playing a role in aiding their evolutionary process.
2. **Altering skills**: The "ecological" inheritance encompasses the alteration of evolution, with parents passing on skills that modify the environment to their offspring.
3. **Adoption**: New characteristics adopted through niche construction become evolutionarily significant, systematically affecting environments.
4. **Adaptation**: Adaptation is essentially the outcome of organisms making their environments more complementary through niche construction.

Ecological niches provide a valuable source of information for understanding species interaction with environmental variables. Ecologists can leverage this knowledge to gain insights into effective species management, conservation strategies, and future development planning. By comprehending the intricacies of ecological niches, ecologists can make informed decisions to promote the well-being of species and ensure sustainable environmental practices.

15.4.8 Wildlife Habitat Improvement and Its Management

Wildlife habitats are expanses distributed both horizontally and vertically across the landscape, catering to the specific needs of a particular wildlife species. These habitats serve as essential areas for fulfilling the basic requirements of food, water, reproduction (nesting), and protection against predators and competitors (cover). Any improvement or modification in such areas (habitats) for management of wildlife is termed as wildlife habitat improvement. Its management is aimed with two objectives:

1. The first goal is to preserve the quality of habitats as they naturally exist within ecosystems, ensuring their integrity and sustainability.
2. The second goal involves creating or restoring quality habitats in areas where they have deteriorated or lack specific components essential for wildlife, such as water, food, or shelter. This aims to enhance and support biodiversity in those environments.

The planning and implementation of habitat management should incorporate the following fundamental principles:

1. It should be substantiated by a thorough examination of biological requirements.
2. The impact of suggested methods on additional natural resources and land utilization needs assessment.
3. Economic viability is essential, and the stated goal, whether it be preservation, enhancement, or complete transformation of the current habitat, should be clearly outlined.
4. Enhancements should imitate natural conditions, with a general emphasis on sustaining native flora and fauna.
5. Projects involving manipulation should adhere to natural topographical features rather than adopting geometric squares or strips.
6. Periodic evaluations are necessary to assess the achievement of set objectives.

15.4.8.1 Food and Cover Production

Here, the goal is to improve habitat by providing food and/or cover for a particular species, or group of species. Over all, there are three major methods:

Propagation The cultivation of food and cover species for wildlife can be costly, with outcomes often unpredictable. Economically speaking, there is no viable substitute for the natural regeneration of native species. Whenever feasible, management efforts should focus on preserving or enhancing existing native species, as they are well-adapted to the environment and have a higher likelihood of thriving. Exotic species typically exhibit lower initial survival rates, frequently necessitate more intensive cultural care, tend to grow slowly, and may produce fewer seeds. When the introduction or restoration of species becomes necessary, this can be achieved through direct seeding or the use of transplants. Key factors contributing to a successful plantation include site selection, site preparation, planting depth, and soil moisture. Optimal results are anticipated on sites with a historical record of supporting the relevant species.

Direct seeding/planting: Certain native and exotic woody species suitable for habitat improvement can be acquired from commercial nurseries, while others may be found growing in the wild or propagated. Regardless of their source, transplants must be kept adequately moist until they are planted. Planting procedures should adhere to accepted nursery practices. Some land management activities that present cost-effective opportunities to establish forage or cover include forestry practices like thinning, harvesting, and post-harvest treatments, utility transmission corridors, soil stabilization projects (especially after events like fires, road and ski-slope construction, and surface mining), range improvements such as reseeding and brush control, and any other project that alters the vegetative cover. It is often possible to select species that best meet the wildlife habitat requirements.

Regeneration: Effective coordination between timber management and wildlife conservation hinges on the strategic scheduling of timber harvests using small land units. Clear-cuts should strike a balance—being sufficiently large to limit consumption of tree reproduction by deer and other wildlife, yet small enough to avoid significant harm to wildlife adapted to old-growth forests, such as squirrels and wild turkeys. Many procedures employed to free desirable browse plants from the competition of less desirable species mirror those used for the complete removal of existing vegetation. The outcomes of such treatments are contingent upon the intensity of application. For instance, chemical sprays may be employed to desiccate the crowns of woody species or to completely eliminate plants, depending on the strength and number of mix and applications, respectively.

Rejuvenation: Numerous shrubs and trees have the ability to regenerate through the sprouting of adventitious buds on the stem or from the root crown. Seeds of chaparral and certain other species display heat resistance, leading to abundant germination following a fire. In instances where these species become too tall, dense, or decadent to provide accessible browse, rejuvenation can be achieved through controlled burning. For many species, it has been observed that the sprouts and young plants exhibit significantly higher levels of protein and other nutritional values for several years post-burning compared to older growth stages. However, some shrubs and trees are susceptible to fire and may not naturally re-establish for decades after a hot burn.

Treatment Various physical treatments, such as crushing, cutting, or mowing, can be employed for vegetation management. Similarly, chemical spraying that burns back the tops without killing the shrubs has a comparable impact on various types of woody plants. In such cases, the root systems generally remain largely undamaged, prompting the plants to respond with rapid and extensive root and leaf development after the reduction of aerial growth. However, there is evidence suggesting that deer may be hesitant to heavily browse on plants where numerous dead stems are intermingled with new growth. The presence of dead stems may pose a disadvantage if moderate-to-heavy browsing is necessary to maintain growth at heights suitable for browsing.

Release At different localities, the growth and seed production of certain species can be enhanced by eliminating surrounding competition. Strategic measures like small group selection cuts or row thinning can be implemented to increase the availability of light, moisture, and nutrients for potential understory species that produce seeds or serve as browse. In stands where seeds are not mature enough for commercial timber sales, specific stems of vital seed and browse species can be freed from competing vegetation through methods such as fire (with species like sassafras and flowering dogwood displaying vigorous re-sprouting after burning), herbicide application, or cutting. When it comes to release cutting of trees in older forest stands, the most effective approach is often in conjunction with some form of commercial timber harvest.

In oak-dominated ecosystems, the selection of trees for release should be done in the fall, particularly during a year with abundant seeding, to ensure that the released trees will produce seed crops. Additionally, certain species crucial for wildlife, like hickories and American beech, tend to grow slowly and are often overshadowed by the robust growth of intolerant tree species that typically thrive after clear-cutting. To allow these slow-growing species to attain seed-bearing size in clear-cut stands, it is essential to release them from competition.

Another approach to boost fruit production involves selecting specific trees, such as wild or abandoned orchard apple trees, wild cherry, hackberry, oak, or hickory, and applying orchardist techniques to increase fruit yield. This not only allows the tree to produce more fruit, but it also opens up the interspaces for enhanced production of grasses and shrubs. In the Western regions where many fruit-producing chaparral species like manzanita, California redberry, and toyon are found, decadent stands can be revitalized through methods such as mechanical crushing, chemical spraying, and controlled burning.

15.4.8.2 Cover Management

Cover serves diverse habitat requirements for wildlife. For instance, a hedgerow might offer escape cover for quail against predators, while the same shrubs can serve as nesting cover for songbirds. Various items, such as rock piles, ground burrows, brush piles, or trees (including cavities), can contribute to providing cover. The absence, scarcity, or poor distribution of cover can limit the utilization of an area by wildlife. Wildlife managers can enhance wildlife numbers or expand the area of use by improving the quality or quantity of cover. When manipulating food or water, it's crucial for managers to ensure that sufficient cover of different types is retained to meet wildlife needs. Cover includes areas for escape, nesting, and refuge from harsh weather conditions. It often consists of vegetation in the form of herbaceous, shrubs, or trees, offering protection against hunters or predators, as well as mechanical or thermal protection from winter storms or summer sun, or a combination of these factors that create a secure nest site.

15.4.8.3 Water Management

Enhancing the quantity, accessibility, and year-round presence of water can be instrumental in increasing wildlife populations or broadening habitat utilization. Conversely, water can be strategically removed to control animal numbers and feeding in areas where their presence is undesirable.

Water development often serves purposes beyond wildlife-specific objectives. When planned thoughtfully, water projects can yield benefits for wildlife. Therefore, habitat managers should be well-versed in diverse techniques for water development, encompassing natural features like springs, seeps, and water holes, as well as man-made structures such as reservoirs, guzzlers, and wells. This comprehensive understanding allows managers to optimize water resources for the advantage of both wildlife and other land uses.

15.4.8.4 Water Holes

Water holes, whether natural or artificial, are open water storage basins. In certain areas, water is a principal necessity for wildlife, and as a result, water holes often serve as focal points for wildlife activities. Consequently, it is crucial to design and maintain these water holes to be accessible and beneficial for all species of wild animals.

15.4.8.5 Springs and Seeps

Springs share common developmental needs, but various planning techniques can be applied. Before developing a spring or seep, it's essential to assess the reliability and quantity of its flow. Typically, installing a protective box to capture and store the water is necessary. In cases where water flow is intermittent, it may be advisable to establish large-capacity storage so that stored water remains available when the spring or seep stops flowing. These water sources are best excavated from firm ground, hardpan, or rock to maximize flow. Regardless of the number of sources, the water should be directed to a collection basin and then piped to a trough. Fencing the water source and collection basin from human or livestock use is usually necessary and desirable.

15.4.8.6 Reservoirs and Small Ponds

In this context, the term "reservoir" refers to water impounded behind a dam, created either by constructing a dam directly across drainage or by enclosing a depression to one side of drainage and constructing a diversion ditch into the resulting basin. Reservoirs should be strategically designed to maximize storage while minimizing surface area to mitigate evaporation loss. Several key considerations when selecting reservoir sites are:

- Wildlife area where water demand is minimum, an excavated pond must be dug with reasonably ground flat. The design will allow low risk damage from excess water.
- Areas with steep slopes must have embankments, which although complex would ensure protection from erosion.
- Pond sedimentation and excessive soil erosion must be avoided by excavating the upland lands.
- Areas which support aquatic wildlife must have shallow waters at least 2–6 ft. deep for feeding purpose and 10 ft. for movement.
- Water birds like water fowl prefer a variety of depths for nesting and feeding, while amphibious and young fish are comfortable in shallow, plant-rich waters.
- Shapes vary from rectangular to rounded. However, generously indefinite shapes are more appealing to wildlife.
- If multiple objectives for wildlife management are sought, deeper water levels take precedence.
- The depth is subject to season and varies between winter and summer.

15.4.8.7 Catchment Water

Over the past two decades, various self-filling watering devices have been planned for the benefit of wildlife. While a significant number have been primarily constructed for quail, such structural designs have also been specifically built to aid other wild animals, including antelope, bighorn sheep, deer, sage grouse, and turkeys. The California Department of Fish and Game, for example, implemented over 2000 catchments for quail between 1943 and 1974. Given that a substantial portion of these devices were installed for upland game birds (*Galliformes*), they have been commonly referred to as Gallinaceous Guzzlers or, more recently, simply as Guzzlers.

Development of Wetland

A wetland is typically an area of land that is either submerged in water or consistently saturated with water, whether throughout the entire year or only during specific seasons. The depth and duration of this seasonal flooding can vary. Wetlands are often referred to as transition zones because they exhibit characteristics of both dry land and underwater environments (Mitsch and Gosselink 2007). However, the most widely accepted definition of wetlands is of by the Ramsar Convention. Established in 1971 in Ramsar, Iran, this convention is the sole global treaty specifically dedicated to wetlands. Currently, 170 nations are signatories to this convention. According to the convention on wetlands, wetlands are classified as "*areas of marsh, fen, peat, and/or water, whether natural or artificial, permanent or temporary, with water that is static or flowing, fresh, brackish, or salt, including areas of marine water the depth of which at low tide does not exceed six meters.*"

As of 2020, India has 42 recognized Ramsar Sites, which include wetlands considered of international importance under the Ramsar Convention. According to WWF India (World Wide Fund for Nature), wetlands are among the most endangered ecosystems in the country. Factors such as loss of vegetation, salinization, excessive inundation, water pollution, invasive species, and extensive development and road construction have collectively contributed to the deterioration of India's wetlands.

Wetlands are characterized by any one of the following three attributes:

1. Periodically, the land supports predominantly hydrophytes (aquatic plants).
2. The substrate is predominantly undrained hydric soil.
3. The substrate is saturated with water or covered by shallow water at some point during the growing season of each year (Balwan and Kour 2021).

The development of water areas and techniques for improving wetlands can vary, contingent on factors such as the existing structural development of the area, water quality, water level management, soil composition, climate, topography, and plant succession. Enhancing wetlands for wildlife can involve the manipulation of biological and physical forces to create an improved environment.

Before initiating any plans for wetland improvements, it is essential to establish a biological need. The primary consideration in judging the potential development of an area should be its chief use or uses, which are largely determined by the location and physical characteristics of the area. Some areas may be best developed for waterfowl, while others may be more suitable for muskrat or other fur production. In certain cases, these features can be combined.

Habitat managers can employ various practices to achieve the interspersion of open water with marshland, interlace ditches and high spoil lands, and plant vegetation for food and cover. This approach creates wetlands favorable to a variety of wildlife, including ducks, geese, beaver, muskrats, mink, and warm-water fishes. For habitat managers focused on the preservation, management, or manipulation of wetlands, careful consideration of these factors is crucial, we recommend the Techniques Handbook of Waterfowl Habitat Development and Management published by the Atlantic Waterfowls Council (1972).

15.4.8.8 Potholes, Sumps, Ponds

Potholes are defined as small, shallow, open water retention areas or basins with surface areas typically under 0.016 km^2. When developed alongside larger, permanent water areas, potholes become a valuable tool in waterfowl management. The purpose of creating potholes is to counteract the loss of water area due to geological changes and plant succession. An ideal wetland for waterfowl typically consists of one-third open water and two-thirds marsh.

Ponds are a widely used system for water management, constructed in various ways. Watershed ponds result from building a dam to confine runoff, either from rainfall overland flow or an existing water source. Alternatively, ponds may be excavated or formed by building an earthen embankment, often requiring an external water supply from sources such as streams, wells, or irrigation systems (Boyd and McNevin 2005).

Coastal plain ponds, considered globally significant water resource habitats, feature gently sloping shorelines and fluctuating water levels seasonally and yearly with water table's height as base. These fluctuations are critical for the survival of rare pond shore vegetation communities, hosting diverse flora including carnivorous plants like bladderworts (*Utricularia* spp.), sundews (*Drosera* spp.), and various rush species (*Juncus* spp.).

In context of wildlife and its management, a large Heron and Egret colony in the Stillwater Wildlife Management Area in Nevada, USA, benefits from a marshy waste water sump supplied by drain water from the Carson and Truckee rivers. This has improved food and water quality, benefiting the bird populations (Giles and Marshall 1954).

However, in contrast to water management through artificial practices like reservoirs, holes, and sumps, there is a guiding principle that natural population dynamics in conservation areas should not be interfered with. Wildlife populations must undergo natural stressors such as droughts and diseases, and should be impacted by cycles of lower survival rates and densities (Karanth and Kumar 2013).

15.4.8.9 Green Tree Reservoirs

Green tree reservoirs refer to bottomland hardwood areas that are intentionally shallowly flooded for short periods during the dormant growth period. This practice is specifically designed to attract avian populations, especially waterfowl. These reservoirs provide a habitat that meets the needs of birds, particularly during certain phases of their life cycles, for their contribution to the overall conservation and management of avian species in bottomland hardwood ecosystems.

15.4.9 Wildlife Corridors

The increasingly significant aspect of conservation of wildlife involves extending its focus beyond the boundaries of protected areas (PAs) to encompass the larger landscape in which such areas are situated. A landscape, in this context, is defined as "a large tract of land constituted by a mosaic of interacting land uses with people and the impacts of their activities as the cornerstone of its management" (MoEFCC 2017). While PAs, along with reserve and protected forests, constitute 20% of India's bio-resource management, the surrounding areas, often designated as production and human habitation areas, frequently lack dedicated biodiversity management interventions.

Therefore, adopting a landscape approach becomes crucial, aiming for an integrated and comprehensive management method that covers the entire matrix of land use forms, including PAs. This approach considers multiple competing demands, ensuring the viability of species populations into the future. In terms of management planning for PAs, a landscape approach enables ecosystem-level conservation actions at various nested spatial scales, including PAs and territorial forest divisions, as well as larger units. This strategy aims to achieve conservation goals at the broadest spatial scale feasible in practical terms. Landscape-level conservation of species is viewed as maintaining or enhancing genetic exchanges between meta-populations, significantly improving the prospects for their long-term persistence (MoEFCC 2017).

Wildlife corridor is thus defined as a patch of habitat that connects two or more isolated habitats. This includes landscape, hedge grows/fence grows, and buffer strip vegetation. These corridors provide some important types of wildlife corridors used for providing space and food. The principal aim of wildlife corridor management is interconnection of wildlife population. The wildlife corridor is thus able to provide the following connectivity within the landscape of diverse land use land cover (LULC) attributes:

1. **Structural connectivity**: A metric assessing the extent to which specific landscape features are connected or adjacent to one another, devoid of any considerations related to species movement or direct links to the behavioral attributes of organisms.
2. **Functional connectivity**: A metric of connectivity that takes into account the species-specific reactions to various landscape elements and considers the real movement of individuals through the matrix connecting habitat patches (Taylor et al. 2006).

Depending on the degree to which species-specific responses or behavior are integrated, it can be further classified into follows:

1. **Potential functional connectivity**: A metric of connectivity that relies on the physical characteristics of the landscape to anticipate connectivity among patches for a species, often with restricted information regarding dispersal ability or specific animal movement data.
2. **Realized functional connectivity**: A metric of connectivity that integrates observed data, such as radio tracking or molecular genetic data, to represent the real rates of movement of individuals (or their genes) between specific patches or throughout a landscape. This approach is used to estimate species-specific connectivity between various landscape elements or habitat patches (Fletcher et al. 2016).

15.4.9.1 Wildlife Corridor Connectivity Issues

The primary challenge in achieving wildlife corridor connectivity is coordinating among the various agencies tasked with regulating or promoting land use within these landscapes. The nature of land use and land cover practices in such areas not only involves competition but may also inadvertently conflict with biodiversity management and conservation goals. The matrix within which protected areas (PAs) exist is progressively degraded and fragmented due to land use changes and development pressures, including roads, railways, linear infrastructure development, mining, and urbanization. Large-landscape dependent species such as tigers and elephants navigate through these adjoining matrix landscapes, which are often inhospitable and contribute to situations of human–wildlife conflicts. The consequences of these conflicts result in a loss of connectivity between wildlife habitats, posing a significant threat to wildlife vulnerability.

Another key issue of landscape connectivity is the anthropogenic resistance resulting in the impact of human behavior on species movement. These anthropogenic behaviors are governed by either psychological (individual) or social (group) factors (Lischka et al. 2019) and/or policy decisions (legality); these together can affect the movement of wildlife in different ways. The above anthropogenic behavior can further be classified into the following (Ghoddousi et al. 2021):

1. Attitude: It can be positive or negative evaluation(s)
2. Social identity: The hierarchical grouping of the persons
3. Norms: Also called as social norms, that is, activities that is socially desirable
4. Values: The core belief that is rooted in their culture
5. Risk generated: It is the risk perceived or real to the loss of humans
6. Economic gain: Consumptive gain from legal or illegal harvest

However, anthropogenic resistance is not entrenched in livelihood generation only. For example, despite recognized livestock losses and human fatalities, in the Khata Wildlife Corridor in Nepal, local communities generally hold positive attitudes toward tigers, *Panthera tigris* (Wegge et al. 2018).

Economic incentives, including compensation and community development efforts, can play a vital role in facilitating the corridor use by tigers, ultimately enhancing its overall functional connectivity. Similarly, Masai pastoralists in the Amboseli region, bordering Tanzania and Kenya, have supported lion (*Panthera leo*) connectivity through interventions such as compensation and participatory conflict mitigation schemes, despite experiencing considerable livestock depredation events (Hazzah et al. 2014). Beyond economic incentives and awareness, factors like cultural or religious beliefs may contribute to a reverence for wildlife. Consequently, understanding the drivers of human behavior, which go beyond what is detectable or quantifiable through commonly used biophysical indicators like land cover maps derived through remote sensing, can enhance our comprehension of true connectivity dynamics (Norton et al. 2010).

One crucial aspect of wildlife corridor management is understanding and modelling the flow of genetic information, processes, and organisms between the areas of unimpaired habitats to design a network of wildlife corridors, for example, Tigers (*Panthera tigris*) in central India and Eastern Ghats landscape complex (Shanu et al. 2008). The application of the concept of minimum spanning tree obtained from the weighted graph in the focal landscape suggested a viable corridor network for the tiger population of the protected areas (PAs) in the landscape complex. Centralities of the network identify the habitat patches and the critical parameters which are central to the process of tiger movement across the network. As a result, the ecological information propagates expeditiously and even on a local scale in these networks representing a well-integrated and self-explanatory model as a community structure.

Noss and Daly (2006) have recognized the methods to identify and design the wildlife corridors. These are:

1. Shortest or most direct routes.
2. Only remaining routes.
3. Routes incorporating localities of conservation importance.
4. Pathways determined by the expertise of individuals knowledgeable about the specific focal species.
5. Approaches that blend various expert-based methods.
6. Based on empirical and modelling approaches.
7. Least cost path analysis.
8. Population modelling.

Section 380 (g) 38v of the 1972 Wildlife (Protection) Act recommends ensuring the ecological integrity of wildlife corridors. However, in the current context of India's protected area network (PAN), which is embedded with diverse land use land cover practices, socio-economic dynamics, and population pressure, this protection as envisaged in the WPA, 1972 could be further elaborated in the following:

1. Bulk of the Indian protected areas is small in size and susceptible to anthropogenic pressures. There must be uninterrupted processes of evolution and creation of buffer zones around them.
2. Most of the wildlife corridors are under severe degradation and fragmentation due to pressure and unplanned developments impacting the connectivity and mobility of animals.
3. Unplanned tourism and commercial activities in the already established buffer zones.
4. Keeping the above context in mind, considerations for declaring these areas as "eco-sensitive zones" must be implemented.
5. Wildlife conservation strategy (2002) envisages that lands falling within the radius of 10 km of the boundaries of protected areas should be notified as Eco-Fragile Zones.

15.4.10 Management of Migratory Birds

Migratory wildlife often presents difficult management challenges. In such situations, management requires coordination between the political units so that a species is not overexploited when it falls under the temporary jurisdiction of one state or a nation. A treaty signed between the USA and Great Britain in 1916 known as *The Migratory Bird Treaty* effectively protected all migratory birds crossing the borders of both nations. The International Union for Conservation of Nature and Natural Resources also known as IUCN and headquartered at Switzerland advocates for the promotion of wild plants, animals, migratory birds, and habitat conservation. Tracking the bird migration along the different routes across the seasons has been an old science. Traditional knowledge, derived from seasonal calendars, ecological indicators, seasonal variations, and the associated activities of birds, serves as a foundational source of information. This knowledge is crucial for comprehending the patterns of bird migration, particularly in the context of climate change (Yang et al. 2019). Apart from this, documenting the occurrence of disease transmission in migratory birds particularly those in close contact of human settlements is essential in understanding the potential risks to bird health as well as humans. A study in this context was conducted by Haq et al. (2021) and unveiled the presence of two proteobacteria (*Pseudomonas fluorescens* and *P. syringae*), three Firmicutes bacteria (*Bacillus pumilus, B. subtilis, B. licheniformis*), and one avian gastrointestinal Enterobacter (*Escherichia coli*). It is noteworthy that the identified *Escherichia coli* possess the potential to be pathogenic and pose harm to humans.

Among the all different conservation solutions, recognizing the contribution of zoos and aquariums in the conservation of migratory birds is considerable (Hutchins et al. 2018). In addition, migratory birds enroute to their final destinations require habitats for rest and refuel called as stopovers. However, a comprehension is required to establish the facts of determination of these bird stopovers during different seasonal migration. For example, Mexico Gulf and Atlantic coasts of the United States of America are important stopover corridors for migratory birds of North America. Studies have shown that migratory birds tend to concentrate along the coast during spring and in inland forested landscapes during autumn. This suggests seasonal variations in habitat functioning, emphasizing the need to comprehend the ecology and role of stopover habitats in sustaining migratory bird populations (Emily et al. 2021). Another important aspect is the impact of change climate on migratory bird assemblages, their population trends, and changes. Several studies have indicated that climate change-induced phenological attributes including senescence of forest vegetation have a long-term impact on the migratory bird populations (Beresford et al. 2019).

15.5 Habitat Selection

Heterogeneity both spatially and temporally is the foremost attribute of a habitat for most species. A habitat can be defined as a specific type of environment characterized by general physical features, for example, type of vegetation, water, or soil structures. Additionally, a habitat can be classified as the specific place occupied by a species, taking into consideration the general features of the areas used by its individuals, ensuring suitability for the species' activities. At a more detailed level, a habitat can also be characterized by the specific portion of the environment dedicated to a particular activity of individuals, such as breeding or foraging. Typically, the term "habitat" does not include conspecifics, referring to the social components of individuals' activities, in contrast to the term "environment." A habitat patch is defined as a continuous and homogeneous portion of a habitat (Doligez and Boulinier 2008). Spatial and temporal heterogeneity in habitat can be linked to biotic, abiotic, and social factors. Spatial and temporal heterogeneity in habitat can be associated with biotic, abiotic, and social factors. Climatic conditions, soil nature, substratum, and level of salinity are some of the features of abiotic conditions of the habitat. However, food and nesting material, prey–predator dynamics, and parasitism are some of the features of biotic attributes. In addition, social components can also play a major role.

15.5.1 Habitat Selection and Its Implications

The implications of habitat selection are not governed by the human intervention only but it includes the geographical affinity, wildlife population of the area, and suitability matrix. Most importantly, it is important to acknowledge that the changes resulting from succession can create deteriorating conditions for certain species of animals. The suitability of an area which determines the planting of preferable vegetation impacts the overall habitat selection and its management. Following are some of the habitat selections with limitation for their selection:

15.5.1.1 Habitat Selection Based on Foraging

Foraging habitats differ quantitatively (food availability) as well as qualitatively (nutritional values) especially in a heterogeneous environment. Decisions related to foraging habitat selection encompass choosing patches to commence foraging and determining the duration of patch exploitation as resources progressively deplete over time. To address these aspects, simple mechanistic rules for patch-leaving decisions have been suggested and experimentally tested.

1. **Incremental rule:** Likelihood of staying in existing area (patch) is inversely proportional to unsuccessful search time. If given more time, individuals' success of finding the resources gets increased.
2. **Decremental rule**: Probability of staying in current patch is inversely proportional to successful search time.
3. **Rule of giving-up**: The inclination to remain in a patch decreases with the time spent searching for resources, unsuccessfully. However, each time a resource item is found, the inclination to stay is reset to a maximum level. Individuals ultimately leave the patch after a predetermined fixed unsuccessful search time, known as the giving-up time.
4. **Rule of fixed number**: Foraging is done for a fixed number of resources.
5. **Rule of fixed time**: Foraging is conducted for a predetermined period, and the individual departs from the patch regardless of the number of resource items found.

15.5.1.2 Breeding Habitat Selection

Individuals of a species must also make a series of habitat choices for breeding, similar to foraging. Breeding requires the availability of high-quality food resources for offspring development, in addition to other resources that influence breeding success, for example, the presence of mates and the availability of secure breeding sites. In some species, breeding decisions within a habitat are sequential in time and space and are independent of specific environmental attributes. Conversely, in other species, all necessary resources must be secured simultaneously, potentially leading to trade-offs between optimal choices for each resource. Many species have a limited number of breeding attempts over an individual's lifetime due to the longer timescales involved in breeding (up to several years) and the seasonal nature of this activity, implying yearly timescales. Consequently, constraints associated with breeding habitat selection often differ from those related to foraging habitat selection. Habitat choice during breeding may differ depending on the type of breeding site and the species' breeding ecology. This selection can occur at variable spatial scales, both in absolute values and relative values, contingent on the spatial range utilized by a given species.

15.5.1.3 Other Habitat Selections

Habitat selection, aside from foraging and breeding, involves individuals making choices in various situations. This occurs in species where mating transpires in a distinct location separate from the breeding activity, as seen in lekking species. For example, individuals might choose a spot close to a dominant male or within a well-lit area in low-light-intensity environments. Similarly, individuals might alternate between resting habitats or patches, where the primary resource is a secure site from predators or a location that enables them to optimize energy expenditure, such as protection against cold or rain. Migration can be viewed as an extreme form of habitat selection, where individuals change habitat due to seasonal variations in resource availability while their requirements remain constant. However, migration behavior constitutes a consistent habitat selection process, with individuals changing habitat in a similar manner year after year. During migration, individuals will select stopover areas, but this choice can be considered a traditional foraging habitat choice, subject to constraints such as energy optimization.

15.5.2 Management of Range Vegetation

The practice of range management is as diverse as the types of rangelands and animals and plants growing on them. In general, range management addresses the manipulations of treatments on vegetation such as spraying or by regulating grazing pressure. Range management involves many practices designed for improvement and regulation of vegetation both quantitatively and qualitatively for livestock and wildlife. These include mechanical and chemical means of managing dense and undesirable vegetation. Prescribed burning also can be a useful management tool on many rangelands to improve conditions for wildlife. Grazing systems, another important means of livestock replenishing, involves periods of rest for range vegetation. Depending on the length of rest periods and the area, the systems are known as deferred or short duration grazing systems.

15.6 Forest Management and Wildlife

Some methods of harvesting trees can enhance the habitats for varied kinds of wildlife; however, it's challenging to fulfill the requirements of all animals within the management plans of a single forest. For instance, some management plans are formulated to enhance species richness, while others may be specifically tailored for one or a few prominently featured species. Fire, now recognized as a part of the environment, has profound effects on the forest community. The general effect of fire is to set back succession to early stages favoring large mammals such as deer and moose. In some cases, silvicultural treatment (manipulations) of a pure forest stand is carried out for management of a wild animal, for example modification of Aspen (*Populus* spp.) stands is carried out for management of ruffed grouse. At present, forest management includes conflict recommendations from forest managers and other key stakeholders. Resolution of these conflicts will entail not only further biological data, but also imaginative management approaches coupled with social and political acumen. Increasing land use land cover practices and changes thereof leads to decrease in wildlife globally. However, in regions where forest area is constantly increasing, forest management tools are apparent and didn't suffer from habitat degradation, such regions are showing marked mitigation of climate change thereby having positive impact on its wildlife, and this is especially in parts of middle European areas or the north-eastern part of the United States of America. A biodiversity assessment for Germany indicated, for instance, that bird populations remained stable. Climate change, leading to longer seasons and milder winters, along with forest and its management practices that promote tree diversity showed a significant positive correlation with the abundance of non-migratory forest birds. These factors explained 92% of the variation in their abundance in Europe. In northeastern North America, non-migratory forest birds have witnessed long-term increases in abundance, and this upward trend is linked to management efforts. The rise in populations of non-migratory forest birds in both Europe and North America is associated with an increase in structural diversity and disturbances at the landscape level (Schulze et al. 2019).

15.6.1 Management of Endangered Species

These are those species facing extinction. However, threatened species are those potentially en route to endangered status. These species both plants and animals are listed in the Red Data Book by International Union for Conservation of Nature and Natural resources also known as IUCN. Reasons for their alarming status are (a) natural causes, (b) unregulated hunting, (c) introduced predators, (d) non predatory exotics, and (e) habitat modification.

Dealing with uncertainty is essential in the management of threatened and endangered species, thus presents following scenarios:

1. Knowledge of decisions in context to threatened species management
2. Knowledge how to make such decisions

The conservation and its management of species that are endangered involve comprehensive strategies, like intensive initiatives like captive breeding and foster parent programs, habitat preservation, predator and competitor control, and other customized approaches derived from species recovery programs.

15.6.2 Structured Decision-Making

In recent times, structured decision-making has been redefined as the "application of formal decision analysis tools to the management of natural resources." The array of analytical tools involved in this process is extensive and varied (Runge 2011). It includes:

1. Multiattribute utility theory
2. Information-gap theory
3. Expected value of information
4. Expert inference
5. Stakeholder engagement
6. Scientific and traditional

Decision analysis is five step processes including,

1. Context understanding
2. Knowledge of fundamental objectives
3. Sets of alternate goals development
4. Evaluation
5. Recommendation of a preferred action

15.6.3 Adaptive Management

A special case of structured decision-making where recurrent decisions under uncertainty are imperative is known as adaptive management. Many natural resource management settings, including threatened and endangered species management decisions, have two key features:

1. They are recurrent; in that the decision is revisited repeatedly.
2. They are plagued by uncertainty; in that the consequences of the alternatives are not known with certainty.

The first feature gives rise to the need for understanding the dynamic nature of decision actions taken now affect future decisions, along with the outcomes of those future actions. The second feature gives rise to the need to know on how to make decisions in the face of uncertainty.

15.6.3.1 Adaptive Management Applications for Threatened and Endangered Species

Applicability of successful completion of adaptive management application of threatened and endangered species management falls into three categories:

1. **Potential of the adaptive management**: Adaptive management application in natural resource management in general and in particular to threatened and endangered species has implied that it is critical method for management as it contains the basic normative ideas of conservation, especially with regard to such species.
2. **Challenges and impediments to implementation**: Importance in realizing the uncertainty in applying to subsequent decisions, especially in areas of population viability in an adaptive framework.
3. **Realization of actual implementation**: Application is the crux of overall success of this management, particularly to such species. Success stories like adaptive management of endangered species of Colorado river (Smith et al. 1998) and recovery of Trout cod in Basin of Murray Darling in Australia (Moore and Conroy 2006) are few highlights of the successful application of this management program.

15.6.3.2 Wildlife Management in Light of Indian Legislation

The indiscriminate and ill management of wildlife which resulted in the decline of important wildlife populations resulted in a necessity for conservation with a precautionary note. Before the enactment of the legislation for wildlife protection in India, number of rules made under the existing Indian Forest Act of 1927 and forest acts of respective states were in vogue. Certain other acts like Act of 1879 for Elephant Preservation, Act of 1912 for Wild Birds and Animal Protection, State of Tamil Nadu 1878 Act for Preservation of Wild Elephants and Bombay's Act of 1951 for Wild Animals and Birds were some notable legislation related to conservation of wildlife in Indian states. However, apart from Bombay's Act of 1951 for Wild Animals and Birds Protection, all the legislations were incomprehensively enacted with considerable deficiencies for wildlife conservation. They were mostly applicable to reserve forests; further there was no provision for controlling illegal wildlife trade.

The obvious limitations in the old legislations, coupled with decrease of wildlife areas in the country, necessitated the enactment of a comprehensive law applicable to conservation of wildlife and its related habitat throughout country. Thus, National Act of 1972 for Wildlife Protection was passed by the Parliament on 9th of September, 1972. This act was applicable to all Indian states and union territories, except Jammu and Kashmir, but since 2019, its applicability has also been extended to the jurisdiction of this erstwhile state of India.

15.6.3.3 Provisions in the Wildlife Protection Act of 1972

1. Chapter 1: (Preliminary) (Sections 1 and 2)
2. Chapter 2: Appointment of authorities (Sections 3–8)
3. Chapter 3: Hunting of wild animals (Sections 9–17)
4. Chapter 3A: Protection of specified plants (Sections 17A–17H)
5. Chapter 4: Protected areas and closed areas (Sections 18–38)
6. Chapter 4A: Central zoo authority and recognition (Sections 38A–38J)
7. Chapter 5: Trade or commerce (Sections 39–49)
8. Chapter 5A: Prohibition of trade (Section 49A–49C)
9. Chapter 6: Prevention and detection of offences (Sections 50–58)
10. Chapter 7: Miscellaneous (Sections 59–66)

Other notable acts and laws pertaining to forests and environment that are also essential for the protection and conservation of the wildlife resources across country are:

1. Indian Forest Act, 1927
2. Forest Conservation Act, 1980
3. Forest Protection Act, 1986
4. Environmental Protection Act, 1986
5. Indian Biodiversity Act, 2002
6. Forest Rights Act, 2006

Besides these, three national forest policies also envisage for the protection and conservation of wildlife under in situ conditions. These are:

1. National Forest Policy, 1894
2. National Forest Policy, 1952
3. National Forest Policy, 1988

15.7 Human–Animal Conflict and Its Management

The decline in wildlife, its species, and individuals has been a regular feature with the advent of human civilization vis-a-vis development. In fact, decline in the number of wildlife species has kept pace with human progress and technological advance. Numerous reasons including habitat fragmentation and consequent degradation, spread of urban areas at the cost of wildlife habitats, rail, road, dams, and agricultural and industrial operations have caused the decline of wildlife population. Along with the destruction of habitats for human progress, uncontrolled hunting and poaching of wild animals for pleasure, furs, horns, tusks, and other animal products have also caused many animal species either to go extinct or endangered.

Inhabiting of a common shared area between the populations of humans and animals may attract interactions in two different ways. They are as follows:

1. **Mutual existence**: This type of interaction occurs when both human and animals remain neutral to each other. It's a hypothetical assumption because human and animal are known to cross each other's domain since pre-historic times.
2. **Conflict**: This type of interaction occurs when there is competition between the human and animal for the common resource. With the increase in human population and consequently less abundant land resources, both human and the animal encroach upon the habitats.

Specifically, populations of some animals like elephant, wild boar, tiger, and deer are supposed to exceed the optimum level that might be supported by the available resources in their habitats. These results in a spillover of wild animals unable to find adequate food and territory, and leads to livestock lifting, crop raiding, property destruction, and even to human injury and death. Wildlife conservation effort directly and indirectly gears up the human–animal conflict. It should be viewed from two sides. On one hand, the conflict arises from restrictions imposed on the local people's access to forest resources. The restriction for the purpose of preservation of wildlife habitats indirectly affects their benefits accrued from staying near habitats. On the other hand, the conflict stems from the harm inflicted upon them by wild animals. As a result, local people directly suffer from incidents such as the intrusion of an increasing animal population in search of food, space, and mates, leading to crop losses, cattle theft, human fatalities, and other related consequences (Das and Guha 2003).

15.7.1 Classification of Human–Animal Conflict

1. Damage of natural resources
2. Damage of forage
3. Damage of agricultural and horticultural crops
4. Damage to livestock by disease transmission
5. Danger to human health and safety
6. Environmental damage

15.7.2 Management

1. **Direct methods**: Shooting, poisoning, or any biological control through predators or translocation
2. **Preventive methods**: Fencing, scaring devices, use of chemicals, repellents, and trapping
3. **Environmental control**: Less attraction of the damaged area and use of alternate area.
4. **Physiological control**: Modification of animal's physiological ability to survive or reproduce

Methods currently used in India to limit the animal damage (Gopal 2018)

1. Driving
2. Trenching
3. Fencing
4. Repellents
5. Lure crops
6. Corridors
7. Mesh fencing
8. Capturing and translocation

15.8 Wildlife Management Using Modern Technologies

The diverse wildlife on the planet faces endangerment due to a range of human activities. From the early days of human existence, people have exploited nature and animals for their benefit, leading to significant environmental harm. This harm encompasses habitat destruction, illicit wildlife trade, the proliferation of invasive species and diseases, and human interventions affecting Earth's climate.

National parks, often spanning thousands of square kilometers, face challenges in effectively monitoring all areas. Drone technology and surveillance tools offer a solution, empowering human rangers to significantly enhance the scope and efficiency of their surveillance efforts without the need for physical deployment to remote locations. Artificial intelligence (AI) excels in processing vast amounts of information, proving invaluable in identifying and tracking illegal wildlife trade. An illustration of this is the AI Guardian program, which can recognize images of illegal wildlife products with 75% accuracy.

In 2007, a consortium of Chinese Internet giants, including Baidu, Alibaba, and Tencent, formed an alliance to leverage their technology against online illegal wildlife trade. Cloud-based technology plays a pivotal role in rapidly expanding these initiatives without the necessity of provisioning, operating, and maintaining expensive physical infrastructure. Moreover, it facilitates widespread accessibility, allowing organizations dispersed over large geographical distances to share information swiftly and seamlessly. This ease of collaboration makes it simpler to combine resources and capabilities, yielding more substantial outcomes in wildlife conservation efforts.

Data analytics further enhances wildlife management strategies by amalgamating multiple data points and providing authorities with a singular, reliable source of information. This not only increases efficiency but also offers enhanced visibility into on-the-ground activities, contributing to more effective wildlife conservation.

15.9 Concluding Remarks

As one of the 17 mega biodiversity countries globally, India boasts an extraordinary diversity of plant and animal species and is very rich in wildlife and exclusively fascinating and diverse ecosystems. India's wildlife spans a vast range of ecosystems, from the lush evergreen forests in the North East to the arid deserts of Rajasthan, and from the alpine and subalpine forests of the Himalayas, including the cold desert Trans-Himalayas, to the biodiversity rich Western Ghats.

In India, although wildlife management and conservation evolved from the ancient and medieval period, it progressed rapidly with the arrival of British who laid the foundation of scientific management and conservation of wildlife. Game management, the precursor of the wildlife management, modified into a more logical and scientific tool which had its roots in human ethics, culture perceptions, and legal concepts. Encouraged by these pioneering efforts, several national and international organizations were formed for the conservation of natural resources. Wildlife research focused on wildlife management and conservation was far and few. Many of these research studies restricted to quantitative population estimation or census, wildlife biology, habitat management, and behavioral ecology. Simultaneously, as the global travel became more frequent, birding and ornithology started picking up momentum. In the mid-twentieth century, these studies were further advanced to scientific appraisal and legislation for wildlife conservation and strengthening of the protected area Top of Form network. The wildlife conservation received a major attention and impetus in 1970s with the enactment of the Act of 1972 for protection of Wildlife and initiation of Tiger conservation Project in 1973. Conservation of tiger as an umbrella species paved way for the establishment of the Tiger Reserves and many other protected areas across the country for the conservation of wildlife in general and tiger in particular.

The substantial rise in both human and livestock populations in India has exerted immense pressure on natural resources. Consequently, forests, pastures, and wastelands have been converted into cultivated areas to meet the escalating demands for housing, cereals, and other food products. The adoption of unsustainable land use patterns in rural regions, encroachments on wild lands, agricultural expansion, excessive livestock grazing, and various extractive industries have led to the fragmentation, degradation, and disturbance of most protected areas.

These alterations in habitat, coupled with disease transmission from livestock, have had widespread and predominantly negative impacts on wildlife populations. Many wildlife species have experienced ecological displacement and local extinction due to these biotic interferences. The current peak in human–wildlife conflicts can be directly attributed to the intensive disturbances in protected areas. Consequently, the acceptance of conservation ideals by local communities has been significantly compromised.

The Indian Government, local communities, multilateral aid agencies, and Non-Governmental Organizations (NGOs) have made commendable financial and other investments in the conservation of wildlife, both within their natural habitats (in situ) and in controlled environments (ex situ). Effective conservation efforts require continuous monitoring of wildlife management practices, wildlife habitats, and species populations to ensure their well-being and sustainability. Climate change and its potential impacts, coupled with increasing anthropogenic demands on land, sea, and freshwater resources, have elevated the conservation and management of wild areas and animals to a paramount priority. All of this requires highly trained personnel. Contrary to the developed countries, in India, there has been a general lack of professional human resources and trained human power, corps of wildlife biologists and trained managers, who will effectively tackle these issues in the protected areas. There is an urgent requirement to cultivate skilled wildlife biologists, providing comprehensive education and training for individuals aspiring to pursue careers in wildlife and nature conservation. These individuals should be equipped to actively advocate and advance science-based wildlife conservation and habitat preservation within the country.

Besides strengthening the conventional management practices of protection, anti-poaching camps and patrolling, fire control, and management, the science-based management of species and their habitats requires a landscape approach and strengthening the use of modern tools and techniques for effective management. There has been an increasing demand for public participation for the long-lasting conservation of wildlife and other resources in the country.

Lessons Learnt

- Wildlife conservation and management have gained significant importance alongside urgent global challenges like climate change, global warming, and health education.
- Wildlife conservation and management in India trace their roots back to ancient and medieval times, but they gained technical advancements during the British colonial era, which laid the foundation for scientific management practices.
- Wildlife management encompasses various disciplines such as monitoring, surveillance, protection, habitat restoration, and population regulation, all integrated with logical reasoning and scientific interpretation.
- Effective wildlife management and enduring conservation depend on the full participation and support of local communities. Involving people through education, awareness campaigns, and sustainable livelihood options can foster a sense of ownership and support for conservation activities.
- Studying species populations, habitats, and ecological dynamics provides scientific solutions for successful management and long-lasting planning of conservation of threatened species and their natural habitats. Innovative science-based approaches and technological interventions increasingly guide decision-making in wildlife conservation.
- Incorporating conservation objectives into development plans and policies can mitigate harmful effects on wildlife and their habitats. Supporting sustainable livelihoods for local communities in alignment with conservation goals can create positive outcomes.
- Addressing human–wildlife conflicts requires a multidisciplinary approach, including community-based conservation methods, habitat management, and flexible wildlife policies. Effective wildlife conservation necessitates collaboration among governments, institutions, conservation groups, researchers, NGOs, and local communities.
- Wildlife conservation is a dynamic process and adaptive management strategies are crucial to adapt to changing ecological, social, and economic conditions.
- Effective wildlife conservation and management require essential qualities such as adaptability and the capacity to derive valuable lessons from both successes and failures.

Key Questions

1. Explain the concept of wildlife management and its significance in conservation efforts.

2. Discuss the importance and inherent values of wildlife, both ecologically and culturally, and what is their contribution to the overall welfare of ecosystems and human society?
3. What are the fundamental requirements for the survival and well-being of wildlife species, and how do these requirements relate to habitat quality?
4. Define the concept of home range and elaborate on the various techniques employed to estimate and analyze home range patterns in wildlife populations.
5. Explain the terms "Territory" and "Territoriality" in the context of wildlife behavior, and discuss their ecological significance and implications for population dynamics?
6. What is a wildlife corridor, and how does it contribute to wildlife conservation and landscape connectivity? Discuss with examples their effectiveness in facilitating wildlife movement.

References

Balwan WK, Kour S (2021) Wetland—an ecological boon for the environment. East Afr Scholars J Agric Life Sci 4(3):38–48

Beresford AE, Fiona J, Paul SF, Ian D, Burfield J, Butler A, Vickery JA, Buchanan GM (2019) Phenology and climate change in Africa and the decline of Afro-Palearctic migratory bird populations. Remote Sens Ecol Conserv 5(1):55–69

Bolen E, Robinson W (2003) Wildlife ecology and management, 5th edn. Pearson Education Inc., New Jersey, pp 84–87

Boyd CE, McNevin AA (2005) Aquaculture, resource use, and the environment. John Blackwell and Sons

Brown JL (1964) The evolution of diversity in avian territorial systems. Wilson Bull 76:160–169

Burt WH (1943) Territoriality and home range concepts as applied to mammals. J Mammology 24(3):346–352

Chardonet P, Des Clers B et al (2002) The value of wildlife. Rev Sci Tech 21(1):15–51

Chundadwat RS, Merten, Agasti J et al (2017) Value of wildlife tourism for conservation and communities: a study around four tiger reserves in Madhya Pradesh (Part-I, Tourism infrastructure and revenue). Bag Aap Aur Van and TOF Tigers

Corsi F, De Leeuw J (2000) Modeling species distribution with GIS. In: De Leeuw J, Skidmore A (eds) Research techniques in animal ecology. Columbia University Press University of Twente, pp 389–434

Das TK, Guha I (2003) Economics of man-animal conflict: a lesson for wildlife conservation. ACB Publications. 757 p

DeWalt SJ, Maliakal SK, Denslow JS (2003) Changes in vegetation structure and composition along a tropical forest chronosequence: implications for wildlife. For Ecol Manag 182(1–3):139–151

Doligez B, Boulinier T (2008) Habitat selection and habitat suitability preference. Encyclop Ecol 1–5:1810–1830

Emily B, Cohen KG, Horton PP, Marra HL, Clipp AF, Jaclyn AS, Jeffrey DS, Buler J (2021) A place to land: spatiotemporal drivers of stopover habitat use by migrating birds. Ecol Lett 24(1):38–49

Fletcher RJ, Burrell NS, Reichert BE, Vasudev D, Austin JD (2016) Divergent perspectives on landscape connectivity reveal consistent effects from genes to communities. Curr Landsc Ecol Rep 1:67–79

Ghoddousi A, Buchholtz EK, Dietsch AM, Williamson MA, Sharma S, Balkenhol N, Kuemmerle T, Dutta T (2021) Anthropogenic resistance: accounting for human behavior in wildlife connectivity planning. One Earth 4(1):39–48

Giles LW, Marshall DB (1954) A large heron and egret colony on the stillwater wildlife management area, Nevada. The Auk 71(3): 322–325

Gopal R (2018) Fundamentals of wildlife management. Natraj Publishers, New Delhi, India. 1288 p

Gwinner E, Rodi T, Schwabl H (1994) Pair territoriality of wintering stone chats: behaviour, function and hormones. Behav Ecol Soc Biol 34:321–327

Haq IU, Rehman S, Bhat BA, Ahmad K, Rahmani AR (2021) Isolation and molecular characterization of bacteria from Ibis bill (*Ibidorhyncha struthersii*) in River Sindh of Kashmir Himalaya. Acta Ecol Sin 41(6):537–544

Hazzah L, Dolrenry S, Naughton L, Edwards CTT, Mwebi O, Kearney F, Frank L (2014) Efficacy of two lion conservation programs in Maasailand, Kenya. Conserv Biol 28(3):851–860

Hutchins M, Marra PP, Micheal ED, Sheppard KC, Hallager S, Lynch C (2018) The evolving role of zoological parks and aquariums in migratory bird conservation. Zoo Biol 37(5):360–368

Karanth KU, Kumar NS (2013) Replenishing water tanks in wildlife reserves: how scientific? Conservation Articles

Lischka SA, Teel TL, Johnson HE, Crooks KR (2019) Understanding and managing human tolerance for a large carnivore in a residential system. Biol Conserv 238:108189

Mitsch WJ, Gosselink JG (2007) Wetlands, 4th edn. Wiley, New York

MoEFCC (2017) India's National Wildlife Action Plan 2017–31. Ministry of Environment and Forests and Climate Change, Government of India. 160 p

Moore CT, Conroy MJ (2006) Optimal regeneration planning for old-growth forest: addressing scientific uncertainty in endangered species recovery through adaptive management. For Sci 52:155–172

Mountfort G (1956) The territorial behaviour of the hawfinch *Coccothraustes coccothraustes*. IBIS 98(3):490–495

Nielsen SE, McDermid G, Stenhouse GB, Boyce MS (2010) Dynamic wildlife habitat models: seasonal foods and mortality risk predict occupancy-abundance and habitat selection in grizzly bears. Biol Conserv 143(7):1623–1634

Norton GL, Wilson R, Stevens JR, Beard KH (2010) A meta-analytic review of corridor effectiveness. Conserv Biol 24:660–668

Noss RF, Daly KM (2006) Incorporating connectivity into broad-scale conservation planning. In: Crooks KR, Sanjayan M (eds) Connectivity conservation. Cambridge University Press, Cambridge, pp 587–619

Orians GH (1969) On the evolution of mating system in birds and mammals. Am Nat 103:589–603

Ozaga JJ, Verme LJ, Bienz CS (1982) Parturition behavior and territoriality in white-tailed deer: impact on neonatal mortality. J Wildl Manag 46(1):1–11

Pitt WC, Box PW, Knowlton FF (2003) An individual-based model of canid populations: modelling territoriality and social structure. For Ecol Model 166(1–2):109–121

Runge MC (2011) An introduction to adaptive management for threatened and endangered species. J Fish Wildl Manag 2(2): 220–233

Sacks BN, Jaeger MM, Neale JCC, Mc Clough DR (1999) Territoriality and breeding status of coyotes in relation to sheep predation. J Wildl Manag 63(2):593–605

Sargent RC (1982) Territory quality, male quality, courtship intrusions, and female nest—choice in the three spine stickleback, *Gasterosteus aculeatus*. Anim Behav 30(2):363–374

Schulze ED, Craven D, Durso AM, Reif J (2019) Positive association between forest management, environmental change, and forest bird abundance. For Ecosyst 6(3):1–12

Shanu S, Shashankaditya U, Roy A, Chundawat R, Bhattacharya S (2008) Community structures in Simplipal complexes: an application to wildlife corridor designing in central India—Eastern Ghats landscape complex, India. Arxiv

Smith DS (1958) Nest-site selection, pair formation and territory in the house-sparrow *Passer domesticus*. IBIS 100(2):190–203

Smith CL, Gilden J, Steel BS, Mrakovcich K (1998) Sailing the shoals of adaptive management: the case of salmon in the Pacific Northwest. Environ Manag 22:671–681

Sparklin BD, Micheal SM et al (2010) Territoriality of feral pigs in a highly persecuted population on fort Benning, Georgia. J Wildl Manage 73(4):497–502

Taylor P, Fahrig L, With K (2006) Landscape connectivity: are turn to the basics. Conserv Biol 14:29–43

Watson A, Jenkins D (1968) Experiments on population control by territorial behaviour in Red Grouse. J Anim Ecol 37(3):595–614

Wegge P, Yadav SP, Lamichchane BR (2018) Are corridors good for tigers, *Panthera tigris* but bad for people? An assessment of the Khata corridor in lowland Nepal. Oryx 52(1):35–45

Yang H, Ranjitkar S, Zhai D, Zhong M, Goldberg SD, Salim MA, Wang Z, Jiang Y, Xu J (2019) Role of traditional ecological knowledge and seasonal calendars in the context of climate change: a case study from China. Sustainability 11(12):3243

Yarrow G (2009) Habitat requirements of wildlife: food, water, cover and space, forestry and natural resources. Clemson Extension. 5 p

Part III

Silviculture, Resource Assessment and Sustainability

Forest Mensuration

16

Ramakrishna Hegde, P. A. Clara Manasa, and Supriya K. Salimath

Abstract

Forest mensuration involves the measurement and quantification of forests and trees. A comprehensive understanding of the methods and processes employed in these measurements is essential for forest managers. This encompasses measures of linear length, area, volume, and weight. Techniques often involve measuring easily identifiable parts of an object, with the actual object size estimated using intricate mathematical formulas derived from component measurements. Whether measuring something basic or highly complex, precision, bias, and accuracy are critical considerations. Therefore, precision-enhancing methods for different dimensions are pivotal to ensure accurate and reliable measurements. This chapter delves into various tree dimensions, including height, diameter at breast height (DBH), girth, bark thickness, tree form, volume, felled logs, and branch measurements. It also explores factors influencing site productivity, such as site quality and canopy density. These measurements are essential for foresters to establish relationships among dependent tree dimensions. This chapter further highlights the importance of forest inventory in estimating the productivity of larger forest areas. The adage "if you cannot measure it, you cannot manage it" underscores the vital role mensuration plays in effective forest management. Mensuration is crucial for obtaining information about forest resources, with the ultimate goal of providing quantitative data to make informed decisions regarding density, use, and management. The discussion extends to factors influencing site production, with a detailed examination of site quality and canopy density. Establishing relationships between dependent tree components is imperative for foresters. This chapter sheds light on forest inventory, contributing to the estimation of productivity in broader forest regions. Effective forest management relies on accurate data, making mensuration an integral component in the quest for information about forest resources and informed decision-making.

Keywords

Mensuration · Breast height · Volume table · Site index · Stand density

16.1 Introduction

All woodland and forest management are based on estimations and measurements, that is, interpreting numerical data. Evaluating the results of observations, estimates, and measurements provides a basis for the practical experience of particular importance to foresters (Prodan 1968). The basis of mathematical statistics lies in pure mathematics. The mathematical statistics method obtains an objective and effective evaluation of the data resulting from observations and experiments. Mathematics was a well-developed discipline in ancient India. Mensuration, an essential part of this discipline, is derived from the Latin word "mensura," meaning "measure." It is, thus, concerned with determining lengths, area, and volume (Chaturvedi and Khanna 1994). Forest measurements are vital for understanding growth, assessing yields, and ensuring proper management. Challenges in directly measuring aspects like tree volume are met by estimating through ground-level measurements, overcoming obstacles posed by towering tree height and size.

Forest mensuration, also known as Dasometrics or Dendrometrics/Biometrics, is the discipline focused on quantifying forests, trees, and forest products (García 1995). It constitutes the branch of forestry dedicated to determining dimensions (e.g. diameter, height, volume), form, age, and increment of individual trees, stands, or entire woodlands, whether standing or after being felled. This encompasses

R. Hegde (✉) · P. A. Clara Manasa · S. K. Salimath
College of Forestry, Ponnampet, Keladi Shivappa Nayaka University of Agricultural and Horticultural Sciences, Shivamogga, Kodagu, Karnataka, India
e-mail: ramakrishnahegde@uahs.edu.in

A. K. Mandal, A. Nicodemus (eds.), *Textbook of Forest Science*, https://doi.org/10.1007/978-981-97-8289-5_16

measurements of both standing and felled trees, sawn wood, round logs, and various minor forest products such as bamboo, charcoal, bark, fruits, and more. The scope of forest mensuration extends to linear, area, volume, and weight measurements. While volume holds significant importance in the realm of forest management, weight, often referred to as biomass, is gaining increasing prominence and is being employed as a substitute for volume. The metric system of weights and measures has been used throughout. In Indian forestry, the unit of measurement before the passage of the Standards of Weights and Measures Act 1956 was the F. P. S. system. Metric system or C. G. S. system introduced in October 1962 is now a legal necessity. Hence, in India, old records of measurements are in the British system, and the measurements taken and table prepared after the introduction of the metric system are in metric measures. Until all the old tables are reconstructed in the metric system, the measurements of the British system would continue to appear in references to those measurements. However, it is essential to convert F. P. S. units to C.G.S. units in current use.

A primary focus lies in measuring the quantity of wood within tree stems and determining the sizes of logs that can be extracted. This wood is subsequently transformed into timber products, commonly known as lumber in America, serving various purposes including construction and paper-making. The extraction of wood in the form of logs from tree stems remains a valuable commercial practice in forestry. Traditionally, courses on forest measurement have emphasized the techniques employed in measuring this wood resource.

16.2 Measurements

The act of measurement is an integral aspect of any scientifically grounded discipline. When faced with the challenge of measuring something directly or when direct measurement is impractical, approximations or estimates are employed. These approaches often entail measuring accessible parts of the object. Subsequently, intricate mathematical processes are applied to translate these partial measurements into an estimate of the overall size of the object. Whether dealing with a straightforward or highly intricate subject, three key considerations come to the forefront in measurement: accuracy, potential bias, and precision (West 2009).

Accuracy refers to "the variance between a measurement or estimation of something and its actual value." Put simply, it represents the degree of closeness with which one can measure or estimate something, considering the limitations of the available measuring tools or estimation techniques. Accuracy is conveyed by indicating that a measurement or estimate has been conducted to the nearest fraction of a unit of measurement, such as the closest 1/10th of a meter, the nearest hectare, or the nearest microsecond, depending on the nature of the object being measured.

Bias is defined as "the disparity between the average of multiple repeated measurements or estimates of something and its true value." When measuring something challenging, the number of attempts or the diversity of measurement equipment employed may not rectify the issue; the obtained results could consistently deviate from the actual value. In this context, "the wrong answer" signifies that the outcomes from numerous measurement attempts would consistently overstate or understate the actual value of the object being measured. In such instances, the measurement or estimation method is described as biased.

Precision is defined as "the fluctuation in a series of repeated measurements or estimates of something." This variability stems from the constraints inherent in the measurement or estimation technique, considering its application at different times and under diverse conditions, as well as the limitations of the individuals conducting the measurements.

16.2.1 Individual Tree Measurements

16.2.1.1 Stem Diameter/Girth

In forestry, the most straightforward, prevalent, and arguably the most crucial measurement applied to trees is the diameter of their stems. The tree stem diameter holds significance for various reasons:

- It frequently exhibits a close correlation with more challenging-to-measure attributes, such as the volume of wood in a tree's stem or the tree's weight (referred to as biomass).
- The diameter often serves as an indicator of the tree's monetary value, as larger trees yield larger logs, allowing for the production of more valuable timber with higher commercial worth.
- It can indicate a tree's competitive standing within a stand and its potential for robust growth relative to other trees.
- Stem diameter gradually diminishes from the base of the tree as it tapers.

To standardize measurements, a widely adopted practice in forestry involves assessing tree stem diameter at breast height. Breast height serves as the global standard for measuring the diameter, girth, and basal area of standing trees. In India, the USA, Burma (Myanmar), the Union of South Africa, and some other former British colonies, the standard breast height is set at 1.37 m (4 ft. 6 in.) above ground level. In Europe and most other Commonwealth countries, the established breast height is 1.30 m (4 ft. 3 in.), a recommendation also endorsed by the FAO.

Standard Rules Governing Breast Height Measurements

To standardize diameter measurements, specific rules have been established for measuring the diameter or girth at breast height.

1. Mark breast height on standing trees at 1.37 m (4 ft. 6 in.) above ground level using a measuring stick (see Fig. 16.1a).
2. Mark the breast height point using a cross mark, consisting of intersecting vertical and horizontal lines each measuring 12 cm in length and painted with white paint.
3. On sloping terrain, measure the breast height on the uphill side after clearing away debris, leaves, needles, etc. (Fig. 16.1b).
4. For leaning trees, measure breast height along the leaning tree stem rather than vertically (Fig. 16.2a).
5. If the stem exhibits abnormalities at the 1.37 m (4 ft. 6 in.) level, adjust the breast height point by shifting it up or down as minimally as possible to a more normal stem position (Fig. 16.2b).
6. For buttressed trees with potential further growth of the buttress, measure the breast height point at the lowest position beyond which the abnormal formation is not expected to extend (Fig. 16.2c).
7. For trees with forks, designate breast height at 1.37 m (4 ft. 6 in.) if the fork is above this point, treating the tree as a single entity for diameter or girth measurement. If the fork is below, measure each fork separately. In cases of abnormal stem conditions at breast height due to forking, adhere

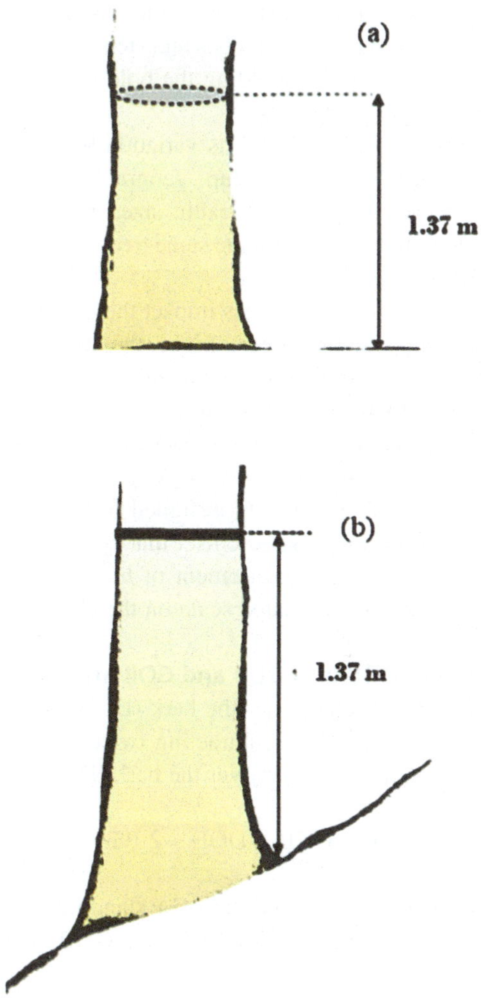

Fig. 16.1 (**a**) Breast height of tree; (**b**) tree on sloping ground

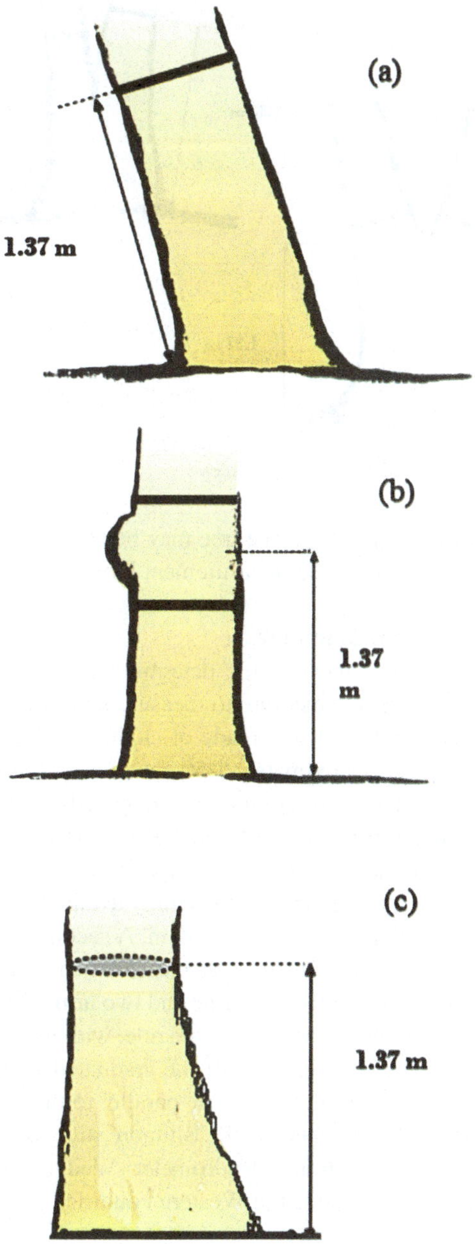

Fig. 16.2 Breast height of (**a**) A leaning tree; (**b**) abnormal stem; (**c**) buttresses tree

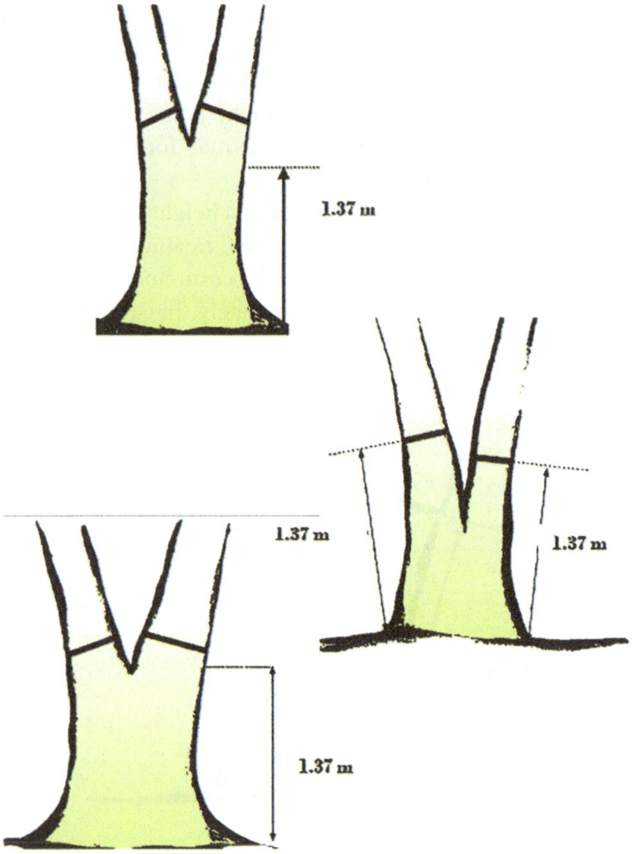

Fig. 16.3 Breast height of forked trees

to the mentioned rule. The tree may be counted as one or two, depending on the measurement location (Fig. 16.3).

Measuring Stem Diameter

The conventional method for determining stem diameter involves using a tape measure to measure its girth. A tape is a graduated band, typically made of cloth, reinforced cloth, plastic, or steel, approximately 1.50 cm wide and of variable length. These tapes are commonly graduated in the metric system (cm and mm, m) and the F.P.S. system (inch and foot). Tree measuring tapes are typically 3 m or 5 m long, while land measuring tapes are generally available in lengths of 5 m, 10 m, 20 m, 30 m, or even 50 m. A second instrument frequently employed for diameter measurement is a caliper. Calipers consist of a graduated rule and two arms. One arm is fixed at right angles to one end of the rule, with its inner edge aligned with the starting point of the graduated scale, while the other arm moves along the rule parallel to the fixed arm. Additionally, instruments like the Biltmore stick, Sector fork, Penta Prism, Barr, Stroud Dendrometer, Wedge Prism, and Tele Relaskop are utilized in Western countries.

Fig. 16.4 Bark of tree

16.2.1.2 Bark Functions

Bark, depicted in Fig. 16.4, serves as the outer protective covering for the stems and roots of woody plants. Woody plants, such as trees, woody vines, and shrubs, feature bark, encompassing all tissues exterior to the vascular cambium. Bark develops around the branches and stems of these plants, playing a crucial role in shielding the bole from desiccation and various forms of damage.

The thickness of bark exhibits variation based on factors such as species, genetic makeup, geographical location or site, age of the tree, overall health, size, growth rate, and position along the bole within the same tree, ranging from the base to the tip.

Differences in bark thickness impact the determination of the volume of standing trees. Since the measurement of diameter or girth is typically conducted over the bark, and it is impractical to measure bark thickness on every tree, it becomes essential to ascertain the bark thickness for trees of various species. The Swedish bark gauge (depicted in Fig. 16.5) is a tool specifically designed for this purpose. It takes the form of a graduated chisel that is inserted into the bark, allowing for the measurement of bark thickness with the assistance of the graduated scale on the chisel.

Conversion of DOB into DUB and GOB into GUB

The estimated diameter under the bark (DUB) of a standing tree can be determined by subtracting twice the bark thickness (BT) from the diameter over the bark (DOB).

$$\text{i.e., DUB} = \text{DOB} - 2\,\text{BT}$$

The girth under bark (GUB) of a standing tree can be estimated using the following formula: GUB $=$ GOB $-$ 2πBT

Fig. 16.5 Swedish bark gauge

Bark Thickness Table

After measuring the bark thickness on a significant number of trees within a species, a plot is created by correlating the thickness of the bark with the tree diameter, and a smooth curve is drawn. The bark thickness is then determined from this curve by reading it against the midpoint of the diameter class interval, and the results are tabulated.

Bark Percentage Table

The table comprises data on the percentage of bark and wood categorized by diameter classes. This table is constructed by measuring numerous trees and determining the percentage of bark and wood for each. The collected data is then graphically represented against tree diameters, and smooth curves are delineated. Subsequently, the percentage of bark and wood is extracted from this curve by reading it against the midpoint of the diameter class interval, and the results are tabulated.

Bark Thickness Relationships

The average bark volume typically falls within the range of 10–20% of the over bark volume in the majority of tree species. Bark thickness demonstrates a regular decrease from the base to the tip, with the taper rate generally (though not invariably) associated with the taper rate of the stem beneath the bark. Despite this variability, there is usually a discernible relationship between average bark thickness at breast height, bark volume, and stem variables such as diameter at breast height (dbhob) and height. The ratio of diameter under bark (dub) to diameter over bark (dob) often exhibits a consistent and predictable pattern with increasing height on a tree trunk. In many hardwoods, the ratio remains relatively constant as height increases. Conversely, in conifers, the ratio of dub/dob tends to rise along the bole, while the bark percentage by volume decreases with height. Some species exhibit a decrease in the ratio of dub/dob with height.

16.2.2 Tree Height

Measuring tree height is critical in forest assessments, serving as a fundamental parameter for estimating tree volume. Tree heights are vital for consulting volume tables, form factor tables, yield tables, and conducting essential evaluations of site productivity.

Total Height The total height of a standing tree is the straight-line distance from the tip of the leading shoot (or from the highest point of the crown where there is no leader) to the ground level usually measured on slopes from the uphill side of the tree (Fig. 16.6a).

Bole Height: Bole height is the distance between the ground level and Crown Point. Crown point is the position of the first crown forming a living or dead branch. Bole height expresses the height or length of the clear main stem of a tree.

Commercial Bole Height: It is the height of the bole that is usually fit for utilization as timber.

Height of Standard Timber: The height of the bole from the ground up to the point where the average diameter over bark is 20 cm.

Stump Height: The height of the top of the stump above ground.

Crown Length: It is the vertical measurement of the crown of a tree from the tip to the point halfway between the lowest green branches forming the green crown all round and the lowest green branch on the bole.

Crown Height: It is the height of the crown as measured vertically from the ground level to the point halfway between the lowest green branch and the green branches forming the green crown all round (Fig. 16.6b).

16.2.2.1 Methods of Measurement of Tree Height

For small trees, height can be directly measured using a pole or rod, while taller trees may require sliding poles or telescopic poles made of wood or metal. Typically, measurements of trees in the forest or stand are conducted through three methods:

1. Ocular estimation
2. Non-instrumental methods
3. Instrumental methods

Ocular Estimation

Ocular estimation visually assesses tree height. To establish a scale, some trees are measured with instruments. Using this

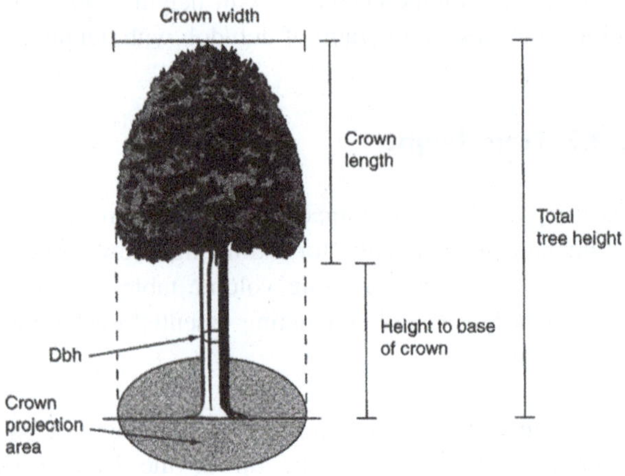

Fig. 16.6 (**a**) Measuring of tree height and other parameters. (**b**) Measuring crown height and related parameters

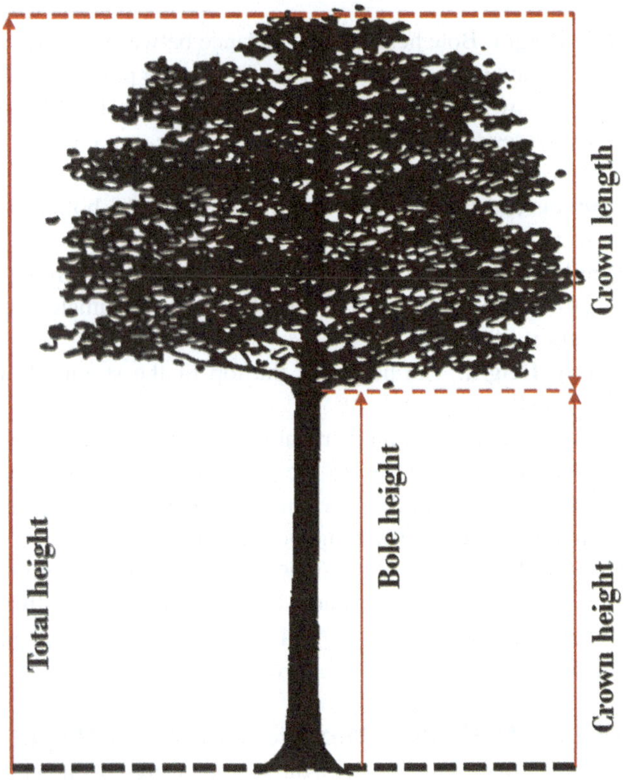

Fig. 16.6 (continued)

standard, the estimator evaluates tree heights. For added reliability, a 3 m pole or a pencil can be used. Place a 3 m pole against the tree and mentally divide it into 3 m sections for height calculation. Alternatively, use a pencil in an outstretched hand to cover the pole against the tree. Imagining the sections marked by the pencil's length, multiplying the number by 3 yields the tree's height in meters. Consistent practice enhances accuracy despite ocular estimation's potential for errors.

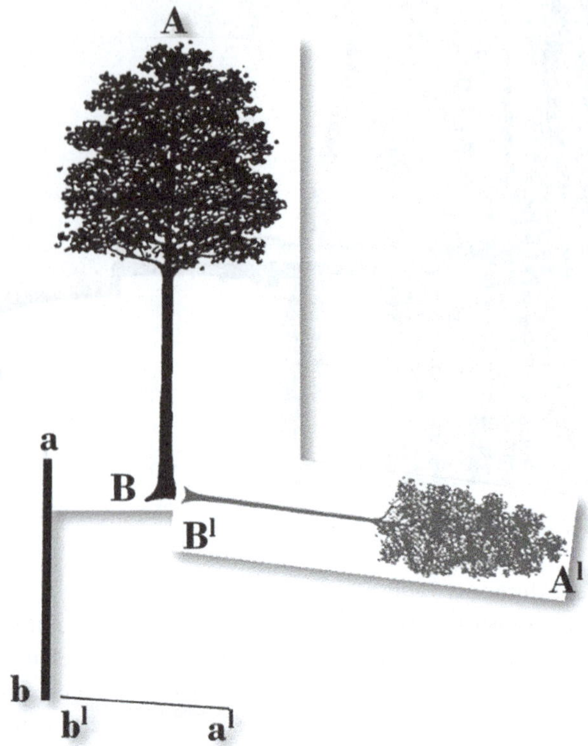

Fig. 16.7 Shadow method on

Non-instrumental Methods

All methods, instrumental and non-instrumental, assume vertical tree alignment, although trees may not be perfectly vertical. Instruments can measure the height of leaning trees, while both methods are applicable to vertical ones. Non-instrumental techniques for vertical trees include:

- Shadow method
- Single pole method

Shadow Method

This technique requires inserting a vertical pole into the ground and measuring its height and shadow, along with those of the tree. Represent the tree's height as AB, the pole's height as ab, the tree's shadow as A^1 B^1, and the pole's shadow as a^1 b^1 (see Fig. 16.7). The tree's height is calculated using a straightforward proportion.

Now,

$$\frac{AB}{A^1B^1} = \frac{ab}{a^1b^1}$$

$$AB = \frac{ab \times A^1B^1}{a^1b^1}$$

This technique is applicable solely on clear, sunny days and yields accurate results when employed during the early morning or late evening, avoiding the period between 11 AM and 2 PM when the tree casts a long shadow. However, in densely stocked forests, especially where obtaining the shadow of the specific tree on the ground proves challenging, this method becomes difficult to apply.

Single Pole Method

In this method, an observer holds a roughly 1.5 m pole vertically at arm's length, ensuring the part above the hand equals the distance from the eye. Without changing the hand's position relative to the eye, the observer moves forward and backward until the line of sight to the tree's tip aligns with the pole's tip, and the line of sight to the tree's base passes through the point where the hand holds the pole. This ensures the section of the pole above the hand entirely covers the tree. Let AB represent the tree, and ac a pole about 1.5 m long, held at b vertically so that the distance from the observer's eye E to b is equal to ab.

Now,

$$\frac{AB}{ab} = \frac{EB}{Eb}$$

$$AB = \frac{EB \times ab}{Eb}$$

Since, Eb = ab and AB = EB.

Instrumental Methods

Instruments for measuring tree height include hypsometers, altimeters, and clinometers. Hypsometers determine tree height from a distance. Altimeters, designed for altitude, have specific adaptations for tree heights. Clinometers, measuring slope angles, can be used for tree height via trigonometric methods. Some clinometers, known as hypsometers, incorporate a scale providing the tangent of an inclination angle as a percentage of horizontal distance. These instruments rely on geometric or trigonometric principles involving similar triangles or right-angled triangles.

Height by Geometric Methods

Two triangles are considered similar when their corresponding angles are equal, and their corresponding sides are proportional. The establishment of the similarity of two triangles can be confirmed by one of the following conditions:

- Each angle of a triangle is equal to its corresponding angle in the other triangle.
- Each side of a triangle is proportional to the corresponding side of the other triangle.
- One angle of a triangle is equal to one angle of the other, and the corresponding sides subtending the equal angles are proportional.

Let ABC and abc be two similar triangles. AB and BC are known in triangle ABC and only bc is known in triangle abc.

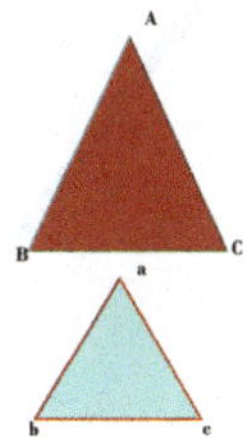

Then ab can be found using the relationships:

$$ab : AB = bc : BC$$

$$\text{or } ab \times BC = bc \times AB$$

$$ab = \frac{bc \times AB}{BC}$$

Certain instruments utilize these principles to directly measure the heights of trees.

Height by Trigonometric Methods

In a right-angled triangle, the trigonometric ratios or functions of angles, excluding the right angle, can be defined based on the sides of the triangle.

Let ABC be a right-angled triangle. The trigonometrical ratios of the $\angle ACB$ are defined as follows:

- $\sin \angle ACB = AB/AC$
- $\cos \angle ACB = BC/AC$
- $\tan \angle ACB = AB/BC$

Therefore, trigonometric ratios can be employed in calculating tree heights using one of the following methods:

Tangent Method

The tangent method calculates tree height using tangents of angles to the tree's top and base, along with the observer's distance. Procedures for height calculation must adjust based on the observer's position relative to the tree, especially in

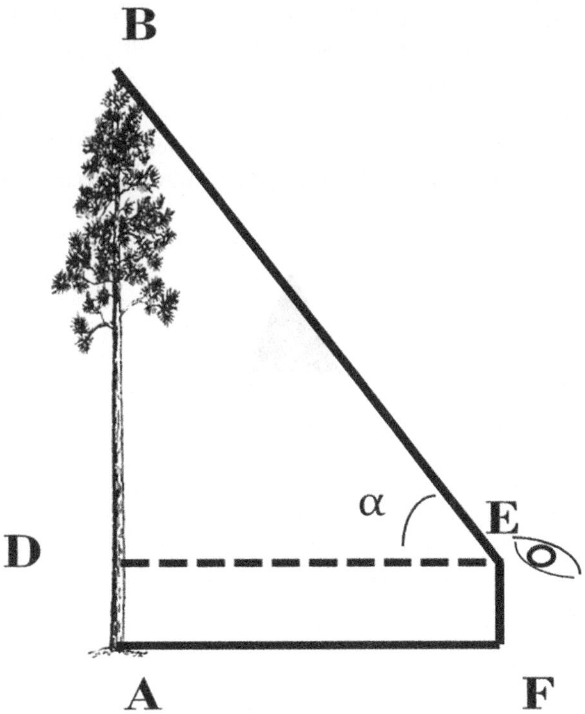

Fig. 16.8 Tree on level ground

sloping terrain. This is crucial for accuracy in various land conditions. The following methods should be followed in such conditions.

Let AB be the tree which is being observed by observer EF standing on the same horizontal plane as the tree (Fig. 16.8). Let angle of elevation (when the observer view the top of the tree) be α and ED the horizontal line of sight making the right angle on the tree at D.

In the right-angle triangle BED

$$\tan \alpha BD/ED$$

$$BD = \tan \alpha \times ED$$

Now AB = AD + BD

$$AB = AD + \tan \alpha \times ED$$

(Since ED = AF and AD = EF)

Therefore, AB = EF + tan α × AF

where observer is standing at such a place that the tree is above the eye level and the base below it. Let AB be the tree, α be the angle to the top and β be the angle to the base of the tree. Let ED be the horizontal line from observer eye E making a right angle on tree at D (Fig. 16.9).

$$AB = AD + BD$$

$$= ED \tan \alpha + ED \tan \beta$$

$$= ED (\tan \alpha + \tan \beta)$$

But if the horizontal line drawn from the eye cuts the tree high making the measurement of ED difficult, it will have to be changed in terms of AE.

In a right-angle triangle ADE,

$$ED = AE \times \cos \beta$$

$$AB = AE \cos \beta (\tan \alpha + \tan \beta)$$

Sine Method

In triangle AEB (Fig. 16.10)

$$\frac{\sin \angle AEB}{AB} = \frac{\sin \angle ABE}{AE}$$

$$AB \times \sin \angle ABE = AE \times \sin \angle AEB$$

$$AB = \frac{AE \times \sin \angle AEB}{\mathrm{Sin} \angle ABE}$$

$$AB = \frac{AE \times \sin (\alpha + \beta)}{\sin (90 - \alpha)}$$

Various instruments, relying on trigonometric principles, are utilized in different countries. In India, commonly used instruments include Brandis hypsometer and Abney's level. Haga altimeter, Topographical Abney's level, Ravi Altimeter, Blume–Leiss hypsometer, and Clinometer are currently in use. Additionally, instruments such as Relaskop, Tele Relaskop, Barr, and Stroud dendrometer are also employed.

16.2.2.2 Tree Stem Form

A casual observation of trees shows that their stems are not perfectly cylindrical. To calculate tree or log volume, one needs the diameter (or length for logs) and an understanding of stem form, which is the taper rate. Taper, the diameter decrease from the base upward, varies with species, age, site, crop density, and even different parts of the same tree. For instance, the basal portion of the tree corresponds to the frustum of a neiloid, the middle portion to the frustum of a paraboloid, and the top portion to a cone (Fig. 16.11).

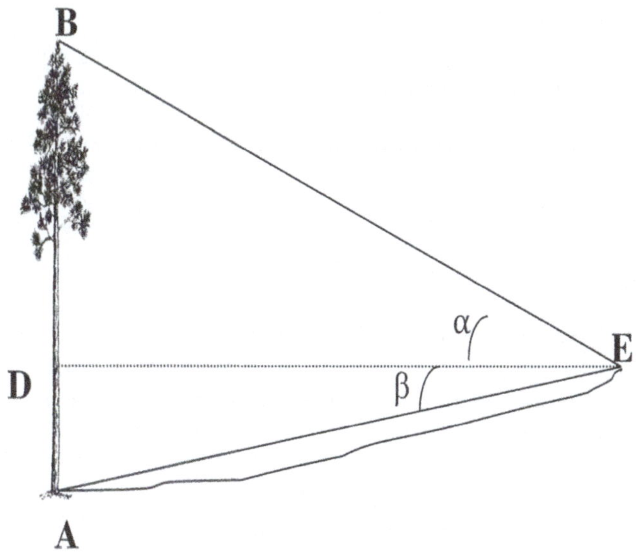

Fig. 16.9 Tree on sloping ground

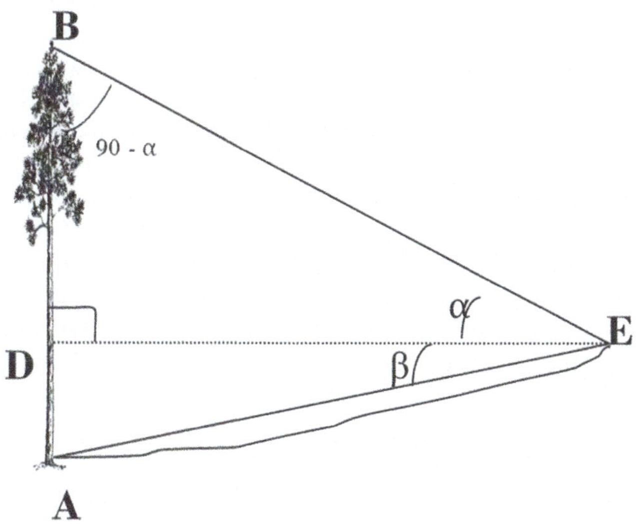

Fig. 16.10 Sine method

An ideally cylindrical tree shape is preferred for efficient timber use. Factors like species, genotype, age, competition, site characteristics (especially wind exposure), silvicultural treatment, and crown size and structure are key in determining tree form. Among these factors, the crown length often decisively shapes the overall tree form (Table 16.1).

Approaches to Studying Tree Form

Analyzing tree form presents challenges as it cannot be fully determined by a few measurements. Form varies not only between trees but also within the same tree, with added complexity from bark thickness variations. Nevertheless, tree form can be studied using the following approaches:

Fig. 16.11 Tree stem form measurement

1. Comparing standard form ratios
2. Classifying form based on form ratios
3. Compiling taper tables

Comparing Standard Form Ratios

There are two form ratios are in vogue

1. Form factor
2. Form quotient

Form Factor

The form factor is the ratio of a tree's or its components' volume to a cylinder's volume with the same length and cross-section as the tree. It represents the ratio between the tree's volume and the product of its basal area and height. Different classes of form factors are distinguished based on basal area measurement height and the tree parts considered.

1. **Artificial Form Factor**: Also known as the breast height form factor, it measures basal area at breast height, including the entire tree above and below the measurement point. While not a reliable indicator of tree form due to inconsistent relationships with tree height, it is widely used for its convenient measurements.

Table 16.1 Different tree situations with its ring area growth and tree ring width

Tree situation	Ring area growth	Tree ring width
Free-growing trees, predominant or vigorous trees growing in gaps, crowns strongly developed	Increases basipetally	Increases basipetally
Stand grown trees not overtopped but side growth of crown hindered, e.g., co-dominants	Approximately equal growth along length of stem	Decreases basipetally
Overtopped stand grown trees, e.g., intermediate and suppressed trees	Decreases basipetally	Decreases basipetally

2. **Absolute Form Factor**: Basal area is measured at any convenient height, and the volume pertains solely to the tree part above this point. It is the ratio between the volume above the measurement point and a cylinder's volume with the same basal area and height as the tree.
3. **Normal (or True) Form Factor**: Basal area is measured at a constant proportion of the total tree height, such as 1/10th or 1/20th, and the volume encompasses the entire tree above ground level.

The standard form factor has drawbacks, including the need to determine the tree's height before fixing the measurement point and potential inconvenience for extremely tall or short trees. Both absolute and standard form factors are no longer in use. In India, unless specified otherwise, "form factor" implies an artificial form factor with basal area calculated at 1.37 m above ground level. The formula for the artificial form factor is provided:

$$F = \frac{V}{Sh}$$

In this context, F represents the form factor, which is a parameter used in calculating tree volume denoted by V in cubic units. S stands for the basal area at breast height, and h represents the height of the tree. These variables are integral in mathematical expressions related to the volumetric assessment of trees.

Uses of Form Factor

1. **To Estimate the Volume of Standing Trees**: Form factors, when tabulated, offer average values for trees of various dimensions categorized by diameter and height classes. These tables aid in estimating the volume of standing trees through diameter and height measurements. While less effective for individual trees, they prove useful for estimating the volume of tree groups, with applicability restricted to conditions resembling the samples used for table creation.
2. **To Study Laws of Growth**: Form factor, alongside form point and form quotient (to be elaborated later), provides insights into the laws of growth, especially pertaining to the stem form, in trees.

Form Quotient

An Austrian forester, A. Schiffel, proposed that taper depends on what he termed the "form quotient" (F.Q.), defining it as the ratio between the mid-diameter and the diameter at breast height (d.b.h.). Therefore, according to his theory,

$$\text{F.Q.} = \frac{\text{Mid diameter}}{\text{DBH}}$$

This assumption was not always accurate because, in the case of a tree with a height of 2 × 1.37 m, the midpoint coincides with breast height, resulting in an F.Q. of 1. To address this limitation, Tor Jonson, a Swedish forester, redefined the form quotient as the ratio between the mid-diameter above breast height and d.b.h. To distinguish it from Schiffel's F.Q. known as the normal form quotient, Jonson termed his form quotient the absolute form quotient. In essence, the normal form quotient is defined as the ratio of the mid-diameter or mid-girth of a tree to its diameter or girth at breast height, while the absolute form quotient is defined as the ratio of the diameter or girth of a stem at one-half of its height above breast height to the diameter or girth at breast height. The term "quotient" now refers to the absolute form quotient, and the normal form quotient is of historical interest only unless specified otherwise.

Classifying Form Based on Form Ratios

There are two ways to classify form based on form ratios

1. Form class
2. Form point ratio

Form Class: Form class is defined as one of the intervals in which the range of form quotients of trees is divided for classification and use. It also applies to the class of trees which fall into such an interval. Trees may be grouped into form classes expressed by form quotient intervals such as 0.50, 0.55 to 0.60 and so on or by mid-points of these intervals such as 0.525, 0.575 and so on.

Form Point Ratio: Form point is defined as the point in the crown at which wind pressure is estimated to be centered. Therefore, the form point ratio (or height) is defined as the relationship, usually expressed as a percentage, of the height of the form point above ground level to the total height of the tree. It is claimed that the form point ratio bears a consistent relation to the form quotient and that by means of a table showing these relations, the form quotient and form class of a tree can be determined if the form point ratio is known.

Compiling Taper Tables

Another method for studying the form of trees involves the compilation of taper tables, illustrating the actual form through diameters at fixed points from the base to the tip of a tree. These tables serve various purposes:

1. The volume of the average tree for each diameter and height class can be easily determined in the office without direct measurement. The only required measurements are the diameter at breast height (o.b.) and the height of the standing tree.
2. Volume tables can be prepared from taper tables in the desired units.

Commonly encountered types of taper tables include:

1. **Ordinary Taper Tables or Diameter Taper Tables**: These tables provide the taper directly for the diameter at breast height without considering the tree's specific form. As a result, they facilitate the computation of volumes without the need for information about the tree's form.
2. **Form Class Taper Tables**: These tables present, for various form classes, the diameters at specific points along the stem, expressed as percentages of diameter at breast height (u.b.).

16.2.2.3 Stem Volume

Stem volume measurement is a labor-intensive and time-consuming task, even for felled trees. In contemporary forestry practices, one of the primary reasons for conducting such measurements is to develop stem volume functions, or taper functions, specific to a particular tree species in a given forest region. Volume functions enable the estimation of the total stem volume of a standing tree based on straightforward measurements, typically its diameter at breast height over bark and its total height.

Volume has traditionally been the predominant measure of wood quantity and remains the most crucial measure, despite the increasing use of weight or biomass as a metric for forest productivity. In the case of trees, volume can be calculated either after the tree has been felled or while it is still standing.

Methods Used for Estimation of Volume of Felled Trees

The volume of felled trees encompasses stem wood, branch wood, and root wood, where the latter may be valuable and permitted for excavation. Given that the stem contributes the majority of the timber in a tree, the calculation of its volume is usually examined in greater detail. Due to the irregular tapering of the tree, it is customary to segment the tree into logs. The lengths of these logs are determined by the rate of taper and market demands. The diameter at the thin end of the log dictates the sawn volume that can be obtained from it, and as the taper rate increases, the log's length decreases.

However, in certain cases, a higher taper rate for a longer timber may be more profitable, leading to the cutting of longer logs despite a reduction in the converted volume. Another factor influencing log length is the mode of transport. Carts can typically transport much longer logs than trucks, so when truck transport is employed, logs are generally limited to lengths of around 4.5 m. For research purposes when calculating the volume of felled trees, all logs, including the first, are typically standardized to a uniform length of 3 meters, except for the top-end log, which may extend up to 4.5 m. If the end section exceeds 1.5 m in length, it is separated and treated as a distinct log.

Calculation of Volume of Logs

The lower portion of a tree corresponds to a frustum of a neiloid, the middle portion to a frustum of a paraboloid, and the top portion to a cone if uncut, and to a frustum of a cone if crosscut at some point (Fig. 16.12). A generally cylindrical tree is the most desirable from the perspective of timber utilization. Critical factors such as species, genotype, age, competition, site characteristics (especially wind exposure), silvicultural treatment, and crown size and structure are essential in determining the tree's form. Among these factors, the crown, particularly its length, often plays a decisive role in shaping the overall form of the tree.

The formulae for calculating the volume of three solids of revolution, along with the formula for a cylinder for comparison, are presented in Table 16.2.

In the provided formula, the variables are defined as follows:

- s is the sectional area at the base
- s_1 is the sectional area at the thick end
- s_{m} is the sectional area at the middle
- s_2 is the sectional area at the thin end
- l is the length of the log or height of the solid in linear units.

Volume by Quarter Girth Formula

The quarter girth formula, also known as Hoppus' rule, is commonly used in India and Britain for calculating the volumes of logs. The formula, named after Hoppus, involves measurements related to the girth of the log.

$$\text{Volume} = \left(\frac{g}{4}\right)^2 \times l$$

where, g—girth of the log at the middle and l—length of the log.

Methods Used for Estimation of Volume of Standing Trees

The processes of management and sales often require estimating the volume of trees without felling them. Therefore, it is essential to be familiar with methods for estimating the volume of standing trees. The following are some of the methods used for volume estimation:

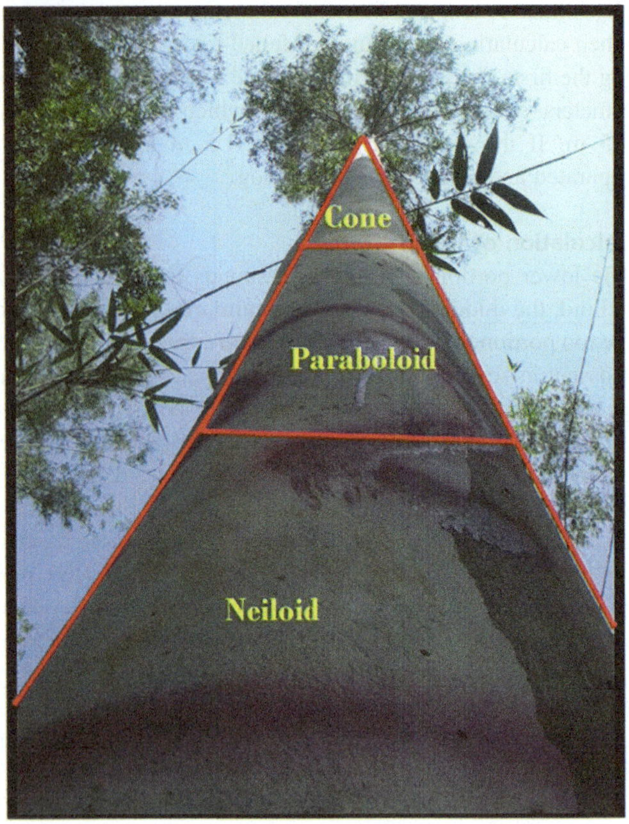

Fig. 16.12 Calculation of volume of logs

1. Ocular estimate
2. Partly ocular and partly by measurement
3. Direct measurement
4. Indirect measurement

1. **Ocular Estimate**: The ocular estimate is the initial method for estimating the volume of standing trees. Experienced individuals can accurately estimate the volume of standing trees marked for felling by carefully inspecting records of past fellings. However, this method is highly subjective, yielding varied results not only among different workers but also for the same worker under the influence of factors such as fatigue and hunger. While every forester should train their eye to make a fair estimate of the cubical contents of standing trees, achieving approximate accuracy requires extensive practice and occasional verification through measurements after trees have been felled. Even with these precautions, the results are still subject to considerable errors.
2. **Partly Ocular and Partly by Measurement**: To overcome the uncertainties associated with purely ocular estimates, one can measure the diameter and height of the tree and then estimate the volume, considering the tree's taper. In this method, the estimator estimates the diameters of successive logs after the basal log, and from these, calculates the volume for the entire length of the tree. This approach also demands a significant amount of experience and practice, as accurate estimates rely on the estimator's familiarity with tree characteristics and taper patterns. Without sufficient experience, estimates may not be accurate.
3. **Direct Measurement**: The unreliability associated with ocular estimates can be eliminated by measuring the diameters of the tree at different heights. This involves a person climbing the tree with the assistance of a ladder up to a certain height and then measuring diameters through their effort. This method is akin to calculating the volume of felled trees, with the distinction that the tree remains standing. While this approach provides accurate measurements, it is labor-intensive, time-consuming, and impractical for scenarios involving a large number of trees.

Table 16.2 Formulae for calculation of volume of three solids with the formula for cylinder for comparison

Sl. No.	Form of solid	Volume of full solid	Volume of a frustum of solid	Remarks
1.	Cylinder	$s\,l$	$s\,l$	–
2.	Paraboloid	$\frac{s\,l}{2}$	1. $\frac{s_1+s_2}{2} \times l$	Smalian's formula
			2. $s_m \times l$	Huber's formula
3.	Cone	$\frac{s\,l}{3}$	$\frac{(s_1+s_2+\sqrt{s_1\,s_2})}{3} \times l$	
4.	Neiloid	$\frac{s\,l}{4}$	$\frac{(s_1+4s_m+s_2)}{6} \times l$	Prismoidal or Newton's formula

4. **Indirect Measurement**: It is no longer necessary to climb trees to measure volume. Instruments such as Spiegel Relaskop, Tele Relaskop, Wheeler Pentaprism Caliper, Barr, and Stroud dendrometer have made it possible to measure upper stem diameters from the ground. These instruments are based on specific principles and methods that enable accurate volume estimation for standing trees. The volume (*V*) of a standing tree can be estimated using the following formula:

$$\text{Volume} = \frac{\pi \times d^2 \times h \times F}{40,000}$$

or

$$\text{Volume} = \text{Basal Area} \times h \times F$$

where *d*—diameter, *h*—height, *F*—Form Factor

Volume Table

The volume of standing trees can be estimated through the use of volume tables, which provide average contents of trees based on specified dimensions. These tables are developed from actual measurements of a substantial number of trees and assume that trees of the same species with similar dimensions will have the same volume. However, their applicability depends on the locality and the source of measurements on which the averages are based. Volume tables typically consider three main variables: diameter, height, and form. The choice of variables depends on factors such as the area of application, desired accuracy, and practicality. Tables based on a single variable, usually diameter at breast height (DBH), are simpler to use but may lack accuracy, especially over larger areas. For broader applications, tables based on at least two variables, such as DBH and height, are more accurate. Tables incorporating three variables (DBH, height, and form) offer greater precision but require more extensive measurements in the field.

The diameter at breast height is considered the most crucial variable, as errors in diameter measurement have a more significant impact on volume estimation due to the squared relationship during sectional area calculation. Height and form follow in importance, with an error in diameter having a more substantial impact on volume estimation than errors in height or form. In summary, the use of volume tables involves a trade-off between simplicity and accuracy, with the choice of variables depending on the specific requirements of the application and the extent of the area being considered.

Classification of Volume Tables

Volume tables can be categorized in three ways: (1) based on the number of variables, (2) according to their application scope, and (3) based on the type of output they provide.

Classification Based on the Number of Variables

The categorization of volume tables can be determined by the number of independent variables upon which they are founded:

1. **Volume Tables Based on One Variable viz., Diameter Alone**: In this type of volume table, trees are categorized solely based on their Diameter at Breast Height (o.b.). Displaying average volumes for individual diameters can be cumbersome and less practical, hence the use of diameter classes.
2. **Volume Tables Based on Two Variables**: Trees with the same diameter can exhibit varying heights and volumes in different locations. Volume tables that incorporate both diameter and height variables are designed for broader geographical areas. Consequently, these tables provide tree volumes categorized by diameter and height classes corresponding to the total height of the trees.
3. **Volume Tables Based on Three Variables**: These tables, known as form class volume tables, are established on the basis of diameter, height, and form quotient. While they offer increased accuracy, they are characterized by high costs, complexity in preparation, and inconvenience, making their application time-consuming.

Classification Based on the Scope of Application

Volume tables are categorized based on the extent of their applicability as follows:

1. **General Volume Tables**: These tables rely on the average volume of trees across a broad geographic region, making them suitable for a wide range of species distributions. Typically based on two variables—diameter at breast height and total tree height—these tables present volumes of trees categorized by diameter classes, and within each diameter class, by height classes.
2. **Regional Volume Tables**: These tables are assembled from measurements of trees within a specific region, resulting in a more limited applicability compared to general volume tables.
3. **Local Volume Tables**: These tables are created from measurements of trees in a confined area. Typically, they are based on a single independent variable, namely Diameter at Breast Height (o.b.). As a result, they find application in restricted localities such as a coupe or

compartment, where the assumption that trees of the same diameter share similar heights is valid in small areas. Local volume tables serve the purpose of (a) estimating the volume of standing trees before felling to regulate yield in line with working plan prescriptions, or (b) making confidential estimates of coupe volume.

Classification Based on the Kind of Outturn

1. **Standard Volume Tables**: These volume tables provide distinct estimates for standard timber, encompassing the portion of the tree stem or branch from ground level to the limit where the diameter is 20 cm measured over the bark, and small wood, representing the volume between diameter limits of 20 cm and 5 cm, both measured over bark. In this table, the volume is expressed in terms of round timber and encompasses the stump volume. The calculation is based on a full basal area basis using over bark diameters, but it's important to note that standard stem timber excludes the volume of bark, while standard small wood includes it.
2. **Commercial Volume Tables**: These volume tables present measurements of round timber content, specified as volume measured down to a slender end diameter for conversion, excluding the stump volume. The applicability of these tables is restricted to areas with consistent conversion practices and specific periods when the conversion standard remains unchanged, as the diameter level for conversion may vary with both time and location.
3. **Sawn Outturn Tables**: These volume tables provide measurements of sawn timber content, expressed as volume measured down to a slender end diameter for conversion, with the exclusion of stump volume. Essentially, these tables resemble commercial volume tables, differing only in that they focus on the volume of sawn timber rather than the volume in the round.
4. **Assortment Tables**: Assortment tables provide volume measurements for round timber down to various specified thin end diameters. For instance, these tables enable the determination of the volume of a tree with given linear dimensions when the conversion was carried out up to specific diameter limits, such as 25 cm, 20 cm, or 15 cm. It's worth noting that both standard and commercial volume tables can be considered special cases of assortment tables.
5. **Sawn Outturn Assortment Tables**: These tables resemble assortment tables, but instead of providing volume measurements in the round, they present the sawn outturn in terms of standardized pieces.

16.2.2.4 Branch Wood

In the past, most branches, except for some thick-sawn ones, were often converted into firewood. The extraction of firewood became more prevalent after World War II, driven by a steep increase in fuel demand. The construction of motor roads in the forest facilitated the export of firewood. The measurement of firewood posed a challenge, considering the impracticality of assessing the volume of each individual piece, known as a billet. Consequently, a customary practice emerged: stacking the fuel in the form of rectangular parallelepipeds and calculating the cubical contents of the stack. This measured quantity is referred to as stacked volume, representing the space occupied by all the billets rather than their actual solid volume. The measurement of stacked volume involves assessing the space occupied by the entire stack, distinct from the cubic contents of the individual billets. When building a stack on sloping ground, it is essential to measure its lengths horizontally, not along the slope. The volume of the stack is then determined by multiplying the length, breadth, and height of the stack. The sizes of these stacks vary across different regions of the country and even within different parts of states. In Western countries, the term "cord" is commonly used to express the volume of stacked wood. The size of a cord can differ between countries; for example, the American cord contains 128 $ft.^3$ of stacked wood and is typically sized at 4 ft. × 8 ft. × 4 ft. Fuel wood is generally stacked immediately after being cut, while it is still green, even if it is sold later. However, the time lapse between cutting and sale can result in shrinkage, weight loss due to moisture evaporation, and rotting of bark and thinner pieces.

Solid Volume of Firewood

The stacked volume of firewood, while a convenient measure for labor payment in forest areas lacking weighment arrangements, does not represent the actual volume of the firewood. The solid volume of the firewood in a stack is influenced by various factors, including the care taken in stacking, the form of billets, the length of billets, and their diameter. Since the value of fuel is contingent on the solid content, understanding the methods for determining the solid contents of stacked wood becomes essential. This can be achieved through any of the following methods:

1. **Xylometric Method**: The volume of billets can be determined using a Xylometer, which comprises a graduated vessel. The calculation is based on the principle of water displacement. Initially, water is poured into the vessel, and a reading of the water level is recorded. Subsequently, the pieces of wood are submerged in the water, and a second reading is taken. The difference between these two readings provides the volume of the submerged wood pieces.

   To avoid the impracticality of submerging large quantities of wood, an alternative method involves weighing the entire stack first. Then, only a portion of the stack is

submerged, and the calculations are adjusted accordingly. This approach facilitates a more manageable and efficient determination of the volume of the stacked wood.
Let W be the weight of the whole stack of wood and w the weight of submerged pieces, V the volume of the whole stack of wood and v volume of submerged pieces, then

$$W : w : V : v$$

$$W \times v = V \times w$$

$$V = \frac{v \times W}{w}$$

This method, despite its accuracy, is often regarded as cumbersome and is rarely employed in practical applications.

2. **Specific Gravity Method**: If the specific gravity of wood is known, the volume of a billet can be calculated from its weight. The specific gravity (Sp. Gr.) of a piece of wood is determined by:

$$\text{Sp.Gr.} = \frac{\text{Weight of Wood}}{\text{Weight of same volume of water}}$$

$$\text{Sp.Gr.} = \frac{\text{Density of wood}}{\text{Density of water}}$$

The density of pure water is 1 g per cc. Therefore, the density of wood in g per cc is equivalent to its specific gravity without units.

$$\text{Volume} = \frac{\text{Weight (g)}}{\text{Sp.Gr.}}\text{cc}$$

16.2.2.5 Crown Width

The development of a tree is intricately linked to its crown, and foresters prioritize the measurement of this aspect. Typically, crown assessment involves evaluating both the length and width. Crown width refers specifically to the widest diameter, representing the maximum spread of the crown. This measurement plays a crucial role in decisions related to thinning, particularly concerning the tree's overall height. The importance of crown width lies in its capacity to signify the functional growing space utilized by the tree. An understanding and quantification of this factor are imperative for the implementation of effective forest management practices.

16.2.2.6 Age of the Tree

Determining the age of individual trees is a crucial aspect to study, especially following the assessment of volume or biomass. This information is vital as it provides insights into the time required for a tree to achieve a particular volume or biomass. The ages of individual trees serve as the foundation for determining the age of woodlands, and this study becomes particularly important. Without knowledge of the age, it becomes challenging to ascertain the rate at which the wood capital is growing. Moreover, the capacity to juxtapose the financial outcomes of forestry against alternative land uses, like agriculture or horticulture, hinges on grasping the age of trees. Consequently, examining the individual tree ages becomes pivotal for proficient forest management and for forming knowledgeable comparisons across diverse land utilization approaches.

Methods of Determination of Age

The method for determining the age of trees depends on the situation, specifically whether the tree is upright or has been felled.

Determination of the Age of Single Tree When Standing

1. **Existing Records**: For trees that have been sown or planted, the records indicating the year of these operations prove highly beneficial for determining tree age. The age of the tree is derived by calculating the difference between the year of age determination and the year of sowing or planting. In the case of trees grown through natural regeneration under concentrated regeneration systems, the age of the trees is determined by subtracting the year of seeding or felling from the year of age estimation. It's important to note that this method may lack precision, given that records of artificial and natural regeneration are typically associated with entire crops rather than individual trees.
2. **General Appearance**: The age of standing trees can also be determined through visual estimation, considering the following factors:
   (a) Stem dimensions and taper: Young trees typically exhibit more pronounced tapering in their boles, while older trees tend to have less tapering.
   (b) Crown size and shape: In certain tree species, the size and shape of the crown undergo changes as the tree ages. An observation of these developments can assist in estimating the age of trees.
   (c) Bark color and condition: The color and condition of the bark undergo changes with age. Generally, older trees have smoother and lighter-colored bark, whereas younger trees, such as sal, display rough, cracked, and darker-colored bark.
   However, achieving accurate age estimates through this method demands considerable practice and experience. These conditions, nevertheless, can be influenced by local factors.

3. **The Number of Annual Shoots or Whorls of Branches**: In species with distinct annual shoot marks, determining the age of a tree involves counting these shoots from top downward. Additionally, a proportionate number of years are added for the part where these marks are not visible. In contrast, certain species produce only one whorl of branches each year. Counting the whorls or remnants thereof and incorporating the proportional number of years for portions without visible whorls provide an approximate estimation of the tree's age. For instance, Semal (*Bombax ceiba*) exhibits annual whorls at an early stage, aiding in age determination.
4. **Pressler's Increment Borer**: Trees with annual rings can be accurately determined using Pressler's increment borer. This tool is designed for extracting a narrow cylinder of wood, known as a core sample. Trees can be bored to the pith for ring counting purposes to ascertain their age. The extracted core is then taken to the laboratory for a detailed ring analysis study. The number of annual rings up to the pith of the core sample is meticulously counted. To determine the tree's age, the age corresponding to the height at which the boring is conducted must be added to the ring count.
5. **Periodic Measurements**: The age of trees lacking annual rings can be determined through three periodic measurements. Let $d1$ represent the initial diameter of the tree, and $d2$ and $d3$ denote the diameters at subsequent periodic measurements. Define $p1$ as the growth per unit diameter per year during the first two periodic measurements, and $p2$ as the growth per unit diameter per year during the period between the second and third periodic measurements. The age of the tree can then be calculated using the following formula:

$$\text{Age} = {}^{1}/_{p_1 s}$$

where s is a constant which is determined

$$s = \frac{\log p1 - \log p2}{\log d2 - \log d1}$$

6. **Diameter Age Curve Prepared from Periodic Measurements of Trees in an Area**: Diameters of trees in the area are initially recorded class-wise, and the average diameter for each class is determined. Subsequently, the same trees are measured at intervals of 5 or 10 years, and the average diameter for each diameter class is recalculated. The difference between these two averages provides the periodic diameter increment, which is then plotted against Diameter at Breast Height (DBH). A smooth curve is drawn to represent the increment pattern. To transform the increment curve into a diameter-age curve, the following steps are typically undertaken:
   (a) To transform the increment curve into a diameter-age curve, the process involves taking the lowest diameter on the increment curve. The increment corresponding to this diameter is directly read off from the curve. This increment is then added to the original diameter, resulting in the final diameter at the end of the specified period. The increment against this new diameter is read off once again, and the process is iteratively repeated for the entire range of values available from the curve. This series of calculations generates a diameter-age curve, providing insights into the growth pattern of the trees over time.
   (b) The diameter values obtained through the iterative process are then plotted against a series of equidistant points, spaced at intervals corresponding to the number of years in the specified period. A diameter growth curve is drawn through these plotted points, depicting the relationship between diameter and age over the given timeframe. This curve provides a visual representation of the growth pattern of the trees, offering insights into how their diameters change over successive years.
   (c) To convert the time axis of the curve into age, a correction is applied by shifting the zero point to the left. This shift is determined by the necessary number of units corresponding to the estimated time required to reach the lowest diameter plotted on the curve. If the time required to reach the lowest point cannot be estimated or is unknown, the curve may be produced backward based on the experience of similar curves. The age is then corrected from where the curve intersects the x-axis. By referencing the diameter on this corrected curve, the tree's age can be read. This approach provides a more accurate representation of the tree's age based on its diameter.
7. **Mathematical Relationship**: Determining a tree's age can be achieved by employing an equation that captures the relationship between age and diameter or girth, specifically tailored for the given species. Notably, researchers like Mishra and others have constructed equations designed for estimating the age of tropical deciduous trees, including sal in the Chakia, UP, forests. These equations rely on girth measurements and their increment over a defined period. Moreover, they have introduced equations that incorporate biomass data during the initial period, offering additional insights beyond relying solely on girth measurements. Nevertheless, it is crucial to acknowledge the potential limitations of this method, as factors beyond age, such as environmental conditions, can influence diameter growth. The precision of these equations is contingent on the unique characteristics of the species and the specific conditions prevailing in the forest environment.

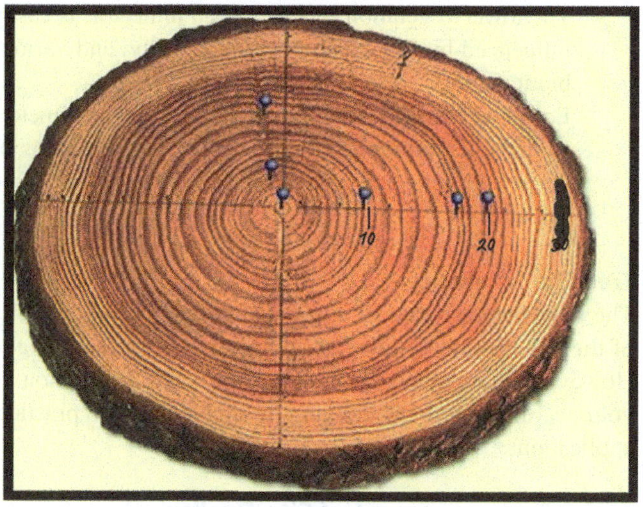

Fig. 16.13 Annual rings

Determination of the Age of Felled Trees

1. **Stump Analysis**: Determining the age of a felled tree is possible if the stump exhibits annual rings (Fig. 16.13). The process involves counting the rings on the stump and adding the estimated period it would have taken for the tree to grow to stump height. However, this method is fraught with the following difficulties:
   (a) Incidence of false rings—False rings, which do not complete a full circle around the tree, are common. Therefore, counting rings must be done meticulously to avoid including false rings in the total count.
   (b) Closed-formed rings—Some trees that experience slow growth or suppression may have rings that are closely spaced. Care must be taken not to overlook any of these closely formed rings during the counting process.
   (c) Absence of growth rings in certain years—During years when a tree is heavily attacked by defoliators, ring formation may be absent. This absence of growth rings in specific years can complicate the accurate determination of the tree's age. Despite these difficulties, it is the best and quickest method of determining the age of a tree and can be used even several years after the felling of the tree.

16.3 Stand Measurements

The concept of a stand, defined as a uniform forest patch, exhibits some flexibility without a strict area requirement. The designation of a stand's size can vary considerably, influenced by the evaluator's interpretation of "homogeneous" in the context of the specific forest. This determination is subjective, relying on individual judgment and influenced by the nature and purpose of the measurements being conducted. The structure of a stand undergoes annual transformations driven by tree growth, mortality, and harvesting activities. Recognizing the stand's growth as a significant and renewable natural energy resource, several studies aim to predict its growth. These studies consider the stand both as a collective unit and as a population of individual trees, allowing for diverse approaches to the study of stand growth.

- Determination of past growth of stand
- Predicting the future growth of stand

16.3.1 Methods of Determining Past Growth of Stand

The conventional approach to determine the past growth of stands involves the repeated measurement of sample plots, as measuring entire stands is often impractical. This can be accomplished either by establishing temporary sample plots each time growth is assessed or by implementing permanent sample plots. Laying out temporary sample plots for each growth estimate involves measuring different sampling units each time, resulting in lower precision and generally reduced accuracy. On the other hand, the method of repeated measurement in permanent sample plots is more effective. Even when such sample plots may not be entirely representative of the stand, the precision tends to be high, although the growth estimate may be biased.

16.3.2 Method of Predicting the Future Growth of Stands

The prediction of future stand growth can be achieved through the use of yield tables derived from growth studies conducted in sample plots systematically distributed across the entire forest. Stand growth is influenced by factors such as site quality and stand density. Yield tables serve as valuable tools in forecasting future growth patterns, providing insights into how these factors impact the development of the stand over time.

16.3.3 Site Quality

A **site** refers to the intricate combination of physical and biological factors within an area that determine the type of forest or vegetation it can support. **Site quality**, in turn, assesses the relative productive capacity of a site for a specific species or group of species. It serves as a measure of the site's ability to support and sustain the growth and development of particular plant life, providing valuable information for forest management and land use planning.

Site Quality Evaluation Every site exhibits a unique growth response to various species. The evaluation of site quality can be accomplished through the measurement of either site factors or vegetative characteristics. Understanding how different species respond to the specific conditions of a site allows for an assessment of the site's quality and its suitability for supporting particular vegetation. The measurements of site factors and vegetative characteristics contribute valuable information for forest management decisions and ecological assessments.**16.3.3.1**

Peterson's CVP index (Climate Vegetation Productivity Index) has been used extensively to calculate the potential productivity of the site

$$\text{CVP index } (I) = \text{Tv}/\text{Ta} \times (P \times G/12) \times E/100$$

where

Tv and Ta are maximum and minimum temperatures
P is precipitation in mm
G is the growing period in months and
E is a measure of evapotranspiration

16.3.3.2 Limitations

This index is overly generalized since it overlooks the specific attributes of soil and other environmental factors.

Vegetative Characteristics

The characteristics of vegetation that could be indicative of site quality include:

- Types of plants occurring naturally in the area
- The volume
- Basal area
- Diameter
- Height of trees

Plant Indicators

- This approach is grounded in the theory that certain species of lower vegetation act as clear indicators of site quality for a particular species or type of forest. While this method has been employed for site quality classification in specific undisturbed forests in Finland and Canada, its universal application faces challenges for several reasons:
  - It is applicable primarily to forests of simple composition, as found in northern latitudes.
  - Implementation requires substantial ecological knowledge.
  - The lower vegetation, serving as an indicator, is easily influenced by forest composition, density, and various biotic factors.
  - Lower vegetation, residing in the top few centimeters of the soil, cannot accurately reflect the conditions of deeper soil layers, which significantly impact tree growth.

Tree Characteristics

The key characteristics of a tree that mirror the productivity of the site include volume, basal area, diameter, and height. However, this method necessitates the prior calculation of volume, making it challenging to implement in practical applications.

16.3.3.3 Measurement of Site Quality

Site quality can be assessed by measuring the volume of desired materials produced, thereby expressing the integrated net effect of all site factors in terms of the product itself. However, determining site quality directly through standing volume is rarely feasible due to the influence of the stand's life history (past cutting, disease, insect infestations, and severe weather) on the standing volume, which may be unknown.

To address this challenge, several indirect methods have been developed to gauge timber site productivity. These methods aim to identify a few easily measurable properties of the vegetation or the land, which represent all factors crucial to the growth of a particular species on a given site. Following Jones (1969), three indirect approaches for estimating site quality are discussed here:

1. Site index
2. Non-tree vegetation (Vegetation typing)
3. Basic environmental and land attributes (environmental factors)

Site Index

Among the various indirect methods investigated, the rate of tree height growth emerges as the most practical, consistent, and useful indicator of forest site quality. While not flawless, it serves as the standard against which other measures, such as soil properties, are compared. The height growth of dominant, free-growing trees in the upper forest canopy proves sensitive to differences in site quality, strongly correlated with volume growth, weakly correlated with density, and species composition. In contrast, diameter growth is strongly correlated with density.

Standard practice defines site index in terms of the total height of the largest, full-crowned trees in a stand, representing the strongest competitors for light, moisture, nutrients, and growing space. Numerically, the site index is defined as the total height at specified base ages, typically 25, 50, or 100 years. The base age selection depends on the stand's maturation characteristics. For example, stands maturing or primarily utilized at a young age often use a 25-year base. Fast-growing species like pines and certain eucalyptus use base ages of 25, 30, or 50 years. Western conifers and northern species typically use 50- and 100-year base ages.

The site is expressed as a site index number, where Site Index 60 on a 50-year base means that dominant trees have an average total height of 60 ft. at 50 years. Similarly, Site Index 140 on a 100-year base indicates that the trees will be 140 ft. tall at 100 years. Site index values on 50- or 100-year bases are commonly grouped into classes of 5 ft., 10 ft., 20 ft., or broader increments. When expressed in broader classes, such as 30 or 40 ft., roman numerals are used to denote the site quality, with Site I being the best, followed by Site II, and so on.

Site Index Curves

Site index curves, depicting the average height of full-crown competitive trees at different ages, have been created for most species in specific regions or locations. These curves have their origin in "normal yield tables." In this process, trees from stands of a particular species or species group, featuring fully developed crowns, are measured for age, diameter, and height. Subsequently, curves are generated by fitting an average height-over-age guide curve to this data. A series of higher or lower curves with the same shape as the guide curve are then constructed through a process known as anamorphic curving (Fig. 16.14).

These curves, when compared to volume-over-age curves of a similar nature, enable the determination of maximum net natural stand productivity as a function of the site index. The data derived from these relationships provide forest managers with insights into how stand growth and development are influenced by site quality.

However, the traditional anamorphic guide curve method has its limitations, especially when it comes to detailed stem analysis of trees on different growth curves. Consequently, many recent site curve development studies have adopted a polymorphic technique that assigns a distinct shape to each site index level curve. This technique involves three steps:

1. A curve is fitted separately to the height growth pattern of each individual tree.
2. The fitted curve assigns a site index value to each tree.
3. A characteristic, multiple-parameter curve form is fitted to each site index value of interest.

This polymorphic approach allows for a more nuanced representation of site quality, considering the unique growth patterns of individual trees and providing a more accurate understanding of how different site index levels impact stand development.

Choosing and Measuring Trees

The accuracy of site index estimates is influenced by the number of trees measured and the method of selecting those trees. The criterion of crown class is commonly employed, with the average height of a sample of dominant or co-dominant trees being the most prevalent selection protocol. However, using the average of dominant trees alone is also a viable option. The quantity of trees chosen for measurement can vary significantly. To enhance precision, more refined selection methods have been developed. Two frequently employed methods are:

- Predominant height, defined as the average height of the tallest 100 trees ha^{-1}.
- Top height is the average of 100 trees ha^{-1} with the largest breast-height diameters.

The application of site index and site curves for natural stands and naturally regenerated site trees often results in a significant underestimation of the productivity observed in intensively managed plantations. This discrepancy is particularly notable when the planted trees have been genetically engineered for rapid height growth and are cultivated with meticulous density control and vegetation management. In such managed plantation settings, the conventional site index and curves may not accurately capture the enhanced productivity achieved through intensive management practices and genetic advancements.

Interpreting and Using Site Index

When interpreting and using site index, especially for good site trees, it is crucial to consider the following points:

1. Age Basis Clarification: When comparing site indices, identify the age on which each index is based and whether it is age at Diameter at Breast Height (DBH) or total age.
2. No Simple Conversion: A site index based on one age cannot be easily converted to one based on another age through a simple numerical relationship.
3. Age at Breast Height: Many contemporary site curves define the age at breast height. This means the tree is chronologically several years older by the number of years it took from seed to breast height.
4. Species Comparison Caution: Similar site indices for different species do not necessarily indicate similar site productivity in an absolute or relative sense.
5. Time-Dependent Measurements: Site index measurements taken at different times in a stand's life can indicate different site qualities for the same stand.

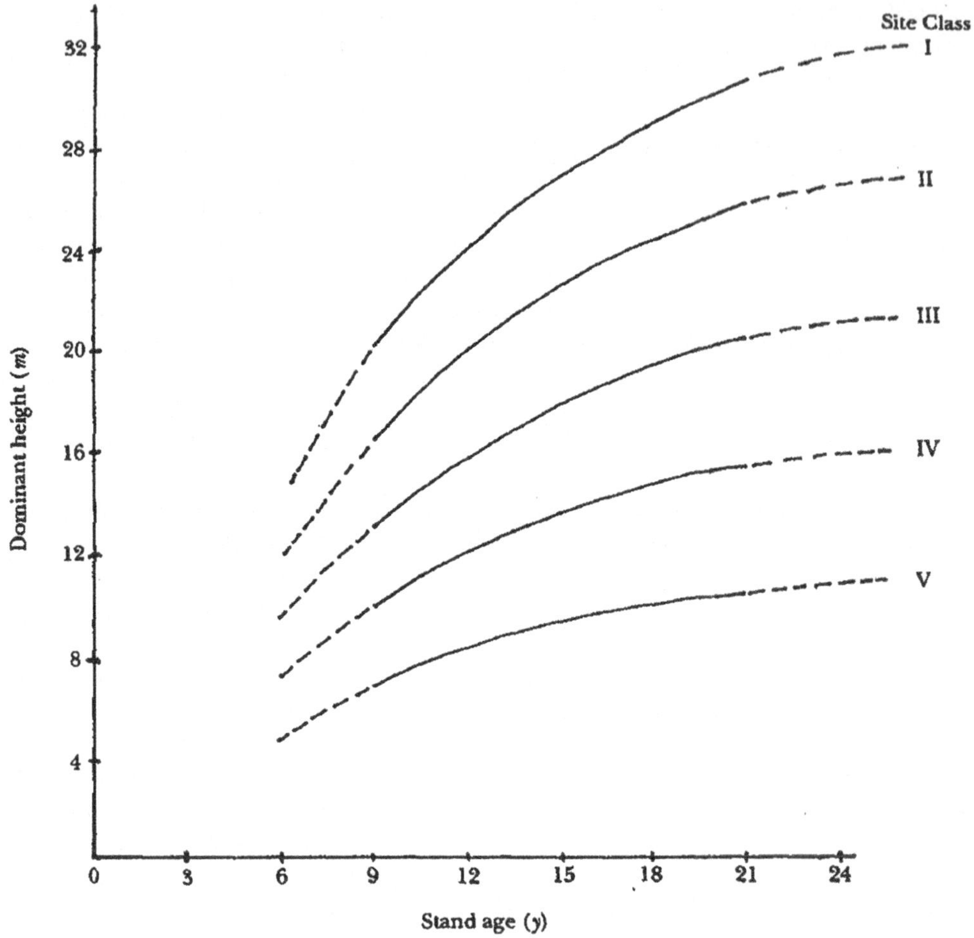

Fig. 16.14 Site index curves for teak

6. Species Range Consideration: Productivity estimates for a species provided by yield tables based on the height–age relationship may be flawed, especially near the fringes of the species range.

Vegetative Typing

Ecosystem management adopts a more holistic perspective, considering non-tree plants not merely as indicators but as integral ecological elements to be preserved for their intrinsic value. Vegetation and wildlife habitat classifications, particularly those with well-documented characteristic species lists for each class, offer valuable data for developing prescriptions and managing for diversity and broader ecological objectives.

The concept of habitat typing, a land classification approach encompassing various habitat types, integrates dominant cover and ground cover. A habitat type is a land class characterized by homogeneity in its growing environment, incorporating variations in elevation, soil, moisture, and other factors to support the same climax overstory and understory vegetation. Utilizing habitat types aids in achieving ecological goals by enhancing the identification and mapping of ecosystems, including the habitats of non-commercial tree plant species. While correlating understory vegetative types with site quality demands substantial effort, and the technique may not be universally tested, such an approach holds the potential to establish predictive plant-cover site quality relationships of useful accuracy.

Environmental Factors

The challenge in correlating understory vegetation with site index has led to a focus on physical environmental factors, particularly soil attributes. The most common approach involves linking soil characteristics with types of site index. Soil exerts a substantial and often controlling influence on tree growth, especially concerning moisture availability and nutrients. Estimating site based on measurable soil factors presents numerous advantages.

Soil properties change gradually over time, allowing for the application of soil measures in various areas, including cutover, deforested, non-forested, and timber stand-present regions. However, the correlation of site with a U.-S. Department of Agriculture (USDA) soil survey types has not proven to be high. This is partly due to these types not being classified according to several factors influencing tree growth, such as drainage class, thickness of surface horizon, and subsurface horizon depth.

The productivity of timberlands exhibits significant variation based on site index. Studies of yields achieved from fully stocked natural stands at different ages measure relative differentials, and these differences typically persist under management. While land treatment, such as fertilization or drainage, can alter site quality, only drastic interventions can transform a poor site into a good one. The timber-related economic value of a high site would be expected to surpass that of a low site.

16.3.4 Stand Density and Stocking

Stand density is a metric that quantifies the extent of tree vegetation within a specific unit of land area. This measurement can be expressed in various forms, including the number of trees, basal area, wood volume, leaf cover, or other less conventional parameters. **Stocking**, on the other hand, represents the ratio of any measure of stand density to a diverse set of norms, also expressed in the same units, selected for different purposes.

The concept of density informs about the composition of the stand, indicating the quantity of tree vegetation present. In contrast, stocking provides a perspective on how this density aligns with a forester's idealized notion of what should be, typically expressed as a percentage. For instance, full stocking could be defined as the basal area of trees larger than 63 cm DBH ha^{-1}, considered necessary for producing sawn boards. A thinning schedule serves as a guide outlining the desired stocking at each stage of development, expressed in terms of a specific measure of stand density. This schedule helps foresters make informed decisions about when and how to thin the stand to achieve the desired composition and structure over time.

16.3.4.1 Measures of Stand Density

Density measurements play various roles in assessing and forecasting outcomes and conditions within a forest. Traditionally, they are employed to ensure precise estimates of timber production. It is crucial for the density measure to consistently gauge the extent to which the trees in a given stand occupy and encapsulate the growth potential of the site.

This section addresses several frequently employed density measures:

1. Number of trees per hectare
2. Volume
3. Basal area
4. Relative density
5. Crown closure
6. Stand density index
7. Relative stand density

1. **Trees per Hectare**
   In homogeneous even-aged stands with identical age, site conditions, and history, the number of trees per hectare serves as a valuable stand density measure for short-term projections. Numerous growth and yield models for plantations utilize this metric as the input variable for stand density. Typically, the count of trees per hectare decreases with age as the trees undergo growth and competitive interactions.
2. **Volume**
   This measure is logical due to the direct or indirect association of numerous objectives with volume. However, to interpret volume effectively, it is often necessary to establish a standard, such as the volume in a stand for the yield table, and express it as a percentage of stocking.
3. **Basal Area**
   The basal area, a measure devised by foresters, represents the total cross-sectional area of trees in a stand, measured in square meters per hectare, and serves as the prevalent density variable in whole stand yield models. Basal area assumes a pivotal role in silvicultural prescriptions, as cutting rules are frequently expressed in terms of thinning or selective harvesting down to a specified residual basal area.
   The quadratic mean stand diameter D_q is a useful parameter characterizing stand and represents the diameter of the tree of mean basal area. It is estimated from N sample trees as

$$D_q = \sqrt{\frac{\overline{\mathrm{BA}}}{0.00007854}}$$

   where $\overline{\mathrm{BA}}$ = Average basal area

$$\overline{\mathrm{BA}} = \frac{\sum \mathrm{BA}}{N}$$

   Several growth models and some stand density indices use quadratic mean diameter as a variable.
4. **Relative Density**
   Curtis (1982) integrated basal area and quadratic diameter to create a composite measure of density.

$$\mathrm{RD} = \frac{\mathrm{BA}}{\sqrt{D_q}}$$

   where
   RD = Relative density
   BA = Basal area
   Dq = Quadratic mean stand diameter
   Relative density, as per DFSIM (Curtis 1982), is the designated density measure in this widely used Douglas-fir stand growth simulator.

5. **Crown Closure**

A diameter-based density measure developed by Krajicek et al. (1961) quantifies the percentage of an area covered by the vertical projection of tree crowns. This measure is determined through three steps:

(a) In a separate study associate the maximum crown width (CW) of open growth trees with their diameter (CW = *f*(DBH)).

(b) Create an average representative tree list representing one hectare of the subject stand from inventory data. Calculate the maximum crown area (MCA) in square meter of trees as if they were all open-grown.

(c) The crown competition factor (CCF) for the subject stand is calculated by summing the maximum crown area for all trees on the representative hectare, dividing the result by 43,560, and then multiplying by 100. This yields the percentage of the hectare covered by the vertical projection of the tree crowns.

As not all trees are open-grown, Krajicek's CCF may somewhat overestimate crown coverage. It is important to note that a stand can have a CCF greater than 100%, especially if there are a high number of stems per hectare and the crowns overlap. This anomaly occurs because the CCF assumes that all trees have crown widths equivalent to open-grown trees of a specified diameter. However, competition among trees can lead to a reduction in crown width for even the strongest dominant trees.

6. **Stand Density Index**

Relationships have been identified between the number of trees, average diameter, basal area, and volume, providing the basis for constructing indices and ratios as measures of stand density. Reineke's stand density index, introduced by Reineke in 1933, is one such measure designed for pure, even-aged stands that remains independent of site index and age. Reineke utilized two elements from normal yield tables to formulate a stocking measure:

(a) Various fully stocked even-aged stands of a species with identical quadratic mean diameter were noted to exhibit an approximate equivalence in the maximum number of trees per hectare.

(b) When plotting a curve of the maximum number of trees per hectare against different quadratic mean stand diameters and fitting it with a log-linear regression using data from normal yield tables, a consistent slope of 1.650 was observed for most species. This uniformity in slope suggests the emergence of a common function across various species.

$$\log N = -1.650 \log D_q + k$$

where

N = number of trees per hectare

D_q = quadratic mean stand diameter

K = a constant for each species

Reineke's stand density index (SDI) is calculated by converting the number of trees at any quadratic mean diameter into an equivalent density at a quadratic mean diameter of 10 in. The process involves using a reference curve based on red fir and redwood, where 24,700 trees per hectare correspond to a quadratic mean diameter of 25.40 cm. Reineke drew a series of parallel lines on a graph (FIG) to relate the maximum number of trees to the quadratic mean stand diameter. These curves permit converting the number of trees at any quadratic mean diameter into an equivalent density at a quadratic mean diameter at 10 in. The number of trees at a quadratic mean stand diameter at 10 in. is Reineke's stand density index (SDI). It is calculated as:

$$\mathrm{SDI} = N\left(\overline{D}_q/10\right)^{1.605}$$

Reineke's stand density index encompasses several valuable attributes. The required data can be readily acquired through inventory. Estimating site and age is necessary, enabling the comparison of stand densities across various sites and ages. When provided with a target number of trees at a specific diameter, it becomes possible to estimate the number of trees at all diameters up to the target with the same density. Primarily applicable to pure, even-aged stands, Reineke's stand density index, along with other stand density indices (SDIs), serves as a useful tool. These indices play a pivotal role in guiding thinning decisions by aiding in the determination of the target number of trees per hectare.

7. **Relative Stand Density**

Drew and Flewelling (1979) formulated a stand density index that shares similarities with Reineke's concept. However, they acknowledged the correlation between tree volume growth and both tree height and diameter. Irrespective of site quality, they identified a consistent association between the maximum mean tree volume and the number of trees per hectare present in natural stands. Their density measure is derived from this relationship and is defined as:

$$\text{Relative Stand Density} = \frac{\text{Density in trees per hectare of subject stand}}{\text{Maximum density in trees per hectare attianable in a stand with the same tree volume as the subject stand}}$$

As tree volume is influenced by both height and diameter, Drew and Flewelling (1979) illustrated the connection between stand density, stand diameter, height, and growth through a series of transformations. This depiction is termed a "density management diagram." The diagram is designed to be locally calibrated, specifically for a particular species.

8. **Canopy Density**
   Canopy density, represented as a decimal coefficient, serves as an indicator of the canopy's relative completeness, with a closed canopy designated as unity (1). It focuses exclusively on the fullness of the canopy and does not correlate with crop volumes, basal area, or the quantity of trees in the stand. This metric offers insights into the extent to which the available space in the canopy is occupied, where higher values signify a more complete or closed canopy.
   Classification:
   Closed: When density is 1.0
   Dense: When density is between 0.75 and 1.0
   Thin: When density is between 0.5 and 0.75
   Open: When density is under 0.5

16.3.4.2 Crown Competition Factor (CCF)

The CCF is an alternative density measure purportedly independent of site quality and stand age. To calculate CCF, the relationship between crown width and Diameter at Breast Height (DBH) is established using a sample of open-grown trees. For instance, an equation like CW = −0.6117 + 20.812D may be used, where CW represents crown width in meters, and D is the DBH in meters. The maximum crown is determined as a percentage of one hectare, indicating the maximum proportion of one hectare that crowns of trees with a given DBH can occupy. This factor provides insights into the extent of crown competition, offering a perspective on how much canopy space is potentially covered by trees of a specific size. Thus,

$$\text{MCA} = \left[\pi\,(\text{CW})^2\right]/(4 \times 100/10{,}000)$$

$$= 0.007854\,(\text{CW})^2$$

By summing the Maximum Crown Areas (MCAs) for all trees on an average hectare of forest land, the Crown Competition Factor (CCF) is derived. In essence, CCF serves as an expression of stand density. It provides an estimation of the area available to the average tree within the stand relative to the maximum area it could utilize if it were in an open environment. This metric offers valuable insights into the competitive dynamics within the stand, helping to assess how closely trees are spaced and how much growing space each tree has in relation to its potential maximum.

16.3.4.3 Point Density and Competition Indices

Different measures of density are commonly used to assess the overall density of a stand, often on an average basis. In addition to these general measures, more specific indicators of density have been developed to characterize the level of competition at a particular point or for a specific tree within the stand. These specific measures are known as point density or competition indices. The fundamental idea behind these indices is to articulate the extent to which growth resources, including light, water, nutrients, and physical growing space, available to an individual tree are constrained by the presence of neighboring trees. With the growing interest in modeling individual tree growth and yield, several competition indices have been developed. These indices can be broadly classified into two categories:

1. Distance-independent measures
2. Distance-dependent measures

Distance-independent measures are designed to portray the competitive status of a tree or a class of trees in relation to all trees within a stand. These measures often involve ratios such as tree height to mean height, tree height to dominant height, and tree volume to mean volume. One significant advantage of distance-independent measures is that they do not necessitate the time-consuming determination of tree location. However, a notable disadvantage is that these indices gauge a tree's status in relation to the average conditions of the entire stand rather than the specific conditions surrounding the tree.

On the other hand, distance-dependent measures aim to describe a tree's competitive status based on the immediate conditions surrounding the tree. These measures can be broadly categorized into three classes:

1. Area overlapping indices
2. Distance weighed size ratio
3. Area potentially available indices

Area overlap indices operate on the principle that each tree has a potential area of influence, within which it acquires or competes for site factors. In these indices, trees whose area of influence overlaps with that of a subject tree are regarded as competitors. An example of an area overlap index is Spurr's (1962) point density, which is derived from adapting Bitterlich's point basal area (Bitterlich 1948) for use as a point density measure. In Spurr's point density, all trees encompassed in a basal area sweep are considered competitors. This approach helps to quantify the competitive influence of neighboring trees on a particular tree, taking into account the spatial arrangement and area of influence of each tree in the stand.

Distance-weighted size ratios are determined by calculating the sum of ratios between the dimensions of each competitor and the subject tree. These ratios are then weighed by a function of the inter-tree distance. This method accounts for both the size of competitors and their proximity to the subject tree, giving a nuanced view of competitive interactions. The area potentially available index takes into consideration the polygons formed by the intersections of the perpendicular bisectors of the distance between a subject tree and its competitors. The area of these polygons, calculated from the coordinates of the vertices, represents the space potentially available for tree growth. This index offers a spatially explicit measure of the available growing space, acknowledging the influence of neighboring trees on the subject tree's competitive environment.

16.4 Forest Inventory

Forest Inventory is the systematic compilation of accurate and reliable tree-related information, organized according to the necessary hierarchical units of assessment. In essence, it seeks to comprehensively depict the quantity, quality, and diameter distribution of forest trees, along with various characteristics of the land on which these trees thrive. Synonymous with the term 'Cruse' in North America or 'Enumeration' in India, forest inventory, as defined by foresters, involves the counting and classification of individual trees based on criteria such as species, size, and condition. As the significance of forest areas extends beyond timber production to encompass non-wood values like recreation, watershed management, and wildlife habitat, the concept of forest inventory has broadened. The primary objective remains the determination of timber volume in the forest, crucial for assessing yield.

Occasionally, inventory is essential for evaluating the value of forests for sale or exchange and estimating returns from clear-felled areas. Additionally, forest inventory aids in determining current and periodic annual increments and may involve mapping regions of high or low volume production per unit of area, facilitating decision-making for industrial development. In summary, the purpose of forest inventories is to furnish information for effective forest management and planning, support pre-investment decisions related to the establishment or expansion of forest industries, and assess project feasibility from economic, social, and other perspectives.

16.4.1 Kinds of Enumeration/Inventory

1. Total or Complete inventory
2. Partial or sample inventory

Total or complete enumeration in forest inventory involves the exhaustive counting and measurement of the desired species above a specified diameter limit across the entire area of the forest unit being considered. This approach is resource-intensive and time-consuming, and it is typically employed in relatively smaller areas of valuable and intensively managed forests, such as Primary Belt Inventory (P.B.I.) areas within uniform or other shelterwood systems. Total enumeration is also chosen in miscellaneous forests with a large number of species or where economically important species are limited to specific favorable locations. It is sometimes used for statistical studies as a check on the results of partial enumerations. However, due to the vast areas involved and the associated time, labor, and financial costs, complete enumeration is generally avoided in forestry.

On the other hand, partial or sample inventory entails conducting the enumeration in only a representative portion of the entire forest. A sample is a part of the population, selected and examined as a representative of the whole. Sample enumeration involves surveying a representative part of the entire population, and the ratio of the sample to the whole population is termed the sampling fraction or intensity of sampling, expressed as a percentage (e.g., 5%, 10%, or 20%). Actual enumeration is then carried out in a few sampling units, which are subdivisions of the sampled area. This method is more practical for large forested areas, allowing for efficient and cost-effective assessment while still providing reliable estimates for the entire population.

16.4.2 Kinds of Sampling

1. Random sampling and
2. Non-random sampling

16.4.2.1 Random Sampling

Random sampling is a method of sampling in which the sampling units that make up a sample are chosen in a way that ensures all possible units of the same size have an equal chance of being selected. This method aims to eliminate bias and provide a representative subset of the entire population. In random sampling:

1. Equal Chance: Every sampling unit within the population has an equal probability of being included in the sample. This ensures that the sample is not skewed toward any particular subset.
2. Random Selection: Sampling units are chosen through random processes, such as using a table of random numbers, drawing unbiased cards from a well-shuffled pack, or employing a random drawing of lots. These methods ensure unpredictability and eliminate any systematic patterns in the selection.

Random sampling is particularly valuable in statistical analysis because it allows researchers to make inferences about the entire population based on the characteristics observed in the randomly selected sample. This method is widely used in various fields, including scientific research, market research, and social sciences, to ensure the generalizability of findings to the larger population.

Kinds of Random Sampling

1. Unrestricted or simple random sampling
2. Stratified random sampling
3. Multistage sampling
4. Multiphase sampling
5. Sampling with varying probability
6. List sampling

Kinds of Non-random Sampling

1. Selective sampling
2. Systematic sampling
3. Sequential sampling

Unrestricted or Simple Random Sampling: The Unrestricted or Simple

Random sampling involves selecting sampling units for the sample through a strictly random process from the entire population or area, without the prior division into homogeneous blocks. To illustrate, consider a scenario where the sampling unit is 0.1 hectares, and the total population area is 1000 hectares. The population is then divided into 10,000 sampling units, each assigned a unique number ranging from 0000 to 9999. Subsequently, for a 5% unrestricted random sampling, 500 different four-digit reference numbers are chosen using a table of random numbers. This ensures that every possible set of 500 units has an equal chance of being selected for the sample.

Stratified Random Sampling

Stratified random sampling is a method in which the population is initially divided into subpopulations or strata, and then sampling units are selected from each stratum in proportion to its size. This approach is a modification of simple random sampling and is particularly essential when unrestricted random sampling is not sufficient. Unrestricted random sampling tends to provide accurate results only when the forest crop is uniform, but in reality, forest populations often exhibit considerable heterogeneity. In the case of stratified random sampling, the precision of estimating the sampled mean is significantly improved by recognizing and addressing these variations. The process involves subdividing, either physically or conceptually, the area to be sampled into homogeneous groups known as strata. From each stratum, sampling units are then selected. This method of dividing the area into homogeneous groups is termed stratification. By accounting for the differences in the forest population through stratification, the sampling design is modified accordingly, leading to more accurate and representative results in the estimation process.

Multi-stage Sampling

In the previously described random sampling, sampling units are drawn from a population all at once, making it a single-stage sampling method. However, in some situations, it may be more cost-effective to take samples at one stage and then subdivide them to obtain additional sampling units in subsequent stages. When sampling units are drawn in two or more stages, the method is referred to as multi-stage sampling. In this approach, the size of the sampling units becomes smaller at each successive stage, but the fundamental principle of random sampling can be maintained at each stage of selection. This method allows for a more flexible and economical way of gathering samples, especially in scenarios where obtaining all samples in a single stage may be impractical or resource-intensive.

Multiphase Sampling

In two-phase sampling, the same sampling units are employed in different phases to collect either distinct information or the same information using different methods. This method is commonly utilized in forest inventories and is often referred to as double sampling due to its two distinct phases. For instance, consider estimating the number of bamboo culms in a forest. In the first phase, the number of clumps per hectare is determined through extensive systematic surveys covering a large area. Subsequently, in the second phase, the number of culms per clump is ascertained through a smaller enumeration survey, covering a much smaller area than the first phase. By utilizing the same sampling units in these two phases, double sampling allows for more precise and efficient estimation of the desired information, making it a valuable technique in various research and inventory applications.

Sampling with Varying Probability

In the earlier described random sampling, the chances of selecting all sampling units are always the same. However, in certain populations, the probabilities of selection may vary as the sampling process proceeds. To address this variation, a modified method known as sampling with varying probability is employed. This method has gained increasing importance in recent years, particularly in the context of forest inventories. Several techniques fall under this category, including the "Bitterlich" method, also known as "variable plot," "point sampling," or "p.p.s." (probability proportional to size). Additionally, the "three p" method, or "sampling with probability proportional to prediction," is another application of this aspect of sampling theory. These approaches aim to adjust the probability of selecting sampling units based on certain characteristics or sizes within the population. Such modifications help to enhance the efficiency and accuracy of the sampling process, especially in scenarios where variability in the population may affect the representativeness of a simple random sample.

List Sampling

This is another form of sampling with varying probability. The method consists of making a list of sampling units along with their measure of size in any order.

Non-random Sampling

Non-random sampling is a method in which samples are chosen based on the subjective judgment of the observer, guided by certain rules or guidelines indicating which samples should be selected. The selection is not governed by random processes, and instead, it relies on the observer's judgment and adherence to specific criteria. It is important to note that non-random sampling introduces a potential for bias because the selection process is influenced by the observer's subjective judgment. Unlike random sampling, where each element has an equal chance of being chosen, non-random sampling introduces a level of subjectivity that can impact the representativeness of the sample and the validity of the results. As a result, caution should be exercised when using non-random sampling methods, and efforts should be made to minimize any potential bias introduced by the observer's judgment.

Selective Sampling

Selective sampling involves choosing samples based on the subjective judgment of the observer. In this method, the observer exercises discretion and selects samples according to specific criteria or guidelines, often relying on their knowledge and expertise regarding the population under study. When properly used by an individual with extensive knowledge of the population, selective sampling can yield good approximations of population parameters. The effectiveness of selective sampling depends on the observer's ability to make informed and unbiased choices that adequately represent the characteristics of the population. While this approach allows for targeted sampling in certain situations, it is essential to acknowledge that selective sampling introduces a potential for bias, and results should be interpreted with caution. A careful consideration of the observer's expertise and the appropriateness of the chosen samples are crucial to ensure the validity of the findings.

Systematic Sampling

Systematic sampling is a method of non-random sampling in which sampling units are selected according to a predetermined pattern without resorting to random selection. Typically, this pattern involves regular spacing of units. The first unit's selection can be either random or based on a fixed arbitrary rule, leading to two types of systematic sampling. When the first unit is selected randomly, and subsequent units are chosen according to a fixed pattern, it is referred to as systematic sampling with a random start. Despite limitations in estimating variances, systematic sampling offers advantages over random sampling, making it a preferred choice in forest inventories. These advantages include simplicity and ease of implementation, a straightforward location of sampling units, a cost reduction in reaching those units, an increased certainty of objectivity in sample selection, and a more uniform distribution of the sample. These benefits are particularly pronounced in tropical forests where challenging environmental conditions can impede fieldwork.

Sequential Sampling

Sequential sampling is a method characterized by the absence of a predetermined number of observations in the sample. Instead, sampling units are taken successively from a population, with each new sample encompassing all the units from the preceding sample. The key feature of this method is that the number of observations is not fixed in advance. In sequential sampling, the confidence interval at each stage decreases, and sampling continues until the desired precision is achieved. This method is commonly employed in hypothesis testing. At each stage, a new observation is made, and the hypothesis is tested to determine whether it should be accepted or rejected. The process repeats until a decision is reached, and the hypothesis is either accepted or rejected. Once this decision is made, the sampling procedure is concluded, and no further sampling units are selected beyond what is necessary to reach a decision. Sequential sampling allows for adaptive decision-making and an efficient use of resources in testing hypotheses.

Point Sampling

Point sampling is a sampling technique where sample units are singular points, a method widely employed in forest inventory. Bitterlich demonstrated that counting trees with breast height cross-sections exceeding a certain critical angle from a random point, when multiplied by a constant factor, provides an unbiased estimate of basal area per hectare. This technique is known by various names, such as "angle count cruising," "pointless cruising," "point sampling," "variable plot cruising," "PPS (probability proportional to size) sampling," and "polyareal plot sampling." Point sampling selects trees based on their size rather than their frequency of occurrence and can be horizontal or vertical depending on whether diameter (or basal area) or height needs estimation.

Horizontal point sampling has significantly influenced forest inventory practices. The process involves selecting random or systematically distributed sampling points across the entire area to be inventoried. Trees around each point are viewed through an angle gauge at breast height, and those forming an angle larger than the critical angle of the instrument are counted. The number of tallied trees, when multiplied by a constant factor dependent only on the angle size, yields the basal area per hectare. Japanese forester Hirata introduced a method for determining the mean stand height through vertical point sampling, utilizing an instrument called a Conimeter. In a complete 360° sweep around the sample point, all trees appearing taller than a critical angle C are counted, providing valuable information about vertical structure.

16.5 Conclusion

Measuring trees and forests is crucial in the field of forestry and forest science. These measurements help in comprehending the growth and development of forests, assessing the quantity of products derived from them, and ensuring appropriate management practices. This chapter provides an overview of forest measurement at two scales—initially focusing on individual trees and expanding to stands of trees. It is noteworthy that much of the measurement at larger scales involves conducting measurements at a smaller scale and subsequently employing mathematical approaches to extrapolate those measurements to a larger scale. The current chapter presents methods for measuring stand quality, stand density, stocking, and other parameters using smaller units. There is a growing inclination toward adopting more sophisticated equipment and techniques in forest measurement. Computers are commonly employed to assist in storing data collected from the forest and performing the necessary mathematical computations to transform raw data into valuable information about the forest. However, even with the utilization of highly advanced measuring technologies, possessing knowledge of the fundamental units in forest measurement remains essential.

Lessons Learnt

- Forest mensuration, the scientific discipline dedicated to measuring and analyzing various attributes of forests, plays a crucial role in sustainable forest management and resource assessment.
- Its significance extends across diverse domains, ranging from ecological research to sustainable resource management, making forest mensuration indispensable for understanding, managing, and conserving forests.
- Forest mensuration yields precise data on parameters such as tree density, species composition, stand structure, biomass, and volume. This information is essential for estimating available timber, carbon storage, and other critical forest resources.
- The accurate measurement of biomass and carbon storage, facilitated by forest mensuration, is vital for evaluating the carbon sequestration potential of forests.
- Beyond resource assessment, forest mensuration contributes valuable data for ecological research and education. It supports studies on forest dynamics, species interactions, growth patterns, and overall ecosystem functioning.
- Governments and organizations leverage forest mensuration data to formulate policies and regulations guiding forest conservation, sustainable development, and land use planning.
- The forest industry relies on precise measurements of tree dimensions and volumes to optimize timber harvesting, processing, and marketing practices. Nevertheless, this field faces its own set of challenges.
- Forest heterogeneity poses a challenge, making accurate and representative sampling a complex task. The selection of sample plots that faithfully represent the entire forest is critical, and biased or non-random sampling may compromise the accuracy of measurements.
- Some forests, particularly those in remote or rugged terrains, present difficulties in accessibility and tree measurement. This limitation can result in incomplete data collection and potentially skewed results. Addressing these challenges necessitates robust study designs, a thorough training of field staff, a calibration of measurement instruments, meticulous data quality control, and ongoing research to enhance methodology accuracy.

Key Questions

1. Discuss the methodology for measuring tree height and elucidate its significance in the estimation of tree volume.
2. Provide a definition of a "stand" in forestry and delve into the methods used to estimate stand growth.
3. Justify the assertion that stand growth is influenced by both stand quality and density.
4. Provide an explanation of the different measures of stand density.
5. Briefly explore the principle involved in controlling growing stock to attain specific management objectives, with a special focus on stand management diagrams.

References

Bitterlich W (1948) Winkelzahlprobe. Algemeine Forst Hoizwirtschafts Zeitung 59(1/2):4–5. (Translated and abstracted in Forestry Abstracts 10(2):423–424, 1949)

Chaturvedi AN, Khanna LS (1994) Forest mensuration. International Book Distributors, Dehradun, India. 406p

Curtis RO (1982) A simple index of stand density for Douglas-fir. For Sci 28(1):92–94

Drew TJ, Flewelling JW (1979) Stand density management: an alternative approach and its application to Douglas-fir plantations. For Sci 25(3):518–532

García O (1995) Notes on forest mensuration. Institute of Forest Management Chair of Forest Mensuration, Universidad Austral de Chile, Faculty of Forestry Sciences
Jones JR (1969) Review and comparison of site evaluation methods, vol 51. US Rocky Mountain Forest and Range Experiment Station
Krajicek JE, Brinkman KA, Gingrich S (1961) Crown competition—a measure of density. For Sci 7(1):35–42
Prodan M (1968) Forest biometrics. Pergamon Press, Oxford
Reineke LH (1933) Perfecting a stand-density index for even-aged forests. J Agric Res 46:627–638
Spurr SH (1962) A measure of point density. For Sci 8(1):85–96
West PW (2009) Tree and forest measurement, 2nd edn. Springer

Contemporary Silviculture

17

Ramakrishna Hegde, P. A. Clara Manasa, and Supriya K. Salimath

Abstract

The genesis of contemporary silviculture can be traced back to the imperative to find sustainable and efficient methods for wood and forest product generation as forested lands diminished and exploitation faced increasing restrictions. Silviculture, situated at the intersection of social and natural sciences, is defined as the theory and practice of controlling forest establishment, composition, structure, and growth. While articulating management objectives is crucial for silviculture's success, it is essential to dispel the misconception that its sole purpose is wood production. Silviculture is employed to manage forests for various uses, encompassing wildlife, water, recreation, and aesthetics, ensuring the vitality and productivity of wooded ecosystems and the continuity of key ecological processes. At the core of silviculture is the concept of a silvicultural system, a planned series of treatments implemented throughout a stand's lifespan to achieve specific structural goals. This treatment program, involving harvesting, regeneration, and stand tending techniques, aims to provide consistent benefits over time. Recognizing that trees have a finite lifespan, actions during regeneration significantly influence a stand's future evolution. Regeneration, whether natural or artificial, is the cornerstone of forest crops, with various disturbances leading to the emergence of natural vegetation. Understanding the interplay between species and events is crucial for effective management. Silviculture, with its focus on timber production, shapes the development and growth of trees across various phases. Despite its historical success in adapting to environmental shifts, silviculture faces unprecedented challenges in the current global transition. Modern silvicultural planning necessitates adaptability to new circumstances and sustainable forest resource use. Proactive silviculturists have a spectrum of choices, including intensive silviculture, precision silviculture, and considerations within the context of climate change and sewage silviculture. These diverse approaches reflect the dynamic nature of silviculture as it responds to evolving societal expectations and ecological insights in the pursuit of resilient and sustainable forest ecosystems.

Keywords

Silviculture · Stand structure · Silvicultural system · Natural regeneration · Growth curve

17.1 Introduction

Silviculture encompasses diverse definitions, including its characterization as both the art and science of cultivating and nurturing a forest to meet the social and economic needs of individuals and society. It is alternatively viewed as the practical application of autecology or silvics knowledge in forest management. Another perspective defines silviculture as the theory and practice of regulating the establishment, composition, structure, and growth of forests. As a practice rooted in applied forest ecology, silviculture plays a pivotal role in biological technology, translating ecosystem management principles into tangible actions. **Contemporary silviculture**, as articulated by Smith et al. (1997), is succinctly described as the "art and science of managing forests." Here, the term "art" reflects the skillful application of knowledge acquired through meticulous observation and prolonged practical experience. Silviculturists leverage this accumulated knowledge to adapt successful practices to new management conditions, thereby ensuring the effective stewardship of forests. The essence of silvicultural practice lies in the diverse treatments employed to care for and enhance forests, optimizing their utility for various purposes. The foundational underpinning of silviculture in the natural sciences is silvics, a discipline addressing the

R. Hegde (✉) · P. A. Clara Manasa · S. K. Salimath
College of Forestry, Ponnampet, Keladi Shivappa Nayaka University of Agricultural and Horticultural Sciences, Shivamogga, Karnataka, India
e-mail: ramakrishnahegde@uahs.edu.in

A. K. Mandal, A. Nicodemus (eds.), *Textbook of Forest Science*, https://doi.org/10.1007/978-981-97-8289-5_17

growth and development of individual trees, as well as other forest species and entire forest ecosystems.

17.2 Importance of Silviculture

Silviculture is crafted to establish and uphold forests that align with the goals of both the landowner and the wider society. While most silvicultural methods are applied during timber harvesting due to a significant reduction in operational costs associated with wood removal, it is crucial to recognize that silviculture extends beyond a singular focus on wood production. This discipline encompasses diverse facets, including the management of seminatural woodlands and protection forests. The manipulation of growing space serves as a key strategy in silviculture to achieve various objectives such as enhancing wildlife habitat, creating scenic vistas, or fostering a robust ground layer for surface watershed protection.

Silvicultural practices, irrespective of the specific management objective, strive to govern the establishment, composition, structure, growth, and roles of trees within managed forests. The establishment of preferred tree species is achieved through natural regeneration, direct seeding, or planting. The term "composition" refers to the diversity of tree species and their relative abundance, while "structure" encompasses internal characteristics like tree crowns, vitality, diameter and height distributions, the presence of dead trees (snags), wood on the ground, and understory vegetation. Implementing a range of individual practices, such as site preparation, promoting natural regeneration, planting, fertilization, thinning, and final harvest based on diameter or age, silvicultural activities work cohesively to cultivate the desired species and structural attributes within and across managed areas in a forested landscape.

17.3 History of Silviculture

Indigenous communities worldwide have historically relied on forests for their sustenance, drawing upon these ecosystems for food, fuel, tools, and shelter. The utilization of wood from forests spans diverse applications, including the crafting of tools, implements, bridges, boats, and dwellings. Archaeological evidence, such as the discovery of extensive caches of wooden tools dating back approximately 5000 years in northern India, attests to the rich history of forest resource utilization by early inhabitants. The variety of tree species used in crafting tools suggests a deep understanding of the distinct properties of different woods. In ancient times, prehistoric cultures exhibited an extensive knowledge of forests, influencing the production of lumber, thatching, staves, weapons, and more. Despite this utilization, the impact on forests by aboriginal cultures was minimal or localized, owing to low population densities and expansive forested areas. Sustainable resource management was not an imperative need, as new, untouched areas could be discovered, colonized, and utilized when the productivity of a particular forest area declined.

However, with the growth of human populations, the availability of untouched forests diminished. Societies transitioned to more sedentary lifestyles, relying heavily on agriculture, leading to concentrated forest exploitation near populated areas. This proximity resulted in unregulated pressure on forests, often leading to overexploitation—a phenomenon commonly known as the "tragedy of the commons." As a response to this challenge, many societies developed rules and regulations governing forest exploitation, including limitations on tree harvesting, grazing, foraging rights, and the collection of minor forest products. The timing of this transition to regulated forest management varied among cultures. In India, historical records from the Gupta period (320–800 AD) described detailed measures to manage forests and wildlife. However, subsequent invasions and colonial periods, such as Mughal rule and British colonization, contributed to renewed forest destruction. By the mid-seventeenth century, the adverse impacts of slope erosion and wood scarcity prompted the urgent need for a well-developed forest management system. Local and provincial governments swiftly implemented regulatory measures to counter the effects of overexploitation. Simultaneously, a rich body of silvicultural techniques was developed and disseminated across the country to address the socio-economic challenges associated with forest resource depletion.

17.3.1 Origins of Modern Silviculture

With the diminishing availability of forested lands and the increasing regulation of forest exploitation, efforts turned toward identifying and implementing more sustainable and efficient methods for wood and other forest product production. This marks the inception of modern silviculture. Among the earliest documented silvicultural techniques was the coppice system, systematically practiced in France since the ninth century. Over subsequent centuries, the coppice working method extended to hardwood forests across much of Europe. In the preceding two centuries, forest science and management primarily concentrated on optimizing commodity production, particularly wood for timber, pulp, and fuel. Driven by the belief that homogeneous products are more cost-effective to produce and manage, management practices often resulted in even-aged, mono-specific, or species-poor stands (Puettmann and Ek 1999; Paquette and Messier 2009; FAO 2010; West 2014). These practices encompass activities such as planting or natural regeneration, including coppicing or natural/sown seed approaches, coupled with weed control and timber stand improvement practices to regulate species composition, tree size, and quality—referred to hereafter as "conventional" forest management. Simultaneously, alternative silvicultural approaches have been developed in various regions globally. However, these alternatives have received comparatively less attention on a global scale, highlighting a need for greater consideration in contemporary silviculture.

17.3.2 Scope of Silviculture

Silviculture serves as a scientific discipline that seeks to bridge the gap between natural sciences and societal disciplines. While many principles of silviculture applicable in temperate ecosystems generally hold true in tropical regions, a notable distinction arises due to the exceptional biodiversity found in most tropical forest ecosystems. This diversity adds complexity to silvicultural planning and interventions. Notably, ensuring compatibility with biodiversity conservation objectives becomes crucial. The reliance of many people in the tropics on forests for subsistence and livelihood further emphasizes the need for silvicultural goals to align with the aims of rural development and poverty reduction. Traditionally, the primary focus of tropical silviculture has been to provide practical methods for sustaining a consistent yield of wood from economically viable species. However, recent years have witnessed substantial shifts in forest management and silviculture due to changes in global and regional societal and environmental conditions. In response to these shifts, silvicultural concepts and techniques must not solely cater to the objectives of forest owners but also consider the diverse societal demands and ecological requirements arising from environmental changes. As outlined by Weber (2011), the following sections will delve into some of the new aspects with paramount impact on forest management in tropical forests.

17.3.2.1 New Perspective on Biodiversity and Conservation Management

Biodiversity's global importance for sustainable forest management and certification schemes is now recognized, extending to technical aspects like tropical silviculture. Historically, biodiversity's role was confined to nature conservation, but recent growth in awareness has broadened its scope. Protected areas were long considered the primary solution for preserving natural landscapes, ecosystems, and genetic diversity in tropical forests. Integrating silvicultural treatments with logging not only reduces costs but also promotes the idea that logging can be silviculturally beneficial. To address escalating soil effects and damages with logging intensity, silvicultural interventions should be moderate and concentrated in small areas, minimizing costs. Advances in forest planning, monitoring, and logging technology provide new tools at the management level to mitigate the damaging effects of silvicultural interventions.

17.3.2.2 Climate Change

Effective forest management influences carbon flows between forests and the atmosphere, impacting both the acceleration and mitigation of atmospheric CO_2 accumulation, a major factor in climate change. Tropical forests, contributing 40% of terrestrial biomass carbon and 30–50% of terrestrial productivity, are pivotal. Destruction and degradation, mainly in the tropics, constitute about 17% of total anthropogenic greenhouse gas emissions. Given forests' role as a carbon sink or source, any activity altering biomass directly affects the carbon budget. Thus, managing tropical forests becomes crucial in political efforts to mitigate climate change. The following sections address key topics aligned with diverse management objectives.

Management for Mitigation

The primary strategies for forest mitigation involve maintaining high carbon stocks in natural and managed forests, reducing losses from management interventions. To preserve high carbon stocks, conserving standing forests, especially old-growth ones, is crucial. Various silvicultural treatments, applied before or after logging, can enhance carbon storage and overall forest health. However, a comprehensive knowledge of stand status and intervention impacts on biomass and soil carbon, as well as other greenhouse gases (e.g., CH_4 and N_2O), is essential. When managing tropical forests for carbon, minimize treatments that affect soil carbon and encourage those creating space for desired species. The greatest carbon losses result from poorly planned or executed logging operations. Despite forest management's mitigation potential, the key opportunity in tropical regions lies in reducing carbon emissions from deforestation and degradation.

Management for Adaptation of Forests to Climate Change

In mitigating tropical forests, the impact of climate change, such as regional drying and warming, extreme weather events, and biodiversity loss, must be considered. Forest management must be highly adaptive, relying on current information about forest conditions. Given the uncertainties of climate change in the tropics, silvicultural adaptation strategies should resemble other risk-reducing approaches.

Prioritizing forest health and ecosystem diversity aims to enhance resilience, crucial for recovery after disturbances.

17.3.2.3 Increasing Importance of Goods and Services Other Than Timber

Non-wood Forest Products

Since ancient times, humans have gathered non-wood forest resources like berries, mushrooms, fruits, herbs, and bushmeat for food, energy, and construction. Silviculture faces a new challenge managing diverse products, requiring expertise in both timber and non-wood forest products (NWFPs). To address this, managers must expand their knowledge from forest management to include agroforestry and farming practices. Therefore, contemporary tropical silvicultural education and training should encompass these aspects.

Biofuels

Fuelwood holds traditional significance in many tropical countries, particularly in Africa, where it dominates heat energy sources for cooking and heating. Given that fossil fuels contribute 57% to total greenhouse gas (GHG) emissions, a shift from fossil fuels to biomass fuels is increasingly vital in GHG reduction strategies. The tropics, with their high production rates of up to 40 tons ha^{-1} $year^{-1}$, merit special attention. While a significant portion of "modern" fuelwood comes from agricultural areas or nonwoody biomass, this situation presents new opportunities and challenges for silviculture.

Ecosystem Services

In addition to yielding marketable goods like timber, non-wood forest products (NWFPs), and biofuels, natural and managed forest ecosystems offer essential services. However, these services are often challenging for forest owners to directly sell on local markets. Over recent decades, a shift from mono-structured silviculture, focused solely on high wood yields, has evolved toward multiple-use forest management. This approach considers timber, NWFPs, and environmental services. Persistent deforestation and degradation, coupled with increasing resource demand, necessitate a comprehensive approach to tropical forest management. Therefore, promoting sustainable extractive uses, alongside the conservation of ecosystem services, particularly biodiversity, is crucial.

Ecotourism

Numerous tropical forests serve as vital repositories of biodiversity, providing sanctuary for endangered species and indigenous communities. This increased significance for conservation and local recreation has also captured the attention of ecotourism. Managed forests in this context demand exceptionally sensitive silvicultural interventions, preserving their natural, biodiverse, and appealing character, as any evident logging damage is incompatible with ecotourism. In the ecotourism framework, a careful consideration of landscape effects and compatibility with attractive animal presence are essential. Forests are integral elements within human-populated landscapes, necessitating the involvement of local communities in participatory planning. Adequate compensation allows for specific activities, such as managing ornamental species, creating openings and viewpoints, establishing scenic roads or hiking corridors, and extending rotation lengths to enhance a forest's value for recreation and ecotourism.

17.4 Principles of Silviculture

Silviculture adheres to essential principles, with the first two focusing on respecting nature's limitations and conserving site productivity. The subsequent four principles serve as checks against the unintended consequences of poor silviculture judgment.

Principle 1: Imitating Nature Through Silviculture

Recognizing the magnificence of natural forests, shaped by millions of years of exposure to climate, disease, and disturbances, underscores the wisdom of imitating nature in silviculture. This discipline intentionally guides natural processes to enhance forest utility efficiently and expeditiously, aligning with anthropocentric ecological constraints.

Principle 2: Conservation of Site Productivity

Forestry, particularly silviculture, places paramount importance on maintaining the productivity of living forests. Soil, a less renewable resource, faces lasting harm. Although damage to the living component of the site may be severe, it is typically not permanent. Soil erosion, especially due to careless treatments associated with timber extraction, poses a significant risk. Permanent forestry without degradation is achievable by prioritizing soil health and employing the four guiding principles described below.

Principle 3: Control of Stand Structure and Process

Silviculture serves as dynamic process engineering, creating structures or developmental sequences that are harmonious with the environment. Managing stands, evolving with time, necessitates a sophisticated design considering environmental influences. The forester, partly creating a new ecosystem, must also adapt to the existing one.

Principle 4: Control of Composition

Silviculture aims to limit stand composition based on ecological and biological standpoints, often resulting in fewer species than the natural stand site. Species composition is regulated through disturbance control during stand establishment, favoring desirable species through planting or artificial seeding, and introducing non-native species when appropriately adapted.

Principle 5: Control of Stand Density

Effective silviculture balances stand density to avoid issues like branchy or malformed trees or excessive competition. Managing trees at optimal density, either too dense or too sparse, ensures ideal growth conditions. In many regions, restocking deforested areas is a common silvicultural goal.

Principle 6: Control of Rotation Length

Recognizing that stands of trees are not immortal, controlled reductions in stand density, fertilization, and drainage can influence rotation lengths. Managing forests efficiently involves planning the growth of stands to reach desired sizes at earlier ages, optimizing economic and predictable use. Planned reductions in tree numbers facilitate efficient timber extraction during partial cutting, keeping transport costs in check.

17.5 Basic Units of Silviculture

A **stand** represents a contiguous assembly of trees, uniform in species composition, an arrangement of age classes, site quality, and condition, forming a distinguishable unit. This unit is fundamental and typically the most detailed management entity suitable for applying silvicultural treatments. The internal structure of the stand primarily varies based on the intermingling of different species' age classes. The most straightforward structure is that of the pure, even-aged plantation, displaying a developmental pattern. Complexity extends to various combinations of age classes and species in diverse vertical and horizontal arrangements, known as **stand dynamics**. The benefits produced by forest stands, whether wood, wildlife, water, forage, or scenery, are controlled by the stand development processes. This developmental process commences with the birth of the stand, involves competition between trees, and concludes with the death of the trees and their replacement. The simplest stand development process is the **pure even-aged stand**, where the trees of a single species start together after removing the previous stand, often seen in planted stands. **Uneven-aged stands** (two to three age classes considered multi-aged; more than three age classes considered all-aged) involve trees or groups of trees of different ages, presenting more intricate developmental patterns. **Mixed stands** comprise more than one tree species, intensifying complexity, particularly when having more than a single age class of trees.

17.6 Fundamentals of Silviculture

17.6.1 Silvicultural Systems

The success of silviculture hinges on well-defined management objectives. It's a misconception to associate silviculture solely with timber management; it encompasses various objectives such as wildlife, water, recreation, aesthetics, or a combination of these. Silviculture ensures the sustained ecological functions, health, and productivity of forested ecosystems. A silvicultural system is a planned treatment program throughout a stand's life, integrating harvesting, regeneration, and tending methods to achieve specific structural goals and predictable benefits over time. Historically focused on maximizing timber production, modern silvicultural systems now incorporate ecological considerations. Silviculturists often work with less-than-ideal stands, aiming for improvement while acknowledging imperfections. Misguided practices can harm a stand's ability to meet the landowner's objectives. Silvicultural systems emphasize conservation, providing predictable yields while ensuring long-term sustainability—an essential concept in forestry practice, aligning sustainability with individual stands.

17.6.1.1 Continuous Cover Silviculture

This approach to forest management, known as continuous cover forestry, embraces a close-to-nature method by keeping the forest ecosystem intact. This method, particularly favored for broadleaves, differs from clear-felling systems. It maintains an unbroken forest cover, with tree harvesting occurring as they reach required sizes. Encouraging natural regeneration, the canopy is carefully opened to allow ample light for seedlings to thrive on the forest floor. Older trees are gradually removed, enabling younger ones to reach the upper canopy. This results in a multi-storied, multi-aged stand structure, sustaining a perpetual forest cover. Continuous cover forestry is more than a silvicultural system; it is an approach to forest management. Terms like "close-to-nature," "holistic," or "ecological" are vague, so focusing on guiding principles is more practical. These principles include managing the entire forest ecosystem, utilizing natural processes for stand management, working within site limitations, and creating a diverse stand structure with various species.

Advantages of Continuous Cover Forestry

- Less Visual Impact: Reduced compared to clear-felling.
- Increased Diversity: Enhances structural and species diversity within stands.
- Wildlife Benefits: Greater structural diversity supports wildlife and regenerating seedlings.
- Cost Savings: Potentially lowers restocking costs, assuming successful natural regeneration.
- High-Quality Saw logs: Production of large diameter, high-quality saw logs.
- Resilience Against Windthrow: Structural diversity provides resilience against windthrow.

Disadvantages of Continuous Cover Forestry

- Complex Stand Management: Requires skilled personnel for more intricate management.

- Difficulty in Yield Prediction/Regulation: Prediction and regulation are more challenging.
- Higher Harvesting Costs: Increased costs due to small, dispersed felling sites.
- Site Damage on Heavy Soils: More significant damage on heavy soils due to less brash for brash-mats.
- Dependency on Natural Regeneration: Cost-effectiveness relies on natural regeneration, less suitable for fertile sites with weed competition or high browsing pressure.
- Risks of Wind Damage: Particularly during the transformation of regular stands, especially on unstable sites.
- Time-Consuming Success Determination: Requires time to assess success.

17.6.2 Forest Regeneration

Forest regeneration involves renewing crops through natural or artificial methods, encompassing both natural and artificial regeneration processes.

17.6.2.1 Natural Regeneration

Natural regeneration involves renewing forest crops through self-sown seeds, coppice, or root suckers. Under favorable conditions, tree seeds germinate and develop into seedlings, ultimately forming a mature crop. There are three ways to achieve natural regeneration:

1. Natural regeneration by seed.
2. Natural regeneration by vegetative plants.
3. Natural regeneration supplemented by artificial regeneration.

Natural Regeneration by Seeds

Successful natural regeneration from seeds relies on the effectiveness of four key processes:

Seed Production

- Influenced by factors like species, tree age, site conditions, weather, and pest attacks.
- Species like *Tectona grandis*, *Dalbergia sissoo*, and *Acacia nilotica* seed annually, while others like *Shorea robusta* seed biennially or at longer intervals.
- Bamboos and *Strobilanthes* undergo gregarious flowering, leading to abundant seed production.

Seed Dissemination

- Essential for species survival as seeds need to be carried away from the parent plant to establish successful germination.
- Dispersal mechanisms involve wind, water, animals, and explosive mechanisms within fruits.

Seed Germination

- Depends on internal factors like seed coat permeability, moisture availability, embryo nature, ripening processes, viability, and seed size.
- External factors include temperature, moisture, oxygen, soil type, and light.

Establishment of Seedling

- Success of natural regeneration relies on a chain of processes, including sufficient seed production, efficient dissemination, successful germination, and establishment of seedlings.
- Factors influencing seedling establishment include soil conditions, climate, weed presence, grazing, forest fires, rainfall impact (drip), competition among crops, and root development.

Natural Regeneration by Vegetative Means

Certain plants exhibit regeneration through vegetative parts, either coppices or root suckers, presenting advantages and limitations.

Natural Regeneration by Coppice

Seedling Coppice: Arises from seedlings cut or burnt back, with multiple shoots produced at the base.

Stool Coppice: Emerges from a living stump, with numerous coppices usually produced, allowing for selection and clear pole development.

Regeneration factors include species, tree age, stump height, cutting method, and rotation.

Natural Regeneration from Root Suckers

- Secondary shoots from tree or shrub roots, termed root suckers, can grow into mature individuals.
- Less commonly observed naturally in species like *Diospyros tomentosa* and *Dalbergia sissoo*.

Advantages of Natural Regeneration

- Lower initial establishment costs, especially without site preparation.
- Reduced need for heavy equipment and labor.
- Naturally shaped root systems in seedlings without nursery growth.
- Aesthetic appeal of uneven, naturally spaced forest stands.

Disadvantages of Natural Regeneration

- Requires available seed crop and timed dispersal for suitable seedbed.
- Soil moisture essential for germination, with dry years impacting success.

- Risk of seed consumption by insects or small seed-eating animals.
- Competing vegetation may challenge survival and growth for an extended period.
- Pre-commercial thinning might be needed for dense stands resulting from abundant seed.
- Genetically unimproved seed from seed trees.
- Initial cost savings may lead to higher long-term expenses for site preparation or thinning.
- Unsuitable for naturally regenerating open sites without trees.
- Lack of control over tree spacing and stocking levels, resulting in uneven stands.
- Longer time to reach harvest compared to direct seeding or planting.

Natural Generation Supplemented by Artificial Regeneration

Natural regeneration varies, with success in some forest areas and failure in others. Artificial regeneration involves sowing, planting, or other methods, creating plantations for reforestation and afforestation.

17.6.2.2 Artificial Regeneration

Artificial regeneration is the renewal of the forest crop by sowing, planting, or other artificial methods. Normally, such crops are called "plantations" which is defined as a forest crop raised artificially, either by sowing or planting.

Objectives of Artificial Regeneration

Artificial regeneration is mainly carried out for the following two objects:

Reforestation: Restocking cleared woodland using artificial means, supplementing natural regeneration or restocking after destruction.

Afforestation: Establishing a forest where vegetation has been absent, focusing on productive, protective, and bio-aesthetic purposes.

Advantages of Artificial Regeneration

- Precise control over spacing and initial stocking.
- Use of genetically improved planting stock.
- Independence from natural seed crops.
- Fewer entries for stand preparation.
- Allows for species conversion.

Disadvantages of Artificial Regeneration

- High establishment costs.
- Requires intensive labor and equipment.
- Some species face severe insect and disease issues.

Artificial regeneration aims to fulfill objectives like productive, protective, and bio-aesthetic purposes, offering precise control but with higher costs and labor demands.

The decision between natural and artificial regeneration hinges on various factors. Considerations include the risk of soil loss and deterioration, crop composition, genetic factors, pest damage susceptibility, operational flexibility, stocking density, yield, time constraints, and cost implications. Generally, natural regeneration is preferable if achieved satisfactorily within reasonable time and cost. However, if natural regeneration proves inefficient, artificial regeneration becomes the preferred option. Recent trends lean toward man-made forests, driven by objectives such as increasing timber yield to meet growing demand, shortening rotation through fast-growing species, strategic forest placement relative to industries, meeting rural needs for implements and fuel, enhancing agro-ecosystems, erosion control, countryside beautification, concentrated work for easier supervision and mechanization, cost-effective logging, and increased employment opportunities. These considerations underscore the evolving emphasis on human-driven forest development in recent years.

17.6.2.3 Ecological Basis for Regeneration

Even trees have a finite lifespan, and there comes a point when they must be naturally or intentionally replaced by new ones. The plasticity of young regeneration is such that the actions taken during the establishment phase significantly shape the future development of trees and stands. The treatments applied in the initial weeks or months exert a profound influence on future characteristics, often more so than subsequent tending efforts. The success or failure of silvicultural treatments is, to a large extent, predetermined during stand establishment. Regenerating a stand is an ultimate and critical act that warrants continuous consideration throughout the entire rotation.

Natural Disturbances

Natural vegetation typically emerges following various types of lethal disturbances. No single species is adapted to all types of disturbances, making it essential to understand which species emerge after specific events. In many regions worldwide, fire stands out as the most common disturbance, having been ignited by lightning and volcanic eruptions long before human intervention. In forest climates where fires recur frequently, the result is often stands of sprouting shrubs or perennial grasses. Another group of species that thrives in the aftermath of fire consists of those germinating from the typically small seeds of wind-dispersed species well-suited to sprouting on mineral soil surfaces left bare by fires. This category of post-fire seeders includes many commercially valuable species, such as most pines and other significant trees.

1. Severe Disturbances

Erosional Geological Events: The most extreme natural disturbances involve erosional geological events, initiating true primary succession. Landslides, glacier melting, and the formation of new land through water, wind, or volcanism mark the starting points. Human-induced erosion or earth-moving can expose nutrient-deficient soil materials lacking organic matter. While deliberately stimulating such disturbances in silviculture is seldom wise, foresters may need to afforest exposed parent material resulting from erosion, road building, strip-mining, or similar events. Successful silviculture in these challenging situations requires a clear understanding of how natural pioneer vegetation develops on similar sites, relying on seed or spore dispersal by wind, water, or animals.

2. Releasing Disturbances

Top-Down Killing by Wind or Pests: Another category of regenerative disturbance involves agencies like wind or pests killing large trees from the top-down while sparing lower strata plants. Species adapted to such conditions possess foliage tolerant to shade. However, their seedlings are usually not suited to exposed microclimates, and their juvenile growth tends to be slow.

3. Patterns in Floristics Concerning Disturbance

Non-uniform Natural Disturbances: Natural disturbances vary in severity, type, and extent. Even in a seemingly fire-laid hectare, numerous tiny spots create distinct regeneration microenvironments. Key disturbance factors include the degree of mineral soil exposure, freedom from root competition, and solar radiation intensity. These elements determine the vegetation in small vacancies, each a few square centimeters in area.

4. Role of Fire as a Disturbance

Late successional: communities are the result of a long series of small, light disturbances, whereas pioneer stages are the product of catastrophe. Fire is one of the most common kinds of natural disturbance, and many species including some of the most valuable forests represent the natural adaptations of the variation to fire. Some species are adapted to reproduce after severe fires with effects akin to those of broadcast burning and others too much gentler disturbances like those of prescribed burning. Occasional fires in natural forests were beneficial to the maintenance of some species as they arrested natural succession, exposed favorable seedbeds, and prevented more destructive fires. Another way of viewing different degrees of adaptations to fire is to note that a given locality may have species with adaptation to different frequencies of fire. In forested climates, sprouting perennial grasses often go with annual burning and sprouting shrubs with less frequent burning. Sprouting tree species are next in line if fires are too frequent for seed production.

5. Disturbances and the Environment of the Microsite

Non-uniform Natural Disturbances: Late successional communities result from the prolonged series of small, light disturbances, fostering diverse ecological developments over time. These communities reflect gradual adaptations to various environmental changes. In contrast, pioneer stages emerge abruptly in response to catastrophic events. Catastrophes like fires play a significant role, and numerous species, including valuable forests, exhibit natural adaptations to fire-induced variations.

Adaptations to Fire: Many species have evolved adaptations to fire, a prevalent natural disturbance. Some thrive in the aftermath of severe fires, displaying responses similar to broadcast burning, while others prefer gentler disturbances like prescribed burning.

Beneficial Role of Occasional Fires: Occasional fires in natural forests play a beneficial role in maintaining specific species. They interrupt natural succession, expose favorable seedbeds, and serve as a preventive measure against more destructive fires.

Diverse Adaptations to Fire: Different degrees of adaptations to fire are observable within a locality, where various species may exhibit adaptations to different fire frequencies. Forested climates often witness the coexistence of sprouting perennial grasses with annual burning and sprouting shrubs with less frequent burning. If fires occur too frequently for seed production, sprouting tree species become the next adaptation in line.

Water Loss Mechanisms in Soil Layers

In the uppermost strata of soil and organic material, water undergoes a process of direct evaporation, dissipating into the atmosphere. The pace of this loss accelerates when surfaces are exposed to substantial amounts of direct solar radiation. Forest floor litter, composed of loose materials, experiences rapid water loss, rendering it inhospitable to plant roots. Although the upper layers of mineral soil or finely divided organic matter offer a more favorable environment, they, too, are prone to water loss through evaporation. Surface tension within the slender water columns in these materials contributes to this phenomenon. As water evaporates from the top ends of these columns, it pulls more water upward, continuing the cycle. The extent of the capillary fringe deepens with finer materials. This direct evaporation distinguishes itself from the transpiration process in plants, wherein water is absorbed through roots. Notably, direct evaporation elucidates water loss from the uppermost soil levels, even when devoid of roots.

Air Dynamics and Thermal Processes Near the Forest Floor

In the initial few centimeters above the forest floor, the air experiences similar dynamics. Frictional effects play a crucial role in impeding the turbulent transfer, which is the primary mechanism for the vertical movement of heated air, water vapor, carbon dioxide, and other airborne substances. This sluggish movement restricts the upward transport of heated air during the day and its downward movement at night, countering radiational cooling heat loss.

Silvicultural manipulation often focuses on shading, involving the reflection or absorption of solar energy by leaves or other objects. While large amounts of radiation are reflected without absorption by surface materials, soil and dead organic materials exhibit relatively consistent reflectivity or albedo. Consequently, modifying surface reflectivity to enhance seedling growth proves challenging. Vertical heat movement through the air relies on turbulent transfer, known as convection, wherein large clumps of air molecules undergo chaotic movement propelled by wind energy from the atmosphere above. The low heat conductivity of air leads to a phenomenon where the stratum immediately above the surface is tightly held by friction, limiting the impact of convection.

Light Dynamics in Forest Ecosystems

Light plays a crucial role in the energy dynamics of photosynthesis and significantly influences seedling establishment and survival. The visible and infrared components of radiation are particularly important for seedling development. Forest canopy conditions can be altered by changing the quality, intensity, and balance between direct and diffuse light. Factors such as moisture and temperature also impact the timing of leaf emergence in deciduous forest canopies, affecting solar radiation at the forest floor. Some seedlings thrive by exploiting temporal mismatches. Canopy openings result in varied radiation quality beneath the forest floor, with side-shade, where slanting direct sunlight is blocked, creating conditions known as "**blue shade**" due to the blue wavelengths from the scattered diffuse light. Conversely, "**green shade**" occurs in full shade with light passing through multiple canopy layers, providing photosynthetically less useful green light, which supports the most shade-tolerant species.

17.7 Tree Growth and Increment

Silviculture focused on timber production presents a unique challenge as it places a greater emphasis on the species and quality of trees compared to other forest uses. The overarching management goal of optimizing timber production paints a clear vision of a well-managed stand that efficiently yields homogeneous, high-quality timber. Consequently, silviculture plays a crucial role in influencing the growth and increment of trees at various stages of their development. Growth, a biological phenomenon, encompasses the processes of forming, differentiating, and expanding new cells, tissues, or organs, resulting in an increase in size over time. It is important to note that the sudden increase in tree diameter observed after rain is not a result of growth but rather reflects the effects of bark swell. On the other hand, increment refers to the quantitative increase in size within a specified time interval, directly attributable to growth. It is essential to recognize that the terms growth and increment are not interchangeable.

17.7.1 Factors Affecting Growth

Several factors play crucial roles in influencing the growth of trees in both plantations and natural forests. In plantations, initial spacing and treatment, silvicultural practices, artificial thinning, pruning activities, site conditions (including nutritional aspects), and climatic conditions collectively shape the growth trajectory. The arrangement of trees, the treatments applied, and interventions such as thinning and pruning all contribute to the overall growth outcomes. Additionally, site conditions, including soil nutrition, and climatic factors play pivotal roles in determining the growth patterns in plantation settings.

In natural forests, growth is intricately linked to factors such as regeneration density and treatment, spatial distribution of trees, silvicultural practices, artificial thinning, site conditions, and climatic influences. The density and treatment of regeneration, along with spatial distribution, significantly impact the natural forest's growth dynamics. Silvicultural practices and artificial thinning continue to be influential, while site conditions and climatic factors also remain pivotal determinants of growth in natural forest ecosystems. Understanding and managing these diverse factors are essential for fostering healthy and sustainable tree growth in both plantation and natural forest environments.

17.7.2 Fundamental Principles of Tree Growth

Wood represents the culmination of biochemical processes, forming relatively stable materials through intricate chemical reactions and energy transformations. During the early stages of cambial growth each growing season, one encounters enlarged, thin-walled, highly vacuolated cells that, apart from their fusiform shapes, bear little resemblance to the eventual woody elements. In the diverse field of dendrology, the starting point often involves the unstated premise that the genome of a vascular plant contains encoded proteins, facilitating life processes essential for the tree's development. Wood, in reality, encompasses a diversity of biomaterials, each exhibiting distinct characteristics, and the variations between species find their explanations in the realm of genetic differences.

There are seven principles to explain growth in trees:

1. ***Uniformitarian Principle***: This principle asserts that the physical and biological processes linking the current environmental conditions with the existing patterns of tree growth must have been consistent throughout history.
2. ***Principle of Limiting Factors***: Emphasizing that rates of plant processes are restricted by the primary environmental variable posing the greatest limitation.
3. ***Principle of Aggregate Tree Growth***: This principle suggests that individual tree-growth series can be analyzed as an aggregate of environmental factors, encompassing both human and natural influences shaping tree growth patterns over time.
4. ***Principle of Ecological Amplitude***: This principle posits that a tree species is more responsive to environmental factors at the latitudinal and elevational extremities of its range.
5. ***Principle of Site Selection***: This principle states that sites suitable for dendrochronology can be identified and chosen based on specific criteria, ensuring the production of tree-ring series sensitive to the examined environmental variable.
6. ***Principle of Cross Dating***: This principle relies on matching patterns in ring widths or other characteristics among various tree-ring series to pinpoint the exact year in which each tree ring was formed.
7. ***Principle of Replication***: This principle advocates for maximizing the environmental signal under investigation and minimizing "noise" by sampling more than one stem radius per tree and more than one tree per site.

17.7.3 Developmental Phases in Individual Tree Growth

Integral to gauging trees and forest stands is a comprehension of how individual trees evolve in diverse settings. For a forester, understanding the correlations between tree size, age, and increment to age is pivotal for predicting future growth. There are nuances to consider within and between growing seasons.

Intra-Seasonal Dynamics: Within a single growing season, various parts of a tree initiate growth at different times, influenced by species-specific reactions. Typically, height growth takes precedence over diameter growth or needle flush. Terminal bud extension (main stem and branches) follows a basipetal progression, starting from the tree's top and advancing downward.

Height Growth Factors: The extent of height growth in a season is contingent on hereditary traits, recent environmental conditions, and prevailing environmental factors.

Diameter Growth Patterns: Diameter growth also exhibits a basipetal trend, closely tied to the current foliage conditions and the existing environmental milieu.

17.7.4 Size/Age Dynamics

Understanding the progression of size in relation to age involves interpreting the Cumulative Growth Curve (CGC), typically sigmoidal for biological organisms. This curve delineates three distinct phases.

1. **Juvenile Phase**: Characterized by an accelerating growth rate, the juvenile phase marks the initial burst of vigor as the organism establishes its formative foundation.
2. **Full Vigor Phase (Maturity)**: Representing a period of constant growth rate, the full vigor phase epitomizes the peak developmental stage where the organism achieves optimal size and functional capacity.
3. **Senescent Phase**: In this stage, the growth rate decelerates, signifying the onset of aging. The senescent phase reflects a decline in the organism's growth dynamics as it gradually transitions toward the end of its life cycle.

17.7.5 Height Dynamics in Trees

Monitoring changes in tree height holds paramount importance in anticipating the future composition of stands and selecting optimal crops, particularly in pure stands. Height growth unfolds gradually once seedlings are well-established, transitioning into a rapid growth phase lasting 20–30 years, contingent on species and site conditions. This cumulative growth encompasses increases in height, diameter, volume, and basal area. In even-aged plantations, the relationship between height growth and age typically exhibits a sigmoid pattern. Early and late stages manifest slow growth, while the intermediate period witnesses a surge in growth rate. In irregular forests, especially with shade-tolerant species, the initial phase of height growth can be remarkably sluggish, only accelerating when trees emerge from overhead shade, showcasing the nuanced intricacies of tree development.

17.7.6 Tree Height and Age Dynamics

The intricate relationship between a tree's height and age serves as a reliable indicator of its inherent vitality and the prevailing environmental conditions. This relationship often forms the foundation for site classification due to its reflection of the tree's vigor and adaptability. While no universal standard governs the height–age relationship, a prevailing sigmoidal pattern captures its essence. This intricate dance of growth is shaped by a medley of internal and external factors.

17.7.7 Tree Girth Expansion (Diameter Growth)

Unlocking the enigmas of an individual tree's diameter growth involves meticulous measurements at the commencement and conclusion of a designated period. The disparity between these measurements unveils the tree's incremental expansion. Given the minuscule nature of annual growth, tools like calipers and diameter tapes are deployed, with measurements typically spaced over several years to capture a meaningful trajectory. For pinpoint accuracy, recurrent measurements within permanent plots prove invaluable, offering a precise lens into the nuanced world of diameter growth. In research pursuits, where the need arises to scrutinize infinitesimal increments, an even finer measurement approach becomes requisite.

17.7.8 Diameter Growth in Even Aged Plantation of One Species

Early Immature Stage: In the initial phases preceding canopy closure, the diameter growth experiences minimal interference from competition, setting the foundation for the tree's developmental journey.

Responsive Middle Stage: As the canopy reaches closure, the ring width undergoes a reduction. However, this stage showcases heightened responsiveness to strategic interventions like thinning and fertilization, offering opportunities for targeted management practices.

Final Mature Stage: In the mature phase, the ring width becomes narrow, exhibiting a less pronounced response to treatments. This stage marks the culmination of the tree's growth trajectory within the even-aged plantation.

17.7.9 Diameter Growth in Natural Forests

An **early stage** of extremely slow diameter (and height) growth, while the sapling is dominated by over wood. A **middle stage** when the tree is growing more rapidly but is still severely affected by its larger neighbors. The **final stage** when the tree is dominant with a large, free, well-developed crown and neighboring trees of same size are a few and distant. Whatever the structure of forest, the rate of diameter growth depends on degree of competition. The width of ring of trees normally decreases as the tree becomes older and thinner wood layer added over a larger diameter or bole surface. So the volume is greater than previous years. The age and rate of diameter growth of a tree depend on soil moisture and the efficiency of leaf functioning in the photosynthesis process. Wider spacing among trees results in more root growing space and larger crowns which lead to faster diameter growth.

Diameter/Age Dynamics: The pivotal point in evaluating tree growth is the diameter at breast height (DBH). This parameter serves as a key indicator of a tree's developmental trajectory and is instrumental in understanding the intricate relationship between diameter and age. The overarching relationship between diameter and age manifests in diverse patterns, ranging from linear to concave curves. This variability hinges on the interplay of species-specific traits, environmental conditions, and the nuances of silvicultural interventions. In instances where the diameter–age relationship unfolds as linear, future projections of diameter growth become more predictable with a higher degree of certainty. This linear alignment offers a reliable basis for anticipating the tree's developmental course. Conversely, when the diameter–age relationship takes on a concave form, the corresponding basal area (G)/age graph may exhibit linearity or near linearity. This unique characteristic facilitates the estimation of future diameter growth through the strategic projection of basal area (G). This nuanced approach provides valuable insights into the complex dynamics of tree development over time.

Volume/Age Dynamics: The volume–age relationship, characterized by its sigmoidal nature, unveils a complexity that may exhibit some degree of unpredictability. This inherent variability stems from the influence of climatic fluctuations and the impact of silvicultural interventions

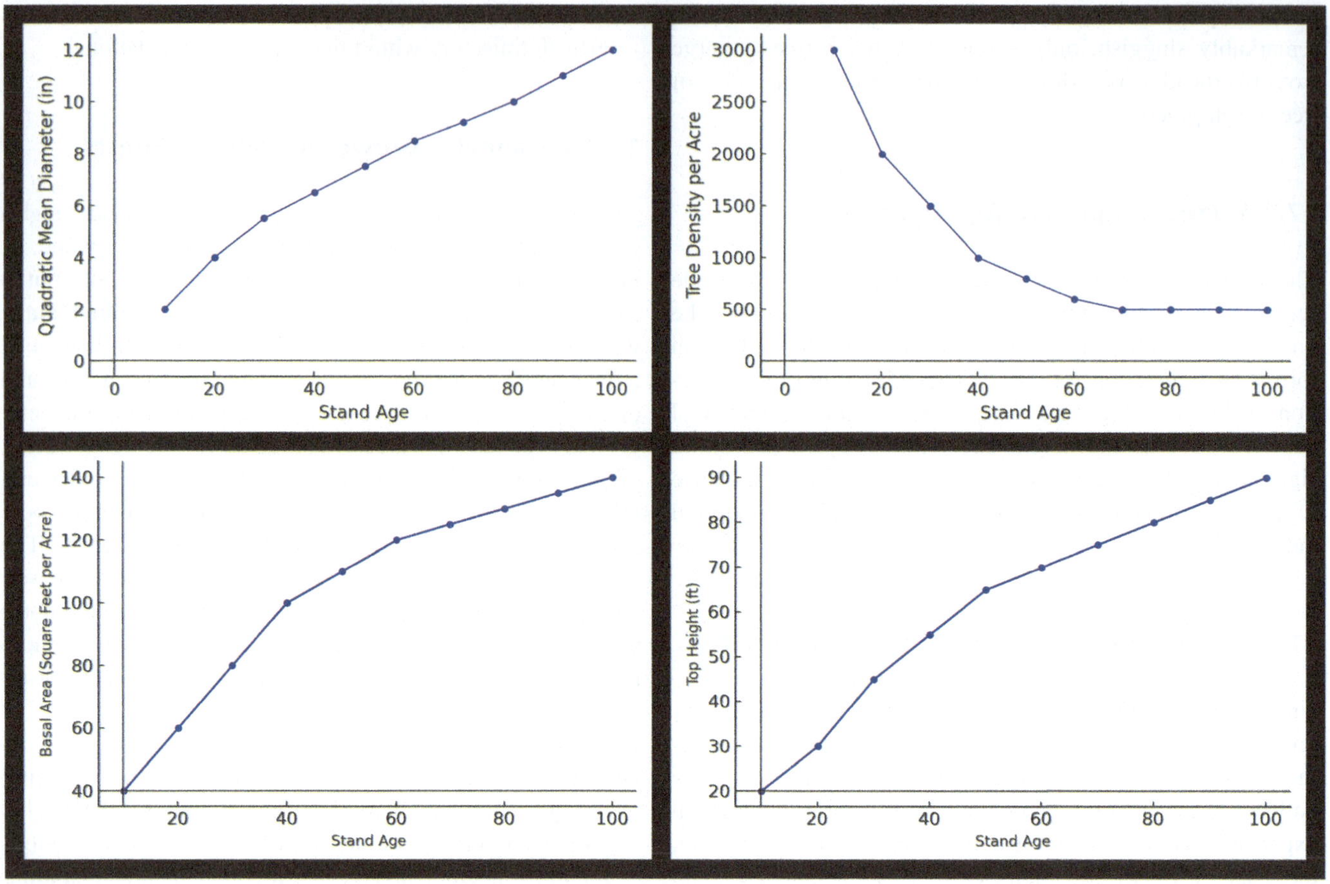

Fig. 17.1 Relationship between tree growth increment and age. (Source: Schnur 1937)

on forest ecosystems. The Cumulative Growth Curve (CGC) often unveils an extended early phase marked by curvilinearity, setting it apart from the more immediate trends observed in the diameter–age and height–age relationships. The emergence of a linear trend in the CGC tends to occur later in the developmental timeline, adding to the nuanced nature of this intricate relationship.

Increment/Age Dynamics: The increment–age relationship finds its representation through the True Growth Curve (TGC) or increment curve (Fig. 17.1). This paradigm serves as a pivotal lens through which the natural increment phases—juvenile, mature, and senescent—come into sharp focus, offering invaluable insights into the science of forest yield. In species thriving in light-demanding environments, the culmination of the increment curve is an early-life phenomenon. Conversely, shade and semi-shade tolerant species depict a more gradual rise, delaying the culmination point and ensuring a more moderate decline in the curve. This contrast underscores the diverse strategies employed by different species in response to their ecological niches.

17.7.10 Primary Influences on Tree and Stand Growth

Several factors wield substantial influence over the growth trajectories of individual trees, shaping the dynamics of forest ecosystems.

1. Site quality
2. Species
3. Stand density

1. **Site Quality**: Site quality, as indicated by the Site Index (SI), experiences an upward trajectory, several key transformations unfold within the forest ecosystem. Height growth in trees accelerates, showcasing a more

rapid vertical development. Simultaneously, stands exhibit an expedited progression toward a closed canopy, indicating denser and more mature forest cover. Additionally, the onset of competition-induced mortality initiates at an earlier age, reflecting the heightened competitiveness and interaction among trees as site quality improves. Enhanced stand development on high-quality sites yields lower densities, larger average diameters, and increased volume compared to low-quality sites. This accelerated volume accumulation is attributed to the presence of taller and larger trees at a given age on sites of higher quality.

2. **Species**: Growth traits exhibit variation within and between species due to various factors. Rainforest tree species, for instance, are frequently categorized into two functional groups, primarily determined by their germination and establishment needs. Shade-tolerant species demonstrate the capacity to germinate, grow, and thrive in low-light conditions, while light-demanding species necessitate a high-light environment for successful establishment. These groups are perceived as representing opposite ends of a spectrum in response to light. Comprehending these relationships enables foresters to make informed decisions when selecting species for specific locations, considering their adaptability and persistence in the given area.
3. **Stand Density**: The height growth of overstory trees is predominantly influenced by extreme stand densities. Trees in open-grown conditions at very low densities and overstory trees in extremely high densities typically exhibit reduced heights compared to other trees in similar-quality sites. The shading effects of the overstory lead to a reduction in height growth for intermediate and overtopped crown class trees. Stand density plays a crucial role in determining tree diameter growth. However, under a specific stand density, diameter growth tends to be higher on better quality sites. Density-dependent mortality rises with a decrease in the available growing space.

17.7.11 Yield

Yield, in forestry, signifies the quantity of harvestable material or attributes generated within a defined land area. This measurement is typically expressed as a rate, indicating the quantity produced per unit of time and area. The fundamental calculation for forest yield establishes a connection between solar energy input and crop output.

Gross yield: Gross yield represents the total amount produced on a specified site at a particular age. It encompasses the volume of living trees along with the volume of mortality. Gross yield tends to increase during stand development but experiences a decline during the complex phase.

Net yield: Net yield, on the other hand, refers to the volume or biomass available for removal at any given age. This measure generally rises over a significant portion of a stand's life but gradually falls below gross yield as mortality accumulates. Understanding both gross and net yield is essential for effective forestry management and sustainable resource utilization.

17.7.11.1 Growth Increment Curves

Traditionally, forestry employs three key measures to evaluate the growth of stem wood in both individual trees and stands. These metrics play a crucial role in understanding and managing the growth dynamics of forest resources.

1. Mean annual increment (MAI)
2. Periodic annual increment (PAI)
3. Current annual increment (CAI)

Utilizing the periodic and current annual increment area proves to be a more effective measure of growth compared to diameter increment. This is attributed to the consideration of the expanding bole diameter over time.

CAI represents the growth observed in the most recent year. The point at which MAI and PAI intersect is identified as the biological rotation age, consistently occurring at the peak of MAI. This age signifies when the tree or stand would be harvested, aligning with the management objective of maximizing long-term yield. It serves as a pivotal value in determining the annual harvestable quantity if a forest is managed for sustained timber yield and establishing the optimal rotation length to enhance production. It's essential to note that stand-level averages of MAI and PAI provide an overview of the entire stand rather than individual variations. Different growth patterns exist among individuals, influenced by species, crown characteristics, and rooting positions within the stand.

Determining MAI for uneven-aged stands poses significant challenges due to their lack of a single age. Nonetheless, in a balanced all-aged forest scenario, the CAI annually approximates the peak MAI observed in an even-aged stand of the same species, cultivated to the identical rotation age on a comparable site.

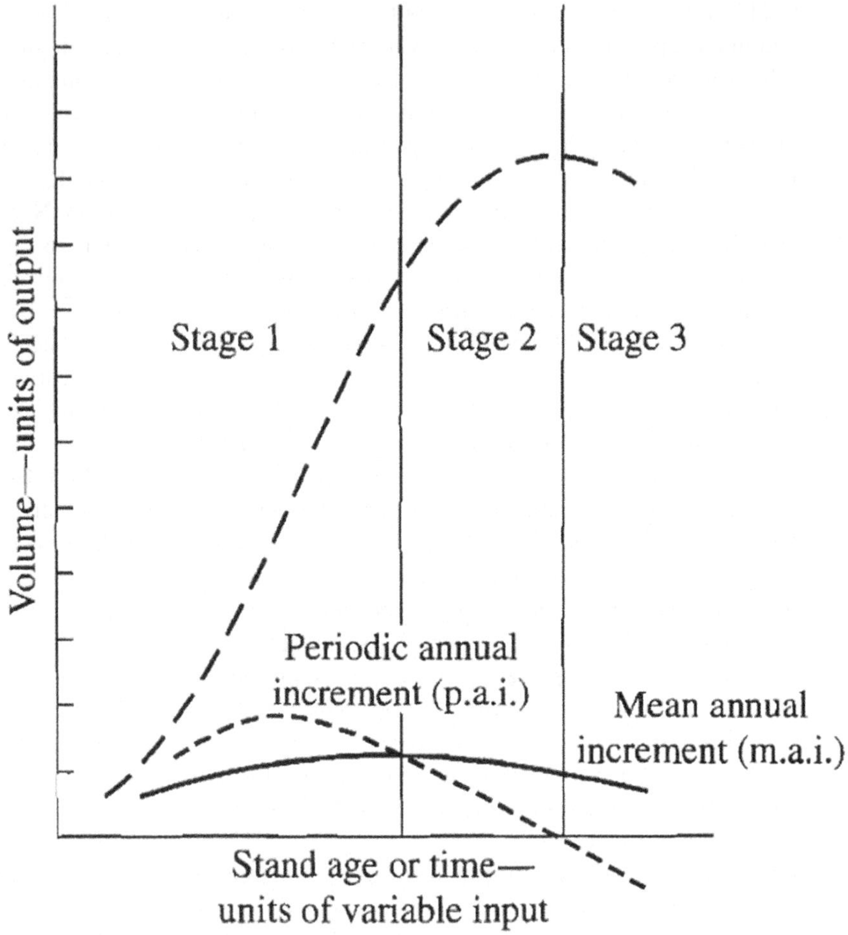

Fig. 17.2 Relationship between PAI and MAI. (Source: Ashton and Kelty 2018)

17.7.11.2 Relationship Between PAI and MAI

Stage 1: Rotation for maximum fiber production ends when PAI = MAI

Stage 2: Rotation for sawn timber production with exact length determined by economic criteria

Stage 3: Understory reinitiation phase where mortality exceeds production and standing volume declines progressively (Fig. 17.2)

17.8 Trends in Silviculture

17.8.1 Precision Silviculture

Precision silviculture, a term echoing concepts in agricultural production like "precision agriculture" or "precision farming," is gaining prominence in contemporary tree farming. This approach necessitates foresters to formulate site-specific prescriptions tailored to distinct sections within individual stands. Operating on a site-specific management model, precision silviculture enables the cost-effective implementation of various site-specific operations, even without full automation. Essentially, it emphasizes making decisions for the smallest practical management unit or a number of units within a given area. This could entail determining precise applications of fertilizer or specific herbicides at distinct locations on a tract. The essence of precision silviculture lies in utilizing diverse tools and technologies to plan, execute, and acquire site-specific forest management activities. The goal is to enhance wood product quality, minimize waste, boost profits, and ensure environmental compliance in timber management practices. Maximizing productivity in plantations or tree farming involves aligning species and genotypes with site conditions. Leveraging the inherent aggressive growth characteristics of species and developing optimal silvicultural practices are the key components. Activities like site preparation, weeding, fertilizer application, intercultural operations, and protection create favorable growing conditions, emphasizing the significance of a well-sequenced approach to plantation establishment and management for heightened productivity.

17.8.2 Intensive Silviculture

In recent decades, the intensive management of forest plantations has undergone significant evolution, driven by advances in comprehending the environmental and silvicultural influences on forest productivity, coupled with strides in information technologies. This evolution is evident in the increasing integration of large-scale precision silviculture into intensive forest plantation management, aiming to estimate site-specific effects on silviculture, biotic factors, and abiotic conditions influencing forest productivity. The utilization of remote sensing measurements, coupled with strategically placed ground information, facilitates the development of spatial modeling tools crucial for this type of silviculture. The incorporation of long-term field experiments within this methodology contributes to a mechanistic understanding of environmental and silvicultural impacts on forest production, essential for the models guiding silvicultural decisions. While the focus on maximizing production presents challenges related to intensive land use and water conflicts, maintaining long-term productivity and sustainability remains a key objective. Future endeavors must deepen our understanding of genetic × environment × silvicultural (G × E × S) interactions to enhance productivity and concurrently optimize ecosystem services. The integration of new silvicultural technology, combining remote sensing data with ground information, allows for modeling resource availability and limitations affecting forest productivity. The successful implementation of these technologies hinges on a refined understanding of G × E × S, offering practical tools for incorporation into scientific and technical models while ensuring robust economic sustainability.

The integration of geospatial and temporal information from remote sensing marks a significant advancement in the silviculture of intensively managed plantations. To harness the full potential of this dynamic information, a comprehensive understanding of soil site characteristics, the dynamics of competing vegetation, and stand nutrient dynamics is crucial. The effective merging of technological advancements with a biological understanding of the system's functioning is pivotal. This synergy is poised to offer substantial support for making strategic, tactical, and operational silviculture decisions, both presently and in the future.

17.8.3 Silviculture and Climate Change

In the face of an uncertain climate future, land managers grapple with uncharted forest dynamics, as the science supporting decision-making continues to evolve. Navigating this uncertainty requires adaptability and a willingness to manage forests in tandem with developing scientific insights. The anticipation of increased temperatures and more frequent summer droughts under various climate change scenarios may lead to heightened stress in forest environments, attributed to increased evaporative demand and limited soil moisture (Spittlehouse 2003). The potential consequences of this heightened moisture stress encompass reduced growth and productivity, along with diminished vigor in forest stands. Vigor declines could render forests more susceptible to large-scale pest attacks and increasingly severe fires. Additionally, existing plant species or genotypes may prove poorly adapted to future climate conditions at different stages of their life cycles, elevating the risk of regeneration failures and altering the trajectories of forest growth, development, and productivity. Identifying effective management actions is crucial to enhancing forest ecosystem health and productivity amidst the challenges posed by a changing climate.

17.8.3.1 Mitigative Options to Increase Carbon Sequestration

The Intergovernmental Panel on Climate Change (IPCC) has determined that implementing a sustainable forest management strategy, with a focus on maintaining or enhancing forest carbon stocks and simultaneously yielding timber fiber or energy annually, can yield substantial and sustained carbon mitigation benefits (Nabuurs et al. 2007). Sustainable management practices contribute to the prolonged and enhanced growth of forests, resulting in net sequestration benefits compared to unmanaged forests (Ruddell et al. 2007). Forest management aligned with carbon sequestration goals offers a dual advantage by facilitating timber production and biodiversity conservation while mitigating soil erosion. Therefore, forest management practices play a crucial role in reducing atmospheric CO_2 through two distinct approaches:

- Preservation and maintenance of existing forests, thereby addressing emissions sources from the forest sector.
- Acceleration of the carbon accumulation rate in the forest area.

17.8.3.2 Management Strategies

Various management strategies can be employed to enhance forest resilience and productivity in the face of evolving climate conditions. Thinning practices should be implemented to prevent overstocked stands, reducing susceptibility to heightened mortality from factors like drought, insects, disease, and wildfires. Maintaining lower canopy foliage densities becomes crucial as site resources, particularly soil moisture, become more constrained. Additionally, underplanting thinned stands with species or genotypes adapted to anticipated future conditions can provide a strategic approach to regeneration. This is especially important when advanced regeneration is deemed unsuitable. Creating opportunities for a diverse range of habitat structures at both stand and landscape scales is vital to meeting the varying requirements of plant and animal species. The use of vegetative buffers helps mitigate the impacts of harvest on stream and riparian microclimates and habitats. Safeguarding instream and riparian habitats involves preserving riparian functions like energy and nutrient exchange, filtration, bank stabilization, and wood production. Managing for swift revegetation with adapted plant communities is essential, ensuring site occupancy by desired species or genotypes before invasive species establish. Exploring options for establishing species or genotypes better adapted to future environments and disturbance regimes is also a valuable strategy. Promoting the development of mixed-species or multi-provenance forests reduces risks associated with major pest outbreaks, enhances genetic diversity, and fosters populations with broader adaptability. A mixture of seedlings, including some provenances adapted to more stressful environments, can be deployed to encourage diversity. Finally, utilizing intensively managed plantations as an opportunity to facilitate the migration of adapted genotypes or species can be beneficial. The regeneration cycles characteristic of commercial timber production offers a relatively rapid means of distributing commercial species and genotypes better adapted to future environments.

17.8.3.3 Adaptive Options

Adaptation options encompass a spectrum of management goals, each associated with varying degrees of desired change. The resistance approach seeks to maintain conditions relatively unchanged over time. In contrast, resilience allows for some change in current conditions but aims for an eventual return to the original state. The transition strategy actively facilitates change to encourage adaptive responses in the face of evolving environmental conditions.

17.8.4 Restoration Silviculture

Restoration Silviculture, a globally emerging trend, is employed for the restoration or monitoring of degraded forests. It adopts a natural disturbance-based silviculture approach, utilizing techniques analogous to natural disturbance processes in attempts to restore forests worldwide. Some proponents of restoration silviculture advocate methods that aim to restore the characteristics of late successional (old-growth) forests. This strategy focuses on recreating an old growth environment with increased biological diversity, seeking to reintroduce the community and processes removed through logging. Aligned with the concept of retention forestry, originating in the northwestern US and southwestern Canada, this approach retains structures and organisms for the long term, fostering a more heterogeneous landscape compared to conventional forest practices. The goal is to enhance ecosystem resilience, potentially minimizing damage from pests. This approach has gained global traction, with notable research activities in Australia, Brazil, Scandinavia, Indonesia, Africa, the US, and Canada. However, the carbon-centered restoration approach poses risks to biodiversity conservation in various regions, including the use of exotic species, an afforestation of native grasslands, an alteration of natural disturbance regimes, and a clearing of native vegetation for tree plantations. The impact of carbon-centered intensive silviculture on the abundance and species diversity of spontaneous regeneration in native tropical plantations remains unexplored as of our knowledge cutoff in 2022 (Muzika 2017).

17.8.5 Sewage Silviculture

People consume substantial amounts of water daily for drinking, cleaning, and various cultural activities, subsequently releasing it as wastewater through sewage systems. The escalating population has resulted in a substantial increase in the volume of this waste, surpassing the recycling capacity of local ecosystems, posing a significant health and environmental threat. The practice of reusing wastewater for afforestation, known as sewage silviculture, offers a dual advantage of "water conservation" and "environmental sanitation." Sewage silviculture contributes to three key ecological benefits: water conservation, environmental sanitation, and afforestation.

17.8.5.1 Fertilizer Value of Sewage Sludge

Undigested Sludge: This by-product of sewage treatment contains nitrogen, phosphorus, and organic matter. With a dry solids content of 2–7%, it acts as a slow-release nitrogen fertilizer due to the high organic matter content.

Digested Sludge: Created through treatment in a digester, it has a dry solids content of 2–5%. The digestion process reduces organic matter and converts nitrogen to ammonium ions, resulting in a quick-acting nitrogen fertilizer with some slow-release properties. It is less odorous than undigested sludge.

Cake Sludge: Produced by pressing or centrifuging liquid sludge, it has a dry solids content of 20–35%, increasing to over 50% through air-drying. While most soluble nutrients are lost, it serves as a good source of phosphate and organic matter.

17.8.5.2 Sludge Application in Forest Sites

Preference for Digested Sludge: Due to reduced odor and pathogen content, digested sludge is preferred, especially in sites frequented by the public. Undigested sludge may be suitable for more remote forest sites.

Application Methods: Sludge can be ploughed in before planting. Liquid sludge is suitable after tree planting, while cake sludge is valuable in restoring sites post-mineral extraction, offering high organic matter input.

Growth Responses: Research indicates that sludge application can yield growth responses comparable to conventional fertilizers, potentially eliminating the need for herbicides in some cases. However, an increased growth of herbaceous vegetation may require additional weeding.

17.8.5.3 Cost-Benefit Considerations

Weeding vs. Sludge Value: While sludge application may lead to an increased growth of grassy vegetation, the cost of additional weeding is unlikely to outweigh the value of sludge as a replacement for conventional fertilizers.

Heather Control: Sludge has shown effectiveness in eradicating heather from sites, reducing or eliminating the need for herbicide use in pure spruce crops.

17.9 Challenges for Silviculturists

Silviculture has a rich history of successful adaptation to changing conditions, yet the challenges posed by global change are formidable. Proactive silviculturists, however, have a range of options that can be customized to specific situations. Recognizing that no single practice operates in isolation, the application of practices within an adaptive forest management framework, accompanied by robust monitoring, emerges as a promising strategy. The diversity of conditions necessitates a suite of silvicultural treatments tailored to respond to global change. Organizing treatments based on objectives, silviculturists can focus on practices that enhance resistance to change, promote resilience, and facilitate the ecosystem's adaptability to evolving conditions. Certain treatments, such as fire breaks, thinning operations, and weed/pest control, can simultaneously enhance resistance and resilience. Practices like assisted migration and managing forests as complex adaptive systems present innovative possibilities.

While some argue that no immediate changes are necessary, embracing these practices may escalate costs and induce short-term environmental impacts. Furthermore, their effectiveness may be limited if ecological conditions persistently shift. Prioritizing the augmentation of forests' adaptability seems to be the most viable long-term solution. Although managing forests as complex adaptive systems is still a theoretical concept, initial research into practical applications has commenced. Shifting the focus within current management practices involves a formal assessment of silvicultural treatments, emphasizing their impacts on adaptability and resilience in decision-making criteria. Certain changes may incur increased costs or short-term reductions in productivity. Yet this could be a worthwhile investment to enhance management flexibility, create options for ecosystem adaptability, and secure long-term forest productivity in an ever-changing and uncertain world. Managing forests as complex adaptive systems might even uncover opportunities to reduce management inputs and associated costs. Identifying these opportunities and striking a balance between current productivity and future adaptability will demand concerted efforts from both researchers and practitioners. History suggests that silviculturists, known for their adaptability, are well equipped to face this challenge head-on.

17.10 Conclusion

Silviculture, with its centuries-old practices, stands at a crossroads where adaptation to societal expectations and ecological insights is paramount to prevent marginalization in broader forest management decisions. While historically focused on objectives tailored to varied conditions, silvicultural management significantly influences the growth, yield, and properties of tree species through intricate interactions between biological processes and environmental conditions. Traditionally geared toward sustaining a consistent wood yield from economically valuable species, silviculture has primarily been tree-centric in its application and evaluation. However, the contemporary landscape demands a paradigm shift. Silvicultural planning must now encompass forest management for adaptation to new conditions and adhere to sustainable resource use requirements. This shift entails promoting a forest's ability to confront and adapt to change, fostering pliability, and offering response options that accommodate substantial shifts in forest management.

Conservation strategies increasingly recognize the value of biodiversity-sensitive silviculture in managed forests, emphasizing actions that restore or sustain compositional, structural, and functional diversity in stands. Consequently, modern silviculture faces the challenge of incorporating a new perspective on biodiversity and conservation management. This has led to managed forests displaying a broader range of tree species composition and stand structures, aligning with diverse core objectives. Recent advances in silviculture cannot be viewed in isolation from traditional practices but should be seen as part of a broader shift, often characterized by the adoption of "close-to-nature silviculture." This evolution is contingent on how the day-to-day requirements of silviculture have influenced research, education, and professional practices. Silviculture's dynamic role in modern forest management is an ongoing narrative shaped by the interplay of societal expectations, ecological knowledge, and the pursuit of sustainable and resilient forest ecosystems.

Lessons Learnt

- Silviculture risks fading into the background of broader forest management decisions unless it responds to public expectations and emerging ecological insights.
- Silvicultural approaches have adapted over time to address diverse objectives and scenarios.
- The influence of silvicultural management on the growth, yield, and properties of tree species is well-established, shaped by interactions between biological processes and environmental conditions.
- In recent years, silvicultural planning has extended to encompass forest management for adaptation to changing circumstances and the imperative of sustainable forest resource utilization.
- This shift has introduced challenges for silviculture in adopting a new perspective on biodiversity and conservation management, resulting in managed forests characterized by diverse tree species compositions and stand structures with varied primary purposes.
- Integrating modern restorations or "close-to-nature silviculture" with current advancements is essential, emphasizing the interconnectedness of contemporary and traditional silviculture.
- The trajectory of silviculture's evolution hinges on how changes in its daily requirements have influenced prior research and impacted teaching methods for students and the actions of professionals in the field.

Key Questions

1. Explore the inception and fundamental tenets of contemporary silviculture.
2. Substantiate the assertion that successful silviculture is contingent on well-defined management objectives, employing relevant foundational concepts and units of silviculture.
3. Delve into the ecological underpinnings of silviculture, with a particular focus on disturbances and their pivotal role in natural regeneration.
4. Define tree growth and expound upon the principles and factors governing both individual tree and stand growth.
5. Unpack the relevance of contemporary silviculture in the context of climate change, elucidating its role in both mitigation and adaptation strategies.
6. Justify the claim that silviculture faces marginalization without responsiveness to societal expectations and new ecological insights. Shed light on how traditional silviculture can adapt to align with current global concerns.

References

Ashton MS, Kelty MJ (2018) The practice of silviculture: applied forest ecology. Wiley

FAO (2010) Global forest resources assessment. Food and Agriculture Organization of the United Nations, Rome

Muzika RM (2017) Opportunities for silviculture in management and restoration of forests affected by invasive species. Biol Invasions 19: 1–17

Nabuurs GJ, Masera O, Andrasko K, Benitez-Ponce P, Boer R, Dutschke M, Elsiddig E, Ford-Robertson J, Frumhoff P, Karjalainen T, Krankina O, Kurz WA, Matsumoto M, Oyhantcabal W, Ravindranath NH, Sanz Sanchez MJ, Zhang X (2007) Forestry. In: Metz B, Davidson OR, Bosch PR, Dave R, Meyer LA (eds) Climate change 2007: mitigation. Contribution of working group III to the fourth assessment report of the Intergovernmental Panel on Climate Change. Cambridge University Press, Cambridge

Paquette A, Messier C (2009) The role of plantations in managing the world's forests in the Anthropocene. Front Ecol Environ 8:27–34

Puettmann KJ, Ek AR (1999) Status and trends of silvicultural practices in Minnesota. North J Appl For 16:203–210

Ruddell S, Sampson RN, Smith M, Giffen RA, Cathcart J, Hagan JM, Sosland DL, Godbee JF, Heissenbuttel J, Lovett SM, Helms JA, Price WC, Simpson RS (2007) The role for sustainably managed forests in climate change mitigation. J For 105(6):314–319

Schnur GL (1937) Yield, stand and volume tables for even aged upland oak forests. US Department of Agriculture, Technical Bulletin No. 560

Smith DM, Larson BC, Kelty MJ, Ashton PMS (1997) The practice of silviculture: applied forest ecology, 9th edn. Wiley, New York

Spittlehouse DL (2003) Water availability, climate change and the growth of Douglas-fir in the Georgia basin. Can Water Resour J 28:673–688

Weber M (2011) Review new aspects in tropical silviculture. Springer, Heidelberg, p 63

West PW (2014) Growing plantation forests. Springer, Dordrecht

Economic Valuation of Ecosystem Goods and Services

18

Samrat Goswami

Abstract

The importance of sustainable development in its truest sense is to recognize the significant contribution of natural resources toward the welfare of the human being and the betterment of human civilization. Natural environment has never been considered in development literature as an important part of the production process due to the lack of clearly defined property rights. It gave birth to a completely new branch, namely Ecological Economics. Based on the structure and openness, ecosystems have features such as ecosystem is a well-designed structural and serviceable unit of ecology and is defined by species diversity. Further, the energy budget controls the operation of an ecosystem and the ecosystem controls both material cycles and energy flow. In the process, further complex ecosystems are formed. Natural ecosystems are divided into terrestrial ecosystem and aquatic ecosystem. In addition, there can be artificial ecosystems, which are shaped and maintained by the human being with certain pre-determined objectives. Gradually the importance of ecosystem services started to attract social scientists and development practitioners due to its contributions that generate goods, valuable to people and, largely, society. To capture the benefits provided by nature to society, an identification of ecosystem goods and services is of immense importance. Provisioning, regulating, socio-cultural, and supporting services were mentioned, followed by the ecosystem services paradigm, started from Millennium Ecosystem Assessment (MA) in 2003 to Common International Classification of Ecosystem Services (CICSE) in 2013, through The Economics of Ecosystem and Biodiversity (TEEB) in 2010. In the fifth section, emphasis has been given to economic valuation of ecosystem services, where monetary valuation method and a brief idea of total economic value method have been discussed in detail. Section 18.5.1 elaborates three important techniques used in the valuation of ecosystem services, namely direct market valuation approach, revealed preference approach, and stated preference approach. The difference between revealed preference and stated preference has also been reflected in this section. It is the fact that after the introduction of the concept of sustainable development, the incorporation of non-marketed services, especially ecosystem services, has gained worldwide importance.

Keywords

Goods and services · Economic valuation · Total economic value · Ecological economics

18.1 Introduction

The important economic functions of forest and other renewable natural resources can be identified as a supplier of resources for economic activities, a reservoir of wastes produced by the production and consumption process and can play an important role in providing well-being to society and can finally be the source of satisfaction. The problem is most of these functions cannot be apprehended through standard market mechanism. This is due to inability to attach clearly defined property rights to them, and, therefore, market valuation fails to put a price on these goods and services. The proper valuation of these ecosystem goods and services through various market and non-market mechanisms has become a real challenge to contemporary researchers. Economic literature has crossed the barrier of traditional economic arguments and has started to take interest in contemporary issues. As a result, a new branch, Ecological Economics, unveiled itself.

The main proposition of ecological economics is about the ecological aspects of environment and is different from the traditional neoclassical approach. Keeping efficiency in

S. Goswami (✉)
Department of Rural Studies, Tripura University, Agartala, Tripura, India

A. K. Mandal, A. Nicodemus (eds.), *Textbook of Forest Science*, https://doi.org/10.1007/978-981-97-8289-5_18

mind, ecological economics emphasizes upon capturing the importance of such functions which nature provides to humans and to maintain the carrying capacity, the level of human or animal population that probably can be supported in the long run (Kadekodi 2001). The story of ecosystem services and their valuation finds its justification through the fact that the entire welfare measure, enjoyed and discussed by and for human civilization, not only talks about the efficiency of allocation of marketable resources but also talks about the sincere uses of ecological resources so that the carrying capacity can be maintained. The present chapter is an effort in that direction, where discussions will be based on various ecosystem services offered by forest resources, importance of their economic valuation, basic ideas on valuation techniques, and discussions on some future scopes of research.

18.2 Characteristics of Ecosystem

The significant characteristic of an ecosystem poses a question whether different ecosystems are independent to each other. For example, whether the ecosystems of forest, lake or river-basin are close, or are interrelated to each other. Every ecosystem receives energy flow in the form of light or primary chemical energy from outside the system and has the mechanism of disposing of heat due to the transpiration of living organisms. Moreover, an ecosystem can gain or lose material to other ecosystems too. For example, a downstream ecosystem can gain water, nutrients, and living organisms from another ecosystem located in the upstream. On the other hand, it loses the same, or less or greater amount to the sea where it is ultimately destined. Though an ecosystem generally maintains the stability of its biotic and abiotic components, individual elements within it would become different from time to time. The ecosystem of any region influences one another to a higher or lower extent depending on the extent of the relative openness of composing individual ecosystems. On the basis of structure and relative openness, ecosystems are characterized as follows:

- Ecosystem is a significantly well-designed structural and serviceable unit of ecology.
- The structure of the ecosystem is defined by the species diversity.
- The specific energy budget controls the operation of an ecosystem.
- Material cycles and energy flow are well controlled by the ecosystem, both inside and outside the ecosystem.
- The progression process leads to the formation of a further complex ecosystem over time.

Ecosystems differ from place to place depending on geographical location and resource type (such as the forest type, coastal belt, inland, and wetland). The types of ecosystems have been discussed in the next section.

18.3 Types of Ecosystems

Different types of ecosystems help human society and extend multiple supports for the welfare of a particular society. The first broad categorization of ecosystem is natural ecosystem on the one hand and artificial or man-made ecosystem on the other hand. The natural ecosystem can be divided in two sub categories, Terrestrial and Aquatic ecosystems. A brief description about the above-mentioned categories of ecosystems has been mentioned below (Fig. 18.1).

18.3.1 Natural Ecosystem

Natural ecosystems are completely natural, and their formation is not predisposed by human activities. These ecosystems are capable enough in operating and maintaining themselves without any major intervention by anthropogenic factors. Natural ecosystems are further classified into terrestrial and aquatic ecosystem based on their habitat.

18.3.1.1 Terrestrial Ecosystem

A terrestrial ecosystem is known as a land-based community of organisms along with the interactions of biotic and abiotic components, in a given area. Examples of terrestrial ecosystems include taigas, temperate deciduous forests, tundra, tropical rainforests, grasslands, and deserts. The type of the terrestrial ecosystem found in a particular place depends on the temperature range, the average amount of precipitation received, the soil type, an amount of light it receives, etc.

18.3.1.2 Aquatic Ecosystem

Ecosystem (various plant species and life forms) that exists in water forms the aquatic ecosystem. Aquatic ecosystems include oceans, lakes, rivers, streams, estuaries, and wetlands. Within these aquatic ecosystems are living things that depend on water for survival, such as fish, plants, and microorganisms. These ecosystems are very fragile and can be easily disturbed by pollution. Marine ecosystems (such as the sea, divergent plant species, and rock pools) and freshwater ecosystems (such as rivers or freshwater lakes) are the two main categories of the aquatic ecosystem.

In fact, the entire biosphere can be considered as an ecosystem, due to the interaction between and among all the elements. On the basis of spatial delimitation and several

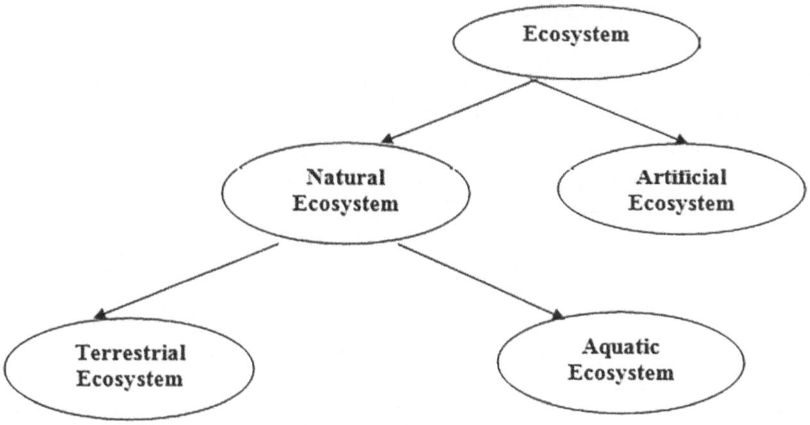

Fig. 18.1 Broad types of ecosystems

other related factors like biophysical surroundings, distribution of organisms and spatial communications, largely ten categories of eco-systems have been identified such as Coastal, Forest, Island, Cultivated Inland, Water, Urban, Marine, Polar, Dry land, and Mountain ecosystem. These ecosystems extend benefits separately to human society as well as they together extend benefits to the larger society, known as human civilization.

18.3.2 Artificial Ecosystem

Artificial ecosystems have few features of natural ecosystems but are shaped and maintained by human beings with certain pre-determined objectives. Such ecosystems are to a great extent simpler than natural ecosystems and are by far the most recognizable surroundings in human experience. "Cultural landscapes" are large-scale ecosystems, generally modified by human beings. Agriculture is an example of artificial ecosystem indispensable for sustaining human population (Guidotti 2015).

18.4 Ecosystem Goods and Services

The concept of ecosystem has been neglected by social scientists for a prolonged period of time, and lately accepted, once they have started to understand the importance of nature, natural resources, natural environment and its importance in social well-being. The absence of clearly defined property rights and public good character, along with inability to apply successful market mechanism, has created a space for estimating the importance of ecosystem services and the benefits generated from them. In this regard, it is imperative to understand an ecosystem and the services provided by it for the benefit of human civilization. An ecosystem is a community of living (biotic) organisms such as animals, plants and microorganisms, interacting with physical environment, such as air, water, and minerals, as an independent system (Odum 1971). Ecosystems can be differentiated as inland or coastal, terrestrial or marine and urban or rural. Ecosystems can be local, or can even be global, in scale, and they often interact and overlap.

Ecosystem services are the "multiple benefits provided by ecosystems to humans" (Ma and Swinton 2011) and the "direct and indirect contributions of ecosystems to human well-being" (TEEB 2010). Following the economic perspective, ecosystem services are described as the "contributions of the natural world which generate goods which people value" (Bateman et al. 2011). Hence, ecosystems and their functions create services that produce goods and services that are valuable, as they derive benefits from them (Haines-Young and Potschin 2011). This implies that ecosystem services include both "final" ecosystem services, which greatly contribute to economic and social well-being (e.g., agricultural crops), and "intermediate" ecosystem services (e.g., pollination) that support "final" ecosystem services (Haines-Young and Potschin 2011).

Hence, ecosystem services consist of all goods and services provided by an ecosystem (such as forest) that directly or indirectly benefit people. Examples of ecosystem services include the production of fuel wood, carbon sequestration, water regulation, and provision of habitat for biodiversity and many more to mention. Some of the services are tangible and well-known in everyday life (e.g., the provision of food, fiber, fuel, and other material for basic life support to the financially marginal community). There are many other intangible services, less known for their inadequately visible direct impacts (e.g., water filtration, climate regulation, protection against extreme weather events and natural hazards) but in no way unimportant.

18.4.1 Ecosystem Services

Biological system administrations are the conditions and cycles through which environment and biodiversity provide benefits, which are essential for human existence. Ecosystem services are the advantages emerging out from the environmental elements of the biological systems. They benefit all living life forms in the specialty, including creatures, plants, and people. Chasing life and through their ability to replicate, organic entities use energy; plants get the vast majority of their supplements from soil or water, while animals will in general get their supplements from different organic entities. Microorganisms are profoundly flexible acquiring supplements from soil, water, food, or different creatures. Ecosystem services are the outcomes, conditions or cycles of normal frameworks that straightforwardly or by implication advantage people or improve social government assistance.

The idea of environment administrations was given expanded public acknowledgment through the Millennium Ecosystem Assessment (MEA) launched in 2001 by the UN Secretary General and completed in 2005. A theoretical structure was created to feature the genuine effects of environment services on human well-being, security, social relations, and actual prosperity to clarify the incorporated angles coordinated into four classes. The characterization of biological system services created Millennium Ecosystem Assessment is given beneath:

- **Provisioning Services**: The substantial/tangible goods that individuals get from environments. For example, food, fresh water, fuel wood, fiber, bio-synthetic substances, and genetic resources.
- **Regulating Services**: Benefits acquired from the regulating process of biological system measures. For example, climate regulation, water regulation, disease regulation, pollination, and water purification.
- **Socio-Cultural Services**: Non-material advantages got from biological systems and they reflect esteem got from the presence of the wetland, including entertainment, style, schooling, natural life review, and ecotourism. For example, recreational, esthetic, educational, heritage, spiritual and religious, inspirational.
- **Supporting Services**: Indirect administrations essential for the creation of any remaining environment administrations. Incorporate capacities and cycles that give the establishment to provisioning, regulatory and cultural administrations. For example, soil formation, nutrient cycling, and primary production.

These services were categorized and overtime modifications have been made. The paradigm of different ecosystem services can be captured through different classifications at different points of time, starting with MA (2003), followed by TEEB (2010) and CICSE (2013).

The paradigm of ecosystem services is profoundly important. It has already been mentioned that ecological economics deals with ecosystem services and their impact on human society changes over time. The ecosystem services classification in the Millennium Ecosystem Services Assessment, adopted in 2003, where the major groups were provisioning, regulation, cultural, and supporting services is followed by the second set of classifications, which portrays ecosystem services as per The Economics of Ecosystem and Biodiversity (TEEB) in 2010. In this structure, major groups were provisioning, regulating, habitat, and cultural and amenity services, and no provision was made separately for supporting services. Later, a scientific ecosystem services classification was done by the Common International Classification of Ecosystem Services (CICES) in 2013. It has grouped all the services in three broad categories, namely, provisioning, regulation and maintenance, and cultural services. In light of sustainable development, there was a felt need to value ecosystem services.

18.5 Economic Valuation of Ecosystem Services

The paradigm shift from development toward sustainable development puts emphasis on the incorporation of natural resources as natural capital in the accounting process. The objective was to capture benefits obtained from the natural environment that keeps on giving support to every economic activity. However, as natural resources are mostly of public good in nature and do not have clearly defined property rights, they suffer from various problems. Perhaps, the most important issues it faces are the problems of missing market and market with imperfections. Hence, resources cannot reflect its scarcity in terms of prices, which is a general signal for products traded in standard commodity markets. Another major problem is related to uncertainty about the availability and accessibility of resources. The supply policy depends on the basis of the available resources stock, and newly invented stocks do not readily be incorporated into the existing resource stock. For some biodiversity products and enterprises, it is fundamental to comprehend and appreciate its other options and elective employments. Valuing environmental services is a challenging task, as environment and natural resources do not have clearly defined property rights, and, hence, there is no structured market for the transaction of environmental goods and services. However, valuing natural resources, especially forest (which provides a number of ecosystem goods and services in terms of benefits to human), has become an essential policy issue for the preservation and conservation of forest stand. It is to be remembered that economic value alludes to the value of an asset, which lies in its job in achieving human goals, be it spiritual edification, esthetic joy or the creation of some marketed item (Koch et al. 2009). As opposed to being an inborn property of an asset, for example, a natural asset value is credited by economic specialists through their willingness to pay for the services that stream from the asset. While this might be determined by the goal (e.g., physical or ecological) of the asset, the willingness to pay relies extraordinarily upon the financial setting where valuation happens on human inclinations, organizations, culture, etc. Environment and resources provide that platform to human society that ultimately helps society to flourish.

In case of natural resources, policy analysts can measure the values of ecosystem goods and services, obtained from forest stands and important for human well-being, broadly using three different approaches, economic valuation approach, quantitative approach, and qualitative approach. In the economic valuation approach, also known as monetary valuation approach (as policies are framed considering money value), money values are being placed on ecosystem goods and services. For example, the NTFPs, also known as Minor Forest Products (MFPs), play a very important role in the life of the rural poor, who are dependent on forest resources for generations. Monetary and quantitative methodologies are discussed in detail below.

18.5.1 Monetary Valuation Methods

Monetary valuation approaches are important from the policy context. There are two main categories of monetary valuation techniques—market-based and non-market-based techniques. The common denominator is their respective economic values, and an associated range of estimates reflects the impact of changes to ecosystem services on human well-being. Non-market-based methods are mainly of two types—Contingent Valuation Technique (CVM) and Choice Experiments. These two methods are used to capture the preferences of individuals, which are not possible to measure through market mechanism. The primary information is collected using a structured questionnaire. Apart from standard socio-economic questions, specific questions are designed to capture the responses on the ecosystem services under valuation. These methods are also considered as stated preference approaches. In addition, non-market benefits, obtained from ecosystem services, can also be captured using value transfer techniques. In this technique, the obtained monetary values in some past valuation study are transferred to the area under consideration. Further, monetary valuation methods are of different categories such as demand-based valuation (both market and non-market situations) techniques, supply-based valuation techniques, cost-based valuation techniques, and value transfer methods. In market-demand-based valuation techniques, market price, travel cost methods, and hedonic pricing methods are important techniques. In the category of non-market demand-based methods, Contingent Valuation and Choice Experiments are significantly important in terms of valuing ecosystem services. In the cost-based method, Replacement cost, Avoided Damage cost, and Averting Expenditure are the most used techniques. Production function approach comes under the supply-based approach, whereas Unit Value approach, WTP Function, and Meta-Analysis are the most used valuation techniques of ecosystem services.

18.5.2 Total Economic Value (TEV) Method

Total economic value (TEV) of the ecosystem is defined as the sum of the values of all service flows that natural capital (Ecosystem) generates both now and in the future—appropriately discounted. These service flows are valued for marginal changes in their provision. TEV encompasses all components of utilities derived from ecosystem services using a common unit of account: money or any market-based unit of measurement that allows comparisons of the benefits of various goods.

18.5.2.1 Components of TEV

Total Economic Value can be considered as the sum of Actual User Value, Option Value, and Existence Value.

1. **Use Values**: These types of values are derived after the actual use of environment or environmental resources, for example, pollution-free air and safe drinking water. This is possible from the current use or future use. Hence, use values can further be segregated into the following components:

   (a) **Direct Use Value**: Results from the direct human use of biodiversity (consumptive or non-consumptive).
   (b) **Indirect Use Value**: Derived from the regulation services provided by species and ecosystems.
   (c) **Option Value**: Relates to the importance that people give to the future availability of ecosystem services for personal benefit (the option value in a strict sense).

Use esteems can be related to private or quasi-private goods, for which market costs generally exist. Use values are partitioned further into two classes: (a) Direct use values, identified with the advantages acquired from the direct utilization of environment administration. Such use might be extractive, which involves utilization (e.g., food and crude materials), or non-extractive use (e.g., tasteful advantages from scenes), (b) Indirect use values are normally connected with controlling administrations, for example, air quality guideline or disintegration avoidance, which can be viewed as open administrations which are by and large not reflected in market exchanges. The concept of option value can be considered as an insurance against ignorance-based uncertainty (Banerjee 2001). The option value can further be segregated as:

- **Bequest Value**: Value attached by individuals to the fact that future generation will also have access to the benefits from species and ecosystems (intergenerational equity concerns).
- **Vicarious Value**: Value attached by individuals to the fact that other people of the future generation can access to the benefits provided by species and ecosystems (inter-generational equity concerns).
- **Existence Value**: Value related to the satisfaction that individuals derive from the mere knowledge that species and ecosystems continue to exist.

The different groups of ecosystem services, along with the types of uses, such as direct use, indirect use, optional use, and non-use, in the total economic value framework are important for calculating total economic value. Provisioning services have direct and optional uses, whereas regulating services have indirect and optional uses. In case of cultural services, direct, optional, and non-use values can be calculated. For a habitat group of services, habitat administration is valued through various classifications of biological systems. Hence, TEV methods try to establish direct or indirect markets through which it tries to estimate the values of different ecosystem services provided for human welfare.

Current markets provide data about the value of a small subset of ecosystem cycles and segments that are estimated and consolidated in exchanges as items or services. These postures structural restrictions on the capacity of markets to give thorough photos of the ecological values engaged with choice cycles. In addition, a data disappointment emerges from the trouble of evaluating most ecosystem services in wording that are similar with services from human-made assets. From this viewpoint, the rationale behind ecosystem valuation is to disentangle the complexities of socio-ecological connections, settle on unequivocal what human choices would mean for ecosystem administration values, and to communicate these value changes in units (e.g., monetary) that allow for their consolidation in open dynamic cycles (Reid et al. 2005).

Economic decision-making should be based on understanding the progressions to financial government assistance from little or minimal changes to ecosystems. Worth in this manner is a negligible idea insofar that it alludes to the effect of little changes in the condition of the world, and not simply the condition of the world. In such manner, the estimation of natural resources, similar to the estimation of different resources, is individual-based and emotional, setting ward, and state-subordinate.

18.6 Valuation Methods Under the TEV Approach

The introduction of sustainable development in development literature was primarily due to the benefits nature provides to society. Among different methods to calculate economic values of ecosystem services, a significantly important method is calculating the total economic value. The available techniques used in valuing ecosystem services are:

- Direct market valuation approaches
- Revealed preference approaches
- Stated preference approaches

18.6.1 Direct Market Valuation Approaches

Direct market valuation approaches are divided into three main components: (a) market price-based approaches, (b) cost-based approaches, and (c) approaches based on production functions. The main advantage of using these approaches is that they use data from actual markets, and thus reflect actual preferences or costs to individuals. Moreover, such data prices, quantities and costs exist and thus are relatively easy to obtain.

18.6.1.1 Market Price-Based Approaches

Market price-based approaches are most often used to obtain the value of provisioning services, since the commodities produced by provisioning services are often sold on, for example, agricultural markets. The efficiency (shadow) prices method is used to capture market prices but is adjusted for transfer payments, market imperfections and policy distortions. It may also incorporate distribution weights, where equality concerns are made explicit. Shadow prices may also be calculated for non-marketed goods.

18.6.1.2 Cost-Based Approaches

Cost-based approaches are based on estimations of the costs that would be incurred if ecosystem service benefits needed to be recreated through artificial means. Different techniques exist, including (a) the avoided cost method, which relates to the costs that would have been incurred in the absence of ecosystem services, (b) the replacement cost method, which estimates the costs incurred by replacing ecosystem services with artificial technologies, (c) the mitigation or restoration cost method, which refers to the cost of mitigating the effects caused by the loss of ecosystem services or the cost of getting those services restored, and (d) preventive expenditure (PE) approach uses the costs of preventing damage or the degradation of environmental benefits.

18.6.1.3 Production Function-Based Approaches (PF)

Production function-based approaches estimate how much a given ecosystem service (e.g., regulating service) contributes to the delivery of another service or commodity which is traded on an existing market. It is based on the contribution of ecosystem services to the enhancement of income or productivity. It generally uses scientific knowledge on cause-effect relationships between the ecosystem service(s) being valued and the output level of marketed commodities.

Direct market valuation approaches fail to work if there is an absence of market system for ecosystem goods and services and in the presence of distorted markets.

18.6.2 Revealed Preference Approaches

Revealed preference techniques are based on the observation of individual choices in existing markets that are related to the ecosystem service that is subject of valuation. Here economic agents "reveal" their preferences through their choices (Banerjee 2001).

18.6.2.1 The Travel Cost Method (TC)

It is mostly relevant for determining recreational values related to biodiversity and ecosystem services. It is based on the rationale that recreational experiences are associated with a cost (direct expenses and opportunity costs of time). The value of a change in the quality or quantity of a recreational site (resulting from changes in biodiversity) can be inferred from estimating the demand function for visiting the site that is being studied.

18.6.2.2 The Hedonic Pricing (HP)

The HP approach utilizes information about the implicit demand for an environmental attribute of marketed commodities. For instance, houses or property in general consist of several attributes; some of which are environmental in nature, such as the proximity of a house to a forest or whether it has a view on a nice landscape. Hence, the value of a change in biodiversity or ecosystem services will be reflected in the change in the value of property. By estimating a demand function for property, the analyst can infer the value of a change in the non-marketed environmental benefits generated by the environmental good.

18.6.3 Stated Preference Approaches

Stated preference approaches simulate a market and demand for ecosystem services by means of surveys on hypothetical (policy-induced) changes in the provision of ecosystem services. Stated preference methods can be used to estimate both use and non-use values of ecosystems and/or when no surrogate market exists from which the value of ecosystems can be deduced. The main types of stated preference techniques are:

Contingent Valuation Method (CV): It uses questionnaires to ask people how much they would be willing to pay to increase or enhance the provision of an ecosystem service, or alternatively, how much they would be willing to accept for its loss or degradation.

Choice Modelling (CM): CM attempts to model the decision process of an individual in a given context. Individuals are faced with two or more alternatives with shared attributes of the services to be valued, but with different levels of attribute (one of the attributes being the money people would have to pay for the service).

Group Valuation (GV): Group valuation combines stated preference techniques with elements of deliberative processes from political science, and are being increasingly used as a way to capture value types that may escape individual-based surveys, such as value pluralism, incommensurability, non-human values, or social justice.

Moreover, there exist pertinent differences between the revealed preference approach and the stated preference approach. The revealed preference theory portrays the real world, whereas the stated preference theory describes the context of hypothetical decision. The revealed preference theory consists the inherent relationship between attributes, whereas the stated preference theory controls relationships between attributes. The revealed preference approach represents the market and personal limitation on decision-maker, whereas the stated preference approach does not represent changes in market and personal limitation effectively. The revealed preference approach has high reliability, but the stated preference approach is reliable when respondents understand, commit and respond to task. In addition, the stated preference approach yields multiple responses per respondent. In this regard, it is worth mentioning that different valuation methods have become important tools to capture the values of many ecosystem goods and services which were important but unaddressed in the accounting process. However, there are many ecosystem services those extend important contributions to social welfare but difficult to address. The constant research in this field will be of immense help in this regard to develop the valuation process.

18.7 Concluding Annotations

Different valuation techniques, along with their possible applications, have been discussed in brief in the prior sections. The concept and features of ecosystems have been discussed at the basic level to have proper understanding. The concepts have been supported by the types of ecosystems along with ecosystem goods and, most importantly, ecosystem services. The new concept of sustainable development emphasizes on the role environment and resources play in shaping up of human civilization, with its support in terms of environmental services, stock of natural resources and acting as receptacle for waste products. Hence, the importance is captured through economic valuation of such important ecosystem services.

Lessons Learnt

- The basic concepts of ecosystems and the types of ecosystems.
- The basic knowledge on ecosystem goods and services along with broad categories.
- Basic concepts of economic valuation of ecosystem services.
- Fundamental concept of Total Economic Value Framework.
- Basic knowledge on Direct Market Valuation approach.
- Basic idea of Revealed Preference and Stated Preference approach.

Key Questions

1. Write the main proposition of ecological economics.
2. What are the major characteristics of ecosystem? Explain different types of ecosystems.
3. Do you think that ecosystems services are remaining same over time? Explain.
4. Suppose you went to visit a reserve forest as a tourist. How would you like to measure the benefits provided by the reserve forest to the local community? Explain.
5. If you need to construct a hypothetical market to capture policy-induced changes in ecosystem services, which broad approach you would like to follow? Mention the broad types of measures of such approach.

References

Banerjee S (2001) Economic valuation of environmental benefits and costs. In: Bhattacharya RN (ed) Environmental economics: an Indian perspective. Oxford University Press, India

Bateman IJ, Mace GM, Fezzi C, Atkinson GR, Turner K (2011) Economic analysis for ecosystem service assessments. Environ Resour Econ 48(2):177–218

CICSE (2013) Common international classification of ecosystem services, V43. European Environment Agency

Guidotti TL (2015) Artificial ecosystems—health and sustainability: an introduction. Oxford University Press

Haines-Young RH, Potschin MB (2011) Ecosystem services. Prog Phys Geogr 35(5):575–594

Kadekodi G (2001) Valuation of natural resources: what have we learnt from Indian experience. Indian J Agric Econ 56(3):285–312

Koch EW, Barbier EB, Silliman BR, Reed DJ, Perillo GME, Hacker SD, Granek EF, Primavera JH, Muthiga N, Polasky S, Halpern BJ, Kennedy CJ, Kappel CV, Wolanski E (2009) Non-linearity in ecosystem services: temporal and spatial variability in coastal protection. Front Ecol Environ 7(1):29–37

Ma (2003) Ecosystem and human well-being: a framework for assessment, Chapter 2. Island Press

Ma S, Swinton SM (2011) Valuation of ecosystem services from rural landscapes using agricultural land prices. Ecol Econ 70(9): 1649–1659

Odum EP (1971) Fundamentals of ecology, 3rd edn. W.B. Saunders Co., Philadelphia

Reid WV, Mooney HA, Cropper A, Capistrano D, Carpenter SR, Chopra K, Dasgupta P, Dietz T, Duraiappah AK, Hassan R, Kasperson R, Leemans R, May RM, McMichael AJ, Pingali P, Samper C, Scholes R, Watson RT, Zakri AH, Shidong Z, Ash NJ, Bennett E, Kumar P, Lee MJ, Raudsepp-Hearne C, Simons H, Thonell J, Zurek MB (2005) Ecosystems and human well-being—synthesis: a report of the Millennium Ecosystem Assessment. World Resource Institute

TEEB (2010) Integrating the ecological and economic dimensions in biodiversity and ecosystem service valuation. http://www.teebweb.org/EcologicalandEconomicFoundation/tabid/1018/Default.aspx

19 Principles and Applications of Geospatial Technology in Forestry

Sudip Kumar Saha

Abstract

Forests contribute to vital environmental and ecosystem processes like climate variation, biogeochemical cycle, hydrological cycle and soil conservation. The livelihood many people depends on food and raw materials produced from forests. A huge aerial extent of forests is undergoing degradation, depletion and loss of productivity due to natural and human activities such as the increase of human and bovine population pressures and non-sustainable developmental activities in forest areas. Therefore, spatial and temporal data on the quality, area and spatial variability of forests are vital for the sustainable management of forests. Geospatial technology consists of aeial and satellite remote sensing (RS), geographic information system (GIS) and global navigation satellite system (GNSS) offer cost-effective, rapid and reliable tools for inventory, change analysis and scientific management of forests compared to traditional time-consuming ground-based forest management techniques.

The initial sections of this chapter document the definition, principles, components and techniques of visual and digital interpretation and analysis of aerospace RS, GIS and GNSS (e.g. GPS, global positioning system). The section dealing with applications of geospatial technology in forestry covers approaches and methodologies of use of RS, GIS and GPS in various areas of forestry. First, the details of techniques and methods of applications of aerial RS using black and white and colour photographs, and digital multi-spectral data in various areas of forestry such as forest mapping, inventory and management are discussed. The principles and both visual and digital analysis approaches using satellite data from different optical sensors in several areas of forestry such as forest type mapping, crown canopy density mapping, forest biomass estimation and modelling, forest fire (active fire detection and burnt area and burnt severity assessment), forest damage by diseases and insects, etc. are discussed. GIS and GPS applications deal with principles and integrated analysis methods using different kinds of geospatial data in various areas of forestry, namely forest working plan preparation and revision, forest fire management (fire risk zonation, fire growth simulation and fire information system), wildlife habitat evaluation, Indian biodiversity information system (BIS) and Indian Bio-resources Information Network (IBIN) are included. Applications of advanced RS techniques, namely microwave, hyperspectral and Lidar (light detection and ranging) in forestry are also covered in this chapter. Conclusions and lessons learnt are included in the end section of the chapter.

Keywords

Remote sensing · Geospatial · GIS · GPS · Satellite

19.1 Introduction

Among the natural resources, the forest resources covering large areas of the terrestrial ecosystem play a vital role in supplying products and ecosystem services. Forests provide vital environmental and ecosystem services such as regulations of climate, hydrological and biogeochemical cycles and soil conservation. By producing food and raw materials, forests provide livelihood to hundreds of millions of people in the world and support the economy. Large areas of the dynamic forest ecosystem are affected by several agents/factors such as natural and human activities. The decline in spatial extent and quality of forest cover by anthropogenic influences are major concerns in sustainable forest management. Scientific management of forests requires a detailed understanding of factors affecting the forest ecosystem.

S. K. Saha (✉)
Formerly with Indian Space Research Organization (ISRO), Indian Institute of Remote Sensing, Dehradun, India

A. K. Mandal, A. Nicodemus (eds.), *Textbook of Forest Science*, https://doi.org/10.1007/978-981-97-8289-5_19

Sustainable management based on inventory and monitoring states of forest requires information on the nature, extent and spatial distribution of forest resources. Managing forest resource bases both spatially and temporally dynamic can be a difficult task only by using conventional ground-based techniques without the utilization of effective geospatial tools. Fast-developing geospatial technology has opened new vistas in forest resources inventory, monitoring and creation of geospatial digital databases and this information is crucial to the preparation of forest working plans, wildlife conservation, soil conservation etc.

Geospatial technology includes several mapping and surveying technologies, and the major technologies are remote sensing (RS), geographic information system (GIS) and global navigation satellite system (GNSS), such as global positioning system (GPS). The RS has revolutionized the techniques of systematic inventory, temporal change analysis and protection of forest resources using aerospace-borne sensors with different spatial and spectral resolution capabilities. During the past one and half decade, GIS technology progressed very fast as an effective tool for natural resources management. GIS is a computerized system designed for data inputting, organizing, storing, analysing, retrieving, displaying and disseminating both geospatial and non-geospatial data, either collected through RS or other techniques. GIS with its multi-source data integration and analysis capabilities helps forest managers to generate forest management plans with varying alternate scenarios. The advent of GPS technology contributed to the widespread acceptance of geospatial technology in geo-web services provided by various geo-mapping portals such as Google Earth/Map; Bhuban (Indian Space Research Organization, ISRO, India) and smartphone devices. GPS is a positioning tool, and part of GNSS provides accurate geographic location anywhere on the earth's surface. GPS is a very effective tool for cost-effective in situ data collection as well as monitoring the changes in features and phenomena in a particular position by repetitive GPS observations.

19.2 Overview of Principles of Geospatial Technologies

19.2.1 Remote Sensing

19.2.1.1 Definition

RS forms the primary data source in GIS analysis. Generally, RS is known as a technique of gathering multi-spectral data and characterize the objects or phenomena without physically contacting them. 'RS is defined as the practice of deriving information about the earth's land and water surfaces using images acquired from an overhead perspective, using electromagnetic radiation in one or more regions of the electromagnetic spectrum, reflected or emitted from the earth's surface' (Cambell and Wynne 2011).

'Remote Sensing' term first known to the people in the 1960s in the USA, particularly applied to aerial photography used for photogrammetry, photo interpretation and photogeological applications. The term 'Remote Sensing' gained popularity after the launch of the maiden earth resource observations spacecraft, Landsat-1 (formerly named Earth Resources Technology Satellites (ERTS-A and ERTS-1) on July 23, 1972, as part of the US Landsat program.

19.2.1.2 Components of Remote Sensing

Remote sensing consists of several components (Fig. 19.1) and the components are briefly described in below sections:

1. An energy source (e.g. the sun) produces EMR) which propagates downwards towards the earth and interacts with objects on the earth's surface. Sun is a common source of energy. EMR while propagating upwards and downwards interacts with the atmosphere (Fig. 19.1).
2. Earth's surface consists of natural and manmade features. These features based on physical and chemical compositions reflect and emit a part of incoming energy back towards the remote sensing sensor. The atmosphere depending on its gas and particulate concentrations scatters some fraction of the earth's reflected EMR and also absorbs part of the earth's emitted EMR.
3. Sensor and platform: The sensor measures radiation coming from the object and converts it to electrical signals which are displayed in the form of digital data or an image after recording. The platform either airborne and/or spaceborne is a vital component of remote sensing which supports sensor operation by providing various services such as orbit and attitude control, power supply and communications with ground tracking and data receiving systems.
4. The ground receiving system collects the sensor-recorded and measured raw digital data and stores the data after proper formatting. Prior distribution of remotely sensed data to the user, several preprocessing corrections, namely radiometric, geometric, atmospheric, etc. are performed by the ground system.
5. Data analysis involves visually interpreting analogue, RS image-applying interpretation devices or computer-aided digital analysis of digital RS sensor recorded data and suitable image processing software. Reference data in the forms of field checks and ancillary spatial and non-spatial data as examples of the topographic map, forest type map, soil map, forest biomass data etc. are used by the analyst to extract thematic information from the RS data. The generated thematic map can be combined with multi-source available other spatial and non-spatial thematic information/data in a GIS.

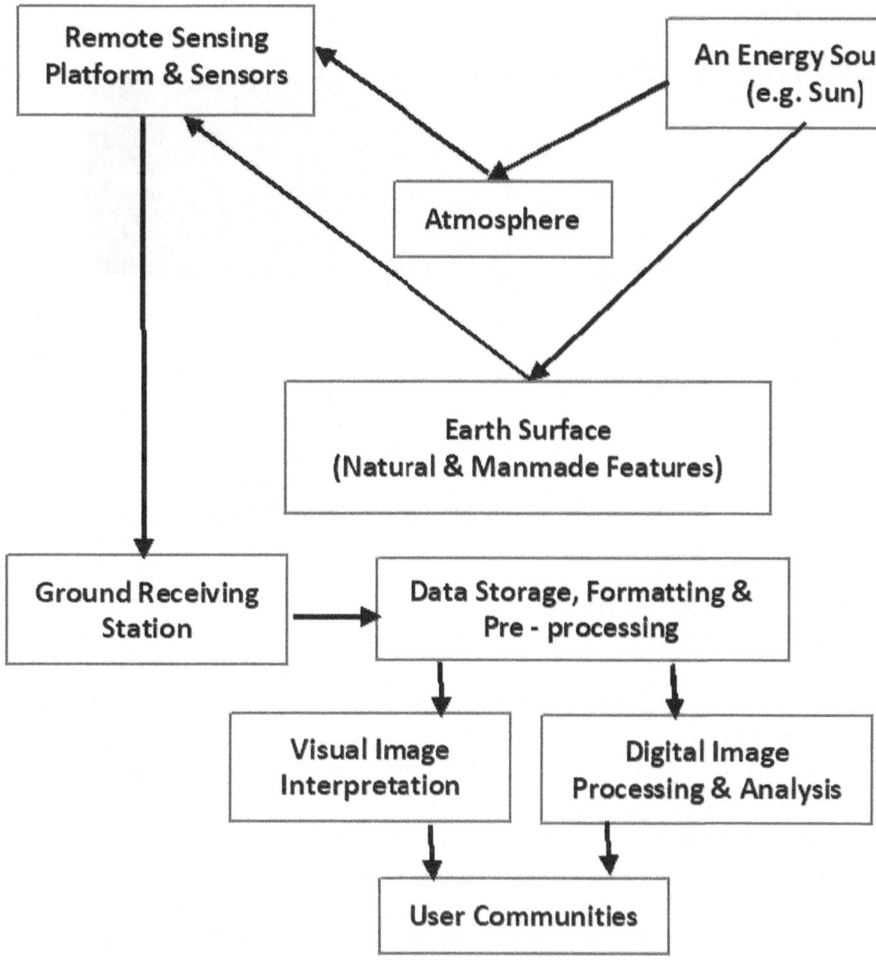

Fig. 19.1 Components of remote sensing

6. Lastly, the analysed thematic map is presented to user communities for appropriate applications in different fields.

19.2.1.3 Classification of Remote Sensing

RS is classified into two groups: passive RS and active RS. Sun forms the natural energy source for RS. Sun's energy in visible (VIS) and short wavelength are reflected and emitted in the thermal infrared and microwave wavelength regions after absorption by the objects. Passive RS deals with the detection and characterization of the EMR of reflection or emission coming from earth features. Active RS uses its own source of EMR to interact with the terrain. Examples of important active RS devices are radio detection and ranging (RADAR), light detection and ranging (LiDAR) and sound navigation and ranging (SONAR).

19.2.1.4 Electromagnetic Spectrum and Radiation Laws Employed in Remote Sensing

Three parameters, namely wavelength (λ), frequency (ν) and velocity (C) characterized EMR which propagates in the form of waves. The EMR after interacting with the objects, is reflected emitted or backscattered from the objects, and this information forms the basis for characterizing the objects. The EMR spectral regions commonly used in remote sensing are illustrated in Fig. 19.2. The EMR wave equation relates these parameters and is expressed as $C = \nu\lambda$. EM spectrum depicts the continuum variations of EMR with respect to wavelength or frequency. Five EMR spectral regions are commonly used for different kinds of RS and the salient features of the five spectral regions are described below:

1. Visible spectral region (wavelength ranges from 0.4 to 0.7 μm): This spectral region covers the eye's sensible spectral wavelengths, and the earth receives maximum energy. The visible region is further divided into

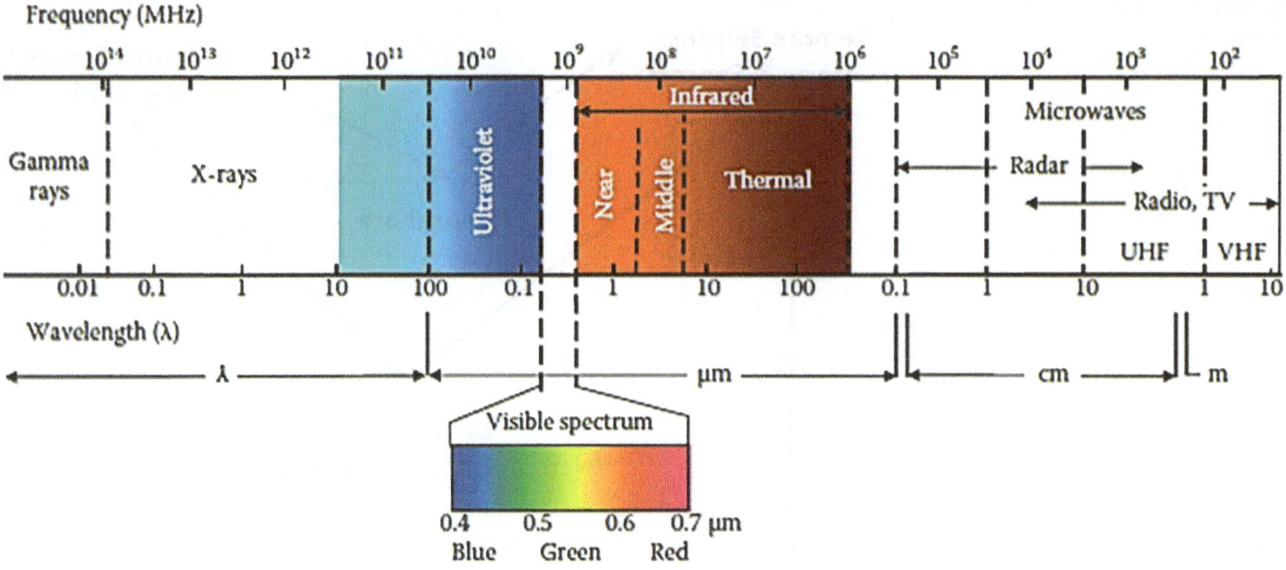

Fig. 19.2 Major spectral bands within the EM spectrum (Source: Chuvieco 2016a)

three primary colour spectral sub-regions: blue (wavelength ranges from 0.4 to 0.5 μm), green (wavelength ranges from 0.5 to 0.6 μm) and red (wavelength ranges from 0.6 to 0.7 μm). The red wavelength region is useful for assessing vegetation vigour as it is sensitive to photosynthetically active radiation (PAR) absorption by plant pigments.

2. Near-infrared (NIR) spectral region (wavelength ranges from 0.7 to 1.2 μm). This spectral region is not sensitive to the human eye's spectral portion of EMR. This region is also known as reflective or photographic infrared because the 0.7–0.9 μm wavelength portion of the near-infrared region can be detected and imaged by special photographic films (e.g. colour infrared known as CIR). The NIR wavelength is very useful for the assessment of plant growth and development.
3. Middle infrared spectral region (wavelength ranges from 1.2 to 8 μm). The wavelength region of 1.2–2.5 μm spectral region is called as short wave infrared (SWIR) region. In the SWIR spectral region, solar radiation is still effective for characteristic reflection/absorption from the objects. SWIR spectral band is effectively utilized for the estimation of vegetation water content. The mixed spectral region of 3–8 μm, which consists of solar-reflected and surface-emitted energy, is generally used for detecting emitting radiation from the objects. The 3–8 μm portion of MIR is specifically used for detecting and characterizing high-temperature phenomena such as forest fires.
4. Thermal infrared spectral region (wavelength ranges from 8 to 14 μm). The spectral RS data from this region is used for the estimation of land surface temperature (LST) by detecting emitted energy from the earth's surface features. Estimation of vegetation evapotranspiration is one of the popular applications of TIR spectral region.
5. The microwave spectral region (wavelength ranges from 1 cm to 1 m). This spectral region is least affected by atmospheric conditions and can penetrate within the volume of the object as well as the sub-surface. Imaging RADAR and microwave radiometer systems use the microwave region for remote sensing. Because of canopy penetration capability, MW remote sensing is an efficient tool for vegetation structure and biomass assessment.

Biophysical and biochemical properties of earth features primarily control the magnitude of energy detected and recorded by the RS payload (sensor).

Four basic EMR laws explaining the physical basis of remote sensing are described below:

Kirchhoff's law quantifies and relates the reflection coefficient, transmission coefficient, absorption coefficient and emission coefficient.

Planck's law describes the relationship among emitted radiation energy from a surface, wavelength of EMR and surface temperature.

Stefan–Boltzmann's law provides the empirical relationship for the estimation of the sum of the surface emitted energy computed over the entire EMR spectrum and the surface temperature.

Wien's displacement law finds the particular wavelength emitting maximum EMR energy from a surface at a given temperature of the surface.

19.2.2 Remote Sensing Sensors and Satellite Systems

19.2.2.1 Remote Sensing Sensors

Based on the mechanism of detection of EMR energy, the remote sensing sensor systems are classified into passive and active sensors. Passive sensors detect radiations coming from the object which is irradiated by external radiation sources. The active sensor emits its own energy to irradiate the object and then detect the energy coming from the object. Passive sensors are further classified based on the method used to record the incoming energy from the object.

19.2.2.2 Photographic Cameras

Photographic cameras (Fig. 19.3a) use photosensitive film to record radiation coming from the object. The lens and shutter optic systems of the camera control the radiation collection conditions and duration of film exposure respectively. By choice of suitable film types and camera optic systems, photographs with high spatial resolution can be collected from long aerial distances. Commonly, the most popular, black and white (grey tones) aerial photographs are collected using panchromatic (PAN) film with one emulsion layer sensitive to the visible portion of EMR. For colour aerial photography, three layers of photochemical film with each layer are sensitive to one of the spectral bands of the visible portion of EMR. The first layer is sensitive to the blue band, the second layer is sensitive to the green band, and the third layer is sensitive to the red band. Film sensitive to 0.7–0.9 μm spectral region of EMR is used for the black and white infrared photograph. CIR photograph specifically used for vegetation studies is generated by using film sensitive to spectral bands green, red and NIR. During recent decades, photographic cameras, which are common analogue remote sensing data collection systems, have been replaced by electronic digital systems that provide imagery with comparable characteristics of aerial cameras.

19.2.2.3 Cross-Track (Wisk Broom) Scanners

This digital imaging technique is common in most satellite remote sensing missions. The terrain is scanned parallel to the scan lines which are at 90° of the satellite track, by the use of an oscillating or rotating mirror and set of lenses (Fig. 19.3b). Forward movement of the remote sensing platform generates the subsequent scan lines and finally a 2D image data of rows (scanlines) and columns collected. A series of detectors as components of the sensor amplify and convert the focused incoming radiance signal from the object to digital numerical value that can be stored onboard or transmitted to a ground-receiving antenna. The EMR coming from the object is separated into several spectral bands which are recorded by special detectors. In the early 1970s, this multi-spectral scanning system (MSS) was adopted by several earth-observing satellite missions such as Landsat MS, and Nimbus Coastal Zone Colour Scanner (CZCS). These RS satellite sensors first time provided spectral data in nonvisible wavelengths (NIR, SWIR and TIR). Several historical satellites used this technique of RS data acquisition, and examples are the Landsat series of satellites [thematic mapper (TM) and enhanced thematic mapper (ETM)] and defence meteorological satellite program-operational line-scan system (DMSP-OLS). Even current satellite missions, namely moderate resolution imaging spectroradiometer (MODIS) sensors of Terra and Aqua, advanced very high-resolution radiometer (AVHRR) of national oceanic and atmosphere administration (NOAA)

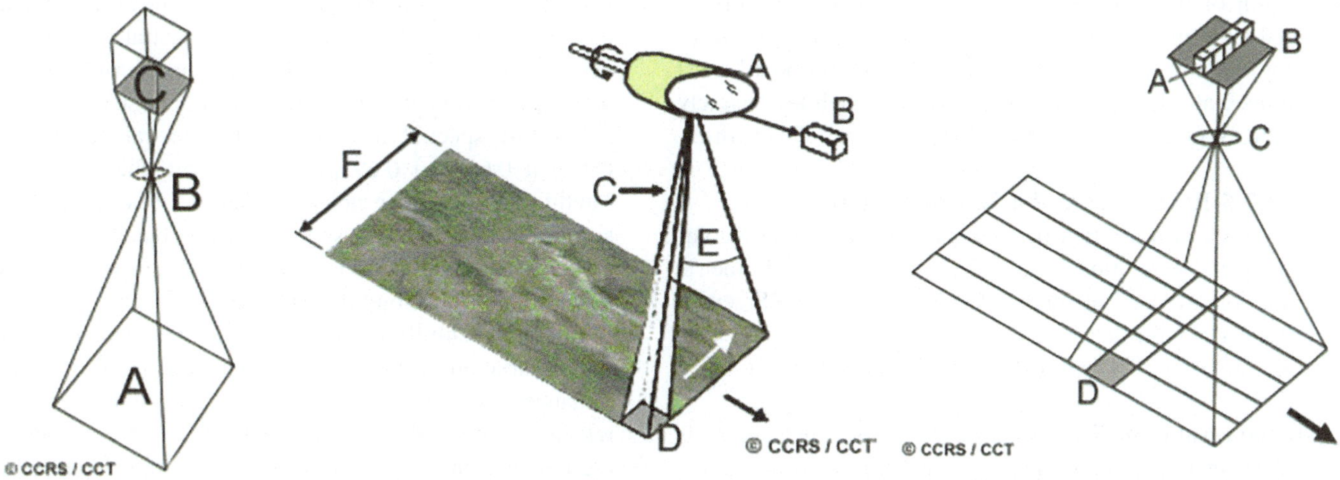

Fig. 19.3 Major scanning systems used for remote sensing data collection. (**a**) Aerial photography (A: ground area, B: lens and C: focal plane); (**b**) cross-track (whisk broom) scanning (A: rotating mirror, B: detectors, C: IFOV, D: viewed ground cell, E: angular field of view F: swath/width of the image); and (**c**) along-tract (push broom) scanning (A: detectors, B: focal plane of the image, C: lens systems, D: viewed ground cell (sources: CCRS and NRC n.d., http://pcmas1.ccrs.nrcan.gc.ca/fundam/chapter1/chapter1_1_e.htm)

and METOP satellites used this technique of remote sensing data collection.

19.2.2.4 Along-Track (Push Broom) Scanners

In 1986, this scanning technology was first tested with the French remote sensing satellite Systeme Pour l'Observation de la Terre (SPOT). Instead of an oscillating mirror, using linear array detectors along-track scanner, scan the entire scan line of the terrain and generate image along with the satellite orbital track (3c). The forward motion of the RS sensor onboard aircraft or satellite sensor by moving along the direction of the flight line or track of the satellite and sequential scanning generates a 2D image. This is the reason the method of scanning is named push broom. Linear arrays consist of many side-by-side-placed semiconductor solid-state devices known as charge-coupled devices (CCD). Detectors of the sensors of the cross and along tracks segregate the radiation coming from the objects into several spectral bands and convert them into digital values before recording. This scanning technique offers two major advantages over cross-tract scanning: (1) improvement in spatial resolution of the image and (2) reduction in geometric distortion of the image caused by misalignment of the mirror oscillation and movement of the remote sensing platform. Presently, push-broom digital scanning is the most common method of earth observation (EO) satellite or aerial remote sensing data collection. Since the late 1980s, the ISRO successfully adopted this technology for various kinds of Indian remote sensing (IRS) satellite sensors such as linear imaging self-scanning (LISS), wide-field sensor (WiFS), advanced wide field sensor (AWiFS) and PAN. French SPOT satellite sensors such as VEGETATION and the haute resolution visible (HRV) also used this imaging technology. This imaging technology was also incorporated in the NASA EO mission of the Landsat-8 satellite sensor called the operational land imager (OLI) satellite sensor. This imaging technology is also very common for the sensor on board most commercial high spatial resolution RS satellites, namely IKONOS, GeoEye, WorldView, RapidEye and QuickBird.

19.2.2.5 Resolutions of a Sensor System

Four types of resolution parameters such as spectral resolution, spatial resolution, temporal resolution and radiometric resolution are used for the characterization of an RS sensor system. The information distinguishing power determines the resolution of an RS sensor (Estes and Simonett 1975).

Spatial resolution: The spatial resolution of an RS sensor is determined by the capability of the sensor to detect the tiniest object on an image generated by RS. The spatial resolution of an RS system is measured by the ability of a sensor to resolve the separation of the smallest angular or linear between two objects (Jensen 2014). The sensor's instantaneous field of view (IFOV) denotes the angle of the illuminated area on the ground observed by a sensor, determining the spatial resolution of an RS system. IFOV covered each ground area from a pixel on the RS image. The height of the RS observing sensor from the ground and the angle (radian) of IFOV values are used for the computation of spatial resolution of an optoelectronic RS sensor system. In general, high spatial resolution RS sensor covers small ground areas with very large recorded data volumes. A wide range of spatial resolution data is collected by operational EO satellite systems such as geostationary satellites ($\approx$5 km^2), polar-orbiting meteorological satellites ($\approx$1 km^2), polar-orbiting satellites: regional scale natural resources studies ($\approx$1 ha), polar-orbiting satellites: local scale natural resources studies ($\approx$0.1 ha) and commercial satellites ($\approx$1–0.1 m^2) (Chuvieco 2016a). Spatial resolution is an important parameter for image interpretation because it provides the levels of details available in an image. The requirement of optimum spatial resolution RS data for thematic mapping depends on the scale of the mapping.

Spectral resolution: Spectral resolution is associated with the number and spectral bandwidths of multi-spectral band data recorded by an RS sensor. The spectral bandwidth refers to the RS sensor-sensitive wavelength range in the EMR. High spectral resolution with more number of bands contributes better discrimination of earth features. Quantification of geophysical and biochemical properties of earth's features can be done efficiently with the RS sensor having spectral bands with narrow bandwidth which is enough to identify characteristics spectral absorption features of the objects. Bands with wider spectral bandwidth are termed as coarse spectral resolution and bands with narrow bandwidth are called fine spectral resolution. For example, the SPOT satellite PAN band has coarse spectral resolution with spectral bandwidth between 0.51 and 0.73 μm. Advanced spaceborne thermal emission and reflection radiometer (ASTER) satellite band 2 has fine spectral resolution with spectral bandwidth between 0.63 and 0.69 μm. At present, the EO-1 RS satellite with a sensor called 'Hyperion', has the highest spectral resolution (220 spectral bands, bandwidth with 10 to 20 nm and covering 0.40 and 2.5 μm wavelength region of EMR) among the EO satellites.

Radiometric resolution: The discrimination capability of minor variations in the sensor recorded incoming spectral radiance from the object is known as the radiometric resolution of an RS sensor. Generally, it is denoted as the dynamic range of values used for coding input radiance in each spectral band. Dynamic range is expressed as quantization level which is denoted by bits used to store the input signal. For example, an 8-bit sensor ($2^8 = 256$, dynamic range of 0–255) may distinguish 256 levels of different radiances per pixel of an image. In past, commonly 8-bit per pixel coding was used in the RS sensor. With the improvement of data transmission speed and data storage, currently, several EO satellite sensors have 11–16 bits per pixel (e.g. MODIS, Landsat 8-OLI, GeoEye, WorldView). IRS satellite sensors have varying radiometric resolutions such as 6 bits (WiFS), 7 bits (PAN and LISS-III) and 10 bits (AWiFS and Cartosat-1). A sensor having higher radiometric resolution can discriminate the minor variations of radiation in the form of reflection or emission incoming from different earth features. Radiometric resolution is very important for digital than visual analysis of remote sensing data.

Temporal resolution: It refers to revisiting the capability of an area by the sensor and determining how frequently the remote sensing sensor collects data over an area. The repeat cycle depends on the orbital parameters of the satellite such as height, speed and declination and IFOV of the sensor. IRS Cartosat-1 PAN sensor acquires images of an area in 5 days repeatability, while IRS LISS III collect data in 24-day intervals. In general, coarse spatial resolution EO satellites have high temporal resolutions and higher spatial resolution satellites offer lower temporal resolutions. Based on the objectives of applications, the temporal resolution of the EO satellite mission is set. Dynamic phenomena like forest fires, floods and volcanos can be better studied with high temporal resolution RS data.

19.2.2.6 Multi-Spectral Earth Resources Satellite Missions

Based on orbital characteristics, EO satellites are grouped into two categories such as sun-synchronous polar and geostationary orbits satellites. An RS polar orbit satellite all time passes the equator at the local time. Sun-synchronous polar orbits with heights ranging between 700 and 900 km are commonly chosen for EO RS satellites. The same region of the earth is observed at all times by the geostationary satellites placed at very high equatorial orbits (36,000 km over the earth). Meteorological remote sensing satellites come under the geostationary orbital satellite class.

Landsat satellites: Worldwide Landsat satellite series EO data were successfully used for the earth's natural resources and environmental inventories and management. ERTS launched on July 23, 1972, was the first satellite of the Landsat series. After the completion of the initial three Landsat missions, a series of satellites named TM as part of the Landsat program with improved multi-spectral imaging sensors were launched between 1982 and 1999. Landsat 4 was launched in 1982, Landsat 5 was launched in 1984, Landsat 6 failed in 1993 and Landsat 7 known as Enhanced TM + launched in 1999 belonged to the improved imaging sensors TM satellites series. MSS (multi-spectral scanner) and video cameras (RBV) were the sensors of the initial three missions of Landsat. MSS included four spectral bands covering the VIS and NIR wavelengths region of EMR, namely blue, green, red and NIR. Landsat 4 and 5 satellites belonging to the TM satellite series provided multi-spectral remote sensing data with improved all three types of sensor's resolutions except temporal resolution than Landsat MSS. TM series satellite data is extensively used for a wide variety of thematic applications. The TM and ETM + sensors included additional PAN (15 m spatial resolution), SWIR and thermal infrared bands that were not contained in the MSS sensor. Landsat 8 satellite latest in the Landsat program includes a unique sensor called an OLI sensor with four, one and six spectral bands respectively covering VIS, NIR and SWIR wavelengths of EMR. The technical specifications of various sensors of Landsat satellite systems are included in Table 19.1.

IRS satellites: The ISRO RS satellite program consists of varieties of operational EO satellite systems. IRS 1A launched on March 17, 1988, was the first satellite of the ISRO's operational EO program. Subsequently, over the years series of satellites have been added named IRS-1B, 1C and 1D; Cartosat and Resourcesat.

IRS-1A, 1B, IC and 1D: IRS-1A and 1B satellites carrying linear imaging self-scanning (LISS)-1 and 2 sensors collect four multi-band spectral data in visible (blue, green and red) and near-infrared wavelengths which are identical to Landsat TM with spatial resolutions of 72.5 m and 36.25 m. IRS-1C and 1D satellites included three sensors with unique capabilities such as LISS-III with four bands [green, red, NIR and SWIR (Shortwave Infrared)]; PAN (black and white) band and two bands (red and near-infrared) WiFS. The spatial resolutions of these three sensors are 36.25 m and 72.5 m (LISS III), 5.2 m (PAN) and 188 m (WiFS).

Cartosat: ISRO launched several high spatial resolution remote sensing satellites named Cartosat for cartographic and large-scale mapping purposes. The number of satellites such as Cartosat-1, Cartosat-2, Cartosat-2A and 2B and Cartosat-3 and 3A belong to the Cartosat series. Two PAN cameras onboard Cartosat-1 generate stereoscopic imagery by simultaneously imaging the same terrain area from two different angles. The sensor onboard the Cartosat-1 satellite has a spatial resolution of 2.5 m and covers a ground area (swath) of 30 km. Cartosat-2 carries a single PAN camera sensible to the visible

Table 19.1 Characteristics of Landsat sensors

Wavelength range of spectral bands (μm)				
Landsat MSS	Landsat RBV	Landsat TM	Landsat ETM+	Landsat OLI and TIR
Landsat 1–3	**Landsat 1 and 2**	**Landsat 4 and 5**	**Landsat 7**	**Landsat 8**
Band 4 (0.5–0.6)	Band 1 (0.475–0.575)	Band 1 (0.45–0.52)	Band 1 (0.45–0.52)	Band 1 (0.43–0.45)
Band 5 (0.6–0.7)	Band 2 (0.580–0.680)	Band 2 (0.52–0.60)	Band 2 (0.52–0.60)	Band 2 (0.45–0.51)
Band 6 (0.7–0.8)	Band 3 (0.690–0.830)	Band 3 (0.63–0.69)	Band 3 (0.63–0.69)	Band 3 (0.53–0.59)
Band 7 (0.8–1.1)		Band 4 (0.76–0.90)	Band 4 (0.76–0.90)	Band 4 (0.64–0.67)
Landsat 3	**Landsat 3**	Band 5 (1.55–1.75)	Band 5 (1.55–1.75)	Band 5 (0.85–0.88)
Band 8 (10.4–12.6)	Band 1 (0.505–0.750)	Band 6 (10.40– 12.50)	Band 6 (10.40– 12.50)	Band 6 (1.57–1.65)
		Band 7 (2.08–2.35)	Band 7 (2.08–2.35)	Band 7 (2.11– 2.19)
			Band 8 (0.52–0.90)	Band 8 (0.50–0.68)
				Band 9 (1.36–1.38)
				Band 10 (10.60–11.19)
				Band 9 (11.50–12.51)
Spatial resolution (m)				
Landsat 1–3	**Landsat 1 and 2**	**Landsat 4 and 5**	**Landsat 7**	**Landsat 8**
Bands 4–7: 79	Bands 1–3: 80	Bands 1–5 and 7: 30	Bands 1–5 and 7: 30	Bands 1–7 and 9: 30
		Band 6: 120	Band 6: 120/60	Band 8: 15
Landsat 3	**Landsat 3**		Band 8: 15	Bands 10 and 11: 100
Band 8: 240	Band 8: 240			

region of EMR, and its spatial resolution is less than 1 m. The camera can obtain a stereoscopic image by steering the camera up to 45° both in along and across track directions. Cartosat-2C provides four bands of multi-spectral data sensitive to 0.40–1.3 μm wavelengths with 1–2 m spatial resolution in addition to the PAN band with <1 m, Cartosat-3 is the latest addition in the Cartosat series with sensors providing data with PAN band of 0.25 m and multi-spectral bands with 1 m spatial resolutions.

Resourcesat: Resourcesat-1 and 2 are the very popular remote sensing satellites for the Indian user community for natural resources and environmental management. Resourcesat-1 carries three sensors: (1) high spatial resolution (2.5 m) LISS IV with monochromatic and three multi-spectral bands; (2) LISS III with three multispectral bands in green, red and near-infrared wavelengths and one in SWIR wavelengths of EMR with a moderate spatial resolution (23.5 m); and (3) AWiFS also with three multi-spectral bands in green, red and near-infrared wavelengths and one in SWIR wavelengths having spatial resolution of 56 m. The sensors of Resourcesat-2 have the same technical specifications as Resourcesat-1 except for the enhanced radiometric resolution and LISS IV sensor with increased swath coverage from 23 to 70 km. The salient technical specifications of IRS/Resourcesat sensors are summarized in Table 19.2.

SPOT satellites: Next to Landsat, a series of satellites under the EO program known as SPOT (system for EO) were launched from 1990 to 2014. By dual modes operation of two HRV (haute resolution visible–high resolution visible) sensors, SPOT satellites acquire both PAN and multi-spectral bands (green, red and NIR) in the first three satellites. The sensor's spatial resolutions are 10 m (PAN) and 20 m (multi-spectral bands). The high-resolution sensors onboard SPOT 4 and 5 enhanced imaging capabilities by the addition of an extra SWIR band and improving spatial resolution from 20 m to 10 m and 10 m to 2.5 m of multiband and PAN sensors respectively. The high-resolution PAN camera also acquired stereoscopic images by off-nadir acquisitions by using a pointable mirror. A special kind of sensor named 'VEGETATION' with a coarse spatial resolution (1 km), four bands covering blue-green, red and near-infrared spectral wavelengths regions of EMR, and global coverage (2250 km swath of an image) also installed on SPOT 4 and 5 for global assessment and monitoring of crops and natural vegetation.

Other medium and coarse resolutions EO satellites: There are other several EO satellite missions developed by space agencies of different countries used for natural resources inventory and monitoring. A brief description of selected missions relevant to vegetation studies is as follows.

MODIS (Moderate Resolution Imaging Spectrometer): Terra and Aqua satellite missions under NASA's Earth Observation System (EOS) had MODIS as an RS sensor. MODIS sensors collect spectral data in thirty-six multi-spectral bands in the wavelength region of 0.4 to 14.4 μm of EMR with varying spatial resolutions in different spectral bands such as 2 m to 250 m (band 1 and 2), 7–500 m (band 3) and 1 km (bands 8–36). It has temporal resolutions of 1–2 days. MODIS data is well calibrated

Table 19.2 Salient technical specifications of major IRS/Resourcesat sensors

IRS/resourcesat sensors				
LISS I	LISS II	PAN	LISS III	LISS IV
Wavelength range of spectral bands (μm)				
Band 1 (B1): 0.45–0.52	Band 1 (B1): 0.45–0.52	PAN: 0.50–0.75	Band 2 (B2): 0.52–0.59	Band 2 (B2): 0.52–0.59
Band 2 (B2): 0.52–0.59	Band 2 (B2): 0.52–0.59		Band 3 (B3): 0.62–0.68	Band 3 (B3): 0.62–0.68
Band 3 (B3): 0.62–0.68	Band 3 (B3): 0.62–0.68		Band 4 (B4): 0.77–0.86	Band 4 (B4): 0.77–0.86
Band 4 (B4): 0.77–0.86	Band 4 (B4): 0.77–0.86		Band 5 (B5): 1.55–1.77	
Name of the spectral bands				
B1: Blue (B),	B1: Blue (B),	PAN:	B2: Green (G),	B2: Green (G),
B2: Green (G),	B2: Green (G),	Panchromatic	B3: Red (R),	B3: Red (R),
B3: Red (R),	B3: Red (R),		B4:Near-infrared (NIR)	B4:Near-infrared (NIR)
B4:Near-infrared (NIR)	B4:Near-infrared (NIR)		B5: Short wave infrared (SWIR)	
Spatial resolution (m)				
B1–B4: 72.5	B1–B4: 36.25	PAN: 5.8	B2–B4: 23.5	B2–B4: 5.8
			B5: 70.5 B5: 23.5 (Resourcesat)	
Radiometric resolution (quantization level–bits)				
B1–B4: 7	B1–B4: 7	PAN: 6	B2–B4: 7 B5: 10 (Resourcesat)	B2–B4: 10
Swath of the image–image area coverage (km)				
148	74	70	140 and 148 (B5: SWIR)	23 (Resourcesat-1) 70 (Resourcesat-2 and 2A)

and processed for the creation of several global datasets such as terrestrial ecosystem variables like spectral vegetation indices (SVI), leaf area index (LAI), PAR and vegetation of net primary productivity (VNPP).

AVHRR (Advanced Spaceborne Thermal Emission and Reflection Radiometer): AVHRR sensors are placed on the NOAA family of polar remote sensing and European MetOp satellites. This sensor provides data in five spectral bands/channels and the central wavelengths of the channels are as follows: 0.6 μm (channel 1), 0.9 μm (channel 2), 0.9 μm (channel 3), 11 μm (channel 4) and 12 μm (channel 5). It collects images of the entire globe two times each day with a 1.1 km spatial resolution. NOAA AVHRR-derived normalized difference vegetation index (NDVI) is worldwide utilized for regional as well as global vegetation monitoring.

ASTER (Advanced Spaceborne Thermal Emission and Reflection Radiometer): The ASTER satellite was launched in 1999, and this satellite mission is a joint venture of NASA and Japan's Ministry of International Trade and Industry. ASTER sensor provides 14 multispectral bands located in different wavelength regions of EMR with varying spatial resolutions digital images such as visible to NIR wavelengths and 15 m spatial resolution (bands 1–4); SWIR wavelengths and 60 m spatial resolution (bands 4–9); and thermal wavelengths and 90 m spatial resolution (bands 10–14). LST and digital elevation model (DEM) of the world are also created from ASTER thermal and stereo remote sensing data respectively.

High spatial resolution satellites: In the late 1990s, different private consortiums, namely Space Imaging (Digital Globe), Orbimage and Earthwatch Inc. developed high spatial resolution (0.5 to 4 m) commercial EO satellites and made them available to various user communities. The data from these satellites are being used for detailed inventories of natural resources, urban studies, national security, precision farming etc. In 1999 Digital Globe launched IKONOS-2 after the failed launch of the first satellite and its sensor provided data of one 1 m spatial resolution black and white (PAN) band and 4 m spatial resolution four multi-spectral bands (blue, green, red and NIR). Preview-3 was launched in 2003 by Orbinage (later renamed Geoeye). The Orbview sensor provided similar types of data of IKONOS-2. GeoEye-1 launched in 2008 provided 41 cm PAN and 1.65 m multi-spectral images. Quick Bird satellite launched by DigitalGlobe acquired data of PAN bands with 61 cm and multi-spectral bands with 2.5 m spatial resolutions. Digital Globe launched several high-resolution satellites such as WorldView-1, 2 and 3 from 2007 to 2014. The latest WorldView-3 provided images with different spectral bands and spatial resolutions such as one 0.31 m black and white (PAN) band, eight 1.24 m visible and near-infrared multi-spectral bands, eight 3.7 m SWIR bands, and twelve 30 m bands. German company's RapidEye constellation of satellites provides global daily coverage with five 6.5 spatial resolution multi-spectral bands data.

19.2.2.7 Fundamentals of Remote Sensing Image Interpretation and Analysis

Interpretation of remote sensing captured photo or image involves finding out and assessing of significance of an object and the patterns on the image based on feature detection and identification (Sivakumar 2010). Aerial photographs were the first remote sensing data which is subjected to photo or image interpretation. Now it is applied to all kinds of processed and unprocessed visual remote sensing data products. Both visual as well as digital interpretation and analysis methods alone and a combination of RS data are used for the extraction of information and thematic mapping. Both photo or image interpretation techniques have advantages and disadvantages.

Visual interpretation: When dealing with photo/image data, visualized as pictures, a set of elements of photo/image interpretation is required to define characteristics present in pictures. The elements of image/photo interpretation are brightness or tone, texture, colour, size, shape, shadow, pattern and site or association. Based on the degree of complexity, the elements of photo/image interpretation can be grouped into four criteria, namely spectral criteria, simple spatial criteria, complex spatial criteria and temporal criteria.

Spectral criteria consisting of brightness and colour are the most primary elements because they control the spectral characteristics of surface features. Spatial properties of objects on the image/photo related to shape, size and texture elements form simple spatial criteria. Complex spatial criteria consist of shadows, context and association. Temporal criteria deal with seasonal changes in dynamic features under study.

Brightness or tone: The relative brightness of objects in an RS image in shades of grey is measured based on brightness or tone. Brightness or tone attributed to the reflectance of solar reflective region radiation or emitted radiation from objects in thermal infrared. The human eye can only discriminate 64 or fewer grey shades. Therefore, by visual image interpretation full radiometric resolution (256–2048 digital values) of an image cannot be exploited. Most land covers reflectance changes with different wavelengths (Fig. 19.4). So, the brightness of objects is different in wavelengths.

Colour: Colour is an important element of image interpretation because of the higher sensitivity of the human eye towards colour variations as compared to the brightness of tonal variations. For colour perception, the human eye is sensitive to only three wavelengths, namely red, green and blue named RGB of the visible region of EMR. These three components refer to 'primary colours' and any colour can be derived by combining these primary colours. By adopting two processes, namely additive and subtractive three spectral bands can be combined to generate colour composite images. In the additive process, any colour is generated by combining primary R, G and B colours and three colour guns are used to assign R, G and B colours to three individual spectral bands. The additive

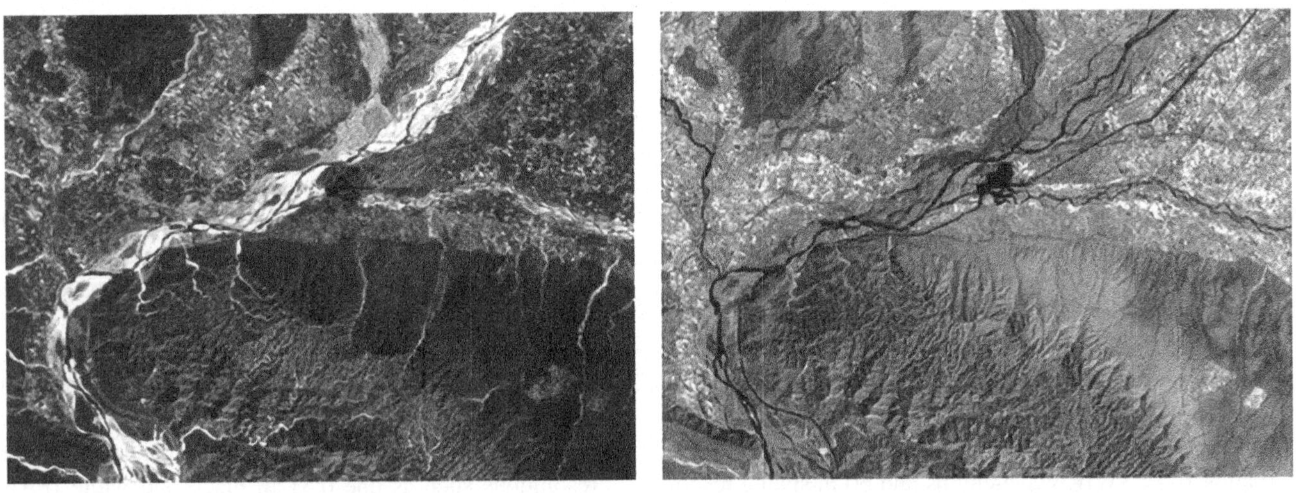

Fig. 19.4 The left image (IRS LISS III, red band) tonal variations represent dark grey represent vegetation (forest and croplands). Vegetation absorbs more in the red band due to plant pigments and reflects less. The dry riverbed shows a bright white tone due to high reflectance in the red band. The right image (IRS LISS III, near-infrared band) tonal variations depict bright white-crop lands and light-grey forests. In the near-infrared band vegetation, reflectance is high. The canal and river course with the presence of surface water appears very dark grey because of high absorption in the near-infrared band by water (image: part of Doon valley, Uttarakhand, India; image courtesy: Indian Institute of Remote Sensing, Dehradun, India)

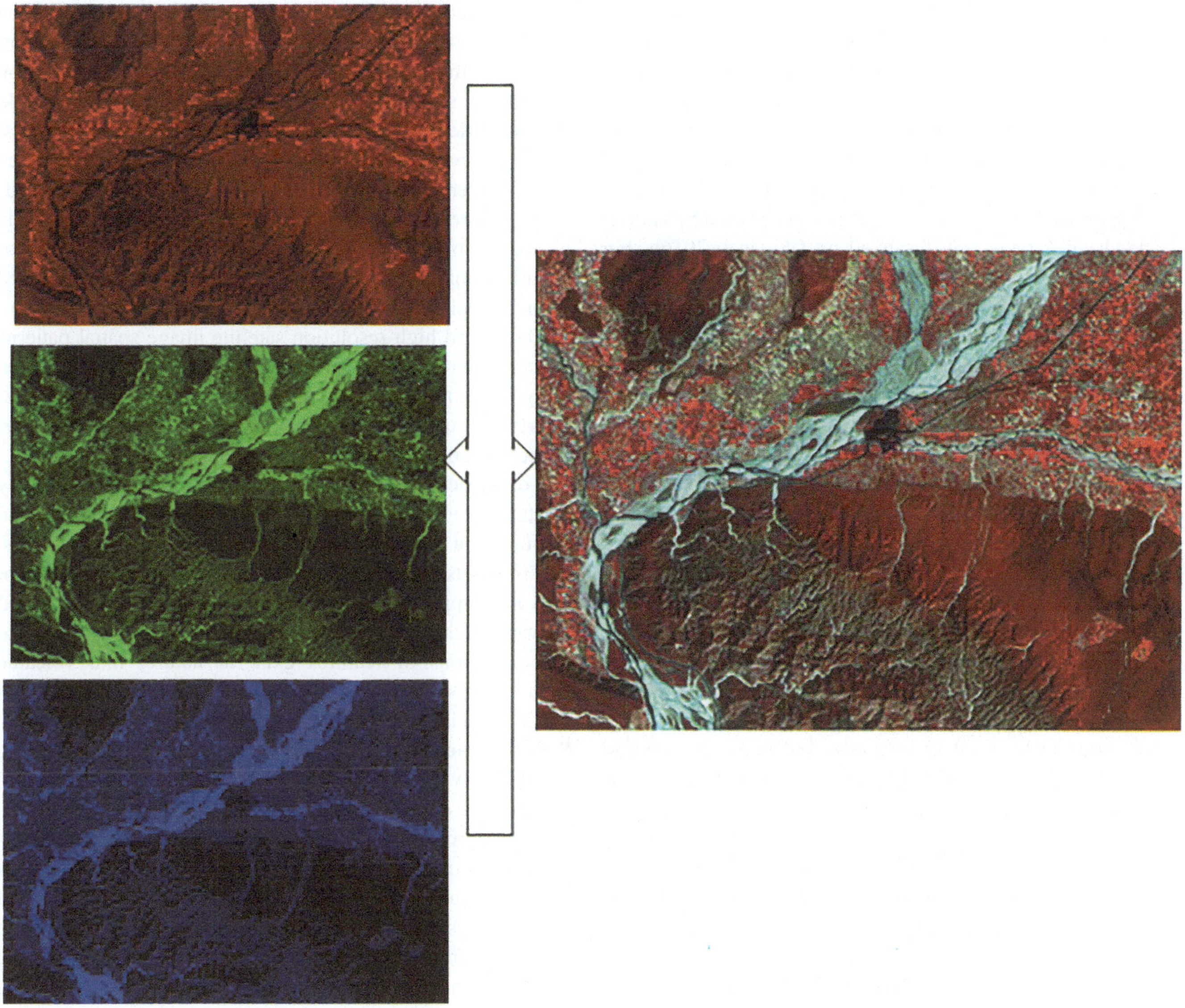

Fig. 19.5 Standard FCC generation process by combining and assigning R, G and B colours to three spectral bands: top image NIR band (R), middle image red band (G) and lower image green band (B). In FCC vegetation, forests and croplands appear in variations of red colours. (image: part of Doon Valley, Uttarakhand, India. Image courtesy: Indian Institute of Remote Sensing, Dehradun, India)

process is commonly used for obtaining colour composite in the electronic display process and popular method for digital remote sensing image processing systems. The subtractive process is generally used in the mechanical reproduction of colour in print media. By mixing yellow, magenta and cyan colour inks, the printed colours are obtained. A colour composite can be generated by combining any three spectral bands. However, standard false colour composite (FCC) generated by combining three spectral bands data such as NIR, red and green and assigning R, G and B colours, in the sequence is the popular one for image interpretation and analysis (Fig. 19.5). This colour composite is called false colour because vegetation appears in red instead of natural green colour due to high reflectance of NIR band by vegetation and assigning red colour in a composite generation. Almost all EO satellite sensors include these three spectral bands which made standard FCC widely used for visual interpretation and analysis of RS data.

Texture: Texture refers to characteristics of tonal or colour arrangement and changes in an image. The texture depends on the unit features an aggregation of an object that is difficult to separate on the image due to small dimension. Homogeneous land cover generates smooth image texture. Rough texture forms due to heterogeneity of cover attributed to reflectance variations in the vicinity.

The size of the constituent objects in a particular cover, shadow effect, spatial resolution and wavelength of the sensor control the texture of an image. Angular orientations of the sun and sensor contribute to shadow effects. A surface smooth at longer wavelengths can appear rough at short wavelengths. Texture is commonly used to discriminate different covers with similar spectral responses. Different image textures contributed by various land cover types are discussed by Chuvieco 2016b. For example, in the Landsat 8 OLI satellite image, urban areas with diverse small dimension covers appear in a coarse rough texture. While homogenous irrigated crops give smooth texture, shrubs and grasslands give medium to low smooth texture. Heterogeneous forest areas appear in rough texture (see details in Fig. 5.9 of Chuvieco 2016b).

Size and shape: Size is a significant parameter in image interpretation because it helps in identifying objects based on relative sizes and measurement of dimensions parameters such as width, length, area and perimeter of an object. For absolute measurement of object size, the image scale must be known. Shape is also useful for identifying objects by examining the shape contours, which can be associated with familiar objects. Based on the shape and size we can identify various geomorphic and structural features such as volcanos, domes and lineaments. Forest plantations appear with geometric borders while natural forests appear in irregular shapes. Shape and size elements can be better extracted using high spatial resolution images. The shape and size image elements of high spatial resolution remotely sensed image (RapidEye satellite) of part of Madrid city Spain used for discriminations of natural and made land cover types: (1) urban park, (2) natural forest and (3) golf courses and hippodrome were demonstrated and shown by Chuvieco (2016a, b) (see details in Fig. 5.11 of Chuvieco 2016b).

Shadow: Shadow in the image is formed due to variations in illumination conditions. In image interpretation, shadows contribute both positive as well as negative effects. Object features lying within shadows are difficult to discern on an image due to very low reflection. However, the shadow is useful for the detection of objects by estimating the height and depth of objects and separating targets from the background. In passive remote sensing, the shadow created by low solar angle illumination enhances relief features of the terrain, which helps in the interpretation of landforms or geomorphology. Shadow is also useful for better discrimination of forested areas by enhancing image texture.

Spatial pattern: Spatial arrangements of objects constituting cover form the spatial pattern element of image interpretation. The spatial pattern is important for the identification of features because of the presence of unique inherent patterns and characteristics of an object. Different objects from different types of characteristic patterns on remote sensing images such as agricultural field boundaries, roads, forest areas and human settlements. Spatial resolution and scale of the image are the key parameters that contribute to spatial patterns. The individual pattern may be visible in high resolution/large scale images, but the patterns merge in coarse or medium resolution/small scale images. Identifications of features based on spatial patterns were discussed by Chuvieco (2016b). QuickBird high-resolution satellite image spatial patterns were used for discriminating land cover features of an agricultural region in Salamanca, Spain crop fields linearly arranged and natural vegetation with scattered trees (see details in Fig. 5.15a of Chuvieco 2016b).

Site or association: Particular features in nature are mutually attached to other features, and this characteristic forms an important image element for the identification of features. Croplands are generally associated with water available or water-supplying devices like irrigation canal networks and water bodies. In geosciences, several examples of associations of features exist in nature, like volcanic landforms such as cones, calderas, dykes and lava flows are associated with extrusive rocks (Gupta 2017).

Period of acquisition: The multi-temporal data acquisition capability in remote sensing is important for the identification or change detection of features by changes in seasonal or multiannual spectral characteristics of the features under study. Time series analysis of aerial/spacecraft photos/images is very effective for the delineation of deciduous forests which shed leaves during the spring season from evergreen forests. Dynamic growth and development patterns of crops can be detected using time-series remote sensing data. Natural disasters like floods and draught are monitored and assessed using temporal remote sensing data.

Stereoscopic view: Stereoscopic remote sensing images/ photographs are acquired through overlapping coverages of the same area of a terrain by the sensor. Stereo images are generally captured by using dual sensors or by tilting sensors to create two different view angles. Stereoscopic view by analogue or digital photogrammetry is an important image analysis technique for the extraction of 3D features of objects. Stereoscopic image interpretation is common for geomorphologic analysis. This technique could be used for tree height estimation. Currently available several satellites such as SPOT-HRV, Terra-Aster and IRS-Cartosat have stereoscopic image acquisition capability.

19.2.2.8 Fundamentals of Digital Image Processing

A remotely sensed digital image consists of a 2D matrix of numbers which represent original incoming radiances recorded by the RS sensor. A specific row and column in the 2D digital image matrix are associated with each digital value in the matrix. The instantaneously observed ground area by the remote sensing sensor is named pixel (picture element). Pixel denotes the smallest non-dividable element of a digital RS image. A digital number (DN) represents the brightness value of an associated pixel. The multi-spectral remote sensing dataset consists of several spectral bands. Therefore, each pixel of an image is characterized by a 3D matrix with row and column location and DN values constituting spectral bands of the RS data. Digital images are organized in a series of files with different formats. Processing, manipulation and analysis of digital images are performed using a computer system and both commercial and open-source software. The major types of digital image processing performed in remote sensing can be grouped into several categories such as image preprocessing; image enhancement, image classification; change detection; biophysical modelling; and GIS integration. Brief descriptions of these processes are as follows:

Image preprocessing: These operations aim to remove radiometric and geometric errors associated with raw remotely sensed data. Radiometric errors are due to sensor system noises associated with optoelectronic components) and the scattering of radiation by the atmosphere into the IFOV of the sensor. Simple image normalization and relative and advanced radiative transfer model-based radiometric calibration techniques are in use for radiometric correction of digital RS data. Radiometrically calibrated remotely sensed data of varying periods/dates are essential for digital change detection analysis and comparison derivative products (e.g. spectral vegetation index, SVI). Remote sensing image exhibits internal and external geometric errors. Sensor scanning system itself and combinations with earth rotation and curvature characteristics introduce internal geometric errors. Geometric errors in digital RS data are contributed by changes in height and roll, pitch and yaw attitudes of remote sensing platform. The placement of individual pixels in their exact x and y planimetric positions in a standard map projection is performed by geometric correction. Geometric rectification using control points in reference image/reference map is performed to correct geometric errors of remote sensing image. Digital preprocessing analysis is generally carried out before digital enhancement, modification and classification of digital RS data for extraction of thematic information.

Image enhancement: These techniques are applied to improve visual distinctions among the objects/features in an image and finally improve information content for interpretation. Several image-processing techniques are in use for image enhancement and these include image contrast manipulation (e.g. contrast stretching, density slicing); manipulation of spatial features (e.g. filtering, edge enhancement, texture analysis); linear and nonlinear transformations (e.g. principal components analysis (PCA), spectral vegetation indices (SVI), hue intensity saturation (HIS) colour space transformation).

Image classification: Digital classification of an image is a quantitative technique for automatic information extraction from multi-spectral/multi-temporal remotely sensed digital images. In digital classification based on statistical decision rules individual pixel of the image is classified to a particular land cover/feature class. Supervised and unsupervised classifications are the two types of digital image classification. In digital supervised classification, the image analysis specialist provides field sample (training) information, which is used in the classification algorithm for the thematic classification of each pixel. While in unsupervised classification thematic information of a pixel is extracted using a classification algorithm only with no instructions from the image analyst. Maximum likelihood classification and ISODATA clustering algorithm are the most widely used supervised and unsupervised classification methods respectively. Supervised and unsupervised classifications are very effective When spectral training data consist of mixed pixels, applications of special digital supervised classification algorithms such as fuzzy, support vector machine (SVM) are found effective. The image segmentation approach using spectral characteristics of contextual surrounding pixels is the basis of the object-based digital image classification method. This classification technique is being successfully applied for automated feature extraction from high spatial resolution remote sensing data. Currently, artificial intelligence (AI) and machine learning (ML)-based advanced classification methods are gaining importance for classifying remote sensing data of complex terrain.

Change detection: Using time series remotely sensed data, the change detection analysis to assess the extent and nature of changes in land use/land cover or features over time has become a very important study objective in most remote sensing-based projects. Digital change detection is being used for seasonal, inter-annual and long-term change analysis using temporal remote sensing data. Several digital change detection algorithms are used for 'change/no change' or more detailed 'from-to' change information of land covers/features over time intervals. The change information is useful for finding out the diver(s) of changes.

Biophysical modelling: Spectral band radiance data as well as digitally processed spectral indices are used as inputs in empirical, semi-empirical and process-based models to estimate biophysical and biochemical properties of land covers/features.

GIS integration: The accuracy of land covers/features identified and mapped using remote sensing data following digital image processing can be significantly improved by the inclusion of other spatial information such as DEM, forest working plan map and soil map. Commonly for the development of expert systems and artificial neural network (ANN)-based digital analysis for feature extraction, in addition to multi-spectral RS data, multi-sources gather spatial data used.

19.2.3 Geographic Information System

'Geographic' because deals with data which is associated with geographic location in space with reference system (latitude/longitude or other spatial coordinates). 'Information' because we want to learn the attributes or characteristics of the data about the space. 'System' because it handles various interlinked and interrelated components having different functions. The database of GIS is a unique information system with geo-location tagged features represented by various types of geospatial data such as points, lines, polygons and areas. Burrough the pioneering researcher in geospatial technology in 1986 defined 'GIS as a computer-based system, which allows capturing, storing, analysing and displaying geographically referenced information' (Burrough 1986). GIS technology gained importance for creating spatial decision support systems (SDSS) and management of earth resources and the environment by combining geospatial and non-geospatial data collected from multiple sources (Fig. 19.6).

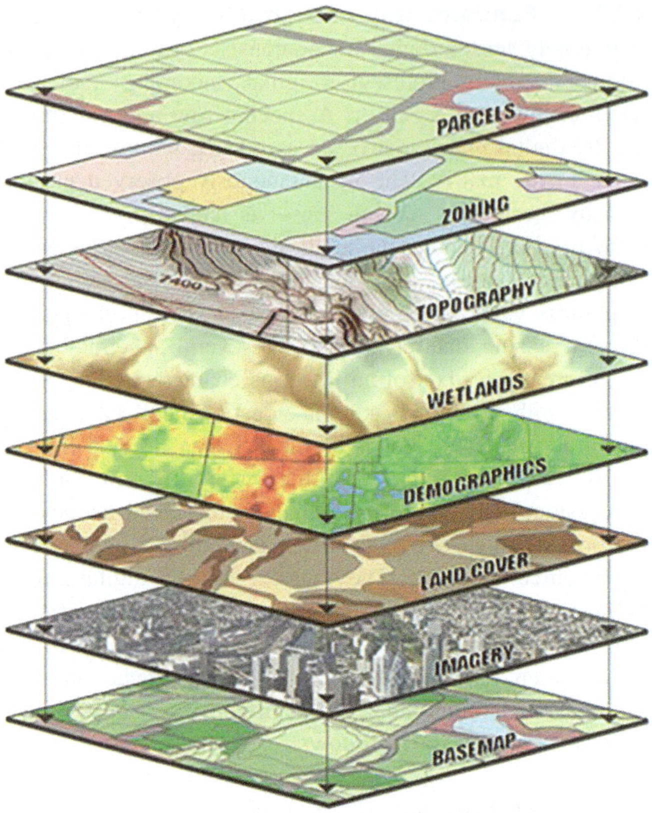

Fig. 19.6 GIS is a data integration technology (source: https://www.usgs.gov/media/images/gis-data-layers-visualization)

The evolution of GIS technology is linked parallel to the history of the development of digital computers and databases. Progress of GIS technology is also contributed by several allied disciplines, namely geography, cartography, remote sensing, photogrammetry, statistics and mathematics (geometry and graph theory). The first operational GIS was Canada GIS (CGIS) designed in the mid-1960s as a computerized map measuring and a producer of tabular land information system. In 1964, the foundation of modern GIS with respect to the development of the theory of geospatial data processing and manipulation was the research contributions by the Laboratory of Computer Graphics and Spatial Analysis, Harvard Graduate School, USA. In the early 1980s, two private companies, namely ESRI (Environmental Systems Research Institute) and Intergraph immensely contributed in the development of GIS software for commercial uses. Although, ESRI's ArcGIS is the most popular and widely used GIS software, currently free of cost and without licence known as open-source GIS software developed by various universities and research organizations is easily accessible to GIS users.

19.2.3.1 Definitions of GIS

Based on the functionality, capabilities and purposes of applications of GIS, several definitions of GIS are available. The following section presents well-recognized GIS definitions:

GIS is a 'Set of tools for collecting, storing, retrieving at will, transforming and displaying spatial data from the real world for a particular set of purposes' (Burrough 1986).

GIS is a ' computer-assisted system for the capture, storage, retrieval, analysis and display of spatial data, within a particular organization' (Stillwell and Clarke 1987).

GIS is a 'computer-based system that provides four sets of capabilities to handle georeferenced data,viz., data input, data management (data storage and retrieval), manipulation analysis, and data output' (Aronoff 1989).

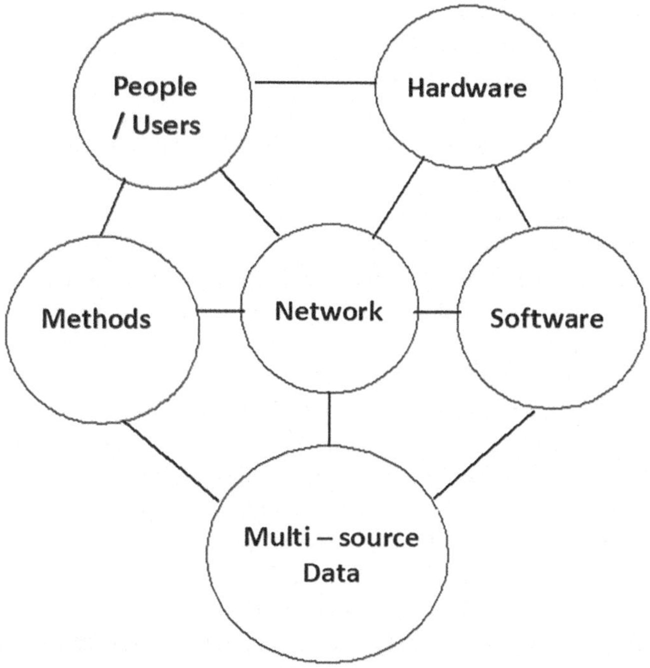

Fig. 19.7 The six components of a GIS

GIS is a 'computer based system for the efficient input, storage, manipulation, analysis, representation and retrieval of all forms of spatially indexed and related descriptive data' (Jackson 1992).

19.2.3.2 Components of GIS

Hardware, software, data, procedure/methods, people and network are the six components which are integrated into a working GIS (Fig. 19.7).

Hardware: The computer system which runs GIS software forms the hardware component. Different types of hardware configurations such as stand-alone desktop computers, laptops and central server network-connected desktop computers are used for running GIS software. Computers with efficient processors for running the software and sufficient memory for storing enough data are the main hardware requirements. A digitizer/scanner is also formed part of GIS hardware for data inputting by converting analogue maps and documents into digital form.

Software: GIS software consists of programs and user interfaces for driving the hardware for performing functionalities of GIS such as inputting, storing, analysing, manipulating and presenting geospatial data. The software program of GIS is called a GIS application. Both commercial (ArcGIS, GeoMedia, MapInfo, etc.) and open-source software (QGIS, ILWIS, SAGA GIS, GRASS, etc.) are used for GIS analysis.

Data: Geo-referenced digital spatial and non-spatial linked tabular data are the key data of GIS. The tabular data are related to the map objects and linked with the digital map. The sources of data can be collected from the organization collected or data can be purchased from the commercial data provider. Using DBMS (Data Base Management System), GIS integrates geo-spatial with other sources of data.

Procedures/methods: The methods of a successful project consist of well design plan with appropriate models and operating practices unique to each task. The GIS project executing organization must establish procedures and mechanisms to monitor the budget and quality of the project and meet the needs of the organization. The procedures look into the various aspects of the project starting from the formats of data received and how to incorporate it into the system, to geo-processing to be performed to get final output (s).

People: GIS users consist of several technical specialists and analyst persons having various GIS skills such as configuring the system, performing GIS analysis, maintaining the system, and helping end users to do analysis day-to-day basis. All the stakeholders of GIS users together plan, implement and participate in the decision-making process.

Network: The development of the ICT (Information Communication Technology) network and the internet, nowadays made it possible to rapidly acquire multi-source data and share large GIS-processed data sets.

19.2.3.3 Functionality of GIS

Data inputting, data storing and managing, analysis and data manipulating, visualization and generation of data output are the five major tasks performed by a typical GIS. Brief descriptions of these functions are as follows:

Data input: Spatial and non-spatial GIS data are obtained from a combination of hard copy maps, remotely sensed images (aerial photographs and satellite images), GPS surveys, reports, survey documents, etc. Analogue hard-copy maps are encoded into digital data by digitization process. Digitization is only an effective technique when few maps are digitized and the quality of encoded data depends on the accuracy of the digitization process. Scanning is an additional technique of conversion of analogue maps to digital data and scanning resolution is a critical factor controlling quality of data. Aero-space RS contributes major GIS data. Raw remote sensing digital data are processed before inputting to GIS. Another geospatial technology, an important source of data for

GIS is the global positioning system (GPS). GPS-derived locational data are increasingly being used together with spatial data in GIS. Recently, large amounts of spatial data are routinely accessed for GIS analysis from various websites maintained by various organizations/universities through the internet.

Data storage and management: These components of GIS perform functions of storing and retrieving to and from the database. The database organizes collected data in such a way that data can be retrieved, updated, expanded and shared by multiple users. The GIS database stores real-world geographic data in the form of objects (point, line, polygon) and attribute or non-spatial data which are described by the locations and characteristics of the object. There are different models of database management systems (DBMS), but because of efficient data storing, management and simple structure, relational database model systems (RDBMS) are widely used in GIS.

Analysis and data manipulation: The unique capability of GIS is the integrated analysis using both spatial and attribute data. A wide range of analyses can be carried out using GIS useful for particular applications. GIS analysis functions are classified into: basic and advanced functions. The basic analysis functions include measurements, reclassification, multi-layer overlay, connectivity and neighbourhood operations. Advanced functions are used for spatial decision-making. Several theoretical geospatial models based on mathematical and statistics developed which are referred to as advanced functions of GIS. Two types of mathematical GIS modelling are in use for decision-making: optimization and simulation. The best solution of either maximum or minimum for a particular problem on the decision can be identified using the Optimization method. Impacts of varying policies under the influence of several factors in a given system are tested with a simulation model.

Visualization and data output: In general, three types of outputs/visualization products are generated using GIS analysis: a) such as user query responses in the form of text output (different kinds of tables, listing of analysed data, numerical values or text, etc.); b) black and white and colour graphic analogue outputs (thematic maps, analysed or data display on the screen, schematic diagrams, different types of graphs, etc.); and c) analysed digital outputs data which can be achieved in local digital storing devices such as hard disk or transfer to a computer network or geo web service networks.

19.2.3.4 Data Types and Representation of Data in GIS

GIS handles and processes geographically referenced data, which consists of location and spatial features on the earth's surface. GIS uses two types of geo-located geographic data: spatial and attribute. GIS data types and their characteristics are given in ESRI (1990).

Spatial data: It refers to the relative and actual geolocation of features and describes the geometry of spatial features. Primary two types of formats of spatial data, namely vector and raster formats are stored in the GIS database.

Vector data: Points, lines and polygons are three basic types of vector data. Points are 0D objects consisting of pairs of x and y coordinates. Lines are 1D objects which have length but no area represented by sets of coordinate pairs (beginning and ending points). Polygons are closed mathematical figures of any shape or size consisting of sets of coordinate pairs enclosing the area. In addition, a topology database depicting the spatial relationship between the lines is created by GIS. Spatial polygon data are stored in GIS by creating databases of identification codes (IDs), arcs/lines make the polygon, perimeter, area and topology. Shapefile, geodatabase, DLG (Digital Line Graph), DXF (Digital Exchange Format), ArcInfo Interchange (e00), TIGER (Topologically Integrated Geographic Encoding and Referencing) etc. are the popular formats used for representation of vector data in GIS.

Raster data: In raster data, a geographic feature is represented as regular sets of cells on a grid. Generally, the regular grid is square cells called pixels. The dimension of cells varies from large or small depends how accurately to represent features within the area. The smaller the pixel size of raster data of an area, the larger be data volume and it will take more storage space and processing time. Digital satellite images and aerial photographs, scanned/digitized hard copy maps etc. are the major raster data used in GIS processing. TIFF (Tagged Image File Format), Geo TIFF (Geographic Tagged Image File Format), GIF (Graphic Interchange Format), JPEG (Joint Photographic Experts Group), BIL (Band Interleaved by line), BIP (Band Interleaved by pixel), BSQ (Band Sequence), ArcInfo GRID etc. are the common raster data formats handle by GIS.

Attribute data: This data represents properties of spatial entities consisting of spatial features. Attributes data are linked with vector and raster through the GIS database. The attribute data is also called non-spatial data as it does not represent location information. A patch of forest cover with a geographic location is an example of spatial data. But the various properties of the forest cover such as type of forest, species composition, crown density and tree height are examples of attribute data.

19.2.3.5 Global Positioning System

Global positioning system (GPS) is one of the global navigation satellite systems (GNSSs) used for operational global

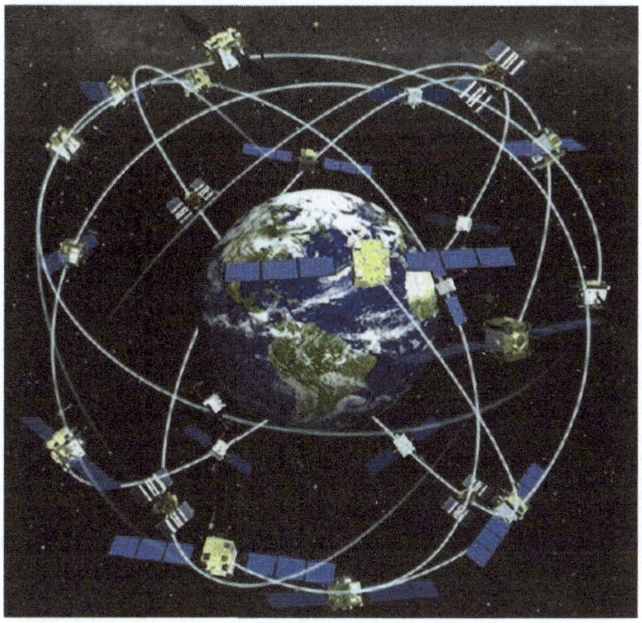

Fig. 19.8 Constellation of GPS satellites orbit the earth on six orbital planes (source: https://spaceplace.nasa.gov/gps/en/)

navigation and geopositioning of any location on the earth's surface. Radio navigation with precise time measurement is the key technology for the function of GPS. Several US government organizations, namely the Department of Defense (DoD), NASA and the Department of Transportation (DoT) are involved in the development and control of the GPS. In the 1960s, GPS was developed and initially; it was known as Navigational Satellite Timing And Ranging-Global Positioning System (NAVSTAR-GPS). GPS provides geolocation and time information worldwide to various users such as military, civil and commercial. Major applications of GPS are navigation, surveying and map-making. In forestry, GPS is very useful for finding geolocation and mapping areas and the extent of different forest ecosystem parameters.

The three major components of GPS are the space segment (satellite constellation), ground segment (satellite control station) and user segment (receivers). 30+ satellites of the GPS space segment are placed at an altitude of 12,000 miles which are transmitting signals and orbiting the earth at an interval of 12 h (Fig. 19.8). The control segment consists of ground facilities involved in tracking the satellites and provides the satellites with precise time and corrected orbital information. The user segment comprises limitless simultaneous users with GPS receivers. The GPS satellites are fitted with atomic clocks that transmit radio signals that contain precise location, time and other locations. Control stations receive, monitor and correct the signals transmitted by the satellites and the user receivers pick the corrected signal to find out the position. Ideally, a GPS receiver needs four satellites to compute the 3D (2 horizontal: x, y and vertical: z) position more accurately.

GPS receiver downloads orbital information (Almanac data) of all visible satellites when it turns on. The GPS receiver calculates the distance between the receiver and each visible satellite using the relationship-transmit radio signal velocity (186,000 miles/s) multiplied by the time value. The time value is determined by how long the time takes the satellite-transmitted signal to arrive at the GPS receiver. The receiver determines position using mathematical triangulation. The position is reported in different coordinate systems such as latitude/longitude and UTM (Universal Transverse Mercator).

The GPS positional accuracy is affected by several factors such as satellite clock error, ephemeris (satellite orbital parameters) error, atmospheric effect, ionospheric effects, multipath and shadowing effect and selective availability. Common GPS receivers provide 5–20 m accuracy. Currently, GPS positional accuracy up to sub-metre to a few metres level is improved by adopting two advanced techniques, namely DGPS (Differential Global Positional System) and RTK (Real Time Kinematic) GPS. DGPS improves the positional accuracy of GPS by using one or more GPS reference stations with known geo-location and post-processing corrections are applied. RTK GPS survey improves accuracy by real-time corrections of GPS receiver position.

19.3 Applications of Geospatial Technology in Forestry

19.3.1 Applications of Aerial Remote Sensing in Forestry

Before the 1970s, various types of aerial remotely sensed data products; namely black and white and colour photographs were widely used for forest resources assessment and management. Forestry applications of aerial remote sensing applications are a century old. In 1963, first time in India, J. A. Jones, a Canadian expert used aerial remote sensing in the form of black and white aerial photographs to make a quick forest inventory of the Kullu and Seraj valleys. After the success of the first study, in 1965 onwards Indian government under the Pre-Investment Survey of Forest Resources (PISFR) project used different aerial remote sensing data products for assessing available commercial bamboo forest resources for the paper industry. From 1965 to 1982, the Forest Survey of India (FSI) undertook several projects on forest mapping and growing stock assessment using aerial photographs (Unni 1983). In general, aerial photographs are used in three distinct areas of forestry such as forest mapping, forest inventory and forest management.

19.3.1.1 Forest Mapping

Forest mapping using aerial photo-interpretation involves the detection and composition of tree species and tree structures. Photo/image analysis based on photo/image elements, namely tone, colour, texture, shape, size pattern and associated is applied for identification of tree species. Other factors such as the scale of the photograph, the type of aerial film used, the photo quality and the season of photography control the effectiveness of identification of types of tree species and forest types.

The photograph's scale to be used for analysis is a function of details of information to be extracted. Large-scale photographs on 1:5000–1:10,000 scales are useful for small-area detailed forest inventory with respect to the identification of individual trees, and canopy density, small-scale photographs on the scale of 1:40,000–1:60,000 are useful for large-area broad forest cover type and broad classes of canopy density mapping. Generally, trees with a full foliage cover period are ideal time of aerial photography.

Aerial remote sensing data of true colour photographs was observed to be more effective than black and white aerial photographs (Unni 1983). Because vegetation highly reflects near-infrared radiation, CIR is found useful for the discrimination of vegetation from other land cover features. Vegetation appears in red in the CIR photograph. Using a CIR photograph, Unni et al. (1983) identified and mapped several tree species in the deciduous forests of Andhra Pradesh, India. A CIR photograph is superior to a true colour photograph for forest mapping.

Forest structure classification is making groups of forest vegetation based on species composition stand or tree height, crown density and crown size. Tree height can be estimated with stereoscopic parallax photogrammetric interpretation using overlapping stereo aerial photographs by using a stereoscope instrument.

19.3.1.2 Forest Inventory

Tree height and crown diameter extracted from stereoscopic interpretation of aerial photographs are used in the computation of the volume of a single tree required for forest inventory assessment. Finally, an aerial photograph estimated stand crown density percentage is used for the estimation of volumes of stands. The crown diameter estimated from the photograph is equivalent to the visible crown diameter or crown width as all parts of the crown are not visible by aerial photography. Crown closure also known as crown density refers to the percentage of ground area covered by tree crown. Forest inventory requires information on two important parameters, namely total growing stock and mean volume per unit area of forest. Aerial photograph-derived parameters, namely visible tree height, crown diameter, crown density and crown count have been successfully used for rapid estimation of growing stock and mean volume of the forest.

Several studies on forest inventory in diversified forested areas concluded that for the classification of timber volume by stands, aerial photographs with scales of 1:15,000–20,000 are suitable. However, individual tree volume estimation, and photographs with scales of 1:5000 or larger are recommended.

19.3.1.3 Forest Management

Aerial photography can also provide valuable information in several areas of forest management such as forest land suitability analysis, timber harvest planning, monitoring logging, afforestation, wildlife management, pests and disease infestation monitoring and forest road alignment and monitoring.

19.3.1.4 Principles of Photo-Interpretation in Forestry

A brief description of approaches to the use of photo/image elements in forest resources information extraction follows:

Tone: Tone is one of the important photo elements for tree species identification. Tonal variation of aerial photographs of tree species is controlled by vegetation phenological changes such as shedding of leaves, fruiting, flowering and new flash of leaves. For example, teak trees appear in white tones in aerial photographs taken during the flowering period and sal trees can be easily identified with silvery grey tones in photographs captured during new flashes of leaves. Infrared B and W aerial photographs are very effective for the discrimination of conifers and broad-leaved species based on their grey and white tones respectively.

Texture: Image texture is characterized by coarseness/smoothness parameters. In general, in aerial photographs younger and older trees appear in smooth and coarse textures respectively. The image texture of trees depends on branching pattern and age.

Shape: The shape of a tree crown is an important parameter for the identification of species. By interpreting the shapes of tree crowns on aerial photographs, many important species can be identified. Young broadleaved and most conifers have an ovate-shaped crown. The dome-shaped crown is found in older hardwoods.

Size: The size of the tree crown and height are the key variables used for the discrimination of tree species. The scale of aerial photographs is a critical factor for interpreting the shape of an object, crown size relating to the basal area of a tree is an important parameter for the assessment of tree volume.

Pattern: Spatial associations of objects in photos/images are called patterns. Patterns can be either manmade or natural. In aerial photographs, natural and forest plantations as well as young and old plantations can be discriminated against based on unique image patterns.

Fig. 19.9 Photo-interpreted area in the central part of the study area with different classes of forest canopy density: 1 closed, 2 semi-closed, 3 semi-open, 4 open and 5 deforested (source: Garcia et al. 2016)

Shadow: Shadow helps in identifying the tree species by its contribution to the image texture. The shadow cast by an individual tree on the ground along stand edges is also useful for the identification of species. Measurement of tree shadow on aerial photographs can be used for obtaining tree height. However, excessive shadows on the photo hinder the identification of objects.

Location and association: It has been observed that certain tree species occupy characteristics of topographic sites and are associated with specific types of land covers. While interpreting aerial photographs these photo elements are used as one of the features for identification of species. Silver oak and Erythrina species are associated with tea plantations in the Nilgiris.

A case study example of the use of photo-interpretation in temperate forest canopy density classification in Monarch Butterfly Biosphere Reserve (MBBR, near Mexico City) was demonstrated by Garica et al. (2016). Forest canopy density was classified into five categories based on the stereoscopic interpretation of a stereo pair of aerial photographs. Five forest density categories identified and mapped in this study were as follows: 1, closed >75%; 2, semi-closed 51–75%; 3, semi-open 26–50%; 4, open 10–25%; 5, deforested <10% (Fig. 19.9) (Garcia et al. 2016).

19.3.2 Satellite Remote Sensing Applications in Forestry

Since 1972, the availability of data from the first remote sensing satellite, Landsat-1, USA, space remote sensing has become a popular and useful tool for forest resources assessment and management. Forest types and characteristics information related to vegetation phenological types, gregarious formations and communities are very effectively derived from satellite multi-spectral data. The availability of multi-spatial resolution satellite data enables multi-level forest information extraction (Fig. 19.10). Multi-temporal satellite data provide an opportunity for forest dynamics study, landscape change detection analysis, phenological studies, etc. Biophysical modelling using digital satellite data derived spectral vegetation indices as input useful for various applications in forestry such as biomass/volume, canopy density and forest condition and degradation processes. Satellite remote sensing data available in digital format enables time-saving, cost-effective computerized mapping of land cover, land use and vegetation types in multi-levels, namely local, regional and global with reliable accuracies regularly.

19.3.2.1 Spectral Characteristics of Vegetation

Understanding spectral characteristics called spectral signatures is vital for the identification of the features. Spectral signatures are the major inputs for the visual and digital processing of remotely sensed data for feature identification. The spectral signature of an object depicts variations of spectral absorption and reflectance with respect to the wavelength of EMR. Several vegetation factors such as pigments of plants, leaf water, plant cell structure, cellulose, lignin and other biochemical constituents control spectral reflectance characteristics of vegetation (Fig. 19.11) (Pu 2017). A typical reflectance spectrum of a green leaf is divided into three regions: 1. 0.4–0.7 μm (visible region), 2. wavelength between 0.7–1.3 μm (near-infrared region) and 3. wavelength between 1.3 and 2.5 μm (middle-infrared region) (Fig. 19.11).

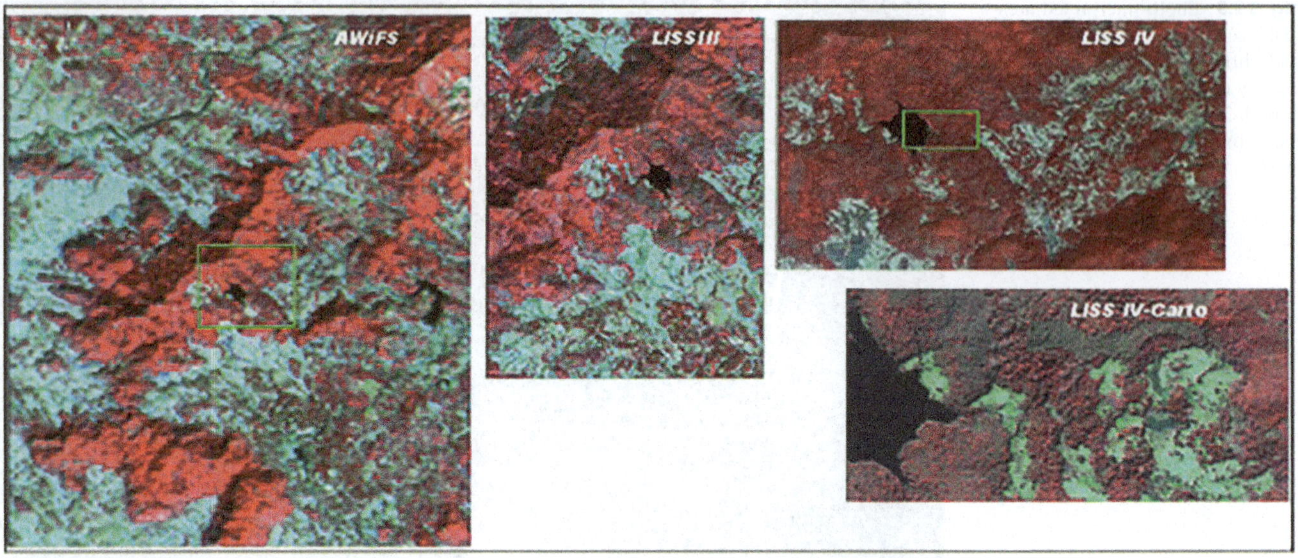

Fig. 19.10 Comparative example of different spatial resolutions of IRS satellite data (AWiFS, LISS-III, LISS-IV and LISS IV + CARTOSAT) (source: Murthy and Jha 2010)

Fig. 19.11 A typical reflectance spectrum of an oak leaf showing characteristics of absorption and reflectance features and the vegetation factors causing variations of reflectance in three spectral regions (source: Pu 2017)

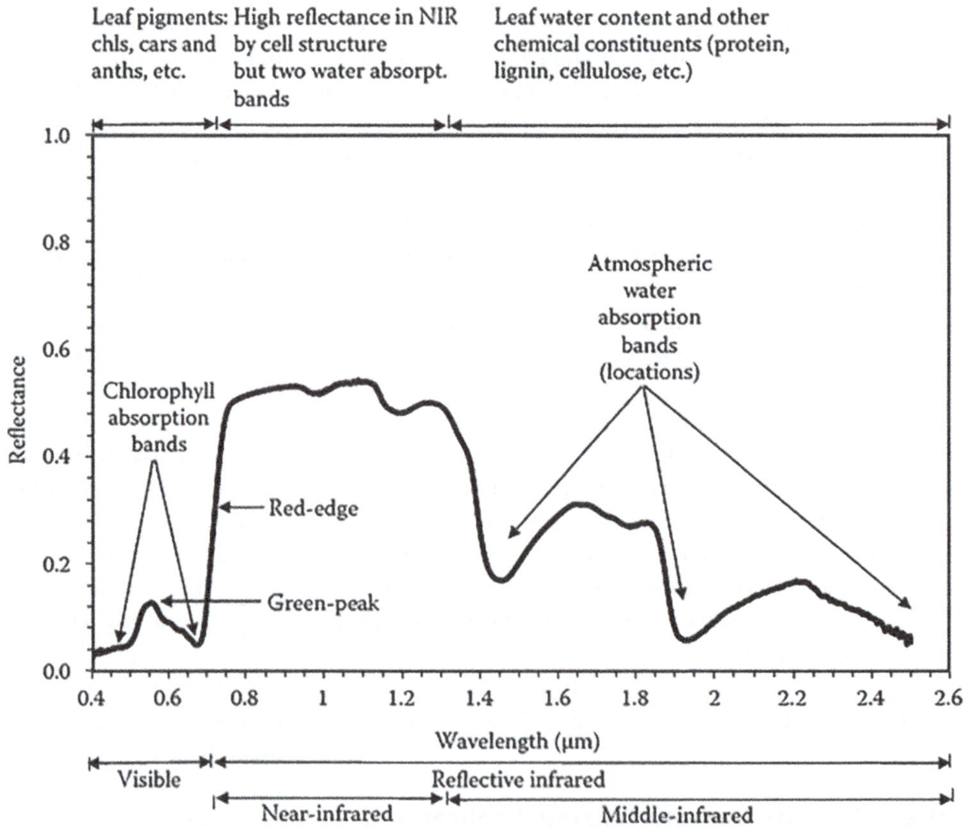

Visible (VIS region): Low reflectance is observed in the visible spectral region of the green plant reflectance spectrum. Low reflectance is due to the high absorption of visible radiation by different pigments of leaves such as chlorophylls, carotenoids and anthocyanins. Chlorophyll is the dominant contributing pigment in spectral absorption with spectral absorption peaks located near 420 nm, 490 nm and 660 nm wavelengths. The absorbed radiation in the visible spectral region is used for the photosynthetic process by the green plant. In green wavelength, there is a small reflectance peak and because of this most plants appear green. In the case of high vigour vegetation, the spectral reflectance increases sharply in the transition zone of red to near-infrared wavelength (660–780 nm). Generally, the inflexion point of the sharp increase of spectral reflectance lies between 700 nm and 725 nm wavelength this inflexion point is called red-edge position and it is very effective for vegetation stress detection.

Near-infrared (NIR region): The healthy green vegetation reflectance is high in the NIR region with a spectral wavelength range from 0.7 to 1.3 μm. The reflectance in the NIR region depends on the internal structure of leaves and varies among plant species. For monitoring plant growth and development, the reflectance of this spectral region is very useful. Volume scattering of NIR radiation within green leaf internal intercellular spaces of spongy mesophyll layer and air causes variations of spectral reflectance in the NIR region.

Middle infrared (MIR region): Like the VIS spectral region, the MIR spectral region also shows high spectral absorption. The vegetation factors responsible for strong absorption are the moisture content of plants and biochemical constituents of leaves such as pectin, lignin and cellulose, lignin. Plant leaf water content is strongly related to the reflectance of MIR reflectance in the 1.3–2.5 μm spectral region. Plant biochemical constituents contribute to the reflectance in MIR in the spectral region of 1.8–2.5 μm, when the green leaf dries up.

19.3.2.2 Spectral Vegetation Index

An SVI is computed by combining reflectance data of several remotely sensed spectral bands using simple mathematical formulae. The significance of the use of SVI is to discriminate vegetation features from other features by exploiting differences in the characteristics of spectral signatures of green vegetation and other features. SVI is related to several vegetation parameters such as per cent vegetation cover, LAI, green biomass, fractional absorption of PAR and carbon dioxide fluxes.

The simple ratio index and NDVI are the two popular basic ratio vegetation index (RVI). Particularly, NDVI is used worldwide for the assessment and monitoring of vegetation conditions. The empirical formulae for calculating RVI and NDVI using near-infrared (NIR) and red (RED) spectral reflectance bands are given below:

$$\mathrm{RVI} = \mathrm{NIR}/\mathrm{RED};\ \mathrm{NDVI} = (\mathrm{NIR}\text{–}\mathrm{RED})/(\mathrm{NIR} + \mathrm{RED})$$

RVI value increases (>1) with an increase in vegetation greenness. Bare soils RVI show values closer to 1 due to similar spectral reflectances in both RED and NIR bands. NDVI is used to minimize the effects of variable illumination levels by normalizing reflectances of RED and NIR bands. −1 to +1 is the range of values of NDVI. Positive values of NDVI are observed in the case of vegetation and NDVI value increases with an increase in vegetation greenness.

Other number of SVIs are also used for vegetation assessment and monitoring. Empirical formulae for calculating these SVIs are presented below:

$$\text{Green difference vegetation index (GDVI)} = \mathrm{NIR}\text{–}\mathrm{green}$$

$$\text{Green normalized difference vegetation index (GNDVI)} = (\mathrm{NIR}\text{–}\mathrm{green})/(\mathrm{NIR} + \mathrm{green})$$

$$\text{Vegetation index green (VIG)} = (\mathrm{green}\text{–}\mathrm{red})/(\mathrm{green} + \mathrm{red})$$

$$\text{Global environmental monitoring index (GEMI)} = \acute{\eta} \times (1\text{–}0.25 \times \acute{\eta})\text{–}[(\mathrm{red}\text{–}0.125)/(1\text{–}\mathrm{red})]$$

$$\text{Green atmospherically resilient index (GARI)} = \mathrm{NIR}\text{–}[\mathrm{green}\text{–}(\mathrm{blue}\text{–}\mathrm{red})]/\mathrm{NIR} \times [\mathrm{green}\text{–}(\mathrm{blue}\text{–}\mathrm{red})]$$

$$\text{Vegetation atmospherically resilient index green (VARIG)} = (\mathrm{green}\text{–}\mathrm{red})/(\mathrm{green} + \mathrm{red}\text{–}\mathrm{blue})$$

$$\text{Soil} - \text{adjusted vegetation index (SAVI)} = [(\mathrm{NIR}\text{–}\mathrm{red})/(\mathrm{NIR} + \mathrm{red} + L)] \times (1 + \mathrm{L})$$

$$\text{Modified soil} - \text{adjusted vegetation index (MSAVI2)} = \left[2 \times \mathrm{NIR} + 1\text{–}\sqrt{}(2 \times \mathrm{NIR} + 1)^2 8\mathrm{NIRred}\right]/2$$

$$\text{Green soil} - \text{adjusted vegetation index (GSAVI)} = [(\mathrm{NIR}\text{–}\mathrm{green})/(\mathrm{NIR} + \mathrm{green} + \mathrm{L})] \times (1 + L)$$

$$\text{Transformed soil} - \text{adjusted vegetation index(TSAVI)} = \frac{[a \times (\mathrm{NIR}\text{–}a \times \mathrm{red}\text{–}b]/[a \times \mathrm{NIR} + \mathrm{red}\text{–}(a \times b) + X}{\times (1 + a^2)]}$$

where, NIR, red, green and blue are spectral reflectances of respective bands. $\acute{\eta} = (2 \times (\mathrm{NIR}^2 - \mathrm{red}^2) + 1.5 \times \mathrm{NIR} + 0.5 \times \mathrm{red})/(\mathrm{NIR} + \mathrm{red} + 0.5)$, $L = 0.5$, a: the slope of the soil line, b: the intercept of the soil line, X: adjustment factor.

GEMI, GARI and VARIg are SVIs that minimize atmospheric effects and are suitable for change detection studies in vegetated landscapes. Whereas, SAVI, MSAVI2, GSAVI and TSAVI remove soil background effects in SVIs and are effective for assessing vegetation greenness in sparse/open vegetation cover conditions.

19.3.2.3 Forest Type Mapping

Both visual and digital analysis methods utilizing remotely sensed data are used for mapping and monitoring forest/ vegetation and both methods have advantages and limitations. Remote sensing analysis approaches for classifying vegetation consider several criteria: identification of plant formation with respect to microclimate zone, characterize each plant formation based on its physiognomy including life forms (without considering floristics), and division of each plant formation into sub-formations according to the physiognomy of the sub-dominant vegetation by using field sampling data.

Visual method: In the past remotely sensed satellite data of enlarged hard copy printout of FCC was used. Nowadays, computer display-on-screen visual interpretation is adopted. Image elements of tone/colour, texture, size, shape, shadow, pattern association and location are applied to discriminate tree species. A tree crown in a forest stand produces a pattern, which is unique to many species. Image shadow is useful for the identification of species based on the profile of trees derived from the shadow cast on the image. Temporal remote sensing data helps in the identification of species by studying the variation of tree appearance due to phenological changes in different seasons in the year.

Several forest vegetation parameters such as stand height, the diameter of the tree crown, the density of stocking and the area of the stand can be extracted by visual interpretation of remotely sensed data. These parameters are being used for the quantification of different parameters of forest stands such as growing stock, carbon stock and vegetation biomass. Remote sensing estimated volume of timber of individual trees or stands are empirically related with ground measured sample plot level tree volume, biomass and carbon mass. The developed empirical relationships are used for extrapolation to the large areas using remotely sensed data.

Forest Survey of India (FSI) in 1982 adopted a visual interpretation technique using remote sensing data to map large areas of forests in India into various categories like closed and open forests, shifting cultivation/biotic factors affected forests and non-forests. Several researchers (cited by Unni 1983) found the usefulness of the visual method using FCC of Landsat MSS data for forest resources studies. NRSA (National Remote Sensing Agency) in 1983 prepared a forest/vegetation cover map of India using a 1:one million Landsat following visual method (Unni 1983).

Digital method: Digital multi-spectral remote sensing data are digitally classified using both supervised and unsupervised approaches for forest-type mapping. Supervised digital classification uses limited ground truth data to derive training area statistical parameters for classification. Commonly, the maximum likelihood algorithm is used for digital supervised classification. Research studies showed that a hybrid digital classification approach combining both unsupervised and supervised techniques improves classification accuracy considerably (Kushwaha and Unni 1989). In India, the first major forest cover types such as temperate, tropical and tropical semi-evergreen forests; tropical deciduous; pure and mixed bamboo; shifting cultivation; cultivated land and water bodies were mapped using Landsat digital data following digital classification in Nagaland (Unni 1983).

The choice of a suitable classification method depends on the complexity of the study area, type of RS data and forest classification system. The appropriate seasons for selecting satellite data for different forest regions of India are presented in Table 19.3 (Roy 1986).

The incorporation of digital image-derived spatial features such as texture or segmented images along with spectral data also improved classification accuracy (Lu et al. 2014a, b). In recent years, object-based image classifiers have been successfully used for detailed forest resource information extraction using higher spatial resolution satellite data (Machala and Zejdova 2014). Currently, for forest tree species identification and detailed land cover/ land use inventories using high spatial resolution satellite data, several advanced machine learning digital information extraction algorithms are gained popularity and these algorithms are decision tree (DT), artificial neural

Table 19.3 Appropriate seasons in different vegetation zones and bio-climatic regions for studying vegetation covers (source: Roy 1986)

Sl. no.	Region/vegetation zone	Proper season
1.	Humid and moist evergreen and semi-evergreen vegetation in western and eastern ghats	January–February
2.	Humid and moist evergreen and semi-evergreen vegetation of north eastern region	February–March
3.	Tropical moist deciduous vegetation of northern and central India	December–January
4	Temperate evergreen vegetation of western Himalayas	March–May
5.	Temperature, subalpine, alpine evergreen and deciduous vegetation of Jammu & Kashmir	September–October
6.	Arid and semi-arid dry deciduous and scrub vegetation	October–December
7.	Tropical coastal mangrove vegetation	February–March

networks (ANN), support vector machine (SVM), random forest (RF) rule-based expert system, etc. (reviewed by Xie et al. 2019).

19.3.2.4 Forest Crown Canopy Density Mapping

For quantification and monitoring of growing stock and forest cover, forest crown canopy density forms a vital parameter. Remotely sensed satellite data has been successfully used in India for the estimation of crown density from 1984 onwards. Since 1987, operationally FSI using remote sensing techniques for the biannual cycle of forest cover mapping and monitoring changes across the entire country. The wall-to-wall mapping of forest cover is carried out following a hybrid approach of unsupervised digital classification of satellite data. The technical persons from state forest departments are involved in the visual interpretation of RS data, evaluation of post-classified thematic maps, field data collection in training areas (ground truth sites) and accuracy assessment (Fig. 19.12). Digital IRS-LISS III sensor data onboard Resourcesat 1 and 2 satellites are used for forest cover mapping.

The criteria for classification and forest crown/canopy density classes are presented in Table 19.4 (FSI 2019). Indian national forest cover map of 2019 prepared by FSI is presented in Fig. 19.13.

IRS black and white (PAN) and colour multi-spectral (LISS IV) high spatial resolution sensors data were used for forest crown density mapping with 20% intervals for forest division and micro level planning (NRSC). Currently, sub-metre spatial resolution (<1 m) satellite data are being used for detailed forest crown-based information such as crown diameter, number of crowns and degree of overlap and this information is valuable for micro-level monitoring and forest condition assessment. Both common parametric and non-parametric digital classification techniques are used for forest crown canopy density. Commonly, parametric supervised and unsupervised digital classification techniques and non-parametric classification methods, namely Nearest neighbour (kNN), SVM and RM are adopted for mapping various forest attributes including forest crown canopy density (reviewed by Abdollahnejad et al. 2017). The use of several spectral vegetation indices, namely NDVI, RVI, atmospherically resistance and soil-adjusted vegetation index were investigated for forest crown canopy density mapping (reviewed by Abdollahnejad et al. 2017).

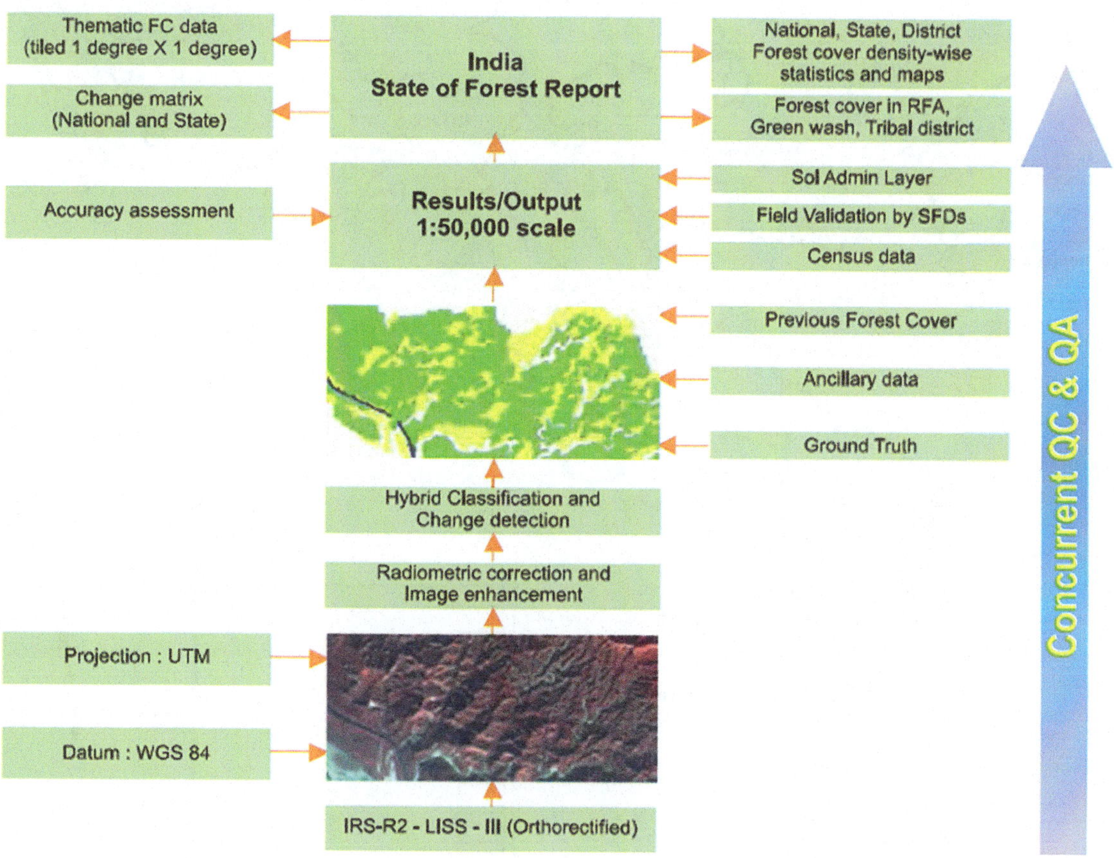

Fig. 19.12 Broad approach followed in forest cover mapping (source: FSI 2019)

Table 19.4 Criteria of classification of forest cover based on canopy density (source: FSI 2019)

Class	Description
Very Dense Forest	All lands with tree canopy density of 70% and above
Moderately Dense Forest	All lands with tree canopy density of 40% and more but <70%
Open Forest	All lands with tree canopy density of 10% and more but <40%
Scrub	Forest lands with canopy density <10%
Non-forest	Lands not included in any of the above classes (includes water)

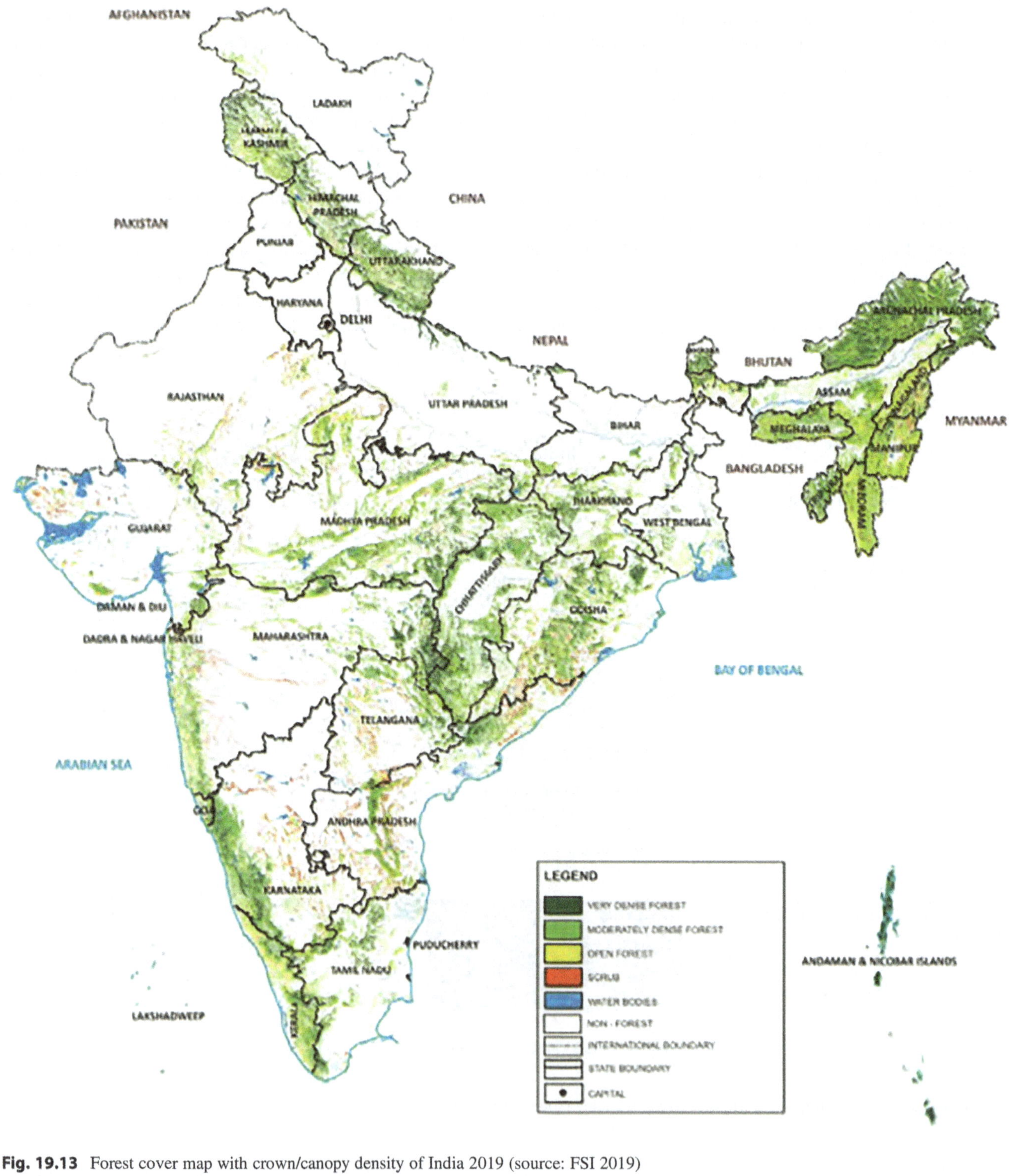

Fig. 19.13 Forest cover map with crown/canopy density of India 2019 (source: FSI 2019)

Table 19.5 Spectral indices of the FCD model

	Formula	Explanation and practical uses
Advanced vegetation index (AVI)	$\mathrm{AVI} = [(B4+1)(256-B3)(B4-B3)]^{\frac{1}{3}}$	The use of power degree on NDVI enables AVI to be more sensitive to forest density and physiognomic vegetation indices
Bare soil index (BI)	$\mathrm{BI} = \frac{(B4+B2)-B3}{(B4+B2)+B3}$	It is the index prepared for analysing soils, in other words it can be used to identify the difference between agricultural and non-agricultural vegetation
Canopy shadow index (SI)	$\mathrm{SI} = \sqrt{(256-B2)(256-B3)}$	Evaluates the different shadow patterns, based on the structure, age, species distribution etc., by affecting the spectral responses each time
Thermal index	–	Source of info is the thermal band of thematic mapper sensor (band 6)

B2 green band, *B3* red band, *B4* near infra-red band, *NDVI* normalized difference vegetation index

Biophysical spectral response modelling (forest canopy density: FCD model): A semi-expert system: FCD uses a linear multi-parametric approach for the classification/grouping of forest crown canopy density. Four satellite-derived spectral indices, namely advanced vegetation index (AVI), bare soil index (BI), canopy shadow index (CSI) and thermal index (TI) are used in FCD (Abdollahnejad et al. 2017, Table 19.5).

Chandrashekhar et al. (2005) compared the performance of the FCD model-derived forest crown canopy density map in part of Doon Valley, Uttarakhand with conventional visual interpreted and object-oriented image analysis maps. FCD gave comparable results as of other two methods and effective for automated mapping of forest canopy density without ground truth information. Spatial map of five forest canopy density classes (>80%, 60–80%, 40–60%, 20–40% and <20%) and the non-forest class was generated using the FCD model and Lansat 7 ETM+ satellite data (for details, see Fig. 6, Chandrashekhar et al. 2005).

19.3.2.5 Forest Biomass Estimation

The forest ecosystem acts as an environmental regulator by reducing atmospheric greenhouse gas concentration transforming it into biomass and emitting oxygen through the photosynthesis process. The majority of NPP accumulates as biomass. The contribution of forests is very vital in the terrestrial C (carbon) cycle and storing of C as biomass. Biomass constitutes the total quantity of organic matter of living as well as dead plant parts above and below ground (FAO 2005). Spatial estimation of forest biomass is useful for the calculation of terrestrial carbon flux, the productivity of the forest, the quantity of energy available from biomass, terrestrial carbon stock and sustainable forest planning and management (Pizana and Nunez 2016). Generally, above-ground forest biomass (AGB) is estimated due to difficulty in measurement and estimation of underground forest biomass.

Conventionally, AGB is accurately estimated using allometric equations developed for individual tree types and field sample plots observations data such as tree height and trunk diameter at breast height (DBH). However, this technique is time-consuming, labour-intensive, non-economical and more applicable for relatively small areas. Multiple spectral bands, different spatial resolutions and repetitive availability of remote sensing data are very effective and cost-saving techniques for spatial estimation, mapping and change analysis of forest AGB ranging from small to large both accessible and non-accessible areas. However, field data of sample plots are essential for training and validation of remote sensing-based AGB estimation techniques.

Biomass modelling: Several researchers developed diverse modelling approaches for AGB estimation using a variety of EO satellite data. Both direct and indirect modelling approaches are applied for the estimation of forest AGB biomass. Two major types of modelling approaches, namely are used for AGB estimation. In the direct approach, remote sensing derived selected variables are related to field-measured biomass. Whereas, the indirect approach, remote sensing derives forest biophysical parameters using biomass estimation.

Uses of field data in biomass estimation: All remote sensing-based AGB estimation models require accurate field biomass data of ground truth sites. Field ground truth estimated biomass data are used in various ways in remote sensing-based AGB modelling:

1. Field-measured reference biomass data is used for the identification of remote sensing-derived variables by studying relationships between the variables and reference biomass data,
2. Development of AGB estimation models using selected remote sensing derived variables and reference field biomass data,
3. Evaluation of model estimate with other field-measured biomass reference data set.
4. Assessment of error, uncertainty and identification of factors controlling the accuracy of estimation (Nandy et al. 2019).

Conventionally, three approaches, namely destructive field sampling, development of allometric equations for individual tree types and conversion of tree volume to biomass

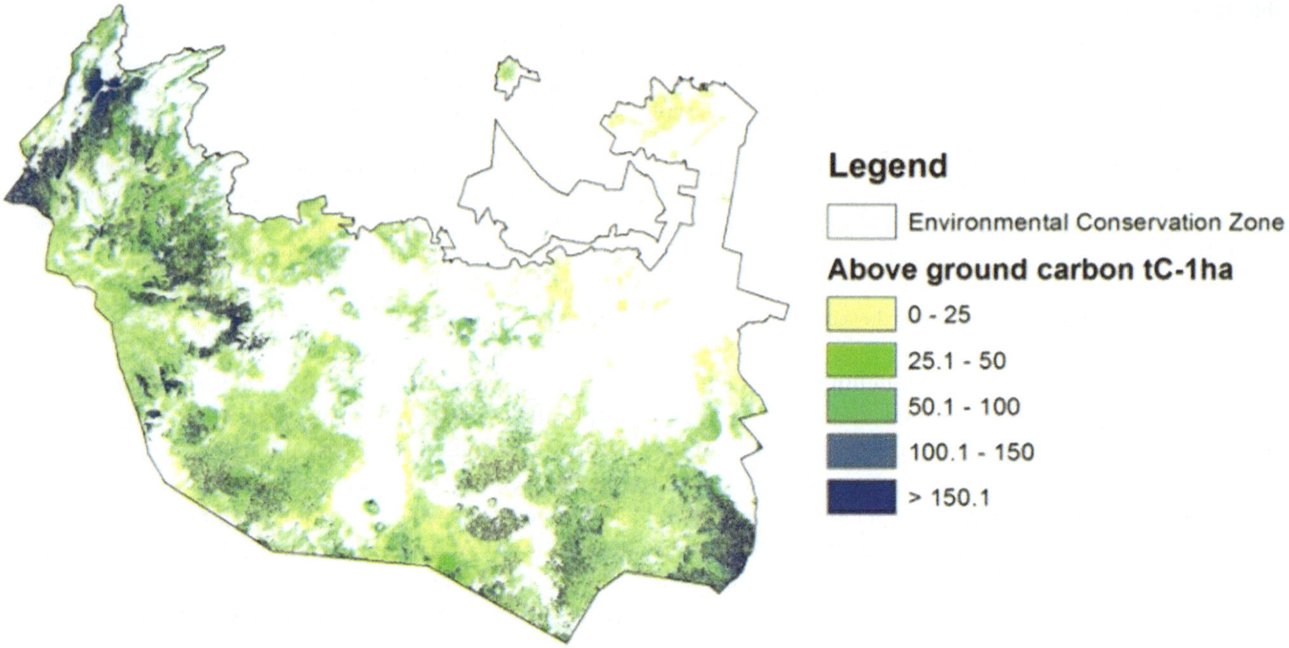

Fig. 19.14 Spatial map of aboveground forest biomass carbon (C) of ECZ of Mexico City (source: Pizana et al. 2016)

are used for the computation of field biomass (details of methodologies are given in Lu 2006). Stratified random sampling is adopted for the selection of ground truth sites in remote sensing-based biomass studies. Remote sensing-derived maps of forest cover type and forest crown canopy density are used for stratification.

Regression model: Regression analysis using simple linear regression (SLR) and multiple linear regression (MLR) are the most common and widely used methods to estimate biomass. Regression analysis involves the establishment of the empirical relationship between remote sensing derived, namely reflectance of spectral bands, spectral vegetation indices and image textural parameters as independent variables and the field sample plots measured biomass as dependent variable. MLR uses more than one independent variable with less correlation between independent variables.

Regression analysis of AGB estimation consists of three steps: measurement and estimation of field biomass in sample plots; development of regression equation (model) between field estimated biomass as the dependent variable and remote sensing derived spectral independent variables extracted from the image pixels of field plots; and preparation of a map of biomass based on regression model spatial predicted values of each pixel.

Spectral vegetation indices, NDVI particularly have been often used in regression models as independent variables with results of varying successes. (Calvao and Palmeirim 2004; Gomez et al. 2014; Kushwaha et al. 2014; Yadav and Nandy 2015). In a Mediterranean holm oak forest region, Ogaya et al. (2015) found a good correlation between MODIS satellite-derived NDVI and EVI (Enhanced vegetation index) and changes (increase or decrease) of AGB. Yadav and Nandy (2015) showed the superiority of the use of NDVI than spectral band reflectance in SLR analysis for the estimation of forest AGB in part of Swalik Himalaya and the prepared biomass map. Pizana et al. (2016) used a multiple regression model and SPOT 5-HRG satellite-derived spectral vegetation indices as inputs and prepared a spatial map of the aboveground forest biomass carbon (C) (Fig. 19.14) in ECZ (Environmental Conservation Zone) of Mexico City covered by sacred fir and Mexican mountain pine forests.

In several studies, image textural indices obtained from digital processing of remotely sensed data were successfully used as independent variables in regression models for AGB estimation (Sarker and Nicole 2011; Dube and Mutanga 2015; Nandy et al. 2017). Grey-level-co-occurrence matrix (GLCM) is one of the popular image textural indices used for AGB modelling. GLCM of a digital image is computed by a moving window using the brightness value of the individual pixel and the mathematical relationships exit with the brightness values of neighbouring pixels.

The passive optical remote sensing-based biomass model performs well in low biomass forested areas because at higher biomass the vegetation canopy spectral reflectance gets saturated. So, linear regression models fail to predict biomass beyond the canopy reflectance saturation point. Therefore, for accurate estimation of biomass, several nonlinear regression models such as logistic regression, geographically weighted regression and power models were often used (reviewed by Pizana and Nunez 2016).

Nonparametric models: These are complex nonlinear models suitable for biomass estimation at higher biomass levels. These models are alternative to parametric regression models and use the same remote sensing variables of regression analysis. Widely used nonparametric models are k-nearest neighbour (k-NN), artificial neural network (ANN), random forest (RF) and support vector machine (SVM).

k-Nearest neighbour (k-NN): In this approach, weighted mean values computed from a weighted distance of k-spectrally nearest neighbouring observations are applied for the estimation of the target variable value (Lu et al. 2012). The accuracy of the k-NN approach depends on the selected value of k and the choice of distance measuring algorithms such as the Weighted function, Mahalanobis and Euclidean. k-NN method is widely employed for the quantification of forest AGB, stock volume and different forest attributes using remote sensing (Nandy et al. 2019).

Artificial neural network (ANN): ANN-based models can solve complicated and nonlinear problems when analysis data is not normally distributed and the dependable variable forest AGB and independent variables remotely sensed spectral indices are not linearly related. It simulates the human brain's thinking process using interconnected neurons and processes the inputted information. Commonly, ANN consists of one data inputting layer, data processing one or more than one layer and one layer for output generation. Using complex mathematical functions, ANN is capable of handling data from multiple sources, namely remotely sensed spectral data and indices and other non-remote sensing data as inputs for improved quantification of forest AGB (Lu et al. 2014a, b).

In Barkot forest, Uttarakhand, the application of ANN-based optimization technique was demonstrated by Nandy et al. (2017) for the selection of a number of remote sensing derived spectral and textural indices as independent variables for quantification of AGB (Fig. 19.15). IRS-LISS III (Resourcesat-1) digital data was used to derive numerous spectral and textural indices. A high correlation ($R^2 = 0.74$) was observed between ANN-predicted biomass and field-estimation biomass.

Support vector machine (SVM): It is an important remote sensing-based method for the quantification of biomass and forest attribute parameters (Mountrakis et al. 2011, Marabel and Alvarez 2013). SVM is an ML algorithm capable of producing higher accuracy than other nonparametric approaches with small training sample data.

Random forest (RF): The RF approach uses the classification and regression tree (CART) digital classification method which generates a large spectral group using a random and number of iterations of selected sample data. A large group for classification using a large number of decision trees is generated by RM through random and iterative selection of sample data. Now RF algorithm is extensively investigated for biomass estimation (Pizana and Nunez 2016).

Models using biophysical variables: Several studies used remote sensing derived canopy structural variables like canopy fraction and crown cover for AGB assessment. Spectral mixture analysis (SMA) was employed for the generation of green vegetation (such as leaves) fraction images using multi-spectral remotely sensed data. In a case study from a Mediterranean pine forest, Manso et al. (2014) showed improved estimation of biomass using SMA to derive green vegetation fraction than spectral bands as an independent variable in a regression model utilizing ASTER satellite multi-spectral data. Object-based image analysis (OBIA) was performed on high spatial resolution satellite data (Quickbird) in a Mediterranean oak forest for the information extraction of crown cover. The crown cover was regressed with field-measured biomass ($R^2 = 0.72$) for spatial estimation of forest biomass with higher accuracy (Sousa et al. 2014).

19.3.2.6 Forest Fire

Forest fire adversely affects ecosystem dynamics by changing vegetation composition, altering soil quality and modifying hydrological regimes. Losses of human life and property are also associated with forest fires. High frequency of forest fires contributes to climate change due to increased emission of CO_2, shoot and aerosols and destruction of CO_2 sink vegetation areas. Satellite RS is a very effective technique for the prevention and management of forest fire through the creation of early warning for fire risk zonation, mapping of fire contributing fuel, real-time monitoring of active fire propagation, assessment of burnt scars and monitoring of vegetation recovery after fire event.

Fuel mapping: Forestland fuel is an important factor that influences the risk, intensity and burn severity of fire. Information on spatial variation of fuel is helpful in the identification of areas with high fire risk and control of fire. Fuel is mapped by fuel loading which is characterized

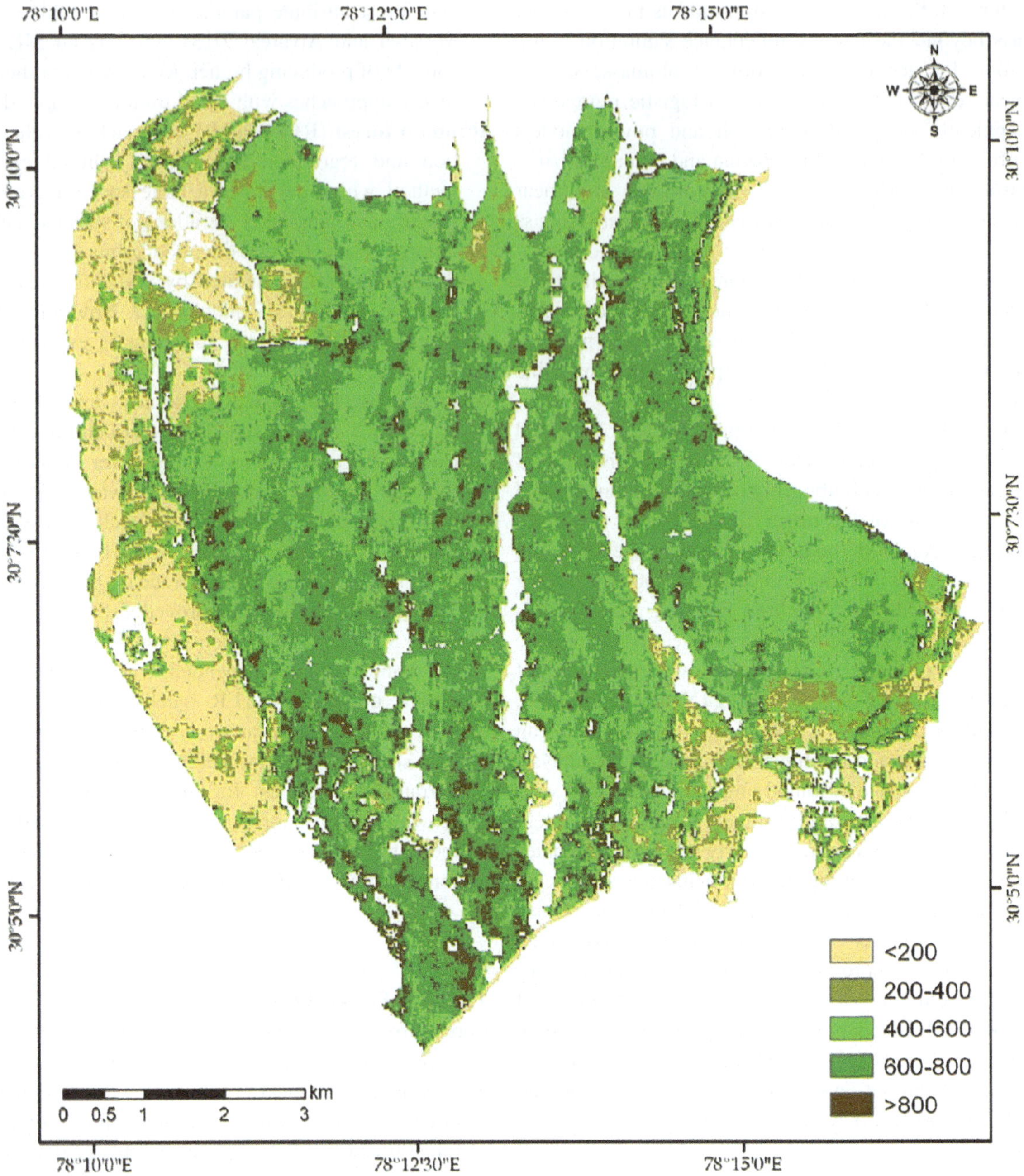

Fig. 19.15 ANN predicted spatial forest biomass (Mg ha^{-1}) map (source: Nandy et al. 2017)

by dry weight of biomass per unit area. Remote sensing-derived inputs such as vegetation type, height, canopy cover and canopy live fuel moisture content are used for fuel mapping. Digital classification (supervised and unsupervised) and spectral vegetation (NDVI) are commonly used for fuel types mapping in forested areas (Riano et al. 2002. Van Wagtendonk and Root 2003). Ji et al. (2020) used spectral mixture analysis of satellite data to derive photosynthetic and non-green/dry vegetation and barren soil surface fractions for mapping and characterizing fuel conditions in Pike National Forest, Colorado. A rule-based approach using satellite data was employed to classify fuel types by combining parameters of type of vegetation, crown canopy cover, tree height and environmental site potential in the US LANDFIRE program. In different vegetation systems in south-eastern Australia, using MODIS satellite-derived visible atmospherically resistant index (VARI) and NDVI, regression models were developed for quantification of real-time fuel moisture content (Caccamo et al. 2012).

Active fire detection: Real-time active fire detection is vital for fire risk assessment, and containment and provides alerts to minimize losses. Remote sensing is effective for providing information on spatial distribution and development of fire. Various satellite remotely sensed data are being used for the detection of active fire. Remote sensing data from the AVHRR sensor onboard NOAA and sensor onboard Terra and Aqua are used worldwide for detection and monitoring real-time active fire detection. Since 1981, AVHRR sensor bands 3 SWIR (3.5–4 μm) and 4 TIR (10.5–11.5 μm) have been used for active fire detection and monitoring until the launch of MODIS sensors in 1999.

MODIS provides improved and more accurate fire detection than AVHRR by using algorithms capable of removing various false alarms other than fire. Recently available 1 km 'fire pixels' 6 MODIS fire product uses several spectral bands such as red (band1) NIR (band 2), SWIR (band 7), TIR (band 21, 4 μm), TIR (band 22, 4 μm), TIR (band 31, 11 μm) and TIR (band 32, 12 μm). Removal of cloud cover, rejection of sun glint and coastal false alarms in MODIS images are performed by use of spectral data of bands 1, 2 and 7. Fire detection is done by use of spectral data of bends 21, 22 and 31. While cloud masking is performed by use of spectral data of bands 31 and 32. Rejection of false alarms of cleared forest areas is done by use of spectral data of band 31. A large number of countries in the world are using AVHRR and MODIS satellite sensors derived from active fire products in routinely monitoring forest fires daily basis (reviewed by Szpakowski and Jensen 2019).

Burnt area and burn severity assessment: Reliable estimates of the spatial extent of burnt area and severity are needed for rehabilitation planning, assessment of gas and particulate matter emissions and calculation of economic and environmental cost of fire. Multi-level (local, regional and global) detection and mapping of fire burnt areas are efficiently done with the use of multi-spatial remotely sensed data. Digital change detection techniques such as NDVI and image differencing using pre and post-forest fire periods high-to-moderate spatial resolution (<100 m) satellite data are found effective in local level burnt area mapping (Silva et al. 2005; Bastarrika et al. 2011) Coarse spatial resolution satellite data derived active fire products are found suitable for detection, mapping and monitoring of burnt areas at regional and global scales. Several global satellite-derived burnt area products are currently available, and the MODIS-MCD64 product was found to be the most accurate (Szpakowski and Jensen 2019). Giglio et al. (2009) developed a spectral normalized burn ratio (NBR) index for burn area mapping using MODIS 1 km data. NBR is expressed as:

$$NBR = (NIR - SWIR)/(NIR + SWIR)$$

Bare ground and recently burnt areas show low values of the NBR index. The high values of the NBR index indicate healthy/highly vigorous vegetation. Normally, NBR index values close to zero indicate non-burnt areas. This index was tested for burnt area assessment in various forested areas in the world (Szpakowski and Jensen 2019). Conventionally, burn severity is measured using a field data-based composite burn index (CBI). CBI estimation is time-consuming and requires physical visits to burnt areas. Remote sensing-based assessment can overcome these problems. Commonly two satellite-derived spectral indices, namely NDVI and NBR have been employed. Widely, changes in values of NBR (dNBR) during pre and post-fire are applied for the assessment of the severity of the burn (Fig. 19.16, reviewed by Szpakowski and Jensen 2019). MODIS satellite-derived land surface temperature (LST) and land surface albedo have been successfully used by several researchers to estimate burn severity (Jin et al. 2012; Veraverbeke et al. 2012; Quintano et al. 2015). The effective use of the digital spectral mixture analysis (SMA) technique was investigated by Quintano et al. (2013) and Tane et al. (2018) for the detection and mapping of the severity of the burn.

Post-fire vegetation recovery assessment: Assessment and monitoring of post-fire recovery of vegetation are important for studying the long-term impacts on ecosystem functions and services. Conventionally, field-based monitoring and assessment are carried out for post-fire vegetation recovery. RS from a satellite platform is a rapid and economical technique for assessing vegetation recovery in large areas compared to the field-based method. Three remote sensing-based approaches are in use for the assessment of post-fire vegetation recovery and these are (1) supervised and supervised digital image classification, (2) spectral vegetation indices and (3) digital spectral mixture analysis (SMA).

Both supervised and unsupervised classification methods using proper ground training data were used for studying the effects of severity of burn on post-fire recovery of vegetation. Application of the object-based image analysis (OBIA) method using high spatial resolution satellite data was also attempted for the assessment of post-fire recovery of vegetation (Szpakowski and Jensen 2019). Spectral vegetation indices such as NDVI, SAVI and MSVI (modified SVI) have been commonly used for assessing vegetation recovery via estimate of ecological variables such as LAI, NPP and proportion of vegetation cover (reviewed by Szpakowski and Jensen 2019). Several researchers found high agreement with SMA-mapped

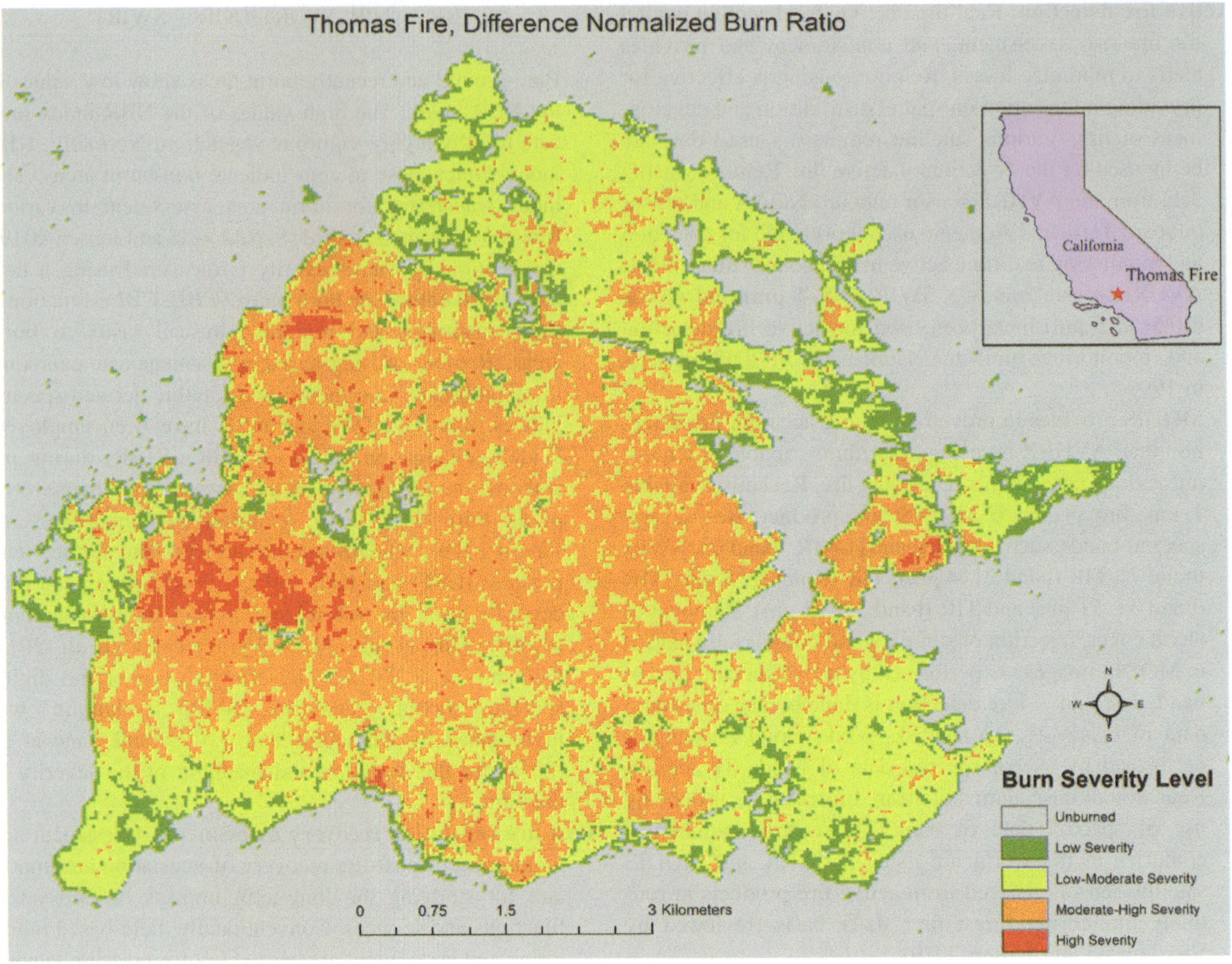

Fig. 19.16 Forest fire burn severity map prepared by using satellite-derived dNBR data (source: Szpakowski and Jensen 2019)

vegetation recovery using various kinds of satellite data and field sampling information (reviewed by Szpakowski and Jensen 2019).

19.3.2.7 Forest Damage by Diseases and Insects

Various pathogens such as bacteria, fungi, viruses and helminths cause diseases in forests. Defoliators feeding on leaves and barks, and bark/wood borers are the dominant insects that attack forests. The frequency and intensity of infestation of diseases and insects causing forest disturbances have extensively increased in the past few decades due to climate changes. Mitigating the destructive impacts of diseases and insects is vital for sustainable forest management. Temporal satellite remote sensing data is a very useful tool for monitoring varying stages of outbreaks of diseases and insects by assessing forest biophysical and biochemical parameters.

Infected damaged trees have characteristic symptoms, namely discoloured foliage, defoliated canopy and reduced moisture, compared to healthy trees. Multi-spectral satellite-derived various spectral vegetation indices, namely disturbance index (DI), NDVI, EVI, normalized difference moisture index (NDMI) and enhanced wetness difference index (EWDI) were used by researchers to assess and monitor diseases and insect infestation damages of forest (reviewed by Chen and Meentemeyer 2016). For example, Coops et al. (2009) applied digital thresholding of Landsat-derived NDMI for mapping tree damage due to beetle attacks and post attacks regrowth of forest vegetation. MODIS-derived 1 km spatial resolution DI change was also used by Coops et al. (2009) to detect infested forested areas. EWDI and NDVI derived from Landsat TM satellite data were successfully utilized for the identification and mapping of infested forest patches respectively (Skakun et al. 2003 and Olsson et al. 2012).

Digital image classification methods using a variety of satellite data and algorithms such as supervised maximum likelihood classification (MLC), SMA and machine learning (SVM and RF) were attempted by several researchers to map disease and insect outbreaks in the forest (Radeloff et al. 1999; Walter and Platt 2013; Adelabu et al. 2014; Long et al. 2023). Landsat 8 OLI satellite-derived time series normalized difference forest index (NDFI) and random forest digital classification were employed for monitoring pine wilt disease in the Anhui province of China (Fig. 19.17, Long et al. 2023).

Statistical regression models, namely multiple linear regression and logistic regression have been shown useful for identifying and mapping damaged forests. In several research studies in regression models variety of satellites derived parameters, namely reflectance of spectral bands, spectral vegetation indices and terrain variables were statistically related with ground truth sites measured with forest diseases and pests damage parameters, namely intensity of defoliation, basal area, LAI, leaf nitrogen and plant growth vigour, chlorophyll and water content of foliage (reviewed by Chen and Meentemeyer 2016).

19.3.3 Geographic Information System Applications in Forestry

19.3.3.1 Forest Working Plan Preparation and Revision

Scientific forest management is done through working plans. Working plans are important for the evaluation of current conditions of different attributes of the forest, studying the effects of past practices of forest management and making decisions on the adaptation of suitable planning strategies to achieve the objective of sustainable forest management. In India, forest division is the basic unit for the preparation of forest working plans. A forest working plan is prepared with forest division-wise collected 5–10% of field data within the time frame of 1–2 years and revised once in 10 years.

Various spatial inputs of forest parameters required for the preparation of a forest working plan can be derived by integration of data from GIS, remote sensing and forest mensuration data measured in the field. GIS is very useful for the creation of spatial and non-spatial digital databases of forest working plan entities and the development of a decision support system (DSS) which could make the job of forest manager very easy and efficient. Presently, GIS is being used with cost-effective, time-saving and greater accuracy than the conventional method in various aspects of forest working plans such as updating administrative boundaries of the forest; preparation of forest stock map; zoning working circles; preparation of treatment plan; treatment based plan for felling of series and coupes; finding suitable sites for coupe operations and specific treatment; marking grazing closure zone and identification of fire control areas, etc. (NRSC 2010).

GIS-based spatial modelling is used to prepare treatment type maps by integration of spatial maps of slope, drainage, forest density, forest type and forest plantation based on weightage given based on specific criteria of each treatment. Generally, forest working plans require several kinds of suitability analysis for identifying potential areas of plantation, afforestation, reforestation, meadow development, gap filling, water harvesting soil conservation structures, etc. GIS also helps in the preparation of a suitability map based on the potentials and limitations of soil, terrain and vegetation qualities of each forest compartment (Fig. 19.18). Commonly, GIS-aided multi-criteria decision analysis (MCDA) technique is applied for suitability analysis following the guidelines included in a working plan.

MCDA is a spatial data analysis method which handles several types of available criteria while making a decision. Based on prior set objectives and criteria, MCA provides multiple solution options to be chosen based on ranking order of suitability (gispeople.com 2020). In the east Melghat forest division, Maharastra, Varghese and Krishna Murthy (2006) used IRS LISS III satellite data and GIS for site suitability analysis as one of the tasks of preparing a working plan (Fig. 19.18).

19.3.3.2 Forest Fire Management

Forest fire risk/hazard zonation, simulation of fire growth/spread and development of fire information systems are the major areas of applications of GIS in forest fire management.

Forest fire risk zonation: Various GIS-based simple and advanced spatial modelling approaches such as index weighted overlay, AHP (Analytical Hierarchy Process), logistic regression, fuzzy logic, ANN and data mining are being used in forest fire risk mapping/zonation. Several parameters, namely fuel and vegetation conditions, topography (slope, aspect, elevation), weather (temperature, humidity, wind speed, rainfall) and proximity to settlement/infrastructure are required as inputs in the fire risk model.

The spatial model result is a map showing fire risk zones with low to very high fire risks. Forest fire risk modelling satellite remote is primarily used to derive land cover and DEM, assessment of vegetation condition using spectral vegetation index (NDVI) and validation of the model to derive a fire risk zonation map. Vegetation types are related to fuel types and characteristics. GIS-based spatial modelling approaches (logistic regression, AHP, index weighted overlay) and satellite-derived parameters were

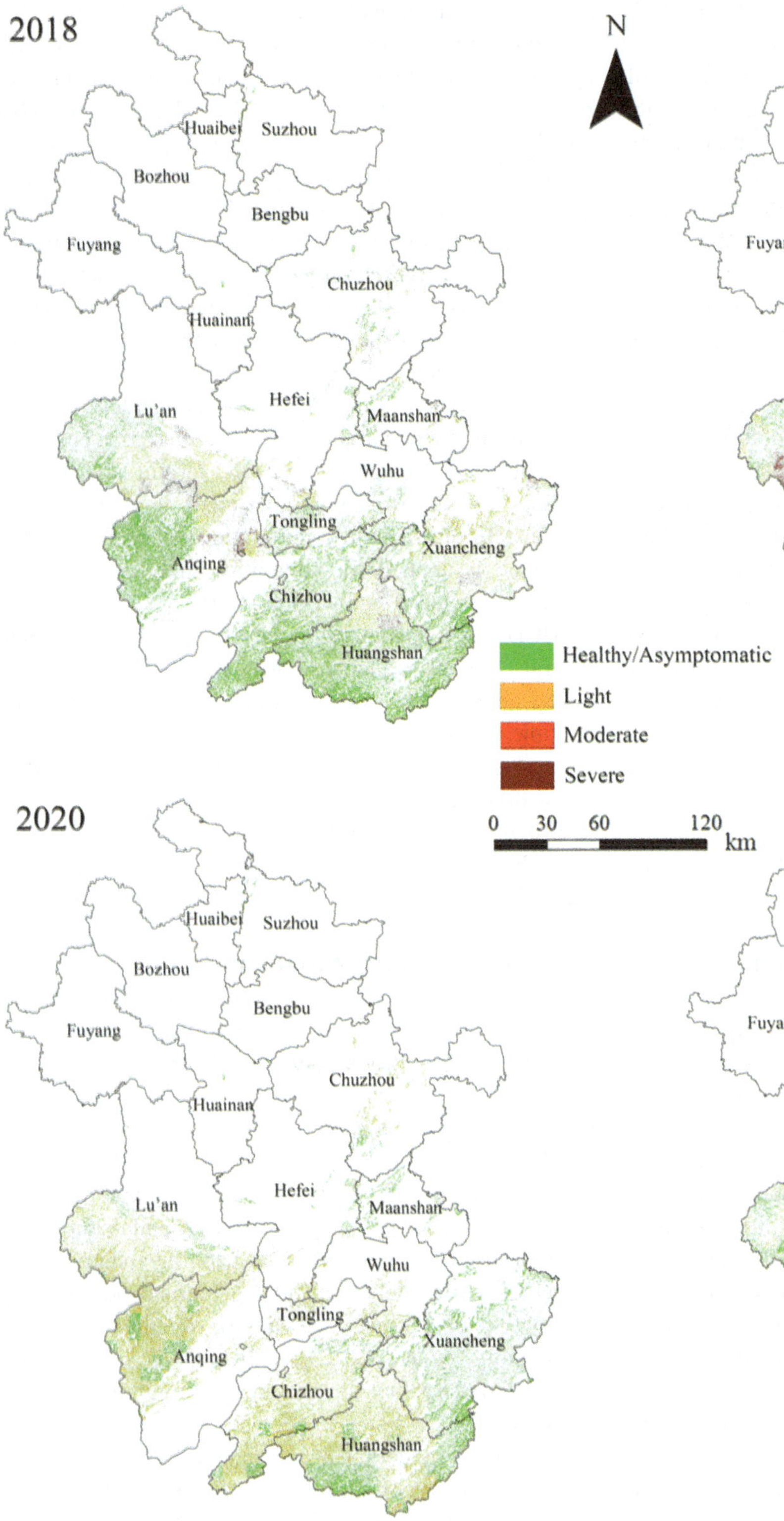

Fig. 19.17 Pine wilt disease severity map of 2018 to 2021, Anhuli province of China (source: Long et al. 2023)

used by Yathish et al. (2019) to generate a map showing varying potential fire risks of Kudremukh National Park. Szpakowski and Jensen (2019) prepared a forest fire risk map of Grand Teltom National Park, WY, USA by GIS-aided integrated analysis using inputs of fuel,

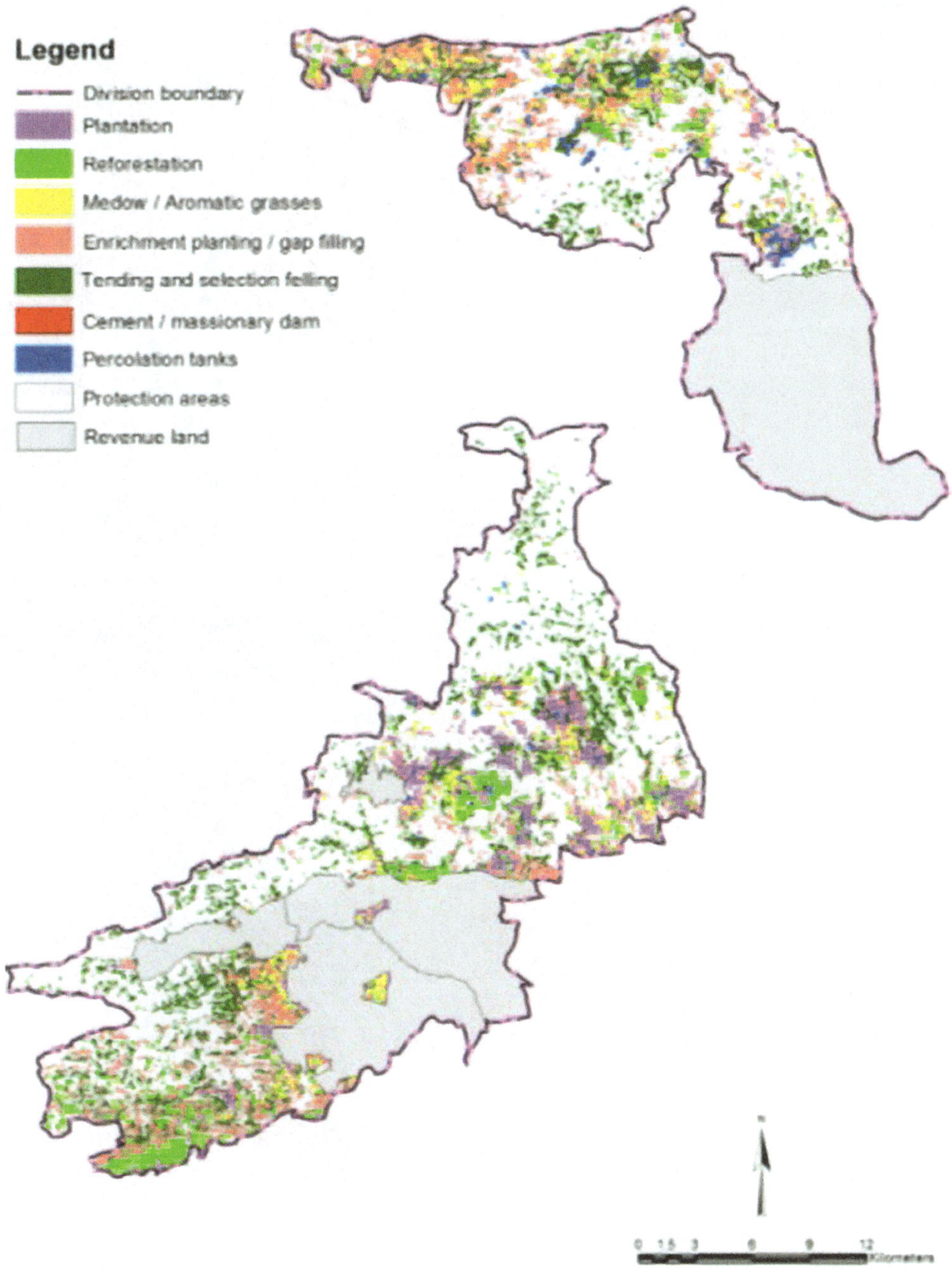

Fig. 19.18 Site suitability map of East Melghat forest division (source: Varghese and Krishna Murthy 2006)

moisture, slope, aspect, distance from roads, proximity to settlements and elevation (Fig. 19.19).

Forest fire growth simulation modelling: Currently, FIRESITE a fire area simulator model is commonly used for simulating fire spread/growth (Szpakowski and Jensen 2019). The various kinds of input data used in this model are terrain parameters, weather variables including wind speed and direction, and fuels. This 2D fire growth model integrates several sub-models such as forest ground surface fire, forest crown fire, spotting, post-frontal combustion, and forest fire acceleration (www.firelab.org/project/farsite 2020). The model uses a variety of inputs such as terrain parameters (slope, aspect and elevation), forest understory fuel conditions, forest vegetation parameters (canopy height, crown bulk density, crown base height and canopy cover) and weather parameters (humidity, wind speed, cloud cover, minimum and maximum temperatures and wind direction).

Forest fire information system: Forest Survey of India (FSI) developed a forest fire web GIS geoportal named 'Van Agni 1.0' for user-friendly interactive visualization and extraction of forest fire-related information (Fig. 19.20). This portal provides information on near real-time forest fire data and large forest fire events along with other

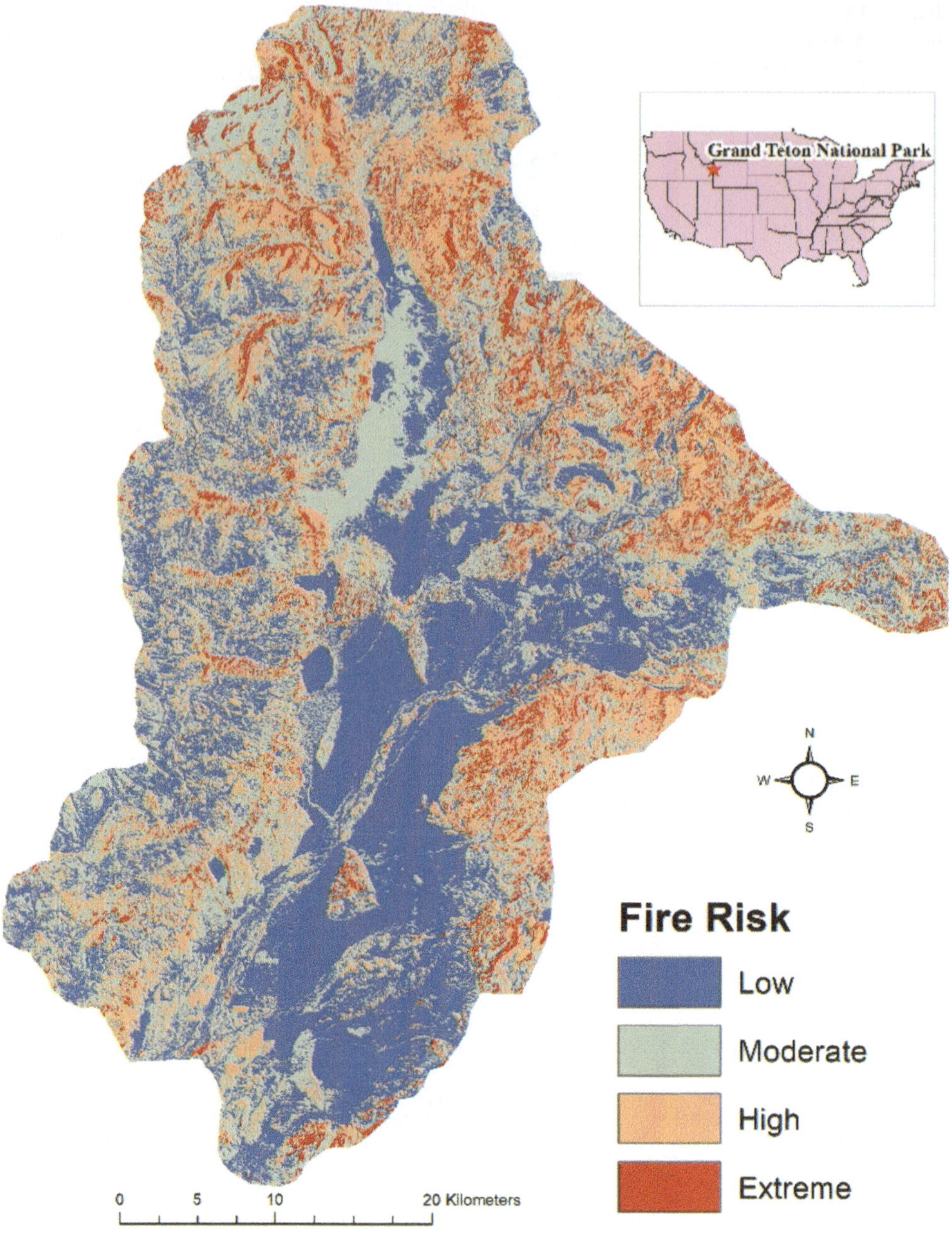

Fig. 19.19 Map showing different zones of potential fire risk (Grand Teltom National Park, WY, USA) (source: Szpakowski and Jensen 2019)

thematic layers such as forest administrative boundaries, forest cover and forest type.

19.3.3.3 Wildlife Habitat Evaluation

Wildlife survival resources, namely food, cover, water and terrain are provided by the habitats. Wildlife habitat evaluation helps in various conservation practices such as habitat restoration conservation and endangered species reintroduction, prediction of threat to invasive species, scientific conservation planning, assessment of risks and identified priorities of conservation (Guiming 2019). GIS-based habitat suitability analysis is a widely used approach for the assessment of conditions of habitat resources. All the wildlife habitat resource parameters spatially could be derived using remote sensing techniques. These parameters along with additional information on the closeness of disturbances sources and grazing pressure are used in GIS-supported spatial modelling for habitat suitability analysis of specific species of wild animal. Several GIS-based spatial modelling methods such as weighted index overlay, AHP, logistic regression and ecological niche factor analysis were evaluated by several researchers for habitat suitability analysis for a variety of wild animals (Porwal et al. 1996; Kushwaha and Roy 2002; Dash et al. 2015; Sanare et al. 2015; Ahmad et al. 2018). AHP-based GIS modelling technique was applied in the Sarnanda forest division, Jharkhand state for the generation of a wildlife habitat suitability map for the dominant wild animals available in the forest division

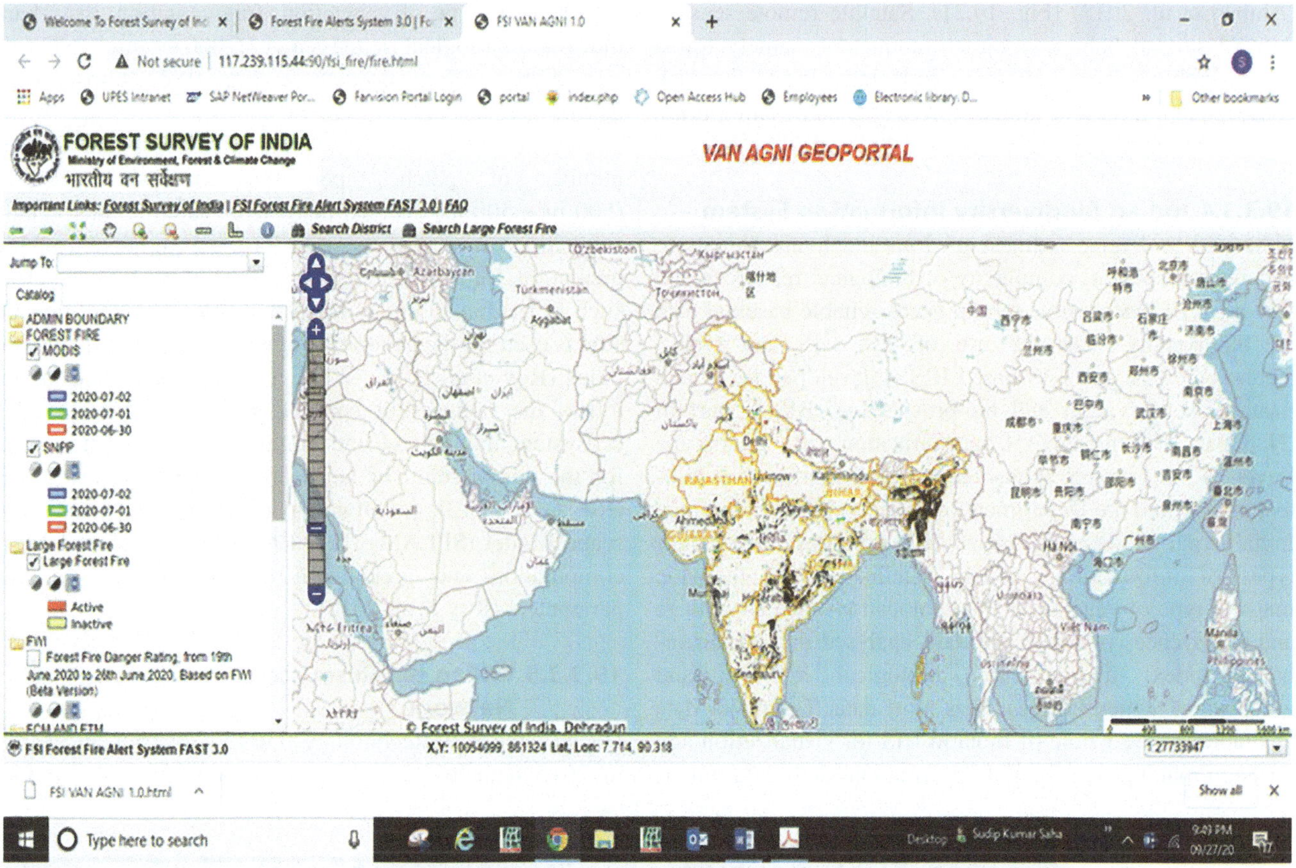

Fig. 19.20 FSI forest fire web GIS geoportal (source: fsi.nic.in/forest-fire-activities 2020)

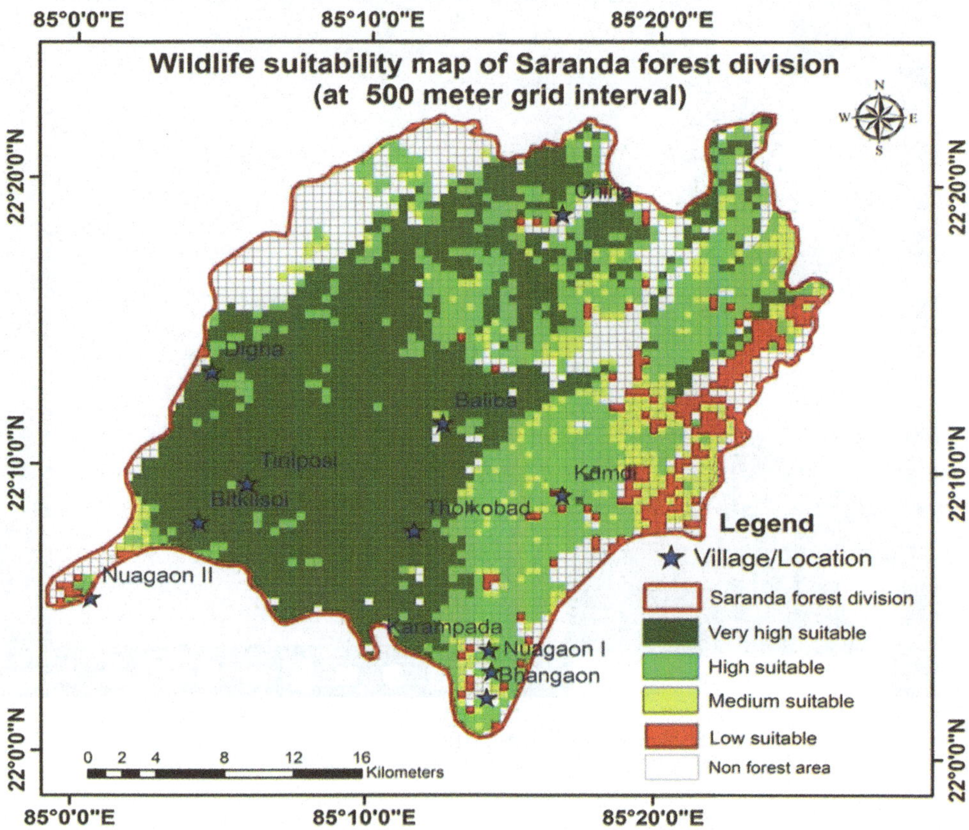

Fig. 19.21 Wildlife habitat suitability map of Saranda forest division (source: Ahmad et al. 2018)

(Ahmad et al. 2018) (Fig. 19.21). Satellite remote sensing derived several habitat resource parameters, namely forest cover, cultivated land, human settlement, forest fire risk zones, roads, river and drainage networks and mining areas used in this analysis.

19.3.3.4 Indian Biodiversity Information System

India with a diverse richness of flora and fauna, threats to endangered species, complexity of biological resources and vast areas of inaccessible terrain needs reliable baseline data on biodiversity with the use of RS, GIS and GNSS technologies. A web GIS-based BIS is developed for prioritization, conservation and bio-prospecting. BIS is part of 'National Biodiversity Characterization at Landscape Level', a joint project sponsored by the Department of Biotechnology and the Department of Space to identify and map India's rich biodiversity areas. BIS provides three major types of information, namely satellite-derived thematic maps of type of vegetation and geolocations of road, village and occurrences of forest fire; GIS analysed data on disturbance index, fragmentation, biological richness; and geolocation tagged field samples plots data. The major data available for the whole of India in BIS for visualization and decision-making are spatial information and data on 1: 50,000 scale-type of vegetation, fragmentation, disturbance index and biological richness and 16,000+ sample plots species phytosociological database (Fig. 19.22, https://bis.iirs.gov.in/).

Several patches of forest and non-forest and the total number of patches per GIS analysis grid cell (500 m × 500 m) data extracted from satellite interpretation vegetation cover map are used for the computation of fragmentation. Linear combination of several landscape matrices such as fragmentation, porosity, juxtaposition and interspersion is used for the generation of a spatial map of disturbance index (Roy et al. 2012). While a linear function disturbance index, the richness of species, the value of biodiversity, ecosystem uniqueness and complexity of terrain are used for the computation of landscape-level biological richness (Roy et al. 2012). A software package named Spatial Landscape Model (SPLAM) (Fig. 19.23) was developed for the computation and preparation of spatial maps of these parameters (Roy et al. 2012).

19.3.3.5 Indian Bio-Resources Information Network

Department of Biotechnology (DBT), Government of India, in 2012 launched a geo-portal named IBIN to provide a

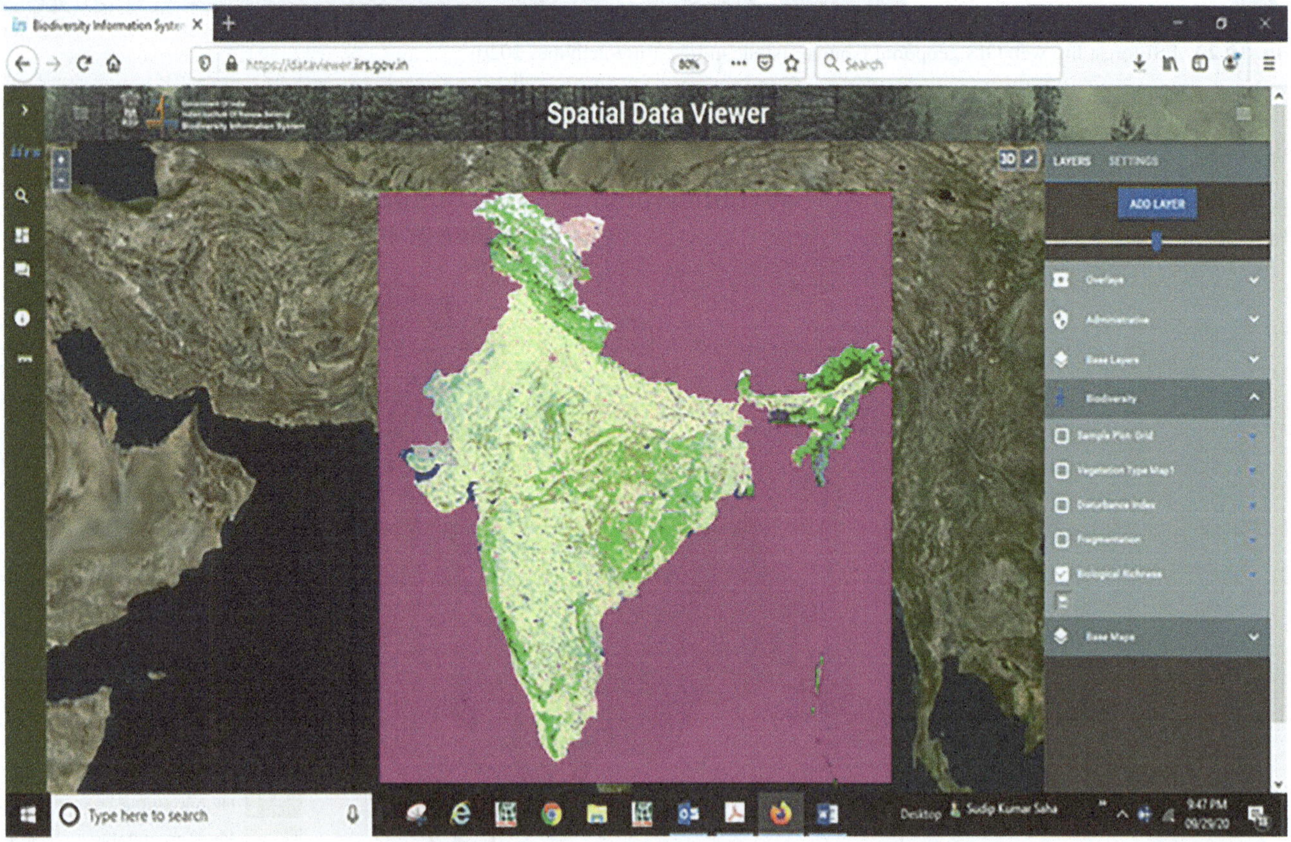

Fig. 19.22 Snapshot of BIS (https://bis.iirs.gov.in/, 2020)

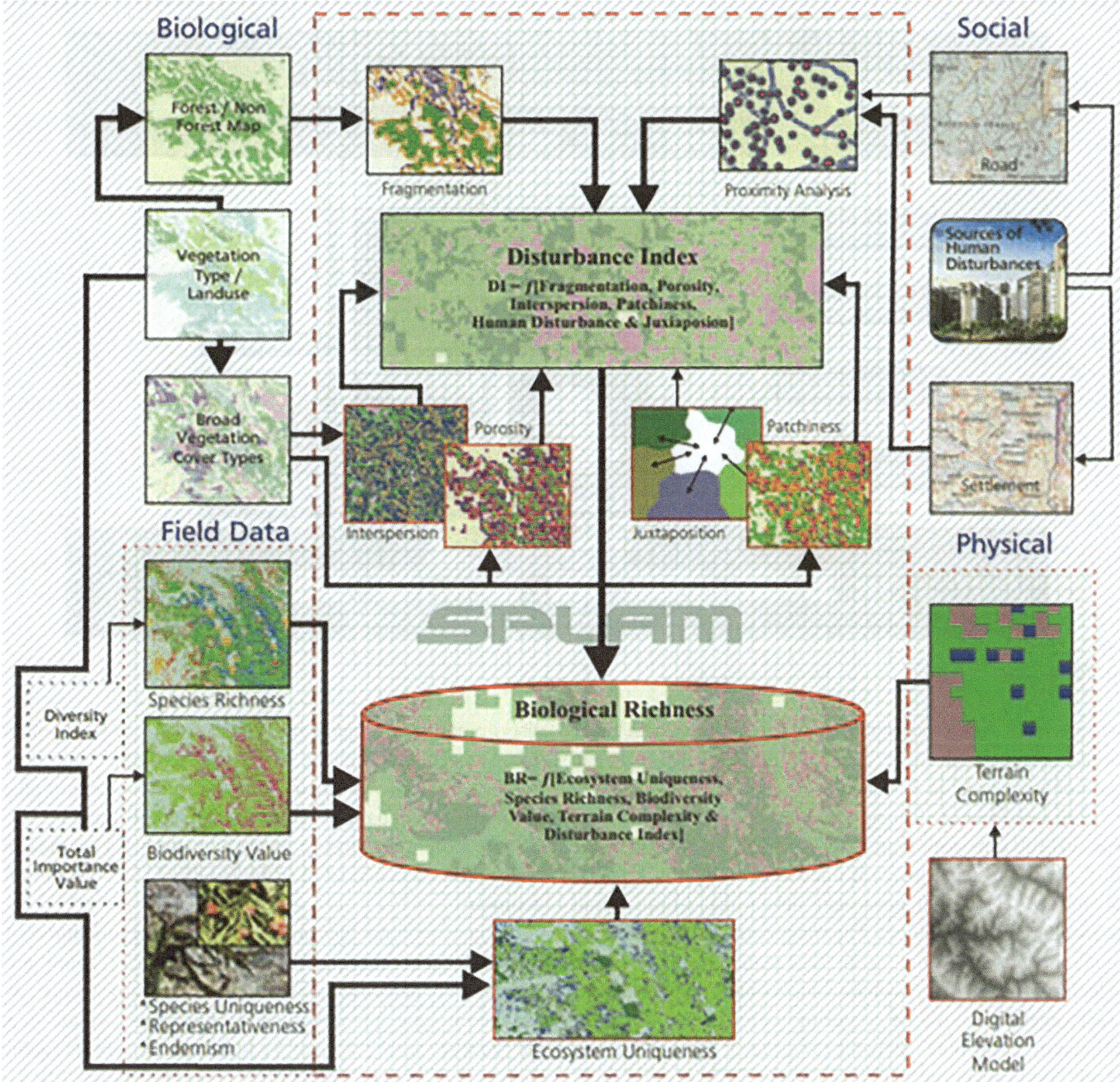

Fig. 19.23 Geospatial model for biodiversity characterization using SPLAM (source: Roy et al. 2012)

digital database of the bio-resource of the country. The IBIN is a single-window digital portal (www.ibin.gov.in) for accessing a variety of India's bio-resources: plant, animal, marine and microbial (Fig. 19.24).

The scientific information/data collected by a large number of researchers representing 150 institutions are collated inputted and organized in the IBIN portal. IBIN is planned to provide India's bio-resources information to the user communities involved in bioresources conservation, biological prospecting, protection of biopiracy and marketing (Saran et al. 2019).

19.3.4 Applications of Advanced Remote Sensing Techniques in Forestry

19.3.4.1 Microwave Remote Sensing

In the past decades, microwave RS with active RADAR imaging systems operating in the wavelength range from 1 cm to 1 m of EMR has gained importance in applications in forestry. Microwave RS offers several benefits compared to visible and infrared RS such as all-weather imaging and surface penetration capability, sensitivity to objects' moisture content and surface conditions (roughness). Synthetic

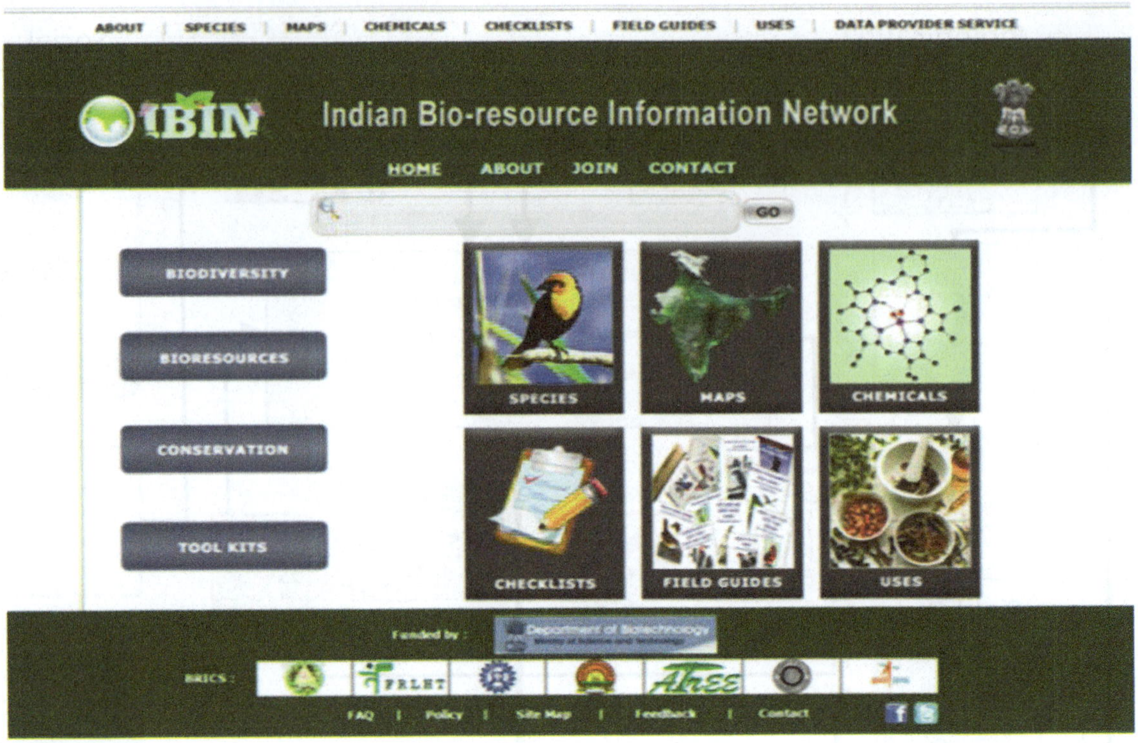

Fig. 19.24 IBIN web portal (source: www.ibin.gov.in)

aperture radar (SAR) sensor system is widely used in radar remote sensing. Airborne and satellite SAR RS systems commonly operate in C-band (λ = 3.75–7.5 cm), X-band (λ = 2.4–3.75 cm), L-band (λ = 15–30 cm) and P-band (λ = 30–100 cm) with increasing order of penetration capability. Radar systems transmit microwave radiation towards the surface area of interest and detect the return signal backscattered by the object. The magnitude of backscattered signal depends on surface properties, namely roughness and dielectric constant (vary with moisture content). In the case of forest vegetation, volume scattering within the forest stand dominantly contributes to radar backscatter. During the last decade, both air-borne and space-borne SAR systems with different frequencies have been used for various applications in forestry. Recently, emerging techniques of Polarimetry and Interferometry enhanced SAR applications in forestry.

A polarimetric SAR (PolSAR) system measures the electrical field as well as the polarization state of the return backscattered signal from the scene. The polarization of the transmitted wave transforms after interaction with objects of the scene. PolSAR remote sensing systems depending on the sensor design and mode of acquisition, collect surface backscatter data in two same polarizations (HH and VV) and two cross polarizations (HV and VH) combinations (H: horizontal and V: vertical polarizations). The polarization of the backscattered wave depends on the polarization of the transmitted wave as well as the scattering characteristics of the imaged objects. PolSAR technique enhances the discrimination of objects by detecting various kinds of scattering mechanisms such as surface, volume and double bounce. SAR interferometry records the difference of phases of microwave radiations recorded by the SAR imaging sensor system from two closely spaced positions. The measured phase difference is related to the height of the object through geometry.

Foliage and branches are the dominant factors that control the SAR backscattering from forested areas. Canopy architecture and dielectric constant of the canopy change with vegetation type, phenology and conditions, which change the radar scattering. Use of SAR data from various satellites such as Seasat SAR, shuttle imaging radar (SIR)-B, SIR-C/X, earth resource satellite (ERS)-1 and 2, Japan earth resource satellite (JERS), environmental satellite (ENVISAT), Advance SAR (ASAR), Radar satellite (RADARSAT)-1 and 2, and recently advance land observing satellite (ALOS), phase array type L band SAR (PALSAR), Sentinel, etc. was investigated in boreal, temperate and tropical forest ecosystems.

Vegetation parameters such as biomass, tree density and tree volume are significantly related to radar backscatters at P and L longer wavelengths having tree crown penetrating capability. Shorter wavelength radar such as C and X bands do not penetrate tree crowns due to dominant surface scattering and are therefore suitable for forest cover types and deforestation mapping (Fig. 19.25). Recently, PolSAR microwave remote sensing became popular in forest biomass

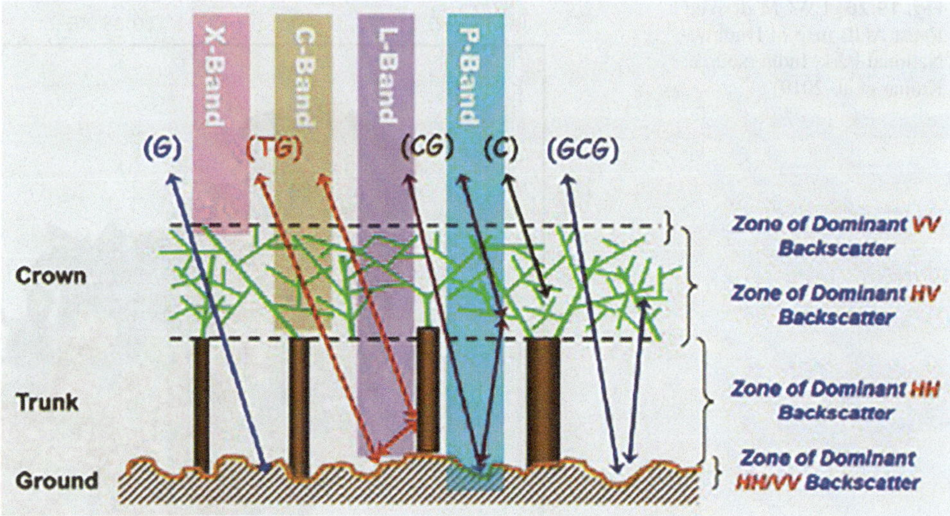

Fig. 19.25 Different scattering components from a forest (C: backscatter from the crown; CG: interaction of crown-ground; G: direct backscatter from the ground; GCG: multiple scatters from ground-crown-ground; TG: interaction of trunk-ground) (source: Das 2017)

estimation. Several research studies found that PolSAR cross polarizations (HV and VH) backscatters from all microwave wavelengths or frequencies are highly correlated with forest biomass in comparison to backscatters of the same polarization bands (VV, HH) (Mohan 2013).

ALOS-PALSAR L band SAR, HV polarization backscatter significantly related with the field aboveground biomass (AGB) in central Indian deciduous forest and the developed empirical model used spatial mapping of AGB (Thumaty et al. 2016). The extended water cloud model (EWCM) using ALOS-PALSAR L band PolSAR satellite data as input was successfully employed by Kumar et al. 2019 for the retrieval of forest AGB of Dudhwa National Park, India and the spatial forest AGB map is shown in Fig. 19.26.

19.3.4.2 Hyperspectral Remote Sensing

Hyperspectral RS simultaneously acquires images with a large number of spectral bands and high spectral resolution. Continuous spectra of objects are generated from the hundreds of spectral bands data with a narrow bandwidth (5–10 m) recorded by the hyperspectral RS sensors. Identification/discrimination of objects and quantification of bio-physical/biochemical/geophysical parameters can be effectively done with the spectral signatures derived from hyperspectral RS data. Most of the hyperspectral RS sensors detect and record spectral data in VIS, NIR and SWIR regions of EMR with wavelengths ranging from 350 nm to 2500 nm. For example, hyperspectral spectral signatures of the dominant tree species of the USA Harvard temperate forest area are shown in Fig. 19.27 (Chaity and Aardt 2024).

Several research studies reported the usefulness of hyperspectral RS in improved identification and mapping of forest species. The sensitivity of hyperspectral indices to foliar pigments forms the basis for discriminating between canopy species in forested landscapes (Thomas et al. 2019). Spectral angle mapper (SAM): The spectral matching technique is a popular method of forest species discrimination and mapping. SAM match spectral patterns of pixels with reference spectra of forest species as end members. The spectral mixing analysis method is also used for quantifying the fraction of each species within the hyperspectral image pixel (Darvishefat et al. 2002).

The boreal species signatures were acquired from a CASI hyperspectral sensor with 56 bands from 498 to 916 nm. The temperate forest signatures were acquired with the European Space Agency's HyPlant instrument, with 629 bands from 380 to 2537 nm (Buddenbaum et al. 2005). Aerial hyperspectral remote sensing AISA sensor data was used for tree species classification and mapping (Fig. 19.28) using two non-parametric classification methods, namely SVM (Support Vector Machines) and RF (Random Forest) in forests of Muir Woods National Monument and Kent Creek Canyon, California, USA. More than 90% accuracies were achieved in tree species classification and mapping (Ballanti et al. 2016).

Quantification of forest foliar chlorophyll content is a very popular application of the use of hyperspectral narrowband spectral indices. Commonly two hyperspectral narrowband spectral indices, namely TCARI/OSVAI (Transformed chlorophyll absorption in reflectance index/optimized soil-adjusted vegetation index) and MTCI (MERIS terrestrial chlorophyll index) are used for chlorophyll content estimation. Physically based radiative transfer modelling approaches were also applied for the quantification of foliar chlorophyll content. Canadian hyperspectral RS sensor called, Compact Airborne Spectrographic Imager (CASI), 72 narrow bands spectral data were used as inputs in canopy radiative transfer model for estimation of canopy level chlorophyll content in boreal forest, Ontario. The model-

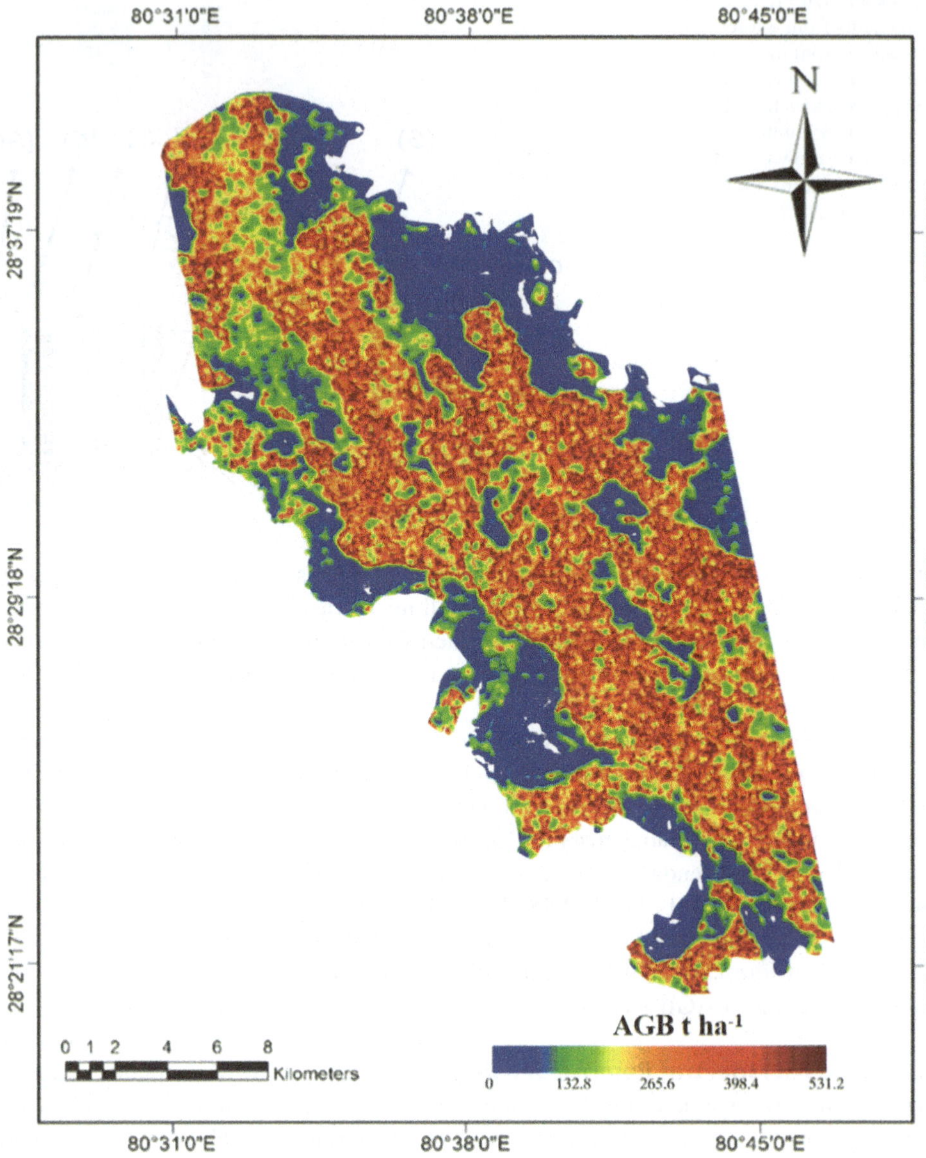

Fig. 19.26 EWCM-derived forest AGB map of Dudhwa National Park, India (source: Kumar et al. 2019)

estimated map of foliar chlorophyll content is shown in Fig. 19.29 (Zhang 2012).

19.3.4.3 Lidar Remote Sensing

Lidar is an acronym for light detection and ranging. Lidar performs remote sensing with the transmission and detection of laser light. Lidar is an active RS technique like radar and its sensors record the two-way travel times for a laser pulse from sensor to target and from target to sensor. The incident pulse of energy of lidar in the NIR wavelength reflects from the canopy, branches, trunks and understory and the return energy is detected by the sensor. The distance or range of the object from the sensor is determined by the estimation of the return trip travel of the pulse of energy from the target to the sensor. So, 'Laser altimetry' is also synonymous with lidar.

Several forest structural parameters, namely the height of the canopy, AGB, sub-canopy topography and the vertical distribution of canopy structure can be estimated using Lidar RS data (Rai 2022). By calculating the difference in elevations of the first and last returns of the lidar signal, the canopy height is calculated. Canopy height is an important biophysical variable because it is strongly related to vegetation height. Successful use of lidar RS for retrieval of canopy heights in different forests such as temperate deciduous, tropical wet, pine and Douglas fir reported by Dubayah and Drake (2000). Waveform images related to the vertical distribution of the canopy structure can be generated by recording both the return signals from the top of the canopy and the ground with large-footprint lidar systems. Vertical distribution of canopy structure is an important parameter for the

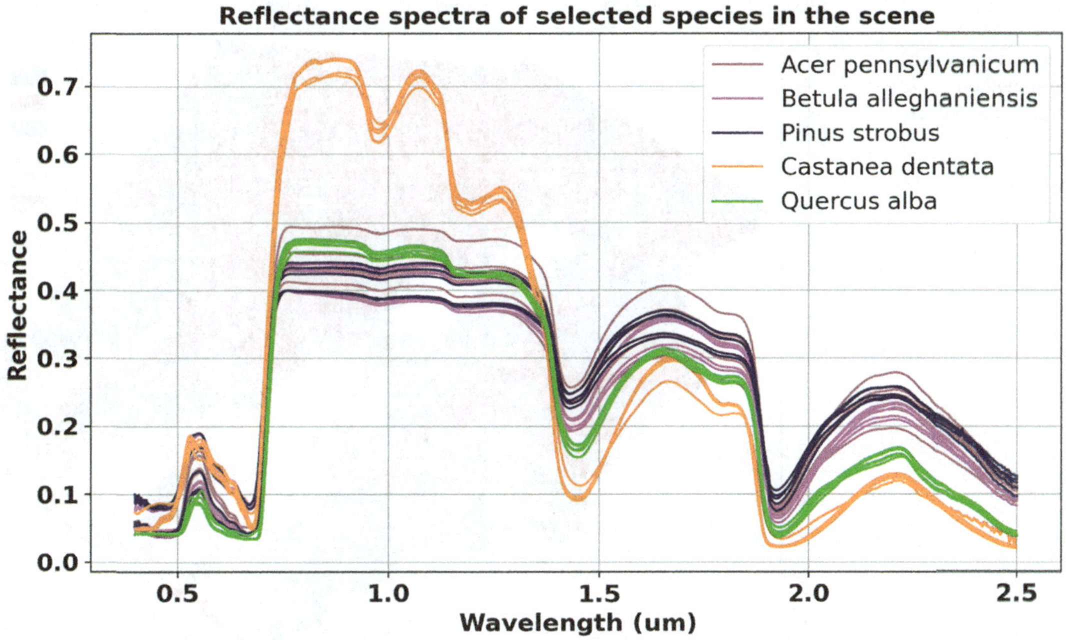

Fig. 19.27 Hyperspectral spectral signatures for selected forest species of the Havard forest (source: Chaity and Aardt 2024)

quantification of AGB and prediction of the successional state of a forest. Lidar-derived heights are highly related to AGB for various types of forests (Dubayah and Drake 2000). Lidar-derived canopy height is also useful for the estimation of tree mean stem diameter and basal area (Drake et al. 2002). Dhanda (2017) used ICESat/GLAS lidar satellite data-derived parameters for the computation of forest height and GB. The height of the forest and AGB were predicted with RMSE of 1.35 m and 13.9 Mg ha^{-1}. Airborne lidar data was evaluated by Lee and Lee (2018) for spatial estimation of forest height (Fig. 19.30).

19.4 Conclusions

Applications of geospatial technology consist of remote sensing; GIS and GNSS are rapidly growing in various areas of forestry. This trend is attributed to the mentioned factors:

1. increase the availability of multi-sensors (optical, microwave, hyperspectral, lidar) satellite data with enhanced multi-spatial and multi-spectral, high radiometric and multi-temporal resolutions;
2. availability of new spectral indices using multi-sensors remote sensing data related to different forest biophysical and structural parameters;
3. development of advanced large remote sensing data (spatially and/or temporally) processing techniques, namely data mining/object-based classification/artificial intelligence (AI)/machine learning (ML) etc.;
4. availability of GIS-aided multi-sources data integration and spatial analysis methods for decision-making and management planning.

Operationally and successfully classification and mapping of various forest types and densities are being carried out using moderate spatial resolution optical remote sensing data (e.g. IRS LISS III, Landsat TM/ETM+) following common digital supervised, unsupervised and hybrid classification approaches. However, for forest species and detailed crown density mapping and crown-based information extraction with reliable accuracies, high spatial resolution satellite data and advanced object-based and machine learning digital information extraction algorithms are to be employed.

Remote sensing technique established as an effective tool for estimation of aboveground ground forest biomass with reasonable accuracy compared to labour-intensive, costly and time-consuming conventional ground-based methods. Multi-spectral remote sensing derived spectral vegetation indices (SVI) have been used as independent variables in simple linear parametric models for spatial estimation of AGB with results of varying successes. However, research studies showed that at high forest biomass levels, simple parametric models are not suitable, instead use of nonparametric nonlinear models using the same SVI variable was found useful and more accurate.

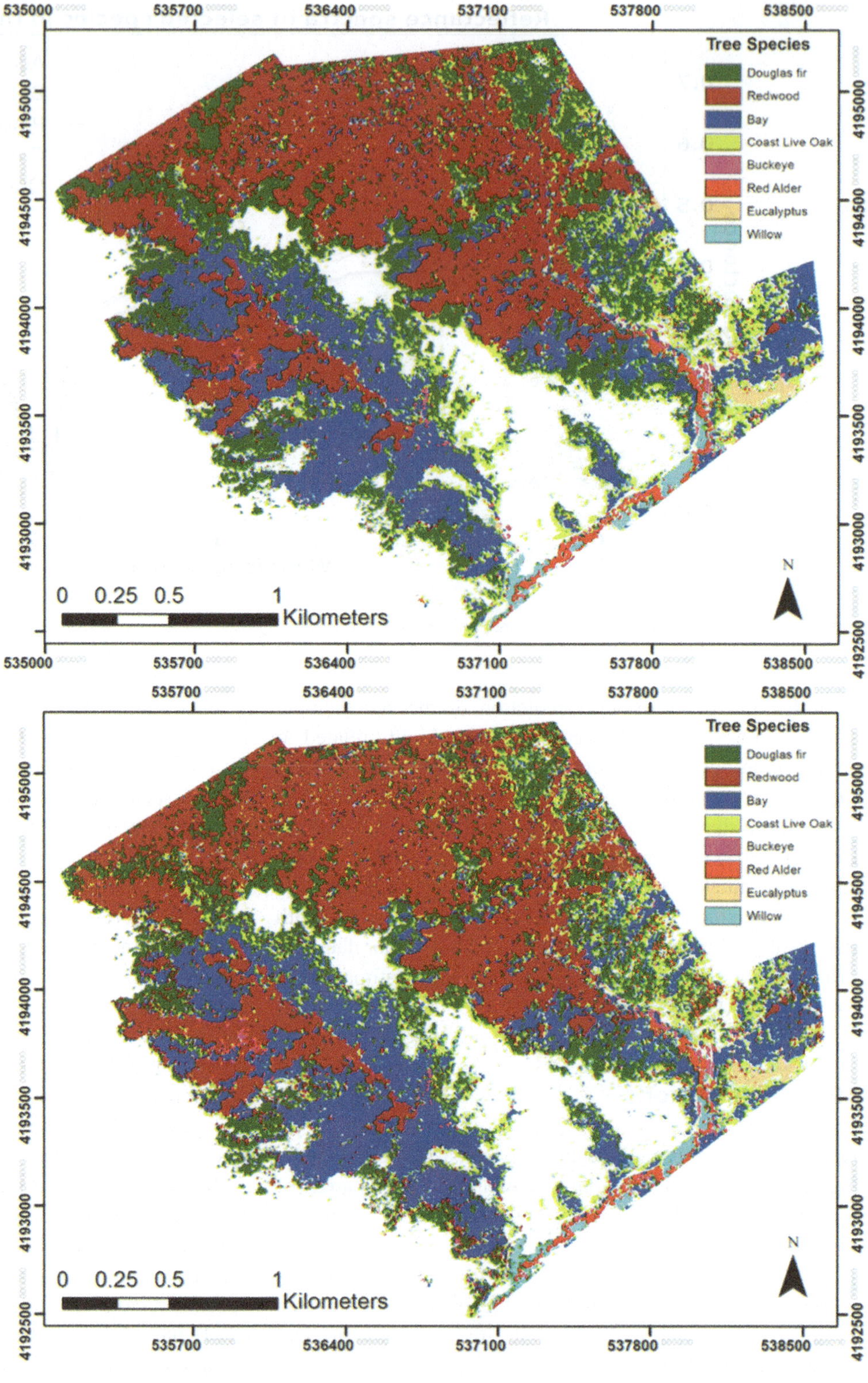

Fig. 19.28 Tree species distribution maps: SVM classification method (top) and RM classification method (bottom) (source: Ballanti et al. 2016)

Currently, high temporal and coarse spatial resolutions satellite data (e.g. MODIS, NOAA) are being operationally used for real-time active forest fire detection, mapping and monitoring at regional and global levels. However, real-time local-level detection and monitoring of small-scale forest fires require geostationary satellites with high spatial resolution and at present this kind of satellite remote sensing system is not available. Multi-spectral satellite-derived spectral indices like NBR and NDVI were found useful for the estimation and assessment of burnt area and post-fire vegetation recovery with desirable accuracies. Methodologies of other important information required for effective forest fire

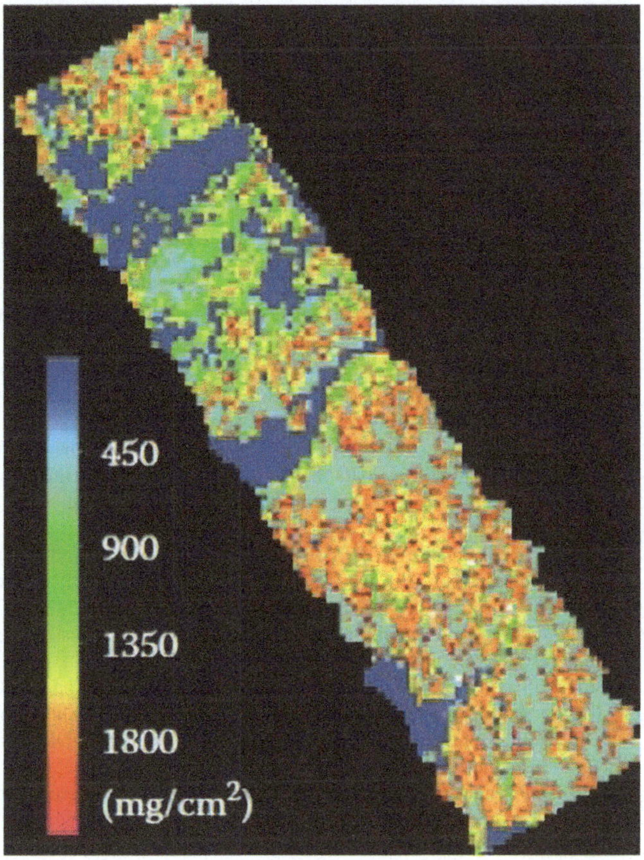

Fig. 19.29 Spatial map of forest canopy chlorophyll content for three vegetated cover types (source: Zhang 2012)

management such as mapping of fuel types and burnt severity assessment are evolving.

Multi-spectral satellite-derived vegetation indices have shown potential for the identification and mapping of insect/disease-infested forest patches. Although, several advanced digital classification techniques such as SMA, SVM and RF were attempted using multi-spectral satellite data to map disease and insect outbreaks in forests, effective operational technique(s) are yet to be developed.

Presently, GIS has been successfully used for the creation of digital databases (spatial and non-spatial data) required for forest working plans and the development of decision support systems (DSS) through integrated spatial analysis using multi-source data including remote sensing-derived information. GIS-based various spatial modelling approaches showed potential applicability for forest fire risk zonation, simulation of fire growth/spread and wildlife habitat evaluation.

GIS is the only well-established tool for the development of geoportal for geo-web services for efficient access to forest resources and management information. In India, several geoportals such as the biodiversity information system (BIS), Van Agni (forest fire) and Indian Bio-resource Information System (IBIN) related to forestry sectors are in operation. These digital information systems are developed using geoinformatics technologies.

Advanced remote sensing techniques, namely microwave, hyperspectral and lidar have shown usefulness in improved

Fig. 19.30 Forest height map derived by airborne Lidar (source: Lee and Lee 2018)

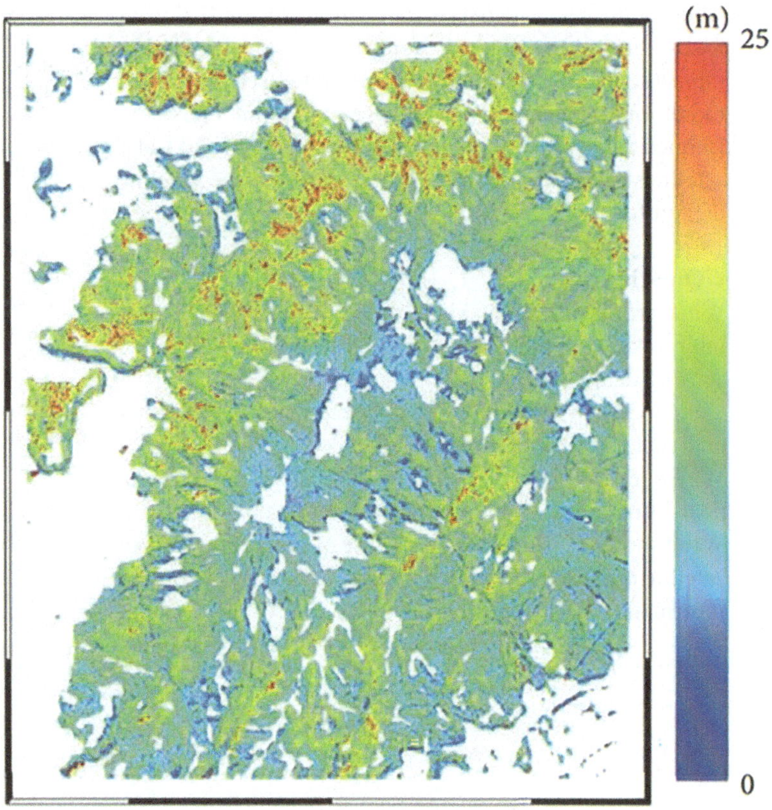

forest resources inventory and management. Synthetic aperture radar (SAR) microwave remote sensing has shown potential for various areas of forestry applications due to its sensitivity to canopy moisture content, canopy surface roughness and forest canopy penetration capability. Various satellite SAR data were successfully used for forest structural and biophysical parameters assessment such as forest types, biomass density and forest canopy height. Recently, several studies demonstrated the usefulness of airborne hyperspectral remote sensing in improved forest species mapping and spatial estimation of forest canopy chlorophyll content. The use of airborne Lidar remote sensing data rapidly growing for individual tree-level information extraction on tree stand and volume, height and structure and this information revolutionizing forest planning and operations.

Lessons Learnt

- Retrieval of forest biophysical and biochemical parameters and mapping of forest species using multispectral high spatial resolution and hyperspectral remote sensing data.
- Vegetation stress analysis and pests and disease assessment and forewarning particularly using hyperspectral remote sensing and other ancillary data using GIS.
- Development of operational techniques for assessment of forest biomass with desirable accuracy using integrated use of multi-sensors (Optical, microwave and Lidar) remotely sensed data.
- Development of useful GIS-based spatial modelling techniques using remote sensing-derived information along with other ancillary data sources for prediction of forest fire growth/spread, forecasting climate change impacts and anthropogenic activities in forest ecosystems, monitoring biodiversity and change modelling, etc.
- Limited availability of operational hyperspectral and Lidar satellites added constraints on the use of these satellite data in forestry.

Key Questions

1. What are the different kinds of resolutions used in the interpretation and processing of satellite remote sensing data?
2. Define GIS. Write briefly the characteristics of the raster and vector format of geospatial data.
3. Why does forest vegetation reflect less in red and more in near-infrared spectral bands in satellite data?
4. Write short notes on principles of use of image elements of satellite data used for extraction of various forest resources information.

References

Abdollahnejad A, Panagiotidis D, Surovy P (2017) Forest canopy density assessment using different approaches. Rev J Forest Sci 63:107–116

Adelabu SA, Mutanga O, Adam EMI (2014) Evaluating the impact of red-edge band from Rapideye image for classifying insect defoliation levels. ISPRS J Photogram Remote Sensing 95:34–41

Ahmad F, Goparaju L, Qayum A (2018) Wild life habitat suitability and conservation hotspot mapping: remote Sensing and GIS based decision support system. AIMS Geosci 4(1):66–87

Aronoff S (1989) Geographic information systems: a management perspective. WDC Publications, Ottawa

Ballanti L, Blesis L, Hines E, Kruse B (2016) Tree species classification using hyperspectral imagery: a comparison of two classifiers. Remote Sens 8(6):445–462

Bastarrika A, Chuvieco E, Martin MP (2011) Mapping burned areas from Landsat TM/ETM+ data with a two phase algorithm: Balancing omission and commission error. Remote Sens Environ 115(4): 1003–1012

Buddenbaum H, Schlerf M, Hill J (2005) Classification of coniferous tree species and age classes using hyperspectral data and geostatistical methods. Int J Remote Sensing 26(4):5453–5465

Burrough PA (1986) Principles of geographical information systems for land resource assessment. Clarendon Press, Oxford, p 193p

Caccamo G, Chisholm L, Bradstock RA, Poutinen M, Pippen BG (2012) Monitoring live fuel moisture content of heathland, shrubland and sclerophyll forest in south-eastern Australia using MODIS data. Int J Wildland Fire 21(3):257–269

Calvao T, Palmeirim JM (2004) Mapping Mediterranean scrub with satellite imagery: biomass estimation and spectral behavior. Int J Remote Sensing 25(16):3113–3126

Cambell JB, Wynne RH (2011) History and scope of remote sensing. In: Cambell R, Wynne H (eds) J.B. The Guiford Press, Introduction to remote sensing, pp 3–28

CCRS and NRC (n.d.) Canada centre for remote sensing and natural resources Canada. Remote Sensing Tutorials, pp 45–59

Chaity MD, Aardt JV (2024) Exploring the limits of species identification via a convolutional neural network in a complex forest scene through simulated imaging spectroscopy. Remote Sens 16(3): 498–528

Chandrashekhar MB, Saran S, Raju PLN, Roy PS (2005) Forest canopy density stratification: how relevant is biophysical spectral response modelling approach? Geocarto Int 20(1):15–21

Chen G, Meentemeyer RK (2016) Remote sensing of forest damage by diseases and insects. In: Weng Q (ed) Remote sensing for sustainability. CRC Press, pp 145–158

Chuvieco E (2016a) Physical principles of remote sensing. In: Chuvieco E (ed) Fundamentals of satellite remote sensing: an environmental approach. CRC Press, pp 23–66. (Chapter 2)

Chuvieco E (2016b) Visual interpretation. In: Chuvieco E (ed) Fundamentals of satellite remote sensing: an environmental approach. CRC Press, pp 149–170. (Chapter 5)

Coops NC, Waring RH, Wulder MA, White JC (2009) Prediction and assessment of bark beetle-induced mortality of lodgepole pine using estimates of stand vigor derived from remotely sensed data. Remote Sensing Environ 113:1058–1066

Darvishefat A, Kellenbarger T, Itten K (2002) Applications of hyperspectral data for forest stand mapping. In: Proceedings of the symposium on geospatial theory, processing and applications. ISPRS Comission IV Symposium, Ottawa, Canada, IAPRS, vol. XXXIV, Part 4, 1–5 p

Das AK (2017) Advanced SAR techniques for forests. In: ERTD TREES training lecture notes on "SAR and Hyperspectral Data Analysis for Forest Applications". Lecture no. 6, pp. 1–10

Dash PP, Joshi PK, Roy PS, Tillu AD (2015) Predictive habitat suitability modeling for prey and predator (Tiger) in Palamau Tiger Reserve. Int J Int Sci 6

Dhanda P, Nandy S, Kushwaha SPS, Ghosh S, Krishna Murthy YVNK, Dadhwal VK (2017) Optimizing spaceborne LiDAR and very high resolution optical sensor parameters for biomass estimation at ICESat/GLAS footprint level using regression algorithms. Prog Phys Geogr Earth Environ 41(3):247–267

Drake JB, Dubaya R, Knox RG, Clark DB, Blair JB (2002) Sensitivity of large-footprint LiDAR to canopy structure and biomass in a neotropical rainforest. Remote Sensing Environ 81(2–3):378–392

Dubayah R, Drake J (2000) LIDAR remote sensing for forestry. J Forest 98:44–46

Dube T, Mutanga O (2015) Investigating the robustness of the new landsat-8 operational land imager derived texture metrics in estimating plantation forest aboveground biomass in resource constrained areas. ISPRS J Photogram Remote Sensing 108:12–32

ESRI (1990) Understanding GIS: The ARC/INFO method. Redlands, ESRI

Estes JE, Simonett DS (1975) Fundamentals of image interpretation. In: Reeves RG (ed) Manual of remote sensing. American Society of Photogrammetry, Falls Church, pp 869–1076

FAO (2005) Global forest resources assessment update, terms and definitions (final version). Forest Resources Assessment Program, Rome. Working Paper 83, Forest, 33p

FSI (2019) Indian State of Forest Report, Forest Survey of India, Ministry of Environment, Forest & Climate Change, Govt. of India, 16p

FSI (2020). fsi.nic.in/forest-fire-activities

Garcia JL, Molina JP, Delgado LM, Higuera AP (2016) Monitoring changes of forest canopy density in a temperature forest using high resolution aerial digital photography. Investigaciones Geogr Bolet 90:59–74

Giglio L, Loboda T, Roy DP, Quayle B, Justice CO (2009) An active-fire based burned area mapping algorithm for the MODIS sensor. Remote Sens Environ 113:408–420. Gispeople.com. 2020. https://www.gispeople.com.au/geospatial-consulting/multi-criteria-analysis/

Gomez C, White CJ, Wulder M, Alejandro P (2014) Historical forest biomass dynamics modelled with Landsat spectral trajectories. ISPRS J Photogram Remote Sesning 93:14–28

Guiming Z (2019) Integrating citizen science and GIS for wildlife habitat assessment. In: Ferretti M (ed) Wildlife population monitoring, IntechOpen, pp 1–18

Gupta RP (2017) Remote sensing geology, 3rd edn. Springer, pp 107–114

Jackson MJ (1992) Integrated geographical information systems. Int J Remote Sens 13(6–7):1343–1351

Jensen JR (2014) Remote sensing of environment. In: Jensen JR (ed) Remote sensing of the environment: an earth resource perspective. Pearson Education Limited, pp 1–38

Ji C, Li X, Wei H, Li S (2020) Comparison of different multispectral sensors for photosynthetic and non-photosynthetic vegetation-fraction retrieval. Remote Sens 12(1):115–132

Jin Y, Randerson JT, Goetz SJ, Beck PS, Loranty MM, Goulden ML (2012) The influence of burn severity on postfire vegetation recovery and albedo change during early succession in North American boreal forests. J Geophys Res Biogeosci 117(1):1–15

Kumar S, Garg RD, Govil H, Kushwaha SPS (2019) PolSAR- decomposition based extended water cloud modeling for forest aboveground biomass estimation. Remote Sens 11(19):2287–2314

Kushwaha SPS, Roy PS (2002) Geospatial technology for wildlife habitat evaluation. Trop Ecol 43(1):137–150

Kushwaha SPS, Unni MNV (1989) Hybrid interpretation for tropical forest classification. Asian-Pacific Remote Sensing J 1(2):69–75

Kushwaha SPS, Nandy S, Gupta M (2014) Growing stock and woody biomass assessment in Asola-Bhatti Wildlife Sanctuary, Delhi. India Environ Monitor Assess 186(9):5911–5920

Lee WJ, Lee CW (2018) Forest canopy height estimation using multiplatform remote sensing dataset. J Sensors 2018:1–9

Long L, Chen Y, Song S, Zhang X, Jia X, Lu Y, Liu L (2023) Remote sensing monitoring of pine wilt disease based on time-series remote sensing index. Remote Sens 15(2):360–378

Lu D (2006) The potential and challenge of remote sensing-based biomass estimation. Int J Remote Sensing 27(7):1297–1328

Lu D, Chen Q, Wang G, Moran E, Batistella M, Zhang M, VaglioLaurin G, Saah D (2012) Aboveground forest biomass estimation with landsat and LiDAR data and uncertainty analysis of the estimates. Int J Forestry Res 2012:1–16

Lu D, Li G, Moran E, Kuang WA (2014a) Comparative analysis of approaches for successional vegetation classification in the Brazilian Amazon. GISci Remote Sens 51:695–709

Lu D, Chen Q, Wang G, Liu L, Li G, Moran E (2014b) A survey of remote sensing-based aboveground biomass estimation methods in forest ecosystems. Int J Digital Earth 9(1):63–105

Machala M, Zejdova L (2014) Forest mapping through object-based image analysis of multispectral and Lidar aerial data. European Journal of Remote Sensing 47(1):117–131

Manso OF, Manso AF, Quintano C (2014) Estimation of above ground biomass in Mediterranean forests by statistical modelling of ASTER fraction images. Int J Appl Earth Observ Geoinform 31(1):45–56

Marabel M, Alvarez TF (2013) Spectroscopic determination of above ground biomass in grasslands using spectral transformations, support vector machine and partial least squares regression. Sensors 13(8): 10027–10051

Mohan S (2013) Radar remote sensing and planetary sciences. Int J Sci Eng Res 2229–5518:212–221

Mountrakis G, Im J, Ogole C (2011) Support vector machines in remote sensing: a review. ISPRS J Photogram Remote Sensing 66:247–259

Murthy MSR, Jha CS (2010) Forest and vegetation. In: Roy PS, Dwivedi RS, Vijayan D (eds) Remote sensing applications. National Remote Sensing Centre, Hyderabad, pp 49–80

Nandy S, Singh RP, Ghosh S, Watham T, Kushwaha SPS, Senthil Kumar A, Dadhwal VK (2017) Neural Network-based modelling for forest biomass assessment. Carbon Manage 8(4):305–317

Nandy S, Gosh S, Kushwaha SPS, Senthil KA (2019) Remote sensing based forest biomass assessment in Northwest Himalayan landscape. In: Navalgund RR, Senthil Kumar A, Nandy S (eds) Remote sensing of Northwest Himalayan ecosystems. Springer Nature, Singapore, pp 266–285

NRSC (2010) Remote sensing and GIS inputs for management plan preparation of Akola forest division, Technical document published by Regional Remote Sensing Center—Central. NRSC—ISRO, Nagpur

Ogaya R, Barbeta A, Basnou C, Penuelas J (2015) Satellite data as indicators of tree biomass growth and forest dieback in a Mediterranean holm oak forest. Ann For Sci 72:135–144

Olsson P-O, Jönsson AM, Eklundh L (2012) A new invasive insect in Sweden—Physokermes inopinatus: Tracing forest damage with satellite based remote sensing. For Ecol Manag 285:29–37

Pizana JMG, Nunez JM (2016) Remote sensing based biomass estimation. In: Marghany M. (ed.) Environmental applications of remote sensing. ExLi4EvA, pp 1–40

Porwal MC, Roy PS, Chellamuthu V (1996) Wildlife habitat analysis for "Sambar" (Cervus unicolor) in Kanha National Park using remote sensing. Int J Remote Sensing 14

Pu R (2017) Characteristics of hyperspectral remote sensing. In: Hyperspectral remote sensing—fundamental and practices. CRC Press, Taylor & Francis Group, pp 1–23. (Chapter 1)

Quintano C, Manso AF, Roberts DA (2013) Multiple Endmember Spectral Mixture Analysis (MESMA) to map burn severity levels from Landsat images in Mediterranean countries. Remote Sens Environ 136:76–88

Quintano C, Manso AF, Calvo L, Marcos E, Valbuena ML (2015) Land surface temperature as potential indicator of burn severity in forest Mediterranean ecosystems. Int J Appl Earth Obs Geoinf 36:1–12

Radeloff VC, Mladenoff DJ, Boyce MS (1999) Detecting jack pine budworm defoliation using spectral mixture analysis: separating effects from determinants. Remote Sensing Environ 69:156–169

Rai N (2022) Mapping forest structure in Mississippi using LiDAR remote sensing, vol 5688. Mississippi State University, USA (Theses and Dissertations)

Riano D, Chuvieco E, Salas J, Palacios-Orueta A, Bastarrika A (2002) Generation of fuel type maps from Landsat TM images and ancillary data in Mediterranean ecosystems. Can J For Res 32:1301–1315

Roy PS (1986) Satellite remote sensing data interpretation using visual and computer aided techniques for wildlife habitat analysis. In: Proceedings of the seminar-cum-workshop wildlife habitat evaluation using remote sensing, 22–23 October, Dehradun, pp 193–208

Roy PS, Kushwaha SPS, Murhty MSR, Roy A, Kushwaha D, Reddy CS, Behera MD, Mathur VB, Padalia H, Saran S, Singh S, Jha CS, Porwal MC (2012) Biodiversity characterisation at landscape level: national assessment. Indian Institute of Remote Sensing, Dehradun, India, p 140p

Sanare J, Ganawa ES, Salih A (2015) Wildlife habitat suitability analysis at Serengeti National Park (SNP), Tanzania- case study Loxodonta sp. J Ecosyst Ecogr 5(3):1–8

Saran S, Padalia H, Ganeshaiah KN, Oberai K, Singh P, Jha AK, Reddy KS, Verma A, Uniyal S, Senthil Kumar A (2019) Indian Bioresources Information Network (IBIN). In: Navalgund RR, Senthil Kumar A, Nandy S (eds) Remote sensing of Northwest Himalayan ecosystems. Springer Nature, Singapore, pp 251–264

Sarker MLR, Nicole J (2011) Improved forest biomass estimates using ALOS AVNIR-2 texture indices. Remote Sensing Environ 115(4): 968–977

Silva JMN, Sa ACL, Pereira JMC (2005) Comparison of burned area estimates derived from SPOT VEGETATION and Landsat ETM+ data in Africa: influence of spatial pattern and vegetation. Remote Sensing Environ 96(2):188–201

Sivakumar R (2010) Image interpretation of remotely sensed data. Geospatial World

Skakun RS, Wulder M, Fraklin SE (2003) Sensitivity of the thematic mapper enhanced wetness difference index to detect mountain pine beetle red-attack damage. Remote Sensing Environ 86(4):433–443

Sousa A, Goncaives AC, Mesquita P, Da Silva JM (2014) Biomass estimation with high resolution satellite images: a case study of Quercus rotundifolia. ISPRS J Photogram Remote Sensing 101: 69–79

Stillwell J, Clarke G (1987) Applied GIS and spatial analysis. Wiley, West Sussex

Szpakowski DM, Jensen JLR (2019) A review of the applications of remote sensing in fire ecology. Remote Sens 11:2683–2714

Tane Z, Roberts D, Veraverbeke S, Casas A, Ramirez C, Ustin S (2018) Evaluating endmember and band selection techniques for multiple endmember spectral mixture analysis using post-fire imaging spectroscopy. Remote Sens 10(3):1–21

Thomas M, Adeline K, Huesca M, Ustin S, Briotiet X (2019) Monitoring LAI, chlorophylls, and carotenoids content of a Woodland Savanna using hyperspectral imagery and 3D radiative transfer modeling. Remote Sens 12:1–26

Thumaty TC, Fararoda R, Middinti S, Rajashekar G, Jha CS, Dadhwal VK (2016) J Indian Soc Remote Sensing 44:31–39

Unni NVM (1983) Forest survey and management using remote sensing. Proc Indian Acad Sci 6(3):209–231

Unni NVM, Roy PS, Parthasarathy V (1983) Feasibility of mapping economically important forest species using Landsat data–a case study. J Indian Soc Remote Sensing 11(1):37–42

Van Wagtendonk JW, Root RR (2003) The use of multi-temporal Landsat Normalized Difference Vegetation Index (NDVI) data for mapping fuel models in Yosemite National Park, USA. Int J Remote Sensing 24(8):1639–1651

Varghese AO, Krishnamurthy YVNK (2006) Remote sensing and GIS inputs for working plan preparation of east Metghat forest division. RRSSC, Nagpur and Department of Forests (Working plan), Maharashtra, pp 35–49

Veraverbeke S, Verstraeten WW, Lhermitte S, Van De KR, Goossens R (2012) Assessment of post-fire changes in land surface temperature and surface albedo, and their relation with fire—burn severity using multitemporal MODIS imagery. Int J Wildland Fire 21(3):243–256

Walter JA, Platt RV (2013) Multi-temporal analysis reveals that predictors of mountain pine beetle infestation change during outbreak cycles. For Ecol Manag 302:308–318

Xie Z, Chen Y, Lu D, Li G, Chen E (2019) Classification of land cover, forest and tree species classes with Ziyuan—3 multispectral and stereo data. Remote Sens 11:1–27

Yadav BKV, Nandy S (2015) Mapping aboveground woody biomass using forest inventory, remote sensing and geostatistical techniques. Environ Monitor Assess 187(5):1–12

Yathish H, Athira KV, Preethi K, Pruthviraj U, Shetty A (2019) A comparative analysis of forest fire risk zone mapping methods with expert knowledge. J Indian Soc Remote Sensing 47(12):2047–2060

Zhang Y (2012) Forest leaf chlorophyll study using hyperspectral remote sensing. In: Thenkabail PS, Lyon JG, Huete A (eds) Hyperspectral remote sensing of vegetation. CRC Press, pp 167–182

Forest Certification for Promoting Responsible Forest Management

20

Manmohan Yadav and Dharmendra Dugaya

Abstract

The criteria and indicators (C&I) approach to sustainable forest management (SFM) has emerged since the climate change debate began in the late 1980s and early 1990s. The International Tropical Timber Organization (ITTO) was the first to introduce 8 criteria and about 45 indicators for the sustainable management of tropical forests in 1992. Since then, some 11 processes have been developed globally, each catering to the contextual relevance of a specific region or forest type. One such initiative is that of the dry forests in Asia regional initiative for developing and implementing national level C&I for the sustainable management of dry forests in Asia. Also known as the Bhopal-India Process, this initiative of nine countries in Asia developed 8 criteria and 49 indicators for sustainable management of dry forests in Asia.

Forest certification emerged as an eco-label to provide a market link between responsible/sustainable forest management practices and green consumers. Forest certification as a possible market-based tool has its foundations in controlling deforestation on the one hand and enhancing the social and environmental values of forests, mainly those of tropical rainforests, on the other hand.

Forest Stewardship Council (FSC) emerged as the first international forest certification in 1993 for responsible forestry practices aimed at promoting environmentally sound and socially beneficial forestry. In this process, forest management practices are evaluated and verified against a set of standards by an external independent certification agency. Thus, FSC certification provides confidence to environmentally conscious consumers that the forest products they are buying have been sourced from sustainably managed forest areas with documented proof of sustainability. When combined with a chain-of-custody (CoC) verification, the certification allows the establishment of traceability of the product, confirming it has been derived from a particular certified forest area. Such products can carry an eco-label or the logo of the FSC as proof of such a claim.

The second global forest certification system which is in practice is the Program for Endorsement of Forest Certification (PEFC), which mutually recognizes the national forest certification schemes that aim to harmonize the global forest certification standards with those of national forest certification standards. PEFC has accorded mutual recognition to about 50 national forest certification schemes, the majority of which are in Europe. Globally about 13% of forest area is certified under these two certification schemes.

Non-timber forest products (NTFPs) are very important forest resources for the local forest-dwelling communities as a source of food, medicine, and income. NTFPs are also a rich biodiversity pool for the global community and important raw materials for natural products such as consumer personal care products, food supplements, and traditional health remedies. Forest certification for NTFPs, hence, is understood as a means of protecting NTFP resources by promoting responsible and improved management practices. It provides third-party assurance that the NTFP management operations and prescriptions meet standards set by the certification system. Forests with NTFP resources are, therefore, evaluated based on a defined set of standards for NTFP sustainable management and harvesting by an independent, qualified third-party auditor.

Keywords

Forest resource assessment · Forest certification · Criteria and indicators · Eco-label · Chain-of-custody

M. Yadav (✉)
Indian Institute of Plantation Management, Jnana Bharathi Campus, Bengaluru, Karnataka, India
e-mail: manmohan@iipmb.edu.in

D. Dugaya
Indian Institute of Plantation Management, Jnana Bharathi Campus, Bengaluru, Karnataka, India

Indian Institute of Forest Management, Bhopal, India

A. K. Mandal, A. Nicodemus (eds.), *Textbook of Forest Science*, https://doi.org/10.1007/978-981-97-8289-5_20

20.1 Forests as Lifeline for All Living Things

A forest is an important natural resource for humans, and it plays an essential role in providing multiple benefits to all living beings and also to commercial enterprises. The forest can be described as an area with a high density of trees. They support a host of different species and play an important role in the livelihoods of forest fringe communities. One-third of the earth's land area is occupied by forests, which account for around 80% of the world's terrestrial biodiversity. Human beings have been dependent on forests for clean air, food, water, fuel, shade, and shelter. Gradually, their dependency on forests has extended to economic gains as well. Due to the overexploitation of forest resources because of the increasing population, this disturbance is now becoming unfavorable to human beings themselves.

Forests maintain the carbon cycle, feed rivers, and also contribute to the nature-based tourism industry known as eco-tourism. With the growing knowledge, it is established that for a healthy planet, it is important to have healthy forests. However, the growing population has been exerting immense pressure on forests and forest products, resulting in deforestation and forest degradation.

According to the Food and Agriculture Organization's (FAO) Global Forest Resources Assessment 2020, the total global forest area is 4.01 billion ha, or 31% of the total global land area. Just ten countries account for two-thirds of the world's forest area (Table 20.1). Russia alone accounts for about 20% of the global forest resources, followed by Brazil, Canada, the USA, China, Australia, the Congo, Indonesia, Peru, and India.

Tropical forests account for 45% of the world's forests, followed by boreal forests (27%), temperate forests (16%), and subtropical forests (11%). In total, 93% (3.75 billion ha) of the forest area worldwide consists of naturally regenerated forests, while 7% (290 million ha) consists of planted forests.

The world has lost 178 million ha of net forest area since 1990. However, the rate of net forest loss has declined during the last three decades, from 7.8 million ha per year during 1990–2000 to 5.2 million ha per year in 2000–2010 and 4.7 million ha per year in 2010–2020. This positive change is the result of the efforts of the global community toward the protection of forests and forest resources.

Plantations of tree species are one of the potential measures for converting barren land into productive green cover and rehabilitating degraded natural forests. Although plantations cannot match the natural forests in terms of multiple functions and services, they still contribute to fulfilling the esthetic, environmental, and industrial requirements of an ever-increasing population. They account for only 7% of the global forest area but contribute to around 40% of industrial requirements. When well-managed with crop tending operations, plantations produce more desirable timber and fiber in perpetuity than natural forests.

Forestry and agriculture are two independent land-use patterns. The latter competes with the former under the relentless pressure of an ever-increasing demand of modern society. The decreasing trend of natural forests calls for a holistic approach toward sustainable development by balancing the existence of both. Various global initiatives have emerged for the sustainable management of forests through a system of criteria and indicators (C&I) approaches.

20.2 Criteria and Indicators for Sustainable Forest Management

The concept of Sustainable Forest Management (SFM) is not new but has become a buzzword and a widely used concept in recent decades. Today, SFM is an integral part of the Sustainable Development concept. Sustainable Development is a way of dealing with the environmental dilemmas that confront the world, discussed and coined at the United Nations Conference on the Human Environment held in Stockholm in 1972.

Table 20.1 Top ten countries with the largest forest area

Rank	Country	Forest area (million ha)	Percent land area as forest
1	Russia	815	49.78
2	Brazil	497	59.42
3	Canada	347	38.15
4	USA	310	35.30
5	China	220	23.30
6	Australia	134	17.44
7	Democratic Republic of the Congo	126	55.65
8	Indonesia	92.1	49.07
9	Peru	72	55.60
10	India	71.2	21.60
	Total of the top ten countries	2684.3	

Source: Global Forest Resources Assessment 2020 (FAO 2020)

The United Nations Conference on Environment and Development (UNCED) held in Rio de Janeiro in 1992 was a landmark event as regard to managing the world's forests, resulting in the publication of five non-legally binding documents. These were milestones in terms of achieving international recognition for the need to deal with a wide range of environmental issues, including the need for sustainable development.

Following the UNCED, the criteria and indicators C&I for sustainable forest management (SFM) were developed as a potent tool for assessing, monitoring, and reporting trends and changes in forest conditions and management systems at the regional, national, and forest management unit (FMU) levels.

Globally, international organizations such as the International Tropical Timber Organization (ITTO), Food and Agriculture Organization (FAO), and Center for International Forestry Research (CIFOR) have led the process of developing C&I as an approach to SFM. As a result, 11 major regional and international C&I processes have evolved for different forest types across the world. These are as follows (Adopted from Stefanie et al. 2018):

1. The International Tropical Timber Organization's (ITTO) C&I for the sustainable management of tropical forests, representing 75% of the world's tropical forests (since 1986, the first set of C&I was developed in 1992).
2. The Pan-European Process on C&I for SFM, under the Ministerial Conference for the Protection of Forests in Europe, known also by its acronym FOREST EUROPE (since 1990, the first set of C&I was developed in 1994).
3. The Montréal Process on C&I for the conservation and sustainable management of temperate and boreal forests, covering 90% of the world's temperate and boreal forests (since 1993, the first set of C&I was developed in 1995).
4. The Amazon Cooperation Treaty Organization (ACTO) Tarapoto Process on C&I for the sustainability of Amazon forests (since 1995, the first set of C&I was developed in 1995). The harmonization process with ITTO started in 2012, which led to the renaming as the "process of harmonized C&I of ITTO–ACTO (Tarapoto)" for the sustainability of the Amazon forests.
5. The Association of Southeast Asian Nations (ASEAN) C&I for the sustainable management of tropical forests in Southeast Asia (since 1998, the first set of C&I was developed in 2000).
6. The Low-Forest-Cover-Countries Process, also known as the Tehran Process (TP for LFCCs) (since 2000, the first set of C&I was developed in 2011).
7. The Dry-Zone Africa Process on C&I for the sustainable management of dry-zone forests in sub-Saharan countries (since 1995, the first set of C&I was developed in 1995).
8. The Near East and North Africa (NENA) Process on C&I for sustainable management of dry-zone forests (since 1996, the first set of C&I was developed in 1996).
9. The Lepaterique Process of Central America on C&I for SFM (since 1997, the first set of C&I was developed in 1997).
10. The Dry Forests in Asia Regional Initiative for the development and implementation of national-level C&I for the sustainable management of dry forests in Asia (also known as the Bhopal-India Process) (since 1998, the first set of C&I was developed in 1999, revised in 2005).
11. ATO/ITTO C&I Process (African Timber Organization (ATO) C&I framework, which was established in 1994 with the support of ITTO and led to the development of the first set of C&I in 2001).

20.2.1 The Bhopal-India Process

The Indian initiative for the development of Criteria & Indicators for sustainable forest management was undertaken by the Indian Institute of Forest Management (IIFM), Bhopal, in 1998. Eight criteria and 51 indicators were developed for SFM in India through a multi-stakeholder consultative process (IIFM 2008). This has been refined over the years and is now named the National Set of Criteria & Indicators for SFM in India, with 8 criteria and 37 indicators.

These criteria and indicators have been included in the National Working Plan Code (NWPC) 2023, a document that provides guidelines for preparing working plans in the country. A working plan is the key guiding document that provides a link between past management practices and future management prescriptions for forest managers. It is a 10-year plan for management interventions at the division (FMU) level. After 10 years, working plans are revised to keep pace with the trends emerging at the forest-people interface and to address other related objectives. The working plans, which are due for revision after the year 2023 are being prepared following the NWPC 2023.

20.2.2 United Nations Global Strategy for Forests

The Intergovernmental Panel on Forests (IPF, 1995–1997) and the Intergovernmental Forum on Forests (IFF, 1997–2000) culminated in over 270 IPF/IFF proposals for action and the establishment of the United Nations Forum on Forests (UNFF) (Desai 2011). The UNFF was established as part of the international arrangement on forests to carry on the work of the IPF and the IFF processes. In 2006, the UNFF and ECOSOC agreed on four Global Objectives on Forests, which aimed at reversing the loss of forest cover, improving the contribution of forests to local livelihoods, increasing protected areas, and enhancing financial support for sustainable forest management. These Global Objectives on Forests were adopted by the UNFF as a Non-Legally Binding Instrument on 28 April 2007.

At a special session of the UN Forum on Forests in January 2017, a historic agreement on the Strategic Plan was achieved, and in April 2017, the UN General Assembly accepted the plan. Member States reaffirmed their commitment to six global goals and 26 associated targets on forests to be achieved by 2030 (UN 2019).

The six global forest goals are as follows: (1) reverse the loss of forest cover worldwide through sustainable forest management, including protection, restoration, afforestation, and reforestation, and increase efforts to prevent forest degradation and contribute to the global effort to address climate change; (2) enhance forest-based economic, social, and environmental benefits by improving the livelihoods of forest-dependent people; (3) significantly increase the area of protected forests worldwide and other areas of sustainably managed forests, as well as the proportion of forest products from sustainably managed forests; (4) mobilize significantly increased, new, and extra funding from all sources to implement sustainable forest management and to fortify linkages and collaborations in science and technology; (5) promote governance frameworks to implement sustainable forest management through the United Nations forest instrument and enhance the contribution of forests to the 2030 Agenda for Sustainable Development; (6) enhance cooperation, coherence, and synergies on forest-related issues at all levels within the UN system and across member organizations of the collaborative partnership on forests, as well as across sectors and relevant stakeholders.

The member states have gradually imbibed these principles of sustainable forest management in their national policies. For example, India has adopted the national set of criteria and indicators for sustainable management of forests as the Indian Forest Management Standard as part of the National Working Plan Code 2023.

20.3 Introduction of Forest Certification

To ensure that the principles of forest management are developed and implemented responsibly and sustainably, the need for some market-based mechanisms was felt.

Forest certification emerged as a market tool to ensure that the products and services are obtained from a forest that is managed by following the principles of sustainable forest management. If forest products are harvested from such forest areas that practice sustainable management principles, then a third-party certification mechanism can verify such practices and the trail of products coming from such forest areas and issue a certificate of compliance to the requirements of the certification scheme. Such a process is called forest certification. Once issued, the forest certification allows the business firm to formally label all such products as certified products with a seal of the certification scheme to create differentiation for its products in the marketplace.

Forest certification is thus a market mechanism to ensure that forests are being managed in accordance with international commitments, national mandates, and universally accepted principles of responsible forest management.

20.3.1 Forest Certification as an Eco-label

Environmental labels, or eco-labels, are market tools that enable consumers to use their purchasing power to promote and incentivize environmentally friendly products. In addition, they provide producers with a rationale for adhering to environmental standards that are put in place to satisfy government mandates or market-driven voluntary obligations. The term "eco-labeling" denotes a process that attempts to provide a credible indicator as to what extent a product is environmentally adapted.

Public concern for the environment has grown remarkably during the last four decades, both in developed and developing countries. As a result, environmental issues have acquired center stage in the economic and trade policies globally.

During the late 1960s, environmental problems and pollution became a concern for the developed countries. The United Nations agencies also responded to such concerns first at its Biosphere Conference in Paris in 1968. The conference recognized the need for more and improved research on ecosystems, human ecology, pollution, and genetic and natural resources, and further catering to inventory assessment for monitoring of resources. As seen by the numerous anti-war and nuclear protests, social concerns related to the environmental balance were significant issues that compelled political commitments and actions in society. The observance of the first Earth Day in the United States in April 1970 serves as an example of environmental action (McCormick 1989; Shabecoff 1993).

To address these social demands, the United Nations Conference on the Human Environment was held in Stockholm in 1972 wherein the sustainable development paradigm was introduced, which was later published as Brundtland's Report. Next in the array of developments, resulting from people's pressure, the United Nations created the United Nations Environment Program (UNEP) aiming at creating legal and policy frameworks to bring environmental issues to the international forum. In 1983, the Food and Agriculture Organization's World Forest Appraisal Program (WFAP) compiled a report on the inventory of forestry resources incorporating the human impact on forests. This report initiated a global debate on the rate of forest loss and interventions, and the need for developing systems for implementing sustainable forest management.

Consumer interests have spurred information systems aimed at differentiating the environmental appropriateness of goods and services. One modern example of this is the advent of "eco-labeling," a process that aims to provide an indicator of how well a product is adapted to the environment. The basic concept of "eco-label" is derived from the word "eco," which means natural environment, and "label," which signifies how a particular product differs from other products.

Eco-label provides information on the environmental characteristics of a product, allowing consumers to use their purchasing power to promote environmentally friendly products. The practice of labeling wood items with a quality mark originated in Europe in 1637 when a royal decree from France mandated that cabinet manufacturers in the guild mark the furniture they produce (Pradère 1989). Relying on this market-driven mechanism, the world's first eco-labeling program, the "German Blue Eco Angel," was created in 1978 (Rametsteiner 2000).

Eco-labels are part of the environmental policy "second generation," which replaced some, but not all, of the earlier environmental policy instruments. These older approaches were developed in the 1970s when the regulation of environmental media became dominant. By contrast, eco-labels aim directly to change the behavior of consumers (sustainable consumption) who are enabled to assess the impacts of a product throughout its entire life cycle. Environmental labeling programs can provide consumers with an immediately available, objective, and accurate evaluation of a product's environmental impact. In addition, they provide manufacturers with a reason to comply with environmental regulations (Sitarz 1998).

"Blue Angel," the first national eco-label, introduced in Germany in 1978, remained the only label used until a 10-year-later, comparable program was developed in Canada. In 1989, the Nordic Council of Ministers (Ministers for Consumer Affairs) decided to create the first multinational eco-label system, which was known as the "Swan Label." Between 1989 and 1992, this policy innovation spread quickly. These eco-labels were introduced in almost all Organization for Economic Co-operation and Development (OECD) countries and even in some newly industrialized and developing countries, such as Singapore, Korea, Taiwan, and India. Crucial to this dynamic development was the introduction of a European eco-label in 1992 named "European Flower." "Eco-mark" was initiated as an environmental label under the aegis of the Ministry of Environment, Forest and Climate Change (MoEFCC) in India. However, the use of Eco-mark in India has been very limited, primarily because of a lack of consumer response.

Forest certification can also be considered as an eco-label that differentiates a forest product sourced from a sustainably managed forest area from the rest of the products. Thus, forest certification as an eco-label placed on product links responsible/good forestry practices with environmentally conscious consumers.

Ghazali and Simula (1994) defined forest certification "as a process which results in a written statement which is a certificate attesting the origin of wood raw material, and its status and/or qualifications following validation by an independent third party." Forest certification is actually a single-issue eco-label or certification that acknowledges the environmental quality of a product at a certain point in its production life cycle or for a specific component of the product. It entails evaluating the effectiveness of forest management in light of a predetermined set of principles and criteria. Forest certification also gives consumers a credible guarantee that the product comes from a forest that is under an environmentally responsible, socially beneficial, and economically viable management regime.

20.3.2 Evolution of Forest Certification

Certification programs have long existed in other economic sectors, such as appliance manufacturing, quality control, and healthcare services. The rise of certification programs in the forestry sector has been led, primarily by the non-governmental actors taking up functions traditionally claimed by the agencies and ministries of nation states. Despite the state predominance in the forestry sector in most countries, forest certification programs did not emerge from the government systems but rather from the markets (markets having a limited role in forest management as compared to other sectors). Instead, they were able to draw upon models and techniques that were developed and standardized for various programs performing similar functions in other sectors. Therefore, forest certification is inherently linked to developments in other sectors.

The genesis of the modern concept of forest certification can be attributed to society's concern for the social and environmental significance of forests, especially in developed countries, furthered by the increased environmental awareness in the 1960s and 1970s (Granholm et al. 1996; Hansen 1997). Because of high societal awareness and concern toward the increased rate of deforestation in tropical forests, the concept of forest certification emerged in the late 1980s as a possible tool to control deforestation and enhance the social and environmental importance of forests (Vogt et al. 2000). The foundation of forest certification has been built as a market-based tool to address public concerns related to deforestation, mainly that of tropical rainforests (Perera and Vlosky 2006).

The 1992 United Nations Conference on Environment and Development (UNCED) can be identified as the watershed event (Finger and Princen 1994) for the environmental movement as a whole and as the point at which certification became a reality. The UNCED 1992 was the first attempt to reach a consensus on forestry issues. Following ECO 92 (UNCED), forestry issues became a part of the international political agenda when efforts were increased to determine how forests should be managed worldwide on a sustainable basis. After the UNCED, the first set of principles, criteria, and indicators (PCI) was released on how SFM should be practiced and evaluated for the certification of forests.

The concept of timber certification emerged in 1990 when a group of stakeholders in California recognized the need for a credible system to identify acceptable sources of forest products (Perera and Vlosky 2006).

In 1990, the Rainforest Alliance launched the Smart Wood Program as a tool to curb deforestation in the tropics. Its premise was that several independent groups would scientifically evaluate and certify forests that were appropriately managed for sustainability and biodiversity. Later, the Forest Stewardship Council (FSC), spearheaded by the World Wide Fund for Nature (WWFN), supported by other environmental non-governmental organizations, social activists, and many governments, was established in 1993, with its head office in Mexico. FSC was formed to prevent forest deterioration globally and to help tropical timber producers avoid environmental boycotts and meet the demand by importing timber from other countries for environmentally sensitive wood products from well-managed forests.

20.3.2.1 Emergence of Certification in America and Europe

The American Tree Farm System (ATFS), established in 1941 to raise public awareness of the active management and investment made in private forests, gave rise to the idea of certification in the United States. The ATFS certification was based on a set of forest management principles and required an on-the-ground inspection of forest practices every 5 years (Bettinger et al. 2009).

In the 1990s, the American Forest and Paper Association (AF&PA), an industry trade group, initiated its own certification system and later added an optional third-party auditing component.

In 1996, the Canadian Standards Association (CSA) developed standards for SFM, which were updated in 2002. The sustainable forestry initiative (SFI), originally an industry self-regulation program, recognized the CSA system as the functional equivalent of the SFI system (Perera and Vlosky 2006). The National Forestry Association (NFA) in the USA developed a system called "Green Tag" for woodland owners (Perera and Vlosky 2006).

In 1999, forest owners in Europe created an alternative to the FSC, naming it the Program for Endorsement of Forest Certification (PEFC) scheme. In 2003, this was renamed as the Program for the Endorsement of Forest Certification Schemes.

The International Organization for Standardization (ISO) 14001 Environment Management Standard is a generic standard for continual improvement in environmental performance. However, forestry operations can use its environmental management system framework as a foundation to implement a forest management standard (Hansen et al. 2006). This framework was used only by the Canadian Standards Association before its adaptation to the PEFC standard. National forest certification schemes were started in many countries, including Europe, Malaysia, Brazil, Chile, and, of late, China and Africa.

20.4 Relevance/Rationale of Forest Certification

The international consensus on better management of forests and forest resources for their sustainability and enhancement of related social and ecological benefits paved the way toward a global commitment for SFM. As a logical next step, forest certification incentivizes forest managers for practicing SFM by linking products from such forests to environmentally conscious consumers.

20.4.1 Objectives and Benefits of Forest Certification

Important questions that forest managers always ask themselves when considering certification are whether or not to get the forest certified at all and how they benefit if they achieve forest certification. Certification is the process of verifying that forest management meets the requirements of a responsible forest management standard.

Forest certification can provide different benefits to multiple stakeholders. More access to international markets; enhanced planning and management practices that incorporate conservation measures like biodiversity conservation, protection of riparian zones, water catchments, and wildlife; adoption of reduced impact logging; and the development of monitoring capacity are just a few advantages that come with certification for the forest products industry.

For forest communities, forest certification ensures their share of benefits because of enhanced external scrutiny that results in better transparency.

Table 20.2 Objectives and benefits of forest certification

Type of objective/benefit	Objective/benefit	Stakeholder(s)
Regulatory	Condition for the allocation of forest resources/management rights	Governments and business
	Reduce the risk of environmental non-compliance	Business and value chain members
	Source-of-origin and legality	Governments, business, and value chain members
Market	Consumers receive certified products	Consumers and value chain members
	Certification as a means to access new markets	Value chain members
	Credibility of sustainability claim	Consumers, business, and value chain members
	Promote responsible resource sourcing	Business and value chain members
	Resource sustainability/availability in the long run	Business, value chain members, consumers, and society
Social	Community rights assured	Local community
	Fair share of benefits to local communities	Local community
	Improvement in work conditions: EHS	Workers
	Equity	Intragenerational and intergenerational
	Maintenance and concern of social, cultural, traditional, and spiritual values of the communities related to forest	Local community
	Using traditional ecological knowledge in decision-making processes	Business, value chain members, and local community
Economic	Tool to achieve management goals	Forest owners and shareholders
	Control over resources	Forest owners and shareholders
	Demonstrate/benchmark best practices	Forest owners and shareholders
	Maintenance and enhancement of forest resource productivity	Forest owners and value chain members
	Sustainable production and consumption	Consumers, governments, society, value chain members, and planet earth/ecosystem
	Price premium	Forest owners, business, and local community
Ecological/environmental	Environmental health and ecological values conserved through holistic development	Planet earth/ecosystem, intergenerational equity
	Sustainable resource extraction and use	Planet earth/ecosystem, intergenerational equity, business, value chain members, and consumers
	Maintenance, conservation, and enhancement of biodiversity	Planet earth/ecosystem, intergenerational equity, and global community
	Maintenance and enhancement of ecosystem function and vitality	Forest owners, planet earth/ecosystem, intergenerational equity, and global community
	Maintenance and conservation of soil and water resources	Forest owners, planet earth/ecosystem, intergenerational equity, and global community

Researchers have made efforts to identify the most common reasons why forest managers may decide to pursue certification. Some of the benefits of forest certification are summarized in Table 20.2.

Each of the above reasons will have different implications when selecting a particular certification scheme. If customers are demanding a specific type of certification, there is little point in achieving forest certification under a different scheme.

Certification has the potential to benefit forest owners economically by improving the return on investment of their goods and assuring customers that the raw materials derived from the forest adhere to management-specific standards (Washburn and Block 2001). Thus, certification is an important market tool that benefits both the timber industry and the consumers. Timber certification schemes offer purchasers the credibility that the origins, processing, and distribution of the timber they are purchasing adhere to a set of predetermined management standards.

Certification of timber can be an application of market tools for achieving widespread SFM. The usual motivation for a producer or supplier is continued access to markets for timber and wood products, which is partially provided by the certification (Hansen et al. 2006). Certification is also a tool to improve forest harvesting practices and forest management around the world. The philosophy is that the certification label will result in an increased market share for forest owners (Vogt et al. 2000).

Certification also provides a mechanism for independent validation of sustainability in markets where forest products are subject to consumer resistance on environmental grounds. Tropical forests were the initial focus of certification, but as the processes rapidly expanded, this resulted in the inclusion of all forest types in the domain of forest certification (Durst et al. 2006).

Forest certification, of late, has been used as a trade tool to stop illicit logging. It was estimated that more than 100 million m^3 of timber are cut illegally each year worldwide (Lawson and McFaul 2010). Illegal logging and the associated trade in illegally sourced wood products are important causes of deforestation and forest degradation all over the world. Forest destruction contributes up to 20% of the global carbon dioxide emissions. It is further reported that in addition to undermining government revenue, illicit logging, and its related industries also encourage corruption, jeopardize the rule of law, and fund armed conflicts (Lawson and McFaul 2010).

Although forest certification emerged as a private sector-based voluntary initiative to address deforestation, later different types of initiatives evolved to combat illegal wood trade. van Dam and Savenije (2011) listed about 122 initiatives, a few of which are listed below (the number indicates the total number of initiatives): forest law, enforcement and governance, forest law enforcement, and governance and trade (8); knowledge development and capacity-building (14); bilateral initiatives (12); global and multilateral agreements and declarations (19); public procurement policies (17); national policies and legislation (5); regional and country support facilities (3); NGOs for advocacy (2), etc.

20.4.2 Types of Forest Certification Schemes

As of date, there are a handful of international forest certification schemes and many regional and national certification schemes setting up standards for measuring practices toward sustainable management of forests. There is also growing competition among various certification schemes to get more area of forest certification under them.

Globally, two forest certification systems are in practice. One is the Program for Endorsement of Forest Certification (PEFC), which mutually recognizes the National Forest Certification Schemes that aim to harmonize the global forest certification standards with those of national forest certification standards.

Phases in developing a PEFC-endorsed national forest certification system are as follows: Phase 1: your organization; Phase 2: developing your system; Phase 3: PEFC council membership; Phase 4: PEFC endorsement; and Phase 5: running your system.

A national entity has to first prepare itself to meet the requirements of PEFC recognition, then it must develop a national standard for forest management and chain-of-custody (CoC) certification, a certification process, an accreditation process for certification bodies (CBs), seek membership in the PEFC council, and then apply for mutual recognition of its certification scheme. The PEFC council will then conduct a peer review of the national forest certification scheme with respect to meeting the requirements of PEFC mutual recognition before taking a decision for mutual recognition.

The second global certification system is the Forest Stewardship Council's (FSC) standards for forestry practices that aim to promote environmentally responsible and socially beneficial forestry. In this process, forest management practices are evaluated and verified against a set of standards by an external independent certification agency. Thus, FSC certification provides confidence to environmentally conscious consumers that the forest products they are buying have been sourced from sustainably managed forest areas with documented proof of sustainability.[1] Whereas in some cases, when combined with a chain-of-custody certificate, the certification allows the establishment of traceability that the product was derived from a particular certified forest area and carries an eco-label.

The FSC has a set of 10 principles, 70 criteria, and several verifiers within each criterion for certifying forests. FSC certification is one of the most widely accepted certification schemes, which is mostly favored by environmental NGOs but is probably less popular with industry (Bettinger et al. 2009). The unique concept and part of the certification plan is to effectively use consumer purchasing of forest products as the catalyst for responsible forest management around the world and, consequently, harness market forces to maintain a healthy forest with little or no need for regulation.

20.4.3 Forest Certification Process

Over the years, two main policy approaches, namely, top-down and bottom-up, have been adopted to protect forest resources. In the top-down approach, the fundamentals of policies are formulated at higher levels of government and implemented under its authority. The success of these command and control methods heavily depends on the strength of the governing body. The bottom-up approach, on the other hand, relies more on a participatory approach where the people/local community agrees on the needs and forms of the policy and implements it by tradition, cooperative agreement, or local rule. However, in modern, complex societies, common interests binding the members of smaller communities are lacking, which hinders the success of this approach. The ineffectiveness and flaws of these two approaches have led to the development of a third approach

[1] Details available at www.forestry.about.com/od/certification/Forest_Certification.html

Table 20.3 Classification of forest products certification

Product certification type	Criteria/assurance to the customer	
	Properties	Characteristics
Product quality	Quality of timber/NTFP products	Appearance grading
	Mechanical properties of timber	Stress grading
Organic in process	Ensures the consumer that the product is herbal/organic in nature and does not contain harmful chemicals	No chemical fertilizers or pesticides were used. Only organic manure was used
Fair labor enforcement	Assures that no child labor was involved in the production process and no exploitative labor practices were followed	Fair treatment of workers
Legally sourced	Assures that the timber/NTFP has come from a known and legal source	Known and legal sourcing
Fair trade practices	All the responsible trade practices have been incorporated	Fair share of benefit to the producers/ collectors
Sustainably managed	Assurance that the timber came from a forest that has been managed on sustainability principles (principles, criteria, and indicators)	The certified product is extracted/harvested based on sustainability principles/standards

called "certification." It introduces policy changes through commercial rather than central or local power and uses market acceptance in place of regulatory compliance as an enforcement mechanism (Naka et al. 2000).

Certification of forest products can be categorized based on many aspects, such as principles, criteria, indicators, purpose, and assurance to customers. Some of these are illustrated in Table 20.3. However, in this chapter, certification of forest implies certification with respect to conformance to SFM practices as per the selected standard/scheme.

Forest management certification is a process by which forest management practices are evaluated against a set of forest management standards. The term "forest certification" is now commonly understood as an "independent verification" (Mater 1999; SAF 1999) of conformity to established standards of SFM, generally done by a third party. It is a market tool designed to document and reward specific forest management practices and to assure consumers of forest products that their purchase comes from a forest whose management meets certain standards.

Forest certification is thus a mechanism for ensuring that forests are managed by following the universally agreed principles of SFM. In this process, forest management practices are evaluated and verified against a set of standards by an external, independent certification organization. Thus, consumers can be confident that their wood purchase is the product of a sustainably managed forest with documented proof of sustainability. In some cases, when combined with a chain-of-custody (CoC) certificate, certification allows products from a particular certified forest area to carry an eco-label on it.

The verification process is conducted through an audit system directed by an external and internal forest management team. Internal auditing is carried out to obtain assurance that the forest management unit fulfills the minimum requirement, which can be then attested by an independent third-party external auditor.

This process ensures the "sustainability of forests" and may induce the market to offer a "green premium," which in turn invigorates the sustainability of forests that are managed under this system. Theoretically, certified forests would be environmentally, socially, and economically adjudged better. In addition, they would also be able to yield timber products on a sustained basis while maintaining adequate wildlife habitats and protecting water quality. Forest certification claims to maintain the long-term viability of commercial forests, protect biodiversity, and provide a continuous stream of social and economic benefits. Thus, forest certification provides economic incentives to achieve and maintain high standards for forest management worldwide.

20.4.4 Understanding the Process of Forest Certification

The process of certification starts with the choice of forest enterprise's owners/managers to receive certification from a preferred certification system/scheme, which concludes with the issuing of a certificate of verification that the management practices followed in the enterprise are reliable and in compliance with the requirements of the selected standard. The forest managers' decision to seek certification is based on one or a combination of objectives/benefits as mentioned in Table 20.2, and accordingly, the certification scheme is selected.

In countries where a national forest certification system exists, the choice could be in its favor. However, in countries where there is no such national scheme, the default choice becomes the Forest Stewardship Council (FSC), which is globally recognized and has a presence in most countries, including India.

In countries where there is a national scheme, which is also mutually recognized by the Program for Endorsement of Forest Certification Scheme, the choice expands to three schemes, that is, a national scheme, PEFC, and/or FSC. This is the case in countries, such as Malaysia, Brazil, Chile, the USA, China (country schemes), Canada, and European countries that have their own national certification schemes, which are mutually recognized by the PEFC.

Having made a decision for certification, the forest manager prepares/adapts the documentation, including the management plan, as per the requirements of the selected scheme or appoints consultants to help develop these documents. Thereafter, the forest manager appoints a certification body out of the available certification bodies that are accredited to the certification scheme of choice. For example, the FSC has accredited 52 certification bodies (CBs) that can undertake audits of the forest management practices of the client organization.

20.4.5 Elements of Forest Certification

Forest certification must consist of following elements: (1) standard (acceptable levels of forest management performance); (2) certification process: assessment and audit (objective assessments of whether the requirements of the standard are fulfilled for FM Certification and/or CoC (identifying a product's origin and tracking it across the value chains); (3) accreditation of certification body; and (4) issuance of certification and label permissions (a labeling process that indicates the previous three elements are met satisfactorily).

Standard: Any forest certification scheme needs to develop a standard that lays out the level of management, which must be implemented by the forest enterprise to be eligible to achieve forest certification.

Certification process: A clearly defined procedure must be followed by the certification body to assess whether a particular enterprise has met the requirements of the certification standard.

Accreditation: There must be a process in place to check the certification body in terms of experience, competence, and system of due diligence in its operations. The accreditation agency must have the authority to derecognize/stop incompetent organizations from awarding certifications.

Certification and labeling: On successful completion of certification assessment with respect to meeting the requirements of the standard, the certification body issues certification to the client organization. Once certified, the client organization can obtain the labeling permission for the certified material as per its need.

FSC does not carry out certification audits and issue certificates itself; rather, its independent certification bodies carry out the forest management and chain of custody assessments that lead to FSC certification. FSC certification bodies are held to account by the FSC through Assurance Services International (ASI), which is an accreditation body for certification bodies.

FSC sets the standards for forest management and chain of custody certification and defines procedures that the certification bodies should follow in their certification assessments through a multi-stakeholder consultative process.

FSC-accredited certification bodies are checked regularly by Assurance Services International to make sure they operate in accordance with the FSC rules and must meet FSC accreditation requirements (FSC website).

The FSC-accredited certification body verifies whether forest resources are managed on a sustainable basis as per the requirements of the FSC standard. In this assessment, the certifiers should give due importance to the overall health and vitality of the entire forest ecosystem, which includes adequate wildlife habitat and watershed protection, and the economic and cultural impacts of the forestry operations on local communities.

20.4.6 Types/Stages of Forest Certification

The certification of forests has two sequential stages that must be completed before the market benefit can be obtained through the use of a label on the finished product. The first stage is the certification of the forests or plantations against the standards of sustainability of forests (principles, criteria, and indicators) and verifiers as prescribed by the certification scheme. For instance, FSC has 10 principles and 70 criteria, whereas PEFC has 6 criteria and 66 indicators. When the first-stage certification confirms that forests are managed sustainably, it is called "Forest Management" certification, or in short, FM certification.

Verifying the path of wood material from the point of harvest or logging to the point of finished product sale at the retail level is the second step, called the chain of certification. This is important as wood products pass from the stump to the processing unit, furniture assembly unit, wholesaler, retailer, and finally to the customer. Therefore, the customer must be assured that the wood-based products purchased indeed are made from the wood material that came from a certified forest. When the second stage of certification confirms that forest products are made with wood harvested from sustainably managed forests, it is termed chain-of-custody or CoC certification. The issuance of FM certification is a prerequisite for CoC certification. CoC is therefore a method of tracing back the source of materials from the forest, all the way from the production process to the final product. It is also a means of controlling the use of labels and claims made to ensure that they are clear and accurate. Combined, these are called "forest certification," and both certifications are needed to receive the market benefit from certification with the legitimate use of the certification scheme "logo" on the finished products.

A related but equally significant component of forest certification is the authorization of "product labeling and logo use." In addition to having received the above two stages of certifications (FM and CoC), the business entities desirous of market/brand benefits need to seek explicit permissions for product labeling as being a certified product and use of the logo of the certification scheme on the products, packaging, stationery, and any promotional material including advertising. Such permissions are granted for a specific purpose(s) and their use requires adherence to the prescribed guidelines of the certification scheme.

Controlled wood certification is the certification of assessment of various kinds of risks in sourcing of wood raw material by a forest enterprise. These risks include legality, source-of-origin, labor laws, business risks, and adherence to the rules and regulations of the country. However, it does not require adherence to sustainable forest management principles, criteria, and indicators. Thus, SFM is not a prerequisite for CS certification.

20.4.7 Steps in Forest Management Certification

Forest certification is applied at the operational forest management level. It is carried out for a particular forest management unit (FMU), which is defined as an area of forest under a single or common system of forest management. This might be a privately held area of forestland or plantation or a series of small areas of forestland/plantation owned by different farmers/owners but managed under a common system through a legal agreement.

A typical FSC certification process involves working through the steps of pre-assessment/scoping, forest manager's decision to proceed, closing the gaps, stakeholder's consultation, development of an interim standard, main assessment, reporting and peer review, and certification decision. Following a positive certification decision, a certificate is issued and annual surveillance visits are undertaken to verify continued compliance with the standard. The certification once issued is valid for a period of 5 years with an annual review for compliance. Figure 20.1 provides the sequence of stages in the forest certification process.

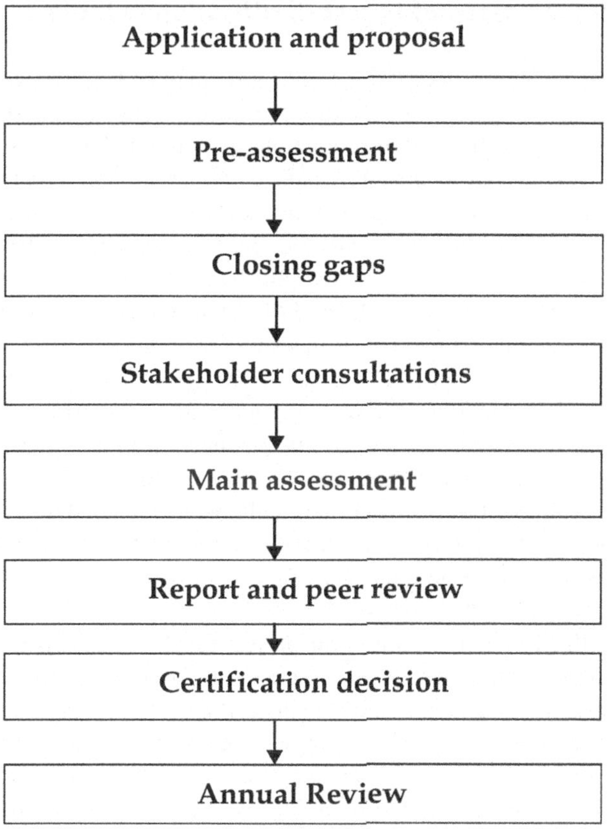

Fig. 20.1 Outline of FSC certification process

Other forest certification schemes also use similar process. However, differences may occur on account of the country context and the type of value chain partners.

20.5 Status of Forest Certification

20.5.1 Global Status of Forest Certification

As of January 2024, around 10.7% of the total forest area reported certified globally. Of which, FSC has certified 157.92 million ha with 1510 FM/CoC certification and 57951 CoC certification. Whereas, 280 million ha of forest area has been certified under the Program for the Endorsement of Forest Certification (PEFC), which includes certifications done by the national certification schemes endorsed by the PEFC.

FSC with around 30 years of experience has a presence in 89 countries; whereas, the PEFC has 55 counties as national members. The share of Forest Certification by FSC is around 36.06% and PEFC accounts for 64.94% share in the total certified area on the globe as of January 2024.

20.5.2 Status of Forest Certification in India

The development of forest certification in India has been very erratic. It started with inquiries from the handicraft export sector led by the Export Promotion Council of India, Ministry of Textiles, Government of India in the year 2001. The first certification in India was an FSC-CoC certification, issued in January 2001 to a toy manufacturer in Saharanpur, UP for Babul (*Acacia nilotica*) and Shisham (*Dalbergia sissoo*) trees. However, there was no FM certification for this CoC. The second certification in India was for both FM and CoC certification under FSC issued in 2002 for Kadam (*Ailanthus grandis*) plantations in West Bengal (432 acres). However, both these certificates were discontinued within 1 year of issuance for unknown reasons (Yadav 2016). The first successful FSC FM certification in India was concluded in 2007 for a rubber (*Hevea brasiliensis*) plantation in a 687.85 ha area in Tamil Nadu, namely, New Ambadi Estates Pvt. Ltd. (IMO-FM/COC-026594).

IIFM undertook a pioneer project for the certification of bamboo resources in the States of Tripura, Arunachal Pradesh, Nagaland, and Madhya Pradesh. In Tripura, a large number of farmers in Katlamara block have been practicing bamboo (*Bambusa affinis*) cultivation using a traditional management system. This was documented by IIFM in the form of a traditional but sustainable management plan in line with international certification standards (such as FSC) and a complete value chain was designed to link this sustainably managed bamboo resource-based handicraft products to environment-friendly consumer in the developed markets in the year 2003. A similar sustainable bamboo management plan was prepared for Apatani bamboo in Arunachal Pradesh. However, these efforts could not result in the achievement of FSC certification in the absence of budgetary support.

As of January 2024, there are 12 valid FSC forest management certificates in India (Table 20.4), of which 6 certificates are in the paper industry and 5 in plywood and woodcraft industry. All these certificates are in the agroforestry/farm-forestry owned by the individual farmers. Whereas, Telangana State Forest Development Corporation owns one valid FM/CoC. The total certified area under FSC FM certification in India is 111918.5 ha. Also, there is one PEFC FM certification owned by the UP Forest Corporation.

20.6 NTFP Certification

Non-timber forest products (NTFPs) also known as minor forest produce (MFPs) in central India are important resources for the forest-dwelling communities for meeting their health, socio-cultural-spiritual, and economic needs. The NTFPs not only provide sustenance to a large community (Yadav and Basera 2013) but also offer opportunities for the NTFP-based enterprises (Subedi 2006), contribute to foreign exchange earnings (Shiva and Verma 2002), provide alternatives for livelihood (Fu et al. 2009), enhance human capital (Fisher et al. 2010), and influence forest policy (McSweeny 2005). However, the burgeoning population and ever-increasing pressure on forests for goods and services are resulting in the deteriorated condition of forests in general and on availability of non-timber forest produces (NTFPs) in particular.

Forest certification of NTFPs can bring about a transformational positive change in ensuring their sustainable

Table 20.4 Status of FM certifications in India (as of January 2024)

S. no.	Certificate holder	Certificate number and year	Area (ha)
FSC			
1	Society for Afforestation, Research and Allied Works (SARA)-Dandeli (Karnataka)	SCS-FM/COC-00143P-2021	2655.92
2	ITC Ltd. PSPD Unit (AP)	SCS-FM/COC-006945-2021	40,111.31
3	International Paper-APPM Ltd. (Telangana)	SA-FM/COC-010151-2021	26,256
4	JK Paper Ltd. Rayagada (Odisha)	SCS-FM/COC-005124-2021	16,375
5	JK Paper Ltd. Unit-CPM (Gujarat)	SCS-FM/COC-007694-2021	3454
6	Tamil Nadu Newsprint and Papers Limited (TN)	NC-FM/COC-006000-2021	2459.28
7	Shivangi Industries (Rajasthan)	SCS-FM/COC-008105-2021	1002
8	Greenply Industries Limited (Nagaland)	SCS-FM/COC-007306-2021	605
9	Pooja wood House	SCS-FM/COC-009232-2022	1000
10	Telangana State Forest Development Corporation	CU-FM/CoC-885729-2023	18,000
11	Reyansh The Wood Art	SCS-FM/COC-009897-2023	
12	Saraf Lumbers Pvt. Ltd	SCS-FM/COC-010015	
		Sub-total	111,918.5
PEFC			
1	Uttar Pradesh Forest Corporation (Uttar Pradesh)	GIPL-FM-0001-2020	404,104
		Sub-total	404,104
		Grand total	516,022.51

utilization (and thus long-term availability), enhanced market opportunities, and sustainable value chains for businesses. This can create a win-win situation for poverty alleviation as well as for biodiversity conservation (Ahenkan and Boon 2010; FAO 2006; Golam et al. 2008).

The traceability of NTFPs from the source of collection (region/local) to final use in product manufacturing is difficult, thus posing serious challenges to establishing the source of origin and hence the credibility of the genuineness of the material.

The forest certification framework provides an opportunity to establish a mechanism for traceability of the NTFPs. Moreover, it addresses the environmental concerns of consumers buying products coming from sustainably managed forests and the ability to support rural livelihoods while contributing to environmental objectives, including biodiversity conservation (Yadav and Dugaya 2013).

The harvesting practices of non-timber forest products have also come under increasing scrutiny from certification programs because of the key role that they play in the local economy and sustainable management of forest resources worldwide.

Forest certification, hence, is understood as a means of protecting forests by promoting responsible and better management practices. It provides a third-party assurance that the forestry operations meet standards set by the certification scheme.

20.6.1 NTFP Certification: Global Status

In addition to the Forest Stewardship Council, the International Federation of Organic Agriculture Movements (IFOAM) and Fairtrade Labeling Organization (FLO) also play significant roles in ensuring market access, sustainability of NTFP resources, and the livelihoods of the local communities dependent on NTFP resources. These organizations have a good degree of experience and capacity in certification of NWFPs, including chicle latex, Brazil nuts, palm hearts, and maple syrup (FSC); berries, tea, honey, coffee, mushrooms, ginseng, and others (IFOAM); coffee, tea, honey, bananas, cocoa (FLO).

20.7 Conclusion

Worldwide, forest certification has become an important marketing tool for various stakeholders involved in the management, conservation, and exploitation of forest resources. SFM is a prerequisite for achieving forest management certification irrespective of the type of forest certification scheme. However, controlled wood certification under FSC does not require SFM principles, but it assesses various kinds of risks associated with the sourcing of raw wood material by the forest enterprise. FM certification for timber provides a credible system of verification of origin (and thus the legality of the resource) as well as the commitment of all stakeholders in responsible forestry. For governments and civil society, it is an assurance of sustainability and legal logging operations. For farmers and small businesses, it provides access to environmentally sensitive markets. The progress in recent years with enhanced areas under certification and involvement of increasing businesses in terms of CoC certificate holders is a testimony to the acceptance of this new market mechanism for ascertaining the sustainability of forest/wood resources. Initially, the concept of forest certification was popular only in the USA and Europe, but of late increasingly expanding in tropical forests and developing countries. Forest certification to receive market incentives must complete the two components: FM and CoC certifications. While CoC certification serves as evidence to green customers that only certified materials were used in the production of the final products they have purchased, FM certification is the attestation of forest management practices against the certification standards. The use of the logo of a certification system adds to the brand value of the company for it being environmentally friendly. The credibility of the certification scheme lies in the development of standards through a consultative process, the competence and due accreditation of the certification body/auditors, transparency in the procedure followed, and stakeholder consultation. Forest certification of NTFPs can enhance their market access, ensure fair prices to the forest-dwelling communities, and the conservation of resources along with enhanced credibility of the product/brand for the consumers and the marketers.

Lessons Learnt

- Understanding the role of forest certification in promoting sustainable forest management
- Grasping the forest certification concept as an environmental label or eco-label
- Learning the benefits of forest certification to various stakeholders in forest product value chains
- Learning the requirements and process to achieve forest certification
- Learning the importance of forest certification for non-timber forest products (NTFP)

Key Questions

1. How forest certification helps in promoting sustainable forest management?
2. Why forest certification is termed as an environmental or eco-label?
3. Explain briefly the crucial elements of the forest certification process.

4. Which are the global forest certification schemes and how do they differ from each other?
5. What are the benefits of forest certification for non-timber forest products (NTFP)?

References

Ahenkan A, Boon E (2010) Commercialization of non-timber forest products in Ghana: processing, packaging and marketing. J Food Agric Environ 8(2):962–969

Bettinger B, Boston K, Siry JK, Grebner DL (2009) Forest management and planning. Academic Press

Desai BH (2011) Forests, international protection. In: Wolfrum R (ed) Max Planck encyclopedia of public international law. OUP Oxford. http://www.mpepil.com/

Durst PB, McKenzie PJ, Brown CL, Appanah S (2006) Challenges facing certification and eco-labelling of forest products in developing countries. Int For Rev 8(2):193–200

FAO (2006) Responsible management of planted forests: voluntary guidelines. Planted forests and trees working paper 37/E. FAO, Rome

FAO (2020) Global forest assessment report. Food and Agriculture Organization, Rome

Finger M, Princen T (1994) Environmental NGOs in world politics: linking the global and the local. Routledge

Fisher M, Chaudhury M, McCusker B (2010) Do forests help rural households adapt to climate variability? Evidence from Southern Malawi. World Dev 38(9):1241–1250

Fu Y, Chen J, Guo H, Chen A, Cui J, Hu H (2009) The role of NWFPs during agro ecosystem shift in Xishuangbanna, South-western China. For Policy Econ 11:18–25

Ghazali BH, Simula M (1994) Certification: who will pay the price. ITTO Trop For Update 4(5):3–5

Golam R, Madhav K, Ram PS (2008) The role of NTFP in poverty reduction in India: prospects and problems. Dev Pract 6:779–788

Granholm H, Vahanen T, Sahlberg S (1996) Background document: intergovernmental seminar on criteria and indicators for sustainable forest management. Ministry of Agriculture and Forestry, Helsinki

Hansen E (1997) Forest certification and its role in marketing strategy. For Prod J 47(3):16–22

Hansen E, Fisher R, Cashore B, McDermott C (2006) Forest certification in North America. Oregon State University Extension Service, Corvallis, OR

IIFM (2008) Manual: criteria and indicators for sustainable forest management in India - 2008. IIFM, pp 1–108

Lawson S, McFaul L (2010) Illegal logging and related trade: indicators of the global response. Chatham House, London

Mater CM (1999) Understanding Forest Certification: answers to key questions. Pinchot Institute for Conservation MEA (Millennium Ecosystem Assessment), Washington, DC

McCormick J (1989) Reclaiming paradise: the global environmental movement. Belhaven Press, London

McSweeny K (2005) Natural insurance, forest access and compounded misfortune: forest resources in smallholder coping strategies before and after hurricane Mitch, North-eastern Honduras. World Dev 33: 1453–1471

Naka K, Hammet AL, Stuart WB (2000) Forest certification: stakeholders' constraints and effects. Local Environ 5(4):475–481

Perera P, Vlosky RP (2006) A history of forest certification. Louisiana Forest Products Development Centre working paper no. 71. Louisiana State University, Baton Rouge, LA

Pradère A (1989) French furniture makers: the art of the Ébéniste from Louis XIV to the revolution. The J. Paul Getty Museum

Rametsteiner E (2000) Sustainable forest management certification. Frame conditions, system design and impact assessment. Ministerial Conference on the Protection of Forests in the Europe, European Commission

SAF (1999) Task force on forest management certification programs. Society of American Foresters. http://www.safnet.org/policy/fmcp1999.html

Shabecoff A (1993) Community job initiatives: readiness, training, creation and retention. Strategy Alert

Shiva MP, Verma SK (2002) Approaches to SFM and biodiversity conservation: with pivotal role of NWFPs, 1st edn. International Book Distributors, Dehradun

Sitarz D (1998) Sustainable America: America's environment, economy and society in the 21st century. Nova Publishing Company

Stefanie L, Bernhard W, Fady A, Bridge SRJ, David G, Vicente G, Mostafa J, Steven J, Pablo L, Guy R (2018) 25 Years of criteria and indicators for sustainable forest management: why some intergovernmental C&I processes flourished while others faded. Forests 9(515):1–23. https://www.researchgate.net/publication/327230923

Subedi BP (2006) Linking plant-based enterprises and local communities to biodiversity conservation in Nepal Himalaya. Akhil Book Distributors, Nepal

UN (2019) Global forest goals and targets of the UN strategic plan for forests 2030. United Nations, Department of Economics and Social Affairs

van Dam JV, Savenije H (2011) Enhancing the trade of legally produced timber. A guide to initiatives. Tropenbos International, Wageningen

Vogt KA, Larson BC, Gordon J, Vogt DJ, Fanzeres A (eds) (2000) Forest certification: roots, issues, challenges, and benefits. CRC Press, Boca Raton, FL, pp 1–10

Washburn MP, Block NE (2001) Comparing forest management certification systems and the Montreal Process criteria and indicators. Pinchot Institute, Washington, DC. http://www.pinchot.org/publications/discussion_papers.htm

Yadav M, Basera K (2013) Status of forest products production and trade. IIFMWP-13-10-01. Working paper. Indian Institute of Forest Management, Bhopal. http://www.iifm.ac.in/node/21

Yadav M, Dugaya D (2013) Non-timber forest products certification in India: opportunities and challenges. Environ Dev Sustain 15:567–586. https://doi.org/10.1007/s10668-012-9393-1

Yadav M (2016) Handbook on forest certification. TERI, New Delhi

Forest Genetic Resources: Conservation, Management and Utilization

21

Bhavani Gurudev Singh

Abstract

India being a tropical country harbours rich biodiversity and has ample forest genetic resources. These genetic resources need to be utilized for the betterment of humankind and also need to be conserved for posterity. A majority of Indian household depends on agrarian economy; however, in the recent years, the development of forest-based industries has contributed substantially to the farm income. The overall increase in income levels has fuelled demand for wood and wood-based products. The boom in construction industry has necessitated increased import of timber. Likewise, the paper and pulp industry, ply wood, matchwood, pencil-making, incense sticks and ayurvedic industry require a large volume of wood and other forest produce. In this situation, sustainable utilization of forest genetic resources and their conservation is of paramount importance. This chapter discusses about status and utilization of resources for various end uses, the need and strategy for conservation by involving various stakeholders.

Keywords

Genetic resource · Conservation · Seed bank · Disjunct distribution · Endemic species

21.1 Introduction

Forest genetic resources (FGR) are important assets of a country, contributing to the welfare of its people and economy. Apart from providing the basic needs of the local communities, they aid in agricultural and industrial growth. Unlike the agricultural crop genetic resources, the FGR did not get the required attention in the past. However, there is increased concern on the FGR in the recent years around the globe. The vastness of FGRs particularly in tropical countries made it difficult to comprehend the method and modes to handle them.

The global forest area accounts for 3870 million ha of which nearly 548 million ha is found in Asia. India ranks third among Asian countries and foremost among the South Asian countries in having the largest area under forests (64 million ha) with 21.6% of land area under forest cover. India with 2.5% of the world's land surface and 1.8% of the world's forest area holds a wide array of biodiversity within the forest ecosystem. Forest areas hold 15,000 species of higher plants of which 33% (i.e. 4900) species is endemic (FSI 2009). The country has 10 biogeographic regions and 15 agroclimatic zones. About 60% of the country's forest is located in ecologically sensitive zones like the Western Ghats and Himalayas.

The contribution of FGR to welfare of humankind was never assessed and quantified because of complexities involved. They have contributed substantially to the growth of indigenous health traditions like Ayurveda, timber production, paper industry apart from innumerable small-scale industries such as wood crafts, toy making, agarbathi (incense) industry, essential oils, rayon, silk, basket making, cosmetics and pesticides. In the past, indiscriminate and unsustainable harvest has resulted in dwindling of these resources. On the contrary, some of the FGRs are yet to be put into the optimum utilization. Rapid reduction of forest cover around the globe coupled with anthropogenic pressure has necessitated the initiation of regional and international network on FGR.

Realizing the importance of these resources, the local communities followed many traditional ways like reserving forest patches as sacred groves. In the recent decades, the *Chipko* movement in Uttarakhand (Shiva 1988), The Joint Forest Management (JFM) in West Bengal and Orissa (Saxena 1999) and the shelving of Silent Valley Hydroelectric Project in Kerala are some of the examples of people's

B. Gurudev Singh (✉)
ICFRE-Institute of Forest Genetics and Tree Breeding, Coimbatore, India

Mysore, India

A. K. Mandal, A. Nicodemus (eds.), *Textbook of Forest Science*, https://doi.org/10.1007/978-981-97-8289-5_21

involvement in conservation of forest and natural resources. The concept of 'conservation through utilization' is getting importance in the recent times. *Gloriosa superba*, a medicinal plant which was considered to be rare in nature three decades ago, is successfully conserved by large-scale cultivation involving local farming communities in the state of Tamil Nādu. There are efforts on conservation and sustainable utilization of FGR at global level by regional and international cooperation. The global database on FGR by food and agriculture organization (FAO) through a number of networks, created in recent decades to address issues related to FGR, is an important step in right direction.

21.2 Characteristics of Tree Genetic Resources

Forest genetic resources have been defined as heritable materials maintained within and among tree and other woody species that are of actual or potential economic, environmental, scientific or societal value. Forest trees exhibit a large amount of variability due to their unique characteristics. Some of the characteristics are:

- Long lifespan making them to expose to changing environmental conditions
- Mostly out breeders
- High fecundity, produce large quantities of seeds every year
- Their existence in nature for long period in geological timescale resulting in adaptation to varied climatic conditions

Further, trees with a wide distribution range (e.g. teak and neem) exhibit a high amount of variability because of differences in edaphic and climatic factors. Since most of the trees are outbreeding in nature, each standing tree is genetically different than others. They exhibit variation in adaptation, resistance to disease and insect attack, form, flowering, fruiting, germination etc. It is prudent to focus on conservation of only those FGR that exhibit variability which favour selection for characteristic such as resistance to disease and pests, higher productivity etc. The conservation of FGR requires strategies which are entirely different from agricultural crops because of the characteristic differences between these two crops (Table 21.1).

21.3 Prioritization, Characterization and Documentation of FGR

Considering the vast number of forest tree species and their wide distribution range harbouring an even larger number of populations, the critical issues in gene conservation planning are (1) how best to identify species to be included in the priority list and (2) how to select the populations to be conserved (Graudal et al. 2001). The following are a few different but complementary criteria that can be considered in setting priorities:

- Current local importance, even if the species currently contribute only a little to the economy.
- Economic value, such as a species that may be of major importance for the subsistence of local populations in rural areas.
- Ecological and geographic considerations, such as status in typical stands, geographic range and capacity for natural regeneration.
- The current conservation status of species and their populations may be used as an indicator for prioritization.

Table 21.1 Differences between genetic resources of crop plants and trees

Feature	Crop plant	Trees
Duration	Annual/biennial	Perennial, several years for a cycle
Periodicity	Complete life cycle in three to 6 months—Grow, produce flowers and fruits and end their life	Leaf shedding, flowering, fruiting is influenced by annual rains and photo and thermocycles—An annual phenomenon
Pollination	Mostly self-pollinated/cross-pollinated. Pollinated by wind and insect visitors	Generally cross pollinated, self-incompatible
Seed setting	Prolific seed producers	Shy seed producers as fruit, nut, capsules—Not domesticated for seed production
Dormancy	Many do not have dormancy. Hard seed is rare	Dormancy is common; pretreatment is needed
Propagation	Mainly by true seed	By seed, cuttings, grafts
Genetics	Inheritance of most of the economically important characters is well understood	Inadequate knowledge, 2n number unclear
Breeding	Traits controlled by single gene	Wood traits mostly controlled by many genes
Seed production	Every year affair	Seed production starts after reaching the maturity phase, which may be 7–8 years. Once orchards are established, seed production is a routine affair
Conservation	Predominantly ex situ	It is in situ supported by ex situ

(Modified after Nagarajan 2013)

Several thorough priority-setting approaches have been developed and described (Franzel et al. 1996; Yanchuk and Lester 1996; Koshy et al. 2002), but any particular set of criteria that are ultimately chosen to rank species for conservation activities need to consider local conditions and factors. Priority-setting processes can, of course, consider factors as equally important, or differentiate them by weighting the criteria. The net results of the priority analysis and weighting system ultimately used must make logical sense. Based on these considerations, countries have prepared the priority list of trees taken up for conservation. Forty-eight tree species have been enlisted as priority species for India (Sharma et al. 2002). Country report on the status of FGR in India enlists 272 species as priority species for the country (Anonymous 2012).

The characterization of forest genetic resources depends critically on the contributions of three scientific disciplines. Taxonomy provides the reference system and depicts the pattern of diverse range of species. Genetics gives a direct knowledge of the genetic variation found within and between species. Ecology provides knowledge of the varied ecological systems in which taxonomic and genetic diversity is located and of which it provides the functional components.

Assessing the genetic variation is fundamental for developing strategies for conservation and utilization. Reliable authentic information on the distribution of genetic variation within and between geographic regions is important in order to establish an effective network of conservation populations. Genetic variation can be assessed by different techniques. It is possible to study morphological and metric characters in field trials, or with biochemical and molecular markers using laboratory techniques.

An array of biotechnological tools is contributing to the knowledge of forest genetic resources. Biotechnology has contributed to the knowledge of genetic variation within and between species and populations, understanding pollination mechanism and establishing clonal identity. In the last two decades, there is a vast number of publications on molecular markers far exceeding the quantitative data. Transgenic research is being carried out in trees such as Pinus, Populus, Picea, Larix and Eucalyptus to develop trees conferring pest resistance, herbicide resistance, abiotic stress resistance, hormone regulation, lignin, cell wall biosynthesis and growth (Flowchart 21.1).

Collection of data on natural range of species, present status, potential threats, distribution pattern, habitat characterization, demographic structure, population dynamic, interspecific competition, genetic variation within and between populations, breeding system, pollen and seed vectors, changes of climate and environment is required for effective conservation and utilization of FGR. Each one of the above aspects is described elsewhere (FAO 2012). In a tropical country like India, it is necessary to study all the above aspects on the species prioritized. For example, various patterns of distribution are exhibited by tree crops (Table 21.2).

A common, uniform approach in conservation of these species is not feasible because of diverse ecological conditions, diversity of species and anthropogenic pressure.

21.4 Factors Driving the Dwindling of FGR

Change in land use patterns and exploitation of the species are the two main aspects responsible for dwindling of forest genetic resources. In the name of development, large forest areas in India as well as around the globe are diverted for mining, hydroelectric projects, construction of roads. In the past, especially during the colonial period, large-scale planting of tea, coffee, rubber, etc. has changed the landscape pattern from natural forests to plantations.

Overexploitation of the species of commercial importance has resulted in dwindling of resources and in some cases has threatened the very existence of species. *Tectona grandis, Dalbergia latifolia* and *Santalum album* are some of the species overexploited for their commercially valuable wood. Large-sized teak trees were removed from forests of India for railway sleepers, ship buildings, etc. It is said the best teak wood is found in the castles and palaces of Britain. Dysgenic felling, removal of best trees from population in the past has resulted in a skewed population structure, not allowing the natural process of evolution. But on a conciliatory note, the Britishers are also credited with initiating the large-scale plantations of teak. *Dalbergia latifolia* (Indian rosewood) was among the most preferred wood for furniture making with great export value. The species with scattered distribution in the plains of Chotanagpur, central India and Western Ghats of south India had population of big size trees exceeding 10 feet girth in the past. Now the population has dwindled in number and it is a rare sight to see big trees. The conservation efforts made for its close cousin *Dalbergia cochinchinensis* in Cambodia, Laos, Vietnam and Thailand are also need to be adopted for *Dalbergia latifolia* in India.

Garcinia gummi-gutta, a dioecious species, is a recent example of overexploitation in the Western Ghats region. The seeds which are used to extract butter are collected by local communities. About 90–95% of fruits are harvested/removed by local people and have resulted in poor/no regeneration in the natural habitat. Another example is *Coscinium fenestratum,* a woody climber of Western Ghats, harvested indiscriminately for its medicinal uses.

Apart from the above threat status, various types of threats to forest genetic resources have been enlisted (Box 21.1). It is a general perception that any developmental activity is considered as antithesis of conservation. Hence, any

Flowchart 21.1 Conservation and management of forest genetic resources

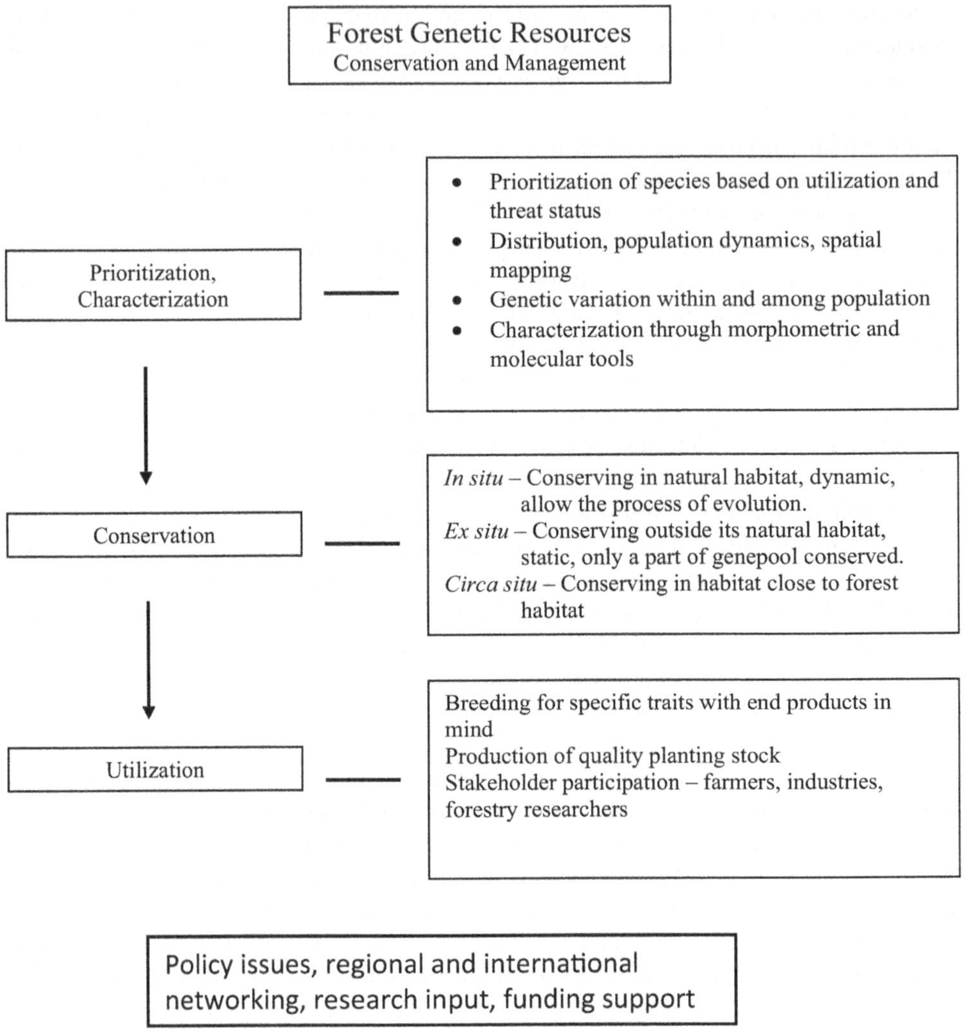

Table 21.2 Occurrence and distribution of tree species in India

Distribution pattern	Example species
Rare	*Vateria macrocarpa*
Only native conifer of Western Ghats	*Podocarpus wallichianus*
Narrow endemic	*Gluta travancorica*
Endemic with gregarious occurrence	*Pterocarpus santalinus*
Endemic with sporadic distribution	*Santalum album*
Distribution with wide geographic range	*Azadirachta indica*
Distribution extending to neighbouring countries	*Tectona grandis*
Disjunct distribution	*Gmelina arborea* *Melia dubia* *Ailanthus excelsa*
Complex species with wide distribution	*Acacia nilotica*
Countrywide common distribution	*Dendrocalamus strictus* *Bambusa bamboos*

developmental activity should also have conservation of natural resources in its centre of focus.

Box 21.1 Type of Threats to FGR

1. Forest cover reduction and degradation
2. Forest ecosystem diversity reduction and degradation
3. Unsustainable logging
4. Management intensification
5. Competition for land use
6. Urbanization
7. Habitat fragmentation
8. Uncontrolled introduction of alien species
9. Acidified soil and water
10. Pollutant emissions
11. Pests and diseases
12. Forest fires
13. Drought and desertification
14. Rising sea level

21.5 Conservation

The future of humankind depends on the effective conservation approaches adopted now. The process of conservation starts with prioritizing the species based on stakeholders' needs and threat status. Various strategies like conserving the ecosystem rich in diversity, species conservation and gene conservation are recommended by the researchers. Conservation is not possible without involving the participation of people and change in policy regime. There are two major approaches followed in conservation strategy. One is in situ conservation, where the natural habitat is conserved by giving adequate protection. It is considered as 'Dynamic Conservation' as the natural process of evolution is allowed to continue with its relative components unaltered. The ex situ conservation involves removal of species from its natural habitat and growing them in another location. It is considered as 'Static Conservation' as only a part of genepool is conserved through human intervention. A comparison of in situ and ex situ conservation is given in Table 21.3.

21.6 In Situ Conservation

In situ conservation is generally considered as easier and cost-effective. There are a lot of deliberations on what should be the size of the conservation area and in case of widely distributed species how many populations need to be conserved. Hamrick (1994) suggested that for tropical tree species, if 80% of the total genetic diversity resides within a population, five strategically placed populations should capture 99% of their total genetic diversity. A successful in situ gene conservation programme must fulfil the following basic requirements:

- Regeneration of the population must be assured and the new generation of trees must predominantly result from mating within the conserved population
- The number of genotypes in the conserved population must be large enough to include most of the common alleles
- The network of conserved stands must be distributed in such a way as to cover the spatial genetic variation present in the species

Protected areas are 'areas especially dedicated to the protection and maintenance of biological diversity and associated cultural resources, and managed through legal or other effective means' (IUCN 1994). They cover various situations, ranging from managed resource areas, protected watersheds, national parks and strictly protected reserves, to sacred forest groves. Successful conservation of threatened tree species in protected areas have been reported from many countries. Some examples are *Cotylelobium melanoxylon* in Thailand, *Eucalyptus nitens* in Australia, *Shorea bakonensis* in Malaysia, *Caesalpinia echinata* in Brazil, *Lodoicea maldivica* in Seychelles (Thomson and Theilade 2001).

Involvement of local communities and local government is important for effective in situ conservation. Creation of protected areas as a policy of Government of India has not only safeguarded forests from degradation but also ensured in situ conservation. In India, there are 14 biosphere reserves, 97 national parks, 508 wildlife sanctuaries, 31 tiger reserves, 7 conservation reserves and 2 community reserves providing niche for FGR (Rawat and Ginwal 2009).

Table 21.3 In situ versus ex situ conservation

In situ conservation	Ex situ conservation
• On-site activity	• Off-site activity
• Conservation of species within its natural habitat with all associated components	• Conservation of species outside its natural habitat with only a part of genepool reserve
• Dynamic process, facilitates natural evolution	• Static process with facilitation of artificial selection
• Perpetuation of natural variability	• Creation of artificial hybrids with desired traits by assemblage of genotypes from different environment
• Facilitates conservation forestry	• Facilitates production forestry with improved yield

Table 21.4 Regeneration status of local species (RET) in Silent Valley MPCA

Name of species	Seedlings per ha	Saplings per ha	Mature trees per ha
Myristica dactyloides	759	352	143
Garcinia morella	704	222	70
Persea macrantha	167	56	45
Symplocos racemosa	148	0	9
Aphanamixis polystachya	111	0	27
Embelia ribes	56	0	10
Canarium strictum	19	0	23
Nothapodytes nimmoniana	0	0	1
Garcinia gummi-gutta	0	0	2
Cinnamomum sulphuratum	0	0	10
Hydnocarpus alpina	0	56	18

Studies carried out in the Silent Valley National Park, Kerala, revealed that there are three patterns of regeneration (Kunhikannan et al. 2011). In the first case, the regeneration of seedlings, the status of saplings and the status of mature trees is adequate for the population to sustain. In the second case, there is a good number of seedlings and mature trees but no saplings recorded. In the third case, there are only a few mature trees with no seedlings and saplings. There are concerns about the sustenance of population in the latter two cases (Table 21.4). Whether this state has arisen due to poor seed set, lack of pollinators, poor regeneration, predation due to insects and animals, trampling by wild animals needs to be studied. Such studies are required for an effective implementation of in situ conservation measure.

Some researchers stress on identifying the genetic hotspots or the sites of diversity as effective in situ conservation measure. Studies conducted on allozyme variation in sandalwood populations of Karnataka have revealed high diversity in the Deccan plateau as well in the Western Ghats, suggesting such areas as ideal in situ conservation sites (Nageswara Rao et al. 2007). Similar studies conducted on two most important bamboo species in India, namely *Bambusa bamboo* and *Dendrocalamus strictus* have revealed Western Ghats and Eastern Himalayas as hotspots and suggested as in situ conservation sites (Chaluvaraju et al. 2001).

However, in situ conservation in tropical ecosystem, particularly like those of the Indian subcontinent, is difficult, not only owing to environmental disasters, landslides, unpredictable rainfall, flood etc. but also due to manmade disasters and pressures like forest fire, illicit felling and overexploitation of wild gene resources for commercial purpose (Box 21.2).

Box 21.2 Circa Situ Conservation

Circa *situ* conservation is a concept involving conservation of forest genetic resources in areas close to the forest habitat such as homesteads in Western Ghats,

(continued)

Box 21.2 (continued)

vested forests in Kerala, remnant forest patches in coffee and tea estates. Some of the genetic resources like Mango, *Artocarpus, Zanthoxylum, Hopea* and many more species are conserved for generations in such habitat. They also supplement the pollen to the wild population in adjoining forest areas and perpetuate the diversity. The genetic resource of 'appemidi' a flavoured variety of mango where whole raw mango is used to prepare pickle, a delicacy in Karnataka, was depleting in the wild. It is conserved in homestead of Uttara Kannada according to a study conducted by College of Forestry, Sirsi.

21.7 Ex Situ Conservation

Ex situ conservation is an important complement to conservation, especially for forest tree species threatened by loss of habitat. These collections are generally easily accessible, well documented and protected from natural disasters, but they do not continue to evolve as it does in situ. The methodologies for selection of sites, sampling strategies and the establishment of ex situ stands have been well discussed (Weber et al. 2004).

Some possible approaches of ex situ conservation of trees are mentioned below (FAO 2012).

- Provenance trials—comparing trees grown from seeds collected from different sites of a species range.
- Seed orchards—plantations established for the production of tree seed.
- Clonal repositories—collection of clones of a species.
- Botanical gardens—plants, especially ferns, conifers and flowering plants, are grown and displayed for the purposes of research and education.

- Arboreta—trees are grown and displayed for the purposes of research and education.
- Herbal gardens—plants of known medicinal values are grown and displayed for the purposes of research and education.
- Seed and pollen banks—storing seeds as a source for planting in case seed reserves elsewhere is destroyed and pollen for controlled pollination. It facilitates germplasm exchange (Box 21.3).
- Vegetative propagules—stored under aseptic conditions.

Box 21.3 Seed Bank—Need for India Initiative

Seeds are generally categorized into three groups: (1) orthodox seeds are characterized by low moisture in the seeds at the time of seedfall. They can be stored comfortably at low temperature for a long period, invariably they suffer from dormancy, e.g. *Tectona grandis;* (2) Recalcitrant seeds have a high moisture content at the time of seed fall, sensitive to reduction in moisture content and low temperature, hence pose problems during storage, e.g. *Artocarpus;* (3) Intermediate seeds have a high moisture content at the time of seed fall, amenable for moisture reduction and can be stored for some period, e.g. *Azadirachta indica.* Subsequently many more categories of seeds have been added based on the sensitivity to drying regime and temperature. But the behaviour of tropical tree seeds poses much more challenges in handling them and there are gaps in knowledge about them.

The Millenium Seed Bank Partnership established in 1996 in the Royal Botanical Garden, Kew of United Kingdom is the largest ex situ conservation programme based on seed storage. This global initiative involves collaboration of 50 countries and agencies like FAO and Bioversity International. Its purpose is to ensure availability against extinction of plants in the wild.

In India, there are many seed centres operated by state forest departments, to name a few, Seed Centre at Nagpur, Maharashtra; Seed Centre at KFRI, Peechi, Kerala; Seed Centre at Kanpur, Uttara Pradesh; Seed Centre at Coimbatore, Tamil Nadu; Seed Centre at Bengaluru, Karnataka. Every year these centres collect large quantity of seeds of over 100 species and supply to the farmers and others for raising the seedlings.

India harbours over 14,000 angiosperm plant species with equal array of seeds. Seed banks are considered as efficient ex situ conservation measure against extinction from wild. The requirement of manageable space in storage of seeds, protection against natural calamities and retrieval of germplasm at any time are the greatest advantages of seed bank. India with rich

(continued)

Box 21.3 (continued)

genetic resources has a challenge to meet the wood requirement of ever-growing population through increased forestry activities in the future enabled by concerted effort to have its own Seed Bank. It will also help in biochemical profiling of tropical seeds and provide the scope for patenting the novel compounds. It is also pertinent to mention the statement made by the then Honourable Minister of Environment and Forests Shri Jayaram Ramesh that 'Seeds are as strategic to India as space and nuclear issues'.

21.8 Tree Improvement Activities and Ex Situ Conservation

Ex situ conservation strategies such as establishing field gene bank, seed bank are considered to be expensive, time-consuming strategies require a great amount of effort. Hence field gene banks are established for those species which are planted in large areas, have a high commercial value, supported by an elaborate tree improvement programme (TIP). They also need to have a broad genetic base, ensure gene flow and take care of genetic drift. Tree improvement programmes are implemented for only those species which are commercially important and planted in large areas. TIP is important component of production forestry backed by conservation of FGR. TIP activities focus on judicious use of resources in conservation and utilization (Christel Palmberg-Lerche 1998). The aim of TIP is production of genetically improved seeds with increased productivity. The process involves selection of trees for desired traits, establishment of orchards as a source of improved seeds, establishment of progeny trials and development of clones. The process of selection and breeding continues for many generations with a focus on increasing productivity. Continuous selection and breeding for many generations results in loss of variability and therefore fresh populations are infused into breeding programme. Thus, TIP is a long-term continuous activity requiring management of the genetic resources for a few decades. Tree improvement in different tree species being carried out by various agencies in India is given in Table 21.5.

The successful conservation and management of FGR involves characterization and documentation of these resources. Characterization of FGR is important to understand the variability existing and the scope for improving the traits of importance. An understanding of morphology, seed technology, biochemistry, wood anatomy and knowledge of application of biotechnological tools is a prerequisite for characterization of FGR.

Table 21.5 List of institutes engaged in TIP

Species	Institution
Azadirachta indica	AFRI, Jodhpur
Tectona grandis	Teak center, Chandrapur TFRI, Jabalpur IFGTB, Coimbatore
Dalbergia sissoo	FRI, Dehradun
Bamboos	RFRI, Jorhat KFRI, Peechi
Melia dubia	FRI, Dehradun IFGTB, Coimbatore
Pinus	HFRI, Shimla
Eucalyptus	IFGTB, Coimbatore
Casuarina	IFGTB, Coimbatore
Leucaena	BAIF, Pune. IFGTB, Coimbatore
Santalum album	IWST, Bangalore
Gmelina arborea	RFRI, Jorhat IFGTB, Coimbatore
Tamarindus indica	TNAU, Periakulam IFGTB, Coimbatore

The above list is indicative. There are many horticultural colleges across India that maintain the germplasm of many wild edible fruits like *Artocarpus, Garcinia, Syzygium, Emblica*

21.9 Status of Forest Genetic Resources in India

21.9.1 Genetic Resources of Medicinal Plants

Ayurveda, the traditional way of healing diseases in India, is mostly based on plants and their parts. Currently more than 500 species of plants including herbs, shrubs, climbers, trees are used by Ayurvedic industry. As most of the plant species are sourced from wild, there is depletion of these resources in the wild. Since roots and bark are the main source of active ingredient, destructive harvesting of plants in large quantities has resulted in severe depletion of these resources. *Saraca asoca*, the Ashoka, is one of the important trees used in Ayurvedic medicine. The demand for bark of this tree is over 2000 tonnes per year. Removing of bark from standing trees has resulted in death of trees and dwindling of the population in its natural range. International Union for Conservation of Nature and Natural Resources (IUCN) has red-listed this species under the threat category 'globally vulnerable'. Some of the trees which are harvested in this way are, *Oroxylum indicum, Gmelina arborea, Aegle marmelos* and *Premna serratifolia.* Continuous unsustainable harvest of these resources has also put the future of Ayurvedic industry at stake.

In order to support sustainable utilization of these resources, the Department of Ayush (erstwhile NMPB) is encouraging the cultivation of medicinal plants by providing subsidy of 30% to 70% of cultivation cost for about 95 medicinal plants including 26 tree species. Foundation for Revitalization of Local Health Tradition (FRLHT), a Bengaluru-based NGO, has established 54 Medicinal Plant Conservation Areas (MPCA) in collaboration with state forest departments of Karnataka, Kerala, Tamil Nadu and Maharashtra.

21.9.2 Genetic Resources of Timber Species

India is a timber deficient country and its import bill is to the tune of USD 1.25 billion a year. Teak is the major timber in the households of the country and is known as 'King of timber'. India imports wood from around 30 countries of which except Myanmar the rest do not have natural teak and the source of planting material is mostly from India. Though India harbours the largest genetic diversity of teak (more than the combined diversity of all other countries where teak occurs naturally), its genetic potential has not been exploited. The recorded teak plantation in the world is 5.28 million ha, out of which India has 1.50 million ha. The pre-independence era Indian Forest Records show that the British knew the importance of teak not only in making warships but also as a timber in house construction and furniture making. Large-scale plantations were established during that time. Even before the colonial rule, teak was a preferred species for shipbuilding, a thriving business in West Bengal and Kerala. The interest in teak cultivation lost focus gradually post-independence. During 1990s, teak was promoted for cultivation by many private companies in India with floating of teak bonds, but these ventures did not succeed because of the use of poor genetic quality of planting material, improper site selection and poor management practices. However, in the recent years scientific intervention has resulted in establishment of large-scale plantations by private companies in the continents of Africa and South America. In countries like Malaysia and Thailand, clonal plantations of teak are harvested under rotation age of 18–20 years. A good performing clone is promoted as 'Super Teak' by Sabah-based company YSG Biotech (Goh and Monteuuis 2009).

Realizing the importance of teak, tree improvement activities started in 1960s. Over the past half century around 250 plus trees have been identified and 5000 ha of seed production areas (SPAs) and 1000 ha of clonal seed orchards (CSOs) established in different teak growing states (Katwal 2003). A national germplasm bank of teak was established at Chandrapur, Maharashtra and an international provenance trial of teak was established at Maredumilli in Andhra Pradesh. The seed orchards established with an intension of supplying quality seeds for plantation activities were bogged down by unperceived low seed production. With the passage of time these resources which were created with a lot of effort, money and time have been pushed to oblivion. The farmers of India are ready to plant teak. There is a demand for wood with increased activities in house construction, it is time to revive these resources and reduce the burden of importing wood. There is untapped genetic variation in teak still to be exploited with scope for extending the plantations in wet zone as well as semi-wet zones of the country.

Dalbergia sissoo and *Shorea robusta* are the two other most widely grown species for which tree improvement activities are in progress albeit in a lesser way. These two species are also prone to serious disease and pest attack, namely die back and borer attack. The priority of research for these trees is breeding for disease and insect resistance. *Dalbergia latifolia* (Rosewood), a wood preferred over teak in furniture making and most sought after species for inlay works and artifacts, has a high export potential. Due to overexploitation in the past and inherent slow growth, the genetic resources of this species have dwindled and the non-availability of wood for artisans and furniture making industry has put them in disarray. A study conducted by IFGTB on the natural regeneration of the species in Tamil Nadu and Kerala revealed the natural population is fast depleting in the wild with poor regeneration. The species has been categorized as 'Vulnerable' by IUCN Red List of Threatened species and overexploitation of the species from wild for timber has been attributed as the major threat factor encountering the species. *Dalbergia sissoides* is another species native to Southern region of Western Ghats with a wood harder than *D. latifolia* and fetches a higher value as premium wood in cabinet and furniture making. It is a distinct species, closely related to *D. latifolia* than *D. sissoo*, but less understood.

Other important wood producing tree species are *Pterocarpus marsupium*, *Terminalia tomentosa* and *Lagerstroemia lanceolata* for which seed production areas have been established as source of quality seeds by many state forest departments.

21.9.3 Genetic Resources of High-Value Wood Producing Species

Santalum album (Sandalwood) and *Pterocarpus santalinus* (Red sanders) are the two high value wood producing species endemic to Peninsular India and deeply associated with the culture of local people. *Santalum album* is native to Karnataka, Tamil Nadu and Kerala and *Pterocarpus santalinus* is restricted to the hills of Cuddapah region of Andhra Pradesh in Eastern Ghats. They have been exploited for over two centuries for revenue generation by the state and illicit felling and smuggling of wood were the major reasons threatening their genetic resources. As a result, farmers and entrepreneurs who grow them are highly concerned since there is a threat looming over illicit cutting of trees by smugglers for their high value wood. It is also evident by the frequent seizure of large volume of wood by enforcement agencies not only in the region of their growth but also from whole of India.

Santalum album grows naturally in forests extending over 8000 km^2. Because of its legal status as 'Royal Tree', there were restrictions in cultivation and trading of sandalwood and used to be the monopoly of the state. Because of slow growth in its natural habitat, the rate of extraction of wood from the forests was much higher than its production potential. The annual production of sandalwood in Karnataka was 2800 tonnes during 1950s, reduced to 1800 tonnes during 1985, dropped to 76 tonnes in 1999 and further dwindled to 3.52 tonnes in 2010. It is interesting to note that the extraction of a large volume of wood from forests is happening for over 200 years, during the precolonial period, colonial period and post-independence (Ezra 2014). Extraction of wood for such a long period without augmenting plantation activities has threatened the very existence of the species. It is an ideal case to study how a highly valuable species with restricted distribution was pushed to the edge of extinction. Concerted efforts in framing the policies, improving the cultivation practices backed by research support are required to regain the past glory of sandalwood in India. The tree is also plagued by sandalwood spike disease, frequent fire and illicit felling of good trees. Sandalwood has been categorized as 'vulnerable' by International Union for Conservation of Nature and Natural Resources. Sandalwood regeneration with bird droppings has resulted in the growth of sandal in farmlands and urban landscapes. Srimathi et al. (1983) reported three phenotypes on the basis of heart wood, sap wood and certain parameter of the tree. Institute of Wood Science and Technology (IWST), Bengaluru, has selected plus trees of *S. album* from southern states, based on the growth, heart wood and oil content (Arunkumar et al. 2011) and established Clonal Germplasm Bank at Gottipura, near Bengaluru. During 1982, Clonal Seed Orchard (CSO) of 25 clones was established at Nallal, also near Bengaluru (Srinivasan et al. 1992; Srimathi et al. 1995). A Vegetative Multiplication Garden and seed orchards at four locations were established in Tamil Nadu. IWST has also developed refined protocols for in vitro cloning of *S. album* through axillary shoot proliferation and somatic embryogenesis of mature trees and clones.

The lifting of the ban on felling of sandalwood trees during 2010 in Karnataka and other states coupled with the awareness on the high returns from its cultivation have encouraged many farmers to take up large-scale plantation of sandalwood planting in India. This increased plantation activity necessitates supply of quality seedlings with fast growth, high heart wood content and disease-free stock.

Pterocarpus santalinus naturally occurs in the rocky out crops of the hill regions of five districts in Andhra Pradesh, namely Cudappa, Chittoor, Nellore, Ananathapur and Karnool (Fig. 21.1). Though the population density is high, the narrow endemism and clandestine removal of large number of trees for a long period have made the species vulnerable and the species is placed in Appendix-II of CITES. Though it coppices well, annual fires and grazing have resulted in poor regeneration. A study on non-detrimental findings (NDF) has shown poor regeneration, slow growth, absence of trees with higher girth class, absence of seed producing mother trees are the reasons of depletion of genetic stock (Hegde et al. 2012).

The tree is smuggled out of the country because of the high value it fetches in China, Japan and other countries. Not only wood, but also wood chips and wood dust are smuggled. The unusual high price of smuggled wood is ascribed to the coloured wood used in making musical instrument, as colorant in drinks, medicinal use, treating the heavy water but knowledge of the actual use of this species is elusive and yet to be unearthed. In the recent years, the species is being planted outside the natural range and reported to have good growth (Fig. 21.2).

21.9.4 Genetic Resources of Pulp Wood Species

There are over 600 paper mills (Box 21.4) in India, of which 12 are major paper mills. These mills require a large volume of wood for pulp production. The major species utilized in

Fig. 21.1 *Pterocarpus santalinus* (red sanders), a narrowly endemic species to the Southern Eastern Ghats

Fig. 21.2 A plantation of red sanders raised by forest department in Rajampet, Andhra Pradesh

these paper mills are short rotation exotic species, namely Eucalyptus, Casuarinas, Phyllodinous Acacias and Leucaena. The rotation of these crops is invariably fixed at 3–6 years. Large-scale plantations of these crops are being raised in farmers' field with incentives provided by paper mills in the form of subsidized seedlings and buy back guarantee. India is considered to have the largest plantations of these pulpwood crops under cultivation. But these tree crops are also prone for serious pest and disease problems. High yielding clones and hybrids have been released by research organizations and paper mills. As a result, clonal plantations have become popular among farmers. The narrowing genetic base is a serious concern in these crops due to repeated selection and breeding. As a result, new populations are infused into the breeding programme at regular interval.

When exotics are being introduced, many precautions need to be taken. The first set of provenances of *Acacia auriculiformis* introduced to India had the trees with forked and heavily branched main stem (Fig. 21.3), a highly undesirable attribute for timber and pulpwood production. Hence, introduction of species should prelude provenance testing. Many exotics have also notoriously known to become weeds. *Acacia mearnsii* (blue wattle) in the Nilgiri mountains, Leucaena in the farmlands of East coast and Phyllodinous Acacias in the Western Ghat region are some of the examples of exotics turning into weeds. It is important in understanding the biology of the species and the management before they are introduced in new locations.

Box 21.4 Paper and Pulp Industry in India

- The pulp and paper industry is one of the key industrial sectors contributing to the Indian economy. There are 861 paper mills in India with an operating capacity of 25 million tonnes (2019–20).
- Paper mills in India continue to face challenges with forest-based raw material. Out of the annual paper production capacity of nearly 25 million tonnes, around 31% (7.75 million tonnes) is produced by 26 major wood-based mills and the rest 69% by waste paper and agro-based mills.

(continued)

Box 21.4 (continued)

- The present annual requirement of wood is 11 million tonnes. The demand for pulp wood in India expected to increase by 70% in the coming decade.
- The Tamilnadu Newsprint and Papers Limited (TNPL) requires about 1800 MT of pulp wood per day.
- The yield from clonal plantations of Eucalyptus (Fig. 21.4) and Casuarinas is about 120 tonnes/ha on a 3-year rotation cycle. At a price of Rs 2500 per tonne, famers will get Rs 3 lakh per ha (2015).

Farmers prefer to grow these trees because of fluctuating and non-remunerative agricultural produce, shortage of labour to work in the field, ease of maintenance and a definite income at the harvest age. Since these are short rotation trees, harvested at the age of 5–6 years, there is a huge demand for seeds and seedlings to raise new plantations every year. There are many private nurseries which supply seeds and seedlings, but the quality of the planting stock supplied by these nurseries is a matter of concern. There is an urgent need to augment these supplies.

21.9.5 Genetic Resources of Minor Wood Producing Species

There are various small- and medium-scale wood-based industries in India which require wood in substantial quantities. They are also encouraging farmers to grow tree crops in their farmland. Some of the species used by industry are *Anthocephalus* for pencil making, *Ailanthus* and poplar in match wood industry, *Melia dubia* in plywood industry, *Grevillea* in truck building, bamboos in incense stick industry. In the recent years, farmers have taken up cultivation of

Fig. 21.3 Forked and highly branched *Acacia auriculiformis*, an early introduction to India

Fig. 21.4 Clonal plantation of Eucalyptus in farmers' field

these tree crops because of better remuneration. They are harvested at the age 10–15 years and fetch a market price better than the pulp wood species. *Ailanthus excelsa* and *Ailanthus triphysa* are the two species grown in south India. *A. excelsa* is grown in river tracts of Tamil Nadu and *A. triphysa* in wetter regions of Kerala. The matchwood industry prefers *A. triphysa* because of light coloured wood.

These crops are either grown as block plantations or a few trees are grown on the bunds of agricultural field. It is a custom in Tamil Nadu to plant few trees of *Ailanthus* on the bunds of agricultural field and sell the trees and utilize the money obtained for meeting different family needs.

21.9.6 Genetic Resources of Wild Fruits

India being a tropical country is also rich in wild animals. There are many trees which support the wildlife by providing them fodder and fruits. It is estimated that there are more than 235 species of wild fruit yielding plants in the Western Ghats. Some of the species have been domesticated and are being grown as horticultural crops or in homestead. *Garcinia, Artocarpus, Syzygium, Emblica* and *Mangifera* are some of the species successfully domesticated. There are also a large number of fruits used by local communities in traditional medicine, whose nutritive value is yet to be exploited and they are the potential source of novel compounds.

21.9.7 Genetic Resources of Oil Yielding Species

There are many trees species in India that support the energy needs of the local communities. Trees like *Azadirachta, Pongamia, Calophyllum, Madhuca* yield oil used by local villagers to run their generator sets and lit the lamps in the night. Most of these oils are also used in local therapy and also as the base for application of pesticides on agricultural crops. In the recent years, these oils are being promoted as biofuel by Government of India and other state governments. The Government of Karnataka has established a 'Biofuel Park' in the district of Hassan and helping the farmers in cultivation, collection, procurement and marketing of oil seeds.

These seeds are also excellent source of pesticides. Azadirachtin extracted from the seeds of *Azadirachta indica* (Fig. 21.5) is found to be effective in controlling over

Fig. 21.5 Variation in seeds of *Azadirachta indica* (neem)

200 insect pests. As a result, there is global market opened for neem-based pesticides. It is estimated that about 4 lakh tonnes of neem seeds can be produced every year from the naturally grown trees (IARI 1999).

21.10 Legal Framework and Relevant Acts on FGR

The Indian Forest Act of 1927 regulates management of forests by the states. The Forest Conservation Act 1980 is considered as a stringent act which has placed strict control over diverting forest land for non-forestry purposes. Following the CBD, Biodiversity Act 2002 was enacted (Mandal 2006) which regulates the use of bioresources for commercial purposes. A 'Scheme for Certification of Forest Reproductive Material in India' was proposed under the Indo-Danish Project on Seed Procurement and Tree Improvement in India. This scheme was based on the OECD scheme and recommended its implementation in India (Madan Gopal 1979). The OECD scheme was basically evolved to certify the tree seeds and plants moving in international trade, particularly in Scandinavian countries and some of the other countries in Europe. The same scheme is followed in Canada, Japan and the United States. The old 1974 Scheme has been updated in order to include new progress in forest tree improvement (Nanson 2001). It is time to revive the scheme in India to ensure the supply of planting stock of proven genetic quality. The forestry crops have already been notified under PPVFRA Act, 2001. The DUS descriptors and testing guidelines have been notified for *Eucalyptus, Casuarina, Pongamia, Melia dubia,* Poplars and Teak. It is also planned to include the forestry crops in proposed Seed Act 2020. The advancement made in tree improvement has resulted in release of clones of Eucalyptus, Casuarina, *Dalbergia sissoo*, Calophyllum by research institutes following the guidelines issued by ICFRE, the national body for forestry research in the country.

21.11 Future Strategies

The genetic resources in India, though rich and diverse, are under constant threat because of its population related pressure to divert forest areas to developmental activities. The demographic change has also led to increased demand for wood and other forestry products. In order to address these issues, there need to be a sound strategy on conservation and utilization of forest genetic resources.

Prioritization of Species Prioritization of tree species is the first step in initiating the process of conservation and utilization of forest genetic resources. While prioritizing, it is important to involve all the stakeholders. But, it is observed that each set of stake holders will have their own list of priority species. The list of trees prioritized by timber industry is different than the species prioritized by artisans, and in turn they differ with the farmers, researchers, forest dwellers, forest department etc. As a result, the list ends up including a large number of species. An exercise done at national level has prioritized 272 tree species for India. Hence, a pragmatic list of prioritised species is required to be prepared for different regions of India, namely North, North-East, East, West, Central and Southern India.

Establishing Linkages The farmers in India are ready to grow tree crops, when there is an assured market for the crop. But getting quality planting stock is one of the constraints. There are many farmers who have planted teak trees in their farmland expecting a good return after few years. It is observed that most of these trees started forking at early age with a short clear bole of no commercial value. There are seed orchards and seed production areas of teak established by state forest departments meant for supplying quality seeds. They need to be put into proper usage. There are many research organizations working on improvement of tree crops. The timber and paper industries require a large volume of wood. It is necessary to develop linkages among these stakeholders in order to make optimum utilization of resources.

Research Requirement Tropical trees exhibit a large amount of variability because of their evolution in the complex ecosystem. Taking up research work on these trees is expensive, time consuming, limited to few species, seldom on population. There is a limited knowledge on a few species and a vast number of species are yet to be investigated. In order to develop a sound conservation strategy, it is important to understand the variability due to genetic and environmental factors, the pollination mechanism, mode of pollen and seed dispersal, threat perception. In the recent years, molecular markers are being extensively used to understand the genetic variability and pollination mechanism. Most of the tropical tree seeds are recalcitrant in nature. It is easy to store hard coated orthodox seeds comfortably in a seed bank, whereas handling recalcitrant seeds is difficult as they lose viability quickly in spite of providing all the conditions necessary for storage. Research is required in handling of these resources.

Field Activities Most of the activities concerned with the in situ and ex situ conservation are field oriented and involvement of field staff of state forest department and local communities plays a very important role in successful implementation. 'Hands off' approach has resulted adversely on the conservation programme; on the other hand, involving the local communities has resulted in sustainable utilization of resources. What is required is a well thought management plan for conservation and utilization of forest genetic resources.

Institutionalization of FGR Programmes on conservation and utilization of forest genetic resources are best undertaken and coordinated by a designated national agency working in close collaboration with state agencies and other stakeholders. This national agency can take the best examples around the globe and coordinate the programmes at national and international level. The natural forests are treasure house of the country and backbone of the conservation programme. Reducing adverse interference and enhancing protection ensure the sustenance of these forests in generations to come. These natural resources can act as 'Permanent Fixed Assets', at the same time provide basic resource material for creation of '*ex situ*' seed sources in the form of seed production areas, seed orchards and multiplication garden. These seed sources, so established can supply the seeds for large-scale plantation programmes. As these forest genetic resources are assets of the country, they need to be 'registered' under National Register of Forest Genetic Resources. Though there are 272 tree species prioritised for the country, there is no comprehensive document available on status of these tree species. There is good amount of information available on the taxonomic features of these species, but there is an urgent need to prepare a comprehensive document on profile of each species with reference to FGR comprising the information on distribution, population structure, number of populations, variability existing, regeneration pattern, extent of utilization, economics of the species, programme of planting, threat status, conservation measures undertaken.

21.12 Conclusion

It is generally believed that we are losing one species before we discover another species indicating the rate of degradation happening around the globe. It is disheartening to know that it takes millions of years to evolve a species in its natural habitat but they get exterminated from nature before they are discovered lest utilized. For a biodiversity rich country like India with a large human population, there is a threat looming large on the future of these natural resources. Large chunks of forest areas have been diverted for mining for bauxite, iron and other minerals. Construction of dams for irrigation and agriculture also has engulfed large forest areas. Grow more food campaign in the 1960s, construction of roads and other activities have reduced the natural forest areas in the last few decades.

Social forestry programme implemented by many state forest departments in the country is one of the main reasons in reducing the pressure on natural forests. The involvement of local people in managing the fringe forest areas and planting of trees in the community land and wasteland have become successful model for extending the tree cover in the country. The policy of Government of India to secure wood from outside forest areas for industrial purpose has made many wood-based industries to encourage farmers to grow the trees in the farmland with buyback arrangement. The supply of wood for paper mills is sourced from short rotation crops such as Eucalypts, Casuarinas, Leucaena which are exotics. These crops have changed the landscape of the country and helped increasing the tree cover. The conservation of the genetic resources of exotic tree crops is equally important for sustenance of the industry. The industries such as plywood, matchwood and pencil industry source their wood mostly from native tree crops grown in farm land. The most successful conservation programmes are those which involve partnership between researchers, farmers and industries.

Conservation of genetic resources through in situ is considered as the most effective strategy. But there are gaps in research inputs required for conservation such as prioritization of species, its distribution, the variability, how much to conserve and how many populations to conserve. For NWFP species, the in situ conservation is recommended as an effective strategy because of vast number of species and their disjunct distribution. Successful conservation is possible only with the involvement and support of local people and favourable government policies. Many forest areas and genetic resources have been well conserved in situ because of declaration of vast areas as protected areas by many countries as a matter of policy. Ex situ conservation strategies are adopted in conjunction with tree improvement programmes.

Networking with global and regional agencies, adoption of policies backed by a strong research support and involvement of people are the key issues in conservation and utilization of forest genetic resources.

Lessons Learnt

- Forest resources are not inexhaustible and the loss of the forests and degradation in the recent years had brought untold misery to humankind and unprecedented ecological disaster in the form of landslides, floods, increase in temperature and erratic rainfall patterns. The wisdom lies in protection, conservation and sustainable utilization of these resources. India is known to have a tradition of conserving the forests in the form of 'Sacred Groves' and utilizing the resources existing outside the sacred groves, thus balancing harmony with nature.
- Conserving the forests as well as utilizing the resources is equally important for the ecology and economy of any country. Thus, conservation forestry and production forestry play pivotal roles. But production forestry cannot survive on its own, without depending on conservation forestry. The conservation forestry focuses on conserving the natural forests in their pristine condition in the form of in situ conservation. These forests apart from serving as source for genetic resources of important crops, they also perform many ecological services such as the source of origin of many rivers, perform the function of soil conservation, water conservation. The production forestry focuses on production of wood and other products outside the forest areas to meet the requirement of local communities and industries apart from increasing the green cover and locking the carbon in the form of wood. The production forestry requires large quantities of planting propagules every year. This is achieved by creation of seed orchards, seed production areas, provenance resources stands, etc. These seed sources also function as ex situ conservation areas; the base material for establishing these seed sources essentially comes from conservation forestry.
- The successful implementation of both in situ and ex situ conservation programmes is possible by involving the local communities, farmers, researchers and industries. The farmers are interested in growing tree crops in their farmland. There are many varieties/clones developed by researchers over a period apart from creation of sources for quality seed production. Wood is required in large quantities by paper and pulp, plywood industries. Coordination among these stake holders is key to improving the economy of the country. It is also necessary to amend the archaic policies from time to time for sustainable utilization of forest genetic resources.

Key Questions

1. What is the significance of forest genetic resources?
2. What are the strategies required for conservation of forest genetic resources?
3. How are forestry species different from agricultural crops with reference to handling of genetic resources?
4. Whether there are enough efforts in conservation and utilization of forest genetic resources in India, if so enlist those efforts.
5. What are the research requirements to make conservation of forest genetic resources a successful programme?

References

Anonymous (2012) State of forest genetic resources in India: a country report - 2012. Institute of Forest Genetics and Tree Breeding, Indian Council of Forestry Research and Education, Coimbatore, 129p

Arunkumar AN, Srinivasa YB, Joshi G, Seetharam A (2011) Variability and relation between tree growth, heartwood and oil content in sandalwood (*Santalum album L.*). Curr Sci 100:827–830

Chaluvaraju, Deepali Singh BS, Nageswara Rao M, Ravikanth G, Ganeshaiah KN, Uma Shaanker R (2001) Conservation of bamboo genetic resources in Western Ghats: status, threats and strategies. In: Uma Shaanker R, Ganeshaiah KN, Bawa KS (eds) Forest genetic resources: status, threats and conservation strategies. Science Publishers, Enfield, NH, pp 97–113

Christel Palmberg-Lerche (1998) Conservation and management of forest genetic resources. J Trop For Sci 11(1):286–302

Ezra DR (2014) Perfumed the axe that laid it low: the endangerment of sandalwood in southern India. Indian Econ Soc Hist Rev 51(1): 41–70

FAO (2012) State of forest genetic resources in India: a country report. Institute of Forest Genetics and Tree Breeding, Indian Council of Forestry Research and Education, Coimbatore

Franzel S, Jaenicke H, Janssen W (1996) Choosing the right tree: setting priorities for multipurpose tree improvement. ISNAR research report no. 8. International Service for National Agricultural Research, The Hague

FSI (2009) State of Forest report 2001. Forest Survey of India, Ministry of Environment and Forests, Dehradun

Goh DKS, Monteuuis O (2009) Status of the 'YSG BIOTECH'-program of building teak genetic resources in Sabah. Bois et Forêts des Tropiques 301:33–49

Graudal L, Thomson L, Kjær E (2001) Selection and management of *in situ* gene conservation areas for target species. In: FAO, DFSC, IPGRI forest genetic resources conservation and management. Vol. 2: managed natural forests and protected areas (*in situ*). International Plant Genetic Resources Institute, Rome

Hamrick JL (1994) Genetic diversity and conservation in tropical forests. In: Proceedings of the international symposium on genetic conservation and production of tropical forest tree seed. ASEAN-Canadian Forest Tree Seed Centre Project, Muak-Lek, Saraburi

Hegde M, Gurudev S, Krishnakumar N (2012) Non-Detriment Findings (NDF) study for *Pterocarpus santalinus* L.f. (Red Sanders) in India. Institute of Forest Genetics and Tree Breeding, Coimbatore, 28p

IARI (1999) Neem News letter. IARI, New Delhi

IUCN (1994) Guidelines for protected areas management categories. IUCN – WCPA and WCMC, Gland

Katwal RPS (2003) Teak in India: status, prospects and perspectives. In: Bhat KM, Nair KKN, Bhat KV, Muralidharan EM, Sharma JK (eds) Quality timber products of teak from sustainable forest management. Kerala Forest Research Institute, Peechi, pp 1–18

Koshy MP, Namkoong G, Kageyama P, Stella A, Gandara FB, Amaral WAN (2002) Decision-making strategies for conservation and use of forest genetic resources. In: Ramantha Rao V, Engels JMM, Brown AHD, Jackson MT (eds) Managing plant genetic diversity. IPGRI, Rome and CABI, Wallingford

Kunhikannan C, Venkatasubramanian N, Sivalingam R, Pramod Kumar N, Salvy T, Sibin T (2011) Diversity of woody species in the Medicinal Plant Conservation Area (MPCA) of Silent Valley National Park, Kerala. Biodiversity 12(2):97–107

Madan Gopal (1979) Certification of forest reproductive material in India (Revised scheme). Indo-Danish Project on Seed Procurement and Tree Improvement, Hyderabad

Mandal AK (2006) Forest genetic resources conservation and management in India – an update on activities, challenges and needs since APFORGEN inception in 2003. In: Hong LT, Sim HC (eds) Forest genetic resources conservation and management. Proceedings of the Asia Pacific Forest Genetic Resources Programme (APFORGEN). Dehradun, India

Nagarajan S (2013) National forest genetics resources conservation and usage plan. Institute of Forest Genetics and Tree Breeding, Coimbatore, 88p

Nageswara Rao M, Ganeshaiah KN, Uma Shaanker R (2007) Assessing threats and mapping sandal resources to identify genetic 'hot-spot' for *in-situ* conservation in peninsular India. Conserv Genet 8:925–935

Nanson A (2001) The new OECD scheme for the certification of forest reproductive materials. Silvae Genetica 50(5–6):181–187

Rawat GS, Ginwal HS (2009) Conservation and management of forest genetic resources in India. In: Jalonen R, Choo KY, Hong LT, Sim HC (eds) Forest genetic resources conservation and management. APFORGEN

Saxena NC (1999) The new forest policy and joint forest management in India. National workshop on JFM, 25–26 Feb. organized by VOKSAT, SPWD, AKF and Gujrat Forest Department, Ahmedabad, India

Sharma MK, Singhal RM, Sudhir Kumar, Jeeva V (2002) Regional update for Indian forest genetic resources. Working papers for the twelfth session of the FAO Panel of Experts on Forest Gene Resources, Rome

Shiva V (1988) Staying alive: women, ecology and development. Zed Press, London

Srimathi RA, Kulkarni HD, Venkatesan KR (1983) Phenotypes of sandal. J Bombay Nat Hist Soc 80:245–246

Srimathi RA, Kulkarni HD, Venkatesan KR (1995) Recent advances in research and management of sandal (*Santalum album L.*) in India. Associated Publishing Co., New Delhi

Srinivasan VV, Shivaramakrishnana VR, Rangaswamy CR, Anathapadmanabha HS, Shankaranarayana KH (1992) Sandal. ICFRE, Dehradun

Thomson L, Theilade I (2001) Protected areas and their role in conservation of forest genetic resources. In: FAO, DFSC, IPGRI forest genetic resources conservation and management. Vol. 2: managed natural forests and protected areas (*in situ*). International Plant Genetic Resources Institute, Rome

Weber A, Yanchuk A, Kjær E (2004) Methodologies for *ex situ* conservation in FAO, FLD, IPGRI forest genetic resources conservation and management. Vol. 3: plantations and gene banks (*ex situ*). International Plant Genetic Resources Institute, Rome

Yanchuk AD, Lester DT (1996) Setting priorities for conservation of the conifer genetic resources of British Columbia. For Chron 72:406–415

Part IV

Production Forestry

Plantation Forestry

22

Uday Kumar Lodh, Vimal Chauhan, Sachin Verma, and Krishan Chand

Abstract

Plantation forests span around 131 million hectares (ha) worldwide, accounting for 3% of the world's forests and 45% of total planted forest area. Plantation woodlands are highly managed forests that are formed between both the first and second genera, are evenly ageing and spaced apart in a consistent manner and are primarily maintained for beneficial ends. Plantations are far more noteworthy in the context of wood output than their percentage of forest area, and their importance is likely to grow over time. In 2000, a quarter of the worldwide consumption of commercial round wood was generated from plantations. According to some projections, plantations and planted forests will produce over half of the world's industrial round-wood supply by 2040. Beyond producing timber and fiber, woodland plantings also produce clean water, regulate the ecological cycle, trap carbon, and promote the interrelationships of terrain mosaics to preserve ecology and lessen the outcome of degradation. The service provision marketplaces have historically been expanding swiftly and planting trees will be important for both the local reimbursements for environmental care networks that manage water and soil assets as well as the world's carbon trading. Plantation forestry in the developing nations of tropical areas is strongly nested on the political agenda due to the planted forest's carbon storage capacity as noticed in the climate change debate. As the forthcoming model regarding environmentally friendly plantation development includes adaptation and mitigation of climate change, it is anticipated that both of the aforementioned areas would grow in significance. The loss of forest cover in many nations has resulted in increased floods, fast desertification, siltation, and accelerated soil erosion. Extensive re-establishment of tree cover by plantations is the most effective means of reversing these processes. The other important environmental role of plantations or systems of agroforestry, where wood and food crops are grown together on the same piece of land, is the key to sustainable land management in many fragile ecosystems in the tropical regions. As the population around all over the world increases and its economies grow, so does number of wood people use, for firewood, for building, for paper, and many other purposes. Plantation forests now play an essential role in wood production, particularly in tropical nations, due to numerous crucial traits such as high yield, short rotations, and accessibility. The subsequent supply of wood for the globe will also depend heavily on forest growth. If afforestation rates significantly improve, plantations can begin to relieve pressure on the world's depleting supplies of wild forest, which is the most compelling reason to pursue plantations. Afforestation/reforestation techniques, the Clean Development Mechanism (CDM) and the controversial Reducing Emissions from Deforestation and Degradation (REDD) mechanism, are some of the new incentives that emphasize forestry plantations as a reasonably inexpensive, safe, and clean source of emission compensation.

Keywords

Plantation · Round-wood · Forest degradation · Fire wood · Exotic species

22.1 Introduction

Plantation forests encompass over 131 million ha of land or 45% of all established forest land globally, accounting for 3% of the globe's total area of forests. Plantation forests are made up of one or two species that are evenly aged, planted with uniform spacing, and are primarily created for agricultural

U. K. Lodh · V. Chauhan (✉) · K. Chand
College of Forestry, Dr. Y. S Parmar University of Horticulture and Forestry, Nauni, Solan, India

S. Verma
Department of Silviculture, Forest Management and Agroforestry, ICFRE-Tropical Forest Research Institute, Jabalpur, India

A. K. Mandal, A. Nicodemus (eds.), *Textbook of Forest Science*, https://doi.org/10.1007/978-981-97-8289-5_22

purposes. They're also under strict control. Since they are not subject to extensive management, other planted forests, which comprise 55% of all cultivated trees, may look like natural forests at stand maturity. Since 1990, the area covered by planted forests has grown by 123 million ha, reaching 294 million ha; however, since 2010, the rate of expansion has decreased. Other planted woods may be used for the preservation of soil and water quality as well as for the restoration of ecosystems (FRA 2020).

Over 44% of planted forests globally consist mostly of introduced species. The species composition of plantations differs substantially across sites. While 97% of the species in the continent of South America are introduced, a great number of species that inhabit forests from plantations in both North and Central America seem native (Fig. 22.1). Contrary to their share of the overall forest area, plantations contribute significantly more to the production of wood and this contribution is expected to increase with time. In 2000, plantations supplied one-third of the industrial round wood market worldwide. Some forecasts indicate that by 2040, almost half of all of the world's corporate round wood supply will come from plantations and planted forests.

The increasing globalization of marketplaces for forest products and services brings with it both new possibilities and challenges facing plantation-based forestry and forest firms. Value-added wood products not only meet the growing demands of the pulp and paper industry by providing fiber from fast growing plantations, but they also have the potential to develop into new domestic and international markets. However, substantial silvicultural interventions are required on both efficiently farmed and sustainably treated plantations of timber in order to fully optimize the cultivation and manufacturing of high-quality products.

Apart from generating wood and fiber, forest plantations offer various other ecosystem services like sequestering carbon, producing clean water, controlling the hydrological cycle, and enhancing the interconnectivity of landscape mosaics to preserve biodiversity and mitigate desertification. It is anticipated that in the future, the proportional value of these services given by forest plantations would rise.

South America has the largest proportion of plantation forests, accounting for 99% of the region's total planted forest area and 2% of its overall forest area. Europe has the lowest percentage of plantation forest, making up 6% of the planted forest estate and 0.4% of the overall forest area. Total 294 million ha are covered by planted forests, an increase of 123 million ha since 1990; nevertheless, the rate of spread has slowed since 2010.

Worldwide, forests that are naturally renewing make up 93% of all forests, with 7% being planted (Fig. 22.2). It's interesting to note that, with the exception of Asia, regional disparities are negligible. For the years 1990–2020, several classifications of introduced species and new forests in India plainly show an upward tendency (Table 22.1).

22.2 Defining a Plantation

A forest crop or stands that are intentionally grown by planting or sowing are referred to as plantations (Ford-Robertson 1971). Simple enough, right? However, the term artificial, or "man-made," as it is now used in the majority of contemporary forestry literature, has to be defined more. For instance, afforestation—also known as plantation—occurs when humans create a new forest on grassland. A comprehensive

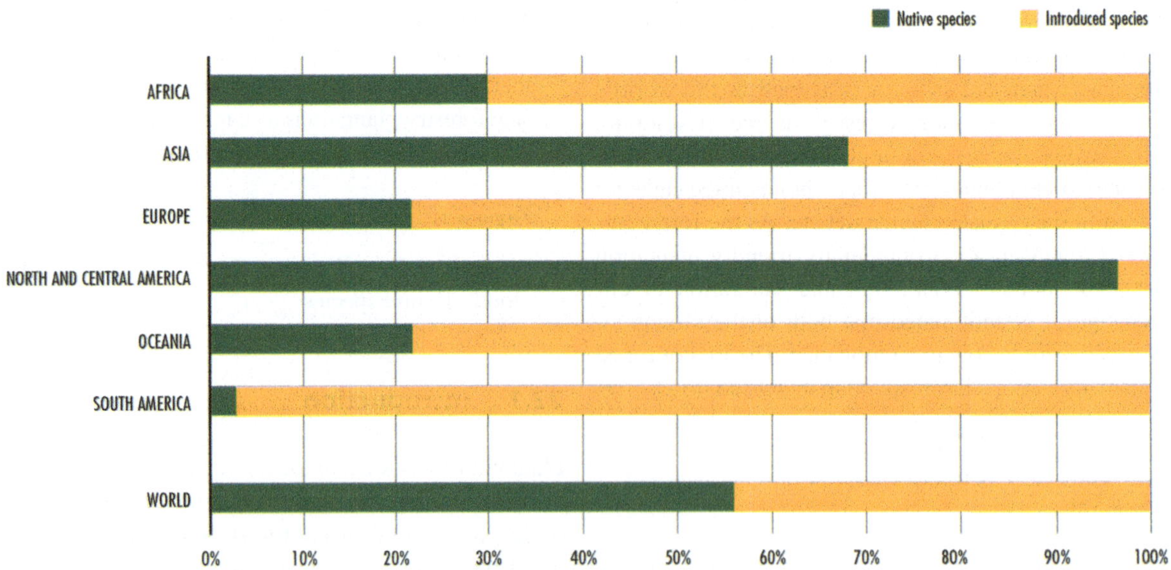

Fig. 22.1 Global status of the proportion of native and introduced species in planted forests (FRA 2020)

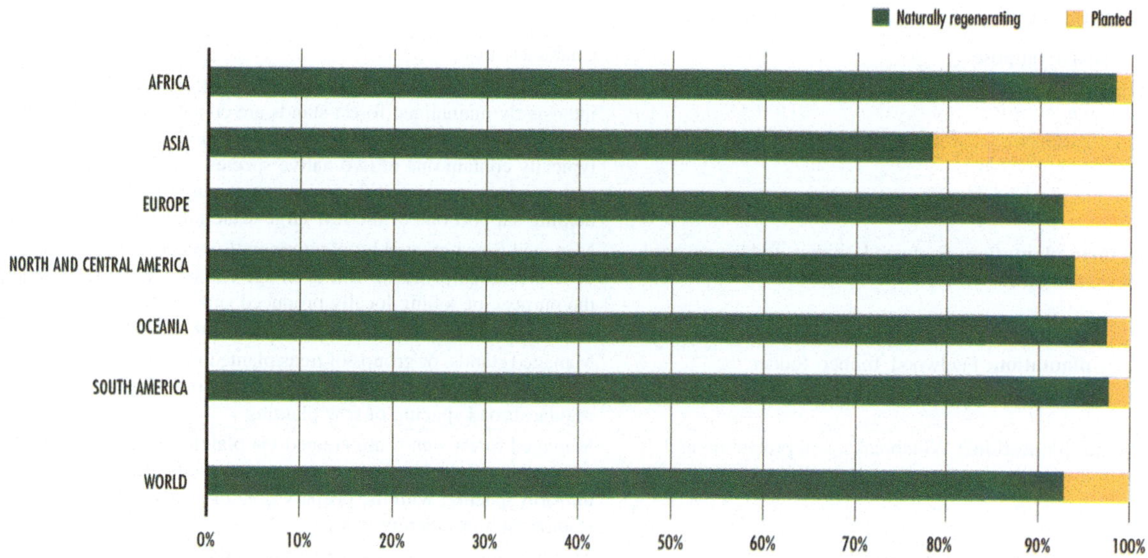

Fig. 22.2 Global status of the proportion of forests regenerated naturally and by plantations (FRA 2020)

Table 22.1 Changes in different categories of planted forests area in India (1000 ha) (Source: FRA 2020)

	Changes in different categories of planted forests area in India (1000 ha)								
	1990	2000	2010	2015	2016	2017	2018	2019	2020
Total planted forest	5715	9368	12,778.69	13,024	13,072	13,121.59	13,170.57	13,219.6	13,269
Plantation forest	3514	7167	9668.06	9854	9890	9927.49	9964.55	10,002	10,039
Introduced species	460	939	1266	1291	1296	1300	1305	1310	1315
Other planted forests	2201	2201	3110.63	3170	3182	3194.1	3206.02	3217.6	3230

list of planted forests is provided by CIFOR (2001) (Table 22.2).

22.3 Trends in Plantation Forest

The interest in tropical farms has grown quickly. Between 1965 and 1980, the plantation's acreage grew by roughly three times. Between 1990 and 2020, the global area of plantation forests expanded by 55.8 million ha, with the largest growth (21.2 million ha) taking place between 2000 and 2010. From 1.98 million ha in 1990–2000 to 2.12 million ha in 2000–2010, the average annual rate of growth rose. However, in the last 10 years, it decreased to 1.48 million ha annually. Between 1990 and 2020, the area covered by other planted forests rose by 66.8 million ha. From 2.08 million ha in 1990–2000 to 3.01 million ha in 2000–2010 and finally to 1.59 million ha in 2010–2020, the average annual rate of gain dropped.

Over the course of the previous three decades, Asia's plantation forest area has grown significantly, but at a decreasing average annual rate of expansion, from 1.26 million ha in 1990–2000 to 975,000 ha in 2000–2010 to 735,000 ha in 2010–2020. The average annual rate of increase in other planted forests more than doubled from 717,000 ha in 1990–2000 to 1.59 million ha in 2000–2010, but it then declined to 824,000 ha in 2010–2020. The major cause of the significant growth between 2000 and 2010 was China's adoption of extensive afforestation programs. Between 1990 and 2020, plantation forests retained a smaller share of Asia's planted forest area (Fig. 22.3).

In the most recent 10 years, the area covered by plantation forests in North and Central America grew at an average annual rate of 164,000 ha, decreasing from 270,000 ha in 1990–2000 and 420,000 ha in 2000–2010. In contrast to the prior growing trend of 94,200 ha in 1990–2000 and 38,200 ha in 2000–2010, the area of planted forests in Europe shrank by an average annual rate of 17,700 ha between 2010 and 2020. Sweden was the primary cause of the change from gain to loss, with an average yearly loss of 30,100 ha between 2010 and 2020. Africa's plantation forest area increased at a slower average yearly pace between 2010 and 2020 (totaling 55,300 ha) than it did between 2000 and 2010 (totaling 89,800 ha); the rate of increase between 1990 and 2000 was 25,300 ha. In 1990, plantation forests accounted for 70% of Africa's planted forest area; by 2020, that percentage had dropped to 67%.

Table 22.2 Different types of planted forests (CIFOR 2001)

Plantation type and purpose	Characteristics
Industrial plantation: Timber, biomass, food	Planting and/or seeding in the process of afforestation or reforestation, intensively maintained forest stands are created to supply material for sale locally or outside the immediate region. Individual stands or compartments typically contain one or two native species, one or more imported species, and an even age class with consistent spacing usually on a huge scale or helping support one of the few large-scale industrial businesses in the area
Home and farm plantation: Fuelwood, timber, fodder, orchards, forest gardens, and other	Managed forest created by planting and/or seeding during the afforestation or reforestation process, with even age classes and uniform spacing, with the purpose of selling locally produced subsistence wood. Typically, on a small scale and selling in a scattered market, if at all
Agroforestry plantation: Fuelwood, timber, fodder	Managed stands, or groups of trees planted in a local market or for personal use, have a positive impact on agricultural output since they often have regular, broad spacing or row planting
Environmental plantations: Windbreaks, soil protection and erosion control, wildlife management, site reclamation or amenity	Managed forest stands are created via planting and/or seeding during the afforestation or reforestation process, generally with even age classes and uniform spacing, with the primary goal of providing environmental stabilization or amenity value
Managed secondary forests with planting	Managed forest, where extra planting and/or seeding is used to preserve the productivity and composition of the forest

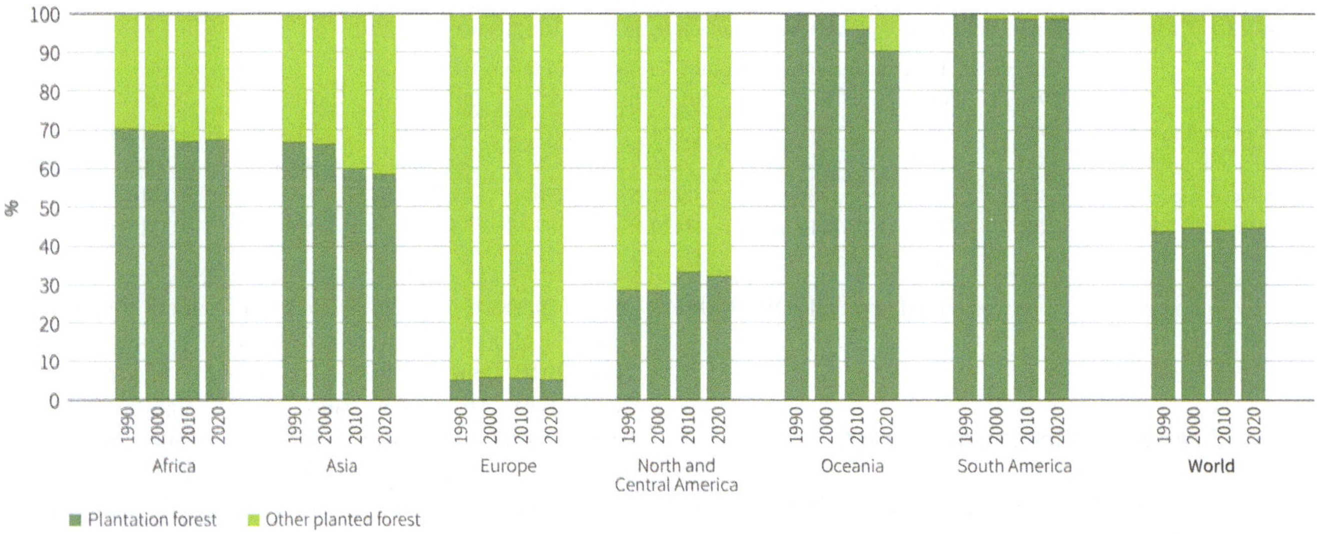

Fig. 22.3 Plantation forest and other planted forest as a proportion of total planted forest area, by region, 1990–2020 (source: FRA 2020)

Plantation forests accounted for 99.9% cent of the planted forest area in South America in 1990 and for about 99% in 2020; in Oceania, they comprised 99.7% in 1990 and 90.5% in 2020.

22.4 Factors Favoring Plantation Forests

22.4.1 Past and Continuing Destruction of Natural Forest

The natural forest has long been damaged by human activity, cut for use, and its extent has been steadily declining. While the loss of forests has been a recent phenomenon in some nations, it has been happening continuously for thousands of years in others. However, net deforestation has only occurred in the last 150–200 years in practically every nation, and the pace of disappearance has significantly accelerated recently. About 420,000,000 ha of forest were allegedly lost since 1990 as a result of converting to other land uses, despite the fact that the rate of deforestation has slowed during the preceding three decades. Between 2015 and 2020, deforestation has declined to ten million ha annually from 16 million ha per year in 1990s (Fig. 22.4). The area of primary forests worldwide has decreased by a total of over 80 million ha since 1990 (FRA 2020).

Net forest loss declined from 7.8 million ha during 1990–2000 to 5.2 and 4.7 million ha during the decades

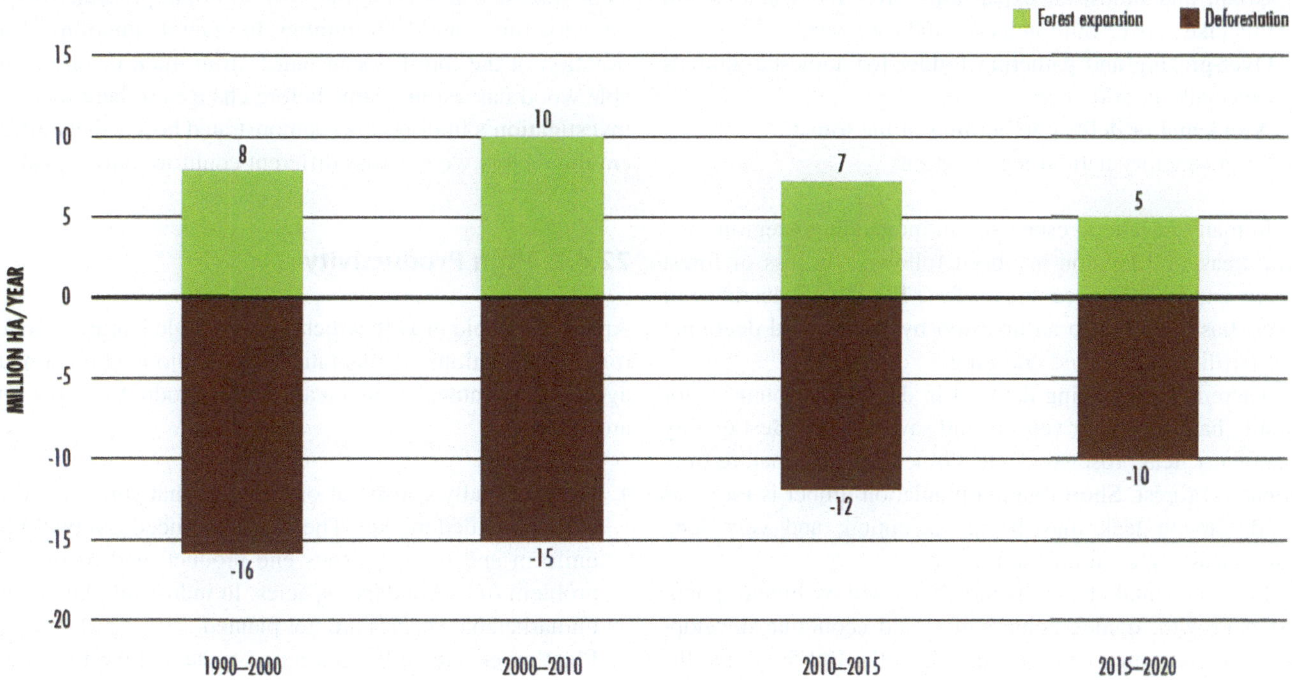

Fig. 22.4 Global forest expansion and deforestation, 1990–2020 (million ha per year). Source: FRA 2020

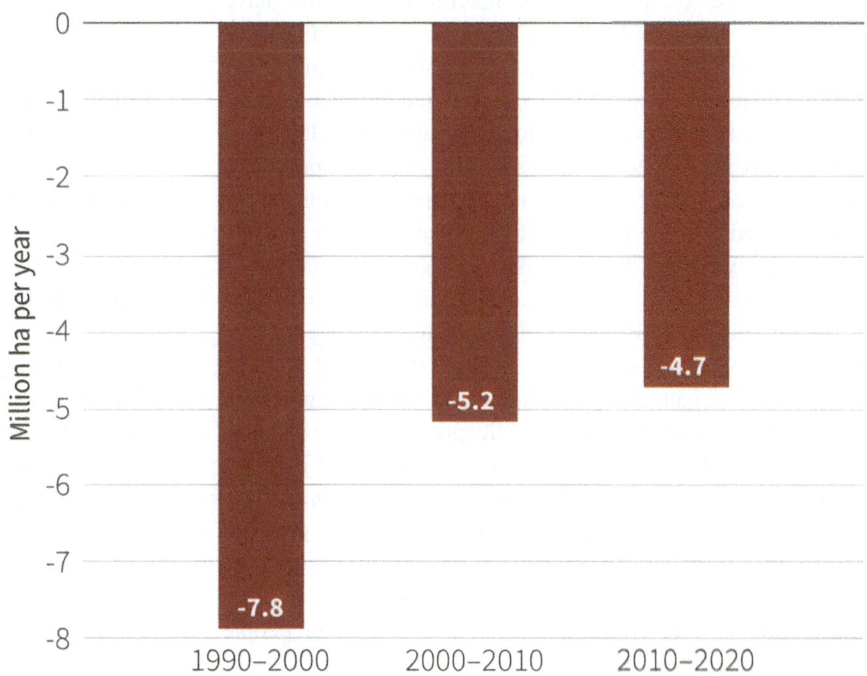

Fig. 22.5 Global annual forest area net change, by decade, 1990–2020. (Source: FRA 2020)

2000–2010 and 2010–2020 (Fig. 22.5) due to reduction in the rate of forest expansion.

The main causes of forest destruction are:

1. Agricultural expansion continues to be the main driver of deforestation and forest fragmentation and the associated loss of forest biodiversity.
2. Intensive logging for veneer, sawn timber and more recently for chip wood.
3. Exploitation for charcoal and firewood.
4. Shifting cultivation on very short cycles—250 million ha of tropical forests are presently affected by shifting agriculture.

5. Urban and industrial expansion—over the whole world this takes up 12 million ha of land each year.
6. Over-grazing and gathering fodder for domestic animals especially in arid areas.
7. Accidental or deliberate burning of the forest.
8. Unsatisfactory natural regeneration.

In many instances, especially in mountainous regions and arid areas, soil erosion has been followed by loss of forest cover, and land has become useless. Elsewhere clearance of forest has often led to an invasion by grasses and declining soil fertility on repeated cropping.

There is a continuing demand in developed countries for quality hardwoods for veneers and sawn timber. Best quality mahogany, teak, rosewood, etc. come from old, mature trees in natural forest. Short rotation plantation timber is rarely as good, though teak may be an exception, and very long rotations are uneconomic at present.

Environmental change is mostly caused by human population growth, demographic shifts, and economic development, as has long been recognized. In the last 50 years, the world population has doubled and the economy has expanded by over four times. In many nations, economic prosperity has allowed billions of people to escape poverty. But in the process, most of the world's natural ecosystems have undergone substantial change, mostly with detrimental effects on biodiversity and frequently also on the most vulnerable members of society, such as indigenous peoples. The recognized key stressors include loss and degradation of habitat, invasive species, unsustainable farming practices, low resource utilization efficiency, overexploitation, illicit logging, and wildlife trading. These forces have an increasingly negative effect due to climate change and variability.

The demand for agricultural and forest products is driven by factors such as food choices, loss and waste along agricultural value chains, and pressures from the global market, which in turn causes deforestation and forest degradation (IPCC 2019). Generally speaking, the primary driver of forest and forest biodiversity loss is the need to feed and power a growing world population. The primary risks to forest conservation in Africa are population pressure and poverty, which push impoverished farmers to convert forests to agriculture and extract wood fuel at unsustainable levels. In other places, the shifting consumption habits of wealthier people are the primary cause of deforestation.

According to Hosonuma et al. (2012), the most common cause of deforestation, accounting for 40% of it, is large-scale commercial agriculture, primarily cattle ranching, and the cultivation of soybean and palm oil. Approximately 78% of the forest area in such climatic regions is represented by the national data analysis covering 46 tropical and subtropical nations. An estimate of 33% of deforestation is attributed to local subsistence agriculture, 10% to urban growth, 10% to infrastructure, and 7% to mining. In several situations, deterioration of the forest, for instance, from illicit or unsustainable wood harvesting, came before changes in land use. The investigation's findings also demonstrated how widely different the factors were across different countries and regions.

22.4.2 High Productivity

Apart from rapid growth which is the notable feature of many tropical plantations, all plantations have following other management advantages which lead to high productivity per unit area.

1. Stands usually consist of one species that suits the needs of the intended market. The wood produced is a relatively uniform and homogeneous end-product, and there is no problem of "secondary" species. In industrial plantations, unmarketable species are not planted.
2. Planting ensures full stocking on a site and the fullest use of its potential. This is efficient land use and the high volumes per hectare lead to cheaper harvesting costs per cubic meter.
3. Initial spacing of trees, thinning regime, and rotation length can be manipulated to produce the desired mix of timber sizes—poles, pulpwood, and sawn timber.
4. A careful pre-planting survey can help match species with site conditions for optimum growth and identify all land unsuitable for planting. Only suitable species and suitable land need ever be planted.

22.4.3 Fast Growth

The most striking advantage of many plantations is rapid growth; and as mentioned above, fast growth is not only in relative volume but in relative value as well. All of the fast-growing crops consist of a "valuable," that is, usable species.

Generally speaking, mean yearly increments gained decrease with longer spins or drier circumstances. In India, teak (*Tectona grandis*) planted for 60–80 years typically grows at a mean each year increment of 4–8 m^3/ha/year. On 10–20 year rotations, *Eucalyptus camaldulensis* plantings in the dry tropics typically yield between 5 and 10 m^3/ha/year, but in moister locations, yields of up to 30 m^3/ha/year may be attained.

22.4.4 Environmental Forestry

Plantations are important not just as a source of wood but also in protecting the environment. In many countries, loss of forest cover has led to increased flooding, rapid desertification, siltation, and accelerated soil erosion. Extensive re-establishment of tree cover by plantations is the most effective means of reversing these processes. The other important environmental role of plantations or any tree planting is integration with agriculture-supportive forestry. It is becoming widely recognized that systems of agroforestry, where wood and food are grown together on the same piece of land, are the key to sustainable land management in many fragile ecosystems in the tropics.

These factors highlight the need for plantations if wood supplies are to be assured. Several positive factors favor plantation development. However, it would be a mistake to assume that plantations are the sole answer to future world wood supplies. Research must continue into the complex ecology of the mixed tropical forest so that their management is more than just selective cuttings, neglect, and hoping for regeneration.

Therefore, in conclusion forest plantations will play a very important role in future world wood supply. Moreover, suppose rates of afforestation substantially increase, in that case, plantations can begin relieving the pressure to exploit the world's dwindling reserves of natural forest, and this is the best reason why plantations should be taken up. As recommended by the Eleventh Commonwealth Forestry Conference (1980) "productions of wood for industrial purposes and fuel will have to increase in the coming decade and will increasingly rely on plantations and other intensive forestry practice."

22.5 Significance of Forest Plantations

The scientific community offers a variety of definitions for plantations of trees and even the standardized definition adapted by FAO has significantly changed over time. The main shared characteristics of these definitions point to forested areas artificially established by planting of seedling. Other features are that the trees usually belong to the same species (native or introduced), have the same age structure, and regularly spaced (FAO 2006; FRA 2010). During the context of greening or planting trees, forest stands created by transplanting and/or sowing is referred to as woodland plantations. They are either introduced or indigenous species that meet the minimum area requirement of 0.5 ha, tree crown cover of at least 10% or more of the total land area and total height of adult trees above 5 m (FRA 2000).

A variety of forest kinds are included in plantation-based forests. The majority of the trees on the property were either planted or seeded (sown), which is their one common characteristic. There is no internationally agreed definition of forest plantations. However, the expression "planted forest" is now widely used to embrace the continuum of forest types where forest origin is known to be by planting. Mostly plantations contain a single tree species, although plantations of a mixture of species are important in some parts of the world. Plantation forestry is often much like any other agricultural enterprise, aiming to grow highly productive forests on relatively small land areas, so the land is being used most efficiently. To achieve this, much attention is to be paid to the silviculture of plantations, that is, the tending of the trees to achieve some desired objectives.

The globe's inhabitants and economy are growing, and with them so is the total quantity of wood consumed for fuel, construction supplies, paper products, and many other purposes. Commercial forestry will play a bigger part in meeting the increasing requirement for wood. These days, owing to their high productivity, quick rotation periods, and mobility, woodlands from plantations are crucial for the growing amount of wood, especially in tropical countries. While some plantings are for conservation, the majority are for production (FAO 2006), and the proportion of plantations supplying round wood to the industry is increasing. Large-scale investments in monocultures are gradually giving way to small-scale investments in plantation forestry where the primary owners of the means of production are the local households and communities. Every continent has sizable plantation areas where wood is cultivated for industrial purposes. They often cover millions of hectares and are major commercial enterprises. Some are publicly owned and some are privately owned. They are important economically to the countries concerned, often earning export income and providing employment, both in their growth and processing of the wood they produce. Many such plantations are grown for only 7–15 year rotations before harvesting. Usually, they produce wood for paper making. Others are grown on 20–30 year rotations, by which time the trees will be large enough for their stems to be sawn to produce timber.

An emerging type of industrial plantation is being grown to produce bioenergy. The wood from such plantations is used to make energy, either through fermentation to produce ethanol as motor fuel or by burning the wood to generate electricity. The primary stimulus for establishing bioenergy plantations comes from concerns about the global warming of the atmosphere, which is believed to be caused by the release of greenhouse gases including CO_2. Bioenergy plantations are usually grown on very short rotations (3–5 years). The small size of trees at harvest is of no importance because they will simply be burnt or used for fermentation. Additionally, plantation forestry in the developing nations of tropical areas is very strongly nested on the political agenda due to the planted forest's carbon storage capacity as noticed in the climate change debate. New incentives such as the Clean Development Mechanism (CDM), afforestation/reforestation methodologies, and the debated mechanism of Reducing Emissions from

Deforestation and Degradation (REDD) emphasize plantations as a relatively cheap, clean, and harmless means of emission compensation.

22.6 Exotics in Plantation Forestry

The majority of the industrial plantations in most countries are being raised with exotic tree species like *Eucalyptus* spp., *Acacia* spp., and *Pinus* spp. The reasons for greater preference for the selection of exotic species include:

- There is generally a lack of adequate knowledge on indigenous species' propagation and silvicultural management.
- There is generally a plentiful supply of seeds of the exotic species.
- The exotic species are easy to handling.
- The exotics are fast-growing and high yielding.

In many countries, the use of indigenous species to grow forest plantations is undervalued. Promoting the identified priority indigenous tree species for national level plantations, along with addressing forestry-related emissions, is considered essential for coordinated research and reliable data. Many countries have already undertaken studies to identify suitable indigenous trees.

22.7 Trends in Introduced Species of Plantation Forests

For 170 nations, or 85% of the global forest area, data on trends in the amount of plantation forests made up of imported species was available. Between 1990 and 2020, the global area of plantation forests made up of imported species grew by 26.7 million ha. The percentage of imported species in the entire plantation forest area grew from 34% in 1990 to 44% in 2020. In Asia, this percentage nearly doubled over that time, rising from 17 to 32%; in all other areas, it decreased. East Asia grew at a faster rate than the rest of Asia; in China, for example, forests that were planted including immigrant species expanded via a score of 4.39 million ha in 1990 to a total of 14.2 million ha by 2020.

22.8 Future Trends in Plantations

Over the past 10 years, consumer appetite for timber and timber products has increased dramatically on a global scale, and this trend will probably continue in the coming years. As enormous geographic regions of tree plantations achieve maturity, the supply of industrial wood from plantations expands quickly on a global scale. In the last two decades, the vast majority of new plantation growth in the southern hemisphere has occurred in temperate and tropical zones. The output of industrial lumber from natural forests has decreased in Asia's top producing countries for tropical forest wood, and this has coincided with a notable drop in the supply of large-diameter timber at a reasonable price in recent years.

According to Spek (2006), a report on investments in the pulp business, Asian countries, especially China, today dominate new investments. This is despite the fact that in the 1990s, over 70% of new investments came from South America (specifically, Brazil and Chile) as well as Indonesia.

In the Asia-Pacific region alone, the yearly usage of hardwood pulp is expected to expand by 73 million cubic meters (Mm^3), while the yearly use of softwood pulp is expected to increase by 32 Mm^3. Forest products are being produced and used in many ways. Recent estimates indicate that the current demand from the Organization for Economic Cooperation and Development (OECD) member countries for forest sector goods would grow more slowly than it did in the past. At the same time, demand will continue to climb sharply in many economies that are developing or transitioning. This suggests a change in the consuming of wood-based products from North America, Japan, and Western Europe to the rest of Asia, Eastern Europe, and Russia (Barr and Cossalter 2004; Seppala 2007). A recent study by Carle and Holmgren (2003) found that, contingent upon the scenario being examined, the area of planted woodlands will increase from 261 million ha as it is now to approximately 303–345 million ha in 2030. This increase equates to 16–32% or 0.66–1.20% each year, during a 25-year period. In light of this, the annual production of wood from trees planted would increase from 1.4 billion m^3 in 2005 to 1.6–2.1 billion m^3 in 2030. The data above refer to forests that have been planted instead of plantations of timber; therefore, it is difficult to compare them with earlier research. Varmola et al. (2005) estimate that the annual commercial supply of timber from forest plantings will be 1.1 billion m^3 (46% of total) in 2040 and 970 Mm^3 (44% of total) in 2020. Informal carbon markets accounted for 72% of trades in the initial two quarters of 2009 and the majority of sales involving forest carbon in the last two quarters of 2008. In 2008, 5 Mt CO_2 quantities of credits were swapped in the voluntary markets, in accordance with Hamilton et al. (2009).

Notwithstanding the truth that forest restoration and afforestation contribute very little to the total Certified Emission Reductions (CERs) obtained according to the Clean Development Mechanism (CDM), ten novels A/R CDM initiatives were registered in 2009 as a result of the recent adoption of innovative project approaches. Predictions regarding the availability of lumber and carbon allowances outlined above indicate that plantations will become more prevalent in various areas. The world's plantations will not expand at the same rate in every location; further new estates will be constructed in Eastern and South Eastern Asia and, to some extent in South America.

22.9 Challenges for the Future

Planting forests can be a profitable way to acquire commercial round lumber. Their proportion of the worldwide manufacturing of industrialized hollow wood has increased from 5% to over 30% in the last 25 years. It is expected to generate half of the world's industrial round wood in the 25 years to come. This accomplishment is typically seen in the context of the numerous social and environmental issues that large-scale plantation enterprises have brought about. The overwhelming numbers of farmers with small holdings and forest dwellers have either contributed very little or nothing at all to the profitable endeavors started by massive or small- to medium-sized enterprises. In the way of wealth or other benefits, they have not gained anything from these firms' establishing, caring for, and harvesting of timber. In other instances, companies have established themselves consistently in large areas of land, displacing the natives from their small-scale farms or rural villages of origin. To address the issue, authorization programs together with additional safety measures—like promoting FAO's (2006) informal recommendations towards responsible stewardship of planted forests—have been put into place. To guarantee their widespread implementation in practice, there is still a lot of work to be done. Biodiversity has suffered as a result of large-scale plantings. This has, however, been heavily reliant on the land usage that the crops replaced. Plantations can contribute to the creation of connection in fragmented landscapes if they are designed and maintained appropriately (Kanowski et al. 2005; Marjokorpi 2006). However, fixed, single-objective plantation management will probably become less appealing in the future as biodiversity continues to disappear.

In order to effectively manage multifunctional landscapes, the plantation management paradigm must adapt. This may be achieved by including a variety of ecosystem services into the overall production function of forest planting schemes. The market places for natural resources are growing rapidly, and forest plantings will be important to the global carbon markets as well as the community reimbursements for ecological amenities regimes that manage water resources. Both of these fields are expected to grow in significance as the contemporary model for environmentally sound plantation development includes efforts to mitigate and adapt to changes in the climate.

Lessons Learnt

- Plantation produces clean water, regulates the hydrological cycle, and plays a significant part in developing ecological services, among them the preservation of carbon dioxide and opening of worldwide carbon markets.
- Natural forests have been damaged by humans, destroyed for development, and exploited. An estimated 420 million ha of forests have been destroyed as a result of other land uses taking over.
- Agricultural expansion continues to be the main driver of deforestation, forest fragmentation, and loss of forest biodiversity.
- In many delicate tropical ecosystems, agroforestry systems—which cultivate food and wood on the same plot of land—are essential to sustainable land management.
- Plantations in forests have demonstrated their capacity to provide industrial round wood. In the previous 25 years, the amount of round wood produced on plantations has increased from 5% to over 30%.

Key Questions

1. Describe the significance of plantation forestry in today's scenario.
2. How does the plantation forestry help in combating climate change?
3. What is the role of plantation forestry in securing the livelihood in areas with a fragile ecosystem?
4. How does the carbon sequestration and carbon trading be achieved through plantation forestry?
5. How does the plantation forestry help in enriching the natural resources?

References

Barr C, Cossalter C (2004) China's development of a plantation based wood pulp industry: government policies, financial incentives and investment trends. Int For Rev 6:3–4

Carle J, Holmgren P (2003) Definitions related to planted forests. UNFF Intersectional Expert Meeting on the Role of Planted Forests in Sustainable Forest Management, Wellington, pp 329–343

CIFOR (2001) Typology of planted forests. CIFOR Info Brief, Center for International Forestry Research (CIFOR), Bogor

FAO (2006) Responsible management of planted forests: voluntary guidelines. Planted forests and trees working paper 37/E, Rome

Ford-Robertson FC (1971) Terminology of forest science, technology, practice and products. IUFRO/Society of American Foresters, Washington, DC

FRA (2000) Global forest resources assessment 2000. Food and Agricultural Organization of the United Nations, Rome

FRA (2010) Global forest resources assessment 2010. Food and Agricultural Organization of the United Nations, Rome

FRA (2020) Global forest resources assessment 2020. Food and Agricultural Organization of the United Nations, Rome
Hamilton K, Sjardin M, Shapiro A, Marcello T (2009) Fortifying the foundation: state of the voluntary carbon markets 2009. New Carbon Finance, New York
Hosonuma N, Herold M, De Sy V, De Fries RS, Brockhaus M, Verchot L, Angelsen A, Romijn E (2012) An assessment of deforestation and forest degradation drivers in developing countries. Environ Res Lett 7(4):044009
IPCC (2019) Climate change and land: an IPCC special report on climate change, desertification, land degradation, sustainable land management, food security, and greenhouse gas fluxes in terrestrial ecosystems. https://www.ipcc.ch/srccl
Kanowski J, Catterall CP, Wardell-Johnson GW (2005) Consequences of broadscale timber plantations for biodiversity in cleared rainforest landscapes of tropical and subtropical Australia. For Ecol Manag 208(1–3):359–372
Marjokorpi A (2006) Biodiversity management in fast-growing plantations: a case study from West Kalimantan, Indonesia. Annales Universitatis Turkuensis, Serie AII, Part 200
Seppala R (2007) Global forest sector: trends, threats and opportunities. In: Freer-Smith PH, Broadmeadow MSJ, Lynch JM (eds) Forestry and climate change. CAB International, pp 25–30
Spek M (2006) Financing pulp mills: an appraisal of risk assessment and safeguard procedures. Center for International Forestry Research (CIFOR), Bogor
Varmola M, Gautier D, Lee DK, Montagnini F, Saramaki J (2005) Diversifying functions of planted forests. In: Mery G, Alfaro R, Kanninen M, Lobovikov M (eds) Forests in the global balance – changing paradigms. IUFRO world series, vol 17. International Union of Forest Research Organizations (IUFRO), Helsinki, pp 117–136

Urban Forestry: Scope and Prospects

23

C. Buvaneswaran and A. Balasubramanian

Abstract

It is estimated that one-third of human population will be living in urban cities by end of 2030, which will lead to emission of substantial air contaminants. Increasing the urban green cover/tree cover as 'Urban Forests' only will be able to minimize the deleterious degradation of urban environment. Further, in the context of climate change, urban green will act as potential sink for CO_2, thereby mitigating global warming. In a broad sense, the benefits from urban forestry can be grouped into (1) air quality improvement, (2) moderation of temperature, (3) abatement of noise pollution, (4) sequestration of carbon, (5) maintenance of biodiversity, (6) providing social benefits of people's health and recreational values and (7) supply of forest products to poor urban dwellers and fodder to cattle. Besides these benefits, urban forests will also aid in achieving some of sustainable development goals (SDGs) of United Nations.

To harness such regional and global benefits from urban forests, a scientific evaluation of species is required to aid in choosing appropriate tree species for the given environmental conditions, particularly for the release of organic compounds of volatile nature (VOCs) and for their air pollution tolerance. Once suitable tree species are selected, the major activities to be followed are (1) having urban tree nurseries to supply certified quality planting materials produced from superior trees, (2) adopting innovative planting techniques suiting to the urban environment in which seedlings will be planted and (3) having definite planning for management of planted tree seedlings particularly for watering and monitoring and managing pests and diseases.

Community participation is an important dimension of any successful urban greening programme. Some of the most successful community-lead urban forestry programmes are Tree City USA, Tree Canada, Greening Chandigarh, *Vanathukkul Tirupur* (Tirupur city within forest) which can be adopted to achieve a minimum green space of 9 m^2 per person as prescribed by Food and Agriculture Organization of the United Nations (UN-FAO). A sound legal framework complimented with appropriate powers to the designated authorities of urban planning would foster healthy and people friendly urban surroundings and green spaces.

Keywords

Sustainable development goals (SDGs) · Urban forestry · Urban landscape · APTI · Urban tree cover

23.1 Introduction

Urbanization It estimated that one-third (UNDESA 2018) to two-third (Schell and Ulijaszek 2018) of human population will be living in urban cities between 2025 and 2030. This proportion of urban population is still on higher end in developed nations like 80% in USA (Wolf 2018) and 85% in Australia (Brack 2002). This increasing trend of urbanization will have bearing on successful implementation of sustainable development goal 11 (SDG11)—relating to bringing cities as inclusive, safe, resilient and sustainable (Fig. 23.1).

This trend of urbanization in India has witnessed an unexpected peaking from urban population of 6 crores in 1947 to 40 crores over the period of 50 years (Chakrabarti 2001) and expected to be 70 crores by 2050 (CII 2010). According to Census India, the proportion of urban population is anticipated to be 40% by 2030 with the present rate of urbanization (GoI 2015). The count of megacities in India is also in increasing trend and there are more than 35 cities with around 10 crore population as per GoI (2011). These

C. Buvaneswaran (✉)
Institute of Forest Genetics and Tree Breeding, Indian Council of Forestry Research and Education, Coimbatore, Tamil Nadu, India
e-mail: buvanesc@icfre.org

A. Balasubramanian
Forest College and Research Institute, Tamil Nadu Agricultural University, Mettupalayam, Tamil Nadu, India

A. K. Mandal, A. Nicodemus (eds.), *Textbook of Forest Science*, https://doi.org/10.1007/978-981-97-8289-5_23

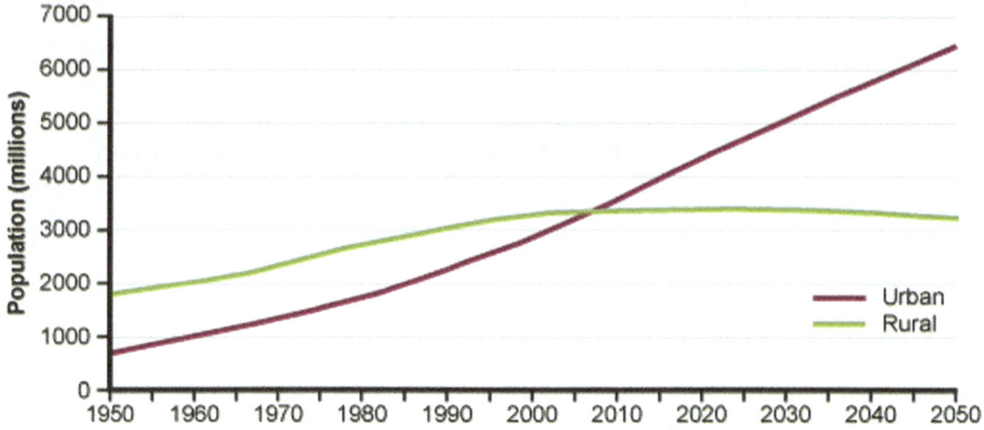

Fig. 23.1 Trend in urban and rural population growth from 1950 to 2050 (UNDESA 2018)

megacities will have 14% of the urban population globally in the year 2025 (McKinsey Global Institute 2010). The pronounced increase in population eventually results in over exploitation of energy and other natural resources, which is a major contributor of urban pollution. A greater concentration of particulate matter, aerosols, SO_2 and NO_2, GHGs, and ozone depleting compounds is being recorded in urban vicinities in India. In comparison to other cities in the world such as Vienna and Tokyo, cities in India are termed as 'most polluted' and 'unsustainable' Further, green cover is also getting affected by increasing population in urban areas. For example, a very less green area of per person is being recorded in Chennai (0.46 m^2) and Mumbai (0.12 m^2) (Srivathsan 2013; FAO 1998), as compared to a minimum green cover of 9 m^2 per person which is recommended by Food and Agriculture Organization of the United Nations (UN-FAO). Considering the importance of tree cover in urban environments, Government of India (GoI) has implemented a mega urban green cover programme through 'Green India Mission' and envisioned it in its ambitious project of 'Smart Cities' for increasing and safe-guarding open spaces, for enhancing quality of urban life, for evading the heat effects in urban areas and for assuring ecological balance.

Problems of Urbanization Manifold environmental problems like increasing levels of pollutants are being observed owing to peaking in trend of urbanization and industrialization (CPCB 2010). The escalating levels of pollution can also be attributed to massive vehicular pollution, improper solid waste management, conversion of lands for real estates, generation and poor processing of e-wastes, deforestation and degradation of land. Mushrooming metal smelting industries, refineries, cement factories, distilleries, pulp and paper industries in urban cities as well are causative to increasing pollution levels in air, water and soil. Two major pronounced problems in many parts of the urban cities across the world facing today are poverty and environmental degradation. (https://www.nationalgeographic.com/environment/habitats/urban-threats/).

Air Quality Assessment Impacts of various developmental activities on air quality are a vital focal point in many of environmental impact assessments (EIAs). Substantial air contaminants may be emitted from such developmental activities and the studies related to these issues are termed 'air quality impact assessments' (DiGiovanni and Coutinho 2017). By considering both regional and global levels of health of human population vis-a-vis impact of air pollution, it is imperative to take up local actions towards controlling and monitoring of air quality (Koch et al. 2018). Numerous establishments are carrying out assessment and monitoring of air quality at regional levels. One such air quality assessment done by Department of Silviculture and Natural Resources Management, Forest College and Research Institute, Tamil Nadu Agricultural University, Mettupalayam, is presented here as an example. In this study, a mobile monitoring facility (air quality monitoring of pollution levels in real time) was utilized for monitoring ambient air quality in real time (NO_x, SO_2, O_3, PM 2.5 and 10.0) at different zones, namely forest area, tribal villages, town (municipality), state highways and brick kiln areas (Fig. 23.2). The town (municipality) area nearer to the bus stands experiencing greater amount of pollution from vehicles (carbon monoxide). It is essential to note that sulphur dioxide (SO_2), nitrogen dioxide (NO_2) and particulate matter (PM 2.5) were observed maximum in brick kiln area (Table 23.1). Hence, the air quality status in terms of SO_2, NO_2, O_3, CO and PM 2.5 in the town area was low due to maximum tree population around the buildings. This clearly indicates that trees offer greatest scope for pollution abatement (Balasubramanian and Hariprasad 2020).

Urban Forestry: As Solutions to the Problem of Urbanization Proper provisions for introduction of trees in city and town planning will help to abate many environmental impacts, moderate climate, reduce carbon

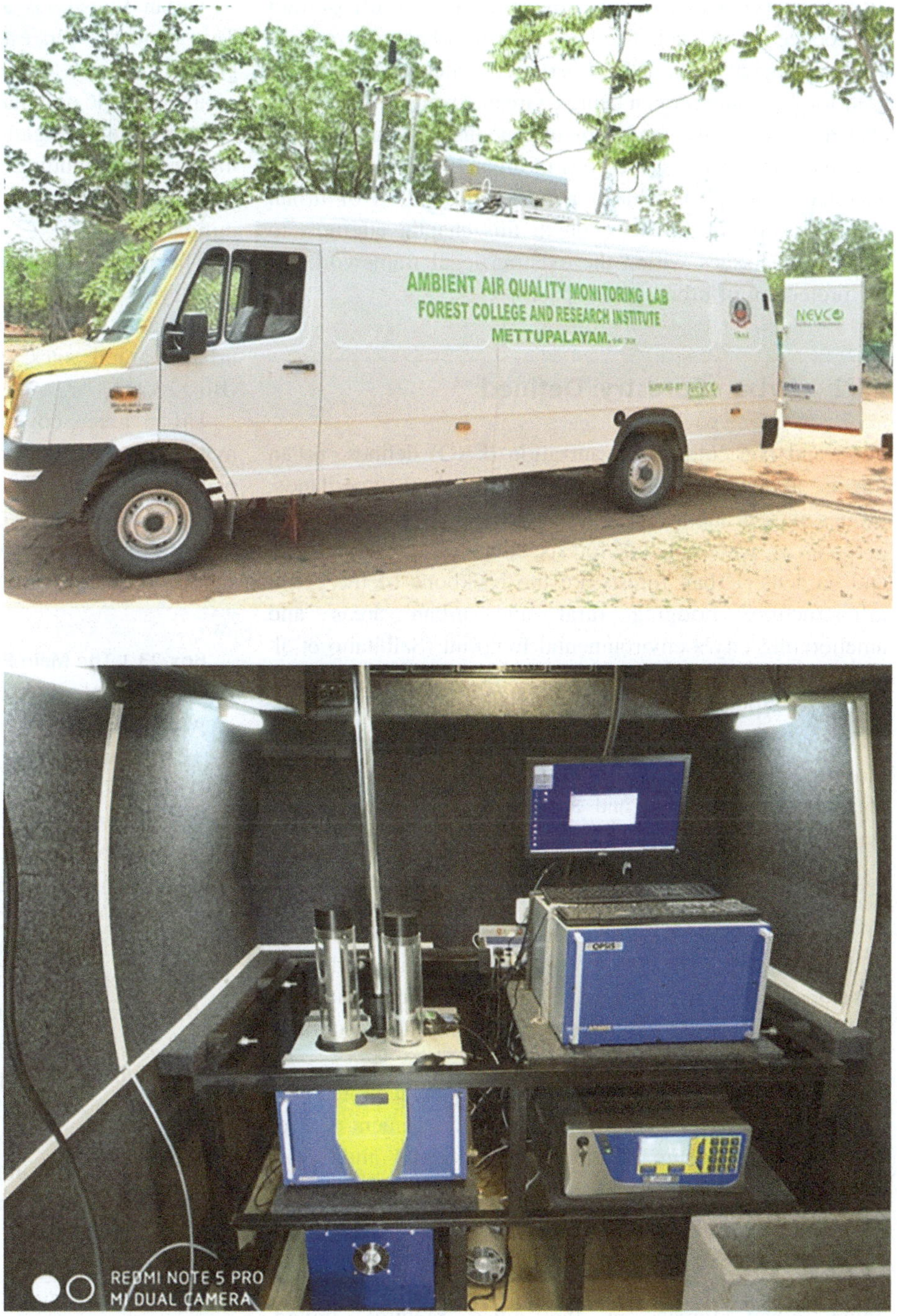

Fig. 23.2 Mobile van-based air quality monitoring facility (real-time air quality monitoring of pollution) used for air quality assessment by Forest College and Research Institute, Mettupalayam, Tamil Nadu, India

Table 23.1 Ambient air pollutant status in different zones (Real time monitoring)

S. no.	Name of the pollutant	Forest area	Tribal village	Town (Municipality)	State highways	Brick kiln area
1	Sulphur dioxide (SO_2) μg/m^3	6.43–25.40	7.15–16.02	6.99–31.01	10.65–43.17	10.05–155.22
2	Nitrogen dioxide (NO_2) μg/m^3	1.96–28.08	4.43–19.85	0.15–33.45	5.21–74.98	3.15–117.85
3	Ozone (O_3) μg/m^3	6.08–19.73	12.02–25.62	6.55–21.70	8.65–25.35	13.72–52.20
4	Particulate matter (PM 2.5) μg/m^3	7.80–22.40	7.30–21.20	18.90–54.70	51.00–66.10	70.00
5	Carbon monoxide mg/m^3	0.13–1.81	0.080–0.770	0.220–2.220	0.460–1.940	0.130–1.950
National Ambient Air Quality Standards		SO_2 μg m^{-3}	NO_2 μg m^{-3}	O_3 μg m^{-3}	PM 2.5 μg m^{-3}	CO mg m^{-3}
Industrial, residential, rural and other areas		80	80	100	60	2

Source: Balasubramanian and Hariprasad (2020)

dioxide, improves air quality, lower surface run off by rain and reduce noise level. The tree cover in urban area renders aesthetic appeal. Avenue trees serve as shelters/resting places both for birds and human beings. Green cover in urban areas also ensures the city's infrastructure against natural disasters. The trees also serve as reservoirs of urban biodiversity and an excellent source for education and conservation of urban natural resources. In this context, this chapter attempts to analyse the issues within the existing legal framework for the protection of urban green spaces.

23.2 Urban Forestry: Defined

Food and Agricultural Organization (FAO) defines 'urban forests' as networks or systems comprising all woodlands, groups of trees and individual trees located in urban and peri-urban areas, which include forests, street trees, trees in parks and gardens. Urban forests are the backbone of the green infrastructure, bridging rural and urban areas and ameliorating city's environmental footprint (Salbitano et al. 2016).

Helms (1998) has defined it as the art, science and technology of trees and forest resource management in and around urban community ecosystems for the physiological, sociological, economic and aesthetic importance that the trees provide to the society.

Ismail and Jusoff (2004) defined 'urban forestry is a practice of developing and managing scientifically suitable types of woody plants which thrive in the populated environment of the cities and are influenced by urban population and urban development, which provide a sustainable environment along with other physical, ecological, recreational and economic benefit'.

Miller (1988) defined urban forestry as 'an integrated, city wise approach to the planting, care and management of trees in the city to secure multiple environmental and social benefits for urban dwellers'.

Urban forestry is the practice of managing urban forests to ensure their optimal contributions to the physiological, sociological and economical well-being of urban societies. Managing urban forests in the cities requires united, multidisciplinary, participatory and planned approach. The scale of operation of urban forests can be from single trees to landscapes (Salbitano et al. 2016).

Creation of urban greenery can be a way to connect people with nature. More than for direct benefits, urban greenery is commonly known for intangible benefits like pollution abatement, energy saving in households, recreational values and carbon capturing, etc. In simple term, urban forestry can be defined as planting and caring of trees in cities, including protecting and managing of trees—the communal resource.

Urban forestry is a new and emerging science subject. This subject covers establishment and management of urban greenery mainly for their ecosystem services like abatement of atmospheric pollutants, generation of oxygen, biodiversity conservation, mitigation of noise pollution, microclimate moderation including reducing of urban heat island effect, recharging of groundwater table and climate change mitigation.

The most common examples for the idea of urban forestry—which focus on all tree resources in and around urban cities—are woodlots, parks and botanical gardens, urban natural sites, avenue trees and block plantations (Box 23.1).

Urban green covers are essential both for their above-mentioned ecosystem services and for social interconnection, psychological and physical well-being of urban population. Urban forest such as 'Delhi Ridge'—the Reserve Forest in New Delhi, India, can also act as green lungs of the city besides providing extremely valuable ecosystem services.

Box 23.1 The Main Principles of Urban Forestry

The subject of urban forestry has several principles, which include:

Urban forestry is integrative and all-inclusive: 'Urban Forest' is a holistic approach to incorporating various green-space elements into a unified system within urban areas. This approach is designed to promote a comprehensive perspective on urban planning and land use, emphasizing the integration of diverse elements for sustainable and multifunctional outcomes. The concept recognizes that sustainability isn't limited to traditional forest resources alone. It extends to other systems like agroforestry and lined tree plantings, emphasizing the importance of diverse and sustainable land-use practices. This principle of integration also extends to land ownership considerations by having a comprehensive understanding of tenure aspects to ensure responsible and equitable management (Mock 2004; Pauleit et al. 2005). This approach aligns with the broader goals of sustainable development and resilient urban ecosystems.

Urban forestry is tactical: The concept of urban forestry extends beyond immediate urban development to encompass long-term policies and plans. This forward-looking perspective aims to address the evolving needs for tree resources and the challenges associated with urbanization. This could involve cooperation with urban planning, environmental conservation, public health, and other relevant domains to create a more comprehensive and joint approach to urban

(continued)

Box 23.1 (continued)
development. Urban forestry also seeks to address the challenges posed by forced migration of rural people, including increased pressure on urban resources and the necessity for sustainable urban development (Mock 2004; Ottitsch and Krott 2005; UN Habitat 2004).

Urban forestry is interdisciplinary: Urban forestry thrives on the active involvement of professionals from diverse disciplines, fostering interdisciplinary collaboration to address the complex challenges of urban development. This inclusive approach is essential for creating well-rounded strategies and solutions that go beyond traditional sectoral boundaries (Miller 1997; Nilsson et al. 2005).

Urban forestry is for social inclusiveness: Urban forestry is not the mere management of trees; it involves creating collaborative partnerships, fostering community involvement, and establishing organizations which stand for transparency, responsibility and unbiased access to benefits. This approach recognizes the interconnectedness of environmental, social and governance aspects in sustainable urban development (Mock 2004; Van Herzele et al. 2005; Jones et al. 2005).

Urban forests are multifunctional: Urban forestry recognizes the diverse needs of urban societies and responds by providing a variety of economic, environmental and socio-cultural benefits. This integrated approach underscores the need of urban forests in enhancing the quality of life for residents and promoting sustainable and resilient urban environments (Mock 2004; Tyrväinen et al. 2005).

Source: Konijnendijk and Gauthier (2006)

23.3 Benefits of Urban Forestry

Urban greening and urban forests offer a comprehensive set of benefits that span environmental sustainability, economic viability and societal well-being (IADB 1997). Recognizing the multifaceted advantages, these approaches play a crucial role in creating healthier, more resilient and more liveable urban environments by providing a list of benefits as detailed below;

1. **Environmental benefits**
   - **Air quality improvement:** Urban greening helps absorb air pollutants, contributing to improved air quality. As of 2016, the World Health Organization (WHO) reported that a substantial proportion of the population in the globe (91%) was residing in areas with poor air quality by not meeting the guidelines of the WHO. The estimate for 2016 indicated that approximately 42 lakhs deaths of infants could be ascribed to the effects of outdoor air pollution. Mexico City experienced a substantial increase in the average level of particulate suspension in the atmosphere over a period of 16 years, from 1974 to 1990. The rise from 65 mg/m^3 to 400 mg/m^3 reflects a notable deterioration in air quality during that time (Carter 1993). The adverse effects of such detrimental air contaminants are particularly pronounced among vulnerable populations, including children, the elderly and individuals with respiratory problems, as they are more susceptible to the health impacts of poor air quality. Urban greening, characterized by the introduction and presence of urban greenery, can play a crucial role in mitigating such detrimental air pollution. To voucher this benefit of urban forests in reducing air pollution, the city Santiago, Chile, has named its urban parks programme as the 'Lungs of Santiago', emphasizing the role of green spaces in reducing air pollution and enhancing overall urban environmental quality. Nowak (2006) also emphasizes on institutionalizing urban forestry as a form of 'biotechnology' and highlights the scope of urban forests to contribute significantly to environmental quality, particularly to air quality.
   - **Water management:** Green spaces increase water catchment areas and contribute to flood plain surfaces, assisting in better water management.
   - **Soil stabilization:** Urban forests play a role in stabilizing soils, preventing erosion and enhancing the overall health of the land.
2. **Climate regulation**
   The positive influences of urban vegetation on moderating climate extend beyond aesthetics. They include tangible benefits for human comfort and energy efficiency in buildings. Effective urban design and planning, considering the size, spacing, and design of vegetated areas, can optimize these influences, making urban environments more liveable and sustainable. The research findings from Baris et al. (2009) also emphasize the pivotal role of trees and open areas in mitigating the impacts of urban development on the local climate. The characteristics of vegetation and its distribution throughout the city are critical factors and even parks of varying scales can play a substantial role in regulating heat and contributing to a more favourable urban climate. This reinforces the importance of incorporating green infrastructure and thoughtful urban planning to enhance the resilience and sustainability of cities.
3. **Noise reduction**
   Urban forests contribute to noise reduction, providing a quieter and more pleasant living environment. Carter

(1993) and Miller (1988) highlight the valuable role of trees and vegetation in mitigating noise pollution, particularly in urban areas with poor populations facing high noise exposure. The key ways in which trees and vegetation can contribute to reducing noise pollution are:

- **Sound absorption:** Leaves, twigs, branches and other parts of herbaceous plants contribute to sound absorption. They absorb sound energy, converting it into other forms and reducing the overall noise levels in the surrounding environment.
- **Deflection:** Tree and plant barriers have the ability to deflect sound away from listeners. This means that the physical presence of vegetation alters the path of sound waves, directing them away from sensitive areas or residences.
- **Reflection:** When vegetation is strategically placed, such as at right angles to the source of noise, it can reflect sound back to its source. This can be an effective means of addressing noise issues, particularly in areas where sound reflection is beneficial.
- **Refraction:** Vegetation can cause refraction, where sound waves bend around the objects (leaves, branches) and are thereby dissipated. This bending or redirection of sound contributes to a reduction in overall noise levels.
- **Masking:** Vegetation contributes to masking by introducing more pleasing sounds that can cover up or distract from unwanted noise. Natural sounds such as bird songs or rustling leaves can serve as a positive auditory backdrop, helping individuals filter out unwanted urban noise.
- **Filtering out unwanted noise:** People have the ability to selectively listen to sounds they find more pleasant. In the presence of vegetation, individuals may consciously or subconsciously choose to focus on the sounds of nature, helping to minimize the impact of city noise.

4. **Economic benefits**
   - The material benefits derived from urban greening, especially in terms of providing poles, firewood, and fodder, contribute to the sustainability and resilience of communities, particularly in regions with specific demand for such resources.
   - Marketable products: Urban forests can yield marketable timber and agricultural products, contributing to the local economy.
   - Non-timber forest products: Besides timber, urban forests provide non-timber forest products like artisan wares and honey, offering additional economic opportunities.
5. **Societal benefits**
   - Health and well-being: The presence of green spaces positively influences the mental and physical health of urban residents.
   - Recreation and education: Urban forests serve as recreational spaces and outdoor classrooms for environmental education.
   - Aesthetic improvements: Greenery in urban areas enhances the overall aesthetic appeal, providing a welcome contrast to the often-dominant concrete and asphalt landscape.
   - Incorporating greenery into urban planning not only enhances the aesthetic value of cities but also contributes to the well-being and satisfaction of residents. It creates a more liveable, sustainable, and enjoyable urban environment (Box 23.2).

Box 23.2 Valuing Recreational Benefits of Urban Forestry: A Case Study of Chandigarh, India

The study of Chaudhry (2013) presents more understandings into the intangible value of recreational benefits derived from urban greenery in Chandigarh, India. The research focuses on various green spaces within the city, including parks, gardens, tree avenues, forests and the Sukhna Wildlife Sanctuary. The primary objectives were to quantify the recreational value and in turn for estimating willingness to pay for greenery and a liveable environment.

Two methods, namely the contingent valuation method (CVM) and the travel cost method (TCM), were employed to calculate the willingness to pay. The study found that willingness to pay for the improvement of present green landscape character and the creation of new parks and gardens by every family in the city was Rs. 153 per annum for 5 years. This is converted to recreational value of the urban forestry capitals of Rs. 275 lakhs per year in the period 2002–2003.

The CVM method was adopted to collect primary information from 2358 peoples in the city. Additionally, CVM and TCM were used to estimate the annual recreational value for tourists. The TCM estimate was more reliable in the Indian context and was calculated to be Rs. 924 lakhs.

Therefore, the annual recreational value of the parks, gardens, avenues, forests, wildlife sanctuary and other landscapes in the city was approximately Rs. 1200 lakhs in the year 2002–2003. It's important

(continued)

Box 23.2 (continued)
to observe that this value corresponds to the recreational value only and not representing the environmental or ecological value of these assets.

The study highlights the dual role of urban forestry in Chandigarh, playing a crucial part in both attracting tourists and satisfying the preferences of residents. The emphasis on greenery in the city's planning and the recognition of its importance by residents demonstrate the power of well-designed and maintained urban green spaces in enhancing the overall attractiveness and liveability of a city.

6. **Equity Considerations**
   - Benefits for low-income residents: Urban forests contribute to the well-being of low-income residents by offering recreational opportunities, access to nature and environmental education.

In the current scenario of global warming, urban forests have a crucial role to play as a sink for CO_2 and mitigating global warming. In a broader sense, the benefits from urban forestry can be grouped into (1) air quality improvement, (2) moderation of temperature, (3) abatement of noise pollution, (4) sequestration of carbon, (5) maintenance of biodiversity, (6) providing social benefits of people's health and recreational values and (7) supply of forest products to poor urban dwellers and fodder to cattle. These benefits of urban forestry are briefly explained below (Table 23.2 and 23.3).

23.4 Evaluating Tree Species for Urban Forestry

Species vary in their ability to tolerate different abiotic stresses in urban environment. Hence, a scientific evaluation of species will aid in selecting tree species for the given environmental conditions. Also, tree height, canopy form, branching traits, flowering behaviour and leaf texture determine their selection for the particular planting sites for the specified functions in urban forests.

Sahu and Panda (2001) conducted a study examining the morphological characteristics of trees within the urban landscape of Calcutta, West Bengal, India, with a particular focus on their effectiveness as pollutant sinks. The investigation delved into the traits of leaves, tree canopy and bark surface. Through the evaluation of these characteristics, the study assessed the efficiency of these trees in acting as sinks for pollutants. The emphasis was placed on understanding how the observed traits, especially those of leaves, bark surface, and tree canopy, contribute to their suitability for pollution abatement and the strategic design of urban forestry initiatives.

Trees within urban environments face various stresses, and the selection criteria for urban tree improvement programmes should be based on the specific stresses these trees are going to encounter. Fundamental tree traits such as climatic adaptation, disease resistance and phenotypic plasticity are essential. Urban-related stresses are arising from social factors, confined soil volume, limited crown space, soil and air pollution, wind, and drought. Additionally, criteria such as aesthetic considerations, growth form, growth potential and resistance to limb breakage play crucial roles in the selection process. The prioritization of these criteria depends on the specific urban environment where the plants are intended for use, as indicated in a study conducted by Saebo et al. (2003).

Drunasky and Struve (2005) investigated the morphological and physiological reactions to substrate moisture stress in two oak tree species, namely Bur oak (*Quercus macrocarpa* Michx.) and chestnut oak (*Q. michauxii* L.). The researchers exposed container-grown seedlings to no moisture stress, a single stress cycle or two stress cycles. Notably, the response to these stress cycles was more pronounced between the two species rather than within each species. The researchers suggested that *Q. michauxii* may be better suited for urban planting sites compared to *Q. macrocarpa*. This preference was attributed to more effective water-absorbing root system of *Q. michauxii* and its apparent ability to thrive in shallow soils.

Leaf Area as Selection Criteria for Urban Forestry Management Tool Numerous advantages obtained from urban forests are linked, either directly or indirectly, to the leaf area of these urban green spaces (Kenney 2000).

Lacan and McBride (2009) propose that when choosing trees for urban areas, consideration should be given to minimizing potential disruptions to city infrastructure. The conflict between city trees and infrastructure remains a challenge in urban forestry. The researchers observed that municipal Wi-Fi, constituting a city-wide wireless computer network, might become an integral part of urban infrastructure. However, they noted that trees could potentially attenuate Wi-Fi signals, leading to a possible clash between urban trees and municipal Wi-Fi. They suggested that by factoring in the impact of urban trees during the planning of Wi-Fi networks, any interference with the operation of municipal Wi-Fi could be effectively avoided.

Urban Green Zones and Related Pollen Allergy Carinanos and Porcel (2011) examined the primary factors contributing to pollen allergenicity. Their analysis underscored a distinct requirement for guidelines in

Table 23.2 Various benefits of urban forestry reported in different countries

S. no.	Benefits of urban forestry	Source
1.	**Air quality improvement**	
	Plants act as natural air purifiers by absorbing and filtering out pollutants, including toxic gases from vehicle exhausts—The principal constituent of smog in urban cities	Nowak et al. 1996
	The cumulative impact of urban trees on removing PM10 can have significant benefits in future for public health and overall air quality	Tallis et al. 2011
	The combined effects of vegetation in trapping and mitigating airborne pollutants, including PAHs, assist in providing improved air quality	Peng et al. 2012
	The vegetation, through various mechanisms, captures fine, coarse and larger particles from the air	Dzierzanowski et al. 2011
	Urban vegetation, especially in parks, plays a crucial role in removing airborne pollutants, including total suspended particles (TSP), Sulphur dioxide (SO_2), and nitrogen dioxide (NO_2)	Yin et al. 2011
	The presence of sufficient vegetation cover in urban areas is often associated with improved air quality, including lower aerosol intensities	Kumar and Pandey 2013
	Urban parks capture SO_x, NO_x, CO_x and particles and thereby improve air quality	Konijnendijk et al. 2013
	54–109 kg of particulate pollution can be absorbed by a mature tree each year	
2.	**Moderation of temperature**	
	Incorporating green areas within residential neighbourhoods helps in regulating temperature, humidity, and overall atmospheric conditions	Baris et al. 2009
	Every increase of 10% tree cover, there will be a reduction of 0.62 °C in land surface temperatures	Caccetta et al. 2015
	The results from 16 studies proved that there will be a reduction of temperature in urban parks by 0.94 °C in day time	Bowler et al. 2010
	Urban tree cover can contribute to temperature reductions, and the relationship between canopy cover and temperature is often proportional, with approximately 1 °C of temperature difference is related to every 10% canopy cover difference	Dixon and Wolf 2007
3.	**Abatement of noise pollution**	
	The more extensive the greenery, the greater the potential for reducing noise levels, especially in the moderate to moderately high range of noise pollution (50–65 dB)	Kumar and Pandey 2013; Firoz and Laxmi 2016
	The dense vegetation creates a physical barrier between the park and the source of noise, acting as a buffer zone. This separation helps in reducing the impact of traffic noise on the park environment	Papafotiou and Stilianaki 2010
	Trees absorb sounds and thereby reduce noise pollution. A 30 m width row of trees having 15 m height can reduce 6–10 decibels of highway noise	New Jersey state Forest Service report
	Natural sounds, such as birdsong, rustling leaves, and flowing water, can effectively mask or camouflage unwanted urban noises amidst vegetation cover present in the parks	Miller 1988
4.	**Carbon sequestration**	
	There are two important ways in which urban vegetation can contribute to the reduction of carbon dioxide (CO_2) levels: (1) photosynthesis and carbon sequestration and (2) mitigation of Heat Island effects in cities and in turn leading to reduced energy consumption and fossil fuel use	McPherson et al. 1994
	Urban trees sequester to the tune of 18 kg CO_2 tree^{-1} year^{-1}	Surya Prabha et al. 2013
	Carbon sequestration can range from16 to 360 kg year^{-1} for small and larger trees respectively	McPherson and Simpson 1999
5.	**Maintenance of biodiversity**	
	Urban forest can give abode to a many of the species which are in the contiguous natural habitation, together with endangered species as well	Alvey 2006
	Urban forest is a 'created biodiversity' built by the constant establishment of new tree species. Urban forest in the cities of USA recorded greater species richness (47–82 species) than that of in the adjoining natural forests (18–23 species)	Wade 2017
	Urban parks help in maintaining biodiversity in cities. The presence of 77% introduced species among the trees in urban parks indicates a high level of non-native vegetation in Bangalore city in India	Nagendra and Gopal 2011.
	Urban parks can evidently be denoted as biodiversity hotspots owing to the species richness in urban greenery	Konijnendijk et al. 2013
6.	**Social benefits—Health and recreation values**	
	Urban forests provide pleasing and relaxing aesthetic environment and in turn reduce stress and improve health of people	Nowak et al. 1996
	The positive impact of natural views, such as trees and outdoor settings, on the health and well-being of patients has been observed in various studies, and this phenomenon can be called as the 'healing power of nature' or 'biophilia'	Ulrich 1990

(continued)

Table 23.2 (continued)

S. no.	Benefits of urban forestry	Source
	Visiting to urban parks can reduce obesity, a main global problem, more particularly in children	Konijnendijk et al. 2013
	Childhood asthma rates and city tree density were highly correlated. With every increase of 340 trees km^{-2}, there was a fall of 25% in rate of asthma	
	The beautification and urban development initiatives in Singapore and Kuala Lumpur, Malaysia, have indeed played significant roles in attracting foreign investment and fostering rapid economic growth	Braatz 1993
	The urban greenery in Chandigarh city plays a significant role in both attracting tourists (to the tune of 87.67%) and influencing residents' decisions to live and work in the city (to the extent of 55.65%)	Chaudhry 2006
	The positive impact of urban trees on rental prices in Portland, Oregon in USA reflects the recognition of the various benefits that trees in residential areas	Donovan and Butry 2011
7.	**Material benefits—Forest products and fodder**	
	Trees in urban and suburban areas provide high-quality fodder for livestock, firewood for primary cooking and heating fuel, for poor urban dwellers, besides giving fruits, nuts and fibre	IADB 1997
	Slum residents of Nairobi, Kenya uses wood extraction from trees in an urban area for domestic fuelwood	Furukawa et al. 2011

Table 23.3 Contribution of urban forests to United Nations' sustainable development goals

Sustainable development goals number	Goals	The role of urban forests
1	No poverty	Urban forestry creates jobs in various sectors, including tree planting and maintenance, landscaping, ecological restoration and environmental education. This helps stimulate the economy by providing employment locally to a diverse range of individuals
2	Zero hunger	Urban forests make available diverse and nutritious food resources, both directly and indirectly, by directly providing edible resources and indirectly supporting elements of a healthy food ecosystem
3	Good health and well-being	Urban forests play a multifaceted role in promoting public health. Their contribution extends beyond physical fitness to encompass mental health, stress reduction and the prevention of non-communicable diseases
6	Clean water and sanitation	Urban forests provide valuable ecosystem services related to water management, enhancing water quality, reducing the risk of floods and erosion and contributing to the sustainable use of water resources in urban environments
7	Affordable and clean energy	The sustainable management of urban forests can play a vital role in providing renewable energy for urban communities, especially in lower-income countries
8	Decent work and economic growth	Investments in urban forests and green infrastructure can have far-reaching economic benefits, contributing to green economic growth. The multiple benefits, ranging from increased property values to job creation and energy savings, highlight the importance of urban greenery
11	Sustainable cities and communities	The careful planning and management of urban forests bring about a host of environmental, economic, and social benefits
13	Climate action	Trees and forests in urban areas contribute directly to carbon sequestration and emissions reduction, and indirectly through energy savings by reducing heat islands in urban areas
15	Life on land	Urban forests play a vital role in creating and sustaining ecosystems within urban areas. Their contributions to biodiversity, soil health and land restoration make them essential components of ecologically sustainable urban development

Adopted from Salbitano et al. (2016)

designing urban greenery that have a minimal allergy impact. Their recommendations encompassed strategies such as enhancing biodiversity, exercising caution when introducing exotic species, opting for species with low pollen production, implementing effective management and maintenance practices and engaging in consultations with botanists to judiciously select tree species for particular green spaces.

Selection of Cold Resistant Species The establishment of green roof stands as a method for urban afforestation, offering numerous advantages for the environment, economy, and urban aesthetics. Despite the environmental stresses such as heat, strong wind, sunlight and cold posing challenges to plant growth in extensive green roof systems, these initiatives remain valuable. In the context of implementing rooftop gardening in Beijing, Fan and Wang (2011) assessed 13 ornamental species for greening flat rooftops. It is observed from their studies that six species exhibited excellence in rooftop gardening by having robust cold resistance. This underscores the importance of considering the specific environmental conditions where the plants are intended to be cultivated when establishing selection criteria.

Evaluation of Trees for the Emission of Volatile Organic Compounds (VOCs) Numerous trees have the capacity to release volatile organic compounds (VOCs), which can play a role in the generation of ozone and particles, particularly within heavily polluted urban atmosphere. Baraldi et al. (2010) explored the carbon sequestration effectiveness and volatile organic compound (VOC) emission capacities of 11 commonly found ornamental broadleaf species. This was achieved using a dynamic leaf enclosure system. The light-saturated CO_2 assimilation (A_{max}) varied within a range of 7.2–22.6 micro mol m^{-2} s^{-1}. The emission rates of monoterpenes from the plant species ranged from negligible (*Fraxinus ornus*) to 12.4 micro g gDW^{-1} h^{-1} (*Liquidambar styraciflua*). To offer insights into the suitability of these trees for enhancing air quality in urban environments, the species were categorized as either low or moderate emitters. Such evaluation will be useful in identifying tree species for planting programme in urban cities to obtain maximum benefits of air quality improvement. Bowler et al. (2010) evaluated the available evidence on whether urban greening interventions, such as tree planting or the creation of parks, affects temperature, ground-level O_3 and its precursors (volatile organic compounds, VOCs, or nitrogen oxides, NO_x) or ultraviolet (UV) within the surrounding urban area. Studies on O_3 indicated that any attempt to use greening to improve air quality would need to consider the biogenic emission of VOCs shown for some species, in order to estimate net air quality benefits.

Bowler et al. (2010) assessed existing evidence to determine the impact of urban greening interventions, such as tree planting or park creation, on temperature, tropospheric ozone (O_3), VOCs and nitrogen oxides (NO_x), within the adjacent urban environment. Findings related to O_3 highlighted the importance of considering the biogenic emission of VOCs observed in certain species when attempting to enhance air quality through greening initiatives, thereby ensuring accurate estimation of net air quality benefits.

Evaluation of Trees for Air Pollution Tolerance Air pollution is the most outstanding of environmental problem in the developing urban areas. In developing urban areas, air pollution stands out as the predominant environmental issue and urban forests aids in enhancing the environment quality in urban areas. Of the various benefits offered by urban forests, the most important is their capacity to mitigate air pollutants from the atmosphere. Urban tree species play a crucial role in directly reducing air pollution by absorbing gaseous pollutants such as sulphur dioxide (SO_2), nitrogen dioxide (NO_2), and ozone (O_3) through their leaves. Additionally, their canopies have the capacity to intercept particulate matter in urban areas. However, the tolerance of trees to the air pollutants varies with species. Several studies have been conducted to assess tolerance in trees to air pollution in terms of 'Air Pollution Tolerance Index (APTI)' and also as 'Anticipated Performance Index (API)' (Tripathi et al. 2009; Nayak et al. 2015; Joshi et al. 2016; Sahu et al. 2020). Panda et al. (2018) reviewed the studies related to APTI and API vis-a-vis air pollution abatement. It is reported that APTI is an inherent quality of trees to abate air pollution, which is an important concern in urban areas. The trees tolerant to air pollutants will have a high value of APTI and such trees can be recommended for controlling air pollution. On the contrary, the trees having low values of APTI may act as indicator plants to assess the levels of air pollutants. API is prepared by adding biochemical and aggregate factors. Panda et al. (2018) have given the grading of plants based on APTI (Table 23.4).

A study was conducted to estimate the tolerance index in trees for air pollution and net pollution mitigation potential of 25 tree species that are generally growing in different polluted zones of Coimbatore and Chennai in Tamil Nadu. The tree species studies were *Albizia lebbeck, Samanea saman, Alstonia scholaris, Annona squamosa, Azadirachta indica, Bauhinia racemosa, Senna siamea, Delonix regia, Ficus racemosa, Ficus religiosa, Holoptelea integrifolia, Leucaena leucocephala, Mangifera indica, Melia azedarach, Magnolia champaca, Muntingiacalabura, Peltophorum pterocarpum, Monoon longifolium, Pongamia pinnata, Spathodea campanulata, Syzygium cumini, Tamarindus indica, Tectona grandis, Terminalia catappa* and *Thespesia populnea.*

To assess the air pollution tolerance index and net air pollution mitigation potential of the 25 tree species, the biochemical parameters, namely ascorbic acid content, total chlorophyll content, leaf extract pH and relative water content (RWC) and dust collecting efficiency, total nitrogen content and total sulphur content in leaves, were estimated in the polluted zone (Tables 23.5, 23.6 and 23.7). Estimation of biochemical parameters revealed that there was an increasing level of ascorbic acid in all species. The study was conducted during summer season and compared in all the polluted zones. Zonation of sampling was done in trees by categorizing Coimbatore city into four zones, namely industrial zone, residential zone, high traffic zone and commercial zone (Balasubramanian and Hariprasad 2020). *Thespesia populnea* showed high performance in terms of dust collecting efficiency, and high nitrogen and sulphur absorption by leaves. The highest mean air pollution tolerance index in *Thespesia populnea* was followed by *Bauhinia racemosa*,

Table 23.4 Gradation of plant species based on APTI

Grading character	Index value	Grade
Air pollution tolerance index (APTI)	8.5–9.0	Positive
	9.1–9.5	Two positive
	9.6–10.0	Three positive
	10.1–10.5	Four positive
	10.6–11.0	Five positive

Table 23.5 Comparison of pollution mitigation potential of tree species growing in metropolitan cities

S. no.	Species name	Mean dust collecting efficiency (mg/cm^2)	Mean total nitrogen content (%)	Mean total sulphur content (%)
1	*Albizia lebbeck*	0.126	3.223	0.266
2	*Samanea saman*	0.146	3.499	0.260
3	*Alstonia scholaris*	0.132	3.410	0.208
4	*Annona squamosa*	0.137	3.458	0.242
5	*Azadirachta indica*	0.110	3.891	0.189
6	*Bauhinia racemosa*	0.122	4.035	0.247
7	*Senna siamea*	0.115	3.910	0.230
8	*Delonix regia*	0.175	4.124	0.162
9	*Ficus racemosa*	0.120	4.217	0.212
10	*Ficus religiosa*	0.154	4.098	0.209
11	*Holoptelea integrifolia*	0.151	4.374	0.221
12	*Leucaena leucocephala*	0.137	5.374	0.213
13	*Mangifera indica*	0.176	4.225	0.166
14	*Melia azedarach*	0.147	4.350	0.267
15	*Magnolia champaca*	0.128	4.593	0.202
16	*Muntingia calabura*	0.136	4.435	0.175
17	*Peltophorum pterocarpum*	0.135	4.906	0.237
18	*Monoon longifolium*	0.130	4.516	0.191
19	*Pongamia pinnata*	0.168	5.176	0.172
20	*Spathodea campanulata*	0.146	5.075	0.192
21	*Syzygium cumini*	0.160	5.468	0.245
22	*Tamarindus indica*	0.165	5.386	0.240
23	*Tectona grandis*	0.168	5.483	0.201
24	*Terminalia catappa*	0.192	5.834	0.237
25	*Thespesia populnea*	0.209	7.216	0.208

Source: Balasubramanian and Hariprasad (2020)

Monoon longifolium and *Samanea saman*. The study also explained that *Terminalia catappa, Albizia lebbeck* and *Magnolia champaca* have shown their susceptibility with lower APTI value under pollution stress condition.

Among the 25 tree species, *Thespesia populnea, Tectona grandis, Terminalia catappa, Melia azedarach, Albizia lebbeck, Samanea saman* and *Bauhinia racemose* exhibited high sulphur content in leaves which can be denoted that these trees are potential sink for SO_2 pollutants. It is concluded that *Thespesia populnea, Tectona grandis, Terminalia catappa, Melia azedarach, Albizia lebbeck, Samanea saman* and *Bauhinia racemose* showed very high absorption of nitrogen dioxide and sulphur dioxide pollutants from the urban polluted air. Besides, they are very efficient in accumulating dust in their leaves. Hence, the above seven species can be recommended for urban pollution abatement.

The physiological attributes, namely photosynthesis rate, stomatal conductance, ascorbic acid content, total chlorophyll, ICNT have the significant positive correlation for pollution tolerance for the above tree species. For photosynthetic rate, transpiration rate, stomatal conductance, leaf temperature and intercellular carbon dioxide concentration trees such as *Mangifera indica, Cassia siamea, Alstonia scholaris, Delonix regia* and *Thespesia populnea* are highly suitable for the urban settings. But the trees like *Neolamarckia cadamba, Syzygium cumini, Lannea coromandelica* and *Muntingia calabura* also demonstrate metabolic strategy to cope with the stress effect.

23.5 Selection Criteria for Trees in Urban Planting

Selecting the most appropriate and often fruit yielding trees for various sites/avenues is important for successful urban planting. Good trees planted will be cherished for long time. An inappropriate tree planted may not survive and will result in conflicts with infrastructure laid on the roads like wires above and walk paths. The trees planted should be healthy, attract birds and butterflies and should add life to the avenue. Participative urban forestry with involvement of immediate neighbours who walk and drive past has to be involved in the task. The selection process should be guided by purpose, the condition of the site, the role the tree has to play in the landscape, etc. Therefore, a vision plan for avenue planting should guide all activities. Computer-aided designing for

Table 23.6 Comparison of air pollution tolerance index of selected trees growing in different zones of metropolitan cities

		Industrial zone	High traffic zone	Commercial zone	Residential zone	Mean
S. no.	Species name	APTI	APTI	APTI	APTI	APTI
1	*Albizia lebbeck*	14.05	12.39	12.05	11.51	12.50
2	*Samanea saman*	14.35	15.35	12.96	13.90	14.14
3	*Alstonia scholaris*	15.39	13.06	12.93	13.45	13.71
4	*Annona squamosa*	12.17	12.49	12.48	12.70	12.46
5	*Azadirachta indica*	13.03	14.20	13.84	13.37	13.61
6	*Bauhinia racemosa*	14.25	15.33	13.31	14.07	14.24
7	*Senna siamea*	13.01	12.93	12.48	12.27	12.67
8	*Delonix regia*	15.33	13.64	13.92	13.48	14.09
9	*Ficus racemosa*	15.25	14.51	13.12	13.23	14.03
10	*Ficus religiosa*	15.31	12.59	12.56	12.17	13.16
11	*Holoptelea integrifolia*	14.76	14.69	13.69	12.97	14.02
12	*Leucaena leucocephala*	13.39	12.57	12.56	12.15	12.67
13	*Mangifera indica*	14.93	13.75	12.73	12.07	13.37
14	*Melia azedarach*	14.08	11.93	12.48	12.17	12.67
15	*Magnolia champaca*	11.65	12.97	12.92	12.52	12.52
16	*Muntingia calabura*	13.85	14.63	13.94	13.88	14.08
17	*Peltophorum pterocarpum*	12.37	12.05	12.39	12.51	12.33
18	*Monoon longifolium*	15.36	14.88	13.53	13.17	14.24
19	*Pongamia pinnata*	13.69	13.13	14.30	13.96	13.77
20	*Spathodea campanulata*	13.95	13.92	12.30	11.62	12.95
21	*Syzygium cumini*	12.02	13.04	13.20	12.76	12.75
22	*Tamarindus indica*	13.11	12.64	12.66	11.96	12.59
23	*Tectona grandis*	14.93	13.48	12.31	11.97	13.17
24	*Terminalia catappa*	12.68	11.91	12.78	12.35	12.43
25	*Thespesia populnea*	17.09	16.37	15.01	15.14	15.90

Source: Balasubramanian and Hariprasad (2020)

each landscape and avenues should be visualized in advance. The trees to be planted in avenues definitely have to withstand urban stresses like compacted soils, pollutants, vandalism, excess heat and temperature from vehicles and buildings and thus influence the growth of the tree. Trees planted should add aesthetic value and attractiveness to the site besides functional value.

In urban planting site selection particularly streets/parks/lanes/by-lanes/islands/campuses should gage the space available. The soil underground structures such as sewers, aboveground structures like street lights and electric lines, which may result in controversy with the interest of the local people. Soil analysis for urban planting is never given importance in our conditions. Soil nutrition, the likely injuries to the planted trees from disease, insects, animals and people are elements to be taken care off. Trees with good crown, strong trunk and that can tolerate stress have to be selected. Pruning of trees is important and, therefore, branching habit of trees and the root architecture should also guide tree selection. Flower colours, foliage colour, height of the tree are other selection criteria. The table of trees for different landscape should be prepared in advance. The genetic makeup of the tree and the propagated material should also guide selections.

Several types of software are being used in United States which select appropriate trees and shrubs based on criteria and site specifications. Such tree selection programmes should guide our selections too. Today there are innumerable books and references are available to guide urban planting and landscape and urban landscaping. However, local knowledge and experience should guide the tree selection.

Criteria for selection of trees species for urban landscape can be based on their use as mentioned below:

Trees for Shade and Dense Foliage

Albizia lebbeck, Alstonia scholaris, Azadirachta indica, Ficus sp., *Peltophorum pterocarpum, Syzygium nervosum, S. cumini, Mangifera indica, Albizia saman, Mimusops elengi, Monoon longifolium, Grevillea robusta, Thespesia populnea, Pongamia pinnata. Bauhinia purpurea, Ixora microphylla, Calophyllum inophyllum, Parkia timoriana, Kigelia pinnata, Manilkara kauki, Saraca indica, Sterculia guttata, Pterospermum acerifolium, Grewia orientalis, Tamarindus indica, Filicium decipiens, Putranjiva roxburghii, Santalum album, Swietenia mahagoni.*

Table 23.7 Selecting tree species based on the type of air pollutants they can mitigate

Species suitable for mitigating Suspended particulate matter	Species suitable for mitigating Sulphur dioxide
Golden apple *Aegle marmelos* SPM—0.581 mg/cm^2 CO_2—1.02 μmol m^{-2} s^{-1}	Kadam tree *Neolamarckia cadamba* SO_2—0.354% CO_2—4.59 μmol m^{-2} s^{-1}
Fry wood tree *Albizia lebbeck* SPM—0.163 mg/cm^2 CO_2—1.01 μmol m^{-2} s^{-1}	Flame of the forest *Butea monosperma* SO_2—0.211% CO_2—0.56 μmol m^{-2} s^{-1}
Milkwood pine *Alstonia scholaris* SPM—0.592 mg/cm^2 CO_2—2.26 μmol m^{-2} s^{-1}	Indian banyan *Ficus benghalensis* SO_2—0.343% CO_2—0.41 μmol m^{-2} s^{-1}
Pride of India *Lagerstroemia speciosa* CO_2—1.1 μmol m^{-2} s^{-1}	Jamun *Syzygium cumini* SO_2—0.324% CO_2—1.18 μmol m^{-2} s^{-1}
Indian mast tree *Monoon longifolia* SPM—0.253 mg/cm^2 CO_2—1.89 μmol m^{-2} s^{-1}	Arjuna tree *Terminalia arjuna* SO_2—0.321% CO_2—1.55 μmol m^{-2} s^{-1}
Mango *Mangifera indica* SPM—0.271 mg/cm^2 CO_2—0.89 μmol m^{-2} s^{-1}	White-cedar *Thuja occidentalis* CO_2—0.9 μmol m^{-2} s^{-1}
Species suitable for mitigating Oxides of nitrogen	**Species suitable for mitigating Air pollution**
Indian rosewood *Dalbergia sissoo* NO_x—6.53% CO_2—6.35 μmol m^{-2} s^{-1}	Pacific rosewood *Thespesia populnea* APTI—16.4 CO_2—2.51 μmol m^{-2} s^{-1}
Indian gooseberry *Phyllanthus emblica* NO_x—6.26% CO_2—0.85 μmol m^{-2} s^{-1}	Bengal almond *Terminalia catappa* APTI—13.8 CO_2—2.08 μmol m^{-2} s^{-1}
Crape myrtle *Lagerstroemia indica* CO_2—0.9 μmol m^{-2} s^{-1}	Yellow flame tree *Peltophorum pterocarpum* APTI—14.6 CO_2—0.87 μmol m^{-2} s^{-1}
Bullet wood *Manilkara kauki* NO_x—5.26% CO_2—0.76 μmol m^{-2} s^{-1}	African tulip tree *Spathodea campanulata* APTI—12.1 CO_2—1.28 μmol m^{-2} s^{-1}
Indian hog plum *Spondias pinnata* NO_x—3.21% CO_2—1.19 μmol m^{-2} s^{-1}	Pongamia *Pongamia pinnata* APTI—14.5 CO_2—2.92 μmol m^{-2} s^{-1}
Jack tree *Artocarpus heterophyllus* NO_x—5.23% CO_2—2.56 μmol m^{-2} s^{-1}	

Source: Balasubramanian and Hariprasad (2020)

Flower Colour

Red flower: *Delonix regia, Butea monosperma, Bombax ceiba, Erythrina variegata, Spathodea campanulata, Saraca indica, Cordia scabra, Amherstia nobilis, Melaleuca citrina, Colvillea racemosa, Firmiana colorata.*

Pink flower: *Cassia renigera, C. roxburghii, C. grandis, C. nodosa, C. javanica, Kleinhovia hospita, Gliricidia maculata, Bauhinia purpurea, Bauhinia variegate, Lagerstroemia speciosa, L. duperreana, Tabebuia aurea. T. heterophylla, T. rosea, Couroupita guianensis, Ceiba speciosa.*

Yellow flower: *Peltophorum pterocarpum, Cassia fistula, Senna siamea, Senna spectabilis, Tabebuia spectabilis, Tabebuia aurea, Tecomella undulata. Cochlospermum religiosum, Markhamia lutea, Heterophragma* sp., *Thespesia populnea.*

Blue flower: *Jacaranda mimosifolia, Solanum crinitum, Derris ovalifolia, Clitoria arborea, Guaiacum officinale.*

White flower: *Millingtonia hortensis, Plumeria* sp., *Nyctanthes arbor-tristis, Manilkara kauki, Azadirachta indica, Melia azedarach, Erythrina alba, Crateva religiosa, Gustavia augusta, Bauhinia variegata* var. *candida, Dillenia indica. Ixora microphylla, Alstonia scholaris, Pterosopermum acerifolium, Calophyllum inophyllum, Wrightia tinctoria, Holarrhena pubescens, Neolamarckia cadamba, Citharexylum spinosum.*

Height of Tree

Large trees having height growth around 15–22 m: *Adansonia digitata, Neolamarckia cadamba, Azadirachta indica, Alstonia scholaris, Ficus* sp., *Casuarina equisetifolia, Couroupita guianensis, Peltophorum pterocarpum, Polyalthia longifolia, Samanea saman, Sterculia foetida, Terminalia catappa, Tabebuia rosea, Terminalia. arjuna, Cassia* sp., *Acacia auriculiformis, Tectona grandis.*

Intermediate trees having height growth of 12–15 m: *Butea monosperma, Calophyllum inophyllum, Delonix regia, Kigelia pinnata, Jacaranda mimosifolia, Brassia actinophylla, Tabebuia aurea, Thespesia populnea, Madhuca indica, Dalbergia sissoo, Mangifera indica, Azadirachta indica, Bauhinia purpurea, Cassia fistula, Cassia nodosa, Cassia renigera, Senna siamea.*

Small trees having height growth of 5–8 m: *Brassia actinophylla, Cochlospermum religiosum, Calophyllum inophyllum, Plumeria alba, P. rubra, Lagerstroemia indica, Nyctanths arbour-tritis, Saraca indica, Ixora microphylla, Tabebuia avalandi.*

Canopy Form

Oval shaped canopy: Thuja, middle aged Eucalyptus and Casuarina.

Columnar canopy: *Casuarina equisetifolia, Polyalthia longifolia, Eucalyptus* sp.

Spreading canopy: *Samanea saman, Delonix regia, Anthocephalus cadamba, Bombax ceiba, Terminalia catappa.*

Round canopy: *Mimusops elengi, Azadirachta indica, Mangifera indica, Neolamarckia cadamba.*

Pyramidal or conical: Christmas tree, Deodar, Juniperus.

Weeping canopy: *Callistemon chinensis, Melaleuca citrina, Salix babylonica, Pongamia glabra, Ficus nitida.*

Umbrella canopy: *Albizia saman, Albizia lebbeck, Delonix regia.*

Plant Texture

Coarse texture: *Kigelia pinnata, Butea monosperma, Calophyllum inophyllum, Madhuca indica, Tectona grandis, Neolamarckia cadamba.*

Medium texture: *Spathodea campanulata, Tecoma argentea, Saraca indica, Manilkara kauki, Cassia fistula.*

Fine texture: *Senna siamea, Delonix regia, Jacaranda mimosifolia, Pink cassia, Peltophorum, Casuarina, Albizia saman.*

Fruit Bearing Avenue Trees

Syzygium cumini, S. nervosum, Mangifera indica, Morus alba, Cocos nucifera, Averrhoa carambola, Phyllanthus emblica, Tamarindus indica, Emblica officinalis, Artocarpus heterophyllum, Punica granatum, Citrus sp., *Annona reticulata, A. squamosa, Areca catechu.*

Pollution Tolerant Trees

Albizia lebbeck, Alstonia scholaris, Butea monosperma, Ficus benjamina, Madhuca indica, Pongamia glabra, Lagerstroemia speciosa, Ficus religiosa, Peltophorum pterocarpum and Terminalia arjuna.

Trees for Noise Abatement

Mangifera indica, Butea monosperma, Madhuca indica, Terminalia arjuna, Alstonia scholaris and Azadirachta indica.

Tolerance to Abiotic Stress

Trees with drought tolerance suited for very dry localities: *Dalbergia sissoo, Butea monosperma, Cochlospermum religiosum, Crataeva religiosa, Albizia lebbeck, Tecomella undulata, Azadirachta indica, Tamarindus indica, Ficus* sp., *Cassia fistula, Senna siamea.*

Trees suitable for wetland: *Dillenia indica, Barringtonia sp, Ficus elastica, Syzygiumcumini., Pongamia pinnata, Magnolia champaca, Terminalia arjuna, Holoptelea integrifolia, Thespesia populnea, Acacia nilotica.*

Salt tolerant trees: *Peltophorum pterocarpum, Thespesia populnea, Ficus* sp., *Acacia auriculiformis, Eucalyptus* sp., *Casuarina equisetifolia, Salvadora persica, Prosopis cineraria, Prosopis juliflora.*

Source: Alka Singh and Dhaduk (2012)

Trees Suitable for House Fronts

Thevetia neriifolia, Bauhinia monandra, Magnolia grandiflora, Caryota urens, Millingtonia hortensis, Tecoma stans, Magnolia champaca, Manilkara kauki, Plumeria rubra.

Trees for House Backs

Averrhoea carambola, Emblica officinalis, Carica papaya, Psidium guajava, Annona squamosa, Achras sapota, Mangifera indica, Citrus nobilis.

Trees Suitable for Small Compounds

Acacia auriculiformis, Bauhinia purpurea, Bauhinia variegata, Butea monosperma, Cassia fistula, Cassia marginata, Gliricidia maculata, Crateva religiosa, Thespesia populnea, Pongamia pinnata.

Trees Suitable for Big Compounds

Lagerstroemia speciosa, Peltophorum pterocarpum, Bombax ceiba, Neolamarckia cadamba, Firmiana colorata, Dillenia indica, Delonix regia, Terminalia catappa.

Trees Suitable for Road Avenue

Tamarindus indica, Albiza lebeck, Peltophorum pterocarpum, Samanea saman, Spathodea campanulata, Azadirachta indica, Alstonia scholaris, Cassia fistula, Ficus religiosa, Ficus benghalensis, Syzygium cumini, Terminalia arjuna, Delonix regia, Delonix elata, Monoon longifolia, Albizia procera, Millingtonia hortensis, Lagerstroemia speciosa, Butea monosperma, Jacaranda mimosifolia, Bassia latifolia, Calophyllum inophyllum, Senna siamea, Kigelia pinnata, Thespesia populnea, Crateva religiosa, Ficus lacor, Tabebuia rosea, Tabebuia spectabilis, Cochlospermum religiosum, Cassia javanica, Cassia nodosa, Amherstia nobilis, Bauhinia purpurea, Bauhinia tomentosa, Bauhinia acuminata, Cheniella corymbosa, Bauhinia variegata, Saraca indica, Sterculia guttata, Sterculia foetida, Guaiacum officinale, Hibiscus platanifolius.

Shrubs and Climbers for Houses

Jasminum sp., *Lawsonia inermis, Nerium oleander, Nyctanthes arbor-tristis, Asparagus racemosus, Clitoria ternatea, Jasminum grandiflorum, Bougainvillea* sp., *Passiflora princeps, Gymnema sylvestre.*

Source: Krishnakumar and Shanmughasundaram (2012)

23.6 Major Activities in Urban Tree Planting

23.6.1 Urban Tree Nursery

Historically, the Kings and other rulers selected and planted fruit and shade bearing trees which even today offer the best goods and services. Tree replacement on the new roads is becoming a common feature in many fast-developing cities and towns. Such plantings should be based on high-quality planting stock from identified nurseries. Best nursery practises have been put in place through the Western Ghats development programme, Hill Area Development Programme and various other state sponsored programmes. The forest department urban planting nurseries are worth visiting to learn the techniques and good practises. Tall nursery seedlings of over 1 m and above are being raised by adopting good nursery practises. Planting high quality tall trees also adds to the success of the programmes. Selection of trees from the nursery may look into following criteria, namely form of the seedling, collar diameter, branches, the health of the seedling, the health of the tree which is based on bark, foliage colour, etc. Selection of the tree is also guided by the root architecture. Transport of trees to the planting site also needs proper planning. Shifting, grading, watering/irrigation practises, fertilizer application and weeding should be the planned operations in nursery in order to get quality seedlings. Tall seedlings well-nourished in the nursery will contain high energy reserves that will enable production of new roots, branches and leaves in the planted sites. During nursery growth, the plants should be well spaced to provide good opportunity for photosynthesis and develop energy reserves. Seedlings grown together may not produce enough roots and develop height. All urban nurseries should certify the seedlings produced to ensure its quality. There should be enough laws to guide nursery practises. The seedlings raised should be governed by good forest practices and should be from superior parent material. Thus, the aboveground quality, below ground quality, nursery practices should be very sound for the success of urban planting (Krishnakumar and Shanmughasundaram 2012).

23.6.2 Planting Techniques in Urban Forestry

Trees planted in urban areas often struggle for survival due to various reasons like human pressures and demands of sunlight, oxygen, water, soil volume, nutrients, etc. There are many challenges in urban forestry and, therefore, innovative techniques are called for. Any new housing facility and/or developing township should have a management plan to ensure healthy tree assemblage and sound budget for tree planting should be provided for such landscape planning. The site selection is important as exposure to various factors like temperature, light, water, soil, air quality and other factors including manmade influences affect the planting site. There are also many urban factors like buildings, drainage systems, street lights, roads, oil spills, air conditioning impact, heating, ventilation, accidental scars, smokes and fires which influences plants. Excessive heat and temperature create artificial environment and may kill a plant. Annual rainfall should also be a consideration in plant selection. Drought tolerance and low oxygen level are important for long-term survival of trees. Many trees also respond to presence and absence of light. Night lighting and long days can result in elongation of the plant. Reflected lights can scorch leaf tissue and result in distorted growth. Air quality, particularly wind, salt sprays, carbon dioxide levels, exhaust fans, branching patterns may result in plant pest and diseases. Soil conditions like pH, sand, silt, clay and drainage composition also impact plant. Tree planting is carried out after the soils have worked and soil tested. Native soils with micronutrients and microorganisms are important for root development. The soil should also be tested against harmful soil organisms. Pitting well before planting and preparing the area is important. The depth of the pit should be based on the bag size and the soil volume in the bag. The current trend is to use tall seedlings and deep pits invariably in all the urban planting. Pruning branches periodically as the plant grows is important. Fertilizer application of newly planted trees using organic materials and addition of mycorrhiza including Vesicular Arbuscular Mycorrhizae (VAM), vermicasting, water absorbent gels enrich the plant growth (Krishnakumar and Shanmughasundaram 2012).

23.7 Innovative Urban Tree Planting in Public Lands: Success Stories

Case I: Urban Greening by Tamil Nadu Public Works Department The environmental division of Tamil Nadu Public Works Department (TNPWD) has implemented many innovative schemes in Tamil Nadu for solving pollution issues. At the same time, the division also promoted the herbal gardens in urban areas and school campuses. They have also developed gardens in urban and semi-urban areas with indigenous medicinal plants. The following case studies have successfully been implemented: (1) formation of herbal garden in PWD office premises; (2) formation of green cover area using the semi-treated dyeing waste water at Kasipalayam, Tirupur district; (3) formation of herbal garden in many school premises; (4) formation of garden on islands in tanks and (5) formation of tree gardens using the treated waste water at all polluted area. Out of these urban greening programmes, two are innovative and novel ideas, that is, (a) green cover in Kasipalayam common effluent treatment plant (CETP) area, and (b) green cover inside the tanks by formation of islands using desilted soil (Ilangovan 2012).

Green Cover Development in Common Effluent Treatment Plant (CETP) This project area is located in Kasipalayam, Tirupur district. The CETP is meant for treating the dyeing effluent. But generally, they treat only colour, Biological Oxygen Demand (BOD) and Chemical

Fig. 23.3 Green cover development in Common Effluent Treatment Plant (CETP) in Kasipalayam, Tirupur district of Tamil Nadu by Public Works Department (1 year after plantation)

Oxygen Demand (COD). The Total Dissolved Solids (TDS) portions will not be treated under CETP and in turn, this high TDS water is not useful for the crop cultivation. Hence, the environmental division of PWD, genetics division of State Forest Department, Coimbatore and CETP Kasipalayam unit have implemented the High Rate Transmission System (HRTS) near Kasipalayam CETP. This treated water with less TDS was used to raise the trees. The tree species tried were *Acacia auriculiformis, Alstonia scholaris, Ailanthus excelsa, Khaya senegalensis, Albizia lebbeck, Azadirachta indica, Albizia procera, Casuarina equisetifolia, Casuarina junghuhniana, Terminalia chebula, Bauhinia purpurea, Acacia nilotica, Acacia chundra, Samanea saman, Terminalia arjuna, Derris indica, Gmelina arborea, Holoptelea integrifolia, Adina cordifolia, Chlorxylon swietenia, Dalbergia latifolia, Simarouba glauca, Terminalia bellirica, Swietenia macrophylla, Swietenia microphylla, Alangium lamarkii, Eucalyptus tereticornis, Dalbergia sissoo, Melocanna baccifera, Acacia auriculiformis* and *Acacia* hybrid. Figure 23.3 shows the growth of trees at 3 years after planting (Ilangovan 2012).

Green Cover Development in Tanks by Formation of Islands Using Desilted Soil This project was done in Mookanery lake which is situated at Kaanankuruchi in Salem district of Tamil Nadu. The tank has around 100 acres water spread area. The ayacut area of this tank has been encroached for urbanization and hence the tank water is not used for irrigation fully. Further, owing to dumping of urban wastes in this lake areas, as the case of all urban lakes, this Mookanery lake has also become sink of the city. However, the Salem citizens group has taken challenging works of de-silting and cleaning of lakes. Normally when any tanks are desilted, the major cost shall go for transportation of silts. So, in this case it was decided to form islands inside the water spread of lake with the support of water resources department of PWD and Corporation of Salem, so that the cost could be reduced. In such a way, 50 number of islands were formed. The innovation here is that these islands are put to use for planting of trees. The trees have grown well after the monsoon and have become home for the aquatic life and the tank biodiversity has improved now (Fig. 23.4). (Ilangovan 2012).

Fig. 23.4 Green cover development in tanks by formation of islands using desilted soil in Mookanery lake at Kaanankuruchi in Salem district of Tamil Nadu (2 years after plantation)

23.7.1 Large Tree Transplanting

Planting larger trees is gaining significance. Lot of time, labour and equipment are saved in this process if carefully planned. In urban landscape planning, many cities are involved in large tree transplanting. Large trees should be supported by adequate stakes and wires. Trunk protection as the tree grows should be taken care off. Mulching of the trees planted for water retention is an important action in tree planting. Three to four inches of mulch with leaves and twigs have been found to trap rainfall and prevent run off. It also reduces evaporation of soil moisture. Over use of mulches may have a deleterious impact, that is harbouring wood decaying fungus which can damage bark and collar tissues. Mulch may also invite rodents that can nibble the bark. Therefore, it is suggested that mulches should be kept away from the trunk (Krishnakumar and Shanmughasundaram 2012).

23.8 Maintenance of Trees Planted in Urban Areas

Watering plants is a challenge. Water regimes should be fixed. Watering would be effective if done early in the morning or late in the day to prevent wilting. Fertigation by adding nutrients in the water will also enhance growth. Trees planted should be inspected as often as possible to ensure survivability. Total tree healthcare is often neglected in municipal and corporation plantings. Plant health monitoring, therefore, becomes important for the success of urban planting. At least 5 years of maintenance is required for the success of urban planting. Maintenance also includes tree straightening, removing the tree guard from overgrown trees as they can choke the tree growth, checking stakes, weed control, pruning, fertilizing, mulching, watering, insect and pest care including causality replacement. In many instances, adequate post planting care is not provided to the trees (Krishnakumar and Shanmughasundaram 2012).

Pest and Disease Management Percival (2001) documented that systemic acquired resistance (SAR) is a phenomenon wherein a plant's innate defence mechanisms are triggered through prior treatment with either a biological or chemical agent. The concept of synthetic aperture radar (SAR) has been extensively acknowledged and researched over the last century concerning its role in enhancing resistance against fungal, bacterial and viral pathogens affecting economically significant crop plants. While there is potential for SAR (synthetic aperture radar) in controlling tree pathogens, its application in disease management within the realm of urban forestry has not been extensively investigated in scientific studies. SAR induction can occur through both biological organisms and chemicals, including the promotion of SAR response through the interplanting of flowering woody plants, as well as SAR in woody plants.

Insect Baits According to Dhang (2011), insect baits are formulations that incorporate information from insect sensory physiology, gustatory chemistry, semi-chemicals and various sources. These baits are specific to certain targets and facilitate easy application, making them a recognized and safer alternative to liquid sprays. Particularly advantageous for pests residing in colonies or groups, such as cockroaches, ants and termites, baits have become a more practical and popular choice. Manufacturers have focused on developing bait formulations primarily for these pests. Successful bait and baiting system development requires key factors, including a suitable active ingredient, an easily accepted bait matrix and a thorough understanding of the pest's biology. Advances in bait development have significantly contributed to supporting integrated pest management (IPM) practices. Furthermore, baits serve as excellent tools for monitoring and managing pests, offering the added benefit of reducing the overall use of insecticides in urban areas.

Improving Drought Tolerance of Trees Creg (2004) reported that the selection of trees with superior drought tolerance is a longstanding interest for those engaged in landscape horticulture and urban forestry. Nevertheless, several challenges need to be tackled when formulating studies aimed at selecting drought-tolerant trees. A significant issue in both field and pot studies is the standardization and quantification of the amount of drought stress applied to trees. In field studies, fluctuations in rainfall can impede the application of a known amount of drought stress. Container studies face the challenge of differing growth rates among genotypes, leading to varying levels of drought stress even with comparable irrigation levels. Two effective techniques for identifying drought tolerance variation in seedlings are (1) stress-bed drought screening and (2) the target water potential concept. Both approaches prove useful for screening a relatively large number of genotypes for drought tolerance. Another substantial challenge in screening trees for drought tolerance is determining the appropriate response variables to measure.

23.9 Role of Community in Urban Planting Programme

Involving community through volunteers for creating awareness has been a challenge to urban forestry. Many urban planting initiatives without the stakeholders' involvement have resulted in failures. Public awareness and involvement should be through participation of residential colonies and sharing the programme with the local residents. Participatory urban planting programme seeking the public opinion on trees to be planted will result in the success of the effort. The local press and media particularly the local television and elected representatives should also be involved in the effort. Involving volunteers particularly the local youth will ensure the success of the programme.

Trees and wires often conflict with each other. Wires are broken due to tree branch falling and, therefore, they require line clearance using trimmers. This also requires strict regime of training for electricity personnel. There are qualified tree trimming companies in the developed countries with much experience. Sufficient safe guard should be taken while trimming trees. Fast growing trees and high maintenance trees can be avoided along electric wires. Planting the right tree in the right location, therefore, becomes important.

Similarly, trees are also blamed for damaging side walls and drainages. Shallow rooted trees with huge crown along the road ways have caused immense damage to the parked vehicles. Understanding the root system particularly the nature of the surface and the tap root is very important in any urban planting programme. Healthy roots are critical for the survival of the trees. The health of the root is governed by many factors like soil, weather and nutrients. The nutrients required for proper tree growth are nitrogen (N), phosphorous (P), potassium (K), calcium (Ca), sulphur (S) and magnesium (Mg) in larger quantities and boron (Bo), zinc (Zn), manganese (Mn), copper (Cu), iron (Fe), molybdenum (Mo) and chlorine (Cl) in smaller doses. Hydrogen (H), oxygen (O_2) and carbon (C) obtained from air and water also enhance root health. The pH, soil structure, soil particles, organic matter and depth enable comfortable penetration of the soil. Selection of the tree species should be guided by the science of urban forestry, good technology, knowledge on root systems and crown occupancy. Appropriate planting and maintenance techniques are already available. Properly constructed side walls and drainages prevent root penetration. The planting escapement is also important.

23.9.1 Successful Community Urban Forestry Programmes

Tree City USA Tree City USA, a programme developed in collaboration with the USDA Forest Service by the Arbor Day Foundation, operates as a non-profit organization. With the support of a million members, donors and partners, the Arbor Day Foundation is dedicated to promoting a greener and healthier world. Since 1976, the Tree City USA is instrumental in enhancing the greenery of cities throughout America. This nationwide initiative offers a framework that enables communities to effectively manage and enhance their public tree resources. https://www.arborday.org/programs/treeCityUSA/about.cfm.

Tree Canada Since 1992, Tree Canada has been the national non-profit organization committed to planting and fostering trees in both rural and urban settings across Canada. Supported by funding from the Canadian Forest Service, this organization has played a pivotal role in re-establishing tree cover in regions affected by natural disasters. Moreover, Tree Canada has provided guidance to communities in the effective management of their urban forests, facilitated the greening of schoolyards and coordinated conferences focusing on urban forestry. With the collaborative efforts of community partners and sponsors, Tree Canada has successfully planted over 83 million trees to date.

Greening Chandigarh Department of Forests and Wildlife, Union Territory of Chandigarh has been committed towards greener and cleaner Chandigarh. Greening Chandigarh programme was supplemented by sustained campaign and much needed technical inputs which in turn have yielded commendable results. Greening Chandigarh programme was supported by many institutions, namely Forest Department, Municipal Corporation, Engineering Department, Punjab University, NGOs, Resident Welfare Associations, Educational Institutions and farmers. Chandigarh administration formed 'Greening Chandigarh Task Group' in 2001 and continued to work till date. Another legacy of this Greening Chandigarh programme is that having 'Greening Chandigarh Action Plan' and it is being prepared annually since 2008–09. The highlights of achievements made under Greening Chandigarh programme are (1) the establishment of The City Forest also called 'Nagar Van' in 150 ha area, (2) Shivalik Arboretum in 5 ha, (3) Butterfly Park in 3 ha area, (4) Peacock Park in 3 ha area, (5) City Bird Sanctuary in 2 ha area, (6) Botanical Garden in 70 ha area and (7) Sukhna Wildlife Sanctuary—a manmade lake in 4207 ha.

Vanathukkul Tirupur This is a successful urban greening done by an NGO—'Voluntary Organization for People Empowerment of Rural Areas by Youth (VETRY)' in Tirupur district in Tamil Nadu. The method adopted in this greening programme is that nursery period for plants is 6 months in order to ensure minimum height of 1 m for planting. Barren lands of farmers are identified to make a fencing protection, with an access gate for water tanker movements. VETRY team will make pitting and planting. The main reason for the success of the planting programme is that watering to the planted seedlings is compulsorily carried out for a 2 years period at least. The concept in this planting programme is that 'Trees belong to the land owners, while pure air and increased rains belong to the society'. Indigenous tree species are selected for planting programme such as Neem, Pungam, Jamun which are drought resistant and also provide food to birds and other animals through which a biodiversity cycle is also rebuilt. In case of a better water source with the land owner, high timber value species are selected such as Teak, Rose Wood, Senegal Khaya, so that after 15 years the land owner also gets a commercial return. Around 5% of plantation is done in public roads/parks where people undertake watering/maintenance responsibilities. This VETRY organization completed planting of one million trees by the end of 2020 and received the prestigious 'Tamil Nadu Government Environmental Award 2016' for their successful greening programme.

Kalam Vanam MIYAWAKI plantations established in the campus of Anna University in Coimbatore district of Tamil Nadu and named after ex-President of India Dr. A. P. J. Abdul Kalam as 'Kalam Vanam'. A local NGO called Siruthuli was associated in establishing this urban planting programme. These Miyawaki plantations were established in 50 acres of land by planting 16 different tree species under very close spacing of 1 × 1 m. Among the 16 species, *Spathodea campanulata* exhibited the fastest growth, high productivity and also superior in eco-physiological traits which suit better for Miyawaki afforestation. *Cassia siamea* performed well next to *Spathodea campanulata* in terms of biometric, productivity and eco-physiological parameters. *Holoptelea integrifolia*, *Swietenia macrophylla* and *Limonia acidissima* have recorded the lowest parameters.

Box 23.3 PPP for the Management of Urban Forests
The involvement of both public and private sectors in urban forest management typically revolves around green space planting and maintenance, with private contractors undertaking tasks on behalf of public authorities. In the United States, a more strategic approach to public–private partnerships has emerged in several cities. In these instances, private conservancies or trusts actively participate in the co-management of extensive urban parks, as observed in iconic locations like Central Park in New York and Golden Gate Park in San Francisco. While such partnerships are less prevalent in Europe, there are exceptions, like the Woodland Trust in Great Britain, UK and Northern Ireland, along with nature conservation organizations in the Netherlands, that engage in the ownership or management of woodland areas.

Sources: Drayson and Newey (2014); Buijs et al. (2016)

23.10 Urban Tree Cover and Green Space

Tree cover encompasses all patches of trees smaller than 1 ha outside the officially recorded forest area. This includes trees in various formations, including scattered trees. It's important to note that while the terms 'Trees Outside Forests' (TOF) and 'tree cover' may seem similar, they denote distinct entities according to the Forest Survey of India (FSI) in 2019. TOF encompasses all trees growing outside the recorded forest area, regardless of patch size, which could exceed 1 ha.

The FSI, in its methodology, calculates the tree cover of urban areas based on the inventory of TOF (urban). Urban centres are defined by the Registrar General of India, and these definitions align with those used in the FSI inventory. However, it's noteworthy that the FSI report provides only the total tree cover of states and union territories (UTs) by aggregating the estimated tree cover of both rural and urban areas. Consequently, the FSI report does not explicitly detail exclusive information on urban tree cover.

Similarly, information on tree cover in urban areas is in decimal, particularly for tier 2 and tier 3 cities and urban agglomerations, from secondary sources. This is because there hasn't been a systematic effort by any agency to undertake this work. However, there is some available information about certain cities and towns, providing a limited perspective on the extent of tree cover in the country's urban areas.

According to the Indian State of Forest Report (ISFR) in 2019, the Union Territory of Chandigarh boasts the highest percentage of tree cover in the country at 22.34%, followed by Delhi at 8.73%, relative to their respective geographic areas (FSI 2019). Hyderabad's tree cover is reported to be 5% of its geographical area, while Bengaluru urban district has an estimated tree cover of 6.85%. A specific study notes that Gandhinagar's tree cover is an impressive 53.9% of its geographical area, contrasting with 4.6 and 16.29% for the municipal areas of Ahmedabad and Vadodara, respectively. However, the tree cover in the Nagar Palika areas of Gujarat is only 8.8% of the geographical area (Table 23.8).

23.10.1 Standards for Urban Tree Cover

Globally countries have varying standards for tree cover. In the United States, the general recommendation for urban tree cover across all zones is 40% of the geographical area. However, specific recommendations vary, with suburban residential zones aiming for 50%, urban residential zones targeting 25%, and central business districts seeking 15%. On average, the tree cover in the top 20 metropolitan areas in the U.S. is estimated to be 27.1% of the geographical area.

In contrast, in the Indian context, the National Forest Policy of 1988 envisions an average forest and tree cover of 33% of the geographical area for the entire country and a higher goal of 60% forest and tree cover in hills. This policy also promotes the planting of trees alongside roads, railway lines, rivers, streams and canals, as well as the establishment of 'green belts' in urban and industrial areas and arid tracts.

The Municipal Acts of various states and Union Territories specify minimum areas for parks and gardens, but achieving these standards is often impractical. Redevelopment and allocating exclusive spaces for parks and greenery are particularly challenging in the older parts of cities. However, such endeavours are more feasible in newly developed or developing areas. Furthermore, the creation of green areas is not consistent across cities, with some having forests within their limits and others experiencing limited tree growth. Establishing new forest areas within city limits may not be feasible, making the increase in green cover in urban areas dependent on trees growing in non-forest land, regardless of ownership. While some states and cities have Tree Preservation Acts regulating tree felling and protection, these regulations typically lack specific minimum area requirements for trees and greens in their management plans (Source: Firoz and Laxmi 2016).

Studies indicate that in developed countries, the green space, urban forest, and woodland per inhabitant in Europe amounted to 104 m^2 per inhabitant (Konijnendijk 2001, 2003) (Table 23.6). In France, this figure was 80 m^2 per inhabitant (Moigneu 2001; Konijnendijk 2003), in the Netherlands, it was 228 m^2 per inhabitant (CBS 1998; Konijnendijk 2003), and in Australia, it was 80 m^2 per inhabitant (Brack 2002). In India, research has found that

Table 23.8 Woodland cover and green spaces in urban areas across various regions

Region/country/ city	Estimated size of urban green space/woodland resource
Europe	The municipal limits of 26 major European cities exhibit an average woodland cover of 18.5%, equivalent to 104 square meters per person (Konijnendijk 2001, 2003)
France/Paris	The Greater Paris region boasts approximately 80 square meters of urban forest per person. (Moigneu 2001; Konijnendijk 2003)
Great Britain	Green areas constitute approximately 14% of urban spaces, encompassing 120,000 hectares of parks and gardens. (DLTR 2002; Konijnendijk 2003)
The Netherlands	The 22 largest cities in the Netherlands have an average green space cover of approximately 19%, providing around 228 square meters per person. (CBS 1998; Konijnendijk 2003)
Australia/ Canberra	The estimated crown cover is around 24 million square meters, equivalent to about 80 square meters per person (Brack 2002)
India/Delhi	The average tree and forest cover amount to approximately 20% of the geographical area, translating to about 21.43 square meters per person (FSI 2009)
India/ Chandigarh	The average tree and forest cover constitute about 35.7% of the geographical area, providing approximately 56 square meters per person (Action Plan 2009–2010)
India/ Gandhinagar	The city's tree and forest cover span 3256 ha, equating to about 164 square meters per person (FDG 2008)
India/Bangalore	The estimated crown cover of the city encompasses about 19.9% of the geographical area, averaging around 16.8 square meters per person (Behera et al. 1985)

Source: Chaudhry and Tewari (2011)

green space, urban forest and woodland in Delhi were 21.43 m^2 per inhabitant (FSI 2009). In Chandigarh, it was 56 m^2 per inhabitant (Action Plan 2009–2010). In Bangalore, it was 16.8 m^2 per inhabitant (Behera et al. 1985), and in Gandhi Nagar, it was 164 m^2 per inhabitant (FDG 2008). The World Health Organization (WHO) proposed an international minimum standard, later endorsed by the United Nations Food and Agriculture Organization (WUP 2014), suggesting that a minimum of 9 m^2 of green open space per city dweller is necessary (Kuchelmeister 1998). Per capita green space available in Indian cities ranged from 147.6 in Gandhinagar to 0.5 sq m in Hyderabad. In cities like Delhi, Bangalore, Jaipur and Chandigarh this figure is 5.5, 17.7, 20.0 and 54.4 sq m respectively (Govindarajulu 2014).

23.11 Legal Framework and Conservation of Urban Trees (Adopted from Rajesh 2020)

Though many states have tree conservation measures, they are less foolproof and effective. The development, protection and management of urban trees are the responsibilities of various agencies such as local bodies, revenue and forest departments, etc. From a legal standpoint, urban spaces are governed by a variety of Acts and Rules, including Municipal Acts, the Indian Forests Act, Land Revenue Code, Master Plans, Bye Laws, etc. However, the actual implementation and specifics may vary from case to case or state to state. Some states have enacted Tree Preservation Acts and established statutory authorities, like the Tree Officer, to safeguard trees and address related issues. In contrast, other states delegate these functions to Revenue and Municipal officers. Certain states have specific acts that cover the felling of trees in non-forest areas, private lands and urban areas. Additionally, Country and Town Planning guidelines dictate the allocation of space for green areas when formulating town plans. These green areas typically encompass neighbourhood parks, district and regional parks, avenues, etc. However, the determination of the allocated area for greenery in accordance with town planning rules and guidelines has predominantly relied on empirical standards rather than being rooted in actual scientific findings and necessities. Consequently, these guidelines frequently lack the precision needed to designate sufficient land cover that could potentially impact the environment, the health of citizens and other related concerns. Another gap in the conservation and preservation of trees in urban spaces is the absence of participation by regulatory authorities for tree conservation in the planning of urban greenery.

The terminology 'Forest' is found in the Concurrent List of the Indian Constitution, allowing both the central and state governments to enact on this subject. Interestingly, there is no explicit mention of 'trees' in the constitution. The fundamental duties outlined in Article 51 A (g) specifically emphasize that protecting forests, lakes and rivers is the fundamental duty of all Indian citizen (Ritwick Dutta 2019). Unlike forests, which have a central law as Forest (Conservation) Act, 1980, there is a lack of centralized legislation specifically focused on the preservation of trees. Few state legislatures have passed acts pertaining to the protection of trees.

The legislation specifically designed for the protection of trees outside forest areas in the country includes the Preservation of Trees Acts of Karnataka (1976), Delhi (1994), Goa, Daman and Diu (1984), Notably, despite being enacted by different states at various points in time, these laws share identical content, broadly outlining key components such as the role of a tree officer, the establishment of a tree authority, obtaining permission for tree felling, and the circumstances under which such permissions must be granted. The establishment of a tree authority and the designation of a tree officer are common components in nearly all tree preservation laws (Ritwick Dutta 2019).

Tree preservation acts adhere to the typical structure of criminal act and grant extensive powers, including the authority to make arrests with no need of a warrant vested in the tree officer. The pivotal authority bestowed upon the tree officer is the ability to arbitrate on applications submitted by persons looking for consent for the cutting down of trees. In accordance with the laws enacted by the states, the tree officer and the tree authority wield significant decision-making authority.

23.11.1 Role of Authorities and Legal System: Case Studies

Ritwick Dutta (2019) in his '*A National law for urban trees*' discusses relevant cases to bring out role of designated authorities and legal system in protecting the urban trees and its environment. Some of the important ones are discussed as case studies in the following paragraphs:

Case 1 The High Court of Bombay lengthily analysed the role of the tree officer under the Tree Preservation Act of Goa, Daman and Diu (1984) in a case connected to the building of the second international airport in Goa (Federation of Rainbow Warriors Case against Goa Government, 2018).

The argumentative project encountered resistance from local communities and environmental organizations. Although Justice N. M. Jamdar's ruling from the High Court of Bombay specifically relates to the Mopa case in Goa, the conclusions and remarks provide valuable perspectives on the application of tree preservation laws nationwide and the responsibilities of tree officers.

Delving into the intent behind the preservation of tree act, the high court noted, 'During the colonial era, trees were looked mainly as timber resource. As concerns regarding environmental degradation grew, the key role of trees in maintaining a healthy ecosystem was recognized, leading to the adoption of various measures to protect and safeguard tree cover by regulating tree cutting'.

Another critical aspect addressed in the Federation of Rainbow Warriors case in the High Court of Bombay was the nature of examination steered by the tree officer when asking permission for tree felling. The court emphasized the necessity for the tree officer to issue a reasoned order, stating, 'In cases where approvals of such magnitude are in request, reasons bring transparency to decision-making'. The tree officer must consider various factors, including the quantity and types of trees and the rationale for cutting the specified number of trees. The high court underscored that the basic principle of administrative law, requiring a decision-making body to issue a well-reasoned order, is applicable to the tree officer as well.

The Karnataka High Court, in the case of Paul D′ Silva and Others vs State of Karnataka and Others (1999), echoed a similar perspective on the role of the tree officer. As per the court's decision, "the tree officer must exercise judgment and assess whether the request for permission to cut down trees should be approved based on factors like the nature of the trees, their location and other pertinent considerations, rather than the applicant's status. The tree officer must consider public interest and the likely consequences on the environment or tree preservation in the area while granting such permission".

The prevailing practice of tree officers often involves issuing a 'no objection', which means granting orders for tree felling without providing any justification or reasons. Despite numerous judicial pronouncements emphasizing the 'duty to give reasons', administrative authorities, including those performing quasi-judicial functions, frequently neglect to record reasons for their decisions (Ritwick Dutta 2019).

Case 2 The recent case of Aarey forest in Mumbai is another case study, which details on the role of tree officer. The Maharashtra (Urban Areas) Protection and Preservation of Trees Act, 1975 requires, vide Section 3 thereof, a Tree Authority to be constituted for the area comprising Municipal Corporation of Greater Mumbai; the authority was constituted comprising 19 members, 13 corporators, 5 nominated experts and 1 municipal commissioner who is the Chairperson of the Tree Authority. The duties and functions of the Tree Authority are as per Section 7 of the Trees Act, 1975 have been well defined. As per the said act if a large number of trees have to be felled, the procedure to be followed is that the persons/body desirous of felling the trees has to make an application to the Tree Officer who has to visit the site, identify the trees proposed to be felled and prominently display on the trees that permission is required to fell the tree. The Tree Officer also has to publish a notice in the locality where the tree is standing. Objections if any received have to be considered by the Tree Officer and the proposal prepared by the Tree Officer has to be placed before the Tree Authority for a decision to be taken. Metro Rail Corporation Ltd. ('MMRCL') applied for permission to fell trees at Aarey on a plot of land admeasuring 33 hectare reserved in the development plan for construction of a metro car shed. The

Tree Officer, after visiting the site and identifying the trees proposed to be felled and affixing the notice on each tree, issued the public notice to conduct a public hearing. The public notice was that 2238 trees were proposed to be cut and 464 trees were to be transplanted. Nearly one lakh written objections was filed online. At the public hearing, the view of the citizens was taken on 2646 trees, that is, 2185 to be felled and 461 to be transplanted. Hon'ble High Court dismissed the petition noting that the 'project proponent has already planted 20,900 trees with GPS tagging on each plant in Sanjay Gandhi National Park and the survival rate is 95%'. The judgement noted that the project proponents 'established about 7 times the number of trees to be felled and have replaced the lost trees by planting saplings of trees, the process which commenced two years ago'. However, subsequently, the Hon'ble Supreme Court set aside the orders of Hon'ble High Court and case is under *sub judice*.

23.12 Way Forward

In 2015, the global community embraced a set of objectives known as the sustainable development goals (SDGs) with the aim of eradicating poverty, safeguarding the planet and ensuring prosperity for all. This marked the inception of a new sustainable development agenda, underscoring that sustainable development is the sole path to securing a better future for the planet and its inhabitants. The unchecked expansion of urban areas poses a significant threat to the attainment of the SDGs, as cities contribute increasingly to carbon emissions, resource depletion, income inequality and other challenges.

Urban forests play a crucial role in advancing the SDGs. They serve as a cost-effective measure to enhance human health and quality of life, providing nature-based solutions to various social and cultural needs. Incorporating urban and peri-urban forests into planning and management strategies represents an intelligent approach to addressing the adverse effects of urbanization. A strong legal framework, along with sufficient powers granted to designated authorities in urban planning and management, would play a role in promoting vibrant and people-friendly urban environments, characterized by ample green spaces.

Lessons Learnt

- Due to rapid urbanization, associated environmental degradation including emission of uncontrolled pollutants is increasing to unprecedented levels. However, urban forestry can be a part of solutions to this massive problem of urbanization.
- As per the main principles of the concept of urban forestry, it is strategic, multisectoral, multidisciplinary, multifunctional and most importantly it emphasizes social inclusiveness.
- In a broader sense, the benefits from urban forestry can be listed as: (1) air quality improvement, (2) moderation of temperature, (3) abatement of noise pollution, (4) sequestration of carbon, (5) maintenance of biodiversity, (6) providing social benefits of people's health and recreational values and (7) supply of forest products to poor urban dwellers and fodder to cattle.
- Species vary in their ability to tolerate different abiotic stresses in urban environment. Hence, a scientific evaluation of species will aid in selecting suitable tree species for the given environmental conditions, particularly for the emission of volatile organic compounds (VOCs) and for their air pollution tolerance.
- Trees planted in urban areas often struggle for survival due to various reasons like human pressures and demands for sunlight, oxygen, water, soil volume, nutrients, etc. There are many challenges in urban forestry and, therefore, innovative techniques are called for.
- Another important dimension of any successful urban greening programme is community participation. There are successful community urban forestry programmes, like Tree City USA, Tree Canada, Greening Chandigarh, Vanathukkul Tirupur, which can be adopted in other countries/cities.

Key Questions

1. Define urban forestry and elaborate the concept of urban forestry.
2. Describe the benefits of urban forestry and valuing of recreational benefits of urban forestry.
3. Which criteria are mainly used for the evaluation of tree species for urban forests? Describe Air Pollution Tolerance Index (APTI) of tree species and gradation of tree species based on APTI.
4. Write in detail the major activities in tree planting in urban localities and scope for innovative techniques in urban tree planting.
5. Explain urban green space and what are the different standards for urban green space?

References

Alvey AA (2006) Promoting and preserving biodiversity in the urban forest. Urban For Urban Green 5(4):195–201. https://doi.org/10.1016/j.ufug.2006.09.003

Balasubramanian A, Hariprasad CN (2020) Urban forestry and pollution abarement. In: Buvaneswaran C, Senthilkumar S, Murugesan S (eds) Manual for training on urban forestry. Institute of Forest Genetics and Tree Breeding, Indian Council of Forestry Research and Education, Coimbatore

Baraldi R, Rapparini F, Tosi G, Ottoni S (2010) New aspects on the impact of vegetation in urban environment. Acta Hortic 881:543–546

Baris ME, Uckac L, Uslu A (2009) Exploring public perception of urban identity: the case of Ankara, Turkey. African Journal of Agricultural Research 4(8):724–735

Behera G, Nageshwara R, Dutt PP, Manikiam B, Balakrishna P, Krishnamurthy J, Jagadeesh KM, Ganesha RK, Diwakar PG, Padmavathy AS, Parvathy R (1985) Growth of Bangalore city since 1980 based on maps and satellite imagery. Technical report, ISRO-EOS-TR-55-85. Indian Space Research Organisation, Bangalore

Bowler DE, Buyung-Ali L, Knight TM, Pullin AS (2010) Urban greening to cool towns and cities: a systematic review of the empirical evidence. Landsc Urban Plan 97(3):147–155. https://doi.org/10.1016/j.landurbplan.2010.05.006

Braatz SM (1993) Urban forestry in developing countries: status and issues. In: Cheryllkollin (ed) Proceedings of the sixth national urban forest conference. Panel on Urban Forestry across the Globe, Minneapolis-St. Paul, MN

Brack CL (2002) Pollution mitigation and carbon sequestration by an urban forest. Environ Pollut 116(1):195–200

Buijs A, Elands B, Havik G, Ambrose-Oji B, Cvejic R, Debellis Y, Davies C, Delshammar T, Erlwein S, Gerőházi E, Goodness J, Hansen R, Fors H, van der Jagt A, Luz A, Mattijssen T, Nastran M, Steen Møller M, Otten R, Rall E, Santos A, Spanò M, Száraz L, Tosic I, Vierikko K, Železnikar S (2016) Innovative governance of urban greenspaces. Collaborative report as part of the European Union FP7 (ENV.2013.6.2-5-603567) GREEN SURGE project (2013–2017). European Union, Brussels

Caccetta P, Collings SC, Drew D, Kassel H, Don M, Anthony T, Xiaoliang W, Zhou Z (2015) Monitoring land surface and cover in urban and peri-urban environments using digital aerial photography. Int J Digit Earth 9(5):457–475. https://doi.org/10.1080/17538947.2015.1046510

Carinanos P, Porcel MC (2011) Urban green zones and related pollen allergy: a review. Some guidelines for designing spaces with low allergy impact. Landsc Urban Plan 101(3):205–214

Carter J (1993) The potential of urban forestry in developing countries: a concept paper. Forestry Department of the Food and Agriculture Organization of the United Nations, Rome

CBS (1998) Statistisch bestand Nederlandse gemeenten, vol 2. Centraal Bureau voor de Statistiek, Voorburg/Heerlen

Chakrabarti D (2001) Urban crisis in India: new initiatives for sustainable cities. Dev Pract 11:260–272. https://doi.org/10.1080/09614520120056397

Chaudhry P (2006) Recreational use value of Chandigarh city's urban forestry. Curr Sci 91(11):1440–1441

Chaudhry P (2013) Valuing recreational benefits of urban forestry-a case study of Chandigarh city of India. Int J Environ Sci 3(5):1785–1789

Chaudhry P, Tewari VP (2011) Urban forestry in India: development and research scenario. Interdiscip Environ Rev 12(1):80–93

CII (2010) A report on intelligent urbanization: roadmap for India. Confederation of Indian Industry

CPCB (2010) Status of the vehicular pollution control program in India. Program objective series. PROBES/136/2010. Central Pollution Control Board, Ministry of Environmental and Forests, Government of India, New Delhi

Creg BM (2004) Improving drought tolerance of trees: theoretical and practical considerations. Acta Hortic 630:147–158

Dhang P (2011) Insect baits and baiting: novel technology for managing urban pests with less insecticide. In: Urban pest management: an environmental perspective. CABI, London, pp 187–206

DiGiovanni F, Coutinho M (2017) Guiding principles for air quality assessment components of environmental impact assessments. International Association for Impact Assessment, USA. https://www.iaia.org/downloads/Guiding%20Principles%20for%20Air%20Quality_2_1.pdf

Dixon KK, Wolf KL (2007) Benefits and risks of urban roadside landscape: finding a livable balanced response. In: 3rd urban street symposium, Seattle, Washington. http://www.urbanstreet.info/3rd_symp_proceedings/Benefits%20and%20Risks.pdf

DLTR (2002) 'Green spaces, better places.' Final report of the urban green spaces taskforce. Department of Transport, Local Government and the Regions, London

Donovan GH, Butry DT (2011) The effect of urban trees on the rental price of single-family homes in Portland, Oregon. Urban For Urban Green 10(3):163–168

Drayson K, Newey G (2014) Green society: policies to improve the UK's urban green spaces. Policy Exchange Report

Drunasky N, Struve DK (2005) *Quercus macrocarpa* and *Q. prinus* physiological and morphological responses to drought stress and their potential for urban forestry. Urban For Urban Green 4(1):13–22

Dzierzanowski K, Popek R, Gawronska H, Saebo A, Gawronski SW (2011) Deposition of particulate matter of different size fractions on leaf surfaces and in waxes of urban forest species. Int J Phytoremediation 13(10):1037–1046

Fan P, Wang K (2011) Evaluation of cold resistance of ornamental species for planting as urban rooftop greening. For Stud China 13:239–244. https://doi.org/10.1007/s11632-011-0305-5

FAO (1998) Urban forestry in the Asia-Pacific region: status and prospects. Food and Agricultural Organisation. http://www.fao.org/docrep/003/x1577e/x1577e06.htm

FDG (2008) Tree cover assessment in Gandhinagar capital project area using remote sensing technology. Report prepared by GIS Centre, Forest Department, Gujarat

Firoz A, Laxmi G (2016) Geospatial technology in urban forest suitability: analysis for Ranchi, Jharkhand, India. Ecol Quest 24:45–57. https://doi.org/10.12775/EQ.2016.011

FSI (2009) State of forest report. Forest Survey of India, Ministry of Environment and Forests, Dehradun. http://www.fsi.nic.in/sfr_2009/delhi.pdf

FSI (2019) India state of forest report 2019. Forest Survey of India, Government of India

Furukawa TF, Kiboi SK, Mutiso PBC (2011) Can stumps tell what people want: pattern and preference of informal wood extraction in an urban forest of Nairobi, Kenya. Biol Conserv 144(12):3047–3054

GoI (2011) List of million plus cities. Ministry of Home Affairs, India. http://censusindia.gov.in/Census_Data_2001/Census_Data_Online/Population/List_of_Million_Plus_Cities.aspx

GoI (2015) Smart cities mission statement & guidelines. Ministry of Urban Development, Government of India

Govindarajulu D (2014) Urban green space planning for climate adaptation in Indian cities. Urban Clim 10:35–41. https://doi.org/10.1016/j.uclim.2014.09.006

Helms JA (ed) (1998) The dictionary of forestry. Society of American Foresters, Bethesda, MD

IADB (1997) Good practices for urban greening. Inter-Am Dev Bank Rep CENV 109:8–18

Ilangovan R (2012) Waste land development using tree sapling plantations - a case study. In: Buvaneswaran C, Shanmughasundaram SK, Jayaraj RSC, Krishnakumar N (eds) Training manual on innovative techniques in urban forestry. Institute of Forest Genetics and Tree Breeding, Indian Council of Forestry Research and Education, Coimbatore

Ismail MH, Jusoff K (2004) Urban forestry planning using remote sensing and GIS technique. J Sci Technol 12:21–32

Jones N, Collins K, Vaughan J, Benedikz T, Brosnan J (2005) The role of partnerships in urban forestry. In: Konijnendijk CC, Nilsson K, Randrup TB, Schipperijn J (eds) Urban forests and trees. Springer, Berlin, pp 187–205

Joshi N, Joshi A, Bist B (2016) Air pollution tolerance index of some trees species from the industrial area of Tarapur. Int J Life Sci 2(2): 173–182

Kenney WA (2000) Leaf area density as an urban forestry planning and management tool. For Chron 76(2):235–239

Koch NM, Lucheta F, Käffer MI, Martins SMA, Vargas VMF (2018) Air quality assessment in different urban areas from Rio Grande do Sul state, Brazil, using lichen transplants. An Acad Bras Cienc 90(2): 2233–2248. https://doi.org/10.1590/0001-3765201820170987

Konijnendijk CC (2001) Urban forestry in Europe. In: Palo M, Uusivuori J, Mery G (eds) World forests, vol 3. Kluwer Academic Publishers, Dordrecht, pp 413–424

Konijnendijk CC (2003) A decade of urban forestry in Europe. Forest Policy Econ 5(2):173–186

Konijnendijk CC, Annerstedt M, Nielsen AB, Maruthaveeran S (2013) Benefits of urban parks: a systematic review - a report for IFPRA. The International Federation of Parks and Recreation Administration Publication

Konijnendijk CC, Gauthier M (2006) Urban forestry for multifunctional urban land use. In: van Veenhuizen R (ed) Farming for the future, urban agriculture for sustainable cities. RUAF Foundation, IDRC and IIRR, pp 414–434

Krishnakumar N, Shanmughasundaram SK (2012) Urban forestry - an overview. In: Buvaneswaran C, Shanmughasundaram SK, Jayaraj RSC, Krishnakumar N (eds) Training manual on innovative techniques in urban forestry. Institute of Forest Genetics and Tree Breeding, Indian Council of Forestry Research and Education, Coimbatore

Kuchelmeister G (1998) Urban forestry: present situation and prospects in the Asia and Pacific region, FAO Asia-Pacific forestry sector outlook study. FAO working paper no: APFSOS/WP/44. Food and Agriculture Organization of the United Nations, Rome

Kumar A, Pandey AC (2013) Spatio-temporal assessment of urban environmental conditions in Ranchi Township, India using remote sensing and geographical information system techniques. Int J Urban Sci 17(1):117–141

Lacan I, McBride JR (2009) City trees and municipal Wi-Fi networks: compatibility or conflict? Arboric Urban For 35(4):203–210

McKinsey Global Institute (2010) India's urban awakening: building inclusive cities, sustaining economic growth. McKinsey & Company

McPherson EG, David JN, Rowan AR (1994) Chicago's urban forest ecosystem: results of the Chicago urban forest climate project. Northeast Forest Experiment Station, Radnor, PA

McPherson EG, Simpson JR (1999) Carbon dioxide reduction through urban forestry: guidelines for professional and volunteer tree planters. General technical report PSW-GTR-171. USDA Forest Service, Pacific Southwest Research Station, Albany, CA

Miller RW (1988) Urban forestry: planning and managing urban greenspaces. Prentice Hall, Englewood Cliffs, NJ

Miller RW (1997) Urban forestry: planning and managing urban greenspaces, 2nd edn. Prentice Hall, Englewood Cliffs, NJ

Mock T (2004) Building a sustainable urban forest. Urban Agric Mag 13:29–30

Moigneu T (2001) Recreation in the Paris region national forests: a priceless service which should be costless? In: Konijnendijk CC (ed) Communicating and financing urban woodlands in Europe, Proceedings of the second and third IUFRO European urban forestry. Aarhus, Denmark and Hungary, pp 201–206

Nagendra H, Gopal D (2011) Tree diversity, distribution, history and change in urban parks: studies in Bangalore, India. Urban Ecosyst 14:211–223. https://doi.org/10.1007/s11252-010-0148-1

Nayak D, Patel DP, Thakare HS, Satashiya K, Shrivastava PK (2015) Evaluation of air pollution tolerance index of trees. Res Environ Life Sci 8(1):7–10

Nilsson K, Konijnendijk CC, Randrup TB (2005) Research on urban forests and trees in Europe. In: Konijnendijk CC, Nilsson K, Randrup TB, Schipperijn J (eds) Urban forests and trees. Springer, Berlin, pp 445–463

Nowak DJ (2006) Institutionalizing urban forestry as a "biotechnology" to improve environmental quality. Urban For Urban Green 5(2): 93–100

Nowak DJ, Dwyer JF, Childs G (1996) The benefits and costs of urban greening. In: Krishnamurthy L, Nascimento JR (eds) Proceedings of seminar on urban greening. Universidad Autonoma de Chapingo, Mexico

Ottitsch A, Krott M (2005) Urban forest policy and planning. In: Konijnendijk CC, Nilsson K, Randrup TB, Schipperijn J (eds) Urban forests and trees. Springer, Berlin, pp 117–148

Panda LRL, Aggarwal RK, Bhardwaj DR (2018) A review on air pollution tolerance index (APTI) and anticipated performance index (API). Curr World Environ 13(1):55–65

Papafotiou MC, Stilianaki G (2010) The effect of design of an urban park on traffic noise abatement. Acta Hortic 881:331–334

Pauleit S, Jones N, Nyhuus S, Pirnat J, Salbitano F (2005) Urban forest resources in European cities. In: Konijnendijk CC, Nilsson K, Randrup TB, Schipperijn J (eds) Urban forests and trees. Springer, Berlin, pp 49–79

Peng C, Wang M, Ouyang Z, Jiao W, Chen W (2012) Vegetative cover and PAHs accumulation in soils of urban green space. Environ Pollut 161:36–42

Percival GC (2001) Induction of systemic acquired disease resistance in plants: potential implications for disease management in urban forestry. J Arboric 27(4):181–192

Rajesh G (2020) Urban forests and urban environments—analysing issues in protection and legal framework. In: Buvaneswaran C, Senthilkumar S, Murugesan S (eds) Manual for training on urban forestry. Institute of Forest Genetics and Tree Breeding, Indian Council of Forestry Research and Education, Coimbatore

Ritwick Dutta (5 Jan 2019) A national law for urban trees - commentary. Econ Polit Wkly 54(1)

Saebo A, Benedikz T, Randrup TB (2003) Selection of trees for urban forestry in the Nordic countries. Urban For Urban Green 2(2): 101–114

Sahu AK, Panda S (2001) Studies on pollutant sink function of trees of Calcutta metropolis with emphasis on urban forestry. J Econ Taxon Bot 25(3):635–644

Sahu C, Basti S, Sahu SK (2020) Air pollution tolerance index (APTI) and expected performance index (EPI) of trees in Sambalpur town of India. SN Appl Sci 2:1327. https://doi.org/10.1007/s42452-020-3120-6

Salbitano F, Borelli S, Conigliaro M, Chen Y (2016) Guidelines on urban and peri-urban forestry. Report. FAO forestry paper no. 178. Food and Agriculture Organization of the United Nations, Rome

Schell LM, Ulijaszek SJ (2018) Urbanism, health and human biology in industrialized countries. Cambridge University Press, Cambridge

Singh A, Dhaduk BK (2012) Green-scaping with trees in urban vicinity. In: Jadeja DB et al (eds) Proceedings of the national conference on agroforestry: an evergreen agriculture for food security and environment resilience. Navsari Agricultural University, Gujarat

Srivathsan A (2013) Where is our patch of green, Mr. Mayor? http://www.thehindu.com/news/cities/chennai/where-is-our-patch-of-green-mr-mayor/article3223739.ece

Surya Prabha AC, Senthivelu M, Krishnakumar N, Nagendran S (2013) Urban forests and their role in carbon sequestration: a review. Int J For Res 16(1):23–29

Tallis MJ, Gail T, Sinnett D, Smith PF (2011) Estimating the removal of atmospheric particulate pollution by the urban tree canopy of London, under current and future environments. Landsc Urban Plan 103(2):129–138

Tripathi A, Tiwari PB, Mahima, Singh D (2009) Assessment of air pollution tolerance index of some trees in Moradabad city, India. J Environ Biol 30(4):545–550

Tyrväinen L, Pauleit S, Seeland K, de Vries S (2005) Benefits and uses of urban forests and trees. In: Konijnendijk CC, Nilsson K, Randrup TB, Schipperijn J (eds) Urban forests and trees. Springer, Berlin, pp 81–114

Ulrich RS (1990) The role of trees in well-being and health. In: Proceedings of the fourth urban forestry: make our cities safe for trees held in St. Louis, MO. Washington, D.C. American Forestry Association

UN Habitat (2004) Dialogue on the urban poor: improving the lives of slum dwellers. Pre-session document for the World Urban Forum, 13–17 September, 2004, Barcelona, Spain. www.unhabitat.org/wuf/2004/documentation.asp. Accessed Jun 2005

UNDESA (2018) World urbanization prospects. https://population.un.org/wup/

Van Herzele A, Collins K, Tyrväinen L (2005) Involving people in urban forestry: a discussion of participatory practices throughout Europe. In: Konijnendijk CC, Nilsson K, Randrup TB, Schipperijn J (eds) Urban forests and trees. Springer, Berlin, pp 207–228

Wade CA (2017) The urban forest: created biodiversity. J Ecosyst Ecogr 7(1). https://doi.org/10.4172/2157-7625-C1-025

Wolf KL (2018) Urban forest values: economic benefits of trees in cities. Center for Urban Horticulture, University of Washington. https://wrrc.arizona.edu/sites/wrrc.arizona.edu/files/Urban%20Forest%20Values.pdf

WUP (2014). https://population.un.org/wup/Publications/Files/WUP2014-Report.pdf

Yin S, Shen Z, Zhou P, Zou X, Che S, Wang W (2011) Quantifying air pollution attenuation within urban parks: an experimental approach in Shanghai, China. Environmental Pollution 159(8–9):2155–2163

Agroforestry: A Climate Resilient and Sustainable Land Use

24

Inder Dev, Asha Ram, Naresh Kumar, Ramesh Singh, S. N. Meena, Sushil Kumar, Kamini, Roomi Devi, Babli Joshi, and A. Arunachalam

Abstract

Agroforestry (AF) means combining shrubs, trees, crops, and livestock to manage rural land resources. It generates economic, environmental, and social benefits. Traditionally, agroforestry systems have been mixed farming systems. In India, agroforestry has been around for generations as a way of life. Approximately 10% of all agricultural land worldwide is believed to be covered by agroforestry. Nearly half of the needs for firewood, small timber (65%), wood for plywood (70–80%), the base material for paper pulp (60%), and nutritious green food for animals (9–11%) are met by it. Agroforestry has become of greater significance in tackling numerous challenges as well as offering a broad range of socioeconomic and environmentally friendly advantages, especially in the wake of the Kyoto Protocol to the United Nations Framework Convention on Climate Change (UNFCCC). When combined with field crops, trees have the ability to significantly boost the economy by diversifying the land, generating sustainable income, and enhancing the security of food, fuel, and fodder. Because agroforestry simultaneously reduces the amount of greenhouse gases produced from the soil by storing carbon in topsoil and biomass from trees, it increases the likelihood of minimizing and adapting to the effects of changing the climate. By 2030, India aims to reduce the amount of greenhouse gases, it emits by 33–35% from 2005 levels. India has launched a landmark National Agroforestry Policy 2014 that promotes agroforestry. Also outlined is a recommendation to upscale and promote agroforestry research at the national level through institutional mechanisms. The policy aims to foster collaboration across numerous projects, plans, and organizations that incorporate agroforestry features to improve the livelihoods, revenue, and productivity of small-scale landowners. The policy also seeks to increase awareness about agroforestry and its benefits among farmers, stakeholders, and the public. It also encourages the use of agroforestry practices for sustainable land management. Lastly, the policy seeks to create an enabling environment for agroforestry development.

Keywords

Agroforestry system · Alley cropping · Silvipasture · Bioenergy plantation · Natural resource management

24.1 Introduction

The practice of raising crops, livestock, shrubs, and trees to achieve social, economic, and environmental benefits is known as agroforestry (AF). Nair (1979) defined AF as a land utilization method that combines livestock, crops, and trees harmonically. It is not only ecologically beneficial and supported by science but also practically viable and socially permissible. AF systems developed over centuries are mixed farming systems. Most AF systems are long-term approaches to land management with a duration of greater than a year. Agroforestry systems are a complex form of land management practices both ecologically and economically. Lands that are not even suitable for growing crops can be utilized more effectively with AF. India has several AF systems (Fig. 24.1).

I. Dev (✉) · A. Ram · N. Kumar · S. Kumar · A. Arunachalam
ICAR-Central Agroforestry Research Institute, Jhansi, India

R. Singh
International Crops Research Institute for the Semi-arid Tropics, Hyderabad, India

S. N. Meena
Agriculture University, Kota, India

Kamini
ICAR-Indian Grassland and Fodder Research Institute, Jhansi, India

R. Devi
CCSHAU, College of Agriculture, Bawal, Rewari, India

B. Joshi
Central University of Himachal Pradesh, Dharamshala, India

A. K. Mandal, A. Nicodemus (eds.), *Textbook of Forest Science*, https://doi.org/10.1007/978-981-97-8289-5_24

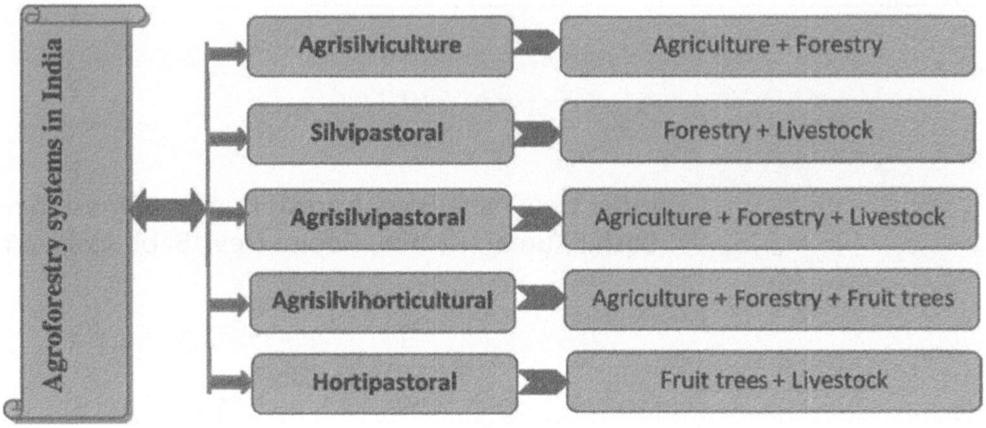

Fig. 24.1 Agroforestry systems prevalent in India

Farmers all around the world engage in agroforestry, which has been a fundamental component of agriculture for thousands of years. On farms and in rural landscapes, AF involves a wide variety of trees. It also entails creating scientifically based tree entrepreneurship options that have the potential to be important in the upcoming years. The agricultural and forestry systems in diverse agro-climatic regions across India have demonstrated their efficacy in terms of productivity and environmental compatibility. It has been estimated that agroforestry constitutes around 10% agricultural area of the world (Zomer et al. 2007). Ravindranath and Hall in 1995 reported that there were 14,224 million trees outside forests, while Prasad et al. reported a total of 24,602 million trees (Ravindranath and Hall 1995; Prasad et al. 2000). According to Rai and Chakrabarti (2001), the country's fuelwood and timber demands are met by the trees that cover an area of approximately 17 M ha (GOI 1999). Agroforestry spans 25.32 M ha in India, which makes up 8.2% of the nation's total land area (Dhyani et al. 2013). This includes cultivated land, of which 7 M ha are in irrigated areas and 13 M ha are in rainfed regions; it additionally encompasses home gardens, 2.28 M ha of shifting cultivation, and 2.93 M ha of restoration of problematic soils.

Agroforestry land use systems may be a preferable choice for mitigating climate change due to their great potential to reduce carbon emissions. As well, it also enhances farm income, and food security, restores and maintains biodiversity above and below ground, maintains watershed hydrology, and conserves soil (Meena et al. 2023a; Ram et al. 2017). It is possible for agroforestry to modify the microclimate under the canopy of trees using capturing carbon from the atmosphere, phytoremediation, watershed maintenance, and ensuring the preservation of biodiversity. AF not only modifies the microclimate but also contributes significantly to increased farm productivity, organic matter, and climate change mitigation (Raj et al. 2014). Agroforestry provides an extensive array of cultural services, including spiritual enrichment, introspection, recreation, and aesthetic values. Additionally, it contributes to supporting services such as soil formation and oxygen production, in addition to its provisioning functions (food, fuel, fiber, fresh water, and genetic resources). Furthermore, as noted by Dev et al. in 2020, agroforestry helps contribute to controlling the climate, maintaining the quality of water, and managing erosion (Dev et al. 2020). About half of the wood for fire needs, two-thirds of the small timber needs, 70–80% of the wood used to make plywood, 60% of the raw materials needed to make pulp for paper, and 9–11% of the green cattle feed needed are met by agroforestry. Agroforestry systems with trees supplement farming by providing the farmer with fuel, fodder, and manure. The integrated farming systems (IFS), which are essential for both sustainable and self-sustaining agriculture, are supported by trees in agroforestry systems. Because of their variety of possibilities and products, AF systems offer rural communities the chance to create jobs (Fig. 24.2). In recent years, there has been a noticeable increase in the number of small- and medium-sized enterprises that deal with timber and timber-derived products due to improved market availability of timber through agroforestry systems. AF systems are more productive and sustainable than forestry or agricultural monocultures, which has major benefits for the environment and the economy.

AF systems reduce water, soil material, organic matter, and nutrient losses according to the Agroforestry Research Trust. Through organic matter maintenance and tree roots, AF systems support soil health through biological activity and favorable soil physical properties. Nutrients are used more efficiently in AF systems due to their closed nutrient cycling. AF systems can reduce soil acidification and salinization, and check the development of soil toxicities. Polluted soil can be reclaimed by planting trees. Compared to monoculture systems, AF systems use solar energy more efficiently, which reduces the number of illnesses and insect pests. The decomposition and pruning of trees contribute significantly to soil fertility maintenance. It is feasible to

Fig. 24.2 Poplar-based agroforestry system (photo: ICAR-CAFRI/Inder Dev)

align the release of nutrients from tree breakdown with the nutrient requirements of neighboring crops. Typically, a significant level of nutrient synchrony is attained by incorporating high-quality pruning into the soil during crop planting. Despite variations in the requirements of different trees and crops, achieving perfect balance may be challenging, but strategic timing helps optimize nutrient utilization. By diversifying farms and stimulating rural economies, AF can make farms and communities more stable. As a result of multiple product production, AF systems also reduce economic risks.

A total of 120.72 M ha, or 37% of India's land area, are impacted by soil degradation due to one or more of the following factors: open forests (16.53 M ha), wind erosion (11 M ha), water erosion (93 M ha), and salt-affected areas (6.74 M ha). There is still a significant amount of land that has to be treated, although up to 2010–2011, about 56.96 M ha had been treated under various watershed development schemes. As a result, AF can contribute substantially to such endeavors, while fulfilling the diverse needs of people through the production of crops, trees, and livestock. In addition, AF can produce multiple products (food, fruit, fuel, fodder, fertilizer, and fiber). The numerous advantages and amenities that trees offer to ecosystems and society make them essential to our traditional farming systems. Various agroforestry practices have thus been incorporated into various national developmental programs/schemes, such as the Multipurpose River Valley Projects, Flood Control/Management Programs, National Watershed Development Program for Rainfed Areas, Integrated Rural Development Program, Agriculture Development Program, Forestry Development Scheme, Drought Prone Area Development Program, and Desert Development Program, in recognition of the importance of agroforestry. Agroforestry has been recognized as a possible crop for cultivation on 10 M ha of agricultural land with irrigation and 18 M ha of rainfed area of land by the Task Force on Greening India (Planning Commission 2001). Sustainable development can be achieved through agroforestry because it increases production, conserves natural resources, provides environmental security, and provides livelihood security.

24.2 Historical Perspective of Agroforestry in India

The forests are an important part of the Indian culture, and farmers have a long tradition of practicing agroforestry for multiple benefits and products. There are evidences which shows that in India agroforestry has been an integral facet of

Fig. 24.3 Coconut tree "Kalpavriksha"

life since the Mesolithic period. The Brhat Samhita, which describes the relationship between irrigation tanks and trees, is one piece of Vedic literature that highlights the importance of agroforestry techniques (Kumar and Sikka 2014). The use of pasturelands by human beings within the forested areas stands as one of the earliest land-use traditions in India. Moreover, the genesis of civilization itself occurred in forests, where the Rishis, instrumental in shaping Hinduism, coexisted harmoniously with nature. The environment around Rishi's ashram reflected a synergistic blend of agriculture, pastures, trees, birds, and animals. Post the Harappan and Mohenjo-Daro civilizations, worship of specific tree species became prevalent in various rituals. The growing of fruit trees such as aonla, ber, banana, pomegranate, coconut, date palm, and various crops, alongside livestock rearing, can be traced back to the Chalcolithic period of civilization. The sacred literature such as ancient Vedas also underscore the religious significance of specific trees like peepal, khejri, and palash. The hymn (Bhatla et al. 1984) "May plants, water, and the sky preserve us, and woods and mountains with their trees for tresses" (Rigveda V. 41.11) is a prime instance of how the Rigveda provides insight regarding the reverence for trees during the period when it was written.

India has a rich historical tradition of different agroforestry practices across diverse agroecological conditions. Archaeological observations from the Harappa excavations advocate that ancient inhabitants had rich knowledge about different fruit species of trees that included coconut, date palm, pomegranate, citrus, and melon. (Fig. 24.3). Further, Emperor Ashoka (274–237 BCE) also played a key role in introducing and boosting the cultivation of different fruit crops like grapes, jackfruit, and mango, thus promoting agri-horticultural agroforestry practices during his reign.

24.3 Major Components of Agroforestry

Agroforestry's major components are described in the sections that follow:

24.3.1 Alley Cropping

The technique of growing food, fodder, or specialty crops in the spaces that exist between tree rows is generally referred to as alley cropping in agroforestry. It includes the strategic planting of trees, either individually or in grouped rows, alongside the crops cultivated between the tree rows. This agroforestry practice can create significant benefits for farmers by modifying the microclimate and improving the soil nutrient status and yield. Alley cropping has many advantages over monocropping practice as it provides diversified and more total farm yield. Moreover, this practice helps in reducing wind and water erosion, reduces nutrient leaching to groundwater, and also provides habitat to wildlife.

24.3.2 Riparian Buffer Strips

Perennial strips made up of grasses, trees, and shrubs are placed between crops or grasslands and bodies of water, which include drainage ditches, lakes, ponds, wetland areas, and streams, to generate riparian buffers. These perennial strips play a critical role in preventing the intense growth of aquatic macrophytes by providing shade from canopies, filtering contaminated air from nearby sources of pollution, and removing pollutants from the overland flow from intensively managed adjacent agricultural fields. In addition, they also stabilize stream banks, for wildlife, provide harvestable products, and also enhance the landscape's beauty.

24.3.3 Windbreaks

Planting shrubs and trees in a planned pattern to improve farm productivity, safeguard livestock, and manage soil erosion is known as a windbreak. It also provides different harvestable products from trees and shrubs and thus increases the farm income. Windbreaks protect wind-sensitive crops and livestock and also provide benefits like improved bee pollination and wildlife habitat.

24.3.4 Forest Farming

The practice of "forest farming" involves using forested areas to grow crops that can withstand the shadow, like shiitake mushrooms, decorative ferns, and ginseng. Forest trees provide the necessary shade levels for these crops. Understory crops suitable for forest farming either naturally grow in forest conditions or adapt well to the site's soil and microclimatic conditions. This approach presents an option for a consistent income stream in addition to or instead of harvesting the trees for wood merchandise.

24.3.5 Silvipasture

Silvipasture is a specific kind of land use system that combines grassland, farm animals, trees, and shrubs in an appropriate combination to produce both fuelwood and fodder on the same piece of agricultural land. The best mix of trees, grasses, and legumes maximizes productivity, preserves plants, soil, and nutrients, and produces sustainable fodder, fuel wood, lumber, and other products. The trees and bushes, particularly those that flourish in semi-arid and dry areas, frequently provide silvipastoral systems with significant volumes of leafy forage. These extra feeds are essential for mitigating livestock feed shortages at times when fodder is scarcely available. In addition, it also provides a habitat for wildlife.

24.4 Agroforestry in Different Agro-Climatic Zones

Various specialized agroforestry systems and techniques that have been created over time and are being adopted in diverse agro-climatic zones of India include energy plantations, home gardens, shelterbelts, and alternatives to shifting farming.

24.4.1 Important Agroforestry Techniques Used by Indian Farmers

Boundary plantations are found in every agroecological zone in the country, while farm woodlots are also found in all agroecological regions. Home gardens are common in Kerala, the West Coastal region, South Karnataka, and the Andaman and Nicobar Islands. Industrial plantations, also known as commercial agroforestry, are prevalent in Haryana, Uttar Pradesh, Punjab, Tamil Nadu, Andhra Pradesh, Karnataka, and Kerala. Agroforestry with plantation crops for shade is practiced in the tea and coffee-growing areas of southern India, West Bengal, and Assam (Puri and Nair 2004).

Shelterbelts and windbreaks are used in coastal and arid regions, and trees scattered across fields and parkland are typical in semiarid and arid regions. Fodder trees are grown across the entire country. The hortipasture system is suited for subtropical and tropical climates, particularly in orchards situated in hilly regions. Trees for salt-affected soils are found in canal-irrigated areas of semi-arid regions, while trees for soil conservation are important in hilly zones,

Fig. 24.4 Apple-based agroforestry system (photo: ICAR-CAFRI/Inder Dev)

seacoasts, and ravine lands. Seasonal grazing in the forest occurs in semi-arid and hilly ecosystems, and the Taungya system is practiced in the states of the Northeast (Puri and Nair 2004).

Additional 20 prevalent AF systems, including agri-silviculture, agri-horticulture (Fig. 24.4), alley cropping, energy plantation, agri-silvi-horti, agri-silvi-pastoral, horti-pastoral, silvi-pastoral, shelter-belts, forage forestry, horti-olericulture, horti-apiculture, aqua-forestry, homestead, wind-breaks, boundary plantation, live fence, block plantation, are currently practiced across many agro-ecological areas of the nation (Dhyani et al. 2009).

Agroforestry is also being studied from a variety of angles by the All-India Co-ordinated Research Program on Agroforestry (AICRP-AF) and the ICAR-Central Agroforestry Research Institute (ICAR-CAFRI), both in Jhansi. AICRP on AF and ICAR-CAFRI Jhansi have created 35 distinct agroforestry models with possible tree-crop combinations based on the nation's various agro-climatic zones (Chaturvedi et al. 2016). The Indo-Gangetic Plain (IGP) is covered by poplar systems; the western Himalayas are covered by morus and grewia systems; the tropical regions are covered by teak systems; the humid and sub-humid regions are covered by gmelina and acacia systems; and the arid and semi-arid regions are covered by aonla and khejri systems. Table 24.1 outlines the significant ecosystems of agroforestry found in the various regions of the nation (Fig. 24.5).

24.5 Agroforestry for Climate Resilience

Changes in the normal climatic conditions are the most important environmental challenges disturbing the natural ecosystems, agriculture, and health globally. The extremes of temperature and precipitation patterns in poor and developing countries are predicted to have negative effects on crop yields, increased infestation of pests diseases in agriculture, and decreased livestock fodder quality. Agriculture as a sector provides a vital source of income for many developing nations; as such, changes in climate effects on the output of agriculture are a major worry in many nations (Dhyani et al. 2016, 2020; Ram et al. 2018; Meena et al. 2023b). As per IPCC (2007), out of total global anthropogenic greenhouse gas (GHG) emissions, around 25–30% is contributed by different production systems like agricultural production systems, forestry, and other land utilization practices. Agriculture alone is responsible for approximately 12% of total non-CO_2 anthropogenic greenhouse gas releases (Gerber et al. 2013).

The lives of farmers who live in tropical and sub-tropical countries may be threatened by global warming, which may cause droughts, floods, strong storms, a rise in disease incidence, and a drop in crop output in certain of these places (Dhyani et al. 2020). Numerous parts of the world are already experiencing these changes, which have an impact on

Table 24.1 Notable agroforestry systems across various Indian states

Agroclimatic zone	Agroforestry systems	Tree component	Crop/Grass
Western Himalayas	Silvipasture	*Celtis australis*, *Grewia optiva*, *M. alba*, *Bauhinia variegata*, *Ficus roxburghii*, *A. lebbeck*	*Cenchrus ciliaris*, *Minor millets*, *Lolium perenne*, *Setaria anceps*, *Chrysopogon fulvus*, *Paspalum* spp.
	Agri-horticulture	*Malus pumila*, *Prunus persica*, *Prunus armeniaca*, *Prunus domestica*	Wheat, millets, *Setaria anceps*, maize, *Lolium perenne*, soybean, *Chrysopogon fulvus*, *Paspalum* spp
Eastern Himalayas	Agri-silviculture	*Anthocephalus cadamba*, *Leucaena leucocephala*, *Melia azadirach*, *Alnus nepalensis*, *Moringa oleifera*	Rice, ragi, jowar, maize
	Agri-horticulture	*Ananas comosus*, *Passiflora edulis*, *Artocarpus heterophyllus*, *Citrus* sp., *Malus pumila*, *Prunus persica*, *Prunus armeniaca*, *Prunus domestica*	Large cardamom/coffee para grass, BN hybrid, guinea grass, white and red clover
	Silvipasture	Bamboos, *Morus alba*, *Parkia roxburghii*, *Kachnar*, *Ficus* spp., *Morus* spp	Para grass, BN hybrid, guinea grass, Napier
Lower Gangetic Plains	Agri-silviculture	*Albizia lebbeck (Siris)*, *eucalyptus* spp., *Acacia* spp, *Albizia* spp, *Leuceana leucocephala*, *Azadiracta indica*, *Dalbergia sissoo*, *Populus deltoides*	Rice, wheat, pulses, mustard, groundnut, jute, sugarcane
	Agri-horticulture	*Litchi*, *Carica papaya*, *Mango*, *Guava*	Maize, paddy, wheat, pulses, mustard
	Silvipasture	*Albizia lebbeck*, *Morus alba*, *Leuceana leucocephala*, *Madhuca latifolia*, *Ziziphus mauratiana*, *Grewia optiva*	*Pennisetum* spp., *Cenchrus ciliaris*, *Panicum maximum*, *Hetropogan contortus*, *Dichanthium annulatum*, *Apluda mutica*, *Brachiaria mutica*
Middle Gangetic Plains	Agri-silviculture	*Populus deltoids*, *Dalbergia sissoo*, *Eucalyptus* spp.	Sugarcane-wheat, rice-wheat, sesame
	Agri-horticulture	*Litchi*, *Carica papaya*, *Mango*, *Guava*, *Citrus* spp.	Rice-wheat, sugarcane-wheat
	Silvipasture	*Albizia lebbeck*, *Ficus* spp. *Leucaena leucocephala*, *Azadiracta indica*, *Madhuca latifolia*, *Ziziphus mauritiana*, *Grewia optiva*, *Morus alba*	*Cenchrus ciliaris*, *Panicum maximum*, *Hetropogan contortus*, *Dichanthium annulatum*, *Apluda mutica*, *Brachiaria mutica*
Trans Gangetic Plains	Agri-horticulture	*Aonla*, *Mango*, *Psidium guajava*	Black gram/green gram
	Agri-silviculture	*Neem*, *L. leucocephala*, *Dalbergia sissoo*, *eucalyptus* spp.	Black gram, wheat/mustard
	Silvipasture	*Albizia lebbeck*, *B. variegata*, *Morus alba*	*Cenchrus* spp., *Pennisetum* sp., *Panicum maximum*, *Hetropogan contortus*
Upper Gangetic Plains	Agri-silviculture	*P. deltoids*, *Eucalyptus* spp., *Leucaena leucocephala*	Wheat, bajra fodder, rice-wheat
	Silvipasture	*Albizia lebbeck*, *Baushinia variegata*	*Chrysopogon*, *Poa* spp. *Panicum maximum*, *Hetropogan contortus*
Eastern Plateau and Hills	Agri-silviculture	*Gmelina arborea*, *Acacia nilotica*, *Tectona grandis*	Rice, linseed
	Silvipasture	*Acacia mangium*, *A. nilotica*, *bamboos*, *L. leucocephala*	*Chrysopogon fulvus*, *Dicanthium* spp., *Pennisetum* spp.
Central Plateau and Hills	Agri-horticulture	*Emblica officinalis* and *Psidium guajava*	Chickpea/groundnut, black gram/green gram
	Agri-silviculture	*A. nilotica/Albizia lebbeck/Azadirachta indica/L. leucocephala*	Soybean, Black gram, mustard/wheat
	Silvipasture	*Albizia amara*, *Dichrostachys cinerea*, *L. leucocephala*, *Ficus* spp.	*Cenchrus ciliaris*, *Chrysopogon fulvus*, *S. hamata*, *S. scabra*
	Tree borne oilseeds	*Jatropha curcas*	–
Western Plateau and Hills	Agri-horti-silviculture	*Achras zapota*, *Tectona grandis*	Rice, maize
	Agri-horticulture	*Areca catechu*	Black pepper, cardamom
	Silviculture	*Tectona grandis*, *Ailanthus excelsa*, *Prosopis juliflora*	–
	Silvipasture	*Acacia mangium*, *Albizia amara*	*Cenchrus* spp. *Chrysopogon fulvus*

(continued)

Table 24.1 (continued)

Agroclimatic zone	Agroforestry systems	Tree component	Crop/Grass
Southern Plateau and Hills	Agri-silviculture	*Ailanthus excelsa*, *Casuarina equisetifolia*, *Eucalyptus tereticornis*, *Melia dubia*	Cotton, groundnut
	Silviculture	*Eucalyptus* spp., *L. leucocephala*, *Casuarina equisetifolia*	–
	Agri-horticulture	*Tamarindus indica*, *Psidium guajava*, *Emblica officinalis*	Chilli, cotton, groundnut
	Tree borne oilseeds	*P. pinnata*	–
East Coast Plains and Hills	Agri-silviculture	*Ailanthus excelsa*, *L. leucocephala*	Cowpea, groundnut, mustard, sorghum, maize, other millets, pulses, vegetables
	Silviculture	*Leucaena leucocephala*, *Casuarina equisetifolia*, *Moringa olelifera*, *Acacia auriculiformis*, *A. indica*	–
	Tree borne oilseeds	*P. pinnata*	–
	Silvipasture	*Artocarpus* spp., *Moringa olelifera*, *Acacia auriculiformis*, *A. indica*	*C. fulvus*, *Napier*, *Cenchrus ciliaris*
West Coast Plains and Hills	Agri-silviculture	*Acacia auriculiformis*, *Casuarina equisetiofolia*	Black pepper, rice
	Agri-horticulture	*Artocarpus heterophyllus*, *Areca catechu/Cocos nucifera*	Black pepper, rice
	Silvipasture	*Hardwickia binata*, *Albizia lebbeck*, *Gliricidia sepium*, *Ficus*, spp., *Moringa olelifera*	*Cenchrus ciliaris*, *Bracharia* spp., *Panicum maximum*
Gujarat Coast Plains and Hills	Agri-silviculture	*Ailanthus excelsa*, *A. indica*, *H. binata*,	Cowpea, mung bean, millet, mustard
	Silviculture	*A. nilotica*, *Prosopis juliflora*, *L. leucocephala*, *H. binata*	–
	Silvipasture	*L. leucocephala*, *Ailanthus excelsa*, *A. indica*	*Lasiurus sindicus C. ciliaris*, *C. setigerus*, *Dichanthium annulatum*, *Hetropogan contortus*
Western Dry Region	Agri-silviculture	*A. nilotica*, *Azadirachta indica*, *Prosopis cineraria*, *Tecomella indica*	Pearl millet, gram, mustard, groundnut
	Tree borne oilseeds	*Jatropha curcas*	–
	Silvipasture	*Hardwickia binata*, *Albizia lebbeck*, *Leucaena leucocephala*, *Ailanthus excelsa*	*Dichanthium annulatum*, *C. ciliaris*, *Lasiurus sindicus C. setigerus*
All Islands	Agri-horticulture	*Cocos nucifera*, *Anacardium occidentale*, *Areca catechu*	Rice, pulses
	Silvipasture	*Bauhinia* spp., *Erythrina indica*, *L. leucocephala*,	*C.* spp., *Pennisetum* spp. *Paspalum* spp.

Source: Dhyani et al. (2009, 2016)

millions of people's lives overall and on marginalized and small-scale farmers especially.

The amount of CO_2 in the earth's atmosphere has amplified at an average rate of 2.07 ppm/year during the last decade, accounting for approximately 77% of all anthropogenic greenhouse gas emissions (Dhyani et al. 2020). There is growing pressure to cut GHG emissions by 50–80% by the year 2050 to avert the most severe consequences of climate change. There is a possibility that various land-use regimes could help stabilize the amount of CO_2 in the atmosphere.

According to Lin (2007), the agroforestry land-use system provides a special chance to combine the dual goals of achieving food security, reducing impacts, and adapting to the effects of climate change. This is because agroforestry increases output compared to annual production systems, reduces GHG emissions from soils, and sequesters carbon in soil and biomass produced by trees (Dhyani et al. 2016, 2020). Agroforestry systems have been shown to have a higher capacity for sequestering carbon dioxide, in addition to their significant role in lowering climate change susceptibility and ensuring livelihood security (Dev et al. 2016). India wants to reduce its GDP's emission intensity to the tune of 33–35% by 2030 compared to 2005 levels. Afforestation programs through agroforestry and a greater percentage of energy produced from renewable sources (such as solar energy, bioenergy, as well as biofuel), would help achieve this goal.

The IPCC's land-use change and forestry report states that the most potential for sequestering carbon is provided by agroforestry land-use systems. Small-scale agricultural communities are less vulnerable thanks to enhanced agroforestry systems, which also aid in their ability to adjust to shifting environmental conditions.

Fig. 24.5 *Ailanthus excelsa*-based AFS (photo: AICRPAF, SK Nagar, Gujarat/Inder Dev)

24.5.1 Measures to Adapt to the Consequences of Climate Change Through Agroforestry

Agroforestry plays a significant role in various climate change activities. It contributes to carbon sequestration both above and below ground using trees and shrubs. Agroforestry also helps in reducing the emission of greenhouse gases by providing alternative timber and other forest products, thereby conserving the existing carbon stock of available forests. It reduces methane emissions through improved forage and fodder quality, decreases fossil fuel use, and lowers farmstead heating and cooling needs. In addition, agroforestry acts as a carbon sink, reducing CO_2 emissions, and decreasing nitrous oxide (N_2O) emissions by minimizing the use of N fertilizers, thanks to nitrogen-fixing trees. The practice further reduces the need for chemical fertilizers due to better nutrient recycling. In terms of climate resilience, agroforestry enhances resilience against extreme weather events affecting crop productivity, increases floral and faunal diversity, and protects natural resources such as soil moisture. It also offers diversified produce. Agroforestry also creates favorable conditions for species during aberrant weather conditions by providing habitat corridors for species migration (Schoreneberger et al. 2012).

The CSP of AFS varies significantly in India since it heavily relies on several variables, which encompasses area, site quality, species kinds, and preceding land use. The estimated values for the tree and crop components of agroforestry are 0.25 to 19.14 and 0.01 to 0.60 Mg C/ha/year, respectively. According to Dhyani et al. (2016), agroforestry is also crucial for soil C sequestration, which varies depending on the system and ranges from 0.003 to 3.98 Mg C/ha/year.

24.6 Agroforestry Network in the Country

Despite being a long-standing practice in India, systematic, targeted research on various aspects of agroforestry has accelerated since the AICRP on Agroforestry was launched in 1983 and the National Research Centre for Agroforestry (NRCAF) was established in 1988 (Dhyani and Handa 2014). The 37 centers spread across the country's agro-climatic regions that are dedicated only to agroforestry research, as well as the AICRP's headquarters in Jhansi, are now running the program.

Due to diversified agro-climatic conditions in India, there are a lot of variations in agroforestry systems, their composition, production potential, and socioeconomic conditions. Agroforestry practices range from simple shifting cultivation to intricate home gardens; from low-tree densities on farmlands in Rajasthan to high-density, intricate, multi-story home gardens in Kerala; from systems where trees primarily serve "service" functions (such as windbreaks and shelter belts) to those where they produce the majority of the plantation crops that are sold. Agroforestry spans 25.32 M ha in India, which makes up 8.2% of the nation's total land area (Dhyani et al. 2013). But neither a measurement nor income records serve as the foundation for these agroforestry area estimates. Dhyani (2014) suggested a GIS tool for reaching near to an accurate estimate of agroforestry in India.

24.7 Agroforestry for Energy Production

24.7.1 Bio-Fuel Production

A large population in India depends mainly on conventional sources of energy to fulfill their energy requirements, and these conventional sources of energy are known to have negative impacts on environmental quality. Hence, renewable sources of energy are being explored worldwide as a possible alternative energy option to conventional sources. Biofuels are widely recognized as a viable alternative to address issues related to environmental deterioration, energy security, import limitation, rural job creation, and bolstering the agricultural sector. In India, the major source of bio-fuels is tree-borne oilseeds (TBOs), viz. *Pongamia pinnata*, *Azadirachta indica Simarouba*, *Jatropha curcas*, and *Madhuca* spp. The integration of such TBOs with annual crops (French bean, green gram, black gram, cowpea, sunflower, sesame, and groundnut) in agroforestry systems can achieve energy as well as food security simultaneously. Whereas, sole plantation of these TBOs on waste and marginal land in rural areas can alleviate poverty and help to protect ecosystems besides increasing green cover. Oils from these sources are CO_2 neutral, which may help to mitigate the GHG effect. The National Oilseeds and Vegetable Oils Development Board reports that compared to our current requirement of more than 5 Mt, we are only receiving 0.8–1.0 Mt of TBOs. According to Dhyani et al. (2015), TBOs are gathered from many different kinds of species, including *Shorea robusta*, *Jatropha curcas*, *Madhuca longifolia*, *Azadirachta indica*, *Pongamia pinnata*, and *Schleichera oleosa*. Wasteland is an excellent place to cultivate several of the TBO species due to their natural hardiness. Such species need to be promoted to harness their full potential for the production of TBOs as an energy substitute and for rehabilitating the wastelands as well as increasing the green cover. But to increase seed and oil yield, more research is required on the genetic modification of trees yielding TBOs.

24.7.2 Bio-Energy Production

Biomass from trees and bushes provides a significant portion of the energy produced in rural areas. Agroforestry projects have led to the production of massive amounts of wood biomass from trees that were previously only found in traditional forest areas. The needs of wood biomass are being met in large part by species of short-rotation trees. Andhra Pradesh (AP) power generation plants mostly rely on the perennial shrub *Prosopis juliflora* as their fuel supply for their boilers. A comparison of the fuel wood potential of exotic (*Acacia. Tortilis*, *A. auriculiformis*, *E. tereticornis*, and *Casuarina equisetifolia*) and indigenous (*A. indica*, *D. sissoo*, *A. nilotica*, *P. cineraria*, and *Z. mauritiana*) trees showed that the range of calorific values for these tree species was 18.7–20.8 and 17.3–19.3 MJ DM/kg, respectively. *L. leucocephala*, *Casuarina equisetifolia*, and *Calliandra calothyrsus* are well-known tree species for producing wood energy with maximum efficiency because of their shorter rotation and excellent adaptability to a variety of habitats and temperatures. There is a huge possibility for producing wood energy from live fences on farm boundaries and agroforestry plants on common property. For improved fuel wood consumption and socioeconomic status, plantations of locally available multi-purpose trees (MPTs) can be encouraged on wasteland, community land, and private property in the Bundelkhand region of central India (Chauhan et al. 2015).

The utilization of biomass from diverse agroforestry systems as an energy source has not only enhanced livelihood conditions but also positively impacted ecological health in the semi-arid Bundelkhand region. According to Chavan et al. (2016), the finest quality different fuelwood trees for agroforestry plantations in the region of Bundelkhand are *Prosopis juliflora*, *Acacia catechu*, and *Anogeissus pendula*. The AICRP on Agroforestry's biofuel research has shown that bioenergy plantations are a cost-effective solution for degraded lands. Because of their higher energy efficiency, shorter rotation times, and greater adaptability to a variety of environments and climates, native tree species like *D. sissoo*, *A. indica*, *A. nilotica*, *Z. mauritiana*, and *P. cineraria*, as well as invasive species like *Casuarina equisetifolia*, *A. auriculiformis*, *A. tortilis*, *E. tereticornis*, and *E. camaldulensis* have a great deal of potential for use as energy producers. According to Raizada et al. (2014), densified plantations of *Bahunia variegata* (4 m × 2 m), *Dalbergia sissoo* (3 m × 2 m), and *Grewia optiva* (2 m × 2 m) on former riverbank lands in the northwest Himalayas were found to be highly beneficial for producing fuel and fodder.

24.8 Agroforestry in Relation to Industry

Agroforestry plays a vital role in maintaining a steady supply of raw materials to both wood and non-wood forest product industries. The demand for wood/timber, and additionally non-wood forest commodities is expanding quickly due to urbanization and rising populations. According to Shrivastava and Saxena (2017), India has seen a growth in wood imports of 0.9 M m^3 every year since 2001. By the end of 2020, 2025, and 2030, the country's wood imports are projected to reach 22.51, 27.01, and 31.5 M m^3. However, the World Wide Fund for Nature (WWF) and the erstwhile Planning Commission (now NITI Aayog) of India have estimated a significant decrease in wood/timber supply compared to national as well as international sources in the coming future owing to declining wood productivity of plantations and conventional forests and ever-rising forest products demand (Manoharan 2011).

Indian forests, which were the prime source of raw material for forest-based industries, now owing to their diminishing area, overexploitation, and low productivity (0.77 m^3/ha/year) in comparison to the world average (2.1 m^3/ha/year), are unable to meet out solely the demand of wood industries (Agarwal and Saxena 2017). As per the FSI (2019) report, the total growing stock of India was 5915.76 M m^3 of which 4273.47 M m^3 is contributed by forests and 1642.29 M m^3 by trees outside forests, which is not enough to meet the demand for forest products in India. Hence, India must develop high-yielding sources of timber and other forest-based goods to cater to the nation's growing requirements and declining availability. This can only be accomplished by making use of wastelands/marginal areas, fallow lands, and agricultural land. Among each of the possible land resources that have been identified, utilizing agricultural land via the establishment of a commercial or industrial agroforestry model would be the best option to ensure livelihood security and environmental safety in the event of catastrophic climate change, in addition to catering to the rising demand for forest products. According to Agarwal and Saxena (2017), the productivity of woody growing stock under agroforestry and farm forestry, and in trees outside of forests is estimated to be 3.06 m^3/ha/year. This productivity significantly helps meet the demand for raw materials for industries based on both wood and non-wood forests. About 80% and 60% of demand for the plywood and pulpwood industries, respectively, is being met by trees raised under agroforestry (Agarwal and Saxena 2017). The increasing disparity between supply as well as demand for wood in addition to forest products other than wood is making it more and more challenging for enterprises that deal with forest products to get raw materials, as noted by Agarwal and Saxena (2017). Therefore, the amount of land that is now covered in trees outside of forests is insufficient to close the disparity between the availability and demand of unprocessed products for enterprises that are reliant on forests for their resources.

There is a pressing need to intensify the area under agroforestry in India to meet ever ever-growing demand for wood products and this will create many opportunities for millions of marginal and smallholders to secure their livelihood by growing commercially high-value trees. Adoption of commercial/industrial agroforestry among farmers would require quality planting material, high-yielding genotypes, technical assistance, financial support for establishing and managing plantations, insurance, and marketing assistance, etc. Among all the above-mentioned requirements of industrial agroforestry, the availability of high-yielding genotypes of commercial tree species is the need of the hour. The connection between FARM and FIRMS must be established to meet the growing demand in commercial wood markets, ensure rural development and poverty alleviation, and sustain ecosystem services while mitigating climate change.

24.9 Role of Private Sector and Forest-Based Industries

As expansion of area under commercial/industrial agroforestry is required to meet the demands of the forest-based industries, it is crucial to encourage farmers to take up commercial/industrial agroforestry. In this regard, private sector and forest-based industries can play important roles in the promotion of industrial agroforestry. The private sector and industries can provide quality planting material, insurance, finance support, market information, and marketing assistance to the farmers. Moreover, the private sector and industries can create an assured market for the farmers venturing into industrial agroforestry (Chavan et al. 2015).In India, the private sector has already begun to play a key role in developing high-yielding clones of commercial tree species and raising quality planting material for agroforestry tree species (Dhiman and Gandhi 2012; Dhiman and Gandhi 2014; Kulkarni 2013). WIMCO is known for its poplar, eucalyptus, and kadamba clones (Dhiman and Gandhi 2015; Dhiman et al. 2010). The private sector is also carrying out research trials to study tree crop interaction under agroforestry to find suitable clones for the agroforestry sector and also on developing diseases and pest-resistant clones (Kumar and Dhiman 2016; Kulkarni 2004). New emerging diseases and pests have been identified for tree species and their control measures have been worked out (Dhiman et al. 2010). Commercial agroforestry models have been developed by the private industries for pulp and paper by Tamil Nadu Newsprint Ltd.; by ITC Paper Board in Andhra Pradesh; by JK Paper Mills in Gujarat and Odisha; by West Coast Paper Mill in Uttar Pradesh; by plywood industries in Haryana, Punjab and Uttarakhand; by WIMCO Ltd. in the Indo-Gangetic region; by Gums and Resins Industries in Jharkhand, Chhattisgarh through contract farming mode (ICFRE 2012; Luukkanen and Appiah 2013; Chavan et al. 2015). Commercial agroforestry is practiced in India over 5 M ha exploiting the potential of *Acacia* spp., *Ailanthus* spp., *Anthocephalus cadamba*, *Bombax* spp., *Casuarina*, *equisetifolia*, *Eucalyptus* spp., *Leucaena leucocephala*, *Melia* spp., *Populus* spp. (Chavan et al. 2015).

Initiatives of private players have opened up avenues of income and livelihood generation to the farmers by establishing commercial agroforestry models with the involvement of local farmers. The private sector needs to establish and support the commercial agroforestry model in various agro-climatic zones of the country to become self-sufficient in meeting the demand of forest-based industries and to support farmers economically. As a further step toward mitigating the consequences of climate change, this will make it easier to accomplish the target of covering 33% of the nation's land area with trees and forests.

The main consideration for the private sector to make industrial agroforestry profitable would be to establish a market in close proximity to the commercial agroforestry models of the specific region and provide farmers with financial and insurance support for forest products. Moreover, establishing industry-driven minimum support prices to support farmers is required from private firms as there is no minimum support price in the case of forest/agroforestry produce in India. Value chain development is another yet important sector that has been ignored and is required to make agroforestry a successful venture for farmers in which private players and industries can play a major role.

24.10 Agroforestry with Reference to Natural Resource Management

In dry and semi-arid regions, insufficient water availability is a primary obstacle to food production. Since the water use efficiency (WUE) of a rain-fed ecosystem is low, raising WUE in this ecosystem is essential to raising the standard of living of farmers. Floods and droughts are predicted to rise in many places due to altered precipitation patterns, which will be extremely difficult for small-scale farmers that rely on rain-fed farming (Verchot et al. 2007). Low crop productivity in arid and semi-arid parts of the country is primarily triggered by land degradation. The low infiltration capacity of degraded soils makes degraded land more susceptible to flooding (Wani et al. 2009). Compared to annual crops, perennial plants withdraw the water from more volume of

Fig. 24.6 Agroforestry-based natural resource management at Parasai-Sindh watershed, Jhansi (photo: ICAR-CAFRI/Ramesh Singh)

soil; therefore, perennials can survive even in long-lasting droughts (Verchot et al. 2007). Due to their larger ecological niches, agroforestry systems are able to use water more effectively (Fig. 24.6). As a result of the trees using water not available for the crops between growth seasons, agroforestry systems are able to double the use of rainwater (Pandey 2007).

Livesley et al. (2004); Radersma and Ong (2004); Odhiambo et al. (2001), and other studies have demonstrated that trees lower soil moisture content and lower yields; yet, other studies have shown encouraging results (Sinare and Gordon 2015; Radersma and Ong 2004; Siriri et al. 2013). According to several researchers (Radersma and Ong 2004; Pandey 2007; Kuyah et al. 2016), trees are more likely to compete for water resources and have a good (in 51% of studies) or negative (in 35% of studies) impact on water availability. According to them, reduced evapotranspiration and improved infiltration were responsible for the positive effects.

24.11 Agroforestry Policy: Need and Potential to Expand Area Under Agroforestry

There is a need for independent policies that can address the numerous bottlenecks connected with the expansion of agroforestry and promote the growth of agroforestry on a larger scale. The National Forest Policy 1988, the National Agriculture Policy 2000, the Planning Commission Task Force on Greening India 2001, the National Bamboo Mission 2002, the National Policy on Farmers 2007, and the Green India Mission 2010 highlight the importance of agroforestry for nutrient cycling, soil fertility, and other ecosystem services. However, various factors have prevented agroforestry from gaining the importance it deserves as a tool for resource development. The absence of a well-thought-out national policy, a comprehensive integrated farming system approach, restrictive regulatory regimes, insufficient attempts to liberalize restrictive regulations, inadequate research, extension, and capacity building, inferior planting materials, a lack of institutional financing and insurance coverage, and limited market access for agroforestry produce are some of the main hindrances to promote agroforestry (National Agroforestry Policy 2014). To overcome the various issues/obstacles related to agroforestry adoption and promotion, India realized that it needed a national policy for agroforestry. The National Agroforestry Policy was created by the Indian government to make agriculture less susceptible to climatic irregularities. India is the first country in the world to have a comprehensive agroforestry strategy, having had the National Agroforestry Strategy approved by the Cabinet in 2014.

The goal of an agroforestry policy is to combine several agencies, programs, and plans that include agroforestry aspects. It aims to increase smallholder farmers' productivity, income, and livelihoods. Through this policy, tree plantations will be encouraged and expanded, existing pressure on forests will be minimized, and a national forest cover target will be achieved. As a result, rural households, especially smallholder farmers, will have more employment opportunities and livelihood security. Various laws and procedures related to harvesting and transporting agroforestry products have also been simplified under this policy. The policy guidelines require different states in the country to identify about 20 commonly grown tree species that can be planted on farmlands to benefit the farming community economically and ecologically. State regulations regarding growing, harvesting, and transportation must be notified for these species. In addition to safeguarding the environment and natural forests, the strategy will help meet the increasing need for agroforestry products like food, fuel, lumber, and other items. As a result, the wood-based industries will benefit from increased raw material availability and reduced imports of wood and wood products, thereby saving foreign exchange.

24.11.1 Exempted and Restricted Agroforestry Tree Species in Different Indian States

In Tamil Nadu, the exempted species include *Eucalyptus* spp., *Grevillea robusta*, *Leucaena leucocephala*, and *Casuarina equisetifolia*, while the restricted species are *Diospyros melanoxylon*, *Santalum album*, *Pterocarpus santalinus*, *Dalbergia latifolia*, and *Tectona grandis*. In Kerala, the exempted species are *Hevea brasiliensis*, *Pongamia pinnata*, *Mangifera indica*, *Garcinia cambogia*, *Ailanthus excelsa*, and *Terminalia* spp. The restricted species include *Santalum album*, *Dalbergia latifolia*, *Michelia champaca*, *Hopea parviflora*, *Xylia xylocarpa*, *Grewia tiliifolia*, *Toona ciliata*, and *Tectona grandis*. In Uttar Pradesh, the exempted species are *Ailanthus* spp., *Anthocephalus cadamba*, *Albizia* spp., *Acacia* spp., Poplar, *Eucalyptus* spp., Subabul, *Gmelina arborea*, *Grevillea robusta*, Casuarina, *Melia composita*, *Morus alba*, *Borassus flabellifer*, *Butea monosperma*, *Tamarindus indica*, and *Grewia oppositifolia*. The restricted species are *A. indica*, *Mangifera indica*, *Ficus bengalensis*, *Carpinus viminae*, and *Madhuca indica*. In Punjab and Haryana, the exempted species are Poplar, *Eucalyptus*, and *Melia composita*. The restricted species are *Acacia catechu*, Bamboo, and *Emblica officinalis*. In Gujarat, the exempted species are *Eucalyptus* spp., *Casuarina equisetifolia*, and *Prosopis juliflora*. The restricted species include *Acacia catechu*, *Dalbergia latifolia*, *Tectona grandis*, *Mangifera indica*, *Madhuca latifolia*, *Santalum album*, *Gmelina arborea*, *Anogeissus pendula*, and *Madhuca indica*. In Maharashtra, the exempted species are *Moringa*, *Phoenix*, *Poplar*, *Acacia auriculiformis*, *Acacia nilotica*, *Leucaena leucocephala*, *Eucalyptus*, and *Sapota*. The restricted species are *Adina cordifolia*, *Tamarindus indica*, *Artocarpus integrifolia*, *Terminalia chebula*, *Madhuca latifolia*, *Tectona grandis*, *Mangifera indica*, *Santalum album*, *Acacia catechu*, *Pterocarpus marsupium*, *Syzygium cumini*, *Terminalia tomentosa*, *Hardwickia binata*, and mangroves. In Bihar, the exempted species are *Litchi*, *Mangifera indica*, *Poplar*, *Eucalyptus* spp., *Phoenix*, *Bombax ceiba*, *Anthocephalus cadamba*, *Gmelina arborea*, and bamboo. The restricted species include *Tectona grandis*. In Madhya Pradesh, the exempted species are *Eucalyptus*, Subabul, *Poplar*, *Casuarina*, and *Pithocellobium dulce*. The restricted species include *Ziziphus mauritiana*, *Albizia lebbeck*, *Butea monosperma*, *Azadirachta indica*, *Acacia nilotica*, *Acacia leucophloea*, *Syzygium cumini*, and bamboo (Chavan et al. 2015).

In West Bengal, no tree species are exempted, while the restricted species are *Madhuca indica*, *Swietenia mahagony*, *Bombax ceiba*, *Acacia catechu*, *Diospyros melanoxylon*, *Michelia champaca*, *Dalbergia sissoo*, *Gmelina arborea*, *Shorea robusta*, mangroves, and *Tectona grandis*. In Odisha, the exempted species are *Acacia auriculiformis*, *Bambusa vulgaris*, *B. tulda*, *B. nutan*, *Eucalyptus* hybrid, *Cassia siamea*, *Casuarina equisetifolia*, *Samania saman*, and Silver oak. The restricted species include *Schleichera oleosa*, *Tamarindus indica*, *Artocarpus heterophyllus*, *Madhuca indica*, *Santalum album*, and *Mangifera indica*. In Karnataka, the exempted species are Areca nut, *Casuarina*, Subabul, Rubber, Eucalyptus, Coconut, Erythrina, Orange, Sesbania, Gliricidia, and Silver oak. The restricted species are *Acacia catechu*, *Santalum album*, *and Dalbergia latifolia*. In Jharkhand, the exempted species are *Terminalia elliptica*, *Boswellia serrata*, Bamboo, Subabul, Cane, *Tectona grandis*, *Acacia nilotica*, *Shorea robusta*, and *Pterocarpus marsupium*. The restricted species include *Syzygium cumini*, *Artocarpus heterophyllus*, *Gmelina arborea*, *Dalbergia sissoo*, *Madhuca indica*, and *Mangifera indica*. In Andhra Pradesh, the exempted species are *Acacia nilotica*,

Anacardium occidentale, *Casuarina equisetifolia*, *Mangifera indica*, *Syzygium cumini*, *Psidium guajava*, *Ficus religiosa*, *Eucalyptus*, *Azadirachta indica*, *Cocos nucifera*, and Subabul. The restricted species include *Tectona grandis*, *Pterocarpus santalinus*, and *Santalum album* (Chavan et al. 2015).

The National Agroforestry Policy (2014) recommended the setting of an agroforestry mission to develop or promote agroforestry in an organized manner. To achieve this, the Sub-Mission on Agroforestry was established under the National Mission on Sustainable Agriculture. Farmland tree cover is to be increased as part of the sub-mission. To achieve quantifiable benefits, the mission aims to increase agroforestry areas to enhance carbon sequestration, enrich soil fertility, provide quality planting materials, enhance livelihoods, improve crop productivity, develop an information system, etc.

24.12 Economic Analysis of Agroforestry Systems

Agroforestry has proven to be more lucrative for farmers than forestry or agriculture in several studies conducted throughout the country. According to Newaj and Rai's (2005) research, a 13-year-old aonla-based agroforestry system that was built on marginal lands within a rainfed ecosystem had a 3.28 benefit-cost ratio (B:C), indicating significant profitability. These systems are also profitable, according to several researchers. The Planning Commission (2001) found a B:C ratio in the range of 1.01–4.17 by analyzing 24 agroforestry systems located in different agroclimatic zones of the country. Farmers in Western Uttar Pradesh, Punjab, and Haryana are drawing a lot of attention because they are using agroforestry concepts, which are quite profitable. In these places, poplar has been seen to be harvested every 6–8 years, and poplar-based agroforestry systems have an average economic return that is significantly higher than that of purely agricultural enterprises. Another example of a eucalyptus-based agroforestry system that was economically more profitable than annual crops in Haryana was worked out by Dhillon and Bangarwa (2016). Research findings suggest that eucalyptus (6 m × 1 m)-based agroforestry, coupled with an eight-year rotation cycle, and intercropping with crops, is a great way to expand agroforestry on farmlands. With a net present value (NPV) and B:C ratio of 2.28, this method produced the highest results. Furthermore, poplar-based agroforestry has proven to be highly successful in northern states. The study examined the financial aspects of a seven-year agroforestry system based on bamboo. The results showed that the configuration of 12 m × 10 m bamboo + sesame – chickpea produced the best results in terms of NPV, B:C ratio, annual equivalent value (AEV), and land expectation value (LEV). This configuration was followed by the arrangement of 10 m × 10 m bamboo + sesame – chickpea (Dev et al. 2020).

24.13 Conclusion and Way Forward

The National Agroforestry Policy of 2014 has made it easier for many of the stakeholders to advocate agroforestry. Government policies pertaining to resource management, infrastructure, and climate change should all take agroforestry into consideration. Agroforestry-related research and education should also get more support. Unfavorable laws must be changed, and forestry and agroforestry rules need to be made simpler. The need for a public–private partnership is essential. It is necessary to create financially sustainable and geographically specific systems of agroforestry. Since agroforestry land use has a lot of possibilities to provide services for the environment, it is essential to measure all of these advantages. Agroforestry has the potential to meet the sustainable development goals.

Lessons Learnt

- Agroforestry provides an opportunity for climate resilience and food security as it stores carbon in above- and below-ground biomass reduces GHG emissions from soils, and gives higher biomass production than annual systems.
- Agroforestry plays a pivotal role in addressing manifold problems and providing a wide range of socioeconomic and environmental benefits.
- Agroforestry is an efficient system for recycling nutrients and a better protector of ecological systems.

Key Questions

1. What are below and above-ground interactions in agroforestry and how to minimize the impact of negative interactions?
2. What are the ecosystem services of agroforestry and how they can be quantified to harness their economic potential?
3. How does the National Agroforestry Policy (2014) help in the convergence of different government schemes and the expansion of tree cover outside the forest area?
4. How does the National Agroforestry Policy (2014) ease the felling and transit rules for different tree species?
5. How agroforestry plantations are supporting and sustaining the diverse wood-based industries in the country?

References

Agarwal S, Saxena AK (2017) The puzzle of forest productivity: are forest development corporations solving it right? Centre for Science and Environment, New Delhi

Bhatla N, Mukherjee T, Singh S (1984) Plants: traditional worshiping. Indian J Hist Sci 19:37–42

Chaturvedi OP, Sikka AK, Handa AK, Bajpai CK (2016) Agroforestry technologies for different agro-climatic zones of the country. ICAR-CAFRI Publication, Jhansi

Chauhan D, Chavan SB, Keerthika A, Jha A, Dwivedi RP, Newaj R, Kumar A (2015) Socio-economic fuelwood survey in Bundelkhand region of central India. In: Mishra GC, Yadav SK (eds) Energy research and environmental management: an innovative approach. Krishi Sanskriti Publications, New Delhi, pp 124–129

Chavan SB, Keerthika A, Dhyani SK, Handa AK, Newaj R, Rajarajan K (2015) National Agroforestry Policy in India: a low hanging fruit. Curr Sci 108:1826–1834

Chavan SB, Uthappa AR, Sridhar KB, Keerthika A, Handa AK, Newaj R, Kumar N, Kumar D, Chaturvedi OP (2016) Trees for life: creating sustainable livelihood in Bundelkhand region of central India. Curr Sci 111(6):994–1002

Dev I, Newaj R, Prasad R, Handa AK, Ram A, Tewari R, Bajpai CK, Pandey AK (2016) Agroforestry for environmental challenges, sustainable land use, biodiversity conservation and rural livelihood options - abstracts book. Indian Society of Agroforestry, ICAR-CAFRI, Jhansi

Dev I, Ram A, Singh R, Kumar N, Kumar D, Chand L, Tewari RK, Kumar S, Venkatesh (2020) Agroforestry based conservation agriculture for resource conservation and climate change mitigation. In: Sharma NK, Mishra PK, Yadav RS, Mandal D, Kumar G, Roy T, Ranjan R, Meena MK, Kumar D, Narayan D, Ojasvi RP (eds) National seminar on resource conservation for soil security and jalshakti: farmers perspective in Bundelkhand. ICAR-IISWC RC, Datia, Madhya Pradesh

Dhillon RS, Bangarwa KS (2016) Agroforestry practices in Haryana. Dep. For. CCSHAU, Hisar, Haryana

Dhiman RC, Gandhi JN (2012) Clonal development and diversity in WIMCO's poplar programme. ENVIS For Bull 12:40–48

Dhiman RC, Gandhi JN (2014) Status of eucalypts clonal culture in India. In: Bhojveid PP, Kaushik S, Singh YP, Kumar D, Thapliyal M, Barthwal S (eds) Eucalypts in India. ENVIS Centre on Forestry, National Forest Library and Information Centre, FRI, Dehradun, pp 60–116

Dhiman RC, Gandhi JN (2015) Cultivation of high value crop-Sweet corn (*Zea mays* Linn.) in poplar based agroforestry. Indian J Agrofor 17:70–75

Dhiman RC, Kulkarni HD, Gandhi JN (2010) Spreading infestation of gall wasp (*Leptocybe invasa* gen. & sp. N.) on eucalypts in North India. Indian For 136:791–803

Dhyani SK (2014) National Agroforestry Policy 2014 and the need for area estimation under agroforestry. Curr Sci 107:9–10

Dhyani SK, Handa AK (2014) Agroforestry in India: current scenario. Indian Farming 63:6–8

Dhyani SK, Handa AK, Uma. (2013) Area under agroforestry in India: an assessment for present status and future perspective. Indian J Agrofor 15:1–11

Dhyani SK, Newaj R, Alam B, Dev I (2015) Agroforestry: present status and way forward. Biotech Books, New Delhi

Dhyani SK, Newaj R, Sharma AR (2009) Agroforestry: its relation with agronomy, challenges and opportunities. Indian J Agron 54:249–266

Dhyani SK, Ram A, Dev I (2016) Potential of agroforestry systems in carbon sequestration in India. Indian J Agric Sci 86(9):1103–1112

Dhyani SK, Ram A, Newaj R, Handa AK, Dev I (2020) Agroforestry for carbon sequestration in tropical India. In: Ghose PK, Mahanta SK, Mandal D, Mandal B, Ramakrishnan S (eds) Carbon management in tropical and sub-tropical terrestrial systems. Springer India, pp 313–331

FSI (2019) State of Forest Report 2019

Gerber PJ, Steinfeld H, Henderson B, Mottet A, Opio C, Dijkman J, Falcucci A, Tempio G (2013) Tackling climate change through livestock - a global assessment of emissions and mitigation opportunities. Food and Agriculture Organization of the United Nations (FAO), Rome

GOI (1999) National forestry action programme, vols. 1/2. https://envisjnu.tripod.com/envnews/nov99/action.html

ICFRE (2012) Country report on poplars and willows, period: 2008 to 2011. Curr Sci 92:455–463

IPCC (2007) Climate change 2000. Oxford University Press, Oxford

Kulkarni HD (2004) Clonal forestry for industrial wood production. In: Parthiban KT, Paramathma M, Neelakantan KS (eds) An ITC experience in compendium on clonal forestry. NATP-FCRI, Coimbatore, pp 92–113

Kulkarni HD (2013) Pulp and paper industry raw material scenario - ITC plantation a case study. J Indian Pulp Pap Tech Assoc 25:79–89

Kumar BM, Sikka AK (2014) Agroforestry in South Asia: glimpses from Vedic to present times. Indian Farming 63:2–5

Kumar D, Dhiman RC (2016) Country report on poplars and willows, 2012–2015. National Poplar Commission of India. Forest Research Institute, Dehradun

Kuyah S, Öborn I, Jonsson M, Dahlin AS, Barrios E, Muthuri C, Malmer A, Nyaga J, Magaju C, Namirembe S, Nyberg Y, Sinclair FL (2016) Trees in agricultural landscapes enhance provision of ecosystem services in Sub-Saharan Africa. Int J Biodivers Sci Ecosyst Serv Manag 12:255–273

Lin BB (2007) Agroforestry management as an adaptive strategy against potential microclimate extremes in coffee agriculture. Agric For Meteorol 144:85–94

Livesley SJ, Gregory PJ, Buresh RJ (2004) Competition in tree row agroforestry systems. 3. Soil water distribution and dynamics. Plant Soil 264:129–139

Luukkanen O, Appiah M (2013) Good practices for smallholder teak plantations: keys to success. World Agroforestry Centre (ICRAF) Southeast Asia Regional Program, Bogor

Manoharan TR (2011) Supply determinants of timber trade in India. A study supported by the planning commission, Gov. of India, WWF 1-83

Meena SN, Sharma SK, Singh P, Ram A, Meena BP, Jain D, Singh D, Debnath S, Yadav S, Dhakad U, Verma P, Meena JK, Nandan S (2023b) Tillage-based nutrient management practices for sustaining productivity and soil health in the soybean-wheat cropping system in Vertisols of the Indian semi-arid tropics. Front Sustain Food Syst 7: 1234344. https://doi.org/10.3389/fsufs.2023.1234344

Meena SN, Sharma SK, Singh P, Ram A, Prakash B, Prajapat K, Sharma NK, Kumhar BL, Meena BS (2023a) Conservation and organic management practices influenced wheat (*Triticum aestivum*) productivity, profitability and weed dynamics. Indian J Agric Sci 93:501–505

Nair PK (1979) Agroforestry research: a retrospective and prospective appraisal. In: Int. conf. on international cooperation in agroforestry. ICRAF, Nairobi, pp 275–296

National Agroforestry Policy (2014) Government of India, New Delhi

Newaj R, Rai P (2005) Aonla-based agroforestry system: a source of higher income under rainfed conditions. Indian Farming 55:24–27

Odhiambo HO, Ong CK, Deans JD, Wilson J, Khan AAH, Sprent JI (2001) Roots, soil water and crop yield: tree crop interactions in a semi-arid agroforestry system in Kenya. Plant Soil 235:221–233

Pandey DN (2007) Multifunctional agroforestry systems in India. Curr Sci 92:455–463

Planning Commission (2001) Report of the task force on greening India for livelihood security and sustainable development. Planning Commission, New Delhi

Prasad R, Pandey D, Kotwal NPC (2000) Trees outside forests in India: a national assessment. Indian Institute of Forest Management, Bhopal

Puri S, Nair PK (2004) Agroforestry research for development in India: 25 years of experiences of a national program. Agrofor Syst 61:437–452

Radersma S, Ong KC (2004) Spatial distribution of root length density and soil water of linear agroforestry systems in sub-humid Kenya: implications for agroforestry models. For Ecol Manag 188:77–89

Rai SN, Chakrabarti SK (2001) Demand and supply of fuel wood and timber in India. Indian For 127:263–279

Raizada A, Ghosh BN, Jayaprakash J, Singh C (2014) Biomass production, carbon sequestration and water transmission properties as influenced by densified plantations raised on old river bed lands in the North West Himalayas. Indian J Soil Conserv 42:85–97

Raj A, Jhariya MK, Pithoura F (2014) Need of agroforestry and impact on ecosystem. J Plant Dev Sci 6:577–581

Ram A, Dev I, Uthappa AR, Kumar D, Kumar N, Chaturvedi OP, Dotaniya ML, Meena BP (2017) Reactive nitrogen in agroforestry systems of India, the Indian nitrogen assessment: sources of reactive nitrogen, environmental and climate effects, management options, and policies. Elsevier

Ram A, Kumar D, Shivay YS, Anand A, Singh N, Dev I (2018) Effect of sulphur on growth, productivity, and economics of wheat (*Triticum aestivum*) and residual soil fertility under aerobic rice (*Oryza sativa*)-wheat cropping system in Inceptisols. Indian J Agron 63:271–277

Ravindranath NH, Hall DO (1995) Biomass, energy and environment: a developing country perspective from India. Oxford University Press, New York, xiv, 376p

Schoreneberger M, Bendrub G, Gooijer H, Soolanayakanaha R, Sauer T, Brandle J, Zhan X, Den C (2012) Branching out agroforestry as a climate change mitigation and adaptation for agriculture. J Soil Water Conserv 67:128–136

Shrivastava S, Saxena AK (2017) Wood is good: but is India doing enough to meet its present and future needs? Center for Science and Environment, New Delhi. https://cdn.cseindia.org/attachments/0.39269300_1505297987_wood-is-good.pdf

Sinare H, Gordon LJ (2015) Ecosystem services from woody vegetation on agricultural lands in Sudano-Sahelian West Africa. Agric Ecosyst Environ 200:186–199

Siriri D, Wilson J, Coe R, Tenywa MM, Bekunda MA, Ong CK, Black CR (2013) Trees improve water storage and reduce soil moisture evaporation in agroforestry systems on bench terraces in SW Uganda. Agrofor Syst 87:45–58

Verchot LV, Noordwijk VM, Kandji S, Tomich T, Ong C, Albretch A, Mackensen J, Bantilan C, Anupama K, Palm C (2007) Climate change: linking adaptation and mitigation through agroforestry. Mitig Adapt Strateg Glob Chang 12:901–918

Wani SP, Rockström J, Oweis T (2009) Rainfed agriculture: unlocking the potential. CAB International, Wallingford, UK

Zomer RJ, Bossio DA, Trabucco A, Yuanjie L, Gupta DC, Singh VP (2007) Trees and water: smallholder agroforestry on irrigated lands in northern India. International Water Management Institute

Good Agricultural and Collection Practices of Important Medicinal Plants

25

A. K. Pandey

Abstract

Medicinal plants play an important role in primary healthcare and are an important source of income in rural areas. Herbal medicines are gaining popularity in both developing and developed countries. This trend is attributed to a growing awareness of the harmful side effects associated with many modern drugs. The current global trade value of medicinal and aromatic plants (MAPs) stands at approximately US$ 62 billion. There is a projection that this trade could reach an impressive US$ 5 trillion by 2050. The rising global demand for herbal drugs has led to large-scale collection from the wild, resulting in the overexploitation of many medicinal plants. Some of these plants are now facing the threat of extinction due to this increased demand.

The concept of good agricultural and collection practices (GACP) is introduced as a set of guidelines aimed at ensuring the sustainable and environmentally friendly cultivation and harvesting of medicinal plants. These guidelines are crucial for maintaining the quality and efficacy of the final herbal products. GACP is presented as a way to meet the increasing demand for medicinal plants while safeguarding their populations and ecosystems.

The chapter focuses on seven commercially important medicinal plants, providing an account of GACPs for *Andrographis paniculata* (Kalmegh), *Asparagus racemosus* (Shatavar), *Bacopa monnieri* (Brahmi), *Gymnema sylvestre* (Gudmar), *Rauvolfia serpentina* (Sarpagandha), *Tinospora cordifolia* (Giloe), and *Withania somnifera* (Ashwagandha). The implementation of GACPs is highlighted as a means to ensure the conservation and sustainable utilization of these plants, maintaining quality, involving communities, and optimizing returns for all stakeholders, including farmers and collectors. Overall, the chapter underscores the importance of GACP in the medicinal plant industry to balance the increasing demand for herbal medicines with environmental conservation and sustainable resource management.

Keywords

Medicinal plants · GACP · Phytochemicals · Variety · Yield

25.1 Introduction

India is recognized as one of the top 12 mega-diversity countries globally. The country's diverse landscape includes 15 different agro-climatic zones, 10 vegetarian zones, 25 biotic provinces, and 426 biomes. As a subcontinent, India experiences a wide range of climates, covering almost all shades of environmental conditions. From the hottest Thar desert to the arctic environment in the Himalayas, India showcases diverse climatic zones with intermediate gradations.

Medicinal plants have been a historical mainstay in traditional healthcare practices across major societies for centuries. Even today, 75–80% of the world's population relies on medicinal plants for their primary healthcare needs (Patwardhan et al. 2005; Pandey and Shukla 2008). The World Health Organization (WHO) has a target of "health for all," which can only be realized through the utilization of herbal medicine. The demand for herbal-based drugs in developing countries is huge with some consumers becoming increasingly skeptical about the use of modern medicine in developed countries.

Medicinal plants have been an integral part of India's healthcare system for centuries. India is home to approximately 18,000 flowering plant species. About 8000 of these plant species are used in traditional Indian medicine systems, including Ayurveda, Unani, Siddha, and Homeopathy (AYUSH). Medicinal plants find application in various

A. K. Pandey (✉)
Forest Research Institute, Dehradun, India

A. K. Mandal, A. Nicodemus (eds.), *Textbook of Forest Science*, https://doi.org/10.1007/978-981-97-8289-5_25

sectors, including herbal industries, Ayurvedic industries, pharmaceutical industries, household usage, and export. According to the National Medicinal Plants Board (NMPB), 1178 species of medicinal plants are active in trade in India. Among these, 242 species have high annual consumption levels, exceeding 100 metric tons per year. The breakdown of the high-consumption species reveals that 42% are herbs, 27% are trees, and 31% are shrubs and climbers (Ved and Goraya 2007). A study conducted by Ravikumar et al. (2018) reveals that nearly 90% of the medicinal plants used by local communities in India are directly sourced from the wild. This study also highlights that approximately 72% of the medicinal plant species and 50% of the annual quantities consumed as herbal raw materials by the domestic herbal industry are also sourced from the wild. This high demand for herbal drugs has led to the rampant collection of medicinal plants from the wild, leading to the depletion of biodiversity. Sustainable harvesting is the only way to conserve our precious resources for future generations (Pandey 2015).

There is a rising global demand for phytochemicals and products derived from medicinal plants. This trend is driven by increased awareness of the potential health benefits associated with natural compounds found in these plants and has opened up opportunities for entrepreneurs to add value to these plants and products through cultivation, harvesting, processing and thereby generating enormous employment avenues. An upward trend has been recorded in the exports of medicinal and aromatic plant products in recent years, which has encouraged government and private organizations to develop processed products with medicinal plants. Different biotechnological approaches have been developed for enhanced production of secondary metabolites from plants and their conservation in natural habitats (Pandey and Tripathi 2017). Conservation of medicinal plants in their natural habitat is crucial to maintain ecological balance and preserve biodiversity. Balancing increased cultivation and harvesting with sustainable good agricultural and collection practices (GACP) help to ensure the long-term availability of these valuable resources.

The geographical origin, cultivar, and growth stage of the plant at the time of collection impact the quality of medicinal plants. Post-harvest handling is a critical factor in maintaining the quality of medicinal plants. The quality of medicinal plants is also dependent on the use of quality planting material, such as seeds or propagules. The right seed source is crucial for planting and growing a diverse range of medicinal plants. World Health Organization (WHO) prescribed general guidelines for Good Agricultural and Collection Practices (GACP) for the supply of quality medicinal herbs (WHO 2003). GACPs for medicinal plants are a set of guidelines for medicinal plant producers on how to improve the safety, efficacy, and quality standards of raw materials used in the preparation of herbal medicines. Although the GACP guidelines only cover production up to the point of selling the raw material, the buyers of medicinal plant materials and manufacturers of herbal medicines also need to learn about the principles of GACPs. Many elements of GACPs can be implemented by farmers and collectors without any additional cost, yet to significantly improve the quality of medicinal plant material some investment may be required, especially to improve post-harvest processing facilities such as for drying and storage. The National Medicinal Plants Board and Quality Council of India (QCI) have initiated a Voluntary Certification Scheme for Medicinal Plant Produce. This scheme aims to create awareness, implement GACPs, and improve the quality and certification of medicinal plant material, making high-quality raw material available to the AYUSH industry (Guidelines on Good Field Collection Practices for Indian Medicinal Plants 2009).

GACP provides suitable cultivation and collection practices for important medicinal plant species and plant parts used (Pandey and Das 2014). The general guidelines for the GACPs of some important medicinal plants are discussed hereunder.

25.2 *Andrographis paniculata* (Kalmegh)

25.2.1 Description and Distribution

Andrographis paniculata (Burm. f) Nees commonly known as Kalmegh belongs to the Acanthaceae family and is indeed an important medicinal plant with a rich history of traditional uses in various Asian countries. This is also called as "King of Bitters" due to the bitter taste of its leaves and stems. It is distributed southwards through Thailand and Peninsular Malaysia to Indonesia, and in India it is distributed in the states of Madhya Pradesh, Chhattisgarh, Odisha, Jharkhand, Maharashtra, Assam, Bihar, West Bengal, Uttar Pradesh, Uttarakhand, Rajasthan, Andhra Pradesh, Tamil Nadu, Karnataka, and Kerala (Ghosh et al. 2012).

A. paniculata holds significant importance in traditional medicine systems, particularly in Ayurveda. The plant is being prominently used in at least 26 Ayurvedic formulations. The Panchang (stem, leaf, flowers, seed, and root) of the plant is utilized in various Ayurvedic formulations. The plant has been traditionally used in Indian medicine for the treatment of fever, malaria, sore throat, and wounds (Chopra et al. 1956). Clinical information suggests the efficacy of Kalmegh in the treatment and prevention of common colds, tonsillitis, and respiratory issues. The plant exhibits immunosuppressive properties, which may be beneficial in conditions where immune system modulation is required. In addition, it possesses alexipharmic properties, indicating its potential to counteract toxins (Puri et al. 1996). The decoction of the plant is considered a blood purifier and is used to treat liver disorders, jaundice, and dermatological diseases. This aligns with its reported hepatoprotective properties (Prathanturarug et al. 2007). The plant is known for its effectiveness against malaria. *A. paniculata* is reported for hepatoprotective, immunomodulatory, anti-cancer, and anti-HIV activities (Kumar et al. 2004; Jayakumar et al. 2013).

25.2.2 Phytochemicals

A. paniculata has been extensively studied for its phytochemical composition, and several bioactive compounds have been identified. Some of the key bioactive compounds found in *A. paniculata* include andrographolide, neoandrographolide, panaculoside, flavonoids, andrographonin, panicalin, and apigenin-7, 4′-di-O-methyl ether. The plant contains diterpenoids: 14-deoxy-11-oxoandrographolide, 14-deoxy-11, 12-dehydroandrographolide, and 14-deoxyandrographolide. These compounds collectively contribute to the medicinal properties of *A. paniculata*, making it a subject of interest in both traditional medicine and modern pharmacological research. The major bioactive constituent "Andrographolide" constitutes a group of diterpene lactones mainly found in leaves whereas stems contain the compound in traces (Saxena et al. 1998).

Among the various medicinal plants, *A. paniculata* is in high demand because of its multiple uses. The estimated demand for this plant in India is more than 1000 tons per year (Goraya and Ved 2017). The plant is mainly collected from wild sources (Pandey and Mandal 2008). However, cultivation has also commenced in the plains of Uttar Pradesh, Madhya Pradesh, Bihar, Jharkhand, Odisha, Gujarat, and Andhra Pradesh. In India, it is mainly cultivated as a Kharif (rainy season) crop.

25.2.3 Cultivation

Climate *A. paniculata* grows well in tropical and subtropical regions under hot and humid conditions. It grows luxuriantly with the onset of monsoon and starts flowering within 100–120 days in the months of September–October. In central India, flowering and fruiting continue up to November–December until the temperature falls drastically. In the southern region, flowering starts in the middle of October and fruiting continues up to February end. The optimum mean temperature for *A. paniculata* is between 30 and 40 °C during the major growth period and grows well in places that receive 1000–1100 mm of well-distributed rainfall. It also grows as an understory crop in forest areas and is well-suited to the agroforestry system, where it can be planted beneath the canopy of larger trees.

Soil Kalmegh may be cultivated in a wide range of soils from loam to lateritic soils having a normal pH range with moderate fertility. It can also be cultivated on shady wastelands. It cannot tolerate waterlogged conditions and high soil sodicity. It could be better grownup as a rainfed crop in culturable wastelands. Although Kalmegh can be grown in moderate to less fertile soils. Sandy loam soil is most suitable for the cultivation of Kalmegh.

Plant Propagation Normally Kalmegh is propagated/cultivated through seeds. In special cases, vegetative propagation can also be done as the nodes have rooting capacity. Seeds are very small and remain dormant for 5–6 months. Kalmegh is mostly cultivated by direct sowing of seeds or by transplanting nursery-raised seedlings. Farmyard manure (FYM) is added to nursery beds for raising healthy seedlings. The mature ripe seeds of Kalmegh are sown in well-prepared nursery beds in May. About 750 g of seeds are sufficient for 1 ha land. Seeds mixed with sand (five times) or fine FYM/compost are manually broadcasted in each bed and covered with 0.5–1.0 cm fine mixture of FYM/compost and sandy soil. Applying a layer of mulch on the nursery beds helps retain soil moisture, reduce evaporation, and suppress weed growth. Mulching promotes germination and prevents newly germinated seedlings from hot sunstroke. However, the mulches need to be removed immediately after germination starts to avoid elongation of the seedlings. The nursery beds should be irrigated lightly to provide consistent moisture to the soil without causing waterlogging or other potential issues. For raising seedlings to be planted in 1 ha of land, six beds of 10 × 1 m size are sufficient. It is advisable to prepare a nursery in partial shade to avoid excess temperature during summer months for proper growth of the seedlings. The seedlings are ready for transplantation after 45 days of sowing.

Varieties Availability of released varieties of *A. paniculata* is limited; however, the Central Institute of Medicinal and Aromatic Plants (CIMAP), Lucknow has developed and released a variety named CIM-Megha (Bahl et al. 2018). AK-1 (Anand Kalmegh-1) and IIIM (J)-90 are some of the other important varieties/cultivars released for commercial cultivation.

Sowing Kalmegh can also be cultivated by directly broadcasting seeds in the well-prepared field. The land should be thoroughly plowed to fine tilth, and weeds and debris should be removed manually. About 7–10 tons/ha of well-discomposed FYM should be properly mixed with the soil for proper growth. In the case of direct sowing, furrows were drawn at about 30 cm distance and seeds mixed with soil/FYM/compost were manually sown in the rows during the last week of June. The germination begins within 15 days after sowing. Seedlings should be thinned to maintain a plant-to-plant distance of about 20–30 cm in the rows. For the direct sowing method, 2.0–2.5 kg seeds are required for 1 ha of land. In the direct sowing method, control of weeds may be a problem. Therefore, transplanting is preferred over direct sowing. Both direct sowing and transplanting have their merits, and the choice depends on the specific needs of the plants, environmental conditions, and available resources. While transplanting may offer an early advantage in weed control, it is important to implement good cultural practices and management strategies for both methods to ensure successful crop growth.

Transplanting About 40–45-day-old seedlings are suitable for transplanting. However, before transplanting the land should be properly prepared following plowing, disking, harrowing, and planking. In a well-prepared field, nursery-raised seedlings are transplanted during the last week of June maintaining a spacing of 30 × 30 cm. Seedlings grown in a nursery are being transplanted into a well-prepared field at the end of June. The spacing between each plant is set at 30 × 30 cm. Field/beds are irrigated immediately after planting.

Weeding Kalmegh being a rainy-season crop is infested by several dicot and monocot weeds. Weed infestation is a problem, especially during the early period of crop establishment. Hand weeding or that done with the help of mechanical hoes is very necessary within the first 4 weeks of planting. This process can be repeated once or twice at an interval of about 2–3 weeks. To begin with, one or two weeding or hoeing sessions are necessary to establish the crop. Once established, the crop thrives during the monsoon without facing much weed competition. The first weeding can be performed around 15 days after planting to maintain optimal growing conditions.

Irrigation Kalmegh is mostly grown as a rainfed crop. Fairly well-distributed rains during monsoon are adequate to get a good harvest in the tropical climate of central India. However, when the monsoon is delayed, 2–3 irrigations are required after transplanting. The requirement for irrigation is very low at the later part of the growth stage. Proper drainage facilities are essential as the crop does not withstand water logging.

Manure and Fertilizer Application The cultivation of medicinal plants requires a chemical-free approach, avoiding the use of synthetic fertilizers and pesticides. Instead, organic manures such as FYM (farmyard manure), vermicompost, and green manure can be applied as needed for specific plant species. To prevent diseases, bio-pesticides can be prepared from natural sources like neem (kernel, seeds, and leaves), chitrakmool, dhatura, either individually or in combinations. This method aligns with organic and sustainable farming practices, promoting the growth of plants in an environmentally friendly manner.

Yield The crop yield is influenced by several factors such as climate, soil properties, planting time, irrigation, intercultural operations, manure application, and harvest timing. Single-cut sole cropping of Kalmegh yields about 3.5–5.0 tons/ha; dried herb yield up to 7.5 tons/ha can be obtained in the case of two harvests of the crop. When practicing intercropping, a dried herb yields of 1.5–2.0 tons/ha can be achieved (Fig. 25.1).

Harvesting The productivity of this crop depends significantly on the timing of planting and harvesting, as it directly impacts both the yield and quality of the crop. The active principle in the crop, andrographolide, varies with the time interval between planting and harvesting, the growth stage of the plant, and the prevailing climatic conditions (Kumar et al. 2002; Pandey et al. 2019). Harvesting at the right time is crucial for maximizing the yield and ensuring the potency of bioactive compounds, such as andrographolide, which contribute to the medicinal properties of the plant. (Pandey and Mandal 2010a; Pandey and Savita 2017). Earlier reports indicate that the quality of Kalmegh diminishes with delayed harvesting. To ensure high-quality produce, it is crucial to adhere to proper planting distances and harvest the crop at the optimal time. The production of andrographolide, a key bioactive compound in Kalmegh, can be influenced by various climatic and environmental conditions (Pandey et al. 2019). Non-destructive harvesting practices of Kalmegh have been

Fig. 25.1 A healthy crop of *Andrographis paniculata* (Kalmegh)

standardized by Pandey and Mandal (2010a) in which plants are cut from above the ground instead of uprooting from the forest areas.

25.3 *Asparagus racemosus* (Shatavar)

25.3.1 Description and Distribution

Asparagus racemosus (Willd.) commonly known as Shatavar is an important medicinal plant recognized for its medicinal properties, especially in traditional medicine systems. The plant is valued for its tuberous roots, which are used in various herbal preparations. It is recognized for its potential health benefits, particularly in women's health, and as an adaptogen. *A. racemosus* is indigenous to tropical and subtropical regions of India at up to 1200 m elevation, commonly found in plains, scrub jungles, hill slopes, forest borders, and fallow lands. It is a woody perennial climber with extensive scandent spinous branches. Roots are numerous, fusiform, succulent, and tuberous, clustered at the basal end of the stem. The stem is woody and sparsely covered with recurved spines. Reduced to small scales (cladode) in tufts of 2–6 in a node, finely acuminate. Inflorescence is a branched raceme. Flowers are white, fragrant, and solitary. Fruits are red berries, globose, or obscurely 3-lobed. Seeds are black and hard, with brittle testa (Goyal et al. 2003).

The roots are recognized with immense medicinal value and find several therapeutic uses in traditional medicine systems. The medicinal properties of the roots are accredited to their steroidal saponins called shatavarins. The plant is widely used in about 64 Ayurvedic formulations that include traditional formulations such as "Shatavari Kalpa," "Phalaghrita," and "Vishnu taila" (Bopana and Saxena 2007).

Asparagus is a nutritious plant with many health-promoting qualities. It is commonly used to boost lactation and treats conditions like diarrhea, diabetes, and rheumatism. The plant's tuber possesses various properties, including being a demulcent, diuretic, aphrodisiac, tonic, alterative, antiseptic, anti-dysenteric, and antispasmodic (Chopra et al. 1956). In Ayurveda, Asparagus is recognized for its cooling properties. The root juice serves as a cooling agent, effectively alleviating the effects of summer heat, quenching thirst, and providing relief from hyper-acidity and peptic ulcers. Notably, it acts as a relaxant for uterine muscles, particularly beneficial during pregnancy, where it aids in preventing abortion and pre-term labor as an alternative to progesterone preparations. As a remedy, a powder of Asparagus boiled with milk is commonly used to prevent abortion. In addition, Asparagus proves effective in addressing various women's health issues, including vaginal discharges such as leucorrhea, uterine disorders, excessive bleeding, and colic pain. Its versatility in addressing multiple health concerns makes it a valuable component in Ayurvedic practices (Nishritha and Saxena 2007). Its preparations help in increasing breast milk in lactating women and avoiding excessive blood loss during periods. It contains a good amount of mucilage that soothes the inner cavity of the stomach. It relieves the burning sensation while passing urine and is used in urinary tract infections (Anonymous 2007).

25.3.2 Phytochemicals

A. racemosus contains a diverse array of bioactive and medicinally important molecules, with steroidal saponins being the major constituent. Alongside steroidal saponins, the plant also contains alkaloids, flavonoids, dihydrophenanthrene derivatives, furan derivatives, and volatile constituents (Alok et al. 2013). Twenty-nine steroidal saponins were reported from *A. racemosus* (Hayes et al. 2008). An oligospirostanoside named 3–0-α-L-rhamnopyranosyl-(1→2)-α-L-rhamnopyranosyl(1→4)-O-βD-glucopyranosyl]-25(S)-spirosta-3βoil is obtained from *A. racemosus*, which on oral administration potentiated antibody synthesis and enhanced cell-mediated immune response in immune cooperated animals (Handa et al. 2003). It contains an anti-cancer agent, asparagine that is useful against leukemia. It also contains active antioxytocic saponins that have an antispasmodic effect and specific action on uterine musculature.

Shatavarin I and IV are characterized using high-performance liquid chromatography (HPLC) with an evaporative light scattering detector (ELSD) using a solvent system of 57.3% ethyl acetate in methanol (Penumajji et al. 2010). Shatavarin V, a key component in *A. racemosus* root extract, was analyzed using reverse-phase high-performance liquid chromatography (RP-HPLC) (Negi et al. 2011).

In India, the demand for Shatavari roots surpasses 500 tons annually, driven by its extensive use in various medicinal preparations (Goyal et al. 2003; Goraya and Ved 2017). *A. racemosus* is now considered endangered in its natural habitat, and the need for its conservation is urgently needed (Nishritha and Saxena 2007).

25.3.3 Cultivation

Soil and Climate Shatavar typically thrives in various soil types, including medium black soil with a pH ranging from 7 to 8, electrical conductivity at 0.15 mS/m, organic carbon content of 0.79%, and phosphorus at 7.3 kg/acre. It adapts well to subtropical and sub-temperate agro-climatic regions, flourishing at elevations up to 1400 m. The most conducive soil for optimal Shatavari root production is identified as an alluvial or sandy loam (Vijayraghavan et al. 2005).

Land Preparation For Shatavari cultivation, the soil undergoes deep plowing (20–30 cm) followed by 2–3 harrowing sessions. Weeds are removed, and the land is leveled. Plantation ridges, 40–45 cm broad, are created with 15–20 cm furrow spaces for irrigation channels.

Manures, Fertilizers, and Pesticides Shatavari is ideally cultivated organically, avoiding the use of chemical fertilizers and pesticides. The best root yield is obtained by applying 15 tons/ha of farmyard manure (FYM) and vermicompost. Precautions are taken to protect against rodents that may consume tender shoots. For disease prevention, the application of bio-pesticides like neem, chitrakmool, and dhatura, has proven effective. This organic cultivation approach ensures the growth of healthy Shatavari plants without relying on synthetic chemicals.

Plant Propagation Asparagus plant is best grown from its stem disc but commercially cultivated by the saplings raised through seeds (Purohit and Vyas 2004). Farmers are advised to obtain seeds from known sources before the start of cultivation.

Varieties CIMAP, Lucknow has released two varieties, CIM-Shakti and CIM-Sunehari (Bahl et al. 2018).

Nursery Raising and Planting Shatavar seeds are soaked in water for 12 h and sown 5 cm apart in raised nursery beds in April. Germination starts around 40 days with a 70% success rate. Planting on ridges with 60 cm spacing and 30 cm proximity is beneficial, while in beds, the ideal spacing is 45 × 45 cm (Shrivastav and Pahapalkar 1997). For planting on mounds, pits of 30 cm are dug at a spacing of 60 × 60 cm, filled with soil and FYM. Alternatively, seedlings can be transplanted on ridges at 60 × 60 cm intervals. Vegetative propagation is done by dividing the rhizomatous disc at the base of the aerial stem, retaining two buds and 2–3 tuberous roots. One-year-old seedlings were planted during April–May followed by mulching with available leaf litter for moisture conservation and weed suppression. These adaptable approaches cater to different preferences in Shatavar cultivation.

Irrigation Irrigation is done once the rainy season is over. During the winter season, it is recommended to provide one irrigation, and in the summer season, one irrigation per month is advised. However, adjustments may be necessary based on specific local climatic conditions. This watering schedule helps to maintain optimal moisture levels for Shatavar cultivation throughout different seasons.

Weeding Weeding is done twice during rainy months, thereafter once in the next 2–3 months. Regular weeding is required to realize higher root yields.

Cropping Systems Shatavar, being a climber, requires support for its growth. In the context of intercropping, Shatavar has demonstrated notable advantages. When planted as an intercrop in plantations, Shatavar exhibits higher root yields and a more favorable benefit-cost ratio (BCR) of 1.25, compared to the BCR of 0.88 in pure crop conditions. The intercrop also shows elevated levels of total saponins, contributing to its therapeutic value, along with increased soluble and insoluble sugar content. On the other hand, the pure crop exhibits higher levels of total free amino acids.

Harvesting and Post-Harvest Management In forest areas, Shatavar harvesting is recommended from January to February. Optimal tubers for collection are white and fair, measuring 15–30 cm in length, and 1.5 cm in thickness. To ensure sustainability, one-third of the tubers should be left during harvesting for regeneration, and the collection of damaged or diseased tubers is to be avoided. For plantations, the recommended harvesting time is in winter after 20 months of planting, and for those grown from one-year-old seedlings, harvesting is suitable after 8 months of planting. These practices emphasize acquiring high-quality Shatavar tubers while considering the long-term health and vitality of the plant.

Harvesting the Shatavar crop after 20 months of planting shows a higher root yield (Fig. 25.2). Pre-harvest irrigation facilitates easier extraction of root tubers. The average yield ranges from 10 to 15 tons/ha of fresh root tubers, although reports indicate yields exceeding 40 tons/ha. Shatavar roots are marketed either fresh or dried in 5–15 cm pieces with a 2 cm thickness. Good quality roots are silvery-white or light ash in color. While fresh roots are relatively smooth, they develop longitudinal wrinkles when dried. Despite lacking distinct odors, the roots possess a sweet and bitter taste. Proper storage of dried roots is crucial to prevent fungal contamination.

When harvesting or collecting from the forest, a sustainable approach involves spreading seeds on the ground during harvesting. To support the plant's regeneration, it is recommended to leave 2–3 roots in the ground and cover them with soil. This method aligns with the plant's rotational period, allowing for a sustainable harvest cycle of 3 years. Such practices ensure the continual growth and propagation of Shatavar in its natural habitat.

Fig. 25.2 *Asparagus racemosus* (Shatavar) plant with profuse roots

25.4 *Bacopa monnieri* (Brahmi)

25.4.1 Description and Distribution

Bacopa monnieri (L.) Pennell (Brahmi or Indian pennywort), commonly known as Brahmi, belongs to the Scrophulariaceae family and is an aquatic herb with creeping stems, small elliptical leaves, and light purple or white flowers with four or five petals. It typically attains a height of 2–3 feet, with branches extending 10–35 cm. Brahmi thrives in wetlands across the Indian subcontinent, favoring moist or boggy areas near streams. It primarily grows in warmer and humid conditions, found in both hills and plains, up to 1500 m altitude. In India, major Brahmi-producing regions include West Bengal, Tamil Nadu, Uttar Pradesh, Bihar, Jharkhand, Kerala, Karnataka, Madhya Pradesh, Assam, Punjab, Haryana, and the foothills of Himachal Pradesh and Uttarakhand. The plant flourishes in hot and humid climates, requiring temperatures between 33 and 40 °C, with a relative humidity of 60–80% (Anonymous 2003).

Brahmi has been used in traditional Indian medicine for a long time. This herb has played a significant role in diverse therapeutic applications since the inception of Ayurveda. It is valued for improving memory and has various health benefits, acting as an anti-inflammatory, pain reliever, fever reducer, calming agent, and anti-seizure remedy. Its role in Ayurveda showcases its versatility in promoting well-being (Stough et al. 2001). The entire herb, including stems, roots, leaves, rhizomes, and other parts, is utilized for preparing various herbal medicines (Fig. 25.3).

Bacopa monnieri is well-regarded for its scientifically proven ability to enhance memory. In addition, it is acknowledged as a brain tonic that supports longevity. Research has substantiated its positive impact on memory improvement, reduced anxiety levels, enhanced social adjustment, and alleviation of mental fatigue (Thorat et al. 2018). Research studies have established Brahmi as a promising herb in the treatments for various diseases such as anxiety, depression, epilepsy, asthma, bronchitis, cardiac problems, hyperthyroidism, gastrointestinal disorders, and, in certain cases, even

Fig. 25.3 A healthy crop of *Bacopa monnieri* (Brahmi)

some cancers (Das et al. 2010; Mishra et al. 2013). *B. monnieri* supplementation is effective in preventing colchicine-induced dementia through anti-inflammatory and antioxidant action. The ethanolic extract proved to have a significant effect in the treatment of diabetes, which poses a serious threat in the world today; further studies on this extract can prove to be beneficial for humans (Rai et al. 2017). Brahmi has been clinically proven to offer functional and therapeutic benefits. Its positive attributes have led to its widespread incorporation as a functional ingredient in many commercial food products.

25.4.2 Phytochemicals

B. monnieri is rich in various active phyto-constituents. The major compounds reported from this plant include two steroidal sapogenins (bacosides), alkaloids (herpestine and brahmine), flavonoids, glycosides, betulic acid, and phytosterols. These compounds contribute to the diverse therapeutic properties and health benefits associated with the herb (Jain and Kulshrestha 1993). The diverse pharmacological features of *B. monnieri* are due to the presence of characteristic saponins which mainly constitute bacosides. Bacosides are a complex mixture of structurally related compounds, glycosides of either jujubogenin or pseudojujubogenin. The saponins differ only in the nature of the sugar units in the glycosidic chain and the position of the olefinic side-chain in the aglycone (Rajani 2008). A quantitative estimation of bacoside, a key compound in *B. monnieri*, was conducted using high-performance liquid chromatography (Deepak et al. 2005). The entire Brahmi herb is commercially utilized for extracting bacosides, the memory-enhancing molecules found in the plant. (Mathur et al. 2002).

Increasing demand for *B. monnieri* is primarily met by harvesting/collecting from natural populations/habitats, straining existing habitats, and leading to the slow depletion of this already threatened herb (Gohil and Patel 2010). The estimated consumption of this herb in India surpasses 1000 MT per year, highlighting a significant gap between production and demand (Ved and Goraya 2007). Generally, *B. monnieri* has been collected from the wild. However, recognizing its importance, cultivation initiatives have been started in some parts of the country. This shift toward cultivation aims to address the challenges posed by over-reliance on wild harvesting and contributes to the sustainable management of *B. monnieri* resources.

25.4.3 Cultivation

Brahmi, known for its adaptability, thrives in diverse soil and climatic conditions. This common perennial creeper is frequently found in damp or marshy areas near streams and on the banks of ponds and lakes across India. While cultivation efforts are underway in many states, there is a notable gap between demand and supply. This gap can be addressed by expanding cultivation to more suitable areas, ensuring a balanced supply to meet the increasing demand.

Soil Though Brahmi can be cultivated in any type of soil, it proliferates well on sandy, sandy loam, and light black cotton soils. It can even tolerate a poor drainage system. It gives the best result in alluvial marshy soil. It can be grown in marshy areas, canals, and near water bodies. It prefers acidic soil for its good growth.

Land Preparation Successful Brahmi cultivation requires well-pulverized and leveled soil. To achieve fine tilth, deep plowing and harrowing is essential. During plowing, it is crucial to mix in farmyard manure (FYM) at a rate of 5 MT/ha. To ensure the successful establishment of plant cuttings, the field should be irrigated a day before planting.

Varieties The CIMAP, Lucknow has released two varieties, Pragyashakti and Subodhak, which have a high herbage yield and Bacoside A content (Bahl et al. 2018).

Propagation Brahmi is a self-propagating plant and, in a way, invasive with a high degree of resilience if ideal conditions prevail. The cuttings of its branches (5–7 cm long) containing a few leaves, nodes, and roots taken from the mother plants are transplanted in the wet soil of the field at a spacing of 20 × 20 cm preferably during June and July or on onset of monsoon in areas that depend on rains for moisture. Rooted saplings prepared in the nursery may also be transplanted at 20 × 20 cm spacing. About 25,000 cuttings are required for planting in a one-acre area. Plantation should immediately be irrigated and allowed to grow and proliferate through hot and humid months of monsoon till the end of September. In areas where sufficient irrigation facilities are available, it should preferably be transplanted during March and April.

Irrigation The plantation may be irrigated by flooding as per requirement, usually at an interval of every 7–10 days. It is a rainy-season crop; therefore, it requires irrigation immediately after the end of the rainy season. In the winter season, irrigation should be done at regular intervals of 15 days, and in the summer season, irrigation should be done at intervals of 10 days. There is no need for irrigation during the monsoon period if there is sufficient rainfall.

Weeding As it is a rainy season crop, hand weeding is to be done to keep the field free from weeds. The first weeding is done in 15–20 days after transplanting and the second weeding at the interval of 2 months.

Harvesting First harvesting is done approximately 100–110 days after planting. If the crop is planted from June to July, then the first harvesting is done in October–November. Harvesting in October yielded better herbage and bacoside content (Gulati et al. 2019). The upper portion from the base (i.e., 4–5 cm from the base) is cut for harvesting. In total, 2–3 harvestings are done in 1 year.

Post-Harvest Processing After harvesting, Brahmi herb undergoes a drying process in the shade. The dried material is then carefully packed in airtight bags to facilitate long-distance transportation. The dried material serves as the raw ingredient for making various products, including Brahmighrtam, Sarasvataristam, Brahmitailam, Misrakasneham, Memory Plus, and Megamind Plus.

25.5 *Gymnema sylvestre* (Gudmar)

25.5.1 Description and Distribution

Gymnema sylvestre (Retz.) R.Br. ex Sm., commonly known as Gudmar, is an important medicinal woody climber belonging to the family Asclepiadaceae. It is found in various regions of Asia, Africa, and Australia. It is found growing in the forests of various states in India, including Andhra Pradesh, Bihar, Chhattisgarh, Karnataka, Kerala, Madhya Pradesh, Maharashtra, Odisha, Tamil Nadu, Uttar Pradesh, and West Bengal. In Sanskrit, it is referred to as Meshashringi or Merasingi, while in English, it goes by the name Periploca of the woods. In addition, it is recognized as "Gurmur" due to its property of lowering sugar levels (Tiwari et al. 2014). *G. sylvestre* is a highly branched, woody climber that can ascend to the tops of trees in the dry forests of central and southern India (Kapoor 2017). The stems are cylindrical, hard, and twining, displaying branching with internodes measuring 0.7–17.2 cm in length and 2–10 mm in diameter. The leaves are simple, opposite, elliptic, or ovate, and possess a hairy texture. Small yellow flowers are arranged in umbellate cymes, while the follicles are terete and lanceolate, reaching lengths of up to 7.5 cm. Seeds are 1.3 cm long, flat, with a thin marginal wing and narrowly ovoid oblong.

G. sylvestre finds extensive use in various Indian systems of medicine to address conditions such as rheumatism, cough, ulcer, eye pain, dyspepsia, jaundice, hemorrhoids, asthma, bronchitis, conjunctivitis, and leukoderma (Nadkarni 1993). Its roots are employed as a remedy for snakebites, and it is recognized for its stomachic and diuretic properties. In addition, it is utilized to control diabetes mellitus (Warrier et al. 1995). The leaves of *G. sylvestre* have been employed for over 2000 years to treat diabetes, earning it a significant place in indigenous medical systems. It is considered a key botanical for diabetes treatment in Ayurveda and is included in the Indian Pharmacopoeia as an anti-diabetic plant (Singh et al. 2008). Popularly known as "Gudmar," it possesses the unique ability to temporarily suppress the taste of sweetness. Administration of its leaves has demonstrated a reduction in blood glucose levels in diabetic patients (Shanmugasundaram and Panneerselvam 1981). In addition, anti-allergic, antiviral, lipid-lowering, and other effects have been reported (Porchezhian and Dobriyal 2003). Gudmar leaves are also employed as food additives against obesity and dental caries (Nakamura et al. 1997). The multifaceted uses have resulted in a continuous demand for the species, rendering it highly vulnerable in nature.

25.5.2 Phytochemicals

G. sylvestre is characterized by the presence of major chemical constituents, namely gymnemic acids and gymnema saponins, both classified as oleanane saponins. The leaves of *G. sylvestre* contain a combination of oleanane and dammarane types of saponins (Khramov et al. 2008). In addition, the leaves exhibit the presence of saponins, anthraquinones, cardiac glycosides, tannins, quinones, flavonoids, and phenols. The plant's anti-diabetic properties are attributed to the presence of a mixture of triterpenes and saponins, including gymnemic acids, gymnemagenin, and gurmarin, found in the leaves (Porchezhian and Dobriyal 2003; Spasov et al. 2008).

25.5.3 Cultivation

G. sylvestre, the second best-selling medicinal plant globally, is in high demand. The species is facing a significant decline in natural forests due to unsustainable harvesting/collection. In India, the current demand, exceeding 100 MT per annum, is predominantly fulfilled through wild collection. The need for sustainable cultivation methods becomes crucial to address these challenges, ensure the continued availability of this valuable medicinal plant, and alleviate the pressure on wild populations.

Soil *G. sylvestre* demonstrates adaptability to various soils but thrives particularly well in well-drained red loamy or medium black soils. It is not suitable for waterlogged areas and flourishes in dry regions. The plant thrives best in humus-rich loamy soils. To enhance growth and increase production, it is recommended to apply farmyard manure (FYM) at a rate of 4000 kg/ha.

Propagation Successful cultivation of Gudmar can be achieved through seeds or rooted cuttings. However, seed propagation poses challenges due to low seed viability. As an alternative approach, successful cultivation can be done by planting rooted cuttings at a spacing of 50 × 50 cm in June and July. Alternatively, terminal cuttings can be planted in February and March to establish the plant.

Nursery To enhance germination, *G. sylvestre* seeds should undergo pretreatment with cold water (soaking) for 24 h. They (soaked seeds) should then be sown in nursery beds with a mixture of sand, soil, and FYM in a 2:1:1 ratio during May. For vegetative propagation, optimal results are achieved using hardwood cuttings with a diameter of 10–15 mm and three nodes. March and July are identified as the best times for vegetative propagation. Among hormonal treatments, dipping cuttings in a 500 ppm IBA solution for 30 min prove effective in promoting maximum rooting (Pandey and Yadav 2010).

Irrigation and Weeding Regular weeding, hoeing, and irrigation should be done for proper management of plants.

Support System *G. sylvestre*, being a climbing plant, requires support system/structures. Bamboo sticks or stone pillars with a height of 2 m can serve as stakes. Two plants per stake are trained to grow in opposite directions. This training aids in easier harvesting since leaves are the harvested part. Lateral branches can be directed upward with additional wires, forming a two-tier system for convenient harvesting. Using wire fences as support not only aids in plant growth but also serves as a fence for other crops, allowing intercropping. Establishing a proper support system, preferably using locally available materials, helps in cost-effective cultivation.

Harvesting *G. sylvestre* is typically ready for harvest 2 years after plantation, with harvesting conducted twice in the year in June and October. The harvesting process involves hand-plucking the leaves. To ensure the plant's normal physiological processes, it is recommended to leave a certain percentage of leaves (approximately 40%) (Pandey 2012). The branches should not be chopped to facilitate the collection of otherwise inaccessible leaves. This harvesting approach promotes the sustainable development of important species.

Post-Harvest Processing After the harvest of Gymnema leaves, it is essential to process them properly to retain their medicinal properties. The initial step involves shade-drying the leaves by spreading them thinly on a clean and clear surface. This drying process typically lasts for about 7–8 days. The purpose of shade drying is to remove moisture from the leaves while protecting them from direct sunlight, preserving their beneficial compounds. Once the leaves are adequately dried, they are carefully packed into storage containers, often polythene-lined jute bags. This packaging method ensures that the dried leaves are protected from environmental factors that could degrade their quality during storage and transportation. Proper processing is crucial to maintaining the effectiveness of Gymnema leaves for their various medicinal applications.

25.6 *Rauvolfia serpentina* (Sarpagandha)

25.6.1 Description and Distribution

Rauvolfia serpentina (L.) Benth. ex. Kurtz commonly known as Sarpagandha, belonging to the family Apocynaceae, is a woody perennial plant found in the Himalayan foothills and Peninsular India. *R. serpentina* thrives in well-drained soil and is often found in the understory of forests or along the edges of wooded areas. It is characterized by its slender, erect stems, elliptical leaves, and small, fragrant white flowers. Sarpagandha has been traditionally used in the treatment of cardiovascular conditions, hypertension, and various psychiatric ailments. In India, its roots have a historical significance in treating mental illness and snakebites, earning it the names "Insanity herb," "snake root," or "Sarpagandha" (Sahu 1983).

The large-scale indiscriminate harvesting/collection of *R. serpentina* from its natural habitats has resulted in a depletion of plant resources. The availability of *R. serpentina* in the wild has significantly decreased to the extent that it has been included under the category of endangered plants. As a response to the critical situation, *R. serpentina* has been included in the red list of species prohibited for export. Urgent conservation measures are necessary to protect this vital endangered medicinal plant.

25.6.2 Phytochemicals

R. serpentina contains more than 50 alkaloids, primarily belonging to the indole group and mainly concentrated in the plant's roots (Phillipson and Zenk 1980). Notable alkaloids extracted from the roots include reserpine, ajmaline, ajmalicine, ajmalinine, serpentine, alloyohimbine, chandrine, deserpidine, isoajmaline, yohimbine, r-yohimbine, isoyohimbine, 11-methoxy-delta-yohimbine, methylreserpate, neoajmaline, papaverine, corynanthine, isorauhimbine, 3-epi-alpha-yohimbine, raunatine, rauvolfinine, rauwolscine, reserpiline, rescinnamine, reserpinine, reserpoxidine, sarpagine, serpenine, and serpentinine (Virmani et al. 1992). These alkaloids demonstrate effectiveness against snake bites and scorpion stings (Anonymous 1969).

R. serpentina roots are reported to contain 0.7–3% of total alkaloids in the dry mass, and the amount varies depending upon the time and source of collection (Kokate et al. 1998). Chatterjee et al. (1956) studied the seasonal variation in the total alkaloid content of *R. serpentina*. The highest alkaloid content (1.69%) was found in the roots harvested/collected from October to January. Sobte et al. (1951) reported that the yield of roots and alkaloid content increases with the age of the plant. Rajagopalan reported that *R. serpentina* roots collected from the Western Ghats of Bombay, Madras, and Travancore-Cochin contained alkaloids ranging from 1.17 to 1.73%. Biswas (1956) in West Bengal and Ahluwalia (1963) in Saurastra also reported differences in the alkaloid content of roots. The alkaloid content in the plants grown in Bangladesh was 1.43% (Ahmad et al. 2002).

Generally, Sarpagandha is gathered from the wild, and the unregulated and indiscriminate collection has resulted in its inclusion in the list of endangered plant species. The pharmaceutical industry has a high demand for Sarpagandha roots and related products. Although reserpine can be synthesized in a laboratory, the cost is higher compared to extracting it from natural sources (Farooqi and Sreeramu 2001).

Cultivation of Sarpagandha is currently limited in scale and is being encouraged by various organizations and institutes. This is a response to the plant's diminishing availability in numerous accessible regions due to excessive exploitation.

25.6.3 Cultivation

R. serpentina grows well in tropical to subtropical climates, particularly in regions experiencing substantial rainfall between June and August. Areas with more consistent climatic conditions appear to be better suited than those with greater variations (Pandey et al. 2001). To meet the increasing demand for *R. serpentina* roots, a farmer-friendly cultivation package needs to be practiced. This approach ensures the production of high-quality roots on a sustainable basis.

Varieties The following varieties of *R. serpentina* have been developed by different institutes/universities:

- TFRI-RS-01, TFRI-RS-02 (Tropical Forest Research Institute, Jabalpur in 2017)
- CIM-Sheel (CIMAP, Lucknow) (Bahl et al. 2018)
- RS-1 (JNKVV, Jabalpur)
- RI-1 (JNKVV, Indore)

Soil *R. serpentina* prefers soil that is rich in humus, nitrogen, and organic matter, with excellent drainage. Commercial cultivation is not suitable in alkaline soils. The ideal soil types are sandy loam to medium black cotton soils, characterized by rich in organic matter, pH levels between 6 and 8, and good drainage. This plant can grow in a diverse range of climatic conditions but shows optimal growth in hot, humid tropical climates. Elevations up to 1300 m, with a temperature range of 10–38 °C and an annual rainfall of 2500 mm, are conducive to its growth. Areas with less severe winters tend to yield better results.

Land Preparation The land is deeply plowed in April–May and left for weathering. After the pre-monsoon showers, FYM is added, followed by second plowing and two cross-harrowings to break the clods. The land is finally dressed by planking and beds are laid out. The nursery should be raised in a partially shaded area with adequate irrigation facilities. Each bed should be about 1.0 m wide, 15–20 cm high, and of a convenient length. Beds with shallow furrows 8–10 cm apart are prepared and irrigated.

Manure/Fertilizer Fifty days after planting two-thirds of the nitrogen is applied and the remaining nitrogen is top-dressed in the next rainy season. Generally, *R. serpentina* is grown using organic practices. Before planting, 10–15 tons of FYM per hectare are applied. For nutrients, 30 kg of nitrogen, along with 30 kg each of phosphorus and potash per hectare, are used. During planting, one-third of the nitrogen and the entire amount of phosphorus and potash are applied. After 50 days, two-thirds of the nitrogen is given, and the rest is added during the next rainy season. This approach aims to provide the necessary nutrients for the optimal growth of the plant.

Propagation The farmers are practicing different propagation methods for its cultivation. It is a 1.5 to 2-year crop and can be grown by transplanting seedlings raised through seed and vegetatively by root and stem cuttings.

Seeds About 5–7 kg of seeds are required for establishing a 1 ha plantation. It is advisable to use fresh seeds as their viability lasts for only 6 months. Seeds stored for more than a year are challenging to germinate. Therefore, it is crucial to use seeds collected between September and December for planting in the following season. The seeds are treated with Thiram (2–3 g/kg seed) after soaking in water for 24 h. They are then sown from the end of April to the first week of May at a depth of 1–2 cm and a distance of 8–10 cm. The seeds are covered with a mixture of FYM and soil, and regular irrigation is provided. Germination is typically complete in 30–35 days, with a germination rate varying from 20 to 60%.

Seedlings, aged 40–50 days and bearing 4–6 leaves, are suitable for transplantation in the already prepared field in the first week of July. Before transplantation, these seedlings are uprooted and treated with Bavistin 0.1% for 30 min. They are then transplanted at a spacing of 45 × 30 cm in the main field, followed by light irrigation. Approximately 10–15% of the seedlings are kept for gap filling, to be done 10–15 days after the initial planting.

Vegetative It can also be propagated by vegetative means using stem and root cuttings and root stumps. Root cuttings 3–5 cm long are planted in June–July and are covered completely with the soil and FYM. The cuttings sprout within 3 weeks if there is good moisture. The success rate is 50–80%, and around 100 kg of root cuttings are required to plantation in 1 ha land. Stem cuttings 15–20 cm long with 3–4 nodes are planted in the nursery in June and kept moist until they sprout. Cuttings treated with Indole acetic acid (IAA) (30 ppm) initiate rooting in 15 days. The success rate obtained in stem cuttings is about 65%. In the case of root stumps, approximately 5 cm long roots with a portion of the stem above the collar are planted in May–June in irrigated fields. Though about 90–95% of success is obtained in this method, only one plant can be raised from a single stump.

Irrigation Irrigation is necessary after planting, then twice a month during the hot dry season, and once a month in winter for Sarpagandha cultivation. As Sarpagandha is a long-duration crop and exhibits slow growth in the initial stages, it can be intercropped. Vegetables such as brinjal, cabbage, okra, and soybean can be planted during the kharif season.

Weeding Two weedings in the first year and one weeding in the second year followed by one hoeing usually at the beginning of the growing season are required. Flowers appearing on very young plants should be nipped to promote root growth.

Harvesting The optimal time for harvesting Sarpagandha is November–December to achieve higher alkaloid content. The crop is irrigated 8–10 days before uprooting. During harvesting, the above-ground foliage is cut, and the roots are extracted. The highest root yield and alkaloid contents were observed in crops grown from seeds harvested after 18 months of planting. Harvesting time significantly influenced quality and root yield. However, different propagation techniques did not have a notable effect on the total alkaloid content (Pandey and Mandal 2010b). The highest root yield and alkaloid content were obtained in the crop harvested after 30 months of planting.

Drying The roots are cleaned after harvest, washed, and dried in the shade till the moisture content reduces to about 8%. It is crucial to handle the roots carefully during cleaning, as the outer skin contains approximately 80% of the total alkaloid content. Brown to black-colored fruits are collected from August to December. These fruits are soaked in water for 15–20 h, manually rubbed to remove the pulp, washed three times, and then dried. The dried seeds are stored in a moisture-free container for the next sowing.

Yield On average, 1 ha of land can yield about 15–20 quintals of dry roots and 200 kg of seeds (Fig. 25.4).

Fig. 25.4 *Rauvolfia serpentina* (Sarpagandha) plant with fruits

25.7 *Tinospora cordifolia* (Giloe)

25.7.1 Description and Distribution

Tinospora cordifolia (Willd.) Miers ex Hook. f. and Thoms., a member of the family Menispermaceae, is extensively utilized in folk and Ayurvedic systems of medicine. It is considered an important medicinal plant with promising potential for the future. It is distributed across the tropical Indian subcontinent and China, growing up to an altitude of 1200 m (Chopra et al. 1956). This is a glabrous, deciduous climbing shrub and is valued for its medicinal properties (Anonymous 2003).

T. cordifolia typically grows as a climber, featuring simple, alternate leaves with long petioles. The leaf lamina is broadly ovate, 7-nerved, and deeply cordate at the base. Flowering occurs in the summer, while fruiting takes place in winter (Anonymous 1994). The stem of *T. cordifolia* exhibits varying thicknesses, with young stems appearing green and smooth, and older ones adopting a light brown color with warty protuberances at the surface due to circular lenticels. The transversely smoothened surface displays a radial structure with conspicuous medullary rays traversing porous tissues, and it imparts a bitter taste (The Ayurvedic Pharmacopeia of India 2001). *T. cordifolia* is employed in diverse traditional medicines to address a spectrum of health concerns. Apart from its medicinal applications, this plant is occasionally cultivated as an ornamental plant, often propagated through cuttings.

The plant is found in various parts of India, Myanmar, Sri Lanka, China, Thailand, the Philippines, Indonesia, Malaysia, Borneo, Vietnam, Bangladesh, North Africa, and South Africa. In India, it is found in tropical regions across India, ranging up to 1200 m above sea level. It is native to places like Kumaon, Assam, West Bengal, Bihar, Deccan, Konkan, Karnataka, and Kerala.

In Hindi, the plant is commonly known as Giloe or Guduchi, which translates to the heavenly elixir that has saved celestial beings from old age, keeping them eternally young. The term "Guduchi" suggests its role in safeguarding the entire body. The term "amrita" is associated with its reputed ability to bestow youthfulness, vitality, and longevity (Chopra et al. 1982). It is considered useful in the promotion and restoration of health and in the treatment and curing of many diseases, often referred to as a panacea for various ailments. *T. cordifolia* is notably highlighted in tribal or folk medicine across different regions of the country. Ethnobotanical surveys reveal the plant's various useful properties, making it a valuable resource in traditional healing practices (Singh et al. 2003a, b).

T. cordifolia is reported for its antispasmodic, anti-inflammatory, and antiallergic properties. Various medicinal attributes have been reported, including antidiabetic, antiperiodic, anti-spasmodic, anti-inflammatory, anti-arthritic, antioxidant, antistress, anti-leprotic, antimalarial, antipyretic, anti-allergic, hepatoprotective, immunomodulatory, and antineoplastic activities (Khosa and Prasad 1971; Pendse et al. 1977; Nayampalli et al. 1986). It acts as a tonic and proves beneficial in treating chronic diarrhea and dysentery. As a hepatoprotectant, it safeguards the liver from damage, particularly when exposed to various toxins. According to Ayurvedic principles, giloe is classified as a rasayana herb, promoting longevity, enhancing intelligence, and preventing diseases. It stands out as an excellent immune modulator (Dhama et al. 2017). Giloe is utilized in conditions such as dengue, swine flu, bird flu, fever of unknown origin, throat infections, sneezing, coughing, and body aches.

25.7.2 Phytochemicals

The chemical constituents of *T. cordifolia* include glycosides, steroids, terpenoids, flavonoids, saponins, starches, and more, contributing to its diverse therapeutic potential (Chopra et al. 1982; Tiwari et al. 2018). The starch obtained from the stem known as Guduchi Satva is highly nutritive and digestive and is used in many diseases. Starch is present throughout the parenchyma of the stem. *T. cordifolia*, is nutritionally rich, with high fiber (15.8%), protein (4.5–11.2%), carbohydrates (61.66%), and low fat (3.1%). It provides 292.54 calories per 100 g. In addition, it is a good source of potassium (0.845%), chromium (0.006%), iron (0.28%), and calcium (0.131%), and contributes to its nutritional value and potential health benefits (Singh et al. 2003a, b). Pradhan et al. 2013 analyzed *T. cordifolia* stems of different diameters for their phytochemical constituents. Geo-climatic conditions and the age of the plant (stem diameter) are known to influence the levels of secondary metabolites in *T. cordifolia* (Pradhan and Pandey 2013).

25.7.3 Cultivation

T. cordifolia, renowned for its significant medicinal value, stands as the foremost recommended natural herb in the Indian system of medicine (ISM). This hardy plant exhibits adaptability to various climates, with a preference for warmer conditions. *T. cordifolia* thrives particularly well in tropical and subtropical climates. Due to the escalating demand, the cultivation of *T. cordifolia* has become imperative, and fortunately, its cultivation is relatively easy.

Soil *T. cordifolia* grows in almost any type of soil, but light medium sandy loam soil rich in organic matter with adequate drainage is suitable for its cultivation (Kumar and Jnanesha

2017). It also grows well in medium black soil or red soil. It does not tolerate high rainfall or waterlogged conditions.

Land Preparation The land is plowed, harrowed, and made weed-free before planting. Organic manures like FYM, vermicompost and green manure may be used to increase the nutritional status of the field. Generally, the basal dose of 10 tons of FYM per hectare and a half dose of nitrogen (75 kg) are applied at the time of land preparation.

Propagation *T. cordifolia* can be cultivated using either seeds or stem cuttings, but for commercial purposes, stem cuttings are the more favorable option. Cuttings approximately 1 cm in thickness and 15–20 cm in length with 2 or 3 nodes are obtained from healthy mother plants in February–March. These stem cuttings are directly planted in the field, and within 15–25 days, they develop roots. Alternatively, the rooted cuttings, initially grown in a nursery, can also be transplanted in the field around July or at the beginning of the monsoon season for better chances of survival.

Nursery Cuttings of 15–20 cm in length and about 1 cm thickness can be planted in nursery beds or poly bags of 10 × 15 cm size filled with soil, sand, and FYM or vermicompost in the ratio 1:1:1 during February–March. The rooting of cuttings takes 15–25 days, and they are ready for planting in the main field in June–July. Promotion of rooting of shoot cuttings by exogenous auxins application has been reported (Hartmann et al. 1997).

Planting Approximately 2500 rooted cuttings are needed for planting in 1 ha of land. An optimal spacing of 2 × 2 m is recommended for better yield. In addition, it can be planted near existing trees in the field or homestead.

Support System The plant requires support for growth, and this can be facilitated by erecting wooden stakes or a trellis. Alternatively, already established shrubs or trees can serve as natural support, believed to enhance the medicinal properties of the plants.

Irrigation The crop is grown under rain-fed conditions. However, occasional irrigation during extremes of cold and hot weather may help the crop survive adverse conditions.

Harvesting and Post-hHarvest The stem is harvested during autumn when it develops to a diameter of more than 2.5 cm. The basal part is left for further growth in the next season. Care should be taken during harvesting/cutting, as the stem bark easily peels off even with a light touch. The harvested stem should be cut into small pieces and dried in the shade (Fig. 25.5). Well-dried crop (stem) can be stored in gunny bags and kept in a cool and airy storage facility. The average yield is 8–10 q/ha.

25.8 *Withania somnifera* (Ashwagandha)

25.8.1 Description and Distribution

Withania somnifera L. Dunal, belonging to the Solanaceae family, is commonly known as Ashwagandha, Indian ginseng, or winter cherry. It has been a significant herb in the Ayurvedic system of medicine for over 3000 years (Sandhu et al. 2010). Ashwagandha is characterized by its erect, greyish, slightly hairy evergreen shrub with long tuberous roots (Fig. 25.6). The entire plant is covered in short, small, fine, branched hairs, giving it a silver-grey color. The stems are brownish and erect, with leaves occasionally absent or fewer on the lower part of the stem. The leaves, measuring 5–10 cm in length, are simple, alternate, ovate, entire, thin, with a cuneate/connate base, and densely hairy with reticulate venation. Near the inflorescence, leaves become opposite with an adnate arrangement. The flowers are small and greenish, either single or in small clusters in the leaf axils. The smooth, round fruit is fleshy, containing many seeds, and turns orange-red when ripe, enclosed in a membranous covering (Khare 2007).

It is distributed in the drier regions of India and is now being cultivated in various parts of the country due to its medicinal significance. The ideal regions for cultivation of this crop are semi-arid tropical areas with an annual rainfall between 500 and 750 mm. Crucially, Ashwagandha thrives when there is a dry season during its growing period. The plant thrives in dry and arid climates (Patra et al. 2004) and shows a preference for acidic soil (Obidoska and Sadowska 2003) in subtropic and semiarid regions. With its ability to tolerate drought, it is grown as an annual crop under rainfed conditions in marginal soils, primarily by small and marginal farmers in states like Madhya Pradesh, Rajasthan, Andhra Pradesh, Karnataka, and others. The plant's ease of cultivation, coupled with the high market value of its roots, attracts farmers to engage in large-scale cultivation. Furthermore, the leaves and seeds are also marketable, contributing to increased profits for farmers.

Ashwagandha is a well-known medicinal plant in Indian traditional medicine systems, featuring over 100 herbal and nutraceutical formulations. With a history dating back to ancient times, it holds a significant place in Ayurveda, Siddha, and Unani practices (Sangwan et al. 2004). Renowned as a "Rasayana," it has been traditionally used to enhance longevity and vitality (Singh et al. 2001). Ashwagandha is widely recognized for its diverse properties, including being an aphrodisiac, sedative, anti-inflammatory,

Fig. 25.5 *Tinospora cordifolia* (Giloe) stem cut into pieces

anti-tumor, anti-stress, antioxidant, cognition-boosting, immune-enhancing, and rejuvenating agent (Rastogi and Mehrotra 1998). In addition, it has been recommended for various health conditions such as polyarthritis, rheumatoid arthritis, lumbago, painful swellings, spermatorrhea, asthma, leukoderma, general debility, sexual debility, amnesia, anxiety neurosis, and leucorrhea (Anonymous 2007; Khare 2007).

25.8.2 Phytochemicals

The roots of *W. somnifera* contain a variety of biochemically diverse alkaloids. Basic alkaloids found in the roots include withananine, withananinine, pseudo-withanine, somnine, somniferine, somniferinine, cuscohygrine, anahygrine, tropine, pseudotropine, anaferine, and isopelletierine. Neutral alkaloids present are 3-tropyltigloate and an unidentified alkaloid. Other alkaloids documented are withanine, withasomnine, and visamine. A key steroidal lactone, withaferin A, is the most significant withanolide identified from the extracts of leaves and dried roots of *W. somnifera* (Khare 2007). Withanolides, particularly withaferin A, exhibit notable anti-cancer activity against various cancer cell lines, including those associated with breast, colon, prostate, lung, and liver cancers (Siddique et al. 2014). Withanolides are predominantly concentrated in leaves and roots, typically ranging from 0.001 to 0.5% of dry weight (Anonymous 2004). Quantitative analysis of withanolides has been conducted using HPLC (Chaurasiya et al. 2008).

25.8.3 Cultivation

W. somnifera is mostly cultivated as a rain-fed crop in semi-arid regions of India. At present, it is mainly cultivated in India in the states of Madhya Pradesh and Rajasthan (Jat et al. 2015). The optimization of growth, yield, and quality in Ashwagandha involves several environmental and agronomic factors, including the timing and method of sowing, plant population/spacing, weed management, harvesting schedules, and post-harvest practices (Pandey and Chowdhry 2006). Among these factors, achieving an adequate plant population per unit area is crucial for realizing the crop's full yield potential.

Fig. 25.6 *Withania somnifera* (Ashwagandha) with emerging flower buds

Varieties CIMAP, Lucknow released the following varieties (Bahl et al. 2018).

CHETAK (Naguri *Withania* variety) or CIMAP Chetak is a semi vigorous with small leaves and a whitish green stem, high dry root yield (11.77 q/ha vs check 5.45 q/ha), withanolide content (0.40 vs 0.20% in check). The fresh and dry leaf yield was also high (1.722 and 0.453 q/ha vs 0.872 and 0.147 q/ha in check) with high withaferin content (1.223 vs 0.788% in the check.)

PRATAP (CIMAP Pratap) has highly vigorous, dark green medium-sized leaves and dark green stem; high dry root yield (34.95 q/ha vs Poshita 21.99 q/ha) with high total withanolide content (0.31 vs 0.25% in Poshita). The fresh and dry leaf yield was also high (5.39 and 0.87 q/ha vs 2.83 and 0.50 q/ha in Poshita) with high withaferin content in dry leaves (0.720 vs 0.528% in the check variety Poshita).

POSHITA is a medium tall, semi-broad, medium dark color leaf with red colored berries. The estimated dry root yield was 14 q/ha vs check which is 8 q/ha. Total withanolides is 0.25%.

Varieties developed by other universities and institutes are Jawahar Ashwaganda-20, Jawahar Ashwaganda-134, Raj Vijay Ashwagandha-100, and Nagori.

Soil Ashwagandha's ability to flourish in different soil types makes it a versatile crop suitable for cultivation in various regions. However, it thrives well in soils that are well-drained, such as sandy, sandy loam, or light-textured red/black soils. The plant prefers soils with a pH ranging from 7.5 to 8.0, indicating a liking for slightly alkaline to alkaline conditions.

Climate Ashwagandha, known for its hardiness and drought tolerance, thrives in conditions characterized by a relatively dry growing season. It is typically cultivated as a late rainy season (Kharif) crop. Regions with semi-tropical climates and receiving 50–75 cm of rainfall are considered suitable for cultivation. The temperature range of 20–35 °C is optimal for its growth. In addition, late winter rains play a crucial role in fostering the proper development of the plant's roots.

Land Preparation In Ashwagandha, the roots constitute the primary economic component. Therefore, it is essential to prepare the land in a manner that promotes the development of roots with increased length and girth, ensuring higher quality. The land is initially plowed using a moldboard plow and then harrowed twice to achieve a fine tilth after receiving pre-monsoon rain. Adequate organic matter is added to the soil during land preparation to enhance its fertility. The field is subsequently leveled through planking, creating an optimal environment for the cultivation of Ashwagandha.

Manure Optimum crop nutrition should be ensured as excess or deficit of any essential plant nutrient may decline the production as well as the quality of the produce. Soil testing should be done before applying the nutrients. Ashwagandha should be grown without chemical fertilizers and use of pesticides. Organic manures like farmyard manure (FYM), vermicompost, and green manure may be used as per the requirement of the species. The manure or compost should be well-decomposed and not made from city waste. About 10–20 tons of FYM per hectare should be mixed into the soil at the time of the last plowing to obtain a good yield.

Propagation Method It is better to cultivate Ashwagandha by broadcasting seeds directly in the field instead of transplanting seedlings (Pandey and Shukla 2007). Broadcasting with higher seed rates at 15–20 kg/ha is the most common method for sowing Ashwagandha in rainfed areas. However, line sowing and raised bed sowing are also gaining importance in recent times and have been reported to yield a higher quantity of roots and also help in performing intercultural practices properly. The seeds are sown in lines 1–3 cm deep in soil. A light shower is applied after sowing seeds to ensure good germination. Seeds can be treated with Thiram Indofil or Dithane-45 (@ 3 gm/kg seed) before sowing to protect seedlings from seed-borne diseases (Chandra and Pandey 2011). It may be noted that since Ashwagandha is a rainy season Kharif crop, the time of sowing is decided by the date of arrival of monsoon in that area.

Planting Material For sowing Ashwagandha, seeds that were harvested during the previous season and of good quality and free of pests should be used. Seeds should be procured from a reliable source with detailed information about the seed. A seed rate of 10–12 kg is sufficient for the sowing of 1 ha of crop. Sowing should be done at the right spacing at 30 cm row-to-row and 10 cm plant-to-plant spacing in line sowing method.

Irrigation Ashwagandha is usually grown as a rainfed crop where irrigation facilities are not available. One or two late winter rains are conducive to the proper development of roots. Excessive rainfall or water is harmful for this crop and irrigation is not required if monsoon is well distributed throughout the growing season. One or two lifesaving irrigations can be given if required. Under irrigated conditions, the crop can be irrigated once in 15 days depending on soil type. Organic mulches such as wheat straw or Ashwagandha straw of previous crops should be spread in between the rows to conserve soil moisture, facilitate better water infiltration during excess rains, and control weeds.

Intercultural Operations and Weeding The seeds sown by broadcasting or inline in furrows should be thinned out by hand 25–30 days after sowing to maintain a plant population of about 60–90 plants per square meter (about 3–6 lakh plants per hectare). The plant density to be used may depend on the nature and fertility of the soil. On the marginal land, the population is kept high. The optimization of the plant population for Ashwagandha under different agro-climatic conditions and varying levels of input supply is an important factor for quality and productivity cost effectively.

Weeds should be managed before they start competing with the crop for nutrients and light. One-hand weeding at an early stage is sufficient to enable the Ashwagandha plants to take over the growth of weeds. At later growth stages, the weeds are suppressed by its smothering effect. Care should be taken not to damage the roots during weeding. The use of chemical herbicides is restricted for weed control in medicinal crops; hence, alternative methods of weed control such as the use of organic mulches to control weeds should be preferred as they inhibit weed growth as well as conserve the soil moisture.

Harvesting Harvesting should be done at the right stage to ensure maximum levels of active ingredients and better quality. Ashwagandha plants start flowering and bearing fruits from December onwards. The crop is ready for harvest in January–March at 150–180 days after sowing. The maturity of the crop is judged when leaves start drying and berries become yellow-red. Root size, root, shoot biomass, and alkaloid content were found maximum in a 180-day crop, which should be considered the best harvesting time for Ashwagandha. The crop should be harvested in the dry weather and not in rain or in the early morning when there is dew on the ground. Harvesting is done by uprooting the whole plant without damaging the roots. There should be sufficient moisture in the soil at the time of harvesting for easy uprooting of the plants. Weed plants or any inert material should not be harvested with the crop plants. The roots are cleaned and either cut into 7–10 cm long pieces and dried or dried as a whole in the sun and stored. Berries are hand-plucked, dried, and threshed, and the seeds are stored for the next crop.

Post-Harvest Management The post-harvest processing stage is pivotal in determining the quality of the final product. To ensure high quality, it is imperative to safeguard the harvested roots from contamination, degradation, or damage throughout the processing stages. During transportation to the processing site, one should use clean vehicles, protecting the material from heat and rain. One should choose a clean processing site, shielded from direct sunlight and rain, and equipped with water access. For processing, one should use a clean surface, preferably a cemented floor or a well-maintained tarpaulin sheet. Before processing, one should eliminate weeds, extraneous physical matter, and substandard material.

The roots are separated from the aerial portion by cutting the stem 1–2 cm above the ground. After digging, the roots are washed, cut into 7–10 cm small pieces, and dried in the sun or shed. Roots should be dried to 10–12% moisture content. The dried roots are beaten with a club to remove adhering soil and to break off thin, brittle, lateral rootlets. Lateral branches, root crowns, and stem remains are carefully trimmed with a knife. Root pieces are then sorted out into the following grades:

A grade: Root pieces up to 7 cm in length, 1–1.5 cm in diameter, solid cylindrical with a smooth external surface and pure white from inside.
B grade: Root pieces up to 5 cm in length, 1 cm or less in diameter, solid, brittle, and white from inside.
C grade: Solid root pieces up to 3–4 cm in length, 1 cm or less in diameter.
Lower: Small root pieces, semisolid or hollow, very thin, yellowish from inside, and <1 cm in diameter.

The superior grade has stout and long roots, which fetches a premium price. To avoid moisture and fungal attack on the dried roots, it should be stored in dry sacks or tin containers. One should pack the harvested produce in a clean, dry sack with a clear label. The sacks should be stored in a clean, dry room, elevating them off the ground and keeping them away from walls or any contact with fertilizers or pesticides. It is advisable to refrain from using rat poison or engaging in fumigation within the storage rooms.

Yield The yield from 1 ha of commercial cultivation is approximately 3–5 q of dry roots and 50–75 kg of seeds (Shrivastava and Sahu 2013). A maximum yield of up to 6.5–7.0 q/ha can be obtained. There are instances where farmers have achieved root yields as high as 1 ton/ha. Commercially, 6–15 mm diameter and 7–10 cm length root species are better. The alkaloid percentage in roots ranges from 0.13 to 0.31%.

25.9 Conclusion

Medicinal plants play a crucial role in providing affordable healthcare and livelihood security for over 70% of the Indian population. Moreover, India has emerged as the second-largest exporter of medicinal plants. Despite the significance of medicinal plants, there is a lack of awareness among farmers and collectors regarding good agricultural and collection practices (GACP) for herbal raw materials. It is important to note that specific guidelines (GACP) may vary for different medicinal plant species due to their unique requirements. While GACPs are available for some important medicinal plants, there is a need to develop species-specific GACPs for others. This would ensure the sustainable supply of quality raw materials and maintain the overall quality of the produce. Quality, sustained availability, authenticity, organic certification, and appropriate pricing are important criteria essential for gaining consumer confidence and ensuring a thriving market for medicinal plant products. With respect to conservation, efforts should focus on the protection, skill development, and capacity building of medicinal plant cultivators and gatherers. Furthermore, techniques for sustainable harvesting and management are crucial, especially for conserving threatened species. Local communities, farmers, women, and youth should be actively involved in the decision-making process. This participatory approach ensures that the perspectives of those directly engaged in cultivation and harvesting are considered.

In conclusion, addressing awareness gaps, meeting consumer expectations, and prioritizing conservation efforts are essential for the sustainable development of the medicinal plant sector in India. The involvement of local communities is crucial for the success of initiatives aimed at ensuring both economic benefits and environmental sustainability.

Lessons Learnt

- A suitable planting site, characterized by favorable soil fertility, ample water availability, and adequate sunlight exposure, contributes to optimal plant growth, and these requirements may vary among different species.

- Sustainable cultivation practices, such as the use of organic fertilizers and employing natural methods for pest and disease control, are crucial for ensuring the prolonged well-being of both plants and the environment.
- Harvesting must be done at the appropriate time to ensure maximum potency and quality of the medicinal plants.
- Post-harvest handling including proper drying, storage, and packaging of medicinal plants is essential to maintain their quality and potency.
- Proper documentation and record keeping of cultivation practices such as planting dates, fertilization schedules, and pest and disease control measures can help to identify potential issues and improve future crop yields.
- Understanding market demand and having a reliable distribution network can ensure a sustainable income for the cultivator.
- Cultivation practices and knowledge related to medicinal plants are continually advancing. It is important to stay up-to-date with new research and techniques to continuously improve cultivation practices and ensure the quality of medicinal plants.

Key Questions

1. What is the significance of good agricultural and collection practices (GACP) of medicinal plants?
2. Give a brief account of lessons learnt by medicinal plant growers.
3. What is the importance of post-harvesting practices on the quality of herbs?
4. Does GACP contribute significantly to the quality, safety, and sustainability of medicinal plants?
5. What is the optimum time to harvest roots of *Rauvolfia serpentine* and *Withania somnifera*?

References

Ahluwalia KS (1963) Observation on cultivation and alkaloid content of *Rauvolfia serpentina* in Saurashtra. Indian Forester 89:373–378

Ahmad S, Amin MN, Mosaddik MA (2002) Studies on moisture content, biomass yield (crude and plant extract) and alkaloid estimation of *in vitro* and field grown plants of *Rauvolfia serpentina*. Pak J Biol Sci 5(4):416–418

Alok S, Jain SK, Verma A, Kumar M, Mahor A, Sabharwal M (2013) Plant profile, phytochemistry and pharmacology of *Asparagus racemosus* (Shatavari): a review. Asia Pac J Trop Dis 3(3):242–251

Anonymous (1969) The wealth of India. CSIR, New Delhi, 377p

Anonymous (1994) Indian medicinal plants, vol 5. Kottakkal Arya Vaidyasala, Kottakkal

Anonymous (2003) Wealth of India: a dictionary of Indian raw materials and industrial products. CSIR, New Delhi

Anonymous (2004) Monograph: *Withania somnifera*. Altern Med Rev 9:211–214

Anonymous (2007) Pharmacopoeia I. The Indian pharmacopoeia commission. Central Indian Pharmacopoeia Laboratory, Ministry of Health and Family Welfare, Govt of India

Bahl JR, Singh AK, Lal RK, Gupta AK (2018) High-yielding improved varieties of medicinal and aromatic crops for enhanced income. In: Singh B, Peter KV (eds) New age herbals. Springer Nature, Singapore

Biswas K (1956) Cultivation of *Rauvolfia serpentina* in West Bengal. Indian J Pharm 6:177–182

Bopana N, Saxena S (2007) *Asparagus racemosus*—Ethnopharmacological evaluation and conservation needs. J Ethnopharmacol 110:1–15

Chandra V, Pandey MC (2011) Jadi butiyon ki kheti. Indian Agriculture Research Institute, New Delhi

Chatterjee A, Parkashi SC, Werner G (1956) Some recent developments in chemistry and pharmacology of Rauvolfia alkaloids. Forstschr Chem Org Natirst 10:416

Chaurasiya ND, Uniyal GC, Lal P, Misra L, Sangwan NS, Tuli R, Sangwan RS (2008) Analysis of withanolides in root and leaf of *Withania somnifera* by HPLC with photodiode array and evaporative light scattering detection. Phytochem Anal 19(2):148–154

Chopra RN, Chopra LC, Handa KD, Kapur LD (1982) Indigenous drugs of India, 2nd edn. M/S VN Dhar & Sons, Kolkata

Chopra RN, Nayar SL, Chopra IC (1956) Glossary of Indian medicinal plants, 1st edn. Publication and Information Directorate, CSIR, New Delhi

Das M, Shrestha B, Datta S, Das S, Deb J (2010) Phyto-pharmacological review of *Bacopa monnieri* Linn. Nat Prod 6:1–4

Deepak M, Sangli GK, Arun PC, Amit A (2005) Quantitative determination of the major saponin mixture bacoside A in *Bacopa monnieri* by HPLC. Phytochem Anal 16(1):24–29

Dhama, K., Sachan, S., Khandia, R., Munjal, A., Hmn, I., Latheef, S.K., Karthik, K., Samad, H.A., Tiwari, R., Dadar, M. 2017. Medicinal and beneficial health applications of *Tinospora cordifolia* (guduchi): a miraculous herb countering various diseases/disorders and its immunomodulatory effects. Recent Pat Endocr Metab Immune Drug Discov 10(2):96–111

Farooqi AA, Sreeramu BS (2001) Cultivation of medicinal and aromatic crops. University Press Ltd., India, pp 210–211

Ghosh KB, Datta KA, Mandal A, Dubey PK, Halder S (2012) An overview on *Andrographis paniculata* (Burm. F.) nees. Int J Res Ayurveda Pharm 3(6):752–760

Gohil KJ, Patel JA (2010) A review on *Bacopa monnieri*: current research and future prospects. Int J Green Pharm 4(1). https://doi.org/10.4103/0973-8258.62156

Goraya GS, Ved DK (2017) Medicinal plants in India: an assessment of their demand and supply. National Medicinal Plants Board, Govt. of India, New Delhi and Indian Council of Forestry Research & Education, Dehradun

Goyal RK, Singh J, Lal H (2003) *Asparagus racemosus*--an update. Indian J Med Sci 57(9):408–414. http://www.indianjmedsci.org/text

Guidelines on Good Field Collection Practices for Indian Medicinal Plants (2009) National Medicinal Plants Board, Department of AYUSH, Ministry of Health and Family Welfare, Government of India

Gulati S, Pandey AK, Gupta A (2019) Influence of different storage conditions on quality of important medicinal herb Brahmi (*Bacopa monnieri*). International Conference on Cultivation and Sustainable Use of High Altitude Medicinal and Aromatic Plants for the Socio-economic Development held at Forest Research Institute, Dehradun

Handa SS, Suri OP, Gupta VN, Suri KA, Satti NK, Bhardwaj V, Parikh GG (2003) Process for the isolation of novel oligospirostanoside. Google Patents

Hartmann HT, Kester DE, Davies FT, Geneve RL (1997) Plant propagation principles and practice. Prientice Hall of India, New Delhi

Hayes PY, Jahidin AH, Lehmann R, Penman K, Kitching W, De Voss JJ (2008) Steroidal saponins from the roots of *Asparagus racemosus*. Phytochemistry 69:796–804

Jain P, Kulshrestha DK (1993) Bacoside Al, a minor saponin from *Bacopa monnieri*. Phytochemistry 33:449–451

Jat RS, Reddy RN, Bansal R, Manivel P (2015) Good agricultural practices for ashwagandha. ICAR – Directorate of Medicinal and Aromatic Plants Research, Boriavi, Anand, Gujarat

Jayakumar T, Hsieh CY, Lee JJ, Sheu JR (2013) Experimental and clinical pharmacology of *Andrographis paniculata* and its major bioactive phytoconstituent andrographolide. Evid Based Complement Alternat Med 2013:846740

Kapoor LD (2017) CRC handbook of ayurvedic medicinal plants. CRC Press, Boca Raton, FL. https://doi.org/10.1201/9780203719473

Khare CP (2007) Indian medicinal plants–an illustrated dictionary. First Indian Reprint, Springer India, New Delhi

Khosa RL, Prasad S (1971) Pharmacognostical studies on Guduchi, *Tinospora cordifolia* (Miers.). J Res Ind Med 6:261–269

Khramov VA, Spasov AA, Samokhina MP (2008) Chemical composition of dry extracts of *Gymnema sylvestre* leaves. Pharm Chem J 42(1):30–32

Kokate CK, Purohi AP, Gokhale SB (1998) In: Furia DK (ed) Pharmacognosy. Nirali Prakashan, Pune

Kumar A, Amit A, Sujatha M, Murali B, Anand MS (2002) Effect of aging on andrographolide content in Kalmegh. J Nat Rem 292:179–181

Kumar A, Jnanesha AC (2017) Cultivation, utilization and role of medicinal plants in traditional medicine in deccan eco-climate. Int J Agric Sci 8(1):98–103

Kumar RA, Sridevi K, Kumar NV, Nanduri S, Rajagopal S (2004) Anticancer and immunostimulatory compounds from *Andrographis paniculata*. J Ethnopharmacol 92:291–295

Mathur S, Gupta MM, Ram M, Sharma S, Kumar S (2002) Herb yield and bacoside a content of field grown *Bacopa monnieri* accessions. Int J Geogr Inf Syst 9(1):11–18

Mishra A, Mishra AK, Ghosh AK, Jha S (2013) Standardization of a traditional polyherbo-mineral formulation - Brahmi vati. Afr J Tradit Complement Altern Med 10(3):390–397

Nadkarni KM (1993) Indian Materia Medica, vol 1. Popular Prakashan, Bombay

Nakamura Y, Tsumura Y, Tonogai Y, Shibata T (1997) Contents of gymnemic acid in health foods using *Gymnema sylvestre*. J Food Hyg Soc Jpn 38:178–184

Nayampalli SS, Desai NK, Ainapure SS (1986) Anti-allergic properties of *Tinospora cordifolia* in animal models. Indian J Pharm 18:250–252

Negi JS, Singh P, Pant GJ, Rawat M (2011) High performance liquid chromatographic analysis of derivatized sapogenin of Asparagus. J Med Plants Res 5:1900–1904

Nishritha B, Saxena SB (2007) *Asparagus racemosus*: ethnopharmacological evaluation and conservation needs. J Ethnopharmacol 110(1):1–15. http://cat.inist.fr

Obidoska G, Sadowska A (2003) Effect of pH and nitrogen content of the substrate on the yield of *Withania somnifera* (L.) Dun. Biulentyn Instytutu Hodowli Aklimatyzacji Roslin 288:373–381

Pandey AK (2012) Cultivation technique of an important medicinal plant *Gymnema sylvestre* R. Br. (Gurmar). Acad J Plant Sci 5(3): 84–89

Pandey AK, Banerjee SK, Shukla PK (2001) Oushdhiya Poudhon Ki Kheti, Sarpagandha (*Rauvolfia serpentina*). ICFRE Br-14, CFRHRD Br.3, Chhindwara

Pandey AK, Chowdhry PK (2006) Propagation techniques and harvesting time on productivity and root quality of *Withania somnifera*. J Trop Med Plants 7(1):79–81

Pandey AK, Das R (2014) Good field collection practices and quality evaluation of medicinal plants: prospective approach to augment utilization and economic benefits. Res J Med Plant 8:1–19

Pandey AK, Gulat S, Gupta A, Tripathi YC (2019) Variation in andrographolide content among different accessions of *Andrographis paniculata*. Pharma Innov J 8(4):140–144

Pandey AK, Mandal AK (2008) Cultivation of *Andrographis paniculata*. J Trop Med Plants 9(1):178–182

Pandey AK, Mandal AK (2010a) Variation in morphological characteristics and andrographolide content in *Andrographis paniculata* (Burm. f.) Nees of central India. Iran J Energy Environ 1(2):165–169

Pandey AK, Mandal AK (2010b) Influence of propagation and harvesting time on root yield and alkaloid contents of *Rauvolfia serpentina*. J Nat Rem 10(1):44–49

Pandey AK, Savita (2017) Harvesting, post harvest processing and value addition of medicinal plants. Pharma Innov J 6(12):229–235

Pandey AK, Shukla PK (2007) Influence of cultivation techniques on root quality and productivity of ashwagandha (*Withania somnifera* Dunal). Recent progress in medicinal plants, vol 18, Natural Products II. Studium Press LLC, pp 465–471

Pandey AK, Shukla PK (2008) Role of medicinal plants in health care and rural economy in the tribals of Central India. Indian Forester 134(11):1438–1446

Pandey AK, Tripathi YC (2017) Biotechnological approach for medicinal plant conservation and enhanced production of secondary metabolites. Int J Pharmacogn 4(11):352–362

Pandey AK, Yadav S (2010) Variation in gymnemic acid content and non-destructive harvesting of *Gymnema sylvestre* (Gudmar). Pharmacognosy Res 2(5):309–342

Pandey AK (2015) Sustainable bark harvesting of important medicinal tree species, India. In: Shackleton CM, Ticktin T, Pandey AK (eds) Ecological sustainability of non timber forest products: dynamics and case studies of harvesting. People and Plants International Conservation, New York, pp 163–178

Patra DD, Singh K, Mishra HO, Gupta AK, Singh J, Singh SC, Khanuja SPS (2004) Agrotechnologies of ashwagandha. J Med Aro Plant Sci 26:332–335

Patwardhan B, Warude D, Pushpangadan P, Bhatt N (2005) Ayurveda and traditional Chinese medicine: a comparative overview. Evid Based Complement Alternat Med 2:465–473

Pendse VK, Dadhich AP, Mathur PN, Bal MS, Madam BR (1977) Anti-inflammatory, immunosuppressive and some related pharmacological actions of the water extract of Neem and Giloe (*Tinospora cordifolia*) - a preliminary report. Indian J Pharm 9:221–224

Penumajji S, Bobbarala V, Naidu KC (2010) Extraction and optimization of molecular marker compounds from *Asparagus racemosus* using HPLC with ELS detector. J Pharma Res 3:159–162

Phillipson JD, Zenk MH (1980) Indole and biologically related alkaloids. Academic press, London, 298p

Porchezhian E, Dobriyal RM (2003) An overview on the advances of *Gymnema sylvestre*: chemistry, pharmacology and patents. Pharmazie 58:5–12

Pradhan D, Ojha V, Pandey AK (2013) Phytochemical analysis of *Tinospora cordifolia* (Willd.) Miers ex Hook. f. & Thoms stem of varied thickness. Int J Pharm Sci Res 4(8):3051–3056

Pradhan D, Pandey AK (2013) Estimation of starch extracted from different diameter thickness stem of Giloe (*Tinospora cordifolia* Willd. Miers). J Pharmacogn Phytochem 2(4):162–163

Prathanturarug S, Soonthornchareonnon N, Chuakul W, Saralamp P (2007) Variation in growth and diterpene lactones among field cultivated *Andrographis paniculata*. J Nat Med 61(2):159–163

Puri AR, Saxena RP, Saxena KC, Saxena V, Srivastava V, Tandon JS (1996) Immunostimulant agents from *Andrographis paniculata*. J Nat Prod 56(7):995–999

Purohit SS, Vyas SP (2004) Medicinal plant cultivation- a scientific approach including processing and financial guidelines. Agrobios, Agro House, Jodhpur

Rai K, Gupta N, Dharamdasani L, Nair P, Bodhankar P (2017) *Bacopa monnieri*: a wonder drug changing fortune of people. Int J Appl Sci Biotechnol 5(2):127–132

Rajani M (2008) *Bacopa monnieri*, a nootropic drug. In: Bioactive molecules and medicinal plants. Springer, Berlin, Heidelberg, pp 175–195

Rastogi RP, Mehrotra BN (1998) Compendium of Indian Medicinal Plants, vol 6. Central Drug Research Institute, Lucknow

Ravikumar K, Noorunnisa Begum S, Ved DK, Bhatt JR, Goraya GS (2018) Compendium of traded Indian medicinal plants. Foundation for Revitalization of Local Health Traditions (FRLHT), Bengaluru

Sahu BN (1983) Rauvolfias, vol II. Today & Tomorrow Printers and Publishers, New Delhi

Sandhu JS, Shah B, Shenoy S, Chauhan S, Lavekar GS, Padhi MM (2010) Effects of *Withania somnifera* (Ashwagandha) and *Terminalia arjuna* (Arjuna) on physical performance and cardiorespiratory endurance in healthy young adults. Int J Ayurveda Res 1(3):144–149

Sangwan RS, Chaurasiya ND, Misra LN, Lal P, Uniyal GC, Sharma R, Sangwan NS, Suri KA, Qazi GN, Tuli R (2004) Phytochemical variability in commercial herbal products and preparations of *Withania somnifera* (Ashwagandha). Curr Sci 86:461–465

Saxena S, Jain DC, Bhakuni RS, Sharma RP (1998) Chemistry and pharmacology of *Andrographis* species. Indian Drugs 35:458–467

Shanmugasundaram KR, Panneerselvam C (1981) The insulinotropic activity of *Gymnema sylvestre* R. Br. An Indian medicinal herb used in controlling diabetes mellitus. Pharmacol Res Commun 13:475–486

Shrivastav JL, Pahapalkar MK (1997) Cultivation trials on satavar in Mandla district of Madhya Pradesh. Vaniki-Sandesh 21(2):24–23

Shrivastava AK, Sahu PK (2013) Yield and quality parameter of alkaloids of *Withania somnifera* (L) Danal. Int J Agron Plant Prod 4(12):3246–3254

Siddique AA, Joshi P, Misra LN, Sangwan NS, Darokar MP (2014) 5,6-De-epoxy-5-en-7-one-17 hydroxy withaferin A, a new cytotoxic steroid from *Withania somnifera* L. Dunal leaves. Nat Prod Res 28: 392–398

Singh B, Saxena AK, Chandan BK, Gupta DK, Bhutani KK, Anand KK (2001) Adaptogenic activity of a novel, withanolide free aqueous fraction from the roots of *Withania somnifera*. Phytother Res 15: 311–318

Singh J, Sinha K, Sharma A, Mishra NP, Khanuja SPS (2003a) Traditional uses of *Tinospora cordifolia* (Guduchi). J Med Aromat Plant Sci 25:748–751

Singh SS, Pandey SC, Srivastava S, Gupta VS, Patro B (2003b) Chemistry and medicinal properties of *Tinospora cordifolia* (Guduchi). Indian J Pharmacol 35:83–91

Singh VK, Umar S, Ansari SA, Iqbal M (2008) *Gymnema sylvestre* for diabetics. Int J Geogr Inf Syst 14(1–2):88–106

Sobte SN, Abrol BK, Handa KL, Chopra IC (1951) Cultivation of *Rauvolfia serpentina* Benth. J Sci Indus Res 13:77

Spasov AA, Samokhina MP, Bulanov AE (2008) Antidiabetic properties of *Gymnema sylvestre* (a review). Pharm Chem J 42: 626–629

Stough C, Lloyd J, Clarke J, Downey L, Hutchison C, Rodgers T, Nathan P (2001) The chronic effects of an extract of *Bacopa monniera* (Brahmi) on cognitive function in healthy human subjects. Psychopharmacology 156(4):481–484

The Ayurvedic Pharmacopoeia of India (2001) Part 1, volume I. Govt. of India, Ministry of Health & Family Welfare, Department of Ayush, New Delhi

Thorat BS, Bagkar TA, Patil RR (2018) Brahmi, the memory booster medicinal herb. J Med Plants 6(1):185–187

Tiwari P, Mishra BN, Sangwan NS (2014) Phytochemical and pharmacological properties of *Gymnema sylvestre*: an important medicinal plant. Biomed Res Int 2014:1–18

Tiwari P, Nayak P, Prusty SK, Sahu PK (2018) Phytochemistry and pharmacology of *Tinospora cordifolia*: a review. Syst Rev Pharm 9: 70–78

Ved DK, Goraya GS (2007) Demand and supply of medicinal plants in India. National Medicinal Plants Board, Govt. of India, New Delhi

Vijayraghavan A, Meshram PB, Nandeshwar DL, Pandey AK (2005) Root production of *Asparagus racemosus* in sandy loam soil of Satpura Plateau of Madhya Pradesh. Indian J Trop Biodivers 13(2): 101–103

Virmani OP, Popli SP, Misra LN, Gupta MM, Srivastava GN, Abraham Z, Singh AK (1992) Dictionary of Indian medicinal plants. CIMAP, Lucknow

Warrier PK, Nambiar VPK, Ramankutty C (1995) Indian medicinal plants: a compendium of 500 species, vol 3. Orient Longman, Hyderabad

World Health Organization (2003) WHO guidelines on good agricultural and collection practices (GACP) for medicinal plants. WHO, Geneva

Clonal Forestry

26

R. C. Dhiman

Abstract

Clonal forestry is the deployment of selected and proven clones in plantation forestry. Operational clonal forestry, in India, is now over four decades old, although multiplication of venerated trees like the Bodhi tree (*Ficus religiosa*) using stem cuttings has been practiced for centuries. Later the practice was used as a management tool for in situ cloning of some tree species. The current level of clonal forestry of selected trees is based on better understanding and exploitation of juvenile-maturity phase, genetically improved clones, sophisticated infrastructure for cloning and skill developed in cloning techniques. Operationally, clonal forestry is a complete package interlinking activities from the collection of suitable starting propagules/explants to their final deployment in plantations. For timber production, mature trees of selected clones are first rejuvenated by serial propagation or micropropagation before being inducted in propagation facilities. Use of macro-propagation methods especially stem cutting is the most common propagation method deployed in most propagation facilities, whereas micropropagation methods have yet to be picked up for commercial use. Planting stock production uses a standard operating procedure (SOP) for cutting production, cutting preparation, conditioning, treatment, rooting media, containers, rooting in mist chambers, conditioning in hardening chambers and transportation to plantation sites.

This chapter describes the evolution of clonal forestry with examples drawn from mass multiplied species like poplar, eucalyptus and melia. It also cites example of two pest epidemics, namely gall wasp infection on eucalyptus clones and leaf blight on G3 clone of *Populus deltoides* which infested clones from production system and quickly replaced by newly developed clones. Clonal forestry under Indian conditions is practiced by thousands of farmers on small farm holdings which exhibit a mosaic of multiple clones that has had better opportunity in handling pest epidemics on narrow genetic base of selected clones on isolated land patches. Of recent, the awareness on registration of newly developed clones has significantly increased and newly developed clones are increasingly registered with national and international agencies. Field testing of genetically modified rubber tree in eastern India has also been attempted which may open up new scientific and operational opportunities in clonal forestry. Approximately 200 million eucalyptus and 20 million of poplar clonal plants are currently planted annually which are effectively managed with standard silvicultural operations for attaining high productivity that indicates the evolving strength of clonal forestry in the country.

Keywords

Clonal forestry · Nursery production · Registration of clone · Mist chamber · Rooting media

26.1 Introduction

Clonal forestry is the deployment of selected and proven clones in plantation forestry. Carson (1986) defined it as 'the establishment of plantations using tested clones'. Cloning trees has been in practice for centuries. Initially, it was restricted to planting a few religiously, socially and ethnically important trees like Bodhi tree (*Ficus religiosa*) under which 'Buddha' got enlightenment and this sacred tree has been reportedly reproduced using cuttings for many reproduction cycles (Joshi and Dhiman 1994). It was further practised in planting of some easy to clone trees, like poplars and willows, for meeting tree-based domestic needs, in situ cloning through coppicing of certain trees like oaks as forest management prescription and finally the deployment of selected and improved clones of certain trees under

R. C. Dhiman (✉)
CIFOR-ICRAF Asia Continental Programme, World Agroforestry, New Delhi, India

A. K. Mandal, A. Nicodemus (eds.), *Textbook of Forest Science*, https://doi.org/10.1007/978-981-97-8289-5_26

production forestry. It has been under debate and discussion for ecological, social, economical and industrial applications in operational forestry. Tree growers, wood-based industry, foresters, nurserymen and wood users have been fascinated by its potential to adapt to varied soil and site conditions with fast and uniform tree growth, increased wood production and capability of generating economical returns to its growers.

The interest in clonal forestry is now significantly increasing in production forestry. With every passing time, new knowledge is added on its different aspects including cloning trees that is making the subject more operationally feasible, economically viable and silviculturally sound. The main revolution in its large-scale adoption in different countries is encouraged due to new understanding and manipulation of juvenile–maturity relations in trees, increased emphasis on long-term tree improvement programmes providing elite material, infrastructure development for cloning facilities, development and cloning of new genotypes including genetically modified (GM) trees and skill developed on using smaller and easy to root propagules in propagation facilities.

The operational level clonal forestry, with limited imported clones, started taking shape during 1970s when large-scale planting of poplars (*Populus deltoides*) was promoted for matchwood production in north-western India (Dhiman 2012). The success of these plantations encouraged many others engaged in raising/promoting plantations and wood consumption to enlarge clonal forestry with other commercially important trees. Eucalyptus-based clonal forestry was started by ITC, a paper making company during early 1990s, and it soon became the leading tree grown in clonal forestry because of its wide adaptability and multiple utility. Other trees subsequently included in clonal forestry were *Casuarina* spp., *Leucaena leucocephala*, *Melia* spp. and *Acacia* spp., whereas many others are in the testing phase for their induction in production forestry. Clonal forestry, thus, picked up with the active participation of wood-based industry for securing tailor-made raw material for their manufacturing factories. Clonal forestry is now fast changing the landscape and occupying a significant space in agriculture and forest land in different states and regions of the country. Unlike many other countries where clonal plantations cover compact large acreage, the majority of these plantations, in India, are grown by farmers integrated with or without agricultural crops on their fields.

Clonal forestry has two independent but interdependent production-related activities, namely clonal planting stock production of selected clones and their deployment in raising clonal plantations. It is a complete package involving a number of interlinked activities beginning from the collection of suitable starting propagules/explants from finally selected and tested genotypes to their final deployment in plantations. The main ingredients of a systematic clonal programme include rejuvenation of selected and tested mature genotypes, their induction in production system, fast multiplication, field planting, tending and management of raised plantations. Each selected genotype may need some refinements in different operations and handling of its propagation material. The main focus of clonal forestry, so far, has been for wood production; yet, some clonal plantations for wild fruit production and environmental purposes are also being raised. The following text discusses in brief the current status of clonal forestry in India.

26.2 Induction of New Genotypes in Propagation Facilities

26.2.1 Rejuvenation

Rejuvenation of mature selected trees to enable their mass cloning, which do not show maturity related abnormalities in the trees grown from such planting stock, is extremely important and a crucial step. Understanding of mature and juvenile phases in trees is of significant importance in clonal forestry. Maturation in trees is a developmental process which is associated with reduced rate of growth, reduced efficiency in rooting of cuttings, reduced ability to propagate mature plants in tissue culture, early onset of flowering and fruiting, loss of vigour and changes in foliar morphology. Rejuvenation of mature trees of proven worth is, thus, necessary so that the produced planting stock starts behaving like a juvenile propagule in field plantations. Trees retain juvenility at their base of trees and at apical meristemic cells located at certain positions of both above- and below ground parts (Bonga 1982; Del Tredici 1995, 1999). Aitken-Christle and Connett (1992) proposed to either collect juvenile tissues near the base and/or buds near the top of the tree those being more juvenile parts. According to Franclet (1983) and Tefsumura et al. (2001), roots in general are more juvenile than the aboveground parts. Rejuvenation of mature trees could, thus, be achieved by using starting propagules from roots and also from shoots produced on stumps on cutting back the selected genotypes. Propagules in the form of branches of difficult to root species can be collected from mature trees and grafted/budded on juvenile root stock. Successful grafting and serial propagation of such shoots help in achieving a fair degree of juvenility which has been effectively followed in some species (St Claire et al. 1985; Dhiman and Gandhi 2014). Of late, the micropropagation methods especially embryogenesis are increasingly used for the rejuvenation of mature trees (Bosela and Ewers 1997). Once the propagules from a mature tree are rejuvenated, they can be maintained at juvenile stage by serial propagation and hedging (Bolstad and Libby 1982; St Claire et al. 1985). This has led to the development of a concept of stool beds or hedging for shoot production for making cuttings for rooting.

26.2.2 Assembly of Selected Clones for Further Processing

A simple model followed in India for inducting new genotypes in propagation facility is given in Fig. 26.1. It highlights three options, A, B and C, any of which could be applied based on the biological nature of genotype and its species. Option A has been extensively applied in many tree species, including eucalyptus, having good coppicing ability. It has an advantage that trees on cut back quickly produce multiple juvenile shoots from stumps which are used to make cuttings having easy rootability compared to those collected from mature branches of old aged trees. It, however, carries a very high risk of losing the best of selections if sprouting from stumps or rooting of cuttings from such trees is inadequate. Even in some cases, the cloning of trees of strategic importance or existing in locations where felling is not allowed needs to be handled through alternate routes as given in option C. This option was used for assembling the eucalyptus plus trees during 1980s by Wimco Seedlings in its

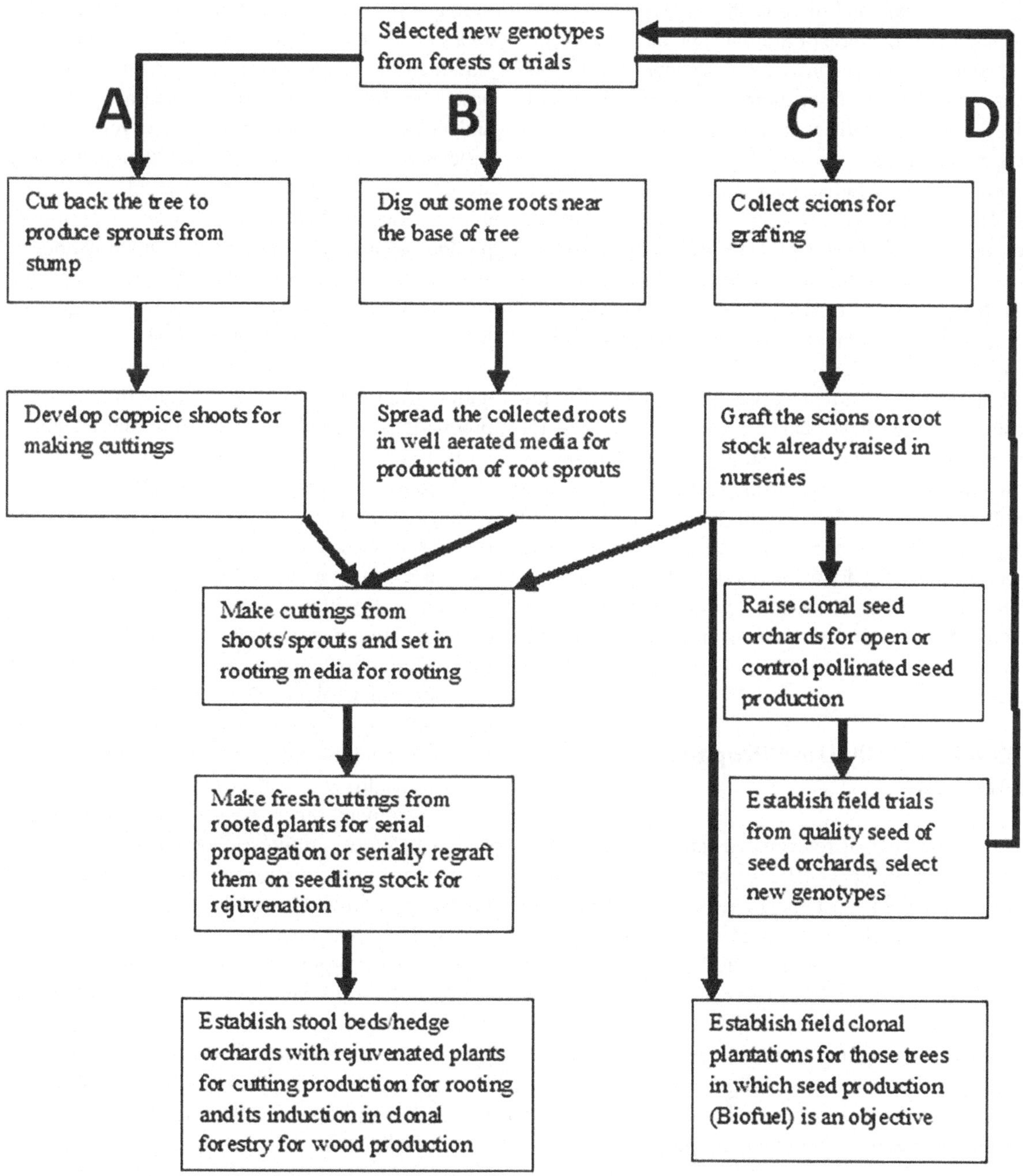

Fig. 26.1 A simple approach for inducting new genotypes in clonal forestry

first ever programme implemented on eucalyptus improvement and cloning outside government sector, in India (Dhiman and Gandhi 2014). Option C is also recommended for establishing clonal seed orchards in a number of species for early and quality seed/fruit production and hybridization to produce new genotypes which could be included in the clonal programme, such as fruit production in *Terminalia chebula* and *Emblica officinalis*, which is separately discussed latter.

The B option is increasingly attempted in species like *Dalbergia sissoo*, *Melia dubia*, *Populus* spp. and *Paulownia* sp. which have dormant buds in roots that give rise to sprouts. Roots collected near the base of trees are spread in moist and aerated rooting media produce new sprouts. Fresh root sprouts or cuttings made from them are juvenile in nature and could be easily rooted for inducting such genotypes in the propagation facilities.

Micropropagation including micrografting is also a potential mean for the rejuvenation of mature trees in some species (Assis et al. 2004; Aitken-Christle and Connett 1992). Embryogenesis using immature and mature embryos from control pollinated seeds is used for production of juvenile plants for induction in clonal forestry (Menzies et al. 1991). Clones so produced are field planted for testing and embryo culture of the same is stored in cryopreservation for the duration during which field planted ones get shortlisted for induction in clonal forestry. The stored embryo culture of shortlisted clones is taken out from cryopreservation and mass multiplied for induction in clonal forestry. In New Zealand, 200 clones of *Pinus radiata* were cryopreserved for 6–8 years before starting multiplication of selected ones performing better in field trials (Horgan 1987). Such an approach effectively combines long-term tree breeding programme with clonal forestry.

26.3 Operationally Used Propagation Methods

Once a limited number of rejuvenated ramets of a selected ortet are developed, the next challenge is to build up the stock to appropriate level to meet the increased requirement of planting stock of such clones. There is a long list of propagation methods that have been tried, tested and applied in cloning the trees. Choosing an appropriate method for cloning is a key to the success of any clonal forestry programme. Initial efforts on attempting vegetative propagation of tree species especially those not responding favourably to cloning remained largely a thematic subject with a lot of research on technological interventions but gradually gained recognition for raising clonal plantations of even some unorthodox species like eucalyptus, acacias, melia and a few others. Repeated experimentations on refinement of methods and appropriately handling the genotypes made it possible to select specific propagation methods under appropriate propagation facilities specific to species and clones.

Propagation methods used in cloning trees include simple coppice growth obtained on tree stumps on their cut back and natural or induced root suckers; to rooting of propagaules collected from aboveground tree parts may include cuttings, bud sprouts, simple leaf cuttings, needle/fascicles etc.; grafting/budding of scions on root stock; layering; singling of macro-proliferated shoots; splitting of seedlings; use of root cuttings and rhizomes in case of bamboos; and in-vitro propagation methods, namely tissue culture, organogenesis and somatic embryogenesis. These methods are sometimes classified into micropropagation and macropropagation based on the size of starting material (propagule/explants) and propagation conditions or auto-vegetative and hetero-vegetative methods based on whether collected propagules are developed into independent plants or propagule of one tree is fused with other plant through grafting/budding to produce another clonal plant. No single method suits all the species and conditions and each of them has its own advantages and disadvantages for mass propagation and tree improvement. Propagation facilities select a method or a combination of them based on their long-term clonal propagation programme, biological nature of tree species, scale of production, available infrastructure and technical skills and costs involved in production of planting stock.

Among different methods of cloning, the use of stem cuttings is the most extensively used method for a number of tree species throughout the world. This method has been in use since long for propagation of easy to root species like poplars and willows in which even simple sets (branch portions) are cut, part of thicker end inserted into the ground to develop roots, whereas aboveground parts develop shoots to produce new plants. Developments in rooting of cuttings, even with some hard to root species, proved highly effective for several hardwoods, in particular for several eucalyptus species (Ondro et al. 1995) which have revolutionized clonal forestry in many other species. Based on species and propagation facilities, cuttings made from aboveground parts are known by numerous names, namely set, hardwood, semi-hardwood, softwood, nodal, mini, apical mini, root sprouts, bud sprouts, leaf buds and leaves, some of which are now used in mass propagation for cloning certain species. One major shift in propagations facilities is from using a large-sized propagule to smaller ones for optimization of infrastructure and resources for their production, processing, rooting and handling of both rooted cuttings and rooted plants. For example, eucalyptus cloning in many countries was initially started by rooting nodal cuttings made from coppice shoots collected from stumps on cutting back the candidate plus trees, then nodal, mini cuttings, micro cuttings and even bud sprouts collected from shoots of stool beds

Fig. 26.2 Multiple grafts made on 2-year cut back eucalyptus to produce multiple shoots for making cuttings

(Dhiman and Gandhi 2014). Similarly, poplars and willows have traditionally been multiplied by planting sets directly in the field, then stem cuttings collected from selected trees initially but latter on from previous year (1 year old) grown saplings to maintain juvenility in the planting stock and finally using nodal and binodal cuttings, root suckers and even leaves (Dhiman 2013a, 2014).

There have been efforts to combine more than one propagation method for mass scale production of cutting material for rooting in some species. Grafting/budding, that involves fixing together parts of two plants to produce a new one, has been extensively used in tree improvement in the past. This method in combination with cuttings is still in use for mass production of cutting material of some species. The example of mass grafting of shoots or single buds on root stock to produce maximum number of shoots from limited stock in case of eucalyptus is demonstrated in Fig. 26.2. Even each axial bud from branches/shoots could be grafted on seedling stock and the grafted plant is used for establishing multiplication hedges near the cutting propagation site. This again has been effectively applied in eucalyptus and poplars for building up the planting stock (Dhiman 2013a; Dhiman and Gandhi 2014). Some hard to root species like *P. gamblei* could be mass multiplied by grafting multiple buds on each sapling, of easy to root species like *Populus deltoides*, to make cuttings for normal nursery stock production (Dhiman 2013a). Root system of open bed nursery plants on their lifting for field planting can be effectively used to produce shoots which are easily multiplied in large numbers in some species like poplar, shisham, albizias, paulownia and a few others (Joshi and Dhiman 1994; Dhiman 1999, 2013a). Normal size root cuttings can directly be planted in the nurseries to get additional planting stock. Similarly, singling of macroproliferated shoots of different bamboo species (Adarsh Kumar 1991, Adarsh Kumar et al. 1992), splitting and planting of seedlings (Chandra 1976) in some other species have been demonstrated. Even, leaves or their segments could be made as simple cuttings for rooting in some tree species which have started finding a space in mass propagation for building up the limited planting stock in selected clones (Dhiman 2013a).

Micropropagation methods have also been explored for mass multiplication of many tree species, while for some others, it is still being researched to exploit their true potential. From the first attempt of tissue culture of trees attempted in Delhi University with *Pinus* species during 1960s (Konar and Guha 1968) to date, much progress has been made in developing infrastructure and protocols for mass multiplication. During late 1960s, micropropagation facility was created in National Chemical Laboratory, Pune and then in many other locations across the country in which work on micropropagation was extended to other tree species (Mascarenhas and Muralidharan 1989). Tree species covered under these facilities were teak (Gupta et al. 1980), eucalyptus (Gupta et al. 1981, 1983), acacias (Mittal et al. 1990), poplar (Chaturvedi et al. 2004), sandal wood (Rao and Bapat 1998), casuarinas (Parthiban et al. 1997), leucaena (Dhawan and Bharwani 1995) and some others which are now being clonally propagated. Biotech Consortium India Limited has identified priority forest trees to be micropropagated (MP) for different states (Anon 2005). These included sandal wood, bamboos, ailanthus, soapnut, teak, eucalyptus, albizias, canes, poplar, sissoo, amla, gmelina, nypa palm, pterocarpus, thyrsostachys, xylocarpus and machilus. The private business on production and supply of MP plants of many species including some of forest trees is now increasing and a number of such links could be located on the internet. Among various micropropagation methods, greater emphasis is now laid on embryogenesis for both mass multiplication, production of mini cuttings for their rooting in mist chambers and rejuvenation of trees (Assis et al. 2004). The cost of production of planting stock is high through micropropagation methods compared to rooting of cuttings (Kaosa-ard 1990). In India, most of the micropropagation research has centred on developing methods for culture, media and techniques to induce juvenility in trees (Dhawan et al. 1993). Most reports on MP trees are restricted to embryos and/or young seedling material (Bonga and Durzan 1987). Many studies also report subnormal performance of MP trees in field plantations. These factors indicate that most of the biotechnological tools are still under experimentation and it may still take some time in their refinement for large-scale application in clonal forestry in India.

26.4 Production and Handling of Planting Stock

Planting stock for clonal forestry is grown both in open nursery beds and containers (also called root trainers in India) under controlled and open conditions depending on tree species. In some temperate species like poplars and willows, it is raised from hardwood stem cuttings in open nursery beds, while for most others it is grown in containers of varying sizes depending upon the species and targeted size of the planting stock. There is need for a standard operating procedure (SOP) for different activities right from cutting production, their handling from mother stool beds (status of mother stock in terms of age, size and eco-physiological condition of parent material, cutting preparation, conditioning, treatment with fungicides and growth promoting hormones etc.), rooting media, types and design of rooting containers, retention period of cutting on rooting in mist chambers, taking out the rooted cuttings from mist chambers to conditioning chambers, maintenance in open conditions, irrigation, fertilization, packing, transportation and field planting. In clonal forestry, only selected clones are exploited for which SOPs vary from clone to clone and species to species. The integrated package of SOPs for different cultural operations and handling should target to produce quality planting stock.

Most nursery production facilities now make heavy investment in term of land, human and monetary resources. Keeping operational controls at each level of pre- and post-nursery production phase before field planting is necessary to keep costs under control. For example, the traditional poplar nursery production involves preparing nursery fields in the month of January–February, make around 20 cm long and 2–3 cm thick stem cuttings from 1 year grown saplings preferably during the month of February, culture them for the entire year and lift them for field planting after 1 year production cycle. In poplar nurseries, production occupies fields for a period equivalent to production cycle of two winter agriculture crops and one summer crop which is highly taxing on land and time resources. An alternate method developed is to plant small mini-nodal stem and root cuttings in cavities (Fig. 26.3), during lifting of saplings from nursery beds, culture them for some time to make them sprout, root and form effective plug within the cavity and shift them anytime up to June month to produce plantable saplings by the following winters (Dhiman 2014). This new method has helped in curtailing expenditure on human, material and land resources and making the nursery production economically viable in spite of rising costs. In other species like eucalyptus, many such operations may save costs and check them on nursery production system and their transportation to field centres established for maintaining and distributing such stock. Rooted cuttings are packed in jute bags, transported in bulk over long distances and shifted in cavities in the field location for further culture, maintenance and field planting. A small truck which could transport around 10,000 plants along with cavities could carry five to six times more such stock and save time and resources in retrieving empty cavities which otherwise are very cumbersome and costly operation.

Clonal planting stock grown under exacting and controlled environmental conditions in mist chambers need to get acclimatized to face the matching conditions existing on planted sites. Handling of plants immediately on taking out of mist chambers is critical for their outplanting performance. If the planting sites are appropriately prepared and planting stock is properly handled, plants from mist chambers can directly be field planted with a reasonable success (Fig. 26.4). However, under normal circumstances, they need to be conditioned for a specific duration before being dispatched for field planting. The stock needs to be of appropriate size (collar thickness and height) along with effective plug formation in containers in which they have been grown. Over retention of stock in nurseries may block the bottom hole of containers (root trainers) with self-pruned roots, which may check growth and create physiological stress that may stagnate such stock on field planting (Dhiman and Gandhi 2014).

Planting stock ready for dispatch for field planting far away from propagation facilities must possess some minimum standards in terms of physical and physiological traits those have been worked out for specific species and clones for achieving desired field performance. Planting stock is evaluated for its quality based on certain performance attributes, namely maximum survival on field planting and quick growth thereafter. Production facility may have taken utmost care in maintaining such traits in planting stock during production process, yet the plants need to maintain the *status quo* till they are finally planted in the field. Physical and physiological conditions of clonal planting stock are more important with a rider to maintain a minimum size for each species and planting site. The size of the clonal planting stock may vary from around 15 cm shoot height with effective root plug formation in cavities for eucalyptus and some other species to around 4–5 m with a small portion of root system for poplars grown in open nursery beds. Physical appearance of poplar saplings in term of their freshness/shining and

Fig. 26.3 Plants grown from poplar nodal and mini hardwood stem cuttings

Fig. 26.4 High survival of eucalyptus rooted cutting immediately on taking out from mist chambers in freshly planted field (left) and 2 weeks after planting (right)

increased number of live buds on stem with high turgidity give better out-planting performance and are attempted through nursery culture and by conditioning in freshwater on lifting from beds. Containerized stock needs balanced internal mineral and moisture contents which are maintained by regulating nutrient and irrigation applications in nursery production phase. A lush green looking planting stock with high internal moisture content and some nutrients may lead to higher mortality and frost damage in the field plantations (Dhiman and Gandhi 2014).

An effective transportation system is needed for long distance transfer of planting stock from production facilities to planting sites and supply centres. A large number of locally made systems are currently used. Planting stock needs to be adequately irrigated before loading and dispatch to field. If the distance and duration of transport are long, such plants need additional sprinkling of water en-route for maintaining adequate turgidity and avoid dehydration of transported planting stock. Plants on retrieving from vehicles need proper handling and immediate irrigation to avoid any stress built in during transportation.

26.5 Clonal Diversity

26.5.1 In Plantations

Operational clonal forestry in India is now over four decades old. Poplar was the first operationally planted tree with selected clones introduced in 1970 onward and later on expanded with eucalyptus and some other tree species. Numerous clones are now available for a large number of tree species grown in clonal forestry. Some initial clones mainly in case of poplar were imported from other countries, while many new ones have been developed by various governmental and private sector players engaged in raising and promoting clonal forestry of such trees. Initial poplar plantations were raised with Forest Research Institute (FRI) introduced clones, from other countries, on forest land around the Tarai region by the then Uttar Pradesh Forest Department (UPFD) had shown some potential with a couple of introduced IC series clones (Chaturvedi 1982). Performance of WIMCO's introduced two clones, namely G3 and G48, were grown by the Company in its trials was better than IC clones. Soon UPFD introduced these clones in its field plantations. During 1990s, introduced 'S' series clones especially S7C15 and S7C8 also found favour among some farmers. Indigenously developed clones like Udai, Kranti and Bahar by Wimco Seedlings and L34 developed by UPFD made inroads in poplar culture in some locations during 1990s. Poplar improvement programme in the country is currently undertaken by the Wimco Seedlings; FRI, Dehradun and some state agricultural universities. By now around a dozen series clones, namely WSL, WIMCO, FRIAM, FRIS, L, EL, PL, Pant, PAU, UFC, UCM, UD and unnamed were promoted for field plantations by these organizations (Dhiman and Gandhi 2012). The major share in the current poplar-based clonal forestry is that of WSL and WIMCO series clones. The present-day culture is largely based on fourth stage developed WIMCO series clones followed by third phase developed clones with some notional presence of second and first phase clones (Table 26.1). Clones G3, St 121, Kranti and Bahar are no more under cultivation, though some of them might be inadvertently multiplied by local nursery growers from old standing trees. Wimco Series clones especially Wimco 110, Wimco 109, Wimco 108, Wimco 81 and Wimco 83 are now more popular than the old ones.

Like poplar, eucalyptus clonal forestry too is based on around two dozen clones mainly developed by ITC. The company has developed 107 eucalyptus, 10 leucaena and 15 casuarina clones (Kulkarni 2008). Three *Melia dubia* clones have been developed by TNAU (Parthiban et al. 2021). Ten varieties of *M. dubia* released by FRI, Dehradun, are also finding space among some farmers (Kumar 2020) and a few candidates plus trees from this germplasm have been cloned and their planting stock is being supplied to the tree growers (Dhiman and Gandhi 2017). There is a very

Table 26.1 Sustaining poplar-based clonal forestry by regularly inducting new clones—a case study of WIMCO poplar programme

Integration of poplar improvement with its clonal culture			
Year	Activity	Clones	Comments
1974	Initiation of the programme	Introduced G3, G48, and St121	Clonal poplar culture established with these clones
1995	Release of first series of indigenously developed clones	Udai, Kranti, Bahar	Udai became preferred clone replacing G3
2000	Second series clones	WSL 22, WSL A/26, WSL 27, WSL 32, WSL 39, WSL A/49	WSL 22 and WSL 39 became preferred clones
2010	Third series clones	WIMCO 62, WIMCO 81, WIMCO 83, WIMCO 108, WIMCO 109, WIMCO 110	WIMCO 110 is the dominant clones presently
2018	Fourth series clones	WIMCO 111, WIMCO 112	WIMCO 110 still dominate clone. New clones recently released are being accepted by farmers

Source: Dhiman and Gandhi (2012) and Dhiman (2018a)

large list of clones indigenously developed by government and private sector, many of which may be of academic interest. For example, Wimco Seedlings has been maintaining a germplasm of over 500 poplar clones (Dhiman and Gandhi 2012) and ITC is having equally rich germplasm of eucalyptus, casuarina and leucaena clones.

Clones have also been reportedly developed for some other species like *Anthocephalus cadamba*, *Dalbergia sissoo*, *Gmelina arborea*, *Tectona grandis*, *Toona ciliata*, *Santalum album*, red sandars and some others by various research establishments in the country. Clonal plantations of these species may pick up in due course of time. Indian Council of Forestry Research and Education (ICFRE) Dehradun has issued guidelines for testing and releasing tree varieties and clones during 2008 which were approved by the MOEF&CC (https://icfre.gov.in/bulletin_board/bulletin72.pdf). Over 70 clones of casuarina, eucalyptus, melia, sissoo, poplar and calophyllum have been released by different ICFRE institutes from 2010 to 2021. Similarly, state agriculture universities under ICAR system have also their own Variety Releasing Committees and some of them have released clones of trees used in clonal forestry.

26.5.2 Risk Aversion by Replacing Old Clones

Clonal forestry is much debated for creating a narrow genetic base in production forestry and exposing such plantations to biotic agents that may cause a catastrophic damage to certain susceptible clones (Burdon and Aimers-Halliday 2003). Many good and favourable clones tend to lose vigour through successive and multiple reproduction cycles. Such poor performance of a few good clones over the years is also reportedly associated with purging of some viruses, nematodes and other deleterious parasites inside plant tissues (Ahuja and Libby 1993a). Clonal forestry, thus, needs a regular development and induction of new productive and stable clones from tree improvement programmes to sustain its culture. Use of relatively a large number of clones is advocated to avoid such threats. A safer number is suggested to be 30–40 clones by Bishir and Roberts (1999), 30 by Yanchuck et al. (2006), 20 by Burdon (1982) and 7–30 by Foster (1993). Inducting too many clones in production forestry may not be preferred from wood quality perspective. Factory managers in wood processing industry need uniform wood quality in terms of anatomical features and log size for better processing and product manufacturing. As the number of clones increases, wood produced, therefore, becomes variable and start affecting its processing for quality product manufacturing in mills. Clonal forestry, therefore, needs to maintain a balanced approach to address both the concerns of narrow genetic base in production forestry and preference for highly uniform wood quality from a few selected clones for product manufacturing.

In India, there have been a couple of instances when a large-scale infection of insects and diseases caused alarm for sustaining clonal plantations. Table 26.1 gives details of poplar clones developed and released by Wimco, the main promoter of poplar plantations, for production at different phases to avoid such a threat. Poplar culture in its present region of culture suddenly developed heavy infection to leaf blight (*Bipolaris maydis*) in the most dominant clone G3 (with over 90% share) during mid 1990s which was gradually replaced with introduced G48 and indigenously developed Udai and Kranti clones. The main focus thereafter shifted to other indigenously developed and released clones (Dhiman and Gandhi 2012; Dhiman 2014). Currently, Wimco 110 is the most dominant clone in poplar culture but has now started showing heavy infection to leaf rust caused by *Melampsora* sp. Two recently developed and released clones, namely Wimco 111 and Wimco 112, are now being tried (Dhiman 2018a). A similar trend has been observed in eucalyptus clonal culture. There are about two dozen series clones released by different organizations. The initially released clones by ITC, namely BCM 3, BCM 7 and BCM 10 were the base for establishment and expansion of clonal eucalyptus in India. These suddenly became susceptible to gall wasp (*Leptocybe invasa*) which started appearing in epidemic form during the first decade of this century and such susceptible clones were gradually replaced with new clones, namely BCM 411, BCM 413 and more recently with BCM 288 (Kulkarni 2004, 2013; Dhiman et al. 2010; Roychoudhury 2016). Similar threats are invariably posed by *Cylindrocladium* leaf blight and stem canker to some eucalyptus clones in certain locations (Dhiman and Gandhi 2014; Mohanan 2014). Despite these isolated cases, clonal forestry has progressed aggressively over the last four decades with regular induction of new species and clones. The experience gained from operational clonal forestry using fast growing trees, so far, indicates that if it is adequately supported with the long-term tree improvement programmes, the fear of narrow genetic base in plantations could be properly handled. There is another unique angle of Indian clonal forestry that the clone mix is better managed by farmers as each farmer may plant different clone on their small land holdings creating a genetical divergent clones across the landscape. For example, in case of poplar clonal forestry based on around a dozen clones, around 40,000 farmers plant on an average of 500 saplings/farmer each year on their respective fields. This forms a mozaic of multiple clones in small areas unlike a continuous plantation of a few selected clones over vast area in some other countries.

Park et al. (1998) suggested planting a mix of clonal plants and seedlings together and cull/thin underperforming ones at regular intervals. This model may not be suitable in clonal plantations on farmers' fields where planting is made at final harvestable density. This, however, could be tried on forest land where such plantations are established with higher initial planting density with an objective to thin such stands during the production cycle. There is also a need to maintain a balance between genetic diversity and expected gain in clonal plantations of native species planted on forest land (Cyr and Klimaszewska 2002).

26.5.3 Regulations for Clonal Forestry and Registration of Clones

Clonal forestry is now entering a new era of regulations. Some countries have enacted legislations to regulate the deployment of clones regarding their numbers and area (Burdon and Aimers-Halliday 2003; Lelu-Walter et al. 2013). Cloning methods are now getting patented. A Patent No. 4353184 of United States patent (19) on method for asexual reproduction of coniferous trees is registered with the United States Patent and Trademark Office (https://www.freepatentsonline.com/4353184.html).

Development of new productive, disease and pest resistant clones is essential for sustaining the culture of clonal forestry. Forestry research establishments across the globe are implementing tree improvement programme and developing new clones/varieties. These establishments are now showing increased interest for registration of their developed clones and varieties for getting their research work recognized. There are numerous bodies at international, national and local levels which are involved in recognition and registration of clones grown in clonal forestry. At international level, the International Society for Horticulture Science created an International Cultivar Registration Authority (ICRA) which provides guidelines for registration of cultivars/clones of different tree species. ICRA has appointed the International Poplar Commission (IPC) for registration of cultivars/clones of two tree genera, namely *Populus* and *Salix*. Wimco Seedlings (ITC-PSPD) has over 500 clones in its germplasm out of which 14 commercially grown clones are registered and three others find recognition with the IPC (Dhiman and Gandhi 2012; Dhiman 2018a). At national level, Government of India has created the Protection of Plant Varieties and Farmers' Right Authority which undertakes registration of plant varieties/clones. The authority has recognized and appointed Institute of Forest Genetics and Tree Breeding (IFGTB), Coimbatore (under ICFRE) as DUS (Distinctness, Uniformity and Stability) Testing Centre for casuarina and eucalyptus in the country (http://www.plantauthority.gov.in/dus-center.htm). Six clones of casuarina and one clone of eucalyptus have been registered under the Protection of Plant Varieties and Framers' Right Act which is a first instance for forestry species in the country.

There are many other introduced and indigenously developed clones of different species which have been developed by various government and private sector research establishments and are being grown in clonal forestry but not registered with any authority. The Plant Variety and Farmers Right Act in India provides a mechanism for using developed clones by nursery growers at some agreed terms and conditions with breeder/breeding institutes and this mechanism is now being explored as a way forward in clonal forestry. Though the Plant Variety and Farmer's Right Act has provisions for recognition of unique varieties and cultivars being preserved and grown by farmers, their registration is not mandatory under the provisions of this act. Similarly, clones of some fast-growing trees developed by many research establishments are not registered with the concerned national authorities but are highly preferred and grown by numerous growers. The existing regulations do not provide uniform code for naming the clones. Some organizations are writing the abbreviated alphabets and numerical characters prefixed with clones with or without space, dash, underscore etc. For example, the names of the clone are sometimes prefixed with abbreviated code of company or institute like FRI by the Forest Research Institute, Dehradun; IFGTB by the Institute of Forest Genetics and Tree Breeding, Coimbatore; ITC-BCM by ITC; WSL/WIMCO by Wimco Seedlings/WIMCO Ltd., MPT by Tamil Nadu Agriculture University, Tamil Nadu etc. Clonal forestry is opening up to keep with market forces not only in term of wood production but for other associated activities related to patenting in propagation, registration and others.

26.5.4 Genetically Modified Trees

Genetically modified (GM) trees have been developed for some tree species which are being routinely planted in some countries (Ledford 2014; Wang 2004; Sedjo 2005), whereas field trials for monitoring and evaluation of many others are in progress in many other countries (Pilate et al. 2002). Field trials of GM trees are only allowed in selected species and locations, once the approval of competent authority is obtained. India joins the select list of countries for field testing the GM trees when first GM rubber (*Hevea brasiliensis*) (Arokiaraj 2000) was recently planted in Assam (Hindu 2021). The tree is reported to be cloned through somatic embryogenesis to produce plantlets (Arokiaraj 2000; Jayashree et al. 2003). GM rubber is the second genetically modified crop to start field trial in India after Bt cotton. In 2010, the Genetic Engineering Appraisal Committee (GEAC) had given permission to initiate field trials of GM rubber at Chetchackal, Thombikandom, in Kottayam. China has approved planting of GM poplar long back in 2002 (Lang 2004) and subsequently around 1.4 million such trees were planted for wood production and for controlling desertification (Then and Hamberger 2010).

Similarly, GM eucalyptus is now field planted for pulpwood production in Brazil (Ledford 2014). There are reports that GM poplars have spread across the areas beyond the originally planted sites and contamination of native poplars with the Bt gene is occurring (Carman et al. 2006). It is expected that more tree species and trial plantations of such modified trees would be raised in many countries in due course. There is an urgent need for a very rigorous monitoring and evaluation of such plantations to avoid unwarranted damage to ecology and diversity of native forests and species.

26.5.5 Field Performance of Micropropagated (MP) Trees

Published information on the performance of MP trees in field trials is a mixed one. Most reports are based on data collected from young trees/plantations, whereas a very few studies mention growth and productivity of MP trees for more than half the rotation. A few studies reported better or comparable growth of MP plants in comparison to same genotype produced using other propagation methods or seed obtained from the same genotype. At the same time, there are many other studies which documented subnormal growth and productivity of MP planting stock in field trials.

Large-scale field plantations and research trials of MP origin poplars were raised in north India during 1990s. Hundreds and thousands of saplings were supplied by commercial nursery growers to farmers for raising plantations on their fields. The field performance of these plantations was subnormal when compared with those raised form traditionally grown saplings from cuttings of same clones (Fig. 26.5). Poor performance of MP poplar for full production cycle was reported from statistically designed field trials (Dhiman and Gandhi 2010). Results of a poplar clonal trial with traditionally propagated (TP) ETPs (entire-trans-plants) of five clones, namely G48, WSL 22, L 34, L 49 and S 7C15, were compared with MP origin S7C15 clone. Height and diameter

Fig. 26.5 Field trial of micropropagated (MP) and traditionally propagated (TP) poplar from stem cuttings. Right side of visitors is a replicated plot of MP and left is TP poplar of same clone

growth both in nursery and field trials for each year recorded lower values for plantlet origin poplar when compared with that of TP poplar. Height and diameter values for 8 years MP S7C15 clone were 23.7 m and 20.1 cm respectively which were significantly lower than for height of 30.3 m and diameter of 22.7 cm for TP S7C15. The final reduction in MP S7C15 was approximately 20% for height and 13% for diameter when compared with TP poplar. In another unpublished study, the tree wood volume (above 45 cm mid girth logs) was 0.5906 cum for TP and 0.1894 for TC poplar. MP origin poplar wood had increased defects of 34.7% in match splint manufacturing compared to 19% from TP poplar. In yet another study, three clones, namely G3, S7C15 and L 34 at 6 years of growth, showed variability in wood anatomical properties and specific gravity in the woods grown from planting stock produced by employing these two propagation methods (Pande and Dhiman 2010). Such variations may be associated with somaclonal variation that may occur due to rapid multiplication of callus cells in the culture media with high level of hormones leading the callus to become a heterogenic mass (Larkin and Scowcroft 1981; Das 1991).

Like poplars, there have been a large number of studies on field trials for MP eucalyptus in India (Dhawan et al. 1993; Gupta et al. 1980, 1981; Muralidharan and Pandalai 2000; Solanki et al. 2000; DBT 2001). A few of them reported comparable or slightly better growth of TC eucalyptus over seedlings origin eucalyptus; some others reported poor performance in term of survival, growth and diseases susceptibility of MP eucalyptus plants over the rooted cuttings. Mascarenhas et al. (1987) reported that micropropagules raised from seedlings of plus trees of teak (*Tectona grandis*) flowered and produced fruits and seed within 2 years of planting in the field, indicting such plants attaining early maturity than seedlings. Such mixed response of MP plants is not restricted to India but also reported from some other countries (see Ahuja and Libby 1993b). The existing information on MP trees from India largely indicated that this biotechnological tool yet to produce encouraging results in large number of cases. It is repeatedly argued by its proponents that more research inputs especially on embryogenesis may make it feasible for mass multiplication of selected genotypes for clonal forestry.

26.6 Area Under Clonal Plantations

There is no authentic estimate on total area covered under clonal forestry in the country. Some isolated studies reported plantation area under fast growing tree species in which some are totally or partially raised using clonal planting stock. According to an estimate, around 15 million ha eucalyptus clonal plantations have been created in the country (Outreach 2019). During the year 2014, Dhiman and Gandhi (2014) estimated production of 200 M eucalyptus clonal plants in the country, out of which around 90% was planted in the field. Twenty-seven paper mills alone have created plantations of over 657,093 ha with a potential wood production of 39,425,580 t/annum by 2013 (CSE 2016); the major share of them was of clonal origin. These mills are increasingly promoting clonal forestry to make tailor-made plantations for paper pulp. It is estimated that a total area of 8.02 lakh ha was planted by ITC till last year (https://www.itcportal.com/), over 3 lakh ha by Wimco Seedlings/Wimco till 2017 (Dhiman 2018b), 173,000 ha by JK papers (https://www.jkpaper.com/), 535,000 acres by IPAPPM (http://www.ipappm.com/contact-us.html) alone. Rizvi et al. (2020) reported area under poplar to be 481,000 ha in two states (276,000 ha in Punjab, and 205,000 ha in Haryana) alone. Major poplar planting state in the country is Uttar Pradesh with around 36% of total poplar plantations (Dhiman 2012). Extrapolation of these figures to the total poplar planted area works out to be 1.6 M ha under poplar culture in the country.

The share of other tree species especially casuarinas, leucaena, melia and some acacias is gradually increasing in clonal plantations. According to Dhiman (2013b), around 5 M ha was under commercial agroforestry which had maximum component of clonally planted fast growing tree species like poplar, eucalyptus, leucaena, casuarinas, acacias, melia and some others. The area under clonal forestry with different species and clones has significantly increased since these estimates were made. Of late, there is also surge in planting MP plantlets in number of tree species. TERI has supplied around 12.8 M plants of different species including that of *Anogeissus pendula*, *A. latifolia*, *Bambusa bambos*. *Dendrocalamus asper*, *D. strictus*, *Eucalyptus tereticornis*, *E. camaldulensis*, *Paulownia fortunei* and *Populus deltoides* to different state forest departments, NGOs, agro-based companies and private growers (https://www.teriin.org/technology/micropropagation-technology-park). According to DBT (2001), over 1.57 M tissue cultured plants were supplied from its promoted production facilities to different states forest departments, NGOs and other private growers. Initially, most of the clonal plantations were raised on farmers' fields, but now state forest departments and forest corporations are also increasingly planting clonal origin planting stock.

26.7 Silviculture for Clonal Plantations

26.7.1 Plantation Activities

Clonal plantations are raised in different soil-sites, ecological regions, production systems and land uses. Majority of such plantations are raised on farmland in compact blocks, boundary and row plantations integrated with or without agriculture crops. Site preparations are excellent with repeated deep ploughing the soils immediately before sowing the agriculture crops. Tree plantations are made on such well-prepared fields any time before, during or after sowing the agricultural crops. During ploughing and levelling the fields, many a times provisions are made for channelizing water in small patches by raising ridges to irrigate fields for both agriculture crops and trees. After each crop harvest, fields are again ploughed for sowing next agriculture crops. Cultural and agronomical operations of repeated soil working, irrigations and nutrient applications for crops also provide excellent conditions for tree growth. Spacing, density, pattern of planting and rotation of trees are decided by farmers themselves based on their past experience and knowledge, though at times they also follow some recommendations and suggestions of professionals and promoters of such plantations. Poplar is recommended for planting at 5 × 4 m, 5 × 5 m, 7 × 3 m, 8 × 3 m in blocks with 400–500 saplings/ha and at varying spacing of 1.5–3 m on boundaries and rows. Now there is a tendency of planting a little more number of saplings per unit area with reduction in production cycles. Similarly, melia and a few other trees are planted at wider spacing for log production. Eucalyptus, leucaena, casuarina and some others are planted, for pulpwood, medium density board (MDF), particle boards, and poles at higher planting densities between 1200 and 2500 plants/ha, even sometimes more. A pair row concept is now being followed for raising eucalyptus by many farmers which allow wider space for agriculture crop production in wider space between two consecutive close spaced rows (Outreach 2019).

Plantations are made at final harvest densities with some replacement soon after planting but almost no replacement during the following season. Saplings replanted in the following year remain suppressed among the grown-up saplings of the preceding year and hence this is considered a wasteful activity. Poplar is planted deep in auger made holes to avoid their dislodging, whereas others are planted in small pits or in crow bar or augor made holes depending upon the species and timing of soil preparation before planting. Fields are, in general, irrigated soon after planting. Poplars and melia are planted with open bed nursery raised saplings in only winter months when they are leafless, whereas other species including melia grown in containers are preferably planted during rainy season over a greater part of the country. With assured irrigation facilities, the activity of planting with containerized seedlings of many tree species is taking place around the year. In termite-infected areas, anti-termite treatment of chlorpyriphos either in pits or in irrigation water is applied. Some fast grown trees and their clones with lanky stems get bended with strong winds. Such poplar trees are straightened by stretching with force and thereafter anchoring them with ropes. At times, dislodged trees are heavily pruned, straighten and tide with ropes. Dislodging is also a common phenomenon in some eucalyptus clones. Sometimes, farmers avoid their straightening as they are grown for pulpwood/MDF/particle board where higher log/billet thickness and straightness do not matter much. However, those grown for log production are straightened, the way it is done for poplar.

The interest of growing clonal plantations on forest land is gradually increasing. Some forest departments and forest corporations are now raising clonal plantations of select trees mainly eucalyptus and casuarina on forest land by following some well-established guidelines for each activity beginning from site clearance and preparation by removing unwanted vegetation, pit digging, spacing, pattern of planting and harvesting age. Plantations are established mainly at 3 × 3 m spacing with some deviation with species, location and objective of management. Most plantations are established during rainy season due to lack of assured irrigation facilities for planted trees. For longer rotation trees, like clonal teak, more plants are planted per unit area with a provision for thinning during the production cycle. Mortality replacement is done during the following planting season along with some maintenance in term of removal of competing vegetation. In general, only a limited care and maintenance is provided once the trees get established.

26.7.2 Rotation

Among all plantation aspects, the rotation of harvesting trees is highly variable on farmland. The timing of harvesting, on attaining some minimum tree size, is mainly decided by the market conditions including log size, wood prices and economical conditions of tree growers. Farmers with larger land holdings prefer harvest for timber production and retain trees for relatively longer rotation. At times, when farmers need some consolidated money, they prefer to harvest and sell the trees even if not grown to full size or even if the current market conditions are not conducive in terms of wood prices. In some locations where there is good market for small sized wood, harvesting is done at young age to take advantage of fluctuating wood prices and good income from intercrops like sugarcane which can be economically grown during the first 2 years of tree growth. Poplar was initially introduced for a production cycle of 20 years, which was reduced to 12 years, then 8 years, 6 years and now some farmers are harvesting poplar before attaining 4 years age for production of peeling logs. Similarly, when there is a scarcity of wood and market conditions are good in term of wood prices farmers tend to harvest trees at younger age, as low as 2 years in case of leucaena for pulpwood production in some locations (Dhiman 2007). Eucalyptus in many locations is harvested at 4–5 years for pulpwood, poles, MDF and particle board, up to 8 years for peeling logs and still longer for timber production. Most plantations raised in agroforestry and some on field boundaries are harvested with roots, and fresh plantations are made in the following planting season. While many others grown in compact blocks as farm forestry without intercrops and some on field boundaries especially those having good coppicing behaviour are cut back to develop new shoots for coppice rotation during the following two production cycles. In such cases, singling of shoots is attempted a couple of months after harvesting to allocate tree resources among the retained shoot(s) to develop them as the next crop. These plantations are harvested, after two coppice rotations, roots are dug out and removed to make fresh plantations or sow agriculture crops. Main crop is harvested at 4–5 years production cycles followed by two rotations of 3 years for each coppice production cycle. Trees are raised for longer rotation on forest land where there is a tendency of taking one main crop and two coppice rotations. Roots from forest land on tree harvesting are hardly removed. Some trees like clonal teak and shisham are managed under longer rotation to produce thick and quality logs.

26.7.3 Pruning

Pruning is an essential tending operation for improving tree form and wood quality in plantations grown for timber production. It also helps in shaping the crown to face the wind storm and avoid likely damage of breaking the stem or trees getting dislodged with strong winds. Clonal poplar, melia and some others are currently grown for peeling logs for which clean, straight and knot-free logs are highly desirable traits. At closer spacing, trees invariably get self-pruned due to increased inter tree competition, whereas those at wider spacing tend to develop branches towards lower portion of main stem which invariably grow in thickness, sometimes with forking on the main stem which results in heavy knot formation and bending of main stem. Branch thickness and crown development are under genetic control and some poplar clones especially those belonging to 'S' series form heavy branches, while those of WIMCO 110, G48, WSL 22 etc. have narrow crown with well-distributed thin branches on the main stem. Pruning is, thus, repeatedly attempted in plantations by removing all less desired branches and even co-leaders to channelize the growth on the main stem to produce extra tradable log.

In melia, the pruning is done after first year growth by selectively removing multiple emerging shoots that form a whorl of branches at the top of the tree. This is performed either with a hook attached with a pole by gently pushing them or by cutting with pruning saw or secateurs. The pruning is attempted during winter months to lift the crown of trees. A detailed pruning protocol has been worked out for poplar plantations (Fig. 26.6) (Dhiman 2012, 2014). Two types of pruning, namely lateral pruning (Fig. 26.6a) and vertical pruning (Fig. 26.6b, c) are attempted on trees based on their size and age. Lateral pruning is carried out on young trees to save trees from being lodged or broken with wind, whereas vertical pruning for lifting the crown to a height size (30 cm girth) which produces tradable log. It is also carried

Fig. 26.6 Pruning in poplar plantations. (**a**) Lateral pruning to half cut the lower branches to control crown spread, (**b**) vertical pruning for lifting the crown by removing lower branches to avoid heavy knot formation on peeling logs, and (**c**) pruning co-leader to develop more tradable logs on main stem

out to remove weaker co-leaders to allocate growth only on the main stem for production of more saleable logs.

26.7.4 Harvesting

Most clonal plantations grown inside fields in agroforestry are harvested by exposing the base of trees, cutting the anchoring side roots with axes, and making the trees to fall. These trees are also sometimes felled with the help of JCBs which are used to excavate the side roots, and making them to fall by pushing the trunk. Removal of most of tree roots with this method helps to prepare the fields for sowing agriculture crops and/or making plantations in the following season. Trees like eucalyptus, casuarina and leucaena grown for pulpwood, MDF and paricle boards on both the foreet and farmland are, many a times, cut back to facilitate coppice shoot production to grow two subsequent coppice crops after the main harvest. The felled/fallen trees are then converted into logs of standard length with cross-cut saws and power chain saws, segregated into tradable lots, loaded on to trucks and dispatched to market for sale the same day or by the next day. Longer period between harvesting and sale of logs reduces their weight by losing moisture resulting in lesser sale money. Immediate dispatch of wood of trees grown on farm land is, thus, preferred as it is sold by weight in most locations. Trees harvested on forest land, converted into logs, segregated into tradable sizes are transported to depots, staked in lots and auctioned to the highest bidder. The inventory in forest depots is maintained by volume. Singling of shoots in coppice crops is attempted a couple of months of harvesting when multiple shoots appear on the stumps.

26.7.5 Growth and Productivity

Growth and productivity of forest plantations including that of clonal plantations are a function of multiple variables related to plant, site conditions and their interaction. Plant-site variables which affect growth and productivity include plant conditions like genotype; climatic conditions like rainfall, temperature etc.; soil-site conditions like soil type, depth and its fertility, associated biotic and abiotic factors; management interventions like cultural, tending and thinning operations and land coordinates like latitudinal and longitudinal synchronizing with the native range geo-cordinates of the species being planted. Clonal plantations with tested and site matched clones under well prepared soil-sites on farm land with best possible management interventions grow to their optimum potential in most cases. This results in manifold increase in growth and productivity of such plantations on farm fields compared to those grown on wild lands with low management prescriptions. Clonal eucalyptus plantations

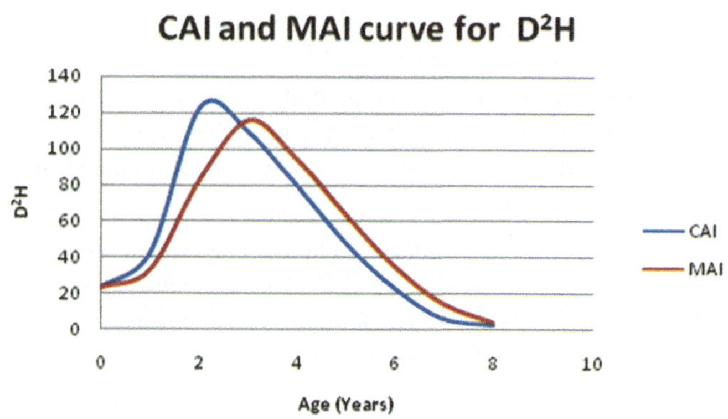

Fig. 26.7 CAI and MAI curve for D^2H for 8 years rotation poplar

are now reported to grow four to six times faster compared to those grown with seedlings from unimproved sources even at half the rotation of 4 years compared to 8 years for the latter one (Outreach 2019). Poplar under good conditions in agroforestry grows at an average rate of 5 m in height and 5 cm in diameter annually. CAI and MAI for tree height culminate earlier than that for diameter growth (Dhiman 2012). The CAI and MAI curve for D^2H (diameter × diameter × height) culminates at the age of around 3 years (Fig. 26.7). A very high timber (up to 15 cm diameter) productivity figure of 58.4 m^3/ha/year is reported for poplar grown in agroforestry by Dhanda and Verma (1995), up to 58 m^3/ha/year in case of eucalyptus for pulpwood production (Kulkarni 2004) and up to 55.83 m^3/ha/year for melia grown for timber by Kumar et al. (2017). Kumar et al. (2017) further claim that each melia tree producing 0.10–0.25 m^3 timber which when extrapolated to per unit area touches this productivity figures per hectare. *E. urophylla*-JKSC U283 clone has been reported to produce 50–60 m^3/ha/year pulpwood (Shukla et al. 2018). These higher productivity figures may be considered as potential productivity obtained in some isolated cases and under extremely well managed plantations. The average productivity of clonal plantations of most species could be considered between 20–30 m^3/ha/year.

Good clones grow fast and give high productivity subject to best use of soil site conditions under good management. If the same material is planted on less attended land with low inputs, the productivity significantly declines. It has been repeatedly noticed that poplar planted on forest land with low inputs hardly grows during first year, whereas same material planted on farm fields getting high inputs grows to 4–5 m in height in the first year itself. As a result, poplar planted on farm fields produces more timber at 6 years rotation compared to almost half the volume by the same clone planted on forest land at double (12 years) the rotation. Eucalyptus also shows similar performance on both the well and poorly managed soil sites. Raina et al. (2011) reported a case of planting 430 plants on 2 acre farmland in Haryana and harvested 18 Mt/ha peeling logs when harvested at 8 years, whereas growing eucalyptus for 8 years on forest land in Kerala produced an average MAI of 6.432 m^3/ha/year based on assessment of 571 plantations covering a total of 39,084 ha plantations (Chundamannil 2001) indicating a significant decline on site quality with low management. Poplar shows good growth and yield in locations near the foot hills of the Himalayas and poor growth in locations away from it towards Bihar in east (Dhiman 2012) and other parts of the country like Chhattisgarh and Maharashtra (Gogate and Deshpande 1994; Mishra et al. 2004).

26.7.6 Root System

Root system of clonal plantations is invariably debated for its form and spread inside the soil. It is argued that clonal trees do not form tap root with the possibility of its spread remaining near the ground surface compared to deep vertical spread of seedling origin plantations. Shallow root system may not be a desirable character as such trees may easily get uprooted due to wind and exploit moisture and nutrients from upper layers of soil by having shallow rootedness thereby competing with other vegetation especially in agroforestry system. The tap root originates from the radicle of germinating seed and is believed to have a tendency to penetrate deep into the soil compared to adventitious root system of clonal origin trees having a shallow rootedness inside the soil. The parallel studies on root system of seedling origin versus clonal origin in fully grown trees are rarely available in support or contradiction to this perception. Investigations on root systems in large sized trees have inherent difficulty as they are in inaccessible depths inside soil. In a resistant free-soil ecosystem, the roots grow and develop to their natural shape and form. Such an ideal condition is rarely available in the underground soil profiles where a large number of abiotic and biotic elements affect the shape, form, spread, growth and development of roots. There is a direct correlation between tree size and its root system (Fayle 1968), and hence the root system grows, complements and supplements aboveground parts, and also meets its functional role to provide stability to aboveground parts in a highly complex manner. Root systems in trees are strongly influenced by genetic and environmental factors (Sutton 1969; Eshel and Waisel 1996). Root system of some clonal plantations has been studied. Kulkarni (2013) gave evidences of anchoring type root system in eucalyptus clonal plantations of 1–4 years age (Fig. 26.8). Dhiman et al. (2020) provided some evidences of vertical spread of root system in eucalyptus plantations. It did not differ between the trees of seedling (40 years old) and cutting origin (14 years old) trees. Root system in clonal poplar plantations has a horizontal spread with a few anchoring roots to support aboveground parts. However, in locations of repeated flooding, it tends to redevelop spreading root system near the surface with additional silt deposition with repeated floods every now and then (Fig. 26.9).

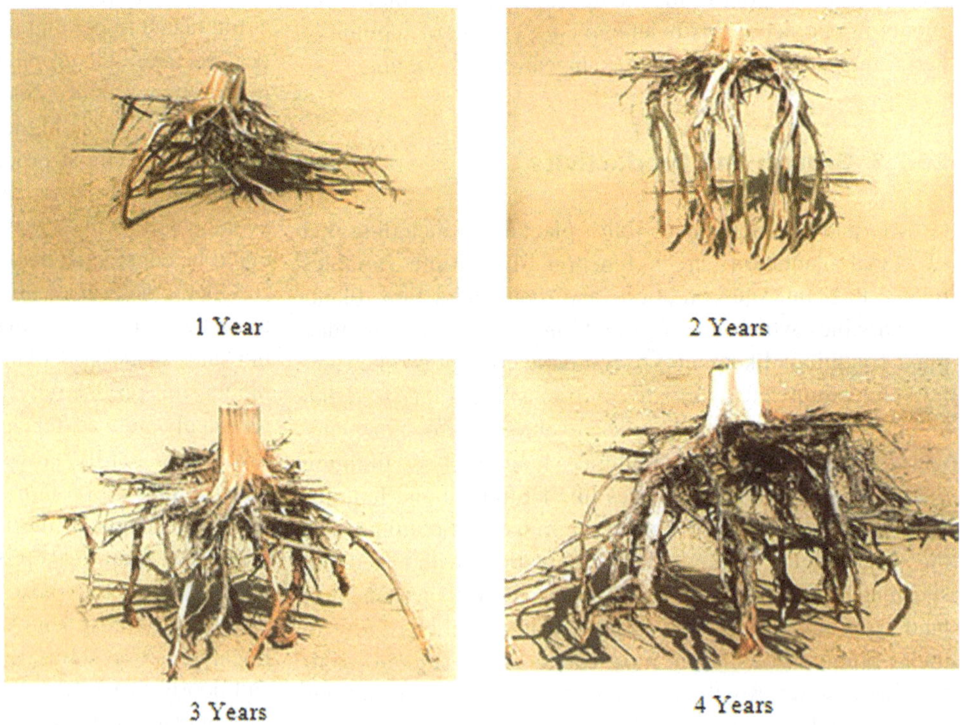

Fig. 26.8 Root system of clonal eucalyptus

Fig. 26.9 Changing root system in poplar with soil deposition along river bed. (**a**) Original depth of planted sapling, (**b**) after 1st flooding (2nd year), (**c**) after 2nd flooding (3rd year), (**d**) after 3rd flooding (4th year), and (**e**) after 4th flooding (5th year)

26.8 Clonal Forestry for Non-wood Products

Clonal forestry is making an overall impact in a number of sectors. It is directly helping the wood-based industry in sustaining existence by ensuring wood production and supply and expansion in more wood-based units on increased wood availability. Its other contributions are help in protecting forests by diverting pressure to farm grown wood, increasing forest and tree cover, better realization of the potential of land resource for productive purposes, enforcing some check on ever increasing wood imports, improving income of tree growers, enhanced revenue collection to states, self-employment to numerous nursery growers and logistic teams, diversification of agriculture to integrated land use; generating employment in nursery and plantation production, wood harvest and processing etc. It is also providing a number of other indirect benefits, especially on farm land, through soil amelioration by adding nutrients (Pathak and Gupta 1994; Thind 2004), amelioration of soil pH (Singh et al. 1997), reclamation of sandy river beds and stressed sites (Dhiman et al. 2007), improving physical and chemical properties of soil by addition of organic matter (Dhiman 2009) and acting as sink for atmospheric carbon (Gera et al. 2006; Kulkarni 2013). In many countries, clonal forestry is still considered as an emerging means for quality wood production. Its use in India for other than wood production is discussed in the following section.

26.8.1 Wild Fruit Production

In trees, the vegetative and reproductive phases are well marked. Trees on attaining reproductive phase slow down the rate of vegetative growth and divert some resources towards reproductive processes. The cloning through grafting/budding of mature trees for reproductive products has been increasingly attempted for establishing clonal seed orchards. These have also been successfully attempted in breeding programmes of some trees in which branches bearing reproductive buds are collected from selected mature trees, grafted/budded on already established root stock and hybridization attempted among selected clones in the flowering season(s) when flowers appear on grafted shoots. Of late, grafting/budding is effectively used in some wild fruit or seed-bearing trees for developing clonal plantations for fruit/seed production. Clonal plantations of grafted *Emblica officinalis*, *Terminalia chebula*, *Ficus* species and *Syzygium cumini* are now increasingly raised for early and abundant fruit production. This approach is now further explored for seed yielding trees for biofuel generation.

26.8.2 Clonal Forestry for Fruit/Seed Production: A Case Study

In India, many wild fruit trees have been used by people for a variety of purposes since ages. Some species belonging to genera like *Ficus*, *Myrica*, *Emblica*, *Syzygium*, *Terminalia*, *Pyrus*, *Prunus*, *Morus*, *Cordia*, *Sapindus*, *Aegle*, *Artocarpus*, *Punica*, *Annona*, *Diospyros*, *Pyracantha*, *Hippophae* and some others have been preferred for fruit/seed production in addition to wood and some other products. Many of them remained unexplored for scientific, nutritional and trade values. Of late, some of them attracted the attention for their improvement and growing their clonal plantations for this purpose. A case study of *Terminalia chebula* (harad) clonal plantations is discussed below.

T. chebula is a large-sized forest tree growing up to 25 m in height and 2.5 m in diameter. The tree in addition to producing quality wood for multiple uses is also highly valued, in pharmaceutical and ayurveda industry for the high medicinal value of its fruits. Realizing its immense social, anthropogenic, economical and medicinal value, efforts were made to identify plus trees for fruit production (Sharma et al. 1995) and till date eight clones have been selected, germplasm collected and preserved at different centres; some grafted on root stock and their clonal plantations raised (Sharma 2021). Many of such plantations are now producing fruits which are high in demand and generating lucrative returns to its growers. Grafted plants start producing good size fruits from the 4th year onwards (compared to 10th year in wild trees) and expected to continue up to 40th year. Economic analysis made from a 12-year-old plantation (8 × 8 m spacing) in the College of Horticulture and Forestry, Neri, Hamirpur, H.P. by considering a net return at 10% rate of interest shows very promising results (Singh et al. 2021). Returns from the fifth year fruit sale were Rs. 9600, which increased to Rs. 18,500 by sixth year, Rs. 32,000 by seventh year, Rs. 60,000 by eighth year, Rs. 82,000 by ninth year and Rs. 202,100 by tenth year. B:C ratio worked out to be 1.41 and 2.93 for fifth and tenth year respectively. Some of the clonal plantations of this tree on farmers' fields are now generating up to Rs. 200,000/ha/year. from the sale of fruits alone (Sharma 2021). These figures clearly establish that grafted harar clonal plantations are now a successful venture. The trees on final felling, when fruit bearing declines, would provide additional income from wood. The returns from sale of fruits are available each year compared to one time returns from wood harvest from other trees grown for that purpose. Clonal plantations of other wild fruit trees like *Emblica officinalis* (Singh and Singh 2016) and *Syzygium cumini* (Sekhar et al. 2015) have also been found high-yielding from early ages and providing handsome income to growers. New clones of wild fruit trees like *Sapindus mukorossi* (soapnut) and *Cordia myxa* (lasoora) have now been selected and their clonal plantations are being developed and tested for economic feasibility (Sharma 2021).

26.8.3 Ecological and Environmental Purposes

26.8.3.1 Averting Floss Menace

One of the potential negative impacts of clonal forestry by planting selected clones is being currently realized in Kashmir valley and other hill cities where poplars have created an environmental crisis by releasing massive floss with seed that remains suspended in the air for some time on release from female trees. This is also sometimes referred as white floss storm and is inadvertently claimed to have connected with human sufferings related to allergy, respiration etc. This matter even attracted public interest litigation (PIL) in Hon'ble J&K High court to stop poplar planting and felling all the existing trees those releasing poplar floss. The felling of poplar trees, once ordered by the Hon'ble Court, was initially stayed and final judgment passed to plant poplars with only male clones to avert this menace. As such the problem is being effectively handled by promoting planting of only male clones in such locations (Dhiman 2014). A large number of male clones have been supplied during the last one decade by Wimco Seedlings to J&K forest department and research establishments which have been multiplied in large numbers for planting on such sites.

26.8.4 Rehabilitating Sandy Riverbanks

Indo-Gangetic plains are subjected to repeated and heavy floods during monsoon rains almost every year. Such floods cause heavy losses to property and lives by submerging land and eroding soils around riverbanks. Cost-intensive engineered structures are raised along banks every year to avert flood threats. However, such structures are washed away following heavy floods and are not a permanent solution to this problem. Indian Council of Forestry Research and Education (ICFRE) has been entrusted with the task to prepare detailed project reports for the major Indian rivers through forestry interventions (ICFRE 2019). These reports are detailing the importance of forestry including agroforestry to be incorporated for actions to avert the high pollution load and floods in these rivers and their catchment areas. While, the forest land, around such rivers being government controlled and could be effectively treated by government agencies, the real challenge is to tackle the private land belonging to individuals. Any strategy needs to incorporate support from local farming community to get them engaged in forestry and agroforestry activities for a long-term solution. Vegetative barriers, including plantations, along riverbanks are more stable than engineering structures in averting such threats. An attempt was made to rehabilitate riverbanks of Gola River that flows through Udham Singh Nagar in Uttarakhand (Dhiman et al. 2007).

A sequence of rehabilitating the riverbank of Gola River, which is a tributary initially to Ram Ganga and subsequently to River Ganga in Uttarakhand, was attempted with clonal poplar along with intercropping of vegetable and fruit crops during the first 2 years after planting is depicted in Fig. 26.10. The sequence includes making of trenches in pure sand before winter, filling them with good manure (organic matter), erecting saccharum (cana) thatch on northern side of trenches to protect germinating seedlings from winter frost, planting poplar saplings at appropriate spacing, levelling the thatches on the ground to provide mulching once the

Fig. 26.10 Rehabilitation of Gola Riverbanks with clonal poplar. (**a**) Vegetable/fruit crop production in trenches made on sandy river bed, (**b**) crops and poplar growing together, (**c**) flooded water leaving debris and soil inside plantation, (**d**) stabilized riverbanks with poplar plantations and (**e**) knee height soil deposit just after receding the water

Table 26.2 Average crown size of poplar trees under 5 × 5 m spacing

Age (years)	Crown diameter (m)	Crown cover/tree (m^2)	Cumulative crown cover % with 400 trees/ha	No. of trees required for making 10% crown cover
1	2.6	5.31	21.24	190
2	3.2	8.05	32.22	124
3	4.1	13.21	52.84	75
4	5.0	19.64	78.56	50
5	5.5	23.77	95.08	42
6	5.8	26.43	105.72	37
7	6.0	28.29	113.16	35
8	6.4	32.18	128.72	31

vegetable plants attain reasonable size and start bearing fruits and allowing the trees to grow after harvest of vegetable and fruits like pumpkin and water melon. Since this activity is taken on pure sand, only selected poplar clones mainly, WSL 81, WSL 83, WIMCO 109 and a few others, could grow better than other commercially grown clones. This in addition to stabilizing riverbanks also generated money from sale of vegetables/fruits and poplar trees on their harvesting (Dhiman 2023).

26.8.5 Improving Area Under Forest and Tree Cover

Clonal forestry when practiced with fast grown species results in quick expansion of crown which covers the ground and attains 10% crown cover for being qualified as forest as per the regulatory parameters fixed for the definition of forest (Table 26.2). For example, around 190 poplar samplings per ha in blocks, rows or boundary fields attain this limit of 10%

crown cover in the first-year growth itself and such areas are picked up as forest cover in biannual inventory monitored by the Forest Survey of India (Dhiman 2005). A large acreage under clonal plantations is indirectly helping in protecting natural forests by diverting pressure on them and hence serving a great environmental purpose.

26.8.6 Others

Some trees especially eucalyptus are increasingly planted in waterlogged areas particularly along the water canals where water seepage and stagnation make such lands unsuitable for agriculture and other productive land use. With increased availability of multiple clones, many of them are now planted to uptake excessive water from upper layers of the water saturated soils with their shallow rootedness restricted to a couple of metre below ground (Outreach 2019; Dhiman et al. 2020). This way, the clonal trees with fast growth are helping in draining out excessive water from upper layers and reclaiming such land for other productive uses. There were also large-scale reclamation of coastal areas with planting Casuarina clonal and seedling plants (around 90% casuarinas) after 2004 tsunami between villages and sea. A large number of clones tested on stressed sites lead to screening many of them specifically for black, red, alkaline and saline soils, sandy river beds, wet and semi dried locations which could now be used for early reclamation of such sites with fast growth of trees.

With the first Forest Stewardship Council Forest Management (FSC-FM) certified project for pulpwood plantations by ITC during 2009, many such projects were initiated mainly by the wood-based industry using clonal and seedling plants. ITC is currently growing over 1 lakh ha under FSC-FM certification which mainly has eucalyptus clonal plantations. This indicates that clonal plantations are now raised as per international norms and responsibly managed to protect the natural forests following all possible regulations. Poplar culture in India is estimated to sequester approximately 3 million tonnes of carbon annually in plantations (Dhiman 2009).

26.9 Way Forward

Around 270 commercial timber tree species are found in Indian forests, out of which 30 are of outstanding forestry value. However, we have been using only a few of them in production forestry including clonal forestry. The current clonal forestry is largely based on exotics like eucalyptus, poplar, leucaena, casuarina and a few others. Some Indian species like teak, gmelina, neem and sandalwood are now increasingly planted in many other countries for their quality wood production. Vegetative propagation methods have been worked out for most of them; yet, operational level plantations of these species have not picked up in the country. However, clonal plantations of some of them like teak are widely raised in other countries. There is a need to recognize indigenous tree resources in clonal forestry for which long-term tree breeding programmes integrated with developing vegetative propagation techniques should be taken up.

Many clonal plantations are now harvested much earlier before they enter reproductive stage. They, therefore, do not interfere with native species even if they are having a threat to cross breed with natural populations. The long-term clonal trials of such trees need to be avoided around the habitat of native species as there may be a genuine threat of interfering with genetic resources of native species. We have already recorded some evidences that some of the poplar clonal trials established long back in hills produce natural hybrids with native species.

Unpredictability of clones especially for developing susceptibility to diseases and insects of both native and invasive pests, years after their deployment in production systems, is one of the serious problems in clonal forestry. Epidemic spread of gall insect *Leptocybe invasa* on eucalyptus, large-scale spread of stem canker and leaf blight on certain eucalyptus clones and epidemic infection of *Bipolaris maydis* on poplar G3 clone during mid 1990s are a couple of examples that lead to removing many useful clones out of production system. These clones developing susceptibility to such insects and diseases after many years of regular culture are a real concern for losing useful genotypes and also for premature harvest of infected plantations. The country needs a long-term strategy for not losing such valuable clones.

Most clones used initially in clonal forestry were selected and shortlisted for growth and productivity from forests, plantations and field trials. Some clones recently developed through controlled crosses and introduced in plantation forestry are also selected based on mainly growth and productivity. Many of them were tested and shortlisted for production for different sites and product manufacturing. Breeding and production of new tailor-made clones for specific sites and products from long-term breeding programmes specially targeting for incorporating wood traits suitable for quality and specific product manufacturing are yet to be started in the country.

Our clonal forestry drew inspiration from poplar plantations raised by safety match industry in Sweden and eucalyptus plantations raised by paper industry in Brazil and Congo. These countries still have long-term breeding programmes well integrated with clonal forestry. The continuity of old tree breeding programmes of some species is presumably under pressure as next-generation plantations are now being encouraged with seed origin planting stock rather than with already proven clones of some species. Our huge economy is based on these plantations and the very sustenance and survival of wood-based industry are exposed to a very high risk. Long-term tree breeding programmes in most research institutions suffer due to difficulties like inadequate funding and continuity of trained personnel. Private sector will only support such a programme till it makes economic sense to their business. It is timely that the country takes a call, review the old work, plan long-term tree breeding programmes and integrate them with clonal forestry using the latest adaptive research followed in some countries to sustain the activity. For example, *Pinus radiata* in New Zealand has an advanced tree improvement programme and its over 60% plantations are reportedly of clonal origin. The limited seed produced through control crosses in clonal seed orchards, is grown to seedlings, stool beds established for producing cuttings used for multiplication in large numbers from genetically improved seed. Of late, the country is effectively using somatic embryogenesis to mass multiply clones from immature and mature embryos produced from controlled crosses. With every passing day new techniques and methods are opening up new avenues in clonal forestry. It is essential to catch up such developments and apply some of promising ones in programmes implemented at operational scale.

The true potential of clonal forestry can be realized by deploying some unique, tested, proven and reliable clones. The published reports, so far, mention around a 1000 clones of different species developed by various private and government sector research establishments in the county. Practically, around three dozens of them find space in plantation forestry. There is a need for a well-established protocol for releasing of tested clones and preventing any inferior/untested clone reaching the market.

The experience gained on clonal forestry has so far largely been from productive farm land with high input and better cultural operations. The practice has not expanded with same rate and zeal to low productive land especially on degraded land which is huge in acreage and holds high potential for expansion. The next revolution is needed on this land use.

Clonal forestry exploits the existing genetic makeup of selected clones as they are mass multiplied through asexual reproduction and as such is accused for being dead end to breeding. Biotechnological tools are now increasingly used to create new genotypes including genetic engineering or by creating somaclonal variation. Our experience with micropropagated poplar and a few other trees has not been good both in exploiting these techniques for mass multiplication and for creating favourable variation for developing new clones.

Adoption of large-scale clonal forestry of fast-growing species especially on farmlands has demonstrated that the practice has been economically remunerative to even small growers who intentionally diversified their cropping system. Over the years the production costs have been significantly increased which could not keep pace with realization of wood sale price. The price of planting stock varies widely based on species, propagation method used, brand name and locations. For example, the market sale price of poplar saplings ranged between Rs. 8 and 36 per sapling in the 2020–21 planting season in north India. The actual cost of production varied between Rs. 7 to 15 per sapling. Tissue culture plants of some species (like teak) are sold at a very exorbitant price by some producers, though they are not performing well in plantations as per existing experience. Similar to ensuring the deployment of only tested clones, the cost of the clonal plants also needs to be regulated to make them accessible and affordable to all tree growers particularly farmers of smallholding.

26.10 Conclusion

There are many arguments advocated in favour and against the practice of clonal forestry. Many of the claimed threats may be true and could be effectively managed by developing additional scientific knowledge to find workable solutions to overcome them. Clonal forestry is proved to offer numerous short-, middle- and long-term advantages (Libby and Rauter 1984; Carson 1986; Burdon 1988, etc.). Some of the significant and meaningful achievements of clonal forestry, in Indian perspective, have been the realization of many of the claimed short-term advantages. The arguments in favour of its culture on operational scale weigh far higher than against it and are the basic reasons for its rapid expansion. There is a very high participation of local farming community and private sector in making it successful. Participation of forest departments and forest corporations is fast increasing to realize its true potential by expanding its culture on forest land. There is an increasing realization that the conventional production forestry of seedling origin is now in transition phase to clonal forestry. Use of biotechnological tools including developing and exploitation of genetically engineered trees in clonal forestry appears to be a reality now. The current need is to move aggressively in scientific pursuits and cautiously in its expansion especially with new biotechnological products like genetically engineered trees with their rigorous and constant monitoring and evaluation of newly developed products and methods.

Lessons Learnt

- Clonal forestry is highly beneficial if adequately supported with tree improvement programme. The genetic potential of improved source material could effectively be utilized by cloning. The genetic gain in clonal forestry has been high with highly improved genotypes.
- In India, clonal forestry occupies a significant space in agroforestry where thousands of farmers grow multiple clones of some select trees. A couple of pest epidemics, namely gall wasp infection on eucalyptus and leaf blight

on poplars could eliminate a couple of susceptible clones, yet a large number of other available clones readily substituted them to sustain the commercial clonal forestry.

- The method used for cloning is specific to the end use of the plantations for which the planting stock is produced. For example, timber production encourages use of the juvenile vegetative propagules and those raised for fruits/ seed use mature vegetative propagules for production of planting material.
- There is space for integrating clonal forestry with improved seed production. Hybrid seed produced by making controlled crosses could be fast multiplied by cloning juvenile seedlings for meeting the large demand of improved planting material.
- The potential of clonal forestry has been adequately realized for increasing the productivity of short rotation, fast growing trees outside forests. It is yet to be realized in many other tree species like timber yielding species widely grown on forest land.

Key Questions

1. What is the safe number of clones for averting danger of epidemics due to narrow genetic base of clones used in clonal forestry?
2. Why clonal forestry could not be taken up so extensively on forest land compared to fast growing and commercially important tree species in agroforestry?
3. Why clonal forestry has limited use without tree improvement?
4. Why tissue culture could not be used on a large scale in clonal forestry?
5. Discuss with examples specific cases where monoclonal plantations were subject to epidemics of pests

References

Ahuja MR, Libby WJ (1993a) Genetics, biotechnology and clonal forestry. In: Ahuja MR, Libby WJ (eds) Clonal forestry I. Genetics and biotechnology. Springer-Verlag

Ahuja MR, Libby WJ (1993b) (eds) Clonal forestry II. Conservation and application. Springer-Verlag

Aitken-Christle I, Connett M (1992) Micropropagation of forest trees. In: Kurata K, Kozai T (eds) Transplant production systems. Kluwer, pp 163–194

Anon (2005) Summary report on market survey of tissue cultured plants for DBT and Small Farmers' Agribusiness Consortium. Biotech Consortium India Limited, 14 p

Arokiaraj P (2000) Genetic transformation of *Hevea Brasiliensis* (rubber tree) and its applications towards crop improvement and production of recombinant proteins of commercial value. In: Jain SM, Minocha SC (eds) Molecular biology of woody plants, vol 66. Kluwer, Dordrecht. https://doi.org/10.1007/978-94-017-2313-8_14

Assis TF, Fett-Neto AG, Alfenas AC (2004) Current techniques and prospects for the clonal propagation of hardwoods with emphasis on *Eucalyptus*. In: Walter C, Carson M (eds) Plantation forest biotechnology for the 21st century. Research Signpost, Kerala, pp 303–333

Bishir J, Roberts JH (1999) On numbers of clones needed for managing risks in clonal forestry. For Genet 6:149–155

Bolstad PV, Libby WJ (1982) Comparison of radiata pine cuttings of hedge and tree-form origin after seven growing seasons. Silvae Genet 31:9–12

Bonga JM (1982) Vegetative propagation in relation to juvenility, maturity and rejuvenation. In: Bonga JM, Durjan DJ (eds) Tissue culture and forestry. Nijhoff, Boston, 420 p

Bonga JM, Durzan DJ (1987) Cell and tissue culture in forestry, General principles and biotechnology, vol 1. Martinus Nijhoff Publishers, Dordrecht

Bosela MJ, Ewers FW (1997) The mode of origin of root buds and root sprouts in the clonal tree *Sassafras albidum* (Lauraceae). Am J Bot 84(11):1466–1481

Burdon RD (1982) The role and optimal place of vegetative propagation in tree breeding strategies. In Proc. IUFRO joint meeting working parties genet breeding strategies including multiclonal varieties, Sensenstein, 6–10 Sept 1982

Burdon RD (1988) When is cloning on operational scale appropriate? In: IUFRO meet breeding tropical trees, Pattaya, Thailand, pp 9–27

Burdon RD, Aimers-Halliday J (2003) Risk management for clonal forestry with *Pinus radiata*—analysis and review. 1. Strategic issues and risk spread. N Z J For Sci 33(2):156–180

Carman N, Langelle O, Perry A, Petermann A, Danna Smith JD, Tokar B (2006) Ecological and social impacts of fast growing timber plantations and genetically engineered trees. Dogwood Alliance. https://media.dogwoodalliance.org/wp-content/uploads/2012/08/manuscript_final.pdf

Carson MJ (1986) Advantages of clonal forestry for *Pinus radiata*—real or imagined? N Z J For Sci 16(3):403–415

Chaturvedi AN (1982) Poplar farming in U.P., India. U.P. Forest Bulletin No. 45

Chaturvedi HC, Sharma AK, Agha BQ, Jain M, Sharma M (2004) Production of cloned trees of *Populus deltoides* through *in vitro* regeneration of shoots, leaf, stem and root explants and their field cultivation. Indian J Biotechnol 2:203–208

Chundamannil M (2001) Economics of forest plantations in India. KFRI Research Report No. 201. Kerala Forest Research Institute, Peechi, Kerala

CSE (2016) Fumbling with forests-why we should not handover forests to the private sector. Centre for Science and Environment, New Delhi

Cyr DR, Klimaszewska K (2002) Conifer somatic embryogenesis. II. Application. Dendrobiology 48:41–49

Das CS (1991) Tissue culture technique in tea. In: Field management in tea. Tea Research Association, Tocklai Experimental Station, Jorhat, Assam, pp 14–17

DBT (2001) Brain storming session to discuss the future strategies for tree culture. Department of Biotechnology, MoST, GoI, New Delhi

Del Tredici P (1995) Shoots from roots: a horticultural review. Arnoldia 55(3):11–19

Del Tredici P (1999) Redwood burls: immortality underground. Arnoldia 59(3):14–22

Dhanda RS, Verma RK (1995) Growth performance of *Populus deltoides* Bartr. in agroforestry plantations in Punjab. In: Khurana DK (ed) Poplars in India—recent research trends. IDRC-UHF, Solan, pp 41–56

Dhawan V, Bharwani SS (1995) *In vitro* vegetative propagation of *Leucaena leucocephala* Landewit. Plant Cell Rep 4:315–318. https://doi.org/10.1007/BF00269887

Dhawan V, Ganapathy PM, Khurana DK (1993) Tissue culture of forest tree species: recent researches in India. In: Proceedings of the national workshop on mass propagation of tree species through *in vitro* methods. IDRC-TIFNET

Dhiman RC (1999) Paulownia: UNDP-ICFRE forestry research and education project. ICFRE, Dehradun, UP

Dhiman RC (2005) Forestry mitigation projects-expectations of project developers. In: Paper presented in the workshop on facilitating forestry mitigation projects in India: promoting stakeholder dialogue and capacity building, 15–17 Jun 2005. ICFRE, Dehradun

Dhiman RC (2007) Important stimulators in the success of commercial agroforestry. In: Singh J, Kumar V, Singh MP, Sharma KD (eds) Proceedings of workshop on status and procedures of agroforestry in North Western Himalayas. Himalayan Forest Research Institute (ICFRE), Shimla, pp 123–131

Dhiman RC (2009) Carbon footprint of planted poplar in India. ENVIS For Bull 9(2):70–81

Dhiman RC (2012) Status of poplar culture in India. ENVIS For Bull 12(1):15–32

Dhiman RC (2013a) Unorthodox asexual means for mass multiplication of poplars. IPC Newsletter, Jun 2013, pp 7–8

Dhiman RC (2013b) Status and impact of commercial agroforestry in India. Indian J Agrofor 15(2):55–67

Dhiman RC (2014) Poplars in India: past, present and future. In: Bhojvaid PP, Khandekar N (eds) Sustainable forest management for multiple values: a paradigm shift. Forest Research Institute, Dehradun, pp 183–206

Dhiman RC (2018a) Sustaining poplar culture and utilization in Haryana. In: Proceedings of the workshop "growing of poplars: bottlenecks and issues", 8 Jun 2018. Haryana Kissan Ayog, Govt. of Haryana, pp 37–38

Dhiman RC (2018b) Development of synergy between private sector and farmers for wood raw material production. In: Souvenir of NAU-ISTS national seminar on forest and tree based land use system for livelihood, nutritional and environmental security. NAU, Navsari, pp 157–167

Dhiman RC (2023) Ecological and economic impact of riverside stabilization with poplar based agroforestry. Indian For 149(7):710–715

Dhiman RC, Gandhi JN (2010) Field performance of traditionally propagation vis-à-vis tissue cultured poplar. Ann For 18(2):223–230

Dhiman RC, Gandhi JN (2012) Clonal development and diversity in WIMCO's poplar program. ENVIS For Bull 12(1):40–48

Dhiman RC, Gandhi JN (2014) Status of eucalypt clonal culture in India. In: Bhojvaid PP, Kaushik S, Singh YP, Kumar D, Thapliyal M, Barthwal S (eds) Eucalypts in India. ENVIS Centre on Forestry, National Forest Library and Information Centre, FRI, Dehradun, pp 60–116

Dhiman RC, Gandhi JN (2017) Performance of *Melia dubia* Cav. under agroforestry in the Tarai Region of Uttar Pradesh. J Trop For 37(11): 73–79

Dhiman RC, Sekhar C, Gandhi JN (2007) Rehabilitation of sandy river beds of Gola River, Uttrakhand with poplar based agroforestry. In: Paper presented in the conference on environmental and livelihood security through resource management in Northern India, 29–31 Oct 2007. Central Soil and Water Conservation Research and Training Institute, Research Centre, Chandigarh

Dhiman RC, Kulkarni HD, Gandhi JN (2010) Spreading infestation of gall wasp (*Leptocybe invasa* gen. & sp. n.) on eucalypts in North India. Indian For 136(6):791–803

Dhiman RC, Kulkarni HD, Kumar J (2020) Some unusual observations on the root system of clonal eucalypts. Wood Good 1(3):73–77

Eshel A, Waisel Y (1996) Multiform and multifunction of various constituents of one root system. In: Waisel Y et al (eds) Plant roots—the hidden half, 2nd edn. Marcel Dekker Inc., New York, pp 231–245

Fayle DCF (1968) Radial growth in tree roots. Faculty of Forestry, University of Toronto, Toronto, pp 25–40

Foster GS (1993) Selection and breeding for extreme genotypes. In: Ahuja MR, Libby WJ (eds) Clonal forestry I & II. Conservation and application. Springer-Verlag, pp 50–67

Franclet A (1983) Rejuvenation: theory and practical experiences in clonal silviculture. In: Zsuffa L, Rauter RM, Yeatman CW (eds) Clonal forestry: its impact on tree improvement and our future forests. Proc. can. tree. improv. assoc. 19th meet. Part II, pp 96–134

Gera M, Mohan G, Bisht NS, Gera N (2006) Carbon sequestration potential under agroforestry in Rupnagar district of Punjab. Indian For 132(5):543–555

Gogate MG, Deshpande MD (1994) Poplar introduction in Maharashtra. In: Proceedings of the multipurpose tree species for agroforestry in Maharashtra, 6–9 Apr 1994, Pune. BAIF Development Research Foundation, 'Kamdhenu', Pune

Gupta PK, Nadgir AL, Mascarenhas AF, Jagannathan V (1980) Tissue culture of forest trees: clonal multiplication of *Tectona grandis* L. (teak) by tissue culture. Plant Sci Lett 17:259–268

Gupta PK, Mascarenhas AF, Jagannathan V (1981) Tissue culture of forest trees—clonal propagation of mature trees of *Eucalyptus citriodora* Hook. by tissue culture. Plant Sci Lett 20:195–201

Gupta PK, Mehta UJ, Mascarenhas AF (1983) A tissue culture method for rapid clonal propagation of mature trees of *Eucalyptus torelliana* and *Eucalyptus camaldulensis*. Plant Cell Rep 2(6):296–299

Hindu (2021) First-ever genetically modified rubber planted in Assam, Hindu, Guwahati edn., 24 Jun 2021

Horgan K (1987) Pinus radiata. In: Bonga JM, Durjan DJ (eds) Tissue culture in forestry, vol 3. Martinus Nijhoff Publisher, Dordrecht, pp 128–145

ICFRE (2019) Annual report 2018–19. Indian Council of Forestry Research and Education, Dehradun

Jayashree R, Rekha K, Venkatachalam P, Uratsu SL, Dandekar AM, Kumari Jayasree P, Kala RG, Priya P, Sushma Kumari S, Sobha S, Ashokan MP, Sethuraj MR, Thulaseedharan A (2003) Genetic transformation and regeneration of rubber tree (*Hevea brasiliensis* Muell. Arg) transgenic plants with a constitutive version of an anti-oxidative stress superoxide dismutase gene. Plant Cell Rep 22(3): 201–209

Joshi NK, Dhiman RC (1994) Vegetative propagation in operational forestry-problems and perspectives. In: Joshi NK (ed) Indian forestry—new trends. Forest Research Institute, Dehradun, pp 1–31

Kaosa-ard M (1990) Teak tissue culture: cost analysis report RFD/CMU/PSTC, 26 p

Konar RN, Guha C (1968) Culture of gymnosperm tissue *in vitro*. Acta Bot Neerland 17(4):258–266

Kulkarni HD (2004) Clonal forestry for industrial wood production—an ITC experience. In: Prathiban KT, Pramathma M, Neelkantan KS (eds) Compendium on clonal forestry. NATP, Forest College and Research Institute, Tamil Nadu Agricultural University, Mettupalayam

Kulkarni HD (2008) Private farmer-private industry partnerships for industrial wood production: a case study. Int For Rev 10(2):147–155

Kulkarni HD (2013) Pulp and paper industry raw material scenario-ITC plantation, a case study. IPPTA 25(1):79–89

Kumar A (1991) Mass production of field planting stock of *Dendrocalamus strictus* through macro-proliferation—a technology. Indian For 117(12):1046–1052

Kumar A (2020) *Melia dubia* Cav.: a multipurpose tree species and quality industrial raw material. Wood Good 1(2):54–61

Kumar A, Pal M, Kumar S (1992) Mass production of field planting stock of *Dendrocalamus hamiltonii* vegetatively through macro-proliferation. Indian For 118(9):638–646

Kumar A, Savita, Shrivastava P, Sharma S, Dobhal S, Rana A, Kumar R (2017) Development of high yielding varieties of *Melia dubia* Cav. (Syn. *Melia composita* Benth). Indian For 143(11):1203–1206

Lang C (2004) China: genetically modified madness. The world rainforest movement. Archived from the original on 3 Feb 2014. Retrieved 29 Jan 2014

Larkin PJ, Scowcroft WR (1981) Somaclonal variation—a novel source of variability from cell cultures for plant improvement. Theor Appl Genet 60:197–214. https://doi.org/10.1007/BF02342540

Ledford H (2014) Brazil considers transgenic trees. Nature 512:357. https://doi.org/10.1038/512357a

Lelu-Walter MA, Thompson D, Harvengt L, Sanchez L, Toribio M, Pâques LE (2013) Somatic embryogenesis in forestry with a focus on Europe: state-of-the-art, benefits, challenges and future direction. Tree Genet Genomes 9:883–899

Libby WJ, Rauter RM (1984) Advantages of clonal forestry. For Chron 60(3):145–149. https://doi.org/10.5558/tfc60145-3

Mascarenhas AF, Muralidharan EM (1989) Tissue culture of forest trees in India. Curr Sci 58(11):606–613. www.jstor.org/stable/24094311

Mascarenhas AF, Kendurkar SV, Gupta PK, Khuspe SS, Agrawal DC (1987) Teak. In: Bonga JM, Durzan DJ (eds) Cell and tissue culture in forestry, Case histories: gymosperms, angiosperms and palms, vol 3. Martinus Nijhoff, Dordrecht, pp 300–315

Menzies MI, Klomp BK, Holden DG (1991) Optimal physiological age of propagules for use in clonal forestry. In: Miller JT (ed) Proceedings FRI/NZFP Forests Ltd clonal forestry workshop. New Zealand Ministry of Forestry, FRI Bulletin No. 160, Rotorua, pp 142–145

Mishra A, Swamy SL, Puri S (2004) Growth and productivity of soyabean under five promising clones of *Populus deltoides* in agrisilviculture system. Indian J Agrofor 6(2):9–13

Mittal A, Agarwal R, Gupta SC (1990) *In vitro* development of plantlets from axillary buds of *Acacia auriculiformis*—a leguminous tree. Plant Cell Tissue Org Cult 19(1):65–70

Mohanan C (2014) Diseases of eucalypts: status and management. In: Bhojvaid PP, Kaushik S, Singh YP, Kumar D, Thapliyal M, Barthwal S (eds) Eucalypts in India. ENVIS Centre on Forestry, National Forest Library and Information Centre, FRI, Dehradun, pp 280–314

Muralidharan EM, Pandalai RC (2000) Assessment of field performance of micropropagated teak and eucalypt. KFRI Research Report 182

Ondro WJ, Couto L, Betters DR (1995) The status and practice of forestry in Brazil in the early 1990s. For Chronicle 71:106–119

Outreach (2019) Farmer and planet friendly NewGen Eucalyptus. Outreach, Bangalore, 338 p

Pande PK, Dhiman RC (2010) Variations in wood traits in micro and macro propagated plantation woods of *Populus deltoides* Bartr. ex Marsh. Adv Biosci Biotechnol 1:263–275

Park YS, Barrett JD, Bonga JM (1998) Application of somatic embryogenesis in high-value clonal forestry: deployment, genetic control, and stability of cryopreserved clones. In Vitro Cell Dev Biol Plant 34:231–239

Parthiban KT, Narayanan R, Rai RSV, Surendran C, Ravichandran VK (1997) Callogenesis and organogenesis in *Casuarina equisetifolia*. F.R. & G. Forst. Indian J For 20:227–230

Parthiban KT, Fernandaz CC, Sudhagar RJ, Sekar I, Kanna SU, Rajendran P, Devanand PS, Vennila S, Kumar NK (2021) Industrial agroforestry—a sustainable value chain innovation through a consortium approach. Sustainability 13:7126. https://doi.org/10.3390/su13137126

Pathak PS, Gupta VK (1994) Evaluation of some new selections of *Leucaena leucocephala* on dry degraded lands. In: Singh P, Pathak PS, Roy MM (eds) Agroforestry systems for degraded lands, vol 1. Oxford and IBH Publishing House, New Delhi, pp 302–307

Pilate G, Guiney E, Holt K, Petit-Conil M, Lapierre C, Leple JC, Pollet B, Mila I, Webster EA, Marstorp HG, Hopkins DW, Jouanin L, Boerjan W, Schuch W, Cornu D, Halpin C (2002) Field and pulping performances of transgenic trees with altered lignification. Nat Biotechnol 20(6):558–560

Raina R, Hara SS, Irland LC (2011) Twenty-first century forest plantations: development of agroforestry in Haryana. J Resour Energ Dev 8(2):67

Rao PS, Bapat VA (1998) Vegetative propagation of sandalwood plant through tissue culture. Can J Bot 56:1153

Rizvi RH, Handa AK, Sridhar KB, Kumar A, Bhaskar S, Chaudhari SK, Arunachalam A, Thomas N, Ashutosh S, Sapra RK, Pujar G, Singh RK, Londhe S, Nayak D, Dogra A, Choudhary R, Dhyani SK, Rizvi J, Vagen T-G, Ahmad M, Prabhu R, Dongre G (2020) Mapping agroforestry and trees outside forest. Jointly published by the ICAR, Central Agroforestry Research Institute (CAFRI), Jhansi and World Agroforestry (ICRAF), South Asia Regional Programme, New Delhi

Roychoudhury N (2016) Gall insect menace in *Eucalyptus*. Indian J For 39:349–357

Sedjo RA (2005) Will developing countries be the early adopters of genetically engineered forests? Resources for the future. AgBioForum 8(4):205–211

Sekhar C, Rajesh R, Vani V (2015) Role of jamun (*Syzygium cuminii*) in enhancing the agripreneurial status of a farmer—a case study in Dindigul District of Tamil Nadu. J Int Acad Res Multidiscip 3: 270–290

Sharma K (2021) Personal communication. Dean, College of Horticulture and Forestry (UHF), Neri, Hamirpur, HP

Sharma K, Thakur S, Badiyala SD, Sharma NK (1995) First report on the propagation of *Terminalia chebula* Retz. through patch budding. Indian For 121(8):760–761

Shukla OP, Khanna NK, Narkhde SL (2018) Developments of high yielding *Eucalyptus urophylla* clone for paper making-JKSC U286 Turbo. In: IUFRO Eucalyptus 2018. Le Corum, Montpellier, 168 p

Singh PP, Singh AK (2016) Variability studies in aonla wild genotypes for fruit character from the north-eastern region of India. Int J Basic Appl Biol 3:170–172

Singh G, Singh H, Singh J (1997) Effects of *Populus deltoides* litter on its saplings, associated agricultural crops, and the properties of an alkali soil. J Trop For Sci 9(4):536–545

Singh S, Sharma K, Bhatia AK, Chaudhary N (2021) Feasibility and economical returns from Harad cultivation. Int J Econ Plants 8(2): 62–65

Solanki KR, Singh G, Dadhwal KS (2000) Transferable agroforestry technologies. NRCA Jhansi, Uttar Pradesh

St Claire JB, Kleinschmit J, Svolba J (1985) Juvenility and serial vegetative propagation of Norway spruce clones (*Picea abies* Karst.). Silvae Genet 34(1):42–48

Sutton RF (1969) Form and development of conifer root systems. Commonwealth Agricultural Bureaux, Tech. Comm. No. 7

Tefsumura T, Tao R, Sugiura A (2001) Some factors affecting the rooting of softwood cuttings of Japanese Persimmon. J Jpn Soc Hort Sci 70(3):275–280

Then C, Hamberger S (2010) Genetically engineered trees—a ticking "time bomb". Testbiotech.de. (n.s.)

Thind HS (2004) Tree growth and nutrient dynamics in wheat-poplar agroforestry system under various management options. In: Book of abstracts. First world congress of agroforestry, 27 Jun–2 Jul 2004, Orlando, FL, USA, p 404

Wang H (2004) The state of genetically modified forest trees in China. Preliminary review of biotechnology in forestry including genetic modification. Forest Genetic Resources Working Paper, Forest Resources Development Service, Forest Resources Division. FAO, Rome, 96 p

Yanchuck AD, Bishir J, Russell JH, Polsson KR (2006) Variation in volume production through clonal deployment: results from a simulation model to minimize risk for both a currently known and unknown future pest. Silvae Genet 55:25–37

Part V

Tree Improvement and Biotechnology

Tree Improvement for Enhancing Productivity of Plantation Forests

27

Ashok Kumar and Anchal Rana

Abstract

In order to enhance productivity of planted forests in a short period, it is essential to adopt tools of genetic improvement followed by commercial deployment of improved planting material to harness maximum gain. The improved planting stock will not only increase the growth and productivity but will also influence survival, disease resistance and wood quality. The existing variation in the wild plays a crucial role in screening trees with desirable characteristics, as it is the platform for other components of the improvement cycle, namely breeding populations, propagation populations and production populations. One of the most important strategies would therefore be to deploy only commercial clones in production populations for increasing productivity and economic gain. The approaches of genetic evaluation and testing will be of immense use in the operationalization of commercial forestry. It is essential to subject the selected germplasm to multilocation testing (MLT) before deploying them in commercial forestry to ensure high adaptability and productivity.

Keywords

Plus tree · Progeny trial · Multilocation testing · Heritability · Wood genetics

27.1 Introduction

The ever-increasing population and rapid decrease in the forest cover are creating environmental, ecological and economic disturbances and irreversible consequences. The trend of higher annual wood consumption (68.9 million m^3) than supply (47.51 million m^3) reported for the year 2020 (Kulkarni 2020) is continuing. The demand is likely to increase further with a higher rate of literacy, liberalization of economic policies and advancement of technology utilization. Meeting the wood demand is further affected by the lower productivity of Indian forests of 0.045 m^3ha^{-1} $year^{-1}$ compared to the world average of 2.1 m^3ha^{-1} $year^{-1}$ (FAO 2009). Due to various restrictions, natural forests supply as low as 3.175 million m^3 of wood annually whereas trees outside forests (ToF) contribute 44.34 million m^3 of wood. It is therefore essential that different afforestation and reforestation activities are carried out only with genetically improved planting material coupled with good silvicultural practices. Short-rotation forest tree species like Acacia, Casuarina, Eucalyptus, Gmelina, Melia, Populus and others have great potential in bridging the gap between demand and supply.

A tree improvement/breeding programme primarily consists of the selection of phenotypically superior trees and assembling them in breeding populations, promoting mating among selected individuals to get recombination of desirable genes, mass multiplication of improved planting materials and maintaining a broad genetic base for advanced generation breeding. Maximum gain can be achieved by the judicious deployment of different types of planting stock derived from selected trees in operational forestry programmes. Most of the tree improvement programmes have objectives similar to the ones given below:

(a) Genetic improvement in terms of fast growth, stem form and better wood quality to increase plantation productivity.
(b) Practising recurrent selection to increase genetic gain across generations.
(c) Maintaining a broad genetic base for continued improvement over many generations.

A. Kumar (✉) · A. Rana
ICFRE-Forest Research Institute, Dehradun, Uttarakhand, India

A. K. Mandal, A. Nicodemus (eds.), *Textbook of Forest Science*, https://doi.org/10.1007/978-981-97-8289-5_27

27.2 Variability in Base Population

The success of a breeding effort largely depends on the extent of variation present in the wild. If the variability present for target traits is low, the expected genetic gain through breeding would also be low. But generally, a large amount of variation exists for most of the economically important characters among different forest tree species. Zobel and Talbert (1984) suggested that there could be three main causes for differences among the trees: (a) the differing environment in which trees are grown, (b) genetic differences among the trees and (c) interactions between genotypes and environment. Genetic variations when predictable, are useful for tree breeders, whereas the random ones are difficult to be employed in tree improvement programmes. In forest trees, the existing variation can be grouped into geographic sources such as provenances, stands, sites and races. Geographic variations often impact traits like survival and adaptability (e.g., drought/cold hardiness); however, for economic gain, traits like individual tree variability, growth parameters, stem straightness and wood-specific gravity have greater importance.

27.3 Selection of Plus Trees

The selection of plus trees (Fig. 27.1) is to be carried out in a base population possessing a wide genetic base. Advanced generation selection is carried out in populations that are raised from genetic tests. Intra- and inter-family selections are usually made to choose superior individuals for advanced-generation breeding programmes. The wild populations and natural stands or unimproved plantations are used to select first-generation plus trees.

For the selection of superior individuals, there are two main kinds of forest stands generally used in the first-generation selection system.

Selecting individuals from a foundational population is advantageous in addressing the phenotypic effects influenced by an individual's genes. In stands of the same age consisting of a single species, responses are predominantly driven by genetic factors, preventing confusion with age-related effects in traits such as growth, form, disease resistance and adaptability. Trees in such stands experience conditions similar to those encountered in commercial plantations where improved planting stock is used.

Plantations featuring uniform age, spacing and cultural practices are considered ideal base populations for selecting individual trees. The absence of intense competition in these settings enhances the prominence of genetic effects (heritability). Consequently, plantations are favoured over natural stands for identifying superior trees in the initial stages of genetic improvement, provided they originate from known seed sources. The search for superior trees should focus on stands and plantations exhibiting average to superior characteristics in growth, stem form, pruning ability, bole straightness, branch angle and other important traits.

Before selection, it is crucial to gather information about the adaptability and other qualities of the seed source used to establish the plantations. As a general guideline, individuals should not be chosen from plantations originating from seed sources known for poor adaptability. The selected trees should be reproductively mature and have reached at least half of the rotation age. However, caution is necessary to avoid selecting over-matured trees or those that have surpassed the rotation age.

Selecting trees should be avoided from stands that have undergone logging for poles, piling, grading or thinning for such purposes. In addition, it is advisable to steer clear of stands with impure species compositions whenever possible. There is no specified minimum size for a stand or plantation in which a potential candidate tree is to be identified. Nevertheless, the size should be sufficient to locate the necessary minimum number of comparison trees in addition to the candidate tree. To reduce the risk of choosing candidate trees with shared descent, it is recommended to select only one tree from any small natural stand. This restriction, however, does not apply in the case of plantations.

27.4 Selection Methods

The methods that are generally used for the selection of phenotypically superior trees under different situations are mentioned below:

Comparison tree method: This approach, also referred to as the point grading method, involves comparing a candidate tree with a minimum of five reference trees based on various characteristics considered for selection. Points are allocated to the candidate tree for each character accordingly. In this method, certain traits such as height, clear bole height and diameter at breast height are quantitatively measured on both candidate and reference trees. Meanwhile, subjective scoring based on visual observations is applied to characteristics like branching habit, apical dominance, crown formation and pruning ability in relation to the reference trees. Bole straightness, disease and insect incidence are subjectively scored exclusively on the candidate tree.

Observations on economically important traits are recorded for both candidate and reference trees. The reference trees, which are the next best in the vicinity of the candidate, must possess dominant or co-dominant characters similar to the candidate tree. The same method employed for selecting the candidate tree is used for choosing reference trees. These reference trees may be located at varying

Fig. 27.1 Plus trees of *Gmelina arborea* (left) and *Eucalyptus tereticornis* (right)

distances from the candidate, typically reported as 25–50 m. Reference trees must be selected in a site and under environmental conditions similar to or better than those of the candidate tree. If the candidate tree is on sloped land, reference trees should also be selected on a similar contour.

Certain traits like plant height, clear bole height and diameter at breast height are objectively measured for both candidate and reference trees. Conversely, subjective scoring is applied to traits such as bole straightness, disease and insect resistance, pruning ability, flowering and fruiting and crown formation. Candidate trees can also be scored based on their percent superiority over the average of reference trees for important quantitative characters. In later stages, wood traits may be considered for evaluating candidate plus trees.

Trees found to be infested with serious diseases or insects should not be selected. For analysing wood characteristics, a 2 mm bark-to-bark increment core is extracted. For the analysis of tracheary elements and fibre length, if included for selection, a larger core of about 8–10 mm in diameter should be taken. This comparison tree or grading method is suitable for selecting plus trees in timber-yielding species. The criteria for selection and the relative importance assigned to different traits may vary depending on the breeding objective and the species undergoing improvement. The impact of meticulous visual selection alone is substantial, but when conducted scientifically, the improvement becomes significantly positive and noteworthy (see Figs. 27.1 and 27.2).

Regression method: This approach is employed to choose individuals in stands with uneven age distribution or a mix of species. The procedure involves creating a regression line by plotting recorded observations of economically significant traits against the age of the tree. A regression curve is established for height or volume, typically utilizing data from 50 trees. Occasionally, determining the age can be

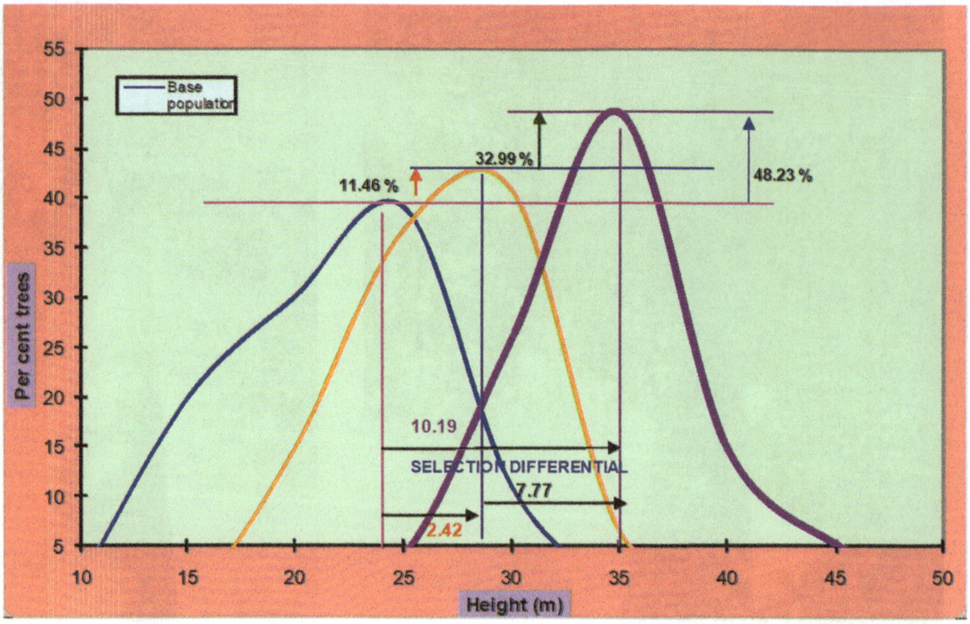

Fig. 27.2 Gains obtained through selecting candidate and plus trees from the base population

challenging, especially when some species lack distinct growth rings. Regression lines are constructed for each trait, stand and region.

The selection or rejection of a candidate tree is determined by its position on the regression line. If the candidate tree falls within a specified distance on the regression line, it is selected. Conversely, if the trait value falls below the minimum acceptable level, the tree is rejected.

Base value method: The base value method is applied in uneven-aged stands for traits that are minimally influenced by environmental factors, including age differences among trees. This is due to the robust genetic control of characteristics such as stem straightness, branching habit, disease resistance, and wood density. A base value, representing the average value for these traits, is established for each stand. The values of candidate trees for various traits are then compared and a candidate tree is either selected as a plus tree or rejected based on these comparisons.

Individual tree method: In this approach, the candidate tree is assessed and selected without comparative evaluation. However, the effectiveness of this method is limited since it fails to account for the confounding effects of both the environment and genetics in character development. There are three variations of this method described below:

1. **Total score method:** The scores assigned to different traits are aggregated and the total score serves as a guide for selecting or rejecting a candidate tree. Trees with the highest scores are chosen as plus trees.
2. **Independent culling method:** Minimum standards are set for each trait and a candidate tree is selected as a plus tree only if it meets these predetermined standards for every trait; otherwise, the candidate tree is rejected.
3. **Selection indices method:** The development of selection indices is a challenging task that demands a comprehensive understanding of the economic value of traits, along with knowledge of the genotypic and phenotypic covariances associated with these traits. When information on these aspects is available, selection indices can be formulated for choosing plus trees. The individual traits are amalgamated into a score or index and selection is then applied to the index as if it were a single trait. The index is derived through a multiple regression equation.

For two traits, X and Y, the index appears as $I = \mathrm{Px} + \mathrm{WPy}$, where I represents the index used for choosing individuals, W is a factor by which the phenotypic value of trait Y is multiplied and Px and Py are phenotypic values measured as deviations from the population mean. Despite being a complex method, selecting plus trees through this approach is highly effective as it incorporates both genetic information and the economic value of a trait. However, the use of a selection index without proper assignment of economic weights can lead to erroneous selection of individuals.

27.5 Marking and Numbering of Trees

Once comparison trees and candidate trees are located, the candidate tree is marked by a band of yellow paint at breast height. After grading, a second band of yellow paint or a plus (+) sign is marked on the candidate tree. The selected plus trees are also given registration numbers by denoting the name of the state, name of the locality and serial number of the trees. For example, in the name of a teak plus tree MHAL-

1, MH denotes Maharashtra state, AL denotes the locality Allapalli and 1 (one) denotes the plus tree number.

27.6 Categories of Trees

The selected plus trees form the base population of a first-generation breeding programme and are used for establishing first-generation seed orchards. After progeny testing, they are used in the establishment of an advanced generation production population and form a part of the germplasm collection and breeding population. Progeny-tested trees (elite trees) with complementary characteristics are used as parents in the controlled breeding. In view of the multipurpose use of plus trees, utmost care should be taken to select them.

- **Candidate plus tree:** The tree is chosen for evaluation based on its favourable phenotypic attributes, such as superior growth, good form and better wood quality. However, it has not been compared with other superior trees (check trees) for the final selection process.
- **Comparison or check tree:** The tree is used for comparison with the candidate plus tree, situated in the same stand, nearly the same age and growing on the same or better site. These trees are considered the next best in comparison to the candidate tree.
- **Select, superior, or plus tree:** This tree has undergone comparison and selection and is recommended for inclusion in production and breeding populations. However, it has not been tested for its genetic merit.
- **Elite tree:** This is a plus tree identified as genetically superior through progeny testing (genetic testing). Such a tree is highly desired for use in a breeding programme and for mass-producing seeds and vegetative propagules.

27.7 Progeny and Clonal Trials

Progeny and clonal trials are an essential part of the tree improvement programme as they help in assessing the extent of genetic control over the desired characters. Important issues are detailed below:

- Mass selection is carried out purely based on phenotypic values and does not guarantee breeding value as a parent. However, progeny and clonal tests enable the evaluation of the performance of selected trees for their breeding value.
- Since the objective of any breeding programme is to develop individuals that have desired genotypic constitution as parents, the progeny test (Fig. 27.3) facilitates the identification of different clones or parents with high general combining ability (GCA).
- Progeny and clonal tests are used for estimating genetic parameters other than breeding value including heritability, genetic advance and correlations among the characteristics. With the knowledge of these parameters, the effectiveness of the selection methods used for the improvement of different characteristics can be estimated properly.
- Selections made through progeny and clonal testing always record higher gains than that of the selections made from the plantations. Progeny and clonal tests are also very useful tools for making strategies for further multi-generation tree improvement programmes, and through these tests desired individuals will be selected and established in advanced generation seed orchards.
- Progeny and clonal tests are integral parts of the management of seed orchards. The first generation of seed orchards, which have already been established, need to be upgraded for quality seed production by culling inferior clones or progenies. This can only be done when all the clones or progenies are tested/evaluated properly.
- The trials established for progeny tests can easily be converted into seedling seed orchards for quick mass production of quality seeds by rouging out inferior families. Only the superior trees within desired families are retained and allowed to inter-breed to produce quality seeds.
- Trees are long-lived and therefore a trial for the required length of time needs to be established so that the actual trend for different characteristics among families is properly determined. However, knowledge of the nature of the relationship between juvenile and mature growth stages helps in deciding the duration of the field trial. Where such information is not available, it is desirable that progeny and clonal trials are designed for longer duration so that both the trend of characteristics and juvenile-mature correlation are estimated.
- The high potential of clonal forestry is now well recognized, particularly for species like Eucalyptus and Casuarina. However, the success of production forestry mainly depends on deploying clonal planting stock that has been adequately field-tested. Field testing ensures the availability of information about the comparative performance of different genotypes, variability, assessing genotype × environmental ($G \times E$) interaction, estimating quantitative genetic parameters and demonstration of potential in production forestry. The genetic worth of an individual line can only be assessed through genetic test-

Fig. 27.3 Progeny trial of *Gmelina arborea* (above) and *Tectona grandis* (below)

ing in the field. In genetic testing, it becomes imperative to test the progeny of selected materials (trees/parents) to evaluate genetic differences among the progenies. The phenotypic value (P) is the sum of genetic effect (G) and the environment in which it is grown (E) and described as:

$$P = G + E$$

Seeds of forest trees are used for afforestation programmes, which usually come from different sites with diverse environmental conditions. To select genotypes that can be used successfully in such diverse conditions, progeny tests need to be carried out in different environments to test $G \times E$ interactions. $G \times E$ interactions are used to measure the relative performance of inconsistency over the environments. It is also necessary that field tests be conducted over the years and environments, particularly for characters that show large year-to-year variations such as growth and resistance to diseases and pests. The results of such trials will help the breeders select and recommend the best progeny or genotype for large-scale deployment on regional and national bases.

Progeny tests are established either by using open-pollinated seed from plus trees through which additive gene effects are estimated or from controlled pollinated seed which gives information both on additive and non-additive components of inheritance. Progeny trials are extremely useful for providing information on the half-sib performance of various genotypes used in production forestry and establishment of seed orchards. It also proves valuable in evaluating the offspring of a set of parents or their crosses to provide estimates for various genetic parameters, including heritability and genetic advancement. These genetic parameters serve as crucial tools in predicting the anticipated gains from progeny trials. The variability among progenies commonly serves as an estimate for total genetic variation and is employed to determine the degree of genetic control over specific traits (Hood and Libby 1980). Likewise, among progenies and clones, covariance estimates between traits can offer insights into the genetic correlation between these traits (Foster 1986). Even though the selection of plus trees involves high selection intensity, it is essential to estimate genetic superiority itself through analytical tools such as heritability and genetic advancement.

Genetic divergence studies are essential to determine the genetic distance among the selected genotypes for establishing clonal/seedling seed orchards of diverse genotypes. After obtaining information about genetic distance, genetically diverse progenies and clones can be planted in such a way that it facilitates cross-hybridization among the genotypes, which will provide quality seed with good hybrid vigour. Estimation of broad sense heritability for various characters in *Casuarina equisetifolia* showed that heritability for height was the highest (86.85%) followed by main bole volume (85.96%) (Kumar 1996). Nelson and Tauer (1987) reported moderate clonal heritability (broad sense) in poplars for juvenile traits like height, diameter, growth and leaf size. However, narrow-sense heritability estimates for individual growth components are widely used in open-pollinated breeding programmes as a measure of additive genetic variance components. These estimates usually focus on the selection of those traits that have a positive genetic response at reasonable selection intensity.

The estimates of heritability and genetic advancement are ideal once the log phase starts. However, estimates at early stage can serve as a pointer to be compared with the results obtained at later stages and also to establish genetic correlation. This type of study would perceptively determine whether genetic analysis at an early stage is reliable. Wright (1976) noted that in clonally propagated plants, heritability would be a more appropriate estimate since any combination of genetic factors obtained can be maintained by clonal propagation. Borges et al. (1980) studied the heritability of diameter at breast height (DBH) in 124 half-sib families of *Eucalyptus grandis* at 18 and 30 months and stated that the values varied from 0.37 to 0.59. The heritability of wood density and number of shoots per seedling in the young progeny of *Eucalyptus citriodora* were estimated to be 0.91 and 0.73 respectively (de Almeida et al. 1981).

27.8 Multilocation Testing

Multilocation testing (MLT) for genetic evaluation of various genetic materials needs to be carried out in diverse agroecological regions to facilitate the screening of elite genotypes. Some of the genotypes may do well in many of the ecological regions whereas others may do so only in a few regions. All genotypes should therefore be subjected to MLT by adopting an appropriate experimental design. Though such evaluation is necessary for analysing the stability of different genotypes across locations, it becomes much more important when clonal forestry is practised with superior genotypes that are screened through such tests.

Though the potential of producing clonal planting stock is well documented, if the planting stock is not of good quality and mass multiplied before field-testing, the whole purpose of clonal forestry may fail. Thus, systematic field-testing of vegetative propagules is necessary to assess the real benefits of utilizing clonal stock. The field-testing of clonal material will give a comparative assessment of different clones for various traits and productivity and genotype × environment interaction. Thus, clones must be tested in an appropriate statistical design on a multi-location basis (three or more contrasting sites). The specific design to test the stability and productivity of the clone functions for the trait(s) of interest and environment(s) in which the clones are tested and grown is known. The following important factors need to be considered for successful clonal forestry:

27.8.1 Number of Clones per Location

As clonal forestry attracts increasing interest and begins to gain wide acceptance, the safety aspect is becoming a major concern to researchers and managers. The approach to finding a solution to disasters like epidemics is to prescribe a minimum number of clones per location (Hedstrom and Krutzsch 1982). Heybroek (1978) reported that in large crowned and long-lived organisms like trees, rules may be different from those that govern the race between pathogen epidemics and host harvest in short-rotation plants like wheat and barley. Libby (1987) developed three general guidelines for a minimum number of clones per location: (i) a mixture of a large number of clones is as safe as a mixture of genetically diverse seedlings, (ii) a mixture of a few clones is not safe, in fact, the mixture of 2–4 clones is often worse than monoclonal deployment and (iii) a mixture of 7–20 clones is as safe as large number of clones.

27.8.2 General Purpose and Interactive Clones

It is reported that the overall test may record a large and highly significant clone-by-location interaction. However, when the performance of the clones is analysed individually, it is found that the relative rank of many clones stays approximately constant in many of the test locations. Further, the ranking of some of the clones may change with regard to growth performance from location to location. This leads to two contrasting selection and deployment strategies: firstly, to select the high-performance clones that are relatively stable over many locations and deploy them as general-purpose clones to the client areas and secondly to select interactive clones for deploying only in those locations where their performance was outstanding. The best performance in each location is often achieved by selecting a few of these interactive clones and planting them on a large scale. In this way, greater genetic diversity can be maintained for future improvement and the danger of epidemics can also be avoided.

In 2007, Kumar evaluated 70 clones of Gmelina arborea and identified the top 20 performing clones for a specific purpose. These top 20 clones, based on height growth, were further assessed for their rankings at the ages of 12 and 24 months. While Clones 106 maintained the first rank consistently, Clones 9, 17 and 79 displayed an upward trend, whereas Clones 3, 7, 11, 16, 27, 39 and 101 exhibited a decreasing trend. Despite an average increment of 190% for these clones, Clones 79, 17 and 9 demonstrated increments of 242, 230 and 227%, respectively. These results strongly suggest that the testing of clones should extend over longer durations to identify and deploy elite genotypes through plantation programmes. In addition, it is essential to analyse the trend of all clones over the years and categorize them for various end uses. Fast-growing clones may be suitable for establishing pulpwood plantations for the paper industry, while slower-growing clones may be recommended for the timber and furniture industry. However, individual clones must be thoroughly analysed for different wood properties before recommending their deployment in plantation programmes. Nevertheless, the results described above only indicate clonal behaviour for height at a single location, emphasizing the need for testing real superiority at different locations.

27.9 Genotype × Environment Interactions

The $G \times E$ interaction is an effective genetic tool to test various individuals over a wide range of environments (locations/years/growing seasons). The process enables understanding genotypic and environmental impacts independently as well as the influence of $G \times E$ interaction. In the absence of $G \times E$ interaction, individuals performing better in one environment are expected to perform similarly in another environment as well. But in the presence of $G \times E$ interactions, the superiority of a genotype is based on a particular environment only. The additive main effects and multiplicative interaction (AMMI) model is an appropriate statistical tool to analyse G, E and $G \times E$ interaction, in which principal components analysis (PCA) for non-additive residuals combines with conventional analyses of variance for additive main effects (Bondari 2003).

Kumar et al. (2012) analysed adaptability and stability in *Dalbergia sissoo* Roxb. for 36 clones selected by indexing method from seed orchards and gene banks. These clones were evaluated at three locations in the state of Punjab (Ludhiana, Hoshiarpur and Patiala) adopting a completely randomized blocks design (RBD). Height, collar diameter, DBH and volume of all the genotypes were studied, which concluded that Clones 124, 36 and 1 were the most stable, Clone 124 ranked on top in adaptability and stability with an average bi value of 1.04. This clearly strategized that the most adaptable and promising clones can be used in future for large-scale deployment, inspiring breeders to produce more such stocks and enhancing the income of the planters.

A comprehensive estimation of genetic parameters and stability analysis is carried out using the following biometrical and statistical tools of analysis.

27.9.1 Analysis of Variance

Analysis of variance (ANOVA) is carried out to split total variation into different components (Sukhatme and Amble 1989). The ANOVA enables the estimation of a series of other genetic parameters, which can be summarized as follows:

1. **Variance:** The magnitude of variance in a population is estimated as a component of variance derived through multi-locational testing and ANOVA. The following genotypic and phenotypic components of variance are calculated as per Burton (1952):

   (a) Genotypic variance (GV)
   $(\sigma^2_g) = (\sigma^2_e + R\ \sigma^2_T) - (\sigma^2_e)/r$ (r = no. of replications)

   (b) Phenotypic variance (PV)
   $(\sigma^2 p) = \sigma^2_g + \sigma^2_e$

   (c) Genotypic coefficient of variation (GCV)
   $\text{GCV} = (\sqrt{\sigma^2_g}/\text{mean}) \times 100$

   (d) Phenotypic coefficient of variation (PCV)
   $\text{PCV} = (\sqrt{\sigma^2_p}/\text{mean}) \times 100$

   (e) Environmental coefficient of variation (ECV)
   $\text{ECV} = (\sqrt{\sigma^2_e}/\text{mean}) \times 100$

(ii) **Heritability (h^2):** In tree improvement, genetic components of variation play an important role as these components are transferred from generation to generation. The extent of a genotype to phenotypic variation for a particular trait in a population is defined as the ratio of genetic variation to the total variation and is known as heritability. Hence, heritability denotes the proportion of phenotypic variance due to genotype. The broad sense heritability is accordingly calculated as per Lush (1949).

$$h^2 = \sigma^2_g / \sigma^2_p$$

(iii) **Genetic advance (Gs):** The selection of a plus tree is usually carried out based on phenotypic performance, which is a joint action of a genotype and the environment in which it exists. It is therefore obvious that the superiority of a plus tree over the base population is not exclusively due to genotypic superiority, but the environment plays a crucial role as well. The improvement in the mean genotypic value of a plus tree over that of a base population is called genetic advance under selection. The genetic advance depends on two major factors (i) the existence of phenotypic variability in the base population and (ii) the heritability of the trait under consideration. The genetic advance under selection is calculated as described by Johnson et al. (1955).

$$Gs = K\,h^2\sqrt{\sigma^2}_p, \text{ where, K} = \text{selection differential at 5\%selection intensity} = 2.0$$

(iv) **Genetic gain (Gg):** Genetic gain, also referred to as genetic improvement, is defined as the enhancement in the average genetic value within a population or the improvement in the average phenotypic value resulting from selection within a population across successive breeding cycles. The anticipated genetic gain, expressed as a percentage of the mean, is calculated according to the method proposed by Burton and de Vane in Burton and de Vane 1953.

$$Gg = (Gs/\text{mean}) \times 100$$

27.9.2 Stability Analysis

Though there are various models suggested to analyse stability, Eberhart and Russell (1966) model considers regression coefficient (b_i) as the first measure of stability and further, it calculates the second measure of S^2D_i as a function of deviation from regression mean square. The progenies (t) are tested for varying environment(s) and considering Y_{ij} as mean observation of ith progeny in jth environment. The stability of genotypes is calculated using the following equation:

$$Y_{ij} = \mu_i + \beta_i I_j + \delta_{ij}$$

where Y_{ij} = mean growth performance of ith progeny at jth geographical location (i = 1, 2,... ..,t and j = 1,2,..,s); μ_i = mean of all the progenies over all the sites; β_i = regression coefficient of ith progeny on an environmental index that measures the response of varying geographical locations.

$$I_j = \frac{\sum_i Y_{ij}}{t} - \frac{\sum_i\sum_j Y_{ij}}{ts} \quad \text{with} \sum\nolimits_j I_j = 0,$$

I_j = environmental index that is defined as the deviation of the mean of all the progenies at a given geographical location from the overall mean, where t = progeny, s = growth period and δ_{ij} = error effect.

The model is preferred as it further calculates the function of deviation for regression mean square as a second measure. Therefore, two parameters of stability are calculated in the following manner:

27.9.2.1 Regression Coefficient

The regression coefficient of ith progeny on an environmental index that measures the response of this progeny to varying environments is estimated as follows:

$$b_i = \frac{\sum_j Y_{ij}\, I_j}{\sum_j I_j^2}$$

where $\sum Y_{ij}\, I_j$ = sum of products.
$\sum I^2_j$ = sum of square.

27.9.2.2 Mean Square Deviations ($\overline{S}^2d$)

Mean square deviations ($\overline{S}^2d$) from linear regression are evaluated using the following equation:

$$\left(\overline{S}^2 d\right) = \frac{\sum_j \delta_{ij}^2}{(s-2)} - \frac{S_e^2}{r}$$

where

$$\sum_j \delta_{ij}^2 = \left[\sum_j Y_{ij}^2 - \frac{Y_i^2}{t}\right] - \frac{\left(\sum_j Y_{ij}\, I_j\right)^2}{\sum_j I_j^2}$$

and S^2e = estimate of pooled error
$\sum_i Y_{ij}^2 - \left(Y_i^2/t\right)$ = The variance due to dependent variable

and
$\left(\Sigma Y_{ij}\, I_j\right)^2 / \left(\sum_j I_i^2\right)$ = The variance due to regression.

27.10 Genetic Divergence

Genetic divergence is a phenomenon wherein two or more populations of a common ancestral species accumulate independent genetic changes over time, typically occurring after these populations have been reproductively isolated for a certain duration. In some instances, subpopulations residing in ecologically distinct peripheral environments may also display genetic divergence from the rest of the population, particularly when the population's range is extensive. The genetic disparities among divergent populations can involve mutations and/or significant morphological and/or physiological alterations. Reproductive isolation, resulting from either novel adaptation through selection and/or genetic drift, always accompanies genetic divergence and serves as the primary mechanism driving speciation. In addition, genetic divergence can occur due to a bottleneck effect, such as a natural disaster causing a substantial portion of the population to perish. Such bottleneck effects induce specific changes in genetic patterns, which become overrepresented in the remaining population, resembling the founder effect.

The D^2 statistic, first presented by Mahalanobis (1928) is used to measure group distance based on multiple characters. With $x_1, x_2, x_3 \ldots\ldots, x_p$ as the multiple measurements on different progenies, and $d_1, d_2, \ldots, d_p$ as $\bar{x}^1{}_1, \bar{x}^1{}_2, \bar{x}^2{}_1, \bar{x}^2{}_2, \ldots\ldots \bar{x}^1{}_p - \bar{x}^2{}_p$ respectively are the difference in means of among populations, Mahalanobis D^2 statistics is defined as follows:

$$_pD^2 = b_1d_1 + b_2d_2 + \ldots\ldots\ldots + b_pd_p$$

Here, the b_i value for different progenies is estimated such that the ratio of variance between populations to variance within the populations is maximized. In terms of variance and covariance, the D^2 value is obtained as per the following formula:

$$_pD^2 = W^{ij}\left(\bar{x}^1{}_i - \bar{x}^2{}_i\right)\left(\bar{x}^1{}_j - \bar{x}^2{}_j\right)$$

where W^{ij} was the inverse of the estimated variance and covariance matrix.

The steps involved collection of field data, processing and analysis for estimation of D^2 values are carried out as follows:

27.10.1 Collection of Data

The growth data is collected in the field for various traits across the geographical locations in which evaluation trials are established. Secondly, data for different wood traits among the progenies is collected and measured in the laboratory as per standard tools of evaluation. The field and laboratory data are arranged appropriately and analysed for D^2 statistics.

27.10.2 Test of Significance

The collected data is calculated for variance and co-variance to determine dispersion values to be utilized in Wilk's criteria for simultaneous tests of differences between mean values of a number of correlated variables (Rao 1948). Using the pivotal condensation method, the determination of error and error + clonal matrix is calculated. The pivotal condensation element is calculated in the following fashion:

$$Л + \frac{|W|}{|S|} = \frac{|\text{ Determinant of error matrix }|}{|\text{ Determinant of error} + \text{clonal matrix }|}$$

But $V\,(\text{stat}) = m \log_e Л = -(n - p + q + 1)/2) \log_e Л$ where $m = n - (p + q + 1)/2)$.

p = number of variables or characters.

q = number of progenies -1 (or degree of freedom for populations).

n = degree of freedom for error + progenies.

$e = 2.7183$, $\log_e л = 2.3026 \log_e л$.

27.10.3 Transformation of Correlated Variables

Analysing D^2 values for a large number of characters can be intricate, leading to the preparation of a higher-order matrix inversion. This approach streamlines the computation of D^2 values, reducing it to a straightforward summation of differences in mean values for various characters, denoted as $\sum d^2 i$. To achieve this, the transformation of correlated variables into uncorrelated values is initially conducted, followed by the calculation of D^2. The transformation process involves employing the pivotal condensation method, where the coefficients for transformation are obtained by dividing the first row of the reduced matrix by the square root of the corresponding pivotal condensation element. Only the absolute values of these elements are considered for subsequent analysis.

27.10.3.1 Computation of D^2 Values

In each combination, mean deviation i.e. $Y^1{}_i - Y^2{}_i$ with $i = 1, 2, \ldots p$ is computed and D^2 is calculated as sum of squares of these deviations i.e. $\sum(Y^1{}_i - Y^2{}_i)^2$.

- **Testing the significance of D^2 values**
  The D^2 value obtained for a pair of populations is taken as the calculated value of X^2 and is tested against the tabulated value of X^2 for p degree of freedom, where p is the number of characters considered.
- **Contribution of individual characters towards divergence**
  In all combinations, each character is ranked based on $d_i = Y^1{}_i - Y^2{}_i$ values. Rank 1 is given to the highest mean difference and rank p to the lowest mean difference (where p is the total number of characters).

27.10.4 Grouping of Progenies into Various Clusters

- The Tocher Method is employed to group various progenies into clusters. In this approach, the relative distance between each pair of progenies is taken into account to sequentially calculate the two nearest progenies until all progenies are arranged accordingly. Following this arrangement, D^2 values, as suggested by Rao (1952), are utilized for cluster formation.

Initially, two progenies with the smallest distance from each other are considered, to which a third progeny with the smallest average D^2 value from the first two progenies is added. This process continues with the inclusion of the nearest fourth progeny and so forth. However, at a certain stage, when there is a sudden increase in the average D^2, the progeny causing the increase is not added to that cluster. Subsequently, a second cluster is formed and the process continues until all progenies are included in one or more clusters.

- **Average intra-cluster distances**
  The intra-cluster distances were measured by adopting the following formula:

$$D^2 = \sum D^2{}_{\mathrm{i}}/n$$

  where $\Sigma D^2{}_{\mathrm{i}}$ is the sum of the distance between all possible combinations (n) of the population included in a cluster
- **Average inter-cluster distances**
  The procedure for calculating inter-cluster distance is first to measure the distance between clusters I and II, between I and III, between I and IV, between II and III, between II and IV and so on. Likewise, clusters are taken one by one and their distances from other clusters are calculated as per the following equation:

$$D^2{}_{1\to 6} + D^2{}_{1\to 8} + D^2{}_{4\to 6} + D^2{}_{4\to 8} + D^2{}_{7\to 6} + D^2{}_{7\to 8}$$

The success of a tree improvement programme ideally depends on the identification of germplasm that is recommended for large-scale deployment. Germplasm is called superior if it produces potential results for greater productivity under favourable environmental conditions and demonstrates consistency in stability. Stability therefore refers to the performance of a genotype with respect to changes in environmental factors within a given location. Adaptability on the other hand is a concept that refers to the stability of different genotypes with respect to change across locations (Prabhakaran and Jain 1994). It has been well understood that genotype and environment interact appropriately. Environmental factors could induce a selection process that affects the genetic constitution of a given population to distorted segregation and recombination as per Mendelian principles, which could even be termed a conditioning effect.

In a tree improvement programme, both genotype and variation play important roles and lead to the effective selection of a genotype. The variability among different genotypes of a species is known as genetic diversity or divergence (Singh 1993; Pande et al. 2013). The divergence study analyses the degree of diversification and relative proportion of each component character to the total divergence. Genetic divergence studies are one of the most effective tools for establishing seed orchards of forest tree species with diverse parents. The approach therefore allows the production of improved seeds in a highly effective and economical manner as diverse parents have equal opportunities for hybridization and production of quality seeds (Kumar and Gurumurthi 2000; Vishnu et al. 2018). Assessing all available genetic variations or divergence within/between species and delineation of the most appropriate genotype(s) matching to the site for increased productivity would be of prime importance in any tree improvement programme (Sharma and Bakshi 2014). A better understanding of genetic variation among the species assists in producing superior germplasm. Hybrids between individuals of diverse origin display greater heterosis than those between closely related individuals. Screening and evaluation of genetically divergent progenies or hybrids can even be useful for further tree improvement programmes.

27.11 Wood Genetics

Enhancing wood quality is acknowledged as a vital component of tree breeding programmes, recognizing that wood volume, growth and wood quality are interconnected traits rather than independent entities (Zobel 1984). Anatomical properties such as cell size, proportion and arrangement significantly influence wood properties (Burley and Palmer 1979). Fibre length, an essential parameter for quality paper production, holds particular importance (Jorge et al. 2000). Specific gravity serves as a broad indicator encompassing anatomy, strength, stiffness and properties related to drying and machining (Panshin and deZeeuw 1980). Additional wood properties, including heartwood content and extractives, are deemed crucial for tree improvement programmes. Considerable variability in certain physical properties of plantation-grown wood has been documented (Kumar et al. 2018). The understanding of wood anatomy aids tree breeders in developing superior varieties suited for diverse wood-consuming industries.

To use wood efficiently and effectively, variation patterns for wood traits within trees as well as among trees need to be understood (Zobel and van Buijtenen 1989). Considerable differences in wood properties exist from one tree to another within a species and these variations may manifest in differences in both anatomical and physical characteristics The strength, bulk, opacity and smoothness mainly depend on morphological characteristics of the fibres, in particular its high Runkel ratio and relatively low fibre width. Though most of the tree breeding programmes have aimed at improving growth (Yanchuk and Kiss 1993), the importance of wood parameters including wood quality was understood to make wood genetics an integral part of tree breeding (Zobel and Talbert 1984).

27.12 Release of Tree Varieties

The contribution of tree improvement has significantly enhanced global plantation productivity. Despite several tree improvement programmes underway in India, their integration with operational forestry through the deployment of tested and genetically superior clonal and seedling varieties is weak. Notably, there were no explicit guidelines for tree breeders/geneticists to systematically conduct progeny/clonal trials before declaring new varieties as elite planting stock and releasing them for large-scale deployment.

The goal of forest geneticists and tree breeders is to develop superior genotypes of various forest species with increased merchantable biomass, resistance to diseases and insect pests and broader adaptability to diverse soils and environmental conditions. It is widely acknowledged that gains from tree improvement programmes are realized through the use of improved seeds and clones from production populations. Forest geneticists and tree breeders can significantly influence the genetic variation of future forests and plantations, thereby reducing costs.

An examination of tree improvement programmes in India revealed that several divergent lines of Eucalyptus tereticornis, Eucalyptus camaldulensis (Verma et al. 1994), Casuarina equisetifolia (Kumar and Gurumurthi 2000), Tectona grandis (Dakshindas and Gogate 1995), Gmelina arborea (Kumar 2007; Kumar et al. 2003) and Populus deltoids (Jones and Lal 1989) still exist but are not utilized in breeding and tree improvement. Substantial improvements in yield could be easily achieved in these species by synthesizing new recombinants using genetically distinct parents and clonally propagating them. However, it is imperative to field evaluate the recombinants before introducing and deploying them in afforestation programmes.

However, in India, Eucalyptus and Poplars have undergone considerable genetic improvement, and the clonal plantations have been adopted in farm/agroforestry systems on a commercial scale (Lal et al. 1994). Screening genetically superior planting stock and its clonal multiplication involves high cost and technical input during production. It is, therefore, necessary not only to release the promising varieties but also to ensure that they are registered with the appropriate authority and made available to the tree growers at an affordable cost. In the absence of any regulation, there have been instances of large-scale marketing of low-quality Poplar planting material with doubtful identity in the states of Haryana, Punjab, Uttarakhand and Uttar Pradesh (Chandra 1998). An approach to systematic testing, verification and release of tree varieties and a framework for providing legal protection to tree varieties was first proposed by Kumar et al. (2001). The Indian Council of Forest Research and Education subsequently framed guidelines for testing and releasing quality planting materials for commercial cultivation, and the same were approved by the Ministry of Environment, Forests and Climate Change, Government of India in 2008. Varietal releases have been made through these guidelines since 2010 and varieties/clones were released till 2022. The verification and approval for variety release is carried out through a three-tier process each handled by a committee of technical experts and forest managers. The composition and functions of each of the committees are discussed below.

27.12.1 Implementation Team

The first level of scrutiny in the varietal release process is undertaken by the implementation team. This team has five members representing research institutions, universities, forest department and the wood-based industry with expertise in forestry, tree breeding and biometrics. A tree breeder from the ICFRE Institute functions as the coordinator of the team. When applications are received from breeders for the release of new varieties, the implementation team makes field visits and assesses the superiority of the new accessions in comparison with the check/control varieties. A detailed report of varietal performance is prepared and submitted to the Regional Variety Testing Committee (RVTC) for further verification.

27.12.2 Regional Variety Testing Committee

The Regional Variety Testing Committee (RVTC) functions at the level of agroclimatic regions and is facilitated by different Institutes of ICFRE. The Director of the regional ICFRE Institute is the chairperson of RVTC, which has eight other members having expertise in forestry, breeding and biometrics and representing research institutions, universities, forest department and the wood-based industry. The RVTC reviews the report submitted by the implementation team and also interacts with the breeders who proposed the new varieties for release. Based on the report of the implementation team and the additional information provided by the breeders concerned, RVTC recommends to the variety releasing committee (VRC) for the release of new varieties.

27.12.3 Variety Releasing Committee

The final approval for the release of a tree variety is provided by the VRC headed by the director general of forest and special secretary, Ministry of Environment, Forest and Climate Change, the government of India. The director general of ICFRE is the co-chair of the committee, which also has 13 other members representing ICFRE's regional Institutes

(Chairpersons of RVTCs), forest department, university and industry. The VRC considers the recommendations made by the RVTC and if necessary, interacts with the breeders proposing the release of new varieties and gives its final approval for public release of the new varieties.

27.13 Conclusion

The forests play an important role in human welfare as the world's most important and valuable renewable natural resource and repository of biological diversity. Globally a large human population depends on forests for energy, housing, timber, fodder, medicine etc. and also derives innumerable ecological services. The demand for wood and wood products is increasing on an exponential scale and wood processing industries require sustainable wood production of a large number of wood-based products, including sawn wood, pulp/paper, veneer, plywood, panel products, parquets, shingles, toys, bobbins, sports goods, boats, carts, musical instruments and pencils. The large-scale import of wood costs valuable foreign exchange that could be saved by planting genetically improved tree species under commercial, social and agroforestry plantations. In tree improvement, the assessment of genetic diversity is an important aspect, which helps in the continuous improvement of productivity and other economically important traits. Though morphological evaluations are direct, easy and inexpensive, they tend to be influenced by environmental conditions and therefore it is essential to practice robust molecular genetic approaches to assess genetic diversity and its efficient utilization.

Lessons Learnt

- Genetic improvement of tree species is the most cost-effective sustainable method of increasing wood production to meet the demand and reduce the pressure on natural forests.
- The genetic and tree breeding strategies involve selection, random mating in breeding populations, hybridization establishment of seed orchards and vegetative multiplication of elite trees.
- Forest trees, being predominantly cross-pollinated species, respond to genetic improvement with ease, provided the tree improvement programme focuses on the incorporation of desired traits into new recombinants.
- Control pollination techniques coupled with clonal propagation could be an important tool to capture the benefits of certain specific species combinations through hybrid vigour. Using these improved planting stocks, the expected growth rate, stem straightness, wood quality, coppicing ability, tolerance of salinity etc. can be achieved.

Key Questions

1. What is tree improvement, and how does it help in increasing the productivity of commercial forestry?
2. What are the methods of selection? Describe the various categories of trees including broad differences among candidate, plus and elite trees.
3. What is stability? Describe the steps of such approaches systematically. How is stable genetic material useful in production forestry?
4. What are the steps involved in the release of genetically improved planting stock and how would such genotypes be useful in increasing productivity?
5. Describe the step-by-step approach for adopting commercial forestry with genetically improved planting stock.

References

Bondari K (2003) Statistical analysis of genotype × environment interaction in agricultural research. Paper SD15, SESUG, Proceeding of the Southeast SAS users group, St Pete beach

Borges RCG, Brune A, Silva JC, Regazzi AJ (1980) Estimativa de parâmetros genéticosem *Eucalyptus grandis* W. Hill ex Maiden. Revista Árvore Viçosa 4(2):134–145

Burley J, Palmer RR (1979) Pulp and wood densitometric properties of *Pinus caribaea* from Fiji. CFI, Oxford. Occasional Paper, p 66

Burton GW (1952) Quantitative inheritance in grasses. Proceeding of the 6th International Grassland. Congress 1: 277–283

Burton GW, de Vane DE (1953) Estimating heritability in tall fescue *Festuca arundinacea* from replicated clonal material. Agron J 45(10):478–481

Chandra JP (1998) Registration of clones and certification of clonal planting material. A letter to the Secretary, MoEF, GOI, New Delhi dated September 22, 1998

Dakshindas SD, Gogate MG(1995) Phenological performance of teak (*Tectona grandis)* seed orchards in Western Ghats of Maharashtra. In: UNDP/ICFRE International Workshop on Forestry Research and Methods, Jabalpur

de Almeida CMVC, Brune A, Silva JC, de Oliveira LM (1981) Estimatives de herdabilidades e correlacoesem progenies jovens de *Eucalyptus citriodora* Hook. Revista Arvore 5(2):250–268

Eberhart ST, Russell WA (1966) Stability parameters for comparing varieties. Crop Sci 6(1):36–40

FAO (2009) India forestry outlook study, Ministry of Environment and Forests, Government of India, Working Paper No. APFSOS II/WP/2009/06

Foster GS (1986) Provenance variation of eastern cottonwood in the lower Mississippi Valley. Silvae Genetica 35:32–38

Hedstrom BS, Krutzsch P (1982) Regulations of clonal forestry with *Picea abies*. In: Proceeding of the , IUFRO joint meeting of working parties on genetics about breeding strategies including multiclonal varieties. pp 109–112

Heybroek HM (1978) Primary considerations: multiplication and genetic diversity. Unasylva 30:27–30

Hood JV, Libby WJ (1980) A clonal study of intraspecific variability in radiata pine. I: cold and animal damage. Aust For Res 10(1):9–20

Johnson HW, Robinson HF, Comstock RE (1955) Estimates of genetic and environmental variability in soybeans. Agron J 47:314–318

Jones N, Lal P (1989) Commercial poplar planting in India under agroforestry systems. Commonwealth For Rev 68(1):19–26

Jorge F, Quilho T, Pereira H (2000) Variability of fibre length in wood and bark in *Eucalyptus globulus*. IAWA J 21(1):41–48

Kulkarni HD (2020) Agroforestry for wood based industry: a new proposal. Wood Good Grow More Use More 1(1):53–62

Kumar A (1996) Genetic improvement of *Casuarina equisetifolia*. Ph. D. Thesis,. Forest Research Institute, Deemed University, Dehradun

Kumar A (2007) Growth performance and variability in different clones of *Gmelina arborea* (Roxb). Silvae Genetica 56(1):32–36

Kumar A, Gurumurthi K (2000) Genetic divergence studies on clonal performance of *Casuarina equisetifolia*. Silvae Genetica 49(2): 57–59

Kumar A, Prasad KG, Mathroo AK (2001) Strategies for testing, registering and releasing genetically improved forestry planting stock. Indian Forester 128(6):619–634

Kumar A, Chawhaan PH, Mathroo AK (2003) Improvement through selection of plus trees in *Gmelina arborea*. J Trop For Sci 15(3): 441–449

Kumar A, Bhatt A, Ravichandran S, Kumar V, Dobhal S (2012) Genotype x environmental interactions for analyzing adaptability and stability in different clones of *Dalbergia sissoo* Roxb. J For Res 23(1):65–74

Kumar S, Kelkar BU, Mishra AK (2018) Variability in physical properties of plantation grown progenies of *Melia composite* and determination of kiln—drying schedule. J For Res 29(5):1435–1442

Lal P, Kulkarni HD, Srinivas K (1994) Four years of eucalyptus improvement through clonal technology by ITC Bhadrachalam Paperboards Ltd. Proceeding of the Second Asia-Pacific Conference on Agricultural Biotechnology, Madras, India

Libby WJ (1987) Testing and deployment of genetically engineered trees. In: Bonga JM, Durzen DJ (eds) Cell and tissue culture in forestry. Martinus Mijhoff Publishers, Lancester, pp 167–197

Lush JL (1949) Heritability of quantitative characters in farm animals. Proc Int Congress Genet Heriditas (Suppl) 35:356–387

Mahalanobis PC (1928) A statistical study at Chinese head measurement. J Asiatic Soc Bengal 25(3):301–377

Nelson CD, Tauer CG (1987) Genetic variation in juvenile characteristics of *Populus deltoides* from southern Great Plains. Silvae Genetica 36:216–221

Pande PK, Kumar A, Ravichandran S, Naithani S, Kothiyal V, Kishore PBK, Raturi A, Gautam P, Dobhal S, Sharma S (2013) Genetic analysis of growth and wood variations in *Leucaena leucocephala* (Lam.) de Wit. For Res 24(3):485–493

Panshin AJ, deZeeuw CE (1980) Textbook of wood technology. McGrawhill, New York, NY, p 643

Prabhakaran VT, Jain JP (1994) Statistical techniques for studying genotype environment interaction. South Asian Publishers Pvt. Ltd., New Delhi, p 320

Rao CR (1948) The utilization of multiple measurement in problems of biological classification. J R Stat Soc 10(2):159–193

Rao CR (1952) Advanced statistical methods in biometric research. John Wiley and Sons, Inc., New York, NY

Sharma A, Bakshi M (2014) Variability in growth, physiological, and biochemical characteristics among various clones of *Dalbergia sissoo* in a clonal seed orchard. Int J For Res 2014:829368

Singh NB (1993) Estimation of variance, heritability, genetic gain and correlation among some growth characters in *Bambusa pallida* Roxb. Indian J For 16(1):33–38

Sukhatme PV, Amble VN (1989) Statistical methods for agricultural workers. Publication and Information Division, ICAR, New Delhi, p 359p

Verma RK, Ravi N, Jayachandran CK, Ganesan M, Kumar A, Balasubramanian A, Gurumurthi K (1994) Hedge orchards for clonal forestry. ICFRE technical bulletin. IFGTB, Coimbatore

Vishnu R, Anoop EV, Warrier KCS, Anish MC (2018) Genetic divergence for growth and wood parameters in different clones of *Casuarina equisetifolia*. J Trop For Sci 30(4):615–621

Wright JW (1976) Introduction to Forest genetics. Academic Press, New York, NY/San Francisco, CA/London

Yanchuk AD, Kiss GK (1993) Genetic variation in growth and wood specific gravity and its utility in the improvement of interior spruce in British Columbia. Silvae Genetica 42:141

Zobel B (1984) The changing quality of the world wood supply. Wood Sci Technol 18(1):1–17

Zobel BJ, Talbert J (1984) Applied Forest tree improvement. John Wiley and Sons, New York, NY, p 505p

Zobel B, Van Buijtenen JP (1989) Wood variation: its causes and control. Springer-Verlag, Berlin, p 363p

Genetic Improvement and Conservation of Medicinal and Aromatic Plants

28

K. A. Geetha and Satyabrata Maiti

Abstract

India is home to a rich plant biodiversity of medicinal and aromatic plants (MAPs). About 7000–8000 species of MAPs are reported from India. About 242 species are traded in large volumes, which is more than 240 MT per year. About 90–95% of the raw drugs are collected from the wild since majority of MAP species are not yet under cultivation.

Collection from the wild is destructive in nature, which causes a great challenge to the survival of these species. To cope with the situation, different conservation strategies are adopted using in situ and ex situ methods. The required level that would assure sustainable conservation of MAP resources is yet to be achieved. Popularization of MAP cultivation thus becomes the top priority in the agenda of commercialization. However, the lack of high-yielding elite cultivars for a majority of MAP species prevents them from achieving the target. Crop improvement programmes of MAPs have been initiated by various national research organizations and universities. Breeding for improved cultivars in MAPs is targeted with specific objectives of higher yield without compromising the quality of bioactive ingredients. The bioactive ingredients in MAPs are secondary metabolites which are highly influenced by the environment. Hence, while working for the improvement of MAPs, all the breeding objectives should be focused on improving the total targeted secondary metabolite content along with crop yield. Setting breeding objectives is made more challenging and complicated due to the demands of a large number of diversified end users. The present chapter deals with various breeding methods to be adopted for developing cultivars having high active ingredients as well as total biomass yield per unit area to fulfil the raw drug/raw material requirements of each client.

K. A. Geetha (✉) · S. Maiti
ICAR-Directorate of Medicinal and Aromatic Plants Research (ICAR-DMAPR), Boriavi, Anand, Gujarat, India

Keywords

Medicinal plant · Aromatic plant · Breeding objective · Germplasm collection · Secondary metabolite

28.1 Introduction

India is gifted with a vast plant biodiversity of medicinal and aromatic plants (MAPs).

Since time immemorial, plants and plant-based products have been in use traditionally by the local people and therefore, India has one of the oldest, richest and most diverse cultural traditions of medicinal plant use.

Ayurveda, Siddha, Yoga, Folklore medicines, etc. are some of the important local healthcare systems that originated in India. Now these local healthcare systems are popularly known by the blanket term 'Indian Systems of Medicines (ISM)'. However, commercial utilization, including cultivation of MAPs, has been in practice for only about 50 years. The World Health Organization (WHO) reported that about 80% of populations in developing countries depend upon herbal drugs for their healthcare. According to their forecast, a similar proportion of the world population may also depend upon herbal-based drugs in the coming years. The world is now witnessing a revitalization and transformation in the traditional system of medicine, causing a quantum upsurge in the use of plant-based medicines across the world. According to the estimate of the WHO, the global herbal market will rise to $5 trillion by 2050 from the current market level of $62 billion (Gautam et al. 2020).

In India, about 1178 species are estimated to be traded out of 7000–8000 medicinal plants reported (Goraya and Ved 2017). According to the National Medicinal Plants Board (NMPB), New Delhi, about 242 species have annual consumption of more than 100 metric tons per year. The domestic demand for medicinal plants was about 1,95,000 MT, and the export demand for medicinal plants was 1,34,500 MT during 2014–2015. Total consumption of herbal raw drugs in

A. K. Mandal, A. Nicodemus (eds.), *Textbook of Forest Science*, https://doi.org/10.1007/978-981-97-8289-5_28

the country was estimated at 5,12,000 MT for the year 2014–2015 with a corresponding trade value of ₹5500 crore. India exported raw herbs worth US$ 329.83 and 319.33 million during 2017–2018 and 2018–2019, respectively with a growth rate of 10–14% over the previous years. The export of value-added extracts of medicinal herbs/herbal products from India during 2017–18 stood at US$ 456.12 million, recording a growth rate of 12.23% over the year before (https://pib.gov.in/PressReleasePage.aspx?PRID=1594929; Gautam et al. 2020). Nearly 90–95% of the medicinal plants in trade in the commercial market of India are obtained from the wild. The major reasons for this are (1) the non-availability of most of the MAP species under cultivation, (2) even if available, the price of the raw drug is comparatively higher than those obtained from the wild, (3) the raw drug collected from the wild is sometimes treated as organically grown; hence, the industry uses the material under the organic label.

The increasing demand for natural product-based medicines in the national and international markets is causing indiscriminate harvest from the wild. High consumption of these wild flora collected mostly by destructive methods (collection of roots, barks or the entire plant) has been a great challenge to the survival of this invaluable wealth in nature. It is estimated that about 70% of the plant collection involves destructive harvesting. Too often, collected material from the wild becomes a focus of criticism due to some drawbacks which are listed below:

1. Inferior quality in terms of active ingredients due to a mixture of different genotypes of the same species.
2. Mechanical mixture with other species or due to failure in identification of the targeted species, sometimes mixed with toxic species.
3. Regular supply in the desired volume cannot be assured from the wild habitat.
4. Irrational collection of MAPs from wild habitats causes damage to the natural plant biodiversity.

28.2 Plant Genetic Resources

Plant genetic resource (PGR) is the foundation of genetic improvement in a crop and also plays an important role in agricultural development. It is also an indicator as well as the reservoir of genetic adaptability and often the buffer against environmental changes. PGR is the source material for any kind of genetic improvement and varietal development. A breeder gets an opportunity to play with the variability to breed a desired variety if the PGR base of a species is wide enough for capturing genes of desired characters. Wild species also play an important role in medicinal plants. About 90–95% of species used are not cultivated but collected from either forests or wild habitats. Utilization of medicinal plant resources should, therefore, be planned in such a way that it should bring out the maximum potential of the resources and ensure the health and prosperity of future generations as well. India may lead the world in this sector if a programme is organized for the proper management of Medicinal Plant Genetic Resources in a sustainable manner.

28.3 Conservation

Convention on Biological Biodiversity (CBD) in 1992 has brought a new awakening in all the facets of biodiversity. The CBD recognizes the sovereign rights of the respective nations over the biodiversity within their domain. India as a signatory to the convention is obliged to take legislative, administrative and policy measures to conserve/protect biodiversity. Accordingly, the national government has passed two acts (PPVFR Act 2001 and the Biodiversity Act 2002) for the protection and efficient utilization of biodiversity.

At present, biodiversity loss is an utmost concern throughout the world. There are many causes of biodiversity loss, such as industrialization, population explosion, and expansion of agricultural land. Medicinal plant diversity is worst affected since most of the gene pool still exists within forest areas, which is shrinking fast throughout the world. There are two main reasons for the loss of medicinal plant biodiversity:

1. Habitat loss which is a global phenomenon now.
2. Overexploitation and unsustainable destructive harvesting from its natural habitat due to ever-increasing demand for increasing population.

The current pace at which the forests are destroyed due to increased developmental activities is inconceivable and shall reach an alarming critical condition of natural biodiversity reserve sooner than the present calculation where reversal will not be possible. Some of the important species in nature will be lost even before their medicinal properties are fully explored. Plant diversity already documented to be in use for medicinal purposes suggests that 'for every two species going to extinct, one will be of the medicinal category'. According to the 1997 IUCN Red List, about 12.5% of the world's flora is facing the threat of extinction. Walter and Gillett (1998) estimated that about 8% of the world's flora is threatened at different levels. If these ratios are extended to the MAP flora of India, about 600–1000 MAP species are feared to fall into the threat category.

Out of 627 traded species, only 70 are in cultivation (Goraya and Ved 2017). Therefore, a large number of plant species require immediate attention for their survival. However, because of resource constraints, it is practically impossible to take care of all these plant species in one go. Fixing research priority for the conservation and sustainable utilization of the existing medicinal plant biodiversity gained attention in this context.

Several government and nongovernmental organizations are engaged in the conservation of MAPs by using various strategies of conservation. All the conservation strategies can be widely divided into two:

1. In situ conservation (conservation of biodiversity in its natural habitat).
2. Ex situ conservation (conservation away from the natural habitat).

In situ conservation is achieved by protecting natural habitats from human interference. Popular examples are natural parks, biosphere reserves, sacred groves, and sanctuaries.

An in situ conservation project was first initiated for the conservation of medicinal plants in India in the States of Karnataka, Kerala and Tamil Nadu involving the State Forest Departments, leading NGOs and research institutes in 1993 in collaboration with the Foundation for Revitalization of Local Health Traditions (FRLHT), a Bengaluru-based NGO group. This project developed a network of Medicinal Plant Conservation Areas (MPCAs), Medicinal Plant Conservation Parks (MPCPs) and Medicinal Plant Development Areas (MPDAs).

A separate Ministry of AYUSH, Government of India, promotes in situ conservation through its programmes on survey, inventorization and documentation of important MAPs in their natural habitat as well as by developing Medicinal Plants Conservation and Development Areas (MPCDAs) by involving state forest/wildlife department/forest development corporation/federations/national and state level research organization/universities and non-government/voluntary organizations.

Indigenous communities develop their own traditional methods for the conservation of natural resources. One such conservation practice is the development of 'sacred groves'. Sacred groves are devoted to gods and goddesses and are strictly protected from human interference. They can be considered the first human effort for the in situ conservation of biodiversity, which are treasure homes of valuable MAP wealth. But at present, these great traditions are also facing a lot of challenges for their existence because of changes in the value system, lifestyles, etc. As a result, the number of such in situ conservation sites is shrinking. This type of our traditional strengths must be recognized, valued and harnessed for the continued preservation of valuable MAPs and ecosystems.

Ex situ conservation methods include seed gene banks, in vitro gene banks, DNA gene banks, field gene banks, botanical and herbal gardens, etc., in addition to the cultivation of MAPs. Rare and unique native plants can be preserved as living collections in this program. For species with adequate seed production, preservation through seeds is useful. In vitro methods are important for the maintenance of a pathogen-free germplasm. The major shortcoming of ex situ conservation is the absence of natural evolution within the conserved materials. This difficulty can be partially addressed by connecting ex situ intervention to in situ program, wherever possible.

The following strategies could be taken up for sustainable conservation and utilization of medicinal plant wealth.

1. Sustainable harvesting from the wild: The collection and trade of critically endangered species should be strictly banned until their status is upgraded. It requires the full-hearted cooperation of the forest departments and the local communities. All harvesting from the wild should be sustainable in nature.
2. Assessment of the threat status of enlisted medicinal plants: This should be done to prioritize the plants to be selected for immediate care. People's participatory programme can be integrated into this type of study as the local people are the factual information source for getting a clear idea of the present and past distribution of a particular species at a particular locality.
3. Medicinal plants under high stable demand in the threat category have to be given prime importance for future research studies.
4. Wherever feasible, integration of food crops and medicinal plant cultivation may be given priority.
5. Implementation of a networking programme partnering with public and private institutions for establishing better linkage.
6. Large-scale cultivation projects funded by industries should be diverted to marginal and sub-marginal lands with appropriate agro-techniques for stabilizing the supply of quality raw drugs in the market.
7. Launching of the research programme for the species having poor regeneration capacity and limited knowledge about species biology and their life cycle.

Cultivation of MAPs has been slowly gaining popularity among farmers, and at present, about 70 MAPs are being cultivated. However, the availability of high-yielding elite cultivars is the main bottleneck for improving profitability and assuring fair prices in the market. Crop improvement programmes of MAPs have been thus initiated by the Indian Council of Agricultural Research (ICAR), Council of Scientific and Industrial Research (CSIR) and State Agricultural Universities (SAUs), which ultimately resulted in the development of some elite cultivars in MAPs.

28.4 Breeding Objectives of MAPs

The role of crop improvement programmes to produce new improved varieties has increased considerably due to the commercialization of MAP cultivation. In traditional crops, namely cereals, pulses, fruits and vegetables, the productivity potential has been increased by adjusting certain morphological characters such as number of branches, number of fruits, seeds and test weight of seeds. However, in MAPs, the approach needs an additional dimension. The quality of bioactive therapeutic compounds (alkaloids, steroids, glycosides, phenolics, essential oil, etc.) are not dependent on morphological characteristics. These quality-determining compounds are secondary metabolites and are not involved in the basic metabolic process of living beings; however, they take part in the various interaction processes of the organism for survival in various adverse environments. These compounds are known to be responsible for attracting pollinators (essential oils in the form of aroma, colouring pigments, namely carotenoids and xanthophylls as colouring agents), providing resistance against insect and disease pests, etc. Each species or genera or family has its own specific set of secondary metabolites, and more than 100,000 compounds are already known as plant-originated, and about 4000 new compounds are discovered every year (Verpoorte et al. 1999).

Because of the dynamic nature of plant environmental interactions, the species growing under diverse ecological conditions differ in the synthesis and accumulation of the metabolites (Wink 1988; Bennett and Wallsgrove 1994; Theis and Lerdau 2003; Edreva et al. 2008; Oh et al. 2009; Ramakrishna and Ravishankar 2011; Gutbrodt et al. 2012; Pavarini et al. 2012). The factors which cause the fluctuations of these secondary metabolites include the plant's requirement for acclimatization or adaptation to newer areas; climatic changes; environmental changes including biotic and abiotic factors; geographical variations involving populations; genetic variations within a species; and environmental conditions during the growing phases (Telascrea et al. 2007; Rahimmalek et al. 2009).

To cope with the plant environment interactions, changes in respective groups of secondary metabolites production/assimilation occur within the plant, which would normally help to interact with the dynamic environments accordingly. Hence, while working for the improvement of MAPs, all the selection criteria of breeding objectives should be adjusted in such a way that it would enhance the total secondary metabolite content and total production in a particular crop species. Just introduction of a variety for successful cultivation in a new environmental condition may sometimes be futile due to chemical changes and infra-specific chemical modifications, which may take place due to altered ecological and geographical conditions (Tetanyi 1992).

Medicinal plants are used by various clients, such as vaidyas, local healers, ayurvedic industries and extraction units. Therefore, quality parameters defined by the various users are not the same. For example, for vaidyas and ayurvedic industries, medicinal plant quality should be as per the Ayurvedic Pharmacopoeia of India (API). On the contrary, extraction industries require a higher percentage of certain chemical compounds for high extraction efficiency as well as more harvest of compounds from low volume of raw drugs. This situation will demand different varieties for different clients. Hence, breeding objectives also should be different for catering the needs of various industries and a new variety must be defined for whom it has been developed.

Hence, breeding for MAPs mainly should be focussed upon the increased stable yield of bioactive therapeutically important compounds in addition to enhanced harvest index and resistance/tolerance against biotic and abiotic stresses. Some important breeding objectives for MAPs are listed below:

1. Enhanced and stable yield with a higher harvest index.
2. Higher content of therapeutically important compounds (active ingredients).
3. Improved agronomic traits, namely uniform flowering/maturity, rapid initial growth to suppress weed growth, non-lodging, non-shattering, etc.
4. Resistance/tolerance to biotic (diseases and insect pests) and abiotic (drought tolerance/saline/frost/climatic changes) stresses.

Most of the harvested products (raw drugs) are not directly used by the end users. The raw drugs pass through different processing industries including the pharmaceutical industry. Different primary processing interventions include cutting, chopping, powdering, and making paste forms. Hence, all the cultivars should yield raw drugs amenable to easy processing as per the prevailing industries' demand. For example, in the case of ashwagandha (*Withania somnifera*), the harvested roots are powdered and used for different formulations or direct consumption. Hence, the demand is for a cultivar with unbranched roots of less fibre. Hence, the development of processable cultivars is very important in MAPs, and species-specific traits have to be defined beforehand.

Truly speaking, in the present MAP scenario, a large number of organizations and universities along with different funding agencies are working in isolation but most of the time unaware of the demand of the industry. Industries are playing hide-and-seek games since quality standards in MAPs are at a very primitive stage. There is no compulsory

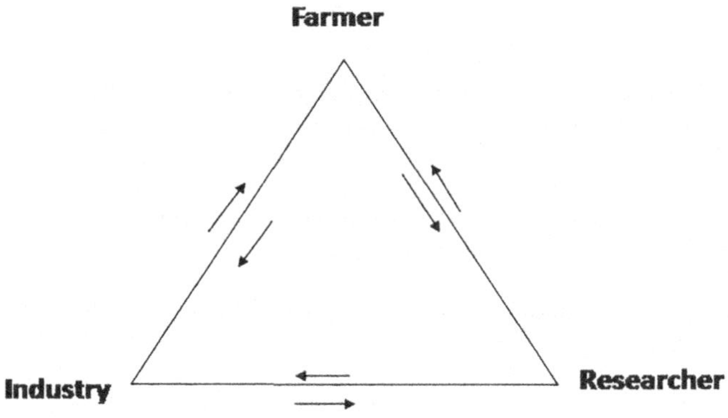

Fig. 28.1 Two-way linkage among researchers, farmers and industry

legal mechanism (third-party certification) to check the quality of each herbal raw drug used in different formulations. Hence, the industry is using raw drugs available at the lowest price, many a time it would not have a direct correlation with the quality. Farmers are also ignorant of the right quality of the raw drug to be produced. Hence, a two-way connected linkage (Fig. 28.1) among researchers, farmers and industry is required for translating the need of the industry in terms of variety, which is to be developed by the breeders, which in turn should reach the farmers for cultivation. The quality raw drug thus produced should reach back to the industry.

Indian Pharmacopoeia (IP) and Ayurvedic Pharmacopoeia (API) have set some standards for the identity, purity and strength of single drugs for medicinal purposes in which chemical content is an integral part. Some of the key components of IP standards of important MAPs are listed in Table 28.1.

28.5 Germplasm Introduction

An introduction is a process in which a new species, superior genotype or its wild relatives from its growing environment are introduced to a new area where they were not growing before. It can be between countries and within countries. Even though India is rich in biodiversity, many of the popular MAP species are introduced intentionally or spontaneously from other countries. The introduced germplasm is sometimes used directly as a cultivar (primary introduction) or used in a crop breeding programme for developing new improved varieties (secondary introduction).

Cinchona cultivation in India is a popular example of plant introduction in MAPs. Seven species of *Cinchona* were introduced from South America for cultivation across West Bengal, Tamil Nadu and Assam in 1859 to fight malaria (The Wealth of India 1992). Isabgol/psyllium (*Planatgo ovata*) is another example of plant introduction in the MAP sector. The species was naturally distributed in Iran, Iraq and Mediterranean areas from where it was introduced to India, and now India has a monopoly in isabgol cultivation in the world, and 95% of psyllium husk is exported from India to the international market. The area under its cultivation spreads over about 1.5 lakh ha, and the isabgol husk trade earned about 201.22 million US dollar worth of foreign exchange in the year 2018–2019 (http://commerce.nic.in/eidb/ecom.asp). Similarly, senna (*Cassia angustifolia, C. acutifolia),* aloe (*Aloe barbadensis*), stevia (*Stevia rebaudiana*), pudina (*Mentha arvensis*), rose-scented geranium (*Pelargonium graveolens*), foxglove (*Digitalis purpurea*), chamomile (*Matricaria chamomilla*), salvia (*Salvia sclarea*), celery (*Apium graveolens*), mulethi or liquorice (*Glycyrrhiza glabra*) and patchouli (*Pogostemon cablin)* are some of the other examples of species introduced from the other countries to India, and now the species are in commercial cultivation in India.

Rose-scented geranium or scented geranium was introduced from South Africa to Shevaroy Hills, TN, India in 1915. KKL 1, Sel 8, Kunti, Hemanti and Bipuli are some of the varieties developed from the introduced germplasm suitable for Indian conditions. Another successful introduction was the Japanese Mint by Regional Research Laboratory, Jammu (Kapoor et al. 1955). Similarly, Chamomile, Salvia, Celery etc. were also introduced to India from France during the 1950s, and the crop has now become a commercial crop in Himachal Pradesh, Uttar Pradesh and Haryana.

Similarly, some of the Indian species that are grown in one part of India can be introduced to another part for cultivation. Safed musli (*Chlorophytum borivilianum*), naturally distributed in Central India (Maharashtra, Madhya Pradesh, Gujarat and Rajasthan), which was introduced to southern parts of the country during the 1990s due to the popularity of the species is an example of plant introduction within the country.

In India, for the introduction of plants including MAPs from other countries, ICAR-National Bureau of Plant Genetic Resources (ICAR-NBPGR), New Delhi is the authorized agency. All the introductions indented by individual scientists, universities or any research organizations must be routed through ICAR-NBPGR.

The major drawback in plant introduction is the narrow genetic base, since the germplasm in most cases may not sufficiently represent the genetic variability available in the primary centre of origin. Hence, targeted reintroductions for particular traits may be conducted through ICAR-NBPGR.

28.6 Germplasm Collection

The collection of a vast diversity of germplasm is the key to the success of developing superior varieties in MAPs as in the case of other crops. The full potential of germplasm has not yet been fully utilized in MAPs. Similar to other commercial

Table 28.1 Components of IP standards of MAPs (The Indian pharmacopoeia 2007)

Plant	Parts used	Description	Quality
Amlaki (aonlas) (*Phyllanthus officinalis*)	Dried fruit pericarp	The dried fruit has a highly shrivelled and wrinkled external surface. The taste is sour and astringent followed by a delicately sweet tinge	It contains not less than 1.0% w/w gallic acid calculated on a dried basis
Anantmool (*Hemidesmus indicus*)	Root	The roots are cylindrical and yellowish-brown in colour. They have vanillin-like odour, and the taste is acrid	Anantmool contains not less than 0.02% iso-vanillin calculated on a dried basis.
Arjuna (*Terminalia arjuna*)	Bark	A flat or minutely curved thick piece of bark with a reddish-grey colour and astringent taste	The dried stem bark contains not less than 0.02% of arjungenin
Artemisia	Dried leaf and flowering top	Slightly camphoraceous odour and bitter taste	Contains not less than 0.8% of artemisinin calculated on a dried basis
Ashwagandha (*Withania somnifera*)	Root	Buff to greyish-yellow, with a slightly mucilaginous bitter and acrid taste	Not less than 0.02% of total withanolide A and withaferin A on a dry weight basis
Belladonna leaf	Dried leaf and flowering top	Leaf: Green to greenish-brown leaf, elliptical to ovate; flowering top: Purple or yellowish-brown in colour	Belladonna herb contains not less than 0.03% of total alkaloids, calculated as hyoscyamine with reference to the material dried at 100–105 °C
Bhibhitaki/ Behda (*Terminalia bellirica*)	Dried fruit pericarp	The dried pericarp appears as curved pieces of irregular shapes, the external surface is velvety, wrinkled grey to brown in colour and has an astringent taste	Contains not less than 0.3% w/w of ellagic acid and 0.75% w/w of gallic acid, calculated on a dried basis
Bhringraj (*Eclipta alba*)	Dried whole plant	A green to greenish-brown colour when completely dried	Contains not less than 0.1% of wedelocatone calculated on a dried basis
Bhuiamla (*Phyllanthus amarus*)	Dried aerial part	A green to greenish-yellow in colour, tastes slightly bitter	Contains not less than 0.25% of total phyllanthin and hypophyllanthin calculated on a dried basis
Brahmi (*Bacopa monnieri*)	Dried whole plant	Colour brown to reddish-brown when completely dried	Bacoside A content should be not less than 2.56% on a dry-weight basis
Coleus (*Coleus forskohlii*)	Dried root	Colour: Light brown, long radially spread; odour: Aromatic; taste: Slightly pungent	Forskolin not less than 0.4% w/w calculated on a dried basis
Garcinia	Dried deseeded fruit	Dark brown to blackish-brown fruits; taste: Acidic	Hydroxycitric acid and hydroxycitric acid lactone not less than 12.0% (calculated on a dried basis)
Gokhru (*Tribulus terrestris*)	Dried fruit	Fruit: Pedicellate and globose, having wedge-shaped cocci, covered with short and stiff spines; odour: Faintly aromatic smell; taste: Acrid	Diosgenin not less than 0.5% (calculated on a dried basis)
Gudmar (*Gymnema sylvestre*)	Dried mature leaf	Colour: Greenish-yellow; surface: Pubescent on both sides; odour: Characteristics; taste: Extremely bitter and acrid	Gymnemic acid not less than 1.0% (calculated on a dried basis)
Guduchi (*Tinospora cordifolia*)	Dried stem piece	Colour: Greyish-black; fibrous fracture; no distinct odour; taste: Bitter	Cordifolioside A not less than 0.02% w/w (calculated on a dried basis)
Guggul resin (*Commiphora wightii*)	Gum (oleoresin exudation)	Colour: Light to dark brown; shape: Rounded/ irregular; odour: Slightly balsamic; slightly sticky to touch	Guggulsterones (Z and E) not less than 1.0% and not more than 1.5%
Guggulipid	Ethyl acetate extractive of Guggul resin	A brown viscous liquid	Guggulsterones (Z and E) not less than 4.0% and not more than 6.0%
Haridra (*Curcuma longa*)	Dried rhizome	Externally yellowish to yellowish-brown with roots scars and annulations; odour: Aromatic; taste: Warmly aromatic and bitter	Curcumin not less than 1.5% (calculated on a dried basis)
Haritaki (*Terminalia chebula*)	Pericarp of dried fruit	The fruit has a shine on its external part with longitudinal ridges; colour: Yellowish-brown to light black; taste: Bitter and slightly astringent	Chebulagic acid not less than 5.0% and chebulinic acid not less than 12.5% (calculated on a dried basis)
Isabgol (*Plantago ovata*)	Seed husk	Pale buff brittle flakes; shape: More or less lanceolate; size: Up to 2 mm long and 1 mm wide in the Centre; the flakes have a small brownish oval spot in the Centre (about 0.8–1.0 mm long); it swells rapidly in water and forms a stiff mucilage	Mucilage

(continued)

Table 28.1 (continued)

Plant	Parts used	Description	Quality
Kalmegh (*Andrographis paniculata*)	Dried stem and leaf	Taste: Intensely bitter	Andrographolide not less than 1.0% (calculated on a dried basis)
Kutki	Dried root	Rhizomes are sub-cylindrical, straight or slightly curved, externally greyish with a wrinkled surface, circular scars on roots and bud scales, with cork exposed at places	Kutkin not less than 5.0% (calculated on a dried basis)
Mandukaparni (*Centella asiatica*	Dried aerial part	Colour: Green to greenish-yellow; taste: Slightly bitter and sweet	Asiaticoside not less than 0.5% (calculated on a dried basis)
Manjishta	Dried root	Cylindrical, slightly flattened, wiry pieces of brown to purple colour; taste: Bitter	Rubiadin not less than 0.02% (calculated on a dried basis)
Mentha oil	Aerial part of the whole plant	Mentha oil is volatile; colour: Colourless or yellowish; clear liquid; odour: Characteristic and pleasant	Menthol not less than 50.0% w/w
Opium (*Papaver somniferum*)	Air-dried latex of unripe capsule	The capsule is soft and shiny. After drying it becomes hard and brittle; colour: Blackish-brown; odour: Strong and characteristics	Morphine not less than 10% and codeine not less than 2.0% (both calculated on a dried basis)
Peppermint oil	Aerial part of the flowering plant	Oil is colourless, pale yellow or pale greenish-yellow; odour: Characteristics; taste: Characteristics followed by sensation of cold	Esters not less than 4.5 w/w% and more than 10.0% w/w (calculated as methyl acetate); free alcohols not less than 44.0% w/w (calculated as menthol); ketones not less than 15.0% w/w and more than 32.0% w/w (calculated as menthone)
Pippali large	Fruiting spike	Spike colour: Blacking green to green; shape: Cylindrical erect and blunt; size: 2–4 cm long and 0.4–0.7 cm in diameter; taste: Pungent; odour: Aromatic and characteristics	Piperine not less than 1.0% w/w (calculated on a dried basis)
Pippali (small) (*Piper longum*)	Fruiting spike	Greenish-black to black in colour, cylindrical, erect, pungent and blunt. Pungent taste with aromatic odour. Spikes are of 1.0–1.9 cm long and 0.2–0.3 cm diameter	Not less than 0.4% w/w of piperine on a dry weight basis
Punarnava (*Boerhavia diffusa*)	Dried root	Colour: Greyish-brown; short and fibrous fracture; no distinct odour; taste: Bitter	Boeravinone B not less than 0.005% (calculated on a dried basis)
Senna leaf	Dried compound leaf of *Cassia angustifolia* or *Cassia senna* Vhal	Pale yellowish-green colour leaflets with mucilaginous and faint odour	Not less than 1.0% w/w of sennosides A and B, calculated on a dried basis.
Senna pod	Fruit of *Cassia angustifolia* or *Cassia senna* Vhal	Pale yellowish colour with a slight odour	Senna pods contain not less than 1.0% w/w of sennosides A and B, calculated on a dried basis
Sarpagandha (*Rauvolfia serpentina*)	Root	Bitter taste; odour indistinct	Not less than 0.15% reserpine and ajmalicine on a dry weight basis
Satavari (*Asparagus racemosus*)	Tuberous root	Dirty white colour, longitudinally wrinkled with yellow hard central colour, starchy and slightly bitter followed by sweet taste	Not less than 0.01% shatavarin IV on a dry weight basis
Tulsi (*Ocimum sanctum*)	Leaf	Greyish-black with a characteristic odour, slightly pungent and aromatic taste	Not less than 0.40% eugenol on a dry weight basis
Vasaka (*Adhatoda vasica*)	Dried mature leaf	Taste: Bitter	Vasaka contains not less than 0.6% of vasicine (calculated on a dried basis)
Yasti (*Glycyrrhiza glabra*)	Dried unpeeled root	Odour characteristics and slightly aromatic; taste: Very sweet and faintly astringent. The bark is not bitter	Yasti contains not less than 3.0% of glycyrrhizic acid

crops, cultivation of MAPs was started with the direct introduction of the species from the wild habitat as domestication to agriculture. Hence, the collection of maximum variability present in any MAP species is the first and the most important step to start with any crop improvement programme. An ideal germplasm collection of a crop should represent all the available variability present in that species and such collection is termed a gene bank. Each entry in the gene bank should be different from the other. Even though, there are two objectives for the collection of any germplasm, namely collection for specific traits (which are mainly going on in the major crops where maximum variability present in the gene bank is already utilized) and collection of maximum variability from the natural habitats of a species (This is going on in MAPs, since most of the species are new to cultivation and related research). One of the difficult tasks in the collection of MAP germplasm is defining the appropriate sampling strategy for a particular species since the breeding behaviour of most of the species is not studied well. This step is very crucial since this important task is considered the interface between the genetic diversity that was created by nature and what is going to be conserved in the gene bank for immediate or future use (Namkoong 1988). A general benchmark for sampling is judicious division of the target area into different eco-geographic sub-division and collecting an optimum number of about 50 random individuals from populations in each sub-division (Allard 1970; Brown and Marshall 1975; Krusche and Geburek 1991). In the case of cross-pollinating species, the number of populations in an eco-geographic sub-division can be reduced and the number of individuals per site/population can be increased since we can expect maximum variability within a population in cross-pollinated species. However, in the case of self-pollinating species, a large number of populations from each eco-geographic sub-division have to be targeted for germplasm collection since variability will be less within a population and to represent maximum variability, one has to survey more populations (Brown and Marshall 1995).

Maintenance of the germplasm of different MAP species requires a considerable amount of resources in terms of space, money and human power. In recent times, a large number of institutes, universities and colleges, are collecting germplasm for their short-term programmes, and after the study period, they are not making any effort to maintain this valuable germplasm. In India, the NBPGR is the nodal agency that collects and maintains large collections of crops including MAPs under the aegis of ICAR. The NBPGR also provides germplasm of different MAP species for research and related purposes. Hence, it is advisable for the researchers of MAPs to send the collected germplasm once their projects are over for its maintenance in the national facility of NBPGR for long-term conservation.

28.7 Characterization and Evaluation

The value and potential of the germplasm are known and can be utilized once it is characterized and evaluated. Characterization and evaluation of genetic variability present in MAP species is being taken by different research organizations for utilization of the useful variability efficiently.

For giloe (*Tinospora cordifolia*), studies were conducted to evaluate the variability present in available germplasms collected from different parts of India based on morphological characteristics. Molecular markers and chemical analysis reported considerable genetic variability in the studied traits especially in leaf shape, size, leaf lamina base and chemical contents, which can be naturally expected in a cross-pollinating dioecious species (Ahmed et al. 2006; Rout 2006; Ishnava and Mohan 2009; Rana et al. 2012; Paliwal et al. 2013; Sharma and Geetha 2014; Singh et al. 2014; Bajpai et al. 2016; Haque et al. 2017; Gargi et al. 2017; Lade et al. 2018, 2020; Nazneen et al. 2019).

In Safed musli (*Chlorophytum borivilianum),* germplasm collections made from different natural habitats of India showed the presence of wide genetic variability in terms of crop maturity, plant type, plant growth characters, size and shape of fleshy roots and yield traits (Jat and Sharma 1996; Aundhe and Deokule 2001; Kothari and Singh 2001; Geetha and Maiti 2001; Bhagat and Jadeja 2003; Maiti and Geetha 2005; Kumar et al. 2008; Singh et al. 2008).

Similarly, in kalmegh (*Andrographis paniculata*), isabgol (*Plantago ovata*), ashwagandha (*Withania somnifera*), sarpagandha (*Rauvolfia serpentine*), guggul (*Commiphora wightii*), opium poppy (*Papaver somniferum*), senna (*Cassia angustifolia*), lemongrass (*Cymbopogon flexuosus*), etc., a wide spectrum of variability was reported by different researchers, which has to be utilized for initiating the breeding programme in these species.

28.8 Reproduction and Pollination Control

As in any other crop, in MAPs also, the knowledge of the reproductive system, i.e. sexually propagated or asexually propagated, is the foremost requirement for handling the breeding programme. In asexual reproduction, the plant is regenerated through vegetative parts (vegetative reproduction) or from the unfertilized gametes, ovular tissues etc. (apomixis).

- **Vegetative Reproduction**
  In vegetative reproduction, plants are regenerated through aerial parts, namely stem (*Tinospora cordifolia*, *Commiphora wightii*) or through underground stems such as bulbs (tuberose), suckers (*Aloe*), tubers, runners

(*Centella*), rhizomes (ginger, *Acorus calamus*), corms (*Typhonium*). Apomixis is another way of asexual reproduction (*Commiphora wightii* and *Garcinia cambogia*). However, in some species, there may be both ways of reproduction. Crops like aloe, safed musli, giloe, lemongrass, palmarosa, madhunashini and guggul are mainly multiplied by vegetative methods, and seeds are also produced, which can be used for regeneration. From breeders' points of view, this property is an added advantage of the species that can be utilized for crop improvement programmes as well as maintaining the variability created by different breeding programmes.

- **Sexual Reproduction**
  In sexual reproduction, the crops can be grouped into (a) self-pollinated crops, (b) cross-pollinated crops and (c) often cross-pollinated crops based on pollination nature. In self-pollinated crops, a limited degree of cross-pollination of up to 5% may occur. Similarly, in cross-pollinated species, a small amount of self-pollination (5–10%) may occur. In often cross-pollinated crops, cross-pollination may range from 5 to 30%.

Aparajita (*Clitoria ternatea*), ashwagandha (*Withania somnifera*), kalmegh (*Andrographis paniculata*), kaucha (*Mucuna pruriens*) and salaparni (*Desmodium gangeticum*), are some examples for self-pollinated MAP species; aloe (*Aloe barbadensis*), asalio (*Lepidium sativum*), basil (*Ocimum basilicum*), chirayita (*Swertia chirayita*), giloe (*Tinospora cordifolia*), sarpagandha (*Rauvolfia serpentine*), guggal (*Commiphora wightii*), lemongrass (*Cymbopogon flexuosus*), palmarosa (*Cymbopogon martinii*), safed musli (*Chlorophytum borivilianum*), satavari (*Asparagus racemosus*), senna (*Cassia angustifolia* and *C. acutifolia*), shankupushpi (*Convolvulus microphyllus*), valerian (*Valeriana jatamansi*) and tulsi (*Ocimum sanctum*) are some of the important cross-pollinated species; while isabgol (*Plantago ovata*) is an often cross-pollinated species.

However, in MAPs, information on the pollination behaviour is lacking in most of the species since they are recently domesticated. Considering the vast number of MAP species, generating detailed data on these aspects is time consuming, and work is under progress at different research institutes. Knowledge of floral biology is very essential to find out the pollination behaviour based on which different breeding programmes have to be formulated for each species including hybridization programmes. Floral biology or reproductive biology includes information on floral structures, particularly about androecium and gynoecium, flower opening pattern, time of flower opening, duration of flower opening, maturation schedule of androecium and gynoecium including stigma receptivity and its duration, anther dehiscence and pollen dispersal.

The determination of the mode of pollination and the extent of cross-pollination in MAPs are neglected fields where research has to be focussed. The first step to finding out the mode of pollination is to critically study the structure of flowers. If the flowers have dioecy, protogyny, protandry, cleistogamy, self-incompatibility, male sterility, etc. the species is adapted for cross-pollination (Fig. 28.2). Guggal (*Commiphora wightii*), *Valeriana jatamansi*, betelvine and giloe are some of the examples of dioecious species. Protogyny is reported from isabgol. In aloe (*Aloe barbadensis*), the flowers are protandrous. Male sterility is reported from isabgol.

If individual plants fail to set seed in isolation, it indicates cross-pollination. Cross-pollination can further be confirmed by the reduced vigour of repeated selfing. If individual plants in isolation set seeds, it is an indication of self-pollination or there may be chances of apomixis. Apomixis can be ruled out by histological studies or by crossing a recessive genotype to a dominant genotype and studying the progenies. The appearance of high-frequency recessive offspring in the progenies proves apomixes. Normally apomixes are associated with polyembryony also (more than one seedling per seed).

The extent of cross-pollination in a species is worked out by growing a homozygous recessive genotype surrounded by its corresponding dominant genotype. After allowing natural pollination, seeds will be collected from the recessive genotype, and the number of progenies showing the dominant character in the next generation can be used for calculating the percentage of natural cross-pollination in the species. In most MAPs, information on the percentage of cross-pollination is lacking except in isabgol, periwinkle, opium poppy etc. Information on reproductive biology is available in detail or partly in some species, namely isabgol, safed musli, glory lily, gentian, Ashoka, palas, opium poppy, khasi kateri; psoralea, periwinkle and lemongrass (Agrawal and Tayyab 1958; Murthy and Abraham 1975; Patel et al. 1980; Pawar 1981; Handique 1986; Sridhar 1986; Chadha and Gupta 1995; Geetha and Maiti 2002; Tandon et al. 2003; Smitha and Thondaiman 2016).

28.9 Selection

Selection is the most important activity in any crop improvement programme in MAPs. It is the widely used method for the identification of superior cultivars since commercial cultivation of MAPs is comparatively a new phenomenon. Hence, selection itself will provide unlimited opportunities to identify superior genotypes of breeders' choice. In most MAPs, even the available variability has not been fully explored from the wild habitats. Most high-yielding MAP varieties are selected from the available variability present in the germplasm.

Fig. 28.2 Plants have different floral adaptations for pollination control like male sterility favouring cross-pollination. (**a**) Normal spike of isabgol, (**b**) male sterile spike of isabgol showing shrivelled male sterile anthers

There are various selection methods, such as mass, progeny, pedigree and pure line which have to be adopted in each species based on the mode of pollination, the predominance of gene action and the breeding objective. The efficiency of selection depends upon the extent of genetic variability present in the germplasm and heritability (phenotypic variation due to genotypic variability) of the targeted trait. In MAPs, as mentioned earlier, quality yield (higher harvest index along with higher active ingredient content) is more important. For example, in the case of kalmegh, genotypes have to be selected based on higher dry herbage yield and andrographolide content. Hence, the higher harvest index of total herbage and higher andrographolide content are to be targeted in kalmegh. Both traits are highly influenced by many components. In addition, due to low heritability, the direct selection for these characters is not effective.

Biometric techniques, namely correlation coefficient, path coefficient analysis and discriminate function analysis (Wright 1921; Dewey and Lu 1959) are adopted for determining the extent of association between yield and yield-attributing traits and to collect information about the inter-relationship of different characters with yield and quality traits. The correlation coefficient offers basic information about the association between pairs of characters. However, it cannot be used to illustrate the relative importance of the direct and indirect effects of each of these component traits towards the final trait (yield and quality). The path coefficient analysis, designed by Wright (1921), takes into account the cause-and-effect relation between the variables by partitioning the association into direct and indirect effects through other independent variables. The path coefficient analysis also measures the comparative significance of the causal factors involved. This is simply a standardized partial regression analysis, wherein the total correlation value is subdivided into the causal scheme.

In ashwagandha, (*Withania somnifera*), Kubsad et al. (2009) reported a highly significant and positive correlation between harvest index, plant height and dry matter per plant with dry root yield.

Khatak et al. (2013) reported that root yield had a highly significant and positive direct correlation with plant height and root length. Sangwan et al. (2013) studied path analysis, which revealed a positive direct effect of total alkaloid content on fresh root yield per plant followed by biomass yield at maturity, seed yield per plant, root diameter and number of berries per plant. Gami et al. (2016) reported a positive correlation between dry root weight per plant and stem diameter, main root diameter, plant height, days to maturity, main root length and days to flowering, suggesting these traits were major influencing factors for increasing the dry root yield in ashwagandha. Sundesha et al. (2016) revealed dry root yield per plant had a strong positive association with root diameter, root length, plant height, root branches and days to maturity. A positive direct effect of plant height and root branches per plant was observed on dry root yield per plant. Hence, plant height and root branches per plant may be good criteria for increasing dry root yield per plant through simple selection in ashwagandha. Patel and Desai (2017) observed the existence of a positive correlation between plant growth, root yield and quality components. Srivastava et al. (2018) observed the highest positive direct effect of fresh root weight, a total alkaloid (%) in leaves and 12 deoxy withastramonolide (%) in roots on dry root weight.

Similar studies were conducted in a number of MAP species, namely aloe, basil, isabgol, kalmegh, opium poppy, safed musli and sarpagandha for detecting genetic variability, heritability, an association of traits and direct and indirect effect of different morphological traits to the final crop yield as well as quality (Patel and Desai 2017; Disha and Tirkey 2016; Kumar et al. 2014; Misra et al. 2000; Dayana et al. 2018; Ibrahim et al. 2011; Maison et al. 2005; Ozturk and Gunlu 2008; Mishra et al. 2010; Nejatzadeh-Barandozi et al. 2012; Pandey and Mandal 2010; Rajnish et al. 2013; Sharma and Singh 2012; Sharma et al. 2009; Singh et al. 2014, 2003; Usmani et al. 2014). These studies will help the breeder select better genotypes for direct use as high-yielding varieties and define selection criteria for the enhancement of harvest index along with chemical content and for the selection of parents for hybridization programmess, etc.

Considering the pattern of genetic variation for various target traits discussed above, different modes of selection can be adopted for selecting superior genotypes of any MAP species for direct use as a cultivar or for the selection of parents for hybridization programmes (Fig. 28.3). As discussed earlier, selection will be effective if:

- Higher heritability of the trait under selection will yield more stable and superior genotypes.
- Information on the association of independent variables (e.g. number of branches, plant height and leaf size) can be used as reliable selection criteria.
- Since selection creates no new variation, the breeder should have germplasm with sufficient genetic variability.

In self-pollinated species, the common selection methods are mass selection, pure line selection and single plant selection. However, for segregating populations, obtained after hybridization, selection may be done through the pedigree method, bulk method and back cross method. In cross-pollinated species, populations collected from natural habitats include heterozygous lines and upon continuous selfing, they show inbreeding depression (loss of vigour). Hence, in cross-pollinated species, selection methods may be adopted either to avoid or minimize selfing or inbreeding depression. Progenies of each heterozygous line will be heterogeneous (a mixture of heterozygous lines). Hence, one of the important selection criteria will be to accumulate desirable gene or allele combinations in a variety without causing inbreeding depression. To achieve this, mass selection or its modification, progeny selection, recurrent selection, etc. can be adopted (Figs. 28.4 and 28.5).

In the case of asexually propagated plants, a single plant based on the breeding objective, namely higher yield and quality may be selected from the available variability, multiplied and directly used as a variety. The procedure for selection used in asexually propagated crops is known as clonal selection. The method can be effectively utilized in crops, namely *Aloe*, *Mentha*, *Chlorophytum borivilianum*, *Centella asiatica* and *Bacopa monnieri*. If natural variability is lacking in the available germplasm, especially in crops that are not indigenous, namely *Aloe*, *Stevia*, *Geranium* and *Patchouli*, induced mutation followed by clonal selections can be practised. Some of the important varieties identified through selection methods are presented in Table 28.2.

28.10 Hybridization

The main objectives of hybridization are (1) creation of variability, (2) combination breeding for bringing desirable traits (both qualitative and quantitative) into a variety from one or many other varieties or genotypes and (3) developing hybrid varieties and utilization of hybrid vigour. In MAPs, only limited efforts have been made to increase productivity in terms of plant yield and secondary metabolites through hybridization.

'Clocimum' is an interspecific hybrid of *Ocimum sanctum* and *O. gratissimum*, which is a rich source of eugenol (75%). An elite evergreen hybrid has been developed from *Ocimum kilimandscharicum* and *Ocimum basilicum*, which demonstrated adaptive behaviour towards cold stress (Dhawan et al. 2016).

Maximum work on hybridization and related work including interspecific hybridization was conducted in opium poppy (*Papaver somniferum*) (Ojala and Rousi 1986; Singh et al. 1998; Yadav et al. 2009; Kumar and Patra 2010, 2012; Matyasova et al. 2011; Nemeth-Zambori et al. 2011; Yadav and Singh 2011). Varieties developed through hybridization are presented in Table 28.3.

Fig. 28.3 Genetic variability present in the germplasm is very important for developing high-yielding varieties. The figure presents different morpho-types of *Andrographis paniculata* (kalmegh). (**a–f**) Leaf types. (G and H) Branching pattern. (**i–k**) Canopy types. (**l** and **m**) Flowering branches

28.11 Induced Mutation Breeding

Beneficial mutations are artificially induced by physical agents (X-rays, gamma rays, neutrons, beta particles, alpha particles, proton or deuterons, UV rays, etc.) or chemical agents (Alkylating agents: EMS, MMS, nitrogen mustard, etc.; Base analogues: 5BU, 2-AP, TMU, etc.; and Acridine dyes: acraflavin, acridine orange, ethidium bromide, etc.) for creating new variabilities. These are utilized for crop improvement. In MAPs, the use of induced mutations is important in crops where there is a narrow genetic variability, which is commonly found in species that are not native to India and are introduced to India. Isabgol, aloe, liquorice, geranium, artemisia, patchouli, senna and stevia are important cultivated species where there is a narrow genetic base due to their exotic origin.

In certain cases, as in traditional crops, mutations can be induced to improve the existing varieties or genotypes. For example, in isabgol, the fruit (capsule) is the shattering type and if unseasonal rain occurs at the crop maturity (summer rains), almost 90–95% seed shattering occurs. In this case, induced mutagenesis can be utilized for the development of a seed-shattering resistant variety.

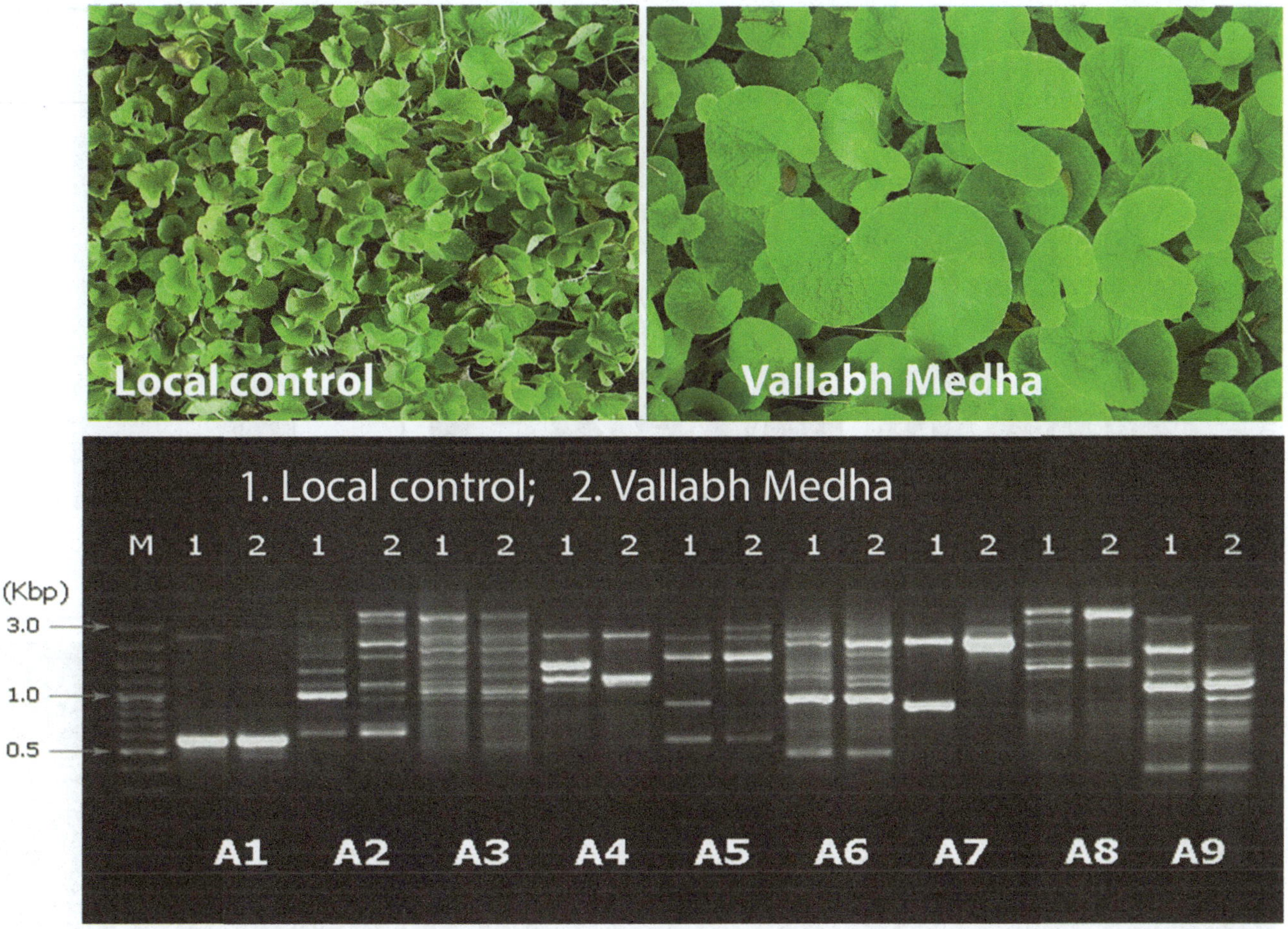

Fig. 28.4 A local and a high-yielding variety (Vallabh Medha) of *Centella asiatica* developed by selection and characterization of the two varieties using RAPD markers (A1–A9: different decamer primers)

In kalmegh (*Andrographis paniculata*), seeds are used for propagation, and since the flowering is not synchronous, fruit (dehiscent capsule), staggered fruit maturity followed by shattering occurs. Hence, also, in this case, mutations can be attempted to develop a synchronous flowering type line.

Similarly, induced mutations can also be attempted in MAPs to enhance the production of a particular active ingredient or aroma of commercial interest. However, for attempting this, information on details of the secondary metabolite pathway should be known so that intervention in terms of mutagen treatment can be attempted at the right metabolic production stage.

Varieties developed through induced mutation breeding are presented in Table 28.4.

28.12 Polyploid Breeding

Polyploid has played a significant role in crop improvement in many crops, especially in crops where plant biomass is important economically. In the case of MAPs, if the increase in ploidy level results in an increased harvest index along with enhanced active ingredient content also, it will be useful. Bach (*Acorus calamus*) is a popular example wherein ploidy is related to the level of phytoconstituent. In Bach, diploid, triploid, tetraploid and hexaploid are reported and β-asarone, a carcinogenic compound, is absent or negligible in diploids and its content varies drastically from 5 to 96% of the essential oil depending upon triploid, tetraploid or hexaploid (Janaki Ammal et al. 1964; Lander and Schreier 1990; Todorova and Ognyanov 1995; Raina et al. 2003; Ogra et al. 2009; Verma et al. 2015). In isabgol, as there is limited scope for natural variability due to a narrow genetic base, induced polyploidy was tried. Tetraploid isabgol was developed in the US for the first time in which seed size and mucilage content were increased; however, there was an associated reduced fertility. Mital et al. (1975) suggested the role of selection for further improvement of the Isabgol polyploid. Induced tetraploidy was also attempted in periwinkle, but due to high level of seed sterility and low seed germination, it could not be used for commercial purposes (Janaki Ammal and Bezbaruah 1963; Dnyansagar and

Fig. 28.5 Varietal development and their characterization in *Andrographis paniculata* (kalmegh). (**a**) Variety DMAPR-AP35 (Vallabh Kalmegh 1); (**b**) Distinct close branching pattern of DMAPR AP35; (**c**) Characterization of two kalmegh varieties using SRAP markers; (**d**) a mature crop of DMAPR AP35

Table 28.2 Superior varieties developed for cultivation through selection

Species	Name of the variety	Characters
Aloe (*Aloe barbadensis*)	Anand Aloe 1	High leaf and gel yield
	CIM Sheel	High leaf and sap yield contains about 40% sap, suited for rain-fed conditions, yield: 40–50 tons of leaves per hectare
Asalio (*Lepidium sativum*)	GA 1	High seed yield
	RVA 1007	Higher seed yield (1127–1827 kg/ha) and higher oil yield (321 kg/ha), resistant to *Alternaria* leaf blight
	HLS 4	Higher seed yield (1244–1725 kg/ha) and higher oil yield (306 kg/ha)
Ashwagandha (*Withania somnifera*)	GAA 1	Higher dry root yield (537–748 kg/ha), resistant to *Alternaria* leaf blight
	JA 20	Higher dry root yield
	JA 134	Higher dry root yield
	RVA 100	Higher dry root yield
	CIM Pratap	Long tap root with less fiber, dry root yield: 34.95 q/ha, withanolide content: 0.31%, withaferin A content: 0.720%
	Poshita	Dry root yield: 14 q/ha, alkaloids: 1.3 kg/ha, withaferin content: 0.538% in dry leaves
Bach (*Acorus calamus*)	CIM Balya	Medium long and broad leaves, light cream-coloured cylindrical rhizomes, rhizome yield: 9–11 t/ha, oil content: 0.60–0.65%
Basil (*O. basilicum*)	CIM Saumya	Short duration, dwarf, early flowering, herb yield: 290 q/ha, oil yield: 197.2 kg/ha, oil content: 0.68%, oil quality: Methyl chavicol 62.54%, linalool 24.61%
Brahmi (*Bacopamonierri*)	CIM Jagriti	Dry herb yield: 40 q/ha, bacoside-A content: 1.94–2.07%
Citronella (*C. winterianus*)	Manjari	Leaf medium broad, leaf sheath purple, herb yield: 145 q/ha, oil yield: 275 kg/ha, citronellal content: 45.07%
Indian valerian (*Valeriana jatamansi*)	Dalhousi clone	3.8% valepotriate
Isabgol (*Plantago ovata*)	GI 1	Higher seed yield
	Hi 5	Profuse tillering, long and compact spikes and moderate resistance to downy mildew
	JI 4	Higher yield
	GI 3	Higher yield
Kalmegh (*Andrographis paniculata*)	Vallabh Kalmegh 1 Anand Kalmegh 1 (AK 1)	Dry herbage yield 3.6 t/ha, high andrographolide content
	CIM Megha	High yield of dry biomass: 32–40 q/h andrographolide content:1.90%, plant height: 58–69 cm, leaf length: 5.0–5.9 cm, leaf width: 1.48–1.64 cm, plant spread: 48.3–50.0 cm, canopy shape: Open, leaf colour: Dark green
Kaucha (*Mucuna pruriens*)	CIM ajar	Trichomeless, early maturing: 120–125 days, L-DOPA content: 6.18%
Khus (*Vetiveria zizaniodes*)	CIMAP Khus 15	Erect growth habit; dry root yield: 9–22 q/ha; oil yield: 35–40 kg/ha; oil content: 2.1%.
Lemon basil (*Ocimum africanum*)	CIM Jyoti	Dwarf; early maturing; white flower; herb yield: 200 q/ha; oil yield: 150 kg/ha; citral content: 68–75%
Lemongrass (*Cymbopogon flexuosus*)	NLG 84	100–110 cm tall, broad leaves, dark purple sheath
	Nima	Dwarf; dark green leaf; leaf sheath dark purple in colour; herb yield: 289 q/ha; oil yield: 261 kg/ha; citral content: 89%
	Chirharit	Frost resistant; herb yield: 256 q/ha; oil yield: 260 kg/ha; citral content: 80%
	CIMAP Sawarna	Very fast growing; suitable for drought prone areas and marginal lands; herb yield: 549 q/ha; oil yield: 208 kg/ha; citral content: 80%
Long pepper (*Piper longum*)	Viswam	2.8% alkaloid content, 25% dry matter content of dry fruits
Mandukaparni (*Centella asiatica*)	Vallabh Medha	Leaf size is 4–5 times bigger, 3 times more herbage and quality yield than the local type
Mint (*Mentha arvensis*)	CIM Kranti	Stem purple green; semierect; 0.57% essential oil; 60–70% carvone; cold tolerant; oil yield: 100 kg/ha (winter) along with suckers (250–300 q/ha); 170–210 kg/ha (summer); menthol content: 80%
	Kosi	Tolerant to leaf spot, rust and powdery-mildew diseases; early maturing: 90–100 days; herb yield: 260 q/ha; oil yield: 200 kg/ha; oil content: 0.75%; menthol content: 75–80%
	MAS 1	Medium tall with broad dark green leaves; highly branched; early maturing; oil content: 0.8%; oil quality: Menthol (84%) and menthone (5.8%)

(continued)

Table 28.2 (continued)

Species	Name of the variety	Characters
Peppermint (*M. piperita*)	CIM Indus	Oil yield: 70–75 kg/ha; oil content: 0.35%; oil quality: Menthofuran (27%) and pulegone (15%)
	CIM Madhuras	Sweet smelling; oil yield: 120–125 kg/ha; oil quality: Limonene 3.5%, menthone 24.3%, menthol 31.2%, menthofuran 6%
Spearmint (*M. spicata*)	Punjab Spearmint 1	Stem purple green; semi-erect; 0.57% essential oil; 60–70% carvone
	Arka	Broad leaves; erect plant habit; early flowering and early maturing; suitable for flavour industry; oil yield: 139 kg/ha; oil content: 0.60%; carvone content: 68%
M. spicata var. viridis	Ganga	Oil has insecticidal properties against stored grain pests, anti-microbial and potato sprouting inhibition properties, oil contains piperitenone oxide in higher concentrations (73.13%)
Mentha x gracilis cv. cardiaca	MCAS 2	Suitable for flavour industry; sweet smelling; herb yield: 482 q/ha; oil yield: 225 kg/ha; carvone content: 64%
Opium poppy (*Papaver somniferum*)	Jawahar Aphim 16	12.5% morphine content; white flowered with deeply serrated petals; 2–3 capsules/plant; latex yield 60–65 kg/ha
	Kirtiman	White-flowered; serrated broad leaf; flat capsule. Latex yield 45–50 kg/ha
	Jawahar opium 539	Deeply serrated leaf with two capsules/plant; latex yield 65 kg/ha; 14.85% morphine content
	Jawahar opium 540	Serrated leaf; latex yield 75 kg/ha; 13% morphine
	Chetak Aphim	Leaves large sized round and smooth; capsule12.6% morphine content
	Subhra	Medium-sized capsules; straw yield: 8 q/ha; morphine content: 0.86% in straw
	Shweta	Very big and bold capsules; seed yield: 8 q/ha; latex yield: 80–90 kg/ha; morphine content: 22% in latex
	Sapna	Tall variety with fringed leaves; white peduncle with flowers; 66.5 kg/ha opium yield and about 18% morphine content in latex
Palmarosa (*Cymbopogon martinii*)	Rosh grass 49	Tall, long and dense inflorescence, long and broad leaves with thick stems, 90% geraniol
	CIMAP Harsha	Medium tillering capacity, 1–1.5% essential content Drought resistant, herb yield: 450 q/ha, oil yield: 300 kg/ha, geraniol content: 94%
Patchouli (*Pogostemon cablin*)	CIM Samarth	Faster regeneration capabilities, performs well both under open and shaded conditions, fresh herb yield: 21.6 t/ha, oil yield: 80.7 kg/ha, oil content: 2.55%
Periwinkle (*Catharanthus rosea*)	Prabhat	High yield
	Prabhal	Winter hardy, leaf yield: 1.1–2.6 t/ha, total alkaloids content: 1.65–1.95%, root yield: 0.8–1.0 t/ha, total root alkaloids content: 1.5–1.85%
Rose (*Rosa damascene*)	Noorjahan	Suitable for north Indian plains and hills, flower yield: 35 q/ha, oil quality: Citronellol 20.8%, geraniol 25.3%, a-damascenone 0.1%
Rosemary (*Rosemarinus officinalis*)	CIM Hariyali	Essential oil is rich in camphor and 1-cineol, herb yield: 350 q/ha, oil content: 1.03%. Oil yield: 350 kg/ha
Sarpagandha (*Rauvolfia serpentina*)	TFRI-RS1	1.4–1.7% alkaloid content, medium plant height 60–80 cm, profuse branching
	TFRI-RS2	High reserpine and alkaloid content, more primary root, longer root
Salvia sclarea	CIM Chandni	Fresh flower spike yield: 170–175 q/ha, oil yield: 19–21 kg/ha, oil content: 0.12–0.14%, oil quality: Linalool 26.19%, linalyl acetate 52.41%
Safed musli (*Chlorophytum borivilianum*)	Jawahar safed musli 405	Higher fleshy root yield
	Anand safed musli −1 (ASM-1)	Higher root yield with 0.62% sapogenin content
	RVSM-414	Anther colour greenish-yellow with higher root yield
	CIM Oj	Tolerant to leaf blight in field conditions, root yield: 8 q/ha
Senna (*Cassia angustifolia*)	Anand late selection	High leaf yield
	Sona	Dry leaf yield: 11 q/ha, seed yield: 4 q/ha, sennoside content:3.51%
Shatavari (*Asparagus racemosus*)	CIM Shakti	Dry root yield: 8–10 t/ha, saponin content: 21%
Sylibum marianum	CIM liv	More number of capitulum, seed yield: 9 q/ha, silymarin content: 2.8%
Tagetes minuta	Vanphool	Oil yield: 61 kg/ha, dihydrotagetone: 32%, (Z)-tagetone: 16.7%
Tusli (*O. sanctum*)	CIM Angana	Tall, leaves light to dark purple in colour, herb yield: 181 q/ha, dry leaf yield: 14.33 q/ha, oil yield: 91.71 kg/ha, oil quality: Eugenol 40.42%, β-elemene 14.11%, β-caryophyllene 9.07%, Germacrene-D 16.65%

Table 28.3 High-yielding varieties developed by hybridization

Species	Name of the variety	Breeding method	Characters
Opium poppy (*Papaver somniferum*)	CIM Ajay	Developed through hybridization followed by selection	White peduncle, seed yield: 9.5 q/ha, straw yield: 8.5 q/ha, morphine content: 1.036% in straw
	Rakshit	Developed through selection in intra-specific hybrids for diverse resistance	Highly resistant to downy mildew disease caused by *Peronospora arborescens*, seed yield: 12–14 q/ha, straw yield: 9–11 q/ha, morphine content: 0.5–0.87% in straw
	Madakini	Hybridization followed by selection	High opium yield (40–42 kg/ha, seed yield (up to 12 q/ha), resistant to diseases
M. spicata	Neerkalka	Developed through hybridization between *Mentha arvensis* and *Mentha spicata*	Presence of underground suckers for vegetative reproduction, unique odour and flavour, herb yield: 128 q/ha, oil yield: 270 kg/ha, oil quality: Carvone (47%), limonene (35%) and menthol (0.6%)
Basil (*O. basilicum*)	Vikarsudha	Developed through intra-specific hybridization	Longs late maturity, herb yield:335 q/ha, oil yield: 167.50 kg/ha, oil content: 0.5%, oil quality: Methyl chavicol 78%, linalool 0.16%

Table 28.4 MAP varieties developed through mutation breeding

Species	Variety name	Traits developed
Periwinkle (*Catharanthus rosea*)	Dhawal	The high alkaloid-producing variety produced by induced chemical mutation followed by rigorous selection in the cultivated variety 'Nirmal'
Opium poppy (*Papaver somniferum*)	Sujata	Non-narcotic (opium-less seed cultivar)
	Vivek	Tall variety with white petals, big sized capsule, seed yield of 9 q/ha, straw yield of 8 q/ha with morphine content of 0.9% in straw
	CIM Sampada	Black peduncle, with seed yield of 7–8 q/ha, latex yield of 78–80 kg/ha and morphine content of 18% in latex
Henbane (*Hyoscyamus* niger L.)	Aela	Vigorous growth with a yield of 73 q/ha, with an alkaloid level of 0.55%, yellow flowers with a slight purple tinge at the base of the petals
Coleus (*Coleus forskohlii*)	Suphala	High yielding (15.93 t/ha) and can be cultivated throughout the year
Khasi kateri (*Solanum viarum*)	Glaxo	Almost spineless
	BARC	Less and curved spines
	Arka Sanjeevani	Spineless
	Arka Mahima	Tetraploid
Isabgol (*Planatgo ovata*)	GI-2	Seed yield 13 q/ha
	Niharika	Seed yield 10–11 q/ha
	Mayuri	Seed yield 11 q/ha
	Vallabh Isabgol-1	Seed yield (923–1040 kg/ha), mucilage yield (9.21 g/kg seed
	Vallabh Isabgol-2	Higher seed yield (916–1360 kg/ha) and early maturity (100–105 days)
Bhui amla (*Phyllanthus niruri*)	CIM Jeevan	0.75–0.77% phyllanthin content
Sarpagandha (*Raovolfia serpentina*)	CIM Sheel	Dry root yield 25–30 q/ha and reserpine content 0.030–0.035% in dry root
Chamomilla recutita	CIM Sammohak	Dry flower yield 7.53 q/ha, oil yield 6.63 kg/ha, chamazulene content: 12.98%
Cymbopogon winterianus	CIM Jeeva	Herb yield 215 q/ha, oil yield 285 kg/ha
Mentha citrata	Kiran	Herb yield 508 q/ha, oil content 0.5%
Mentha piperita	Pranjal	Oil yield 90 kg/ha
	Tushar	Oil quality: Menthol (33.3%), menthone (27.3%)
	Kukrail	Oil yield 90 kg/ha
Mentha spicata	Neera	Herb yield 80 q/ha and oil content 0.40% content0.5%)
Mentha x gracilis cv. cardiaca	Pratik	Vigorous growth and larger leaves, suitable for flavour industry, herb yield of 212 q/ha/harvest, oil yield 81 kg/ha/harvest, oil content 0.4%, carvone content 66%, limonene content 21%

Sudhakaran 1970; Krishnan 1995; Dhawan and Lavania 1996). In safed musli also the indication of the presence of polyploid series (Geetha and Maiti 2004) has to be exploited further. Even though they do not perform superior to the diploids, attempts were made in sarpagandha also to develop polyploids with superior alkaloid content (Mittal et al. 1979; Parimoo 1974; Guniyal et al. 1988). In Henbane, a tetraploid (1 + MT-1) was developed by the Central Institute of Medicinal and Aromatic Plants (CIMAP). *Solanum viarum*, an autotetraploid developed by the Indian Institute of Horticultural Research (IIHR), Bengaluru, was released as a variety (Arka mahima) for its commercial cultivation.

28.13 Conclusion and Future Prospects

Among the 7000–8000 medicinal plants reported from India, about 1178 species are estimated to be in commercial trade. In the Indian market, about 90–95% of the medicinal plants traded are obtained from the wild. The cultivation of more medicinal plants has thus become a top priority in the agenda of commercialization. Organized cultivation is practised only in 65–70 medicinal plants and superior-performing varieties are lacking in most cultivated medicinal species. However, utmost care should be taken to prioritize the number of species to be brought under cultivation. Profitability vis-a-vis volume of the raw drug material required should be some of the important factors for selection of species for cultivation. A wide spectrum of variability was reported by different researchers in most MAP species distributed in India, which have not yet been utilized fully. Hence, crop improvement programmes are mostly led by classical breeding methods to harness this variability. The use of biotechnological tools for revealing the biosynthetic pathway for understanding the genes responsible for control of the secondary metabolite pathway is also a fast-growing area of research in MAPs. Understanding of genes responsible for different active ingredients of interest can be exploited by linking this information with the genetic variability which will cut short the time frame required for conventional breeding programmes in MAPs.

However, the problems faced in MAP breeding are as follows:

1. The number of MAP species is huge, and trained manpower in the sector is limited.
2. The same species is used for different therapeutic purposes.
3. A single species will have multiple active ingredients, and targeting higher multi-compounds with a high harvest index is highly complicated.
4. Information is lacking in most species regarding breeding behaviour, a spectrum of genetic variability, the inheritance pattern of traits and their association with an active ingredient.
5. Information on the active ingredient responsible for therapeutic action and its inheritance pattern is lacking.
6. Lack of efficient phytochemical methods to screen the active ingredient of therapeutic interest at early generation.
7. Duplication of research by different national institutes, universities and state-level agencies.
8. Lack of communication among the MAP researchers, industries and farmers.
9. Limited funding support for research and absence of coordination among the funding agencies.

Future research thrust should therefore be addressed by taking up on priority some of the problems mentioned above in a phased manner as follows:

1. National policy for prioritization of species to be attended in a phased manner.
2. MAP germplasm collection and conservation from natural habitats should be monitored under one national organization.
3. Undertaking basic research on MAPs, namely floral biology, breeding behaviour and synthesis pathway of active compounds.
4. Efficient analytical facilities for phytochemicals for identification of active ingredients and their screening in the breeding material.
5. Direct linkages of farmers and industry with the MAP research groups.
6. Clinical study for prioritizing or limiting the targeted active ingredient/s in the breeding material.
7. Utilization of biotechnological tools for utilizing genetic variability in a faster way.
8. Compulsory certification of raw drugs so that the quality of raw drugs can be assured, which would improve the profitability of MAP cultivation.

Lessons Learnt

- MAP biodiversity utilization in India is massive and mainly from nature. Hence, to relieve the pressure on nature there is a need to bring more species under cultivation for sustainable supply to various industries.

- Conservation of MAPs by employing in situ and ex situ methods must be in place to ensure sustainability.
- Promoting cultivation of MAPs needs a steady supply of high-yielding cultivars in terms of active ingredients and also biomass per unit area.
- Breeding of MAPs is more challenging and complicated compared to crops because secondary metabolites are highly influenced by the environment in addition to the diversified requirements of utility industries.

Key Questions

1. What are the different conservation steps for the sustainability of MAPs?
2. What is the role of cultivation in MAPs?
3. What are the major breeding objectives in MAP improvement? In what way breeding objectives in MAPs are different from traditional crops?
4. What are the different methods employed for the breeding of MAPs?
5. What should be future research thrust in MAP breeding?

References

Agrawal JS, Tayyab MA (1958) Chromosome number in *Psoralea corylifolia* Linn. (Papilionaceae). Sci Cult 23(12):652–653

Ahmed SM, Verma V, Qazi PH, Ganaie MM, Bakshi SK, Quazi GN (2006) Molecular phylogeny in Indian *Tinospora* species by DNA based molecular markers. Plant Syst Evol 256:75–87

Allard RW (1970) Population structure and sampling methods. In: Frankel OH, Bennet E (eds) Genetic resources in plants- their exploration and conservation. Blackwell Scientific Publications, Oxford, pp 97–107

Aundhe DJ, Deokule SS (2001) Studies on root morphology of *Chlorophytum* Ker-Gawl. J Phyto Res 14:195–197. https://scholar.google.com/scholar_lookup?title=Studies+on+root+morphology+of+Chlorophytum+Ker-Gawl&publication+year=2001&author=Aundhe+DJ&author=Deokule+SS&journal=Journal+of+Phytological+Research&volume=14&pages=195-197

Bajpai V, Kumar S, Singh A, Singh J, Negi MPS, Bag SK, Kumar N, Konwar R, Kumar B (2016) Chemometric based identification and validation of specific chemical markers for geographical, seasonal and gender variations in *Tinospora cordifolia* stem using HPLC-ESI-QTOF-MS analysis. Phytochem Anal 28:277–288. https://doi.org/10.1002/pca.2673

Bennett RN, Wallsgrove RM (1994) Secondary metabolites in plant defence mechanisms. New Phytol 127:617–633

Bhagat C, Jadeja GC (2003) Variation and correlation in root yield and biochemical traits of Safed musli (*Chlorophytum borivilianum*). J Med Arom Plant Sci 25:33–36. https://scholar.google.com/scholar_lookup?title=Variation+and+correlation+in+root+yield+and+biochemical+traits+of+Safedmusli+(Chlorophytum+borivilianum)&publication+year=2003&author=Bhagat+C&author=Jadeja+GC&journal=Journal+of+Medicinal+and+Aromatic+Plant+Sciences&volume=25&pages=33-36

Brown AHD, Marshall DR (1975) Optimum sampling strategies in genetic resources. In: Frankel OH, Hawkes JG (eds) Crop genetic resources for today and tomorrow. Cambridge University Press, Cambridge, pp 53–80

Brown AHD, Marshall DR (1995) A sampling strategy: theory and practice. In: Guarino L, Rao VR, Reid R (eds) Collecting plant genetic diversity: technical guidelines. International Plant Genetic Resources Institute, Rome, pp 75–91

Chadha KL, Gupta R (1995) Advances in horticulture—medicinal and aromatic plants. Malhotra Publishing House, New Delhi

Dayana SN, Himabindu K, Suryanarayan MA, Umeha K, Rao VK, Shivapriya M (2018) Correlation and path analysis in kalmegh. J Pharmacog Phytochem 3:504–507

Dewey DR, Lu KH (1959) A correlation and path coefficient analysis of components of crested wheat grass and seed production. Agron J 51: 515–517

Dhawan OP, Lavania UC (1996) Enhancing the productivity of secondary metabolites via induced polyploidy: a review. Euphytica 87:81–89

Dhawan SS, Shukla P, Gupta P, Lal RK (2016) A cold-tolerant evergreen interspecific hybrid of *Ocimum kilimandscharicum* and *Ocimum basilicum*: analyzing trichomes and molecular variations. Protoplasma 253:845–855

Disha N, Tirkey A (2016) Studies on genetic diversity in various quantitative characters in kalmegh (*Andrographis paniculata*) germplasm. Adv Res J Crop Improv 7(1):60–64

Dnyansagar VR, Sudhakaran IV (1970) Induced tetraploidy in *Vinca rosea* Linn. Cytologia 35:227–241

Edreva A, Velikova V, Tsonev T, Dagnon S, Gürel A, Aktaş L, Gesheva E (2008) Stress-protective role of secondary metabolites: diversity of functions and mechanisms. Gen Appl Plant Physiol 34:67–78

Gami RA, Solanki SD, Patel MP, Tiwari K, Bhadauria HS, Kumar M (2016) Correlation study in ashwagandha [*Withania somnifera* (L.) Dunal] and identify better genotypes for North Gujarat. Adv Life Sci 5(7):2844–2848

Gargi M, Thakur S, Anand SS, Choudhary S, Bhardwaj P (2017) Development and characterization of genomic microsatellite markers in *Tinospora cordifolia*. J Genet 96:25–30

Gautam RJ, Mishra M, Prasad R (2020) Working paper on development and trade of medicinal and aromatic plants (MAPs): learning from comparative analysis of MAPs export of India and China. Rajiv Gandhi Institute for Contemporary Studies, New Delhi

Geetha KA, Maiti S (2001) Reproductive biology of *Chlorophytum borivilianum*. J Trop Med Plant 2:287–290

Geetha KA, Maiti S (2002) Biodiversity in *Chlorophytum borivilianum* Santapau and Fernandes. Plant Genet Resour Newsl 129:52–53

Geetha KA, Maiti S (2004) New basic chromosome number in *Chlorophytum borivilianum* Santapau and Fernandes. J Spices Aromat Crops 13:121–123

Goraya GS, Ved DK (2017) Medicinal plants in India: an assessment of their demand and supply. National Medicinal Plants Board, Ministry of AYUSH, Government of India, New Delhi and Indian Council of Forestry Research & Education, Dehradun

Guniyal AK, Kapoor A, Virmani OP (1988) Rauvolfia serpentina. CROMPA 10(3):113–137

Gutbrodt B, Dorn S, Unsicker SB, Mody K (2012) Species-specific responses of herbivores to within-plant and environmentally mediated between-plant variability in plant chemistry. Chemoecology 22:101–111

Handique AK (1986) Breeding behaviour of *Solanum khasianum* Clarke. Euphytica 35:631–632

Haque MA, Jantana I, Bukhari ANA (2017) *Tinospora* species: an overview of their modulating effects on the immune system. J Ethnopharmacol 207:67–85. https://pib.gov.in/PressReleasePage.aspx?PRID=1594929

Ibrahim MM, Aboud KA, Hussein RM (2011) Genetic variability and path coefficient analysis in sweet basil for oil yield and its components under organic agriculture conditions. Am J Sci 7(6): 150–157

Indian Pharmacopoeia (2007) The Indian Pharmacopoeia Commission, vol 3, Ghaziabad, pp 2463–2552
Ishnava K, Mohan JSS (2009) Assessment of genetic diversity in medicinal climber of *Tinospora cordifolia* (Willd.) Miers. (Menispermaceae) collected from Gujarat, India. Asian. J Biotechnol 1(3):93–103
Janaki Ammal EK, Bezbaruah HP (1963) Induced teteraploidy in *Catharanthus roseus* (L.) Don. Proc Natl Acad Sci India Sect B 57:339–342
Janaki Ammal EK, Sobti SN, Handa KL (1964) The interrelationship between polyploidy, altitude and chemical composition in *Acorus calamus*. Curr Sci 33:500
Jat RD, Sharma GS (1996) Genetic variability and correlation in Safed musli (*Chlorophytum borivilianum*). J Med Arom Plant Sci 18:496–498. https://scholar.google.com/scholar_lookup?title=Genetic+variability+and+correlation+in+Safed+musli+(Chlorophytum+borivilianum)&publication+year=1996&author=Jat+RD&author=Sharma+GS&journal=Journal+of+Medicinal+and+Aromatic+Plant+Sciences&volume=18&pages=496-498
Kapoor LD, Handa KL, Chandra I, Abrol BK (1955) Cultivation of *Mentha arvensis* in Jammu and Kashmir. J Sci Ind Res 14A:374–378
Khatak S, Dhillon S, Yadav OP, Grewal A, Sheokand RN (2013) Morphological and rapid marker-based characterization of genetic diversity in different genotypes of *Withania somnifera* L. Dunal. Int J Adv Biotechnol Res 3(4):1–16
Kothari SK, Singh K (2001) Evaluation of Safed musli (*Chlorophytum borivilianum* Santapau and Fernandes) germplasm. J Spices Aromat Crops 10: 147–149. https://scholar.google.com/scholar_lookup?title=Evaluation+of+Safed+musli+(Chlorophytum+borivilianum+Santapau+and+Fernandes)+germplasm&publication+year=2001&author=Kothari+SK&author=Singh+K&journal=Journal+of+Spices+and+Aromatic+Crops&volume=10&pages=147-149
Krishnan R (1995) Periwinkle. In: Chadha KL, Gupta R (eds) Advances in horticulture, vol 11. Malhotra Publishing House, New Delhi, pp 409–428
Krusche D, Geburek T (1991) Conservation of forest gene resources as related to sample size. Forest Ecol Manag 40:145–150
Kubsad VS, Palled YB, Mansur CP, Alagundagi SC (2009) Correlation and path coefficient analysis in ashwagandha [*Withania somnifera* L. (Dunal)]. Madras Agric J 96(7–12):314–315
Kumar B, Patra NK (2010) Genetic analysis of capsule and its associated traits in opium poppy (*Papaver somniferum* L.). J Hered 101(5):657–660
Kumar B, Patra NK (2012) Inheritance pattern and genetics of yield and components traits in opium poppy (*Papaver somniferum* L.). Ind Crop Prod 36(1):445–448
Kumar B, Verma AK, Singh HP, Mishra HO, Lal RK (2008) Genetic divergence and correlations study in *Chlorophytum borivilianum*. J New Seeds 9:321–329
Kumar A, Semwal DP, Bhatt KC, Raina A (2014) Characterization of indigenous germplasm of *Andrographis paniculata* (Burm. f.) Wall. ex Nees for variability analysis. Intl J Phytomed Rel Ind 6(4): 277–281
Lade S, Sikarwar PS, Ansari MA, Khatoon S, Kumar N, Yadav HK, Ranade SA (2018) Diversity in a widely distributed dioecious medicinal plant, *Tinospora cordifolia* (Willd.) Miers ex Hook f and Thomas. Curr Sci 114(7):1520–1526
Lade S, Pande V, Rana TS, Yadav HK (2020) Estimation of genetic diversity and population structure in *Tinospora cordifolia* using SSR markers. J Biotech 10:310. https://doi.org/10.1007/s13205-020-02300-7
Lander V, Schreier P (1990) Acorenone and γ-asarone: indicators of the origin of calamus oils (*Acorus calamus* L.). Flav Fragr J 5:75–79
Maison T, Volkaert H, Boonprakob U, Paisooksantivatana Y (2005) Genetic diversity of *Andrographis paniculate* wall. ex Nees as revealed by morphological characters and molecular markers. Agric Nat Resour 39(3):388–399
Maiti S, Geetha KA (2005) Characterization, genetic improvement and cultivation of *Chlorophytum borivilianum* an important medicinal plant of India. Plant Genet Resour 3(2):264–272
Matyasova E, Novak J, Stranska I, Hejtmankova A, Skalicky M, Hejtmankova K, Hejnak V (2011) Production of morphine and variability of significant characters of *Papaver somniferum* L. Plant Soil Environ 57(9):423–428
Mishra BK, Shukla S, Rastogi A, Sharma A (2010) Study of heritability and genetic advance for effective selection in opium poppy (*Papaver somniferum* L.). Indian. J Agric Sci 80(6):470–476
Misra HO, Sharma JR, Lal RK, Shukla N (2000) Pattern of genetic variability for different traits in a collection of kalmegh *Andrographis paniculata* genotypes. J Med Arom Plant Sci 22(4a):348–351
Mital SP, Bhagat NR, Maheswari ML (1975) Improvement of Isabgol (*P. ovata* Forsk.) through polyploidy and mutation breeding. Indian J Agric Sci 45:426–429
Mittal SP, Issar SC, Bhagat NR, Maheswary ML (1979) Selection of superior genotypes in Sarpagandha (*R. serpentina*). IDMA Bull 15: 175–178
Murthy UR, Abraham K (1975) Spineless mutant in *Solanum khasianum* Clarke. Curr Sci 44:718–719
Namkoong G (1988) Sampling for germplasm collections. Hort Sci 23: 79–81
Nazneen HF, Naik A, Ramesh P, Sekhar AC, Shankar PC (2019) Genetic diversity using random amplified polymorphic DNA (RAPD) and inter-simple sequence repeats (ISSR) markers in *Tinospora cordifolia* f rom the Rayalaseema region in Andhra Pradesh. Afr J Biotechnol 18(10):231–241
Nejatzadeh-Barandozi FN, Naghavi MR, Enferadi ST, Mousavi A, Mostofi Y, Hassani ME (2012) Genetic diversity of accessions of Iranian *Aloe vera* based on horticultural traits and RAPD markers. Ind Crop Prod 37:347–351
Nemeth-Zambori E, Jaszberenyi C, Rajhart P, Bernath J (2011) Evaluation of alkaloid profiles in hybrid generations of different poppy (*Papaver somniferum* L.) genotypes. Ind Crop Prod 33(3):690–696
Ogra RK, Mohanpuria P, Sharma UK, Sharma M, Sinha AK, Ahuja PS (2009) Indian calamus (*Acorus calamus* L.): not a tetraploid. Curr Sci 97(11):1644–1647
Oh MM, Trick HN, Rajashekar CB (2009) Secondary metabolism and antioxidants are involved in environmental adaptation and stress tolerance in lettuce. J Plant Physiol 166:180–191
Ojala A, Rousi A (1986) Interspecific hybridization in *Papaver*: 1. F1 hybrids of *P. Somniferum* with perennial species of sect. Oxytona Ann Bot Fenn 23(4):289–303
Ozturk O, Gunlu H (2008) Determining relationships amongst morphine, capsule and oil yield using path coefficient analysis in poppy (*Papaver somniferum* L.). Asian J Chem 20(4):2979–2988
Paliwal R, Singh R, Singh AN, Kumar S, Kumar A, Majumdar RS (2013) Molecular charcaterization of Giloe (*Tinospora cordifolia* Willd. Miers. Ex. Hook. F. and Thoms.) accessions using start codon targeted (SCoT) markers. Int J Med Aromat Plan Theory 3(4):413–422
Pandey AK, Mandal AK (2010) Variation in morphological characteristics and andrographolide content in *Andrographis paniculata* (Burm. f.) Nees of Central India. IJEE 1:165–169
Parimoo RD (1974) Induced tetraploid in geographical races of *Rauvolfia serpentina* Benth. Proc Indian Acad Sci 80(B):92–96
Patel AI, Desai BS (2017) Genetic divergence in ashwagandha [*Withania somnifera* (L.) Dunal.]: a review. J Med Plants 5(3): 119–122
Patel NH, Sriram S, Dalal KC (1980) Effect of seed age and temperature on germination of isabgol seed. III All India Co-ordinated Workshop on Medicinal and Aromatic Plants held at Indore

Pavarini DP, Pavarini SP, Niehues M, Lopes NP (2012) Exogenous influences on plant secondary metabolite levels. Anim Feed Sci Technol 176:5–16

Pawar VA (1981) Pollen viability, duration of stigma receptivity and stigma emergence schedule in isabgul (*Plantago ovata* Forsk.). M. Sc. thesis,. Gujarat Agricultural University, Anand

Rahimmalek M, Tabatabaei BES, Etemadi NE, Goli SAH, Arzani A, Zeinali H (2009) Essential oil variation among and within six *Achillea* species transferred from different ecological regions in Iran to the field conditions. Ind Crop Prod 29:348–355

Raina VK, Srivastava SK, Syamsunder KV (2003) Essential oil composition of *Acorus calamus* L. from lower Himalaya. Flav Fragr J 18: 18–20

Rajnish K, Dodiya NS, Khatik CL (2013) Assessment of genetic variability and magnitude of correlation coefficient among different traits in isabgol (*Plantago ovata* L. Forsk). Intl J Plant Sci 8(1): 193–196

Ramakrishna A, Ravishankar GA (2011) Influence of abiotic stress signals on secondary metabolites in plants. Plant Signal Behav 6: 1720–1731

Rana V, Thakur K, Sood R, Sharma V, Sharma TR (2012) Genetic diversity analysis of *Tinospora cordifolia* germplasm collected from north-western Himalayan region of India. J Genet 91:99–103

Rout G (2006) Identification of *Tinospora cordifolia* (Willd.) Miers ex Hook F. & Thomas using RAPD markers. Z Naturforsch 61c:118–122

Sangwan O, Avtar R, Singh A (2013) Genetic variability, character association and path analysis in ashwagandha [*Withania somnifera* (L.) Dunal] under rainfed conditions. Res Plant Biol 3(2):217–220

Sharma A, Geetha KA (2014) Variability in seed characters of *Tinospora cordifolia*. Open Access J Med Aromat Plants 5(1):1–4

Sharma MM, Singh OP (2012) Heritability and genetic advance for different morphological and quality traits in germplasm of kalmegh (*Andrographis paniculata*). Adv Plant Sci 25(2):681–683

Sharma SN, Sinha RK, Sharma DK, Jha Z (2009) Assessment of intra-specific variability at morphological, molecular and biochemical level of *Andrographis paniculata* (kalmegh). Curr Sci 96:402–408

Singh SP, Shukla S, Dixit BS, Banerji R (1998) Variation of major fatty acids in F8 generation of opium poppy (*Papaver somniferum* L.). J Sci Food Agric 76:168–172

Singh SP, Dubey T, Tiwari RK (2003) Correlation and path analysis in senna (*Cassia senna*). Indian J Genet 63(4):356

Singh AK, Singh HP, Singh SP, Kalra A (2008) Genetic variability and correlation studies for selection criteria in Safed Musli (*Chlorophytum borivilianum*, Santapau). Commun Biometry Crop Sci 3(2):67–71

Singh K, Kadyan S, Panghal M, Yadav JP (2014) Assessment of genetic diversity in *Tinospora cordifolia* by inter simple sequence repeats (ISSR) and expressed sequence tagged simple sequence repeats (EST-SSR). Int J Pharm Sci 6:520–524

Smitha GR, Thondaiman V (2016) Reproductive biology and breeding system of *Saraca asoca* (Roxb.) De Wilde: a vulnerable medicinal plant. Springerplus 5:2025

Sridhar B (1986) Floral bilogy and seed character in diploid and induced autotetraploids of *Solanum viarum* Dunal. M.Sc. thesis,. University of Agricultural Sciences, Bangalore

Srivastava A, Gupta AK, Shanker K, Gupta MM, Mishra R, Lal RK (2018) Genetic variability, associations, and path analysis of chemical and morphological traits in Indian ginseng [*Withania somnifera* (L.) Dunal] for selection of higher yielding genotypes. J Ginseng Res 42(2):158–164

Sundesha DL, Tankii CJ, Tulsaniiii NJ (2016) Correlation and path analysis for dry root yield in Ashwagandha [*Withania somnifera* (L.) Dunal]. Electron J Plant Breed 7(2):448–453

Tandon R, Shivanna KR, Mohanram HY (2003) Reproductive biology of *Butea monosperma* (Fabaceae). Ann Bot 92(5):715–723

Telascrea M, de Araúlp CC, Marques MOM, Facanali R, de Moraes PLR, Cavalheiro AJ (2007) Essential oil from leaves of *Cryptocarya mandioccana* Meisner (Lauraceae): composition and intraspecific chemical variability. Biochem Syst Ecol 35:222–232

Tetanyi P (1992) Plant breeding and variety control of medicinal plants. In: John HL (ed) Cultivation and processing of medicinal plants. Wiley, New York, pp 10–18

The Wealth of India (1992) Indian Council of Scientific and Industrial Research, New Delhi, Volume V, pp 408

Theis N, Lerdau M (2003) The evolution of function in plant secondary metabolites. Int J Plant Sci 164:S93–S102

Todorova MN, Ognyanov IV (1995) Chemical composition of essential oil from Mongolian *Acorus calamus* L. rhizomes. J Essent Oil Res 7: 191–193

Usmani G, Chawhaan PH, Mishra Y, Mandal AK (2014) Variability, heritability and character association analysis of chemoagronomic traits in *Rauwolfia serpentina*. Euphytica 200(2):259–268

Verma RS, Padalia RC, Chauhan A (2015) Chemical composition of root essential oil of *Acorus calamus* L. Natl Acad Sci Lett 38(2): 121–125

Verpoorte R, van der Heijden R, ten Hoopen HJG, Memelink J (1999) Metabolic engineering of plant secondary metabolite pathways for the production of fine chemicals. Biotechnol Lett 21:467–479

Walter KS, Gillett HJ (1998) 1997 IUCN red list of threatened plants. IUCN Publication Services Unit, Cambridge

Wink M (1988) Plant breeding: importance of plant secondary metabolites for protection against pathogens and herbivores. Theor Appl Genet 75:225–233

Wright S (1921) Correlation and causation. J Agri Res 7:557–585

Yadav HK, Singh SP (2011) Inheritance of quantitative traits in opium poppy (*Papaver somniferum* L.). Genetika-Belgrade 43(1):113–128

Yadav HK, Shukla S, Singh SP (2009) Genetic combining ability estimates in the F1 and F2 generations for yield, its component traits and alkaloid content in opium poppy (*Papaver somniferum* L.). Euphytica 168:23–32

Seed Production Systems for Forest Tree Species 29

Abel Nicodemus

Abstract

An efficient seed production system is essential for both natural and artificial regeneration of forests. Annual new recruitments ensure continuous forest growth and the evolution of different species. A large quantity of seeds is also required to establish plantations outside the forest area for fulfilling the need for wood and other forest produce. Seeds are also vehicles for channelling genetic improvement of different traits to the plantations. Any breeding programme of a forestry species has a dedicated propagation or production population which supplies the genetically improved planting material either as seed or vegetative propagule. Seeds from the seed production areas (SPA) meet the immediate seed demand till such a time the production populations become functional. Although the genetic gain obtained from SPAs is not high, they can be established with low cost and limited technical expertise. Throughout the world seed orchards are planted for both tropical and temperate species for cost-effective production of genetically improved seeds. They are plantations established with seeds or clonal material from trees selected for desirable characters. They are intensively managed to produce an abundant seed crop every year which is genetically improved and possessing high level of genetic diversity. Two types of seed orchards are recognized based on the kind of planting material used for establishing them. The clonal seed orchards (CSO) are raised with vegetatively propagated material of the selected trees and preferred for those species which take a long time to become reproductively mature. Seedling seed orchards (SSO) are planted using seedlings obtained from seeds of selected trees and best suited for species which commence flowering early. The SSOs are usually serve as a progeny trial in the first few years (testing phase) and after thinning inferior accessions start functioning as an orchard (production phase). Seed orchards are further improved through thinning inferior families and clones identified through progeny testing. Several technical and financial aspects have to be considered while planning for the development of seed orchards like choice of orchard type, location, orchard size and requirements for the long-term management. Similarly, the factors like low or lack of flowering, low fruit set, fertility variation among orchard trees, reduced pollination, pollen contamination from inferior sources have to be constantly addressed to ensure the expected seed production and genetic gain. In general, a gain of 5–10% in wood production is realized form SPA seeds, whereas seed orchards may return higher gains of 15–30% depending on the species, stage of genetic improvement and benchmark used. Similar levels of genetic gain are expected from the two types of orchards for those traits which have high heritability. Case studies of orchards of three tropical trees with contrasting breeding systems are discussed to explain the various biological, silvicultural and environmental factors involved in establishing and managing successful seed production systems.

Keywords

Seed production · Clonal seed orchard · Seedling seed orchard · Genetic thinning · Pollen dilution zone

29.1 Introduction

Seed is the most preferred medium of forest regeneration throughout the world. Seed production and germination ensure that new members are added to the forest annually marking the continuous growth of the forest. Seeds, being the output of sexual reproduction, also help in the evolution of different species and influence the composition of the forest. In the natural ecosystem, seed production and their dispersal

A. Nicodemus (✉)
ICFRE-Institute of Forest Genetics and Tree Breeding, Coimbatore, Tamil Nadu, India
e-mail: nico@icfre.gov.in

A. K. Mandal, A. Nicodemus (eds.), *Textbook of Forest Science*, https://doi.org/10.1007/978-981-97-8289-5_29

involve a complex plant–plant and plant–animal interactions particularly in the tropical forests. In case of trees, such adaptations have helped them to gain fitness for withstanding the changes in soil and climatic conditions. Indeed, an efficient system of seed production is considered as a reliable indicator of healthy forests.

Seed is the major propagule for the artificial regeneration of trees also. A large number of trees belonging to different species are planted throughout the world. The planting within the forest area supplements the low natural regeneration in some species. Outside the forests trees are planted for both wood production and environmental services. Increasing green cover, windbreaks, shelterbelts, sand dune stabilization, reclamation of mined areas, saline and waterlogged areas and avenues are some of the services for which plantations are raised. Trees are also planted in combination with various agricultural and horticultural crops to maximize land productivity and income from farming operations. A majority of tree planting is taken up as plantations to produce wood and wood products like timber, pulpwood for papermaking, poles for construction, charcoal making and fuelwood. Such organized large-scale plantation programmes require suitable sources for producing the seeds needed to raise seedlings.

The perennial nature of trees facilitates collection of seeds every year and use them for establishing plantations. Seeds can be handled with least technical expertise and costs during collection, processing, storage and transport. Planting stock derived from seeds invariably is less expensive than any means of vegetative propagation. In fact, seedling cost constitutes only a miniscule of overall costs involved in growing a plantation. From the breeding point of view, seeds indicate a generation turnover and with appropriate control over mating between desirable trees, the quality of the seeds can be improved.

The extent of forest tree plantations has steadily increased during the past half a century. In many countries, natural forests are protected from harvesting timber as envisaged by different forest policies and legal provisions. As a result, the area under plantations increased rapidly during the last three decades for meeting the demand for wood. India, for example has the largest areas under plantations for species like eucalyptus, teak, casuarina and others. They are owned by different agencies like forest departments, industries and farmers. Since a sizeable part of these plantations is harvested every year, a continuous supply of large quantities of seeds is required to replant the harvested areas.

Faced by low productivity of plantations, countries like - India are left with the only option of importing timber and other wood products to fulfil the domestic demand. Most of the tropical countries export timber to India (ITTO 2017). The low plantation productivity is mainly due to the use of seeds and other planting material from unknown and unimproved sources. The twin requirements of continuous supply of large quantity of seed and ensuring their quality can be met by creating suitable seed production systems for the important tree species. During past 50 years, productivity of plantations of many tree species have been substantially increased through tree improvement activities. They include fast growing varieties and improved silvicultural techniques aimed at maximizing the wood production and optimizing input requirements. Production of genetically improved seeds is an important component of plantation development programmes. In this chapter, the various methods and techniques for producing seeds of forest trees are discussed.

29.2 Seed Production Systems—The Underlining Principle

Production of genetically improved seeds can be achieved through various methods ranging from the low-input and low-genetic gain seed production areas (SPAs) to sophisticated control pollination between parents known to have high breeding value for the desirable traits. A suitable system is selected based on the biology, the extent of its cultivation, the extent of genetic gain intended to be achieved and the costs involved.

Seed production systems can be classified based on different aspects. The most common way of looking at them is whether they are needed for a limited period of time or will be used for a long time. The immediate need for seed is usually met from reproductively mature plantations after thinning out trees with poor growth and other traits. Production of genetically improved seeds is usually obtained through seed orchards established by assembling superior trees for different end-use characters. Seeds from seed stands and orchards are produced under open-pollinated (OP) conditions, that is, pollination occurs through natural vectors among the trees. In general, the open-pollination systems involve low costs for seed production. In sophisticated seed production systems, seeds are produced through artificial pollination between selected parents to produce progeny with specific combination of characters. Since such sophisticated systems need a good understanding of the species' biology and technical expertise, controlled pollinated seed production is limited among forestry species. The genetic gain may be high with the use of seed produced with parents known to have the desirable characters but the costs involved are usually high which forces its use for mostly breeding purpose only.

29.3 Seed Production Areas

Seed production area (SPA) is an interim and cost-effective arrangement for obtaining genetically improved seed although only a limited genetic gain is realized. SPAs are generally established while beginning the long-term genetic improvement of a species for sourcing seeds until the more advanced seed production systems like seed orchards become productive. They are also a preferred seed production system for exotic species for which getting seeds regularly from the natural range may be cost-prohibitive. If the seed source of the plantations converted into SPA is well-adapted to the planting environment, SPA seeds guarantee successful establishment of plantations even though the expected genetic gain may be low.

29.3.1 Selecting Stands for SPA

SPAs are established by thinning a natural stand or an even-aged plantation and subsequently maintaining it for abundant seed production. The major criterion for short listing stands for conversion into SPA is possessing higher preponderance of dominant and co-dominant trees than other stands in the target area. In addition, they are also expected to have as many of the following features to facilitate regular maintenance and seed collection.

(i) A minimum area of 5 ha to justify the investments made for establishment and seed collection. An optimum extent would be 10 ha. Small areas increase the cost of developing and maintaining per unit area.
(ii) A stand established through a known seed source with a history of producing good seed crop in the previous years.
(iii) Even-aged stands are preferred with a stocking appropriate to the species. Low stocking will cause problems for thinning and subsequent mating in the SPA.
(iv) Easily accessible to undertake regular maintenance and seed collection activities.

29.3.2 Thinning in SPA

Stands identified for conversion to SPA are subjected to one or more thinning to remove inferior trees and leaving only dominant and co-dominant trees. Thinning achieves the dual objectives of increasing mean growth of the plantation and to open-up the canopy so that crown development takes place leading to more flower and fruit production. Inferior trees are anyway removed even if their removal causes large gaps because they will further reduce the minimum genetic gain expected from SPAs. The maintenance works include removal of all residues of the felled trees and regular weeding to facilitate seed collection. This is essential particularly for trees like teak, where the woody drupes are usually collected from the ground which reduce the collection costs.

29.3.3 Improved Versions of SPA

29.3.3.1 Extensive Seedling Seed Orchards/Seedling Seed Production Areas

The genetic gain expected from SPAs can be enhanced by establishing plantations with well-adapted and selected seed sources and thinning them when they commence reproduction. Such seed production systems are useful for short-rotation species since both economic rotation and reproductive maturity are reached early. In fact, if most suitable seed sources are used, for example, those obtained from an ongoing breeding programme, their output can be comparable to that of more advanced seed production systems like seed orchards (Hai et al. 2008; Varghese et al. 2009a). Seeds from several dominant trees of seed sources/provenances known for fast growth and other desirable traits through earlier field testing are mixed and planted in an area suitable for the species. Plantations are subjected to one or more thinning to remove inferior phenotypes to increase crown development and seed production. Sometimes, provenance trials are thinned to remove less adapted provenances and inferior trees for conversion into SPA. These unpedigreed seed production systems have been popular in widely planted short rotation species like phyllodinous acacia (e.g. *Acacia auriculiformis, A. mangium*), casuarina and eucalyptus. The genetic gain from SPAs can further be enhanced by collecting seeds from only the best trees (say the top 25%) in the SPA. All trees function as male trees, while only an elite set functions as female trees. Biological (phenology, flower biology and pollination limitation) and financial constraints have limited the extent of seed orchard use in developing countries in the tropics. If properly planned, the seed production areas can function as excellent unpedigreed seed orchards for the extensively cultivated short rotation trees like acacia, casuarina and eucalyptus. In some species like eucalyptus, second-generation seed production areas were developed through progeny of earlier seed production area which showed growth comparable to clonal seed orchard (CSO) (Suraj et al. 2019).

29.3.3.2 Low-Budget Genetically Improved Seed Collection

SPAs are a good means of performing low-budget breeding or breeding without breeding approaches (Lindgren and Wei 2007; El-Kassaby and Lstiburek 2009). They are ideal for the situations when (i) the financial resources are limited and

(ii) genetic improvement in the species (like many tropical trees) is in the early stages and its biology is not well understood. The strategy is to get seeds of many trees from three or more existing seed production areas and use them to raise the next-generation seed production areas. If the budget is permitting, the family identity may be retained while planting or otherwise it can be raised as an unpedigreed seed production area. A compromise between these two options is to establish 1 or 2 ha of pedigreed and the rest unpedigreed SPAs. The best aspect of this approach is that the SPAs are primarily raised as wood-producing plantations and the seed is considered only as a by-product and thus the cost of seed is low. In case of teak, the average rotation age in government owned plantations is 50 years. Assuming that the regular seed production starts at the age of 20 years (by that time the plantation would have undergone 3 to 4 thinning), it is possible to collect seeds for the next 30 years. If the next-generation SPAs are raised with seeds collected from 20 to 25-year-old existing SPAs, seeds of advanced-generation SPA will be available when the older ones reach the age of commercial timber harvest. Thus, both wood and seed production functions can easily be achieved in the same plantation without the requirement of large funding and technical expertise. Considering that seed orchards of teak are not meeting demand for genetically improved seeds, such low-cost options ensure both seed availability and fairly good genetic gain.

29.4 Seed Orchards

A seed orchard is an assemblage of selected clones or families and then managed to obtain genetically improved seed for operational forestation (White et al. 2007). It is a plantation established with propagules from selected trees primarily for seed production, isolated from other inferior pollen sources and managed for maximum seed production. It is the most rapid and cost-effective way to mass produce seeds with substantial genetic gain. Seed orchards are easy to plant, manage and achieve desired levels of genetic improvement without the need of expensive infrastructure and human resources (Varghese et al. 2000).

It is reported that seed orchards were planted in Java during 1880 to produce seeds of *Cinchona ledgeriana* by the Dutch colonists to fight malaria (Schreiner 1961). However, establishing orchards of important forestry species started much later. Predictably the first seed orchard in India was established for teak considering its economic importance and the large area under cultivation. Since then, every teak growing state in the country established teak seed orchard following more or less a similar protocol. Many of these orchards are still being maintained well and some of them have also been subjected to progeny testing (Nagarajan et al. 1996a; Swain et al. 1996; Kumar et al. 1997; Mandal and Chawhaan 2003). Teak seed orchards aged 40–50 years with information on the performance of their 20+ year old progeny constitute an important genetic resource for both forward and backward selections to achieve maximum genetic gain.

29.4.1 Types of Seed Orchards

Generally, seed orchards are distinguished into two major types on the basis of the kind of planting material used for establishing the orchard.

1. A clonal seed orchard (CSO) is established with clonally propagated plants of selected trees.
2. A seedling seed orchard (SSO) is a seed orchard established with seed propagules of selected trees.

The choice of orchard type is mainly based on the suitability for the given species and the costs involved. Each of them functions the best in certain situations and returns the maximum genetic gain. Despite the several interconnected factors involved in choosing the type of orchard, there are certain situations where a particular type will work better than the other.

Clonal seed orchards are preferred under the following situations.

(i) The species is amenable for easy and cost-effective vegetative propagation. Producing enough number of propagules (e.g. rooted cuttings, grafts or tissue cultured plants) is a key factor for establishing CSO. This requires that the species allows itself for hassle-free propagation of trees selected for various characters.
(ii) The species requires many years to attain reproductive maturity and produce seeds. Many tree species take several years to flower and set fruits which delay deployment of genetically improved seed into planting programmes. Propagating reproductively mature trees and planting them in CSO ensure early flowering and seed production making quick returns from the investments made in tree improvement.
(iii) The intended purpose of the orchard is only seed production and no additional functions like genetic testing of accessions is envisaged. CSOs are predominantly established using grafts, that is, scion material from the selected trees grafted on to root stocks of unknown sources (Zobel and Talbert 1984). This combination of two genotypes does not permit using the orchards as genetic tests. In rare cases, clonal testing and CSO are combined where rooted cuttings attain reproductive maturity quickly (e.g. casuarina and eucalyptus).

SSOs work well for the conditions listed below.

(i) The species is difficult for any means of cost-effective vegetative propagation. The only option to capture the superiority of the selected trees is through their seedling progeny. Seeds from each of the selected trees are collected separately and seedlings are grown family-wise. Collection of open pollinated seeds (assumed half sib families) is the most common method, although full sib families produced through control pollination are also used. The families are planted as a progeny trial and then converted into SSO by removing inferior families and individuals.
(ii) SSOs work well for species that attain reproductive maturity early and produce a reasonable seed crop within a short time of 5–7 years (e.g. casuarina, eucalyptus, phyllodinous acacias). Most of these species also have short commercial rotations which help in ranking the families and individuals for different traits and undertake thinning of inferior accessions.
(iii) Wherever possible SSOs are used to combine many functions like breeding, testing and seed production in a single plantation to save cost and time.

In the early decades of orchard development, there was an intense debate on the advantages of choosing one orchard type over the other (Toda 1964; Zobel and Talbert 1984). The conclusion of such debate was tended to be in favour of CSOs in the beginning. It is also a fact that the CSOs are the most widely planted type of orchard for a range of species throughout the world. But over a period of time with the accumulation of knowledge, now it is widely regarded that the genetic gain obtained from two types of orchards is usually the same particularly when the heritability is high for the trait under improvement (Giertych 1975; White et al. 2007). Now it is common to see both CSO and SSO are used in the same tree improvement programme of a species under different circumstances. It is possible by suitably structuring orchard development and management; the advantages of one type of orchard can be obtained from other type as well. According to Shelbourne (1969), a CSO and associated progeny testing and a SSO established with cross pollinated families are expected to yield comparable genetic gain. Similarly, an untested CSO is as valuable as a SSO based on plants derived from open pollination.

29.4.2 Structure, Composition and Design of CSO

While discussing CSO, three terms with regard to the selected trees are used: clone, ortet and ramet. The tree selected in the field is called ortet and the vegetative propagules derived from that tree are called ramets. The ramets obtained by vegetative means are collectively referred to as clone. The first consideration for designing a CSO is to decide how many clones to be planted. Seed orchards are typically established with 20–60 clones (White et al. 2007). In early stages of tree improvement programmes, inclusion of a larger number of clones is considered to allow reduction in the number by thinning and other factors. Clones of CSO will undergo progeny test and the ones whose progeny did not perform well will be removed from the orchard to improve the pollen pool. A few others will also be eliminated if they show no or low flowering and have asynchronous phenology. Planting too few clones will result in low genetic diversity in seed crop and high level of pollen contamination, whereas deploying too many clones will make planting and management difficult and also likely to reduce genetic gain (because some of the clones may be inferior).

Once the required number of ramets for all the clones to be planted in a CSO are produced, a suitable field design is prepared to decide the position each ramet will take. Using the most suitable design helps in positioning ramets of the same and related clones farther from each other to avoid/minimize mating between them. A number of studies have been undertaken to propose different kinds of seed orchard designs. They include pure rows, chessboard, completely random, randomized complete block, fixed block, rotating block, reversed block, unbalanced incomplete block, balanced incomplete block, directional cyclic balanced block, cyclic balanced incomplete-block, balanced lattice, permuted neighbourhood and systematic designs (Giertych 1975; Bell and Fletcher 1978; El-Kassaby 2003; El-Kassaby et al. 2007; White et al. 2007). Each of these designs work under specific conditions and requirements like the chessboard design may be suitable for bi-clonal orchards and balanced lattice design can be used only when the number of clones is a square of a whole number. The types of designs that are widely used for establishing seed orchards are briefly discussed below.

(i) Completely randomized design: All the ramets are distributed randomly in the given area without keeping any blocks. However, suitable adjustments are made to avoid trees of related clones that are placed apart. This design is suitable to accommodate unequal number of ramets of different clones.
(ii) Modified randomized block design: In this design one ramet of each clone is randomly allocated in each block in such a way that related accessions are placed at a distance either within or between the blocks.
(iii) Permuted neighbourhood design (La Bastide 1967): Two rings of trees of other clones protect each ramet from the other ramets of the same clone. Even within the rings the combination of any two clones is kept as minimum as possible. A computer programme called

COOL (coordinated orchard layout) was developed to generate efficient neighbourhood designs for many species (Bell and Fletcher 1978).

(iv) Systematic design: In every block of the orchard, the clones take similar positions. Thus, it is easy to establish and maintain orchards using systematic design. The main drawback of the design is removal of two or more clones (placed together) during thinning leaves large gaps in every block of the orchard. But if clones from advanced generation with known breeding values are used, the systematic design has the advantage of deploying higher number of ramets for the best clones to increase the genetic gain. Models like linear deployment of clones can also be adopted with this design (Lindgren and Matheson 1986; Lindgren et al. 2009). Another situation where systematic designs may be preferred is establishing orchards of dioecious species. Entire rows of female and male clones can be placed alternatively for easy establishment, maintenance and seed collection. In order to balance seed production and diversity, equal number of female and male clones may be included, but more trees of female clones are preferred to increase the seed production.

29.4.3 Steps Involved in Establishing CSO

(i) Identifying trees with desirable characters from natural population and breeding populations.
(ii) Vegetative propagation of the selected tree by adopting the most cost-effective and easy method suitable for the species. Often a clone bank is established in an area close to the nursery by assembling the first set of propagules of trees selected from different areas and shoots collected from the clone bank is subsequently used for raising the required number of plants for establishing the orchards.
(iii) Field planting the CSO in a suitable area by adopting the most appropriate genetic/field design.
(iv) Since the only function of the trees in a CSO is seed production, they are usually planted in a wider spacing than what is followed in a regular plantation of the species. A wide spacing between trees promotes early and large crown development and adequate sunlight falling on them which results in abundant flowering and seed production. At the same time how wide the initial spacing can be is also decided considering the thinning that will take place in future. The usual range of spacing is 3 × 3 m to 10 × 10 m depending on the size and time taken by a species to commence flowering.
(v) Maintenance of orchard by adopting species- and location-specific cultural operations to promote flowering, pollination and seed production in the orchard.
(vi) Seeds collected for commercial planting programmes when at least 50% of the clones flower and produce seeds. Progeny tests are established with seeds collected from individual clones when at least two-thirds of the clones are flowering.
(vii) Thinning of clones whose progeny were found inferior through progeny testing. Since progeny tests results are obtained fast for short rotation species, one or two thinning may be sufficient. In case of long rotation species, a series of thinning will be undertaken as and when the latest data is obtained from the progeny tests.

29.4.4 Genetic Thinning in CSO

Thinning in a CSO is both a genetic (backward selection) and silvicultural intervention to increase seed production and increase genetic quality of seeds in terms of genetic gain and genetic diversity. The ranking of the progeny in the progeny test determines which clones to be retained and which one to be removed. Clones whose progeny has shown good performance (i.e. possessing high breeding values) are retained, while those with low breeding values are removed. Thinning may be taken in one or more phases depending on the rotation age of the species and nature of characters under improvement. Although the main benefit of removing inferior clones is to increase the quality of pollen pool of the orchard (i.e. paternal parent), additionally the space released by them is utilized by the retained trees to increase their crown area and thereby increase flowering and seed production as well.

A CSO with a unique composition of having many copies of a clone provides an opportunity to retain varying numbers of ramets per clone depending on their genetic superiority and seed production capacity. Lindgren (1974) proposed to gainfully utilize this situation by retaining more ramets of clones with high breeding value and less for those with low breeding values. In other words, parental representation during orchard establishment or thinning would be linearly related to their respective breeding values (Lindgren and Matheson 1986; Bondesson and Lindgren 1993).This strategy aims at deriving the maximum genetic gain through higher representation of high breeding value accessions and broadening the genetic diversity in seed crop through increasing the total number of clones retained in the orchard. If the seed demand is high, more orchards can be planted using only those clones having a high breeding value which are sometimes called 1.5-generation orchards. But the term only emphasizes the additional genetic gain obtained within a single generation since no generation turnover has taken place (Zobel and Talbert 1984). They are now referred as tested orchards or half-generation orchards (i.e. 0.5 generation over the current generation) (White et al. 2007).

29.4.5 Structure, Design and Composition of SSO

Seedling seed orchards are formed using seedlings from trees with desirable characters. They are predominantly developed with seeds formed generally through open-pollination and rarely full sib seeds produced through controlled mating between selections. As discussed earlier, SSOs are preferred where the species attains flowering phase early and both testing and seed production requirements are met from the same plantation. Seedlings with family identity are first planted as a progeny trial and converted into a functional SSO through one or more thinning to remove inferior trees within the families and all trees of the poorest families. SSOs are the first choice in case of early stages of tree improvement involving exotic species where it is easier to obtain family-wise seeds from different countries than to get scion or any other vegetative material. Other advantages of SSO are unlike the CSO they are established with the progeny of the selected trees, that is, one generation ahead of CSOs (forward selection). Seed collection from the selected trees inherently helps in selecting for reproductive characters in the progeny considering the fact that lack of or low flowering is a major problem faced in CSOs across many species.

The structuring and designing of SSOs are done keeping in mind multiple functions they are expected to perform (testing, seed production, base for future selection). Usually, a large number of families are included (50–300) in the progeny trial to keep a broad genetic base and to have more choices for selection (White et al. 2007). If the progeny testing/SSO is part of a long-term tree breeding programme, inclusion of 100 open-pollination families is considered as minimum (Lindgren and Wei 2014). Each family may be represented by 15 to 50 trees planted in a suitable statistical design. Usually, they are planted in replicated blocks; the families are randomly distributed within a block. The size of the orchard depends upon the number of blocks which is arrived at based on the amount of seed needed in the future.

SSOs have two distinct phases, the first phase is testing to determine the merit of individuals and families and the second phase is seed production. The field design of the orchard is decided to balance these distinct functions which may affect each other at some point of time. Being a test plot, it will be located within the species' planting zone adopting spacing and cultural practices similar to the routine plantations. This may affect the flowering and seed production in the trees. On the other hand, if wider spacing and higher cultural inputs are given, the precision of testing may be reduced. The most widely used designs for SSOs are completely randomized, randomized complete and incomplete block designs. The completely randomized design with single-tree plots is efficient in tackling the field heterogeneity and also ensures that trees of the same family (siblings) do not occur together leading to related mating. Single tree plots provide the smallest blocks but cause difficulties in analysis in case of many casualties and would be very costly and difficult to keep identity of the trees when there are many families. The randomized complete block design involves replicated blocks each having all the families planted randomly. This design works better when the block size does not exceed 0.1 ha. Large block size means higher soil heterogeneity within the block which reduces the effect of randomization. In such cases, the incomplete block design is used so that the effects of small-sized incomplete blocks can also be accounted within each block to arrive at precise estimation of individual and family performance. In case of complete and incomplete block design, each family is represented by a row-plot of three to five trees initially and once they are assessed for different characters at an appropriate age, only the best tree is retained to avoid inbreeding among the siblings. Within-family thinning also releases space for the retained tree to develop wider crown and more seed production in the second phase of the SSO.

29.4.6 Steps Involved in Establishing an SSO

(i) Seeds collected from selected trees keeping the family identity and seedlings raised separately from them.
(ii) Seedlings are planted using a suitable design (e.g. completely randomized design, randomized complete block design and incomplete block design) in areas similar to the species planting zone adopting the same spacing between trees as that of routine plantations.
(iii) The progeny trial is maintained by adopting the silvicultural practices similar to that of commercial plantations during the testing phase.
(iv) Usually, trials are planted in contrasting locations within the species' planting zone using the same set of families to assess the G x E interactions for different traits.
(v) At the end of the testing phase (at half- or two-third rotation age), the trees are assessed for different characters and breeding values of the families and the extent of G × E interactions is determined.
(vi) Genetic thinning is undertaken based on the breeding values of the families: a large number of trees from the families with best breeding value and the best trees of

the families with intermediate values are retained, while most of the trees of the poorest families are removed.

(vii) Once the thinning process is completed, the orchard is intensively managed to promote abundant flowering and seed production.

Thinning in a progeny trial is undertaken based on the performance of both families and individual trees within a family. Often the thinning process is taken up in at least two phases. The first phase is almost entirely within-family in which the inferior trees of a family are removed. In case of multi-tree plots, the plot size gets reduced to one or two trees, whereas the number of families left more or less untouched. The second phase of thinning is among the families based on the mean performance of all trees of a particular family. The simplest method is truncated selection in which all trees of families whose mean performance is below a cut-off value are removed. While this method helps in increasing genetic gain from the orchard, the genetic diversity in the seed crop will be drastically reduced. The linear deployment option helps in balancing the genetic gain and diversity by retaining higher number of trees from the best families and lower number of trees from intermediate and poor performing families (Lindgren 1974). This is particularly useful if the SSO also has long-term function like providing families for establishing the advanced-generation breeding populations which should address both high genetic gain and genetic diversity. Selecting the trees with high breeding value from the outstanding families ensures capturing the genetic gain from parents especially for the traits that have low heritability.

29.5 Silvicultural Requirements for Seed Orchards

Seed orchards being special plantations formed for a specific function require silvicultural practices suitable for the given species and climatic and edaphic conditions of the site where the orchard is located. The planting materials with which the seed orchards are established are developed by investing substantial financial and human resources. Planting and maintaining them by adopting the cultural techniques are necessary to ensure efficient functioning of orchards and to derive the intended genetic gain through the seeds collected from them. The important silvicultural requirements for establishment and management of seed orchards are discussed below.

A seed orchard houses selected trees expected to produce high quality seed that will improve productivity in future plantations significantly. In this context, an orchard should not be seen in terms of area of plantation but as a recombination centre. In several cases, the seed orchards are also clone banks, clonal/progeny tests, breeding populations and germplasm assemblage. Hence, intensive management of orchard is essential to gain maximum output in terms of seed, genetic gain and information. A compact but intensively managed orchard is more preferable than a large one poorly looked after.

29.5.1 Ensuring Survival

High survival (90% and above) is necessary in the initial years of an orchard. In case of low levels of casualties, avoid filling the gaps at a later time which may result in uneven growth and interfere with ranking of different accessions. In particular, high survival is essential during the 'testing phase' of seedling seed orchard for accurately assessing and ranking families and individuals for different characters. Every tree in an orchard earned its place after a series of selection and elimination process and is highly valuable. Losing them due to factors that are well within a manager's control should be avoided.

29.5.2 Ground Management

Weed control in the initial years is essential for the seedlings to establish fast and grow vigorously. The ground should be free from weeds and residual stumps. Apart from helping the young trees to grow without competing with weeds, ground management also facilitates fertilizer application, prevents fire incidence and reduces possibility of insect attack. In older orchards, weed control helps in easy assessment and seed collection. Compaction of soil is frequently observed in orchards especially in dry locations. Hardening of top soil results in increased run-off of rain water severely reducing tree growth and seed production and dying of trees in serious cases. Loosening of soil just before the beginning of rains is often found to promote growth in orchards during first few years. If the spacing in an orchard allows movement of tractor, this operation can easily be carried out.

29.5.3 Fertilizer Application

Application of fertilizers promotes vigour, growth and flowering in seed orchards. This is usually done based on results from soil analyses. It is always desirable to carry out a detailed soil testing right at the time of selecting the site for an orchard. Fertilizers particularly nitrogen and phosphorous have promoted flowering in almost all the hardwood species for which trials have been established (Varghese et al.

2009b). The timing of fertilizer application is critical if satisfactory results are to be obtained. It should be applied before the initiation of floral buds if immediate increased flowering is to result. Fertilizers also keep the trees healthy and to grow to a large size resulting in more reproductive bud locations. The emphasis on vigorous vegetative growth in the early years and shifted to flowering and fruiting once reproductive maturity is reached. Application of growth regulators like paclobutrazol dramatically increases flowering in orchards but should be used with great caution as overdose and inappropriate time of application may kill trees (Griffin 1989, 1993; Williams et al. 2003; Varghese et al. 2009b).

29.5.4 Control of Pests

Managing pest attack (insect and diseases) is important both during the establishment and reproductive phases of the trees in the orchard. In particular, pests attacking the flowers, fruits and seeds have to be effectively tackled to ensure high seed production with acceptable quality (like seed filling). Heavy losses in seed production and seed quality have been reported for many species. The annual seed yield in a 20-year-old orchard of *Pinus elliottii* was 30 kg with insect control but only around 20 kg when no control measures were taken (Powell and White 1994). Reduction in fruit and seed yield was also reported from seed orchards of many conifers, teak, eucalyptus, acacia and casuarina (Balu et al. 2016). While undertaking chemical control of insects damaging flowers in entomophilous species, care has to be taken not to affect the pollinating insects.

29.5.5 Assessment and Record Keeping

Seed orchard development, management and deployment of seeds are long-term activities and keeping a good record of all events will be useful in multiple ways. The foremost need is to maintain the identity of entries (clones and families) in the orchard both in the field and in records. Orchard trees are planted in a specific field design by adopting uniform spacing between them. It is easy to identify the location of a particular tree during the initial years when the stocking is high and planting rows are continuous. During the later years when trees are removed either by casualty or genetic thinning, the spacing between trees becomes irregular and locating a tree as per the planting design will be difficult. It is necessary to periodically update the field design like once in a year or after every thinning and also mark the identity of the trees in the field as far as possible. Additional benefits are derived when seeds are collected clone or family-wise for effective deployment in plantations and also to conduct progeny tests. Those trees identified to possess high breeding value estimated through the performance of their progeny in field tests many years later will have to be visited for more seed collection or for producing new grafts. It is essential to be able to return to the correct accession which keeps the orchard and the progeny test connected.

It is useful to measure important characters like height, girth, stem straightness and health of trees even if the orchards are not planted as trials. Once the trees reach flowering stage, data on flower initiation, peak flowering, pollinator visits and cessation of flowering may be recorded. The appropriate time for seed collection can also be determined by continuous observation of fruit development. Information gained through these assessments will be useful in determining whether the growth and flowering of orchard trees are optimum in the given site, to initiate appropriate management practices and also to make decisions on future establishment of orchards in that and similar sites. Similarly understanding the flowering initiation, peak and cessation, time of fruit maturity will help in estimating seed production and undertaking timely seed collection. Records of the quantity of seed collected from individual clones/families, their seed filling and germination characters and details of the plantation raised with them will facilitate long-term monitoring of orchard dynamics and the performance of its progeny. A robust data base on the seed orchard is also necessary to make decisions on continuous investments in the orchard management, fixing the cost and marketing of the seed crop. It is equally important to record the movement of seeds to various planting agencies and information on their field performance.

29.6 Factors Affecting Orchard Output and Quality

29.6.1 Orchard Location

Locating a seed orchard should be decided according to the species, and planted within the species' range of adaptation. Further locations where the species is known to flower abundantly and set seed consistently across years will be more preferable within the range. Areas where the species recorded good wood volume need not always be ideal for abundant seed production. The site is also expected to be easily accessible and be available for long-term without any developmental projects. The other requirements are areas not prone to natural disasters like cyclone, prolonged waterlogging, pest attack and damage by wildlife. Soils with both low and high nutrient status should be avoided. While the growth is slow in the former, the latter tends to promote more of vegetative growth. Soils of average fertility are preferred which give the orchard manager to meet any nutrient requirement thorough fertilizer application (Zobel and Talbert 1984). Another preference would be to have the orchard as close to the managing agency's headquarters as possible. Since orchards need intensive care, easy access will help in efficient management. Though meeting all of these requirements is nearly impossible, all available options have to be considered carefully before selecting a site.

29.6.2 Orchard Size

Size of the orchard is determined mainly by the seed requirement. This in turn depends on the species and the amount of seed production per tree. In terms of physical space, different species will need varying area to have a well-developed crown so the area needed to house the same number of trees will vary across species. Small-sized (0.5–1.0 ha) orchards work well for short-rotation species like casuarina and eucalyptus where individual trees produce large quantity of seed. Since the seed rate is also very high for these species, concise orchards are preferred for them to be cost-effective and easy to plant and manage. But it is also a fact that small-sized orchards are more prone to pollen contamination from outside sources especially if they are surrounded by plantations originated from unimproved sources. Long rotation tropical hardwood species like teak take long time to commence flowering and require large space to develop crown. Further seed production in teak orchards is generally low. Only 16 ha of plantations can be raised from seeds produced by 1 ha of teak CSO necessitating large-sized orchards (Nicodemus et al. 2009). Species of single or few seeded fruits need more trees to be planted in an orchard than those with multiple-seeded fruit. Teak fruits have only four ovules and rarely all of them develop into viable seeds. But in case of *Eucalyptus tereticornis*, more than 50 seeds present in a fruit. All these factors have to be considered before arriving at the size of the orchard.

29.6.3 Phenology and Fertility Variation

The genetic gain expected from the seed orchards is based on the assumption that different accessions are unrelated and they have equal chance of mating with each other (panmixis). If the flowering phase of a member is drastically different from others, it may not contribute to the orchard pool by being reproductively isolated. Generally, those members of the orchard which flower much earlier or later than most of others reduce the effective population size of the orchard. In species where orchards established with selections from different provenances, they tend to have differing phenophases corresponding to their origin (Spencer et al. 2020). In such cases, mating is possible only with accessions from similar origin with overlapping flowering period which reduces both seed production and genetic diversity in the seed crop. Seeds of an early flowering clone in a teak seed orchard showed low seed filling and germination (Nicodemus et al. 2009).

Apart from flowering synchrony among the members of the orchard, it is also necessary that their flower and fruit production is balanced to achieve panmixis. But invariably the clones/trees show large variation in their flowering and fruiting behaviour. A few accessions tend to produce a higher proportion of gametes (pollen and ovules) through abundant flowering than the others. The effective population size of the orchard and gene diversity in the seed crop will be seriously affected by differential gamete production (Lindgren and Mullin 1998). In particular, if the clone/family that contributes the major proportion of seed is having a low breeding value, it will have serious impact on the genetic gain expected from the orchard progeny (Nagarajan et al. 1996a; Nicodemus et al. 2009). Fertility variation has been widely reported from seed orchards of both conifers and broad-leaved trees (Kang 2001). Seed orchards of the most widely planted tropical trees like teak, eucalyptus and casuarina showed high levels of fertility variation severely affecting the quality of the seed crop produced from them (Varghese et al. 2004, 2009a; Nicodemus et al. 2009).

While it is not possible to avoid variation in gametic and progeny contribution of seed orchard accessions, it can be reduced to acceptable levels by resorting to suitable management practices. Low level of variation was observed in orchards located in areas conducive for flowering of the species. Casuarina trees in coastal locations and teak orchards in medium-to-high rainfall (1500–2000 mm) showed less fertility variation compared to orchards established with similar composition in other locations (Varghese et al. 2004). Assembling accessions selected from the same provenance/population is expected may lead to similar flowering output. Since most of the forest tree species possess a major part of genetic variation within the population, selecting candidates for seed orchards from within a population is likely to capture both the diversity and adaptability to the orchard site. High levels of fertility differences were found during poor flowering years. Flowering can be improved through appropriate silvicultural interventions, or altogether avoid seed collection during scant flowering years. Alternatively constrained seed collection, that is, collecting equal amount of seed from each accession will result in all gene combinations equally represented in the orchard seed crop.

29.6.4 Pollen Contamination

Pollen contamination from unimproved sources is a major problem faced in seed orchards which reduces the genetic gain. Further if the source of contamination has poor adaptability, it will also affect the seed crop. If the orchard is located in the midst of routine plantations, it will increase pollen contamination from external sources. Higher level of contamination is witnessed in small-sized orchards compared to large orchards due to smaller orchard pollen pool. The extent of overlapping of flowering phenology between the orchard trees and the surrounding population will also influence the level of pollen contamination. Estimates of pollen contamination levels are available predominantly for the conifers, ranging from 2 to 74% for *Pinus sylvestris*, 43 to 71% for *Picea abies* and 30 to 40% is considered as normal for mature conifer seed orchards in Sweden with an average range of 30 to 50% (Kang 2001; Prescher 2007). Estimation of pollen contamination in 21 orchards of conifers including *Picea glauca* and *Pseudotsuga menziesii* in USA and Canada showed two thirds of them receiving 33% pollen from other sources (Adams and Burczyk 2000). Pollen from an exotic plantation of *Eucalyptus loxophleba* subsp. *Lissophloia* has been reported to be pollinating a native population of *E. loxophleba* ssp. *supralaevis*, located 1940 m away (Sampson and Byrne 2008). The extent of economic loss from pollination by inferior sources can be realized from the fact that a 50% pollen contamination in an orchard will reduce the genetic gain by 25%. It is important to take up every possible intervention available with the seed orchard manager to reduce the pollen contamination and realize the expected genetic gain.

29.6.5 Pollen Dilution Zones

The following are the major interventions taken to prevent/reduce pollen contamination.

(i) Keeping an isolation distance from the plantations of the same species and the orchard is important. This distance is called pollen dilution zone which varies depending on the species (wind or insect pollinated) and the location of the orchard. Small-sized orchards may need longer isolation distance. For wind pollinated species, the minimum distance is 150 m but normally distances of 500–1000 m were found to be more effective since it was reported that external pollen sources can be even 50 km away from orchards (Adams and Burczyk 2000; Di-Giovanni et al. 1996). There is scant information on the isolation distances required for insect pollinated species where the pollen transfer is mainly dependent on the type of insect and its behaviour (Zobel and Talbert 1984). It is reported that eucalyptus and acacia orchards with insects as major pollinators require an isolation distance between 100 and 200 m (van Wyk 1981; Griffin 1989; Eldridge et al. 1994; White et al. 2007; Spencer et al. 2020).

Setting aside areas for pollen dilution zones and their maintenances involves cost and continuous efforts. Leaving a minimum distance of 150 m around a square shaped 10 ha orchard will increase the orchard area nearly by four times (White et al. 2007). Even if the high costs are accepted owing to the anticipated benefits of reduced pollen contamination, the required extent of land may not be available in an area otherwise suitable for orchard establishment. Such limitations have made orchard managers to develop alternative methods for reducing or preventing pollen contamination from inferior sources.

Planting of a few rows of non-interbreeding species around the orchard is commonly practiced to reduce the entry of external pollen into the orchard. In cases where orchard trees are pruned, those in the peripheral rows are retained without pruning so that they function as barriers for pollen from outside the orchard. It was shown that the pollen contamination was more in the edges compared to the lee side and central portion of the orchard (Yazdani and Lindgren 1991; Nikkanen 2002).

(ii) As far as possible establish plantations using known and genetically superior sources in the areas surrounding the orchards as pollen enrichment zones. Alternatively locate the orchards in an area where the origin of the existing plantations is known to be an improved source. The expectation from having pollen enrichment zones is that even if pollen contamination occurs, it will not affect genetic gain substantially.

(iii) The decision of having the plantations surrounding the orchard may not be entirely with the orchard mangers. In such cases, keeping the pollen pool of the orchard larger than that of other sources will increase mating among orchard trees. Larger-sized orchards will have larger pollen pool compared to small-sized orchards. It is also possible to take suitable cultural interventions which increase flowering in trees. If the fecundity of different accessions (clones/families) is already known, having the most fecund entries in the peripheral rows and seed collection can be optional from these rows. In conifers which are wind pollinated, supplemental mass pollination method is adopted to increase the pollen pool in the orchards and thereby improve overall seed production and genetic gain in orchards. Pollen from a known and genetically improved source is collected and sprayed in the orchard when majority of the trees are in flowering phase. Supplemental mass pollination has been reported to reduce pollen contamination (Askew 1992; El-Kassaby and Ritland 1986; Stoehr et al. 1998). In case of insect, pollinated species apiaries are set up in orchards to increase pollination services within the orchard (Moncur et al. 1993, 1995; Hardi and Kanso 1995).

(iv) In case of conifers, the onset of flowering in an orchard is delayed by spraying water over the canopies to keep the temperature low. This method of influencing the flowering phenology is called bloom delay which prevents mating between trees in the orchard and those outside. Although blooms delay involves high costs, it was proven to be effective in reducing contamination in conifers like *Pseudotsuga menziesii* (El-Kassaby and Ritland 1986; Wheeler and Jech 1986).

29.6.6 Seed Collection and Deployment Options for Seed Orchards

Seed is collected from all flowering trees in the seed orchard and bulked and supplied to planting programmes. There are many options available to collect seeds from orchards which can improve genetic gain and/or genetic diversity in the seed crop. Seeds can be collected from only the best clones/families in an orchard to increase genetic gain provided the entire seed requirement can be met from such accessions. Collecting seeds in equal quantity from each of the seed orchard accession ensures high diversity within the seed crop. This method of seed collection will be particularly effective when there is high variation in the fertility among seed orchard trees. Capturing maximum diversity will also be useful in cases where the location of the prospective plantations is not known and there is a need to have maximum diversity within the seed crop to increase its adaptability.

In a CSO, seeds may be collected clone-wise, that is, seeds from all ramets of a clone are pooled and deployed as open-pollinated (OP) family seedlot. Such a deployment has advantages like (i) the high level of uniformity within a family in terms of nursery development, uniform early establishment in the plantation and quality of wood harvested have significant impact on the costs involved and returns obtained in the forest plantation programme; (ii) OP families suitable for different growth and end-use characters can be deployed and (iii) with prior information on the G x E interactions of OP families, the site-family matching can be effectively taken up to increase adaptability and wood production. Use of OP families from orchards is widely practiced for raising large areas of plantations of *Pinus taeda* and *P. elliottii* in south-eastern USA where large blocks of single OP families are raised. In order to keep diversity, many families are used which may vary from one location to another depending on the site quality and other requirements (McKeand et al. 2003; White et al. 2007).

29.7 Genetic Gain from SPA and Seed Orchards

Genetic gain obtained from breeding programme is calculated as the difference between wood production and quality of plantation raised with improved seed and those established with unimproved sources. It can be either predicted from the genetic tests on the basis of parental performance or directly measured from field tests conducted with improved and unimproved seed sources (realized gain). While predicted gains are widely reported, information on realized gains is scant. The genetic and economic gain actually realized from planting the seed orchard seeds is determined through genetic gain trials. These trials are raised with the seed orchard bulk seedlot along with an unimproved seed source and any other available alternate seed source. They are located in a typical plantation environment adopting only those silvicultural practices taken up in the plantations. The value of the additional wood produced by the trees from orchard seeds minus the difference in the cost of seed is the economic gain obtained from the tree improvement programmes which justifies the investments made in the programme. The genetic gain reported for even those traits with low heritability (0.1–0.3) were 5–25% in the first generation both for conifers and angiosperms which justify the investments made in tree improvement. It has also been amply demonstrated that even a few percentages of genetic gain in perennials results in reasonable economic returns at the time of harvest. Economists have determined that forest seed orchards are sound investments if they are productive. On the other hand, any activity of breeding brings societal benefit only when the trees raised from genetically improved variety has been harvested by the growers and obtaining a higher economic return (Talbert et al. 1985; White et al. 2007).

Information available on genetic gain obtained through use of seed orchard seeds is mostly for conifers planted in the temperate regions of North America, Europe and New Zealand. These trees are in their advanced-generation breeding and seed orchards are the main propagation populations of these programmes. In contrast to the tropics (South America, Asia), the share of clonal forestry is negligible (Sonesson et al. 2001). Seeds from orchards of scots pine (*Pinus sylvestris*) produced 10–25% higher wood volume than unimproved sources (Rosvall et al. 2002; Andersson et al. 2007). The gains realized through subsequent generations of breeding of *Pinus radiata* has been denoted by ratings starting from GF1 (growth and form; unimproved source). The improved variety GF22 had mean annual increment of 24.9 m^3 ha^{-1} $year^{-1}$ compared to 19.9 m^3 ha^{-1} $year^{-1}$ of unimproved variety, a genetic gain of over 25% (Carson et al. 1999). The temperate species *Eucalyptus globulus* yielded 60% genetic gain over unimproved variety

Table 29.1 Current extent of seed orchards of different tropical tree species and gains realized

Species	Annual planting area (000 Ha)	Seed orchard area (Ha)	Realized genetic gain (%)	Proportion of plantations from orchard seed (%)	References
Tectona grandis	40	1648	10–68	>5	Nagarajan et al. (1996a), Mandal and Chawhaan (2003), Goh et al. (2013)
Eucalypts camaldulensis	300	43	Height-17; DBH-14	5–10	Kumar et al. (2016)
Casuarina equisetifolia and *C. junghuhniana*	100	18	Stem volume-13–28	2–25	Nicodemus et al. (2011)
Acacia auriculiformis	10	15	Stem volume-59–129	>5	Hai et al. (2008)

for Kraft pulp production contributed by improvement in growth rate, wood specific gravity and pulp yield (Cotterill 2001).

Reports on realized gain by using genetically improved seeds are far less for tropical trees compared to conifers as they are in their early stages of domestication and breeding. However, the extent of gains realized from even the first-generation orchards is encouraging. Table 29.1 shows the details of genetic gain obtained for the most extensively planted tropical tree species of acacia, casuarina, eucalyptus and teak. Conical stem volume of 4-year-old trees of *Acacia auriculiformis* raised with seedling seed orchard seeds was 22.3–27.7 m^3 ha^{-1} compared to 12.1 for the unimproved commercial seedlot. Even the seeds from an unpedigreed SPA yielded 23 m^3 ha^{-1}, returning a gain of about 90% (Hai et al. 2008). The huge difference between the seed sources indicates the inferior nature of the local seeds used for raising plantations which lead to suboptimal utilization of potential of the planting site. In South Africa, the average gain in wood volume per generation was 14% for three generations of breeding *Eucalyptus grandis* (Verryn et al. 2009; Harwood 2014). SSO families of *E. globulus* showed 22.6% increase in volume over the check seedlots (Callister et al. 2013). The realized genetic gain from a second-generation SSO of *E. pellita* in Indonesia was 16% for height, 19% for diameter and 21% for stem form (Leksono et al. 2008). In China, SSO of *E. urophylla* yielded a gain of 5.2% over seed stand and 16.1% over imported commercial seed for individual tree volume. Advanced-generation CSO of the same species produced a 10.4% gain over seed stand and 31.3% over commercial seed (Huang et al. 2003). Plantations raised with seeds from unpedigreed orchards of *E. camaldulensis* showed 17% gain in height and 14% for diameter in India. Similarly seeds of SSO of nitrogen fixing species, *Casuarina equisetifolia* produced 50% higher conical volume and that of *C. junghuhniana* 112.5% more than unimproved seed sources in south India (Nicodemus et al. 2011).

Seed from teak orchards has not been subjected to systematic testing to quantify the genetic gain that can be realized in operational planting programme. A wide range of 10–68% genetic gain for single tree volume/basal area has been reported from progeny tests established with limited number of trees and evaluated at varying ages from 5 to 20 years (Nagarajan et al. 1996a; Mandal and Chawhaan 2003; Goh et al. 2013). Considering the heavy demand for teak timber and the high prices prevailing in the international timber market, even the lowest genetic gain of 5% is sufficient to justify the investments made in teak improvement (Kjaer and Foster 1996).The current extent of seed orchards of all widely cultivated trees is inadequate to meet the annual planting target in India. Although clones fulfil considerable portion of the demand for improved planting material, they have limitations like cost and risks posed by low genetic diversity. The data presented in Table 29.1 indicate that vast areas of plantations are still raised with seeds from unknown and probably unimproved sources leading to under realized potential of the land and the crop. In contrast, in the USA more than 2000 ha of orchards of *Pinus taeda* and *P. elliottii* produce seeds sufficient to plant approximately 700,000 ha annually which is about 90% of the total planting area in the country (White et al. 2007).

29.8 Case Studies of Tropical Trees

29.8.1 Teak

Teak is a high-value timber species preferred world over for its durability, suitability and aesthetics for various uses. It is naturally distributed in India, Thailand, Myanmar and Laos and planted as exotic in almost all tropical countries. Teak timber was harvested from natural forests until mid-nineteenth century and when the supplies started dwindling, large-scale plantations were raised. Now the estate of teak-bearing forest and plantations is among the largest in the world (Graudal and Moestrup 2017). A large quantity of seeds is required to meet the annual planting target throughout the world which was usually met by collecting seeds from the existing plantations. But the productivity of teak plantations gradually declined with continuous harvesting and replanting. Indian teak plantations showed a low productivity ranging from 0.97 to 5.64 m^3 per ha per year with an average of 2.85 m^3 per ha per year in a 53-year rotation period (Subramanian et al. 2000). Since the slow growing plantations could not meet the rapidly increasing timber demand, large-scale imports were made to bridge the gap. India is among the largest importers of timber (mainly that of teak; Kollert and Walotek 2017).

29.8.1.1 Seed Production Systems

Establishment of seed orchards of teak began almost at the same time when modern tree improvement activities were started throughout the world during the mid-twentieth century. They were designed similar to the clonal seed orchards of conifers in Europe and North America. Plus trees were selected from natural populations and plantations mainly based on traits that determine the timber volume, that is, growth, clear bole (axis persistence) and stem form. Bud grafts of the selected trees were used to establish clonal seed orchards. CSO generally included 15–25 ramets each of 20–40 clones planted in a completely randomized design. A few rows of fast-growing species like eucalyptus were planted around the orchard as a pollen dilution zone. Large areas of CSOs were established in India (1022 ha) and Thailand (1830 ha) during 1950–2000 (Kaosa-ard et al. 1998; Katwal 2005). SPAs were also established simultaneously to meet the current demand till the time CSOs commence seed production. Reproductively mature plantations (20–30 years) were subjected heavier thinning than commercial level and used for seed collection. SPAs occupy an area of 2185 ha in India (Katwal 2005).

29.8.1.2 Status of Seed Production

A vast majority of the teak seed orchards did not function well in terms of seed production. The seed production was generally low, 18 kg per ha per year in India and 50 kg in Thailand in 15–30-year-old orchards (Wellendorf and Kaosa-ard 1988; Nicodemus et al. 2009). The inadequacy of such low seed output needs to be understood in the light of reproductive biology and breeding system of teak. Teak inflorescences are borne terminally and the formation of first inflorescence results in forking of the main stem (Kaosa-ard 1996; Nicodemus et al. 2009). Teak inflorescences are large in size having a few thousand flowers each, but the fruit to flower ratio is very low (<1%). Teak is an outcrossing species pollinated mainly by insects (Kjaer and Suangtho 1995; Nagarajan et al. 1996b). The fruit which is botanically a drupe is the unit of sowing in the nursery and is normally referred to as 'seed'. Although a fruit has four locules, many of them possess a single seed at the time of maturity. A significant proportion of fruits may also be empty without any filled seed. These factors contribute to low and staggered germination of teak seeds which seldom exceeds 40% even with the best nursery practices. In Thailand, 12 kg of seed is needed to raise seedlings sufficient to plant 1 ha of plantation (Wellendorf and Kaosa-ard 1988). Considering such heavy seed requirement, it is nearly impossible for the CSOs with low output to meet the seed demand and make an impact on the productivity of plantations.

The reasons for low reproductive output and success in teak in general and seed orchards in particular have been the subject of several studies undertaken during the last three decades. Lack of or low flowering, lack of sufficient pollination by insects, asynchronous flowering phenology and high level of fertility variation among clones and damage of flowers and fruits by insects were reported as major factors limiting both seed production and the intended genetic gain from the teak orchards. A list of such studies is provided in Table 29.2.

In teak the age at which the first flowering occurs is reported to be under genetic control (Nanda 1962). The apical meristem is used for producing the terminal inflorescences which results in branching of the main stem. Thus, the clear bole height which is an important character influencing timber yield is determined by the age at which the first flowering occurs. A strong inverse relationship between clear bole height and the number of flowers produced was found among clones in an orchard (Nicodemus et al. 2009). The clear bole height is one of the primary selection criteria for identifying plus trees which has led to selection for apical dominance and delayed flowering when assembled in

Table 29.2 Factors affecting seed production, genetic gain and gene diversity in teak seed orchards

Limiting factor	Effect on seed production and outcrossing	Reference
Clonal variation for flowering	Unbalanced gene representation in seed crop	Gunaga et al. (1999), Nicodemus et al. (2009)
Non-overlapping of flowering phenology	Reduction in panmixis; preferential mating	Nagarajan et al. (1996b);, Vasudeva and Gunaga (2012)
Pollinator behaviour/efficiency	Increase in geitonogamy	Tangmitcharoen et al. (2006)
Fertility variation: Clonal, annual	Reduction in effective population size and genetic diversity in seed crop	Varghese et al. (2004, 2006) Nicodemus et al. (2009)
Poor germination and fitness	Low nursery output and reduced genetic gain	Indira and Basha (1999)

orchards. Most of the orchards have not been thinned and the trees could not develop a large crown for producing inflorescences. Assembling clones from populations under diverse environmental conditions caused variation in their flowering phenology. The effective population size is reduced when some clones flowered well before other clones in the orchard. Inclusion of clones selected from similar environmental conditions and locating the orchard close to the origin of selected trees may help trees retain their adaptability to flower. The spacing between trees in teak orchard needs to be increased either at the time of planting or through periodic thinning to facilitate adequate crown development and reduce fertility variation among clones. Many silvicultural and chemical interventions were attempted to increase flowering but none could significantly increase flower and seed production. Seed collection from clones having flowering phenology highly non-overlapping with rest of the orchard may be avoided to improve genetic gain and diversity in the seed crop.

Teak is predominantly pollinated by insects although pollination by birds was also reported (Hedegart 1973; Nicodemus et al. 2005). Thirty-five insect species belonging to different orders were found visiting teak flowers in south India (Mathew et al. 1987). Both solitary and social bees were found to be the most frequent pollinators. The implications of these observations are maintaining adequate ground vegetation in and around the orchard as refugia for the bees and introduction of apiaries to increase pollinator visitation. But it is difficult to come across reports on actual implementation of such recommendations and their impact on seed production and quality in orchards.

Although varying levels of loss of seed produced in an orchard due to pest attack have been reported for almost all species throughout the world, it assumes more significance for teak since there are two major insect pests (defoliator *Hyblaea purea* and skeletonizer *Eutectona machaeralis*) attacking teak which also damage flowers and young fruits. In addition, teak flowers and fruits were found to be predated by fruit borer and inflorescence feeder *Pagyda salvalis*, sap sucking bug, *Leptocentrus vicarius* and fruit borer and *Dichocrocis punctiferalis* (Neelay et al. 1983). Heavy dropping of developing fruits was reported to occur due to infection by the fungus *Phomopsis* and *Colletotrichum* sp. in Kerala, India (Mohanadas et al. 1999; Indira and Mohanadas 2002). Chemical control through spraying of insecticides and fungicides has been found controlling the pest damage and increasing seed production. However, such practices are not widely followed due to logistical constraints.

29.8.1.3 Realized Gain from Orchards and Alternate Options

Despite the factors limiting seed production in teak CSOs, many tests have demonstrated the superiority of CSO seeds over unimproved seed sources. Kjaer and Suangtho 1997 predicted a genetic gain of 12–17% from teak CSOs over unclassified seed sources and results from different field tests have confirmed that prediction (Nagarajan et al. 1996a; Mandal and Chawhaan 2003; Goh et al. 2013).

The sustained low seed production and the absence of robust interventions to reverse it have kept the teak orchard not connected with the commercial planting activities. The demand for improved planting material is mainly met from seed production areas particularly in India and Thailand. A minimum of 8% gain in wood production is expected from plantations raised with seeds from classified seed stands through improvement in volume and stem form (Kjaer and Suangtho (1997). This gain is still profitable considering the value of the additional timber produced and the steady increase in timber prices in the international market (Kjaer and Foster 1996). SPAs are primarily timber-producing plantations and the seed production is only an additional function derived from them which makes the seed production cost low. Teak SPAs were found to function well in terms of fertility variation among trees and effective population number (Varghese et al. 2008). Seeds from SPAs showed better germination and seedling quality compared to the progeny of CSOs (Indira and Basha 1999; Jayasankar et al. 1999). Considering these factors, Lindgren (2005) proposed a low input breeding strategy for teak involving successive generations of seed production areas. It proposes to have a sufficiently large SPA of around 25 ha and collect seeds from the best trees and establish the next-generation SPAs. Once the seed is available from the new SPA, seed collection is stopped from the earlier area and it is harvested when the commercial rotation period is completed. The advantage of this strategy is that it provides continuous genetic improvement with low costs and technical expertise and land requirement and it overcomes the limitations of seed production through keeping a large number of trees in the SPA.

The challenges faced at every stage of sexual reproduction in teak have shifted the focus on vegetative propagation of selected material for commercial planting in the recent years. The argument for large-scale multiplication of selected clones is based on the facts (i) overcoming the problems faced in seed germination and making available the required amount of planting stock across seasons and years, (ii) capturing the full genetic potential of the selected trees and ensuring high level of uniformity of characters in the plantations, (iii) benefitting from the apical dominance which increases the clear bole height which otherwise delays sexual reproduction (Monteuuis and Goh 2017). Progeny of seeds collected from seed orchards or plus trees is raised in suitable trials and the outstanding trees in terms of growth, stem form and wood properties are selected for vegetative propagation, clonal testing and deployment for plantation development. Highly refined tissue culture techniques (e.g. micro-grafting) are used to mass produce the selected germplasm overcoming the limitations like season and space requirement usually faced in seedling nurseries. The share of plantations raised with such clonal planting material has been steadily increasing especially in the exotic areas of West Africa and central and South America. Clonal forestry in teak is a welcome development to realize the genetic gain from selection and breeding but it can only complement the sexual breeding programmes to achieve the breeding objective. In order to make more improved clones with new characters in terms of adaptability and timber production, the sexual breeding programme has to continue. The new recombinants generated for various characters may be subjected to mass multiplication through vegetative means to meet the demand of genetically improved planting material.

29.8.2 Eucalyptus

Eucalypts are the most widely planted hardwood trees throughout the world. The total plantation area is estimated to be over 20 million ha. India and Brazil have around four million ha each followed by China (GIT Forestry 2008). Eucalyptus plantations have a high socioeconomic importance wherever planted by meeting the domestic, environmental and industrial demand for wood and wood products. They are highly adaptable to a range of climatic and soil conditions, possess a fast growth rate, relatively free from insect and disease attack and produce wood suitable for many end-uses. The major industrial use of eucalyptus wood is for papermaking followed by plywood, solid timber products, charcoal production and fuelwood for domestic and industrial purposes. Essential oil and honey production are some of the niche end uses of eucalyptus trees. Eucalyptus plantations is considered as an important source of livelihood in many developing countries and widely regarded as helpful in easing the pressure on the natural forests for wood and wood products.

The eucalyptus group of plants has two genera, *Eucalyptus* and *Corymbya* having around 900 species naturally occurring in Australia, Papua New Guinea and parts of Indonesia. Of these around 20 species are under large-scale commercial cultivation throughout the world. They are *Eucalyptus grandis, E. urophylla, E. saligna, E. pellita, E. globulus, E. nitens, E. dunnii, E. benthamii, E. viminalis, E. smithii, E. camaldulensis, E. tereticornis, E. bassiana* (subgenus Symphyomyrtus), *E. cloeziana* (subgenus Idiogene), *E. pilularis* (subgenus Monocalyptus) belonging to the *Eucalyptus* genus and *Corymbia citriodora, C. torelliana* and *C. maculata* belonging to the genus *Corymbia*. All the plantation species have a chromosome number of $2n = 22$ but inter-breeding takes place only within the subgenus. Eucalyptus trees have a mixed mating system, that is, selfing takes place in the absence of outcrossed pollen. The extent of self-pollination is reported to be in the range of 10–35%. Insects are the major pollinators in many species, although other animals like birds also participate in pollination.

29.8.2.1 Structure of Seed Production Systems

The large area of planting, short rotations (particularly for pulpwood production) and the presence of large variation at various taxonomic levels have generated willingness to invest in genetic improvement of eucalyptus long ago. Systematic long-term breeding programmes were started in many countries to increase genetic gain for various commercially important characters. Expectedly seed orchards were the most preferred form of production populations in these programmes meeting the demand for genetically improved planting material till the 1980s when clonal forestry was adopted to maximize genetic gains in plantations.

Both types of seed orchards (SSO and CSO) have been reported for almost all widely planted species. SSO came to be established first, because progeny trials with selections from the natural populations are established in the beginning of a breeding programme. They underwent one or more thinning to remove inferior trees within average and above average families and all trees of poor families and get converted into seedling seed orchard. Since eucalyptus trees attain reproductive maturity early (4–8 years), SSO is an easy and cost-effective option of creating propagation population. In particular creating SSO in regions outside the natural distribution range of species is facilitated by easy movement of family seedlots. Most of the first-generation progeny tests that eventually served as SSOs were established with 50–100 families drawn from different provenances. The testing phase of these plots lasts up to half or two-thirds of the rotation period (4–6 years) after which they undergo genetic thinning to remove inferior trees and families. The multi-tree family plot is thinned to retain only the most outstanding tree to prevent mating between siblings and also to free space for the retained tree to develop a large crown for increased flower bud production.

Clonal seed orchards are usually established with the best subset of trees from the seedling seed orchards to further increase the genetic gain. Outstanding trees with high breeding values for the desirable characters from the best families are selected for inclusion in CSO. Such trees are propagated through grafting of scion of the selected tree on a root stock. Use of scion material from the reproductively mature trees results in early flowering of the trees in CSO. Since the number of clones is usually in the range of 20–40, the randomized complete block design is used to plant eucalyptus CSOs. A wider space than progeny tests is given between

Fig. 29.1 A potted hybridization orchard of *Corymbia* spp.

trees for early crown development and subsequent flowering. CSOs with grafts are also established with potted plants to facilitate easy access to crown for performing control pollination and other experiments like chemical induction of flowering (Fig. 29.1).

Unpedigreed seed orchards of *E. camaldulensis* (EC) and *E. tereticornis* (ET) were established in India with a broad genetic base of seeds collected from 506 trees belonging to 21 provenances that were found to be superior to local landrace in provenance tests. The bulked seedlot was used to plant an extensive seedling seed orchard or unpedigreed seedling seed orchard. Inferior trees were thinned at the age of 4–5 years before seed collection and supply to planting programmes (Varghese et al. 2004). A progeny trial established with around 200 families of identified provenances was converted to a seedling seed orchard. The gain from a thinned seed production area was found to be the same as that obtained from a pedigreed first-generation seedling seed orchard (Varghese 2014).

Seed supply from seed orchards of EC and ET began during 2006 in India. The demand for seeds has mainly been from paper and pulpwood industries, state forest departments, forest development corporations, farmers and NGOs. The annual supply of eucalyptus seeds ranges from 5 to 60 kg. This demand is influenced by various factors like planting targets, programmes and policies.

29.8.2.2 Functioning of Seed Production Systems and Genetic Gain

A vast majority of eucalyptus orchards have been reported to be functioning well in different parts of the world. They are connected with the operational planting programmes in different countries. A study conducted on fertility variation, outcrossing and genetic gain testing involving four unpedigreed orchards (two each of EC and ET) and one SSO of ET revealed new insights on orchard establishment and management particularly in exotic environments. An orchard of each species was planted in a high (1500 mm) and low rainfall (800 mm) areas in south India. At the age of

5 years, the proportion of flowering was the highest in the EC orchard in moist area with 73% of the 182 trees flowering. In the other three orchards, the flowering trees were lower ranging from 23 to 30%. An SSO of ET located in medium rainfall area (1100 mm) also showed low proportion of fertile trees at the age of 4 years. The flowering trees in turn showed a high level of variation in their fecundity. In the SSO where the provenance-family identity was retained, large variation was found in fertility and fecundity among provenances and families within provenance. The parameter called sibling coefficient (Ψ) is used to quantify the variation in the fertility among orchard trees. It is the probability that two genes originate from the same parent, compared to a panmictic situation and derived from the number of fertile trees in the orchard (N) and individual fertility (p_i) of each tree (Kang et al. 2003). The EC orchard in the moist area had the lowest fertility variation ($\Psi = 2.2$), whereas all other orchard had high variation ($\Psi = 6.7$ to 17.4). A few genotypes producing a major share of the orchard seed crop will increase related progeny resulting in lower genetic gain and genetic diversity (Varghese 2014).

The differences of fertility level and sibling coefficient observed across the species and orchard locations indicate that choosing a species and locating its orchard have to be decided after a careful study of various factors. Even within the species the large provenance variation, usually present in eucalyptus, will contribute towards high fertility differences. In the ET SSO, 12% of the trees belonging to the Indian landrace (Mysore gum) produced 52% of the fruits before thinning. The contribution increased to 81% after thinning to retain the best 20% of the trees. Expectedly the fertility variation among the retained trees was very high ($\Psi = 17.4$) (Varghese et al. 2009a). Such skewed contribution by different orchard trees will result in many entries with good phenotypes for desired characters may not be represented in the orchard's bulk seed collection leading to reduced gain and diversity. *Eucalyptus tereticornis* is generally reported to be a shy flowering species in tropical moist environments of Asia particularly in the first-generation orchards (Arnold 1996; Pinyopusarerk and Harwood 2003a, b). Kang et al. (2003) reported an average value of sibling coefficient 2.62 for seed orchards of broad-leaved species and in general, the values may be high in young orchards and during poor flowering years. The EC orchard in moist location functioned well as indicated by the low fertility variation of 2.2 which is lower than the mean values reported for broadleaved orchards. It is recommended to plant a larger number of trees if the orchard has to be located in a less optimal place for flowering. Large-sized orchards can offset low fertility up to an extent by retaining a high-status number. Pinyopusarerk and Harwood (2003b) recommended that seed collection may be made only when a minimum of 50% of trees in an orchard are fertile.

This variation in fertility among orchards of different species located in different environments had shown impact on the outcrossing in the orchard and the performance of their progeny. Eucalyptus trees have a mixed mating system, that is, both self- and cross-pollination occur in an individual depending upon the conditions prevailing in the population. The extent of selfing in eucalyptus is reported to be between 10 and 35% (Rezende et al. 2014). The progeny arising out of selfing displays low adaptability and vigour due to inbreeding depression (Eldridge et al. 1994). High level of outcrossing ($t_m = 0.95$) was reported for the Petford provenance in Queensland, Australia (Butcher and Williams 2002). This provenance has consistently performed the best in a range of trial sites including India. The outcrossing rate estimated from 20 progeny each of 15 trees in the EC unpedigreed orchard in a moist site was fairly high ($t_m = 0.86$). The individual values for the 15 mother trees ranged from 0.41 to 1.00 and selfing was detected among one-third of the trees. Higher level of outcrossing was reported from an EC progeny trial in Thailand ($t_m = 0.95$) and a seed orchard with bee hives in Queensland, Australia ($t_m = 0.94$) (Butcher and Williams 2002; Moncur et al. 1995). Although 73% of the trees were fertile in the Indian unpedigreed orchard of EC, 22% of 300 offspring from the 15 trees studied for outcrossing were full sibs, that is, some or all of the 15 open pollinated families received pollen from a few trees (Varghese et al. 2009a). The bulk seedlot used to establish the unpedigreed orchard was drawn from a large number of trees (506) belonging to 21 provenances. Expectedly there would have been variation among trees of different provenances for flowering phenology affecting the outcrossing in the orchard.

The flowering incidents were highly synchronous within provenances but differed between provenances in *E. nitens* and *E. Loxophleba* orchards (Gore and Potts 1995; Jones et al. 2011; Spencer et al. 2020).There was a 100-day difference between two races of *E. globulus* which means that the mating between the races is meagre. A moderate-to-strong heritability for many flowering traits including starting, peak and end of flowering period has been reported for *Eucalyptus* species (Gore and Potts 1995; Jones et al. 2011; Mora et al. 2009; Cané-Retamales et al. 2011; Spencer et al. 2020). These studies reiterate that an understanding of the flowering phenology at provenance, family and individual levels is essential for designing of orchards so that the flowering synchrony is maximized to increase outcrossing and genetic gain for the desired traits of commercial interest.

Silvicultural and ecological interventions have been widely adopted in eucalyptus orchard to increase flowering, pollination and seed set all of which eventually led to high outcrossing and genetic gain. Use of plant growth regulators like paclobutrazol has been widely practiced in eucalyptus to induce/increase flowering and to shorten generation time (Hasan and Reid 1995; Williams et al. 2003). Soil drenching of seed orchard trees of *E. camaldulensis* and *E. tereticornis* increased the proportion of fertile trees up to four times in different locations. Treated trees continued to produce more flowers than the untreated trees (control) even 3 years after application (Varghese et al. 2009b). Insects are the major pollinating agents in eucalyptus. Even in a highly fertile orchard of eucalyptus, the services of pollinator insects are essential to effect maximum seed set and outcrossing. The location of the orchard needs to support the presence of bees throughout the year so that their services are available during the eucalyptus flowering season. Knowledge on the behaviour of the pollinating insects is also essential to determine isolation distance for orchards particularly in areas surrounded by plantations of unknown/unimproved sources. Although the longest distance to which pollen can be transferred differs among eucalyptus species, most of the transfers occur locally (Potts and Wiltshire 1997). Increasing the pollen isolation distance from 400 to 800 m in a CSO reduced the pollen contamination from 14.2 to 2.8% (Campinhos et al. 1998; Junghans et al. 1998).

A genetic gain trial was conducted with bulk seedlots from four unpedigreed orchards of *E. camaldulensis* along with a local seedlot and an Australian provenance as control entries. At the age of 3 years, the height and diameter growth of progeny of the four orchards were comparable but were faster growing than the local seedlot. Overall, the orchard progeny showed 17 and 14% gain for height and diameter respectively over the local seed lot. Although many long-term breeding programmes have been implemented for different species of eucalyptus, only a few published reports are available on quantification of increase in plantation productivity (White et al. 2007; Harwood 2014). In South Africa, seed from first-generation SSO of *E. nitens* delivered an average volume increase of 62% over the unimproved planting material (Swain et al. 2013). The realized average genetic gain in volume production from three generations of *E. grandis* breeding programme in South Africa was 14% per generations (Verryn et al. 2009).

29.8.3 Casuarina

Casuarina trees belong to the family Casuarinaceae naturally occurring in Australia, Pacific, Oceania, Melanasia and SE Asia. Around 90 species of this family are distributed among four genera, namely *Allocasuarina, Casuarina, Gymnostoma* and *Ceuthostoma*. Among them *Casuarina equisetifolia, C. junghuhniana, C. cunninghamiana* and *C. glauca* are widely planted throughout the tropics for various services and end-uses. It is estimated that around 800,000 ha are planted with casuarina mainly in China and India (Pinyopusarerk and Williams 2000). They are nitrogen-fixing species through actinorhizal symbiotic association with the actinomycete Frankia. It confers them the ability to grow in a range of climatic conditions and soil types. The principal uses of the wood are fuel wood, pulpwood for papermaking, core veneer, charcoal and minor implements, while the major environmental services include windbreak, shelterbelts, reclamation of salt-affected and mined areas and bioenergy. All these benefits coupled with the ease of growing casuarina trees in short-rotation high density plantations made their extensive cultivation by farmers of smallholding in China and India. A large quantity of seed and other types of planting material is required for raising the plantations.

Casuarina trees are dioecious; that is, female and male flowers are borne in different individuals and thus obligatory outbreeders. In general, the proportion of the two sexes is equal in a population, although a small percentage of monoecious (male and female flowers borne on the same individual) is also reported for *Casuarina equisetifolia*. Casuarinas are wind-pollinated and floral structures are highly adapted to send and receive pollen. Flowers lack perianth, while stigma and stamens constitute the female and male flowers respectively. Female inflorescenses are borne on the axils of woody permanent branches, whereas the male inflorescences are terminal in the deciduous branchlets (commonly called 'needles'). Pollen which is dry and light is produced in large quantities and carried by wind. Upon pollination, the entire female inflorescence develops into a woody infructescence (commonly called 'cones'). Each valve of the infructescence has a winged fruit called samara. The samara is generally referred as the 'seed' in casuarina nursery (Pauldasan et al. 2022, 2023).

Till recently almost the entire planting of casuarina was done through seed raised plants. Use of clones has been in practice in China and India for the past two decades and is rapidly increasing. Still the quantity of seed required for planting programmes is large, for example, around two tonnes a year for south India. Seeds are traditionally collected from the existing plantations especially when the trees are harvested for wood. The presumably narrow genetic base of original introduction and repeated use of seeds from unimproved sources to raise new plantations have resulted in low productivity and large variation among trees for various characters.

29.8.3.1 Seed Production Systems and Genetic Gain

The earliest tree improvement activities of *Casuarina equisetifolia* included selecting phenotypically superior trees from the plantations, propagating them vegetatively and establish clonal seed orchards. The ease of rooting soft wood cuttings from reproductively matured trees of casuarina was utilized to clone the plus trees selected for fast growth, straight stems and freedom from pests. The CSOs were established with around 80 clones with a ratio of 1:4 for male and female clones. The ramets were distributed in the orchard using a permuted neighbourhood design in which each male tree was surrounded by a ring of six female trees (Kumaravelu and Paramathma 2001). Randomized complete block design was also used to plant a CSO with eight female and four male clones (Warrier et al. 2001). These orchards became reproductively matured early (often from the second year after planting) due to the origin of the planting material used. They generally functioned well with adequate flowering and seed production probably due to the fact that all clones were selected from the locally adapted plantations. The major problem encountered in CSOs of *C. equisetifolia* was heavy mortality due to the blister bark disease caused by the fungal pathogen *Trichosporium vesiculosum*. Since the clones assembled in the orchard were drawn from a narrow genetic pool (local plantations), they were equally susceptible to the pathogen leading to the complete loss of the orchards (Soni et al. 1989; Warrier et al. 2001).

Breeding programmes involving long-term activities for *C. equisetifolia* and *C. junghuhniana* were begun in China and India during the 1990s starting with the testing of range-wide provenances and landraces. Many natural provenances showed faster growth than local landraces in all test locations. Large breeding populations were established as progeny trials involving open-pollinated families of superior provenances. These progeny trials had 100+ families drawn from 10 or more provenances and planted using incomplete block design to effectively handle the heterogeneity within the block. A row of four to five trees were planted for each family. The progeny trials were subjected two thinning to retain only the best tree in each plot so that mating between siblings is reduced. An overall ratio of 1:1 was ensured between female and male trees retained. But in practice, ensuring the presence of such gender balance took longer time than expected especially in the first-generation progeny trials. Originating from diverse geographical areas experiencing different soil and climatic conditions, there is a significant provenance variation for gender ratio, age of flowering, flowering phenology and fecundity. Casuarina trees generally start flowering from third year onwards and some landraces flower even earlier than that (Pinyopusarerk and Williams 2000; Nagarajan et al. 2006). But many of the natural provenances did not commence flowering at the age of 3 years when the first within-family thinning was planned in the breeding programme. The proportion of flowering trees of provenances from Thailand was less than 5% at this age, whereas Chinese and Indian land races had more than 50% of trees that were found flowering (Nicodemus et al. 2001). In the absence of information on their gender, thinning process was deferred till their reproductive maturity. Some of the provenances outstanding in terms of growth and stem form were found to be shy fruit bearers limiting their contribution to the expected genetic gain from the orchard (Nagarajan et al. 2006). The proportion of flowering trees increased to 80% at the age of 6 years and the final thinning in the orchards took place after that. As a result, the first-generation breeding cycle was completed in 10 years instead of 6 years as envisaged at the beginning of the breeding programme (Pinyopusarerk 1996; Nicodemus et al. 2011).

CSOs of casuarina provide opportunity to optimize the representation of male and female trees and their position in the orchard since the gender of the tree is already known. Since selfing and mating between siblings within the plot are prevented by dioecious nature, more freedom is available to choose the orchard design and the number of ramets deployed per clone. Since only the female clones will produce seeds, keeping a ratio of 1:1 for the two sexes will lead to seed production in only 50% of the trees in the orchard. On the other hand, reducing the male accessions will reduce diversity in the seed crop leading to low adaptability (White et al. 2007). A judicious way to balance seed production and genetic diversity is to have equal number of genotypes (clones) for both the sexes but twice the number of ramets for female clones (a ratio of 1:1 for clones and1:2 for male and female ramets). Since *C. equisetifolia* has two flowering seasons in a year (Nagarajan et al. 2006) which compensate reduction in seed production by the presence of male trees in the orchard.

A field layout for a CSO with 30 clones (15 clones each sex) is given in Fig. 29.2. Each row has ramets of a single female clone and every two rows of female clone alternates with a row of male clone. Within the male row, each clone is represented by a single ramet and they are randomly distributed. This arrangement helps every female clone having equal chance to be pollinated by all the male clones provided there is sufficient overlapping between their phenology. A few additional rows of male clones planted around the perimeter of the orchard to increase the orchard pollen pool and to function as barrier for pollen from other sources. These orchards are easy to plant, ensure clonal identity, silvicultural management (e.g. similar dose of fertilizer in a row with same clone) and seed collection (seeds from all trees in a row can be pooled as open-pollinated family). They are also suitable for production of interspecific hybrids through open pollination. In such cases, the clones of the two sexes belong to different species known to have sufficient synchronicity in their flowering phenology. It is also possible to raise these CSOs using potted plants (Fig. 29.3).

Although both SSOs and CSOs of casuarina were found to function well in terms of overall flowering and seed production, there are still concerns in terms of genetic gain and relatedness in seed orchard progeny. The large variation in fecundity of different provenances of families within an orchard impacts the quality of the orchard seed crop. Natural provenances were faster growing than the land races but produced on average just one fourth of the fruits per tree

	1	2	3	4	5	6	7	8	9	10	11	12	13	14	15	16	17	18	19	20	21	22	23	24	
1	B	B	B	B	B	B	B	B	B	B	B	B	B	B	B	B	B	B	B	B	B	B	B	B	1
2	B	F1	F2	M1	F3	F4	M12	F5	F6	M14	F7	F8	M4	F9	F10	M12	F11	F12	M6	F13	F14	M9	F15	B	2
3	B	F1	F2	M2	F3	F4	M1	F5	F6	M4	F7	F8	M3	F9	F10	M9	F11	F12	M5	F13	F14	M8	F15	B	3
4	B	F1	F2	M3	F3	F4	M13	F5	F6	M12	F7	F8	M9	F9	F10	M13	F11	F12	M12	F13	F14	M13	F15	B	4
5	B	F1	F2	M4	F3	F4	M11	F5	F6	M5	F7	F8	M10	F9	F10	M10	F11	F12	M2	F13	F14	M7	F15	B	5
6	B	F1	F2	M5	F3	F4	M9	F5	F6	M2	F7	F8	M5	F9	F10	M8	F11	F12	M15	F13	F14	M2	F15	B	6
7	B	F1	F2	M6	F3	F4	M3	F5	F6	M6	F7	F8	M8	F9	F10	M11	F11	F12	M3	F13	F14	M3	F15	B	7
8	B	F1	F2	M7	F3	F4	M8	F5	F6	M15	F7	F8	M13	F9	F10	M6	F11	F12	M11	F13	F14	M10	F15	B	8
9	B	F1	F2	M8	F3	F4	M14	F5	F6	M7	F7	F8	M14	F9	F10	M7	F11	F12	M1	F13	F14	M4	F15	B	9
10	B	F1	F2	M9	F3	F4	M6	F5	F6	M11	F7	F8	M1	F9	F10	M1	F11	F12	M13	F13	F14	M6	F15	B	10
11	B	F1	F2	M10	F3	F4	M15	F5	F6	M10	F7	F8	M12	F9	F10	M2	F11	F12	M10	F13	F14	M11	F15	B	11
12	B	F1	F2	M11	F3	F4	M4	F5	F6	M13	F7	F8	M15	F9	F10	M3	F11	F12	M14	F13	F14	M1	F15	B	12
13	B	F1	F2	M12	F3	F4	M2	F5	F6	M9	F7	F8	M6	F9	F10	M5	F11	F12	M9	F13	F14	M15	F15	B	13
14	B	F1	F2	M13	F3	F4	M5	F5	F6	M3	F7	F8	M11	F9	F10	M4	F11	F12	M7	F13	F14	M14	F15	B	14
15	B	F1	F2	M14	F3	F4	M7	F5	F6	M1	F7	F8	M7	F9	F10	M15	F11	F12	M8	F13	F14	M12	F15	B	15
16	B	F1	F2	M15	F3	F4	M10	F5	F6	M8	F7	F8	M2	F9	F10	M14	F11	F12	M4	F13	F14	M5	F15	B	16
17	B	B	B	B	B	B	B	B	B	B	B	B	B	B	B	B	B	B	B	B	B	B	B	B	22
	1	2	3	4	5	6	7	8	9	10	11	12	13	14	15	16	17	18	19	20	21	22	23	24	

Fig. 29.2 A field lay out of CSO of a dioecious tree, *Casuarina equisetifolia* (*F*—female clones, *M*—male clones, *B*—border row, male clones)

Fig. 29.3 A potted plant of *Casuarina equisetifolia* with mature fruits

which limit the transfer of their breeding values to next generation (Nagarajan et al. 2006). Location of the orchard also impacted the fecundity of seed orchards of *C. equisetifolia* and *C. junghuhniana*. Seed orchards located in the coastal sites of India performed better than those planted in the inland sites.

Fertility variation assessed through sibling coefficient was less than 3 in the coastal orchards of both the species which is reported as acceptable for broadleaved species by Kang et al. (2003). Due to high variation in flowering and fruit production among different provenances, the inland orchards showed a high sibling coefficient of 5.83 and 7.06 for *C. equisetifolia* and *C. junghuhniana* respectively (Varghese et al. 2004). Locating the orchards in sites favourable for uniform flowering of all orchard trees and targeted seed collection from all genetic entries of the orchard are recommended to ensure high genetic gain and diversity in the seed crop. The volume production of progeny of the four orchards did not differ significantly among themselves and with that of a natural provenance from Thailand when tested in genetic gain trials (Varghese et al. 2011). But seeds from orchards produced 50 and 112% gain in wood volume in plantations of *C. equisetifolia* and *C. junghuhniana* respectively over the unimproved local seed source (Nicodemus et al. 2011).

29.9 Future Research Needed to Improve Seed Production and Deployment

Seed orchard development is a long-term activity. Even for short rotation trees like casuarina and eucalyptus, a minimum of 5 years is required for normal seed production to commence and the orchard should be maintained for at least another 5 years for seed collection. Unless an organization is committed to provide the required financial and human resource needs for 10 years, deriving the intended benefits from the orchards will be difficult. In case of long rotation species like teak, the commitment period may be 30 years and more. Considering this factor, it is essential to revisit all the existing orchards, recording their current status and potential to meet the current seed demand and document them to increase accessibility for the prospective seed users. Seeds from orchards producing consistent and commercially viable quantities of seed should be subjected to genetic gain testing and the realized gains should be utilized for justifying the investments to be made on the orchard maintenance and for fixing the seed price. Forward and backward selections need to be carried out for thinning and to identify next-generation orchards.

In order to effectively design, establish and manage productive seed orchards, basic research on the reproductive biology and mating system of the trees under genetic improvement has to be stepped up. Information on flowering phenology, fertility variation, pollinator behaviour, mating pattern and seed germination are still inadequate for overcoming problems that affect seed production and realizing genetic gain from orchards. Seed orchards provide the best opportunity to study the basic aspects of biology of tree species and apply them for the larger benefits of society (Kang 2001).

Lessons Learnt

- A healthy seed production is essential for both natural and artificial regeneration of forests.
- Sourcing the seeds from known and suitable seed producing population is a perquisite for successful plantation development.
- Immediate seed requirement is met from seed production areas which are matured plantations thinned heavily.
- Seed orchards are specialized plantations established with propagules obtained from trees selected for desirable characters.
- Clonal seed orchards are established with vegetatively propagated planting material and suitable for trees that take long time to attain reproductive maturity.
- Seedling seed orchards are planted with seedlings raised from seeds of selected trees and work well for species that flower at young age.
- Orchards have to be located in sites suitable for the species to flower and fruit and planted using statistical designs that minimize mating among related individuals.
- Several factors like low flowering, asynchronous phenology, limited pollination and pollen contamination from inferior sources affect both quantity and quality of seed production.
- Appropriate silvicultural practices have to be followed to promote seed production and increase the quality of seed crop.
- Seed production systems are dynamic in nature and have to advance constantly to meet the demands of changing edaphic and climatic conditions and emerging socioeconomic needs.

Key Questions

1. What are the functions of seed production systems of forestry species?
2. What are the types of seed orchards and explain the how to choose them?
3. Explain the factors that affect seed production in seed orchards.
4. What are the silvicultural practices to be followed for increasing seed production in seed orchards?
5. What is the extent of genetic gain obtained for different traits from seed orchard seeds? Explain the methods to balance genetic gain and genetic diversity in seed crops.

References

Adams WT, Burczyk J (2000) Magnitude and implications of gene flow in gene conservation reserves. In: Young A, Boshier D, Boyle T (eds) Forest conservation genetics: principles and practice. CSIRO Publishing, Collingwood, VIC, pp 215–244

Andersson B, Elfving B, Persson T, Ericsson T, Kroon J (2007) Characteristics and development of improved *Pinus sylvestris* in northern Sweden. Can J For Res 37(1):84–92

Arnold RJ (1996) Eucalyptus genetic improvement. CSIRO, Canberra. [Unpublished report]

Askew GR (1992) Potential genetic improvement due to supplemental mass pollination management in conifer seed orchards. For Ecol Manag 47:135–147

Balu A, Nicodemus A, Murugesan S, Asokan M, Pauldasan A (2016) The extent of infestation by the seed pest *Bootanelleus orientalis* Mathur and Hussey among family seedlots from breeding orchards of *Casuarina equisetifolia* and *C. junghuhniana*. In: Nicodemus A, Pinyopusarerk K, Zhong CL, Franche C (eds) Casuarina improvement for securing rural livelihoods. Institute of Forest Genetics and Tree Breeding, Coimbatore, p 260

Bell GD, Fletcher AM (1978) Computer organized orchard layouts (COOL) based on the permutated neighborhood design concept. Silvae Genetica 27:223–225

Bondesson L, Lindgren D (1993) Optimal utilization of clones and genetic thinning of seed orchards. Silvae Genetica 42:157–163

Butcher PA, Williams ER (2002) Variation in outcrossing rates and growth in *Eucalyptus camaldulensis* from the Petford Region, Queensland: evidence of outbreeding depression. Silvae Genetica 51(1):6–12

Callister AN, England N, Collins S (2013) Predicted genetic gain and realized gain in stand volume of *Eucalyptus globulus*. Tree Genet Genom 9(2):361–375

Campinhos EN, Peters-Robinson I, Bertolucci FL, Alfenas AC (1998) Interspecific hybridization and inbreeding effect in seed from a *Eucalyptus grandis* × *E. urophylla* clonal orchard in Brazil. Genet Mol Biol 21(3):369–374

Cané-Retamales C, Mora F, Vargas-Reeve F, Perret S, Contreras-Soto R (2011) Bayesian threshold analysis of breeding values, genetic correlation and heritability of flowering intensity in *Eucalyptus cladocalyx* under arid conditions. Int J Plant Breed 178:177–183. https://doi.org/10.1007/s10681-010-0292-y

Carson SD, Kimberley MO, Hayes JD, Carson MJ (1999) The effect of silviculture on genetic gain in growth of *Pinus radiata* at one-third rotation. Can J For Res 29:1979–1984. https://doi.org/10.1139/x99-152

Cotterill PP (2001) Enterprise and leadership in genetic project breeding: nucleus and cluster strategies. In Proceedings of the International Union of Forest Research Organizations (IUFRO) Congress on Developing the Eucalypt of the Future. Valdivia, Chile, pp 12–16

Di-Giovanni F, Kevan PG, Arnold J (1996) Lower planetary boundary layer profiles of atmospheric conifer pollen above a seed orchard in northern Ontario, Canada. For Ecol Manag 83:87–97

Eldridge K, Davidson J, Hardwood C, van Wyk G (1994) Eucalypt domestication and breeding. Oxford University Press, Oxford

El-Kassaby YA (2003) Clonal-row vs. random seed orchard designs: mating pattern and seed yield of western hemlock (*Tsuga heterophylla* (Raf.) Sarg.). For Genet 10:121–127

El-Kassaby YA, Lstiburek M (2009) Breeding without breeding. Genet Res 91:111–120

El-Kassaby YA, Ritland K (1986) The relation of outcrossing and contamination to reproductive phenology and supplemental mass pollination in a Douglas-fir seed orchard. Silvae Genetica 35:240–244

El-Kassaby YA, Stoehr MU, Reid D, Walsh CG, Lee TE (2007) Clonal-row versus random seed orchard designs: interior spruce mating system evaluation. Can J For Res 37:690–696

Giertych M (1975) Seed orchard designs. In: Faulkner R (ed) Seed orchards. UK Forestry Commission, London, pp 25–37

GIT Forestry (2008) Global cultivated eucalypts forest map 2008. http://www.git-forestry.com

Goh DKS, Bacilieri R, Chaix G, Monteuuis O (2013) Growth variations and heritabilities of teak CSO-derived families and provenances planted in two humid tropical sites. Tree Genet Genom 9:1329–1341

Gore PL, Potts BM (1995) The genetic control of flowering time in *Eucalyptus globulus*, *E. nitens* and their F_1 hybrids. In: Potts BM, Borralho NMG, Reid JB, Cromer RN, Tibbits WN, Raymond CA (eds) Eucalypt plantations: improving fibre yield and quality. CRC for Temperate Hardwood Forestry, pp 241–242

Graudal L, Moestrup S (2017) Genetic resources conservation and management. In: Kollert W, Kleine M (eds) The global teak study. Analysis, evaluation and future potential of teak resources, vol 36. IUFRO World Series, pp 18–29

Griffin AR (1989) Sexual reproduction and tree improvement strategy—with particular reference to eucalyptus. In: Gibson GI, Griffin AR, Matheson AC (eds) Breeding tropical trees: population structure and genetic improvement strategies in clonal and seedling forestry. Oxford Forestry Institute, Oxford, pp 52–67

Griffin AR (1993) Potential for genetic improvement of eucalyptus in Chile. In: Barros S, Prado JA, Alvear C (eds) Proceedings of the Los Eucaliptosen el Desarrollo Forestal de Chile, Pucon, Chile, pp 1–25

Gunaga RP, Vasudeva R, Hanumantha M, Swaminath MH (1999) Blooming variation among clones of different provenances in teak. Myforest 35:237–247

Hai PH, Harwood C, Kha LD, Pinyopusarerk K, Thinh HH (2008) Genetic gain from breeding *Acacia auriculiformis* in Vietnam. J Trop For Sci 20:313–327

Hardi T, Kanso A (1995) The role of European honey bees (*Apis melliferaL.*) in the pollination of a eucalypt plantation in Pati. Honeybee Sci 16:24–26

Harwood CE (2014) Classical genetics and traditional breeding. In: Henry R, Kole C (eds) Genetics, genomics and breeding of eucalypts. CRC Press, New York, NY, pp 11–33

Hasan O, Reid JB (1995) Reduction of generation time in *Eucalyptus globulus*. Plant Growth Reg 17:53–60

Hedegart T (1973) Pollination of teak (*Tectona grandis* L.). Silvae Genetica 22:124–128

Huang S, He P, Chen B, Huang K, Chen Y, Wang S, Luo Z, Zhou Y (2003) Genetic gains in growth of *Eucalyptus urophylla* from seed orchards in West Guangdong. Eucalyptus Plantations, pp 186–191. https://doi.org/10.1142/9789812704504_0015

Indira EP, Basha SC (1999) Effect of seeds from different sources on germination and growth in teak (*Tectona grandis* L.f.) nursery. Ann For 7:39–44

Indira EP, Mohanadas K (2002) Intrinsic and extrinsic factors affecting pollination and fruit productivity in teak (*Tectona grandis* Linn. f.). Indian. J Genet 62(3):208–214

ITTO (2017) Biennial review and assessment of the world timber situation 2015–2016. International Tropical Timber Organization, Yokohama, p 223

Jayasankar S, Babu LC, Sudhakara K, Kumar PD (1999) Evaluation of provenances for seedling attributes in teak (*Tectona grandis* Linn. f.). Silvae Genetica 48:115–122

Jones RC, Vaillancourt R, Gore P, Potts BM (2011) Genetic control of flowering time in *Eucalyptus globulus* ssp. *globulus*. Tree Genet Genom 7:1209–1218. https://doi.org/10.1007/s11295-011-0407-1

Junghans TG, Peters-Robinson I, Bertolucci FL, Alfenas AC (1998) The use of self-incompatibility in the production of hybrid *Eucalyptus* seed by 'Aracruz Celulose' in Brazil. Genet Mol Biol 21(3):375–379

Kang KS (2001) Genetic gain and gene diversity of seed orchard crops, vol 187. Acta Universitatis Agriculturae Sueciae, Silvestria, p 75

Kang KS, Bila AD, Harju AM, Lindgren D (2003) Estimation of fertility variation in forest tree populations. Forestry 76(3):329–344

Kaosa-ard A (1996) Teak (*Tectona grandis* Linn. f.) domestication and breeding. UNDP/FAO, Los Banos, Phillipines, RAS/91/004

Kaosa-ard A, Suangtho V, Kjær ED (1998) Experience from tree improvement of teak (*Tectona grandis*) in Thailand. Technical note no. 50. Danida Forest Seed Centre, Humlebæk

Katwal RPS (2005) Teak in India: status, prospects and perspectives. In: Bhat KM, Nair KKN, Bhat KV, Muralidharan EM, Sharma JK (eds) Quality timber products of teak from sustainable forest management. Proceedings of the IUFRO International Conference held between 2–5 December 2003 at Peechi, India. Kerala Forest Research Institute, Peechi, Kerala, India and ITTO, Yokohama, Japan, pp 1–18.

Kjær ED, Foster GS (1996) The economics of tree improvement of teak (*Tectona grandis* L.). Technical note no. 43. Danida Forest Seed Centre, p 22

Kjaer ED, Suangtho V (1995) Outcrossing rate of teak (*Tectona grandis* (L.)). Silvae Genetica 44:175–177

Kjaer ED, Suangtho V (1997) A review of the tree improvement plan for teak in Thailand. Report, Royal Forest Department, Thailand and Danida Forest Seed Centre, Denmark

Kollert W, Walotek PJ (2017) World teak resources, production, markets and trade. In: Kollert W, Kleine M (eds) The global teak study. Analysis, evaluation and future potential of teak resources, IUFRO World Series, vol 36, pp 83–89

Kumar A, Gogate MG, Sharma R, Mandal AK (1997) Genetic evaluation of teak clones of Allapalli region, Maharashtra. Indian Forester 12(3):187–189

Kumar A, Rathore TS, Palanisamy K, Viswanath S (eds) (2016) Advances in improvement and forest genetic resources conservation and management, ICFRE State of Knowledge Series III. Greenfields Publishers, Dehradun, p 459p

Kumaravelu G, Paramathma M (2001) Stability of open pollinated families of *Casuarina equisetifolia*. In: Gurumurthi K, Nicodemus A, Siddappa (eds) Casuarina improvement and utilization. Institute of Forest Genetics and Tree Breeding, Coimbatore, pp 51–56

La Bastide JGA (1967) A computer programme for the layouts of seed orchards. Euphytica 16:321–323

Leksono B, Kurinobu S, Ide Y (2008) Realized genetic gains observed in second generation seedling seed orchards of *Eucalyptus pellita* in Indonesia. J For Res 13:110–116. https://doi.org/10.1007/s10310-008-0061-0

Lindgren D (1974) Aspects on suitable number of clones in a seed orchard. In Proceedings of the IUFRO Joint Meeting of Working Parties S20401-3 on Population and Ecological Genetics, Breeding Theory, and Progeny Testing. Department of Forest Genetics, the Royal College of Forestry, Stockholm, Sweden, pp 293–305

Lindgren D (2005) India and low budget breeding. http://daglindgren.upsc.se/Reflections/Default.htm

Lindgren D, Matheson AC (1986) An algorithm for increasing the genetic quality of seed from seed orchards by using the better clones in higher proportions. Silvae Genetica 35:173–177

Lindgren D, Mullin TJ (1998) Relatedness and status number in seed orchard crops. Can J For Res 28:276–283

Lindgren D, Wei RP (2007) Low-input tree breeding strategies. In Proceedings of the IUFRO division 2 joint conference: low input breeding and conservation of Forest genetic resources (Ed. F. Isik). Antalya, Turkey, 9–13 October 2006, pp 124–138

Lindgren D, Wei RP (2014) Low input breeding strategies. https://doi.org/10.1142/9789812704504

Lindgren D, Danusevicius D, Rosvall O (2009) Unequal deployment of clones to seed orchards by considering genetic gain, relatedness and gene diversity. Forestry 82:17–28

Mandal AK, Chawhaan PH (2003) Investigations on inheritance of growth and wood properties and their interrelationships in teak. In: Bhat KM, Nair KKN, Bhat KV, Muralidharan EM, Sharma JK (eds) Quality timber products of teak from sustainable forest management. Proceedings of the IUFRO International Conference held between 2–5 December 2003 at Peechi, India, Kerala Forest Research Institute, Peechi, Kerala, India and ITTO, Yokohama, Japan, pp 506–510

Mathew G, Koshy MP, Mohanadas K (1987) Preliminary studies on insect visitors to teak (*Tectona grandis* Linn f) inflorescence in Kerala, India. Indian For. 113:61–64

McKeand SE, Mullin TJ, Byram TD, White TL (2003) Deployment of genetically improved loblolly and slash pines in the south. J For 101: 32–37

Mohanadas K, Anand G, Florence EJM, Radhakrishnan K, Indira EP (1999) June drop in teak is caused by fungi. Evergreen 43:14

Moncur MW, Charles R, Kleinschmidt GJ (1993) Can honeybees increase seed production in blue gum, *Eucalyptus globulus*? Australian Beekeeper 95:248–253

Moncur MW, Mitchell A, Fripp Y, Kleinschmidt GJ (1995) The role of honey bees (*Apis mellifera*) in eucalypt and acacia seed production areas. Commonwealth For Rev 74:350–354

Monteuuis O, Goh D (2017) Origin and global dissemination of clonal material in planted teak forests. In: Kollert W, Kleine M (eds) The global teak study. Analysis, evaluation and future potential of teak resources, IUFRO World Series, vol 36, pp 30–36

Mora F, Gleadow R, Perret S, Scapim C (2009) Genetic variation for early flowering, survival and growth in sugar gum (*Eucalyptus cladocalyx* F. Muell) in southern Atacama Desert. Int JPlant Breed 169:335–344. https://doi.org/10.1007/s10681-009-9962-z

Nagarajan B, Gireesan K, Venkatasubramanian N, Shanthi A, Mandal AK (1996a) An early evaluation of gene action in teak. Myforest 32: 136–139

Nagarajan B, Varghese M, Nicodemus A, Sasidharan KR, Bennet SSR, Kannan CS (1996b) Reproductive biology of teak and its implication in tree improvement. In: Dieters MJ, Matheson AC, Nickles DG, Harwood CE, Walker SM (eds) Tree improvement for sustainable tropical forestry. Proceedings of QFRI-IUFRO Conference, Caloundra, Australia, pp 244–248

Nagarajan B, Nicodemus A, Sivakumar V, Mandal AK, Kumaravelu G, Jayaraj RSC, Narmatha Bai V, Kamalakannan R (2006) Phenology and control pollination studies in *Casuarina equisetifolia* Forst. Silvae Genetica 55(4–5):149–155

Nanda KK (1962) Some observations on growth, branching behaviour and flowering of teak (*Tectona grandis* L.f.) in relation to light. Indian Forester 88:207–218

Neelay VR, Bhandari RS, Negi KS (1983) Effect of insecticidal and hormonal spray on the production of fruits in teak seed orchards. Indian For 109:829–837

Nicodemus A, Varghese M, Nagarajan B (2001) Genetic improvement of *Casuarina equisetifolia* through selection and breeding. In: Gurumurthi K, Nicodemus A, Siddappa (eds) Casuarina improvement and utilization. Institute of Forest Genetics and Tree Breeding, Coimbatore, pp 3–13

Nicodemus A, Nagarajan B, Jacob JP (2005) Pollination by nectarivorous birds in teak clonal seed orchards. Indian Forester 131:1613–1616

Nicodemus A, Varghese M, Nagarajan B, Lindgren D (2009) Annual fertility variation in clonal seed orchards of Teak (*Tectona grandis* L. f.) and its impact on seed crop. Silvae Genetica 58:85–93

Nicodemus A, Gurudev Singh B, Krishna Kumar N, Pinyopusarerk K (2011) Improved Casuarina for farm forestry in south India and its socio-economic implications for small-holding farmers. In: Zhong CL, Pinyoupusarerk K, Kalinganire A, Franche C (eds) Improving smallholder livelihoods through improved casuarina productivity. Proceedings of the 4th International Casuarina Workshop. Chinese Forestry Publishing House, Beijing, pp 31–37

Nikkanen T (2002) Functioning of a Norway spruce (*Picea abies* (L.) Karst.) seed orchard. Finnish Forest Research Institute, Research papers 850, p 58

Pauldasan A, Vipin P, Durai A, Nicodemus A (2022) Hybridization, pollen-pistil interactions and hybrid seed set among *Casuarina* species. Aust J Bot 70:174–186. https://doi.org/10.1071/BT21120

Pauldasan A, Vipin P, Durai A, Mayavel A, Gideon AV, Nicodemus A (2023) Floral biology, pollen viability and stigma receptivity in three species of *Casuarina*. S Afr J Bot 152:182–191

Pinyopusarerk K (1996) A breeding programme for *Casuarina equisetifolia* in India. CSIRO Forestry and Forest Products, Canberra, p 39

Pinyopusarerk K, Harwood CE (2003a) Flowering and seed production in tropical eucalyptus seed orchards. In: Turnbull JW (ed) Proceedings of International Conference on Eucalypts in Asia, 7–11 April. ACIAR, Canberra, pp 247–248

Pinyopusarerk K, Harwood CE (2003b) Surveys of seed orchards of key Australian tree species. Client Report No. 1242. Australia, CSIRO Forestry and Forest Products, p 58

Pinyopusarerk K, Williams ER (2000) Range-wide provenance variation in growth and morphological characteristics of *Casuarina equisetifolia* grown in Northern Australia. For Ecol Manag 134(1): 219–232. https://doi.org/10.1016/S0378-1127(99)00260-1

Potts BM, Wiltshire RJE (1997) Eucalypt genetics and genecology. In: Williams JE, Woinarski JCZ (eds) Eucalypt ecology: individuals to ecosystems. Cambridge University Press, pp 56–91

Powell GL, White TL (1994) Cone and seed yields from slash pine seed orchards. South J Appl For 18:122–127

Prescher F (2007) Seed orchards—genetic considerations on function, management and seed procurement. Doctoral thesis, Swedish University of Agricultural Sciences, Umea, p 49. https://pub.epsilon.slu.se/1517/1/FPfin0.pdf. Accessed 24 June 2021

Rezende GDSP, de Resende MDV, de Assis TF (2014) Eucalyptus breeding for clonal forestry. Challenges and opportunities for the world's forests in the 21st century (Ed. T. Fenning). For Scinces 81: 393–424. https://doi.org/10.1007/978-94-007-7076-8_16

Rosvall O, Jansson G, Andersson B, Ericsson T, Karlsson B, Sonesson J, Stener LG (2002) Predicted genetic gain from existing and future seed orchards and clone mixes in Sweden. Finnish Forest Research Institute, Helsinki. Research Papers 842, pp 71–85

Sampson JF, Byrne M (2008) Outcrossing between an agroforestry plantation and remnant native populations of *Eucalyptus loxophleba*. Mol Ecol 17:2769–2781

Schreiner EJ (1961) Clonal or seedling seed orchards? In: Ninth northeastern forest tree improvement conference. Syracuse, New York, NY, pp 53–57

Shelbourne CJA (1969) Tree breeding methods. N Z For Serv For Res Inst Tech Pap 55:1–43. Cited in Giertych 1975

Sonesson J, Bradshaw R, Lindgren D, Ståhl P (2001) Ecological evaluation of clonal forestry with-cutting propagated Norway spruce. Uppsala, Skogforsk. Report No. 1

Soni KK, Srivastava KK, Kalyani KB, Rajarishi R (1989) Studies on mortality in *Casuarina equisetifolia* clonal seed orchard at Trichy (Tamil Nadu). Proceedings of a National Seminar on Casuarinas. Institute of Forest Genetics and Tree Breeding, Coimbatore

Spencer B, Mazanec R, Abadi A, Gibberd M, Zerihun A (2020) Flowering phenology in a *Eucalyptus loxophleba* seed orchard, heritability and genetic correlation with biomass production and cineole: breeding strategy implications. Sci Rep 10(1):15303. https://doi.org/10.1038/s41598-020-72346-3

Stoehr MU, Orvar BL, Vo TM, Gawley JR, Webber JE, Newton CH (1998) Application of a chloroplast DNA marker in seed orchard management evaluations of Douglas-fir. Can J For Res 28:187–195

Subramanian K, Mandal AK, Ram Babu N, Chundamannil M, Nagarajan B (2000) Site, technology and productivity of teak plantations in India. In: Enters T, Nair CTS (eds) Site, technology and productivity of teak plantations. FORSPA Publication No. 24, FAO, Bangkok, pp 51–68

Suraj PG, Nagabhushana K, Kamalakannan R, Varghese M (2019) Impact of fertility variation on genetic diversity and phenotypic traits in second generation seed production areas and clonal seed orchards of *Eucalyptus camaldulensis*. Silvae Genetica 68:29–40. https://doi.org/10.2478/sg-2019-0006

Swain D, Mohanty SC, Sharma R, Mandal AK, Gupta BN (1996) Preliminary analysis of quantitative characters in teak. Proc Indian Natn Sci Acad B62:169–172

Swain TL, Verryn SD, Laing MD (2013) A comparison of the effect of genetic improvement, seed source and seedling seed orchard variables on progeny growth in *Eucalyptus nitens* in South Africa. Tree Genet Genom 9(3):767–778

Talbert JT, Weir RJ, Arnold RD (1985) Costs and benefits of a mature first-generation loblolly pine tree improvement programme. J For 83: 162–166

Tangmitcharoen S, Takaso T, Siripatanadilox S, Tasen W, Owens JN (2006) Behavior of major insect pollinators of teak (Tectonia grandis L.f.): a comparison of clonal seed orchard versus wild trees. For Ecol Manag 222:67–74

Toda R (1964) A brief review and conclusions of the discussion on seed orchards. Silvae Genetica 13:1–4

van Wyk G (1981) Pollen management for eucalypts. In: Pollen management handbook. United States Department of Agriculture Handbook No.587, Washington, DC, pp 84–88

Varghese M (2014) Domestication strategies for eucalypts in India. In: Eucalyptus in India. ENVIS, Dehradun, pp 21–59

Varghese M, Nicodemus A, Nagarajan B, Sasidharan KR, Siddappa B, S.S.R. and Subramanian, K. (2000) Seedling seed orchards for breeding tropical trees. Institute of Forest Genetics and Tree Breeding, Coimbatore, p 126

Varghese M, Nicodemus A, Lindgren D (2004) Fertility and effective population size in seedling seed orchards of *Casuarina equisetifolia* and *C. junghuhniana*. Silvae Genetica 53:164–168

Varghese M, Lindgren D, Ravi N (2006) Linear thinning in a clonal test of *Eucalyptus camaldulensis* for conversion to a clonal seed orchard. J Trop For Sci 18(2):102–108

Varghese M, Kamalakannan R, Nicodemus A, Lindgren D (2008) Fertility variation and its impact on seed crops in seed production areas and a natural stand of teak in southern India. Euphytica 160(1): 131–141

Varghese M, Kamalakannan R, Harwood CE, Lindgren D, McDonald MW (2009a) Changes in growth performance and fecundity of *Eucalyptus camaldulensis* and *E. tereticornis* during domestication in southern India. Tree Genet Genom 5(4):629–640

Varghese M, Ravi N, Kamalakannan R, Harwood CE (2009b) Effect of silvicultural treatments on growth, fertility and capsule traits in seedling seed orchards of *Eucalyptus camaldulensis* and *E. tereticornis*. New For 37(1):99–107

Varghese M, Kamalakanna R, Nicodemus A, Pinyopusarerk K (2011) Fertility and progeny performance in seedling seed orchards of *Casuarina equisetifolia* and *C. junghuhniana*. In: Zhong CL, Pinyoupusarerk K, Kalinganire A, Franche C (eds) Improving smallholder livelihoods through improved casuarina productivity. Chinese Forestry Publishing House, Beijing, pp 75–81

Vasudeva R, Gunaga R (2012) Flowering phenology in teak seed orchards—genetic variation, measurement and implications to seed orchard fertility. In: Zhang X (ed) Phenology and climate change. In Tech. https://doi.org/10.5772/35658

Verryn SD, Snedden CL, Eatwell KA (2009) A comparison of deterministically predicted genetic gains with those realized in a South African *Eucalyptus grandis* breeding program. Southern For 71(2):141–146

Warrier KCS, Venkataraman KS, Kumar A, Gurumurthi K (2001) Clonal seed orchards for quality seed production in *Casuarina equisetifolia*. In: Gurumurthi K, Nicodemus A, Siddappa (eds) Casuarina improvement and utilization. Institute of Forest Genetics and Tree Breeding, Coimbatore, pp 21–28

Wellendorf H, Kaosa-ard A (1988) Teak improvement strategy in Thailand. For Tree Improve 21:1–43

Wheeler NC, Jech KS (1986) Pollen contamination in a mature Douglas-fir seed orchard. In: Proceedings of the IUFRO Conference on Breeding theory, progeny testing and seed orchards. Williamsburg, VA, pp 160–1171

White TL, Adams WT, Neale DB (2007) Forest genetics. CABI Publishing, Wallingford

Williams DR, Potts BM, Smethurst PJ (2003) Promotion of flowering in *Eucalyptus nitens* by paclobutrazol was enhanced by nitrogen fertilizer. Can J For Res 33:74–81

Yazdani R, Lindgren D (1991) Variation of pollen contamination in a scots pine seed orchard. Silvae Genetica 40:243–246

Zobel B, Talbert J (1984) Applied forest tree improvement. John Wiley and Sons, New York, NY

Forest Seed Technology: Seed Biology, Collection, Quality Evaluation, Storage and Certification 30

Manisha Thapliyal, Karuna Phular, N. K. Namitha, Shweta Rawat, and Vaisakhy P. Chand

Abstract

The availability of the good-quality seed is the most important requirement for successful plantation programmes. The use of high-quality seed is very important because it increases the chance of early gains by improving the establishment success and forest stand performance due to enhanced genetic gain. Knowledge and information regarding the collection and processing of tree seeds, its physical and physiological quality, types and levels of seed dormancy, breaking of dormancy through different pre-treatments to enhance germination, seed storage physiology and longevity, etc. are important. Forestry species offer good scope for quality seed production owing to seed source variability which is being explored and utilized in tree improvement programmes. The guarantee that the new stock will have good survival, growth and shape as well as be better adaptable to stressful situations or severe climates comes from using seeds from carefully chosen, healthy, well-formed trees. The science of seed technology is multidisciplinary and covers a wide range of topics. Based on contemporary agricultural and forestry sciences, seed technology encompasses, in its broadest meaning, seed biology, production, processing, storage, testing, certification, quality control, marketing, distribution and sale of seeds. For the correct handling of seed as well as the management of seed sources, an understanding of seed biology is essential. The use of any species in the plantation programme is frequently restricted by issues with seed acquisition, handling, processing and lack of technology. This chapter covers various aspects like seed development, morphology and anatomy, seed collection methods and tools, seed extraction, processing and handling, seed quality evaluation, direct and indirect tests of viability, seed dormancy and pretreatments, seed drying and desiccation, storage, seed storage facilities, standards for seed certification and future scope of forest seed technology.

Keywords

Seed technology · Seed development · Maturity indices · Recalcitrant seed · Desiccation

30.1 Introduction

The aim of the draft National Forest Policy (2018) is to protect people's ecological security and means of subsistence, both for the current and future generations, by managing forests sustainably to maintain the flow of ecosystem services. The previous goal of keeping 33% of India's land covered by forests and trees is still in place, and in the country's hilly and mountainous areas, the goal is to have two-thirds of the land covered by these types of vegetation. Additionally, it highlights the 'promotion of trees outside forests and urban greens' and states that this will be the focus of a 'mission mode' initiative. This has increased the demand for quality seed and clonal material to raise new plantations and trees outside forests as well as enrichment plantations in the forest areas. Therefore, there is a pressing need to advance and broaden the current programmes into a mission mode in order to incorporate cutting-edge technologies for the manufacture of high-quality, certified seeds; establish a network of laboratories for testing seeds; assist organizations that produce seeds in implementing new technologies etc. It will increase the quantity of high-quality seeds available to the forest departments, tree growers, nurseries etc. in a time-bound manner.

Another serious concern for the humanity is the destruction and disappearance of natural forests and habitats. This calls for focusing towards conserving the species by stockpiling wild plant species or their seeds in large centralized seed banks which is a kind of ex situ conservation (Khanna and Singh 1991). Seed banks are a way to combat

M. Thapliyal (✉) · K. Phular · N. K. Namitha · S. Rawat · V. P. Chand
ICFRE-Forest Research Institute, Dehradun, Uttarakhand, India

A. K. Mandal, A. Nicodemus (eds.), *Textbook of Forest Science*, https://doi.org/10.1007/978-981-97-8289-5_30

threat over biodiversity by human interference. Seed banks help to conserve seeds in order to save plant genetic diversity. For conservation of seeds in seed banks, they have to be tested for their viability, so that they could be stored for a long period of time. Testing of seeds for their viability and quality requires different methods and technologies. Seed technology is the scientific field that considers seeds as living things and guarantees their productivity and quality. It encompasses knowledge of seed biology, quality, production, storage and certification.

Seed-producing plants are generally known as spermatophytes which comprise gymnosperms and angiosperms. Gymnosperms include conifers, cycads, ginkgo and gnetophytes. The term 'gymnosperm' comes from the word gymnos means 'naked' and sperma means 'seed' which produces seeds directly on sporophylls that are not contained within an ovary or fruit, while in angiosperms, seeds are enclosed within an ovary. Angiosperms are more advanced and adapted to a wide plethora of terrestrial ecosystems. Seeds can be produced by sexual reproduction or asexual reproduction. Sexual reproduction offers genetic diversity of a population, whereas asexual reproduction/apomictic or vegetative reproduction results in clones of genetic uniformity.

30.2 Seed

Seeds are mature ovules that form the dispersal and propagation unit of spermatophytes. The seed is the final product of the process of reproduction initiated by the flower in the plant. Seeds contain necessary cells, tissues, nutrients and hormones for growth and development of the new plant. They consist of three important parts. The embryo, which is a young sporophyte, diploid ($2n$) as a result of fertilization and at maturity consists of cotyledons, hypocotyls and radicle. The endosperm consists mainly of stored food in the form of starch. The seed coat (testa) is the outer most protective layer of the seed that mainly develops from outer integuments of ovule and diploid maternal tissue.

30.2.1 Seed Morphology and Anatomy

The mature angiosperm embryo has either a single cotyledon (monocots) or two cotyledons (dicots) on an embryonic axis. The radicle, hypocotyl, epicotyl and plumule (shoot apex) are all carried on the embryonic axis. Endospermic seeds have thin, flattened cotyledons, while non-endospermic seeds have well-developed cotyledons that act as storage organs for reserves. The embryo can be peripheral, axial or rudimentary. Endosperm acts as storage organ for the growing embryo. The reserves that have been transferred to the embryo during seed development determine the relative size of the endosperm. The ovule's nucellar area gives rise to the perisperm, a storage tissue, in certain species. The embryo, endosperm and perisperm are surrounded by a maternal tissue called the seed coat or testa. It safeguards the internal structures from biotic stresses and against desiccation and mechanical injury, assists in gas exchange and water uptake, provides a conduit for nutrients for the embryo and endosperm during their development and may be a means for seed dispersal (Beniwal and Singh 1985).

During seed development, the integuments, which are an initial seed coat tissue, differentiate into layers of specialized cells. Certain seed coat cell layers have the potential to accumulate significant amounts of substances like pigments, phenolics and mucilage. Certain angiosperm species have thick, colourful coats that draw animals, primarily birds. It is referred to as the sarcotesta. Three main groups of materials make up storage reserves: proteins, carbohydrates and oils (triacylglycerols) (Egli 1998). The growing seedling is dependent on the stored reserves in the seed for supply of nutrients to support its biosynthetic processes and energy requirements until it becomes autotrophic. Other non-storage compounds are present in low concentrations, such as in the testa or pericarp. During seed development, during histo-differentiation to create the embryo and endosperm, these reserves are deposited. The reserves in seeds containing non-persistent endosperms are first stored in this structure, where they are later repurposed for reserve synthesis in the cotyledons (Geneve 1996).

30.2.2 Seed Development

The process of seed formation involves the following steps:

- The formation and maturity of reproductive organs of the plant, that is, stamens and pistil in separate buds or in the same bud in species having hermaphrodite flowers.
- Pollination, in which the stigma of a flower receives male genetic material contained in pollen grains from the anther, through different dispersal agents such as wind, water and insects. Pollination taking place within the same plant is known as self-pollination or autogamy, whereas pollen delivered from a different plant is known as cross-pollination or allogamy. Plants frequently display self-incompatibility in order to inhibit self-pollination and to increase genetic diversity, which causes the rejection of pollen from the same flower or plant. The self-incompatible plants pollinate by cross-pollination.
- During fertilization, male and female gametes combine together to create a diploid zygote. This occurs by discharge of the sperm nuclei from the pollen tube and their delivery into the embryo sac, one fuses with the egg cell

nucleus and the second with the two polar nuclei of the central cell. As a result, a diploid embryo cell (zygote) and a triploid endosperm cell are created (Geneve 2005).

- Seed maturity is accompanied by significant chemical changes. Seed development can be divided into three stages namely, Phase I—histo-differentiation characterized by extensive cell divisions. In this stage, different tissues are formed inside the embryo and surrounding structures. Phase II—cell enlargement and expansion occur (little cell division, dry weight increases due to reserve deposition, water content decline), and Phase III—at physiological maturity, dry matter accumulation slows and eventually stops.
- In many species, the process of maturation drying occurs after physiological maturity, during which the seed loses water to a moisture content of 7–15% and becomes desiccation tolerant (orthodox seeds), making it resistant to harsh climatic circumstances. A seed is quiescent (showing little metabolic activity) at this point, and it might even be dormant. Recalcitrant seeds, on the other hand, are susceptible to desiccation during shedding and do not go through maturation drying. Because these seeds quickly become inactive when the water is removed through drying, they are challenging to preserve.

30.3 Seed Technology

The processes that could be used to improve the genetic and physical properties of seeds are known as seed technology (Feistritzer 1975). It involves such activities as seed production from known sources/varieties, collection, processing, storage and certification. The quality of a seed decides whether a seedling/plantation is good at the expected level or not. Quality of a seed is determined by the physical attributes (seed with minimum damage, without pathogens, without inert matter etc.); physiological attributes (germination percentage, viability, vigour etc.); genetic attributes (adaptations to varying temperature, moisture and other local conditions; disease and pest tolerance; high yielding ability etc.); and storability of a seed (temperature, moisture content, seed treatment etc.). The quality of seed can only be maintained through appropriate procedures. The sequential steps used for the production of quality seed/seedling are seed collection, seed extraction, processing and handling, seed quality tests, seed pre-treatment, seed drying and seed storage (Thapliyal et al. 2018).

30.3.1 Seed Collection

The initial stage in starting a plantation is gathering seeds, and the quality of the seeds directly affects how well each plantation operation goes. The early development and survival, vigour, disease and insect resistance, productivity and quality of a plantation are all influenced by the source of the seed. Plants that have been selected or bred to increase their genetic makeup can produce seed orchards that yield seeds of exceptional genetic quality.

Depending on the availability of seeds at a given time or season, seed collection can be prioritized. Certain tree species are seasonal, meaning they reproduce year-round in a largely continuous manner. A small number of ripe seeds, such as *Lagerstroemia* spp., may be available during each collection because the same tree contains fruits and seeds of different age classes. The seeds of trees in the second group should be harvested within that time frame because they have a short seed maturity period and are very season sensitive. Seeds such as *Mesua ferrea, Diploknema butyracea, Acacia* spp. and *Cassia* spp. are quickly lost or distributed within a short amount of time, despite the fact that a great number of seeds can be collected at once. The third group of trees, which includes *Delonix regia, Schizolobium* spp., *Melia* spp., and others, has a set maturity season but persists on the tree for a long time before dispersing (Bonner and Karrfalt 2008).

Recalcitrant seeds begin to lose viability as soon as they reach maturity. Therefore, it's critical to ascertain when seeds are mature and to harvest them appropriately. After maturing, recalcitrant *Diploknema butyracea* seeds lose viability in 2–3 weeks.

30.3.2 Time of Collection

Fruit harvesting is a crucial process that necessitates understanding when seeds should ripen and when high-quality seeds can be collected. Most tree species have irregular seed crops and in some cases seeding takes place at intervals of over 40 years, for example, bamboos. Seeds should be collected in their good seed years to obtain higher quality and quantity of seeds (Tomar 2016). Good seed year of a few selected species are as follows: *Dalbergia sissoo* 2–3 years, *Shorea robusta* 3–4 years, *Pinus roxburghii* 4–5 years, *Cedrus deodara* 4–5 years, *Abies pindrow* 10–11 years.

30.3.3 Maturity Indices

Generally, seeds should be harvested when fully ripe. Certain characteristics of fruits and seeds are known as maturity indicators because they help us determine when the fruit is ready to ripen. Depending on the species and fruit type, maturity indices can vary. A fruit's colour, size, taste, moisture content and abscission zone growth can all vary (Schmidt 2000). The majority of the time, fruit and seeds reach maturity around the same time. However, some seeds—like those of the *Fraxinus* species and *Ilex opaca*—need to ripen after fruit does. Conversely, still-green Fabaceae (Leguminosae) pods can yield germination seeds. Physical and chemical properties that are crucial for defining the collection method are included in maturity indices (Kumar 2015).

30.3.4 Physical Characteristics

Colour of fruits: Fruit colour change, as they mature is a significant and simple diagnostic indicator of maturity. Fruits of *Pterocarpus* change colour from green to yellow to brown or black; those of *Prunus* change colour from green to red to purple or black; and those of conifers change colour from green to brown.

Moisture content (MC): MC and protein synthesis are connected. It takes a drop in seed moisture below 60% to start protein synthesis. The growth of seedlings depends on this process. The moisture content of orthodox, dry seeds and fruits gradually drops as the seeds mature. In pulpy orthodox fruits, moisture levels initially drop and thereafter rise primarily in the pulp. Moisture levels in recalcitrant seeds rise in the early stages and then slightly falls as the seed reaches maturity.

Specific gravity: Fruit's specific gravity is influenced by internal factors such soluble solids, dry matter and physical problems. In conifers, specific gravity denotes seed maturation and serves as a gauge for moisture content (Schmidt 2000).

Seed should be harvested from a wide number of trees (at least 15–25 trees) to maintain genetic variety. Generally speaking, there should be at least 100 meters separating the chosen seed-bearing trees. Seeds harvested from nearby trees may have similar genetic characteristics, which can lead to inbreeding and less genetic variation in the ensuing generations, lowering predicted yields. A seed-bearing tree's progeny will exhibit good qualities when it is surrounded by many other good quality trees; conversely, if the seed-bearing tree is surrounded by many other inferior quality trees, the progeny will exhibit poor quality. Trees may be chosen using a variety of criteria, depending on their utility. It is best to choose timber species with a straight stem, few or no branches, a long, straight and clean merchantable bole, no buttress and a uniform crown, no disease, no insect pests and large volume of excellent quality timber. For fodder, choose fast-growing, nutritious trees with lots of branches and a high coppicing potential. Fruit trees should have large, healthy fruits that taste well and are abundant. They should also be free of pests and diseases (Thapliyal et al. 2018)

30.3.5 Methods of Seed Collection

30.3.5.1 Collection from Felled Trees

With the help of expert climbers or mechanical access, seeds are harvested from selected trees. In order to avoid wasting seeds and limiting their availability in the same location for future years, trees are not heavily lopped for the purpose of collecting seeds. Nonetheless, as part of routine forest maintenance procedures, seeds are harvested from trees that are cut down. It is possible to gather mature fruits from the top of recently felled trees. They can be manually gathered from tree branches that have been clipped

30.3.5.2 Collection from the Ground

Species that can be harvested from the ground are those with large, heavy seeds or enormous, unbroken fruits. *Quercus, Fagus* and *Castanea* seeds are among the temperate species that are frequently picked from the ground. Many tropical plants, including *Tectona, Gmelina*, *Triplochiton, Cassia, Acacia* and a few dipterocarps have seeds that are commonly collected from the ground.

30.3.5.3 Collection from the Standing Trees

When harvesting fruits from shrubs or low-branched trees, such as *Fraxinus* species, *Hardwickia binata* and *Pterocarpus* species, fruits can be collected while standing on the ground.

30.3.5.4 Collection Using Ladders, Laggi

For collection of fruits from medium-sized trees, the most suitable instruments are ladders and laggi/pole pruner (a long hollow, metal pole with a sickle at one end), for example, *Mallotus philippensis, Michelia champaca, Sapindus mukorossi* and *Enterolobium* spp. (FAO 1985).

30.3.5.5 Tools and Equipment

The following tools and equipment may be used for seed collection from standing trees:

(i) Bag or Basket.
(ii) Ladder.
(iii) Rope.
(iv) Climbing gear (climbing iron and spur).
(v) Laggi (long hollow metal rod with a sickle at one end).
(vi) Canvas or tarpaulin sheet.
(vii) Extension pruner, rake or cutting knife.
(viii) Cone picking knife and hook.

30.3.6 Seed Extraction, Processing and Handling

The process of removing seeds from their enclosing structures is known as seed extraction. In order to prevent quick desiccation during the extraction process, seeds should be mature before being extracted and should be done after fruits have ripened. Fruits that have not fully matured can be kept in cool, well-ventilated containers like trays with holes, wire-mesh bottoms and nylon net bags. Hard or dry fruits are best kept in thin layers in the shade; soft fruits require the storage area/chamber to be kept between 10 and 15 °C with enough humidity to prevent drying.

Extraction of seed from dry dehiscent fruits: By spreading the fruits on a tarpaulin sheet in the shade, seeds can be extracted from dry dehiscent fruits. Strong attachments between seeds with pods can be broken by hand threshing, gently tapping the pods with sticks within the sack or rubbing them against a rough surface. Dry dehiscent fruits include *Bauhinia, Desmodium, Eucalyptus* and *Acacia* spp.

Extraction of seed from dry indehiscent fruits: Small indehiscent fruits can be broken by rubbing them on a rough surface, threshing them by hand or gently pounding them with sticks. Seeds from large fruits can be extracted by hand or by mechanically dividing them. Several rounds of threshing and sporadic drying are necessary for pods containing sticky substance, such as in *Cassia fistula* and *Prosopis cineraria.* Dry indehiscent fruits retain their seeds and do not crack open after ripening, for example, *Acer, Fraxinus, Ulmus* and *Corylus avellana.*

Extraction of seed from fleshy fruits: In case of fleshy fruits, seeds are extracted by cutting the fruit in half or removing distal end and squeezing the contents into a container. By mashing the pulp, combining it with water, letting the seeds settle and then decanting off the pulp, small seeds from pulpy fruits can be recovered. Large seeds from pulp can be removed with hands or forceps, or the seeds can be gently rubbed with wire mesh after being washed in sieves under running water. *Neem, Hippophae, Ficus, Putranjiva, Rubus, Cinnamomum camphora* and *Ilex.* are few examples of fleshy fruits.

Extraction of mucilaginous seeds/fruits: A few fruits or seeds, such *Dillenia indica* and *Sterculia* spp., stand out for having a lot of mucilage. It takes effort to separate the mucilage from seeds, as it is difficult to remove with a simple water wash.

Extraction of pulpy fruits: Fruits are soaked in containers until they soften and are taken out once the fermentation process begins. The pulped seeds, such as *Gmelina arborea,* are properly cleaned, rinsed under running water, and dried in a thin layer on an absorbent sheet with air circulation.

Extraction of stony fruits: Stony fruits are de-pulped manually using a sharp knife and the pulp is removed by washing them under running water. After that, the seeds such as *Prunus* spp. are surface dried (ICFRE 2019).

30.3.7 Seed Processing

Pre-cleaning: The initial stage of seed processing is cleaning the seed lots. Impurities such as inert materials, debris, foreign materials, damaged and infected seeds must be removed to enhance the quality of the seeds. As soon as seed arrives at the processing site, it needs to be cleaned. They are separated from dirt and hollow seeds and chaff are removed by gentle winnowing or using seed blower, aspirator, specific-gravity separators, flotation techniques, absorption techniques etc. Seeds are examined for the insect and fungal damage and if infestation is suspected, the seeds are frozen for 7 days at a sub-zero temperature to kill insects before the diseased seeds are removed. To stop fungi or insects from spreading further, the contaminated samples are separated from the remaining material and seeds are dried to low moisture content and placed in sealed containers with silica gel (Schmidt 2000).

Pre-curing: The fresh fruits collected may not be ripe or matured. Pre-curing is the storage for ripening of seeds. They are also air dried to make them suitable for further processing and storage.

Sun drying: The fruits are exposed to the sun by spreading them on the floor till they open up for the seeds. The cones of many conifers are opened merely by spreading them out in trays in direct sunlight or by heating in kilns. Cones can also be placed in open sheds. After the cones open, they should be shaken to extract seeds. The species that require sun drying are pods and capsules, for example, *Acacia, Ailanthus, Pinus, Adina, Lagerstroemia, Bauhinia, Salix, Alnus, Betula, Casuarina, Toona* and *Eucalyptus.*

Drying under shed or cover: The fruits can be dried by spreading in sheds or ventilated rooms and covered with cloth. The seeds should be placed in trays with bottom wire mesh for air circulation. Seeds should be stirred regularly. The delicate seeds such as *Quercus, Cedrus, Dipterocarpu, Hopea, Vitex* and *Abies* are dried by this method.

Kiln drying: Seeds of conifers and many broadleaved species can be extracted by drying the fruits in kiln. The brick and solar kiln are able to control air, moisture and temperature; therefore, seeds can be dried without affecting their quality.

Threshing: Seeds of dry fruits are extracted by beating the fruits by sticks and then cleaned by hand or by winnowing them through air, for example, *Acacia.*

Parching: Some acacias which have extremely hard seed case require parching to increase the brittleness of the case. The seeds are covered with hot soil, ash and coal from the fire. The hot material is stirred through the seeds for few minutes. After the parching process, seeds are separated from ash.

Dewinging: The wings of fruits/seeds are removed before sowing, for example, *Cedrus, Picea* and *Shorea.*

Grading: Grading is removing weak, empty, damaged, immature or odd-sized seeds from a seed lot in order to increase its potential performance (Schmidt 2000)

30.4 Seed Quality Testing

Seed quality has a major role to play in seed industry. Seed quality is the combination of numerous characters, like physical and genetic purity, viability, germination, vigour, colour, form and condition, etc., which all help to establish planting value of seed. The measurement of these parameters or the evaluation of the seed's planting value is known as seed testing. Seed tests are generally performed immediately after extraction or prior to sowing. It is also done periodically on seed lots kept in long storage (Tomar 2016).

The following goals have been set to be used in the development of seed testing rules (Rao et al. 2006)

- To provide techniques for precisely assessing the quality of seed samples.
- To provide guidelines so that seed analysts working in various laboratories across the globe can obtain uniform findings.
- To relate the laboratory results, in so far as possible, to planting value.
- Completing the test within the shortest period of time possible while adhering to the goals.
- To conduct the tests in the most cost-effective way possible.

There are many advantages of better seed quality, namely

(i) Better storability
(ii) Minimum waste of seed
(iii) High rate of survival
(iv) Uniform nursery plants
(v) Higher certainty in plant production

30.4.1 Seed Lot

A seed lot is any amount of seed that has every portion or bag uniform with respect to percentage of pure seed, inert matter, other crop seed, germination and dormant seed, weed seed, and rate of occurrence of noxious weed seeds, all falling within allowed tolerances. A quantity of seed should not be considered a seed lot if it is not consistent within allowed typical limits.

In forest tree seeds, maximum lot size may range from 250 to 1000 kg and for largest seeds it can be up to 5000 kg. Standard lot size of Indian species is given below:

- Seed lot of 1000 kg where 1 g seed number is 5 and above.
- 5000 kg where 1 g seed number is less than 5.

30.4.2 Sampling

The initial step of seed testing is sampling, which is done to get a uniform and representative sample from the seed lot. A sample is an accurate representation of the quality and features of the entire seed batch that it was taken from. When sampling is done correctly, the findings of laboratory tests will represent the complete seed lot from which the sample was taken. The main problem with seed sampling is that it is almost impossible to get a completely homogeneous seed lot (Simwanza 2012).

Primary samples: Small quantities of the seed randomly obtained from the seed lot are called primary samples.

Composite sample: All the primary samples that were taken from one lot are thoroughly mixed and made homogenous to the extent possible is called as composite sample.

Submitted sample: The required quantity of the seed to be used for various tests to be sent to seed testing laboratory taken out from the composite sample is known as submitted sample.

Working sample: A small part of the submitted sample used for seed testing (purity, germination and other tests) is known as working sample.

30.4.3 Sampling Intensity

The minimum quantity of bags or containers of seed that must be sampled from a particular seed lot is determined by the size of the seed lot. The following is the minimal requirement for seed lots in uniformly sized bags or containers with a capacity of 15–100 kg (inclusive):

- 1–4 containers, take 3 primary samples from each container.
- 5–8 containers, take 2 primary samples from each container.
- 9–15 containers, take 1 primary sample from each container.
- 16–30 containers, take 15 primary samples in total from the seed lot.
- 31–59 containers, take 20 primary samples in total from the seed lot.
- 60 or more containers, take 30 primary samples in total from the seed lot (Simwanza 2012).

30.4.4 Tools Used for Sampling

Hand scoop: The hand scoop is a sampling tool consisting of a sturdy 50–100 cm handle which is stiff and durable. For each sample action, place the sampling tool into the stream at a different location on the stream (left, middle, or right).

Sleeve trier: This is used to extract seed samples from the seed lots that are containerized or packed in bags.

Nobbe trier: This type of trier is used for drawing different kinds of seeds. Its pointed tube extends to the center of the bag and has an oval slot close to the pointy end. This is ideal for sampling seeds in bag not in large quantities.

30.4.5 Purity Test

Tree seed samples are usually contaminated with foreign objects such as weed seeds, seed from other tree species, fragments of leaf debris and other waste. The purity test aims to identify the different species of seeds and inert particles that make up the lot as well as determine the composition of the sample by weight and, consequently, the lot composition.

The purity of a seed lot is the weight of pure seeds divided by the weight of pure seeds plus debris and is presented on a percentage basis.

$$\text{Purity}\ (\%) = \frac{\text{Pure seed weight}}{\text{Pure seed weight} + \text{debris weight}} \times 100$$

30.4.6 Seed Weight

It is the general term used to describe the relative size of tree seeds. The weight of 100 or 1000 seeds is the test used to quantify the weight of tree seeds.

The following formula can be used to convert 1000 pure seed weight to seed per gram or per kilogram.

$$\text{Number of seeds per gram} = \frac{1000}{\text{Weight (in g) of 1000 seeds}}$$

$$\text{Number of seeds per kilogram} = \frac{1000 \times 1000}{\text{Weight (in g) of 1000 seeds}}$$

30.4.7 Moisture Content

The water content of a seed is known as its seed moisture content. The moisture content is the primary determinant of the rate of seed deterioration and a major influencer of seed longevity.

$$\text{Seed moisture content}\ (\%) = \frac{\text{Fresh weight} - \text{Ovendry weight}}{\text{Fresh weight}} \times 100$$

Seed moisture content can be determined by two different methods

1. Oven drying method.
2. Moisture meters.

30.4.8 Oven Drying Method

The oven drying method is the standard for determining the moisture content. According to ISTA (2010), there are two distinct oven-drying techniques for figuring the moisture content depending on the chemical makeup of seeds:

- The low constant temperature oven method for oily seeds.
- The high constant temperature oven method for non-oily seeds.

30.4.9 Low Constant Temperature Oven Method

This technique is advised for seeds with high oil content or volatile compounds. This approach involves keeping pre-weighed moisture bottles and seed material (for drying) inside the oven which is kept at temperature of 103 °C for 17 ± 1 h. When moisture determination is carried out, the ambient air in the laboratory must have a relative humidity (RH) of less than 70%.

30.4.10 High Constant Temperature Oven Method

In this method, seed material and pre-weighed moisture bottles are put in an oven that is kept at 130 °C. Depending on the species, seeds are dried at this temperature for 1–4 h. The relative humidity of the surrounding air in the laboratory during the moisture measurement phase of this procedure is not specifically required (Engels and Visser 2003).

30.4.11 Moisture Meter

It provides an appropriate estimation of seed moisture content in a matter of minutes and does so rapidly that it gives a rough idea about the moisture content of a seed within minutes. But compared to the oven dry approach, the estimation is not as accurate

30.4.12 Testing Seed Viability

30.4.12.1 Direct Test of Viability

Germination Test

The primary goal of a germination test is to determine the maximum number of seeds that can germinate under ideal condition. The germination percentage is only based on normal seedlings; not all germinated seeds result in normal seedlings. There are numerous methods used for testing seed germination. When it comes to light, substrate, temperature and water requirement during germination, seeds from various species have varied needs (Thapliyal et al. 2018).

Germination Using Absorbent Paper

(i) ***Top of paper method:*** This method is convenient for small- and medium-sized seed like *Eucalyptus, Hippophae, Toona ciliata, Desmodium oojeinense, Aegle marmelos, Rhododendron arboreum, Holoptelea integrifolia* and *Fraxinus* spp.
(ii) ***Between paper method:*** This method is useful for medium-sized seed, for example, *Cassia* species, *Aegle marmelos*, *Dalbergia sissoo, etc.*
(iii) ***Germination in sand:*** This method is practical for medium- and large-sized seed like *Terminalia, Tectona, Diploknema*, *Quercus*, etc.

30.4.12.2 Indirect Tests of Viability

A germination test is always the most effective method for assessing the quality of seeds under ideal conditions. However, in certain circumstances germination tests are not feasible, so the quality of the seed must be estimated using so-called rapid tests.

Tetrazolium Test

This technique establishes the proportion of viable seed that could potentially germinate. In this method, seeds are moistened by soaking in water for 24 h. Hard coated seeds should be scarified before placing in water. Seed coat is removed and seed is divided into two parts through the embryo. For very small seeds this is not possible; therefore, they are kept directly in the solution. A 0.25 to 1% solution of 2,3,5, triphenyl tetrazolium chloride which is colourless having pH between 6–7 is made in water. This solution is kept in the dark. The cut embryos are placed in the solution in darkness for 1–18 h. The temperature should be around 30 °C. After that seeds are taken out and washed with cold water. When seen through a magnifying lens/microscope, viable seed will show normal red colour which is due to formation of formazan. Non-viable seeds will fail to show such colour. Non-viability is shown by the quantity and distribution of necrotic areas. The colour depends on formation of formazan in the living cells and the shade of colour may differ among species due to different structure of embryo. This method requires practice and experience for correct judgment (Kumar 2015).

$$\text{Viable seed (\%)} = \frac{\text{Number of half seeds stained red}}{\text{Total number of half seeds taken}} \times 100$$

Excised Embryo Test

The test is used for seeds whose dormant embryos require long periods of after-ripening. Seeds are immersed in water for several days till they are completely swollen. The water is changed two times in a day. Fully imbibed embryos are excised from the seed and germinated alone. Hard coated seeds are cracked or mechanically scarified before soaking. The excision must be performed carefully to protect the embryo. The seed coats should be removed with some instrument like needle or scalpel. These decoated seeds are put in petri dishes on moist filter papers and examined daily for 2 weeks. The colour of healthy embryos remains same or sometimes may become green and start growing. The seeds which are weak or dead show brown or black colouration or get degenerated.

Hydrogen Peroxide Method

In this test, seed coats are clipped and seeds are incubated in 1% dilute H_2O_2 solution at room temperature for a week. At the end of 1 week, seeds having more than 1 mm radicle are counted as viable and the test is terminated. This test is often used in lieu of tetrazolium, X-ray and excised embryo methods.

X-radiography

Among the rapid tests, X-radiography is the most expensive but it's not always the finest. Like other rapid tests, it works incredibly well in certain circumstances. When there is not enough time for a thorough germination test, X-radiography provides a quick assessment of seed quality. This test is based on the principle of semi-permeability. The seeds are given a treatment of barium chloride or sodium iodide; for 1–2 h, the solution is not able to penetrate in the living cells due to semi-permeability of sound cells. But dead cells are easily impregnated by the chemical solution. In order to get rid of barium chloride, the seeds are rinsed for 1–5 min, then dried and exposed to X-ray. If an embryo is found to be impregnation-free, it is viable and the endosperm is impregnated not more than 25%.

Leachate Conductivity

The degree of degradation in seeds can lead to the leaching of chemicals and harm to cellular membranes. There is a correlation between these compounds' measurements and seed quality. Sugars, amino acids and electrolytes are materials that can be measured; these are the most easily measured in terms of time and cost.

30.4.13 Seed Dormancy and Pretreatments

30.4.13.1 Seed Dormancy

There is typically a lag period between the reaching of seed maturity and the seed germination in most forestry species. Even in environments that are normally conducive to germination, such seeds are unable to germinate. We refer to this as 'seed dormancy.' Plants use seed dormancy as a crucial means of survival in their natural environments. Temperature, relative humidity and length of light exposure are among the environmental and genetic elements that determine how much dormancy a seed experiences (Thapliyal et al. 2018).

30.4.13.2 Kinds of Dormancy

(i) ***Exogenous or seed coat dormancy:*** The cause of this type of dormancy is hard seed coat or pericarp which is impermeable to oxygen and moisture. The embryo is quiescent (i.e. non-dormant, germination of the seed is possible soon after soaking in water) but is sealed inside a water impermeable coat. This type of dormancy occurs in *Betula pendula*, teak and several leguminous species. In nature, the seed coverings are softened by mechanical abrasion (scarification), alternate freezing and thawing, action of soil micro-organisms, birds, animal digestive tract passage and forest fires (Kumar 2015).

(ii) ***Endogenous, internal or embryo dormancy:*** In this kind of dormancy, the dormancy aspects are located in the inner tissues of seed. The dormancy is determined by the properties of embryo or endosperm which is due to morphological or physiological conditions in embryo. The reasons attributed to this dormancy are (a) immature embryo, (b) lack of metabolic reaction, (c) inactive enzyme activity which is due to presence of some inhibitory substances. Removal of these substances facilitates germination. The species that show embryo dormancy are *Acer platanoides, Adina cordifolia, Albizia lebbek, Cassia siamea, Dalbergia latifolia, Fraxinus nigra, Ginkgo biloba, Ziziphus jujuba, Picea* spp. *Prunus persica*, etc. (Schmidt 2000).

(iii) ***Combined or double dormancy:*** In this, both seed coat and embryo dormancy occur at the same time, for example, *Fraxinus excelsior*. The dormancy is due to combined effect of embryo, testa and pericarp. Several treatments may be necessary in order to break the double dormancy.

30.4.14 After Ripening

For seeds with a modest degree of embryo dormancy, ripening is a necessary step. These seeds undergo specific chemical and physical alterations as a result of being stored dry. Biochemical composition of seed will change as a result of reduced protein, carbohydrate and lipid storage. Additionally, gibberellin production and protein hydrolysis during dry storage will be encouraged, and this will raise metabolic activity and cause the initiation of seed germination. Due to underdeveloped embryos, seeds of certain species, such as *Schleichera oleosa, Fraxinus*, and *Ginkgo*, must ripen before germination.

30.4.15 Pretreatments of Seed to Overcome Dormancy

'Various treatments applied to seed prior to sowing in order to increase the rate of germination' is the definition of 'pre-treatment,' also known as 'pre-sowing treatment of seed.' The different treatments that can be given to seeds are

Soaking in cold water: In some hard seeded species, there is some water permeability in the seed coat. The germination is improved if seed coat is softened and adequate water is absorbed by the seed. For complete imbibition and subsequent germination, soaking such seeds in room-temperature water for 24–48 h may be sufficient, for example, *Putranjiva roxburghii, Acacia, Albizia* and *Trewia nudiflora*.

Weathering or alternate wetting and drying: For seed of some species mere soaking in water does not help them to germinate, for such seeds several cycles of soaking and drying are required, for example, teak (*Tectona grandis*). Alternate soaking and drying are also used for seeds of *Diospyros melanoxylon, Terminalia bellirica, Terminalia chebula, Terminalia citrina* etc.

Passage through animal body: Some hard-coated seeds when passed through the digestive tract of the animals, for example, goats, poultry and sheep, get treated for dormancy. The seeds of *Acacia arabica, Prosopis cineraria, Prosopis juliflora* and *Santalum album* are treated by this method, which gives quick germination but it may not always be practicable to feed the animals and collect its droppings.

Fermentation: The partial fermentation in organic matter or animal dung paste reduces germination period of many seeds. The seed is spread on the floor and covered with moist grass or peat moss. Alternatively, the seeds are well mixed in animal dung and soil paste and spread on the ground in shade. The treatment of teak takes nearly 2 months whereas the seeds of arjun can be treated within 2 weeks by this method.

Scarification: The process of breaking dormancy in many seeds is achieved by scarification which is cracking, breaking, scratching or altering the seed coat chemically/mechanically or outer covering to enable the seed to imbibe water and uptake of gases. A number of methods are used for scarification.

Acid scarification: Soaking the seeds with hard and impermeable seed coats in concentrated sulphuric acid is also used for breaking the dormancy of seeds. The seed coat experiences a type of wet combustion due to the acid. Because the acid may seep through and harm the embryo, this approach is not suitable for seeds with thin seed coverings that quickly become permeable.

Mechanical scarification: Different methods are deployed to enhance the permeability of hard coated seeds by using a hot wire burner, knife, needle, file, abrasion paper etc. to scratch the seeds, for example, *Sapindus mukorossi.*

Growth promoters/chemicals: By immersing the seeds in hydrogen peroxide, which promotes respiration and speeds up germination, one can break the embryo dormancy of seeds. By immersing the seeds for varied duration of time in a solution of KNO_3, cytokinins, gibberellic acid etc., mild dormancy in seeds can be overcome.

Boiling water/hot water soak: After immersing seeds in boiling or extremely hot water, they are left undisturbed until the water cools. The cooling hot water softens or cracks the seed covering, allowing the seed to absorb water. In this way, seeds from several leguminous species can be handled, for example, *Acacia, Albizia, Cassia* and *Prosopis*.

Hot wire: Using this technique, one can burn tiny holes through seed coats by using a heated needle or an electric wood burning tool like in chestnut and *Albizia*.

Cold stratification: One of the most significant and effective techniques for overcoming dormancy involves keeping moistened seeds at a low temperature and covering them with layers of materials that retain moisture, such as sawdust, peat moss, and vermiculite. The length of the moist stratification pretreatment varies according to the species' level of dormancy, for example, *Acer, Corylus jacquemontii* and *Junipers*.

30.4.16 Seed Storage

After collection, seeds can be stored for varying length of time. The storage of seed becomes necessary when there is long interval between seed ripening and time of sowing. Depending upon requirement seeds can be stored for up to 1 year, 1–5 years or long-term storage for conserving genetic resources. The period of storage depends on seed longevity, storage conditions.

There are three types of seeds, for example, short-lived, medium-lived and long-lived seeds.

Short lived seeds: These seeds may remain viable for a few days, months or for a year, at the most. Seeds of Dipterocarpaceae and Lauraceae are short lived and last only for few days under natural conditions. The short-lived seeds are *Populus, Acer, Salix, Quercus, Aesculus* etc. and some oil seeds such as sal, neem, karanj, mahua and cashew nut. These seeds generally drop to the ground and germinate immediately.

Medium lived seeds: These seeds remain viable for periods between 2 and 15 years provided they are stored at low humidity and low temperature conditions. Most of the conifers, *Acacia, Albizia* etc. come under this group.

Long lived seeds: These seeds remain viable for long time even at warm temperatures. They are usually characterized by hard seed coats which are impermeable to water. In the seeds with undamaged seed coat, they can stay viable for 15–20 years.

Storage means preservation of viable seeds till they are finally required for sowing. Storage preserves physiological and genetic characteristics of the seeds. The storage conditions for greater longevity of seeds should be such that its respiration and other metabolic processes slow down. The most significant factors required are low moisture content of seed and low storage temperature.

Based on storage behaviour, seeds are broadly classified into three groups as discussed below.

30.4.17 Orthodox Seeds

The term 'orthodox' refers to seeds that can be dried to 5% moisture content or less and are able to withstand freezing and can be kept at low temperatures for extended periods of time without losing viability. The majority of tropical forest pioneer species, which represent the initial stages of forest succession, have dormant and orthodox seeds, for example, species of *Acacia, Albizia Bauhinia, Cassia, Eucalyptus, Prosopis, Pinus, Picea, Terminalia* and *Tectona*.

30.4.18 Recalcitrant Seeds

Recalcitrant seeds, depending on the species, have a high moisture content (30–50%) during maturity and are susceptible to desiccation below a specific moisture content threshold (12–30%). Under any storage circumstances, these seeds quickly lose their viability and have a limited shelf life. Conventional seed-banking settings of low humidity and below-freezing temperatures are not suitable for storing recalcitrant seeds. The species that produce these seeds are usually found in humid environments, especially rainforests where the seeds are released in a very moist, metabolically active state. These seeds have a low seed coat ratio (SCR), thin coatings and are frequently spherical and large in size. Recalcitrant seeds have a high moisture content, which makes them susceptible to desiccation and chilling harm from the formation of ice crystals that damage cells when exposed to subzero temperatures.

Species belonging to *Dipterocarpaceae, Lauraceae* and *Quercus* spp., *Artocarpus heterophyllus, Castanea* spp., *Castanospermum australe, Cocos nucifera, Diploknema butyracea, Dysoxylum binectariferum, Mesua ferrea*, etc. have recalcitrant seeds.

30.4.19 Intermediate Seeds

These seeds are not fully orthodox or recalcitrant category. Certain seeds might be able to tolerate a limited amount of desiccation (7–10% to 20% moisture content) through air drying, but they won't survive below freezing temperatures. Examples of intermediate seeds are *Araucaria columnaris, Azadirachta indica, Bixa orellana, Citrus limon, Carica papaya* and *Coffea arabica*.

The process of reducing seed moisture content to levels advised for storage by methods that do not jeopardize seed viability is known as 'seed drying.' When harvested, seeds with a high moisture content can promote the growth and respiration of seed embryos as well as the proliferation of fungi and insects. Therefore, during storage, seeds must be dried to a safe moisture level to avoid damage, heating and infestation. As soon as seeds are collected, they should be dried to prevent degradation (Thapliyal et al. 2018; ICFRE 2019).

30.4.20 Seed Drying and Desiccation

Storing seeds under conditions that enhance their post-harvest lifespans is the most convenient way to conserve plant germplasm ex situ. An essential addition to in situ conservation efforts, seed storage provides a safe, inexpensive means of conserving genetic resources ex situ as well as suitable material for the development of seedlings (Khanna and Singh 1991). Determining the behaviour of seeds for storage and desiccation tolerance is the first step in the storage process. Desiccation tolerance is the capacity of an organism to withstand extreme drying conditions and fully resume normal metabolism following water absorption. Extending the lifespan of seeds is most effectively achieved by drying and storing them at low temperatures. This approach, however, is ineffective for seeds that are susceptible to desiccation since it may lower viability and cause death. The conservation of certain species, especially trees and shrubs from moist tropical habitat, may be hampered due to their susceptibility to seed desiccation (Oliver et al. 2014).

Three categories of seeds are created on the basis of the desiccation tolerance and storage conditions of seeds: (1) Conventional or desiccation-tolerant seeds, which may withstand drying out in a low-water content level and storage at low temperatures; (2) seeds that are resistant to drying or desiccation and cannot withstand this step and (3) intermediate seeds that can withstand drying at temperatures that are similar to those of orthodox seeds but may not be suitable for storage at below-freezing temperatures (Gold and Hay 2014).

30.4.21 Characteristics of Desiccation Tolerant/Sensitive Seeds

Conventional seeds are typically scattered at low water content and become quiescent after dispersion, whereas sensitive (recalcitrant) seeds are dispersed with high water content and have an active metabolism. To determine the best practice for preserving seeds, it is crucial to understand how seeds behave during the drying and storing processes. It matters because only orthodox seeds can be stored at low temperatures and dried with little water content.

The ratio of tegument to dried seed mass, or SCR (seed coat ratio), is another feature linked to seed storage behaviour. In comparison to seeds that are resistant, seeds that are susceptible to desiccation, typically have a larger seed mass and a lower tegument to seed ratio. According to Berjak and Pammenter (2013), desiccation tolerance is correlated with the ecological, physiological and morphological traits of fruits and seeds. These traits indicate that seeds that are susceptible to desiccation are typically large, have a thin tegument, are dispersed with a high-water content, germinate quickly and are native to moist environments.

Desiccation sensitivity is thought to be correlated with the origin of the seed. For instance, desiccation-sensitive seeds are more likely to be produced by tropical species, whereas desiccation-tolerant seeds are more likely to be produced in savannas. It is thought that as the environment gets colder and drier, fewer seeds that are susceptible to desiccation would appear. Hamilton et al. (2013) claimed that a deeper comprehension of how seeds react to desiccation and storage can be attained by utilizing the morphological and ecological features of seed characteristics. As a result, morpho-physiological traits including seed mass, shape and water content at dispersion can be used to infer the classification of seeds based on desiccation tolerance and storage.

30.4.22 Seed Drying Instruments

Desiccator: Seed samples are kept dry and protected from humidity using desiccators. The seed samples are freed of any remaining water by filling the desiccators beneath the platform with desiccants such as calcium chloride or silica gel. To create an airtight seal, grease or another lubricant is used to seal the glass lid of the desiccator.

Seed dryer: A seed dryer can be used to remove extra moisture from seeds in low-temperature, low-humidity environments, extending the seeds' shelf life.

30.4.23 Types of Seed Drying

Maintaining a drying room at very low relative humidity (RH) is more expensive and challenging in tropical areas with high RH. Saturated salts and low-cost technologies like silica gel drying can be employed in conjunction with other strategies to efficiently lower the moisture content of seeds to acceptable levels.

30.4.23.1 Silica Gel-Drying

Silica gel can be used to dry small samples. The dried blue self-indicating silica gel is sealed tightly in an airtight glass container or desiccator. For effective drying, the weight of the silica gel used and the seeds should match. Some gene banks utilize higher gel-to-seed ratios, such 3:1, for quicker drying. The seeds are stored near the silica gel in permeable bags. The desiccator is maintained at a low temperature of about 20 °C. If the silica gel turns from a deep blue colour to a pink or pastel blue colour, it can be changed every day. Recharging the silica gel involves heating it to 100 °C until it regains its deep blue colour. It's ready for reusage after cooling it in an airtight container. The seeds are stored in the container with fresh silica gel until their moisture content reaches the range needed for storage. Once the ideal moisture content or equilibrium seed weight is reached, and the seed germination and health are deemed satisfactory, the seeds are placed into suitable containers.

30.4.23.2 Calcium Chloride Drying

Granules of anhydrous calcium chloride are put in an airtight-sealed glass jar or desiccator with a wire mesh shelf above the chemical. Hastily closing the container prevents moisture from the air from getting inside. The wire mesh or desiccator plate is used to hold the permeable bags containing seeds. The calcium chloride is rolled over such that the bottom portion is at the top when the top layer hardens and shines. It can be used again to create a saturated salt solution once it solidifies. Seeds are kept in the container with fresh calcium chloride until their moisture content reaches the range needed for storage.

30.4.23.3 Self-Defrosting Refrigerator

If mineral salts are unavailable, seeds can be dried in a refrigerator that defrosts on its own. Although precise RH control is challenging, this approach works well in the absence of better options. Many refrigerators have RH levels between 10 and 40%, which correspond to seed moisture values appropriate for medium- or long-term storage. Arrange the seeds in an open container in a thin layer. Put the container in a refrigerator that defrosts on its own and wait for the seeds to come into balance with the interior humidity. To keep moisture from condensing on the seeds, carefully seal the drying container, take it out of the refrigerator and let it get to room temperature before opening it. Place the seeds in storage after sealing them in airtight jars.

30.4.23.4 Shade-Drying

When the relative humidity (RH) is low (less than 40%), shade-drying can be a useful technique for lowering the moisture content of seeds; the less humid the environment, the more successful the drying process will be. Shade-drying works especially well for preliminary drying. Since sun drying is thought to negatively impact some species' long-term seed viability, seeds are not dried in the sun. To allow for optimal air circulation, seeds are evenly distributed in a single layer on either open mesh racks set in the shade or a linen sheet. Any tool (like a fan) that can force more air over the seeds will boost the drying efficiency. A protective net is placed over seeds to keep animals (birds, rats etc.) from eating them. It may take a few days for the seed moisture to reach equilibrium with the surrounding relative humidity.

30.4.23.5 Zeolite Desiccant Beads

Zeolite desiccant drying beads absorb water from seeds, reduce moisture content and extend their storage lives. The ratio of bead weight to seed weight required is estimated according to the air's relative humidity in which the seeds are kept or air-dried. The ratio of beads used for drying the seeds increases with increase in relative humidity for starchy and oily seeds (Table 30.1). Beads can be re-used when they have absorbed approximately 18–20% moisture of their initial weight, by heating for 2 h at a minimum of 250 °C in an oven (Kasemsap 2015).

Table 30.1 Relative humidity as indicated for starchy and oily seed

Seed type	50% RH	75% RH	85% RH	100% RH
Starchy seed	0.3:1	0.5:1	1:1	1.25:1
Oily seed	0.2:1	0.4:1	0.8:1	1:1

Zeolite bead drying technology is standardized for many agricultural crops like beans, rice, spinach, cereals, sugar beet, melons, tomatoes and sunflower.

30.4.24 Types of Seed Storage

The different types of seed storage are:

30.4.24.1 Open Storage

In this type of storage, the moisture and temperature are not controlled. Seeds are kept in cans, baskets, sacks or tins and stored in cool and dry place. The viability of seeds varies with the climatic factors. Warm and humid climates are not good, whereas dry and cold regions are the best. To control insect infestation in seeds suitable insecticide treatment should be given. Seeds having impervious seed coat remain viable for many years by this method, for example, *Acacia* spp., *Albizia* spp., *Eucalyptus* spp. and *Robinia pseudoacacia*.

30.4.24.2 Warm Storage with Humidity Control

In this type of storage, the seeds are dried and stored in humidity-controlled rooms or chambers. The desirable temperature range is 20–28 °C and relative humidity should not be higher than 45–60%. The short-lived seeds are kept at even lower humidity. The various types of containers that can be used for seed storage are aluminium cans, tin cans, glass jars etc. Silica gel which is a desiccant can be used in containers for orthodox seeds, the quantity being one tenth of the weight of seed.

30.4.24.3 Cold Storage with or without Humidity Control

The storage temperature is kept 10 °C or below with relative humidity below 50%. Below freezing point temperatures are maintained for very long storage with 70% or lesser relative humidity. The containers should be sealed so that ice crystals are not formed. Low temperature is best for the seeds that deteriorate rapidly. Before storing, the seeds are dried to 3–8% moisture content. Oak acorns are kept under dry cold storage.

30.4.24.4 Cold Moist Storage

In this storage, un-desiccated seeds are stored in container that maintains high moisture content or seeds are mixed with moisture retaining material. Inside the container relative humidity is 80–90% and temperature between 0 and 10 °C. Seeds that can be stored by this method are *Acer, Aesculus, Juglans, Quercus, Litchi* etc. Various pines, deodar, spruce and firseeds can be stored for 10 years at 3 °C.

30.4.24.5 Cryogenic Storage

The majority of orthodox seeds may be successfully kept in climate-controlled storage facilities; however, regular upkeep and administration of these establishments are expensive. Cryogenic storage is an additional strategy to reduce these challenges. Using this process, seeds are lowered to −196 °C in liquid nitrogen. This method has the benefit of extending storage life because there are less harmful physiological activities occurring at this temperature. While this technique is not feasible for the majority of commercial seeds, it can be beneficial in preserving precious germplasm for extended periods of time.

30.4.25 Factors Affecting Longevity of Seeds

The various factors affecting the seed longevity are:

Kind of seed: Some seeds are naturally short lived such as sal and mahua. The seeds of different species differ in storability, due to different size of seed, thickness and texture of seed coat and chemical composition of the seeds. Recalcitrant seeds cannot be preserved for an extended length of time; in contrast, orthodox seeds can be kept for a long time by reducing their moisture content.

Initial quality of seed: The initial quality of seed has considerable effect on its storability. Poor-quality seeds or seeds that have begun to disintegrate as a result of unfavourable pre-harvest environmental circumstances, damage from harvesting and threshing, heat damage during drying, chemical damage during treatment will lose their viability and vigour rapidly, even in best storage conditions. The injured seeds are more susceptible to the attack of fungus and insect-pests. The impurities in the seeds provide shelter for micro-organisms and insects-pests. So, only high-quality seeds should be stored.

Moisture content in the seeds: The most significant factor influencing the quality of the seeds during storage is the amount of moisture they contain. The lower the moisture content of the seeds longer will the seeds survive. During harvest, seeds have a high moisture content, which is lowered to a safe level before storage. A 'one percent decrease in the safe moisture content of seed roughly doubles the seed's storage capability.'

Temperature: As the temperature rises, so does the biological activity (respiration) of seeds, insects and mould. Even if the relative humidity may be a little high, low temperatures are quite efficient in protecting seed quality.

Relative humidity: Mature seeds are hygroscopic in nature, so their moisture content will vary with the atmosphere as they gain or lose moisture until the equilibrium is established. For most of the seeds, if the relative humidity is more than 80%, it affects the viability of seeds, adversely. So, the storage conditions should be as dry as possible.

Supply of oxygen: Increased supply of oxygen decreases the seed viability due to increase in respiration rate, in which heat, CO_2 and water are produced. So, the quality of seed is well maintained in airtight storage than open storage.

Activity of micro-organisms: The seeds suffer vigour and viability loss due to damage from insect pests and micro-organisms. Temperature, moisture content and humidity all have an impact on the activities of these organisms.

Storage containers: Seeds that are kept in sealed, moisture-resistant containers need to be first dried to a low moisture content to reduce respiration rate, otherwise seed deterioration will proceed rapidly in comparison to the seeds stored in porous or open containers.

Seed maturity: When the seed attains physiological maturity, its quality is the highest and deterioration starts immediately after that. So, it is necessary to harvest seeds at their proper maturity time.

Table 30.2 presents summary of seed related information on some important tree species of India.

30.5 Seed Certification

An officially recognized organization's guarantee of the character and quality of seeds is known as seed certification. This guarantee is typically represented by a certificate that contains details about the origin, year of collection, authenticity of species and variety, purity, soundness and germinative capacity of the seeds. The genetic makeup and provenance of the seed have a major impact on the production, growth, survival, long term viability, adaptability and self-sustainability of tree populations. When introducing a certain species to new areas, the origin of seed will provide the information about topography, climate, edaphic characters and association of the area. A high frequency of inbreeding must be present if seed is collected from a single tree, a small number of trees or a restricted environment, since this will result in low germination and vigour. The risk of inbreeding, which lowers fitness, survival and success rates of succeeding generations relative to the global average, exists in India's forests and plantations. A geographical area defined by administrative and physical borders depending on factors such as climate, topography, altitude, latitude, longitude, temperature, and precipitation is referred to as the 'seed zone,' which is the first step in the seed certification process. The immediate need of forest seed for plantation establishment should be met by creating seed orchards. In order to obtain the desired traits and minimize pollen contamination, these plants have undergone genetic isolation from genetically inferior trees. At the New Forest campus of the Forest Research Institute, Dehradun, the first experimental clonal teak seed orchard was established. After then, on an area of almost 1000 hectares, teak seed orchards were created in several states. In India, more than 3000 hectares of seed stands and seed production areas have been identified and being managed.

When collecting tree seeds for utility-based plantings, the phenotypic characteristics of the tree must be highly relevant. A tree's phenotype results from its genetic makeup as well as its growth environment. While branchiness, forkedness and wood quality are highly heritable traits based on genotype, growth rate is frequently mostly influenced by environmental factors. As a result, collectors ought to choose a location that corresponds with the planting location before gathering seeds from the best trees there.

30.5.1 International Seed Certification Schemes

The two major seed certification schemes are the European Union (EU) and Organization for Economic Co-operation and Development (OECD) (2014) certification schemes. Reproductive materials are classified into four groups under the OECD Scheme. Forest reproductive material of 'Identified,' 'Selected,' 'Qualified' and 'Tested' categories, that is derived from forest basic material of 'Seed-source,' 'Stand' and 'Seed Orchard,' 'Parents of Family(ies),' 'Clone' and 'Clonal Mixture' is subject to the regulations of the OECD Scheme.

30.5.2 Standards for Certification

30.5.2.1 Categories of Reproductive Materials

There are different certification classes for forest reproductive material than for agricultural seeds. The OECD guidelines for forest reproductive material which comprise three more classes of less stringent genetic control and 'tested reproductive material,' which is comparable to the 'certified' category for agricultural seeds, are generally followed by forestry programmes. The prerequisites for the four classes are as follows:

Source—identified reproductive material (yellow tag) originates from stands located inside designated seed collection zone. It is required that:

- The provenance and/or source of the seed must be identified and registered with the appropriate authority.
- The designated authority must inspect the collected, processed, and storage of seeds.

Table 30.2 Seed-related information on some important tree species of India

Sl. no	Species	Common name	Fruiting season	Pre-sowing treatment	Germination percentage	Seed viability	Seed storage temp	Seed kg^{-1} (approx.)	Nature of seed
1	*Abies pindrow*	Silver fir	Sep–Nov	Not required	30–65	1–6 months	Low	27,000	Orthodox
2	*Acacia catechu*	Khair	Jan–Mar	Hot water soaking—1 h	60–80	6–12 months	Room	40,000	Orthodox
3	*Acacia nilotica*	Babul	Apr–June	Acid scarification—H_2SO_4	80–90	>2 years	Room	7000–11,000	Orthodox
4	*Acer caesium*	Maple	July–Oct	Not required	50–80	1–6 months	Low	12,000	Orthodox
5	*Acer oblongum*	Maple	Jan–Apr	Not required	50	6–12 months	Low	20,000–40,000	Orthodox
6	*Adina cordifolia*	Haldu	Jan–Mar	Not required	30–50	6–12 months	Room	1,10,000	Orthodox
7	*Aegle marmelos*	Bel	Mar–May	Not required	50–55	1–6 months	Room	5000	Orthodox
8	*Aesculus indica*	Horse chestnut	Sep–Oct	Not required	80–90	< 1 month	–	30–40	Recalcitrant
9	*Albizia lebbeck*	Siris	Jan–Mar	Hot water soaking—1 h	60–90	>2 years	Room	7000–12,000	Orthodox
10	*Casuarina equisetifolia*	Sura	June–Dec	Soaking in water– 24 h	70	1–2 years	Low	7,60,000	Orthodox
11	*Cassia siamea*	Kassod	Mar–Apr	Soaking in water—24 h	80–90	>2 years	Room	37,000	Orthodox
12	*Cassia fistula*	Amaltas	Mar–Apr	Acid scarification-H_2SO_4	22–60		Room	6500	Orthodox
13	*Cassia javanica*	Java cassia	Dec–Apr	Acid scarification-H_2SO_4	60		Room	6000	Orthodox
14	*Cedrus deodara*	Deodar	Oct–Nov	Stratification	80–90	>2 years	Room	8000	Orthodox
15	*Cinnamomum camphora*	Camphor	Oct–Nov	Soaking in water—24 h	40–50	1–6 months	Room	8000	Orthodox
16	*Dalbergia latifolia*	Rosewood	Jan–Mar	Not required	45–80	6–12 months	Room	24,000	Orthodox
17	*Dalbergia sissoo*	Shisham	Nov–Mar	Not required	90–100	6–12 months	Room	53,000	Orthodox
18	*Dendrocalamus strictus*	Narban	Apr–June	Soaking in water—24 h	25–80	6–12 months	Low	30,000–40,000	Orthodox
19	*Delonix regia*	Gulmohar	Aug–Oct	Hot water soaking—1 h	25–50	1–2 years	Room	2000–3000	Orthodox
20	*Diospyros melanoxylon*	Tendu	Apr–June	Soaking in water—24 h	50–60	6–12 months	Room	800–1500	Orthodox
21	*Emblica officinalis*	Amla	Nov–Feb	Not required	40–50	1–6 months	Room	65,000–85,000	Orthodox
22	*Eucalyptus tereticornis*	Mysore gum	Sep–Oct	Not required	90	1–2 years	Low	3,60,000	Orthodox
23	*Gmelina arborea*	Gamhar	May–June	Soaking in water—24 h	50–90	6–12 months	Room	2500	Orthodox
24	*Grevillea robusta*	Silver oak	June	Not required	20–50	6–12 months	Low	1,00,000	Orthodox
25	*Hardwickia binata*	Anjan	Apr–May	Soaking in water—24 h	50–80	6–12 months	Room	3900	Orthodox
26	*Holoptelea integrifolia*	Kanju	Apr–May	Not required	70–80	1–6 months	Room	27,000	Orthodox
27	*Lagerstroemia speciose*	Lendi	Dec–Feb	Soaking in water—24 h	90	1–2 years	Room	1,20,000	Orthodox
28	*Madhuca indica*	Mahua	June–Aug	Not required	80–90	1–6 months	Room	450	Recalcitrant
29	*Mallotus philippinensis*	Kumkum	Feb–Mar	Not required	20–30	<1 month	Low	25,000	Orthodox
30	*Mesua ferrea*	Nahor	July–Sep	Soaking in water—24 h	90	6–12 months	Room	250	Orthodox

(continued)

Table 30.2 (continued)

Sl. no	Species	Common name	Fruiting season	Pre-sowing treatment	Germination percentage	Seed viability	Seed storage temp	Seed kg^{-1} (approx.)	Nature of seed
31	*Desmodium oojeinense*	Sandan	May–June	Soaking in water—24 h	75	<1 month	Low	28,000–33,000	Orthodox
32	*Pinus roxburghii*	Chir	Feb–Mar	Not required	80–90	6–12 months	Low	9000–12,000	Orthodox
33	*Pinus wallichiana*	Kail	Sep–Nov	Soaking in water—24 h	50–60	>2 years	Low	20,000	Orthodox
34	*Pongamia pinnata*	Karanj	Mar–May	Soaking in water—24 h	80	6–12 months	Room	800–1500	Orthodox
35	*Populus ciliata*	Himalayan poplar	May–June	Not required	50–80	6–12 months	Room	1.5 crore	Orthodox
34	*Prosopis juliflora*	Mesquite	Nov–Dec	Acid scarification-H_2SO_4	80–90	1–2 years	Room	30,000	Orthodox
35	*Prosopis chinensis*	Kikkar	May–June	Scarification	80–90	1–2 years	Room	12,500	Orthodox
36	*Prosopis cineraria*	Jand	June–Aug	Scarification (H_2SO_4—5 min)	65	1–2 years	Room	25,000	Orthodox
37	*Prunus nepalensis*	Saiong	Oct–Nov	Scarification (H_2SO_4—5 min)	70–80	1–6 years	Low	1350	Orthodox
38	*Pterocarpus dalbergioides*	Padauk	Jan–Mar	Soaking in water—24 h	35–40	1–2 years	–	1450	Orthodox
39	*Pterocarpus marsupium*	Bijasal	Dec–Apr	Soaking in water—72 h	40–90	6–12 months	Room	1500–2000	Orthodox
40	*Pterocarpus santalinus*	Red sander	Feb–Apr	Soaking in water—72 h	50–70	6–12 months	Room	1000	Orthodox
41	*Putranjiva roxburghii*	Putranjiv	Jan–Feb	Not required	20–30	6–12 months	Low	3000	Orthodox
42	*Quercus incana*	Ban oak	Nov–Dec	Not required	40–60	<1 month	–	600–700	Recalcitrant
43	*Quercus dilatata*	Moru oak	July–Aug	Not required	50–60	<1 month	–	600–700	Recalcitrant
44	*Quercus semecarpifolia*	Kharsu oak	July–Aug	Not required	50–60	<1 month	–	300–500	Recalcitrant
45	*Sapindus mukorossi*	Soapnut	Nov–Jan	Scarification H_2SO_4–5 min	50–60	6–12 months	Room	640–670	Orthodox
46	*Santalum album*	Chandan	Feb–Mar	Soaking in water—24 h		6–12 months	Room	7000–10,000	Orthodox
47	*Schleichera oleosa*	Kusum	Jun–July	Not required	50–60	<1 month	Low	1400–2000	Orthodox
48	*Shorea assamica*	Makai	Jan–Apr	Not required	50–60	<1 month	–	500–600	Recalcitrant
49	*Shorea robusta*	Sal	May–July	Not required	75–90	<1 month	–	575–600	Recalcitrant
50	*Shorea talura*	Lac tree	Apr–May	Not required	50–60	<1 month	–	575–1000	Recalcitrant
51	*Sterculia urens*	Karaya	April	Not required	60–90	1–6 months	Room	4400–6400	Orthodox
52	*Stereospermum suaveolens*	Paral	Dec–Jan, Apr	Soaking in water—24 h	45	1–6 months	Room	27,000	Orthodox
53	*Strychnos nux-vomica*	Nux – vomica	Dec–Jan	Scarification (hot water—1 h)	25–35	6–12 months	Room	600–650	Orthodox
54	*Syzygium cumini*	Jamun	June	Not required	90	<1 month	Low	1200	Recalcitrant
55	*Tamarindus indica*	Imli	March–Apr	Soaking in water—24 h	66	1–2 years	Room	1800	Orthodox
56	*Taxus baccata*	Yew	Oct–Nov	Warm—Cold—Warm treatment	30–40	1–2 years	Low	1200–2000	Orthodox

(continued)

Table 30.2 (continued)

Sl. no	Species	Common name	Fruiting season	Pre-sowing treatment	Germination percentage	Seed viability	Seed storage temp	Seed kg^{-1} (approx.)	Nature of seed
57	*Tectona grandis*	Teak	Nov–Jan	Alternate soaking and drying	10–60	6–12 months	Room	1800–3000	Orthodox
58	*Terminalia arjuna*	Arjun	Feb–May	Soaking in water—24 h	50–60	6–12 months	Room	700–800	Orthodox
59	*Terminalia bellirica*	Bahera	Feb–Mar	Alternate soaking and drying	80–90	6–12 months	Room	400–500	Orthodox
60	*Terminalia chebula*	Harra	Jan–Feb	Scarification (mechanical)	60	6–12 months	Room	150–250	Orthodox
60	*Toona ciliata*	Toon	Apr–Jun	Not required	60–80	1–6 months	Low	5, 50, 000	Orthodox
61	*Ziziphus mauritiana*	Ber	Feb–Mar	Soaking in water—24 h	31–95	1–2 years	Room	1200–1700	Orthodox

Source: Thapliyal et al. (2018); ICFRE (2019)

Selected reproductive material (green tag) are derived from stands and cultivars that have undergone phenotypic selection. The genetic quality of these stands and cultivars has not been investigated, but they must:

- Be isolated by distance from inferior stands.
- Show the typical difference between trees in a stand.
- Have sufficient size to allow for effective cross-pollination.
- Have sufficient maturity and age for phenotypic assessment.
- Show phenotypic superiority in a desired quality (e.g. volume, wood quality, form or growth habit, resistance to disease, fodder production or fruit production) .

Reproductive material from untested seed orchards (pink tag) derived from parent trees in a seed orchard that have been phenotypically selected or from the progenies of such trees.

Tested reproductive material (blue tag) refers to material derived from stands, cultivars or seed orchards whose genetic superiority in at least one desirable trait has been demonstrated through tests authorized by the designated authority. Test age and test environment are the only two factors that can certify superiority. Forestry certification classes can be applied to both plant parts (cuttings) and seeds.

30.5.2.2 Types of Basic Materials Recognized in the Scheme

Six basic material types—seed source, stand, seed orchard, parents of family (ies), clone and clonal mixture—can be used to obtain reproductive material.

(a) **Seed source:** Trees within an area from which seeds are collected.

(b) **Stand:** A defined population of trees with adequate homogeneity. These are of two types:
- Autochthonous stand: Stand restored by organic regeneration. Artificial regeneration of the stand is possible using reproductive material taken from nearby autochthonous stands or from the same stand.
- Indigenous stand: An autochthonous stand, or one artificially generated from seed with its origin located in the same region of provenance, is called an indigenous stand.

(c) **Seed orchard:** It is a plantation of selected individuals, each identified by clone, family, or origin; the plantation is isolated or controlled to prevent or minimize pollination from external sources, and it is cultivated to yield regular, abundant and readily collected seed crops.

(d) **Parents of family(ies):** Trees that bear fruit through the regulated or controlled pollination of a single, identifiable parent acting as a female, using the pollen of that parent (full-sibling) or several, unidentified parents (half-sibling).

(e) **Clone:** A collection of individuals (ramets) that were first produced via vegetative propagation (e.g. by cuttings, micro-propagation, grafts and layers) from a single individual (ortet).

(f) **Clonal mixture:** A combination of the initially identified clones in predetermined ratios.

30.5.2.3 Origin of Reproductive Materials

The origin of an autochthonous seed source or stand is the location where the trees are growing. The origin of a non-autochthonous seed source or stand refers to the location of the seed or plants' initial introduction. A seed source's or stand's origin might not be known.

Provenance: The area where a tree stand or other source of seeds naturally grows.

Region of provenance: The region of provenance of a species or subspecies refers to the area or collection of places with reasonably consistent ecological conditions where stands exhibiting comparable phenotypic or genetic characteristics are observed.

30.5.3 Seed Certification in India

The National Seed Corporation (NSC) was founded in 1963 in India, marking the beginning of seed certification for agricultural crops. The Seed Act of 1966 also governs this process. After India joined the OECD Seed Schemes in 2008, 109 varieties of 20 different Indian agricultural crops were added to the OECD list of varieties qualified for OECD varietal certification.

However, the country lacks a properly developed administrative framework and sufficient rules for the certification of forest seed. The need for forest seed certification, in contrast to that of agricultural crops, has not grown significantly because of a number of factors, including the long tree rotation period, the slower rate of breeding compared to crop cultivation, the low demand for international trade, the idea of secondary outputs from trees and forests, the absence of appropriate standards regarding the genetic, phenotypic and environmental aspects of tree seeds etc.

The 'Indo-Danish Project on Seed Procurement and Tree Improvement,' which was approved by the Indian government in the 1970s, produced two documents for state-level forest seed certification: 'Certification of Forest Reproductive Material in India' and 'Seed Zonation followed in India' (Gopal and Pattanath 1979). This plan was derived on ISTA (1971) guidelines and OECD standards (OECD 2014), which established a set of principles for the distribution of seed and included its transportation, processing, storage, sampling, labelling and sealing.

The project formulated set of minimum standards for seed collection.

1. **Source-identified reproductive materials:** Seed from indigenous or non-indigenous species collected from demarcated seed zones and approved by designated authority.
2. **Selected reproductive materials:** Derived from stands, cultivars with minimum standards mentioned in the scheme.
3. **Reproductive materials from untested seed orchards:** Originated from untested seed orchards of a particular species that were collected from a particular area or source. Only seed orchards planted to generate provenance hybrids or species hybrids may yield reproductive material that falls under this category.
4. **Tested reproductive materials:** Reproductive resources, such as seed, are derived from seed orchards, stands or cultivars whose genetic superiority to suitable criteria in one or two crucial forestry-related features has been demonstrated by comparative studies carried out in specific environments.

30.5.4 Forest Seed Certification Activities at State Level

Significant advancements have been made in the creation of SPAs, SSOs, and CSOs in states like Andhra Pradesh, Tamil Nadu, Uttarakhand, Uttar Pradesh, Gujarat, Haryana, Kerala, Madhya Pradesh, and Maharashtra. In Kerala, the Kerala Forest Seed Centre managed by the Kerala Forest Research Institute collects seeds from known and improved sources of forestry species and supply them with certificate of source, purity and germination capacity to forest department and other users (Lal et al. 2008)

The Maharashtra Forest Department set up a seed centre in 1992 to identify and manage various seed sources, including seed stands, seed production regions and seed orchards, in order to provide high-quality seeds of forestry species for the planting programme. The facility has all the infrastructure required for processing, storing, and certifying seeds. (ICFRE 2019).

A few other states also have an internal certification system in place in addition to the infrastructure needed to collect, process, grade, and test seed. Still, there has been inconsistent development in this area, with most states not using enough verified or identifiable sources of material for plantation growth. To maintain uniformity in seed certification methods across the nation, a Designated Authority with nationally accepted protocols for seed certification should be implemented.

30.6 Future Scope of Forest Seed Technology

Seed quality comprises both the physiological viability, vigour and the ability to produce healthy offspring, so careful selection of good seeds will lead to production of vigorous nursery stock and successful plantations. However, the quality of seeds being employed in the plantation programme is not evaluated which results in low germination percentage, poor emergence, poor survival and adaptability to site and many a times susceptibility to disease. This is despite continual advocating on use of high-quality seeds from identified superior seed sources. The most probable reason for use of inferior seeds resulting in such failures is lack of seed testing laboratories and certification procedures along with regional seed database pertaining to variation in seed quality parameters, germination and other attributes due to seed sources.

The mission and objective of future research in forest seeds should be to supply highest quality seeds and necessary, useful technologies to forest departments, farmers, plantation owners and other stakeholders. It should include a plan that, using the underlying data and analysis, expands on the conclusions of the forest seed research. Based on the deficiencies found in forest seed research, this can be accomplished through network initiatives that emphasize basic, applied and anticipatory research.

Lessons Learnt

- Seed production has a major role in the dynamics of a forest ecosystem. Seed germination, seedling and sapling stages are crucial processes in the development of natural forest communities. Knowledge of the natural regeneration associated with seed quality and seed production are crucial for developing scientific models, methods and fresh management directives to protect the dwindling natural forests areas and for plantation programmes.
- Emptiness in seeds is one of the major issues occurring in several species. It is associated with pollination–fertilization failures. There can be early and immature seed fall because of it. Incomplete development and partially filled seeds are also observed in species. Studies need to be conducted on this very important aspect of seed biology in various species and problems associated with it.
- Forest seeds need special techniques and care during collection, handling, processing etc. Seed quality may deteriorate by poor handling techniques which then may have to be replaced by inferior seed from unknown sources in order to meet planting targets. Good seed handling techniques are essential for raising quality nursery stock.
- Seeds of many forestry species are characterized by deep dormancy leading to regeneration delay, at times failure in nature due to long time requirement for breaking dormancy in the soil seed banks. So, understanding the type and level of dormancy in seeds and devising suitable pre-treatments for cracking the dormancy and providing suitable conditions for early and fast germination are essential.
- Storing seeds under conditions that enhance their post-harvest lifespan is the most convenient way for conservation of plant germplasm ex situ. Investigating seed physiology for desiccation sensitivity/tolerance is important for every species in order to develop their storage protocols for longevity and further usage.
- Seed-borne pathogens, especially fungi, are associated with seeds of several species. Insect injuries also occur in several species. Seed-borne fungi and insect-pests reduce the quality of seeds. Pathogens may infest during seed development, storage or germination. These need to be managed using eco-friendly products/methods for retaining the quality of seeds.

Key Questions

1. What are the various factors affecting the efficiency of seed production in forestry?
2. What is the importance of quality seed in nursery stock production and in afforestation activity?
3. Elucidate different direct and indirect tests for evaluating viability and vigour of seeds.
4. What are the various storage categories of seed according to their storage physiology? Explain the various factors affecting seed longevity.
5. What do you understand by seed certification? Elaborate on various acts, rules and legislations on seed certification.

References

Beniwal BS, Singh NB (1985) Observation on flowering and germination behavior of some useful forests plants of Arunachal Pradesh. Indian Forester 4:216–225

Berjak P, Pammenter NW (2013) Implications of the lack of desiccation tolerance inrecalcitrant seeds. Front Plant Sci 4:478. https://doi.org/10.3389/fpls.2013.00478

Bonner FT, Karrfalt RP (2008) The woody plant seed manual. USDA Forest Service, Northern Research Station, Hamden, CT

Egli DB (1998) Seed biology and the yield of grain crops. Wallingford, CAB International

Engels JMM, Visser L (2003) A guide to effective management of germplasm collections. IPGRI handbooks for Genebanks no. 6. IPGRI, Rome

FAO (1985) A guide to forest seed handling. FAO forestry paper, 20-2, Rome

Feistritzer WP (1975) Cereal seed technology. FAO, Rome, pp 163–185

Geneve RL (1996) New developments in seed germination. Comb Proc Intl Plant Crop Soc 46:46–49

Geneve RL (2005) Vigor testing in flower seeds. In: McDonald MB, Kwong FY (eds) Flower seeds: biology and technology. Wallingford, CAB International, pp 311–332

Gold K, Hay F (2014) Royal botanical gardens Kew. Seed information database (SID). Version 7.1. http://data.kew.org/sid/. August 2014

Gopal M, Pattanath PG (1979) Certification of forest reproductive material in India. Revised Scheme. Indo-Danish Project on Seed Procurement and Tree Improvement, Hyderabad, Andhra Pradesh

Hamilton K, Offord CA, Cuneo P, Deseo MA (2013) A comparative study of seed morphology in relation to desiccation tolerance and other physiological responses in 71 Eastern Australian rainforest species. Plant Species Biol 28(1):51–62

ICFRE (2019) Forest seed management, ICFRE manual. Indian Council of Forestry Research and Education, Dehradun, p 52

ISTA (1971) Proceedings of International Seed Testing Association. 36:547–576

ISTA (2010) International rules for seed testing. Zurichstr.50, CH-8303 Basserdorf, Switzerland

Kasemsap P (2015). https://horticulture.ucdavis.edu/

Khanna PP, Singh N (1991) Conservation of plant genetic resources. In: Paroda RS, Arora RK (eds) Plant genetic resources conservation and management: concepts and approaches. International Board for Plant Genetic Resources, New Delhi

Kumar V (2015) Nursery and plantation practices in forestry. Scientific Publishers, Jodhpur, p 547

Lal P, Dogra AS, Dabral SL (2008) Need for an institutional and regulatory mechanism for certification of forest reproductive material in India. ICFRE, Dehradun, p 223p

OECD (2014) OECD scheme for the certification of Forest reproductive material moving in international trade. OCED Forest seed and plant scheme-rules and regulations, Paris. http://www.oecd.org/tad/forest

Oliver MJ, Tuba Z, Mishler BD, Ecology SP, Nov N, Oliver J, Tuba Z (2014) The evolution of vegetative desiccation tolerance in land plants. Plant Ecol 151(1):85–100

Rao NK, Hanson J, Dulloo ME, Ghosh K, Nowell D, Larinde M (2006) Manual of seed handling in gene banks. Handbook for gene banks no. 8. Bioversity International, Rome

Schmidt L (2000) Guide to handling of tropical and subtropical forest seed. Danida Forest Seed Centre. Humblebaek, Denmark, pp 263–303

Simwanza M (2012) Establishment of seed testing system. Seed quality control unit. FAO/Germany cooperative programme. For ministry of agriculture, forestry and food security. Rome, p 50

Thapliyal M, Gera N, Singh O, Namitha NK (2018) Manual on forest seed technology. Forest Research Institute, Dehradun, p 85p

Tomar HS (2016) Tree seed technology. Rama Publishing House, Meerut, p 179

31 Vegetative Propagation of Forest Tree Species

Santan Barthwal, Sanjay Singh, Romeet Saha, Ravita Tadiya, and Sunita Rawat

Abstract

Vegetative propagation aids tree improvement programme. Over a period of time the technique has been developed for many species, yet, there is ample scope for research especially addressing maturity-related decine in rooting, physiology, genetics and molecular biology of adventitious root formation in woody perennials. This will enable development of rapid, efficient and cost-effective methods of clonal propagation for tree species of commercial importance. Not only induction of appropriate quantity of adventitious roots, but the quality of the parent material and the produced plantlets are equally important in clonal propagation. The root architecture in case of clonally propagated trees plays an important role in survival and growth of the tree. Inter- and intra-specific variability in adventitious rhizogenesis needs to be studied to generate information which may be of great value for evolving uniform and workable clonal protocols. Eucalyptus can serve as an ideal system for such studies as evident from recent research. Eucalyptus clonal propagation is practised extensively and it serves as a good example for understanding the concept of juvenility, difference between leafy and non-leafy cuttings, hormonal treatments required to induce rooting, environmental requirement, etc. This chapter discusses the technique of propagation of *Eucalyptus* species through leafy stem cutting collected from the juvenile coppice shoots. Mist chamber propagation is a technique that uses high humidity and temperatures to promote adventitious root growth in leafy stem cuttings. The cuttings are hardened under shade house conditions with irrigation and nutrition control. Various rooting techniques and variables are described.

S. Barthwal (✉) · R. Saha · R. Tadiya · S. Rawat
Division of Genetics and Tree Improvement, ICFRE-Forest Research Institute, Dehradun, India

S. Singh
ICFRE-Forest Research Centre for Eco-Rehabilitation, Prayagraj, India

Keywords

Vegetative propagation · Cutting · Adventitious rooting · Root architecture · Root quality index

31.1 Introduction

Vegetative propagation is an important method of tree improvement and conservation. The strategy of vegetative propagation can be discussed using eucalyptus, a valuable plantation species that has been successfully mass-multiplied and planted commercially in the country and elsewhere. The *Eucalyptus* is an extremely important genus of the Myrtaceae family occupying a worldwide planted area of 20 m ha (Oballa et al. 2010). The short-fibre pulp of eucalyptus is credited with its increasing demand in the paper and pulp industries. Their ability to grow in a wide variety of environmental conditions is also a highly desirable quality to planters.

Approximately 700 species of eucalyptus are in existence out of which 37 are of importance in the forestry industries while only 15 have commercial value (Ginwal 2015). First planted by Tipu Sultan in 1790 on Nandi Hills, Karnataka, eucalyptus has now come to thrive and flourish in Indian climatic conditions (Shyam Sundar 1984). Later in 1843 and 1856, plantations were established in Nilgiri Hills, Tamil Nadu to meet the local firewood requirements (Wilson 1973). *Eucalyptus tereticornis* and its hybrids (Mysore Gum) represent more than half of the eucalyptus that are planted in the country (Ginwal 2015). Their origin is considered to be from a small patch in Nandi Hills and insufficient genetic variability resulted in the reduction of productivity (Ginwal et al. 2004a, b). Tree improvement programmes for eucalyptus have been carried out by Institute of Forest Genetics and Tree Breeding (IFGTB), Coimbatore; Forest Research Institute (FRI), Dehradun and paper and pulp companies like Indian Tobacco Company (ITC) along with state forest departments for evaluation and clonal testing. A coordinated

A. K. Mandal, A. Nicodemus (eds.), *Textbook of Forest Science*, https://doi.org/10.1007/978-981-97-8289-5_31

programme was initiated by IFGTB using selections from natural provenances (Ginwal 2015). Provenance-progeny trials were planted in three different areas to create a diverse and pedigreed genetic background of *E. camaldulensis* (Varghese et al. 2000). IFGTB managed to establish a large breeding population to undertake genetic improvement of *E. tereticornis* and *E. camaldulensis* (Ginwal 2015; Varghese et al. 2008). FRI, Dehradun established trials of *E. tereticornis* and *E. camaldulensis* throughout northern India and developed a series of F_1 hybrids from FRI 4 to FRI 16. Some hybrids like FRI 4, FRI 5, FRI 10, FRI 15 and FRI 16 have shown excellent hybrid vigour by consistently outperforming their parents (Ginwal 2015).

The paperboard division of ITC uses eucalyptus wood for pulp and paperboard (Kulkarni 2001). To meet the rising requirements, ITC imported seeds from CSIRO, Australia and established trials in South India. Exceptional clones were selected and established in Clonal Seed Orchards (CSO) (Sekar et al. 1984). But seedling plantations were affected by termite attacks and *Cylindrocladium* spp. causing foliar blight disease. The other reasons for the low productivity of seedling plantations were: hybrid breaking, close spacing, lack of quality seeds, archaic nursery techniques, mismatching of species and provenances to location, and failure to implement an appropriate package of measures (Kulkarni 2001). Financial aid from the National Bank for Agriculture and Rural Development (NABARD) ushered in a new era of tree development programmes, with a focus on root trainer technology and eucalyptus clonal propagation. The most difficult problem was matching genotypes to locations.

Plants reproduce naturally through a process known as propagation, which can be achieved sexually or asexually. Sexual reproduction, which happens through the complex phenomena of flowering and fruiting, is the favoured pathway for gene transfer. Seeds are the most prevalent way for plants to reproduce in nature. The approach is also effective and widely used in tree propagation. Except for few species with solely male blooms, all trees yield seeds. Sexual reproduction has been linked to the development of large levels of genetic variation in many tropical tree species as a result of gene recombination. The seed-based strategy, while easy and inexpensive, has certain limitations because seeds include genes from both male and female parents, and the plants that grow from them are not genetically homogeneous. The seed raised plant approach needs more time for tree improvement. The vegetative propagation approach for tree enhancement is frequently beneficial in terms of obtaining faster results and ensuring economic gains.

The method of producing new plants from existing ones while keeping the same genotype as the mother plant is known as vegetative propagation. Plants, unlike animals, have meristematic, undifferentiated cells that can give rise to various types of tissues and organs required to build a new plant. As a result, a new plant can grow from a cutting of a branch, root, or leaf that contains all of the parent plants' genetic information.

It is possible to propagate plants using a variety of vegetative structures such as segments of stems and roots, leaves and other specialized structures such as rhizomes, bulbs and corms. The genetic uniformity and similarity of plants regenerated through vegetative means are ensured. A clone is a population or a group of genetically related plants derived from a single plant (source) by vegetative processes. The word clone is widely used in forest tree improvement literature. Only genetically superior trees with desirable features (fast growth, good biomass output, straight bole, disease and insect resistance, stress tolerance, etc.) are cloned or vegetatively multiplied to achieve a population of superior trees. Donor plants that are used to collect vegetative parts are referred to as ortet. Plants that are obtained by vegetative means from the donor plants are called ramets and the ramets together constitute a clone.

Plantations raised from clonal plants bring in more genetic and economic gain. The profits realized are substantial since each plant becomes a better tree when properly taken care of. It is also easier and more effective to apply fertilizer, water the plants and manage the disease because the plantation's growth is uniform. Marketing advantages can be gained by having uniform growth of clonal plantations (Fig. 31.1).

Vegetative propagation requires a good understanding of the complex physiological mechanism to grow a complete plant from a severed part by application of hormones along with manipulation of environmental conditions for induction of shoots and roots in cuttings. Vegetative propagation includes different techniques like rooting of cuttings, layering, grafting, budding and also involves specialized plant structures.

31.2 Propagation by Cuttings

Cuttings can be prepared using any of the plant organs like stem, root or leaf. These cuttings are then normally placed in a suitable rooting media under humidified and warm environment until rooting occurs. Using this method of propagation, a high number of plantlets can be grown in a very short period of time.

The most popular and simple method for propagating forest tree species is rooting of stem cutting which is also a very commonly used propagation technique for *Eucalyptus* species. This is the most cost-effective method of propagating forest tree species.

Fig. 31.1 Plantation of eucalyptus—Clonal (above) (Photo credit—V. Sivakumar) and seedling (below) (Photo credit—A. Nicodemus)

31.2.1 Types of Cuttings

Cuttings can be divided into three categories: stem cuttings, leaf cuttings and root cuttings (Fig. 31.2).

31.2.1.1 Stem Cuttings

Stem cuttings need to be collected from the lowermost portion of the stem where the tissues are juvenile and have better-rooting ability. The shoots from the topmost parts of

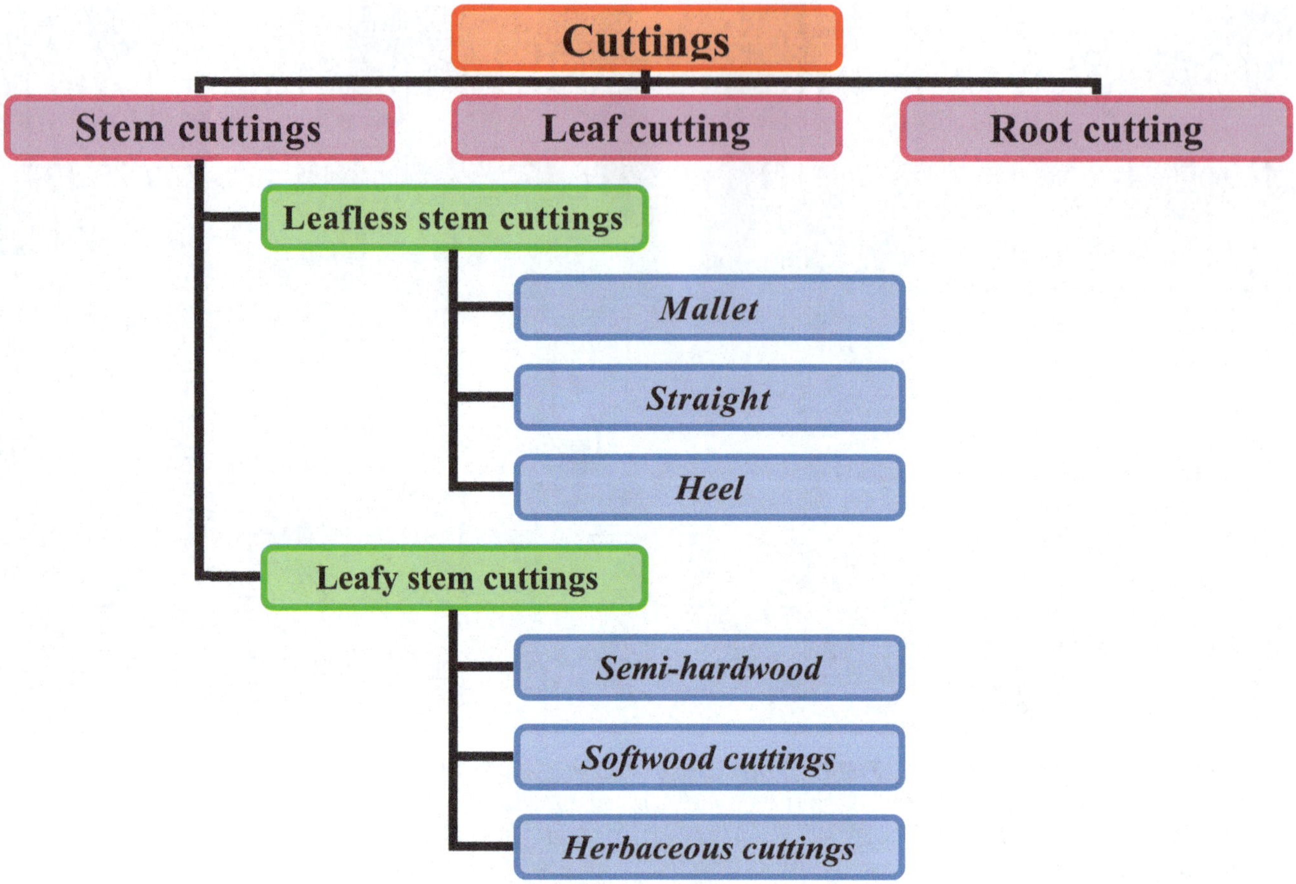

Fig. 31.2 Various types of cuttings used for rooting in clonal propagation

the trees are generally mature and have very poor rooting ability, thus may be avoided for the collection of stem cuttings for propagation purposes. To obtain juvenile cutting clonal multiplication gardens are maintained. In *Eucalyptus*, the tree is cut using a sharp saw leaving some 12–15 inches stump. The cut surface is generally slanted to avoid the accumulation of water leading to any fungal infection. The cut surface is treated with any commercially available fungicidal paste containing copper and lead. After one month the coppice shoots arise and can be used for making 3 nodal leafy cuttings. In other species sometimes the ortet trees are maintained in the form of hedges which are neatly trimmed and maintained to get profuse side branches. Such hedges are common for species like *Casuarina equisetifolia*.

There is a range in the size of cuttings, from a few centimetres as in *Eucalyptus*, *Casuarina* and *Dalbergia sissoo* to 1–2 m in *Cassia siamea*, *Gliricidia sepium* and *Moringa oleifera*. In stem cuttings, the presence of nodes with active buds is critical. The existence of leaves is also significant since they provide vital carbohydrates for root development. The absence or presence of leaves determines whether a cutting is leafless or leafy. Non-leafy stem cuttings or hardwood cuttings are primarily composed of dormant woody branches with at least two nodes. The lower cut is performed close to the lower node, whereas the upper cut is performed somewhat above the upper node. Based on the presence of an old section of the branch or wood at the bottom of the cut, they are further classified as mallet, heel and straight. In the mallet type of cutting a segment of the older branch is present at the base giving the cutting a form of an inverted 'T'. A little chip of wood from an older branch is found at the base of the heel cutting. Furthermore, no part of an earlier branch is present in the straight form of cutting.

Leafy cuttings are, as the name implies, cuttings that have leaves or parts of leaves on them. Most of the time, one or two trimmed leaves are retained with the cuttings as they provide carbohydrates and auxins, both of which are beneficial to root development. The presence of leaves, on the other hand, makes the cutting more susceptible to desiccation due to transpiration. To maintain high humidity conditions for the leafy cuttings, specific propagation facilities such as mist chambers and polytunnels are required for maintaining a humidity level of 85% and above. Leafy cuttings include semi-hardwood, softwood and herbaceous cuttings. Semi-hardwood cuttings are partially developed woody tree branches, softwood cuttings are young branches from the fresh flush and herbaceous cuttings are soft branches or shoot cuttings generally obtained from herbaceous plants.

31.2.1.2 Leaf and Root Cuttings

Leaf-cutting propagation is more common in ornamental plants and has limited utility in forestry. However, the technique can be useful for the multiplication of herbaceous medicinal and aromatic plants. Common examples of leaf-cutting are species like Saintpaulia, *Begonia rex* and *Bryophyllum pinnatum*. Root segments used for propagating plants are another approach which is generally successful in species known to produce root suckers like *Santalum album*, *D. sissoo*, *Populus alba* and *Azadirachta indica*. The root cuttings collected during the active growth phase and kept in a moist aerated medium develop several adventitious shoots followed by root development.

31.2.2 Adventitious Root Development in Cuttings

The roots developed in cuttings are adventitious that can arise from either preformed root initials or wound-induced initials or primordia. Easy-to-root species have preformed root initials and when such cuttings are kept in suitable rooting media under suitable humidity and temperature conditions the root initials quickly develop into adventitious roots as in the case of species like *Populus* and *Salix*. The cut end of the stem cuttings undergoes a process of healing where the xylem vessels are plugged with depositions of suberin followed by repeated cell division in the wounded region with the formation of callus-like growth. Some of the actively dividing cambial cells differentiate into root initials then to root primordia and finally grow into adventitious roots. The carbohydrates stored in the cuttings are translocated downwards and get converted to energy-rich sugars. The process of rooting is regulated by auxins that are either present endogenously or applied exogenously to promote rooting. The auxin hormones are thus the key component of the entire rooting process which is mainly required during the early stages of root initiation for differentiation of actively dividing cells into root initial. The following stage of root primordia generation and elongation into root auxins do not play any significant role. The use of auxin, especially indole butyric acid (IBA), is very extensively used in nurseries for cutting propagation.

Adventitious rhizogenesis is a complicated developmental process impacted by a wide range of endogenous and external factors, as well as the interactions between them. Hormone levels, quality/quantity of light, rooting cofactors and plant maturation have been implicated in adventitious root initiation and development (Hartmann et al. 1997). Adventitious roots (AR) are new root structures that emerge post-embryonically from non-root tissues (such as leaves and stems) in response to several stimuli. This process is highly regulated by a network of hormones (Bellini et al. 2014; Atkinson et al. 2014; Lakehal and Bellini 2019; Gonin et al. 2019; Steffens and Rasmussen 2016). According to Steffens and Rasmussen (2016), adventitious root formation can occur naturally or in reaction to environmental stressors, including injury, flooding or a lack of nutrients for a wide range of species. The appearance of adventitious roots is linked to improved dynamics in plant populations, greater resistance to biotic and abiotic stressors and increased phytoextraction efficacy from polluted soils (Steffens and Rasmussen 2016).

In forestry adventitious rhizogenesis has immense practical utility as propagation through cuttings is an economical and convenient procedure for rapid mass multiplication of elite genotypes to establish plantations of trees having desired commercial characteristics. The ability of the detached cuttings to rapidly develop a large number of adventitious roots, however, varies greatly across the tree species, and most woody perennials have a low capacity for adventitious root formation, especially in the mature stage. This is the main factor determining the effectiveness of this process. Because of this, one of the main obstacles to the clonal multiplication of forest tree elite germplasm is their susceptibility to adventitious root production (Abarca and Díaz-Sala 2009). Until we learn more about the molecular mechanisms that regulate rooting and the maturation-related loss in rooting potential, this problem will not be solved. Despite decades of strenuous efforts by researchers to understand the mechanism, physiology and genetics of adventitious rhizogenesis remain baffling. Therefore, eucalyptus and other economically important trees could benefit from having this trait enhanced if the mechanisms underlying adventitious rooting are better understood.

31.2.2.1 Competence for AR Formation

The mechanisms controlling the ability to develop adventitious roots are not well understood, particularly when it comes to detached cuttings. Determining the mechanisms that let differentiated somatic cells change into root meristematic cells is therefore essential, considering the effects of these mechanisms on the developmental age and maturity of the organism (Vilasboa et al. 2019; Pizarro and Díaz-Sala 2019). Many factors and mechanisms, some of which may function independently, affect cell reprogramming's ability to create adventitious root meristem (Bonga 2016). Cell fate changes dynamically during adventitious rhizogenesis in response to multiple levels of regulatory interactions. Modifications in cell fate are caused by a variety of factors, including the effects of spatiotemporal cellular and tissue signals, the interaction of hormones with important transcriptional regulators and a dynamic cascade of regulatory changes in gene expression linked to organ patterning. A differentiated somatic cell can restart pathways that result in adventitious root meristem development owing to these alterations (Díaz-Sala 2014; Pizarro and Díaz-Sala 2019).

Cells must be dedifferentiated in order to procure rooting competence and react to signalling stimuli, allowing for the recording of the induction phase of AR development, which is characterized by cell cycle activation, primordium development and subsequent cell division. This results in the primordium of the root elongating and the creation of vascular connections in the final phase (Legué et al. 2014). ARs typically form in cells positioned close to vascular structures (Bellini et al. 2014). One of the most significant economic reasons restricting tree species proliferation is the loss of rooting competence, which is critical to the process of upgrading elite genotypes (Bellini et al. 2014). Tissue regeneration in plants is most efficient in the early phases of development and decreases with age in a species- and clone-specific manner. When trees get older and more mature, the cells that produce ARs become less able to undergo de novo root regeneration (Díaz-Sala 2014). Tissue culture and micropropagation are commonly used to rejuvenate plants and regain their ability to establish roots, especially in species that are challenging to establish (Gupta et al. 1981). The cuttings from the difficult-to-root genotypes, hybrid aspen *Populus tremula* and *P. tremuloides*, could root easily after being rejuvenated through in vitro micropropagation methods (Haapala et al. 2004). Determining the genetic and molecular variables that affect the variations in rooting capacity between juvenile and mature cuttings is therefore crucial, especially in genotypes that are challenging to root, like aspens, where softwood cuttings can form adventitious roots while hardwood cuttings typically do not.

31.2.2.2 Phases of Adventitious Root Formation

Shoot cuttings severed from the mother trees represent an unstable metabolic state. Substantial morphological, physiological and biochemical changes subsequently occur within the cuttings towards adventitious root initiation or otherwise, depending upon complex interactive influences of several internal and external factors. The three developmental phases of adventitious root formation are typically referred to as induction, initiation and extension (Li et al. 2009). Through successive interdependent phases of adventitious rhizogenesis that are commensurate with specific physio-biochemical flux, two distinct phases of adventitious root formation—the auxin-insensitive root initiation phase and the auxin-sensitive root induction phase—have been identified in Azukia cuttings (Went 1934; Mitsuhashi-Kato et al. 1978a, b; Shibaoka 1971; Mohammed and Eriksen 1974; Gaspar et al. 1997; Goldfarb et al. 1998; Druege et al. 2019). The former is most likely connected to the target cells' increasing susceptibility to auxins for the beginning of adventitious roots.

Profound changes in endogenous biochemicals were registered in *Gmelina arborea* shoot cuttings during adventitious root formation especially in the first 24 h, indicating the incidence of the consequential process(s), which influence subsequent rhizogenesis in auxin-treated cuttings but not in non-treated ones due to inadequate endogenous auxin (Singh 2006). Marked decline in endogenous moisture was recorded in auxin-treated and untreated semi-hardwood cuttings; more pronounced though in treated ones within 24 h of severance indicating that auxin stimulates rooting via water stress-related signal leading to initiation of rhizogenesis. Total phenol content exhibited an increase in both auxin-treated and non-treated cuttings. An elevation in peroxidase activity in cuttings has been noticed within 24 h of planting. According to Molnar and LaCroix (1972), peroxidase was the first enzyme whose activity increased during root initiation in Hydrangea species cuttings. Gaspar (1981) has also correlated an initiation phase of rooting with the rise of total peroxidase activity of the whole plant. A similar pattern occurs during in vitro rooting of cuttings of *Sequoiadendron giganteum* (Berthon et al. 1989) and *Vitis* (Moncousin et al. 1988). However, the minimum level of peroxidase activity coincides with the advent of rooting in auxin-treated cuttings of *G. arborea*, *Theobroma cacao* (Balasimha and Subra Monian 1983) and *Populus deltoides* (Ansari et al. 1995). A gradual decline occurred in sugar levels with time till the induction of adventitious roots in the cuttings of many Poplars (Okoro and Grace 1976) and *Laryx* (Pellicer et al. 2000). Auxin-treated cuttings exhibit more efficient utilization of soluble sugar than non-treated cuttings as exogenous auxin increases the supply of carbohydrates at the base of cuttings for root induction (Dick and Dewar 1992).

It seems that in tree shoot cuttings two possible phases of adventitious rhizogenesis may be present: (1) the initial phase exhibiting low endogenous moisture, *o*-phenol but elevated peroxidase activity, and (2) the later phase coinciding with higher endogenous moisture and low phenol content and peroxidase activity. Both phases are separated by an intervening unstable phase duration of which possibly varies among different types of cuttings/genotypes (Fig. 31.3).

31.2.2.3 Hormonal Regulation of Rooting

A number of hormones control the complicated process of AR synthesis. Throughout the process, auxin appears to be the key regulator and mediator of hormonal crosstalk (Lakehal and Bellini 2019; Druege et al. 2019). For adventitious root induction in most tree species, exogenous auxin treatments are necessary. Application of anti-auxin during AR phases considerably reduced the development of AR in poplar cuttings (Bellamine et al. 1998). Numerous

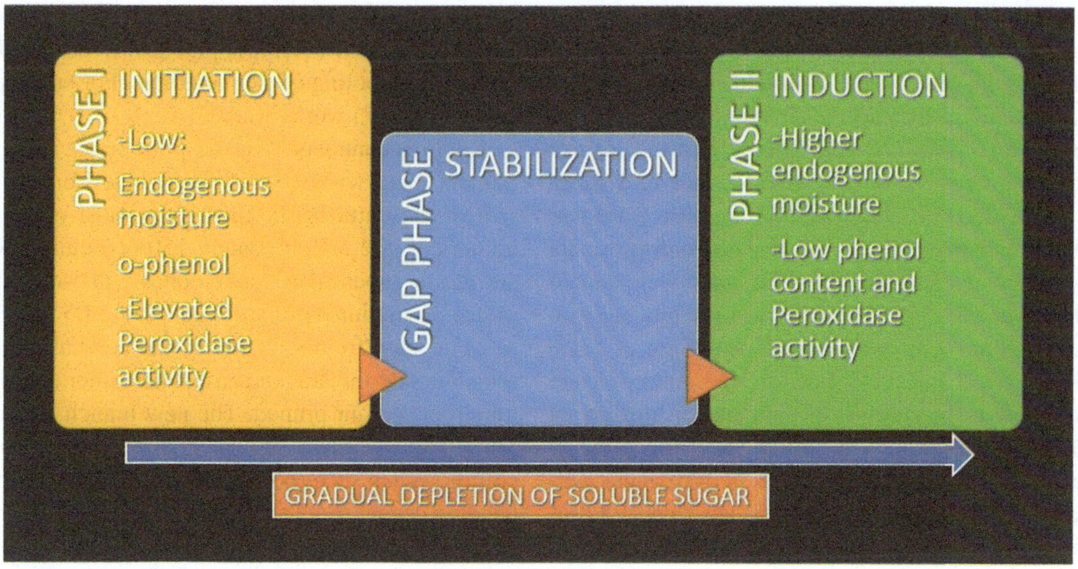

Fig. 31.3 Possible phases of adventitious rhizogenesis in shoot cuttings

experiments have been carried out to ascertain the physiological impact of various hormones on adventitious rooting, and several genes implicated in adventitious rooting have been found in several species. Although the precise role that several hormones play in the formation of AR is still not understood, research has demonstrated a complicated connection. Auxin and cytokinins have opposing functions during adventitious rhizogenesis. A comparison of the amounts of these two hormones in the basal region of cuttings from a variety of woody species, including Populus and Malus, showed that auxin and cytokinin were inversely correlated in the early phases of AR development. De Klerk et al. (1999) revealed contrasting patterns in the auxin and cytokinin concentrations in the basal region of Populus cuttings during the early phases of adventitious rooting, demonstrating their antagonistic roles in the process.

31.2.2.4 Environmental Factors

Environmental conditions at the time of clonal propagation procedures also have a significant effect on its success. Humidity is an essential factor for root initiation. Cuttings being severed part with no root system tend to lose water very quickly leading to dehydration of the tissues. The moment a cutting is removed from its parent plant, it starts to lose water. This process is even faster in the case of leafy stem cutting which loses water more quickly because of transpiring leaves on the cutting. This makes it essential to keep leafy cutting in the mist chamber or other specialized propagation structures (Mist chambers, Polyhouses, Polytents, etc.) which maintain higher humidity through the system of the tank, pump and misting nozzles. The desired humidity of 85% and above (depending on the species) is maintained by controlling the on and off of the pump using a humidity sensor and electronic timer. Advanced systems are now available that can set the humidity level at any value using microprocessor-based controllers.

Like humidity, there are different sets of temperatures and environmental conditions needed for the successful initiation and development of roots. Root initiation requires higher temperatures. A temperature of around 25 °C is needed for most of the temperate species while 35 °C temperature is needed for subtropical species. The temperatures are maintained by the combination of sunlight, cooling pads and shade nets generally. The sunlight warms the mist chamber via the greenhouse effect where the heat of the sunlight gets trapped inside the transparent cover of the mist chamber. Shade nets can be used to cut down the light to 30%, 50%, or 75% shade levels. Cooling pads installed at the end of the mist chamber cool the inside temperature as and when required. The cooling pads generally made up of cellulose pads are kept wet using the small pump and water tank. When air is passed through the pads due to evaporative cooling the temperature is reduced. The exhaust fans fitted on the opposite sides of the cooling pad inside the mist chamber maintains the drift of air through the pads. This system is again connected to a thermostat that operates based on the set values. Low light is typically required by most tree cuttings that need to be adjusted based on the cutting so as not to inhibit photosynthesis. Shade nets of the required grade (30–75%) at a height of about 2–2.5 m generally provide the best light levels for cuttings.

The success rates of clonal propagation in nurseries can be drastically impacted by the usage of salt water. A small amount of salt in commercial nurseries can cause massive financial losses over time. The inability of cuttings to root in a stressed environment and the absorption of toxic levels of Na^+ and Cl^- through the cut ends contribute to the mortality of cuttings subjected to increasing salt stress (Kuligod et al. 1995). The degree of loss depends on how well the species can withstand the negative effects of salt stress. In a study conducted on *Casuarina equisetifolia* cuttings, Barthwal et al. (2005) found that cuttings planted in rooting media with NaCl contents greater than 1% (171 mM) were rooted significantly lower (10–35%). Salinity also inhibits the root length of the cutting, with a maximum of 0.4 cm at 1.4% NaCl. Both untreated and 0.2% NaCl-treated cuttings were rooted successfully. The survival of cuttings was meagre in rooting media with higher salt concentrations. The rooting % was impacted by salt, and the growth of root hairs was prevented as a result. Inadequate root systems (reduced biomass and root hairs) and chlorotic or hypertrophied shoots on rooted cuttings decrease the success of the transplant in the field. With an increase in salinity, it has been reported that the root biomass of *D. sissoo* and the rooting percentage and root length of *P. tremula* decrease (Singh et al. 1996; Evers et al. 1997). Even at 50 mM, *Helianthus annuus* reported complete suppression of roots in MS media when grown in vitro (dos Santos et al. 2000). Rooting media with increased salt contents had a detrimental effect on cutting viability.

31.2.2.5 Rooting Media and Containers

Rooting media should have a good water-holding capacity, porous, well-aerated and sterile. It must be moist but not too wet. Oxygen is required for the developing roots and aeration is very crucial for the success of propagation through cuttings. Some common materials used as rooting media includes sand, sawdust, peat moss, composted coconut coir pith, vermiculite, perlite and pumice. Vermiculite, perlite and pumice are excellent materials that are sterile, lightweight and have good water-holding capacity. A combination of rooting media can also be used to get better results. It is also possible to use even aerated water for plants that are easy to root. However, it is practical to use locally available material for rooting of stem cuttings. For eucalyptus vermiculite (grade IV) works well.

Rooting containers can be pots, trays, polybags or root trainers filled with rooting media based on the species and size of the cuttings. In eucalyptus most nurseries use root trainers of size approximately 150 cc volume. Root trainers are plastic tubes shaped like cones with vertical grooves or ridges on the inner walls (Figs. 31.4, 31.5 and 31.6). When roots develop from the cuttings the vertical ridges guide the roots directly towards the base where after coming out of the tube roots get air pruned. The new branch again follows the same path making a root plug-like structure devoid of any coiling. When such rooted cutting is planted in the field, they develop deeper roots providing better anchorage to the growing tree.

31.2.2.6 Interplay of Multiple Factors

The vegetative propagation of woody perennials has long been thought to depend significantly on light. Even though the rooting period remained constant, light hampered the roots of *P. tremula* and *P. tremuloides* hybrid aspen cuttings, with the average rooting percentage being lower in the light than in the dark (Stenvall et al. 2005). In some tree species, it has been found that the etiolation of the mother plants greatly enhanced the rooting of cuttings (Eliasson and Brunes 1980; Maynard and Bassuk 1985). Auxin treatment, combined with the addition of catechol and sucrose, as well as keeping the cuttings in the dark, was found by Pal and Nanda (1981) to boost the rooting of *P. robusta* cuttings. According to De Almeida et al. (2017) and Klopotek et al. (2016) light is thought to affect the availability and distribution of carbohydrates as well as the regulation of growth regulators like auxin and cytokinins, which in turn affects the development of AR. Peroxidases are a type of enzyme that catalyses the oxidation of a substrate by hydrogen peroxide or other

Fig. 31.4 Root trainer showing internal vertical ridges

Fig. 31.5 Leafy *Eucalyptus* cuttings planted in root trainer

organic peroxides. According to multiple studies (Hand 1994; Hatzilazarou et al. 2006; Rout et al. 2000; Saxena et al. 2000), there is a connection between peroxidase enzyme activity and rooting process. Peroxidase activity variations can be utilized as biochemical indicators for distinct rooting stages in different species. It has been found that during wound-induced rooting, peroxidase activity and endogenous auxin levels are inversely correlated (Gaspar et al. 1992; Nag et al. 2013). The oxidation of IAA (IAA-Os) can be catalysed by certain peroxidase isoenzymes called IAA oxidases. These IAA-Os control IAA levels following the induction phase and promote roots (Fig. 31.7).

31.2.2.7 Role of Beneficial Microbes/Bacteria in Root Promotion

Not always bacteria and microbes are pathogenic to plants. Certain types of bacteria like *Streptomyces* spp., *Bacillus* spp. and *Azotobacter* spp. can synthesize auxin-like IAA (Khalid et al. 2004) which can stimulate plant growth and help in better root growth. The production of IAA by bacteria and also by other microorganisms such as fungi and algae can result in enhanced roots in vegetatively propagated plants. When the plant growth-promoting rhizobacteria (PGPR) was introduced to plants, it increased the number of root hairs and side roots while decreasing root lengths. According to Dobbelaere et al. (2003), auxin, which is produced by bacteria, *Bacillus subtilis*, enhances plant rooting. As the levels of IAA fluctuate, primary and secondary roots are elongated, and new roots are formed. Mycorrhizae inoculation dramatically changed the poplar cuttings' root system by increasing the quantity and length of lateral roots (Hooker et al. 1992). Analogous to this, hybrid aspen (*P. tremuloides* × *P. alba*) cuttings were found to exhibit adventitious root branching when exposed to an ectomycorrhizal fungus (Felten et al. 2009). PIN9-dependent auxin redistribution and IAA-dependent auxin signalling are involved in a model of auxin buildup at the root apex brought on by fungus (Felten et al. 2009). Furthermore, mycorrhizal co-cultivation of rooted *Populus* × *canescens* cuttings boosted root biomass and plant survival (Müller et al. 2013). It has been demonstrated that endophytic bacteria aid in the rooting of cuttings. After 10 weeks of growth, hybrid poplar *P. deltoides* × *P. nigra* hardwood cuttings injected with several bacteria strains showed a considerable increase in the number of roots (Taghavi et al. 2009). Based on the findings of these investigations, the use of microbial inoculum for increasing rooting in hard-to-root poplar genotypes is promising. However, the application of this method in the field is still restricted due to a variety of factors, including interactions between plant genotypes and various microorganisms (Zavattieri et al. 2016).

31.2.2.8 Regulating Gene Expression in Adventitious Root Formation

Transcription factors form a cross-regulatory network that is required for organ formation, and they play an important part in the regulatory modules that control auxin gradients, positional information and polarity field development (Feng et al. 2012). The adventitious root formation process is regulated at critical developmental stages by a variety of transcription factor groups.

These include the HOMEODOMAIN-LEUCINE ZIPPER (HD-ZIP I), WRKY, NAM/ATAF/CUC (NAC), APETALA2 (AP2), GIBBERELLIC-ACID INSENSITIVE, REPRESSOR of GAI and SCARECROW (GRAS) and WUSCHEL families of transcription factors (Fukuda et al. 2018; Li et al. 2017; Liu et al. 2014a; Quan et al. 2017; Ramírez-Carvajal et al. 2009; Rigal et al. 2012; Wang et al. 2019; Wei et al. 2020; Xuan et al. 2014).

One common characteristic of somatic embryogenesis and organogenesis induction is the expression of genes encoding transcription factors linked to stem-cell identity (Díaz-Sala 2019). The capacity of certain tree species to experience adventitious roots may be significantly influenced by the activation of transcription factors associated with meristematic programmes (Legué et al. 2014).

Fig. 31.6 Propagation of *Eucalyptus* through coppice shoot cuttings: emergence of shoots from the coppiced tree (**a**), the shoots ready to be harvested (**b**), three-node leafy cuttings with half trimmed leaves (**c**), leafy cuttings being treated with diluted IBA powder (**d**), treated cuttings planted in root trainers ready to be shifted in the mist chamber (**e**), rooted cutting without root trainer (**f**), rooted cutting washed to show root development (**g**), rooted cuttings along with the root trainers are kept in the shade house for planting (**h**), shade house from where they are shifted to open area under full sun light before planting in the field

A substantial transcriptome remodelling during adventitious root (AR) formation has been revealed by transcriptome analyses conducted on a variety of tree species, including poplar at different stages of AR formation (Brinker et al. 2004), poplar species (Ramírez-Carvajal et al. 2009), and *Eucalyptus grandis* (Rigal et al. 2012). Also, these analyses have revealed information about the species of poplar. During various stages of poplar AR development, there has been reported differential expression of transcripts encoding putative transcription factors (TFs) from 35 TF families (Rigal et al. 2012). However, the roles of these genes are still unknown. The AP2/ERF protein family is one of the well-characterized TFs involved in AR synthesis in trees; it is composed of plant-specific TFs (Riechmann and Meyerowitz 1998). There are 210 members of this superfamily in poplar, and they are all members of the conserved AP2 DNA-binding domain (Jofuku et al. 1994). This superfamily is important in controlling two important plant processes: stress response and developmental regulation.

During AR primordium formation and activation, PtAIL1 transcript levels exhibit highly regulated expression. The number of ARs was also increased in transgenic poplar lines that overexpressed PtAIL1, while it was decreased in lines where the PtAIL1 transcript was downregulated via RNA interference. The Arabidopsis gene AINTEGUMENTA (ANT) homolog AINTEGUMENTA LIKE 1 (PtAIL1), which is a Populus homolog, is expressed in the cambial zone of cells that are undergoing intense cell proliferation (Karlberg et al. 2011; Schrader et al. 2004). Moreover, PtAIL1 has been identified to bind to the D-type

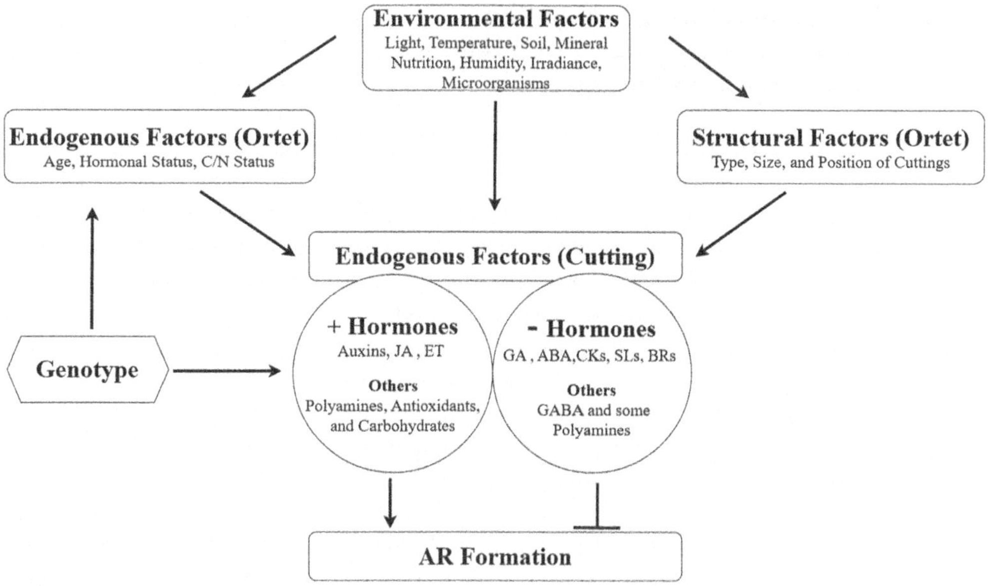

Fig. 31.7 Interplay of various factors in adventitious root formation

cyclin promoter in both Arabidopsis and hybrid aspen trees (Karlberg et al. 2011). This finding suggests that PtAIL1 may play a role in cell proliferation during AR primordium formation. According to the Arabidopsis model, AUX/IAA repress the transcriptional activity of ARF genes, which inhibits adventitious rooting. This observation suggests that this mechanism is conserved in both species. Interactions between AUX/IAA proteins have been shown to lead to the formation of oligomers that repress ARF genes and control adventitious rooting (Korasick et al. 2014; Lakehal and Bellini 2019). It is unclear how these signalling pathways are translated into changes in cell behaviour and how cell identity changes affect developmental signalling, resulting in a rechanneling of developmental memory (Díaz-Sala 2019).

31.2.2.9 QTLs and Genes for Adventitious Rhizogenesis

Attempts have been made to map the regions of the genome which contain the genes responsible for adventitious root formation. The number of roots, the length of the roots, the average diameter of the roots, the area of the roots' surface and the volume of the roots are just some of the adventitious rooting-related traits for which many quantitative trait loci (QTL) have been identified (Ribeiro et al. 2016; Sun et al. 2019; Zhang et al. 2009). A significant QTL can be identified and adventitious rooting is substantially regulated by genes when heritability is medium to high. Many QTL have been found, indicating that adventitious rooting is a quantitative trait regulated by discrete regions. Variations amongst genetic backgrounds are to be expected, even with the use of diverse software, mapping models and markers, which explain variations in the results.

Zhang et al. (2009) reported quantitative trait loci (QTL) controlling two adventitious root (AR) growth parameters within a full-sib family consisting of 93 hybrids that were produced through an interspecific cross between two *Populus* species: *P. deltoides* and *P. euramericana*, which are known to be difficult-to-root and easily rooted, respectively. In accordance with other QTL analyses in forest trees, the maximum root length and total number of AR showed a substantial association and were under strong genetic control (Geiss et al. 2009). The finding of several QTL for related variables highlights the complexity of adventitious root genetics (Ribeiro et al. 2016; Sun et al. 2019; Zhang et al. 2009).

Arabidopsis, as a genetic model, has proven critical in finding multiple genes and signalling processes involved in the regulation of AR development (Gutierrez et al. 2012; Hu and Xu 2016; Lakehal et al. 2019, 2020; Liu et al. 2014b; Sorin et al. 2005). According to studies conducted by Cai et al. (2016, 2019), Ramírez-Carvajal et al. (2009), Rigal et al. (2012), Trupiano et al. (2013), Wuddineh et al. (2015), Xu et al. (2015), Yordanov et al. (2017) and Yue et al. (2020); the genes and gene networks linked to adventitious root production have been identified. Furthermore, pharmacological experiments investigating physiological regulators as well as comprehensive data analysis have been carried out to identify QTL (Gou et al. 2010; Mauriat et al. 2014; Ribeiro et al. 2016; Zhang et al. 2019).

31.3 Miniscule Macro-propagation Propagules and Their Potential in Large-Scale Cloning

31.3.1 Micro-cuttings

To retain the juvenility of plants, de Assis et al. (1992) developed the technique of propagation involving micro-cuttings. They conducted preliminary tests in Aracruz Guaiba, where young mini-cuttings obtained from cotyledonary pair of leaves rooted with 100% success; and the reduction of rooting rate kept increasing with age ranging from a few days to months. In addition, the authors found that different species experienced different drop in rooting ability with age. Based on this, they postulated that 'the rooting ability of *Eucalyptus* clones ex vitro should increase if the physiological distance between the propagule collecting stage and the maximum juvenility stage, obtained in vitro, is reduced'. Using micro-cuttings of highly juvenile micropropagated *E. saligna*, which generated 30% greater rooting than stem cuttings, de Assis et al. (1992) tested their theory. The new method was tested in both *E. saligna* and *E. grandis* and on average 18% higher rooting was observed in the case of micro-cuttings. It was also reported that retention of shoot apex induces a tap root-like system in the cuttings. The micro-cutting has been improved over the years and has become a consistent methodology for the clonal propagation of eucalyptus (Higashi et al. 2000; Wendling et al. 2000; Xavier and Comério 1996).

31.3.2 Mini-cuttings

During 1980s, initial attempts were made at rooting mini-cuttings from shoots obtained after thinning operations, but the results were inconsistent due to nutritional deficiencies of the mother plant (de Assis et al. 2004). After the establishment of micro-cutting as a viable technique for the clonal propagation of eucalyptus, the focus shifted to the development of a more cost-efficient methodology. In the year 2000, Higashi et al. developed a fully functional mini-cutting system which was followed by other researchers who significantly contributed to its improvement (Campinhos et al. 1999; Xavier and Wendling 1998). The main difference between the mini-cutting and micro-cutting systems is in the origin of the initial propagule. Shoot apices of micropropagated plants form the propagules of micro-cuttings while axillary sprouts of rooted stem cuttings form the propagules of mini-cuttings. Subsequent steps after the first rooting cycle are similar in both techniques. Mini-cuttings often require serial propagation to achieve the reactivation of their full rooting potential while micro-cuttings get naturally rejuvenated through monthly sub-culturing of explants (de Assis et al. 2004). Improved nutritional status of mini-cuttings is believed to enhance rooting potential. Labour-intensive practices in maintaining vegetative multiplication garden (VMG) are reduced in the case of mini-clonal hedges which can bring down the overall cost of the setup. Further, the lower lignification of mini-cuttings induces a tap-root-like system as opposed to a more predominant lateral root system in stem cuttings.

The concept of super-intensive management of propagules took shape and gave rise to mini-clonal hedges. In the nascent stages, this technique constituted the setting up of dibble tubes which produced cuttings with good root quality (de Assis et al. 1992; Xavier and Comério 1996). But the earlier systems were still plagued with problems such as diseases, poor nutritional status and unsuitable environmental conditions (de Assis et al. 2004). This led to indoor hydroponic mini hedges which contributed to higher efficiency. The main contribution of this system was in providing higher nutritional status in mini-cuttings which subsequently resulted in a better root system. Two different methods emerged; the first was based on drip fertigated sand beds (Higashi et al. 2000) and the second was based on intermittent flooding (Campinhos et al. 1999).

A comparative study that was conducted on biochemical and morphological changes which take place during adventitious rooting of *Eucalyptus globulus* × *E. maidenii* mini-cuttings obtained from mini-stumps that were grown in sand beds and intermittent flooding trays was conducted by Schwambac et al. (2008). The authors concluded that all the parameters under study were similar in both systems. Kinetic profiles of biochemical parameters were also similar indicating a similarity between results obtained from both systems.

Apart from eucalyptus, mini-cuttings have been successfully utilized in overcoming the maturation-related loss of rooting and mass clonal propagation of some important tropical tree species like Teak (Palanisamy and Subramanian 2001), *G. arborea* (Singh and Ansari 2014) and *Casuarina junghuhniana* (Sureshkumar et al. 2016).

31.3.3 Virtual Clonal Hedges

It was repeatedly observed that the first flush of shoots rooted better than subsequent sprouts and this led to postulating virtual clonal hedges. It used the apices of rooted mini-cuttings of 15 cm height to produce new stock for rooting without losing the quality of the stock material which could still be used for out planting. Sprouts of pruned plants and apices of rooted mini-cuttings are used in a continuous and self-sustained cycle of production representing outstanding technical and operational viability. Extension of sub-fertigation to the newly rooted cuttings enhances the suitability of this system to industrial standards and further enhances practicality. Theoretical concepts like CO_2 enrichment and control of temperature, light intensity and photoperiod help in enhancing rooting predisposition. Indoor clonal hedges also have the added advantage of phytosanitation over outdoor clonal hedges. Ferreira et al. (1997) demonstrated shoot contamination by soil-borne pathogenic fungi.

31.3.4 Productivity of Propagules

Early ventures into clonal propagation of eucalyptus reported 114 cuttings m^{-2} in field cuttings banks (Campinhos and Ikemori 1983). The advent of mini-hedges increased productivity to 1752 cuttings m^{-2} in 1991 (Carvalho et al. 1991) and 24,000 cuttings m^{-2} in 2000 (Higashi et al. 2000). Hydroponic systems have increased the productivity of propagules by about 350-fold compared to the initial techniques (de Assis et al. 2004). This increase in productivity in addition to ease of handling shows great potential for economic exploitation. However, the viability of mini-clonal hedges depends largely on the species. It has been seen that hybrids of *E. grandis* × *E. urophylla* give higher productivity as mini-cuttings compared to those of other species, while *E. saligna* is extremely difficult to root as mini-cuttings and coppicing is preferred in their case (de Assis et al. 2004). A major drawback of mini-clonal hedges is the loss of mother plants due to a varied number of causes ranging from disease incidence, excessive harvesting, excess salinization etc.

31.4 Root Quality Assessment

31.4.1 Evaluation of Root Architecture

A careful study of past research, current trends and futuristic predictions all in unison that a large-scale shift in species distribution will occur as a result of climate change (Hamer et al. 2016). Plant species will be forced to adapt to changing ecological conditions (Hoffmann and Sgró 2011), recede into refugia (Hamer et al. 2016) or face extinction. Below-ground traits can often become extremely important in determining the adaptability of a plant. Water and nutrient uptake are partly determined by the distribution of the root system in the soil profile and show diverse characteristics with species-specific root architecture depending on their functional group and landscape position (Schenk and Jackson 2002a, b, 2005). The ability of plants to invest their assimilates in either shallow soil roots to capture sub-surface water or deep roots to exploit the water table is of extreme importance in plantation forestry within a specific climatic zone (Hamer et al. 2016). The potential implications this has on the functional equilibrium and root system architecture as the proportion of roots distributed in the soil profile and the biomass allocated to them is immense (Hamer et al. 2016). Plants with a deeper root system that can access the water table below can keep its stomata open longer than shallow-rooted plants (Mitchell et al. 2009). The strategy employed by plants to increase its water use efficiency is highly dependent on their root distribution and governs a plant's ability to adapt to changing climatic conditions. Recent periods of drought have caused the crown decline and tree death in many *Eucalyptus* species (Poot and Veneklaas 2013) and models predict a further decline in health in the future, especially for inland species (Hamer et al. 2015).

Based on the commercial significance, Saha et al. (2020a, b, 2021) chose six clones of *Eucalyptus*, that is, 1, 72, 288, 411, FRI 4 and FRI 5 for a study. Clones 1, 72, 288 (*E. tereticornis*) and 411 (*E. camaldulensis*) were arranged from Pragati Biotechnologies in Hoshiarpur, Punjab (Lal et al. 2006) and FRI 4 and FRI 5 were obtained from the Central Nursery, Silviculture Division, Forest Research Institute, Dehradun. The experiment was conducted in the Division of Genetics and Tree Improvement nursery, Dehradun (30°20′40.2″N 78°00′41.5″E, 650 m elevation). A randomized complete block design (RCBD) was used, with six replicates of each treatment in each block.

Six bare-root clonal plants were randomly assigned to each block with a spacing of 3 × 2 m at the beginning of September 2015. The blocks were separated from one another by a 3 m buffer region and displayed the time in months after transplantation (MAT). The common trench profile wall method was used to map the effects of the roots (Böhm 1976).

A critical parameter that has been attributed to spatial root system architecture (Fig. 31.8) and water and nutrient uptake is root length density (RLD) (De Silva et al. 2011). It is widely used in varietal studies in plant breeding programs, agronomic analyses and selection programmes (Chopart et al. 2008). To this end, the researchers evaluated the theoretical model proposed by Noordwijk (1987), which accounted for the relationship between root impact density (RID) and root length distribution (RLD), which is based on the root orientation relative to the counting plane, and also accounted for the geometric model (Chopart et al. 2008), which required the calibration of the model equation based on root impacts observed on the three perpendicular cubical planes and subsequent RLD prediction from a single vertical profile plane. For number of root impacts (NI) and RLD sampling, a three-sided (10 × 10 × 10 cm) partial cube was utilized (Chopart et al. 2008). Based on our preliminary analysis, the mean anisotropy of the genotypes was found to be between

Fig. 31.8 Evaluation of rooting pattern in eucalyptus

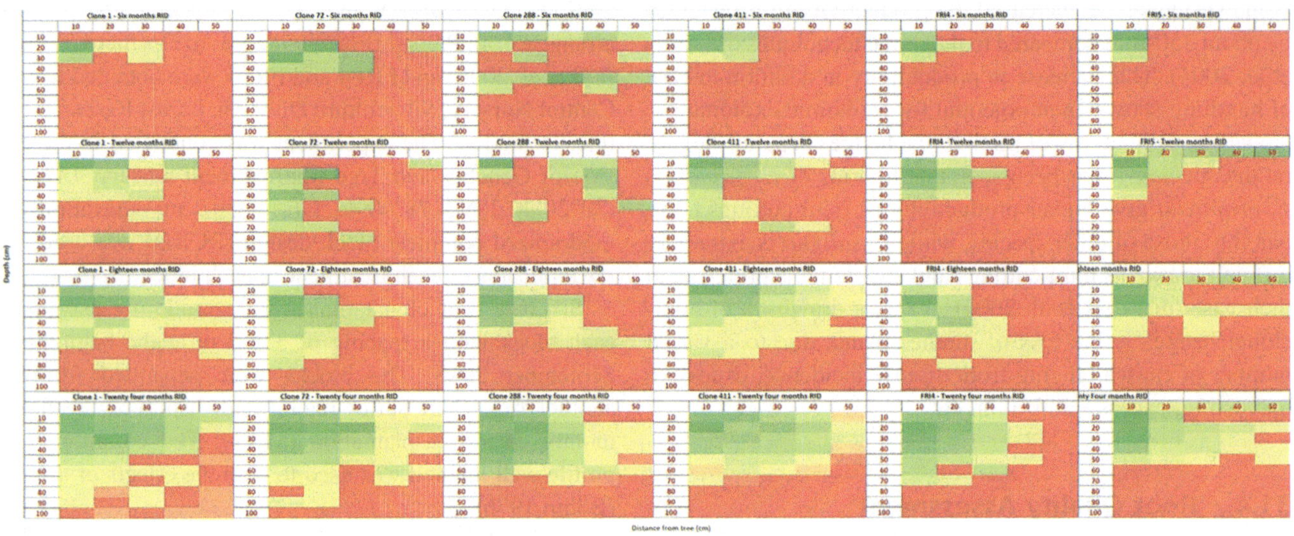

Fig. 31.9 Graph of the average root impacts of clones at 6, 12, 18 and 24 months for genotypes 1, 72, 288, 411, FRI 4 and FRI 5 (Green to red depicts a decrease in root impacts). The columns show the various genotypes, and the rows show the different periods after transplantation in soil

0.2 and 0.25, and the mean preferential direction of the root system for each genotype was found to be <1. This means that the root systems prefer to grow in a direction that is perpendicular to the profile of observation.

The orientation coefficient based on the geometry of root distribution was found to be stable at 1.54 which deviated from the value of 2 which was proposed for isotropic distributions by van Noordwijk et al. (2015). The model was validated by testing it with the genotypes, where it was found to exhibit a similar mean anisotropy. Thus, the model was found to be accurate for this investigation, even though it was applied to a wide variety of species with varying genetic compositions (i.e. *E. tereticornis* and *E. camaldulensis*) (Saha et al. 2020b).

Root systems of the genotypes showed substantial differences in their ability to penetrate the soil, as well as their lateral extent and total RLD in the soil, across all four observations made at various MATs. Seedling origin genotypes displayed much lower RLD than vegetative origin genotypes; additionally, AR demonstrated considerably greater proliferation capacity than the taproot system at the present study site (Fig. 31.9). Clones 1 and 411 exhibited a greater RLD between the surface soils, whereas clones 72, FRI 4 and FRI 5 exhibited a lower RLD between 0 and 20 cm depths. A significant factor in genotypic selection for agroforestry systems, in which crop roots are primarily concentrated on the surface soils, this makes clone 72 and the reciprocal hybrids an excellent choice for any type of mixed-cropping system. When there is less competition between crops and trees, both can flourish, leading to greater productivity. Roots from clones 411 and 1 were more concentrated in the top few inches of soil because the area had been ploughed before the experiments. Similarly, at 24 months, clones 1, 288 and 411 showed a relatively higher RLD even when the roots were spaced 30–40 cm apart from the tree, which was also observed in studies of root horizontal growth. The sandy texture of the soil may have something to do with the roots' ability to spread so widely laterally.

Water use efficiency (WUE) was found to significantly correlate with both lateral root growth and vertical root growth in the current study. The establishment of farm forestry plantations in arid/semiarid soil types would benefit from a species with a higher water use efficiency (Falkiner et al. 2006), which would mean selecting a species that has deeper root penetration capabilities, and this could be helpful in the initial screening for genotypic selection. Clones 72, FRI 4 and FRI 5 were found to have a restricted lateral root system spread in the study, suggesting they could be suitable in agroforestry models. The mathematical model used in this study to simulate the dynamic architecture, which includes both structural and geometric components, was able to reproduce the morpho-genetic and temporal behaviour of the genotypes to a great extent. Such mathematical models are even more important when conclusions have to be made about both the dynamic and static parts of RSA. So, it would be possible to simulate how different species behave in response to environmental factors that are always changing (de Dorlodot et al. 2007). The results from the field experiments were confirmed by the sand-culture experiment which showed that clone 1 had the highest total length of root system (TLRS) and the highest angle of emergence. We already knew that clone 1 had one of the highest lateral distributions. The fact that clone 72 had the lowest angle of emergence was in line with our field experiments, which showed that root system of clone 72 spread less than the other clones. Initial genotype testing at the nursery stage could become a potential indicator of its scavenging capability, assisting in proper selection.

31.4.2 Multiple Attribute-Based Assessments of Rhizogenesis

It is well known that eucalyptus are very well suited to the edapho-climatic conditions of the Indian subcontinent and show average potential productivity of 5 tons ha^{-1} $year^{-1}$ (Palana 2017). Therefore, proper genotypic selections, bearing in mind their ultimate fate and site-specific development become crucial for the successful utilization of the final product. With *Eucalyptus* taking a centre stage in plantation forestry, it becomes important for us to establish successful plantations, the success of which to a large extent rests on a solid root system. The root system has been previously correlated with out-planting success and therefore, proper selection of eucalyptus w.r.t. their RSA can contribute to a much more economically viable plantation. Despite the release of superior clones, research on the assessment of root quality in rooted cuttings has not been adequate. A major focus was laid on the rooting percentage of cuttings and to that end, various modifications like the reversal of ontogenic and physiological maturity of the plant were resorted to. Techniques such as coppicing, micro-cuttings (de Assis et al. 1992) and mini-cuttings (Wendling et al. 2000) were developed to counteract the negative impact of aging. In India, coppicing is widely prevalent as the means of procuring vegetative propagules (Dhiman and Gandhi 1999). Along with ideal propagule, environmental conditions like temperature and humidity have also been optimized by the use of propagation chambers. But despite all efforts low rooting percentage causes economic losses to industries (Husen et al. 2017). Proper regeneration of the root system are essential for successful plantations (Grossnickle 2005). Difficulty in measuring parameters defining root growth has resulted in widely correlating out-planting success and productivity with shoot morphological characters (Davis and Jacobs 2005) but shoot morphological measurements have not been adequate in predicting out-planting performance; thus the assessment of root quality at an early stage is critical (Davis and Jacobs 2005). Previous studies showed use of different parameters to quantify the root system. We take a more integrated approach to evaluate the root system quality. First, we created a comparability sequence through which we evaluated the effects of different treatments on the root system of rooted cuttings. We further simplified the procedure by developing an index that can be used by nursery managers for easy evaluation of propagated material. But, in every selection procedure, there exists a human bias. To eliminate the human bias in the decision-making process and create accurate classification of trees we used machine learning algorithm (MLA), that is, C5.0. We used multiple parameters as predictors to classify our model into different nodes. High accuracy classified trees were used for classifying the selection procedure. Another major problem that we identified and tried addressing stems from the fact that eucalyptus' susceptibility to a plethora of pathogens leading to massive and recurrent losses to nurseries. With the advent of time, the chances of infection increase greatly, and we aimed at modelling the rooting dynamics in an attempt to find the optimal time of permanence of the different clones and decrease the time the propagules are kept in propagation chambers.

Seven different treatments, including IBA, were applied to the cuttings. The Prolonged Dip involved submerging the basal area in a 500–1000 mg L^{-1} IBA solution for an entire day. The basal region has to be submerged in an IBA solution at concentrations of 4000 and 6000 mg L^{-1} for 2 h in order to receive the Pulse Treatment. In addition, there was a 10-s treatment of IBA (4000 mg L^{-1}) + Tannic Acid (2000 mg L^{-1}).

The basal region was immersed in an IBA (HiMedia) solution for 10 s at doses of 4000 and 6000 mg L^{-1} in each instance. Once the cuttings had received the auxin treatment, and inserted 2 cm deep into 150 cc root trainers that had been filled with vermiculite. The root trainers were then kept in a mist chamber for 35 days with a relative humidity of 85 ± 2% and daily maximum and lowest temperatures of 32 ± 2 °C and 28 ± 2 °C. Next, the root trainers were placed in a glasshouse for a 25-day hardening phase. After that, they were evaluated using Expert Vision Labs Pvt. Ltd Biovis Image Plus (Saha et al. 2020a).

Manual assessments of the number of roots, rooted cutting weight, and shoot-to-root ratio were among the measurements that were recorded. Some of the manual measurements that were performed included the following: rooting %, average root diameter, total root system area, number of root tips, total length of the root system (TLRS), segmentation of the root system, frequency of forks (branching frequency), among others. This extensive study sought to evaluate the quality of adventitious rooting in addition to identifying treatments that produced increased rooting rates. According to Bhardwaj and Mishra (2005), this strategy was used to reduce the possibility of higher mortality during the ensuing hardening phase. One way to reduce transplant shock when moving from relatively mild nursery environments to the field is to have a strong root system that has been formed through these parameters. While previous studies have frequently concentrated on one factor to evaluate root development, our work highlights the importance of taking into account different characteristics. The IBA Quick Dip at 6000 mg L^{-1} treatment was selected for *E. tereticornis* clones, according to an analysis of individual parameters. However, in certain instances, alternative treatments produced better results in particular categories. Clone 1 bi-nodal cuttings treated with IBA Pulse at 4000 mg L^{-1}, for example, showed the highest Total Length of Root System (TLRS) at 122.47 cm, exceeding the TLRS of 112.93 cm obtained with the IBA Quick Dip at 6000 mg L^{-1} treatment ($p = 0.769$), while the difference was not significantly different. When each parameter is taken into consideration independently, it becomes difficult to determine the optimal course of action based on those parameters. While clones 1, 72 and 288 (*E. tereticornis*) shown favourable rooting ability with the IBA Quick Dip treatment, clones 411 (*E. camaldulensis*), FRI 4 and FRI 5 (reciprocal hybrids) indicated a preference for the IBA Prolonged Dip treatment. Average root diameters increased with IBA Quick Dip at 4000 mg L^{-1} IBA, however there was significant variation in the parameters between clones and treatments. For example, FRI 4 treated with IBA Prolonged Dip at 1000 mg L^{-1} IBA showed a higher TLRS value. Because of the different responses of each parameter to multiple treatments, selecting parameters individually as indications of root development becomes difficult. Consequently, by taking into account each distinct performance measure and aggregating them into a single grade number for every treatment, using a multiple-attribute decision-making methodology simplifies the process.

31.4.3 Development of Root Quality Index (RQI)

To develop an Adventitious Root Quality Index, a modified version of the 'Dickson quality index' was used (Dickson et al. 1960). The equation was modified in this work (Saha et al. 2020a), although the original quality index took into account the ratio of total mass (TM) to the sum of shoot/root and the sturdiness quotient (SQ), which was thought to be the ratio of shoot length to shoot caliper (Currey et al. 2013; Dickson et al. 1960). The adjustment was made to precisely examine root quality, resulting in the following equation:

$$\mathrm{RQI} = \frac{\mathrm{TM}}{\frac{S}{R} + \mathrm{RSQ}}.$$

In this case, *S*/*R* (gm gm^{-1}) indicates the shoot-to-root ratio, TM is the total mass of the rooted cutting in grams (gm) and RSQ is the root sturdiness quotient. The quality of adventitious roots is evaluated using the Adventitious Root Quality Index (RQI). To assess root sturdiness in this case, SQ has been adjusted. This is measured in centimetres per centimetre (cm cm^{-1}), which is the average root diameters divided by the entire root system length.

Grey-Relational Analysis (GRA) was mainly employed to support Multiple Attribute Decision-Making (MADM), a procedure that takes into account a number of attributes or parameters in order to choose the best course of action from a variety of alternatives. When applying GRA, all performance criteria were taken into account and a consolidated output was produced for evaluating root quality based on different treatments. Variations in responsiveness to different treatments were shown to be genotype- and species-specific based on the GRA scores. While FRI 4 and FRI 5 (interspecific reciprocal hybrids) responded favourably to extended exposure to a low concentration of IBA, *E. tereticornis* clones (1, 72 and 288) showed a favourable response to short-duration treatment to a high concentration of IBA. In both circumstances, Clone 411 or *E. camaldulensis* showed high rooting capacity; however, the ranking suggested that longer exposure to a low concentration of IBA was more beneficial.

An additional way of assessment was developed: the Root Quality Index (RQI). This is useful in cases where determining many characteristics may be impractical or where the goal of the study is to quantify the root quality. A favourable association between RQI and GRA grades was found. As a result, there are situations where employing RQI instead of multiple attribute decision-making methods makes sense. While RQI is easy to define by using relatively simple-to-evaluate parameters, GRA provides a more thorough grasp of integrated selection and functions as a rigorous assessment approach.

According to Kuo et al. (2008), depending on the decision maker's assessment, numerous options can produce varying outcomes, hence it is important to adjust the weights allocated to attributes. In order to overcome this problem and improve the accuracy and performance of our model, we investigated a selection basis that uses intrinsic data from a dataset to give weights to every parameter. As a result, we classified and developed a decision tree-based strategy using the data collected for evaluating adventitious root quality, comparing it with our Multiple Attribute Decision-Making (MADM) model.

Utilizing the evaluation attributes, the C5.0 algorithm was applied to each clone separately in order to determine the most effective treatment. The study's findings demonstrated that new values' class membership could be predicted with a reasonable degree of accuracy using the training features that were available and that they could then be categorized using pattern recognition. After comparing the results obtained from GRA and C5.0, we observed that our initial weight selection was highly accurate based on the predictor importance determined by the information gain ratio in each clone. Treatment segregation was observed in a species-wise allocation in both GRA and C5.0, where the response to IBA concentration and duration was genotype-specific. There were similarities between our comparability sequence and the output pattern that our data mining approach found. In addition to auxin degradation at the wounded end, another significant issue affecting cuttings during rhizogenesis in intermittent misting chambers is the occurrence of diseases owing to the favourable environment created by high relative humidity and temperature (Brondani et al. 2012; Steffens and Rasmussen 2016). Therefore, in order to minimize mortality, it is crucial to optimize the length of time propagules are retained under such circumstances. The Daily Current Increment (DCI) of the clones varied, with the maximum increment being attained between 35 and 40 days, and the DCI and Daily Medium Increment (DMI) curves intersecting between 49 and 57 days. In order to optimize output and reduce mortality, it is therefore becoming more and more important to model the ideal time of persistence for each genotype during mass propagation. Though the best place to remove cuttings would be at the intersection of the DCI and DMI curves (Figs. 31.10 and 31.11), the interval between ideal DCI and the intersection would also be advantageous for proper adventitious root growth.

The application of IBA demonstrated significant heterogeneity, according to Saha et al. (2020a, b, 2021), highlighting the importance of altering the strategy for individual clones in order to improve mass propagation. The economic feasibility and ease of implementation should be considered while choosing a methodology. Our findings show that coppice cuttings of *Eucalyptus tereticornis* responded favourably to a 10-s exposure to 6000 mg L^{-1} IBA. By contrast, after being treated with 1000 mg L^{-1} IBA for 24 h, *E. camaldulensis* and interspecific hybrids showed improved rooting ability. By making it easier to choose cuttings with better rooting potential through vegetative propagation, the evaluation criteria that have been suggested should increase the probability of survival. The favoured strategy for selection is establishing a comparability sequence using GRA (Grey Relational Analysis). But if there are more drawbacks than advantages, we advise applying the Rooting Quality Index, or RQI. It was additionally found that different genotypes had different ideal times for transferring rooted cuttings for hardening. As a result, genotype-specific modelling is critical for increasing output, especially when developing protocols. Studying the dynamics of adventitious rooting also gave us an understanding of a decrease in daily increment due to confinement (Davis and Jacobs 2005) and the non-availability of nutrients for further development. By improving the selection of superior rooting during vegetative propagation, the suggested assessment criteria attempt to increase survival rates. As the favoured selection approach, we recommend using GRA to construct a comparison process; RQI is a good alternative when advantages are exceeded by drawbacks. Different genotypes require different times for transplanting rooted cuttings for hardening, which highlights the significance of modelling unique to a given genotype and developing methods for commercially important clones. This becomes particularly important for clones that are prone to illnesses found in nurseries. Our analysis of adventitious rooting dynamics revealed deficiencies in nutrients for continued development in addition to shedding light on Davis and Jacobs (2005)'s observation that confinement reduces daily increment. It was determined that 35–40 days is a perfect period of time for taking away rooted cuttings from mist propagators. Consequently, in order to reduce losses during the nursery stage, we advise conducting a separate study to determine the ideal permanence time for each commercially significant clone that is extensively propagated. Higher financial returns and the availability of reliable planting material for propagation can be ensured by putting these strategies into practice.

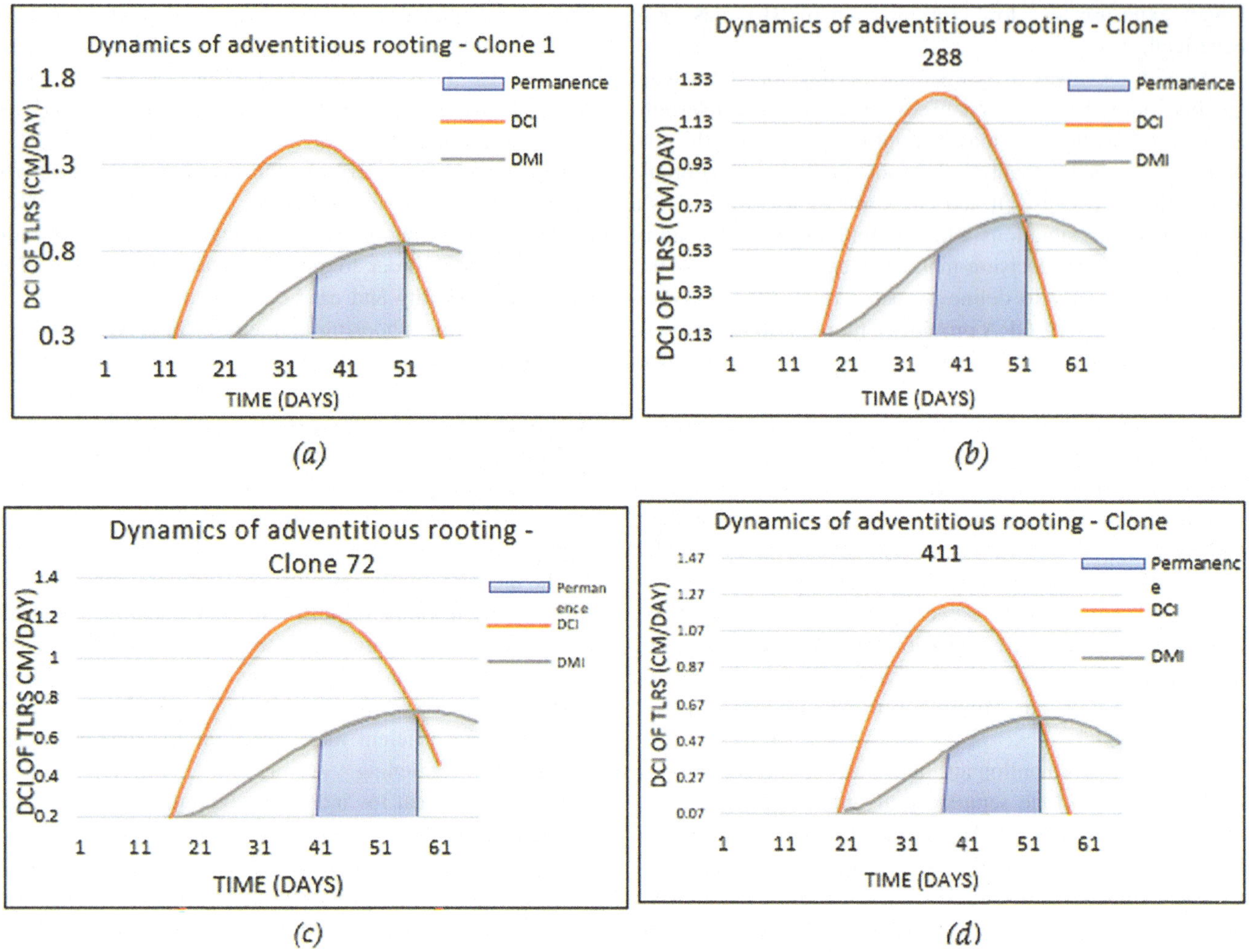

Fig. 31.10 The daily current increment (DCI) and daily medium increment (DMI) of eucalyptus coppice cuttings fluctuate with time for the following clones: (**a**) clone 1, (**b**) clone 72, (**c**) clone 288 and (**d**) clone 411

31.5 Utility of Vegetative Propagation in Tree Improvement

1. Clonal multiplication of trees: It is possible to multiply the selected elite tree, which aids in the enhancement of planting stock. Commercially, vegetative propagation is used to mass multiply Salix, Populus, Casuarina, Eucalyptus, Morus, Teak and other forest tree species. Tissue culture allows for the quick production of a large number of plants in a small space.
2. Establishment of clonal seed orchards (CSO): This is a plantation where clones of selected plus trees are planted to produce genetically improved seeds. Vegetative propagation strategies are critical for the formation of CSOs.
3. Establishment of clone bank: It is possible to create clone banks for several tree species. After multi-locational clonal trials, superior planting stock from the clone bank can be chosen.
4. Genetic improvement: Flowering is essential for a variety of breeding operations. Flowering time can be shortened in many tree species through vegetative propagation.
5. Research work: Because they are genetically homogeneous, vegetatively propagated plants are important materials for research. This reduces experimental errors, thus the results of ramet tests produce more precise results.
6. Other uses of vegetative propagation: Due to limited germination, shelf life, and inadvertent destruction of blooms and seeds due to disease or pest attack, only vegetative growth is sometimes possible. Flowering is quite late in bamboo; thus, only vegetative strategies can be used to multiply immature plants. Genetic variants occur in rapidly dividing cells and developing somatic embryos in tissue culture, particularly in callus cultures (somaclonal variations). These variations can be used to test stress tolerance, disease resistance, and other traits.

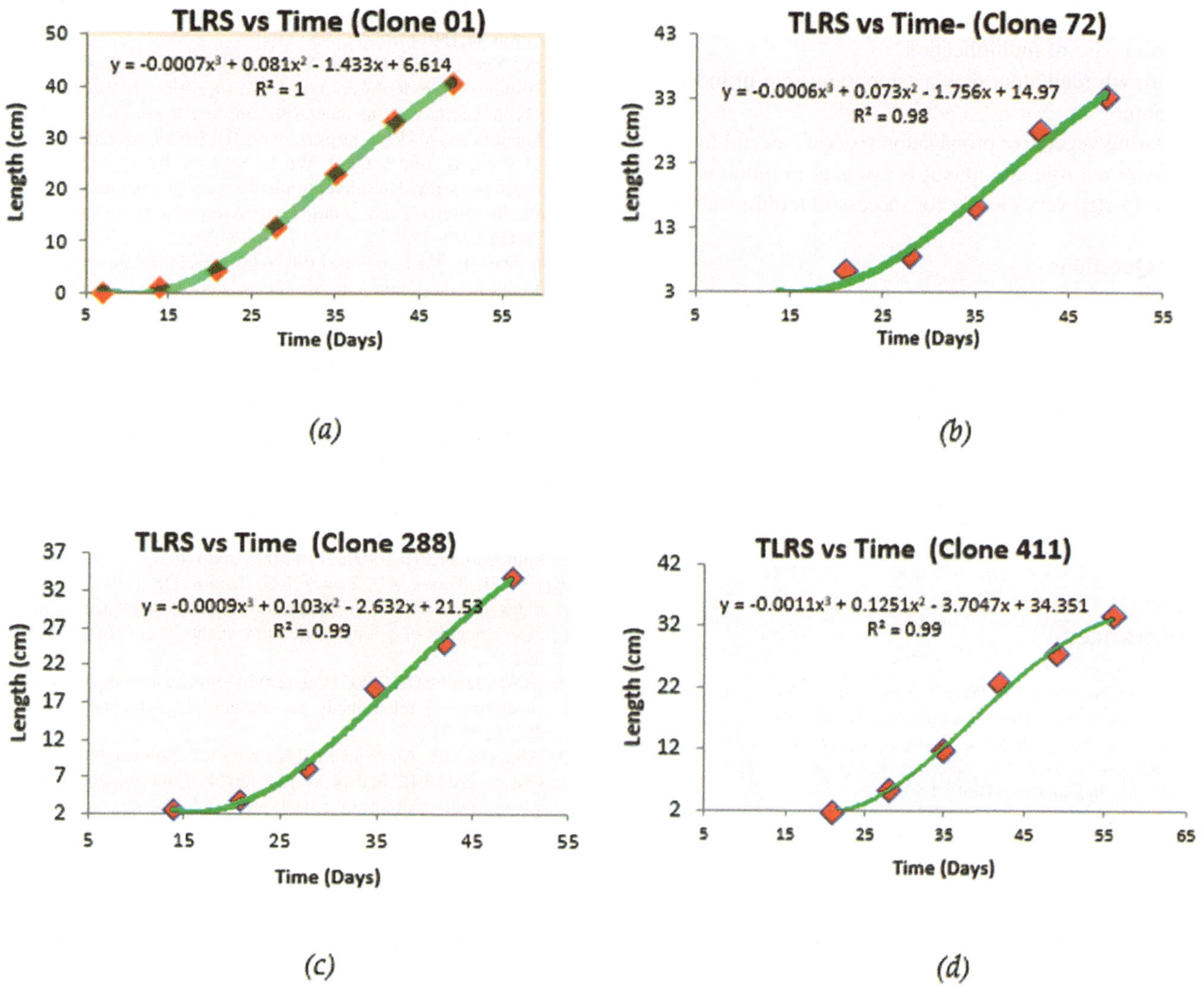

Fig. 31.11 TLRS of eucalyptus clones with evaluation time; (**a**) clone 1, (**b**) clone 72, (**c**) clone 288 and (**d**) clone 411

Vegetative propagation is an important tool for tree improvement purposes. Over a period of time the technique has evolved for many species yet there is ample scope for research especially addressing maturity related decline in rooting, physiology, genetics and molecular biology of adventitious root formation in woody perennials. This will enable development of rapid, efficient and cost efficient methods of clonal propagation for tree species of commercial importance. Not only induction of appropriate amount of adventitious roots the quality of parent material and produced plantlets is also equally important in clonal propagation. The root architecture in case of clonally propagated trees plays an important role in survival and growth of the tree.

Inter and intra-specific variability in adventitious rhizogenesis needs to studied to generate information which may be of great value for evolving uniform and workable clanal protocols. Eucalypts can serve as an ideal system for such studies as evident from recent research. Eucalyptus clonal propagation is practiced extensively and it serves as a good example for understanding the concept of juvenility, difference between leafy and non leafy cuttings, hormonal treatments required to induce rooting, environmental requirement etc. The subject is fascinating to satisfy researchers who want to understand the developmental biology of rooting and also to those who are more interested in developing efficient and smart technologies for clonal propagation of elite germplasm for tree improvement.

Lessons Learnt

- Vegetative propagation is the process of replicating plant material from vegetative organs so that the replicated plant has the same genotypes as the mother plant.
- Cuttings, layering, grafting and budding are common means of vegetative propagation, each having its own importance and usefulness. In forestry rooting of cutting

is the most significant clonal propagation method considering ease of multiplication.

- Growth regulators play an important role in induction of rooting.
- During vegetative propagation process, several factors play important role and thus it is essential to monitor each and every step very closely for successful multiplication.

Key Questions

1. How adventitious root development takes place in cuttings?
2. Describe the role of auxin in vegetative propagation.
3. What do you understand by juvenility in trees? How does it affect rooting in stem cuttings?
4. Describe a mist chamber. How is it different from a green house?
5. What are the advantages and disadvantages of presence of leaves on the stem cuttings?

References

Abarca D, Díaz-Sala C (2009) Reprogramming adult cells during organ regeneration in forest species. Plant Signal Behav 4(8):793–795

Ansari SA, Palanisamy K, Kumar P (1995) Peroxidase and nitrate reductase activity in relation to rhizogenesis in *Populus deltoides* Marsh. Indian J Exp Biol 33:892

Atkinson JA, Rasmussen A, Traini R, Voß U, Sturrock C, Mooney SJ, Wells DM, Bennett MJ (2014) Branching out in roots: uncovering form, function, and regulation. Plant Physiol 166(2):538–550

Balasimha D, Subra Monian N (1983) Role of phenolics in auxin induced rhizogenesis and isoperoxidases in cacao (*Theobroma cacao* L.) stem cuttings. Indian J Exp Biol 21(2):65–68

Barthwal S, Nautiyal R, Ganesan M, Venkataramanan KS, Gurumurthi K (2005) Effects of salt stress on rooting of *Casuarina equisetifolia* cuttings. J Trop For Sci 17(2):312–314

Bellamine J, Penel C, Greppin H, Gaspar T (1998) Confirmation of the role of auxin and calcium in the late phases of adventitious root formation. Plant Growth Regul 26(3):191–194

Bellini C, Pacurar DI, Perrone I (2014) Adventitious roots and lateral roots: similarities and differences. Ann Rev Plant Biol 65(1): 639–666

Berthon J-Y, Maldiney R, Sotta B, Gaspar T, Boyer N (1989) Endogenous levels of plant hormones during the course of adventitious rooting in cuttings of *Sequoiadendron giganteum* (Lindl.) in vitro. Biochemie Und Physiologie Der Pflanzen 184(5–6):405–412

Bhardwaj DR, Mishra VK (2005) Vegetative propagation of *Ulmus villosa*: effects of plant growth regulators, collection time, type of donor and position of shoot on adventitious root formation in stem cuttings. New For 29(2):105–116

Böhm W (1976) In situ estimation of root length at natural soil profiles. J Agric Sci 87(2):365–368

Bonga JM (2016) Conifer clonal propagation in tree improvement programs. In: Park Y-S, Bonga JM, Moon H-K (eds) Vegetative propagation of forest trees, pp 3–31

Brinker M, van Zyl L, Liu W, Craig D, Sederoff RR, Clapham DH, von Arnold S (2004) Microarray analyses of gene expression during adventitious root development in *Pinus contorta*. Plant Physiol 135(3):1526–1539

Brondani GE, Wendling I, Brondani AE, Araujo MA, da Silva ALL, Gonçalves AN (2012) Dynamics of adventitious rooting in mini-cuttings of *Eucalyptus benthamii* × *Eucalyptus dunnii*. Acta Sci Agron 34:169–178

Cai X, Yao L, Gao G, Xie Y, Yang J, Tang X, Zhang M (2016) Responses in root physiological characteristics of *Vallisneria natans* (Hydrocharitaceae) to increasing nutrient loadings. Knowl Manag Aquat Ecosyst 417:4. https://doi.org/10.1051/kmae/20150

Cai H, Yang C, Liu S, Qi H, Wu L, Xu L-A, Xu M (2019) MiRNA-target pairs regulate adventitious rooting in Populus: a functional role for miR167a and its target Auxin response factor 8. Tree Physiol 39(11):1922–1936

Campinhos E, Ikemori YK (1983) Introdução de novas técnicas na produção de mudas de essências florestais. Silvicultura 8(28): 226–228

Campinhos EN, Servin CMI, Almedia MA, Rosa AC (1999) Hidrojardim clonal Champion: uma otimização na produção de mudas de eucalipto. Silvicultura 19(80):42–46

Carvalho P, Moreira AM, Souza AJ, Bertol R, Magnago JM, Buffon JB, Azevedo JA (1991) Jardim clonal como área de multiplicação de estacas na Bahia Sul Celulose S/A. SIMPÓSIO IPEF 2:71–75

Chopart JL, Rodrigues SR, Carvalho De Azevedo M, De Conti Medina C (2008) Estimating sugarcane root length density through root mapping and orientation modelling. Plant Soil 313(1–2):101–112. https://doi.org/10.1007/s11104-008-9683-4

Currey CJ, Torres AP, Lopez RG, Jacobs DF (2013) The quality index—a new tool for integrating quantitative measurements to assess quality of young floriculture plants. Acta Hortic 1000:385–392

Davis AS, Jacobs DF (2005) Quantifying root system quality of nursery seedlings and relationship to outplanting performance. New For 30(2):295–311

De Almeida MR, Aumond M, Da Costa CT, Schwambach J, Ruedell CM, Correa LR, Fett-Neto AG (2017) Environmental control of adventitious rooting in Eucalyptus and Populus cuttings. Trees 31(5):1377–1390

de Assis TF, Rosa OP, Gonçalves SI (1992) Propagação por microestaquia. Congresso Florestal Estadual 7:824–836

de Assis TF, Fett-Neto AG, Alfenas AC (2004) Current techniques and prospects for the clonal propagation of hardwoods with emphasis on Eucalyptus. In: Plantation forest biotechnology for the 21st century. Research Signpost, Trivandrum, pp 303–333

de Dorlodot S, Forster B, Pagès L, Price A, Tuberosa R, Draye X (2007) Root system architecture: opportunities and constraints for genetic improvement of crops. Trends Plant Sci 12(10):474–481

De Klerk G-J, Van Der Krieken W, de Jong JC (1999) Review the formation of adventitious roots: new concepts, new possibilities. In Vitro Cell Dev Biol Plant 35(3):189–199

De Silva ALC, De Costa W, Bandara D (2011) Growth of root system and the patterns of soil moisture utilization in sugarcane under rain-fed and irrigated conditions in Sri Lanka. Sugar Tech 13(3):198–205

Dhiman RC, Gandhi (1999) Paulownia. UNDP-ICFRE Forestry Research and Education Project, Dehradun

Díaz-Sala C (2014) Direct reprogramming of adult somatic cells toward adventitious root formation in forest tree species: the effect of the juvenile–adult transition. Front Plant Sci 5:1–8

Díaz-Sala C (2019) Molecular dissection of the regenerative capacity of forest tree species: special focus on conifers. Front Plant Sci 9:1–10

Dick JM, Dewar RC (1992) A mechanistic model of carbohydrate dynamics during adventitious root development in leafy cuttings. Ann Bot 70(4):371–377

Dickson A, Leaf AL, Hosner JF (1960) Quality appraisal of white spruce and white pine seedling stock in nurseries. For Chronicle 36(1): 10–13

Dobbelaere S, Vanderleyden J, Okon Y (2003) Plant growth-promoting effects of diazotrophs in the rhizosphere. Crit Rev Plant Sci 22(2): 107–149

dos Santos CLV, Gomes S, Caldeira G (2000) Comparative responses of *Helianthus annuus* plants and calli exposed to NaCl: II. Selection of

stable salt tolerant calli cell lines and evaluation of osmotic adjustment and morphogenic capacity. J Plant Physiol 156(1):68–74. https://eurekamag.com/research/003/387/003387622.php
Druege U, Hilo A, Pérez-Pérez JM, Klopotek Y, Acosta M, Shahinnia F, Zerche S, Franken P, Hajirezaei MR (2019) Molecular and physiological control of adventitious rooting in cuttings: phytohormone action meets resource allocation. Ann Bot 123(6):929–949
Eliasson L, Brunes L (1980) Light effects on root formation in aspen and willow cuttings. Physiol Plant 48(2):261–265
Evers D, Schmit C, Mailliet Y, Hausman JF (1997) Growth characteristics and biochemical changes of poplar shoots in vitro under sodium chloride stress. J Plant Physiol 151(6):748–753
Falkiner RA, Nambiar EKS, Polglase PJ, Theiveyanathan S, Stewart LG (2006) Root distribution of *Eucalyptus grandis* and *Corymbia maculata* in degraded saline soils of south-eastern Australia. Agrofor Syst 67(3):279–291. https://doi.org/10.1007/s10457-005-5258-z
Felten J, Kohler A, Morin E, Bhalerao RP, Palme K, Martin F, Ditengou FA, Legué V (2009) The ectomycorrhizal fungus *Laccaria bicolor* stimulates lateral root formation in poplar and Arabidopsis through auxin transport and signaling. Plant Physiol 151(4):1991–2005
Feng Z, Sun X, Wang G, Liu H, Zhu J (2012) LBD29 regulates the cell cycle progression in response to auxin during lateral root formation in *Arabidopsis thaliana*. Ann Bot 110:1–10
Ferreira LA, Santos JL, Farahi F (1997) Pseudoheterodyne demodulation technique for fiber Bragg grating sensors using two matched gratings. IEEE Photon Technol Lett 9(4):487–489
Fukuda Y, Hirao T, Mishima K, Ohira M, Hiraoka Y, Takahashi M, Watanabe A (2018) Transcriptome dynamics of rooting zone and aboveground parts of cuttings during adventitious root formation in *Cryptomeria japonica* D. Don. BMC Plant Biol 18(1):1–14
Gaspar T (1981) Rooting and flowering, two antagonistic phenomena from a hormonal point of view. In: Aspects and prospects of plant growth regulators, pp 39–49
Gaspar T, Kevers C, Hausman JF, Berthon JY, Ripetti V (1992) Practical uses of peroxidase activity as a predictive marker of rooting performance of micropropagated shoots. Agronomie 12(10): 757–765
Gaspar T, Kevers C, Hausman J-F (1997) Indissociable chief factors in the inductive phase of adventitious rooting. In: Biology of root formation and development. Springer, pp 55–63
Geiss G, Gutierrez L, Bellini C (2009) Adventitious root formation: new insights and perspectives. Annu Plant Rev 37(1):127–156
Ginwal HS (2015) Eucalypt improvement: efforts and achievements in India. In: Eucalypts in India. Forest Research Institute, Dehradun, pp 117–138
Ginwal HS, Kumar P, Sharma VK, Mandal AK, Harwood CE (2004a) Genetic variability and early growth performance of *Eucalyptus tereticornis* Sm. in provenance cum progeny trials in India. Silvae Genetica 53(4):148–153. https://doi.org/10.1515/sg-2004-0027
Ginwal HS, Rawat PS, Srivastava RL (2004b) Seed source variation in growth performance and oil yield of *Jatropha curcas* Linn. in central India. Silvae Genetica 53(4):186–192. https://doi.org/10.1515/sg-2004-0034
Goldfarb B, Hackett WP, Furnier GR, Mohn CA, Plietzsch A (1998) Adventitious root initiation in hypocotyl and epicotyl cuttings of eastern white pine (*Pinus strobus*) seedlings. Physiol Plant 102(4): 513–522
Gonin M, Bergougnoux V, Nguyen TD, Gantet P, Champion A (2019) What makes adventitious roots? Plants 8(7):240
Gou J, Strauss SH, Tsai CJ, Fang K, Chen Y, Jiang X, Busov VB (2010) Gibberellins regulate lateral root formation in Populus through interactions with auxin and other hormones. Plant Cell 22(3): 623–639
Grossnickle SC (2005) Importance of root growth in overcoming planting stress. New For 30(2–3):273–294. https://doi.org/10.1007/s11056-004-8303-2
Gupta PK, Mascarenhas AF, Jagannathan V (1981) Tissue culture of forest trees—clonal propagation of mature trees of *Eucalyptus citriodora* Hook, by tissue culture. Plant Sci Lett 20(3):195–201
Gutierrez L, Mongelard G, Floková K, Păcurar DI, Novák O, Staswick P, Kowalczyk M, Păcurar M, Demailly H, Geiss G (2012) Auxin controls Arabidopsis adventitious root initiation by regulating jasmonic acid homeostasis. Plant Cell 24(6):2515–2527
Haapala T, Pakkanen A, Pulkkinen P (2004) Variation in survival and growth of cuttings in two clonal propagation methods for hybrid aspen (*Populus tremula* × *P. tremuloides*). For Ecol Manag 193(3): 345–354
Hamer JJ, Veneklaas EJ, Poot P, Mokany K, Renton M (2015) Shallow environmental gradients put inland species at risk: Insights and implications from predicting future distributions of Eucalyptus species in South Western Australia. Aust Ecol 40(8):923–932. https://doi.org/10.1111/aec.12274
Hamer JJ, Veneklaas EJ, Renton M, Poot P (2016) Links between soil texture and root architecture of Eucalyptus species may limit distribution ranges under future climates. Plant Soil 403(1–2):217–229. https://doi.org/10.1007/s11104-015-2559-5
Hand P (1994) Biochemical and molecular markers of cellular competence for adventitious rooting. In: Biology of adventitious root formation. Springer, pp 111–121
Hartmann HT, Kester DE, Davies FT, Geneve RL (1997) Plant propagation: principles and practices, 6th edn. Prentice-Hall Inc.
Hatzilazarou SP, Syros TD, Yupsanis TA, Bosabalidis AM, Economou AS (2006) Peroxidases, lignin and anatomy during in vitro and ex vitro rooting of gardenia (*Gardenia jasminoides* Ellis) microshoots. J Plant Physiol 163(8):827–836
Higashi EN, Silveira R, Gonçalves AN (2000) Propagação vegetativa de Eucalyptus: princípios básicos e a sua evolução no Brasil. Circ Técnica IPEF 192:1–11
Hoffmann AA, Sgró CM (2011) Climate change and evolutionary adaptation. Nature 470(7335):479–485. https://doi.org/10.1038/nature09670
Hooker JE, Munro M, Atkinson D (1992) Vesicular-arbuscular mycorrhizal fungi induced alteration in poplar root system morphology. Plant Soil 145(2):207–214
Hu X, Xu L (2016) Transcription factors WOX11/12 directly activate WOX5/7 to promote root primordia initiation and organogenesis. Plant Physiol 172(4):2363–2373
Husen A, Iqbal M, Siddiqui SN, Sohrab SS, Masresha G (2017) Effect of indole-3-butyric acid on clonal propagation of mulberry (*Morus alba* L.) stem cuttings: rooting and associated biochemical changes. Proc Natl Acad Sci India Sect B Biol Sci 87(1):161–166. https://doi.org/10.1007/s40011-015-0597-7
Jofuku KD, Den Boer BG, Van Montagu M, Okamuro JK (1994) Control of Arabidopsis flower and seed development by the homeotic gene APEALA2. Plant Cell 6(9):1211–1225
Karlberg A, Bako L, Bhalerao RP (2011) Short day-mediated cessation of growth requires the downregulation of AINTEGUMENTALIKE1 transcription factor in hybrid aspen. PLoS Genet 7(11):e1002361
Khalid A, Tahir S, Arshad M, Zahir ZA (2004) Relative efficiency of rhizobacteria for auxin biosynthesis in rhizosphere and non-rhizosphere soils. Aust J Soil Res 42(8):921–926. https://doi.org/10.1071/SR04019
Klopotek Y, Franken P, Klaering H-P, Fischer K, Hause B, Hajirezaei M-R, Druege U (2016) A higher sink competitiveness of the rooting zone and invertases are involved in dark stimulation of adventitious root formation in *Petunia hybrida* cuttings. Plant Sci 243:10–22
Korasick DA, Westfall CS, Lee SG, Nanao MH, Dumas R, Hagen G, Guilfoyle TJ, Jez JM, Strader LC (2014) Molecular basis for auxin response factor protein interaction and the control of auxin response repression. Proc Natl Acad Sci 111(14):5427–5432
Kuligod V, Doddamani VS, Paul SG (1995) Effect of soil salinity on growth of *Casuarina equisetifolia*. Curr Agric 19:51–53

Kulkarni HD (2001) Eucalyptus hybrid breeding in ITC Bhadrachalam, India. In: Proceedings of the international symposium, International Union of Forest Research Organizations (IUFRO) on Developing the Eucalyptus for the Future (Unit 2.08. 03 Improvement and Culture of Eucalyptus), Held in Valdivia, Chile

Kuo Y, Yang T, Huang G-W (2008) The use of grey relational analysis in solving multiple attribute decision-making problems. Comput Ind Eng 55(1):80–93

Lakehal A, Bellini C (2019) Control of adventitious root formation: insights into synergistic and antagonistic hormonal interactions. Physiol Plant 165(1):90–100

Lakehal A, Chaabouni S, Cavel E, Le Hir R, Ranjan A, Raneshan Z, Novák O, Păcurar DI, Perrone I, Jobert F (2019) A molecular framework for the control of adventitious rooting by TIR1/AFB2-Aux/IAA-dependent auxin signaling in Arabidopsis. Mol Plant 12(11):1499–1151

Lakehal A, Ranjan A, Bellini C (2020) Multiple roles of jasmonates in shaping rhizotaxis: emerging integrators. Methods Mol Biol 2085:3–22

Lal P, Dogra AS, Sharma SC, Chahal GBS (2006) Evaluation of different clones of Eucalyptus in Punjab. Indian For 132(11):1383–1390

Legué V, Rigal A, Bhalerao RP (2014) Adventitious root formation in tree species: involvement of transcription factors. Physiol Plant 151(2):192–198

Li S-W, Xue L, Xu S, Feng H, An L (2009) Mediators, genes and signaling in adventitious rooting. Bot Rev 75(2):230–247

Li S-W, Leng Y, Shi R-F (2017) Transcriptomic profiling provides molecular insights into hydrogen peroxide-induced adventitious rooting in mung bean seedlings. BMC Genom 18(1):1–23

Liu B, Wang L, Zhang J, Li J, Zheng H, Chen J, Lu M (2014a) WUSCHEL-related Homeobox genes in *Populus tomentosa*: diversified expression patterns and a functional similarity in adventitious root formation. BMC Genom 15(1):1–14

Liu J, Sheng L, Xu Y, Li J, Yang Z, Huang H, Xu L (2014b) WOX11 and 12 are involved in the first-step cell fate transition during de novo root organogenesis in Arabidopsis. Plant Cell 26(3):1081–1093

Mauriat M, Petterle A, Bellini C, Moritz T (2014) Gibberellins inhibit adventitious rooting in hybrid aspen and Arabidopsis by affecting auxin transport. Plant J 78(3):372–384

Maynard BK, Bassuk NL (1985) Etiolation as a tool for rooting of cuttings of difficult-to-root woody plants. In: Combined proceedings of International Plant Propagators Society, pp 448–495

Mitchell PJ, Veneklaas E, Lambers H, Burgess SSO (2009) Partitioning of evapotranspiration in a semi-arid eucalypt woodland in southwestern Australia. Agric For Meteorol 149(1):25–37. https://doi.org/10.1016/j.agrformet.2008.07.008

Mitsuhashi-Kato M, Shibaoka H, Shimokoriyama M (1978a) Anatomical and physiological aspects of developmental processes of adventitious root formation in Azukia cuttings. Plant Cell Physiol 19(3):393–400

Mitsuhashi-Kato M, Shibaoka H, Shimokoriyama M (1978b) The nature of the dual effect of auxin on root formation in Azukia cuttings. Plant Cell Physiol 19(8):1535–1542

Mohammed S, Eriksen EN (1974) Root formation in pea cuttings: IV. Further studies on the influence of indole-3-acetic acid at different developmental stages. Physiol Plant 32(1):94–96

Molnar JM, LaCroix LJ (1972) Studies of the rooting of cuttings of *Hydrangea macrophylla*: DNA and protein changes. Can J Bot 50(3):387–392

Moncousin CH, Favre JM, Gaspar T (1988) Changes in peroxidase activity and endogenous IAA levels during adventitious root formation in vine cuttings. In: Kutacek M, Bandurski RS, Krekule J (eds) Proceedings of the symposium on physiology and biochemistry of auxins in plants

Müller A, Volmer K, Mishra-Knyrim M, Polle A (2013) Growing poplars for research with and without mycorrhizas. Front Plant Sci 4:332

Nag S, Paul A, Choudhuri MA (2013) Changes in peroxidase activity during adventitious root formation at the base of mung bean cuttings. Int J Sci Technol Res 2(5):171–177

Noordwijk M (1987) Methods for quantification of root distribution pattern and root dynamics in the field. In: Proceedings of the 20th Colloquium of the International Potash Institute, Baden bei Wein, 1987. International Potash Institute (IPI), Berne, pp 263–281

Oballa PO, Konuche PKA, Muchiri MN, Kigomo BN (2010) Facts on growing and use of eucalyptus in Kenya. Kenya Forestry Research Institute, Nairobi, p 30. https://www.fornis.net/syste

Okoro OO, Grace J (1976) The physiology of rooting Populus cuttings. I. Carbohydrates and photosynthesis. Physiol Plant 36(2): 133–138

Pal M, Nanda KK (1981) Rooting of etiolated stem segments of *Populus robusta*-interaction of temperature, catechol and sucrose in the presence of IAA. Physiol Plant 53(4):540–542

Palana RM (2017) Eucalyptus in India. FAO Corporate Document Directory, pp 1–11

Palanisamy K, Subramanian K (2001) Vegetative propagation of mature teak trees (*Tectona grandis* L.). Silvae Genetica 50(5–6):188–190

Pellicer V, Guehl J-M, Daudet F-A, Cazet M, Riviere LM, Maillard P (2000) Carbon and nitrogen mobilization in *Larix* x *eurolepis* leafy stem cuttings assessed by dual 13C and 15N labeling: relationships with rooting. Tree Physiol 20(12):807–814

Pizarro A, Díaz-Sala C (2019) Cellular dynamics during maturation-related decline of adventitious root formation in forest tree species. Physiol Plant 165(1):73–80

Poot P, Veneklaas EJ (2013) Species distribution and crown decline are associated with contrasting water relations in four common sympatric eucalypt species in southwestern Australia. Plant Soil 364(1–2): 409–423. https://doi.org/10.1007/s11104-012-1452-8

Quan J, Meng S, Guo E, Zhang S, Zhao Z, Yang X (2017) De novo sequencing and comparative transcriptome analysis of adventitious root development induced by exogenous indole-3-butyric acid in cuttings of tetraploid black locust. BMC Genom 18(1):1–14

Ramírez-Carvajal GA, Morse AM, Dervinis C, Davis JM (2009) The cytokinin type-B response regulator PtRR13 is a negative regulator of adventitious root development in Populus. Plant Physiol 150(2): 759–771

Ribeiro CL, Silva CM, Drost DR, Novaes E, Novaes CRDB, Dervinis C, Kirst M (2016) Integration of genetic, genomic and transcriptomic information identifies putative regulators of adventitious root formation in Populus. BMC Plant Biol 16(1):1–11

Riechmann JL, Meyerowitz EM (1998) The AP2/EREBP family of plant transcription factors. Biol Chem 379:633–646

Rigal A, Yordanov YS, Perrone I, Karlberg A, Tisserant E, Bellini C, Busov VB, Martin F, Kohler A, Bhalerao R (2012) The AINTEGUMENTA LIKE1 homeotic transcription factor PtAIL1 controls the formation of adventitious root primordia in poplar. Plant Physiol 160(4):1996–2006

Rout GR, Samantaray S, Das P (2000) In vitro rooting of *Psoralea corylifolia* Linn: peroxidase activity as a marker. Plant Growth Regul 30(3):215–219

Saha R, Ginwal HS, Chandra G, Barthwal S (2020a) Integrated assessment of adventitious rhizogenesis in Eucalyptus: root quality index and rooting dynamics. J For Res 31(6):2145–2161. https://doi.org/10.1007/s11676-019-01040-6

Saha R, Ginwal HS, Chandra G, Barthwal S (2020b) Root distribution, orientation and root length density modelling in Eucalyptus and evaluation of associated water use efficiency. New For 51(6): 1023–1037. https://doi.org/10.1007/s11056-020-09772-8

Saha R, Ginwal HS, Chandra G, Barthwal S (2021) A comparative study on grey relational analysis and C5.0 classification algorithm on adventitious rhizogenesis of Eucalyptus. Trees 35(1):43–52

Saxena C, Samantaray S, Rout GR, Das P (2000) Effect of auxins on in vitro rooting of *Plumbago zeylanica*: peroxidase activity as a marker for root induction. Biol Plant 43(1):121–124

Schenk HJ, Jackson RB (2002a) Rooting depths, lateral root spreads and below-ground/above-ground allometries of plants in water-limited ecosystems. J Ecol 90:480–494

Schenk HJ, Jackson RB (2002b) The global biogeography of roots. Ecol Monogr 72(3):311–328

Schenk HJ, Jackson RB (2005) Mapping the global distribution of deep roots in relation to climate and soil characteristics. Geoderma 126(1–2):129–140

Schrader J, Nilsson J, Mellerowicz E, Berglund A, Nilsson P, Hertzberg M, Sandberg G (2004) A high-resolution transcript profile across the wood-forming meristem of poplar identifies potential regulators of cambial stem cell identity. Plant Cell 16(9):2278–2292

Schwambac J, Ruedell CM, De Almeida MR, Penchel RM, De Araújo EF, Fett-Neto AG (2008) Adventitious rooting of *Eucalyptus globulus* × *E. maidennii* mini-cuttings derived from mini-stumps grown in sand bed and intermittent flooding trays: a comparative study. New For 36(3):261–271. https://doi.org/10.1007/s11056-008-9099-2

Sekar S, Srimathi RA, Kulkarni HD, Venkatesan KR (1984) Computerized design for tree seed orchards. CABI

Shibaoka H (1971) Effects of indoleacetic, *p*-chlorophenoxyisobutyric and 2,4,6-trichlorophenoxyacetic acids on three phases of rooting in Azukia cuttings. Plant Cell Physiol 12(2):193–200

Shyam Sundar S (1984) Some aspects of Eucalyptus hybrid. Workshop on eucalyptus plantation. Indian Statistical Institute, Bangalore

Singh S (2006) Studies on physiomorphological basis of adventitious rhizogenesis vis-à-vis clonal propagation of *Gmelina arborea* Roxb. FRI, Dehradun

Singh S, Ansari SA (2014) Callus formation impedes adventitious rhizogenesis in air layers of broadleaved tree species. Ann For Res 57(1):47–54

Singh YP, Minhas PS, Tomar OS, Gupta RK (1996) Salt tolerance and water use by *Dalbergia sissoo* during the establishment stage. Arid Land Res Manag 10(4):379–390

Sorin C, Bussell JD, Camus I, Ljung K, Kowalczyk M, Geiss G, McKhann H, Garcion C, Vaucheret H, Sandberg G (2005) Auxin and light control of adventitious rooting in Arabidopsis require ARGONAUTE1. Plant Cell 17(5):1343–1359

Steffens B, Rasmussen A (2016) The physiology of adventitious roots. Plant Physiol 170(2):603–617

Stenvall N, Haapala T, Aarlahti S, Pulkkinen P (2005) The effect of soil temperature and light on sprouting and rooting of root cuttings of hybrid aspen clones. Can J For Res 35(11):2671–2678

Sun P, Jia H, Zhang Y, Li J, Lu M, ans Hu, J. (2019) Deciphering genetic architecture of adventitious root and related shoot traits in Populus using QTL mapping and RNA-Seq data. Int J Mol Sci 20(24):6114

Sureshkumar G, Seenivasan R, Chezhian P (2016) Mini-cutting technique for large scale clonal propagation of *Casuarina junghuhniana*. Casuarina improvement for securing rural livelihoods. In: Nicodemus A, Pinyopusarerk K, Zhong CL, Franche C (eds) Proceedings of the 5th international casuarina workshop. Institute of Forest Genetics and Tree Breeding, Coimbatore, pp 130–132

Taghavi S, Garafola C, Monchy S, Newman L, Hoffman A, Weyens N, Barac T, Vangronsveld J, van der Lelie D (2009) Genome survey and characterization of endophytic bacteria exhibiting a beneficial effect on growth and development of poplar trees. Appl Environ Microbiol 75(3):748–757

Trupiano D, Yordanov Y, Regan S, Meilan R, Tschaplinski T, Scippa GS, Busov V (2013) Identification, characterization of an AP2/ERF transcription factor that promotes adventitious, lateral root formation in Populus. Planta 238(2):271–282

van Noordwijk M, Lawson G, Hairiah K, Wilson J (2015) Root distribution of trees and crops: competition and/or complementarity. In: Tree-crop interactions: agroforestry in a changing climate. CABI, Wallingford, pp 221–257. https://doi.org/10.1079/9781780645117.0221

Varghese M, Nicodemus A, Nagarajan B, Subramanian K (2000) Hybrid breakdown in Mysore gum and need for genetic improvement of *Eucalyptus camaldulensis* and *E. tereticornis*. In: Proceedings QFRI/CRC-SPF symposium on hybrid breeding and genetics of forest trees

Varghese M, Harwood CE, Hegde R, Ravi N (2008) Evaluation of provenances of *Eucalyptus camaldulensis* and clones of *E. camaldulensis* and *E. tereticornis* at contrasting sites in Southern India. Silvae Genetica 57(3):170

Vilasboa J, Da Costa CT, Fett-Neto AG (2019) Rooting of eucalypt cuttings as a problem-solving oriented model in plant biology. Progress Biophys Mol Biol 146:85–97

Wang P, Ma L, Wang S, Li L, Wang Q, Yang R, Li Y (2019) Identification and analysis of a candidate WRKY transcription factor gene affecting adventitious root formation using association mapping in Catalpa Scop. DNA Cell Biol 38(4):297–306

Wei M, Liu Q, Wang Z, Yang J, Li W, Chen Y, Lu H, Nie J, Liu B, Lv K (2020) PuHox52-mediated hierarchical multilayered gene regulatory network promotes adventitious root formation in *Populus ussuriensis*. New Phytol 228(4):1369–1385

Wendling I, Xavier A, Gomes JM, Pires IE, Andrade HB (2000) Minicuttings and microcuttings rooting of *Eucalyptus grandis* W. Hill ex Maiden clones as affected by IBA. Revista Árvore 24(2):187–192

Went FW (1934) A test method for rhizocaline, the root-forming substance. Proc Kon Acad Wetensch Amsterdam 37:547–555

Wilson J (1973) The need for a rational utilisation of the montane temperate forests of South India. Indian Forester 99(12):707–716

Wuddineh WA, Mazarei M, Turner GB, Sykes RW, Decker SR, Davis MF, Stewart CN Jr (2015) Identification and molecular characterization of the switchgrass AP2/ERF transcription factor superfamily, and overexpression of PvERF001 for improvement of biomass characteristics for biofuel. Front Bioeng Biotechnol 3:101

Xavier A, Comério J (1996) Microestaquia: uma maximização da micropropagação de Eucalyptus. Revista Árvore 20(1):9–16

Xavier A, Wendling I (1998) Miniestaquia na clonagem de Eucalyptus. 8p. SIF/UVF. Information Tecnico SIF, 11.

Xu Y, Li X, Lin J, Wang Z, Yang Q, Chang Y (2015) Transcriptome sequencing and analysis of major genes involved in calcium signaling pathways in pear plants (*Pyrus calleryana* Decne.). BMC Genom 16(1):1–13. https://doi.org/10.1186/s12864-015-1887-4

Xuan L, Xu M, Chen C, Yang C, Xu L, Huang M (2014) Identification and characterization of three PeSHRs and one PeSCR involved in adventitious root development of Populus. Plant Cell Tissue Organ Cult 117(2):253–264

Yordanov YS, Ma C, Yordanova E, Meilan R, Strauss SH, Busov VB (2017) Big leaf is a regulator of organ size and adventitious root formation in poplar. PLoS One 12(7):e0180527

Yue J, Yang H, Yang S, Wang J (2020) TDIF regulates auxin accumulation and modulates auxin sensitivity to enhance both adventitious root and lateral root formation in poplar trees. Tree Physiol 40(11): 1534–1547

Zavattieri MA, Ragonezi C, Klimaszewska K (2016) Adventitious rooting of conifers: influence of biological factors. Trees 30(4): 1021–1032

Zhang B, Tong C, Yin T, Zhang X, Zhuge Q, Huang M, Wang M, Wu R (2009) Detection of quantitative trait loci influencing growth trajectories of adventitious roots in Populus using functional mapping. Tree Genet Genomes 5(3):539–552

Zhang Y, Xiao Z, Zhan C, Liu M, Xia W, Wang N (2019) Comprehensive analysis of dynamic gene expression and investigation of the roles of hydrogen peroxide during adventitious rooting in poplar. BMC Plant Biol 19(1):1–16

Micropropagation of Forest Tree Species

32

T. S. Rathore, S. S. Rathore, and A. K. Chauhan

Abstract

The genetic variability found in tree species presents an opportunity for enhancing genetic traits and utilizing superior genotypes through clonal forestry. While traditional macropropagation methods have limitations in mass-producing clonal plants for challenging-to-root tree species, plant tissue culture-based biotechnology emerges as a solution to overcome these challenges and complements tree improvement programs. There are three main methods of micropropagation for tree species: axillary shoot proliferation, adventitious shoot regeneration (both direct and through the callus phase), and somatic embryogenesis (both direct and through the callus phase). Among these, axillary shoot induction is the preferred and safest method for producing genetically true-to-type planting material. Direct adventitious regeneration and somatic embryogenesis follow in preference. However, the presence of prolonged cultures (over 1-year old) may pose challenges of genetic variability. Therefore, before large-scale micropropagation for operational planting, it is crucial to conduct genetic fidelity studies using DNA markers to ensure a secure mode of regeneration and the maintenance of genetic uniformity in planting materials.

Several factors significantly influence micropropagation, including (1) the type and source of explant (seedling/mature tree), (2) the management of trees serving as the explant source, (3) the timing of explant collection, (4) surface sterilants, their concentrations, and treatment duration, (5) nutrient media, (6) growth hormones and their concentrations, (7) genotypes/clones, (8) growth adjuvants, (9) gelling agents and media pH, (10) carbohydrate source and its concentrations, (11) incubation conditions of cultures, and (12) hardening conditions and duration. Micropropagation, requiring a small quantity of plant material (explant), offers high production potential for mass multiplication of selected genotypes/clones and plant rejuvenation. It can be carried out in small spaces, allowing year-round production unaffected by external environmental factors. The benefits include uniformity of planting material, high production rates, and germplasm preservation. This chapter provides detailed insights into the micropropagation of tree species through various modes of plant regeneration.

Keywords

Micropropagation · Adventitious shoot regeneration · Somatic embryogenesis · Explant · Genetic fidelity

32.1 Introduction

The presence of genetic diversity within and among tree species presents an opportunity for targeted selection to meet the requirements of commercial forest plantations, farm forestry, and agroforestry. Selections made at various levels, such as species, provenance, and individual trees possessing desirable characteristics, are employed in plantations and further utilized for genetic improvement to enhance productivity and product quality. Traditional methods of improving forest tree species involve selecting superior trees for seed production areas, seed orchards, provenance, progeny, clonal testing, and controlled pollination to develop hybrids (Kumar et al. 2016). While seeds remain the most convenient source for raising planting material, the outbreeding nature of many tree species poses challenges. Seed-based progenies may not be true to the mother plants, limiting the realization of the full potential of superior genotypes/clones. As an alternative, stem cuttings can be utilized for the mass production of superior genotypes,

T. S. Rathore (✉) · S. S. Rathore · A. K. Chauhan
Arid Forest Research Institute, Jodhpur, India

A. K. Mandal, A. Nicodemus (eds.), *Textbook of Forest Science*, https://doi.org/10.1007/978-981-97-8289-5_32

providing a means to overcome the limitations associated with seed-based propagation methods.

Mass multiplication of superior genotypes through leafy stem cuttings is widely practiced for *Acacia*, *Casuarina*, *Eucalyptus*, and *Populus* species. It is easy to rejuvenate by hedging and achieving a high level of rooting in these species. Most of the long-gestation tree species such as *Tectona grandis*, *Pterocarpus* spp., and *Santalum album* are difficult to root and prevent mass production of clonal plants. In addition, factors like seasonal availability of cuttings, requirement of a large space, infrastructure requirement of bulk quantity of cuttings, low production potential, and poor rejuvenation are also limiting macropropagation in forest trees.

To address challenges and complement traditional macropropagation methods, the application of plant tissue culture-based biotechnology, particularly micropropagation, offers a viable solution for the rapid multiplication of a large number of plants from tree species that are difficult to root (Gupta et al. 1980; Bonga 1987; Mascarenhas and Muralidharan 1989; Rathore et al. 1991a, b; Shekhawat et al. 1993; Sanjay et al. 2006; Rathore et al. 2008a, b). Micropropagation involves the in vitro culture of plant cells, tissues, and organs (explants) in a sterile environment on an artificial medium, maintaining desired temperatures and photoperiods, with the goal of producing plants of specific genotypes.

32.2 Explant

Explant can be defined as a part of the plant, namely apical shoot, nodal shoot, leaf segment, inflorescence, anther, pollen, ovary, hybrid embryo, and somatic cells, used for the production of plants. Any living parts of the plant can be used as an explant.

In vitro, seedlings, and field-grown trees can be sources of explant material. The selection of an explant is based on the objectives of micropropagation. The apical or nodal shoot segment is used as an explant for the clonal propagation to get plants genetically similar to the mother plant. Leaf, internode, hypocotyl, and inflorescence are commonly used as explants for adventitious shoot production, callus development, and somatic embryogenesis.

32.3 Aseptic Condition

To maintain a microbe-free environment for in vitro cultures, both explants and nutrient medium undergo sterilization. Explants are surface sterilized using 70% ethanol, sodium hypochlorite, mercuric chloride, chlorine water, or bromine water. The choice of one or two sterilants is flexible to ensure the removal of microbes from the explants. The exposure time to sterilants varies based on the type of explant. Nutrient media are autoclaved at 15 psi for 20 min or 20 psi for 15 min. Similarly, sterilized distilled water is employed for washing explants post-surface sterilization, and Petri plates, as well as surgical items, undergo steam sterilization, mirroring the process applied to the nutrient medium. The maintenance of a sterile environment is facilitated by laminar airflow (LAF), which provides microbe-free air through positive air pressure passing through high-efficiency particulate air (HEPA) filters with a pore size of 0.22 μM. Ultraviolet light installed in the LAF serves to eliminate microbes within the airflow.

32.4 Nutrient Medium

The nutrient medium plays a crucial role in micropropagation as it provides essential elements, vitamins, amino acids, carbohydrates, and growth hormones necessary for the growth and differentiation of cultured explants. In 1902, Haberlandt made the first attempt to culture plant cells on a simple nutrient medium and introduced the term "totipotency" of plant cells, though he did not succeed in plant regeneration (Haberlandt 1902). This pioneering effort laid the foundation for subsequent research. The first successful unlimited growth of callus was achieved by White (1939), and Heller (1953) formulated a nutrient medium containing micro and macroelements essential for cultured tissue growth. White (1963) later developed a simple nutrient medium. Up until the 1950s, there was a limited understanding of nutrient medium and growth hormones, but progress was made by Skoog and Miller (1957) who worked with tobacco tissue culture. They observed that high cytokinin and low auxin induce shoots, while the reverse induces roots, and an equal ratio of auxin and cytokinin induces callus formation. In 1962, Murashige and Skoog (MS) formulated the MS medium, comprising macro and micronutrients, vitamins, amino acids, sucrose, and agar-agar. This medium gained popularity and became widely used in plant tissue culture, including the micropropagation of woody plants (WPs). Table 32.1 provides details of the composition of the commonly used MS medium, while Table 32.2 outlines the details of selected nutrient media commonly used for the propagation of tree species.

32.5 Controlled Incubation Conditions

In order to achieve reproducible and the best result on micropropagation, incubation conditions of culture rooms/ growth rooms are controlled in terms of temperature (generally 25 ± 1 °C), light intensity (2000–4000 lux), photoperiod 12–16 h, and humidity (60%). The requirement of optimum

Table 32.1 Chemical composition, preparation of stock solution of MS medium (Murashige and Skoog 1962), and volumes to be used

Stock solution	Chemicals	Quantity (mg/L)	Quantity (g/L) in stock solution	Strength of stock solution	Volume in mL for 1 L of medium
A	NH_4NO_3	1650	82.5	50X	20
B	KNO_3	1900	95.0	50X	20
C	KH_2PO_4	170	34.0	200X	5
	H_3BO_3	6.2	1.24		
	$Na_2Mo\ O_4.2H_2O$	0.25	0.05		
	KI	0.83	0.166		
	$COCl_2\ 6H_2O$	0.25	0.005		
D	$CaCl_2.2H_2O$	440	88.0	200X	5
E	$MgSO_4.7H_2O$	370	74.0	200X	5
	$MnSO_4.4H_2O$	16.9	3.38		
	$ZnSO_4.7H_2O$	8.6	1.72		
	$CuSO_45H_2O$	0.025	0.005		
F	$FeSO_4\ 7H_2O$	27.8	5.50	200X	5
	$Na_2EDTA.H_2O$	37.3	7.46		
	Thiamine HCl	0.5	1.0	1000X	0.5
	Pyridoxine hydrochloride	0.5	1.0	1000X	0.5
	Nicotinic acid	0.5	1.0	1000X	0.5
	Glycine	2.0	1.0	500X	2.0
	Myo inositol	100			
	Sucrose ($C_{12}H_{22}O_{11}$)		30.0		
	Agar-agar		5.5–7.0		

temperature, light intensity, and photoperiod may vary with the tree species and maintaining such conditions is essential for reproducibility of results.

32.6 Modes of Micropropagation

There are three modes of micropropagation, viz. (1) forced axillary shoot proliferation, (2) adventitious regeneration, and (3) somatic embryogenesis.

32.6.1 Regeneration Through Axillary Shoot Proliferation

Micropropagation (regeneration) through axillary shoot induction stands out as the widely employed method for propagating forest tree species (Fig. 32.1). In this method, apical or nodal shoots are selected as explants. The initial step involves forced axillary shoot proliferation on a shoot initiation medium containing low auxin and high cytokinin. Commonly used media for this purpose include MS (Murashige and Skoog 1962), WP (Lloyd and McCown 1980), and B5 (Gamborg et al. 1968). The selection of growth hormones, such as auxins (indole acetic acid [IAA], indole butyric acid [IBA], and naphthalene acetic acid [NAA]) and cytokinins (kinetin, 6-furfuryl amino purine, benzyl amino purine [BAP], zeatin, 2-iso-pentyl adenine [2ip], and thidiazuron [TDZ]), depends on the plant species. The second step involves further shoot proliferation or bulking through subculturing on the same medium or with modifications in nutrients and growth hormones. Subculturing is typically performed every 3–5 weeks, and it can be carried out for up to 10–12 passages. The multiplication rate varies between two- and four-folds in each passage, depending on the tree species and source material.

The third step focuses on in vitro or ex vitro rooting of the multiplied shoots. This involves the use of a low-nutrient medium or modifications in macro and microelements, with auxins, mostly IBA, playing a crucial role. In vitro rooting takes approximately 2–4 weeks, and rooted plants can be utilized for the subsequent hardening process. Ex vitro rooting, which involves sturdy shoots with a minimum of two to three nodes treated with IBA, reduces both time and production costs. The fourth step entails the hardening/acclimatization of plants, a critical phase for ensuring high survival rates during field transfer. In vitro plants, being partially autotrophic, undergo gradual acclimatization to overcome challenges such as poorly developed wax and cuticle formation and poorly functional stomata due to high humidity within culture vessels. Plantlets are washed and treated with fungicide to prevent microbial attacks. A potting medium, consisting of soilrite, cocopeat, vermiculite, compost, sand, and soil, either alone or in combination, is commonly used. The potted plants initially reside in a greenhouse with low temperature (25–28 °C) and high humidity

Table 32.2 Chemical constituents (in mg/L) of various nutrient media commonly used for micropropagation of tree species

Composition	MS	B5	WPM	SH	HE
Macronutrients					
KH_2PO_4	170.00		170.00		140.00
$Na_2H_2PO_4.H_2O$		150.00			
KNO_3	1900.00	2500.00		2500.00	
NH_4NO_3	1650.00		400.00		
$CaCl_2.2H_2O$	440.00	150.00	96.00	151.00	500.00
Ca $(NO_3)_4.H_2O$			556.00		
$(NH_4)_2SO_4$		134.00		300.00	25.00
Na_2SO_4		150.00			
KCl					
$MgSO_4.7H_2O$	370.00	250.00	370.00	195.05	140.00
Micronutrients					
H_3BO_3	6.2	3.0	6.2	5.00	1.0
$MnSO_4.4H_2O$	16.9				0.01
$MnSO_4.H_2O$		10.0	22.3	10.00	
$ZnSO_4.7H_2O$	8.6	2.0	8.6	1.00	1.0
$Na_2Mo_42H_2O$	0.25	0.025	0.025	0.10	
$CuSO_45H_2O$	0.025	0.025	0.025	0.20	0.03
$CoCl_26H_2O$		0.025		0.10	
KI	0.83	0.75		1.00	0.01
$FeSO_47H_2O$	27.8		27.8		
Na_2EDTA H_2O	37.3	28.0	37.3		
$FeCl_2.6H_2O$					1.0
FeNaEDTA				19.80	
Vitamins					
Thiamine HCl (vitamin B1)	0.5	10.0	1.0		1.0
Pyridoxine HCl (vitamin B6)	0.5	1.0	0.5	0.5	
Nicotinic acid (vitamin B3)	0.5	1.0	1.0	5.0	
Glycine ($C_2H_5NO_2$)	2.0			5.0	
Myo inositol	100	100	100	1000	

B5, Gamborg et al. (1968); HE, Heller's medium (Heller 1953); MS, Murashige and Skoog (1962); SH, Schenk and Hildebrandt (1972); WPM, woody plant medium (Lloyd and McCown 1980)

(80–90%), gradually adjusting the temperature and humidity. Within 4–6 weeks, new growth (roots and shoots) begins, and the hardening period, ranging from 4 to 8 weeks, is species-dependent. This process, known as primary hardening, may be followed by secondary hardening in partial shade or an open nursery. Typically, plants are ready for field transfer at 4–6 months of age.

Axillary shoot proliferation method of recovery of plants has been used in various trees, viz., *Anogeissus acuminata* (Rathore et al. 1993), *Anogeissus pendula*, *Anogeissus latifolia* (Saxena and Dhawan 2011), *Acacia auriculiformis* (Girijashank 2011), *Acacia mangium* (Zhang et al. 1995), *Acacia* spp. (Kaur et al. 1998), *Aegle marmelos* (Parveen et al. 2015), *Aquilaria malaccensis* (Borpuzari and Kachari 2018), *Azadirachta indica* (Arya et al. 1995), *Balanites aegyptiaca* (Anis et al. 2010), *Capparis decidua* (Deora and Shekhawat 1995), *Cassia* spp. (Parveen and Shahzad 2012), *Crateva adansonii* (Sharma et al. 2003), *Eucalyptus resinifera*, *Eucalyptus maculata* (McComb and Wroth 1986), *Ficus carica* (Kumar et al. 1998), *Melia dubia* (Bhimi Ram et al. 2014), *Nothapodytes nimmoniana* (Rai 2002; Isha and Singh 2015), *Osyris lanceolata* (Xavery and Feyissca 2015), *Prosopis cineraria* (Shekhawat et al. 1993), *Pterocarpus marsupium* (Husain et al. 2007), *Santalum album* (Sanjay Muthana et al. 2006; Rathore et al. 2008a, b), *Syzygium travancoricum* (Ajith et al. 1999), *Terminalia arjuna* (Gupta et al. 2014), *Terminalia bellirica* (Ramesh et al. 2005), *Vitex negundo* (Ahmed and Anis 2007), and *Ziziphus* spp. and *Tecomella undulata* (Rathore et al. 1991a, b).

32.6.2 Adventitious Regeneration

This method deals with the production of adventitious shoots directly or callus phase from the leaf, internode, petiole, inflorescence, and root of adult trees or hypocotyl, cotyledon, and leaf of in vitro raised seedlings.

Fig. 32.1 Cloning of teak (*Tectona grandis*) through axillary shoots production: (**a**) a plus tree, (**b**) multiplication of shoot, (**c**) ex vitro root induction and (**d**) acclimatized plants

32.6.2.1 Direct Adventitious Shoot Induction

In this procedure, direct adventitious shoot primordia are induced without a callus phase, originating from leaf epidermal, mesophyll, or cortical cells of internode and root (Fig. 32.2). Permanent cells undergo conversion into dividing cells under the influence of growth hormones. These dividing cells multiply and undergo differentiation into specialized tissues, resulting in the production of shoot primordia. These shoot primordia can be elongated for further shoot bulking or multiplication on a fresh shoot multiplication medium. The frequency of shoots and the number of shoot primordia per explant may vary depending on the plant species, nutrient medium, and growth hormones. Directly induced adventitious shoots can be further multiplied on a fresh multiplication medium, typically requiring less cytokinin compared to the direct shoot initiation medium. Individual shoots, shoot segments, or clumps of two to three shoots can be selected for subsequent shoot bulking. Multiplication of shoots is generally performed up to 10–12 passages, with a subculturing period of 4–5 weeks on a fresh medium. Extended multiplication may increase the chances of somaclonal variation. In vitro, shoots multiplied through direct adventitious shoot induction, measuring 4–6 cm in length with three to four nodes, can be employed for rooting. Low-nutrient media like MS/4 basal salt, woody plant, Heller's, and White (1963) media are favorable for rooting without the need for callus. Depending on the tree species, auxins such as IBA and NAA, either alone or in

Fig. 32.2 Adventitious regeneration from internode of *Santalum album*: (**a**) adventitious shoots, (**b**) shoot multiplication and (**c**) hardened plant

combinations, are used for rooting. Ex vitro root induction follows the earlier mentioned methods. The hardening process is crucial for achieving a high rate of survival and protocol success. As shoots are induced directly without a callus phase, the likelihood of genetic variability is minimal. This method proves valuable for cloning desired clones/genotypes in planting programs, provided the genetic stability of plants is ensured. The direct adventitious shoot regeneration method holds potential for the genetic transformation of desired traits in target tree species.

32.6.2.2 Adventitious Regeneration Through Callus Phase

In this regeneration method, callus is generated from various plant parts such as leaves, internodes, roots, and inflorescences. on a nutrient-rich medium like MS, SH, B5, and WP supplemented with plant growth regulators (PGRs) including auxins (IAA, NAA, and 2,4-D) and cytokinins (Kn, BAP, Zeatin, and TDZ). Callus induction can be initiated by auxin alone or in combination with cytokinin. The callus produced can be further multiplied on the same medium composition with minor modifications, including growth hormones. The nutrient medium and PGR requirements may vary depending on the tree species. Callus multiplication can be conducted for three to four cycles before its differentiation into shoots. In vitro multiplied callus undergoes differentiation into shoot primordia on a nutrient medium such as MS, MS/2 basal salts, WP, B5, and SH with low levels of IAA/IBA/NAA and Kn/BAP/2iP/TDZ. Some of the dividing callus cells differentiate into tissues, giving rise to shoot primordia. These differentiated shoot primordia can be elongated on a low-cytokinin medium with or without gibberellic acid (GA3). Elongated shoots are further multiplied by subculturing shoot segments or shoot clumps on a fresh medium with a new composition to increase the number of shoots before being transferred to a rooting medium.

In vitro differentiated shoots, measuring 4–6 cm in length with three to four nodes, can be isolated and used for rooting in reduced-nutrient levels of MS/2, MS/4 basal salts, WP, and White's media with auxin (IAA/IBA). Ex vitro rooting can be induced in greenhouse conditions using auxins like IAA/IBA/NAA and a rooting medium comprising soilrite, vermiculite, cocopeat, peat-moss, and perlite, either alone or in combinations, placed in containers. Rooting and hardening are conducted in a greenhouse under semi-controlled conditions. The primary hardening duration may vary depending on the species, and secondary hardening can be carried out in shade or an open nursery. There is a possibility of somaclonal variation in plants produced through the callus phase, making it essential to conduct genetic fidelity studies using DNA markers before field planting.

Regeneration through adventitious shoot induction has been reported in tree species like *Acacia auriculiformis* (Das et al. 1993), *Acacia sinuta* (Vengadesan et al. 2003; Shahzad et al. 2006), *Aegle marmelos* (Islam et al. 1995), *Cercis canadensis* (Distabanjong and Geneve 1997), *Cydonia oblonga* (Dolcet et al. 1991), *Dalbergia sissoo* (Singh et al. 2002), *Garcinia mangostana* (Goh et al. 1994), *Gmelina arborea* (Ujjwala et al. 2013), Kentucky coffee tree (Geneve 2005), *Murraya koenigii* (Rajendra and D'souza 1998), *Oroxylum indicun* (Gokhale and Bansal 2010), *Paulownia* spp. (Rao et al. 1996), *Populus ciliata* (Thakur and Srivastava 2006), *Santalum album* (Mujib 2005), *Sesbania grandiflora* (Detrez et al. 1994), *Simarouba glauca* (Lavanya et al. 2016), and *Ziziphus jujuba* (Cui et al. 2017).

32.6.3 Plant Regeneration Via Somatic Embryogenesis

Somatic embryogenesis directly can be induced from the explants like leaf, petiole, internode, root, and inflorescence, whereas indirect somatic embryo production is through the callus phase. Callus is first induced from the explant and later on, callus cells divide within the common boundary and produce somatic embryos. The developmental pathway of somatic embryo production follows is similar to zygotic embryo production pathway. It passes through pre-globular, globular, heart-shaped, torpedo, and bipolar embryo stages.

32.6.3.1 Somatic Embryo Induction Without Callus Phase

The generation of direct somatic embryos involves the use of various plant parts such as leaves, internodes, roots, inflorescences, anthers, pollen, and unfertilized ovaries as explants. Dividing (meristematic) cells are formed from epidermal or mesophyll cells of leaf explants, initiating division within a shared boundary. This division progresses through two cells, four cells, eight cells, multicellular pre-globular, globular, heart-shaped, torpedo, and bipolar stages of somatic embryos. The direct somatic embryo production method boasts a high efficiency in plant production.

Several factors, including explant type, nutrient medium, growth hormones, genotypes, and plant species, play crucial roles in governing direct somatic embryo production. The process of plant regeneration through embryogenesis involves three broad stages: embryo production, maturation, and germination of mature embryos into well-developed roots and shoots. The nutrient medium and plant growth regulator requirements may vary during different stages of embryo development. Abscisic acid (ABA) and polyethylene glycol (PEG) are generally employed to facilitate somatic embryo maturation. A low-nutrient medium without growth hormones or with auxin (IAA/NAA) and gibberellic acid (GA3) supports the germination of mature somatic embryos into plantlets. The primary and secondary hardening stages for these plants are similar to those raised through axillary shoot proliferation.

32.6.3.2 Somatic Embryogenesis Through Callus Phase

Various living plant materials can serve as explants for callus induction, with commonly used ones including leaves, internodes, and inflorescences. These explants are cultivated on a nutrient-rich medium supplemented with auxins and cytokinins, either individually or in combination, to induce callus formation. The induction process typically results in the formation of both embryogenic and non-embryogenic callus. The embryogenic callus is then subcultured onto fresh nutrient media (MS/B5/SH/WP) with auxins and cytokinins, either alone or in combinations, to facilitate further multiplication.

The development of embryos from the embryogenic callus is species-specific and is induced by subculturing on an induction medium (Fig. 32.3). For example, in *Santalum album*, MS medium with 2,4-D is used to induce embryogenic callus, and the same medium can be utilized for subsequent multiplication. However, WP medium with 1.0 mg/L IAA is preferred for achieving a high frequency of globular-shaped somatic embryos within a 3-week period (Rangaswamy 2007; Rathore et al. 2008a). The maturation of somatic embryos to torpedo and bipolar stages is supported by WP medium with ABA + PEG. A high germination rate of mature *S. album* embryos can be achieved on MS/4 basal salts medium with GA3 + IAA + activated charcoal (Rangaswamy 2007).

Somatic embryogenesis has been documented in various tree species, viz., *Acacia* spp. (Ortiz et al. 2000), *Azadirachta indica* (Akula et al. 2000), *Balanites aegyptiaca* (Saharan et al. 2011), *Cydonia oblonga* (Morini et al. 2000), *Dalbergia sissoo* (Das et al. 1997), *Eucalyptus citriodora* (Muralidharan et al. 1989), *Nothapodytes nimmoniana* (Fuzele and Stdive 2003), *Picea*, *Quercus*, *Betula*, *Tilia*, *Robinia*, *Fagus*, *Aesculus* spp. (Chalupa 1987), *Populus nigra*, *Populus maximowiczii* (Park and Son 1988), *Populus* spp. (Michler

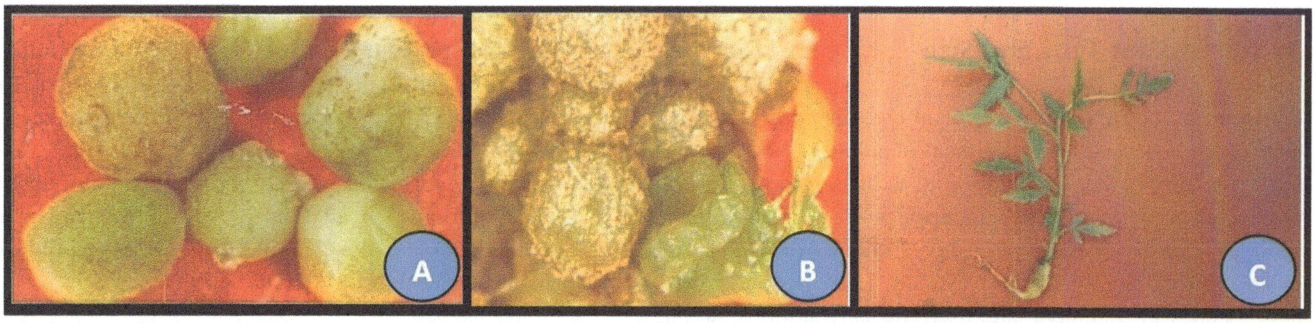

Fig. 32.3 Somatic embryogenesis in *Azadirachta indica*: (**a**) globular embryos, (**b**) torpedo and bipolar, and (**c**) germination

and Baner 1991), *Quercus acutissima* (Kim et al. 2003), *Quercus ilex* (Mauri and Mozanera 2003), *Tilia cordata* (Chalupa 1990), *Santalum album* (Lakshmi Sita et al. 1979; Rangaswamy 2007; Rathore et al. 2008a), *Sesbania bispinosa* (Sinha and Mallick 1991), and *Ulmus minor* (Conde et al. 2004).

32.7 Factors Affecting Micropropagation

32.7.1 Age of Mother Plant/Source of Explants

It is easy to regenerate plants from the in vitro seedlings, field-grown seedlings, and young trees as compared to mature trees. With aging, accumulation of phenolics, inhibitors, complexity in biochemical composition, structural changes, and cytodifferentiation occur, which cause a poor response of mature trees particularly during the first two to three passages of shoot initiation and multiplication. After five to six passages of shoot multiplication of mature trees, the rate of multiplication increases through rejuvenation.

32.7.2 Management of Mother Tree

Age of mother trees, that is, source of plant material, is one of the important factors that plays a crucial role in micropropagation of tree species. Management of the source of explant trees influences the production of new shoot growth with less accumulation of inhibitors (particularly phenolics) and microbial load on the surface of the branches. Pruning and lopping of branches help to obtain new shoots for longer periods and chances of contamination are also reduced apart from low levels of phenolic accumulation (Shekhawat et al. 1993). Periodic irrigation and fertilizer application also improve the vegetative growth of shoots from the mother trees. To reduce the load of microbes, insecticide and pesticide spray can help. Management of the mother tree (source of explants) also helps in the availability of the right stage of explants for a longer period.

32.7.3 Explant Type and Size

Selection of the right kind of explant type and size is important for the successful establishment of cultures. Nodal shoot segments of 1–2 cm in length with axillary shoot bud are widely used for cloning of tree species via forced axillary shoot production. In some cases, apical shoots are also useful as an explanation for in vitro cloning. Leaf, petiole, internode, and root segment are used for adventitious shoot induction and somatic embryogenesis in tree species. Explants collected from managed trees respond better than unmanaged trees. The size of explants is dependent on the type of explants and the objective of their use.

32.7.4 Surface Sterilant and Their Exposure Period

The role of standardization of surface sterilant is very important to establish aseptic cultures for the regeneration of plantlets. Prolonged surface sterilization of explants results in leaching and browning of explants ultimately leading to the death of explants. Whereas, short period of surface sterilization leads to a high rate of microbial contamination. Generally, sodium hypochlorite, mercuric chloride, bromine water, chlorine water, and 70% ethanol are used as sterilant for surface sterilization of explants either alone or in combination. The concentration of sterilant and duration of treatment are important to ensure exposure to microbial load from the explants and minimize contamination load. Selection of sterilant and its exposure time depend on the type of explant and its nature (soft or hard). Sometimes, two types of sterilant are more effective than a single one. Ethanol is useful in waxy and hairy types of explants. Standardization of surface sterilant type, concentration, and exposure duration of treatment to the explants is essential for the successful establishment of aseptic cultures.

32.7.5 Nutrient Medium

Nutrient media have a significant role in fulfilling the nutritional requirement of in vitro cultures for shoot initiation, multiplication, adventitious shoot induction, embryogenic callus induction, and somatic embryogenesis. Nutrient media consist of macro and microelements, vitamins, amino acids, sugar, and gelling agents. Every tree species has its specific requirement of macro and microelements to obtain the best result for shoot initiation, multiplication, and rooting. Most of the tree species respond well to MS medium or its modifications due to its balanced composition of nutrients. A few tree species respond better on B5, WP, and SH media. Shoot initiation and shoot multiplication can be obtained better on nutrient-rich medium, whereas low-nutrient media like half or one-fourth reduced basal salts of MS, White, and Heller's media favor better rooting without callus formation at the base. Nutrient-rich medium is favorable for adventitious shoot induction, callus induction, and callus multiplication for embryogenesis.

32.7.6 Growth Hormones/Regulators

In vitro cultured explants contain a low amount of endogenous growth regulators; therefore, an exogenous supply of growth regulators (PGRs) is essential. The exogenous supply of PGRs plays a significant role in cell division, differentiation, and development of organs. The optimal requirement of PGRs may vary with tree species, explant type, source, and objective of its use in the nutrient media. Commonly used growth regulators for micropropagation are auxins (IAA, IBA, NAA, and 2,4-D), conjugates of auxins, cytokinins, (Kinetin, BAP, 2ip, Zeatin, and TDZ), gibberellic acid (GA3), and ABA, which are used for specific purposes. The GA3 helps in shoot elongation and germination of somatic embryos, whereas ABA alone or with PEG (polyethylene glycol) helps in the maturation of embryos. The optimum concentration of a particular auxin and cytokinin in the medium may vary with the tree species, explant type, and mode of regeneration. The 2,4-D is used for the callus induction either alone or with one of the cytokinins. The IAA is a natural auxin and comparatively weak. The IBA is a preferred auxin for root production from the shoots of tree species. The TDZ is used in small quantities and is very strong. It can be used for shoot induction from difficult-to-respond species as well as for somatic embryogenesis. The kinetin, BAP, and 2ip use in the medium may vary with the plant species. The PGR concentration in the media is dependent on the tree species and the objective of the experiment.

32.7.7 Effect of Period (Season) of Explants Collection

The period (season) of explant collection has a significant role in the establishment of in vitro cultures. The right period of explant collection poses the minimum problem of microbial contamination and ensures a high rate of response of cultures for shoot initiation, callus induction, adventitious shoot induction, and somatic embryo induction. Phenological events like flowering, fruiting, and vegetative growth period vary among the tree species. Explant collection during the new vegetative growth period poses the minimum problem of contamination and a high rate of response under in vitro conditions on shoot proliferation, adventitious shoot induction, and somatic embryogenesis. Response of explants collected during winter poses more problems of contamination and exhibits poor response in culture. During the monsoon period, microbial activities are more and pose the problem of contamination. New shoots produced immediately after the spring season are an ideal period of explants collection with low microbial load and phenolic accumulation. However, the suitable period of explants collection may vary with different species.

32.7.8 Growth Adjuvants

To achieve a better response to micropropagation, the addition of certain adjuvants like ascorbic acid, citric acid, cysteine, polyvinylpyrrolidone, activated charcoal, coconut water, polyamines (putrescine, spermine, and spermidine), yeast extract, and malt extract are useful in the medium for better growth and differentiation. Antioxidants, ascorbic acid, citric acid, cysteine, and activated charcoal are used in the medium to reduce the chances of leaching and browning of explants. Coconut milk, yeast extract, and malt extract supplement nutritional requirements of in vitro cultures.

32.7.9 Genotypes

Apart from tree species, genotypes/clones also influence in vitro regeneration of plants. The difference in response of different genotypes is attributed to differences in their genetic makeup. Genotypes are one of the key factors that play an important role in the proliferation of shoot frequency, shoot number produced per explant, shoot multiplication rate, and rooting percentage. Similarly, genotypes also have a significant effect on adventitious regeneration and somatic embryogenesis.

32.7.10 Gelling Agent and Their Concentrations

The role of the gelling agent is the solidification of nutrient media to support in vitro cultures. Agar-agar is a commonly used gelling agent. Agar-agar is a widely used gelling agent from 0.6 to 0.8% (w/v) depending on the plant species. The purity of gelrite is better than agar-agar and is also transparent. Both are considered inert materials. Gelrite is used at less concentration, 0.2–0.4% (w/v). High and low concentrations of gelling agents are not favorable to the optimum uptake of nutrients. The optimal requirement of the gelling agent may also vary with the tree species.

32.7.11 pH of the Medium

Nutrient uptake is dependent on pH of the medium. Normally, pH of 5.5–5.8 is considered ideal for the optimum uptake of nutrients from the medium. Low pH of the medium leads to poor gelling of the medium whereas high pH (>6.5) results in hardness of the medium leading to poor uptake of the nutrient from the medium. Therefore, optimum pH of the medium is very important for maximum uptake of the nutrient from the medium.

32.7.12 Carbohydrate and Its Concentration

In vitro cultures are partially autotroph, therefore an exogenous supply of energy in the form of carbohydrates is essential for shoot initiation, multiplication, rooting, callus induction, multiplication, and somatic embryogenesis. Glucose, fructose, ribulose, and sucrose are used as a source of energy in the nutrient media. Out of these, sucrose as a source of energy is the most widely used in the media. The concentration of carbohydrates may vary (2–4%) with the tree species. In the majority of the trees, 3% sucrose is used in the medium. The requirement for carbohydrates is comparatively more (3–6%) in the case of somatic embryogenesis. In some cases, combined use of glucose and sucrose is better than sucrose alone for micropropagation. The requirement of carbohydrates is comparatively less (2% w/v) for in vitro rooting from shoots.

32.8 Hardening (Acclimatization)

Micropropagated plants under in vitro conditions are partially autotroph, are poor, or lack cuticle and wax formation on the leaf surface. Stomata are not functional due to high humidity inside culture vessels. Exposing micropropagated plants directly to the outer environmental conditions without acclimatization will lead to a high rate of mortality due to rapid water loss. Therefore, gradual hardening (acclimatization) of micropropagated plants is essential. During the hardening process, potting medium, hardening conditions, and duration of hardening are important factors for the high rate of survival of micropropagated plants. Soilrite, vermiculite, cocopeat, perlite, peat moss, soil, and compost are used singly or in different combinations in a potting medium for the high rate of survival and growth of the micropropagated plants. Initially, low temperature (25–28 °C) and high humidity (85–90%) for 2–3 weeks, followed by high temperature (30–35 °C) and low humidity (70–80%) for 3–4 weeks favor primary hardening. At this stage, cuticle and wax formation and autotropism take place. At a primary hardening stage, new growth of root and shoot is also visible. Secondary hardening is carried out under shade for 3–4 weeks, before keeping plants under open nursery conditions. The condition and duration of hardening vary among tree species. The success of micropropagation depends on shoot multiplication rate, rooting percentage, and hardening conditions and duration.

32.9 Genetic Fidelity

Micropropagated plants raised through axillary shoot proliferation are considered genetically stable and uniform as compared to trees raised through adventitious regeneration and somatic embryogenesis. Chances of genetic variability are higher in the plants raised through the callus phase in case of adventitious regeneration and somatic embryogenesis. DNA markers like RAPD (randomly amplified polymorphic DNA), ISSR (inter single sequence repeat), RFLP (restricted fragment length polymorphism), SSR (single sequence repeat), and microsatellites are commonly used to find out the genetic fidelity of regenerated plants. Genetic stability or instability depends on the tree species, mode of regeneration, and passages of in vitro cultures from which plants are produced. A genetic fidelity test should be carried out for micropropagated plants to ensure a safe mode of regeneration and the number of passages of subculturing from which plants are raised. Genetic stability is essential for micropropagated plants before field planting. Without genetic fidelity studies, if large-scale production is carried out and genetic variation is detected in the field then it may lead to great problems.

32.10 Advantages of Micropropagation

Advantages of micropropagation are:

- Requirement of low quantity of plant material (explant)
- Large-scale production of planting material in a short time
- Production of microbes (disease)-free planting material
- Rejuvenation of micropropagated plants
- Requirement of small space for the production of plants
- No influence of season (summer, winter, and monsoon) on the production of planting material
- Uniformity of planting material
- High production potential of plants
- In vitro preservation of superior clones, and endangered and critically endangered tree species under low and super low (cryopreservation) temperatures

32.11 Applications of Plant Tissue Culture in Forestry

Following are applications of plant tissue culture in tree species:

- Virus-free production of planting material by meristem culture on a defined medium under controlled environmental conditions
- Mass cloning of tree species is difficult through traditional methods of macropropagation
- Clonal production of plants of selected clones, desirable genotypes, and elite clones of commercially important tree species for plantation forestry
- Micropropagation of threatened tree (rare, vulnerable, and endangered) species for in situ and ex situ conservation

- Somatic embryogenesis (direct/through callus phase) for rapid production of planting material of industrially important tree species by stationary, batch culture, or in bioreactor
- Synthetic seeds production from the somatic embryos by encapsulating in calcium alginate sheath
- Genetic improvement by the endosperm culture and production of triploid and polyploid plants.
- Genetic improvement by in vitro induced flowering (useful particularly in bamboo species due to their monocarpic nature
- Genetic improvement by hybrid embryo rescue technique and production of hybrid plants
- Genetic improvement by protoplast fusion within and between species for the production of hybrids and cybrids
- Genetic improvement by somaclonal variation and selection of desired cell line for salt, drought, and disease tolerance/resistance
- Production of secondary products of important tree species through cell suspension cultures in bioreactors for commercial production
- Production of transgenic plants for low lignin, herbicide tolerance, insect resistance, and abiotic (salt, drought, and cold) tolerance
- Medium and long-term conservation of germplasm (cell, tissue, organ, and pollen) under in vitro conditions

32.12 Limitations of Micropropagation

Limitations and problems encountered in micropropagation of tree species are as follows:

- High cost of infrastructure requirement
- Requirement of skilled manpower
- Requirement of more time for development of micropropagation protocols of tree species
- Lack of proper understanding of the physio-chemical requirement for commercial production of planting material of certain tree species
- Some species are recalcitrant and difficult to propagate at large scale
- Leaching and browning of explants of some species are common and difficult to regenerate plants
- Encounter the problem of contamination to raise cultures of mature trees

32.13 Future Prospects

There is a need for a proper understanding of the physio-chemical requirement to establish cultures of tree species for commercial production. Emphasis on the high rate of shoot multiplication and rooting of forestry species for economic feasibility for commercial production is required. Keeping in mind the future requirement of clonal planting material, emphasis on automation for commercial production of clonal planting material is essential. There is a need for a better understanding of somatic embryogenesis and synthetic seed production. To develop hybrids and cybrids for genetic improvement, emphasis on somatic hybridization within and between tree species will be required. There is a need for the development of protocols of micropropagation of tree species that are difficult to clone through traditional methods. To achieve a high rate of survival of micropropagated plants, an emphasis on a better understanding of the primary and secondary hardening of micropropagated tree species will be essential. Thrust on micropropagation of rare, vulnerable, and endangered tree species for their conservation and plantation will be required. To ensure the genetic stability of micropropagated plants, genetic fidelity studies should be mandatory before using micropropagated planting material for field plantation. Better coordination is required between R&D laboratories and commercial units for the industrial production of cloned material of forestry species.

32.14 Conclusion

Plant tissue culture-based biotechnology has the potential to overcome the problems of traditional methods of vegetative propagation and improve the production of high-quality planting material for plantation forestry. Cloning by axillary shoot production, somatic embryo induction, and adventitious shoot regeneration can produce large-scale planting material of tree species around the year without the influence of outer environmental conditions in a small space as compared to traditional methods of propagation. In vitro production of plants by axillary shoot induction is the safest mode to produce genetically uniform clonal planting material. Critical factors like explant type, PGRs, media, incubation conditions, growth adjuvant, and genotypes are some of the important ones that play an important role in the rapidity of micropropagation. Prolonging multiplication of cultures may pose chances of genetic variability; therefore, genetic fidelity studies are mandatory before planting of micropropagated plants in the field to ensure genetic stability. There is a need to focus on better coordination between R&D laboratories and commercial units to produce industrial-level clonal plants of tree species. Emphasis on automation, high shoot multiplication, and rooting rate; somatic embryogenesis; micropropagation of difficult-to-clone tree species and rare, vulnerable, and endangered forestry species; primary and secondary hardening; and mandatory studies on genetic fidelity is essential.

Lessons Learnt

- Micropropagation has the potential to overcome the problems encountered in conventional macropropagation of difficult-to-root tree species.
- Micropropagation through axillary shoot production is the preferred mode of plant regeneration for cloning genetically true to types of plants.
- Direct adventitious shoot regeneration and somatic embryogenesis have a high potential for the regeneration of plants and can also be used for genetic engineering for the production of transgenic plants with desired traits.
- Explant type and source (seedling/mature trees), genotypes, nutrient media, and growth hormones play key roles in plant regeneration through all three modes of regeneration.
- Micropropagation has a high production potential from a small quantity of plant material without the influence of outer environmental conditions.
- Primary and secondary hardening (acclimatization) of micropropagated planting material is essential for the high rate of survival and growth of plants in the field.
- Genetic fidelity studies of micropropagated plants using DNA markers are essential to ensure a safe mode of regeneration and duration of multiplication of cultures under aseptic conditions before planting in the field.

Key Questions

1. What are the different methods of micropropagation? How are they different from each other?
2. What are the important factors that have influence on different stages of micropropagation?
3. How micropropagation is different from macropropagation? Explain.
4. What are the different applications of plant tissue culture in general and micropropagation in particular? Explain in detail.
5. Why hardening (acclimatization) and genetic fidelity studies are essential for micropropagated plants?

References

Ahmed N, Anis M (2007) Rapid clonal multiplication of a woody tree, *Vitex negundo* L. through axillary shoot proliferation. Agrofor Syst 71:195–200

Ajith A, Rao SC, Balakrishna P (1999) *In vitro* propagation of *Syzygium travancoricum* gamble – an endangered tree species. Plant Cell Tissue Organ Cult 56:59–63

Akula C, Akula A, Drew R (2000) Somatic embryogenesis in neem, *Azadirachta indica* A. Juss. and analysis for azadirachtin production. In Vitro Cell Dev Biol Plant 39:304–310

Anis M, Varshney A, Siddique A (2010) *In vitro* clonal propagation of *Balanites aegyptiaca* (L) Del. Agrofor Syst 78:151–158

Arya S, Rathore TS, Arya ID (1995) *In vitro* propagation of neem from seedling and mature trees. Indian J Plant Genet Resour 2:247–252

Bhimi Ram B, Rathore TS, Beena DB, Rajput DS (2014) An efficient protocol for micropropagation and genetic fidelity analysis of *Melia dubia* Cav. - an important multipurpose tree. Int J Curr Microbiol App Sci 3(7):533–544

Bonga JM (1987) Clonal propagation of mature tree: problems and possible solution. In: Bonga JM et al (eds) Cell and tissue culture in forestry. Springer & Dordrecht

Borpuzari PP, Kachari B (2018) Effect of glutamine for high frequency *in vitro* regeneration of *Aquilaria malaccensis* Lam. through nodal culture. J Med Plant Stud 6(2):9–16

Chalupa V (1987) Somatic embryogenesis in *Picea*, *Quercus*, *Betula*, *Tilia*, *Robinia* and *Aesculus*. Common Inst For Ceh 15:133–148

Chalupa V (1990) Plant regeneration by somatic embryogenesis from cultures of immature embryo of Oak (*Quarcus rubber* L.) and Cinden (*Tilia cordata* Mill). Plant Cell Rep 9:398–401

Conde P, Loureiro J, Santos C (2004) Somatic embryogenesis from leaves of *Ulmus minor* Mill. Plant Cell Rep 22:632–639

Cui Y, Hou L, Li X, Huang F, Pang X, Li Y (2017) *In vitro* induction of tetraploid *Ziziphus jujube* Mill. Var. *spinosa* plants from leaf explants. Plant Cell Tissue Org Cult 131:175–182. https://doi.org/10.1007/s11240-017-1274-8

Das P, Samantaray S, Roberts AV, Rout GR (1997) *In vitro* somatic embryogenesis of *Dalbergia sissoo* Roxb. A multipurpose timber yielding tree. Plant Cell Rep 16:578–582

Das PK, Chakravarty VI, Maity S (1993) Plantlet formation in tissue culture from cotyledon of *Acacia auriculiformis* A. Cunn. ex Benth. Indian J For 16:182–192

Deora NS, Shekhawat NS (1995) Micropropagation of *Capparis decidua* (Forsk) Edgew – a tree of arid horticulture. Plant Cell Rep 15:275–281

Detrez C, Ndiaye S, Drefus B (1994) *In vitro* regeneration of the tropical multipurpose leguminous tree *Sesbania grandiflora* from cotyledon explants. Plant Cell Rep 14:87–93

Distabanjong K, Geneve R (1997) Multiple shoot formation from normal and malformed somatic embryo explants of eastern redbud (*Cercis canadensis* L.). Plant Cell Rep 16:334–338

Dolcet R, Mok DWS, Mok MC (1991) Plantlet regeneration from cultured leaves of *Cydonia oblonga* (Quince). Plant Cell Rep 10: 240–242

Fuzele DP, Stdive RK (2003) Somatic embryogenesis, plant regeneration and evaluation of camptothecin. In Vitro Cell Dev Biol Plant 39(2):212–216

Gamborg OL, Miller RA, Ojma K (1968) Nutrient requirement of suspension cultures of soyabean root cells. Exp Cell Res 50:151–158

Geneve RL (2005) Comparative adventitious shoot induction in Kentucy coffee tree root and petiole explants treated with thiadiazuron and benzylaminopurine. In Vitro Cell Dev Biol - Plant 41:489–493

Girijashank V (2011) Micropropagation of multipurpose medicinal tree *Acacia auriculiformis*. J Med Plant Res 5(3):462–466

Goh CJ, Lakshmanan P, Loh CS (1994) High frequency direct shoot bud regeneration from excised leaves of mangosteen (*Garcinia mangostana* L.). Plant Sci 101:173–180

Gokhale M, Bansal YK (2010) Indirect organogenesis in *Oroxylon indicum* (L.) Vent. Indian For 136:804–811

Gupta AK, Harish Rai MK, Phulwaria M, Agarwal T, Shekhawat NS (2014) *In vitro* propagation, encapsulation and genetic fidelity analysis of *Terminalia arjuna*: a cardioprotective medicinal tree. Appl Biochem Biotechnol 173:1481–1494

Gupta PK, Nadgir AL, Mascarenhans AF, Jagannathan V (1980) Tissue culture of forest tree species: clonal multiplication of *Tectona grandis* (Teak) by tissue culture. Plant Sci Lett 17:259–268

Haberlandt G (1902) Kultuversuchemitisolliertenpallazenzellen. Stizungsber. Akad. Wiss. Wien., Math.-Naturewiss. Kl. Abt 111: 69–92

Heller R (1953) Researches surla nutrition minerale de tissue vegetanucultivees *in vitro*. Ann Sci Bot Et Bio Veg 11th Series. 14:1–223

Husain MK, Anis M, Shahzad A (2007) *In vitro* propagation of Indian kino (*Pterocarpus marsupium* Roxb.) using thiodiazuron. In Vitro Cell Dev Biol - Plant 43:59–64

Isha T, Singh A (2015) *In vitro* propagation and camptothea in production in *Nothapodytes nimmoniana*. Plant Cell Tiss Organ Cult 121:1–10. https://doi.org/10.1007/s11240-014-0683-1

Islam R, Hossain M, Karim MR, Joarder OI (1995) Regeneration of *Aegle marmelos* (L.) Corr. plants *in vitro* from callus cultures of embryogenic tissues. Curr Sci 69:494–495

Kaur K, Verma B, Kant U (1998) Plant obtained from Khair tree (*Acacia catechu* Willd.) using mature nodal segments. Plant Cell Rep 17: 427–429

Kim YW, Lee BC, Lee SK, Jang SS (2003) Somatic embryogenesis and plantlet regeneration in *Quercus accutissima*. Plant Cell Rep 13: 313–318

Kumar A, Rathore TS, Palanisamay K, Vishwanath S (2016) Advances in tree improvement and forest genetic resources conservation and management. Published for ICFRE, Dehradun by Greenfield Publishers, Dehradun, 459p

Kumar V, Radha A, Goh CJ (1998) *In vitro* plant regeneration of fig (*Ficus carica* L. Cu Gular) using apical buds from mature tree. Plant Cell Rep 17:717–720

Lakshmi Sita G, Raghava Ram NV, Vaidyanathan CS (1979) Differentiation of embryoids and plantlets from shoot callus of sandalwood. Plant Cell Lett 15:265–270

Lavanya AR, Muthukumar M, Muthukrishnan S, Kumaresan V, Senthil Kumar T, Rao AS, Rao MV (2016) *In vitro* micropropagation of *Simarouba glauca* DC. Indian J Biotechnol 15:107–111

Lloyd G, McCown BH (1980) Commercially feasible micropropagation of mountain laural *Kalmia latifolia* by use of shoot tip culture. Proc Int Plant Propagators Soc 30:421–427

Mascarenhas AF, Muralidharan EM (1989) Tissue culture of forest trees in India. Curr Sci 58(11):606–613

Mauri PV, Mozanera JA (2003) Induction, maturation and germination of holm oak (*Quercus ilex* L.) somatic embryos. Plant Cell Tissue Org Cult 74(3):229–235

McComb JA, Wroth M (1986) Vegetative propagation of *Eucalyptus resinifera* and *E. maculata* using coppice cuttings and micropropagation. Aust For Res 16(3):231–242

Michler CH, Baner E (1991) High frequency embryogenesis from leaf tissue of *Populus* spp. Plant Sci 77:111–118

Morini S, D'onofrio C, Bellochi G, Fisichella M (2000) Effect of 2,4-D and light on production and differentiation from *in vitro* culture of quince leaves. Plant Cell Tissue Org Cult 63:47–55

Mujib A (2005) *In vitro* regeneration of sandal (*Santalum album* L.) from leaves. Turk J Bot 29:63–67

Muralidharan EM, Gupta PK, Mascarehans AF (1989) Plantlet formation through high frequency somatic embryogenesis in long term cultures of *Euclyptus citriodora*. Plant Cell Rep 8:41–43

Murashige T, Skoog F (1962) A revised medium for rapid growth and bio assays with tobacco tissue cultures. Physiol Plant 15:473–497

Ortiz BOC, Reyes MEP, Baltch EPM (2000) Somatic embryogenesis and plant regeneration in *Acacia farnesiana* and *A. schaffneri*. In Vitro Cell Dev Biol - Plant 36:268–272

Park YG, Son SH (1988) *In vitro* organogenesis and somatic embryogenesis from punctured leaf of *Populus nigra* x *P. maximo wiczii*. Plant Cell Tissue Org Cult 15:95–105

Parveen S, Shahzad A (2012) *In vitro* propagation of medicinally important *Cassia* species - a review. J Func Env Bot 2(1):1–11

Parveen S, Shahzad A, Nazim F (2015) Effect of explants origin on clonal propagation of *Aegle marmelos* (L.) Corr.- A multipurpose tree species. Indian J Biotechnol 14:566–573

Rai VR (2002) Rapid clonal propagation of *Nothapodytes nimmoniana* (Wright) Sleumer – a threatened medicinal tree. In Vitro Cell Dev Biol Plant 38(4):347–351

Rajendra K, D'souza L (1998) Direct organogenesis from internodal segments of mature *Murrayako engii*. Phytomorphology 48:427–431

Ramesh M, Vmate P, Rao KV, Sadanandam A (2005) Micropropagation of *Terminnalia bellirica* Roxb. – A sericulture and medicinal plant. In Vitro Cell Dev Biol - Plant 41:320–323

Rangaswamy R (2007) Studies on somatic embryogenesis for rapid and mass production of high quality planting material of Sandal (*Santalum album* L.) and red sanders (*Pterocarpus santalinus* L.), evaluation of genetic fidelity and field performance. Ph.D. thesis, Forest Research Institute University, Dehradun, India, 249p

Rao CD, Goh C, Kumar PP (1996) High frequency adventitious shoot regeneration from excised leaves of *Paulownia* spp. cultured *in vitro*. Plant Cell Rep 16:204–209

Rathore TS, Deora NS, Shekhawat NS, Singh RP (1993) Cloning of *Anogeissus acuminata* through tissue culture. Biol Plant 35(3): 381–386

Rathore TS, Goyal B, Dubey AK, Rao PS (2008b) *In vitro* cloning of sandalwood (*Santalum album* L.) through shoot proliferation and evaluation of genetic fidelity. In: Gairola S, Rathore TS, Joshi G, Arun Kumar AN, Aggarwal PK (eds) Proceeding of national seminar on conservation, improvement, cultivation and management of sandal (*Santalum album* L.). Institute of Wood Science and Technology, Bangalore, pp 85–91

Rathore TS, Rangaswamy M, Goyal B, Dubey AK, Rao PS (2008a) Micropropagation through axillary shoot proliferation and somatic embryogenesis in sandalwood (*Santalum album* L.) from mature trees. In: Ansari SA, Narayanan C, Mandal AK (eds) Forest biotechnology in India. Satish Serial Publication House, New Delhi, pp 215–228

Rathore TS, Singh RP, Deora NS, Shekhawat NS (1991a) Clonal propagation of *Zizyphus* species through tissue culture. Sci Hortic 51:165–168

Rathore TS, Singh RP, Shekhawat NS (1991b) Clonal propagation of desert teak (*Tecomella undulata*) through tissue culture. Plant Sci 79(2):217–222

Saharan V, Yadav RC, Yadav NR, Wiesman Z (2011) Somatic embryogenesis and plantlet regeneration of *Balanites aegyptiaca* Det. (L.): an industrially important arid tree. J Cell Tissue Res 11:2529–2534

Sanjay, Muthana B, Rathore TS, Rai VR (2006) Micropropagation of endangered Indian sandalwood (*Santalum album* L.). J For Res 11:203–209

Saxena S, Dhawan V (2011) Large scale production of *Anogeissus pendula* and *A. latifolia* by micropropagation. In Vitro Cell Dev Biol - Plant 37:586–591

Schenk RU, Hildebrandt AC (1972) Medium and techniques for induction and growth of monocotyledons and dicotyledons plant cell cultures. Can J Bot 50:199–204

Shahzad A, Naseem A, Anis M (2006) An improved method of organogenesis from cotyledon callus of *Acacia sinuata* (Lour.) Merr. using thiadiazuron. J Plant Biotech 8:15–19

Sharma PK, Purnima T, Sharma KC, Kothari SL (2003) Clonal propagation of *Cratevea adansonii* (DC) Prodr: a multipurpose tree. In Vitro Cell Dev Biol - Plant 39:156–160

Shekhawat NS, Rathore TS, Singh RP, Deora NS, Rao SR (1993) Factors affecting *in vitro* cloning of *Prosopis cineraria*. Plant Growth Reg 12:273–280

Singh AK, Chand S, Patnaik S, Chand PK (2002) Adventitious shoot organogenesis and plant regeneration from cotyledons of *Dalbergia sissoo* Roxb. A timber yielding tree legume. Plant Cell Tissue Org Cult 68:203–209

Sinha RK, Mallick R (1991) Plantlet from somatic embryo callus tissue of woody legume *Sesbania bispinosa* (Tacq.) W.B. Wight. Plant Cell Rep 10:247–250

Skoog F, Miller CO (1957) Chemical regulation of growth and organ formation in plant tissue culture *in vitro*. Symp Soc Exp Bot 11:119–131

Thakur AK, Srivastava DK (2006) High frequency plant regeneration from leaf explants of male Himalayan poplar (*Populus ciliata* Wall). In Vitro Cell Dev Biol - Plant 42:144–147

Ujjwala D, Rambabu M, Ramaswamy N (2013) Effect of plant growth regulators on callus induction and organogenesis in *Gmelina arborea* Roxb. Asian J Exp Biol Sci 4(2):203–208

Vengadesan G, Ganapathi A, Amutha S, Selvaraj N (2003) High frequency plant regeneration from cotyledon callus of *Acacia sinuta* (Lour.) Merr. In Vitro Cell Dev Biol - Plant 39:28–33

White PR (1939) Potentially unlimited growth of excised plant callus in artificial nutrient. Am J Bot 26:59–64

White PR (1963) The cultivation of animal and plant cells, 2nd edn. Ronald Press, New York, 239p

Xavery DK, Feyissca T (2015) Prospect of biotechnological approaches for propagation and improvement of threatened African sandalwood (*Osyris lanceolata* Drew, R.A. Hochst & Steud). Am J Plant Sci 6: 1822–1826

Zhang H, Huang X, Fu J, Yang M, Chen C (1995) Axillary bud culture and plantlet regeneration of *Acacia auriculiformis* and *A. mangium*. J Trop Bot 3(3):62–65

Genetic Markers, Genomics and Genetic Modification in Forest Trees: Current Status and Prospects

33

Yasodha Ramasamy, Shanthi Arunachalam, Modhumita Dasgupta, and Mathish Nambiar-Veetil

Abstract

Forest trees still exist in an undomesticated stage and they carry a variety of wild and productive alleles. These genetic variations determine many traits of economic and ecological significance. The development of efficient DNA marker systems for genetic improvement of tree species would benefit from knowledge of genetic variability. Tree breeding projects aim to maximize the genetic gain for a variety of beneficial qualities. Genetic variability is the fundamental resource for the development of disease and pest resistant varieties and to create climate-ready tree types. Sequencing initiatives in woody perennials have increased dramatically as a result of remarkable developments in high-throughput sequencing technology and decreasing sequencing costs. Numerous genes that are activated as a result of stress responses, as well as their related signaling and regulatory networks, are identified and characterized using genomic technologies to comprehend the functional genetic architecture in trees. The development of superior planting stock with multiple traits would be accelerated by the convergence of conventional quantitative genetics and genomics-based selection. Further, integration of precision genome engineering approaches in conjunction with traditional genetic transformation methods are expected to hasten generation and delivery of improved planting stock in a short span of time.

Keywords

Genetic markers · DNA barcode · Linkage map · Quantitative trait loci · Transgenic

All authors contributed equally.

Y. Ramasamy (✉) · S. Arunachalam · M. Dasgupta · M. Nambiar-Veetil
ICFRE-Institute of Forest Genetics and Tree Breeding, Coimbatore, India

33.1 Introduction

About 31% of the land surface is occupied by forests. Forests and trees outside forests (ToFs) play a vital role in rural economy and safeguards poor from major disasters including climate change. Trees on small holder plantations contribute to resilient livelihoods for farmers by assuring a variety of ecosystem goods and services. They have high resilience to adverse weather conditions and depend on forests for food, fodder, and wood products. The improvement in wood production of selected few tree crops in the last century has seen remarkable progress. However, there are many economically important trees need to be considered for genetic improvement and production of designer genotypes becomes inevitable. Tree breeders are at the crossroads because of continually emerging demands for trait-specific varieties growing silviculture problems like climate, pests, and soil conditions. Negative consequences of biotic and abiotic stress such as susceptibility to pests, drought and salinity can be ameliorated by planting genetically improved trees. While most of the progress accomplished so far has been achieved with traditional methods of tree improvement, biotechnology has potential to obtain a greater probability of success in tree genetic improvement. Forest biotechnology indeed had impacts from the tissue culture multiplication of elite individuals since 1990s. Biotechnologies are increasingly applied in forestry to enhance wood production, understand individuals and ecosystem, produce secondary metabolites, adaptation to climate change and long-term management of genetic resources. To ensure that the 9.7 billion people on the planet can sustainably fulfil their wood requirements by 2050, business-as-usual approach needs infusion of several technologies at different stages of production. Evidence-based perspective revealed that wood producers and consumers experience the benefit of biotechnology. Traditional breeding programs integrate biotechnological tools to accelerate the generation of improved planting stock. Biotechnologies greatly contributed to the adoption of

A. K. Mandal, A. Nicodemus (eds.), *Textbook of Forest Science*, https://doi.org/10.1007/978-981-97-8289-5_33

short-rotation woods for the production of paper pulp and biofuel. Methodologies are already available to identify specific sources of genetic variations which influence a particular trait and to predict mature traits at the seedling stage. Many of the traits having economic and ecological significance are genetically determined. Following the genome sequencing of human and rice, the whole genome of poplar was sequenced (Tuskan et al. 2006) which revealed complexity associated with quantitative traits in trees. Biotechnological issues of trees are always unique thus requiring extensive understanding of the tree genome. Next-generation sequencing technology unearths several prospects to discern the distinctiveness of trees. Rapidly evolving fields such as genomics provides unprecedented opportunities to understand the basics of neutral an adaptive plant traits, accelerate the process of domestication as well as safeguarding the tree germplasm for future use. This chapter highlights the major biotechnologies in areas such as genetic diversity, genomics, whole genome and transcriptome sequencing and genetic modifications. Novel studies published in the recent past are summarized and research gaps requiring special attention for tree genetic improvement and conservation are discussed.

33.2 Genetic Diversity

Plant populations which undergo sexual reproduction encompass certain amount of inheritable genetic variation generated through mutations, substitutions, copy number variations, insertion/deletions (InDels) and translocations. These variations can be tracked and quantified with the help of genetic markers. Genetic markers or DNA markers in broad sense represent specific loci revealing variations, which can be used to detect polymorphism between different genotypes or alleles in a population. Further, markers are heritable, can be easily traced and measured in a breeding population, and may be linked to a specific gene or trait of interest (Hayward et al. 2015). DNA markers have several advantages over morphological and protein markers because they are stable, ubiquitous in genome, locus specific, reproducible, generally obey Mendelian segregation rules. In addition, markers exhibit pleiotropic and epistatic effects and are independent of the growing environment. These markers are applied for germplasm characterization, hybrid purity, varietal identification, phylogenetic relationships, and genetic diversity existing in various populations, seed orchards. Detection of allelic variations through genetic linkage analysis, marker-assisted selection and incorporation of genes for resistance of biotic and abiotic stress responses are carried out precisely using DNA markers. The characteristics of ideal DNA markers include reproducibility, multiallelism, cost-effectiveness, fast and easy to practice and reveal polymorphism to differentiate chromosome having mutant gene and normal gene. Majority of these markers exhibit selective neutrality as they reside within noncoding regions of DNA with no specific role on fitness of individuals. Techniques applying the polymerase chain reaction (PCR) brought in variety of novel DNA markers. Methods such as RAPD (random amplified polymorphic DNA), AFLP (amplified fragment length polymorphism), ISSR (inter simple sequence repeat), and microsatellites or SSRs (simple sequence repeats) were used extensively in genetic variability estimation (Porth and El-Kassaby 2014). In the recent past, marker techniques based on sequencing, such as single-nucleotide polymorphisms (SNPs), have become increasingly common. The field of landscape genomics open up possibilities for studying the adaptive variations using common garden experiments (provenance trials), gene expression studies, candidate genes (CG), and whole genomic scans which link the genomic information with adaptive traits (Holderegger et al. 2008).

Understanding the genetic diversity (entire genetic variation within a species) of both common and endangered tree species would aid in developing effective strategies for their genetic improvement and conservation, ensuring their future utility. Compared to agricultural crops, forest trees are still in undomesticated stage harboring the array of wild alleles with them. Genetic diversity is the fundamental asset for tree breeding aimed to increase genetic gains across a variety of useful traits and to develop climate ready tree varieties. The genetic composition of natural populations of tree species indicates the optimal structure to employ in tree breeding, either to reduce susceptibility to pests and diseases or to adapt to challenging environments. Researchers considered the study of genetic variations and structure of population as essential requirement for breeding and conservation and experiments have been carried out in many forest trees such as *Populus* (Chen et al. 2020), *Eucalyptus* (Grattapaglia et al. 2012), *Pinus* (Eckert et al. 2010) *Salix* (Berlin et al. 2014), and *Acacia* (Omondi et al. 2010). Advances made in application of markers to tree improvement, conservation, and sustainable utilization of forest trees have been reviewed thoroughly (Holliday et al. 2017; Changtragoon et al. 2017). A web-based repository, TreeGenes database houses archive for genomic and phenomic information for 2050 tree species (https://treegenesdb.org/species_directory/main. accessed on 03 August 2020).

Tree genetic improvement in a species often reduces the genetic variability (a measure of the genetic differences that exist within a population); hence, it becomes paramount importance to estimate the reduction at different stages of breeding process. Breeding populations of *Eucalyptus dunnii* were subjected to genetic assessment with AFLP and SSR markers to design clonal seed orchard based on the genetic relatedness of the individuals (Poltri et al. 2003). Molecular diversity information confirmed the absence of significant difference in genetic diversity among seed orchard individuals and breeding population and limited individuals would suffice seed orchard to maintain expected variations in progenies. The genetic analysis of *Prunus sibirica* populations using SSR markers revealed no specific variation between the wild and semi-wild germplasm (Wang et al. 2014). This suggests that recent cultivation practices have had minimal impact on the genetic diversity of the species and would serve as suitable base for genetic resources for breeding. Whereas SSR analysis in *P. thunbergii*, a major Japanese pine species, available only as plantation resource showed reduced genetic diversity and population distinctiveness due to genetic drift possibly influenced by seed use and seed transfer patterns (Iwaizumi et al. 2018). In the case of *Jatropha curcas,* very low levels of genetic diversity were recorded with SSR markers in many different populations (Sanou et al. 2015). The SSR allelic information contributed for understanding the population history and laid basis for designing genetic resource conservation units to protect local genetic stock and breeding populations.

Tree breeders choose prospective parents to create populations with desirable characteristics and significant variability for devising breeding strategies with the assistance of markers. A provenance cum progeny trial of *Eucalyptus cladocalyx*, a species with high levels of drought and saline tolerance, was analyzed with nuclear SSR markers to ascertain the presence of highly structured and genetically similar sources (Mora et al. 2017). In breeding programs, when the recurrent directional selection is followed, variability and trait-specific genetic gains are expected to decrease in each cycle. Trait selection increases the frequency of favorable alleles associated with the traits under improvement while reducing the frequency of unfavorable alleles with each selection cycle. SSR markers being the best resources to determine mating pattern and genetic variability, *Eucalyptus grandis* provenance trail consisting of natural populations from Australia, and landraces in Brazil were subjected to genetic analysis (Miranda et al. 2019). It was concluded that landraces tend to preserve significant levels of variability compared to the wild populations through high outcrossing rates, and hence non-native tree breeding programs should have 30% more seed trees to sustain the required level of genetic variability. Breeding populations of *Eucalyptus urophylla* undergoing third cycle of breeding in China were analyzed with SSRs and found that the genetic variability in all three cycles of breeding was equal because of the caution exercised to select provenances/individuals with wide genetic base (Lu et al. 2018).

With the advancement in sequencing technology, a battery of new markers is generated through next-generation sequencing (NGS) approaches. Abundant SSR loci with high polymorphism have been identified as a spin-off of transcriptome and whole genome sequencing projects in a cost-effective manner. These sequences yield both genome wide and gene-specific SSRs. Some of the trees such as *Azadirachta indica* yielding several bioactive compounds used for human welfare attracted less attention in population genetic studies would benefit from NGS approaches. Over the last few years, highly informative and stringent markers like SNPs and SSRs find wider adoption in forestry (Gramazio et al. 2018; Liu et al. 2019; Martin et al. 2010).

33.3 Genotype Identification

DNA markers are used for the development of individual specific fingerprints in forest trees. It involves the display of a unique genetic profile from a specific sample. In tree improvement programs, individual selections and hybrids are deployed through rooted cuttings and micropropagation to produce quality planting stock for commercial clonal plantations. Genotype identity of these operational clones is of paramount importance for genetic purity testing of controlled crosses, seed orchard establishment, large-scale commercial vegetative propagation, and protection of proprietary rights of clones. Clonal identity has a great impact in attaining productivity in commercial plantations and expected gain from the seed production programs (De-Lucas et al. 2008). DNA fingerprinting for genotype tagging using dominate and co-dominant markers is not uncommon in forestry. Pure species and their hybrids of *Picea mariana*, *P. glauca*, and *P. engelmannii*, and their putative hybrids were distinguished with DNA markers for seed certification (Nkongolo et al. 2005). *Pinus contorta* var. *latifolia* and *P. banksiana* and its hybrids were distinguished by SNP markers (Cullingham et al. 2013). The species like *Populus* (De-Lucas et al. 2008), *Prunus avium* (Avramidou et al. 2010), *Eucalyptus* (Keil and Griffin 1994; Kirst et al. 2005), *Acacia* (Le et al. 2017), and *Robinia pseudoacacia* (Dong et al. 2019) had been subjected to fingerprinting analysis for various tree breeding objectives. Similarly hybrid purity and species-diagnostic DNA markers are beneficial to apply quality control standards by assessing contamination level in hybrid seeds and hybrid validation. In *Acacia mangium* and *A. auriculiformis*, multiplexed SSR markers with greater discriminatory power were developed to distinguish pure and hybrid individuals (Le et al. 2019).

33.4 DNA Barcoding

Species-level identification in a cost-effective way is achieved through DNA barcodes generated with organelle genome. Sequences such as chloroplast (cpDNA) or mitochondrial DNA (mtDNA) markers differentiate chlorotypes and mitotypes, respectively, for the inference of demographic structure and phylogenetic and genetic diversity analyses. Taxonomic studies of cryptic taxa or species complexes, identification of cryptic species, and taxonomic revisions were made with the assistance of organelle genome barcodes (Simeone et al. 2013). Authentication of medicinal plant species for their identity is vital to control adulteration and protect consumers and conservation of specific plant species. DNA barcodes become easy to use tools for the authentication of medicinal plants (Kshirsagar et al. 2017). Traditionally used cpDNA genes in barcoding consist of megakaryocyte-associated tyrosine kinase (matK), ribulose-bisphosphate carboxylase (rbcL), photosystem II protein D1-structural RNA-His tRNA (psbA-trnH), and internal transcribed spacer region (ITS) (Thakur et al. 2016). Currently, the entire plastome and large single-copy (LSC) datasets are utilized to discover divergence variables like SNPs, indels, and repeats to resolve the species complexes close to 100% accuracy (Li et al. 2020). Similarly, the haploid mtDNA with maternal mode of inheritance shows hypervariable region in D-loop or control region. The SNP polymorphisms in the sequence of the mitochondrial genome of *Pinus uliginosa* revealed the genetic structure of relict and endangered populations, past population history, and hybridization with closely related species, which led to the development of potential conservation strategies for the preservation of genetic resources (Łabiszak et al. 2019). Variations in mitochondrial markers such as *nad7*, *cox*I, and *syn*31 established colonization history, colonization routes, and locations of refugia of *Pinus sylvestris* (Semerikov et al. 2018).

33.5 Linkage Mapping and QTL Localization

Important silvicultural traits are typically inherited in a quantitative manner, and mapping quantitative trait loci (QTL) proves highly beneficial for identifying distinct alleles governing these significant and intricate traits to accelerate the tree breeding process. One of the most promising applications of DNA markers in forest trees is genetic linkage mapping and quantitative trait loci (QTL) detection followed by marker-assisted selection (MAS). The selection process relies on genotypes in biparental mapping populations, employing DNA markers linked to the desired traits (Butcher and Southerton 2007). It requires development of markers associated with genes governing the traits and applicable to different types of breeding populations. Possibilities of introgression of few or more genes into a single desired genotype exist thus having potential for the production of superior varieties. However, forest trees pose several challenges in applying MAS because of their long generation, traits with very low heritability, unreliable phenotypic evaluation, rapid decay of linkage disequilibrium (LD), diverse genetic background of breeding material, and limited genomic information (Grattapaglia and Kirst 2008). Nevertheless, over the last two decades, significant efforts have been made on linkage mapping and QTL localization for economically important traits. High-density genetic linkage maps not only offer valuable information for marker-assisted selection but also prove useful in positional cloning of genes, anchoring scaffold sequences, genome assembly, and conducting comparative genomics (Hudson et al. 2012). In the beginning, pseudo-testcross mapping strategies (Grattapaglia and Sederoff 1994) with inter-specific crosses were utilized for linkage map generation. These maps involved dominant and co-dominant markers (RAPD, AFLP, SSR) either individually or in various combinations. The drawbacks with such traditional markers include low-throughput, poor reproducibility, and laborious experiments. Therefore to obtain comprehensively covered genetic linkage maps, NGS-based markers are increasingly utilized. Statistical advances have allowed integration of several marker types and linkage phases to enhance the number of markers on the linkage map. In the recent years, array-based SNP genotyping is employed in generation of ultra-high density linkage maps in species like *Pyrus* and about 16,000 SNPs were placed on the map (Montanari et al. 2019). Likewise, in Populus tremula, approximately 14,500 SNPs were utilized to construct a linkage map and evaluate genomic characteristics, including gene density, repeat content, methylation levels, and their impact on the recombination rate (Apuli et al. 2019).

Complex traits like wood volume, tree height, and diameter and pulp yield serve as target for tree breeding and are used to explain the contribution of QTL to phenotypic variation (Sang et al. 2019). QTL mapping has unveiled genetic regions that exert influence on wood properties, climate sensitivity, as well as abiotic and biotic stress tolerance in major forest tree species, such as *Eucalyptus* (Thumma et al. 2010; Subashini et al. 2018; Bartholomé et al. 2020), Corymbia (Butler et al. 2017), *Populus* (Bdeir et al. 2017), *Pinus* (Pot et al. 2006), and *Cryptomeria* (Mori et al. 2020). The interaction between genotype and environment (G × E), estimated by assessing trait values across various genotypes in different environments, plays a crucial role in comprehending genetic composition of the complex traits and genotype deployment decisions. Precise characterization of gene and environment interaction and identification of stable QTL across the environments could maximize breeding efficiency (Mori et al. 2019). Potential and challenges associated with genetic mapping, QTL analysis and MAS in crops and trees are discussed elsewhere (Boopathi 2020).

33.6 Association Mapping

Linkage disequilibrium (LD)-guided association mapping strategy was introduced to circumvent disadvantages associated with linkage map-based QTL detection and to determine the genetic composition of quantitatively inherited traits. This strategy was promoted in forest trees because of its applicability in provenance collections, or individuals with contrasting geographical origin instead of synthetic mapping populations. The decreased genotyping cost and improved statistical methods favored the adoption of candidate gene (CG) and genome-wide association studies (GWAS) in QTL localization. Association mapping relies on the extent of linkage disequilibrium (LD), which is the nonrandom association of alleles at two or more loci in the genome (Abdurakhmonov and Abdukarimov 2008) and hence the information on LD distribution is important. Plantation forest trees such as *Populus* (Slavov et al. 2012) and *Eucalyptus* (Arumugasundaram et al. 2011) showed faster decay of LD at whole genome as well as individual gene level encouraging association mapping strategy.

Genetic associations with stable QTL were reported for various quantitative traits like wood characteristics, cold hardiness, wood metabolites, growth parameters, and bud set for many tree species. In *Eucalyptus nitens*, association mapping proved effective in identifying SNP markers within xylem-related genes that are associated with wood properties such as cellulose and lignin content, pulp yield, and microfibril angle (MFA) (Thumma et al. 2005). Genomic regions linked to natural variation in bark texture in *Populus trichocarpa* were found to be associated with SNPs located within candidate genes (Bdeir et al. 2019). Combined studies on linkage mapping and association mapping analysis were proposed as effective models in natural populations of forest trees which were proved to be effective in *Populus* (Du et al. 2016), *Pinus pinaster* (Bartholomé et al. 2016), and *Salix purpurea* (Carlson et al. 2019). In *Eucalyptus*, an alternative strategy to detect advantageous alleles applying GWAS along with regional heritability mapping (RHM) was proposed by integrating different breeding populations to validate the SNPs associated with target traits (Müller et al. 2019a). Although potentially interesting DNA markers were identified and linked to specific traits, practicing MAS is still continue to face challenges due to undetected causative mutation(s) and markers confer a small fraction of the genetic variability of target traits in the investigated populations thus requiring repeated validation experiments. Due to these reasons, the tree breeding industry has not fully capitalized on most markers, as the integration of each potentially beneficial allelic variation becomes impractical in the process.

33.7 Genomic Selection

The emergence of SNP-chip genotyping technology has brought cost-effective genotyping, introducing the concept of genomic selection (GS) or genome-wide selection. This is regarded as an advanced iteration of marker-assisted selection. Initially proposed by Meuwissen et al. in 2001, genomic selection involves predicting the genetic value of animals based on their genotype at thousands of single-nucleotide polymorphisms (SNPs) spanning the entire genome (Meuwissen et al. 2001). GS does not require pedigree records and genetic estimates of breeding values GEBV are estimated using phenotypes and genomic relationships, which are based on dense marker data generated at genome scale. The nonadditive genetic effects also have an impact in total genetic variation of phenotypes, and genomic estimated genotypic values (GEGV) can be obtained in genomic selection. Trees selected based on GEGV can be propagated vegetatively to capture both additive and nonadditive genetic variations (Grattapaglia 2017).

In GS, the effects of all the SNPs are estimated simultaneously without any significance testing. This approach enhances time efficiency in the analysis. Further, only the training population is phenotyped at high precision and accuracy but future selections are not necessarily recorded for the trait under analysis. Deploying genetic material selected through genomic selection approach can be clones, thus implementing the breeding results is achieved early. In conifers, it has been estimated that a training population comprising approximately 2000 clones is the minimum size required for the effective implementation of genomic selection (Li and Dungey 2018). A substantial number of GS studies have been reported in white spruce (Beaulieu et al. 2014), *Populus* (Evans et al. 2014), *Pinus* (Isik et al. 2016), Douglas fir (Thistlethwaite et al. 2019), Norway spruce (Lenz et al. 2020), and *Eucalyptus* (Ballesta et al. 2019) for diverse traits including growth and disease resistance. Its practical application to forest trees in the form of prediction across multiple environments is getting addressed and validation with unrelated population is in progress (Suontama et al. 2019; Westbrook et al. 2019). The potential of GS in reducing breeding cycles by accurately assessing traits at a young age makes it a valuable tool for enhancing productivity in tree breeding.

33.8 Population Genomics

In the era of climate change, exploration of genes and polymorphisms determining adaptation would support the sustainable management of genetic resources and development of genetically improved germplasm that sustain in climate change conditions. Population genomics integrates genome resequencing and transcriptomic approaches, bioinformatics tools, and statistical methods to conserve adaptive potential and functional diversity of the individuals. High efficiency and inexpensive SNP marker generation methodologies such as restriction site-associated DNA sequencing (RADseq), double-digestion RAD sequencing (ddRAD) (Andrews et al. 2016), genotyping by sequencing (GBS), and associated procedures that can readily be applied to nonmodel species are promising approaches to forest trees (Neophytou et al. 2019; Parchman et al. 2018). In these techniques, a random fraction of the genome is sequenced, eliminating the need for a reference genome or prior knowledge of polymorphisms. Additionally, several thousand markers can be genotyped in a single step for many samples by multiplexing (Andrews et al. 2016). RADseq was advocated for genomic scans of local adaptation when prior knowledge on candidate genes exists for the target species (Lowry et al. 2017). For example, *Castanea sativa* wild populations were examined for the adaptive ability to drought using EST-SSR markers and demonstrated that the presence of specific private alleles for tolerance and susceptibility (Alcaide et al. 2019). Similarly, in Norway spruce, 384 selected SNPs from 285 genes were applied to understand adaptive responses to changing climate, wherein environmental association analysis detected potentially adaptive SNPs associated with climatic variables and indicated potential of the species to genetically adapt to varying climatic factors (Di Pierro et al. 2016). In non-model tree species, inferring SNPs from sequenced data can be challenging, especially when the genome is highly repetitive. This complexity may result in difficulties distinguishing true SNPs from false positives, often caused by issues like allele dropout. Such problems arise during in silico analysis are getting addressed by inclusion of technical replicates and tweaking design of genotyping experiments, sequencing depth and bioinformatic parameters.

Combining the high-throughput efficiency of GBS approach and stringency of SSR markers, recently SSR-GBS (simple sequence repeats-genotyping by-sequencing) was reported for population genetic studies (Curto et al. 2019). This method allows co-amplification of several microsatellites in multiplex PCR reactions, and sequenced in a single flow cell lane by pooling all PCR products. SSR-GBS analysis can overcome the size homoplasy because by amplicon sequencing the difference in unique primer binding sites can be visualized despite same number of repetitions, which is the major constraint in electrophoresis-based SSR fragment length detection. Amplicon sequences not only provided repeat length variations but also information on SNPs and insertions-deletions was obtained. Using this approach, new microsatellite loci for *Ailanthus altissima*, a species with very limited number of molecular markers, were developed and also factors involved in shaping up of population structure, genetic diversity, and genetic differentiation among populations were assessed (Neophytou et al. 2019).

Genetic diversity, population differentiation, and signatures of selection were studied in *Populus deltoides* and uncovered many genes as adaptation candidates (Fahrenkrog et al. 2017). Environmental association analysis is carried out to recognize genes linked to bioclimatic variables that explain the proportion of the environmental variance among the populations. In *Taxus baccata*, signatures of polygenic adaptation in biological pathways was identified using transcriptomes and the findings were used to develop strategies for conservation and reforestation programs (Mayol et al. 2019). In recent times, cost-efficient SNP genotyping method using custom amplicon sequencing approach, genotyping-in-thousands by sequencing (GT-seq) is highly recommended for molecular ecology and conservation of genetic resources (Meek and Larson 2019). As population genomics continues to advance in forestry, researchers must develop a comprehensive roadmap to guide the programs. This includes carefully planning sampling strategies, data production, aggregating global nucleotide variations, analyses, and interpretation of data to minimize potential pitfalls.

33.9 Whole Genome Sequencing in Tree Species

In the last decade due to drastic reduction in sequencing cost, more than 4500 eukaryotic genomes have been assembled. Next-generation sequencing platforms have made it cost-effective and accessible. However, in trees <35 species have been assembled as reference genomes and are predominantly members of Pinaceae, Salicaceae, Myrtaceae, and Fagaceae (Plomion et al. 2016). The first chromosome-level assembly was reported in *Populus trichocarpa* (Nisquallly-1) (Tuskan et al. 2006) and presently, v 3.2 includes 423 Mb genome in 1446 scaffolds with 2585 gaps and 73,013 transcripts. The *Populus* pan genome project has revealed presence of 71 million genomic variations. The genomes are significantly enriched for disease resistance (R) genes and provide insight into genomic control of self-fertilization and possible ancient intra-specific hybridization (Zhang et al. 2019). Comparative genomics studies in poplar lineages including *P. euphratica* and *P. trichocarpa* revealed extensive co-linearity of the genomes and two events of whole genome duplications (Ma et al. 2013).

Eucalyptus is a commercially planted genera providing raw material for paper pulp industries and bioenergy sector. The first 691 Mb reference genome was generated for

Eucalyptus grandis (BRASUZ1) with 34% protein-coding genes and more than 12,500 tandemly duplicated genes. The study also reported diversity in secondary metabolite pathway gene families (Myburg et al. 2014). Additionally, the draft genomes of *E. camaldulensis* (Hirakawa et al. 2011) and *E. globulus* (Rigault et al. 2012) were released. A 594.87 Mb genome of *E. pauciflora* with 61.56% of genome similarity with *E. grandis* was reported by Wang et al. (2020).

The draft genomes in *Salix* spp. were generated for *Salix purpurea* (Wu 2015) and *S. suchowensis* (Wei et al. 2020). *Salix* genomes have revealed evidence of extensive fractionation, chromosomal fission, and fusion events resulting in divergence of lineages. Another hardwood species which has been a focus of intensive research is pedunculate oak (*Quercus robur*) where the genome sequencing has been reported by several groups. The distinctive feature of oak genome is the expansion of gene families related to biotic factors and accumulation and transmission of somatic point mutations (Plomion et al. 2018). The contribution of somatic mutations in evolution of woody perennials was elucidated from this study.

Chinese chestnut (*Castanea mollissima*) is an important species which has been sequenced and assembled (Xing et al. 2019). The genome information was used for breeding resistance against fungal blight disease in *C. dentata* (Westbrook et al. 2019). Draft genomes of *Fraxinus excelsior* (European Ash) (Sollars et al. 2017), *F. americana* (American Ash) (http://www.ashgenome.org/Fraxinus_americana) and birch (*Betula nana*, *B. pubescens*, and *B. pendula*) (Wang et al. 2013) were also reported. In 2019, the genome assembly of *Casuarina equisetifolia* ssp. *incana* was published. The 300 Mb de novo assembled genome had one-third representation of repetitive elements and 29,827 annotated genes. The DNA modification sites across the genome using methylation landmarks like N^6-methyladenine (6 mA) and N^4-methylcytosine (4mC) were also inferred. The study documented lignin, cellulose, and hemicellulose-related genes (Ye et al. 2019).

Sequencing and assembly of conifer genomes have been a major challenge due to genomic "obesity." *Picea abies* with ~20 Gb genome size was the first conifer genome to be released (Nystedt et al. 2013). The enormous genome size was caused by the accumulation of long-terminal repeat transposable elements, introns, long noncoding RNA (lnc RNA), and small RNA (sRNA) due to inefficient mechanism of elimination of these genomic features. The draft genome of haploid seed megagametophyte of loblolly pine, *Pinus taeda* was reported by Neale and coworkers in 2014 (Neale et al. 2014). The genome was 23.2 Gbp size with 50,172 annotated gene models. *P. lambertiana* (Sugar pine) is the largest conifer genome (31 Gbp) assembled and its functional characterization enabled identification of a resistant gene *Cr1* governing tolerance to fungal blister rust caused by *Cronartium ribicola*. The third genus under Pinaceae to be sequenced was *Pseudotsuga menziesii* (Coastal Douglas-fir). The genome harbored lesser number of repeat elements when compared to pine and spruce. This reference genome provided insight into the significant morphological, anatomical, and physiological differences existing between angiosperms and gymnosperms (Neale et al. 2017). Recently, 18.16 Gb genome was assembled and 50,757 genes were functionally annotated in *Abies alba* (Mosca et al. 2019). The draft genome of another conifer, *Larix sibirica,* was also reported by Kuzmin et al. (2019).

The two key tropical species which were sequenced in the recent years are teak and sandalwood. A draft 317 Mb genome of teak (*Tectona grandis*) was assembled into 2993 scaffolds and annotated for 36,172 protein-coding genes. Genes encoding biosynthesis of terpenoids and polyketides was identified from the genome. Additionally, 182,712 microsatellite motifs were mined and 16,252 polymorphic SSRs were predicted across five genotypes through low coverage genome sequencing (Yasodha et al. 2018). Subsequently, chromosomal-scale genome assembly of teak was reported by Zhao et al. (2019) and 17 near-complete pseudomolecules were assembled and 41,155 high confidence gene models encoding 39,930 loci were predicted. The study also highlighted tandem duplication of terpene synthase (TPS) genes resulting in expansion of this gene family in the species.

The draft genome of Indian sandalwood (*Santalum album*) was first published by Mahesh and coworkers in 2018 (Mahesh et al. 2018). An integrated multiomics approach was employed to decode the 221 Mb genome encoding 38,119 protein-coding genes. The second draft genome was released in 2019 (Dasgupta et al. 2019) and the predicted variants between the two genomes were 1,191,838 (0.42%) SNPs and 98,312 (0.034%) InDels, indicating high level of heterozygosity in sandalwood genome (Dasgupta et al. 2019). Additionally, a 147.25 kb chloroplast genome was assembled and phylogenomic analysis grouped the order Santalales with Asterids with no taxonomic ambiguity.

33.10 Transcriptome Resources in Tree Species

Perennial trees in contrast to herbaceous annuals are characterized by delayed expression of several traits like flowering and secondary growth, and in temperate regions, they show an annual behavior of growth and dormancy. Trees being perennial are unique in their response to environmental challenges including abiotic, biotic, and climate adaptation. Hence, gene discovery, expression analysis, and transcriptome resources are mainly generated for these traits in different tree species (Table 33.1).

Table 33.1 Transcriptomic resources in forest trees

Sl. No.	Species	Traits/developmental stages studied	No. of transcripts analyzed	Reference
Wood formation				
1	*Eucalyptus globulus, E. grandis* and *E. urophylla*	Comparative analysis for wood quality traits	53,412	Salazar et al. (2011)
2	*Populus trichocarpa* and *Eucalyptus grandis*	Xylem development and secondary cell wall biosynthesis	15,925 clusters of orthologs and paralogs	Hefer et al. (2015)
3	*Tectona grandis*	Nonlignified (leaf, root, seedling, flower) and lignified (secondary xylem) from two age groups of trees	80,749–139,535 contigs	Galeano et al. (2015)
4	*Eucalyptus urophylla* × *E. grandis*	Lignocellulose formation	–	Shinya et al. (2016)
5	*Populus tremula*	Wood formation	28,294	Sundell et al. (2017)
6	*Casuarina equisetifolia*	Lignin biosynthesis pathway	26,985	Vikashini et al. (2018)
Other developmental processes				
7	*Quercus petraea*	Endodormant and ecodormant buds	6471	Ueno et al. (2013)
8	*Fagus sylvatica*	Quiescent and swelling buds stages	21,057	Lesur et al. (2015)
9	*Eucalyptus grandis*	Floral development	>35,000	Vining et al. (2015)
10	*Tectona grandis*	Vegetative and generative bud tissues	87,365	Diningrat et al. (2015)
Biotic interactions				
11	*Pinus monticola*	Resistant and susceptible trees to white pine blister rust, *Cronartium ribicola*	23,000	Liu et al. (2013)
12	*Eucalyptus melliodora*	Resistant and susceptible trees to herbivory	13,104	Padovan et al. (2013)
13	*Poplar* hybrids	Resistance and susceptibility to leaf spots and stem canker pathogen *Septoria musiva*	550–1499 DEGs	Liang et al. (2014)
14	*E. grandis*	Defense response to stem canker caused by *Chrysoporthe austroafricana*	27,714 and 29,829	Mangwanda et al. (2015)
15	*E. urophylla* × *E. tereticornis*	Response to leaf blight caused by *Calonectria pseudoreteaudii*	8585 differentially expressed genes	Chen et al. (2015)
16	*E. nitens*	Defense response to *Phytophthora cinnamomi*	1400 DEGs	Meyer et al. (2016)
17	*Fraxinus pennsylvanica*	Biotic (Emerald ash borer feeding) and abiotic (heat, drought, cold and ozone) treatments	107,611	Lane et al. (2016)
18	*Picea abies*	Response to needle rust fungus *Chrysomyxa rhododendri*	2444 DEGs	Trujillo-Moya et al. (2020)
Abiotic stress				
19	*Populus euphratica*	Response of callus to salt stress	86,777	Qiu et al. (2011)
20	*Eucalyptus camaldulensis*	Response to water stress	32,474	Thumma et al. (2012)
21	*Eucalyptus urophylla* × *E. grandis* *E. alba*	Response to water stress	14,460	Villar et al. (2011)
22	*Bombax ceiba*	Response to water stress	95,586 unigenes	Zhou et al. (2015)
23	*Abies alba*	Response to drought stress	65,535	Behringer et al. (2015)
24	*Santalum album*	Response to cold stress	80,426	Zhang et al. (2017)
25	*Quercus lobata*	Response to drought stress	68,434 contigs	Gugger et al. (2017)
26	*Casuarina equisetifolia*	Response to cold stress	118,270 unigenes	Li et al. (2017)
27	*Pinus massoniana*	Response to drought stress	197,612 unigenes	Du et al. (2018)
28	*Populus alba* × *P. glandulosa*	Response to salt stress	65,535	Yao et al. (2018)
29	*Tectona grandis*	Response to drought stress	132,647 contigs	Galeano et al. (2019)
Secondary metabolite pathways				
30	*Lindera glauca*	Terpenoid biosynthesis	98,141 unigenes	Niu et al. (2015)
31	*Santalum spicatum*	Sesquiterpene biosynthesis	12,537 and 593, 510	Moniodis et al. (2015), Celedon et al. (2016)

DEG Differentially expressed genes

Secondary growth and wood formation are unique in tree species and are a result of highly regulated process governed by developmental stage and environmental factors. This process results in various forms of wood like early and late wood, juvenile and mature wood, and reaction wood. Hence, understanding the biogenesis of this complex developmental process has received significant focus in recent years. Expression profiling of genes using microarray, transcriptome analysis, and ChIP-Seq assays was employed to comprehend the role of genes regulating wood formation and correlate their expression patterns with wood characteristics. The key genes regulating secondary cell wall biogenesis are reported from different species (reviewed by Zhong et al. 2019). The major transcription factors (TFs) associated with this process included MYB, LIM domain, NAC, HB, MADS-box, zinc finger, ARF, WRKY, AUX-IAA, AS2/LOB AP2-EREBP, bHLH, ARID, and bZIP. The multitier transcriptional regulation of wood development has been extensively reviewed (Zhang et al. 2018).

Active research is in progress to understand the difference between wood formation in gymnosperms and angiosperms. The absence of S units in lignin confirmed by the lack of *F5H* homolog in pine genome (Neale et al. 2014) and the reduced number of vascular-related NAC domain transcription factors in Norway spruce (Nystedt et al. 2013) highlight the different mode of water transport and mechanical support in gymnosperms and angiosperms.

MicroRNAs (miRNAs) are 20–24 nucleotide heterogeneous molecules that mediate post transcriptional gene silencing (PTGS) and chromatin modification. miRNAs are characterized by conservation and divergence resulting in tissue and lineage-specific populations. Several developmental processes in tree species are regulated by miRNAs including development, metabolism, adaptation, and evolution. Regulation of tree processes by miRNAs is reported from Poplars, *Eucalyptus*, Acacias, and gymnosperms like spruce and pines (reviewed by Sun et al. 2012).

One of the emerging fields in genomics is the powerful concept of systems genetics which bridges the knowledge gap by integrating multilevel data from inter-trait relationships and molecular networks to understand phenotypic variations. It attempts to study biological processes by examining all components and their interactions in relation to the context of a whole system. This approach facilitates characterization of complete biological pathways and networks which are unique, essential, and redundant. Genome-scale analysis of secondary development through gene regulatory networks is reported in *Populus*, *Pinus taeda*, and *Eucalyptus* (reviewed by Myburg et al. 2019).

The response of tree tissues to environmental stressors is predominantly reported for drought followed by salinity, elevated CO_2 concentration, hypoxia, heat, cold, metal toxicity, and nitrogen deprivation. Methods like cDNA-AFLP, dot blot, macroarray, microarray, quantitative real-time PCR, and transcriptome sequencing have been employed to catalogue stress inducible genes in pines, poplars, *Eucalyptus*, and oaks (Table 33.1).

Natural forests and plantations are continuously challenged by outbreaks of insect pests and pathogens which affect the population dynamics, ecosystem, and tree health in planted forests. Chestnut blight caused by the fungal pathogen *Cryphonectria parasitica* and outbreak of gall wasp, *Leptocybe invasa* in *Eucalyptus* plantations are examples which demonstrates the severe devastations caused by biotic factors. These outbreaks are expected to be more severe in the changing climate. Understanding the molecular response to biotic stresses and identifying candidate loci/genes imparting resistance can support the breeding programs as demonstrated in chestnut. Transcriptome-based documentation of transcripts and gene networks conferring resistance to chestnut blight fungus caused by *Cryphonectria parasitica* facilitate resistance breeding in American chestnut (Barakat et al. 2012). Transcriptomic approaches including dual-RNA sequencing approach have been used to decipher the tree-pathogen interaction (reviewed by Kovalchuk et al. 2013). Recently, eCALIBRATOR, a web-based tool for comparing the *Eucalyptus* biotic stress interaction, was reported (du Toit et al. 2020).

Transcriptome-wide expression variation in *Eucalyptus* genotypes which are resistant and susceptible to gall wasp *Leptocybe invasa*, was reported by Oates et al. (2015) and an interaction model from chemical and transcriptome profiles were proposed. Similarly, in *Castanea mollissima*, dynamic changes in the transciptome profile was recorded during infestation by *Dryocosmus kuriphilus* and hormone signalling, transcription and Ca^{2+}-mediated signal transduction pathways were differentially regulated (Zhu et al. 2019). The major pest in poplar species are the defoliators and the most devastating is the cottonwood leaf beetle affecting the biomass production. The expression of diverse group of wound and insect-responsive genes was established followed by identification of defense-related genes (Lawrence and Novak 2006) using the technique of differential display of RNA. In a later study on hybrid poplar, transcriptome analyses revealed spatial and temporal expression patterns following application of oral secretions of forest tent caterpillar (*Malacosoma disstria*) and a significant upregulation of galactinol synthases (Philippe et al. 2010). Recently, the response of different poplar species to feeding by *Phratora vitellinae* was reported by deep sequencing of leaf transcriptome and the most enriched domain in the promoters of induced genes was the G-box. Major upregulated transcripts were WRKY transcription factors, cytochrome P450s, chitinases, and protease inhibitors (Müller et al. 2019b). Some of the major studies on biotic interactions of different tree species are presented in Table 33.1.

33.11 Cyber Infrastructure in Tree Genomics

High-throughput sequencing has facilitated understanding biological concepts through multiple levels of data integration. The main bottleneck associated with generation of voluminous data is the cost on storage of sequence datasets which has exceeded the cost of sequencing. This has necessitated curation of data sets to provide value-added context (Falk et al. 2018). Hence, the need for dedicated cyber infrastructure to host databases as a repository of sequences and also enable integrated analysis is essential in tree species.

Some of the online portals which house tree datasets include:

- TreeGenes (previously known as Dendrome) is a clade organism database of forest trees, representing 16 orders and over 1790 species (Falk et al. 2018). It contains 27 reference genomes, 100 genetic maps, transcriptomes of 303 species, 36.7 M genotypes, and 935,596 phenotypic measures (Wegrzyn et al. 2019).
- Hardwood Genomics Project (HWG): This open-source database is a repository of genomic resources of angiosperm trees of ecological and agricultural significance (https://www.hardwoodgenomics.org/). It allows functional annotation of sequence data through integration of several computational workflows.
- Plant Genome Integrative Explorer (PlantGenIE): This platform provides intuitive access to genomic data from model tree species and includes expression networks and associated visualization tools. It allows community annotation and comparative expression analyses to provide biological insight to data resources (Sundell et al. 2017).
- Phytozome: This is a comparative genomics portal for plant species and facilitates retrieving, analyzing, and visualization of genomes sequenced by Department of Energy's Joint Genome Institute (Goodstein et al. 2012). The latest version (v13) hosts 375 assembled and functionally annotated genomes.

Additionally, there are several smaller databases which facilitate comparative genomic analyses including *Quercus* portal, PineDB, ConiferGDB, EuroPineDB, PopulusDB, PoplarDB, EucalyptusDB, Eucanext, Eucspresso, EUCATOUL, EUCAWOOD, EucaCold, EucGenIE, PopGenIE, Fagaceae genomics web, and AspWood.

33.12 Genetic Modification of Forest Trees

The tools for moving genes across kingdoms and edit genome sequences provide new opportunities for the tree breeders to tailor plants for traits not inherently available in them. Since the development of the first transgenic hybrid poplars for glyphosate resistance (Fillatti et al. 1987), several traits have been targeted for genetic modification in trees. Field trials of various transgenic tree species have been promising; however, the regulatory hurdles have stalled the field application of this powerful technology. Genetic engineering techniques are being continually refined to address the regulatory issues. Modern techniques like RNAi and CRISPR/Cas9 now permit minimal modification of genomes, thereby minimizing the concerns of the presence of transgenes, and providing the required impetus for advancing research in this area. Furthermore, these tools provide the only option for validating gene function in the physiological context of the plant, thereby providing the requisite insights for development of applied tools like genetic markers for desired traits. Forestry species are also a resource of genes that code for desired traits like salt/metal/and pest tolerance. Genes that confer fitness in plants that naturally survive in harsh conditions could be co-opted into other plants, as has been done for imparting salt and drought tolerance in transgenic rice expressing a Cu-Zn SOD gene from a mangrove species, *Avicennia marina* (Prashanth et al. 2008).

33.12.1 Tools for Transgenesis in Trees

Genetic transformation of forestry species has mainly relied on *Agrobacterium*-mediated gene transfer approaches. *Agrobacterium tumefaciens,* a soil bacterium that causes crown gall formation in dicots, is, however, relatively less efficient in woody species (De Cleene and De Ley 1977). The *Vir* region in the large Ti (tumor-inducing) plasmid of the *Agrobacterium* encodes for proteins that enable the transfer of the T-DNA into the plant cell. The ability of *Vir* genes to act on T-DNA residing on another plasmid led to the development of the smaller binary Ti vectors. The genes of interest are spliced onto the T-DNA region of the binary vector and introduced into the *Agrobacterium* using techniques like electroporation and triparental mating. The *Agrobacterium* strains are, therefore, defined by their chromosomal background, the helper Ti plasmid carrying the vir genes, and the binary vector carrying the gene of interest. The *Agrobacterium* strain is then co-cultivated with plant cells under appropriate conditions to transfer the genes of interest. The use of *Agrobacterium* cells in early log phase, preincubated injured leaf explants, vir gene inducing compounds like acetosyringone, and sugars in the cocultivation media, appropriate media pH, and temperature conditions enhance transformation efficiency. In order to ensure that only the transformed cells grow into plantlets, the transformation vector also includes selectable marker genes that enable the transformed cells to have a fitness advantage under specific selection pressure applied in the media like the use of antibiotics or herbicides. Screenable markers like GUS and GFPs permit identification of transgenic plants in the absence of a selective agent. Several strategies have been developed to remove selectable markers after they have served their function (Hare and Chua 2002).

Regeneration of plantlets from single transformed cells is a requisite to achieve homogenetically transformed plants. Regeneration methods via organogenesis and somatic embryogenesis are optimized before attempting genetic transformation. The main challenge for genetic transformation of trees is the lack of efficient methods for regeneration from single transformed cells of the most desired genotypes (Ellis et al. 2001). Most of the protocols, therefore, use seedling-derived tissues or are restricted to a few genotypes, mainly in poplars. As most of the regeneration protocols are callus based, possibilities of somaclonal variation pose a serious threat (Ellis et al. 2001; Tripathi et al. 2006). It, therefore, becomes necessary to generate and screen many plant lines to identify lines having minimal genetic changes arising under tissue culture conditions (Birch 1997; Hansen and Wright 1999). Regeneration methods involving minimal use of hormones and avoiding unorganized callus phase can reduce the problems of somaclonal variation (Ellis et al. 2001).

Alternative approaches not involving regeneration via tissue culture have been attempted in model plants to overcome the regeneration bottleneck (Bent 2000). *In planta* vacuum infiltration-based *Agrobacterium* transformation methods in *Arabidopsis* have resulted in a high transformation rates exceeding 1% of the seeds tested (Bechtold et al. 1993). *Arabidopsis* plants at the early stages of flowering were dipped in *Agrobacterium* solution followed by application of vacuum for causing air trapped within the plant to bubble off and be replaced with the *Agrobacterium* solution. The stably transformed progenies were selected using appropriate selection. In cases where *Agrobacterium*-mediated transformation is difficult, direct delivery of DNA has been attempted. Direct gene transfer approaches have mainly relied on DNA delivery via the biolistic approach. Hamada et al. (2017, 2018) targeted mature plant tissues in wheat by bombarding shoot apical meristems. Developing *in planta* approaches in trees can obviate the requirement of regeneration from dedifferentiated cells making it possible to engineer commercially cultivated varieties (Nagle et al. 2018).

33.12.2 Desirable Traits for Modification in Trees

In perennials, engineering for fast growth is one of the most important goals of tree breeding. In 2015, Brazil released the first genetically modified hybrid event of *Eucalyptus grandis* and *E. urophylla*, H421, for commercial plantations. The event engineered to express Arabidopsis *Cell* (β-1,4 endoglucanase) gene enabled faster growth by promoting cell wall relaxation during phases of expansion and elongation of the plant cell (Shani et al. 2004, 2006). Overexpressing the gibberellin biosynthesis gene, GA20ox1, has been shown to significantly improve biomass in poplars (Jeon et al. 2016). While the CaMV 35S promoter-driven constitutive overexpression led to reduced root weight and smaller leaves,these undesirable effects were absent when directed by the developing xylem-specific promoter, DX15.

Increasing concern about environmental pollution by paper and pulping industries necessitates improvement in raw material and processes used for pulping. Desirable traits for pulping industries include reduced lignin, wood density and ash content, and increased cellulose content. Antisense strategy has been used to suppress the expression of lignin biosynthesis genes like caffeate *O*-methyltransferase (COMT) (van Doorsselaere et al. 1995; Tsai et al. 1998) and cinnamyl alcohol dehydrogenase (CAD) (Baucher et al. 1996) resulting in modified lignin composition and content. Hu et al. (1999) reported reduced lignin content and increased cellulose accumulation by the downregulation of 4-coumarate: coA ligase. Long-term field trials in United Kingdom and France have shown that poplars expressing cinnamyl alcohol dehydrogenase (CAD) and caffeate O-methyltransferase (COMT) antisense transgenes were healthy, had stable transgene expression, and had better pulping traits (Pilate et al. 2002). Although transgenic trees with reduced lignin have been generated, field utilization of these genotypes has not been successful because of the associated effects of reduced lignin resulting in stunted growth and increased pest susceptibility (Voelker et al. 2011). Alternative approaches of structurally altering lignin for easier deconstruction instead of reducing lignin content have shown promise. Transgenic poplars overexpressing F5H have been demonstrated to have high S/G ratios (Franke et al. 2000; Huntley et al. 2003; Stewart et al. 2009) with normal growth, improved pulping efficiency, reduced chemical requirements for pulping and bleaching (Huntley et al. 2003; Stewart et al. 2009). These transgenic poplars were also reported to have improved fungal decay tolerance (Skyba et al. 2013). Poplars engineered to express the feruloyl coenzyme-A (CoA) monolignol transferase (*AsFMT*) gene from *Angelica sinensis* could incorporate monolignol ferulate conjugates that resulted in an altered lignin structure with readily cleavable ester bonds (Wilkerson et al. 2014). Transgenic events showed 1–2% higher pulp yield using 5% less chemicals, and were agronomically indistinguishable from wild type. In another approach, pectin biosynthesis gene, Galacturonosyltransferase 4 (*GAUT4*) was targeted for RNAi mediated downregulation resulting in an increase of 44% aerial biomass and 7% sugar release (Biswal et al. 2018).

Clonal plantations of fast growing trees are often grown in large areas, using limited number of genotypes, thereby posing pest and disease issues. Difficulties in pesticide application in tall trees and inaccessibility of the chemicals to the pests at its most susceptible stage make breeding for biotic stress tolerance a major research goal. Strategies to incorporate insect resistance have mainly been through the use of insecticidal *Bacillus thuringiensis* δ-endotoxin genes. Poplars transgenic for *cryIA(a)* and *cryIIIA* have shown resistance to the pests *Lymantria dispar* and *Chrysomela tremula* (McCown et al. 1991; Cornu et al. 1996). Transgenic *Pinus taeda* L. expressing a synthetic version of the *CRY1Ac* gene of *B. thuringiensis* showed enhanced resistance to *Dendrolimus punctatus* Walker and *Crypyothelea formosicola* Staud (Tang and Tian 2003). In China, two lines of *B. thuringiensis* (Bt) toxin-producing poplar were authorized for commercial planting (James 2015). Genes encoding inhibitors targeting insect digestive enzymes have been used to impart insect tolerance. *Populus nigra* transgenic for soybean Kunitz proteinase inhibitor gene (*KTi_3*) showed inhibition of trypsin-like digestive proteinases of *Lymantria dispar* and *Clostera anastomosis* (Confalonieri et al. 1998). Tolerance to the blight fungus, *Castanea dentata* in transgenic chestnut trees was conferred by engineering detoxification of oxalate made possible by overexpression of Oxalate oxidase (Oxo) (Zhang et al. 2013; Newhouse et al. 2014).

As agricultural crops are preferred in fertile lands, tree plantations are grown in less productive lands that are often affected by various abiotic stresses. Poplars transgenic for enzymes involved in abiotic stress tolerance like PrxA1 (Kajita et al. 1994), superoxide dismutase (Arisi et al. 1998) and glutathione synthetase (Herschbach et al. 1998) have been reported. The *Eucalyptus globulus* plants transformed with a bacterial choline oxidase gene (*codA*) showed salt tolerance up to 300 mM NaCl (Matsunaga et al. 2012). Similarly, *Eucalyptus* F1 hybrids (*E. camaldulensis* and *E. tereticornis*) expressing yeast *HAL2* gene encoding 3′(2′), 5′-bisphosphate nucleotidase (DPNPase) resulted in transgenic hybrids showing enhanced salt tolerance of up to 300 mM NaCl (Thananantaet al. 2018). Enhancement of *codA* transcription by the *Arabidopsis thaliana HSP* terminator in *Eucalyptus* resulted in enhanced tolerance to salinity (Tran et al. 2018). Transgenic *E. camaldulensis* plants harboring an RNA-binding-protein (*McRBP*) gene derived from a *Mesembryanthemum crystallinum* showed more than 70% survival at 400 mM NaCl treatment (Tran et al. 2019). Several other studies for engineering salt and drought stress tolerance have been recently reviewed by Polle et al. (2019). Improved freezing tolerance in Eucalyptus by heterologous expression of *AtCBF2* has also been reported (Hinchee et al. 2011).

Trees with their long root system and perennial nature could be used in phytoremediation of soils. Yellow poplar (*Liriodendron tulipifera*) transgenic for mercuric ion reductase gene (*mer A*) was able to volatilize mercury into the atmosphere (Rugh et al. 1998). Poplars transgenic for tadpole ferritin were found to accumulate cadmium and grow well under higher concentrations of cadmium (Choi et al. 1999). A number of valuable medicinal compounds are produced in roots of trees. In vitro roots generated using *Agrobacterium rhizogenes* could be used in the production of secondary metabolites thereby providing a nondestructive system for producing these metabolites that are specifically produced in root systems. Many of the secondary metabolites have undesirable properties. As secondary metabolites are species specific, engineering them would require the knowledge of enzymes and genes involved in their biosynthesis at the species level.

33.12.3 Recent Advances in Gene Silencing and Genome Engineering

While early transformation technologies enabled interkingdom transfer of genes, new tools have been developed for knock-down and knock-out of native genes enabling applications in generating plants with desired traits, and also for reverse genetics approaches to understand the function of genes. RNA interference (RNAi) is a powerful approach for downregulating the expression of a targeted gene. RNAi relies on the cyclic generation of double-stranded RNA (dsRNA) and small interfering RNAs (siRNA) that effect destruction of cognate mRNA molecules in the cytoplasm using cellular components. dsRNA has been shown to induce and direct sequence-specific RNA degradation, and DNA methylation (Jones et al. 1998; Mette et al. 2000; Waterhouse and Helliwell 2003). Double-stranded RNA is expressed in plants using transgenes that encode hairpin RNA. Few examples of RNAi transgenics include generating reduced lignin poplars (Coleman et al. 2008; Zhang et al. 2018), nonopium poppy (Allen et al. 2004), and decaffeinated coffee (Ogita et al. 2003).

Recent years have witnessed the development of gene editing technologies like Zinc Finger Nucleases (ZFNs), TALENS (TALE effector nucleases), and CRISPER/Cas9. The CRISPR/Cas system being directed by simple RNA molecules has gained preference over the protein directed ZFNs and TALENs. While RNAi approach relies on dsRNA-induced downregulation of the genes at the transcription level, CRISPR/Cas9 system relies on Cas9 generated double-stranded breaks at gRNA directed genomic loci (Jinek et al. 2012). Double-stranded breaks that are repaired by the error-prone non-homologous end joining (NHEJ) repair pathway result in the introduction of mutations and loss of function. Gene editing approaches are increasingly being applied in tree species (Bewg et al. 2018). Current gene editing technologies rely on transgenic expression of Cas9 and gRNA. In order to increase public acceptance of gene edited plants, recent approaches in agricultural crops have relied on transgene-free gene editing, in which transiently expressed preassembled Cas9 protein–gRNA ribonucleo proteins (RNPs) are delivered into the plant cell (Svitashev et al. 2016; Liang et al. 2017). Therefore, efficient methods need to be optimized for delivery of the RNPs through the plant cell wall, and regeneration of plants from these tissues (Metje-Sprink et al. 2019). Most of the initial technologies for DNA free editing relied on delivery of CRISPR–Cas9 RNPs into protoplasts as in the case of apple and grape vine (Osakabe et al. 2018). However, regeneration of the transformed protoplasts is challenging in woody species as has been observed in apple or grape vine (Malnoy et al. 2016). In wheat, delivery of CRISPR–Cas9 RNPs into immature embryos was achieved via biolistics for DNA-free genome editing (Liang et al. 2018). Transgene free editing approaches would enable making genome edited crops similar to the conventionally bred varieties, thereby alleviating biosafety concerns.

33.12.4 Functional Genetic Approaches for Understanding Gene Function

Understanding the role of genes in contributing to a particular trait is crucial for the development of molecular marker and genetic modification approaches for a particular trait. The low transformation efficiency and the time required for generating transgenic trees do not permit rapid functional evaluation of genes. To overcome these limitations, induced somatic sector analysis (ISSA) approach has been developed. ISSA involves *Agrobacterium tumefaciens* mediated transformation of the cambial tissues that have been exposed by incising through the bark tissue using a razor blade. Somatic sectors of transformed cells in the woody stem tissue provide a suitable model system for studies on cell fate and pattern formation during secondary growth and xylogenesis (Spokevicius et al. 2005; Van Beveren et al. 2006), and analysis of transgenes and promoters in trees (Hussey et al. 2011; Creux et al. 2013; Baldacci-Cresp et al. 2015).

Another rapid approach involving generation of composite transgenic plants has been developed using *Agrobacterium rhizogenes*. By disrupting gene function, RNAi and CRISPR/Cas9 tools now enable deciphering the precise role of the gene in the genetic and physiological context of the plant. Composite transgenics in which GFP tagged transgenic roots are generated on nontransgenic shoots have been used in conjunction with RNAi for rapid functional analysis of genes involved in root growth and development in difficult to transform plants like casuarina (Gherbi et al. 2008; Svistoonoff et al. 2013). In eucalyptus, the composite transgenic strategy was reported by Balasubramanian et al. (2011), and has been successfully used in functional validation of the lignin biosynthesis gene (Plasencia et al. 2016; Dai et al. 2020). The high transformation efficiency (~ 60%) and a less protracted procedure of around 5 months required to generate hardened composite transgenic enable rapid analysis of the phenotypic effects of the transgene expression in roots.

Gene silencing via dsRNA approaches is increasingly being used for functional analysis of genes in insect pests of agricultural crops (Sivakumar et al. 2007). Silencing genes directing the growth and development of some insect pests have shown some promise as a strategy for their management (Gordon and Waterhouse 2007; Price and Gatehouse 2008). Insect larvae fed on transgenic tissues expressing dsRNA showed larval stunting and mortality (Baum et al. 2007; Mao et al. 2007). However, RNAi responses in lepidopterans have been found to vary, with many insects exhibiting recalcitrance (Terenius et al. 2011).

33.12.5 Adapting Transgenic Technology in Light of Biosafety Concerns

Long-generation time, together with monoculture of tree species transgenic for single resistance genes, may cause strong selection pressure for the development of resistance in the pest population. Ensuring stable tolerance during different developmental stages may require altering more than one genetic mechanism in addition to developing gene delivery systems that ensure targeted single copy integration of multiple genes. Gene pyramiding and planting of nontransgenic trees (refugia) along with the transgenic plantations have been advocated as strategies for preventing the development of resistance in pest populations (Brousseau et al. 1999).

Recent studies in GM eucalyptus have evaluated distances of pollen dispersal and spontaneous seedling establishment (da Silva and Abrahão 2021). In trees which have limited data available, risk assessment of transgenic trees could be simulated based on data from existing nontransgenic populations, modelling of transgene dispersal, and monitoring field releases (James et al. 1998). To minimize ecological risks, various strategies for gene containment and excision of selectable marker genes have been attempted in genetically modified crops (Daniell 2002; Hare and Chua 2002). Chloroplast transformation has been advocated as an alternative to prevent transgene escape, as pollen lacks chloroplasts. However, chloroplast transformation technology has not been standardized in many tree crops. Approaches for engineering reproductive sterility include targeted cell ablation of floral tissues using barnase in *Pinus radiata* (Mouradov et al. 1998) and suppression of a *Populus* homolog of *Arabidopsis* floral meristem identity gene, PTLF (Rothmann et al. 2000). Stable gene editing approaches possible through recent CRISPR/Cas9 approaches (Fritsche et al. 2018) open up avenues in the containment of transgenes. Control of flowering may also have implications in increasing wood production.

Transgenic technology has expanded the plant breeder's repertoire of tools for imparting desired traits and could be a potent tool to breed new varieties of tree species that could be used for climate mitigation strategies, improving pulping efficiency and biofuels. As wood is the final product in case of pulpwood species, consumption-associated biosafety concerns are limited. However, fears of unknown consequences of genetic modification have resulted in regulatory hurdles involving long-drawn biosafety assessments resulting in slow adoption of this technology. The potential of transgenic approaches has, therefore, been harnessed mainly for understanding the contribution of a particular gene towards a desired trait. Addressing societal concerns could see improved acceptance of this powerful technology. The newer approaches of transgene-free gene editing would enable generation of genetically modified crops that are indistinguishable from the conventionally bred varieties. The modern tools of transgene-free gene editing when integrated with the ongoing genetic improvement approaches could potentially revolutionize breeding for desired traits.

33.13 Future Perspectives

The unique biological traits of forest trees make them pivotal structural components within forest ecosystems, endowing them with distinct roles and functions. Forest trees gained significance of molecular research and take advantage of rapid advances in NGS technologies, linkage and association mapping, candidate gene approaches, functional molecular genetics, and genetic transformation. These are crucial tools for understanding the biological interactions occurring on a genome scale and application in conservation and tree improvement. There is a need for advanced research and development in certain areas, particularly regarding reference genomes in tree species. Currently, such genomes are limited to a few economically important species. In the case of other species, such as conifers, draft genomes have been published, but comprehensive annotations remain incomplete. In tropical trees, there is a need for concerted efforts to sequence genomes, aiming to comprehend their adaptive ecological and economic significance. Support for programs focused on cataloging and managing tree genetic resources should include genome-wide marker studies. These studies are essential to understand the dynamics of population differentiation. Identification of adaptive genes (both for biotic and abiotic stress tolerance) from tree species could serve as a future warehouse of functional genes for use in genetic transformation. The huge volume of data generated in these programs necessitates safe management for both current and future utilization. Its demands increased investment in database resources and the development of cyber infrastructure. Given that traits in trees are shaped by a myriad of environmental and genetic factors, having access to whole genome sequences, transcript, protein, and metabolite profiles would significantly enhance our perception of the genetic networks at play. This, in turn, would be advantageous for the genetic manipulation of target traits. New insights can be gained to resolve uncertainties concerning the molecular processes leading to wood development, growth, and plant–environment interactions. Such advances in key molecular pathways will have broad impact on forest trees to provide sufficient goods with desired properties and to meet continuously growing demand for wood products.

Lessons Learnt

- Use of DNA tools and other biotechnological interventions are important components of most forest-tree breeding programs. Application of molecular markers and DNA barcodes was successfully demonstrated in many tree species in varietal identification to timber tracking to control illegal timber harvesting. However, operational deployment of such technologies is limited.
- Modern biotechnologies in tree breeding are frequently a medium- to long-term strategy requiring consistent effort, financial commitment, and coordination, despite the fact that commercially significant and widely cultivated trees can have substantial ecological, social, and economic implications. It has been established in species including eucalyptus, poplars, and some conifers that combining genomic selection techniques with results of bi-parental mapping and genome-wide association studies may speed up reproduction, intensify selection, and improve breeding values.
- The rapidly developing genomic resources have made it much easier for us to comprehend the forest tree genomes for their structure and functions. Likewise, advancements in transgene-free genome editing technologies are expected to simplify onerous regulatory requirements and accelerate the generation of novel variations that confer enhanced resistance to biotic and abiotic stresses and favor geographic adaptations.

Key Questions

1. Write an account on the applications of DNA markers in illegal logging.
2. Genomic prediction is more useful to explain complex traits than genome-wide association analysis. Justify the statement with suitable examples.
3. Explain the economic implications of applying modern biotechnology in forest tree improvement.
4. How will the convergence of artificial intelligence, genomic resources and technologies, genome engineering, and synthetic biology impact the way we breed trees in future?
5. What are the biotechnological tools you would use to conserve threatened species or bring back an endangered species?

References

Abdurakhmonov IY, Abdukarimov A (2008) Application of association mapping to understanding the genetic diversity of plant germplasm resources. Int J Plant Genomics 2008:574927. https://doi.org/10.1155/2008/574927

Alcaide F, Solla A, Mattioni C, Castellana S, Martín A (2019) Adaptive diversity and drought tolerance in *Castanea sativa* assessed through EST-SSR genic markers. Forestry 92:287–296

Allen RS, Millgate AG, Chitty JA, Thisleton J, Miller JAC, Fist AJ, Gerlach WL, Larkin PJ (2004) RNAi–mediated replacement of morphine with the non narcotic alkaloid reticuline in opium poppy. Nat Biotechnol 22:1559–1566

Andrews KR, Good JM, Miller MR, Luikart G, Hohenlohe PA (2016) Harnessing the power of RADseq for ecological and evolutionary genomics. Nat Rev Genet 17:81–92

Apuli RP, Bernhardsson C, Schiffthaler B, Robinson KM, Jansson S, Street NR, Ingvarsson PK (2019) Constructing a high-density linkage map to infer the genomic landscape of recombination rate variation in European Aspen (*Populus tremula*). bioRxiv. https://doi.org/10.1101/664037

Arisi ACM, Cornic G, Jouanin L, Foyer CH (1998) Overexpression of iron superoxide dismutase in transformed poplar modifies the regulation of photosynthesis at low CO_2 partial pressures or following exposure to the prooxidant herbicide methyl viologen. Plant Physiol 117:565–574

Arumugasundaram S, Ghosh M, Veerasamy S, Ramasamy Y (2011) Species discrimination, population structure and linkage disequilibrium in *Eucalyptus camaldulensis* and *Eucalyptus tereticornis* using SSR markers. PLoS One 6(12):e28252

Avramidou E, Ganopoulos IV, Faravanopoulos FA (2010) DNA fingerprinting of elite Greek wild cherry (*Prunus avium* L.) genotypes using microsatellite markers. Forestry 83(5):527–533

Balasubramanian A, Venkatachalam R, Selvakesavan R, Sanu Mary A, Gherbi H, Svistoonoff S, Franche C, Bogusz D, Krishna Kumar N, Nambiar-Veetil M (2011) Optimisation of methods for *Agrobacterium rhizogenes* mediated generation of composite plants in *Eucalyptus camaldulensis*. BMC Proc 5:045

Baldacci-Cresp F, Moussawi J, Leplé JC, Van Acker R, Kohler A, Candiracci J, Twyffels L, Spokevicius AV, Bossinger G, Laurans F, Brunel N (2015) PtaRHE1, a *Populus tremula* × *Populus alba* RING-H2 protein of the ATL family, has a regulatory role in secondary phloem fibre development. Plant J 82:6

Ballesta P, Maldonado C, Pérez-Rodríguez P, Mora F (2019) SNP and haplotype-based genomic selection of quantitative traits in *Eucalyptus globulus*. Plan Theory 8:331

Barakat A, Staton M, Cheng CH, Park J, Yassin NMB, Ficklin S, Yeh CC, Hebard F, Baier K, Powell W, Schuster SC (2012) Chestnut resistance to the blight disease: insights from transcriptome analysis. BMC Plant Biol 12(1):38

Bartholomé J, Bink MC, Van Heerwaarden J, Chancerel E, Boury C, Lesur I, Isik F, Bouffier L, Plomion C (2016) Linkage and association mapping for two major traits used in the maritime pine breeding program: height growth and stem straightness. PLoS One 11(11): e0165

Bartholomé J, Mabiala A, Burlett R, Bert D, Leplé JC, Plomion C, Gion JM (2020) The pulse of the tree is under genetic control: *Eucalyptus* as a case study. Plant J 103(1):338–356

Baucher MB, Cabbert GP, Dorsselaer JV, Tollier MT, Petit-Conil M, Cornu D, Monties B, Montagu MV, Inze D (1996) Red xylem and higher lignin extractability by down-regulation of a cinnamyl alcohol dehydrogenase in poplar. Plant Physiol 112:1479–1490

Baum JA, Bogaert T, Clinton W, Heck GR, Feldmann P, Ilagan O, Johnson S, Plaetinck G, Munyikwa T, Pleau M, Vaughn T, Roberts J (2007) Control of coleopteran insect pests through RNA interference. Nat Biotechnol 25:1322–1326

Bdeir R, Muchero W, Yordanov Y, Tuskan GA, Busov V, Gailing O (2019) Genome-wide association studies of bark texture in *Populus trichocarpa*. Tree Genet Genomes 15. https://doi.org/10.1007/s11295-019-1320-2

Bdeir R, Muchero W, Yordanov Y, Tuskan GA, Busov V, Gailing O (2017) Quantitative trait locus mapping of *Populus* bark features and stem diameter. BMC Plant Biol 17:224

Beaulieu J, Doerksen TK, MacKay J, Rainville A, Bousquet J (2014) Genomic selection accuracies within and between environments and small breeding groups in white spruce. BMC Genomics 15:1048

Bechtold N, Ellis J, Pelletier G (1993) In planta Agrobacterium-mediated gene transfer by infiltration of adult *Arabidopsis thaliana* plants. Comptes Rendus de l Academie des Sciences Paris, Sciences de la vie/Life Sciences 316:1194–1199

Behringer D, Zimmermann H, Ziegenhagen B, Liepelt S (2015) Differential gene expression reveals candidate genes for drought stress response in *Abies alba* (Pinaceae). PLoS One 10(4):e0124564

Bent AF (2000) Arabidopsis *in planta* transformation. Uses, mechanisms, and prospects for transformation of other species. Plant Physiol 124:1540–1547

Berlin S, Trybush SO, Fogelqvist J, Gyllenstrand N, Hallingbäck HR, Åhman I, Nordh NE, Shield I, Powers SJ, Weih M, Lagercrantz U, Rönnberg-Wästljung AC, Karp A, Hanley SJ (2014) Genetic diversity, population structure and phenotypic variation in European *Salix viminalis* L. (Salicaceae). Tree Genet Genomes 10:1595–1610

Bewg WP, Ci D, Tsai C-J (2018) Genome editing in trees: from multiple repair pathways to long-term stability. Front Plant Sci 9:1732

Birch G (1997) Plant transformation: problems and strategies for practical application. Ann Rev Plant Physiol Plant Mol Biol 48:297–326

Biswal AK, Atmodjo MA, Li M, Baxter HL, Yoo CG, Pu Y, Lee YC, Mazarei M, Black IM, Zhang JY, Ramanna H, Bray AL, King ZR, La Fayette PR, Pattathil S, Donohoe BS, Mohanty SS, Ryno D, Yee K, Thompson OA, Rodriguez M Jr, Dumitrache A, Natzke J, Winkeler K, Collins C, Yang X, Tan L, Sykes RW, Gjersing EL, Ziebell A, Turner GB, Decker SR, Hahn MG, Davison BH, Udvardi MK, Mielenz J, Davis MF, Nelson RS, Parrott WA, Ragauskas AJ, Stewart CN Jr, Mohnen D (2018) Sugar release and growth of biofuel crops are improved by down regulation of pectin biosynthesis. Nat Biotechnol 36:249–257

Boopathi MN (2020) Genetic mapping and marker assisted election: basics, practice and benefits. Springer

Brousseau R, Masson L, Hegedus D (1999) Insecticidal transgenic plants: are they irresistible? Agric Biotechnol Net 1:1–10

Butcher PA, Southerton SG (2007) MAS in forestry species. Marker-assisted selection (MAS) in crops, livestock, forestry and fish: current status and the way forward. FAO, Rome, pp 283–305

Butler JB, Vaillancourt RE, Potts BM, Lee DJ, King GJ, Baten A, Shepherd M, Freeman JS (2017) Comparative genomics of *Eucalyptus* and *Corymbia* reveals low rates of genome structural rearrangement. BMC Genomics 18:397

Carlson CH, Gouker FE, Crowell CR, Evans L, DiFazio SP, Smart CD, Smart LB (2019) Joint linkage and association mapping of complex traits in shrub willow (*Salix purpurea* L.). Ann Bot 124:1–15. https://doi.org/10.1093/aob/mcz047

Celedon JM, Chiang A, Yuen MM, Diaz-Chavez ML, Madilao LL, Finnegan PM, Barbour EL, Bohlmann J (2016) Heartwood-specific transcriptome and metabolite signatures of tropical sandalwood (*Santalum album*) reveal the final step of (Z)-santalol fragrance biosynthesis. Plant J 86(4):289–299

Changtragoon S, Jalonen R, Lowe AJ (2017) Use of molecular markers in the conservation management of tropical trees. In: Ahuja M, Jain S (eds) Biodiversity and conservation of woody plants. Sustainable development and biodiversity. Springer, Switzerland, pp 155–195

Chen C, Chu Y, Ding C, Su X, Huang Q (2020) Genetic diversity and population structure of black cottonwood (*Populus deltoides*) revealed using simple sequence repeat markers. BMC Genet 21:2

Chen Q, Guo W, Feng L, Ye X, Xie W, Huang X, Liu J (2015) Transcriptome and proteome analysis of *Eucalyptus* infected with *Calonectria pseudoreteaudii*. J Proteome 115:117–131

Choi YI, Noh EW, Moon HK, Son SH, Noh ER, Yang MS (1999) Development of transgenic poplars carrying a tadpole ferritin cDNA. Poster presented at "Forest Biotechnology 99" conference, Oxford, UK, July 11–16, 1999. Poster number 21

Coleman HD, Park J-Y, Nair R, Chapple C, Mansfield SD (2008) RNAi-mediated suppression of *p*-coumaroyl-CoA 3′-hydroxylase in hybrid poplar impacts lignin deposition and soluble secondary metabolism. Proc Natl Acad Sci USA 105:4501–4506

Confalonieri M, Allgro G, Balestrazzi A, Fogher C, Delledonne M (1998) Regeneration of *Populus nigra* transgenic plants expressing a Kunitz proteinase inhibitor (KTi_3) gene. Mol Breed 4:137–145

Cornu D, Leple JC, Bonade Bottino M, Ross A, Augustin S, Delplanque A, Jouanin L, Pilate G, Ahuja MR (1996) Expression of a proteinase inhibitor and a *Bacillus thuringiensis* delta-endotoxin in transgenic poplars. In: Boerjan W, Neale DB (eds) Somatic cell genetics and molecular genetics of trees. Kluwer Academic Publishers, Dordrecht, pp 131–136

Creux NM, Bossinger G, Myburg AA, Spokevicius AV (2013) Induced somatic sector analysis of cellulose synthase (*CesA*) promoter regions in woody stem tissues. Planta 237:799–812

Cullingham CI, Cooke JEK, Coltman DW (2013) Effects of introgression on the genetic population structure of two ecologically and economically important conifer species: lodgepole pine (*Pinus contorta* var. *latifolia*) and jack pine (*Pinus banksiana*). Genome 56:577–585

Curto M, Winter S, Seiter A, Schmid L, Scheicher K, Barthel L, Plass J, Meimberg H (2019) Application of a SSR-GBS marker system on investigation of European Hedgehog species and their hybrid zone dynamics. Ecol Evol 9:2814–2832

da Silva PHM, Abrahão OS (2021) Gene flow and spontaneous seedling establishment around genetically modified eucalypt plantations. New For 52:349–361

Dai Y, Hu G, Dupas A, Medina L, Blandels N, Clemente HS, Ladouce N, Badawi M, Hernandez-Raquet G, Mounet F, Grima-Pettenati J, Cassan-Wang H (2020) Implementing the CRISPR/Cas9 technology in *Eucalyptus* hairy roots using wood-related genes. Int J Mol Sci 21:3408

Daniell H (2002) Molecular strategies for gene containment in transgenic crops. Nat Biotechnol 20:581–586

Dasgupta MG, Ulaganathan K, Dev SA, Balakrishnan S (2019) Draft genome of *Santalum album L.* provides genomic resources for accelerated trait improvement. Tree Genet Genomes 15(3):34

De Cleene M, De Ley J (1977) The host range of crown gall. Bot Rev 42:389–466

De-Lucas AI, Santana JC, Recio P, Hidalgo E (2008) SSR-based tool for identification and certification of commercial *Populus* clones in Spain. Ann For Sci 65:107

Di Pierro EA, Mosca E, Rocchini D, Binelli G, Neale DB, La Porta N (2016) Climate related adaptive genetic variation and population structure in natural stands of Norway spruce in the South-Eastern Alps. Tree Genet Genomes 12:16

Diningrat DS, Widiyanto SM, Pancoro A, Shim D, Panchangam B, Zembower N, Carlson JE (2015) Transcriptome of teak (*Tectona grandis, Lf*) in vegetative to generative stages development. J Plant Sci 10(1):1–14

Dong L, Sun Y, Zhao K, Zhang J, Zhang Y, Li X, Xun S, Zhang J, Wang S, Li Y (2019) Development and application of EST-SSR markers for DNA fingerprinting and genetic diversity analysis of the main cultivars of black locust (*Robinia pseudoacacia* L.) in China. Forests 10:644

Du M, Ding G, Cai Q (2018) The transcriptomic responses of *Pinus massoniana* to drought stress. Forests 9(6):326

Du Q, Gong C, Wang Q, Zhou D, Yang H, Pan W, Li B, Zhang D (2016) Genetic architecture of growth traits in *Populus* revealed by integrated quantitative trait locus (QTL) analysis and association studies. New Phytol 209:1067–1082

du Toit Y, Coles DW, Mewalal R, Christie N, Naidoo S (2020) eCALIBRATOR: a comparative tool to identify key genes and pathways for *Eucalyptus* defense against biotic stressors. Front Microbiol 11:216

Eckert AJ, Bower AD, González-Martínez SC, Wegrzyn JL, Coop G, Neale DB (2010) Back to nature: ecological genomics of loblolly pine (*Pinus taeda*, Pinaceae). Mol Ecol 19:3789–3805

Ellis D, Meilan R, Pilate G, Skinner JS (2001) Transgenic trees: where are we now? In: Strauss SH, Bradshaw HD (eds) Proceedings of the 1st international symposium on ecological and societal aspects of transgenic plantations. College of Forestry, Oregon State University, pp 113–123

Evans L, Slavov G, Rodgers-Melnick E, Martin J, Ranjan P, Muchero W, Brunner AM, Schackwitz W, Gunter L, Chen JG, Tuskan GA, DiFazio SP (2014) Population genomics of *Populus strichocarpa* identifies signatures of selection and adaptive trait associations. Nat Genet 46:1089–1096

Fahrenkrog A, Neves L, Resende MFR, Vasquez A, De Los Campos G, Barbazuk WB, Kirst M (2017) Genome-wide association study reveals putative regulators of bioenergy traits in *Populus deltoides*. New Phytol 213(2):799–811

Falk T, Herndon N, Grau E, Buehler S, Richter P, Zaman S, Baker EM, Ramnath R, Ficklin S, Staton M, Feltus FA (2018) Growing and cultivating the forest genomics database, TreeGenes. Database 2018: 1–11

Fillatti JJ, Sellmer J, McCown B, Haissig B, Comai L (1987) *Agrobacterium* mediated transformation and regeneration of *Populus*. Mol Gen Genet 206:192–199

Franke R, McMichael MC, Meyer K, Shirley MA, Cusumano CJ, Chapple C (2000) Modified lignin in tobacco and popular plants over-expressing the Arabidopsis gene encoding ferulate 5-hydroxylase. Plant J 22:223–234

Fritsche S, Klocko AL, Boron A, Brunner AM, Thorlby G (2018) Strategies for engineering reproductive sterility in plantation forests. Front Plant Sci 9:1671

Galeano E, Vasconcelos TS, Novais de Oliveira P, Carrer H (2019) Physiological and molecular responses to drought stress in teak (*Tectona grandis* L. f.). PLoS One 14(9):e0221571

Galeano E, Vasconcelos TS, Vidal M, Mejia-Guerra MK, Carrer H (2015) Large-scale transcriptional profiling of lignified tissues in *Tectona grandis*. BMC Plant Biol 15:221

Gherbi H, Nambiar-Veetil M, Zhong C, Félix J, Autran D, Girardin R, Vaissayre V, Auguy F, Bogusz D, Franche C (2008) Post-transcriptional gene silencing in the root system of the actinorhizal tree *Allocasuarina verticillata*. Mol Plant-Microbe Interact 21:518–524

Goodstein DM, Shu SQ, Howson R, Neupane R, Hayes RD, Fazo J, et al (2012) Phytozome: a comparative platform for green plant genomics. Nucleic Acids Res 40:D1178–D1186

Gordon KHJ, Waterhouse PM (2007) RNAi for insect-proof plants. Nat Biotechnol 25:1231–1232

Gramazio P, Plesa IM, Truta AM, Sestras AF, Vilanova S, Plazas M, Vicente O, Boscaiu M, Prohens J, Sestras R (2018) Highly informative SSR genotyping reveals large genetic diversity and limited differentiation in European larch (*Larix decidua*) populations from Romania. Turk J Agric For 42:165–175

Grattapaglia D (2017) Status and perspectives of genomic selection in forest tree breeding. In: Varshney RK, Roorkiwal M, Sorrells ME (eds) Genomic selection for crop improvement: new molecular breeding strategies for crop improvement. Springer, Cham, pp 199–249

Grattapaglia D, Kirst M (2008) *Eucalyptus* applied genomics: from gene sequences to breeding tools. New Phytol 179:911–929

Grattapaglia D, Sederoff R (1994) Genetic linkage maps of *Eucalyptus grandis* and *Eucalyptus urophylla* using a pseudo-testcross: mapping strategy and RAPD markers. Genetics 137:1121–1137

Grattapaglia D, Vaillancourt R, Shepherd M, Thumma B, Foley W, Külheim C, Potts B, Myburg A (2012) Progress in Myrtaceae genetics and genomics: *Eucalyptus* as the pivotal genus. Tree Genet Genomes 8:463–508

Gugger PF, Peñaloza-Ramírez JM, Wright JW, Sork VL (2017) Whole-transcriptome response to water stress in a California endemic oak, *Quercus lobata*. Tree Physiol 37(5):632–644

Hamada H, Linghu Q, Nagira Y, Miki R, Taoka N, Imai R (2017) An *in planta* biolistic method for stable wheat transformation. Sci Rep 7: 11443

Hamada H, Liu Y, Nagira Y, Miki R, Taok N, Imai R (2018) Biolistic-delivery-based transient CRISPR/Cas9 expression enables *in planta* genome editing in wheat. Sci Rep 8:14422

Hansen G, Wright MS (1999) Recent advances in the transformation of plants. Trends Plant Sci 4:226–231

Hare PD, Chua NH (2002) Excision of selectable marker genes from transgenic plants. Nat Biotechnol 20:575–580

Hayward AC, Tollenaere R, Dalton-Morgan J, Batley J (2015) Molecular marker applications in plants. In: Batley J (ed) Plant genotyping: methods and protocols, methods in molecular biology. Springer, New York, pp 13–27

Hefer CA, Mizrachi E, Myburg AA, Douglas CJ, Mansfield SD (2015) Comparative interrogation of the developing xylem transcriptomes of two wood-forming species: *Populus trichocarpa* and *Eucalyptus grandis*. New Phytol 206(4):1391–1405

Herschbach C, Jouanin L, Rennenberg H (1998) Overexpression of gamma-glutamyl cysteine synthetase, but not of glutathione synthetase, elevates glutathione allocation in the phloem of transgenic poplar trees. Plant Cell Physiol 39:447–451

Hinchee M, Zhang C, Chang S, Cunningham M, Hammond W, Nehra N (2011) Biotech *Eucalyptus* can sustainably address society's need for wood: the example of freeze tolerant *Eucalyptus* in the southeastern US. BMC Proc 5:124

Hirakawa H, Nakamura Y, Kaneko T, Isobe S, Sakai H, Kato T, Hibino T, Sasamoto S, Watanabe A, Yamada M, Nakayama S (2011) Survey of the genetic information carried in the genome of *Eucalyptus camaldulensis*. Plant Biotechnol J 28(5):471–480

Holderegger R, Herrmann D, Poncet B, Gugerli F, Thuiller W, Taberlet P, Gielly L, Rioux D, Brodbeck S, Aubert S, Manel S (2008) Land ahead: using genome scans to identify molecular markers of adaptive relevance. Plant Ecol Divers 1:273–283

Holliday JA, Aitken SN, Cooke JEK, Fady B, Gonzalez-Martinez SC, Heuertz M, Jaramillo-Correa JP, Lexer C, Staton M, Whetten RW, Plomion C (2017) Advances in ecological genomics in forest trees and applications to genetic resources conservation and breeding. Mol Ecol 26:706–717

Hu WJ, Harding SA, Lung J, Popko JL, Ralph J, Stokke DD, Tsai CJ, Chiang VL (1999) Repression of lignin biosynthesis promotes cellulose accumulation and growth in transgenic trees. Nat Biotechnol 17:808–812

Hudson CJ, Freeman JS, Kullan ARK, Petroli CD, Sansaloni CP, Kilian A, Detering F, Grattapaglia D, Potts BM, Myburg AA, Vaillancourt RE (2012) A reference linkage map for *Eucalyptus*. BMC Genomics 13:240

Huntley SK, Ellis D, Gilbert M, Chapple C, Mansfield SD (2003) Significant increases in pulping efficiency in C4H-F5H-transformed poplars: improved chemical savings and reduced environmental toxins. J Agric Food Chem 51:6178–6183

Hussey SG, Mizrachi E, Spokevicius AV, Bossinger G, Berger DK, Myburg AA (2011) SND2, a NAC transcription factor gene, regulates genes involved in secondary cell wall development in *Arabidopsis* fibres and increases fibre cell area in *Eucalyptus*. BMC Plant Biol 11:173

Isik F, Bartholomé J, Farjat A, Chancerel E, Raffin A, Sanchez L, Plomion C, Bouffier L (2016) Genomic selection in maritime pine. Plant Sci 242:108–119

Iwaizumi MG, Miyata S, Hirao T, Tamura M, Watanabe A (2018) Historical seed use and transfer affects geographic specificity in genetic diversity and structure of old planted *Pinus thunbergii* populations. Forest Ecol Manag 408:211–219

James C (2015) 20th Anniversary (1996 to 2015) of the global commercialization of biotech crops and biotech crop highlights in 2015. ISAAA brief no. 51. ISAAA, Ithaca

James RR, Difazio SP, Brunner AM, Strauss SH (1998) Environmental effects of genetically engineered woody biomass crops. Biomass Bioenergy 14:403–414

Jeon HW, Cho JS, Park EJ, Han KH, Choi YI and Ko JH (2016) Developing xylem-preferential expression of *PdGA20ox1*, a gibberellin 20-oxidase 1 from *Pinus densiflora*, improves woody biomass production in a hybrid poplar. Plant Biotechnol J 14:1161–1170

Jinek M, Chylinski K, Fonfara I, Hauer M, Doudna JA, Charpentier E (2012) A programmable dual-RNA-guided DNA endonuclease in adaptive bacterial immunity. Science 337:816–821

Jones AL, Thomas CL, Maule AJ (1998) *De novo* methylation and co-suppression induced by a cytoplasmically replicating plant RNA virus. EMBO J 17:6385–6393

Kajita S, Osakabe K, Katayama Y, Kawai S, Matsumoto Y, Hata K, Morohoshi N (1994) *Agrobacterium*-mediated transformation of poplar using a disarmed binary vector and the overexpression of a specific member of a family of poplar peroxidase genes in transgenic poplar cell. Plant Sci 103:231–239

Keil M, Griffin AR (1994) Use of random amplified polymorphic DNA (RAPD) markers in the discrimination and verification of genotypes in Eucalyptus. Theor Appl Genet 89:442–450

Kirst M, Cordeiro CM, Rezende G, Grattapaglia D (2005) Power of microsatellite markers for fingerprinting and parentage analysis in *Eucalyptus grandis* breeding populations. J Hered 96:161–166

Kovalchuk A, Keriö S, Oghenekaro AO, Jaber E, Raffaello T, Asiegbu FO (2013) Antimicrobial defenses and resistance in forest trees: challenges and perspectives in a genomic era. Annu Rev Phytopathol 51:221–244

Kshirsagar P, Umdale S, Chavan J, Gaikwad N (2017) Molecular authentication of medicinal plant, *Swertia chirayita* and its adulterant species. Proc Natl Acad Sci India B 87:101–107

Kuzmin DA, Feranchuk SI, Sharov VV, Cybin AN, Makolov SV, Putintseva YA, Oreshkova NV, Krutovsky KV (2019) Stepwise large genome assembly approach: a case of Siberian larch (*Larix sibirica* Ledeb). BMC Bioinformatics 20(1):37

Łabiszak B, Zaborowska J, Wachowiak W (2019) Patterns of *mt*DNA variation reveal complex evolutionary history of relict and endangered peat bog pine (*Pinus uliginosa*). AoB Plants 11(2): plz015

Lane T, Best T, Zembower N, Davitt J, Henry N, Xu Y, Koch J, Liang H, McGraw J, Schuster S, Shim D (2016) The green ash transcriptome and identification of genes responding to abiotic and biotic stresses. BMC Genomics 17(1):702

Lawrence SD, Novak NG (2006) Expression of poplar chitinase in tomato leads to inhibition of development in Colorado potato beetle. Biotechnol Lett 28(8):593–599

Le S, Hardwood CE, Griffin AR, Son DH, Thinh HH, Ratnam W, Vaillancourt RE (2017) Using SSR markers for hybrid identification and resource management in Vietnamese *Acacia* breeding programs. Tree Genet Genomes 13:102

Le S, Harwood CE, Nghiem CQ, RodGriffin A, Vaillancourt RE (2019) Patterns of hybrid seed production in adjacent seed orchards of *Acacia auriculiformis* and *A. mangium* in Vietnam. Ann For Sci 76:10

Lenz PRN, Nadeau S, Azaiez A, Gérardi S, Deslauriers M, Perron M, Isabel N, Beaulieu J, Bousquet J (2020) Genomic prediction for hastening and improving efficiency of forward selection in conifer polycross mating designs: an example from white spruce. Heredity 124:562–578

Lesur I, Bechade A, Lalanne C, Klopp C, Noirot C, Leplé JC, Kremer A, Plomion C, Le Provost G (2015) A unigene set for European beech (*Fagus sylvatica L.*) and its use to decipher the molecular mechanisms involved in dormancy regulation. Mol Ecol Resour 15(5):1192–1204

Li HB, Li N, Yang SZ, Peng HZ, Wang LL, Wang Y, Zhang XM, Gao ZH (2017) Transcriptomic analysis of *Casuarina equisetifolia L.* in responses to cold stress. Tree Genet Genomes 13(1):7

Li L, Jiang Y, Niu Z, Xue Q, Liu W, Ding X (2020) The large single-copy (LSC) region functions as a highly effective and efficient molecular marker for accurate authentication of medicinal *Dendrobium* species. Acta Pharm Sin B 10(10):1989–2001. https://doi.org/10.1016/j.apsb.2020.01.012

Li Y, Dungey HS (2018) Expected benefit of genomic selection over forward selection in conifer breeding and deployment. PLoS One 13(12):e0208232

Liang H, Staton M, Xu Y, Xu T, LeBoldus J (2014) Comparative expression analysis of resistant and susceptible *Populus* clones inoculated with *Septoria musiva*. Plant Sci 223:69–78

Liang Z, Chen K, Li T, Zhang Y, Wang Y, Zhao Q, Liu J, Zhang H, Liu C, Ran Y, Gao C (2017) Efficient DNA-free genome editing of bread wheat using CRISPR/Cas9 ribonucleoprotein complexes. Nat Commun 8:1426

Liang Z, Chen K, Zhang Y, Liu J, Yin K, Qiu JL, Gao C (2018) Genome editing of bread wheat using biolistic delivery of CRISPR/Cas9 *in vitro* transcripts or ribonucleoproteins. Nat Protoc 13:413–430

Liu F, Hong Z, Xu D, Jia H, Zhang N, Liu X, Yang Z, Lu M (2019) Genetic diversity of the endangered *Dalbergia odorifera* revealed by SSR markers. Forests 10:225

Liu JJ, Sturrock RN, Benton R (2013) Transcriptome analysis of *Pinus monticola* primary needles by RNA-seq provides novel insight into host resistance to *Cronartium ribicola*. BMC Genomics 14(1):884

Lowry DB, Hoban S, Kelley JL, Lotterhos KE, Reed LK, Antolin MF, Storfer A (2017) Striving for best practices in population genomic studies of adaptation. Mol Ecol Resour 17:366–369

Lu W, Qi J, Lan J, Luo J (2018) Genetic diversity of advanced generation breeding population of *Eucalyptus urophylla* in China. J Trop For Sci 30:320–329

Ma T, Wang J, Zhou G, Yue Z, Hu Q, Chen Y, Liu B, Qiu Q, Wang Z, Zhang J, Wang K (2013) Genomic insights into salt adaptation in a desert poplar. Nat Commun 4(1):1–9

Mahesh HB, Subba P, Advani J, Shirke MD, Loganathan RM, Chandana SL, Shilpa S, Chatterjee O, Pinto SM, Prasad TSK, Gowda M (2018) Multi-omics driven assembly and annotation of the sandalwood (*Santalum album*) genome. Plant Physiol 176(4): 2772–2788

Malnoy M, Viola R, Jung MH, Koo O, Kim S, Kim JS (2016) DNA-free genetically edited grapevine and apple protoplast using CRISPR/Cas9 ribonucleoproteins. Front Plant Sci 7:1904

Mangwanda R, Myburg AA, Naidoo S (2015) Transcriptome and hormone profiling reveals *Eucalyptus grandis* defence responses against *Chrysoporthe austroafricana*. BMC Genomics 16(1):319

Mao YB, Cai WJ, Wang JW, Hong GJ, Tao XY, Wang LJ, Huang YP, Chen XY (2007) Silencing a cotton bollworm P450 mono oxygenase gene by plant-mediated RNAi impairs larval tolerance of gossypol. Nat Biotechnol 25:1307–1313

Martin MA, Mattioni C, Cherubini M, Taurchini D, Villani F (2010) Genetic diversity in European chestnut populations by means of genomic and genic microsatellite markers. Tree Genet Genomes 6: 735–744

Matsunaga E, Nanto K, Oishi M, Ebinuma H, Morishita Y, Sakurai N, Suzuki H, Shibata D, Shimada T (2012) *Agrobacterium*-mediated transformation of *Eucalyptus globulus* using explants with shoot apex with introduction of bacterial choline oxidase gene to enhance salt tolerance. Plant Cell Rep 31:225–235

Mayol M, Riba M, Cavers S, Grivet D, Vincenot L, Cattonaro F, Vendramin GG, González-Martínez SC (2019) A multiscale approach to detect selection in non model tree species: widespread adaptation despite population decline in *Taxus baccata* L. Evol Appl 13:143–160

McCown BH, McCabe DE, Russell DR, Robinson DJ, Barton KA, Raffa KF (1991) Stable transformation of *Populus* and incorporation of pest resistance by electric discharge particle acceleration. Plant Cell Rep 9:590–594

Meek MH, Larson WA (2019) Amplicon sequencing and sequence capture usher in the conservation genomics era. Mol Ecol Resour 19:795–803

Metje-Sprink J, Menz J, Modrzejewski D, Sprink T (2019) DNA-free genome editing: past, present and future. Front Plant Sci 9:1957

Mette MF, Aufsatz W, van der Winden J, Matzke MA, Matzke AJM (2000) Transcriptional silencing and promoter methylation triggered by double-stranded RNA. EMBO J 19:5194–5201

Meuwissen TH, Hayes BJ, Goddard ME (2001) Prediction of total genetic value using genome-wide dense marker maps. Genetics 157:1819–1829

Meyer FE, Shuey LS, Naidoo S, Mamni T, Berger DK, Myburg AA, Van den Berg N, Naidoo S (2016) Dual RNA-sequencing of *Eucalyptus nitens* during *Phytophthora cinnamomi* challenge reveals pathogen and host factors influencing compatibility. Front Plant Sci 7:191

Miranda AC, da Silva PHM, Moraes MLT, Lee DJ, Sebbenn AM (2019) Investigating the origin and genetic diversity of improved *Eucalyptus grandis* populations in Brazil. Forest Ecol Manag 448:130–138

Moniodis J, Jones CG, Barbour EL, Plummer JA, Ghisalberti EL, Bohlmann J (2015) The transcriptome of sesquiterpenoid biosynthesis in heartwood xylem of Western Australian sandalwood (*Santalum spicatum*). Phytochemistry 113:79–86

Montanari S, Bianco L, Allen BJ, Martínez-García PJ, Bassil NV, Postman J, Knäbel M, Kitson B, Deng CH, Chagné D, Crepeau MW, Langley CH, Evans K, Dhingra A, Troggio M, Neale DB (2019) Development of a highly efficient Axiom™ 70 K SNP array for *Pyrus* and evaluation for high-density mapping and germplasm characterization. BMC Genomics 20:331

Mora F, Arriagada O, Ballesta P, Ruiz E (2017) Genetic diversity and population structure of a drought-tolerant species of *Eucalyptus*, using microsatellite markers. J Plant Biochem Biotechnol 26:274–281

Mori H, Ueno S, Ujino-Ihara T, Fujiwara T, Yamashita K, Kanetani S, Endo R, Matsumoto A, Uchiyama K, Matsui Y, Yoshida T, Sakai Y, Moriguchi Y, Kusano R, Tsumura R (2019) Mapping quantitative trait loci for growth and wood property traits in *Cryptomeria japonica* across multiple environments. Tree Genet Genomes 15:43

Mori H, Yamashita K, Saiki ST, Matsumoto A, Ujino-Ihara T (2020) Climate sensitivity of *Cryptomeria japonica* in two contrasting environments: perspectives from QTL mapping. PLoS One 15(1): e0228278

Mosca E, Cruz F, Gómez-Garrido J, Bianco L, Rellstab C, Brodbeck S, Csilléry K, Fady B, Fladung M, Fussi B, Gömöry D (2019) A reference genome sequence for the European silver fir (*Abies alba* Mill.): a community-generated genomic resource. G3 9(7): 2039–2049

Mouradov A, Sawbridge T, Hamdorf B, Glassick T, Murphy L, Marla S, Yang Y, Teasdale RD, Drew RA (1998) Genetic engineering of reproductive sterility in *Pinus radiata*. Acta Hortic 461:417–426. https://doi.org/10.17660/ActaHortic.1998.461.48

Müller BSF, de Almeida Filho JE, Lima BM, Garcia CC, Missiaggia A, Aguiar AM, Takahashi E, Kirst M, Gezan SA, Silva-Junior OB, Neves LG, Grattapaglia D (2019a) Independent and Joint-GWAS for growth traits in *Eucalyptus* by assembling genome-wide data for 3373 individuals across four breeding populations. New Phytol 221: 818–833

Müller NA, Kersten B, Fladung M, Schroeder H (2019b) RNA-seq of eight different poplar clones reveals conserved up-regulation of gene expression in response to insect herbivory. BMC Genomics 20(1): 1–10

Myburg AA, Grattapaglia D, Tuskan GA, Hellsten U, Hayes RD, Grimwood J, Jenkins J, Lindquist E, Tice H, Bauer D, Goodstein DM (2014) The genome of *Eucalyptus grandis*. Nature 510(7505): 356–362

Myburg AA, Hussey SG, Wang JP, Street NR, Mizrachi E (2019) Systems and synthetic biology of forest trees: a bioengineering paradigm for woody biomass feedstocks. Front Plant Sci 10:775

Nagle M, Déjardin A, Pilate G, Strauss SH (2018) Opportunities for innovation in genetic transformation of forest trees. Front Plant Sci 9:1443

Neale DB, McGuire PE, Wheeler NC, Stevens KA, Crepeau MW, Cardeno C, Zimin AV, Puiu D, Pertea GM, Sezen UU, Casola C (2017) The Douglas-fir genome sequence reveals specialization of the photosynthetic apparatus in Pinaceae. G3 7(9):3157–3167

Neale DB, Wegrzyn JL, Stevens KA, Zimin AV, Puiu D, Crepeau MW, Cardeno C, Koriabine M, Holtz-Morris AE, Liechty JD, Martínez-García PJ (2014) Decoding the massive genome of loblolly pine using haploid DNA and novel assembly strategies. Genome Biol 15(3):1–13

Neophytou C, Konnert M, Fussi B (2019) Western and eastern post-glacial migration pathways shape the genetic structure of sycamore maple (*Acer pseudoplatanus* L.) in Germany. Forest Ecol Manag 432:83–93

Newhouse AE, Polin-McGuigan LD, Baier KA, Valletta KER, Rottmann WH, Tschaplinski TJ, Maynard CA, Powell WA (2014) Transgenic American chestnuts show enhanced blight resistance and transmit the trait to T1 progeny. Plant Sci 228:88–97

Niu J, Hou X, Fang C, An J, Ha D, Qiu L, Ju Y, Zhao H, Du WZ, Qi J, Zhang Z, Liu G, Lin S (2015) Transcriptome analysis of distinct *Lindera glauca* tissues revealed the differences in the unigenes related to terpenoid biosynthesis. Gene 559(1):22–30

Nkongolo KK, Michael P, Demers T (2005) Application of ISSR, RAPD and cytological markers to the certification of *Picea mariana*, *P. glauca* and *P. engelmannii* trees, and their putative hybrids. Genome 48:302–311

Nystedt B, Street NR, Wetterbom A, Zuccolo A, Lin YC, Scofield DG, Vezzi F, Delhomme N, Giacomello S, Alexeyenko A, Vicedomini R (2013) The Norway spruce genome sequence and conifer genome evolution. Nature 497(7451):579–584

Oates CN, Külheim C, Myburg AA, Slippers B, Naidoo S (2015) The transcriptome and terpene profile of *Eucalyptus grandis* reveals mechanisms of defense against the insect pest, *Leptocybe invasa*. Plant Cell Physiol 56(7):1418–1428

Ogita S, Uefuji H, Yamaguchi Y, Koizumi N, Sano H (2003) Producing decaffeinated coffee plants. Nature 423:823

Omondi SF, Kireger E, Dangasuk OG, Chikamai B, Odee DW, Cavers S, Khasa DP (2010) Genetic diversity and population structure of *Acacia Senegal* (L) Willd. in Kenya. Tropical Plant Biol 3: 59–70

Osakabe Y, Liang Z, Ren C, Nishitani C, Osakabe K, Wada M, Komori S, Malnoy M, Velasco R, Poli M, Jung MH (2018) CRISPR-Cas9-mediated genome editing in apple and grapevine. Nat Protoc 13:2844–2863

Padovan A, Keszei A, Foley WJ, Külheim C (2013) Differences in gene expression within a striking phenotypic mosaic *Eucalyptus* tree that varies in susceptibility to herbivory. BMC Plant Biol 13:29

Parchman TL, Jahner JP, Uckele KA, Galland LM, Eckert AJ (2018) RADseq approaches and applications for forest tree genetics. Tree Genet Genom 14(3):39

Philippe RN, Ralph SG, Mansfield SD, Bohlmann J (2010) Transcriptome profiles of hybrid poplar (*Populus trichocarpa* × *deltoides*) reveal rapid changes in undamaged, systemic sink leaves after simulated feeding by forest tent caterpillar (*Malacosoma disstria*). New Phytol 188(3):787–802

Pilate G, Guiney E, Holt K, Petit-Conil M, Lapierre C, Leplé J-C, Pollet B, Mila I, Webster EA, Marstorp HG, Hopkins DW, Jouanin L, Boerjan W, Schuch W, Cornu D, Halpin C (2002) Field and pulping performances of transgenic trees with altered lignification. Nat Biotechnol 20:607–612

Plasencia A, Soler M, Dupas A, Ladouce N, Silva-Martins G, Martinez Y, Lapierre C, Franche C, Truchet I, Grima-Pettenati J (2016) *Eucalyptus* hairy roots, a fast, efficient and versatile tool to explore function and expression of genes involved in wood formation. Plant Biotechnol J 14:1381–1393

Plomion C, Aury JM, Amselem J, Alaeitabar T, Barbe V, Belser C et al (2016) Decoding the oak genome: public release of sequence data, assembly, annotation and publication strategies. Mol Ecol Resour 16:254–265

Plomion C, Aury JM, Amselem J, Leroy T, Murat F, Duplessis S, Faye S, Francillonne N, Labadie K, Le Provost G, Lesur I (2018) Oak genome reveals facets of long lifespan. Nat Plants 4:440–452

Polle A, Chen SL, Eckert C, Harfouche A (2019) Engineering drought resistance in forest trees. Front Plant Sci 9:1875

Poltri MSN, Zelener N, Traverso RJ, Gelid P, Hopp H (2003) Selection of a seed orchard of *Eucalyptus dunnii* based on genetic diversity criteria calculated using molecular markers. Tree Physiol 23:625–632

Porth I, El-Kassaby YA (2014) Assessment of the genetic diversity in forest tree populations using molecular markers. Diversity 6:283–295

Pot D, Rodrigues JC, Rozenberg P, Chantre G, Tibbits J, Cahalan C, Pichavant F, Plomion C (2006) QTLs and candidate genes for wood properties in maritime pine (*Pinus pinaster* Ait.). Tree Genet Genomes 2:10–24

Prashanth SR, Sadhasivam V, Parida A (2008) Over expression of cytosolic copper/zinc superoxide dismutase from a mangrove plant

Avicennia marina in indica rice var Pusa Basmati-1 confers abiotic stress tolerance. Transgenic Res 17:281–291

Price DRG, Gatehouse JA (2008) RNAi-mediated crop protection against insects. Trends Biotechnol 26:393–400

Qiu Q, Ma T, Hu Q, Liu B, Wu Y, Zhou H, Wang Q, Wang J, Liu J (2011) Genome-scale transcriptome analysis of the desert poplar, *Populus euphratica*. Tree Physiol 31(4):452–461

Rigault P, Spokevicius A, Merchant A, Najera J, Tibbits J (2012) Generation of a *Eucalyptus globulus* reference genome and gene catalogue. In: Plant and animal genome (PAG) 2012 conference, p P0522

Rothmann WH, Meilan R, Sheppard LA, Brunner AM, Skinner JS, Ma C, Cheng S, Jouanin L, Pilate G, Strauss SH (2000) Diverse effects of overexpression of LEAFY and PTLF, a poplar (*Populus*) homolog of LEAFY/FLORICAULA, in transgenic poplar and *Arabidopsis*. Plant J 22:235

Rugh CI, Senecoff RB, Meagher RB, Merkle SA (1998) Development of transgenic yellow poplar for mercury phytoremediation. Nat Biotechnol 16:925–928

Salazar MM, Nascimento LC, Camargo E, Vidal RO, Lepikson-Neto J, Goncalves DC, Marques WL, Teixeira P, Pereira G (2011) Comparative transcriptome analysis of tree *Eucalyptus* species using RNAseq technology: analysis of genes interfering in wood quality aspects. BMC Proc 5(Suppl 7):P175

Sang M, Shi H, Wei K, Ye M, Jiang L, Sun L, Wu R (2019) A dissection model for mapping complex traits. Plant J 97:1168–1182

Sanou H, Angulo-Escalante MA, Martínez-Herrera J, Koné S (2015) Loss of genetic diversity of *Jatropha curcas* L. through domestication: implications for its genetic improvement. Crop Sci 55:749–759

Semerikov V, Semerikova S, Putintseva J, Tarakanov V, Tikhonova I, Vidyakin A, Oreshkova N, Krutovsky K (2018) Colonization history of Scots pine in Eastern Europe and North Asia based on mitochondrial DNA variation. Tree Genet Genomes 14:1–7

Shani Z, Dekel M, Roiz L, Horowitz M, Kolosovski N, Lapidot S, Alkan S, Koltai H, Tsabary G, Goren R, Shoseyov O (2006) Expression of endo-1,4-β-glucanase (cel1) in *Arabidopsis thaliana* is associated with plant growth, xylem development and cell wall thickening. Plant Cell Rep 25:1067–1061

Shani Z, Dekel M, Tsabary G, Goren R, Shoseyov O (2004) Growth enhancement of transgenic poplar plants by overexpression of *Arabidopsis thaliana* endo-1,4-glucanase (*cel*1). Mol Breed 14: 321–330

Shinya T, Iwata E, Nakahama K, Fukuda Y, Hayashi K, Nanto K, Rosa AC, Kawaoka A (2016) Transcriptional profiles of hybrid *Eucalyptus* genotypes with contrasting lignin content reveal that monolignol biosynthesis-related genes regulate wood composition. Front Plant Sci 7:443

Simeone MC, Piredda R, Papini A, Vessella F, Schirone B (2013) Application of plastid and nuclear markers to DNA barcoding of Euro–Mediterranean oaks (Quercus, Fagaceae): problems, prospects and phylogenetic implications. Bot J Linn Soc 172:478–499

Sivakumar S, Rajagopal R, Raja Venkatesh G, Srivastava A, Bhatnagar RK (2007) Knockdown of aminopeptidase-N from *Helicoverpa armigera* larvae and in transfected Sf21 cells by RNA interference reveals its functional interaction with *Bacillus thuringiensis* insecticidal protein Cry1Ac. J Biol Chem 282:7312–7319

Skyba O, Douglas CJ, Mansfield SD (2013) Syringyl-rich lignin renders poplars more resistant to degradation by wood decay fungi. Appl Environ Microbiol 79:2560–2571

Slavov GT, DiFazio SP, Martin J, Schackwitz W, Muchero W, Lipphardt R-ME, M. F., Pennacchio, C. P., Hellsten, U. and Pennacchio, L. A. (2012) Genome resequencing reveals multiscale geographic structure and extensive linkage disequilibrium in the forest tree *Populus trichocarpa*. New Phytol 196:713–725

Sollars ES, Harper AL, Kelly LJ, Sambles CM, Ramirez-Gonzalez RH, Swarbreck D, Kaithakottil G, Cooper ED, Uauy C, Havlickova L, Worswick G (2017) Genome sequence and genetic diversity of European ash trees. Nature 541(7636):212–216

Spokevicius AV, Van Beveren K, Leitch MA, Bossinger G (2005) *Agrobacterium*-mediated *in vitro* transformation of wood-producing stem segments in eucalypts. Plant Cell Rep 23:617–624

Stewart JJ, Akiyama T, Chapple C, Ralph J, Mansfield SD (2009) The effects on lignin structure of overexpression of ferulate 5-hydroxylase in hybrid poplar. Plant Physiol 150:621–635

Subashini V, Bachpai VKW, Mayavel A, Nagarajan B, Sivakumar V, Yasodha R (2018) Quantitative trait loci (QTL) for salinity tolerance traits in interspecific hybrids of *Eucalyptus*. Indian J Plant Physiol 23:822–832

Sun YH, Shi R, Zhang XH, Chiang VL, Sederoff RR (2012) MicroRNAs in trees. Plant Mol Biol 80(1):37–53

Sundell D, Street NR, Kumar M, Mellerowicz EJ, Kucukoglu M, Johnsson C, Kumar V, Mannapperuma C, Delhomme N, Nilsson O, Tuominen H (2017) AspWood: high-spatial-resolution transcriptome profiles reveal uncharacterized modularity of wood formation in *Populus tremula*. Plant Cell 29(7):1585–1604

Suontama M, Klápště J, Telfer E, Graham N, Stovold T, Low C, McKinley R, Dungey H (2019) Efficiency of genomic prediction across two *Eucalyptus nitens* seed orchards with different selection histories. Heredity 122:370–379

Svistoonoff S, Benabdoun FM, Nambiar-Veetil M, Imanishi L, Vaissayre V, Cesari S, Diagne N, Hocher V, Billy F, Bonneau J, Wall L, Ykhelf N, Rosenberg C, Bogusz S, Franche C, Gherbi H (2013) The independent acquisition of plant root nitrogen-fixing symbiosis in Fabids recruited the same genetic pathway for nodule organogenesis. PLoS One 8:5

Svitashev S, Schwartz C, Lenderts B, Young JK, Cigan AM (2016) Genome editing in maize directed by CRISPR–Cas9 ribonucleoprotein complexes. Nat Commun 7:13274

Tang W, Tian Y (2003) Transgenic loblolly pine (*Pinus taeda* L.) plants expressing a modified δ-endotoxin gene of *bacillus thuringiensis* with enhanced resistance to *Dendrolimus punctatus* Walker and *Crypyothelea formosicola* Staud. J Exp Bot 54:835–844

Terenius O, Papanicolaou A, Garbutt JS (2011) RNA interference in Lepidoptera: an overview of successful and unsuccessful studies and implications for experimental design. J Insect Physiol 57:231–245

Thakur VV, Tiwari S, Tripathi N, Tiwari G, Sapre S (2016) DNA barcoding and phylogenetic analyses of *Mentha* species using *rbcL* sequences. Ann Phytomed 5:59–62

Thanananta N, Vuttipongchaikij S, Apisitwanich S (2018) *Agrobacterium*-mediated transformation of a *Eucalyptus camaldulensis* x *E. tereticornis* hybrid using peeled nodal-stem segments with yeast HAL2 for improving salt tolerance. New For 49:311–327

Thistlethwaite FR, Ratcliffe B, Klápště J, Porth I, Chen C, Stoehr MU, El-Kassaby YA (2019) Genomic selection of juvenile height across a single-generational gap in *Douglas-fir*. Heredity 122:848–863

Thumma BR, Nolan MR, Evans R, Moran GF (2005) Polymorphisms in *cinnamoyl CoA reductase* (*CCR*) are associated with variation in microfibril angle in *Eucalyptus* spp. Genetics 171:1257–1265

Thumma BR, Sharma N, Southerton SG (2012) Transcriptome sequencing of *Eucalyptus camaldulensis* seedlings subjected to water stress reveals functional single nucleotide polymorphisms and genes under selection. BMC Genomics 13(1):364

Thumma BR, Southerton SG, Bell JC, Owen JV, Henery ML, Moran GF (2010) Quantitative trait locus (QTL) analysis of wood quality traits in *Eucalyptus nitens*. Tree Genet Genomes 6:305–317

Tran NHT, Oguchi T, Akatsuka N, Matsunaga E, Kawaoka A, Yamada A, Ozeki Y, Watanabe KN, Kikuchi A (2019) Development and evaluation of novel salt-tolerant *Eucalyptus* trees by molecular breeding using an RNA-binding-protein gene derived from common ice plant (*Mesembryanthemum crystallinum* L.). Plant Biotechnol J 17:801–811

Tran NHT, Oguchi T, Akatsuka N, Matsunaga E, Kawaoka Watanabe KN, Kikuchi A (2018) Transcriptional enhancement of a bacterial *choline oxidase A* gene by an *HSP* terminator improves the glycine betaine production and salinity stress tolerance of *Eucalyptus camaldulensis* trees. Plant Biotechnol 35:215–224

Tripathi SB, Mathish NV, Gurumurthi K (2006) Use of genetic markers in the management of micropropagated *Eucalyptus* germplasm. New For 31:361–372

Trujillo-Moya C, Ganthaler A, Stöggl W, Kranner I, Schüler S, Ertl R, Schlosser S, George J-P, Mayr S (2020) RNA-Seq and secondary metabolite analyses reveal a putative defence-transcriptome in Norway spruce (*Picea abies*) against needle bladder rust (*Chrysomyxa rhododendri*) infection. BMC Genomics 21:1–21

Tsai CJ, Popko JL, Mielke MR, Hu WJ, Podila GK, Chiang VL (1998) Suppression of O-methyl transferase gene by homologous sense transgene in quaking aspen causes red-brown wood phenotypes. Plant Physiol 117:101–112

Tuskan GA, DiFazio S, Jansson S, Bohlmann J, Grigoriev I, Hellsten U, Putnam N, Ralph S, Rombauts S, Salamov A, Schein J (2006) The genome of black cottonwood, *Populus trichocarpa*. (Torr. & Gray). Science 313(5793):1596–1604

Ueno S, Klopp C, Leplé JC, Derory J, Noirot C, Léger V, Prince E, Kremer A, Plomion C, Le Provost G (2013) Transcriptional profiling of bud dormancy induction and release in oak by next-generation sequencing. BMC Genomics 14:236

Van Beveren K, Spokevicius A, Tibbits J, Wang Q, Bossinger G (2006) Transformation of cambial tissue *in vivo* provides an efficient means for induced somatic sector analysis and gene testing in stems of woody plant species. Funct Plant Biol 33:629–638

Van Doorsselaere J, Baucher M, Chognot E, Chabbert B, Tollier MT, Petit-Conil M, Leplé JC, Pilate G, Cornu D, Monties B, van Montagu M, Inzé D, Boerjan W, Jouanin L (1995) A novel lignin in poplar trees with a reduced caffeic acid/ferulic acid O-methyl transferase activity. Plant J 8:855–864

Vikashini B, Shanthi A, Dasgupta MG (2018) Identification and expression profiling of genes governing lignin biosynthesis in *Casuarina equisetifolia L.* Gene 676:37–46

Villar E, Klopp C, Noirot C, Novaes E, Kirst M, Plomion C, Gion JM (2011) RNA-Seq reveals genotype-specific molecular responses to water deficit in *Eucalyptus*. BMC Genomics 12(1):538

Vining KJ, Romanel E, Jones RC, Klocko A, Alves-Ferreira M, Hefer CA, Amarasinghe V, Dharmawardhana P, Naithani S, Ranik M, Wesley-Smith J (2015) The floral transcriptome of *Eucalyptus grandis*. New Phytol 206(4):1406–1422

Voelker SL, Lachenbruch B, Meinzer FC, Kitin P, Strauss SH (2011) Transgenic poplars with reduced lignin show impaired xylem conductivity, growth efficiency and survival. Plant Cell Environ 34: 655–668

Wang N, Thomson M, Bodles WJ, Crawford RM, Hunt HV, Featherstone AW, Pellicer J, Buggs RJ (2013) Genome sequence of dwarf birch (*Betula nana*) and cross-species RAD markers. Mol Ecol 22(11):3098–3111

Wang W, Das A, Kainer D, Schalamun M, Morales-Suarez A, Schwessinger B, Lanfear R (2020) The draft nuclear genome assembly of *Eucalyptus pauciflora*: a pipeline for comparing *de novo* assemblies. GigaScience 9(1):giz160

Wang Z, Kang M, Liu H, Gao J, Zhang Z, Li Y, Wu R, Pang X (2014) High-level genetic diversity and complex population structure of siberian apricot (*Prunus sibirica* L.) in China as revealed by nuclear SSR markers. PLoS One 9(2):e87381

Waterhouse PM, Helliwell CA (2003) Exploring plant genomes by RNA-induced gene silencing. Nat Rev 4:29–38

Wegrzyn JL, Staton MA, Street NR, Main D, Grau E et al (2019) Cyberinfrastructure to improve forest health and productivity: the role of tree databases in connecting genomes, phenomes, and the environment. Front Plant Sci 10:813

Wei S, Yang Y, Yin T (2020) The chromosome-scale assembly of the willow genome provides insight into Salicaceae genome evolution. Hortic Res 7(1):1–12

Westbrook JW, Zhang Q, Mandal MK, Jenkins EV, Barth LE, Jenkins JW, Grimwood J, Schmutz J, Holliday JA (2019) Genomic selection analyses reveal trade off between chestnut blight tolerance and genome inheritance from American chestnut (*Castanea dentata*) in (*C. dentata* × *C. mollissima*) × *C. dentata* backcross populations. bioRxiv. https://doi.org/10.1101/690693

Wilkerson CG, Mansfield SD, Lu F, Withers S, Park JY, Karlen SD, Gonzales-Vigil E, Padmakshan D, Unda F, Rencoret J, Ralph J (2014) Mono lignol ferulate transferase introduces chemically labile linkages into the lignin backbone. Science 344:90–93

Wu Z (2015) The new completed genome of purple willow (*Salix purpurea*) and conserved chloroplast genome structure of Salicaceae. J Nat Sci Res 1:e49

Xing Y, Liu Y, Zhang Q, Nie X, Sun Y, Zhang Z, Li H, Fang K, Wang G, Huang H, Bisseling T (2019) Hybrid de novo genome assembly of Chinese chestnut (*Castanea mollissima*). GigaScience 8(9):giz112

Yao W, Zhao K, Cheng Z, Li X, Zhou B, Jiang T (2018) Transcriptome analysis of poplar under salt stress and over-expression of transcription factor NAC57 gene confers salt tolerance in transgenic *Arabidopsis*. Front Plant Sci 9:1121

Yasodha R, Vasudeva R, Balakrishnan S, Sakthi AR, Abel N, Binai N, Rajashekar B, Bachpai VKW, Pillai C, Dev SA (2018) Draft genome of a high value tropical timber tree, Teak (*Tectona grandis* L. f.): insights into SSR diversity, phylogeny and conservation. DNA Res 25(4):409–419

Ye G, Zhang H, Chen B, Nie S, Liu H, Gao W, Wang H, Gao Y, Gu L (2019) *De novo* genome assembly of the stress tolerant forest species *Casuarina equisetifolia* provides insight into secondary growth. Plant J 97(4):779–794

Zhang B, Oakes AD, Newhouse AE, Baier KM, Maynard CA, Powell WA (2013) A threshold level of oxalate oxidase transgene expression reduces *Cryphonectria parasitica*-induced necrosis in a transgenic American chestnut (*Castanea dentata*) leaf bioassay. Transgenic Res 22:973–982

Zhang B, Zhu W, Diao S, Wu X, Lu J, Ding C, Su X (2019) The poplar pangenome provides insights into the evolutionary history of the genus. Commun Biol 2(1):1–8

Zhang J, Xie M, Tuskan GA, Muchero W, Chen JG (2018) Recent advances in the transcriptional regulation of secondary cell wall biosynthesis in the woody plants. Front Plant Sci 9:1535

Zhang X, Da Silva JAT, Niu M, Li M, He C, Zhao J, Zeng S, Duan J, Ma G (2017) Physiological and transcriptomic analyses reveal a response mechanism to cold stress in *Santalum album L.* leaves. Sci Rep 7:42165

Zhao D, Hamilton JP, Bhat WW, Johnson SR, Godden GT, Kinser TJ, Boachon B, Dudareva N, Soltis DE, Soltis PS, Hamberger B (2019) A chromosomal-scale genome assembly of *Tectona grandis* reveals the importance of tandem gene duplication and enables discovery of genes in natural product biosynthetic pathways. Gigascience 8(3): giz005

Zhong R, Cui D, Ye ZH (2019) Secondary cell wall biosynthesis. New Phytol 221(4):1703–1723

Zhou Z, Ma H, Lin K, Zhao Y, Chen Y, Xiong Z, Wang L, Tian B (2015) RNA-seq reveals complicated transcriptomic responses to drought stress in a nonmodel tropic plant, *Bombax ceiba* L. Evol Bioinforma 11(Suppl 1):27–37

Zhu C, Shi F, Chen Y, Wang M, Zhao Y, Geng G (2019) Transcriptome analysis of Chinese chestnut (*Castanea mollissima* Blume) in response to *Dryocosmus kuriphilus* Yasumatsu infestation. Int J Mol Sci 20(4):855

Part VI

Forest Protection

34 Fire Science for Indian Forests: From Theory to Practice

Narendran Kodandapani

Abstract

India has four of the 34 global hotspots of biodiversity; the Western Ghats, being one of these four, is the hottest of the global biodiversity hotspots due to the high level of endemism as well as human pressures. Disturbances such as wildfires have been a part of these ecosystems for several millennia, initially used during slash and burn agriculture and later for maintaining various ecosystem services. Fires are mainly caused by humans, during the collection of minor forest produce; set by pastoralists, and also due to incendiarism. Fires are a natural part of the ecosystem and are important for recycling nutrients back into the soil, checking the proliferation of invasive species and pests, and for maintaining the functioning of ecosystems. Climate change, land use and land cover change, and fire management practices have substantially altered fire regimes in the Western Ghats. Climatic extremes such as droughts and heat waves have exacerbated disastrous fires in different parts of the globe and similar effects have been witnessed in the Western Ghats. Human-caused land use land cover changes and fire suppression have also resulted in changes in fire regimes in most parts of world and similarly in India. However, despite the importance of wildfires in maintaining ecosystem characteristics, and the increasing links between climate change and increasing wildfire activity, little is known about wildfires in the Western Ghats of India, and it continues to be an under studied topic. Here, we provide a synoptic overview of wildfires, its origins, causes, its ecological characteristics, through the lens of the fire regime, in a focal study area, the Nilgiris landscape, in the Western Ghats.

Keywords

Wildfire · Fire effects · Fire climate · Human drivers · Fire management

34.1 Introduction

Global patterns of wildfire activity are characterized by contrasting features with some regions revealing a declining trend in burned areas. Recent changes in the global patterns of fire activity have been pronounced especially in savannas, where a pervasive decline in burned area has been observed during the last two decades (Andela et al. 2017; Kodandapani and Parks 2019; Zubkova et al. 2019; Jiang et al. 2020). Simultaneously, certain other regions of the world are characterized by an acceleration of the various components of the fire regime, including increasing trends in burned areas, such as in western United States, Europe, and Australia (Bowmann et al. 2017; Parks and Abatzoglou 2020). Two fundamental components of this contrasting pattern of global fire activity are climate and humans (Syphard et al. 2007; Syphard et al. 2017; Archibald et al. 2010; Abatzoglou et al. 2018; Jolly et al. 2015). Climate is often considered as the primary driver of fire activity through its effects on fire environment, fuel aridity, and indirectly through its effects on productivity and fuel loads (Abatzoglou and Kolden 2013; Abatzoglou et al. 2018; Vadrevu et al. 2019). In moist forests, drought conditions during the fire season contribute to enhanced fire activity, whereas in arid ecosystems and savannas, antecedent climate conditions contribute to fire activity through their impacts on fuel loads (Riley et al. 2013; Kelley et al. 2019).

Similar to climate, humans also alter fire activity through either dampening or accentuating the fire signal (Lasslop and Kloster 2017). They influence fire activity through changes in the timing, spatial pattern and frequency of ignition sources; likewise, humans contribute to changes in fuel structure, fuel loads, and composition through land use and landcover

N. Kodandapani (✉)
Center for Advanced Spatial and Environmental Research, Bengaluru, India

Centre for Wildlife Studies, CWS, Bengaluru, India

A. K. Mandal, A. Nicodemus (eds.), *Textbook of Forest Science*, https://doi.org/10.1007/978-981-97-8289-5_34

changes (McLauchlan et al. 2020; Guiterman et al. 2019). Importantly, fire suppression and prescribed burning also significantly alter fire regimes through fire exclusion or inclusion. Humans have also significantly increased the length of the fire-weather season by almost 20% between 1970 and 2013 (Jolly et al. 2015). However, synergies between climate and human effects are common, and it is likely that, the magnitude of each control may vary depending on the interactions between the two at various spatial and temporal scales.

Simultaneously, wildfires are annual disturbance events in several ecosystems across the globe and are important for maintaining ecological diversity and various ecological processes such as nutrient recycling and equilibrium between woody encroachment into grass layer in savannas (Bond and Keeley 2005; Abatzoglou and Williams 2016). Although fire is important for maintaining ecosystem function and diversity, continued declines in burned areas, as reported (Kodandapani and Parks 2019) may have negative implications for conservation of biodiversity in savanna ecosystems. Fires occur frequently in many parts of India (Fig. 34.1), especially, along the chir-pine forests, in the north-east, and also in the deciduous forests, in central and southern India. In pine forests, the needles are extremely flammable due to their chemical makeup and since the accumulated leaf litter of pines takes a long time to decompose. In southern India, the leaf litter and grass fuels generated by many grasses such as *Themeda cymbaria* and *T. triandra* contribute significantly to the combustible material of these forests. Over 90% of all fires are anthropogenic in origin, caused due to various land management practices in the region. Some fires also occur due to human negligence, like fires set on farmlands to clear agricultural residues which escape into forests (Kodandapani et al. 2004).

Fig. 34.1 Raging wildfires in the Nilgiris landscape

34.1.1 Fire in Context: Land Use and Burning in the Western Ghats

Humans have modified vegetation in the Western Ghats with fire for several millennia (Gadgil and Chandran 1988). Low population densities and longer fire-rotation intervals, during earlier times, permitted the return of forests compared to the present (Morrison 1994; Chandran 1997; Kodandapani et al. 2008). Fire was the first tool used by humans to significantly alter the ecological history of the Western Ghats (Chandran 1997; Pascal 1986). Fire has been fundamental in shaping the environments of the Ghats, as well as in the development of ecosystem structure, species diversity, and distribution of biomass in various forest types. Frequent burning during pre-colonial times and subsequent fire suppression have significantly altered various ecological attributes of forests in the Western Ghats (Chandran 1997). Currently, in mesic and xeric savannas of India, fire is an almost annual event; fire-return intervals are typically 1–3 years, depending on a number of factors such as drought conditions, ignition sources, and fire management by the forest departments (Kodandapani et al. 2008). Figure 34.2 shows the distribution and frequency of fires in the Western Ghats, obtained from MODIS thermal anomalies dataset (Kodandapani and Parks 2019). Despite the frequent occurrence of fires, information on important aspects of the fire regime is lacking. In this chapter, we present information on the forest fire regime from a savanna ecosystem in India, the Nilgiris landscape (Kodandapani and Parks 2019).

34.2 The Fire Triangle and the Combustion Process

In order to understand fire occurrence, the fundamental aspect to begin with would be the fire process or the combustion process. At the very fundamental level, fire occurrence depends on three very important factors: a fuel source that will provide the energy, a heat source, and oxygen. These three factors constitute what is popularly referred to as the "triangle of combustion" or "fire triangle" (Stott 2000). However, to sustain the chemical reactions in the combustion process it will require the right mix of all the three factors. In order for combustion to occur, there is a need for an ignition source. In Indian ecosystems, there are both natural and human-caused ignitions. Mean flash rate is low and about 4 to 20 $km^{-2}y^{-1}$ in the Western Ghats (see website http://ghrc.msfc.nasa.gov); flash strikes are also accompanied by copious rain, which precludes fire occurrence. For sustained ignition, the fuel moisture, fuel temperature, relative humidity, ambient temperatures, along with oxygenation should be in suitable amounts (Pyne et al. 1996). Lightening-caused fires occur in several parts of the world and are important causes of fire in African savannas and in boreal and temperate ecosystems in the United States, Canada, and Russia. On average, about 11,912 lightning-caused fires were detected in the United States between 2001 and 2008 resulting in an average burned area of 1.8 M ha y^{-1} (see the National Interagency Fire website, http://www.nifc.gov/stats/). However, in the Western Ghats majority of the fires are due to anthropogenic sources (Kodandapani et al. 2008).

Changes in fuel moisture and surface dew are extremely important for the combustion process. Fuel moisture <30% is a critical threshold for the commencement of the combustion process (Pyne et al. 1996). Fine fuels in the Nilgiris landscape easily fall below this threshold during the fire season and hence are extremely vulnerable to fires. The fuel moisture content in savanna fuels can range from 5 to 200% of dry weight (Stott 2000). On applying an external heat source, drying of the fuel occurs and once this process removes the fuel moisture, the temperature of the fuel goes beyond a threshold called ignition temperature (about 300 °C). At the next stage, also called as the pyrolysis stage, the vegetation releases gaseous fuels rapidly which react with oxygen releasing abundance of heat and this stage is referred to as the flaming combustion process.

34.2.1 Components of the Fire Regime

The fire regime is unique to every ecosystem and defined by several parameters including the fire frequency, fire-return interval, seasonality, intensity, size, severity, and type of fire (Agee 2000; Flannigan et al. 2000; Bond and Keeley 2005; Cochrane and Ryan 2009). The frequency, intensity, and seasonality of fires depend on weather and climate apart from factors such as vegetation structure, composition, fuels, and topography (Dale et al. 2001). We describe below these components of the fire regime with reference to the Nilgiris landscape in the Western Ghats.

Fire Frequency It is the occurrence of fire for a given area and for a certain time period of interest (Bond and Keeley 2005). The frequency depends on several factors such as the fire behavior, climate, and sources of ignitions. The fire-return interval is the interval between fires which varies in savanna ecosystems; in tropical savannas of Africa, it is 1.33 years and in Asian savannas it is 2 years (Stott 2000). Although there are variations in the fire-return interval across the landscape, in general most savannas in the Western Ghats have fire-return intervals ≤2 years (Kodandapani et al. 2009). In general, most of the world's savannas burn at intervals of 1–3 years (Hoffmann et al. 2003).

Fire-Rotation Interval In contrast to the fire-return interval, the fire-rotation interval is the time required to burn an area

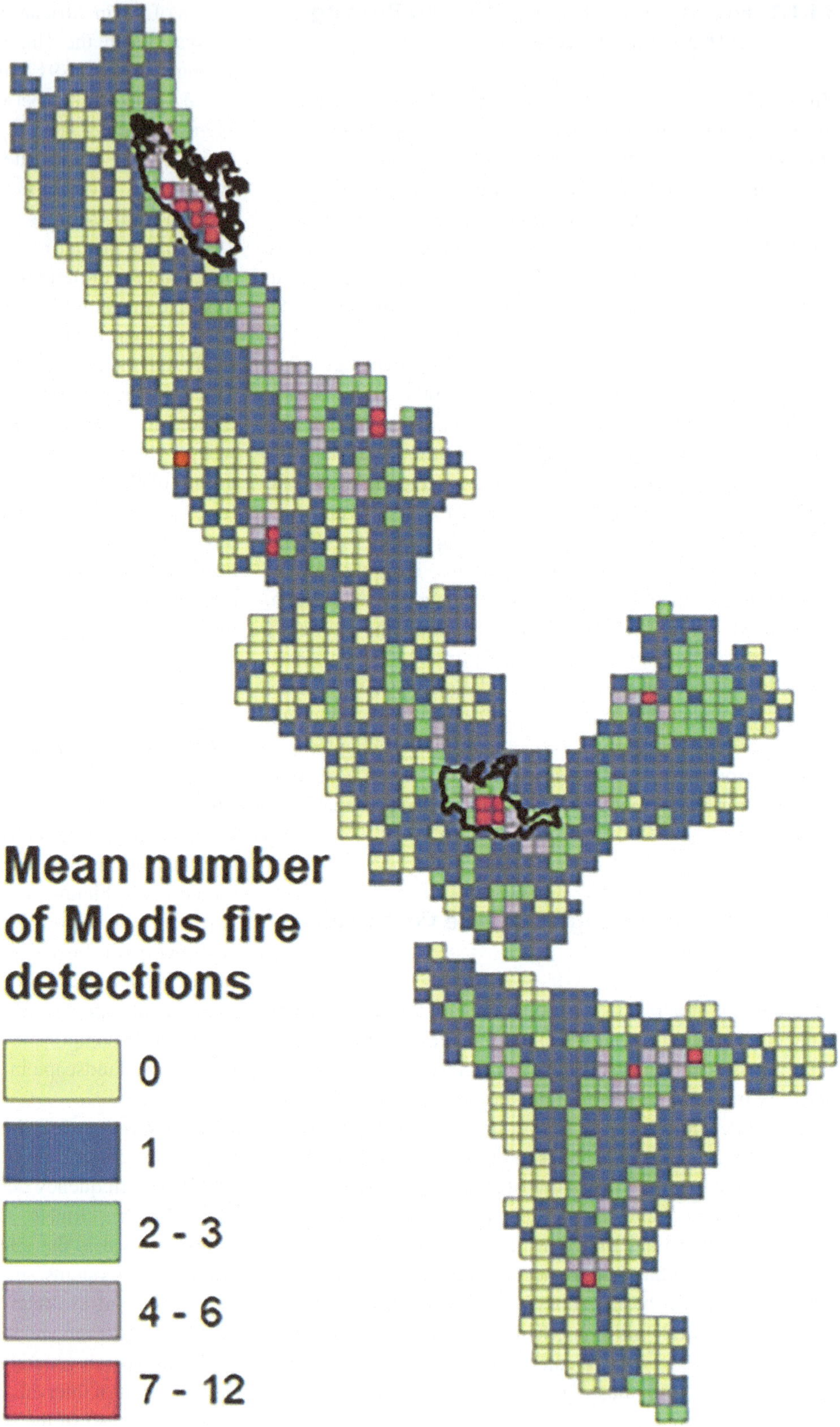

Fig. 34.2 Western Ghats I quarter (Jan–Mar) fire, figure shows the mean number of fires in (5′ latitude × 5′ longitude) pixels from 2001 to 2015

equal to the total forest area with the caveat that some areas would never burn, while some areas would burn more than once (Van Wagner 1978). For the Nilgiris landscape in the Western Ghats, the mean fire-rotational interval is about 7 years (Kodandapani et al. 2008). The fire-rotation interval in certain forests of the Western Ghats has increased three-fold; the mean burned area in 1910s in the Mudumalai wildlife sanctuary was about 10% which has increased to 30% during the 1990s (Kodandapani et al. 2004).

Fire Intensity It is the heat energy release per length of the fire front (Bond and Keeley 2005). Fire intensity in the Nilgiris landscape was estimated to be between 1.44 and 80 kWm^{-1}, with a mean of 25.3 kWm^{-1} (Kodandapani 2006). In the Amazon, studies have shown that the fire intensity ranges from 4 to 55 kWm^{-1} and can sometimes reach as high as 750 kWm^{-1} (Cochrane 2003). Similarly in the deciduous forests of Thailand, fire line intensities are between 11 and 25 kWm^{-1} (Baker and Bunyavejchewin 2009).

Fire Seasonality The fire season in India is spread between January and June; the large window is due to the tremendous spatial variations in climate, forest types, and ignitions in the country. For the Western Ghats, the fire season extends between January and April and fire occurrence depends on sources of ignitions as well as appropriate weather conditions especially fuel moisture conditions. Studies in the Western Ghats have shown a negative effect of seasonality of fire on the fruiting and population growth in a deciduous forest and this study specifically suggests that fire during the budding or flowering stages in the tree species *Phyllanthus emblica* could negatively impact fruit production subsequently (Sinha and Brault 2005).

Fire Severity Apart from fire intensity which is a measure of the heat release of a fire, fire severity can be related to the ecological effects of a fire. It commonly refers to the basal area removal by a fire and has been applied to map severity of fires in the United States (Agee 2000). Fire severity is generally thought to be high within temperate forests; however, it is also high among tropical forests despite the fact that fires are generally slow spreading and of low intensity (Cochrane 2003). The effects of fires in ecosystems are closely tied to several factors such as the tolerance of species to fires and adaptations that they possess to mitigate the effects of fires such as thick barks (Pinard and Huffman 1997).

34.2.2 Fire and Forest Fuels

Fine Fuel Loads In the Nilgiris landscape in India, the dry season begins during January and extends to April and <20% of the annual rainfall is received during these 4 months. The start of the dry season triggers several phenological changes especially the leaf shedding by several tree species in the deciduous forests. Close to about 90% of all leaf litter are completed by the end of January providing substantial leaf fuel load for the ensuing fires. Leaf litter loads can range from 0.2 $Mgha^{-1}$ in xeric savannas to 2.6 $Mgha^{-1}$ in mesic savannas. Once the leaves have fallen, the grass below is exposed to direct solar radiation and increased drying. This time of the year also experiences high temperatures—minimum temperature during the fire season is between 16.1 and 19.8 °C, and maximum temperature during the fire season is between 26.7 and 34 °C. Grass fuel loads can range from 0.34 $Mgha^{-1}$ in xeric savannas to 2.7 $Mgha^{-1}$ in mesic savannas.

Fuel Classes Apart from the fine fuels such as the grass and leaf litter in the forest, fuels can be classified based on their sizes which are in turn related to the ease with which they either lose or gain moisture. Specifically fuels are divided into 1-h, 10-h, 100-h, 1000-h fuel size classes, which correspond to sizes (0–0.62 cm, 0.62–2.54 cm, 2.54–7.62 cm, and >7.62 cm). The number reflects the amount of time required for a fuel particle of a given size to reach 63% of its fuel moisture content and come to equilibrium after a change in ambient moisture conditions (Cochrane 2003). For example, a 1-h time lag fuel at 10% fuel moisture in a drier environment, where the equilibrium fuel moisture content is at 5%, will be at about 7% fuel moisture after 1 h (Agee 1994). In the Western Ghats, there is a considerable variation in the total fuel loads (1-h, 10-h, 100-h, 1000-h) across the different forest types and the mean total fuel load in the mesic savannas ranges from 16 to 54 $Mgha^{-1}$; the mean total fuel load in the xeric savannas ranges from 3.6 to 6.3 $Mgha^{-1}$.

34.2.3 Fire and Forest Regeneration

Fire has a disproportionately large negative effect on the survival of seedlings, saplings, and juvenile individuals in a forest (Cochrane and Schulze 1999). Recent studies suggest declines in the density of seedlings and saplings with increasing fire frequency especially in mesic savannas (Kodandapani et al. 2008). For example in mesic savannas in the Nilgiris landscape, density of seedlings is significantly higher in the 0–5% mortality class compared to the 5–10% and >15% mortality classes. Figure 34.3 shows the differences in seedling density across fire severity classes in mesic savannas. The primary Y-axis (bar) shows the mean density of seedlings, and the secondary Y-axis (line) shows the coefficient of variation. Studies in a tropical dry deciduous forest in the Nilgiris landscape showed that frequent fires were the leading cause of mortality among woody plants,

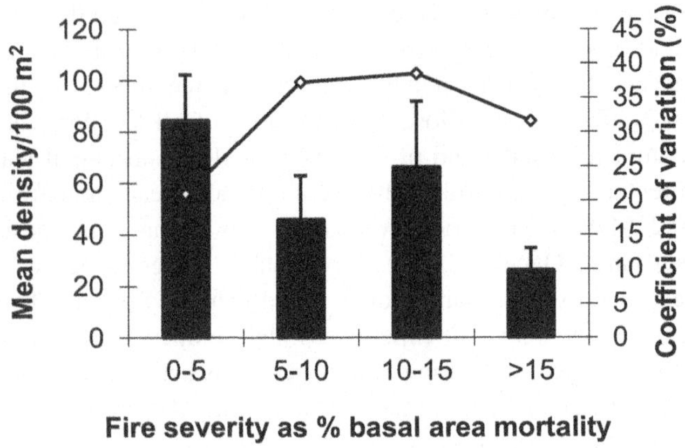

Fig. 34.3 Seedling density in different fire severity classes in the Nilgiris landscape

especially among lower size classes (1–10 cm dbh) (John et al. 2002). The frequent and high intensity fires in mesic savannas are leading to increased mortality of trees, and producing a woody population structure comprising of mature adult trees and an undergrowth of re-sprouting species as shown in other savannas (Whelan 1995).

34.2.4 Fire and Forest Structure

Survival from fire depends on the insulation of the meristematic tissue from heat. Bark thickness is an important characteristic that protects trees from cambial kill and it varies with species, diameter, tree age, distance from the ground, and site characters (Brown and Smith 2000). In certain species such as *Shorea obtusa* and *Dipterocarpus intricatus* distributed in deciduous forests of Thailand, the inner bark temperatures rarely rises to 75 °C, whereas the bark surface temperatures could be as high as 300 °C (Stott 1986). The mere presence or absence of thick bark does not ensure protection from fire because the insulating quality of bark depends on structure, moisture content, composition, and density of bark (Brown and Smith 2000). A few fire-tolerant species proliferate in the forests of the Western Ghats as these species possess thick barks which protect the trees from mortality due to fire (Hegde et al. 1998). The thick bark insulates the cambium from high temperatures, for example, in certain savanna species, the bark forms almost 28.5% of the stem diameter on average (Hoffmann et al. 2003). Certain species such as *Terminalia crenulata* in the deciduous forests of the Nilgiris landscape possess thick barks and this species has low mortality rates in all size classes (Fig. 34.4a). In contrast to the low mortality numbers in all dbh classes in *Terminalia crenulata*, another species *Tectona grandis*, which has a thin bark, has witnessed high mortality in all dbh classes (Fig. 34.4b). Mortality is especially high (>50%) for individuals <20 cm dbh. Apart from the thick bark, another important fire tolerant characteristic is the presence of high moisture content, smooth, and flaking barks; this appears to be an important reason, why species such as *Anogeissus latifolia* found in the Nilgiris landscape, despite a thin bark are able to survive and dominate in these forests. There are different views on re-sprouting as a fire tolerant trait; savanna species survive better as saplings due to higher root to shoot ratios, in fact compared with forest tree species the ratio is 82% greater, ensuring survival from the low intensity fires (Hoffmann et al. 2003). Apart from thick bark, insulation from heat by the presence of subterranean stems and meristematic cells has been shown to confer protection from thermal kill (Whelan 1995). For example during a fire in certain deciduous forests of Thailand, the below (5 cm) ground surface temperature is about 75 °C, probably as low as 35 °C; whereas above the ground the temperature reaches almost 700 °C (Stott 1986).

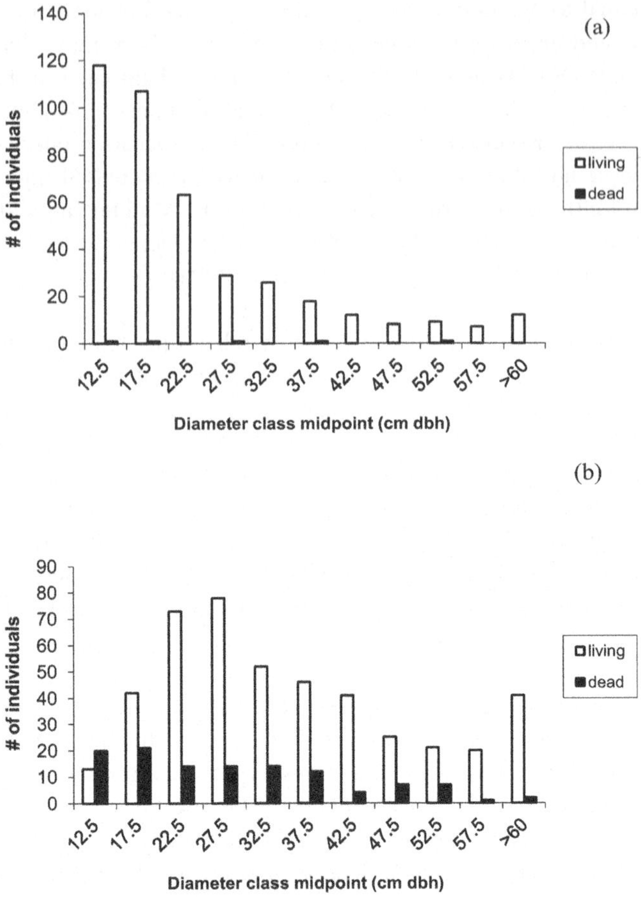

Fig. 34.4 Fire and forest structure in the Nilgiris landscape, Western Ghats. (**a**) Size-class distribution of *Terminalia crenulate*. (**b**) Size-class distribution of *Tectona grandis*

34.2.5 Fire and Climate Interactions in the Western Ghats

Drought plays a key role in driving fire activity in southwestern India (Renard et al. 2012). At regional scales, inter-annual variability in fire is strongly correlated with climate variables, especially precipitation of the antecedent year monsoon season. However, at landscape scales, proxies of forest fuel characteristics such as the type of vegetation, provided more robust estimates of the fire pattern (Renard et al. 2012). Simultaneously, recent studies in the Western Ghats demonstrated that increasing water deficits and drought stress at short time scales (e.g., 3–6 months) enhanced fire activity (Kodandapani and Parks 2019). For example, short-term (3 months) drought exemplified by soil moisture, climatic water deficits, and local weather were found to be important drivers of fire in tropical dry forests in the Western Ghats (Kodandapani and Parks 2019; Kodandapani 2021). Changes in climate would increase the intensity of fires, extent of burned areas, frequency of large fires and most importantly increase the susceptibility of fire immune ecosystems to recurrent fires (Barlow and Peres 2004). The climate of the Nilgiris landscape is characterized by recurrent but nonperiodic droughts which largely depend on the monsoon rainfall (Maréchal et al. 2009). Thus, the solar energy and water budget would largely determine the primary productivity and type of vegetation in the landscape. The vegetation in turn would have impacts on climate through evapotranspiration, thereby accounting for rainfall and changes in relative humidity and temperature in the landscape (Brown and Smith 2000). Vegetation has an important effect on the climate through the water budget. In the deciduous forests of the Nilgiris landscape, evapotranspiration represents almost 87% of the annual rainfall (Maréchal et al. 2009). Changes in climate, especially increasing surface temperatures, increasing fire season length and drought stress would further exacerbate the frequency of fires and also the area burned (Kodandapani and Parks 2019; Kodandapani 2021). Increased burned areas would result in enhanced carbon fluxes into the atmosphere resulting in positive feedbacks in global warming. Declined albedo due to the burned surface also results in a positive feedback in global warming, although albedo would recover within a few months in tropical ecosystems (Running 2008). Apart from these direct feedbacks to global warming, the mortality of trees due to forest fires reduces the opportunity of carbon sequestration and a sink for CO_2 (Fig. 34.5).

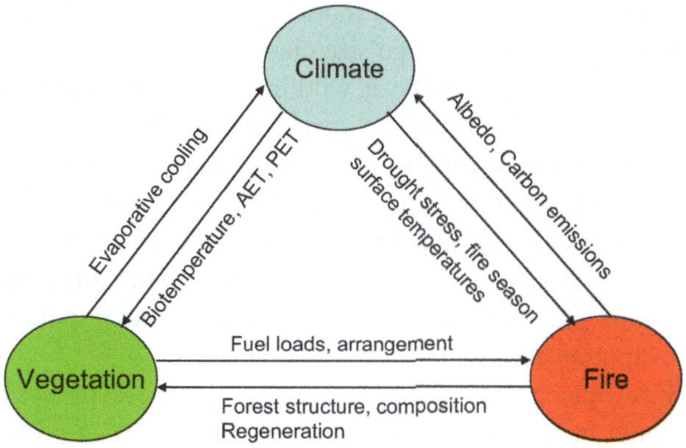

Fig. 34.5 Fire, vegetation, and climate

34.3 Conclusions

This chapter provides information on fire in Indian ecosystems in general, the spatial distribution of wildland fire in India, and the forest types vulnerable to frequent fires. The fire regime has been discussed in much greater detail with respect to the Western Ghats, where several studies have been undertaken to assess the effects of fire in ecosystems. The comparative analyses of fire regimes in different savannas reveal variations in fire characteristics across the globe; future studies would benefit by considering these differences in fire regimes. Thus, forest fire management policies should follow thorough critical analyses of fire regimes in different forest types in the country. In India, the increased use of fire in ecosystems is especially an important factor to consider the perspectives of forest dwellers, forest managers, rural communities and forest scientists should be integrated into future fire management strategies. Impending environmental challenges such as global warming and climate change could be incorporated into the assessment of changes in future fires in forests across the country. While there are several mechanisms available for monitoring fires in Indian ecosystems, integrating other information such as the environmental, socio-economic, and climatic factors would result in the better management of fires in India.

Lessons Learnt

- Wildfires are an almost annual phenomenon in the deciduous and savanna ecosystems of the Western Ghats. The fire regime in terms of its frequency, fire-return interval, intensity, and seasonality exhibits substantial variability across the Western Ghats.
- Predominantly low-to-moderate fire frequencies (1–5 times) and restricted high fire frequencies (>5 times) in a 16 year time window have been observed in the Nilgiris landscape. Fire-return intervals range from 1 to 5 years in the frequently burning areas, especially savannas. The fire intensity is low-to-moderate, between 5 and 50 kWm^{-1} in most parts of the landscape; however, this could increase depending on prior fire activity, the occurrences of invasive species, and accumulation of fuels.
- A few fire tolerant species proliferate in the frequently burning forests of the Nilgiris. The thick bark of these

trees protects them from mortality due to fire and makes them dominate the large size classes.

- Interannual variability in wildfire activity is moderately correlated with several climate metrics such as climatic water deficit, soil moisture, precipitation, maximum temperature, vapor pressure deficit, and palmer drought severity index. Daily fire activity is strongly correlated with meteorological variables such as relative humidity, maximum temperature, wind speed, and potential evapotranspiration.
- Fire management practices, especially fire suppression, could be contributing to fuel build-up in certain parts of the Nilgiris landscape, thereby accentuating fire risk and its effects in these ecosystems.
- Human actions both within and outside protected areas could be driving slow but insidious impacts of fires on various ecological characteristics in the Western Ghats.
- The occurrences of multiyear droughts associated with climate change could amplify fire activity in the Western Ghats. An integrated forest fire management approach should be adopted taking into account past fire practices, ecological characteristics of forests, future climate risk, and changing human drivers of fire.

Key Questions

1. How would natural fire regimes contrast with current fire activity in the Western Ghats? Could paleo fire studies, through the analysis of lake sediments, peat bogs, tree rings provide high-resolution data on past fire activity?
2. Nilgiris landscape is a protected area where fire suppression is the norm, how has the fire pattern altered due to fire management practices? How does this compare with traditional fire management practices?
3. What are the likely impacts of climate change on fire activity in the Nilgiris landscape and the Western Ghats in general?
4. What are the impacts of socio-economic factors and land use land cover change on fire activity in the Western Ghats?
5. Very little is known about the impacts of fires on fauna, assessing the impacts of fire on faunal biodiversity would be critical considering the conservation value of carnivores and mega herbivores in the Western Ghats. In general, large mammalian herbivores are attracted towards recently burned forests and fire is the preferred strategy to maintain these forests open, and tigers indirectly depend on fire-maintained habitats for ungulates as prey and unburnt areas for cover. Understanding the drivers of wildfire both within and outside protected areas along with the effects of fires in maintaining habitat heterogeneity could have implications for the conservation of the Bengal tiger and the Asian elephant in protected areas across the country.

Acknowledgements I thank the forest departments of Kerala, Karnataka and Tamil Nadu for permissions to conduct research in the forest areas. I thank Mr. Khushagra Urs for the pictures of fires.

References

Abatzoglou JT, Kolden CA (2013) Relationships between climate and macroscale area burned in the western United States. Int J Wildland Fire 22:1003–1020

Abatzoglou JT, Williams AP (2016) Impacts of anthropogenic climate change on wildfire across western US forests. PNAS 113(42): 11770–11775

Abatzoglou JT, Williams AP, Boschetti L, Zubkova M, Kolden C (2018) Global patterns of interannual climate fire relationships. Glob Chang Biol 24:5164–5175. https://doi.org/10.1111/gcb.14405

Agee JK (1994) Fire and weather disturbances in terrestrial ecosystems of the eastern Cascades. Gen. Tech. Rep. PNW-GTR-320. Department of Agriculture, Forest Service, Pacific Northwest Research Station, Portland, OR, 52p

Agee JK (2000) Wilderness fire science: a state-of-knowledge review. USDA Forest Service Proceedings RMRS-P-15-VOL-5

Andela N, Morton DC, Giglio L, Chen Y, Van Der Werf GR, Kasibhatla PS, DeFries RS, Collatz GJ, Hantson S, Kloster S, Bachelet D, Forrest M, Lasslop G, Li F, Mangeon S, Melton JR, Yue C, Randerson JT (2017) Ahuman-driven decline in global burned area. Science 80:1356–1362

Archibald S, Nickless A, Govender N, Scholes RJ, Leshten V (2010) Climate and the inter-annual variability of fire in southern Africa: a meta-analysis using long-term field data and satellite-derived burnt area data. Glob Ecol Biogeogr 19:794–809

Baker PJ, Bunyavejchewin S (2009) Fire behavior and fire effects across the forest landscape of continental Southeast Asia. In: Cochrane MA (ed) Tropical fire ecology: climate change, land use, and ecosystem dynamics. Springer, pp 311–334

Barlow J, Peres CA (2004) Ecological responses to El Nino-induced surface fires in central Brazilian Amazonia: management implications for flammable tropical forests. Philos Trans R Soc Lond B359:367–380

Bond WJ, Keeley JE (2005) Fire as a global 'herbivore': the ecology and evolution of flammable ecosystems. Trends Ecol Evol 20:387–394

Bowmann DMJS, Williamson GJ, Abatzoglou JT, Kolden CA, Cochrane MA, Smith AMS (2017) Human exposure and sensitivity to globally extreme fire events. Nat Ecol Evol 1(3):58. https://doi.org/10.1038/s41559-016-0058

Brown JK, Smith JK (2000) Wildland fire in ecosystems: effects of fire on flora. Gen. Tech. Rep. RMRS-GTR-42-vol. 2. U.S. Department of Agriculture, Forest Service, Rocky Mountain Research Station, Ogden, UT, 257p

Chandran MDS (1997) On the ecological history of the Western Ghats. Curr Sci 73:146–155

Cochrane MA (2003) Science for rainforests. Nature 421:913–919

Cochrane MA, Ryan KC (2009) Fire and fire ecology: concepts and principles. In: Cochrane MA (ed) Tropical fire ecology: climate change, land use, and ecosystem dynamics. Springer, Heidelberg, pp 25–62

Cochrane MA, Schulze MD (1999) Fire as a recurrent event in tropical forests of the eastern Amazon: effects on forest structure, biomass, and species composition. Biotropica 31:2–16

Dale VH, Joyce LA, Mcnulty S, Neilson RP, Ayres MA, Flannigan MD, Hanson PJ, Irland LC, Lugo AE, Peterson CJ, Simberloff D, Swanson FJ, Stocks BJ, Wotton MB (2001) Climate change and forest disturbances. BioScience 51:723–734

Flannigan MD, Stocks BJ, Wotton BM (2000) Climate change and forest fires. Sci Total Environ 262:221–229

Gadgil M, Chandran MDS (1988) On the history of the Uttara Kannada forests. In: Dargavel J, Dixon K, Semple N (eds) Changing tropical forests. Center for Resource and Environmental Studies, Canberra, Australian Capital Territory, pp 47–58

Guiterman CH, Margolis EQ, Baisan CH, Falk DA, Allen CD, Swetnam TW (2019) Spatiotemporal variability of human-fire interactions on the Navajo Nation. Ecosphere 10(11):e02932

Hegde V, Chandran MDS, Gadgil M (1998) Variation in bark thickness in a tropical forest community of Western Ghats in India. Funct Ecol 12:313–318

Hoffmann WA, Orthen B, Kielse P, Nascimento VD (2003) Comparative fire ecology of tropical savanna and forest trees. Funct Ecol 17: 720–726

Jiang Y, Zhou L, Raghavendra A (2020) Observed changes in fire patterns and possible drivers over Central Africa. Environ Res Lett 15:0940b8

John R, Dattaraja HS, Suresh HS, Sukumar R (2002) Density-dependence in common tree species in a tropical dry forest in Mudumalai, southern India. J Veg Sci 13:45–56

Jolly WM, Cochrane MA, Freeborn PH, Holden ZA, Brown TJ, Williamson GJ, Bowman DMJS (2015) Climate induced variations in global wildfire danger from 1979-2013. Nat Commun 6:7537. https://doi.org/10.1038/ncomms8537

Kelley DI, Bistinas I, Whitley R, Burton C, Marthews TR, Dong N (2019) How contemporary bioclimatic and humans controls change global fire regimes. Nat Clim Chang 9:690–696

Kodandapani N (2006) Fire regimes and ecological effects in seasonally dry tropical ecosystems in the Western Ghats, India. Ph.D. Dissertation, Michigan State University, East Lansing, MI

Kodandapani N (2021) Meteorological drivers of anomalous wildfire activity in the Western Ghats, India. In: Vadrevu KP, Ohara T, Justice C (eds) Biomass burning in south and Southeast Asia. CRC Press, pp 269–282

Kodandapani N, Cochrane MA, Sukumar R (2004) Conservation threat of increasing fire frequencies in the Western Ghats, India. Conserv Biol 18:1553–1561

Kodandapani N, Cochrane MA, Sukumar R (2008) A comparative analysis of spatial, temporal, and ecological characteristics of forest fires in a seasonally dry tropical ecosystem in the Western Ghats, India. For Ecol Manag 256:607–617

Kodandapani N, Cochrane MA, Sukumar R (2009) Forest fire regimes and their ecological effects in seasonally dry tropical ecosystems in the Western Ghats, India. In: Cochrane MA (ed) Tropical fire ecology: climate change, land use and ecosystem dynamics. Springer, Heidelberg, pp 335–354

Kodandapani N, Parks SA (2019) Effects of drought on wildfires in forest landscapes of the Western Ghats, India. Int J Wildland Fire 28(6):431–444

Lasslop G, Kloster S (2017) Human impact on wildfires varies between regions and with vegetation productivity. Environ Res Lett 12: 115011

Maréchal J-C, Varma MRR, Riotte J, Vouillamoz JM, Kumar MSM, Ruiz L, Sekhar M, Braun JJ (2009) Indirect and direct recharges in a tropical forested watershed: mule hole, India. J Hydrol 364:272–284

McLauchlan KK et al (2020) Fire as a fundamental ecological process: research advances and frontiers. J Ecol 108(5):2047–2069. https://doi.org/10.1111/1365-2745.13403

Morrison KD (1994) Monitoring regional fire history through size-specific analysis of microscopic charcoal: the last 600 years in South India. J Archaeol Sci 21:675–685

Parks SA, Abatzoglou JT (2020) Warmer and drier seasons contribute to increases in area burned at high severity in Western US forests from 1985 to 2017. Geophys Res Lett 47(22):e2020GL089858. https://doi.org/10.1029/2020GL089858

Pascal JP (1986) Explanatory booklet on the forest map of South India. French Institute, Pondicherry

Pinard MA, Huffman J (1997) Fire resistance and bark properties of trees in seasonally dry forest in eastern Bolivia. J Trop Ecol 13:727–740

Pyne SJ, Andrews PL, Laven RD (1996) Introduction to wildland fire. Wiley, New York

Renard Q, Pélissier R, Ramesh BR, Kodandapani N (2012) Environmental susceptibility models to predict wildfire occurrences in the Western Ghats of India. Int J Wildland Fire 21(4):368–379. https://doi.org/10.1071/WF10109

Riley KL, Abatzoglou JT, Grenfell IC, Klene AE, Heinsch FA (2013) The relationship of large fire occurrence with drought and fire danger indices in the western USA, 1984–2008: the role of temporal scale. Int J Wildland Fire 22:894–909

Running SW (2008) Ecosystem disturbance, carbon, and climate. Science 321:652–653

Sinha A, Brault S (2005) Assessing sustainability of non-timber forest product extractions: how fire effects sustainability. Biodivers Conserv 14:3537–3563

Stott P (1986) The spatial pattern of dry season fires in the savanna forests of Thailand. J Biogeogr 13:345–358

Stott P (2000) Combustion in tropical biomass fires: a critical review. Prog Phys Geogr 24:355–377

Syphard AD, Keeley JE, Pfaff AH, Ferschweiler K (2017) Human presence diminishes the importance of climate in driving fire activity across the United States. Proc Natl Acad Sci USA 114:13750–13755. https://doi.org/10.1073/pnas.1713885114

Syphard AD, Radeloff VC, Keeley JE, Hawbaker TJ, Clayto MK (2007) Human influence on California fire regimes. Ecol Appl 17(5): 1388–1402

Vadrevu KP, Lasko K, Giglio L, Schroeder W, Biswas S, Justice C (2019) Trends in vegetation fires in south and southeast Asian countries. Sci Rep 9:7422

Van Wagner CE (1978) Age-class distribution and the forest fire cycle. Can J For Res 8:220–227

Whelan RJ (1995) The ecology of fire. Cambridge University Press, Cambridge, UK

Zubkova M, Boschetti L, Abatzoglou JT, Giglio L (2019) Changes in fire activity in Africa from 2002 to 2016 and their potential drivers. Geophys Res Lett 46:7643–7653

35 Diseases of Tree Species

N. S. K. Harsh

This chapter is dedicated to the fond memories of Late Dr. Ram Keerti Verma, Retired Scientist-G from Tropical Forest Research Institute, Jabalpur, who was initially entrusted to write this chapter but could not complete due to his sad demise. He had successfully completed several research projects on forest pathology, biofertilisers and taxonomy of fungi. He was a member of many academic societies and was awarded Fellow of Phytopathological Society of India (FPSI) by Indian Phytopathological Society. He was awarded with Prof. K.S. Thind medal for the year 2010 by the Association for Plant Taxonomy, Dehradun, for his significant contribution in the field of fungal taxonomy. His major fields of research were Fungal Taxonomy, Fungal Diversity, Fungal Ecology and Biological Control of forest diseases. He authored several research papers and books mostly on the taxonomy of fungi.

Abstract

The stages of forest tree health are different in seeds, seedlings, plantations, forests and after harvesting. Losses occur resulting into seed abortion and deterioration, which directly affect the seed germination and the survival of seedlings. Diseases of leaves decrease the appearance, growth and vigour of the plants and also lead to mortality. Root pathogens kill plants or decrease their growth soon after germination till maturity. Girdling and killing of the branches and trees can be caused by bark and cambial parasites, the latter also producing lesions leading to decay. Wood discolouration and decay decrease the merchantability of wood. During storage and use, fungal attack deteriorates quality and quantity of forest produce and also adds toxic substances making it unsuitable for human consumption. In this chapter some major diseases of forest trees growing in India have been described.

Keywords

Disease · Host · Pathogen · Predisposing factor · Symptom

N. S. K. Harsh (✉)
Forest Research Institute, Dehradun, India

35.1 Introduction

Various stress factors affect trees/plants throughout their life that interfere the normal development. These stress factors may be biotic (living) or abiotic (non-living). Disease is expressed by an abnormal physiology caused by an extended interaction between a plant and a biotic stress factor. In other words, a disturbance in the normal physiologic functioning of a plant results into a disease, which may have many causes, and exhibits different appearances and results. A simple dictionary (Collins) meaning of disease is 'any departure from health, presenting marked symptoms, malady, illness, disorders'. Mostly diseases are caused by biotic agents (pathogens like viruses, bacteria, fungi, nematodes, etc.) and are called biotic plant diseases or infectious diseases. Whereas diseases caused by abiotic (non-living) agents such as extremes of temperatures, excess or deficiency of moisture, nutritional imbalance and phytotoxic gases are called abiotic plant diseases or non-infectious diseases.

Disease is an extended interaction between the pathogen, host and environment, which is always detrimental to the host. The external expression of disease is in the form of symptoms. Diseases are one of the most destructive agents adversely affecting the forest productivity as they attack the planting stock in nurseries and consequently the plantation yield. Any failure in plantations, in the form of partial or total death of trees or when the expected growth rate is not

A. K. Mandal, A. Nicodemus (eds.), *Textbook of Forest Science*, https://doi.org/10.1007/978-981-97-8289-5_35

achieved, is bound to hamper the anticipated ecological and environmental benefits. Sometimes diseases are not the primary cause but are the contributing factors, which attack the weakened plants after they are predisposed by various environmental, edaphic and biotic factors. It is estimated that of all the damages caused to forest crops by diseases, insect pests, fire and weather, diseases rank as the most destructive agents attributing nearly 45% of all losses (Anon 1958).

In a natural ecosystem although a large number of pathogens are known to be associated with the tree species, only a few of them attain a status of serious pathogens in epidemic proportions because of a dynamic equilibrium between the host and its parasites. Their co-evolution produces parasites that are not so aggressive that hosts are destroyed, and equally the host does not become so resistant that the parasite cannot exist. But as soon as such a natural system is disturbed and converted into an artificial one, as in man-made plantations, the relatively non-specific microorganisms may become highly pathogenic to the introduced tree species. The changed situation may greatly influence the virulence of the native pathogens, the susceptibility of the host or both. This situation becomes more intriguing when an introduced host faces an indigenous pathogen against which it has no in-built resistance or a pathogen may be introduced along with the host, which may become threatening for the native hosts.

The stages of forest tree health are different in seeds, seedlings, plantations, forests and after harvesting. Losses occur resulting into seed abortion and deterioration, which directly affect the seed germination and the survival of seedlings. Diseases of leaves decrease the appearance, growth and vigour of the plants and also lead to mortality. Root pathogens kill plants or decrease their growth soon after germination till maturity. Girdling and killing of the branches and trees can be caused by bark and cambial parasites, the latter also producing lesions leading to decay. Wood discolouration and decay decrease the merchantability of wood. During storage and use, fungal attack deteriorates quality and quantity of forest produce and also adds toxic substances making it unsuitable for human consumption (Harsh 2014a).

A uniform crop of hosts in plantations with monocultures of same species and/or clone provide opportunity to a virulent pathogen for causing epidemics during conducive environmental conditions. In agroforestry systems some diseases are common in both agriculture and forestry crops or they may migrate from one crop to another (Jamaluddin 1988). Besides, intensive management practices encourage diseases. Therefore, adequate control measures are absolutely necessary to reduce the losses. According to Gibson (1975), effective and reliable control, however, depends on:

1. Correct diagnosis
2. Knowledge of pathogen, host range, disease cycle and factors favouring disease development
3. Economic feasibility

35.2 Causes of Tree Diseases

As is evident from foregoing discussion, a cause or causes are essentially required for the manifestation of diseases; the cause may be living or non-living entity generally called stress factors, in case of living being as pathogens. A pathogen is a disease-causing agent.

35.2.1 Biotic Agents of Tree Diseases

35.2.1.1 Nematodes as Plant Parasites and Agents of Tree Diseases

Nematodes are unsegmented round worms in the phylum Nemathelminthes, having their body wall composed of cuticle, epidermis and muscles and well-developed digestive, reproductive, nervous and excretory systems. Nematodes can be placed in four groups:

1. Small free-living saprophytic nematodes characterised by a hollow cavity like a mouth for picking up organic matter and feeding it into the oesophagus.
2. Variable-sized nematodes parasitising animals having smooth adaptation for suction attachment to animal cells.
3. Predators on small invertebrates in the soil having mouth parts with modified teeth-like projection.
4. Plant parasitic nematodes 0.5–2.5 mm in length characterised by a hollow dagger-like feeding structure known as stylet.

Nematodes can attack almost any plant part but in tree species the root system, the tips of small feeder roots. Nematodes can be classified into two groups on the basis of their feeding habit, movements and type of injury they cause: ectoparasites, feeding on the surface of the roots, and endoparasites, moving within the roots. Nematodes that move along a root or root to root are known as migratory nematodes while those remaining localised are called sedentary nematodes. Migratory ectoparasites cause stunting of feeder root formation and growth, resulting into blunt or stubby roots. Migratory endoparasites cause death of large number of cells by their feeding and movements, resulting into sunken dead areas or lesions. Sedentary endoparasites cause small swellings or knots to form on the roots.

Symptoms

Symptoms

- Above ground symptoms
  - Stunting
  - Chlorosis
  - Decline
- Below ground symptoms
  - Root lesions
  - Root knots
  - Stubby root

Predisposing agents for root diseases: Nematodes may act as the predisposing factors of root diseases because the injuries caused to the roots act as the infection courts of vascular wilt and root-rot diseases of plants

Other roles: Nematodes may act as the vectors for viruses and help in establishing viral and mycoplasmal diseases in plants. They act as parasites of mycorrhizal fungi and roots and also as contributing factors of decline diseases in trees

35.2.1.2 Viruses as Agents of Tree Diseases

Viruses are infectious agents characteristically composed of a protein coat and a hollow nucleic acid core of either DNA or RNA. Viruses have no cell wall, membranes, cytoplasm, organelles or nucleus. They simply contain genetic material in the form of nucleic acid, for reproduction, which is protected by a protein coat. Viruses may occur in many shapes but most are either rod-shaped or polyhedral as seen through Scanning Electron Microscope. Viruses appear to be on the threshold of life, being more an organic chemical than a living organism but still being capable of causing diseases. The plant viruses are generally simple in structure and function. The infectious nucleic acid material, usually a single strand of RNA, is contained within a protein shell composed of large numbers of identical protein subunits.

Symptoms

Most virus diseases of trees involve one or more of the following general symptoms:

1. Lack of chlorophyll promotion in normally green parts as mottling mosaic and chlorosis either marginal or veinal
2. Dwarfing or growth inhibition (stunting) and witches' broom
3. Distorted growth of part or the entire plant such as curling
4. Necrotic areas or lesions

35.2.1.3 Mycoplasmas/Phytoplasmas as Agents of Tree Diseases

Until recently, all mycoplasma diseases of plants were thought to be caused by viruses. However, there are many differences between these two types of pathogens, not only in their structure but also in the mode of attack and symptoms caused. Mycoplasmas are much larger than viruses and some of them are smaller than bacteria though they are often longer than bacteria. The mycoplasmas attacking plants are usually spherical or occur in long filaments. Unlike viruses, which resemble organic chemicals, mycoplasmas lack a cell wall, but like the bacteria they contain a plasma membrane, cytoplasm and strands of DNA.

Mycoplasmas are parasites of plants or animals or they are saprophytes. Once in a living cell, they may divide by fission, in a manner similar to bacteria, by budding, where a small swelling cut off to form a new mycoplasma, or through constriction along the body in the filamentous forms, which break off and produce several mycoplasmas.

Symptoms

The universal symptoms of mycoplasma diseases appear to be a progressive decline that eventually results in death of the host in a few months to several years. The decline may be associated with any or all of the following symptoms: chlorotic foliage, little terminal growth or stunting, progressive die-back of branches, lack of flowering and necrosis of phloem tissues. Other common symptoms of mycoplasma diseases are reddening of foliage, witches' brooms on branches and on axillary shoots, and proliferation of roots.

35.2.1.4 Bacteria as Agents of Tree Diseases

Bacteria are microscopic, usually single-celled microorganisms that like fungi, lack chlorophyll, but are much smaller than fungi. They have a variety of shapes, such as spherical, spiral or rod-shaped. Some have fine appendages called flagella, which provide a small degree of motility. Bacteria are found singly, in loose masses and in chains. They multiply rapidly by the simple splitting (fission) of one bacterial cell into two cells. Some bacteria are beneficial while some are pathogenic to trees also.

Symptoms

Bacterial diseases exhibit different symptoms on different plant parts. Bacterial pathogens of leaves and shoots produce water-soaking initially, followed by necrosis and maceration of tissues. Bacterial infection of vascular tissue produces wilting. Crown gall bacterial infection is characterised by hypertrophy of stem tissue. Bacterial infection of wood results in increased permeability due to perforation of pit membranes.

35.2.1.5 Fungi as Agents of Tree Diseases

Fungi constitute the most common group of plant pathogens that can attack any part of the plant (roots, stem and leaves), although each species of fungus may attack many parts or only one part. Many plant pathogenic fungi usually live on dead organic matter as saprophytes and only attack a suitable living host tree when the environmental conditions favour them as facultative parasites.

Fungi are often seen as mushrooms on the ground, conks on a tree or mould on cheese, bread or leather items like shoes, but these rather complex structures are made of much smaller structural units of the fungi, the hyphae. Hyphae are thread- or tube-like branching strands of the fungus visible only with the aid of a microscope. Hyphae penetrate the defence barriers of the host and produce diseases. Once the hyphae have gained entry into the host, the fungus begins to derive its nutrition from the plant and to grow within it. When a pathogen derives its nutrition from a plant, it is termed as an infection, and when the pathogen moves within the host, it is called an invasion.

Most fungi produce reproductive structures called spores, the disseminating units, which germinate to produce hyphae capable of entering susceptible host plants.

Symptoms

The fungal diseases produce different symptoms on different plant parts:

- On foliage: total necrosis and shrivelling of leaves or discrete localised necrotic areas/spots, coloured pustules
- On stem and twigs: die-back, cankers, heart-rot, decay
- On roots: plugging of vascular system, wilt, root-rot

35.2.1.6 Flowering Plants (Phanerogams) as Agents of Tree Diseases

Flowering plants also act as parasites of trees. All parasitic flowering plants are dicotyledons. The members of order Santanales have a number of parasitic flowering plants, particularly in the family Loranthaceae (*Dendrophthoe falcata*, *Loranthus* spp., mistletoes), Viscaceae (*Arceuthobium*, *Viscum*—dwarf mistletoe) and Santalaceae (sandalwood), which are partial parasites on host plants. Whereas the genus *Cuscuta* (dodders) of Convolvulaceae family is a total parasite on host plants (Fig. 35.1).

Some of the plant parasites are totally dependent upon the host for preformed photosynthates (holoparasitic) while others are only partially dependent upon the host for photosynthates, because they still maintain some

Fig. 35.1 *Cuscuta reflexa* on a tree

photosynthetic leaf surface (hemiparasitic). Often the leaves are much reduced, so that photosynthetic capacity is minimal. Some parasitic plants are associated with roots while others are with stem.

Symptoms

Effects on parasitised hosts range from excessive hypertrophy and witches' brooms to almost no effect. Reduced reproductive capacity, chlorosis and reduced growth are common symptoms.

35.3 Major Diseases

The two different soil and the sub-aerial environment of root and shoot systems of flowering plants affect the health and vigour of the plants. Some major diseases are explained in the following text for seed-borne infections, nursery diseases, plantation diseases, forest diseases and wood decay.

35.3.1 Seed-Borne Infection

Seeds get infected during development while still on the trees and/or during storage, due to poor collection and storage practices (Fig. 35.2). Seed-borne infections can be of two types:

1. *Externally seed borne*: By fungi causing spoilage resulting into deterioration of seed contents and reducing quality as well as quantity. They also affect seed germination by causing pre-emergence and post-emergence damping-off diseases (explained later).
2. *Internally seed borne*: By fungi causing deep-seated infection, which affect viability by killing embryo and is also source of pre- and post-emergence damping-off diseases.

Species of *Alternaria*, *Aspergillus*, *Botrytis*, *Cladosporium*, *Fusarium*, *Penicillium*, etc. are generally associated with the seed-borne infections.

35.3.1.1 Symptoms of Seed-Borne Infections

- *Change in seed colour*: Can be observed by deviation from normal seed colour. The infected seeds are darker, spotted or lighter.
- *Change in seed size*: The infected seeds exhibit enlargement or reduction in normal seed size.
- *Wrinkled seed surface*: Abnormally wrinkled or folded seed surface/seed coat is also an indicator of infected seeds.

Fig. 35.2 *Fusarium pallidoroseum* causing infection of seeds inside the pods

- *Fungoid seeds*: The infected seeds show superficial fungal growth as white or coloured fungal mycelium and/or spores on the seed surface.
- *Foul smell*: The infected seeds omit foul smell mostly due to bacterial infection.

35.3.2 Soil-Borne Infection

Those pathogens that undertake the major part of their life cycle in the soil are called soil-borne pathogens. Fluctuation in temperature and atmospheric humidity in the sub-aerial environment is not permitted in the soil due to the buffering effect of the soil mass; therefore, the soil remains sufficiently warm and moist throughout the year for microbial activity. Source of energy for this continued microbial activity is provided by dead organic matter of plant and animal origin (Garrett 1970).

Soil-borne pathogens infect roots and/or stem bases and their dispersal and survival mostly remain confined to the soil. They generally produce air- and/or water-dispersed spores for dispersal. Many pathogens, which cause diseases of aerial plant parts, also survive in the soil. Soil-borne pathogens belong to Fungi, Bacteria, Actinomycetes, Nematodes, and Viruses.

35.3.2.1 Soil-Borne Plant Pathogenic Fungi/Root-Infecting Fungi

On the basis of mode of parasitism, they are *specialised* and *non-specialised pathogens*. On the basis of ecology, they are *soil inhibitors* and *soil invaders* (Garrett 1970). Specialised pathogens are mostly soil invaders. Non-specialised pathogens are true soil inhibitors. Soil-inhabiting fungi have the ability to survive indefinitely as a saprophyte. After the death of the host, soil-invading or root-inhabiting fungus consists of an expanding parasitic phase and a declining saprophytic phase. Soil-invaders have a restricted host range because they show more parasitic specialisation than the soil-inhibitors.

35.3.3 Soil-Invading/Root-Inhabiting (Specialised Pathogens)

Generally, they can be classified into three types:

1. *Vascular wilt fungi*: After initial invasion these remain confined to the vascular tissue, e.g. *Verticillium dahliae*, *Fusarium oxysporum*.
2. *Ectotrophic root-infecting fungi*: These grow extensively on the outer surface of the root and can enter the root cortex at many loci. They produce specialised mycelial strands called rhizomorphs that grow along adjacent roots of the adjacent plants and do not depend on soil for survival, e.g. *Armillaria mellea*, *Ganoderma lucidum*.
3. *Obligate zoosporic fungi*: They produce motile zoospores, e.g. *Synchytrium endobioticum*.

35.3.4 Soil-Inhabiting (Non-specialised)

They are mostly facultative parasites and live as saprophytes on dead organic matter in soil but become parasitic under favourable conditions. They are of two types:

1. *Seedling diseases*: 'Damping-off', 'seedling blight', e.g. *Rhizoctonia solani*, *Pythium* spp., *Phytophthora* spp., *Fusarium* spp., *Sclerotium rolfsii*, *Thielaviopsis basicola*, *Macrophomina phaseolina*, etc.
2. *Diseases of older plants*: Occur due to a progressive killing of rootlets and sometimes of the older parts of the root system as well. 'Decline' diseases are caused by an interacting set of factors and not by a single causal agent. Symptoms syndrome such as die-back and blight are commonly but not exclusively used to identify 'Decline' diseases (Tattar 1988). The factors that result into 'Decline' diseases include:

   - Predisposing factors—which are long term, such as climate, soil moisture, genotype of the host, soil nutrients and air pollutants.
   - Inciting factors—which are short term, such as insect defoliation, frost, drought, salt, air pollutants and mechanical injury.
   - Contributing factors—which are long term, such as bark beetle, canker causing fungi, viruses and root-rot fungi.

35.3.5 Seedling Diseases: Nursery Diseases

Nurseries are the suppliers of the planting material for plantations. Success of the future plantations will largely depend on the health of the seedlings raised and maintained at nurseries. Taking a diseased/unhealthy planting material to field for plantation results into failure and can be considered as colossal waste of time, money and manpower. Major diseases in nurseries are root and foliar diseases, which are discussed here.

35.3.5.1 Root Diseases in Nurseries and Young Plants and Their Management

The roots are one of the most important systems that give nourishment to the plant, but may suffer from diseases that remain generally unnoticed until outward symptoms appear on the above-ground parts. The health of the root system determines the total health of a plant and that is why the root diseases are important in nurseries. The most common root diseases in nurseries are damping-off, vascular wilt and root-rot diseases.

Damping-Off Diseases

This is one of the most common diseases of conifer and broad-leaved tree species in forest nurseries. Estimation of loss caused by this disease is difficult. Disease development is governed by the factors like soil texture, soil pH and moisture. The fungi causing damping-off are soil borne in nature. The disease may destroy the entire nursery stock in 1 month and may remain unnoticed in the other.

Pre-emergence Damping-Off

It comes during germination and soon after seed germination but prior to the plumule emergence out of the soil. As a result of pre-emergence damping-off disease, the emergent crop of seedlings is very sparse, which most of the times is wrongly attributed to the poor quality of the seeds. Species of

Fig. 35.3 Post-emergence damping-off in *Moringa pterygosperma* by *Fusarium acuminatum*

Alternaria, Aspergillus, Fusarium, Penicillium, Rhizopus, etc. are common fungi associated with this disease.

Post-emergence Damping-Off

This disease occurs soon after the emergence of seedlings. The seedlings are attacked by the pathogen at the collar region, which gets deteriorated and cannot support upper plant parts and as a result the seedlings topple over on the ground. Species of *Fusarium, Pythium, Phytophthora, Verticillium, Rhizoctonia solani, Sclerotium rolfsii*, etc. are common fungi causing post-emergence damping-off diseases in forest nurseries (Fig. 35.3).

The majority of agroforestry species are susceptible but *Eucalyptus* and *Pinus* spp. suffer more from these diseases than others. The important species in which damping-off disease have been reported are *Acacia* spp., *Albizia* spp., *Azadirachta indica, Cassia* spp., *Delonix regia, Dendrocalamus strictus, Holoptelia intigrifolia, Lagerstroemia* spp., *Leucaena leucocephala, Oroxylum indicum, Tectona grandis*, etc. (Harsh and Gupta 1993).

Collar Rot in Mature Seedlings

Collar rot is a common disease in forest nurseries under moist climatic conditions. The young seedlings are attacked at collar region resulting into toppling over of the seedlings. *Rhizoctonia solani* and *Sclerotium rolfsii* are common collar rot causing fungi in forest nurseries. *Ailanthus triphysa, Bombax ceiba, Eucalyptus* and pines are most susceptible hosts of this disease (Sharma et al. 1985).

Vascular Wilt Disease

In this disease plant becomes yellow and shows withering of leaves at the tip and gradually dying back of seedling, resulting into complete death of plant. Wilt disease is common in forest nurseries particularly in dry climates. This disease is also common in container seedlings. Species causing wilt disease are *Fusarium* spp., *Macrophomina phaseolina, Rhizoctonia solani* and *Verticillium* spp. All the hosts are susceptible when the conditions are favourable for fungal infection.

Root-Rot Disease

It causes die-back of seedlings in which fungi attack the succulent root tips. Normally the damage is not very fatal; if soil moisture conditions are favourable and new roots develop, the seedlings continue to survive though with less vigour. When the initial root tip mortality is followed by drought or excessive soil moisture, which inhibit the formation of new roots, large-scale mortality of seedlings occurs. The root system becomes completely damaged due to rotting. Charcoal root rot caused by *Macrophomina phaseolina* is the most common disease of pines and casuarina.

Root diseases may cause nearly 70% mortality of seedlings in nurseries (Harsh and Gupta 1993); therefore, prevention and control of these diseases become imperative to get the desired quality and quantity of seedlings for various plantations.

Set-Rot of Cuttings

Poplar is the most important species raised vegetatively from cuttings under agroforestry. The cuttings in nursery beds are infected by certain fungal pathogens and could not survive even after sprouting of leaves with well-developed root system. *Botryodiplodia palmarum* (Cooke) Petr. & Syd. is a common fungus, which has been reported to damage more than 50% plantlets in shade conditions (Singh and Singh 1986). The same pathogen causes bark and stem canker disease in poplars (Bakshi et al. 1972) and it is quite likely that the pathogen comes from the mother plant from which cuttings are made.

35.3.5.2 Control of Root Diseases in Nursery

Although it is very difficult to eradicate the diseases completely from the system but certain measures can be adopted to minimise the diseases to some extent.

Cultural Methods

- Always use apparently healthy seeds/cuttings for sowing/planting. Diseased seeds with discolouration, wrinkles, spots, fungal growth and any visible defects such as over-size or under-size than normal seeds should be avoided.
- Use well-drained seed beds as excessive moisture favours the infection and development of root diseases.
- Use proper (porous) potting medium for containerised seedling production system.
- Give adequate watering neither too less nor too much.
- Carefully uproot the plants (two to three leaves stage) at the time of pricking without causing injury to tender roots.
- Remove weeds regularly from the seed beds and containers.
- Eradicate infected parts/plants after early diagnosis.
- Avoid use of same bed for raising same species year after year. The bed should be left fallow or raised with different species.

Chemical Control

- Seed dressing with fungicides like Bavistin (Carbendazim), Topsin-M (Thiophanate methyl), Benlate (Benomyl–Benzimidazole) or Thiram (Dimethyl-dithiocarbamate) @ 0.2% by weight of seeds before sowing.
- Soil drenching with fungicides (Bavistin, Topsin-M, Benlate or Thiram) @ 0.05%. Ten litre solution should be applied for 10 × 1 m bed with repeated doses after the interval of 7–10 days in diseased areas.
- Soil sterilisation with 200 mL commercial-grade formalin diluted in 4.5 L of water can be used in a bed of 1 × 1 × 0.15 m or in potting mixture heaped over ground. After formalin treatment, the bed and soil mix is to be covered with polythene sheet for 3 days and after removal of the sheet the soil should be turned repeatedly for next 3 days. The soil should be used for planting only after 6 days of the treatment.
- Root dip in 0.02% solution of Bavistin, Topsin-M, Benlate or Thiram for 3–5 min prior to pricking out in containers provides protection against root diseases. Dipping of cuttings in 0.3% solution of same fungicides for 5–10 min before placing them in beds or containers has been found to control the set-rot disease.

Biological Control

The use of fungicides for the control of diseases has its disadvantages because of their persistence in the ecosystem, affecting non-target organisms and being toxic to the environment. The alternate is biological control with the application of a biological control agent of fungal or bacterial origin to the plant or soil. In forest nurseries, application of *Trichoderma* spp. (*Trichoderma harzianum*, *Trichoderma viride*, *Trichoderma virens*, etc.) and *Pseudomonas fluorescens* as seed pelleting or soil mixing helps in controlling the root diseases.

35.3.5.3 Foliar Diseases in Nursery

Foliar diseases are serious because they cause premature defoliation apart from checking and reducing photosynthetic activity, which reflect on the vigour of planting stock. The common foliage diseases of nurseries are leaf spot, leaf blight, rusts, powdery mildew, blotch, shot holes, leaf curl and necrosis in the leaf and tip. *Maravalia achroa* and *Olivea tectonae* are two important rusts found in *Dalbergia sissoo* and *Tectona grandis*, respectively. Severe attacks of *Cyclindrocladium quinqueseptatum* on eucalyptus seedlings caused initial leaf spots followed by necrosis of total plants in ideal (high humid) disease conditions. *Cercoseptoria pini-densiflorae* is the prominent fungus, which caused black needle spot in various exotic and indigenous pines. *Alternaria alternata*, *Cercospora populina*, *Cladosporium humile*, *Myrothecium roridum*, *Phyllosticta adjuncta* and *Sphaceloma populi* are the main fungi responsible for leaf spot and blight diseases in poplars.

Some major foliar diseases occurring in nurseries and plantations are discussed below.

Leaf Blight Disease in Teak

Cercospora tectonae causes spots with the onset of monsoon, which appear on both the surfaces of the leaf as light brown specks that turn brown to blackish on maturity. Individual spots develop more or less circularly initially, and later merge together forming larger lesions. In case of severe infection, the leaves drop prematurely, young shoots are infected and the juvenile plants are killed.

Control: Three fortnightly sprays with 0.2% Bavistin or Copper oxychloride can control the disease.

Leaf Rust in Teak

Olivea tectonae attacks old leaves resulting in severe damage to plants in nurseries and plantations (Fig. 35.4). Rust pustules are formed on lower surface of the leaves. Heavily infected plants in nurseries and young plantations usually succumb to rust disease. The rust appears during the month of October–November and persists up to leaf fall.

Control: Fungicidal treatment is recommended as foliar spray with Bayleton (Triadimefon) or Sulfex (Sulphur) at the rate of 0.2% at monthly interval from the month of October to control the disease.

Phyllosticta Leaf Blotch and Blight in Poplars

Phyllosticta adjuncta causes spots with the onset of monsoon, which appear on both the surfaces of the leaf as light brown specks that turn greyish white to dirty white on maturity. Individual spots develop more or less circularly initially, and later merge together forming larger lesions. In case of

Fig. 35.4 Teak leaf rust by *Olivea tectonae*

severe infection, the leaves drop prematurely, young shoots are infected and the juvenile plants are killed. Both G-3 and G-48 clones of poplar are affected by this disease. This disease has been attributed for the disappearance of the popular clone G-3 in North India.

Control: Three fortnightly sprays with 0.2% Bavistin or Copper oxychloride can control the disease.

Leaf Rust in Poplars

Melampsora populnea attacks juvenile leaves and twigs resulting in severe damage to plants in nurseries and plantations. Rust pustules are formed on lower surface of the leaves. Heavily infected plants usually succumb to *Melampsora* infection. The rust appears in the month of September, that is, after rainy season, and persists up to leaf fall.

In *Populus ciliata* the rust is caused by *Melampsora ciliata* in Himachal and Jammu & Kashmir in India. First evidence is noticed by the development of small yellow spots on the upper surface of the leaves. Uredosori develops on the lower surface corresponding with yellow dots, covering almost the entire surface. The disease starts at the end of rainy season.

Control: Fungicidal treatment is recommended as foliar spray with Bayleton or Sulfex at the rate of 0.2% at monthly interval from the month of September till November to control the disease.

Leaf Spots and Blight in Eucalyptus

Cylindrocladium quinqueseptatum causes severe foliar infection in almost all species of Eucalyptus. Initially pinkish-purplish dots appear on leaves, which soon enlarge and change to greyish-brown in colour and cause blight symptoms. In severe infection, premature defoliation and seedling mortality occurs.

Control: Foliar sprays with Bavistin (0.1%) and Copper oxychloride (0.1%) at weekly interval with the onset of disease in July–August check further spread of the disease and maintain healthy plants.

35.3.6 Diseases of Older Plants: Plantation Diseases

Plantations are not free from diseases. Some diseases are causing mortality in plantations while others reduce vigour and form of the trees affecting the productivity directly. Mortality of a tree in a roadside, park and avenue plantations can be understood by the fact that if a tree dies at an age of 15 years it is not only the loss of the tree but also the time it took to grow, and another tree of same dimension will take another 15 years to grow; thus the actual loss is of 30 years. The ecological benefits provided by that particular tree are also lost.

Plantation diseases of tropical tree species have been documented by Jamaluddin et al. (1997). Some major diseases are described below.

35.3.6.1 Root Rot in Trees by *Ganoderma lucidum*

Generally, in drier localities, in North India and Peninsular India, *Ganoderma lucidum* commonly causes root-rot disease in disturbed sites. Symptoms are die-back or top-dying, stag-headedness, mortality, sudden wilting and wind-throw. The pathogen invades root system, kills functional vascular tissue and rapidly decays wood in roots and butt region of the tree. The affected roots show white fibrous rot that kills them and affects the anchorage of the trees and results in uprooting. The disease spreads through root-to-root contact, and once noticed in a pure plantation is likely to affect all the trees of same species sooner or later. The signs of the disease appear as sporophores of the fungus at the basal region and/or underneath roots, which can be easily recognised as stalked brownish maroon shiny upper surface and creamy lower surface. The pathogen has been reported to attack 64 tree species in India of which 38 are leguminous tree species (Kumari and Harsh 2004). It has been associated with the shisham (*Dalbergia sissoo*), khair (*Acacia catechu*) and babul (*Acacia nilotica*) mortality in north, central and western regions of India causing great economic losses. In a khair plantation, mortality of 12% seedlings up to 3 years, 40% by 7th year and 55% by 9th year in Siwaliks was recorded (Bakshi 1976). In an acacia arboretum at Seoni, Madhya Pradesh, mortality in first year, 9%; second year, 15%; third year, 31%; and ninth year, 52% due to *G. lucidum* was reported (Harsh et al. 1993).

In conifers, common root-rot pathogens are *Armillaria mellea* and *Heterobasidion annosum*. Bakshi (1976) has worked out the management of root-rot diseases, which include following measures.

Mechanised plantation: Control of root and butt-rot pathogenic fungi can be effected by removal of residual stumps and roots left after felling. This should be done prior to planting either mechanically or manually. In raising new plantations particularly with susceptible species like *Acacias*, *Cassias*, *Dalbergias* and other legume trees in new sites, the sites should be cleared of tree stumps and residual roots to eradicate the source of infection. If still trees die because of disease, they should be removed along with the roots from the stand.

Isolation trenches: Once a root disease establishes in the forest, it spreads centrifugally from the infection centre to adjacent trees and through root contact or root graft. In such cases, the spread of the diseases can be checked by isolating diseased trees by digging trenches of about 0.30 m wide, 0.7 m deep and 1.5 m long. Bakshi (1976) advised to dig an additional trench immediately after the apparently healthy tree next to the first trench. The trenches are meant to break the root contact between healthy and diseased trees.

The urban forestry programme in big cities face the problem of how to make trenches around trees on footpath and road dividers where making trenches is not feasible. In such cases insertion of vertically perplex glass sheets of size 1 × 1 m between the healthy and diseased trees will break the root contact.

Resistant species and mixed stands: Bakshi (1976) has reported two tree species *Ailanthus excelsa* and *Bombax ceiba* as resistant to *Ganoderma lucidum*. He advised that susceptible species like *Acacias*, *Cassias*, *Dalbergias* and other legume trees are not to be planted in the first rotation but with *Ailanthus excelsa* and *Bombax ceiba*. If the susceptible species are to be planted in a reforested stand, they should be raised with an intimate mixture of resistant species like *Ailanthus excelsa* and *Bombax ceiba*.

Disease resistance: Harsh (2012) identified resistant and susceptible clones of *Dalbergia sissoo* against *Ganoderma lucidum* in field plantations in Haryana in India and recommended to the Haryana Forest Department to use resistant clones for further multiplication.

35.3.6.2 Wilt Disease (Mortality) of *Dalbergia sissoo* (*Shisham* or *sissoo*)

Wilt disease of shisham or shisham mortality is widespread throughout the shisham distribution zone in India, being more severe in eastern Uttar Pradesh and Bihar. It falls under the category of decline diseases. Two pathogens have been found responsible for shisham mortality: *Fusarium solani* f. sp. *dalbergiae* causing vascular wilt and *Ganoderma lucidum* causing root rot (Fig. 35.5). The former pathogen was found spreading throughout shisham-growing area whereas *G. lucidum* is associated in drier localities such as Haryana and Punjab in India. The description of shisham mortality is based on the reports by Bakshi (1976), Shukla and Harsh (2010) and Harsh (2014b).

Symptoms: The symptoms are yellowing of foliage, followed by gradual wilting and drying leading to top dying, oozing of sap (pitch) and ultimately complete death of the plant. Decline/wilt is result of both root rotting and colonisation of vascular system of stem by the fungus, which block water transport though fungal structures and fungal metabolites and byproducts.

Extent of disease: Trees of all the age and diameter classes are affected in the monoculture plantations. Mixed plantations show less signs of mortality. Mature trees of more than 50 years age are highly affected in shisham plantations. Younger plantations of shisham show high mortality in some areas, particularly where biotic stresses, and erratic rainfall and extreme temperature variations/water-logging have occurred during recent years. Dead trees, partially dead trees as well as stumps that are not removed act as source of infection.

Predisposing factors for shisham mortality—edaphic, ecological and biotic: Unsuitable site selection is one of the main reasons of shisham decline; it grows well on well-drained sites and does not grow well in soils of heavy texture (with more clay and silt content and less sand). Improper soil aeration in stiff and clayey soils causes death of fine/feeder roots due to asphyxiation besides water logging. Root disease pathogens enter through these dead roots (infection courts) and then attack living roots. Extended period of low temperature/freezes expose the trees to disease/pathogen.

Biotic pressure: The following biotic pressures were found weakening the tree vigour:

- Injury to roots in agriculture (in fields) and construction activities (near roads, avenues)
- Cleaning and weeding done through harrowing in plantations causing root damage
- Covering of basal area with landfill, garbage and other material, particularly on roadsides and urban areas
- Uncontrolled lopping in agriculture fields
- Blazing of bark for numbering of trees (on roadsides)
- Injuries by cattle
- Intentional girdling (on roadsides for removal)

The above biotic pressure result into opening of wounds, site deterioration and overall weakening of vigour resulting in attack by the pathogens and decline/mortality.

Fig. 35.5 Wilt and mortality caused by *Ganoderma lucidum* in *Dalbergia sissoo*

Integrated Disease Management

- *Selection and preparation of site*: Raise shisham on a site having light textured soil with adequate soil moisture and good drainage; avoid stiff clayey soils with water-logging tendency.
- *Sanitation in the field of dead trees, stumps and roots*: This will not only allow the timely utilisation of the timber from the dead trees but also check the spread of the pathogen to the adjacent trees.
- *Avoid pure plantation of shisham*: Raise mixed plantations preferably with non-leguminous species.
- *Integrated disease management*: Application of biological control agents like *Trichoderma* spp. and *Pseudomonas fluorescens*, and chemicals–fungicides (Propicanojole/ Bavistin), insecticides (chloropyriphos) and fertilisers (after soil testing).
- *Selection of resistant clones and propagation*: Such material has been worked out by Forest Research Institute, Dehradun, and should preferably be used for future plantations. DS-14 is a clone of *Dalbergia sissoo* released by Indian Council of Forestry Research and Education (ICFRE) possessing resistance to *Fusarium* wilt disease. Resistant clones have been identified against *Ganoderma* root rot in Clonal Seed Orchard in Haryana and seeds from these clones are being used by Haryana Forest Department for raising the planting material.
- *Control biotic pressure* to check injuries to root system.

35.3.6.3 Sandalwood Decline

The deterioration of the health of sandalwood (*Santalum album*) trees in Seoni, Madhya Pradesh (India), was categorised as a decline disease (Harsh et al. 2000). Uncontrolled grazing and frequent fires eventually damage the ground flora, natural regeneration and leaf litter, thus contributing to site deterioration. This leads to the drying of roots, shoot and branch tips (die-back) and secondary invasion by the termites, decay fungi (*Flavodon flavus*, *Hexagonia tenuis* and *Schizophyllum commune*) and heart-rot fungi (*Phellinus caryophylli* causing punk knot). These are the series of events identified as the root cause of the decline. Management practices such as protection of the site from fire, grazing, theft and fencing and removal of dead and unsound trees were recommended.

35.3.6.4 Stem Cankers

The death of tissues in stem, branches and twigs is usually associated with a wound caused by broken branch or a mechanical injury, extending radially from the wound. Tattar (1988) termed this localised lesion as canker. Stem diseases occur on the trunk, branches and twigs. Cankers that occur only on shoots and twigs, such as anthracnose or juniper twig blight, could also attack leaves. Many rust fungi also cause cankers like white pine blister rust.

Cankers vary considerably in size and shape. The canker on a tree manifests the host–parasite interaction, which may often continue for many years. A typical sequence for canker development is as follows (Tattar 1988):

1. Entry of the pathogen into the host occurs through a wound, which then invades and healthy bark gets killed, during a dormant period of growth.
2. The host develops a layer of callus over the edge of infected tissue to check further invasion by the pathogen.
3. During the following dormant period, the pathogen invades the callus tissue.
4. Then new callus is formed by the host.

Every year throughout the life of the host, Steps 3 and 4 may be repeated. The type of the canker will be determined by the growth rate of the pathogen and the developed callus.

Most canker fungi remain confined to the bark; however, some fungi can attack the bark as well as the xylem underneath. Canker rots are the stem diseases resulting from a simultaneous canker and wood decay (Tattar 1988).

Fungi causing canker diseases in trees are the parasites in the living cells of phloem, cambium and outer xylem region. Symptoms range from small irregularities in the morphology of the stem, branches and twigs, which are caused by the callus formation around the cankers, to large areas of dead bark and death of the trees. Callus formation around the cankers may close the canker wound and most of the times result into abnormal swelling of the stem, bark and twigs around the canker. Canker fungi may become inciting and contributing factors in decline diseases of older trees.

Fungi causing cankers include Phycomycetes, such as *Phytophthora cactorum*, which causes bleeding canker of maples and other tree species. Basidiomycete fungi also cause cankers. Rust cankers are also there. Some Basidiomycete decay fungi, such as *Inonotus obliquus* (= *Poria obliqua*) and *Phellinus igniarius* (= *Fomes igniarius*), also cause cankers on northern hardwoods. *Inonotus hispidus* (= *Polyporus hispidus*), *Phellinus spiculosa* (= *Poria spiculosa*) and *Spongipellis pachyodon* (= *Irpex mollis*) are canker rots of southern hardwoods (Manion 1981). *Ganoderma applanatum* and *Phellinus pachyphloeus* were found associated with canker rots causing tree breakage in some hardwoods in Dehradun, India (Harsh et al. 2004). Basidiospores of these fungi infect through wounds. Mycelial invasion of sapwood results in decay up to 2 m above and below the canker infection point and cambium death.

Cankers are among the most destructive and hard to manage diseases in trees. Canker fungi cause girdling followed by the killing of the branches of trees and shrubs affecting the landscape. Cankers also act as the infection courts for other fungi, bacteria and insects, eventually hastening the decline and death (Manion 1981).

Canker disease and cankers caused by mechanical injuries can be the sites for invasion by wood-rotting organisms. Wood decay in trees causes the weakening of branches and main stem making them hazardous in high-velocity winds, heavy rains or snow. It is, therefore, imperative to prevent cankers from developing on trees. Proper care and maintenance of trees from cankers can check subsequent damage. Some examples of canker diseases are mentioned here.

Fusarium Canker

A typical example of an annual canker is *Fusarium* canker of sugar maple. *Fusarium solani* and a number of other fungi were isolated from annual cankers on sugar maple in Pennsylvania. Inoculation with pure cultures of the fungi demonstrated the pathogenicity of *F. solani* (Manion 1981). Uniyal et al. (2004) reported canker and die-back caused by *Fusarium sublateritium* f. sp. *acaciae* on *Acacia nilotica* from Dehradun. Harsh et al. (2006) have reported a new canker disease on *Eucalyptus camaldulensis* from Haryana caused by *Fusarium sambucinum.*

Nectria Canker

It is a very common defect of many hardwoods, including aspens, maples, birches, elm, basswood and apple. The red perithecia of *Nectria galligena* that may fruit on the margin of the canker positively identify it. The *Cylindrocarpon* imperfect stage that sometimes fruits on the margin of the young cankers is the stage of the fungus most often seen in culture (Manion 1981). The fungus can invade healthy trees through frost cracks and other wounds. Small injuries that would occur in a branch axil because of heavy snow pushing the branch down are points of infection in the Northeast. In apples grown in the western United States, infection occurs through conidia infecting leaf scars in the fall of the year. Whether leaf scar infection occurs in forest trees is not known. There is no known control for the disease. Canker-infected trees will not produce high-quality wood products and, therefore, should be cut during thinning operations (Manion 1981).

Ceratocystis Canker

It is sometimes diagnosed as the target-like series of concentric rings. *Nectria* canker is sometimes diagnosed based on the similar appearance, so confusion and error sometimes result. There are very few reports of perithecia of *Nectria* fruiting on target cankers of trembling aspen, but perithecia of *Ceratocystis fimbriata* are commonly seen fruiting on these cankers in Colorado and Minnesota. Controlled inoculation experiments with *C. fimbriata* and *Nectria galligena* show that both are capable of producing cankers in some instances. *Certaocysytis* canker has been reported by Pandey et al. (2002) on poplar from Dehradun and nearby areas.

Pestalotiopsis Canker

Harsh and Chandra (2008) have reported a twig canker disease of *Cinnamomum camphora* caused by *Pestalotiopsis versicolor* from Dehradun, India. Initial symptoms of the branches were noticed as that of wilt, leaves showing drooping and turbidity and gradually the leaves drying. The branches appeared prominently dead and brown among the green canopy. On examination of the affected branches, diamond- to irregular-shaped greyish multiple erumpent cankers were noticed scattered all over the surface of the twigs. The size varied from 1–2 mm in length in small cankers to 1–2 × 0.5–1 cm in larger erumpent diamond-shaped cankers. Larger cankers later on developed into pits on the twigs, 1–2 cm in diameter and up to 1 cm deep.

35.3.6.5 Pink Disease

Erythricium salmonicolor (B. & Br.) Burd. (= *Corticium salmonicolor* B. & Br.), the cause of pink disease, is widely distributed in the tropics and sub-temperate regions of Asia, Africa, West Indies, New South Wales and New Zealand. It attacks many plants of economic importance like coffee and cinchona in Java, tea in Ceylon, cocoa in West Indies and rubber in India and Malaysia. It attacks *Eucalyptus* planted as exotic. In India, it is recorded on *Eucalyptus grandis* (Hill) Maiden, *Eucalyptus tereticornis* Sm., *Eucalyptus citriodora* Hook. and *Eucalyptus globulus* Labill and probably other species. The fungus is recorded on *E. grandis* and *Eucalyptus camaldulensis* Dehn. in North East Brazil and on *Eucalyptus kirtoniana* F. v. m. in Costa Rica. The fungus has a wide host range comprising of 141 species of woody plants belonging to 104 genera (Sharples 1936). In India, fungus is recorded on a wide range of forest trees including *Acacia leucophloea* Willd., *Albizia falcataria* (L.) Fosberg, *Aleurites montana* E.H. Wilson, *Anacardium occidentale* L., *Artocarpus hetrophyllus* Lam., *Azadirachta indica* A. Juss., *Butea monosperma*(Lam.) Taub., *Cassia siamea* Lam., *Casuarina equisetifolia* Frost., *Casuarina montana* Jungh., *Eucalyptus alba* Reinw., *E. camaldulensis*, *E. globulus*, *Eucalyptus saligna* Sm., *Eucalyptus multiflora* Poir, *Eucalyptus tereticornis*, *Eucalyptus tessellaris* F. v. M., *Eucalyptus trabuti* Vilmorin, *Grevillea robusta* A. Cunn., *Hevea brasiliensis* Muell.-Arg., *Leucaena glauca* (Linn.) Benth., *Mangifera indica* Linn., *Podocarpus gracilior* Pilger, *Populus* spp., *Salix daphnoides* Will., *Tamarindus indica* Linn. and *Tectona grandis* Linn. f. (Browne 1968; Bakshi et al. 1970; Seth et al. 1978). High summer rainfall above 2500 mm and warm tropical conditions encourage development of different growth forms of the fungus, which is disseminated rapidly resulting in epidemic outbreak of the disease. The severity of the disease becomes greater with increasing rainfall above 2500 mm as in the Kerala, Goa and western districts of Karnataka (Sharma et al. 1985). Pink disease in eucalyptus clonal plantation in Laos PDR (Houydeua Village, Phonhong District, Vientiane Province) was studied by Harsh (2019).

All the three species of eucalyptus, namely *E. tereticornis*, *E. grandis* and *E. globulus*, which are widely planted in India are highly susceptible to the disease, which becomes epidemic under conditions of high rainfall and high temperature during the summer. *Eucalyptus* is susceptible to the disease at a very early stage. The disease appears sporadically during the second year but becomes epidemic in the two subsequent years when the entire plantation may become affected under ideal conditions for development of the disease. Young plants may show repeated die-back and mortality, while those that have escaped infection result in development of cankers.

In young plants, infection results in girdling of branches after canker development and the main stem towards the latter part of the rainy season. The affected plants are killed outright in *E. grandis* and *E. globulus*. In *E. tereticornis*, however, infection on the stem or branches causes girdling and death of parts above the girdled region. Simultaneously below the region of girdling, many epicormic branches develop, one of which becomes the leader, which again becomes infected and killed in the following year, and the process continues resulting in diminution of height growth of plants. The growth increment and productivity become seriously affected in plantations where the disease is epidemic.

If plants escape infection at an early age, they put up good height and diameter growth. However, such trees subsequently become infected on the main bole resulting in cankers splitting and cracking of bark and exposure of wood. Gum eventually exudes from the resultant cankers. The attack is localised and girdling is usually not complete and therefore, the main leader continues to grow and the crown appears healthy. *E. salmonicolor* occurs on the host in four growth forms as described below:

1. **Pustule**: Pustules appear on branches and stem within a few weeks after infection at the beginning of the monsoon as pink- to salmon-coloured sterile cellular bodies up to 1 mm in diameter.
2. **Cobweb**: The infected parts produce on the bark thin, white, arachnoid mycelia originating from pustules or from the edges of the pink incrustation.
3. **Necator**: The fruit bodies are orange red, about 2 mm in diameter and develop on upper side of the attacked branch exposed to sun. Conidia develop after the rains.
4. **Pink incrustation**: This comprises perfect stage of the fungus and develops as thin, light pink incrustation on the bark. It develops during autumn on stem and branches originating at the fork usually on the undersurface of the stem on the shady side, but spreads around the stem later on.

The different stages of fungus develop on the bark, which usually becomes killed. The disease is disseminated by the two sterile and two sporulating forms by wind to infect trees.

Effective control of pink disease of *Eucalyptus* has not yet been developed, though efforts are being made on the following lines. In high rainfall areas in Kerala and Karnataka where the pink disease has limited the success of *E. grandis* and *E. tereticornis*, it may be necessary to replace them with other species of *Eucalyptus* or other fast-growing species suitable for pulpwood and at the same time look for resistance to pink disease. This may be determined from species trials when the performance of the species and their resistance to the disease may be ascertained in the field. It may be mentioned that *E. torlliana* F. v. M., growing in heavily diseased locations, exhibits certain degree of freedom from the disease.

Pink disease in young was also observed in 1–5-year-old plantations of teak (*Tectona grandis*). It was also recorded from Kerala from areas that receive high rainfall (ca. 3000 mm annually) and was prevalent during the monsoon period (June–September) (Sharma et al. 1985).

The disease is characterised by a pink encrustation formed over a canker on the stem. The infection results in killing of inner bark with phloem and cambial tissue and further development of a canker. The portion of the branch and shoot above the canker gets killed outright after complete girdling by the pathogen. Longitudinal splitting of the bark is rampant during the dry weather. Epicormic branches develop soon below the canker.

Development of cobweb stage is the first sign of the infection on the stem during the monsoon followed by small pin-head size white mycelial bodies called pustules that develop over the cobweb. Brown depressions in the bark of this area are visible and leaves show wilting. The perfect stage, which is in the form of pink encrustation, develops over the dead tissue, which produces many club-shaped basidia and air-borne basidiospores as inoculums for fresh infection. Necator stage was not found in teak.

Pink disease of teak has been reported from Karnataka, India (Bakshi 1976), Indonesia (Schwartz 1926) and Laos PDR (Khair Village, Paklai District, Xayaboury Province) (Harsh 2019) (Fig. 35.6). Pink disease of *Bombax ceiba* in 4-year-old trees was recorded during September at the end of South-West monsoon at Mampazhathara (Sharma et al. 1985). The pink disease of *Ailanthus triphysa* was recorded at Kuttampuzha during 1980 and 1981 and in severe cases the top of the diseased tree died due to complete girdling (Sharma et al. 1985). In *Gmelina arborea* low severity of pink disease was recorded in the plantations at Kottappara and Onthupacha.

35.3.6.6 Stem Canker and Die-Back in *Ailanthus triphysa*

This disease caused by *Lasiodiplodia theobromae* (Pat.) Griff. & Maubl. was recorded in a 4-year-old plantation at Pothuchadi, Paravattani, Mala, and in a homestead at Mullakkara in Trichur area (Sharma et al. 1985). Initially small longitudinal cracks up to 30–60 cm in length appeared on the main stem at Paravattany and a dark gummy substance could be seen oozing. Immediately below the canker numerous epicormic branches developed, which got infected later due to spreading canker showing fructifications of the casual organism on the surface (Sharma et al. 1985). The plants showed symptoms of decline disease at Pothuchadi with a sparse crown showing fewer and yellowed leaves. The infection started at the root collar several millimetres deep rapidly spreading upwards, and several epicormic branches emerged. The basal stem canker reached downwards to the root system causing the infected trees to get killed due to complete girdling of the outer bark (Sharma et al. 1985), due to the causal organism of the disease.

35.3.6.7 Eucalyptus Cankers

Cryphonectria cubensis and *Cryphonectria gyrosa* cause serious stem and branch cankers in *Eucalyptus grandis* and other species in Kerala (Sharma et al. 1985) (Fig. 35.7). The pathogens invade through growth cracks at the base, branch stubs and occluded suppressed branches. The pathogens can grow from basal cankers on stumps to attack newly developed coppice stem (Barnard et al. 1987). Ascospores and conidia of *C. cubensis* are found in South America (Hodges 1980), India (Sharma et al. 1989), Indonesia, Thailand and Vietnam (Old et al. 2003) whereas in Africa, only conidia have been found (Wingfield et al. 1989). High rainfall (>2000 mm), high humidity throughout the year, average temperatures 23 °C or higher and the presence of susceptible host species favour the spread of the disease by *C. cubensis*. Information on the pathology of *C. gyrosa* is limited as

Fig. 35.6 Pink disease in teak

compared to *C. cubensis*. Large basal cankers can kill trees during the first 2–3 years of growth, whereas several metres large perennial cankers develop on the stem in older trees. Alfenas et al. (1983) reported that in susceptible species or clones, up to 50% trees in plantations have been killed under favourable climatic conditions. Cankers cause death of phloem, cambium and sapwood with partial girdling of trees and copious flow of kino. Economic effects of *C. cubensis* canker are reduced growth rate (Camargo et al. 1991), reduced coppicing (Hodges and Reis 1976; Sharma et al. 1985; Barnard et al. 1987) and increased mortality (Boerboom and Maas 1970; Hodges et al. 1979). Ferrari et al. (1984) reported significant loss in wood yield when cankers are affecting more than 25% of the utilisable stem length. The canker-affected wood is denser with short fibres and thinner cell walls and contains more extractives and lignin (Foekkel et al. 1976). Apart from loss in pulp yield, an increase in extractives adversely affects bleaching. Stands with more than 50% infected trees are not suitable for pulping (Foekkel et al. 1981).

Control

Some species of *Eucalyptus* are more susceptible to *C. cubensis* such as *E. grandis* and *E. saligna*, which adversely affects the plantations of these species. Control of the disease has been successfully achieved through hybrid plantations of *E. grandis* × *E. europhylla* and selection of resistant genotypes within these hybrids in Brazil. Clonal propagation of resistant genotypes and selection for resistant hybrid clones based on *E. grandis* and many other pure species were also made in South Africa (Van Zyl and Wingfield 1999; Van Heerden and Wingfield 2002).

A bark split and canker disease in Eucalyptus clone No. 413 caused by *Botryosphaeria dothidea* was reported by Chandra et al. (2016) from Haryana, causing as much as 100% mortality in water-logged areas. They found copper oxychloride spray effective in controlling the disease in field.

35.3.6.8 Casuarina Stem Wilt and Bark Blister Disease

The disease is caused by *Subramaniospora vesiculosa* (= *Trichosporium vesiculosum*) and first reported by Butler (1905) from India. Bakshi (1951) reported nearly 50% mortality due to this disease. The disease symptoms include yellowing and drying of needles resulting in wilt. On the stem bark blisters develop due to formation of the spores beneath the bark, which soon rupture to expose black powdery mass of spores. Sporulation is also noticed on the roots and the disease spreads through root-to-root contact showing diseased trees in groups.

Fig. 35.7 Eucalyptus stem canker by *Teratosphaeria zuluensis*

Unscientific lopping and pruning leads to establishment and spread of the disease in plantations. Wound dressing after cutting of branches should be used. Early removal of dead and diseased trees checks the spread of the disease in the plantation.

35.3.7 Diseases of High Forests

In old forests and older plantations established as forests, diseases cause damages and loss of productivity. Some major diseases are mentioned below.

35.3.7.1 Root Rots

Root Rot in Wet *Shorea robusta* (Sal) Forests

In Assam, Odisha, Uttar Pradesh and northern parts of West Bengal, high rainfall aided with fire protection measures adopted during 1940s and 1960s resulted in excess soil moisture and heavy weed growth (Bakshi 1976). This exposed them to the root-rot pathogen *Aurificaria shoreae* (= *Polyporus shoreae*) causing root rot called Partridge root rot because of its peculiar pattern. The symptoms comprised of top dying, mortality, uprooting and wind-throw. The removal of diseased/dead trees helps in inoculum eradication and disease management. Controlled burning to lower soil moisture and checking weed growth is effective in reducing soil moisture and controlling the disease.

Root Rot in Conifers and Hardwoods

The root-rot affected trees showed top-dying symptoms preceded by yellowing of leaves and reduction of crown. Root-rot diseases spread very rapidly from tree-to-tree through root-to-root contact. The affected trees are liable to windthrow as the anchor roots get decayed and thus are unable to hold the trees. The falling trees injure nearby ones also. Root rot is caused by *Armillarea mellea* and *Heterobasidion annosum* in conifer forests and by *Phytophthora cinnamomi* in *Cedrus deodara* in the temperate parts of the country, and by *Ganoderma lucidum* in tropical and subtropical deciduous and dry deciduous forests of hardwoods such as teak, acacias and dalbergias.

Control measures of root-rot diseases have already been mentioned under plantation diseases.

35.3.7.2 Dying of Dry Sal (*Shorea robusta*)

Widespread drying and dying of sal trees in Bihar, Uttar Pradesh and Madhya Pradesh was investigated by Prasad and Jamaluddin (1985). The symptoms include top-dying and mortality and the causes were identified as lessening of soil moisture due to continued droughts and site deterioration by fire, grazing and soil erosion. Protection from fire and grazing and contour trenching to hold water run-off were recommended.

35.3.7.3 Dwarf Mistletoe on Blue Pine

In low rainfall areas in the Himalayas, dwarf mistletoe *Arceuthobium minutissimum* causes heavy damage to blue pine (*Pinus wallichiana*) in Uttarakhand, Himachal Pradesh and Jammu & Kashmir (Bakshi 1976; Shukla and Singh 2007; Rawat et al. 2007). The infection shows spindle-shaped or fusiform swellings on branches or main stem causing deformity resulting into witches' brooms, die-back and death of affected branches. Heavy infection results into stunting of the trees leading to mortality (Bakshi and Puri 1971). *A. minutissimum* is one of the smallest flowering plant parasites. The infection spreads by its fruits (berries), which on ripening develop internal pressure to eject the seeds with great force sometimes up to ten metres in distance. The seeds get adhered to the branches and stems due to their mucilage content and upon germination initiate infection. Management can be done by sanitary cuttings of diseased trees and branches in lightly infected stands whereas clear felling and burning are recommended in heavily diseased stands. New regeneration should be delineated by 20 m width (effective range of dispersal of seeds of dwarf mistletoe is 10 m). Species substitution by *Pinus gerardiana*, which is not the host to dwarf mistletoe and also a valuable cash crop, was also suggested.

Another leafy mistletoe *Taxillus kaempferi* causes severe damage in conifers in neighbouring Bhutan (Yangchen 2017) (Fig. 35.8).

35.3.7.4 Heart Rot: Decay in the Heartwood of Standing Trees

Heart rot is the disease that causes decay in the heartwood amounting to extensive loss of wood volume in living trees. At advance stages of decay in the heartwood, the disintegration of rotten wood leaves a clear hollow bole in the centre of the tree. External indicators like swollen boles, punk knots, branch stubs, dead branches and sporophores are prominent in malformed and heart-rotted trees for marking and extraction. The heart-rot-causing fungi belong to Basidioimycotina, which are responsible for causing white or brown rot in the wood and spoiling the merchantable quality of the timber. In majority of cases the fungi belonging to genera *Fomes*, *Fomitopsis*, *Hymenochaete*, *Inonotus*, *Phellinus*, etc. cause decay in the heartwood. Heart rot in three important species is discussed below

Sal high forests: In sal forests of the Sivaliks, incidence and volume of decay was high and, in an appraisal, nearly 70% trees were found affected resulting 10% loss in wood volume (Bakshi 1976). External indicators exhibited by the affected trees were swollen bole, punk knots, branch stubs and fruiting bodies of fungi. Injuries caused by frost to young plants establish heart rot by *Phellinus caryophylli* causing classical punk knots on the outer side of the bark. Fire injuries establish heart rot by *Hymenochaete rubiginosa.* Protection of trees from injuries due to frost and fire, removal of decayed trees during thinning, improvement and selection felling, lowering of rotation age and eradication of fruiting bodies of fungi were the management practices suggested.

Sal coppice forests: Decay passes to trees when coppice shoots are retained with heart rot in the stool and limitation of rotation age is up to 15–20 years. Coppicing from stools should not to be allowed and plants of seedling origin be allowed to grow.

Dry coppice teak: In dry locations in western parts in Gujarat, teak is worked on a system coppice-with-standard on a rotation of 30 or 48 years. At the end of the rotation, trees exhibit unsoundness/hollowness extending up to 2–4 m in the stem. After felling side and callus shoots arise from the stool, but the practice was to remove the side shoots and retain callus shoots. Due to connection between the heartwood of the stool and callus shoots, the decay passed on to them (Singh et al. 1973). It resulted in large-scale dying. At the end of the rotation or before, dry teak commonly exhibits unsoundness. The incidence was

Fig. 35.8 Leafy mistletoe *Taxillus kaempferi*. (Photo by Tashi Yangchen)

reported as high as 50%. Trees from high coppice and side shoots develop decay from the stool by the fungus *Rigidoporus zonalis* in Gujarat. Timely application of management and silvicultural practices may prevent or reduce losses due to heart-rot disease in trees. Thinning and improvement felling help in stand management. By retaining the side shoots of lower origin, the chances of decay entering from the stool into the coppice shoots could be lowered as they develop their independent root system. The coppicing height of 10–15 cm was found suitable for initiation of low side shoots and discouraging callus and high side shoots, which are more susceptible to heart rot. Removal of coppice and high side shoots during cleaning operations was recommended to reduce the damage.

Teak high forests: Heart rot in dry teak was observed in Madhya Pradesh, which was caused by *Perenniporia tephropora* (Harsh and Tiwari 1995). In the appraisal 38–88% trees were found affected entailing a loss of about 11% in wood volume. The annual monetary loss was estimated to be around 200 million rupees based on the market value of teak wood.

Blue pine (*Pinus wallichiana*): In western Himalaya, blue pine trees were heavily lopped to permit sunlight for agriculture and horticulture crops. The wounds established *Phellinus pini*, causing heart rot in the stem. Restriction of lopping, removal of decayed trees and eradication of fruiting bodies of the fungus were recommended (Bakshi 1976).

35.3.7.5 Spike Disease of Sandalwood

Sandalwood (*Santalum album*) spike disease is unknown outside India. This infectious disease caused by phytoplasma (mycoplasma) is transmitted by insect vectors from plant to plant. No cure could be found for this disease. Every year nearly 1–5% of sandalwood trees are lost due to spike disease. The researchers have warned that the entire natural population of sandalwood is at the threat of elimination if suitable measures are not found to check the spread of the disease. The delay in checking the disease poses serious threat of its spreading to cultivated sandalwood trees. Severe reductions in the size of the leaves and in the length and the breadth are the characteristic symptoms of the disease. Due to shortening of the internodes, the leaves become closely crowded standing out stiffly like spikes from the branch giving a rosette appearance. Leaves stand out like spikes and hence the disease is so named. Leaves become yellowish and may turn reddish tinge later on. Leaf formation and their development are affected. The infected shoots show continuous apical growth and have a drooping habit giving a pendulous appearance to the spike. Infected branches have flowers exhibiting phyllody (retrograde metamorphosis of floral organ). Clustering of branches results into witches' broom.

Removal of infected host trees manually and avoiding monoculture of sandalwood are recommended. Different host plants of sandal impart resistance or susceptibility to spike disease in sandal. The host plants *Azadirachta indica*, *Bambusa arundinacea*, *Cassia siamea*, *Casuarina equisetifolia*, *Dalbergia sissoo*, *Dodonea viscosa*, *Ficus benghalensis* and *Murraya konigii* provide relative resistance whereas *Caesalpinia coriaria*, *Acacias*, *Cajanus cajan*, *Lantana camara* and *Pongamia pinnata* render sandalwood susceptible to spike disease and should be avoided. The selection of host plants should be confined to fast-growing species with a light crown, as sandalwood is essentially a light-demanding species. Apart from the acquired resistance imparted to sandal by host plants, freedom from spike disease could possibly also be due to the fact that the top canopy serves as a protective umbrella against vectors transmitting the disease to sandalwood trees. Phytoplasma diseases in plants are known to be amenable to treatment with antibiotics, which however cause temporary emission in disease symptoms as obtained for spike disease in sandalwood through use of tetracycline or with benlate, a systemic fungicide. The recovery from the disease lasts only for few weeks after which the symptoms reappear. Spike-disease-resistant sandalwood may be developed through selection of apparently resistant sandal trees in a heavily spiked area and testing them for resistance to spike disease through grafting and insect transmission test. Resistant individuals thus screened can be propagated vegetatively in future. Indian sandalwood may be hybridised with other species of Santalum to develop desirable hybrid resistant to spike disease.

35.3.7.6 Diseases of Bamboos

Foliar Diseases in Nurseries

Foliar diseases are serious because they cause premature defoliation apart from checking and reducing photosynthetic activity, which reflect on the vigour of planting stock. There are many foliar diseases occurring in bamboo nurseries but most common are leaf spots caused by *Exserohilum rostratum* and leaf rust by *Dasturella divina*.

Diseases of Culms

Diseases of culms reported by Mohanan (1997) are discussed here.

Rot of New Culms

The new culms soon after emergence are attacked by the fungus *Fusarium verticillioides* and killed. Initially greyish brown lesions cover the culm sheath but later on whole culm tissue gets affected.

Cleaning around the soil and drenching with 0.1% Bavistin and 0.1% Copper oxychloride is useful in controlling the disease.

Rot of Growing Culms

After injuries on culm sheaths and nodes made by the sap-sucking insect *Purohita cervina*, a fungus *Fusarium equiseti* starts infection. Water-soaked greyish-brown spindle-shaped lesions start developing at the base of culm sheaths, which enlarge to form dark necrotic areas on the entire sheath and cause rotting of culm surface below. Severely infected culms cease to grow, become shrivelled and fall off.

Insecticidal spray to control the sap-sucking insects and spray with 0.1% Bavistin and 0.1% Copper oxychloride is useful in controlling the disease.

Die-Back of Branches

Gradual death of the branches from the tip backwards is caused by *Fusarium pallidoroseum*. The repeated die-back results in deformity.

Spray with 0.1% Bavistin and 0.1% Copper oxychloride is useful in controlling the disease.

Sheath and Culm Blight

The disease has been reported from the coastal areas of Orissa and causes severe mortality in *Bambusa bambos*. It is caused by *Sarocladium oryzae*, a common sheath blight pathogen in rice fields (Jamaluddin et al. 1992).

Cleaning of leaf litter and debris and spraying with 0.3% Dithane M-45 and 0.2% Copper oxychloride have been reported to control the disease.

Culm Rot

Initially profuse white mycelial growth at the base of clumps spreading over dried fallen leaves is noticed which is soon followed by circular- to fan-shaped white mycelial patches (subiculum) spreading over the culm sheath. Needle-shaped white to dirty white, branched fruiting bodies arise from the affected culm surface from the subiculum with a clear disc-like attachment at the base of the fruit bodies followed by necrotic lesions on the culm surface and curling of internodes and shortening of nodes, resulting into deformity. Ultimately, drying up of culms from tip backwards results in mortality. The pathogen is *Pterulicium xylogenum*. Harsh et al. (2005) found that the disease can be controlled to an appreciable level by spraying and drenching with a mixture of 0.05% copper oxychloride and 0.05% carbendazim.

Fungal Deterioration in Storage and Use of Bamboos

Bamboos are subjected to fungal deterioration during storage and use, affecting their strength and aesthetic values. Nearly 27% bamboo is lost by weight during 1 year of outside storage due to fungal decay (Harsh and Kapse 1999). Fungal decay especially of white rot type decomposes both lignin and cellulose and affects the pulp and paper quality and quantity and also adversely affects the strength properties.

The common decay fungi are *Coriolopsis telfarii*, *Datronia caperata*, *Flavodon flavus*, *Lenzites acuta*, *Polyporus arcularius*, *P. grammocephalus*, *Schizophyllum commune*, etc. (Harsh and Kapse 1999).

Proper storage conditions to facilitate faster drying and seasoning and early utilisation can help reduce the losses due to decay. Biological control using *Trichoderma* spp. has been reported for the management of fungal decay of stored bamboo.

Stains and Moulds

Being rich in carbohydrates, the cut and split bamboo are soon attacked by various moulds that impart different colours and stains. Common fungi are species of *Alternaria*, *Aspergillus*, *Cladosporium*, *Curvularia*, *Fusarium*, *Monilia*, *Penicillium*, etc. The problem can be resolved by proper drying and seasoning soon after felling and splitting. Besides, spray with 1% Borax or Boric acid soon after felling is also found useful in checking this problem.

35.3.8 Fungal Decay and Discolouration (Stain) of Wood/Timber

35.3.8.1 Decay or Rot

Breaking down of wood and change in its physical and chemical properties is called decay or rot. A wood-decaying fungus utilises sound wood by enzymatic degradation of the wood cell walls. Fan-shaped patches, strands, rhizomorphs, fruit bodies as mushrooms, brackets or crusts are the distinguishing structures of wood-decaying fungi. Lignin-cellulose/cellulose are used as food by the mycelium that enters the wood cell walls. Heartwood in standing trees is attacked by some decay fungi causing heart rot, and adjacent sapwood is also attacked sometimes. Most of the wood-decaying fungi can be seen growing on logs, sawed and structural timber, poles, etc. It is not unusual that some decay fungi continue to grow from trees to converted wood products. The wood having moisture above fibre saturation point (27–30%) but below wetness is attacked by the wood-decaying fungi. Decay and discolouration of wood occur simultaneously. Rots are of following types.

Brown Rot

In brown rot of wood, cellulose is utilised while the lignin remains unaltered or slightly modified. The cuboidal cracking of the wood is the distinguishing feature of brown rotten wood.

White Rot

The lignin and cellulose components of the wood are utilised; wood becomes bleached retaining outward dimensions. It is of white fibrous rot and white pocket rot types in the affected wood.

Dry Rot

Dry and friable condition of rotted wood is known as dry rot, found mostly in buildings. Particular decay of timber is brought about by the activity of fungus *Serpula lacrymans* (= *Merulius lacrymans*). *Lacrymans* means 'weeping'; during active growth, numerous globules of water sparkle in light as tear drops. *Merulius* refers to the bright yellow colouration of mycelium similar to colour of the beak of a male blackbird (*Merula*). Mycelium on timber surface is found as thin sheets of silvery-grey or mouse grey-tinged with bright yellow patches in dry conditions. Rusty red/mustard yellow spores germinate at wood moisture of 20–30%, though hyphae can colonise timber even at a lesser moisture content of 20% and optimum growth occurring at 30–40%. The term dry rot comes as the rot occurs at low moisture content of wood as compared to other wood decay. Fruit body of *M. lacrymans* is thin, pan-cake like, white/yellow round the edges with centre thrown into corrugation. Decay is of brown-rot type, and decayed wood shows deep transverse and longitudinal fissures as cuboidal cracking. Rhizomorphs are brittle when dry and flexible when moist, up to 6 mm in diameter and formed from hyphae to form vein-like structures. They can spread through plaster, brick-work and masonry and can reach out several metres. Rhizomorphs are conducting strands that transport nutrients and moisture. Metabolic breakdown of wood by the mycelium produces water, which is translocated from one timber to another via these thick strands. Dry rot requires a moisture content of timber above 20%, which is essential for dry rot to occur, and it becomes dormant below 20% and eventually die.

Management

- Cut out all timber showing cuboidal cracking, mycelium and sound timber within a radius of 1 m and burn.
- Hack away all plaster and rendering coats and remove any skirtings, linings, studding, panelling and ceilings, over or through adjacent brick, block, concrete or timber surface necessary to trace extent of mycelium.
- Thoroughly clean with a wire brush all surface up to a radius of 2 m; collect dust and debris and burn.
- Subject to heat treatment all bricks, block, concrete and earth surfaces with a heat machine to kill surface mycelium and dry out excess moisture.
- Apply fungicidal fluid to all such brick, block, concrete and earth surfaces.
- Apply two coats of fungicidal fluid to all timber surfaces adjacent to cutting away to a distance of 2 m.
- Select well-seasoned timber for replacement and replace after giving fungicidal treatment.

- Re-render with cement, lime and sand (1:1:6) and then apply two coats of 3-in. oxychloride paint.
- Check and remove causes of dampness.

Wet Rot

Type of decay is white rot, producing a soft, felty or spongy texture; attacks both softwoods and hardwoods and hollow giant beams. Wet rot occurs in persistently damp conditions with optimum moisture content of 50–60%, such as in cellars, mines, etc. It is caused by *Coniophora cerebella* (cellar fungus) and other species like *Coniophora puteana*, *Antrodia* sp., *Phellinus contiguus*, *Pleurotus ostreatus* and *Donkioporia expansa*. Fungal strands are not as thick as that of dry rot, are brownish or black, and yellowish brown when fresh. Fungal strands when growing on the surface of the wood or plaster are vein-like and said to be similar to that of the blood vessels of the cerebellum, hence specific name for *C. cerebella*. These conducting strands do not extend far away from their nutrient source. White mycelium is not found and fruit bodies of fungi are rarely found in buildings though common outdoors as a thin plate, are olive brown in colour, of indeterminate shape covered with small tubercles, and spores are less common.

Management

- Remove timber that already suffered break down.
- Protect timber already attached by retaining adequate strength through prying.
- Protect adjacent timber with possible latent attack.
- Cut out and remove timber that suffered surface or sun-surface breakdown and burn all dust, dirt and debris.
- Select well-seasoned timber for replacement after two brush coats of fungicidal fluids and on adjacent surface of existing timber, brick, block and concrete.
- Detect and remove defective damp courses and treat.
- Re-render with cement, lime and sand (1:1:6) and set in anhydrous plater.

Other wood-decaying fungi found in buildings are *Antrodia serialis*, *Daedalea quercina*, *Gloeophyllum sepiarium*, *Laetiporus sulphureus*, *Neolentinus lepideus*, *Paxillus panuoides*, *Peniophora gigantea*, *Phellinus megaloporus*, *Tyromyces placenta* etc.

35.3.8.2 Sap Stains

The pigmented hyphae grow primarily in parenchyma tissues of sapwood causing sap stains. Shades of blue, black, brown or grey, yellow, pink or green cause sapwood discolouration. The pigments secreted by hyphae cause stains predominantly of bluish discolouration and called 'Blue stain'. Harvested trees with higher percentage of sapwood when seasoned under warm and humid conditions result into stains development. Rapidly colonise ray parenchyma to utilise readily available carbohydrates, as they grow through the wood-pigmented hyphae and discolour the wood. Species of *Ophiostoma*, *Ceratocystis*, *Alternaria*, *Cladosporium*, *Phialophora*, etc. are the common species causing discolouration. Species of *Ophiostoma* and *Ceratocystis* are associated with bark beetles and other wood-inhabiting insects. Sticky spores are transmitted by insect vectors and water splashing. During initial stages of lumber seasoning and storage of logs, sap stains flourish. Other fungi (species of *Alternaria*, *Cladosporium*) are opportunists; their dry spores are disseminated by air and attack wood in wide range of uses. Free water, adequate temperature (4–30 °C), O_2 and a food source are required for the growth of sap stain fungi. Spores are either airborne and/or disseminated by insect vectors (ambrosia beetles) and sawmill machinery from infected wood to other surfaces. On landing on wood surface, the spores germinate within hours and then penetrate wood surface under favourable conditions through open surfaces of tracheids and wood rays. The growing hyphae by direct pit penetration rapidly colonise the parenchyma cells in the wood rays or longitudinal parenchyma surrounding the resin canals. Movement of the infection along the rays from the sides results into wedge-shaped stain on cross-cut surfaces of logs. After some weeks, hyphae penetrate the wood cell walls directly to move between tracheid and fibre walls badly damaging the parenchyma cells. Under suitable moisture conditions after prolonged incubation, a few stain fungi act as soft rot fungi and attack S2 layer of tracheids, for example, *Alternaria alternata*.

Sap stain/discolouration causes aesthetic and physical changes in wood properties and value. Stained wood consumes more bleach thus increasing the cost. Stained wood has aesthetic value as blue pine panels. Stains reduce wood toughness and are not recommended for structural purposes. Stains are conducive for decay initiation. The wood permeability is increased by removing pit membranes so that wood rapidly becomes wetter and drier, absorbs excessive solution during finishing and enhances the risk of fungal decay as it absorbs water.

Management

- Quick removal of the logs from the forests, faster conversion into dry lumber or storage in water ponds can prevent or reduce the fungal staining.
- Creating conditions unfavourable for the stain fungi such as keeping the wood dry, maintaining temperatures above or below optimum and protecting the wood from insect attack can prevent blue stains.
- Moisture content of 20% or less does not allow blue stain fungi to grow in wood. Temperatures above 65 °C are lethal to blue stain.

- Blue stain fungi have sticky spores that are carried into wood by insects, so insect control measures are effective.
- Chemical wood preservatives or biocides make the sap wood unsuitable for blue stain fungi.
- End-coating is the process of applying a wax-like barrier at the log ends or other exposed areas to slow down moisture loss. End-coating of the logs can reduce the risk of stain fungi for a month or 2 if there is delay in transportation of the logs from forest.
- Use of a safe and registered fungicide at the exposed areas of the log kills stain-causing fungi.
- Mixing of logs with stain problems with sound logs is not a good practice.

35.3.8.3 Soft Rot

Savory in 1954 reported an unusual decay in wood by a fungus *Cheatomium globosum* belonging to Ascomycotina, which he termed as soft rot. Macroscopic features include surface softness, and dull grey to brown colour showing surface cracking. Only water-soaked wood such as in cooling towers, jetties and wooden poles put up in soil suffer from soft rot. Soft rots showed the formation of unique longitudinal bore holes also called diamond cavities in the S2 layer of secondary cell wall and are known as Type 1 soft rots. Some eroded the secondary cell wall similar to white rot and are called Type 2 soft rots. In soft rot, all the cell carbohydrates in the S2 layer of the secondary cell wall preferentially attack forming longitudinal cavities (Type 1) or the cell wall is gradually degraded from the lumen surface or S3 layer.

35.4 Conclusion

The diseases and deterioration of forest plants and products cause severe losses to plantation and productivity targets, environment amelioration, ecological goods and services, and landscape development. The presence of a large number of susceptible hosts, an aggressive pathogen and favourable environmental conditions determine the severity of a disease. When all these factors corroborate with each other, outbreak of an epidemic disease occurs. In forest nurseries, planting materials in beds and/or containers provide a continuous crop of seedlings and management practices like over- or under-watering, overcrowding, injuries to root system while shifting, using same bed area year after year for same plant species, inefficient use of fungicides etc. provide congenial conditions for the outbreak of epidemics. Epidemics occur when a continuous crop of hosts in monocultures in plantations and a virulent pathogen are present. Under favourable environmental conditions, the indigenous pathogens and/or the pathogens introduced with the exotic hosts may cause epidemics. Whereas in the natural forests a large number of pathogens exist, however, only a few of them become serious enough to cause epidemic diseases.

Disease diagnosis is the first step in the treatment of a disease. Accurate diagnosis of diseases of forest plants is a complex activity requiring a combination of many disciplines related to plant health, such as botany, entomology, microbiology, plant pathology, plant physiology and soil science as well as broad exposure to forestry. The key features of a good diagnostician (plant doctor) are an inquisitive mind, a reluctance to jump to hasty conclusions, a persistent determination to find the cause of a problem and a willingness to seek additional information from other specialists in the field and reference texts.

Correct diagnosis and timely appropriate treatment can help minimise the losses caused by the diseases in nurseries, plantations and high forests. The cost of control measures in nurseries is bearable whereas in large plantation and forest areas it becomes prohibitive. However, use of drones for spraying the pesticides has made it cost effective. For long-term effective control of diseases in forests and plantations, use of disease-resistant planting material is the answer.

Increasing pressure on forests and forest produce will demand more production, which will dictate the need of protection from diseases and insect pests besides other biotic and abiotic disturbances. With the changing climatic conditions, there will be more outbreak of epidemics in the forests (Harsh 2014a).

Lessons Learnt

- Diseases of tree species are responsible for failure of tree crops in nurseries, plantations and forests.
- Timely diagnosis and control measures for the management of these diseases are, therefore, key to the success of tree crops.
- For large areas of plantations and forests, disease-resistant planting material should be used to meet the desired goals.
- Assessment of biotic and abiotic factors predisposing the trees to diseases will help in controlling them.
- Use of information technology for the assessment of damage caused by diseases to tree crops and their management will provide quicker and cost-effective solutions.
- Studies on climate change influence on diseases of trees will help in addressing the tree health issues in more pragmatic way.

Key Questions

1. Why not the losses due to diseases and insect pests to the forest crops are estimated to justify the cost of control measures?
2. Do we have sufficient number of forest pathologists in the country to address the disease problems in such vast areas under plantations and forests?

3. Do we prepare forest professionals to understand the importance of diseases and insect pests so that plantation managers can approach the specialists in time when problems arise in field?
4. What is our preparedness in the face of challenges being put up by the climate change scenario for diseases and insect pests attack in the forests?
5. Do the policy makers and general public take forests and trees seriously to understand their problems?

References

Alfenas AC, Jeng R, Hubbes M (1983) Virulence of *Cryphonectria cubensis* on Eucalyptus species differing in resistance. Eur J For Pathol 13:197–205

Anon (1958) Timber resources for America's future. Forest Resources Report 14. US Forest Service Publication, USDA

Bakshi BK (1951) Mortality of *Casuarina equisetifolia* Forst. Indian For 77:269–276

Bakshi BK (1976) Forest pathology—principles and practice in forestry. Forest Research Institute & Colleges, Dehradun, 400 p

Bakshi BK, Puri YN (1971) Dwarf mistletoe on blue pine in the Western Himalayas and its control. Indian For 97:547–552

Bakshi BK, Reddy MAR, Singh S, Pandey PC (1970) Disease situation in Indian forests. 1. Stem diseases due to *Corticium salmonicolor* and *Monochaetia unicornis*. Indian For 96:826–829

Bakshi BK, Reddy MAR, Puri YN, Singh S (1972) Forest disease survey (final technical report). Forest Pathology Branch, Forest Research Institute & Colleges, Dehradun, 117 p

Barnard EL, Geary T, English JT, Gilly SP (1987) Basal cankers and coppice failure of *Eucalyptus grandis* in Florida. Plant Dis 71:358–361

Boerboom JHA, Maas PW (1970) Canker in *Eucalyptus grandis* and *E. saligna* in Surinam caused by *Endothia havanensis*. Turrialba 20: 94–99

Browne FG (1968) Pests and diseases of forest plantation trees. Clarendon Press, Oxford, p 1330

Butler EJ (1905) Some Indian forest fungi. Indian For 31:487–496

Camargo LE, Filho AB, Krugner TL, Chaves RAB, do Couto, H.T.Z. (1991) Growth losses in Eucalyptus due to eucalyptus canker caused by *Cryphonectria cubensis*. Phytopathology 81:1235

Chandra S, Chander J, Harsh NSK (2016) First report of *Botryosphaeria dothidea* bark split and canker disease in Eucalyptus clone no. 413 from Haryana. Indian For 142:788–791

Ferrari MP, Krugner TL, Couto HTZ (1984) Avaliacao de perdas em rendimento de madeiras devido ao cancro do Eucalyptus causado por *Cryphonectria cubensis* (Bruner) Hodges. Instituto de Pesquisas e Estudos Florestais Report 27:9–15

Foekkel CEB, Zrinakevicius C, Andrada de Papel JOM (1976) Evaluation of quality of wood of *Eucalyptus saligna* and *Eucalyptus grandis* affected with canker. Abstr Bull Inst Paper Chem 48:1553

Foekkel CEB, Zrinakevicius C, Andrada de Papel JOM (1981) Eucalypt canker and its influence on quality of kraft pulp. Abstr Bull Inst Paper Qual 52:1112

Garrett SD (1970) Pathogenic root-infecting fungi. Cambridge University Press, Cambridge, p 294

Gibson IAS (1975) Diseases of forest trees widely planted as exotics in the tropics and southern hemisphere. Part I. Important members of the Myrtaceae, Leguminosae, Verbenaceae and Meliaceae. Commonwealth Forestry Institute, University of Oxford

Harsh NSK (2012) Disease resistance in genetic material in tree improvement programme—screening disease resistance in trees. Lambert Academic Publishing GmbH & Co., Saarbrücken, 69 p

Harsh NSK (2014a) Health and hygiene of national forests. In: Bhojvaid PP, Khandekar N (eds) Sustainable forest management for multiple values: a paradigm shift, vol 1. Forest Research Institute, Dehradun, pp 133–147

Harsh NSK (2014b) Shisham mortality—finding solutions for future plantations (2010–2014). Project Completion Report submitted to ICFRE, Dehradun

Harsh NSK (2019) Annual progress report of the project "Improved sanitary and phytosanitary (SPS) handling in greater Mekong Sub-region (GMS) trade project in Laos PDR". Submitted to the Asian Development Bank.

Harsh NSK, Chandra S (2008) A new twig canker disease of *Cinnamomum camphora*. Indian J For 31:217–219

Harsh NSK, Gupta BN (1993) Root-diseases in forest nurseries of Madhya Pradesh and their control. T F R I Information Series No. 1, Topical Forest Research Institute, Jabalpur, 31 p

Harsh NSK, Kapse NK (1999) A biocontrol formulation for protection of stored bamboo decay fungi. In: Singh J, Aneja KR (eds) From ethnomycology to fungal biotechnology: exploiting fungi from natural resources for novel products. Plenum Publ. Corp., New York, pp 145–152

Harsh NSK, Tiwari CK (1995) Assessment of damage caused by heart rot in teak in Madhya Pradesh. Indian For 121:540–544

Harsh NSK, Soni KK, Tiwari CK (1993) *Ganoderma* root-rot in an *Acacia* arboretum. Eur J For Pathol 23:252–254

Harsh NSK, Soni KK, Tiwari CK, Verma RK, Jamaluddin (2000) Decline of sandal trees in Seoni district of Madhya Pradesh. J Trop For 16:85–91

Harsh NSK, Chandra S, Singh YP, Shukla AN (2004) Canker rots—cause of tree breakage: a case study. Indian J For 27:257–260

Harsh NSK, Singh YP, Gupta HK, Mishra BM, McLaughlin DJ, Dentinger B (2005) A new culm rot disease of bamboo in India and its management. J Bamboo Rattan 4:387–398

Harsh NSK, Gupta HK, Ginwal HS (2006) A new twig canker disease of *Eucalyptus camaldulensis* from India. Indian For 132:121–124

Hodges CS (1980) The taxonomy of *Diaporthe cubensis*. Mycologia 67: 542–548

Hodges CS, Reis M (1976) A canker disease of Eucalyptus in Brazil caused by *Diaporthe cubensis*. Brazilian Institute for Forestry Development, UNDP and Agriculture Organisation of the United Nations, FO: DP/BRA/71/545 Field Document No. 14

Hodges CS, Geary TF, Cordell CE (1979) The occurrence of *Diaporthe cubensis* on Eucalyptus in Florida, Hawaii and Puerto Rico. Plant Dis Rep 63:216–220

Jamaluddin (1988) The status of diseases in agroforestry. Workshop on technology development for wastelands through agroforestry. State Forest Research Institute, Jabalpur, Madhya Pradesh

Jamaluddin, Gupta BN, Bohidar SC, Dadwal VS (1992) Mortality of bamboo (*Bambusa nutans* Wall.), in the coastal areas of Orissa. J Trop For 8(3):252–261

Jamaluddin, Harsh NSK, Nath V (1997) Handbook of diseases in tropical tree species. T F R I Publication No. 8, Topical Forest Research Institute, Jabalpur, 53 p

Kumari S, Harsh NSK (2004) Host range of *Ganoderma lucidum* in India. Indian J For 27:213–215

Manion PD (1981) Tree disease concepts. Prentice-Hall Inc., Englewood Cliffs, NJ, 399 p

Mohanan C (1997) Diseases of bamboos in Asia: an illustrated manual. INBAR, New Delhi, p 228p

Old KM, Wingfield MJ, Yuan ZQ (2003) A manual of diseases of eucalyptus in South East Asia. Center for International Forestry Research, Bogor, 98 p

Pandey A, Harsh NSK, Chandra S (2002) A new stem canker disease of poplar in Doon Valley. Indian For 128:1033–1036

Prasad R, Jamaluddin (1985) Preliminary observations on sal mortality in Madhya Pradesh. Indian For 111:250–271

Rawat PS, Chandra S, Khaneja A (2007) Mortality of *Pinus wallichiana* by dwarf mistletoe in Uttarkashi. Indian For 133:937–944

Schwartz MB (1926) Cobweb disease of mahogany and teak, vol 2. Korte Meded Inst Plziekt Buitenz, 9 p

Seth SK, Bakshi BK, Reddy MAR, Singh S (1978) Pink disease of Eucalyptus in India. Eur J For Pathol 8:200–216

Sharma JK, Mohanan C, Maria Florence EJ (1985) Disease survey in nurseries and plantations of forest tree species grown in Kerala. KFRI Research Report 36, 268 p

Sharma JK, Mohanan C, Florence EJM (1989) Diseases of forest trees in Kerala: diseases of eucalyptus in plantations. Evergreen (Trichur) 23:4–6

Sharples A (1936) Diseases and pests of the rubber tree. Printed by MacMillan & Co. Ltd., London, pp 267–277

Shukla AN, Harsh NSK (2010) Extent and evaluation of die-back of shisham (*Dalbergia sissoo*) (2006–2010). Project Completion Report submitted to ICFRE, Dehradun

Shukla AN, Singh YP (2007) Status of Dwarf Mistletoe (*Arceuthobium minutissimum* Hook. f.) on blue pine in Jammu & Kashmir. Indian For 133:1179–1184

Singh P, Singh S (1986) Insect pests and diseases of poplars. Forest Research Institute & Colleges, Dehradun, p 74

Singh S, Puri YN, Bakshi BK (1973) Decay in relation to management of dry coppice teak forests. Indian For 99:421–430

Tattar TA (1988) Diseases of shade trees. Academic Press Inc., California

Uniyal K, Harsh NSK, Uniyal DP, Verma S (2004) Canker and dieback of *Acacia nilotica* seedlings in nursery. Indian For 130:95–100

Van Heerden SW, Wingfield MJ (2002) Effect of environment on the response of Eucalyptus clones to inoculation with *Cryphonectria cubensis*. For Pathol 32:395–402

Van Zyl LM, Wingfield MJ (1999) Wound response of Eucalyptus clones after inoculation with *Cryphonectria cubensis*. Eur J For Pathol 29:161–167

Wingfield MJ, Swart WJ, Abear B (1989) First record of *Cryphonectria* canker of Eucalyptus in South Africa. Phytophylactica 21:311–313

Yangchen T (2017) Conifers conservation in Himalayan Country: diversity and infection assessment of plant parasitic-mistletoes in conifer forests in Thimphu District of Bhutan. The Rufford Foundation Final Report, 16 p

Insect Pests of Forest Nurseries, Plantations and Natural Forests and Their Management

36

N. Roychoudhury

Abstract

There is a good number of tree species that always occur in nurseries and plantations, each of which may be a host for numerous insect pests. However, to develop effective control measures the insect must be responsible for economic losses large enough to justify the costs of materials and labour. Insect species, which normally occur in low densities, are of little or no economic consequence. In such cases, control measure would be justified if the costs are minimal. The insects that do cause great damage to forest plants either consistently or sporadically should be the focus for possible implementation of control programmes to develop long-term protection.

A good amount of research has been undertaken on the insect pest problems of forestry species and to develop control measures. The key insect pests found on forest plants are grouped as soil-inhabiting insects, defoliators, sap suckers and borers. To combat these insect pests, safer chemical pesticides have been standardised and are successfully being practiced to prevent insect damage. Emphasis is now being laid on developing biocontrol strategy (screening resistance, biopesticides and classical biological control) against major insect pests. The serious problems of key insect pests of promising forestry species and the strategy of their management are mentioned in this chapter.

Keywords

Insects · Forestry · Defoliators · Sap-suckers · Borers · Biopesticides

36.1 Introduction

Insects constitute the largest class of the living world. Mani (1989) reported that 12,000 species of fossil insects have so far been named. Out of 89,317 species of animals recorded in India (Alfred 1998), about 40,000 insect species are found in forests and described (Beeson 1941). Browne (1968) published an annotated list with brief accounts of the principal forest insect species and their food plants occurring in the British Commonwealth. Later on, many entomologists published books/brochures/bulletins/reviews/research papers from different parts of India on forest insects and their host plants (Joshi 1992; Jha and Sen-Sarma 1994; Nair 2007). This chapter has presented aspects related to insect problems on forest plants, including recent published information available on key insect pests of forestry species and their management with emphasis on biological options.

36.2 Forest Insects

The promising species for plantations in India include *Acacia* spp., *Eucalyptus* species and *Tectona grandis*. *Eucalyptus grandis*, *Eucalyptus camaldulensis* and *Eucalyptus tereticornis* are most common, while among the acacias, *Acacia auriculiformis*, *Acacia catechu*, *Acacia mearnsii*, *Acacia nilotica* and *Acacia tortalis* are common. Other commonly planted species are *Albizia* spp., *Azadirachta indica*, *Casuarina equisetifolia*, *Dalbergia sissoo*, *Gmelina arborea*, *Populus* spp., *Prosopis* spp., *Shorea robusta* and *Terminalia* spp. Among conifers, *Cedrus deodara* and *Pinus roxburghii* occupy a major area; *Pinus patula* and *Pinus caribaea* have been planted to a limited extent. All these species are damaged by various species of insect pests from the day of flowering to seed setting, fruits, seeds, seed germination to young seedlings, saplings, standing trees and stored timbers (Stebbing 1914; Beeson 1941; Browne 1968; Joshi 1992;

N. Roychoudhury (✉)
ICFRE-Tropical Forest Research Institute, Jabalpur, India

A. K. Mandal, A. Nicodemus (eds.), *Textbook of Forest Science*, https://doi.org/10.1007/978-981-97-8289-5_36

Thakur 2000; Joshi and Jamaluddin 2007; Nair 2007; Wylie and Speight 2012; Senthilkumar and Murugesan 2015).

The regular outbreaks of insect pests seriously hamper the growth and productivity of forests. Many groups of insects belonging to orders Coleoptera, Hymenoptera, Lepidoptera and Isoptera are the key pests that cause significant economic loss. The important insects that cause damage to nursery stock are the cutworms, termites and cockchafers; besides, some defoliators, sap suckers and shoot borers are key pests. Forest trees growing in man-made and natural forests in India suffer serious periodical outbreaks of insects, which lead to significant economic loss of quality and quantity of the timber productivity. According to earliest record given by Fletcher (1914), India and Myanmar have a loss of Rs. 12.5 million due to insect damage to the forests and forest products. A good number of insects are found to damage tress in nurseries, plantations and naturally regenerating forests in India, the list of which is presented in Table 36.1.

36.3 Insect Pests in Nurseries, Plantations and Natural Forests

36.3.1 *Acacia nilotica* (L.) Willd. ex Delile

Acacia (family Fabaceae) is well known as the wattles or acacias, a large genus of shrubs and trees. The genus *Acacia* includes nearly 1300 species (Turnbull et al. 1998), mostly distributed in the Australian region. *Acacia nilotica* is commonly known as babul.

36.3.1.1 Insect Pests of *A. nilotica*

Browne (1968) mentioned 40 species of insects belonging to the orders Coleoptera (14 species), Hemiptera (5 species), Isoptera (1 species), Lepidoptera (19 species) and Orthoptera (1 species) associated with *Acacia nilotica* subsp. *indica*. They are defoliators, bark feeders, sap suckers, seed borers and stem and root borers. In general, there is no major threat to *A. nilotica* plantations from pests, although in some places the stem and root borer *Celosterna scabrator* is rated as a serious problem.

***Celosterna scabrator* Fabricius**

Celosterna scabrator (syn. *Cerosterna scabrator*) (Coleoptera: Cerambycidae) is popularly known as babul borer in India, due to its frequent presence in plantations of *A. nilotica*. The beetle is 25–40 mm in length and dull yellowish brown in colour (Fig. 36.1). In India, the life cycle of this insect has been worked out by Stebbing (1914) and Beeson (1931). The life cycle is annual. Emergence of adults from the host plant starts with the onset of monsoon during June–July and feed on the bark of young shoots. Eggs are laid single, one egg per stem, within 15 cm above ground level. After hatching, larva bores into the stem and makes tunnels downwards hollowing out the main root. The larva ejects frass from the borer hole that accumulates just above the ground level of stem. The larval duration is 9–10 months, which then transform into pupa. Emergence of beetle takes place through the hole made on the stem, just below the ground level. The longevity of beetles is around 45 days.

C. scabrator is a polyphagous insect. Alternative plant species attacked by this insect include *Acacia catechu*, *Cassia siamea*, *Casuarina equisetifolia*, *Dipterocarpus alatus*, *Eucalyptus* spp., *Morus alba*, *Pithecolobium dulce*, *Prosopis cineraria*, *Prosopis juliflora*, *Shorea robusta*, *Tamarix indica*, *Tectona grandis*, *Terminalia chebula* and *Zizyphus jujuba* (Stebbing 1914; Beeson 1941; Browne 1968; Chatterjee and Singh 1968; Jain 1996).

Damage due to this borer results in tunnelling of the tap-root and the stem just above the ground, causing arrestation of growth of the sapling and finally death of plant. The length of the tunnel is about 60 cm. Fertile soil helps in survival of healthy damaged plants by healing the wound as growth proceeds (Beeson 1941). The incidence of this insect attack is more where soil is not suitable to grow *A. nilotica*.

Control Measures

Natural enemies of *C. scabrator* are not known and noticed to be practically free of parasitoids and predators and due to this reason, preventive and remedial measures for control of this borer have been suggested (Beeson 1941). The breeding centres from where the beetles spread to other areas are believed to be the young and stagnating crops of *A. nilotica* when grown on poor dry soils. Therefore, as a preventive measure, planting should be done only in the areas exclusively suitable for this species and other preferred plant species of *C. scabrator* should not be allowed. The suggested remedial measures include grubbing-up the damaged saplings by inserting a thick wire pour through the borer tunnel and killing the larvae. Alternatively, the damaged sapling may be cut off at ground level and small quantity of crude oil, petrol or kerosene or 0.1% Dichlorvos (Nuvan) be poured into the open tunnel through the exposed hole and then, it should be sealed with soil (Joshi and Jamaluddin 2007). The beetle collection can also be done from the plants and killed.

36.3.2 *Ailanthus excelsa* Roxb.

Ailanthus (Family Simaroubaceae) is a genus of trees, generally known as tree of heaven and native of Asia and Australia. *Ailanthus excelsa* Roxb., commonly known as mahanimb or maharukh in India, is an important promising plantation species.

Table 36.1 Key insect pests of forests in India

Host	Common name	Scientific name (Order: Family)	Occurrence
Acacia nilotica	Babul borer	*Celosterna scabrator* (Coleoptera: Cerambycidae)	Plantation
Ailanthus excelsa	Ailanthuswebworm	*Atteva fabriciella* (Lepidoptera: Yponomeutidae)	Nursery and plantation
Albizia procera	Albizia foliage feeder	*Spirama retorta* (Lepidoptera: Noctuidae)	Nursery and plantation
Bamboos	Bamboo leaf roller	*Crypsiptya coclesalis* (Lepidoptera: Pyralidae)	Nursery, plantation and natural forest
Casuarina equisetifolia	Australian bug	*Icerya purchasi* (Hemiptera: Margarodidae)	Nursery and plantation
Cedrus deodara	Deodar looper	*Ectropis deodarae* (Lepidoptera: Geometridae)	Plantation
Dalbergia sissoo	Shisham defoliator	*Plecoptera reflexa* (Lepidoptera: Noctuidae)	Plantation
Eucalyptus species	Blue gum chalcid	*Leptocybe invasa* (Hymenoptera: Eulophidae)	Nursery
Gmelina arborea	Khamer defoliator	*Calopepla leayana* (Coleoptera: Chrysomelidae)	Plantation and natural forest
Leucaena leucocephala	Jumping louse	*Heteropsylla cubana* (Homoptera: Psyllidae)	Nursery and plantation
Picea species	Spruce budworm	*Eucosma hypsidryas* (Lepidoptera: Tortricidae)	Plantation
Pinus roxburghii	Chir pine defoliator	*Cryptothelia crameri* (Lepidoptera: Psychidae)	Plantation
Populus species	Poplar defoliator	*Clostera cupreata* (Lepidoptera: Notodontidae)	Nursery and plantation
Salix species	Indian gypsy moth	*Lymantria obfuscata* (Lepidoptera: Lymantriidae)	Plantation
Shorea robusta	Sal borer	*Hoplocerambyx spinicornis* (Coleoptera: Cerambycidae)	Natural forest
	Sal defoliator	*Lymantria mathura* (Lepidoptera: Lymantriidae)	Natural forest
	Sal defoliator	*Trabala vishnou* (Lepidoptera: Lasiocampidae)	Natural forest
Swietenia species	Mahogany shoot borer	*Hypsipyla robusta* (Lepidoptera: Pyralidae)	Nursery and plantation
Tectona grandis	White grubs	*Holotrichia* spp. (Colepoptera: Scarabaeidae)	Nursery
	Teak defoliator	*Hyblaea puera* (Lepidoptera: Hyblaeidae)	Nursery, plantation and natural forest
	Teak leaf skeletoniser	*Eutectona machaeralis* (Lepidoptera: Pyralidae)	Nursery, plantation and natural forest

36.3.2.1 Insect Pests of *A. excelsa*

About 71 species of insects of *A. excelsa* were recorded by Mishra (2001). This includes both nursery and plantation pests, such as defoliators, leaf rollers and miners, tree borers, bag worms, leaf-eating beetles, grass hoppers, sap suckers, midges, hymenopterous insects and termites. Among all, *Atteva fabriciella* is a key insect defoliator of *A. excelsa*.

Atteva fabriciella Swederus

Atteva fabriciella (syn. *Phalaena fabriciella*) (Lepidoptera: Yponomeutidae) is well known as *Ailanthus* webworm due to larval behaviour of webbing the leaves together with silk thread. The impact of defoliation of this insect is very serious (Fig. 36.2). The female moths lay eggs on young tender leaves. The larvae exhibit gregarious feeding habit and stick the leaves together with silken web, making the damage conspicuous. The last instar larva is about 20 mm and greenish grey, with presence of pale longitudinal stripes and scattered short hairs arising from small, whitish warts (Browne 1968) (Fig. 36.3). The larval duration varies from 13 to 28 days. Formation of pupa occurs in transparent boat-shaped cocoons and pupal duration ranges from 7 to 10 days (Beeson 1941). The moth is small and slender and wing expanse varies from 25 to 30 mm. The moth is dark orange

Fig. 36.1 Adult beetle of *Celosterna scabrator*. (Source: Roychoudhury and Mishra 2020e)

in colour, with white spots of variable size on the forewing (Nair 2007). The duration of life cycle varies from 21 to 48 days, depending on weather conditions and food availability. In Central India, this insect shows ten generations per year. The insect population increases during June–July and declines gradually during summer months (Mathur et al. 1970).

Varma (1996) has reported that this insect defoliator is also a key pest of *Ailanthus triphysa*. Besides *Ailanthus* species, *A. fabriciella* is also noticed to be a defoliator of *Boswellia serrata*, *Santalum album* and *Quassia indica* (Beeson 1941; Browne 1968; Mohanadas and Varma 1984).

Severe attack and repeated defoliations of *A. fabriciella* on *A. excelsa* plants cause mortality of seedlings and saplings and significant loss in growth increment, particularly in plantations (Thakur 2000). Older plants when severely damaged lead to die-back and seed formation is badly hampered by larval feeding and boring activities, thus affecting seriously the seed production.

Control Measures

Efficacy of varietal toxins of *Bacillus thuringiensis* (*B.t.*), var. *dendrolimus*, *kurstaki* and *thuringiensis* against larvae of *A. fabriciella* has revealed that 1.5% concentration can be

Fig. 36.2 Damage caused by *Atteva fabriciella* to *Ailanthus excelsa*. (Source: Roychoudhury and Mishra 2020f)

Fig. 36.3 Larvae of *Atteva fabriciella*. (Source: Roychoudhury and Mishra 2020f)

used irrespective of toxins for spraying in nursery and plantation to kill larvae (Joshi et al. 1996). Roychoudhury and Joshi (2009) and Roychoudhury and Mishra (2020a) have suggested to use biopesticide, 0.05% Ivermectin (Ivecop-12) as a foliar spray for management of *A. fabriciella* larvae. Roychoudhury et al. (2009) and Roychoudhury and Mishra (2020b) have recommended to use biopesticide 0.001% Spinosad 45% Suspension Concentrate (SC) (Spintor 45% SC) as a foliar spray for control of *A. fabriciella* larvae. Bioassay of a series of contact insecticides against the third instar larvae of *A. fabriciella* has showed that Formothion (0.004298%) and Chlordimeform (0.004624%) are most effective to manage larval population (Singh and Gupta 1978a). Meshram and Jamaluddin (1989) have carried out a field trial and suggested that control of the present insect may be achieved by applying 0.02% Monocrotophos. Roychoudhury et al. (1995b) studied efficacy of different chemical pesticides against *A. fabriciella* larvae in laboratory and suggested judicious use of 0.004% Cypermethrin or 0.04% Monocrotophos or 0.07% Malathion for larval control in nursery and plantation. Joshi and Jamaluddin (2007) have recommended for dusting with 5% Malathion during May–June for larval management.

Fig. 36.4 Seedlings of *Albizia procera*

36.3.3 *Albizia procera* (Roxb.) Benth.

Albizia (Family Fabaceae) is a genus of about 160 fast-growing species, occurring in Africa, America, Australia and Asia. *Albizia* spp. are commonly known as siris. Around 16 species of *Albozia* are native to India. The most common and popular species of these is *Albizia procera* (Roxb.) Benth., commonly known as white siris and is native to India. This species is distributed throughout the sub-Himalayan tract and Central and South India. Having nitrogen-fixing ability, this tree species is preferred by farmers (Anonymous 1994).

36.3.3.1 Insect Pests of *A. procera*

A. procera is always threatened by insects. About 59 insects have been identified on this species, out of which 33 insects species are recorded on green plant (Browne 1968; Roychoudhury and Sambath 1998; Roychoudhury 2002). Among all, *Spirama retorta* is a devastating key pest in nurseries and plantations of *A. procera* (Fig. 36.4).

Spirama retorta Clerck

Spirama retorta (syn. *Phalaena retorta*) (Lepidoptera: Noctuidae) is known as the Indian owlet-moth, found in India, Bangladesh, China, Korea, Japan, Nepal, Thailand, Myanmar, Sri Lanka, Malaysia, Philippines and Indonesia. Eggs are laid by female moth on new shoots and eggs hatch in 3–5 days. The larvae undergo seven instars in a period of 25–27 days. The mature larva, dull yellowish-brown in colour, reaches 65 mm body length and 5.0 mm head capsule width (Fig. 36.5). The head is brown with white lateral stripes on the vertex. The larva is very active at night. The pupa is dark brown in colour and the pupal stage lasts for 7–13 days during July–October. The wingspan of the male is 64–76 and 66–88 mm in female (Fig. 36.6). Forewings are with arched costa towards apex, which is nearly rectangular. Male is with a fold on inner area of hindwing, containing an erectile ridge of very long hair. Antennae are fasciculate. Male has dark chestnut-brown coloured head and collar. Thorax is paler with dark bands. Abdomen is crimson with triangular black dorsal patches. Wings are fuscous brown. Hindwings are with indistinct antemedial, medial and traces of two postmedial lines along with a submarginal line. Ventral side is suffused with dull red and having two medial lines and single postmedial line to each wing (Hampson 1894). Female is basically ochreous coloured. Forewings have larger stigma than male. Hindwings are with prominent markings. Two antemedial lines and an ochreous submarginal line can be seen along with two crenulate black lines inside the margin. Ventral side is orange-scarlet coloured, each wing with cell spot, medial, postmedial and submarginal lines and series of lunules inside the margin.

Fig. 36.5 Larva of *Spirama retorta*. (Source: Roychoudhury and Mishra 2022a)

Fig. 36.6 Adult moth of *Spirama retorta*. (Source: Roychoudhury and Mishra 2022a)

S. retorta is a leaf feeder of *Albizia* species. This insect is a key pest of *Albizia lebbek* and *A. procera* in forest nurseries and young plantations (Sambath and Joshi 1999). The caterpillars are known to feed on *Acacia mangium*, *Acacia auriculiformis*, *Acacia crassicarpa* and *Falcataria moluccana* (Sajap et al. 1997; Sajap and Samad 2000).

The frequent defoliation of *S. retorta* results in retardation of growth and development of seedlings or trees (Beeson 1941; Sen-Sarma 1987). Larval host range and damage impact of this pest in plantations of *Albizia julibrissin* have been studied by Kim and Lee (1986).

Control Measures

Laboratory evaluation of varietal toxins of a bacterium *Bacillus thuringiensis* (*B.t.*), var. *dendrolimus*, *kurstaki* and *thuringiensis* against *S. retorta* larvae revealed that 1.5% concentration of *B.t.* can be used irrespective of toxins for spraying on seedlings in nursery and saplings in young plantations to kill the larvae (Sambath and Joshi, Unpublished). Roychoudhury and Joshi (2011) and Roychoudhury and Mishra (2020a) have suggested to use biopesticide 0.02% Ivermectin (Ivecop-12) as a foliar spray for control of *S. retorta* larvae. Roychoudhury (2015a) and Roychoudhury and Mishra (2020b) have recommended to use biopesticide 0.0005% Spinosad 45% SC (Spintor 45% SC) as a foliar spray for management of *S. retorta* larvae. Sambath and Joshi (1999) have worked out Lethal Concentration$_{50}$ (LC_{50}) values of different insecticides against the larvae of *S. retorta* and suggested to use of 0.002% Cypermethrin or 0.02% Chlorpyriphos or 0.02% Monocrotophos for larval killing in nursery and young plantation. Joshi and Jamaluddin. (2007) have suggested foliar spraying of Monocrotophos 0.04% for control of *S. retorta* larvae and spraying needs to be repeated after 15 days of interval.

36.3.4 Bamboos

Bamboo (Family Poaceae), the 'poor man's timber' or the 'green gold', is well known throughout the world. Bamboo has always been a material of culture in India (Soderstrom and Ellis 1988). Bamboos are versatile group of monocots consisting of nearly 75 genera and 1250 species and found throughout the globe except Europe. India is endowed with rich diversity of bamboo distributed in different agroclimatic zones of the country and largest reserve of bamboo in the world. There are 136 indigenous and exotic species of bamboo belonging to 23 genera in an area of 10.03 million ha (Mohanan 1990).

36.3.4.1 Insect Pests of Bamboo

Bamboo has a rich complex of insect fauna and insect damage in varying degrees, right from the seed to the finished products (Tewari 1992a). Bamboo entomology reveals about 200 odd insects which infest various bamboos in India and its adjoining countries, belonging primarily to Coleoptera, Homoptera, Isoptera, Lepidoptera and Thysanoptera (Beeson 1941; Singh and Bhandari 1988; Thakur 1988; Singh 1990; Joshi et al. 2008). The insect fauna of bamboo is grouped into seed pests (2 species), nursery pests (5 species), defoliators (48 species), sap suckers (90 species), culm and shoot borers (12 species), borers (44 species) and termites (13 species) of felled and dried bamboos (Tewari 1992a). Among all, leaf rollers cause severe epidemic defoliation in bamboo (Mathur 1943). The most devastating bamboo leaf roller that occurs in Indian subcontinent is *Crypsiptya coclesalis*.

Fig. 36.7 Bamboo leaf roll caused by *Crypsiptya coclesalis*. (Source: Roychoudhury and Mishra 2022b)

Crypsiptya coclesalis (Walker)

Crypsiptya coclesalis (syn. *Pyrausta coclesalis*, *Algedonia coclesalis*) (Lepidoptera: Pyralidae) is well known as a bamboo leaf roller (Tewari 1992a; Roychoudhury and Joshi 2008). It occurs up to 4000 ft elevation, with different seasonal climatic factors, in the Indian subcontinent, Myanmar, Indo-China, Sri Lanka and South East Asia (Mathur 1943; Browne 1968; Tewari 1992a). The light green young larvae of leaf rollers feed in groups and eat out the upper leaf tissues causing transparent blotches on the leaves (Roychoudhury and Joshi 2008). As they attain maturity, they scatter and start webbing and rolling leaves for protection and concealment (Fig. 36.7). The rolled leaves harbour one to five caterpillars, which feed on the inner green leaves (Fig. 36.8). The attacked bamboo clumps show skeletonised and brownish leaves webbed together in branches. The outer leaves of the roll

Fig. 36.8 Larva of *Crypsiptya coclesalis*. (Source: Roychoudhury and Mishra 2022b)

gradually wither, turn pale or straw colour and eventually drop off. Caterpillars abandon the rolls after feeding and migrate to fresh leaves, which are webbed together in a roll. However, leaves on the lower branches are less attacked. Generally, larval population of leaf rollers appear with the onset of rain in June, when leaf flush occurs in bamboo to ensure sufficient food to the larvae. The population outbreak occurs during July and population build-up reaches its peak in August and September and causes severe damage to bamboo on seedlings, saplings and forests.

In India, this insect is found damaging *Bambusa arundinacea*, *Bambusa bambusoidea*, *Bambusa burmanica*, *Bambusa nana*, *Bambusa nutans*, *Bambusa polymorpha*, *Bambusa tulda*, *Bambusa vulgaris*, *Cephalostachyus pergracile*, *Dendrocalamus asper*, *Dendrocalamus giganteus*, *Dendrocalamus hamiltonii*, *Dendrocalamus longispathus*, *Dendrocalamus membranaceous*, *Dendrocalamus strictus*, *Melocanna albociliata*, *Melocanna bacifera* and *Melocanna bambusoidea* (Beeson 1941; Mathur 1943; Browne 1968; Tewari 1992a), whereas in Myanmar, *Cephalostachyus coclesalis* is noticed as a defoliator of *C. pergracile*, in Sri Lanka on *D. strictus* and in Java on bamboo and sugarcane (Mathur 1943). According to Lefroy (1971), it is occasionally found on maize.

Control Measures

To control the population of bamboo leaf rollers, resistant species of bamboo, viz. *Bambusa arundinacea*, *B. nana*, *Bambusa polymoropha*, *Dendrocalamus asper*, *D. giganteus*, *D. membranaceous*, and *B. vulgaris*, var. *green*, which are less attacked, should be preferred for plantations (Roychoudhury 2010). Kalia and Joshi (1995) have suggested for foliar spraying of 1% *Bacillus thuringiensis* var. *kurstaki* against bamboo leaf roller larvae. On the basis of LC_{50} value of a biopesticide, Ivermectin (Ivecop-12), it is recommended to use 0.1% as a foliar spray for control of *C. coclesalis* larvae (Roychoudhury 2012a, b; Roychoudhury and Mishra 2020a). Roychoudhury (2011, 2012b) and Roychoudhury and Mishra (2020b) have suggested to use biopesticide 0.0006% Spinosad 45% SC (Spintor 45% SC) for management of *C. coclesalis* larvae.

Singh (1990) has recommended a solution of 0.2% Fenitrothion or 0.1% Carbaryl in water with a surfactant to kill larval population of bamboo leaf roller. In China, leaf rollers damaged *Phyllostachyus pubescens* in about 7 million ha area as a result of which around 4 million stocks died. Aerial spraying and injection of insecticides into the cavity at the bottom of the plant reduced the extent of infestation (Shi 1980). Joshi and Jamaluddin (2007) have recommended spraying Monocrotophos 0.08% for control of bamboo leaf roller in nurseries and plantations.

36.3.5 *Casuarina equisetifolia* L.

Casuarinas (Family Casuarinaceae) are widely accepted plant species in the tropical countries due to their easy adaptability to a variety of climatic conditions and also for their fast growth. According to Turnbull (1990), among the 96 species, *Casuarina equisetifolia* is a potential species of multiple end-uses. *C. equisetifolia* was introduced to India from Australia (Warrier et al. 2014). This is the most widely accepted plant species of *Casuarina* in the country and used for coastal afforestation programme in coastal area of Tamil Nadu and Andhra Pradesh.

36.3.5.1 Insect Pests of *C. equisetifolia*

C. equisetifolia is nearly free from serious insect attack in its homeland. But this species is facing many major insect problems in countries where it has been planted, including India. About 58 insect species belonging to the orders Coleoptera (19 species), Hemiptera (15 species), Isoptera (5 species), Lepidoptera (15 species) and Orthoptera (4 species) are attached to *C. equisetifolia* (Browne 1968). Sasidharan (2004) has reported that in Tamil Nadu, 40 insect species are recorded, viz. 3 species of stem borers, 2 species of bark feeders, 23 species of needle feeders, 11 species of sap suckers and 1 species of seed feeder. The majority of insects are noticed damaging needles. Among all, *Icerya purchasi* is a key insect of economic concern.

Icerya purchasi Maskell

Icerya purchasi (syn. *Pericerya purchasi*) (Hemiptera: Margarodidae) is generally called as cottony cushion scale or Australian bug. This is a scale insect of Australian origin, which accidentally entered in India. The species can be identified by its conspicuous, fluted egg sac, which is often more than twice as long as the reddish or yellowish colour of adult body (Browne 1968). *I. purchasi* is a hermaphrodite insect and mated individuals give birth to hundreds of offspring. Infestation of this insect is favoured by humid conditions and shade.

I. purchasi is known mainly as a pest of citrus fruit trees and other crops, such as tea and coffee. This insect is commonly associated with *Acacia* spp., and specifically noticed host plants in India include *Acacia dealbata*, *Acacia melanoxylon*, *Albizia julibrissin*, *Cassia* spp., *Casuarina equisetifolia*, *Eucalyptus* spp., *Juglans regia*, *Morus alba* and *Pinus roxburghii* (Beeson 1941; Browne 1968). This insect species is mainly injurious to nurseries, causing stunting of seedlings and saplings.

Control Measures

The population of *I. purchasi* can be controlled by planting resistant source of clones/provenances (Sasidharan 2004). Gokkes et al. (1989) have recorded that insect growth regulator Pyriproxyfen is effective in controlling *I. purchasi* and found highly effective even if it is applied alone or with 0.5% mineral oil (Peleg 1989). Mendel et al. (1991) have noticed that insect growth regulator, Buprofezin, induces 100% mortality of crawlers and spraying on adults can decrease 31% hatching of eggs. Foliar spray of 0.06% Dimethoate or 0.05% Methyl demeton also can control the pest (Senthilkumar and Murugesan 2015).

36.3.6 *Cedrus deodara* (Roxb.) G. Don

Cedrus deodara (Family Pinaceae), commonly known as deodar, is a promising species of Indian conifers (Tewari 1994a). This is a species of cedar native to the western Himalayas. In India, it occurs in Jammu and Kashmir, Himachal Pradesh, Uttarakhand, Sikkim, Arunachal Pradesh and the Darjeeling district of West Bengal (Farjon 2013). Deodar is typically gregarious and is usually found in pure stands.

36.3.6.1 Insect Pests of *C. deodara*

Insect infestations in Deodar have been studied extensively. About 60 species of insects belonging to 22 families of 6 orders, viz. Coleoptera, Hemiptera, Isoptera, Hymemoptera, Lepidoptera and Orthoptera, have been recorded feeding deodar at different stages of plant development (Beeson 1941; Browne 1968; Tewari 1994a). Trees of all age groups are always under threat of defoliation. Repeated defoliation kills the young trees and also affects annual growth increment. Deodar foliage furnishes food for a number of defoliating insect species. Main defoliating insect pest is *Ectropis deodarae.*

***Ectropis deodarae* Prout**

Ectropis deodarae (Lepidoptera: Geometridae) is called as deodar looper. It is a damaging defoliating pest of deodar distributed all along northwestern part of India, including Himalaya. Adult female moth has only very small vestigial wings and the male has broad, white wings in colour mottled with black. Wing expanse varies from 35 to 40 cm (Beeson 1941; Browne 1968; Tewari 1994a). The generation is annual. Emergence of moths takes place during the month of March and female moths lay eggs in the needles of the crown of the host tree. The larvae, which are green, white-haired loopers, attain 25 mm in length. Larvae feed on the foliage until June, when they descend on silken threads and pupae, without forming cocoons in the soil litter.

Large areas of deodar forests in the northwestern and western Himalayan regions are often defoliated completely by *E. deodarae*, causing heavy mortality. Thakur (2000) has compiled the history of epidemics of this insect. The epidemic defoliation of *E. deodarae* occurred four times in Himachal Pradesh during the year 1901–1903, 1906–1907, 1922–1924 and 1961–1962. An epidemic of this defoliator has been reported during 1982–1983 in Jammu and Kashmir (Singh et al. 1989). Epidemics occur at about 10-year intervals and may last for 2 or 3 years (Beeson 1941; FAO 2003). An outbreak has also been noticed in June 1994 in Shimla (Verma and Chander 1995). Larvae prefer to feed on the needles and as a result, the needles turn brown, dry up and fall to the ground prematurely. Beeson (1941) has identified biotic and abiotic factors responsible for the incidence of epidemic defoliation in deodar forests, such as grazing, removal of litter, excessive thinning and, most important of all, the natural climatic conditions and natural enemy complex, which influence to a great extent.

Control Measures

The infestation of this defoliator can be controlled by using grease-bands on the main trunk region, which can restrict wingless female moths from crawling up the trunk and egg laying of moths on the needles (Beeson 1941). Periodical visits can help to collect moths and eggs below the greasy bands for immediate killing. Singh et al. (1989) have effectively controlled the epidemic of *E. deodarae* by aerial spraying of fenitrothion @ 1 L/ha. Use of natural enemies of this deodar defoliator is an efficient method of control (Tewari 1994a). Among the natural enemies, *Calosoma beesoni* Andr. (Coleoptera: Carabidae) is a potential predator, which in the adult form eats the caterpillars and larvae feed the pupae. The main parasites of *E. deodarae* are *Dusona deodarae* Cush (Hymenoptera: Ichneumonidae), *Brachymeria obscurata* Wlk. (Hymenoptera: Chalcidae) and *Campsilura concinnata* Heigh (Diptera: Tachinidae). Many insectivorous birds like Jackdaws, *Corvus monedula* Linn. (Aves: Corviidae), are also recorded to feed on deodar defoliator (Singh et al. 1989).

36.3.7 *Dalbergia sissoo* Roxb.

Dalbergia Linn.f. (Family Fabaceae) is a genus that includes trees, shrubs and woody climbers, and widely distributed in tropical and subtropical regions. In India, around 25 species of *Dalbergia* occur. Among these, *D. sissoo* Roxb. is a promising nitrogen-fixing tree species, popularly known as shisham, found throughout the sub-Himalayan valleys. This is a potential multipurpose tree of economic importance (Tewari 1994b).

36.3.7.1 Insect Pests of *D. sissoo*

About 85 insects have been reported to feed on *D. sissoo*, belonging to the orders Coleoptera (28 species), Hemiptera (20 species), Isoptera (1 species), Lepidoptera (30 species) and Orthoptera (6 species) (Browne 1968). Recent

Fig. 36.9 Defoliation of *Dalbergia sissoo* by *Plecoptera reflexa*. (Source: Roychoudhury and Mishra 2021a)

Fig. 36.10 Larvae of *Plecoptera reflexa*. (Source: Roychoudhury and Mishra 2021a)

information reveals around 130 species of insects are associated with *D. sissoo* (CABI 2005), out of which defoliators and stem borers are the most damaging. Among defoliators, *Plecoptera reflexa* is considered as a serious pest of economic concern.

Plecoptera reflexa Gunee

Plecoptera reflexa (Lepidoptera: Noctuidae) is a major defoliator of *D. sissoo* in India and Pakistan. The caterpillar is a semi-looper, green in colour, and full-grown larvae are around 25 mm in length (Figs. 36.9 and 36.10). The moths are greyish brown in colour with a wing expanse of 30–35 mm. The female moth lays eggs at night on young soft leaves and each female lay up to 400 eggs within 6 days. The young larvae feeds on leaf surface, but maturing laraval instars consume whole leaf including petiole and green shoot. Irrigated plantations of *D. sissoo* are most preferred for outbreaks. Indian rose wood, *Dalbergia latifolia*, is a suitable host plant of *P. reflexa*.

P. reflexa was first reported as a defoliator of *D. sissoo* in 1899 in Changa Manga, Pakistan. The defoliation is annual both in India and Pakistan. This insect mostly attacks plantations aged 3 years and above. Due to repeated attacks of *P. reflexa*, the tree remains leafless during the growing season and epicormic branches are produced. According to Beeson (1941), economic impact of this defoliation is considered serious and poor-quality plantations have often been produced.

Control Measures

Gupta and Joshi (1995) have recommended to use *Bacillus thuringiensis* (*B.t.*) of 90–102 billion spores/g at a concentration of 1.5% for maximum kill of larvae within 72 h of spray, irrespective of varietal toxins and available forms, that is, either in wettable powder or emulsifiable concentrate. Joshi and Roychoudhury (1996) have suggested for spraying Deltamethrin 0.003% otherwise Chlorpyriphos 0.05% against the larvae of *P. reflexa*. According to Joshi and Jamaluddin. (2007), weak seedlings should not be planted and spraying Carbaryl 0.1% or Endosulfan 0.05% can be used for control of *P. reflexa*.

36.3.8 *Eucalyptus* Species

Eucalyptus (Family Myrtaceae) is usually regarded as an Australian tree and undoubtedly is among the most successful world's exotic multipurpose fast-growing tree species. *Eucalyptus* is grown in about 110 countries of the world. Around 600 species and varieties of the genus *Eucalyptus* are available in the world. About 170 species, varieties and provenances of *Eucalyptus* are grown in India (Bhatia 1984), out of which the most promising and suitable has been the *Eucalyptus* hybrid, a form of *Eucalyptus tereticornis*, called as Mysore gum. Other species such as *Eucalyptus camaldulensis*, *Eucalyptus citridora*, *Eucalyptus globulus* and *Eucalyptus grandis* are grown on large-scale plantation.

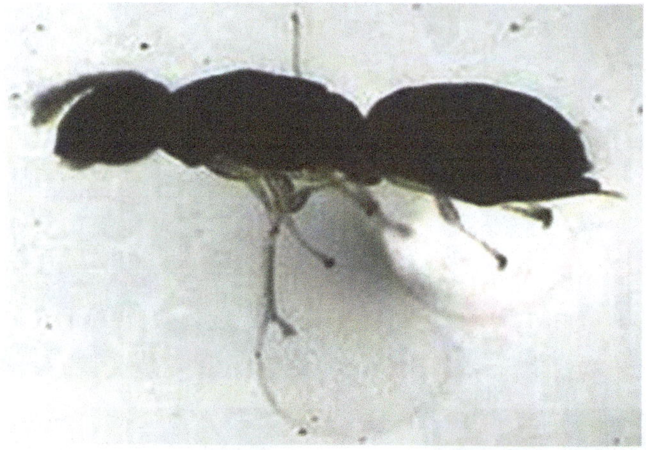

Fig. 36.11 Blue gum chalcid wasp, *Leptocybe invasa*. (Source: Roychoudhury 2013a)

Fig. 36.12 Gall formation in leaves, petioles and stems of *Eucalyptus* due to *Leptocybe invasa*. (Source: Roychoudhury 2013a)

36.3.8.1 Insect Pests of *Eucalyptus*

Insects associated with *Eucalyptus* have been thoroughly reviewed by Ohmart and Edwards (1991). About 400 insects have been found feeding on *Eucalyptus* that includes 160 species of foliage feeders, 110 species of xylophagous borers, 76 species of sap suckers, 32 species of timber borers and 22 species of miscellaneous insects (Nair et al. 1986; Tewari 1992b). Around 920 species of insects are associated with *Eucalyptus* throughout the world (Nair 2007). Most of the insect fauna associated with *Eucalyptus* are minor pests. Recently, only in last decade a serious problem of gall insect, *Leptocybe invasa*, has been encountered in this species throughout the *Eucalyptus*-growing countries, including India, which is described as hereunder.

Leptocybe invasa Fisher & LaSalle

Leptocybe invasa (Hymenoptera: Eulophidae) is generally called as blue gum chalcid wasp. Until the year 2000, *Eucalyptus* gall wasp, *Leptocybe invasa* Fisher & LaSalle (Hymenoptera: Eulophidae), was not known (Mutitu 2003). Mendel et al. (2004) have identified *L. invasa* as a tiny wasp responsible for gall formation in *Eucalyptus* in nurseries and young plantations (Fig. 36.11). This gall insect originated from Queensland, Australia, and then this insect was noticed in the Middle East during the year 2000 (Aytar 2003). Currently, the wasp is reported from Algeria, Brazil, Cambodia, China, Ethiopia, France, Greek, India, Iran, Iraq, Israel, Italy, Jordan, Kenya, Morocco, New Zealand, North America, Portugal, Spain, Syria, South Africa, Tanzania, Thailand, Turkey, Uganda and Vietnam (https://www.cabi.org/isc/datasheet/108923#todistributionDatabaseTable).

According to Mendel et al. (2004), the diagnostic features of *L. invasa* are as follows: length of the body is 1.44 ± 0.16 mm and varies from 1.17 to 1.72 mm; head, thorax and abdomen are dark brown in colour; and antenna, fore coxa, legs and tarsi are yellow in colour. Wings hyaline are very faintly and evenly infumated and veins are brown in colour. Weak head, with distinct groove and weakened area around ocellar triangle is present. Propodeum with a raised lobe of the callus is present that partially overhangs the outer rim of the spiracle, and spiracular depression is open to anterior margin of propodeum. Two longest cercal setae are subequal in length and straight. Postmarginal vein is short and less than length of stigma vein. Mesoscutum is without median line and with two to three small adnotaular setae. Malar sulcus is distinctly curved. Dorsellum is long, medially as long as propodeum. Ovipositor sheaths are short, not reaching to the apex of abdomen.

On the basis of extensive surveys, observations revealed that *L. invasa* is the main cause of gall formation on the mid-ribs, petioles and stems of new shoots both in nurseries and young plantations of *Eucalyptus*, including coppice shoots (Fig. 36.12) (Roychoudhury 2013a). *L. invasa* has been found to damage *E. camaldulensis*, *E. tereticornis* and *Eucalyptus* hybrid. This is a stenophagous insect because it has been reported that it infests many species of *Eucalyptus*, such as *Eucalyptus botryoides*, *Eucalyptus bridgesiana*, *E. camaldulensis*, *Eucalyptus cinerea*, *Eucalyptus dunnii*, *E. globulus*, *E. grandis*, *Eucalyptus gunii*, *Eucalyptus maidenii*, *Eucalyptus nicholii*, *Eucalyptus pulverulenta*, *Eucalyptus robusta*, *Eucalyptus rudis*, *Eucalyptus saligna*, *E. tereticornis*, *Eucalyptus urophylla* and *Eucalyptus viminalis*, and various clones and hybrids (Wylie and Speight 2012). Adult females (1.1–1.4 mm) lay their eggs in the young leaves, in the petioles and in the twigs (Hesami et al. 2005). Larvae develop inside round galls; adults then emerge leaving round exit holes. There are five stages of gall development (Fig. 36.13) (Roychoudhury 2018a). Mendel et al.

Fig. 36.13 Stages of gall development in *Eucalyptus* caused by *Leptocybe invasa*. (Source: Roychoudhury 2018a)

(2004) have observed only females reproducing by parthenogenesis, with the exception of two records describing males in Turkey (Doganlar 2005) and India (Akhtar et al. 2012). In Iran, Israel and Turkey, two to three overlapping generations were recorded in a year (Hesami et al. 2005). Adults of this gall insect are able to fly for short distance but no such data are available on natural spread of this insect (Branco et al. 2006). Movements of plants from one place to another carry this pest over long distances (Nyeko et al. 2009).

L. invasa was first noticed in 2001 in Mandya district of Karnataka, India. Then this species was recorded in 2002 at Marakkanam in Villupuram district of Tamil Nadu, and later on it spread throughout the India (Jacob et al. 2007). The epidemic of this exotic pest in India has concomitantly changed the scenario by damaging vast areas (Jacob 2009; Roychoudhury et al. 2007). Infestation of *L. invasa* harms *Eucalyptus* in nurseries and plantations throughout the country. Serious infestations can lead to production of deformed leaves and shoots and stop growth of the plant. Major damage to nursery seedlings leads to ultimately death of plants. Presently, the insect is infesting *Eucalyptus* all over the country (Roychoudhury 2016a).

Control Measures

Resistant species or clones or hybrids or provenances of *Eucalyptus* can be used to control *L. invasa* (Thu et al. 2009; Kulkarni 2010; Nyeko et al. 2010). Kumari (2009) suggested spraying of *Azadirachtin* in nursery to reduce adult emergence. Foliar spraying of biopesticide 0.05% Spinosad (Spintor), in nursery stage and young plantation is effective to control *L. invasa* (Roychoudhury 2017a; Roychoudhury and Mishra 2018).

Regarding the chemical pesticides, spraying of different insecticides, such as Acephate, Acetamiprid, Carbofuran, Dimethoate, Fenkil, Imidacloprid, Methyl-o-demeton, Methyl parathion, Phorate, Phosphamidon, Thiacloprid and Thiamethoxam, is effective to kill *L. invasa* and subsequent in reduction of adult wasp emergence (Kumari 2009; Javaregowda et al. 2010; Jhala et al. 2010). Roychoudhury (2013a) suggested foliar spraying of 0.5% Monocrotophos or Imidacloprid or Acetamiprid or Dimethoate or Dichlorvos for 6 months at the interval of 15 days can control *L. invasa* (Figs. 36.14 and 36.15).

36.3.9 *Gmelina arborea* Roxb.

Gmelina arborea (Family Lamiaceae) is generally known as khamer. This tree species is native to Asia and occurs in deciduous and moist deciduous forests in India (CABI 2005). *G. arborea* plantations have been raised both in homeland and outside Asia. This species has also been planted in several countries worldwide.

Fig. 36.14 Eucalyptus seedling infested with galls

Fig. 36.15 Pesticide-treated seedlings without gall. (Source: Roychoudhury, Unpublished)

36.3.9.1 Insect Pests of *G. arborea*

In India, about 101 insect species have been noticed in native plantations of *G. arborea* (Beeson 1941; Browne 1968; Mathew 1986). Most of the insects are minor pests, but few are major pests. The key insect species is the chrysomelid beetle, *Calopepla leayana*, responsible for epidemic defoliation of plantations, mainly in northeast India (Beeson 1941).

Calopepla leayana (Latreille)

Calopepla leayana (syn. *Craspedonta leayana*) (Coleoptera: Chrysomelidae) is commonly known as Khamer defoliator. This insect defoliator has been recognised since the 1920s as a major pest of *G. arborea* in northern India. The adult beetle is oblong, 12–16 mm in length and has a shining metallic colouration, with coarsely wrinkled, bluish green to violet blue elytra and pale yellow to reddish brown pronotum and legs. The larva has lateral spines. The excrement, instead of being ejected, is extruded in fine, black filaments, longer than the body, and formed into bunches at the anal end. The moulted exuviae are also carried out at the last abdominal segment. When disturbed, the larva flicks the anal filaments up and down and assumes a defensive posture. Under favourable temperatures, 35–50 days takes to complete the life cycle, but third-generation adults enter hibernation in winter. Infestation period is June–October.

This insect is a key pest of *G. arborea* in plantations within the natural range of the tree. Larvae during the young stage feed khamer foliage, but mid-ribs and main veins remain intact. Adult beetles of *C. leayana* consume foliage by circular holes and also feed growing young buds and shoots. Heavy infestation leads to drying up of plants leading to ultimate death. *C. leayana* was reported on *Gmelina* in Meghalaya during the year 1995, indicating an apparent expansion of its range to the northeast of India (Wingfield and Robison 2004). It is considered as a key insect pest of khamer plantations in Assam.

Control Measures

Trapping of adults in hibernating shelters, hand-picking of beetles returning to the plantation after overwintering and mixed cropping (instead of monoculture) have been suggested in the past (Beeson 1941). The beetles can be attracted to white piece of cloth, then can be collected and killed mechanically. Insecticides, a commercial preparation of *Bacillus thuringiensis* subsp. *kurstaki* and the fungus *Beauveria bassiana*, are found effective against the larvae (Gupta et al. 1989; Sankaran et al. 1989; Sharma et al. 2001). Joshi and Jamaluddin. (2007) have suggested for foliar spraying of 0.05% Chlorpyriphos or 0.04% Monocrotophos or 0.05% Malathion to kill all larval instars and beetles.

36.3.10 *Leucaena leucocephala* (Lam.) de Wit

Leucaena leucocephala (Family Fabaceae) is commonly known as subabul. This is a multipurpose legume plant native to Mexico and Central America. Its true natural distribution is not known (CABI 2005). Plantations of *L. leucocephala* have been raised in agroforestry planting programmes for many purposes. This tree species is planted over 130 countries (CABI 2005).

36.3.10.1 Insect Pests of *L. leucocephala*

Insect pests are not a serious threat to homeland. Nair (2001) has listed about 40 species of native insects that feed on exotic plantations of *L. leucocephala* in different countries.

These include leaf-feeding curculionid and chrysomelid beetles and grasshoppers; sap-sucking coreid, pentatomid, aleurodid, coccid, psuedococcid, eurybrachyid and membracid bugs; root-feeding white grubs and termites; and stem or branch boring cerambycid beetles and a cossid caterpillar. But none of these insects cause major damage. However, the psyllid bug *Heteropsylla cubana* has become a serious pest, since leucaena planting began.

Heteropsylla cubana Crawford

H. cubana (syn. *Heteropsylla incisa*) (Homoptera: Psyllidae) is commonly known as leucaena psyllid. This bug was described in 1914 in Cuba. Adult bug is of a small size measuring 1.5–2 mm in length, usually yellowish green in colour, some with brown shades. The nymphs are yellowish green in colour. Both adults and nymphs feed gregariously on the developing terminal shoot and sucking the sap of leaves. The adult female bug lays about 240 eggs (Rauf et al. 1990) that are usually attached to the tender leaves. Hatching of eggs takes place in 3 days and pass through five instars in 8 days. Adult longevity is 10–15 days. The insect shows several overlapping generations in a year and all developmental stages usually occur together on terminal young shoots.

Other than *L. leucocephala*, *H. cubana* feed on *Leucaena diversifolia*, *Leucaena pulverulenta*, *Leucaena trichodes* and *Leucaena salvadorensis*. Besides these, it can feed on 13 species of *Leucaena* to some extent and also on *Samanea saman* (Geiger et al. 1995; CABI 2005).

L. leucocephala is a tree grown extensively in community forestry and agroforestry ecosystems for fodder and fuel throughout the tropics including India. The tree was almost pest free in India until 1988, when the leucaena psyllid, *H. cubana*, appeared in Chengalpattu, Tamil Nadu, and caused severe defoliation and extensive death of young trees. By 1990, it had attacked all the *Leucaena* plantations in the country (FAO 2005).

Severe infestation of *H. cubana* occurs in terminal shoots. This bug has gregarious feeding habit and consumes developing terminal leaves. These leaves become chlorotic and deformed or the leaflets turn yellow and drop. The petioles turn black and the shoots dry up. About a week of infestation, the damage due to this bug occurs and serious infestation usually results in complete defoliation and stunted growth of the tree. Repeated damage by this bug leads to death of the trees. According to Napompeth (1990), secondary infection by pathogenic microorganisms is possibly responsible for mortality of *L. leucocephala* tree.

Control Measures

Joshi and Jamaluddin. (2007) have suggested to use resistant cultivars, such as K6, K8, K500 and K636, to control psyllid *H. cubana*. According to Senthilkumar and Murugesan (2015), *Menochilus sexmaculatus*, a coccinellid predator, is found to be effective on *H. cubana*. Joshi and Jamaluddin. (2007) have recommended foliar spraying of 0.001% cypermethrin to minimise the infestation or 0.03% monocrotophos for effective control of psyllid. The spraying may be repeated after a week, if required.

36.3.11 *Picea* Species

Picea (Family Pinaceae) is a genus of spruce tree and includes about 35 species of coniferous evergreen trees. Indian spruce, *Picea smithiana* (Wall) Boiss, is commonly known as Himalayan spruce. In the western Himalaya, spruce forests occur at an altitude of 2150–3300 m (Singh and Pusalkar 2020), which grows in close association with silver fir *Abies pindrow* (Bhalla and Parmar 1996). It has a self-supporting growth form.

36.3.11.1 Insect Pests of *P. smithiana*

About 14 species of insects belonging to three orders, viz. Coleoptera (8 species), Hemiptera (2 species) and Lepidoptera (4 species), feed on Himalayan spruce *P. smithiana* (Browne 1968). Among all, *Eucosma hypsidryas* is a major insect pest that causes serious damage of spruce in India.

Eucosma hypsidryas Meyrick

Eucosma hypsidryas (Lepidoptera: Tortricidae) is commonly known as spruce budworm, a small moth available in India and Pakistan (Beeson 1941; Browne 1968). The female moth of spruce budworm lays eggs on the young buds, each of which becomes infested by a single larva. The larva feeds within the bud on the undeveloped needles, except for the outermost needles that are spun together with silk and eventually form the walls of a prominent, spindle-shaped chamber within which the larva pupates. The moths of this generation emerge in June. It is believed that there are at least two generations during a year, but the most serious injury to the host is done by the larvae of spring generation.

The budworm, *E. hypsidryas*, causes considerable damage to *P. smithiana* and mortality of spruce trees in the Himalayas (Bhalla and Parmar 1996). Trees of all ages are attacked and are easily observed because of the brown, aborted tips of their branches (Browne 1968). Heavy and repeated infestation results in stagnation and weakening of the host. The drying-off of spruce in the Himalayas is probably mainly due to *E. hypsidryas* (Beeson 1941).

Control Measures

Thakur (2000) suggested spraying of either 0.18% Azinphosmethyle or 0.025% Fenvelrate or 0.025%

Permethrin can be used for effective control of *E. hypsidryas* attacking cones in seed orchards and seed production areas.

36.3.12 *Pinus roxburghii* Sarg.

The genus *Pinus* (Family Pinaceae) includes more than 90 species and constitutes an important group of conifers. Chir pine, *Pinus roxburghii* Sarg., is an important potential species and commonly called as Indian chir pine (Anonymous 1996). Chir pine is a native of the Himalaya.

36.3.12.1 Insect Pests of *P. roxburghii*

Chir pine is damaged by insects at all stages of its growth, such as cones and seed, seedlings, standing trees and timber. Thirty-two insect species belonging to Coleoptera (14 species), Hemiptera (3 species), Isoptera (2 species), Lepidoptera (6 species) and Orthoptera (7 species) are found feeding on all different plant developmental stages (Beeson 1941; Browne 1968; Anonymous 1996). Among these, *Cryptothelia crameri* is a key insect defoliator that causes serious outbreak.

Cryptothelia crameri Westwood

Cryptothelia crameri (syn. *Clania crameri*) (Lepidoptera: Psychidae) is a polyphagous bagworm pest, found throughout India (Beeson 1941; Browne 1968). Eggs are laid in the case from December to April, according to the climatic conditions of the area. The larva feeds on young leaves and tender bark by making and living within a portable bag of leaves, twigs, barks, spines etc. (Baksha 2000). The wing span of male moth is 6–7 mm, reddish brown in colour streaked with black, and the middle of the forewings are translucent. The female is wingless and lays eggs in bag. The generation is annual. Larval feeding takes place for about 9–10 months and pupal formation takes place in the case suspended from a twig.

Infestation of *C. crameri* is noticed in cacao and tea bushes and recorded as an important pest in *Casuarina equisetifoliai* plantations of southern India. This insect has also been noticed on host plants like *Acacia catechu*, *A. nilotica*, *Albizia* spp., *Bischofia javanica*, *Bobax malabaricum*, *Cassia* spp., *Hevea brasiliensis*, *Santalum album*, *Shorea robusta*, *Tamarindus indica* and *Terminalia chebula* (Beeson 1941; Browne 1968).

The first outburst of *C. crameri* was recorded in 1885 from Tons Valley, Uttarakhand. Subsequently, it was reported from Bilaspur, Himachal Pradesh, in 1928 and also from Kahhula, Pakistan, in 1934. Outbreak of *C. crameri* on *P. roxburghii* was also reported in the state of Jammu and Kashmir during 1989–1990. This outbreak caused a casualty of 5% trees over 2000 ha area and 0.3 million trees were lost resulting in a total loss of 22.5 million rupees (Thakur 2000).

Control Measures

Baksha (2000) observed that *C. crameri* larvae are parasitised by *Brachymeria* sp. This Dipteran parasitoid can be used after mass multiplication by releasing them during the population outbreak of bagworm in chir pine forests. The pest could be collected manually and killed. Foliar spraying of 0.1% Malathion can be used for effective control of larvae. Use of an insect chitin inhibitor, Dimilin, against *C. crameri* has been suggested by Thakur (2000).

36.3.13 *Populus* Species

The genus *Populus* (Family Salicaceae) is known as poplar, which includes around 35 species that are native to the temperate regions, and many species and their clones/hybrids have entered in different countries for large-scale plantations. In India, apart from an indigenous species, *Populus ciliate*, *Populus alba* and *Populus nigra*, many species and clones have been introduced and planted in plains of Haryana, Punjab and the Tarai Region of Uttarakhand.

36.3.13.1 Insect Pests of *Populus*

Around 119 insect species are found in *Populus* species. These insects belong to the orders Coleoptera (28 species), Diptera (2 species), Hemiptera (21species), Hymenoptera (19 species), Lepidoptera (48 species) and Thysanoptera (1 species), which are associated with *Populus* species (Browne 1968). According to Thakur (2000), 112 insect species are found to feed on various species of *Populus*. All of them are native insect fauna, comprising chiefly defoliators (58 species), stem and shoot borers (23 species), sap suckers (27 species) and termites (4 species). Among the defoliators, *Cloetera cupreata* is a key defoliator insect pest in most exotic poplar in nurseries and plantations (Singh et al. 2005), which was reported originally from *Salix tetrasperma*, but has established on *Populus* as new host.

Clostera cupreata Butler

Clostera cupreata (syn. *Pygaera cupreata*) (Lepidoptera: Notodontidae) is popularly known as poplar defoliator and occurs in Jammu and Kashmir, Himachal Pradesh and Uttarakhand. The female moths lay 200–300 eggs on the leaf surface (Thakur 2000). The colour of young larvae is light brown with greenish mid-dorsal portion. Head is light brown in colour, speckled with dark brown spots. Two reddish tufts of hairs are present on each mid-dorsal margin of first and eighth abdominal segments. The length of mature larvae is 25–28 mm. There are five larval instars. The total larval duration is 11–12 days. Pupa formation takes place in thin hairy yellowish-brown colour cocoon either between the leaves or in leaf folds. The pupal duration is 4–6 days. The diagnostic features of moth are pale reddish brown in colour;

head, abdomen and hindwings pale in colour; expanse of wings 25–30 mm; forewing pale reddish brown in colour with two pale oblique antemedian lines, the first line angled to medial nervate and the second line curved near the inner margin; a medial outwardly curved line joining a straight postmedial line at inner margin; and the colour of hindwing paler than forewing. Male moths have a characteristic anal tuft arising on the dorsal side of the anal segment (Hampson 1892). The duration of the life cycle varies between 19 and 20 days and there are overlapping 8–9 generations existing in a year.

The larva of this moth is also found to occur as a defoliator of *Salix alba*, *Salix babylonica* and *Salix tetrasperma* (Beeson 1941; Browne 1968). Since 1966 in the Tarai Region of Uttarakhand and since 1986 in Punjab State, *C. cupreata* has been recorded as an important pest of poplar plantations. After 3 years of plantation establishment of poplar, epidemics of *C. cupreata* develop (Thakur 2000).

Control Measures

Ahmad (1993) has suggested screening and selection of *C. cupreata*-resistant clones of poplar is necessary for plantations. Release of egg parasitoids *Trichogramma achae*, *Trichogramma chilionis* and *Trichogramma perkensi* in poplar plantation can minimise the population of *C. cupreta* (Thakur 2000). Foliar spraying of insecticides, viz. 0.04% Endosulfan, 0.04% Fenithrothion, 0.05% Monocrotophos and 0.05% Quinolphos, showed larval killing of the poplar defoliator (Thakur 2000). According to Sangha et al. (2017), foliar application of insecticides, viz. Quinalphos, Profenofos and Carbaryl @ 0.1% concentration, kept the population of *C. cupreata* under control for 21 days.

36.3.14 *Salix* Species

Willow belongs to the genus *Salix* (Salicaceae) with 330–500 species worldwide (Zhenfu et al. 1999). About 33 species of *Salix* are reported from India and the most popular species is *Salix alba* L., commonly known as white willow.

36.3.14.1 Insect Pests of *S. alba*

Browne (1968) has mentioned about 19 species of insects belonging to five orders, viz. Coleoptera (2 species), Diptera (3 species), Hemiptera (4 species), Hymenoptera (5 species) and Lepidoptera (5 species), which are associated with *Salix alba* in India. Among all, *Lymantria obfuscata* is a key insect defoliator of *S. alba*.

Lymantria obfuscata Walker

Lymantria obfuscate (Lepidoptera: Lymantriidae) is popularly called as Indian gypsy moth, found in the montane and submontane zones of northwestern India and West Pakistan. The eggs hatch in early April and caterpillars feed for about 6 weeks. Feeding takes place during the cooler hours of the day and at night. During the middle of the day the caterpillars congregate in colonies of hundreds of individuals on the bark of the trunk, underside of the branches, in forks, under the shelter of epicornic shoots and the foliage of ground plants, under stones, etc. (Beeson 1941). The full-grown larva is about 45 mm. Pupation occurs in similar positions and under the protection of sticky silk webs. The pupal period lasts for 10–14 days and emergence of moths occurs during June–July. The female has vestigial wings and is unable to move far from her pupation site, close to which she lays a mass of eggs covered with yellow hairs from her body. The eggs remain unhatched until next spring. Hence, the generation is annual.

The species is known principally as a pest of orchards and plantations of *Alnus nitida*, *Malus domestica*, *Populus* spp., *Quercus dilatata*, *Juglans regia*, *Salix alba*, *Salix babylonica* and *Salix fragilis*, in Kashmir and the northwest Himalayas (Beeson 1941; Gupta 1982; Masoodi et al. 1987). *L. obfuscata* is a damaging defoliator of willows and defoliation causes loss of increment. Trees may be killed if they are severely defoliated (FAO 2003).

Control Measures

Gupta (1982) evaluated several insecticides against the eggs and larvae of *L. obfuscata* in apples on hills of Uttar Pradesh and found mineral-oil-based tree spray oils, all at 3%, 6% or 9%, and Sandolin A, a dinitro or the cresol preparation, at 1%, 2% or 3% very effective in reducing the viability of the eggs. Further, application of 5% dust formulation of insecticides like aldrin, Dichlorodiphenyl Trichloroethane (DDT), heptachlor, chlordane or carbaryl in 30 cm wide and 0.5 cm thick bands on the ground around the trunk of the trees or spraying the trees with 0.05% emulsion of insecticides like diazinon, endosulfan, methyl parathion, dichlorvas, quinalphos or fenitrothion resulted in satisfactory control of the larvae. Masoodi et al. (1987) carried out field trials to evaluate the effectiveness of endosulfan, fenthion, fenvalerate and burlap bands against larvae of *L. obfuscata* on walnut in Kashmir. Burlap bands 30 cm wide were stapled around the trunks of experimental trees at waist to breast height and subsequently endosulfan and fenthion were applied at 0.05% and fenvalerate at 0.03%. All treatments significantly reduced the pest population, fenvalerate and endosulfan being the most effective.

36.3.15 *Shorea robusta* Gaertn. f.

Shorea (Family Dipterocarpacae) is a potential genus of timber species of commercial importance, found in the lowland rain forests of India, Indonesia, Malaysia and the

Philippines. *Shorea robusta* Gaertn. f. is known as sal in India. According to Kulkarni (1956), sal is of Indian origin, and the northeast India is considered as homeland of sal. This tree species is a promising timber of India both ecologically and economically. *S. robusta* constitutes an ecosystem of cool and calm environment, rich in biodiversity.

36.3.15.1 Insect Pests of *S. robusta*

Sal has a major insect problem. Out of 346 insects recorded on sal, 155 insects are attached with green tree that includes mainly defoliators (114), seed feeders (19), borers (18) and sap suckers (4) (Stebbing 1914; Beeson 1941; Browne 1968; Sen-Sarma and Thakur 1994; Thakur 2000; Nair 2007; Roychoudhury 2015b). Rest of the insects feed either on the freshly felled timber or dry timber. Other than periodic outbreaks of a cerambycid trunk borer *Hoplocerambyx spinicornis*, and defoliators *Lymantria mathura* and *Trabala vishnou* (Roychoudhury 2017b), there are no major insect problems for *S. robusta* in India.

Hoplocerambyx spinicornis Newman

Hoplocerambyx spinicornis (syn. *Hoplocerambyx relictus*) (Coleoptera: Cerambycidae) is generally known as sal heartwood borer or sal borer. This insect species is the only known Indian representative of its genus and one of the most pernicious pests of sal that commits catastrophic damage to forests (Fig. 36.16). This insect was identified as a destructive pest of sal by Stebbing in 1897 from Singhbhum division in Chhotanagpur, Bihar (Stebbing 1914). Sal heartwood borer and its host tree have received much attention compared to any other tree-insect system in India (Stebbing 1914; Beeson 1941; Mathur 1962; Browne 1968; Roonwal 1976, 1978, 1982; Tewari 1995; Bhandari and Rawat 2001; Joshi et al. 2006; Roychoudhury et al. 2018a).

The life cycle, ecology and management of *H. spinicornis* on sal in India have been the subject of many studies since the 1900s and the literature is vast (Beeson 1941; Roonwal 1978; Thakur 2000; Bhandari and Rawat 2001; Joshi et al. 2006). The beetles start appearing every year with the onset of monsoon rainfall during June–July and continue up to August–September till all beetles have emerged from the tree trunks. After emergence, beetles start mating and lay eggs within a week on bark of sal trees (Fig. 36.17). Normally, the trees chosen for egg laying are dying or weak trees, but during epidemic even healthy trees are attacked. Each female lays 100–300 eggs over a life span of about 30 days. High humidity favours egg laying and fecundity is highly variable, and it may reach up to 465 per female at 91% Reative Humidity (RH) (Beeson 1941).

After hatching, larvae feed under the tree bark for a short period of about 1 h and then in the sapwood and finally enter inside the heartwood. Large sal trees support the development of around 300 larvae. As the internal damage increases, coarse dust is thrown out of holes in the bark and then deposited at the tree base in large heaps. The mature larva is large and reaches up to 9 cm in length (Fig. 36.18). The larval growth and development is completed during the month of November when the larva constructs a chamber in the heartwood with an exit hole for adult emergence and turns into a pre-pupa, then pupa and finally adult beetle during May–June and remains quiescent until it emerges with rainfall initiation. The adult beetle is dark brown in colour and variable in size, measuring 20–65 mm in length (Fig. 36.19). The antennae of the male are much longer than the body. The ontogeny of this borer is completed by 1 year.

H. spinicornis is the most notorious forest pest of India because of its periodic outbreaks, during which millions of sal trees are killed. Extensive galleries in the sapwood made by several larvae cause partial or complete girdling of the tree, leading to its death. Although the tree offers resistance by the outflow of resin, mass attack during epidemics kills even vigorous trees. Both the main trunk and crown branches are attacked. In a typical dead tree, 60–70% of the borer population occurs in the main trunk and 40–30% in the crown branches (Beeson 1941).

The sal borer causes extensive damage to both standing trees and freshly felled timbers (Joshi et al. 2002) because of the kairomonal activity present in sal sap (Roychoudhury 1997). The less vigorous, weak, unhealthy trees or those with reduced vitality are easily attacked or killed due to borer (Beeson 1941). Different abiotic and biotic factors have been the major cause for the recurring epidemics (Joshi 2009; Roychoudhury 2017c; Roychoudhury et al. 2017a, 2018b, 2019). The borer kills trees of all age groups above the girth of 20 cm (Bhandari and Singh 1988); the trees between the girth classes of 91 and 150 cm are the most preferred (Beeson 1941), with maximum mortality recorded in the girth class of 121–150 cm (Roychoudhury et al. 2004). The population density of sal borer also varies with respect to girth class of infested trees (Roychoudhury 2018b).

H. spinicornis has also been recorded to feed on other tree species, such as *Dipterocarpus tuberculatus*, *Duabanga grandiflora*, *Hevea brasiliensis*, *Parashorea robusta*, *Shorea assamica* and *Shorea obtuse* (Beeson 1941; Browne 1968; Nair 2007). However, population outbreaks occur only on *S. robusta*.

The sal borer attack is epidemic when the trees damaged by the insects are above economic threshold level (ETL), that is, the total number of affected trees should be more than 1% (Beeson 1941). So far, 21 epidemics of this borer have occurred in sal forests of different states of the country (Bhandari and Singh 1988; Bhandari and Rawat 2001; Joshi et al. 2006). Severe sal borer epidemics have occurred in Madhya Pradesh and Chhattisgarh during the year 1997–2000. Out of 16.84 lakh ha of sal forests, about 3 lakh ha got damaged during the epidemic, and the seriously

Fig. 36.16 Sal forest damaged by *Hoplocerambyx spinicornis*. (Source: Roychoudhury, Unpublished)

Fig. 36.17 Eggs of *Hoplocerambyx spinicornis*. (Source: Roychoudhury et al. 2018a)

Fig. 36.18 Grubs of *Hoplocerambyx spinicornis*. (Source: Roychoudhury et al. 2018a)

forest-damaged district was Mandla where about 8 lakh trees were damaged (Joshi et al. 2006). Infestations of sal borer were noticed in Dindori and East Mandla Forest Divisions, Kanha and Satpura Tiger Reserves of Madhya Pradesh, and Banhupratapur, Jagdalpur and Kabirdham Forest Divisions of Chhattisgarh (Roychoudhury 2015c, 2016b; Roychoudhury et al. 2017b).

Control Measures

H. spinicornis is a chronic and endemic pest. This insect is found always in small numbers in sal forests infesting fallen, weak, unhealthy or dying trees. During population outbreaks of sal borer, green healthy trees are infested. Pest management of this insect aims at two goals: preventive measures and remedial measures. Preventive measures deal with outbreaks by keeping the infestation below the defined tolerable level (<1%) and remedial measures deal with actions to limit the damage during outbreaks. The control measures of sal borer require the following preventive and remedial measures.

Fig. 36.19 Adult beetles of *Hoplocerambyx spinicornis*. (Source: Roychoudhury et al. 2018a)

Preventive Measures

1. Harvesting of sal trees should be done before the onset of monsoon.
2. Matured and overmatured sal trees lose their resistance against insect pests and diseases and hence, they should be regularly harvested.
3. The borer-attacked trees should be marked and categorised according to the category of the attacks

Fig. 36.20 Types/categorisation of *Hoplocerambyx spinicornis* affected trees in sal forests. (Source: Roychoudhury et al. 2018a)

during the winter months (December–February) as per the recommendations made by Beeson (1941) (Fig. 36.20).

4. The dead and dying trees should be removed after cutting from the forest before the initiation of monsoon.
5. Lops and tops of felled trees should be removed from the forest to maintain the forest hygiene.
6. Thinning of weak, diseased sal trees and transplanting its associate species should be done to reduce the density of pure sal forests.
7. Collection of sal seeds should be suspended in the areas affected by borer till the area gets fully regenerated.
8. In badly affected areas, where sal regeneration is poor or does not occur, the open area created should be filled by introduction of suitable miscellaneous species either through seed sowing or transplanting.
9. Grazing should be kept under control as far as possible.
10. Timber depots should be at least 3 km away from the sal forest.

Remedial Measures

1. *Trap-tree operation (TTO)*: The beetles get attracted to the smell of the sap oozing out from the injured or wounded sal trees. Making use of this phenomenon for monitoring the beetle population and for trapping them in heavily infested forests or in endemic conditions is the only environmentally sound and very effective non-toxic method of control. Many of the outbreaks of this borer have been successfully brought under control by this operation. This exercise has now become a regular activity of maintaining healthy sal forest of the country. Sal trees, which are silviculturally unsound or having borer attack but remaining green, wind fall or broken, with preferably 60–90 cm girth, should be used as trap to attract the beetles during monsoon. About 2–3 m long sal logs are beaten up to loosen its bark. Beetles are attracted to these trap logs due to their odour coming out from oozing sap from cut-end of the logs. Trap-tree operations are required to be carried out immediately with the inception of the first shower of rains and continued till the presence of insect is absent continuously.
2. All attacked trees belonging to categories 1 and 2, that is, dead trees and trees with dead crown having heaps of more than 7 cm high wood dust above the ground, should be felled, debarked and converted into the logs.
3. The borer affected sal logs when stored in stacks should be sprayed with water emulsion of 0.05% chlorpyriphos (10 mL/L water) during the month of June to kill the emerging borer beetles (pharate adults) and for effective results, the sprayed stacks should be covered with hard polythene sheets.

Lymantria mathura Moore

Lymantria mathura (syn. *Porthetria mathura*) (Lepidoptera: Lymantriidae) is generally known as the rosy gypsy moth or pink gypsy moth. This insect occurs in India, China, Nepal, Japan, Korea and Russia.

The adult female of *L. mathura* lays eggs in masses on the tree trunks or stems of sal plant. The length of full-grown larva is about 50 mm in male and 90 mm in female; larval colour is ashy with yellow colour bands across the thorax; abdomen has rows of papules bearing tufts of long hairs, with presence of two long plumes of hair projected on both sides of the head (Beeson 1941). Pupation occurs in a leaf fastened with a few stands of silk. Pupa is buff to dark brown in colour, around 20–36 mm in length. Pupa exhibits sexual dimorphism; the female pupa is paler in colour and larger in size. The moths of *L. mathura* are moderate in size (Figs. 36.21 and 36.22). There is a distinguished sexual dimorphism in size and colour. The female is larger than male in size. The diagnostic characters of adult moths are given by Hampson (1892), Beeson (1941), Browne (1968) and Roonwal (1979). The female moth has a white-coloured forewing with dark markings and edged with pink colour; hindwing is pink in colour with a band of dark brown colour; abdomen is half white, half pink in colour; and legs are pink and black in colour. Wing span of female moth is about 70–90 mm. The male with the forewing is mainly marked by dark brown colour, while hindwing is yellow in colour with a spot and line of black colour. Wing span of male moth is around 40–50 mm. *L. mathura* is univoltine or bivoltine (Beeson 1941; Browne 1968; Roonwal 1979). The first generation is found during April and October. Brood larvae occur during April and moths are found during May after pupal duration of 8–10 days. Larvae occur again during the

Fig. 36.21 Female moth of *Lymantria mathura*. (Source: Roychoudhury et al. 2020b)

Fig. 36.22 Adult male moth of *Lymantria mathura*. (Source: Roychoudhury et al. 2020b)

summer weather and early rains. Pupa formation occurs during July–September and moths are found after around 10 days. Next-generation eggs are laid in October.

L. mathura is one of the key defoliators of *S. robusta* in India. This insect also prefers food plants, such as *Quercus leucotrichophora*, *Quercus serrata*, *Syzygium cumini*, *Terminalia arjuna* and *Terminalia myriocarpa* (Roonwal 1979).

L. mathura larvae are gregarious defoliators. The larvae are able to consume whole leaves and sometimes avoid tough veins in older foliage growth. Larvae may also feed on flowers and tender young shoots (Browne 1968; Roonwal 1979). Damage of this nature can result in decline in overall growth and development of host trees, a reduction in yield or total crop loss in fruit crops, or even tree death (Roonwal 1979). *L. mathura* is a major defoliator of deciduous trees. According to Beeson (1941), larvae of *L. mathura* were found to infest and cause defoliation of *S. robusta* in Assam and Madhya Pradesh. Appanah and Turnbull (1998) have mentioned that in Assam and North India, the successive defoliations of *L. mathura* have been known to kill sal trees.

Control Measures

The management of *L. mathura* can be possible by use of biocontrol options like spraying of water solution of bacterium *Bacillus thuringiensis* (*B.t.*) var. *kurstaki* 1%, that is, @ 10 g wettable powder dissolved in 1 L on larvae-bearing foliage. According to Zheng et al. (1994), infestations of *L. mathura* in China have been controlled in chestnut orchards by the application of pesticides to tree trunks.

***Trabala vishnou* Lefebvre**

Trabala vishnou (syn. *Gastropacha vishnou*) (Lepidoptera: Lasiocampidae) is generally called as the rose-myrtle lappet moth. This insect defoliator occurs in South East Asia, including India, Pakistan, Thailand, Sri Lanka, Myanmar, Java, China, Japan and Indonesia. This insect is a sporadic pest of castor, *Ricinus communis* (Beeson 1941). *T. vishnou* is polyphagous, and feeds on leaves of sal forests (Fig. 36.23). The female lays oval, light yellowish eggs on the tree trunk or surface of the leaves in rows. They turn bluish brown before hatching. Within 12 days hairy, light brown, gregarious, colourful larvae hatch out. They remain assembled in groups in early instars but scatter on the tree in later instars (Fig. 36.24). They undergo five moults resulting in formation of six larval instars. The full-grown larva is hairy, white or yellowish brown, having a row of blue spots along dorsolateral side and grows to about 69 × 8 mm in size (Joshi et al. 1988). The larval period is nearly 45 days in January–March, 27–53 days in April–May, 38–61 days in June–July and nearly 45–53 days in August–September to October–December. The larva constructs a tent-shaped papery cocoon, interwoven with larval hairs, usually fixed along a twig and ultimately turns to pupa. After 11–17 days of pupal period, the moths emerge out. Its female is bulky and yellowish with a wingspan of 60 mm (Fig. 36.25) whereas the male is stout, light green in colour with wing expanse of 40 mm. *T. vishnou* shows four to five overlapping generations in a year (Beeson 1941; Joshi 1992).

Other than sal, *T. vishnou* is found on *Anogeissus latifolia*, *Bischofia javanica*, *Berberis asiatica*, *Butea frondosa*, *Careya arborea*, *Eucalyptus robusta*, *Lagerstoemia specieosa*, *Lantana camera*, *Mallotus philippinensis*, *Psidium guajava*, *Quercus diltata*, *Q. leucotrichophora*, *Quisqualis indica*, *Syzygium cumini*, *Terminalia belerica*, *T. myriocarpa* and *Verbascum thapsus* (Beeson 1941; Browne 1968; Joshi et al. 1988). The species is frequently responsible for severe defoliation to *S. robusta* forests (Beeson 1941; Browne 1968).

Control Measures

Foliar spraying of bacterium *Bacillus thuringiensis* (*B.t.*) var. *kurstaki* 1%, that is, @ 10 g wettable powder dissolved in 1 L water, can be used to control the larvae of *T. vishnou.*

36.3.16 *Swietenia* Species

The genus *Swietenia* (Family Meliaceae) is generally known as mahogany. The species of this genus are native to tropical America (CABI 2005). *Swietenia macrophylla* King is a well-known and widely planted species. This species is a well-accepted planted tree, both in native and exotic places.

36.3.16.1 Insect Pests of *S. macrophylla*

About 21 insect species belonging to the order Coleoptera (10 species), Hemiptera (1 species), Isoptera (2 species), Lepidoptera (7 species) and Orthoptera (1 species) are associated with *S. macrophylla* (Browne 1968). In exotic

Fig. 36.23 Sal defoliation by *Trabala vishnou*. (Source: Roychoudhury et al. 2017b)

Fig. 36.24 Larva of *Trabala vishnou*

Fig. 36.25 Adult moth of *Trabala vishnou*. (Source: Roychoudhury and Mishra 2021b)

plantations of mahogany, the dominant pest is a shoot borer, *Hypsipyla robusta*, which is found in both *S. macrophylla* and *S. mahagoni*.

Hypsipyla robusta Moore

Hypsipyla robusta (syn. *Hypsipyla pagodella*) (Lepidoptera: Pyralidae) is commonly known as mahogany shoot borer. This insect is widely distributed in the old-world tropics and extending into the subtropics (Browne 1968). The female moth lays white eggs on leaves, young shoots or other succulent areas of host. The larva attains a fully grown length of about 30 mm and is of variable colour with longitudinal rows of black and setiferous spots. One larva may infest more than one shoot during the course of its development. The larva is principally destructive as a shoot borer, but also attacks inflorescences and fruits (Beeson 1941). Pupation occurs in a cocoon, either within the larval tunnel or in other sheltered places on the host. The moth is rufous-brown mixed with grey and black, forewing with veins streaked with black and crossed with zigzag black lines and patches; hindwing is whitish, semi-hyaline, with the margin and costal zone darker; and wing expanse is 26–32 mm in male and 28–42 mm in female. In India, the length of ontogeny of this insect is 1–6 months according to season. The species usually undergo overwintering as a fourth instar larva.

About 21 species are found as food plants of *H. robusta*. These food plants are *Canarium schueinfurthii*, *Carapa guianensis*, *Cedrela australis*, *Cedrela mexicana*, *Cedrela multijuga*, *Cedrela odorata*, *Cedrela sureni*, *Cedrela toona*, *Chukrasia tabularis*, *Chukrasia velutina*, *Entandrophragma angolense*, *Entandrophragma utile*, *Khaya anthotheca*, *Khaya ivorensis*, *Soymida febrifuga*, *Swietenia candollei*, *S. macrophylla*, *S. mahagoni*, *Toona ciliata*, *Xylocarpus guianensis* and *Xylocarpus moluccensis* (Ramaseshiah and Sankaran 1994).

Larvae of *H. robusta* bore into the tips and shoots of many potential tree species. It is a key pest of mahogany in India and causes total mortality of seedlings and young plantations. The larvae damage the apical shoot and trunk, which leads to reduction in the timber value. This insect is responsible for destruction of plantations.

Control Measures

Wightman et al. (2008) have suggested the use of resistant clones/provenances that need to be exploited as an element for integrated management of mahogany shoot borer. According to Mishra (1993), the entomogenous fungus *Beauveria bassiana* is a promising pathogen for the biocontrol of this serious shoot borer infesting *S. macrophylla* in India. Joshi and Jamaluddin (2007) have recommended pruning of attacked shoots in young plantations and regular thinning of young attacked trees. Mohanadas (2000) has suggested for foliar spraying of 0.5% Dimethoate or 0.1% Phosphamidon for 100% mortality of mahogany shoot borer.

36.3.17 *Tectona grandis* L.f.

Tectona grandis (Family Verbanaceae) is commonly known as teak. Teak is a paragon among timber species. Teak is truly an Indian species because the Indian region shows as the only known centre of maximum genetic diversity and variability of teak (Hedegart 1975) with distribution over 8.9 million ha. The natural teak forests of India are mainly confined to Peninsular India with a total area of about 1.4 million ha; however, besides Kerala, Karnataka, Tamil Nadu and Andhra Pradesh, other major teak-growing states include Madhya Pradesh, Chhattisgarh, Maharashtra, Orissa, Uttar Pradesh, Gujarat and Rajasthan.

36.3.17.1 Insect Pests of *T. grandis*

Extensive studies have been carried out on teak entomology in India since inception of forestry research, and scientists are engaged in conducting research (both fundamental and applied) on management of insects. About 294 insects have been identified so far on teak, which includes the representatives of the order Coleoptera (147), Lepidoptera (94), Orthoptera (23), Hemiptera (21), Isoptera (7), Diptera (1) and Hymenoptera (1) (Beeson 1941; Browne 1968; Tewari 1992c; Roychoudhury et al. 2001; Nair 2007). These insects attack all the stages of teak growth, from seed to mature trees. Of these, 196 species of insects feed on standing teak in India and its adjoining countries, comprising mainly defoliators (141), sap suckers (17), stem borers (16), root feeders (12), seed feeders (9) and gall former (1) (Shukla et al. 2001; Roychoudhury et al. 2002). Many of these insects are minor or occasional pests and very few are considered as insect pests of economic importance. Among all, *Holotricha* species, *Hyblaea puera* and *Eutectona machaeralis* are considered as major insect pests of teak.

Holotrichia Species

Holotrichia Hope (Coleoptera: Scarabaeidae), the genus of beetles, are generally called as white grubs because larvae (grubs) are white in colour and found under the soil where they feed on the roots and rootlets of teak seedlings. About 50 species of white grubs are found in India, out of which 14 species are identified as key pest attacking different crops (Chandel et al. 2009). Sometimes, in teak nurseries, white grubs become a serious problem (Fig. 36.26). The incidence of white grubs gradually increases with the age of the nursery and after 4–5 years of establishment. Four principal species of *Holotrichia* causing serious damage of teak in nurseries are *Holotrichia consanguinea*, *Holotrichia insularis*, *Holotrichia problematica* and *Holotrichia serrata* (Thakur 2000). About 80% damage in teak nurseries of Madhya Pradesh and Maharashtra was due to white grubs (Vaishampayan and Bhandari 1981). In Kerala, it has been

Fig. 36.26 Teak nursery bed damaged by *Holotrichia* sp. (Source: Roychoudhury and Mishra 2021c)

Fig. 36.27 Grubs of *Holotrichia* sp. (Source: Roychoudhury and Mishra 2021c)

noticed that *H. fissa* causes 20–30% mortality of seedlings in various teak nurseries (Varma 1991).

The adult beetles are called as chaffer beetles or June beetles that emerge from the soil soon after the pre-monsoon showers in June. They feed on the host leaves in the night, and after mating lay the eggs in soil at various depths. The beetles prefer sandy, loose soil for egg laying. The eggs are oval and white in colour. The freshly hatched grub is creamy white, C-shaped, about 4.5 mm in length with a hard chitinised head (Fig. 36.27). The grubs exist up to 8 cm deep in soil. They feed on the semi-decomposed leaves and young living rootlets available in soil during June–July. After a few days, these grubs moult, increase in size and become more injurious to roots and rootlets of teak seedlings in nurseries. The mature grubs move in soil at different depths from 10 to 30 cm. Sometimes they reach up to 100 cm in depth in soil during October. The full-grown grubs construct earthen cocoons and undergo pupation during October–November. The beetles develop after 18 days of pupal period but remain hidden in soil as pharate adults up to pre-monsoon showers in June. Thus, these pests have annual life cycle.

Chafer beetles cause considerable damage by feeding on foliage, flowers and fruits of many economically important plants. Generally, the adult beetle feeds on broad leaf plants. Ali (2001) has reported 143 hosts of beetles in India. Among the forest trees, *Acacia nilotica*, *Azadirachta indica*, *Butea monosperma*, *Dalbergia sissoo*, *D. latifolia*, *Juglans* spp., *Quercus* spp., *Morus* spp., *Populus* spp., *Prosopis cineraria*, *Shorea robusta*, *Tamarindus indica*, *Tectona grandis*, *Terminalia arjuna*, *Toona ciliate* and *Zizyphus jujube* are the important hosts.

The white grubs are considered as national pest in India. The infestations of these insects have been noticed in many localities causing serious damage to a wide range of cultivated plants. Attack of white grubs causes immediate mortality of teak seedlings in nurseries by exhibiting symptom. of drying noticed only when the roots are already eaten-up by the white grubs.

Control Measures

Sandy soil and immature Farm Yard Manure (FYM) should not be utilised for raising seedlings, and during monsoon (June–July) soil working should be carried out (Joshi and Jamaluddin 2007). Light traps may be used to collect beetles and then, beetles should be killed. Meshram et al. (1990) have suggested for soil mixing with phorate 10 G @ 150 g/nursery bed (size 10 × 1 m). Chlorpyriphos 20 Emulsifiable Concentrate (EC) and phenthoate 36 EC have been found effective at 0.04% and 0.1%, respectively, for the control of grub population in forest nurseries (Thakur et al. 1992).

Hyblaea puera Cramer

Hyblaea puera (Lepidoptera: Hyblaeidae), generally called as teak defoliator, is a key insect defoliator of teak in nurseries, plantations and natural forests. Larvae of this insect feed only on soft foliage during the early leaf flushing season, leaving only the major veins intact (Fig. 36.28). Hampson (1894) described this insect in 1794 and included it in the family Noctuidae. This teak defoliator is widely distributed in the tropical regions, viz. India, Myanmar, Sri Lanka, Java, Papua New Guinea, Cape York Peninsula of Northern Queensland in Australia, Solomon Islands, West Indies and Africa (Beeson 1941; Mathur 1960; Browne 1968; Nair 1988).

The moths of *H. puera* are comparatively small, with a wing span of 3–4 cm (Fig. 36.29). The forewing is greyish brown and the hindwing with black and orange-yellow markings. Eggs are laid singly, on the upper or lower surface of tender unexpanded brownish tinge foliage. Each female lays about 400–500 eggs. The freshly hatched larvae moult to second instar after 2–4 days. It moults to third instar that cut outs semi-circular or rectangular flaps after 2–3 days, then moults to fourth instar. The duration of fourth and fifth larval stages exists for 2–4 and 3–8 days, respectively. This insect shows five larval stages with a distinct pre-pupal stage. The mature larvae are about 3.4–4.5 cm in length. There is colour difference in the last instar; the body may be either wholly black, or with longitudinal-coloured bands that may include a dorsal orange or ochreous band and lateral white lines. The mature larvae feed whole leaf and only the major veins remain intact (Fig. 36.30). On completion of feeding, mature larvae descend to the ground on silken threads and pupate under thin layer of litter or soil within a loosely built pupal covering constructed by dry or decayed leaves and silk. The total life cycle is completed within 22–30 days depending mainly on the prevailing temperature and humidity.

H. puera feeds on many plants, but teak is its principal host plant in India (Beeson 1941; Mathur 1960). There are 31 food plants of *H. puera*, including teak (Roychoudhury 1999), out of which 15 species belong to the family

Fig. 36.28 Teak defoliation by *Hyblea puera*. (Source: Roychoudhury and Mishra 2021c)

Fig. 36.29 Adult moth of *Hyblea puera*

Fig. 36.30 Larva of *Hyblea puera*. (Source: Roychoudhury and Mishra 2021c)

Bignoniaceae (*Bignonia megapotanica*, *Catalpa kaempferi*, *Dolichandrone spathacea*, *Dolichandrone stipulata*, *Heterophragma adenophyllum*, *Heterophragma roxburghii*, *Heterophragma sulfureum*, *Kigelia pinnata*, *Markhamia platycalyx*, *Millingtonia hortensis*, *Oroxylum indicum*, *Spathodea campanulata*, *Stereospermum chelenoides*, *Stereospermum suaveolens* and *Tecoma undulate*), 13 species including teak to the family Verbenaceae (*Callicarpa arborea*, *Callicarpa macrophylla*, *Prema latifolia*, *Prema pyramidata*, *Symphorema involucratum*, *Tectona grandis*, *Vitex agnus-castus*, *Vitex altissima*, *Vitex canescens*, *Vitex globrata*, *Vitex negundo*, *Vitex penduncularis* and *Vitex pubescens*), and one species each to the family Araliacae (*Heptapleurum venulosum*), Juglandaceae (*Engelhardtia spicata*) and Oleaceae (*Schrebera swietenoides*). Most of the food plants are trees, others woody climber or shrubs. These food plants may sustain the insect population when teak is not in flush (Beeson 1941).

The outbreak of *H. puera* in Central India, consisting of Maharashtra and Madhya Pradesh, starts from the second half of June when new flush of young, tender, reddish tinged leaves in teak put forth in forest nurseries and plantations and continue till the end of August. On the other hand, outbreak takes place from April to early June and again from August to September at Nilambur, Kerala (Mathur 1960). The trees, thus, remain without foliage for nearly 2 months in Central India and 4 months in South India. It results in a loss of about 13% annual increment during normal infestation to one-third of the total increment in growth during severe infestation (Beeson 1941). Continuous feeding on young teak crop for three successive generations in a season causes a loss of about 65% of the normal increment in growth and sometimes even the death of the defoliated trees (Champion 1934). Nair et al. (1985)

calculated a loss of 44% potential volume growth of teak due to defoliation by *H. puera* in 4–8-year-old plantations in Kerala.

Control Measures

To minimise the incidence of *H. puera*, teak plantation should be subdivided into different patches by raising mixed tree species (Beeson 1941). Desirable plant species such as *Anogeissus latifolia*, *Bauhinia racemosa*, *Carcya arborea*, *Cassia fistula*, *Grewia asiatica*, *Grewia tiliaefolia*, *Helicteres isora*, *Kydia calyeina*, *Shorea robusta*, *Sterculia villosa*, *Terminalia tomentosa* and so on, should be grown preferably in mixture within the teak plantations (Beeson 1941). These species help to maintain the population of the natural parasites and predators. During afforestation programme, teak seeds from relatively resistant genetically superior teak trees should be selected for control of teak defoliator (Ahmad 1987; Nair et al. 1989; Roychoudhury 2012c, 2013b; Roychoudhury and Mishra 2020c). Spraying of bacterium *Bacillus thuringiensis* @ 1% is highly effective in controlling the population outburst of this pest, irrespective of toxins and products (Gupta and Joshi 1995). Application of nuclear polyhedrosis virus (NPV) @ 14 × 106 Purified Inclusion Bodies (PIB)/mL has been proved to be very promising against this pest (Sudheendrakumar et al. 1988). Roychoudhury and Joshi (2010a) and Roychoudhury and Mishra (2020a) have suggested for biopesticide, 0.005% ivermectin (Ivecop-12) application for larval control of *H. puera*. Roychoudhury and Joshi (2010b) and Roychoudhury and Mishra (2019, 2020b) have recommended to use 0.0001% biopesticide spinosad 45% SC (Spintor 45% SC) for management of *H. puera* larvae. Release of potential egg parasitoid *Trichogramma* raoi @ 1.25 lakh/ha in teak forests during the outbreak period (July–September) can reduce the larval population of this pest (Sudheendrakumar 1985; Roychoudhury et al. 2018c, 2020a). The predator beetle, *Chalaenius rayotus*, may be introduced in the area of this pest infestation to minimise the larval population (Sudheendrakumar 1985).

Regarding the chemical pesticides, foliar spraying of either Carbaryl 0.1% or Malathion 0.05% can be used for killing of *H. puera* larvae (Singh 1988). According to Roychoudhury and Joshi (1996a), spraying of Deltamethrin 0.002% (0.6 mL/L) or Alphamethrin 0.002% (0.2 mL/L) or Monocrotophos 0.05% (1.3 mL/L) is proved to be highly effective in killing larval population of this defoliator pest.

Eutectona machaeralis (Walker)

Eutectona machaeralis (syn. *Pyrausta machaeralis*, *Hapalia machaeralis*, *Paliga machaeralis*) (Lepidoptera: Pyralidae), generally known as teak skeletoniser or teak leaf skeletoniser, is a key pest of teak in nurseries, plantations and natural forests. Larvae of this insect feed only green matter of leaves leaving all the veins intact, thus qualifying for the name skeletoniser (Fig. 36.31). Larvae consume all qualities of leaves till natural leaf fall. Teak skeletonisers occur in teak forests of Indo-China, India, Myanmar, Sri Lanka and throughout Malayan region to Australia (Beeson 1941; Mathur 1960; Browne 1968).

The moths of *E. machaeralis* are small, bright yellow, forewings having pink or reddish zig-zag or serrate line and hindwings with an ochreous or reddish marginal line or band (Fig. 36.32). The female moth lays 250–550 greenish white eggs single on the ventral area of leaf. Larvae after hatching are dirty white, which changes to pale green in colour in due course of time. This insect shows five stages of larvae; growth and development differ with respect to stages with distinct pre-pupal stage. The larvae consume the whole fleshy leaf tissues and vein network remains intact thereby skeletonising the leaf (Fig. 36.33). Larvae consume teak leaves at different levels of maturity and quality. However, larvae prefer soft tender leaves (Roychoudhury et al. 1995a). Larvae spend 10–14 days based on the climatic factors. Pupation occurs in small loose cocoon on green or fallen

Fig. 36.31 Teak skeletonisation by *Eutectona machaeralis*. (Source: Roychoudhury and Mishra 2021c)

Fig. 36.32 Adult moth of *Eutectona machaeralis*. (Source: Roychoudhury and Mishra 2021c)

Fig. 36.33 Larva of *Eutectona machaeralis*. (Source: Roychoudhury and Mishra 2021c)

dry teak leaves. The pest undergoes pre-pupal diapause in Central and North India for 4–5 months, whereas presence of diapause in this insect is not known in South India and Myanmar (Beeson 1941).

E. machaeralis is an oligophagous insect and in India, this insect mainly survives on teak as host plant. Other host plants of *E. machaeralis* are *Callicarpa arborea*, *Callicarpa cana*, *Callicarpa macrophylla* and *Tectona hamiltonia* (Beeson 1941; Mathur 1960). This skeletoniser feeds on *Callicarpa cana* in Java and is never found on teak (Beeson 1941).

In Central India, the larvae of leaf skeletoniser start appearing in May and increase to epidemic form only in September whereas in South India the epidemic level is reached in November. The impact of damage seems to be very little in South India in comparison to Central India on current growth increment because during the month of November growth of teak stops (Beeson 1941). However, skeletonisation of leaves presumably reduces the capability of producing enough food, which may hamper the increment and height growth in the following year. In Central India, the plantations defoliated by *H. puera* in June–July suffer from infestation of leaf skeletoniser from September onwards. This hampers the growth of teak. Complete defoliation by these insects stimulates production of epicornic branches and die-back of leading shoots, which sometimes cause death of young trees.

Control Measures

To minimise the outburst of leaf skeletoniser, regeneration areas of pure teak should be subdivided into economically viable patches (Beeson 1941). The desirable plant species acting as food plants of other defoliators should be grown to maintain natural enemies, viz. parasites and predators. Use and selection of teak seeds from relatively resistant genetically superior teak trees should be made for control of leaf skeletoniser (Ahmad 1991; Mishra 1992; Roychoudhury and Joshi 1996b; Roychoudhury et al. 1997a, b, c; Roychoudhury and Mishra 2020c). Spraying of bacterium, *Bacillus thuringiensis*, *B.t.* @ 1% has been found effective to kill larvae of this pest, irrespective of toxins and products (Roychoudhury et al. 1994). Roychoudhury and Mishra (2020a, d) have found biopesticide 0.03% Ivermectin (Ivecop-12) as effective dose for larval control of *E. machaeralis*. Likewise, Roychoudhury and Joshi (2010b) and Roychoudhury and Mishra (2020b) have recorded biopesticide 0.0001% Spinosad 45% SC (Spintor 45% SC) as effective concentration management of larvae of *E. machaeralis*. Release of egg parasitoid *Trichogramma* raoi @ 1.25 lakh/ha in teak forests during July–September has been reported to control the skeletoniser (Patil and Thontadarya 1983, 1984; Joshi et al. 2007; Roychoudhury et al. 2018c, 2020a).

Regarding the chemical pesticides, Singh and Gupta (1978b) and Singh (1988) have suggested spraying of Carbaryl 0.1% or Malathion 0.05% for effective larval killing of this skeletoniser. Roychoudhury and Joshi (2000) have recommended foliar application of either Delthamethrin 0.002% or Fenvalerate 0.003% or Alphamethrin 0.003% or Cypermethrin 0.003% or Monocrotophos 0.03% or Chlorpyriphos 0.05% or Phosphamidon 0.05% or Dichlorvos 0.06% or Methyl demeton 0.06% to kill larvae of this insect.

36.4 Future Scope

Periodical surveys of forest insect damage in nursery, plantation and natural forest are necessary, which is highly limited (FAO 2007). Information on pest infestation is obtained via informal observations by forest scientists, foresters and forest workers. The pest control techniques include silvicultural, biological and chemical, for management of insects in nursery and man-made forests. An example in this regard is integrated method of silvicultural and biological control tactics to control defoliator of teak, *Tectona grandis* (Mathur 1960).

The first step of entomological research is the exploration, inventorisation, documentation and digitalisation of insect fauna, their host plants, major insect pests, quantification of damage assessment, role of climate/weather in different eco-zones, emphasis on biocontrol options and eco-friendly management (Fig. 36.34). According to Zechendorf (1995), biological pesticide is an eco-friendly alternative. The promising biopesticides are plant pesticides, toxins of soil actinomycetes such as Avermectins and Spinosad, microbial pesticides such as *Bacillus thuringiensis* (*B.t.*) and nuclear polyhedrosis virus (NPV), and classical biological control by use of parasitoids and predators. Insecticides have limited prospects in today's pest management practices due to environmental concerns. Eco-friendly approaches such as biocontrol agents including microbial control assume greater importance in this context.

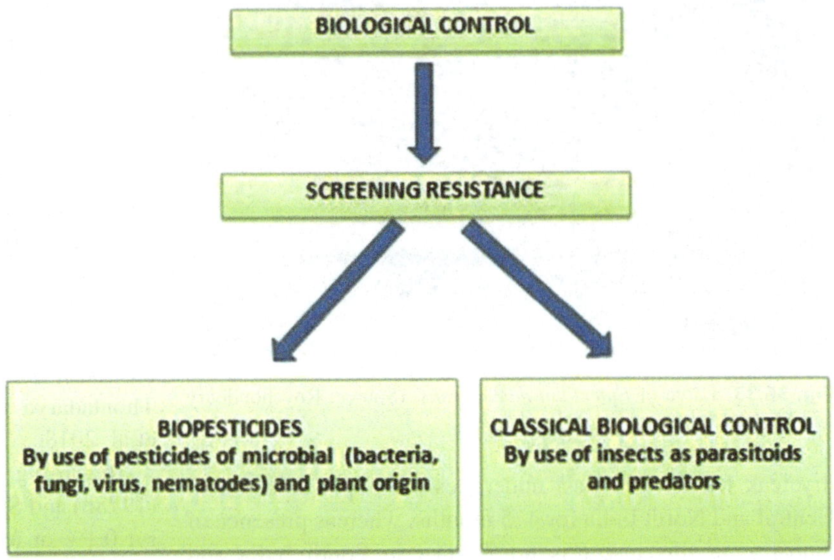

Fig. 36.34 Eco-friendly control measures of insects. (Source: Roychoudhury, Unpublished)

Research on the insect management of forest tree species is not enough. Management of the major insect pests of tree species like *Azadirachta*, *Cedrus*, *Dalbergia*, *Eucalyptus*, *Gmelina*, *Picea*, *Pongamia*, *Pinus*, *Salix*, *Shorea and Swietenia*, particularly in plantations and natural forests, is lacking till today; the plantation growers are unable to safeguard valuable timber species in nursery, plantations and even forests. Many issues are often referred by stakeholders, viz. State Forest Departments, Forest Development Corporations and other non-governmental organisations (NGOs), which need special attention. Thus, wide scope for sorting out the control of various insects of forestry species still exists and remains unexplored and the scientists should fully devote to fill up the gaps.

Lessons Learnt

- The insect pests damaging seeds, seedlings and trees seriously hamper forest produce, both quality and quantity wise.
- Many insects are found to be harmful to forestry species, including naturally regenerating tropical and temperate forests in India.
- Overview of forest insects of India shows that the major key insect pests of forestry species have been identified.
- Insect infestation in natural and plantation forest trees in India causes serious seasonal outbreaks, which lead to substantial economic loss in the timber productivity.
- Pest profile of major insect pests of tree species has been succinctly described.
- Control measures of key insect pests by biocontrol strategy (screening resistance, biopesticides and classical biological control) and safer chemical pesticides have been mentioned for successful control of forest insects.

Key Questions

1. What is the economic injury level of identified key insects of tropical and temperate forests of India?
2. What is the impact of weather in outbreak of key insect pests, and their effects on nurseries, Clonal Seed Orchard (CSO) and Seedling Seed Orchard (SSO) and natural forests?
3. What is the infestation level of five major insect pests of tree species in different eco-zones of India?
4. What method should be followed for screening resistance to adopt biocontrol of pests? Which biopesticides and biocontrol agents can be considered against target insects in field?
5. What should be the Integrated Insect Pest Management (IIPM) strategy to manage major insects of potential forestry species?

Acknowledgement The author is thankful to Dr. Rajesh Kumar Mishra, Chief Technical Officer, Computer and IT Section, ICFRE-Tropical Forest Research Institute, Jabalpur, 482021, Madhya Pradesh, for technical assistance in preparation of manuscript.

References

Ahmad M (1987) Relative resistance of different clones of *Tectona grandis* to teak defoliator, *Hyblaea puera* Cram. (Lepidoptera: Pyralidae) in south India. Indian For 113:281–286

Ahmad M (1991) Natural resistance in teak clones to leaf skeletonizer, *Eutectona machaeralis* (Lepidoptera: Pyralidae). Indian J For 14: 228–231

Ahmad M (1993) Relative resistance in different clones of *Populus deltoides* to poplar defoliator, *Clostera cupreata* (Lepidoptera: Notodontidae). Ann For 1:1–12

Akhtar MS, Patankar NV, Gaur A (2012) Observations on the biology and male of eucalyptus gall wasp *Leptocybe invasa* Fisher and

LaSalle (Hymenoptera: Eulophidae). Indian J Entomol 74:173–175

Alfred JRB (1998) Faunal diversity in India: an overview. In: Alfred JRB, Das AK, Sanyal AK (eds) Faunal diversity in India. Zoological Survey of India, Kolkata, pp 1–9

Ali MTM (2001) Biosystematics of phytophagous Scarabaeidae—an Indian overview. In: Sharma G, Mathur YS, Gupta RBL (eds) Indian phytophagous scarabs and their management: present status and future strategy. Agrobios (India), Jodhpur, pp 5–37

Anonymous (1994) *Albizia procera* (Safed siris). ICFRE Brochure, Dehradun

Anonymous (1996) Chir pine (*Pinus roxburghii*). Indian Council of Forestry Research and Education, Dehradun. http://www.frienvis.nic.in/WriteReadData/UserFiles/file/pdfs/Chir_Pine.pdf

Appanah S, Turnbull JM (1998) A review of dipterocarps: taxonomy, ecology and silviculture. Center for International Forestry Research (CIFOR). http://www.cifor.cgiar.org/publications/pdf_files/Books/Dipterocarps.pdf#search=%oplocerambyx%20spinicornis%20%22

Aytar F (2003) Natural history, distribution and control method of *Leptocybe invasa* Fisher & La Salle (Hymenoptera: Eulophidae) in Turkey. J DOA 9:47–66

Baksha MW (2000) Biology, ecology and management of the bagworm, *Cryptothelea crameri* Westwood (Psychidae: Lepidoptera) in Bangladesh. Bangladesh J For Sci 29:37–40

Beeson CFC (1931) The life history and control of *Celosterna scabrator* Fab. (Coleoptera: Cerambycidae). Indian For Rec (Old Series) (Entomology) 16:1–16

Beeson CFC (1941) The ecology and control of forest insects of India and neighbouring countries. Vasant Press, Dehradun

Bhalla OP, Parmar YS (1996) Binomics of *Eucosma hypsidryas* Meyrik (Lepidoptera: Eucosmidae), a pest of spruce in the western Himalaya, India. In: Nair KSS, Sharma JK, Varma RV (eds) Impact of diseases and insect pests in tropical forests. Kerala Forest Research Institute, Peechi, Kerala, pp 389–391

Bhandari RS, Rawat JK (2001) Sal heartwood borer, *Hoplocerambyx spinicornis* Newn. (Coleoptera: Cerambycidae) and its management. Indian For 127:1387–1393

Bhandari RS, Singh P (1988) Epidemic of sal heartwood borer, *Hoplocerambyx spinicornis* Newm. (Coleoptera: Cerambycidae) and its control in Pachmarhi, Madhya Pradesh. Indian For 114: 152–157

Bhatia CL (1984) *Eucalyptus* in India—its status and research needs. Indian For 110:91–96

Branco M, Franco JC, Valente C, Mendel Z (2006) Survey of Eucalyptus gall wasps in Portugal. Boletin de Sanad vegetal De Plagas 32: 199–202

Browne FG (1968) Pests and diseases of forest plantation trees. Clarendon Press, Oxford

CABI (Commonwealth Agricultural Bureau International) (2005) Forestry compendium (CD version), 2005 edn. CAB International, Wallingford

Champion HG (1934) The effect of defoliation on the increment of teak samplings. Indian For Bull 89:1–6

Chandel RS, Mehta PK, Sharma K (2009) White grubs: polyphagous pests of national importance. In: Singh R, Pal M, Sharma RC (eds) Forest insect pest and disease management in Northwest Himalayas. Himalayan Forest Research Institute, Shimla, pp 45–58

Chatterjee PN, Singh P (1968) *Celosterna scabrator* Fab. (Lamiidae, Coleoptera), a new pest of *Eucalyptus* and its control. Indian For 94: 826–830

Doganlar O (2005) Occurrence of *Leptocybe invasa* Fisher & La Salle, 2004 (Hymenoptera: Chalcidoidea) on *Eucalyptus camaldulensis* in Turkey, with a description of the male sex. Zool Middle East 35: 112–114

FAO (2005) The unwelcome guests. In: McKenzie P, Brown C, Jianghua S, Jian W (eds) Proceedings of the Asia-Pacific forest invasive species conference. RAP Publication 2005/18, Bangkok. http://www.fao.org/docrep/008/ae944e/ae944e00.htm

FAO (2007) Overview of forest pests—India. Working Paper FBS/18E. Forestry Department, Forest Management Division, FAO, Rome

Farjon A (2013) *Cedrus deodara*. In: IUCN red list of threatened species. IUCN. https://doi.org/10.2305/IUCN.UK.20131.RLTS.T42304A2970751.en

Fletcher TB (1914) Some south Indian insects and other animals of importance. Superintendent, Government Press, Madras

Food and Agriculture Organization (FAO) (2003) Global forest resources assessment update 2005—FRA 2005—pilot study for country reporting: India. Forest Resources Assessment Working Paper 78. http://www.fao.org/docrep/007/ae354e/AE354 E00.htm

Geiger CA, Napompeth B, Van Den Beldt RJ (1995) An update on the status of the leucaena psyllid in Southeast Asia. In: Shelton HM, Piggin CM, Brewbaker JL (eds) Leucaena: opportunities and limitations. Australian Centre for Agricultural Research, Canberra, pp 125–128

Gokkes M, Eshel G, Tadmor U (1989) Field trials for the control of chaff scale and Florida wax scale in citrus orchards with Tiger (pyriproxifen). Hassadeh 69:2019

Gupta BP (1982) Control of the Indian gipsy moth, (*Lymantria Obfuscata* Walker) on apple. Indian J Hort 39:279–283

Gupta BN, Joshi KC (1995) Evaluation of some biopesticides against forest insect pests. J Trop For 11:51–57

Gupta BK, Ahmad SI, Veer V (1989) Relative toxicity of some conventional insecticides against adult beetles of *Calopepla leayana* Latr. (Chrysomelidae: Coleoptera). Indian For 115:430–434

Hampson GF (1892) The fauna of British India. In: Moths, vol I, Reprint edition 1976. Today & Tomorrow Printers & Publishers, New Delhi

Hampson GF (1894) The fauna of British India. In: Moths, vol II, Reprint edition 1976. Today & Tomorrow Printers & Publishers, New Delhi

Hedegart T (1975) Breeding systems, variation and genetic improvement of teak (*Tectona grandis* Linn.f.). In: Burley J, Styles BT (eds) Tropical trees: variation, breeding and conservation. Academic Press, New York, pp 109–121

Hesami S, Alemansoor H, Seyedebrahimi S (2005) Report of *Leptocybe invasa* (Hym.: Eulophidae), gall wasp of *Eucalyptus camaldulensis* with notes on biology in Shiraz vicinity. J Entomol Soc Iran 24:99–108

Jacob JP (2009) Invasive insect pest (*Leptocybe invasa*) in *Eucalyptus* plantations in India. ENVIS For Bull 9:65–70

Jacob JP, Devaraj R, Natarajan R (2007) Outbreak of the invasive gall-inducing wasp, *Leptocybe invasa*, on eucalypts in India. Invasives 8: 4–5

Jain NC (1996) Growing pest problems in *Prosopis cineraria* in arid zones of India. In: Nair KSS, Sharma JK, Varma RV (eds) Impact of diseases and insect pests in tropical forests. Kerala Forest Research Institute, Peechi, pp 507–510

Javaregowda PST, Prabhu ST, Patil RS (2010) Evaluation of botanicals and synthetic insecticides against eucalyptus gall wasp, *Leptocybe invasa* (Hymenoptera: Eulophidae). Karnataka J Agric Sci 23:200–202

Jha LK, Sen-Sarma PK (1994) Forest entomology. Ashish Publishing House, New Delhi

Jhala RC, Patel MG, Vaghela MN (2010) Effectiveness of insecticides against blue gum chalcid, *Leptocybe invasa* Fisher & La Salle (Hymenoptera: Eulophidae), infesting eucalyptus seedling in middle Gujarat, India. Karnataka J Agric Sci 23:84–86

Joshi KC (1992) Handbook of forest zoology and entomology. Oriental Enterprise, Dehradun

Joshi KC (2009) Biology, life history of sal heartwood borer, factors responsible for outbreak and its management—an updated information. ENVIS For Bull 9:1–8

Joshi KC, Jamaluddin (2007) Handbook of diseases, insect pests and their control measures in forest nurseries/plantations. Tropical Forest Research Institute, Jabalpur, Madhya Pradesh

Joshi K, Roychoudhury N (1996) Relative toxicity of some insecticides against the larvae of shisham defoliator, *Plecoptera reflexa* Guenee. J Trop For 12:44–47

Joshi KC, Gurung D, Sarma PC (1988) Some observations on the bionomics of *Trabala vishnou* Lefroy—a pest of *Eucalyptus* in Assam. J Trop For 4:285–289

Joshi KC, Roychoudhury N, Sambath S, Humane S, Pandey DK (1996) Efficacy of three varietal toxins of *Bacillus thuriengiensis* against *Ailanthus* defoliator, *Atteva fabriciella* Swed. (Lepidoptera: Yponomeutidae). Indian For 122:1023–1027

Joshi KC, Kulkarni N, Roychoudhury N, Yousuf M, Sambath S (2002) Population dynamics and behaviour of Sal heartwood borer and its control measures. Project completion report submitted to Indian Council of Forestry Research and Education, Dehradun

Joshi KC, Roychoudhury N, Kulkarni N, Sambath S (2006) Sal heartwood borer in Madhya Pradesh. Indian For 132:799–808

Joshi KC, Sambath S, Chandra S, Roychoudhury N, Yousuf M, Kulkarni N (2007) Evaluation of *Trichogramma* spp. to minimize the attack of teak skeletonizer, *Eutectona machaeralis* (Walk.). Indian J For 30:267–271

Joshi KC, Roychoudhury N, Kulkarni N, Chandra S (2008) Insects associated with bamboos and their management. In: Mandal AK, Berry N, Rawat GS (eds) Proceedings of the national conference on bamboos: management, conservation, value addition and promotion. Tropical Forest Research Institute, Jabalpur, pp 143–162

Kalia S, Joshi KC (1995) Efficacy of three commercially available products of *Bacillus thuringiensis* Berliner against bamboo leaf roller, *Crypsiptya coclesalis* Walker. MyForest 31:1–3

Kim KC, Lee TS (1986) Identification of larval host plants, range and damage of the fruit sucking moth to the major fruits in Cheonnaram province. Korean J Plant Prot 24:180–184

Kulkarni DH (1956) Comparative distributional characteristics of sal and teak in Madhya Pradesh. In: Proceedings IX silvicultural conference, Dehradun

Kulkarni HD (2010) Screening eucalyptus clones against *Leptocybe invasa* Fisher and La Salle (Hymenoptera: Eulophodae). Karnataka J Agric Sci 23:87–90

Kumari NK (2009) Bioecology and management of eucalyptus gall wasp, *Leptocybe invasa* Fisher & La Salle (Hymenoptera: Eulophidae). M.Sc. (Agri) thesis, UAS, Dharwad

Lefroy HM (1971) Indian insect life. Today & Tomorrow's Printers & Publishers, New Delhi

Mani MS (1989) Indian insects. Satish Book Enterprise, Agra

Masoodi AM, Koul VK, Bhagat KC, Singh T (1987) Control of *Lymantria obfuscata* Walker on walnut in Kashmir. Indian J Plant Prot 15:15–17

Mathew G (1986) Insects associated with forest plantations of *Gmelina arborea* Roxb. in Kerala, India. Indian J For 94:308–311

Mathur RN (1943) Bamboo defoliators. Indian J Entomol 5:117–128

Mathur RN (1960) Pests of teak and their control. Indian For Rec 10:43–65

Mathur RN (1962) Sal heartwood borer, *Hoplocerambyx spinicornis* and its control by trap tree method. Indian For Bull 238:1–13

Mathur RN, Chatterjee PN, Sen-Sarma PK (1970) Biology, ecology and control of *Ailanthus* defoliator, *Atteva fabriciella* Swed. (Lepidoptera: Yponomeutidae) in Madhya Pradesh. Indian For 96:538–552

Mendel Z, Blumberg D, Ishaaya I (1991) Effect of buprofezin on Icerya purchasi and *Planococcus citri*. Phytoparasitica 19:103–112

Mendel Z, Protasov A, Fisher N, La Salle J (2004) Taxonomy and biology of *Leptocybe invasa* gen. & sp. n. (Hymenoptera: Eulophidae), and invasive gall inducer on *Eucalyptus*. Aust J Entomol 43:101–113

Meshram PB, Jamaluddin (1989) Efficacy of monocrotophos against *Ailanthus* webworm, *Atteva fabriciella* Swed. (Lepidoptera: Yponomeutidae). Indian For 115:113–116

Meshram PB, Pathak SC, Jamaluddin (1990) Effect of some soil insecticides in controlling the major insect pest in teak nursery. Indian For 116:206–212

Mishra SC (1992) Comparative natural resistance of different clones of *Tectona grandis* L.f. to teak skeletonizer, *Eutectona machaeralis* Walk. (Lepidoptera: Pyralidae). Indian For 118:274–278

Mishra RM (1993) *Beauveria bassiana* (Balsamo) Vuillemin a fungal pathogen of *Hypsipyla robusta* Moore (Lepidoptera: Pyralidae). Indian J For 16:236–238

Mishra RM (2001) Insect pests of *Ailanthus excelsa* and their management. M/s Bishen Singh Mahendra Pal Singh, Dehradun

Mohanadas K (2000) Management of the shoot borer *Hypsipyla robusta* (Lepidoptera: phycitidae) in mahogany plantations. KFRI Research Report No. 184. Kerala Forest Research Institute, Peechi, Kerala

Mohanadas K, Varma RV (1984) A new host record for *Atteva fabriciella*, a pest of *Ailanthus*. J Tree Sci 3:128

Mohanan C (1990) Diseases of bamboos in Kerala. In: Rao IVR, Gnanaharan R, Sastry CB (eds) Bamboos—current research. Proceedings of the international bamboo workshop, 14–18 Nov 1988, Kerala, pp 173–183

Mutitu KE (2003) A pest threat to Eucalyptus species in Kenya. KEFRI Technical Report, Kenya

Nair KSS (1988) The teak defoliator in Kerala, India. In: Berryman AA (ed) Dynamics of forest insect populations. Plenum Publishing Corporation, New York, pp 267–289

Nair KSS (2001) Pest outbreaks in tropical forest plantations: is there a greater risk for exotic tree species? Center for International Forestry Research, Bogor

Nair KSS (2007) Tropical forest insect pests: ecology, impact and management. Cambridge University Press, Cambridge

Nair KSS, Sudheendrakumar VV, Varma RV, Chacko KC (1985) Studies on the seasonal incidence of defoliators and the effect of defoliation on volume increment of teak. Research Report No. 30. Kerala Forest Research Institute, Peechi, Kerala

Nair KSS, George M, Varma RV, Sudheendrakumar VV (1986) Insect pests of *Eucalyptus* in India. In: Proceedings national seminar on *Eucalyptus* in India: past, present and future. Kerala Forest Research Institute, Peechi, Kerala, pp 325–335

Nair KSS, Kedharnath S, Koshy MP, Mathew P, Sudheendrakumar VV (1989) Search for natural resistance to the insect pest, *Hyblaea puera* in teak. Research Report No. 62. Kerala Forest Research Institute, 32 p

Napompeth B (1990) Leucaena psyllid problems in Asia and the Pacific. In: Napompeth B, MacDicken KG (eds) Leucaena psyllid: problems and management. Winrock International/IDRC/NFTA/F-FRED, Bangkok, pp 1–11

Nyeko P, Mutitu EK, Day RK (2009) *Eucalyptus* infestation by *Leptocybe invasa* in Uganda. Afr J Ecol 47:299–307. https://doi.org/10.1111/j.1365-2028.2008.01004.x

Nyeko P, Mutitu B, Otieno, Ngae GN, Day RK (2010) Variations in *Leptocybe invasa* (Hymenoptera: Eulophidae) population intensity and infestation on eucalyptus germplasms in Uganda and Kenya. Int J Pest Manag 56:137–144

Ohmart CP, Edwards PB (1991) Insect herbivory on eucalyptus. Annu Rev Entomol 36:637–657

Patil BV, Thontadarya TS (1983) Studies on the acceptance and biology of different *Trichogramma* spp., on the teak skeletoniser, *Pyrausta machaeralis* Walker. Indian For 109:292–297

Patil BV, Thontadarya TS (1984) Efficacy of egg parasite, *Trichogramma* spp. in parasitising the egg of the teak skeletoniser, *Pyrausta machaeralis* Walker. Indian For 110:413–418

Peleg BA (1989) Evaluation of the insect growth regulator pyriproxyfen (Tiger) as a control agent for the California red scale and the cottony-cushion scale. Alon Hanotea 43:681–686

Ramaseshiah G, Sankaran T (1994) Studies on some parasites of the Meliaceous shoot borer *Hypsipyla robusta* Moore in south India. Hexapoda 6:39–46

Rauf A, Hidayat P, Maryana N, Winasa IW (1990) Biology and demography of *Heteropsylla cubana* Crawford (Homoptera: Psyllidae). In: Napompeth B, MacDicken KG (eds) *Leucaena* psyllid: problems and management. Winrock International/IDRC/NFTA/F-FRED, Bangkok, pp 114–118

Roonwal ML (1976) Field ecology of sal heartwood borer, *Hoplocerambyx spinicornis* in sub-Himalayan forests. Part I. Preferred tree size and borer population dynamics. J Ind Acad Wood Sci 7:87–98

Roonwal ML (1978) The biology, ecology and control of the sal heartwood borer, *Hoplocerambyx spinicornis*: a review of recent work. Indian J For 1:21–34

Roonwal ML (1979) Field-ecological studies on mass eruption, seasonal life-history, nocturnal feeding and activity rhythm, and protective behaviour and coloration in the sal defoliator, *Lymantria mathura* (Lepidoptera: Lymantriidae), in sub-himalayan forests. Rec Zool Surv India 75:209–236

Roonwal ML (1982) Population ecology of the two major forest pests of sal (*Shorea robusta*), *Hoplocerambyx spinicornis* (Coleoptera) and *Lymantria mathura* (Lepidoptera). In: Proc. symp. ecol. anim. popul. Zoological Survey of India, pp 37–48

Roychoudhury N (1997) Role of kairomones in sal borer operation. Adv For Res India 17:181–187

Roychoudhury N (1999) Teak defoliator and host plants: an ecological relationship. Adv For Res India 20:182–189

Roychoudhury N (2002) *Albizia procera* (Roxb.) Benth.—a new host record for *Catopsilia crocale* Cramer (Lepidoptera: Pieridae). Indian For 128:101–102

Roychoudhury N (2010) Screening non-preference response in bamboos against key insect pest, *Crypsiptya coclesalis* Walker. In: Nath S, Singh S, Sinha A, Das R, Krishnamurthy R (eds) Conservation and management of bamboo resources. ICFRE, Institute of Forest Productivity, Ranchi, pp 73–79

Roychoudhury N (2011) Spinosad-a potential biopesticide for management of greater bamboo leaf roller, *Crypsiptya coclesalis* (Walker) (Lepidoptera: Pyralidae). In: Singh S, Das R (eds) Productivity enhancement and value addition of bamboos. Excel India Publisher, New Delhi, pp 53–59

Roychoudhury N (2012a) Toxicity of ivermectin in inducing larval mortality in bamboo leaf roller, *Crypsiptya coclesalis* Walker. Bangladesh J For Sci 32:1–9

Roychoudhury N (2012b) Comparative effect of ivermectin and spinosad on larval mortality in greater bamboo leaf roller, *Crypsiptya* coclesalis (Walker) (Lepidoptera: Pyralidae). In: Aggarwal P, Tewari VP, Triveni PR, Joshi SC (eds) Recent advances in bamboo propagation, management and utilization. Institute of Wood Science and Technology, Bangalore, pp 116–123

Roychoudhury N (2012c) Deployment of resistance in teak to key insect pests. Indian For 138:123–130

Roychoudhury N (2013a) Damage assessment of gall making insect species of eucalyptus and its management by pesticides. Project Completion Report No.: 153/TFRI/2010/Ento-3(24). Submitted to Indian Council of Forestry Research and Education (ICFRE), Dehradun

Roychoudhury N (2013b) Resistance in genotypes of teak to key insect pests. In: Singh S, Das R (eds) Assessment and conservation of forest genetic resources. Institute of Forest Productivity, Ranchi, Jharkhand, pp 136–143

Roychoudhury N (2015a) Spinosad toxicity induced larval mortality in *Albizia* foliage feeder, *Spirama retorta* Cramer (Lepidoptera: Noctuidae). Pestology 39:24–30

Roychoudhury N (2015b) Insect pests of *Shorea robusta* Gaertn. f.: an update. Indian J For 38:313–322

Roychoudhury N (2015c) Emergence of sal heartwood borer, *Hoplocerambyx spinicornis* Newman (Coleoptera: Cerambycidae), in Satpura Tiger Reserve, Madhya Pradesh. MyForest 51:33–42

Roychoudhury N (2016a) Gall insect menace in *Eucalyptus*. Indian J For 39(4):349–357

Roychoudhury N (2016b) Status of sal heartwood borer, *Hoplocerambyx spinicornis* Newman and its management. Project Completion Report Submitted to Indian Council of Forestry Research and Education, Dehradun

Roychoudhury N (2017a) Evaluation of biopesticide, spinosad and neonicotinoids, acetamiprid and imidacloprid against *Eucalyptus* gall insect, *Leptocybe invasa* Fisher & LaSalle. Pestology 41:23–32

Roychoudhury N (2017b) Sal defoliation in Knaha Tiger Reserve-search for causative agents. Indian J For 40:267–270

Roychoudhury N (2017c) New record of natural enemies of sal heartwood borer, *Hoplocerambyx spinicornis* Newman (Coleoptera: Cerambycidae). Pestology 41:33–38

Roychoudhury N (2018a) Development of galls in *Eucalyptus* due to infestation of *Leptocybe invasa* and its effects on growth of seedlings. Indian J For 41:79–83

Roychoudhury N (2018b) Determination of population density of heartwood borer, *Hoplocerambyx spinicornis* Newman in relation to girth class of infested logs of sal (*Shorea robusta* Gaertn. f.). Pestology 42:28–33

Roychoudhury N, Joshi KC (1996a) Relative toxicity of certain insecticides against the larvae of teak defoliator, *Hyblaea puera* Cramer. Indian J Entomol 58:253–255

Roychoudhury N, Joshi KC (1996b) Search for natural resistance in teak clones against *Eutectona machaeralis* Walker (Lepidoptera: Pyralidae). Indian J For 19:205–213

Roychoudhury N, Joshi KC (2000) Relative toxicity of some insecticides against larvae of teak leaf skeletonizer, *Eutectona machaeralis* Walker. Pestology 24:80–84

Roychoudhury N, Joshi KC (2008) Leaf roller, *Crypsiptya coclesalis* Walker (Lepidoptera: Pyralidae), a major pest of bamboos in nurseries and plantations. Indian For 134:1229–1235

Roychoudhury N, Joshi KC (2009) Toxicity of ivermectin in inducing larval mortality in *Ailanthus* webworm, *Atteva fabriciella* Swederus (Lepidoptera: Yponomeutidae). World J Zool 4:277–280

Roychoudhury N, Joshi KC (2010a) Toxicity of ivermectin against the larvae of teak defoliator, *Hyblaea puera* Cramer (Lepidoptera: Hyblaeidae). Indian For 136:38–44

Roychoudhury N, Joshi KC (2010b) Response of larvae of teak leaf skeletonizer, *Eutectona machaeralis* (Walker) (Lepidoptera: Pyralidae) to spinosad. Ann For 18:297–302

Roychoudhury N, Joshi KC (2011) Effect of ivermectin on larval mortality and reproductive potential of *Albizia* foliage feeder, *Spirama retorta* Cramer (Lepidoptera: Noctuidae). J Pest Sci 84: 117–122. https://doi.org/10.1007/s10340-010-0334-1

Roychoudhury N, Mishra RK (2018) *Eucalyptus* gall insect, *Leptocybe invasa* Fisher & LaSalle and its management by bio- and chemical pesticides—an update. Pestology 42:11–22

Roychoudhury N, Mishra RK (2019) Spinosad-induced larval mortality in teak defoliator, *Hyblaea puera* Cramer (Lepidoptera: Hyblaeidae). Pestology 43:37–41

Roychoudhury N, Mishra RK (2020a) Toxicity of ivermectin against the defoliator larvae of some major forest insect pests-a critical review. Pestology 44:34–38

Roychoudhury N, Mishra RK (2020b) Toxicity of spinosad against the larvae of some key insect defoliators of forest tree species—a critical review. Pestology 44:31–35

Roychoudhury N, Mishra RK (2020c) Resistance in teak for the management of major insect pests-a review. Pestology 44:21–38
Roychoudhury N, Mishra RK (2020d) Ivermectin-induced larval mortality in teak leaf skeletonizer, *Eutectona machaeralis* (Walker) (Lepidoptera: Pyralidae). Pestology 44:45–50
Roychoudhury N, Mishra RK (2020e) Babul borer, *Celosterna scabrator* and its control measures. Van Sangyan 7:13–16
Roychoudhury N, Mishra RK (2020f) *Ailanthus webworm, Atteva fabriciella* and its control measures. Van Sangyan 7:27–30
Roychoudhury N, Mishra RK (2021a) Shisham defoliator, *Plecoptera reflexa* and its control measures. Van Sangyan 8:30–32
Roychoudhury N, Mishra RK (2021b) Occurrence of sal defoliator, *Trabala vishnoi* in Kanha Tiger Reserve, Mandla Forest Division, Madhya Pradesh. Pestology 45:36–40
Roychoudhury N, Mishra RK (2021c) Major insects pests of teak and their management—an update. Pestology 45:24–31
Roychoudhury N, Mishra RK (2022a) *Albizia* foliage feeder, *Spiroma retorta* and its control measures: an update. Pestology 46:34–36
Roychoudhury N, Mishra RK (2022b) Bamboo leaf roller, *Crypsiptya coclesalis* and its management: an update. Pestology 46:33–38
Roychoudhury N, Sambath S (1998) Pest status, growth and development of *Eurema blanda* Boisduval (Lepidoptera: Pieridae) on *Albizia procera* (Roxb.) Benth. Indian J For 21:160–163
Roychoudhury N, Joshi KC, Sambath S, Pandey DK (1994) Effectiveness of three varietal toxins of *Bacillus thuringiensis* against leaf skeletonizer, *Eutectona machaeralis* Walker (Pyralidae). MyForest 30:57–60
Roychoudhury N, Jain A, Joshi KC (1995a) Alteration of growth and development in leaf skeletonizer, *Eutectona machaeralis* Walker, due to variations in teak leaves of different maturity. Indian J Exp Biol 33:227–229
Roychoudhury N, Sambath S, Humane S, Joshi KC (1995b) Relative toxicity of some insecticides against *Ailanthus* defoliator, *Atteva fabriciella* Swed. (Lepidoptera: Yponomeutidae*)*. Indian J For 18: 309–311
Roychoudhury N, Jain A, Joshi KC, Lal RB (1997a) Natural resistance in teak clones to leaf skeletonizer, *Eutectona machaeralis* Walker: an appraisal. Indian For 123:1027–1035
Roychoudhury N, Jain A, Pal R, Joshi KC (1997b) Antixenosis resistance in teak clones to leaf skeletonizer, *Eutectona machaeralis* Walker, in relation to certain phytochemicals. Malay For 60:7–15
Roychoudhury N, Joshi KC, Gera M, Sharma S (1997c) Selection of teak clones for resistance against leaf skeletonizer, *Eutectona machaeralis* Walker. Malay For 60:183–190
Roychoudhury N, Jain A, Joshi KC (2001) Insect pests of teak and breeding for insect resistance. In: Mandal AK, Ansari SA (eds) Genetics and silviculture of teak. International Book Distributors, Dehradun, pp 187–205
Roychoudhury N, Joshi KC, Chourasia M (2002) Insect pests of *Tectona grandis* L.f.: an update. Adv For Res India 25:196–224
Roychoudhury N, Sambath S, Joshi KC (2004) Girth class of sal trees prone to the attack of heartwood borer, *Hoplocerambyx spinicornis* Newman (Coleoptera: Cerambycidae). Indian For 130:1403–1409
Roychoudhury N, Chandra S, Joshi KC (2007) Infestation of Australian insect, *Leptocybe invasa* on *Eucalyptus* in Madhya Pradesh. Vaniki Sandesh 31:13–15
Roychoudhury N, Singh K, Joshi KC (2009) Insecticidal activity of spinosad against the larvae of *Ailanthus* webworm, *Atteva fabriciella* Swederus (Lepidoptera: Yponomeutidae). Pestology 33:34–36
Roychoudhury N, Gupta DK, Mishra RK (2017a) Role of climatic factors in emergence of Sal heartwood borer, *Hoplocerambyx spinicornis* Newman in Kanha Tiger Reserve. J Trop For 33:1–20
Roychoudhury N, Singh RB, Das AK (2017b) Occurrence of sal heartwood borer, *Hoplocerambyx spinicornis* Newman (Coleoptera: Cerambycidae), in East Baihar Forest range of North Balaghat Forest Division, Madhya Pradesh. Indian J For 40:63–67
Roychoudhury N, Meshram PB, Mishra RK (2018a) Problem of sal heartwood borer, *Hoplocerambyx spinicornis* Newman, in India and its management: a critical review. Pestology 42:23–32
Roychoudhury N, Gupta DK, Mishra RK (2018b) Role of climatic factors in emergence of sal heartwood borer, *Hoplocerambyx spinicornis* Newman, in Dindori Forest Division, Madhya Pradesh and its management. Indian J For 41:149–159
Roychoudhury N, Chandra S, Singh RB (2018c) Biological control of *Hyblaea puera* and *Eutectona machaeralis* by introduction of native egg parasitoid, *Trichogramma raoi* in teak forests. Pestology 42:36–41
Roychoudhury N, Gupta DK, Mishra RK (2019) Sal heartwood borer, *Hoplocerambyx spinicornis* Newman, its recent emergence in Madhya Pradesh and role of climatic factors. Pestology 43:28–43
Roychoudhury N, Meshram PB, Singh RB, Mishra RK (2020a) Native egg parasitoid, *Trichogramma raoi*: a potential biocontrol agent against defoliator and skeletonizer in teak forests of Madhya Pradesh. J Trop For 36:62–73
Roychoudhury N, Singh RB, Mishra RK (2020b) Occurrence of *Lymantria* mathura in sal forests of Odisha. Van Sangyan 10:27–31
Sajap AS, Samad AK (2000) Effects of food plants on development of *Spirama retorta* (Lepidorptera: Noctuidae). Pertanika J Trop Agric Sci 23:97–101
Sajap AS, Wahab YA, Marsidi A (1997) Biology of *Spirama retorta* (Lepidoptera: Noctuidae), a new pest of *Acacia mangium* in Peninsular Malaysia. J Trop For Sci 10:167–175
Sambath S, Joshi KC (1999) Relative toxicity of some insecticides tested against the larvae of *Albizia* foliage feeder, *Spirama retorta*. Indian J Entomol 61:275–278
Sangha KS, Dhaliwal AK, Singh R (2017) Seasonal incidence and management of *Clostera cupreata* (Butler) (Lepidoptera: Notodontidae) on *Populus deltoids*. Indian J Agrofor 19:83–89
Sankaran KV, Mohanadas K, Ali MIM (1989) *Beauveria bassiana* (Bals.) Vuill., a possible biocontrol agent against *Myllocerus viridanus* Fabr. and *Calopepla leayana* Latreille in south India. Curr Sci 58:467–469
Sasidharan KR (2004) Studies on the insect pests of *Casuarina equestifolia* L. in Tamil Nadu and their management. Ph.D. thesis, F.R.I. Deemed University, Dehradun
Sen-Sarma PK (1987) Insect pest problems in social forestry plantations and their management. Indian J For 10:239–240
Sen-Sarma PK, Thakur ML (1994) Pests of Dipterocarpaceae and their management. In: Jha LK, Sen-Sarma PK (eds) Forest entomology. Ashish Publishibg House, New Delhi, pp 165–186
Senthilkumar N, Murugesan S (2015) Insect pests of important trees species in South India and their management information. Institute of Forest Genetics and Tree Breeding, Coimbatore, Tamil Nadu
Sharma G, Dutta BK, Seth R (2001) Potential of *Bacillus thuringiensis* subsp. *kurstaki* in protection of *Gmelina arborea* Roxb. J Entomol Res 25:283–288
Shi Q (1980) Country report on China. In: Lessard G, Chouinard A (eds) Bamboo research in Asia. IDRC, Ottawa, pp 57–62
Shukla PK, Jamaluddin, Roychoudhury N (2001) Diseases and insect pests of teak. ICFRE Brochure No. 68, Tropical Forest Research Institute, Jabalpur
Singh P (1988) Insect pest problems in forest nurseries, natural and man-made forests and their management. In: Gupta VK, Sharma NK (eds) Tree protection. Indian Society of Tree Scientists, Solan, Himachal Pradesh, pp 304–331
Singh P (1990) Current status of pests of bamboos in India. In: Rao IVR, Gnanaharan R, Sastry CB (eds) Bamboos—current research. Proceedings of the international bamboo workshop, 14–18 Nov 1988, Kerala, pp 190–194
Singh P, Bhandari RS (1988) Insect pests of bamboos and their control in India. Indian For 114:670–683

Singh P, Gupta BK (1978a) Laboratory evaluation of insecticides as contact sprays against the forest pests. II-*Ailanthus* webworm, *Atteva fabriciella* Swed. (Lepidoptera: Yponomeutidae). Indian For 104: 696–702

Singh P, Gupta BK (1978b) Laboratory evaluation of insecticides as contact sprays against forest pests. I. Teak skeletonizer, *Pyrausta machaeralis* (Walk.) (Lepidoptera: Pyralidae). Indian For 104:359–366

Singh DK, Pusalkar PK (2020) Floristic diversity of the Indian Himalaya. In: Dar GH, Khuroo AA (eds) Biodiversity of the Himalaya: Jammu and Kashmir state. Springer Nature, Singapore, pp 93–128

Singh P, Bhandari RS, Fasih M, Prasad G, Tyagi BDS, Mishra BM, Mishra RK (1989) Epidemic of deodar defoliator, *Ectropis deodarae* Prout (Lepidoptera: Geometridae) in Kamraj Forest Division (J&K) and its control by aerial spraying of Fenitrothion. Indian For 115: 459–473

Singh AP, Bhandari RS, Verma TD (2005) Important insect pests of poplars in agroforestry and strategies for their management in north-western India. Agrofor Syst 63:15–26

Soderstrom TR, Ellis RP (1988) The woody bamboos (Poaceae-Bambuseae) of Sri Lanka: a morphological—anatomical study. Smithson Contrib Bot 72:1

Stebbing EP (1914) Indian forest insects of economic importance, Coleoptera. Reprint edition 1977. J.K. Jain Brothers, Bhopal

Sudheendrakumar VV (1985) Studies on the parasites of *Hyblaea puera* in teak plantations at Nilambur. In: Joseph KJ, Abdurahima UC (eds) Proceedings of national seminar on entomophagous insects, Calicut, pp 116–122

Sudheendrakumar VV, Ali MIA, Varma RV (1988) Nuclear polyhedrosis virus of the teak defoliator, *Hyblaea puera*. J Invertebr Pathol 51: 307–308

Tewari DN (1992a) A monograph on Bamboo. International Book Distributors, Dehradun

Tewari DN (1992b) Monograph on Eucalyptus. Rohini Printers & Publiishers, Dehradun

Tewari DN (1992c) A monograph on Teak (*Tectona grandis* Linn.f.). International Book Distributors, Dehradun

Tewari DN (1994a) A monograph on Deodar (*Cedrus deodara* (Roxb.) G. Don). International Book Distributors, Dehradun

Tewari DN (1994b) A monograph on *Dalbergia sissoo* Roxb. International Book Distributors, Dehradun

Tewari DN (1995) A monograph on Sal (*Shorea robusta* Gaertn.f.). International Book Distrubutors, Dehradun

Thakur ML (1988) Current status of termites as pest of bamboos and their control. Indian For 114:720–725

Thakur ML (2000) Forest entomology (ecology & management). Sai Publishers, Dehradun

Thakur ML, Kumar S, Negi A, Rawat DS (1992) Chemical control of white grubs in forest nurseries under arid soils. J Trop For 8:244251

Thu PQ, Dell B, Burges TI (2009) Susceptibility of 18 eucalypt species to the gall wasp *Leptocybe invasa* in the nursery and young plantations in Vietnam. Sci Asia 35:113–117. https://doi.org/10.2306/scienceasia1513-1874.2009.35.113

Turnbull JW (1990) Taxonomy and genetic variation in casuarinas. In: El-Lakany MH, Turnbull JW, Breybaker JL (eds) Advances in casuarina research and utilization. Desert Development Center, Cairo, pp 1–11

Turnbull JW, Midgley SJ, Cossalter C (1998) Tropical acacias planted in Asia: an overview. In: Turnbull JW, Crompton HR, Pinyopusarerk K (eds) Recent developments in *Acacia* planting. Australian Center for International Agricultural Research, Canberra, pp 14–28

Vaishampayan SM, Bhandari RS (1981) Chemical control of white grub, *Holotrichia insularis* Blanchard (Coleoptera: Scarabaeidae) in teak nurseries. Pestology 5:15–20

Varma RV (1991) White grub damage and its control in teak nurseries. In: Basha SC, Mohanan C, Sankar S (eds) Proceedings of the international teak symposium, Thiruvananthapuram, 2–4 Dec 1991. Kerala Forest Department and Kerala Forest Research Institute, pp 84–85

Varma RV (1996) Impact of *Atteva fabriciella* on growth of *Ailanthus triphysa*. In: Nair KSS, Sharma JK, Varma RV (eds) Impact of diseases and insect pests in tropical forests. Kerala Forest Research Institute, Peechi, Kerala, pp 311–316

Verma TD, Chander R (1995) Epidemic of deodar defoliator, *Ectropis deodarae* Prout (Lepidoptera: Geometridae) in the north-west Himalaya. Indian For 121:663–664

Warrier KCS, Singh BG, Kumar NK (2014) Twenty-five years of research on casuarinas at IFGTB. Institute of Forest Genetics and Tree Breeding, Coimbatore, Tamil Nadu

Wightman KE, Ward SE, Haggar JP, Santiago BR, Cornelius JP (2008) Performance and genetic variation of big-leaf mahogany (*Swietenia macrophylla* King) in provenance and progeny trials in the Yucatan Peninsula of Mexico. For Ecol Manag 255:346–355

Wingfield MJ, Robison DJ (2004) Diseases and insect pests of *Gmelina arborea*: real threats and real opportunities. New For 28:227–243. http://www.springerlink.com/content/j41r34826g7h0620/fulltext.pdf

Wylie FR, Speight MR (2012) Insect pests in tropical forestry, 2nd edn. CABI, Wallingford

Zechendorf B (1995) Biopesticides as integrated pest management strategy in agriculture. In: Reuveni R (ed) Novel approaches to integrated pest management. Lewis Publishers, Florida, pp 231–259

Zhenfu F, Shidong Z, Skvortsov A (1999) Salicaceae. In: Zhengnys W, Reven PH (eds) Flora of China. Missouri Botanical Garden Press, St. Louis, MO, pp 139–274

Zheng YX, Song JR, Wang YQ (1994) Control of *Lymantria dispar* and *Lymantria mathura* by spreading pesticides on chestnut trunks. China Fruits 4:11–13

Part VII

Forest Products and Utilization

Wood Properties and Utilization

37

S. R. Shukla

Abstract

Wood as a material, types of wood species, wood properties and their distinguishing attributes are discussed. Measurements and evaluation of various physical and mechanical properties of wood as well as wood working quality parameters as per standard procedures are elaborated. The computation of suitability indices and safe working stresses of timbers, interaction of wood with environment and its behaviour and variability in wood quality are presented. A concise list of some important timber species of the country along with their specific gravity, characteristic features and general usage has also been listed at the end of the chapter.

Keywords

Wood properties · Heartwood · Moisture content · Wood density · Working quality

37.1 Introduction

Wood is one of the most valuable and versatile raw materials being used for creating many finished and semi-finished products. This naturally grown material has provided valuable services to humanity since its appearance on the planet and has continued to contribute in many ways to the survival and development of many civilizations. In spite of presence of many other natural and synthetic materials and products, wood is still a source of innumerable products useful to the society. Wood is mainly composed of three main polymeric components, namely cellulose, hemicelluloses and lignin. In addition to these, varying quantities of extractives, gums, resins and some other organic and inorganic chemicals are also found in different wood species (Anon 1970, 1972; Tsoumis 1991). At microscopic level, wood possesses a porous cellular structure which may be modified in different ways providing it (wood) versatility and multi-utility character. Additionally, this natural material is also widely available, renewable, sustainable and eco-friendly.

The usefulness of timber is due to its internal structure, range of properties and chemical composition (Dinwoodie 1981; Panshin and de Zeeuw 1980; Shmulsky and Jones 2011; Tsoumis 1991). Rational and efficient utilization of this precious and multipurpose material for various applications, however, presents a challenge; for that its in-depth and proper knowledge as well as working experience is absolutely necessary. Most people who work closely with wood can easily distinguish between different types of wood. They are also often familiar with the different characteristics that make different types of wood react or appear differently when cut, surfaced, treated, finished, etc. The obvious first step to accomplishing these tasks is to accurately identify different types of wood species. It is also helpful to know that many different tree species are divided into groups based on their similar physical characteristics (Chowdhury and Ghosh 1958; Anon 1972). Furthermore, it may be noted that the use of wood species depends on a number of internal characteristics and external factors, including the basic properties of the material. Different types of wood species may not be useful in all end-use applications. In other words, some types of wood are much better suited for specific purposes than others. Accurate wood identification is also associated with various benefits such as judicious utilization of a suitable wood species for intended application at the appropriate cost, controlling the illegal trade, forensic investigations, etc.

37.2 Wood as a Material

The knowledge of various properties and characteristics is a pre-requisite not only for the efficient and rational utilization but also helps in better understanding the wood as a material.

S. R. Shukla (✉)
ICFRE-Institute of Wood Science and Technology, Bengaluru, India

A. K. Mandal, A. Nicodemus (eds.), *Textbook of Forest Science*, https://doi.org/10.1007/978-981-97-8289-5_37

Non-homogeneous nature and vast variability in wood properties within and between different species restrict indiscriminate utilization for various purposes. Only certain numbers of wood species are found to be more suitable than others for any given application. For efficient utilization of different timber species, detailed information about their microstructure, physical and mechanical properties, durability, effects of varying environmental conditions as well as presence of various defects and abnormalities is very important. The microstructure of wood, in turn, has profound influence on its physical and mechanical properties which are quite important in determining its utility in various applications (Panshin and de Zeeuw 1980; Dinwoodie 1981; Tsoumis 1991).

37.2.1 Comparison with Other Materials

Despite several shortcomings, wood has numerous obvious advantages compared to other synthetic and non-synthetic materials available for identical applications. Wood is natural, versatile and universally available from a variety of trees showing variations in properties, thus, allowing the scope of choosing the suitable wood for a given end use. Being renewable, sustainable, efficient and attractive, it provides cost-effective solutions to several technological designs and applications. Many woods are quite strong, lightweight and present flexibility in designing, making building construction and other engineering products simpler and safer than common alternatives such as steel, aluminium, reinforced concrete, fibre-polymer composites, etc. (Chowdhury and Ghosh 1958; Anon 1970, 2020; Gillie 2004; AEC 2020). It also exhibits insulation behaviour against heat, sound and electricity. The wood also shows good vibration and shock resistance, quite easy to work and design different shapes and sizes using simple tools, convertible to engineered wood products with improved performance and found to be suitable for many residential uses and producing a variety of industrial materials (Anon 1970, 1972).

Gillie (2004), AEC (2020) and Anon (2020) presented various properties of different materials in comparison with wood. It may be seen from the data that for weight to weight, wood is not an inferior material in comparison with other costly and synthetic products. Lighter woods such as Balsa (*Ochroma pyramidale*) show higher specific strength than stainless steel. Similarly, most of the wood properties are either better or at par with the moulded plastic which is a synthetic material. Wood as a structural material possesses many positive features compared to other materials being used in the same field.

Although wood is a versatile material, it also possesses certain inferior qualities which hinder its widespread applicability in different areas. It is highly anisotropic, so different quality parameters change unequally on the three main axes depending on the direction of measurement. Depending on the environmental conditions, the wood material also exhibits hygroscopicity and instability of its dimensions. Differences in various wood properties of the same and different species within and between trees also cause serious problems in their selection, handling, processing and utilization. Being biological by nature, wood is also vulnerable to be attacked by several bio-organisms under suitable conditions such as fungi, termites and insects (Anon 1972; Tsoumis 1991).

Eventually, it is to be emphasized that combination of various good and poor qualities of wood presents it as even more complex material compared to other engineering materials which needs to be understood properly. The efficient, economic and improved utilization of wood requires scientific knowledge of various physical, mechanical, chemical, biological and structural properties of a wide variety of wood species. In addition to converting solid wood into reconstituted wood products, modifications of wood material by different chemical or physical methods as well as modern processing techniques may be applied for enhancing its performance and service life (Tsoumis 1991; Shmulsky and Jones 2011).

37.2.2 Primary and Plantation Timbers

In India, there are close to 1600 identified timber species. Of these, about one third including various secondary lesser known and plantation grown timber species have been studied for various anatomical, physical, mechanical, chemical and biological properties (Rajput et al. 1996). Tremendous amount of data and information have been generated on various aspects of these timbers. Based on the laboratory data collected from testing of various properties as well as field data on different types of trials, a number of timber species have been recommended for many residential and industrial applications (Chowdhury and Ghosh 1958; Anon 1963; Rao and Purkayastha 1972; Sekhar and Gulati 1972; Purkayastha 1982, 1985; Rajput et al. 1996).

Since long time, many wood species are being traditionally used for various applications in the country such as *Tectona grandis* (Teak), *Shorea robusta* (Sal), *Dalbergia sissoo* (Sissoo), *Pterocarpus dalbergioides* (Andaman padauk), *Dipterocarpus indicus* (Gurjan), *Adina cordifolia* (Haldu), *Santalum album* (Sandalwood), *Dalbergia latifolia* (Rosewood), *Lagerstromia microcarpa* (Benteak), *Pterocarpus marsupium* (Bijasal), *Terminalia tomentosa* (Laurel), *Pinus roxburghii* (Chir) and *Cedrus deodara* (Deodar). These and many other region-specific long-established timbers are generally referred to as conventional or primary timbers (Anon 1970, 1972). However, their continuous usage for a long time, lesser availability and higher prices are major concerns making them unaffordable to many. Numerous research projects have evaluated the mechanical characteristics of both naturally occurring and planted trees, including *Tectona grandis* and *Terminalia myriocarpa*. It was discovered that the strength characteristics of trees grown on plantations and those grown naturally do not significantly differ. The plantation grown species were not of inferior quality compared with identical natural grown species. However, in few localities, natural grown teak trees showed some superiority in certain properties over plantation gown tree, while in other localities, the reverse was found valid. Simultaneously, the degree of variation in certain wood characteristics was somewhat greater in naturally grown trees compared to those in plantations (Rajput et al. 1996). There are many lesser known and plantation grown timber species which have been tested and evaluated for various properties. These are also recognized as commercially important and viable alternatives to the primary timber species for various applications. It is to be noted that proper seasoning, preservative treatments and modern processing techniques may be employed for rational utilization and value addition. Some of them have good potential for their exploitation as alternative wood to traditionally used timber species in various applications including building and construction, furniture, packaging, toys and handicrafts, production of plywood and panel products. such as *Acacia auriculiformis* (Bengal jali), *A. mangium* (Mangium), *A. nilotica* (Babul), *Adina cordifolia* (Haldu), *Ailanthus* spp., *Albizia* spp., *Casuarina equisetifolia* (Casuarina), *Dalbergia sissoo* (Sissoo or sheesham), *Eucalyptus* spp., *Gironniera reticulata, Grevillea robusta* (Silver oak), *Gyrocarpus jacquinii, Hevea brasiliensis* (Rubberwood), *Maesopsis eminii, Wrightia tinctoria* (Dudhi), *Gmelina arborea* (Gamari), *Dysoxylum malabaricum* (White cedar), *Melia dubia, Populus* spp., *Protium serratum, Simarouba glauca, Swietenia mahagoni* (Mahogany) and *Tectona grandis* (Teak), *Toona ciliata* (Toon).

37.3 Types of Woods

Wood species are generally divided into two types: hardwoods and softwoods. It's critical to understand that the labels 'hardwood' and 'softwood' don't really refer to how soft or how hard the wood is made. In actuality, the wood from a number of softwoods is stronger and harder than the wood from a number of hardwoods. For example, Chir (*Pinus roxburghii*), a softwood, produces a wood that is harder and stronger than Poplar (*Populus deltoids*) which is a hardwood. Even Balsa wood (*Ochroma pyramidale*), an extremely light-weight wood, commonly used in making model airplanes, is the product of a hardwood. In most of the tropical countries, hardwoods make up the major volume of timbers than the softwoods. Because of the increased extent of forest cover, the global hardwood growing stock is predicted to be about double that of softwoods (Shmulsky and Jones 2011).

37.3.1 Hardwoods

Hardwoods are trees with large leaves that are deciduous in temperate climates. The wood produced by these trees is also called hardwood. In contrast to the simple structure of softwoods, hardwood structure is complex. Surface hardness of wood has nothing to do with this class of timbers. Even some of the hardwoods are very soft in nature and vice versa. At least, four types of principal cells which account for a substantial portion of hardwood volume are vessel (pores), fibres, parenchyma and rays. There is considerable variability of structure between different hardwood species. Vessels are present only in hardwoods, and this is one of the major ways by which the hardwoods are distinguished from the softwoods. In some hardwoods, vessels and fibres may occupy, respectively, about 65% and 30% of the volume while rays may comprise 5%. In another hardwood, there might be 55% fibres, 10% vessels, 5% parenchyma cells and 30% rays. In yet another species, these proportions might be considerably different. Figure 37.1 shows teakwood disc and transverse section showing micro-structural features such as growth rings, large open vessels and radially aligned rays. The complex structure, the presence of vessels and considerable variability within and between species imply that hardwoods look quite different from one another, for example, *Tectona grandis, Shorea robusta* and *D. sissoo*.

37.3.2 Softwoods

The softwoods come from the coniferous trees and refer to those trees with needles, and which, for the most part, remain green throughout the year. The wood produced by such trees is called softwood. It may be noted that many softwood species possess fairly hard surfaces. The cellular composition of softwood is relatively simple. About 95% of the volume of all conifers consists of long cells known as tracheids, which are 3–5 mm long (equivalent to hardwoods fibres which are much shorter in length). The cells are also hollow, pitted and tapered shut at both ends. The remaining 5% or so of volume is essentially accounted for by radially aligned strips of short brick-shaped cells called rays. Disc of pine wood (*Pinus* spp.) and its transverse section with microstructural features such as resin canals, growth ring and ordered arrangement of tracheids are shown in Fig. 37.2. Resin canals are found consistently in only a few softwood genera such as *Pinus* and serve as path for transport of resin in living trees. Resin

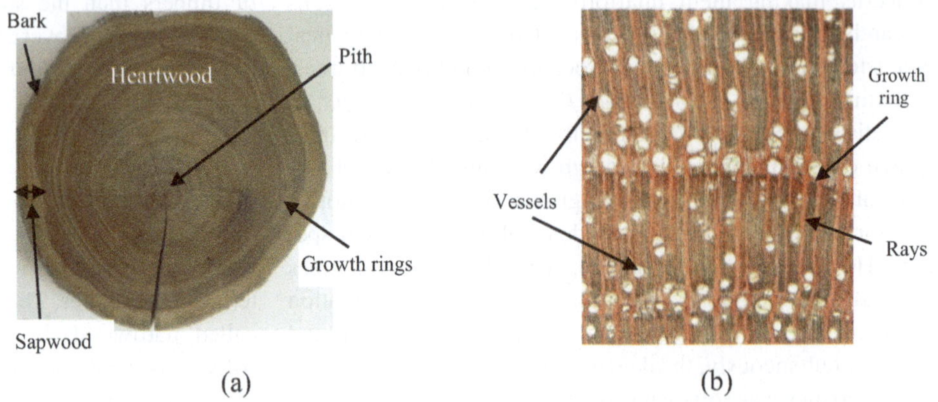

Fig. 37.1 Teak (hardwood) disc (**a**) and transverse section (**b**) showing micro-structural features: growth rings, large open vessels and radially aligned rays

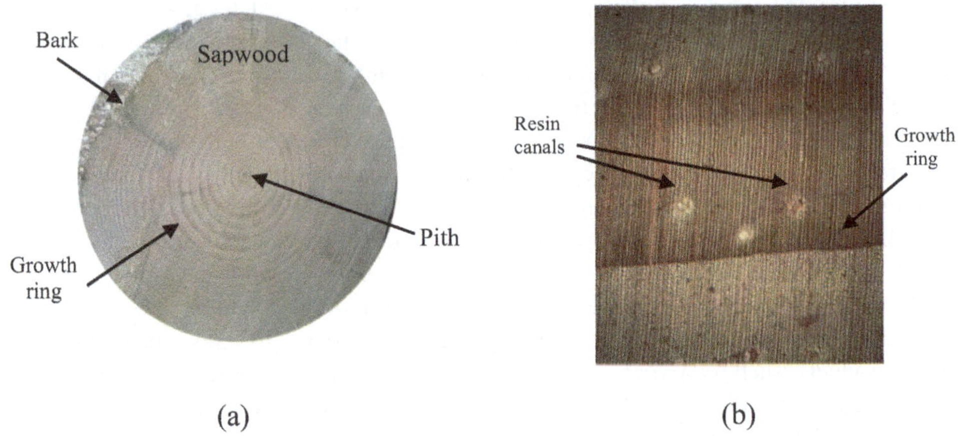

Fig. 37.2 Pine (softwood) disc (**a**) and transverse section (**b**) showing micro-structural features: resin canals, growth ring and ordered arrangement of tracheids

canals are spaces between cells created by the separation of adjacent cells which may be oriented longitudinally or radially (within rays). *Abies pindrow* and *Pines roxburghii* are examples of softwoods.

37.3.3 Heartwood and Sapwood

Observations of cross-section of a tree stem often reveal dark-coloured inner portion called heartwood, which is surrounded by a lighter-coloured outer portion called sapwood. The sapwood part of the tree contains the living cells and responsible for transport of food (mostly sugars) and water. The mechanical support to the tree is contributed by the dead zone of heartwood. It may be noted that sometimes, colour wise distinction between heartwood and sapwood is not very apparent which may be due to paucity of dark-coloured extractives in the heartwood.

Certain important features of heartwood are (1) distinct characteristic odour and colour, (2) high amount of decay resistance against deterioration from certain bio-organisms, (3) difficult to penetrate with liquids like water and preservative chemicals due to blockage of tiny passages in the wood structure, (4) difficult to dry or season due to poor permeability, (5) slightly higher density than the sapwood due to presence of extractives and other infiltrations, (6) comparatively lower hygroscopicity than the sapwood and (7) lower fibre saturation point (FSP) than the sapwood. However, in most of the cases, the strength properties of heartwood are not apparently different than corresponding sapwood of the same tree (Rajput et al. 1996).

37.4 Directions in Wood

The seasonal growth patterns of the wood usually result in the formation of one circular ring annually called growth ring or sometimes, annual ring. They comprise more or less distinct layers of earlywood (forms early in growth season) and latewood (darker coloured portion form later). Rays in the cross-section of wood trunk are visible extending radially from pith to periphery (cambium side) and facilitate the transport of sap in horizontal direction. All types of wood contain rays, readily visible in some, but difficult to visualize with naked eyes in others. The distance between the rays remains relatively constant in growth rings near the pith. The fibres and vessels are oriented vertically in the tree trunk. Depending on the direction of the fibres and the growth

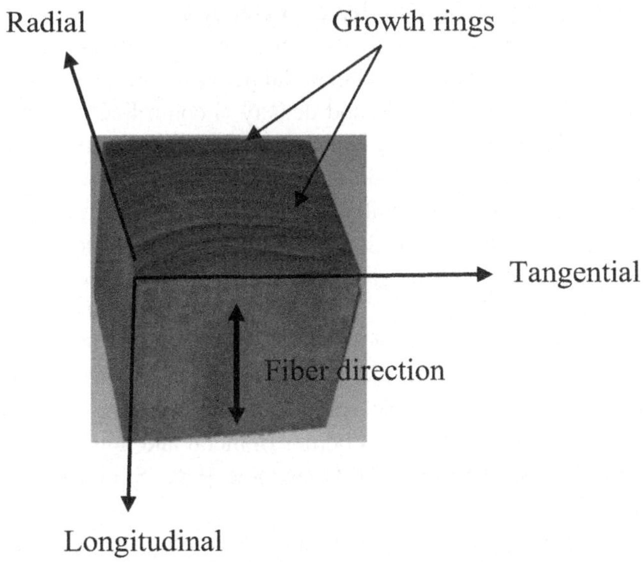

Fig. 37.3 Directions of principal axes in wood

of the tree diameter, the wood properties generally vary along three axes perpendicular to each other: radial, tangential and longitudinal (Fig. 37.3).

- Radial: This direction is perpendicular to both the growth rings and the long axis of the stem.
- Tangential: This direction is tangent to the growth rings.
- Longitudinal: It is parallel to the long axis of the stem and the fibres are oriented in this direction.

37.5 Physical and Mechanical Properties

The timber species exhibit a remarkably wide variation in their wood properties. The suitability of wood for a specific purpose depends on its properties (Anon 1970, 1972; Rajput et al. 1996). Because of inherent variations in the properties of different woods, not all timber species are suitable for all purposes. The selection of appropriate wood species for different applications is the first step towards rational utilization of timber. Several systematic studies have been carried out to evaluate the anatomical, physical, mechanical and other properties of different wood species (Anon 1970; Rajput et al. 1996).

37.5.1 Physical Properties

The term 'physical properties of wood' is generally ascribed to denote such inherent qualities as the wood moisture content, density, shrinkage and swelling characteristics (dimensional stability), appearance, colour etc. (Anon 1970; Tsoumis 1991; Rajput et al. 1996). Many of physical properties are very important to study for proper characterization of wood material and have great influence on the performance (Anon 1970; Shmulsky and Jones 2011). The heartwood colour of timbers is mostly influenced by the nature and relative percentage of extractive chemicals present in the wood. The extractives are also responsible for the natural durability of timber by providing resistance against biodegradability and longer service life. Reactions of wood to different forms of energy such as sound, heat, light and electricity are also broadly classified under physical properties. These properties have sometimes been predominantly responsible for the numerous uses of wood such as for musical instruments, decorative surfaces, insulating media, etc. Study of these aspects has great value in the economic utilization of various species. Sampling of trees and logs for the evaluation of different properties and their conversion for testing is carried out as per standard procedure (Anon 1974). Testing and evaluation of the physical properties of wood are usually done according to IS standards (Anon 1986a). Some important physical properties such as moisture content, specific gravity (SG; basic density) and shrinkage (and swelling) are described in the following paragraphs:

37.5.1.1 Moisture Content (MC)

The cross-section and length of the specimen selected for determining the moisture content of the wood piece are 2 × 2 cm^2 and 2.5 cm respectively. Many times, the specimen for MC is taken from the portions of wood samples of mechanical properties immediately after completing the test. For determining the moisture content, the specimen is first weighed and then dried in an electrically heated hot-air oven maintained at a temperature of 103 ± 2 °C for sufficiently long time, usually in the range of 24–48 h. The weight of specimen is recorded at regular intervals till constant weight is obtained or difference between last two readings should not be greater than 0.002 g. The final weight is recorded as the oven-dry weight of specimen and per cent of moisture content (MC) is computed as follows:

$$\text{MC}\ (\%) = ((W_1 - W_o)/W_o) \times 100, \qquad (37.1)$$

where W_1 is the weight of wood specimen at test (g) before drying and W_o is the final weight of same specimen (g) after oven-drying.

The oven-dry weight 'W_o' of a wood piece may be calculated from its weight at any moisture content 'mc' (W_m) using the following equation:

$$W_o = W_m/(1 + \text{mc}/100). \qquad (37.2)$$

It should be noted that the amount of moisture or water present in the test sample is expressed as a percentage of the weight of the oven-dried wood. Wood dried long enough in a well-ventilated oven at 103 ± 2 °C and maintaining a relatively constant weight is said to reach zero per cent moisture level. Freshly felled trees or 'green' sawn wood from the logs can have a moisture content of up to 150–250% (Tsoumis 1991; Shmulsky and Jones 2011).

Because wood is a naturally hygroscopic material, it absorbs moisture or water vapour in a wet environment and loses moisture in a dry environment. Therefore, the moisture content of wood varies according to atmospheric conditions and depends on the relative humidity (RH) and temperature of the surrounding air. Under conditions of constant temperature and humidity, the wood reaches an equilibrium moisture condition, called equilibrium moisture content (EMC), where the wood does not increase or lose moisture. Here, the EMC value represents the balance point where the wood is in complete balance with the surrounding environment. Thus, the EMC value of wood is always regulated by relative humidity and ambient temperature. In an uncontrolled environment, the EMC of wood varies according to seasonal climatic conditions and different locations. Under normal conditions, the average EMC value of wood should be 10–14% (12 ± 2%). Equilibrium points often vary within the country, from dry deserts to wet oceanic coastal to hills and glacial mountains. In addition, there can be a wide range of relative humidity variations between individual points in the same location. For example, a house by the sea has a higher EMC than a house on land. Different heating/cooling systems can also significantly change the local relative humidity and thus change the EMC of the wood (Panshin and de Zeeuw 1980; Tsoumis 1991).

37.5.1.2 Fibre Saturation Point (FSP)

The FSP is the condition during the increase or loss of wood moisture when the cell walls of the wood are completely saturated and there is no water in the cell cavity. Water that is chemically bound to the wood material of the cell wall is called '*bound water*', while the water filling the cell cavity is called '*free water*'. The FSP for most woods is generally considered to be approximately in the range of 25–30% moisture content, calculated from the oven-dry weight basis. Between 0% moisture (oven-dry) and around FSP, wood shrinks or expands to the maximum extent depending on the loss or gain of moisture. The size of the wood changes as the moisture content drops below the FSP. Wood shrinks when it loses moisture below the FSP and expands when it gains moisture up to the FSP (Panshin and de Zeeuw 1980; Shmulsky and Jones 2011). However, above the FSP, the wood does not shrink or expand significantly due to changes in moisture content, because the cell walls remain fully saturated and the cell cavities are filled with free water without affecting the dimensional changes of the wood sample.

37.5.1.3 Specific Gravity and Density

Both wood specific gravity (SG) and wood density determine the amount of actual wood material per unit volume of wood (Zobel and Jett 1995). Wood density is controlled by many factors, including fibre or cell wall thickness, its diameter, the ratio of early to late wood and the chemical constituents of the wood (Cave and Walker 1994). The specific gravity of wood is one of the most important physical properties. Most physical and mechanical properties of wood are correlated with the specific gravity. The SG value of a piece of wood is determined as the ratio of its mass to the volume at the test. It should be noted that the weight of wood is the sum of different components present such as (1) the weights of the wood material; (2) foreign bodies (mineral and non-mineral) and (3) water (Anon 1970; Dinwoodie 1981; Shmulsky and Jones 2011).

Size of the wood specimen for determining specific gravity or density at test is generally taken as 2 cm × 2 cm in cross-section and 6 cm in length (or bigger size of 5 cm × 5 cm × 15 cm) according to the Indian standards (Anon 1986a). For computing the volume (in cm^3), three dimensions of rectangular specimen (width, height and length) should be measured accurately (in cm) and multiplied. For irregular shaped specimens, the immersion method in a non-adsorbing liquid is adopted or 'mercury volumeter' may be used for determination of the volume. The specific gravity or density at test may be calculated using the following formula:

$$\mathrm{SG} = W_{\mathrm{m}}/V_{\mathrm{m}}, \tag{37.3}$$

where W_m = weight of test specimen (g), V_m = volume of test specimen of wood (cm^3).

Standard or basic specific gravity is mostly estimated from oven-dry weight and volume at test (green or air dry). Specific gravity based on green volume is referred to as standard specific gravity. The standard specific gravity of wood is evaluated using the following formula:

$$S = W_{\mathrm{o}}/V_{\mathrm{g}}, \tag{37.4}$$

where W_o = oven-dry weight of wood specimen (g), V_g = volume of same specimen in green (fully saturated with water) condition (cm^3).

Density of a wood ('weight' or 'heaviness') is determined as the weight divided by its volume at test and measured in kg/m^3 (or g/cm^3). It is inversely proportional to the porosity of the wood (or the ratio of pore volume). Specific gravity, or relative density, also expresses the ratio of the density of wood to the density of water. Density values for green and air-dry conditions are calculated from the weight and volume of the test samples under these two conditions. Numerically, the specific gravity and density of oven-dry wood are the same. It may be noted that the density of wood varies according to its moisture content, which should be recorded for the reporting purpose.

37.5.1.4 Importance of Specific Gravity/Density

The density of wood is a rather complex property that is affected by many factors, such as moisture content, wood microstructure and the presence of extractives. The density of wood is primarily a measure of the amount of cell wall material and other foreign materials per unit volume. Some other important factors influencing the wood density are the ratio of earlywood to latewood, cell (fibre) diameter, fibre wall thickness and chemical constituents of the wood. In general, an increase in the thickness of the cell wall and a decrease in the area of the cell lumen with the age of the tree leads to an increase in the corresponding wood density. The density of perfectly dried wood species varies greatly depending on the air spaces (cell cavities) within the material. The density of the wood substance in the cell wall in all the tree species is ascertained as about 1520 kg/m^3 (Shmulsky and Jones 2011). Non-mineral infiltrations (organic in nature) which are responsible for the various colours of the heartwood in different species constitute about 10% of the dry weight.

Density not only affects the strength properties and processing behaviour of sawn wood, but also affects the yield of wood fibre for pulp production, that is, wood with a lower density generally has lower strength and produces less wood fibre than wood with a higher density. Figure 37.4 depicts microstructure of three wood species namely *Bombax ceiba* (semul), low density (<450 kg/m^3), *Tectona grandis* (teak), a medium density (550–750 kg/m^3) and *Dalbergia sissoo* (sheesham), high density (>750 kg/m^3) wood species. Anatomical structure of low-density wood shows bigger size and large number of pores (vessels), while high density woods have smaller size and a smaller number of pores. In general, the denser wood exhibits greater mechanical properties such as strength, stiffness and hardness (Bootle 1983). In conjunction with other factors, denser woods are often more durable and is less susceptible to collapse (Panshin and de Zeeuw 1980; Tsoumis 1991). However, very dense wood is often more difficult to work and takes a lot of time to dry (Bootle 1983). The wood samples are dried to moisture content in the range of 10–14%, and then the strength properties are adjusted to 12% MC for comparison to other timber species. The classification of wood by density (or weight) and standard specific gravity is generally practiced in commercial fields and presented in Table 37.1.

37.5.1.5 Shrinkage Properties

Shrinkage is the reduction in size or volume of wood specimen when its moisture content changes from higher to lower values. The changes in dimensions of the wood samples are more visible when moisture content falls below the FSP (25–30%). Generally, the amount of shrinkage is proportional to the moisture loss. Some of the important factors affecting the shrinkage and swelling of wood are its moisture content, density, chemical composition, mechanical stresses, etc. Wood shrinks when water (called bound water) is lost from the cell wall and swells after water is absorbed by the cell walls. As explained earlier, the amount of changes in moisture content above the FSP has no significant effect on the size or volume of the wood sample. Shrinkage or

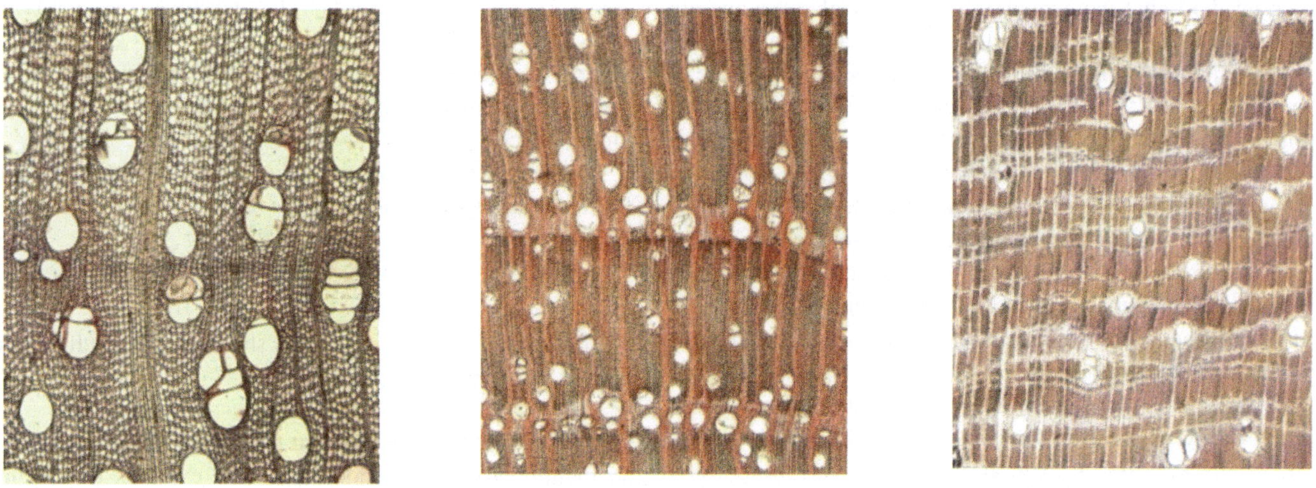

(a) (b) (c)

Fig. 37.4 Microstructures of low (**a**), medium (**b**) and high density (**c**) woods

Table 37.1 Classification of wood based on density and standard specific gravity (Anon 1970; Rajput et al. 1996)

Classification	Limit of density (or weight) at 12% MC (kg/m^3)	Standard specific gravity	Examples
Very light	Up–300	Below 0.28	Balsa, Dhupi
Light	301–450	0.28–0.42	Papita, Gokul
Moderately heavy	451–600	0.42–0.56	Salai, Vellapine
Heavy	601–800	0.56–0.70	Teak, Mundani
Very heavy	801–950	0.70–0.85	Babul, Bael
Extremely heavy	951 and above	0.85 to above	Rohan, oak

Table 37.2 Range of shrinkage values of wood from green to oven-dry conditions (Anon 1970; Rajput et al. 1996)

Shrinkage type	Range of values (%)
Radial (S_R)	2–6
Tangential (S_T)	5–10
Longitudinal (S_L)	0.1–0.3
Volumetric (S_V)	8–18

expansion must occur evenly in the wood to maintain dimensional stability. As such, the degree of contraction or expansion in the wood varies in the three principal directions. Specifically, the dimensional change is smallest in the longitudinal direction (along the fibre direction), much larger in the radial direction (core to edge), and even larger in the tangential direction of the growth rings. Table 37.2 shows the range of values of four types of shrinkage (radial, tangential, longitudinal, volumetric) in typical wood samples dried from green (very high moisture content, above the FSP) to oven-dry conditions.

37.5.1.6 Measurement of Shrinkage

For measuring the amount of shrinkage in the wood, true radial, tangential, longitudinal and volumetric specimens are prepared. According to Indian standard IS:1708 (Anon 1986a), the recommended specimen dimensions for determining the radial, tangential and longitudinal shrinkage are 2×2 cm^2 in cross-section and 5 cm in length. Similarly, for evaluating the volumetric shrinkage, test specimen may be of smaller size (2×2 cm^2 in cross-section and 6 cm in length) or bigger size (5×5 cm^2 in cross-section and 15 cm in length). The wood specimens are weighed in the green condition (water saturated) to 0.001 g accuracy and their radial, tangential and longitudinal dimensions are measured to an accuracy of 0.01 mm using digital caliper. The volume of wood specimen is determined by immersion method with an accuracy of 0.01 cc. The wood specimens under investigation are then allowed to dry slowly under shade and periodically weighed and dimensions (and volume) are measured until no further loss of weight is observed (air-dry condition). The specimens are finally dried in an oven at 103 ± 2 °C and weight and dimensions (and volume) of oven-dried samples are recorded and used for computing the amount of different types of shrinkages and also the corresponding moisture content of wood. As stated above, the changes in dimensions are used to calculate different shrinkage values in wood: radial (S_R), tangential (S_T), longitudinal (S_L) and volumetric (S_V). The amount of shrinkage is determined from initial (green) condition to a required condition (usually air-dry—12% MC or oven-dry—0% MC) by measuring the dimensions or volume of wood specimens in respective conditions. The radial (S_R), tangential (S_T) or longitudinal (S_L) shrinkage value is computed using following formula:

$$S_R \text{ (or } S_T \text{ or } S_L) = ((L_1 - L_2)/L_1) \times 100\%, \tag{37.5}$$

where L_1 = length (radial or tangential or longitudinal) at initial (or green) condition (cm), L_2 = length (radial or tangential or longitudinal) at a given moisture content (air-dry or oven-dry) condition (cm).

Similarly, the volumetric shrinkage (S_V) is calculated as follows:

$$S_v = ((V_1 - V_2)/V_1) \times 100\%, \tag{37.6}$$

where V_1 = volume at initial (or green) condition (cc), V_2 = volume at a given moisture content (air-dry or oven-dry) condition (cc).

The volumetric shrinkage can also be calculated if wood density of the specimen is known:

$$S_V(\%) = \left[\text{FSP } (\%) \times \text{density } \left(\text{kg/m}^3\right) \times 10^{-3}\right]. \tag{37.7}$$

Thus, the high-density wood shrinks and swells more (Rajput et al. 1996).

37.5.1.7 Anisotropic Nature of Shrinkage

As mentioned above, wood shrinks or expands to varying degrees depending on the direction in the wood specimen, indicating the anisotropic nature of this material. As listed in Table 37.3, green to oven-dry longitudinal shrinkage value is one to two orders of magnitude less than tangential

Table 37.3 Mechanical properties of wood

Sl. no.	Properties
1.	Static bending strength (flexural strength and stiffness)
2.	Impact bending strength
3.	Compressive strength parallel to grain
4.	Compressive strength perpendicular to grain
5.	Hardness (under static indentation)
6.	Shear strength parallel to grain
7.	Tensile strength parallel to grain
8.	Tensile strength perpendicular to grain
9.	Cleavage strength parallel to grain
10.	Nail and screw holding powers
11.	Brittleness (Izod and Charpy impact tests)
12.	Torsional strength

shrinkage, and tangential shrinkage is 1.5–2.5 times greater than the radial shrinkage (Tsoumis 1991; Shmulsky and Jones 2011). The volumetric shrinkage is generally the sum of three shrinkage values ($S_R + S_T + S_L$).

37.5.1.8 Importance of Shrinkage and Swelling: Dimensional Instability

The dimensional stability of wood is one of the few properties that, due to its anisotropy, varies significantly in each of the three axes. Dimensional changes in the longitudinal direction between the FSP and the oven-dry are approximately 0.1–0.3% and have little practical significance. The combination of tangential (up to 10%) and radial axis (up to 6%) shrinkage can distort the shape of wood pieces due to differences in shrinkage and annual ring curvature. Wood also shows hysteresis in increasing and decreasing humidity fluctuations. It should be noted that differential shrinkage in the radial and tangential directions and related effects such as checking, deformation and cracking (also called seasoning defects) constitute the most troublesome physical characteristic of wood (Anon 1970). Dimensional changes associated with shrinkage and expansion can cause further deterioration of wood products. As a result of repeated cycles of contraction and expansion during service, various defects can occur, such as the opening or tightening of joints and changes in the cross-sectional shape of wooden objects. Distortions of wood are deviations from straightness that occur after sawing. Distortion affects the utility of the wood by limiting the effective use of the resource but has little effect on overall strength. Different types of distortion include bow, spring and coil. Defects in wooden boards, such as bowing, warping, and warping, are also caused by differences in radial and tangential shrinkage. Similarly, the checking of wooden logs and planks is due to uneven dimensional changes, because there is a certain moisture gradient between the surface and the interior of the wood. Traditionally, wood is graded either visually or measurably according to the level of distortion, which is given an average value. The recovered volume of each class or grade can be used to indicate the severity of the distortion.

37.5.2 Mechanical Properties

The mechanical properties of a material are usually related to the resistance with which it resists external forces. Mechanical properties include strength properties, which refer to the fracture aspects of a material and elastic properties, which measure resistance to deformation and distortion. Both of these parameters will eventually show performance and other related issues related to the material. The strength properties of wood are closely related to its density, which varies widely from species to species. While mechanical properties vary drastically with timber species, they also depend upon the shape of a member, the nature of forces applied and the relative direction in which forces are applied (Anon 1970; Tsoumis 1991; Shmulsky and Jones 2011). For example, a lesser force will be required to bend a piece of wood than to crush it, and yet a different degree of force will be required to pull the piece apart. In addition, the magnitude of forces differs depending upon whether forces are applied parallel or perpendicular to the grain. Table 37.3 lists basic mechanical properties of wood which are mostly evaluated to find the suitability of timber of various end-use applications. The mechanical properties of wood also vary with changes in moisture content. Wood is usually much stronger in its dry state than in its green or wet state. The exception is impact resistance; dry wood tends to shatter and crack under high-impact loads. Due to the comparison of two types of wood with different specific gravity, wood with a higher specific gravity corresponds to higher strength values. It must also be considered that wood is much stronger (7–10 times) when loaded parallel to the direction of the grain than perpendicular to the grain.

37.5.2.1 Testing and Evaluation of Different Mechanical Properties

The widespread application of various timbers and the applicability of evaluating the mechanical properties of different wood species require testing them in some standardized way so that their comparison also becomes easier. Therefore, different specific methods of Indian Standards IS:1708 (Anon 1972, 1986a) some other accepted standards are followed to determine the mechanical properties of wood samples in different conditions (green, kiln-dry, oven-dry). Testing of mechanical properties under static and dynamic loading is done with universal testing machine (UTM) and impact tester shown in Fig. 37.4a, b respectively. The direction of load applied during testing of mechanical properties of wood in UTM is illustrated in Table 37.4 and Fig. 37.5.

Table 37.4 Some basic mechanical properties and direction of applied forces

Properties	Directions of applied forces
Static bending (3-point loading; modulus of rupture-MOR, modulus of elasticity-MOE, fibre strength at limit of proportionality-FS at LP)	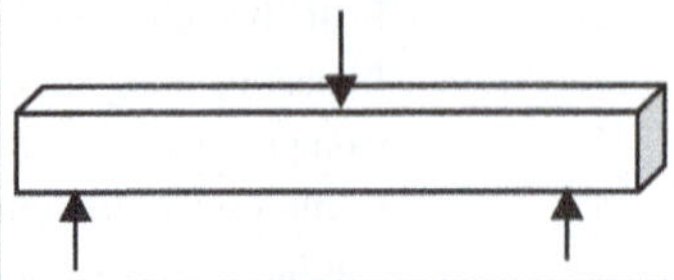
Compressive strength parallel to grain	
Compressive strength perpendicular to grain	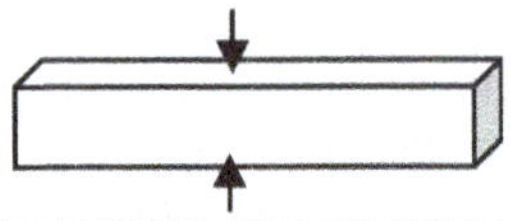
Hardness (surface indentation)	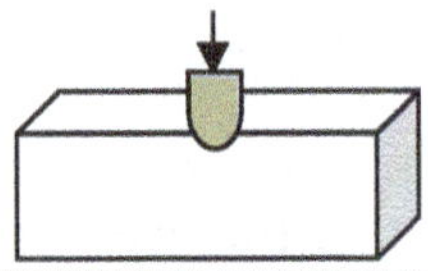
Shear strength parallel to grain	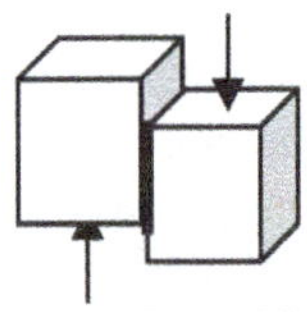
Tensile strength parallel to grain	
Brittleness (Charpy and Izod impact tests)	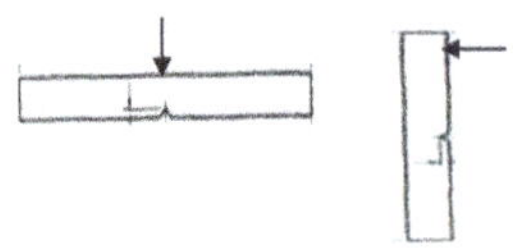
Torsional strength	

37.5.2.2 Static Bending Parameters (Central or 3-Point Loading)

Modulus of Rupture (MOR)

Strength of wood is the load-bearing capacity and shows the ability of the material to withstand the applied forces. Bending (or flexural) strength and compressive strength parallel to grain are commonly used parameters of wood which are considered as very important mechanical properties for structural and other load-bearing applications. The strength of wood can be measured in several different ways using conventional destructive and non-destructive tools. The flexural strength of wood measured using central loading is usually expressed as modulus of rupture (MOR), which refers to the maximum load-carrying capacity of a piece of wood when a load is applied slowly, as shown in Fig. 37.6. The MOR values are usually computed for wood samples equilibrated to green (wet) and seasoned (dry) conditions.

Modulus of Elasticity (MOE)

MOE, which is a direct measurement of the resistance of wood to deflection, has traditionally been used to measure

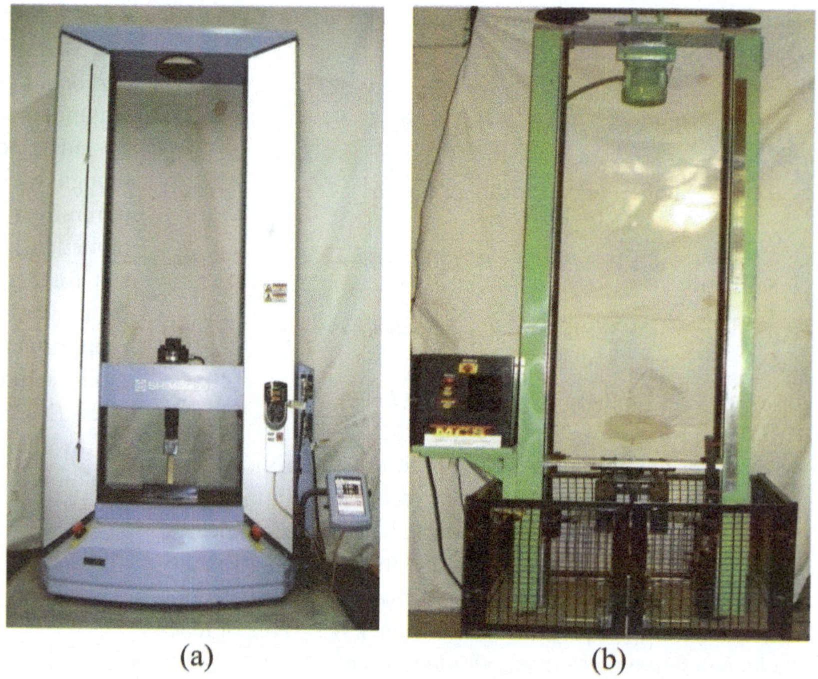

Fig. 37.5 Universal timber testing machine (**a**), impact tester (**b**)

stiffness. It determines the ability of the material to recover when the load applied to it is removed. MOE corresponds to the slope of the linear part of the relationship between applied load (stress) versus deformation (strain) from zero to the limit of proportionality (LP), as shown in Fig. 37.6. It is an important mechanical property of wood that is often considered more powerful in predicting the quality of wood. The high stiffness enables the use of minimal material and lightweight in new furniture and designed products, resulting in more efficient use of this valuable resource. Similar to MOR, MOE is also calculated for small pieces of wood specimens in the green (wet) or seasoned (dry) state.

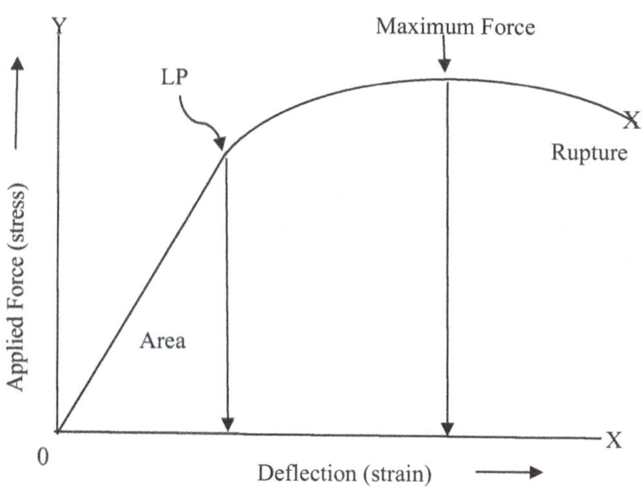

Fig. 37.6 Schematic diagram of applied force versus deflection curve of wood sample

Fibre Stress at Limit of Proportionality (FS at LP)

FS at LP is also an important mechanical parameter that indicates the maximum bending stress the wood specimen can withstand under static load condition without showing any permanent set or distortion in the piece.

Work to Proportional Limit (WPL) and Work to Maximum Load (WML)

WPL is a measure of the work done, that is, the energy used in moving from the unloaded state to the limit of proportionality of the wood material. Similarly, WML measures the amount of work required to split or break wood material. It is a measure of the energy required to fracture a material.

Testing and Evaluation of Static Bending Parameters

The nominal size of clear wood test specimen of 2×2 cm^2 in cross-section and 30 cm in length with a span of 28 cm is selected. Larger specimens of the size $5 \times 5 \times 75$ cm^3 having a span length of 70 cm may also be taken for testing. The test specimen is placed on a suitable rig having central loading arrangement and force is applied through the moving cross-head at a constant speed of 1.0 mm/min (2.5 mm/min for larger size test specimens) on the tangential surface of the wood sample. The applied load through the UTM's cross-head may be measured in kilogram (kg) or Newton (N), while dimensions of the wood specimens are measured in cm or mm giving the stress units as kg/cm^2 or N/mm^2 (MPa) respectively. Different static bending strength parameters, namely MOR, MOE, FS at LP, WPL and WML, are computed from the load-deflection curve using following equations:

$$\mathrm{MOR} = (3 \times P_{\max} \times l)/(2 \times b \times h^2) \quad (37.8a)$$

$$\mathrm{MOE} = (P \times l^3)/(4 \times \Delta \times b \times h^3) \quad (37.8b)$$

$$\mathrm{FS\ at\ LP} = (3 \times P \times l)/(2 \times b \times h^2) \quad (37.8c)$$

$$\mathrm{WPL} = (C \times A)/(l \times b \times h) \quad (37.8d)$$

$$\mathrm{WML} = (C \times A^1)/(l \times b \times h), \quad (37.8e)$$

where P = load at the limit of proportionality (N), P_{max} = maximum load (N), l = span of the test specimen (mm), b = breadth of the test specimen (mm), h = depth of the test specimen (mm), Δ = deflection at the limit of proportionality (mm), C = area constant (N mm), A = area of load-deflection curve up to limit of proportionality (mm^2), A^1 = area of load-deflection curve up to maximum load (mm^2).

37.5.2.3 Impact Bending Strength

The dimensions of the clear test specimen for impact bending strength are same as static bending strength, that is, 2 × 2 × 30 cm^3 with span length of 28 cm for smaller size samples or 5 × 5 × 75 cm^3 with span length of 70 cm for larger size samples. Impact bending strength test is conducted on a suitable machine. A hammer weighing 1.5 kg for smaller size samples or 25 kg for bigger size wood samples is initially dropped from a height of 5 cm at the centre on tangential surface. The height of drop is then gradually increased in steps of 2.5 cm until a height of 25 cm is reached. After that, the height increment is increased to 5 cm till the complete failure of the wood sample occurs. Actual deflection of the test specimen is recorded for each height of drop of hammer. The graph is plotted between sum of height of drop and maximum deflection at that drop versus square of deflection and limit of proportionality is determined. Different impact bending strength parameters such as modulus of elasticity (MOE_d) and fibre stress at limit of proportionality (FS_d at LP) are evaluated by using following equations:

$$\mathrm{MOE_d} = (H^1 \times W \times s^3)/(2 \times b \times h^3 \times \Delta^2) \quad (37.9a)$$

$$\mathrm{FS_d\ at\ LP} = (3 \times H^1 \times W \times s)/(b \times h^2 \times \Delta), \quad (37.9b)$$

where H^1 = height of drop at limit of proportionality (cm), W = weight of hammer (kg), s = span of specimen (cm), b = breadth of specimen (cm), h = height of specimen (cm), Δ = deflection at limit of proportionality (cm).

37.5.2.4 Compressive Strength Parallel to Grain

The size of clear wood specimens for compressive strength parallel to grain test is 2 × 2 cm^2 in cross-section and 8 cm in length (smaller size sample) or 5 × 5 cm^2 in cross-section and 20 cm in length (larger size sample). The rate of loading though moving cross-head of the machine is kept at 0.6 mm/min. A graph between applied load and deflection is plotted till complete failure of the wood specimen under test for determining compressive strength parallel to grain parameters. Compressive stress parallel to the grain at limit of proportionality (CS at LP), compressive stress parallel to the grain at maximum load (CS at ML or maximum crushing stress (MCS) and modulus of elasticity in compression parallel to the grain (MOE) are computed using following equations:

$$\mathrm{CS\ at\ LP} = P/A \quad (37.10a)$$

$$\mathrm{MCS} = P^1/A \quad (37.10b)$$

$$\mathrm{MOE} = (P \times L)/(A \times \Delta), \quad (37.10c)$$

where P = load at limit of proportionality (N), A = area of cross-section of specimen (mm^2), P^1 = maximum crushing load (N), L = gauge length between two points of compression or length of smaller size specimen (mm), Δ = deflection at limit of proportionality (mm).

37.5.2.5 Compressive Strength Perpendicular to Grain

The size of clear wood specimen for evaluating the compressive strength perpendicular to grain is 2 × 2 cm^2 in cross-section and 10 cm in length (smaller size samples) or 5 × 5 cm^2 in cross-section and 15 cm in length (larger size samples). The load is applied on radial surface through a metal plate of 5 cm width kept centrally on the specimen. The cross-head loading speed is kept at 0.6 mm/min. The load versus deformation curve is plotted until deflection of 2.5 mm or maximum load is observed. Different parameters such as compressive stress perpendicular to the grain at limit of proportionality (CS at LP), compressive stress perpendicular to the grain at 2.5 mm compression and modulus of elasticity in compression perpendicular to the grain (MOE) are evaluated using following equations:

$$\mathrm{CS\ at\ LP} = P/A \quad (37.11a)$$

$$\mathrm{CS\ at\ 2.5\ mm} = P^1/A \quad (37.11b)$$

$$\mathrm{MOE} = (P \times h)/(A \times \Delta), \quad (37.11c)$$

where P = load at limit of proportionality (N), P^1 = load at 2.5 mm compression (N), A = area of specimen on which force was applied (mm^2), h = height of specimen (mm), Δ = deflection at limit of proportionality (mm).

37.5.2.6 Hardness Under Static Indentation

Hardness is a measure of the penetration resistance of wood against any object. It is one of the important properties for flooring and other similar applications of timbers. It is traditionally measured by the Janka hardness test which measures the force required to press a small steel ball bearing into a piece of wood. According to standard procedure, the size of clear wood specimen selected for hardness under static indentation test is 5 × 5 cm^2 in cross-section and 15 cm in length. A steel rod with a hemispherical end of 1.13 cm in diameter is penetrated to half of its diameter through the radial, tangential and end surfaces of the test specimen. The rate of loading is kept constant at 6 mm/min. Penetrations are made at the centre of each face and care is taken such that no splitting or chipping occurs in the wood specimen under test. The load required to penetrate the steel bar to depth of 0.564 cm into the wood specimen is recorded for determining the hardness on each radial, tangential and end faces. The average of radial and tangential hardness values is computed as the side hardness.

37.5.2.7 Shear Strength Parallel to Grain

The cross-section of the test pieces is 5 × 5 cm^2 and the length is 6 cm. The wood samples are notched and one end of the test specimen is stepped to produce a shear failure over an area of 5 × 5 cm^2 in the radial or tangential plane. The cross-head loading speed is maintained at 0.4 mm/min. The load versus deformation curve is drawn until the maximum load is observed in shearing the area. The shear force parallel to the grain in radial and tangential planes, called maximum shear stress (MSS), is calculated using following equation:

$$\mathrm{MSS} = P_{\max}/A, \qquad (37.12)$$

where P_{max} = maximum load required for shearing the area (N), A = shearing area of the specimen on which force was applied (mm^2).

37.5.2.8 Tensile Strength Parallel to Grain

Shape and dimensions of the clear wood specimen for this test are taken as dumbbell and 32.5 cm with 7 × 7 cm^2 respectively in length and cross-section. The effective length at the centre of the test specimen remains as 5 cm. Suitable grips are used to conduct the test in the machine and speed of applied load is fixed at 1 mm/min. Elongation of the effective gauge length of the test specimen is recorded using an extensometer and plotted against the applied load. Three properties from tensile strength parallel to grain test, namely tensile stress at limit of proportionality (TS at LP), tensile stress at maximum load (TS at ML) and modulus of elasticity in tension parallel to grain (MOE) are determined using following equations:

$$\text{TS at LP} = P/A \qquad (37.13a)$$

$$\text{TS at ML} = P^1/A \qquad (37.13b)$$

$$\mathrm{MOE} = (P \times L)/(\Delta \times A), \qquad (37.13c)$$

where P = load at limit of proportionality (N), A = cross sectional area of central portion of specimen (mm^2), P^1 = maximum load (N), L = gauge length of the specimen between extensometer (mm), Δ = deformation at the limit of proportionality (mm).

37.5.2.9 Tensile Strength Perpendicular to Grain

The size of wood specimen for the tensile strength perpendicular to grain test is 5 cm × 5 × 5.6 cm^3. The specimen is notched on the two surfaces perpendicular to the grain so as to produce a failure on 5 × 2 cm^2 area. Suitable fixtures are used to conduct the test in the machine. Load is applied at a rate of 2.5 mm/min until failure occurs. Load versus deformation curve is plotted to find maximum load of failure in the concerned plane. Tensile strength perpendicular to grain in radial and tangential planes (TS) is calculated using equation:

$$\mathrm{TS} = P_{\max}/A, \qquad (37.14)$$

where P_{max} = maximum load required for failure in tension perpendicular to grain (N), A = area of the portion of specimen on which force is applied (mm^2).

37.5.2.10 Cleavage Strength Parallel to Grain

The size of the wood sample for cleavage strength parallel to grain is 5 × 5 × 9.3 cm^3. Appropriate grips are used to perform the test and loading speed of the cross-head is fixed at 2.5 mm/min. One end of the specimen has a special notch to break in the radial or tangential plane. The test is performed in both the radial and tangential planes and the maximum load required to cause failure of the test specimen is recorded. Maximum cleavage resistance parallel to grain (MCR in kg/cm) in radial or tangential plane is calculated using the equation:

$$\mathrm{MCR} = P_{\max}/w, \qquad (37.15)$$

where P_{max} = maximum load (N), w = width of the specimen (mm).

37.5.2.11 Nail and Screw Holding Powers

According to standard procedure, the size of clear wood specimen selected for nail and screw-holding power test is 5 × 5 cm^2 in cross-section and 15 cm in length. Nails of 50 mm length and 2.50 mm shank diameter and screws of 50 mm length and 8 gauge are used for testing. Special fixtures are used to conduct the test and UTM cross-head speed to pull out the nails and screws is fixed to 2 mm/min. The maximum load required for pulling out nails or screws is recorded for radial, tangential and end surfaces. 'Side values' are the average of radial and tangential surface loads.

37.5.2.12 Toughness and Impact Strength in Dynamic Bending

Wood specimen of the dimensions of 2 × 2 cm^2 in cross-section and 30 cm in length on grain direction are recommended for testing and evaluation of toughness and impact strength parameters in dynamic bending mode (Anon 1999; Moreira et al. 2017). The wood specimens are tested using an impact bending machine as shown in Fig. 37.7. The pendulum impact is used for the toughness value determination. The energy absorbed in breaking the specimen is recorded after the impact and used for computing the toughness (W, in N m). The impact bending strength (f_{bw}) is determined according to the following equation:

$$f_{\text{bw}} = (1000 \times W)/(b \times h),$$

where b and h are the dimensions of cross-section of the test specimen respectively.

Fig. 37.7 Impact tester for the evaluation of dynamic bending properties (Moreira et al. 2017)

37.5.2.13 Torsional Strength

The test specimen for torsional strength test is of cylindrical shape having 30 cm length and 2.5 cm diameter. The length of the middle part of the wooden sample is taken as 22 cm and the gauge length is 15 cm. Angular twist measuring device is used at the centre. Torque is applied gradually at the rate of about 50 kg cm/min. From the torque and twist data, torque and angular twist at proportional limit is recorded. Torsional strength parameters such as torsional shear stress at limit of proportionality (TSS at LP), torsional shear stress at maximum torque (TSS at MT) and torsional modulus of rigidity (TMR) are computed using following equations:

$$\text{TSS at LP} = (2 \times M)/(\varphi \times r^3) \tag{37.16a}$$

$$\text{TSS at MT} = (2 \times M^1)/(\varphi \times r^3) \tag{37.16b}$$

$$\text{TMR} = (2 \times M \times L)/(\varphi \times r^4), \tag{37.16c}$$

where M = torque at the limit of proportionality (N mm), r = radius of specimen under gauge (mm), M^1 = maximum torque (N mm), L = gauge length of specimen (mm), φ = angular twist of one end relative to other at a distance of L (radians).

37.5.3 Physical and Mechanical Properties

Average physical properties of *Tectona grandis* and *Acacia auriculiformis* such as equilibrium moisture content, specific gravity, weight (in green and air-dry conditions) and shrinkage values from green to oven-dry condition (Sekhar and Rawat 1966; Shukla et al. 2007) are listed in Table 37.5. Table 37.6 shows the average mechanical properties of these two timber species evaluated in green and air-dry conditions (Sekhar and Rawat 1966; Shukla et al. 2007). Average values of dynamic mechanical parameters such as impact bending strength, toughness, strength in compression parallel to grain and modulus of elasticity in compression parallel to grain of a few South American timber species evaluated in air-dry condition and reported by Moreira et al. (2017) are listed in Table 37.7.

37.5.4 Comparative Suitability Indices

Owing to wide differences in the microstructure, homogeneity and various properties, many woods behave differently and do not strictly follow the laws of mechanics. Several factors are generally taken into consideration while computing the suitability of various wood species for different

Table 37.5 Average physical properties of *Tectona grandis* and *Acacia auriculiformis* (Sekhar and Rawat 1966; Shukla et al. 2007)

Properties	*Tectona grandis*		*Acacia auriculiformis*	
	Green	Air-dry	Green	Air-dry
Moisture content (%)	76.60	12.00	44.07	12.00
Weight (kg/m^3)	1056.00	672.00	883.45	729.00
Specific gravity	0.596	0.604	0.625	0.645
Shrinkages				
Longitudinal shrinkage (%)	–	–	0.61	–
Radial shrinkage (%)	2.30	–	2.64	–
Tangential shrinkage (%)	4.80	–	5.43	–
Volumetric shrinkage (%)	6.80	–	8.22	–

Table 37.6 Average mechanical properties of *Tectona grandis* and *Acacia auriculiformis* in green and air-dry conditions (Sekhar and Rawat 1966; Shukla et al. 2007)

Properties	*Tectona grandis*		*Acacia auriculiformis*	
	Green	Air-dry	Green	Air-dry
Static bending				
Modulus of rupture, MOR (MPa)	82.50	94.08	91.63	106.62
Modulus of elasticity, MOE (GPa)	10.76	11.73	10.86	12.96
Fibre stress at limit of proportionality, FS at LP (MPa)	49.93	63.86	63.06	72.67
Compression parallel to grain: maximum crushing stress, MCS (MPa)	40.71	52.19	37.09	49.76
Compression perpendicular to grain: compressive stress at limit of proportionality, CS at LP (MPa)	8.44	9.91	7.90	10.98
Hardness (static indentation)				
Radial (kN)	5.46	4.92	3.63	4.76
Tangential (kN)	5.41	5.14	3.71	4.88
End (kN)	4.77	4.79	3.88	4.01
Maximum shearing stress parallel to grain (MSS)				
Radial (MPa)	8.8	9.5	6.75	6.91
Tangential (MPa)	9.8	10.6	7.65	8.95
Tensile stress perpendicular to grain (TS)				
Radial (MPa)	6.7	5.6	2.72	1.22
Tangential (MPa)	7.9	6.5	3.78	2.01
Nail holding power				
Side (kN)	1.25	–	0.96	0.62
End (kN)	0.89	–	0.71	0.52
Screw holding power				
Side (kN)	3.25	–	3.02	2.99
End (kN)	2.32	–	1.69	2.37

general, engineering and industrial applications. Some factors are given higher importance compared to others depending upon the basic wood property related to the end-use application, condition of its evaluation and choice of timber species. On account of various degrees and factors of importance in different uses, only limited wood species are acceptable for a particular application. Suitability indices for any use may only serve as quantitative guide in absence of reliable data on actual service trials in the laboratory or field conditions. Using standard basic data on various physical and mechanical properties in 'green' and 'air-dry' conditions, suitability indices for various wood species are computed (Sekhar and Gulati 1972; Rajput et al. 1996). Table 37.8 presents average values of physical and mechanical properties of Teak and *A. auriculiformis* in green and air-dry conditions.

The comparative suitability index is generally expressed as percentage of suitability figure of a well-known and widely used wood species for any particular use. Teak (*Tectona grandis*) from 14 different locations across the country is taken as a standard timber and average values of different properties are used for computing the comparative suitability index. The suitability indices may also be used for grouping of various species for commercial classification and their preferences. Following suitability indices are generally

Table 37.7 Average impact bending strength and toughness of few timber species in air-dry condition (Moreira et al. 2017)

Wood species	Impact bending strength (f_{bw}) (kJ/m^2)	Toughness (N m)	Strength (f_c) in compression parallel to grain (MPa)	Modulus of elasticity (E_c) in compression parallel to grain (MPa)
Cedrella spp. (Cedro)	19.46	7.78	31.00	8354.0
Erisma uncinatum warm (Cambará Rosa)	8.38	3.35	35.00	12,967.0
Cedrelinga cateniformis (Cedrorana)	20.02	8.01	31.00	8962.0
Calophyllum spp. (Catanudo)	33.65	13.46	51.00	14,279.0
Goupia glabra (Cupiúba)	15.35	6.14	57.00	12,970.0
Parkia spp. (Angelim Saia)	11.38	4.55	63.00	19,748.0
Bagassa guianensis Aubl. (Tatajuba)	7.52	3.39	60.00	26,723.0
Luetzelburgia spp. (Guaiçara)	44.98	17.99	71.00	15,301.0
Dipteryx odorata (Cumaru)	57.32	22.93	93.00	23,002.0
Dinizia excelsa Ducke (Angelim Vermelho)	49.52	19.81	78.00	16,695.0

Table 37.8 Suitability indices of *A. auriculiformis*, *A. mangium* and *E. tereticornis* in comparison with teak taken as 100 (Sekhar and Rajput 1968; Rao et al. 2004)

	A. auriculiformis		*A. mangium*	
Properties	8-year-old	13-year-old	18-year-old	*E. tereticornis*
Strength as a beam	104	123	96	104
Stiffness as a beam	91	106	84	95
Suitability as a post	71	95	83	95
Shock-resisting ability	089[a]	100[a]	109	125
Retention of shape	085	083	85	47
Shear	072	79	83	134
Surface hardness	085	97	79	126
Splitting coefficient	057	44	100	61
Nail-holding power	077[a]	064	75	84
Screw-holding power	073[a]	082	70	86
Weight or heaviness	099	109	106	141

[a]Values calculated using strength-specific gravity relationship (Rajput et al. 1996)

evaluated for any wood species (Sekhar and Gulati 1972; Rajput et al. 1996):

(a) Strength as a beam
(b) Stiffness as a beam
(c) Suitability as a post
(d) Shock resisting ability
(e) Retention of shape
(f) Shear
(g) Surface hardness
(h) Refractoriness (splitting co-efficient)
(i) Nail holding power
(j) Screw holding power

Variation in structure and non-homogeneity limit the application of simple laws of mechanics on wood without suitable approximation at different stages. Owing to operation of factors of importance of different kinds in different uses, only certain species are more suitable for specific application. Comparative suitability indices of wood for different uses serve as a guide for selection of one species in preference to other. Comparative suitability indices of *A. auriculiformis* of 8 and 13 years old, *A. mangium* and *E. tereticornis* in comparison with *Tectona grandis* are listed in Table 37.8. Teak wood is conventionally selected as a model timber and given the score of 100 for making the comparison to other timbers. The comparative suitability indices of a few representative wood species such as *Dalbergia latifolia* (Rosewood), *Pterocarpus dalbergioides* (Andaman padauk), *Adina cordifolia* (Haldu), *Artocarpus hirsutus* (Aini), *Azadirachta indica* (Neem) and *Mangifera*

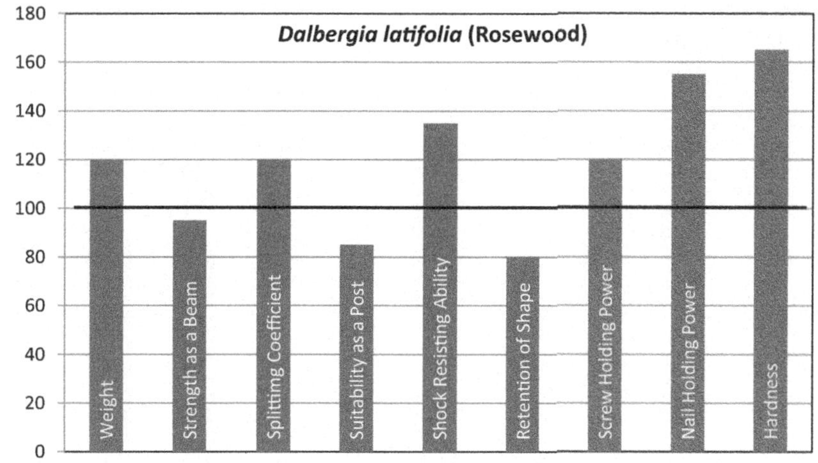

Fig. 37.8 Suitability indices of rosewood in comparison with teak taken as 100

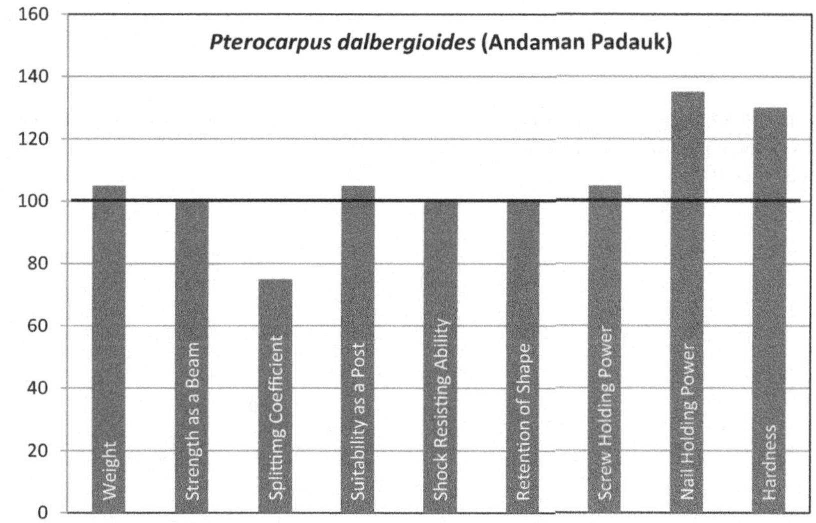

Fig. 37.9 Suitability indices of Andaman padauk in comparison with teak taken as 100

indica (Mango) are plotted as bar diagram (Anon 1970, 1974) and shown in Figs. 37.8, 37.9, 37.10, 37.11, 37.12 and 37.13.

37.5.5 Suitability Figures for Different End Uses

According to standard procedure, basic mechanical properties and suitability indices are used in association with certain adjusting and weightage factors to compute the suitability figures of various timber species for different end use applications such as furniture artefacts, building and construction, door, window shutter and frames, floorings, packaging, wooden pallets, oars and paddles and tool handles (Sekhar and Gulati 1972; Rajput et al. 1996).

Based on MOE and MOR of static bending, different types of wood are structurally classified (Anon 1986b) into three different groups (A, B and C) as shown in Table 37.9. It may be emphasized that all the species cannot be equally suitable for various applications. At the same time, all properties may not be contributing and so important to various end uses. Classification of timbers and their suitability for important end uses is generally based on mechanical criteria. Hence, it is necessary to select certain timbers for different applications according to their specific requirement and wood properties.

37.5.6 Safe Working Stresses

The basic or fundamental strength properties as described above are generally evaluated from the testing of defect free wood specimens according to prevailing standard procedures. However, in actual practice of designing the wooden structures, the timber to be used may not always be clear and defect free and may contain defects and actual conditions of use may be different from the standard laboratory conditions where timber was tested. Moreover, intrinsic variability of different properties also tends to affect the

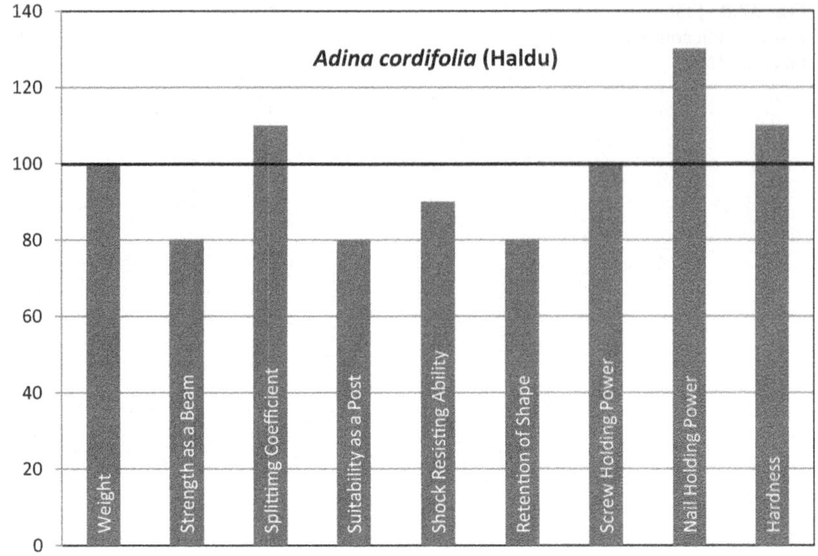

Fig. 37.10 Suitability indices of Haldu in comparison with teak taken as 100

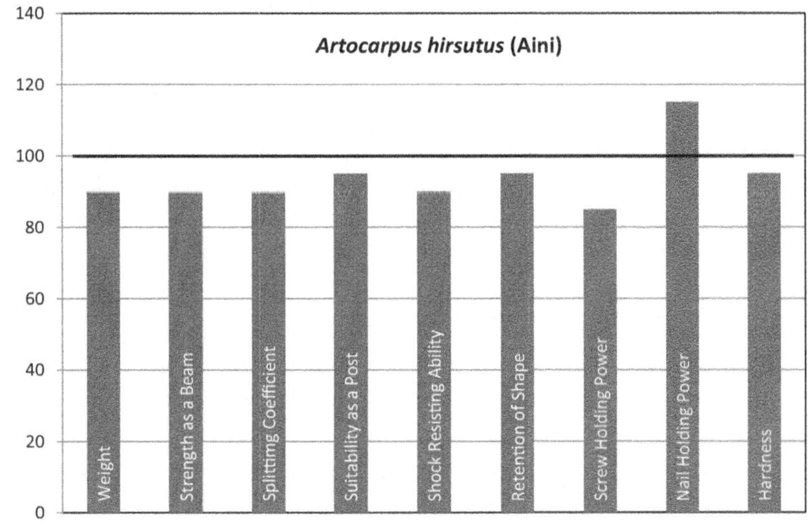

Fig. 37.11 Suitability indices of Aini in comparison with teak taken as 100

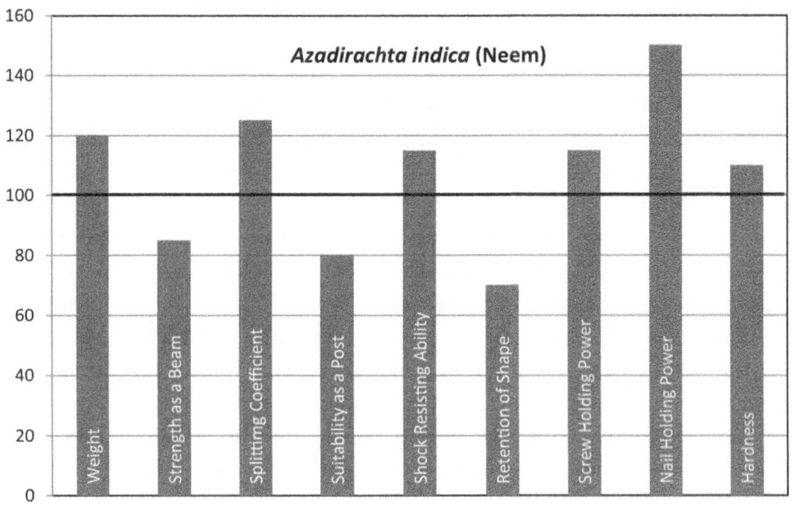

Fig. 37.12 Suitability indices of Neem in comparison with teak taken as 100

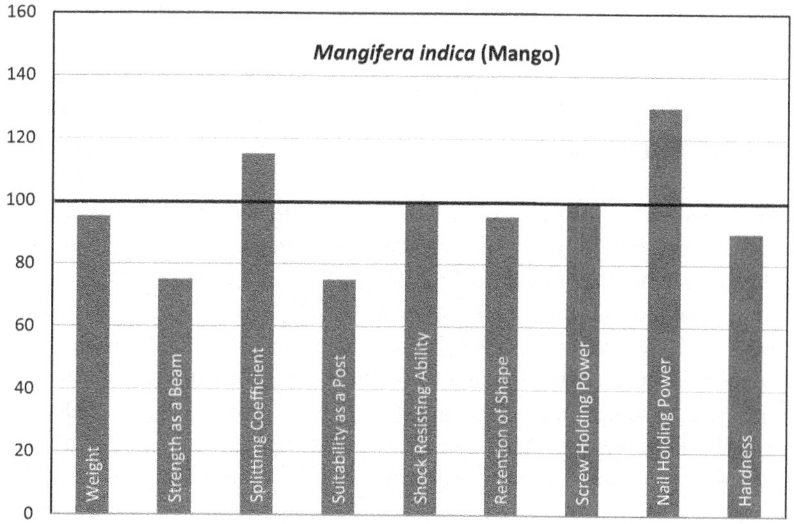

Fig. 37.13 Suitability indices of Mango in comparison with teak taken as 100

Table 37.9 Classification of different timber species for construction purposes (Anon 1986b)

Category	Static bending parameter	Value	Number of timber species
Group A	MOE≥	126 tonne/cm^2 (12.36 GPa)	17 species
	MOR≥	900 kg/cm^2 (88.26 MPa)	
Group B	MOE	98–126 tonne/cm^2 (9.61–12.36 GPa)	70 species
	MOR	600–900 kg/cm^2 (58.84–88.26 MPa)	
Group C	MOE	56–98 tonne/cm^2 (5.49–9.61 GPa)	84 species
	MOR	425–600 kg/cm^2 (41.68–58.84 MPa)	

overall strength and performance of wood product. Therefore, fundamental stresses may not be applicable with confidence in designing of timber products. Hence, more realistic values of different stresses are required to be used. These are termed as 'safe working stresses' which are evaluated by applying certain safety factors on the actual fundamental stresses computed after laboratory testing on clear wood specimens (Sekhar and Gulati 1972; Rajput et al. 1996).

37.6 Environmental Factors

Wood being hygroscopic and having high micro-structural and natural variability, the strength properties of wood are influenced by many external and internal factors such as moisture content, temperature, exposure to different chemicals, inherent and processing defects in the wood material. Besides these, the shape and size of the test specimens as well as rate of loading of the specimens during test also found to affect the strength properties of wood. This information is quite useful for better understanding of timber as material for diverse structural or non-structural applications in varying climatic conditions. Effect of some of these factors on wood properties is described in the following paragraphs.

37.6.1 Moisture Content

The variations in the moisture contents between oven-dry condition and FSP affect strength properties of wood. As explained above, changes in the amount of moisture content of wood influence other important properties such as dimensional stability (due to shrinkage and swelling) as well as durability of wood (against exposure to fungi). As the wood dries, the strength properties are found to increase. In other words, dry wood has greater strength and stiffness than wet wood. The increase in bound water interrupts and reduces the hydrogen bond between the organic polymers of the cell wall, which weakens the strength of the wood. In practice, sometimes it is not possible to dry wood to exactly 12% moisture content (dry state), which is necessary to compare the strength parameters of different wood species. The following formula has been proposed to adjust the strength properties to 12% moisture content, assuming that the actual variation in EMC of seasoned/air-dry wood specimens is between 10 and14% (Sekhar and Rajput 1968; Rajput et al. 1996).

$$S_{12} = (M_d \times S_d)/12, \quad (37.16)$$

where S_{12} = value of strength property at 12% moisture content, M_d = actual moisture content of specimen in dry condition (%), S_d = value of strength property in dry condition.

37.6.2 Temperature

At constant moisture content, the effect of temperature change on strength properties has been investigated by several researchers. A linear decrease in flexural and compressive strength with increasing temperature (up to 70 °C) was observed (Rajput et al. 1996). These changes are usually reversible when the temperature returns to normal. An increase in temperature weakens the strength properties of wood. Permanent effects are also observed at higher temperatures due to certain chemical changes in the wood.

37.6.3 Chemical Exposure

Wood is an organic material and its structural integrity gets disturbed when it is exposed to severe acidic or alkaline conditions. Softwoods are more resistant to chemicals than hardwoods. Wood less permeable to moisture movement is less affected by chemicals.

37.7 Working Qualities of Timbers

In addition to anatomical, physical and mechanical properties, the working quality of wood in processing operations is also important in determining the suitability of wood for the manufacture of various wood products. According to standard procedures, important operations (tests) in evaluating woodworking quality are planning, sanding, turning, forming, grooving, drilling, carving, ease to work and polishing (Anon 1992). To evaluate these working properties, experiments are carried out with seasoned wood with an EMC value of 10–14%, which can be obtained by conditioning the wood in a chamber with a temperature of 27 ± 2 °C and 65 ± 5% RH. A total of 50 clear and defect-free wood samples are selected to evaluate the working quality parameters (Anon 1992). Table 37.10 shows the recommended sample size for different tests.

37.7.1 Different Woodworking Parameters

Following tests are generally carried out under controlled conditions to evaluate the woodworking quality parameters (Anon 1992).

37.7.1.1 Planning

A thickness planner with an auto-feed configuration and four sets of cutter blades with cutting angles of 15°, 20°, 25° and 30° are required. In both directions, four incisions of 2 mm depth are produced at each angle. The wood specimen is visually checked after each pass for planning characteristics and faults such as raised, ripped and fuzzy grains and chip marks.

37.7.1.2 Sanding

The sanding test is performed on each specimen following the planning test. This test requires a 3-drum sander machine with a feed speed of 60 mm per sec and sand papers with grit numbers 60, 80 and 100 respectively. All specimens are run through the sanding machine with a 1 mm depth cut. The specimens are visually evaluated and recorded after each pass for various sanding characteristics and faults such as wooliness, spiral markings and scratching.

37.7.1.3 Turning

For this test, a turning lathe with 3000 rpm and turning tools are used. Defects like fuzzy grains, roughness and torn grains are noted after visual examination.

37.7.1.4 Shaping

The shaping test uses a single-spindle moulder machine with a speed of about 6000 rpm along with a cutter capable of giving 2 mm deep sweep. A band saw may also be needed to make the proper cuts. The samples are visually inspected for defects such as fuzzy and chipped grains, roughness, charring and split grains.

37.7.1.5 Mortizing

This test is carried out with a 12.5 mm hollow chisel type single-spindle electric mortizing machine at a speed of approximately 3000 rpm. Appropriate jigs are required to perform the experiment. Defects such as roughness, tearing, crushing, charring are visually examined on the wood samples.

Table 37.10 Specimen size for evaluating woodworking quality parameters (Anon 1992)

Name of the test for evaluation of the woodworking quality parameters	Size of the wood specimen (length by width by thickness; mm)
Planning and sanding tests	900 × 100 × 25
Turning test	150 × 25 × 25
Shaping, mortising and boring tests	300 × 75 × 25
Moisture content and specific gravity tests	25 × 25 × 25

Table 37.11 Different grades and quality parameters-based total defect value (Anon 1992)

Grade	Quality	Computed value of defect
Grade I	Excellent	Defect free
Grade II	Good	Total defect value 1
Grade III	Fair	Total defect value 2
Grade IV	Poor	Total defect value 3 or 4
Grade V	Very poor	Total defect value 5

Table 37.12 Working qualities of plantation grown timbers

Species	Turning	Boring	Planning	Sanding	Carving	Ease of work	Polishing
Acacia auriculiformis	Turns well	Bores well	Smooth finish	Sands well	Very good	Easy with sharp tools	Takes lacquer polish well
Maesopsis eminii (Musizi)	Turns well	Bores well	Fairly smooth surface	Sands well	Excellent	Very easy to work	Finish with wax and lacquer
Eucalyptus camaldulensis	Turns well	Bores easily	Planes well	Sands well	Excellent	Easy to work	Good glossy finish
Eucalyptus tereticornis	Turns well	Bores well	Planes well	Sands well	Excellent	Easy to work	High glossy finish
Leucaena leucocephala (Subabul)	Turns well	Bores well	Smooth surface	Sands well	Good	Easy to work	Takes lacquer with wax polish
Gyrocarpus americanus	Raised grain and rough surface	Not suitable	Rough surface	Sands well	No sharp edges	Very easy	Can be painted to make imitation
Swietenia mahagoni (mahogany)	Turns well	Easy boring	Smooth surface	Sands well	Excellent	Very easy	Takes lacquer and wax polish well

37.7.1.6 Boring

In this test, two holes are drilled, one in the grain direction and the other across the grain direction. Various types of defects such as fuzziness, roughness, torn grains and crushing are visually identified and recorded.

37.7.2 Grading Based on Working Properties

In each test, each wood sample must be carefully visually inspected for various defects. Each defect detected is given a value between 1 and 5, depending on the severity of the defect. For each sample, these defect values are added to obtain the total defect value. Accordingly, the quality and grade are decided as shown in Table 37.11 (Anon 1992). The working qualities of a few plantations grown timber species are listed in Table 37.12.

37.7.3 Wooden Handicrafts

As known, handicrafts including wooden items are unique expressions and represent the culture, tradition and heritage of a society/country. These products are mainly made either entirely by hand or by hand/power tools. Mechanical tools can be used as long as direct manual input by the craftsman is the most important part of the final product. Crafts can be useful, aesthetic, artistic, creative, cultural, decorative, functional, traditional, religiously and socially symbolic and meaningful. India is rich in history, tradition and culture and has been one of the largest producers and suppliers of handicrafts. Indian handicrafts are known worldwide for their versatile selection and skillful craftsmanship. Before the era of industrial development, this industry was a potential source of economic and local employment. Like ours, in many countries, the share of handicrafts in the national income is equal to the quality and quantity of the processed products, and it has been a considerable source of foreign income. These countries are also focusing on developing craft industries and marketing strategies to earn a better income and strengthen the economy. India is a land of exotic arts, traditions, colours, smells, festivals, languages, etc., and has a rich cultural heritage of which craftsmanship is an integral part.

Woodcraft (wooden crafts) has flourished for centuries and has an important place in Indian history and culture. Skilled and experienced craftsmen carve traditional patterns on wooden objects, enhancing their appearance by painting them with intricate metallic/ivory decorations. Being a tropical land, there is a huge variety of wood species available locally, which become an unlimited source of basic raw materials for woodworking. The important types of wood that are traditionally used for handicrafts in the country are sandalwood, teak, sissoo, sal, walnut, deodar, rosewood, mahogany, red cedar, ebony etc. to mention a few. Due to the good decorative carving and inlay, these types are widely used by craftsmen. Woodcrafts are available in many sizes, shapes and finishes. Indian woodcrafts are known for their beauty and utility and reflect unparalleled artistic virtues. Skilled craftsmen have the expertise to create masterpieces with exotic shapes, designs, patterns and colours. Indian wood industry mainly includes diverse articles, artifacts and objects such as finely carved figures, furniture, and accessories, windows, doors, boxes, ornaments, vases, panels, beads, partitions, screens, etc., available in carved, painted or inlaid wood. The highest qualities of wooden furniture have always been its durability, ethnic taste, style and design, attractive colours, unparalleled finish and exquisite elegance. With growing centres, advanced processing and manufacturing facilities and export possibilities, India's traditional craft culture has reached every corner of the world.

Being very creative, the craft industry has localized segments of domestic and international markets and produces a wide range of products. Indian textile industry is also very labour-intensive, cottage-based and decentralized. Most of the manufacturing units are located in rural areas and small towns and the market potential is huge in all Indian cities and abroad. Although considered a cottage industry, Indian handicrafts have emerged as one of the biggest earners over the years. The domestic craft industry has consistently shown good and commendable growth in recent years and has become one of the most important producers of exports and foreign earnings (Anon 1970). Indian handicrafts are in high demand and have good potential to reach both domestic and international markets. For demand and supply to match quality, greater technological support and innovation with industry uniqueness are needed. The rich cultural diversity and heritage of the country have been a constant source of inspiration, providing a unique and vast resource for the development of traditional and innovative wood crafts.

37.8 Some Important Timbers and Their Usage

India is home to a wide range of tree species. Research work has been done with great effort and many scientists have collected and compiled a huge amount of data and information about Indian forests, Indian woods, their identification and classifications, wood characteristics, quality parameters and various end-use applications (Anon 1963, 1972; Chowdhury and Ghosh 1958; Purkayastha 1982, 1985; Rajput et al. 1985, 1996; Rao and Purkayastha 1972). Table 37.13 presents a concise list of some important timber species being used for various purposes in the country, their scientific and common names, specific gravity, certain characteristic features of the heartwood including colour and their general use for different purposes.

37.9 Variability in Wood Properties

Wood being a natural biomaterial varies in its properties in radial direction from pith to periphery as well as in longitudinal direction along the height of the tree. In general, wood density and all other properties are subject to great variability among genera and species as well as trees grown on the same site. In fact, variations can even be observed within a single tree and may be higher than 20% between the trees of same species (Panshin and de Zeeuw 1980; Zobel and van Buijtenen 1989; Tsoumis 1991; Shmulsky and Jones 2011; Ramagea et al. 2017). Variability is generally caused by differences in the wood anatomical structure, notably the width of annual rings and the varied proportions of early to late wood as well as juvenile to mature wood (Dekort et al. 1991; Guler et al. 2007; Adamopoulos et al. 2010). Moreover, the plantation trees of faster growth exhibit far greater variations in various wood quality parameters compared to slow grown trees of the same species. Fast grown trees from the plantations also have lower age of rotation and harvested early. Hence, such trees are expected to contain higher proportions of juvenile wood. Other types of growth-related defects such as wider growth rings, tension wood and growth stresses etc. as well as natural defects such as knots and spiral grains may be present in the tree, which ultimately affect the wood properties, processing and performance of the products.

37.10 Future Scope

Many pioneering efforts have been made in the past to generate a data base of various wood properties of different timber species of different age plantations including clones and hybrids. There is still not much scientific information available about the performance of different tree species being raised in various agro-climatic regions. Information on growth and wood properties is helpful in evaluating their potential for various end-uses. Such information is also important for making informed investments in large-scale tree species planting programmes. Understanding the wood quality parameters of tree species, the variability due to their age, location, etc., and processing provides a basis for assessing the potential of their value-added applications in various sectors of the wood industry.

Lessons Learnt

- Wood is one of the versatile natural materials suitable for numerous end-use applications. It is also a low-carbon footprint structural material with minimal processing, easy to work and exhibits high specific mechanical, thermal and other properties.

Table 37.13 Important timber species, their specific gravity, characteristic features and general usage (Chowdhury and Ghosh 1958; Anon 1963, 1972; Rao and Purkayastha 1972; Purkayastha 1982, 1985; Rajput et al. 1985, 1996)

Scientific name	Common name	Specific gravity	Important characteristic features of heartwood	General use
Abies pindrow	Himalayan fir	0.447	Light weight, soft, not very durable, to be treated for external use, whitish in colour	Internal work, boxes, crates, packing cases, planking, shingles, match manufacture, wood pulp, treated timber for building purpose
Acacia auriculiformis	Bengal jali	0.645	Strong, heavy, tough, hard, heartwood resembles with teak, brownish yellow	Door and window frames, posts, poles, columns, beams, turnery articles, tool handles crates
Acacia catechu	Khair	0.875	Very heavy, very strong, very hard, tough, durable, kiln seasons well, hard to work, deep red or reddish-brown darkening on exposure	Mainly used for extraction of cutch and Katha, posts and house construction
Acacia chundra	Lal Khair	0.976	Extremely heavy, very strong, very hard, tough, durable, difficult to dry, hard to work, deep red or reddish-brown darkening on exposure	House construction, posts, agricultural implements, hubs and axle of carts, good for underwater usage, chemical extraction etc.
Acacia mangium	Mangium	0.520	Heavy, hard, very strong, tough, liable to crack, glossy and smooth finish, brownish yellow	Furniture: Tables, teapoys, door and window frames and shutters, artifacts, toys etc.
Acacia nilotica	Babul	0.738	Very strong, tough and hard, seasons well, very durable, difficult to saw, takes up a good polish, pinkish brown ageing to reddish brown	Posts, rafters, beams, door frames, rural housing, tool handles, cart bodies, wheels, agricultural implements etc.
Acrocarpus fraxinifolius	Mundani	0.520	Moderately heavy, strong, moderately refractory, non-durable, easy to saw and work, light pinkish or reddish-brown with darker streak	Planking, flooring, furniture, heavy packing cases, crates, boxes, plywood for tea chest, life-boat oars, etc.
Adina cordifolia	Haldu	0.614	Moderately strong, harder than teak, not durable, moderately refractory to season, easy to treatment, easy to saw and work, fresh wood appears deep yellow, brownish or reddish yellow when exposed	Mostly used in artifacts, toys, pen-holders, combs, rulers and other handicraft, picture frames, sports items, light furniture, tool handles, veneers, general purpose plywood, tea-chests, panelling, etc.
Ailanthus excelsa	Maharukh	0.356	Fairly strong, light, soft, easy to dry, easily perishable in open, easy to work, whitish to light yellow or greyish turning brownish yellow on aging	Light packing cases, box shooks, match splints, toys, slate frames, veneering and panel products
Albizia chinensis	Kala siris	0.450	Soft, light to medium weight, not very strong, easy to work, light coloured	Packing boxes, boards, turnery products, furniture, yokes, slate, frames, pulping, etc.
Albizzia lebbeck	Kokko	0.549	Moderately heavy to heavy, strong, hard, moderately durable, moderately refractory, difficult to machine, brown to dark brown, tends to darken with age	Furniture, parquet and strip floorings, railway carriage work, boat building, turnery items, carvings, interior decoration, veneering and other small specialty items
Albizzia procera	White siris	0.579	Moderately heavy, strong, hard, moderately durable, moderately refractory, difficult to work and machine, brown to dark brown	Construction and building such as beams and posts, cabinets, furniture, panelling and veneer, agricultural implements and carts
Anogeissus latifolia	Axle wood	0.828	Very strong, hard and tough, difficult to season, difficult to saw, takes smooth finish	Making wheels, picker arms used in textile industry, railway sleepers, cupboards, bent-wood articles, good for making charcoal
Artocarpus heterophyllus	Kathal	0.505	Moderately strong, hard, heavy, durable, easy to season, refractory to treatment, easy to work, yellowish brown to golden brown, darker streaks turning dark brown on contact	Piles, wooden bridge platforms, door and window panels, other building and constructional works, musical instruments, tool handles, turnery items
Artocarpus hirsutus	Aini	0.516	Strong, moderately hard, durable, season easily, refractory to treatment, close-grained, easy to saw and work, takes polish, yellowish brown to golden brown turning dark brown on aging	Building and construction for beams, rafters, window and door frames, posts, railway sleepers, paving, furniture and cabinet works, tool handles, sports goods, masts, rafts and life boats
Azadirachta indica	Neem	0.693	Hard and strong timber, seasons well, durable, difficult to saw and work, takes good finish, reddish brown	Different tree parts of medicinal value, wood is used for carving, toys, agricultural implements, carts, boards, panels, etc.
Bombax ceiba	Semul	0.329	Loose grained, light weight, inferior quality wood, whitish in colour	Packaging boxes, match industry, well curbs, light furniture
Casuarina equisetifolia	Casua- rina	0.693	Moderately strong, tough, very heavy, very hard, very refractory to season, twist badly, non-durable, difficult to saw and work, pink to light reddish brown, darkening on exposure	Construction work for posts, scaffolding for temporary structures, poles, tent accessories, masts, oars, and paddles, boat building, valuable as fuelwood

(continued)

Table 37.13 (continued)

Scientific name	Common name	Specific gravity	Important characteristic features of heartwood	General use
Cedrus deodara	Deodar	0.476	Softwood, distinct growth rings, easy to work, moderately strong, yellowish brown	Furniture, carriages, packing boxes, structural works
Chloroxylon swietenia	Satin wood	0.771	Very dense, hard and strong, durable, close grained, moderately hard to saw and machine dry wood, light yellow, cream yellow to golden yellow darken with age	Poles, posts, rafters, railway sleepers, bridge constructions, agricultural implements, cabinet, furniture, turnery, inlays, picture frames and ornamental works
Cocos nucifera	Coconut	0.680	Takes good polish, requires preservative treatment; reddish brown	Poles, piles, furniture, formwork, flooring, canoes, rafts, walking sticks, tool handles, turned objects
Dalbergia latifolia	Rose-wood	0.660	Hard, heavy, strong, tough, very durable, seasons well, close-grained, maintains shape well, takes good finish, purplish brown to deep purple with streaks	Decorative wood for superior quality furniture, cabinet work, doors and windows, panelling, ornamental carvings, face veneers for decorative plywood, marine usage
Dalbergia sissoo	Sissoo/ Sheesham	0.692	Strong and tough, durable and beautiful, holds its shape well, easily seasoned, difficult to work, withstands polishing, dark brown in colour	High quality furniture, very good for decorative works and carvings, plywood, bridge piles, sport goods, railway sleepers
Dipterocarpus indicus	Gurjan	0.618	Heavy, strong, very hard, moderately durable, easy to work, yellowish brown to reddish brown	Furniture, flooring, veneering, plywood, panel products, trailer decks, other utility items
Dysoxylum malabaricum	White cedar	0.581	Moderately strong, moderately tough, hard, very durable, not difficult to season, easy to work, takes good finish, straw yellowish turning golden yellow or golden brown on aging	Furniture, cabinet making, turnery articles, toys, oil casks, veneer and plywood, ceiling and floor boards, interior fittings, suitable for match industry
Eucalyptus camaldulensis	Eucalypt	0.639	Moderately heavy, moderately strong moderately tough, moderately hard, non-durable, difficult to season, not easy to work, pale brown to reddish brown	Building and construction, door and window frames, rafters, artifacts, carving, turnery items, tool handles, packing cases, pallets
Eucalyptus citriodora	Eucalypt	0.742	Moderately hard, moderately heavy, not durable, refractory to seasoning, refractory to treatment, light brown to greyish brown	Mainly used as poles, pulp and paper, fuelwood, producing charcoal etc.
Eucalyptus tereticornis	Mysore gum	0.713	Heavy, moderately tough, strong, hard, non-durable, refractory to season, refractory to preservative treatment, difficult to work, takes good finish, pale brown to reddish brown	Mainly used for pulp and paper, rayon grade pulp, power transmission poles, constructional work such as beams and columns, etc.
Ficus benghalensis	Banyan	0.385	Light, soft to moderately hard, not very strong, not durable, non-refractory, very easy to saw and work, greyish white darkening to brownish grey on exposure	Packing cases, match boxes, cheap turnery works, aerial roots utilized for tent poles, cart yokes, carrying shafts, etc.
Gironnera reticulata	Churchi	0.77	Very heavy, strong, hard, moderately steady and hard, needs to be seasoned, greyish brown to golden brown	Structural and engineering works, railway sleepers, furniture, tool handles, packing cases, crates
Gmelina arborea	Gamari	0.445	Strong, moderately hard, moderately heavy, very durable, easy to season, refractory to treatment, easy to saw and work, pale yellow to cream coloured turn yellowish brown on aging	Door and window panels, other joinery and furniture, tool handles, decking, bent wood articles, toys, picture and slate frames, turnery items, veneering, plywood, panelling, etc.
Grevillea robusta	Silver oak	0.572	Moderately hard, tough, strong, moderately durable, coarse texture, straight grain, fairly easy wood to work, light to medium reddish brown with grey to light brown, exhibit figures	Light furniture, packing cases, veneer, plywood, panel products, cabinetry, turnery and other specialty items
Gyrocarpus jacquini	Tanaku	0.303	Light, very weak, not tough, very soft and steady timber	Model toys, craft floats, catamarans, light packing cases, slate frames, handicraft, etc.
Hevea brasiliensis	Rubber wood	0.521	Soft, weak, light, non-durable, dries fast, easy to treat, easy to saw and work, requires treatment, whitish to greyish white turning to dirty creamy-white on aging	Packing cases, second grade door frames, shutters, furniture, light joinery, dunnage pellets, match industry, bent wood articles
Heritiera spp.	Sundri	0.694	Strong, hard, tough, durable, difficult to season and work, close grained, dark red	Used in boat building, piles, poles, tool handles, carriage shafts
Hopea parviflora	Hopea	0.794	Extremely strong and tough, difficult to work. Season easily, durable, light to deep brown	Mostly used for construction, structural work, railway sleepers, piles, boat building

(continued)

Table 37.13 (continued)

Scientific name	Common name	Specific gravity	Important characteristic features of heartwood	General use
Lagerstroemia microcarpa syn. *L. lanceolata*	Benteak	0.603	Heavy, strong, hard, moderately tough, very durable, moderately refractory to air-season, very refractory to treatment, not difficult to saw and work, takes smooth finish, light red to reddish brown darkening on exposure	Building and constructions, rafters, beams, columns, posts, door and windows, lorry bodies, poles, furniture, tool handles, wooden crates, bend items for sports good, boat and ship building, marine structures, railway carriages, etc.
Lagerstroemia parviflora	Lendi	0.620	Heavy, moderately tough, strong, hard, fairly durable, very refractory to treatment, moderately easy to air-season, fairly easy to saw and work, takes good finish, light brown to pinkish brown	Building and constructions such as posts, beams, rafters, bridges, oars, cheap quality furniture, railway sleepers, wooden packing cases and crates, poles, lorry body building
Lagerstroemia speciosa	Jarul	0.521	Heavy to moderately heavy, moderately strong, hard, moderately durable, easy to air-season, very refractory to treatment, easy to saw and work, takes a good finish, reddish brown darkening on exposure	Building construction, beams, doors, windows, cheap quality furniture, parquet flooring, poles, railway sleepers, wooden packing cases, crates, chests, tool handles
Leucaena leucocephala	Subabul	0.601	Hard, strong, moderately heavy to heavy, not durable, dry without seasoning defects, close grained, easy to work, light reddish-brown	Door and window frames, tool handles, turnery, tool handles, poles, posts, props, dunnage pallets, cabinet making, parquet flooring, small furniture, pulp and paper
Madhuca longifolia	Mahua	0.737	Moderately strong extremely hard, highly durable, refractory to season, very refractory to treatment, difficult to saw and work, takes a good finish, dark red or reddish brown	House construction, posts, beams, piles, axels, frames, parts of carts, tool handles, etc.
Maesopsis eminii	Musizi	0.410	Soft, moderately heavy, not tough, non-durable, easy to saw and work, takes smooth lacquer finish, bright yellow green or green brown turning to a golden brown on exposure	Light construction, joinery, turnery articles, toys, light packing cases, wooden crates, plywood, light furniture, millwork light tool handles
Mangifera indica	Mango	0.588	Fairly strong, light to moderately heavy, soft to moderately hard, non-durable, easy to treatment, easy to saw and work, shape well, peels well, whitish yellow, greyish brown to buff in colour	Cheap furniture, lining, floor and ceiling boards, door and window frames, carvings and turnery items, packing boxes, veneering, plywood and panel products, tea chests, etc.
Mastixia arborea	Bolong	0.450	Light, weak, soft, moderately steady, straight grained, medium fine textured, greyish white turning brownish with age	Light packing cases, slate frames, toys, veneering, plywood, etc.
Melia azedarach	Persian lilac	0.550	Moderately strong, moderately heavy, moderately hard and tough, durable, straight grain, seasons well, easy to saw and work, light red to reddish brown on aging	Toys, small articles, turnery items, packing cases, carvings, light furniture, veneering and decorative plywood
Melia dubia	Malabar neem	0.490	Softer and lighter, moderately durable under cover, straight grain, coarse texture, easy to saw and work, light pink to light red on aging	Light packing cases, ceiling planks, musical instruments, veneering, good quality plywood, boxes
Mesua ferrea	Mesua	0.914	Durable, very hard, and not easy to work. Resists penetration of nails, reddish brown	Construction, bridges, agricultural implements, wagons, sleepers, piles
Mimusops elengi	Bakul	0.800	Strong, tough, hard very durable, difficult to season, close-grained, difficult to work, finishes well, light or deep reddish brown with darker streaks	Heavy construction, bridges, piles, crushers, carts, boats, etc.
Morus alba	Mulberry	0.603	Moderately hard to hard, moderately heavy, strong, not durable, easy to season, easy to saw and work, takes up a clean finish, light yellowish brown to dark brown on aging	Baskets, sports goods: hockey sticks, badminton and tennis rackets, cricket bats, stumps, slate frames, packing cases, excellent for steam bending, artificial limbs, tool handles, oars, paddles, furniture, cabinets
Populus deltoides	Poplar	0.419	Light, not strong, non-durable, easy to work, heartwood-light brown, sapwood- yellowish white	Boxes and crates, veneer, plywood, used in match industry, various other utility items
Pinus roxburghii	Chir pine	0.491	Hard and tough, not durable, coarse grained, light greyish-yellow	Door/window frames, paving material, white pine is light and straight grained, match boxes
Prosopis juliflora	Vilayati babool	0.830		Cheap furniture, turnery, tool handles, fence, posts, agricultural implements, construction, etc.

(continued)

Table 37.13 (continued)

Scientific name	Common name	Specific gravity	Important characteristic features of heartwood	General use
			Heavy, strong, hard, tough, requires careful seasoning, durable, difficult to saw, takes fairly smooth finish, dark brown to purplish brown	
Protium serratum	Murtenga	0.735	Fairly strong, tough, heavy, moderately hard, fairly durable, dry slowly, easy to work, red turning dull brown or brick red with age	Structural beams, house posts, plough, carts, photo frames, pencil making, railway sleepers
Pterocarpus dalbergioides	Andaman padauk	0.620	Heavy, strong, hard, extremely durable, seasons well, not difficult to work, yellowish pink to deep pinkish brown or brick red, darkening to a reddish/purplish brown over time	Decorative uses, high class furniture, cabinets, musical instruments, marine grade plywood, ship cabins, veneer, pane/ling, flooring, turnery items, tool handles and other small wood items
Pterocarpus marsupium	Bijasal	0.671	Strong, tough, very durable, seasons well, very refractory to treatment, coarse-grained and difficult to work, golden brown to reddish brown with darker streaks	Building and construction: Doors and window frames, posts, pillars, beams, furniture, agricultural implements, cart building, tool handles, carvings, etc.
Pterocarpus santalinus	Red sanders	0.967	Extremely strong and heavy, very tough, durable, seasons well, dark orange red with darker streaks turning deep red to purplish black	Wood has high export potential for musical instruments, medicinal value, carving idols, toys, ornamental house posts, furniture, tool handles, picture frames and other small wooden items
Santalum album	Sandal wood	0.840	Hard, heavy, strong, durable, moderately refractory to air season, straight grained, very fine textured, takes smooth finish, gives pleasant smell, yellowish brown to dark brown	Wood used for oil extraction in fragrance industry, handicrafts items, carving statues, toys, wood powder for agar bathies, etc.
Salix spp.	Indian willow	0.443	Soft, light weight, tough, non-durable, seasons well with care, easy to saw, glues and finishes well, light brown turning to greyish brown on aging	Cricket bats, posts, planks, handles, baskets, packing cases, crates, carvings, pen holders, artificial limbs and other specialty wood items
Shorea robusta	Sal	0.745	Hard, heavy, very durable, difficult to season, easy to saw, difficult to plane and turn reddish brown	Structures and house construction, ships and railway cars, poles, railway sleepers, posts, agricultural implements, bridges, floors, window frames. Interior finishing
Simarouba glauca	Simaruba	0.380	Light, soft, not tough, easy to work, split during sawing, creamy whitish in colour	Light construction, light furniture, match industry, light packing cases, veneer, plywood core, wood chips, handicraft items, decorative uses
Sterculia urens	Karar	0.528	Moderately heavy, strong, moderately hard, moderately tough, fairly durable under cover, easy to season, easy to saw and work, takes good finish, reddish brown	Door and window frames, furniture, joinery works, packing cases, tool handles, staircase railings, boat building, cheap plywood, etc.
Swietenia mahagoni	Maho-gany	0.540	Hard, strong, very durable, seasons well, takes good polish, easy to saw and work, good peeling properties, reddish brown	Furniture, pattern making, cabinet work, turnery articles, bowls, inlay works, jewellery boxes, carvings, panelling, etc.
Tamarindus indica	Imli	0.747	Hard to very hard, heavy to very heavy, moderately refractory to air-seasoning, durable for indoor, extremely difficult to work, excellent for turning, dark purplish-brown	Internal fittings of buildings, parts of cart wheels, other agricultural implements, tent pegs, chopping blocks, turnery and toys, excellent for charcoal production
Tecomella undulata	Rohida	0.930	Strong, hard, heavy, tough, durable, straight grained, fine textured, polishes well, brownish to dirty grey	Door and window frames, shutters, handicraft, carving work, furniture, parquet flooring, agricultural implements, etc.
Tectona grandis	Teak	0.604	Moderately hard, durable, fire-resistant, stable, season easily, takes good polish, deep yellow to dark brown	Superior quality construction work, furniture, boats/ships, decorative veneers for panel products, etc.
Terminalia arjuna	Arjun	0.686	Heavy, tough and strong, hard to very hard, moderately durable, moderately refractory to seasoning, difficult to saw and work, dark brown	Construction purposes such as beams, columns, rafters, trusses, posts, door and window frames, panelling, transmission poles, agricultural implements and boat building
Terminalia bellerica	Bahera	0.540	Strong and moderately tough, hard, moderately durable, moderately refractory to seasoning, easy to saw and work, yellowish grey turning brownish with age	House building as rafters and boards, agricultural implements such as plough shaft, carts, heavy packing cases, tea chest, commercial plywood and blackboards

(continued)

Table 37.13 (continued)

Scientific name	Common name	Specific gravity	Important characteristic features of heartwood	General use
Terminalia tomentosa	Laurel	0.796	Strong, tough, hard to very hard, moderately durable, very refractory to seasoning, difficult to saw and work, brown to deep brown with darker streaks	Building and construction, door and window frames, furniture and cabinet work, plywood, panelling, poles, carriages, agricultural implements
Toona ciliata	Toon, red cedar	0.424	Light to moderately heavy, strong, hard, moderately durable, easy to work, takes good polish, dark reddish-brown, darkens upon exposure, reddish-brown with darker streaks	Light weight racing boats, door and window panels, match boxes, packing boxes, crates, furniture, joinery, cabinets, musical instruments, veneering, plywood, panelling
Ulmus wallichiana	Elm	0.435	Moderately hard and strong, fine grained, take good polish. Reddish in colour	Door and window frames, furniture, carts, decorative and ornamental uses
Wrightia tinctoria	Dudhi	0.480	Soft, light to moderately heavy, hard, durable under cover, easy to season, straight grained, fine textured, easy to work, white when fresh, turning yellowish grey with age	All classes of turnery articles, toys, cups, plates, frames, spoon, combs, screens, rulers, engineering scales, pen holders, cheap grade pencils, etc.
Xylia xylocarpa	Irul	0.715	Very hard, heavy, strong, durable, difficult to saw and work, requires slow and careful seasoning, reddish brown turning to deep reddish brown on ageing	House construction, agricultural implements, railway sleepers, wagons, heavy constructional works such as bridges, piles, decking, overhead poles and harbour works

- At present, most of the wood resources available are from plantation timbers along with a few primary species. Selection of right wood at right price for a specific application is very important from end-user point of view.
- A variety of wood species having a range of different properties are available for various applications including building and construction, flooring, plywood and panel products, structural wood composites (mass timber), pulp and paper, furniture and cabinetry, packaging, handicrafts, artifacts, agricultural and sports equipment etc.
- Testing and evaluation of different properties of wood and finding its suitability for a specific application is a prerequisite for effective utilization of wood products. Applications of technological innovations in wood processing and finishing not only facilitate in providing the best quality end-products but having competitive and improved performance.

Key Questions

1. Describe wood as a material, how it compares with other structural materials.
2. What are different types of woods? Discuss various end-use applications of wood.
3. Discuss different shrinkage/swelling properties, fibre saturation point and dimensional stability in wood.
4. Discuss the importance of density (or specific gravity) and its relation with physical and mechanical properties of wood.
5. What are different comparative suitability indices and suitability figures for different end uses?

References

Adamopoulos S, Chavenetidou M, Passialis C, Voulgaridis E (2010) Effect of cambium age and ring width on density and fibre length of black locust and chestnut wood. Wood Res 55(3):25–36

AEC (2020) Aluminum Extruders Council. https://www.aec.org/page/material-comparisons-plastic-wood-vinyl?&hhsearchterms=%22molded+and+plastic%22. Accessed 18 June 2020

Anon (1963) Indian woods: their identification, properties and uses, vol 2. Manager of Publications, Delhi, p 430

Anon (1970) Indian forest utilization. Compiled and written at the Editorial Board, Forest Research Institute and Colleges, Dehradun, vol 1. Manager of Publications, Delhi, p 508

Anon (1972) Indian forest utilization. Compiled and written at the Editorial Board, Forest Research Institute and Colleges, Dehradun, vol 2. Manager of Publications, Delhi, p 941

Anon (1974) IS:2455. Methods of sampling of model trees and logs and their conversion for timber testing. Bureau of Indian Standards, New Delhi, p 18

Anon (1986a) IS:1708 (Reaffirmed 2005). Methods of testing of small clear specimens of timber. Parts 1–18. Bureau of Indian Standards, New Delhi, p 64

Anon (1986b) IS:3629 (Reaffirmed 2000). Structural timber in building: specification. Bureau of Indian Standards, New Delhi, p 31

Anon (1992) IS:8292. Evaluation of working quality of timber under different wood working operations: method of test. Bureau of Indian Standards, New Delhi, p 7

Anon (1999) ASTM D143. Standard test methods for small clear specimens of timber. American Society for Testing and Materials (ASTM), Philadelphia

Anon (2020) Wikipedia contributors. Specific strength. In: Wikipedia, The Free Encyclopedia. https://en.wikipedia.org/w/index.php?title=Specific_strength&oldid=945501282. Accessed 12 July 2020

Bootle K (1983) Wood in Australia: types, properties and uses. McGraw-Hill Book Company, Sydney

Cave ID, Walker JCF (1994) Stiffness of wood in fast-grown plantation softwoods: the influence of microfibril angle. For Prod J 44(5):43–48

Chowdhury AK, Ghosh SS (1958) Indian woods: their identification, properties and uses, vol 1. Manager of Publications, Delhi

Dekort I, Loeffen V, Baas P (1991) Ring width, density and wood anatomy of Douglas-fir width different crown vitality. IAWA Bull 12(4):453–465

Dinwoodie JM (1981) Timber: its nature and behaviour. Van Nostrand Reinhold Company, London, England, p 190

Gillie M (2004) Fire resistance of structures. The University of Edinburgh, Edinburgh, pp 1–16. http://civil.iisc.ernet.in/~manohar/Fire/Part-23_Materials.pdf

Guler C, Copur Y, Akgul M, Buyuksari U (2007) Some chemical, physical and mechanical properties of juvenile wood from black pine (*Pinus nigra* Arnold) plantations. J Appl Sci 7(5):755–758

Moreira AP, Silveira E, Almeida DH, Almeida TH, Panzera TH, Christoforo AL, Rocco FA (2017) Toughness and impact strength in dynamic bending of wood as a function of the modulus of elasticity and the strength in compression to the grain. Int J Mater Eng 7(4):61–67

Panshin AJ, de Zeeuw C (1980) Textbook of wood technology, 4th edn. McGraw-Hill Publishing Company, New York

Purkayastha SK (1982) Indian woods: their identification, properties and uses, vol 4. Controller of Publications, Delhi

Purkayastha SK (1985) Indian woods: their identification, properties and uses, vol 5. Controller of Publications, Delhi

Rajput SS, Shukla NK, Gupta VK (1985) Specific gravity of Indian timbers. J Timber Dev Assoc India 31(3):12–41

Rajput SS, Shukla NK, Gupta VK, Jain JD (1996) Timber mechanics: strength, classification and grading of timber, 2nd edn. ICFRE-38, New Forest, Dehradun, p 189

Ramagea MH, Burridgeb H, Busse-Wicherc M, Feredaya G, Reynoldsa T, Shaha DU, Wud G, Yuc L, Fleminga P, Densley-Tingleye D, Allwoode J, Dupreec P, Lindenb PF, Schermane O (2017) The wood from the trees: the use of timber in construction. Renew Sustain Energy Rev 68:333–359

Rao KR, Purkayastha SK (1972) Indian woods: their identification, properties and uses, vol 3. Manager of Publications, Delhi

Rao RV, Shukla SR, Sharma SK, Hemavathi TR, Kumar P, Dubey YM, Singh D, Sudheendra R, Sujatha M, Shashikala S, Maddurappa (2004) Study of anatomical, physical and mechanical properties of plantation grown *Acacia auriculiformis*, *Acacia mangium*, *Tecomella undulata* and *Eucalyptus tereticornis* clones, p 105

Sekhar AC, Gulati AS (1972) Suitability indices of Indian timbers for industrial and engineering uses. Indian Forest Records (N.S.). Timber Mech 2(1):62

Sekhar AC, Rajput SS (1968) A note on strength adjustment for moisture content in wood. Indian For 94(3):263–266

Sekhar AC, Rawat BS (1966) Physical and mechanical properties of teak from different localities in India and neighbouring areas. Indian Forest Records (N.S.). Timber Mech 1(13):197–212

Shmulsky R, Jones PD (2011) Forest products and wood science—an introduction, 6th edn. Wiley-Blackwell, New York, p 477

Shukla SR, Rao RV, Sharma SK, Kumar P, Sudheendra R, Shashikala S (2007) Physical and mechanical properties of plantation grown *Acacia auriculiformis* of three different ages. Aust For 70(2):86–92

Tsoumis G (1991) Science and technology of wood- structure, properties, utilization. Van Nostrand Reinhold, New York, p 494

Zobel BJ, Jett JB (1995) Genetics of wood production. Springer, Berlin, p 337

Zobel BJ, van Buijtenen JP (1989) Wood variation and wood properties. In: Chapter 1: Wood variation. Springer, Berlin

38 Wood Processing

Shakti Singh Chauhan

Abstract

Understanding the primary wood processing is very critical for augmenting throughput and improving the quality of the wooden products which results in economic gains. This chapter covers the basic aspects of three fundamental wood processing stages namely sawing, seasoning and preservation. The first section deals with the sawing process wherein different sawing methods are described based on the material dimensions. The saw blades play a critical role in defining sawing efficiencies. Various saw blade parameters have been discussed in detail. The second section is on wood seasoning which is one of the most important fundamental processing steps essentially required for the efficient utilization of wood. The basic aspects of wood seasoning covered include specific terms like moisture content, equilibrium moisture content and fibre saturation point which are essential parts of wood seasoning. Different seasoning methods are discussed in detail with their pictorial representation. Defects during drying are a major concern in terms of recovery of quality material. Various types of common defects along with their cause and ways to overcome them have been described. In addition to seasoning, treatment with preservative chemicals is required to increase the longevity of wood material. Being a biological material, wood is susceptible to degradation by decay organisms like fungi, termites and many other insects. The section on wood preservation covers the natural durability of wood and ways of testing it, different types of wood preservatives and their efficacies and methods of treatment.

Keywords

Saw-milling · Wood seasoning · Fibre saturation point · Seasoning defect · Wood preservative

S. S. Chauhan (✉)
ICFRE-Institute of Wood Science and Technology, Bengaluru, India

38.1 Introduction

Appropriate wood processing is essential for efficient utilization of woody resources. Wood processing strategies are decided based on the end application of wood. The processing requirements of pulp products are different from those of solid wood applications like timber. This chapter mainly deals with wood processing for solid wood applications. Primary wood processing can be divided into three major categories namely saw-milling, seasoning and preservation.

38.2 Saw-Milling

Saw-milling is the basic step of converting a log to the sawn form to obtain valuable timber. Saw-milling operations primarily depend on the quality and size of raw material or logs to be processed in the mill and the ultimate end-product. The basic process in a sawmill is breaking down of the logs into rectangular cross-section slabs or large dimension timber. Further, the large dimension timber is converted into boards or planks through ripping, edging and trimming. Finally, the planks are cross-cut to desired sizes, and defects, if any, are removed during the process to enhance the quality and grade of the sawn timber. Every sawmill is unique and is characterized by the type of machines used for converting timber and the wood type. Factors like wood species, log quality in terms of shape and size and log type in terms of tree maturity (old-growth or plantation timber) determine the saw-milling operations to optimize the output and recovery of high-grade material without many defects. For a large-scale sawmill, the certainty in the inflow of the round logs with the desired sizes and log quality plays a critical role in the overall operation of the mill and its economic viability. In India, most of the sawmills are in the unorganized sector performing very basic operations due to uncertainty in wood supply and highly variable output requirements. The

A. K. Mandal, A. Nicodemus (eds.), *Textbook of Forest Science*, https://doi.org/10.1007/978-981-97-8289-5_38

size and design of a sawmill mainly depend on the market potential. A small sawmill may only be involved in sizing, trimming and cross-cutting for the furniture sector or a mill may only be involved in converting logs into thick slabs for further processing. Mill locations close to log resources can have a significant role in operational logistics and profitability of saw-milling. As an example, several sawmills are operational near Gandhidham and Kandla in Gujarat since it is one of the major ports for wood import in India. For a large saw-milling operation, volume processing, conversion efficiency and grade recoveries determine the profitability of the operations. Thus, the proximity of saw-milling close to the port makes it much more economically viable.

Each sawmill is defined by its operational scale, types of machinery, operational efficiency and scale of automation. The primary saw-milling operations revolve around two types of saws namely band saw and circular saw. Band saw can have blade orientation either horizontal or in vertical direction and is primarily used to convert logs into boards by cutting along the grains or the length of the log (Fig. 38.1). The process is also termed as ripping. For large dimension logs, a horizontal band saw is the most preferred for converting logs into slabs or thick sections.

A band saw has an endless band of steel saw blade that is mounted between two large wheels. In a vertical band saw, the lower wheel is powered by a motor pulling the blade down and is heavier than the upper wheel. The blade is pre-stressed between the wheels to a specific tension and the pulling of the blade cuts through the log when fed into the saw to get an efficient sawing. Since the blade is pre-stressed between the wheels, it allows a thinner gauge blade suitable for operation. If the saw blade is not stressed adequately, the blade wanders resulting in a wavy cut and poor finish in sawn wood. Band saws are mainly used for the primary conversion of logs to slabs. For the primary breakdown, log carriages are used to hold logs firmly and accurately while sawing.

Cross-cutting is the process of sawing or cutting across the grains and the process is used for down-sizing of logs, cross-cutting of sawn timber, edging and trimming. In a sawmill dealing with large dimension logs, cross-cutting of logs to uniform and desirable lengths is the first operation involving large-size circular saws (Fig. 38.2). In a circular saw, the circular blade rotates on a spindle at a high speed (RPM). Since the RPM of circular saws is very high, a saw blade of thicker gauge is required to maintain stability during its operation.

Several technical parameters in a saw-milling system define the sawing quality, operation of the machine and the amount of sawdust produced. The width of the cut produced by the saw blade is known as the saw kerf (Fig. 38.3). The saw kerf width should be broader than the thickness of the saw blade to provide an adequate clearance between the blade and the wood surface. In general, saw kerfs are kept nearly double the saw blade thickness. Kerf width is an important factor determining the saw dust formation during sawing as thicker kerf generates more sawdust. Band saw has generally lower kerf width due to thinner saw gauge as compared to circular saws which require a thicker gauge blade.

Many factors influence the efficiency of sawing, waste generation, quality of sawing and power consumption during sawing operations. Knowledge of these factors is very much essential for a saw-miller. Blade material, blade thickness, blade width, tooth sharpness which is defined by the tooth angle, rake angle, clearance angle, gullet depth, gullet shape, tooth design, tooth tipping, bite per tooth, etc. Some of the

Fig. 38.1 Horizontal and vertical band saw

Fig. 38.2 Circular saw

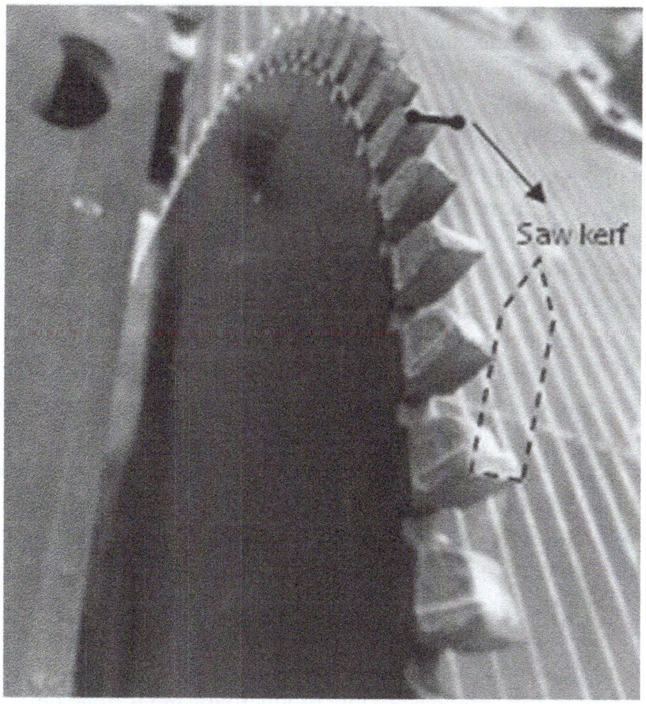

Fig. 38.3 Saw kerf in a circular saw

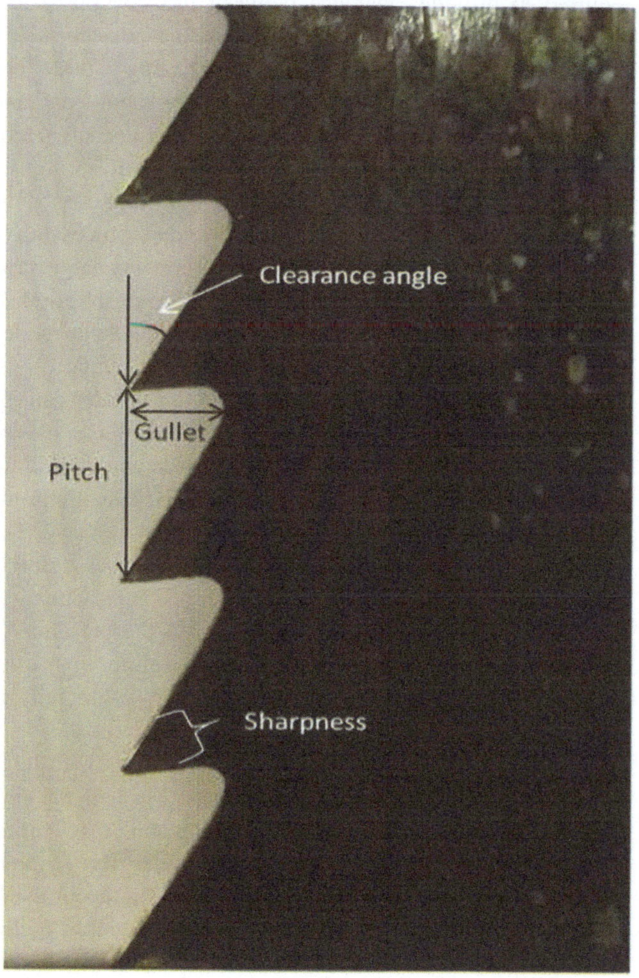

Fig. 38.4 Saw blade parameters

important characteristics of a saw blade are shown in Fig. 38.4. Saw blades require good durability, corrosive resistance, fatigue resistance, the ability to sustain tension and the ability to remain flat and straight during the sawing process. Therefore, the highest quality steel is required to be used for manufacturing quality saw blades. Besides this, blades need to be re-sharpened regularly to increase the service life of the saw.

The pitch of the blade is the distance from the tip of one tooth to the tip of the consequent tooth. The gullet part of the blade collects and takes out sawdust generated during the cut. An increase in the pitch results in increased bite per tooth and therefore increases the power of the saw. Small bite per tooth generates fine sawdust which escapes very easily from the gullet space and falls between the blade surface and wood which results in overheating of the blade due to rubbing friction. This frictional heating can become considerably high resulting in choking of the saw due to compacted sawdust. Large bite per tooth results in a large volume of coarse saw dust beyond the capacity of the gullet which can also result in frictional heating. The gullet capacity can be increased by increasing the gullet depth and that can allow higher feed speed. However, increasing gullet depth increases the length of the tooth which can surge strains on the tooth during the operation leading to crack formation in the gullet. Bigger gullets are used for sawing softwoods with lighter weight. Increasing the rake angle of the tooth increases the sharpness of the tooth which can ultimately help in reducing power consumption but the tooth becomes more prone to accelerated wear and tear and breakage. A large clearance angle prevents the rubbing of the saw with cut wood and smooth sawing operation.

Teeth setting is another important parameter that is required to provide clearance between the saw blade and the wood. Teeth setting is done by bending adjacent teeth, also called spring-set. In band saw, teeth setting is done left-right-raker scheme where the raker tooth is not bent and keeps the cut straight. In the case of a circular saw, left-right bend to the adjacent tooth is a common spring set when the saw is used for cross-cutting. Spring set is done generally in small saw-mills used for secondary processing of wood or handling small dimension materials. For large saw-mill, a swage set is preferred wherein the tip of the tooth is punched. The punched tooth is sharpened and side grounded to result in a broadened tip compared to the gauge of the blade. Further, the wear resistance of the blade can be increased by tipping the teeth and superior sawing is achieved particularly when sawing high-density woods with high mineral contents like silica. Tungsten carbide and cobalt/chrome stellite tips are the prominent ones used in the industry. The tungsten carbide tips are very hard and brittle, whereas stellite tips are not very hard but are corrosion resistant. Chemical corrosion of blades may take place due to the presence of extractives in wood. Among the two, stellite tips offer narrower kerf, better finish and increased saw life but are more expensive.

In general, circular saws are not used for logs with large diameter or thicker cross-section timbers. Most of the mill uses small circular saws mainly for ripping and cross-cutting of small dimension timber. The large circular saw requires thicker gauge metal for the blade stability and correspondingly the kerf width also becomes large (double to the blade thickness). This results in an excessive amount of sawdust production which is highly undesirable in the industry. Only the large-scale sawmills use circular saws for cross-cutting logs into specific lengths as shown in Fig. 38.2.

38.2.1 Sawing Patterns

Depending on the log form, size and type of sawmill requirements, sawing methods have to be adopted to achieve specific quality material having a specific pattern. Generally, four basic sawing methods have evolved namely plain or flat sawing, cant sawing, sawing around and quarter sawing (Fig. 38.5). Among these, plain or flat sawing is one of the easiest ways of sawing wherein successive ripping of the log is carried out in the same plain, mainly in a plane tangential to growth rings. This type of sawing yields high productivity and recovery of sawn timber and therefore is the most popular in the industry. The quarter sawing method and sawing around method are suitable for large girth logs and are currently hardly practiced by the industry. Cant sawing is an attractive option for achieving higher productivity but it requires more skills and understanding of log form.

Flat sawing and cant sawing generally produce boards having grains in multiple directions and result in 'U' or 'V' shapes figures in species having distinct growth rings (Fig. 38.6). However, these boards are more prone to develop cupping, twisting and bowing due to uneven drying pattern in boards. Quarter-sawn boards have almost uniform grain directions and produce boards with line or ribbon patterns. These types of boards are dimensionally more stable and exhibit relatively fewer drying defects. It is very easy to match grains in quarter-sawn boards. However, low productivity and recovery make them an unattractive proposition to saw millers. More so, such a type of sawing is not feasible for small-diameter logs.

Saw-milling industries are highly fragmented and unorganized in most countries including India. The efficiency of the mill depends on throughput, volume recovery and grade recovery. In most of the mills volume recovery is the primary criterion. For improved grade recovery, the selection of a proper sawing method is critical and it may involve frequent turning of logs while sawing as well as careful examination of logs. This may lead to low productivity and is done only in specialized cases where quality is the main concern. The sawmill efficiency is highly variable and the conversion from round log to sawn timber can be anywhere between 50 and 70%. It is generally recommended that the sawdust should not be more than 10%, but in a poorly managed sawmill, the sawdust can be as high as 25% which is highly undesirable. Despite the significant advances in automation, most of the operations in the majority of sawmills are manually carried out on substantially older machines affecting the efficiency and quality of the sawn timber. Smooth flow of material right from the debarking of logs to the packing of

Fig. 38.5 Different sawing patterns

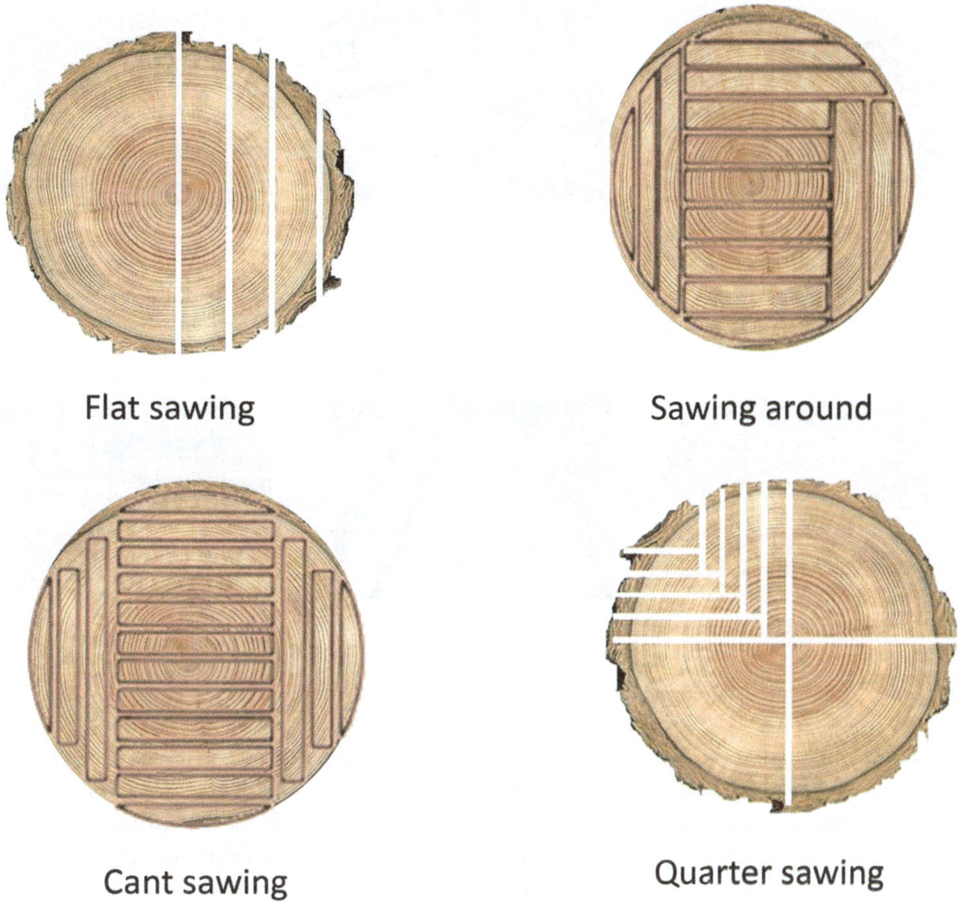

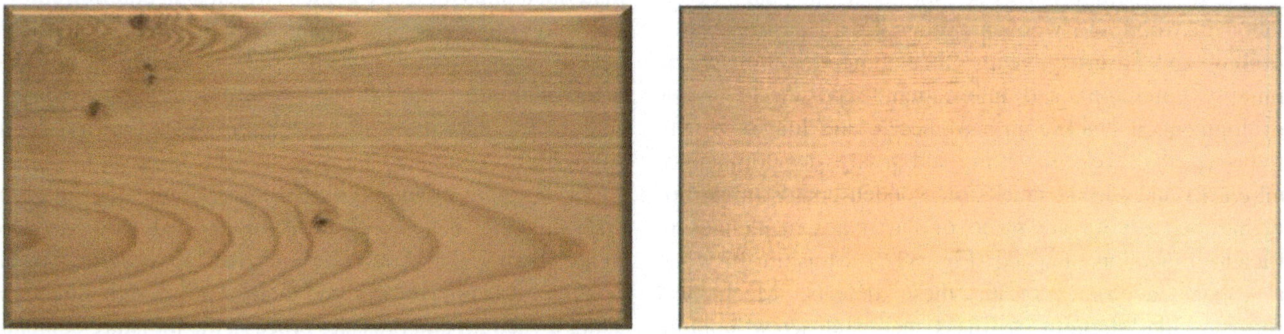

Fig. 38.6 Sawing figure in flat-sawn and quarter-sawn boards

sawn timber along with logs characterization, and standardization in sawing, grading and sorting are some of the basic elements of a highly efficient mechanized sawmill (Fig. 38.7).

38.3 Wood Seasoning

Wood seasoning is one of the most vital processing steps for increasing the longevity of wood mainly sawn wood and its products. Once logs are sawn and planks of different sizes are obtained, the sawn material should be seasoned before being put to use. The process is often overlooked by many wood-based industries which ultimately affects the quality and performance of the final product. Moisture/water is necessary for tree growth and is one of the major factors contributing significantly to the growth of trees. Water is transported from roots to leaves through wood as the conducting channel. Therefore, wood from a freshly cut tree will have a substantial amount of moisture which may vary from 50 to 200% depending on species, location and season. Within a tree also, the moisture content (MC) of wood may vary between

Fig. 38.7 A fully mechanized and efficient sawmill

sapwood and heartwood. Fast-grown trees tend to have high moisture and in many species, the amount of water may be more than the amount of wood (on a weight basis).

Wood is among the most severely affected materials at elevated moisture content when used in sawn form. Microbial growth and decomposition of wood above a certain moisture content; discolouration of wood; shrinkage and shape deformation in wood and wooden products with moisture loss; swelling and shrinkage with fluctuations in atmospheric humidity conditions and intermittent exposure to water; development of cracks, surface checks, and fungal growth on the repeated cycle of wetting and drying; development of end-cracks and surface cracks on wooden boards in case of uncontrolled drying are some of the major challenges in efficient utilization of wood (Fig. 38.8). Moisture in wood is responsible for most of these defects. Mechanical properties like strength, stiffness, nail and screw holding capacity are adversely affected by elevated moisture content in wood. Moisture is the main element for mould and fungus growth on wood. Moulds impart discolouration to the wood surface as yellow, green, orange, black or an array of other colours due to the production of pigmented spores. However, fungus growth leads to structural disintegration of wood. Mould and fungal growth cannot take place on the wood below 20% MC.

To avoid moisture-associated defects in wood as shown in Fig. 38.8, it is essential to season wood. If left unattended, wood tends to lose moisture on its own over a period and reaches a level that is in equilibrium with the immediate environment in which wood is placed. However, during this process, it also tends to develop defects which can affect the recovery of good quality sawn timber. If the seasoning process is not properly carried out, defects generated during the moisture loss can devalue the timber.

Wood seasoning is primarily a process of removing moisture from wood in a controlled manner such that the resulting material attains moisture content which is in equilibrium with the environmental conditions without any undesirable defects. In addition to reducing defects, seasoning also adds a lot of value to the timber in terms of high strength as seasoned timber typically has more than twofold higher strength as compared to wet timber, improved nail and screw holding strengths, dimensional stability, easy gluing and finishing, easy to handle as seasoned wood is much lighter than wet wood, better electrical and thermal insulation, etc. It is also to be noted that for preservative treatment using a vacuum and pressure process, the moisture content of wood needs to be reduced to about 30% (moisture content at fibre saturation point) to allow chemicals to penetrate inside the wood. Considering the numerous advantages the seasoning process offers, it should be mandatorily implemented by the wood-based industries.

However, understanding the fundamental concepts involved in wood seasoning is very important for economical and efficient seasoning. The knowledge of these concepts allows controlling the seasoning parameters desired for specific situations. While wood seasoning, it is very critical to know what level of moisture content and quality in seasoned wood is desired. The seasoning process needs to ensure that the seasoned wood has correct moisture content; planks

Fig. 38.8 Moisture associated defects in wood

within a stack have small variations in moisture content; within a plank moisture content is uniform; seasoned wood is free from surface checks, warping, end-splitting, fungal or chemical stains and there is no colour change. The possible seasoning defects are shown in Fig. 38.9.

The following section deals with the basic principles involved in the wood seasoning process which need to be understood for an economical and efficient process.

38.3.1 Moisture Content

Anatomically, wood is a porous and permeable material. The porous structure allows spaces for air and water inside wood. Additionally, the chemical constituents of wood mainly cellulose, hemicelluloses and lignin are also hygroscopic with abundant hydroxyl groups which have the ability to chemically bond water molecules. The moisture content is defined as the amount of moisture in wood with respect to the dry content of wood and mathematically, it is represented as,

$$\text{Moisture content}(\%) = \frac{\text{Mass in moist condition} - \text{Mass in oven dried condition}}{\text{Mass in oven dried condition}}$$
$$= \times 100$$

38.3.2 Fibre Saturation Point

In wood from freshly felled trees, moisture is present in liquid form within cell cavities (called 'lumen') and also is held within cell wall pores and chemically bonded with the cell wall polymers. The greater part of moisture is present in the cell cavity also called as lumen and is theoretically called 'free water'. However, in a true sense, it is not completely free water as it is subjected to capillary forces. Moisture closely associated with cell wall polymers is called 'bound water'. In general, 25–35% moisture is bound water chemically attached with cell wall polymers and the rest is present in the liquid form. When wood is exposed to the drying process, initially free water evaporates and thereafter bound

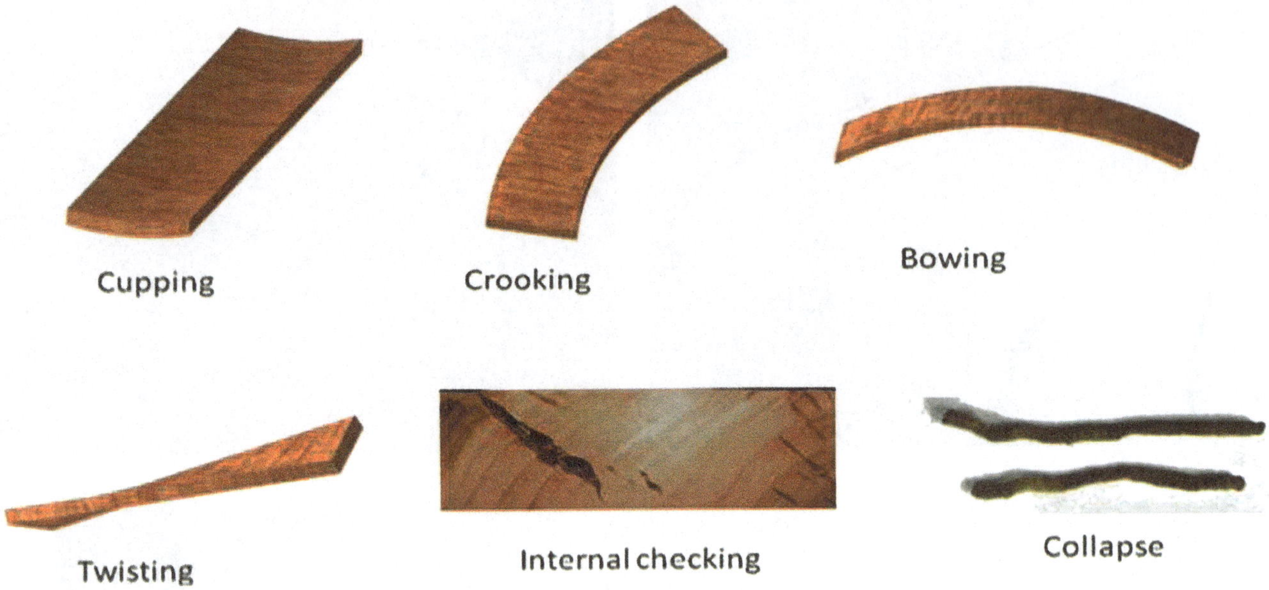

Fig. 38.9 Defects in wood during drying/seasoning

Fig. 38.10 Schematic representation of free water, bound water and FSP

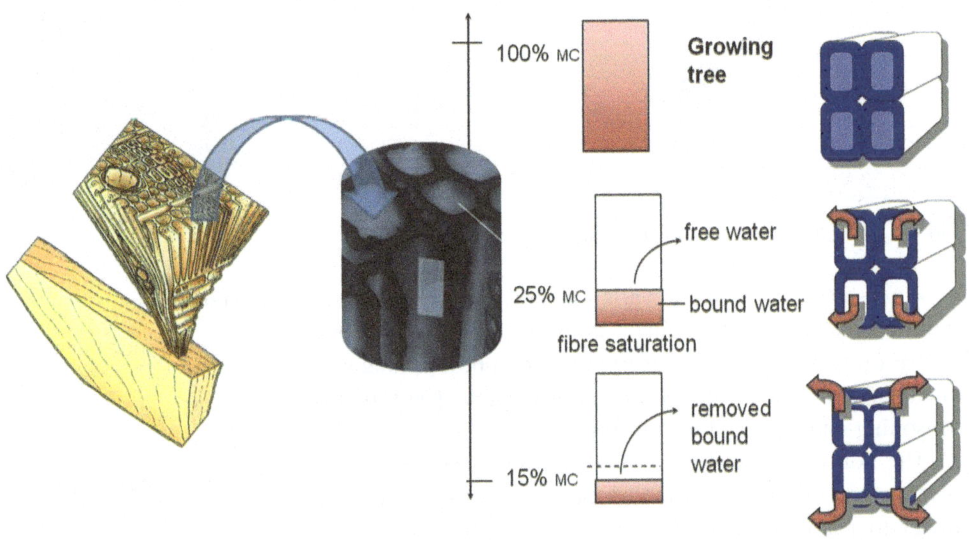

water. The point at which all the water present in liquid form in the cell lumen (free water) is evaporated while cell walls remain saturated with chemically bonded water (bound water) is known as the fibre saturation point (FSP). FSP is very important in the utilization of wood and controlling moisture-related degrades in wood. No dimensional movement takes place in wood until it dries to FSP. When wood continues to dry below FSP, dimensional deformation starts with shrinkage in all directions in the wood. The shrinkage is due to the loss of moisture from the cell walls of wood elements (fibres, vessels, etc.). The concept of FSP has been illustrated in Fig. 38.10.

All dimensional changes (either swelling or shrinkage) related deformation in wood take place with fluctuations in moisture content below FSP. On the other side, strength properties increase with decreasing moisture below FSP. FSP varies between species, denser wood generally has lower FSP as compared to lighter or low-density wood. Understanding of FSP is also necessary for wood preservation as chemical impregnation would be most effective when treatment is given to wood as FSP.

38.3.3 Equilibrium Moisture Content

Wood loses water when exposed to dry conditions and gains in wet conditions. Therefore, the moisture content of a wooden piece is seldom constant. Fluctuations in wood

Table 38.1 Equilibrium moisture content of some Indian wood species at two RH (93% and 33% and 35 °C)

Species	EMC at 93% RH	EMC at 33% RH
Albizia chinensis	22.0	4.1
Gmelina arborea	20.1	4.2
Hevea brasiliensis	19.4	4.4
Lannea coromandelica	25.0	5.1
Mangifera indica	19.6	5.1
Tectona grandis	16.0	4.6
Shorea robusta	20.0	5.6

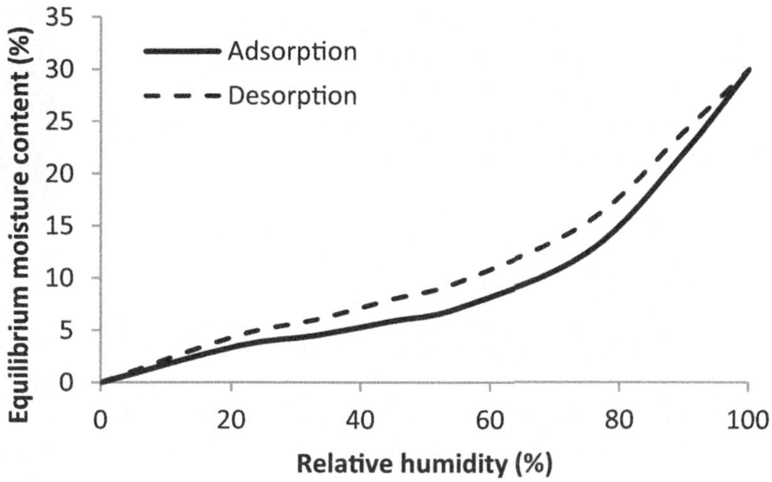

Fig. 38.11 Moisture sorption curves for wood

moisture content depend on the variation in relative humidity (RH) and temperature of the atmosphere. When exposed to a constant temperature and relative humidity, wood attains a stable moisture content which will be in equilibrium with the moisture present in the environment. The moisture content of wood in such conditions is defined as 'Equilibrium Moisture Content (EMC)'. Different wood species exhibit different EMC at a specific atmospheric temperature and relative humidity depending on the presence of extractives, anatomical structure, stress condition, etc. The information on the EMC of species is very important in deciding the final moisture content during seasoning operations depending on the product and its final destination. In general, EMC for a species is inversely proportional to both temperature and relative humidity. EMC of a few Indian timbers is given in Table 38.1 (Chauhan, unpublished).

The relationship between EMC and relative humidity at a given temperature is known as sorption isotherm (Walker 2006). When the moist wood is subjected to drying from high relative humidity to low relative humidity below FSP at a constant temperature, a change in EMC with respect to relative humidity is known as desorption. Subsequently, when the dried wood is exposed from low humidity to high humidity, moisture adsorption takes place where water molecules get attached to the cell wall elements of the wood, and the wood moisture content increases. Generally, moisture adsorption behaviour follows a typical sigmoid curve (Fig. 38.11). However, at a specific relative humidity condition and temperature, the EMC is generally found to be higher during desorption compared to adsorption when the wood is subjected to adsorption-desorption cycle resulting in a hysteresis effect (Skaar 1988).

The relative difference in EMC during adsorption and desorption is mainly attributed to incomplete rehydration of sorption sites and the effect of compressive stresses during swelling (Siau 1984). A similar hysteresis effect is also observed in dimensional changes in many tropical hardwood species (Chauhan and Aggarwal 2004).

Several theories have been proposed to describe the moisture adsorption behaviour in wood. Among these, the surface sorption theory proposed by Brunauer et al. (1938) popularly known as Brunauer-Emmett-Teller (BET) theory and the solution theory proposed by Hailwood and Horrobin (1946) are the most extensively studied and popular theories explaining the moisture sorption phenomenon in wood. BET theory is based on the principle of layers of water molecules that can be adsorbed on the internal surface of cell-wall elements of wood. According to theory, the cell-wall surface and water molecule bonding take place in the initial stages forming a monolayer of water molecules on the adsorbate substrate. Subsequently, bonding between water molecules takes place forming multi-layers of adsorbent. The mathematical derivation of BET theory is as follows (Skaar 1988):

$$\frac{h}{M(1-h)} = \frac{1}{M_m C} + \frac{(C-1)h}{M_m C},$$

where 'h' is the relative vapour pressure, 'M' is the moisture content, 'M_m' is the moisture content of the wood corresponding to the complete monolayer adsorption and 'C' is the constant.

Hailwood–Horrobin theory assumes that the adsorbed water can be divided into two specific components. One component is the water of wood hydration that is similar to chemically adsorbed water and the second component is the dissolved water within the wood material (Skaar 1988). The theory provides a reasonable explanation of the water sorption mechanism for wood. The mathematical derivation of the theory is as follows:

$$M = \frac{1800K_1K_2H}{W(100 + K_1K_2H)} + \frac{1800K_2H}{W\,(100 - K_2H)},$$

where 'M' is percentage of moisture content, H is relative humidity in percent and K_1, K_2, and W are constants. K_1 is the equilibrium constant when hydrate is formed from dissolved water and dry wood, K_2 is the equilibrium constant between dissolved water and external vapour pressure and W is the molecular weight of the substance required to associate one gram mole of water (Chauhan et al. 2001). The first term in the equation represents the water of hydration and the second term represents the dissolved water in wood (Skaar 1988).

38.3.4 Anisotropic Shrinkage

Being a hygroscopic material, wood swells and shrinks with fluctuations in relative humidity and temperature of the surroundings. The swelling and shrinkage phenomenon in wooden products is known as movement in timber. This movement is mainly due to the interaction of moisture within cell walls which internally causes dimensional deformation in wood tissue (Naderi and Hernandez 1997). Proper understanding and knowledge of the extent of shrinkage-swelling in wood is of paramount importance in designing wooden structures and during the processing of timber. Bowing and twisting in door and window frames, nonconformist doors and windows during different seasons slack joints and fittings, shape deformation and development of cracks in wooden articles are some of the commonly observed defects in wooden structure due to dimensional movement. These defects are manifestations of anisotropic shrinkage in wood due to the intrinsic orientation of cells. The three principal directions, longitudinal, radial and tangential (Fig. 38.12), swell and shrink with different magnitudes. Dimensional movement is least in the parallel-to-grain direction and highest in the tangential direction (Walker 2006). Generally, mature wood of most of the hardwood species shrinks about 0.1–0.3% along the longitudinal direction, 2–6% along the radial direction and about 3–10% along the tangential direction. The ratio of tangential and radial shrinkage is also known as the coefficient of anisotropy. A high value of this ratio indicates a high probability of developing deformation in wood during drying. Generally, tangential shrinkage is typically 1.5–2.5 times higher than the radial shrinkage. Some species like eucalyptus exhibit excessive shrinkage and also high coefficient of anisotropy.

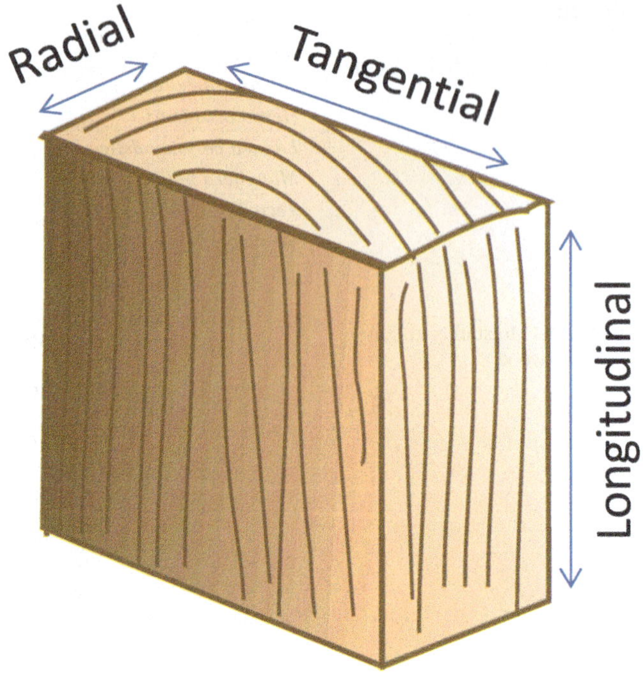

Fig. 38.12 Three principal directions of wood

The shrinkage/swelling anisotropy and the dimensional movement in wood are complex issues and several theories have been postulated to explain the anisotropy. Barber and Meylan (1964) proposed a theoretical model to explain the dimensional movement based on the orientation of microfibrils (crystalline cellulose) and matrix (hemicelluloses and lignin) interaction in the cell walls of softwoods. Differences in density in earlywood and latewood, which are prominent in softwood species, have been attributed to the reason for differences in radial and tangential shrinkage by Pentoney (1953). The restraining effect of ray tissues is also attributed to anisotropic shrinkage in the tangential and radial directions (Harda and Cote 1985). Factors like elastic anisotropy and the presence of extractives have also been considered to contribute significantly in explaining the shrinkage behaviour of wood (Walker 2006). Further, factors like the slope of grain, knots, reaction wood and faulty sawing of timber are known to add to the complexity of movement-related effects in timber. During the seasoning process, all these factors need to be considered to control the dimensional deformities.

38.3.5 Moisture Movement During Wood Drying

Moisture movement in wood involves complex mechanisms of heat and mass transfer. There are two primary steps involved in moisture loss from wood during the drying process (Fig. 38.13)

- Evaporation of moisture from the surface
- Moisture movement within the wood from the inner core to the surface by mass flow and/or diffusion

Both these mechanisms act concurrently and for efficient drying, it is essential to maintain equilibrium in the two processes. Surface evaporation controls the initial rate of drying and mainly depends on the temperature and relative humidity of the surroundings. The moisture movement from the core to the surface can take place through the capillary movement of water from one fibre to another when fibres are filled with free water and pathways are available for capillary movement. This is the prominent mechanism in highly permeable and low-density wood particularly softwoods. The other mechanism involves the diffusion of water vapours through cell walls and the diffusion of bound water from cell walls below FSP. In hardwood species, the diffusion process is the dominating mechanism during drying. In ideal drying conditions, the rate of evaporation must be broadly equal to the rate of moisture movement from the inner core to the surface. Rapid evaporation can result in steep moisture gradients and excessive drying stresses within the wood. Understanding the drying mechanism is of utmost importance in deciding the drying parameters. When the moisture movement rate from the inner region to the surface lags the surface evaporation rate, the surface region dries quickly and the dried surface becomes a thermal insulator reducing the drying rate. As the drying process progresses, the diffusion process becomes dominant with the reduced moisture content, and higher temperatures are required to increase the diffusion rate.

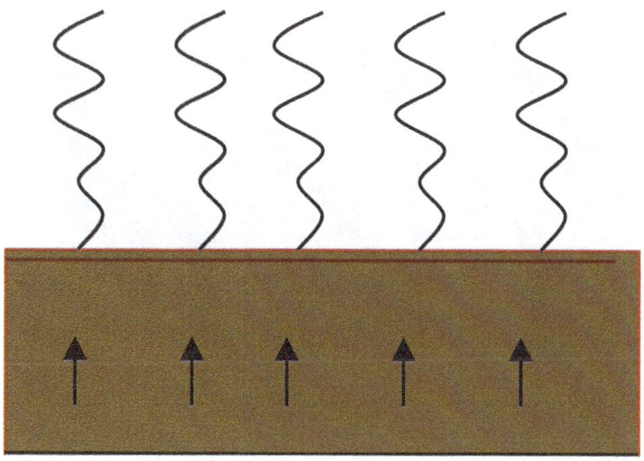

Fig. 38.13 Process of moisture movement in wood

38.3.6 Basic Elements for Wood Seasoning

For any moisture removal process from any substrate temperature of the surroundings, relative humidity and airflow are the most critical factors determining the moisture removal rate. Thermal energy (heat) is required to evaporate moisture from the surface as well as accelerate the moisture diffusion process in the inner core of the wood. Evaporation depends on the heat transfer rate to the wood surface. Proper control of temperature is essential to balance evaporation from the surface and moisture movement in the inner core for quality seasoning. Seasoning at a higher temperature in the early stages of drying can lead to rapid moisture evaporation from the wood surface resulting in a dried surface and moist core region characterized by a steep moisture gradient within wood. The rapid surface drying causes poor heat transfer from air to wood surface leading to reduced moisture loss rates and increased risks of drying stresses. When the surface temperature approaches the ambient temperature, there can be no heat transfer and the drying process can stop. However, low temperature affects the seasoning efficiency and increases the risk of mould and stain growth resulting in discolouration. The amount of energy required to remove bound water far exceeds the energy required to remove free water. Therefore, the energy requirement for removing moisture increases as the moisture content of wood decreases below the fibre saturation point.

Proper control of relative humidity is important in the efficient seasoning of wood. Low relative humidity at the start of the seasoning process may lead to excessively faster drying inducing drying stresses in wood. Airflow within the stack is equally important as temperature and RH in controlling the seasoning process. Proper control of the airflow in terms of air passage and velocity is very much essential during drying at higher and intermediate moisture content for defect-free seasoning. Airflow acts as a carrier for heat through the wood and also absorbs the evaporated moisture thus playing a dual role in efficient seasoning. Inadequate air flow (low air velocity) can result in excessively high humidity inside the chamber causing slow drying. This may lead to warping, discolouration and mould growth in some species. On the contrary, higher air velocities at the initial seasoning stages (when the wood moisture content is high) can result in excessive drying rates and drying degrades. At lower moisture content (<40%) air velocity does not make any significant impact in controlling the drying rate. It is advisable to make sure that the air is moving equally through the wood stack, otherwise uneven drying can occur.

The combination of temperature, relative humidity and air velocity determines the moisture loss or drying rate. These factors can be adjusted optimally to regulate the seasoning process. While fixing these parameters due consideration should be given to the species to be dried, initial moisture content and thickness of the timber. Different species dry at a different rate as the drying behaviour of species depends on the anatomical structure like pit aspiration, the organizational structure of capillary systems (fibres/vessels/rays) of wood and the presence of extractives. A large number of Indian timber species (>200) have been studied for their seasoning behaviour and have been classified into the following three broad categories:

38.3.6.1 Highly Refractory Timbers

These species are very difficult to season without any cracks and splitting and therefore seasoned very slowly. Sal (*Shorea robusta*), Eucalyptus (*Eucalyptus* spp.), *Khair (Acacia catechu),* axlewood (*Anogeissus latifolia*) and Jamun (*Syzygium* spp.) are examples of highly refractory timber species.

38.3.6.2 Moderately Refractory Timbers

These species show a moderate tendency to crack and split during seasoning and therefore should be seasoned with some care. However, species of this category can be seasoned with reasonably fast drying conditions without many seasoning defects. Most of the furniture grade timber species like Sissoo (*Dalbergia sissoo*) and Teak (*Tectona grandis*) are classified in this category.

38.3.6.3 Non-refractory Timbers

These timber species can be dried rapidly without any defects. Packing case timbers such as semul (*Bombax ceiba*), kadam (*Anthocephalus chinensis*), mango (*Mangifera indica*), rubber wood (*Hevea brasiliensis*) and deodar (*Cedrus deodara*) are some of the species in this category.

As a prerequisite of seasoning, sawn timber needs to be stacked properly. For stacking, well-seasoned heartwood stickers of uniform thickness are placed across the timber layer for uniform air circulation (Fig. 38.14). The stickers should be accurately aligned above one another in the vertical direction to prevent bending under an eccentric load and allow uniform drying. The distance between stickers has to be adjusted based on the species and its thickness. Too close spacing of stickers will hamper the free flow of air resulting in slow drying whereas too far spacing may lead to warping in timber. Generally, spacing between the stickers is kept about 60–75 cm. For the seasoning of most of the hardwood species, 19–25 mm thick stickers are used in commercial operations. The length of the stickers should be matching with the width of the stack. The schematic arrangement of planks and stickers is shown in Fig. 38.14 showing front (left) and side (right) views of the stack. For sawn timber with variable thickness, the thicker and heavier sections should be kept at the top. Two consecutive stacks should be separated by enough gaps between the stacks to allow proper airflow. The airflow can be controlled by adjusting the height of the stack and sticker thickness and the design of a shed.

38.3.7 Wood Seasoning Methods

Wood seasoning is an art coupled with scientific principles. Quality seasoning of timber to obtain optimum recovery of high-grade material is one of the most important challenges

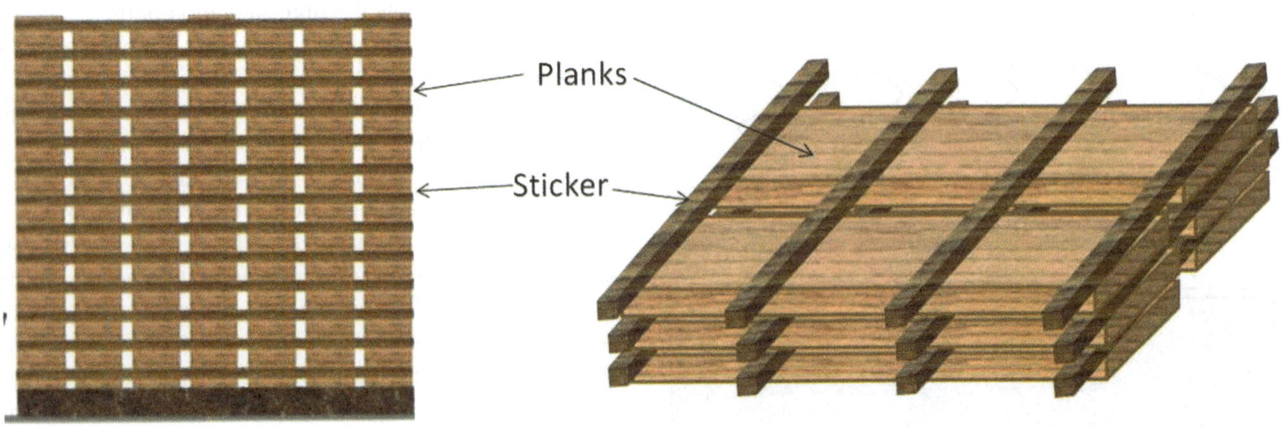

Fig. 38.14 Stacking of timber for seasoning

associated with the processing of wood. It is about going fast but without losing the focus of getting degrade-free material with appropriate moisture content levels. Wood seasoning is an energy intensive operation in wood product manufacturing. It is estimated that about 60–70% energy of the total energy required for timber processing is consumed in seasoning operations. This can have a large bearing on the seasoning cost. The major economic drivers for wood seasoning are seasoning time, energy requirement and extent of degrade in the seasoned wood. Both time and energy requirements have a direct bearing on the cost of seasoning, whereas degrades are linked with the quality of the material. Degrades development during the seasoning process can have significant implications on the quality of the final product and its economic value. Any changes made in the seasoning process to reduce energy consumption and the seasoning time can have a significant impact on the cost of production as a whole. Considering the cost factor in seasoning operations, many times industry overlooks the proper seasoning practices and timber is dried at a faster rate with little consideration to degrade and quality of the dried wood. Historically, air seasoning was extensively practiced by the woodworkers and carpenters. With the advancement in technologies, there have been continuous developments in wood seasoning processes and methodologies. This section discusses major wood seasoning methods.

38.3.7.1 Air Seasoning

Air seasoning depends on the natural process of drying of wood but with certain codes of practices like proper stacking of wood, having effective ventilation for efficient airflow within the stacked timber, protection from direct exposure to sunlight and rain during the seasoning process and prevention of faster drying from the ends (along the grains) of planks. Not long ago, it was one of the most popular methods of wood seasoning and even now it is preferred by many for seasoning high-density hardwood species. The combination of atmospheric temperature, relative humidity and air velocity are the determining factors for the seasoning rate.

A shed with only a roof supported over pillars and open from all sides is the simplest shed providing protection from direct sunlight and rain (Fig. 38.15). However, this type of shed is useful for easy-to-dry timbers like rubber wood, mango wood and semul. Sheds with three side walls (north side open) or a closed shed with a roof and all side walls with proper ventilation at the top and bottom on two side walls are also used for air seasoning.

Capital investment and operating costs for air seasoning in a shed are very low and the drying defects are also generally less provided precautionary measures are taken care of. However, one of the major limitations of air seasoning is that none of the factors influencing drying, that is, temperature, relative humidity and air velocity, are controlled and

Fig. 38.15 Air seasoning of wood

the overall seasoning process is very time-consuming. Air seasoning of 25 mm thick planks of most of the timbers having moderate-to-high density may take up to 2–3 months to season in moderate climatic conditions. Thicker cross-sections (3–4 inches which are generally used for door and window frames) may take anywhere between 6 and 8 months for proper seasoning. This makes it less attractive and uneconomical for industrial-scale operations for mass production.

To fasten the drying process, forced air drying is also adopted in some cases where fans are used to circulate air with a speed of 1–2 m/s to quicken the drying of timber (Fig. 38.16). The forced circulation of air through the stack is effective, particularly in accelerating moisture removal from green condition to around FSP. Below FSP, forced air drying is not very effective (Chauhan et al. 2006). The operational cost of the fans is mainly compensated by the benefit of faster drying.

38.3.7.2 Kiln Seasoning

Commercially, kiln seasoning is practiced extensively by most of the wood processing industries to achieve faster, cheaper and degrade-free timber. In kiln seasoning, timber is dried in a closed chamber in a specified manner by controlling temperature, relative humidity and airflow to get the desired moisture content. A schematic representation of a conventional steam-heated kiln is shown in Fig. 38.17.

The conventional kiln chamber is fitted with reversible fans of the desired capacity depending on the chamber size and heating coils above false ceiling/baffles to circulate warm air inside the chamber. The relative humidity inside the kiln chamber is maintained by operating (opening and closing) the vents available on top or bottom of the chamber. The heat

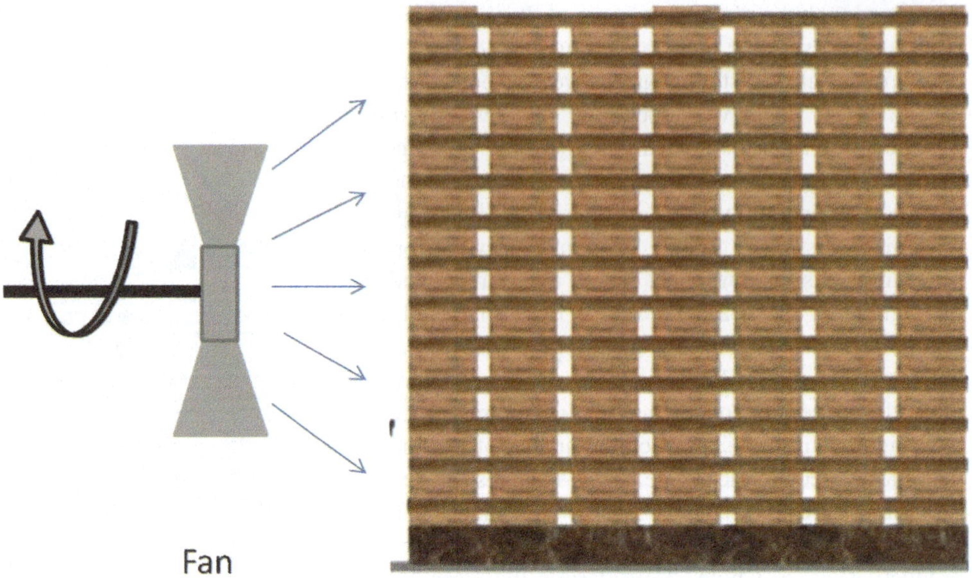

Fig. 38.16 Forced air drying of timber

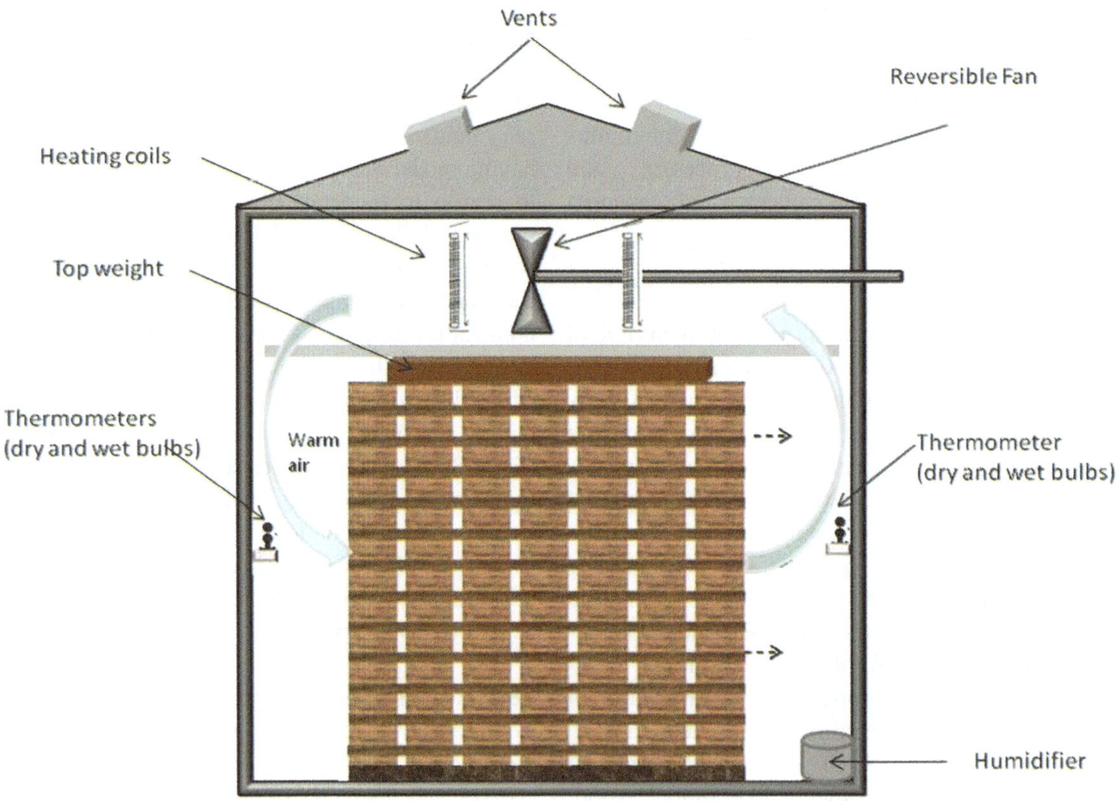

Fig. 38.17 Schematic diagram of wood seasoning kiln

source can be steam, electrical heating, oil heating, gasifier-based heating, solar heating or any other system capable of raising the air temperature inside the chamber to the desired level. The capacity of the required furnace for thermal oil heating or boiler for steam generation would depend on the size of the kiln and the number of chambers.

The timber is pre-stacked as per the procedure described earlier and loaded in the chamber. There should be about a 50–100 mm gap between the false ceiling and the stack. With the heating on, relative humidity builds up inside the chamber with the evaporation of moisture from the wood. Uniform airflow through the stack requires rectangular stacking and using the proper-sized stickers. In practice, 40% of the air by-passes and in an uneven stacking 100% of air can bypass affecting drying efficiencies adversely. During the drying operations, the direction of dry air is regularly reversed to get uniform drying throughout the stack. In some kiln designs, side-mounted fans are used instead of overhead fans. In conventional kilns, drying temperature ranges between 38 and 68 °C, and air velocity through the stack is generally kept between 0.6 and 1.5 m/s. Lower velocities are preferred for refractory species. Relative humidity inside the chamber is monitored either using a dry bulb-wet bulb system or using a humidity controller. Dry bulb and wet bulb systems have been traditionally used in most of the kilns. The depression between dry-bulb and wet-bulb temperatures indicates the relative humidity inside the chamber.

In kiln seasoning, initially, the temperature inside the chamber is kept low (about 38–40 °C) and high relative humidity is maintained to avoid rapid moisture loss from the surface and to maintain equilibrium between surface evaporation and moisture movement from core to surface. Moisture loss from wood depends on the heat transfer from hot air to wood surface. Heat transfer per unit surface is defined as

$$Q = K_1(T_{\mathrm{db}} - T_{\mathrm{wb}}),$$

where K_1 is the heat transfer coefficient, T_{db} is the dry bulb temperature and T_{wb} is the wet bulb temperature. A high temperature at the initial stages of drying may develop surface cracks, end splits and reduced drying efficiencies. As surface temperature approaches the dry bulb temperature there can be no heat transfer and the drying eventually stops. Therefore, dry bulb temperature is increased and relative humidity is decreased gradually based on the moisture content of the stack. For kiln seasoning, specific steps of temperature and relative humidity conditions are optimized for a specific timber species based on its refractoriness. These optimized conditions are called kiln schedules. Seven such kiln schedules have been recommended by the Bureau of Indian Standards (IS 1141-1993). For the satisfactory seasoning of timbers, these schedules have to be followed during kiln drying. The schedules differ for different species. The seasoning schedules have been derived based on the ease of seasoning for a particular species. For refractory timber species, the initial drying temperature is kept at 38–40 °C and raised to 55 °C for the final drying from 15 to 12% MC. In between two to three steaming operations are also carried out to avoid the development of drying defects. Whereas for easy-to-dry species, drying is initiated at 50–52 °C, and the temperature is raised to 68–70 °C in the final stages of drying. During kiln seasoning, it is recommended not to mix species with different seasoning characteristics and also planks of varying thickness.

38.3.7.3 Solar Seasoning Kiln

A solar kiln is an attractive option for seasoning timber with a substantially accelerated rate as compared to air seasoning and at a lower cost as compared to steam/electrically heated kilns. The process is also one of the most energy efficient ways of drying wood operated during daylight hours and effectively be used as a pre-dryer for slow-drying timber species. The selection of the site and direction of the kiln is very critical for efficient operation and harnessing the maximum solar energy for the kiln. Solar kilns are particularly useful at locations where sufficient solar energy can be harvested and are more suitable for small-scale operations. Many designs are available for fabricating a solar kiln; however, the greenhouse type of design is very popular and easy to fabricate. The schematic diagram of a solar seasoning kiln is shown in Fig. 38.18.

A simple design of a solar seasoning kiln consists of a main structure with a timber/metal frame, double sheathed roof and all side walls (except the north side wall) having clear transparent double-layered glass with wide intervening air-gap. The kiln is generally placed east-west direction along its length. The kiln roof is generally slanted towards the south at an angle to the horizontal which is recommended to be 0.9 times the latitude for maximum absorption of solar energy throughout the year (Pandey and Jain 1992). Corrugated galvanized iron sheets are used for false ceilings which are blackened to absorb maximum solar radiation and increase the temperature inside the kiln. The inside structure of the kiln is also painted black to increase energy efficiency. Electrically operated fans of appropriate dimensions and airflow capacity are mounted on a partition wall on the back side (North side wall) of the kiln for uniform air circulation. The kiln chamber generally has double-panelled glass sheets, wooden framed doors for charging timber inside the kiln and an inspection door located on either the eastern or western side of the kiln. The entire structure fittings are mostly heat-resistant gaskets and have fastening arrangements for preventing heat loss. Enough space is provided on the north wall for air flow movement and maintenance of the system. The kilns can also be equipped with energy storage devices to make them functional during the night times and efficiencies can be improved.

38.3.7.4 Dehumidification Drying

Dehumidification-based drying is a popular technology for drying temperature-sensitive products and is also used for drying wood. Dehumidification-based wood drying systems are considered to be energy efficient and dry wood without many defects. Furthermore, the time required for drying in such systems is at par with steam kilns. There are two types of dehumidification kilns. One is based on refrigerant type (heat pump-based dehumidification) and the other is based on the desiccant principle. Heat pump-based dryers have been in

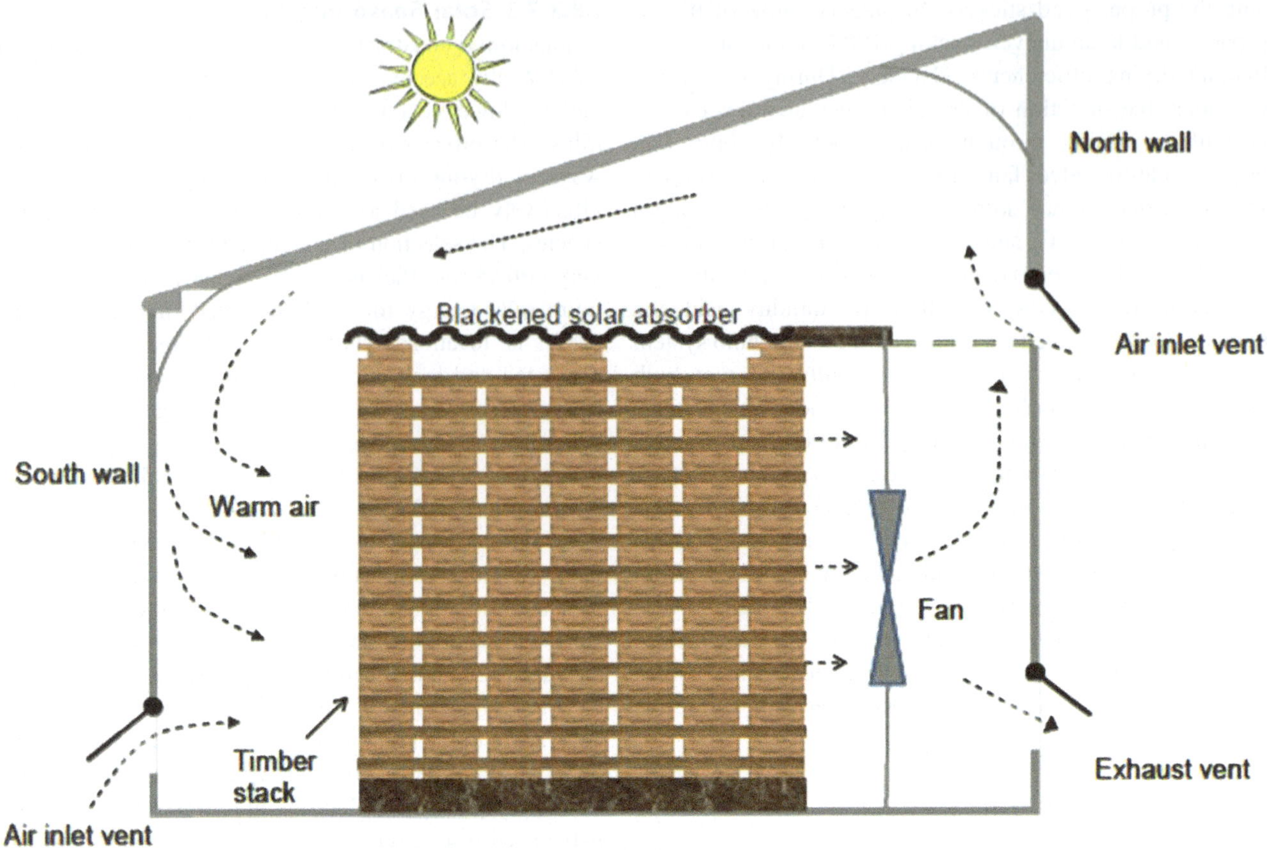

Fig. 38.18 Schematic diagram of a greenhouse-type solar seasoning kiln

industrial applications for the last six decades, particularly in food and timber drying industries. In this type of kiln, water is condensed by passing the moist air through cold coils within a closed structure. The system consists of an evaporator and a condenser both placed inside the chamber. As moisture from the timber is evaporated by the heated circulating air, a part of the warm moist air is continuously sucked through the dehumidifier, and the condensed water is drained out. When water vapours are condensed, the latent heat of condensation is used to heat the air within the kiln. The dehumidification-based drying is considered to be energy efficient as venting energy losses are controlled in such systems and also latent heat is used to maintain the air temperature by condensing moisture from the moist air.

The other type of dehumidifier kiln system works on the chemical dehumidification principle wherein the moisture is trapped by a fluted flat-bed type desiccant dehumidifier (Chauhan and Sethy 2008). The drying system is a self-contained well-insulated chamber, very easy to operate and requires relatively less attention during the drying operations. The desiccant-based dehumidifier wood dryer requires an auxiliary heat source to maintain the temperature, a heat source to reactivate desiccant and energy for maintaining airflow in the chamber. The reactivation of desiccant chemicals is needed to keep it effective for continuous moisture adsorption and therefore consume a significant amount of energy. The efficiency of such kiln mainly depends on total capacity usage during the operation. The cost of drying is directly derived from the electrical energy consumption in reactivating the desiccant and maintaining the temperature inside the chamber. When not used to full capacity, the essential energy consumption by the desiccant and auxiliary heaters makes such drying uneconomical. More so dehumidification drying is not very efficient at lower moisture content (below 20%) as the energy requirements become too high since they operate purely on electricity rather than cheap sources of energy like boilers in conventional kilns.

There are certain misconceptions about the dehumidification drying process particularly being very slow as it operates at low temperatures (~50 °C), uses electricity for entire operations, has poor efficiencies, etc. In general, such kilns are not appropriate for timbers that are dried at high temperatures like pines, rubber wood and semul. These kilns are very suitable to season refractory timber species which require milder drying conditions. The design of the kiln plays a very critical role in dehumidification drying particularly for heat pump-based kilns. These kilns require far superior sealing for doors and vents, better insulation and proper sizing of the equipment (heat pump/condenser/desiccant chamber).

38.3.7.5 Vacuum Drying

The basic principle of vacuum drying is based on the low boiling temperature of water in a vacuum. The boiling point of water at 760 mmHg (one atmospheric pressure) is 100 °C which drops to 88.3 °C at 500 mmHg. The timber is stacked inside a vessel, a vacuum is applied and the material is heated allowing rapid drying at lower temperatures than conventional seasoning. Efficient heat transfer to the wood surface is a major challenge in vacuum drying. Vacuum dryers are classified based on heat transfer mechanisms like conductive heat transfer (hot plate vacuum drying), convection heat transfer (superheated steam, cyclic vacuum application) or radio frequency heating. Espinoza and Bond (2016) have provided an extensive review of the different methods of vacuum drying. Although the process was developed almost a century ago, it is still a niche player in commercial wood seasoning limited to specialty applications. In the current scenario, the technology can be appealing for small-scale operations as small dimension vacuum chambers are more efficient. A well-designed and properly operated vacuum system with an efficient heat transfer mechanism can result in faster and quality drying. The drying time in vacuum kilns is substantially lower than the time required in conventional or dehumidification kilns. However, significantly higher capital and operating costs (three to four times) as compared to conventional kilns are the major hindrances in the large-scale commercialization of the technology.

38.3.7.6 Dielectric drying

Dielectric heating has received significant attention in the recent past for drying agricultural produce, disinfestations of grains, food processing, etc. due to its higher efficiency, easy control and environmentally friendly nature. Dielectric heating includes both radio frequency (RF) waves (frequency range 1–100 MHz) and microwaves (above 300 MHz). Moist wood being a dielectric material when exposed to an electric field, the polar molecules in the wood begin to oscillate depending on the frequency. These oscillations give rise to heating due to friction from the oscillating charges. Since energy is transferred by the interaction between the high-frequency electric field and the responsive components, mainly moisture, in wood, heat is generated instantaneously throughout the volume of the wood material, if moisture is uniformly distributed inside. Because of the volumetric heating, there is a lower probability of developing case hardening and other surface damage in this process, which is a major concern in conventional kiln drying of timbers.

The technology has garnered commercial acceptance in many developed countries. The basic research of the application of high-frequency electro-magnetic waves for drying started way back in the 1940s but high energy costs and technological limitations had constrained the commercial acceptance of the technology. However, in the late 1980s, interest was renewed in dielectric heating of wood with the advancement in technology and since then significant research has been carried out in the developed world and RF drying has received commercial success (mainly in Canada, the US, NZ). In the 1990s, RF heating in combination with vacuum received a lot of commercial attention from the wood industry in the USA. A radio frequency-vacuum dryer was designed with a capacity of 20 m^3 and commissioned at Norton Smith Lumber Co. in 1996 to dry red oak furniture parts (Smith et al. 1996). The technology has been reported to be very effective in drying large dimension timber particularly thicker sections which are difficult to season or dry in conventional seasoning kilns and if done carefully, satisfactory seasoning requires an enormous amount of time and energy. RF dryers of different designs with varying frequencies (ranging from 4 to 40 MHz) and different power ratings (10–300 kW) have been developed in many countries depending upon the specific requirements. RF drying was found to be very effective in drying softwood as well as hardwood species without developing any significant drying degrades (Avramidis and Zwick 1996; Resch and Hansmann 2002) Dielectric heating has advantages in terms of instant heating and very high heating efficiency (60–70%) as compared to heating transfer efficiencies in conventional steam heated kilns which is about 10% (Jiao 2012; Jiao et al. 2015). In addition, no chemical stains or residues on seasoned timber products, high drying quality and little impact on the environment make the technology more appealing (Mitcham et al. 2004).

Microwaves have also been used to dry wood but mainly timber with smaller thicknesses (usually about 25–50 mm thick) as the high frequencies of microwaves are associated with shorter wavelengths and therefore lower penetration depths. Microwave heating at 2.45 GHz is the most practiced process and is found to be efficient mainly for treating smaller and thinner wood sections (Hassan et al. 2015). Aggarwal and Chauhan (2013) also reported faster drying of silver oak boards with a 13 g/min water loss rate with 2 kW microwave power intensity and the initial moisture content of wood had no significant effect on the drying rate. The drying rate of 1′ and 2′ thick planks was almost similar suggesting volumetric heating of wood. Microwave drying is mainly effective at low moisture content. Microwave heating was also found to be very effective in drying bamboo, which is otherwise very difficult to dry in a kiln (Prasad and Pandey 2012).

The design of the system and the power source for microwave heating and radio frequency heating are very different

Fig. 38.19 Radio frequency (left) and microwave wood dryer (right)

Table 38.2 Major seasoning defects and their cause

Defects	Cause
Surface checks, end cracks, internal checks, collapse	Drying too fast
Fungal growth, mould growth, discolouration, chemical stain	Drying too slow, wood kept in a damp environment
Warping in sawn boards	Drying too slow, poor stacking, improper drying practice
Uneven drying	Poor stacking and storage, operational errors in drying
Gluing problem	Improper drying, wet surface, defects
Raised and fuzzy grains and woolly surface	Processing defects, presence of reaction wood
Bad odour	Bacterial growth
Loose or open joints	Variation in moisture content within wood pieces

(Fig. 38.19). In microwave heating, single or multiple waveguides are used to introduce the electromagnetic waves in the chamber and the efficiency of interaction of waves with the wood mainly depends on the waveguide and design of the chamber. A poorly designed chamber may lead to hotspots on wood elements resulting in uneven heating of wood. In the case of radio frequency dryers, an electromagnetic field is generated between a pair of electrode plates, and wood is stacked between the plates. The electrode plates create a unidirectional electromagnetic field on the entire stack of wood.

Whichever method of seasoning is applied to timber, one needs to know to what moisture content levels timber should be seasoned to avoid any moisture-associated problems during product development and in the finished product. Once timber is properly seasoned, it must be stored properly and products to be manufactured at humidity conditions according to the atmospheric conditions where the products are to be used. Not following these basic principles may lead to significant economic implications as the product may develop some defects.

38.3.8 Drying Defects

Though several seasoning methods have been developed, steam heated kiln seasoning is the most practiced and commercially acceptable method worldwide. In kiln seasoning, timber may develop different types of drying defects due to inappropriate kiln schedules, poor or improper stacking, refractoriness of timber, presence of reaction wood, irregular or non-homogeneous grain and anisotropic properties. A thorough understanding of the different drying stages and proper monitoring of material during drying is essential for preventing and controlling degrades. Common defects and their causes in kiln seasoning are given in Table 38.2.

Industries must follow proper seasoning practices to produce quality materials to produce high-quality end-products. Many times, poor seasoning results in the devaluation of the products and also the rejection by the consumer. These also contribute to the short life of the products and considerable economic loss to the manufacturer and consumer.

38.4 Wood Preservation

Wood preservation can be interpreted in terms of protection against both biotic (biological degradation) and abiotic (fire, weathering, chemical degradation) agencies degrading wood. However, in general, wood preservation terminology is closely associated with the biological degradation of wood. Being a biological material, wood is prone to deteriorate by biological organisms like termites, fungi, beetles, ants, borers and many other insects in terrestrial conditions. In aquatic and marine conditions, borers and foulers degrade wooden structures within a short span. The biological degradation of wood is considered to be one of the most undesirable characteristics when compared to other building materials. Biodegradation of wooden structures beyond a critical stage result in a significant loss of strength in the structure and that can be catastrophic. In this scenario, the natural resistance of wood against deteriorating agents plays a crucial role in the durability and longevity of the wood. The heartwood of many tree species exhibits excellent resistance against degrading organisms due to the deposition of specific extractives also called the secondary metabolites that inhibit degradation. This inherent property of resisting biodegradation is known as natural durability. Naturally durable timber species are preferred for outdoor or exterior applications and also environmentally preferable over chemically treated wood. However, irrespective of timber species, sapwood is highly susceptible to biodegradation and therefore needs to be protected. The natural durability of wood is tested with the stack test method also known as 'Timber graveyard test' (Fig. 38.20). The test involves half burial of a timber stack of specific size underground in a test yard (generally termite-prone areas) and the samples are monitored frequently for their degradation. Many tropical timber species like mango, rubberwood and silver oak are found to perish quickly in the test. Timbers are categorized under different durability classes based on the natural durability of their heartwood when in ground contact and specific to only terrestrial conditions. The timber species is classified as naturally durable (class-I) if the stack survives for more than 120 months, moderately durable (Class-II) if the average life is 60–120 months and nondurable (Class-III) if the timber perishes within 60 months in ground contact in the test yard.

Testing of durability in marine conditions is carried out by exposing the wooden stacks in the marine environment and monitoring the same for degradation. The marine degradation depends on the presence of degrading agents in specific locations. Very few wood species like *Syncarpia glomulifera* (Australia) and *Ocotea rodiei* (Guyana) have sufficient natural durability in marine conditions (Walker 2006). Tarakanadha et al. (2005) documented the natural durability of 30 Indian timbers widely used for fishing crafts and marine structures and they were found to get destroyed within 3–7 months in the marine environment.

The philosophy of wood preservation is primarily based on increasing the life of wood and wood products and thereby reducing the burden on forests for durable timbers which have become scarce. The importance of protecting wood from biodegradation was well recognized historically and treatment practices like soaking timber in garlic boiled vinegar, soaking or coating with certain oils like oil of cedar, linseed oil, cashew-nut shell oil, tar, etc., smoking and charring treatment were used to protect timber from biodegradation even in case of durable and moderately durable timber species. There has been a paradigm shift in forestry with more emphasis on growing fast-grown and short-rotation timber species to meet the increasing demand for timber and most of these are non-durable. In such situations, wood preservation becomes more important as it can help in storing the organic carbon for a much longer period and also reduce the burden of timber resources.

Fig. 38.20 Graveyard test for assessing the natural durability of wood with the extent of decay in a few samples

However, a key question in wood preservation is the longevity required for a wooden product and the geographical locations where it is to be used. In the early nineteenth century, impregnating wood with toxic chemicals capable of inhibiting degradation became the scientific method for wood preservation. Consequently, several chemical systems and treatment methods have been developed to enhance the life of wood. The choice of chemicals and treatment method mainly depends on:

- Environmental conditions in which wood is to be exposed and bio-deteriorating organisms present there
- The effectiveness of the chemicals against target organisms
- Fixation of chemicals in wood
- Desired loading, penetration, and distribution of preservative chemicals in wood

38.4.1 Wood Preservatives

Wood preservative chemicals are either impregnated or applied to wood to make it resistant to biodegradation. These chemicals are classified as oil-based, organic solvent-based and water-soluble chemicals.

38.4.1.1 Oil-Based Wood Preservative

Oil-type systems are among the oldest and most effective preservatives commonly used for treating utility poles, bridge timbers, railway sleepers etc. Among the oil-based systems, creosote is one of the most popular preservatives. It is a fraction of coal tar distillate having a blend of polycyclic aromatic hydrocarbons (PAHs) and tar with a boiling point above 200 °C. Creosote is an effective broad-spectrum wood preservative especially suitable for timber used in exterior applications. It may be used as a stand-alone chemical or in combination with other suitable oils to the extent of 50% by weight of creosote. The mixture of oil protects against splitting and cracking in timber. Creosote as a wood preservative is highly toxic to target organisms, non-corrosive and very stable in wood. It provides very good protection against termites and other insects. However, the chemical has an unpleasant odour and is slightly difficult to handle due to its high viscosity. Wood surfaces treated with creosote are also difficult to paint. Hot and cold process is used for the treatment of wood with creosote and in case of very high viscosity grade creosote, appropriate diluents or organic solvents are used to reduce the viscosity for effective penetration.

38.4.1.2 Organic Solvent-Based Preservatives

These preservatives are used by dissolving in suitable organic solvents. The solvent selection depends on the solubility of the preservative chemical and the end-use of the treated material. These preservatives are relatively easy to handle as compared to creosote when lighter and non-toxic or low-toxicity solvents are used. These types of preservatives are non-leachable and are considered to be fixed and permanent. In most cases, wood treated with such types of preservatives can easily be painted, unlike creosote-treated wood. As most of the organic solvents used in these preservative solutions are highly flammable, proper safety precautions are necessary in handling the solutions. Copper naphthenate, zinc naphthenate, pentachlorophenol (PCP), benzene hexachloride, *p,p*-dichloro-diphenyl-trichloro-ethane (DDT) are some of the organic solvent type wood preservatives. However, their application is very limited and the use of chemicals like PCP has raised environmental and health hazard concerns to humans, and has been banned as a wood preservative.

38.4.1.3 Water-Soluble Preservatives

Water-soluble wood preservatives are classified into two categories, that is, leachable and non-leachable (fixed). Non-fixative types of wood preservatives are water-soluble inorganic or organic salts that do not get fixed or bond chemically with any of the chemical constituents of wood. When the wood treated with such preservatives is exposed to water, the chemicals tend to dissolve in water and leach out gradually. This leaching can be prevented to some extent by applying a waterproof coating or painting the wood surface after chemical treatment to avoid any moisture movement within the wood and the coating is maintained properly. Such types of preservatives are easy to transport, generally odourless and generally free from fire hazards. Dry surfaces of wood treated with these preservatives do not pose any significant challenges in painting, varnishing or wax treatment. Such preservatives are suitable for indoor applications where wood is not often in contact with water or ground. Simple salts like zinc chloride and sodium fluoride are simple water-soluble wood preservatives mainly used against fungi and insect borers but are not very effective against termites. Boric acid and borax are effective against lyctus borers, sap-stain fungi and some species of termites. These chemicals are suitable for protecting veneers, plywood glue line protectants and packing case timbers for tea chests and for packing edible articles as it has low toxicity against non-target organisms. These chemicals are used in combination as a prophylactic treatment for freshly sawn timber from mould growth. Boric acid is odourless, colourless and a low-cost wood preservative. In addition to these, chemicals like sodium pentachlorophenate and commercial insecticides like DDT, benzene hexachloride, chlorpyrifos and pyrethroids have been used as wood preservatives.

38.4.1.4 Water-Soluble-Non leachable Wood Preservatives

These preservatives consist of mixtures of various salts that can chemically bond or be fixed within the wood or form an insoluble compound that does not leach out on exposure to water. Materials treated with such types of preservatives may be used for outside locations and unlike creosote these can also be painted over. There are several such formulations developed over the years.

38.4.1.5 Copper-Chrome-Arsenic Composition (CCA)

CCA is one of the most effective and until the recent past was widely used wood preservatives worldwide. CCA wood preservative (AsCu) was invented in 1933, and patented in India by Dr. Sonti Kamesam of Forest Research Institute, Dehradun. This preservative is especially suitable in highly termite-prone areas in terrestrial conditions, borer-infected areas in marine conditions, and for cooling tower timbers where wood is susceptible to soft rot deterioration. It is a combination of copper oxide, sodium dichromate or chromic acid and arsenic pentoxide. Varying proportions of copper, chromium and pentavalent arsenic compounds have been formulated and tested for their efficacy. The formulation commercially used in India has 50% sodium dichromate, 37.5% copper sulphate and 12.5% arsenic pentoxide. CCA is fixative to the wood with chromium being the fixative agent provided a proper treatment schedule is followed. Since the fixation process involves the reduction in hexavalent chromium (Cr^{VI}) to trivalent chromium (Cr^{III}) which is a slow process and therefore it is necessary to complete the fixation process which may require 2–3 weeks. Treatment temperature, initial pH of the preservative solution and concentration of solution play critical roles in the fixation mechanism. With the correct treatment protocol, CCA-treated wood is expected to provide assured service life for a very long period even in ground contact. However, arsenic and chromium compounds are recognized as potential carcinogenic for humans, and therefore the usage of CCA as wood preservatives is being restricted in many parts of the world.

38.4.1.6 Copper-Chrome-Boron Composition (CCB)

This preservative was developed as an alternative to CCA with the arsenic component replaced by boron to make it less harmful to non-target organisms. The composition contains boric acid, copper sulphate and sodium/potassium dichromate in 1:5:3:4 (by weight) proportion respectively. CCB is relatively less efficacious than CCA as more toxic arsenic is replaced by boron. The preservative has become very popular and is recommended for general use including protecting timber in cooling towers.

38.4.1.7 Copper-Zinc-Borax (ZiBOC)

Recently, a combination of copper sulphate, zinc chloride and borax in a fixed ratio of 15:2:9 was formulated and patented as an effective wood preservative (Indian Patent 257393 2013), which is considered to be eco-friendly, easy to handle, economical and non-hazardous. All the ingredients are pre-mixed in a small amount of water and then a 60% concentration paste is made in acetic acid. This paste is diluted in water to a specific concentration for the treatment of wood. Efficacy studies of ZiBOC against fungi and termites have revealed its effectiveness in protecting wood against these agents.

38.4.1.8 Acid Cupric-Chromate Composition (ACC)

This preservative comprises chromic acetate (5 parts), copper sulphate (50 parts) and sodium/potassium dichromate (45 parts). It was developed way back in the 1920s but could not match the efficacy of CCA. Since the formulation contains an enormous amount of chromium, it did not get commercial acceptance for general application and is used in limited quantity for industrial applications where human contact is limited.

38.4.1.9 Chromated-Zinc Chloride

This preservative composition consists of only two chemicals namely zinc chloride and sodium dichromate/potassium dichromate in the proportion of (by weight) 81.5:18.5 respectively.

38.4.1.10 Ammonia-Based Preservatives

There has been a great deal of activity in the evaluation of new biocides as potential preservatives. The most promising new biocides are the alkyl ammonium compounds (AAC). Some of the alkylammonium compounds (AAC) have good fungicidal activity and many have potential as wood preservatives. Didecylmethylammonium chloride emerged as a potential wood preservative. Combining alkylammonium compounds with copper and ammoniacal solutions (ACQ) improved the performance of wood in ground contact. This ACQ formulation gives a protection similar to that of CCA. Ammonia also can soften lignin and swell wood cell walls which helps in easy impregnation of the chemicals inside wood. Similarly, water-soluble ammoniacal copper zinc arsenate (ACZA) is used as a commercial wood preservative in many countries.

38.4.1.11 Eco-Friendly Preservatives of Plant Origin

Copper and zinc abietates prepared from chir pine are toxic to fungi and insects, including termites, and have shown potential as an effective wood preservative even in marine conditions. Treatment with cashew nut shell liquid (CNSL) is found to be effective against termites and indicated toxicity of the oil for termites. Several other plant extracts are being explored as potential wood preservatives.

38.4.1.12 Nano Wood Preservatives

Nanotechnology is an emerging field that has enormous potential in wood preservation. Impregnation of nanoparticles of copper, silver, zinc borate, zinc oxide, titanium dioxide, etc. has been tested as wood preservatives by many researchers and reported to be promising in protecting wood from wood destroying organisms (Borges et al. 2018). Micronized copper systems like micronized copper quaternary and micronized copper azole have already gained commercial acceptance in many developed countries. Nano-scale metallic particle impregnation offers several advantages like the fixation of nanoparticles in the cell wall structure, uniform dispersion of particles, low concentration of chemicals and reduced leaching. Nano-based wood preservatives have been reported to be effective against both termites and fungi (Teng et al. 2018). Nair et al. (2017) reported effective decay resistance in nano zinc oxide and copper oxide treated rubber wood against both white and brown rot fungi. The technology is still evolving and may play a big role in wood preservation in the near future.

38.4.1.13 Wood Modification

Wood modification is an efficient way of making wood decay-resistant. Chemical and thermal modifications are the two major wood modification methods. Chemical modification involves altering the chemistry of cell wall constituents wherein hydroxyl groups present on hemicelluloses, cellulose and lignin are replaced with other functional groups. Acetylation of wood has been one of the most prominent chemical modification methods and has already gained commercial success. The process requires the reaction of wood with acetic anhydride with or without a catalyst. The modified wood exhibits reduced hygroscopicity, increased dimensional stability, improved photo-stability, better weather resistance and superior decay resistance against both termites and fungi (Teng et al. 2018; Giridhar et al. 2017). The modified process generally alters colour or mechanical properties and the modified wood is non-toxic. Other than chemical modification, thermal modification has also been used as a method for imparting fungal resistance in wood. The process involves heat treatment of wood at elevated temperatures (above 160 °C) in the absence of oxygen. Thermal modification technology has also reached to commercial scale with ThermoWood® a well-established product available in the European market. However, loss in mechanical strength, darkening of the colour of wood, rapid weathering and no termite resistance are the major drawbacks of the technology which restricts its application in certain areas.

38.4.2 Treatment Methods

There are different methods of treating wood with preservatives. The choice of treatment method ideally depends on the moisture content of the wood and the end application. Methods like diffusion and sap displacement are applicable when the wood is in green condition (high moisture content). In other treatment methods, material needs to be dried to appropriate moisture content particularly up to fibre saturation point. All woodworking like cutting to sizes and boring has to be completed before chemical treatment to avoid wastage of treated timber. In the case of timbers where heartwood is non-durable but is difficult to treat and when treating thick sections like beams, incision to a depth of 15–20 mm at surfaces except cross-cut surfaces (both ends) helps in proper penetration of the preservatives. The treatment methods are broadly divided into two categories, that is, non-pressure processes and vacuum/pressure impregnation processes.

Non-pressure Processes: Non-pressure methods have been practiced in many applications and different approaches are adopted for dry wood and wood in green condition.

Surface application: The simplest way of treatment is by brush application or spraying of wood preservatives on the wood surface. However, surface application is inadequate in most cases for long-term effectiveness and is used mostly for treating material at the site as prophylactic treatment.

Soaking treatment: This treatment requires wood material without any bark (in logs or poles) and sawn timber and submerging the material in the preservative solution until the required absorption is obtained.

Sap displacement (Boucherie) method: This method is adopted to treat freshly felled poles and bamboo using water-soluble preservatives. In this method, the butt ends of the freshly cut poles or bamboo are immersed in the preservative solution (Fig. 38.21). The preservative replaces sap of the poles/bamboo by wick action throughout the length of the pole or bamboo with time. This method is very simple to adopt, low cost and the process can be carried out in the field without much technical know-how at the felling site. However, the method is useful when there is no end-grain drying and has to be done immediately on felling. The method requires reversing the poles upside down after a few hours to achieve uniform penetration of chemicals. The process may take a considerable amount of time depending on the size of the pole. Alternatively, the sap can be replaced by wood preservatives by pressurizing preservative solution (Boucherie process) which is a faster method for treating poles and bamboo (Fig. 38.22). The preservative solution is filled in a reservoir, and the butt end of the pole/bamboo is connected to a charge cap with proper sealing. The other end of the pole is kept at a lower elevation than the connected

Fig. 38.21 Sap displacement technique to treat green poles and bamboo

Fig. 38.22 Sap displacement using hydraulic pressure—Boucherie treatment units

end. An air pressure of 0.5–2 kg/m^2 is applied to the treating solution in the reservoir. With the exertion of pressure on the solution, preservative displaces the sap which is forced out at the other end within a short period. These methods are extremely useful in treating at remote locations, at felling sites or when on-farm treatment is required as multiple poles/bamboo can be treated.

38.4.2.1 Diffusion Process

This method is also adopted for treatment in the green condition when the moisture content of wood is well above the fibre saturation point. In this process, rough-sawn timber or debarked logs are either dipped in a highly concentrated preservative solution or a preservative paste is applied on the surface and then they are concealed with polythene covers and stacked for different periods. During the holding period, the preservative salt slowly diffuses into wood due to concentration gradient. Timber that is not treated easily by the pressure process can be treated by this method in the green condition. Timber with higher basic density and/or relatively less moisture content needs to be kept in a highly concentrated solution for efficient diffusion. The process is time-consuming and may take 4–6 weeks for 25 mm thick boards and up to 12–14 weeks for 5 mm thick sections. The uptake of salts depends on the initial concentration of the solution, initial moisture content, ratio of surface area to

Fig. 38.23 A vacuum-pressure impregnation chamber

volume and temperature of the treating solution. The method is an effective way of treating green wood with boron salts as it does not raise any health hazards, is simple to operate and can treat wood locally.

38.4.2.2 Vacuum-Pressure Impregnation

Combining vacuum and pressure for treating wood with chemicals is the most accepted commercial method. The process is very efficient in achieving deep penetration of wood preservatives and the chemical loading can be controlled by using specific vacuum-pressure conditions called 'treatment schedules'. The method requires a large size heavy-gauge cylindrical pressure vessel for treatment (Fig. 38.23). Mainly, there are two processes, that is, full cell process and the empty cell process.

38.4.2.3 Full Cell (Bethell) Process

This process was patented by John Bethel in 1838 and is used where high absorption of preservatives is required. In the process, wood material is charged to the cylinder and the door is tightly closed. A vacuum of at least 56 cm of mercury is created and the vacuum is maintained for 15–30 min. The objective of this operation is to remove air from the wood cells so that preservatives can impregnate rapidly inside the wood. At the end of the vacuum period, the preservative is introduced into the cylinder by opening the valve connected to the solution storage tank while maintaining the vacuum. Once the cylinder is filled with the preservative, the solution is subjected to an air pressure of 3.5–12.5 kg/cm^2 using an air compressor or pressure pump. The air pressure to be applied depends on the species, thickness of the timber, treatability of the wood etc. The applied pressure ensures the injection of the preservative into the timber. The pressure inside the chamber is maintained for a specified period as per the treatment schedule to get the desired absorption. Once the treatment is over, the solution is withdrawn from the cylinder back to the storage tank. Finally, the chamber is subjected to a vacuum for about 15 min to remove any excess chemicals from the wood surface. Specified retention of preservatives is achieved using an appropriate concentration of the solution and controlling vacuum and pressure durations.

38.4.2.4 Empty Cell Process

Empty cell process aims at maximizing penetration with minimum net absorption of preservatives. There are two processes namely the Rueping process and the Lowry process used in the empty cell method.

38.4.2.5 Rueping Process

This method is principally used for hot oil-type preservatives such as creosote where low absorption of chemical is desired. In this process, material is charged into the cylinder and air pressure of 2.0–5.0 kg/cm^2 is applied for a specific period. The cylinder is filled with the preservative while maintaining the pressure and when filled, pressure is further increased to 3.5–12.5 kg/cm^2 depending on the species and size. After the desired time, the pressure is released and a final vacuum of 35–56 cm of Hg is drawn to wipe out the excessive preservative from the surface and get a much drier surface.

38.4.2.6 Lowry Process

In this process, the material is loaded and filled with the preservative. A hydraulic pressure of 2–12.5 kg/cm^2 is applied till the desired absorption is achieved and the pressure is slowly released. During pressure release, a part of the injected preservative is expelled due to the expansion of trapped air in the wood cells. Finally, a vacuum of 36–56 cm of Hg is applied for 15 min to remove excess preservatives. The treatment process is appropriate for treating low density and highly permeable timbers which are to be used for exterior joinery and framing applications.

In addition to these methods, oscillation pressure method is used to treat wood in green condition wherein wood material is charged into the cylinder and once the chamber is filled with preservative repeated application of high pressure and vacuum is applied. When the vacuum is applied, some amount of sap is extracted from the wood which gets mixed with the preservatives, and subsequent pressure forces chemicals to penetrate inside the wood. The cycles of pressure and vacuum are repeated for a long duration to achieve the desired absorption of the preservative. However, in most commercial practices, the full cell process is the preferred method for an effective treatment with water-soluble preservatives. The treatment protocols of different species are different and depend on permeability and treatability. Sapwood portion of most of the timbers can easily be treated but heartwood of many species poses serious challenges. The treatability of heartwood depends on anatomical structure and the presence of pathways for impregnation of chemicals inside wood and therefore is different for different species. The heartwood treatability is categorized into five grades namely easily treatable; treatable but complete penetration not always achieved in thick sections; only partially treatable; difficult to treat; very difficult to treat wherein the preservative penetration is negligible or nil even from the ends. Based on the treatability class, the treatment schedules are adopted to achieve satisfactory loading of preservatives.

38.5 Concluding Remarks

Sawing, seasoning and preservative treatment are the basis of secondary wood processing for product development. The unorganized and fragmented nature of the saw-milling industry in the country has been a major reason for technologically being outdated. Many developed countries have adopted considerable automation in the sawing process to improve efficiency like automatic log holding, log scanning for deciding optimum cutting pattern and infeed systems for positioning the logs for optimal sawing. Further, the benefits of optimized sawing practices yield more benefits with small logs. As far as seasoning and preservation are concerned, these essential processing steps are quite often overlooked by industries as well as consumers as both these processes add to the cost of the product considerably. Wood seasoning technologies are well established and benefits are also well known but their adoption is far from satisfactory. Large capital investments, time required for effective seasoning and cost have been major restraining factors. Environmental issues related to wood preservation particularly the use of chemicals, disposal of treated wood and human hazards associated with preservatives during service life have raised concerns about adopting such practices. However, the preservative treatment in applications where wood is in ground contact or exposed to fungal degradation like in building construction, cooling towers and marine structures can have significant economic implications.

It is important to recognize that appropriate processing becomes more crucial in utilizing fast-grown plantation timbers which are generally characterized by high moisture content, a higher proportion of sapwood, a low proportion of mature wood, low natural durability and a steep gradient in wood properties. With the advances in technologies, wood processing technologies are also going to be more effective enabling us to efficiently utilize timber resources. A cluster or cooperative-based approach for seasoning and preservation could be adopted for small-scale wood processing operations like for handicraft and furniture industries. There has been a paradigm shift, globally, with emphasis on growing more wood and using more wood as wood is one of the most environmentally friendly materials. The adoption of appropriate primary processing can play a significant role in increasing the life cycle of wood-based products and in turn, can play a key role in storing carbon, reducing frequent replacements and thereby contributing to the mitigation of climate change.

Lessons Learnt

- Different saw types, saw blade parameters and sawing patterns
- Design of an efficient sawmill
- Advantages of wood seasoning for its efficient utilization
- Theoretical aspects of wood seasoning including moisture sorption and drying mechanism
- Different seasoning methods
- Defects due to improper seasoning and uncontrolled drying of wood
- Natural durability of timbers
- Various types of wood preservatives
- Methods of preservative treatment of wood

Key Questions

1. What are the different saw blade parameters and how do they influence sawing quality?
2. Define the fibre saturation point and discuss its relevance in wood seasoning.
3. A piece of wood is extracted from a freshly felled tree. The initial weight of the piece is 890 g and after oven drying the weight is 423 g. What will be its weight at 12% moisture content?
4. What are different seasoning methods? Describe their relative advantages and disadvantages.
5. What are the different methods for preservative treatment of wood? Is the sap displacement technique suitable to treat air-dried poles?

References

Aggarwal P, Chauhan S (2013) Microwave drying of planks of *Grevillea robusta* A. Cunn. ex R. J Indian Acad Wood Sci 8:84. https://doi.org/10.1007/s13196-012-0050-y

Avramidis S, Zwick RL (1996) Commercial-scale RF/V drying of software lumber. Part 2. Drying characteristics and lumber quality. For Prod J 46(6):27–36

Barber NF, Meylan BA (1964) The anisotropic shrinkage of wood. Holzforschung 18(5):146–156

Borges CC, Tonoli GHD, Cruz TM, Duarte PJ, Junqueira TA (2018) Nanoparticles-based wood preservatives: the next generation of wood protection? CERNE 24(4):397–407

Brunauer S, Emmet PH, Teller EJ (1938) Adsorption of gases in multilayers. J Am Chem Soc 60:309–319

Chauhan S, Aggarwal P (2004) Effect of moisture sorption state on transverse dimensional changes in wood. Holz-als Roh-und Werkstoff. 62:50–55

Chauhan S, Sethy A (2008) Drying behaviour of *Eucalyptus tereticornis* wood in a desiccant based dehumidifying wood dryer. J Indian Acad Wood Sci 5(1&2):21–30

Chauhan S, Aggarwal P, Karmarkar A, Pandey KK (2001) Moisture adsorption behaviour of esterified rubber wood. Holz als Roh- und Werkstoff 59(4):250–253

Chauhan S, Upreti NK, Sethy A (2006) Drying behaviour of some plantation grown timbers in forced air drying. J Indian Acad Wood Sci 3(1):62–67

Espinoza O, Bond B (2016) Vacuum drying of wood—state of the art. Curr For Rep 2016(2):223–235. https://doi.org/10.1007/s40725-016-0045-9

Giridhar N, Pandey K, Prasad B, Bisht SS, Vagadevi H (2017) Dimensional stabilization of wood by chemical modification using isopropenyl acetate. Maderas Cincia Technol 19(1):15–20

Hailwood AJ, Horrobin S (1946) Absorption of water by polymers: analysis in terms of a simple model. Trans Faraday Soc 42B:84–92

Harda H, Cote WA (1985) Structure of wood. In: Higuchi T (ed) Biosynthesis and biodegradation of wood components. Academic Press, London, pp 1–42

Hassan HF, Tola Y, Ramaswamy HS (2015) Radio-frequency and microwave applications: similarities, advantages and limitations. In: Radio-frequency heating in food processing principles and applications. CRC Press, Boca Raton, pp 21–28

Indian Patent 257393 (2013) An eco-friendly, economical and non-hazardous wood preservative. Forest Research Institute 264/DEL/2008

Jiao S (2012) Development of non-chemical post-harvest treatments for disinfecting agricultural products (Ph.D. Thesis). Washington State University

Jiao S, Luan D, Tang J (2015) Principles of radio frequency and microwave heating. In: Radio-frequency heating in food processing principles and applications. CRC Press, Boca Raton, pp 1–19

Mitcham EJ, Veltman R, Feng X, De Castro E, Johnson JA, Simpson TL, Biasi WV, Wang S, Tang J (2004) Application of radio frequency treatments to control insects in in-shell walnuts. Postharvest Biol Technol 33(1):93–101

Naderi N, Hernandez RE (1997) Effect of rewetting treatment on the dimensional changes of sugar maple wood. Wood Fibre Sci 29:340–344

Nair S, Pandey KK, Giridhar BN, Vijayalakshmi G (2017) Decay resistance of rubberwood (*Hevea brasiliensis*) impregnated with ZnO and CuO nanoparticles dispersed in propylene glycol. Int J Biodeterior Biodegrad 122:100–106

Pandey CN, Jain VK (1992) Wood seasoning technology. ICFRE Publication

Pentoney RE (1953) Mechanism affecting tangential and radial shrinkage. For Prod J 3(2):27–33

Prasad BE, Pandey KK (2012) Microwave drying of bamboo. Eur J Wood Wood Prod 70:353–355

Resch H, Hansmann C (2002) Tests to dry thick Eucalyptus boards in vacuum using high frequency heating. Holzforschung 54(3):59–61

Siau JF (1984) Transport processes in wood. Springer, Berlin

Skaar C (1988) Water in wood. Springer, Berlin

Smith WB, Canham HO, Harris J, Neuhauser EF, Smith A (1996) Economic analysis of producing red oak dimension squares with a radio-frequency vacuum dry kiln. For Prod J 46(3):30–34

Tarakanadha B, Raveendra Prasad NR, Satyanarayana Rao K (2005) Natural durability of Indian timbers under marine environment at Krishnapatnam Harbour, East Coast of India, Andhra Pradesh. J Inst Wood Sci 17(2):75–84. https://doi.org/10.1179/wsc.2005.17.2.75

Teng TJ, Arip MN, Kumar S, Nemoikina A, Jalaludin Z, Ng EP, Lee HL (2018) Conventional technology and nanotechnology in wood preservation: a review. Bioresources 13(4):9220–9252

Walker JCF (2006) Primary wood processing- principles and practice. Springer, Dordrecht

39 Sustainable Management of Non-Timber Forest Products

Prodyut Bhattacharya and Rituparna Bhattacharya

Abstract

Traditionally Non-Timber Forest Products (NTFP) provide a significant opportunity for socially and economically weaker sections of society particularly tribal, landless farmers, women and other rural poor for cash and non-cash income for livelihoods. Forests provide various NTFPs which are used for food, fodder, medicines, firewood, mushroom, honey, bamboo and many more. There are more than 200 different types of NTFP available in tropical and temperate forests, which have commercial value and require multiple strategies to manage sustainably in wild areas.

From the perspective of management, there are two categories of NTFPs: (a) products of commercial value and (b) products of subsistence value, which never reach to market. This makes it difficult to estimate the real value of NTFPs, which are under provisional ecosystem services. If collection, processing and value addition activities are sufficiently strengthened with modest investment from the state and central governments, tribal and poor forest living communities will quickly become prosperous in their own habitat. Management of NTFPs covers a range of activities starting from adopting appropriate silvicultural practices to caring regeneration and practicing healthy collection methods. A large number of NTFPs have medicinal properties and are collected by herbal industries in large quantity through the local communities. In many cases premature and over harvesting has made the collection unsustainable for several products from natural forests.

This chapter discusses different issues of NTFP management including the causes of unsustainable harvesting, informal market system, fast depleting resource base, unaccounted inventory and weak institutional structure for collection and marketing of products in remote forest areas. Many recent initiatives by state forest departments and central organizations, Tribal Cooperative Marketing Development Federation (TRIFED) to boost the sector through various supporting services like introduction of Minimum Support Price (MSP), inclusion of NTFP management in forest working plan code as well as conservation practices for rare and threatened species. Many studies indicate the emerging impact of climate change, fire and other vulnerabilities in many NTFP growing areas resulting in low yield. There are huge gaps perceived in awareness, communication and capacity building of people who are stakeholders in this sector. In the past few decades many landmark policy interventions emerged in India, which have direct and indirect link with NTFP management but their impact is yet to be seen. Proper community-based enterprise development linked with sustainable income sources for the tribal and other local communities is still a dream in NTFP collection.

Keywords

NTFP · MFP · Sustainable harvesting · Stakeholder · Van Dhan scheme

39.1 Introduction

In the last seven decades since independence, India has made impressive strides in critical development sectors. Nonetheless, a substantial portion of the population still relies on forest resources for their livelihoods and income generation. Approximately 140 million individuals are forest dwellers, whose primary reliance on forests extends beyond fulfilling various household requirements to ensuring food security as well (Bhattacharya and Hayat 2004). A distinct subset within this forest-dwelling population comprises the indigenous tribal communities, dispersed throughout the nation.

P. Bhattacharya (✉)
University School of Environment Management, Guru Govind Singh Indraprastha University, New Delhi, India

R. Bhattacharya
Centre for Economic and Social Studies, Hyderabad, India

A. K. Mandal, A. Nicodemus (eds.), *Textbook of Forest Science*, https://doi.org/10.1007/978-981-97-8289-5_39

Among the nearly 68 million tribal individuals, half rely directly on forest resources to fulfil their livelihood needs, including food, fodder, medicinal plants, small timber and various other Non-Timber Forest Products (NTFPs) gathered from the wilderness. Non-timber forest produce, often referred to as 'Minor Forest Produce,' plays a significant role in the economic sustenance of tribal societies.

The FAO defines non-timber forest products as 'biological resources of plant and animal origin, harvested from natural forests, plantations, wooded land and trees outside forests'. From this definition, it becomes easy to deduce that all the forest produce minus the timber could be perceived as 'non-timber forest produce' (Bhattacharya and Hayat 2004). However, from the perspective of management, there would be two categories of NTFPs: (a) products of commercial value and (b) products of subsistence value, which never reach a 'market', estimating whose values under the provisional services of the ecosystem become cumbersome. Examples of the category 'b' are usual in many parts of India especially among tribes in Northeast, where the use of birds, insects, mushrooms, leafy vegetables, larvae, fish and other animals and medicinal plants from the forest is integral to their diet and personal use (Bhattacharya and Hayat 2004).

De Beer and Mc Dermot (1989), who coined the term NTFP, define it as 'NTFP encompasses all biological materials, other than timber, which are extracted from forests for human use'. This definition excludes minerals and includes fuelwood, bamboo and animal products. It also excludes non-forest woodlots and TOF (Trees Outside Forests).

According to a report of the sub-group-ii Planning Commission on NTFP and their sustainable management in the 12th Five Year Plan, NTFPs refer to 'all biological materials other than timber extracted from natural forests for human and animal use and have both consumptive and exchange value'. Globally NTFP/NWFP are terms for 'forest products consisting of goods of biological origin other than wood, derived from the forest, other woodlands, and trees outside forests'.

The Scheduled Tribes and Other Traditional Forest Dwellers (Recognition of Forest Rights) Act, 2006 still uses the term Minor Forest Produce (MFP)[1] for a similar concept. The Forest Rights Act tries to define MFP as 'Minor Forest Produce includes all non-timber forest produce of plant origin including bamboo, brushwood, stumps, cane, tussar, cocoons, honey, wax, lac, tendu or kendu leaves, medicinal plants and herbs, roots tubers and the like'. A closer scrutiny of this definition will reveal that in an attempt to define 'MFP' for the limited purpose of this act, it is taking support of the term 'non-timber forest produce'. Further, trying to limit the scope of 'NTFP' also, is adding the condition 'of plant origin' . However, the inclusion clause, also mentions items like tussar, cocoons, honey, wax and lac which essentially are animal products.

Annually, around Rs. 2,00,000 crores worth of NTFP is gathered by the tribal people and traded in the *haat bazaar* (local market). The potential of this huge capital, grossly under-managed, is yet to be fully appreciated by the government and planning agencies (TRIFED 2020). If NTFP activity is sufficiently strengthened, tribals can become prosperous in their habitat, in the shortest period, with moderate investment by the state and central government. Earlier time the Forest Department used to focus mostly on the management of timber trees, considered NTFPs as byproducts of forest, and they had provided little importance to its management at their territorial forest areas. NTFP is primarily gathered from the forests and its primary processing is done by women folk at home. A holistic model of NTFP-led tribal development for the scheduled tribes in forest areas is need of the hour. This requires synergy among various concerned ministries/departments for concerted actions towards NTFP based economic development. The government has progressed significantly in this direction.

NTFPs are estimated to contribute significantly to the income of 50 million tribal households in India, ranging from 10 to 40%, with an additional 200–300 million villagers relying on them to a lesser extent (Shiva 1993). Particularly in the arid tropical forest regions of India where water resources are scarce and agricultural opportunities are limited, NTFPs often serve as the primary means of survival for tribal communities. These forests exhibit a concerning trend of rapid degradation and poor regeneration, posing challenges to the sustainable utilization of NTFPs. The tribal populations dependent on these resources are among the most marginalized, experiencing severe deprivation in terms of health, nutrition and education, as indicated by their low development index (Bhattacharya and Hayat 2004). Against this backdrop, the role of NTFP-based development could be considered as the 'first step survival strategy'. There is a correlation between high tribal community concentration and a high proportion of forest area per state (Fig. 39.1). Consequently, these are also the states where the NTFP collection and allied activities are higher.

Traditionally, Non-Timber Forest Products provide a significant opportunity for the social and economic life of millions of forest dwellers, particularly the tribal, landless, women and other rural poor for cash and non-cash income to local livelihood. Besides, NTFPs yielding biodiversity is of immense ecological significance to the forest ecosystem ensuring ecological security to the nation. Throughout the year, as per the forest types of the region, NTFPs are produced in the Indian forest and the local community collects these products. Although there is no precise estimate of the total amount of NTFPs extracted from the natural forests,

[1] The terms- NTFP and MFP are interchangeably used in India.

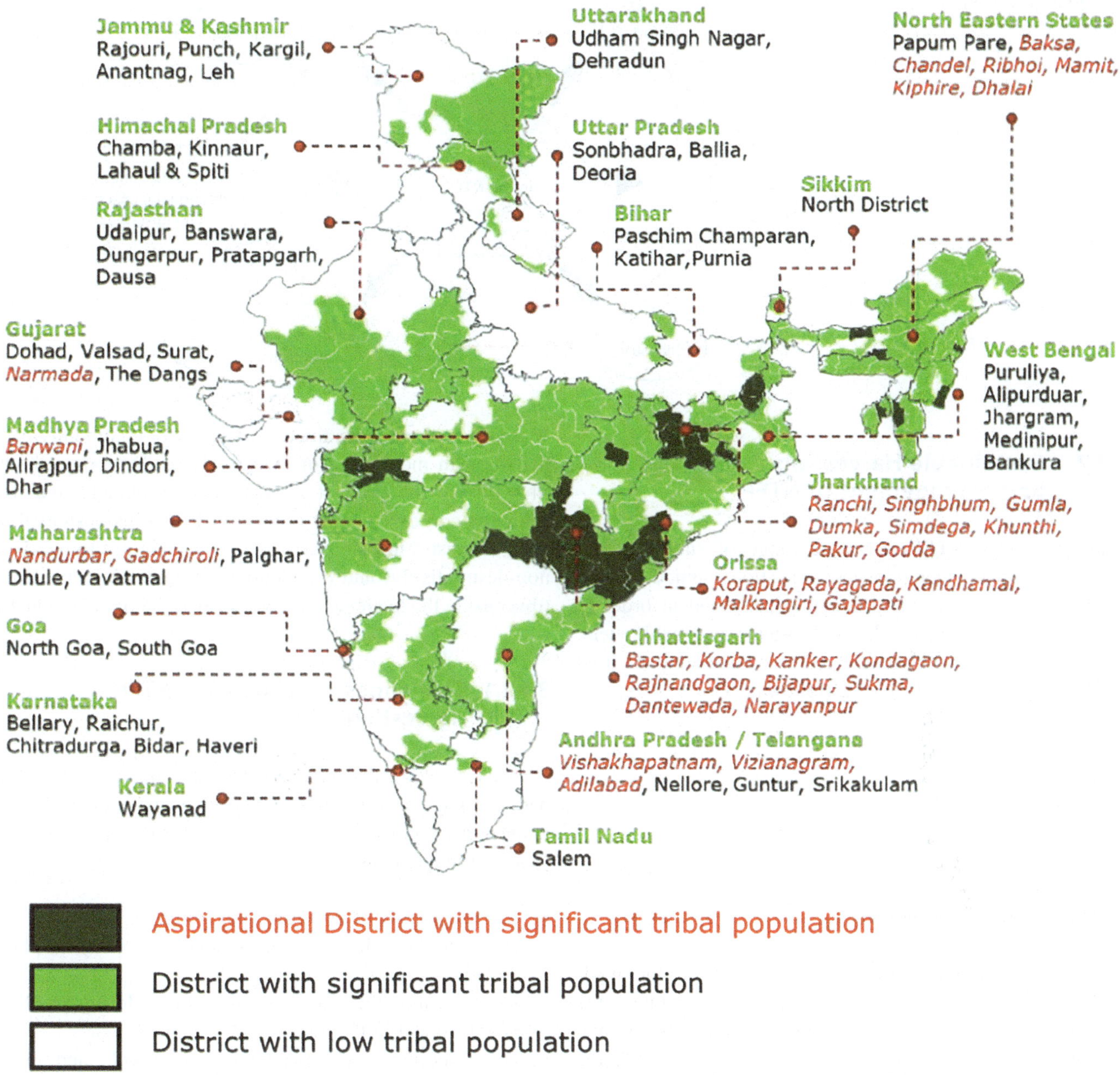

Fig. 39.1 Districts with concentration of Scheduled Tribes in India (TRIFED 2020)

according to an estimate, it could vary between 10,000 and 50,000 tons annually running into billions of Indian rupees. Furthermore, more than 60% of NTFPs go unrecorded and is consumed or bartered by about 15 million population living in and around the forest (Prasad 2007).

Management of NTFPs is a subject of interest for various institutions and researchers working in the field of conservation biology and field foresters who are working as stakeholders of the resources (Fig. 39.2). The complex interaction of ecological and economic aspects of NTFPs has been studied for the past 3–4 decades in India. High demand by the herbal sector has seen many medicinal and aromatic plants (MAP) become extinct in the wild especially in the parts of central and northern India, whereas other reasons of constant pressure are changing land use, poaching and indiscriminate harvesting practices.

This chapter analyses a detailed account of sustainable NTFP management and harvesting practices in India, including conservation and livelihood dependency of tribal communities, marketing and enterprise development initiatives and NTFP-related policy environment in India.

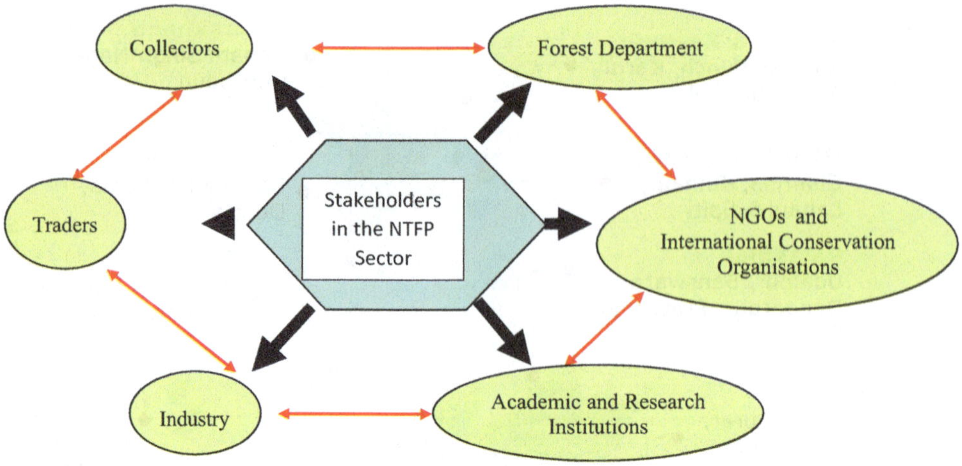

Fig. 39.2 Stakeholders in NTFP sector

39.2 Sustainable Harvesting and Management of NTFP

Management of NTFPs covers a range of subjects from silviculture to regeneration and collection methods in the field. NTFPs need to be sustainably harvested in order to preserve their medicinal value and foster a sustainable yield. Plant species have been categorized based on their parts used for NTFP purposes which are as follows:

- Fruit
- Leaves
- Flowers
- Bark
- Gum/resin
- Root/rhizome
- Other medicinal plant parts

In theory, any harvesting plan should consider the availability of resources, the rate of use and the renewal rate of the resources. Thus, plans that promote harvest below the resource regeneration threshold and that do not modify the natural prevalence of the harvested species should only be used (Lawes and Obiri 2003). Seedling mortality, even if the seed germinate, over 90% of the new seedlings usually die before becoming established in the understory. Only a very small fraction of these seedlings (less than 1 in 10,000) will make it to canopy and produce fruits, so there are many biological factors we need to consider to understand sustainable harvesting.

The protocols of harvesting NTFP at the individual level, population level, ecosystem level and genetic level for fruit, leaf, root, bark, stem and seed require adaptive management strategies based on available information with forest managers and biologists; they get inputs from harvesters and traders as well. Sustainable harvesting protocols are based on ecological parameters (that ensure community participation in monitoring of the resources), harvest methods (that ensure the product is harvested sustainably and the source is conserved), raw produce quality (this ensures that the harvest takes place at the right time and non-destructively) and production and processing standards (that ensure the hygiene, quality and shelf life of the product).

39.2.1 Challenges in Sustainable NTFP Harvesting

Various ecological characteristics can make sustainable harvesting a very difficult objective to achieve, under NTFP management practices, where many actors are involved. Major problem areas are as follows:

- The irregularity of flowering and fruiting
- The high diversity and low population density of plant species
- The high mortality and low success rate during seedling establishment in the field
- The importance and availability of insects and other animals for pollination and seed dispersal
- The sensitivity of population structure to changes in the level of natural regeneration.

  India boasts an impressive botanical diversity, hosting over 46,000 plant species within its borders. Among these, a substantial number—approximately 760—are actively harvested from the wild for their medicinal properties, catering to the extensive herbal medicine industry prevalent in the country. However, this diversity of NTFPs is increasingly threatened by the interplay of factors: rapid deforestation and habitat destruction, indiscriminate collection, and spurious substitutes in the market.

The process of sustainable exploitation may have various steps like (1) species selection, (2) forest inventory, (3) yield condition, (4) regeneration surveys, (5) harvest assessment and (6) harvest adjustments.

39.2.2 Important Indicator of Sustainable NTFP Harvesting

- Monitoring the seedling, sapling and adult tree densities helps in appraising the sustainability of the current harvest.
- Size class distribution of trees is an important indicator of regeneration rate.
- The intensity of harvesting should be reduced as per the approach in the working plan if the density drops while monitoring the size class distribution. For example, if aonla trees are reduced in number in a particular forest patch, we must reduce the quantity of harvest from that area, at least there should be some fruits in the trees for future regeneration.

39.2.2.1 Protocols for Harvesting NTFPs

Plant species should be harvested under the best possible conditions avoiding wet soil, rain, or exceptionally high air humidity. For instance, plants with volatile oils, stems or flowers and leaves, should always be gathered on a sunny day after the dew has evaporated off the plants but before the full heat of the day.

Retention and management protocols: Considering the gatherer's perception, retention principles advised for different plant parts during collection are given below:

- A quarter of the fruits harvested must be left for regeneration.
- Thirty percent must be left for regeneration after harvesting seeds.
- Twenty percent of the population and 30% of each plant's flowers must be left unharvested during flower harvesting.
- Seventy percent of plant leaves must be left on the trees at the time of harvest. Similarly, 80% of the plant population must remain unharvested in order to harvest roots, or bulbs.
- Unless otherwise advised, annual plants' leaves and roots should be harvested after flowering to ensure regeneration and higher product quality when harvesting the entire plant, leaving at least 30% of the population untouched.
- While harvesting roots of perennial plants, some plants from each life stage should be left out. For species that regenerate from portions of roots or root crown a portion of the root should be left in the ground or the whole or divided crowns should be replanted, as appropriate (e.g. *Asparagus* sp.).
- Many herbs, such *as Cymbopogon martinii* and *Mentha piperita*, lose some of their aromatic oils when it rains, thus harvesting should be done 2–3 days following the rain to allow the plant's oil to re-accumulate.
- The harvest areas where the target plant can be easily located and where plant materials are expected to be of high quality and uncontaminated by pollution and other adverse environmental factors should be chosen by the collectors. Unless it is organic, picking herbs along roadsides or in agricultural areas should be avoided.
- To harvest roots (*Asparagus racemosus, Decalepis hamiltonii*), dig a hole that is parallel to the roots and straight down. Remove dirt from one side of the hole gradually along the direction of the root. After that, just slide the root into the hole sideways. Comparatively speaking, this approach will cause less damage to the roots than traditional digging.
- Harvested stems or branches must be left upside down in bunches for a few days if herb leaves are to be collected. As a result, the sap found in the stems and branches will enter the leaves.
- To prevent undue damage to the tree, sharp-cutting instruments such as chopper, sickle, secateurs, scissor and hedge scissor must be used. Using scissors or a sharp knife, harvesting leaves, stems, or blossoms from tender, non-woody stemmed herbs is a simple task. To prevent contamination, cutting instruments need to be sharpened and cleaned in between collections.
- When harvesting a whole plant (like *Andrographis paniculata*), it is important to limit the amount taken from each population so that there is still enough leftover for the population to regenerate. If the species is spread by seed, harvesting should only take place once the fruit has ripened and the seed has been discharged.
- Plant species that are prohibited from being harvested by state regulations because of worries about overharvesting, such as *Taxus baccata* in the Himalayas and Sarpagandha in Madhya Pradesh and Chhattisgarh, or those listed under RET (rare, endangered and threatened) or on the Convention of International Trade in Endangered Species (CITES) (Table 39.1) (Box 39.1).

Prescribed Silvicultural Manipulation for NTFP: The goal of silvicultural interventions is to provide an ideal environment in natural forests for the growth of the most desired NTFP species. The most challenging task in managing non-timber forest products (NTFPs) involves silvicultural manipulations due to a variety of species involved, as well as their differing needs for development and regeneration.

Silvicultural requirement: The conditions that different plant species need to grow, such as light, space and soil

Table 39.1 Examples of commercial NTFP trees, which require an immediate care of germplasm conservation before they deplete from natural forests

S. no.	Species of NTFP	State	Region
1	*Aegle marmelos* (bael) *Terminalia arjuna* (arjun)	Madhya Pradesh and Chhattisgarh	Central India
2	*Garcinia indica* (kokum) *Cinnamomum zeylanicum* (dalchini/cinnamon) *Decalepis hamiltonii* (mahalikizhangu)	Karnataka and Tamil Nadu	South India
3	*Saraca asoca* (ashoka) *Litsea glutinosa* (maidalakdi) *Aquilaria malaccensis* (agarwood)	Tripura and Odisha	Northeast India
4	*Commiphora wightii* (guggal)	Gujarat and Madhya Pradesh	North central India
5	*Emblica officinalis* (aonla)	Madhya Pradesh and Chhattisgarh, Uttar Pradesh, Rajasthan, Karnataka	Central and South India
6	*Buchanania lanzan* (chironji) *Acacia catechu* (Katha)	Madhya Pradesh, Chhattisgarh, Uttar Pradesh, Rajasthan	Central India and Himalaya foothills
7	*Cinnamonum tamala* (tejpatta)	Uttarakhand, Tamil Nadu, Karnataka, Assam, Meghalaya and Tripura	South India, Northeast India and Northern Himalaya
8	*Taxcus baccata* (thuner)	Uttarakhand	Northern Himalaya
9	*Sapindus* sp. (ritha)	Andhra Pradesh, Tamil Nadu	South India

quality, vary. Certain species may be shade carriers, while others may be shade demanders. Some species are light demanders. While certain species may survive in dry environments, others require more water to grow than others. Plants tend to occupy distinct forest stories and different sorts of habitats based on their development habits. Shrubs often form a lower canopy, whereas most trees occupy either the upper canopy or the understorey. The ground flora is made up of grasses and herbs. Shade-bearers are smaller trees. Following operations are the examples of possible silvicultural modifications.

- Seedling adoption and weeding around the desired plant
- Manipulation of ground vegetation
- Manipulation of the middle storey
- Manipulation of the top canopy

Box 39.1 Biodiversity Conservation and Development—For Research on NTFP

National Working Plan Code (NWPC), 2014

For studying the impacts of ecological, phenological and climate change along with the harvesting practices on forests, it was recommended to designate permanent plots of appropriate sizes for the formulation of safe harvesting protocols, emphasizing the standardization of optimal harvest limits to ensure the regeneration of species (NWPC 2014). The exploration of various harvesting regimes is suggested to determine safe limits, including:

- 100% harvest of the marketable parts (by removing the whole number)

(continued)

Box 39.1 (continued)

- 75% harvest of the marketable parts (by leaving 25% of the whole number)
- 50% harvest of the marketable parts (by leaving 50% of the whole number)
- 25% harvest of the marketable parts (by leaving 75% of the whole number)

The plots should be established in triplicate and revisited annually to assess the number of new recruits. The impact of the above removal should be evaluated using standard mathematical formulas. The area under NTFP is shrinking because of the large-scale supplanting of timber species in the decades after independence. Tribal communities are the best conservators of the forest, if only the forest can sustain them. The symbiotic relationship between tribes and forests needs to be appreciated and encouraged. The area surrounding a forest village (at least 5 km around the village) should be reserved only for NTFP forests, and no economically rich timber species plantation should be planted in this area.

39.2.3 Strategy for Sustainable Harvesting of NTFPs

Enrichment planting or seeding: The goal is to manage the region for the conservation and development of a particular medicinal plant or a group of medicinal plants and other NTFP species when an autonomous working group for the conservation and management of NTFPs is established. It is possible that some NTFP species that were once common in great concentrations in some woods have become rare. These spaces could be used for seeding or enrichment planting. *Chlorophytum species, Rauvolfia serpentina, Withania somnifera Andrographis paniculata, Asparagus species, Aloe barbadensis, Centalla asiatica, Curcuma spp.* and many more are some of the key NTFP and medicinal plant species. When planting and seeding for enrichment, bear the following points in mind:

- Several studies have been conducted on seed bank analysis to learn about natural regeneration studies, and it is possible that various species have distinct methods of regeneration. In general, species reproduce by seed; but some species also do so by cuttings or other vegetative parts, while still others do so through seedlings that have been cultivated in nurseries. To increase the number of NTFP species in forests, a thorough understanding of the most effective regeneration method is necessary.
- The species selected for enrichment planting or seeding should be in great demand or whose population has dwindled in the forest due to overexploitation.
- Since a majority of NTFPs grow in the understorey, it is possible to find even ideal forest sites for NTFP planting and enrichment seeding. There is no need for artificial support because the climbers have the support of the existing trees.
- A few square kilometres may be included in the NTFP operating circle. Not a very big space is needed. The area selected should be such that it meets the ecological requirement of the species or a group of species.
- The most desirable seed and other planting material should be gathered for enrichment sowing and planting.

Prescribe deferred or rotational collection: The continuous extraction of non-timber forest products (NTFPs) from a particular forest area may lead to a reduction in the quantity of these plant resources over time. Sustainable harvesting is often not implemented in the case of most NTFPs. Since they are a popular resource, collectors usually try to get as much of them as they can. When roots or rhizomes are harvested, the plant is harmed; fruits are harvested before they are fully ripe. Significant portions of the plant are chopped during the harvesting process, even if only the fruits or flowers are needed. Bark accumulation causes girdling, which ultimately kills the plant. The gum and resin tapping also causes a major harm to the plants. As a result, it is advised that some locations take a short break while collecting in other places continues. Usually, 5 years' rest may help to rejuvenate the NTFP resource. The period of rest can be decided by keeping in view the condition of the resource. The JFMCs may also be motivated to practice the rotational collection of NTFP in their area.

The Madhya Pradesh Biodiversity Rules 2004 formulated a required closure of territory for a certain term, based on the Biodiversity Act 2002. There may be a seasonal shutdown. For one or more seasons, the entire area may be closed. The collection may be restricted in some locations. This restriction might only apply for a few years. The needs of the species and the region can also be taken into account when determining the harvest limit.

Control on fire, grazing and shifting cultivation: Climbers, creepers, annuals and herbs make up a significant component of NTFPs. In forests, fire and grazing seriously harm these plants. However, trees and plants are somewhat protected from the harm. Frequent fires destroy ground flora, including herbs and annuals, permanently. These plants burn up in fires because they cannot tolerate the extreme heat. Because of frequent fires, a number of herbs and ground flora have become locally rare. Annual fires produce arid conditions that lead to the extinction of organisms that require rainfall. Additionally, the flames change the qualities of the soil, which hinders some species' ability to grow.

Numerous species are trampled by the hooves of numerous cattle grazing over the area, and the animals consume the vulnerable sections. There is insufficient regeneration as a result of eating fruits, flowers, etc. Thus, it would seem that a combination of fire and grazing has detrimental effects on the ground flora, which includes climbers, annuals and herbs. In a similar vein, shifting agriculture leads to unfavourable circumstances for these species' growth. To safeguard rapidly declining ground-based NTFPs from open forests, these practices must be regulated.

Establish Preservation Plots in Each Felling Series: At least one preservation plot must be included in every felling series in order to monitor the fast-declining biodiversity of forests caused by strong biotic pressure and unsustainable harvesting practices. These preservation plots can be sites that are off-limits to harvesting. These plots, which will belong to the local population, could range in size from 5–10 to 50–100 ha depending on the amount of forest land available. It is possible to locate these preservation plots both on the ground and on the map. These preservation plots ought to be permanently designated on the ground with clear markings. In order to prevent damage and to enable the long-term monitoring of the effects of ecological, phenological and climate change, the designated area needs to be fenced.

Increase research efforts: Data regarding the management of NTFP plants is scarce. Whatever data is available must be documented, and study must be started on them. This is required in order to manage the resource in a scientific manner. Research topics may include ecological (including reproductive and regenerative ecology), forestry, economics, production, and processing. Researchers and managers can exchange documents and best practices in management. Instead of concentrating solely on the variables affecting the collection and usage of NTFPs, research ought to evaluate the several aspects affecting their administration. Little emphasis has so far been made to developing these habitat traits and situating NTFPs within a typological framework that explains the patterns of NTFPs management (Ruiz 1995; Peters 1996; Bhattacharya 2013).

Include management of NTFPs Working Plan Code: The National Working Plan Code (NWPC) of 2014, the latest iteration of its kind, outlines provisions for the management of non-timber forest products (NTFPs) within forest management units (FMUs). This document serves as a comprehensive guide for the formulation of working plans across the country, ensuring consistency and adherence to standardized procedures among all Indian states (Kumar et al. 2020). The NWPC (2014) has several provisions on NTFP (including MAPs), Estimation (Appendix iii of NWPC), Assessment of NTFPs and Optimization of Forest Resource Utilization (Chap. 7). These sections of the NWPC mandate that the forest department and forest management units (FMUs) document various aspects related to non-timber forest products (NTFPs). This includes recording estimated harvestable quantities, details of plant types and their parts used, utility of the products, geographical distribution, scientific and local names of species, as well as information on demand, supply and import/export dynamics of NTFPs (NWPC 2014).

The Working Plan Officer (WPO) may plan and carry out survey, sampling and evaluation in accordance with the NWPC (2014) in order to estimate the number of prioritized NTFP species that will initially be found inside the chosen grids. Generally speaking, local research, plot enumeration data and accessible historical documents are used by WPO to evaluate possible NTFPs. Plot enumeration, data on shrubs, herbs and climbers and the identification of a few NTFP species can all be used to estimate resource species. It is possible to use data gathered for various studies and/or kept in the JFM locations.

For each species of tree, shrub, climber, grass, herb, lichen, fungus etc., summarized estimated quantities may be recorded for each compartment, village, or other management unit along with the scientific name, local name, type of plant, part and utility, area (ha), quantity per hectare, estimated harvest/hectare, etc. For MAPs, a different estimate could be made.

Based on the potential NTFP calculation of market demand and production, a priority list of the most significant species may be determined for the purpose of optimizing forest resources. To promote the sustainable use, management and value addition of non-timber forest products (NTFPs), additional features for general prescriptions can be developed within the overlapping working circles. To preserve biological diversity's capacity to satisfy the needs and ambitions of both the present and the future, its components should be used in a way and at a rate that does not cause a long-term reduction in biological diversity.

To help issuing certificates for NTFPs, a good practice guide may be created to guarantee the quality and purity of the material that has been collected as well as its proper storage, controlled transit and chain of custody. To promote sustainable management and value addition, knowledge on best times to harvest, grade and store commercially significant species under the NTFP should be imparted in working plan (Bhattacharya 2013). For protecting and managing NTFPs from government forests in a sustainable manner, the state government may establish the necessary regulations. Regulating and conservation services for NTFPs are sadly not being extended by the Forest Department or the Panchayat, the new owners of NTFP resources.

There isn't yet a governance structure in place to guarantee that harvesting is sustainable. The development and application of Good Manufacturing Practices (GMP) and Good Collection Practices (GCP) for numerous commercial NTFPs in various places yields positive outcomes when locals are actively involved. It is necessary to adhere to strong protocols for sustainable harvesting procedures as well as community-based ecological monitoring techniques. Harvesters, local collectors, dealers, forest managers and scientists must all receive a proper training for all parties involved to see the advantages of sustainable harvesting. Hence, the Working Plan Organization (WPO) may recommend suitable measures such as temporarily closing an area for the collection or extraction of specific forest produce for a designated duration (closed area); imposing restrictions or prohibitions on the collection or extraction of any forest produce during certain periods of the year (closed season); setting limits on the quantities of forest produce to ensure sustainable harvesting practices for the future (sustainable harvesting limits); promoting sustainable harvesting and collection methods; requiring collectors to submit returns; and other relevant actions. Improving site production while preserving regrowth, the health and vitality of resources, and the biodiversity of the forests is the best way to guarantee an increased productivity of NTFPs (Box 39.2).

Box 39.2 Seven Step Methodology for Community-Based NTFP Management

(National Working Plan Code 2014)

STEP 1: Identify the NTFPs to be managed, their uses and the people who are to manage them

1.1. Which NTFPs are to be managed?

1.2. What are the uses of these NTFPs?

1.3. Who will manage these NTFPs?

(continued)

Box 39.2 (continued)

STEP 2: *Determine the community's rationale and goals for sustainable NTFP management*

2.1. Rationale for sustainable NTFP management

2.2. Goals for sustainable NTFP management

STEP 3: *Document the community's existing NTFP management system*

3.1. What is the community's existing NTFP management system?

3.2. When to document the existing NTFP management system?

3.3. How to document the existing NTFP management system?

3.4. Questions to ask about the existing management system

STEP 4: *Evaluate the existing system's potential for sustainable NTFP management*

4.1. The potential for sustainability of the existing system

4.2. Ecological predictors of the potential for sustainability

4.3. Social predictors of the potential for sustainability

4.4. Economic predictors of the potential for sustainability

4.5. Political predictors of the potential for sustainability

4.6. Implications for investment of community effort

STEP 5: *Consider incorporating new approaches to NTFP resource management*

5.1. Consider incorporating new approaches to NTFP resource management

5.2. Quantitative approaches used in NTFP resource management

STEP 6: *Develop a plan for sustainable NTFP management introduction*

6.1. What is a management plan?

6.2. How to develop the management plan

STEP 7: *Implement and monitor the management plan*

7.1. Develop a monitoring plan

7.2. Implement monitoring and evaluation activities

7.3. Develop a new management plan

39.3 Livelihood Support From NTFP

39.3.1 Significance of Forest Products and Dependence of Forest-Dwelling Communities on Forest Products

Forest fringe communities gather forest produce not only for personal consumption but also for commercial purposes, contributing significantly to their income. Sales of forest products account for 40–60% of the total income for households residing in and near forested areas. A study conducted in northeast India reveals that tribal communities utilize 343 non-timber forest products (NTFPs) for various purposes, including medicinal use (163 species), consumption as edible fruits (75 species) and as vegetables (65 species) (Saha and Sundriyal 2012). The communities surveyed displayed a complete reliance on forest resources for firewood and materials used in house construction. Additionally, non-timber forest products (NTFPs) accounted for 19–32% of their overall household income.

Forests in India play a crucial role in meeting the country's energy and fodder requirements. They fulfill nearly 40% of the nation's energy needs, with over 80% utilized in rural areas, as well as contribute about 30% to the fodder for the cattle population. Non-timber forest products (NTFPs) are vital for the sustenance of rural communities, constituting around 70% of forest products and contributing about 40% to state forest revenues. Particularly for landless families and marginal farmers, NTFP-related activities often serve as their primary source of income. Around 275 million impoverished rural individuals in India, constituting 27% of the total population, rely on Non-timber Forest Products (NTFPs) to sustain their livelihoods, as per studies by Bhattacharya and Kandya (2016) and Bhattacharya and Hayat (2004). This dependence is especially pronounced among half of India's 89 million tribal population, considered the most marginalized segment of society, residing in forest fringe regions. An estimate suggests that the Non-timber Forest Product (NTFP) sector has the potential to generate approximately 10 million workdays annually in India. This sector engages significant populations of rural and tribal individuals in activities such as collection, processing and marketing of NTFPs. Additionally, an equal number of individuals are likely involved in healthcare practices that utilize these resources. The environmental, social and economic significance of NTFPs, including medicinal plants, to countries like India is substantial.

The Planning Commission report of the sub-group-ii on NTFP and their sustainable management in the 12th five-year plan highlights the importance or role of this natural resource in the following manner:

- Ayurveda, the world's oldest healthcare system and India's heritage, completely depends upon forests as almost 85% of the raw herbs used in this system of medicine are sourced from this resource. Similar is the case with other systems of medicine included in AYUSH (Unani, Siddha and Homoeopathy) also.
- The NTFP sector is one of India's largest unorganized sectors having a dependent population of about 275 million, and with a business turnover of more than Rs. 6000 crore per annum, the NTFP sector has unfortunately been neglected since the pre-independence period. Fortunately, for natural reasons, India remains No. 1 in the case of a few items like lac because it is only here that the best quality lac is produced that too in substantial quantities.
- The collection and utilization or sale of Non-timber Forest Products (NTFPs) are primarily carried out by women, highlighting a significant connection between NTFPs and women's financial empowerment in forest-fringe areas. Particularly for vulnerable groups such as primitive tribal communities and landless individuals, NTFPs play a crucial role in providing critical subsistence during lean seasons.
- NTFPs have a tremendous potential to create large-scale employment opportunities thereby helping in reducing poverty and increasing empowerment of tribal and poor people of the poorest and backward districts of the country.

Role of women in NTFP sector: Strengthening NTFP is a sure way to women's empowerment in tribal areas. Women are mostly dependent on NTFPs due to the nature of their production, quantity, collection procedure, processing and local selling (Bahuguna 2000; Bhattacharya and Hayat 2004). Most NTFPs are seasonal and we observe there are various cultural rituals attached in society which are corresponding with the availability of these NTFPs either fruits, leaves, flowers, twigs, roots/rhizomes, mushrooms and various other animal NTFPs. They use most of NTFPs for their consumption, and they have better knowledge of resources.

There are just a few research-based studies that provide information on the use of 62 different NTFP products in diet, such as vegetables and regular medication, by women in Tamil Nadu, Kerala and Karnataka. Similar to this, women in the north-eastern states of Tripura, Arunachal and Meghalaya frequently visit wetlands and forest areas to gather leafy vegetables. Of the 25–85 items on the list, all of them are startlingly uncultivated. An enormous portion of India's prehistoric tribes' diets still came from a variety of non-timber forest products (NTFPs) found in various forest types.

Current challenges in livelihood support from NTFP: In recent years tribal and forest-dwelling communities have observed a reduction in the level of their food security, health cover, cash income and dependable wage income; these situations have exacerbated further due to the COVID-19 pandemic, in almost all Indian states. The people who depend on such products are also beset with many problems such as the perishable nature of the produce, lack of holding capacity, lack of marketing infrastructure, exploitation by middlemen, lack of government intervention at the required time and scale, volatile nature of markets, etc. Most of the products are sold in raw form, which is not remunerative to tribal gatherers. After walking long distances, they receive low amounts. Thus, they try to collect as much as possible to maximize their earnings from the day's collection by compromising the products' maturity level. As a result, important NTFP species are being depleted, their natural regeneration is almost nil (Murali et al. 2006; Devi and Bhattacharya 2015).

Economic dependence of tribal communities on NTFPs: Processing and value addition at the primary collector's level boosts NTFP profits by 4–5 times, according to several studies. Simple value-adding tasks that may be completed without any investment at the home and community level, including cleaning, washing, air or sun drying, grading, packing, etc., can also pay gatherers two to three times more. Collectors might be encouraged to adopt sustainable harvesting practices by providing institutional support for marketing, value addition procedures and awareness training (Tewari and Campbell 1995).

Conservationists by nature, NTFP gatherers would typically refrain from harming the foundation of natural resources. Because there are no local institutions, credit facilities, or value-added choices, middlemen and contractors that operate in and near forest areas have been taking advantage of the gatherers. Table 39.2 helps to explain how indigenous communities are economically dependent on NTFPs. The distribution of economically significant MFPs by state is seen in Fig. 39.3.

A list of important NTFPs in terms of their production and collection potential besides importance for livelihood is given in Table 39.3.

Table 39.2 Economic dependence of tribal communities on important NTFPs (TRIFED 2020)

Seasons	NTFPs collected	Economic dependence of tribal and forest-dwelling communities
January–March	Lac (resin), mahua flower and tamarind	In Odisha, Madhya Pradesh and Andhra Pradesh, over 75% of tribal households engage in the collection of mahua flowers, earning approximately Rs. 5000 annually. Additionally, approximately 3 million individuals are involved in lac production
April–June	Tendu leaves, sal seeds and chironji	30 million forest dwellers rely on seeds, leaves and resins from sal trees for their livelihoods. Tendu leaf collection alone provides about 90 days of employment to 7.5 million people, while an additional 3 million individuals are employed in bidi processing
July–September	Chironji, mahua fruits, tussar cocoons and bamboo	10 million people depend on bamboo for their livelihood, while 1,26,000 households are involved solely in tussar silk cultivation
October–November	Lac, kullu gum, aonla, hill broom grass, resins used in incense sticks	3 lakh person-days of employment from the collection of gums

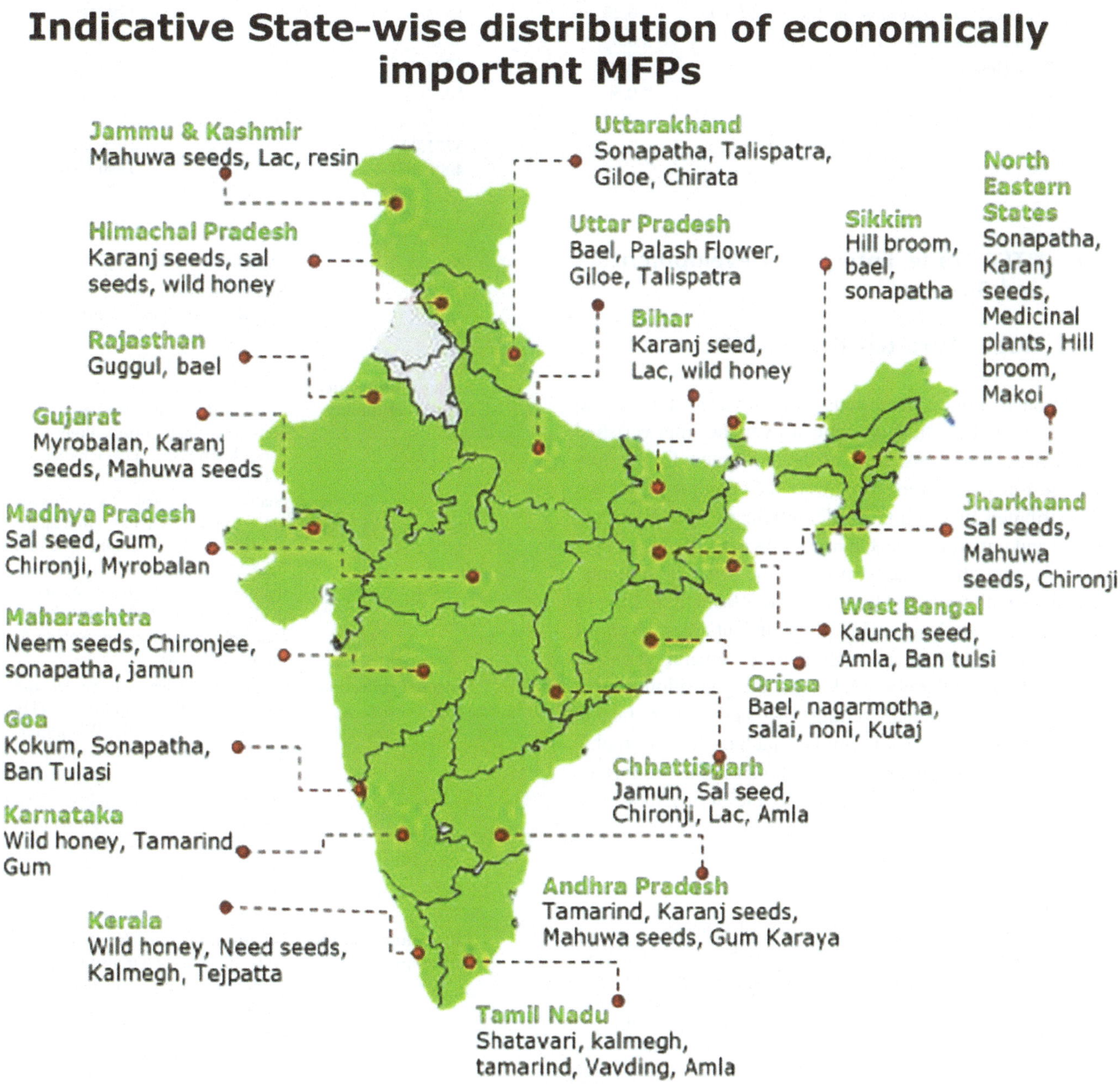

Fig. 39.3 State-wise distribution of economically important NTFPs (TRIFED 2020)

Table 39.3 Estimated production and collection potential of NTFPs (TRIFED 2020)

S. no.	Commodity	Estimated production potential		Estimated collection potential	
		Qty in lakh MTs	Value in Rs. crores	Qty in lakh MTs	Value in Rs. crores
1.	Tamarind (*Tamarindus indica*)	2.00	240.00	2.00	240.00
2.	Mahua Flower (*Madhuca longifolia*)	1.50	122.00	1.00	81.00
3.	Mahua Seed (*Madhuca longifolia*)	1.00	110.00	0.50	55.00
4.	Sal Seed (*Shorea robusta*)	1.60	160.00	1.00	100.00
5.	Tendu	80 (In standard bags)	1040.00	40 (In standard bags)	520.00
6.	Bamboo (*Dendrocalamus strictus, Phyllostachys edulis*)	48.00	12.00	12.00	300.00
7.	Karanj Seed (*Pongamia pinnata*)	0.40	40.00	0.25	25.00
8.	Myrobalan (*Terminalia chebula*)	1.30	78.00	0.75	45.00
9.	Chironji (*Buchnania lanzan*)	0.10	230.00	0.05	110.00
10.	Lac (Stick Lac) (Kusumi and Rangeeni Lac)	0.25	150.00	0.20	120.00
11.	Gum Karaya (*Sterculia urens* gum)	0.05	62.00	0.03	37.00
12.	Wild Honey	0.30	270.00	0.25	230.00
13.	Puwad Seed (*Cassia tora*)	0.50	50.00	0.20	20.00
14.	Neem Seed (*Azadirachta indica*)	0.25	25.00	0.25	25.00
			3777.00 **4000.00**		**1908.00** **1900 Crores**

39.4 NTFP Trade, Enterprise Development and Business

39.4.1 Trade Characteristics of NTFP

In addition to policy distortions resulting from public and private monopolies, traders' control over the impoverished and illiterate forest residents also contributed to the low returns to forest gatherers. They were the outcome of uneven and distributed production by definition, as well as erratic demand and underdeveloped markets. As a result, it is important to examine the unique aspects of the forest inhabitants' interactions with trade and how they differ from the selling of food grains in India's agriculturally abundant areas. Even for those products that are not nationalized, several variables contribute to the gatherers' weak negotiating position against traders in the particular setting of NTFP. The factors influencing the value of the product include its geographic location, unique supply and demand dynamics, as well as the dynamics of interaction between the collectors and purchasers.

Characteristics and nature of NTFPs

S. no.	Issues related to the nature of NTFPs	Explanation and examples
1	**Change in production**	The majority of NTFP commodities has annual output swings that range from 300 to 400%, which causes a large change in supply. This also depends on the annual rainfall. Good seed year yields large crops for many NTFPs. This stands in stark contrast to agricultural goods, wherein the state's production variance seldom surpasses 20% of the average
2	**Lack of uniformity in product quality**	Since NTFPs are natural products, their qualities can never be consistent. Size, form, colour, scent and other physical characteristics are contingent upon variables that change from season to season and place to place, such as temperature, moisture, soil and rainfall. One of the biggest drawbacks of marketing is that consumers desire consistent quality and a regular supply. This is especially important for users in the industrial sector
3	**Seasonal collection**	Even though there may be demand all year round, the majority of NTFPs is collected seasonally. When they are sold locally during the flush season, there is more supply than there is demand, which drives down prices. When the short season falls during India's monsoon, it poses even greater challenges because drying and shipping the goods become extremely difficult
4	**Low volumes of the collection**	NTFPs in the forest are dispersed, and collecting and transporting them are frequently not financially feasible due to the quantity available from one location. Lower returns are the outcome of the producers'

(continued)

S. no.	Issues related to the nature of NTFPs	Explanation and examples
		diminished negotiating power due to the low volume of NTFPs
5	**Fluctuating demand**	The manufacturing of its replacements and the shifting export climate both have a significant impact on the market for these products, which varies greatly. Primary gatherers may benefit from it at times For example, traders who camp in the villages of Tripura and purchase hill brooms at a premium price of 35/kg are in high demand from the neighbouring state. However, the gatherers are frequently at a disadvantage due to a combination of unpredictable demand and unpredictable supply
6	**Exports**	Due to their primary export function, some NTFPs are quite vulnerable to global demand and pricing. For instance, the boom-and-bust phenomena is causing overharvesting or a price collapse in certain products, such as aromatic oils, gums, herbal remedies and cashew nuts. This cycle has recently affected Indian rubber and shellac

Nature of the actors involved in NTFP sector

S. no.	Issues related to nature of actors involved in NTFP trading	Explanation
1	**Poverty of gatherers**	Due to their poverty and ongoing debt to intermediaries or landowners, the majority of NTFP collectors and extractors lacks control over their labour and other terms of exchange. Consequently, the low price at which NTFPs are disposed of is influenced by both underdeveloped rural credit markets and acute poverty (Mott 1998)
2	**Gender dimensions**	For female entrepreneurs, the aforementioned issues become more serious. Burdened with various tasks within the household typically ascribed to women, their ability to explore for far-off markets is curtailed. Their limited production scale exacerbates the issue further, trapping them in a cycle of low output, a tiny market and a tiny excess (Agarwal 1989)

(continued)

S. no.	Issues related to nature of actors involved in NTFP trading	Explanation
3	**Too many intermediaries**	The link between main gatherers and end users is lengthy and vertical. Certain village-level vendors work for commission agents or wholesalers situated in the market, who subsequently provide to other wholesalers based outside the state. In the majority of situations and frequently even down to 10%, the gatherers' portion of the ultimate price for therapeutic plants is less than 33% (Saigal et al. 2002). Gatherers are limited in the number of intermediates they can choose from, even with a high number of them. A competitive and effective system should provide multiple buyers with options
4	**Nature of buyers**	Since they can satisfy the primary gatherers' urgent needs, the local trader's intermediaries are able to hold a prominent position in the marketing network. They have a strong network of procurement right at the tribe collector's doorstep, provide prompt and timely credit and make prompt payments. The local trader's (Kochia) control is further strengthened by a number of issues, including poverty and the impoverishment of the gatherers; bad marketing structure; lack of understanding of market prices by the gatherers; and ineffectual state agencies. In addition, a highly segmented market, inadequate communication and transportation infrastructure and unequal bargaining power between buyers and sellers all contribute to the situation's increased profitability for intermediaries

A major hindrance to the efforts of various government and non-government organizations (NGOs) assisting the collectors is the lack of precise and dependable market data. The indigenous people of India typically gather and market wild medicinal herbs unprocessed, without much added value. Only sales conducted in nearby villages or at the weekly village marketplaces, or haats, will provide the NTFP collector with access to customers. Before they reach the organized sector, a sizeable amount of their collection is sold to local middlemen like commission agents and contractors.

Furthermore, access to the market is restricted for perishable goods and goods containing active ingredients that alter or degrade over time due to regional restrictions. The issue is made worse by the comparatively small volume of collecting, which puts the tribal population in the area in a vicious cycle of poor productivity, a tiny market and a tiny amount of marketable surplus. As their need for the conversion of little production into cash becomes more pressing, this restricted marketable surplus continually erodes their bargaining power, making them more susceptible to exploitation.

Nature of the market and marketing operations

S. no.	Nature of the market and marketing operations	Explanation
1	**Restrictions related to storage, transportation, processing, and marketing**	Different items have different restrictions governing the quantity of NTFPs that producers or gatherers may store. Additionally, growers of specific forest products whose production exceeds a set quantity are required by law to register Similarly, for moving NTFPs, transit licenses granted by the Forest Department are still necessary for most goods for their travels inside and outside the state. There might also be limitations on value addition at the primary level. For instance, Mahul leaf (*Bauhinia vahlii*) plates manufactured of Mahul leaves need a transit authorization. Permission for higher-level processing must be obtained from the Forest Department through registration. The processor/manufacturer is required to turn in the required accounts, refunds and declaration. Enforcing these laws is the responsibility of the Forest Department These restrictions and permits mean that the traders are continually reliant upon and affected by the actions of the Forest Department Farmers find it challenging to stay informed about the most recent changes to the transit regulations, which are subject to frequent changes. The government order that established the new restriction did not include a justification. In a similar vein, when limitations are lifted, harassment of the forest inhabitants persists as the new regulations are not made public Even in cases when NTFPs are not nationalized, laws limiting their freedom of movement create uncertainty in the market and prevent gatherers from optimizing their returns on investment. Government regulations not only encourage corruption but also provide current market participants a stronger grip, making it more difficult for new competitors to enter the market. As a result, a small number of buyers have monopolistic advantages
2	**Harvesting**	The tribal people occasionally have a tendency to gather product before it is ready to be harvested or to utilize extraction techniques that are not scientific and may endanger the trees in an effort to boost their immediate revenue. For instance, similar extractive techniques are frequently used to injure trees when harvesting kullu gum in Madhya Pradesh. The tribes have the freedom to gather NTFPs thanks to the current access rights, but they are not encouraged to make a sustained effort to develop the forests. Similar to this, post-harvesting methods receive minimal attention, which results in significant losses in terms of both quantity and quality
3	**Grading and storage**	Enhancements are also required for the produce's grading and storage. Produce is brought by the gatherers in a mixed state and at the collection site it is graded, incurring additional costs. Lower prices are paid for ungraded commodities. In general, the blended product is priced according to the lowest quality The quality of NTFPs is impacted

(continued)

S. no.	Nature of the market and marketing operations	Explanation
		by temporary storage, when products are kept for many days before being transported. During the monsoon season, warehouses or storage facilities may encounter issues such as insect infestations, rat abandonment, or elevated moisture levels, leading to a decline in product quality. Therefore, the potential returns on NTFPs like tamarind, mahua and aonla could be significantly enhanced by storing them in cold storage for a period of 5–6 months
4	**Market information**	The NTFP collector lacks sufficient knowledge and awareness of purchasers, the going rate in the market, and the regulations. Gatherers have no idea what customers need or desire. The likelihood that the producer/gatherer will have access to this information decreases with the length of the marketing chain. Low pricing and waste arise from a lack of information between what the final buyer wants and the actual product
5	**Lack of infrastructure facilities**	Limited infrastructure facilities prevent the full utilization of forest resources, leading to the collection of NTFPs primarily from the forest periphery

39.4.1.1 Value Addition and Processing of NTFP

Value addition is the process of increasing the worth of a product or service at each stage of production, beyond its initial cost. This enhancement can include additional features, improvements, or modifications that increase its economic value to customers. By adding value, companies can attract more customers, leading to increased revenue and profits. Processing of NTFP involves performing a series of physical or chemical operations on (something) to improve the product (or collected raw material) condition, change it, or preserve it. Processing is usually done in a processing unit, with the help of equipment or machinery structures.

Primary-level processing includes drying, cleaning and grading, and is undertaken by the local community at the village level with the help of self-help groups (SHGs) or Panchayat. Raw NTFP material is semi-processed (secondary processing) with the help of equipment in a processing unit, for example, aonla fruits are de-seeded, sliced and dried. The secondary or final processing includes packaging and labelling of the product, for example, dry aonla converted to powder, or used with coconut hair oil for hair oil, from aonla fruit to aonla murabba. A few more examples of value-added primary forest-based products are given below:

Forest honey: Bitter honey; sweet honey; pepper-, ginger and saffron-infused honey; neem, mahua and karanj medicated honey.

Bee wax: Candles; camphor, eucalyptus, coconut, olive, lavender, geranium and citronella-added balms.

Tamarind: Deseeded tamarind bricks, candy and chutney.

Shikakai: Powder with vetiver, hibiscus and aonla added.

39.4.2 Marketing Issues for NTFPs

According to Yadav and Mishra (2010), the NTFP market is an oligopsony, with a very large number of ill-informed vendors and a comparatively small number of well-informed (and secretive) customers. While the seller's information flow is accidental, the demand side's information flow is purposefully restricted to intentionally influence the market price in their favour. During the last few decades, the majority of NTFP research conducted in India has concentrated on biological aspects to address issues related to useful part, collection time, nutritive efficiency, factors influencing NTFP production, effects of silvicultural practices on the yield of different NTFP, and NTFP inventory methods. The research has been progressively incorporating market participants, marketing processes, future strategy, product analyses and certification as marketing in recent years.

Following features generally characterize the current NTFP market (Saxena 2003):

- Lack of scientific quality parameters or standards—traders go by physical characteristics, giving scope for reducing prices arbitrarily.
- Lack of value chain knowledge and market informally linked.
- Absence of scientific weighing and volume assessment at the harvesters'/haat traders' level, resulting in collectors not getting fair prices.
- Prevalence of adulterants in the market (also called substitutes) which are difficult to identify but put added pressure on forest resources.
- Demand from across the country and abroad is met by informal market players.
- Wide swings in prices depending on season and factors outside the local domain.

- The practice of even Rare and Endangered and Threatened species (banned items) traded and even billed in different names.
- Variable tax and levies (VAT, GST, Transit taxes etc.) between states and restrictions on inter-state transit.

39.4.3 Certification of NTFPs in India

Certification is a voluntary market tool that attracts 'green consumerism' by ensuring the sustainability of the resources. Certification is done based on developed principles, criteria and indicators towards achieving sustainable forest management (SFM). NTFP-based certified products market is still to develop and tends to occupy the niche market for high-quality products, particularly for medicinal plants, honey and other food items (Bhattacharya and Manoharan 2017) (Table 39.4).

Certification is envisaged as a market-driven mechanism for the promotion of sustainable forest management in four main ways:

- By establishing standards for forest practices and management that guarantee a certain level of management performance. Producers and harvesters can use the principle, criteria, indicators and verifiers of sustainable harvesting and NTFP certification anywhere as a model for best practices.
- By ensuring that NTFPs are coming from legal sources and that the resource is sustainably sourced
- By enhancing market opportunities for products from sustainably managed forests.
- By promoting public education about improved forest management for both producers and consumers.

There are, in general, three types of initiatives in NTFP certification (Mallet 1999):

1. Well-managed forests (sustainable forestry) by applying criteria that address ecological, social and economic issues
2. Organic certification
3. Fair-trade system

Like the organically certified product market, the progress of NTFP certification is growing at a slow pace. As per the concept, certified products receive a higher price for sustainable/responsible management. But available evidence is somewhat contradictory and generally suggests that buyers are unwilling to pay a price premium for certified products (Bhattacharya and Manoharan 2017). However, due to an increase in the purchasing power of the Indian middle-class people, and interest in quality products, domestic market opportunities for certified NTFPs are likely to improve in future. Pharmaceutical companies are also trying to understand the prospects of certification and its scope, as in the case of ISO, Bureau of Indian Standards and other food standards; such understanding has helped by marketing quality products.

Table 39.4 Challenges, opportunities, costs and benefits of NTFP certification (Bhattacharya and Manoharan 2017)

Challenges of NTFP certification	Opportunities of NTFP certification
• Unavailability of funds for meeting direct and indirect cost of certification. There is also difficulty in reporting and administrative procedures, which creates heavy workload for unprepared communities • Lack of capacity (technical and ecological knowledge about individual species) • Lack of baseline data, sustainable harvesting levels and resiliency levels • Non-equitable benefit sharing for the local communities. Unclear tenure rights and access • Complex chain of custody • Difficulty in realizing market benefits from certification • Lack of standard policies and difficulty with integrating timber and NTFP certification	• Green public procurement • Foreign direct investment (FDI) in the retail sector by multinational corporations (MNCs) • Participation in corporate and governmental programmes and projects that promote sustainable standards • Foster green public procurement to create demand in the domestic market • Establish policy guidelines for forest certification on government forests and plantations • Provide incentives for small holders to receive forest certification support • Raise awareness of the benefits of certification • Support the development of national standards and indicators with the aim of adapting certification in the national and local contexts
Costs of NTFP certification	**Benefits of NTFP certification**
• The direct costs of certification are the cost of the certification process itself. Direct cost may vary with the size of the area of the enterprise and distance travelled by the certifiers • The indirect cost of the certification process adopts the change of management at field to meet the sustainable forest management standards	• Direct benefits cover sustainable collection of material that secures future availability, improved quality value-added material, ensuring price premium as well as market expansion through international acceptance • Indirect benefits cover conservation of biodiversity by following environmental standards, respecting traditional rights and practices, improvement in social setup, economic development and benefit sharing

39.4.4 Fair Returns to NTFP Collectors

How much should be paid back to the NTFP collectors? Is it the primary collector's pay for obtaining NTFP from the forests, based solely on the worth of their labour? This perspective must be balanced by the understanding that the health of a forest resource is closely related to the health of the dependent populations, even under the assumption that fair wages are provided (Saxena 2003). The national forest policies explicitly acknowledge that these people are the resource's principal stakeholders (Yadav and Dugaya 2013). The recently adopted Scheduled Tribes and Other Traditional Forest Dwellers (Recognition of Forest Rights) Act 2006 guarantees the right of ownership and control of NTFP to the tribal people and forest dwellers. The need for providing the real value of the resource to the primary collectors is of paramount importance in the management of NTFP (Saxena 2003).

One must distinguish between a fair return on labour and the actual price for primary product if the primary goal of a fair price is to increase the incomes of primary collectors. In the context of NTFP management, a higher price for primary products may also have unforeseen consequences, if production is not price responsive. Raising procurement costs as a symptom could result in an unsustainable supply of NTFP resources, endangering the current's future. Offering a 'support price' purely to provide collectors a higher price may have unforeseen implications on the NTFP resource base due to the low productivity of forest resources and the rising demand for items from forests (Saxena 2003).

In this situation, boosting the product's value through methods other than managed raw material pricing is a different strategy to improve earnings for NTFP-dependent communities (Saxena 2003). The goal of the strategy should be to make a supply chain more efficient so that NTFP collectors can extract more value from it. This can be done by reducing waste, improving market knowledge, creating more opportunities for value addition close to the supply chain's initial stages, and providing access to organizational and R&D resources. These days, there are ambiguous, insufficient, or non-existent connections between the value produced by NTFPs and reverse flow inputs to the manufacturing base. The alternative strategy requires a great deal more work and data than what is now accessible. This is also the right strategy to help change people's perspectives and manage NTFP resources responsibly while prioritizing human well-being. It is important to view the shortcomings in the NTFP resource management from a wider angle and avoid reducing them to a simple and transient requirement for 'higher prices' only because they are quick or easy to implement.

39.4.4.1 Challenges of NTFP Pricing

NTFP pricing by itself may not address the structural problems that need to be tackled as follows:

- Inadequate market intelligence, product and market development and market information systems (which identify new markets, new product developments, and regular pricing update mechanisms).
- Poor investments in the NTFP sector and inadequate support to the primary collectors and their organizations for company development and market access.
- Exploitative market mechanism for NTFP leading to inadequate earnings from NTFP trade for the main collectors and producers.
- The main collectors' unsustainable collecting methods have destroyed forests and left a lack of data necessary to calculate sustainable harvest levels.
- Lack of infrastructure for branding, advertising, certification, quality control and standardization of quality.
- Poor supply chain infrastructure (warehousing, cold storage facilities, efficient and transparent trading platforms).
- Inadequate institutional structures for NTFP procurement, value addition, marketing and other technical services;
- Inadequate capacity with stakeholders for NTFP management and connecting it to livelihood security.
- Inadequate public awareness of NTFP trading and processing, as well as a lack of sufficient research and development on NTFP processing and marketing.

The aim of the NTFP market should be to move from being highly unregulated to being a fair and regulated market. Steps, therefore, need to be taken towards market regulation. NTFP societies are playing the role of market regulation to some extent—but since most of them do not function effectively on the ground, their role in paying fair prices to the primary collectors is undermined.

39.5 New Government Initiatives for Fair Price

39.5.1 Minimum Support Price (MSP) to the NTFP Collectors

As per the Haque Committee Report, May 2011, the procurement value of 14 major MFPs is estimated at Rs. 1900 crores (including tendu and bamboo). Based on a survey and data collected by TRIFED, the value of the potential stock of the proposed 55 MFPs is estimated to be about Rs. 20,000 crores.

The development of minor forest products is crucial to the tribe's transformational evolution. 40–60% of tribal members' yearly income comes from the collection and selling of MFPs; these individuals are mostly tribal women who gather, process, utilize and sell MFPs. In 2014, the Indian government introduced the Scheme for Marketing of Minor Forest Produce through Minimum Support Price (MSP) and Value Chain Development in response to these circumstances. The benefits have not trickled down to the anticipated levels of tribal gatherer engagement, even though the MSP for the MFP Scheme was intended to be a comprehensive plan for the growth of MFP commerce to cover non-nationalized/non-monopolized MFPs. MSP has been declared for multiple MFPs, and tribal people have been given an ownership of MFPs in their territories; nevertheless, the ownership is still largely notional.

The Government of India has decided to introduce the scheme of 'Mechanism for marketing of Minor Forest Produce through Minimum Support Price (MSP) and development of value chain'. The scheme is designed as a social safety net for the improvement of the livelihood of MFP gatherers by providing them a fair price for the MFPs they collect.

The scheme has been started with the following objectives:

- The objective of the scheme is to have a huge social dividend for NTFP gatherers; the majority of whom are tribal people.
- To provide a fair price to the NTFP gatherers for the produce collected by them and enhance their income level.
- To ensure sustainable harvesting of NTFPs.

According to the proposal, the Tribal Affairs Ministry will put in place a fair price mechanism under TRIFED, the marketing arm of the ministry. It would fix MSP or a remunerative and fair price for the 49 MFPs. If the price of the produce gathered by the tribal falls below this price, the state's designated agencies would be alerted to procure the MFP.

The centre would share the losses with the state in a 75:25 ratio. The government has earmarked almost Rs. 125 crores for the scheme. Its target group is tribals of nine states—Andhra Pradesh, Chhattisgarh, Gujarat, Himachal Pradesh, Jharkhand, Madhya Pradesh, Maharashtra, Odisha and Rajasthan.

39.5.2 Van Dhan Scheme, 2018

With MFPs available and a sizable tribal population that lives in the forests, TRIFED leads the Van Dhan program's implementation in 27 states and 307 districts. Tribal members receive between 40 and 60% of their yearly income from the collection and sale of MFPs, and further 'Value Addition' helps to triple or quadruple their revenue. The initiative tackles the severe issues that Native Americans confront, including their illegitimate land and housing, limitations on their ability to harvest small-scale forest products, middlemen's exploitation, their forced eviction from national parks and wildlife sanctuaries, a lack of development in their forest villages, etc.

The Van Dhan Yojana, also known as the Van Dhan Scheme, is an integral part of the 'Mechanism for Marketing of Minor Forest Produce (MFP) through Minimum Support Price (MSP) and Development of Value Chain for MFP'. Launched on April 14, 2018, this initiative is overseen by TRIFED, serving as the central agency. The Van Dhan start-up represents a comprehensive strategy aimed at fostering the socio-economic progress of India's tribal communities (TRIFED 2018). It aims to overcome all issues in the NTFP sector especially during the Covid-19 pandemic, such as proper pricing of products, value addition and processing centre development, marketing NTFPs, supporting entrepreneurship and providing training.

Under the Van Dhan Scheme (2018), the primary procurement at haat bazaars will be facilitated through the existing infrastructure of the state/district or by involving primary procurement agencies. These agencies may include cooperative societies, LAMPS, mahila samities, VDCs, JFMCs and self-help groups (SHGs) formed in the state, including those established under the National Rural Livelihood Mission. The members of these groups are predominantly local tribal individuals selected by the District Collector in consultation with the Gram Panchayat. These agencies will receive a commission, not exceeding 7.5% of the value of Minor Forest Produce (MFP) procured through them. Buyers, including government-led Self-Help Groups (SHGs), will also participate in the haat bazaars, oversee by the Gram Sabha in accordance with the provisions of the Panchayats (Extension to Scheduled Areas) Act, 1996 (PESA Act). The establishment of 'Van Dhan Vikas Kendras', aimed at value addition and marketing, will be aligned with the haat bazaars. Additionally, facilities for primary, secondary and tertiary level value addition will be established in villages, tehsils, or talukas, as well as district headquarters, based on the availability of minor forest produce and the tribal population.

People's lives and livelihoods have been significantly affected across the nation over the past year due to the unprecedented crisis caused by the ongoing pandemic, particularly among the impoverished tribal population. A change-making light has emerged in the form of the 'Mechanism for Marketing of Minor Forest Produce (MFP) through Minimum Support Price (MSP) and Development of Value Chain for MFP' (Fig. 39.4). Developed and executed by TRIFED in collaboration with state government agencies in 21 states nationwide, the programme has proven to be a lifeline for tribal gatherers, contributing approximately

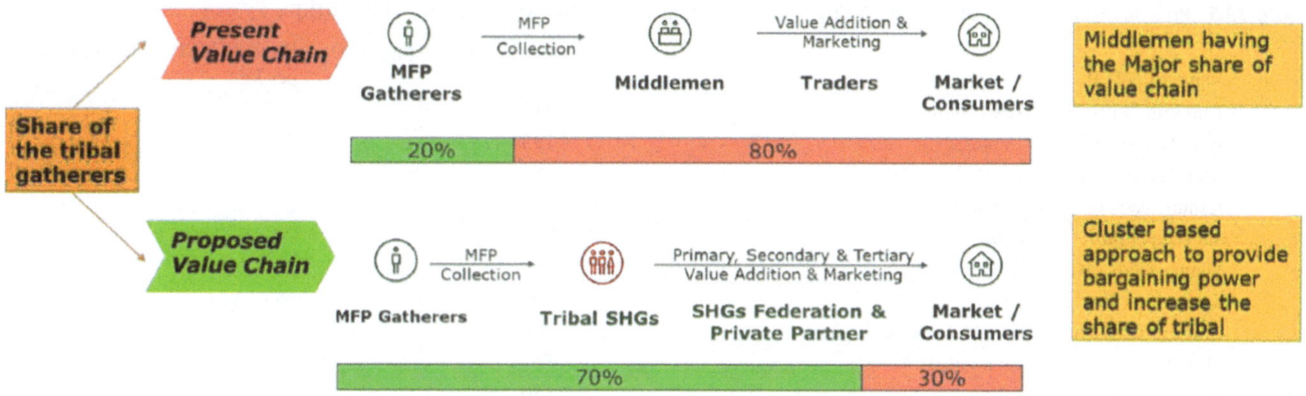

Fig. 39.4 Value chain proposed under the MSP scheme covered by Van Dhan Yojna (TRIFED 2020)

Rs. 3000 crores in direct economic activity to the tribe since April 2020. This has been possible mainly due to the government push and active participation from the states and has provided much-needed liquidity in the tribal economy, particularly in the adverse times like pandemic.

The Ministry of Tribal Affairs has updated the MSP for MFP list and included 14 new MFPs, carrying on its previous efforts to give tribal gatherers of forest products fair and remunerative pricing. Above and beyond the prior notifications of 26 May 2020 (when the list was updated to include 23 MFPs) and 1 May 2020 (when the MSP changes for the MFPs were revealed), this suggestion of extra items is being made. Moreover, state administrations have the authority to set the MSP at 10%. Table 39.5 lists all the price revisions of MSP of different NTFPs.

39.5.3 NTFP Enterprise Development

Enterprise initiatives on NTFP historically started in the pre-independence period through an informal arrangement based on local expertise and resource availability. Whereas in the 1980s, the Forest Research Institute, Dehradun, started research in the field of processing and value addition mechanism of NTFPs such as rubber, pine resin, gum, katha, charcoal and tendu leaf (Fig. 39.5). This was adopted by local traders for setting up small- and medium-scale enterprises in central and north India. This led to the development of enterprises in the last four decades, with the involvement of different stakeholders, change in the policy environment and people-led initiatives. NTFPs, such as honey, tamarind, hill broom, sheekakai, aonla, myrobalans, were some of the first NTFPs with processing and packaging units set up.

There are several key factors responsible for setting up any small-scale NTFP enterprises such as availability of raw material, microfinance, infrastructure, skilled labour, the market for the product and institutional support through policies (Fig. 39.6). However, there are also several challenges in meeting these above-mentioned prerequisites.

- Field models are available with different NGOs and Science and Technology Institutions; they only work seasonally (few products) as per the availability of the raw material.
- Field testing for enterprise model functioning may have been done only at a pilot scale in one/two locations.
- Reliability for further field-level assessment/validation of the technology, field model requires functional data for further strengthening and expansion of the enterprise.
- In most cases NTFP-based technology was set up in isolated manner. Line departments/district-level administrative coordination were mostly absent or inactive.
- Proper training of staff was lacking; technology providers make some occasional visits to enterprise locations, which doesn't ensure sustainability.
- Hand-holding support after the transfer of know-how and troubleshooting is one of the most important components in enterprise development. This was lacking in many cases.

39.5.3.1 Stakeholders and Institutions in NTFP Sector

In most cases, trading is through state-controlled institutions such as state forest development corporations, federations, cooperatives and tribal-led societies. Odisha used this monopoly to grant rights over 29 NTFP items to a private company. However, the whole NTFP trade is now de-regularized (79 items) after PESA Act (Saxena 2003). Now local people who collect NTFPs are not required to sell their materials to the company's agents.

Table 39.5 Price revisions of MSP of different NTFPs [MOTA (Ministry of Tribal Affairs) 2020; TERI 2018]

S. no.	Name of NTFP	MSP (2018) in rupees per kg	Revised MSP (2020) in rupees per kg
1	Tamarind with seeds (*Tamarindus indica*)	31	36
2	Wild Honey	195	225
3	Gum Karaya (*Sterculia urens*)	98	114
4	Karanj seeds (*Pongamia pinnata*)	19	22
5	Sal seeds (*Shorea robusta*)	20	20
6	Mahua seeds (*Madhuca longifolia*)	25	29
7	Sal leaves (*Shorea robusta*)	30	35
8	Chironji pods with seeds (*Buchanania lanzan*)	109	126
9	Myrobalan (*Terminalia chebula*)	15	15
10 A	Rangeeni lac	130	200
10 B	Kusumi lac	203	275
11	Kusum seeds (*Schleichera oleosa*)	20	23
12	Neem seeds (*Azadirachta indica*)	23	27
13	Puwad seeds (*Cassia tora*)	14	16
14	Baheda (*Terminalia bellirica*)	17	17
15	Hill broom (*Thysanolaena maxima*)	30	50
16	Dry Shikakai pods (*Acacia concinna*)	43	50
17	Bael pulp (dried) (*Agele marmelos*)	27	30
18	Nagarmotha (*Cyperus rotundas*)	27	30
19	Shatavari roots (dried roots) (*Asparagus racemosus*)	92	107
20	Gudmad (*Gymnema sylvestre*)	35	41
21	Kalmegh (*Andrographis paniculata*)	33	35
22	Tamarind deseeded (*Tamarindus indica*)	54	63
23	Guggul (Exudates) (*Commiphora wightii*)	700	812
24	Mahua flower (dried) (*Madhuca longifolia*)	17	30
25	Tejpatta (Dried) (*Cinnamomum* sp.)	33	40
26	Jamum (Dried) seeds (*Syzygium cumini*)	36	42
27	Dried Amla pulp (*Phyllanthus emblica*)	45	52
28	Marking nut (*Semecarpus anacardium*)	8	9
29	Soap nut (Dried) (*Sapindus emarginatus*)	12	14
30	Bhava seed/Amaltas (*Cassia fistula*)	11	13
31	Arjuna bark (*Terminalia arjuna*)	18	21
32	Kokum (dry) (*Garcinia indica*)	25	29
33	Giloy/giloe (*Tinospora cordifolia*)	21	40
34	Kaunch seed (*Mucuna pruriens*)	18	21
35	Chiraita (*Swertia chirayita*)	29	34
36	Vaybidding/Vavding (*Embelia ribes*)	81	94
37	Dhavaiphool dried flowers (*Woodfordia floribunda*)	32	37
38	Nux Vomica (*Strychnos nux vomica*)	36	42
39	Van Tulsi leaves dried (*Ocimum tenuiflorum*)	19	22
40	Kshirni (*Hemidesmus indicus*)	30	35
41	Bakul dried bark (*Mimusops elengi*)	40	46
42	Kutaj dried bark (*Holarrhena pubescens*)	27	31
43	Noni/Aal dried fruits (*Morinda citrifolia*)	15	17
44	Sonapatha/Syonak pods (*Oroxylum indicum*)	18	21
45	Chanothi seeds (*Abrus precatorius*)	39	45
46	Kalihari dried tubers (*Gloriosa superba*)	27	31
47	Makoi fried fruits (*Solanum nigrum*)	21	24
48	Apang plant (*Achyranthes aspera*)	24	28
49	Sugandhmantri roots/tubers (*Homalomena aromatica*)	33	38
	New NTFP items under MSP scheme w.e.f. 11.11.2020		

(continued)

Table 39.5 (continued)

S. no.	Name of NTFP	MSP (2018) in rupees per kg	Revised MSP (2020) in rupees per kg
50 A	Tassar Cocoon Reeling Grade I (when average shell weight is 1.55 g and above)		3200/thousand number
50 B	Tassar Cocoon Unreeling Grade (when average shell weight is 1.40 g and above)		1500/thousand number
51	Cashew kernel (*Anacardium occidentale*)		90
52	Elephant apple dry (*Dillenia indica*)		120
53	Bamboo shoot (*Phyllostachys edulis*)		70
54	Malkagni seed (*Celastrus paniculatus*)		100
55	Mahul leaves (*Bauhinia vahlii*)		15
56	Nagod (*Vitex negundo*)		20
57	Gokhru (*Tribulus terrestris*)		60
58	Pipla/Uchithi dry bark (*Piper pedicellatum*)		120
59	Gamhar/Gamari (*Gmelina arborea*)		20
60	Oroxylum indicum dry bark (*Oroxylum indicum*)		40
61	Wild Mushroom dry (*Agaricus* sp.)		400
62	Shringraj (*Eclipta alba*)		18
63	Tree moss (Bryophytes)		350

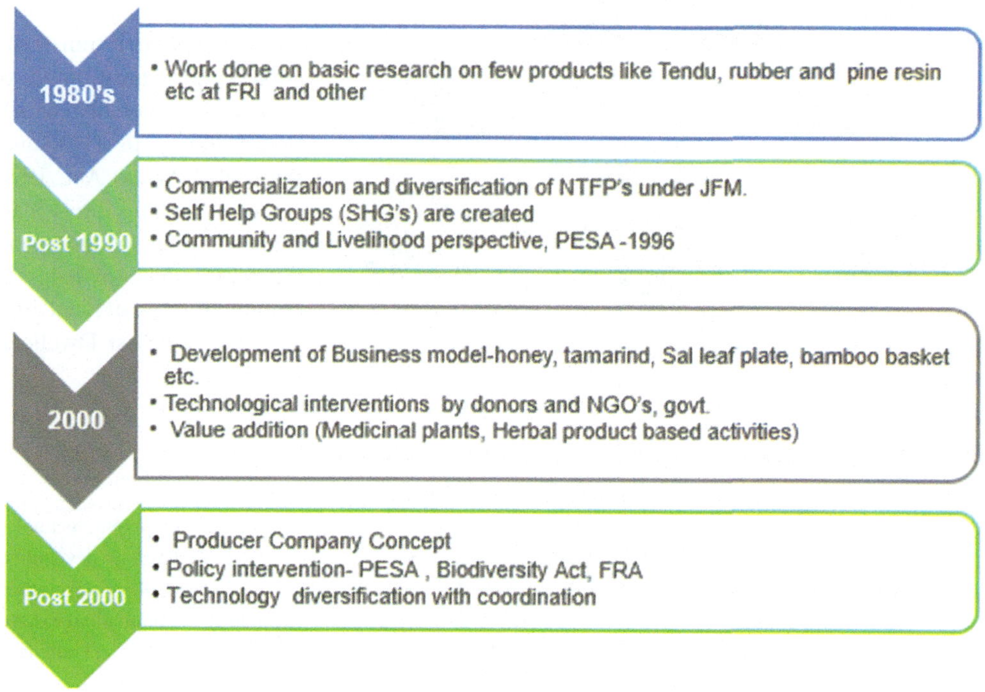

Fig. 39.5 History of NTFP-based enterprise development

39.5.3.2 State Monopoly Over NTFPs

The economic potential of NTFPs has become apparent in recent times, and infrastructure for trade potential and processing has developed, which has raised concerns for the sustainability of the resources and the distribution of benefits derived from them.

In reaction to these concerns, a number of state governments have taken over the control of many NTFPs. Some of the explicit objectives for state monopoly over NTFP trade are:

- To prevent unscrupulous intermediaries and their agents from exploiting NTFP collectors
- To ensure fair wages to collectors
- To enhance revenue for the state
- To ensure quality and quantity
- To maximize the collection of produce

Minor Forest Products (MFPs) include bamboo, cane, tussar, cocoons, honey, wax, lac, tendu or kendu leaves, medicinal plants and herbs—these products were gathered, processed, packaged and sold by monopolistic, state-owned Forest Development Corporations (FDCs) (Tables 39.6 and 39.7). While forest dwellers are allowed to collect, their status remains as labourers of FDCs.

The Government of Kerala has created a monopoly over 120 notified items of Non-Timber Forest Products. The scheduled tribes and forest dwellers had to sell it to cooperative societies only. Problem is that the open market price was more than double the price being offered by the government. Government monopoly was not only inefficient but also exploiting the tribal people (Kerala World Bank PAD 1998). In Andhra Pradesh, state monopoly is carried out by GCC (Girijan Cooperative Cooperation). Maharashtra has also included a large number of NTFPs causing undue hardships to the gatherers. In Madhya Pradesh, state trading is selective in accepting only quality tendu leaves and other NTFPs.

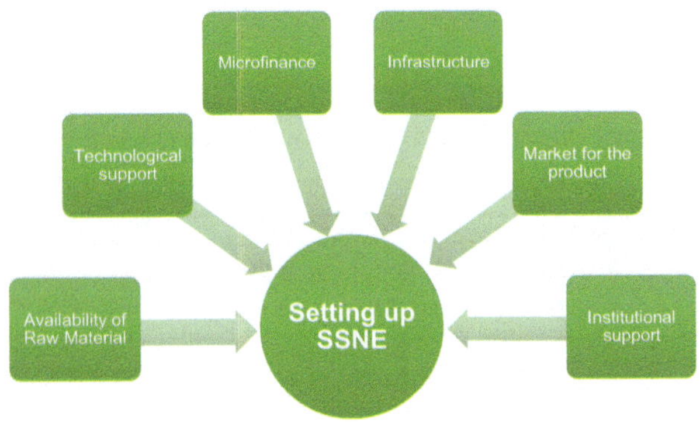

Fig. 39.6 Prerequisites for setting up small scale NTFP enterprise (SSNE)

Although the goals of establishing state monopolies are praiseworthy, the stated goals and the real circumstances imply that the overall effects will not be positive. State forest income has gone up, but wages, socioeconomic circumstances and gender equity do not seem to be improving in the communities that depend on the forests. The agents of the monopoly leaseholders have taken the position of intermediaries, rather than their elimination (Prasad 1999; Khera 2016). As a result, there was increased number of NGOs and CBOs intervening, as these groups assisted the NTFP collectors with value addition, processing and marketing as well as covering the gaps left by government organizations (Tables 39.8 and 39.9).

39.6 NTFP Policy and Legal Issues

39.6.1 Institutional Perspectives of Policy Change for NTFPs

India underwent a transformative shift in forest management approaches, transitioning from regulatory to participatory frameworks, notably with the enactment of the National Forest Policy in 1988 and the Panchayats (Extension to Scheduled Areas) Act, 1996 (PESA). These policies underscore sustainable forest management principles, recognizing forests as vital social and environmental assets. India's commitment to acknowledging the land rights of indigenous communities is evident in the Scheduled Tribes and Other Traditional Forest Dwellers (Recognition of Forest Rights)

Table 39.6 NTFP enterprise support by the government organization

NTFPs	Government organization	Activity
Tamarind, honey, gum Karaya	GCC, Hyderabad	Business development, technological implementation and marketing
Aonla, arjun, honey	MP MFP Federation	Technology development, training and marketing
Herbal, lac, char	Chhattisgarh MFP Federation	Technology development, training and marketing
Gum karya, honey	TRIFED	Technology dissemination, extension, training and market development
Honey, sissal, aromatic oil	KVIC	Training and marketing

Table 39.7 NTFP Enterprise support by the research and academic institutions

NTFPs	Research and academic organization/ institution	Activity
Sal leaf, sabai grass	Indian Institute of Technology, Kharagpur	Product development, market development and institution building.
Asparagus root, aonla	Indian Institute of Forest Management, Bhopal	Product development, market development and institution building.
Jatropha/karanj seed	SUTRA, Indian Institute of Science, Bengaluru	Biodiesel production technology
Lac	Indian Institute of Natural Resins and Gum, Ranchi	Technology development, dissemination of information and extension services
Honey and bee products	Central Bee Research and Training Institute, Pune	Rock bee—Sustainable honey harvesting technology and extension services

Table 39.8 NTFP enterprise support by NGOs

NTFP	Non-governmental organization	Activity
Honey, bee wax, shikakai, aonla etc.	Keystone Foundation, Nilgiris District, Tamil Nadu	Wild honey harvesting, processing, marketing and certification
Siali leaf, harada flower and mahua flower	Vasundhara, Odisha	Collection and plate stitching enterprise and business development
Bamboo incense stick	Tripura Bamboo and Cane Development Centre, Agartala, Tripura	Technology, processing, value addition and marketing
Tassar	Pradhan, Madhya Pradesh	Tassar silk production and processing
Gum karya	Kovel Foundation, Vishakhapatnam, AP	Gum tapping and picking technology and monitoring

Table 39.9 NTFP enterprise efforts by community-based organization

NTFP	Community-based organization	Activity
Herbal products	GMCL, FRLHT, Karnataka	Capacity development, institution building and business development
Broom grass, natural dyes	WIDA, Koraput, Orissa	Technology development, business model and value chain Improvement
Bel juice, honey	MP Vigyan Sabha	Technology development, business model and value chain improvement
Medicinal plants, honey	Vikas Mitra, Bastar, Chhattisgarh	Capacity development, institution building and business development
Lac, herbal products	SPAR, Ranchi	Capacity development, institution building and business development

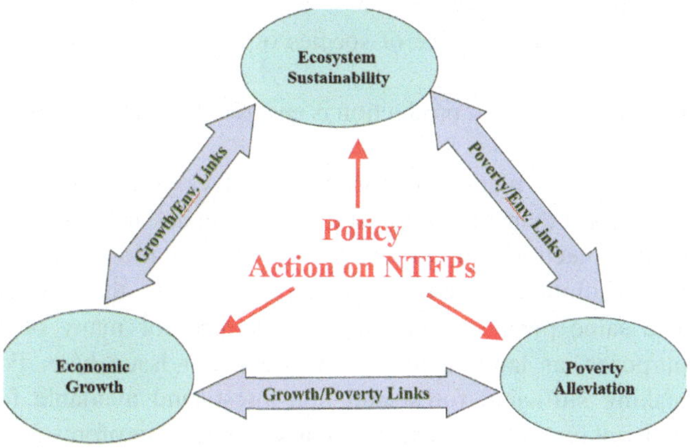

Fig. 39.7 Policy actions on NTFP sector

Act, 2006, demonstrating sensitivity towards forest tenure issues.

Changes in policies from the perspective of livelihood promotion and natural resource sustainability are the key drivers which have helped to pump a good amount of public money through various projects from the Ministry of Rural Development and Ministry of Environment, Forest and Climate Change. With the recent emphasis on environmental conservation, NTFPs have emerged as eco-friendly products as their harvest/gathering is believed to cause less damage to the ecosystem, compared to logging. NTFPs deserve special mention because of their great potential to support economic development and ecological sustainability consistent with the principles of Sustainable Forest Management (SFM) (Fig. 39.7).

39.6.1.1 PESA (Panchayat Extension to the Scheduled Areas) Act, 1996

Forestry in India has undergone significant evolution in recent decades, marked by legislative milestones such as the enactment of the Environment Protection Act in 1986, aimed at enhancing the country's environmental quality. Over the past 25 years, notable progress has been made in leveraging forests to alleviate poverty, particularly through initiatives like the Panchayats (Extension to Scheduled Areas) Act of 1996, which empowered local communities by granting them ownership rights over Non-Timber Forest Products (NTFPs).

To broaden the scope of the 1993 73rd constitutional amendment to encompass the Schedule V Areas of the country, the Indian Parliament enacted the Panchayat (Extension to the Scheduled Areas) Act in 1996. This legislation mandates that state governments confer upon panchayats in these designated regions the requisite authority and powers to function as self-governing bodies. Additionally, this Act gave panchayats the authority to manage natural resources sustainably, to own the rights to these resources and to create value-adding chains that will increase revenue for both primary collectors and processors and PRIs (Panchayat Raj Institutions).

39.6.1.2 Biodiversity Conservation Act, 2002

The Biological Diversity Act was passed by the Parliament in 2002 after consultation among stakeholders. It looks at that conservation of biological diversity, sustainable use of its components and equitable sharing of benefits arising out of the use of biological resources. The National Biodiversity

Authority and the State Biodiversity Board have the responsibility to see the sustainability aspect of those NTFPs which are getting rarer day by day; out of which, many are placed under the threatened category by the International Union for Conservation of Nature (IUCN).

39.6.1.3 Forest Right Act, 2006

The Scheduled Tribes and Other Traditional Forest Dwellers (Recognition of Forest Rights) Act, 2006, was introduced to grant tribal communities and forest-dwelling individuals the right to live and work on forests. It also assigned the communities responsibility for maintaining ecological balance and conserving biological resources. This law gives tribes and forest dwellers the authority to manage the forests they use as common property resources and acknowledges their rights to do so.

During the course of the Act's 15-year implementation, the Ministry of Tribal Affairs has become aware of a number of issues, including the holding of Gram Sabha meetings at the Panchayat level, which has led to the exclusion of smaller habitations that are not formally part of any village, the refusal to acknowledge forest dwellers' unrestricted rights over Minor Forest Produce (MFP), the rejection of claims by requiring specific types of evidence and harassment and eviction of forest dwellers.

The guidelines do little more than reiterate the rules which were set out following the original Act, in January 2008. Section II (a, b, c, d, e) of the guidelines, for example, requires the disbanding of FDCs and the scrapping of transit rules required for the movement of MFPs. They highlight the rights of the unrestricted use of MFPs by federations or cooperatives of forest dwellers, and the abolition of any fees or charges by the Forest Department. Further, the guidelines (Section III {a, b, c, d} and IV {a, b, c, d, e}) call on state governments to recognize community rights of various kinds (nistar, pastoralists, primitive tribal groups).

The key expectation that the FRA would encourage cooperatives and other business entities owned by forest dweller communities dealing in NTFP has not been met. The salient features of the new FRA guidelines and rules are as follows:

- Forest dwellers have unrestricted access to collect Minor Forest Produce
- Do away with the monopoly of forest corporations
- Conversion of declared forest villages to revenue villages
- Recognition of Nistar Rights and rights of Primitive Tribal Groups
- Scrap fees, royalties, charges on trade by forest dwellers
- Honour the constitutional power of the Gram Sabha

39.6.2 Economic Perspective of Policy Change Required for NTFP Production

Expanding the forest resources by planting NTFP species on a big scale could be the solution to producing enough in degraded forest areas. To establish and broaden the base of forest resources, plantations have been started. Forest plantations built under a technique of clear-felling followed by artificial regeneration, or by an afforestation of bare lands, wastelands, grasslands and other degraded lands, have been prevalent in the past. Due to their many benefits, NTFP plantings can also be considered a way to preserve natural forests:

- Vegetation growth and yield can be manipulated using appropriate inputs
- Plantations supplement production from natural forests, thus, reducing the pressure on the latter
- Harvesting is less expensive in the case of a carefully planned plantation
- They allow a choice of species of desired characteristics for an area
- Homogeneity of production is ensured through plantations

However, plantation forestry involves a much larger investment than natural forest management and needs continuous protection and upkeep. Yet NTFP plantations can provide a solution to the resource scarcity faced by the country. The same piece of land may be claimed for many other purposes, as land scarcity is also a major issue here. For making sufficient forest and non-forest land available for undertaking NTFP species plantations under agroforestry or in other models. Following potential areas may be suggested to supplement the natural regeneration:

- Village common lands
- Degraded private lands
- NTFP species plantation in margins of roads (NH), railways, canals and tanks
- Vacant lands belonging to schools, offices and hospitals

By an appropriate selection of species and system (the mixture of species, rotation, input levels, method of planting and maintenance), it will be possible to raise need-oriented plantations or mixed tree crops in dry areas through social forestry programmes.

39.6.2.1 Overall Linkages with Schemes and Present Policy Initiative in the Management of NTFPs

The main policies and related governing process of the NTFP-based livelihood developments are as follows (Bhattacharya 2018):

- Rights and Concession: NCA, 1976 to PESA, 1996; M.P. Tendu Patta Act, 1964; Recognition of Forest Right Act, 2006; MP Laghu Van Upaj Vidheyak, 1999.
- Benefit Sharing Arrangement: JFM, 1990—JFM Present; Nistar, 1996; Tendu Patta Act, 1964, Van Dhan Scheme 2018.
- Minimum Support Price (MSP), 2020: Revised MSP implemented through Ministry of Tribal Affairs, TRIFED and State Forest Department provides a minimum rate for more than 60 NTFP items.
- Conservation and Production: NCA, 1976; NFP, 1988; JFM and its subsequent amendments, NAP, NMPB, BCA.
- Role of Panchayat and gram sabha: JFM, 1998; PESA, 1996; JFM, 2002; Van Dhan Scheme, 2018.
- Sustainable Harvest and Non-Destructive Harvest: NFP, 1988; JFM, 2002; M.P. Gazette Notification, 2005; People's Protected Area, Chhattisgarh.
- Restriction on Negative Trade List: Biodiversity Act, 2002; Forest Protection Act, 1927; Forest Conservation Act, 1980.
- Institutional Framework Strengthening: NCA 1976; NFP 1998; JFM 2002; Van Dhan Scheme, 2018.
- Marketing Linkages: NCA, 1976; Formation of MP MFP Federation, 1984; NFP, 1988; JFM, 2000; JFM, 2002.
- Transit Policies: Madhya Pradesh Transit Policies, 2000; Individual State Policy
- Defining what comes under NTFP (Nationalized, Non-Nationalized): M.P. Forest Produce Trading Act, 1969; JFM, 1998; NAP, 2000.
- Capacity Building: JFM, 2002; NAP, 2001; Van Dhan Scheme 2018.
- Taxation: Sales Tax; VAT; Forest Development Tax; Education Cess; Royalty; Commercial Tax.
- Livelihood and NTFPs: PESA, 1996; Recognition of Forest Rights Act, 2006;
- Domestication and Commercialization: MGNREGA, 2005; SGSY, 2002; FCA, 1980, NAP, 2001; NMPB, 2002;
- Certification: Yet to be accepted by Govt. of India as policy in the NTFP sector, FSC, Fair Trade, PEFC
- Processing and Value Addition: Policy support is available in various forms, but exclusive focus is yet to come.

39.6.2.2 Convergences Under Van Dhan Scheme, 2018

The Van Dhan Scheme, 2018 aims to leverage synergies with several central and state ministries, departments, agencies and PSUs. For instance:

- Niti Aayog will assist in implementation across 39 aspirational districts with a significant tribal population.
- The Ministry of Rural Development will coordinate to utilize the resources, skills and experiences of Aajeevika and tribal SHGs.
- The Ministry of Panchayati Raj will support the implementation of tribal-centric micro-market reforms.
- The Department of Financial Services will collaborate to align Jan Dhan and MUDRA Schemes with Van Dhan.
- The Ministry of Environment and Forests will work on integrating MFP plantation into the National Forest Policy.
- The Ministry of AYUSH will support the value addition of medicinal MFPs.
- The Department of Space (ISRO) will undertake MFP resource mapping initiatives.
- The Department of Home Affairs will promote the scheme in tribal LWE affected districts.
- The Ministry of Food Processing Industries will encourage value addition and training partnerships in MFPs.
- PSUs and other enterprises will promote the scheme as part of their CSR activities.

39.6.3 NTFP Trade-Related Policy in Different States

NTFPs can be broadly classified as nationalized and non-nationalized items. For the nationalized items, state governments have a monopoly over collection, marketing and trade, for example, tendu leaves, kullu gum, harra, sal seeds etc. Non-nationalized items do not have monopolistic control, for example, sal leaves, medicinal plants, mahua flower, neem seeds etc. Indian forests are under great pressure, as 78% of the forest areas are subjected to heavy grazing, adversely affecting productivity and regeneration, which is a concern for NTFP supply. All different states have their own nationalized NTFPs, price fixation methods, transit permit regulations and state marketing institutions (Tables 39.10 and 39.11).

39.6.4 Policy Constraints for NTFP-Based Livelihood Promotion

There are several constraints in NTFP-based livelihood promotion for the actual field-level implementation

1. Depleting NTFP resource base and its conservation/regeneration in the forest areas
2. Lack of actual assessment of resources and their inventorization, digitization of data
3. Differential policy and its solution

Table 39.10 NTFP categorization under trade and policy in different states

Category	Chhattisgarh	Madhya Pradesh	Odisha	Jharkhand	Andhra Pradesh
Nationalized NTFPs	Bamboo, tendu leaf, harra, gums and sal seeds	Bamboo, tendu leaf, harra, gums and sal seeds	Bamboo, sal Seed and kendu leaf	Bamboo, kendu leaf, mahua seed, mahul leaves and harra	Bamboo and kendu leaf. 24 produces including gums, tamarind and mahua leaves are procured and marketed by GCC
Free/notified NTFPs	Apart from nationalized items rest are free to collect	Apart from nationalized items rest are free to collect	69 items handed over to Gram Panchayat	–	28 items handed over to Gram Panchayat
Price fixation method	Nationalized state govt	Nationalized state govt.	Nationalized state govt. and Panchayat Samiti	Nationalized state govt. and cooperatives	Nationalized state govt. and 28 items by GCC
Transit permit (TP)	TP is required	TP is required	TP is required	Not required for 69 items	TP is required
State marketing Institution	Chhattisgarh state MFP Federation	MP State MFP Federation	FD, OFDC, TDCC, ORMAS	JFDC, TDCC	Girijan Cooperative Cooperation, APFDC

Table 39.11 State trading regulations promulgated by different NTFP producing states

State	Acts	
Andhra Pradesh	Andhra Pradesh Minor Forest Produce (Regulation of Trade) Act, 1971, Andhra Pradesh scheduled areas	Trade-in NWFPs is declared state monopoly whether ownership is with government or not
Bihar	Bihar Kendu Leaves (Control of Trade) Act, 1973, Bihar Forest Produce (Regulation of Trade) Act, 1984	Bihar State Forest Development Corporation operates as a state government agent for the collection and marketing of kendu leaves, sal seed, mahua (*Madhuca latifolia*) and harra
Gujarat	Gujarat Minor Forest Produce (Regulation of Trade) Act, 1979	Minor Forest products identified include timru leaves (tendu leaves), mahua flowers, fruits, seeds and gum
Himachal Pradesh	Himachal Pradesh Resin and Resin Product (Regulation of Trade) Act, 1981	Resin, bamboo and *Acacia catechu* (khair) collection through Himachal Pradesh Forest Development Corporation Ltd.
Madhya Pradesh	Madhya Pradesh Vanopaj (Vyapar) Viniyam) Adhiniyam, 1969	Items under monopoly include tendu leaves, sal seed, harra and gums; Madhya Pradesh Minor Forest Product (Trade and Development) Federation acts as agent of state government
Rajasthan	Rajasthan Tendu Leaves Act, 1974	Rajasthan Tribal Area Development Federation collects and markets NWFPs
Odisha	Odisha Forest Product (Control of Trade) Act, 1981, Odisha include Kendu Leaves (Control of Trade) Act, 1981	Collection and sale of NWFPs are monopolized by the Forest Department and leased to the Tribal Development Cooperative Society which in turn delegates to an individual; collection and trade of leaves are handled by Forest Corporation

Source: MoEF 1998

4. Inadequate skill and capacity at various levels on value chain development and interventions for its solution
5. Weak institutionalization, and policy solution is need of the day, partially supported in Van Dhan scheme
6. Poor R&D focus and connection with R&D industries
7. Lack of special attention for critically dependent communities and areas
8. Effective marketing linkages
   (a) The policy decision for pricing, example tendu patta (leaves) does not involve cooperative members at any level for price fixation.
   (b) The minimum support price for tendu (a nationalized product) and the collection rates for the whole state are decided in Bhopal once every year. In this system, the district unions and the societies do not have any leverage regarding the rates of collection.

39.7 Conclusions

NTFPs are a lifeline for rural and tribal households as more than 140 million forest dwellers depend on them for their subsistence and income. It has an immense untapped potential and scope for commercialization. However, over time, there have been instances where the forests have witnessed a rapid fall in the regeneration rates of NTFPs, unsustainable harvesting, premature harvesting and over-harvesting of NTFPs such as aonla, chiroji, baheda, gums and resins. There is limited data on the management of these NTFP species. It is required to document whatever data is available and to initiate research on ecological, social and economic parameters in the NTFP sector. This is needed to manage NTFPs scientifically under sustainability principles. The area of research agenda may include the suitable requirement of natural regeneration of species, silvicultural manipulation in forest stands, economical, production, processing and application of technological innovation and marketing (Gadgil et al. 1993; Ticktin and Johns 2002; Saxena 2003; Bhattacharya 2013).

The best NTFP management practices may be documented and circulated among the field foresters who manage the resources on a day-to-day basis. It is required coordination amongst researchers, managers and industry people and development of improved varieties/hybrids for selected high value species like *Santalum* sp. (Chandan), aonla, agarwood, shatawar (*Asparagus racemosus*), haldi (*Curcuma longa*), aromatic plants and grasses. Detailed and systematic research is required for the conservation of medicinal species under MPCA (Medicinal Plants Conservation Area) and MPDA (Medicinal Plants Development Area) under in situ and ex situ conservation and development. Germination and plantation techniques, good collection practices and non-destructive harvesting, best gum tapping practice, improved processing techniques, better packaging technology, NTFP certification and promoting domestic, national and global trade through tariff regulations are top priority actions (Bhattacharya et al. 2008).

Participatory resource inventory, periodic monitoring of resource availability and sustainable utilization and proper valuation are a few key actions that are required for the NTFP sector development. Furthermore, a thrust towards the implementation of protocols of sustainable harvesting and equitable distribution with special attention to vulnerable people, is also the need of the hour for state implementation agencies.

NTFPs are a major raw material for food, pharmaceuticals and FMCG companies. Annually, around Rs. 2,00,000 crores worth of NTFP in raw form are collected by the tribals and rural forest dwelling community and traded in the weekly *haat bazaars*. However, the income accruing to tribal gatherers in the above business is far below what may be considered a fair price. They are getting around 20% (the rest passes on to middlemen), whereas they deserve more than 60% of the product value. If NTFP and market-led allied activities are strengthened, tribals can prosper in their habitat, in the shortest time span.

The role of gender is crucial in gathering of NTFPs from the forests and other allied activities such as the primary and secondary processing done at homes is mostly by womenfolk. Strengthening the NTFP collection, value addition and marketing is also a sure way to women's empowerment in tribal areas. The government has designed a holistic model of NTFP-led Tribal Development in Mission Mode since 2018. Tribal communities have been made the legal owners of MFP in their areas after the enactment of PESA, 1996 and FRA, 2006; both the acts supported the concept of empowerment of forest-dwelling communities. Following this, Minimum Support Price (MSP) has been implemented for about 70 NTFPs. During the Covid-19 pandemic, such efforts of the government have supported 1.1 crore tribal households and 5.5 crore population dwelling in forest areas (TRIFED 2020). The Ministry of Tribal Affairs has been emphasizing the processing of MFP locally to generate more employment and income for the tribes in their villages. TRIFED is spearheading this program for processing and value-addition to MFP in well-equipped common facility centres, to be called Van Dhan Vikas Kendra.

Capacity development training, awareness building is another thrust area in the NTFP sector. Training to primary collectors, processors and traders, and to the frontline staff in basic and advanced areas to build their confidence. It is important to train them regarding their rights and proper harvesting techniques; and complementary facilitation should be made for forest protecting/forest managing communities too in the form of NTFP management protocols. Furthermore, lack of knowledge regarding value addition and marketing process is one of the biggest gaps, which needs to be addressed immediately.

Climate change is one of the key challenges faced by the rural poor due to uncertainty about the magnitude of these future impacts across different ecosystems and the NTFP sector is certainly facing huge threats, which require adjustment in the ecological, social and economic system as adaptation strategy (Bhattacharya and Prasad 2009). Climate change impact is a huge potential area for NTFP research; a proper planning of research agenda is needed on the reproductive biology of most affected NTFPs, production pattern of high-altitude species, development of coping mechanism and mitigation and adaptation plan of most dependent areas (Thomas and Twyman 2005; Bhattacharya and Prasad 2009; Shrivastava and Bhattacharya 2015).

The National Highways Authority of India (NHAI) has a very aggressive target of constructing about 50 km of roads in a single day; they need to take positive steps for planting NTFP species which can provide medicine and other NTFPs for local people. State- and central-level public sector organizations (CGMFPF, GCC, MPMFP Fed, UFDC, OFDC, ORMAS, TRIFED, etc.) and informal organizations (SHGs, NGOs, primary cooperative societies, etc.) are working concerning procurement, storage, processing and sale need greater coordination and better efficiency. But unsustainable harvesting, inequitable benefits distribution and overall economic inefficiencies characterize the current NTFP practices. Many research studies suggested that there should be an independent NTFPs policy and a separate NTFP board (following the structure and functions of other existing boards such as rubber board, coir board, spices board, etc.) in India. However, this is also a fact that in India there is no dearth of good policies and Acts. The problem is with the execution of these policies with responsibility and for the national interest.

Lessons Learnt

- Historically the NTFP sector in forestry was neglected. Timber was accorded priority over NTFPs by the forest departments. But currently NTFP is a key to connect with local tribal and other poor people who live in and around forests for the livelihood support they provide.
- Tropical forests produce more NTFPs compared to other forest types particularly from the dry deciduous forest which produces maximum NTFPs in quantity and diversity. Factors like forest fire, poor forest regeneration, climate vulnerability and degradation of habitat have reduced the NTFP yield significantly in the last few decades.
- The collection and harvesting practices of NTFPs are mostly unscientific, largely unsustainable and the quantity of products are reducing every passing year. Faulty harvesting methods like over harvesting and premature harvesting are primary reasons for depletion. Maturity times of different NTFPs are round the year and silvicultural requirements vary for different products.
- Due to the paucity of data on production and proper inventory, trade projections are difficult unlike the agricultural sector. Informal trade is most prevalent in the sector. Local collectors of NTFP get very low profit compared to traders at different levels. Marketing support to local people are available for a few products only, which need further extension.
- The primary-level processing and value addition have been tried by state government and NGOs, but they have limited spread and continuity issues are observed. Entrepreneurs require more exposure to understand quality and value chain.
- The policy intervention is successful in a few central Indian states only, where foresters are involved to a great extent with NTFP trade and value addition aspects. Several related Acts (PESA 1996; Biodiversity Act 2002; FRA 2006) and Policy are in place, whereas no separate NTFP policy available in India.
- Local people are very much interested to receive training in capacity building and orientation towards better NTFP management and development. Awareness at village level has increased after better communication facilities including mobile phone. MSP, Van Dhan 2018 and other schemes on nursery development have helped local people to improve their confidence.

Key Questions

1. What is the significance of NTFPs towards supporting the livelihoods of local people?
2. What do you mean by unsustainable harvesting? Discuss various reasons of unsustainable harvesting of NTFPs.
3. What are the various issues involved in NTFP management? How working plan code has suggested to follow the management practices in natural forest?
4. Discuss the pattern and status of NTFP trade in India. Enumerate the nature of NTFP market. What are the various organizations working for NTFP trade?
5. What are the important policy and legal conditions currently available in different Indian states for the NTFP sector?

References

Agarwal B (1989) Rural women, poverty and natural resources: sustenance, sustainability and struggle for change. Econ Political Wkly 24:46

Bahuguna VK (2000) Forests in the economy of the rural poor: an estimation of the dependency level. AMBIO J Hum Environ 29: 126–129

Bhattacharya P (2013) Prescriptions for sustainable NTFP management in forest working plan: addressing new market demand. Int J Usuf Manag 14(2):31–50

Bhattacharya P (2018) Redefining forestry for effective livelihoods. TERI, New Delhi, pp 373

Bhattacharya P, Hayat SF (2004) Sustainable NTFP management for rural development: a case from Madhya Pradesh, India. Int For Rev 6(2):161–168. https://doi.org/10.1505/ifor.6.2.161.38399

Bhattacharya P, Kandya AK (2016) Sustainable forestry. International Publication House, New Delhi

Bhattacharya P, Manoharan TR (2017) Forest certification: benefits and issues of sustainable forestry. In: Bhattacharya P, Kandya AK (eds) Sustainable forestry- emerging challenges. IK International Publishing House Pvt Ltd., New Delhi, pp 28–52

Bhattacharya P, Prasad R (2009) Initial observation on impact of changing climate on NTFP resources and livelihood opportunities in Sheopur District of Madhya Pradesh, XIII World Forestry Congress, Buenos Aires, Argentina, 18–23, October 2009

Bhattacharya P, Prasad R, Asokan A (2008) Towards certification of wild medicinal and aromatic plants in four Indian states. Unasylva 59:35–44

De Beer JH, Mc Dermot M (1989) The economic value of Non Timber Forest Products in southeast Asia. The Netherlands Committee for IUCN, Amsterdam, p 199

Devi S, Bhattacharya P (2015) Population structure and generation study in forests of Barella and Sohagpur, Madhya Pradesh. India. Indian Journal of Tropical Biodiversity 23(1):106–112

Gadgil M, Berkes F, Folke C (1993) Indigenous knowledge for biodiversity conservation. Ambio 22:151–156

Khera A (2016) Poverty alleviation through non wood forest products in Madhya Pradesh. J Trop For 32:19–33

Kumar M, Singh MP, Singh H, Dhakate PM, Ravindranath NH (2020) Forest working plan for the sustainable management of forest and biodiversity in India. J Sustain For 39(1):1–22. https://doi.org/10.1080/10549811.2019.1632212

Lawes MJ, Obiri J (2003) Using the spatial grain of regeneration to select harvestable tree species in subtropical forest. For Ecol Manag 184(1–3):105–114. https://doi.org/10.1016/S0378-1127(03)00215-9

Mallet P (1999) Certifying the harvest: developments in NTFP certification. In: Davidson-Hunt I, Duchesne CL, Zasada JC (eds) Forest communities in the third millennium: linking research, business, and policy toward a sustainable non-timber forest product sector. US Department of Agriculture, St. Paul, pp 124–127

MoEF (1998) Report of Expert Committee on conferring Ownership rights of MFPs on Panchayat. GOI, New Delhi

MOTA (Ministry of Tribal Affairs) (2020) Guidelines for mechanism for marketing of minor forest produce (MFP) through minimum support price (MSP) and development of value chain for MFP. MOTA and TRIFED, New Delhi

Mott J (1998) A framework to consider forest product marketing. In: National workshop on marketing of forest products in India. Oct 30–31, 1998. IGFRI, Dehradun

Murali KS, Shankar U, Shanker RR, Ganeshaiah KN, Bawa KS (2006) Extraction of non-timber forest products in the forests of Biligiri Rangan Hills, India. II. Impact of NTFP extraction on regeneration, population structure, and species composition. Econ Bot 50:252–269

NWPC (2014) National Working Plan Code—2014 (for sustainable management of forests and biodiversity in India). Forest Research Institute, Ministry of Environment and Forests, Government of India. Forest Research Institute, Dehradun

Peters CM (1996) The ecology and management of non-timber forest resources. World Bank

Prasad R (1999) Joint Forest Management in India and the impact of state control over Non Wood Forest Products Management. Unasylva 50:198

Prasad R (2007) Strategy for sustainable NTFP management in India. Indian Institute of Forest Management, Bhopal

Ruiz MR (1995) A conceptual framework for CIFOR's research on non-wood forest products. CIFOR Working Paper No. 6, Center for International Forest Research, Bogor, Indonesia

Saha D, Sundriyal RC (2012) Utilization of non-timber forest products in humid tropics: Implications for management and livelihood. For Policy Econ 14(1):28–40

Saigal S, Arora H, Rizvi SS (2002) The New Foresters: the role of private enterprise in the Indian forestry sector. Echo Tech Services, New Delhi

Saxena NC (2003) From monopoly to de-regulation of NTFPs: policy shifts in Odisha, India. Int For Rev 5(2):168–176

Shiva MP (1993) Technology scenario in Asia-Pacific Forestry Sector. In: APFSOS Working Paper no. 25. Forestry Policy and Planning Division, FAO, Rome, Oct. 1997

Shrivastava K, Bhattacharya P (2015) Lichen as a bio-indicator tool for assessment of climate and air pollution vulnerability: review. Int J Environ Sci 4(12):107–117

TERI (2018) Minimum support price of minor forest produce (MFP) and its sustainable harvest: a social safety measure for MFP collectors in India. The Energy and Resources Institute, New Delhi

Tewari DD, Campbell JY (1995) Developing and sustaining non-timber forest products: some policy issues and concerns with special reference to India. J Sustain For 3(1):53–79

Thomas DSG, Twyman C (2005) Equity and justice in climate change adaptation amongst natural resource dependent societies. Glob Environ Change 15:115–124

Ticktin T, Johns T (2002) Chinanteco management of *Aechmea magdalenae* (Bromeliaceae): implications for incorporating TEK and TRM in management plans. Econ Bot 56:43–57

TRIFED (2018) The Van Dhan Scheme, 2018. TRIFED, New Delhi

TRIFED (2020) 200 Days Report on Pradhan Mantri Van Dhan Yojana. TRIFED, New Delhi

Yadav M, Dugaya D (2013) Non-timber forest products certification in India: opportunities and challenges. Environ Dev Sustain 15:567–587. https://doi.org/10.1007/s10668-012-9393-1

Yadav M, Mishra S (2010) Sustainable development: a role for market information system for NTFP. Sustain Dev 20:128. https://doi.org/10.1002/sd.470

The manufacturer's authorised representative in the EU is Springer Nature Customer Service Centre GmbH, Europaplatz 3, 69115 Heidelberg, Germany. If you have any concerns regarding our products, please contact ProductSafety@springernature.com

Printed and bound by CPI Group (UK) Ltd, Croydon, CR0 4YY
02/10/2026
02226756-0002